to my students

STUDY GUIDE AND SOLUTIONS MANUAL

ORGANIC CHEMISTRY

FIFTH EDITION

PAULA YURKANIS BRUICE
UNIVERSITY OF CALIFORNIA, SANTA BARBARA

PEARSON
Prentice Hall

Upper Saddle River, NJ 07458

Assistant Editor: Jennifer Hart
Executive Editor: Nicole Folchetti
Editor-in-Chief, Science: Dan Kaveney
Executive Managing Editor: Kathleen Schiaparelli
Assistant Managing Editor: Karen Bosch
Production Editor: Gina M. Cheselka
Supplement Cover Manager: Paul Gourhan
Supplement Cover Designer: Christopher Kossa
Manufacturing Buyer: Ilene Kahn
Manufacturing Manager: Alexis Heydt-Long
Cover Image: Neo Vision/Photonica/Getty Images, Inc.

PEARSON Prentice Hall

Pearson Prentice Hall
Pearson Education, Inc.
Upper Saddle River, NJ 07458

Printed in the United States of America

10 9 8 7 6 5 4 3

ISBN 0-13-196328-7

Pearson Education Ltd., London
Pearson Education Australia Pty. Ltd., Sydney
Pearson Education Singapore, Pte. Ltd.
Pearson Education North Asia Ltd., Hong Kong
Pearson Education Canada, Inc., Toronto
Pearson Educación de Mexico, S.A. de C.V.
Pearson Education—Japan, Tokyo
Pearson Education Malaysia, Pte. Ltd.

TABLE OF CONTENTS

CHAPTER 1
Electronic Structure and Bonding • Acids and Bases

Important Terms

acid	a substance that donates a proton.
acid-base reaction	a reaction in which an acid donates a proton to a base.
acid dissociation constant	a measure of the degree to which an acid dissociates.
acidity	a measure of how easily a compound gives up a proton.
antibonding molecular orbital	a molecular orbital that results when two atomic orbitals with opposite phases interact. Electrons in an antibonding orbital decrease bond strength.
atomic number	tells how many protons (or electrons) the neutral atom has.
atomic orbital	an orbital associated with an atom.
atomic weight	the average mass of the atoms in the naturally occurring element.
aufbau principle	states that an electron will always go into the available orbital with the lowest energy.
base	a substance that accepts a proton.
basicity	describes the tendency of a compound to share its electrons with a proton.
bond dissociation energy	the amount of energy required to break a bond homolytically (each of the atoms retains one of the bonding electrons) or the amount of energy released when a bond is formed.
bonding molecular orbital	a molecular orbital that results when two atomic orbitals with the same phase interact. Electrons in a bonding orbital increase bond strength.
bond length	the internuclear distance between two atoms at minimum energy (maximum stability).
Brønsted acid	a substance that donates a proton.
Brønsted base	a substance that accepts a proton.
buffer solution	solution of a weak acid and its conjugate base.
carbanion	a species containing a negatively charged carbon.
carbocation	a species containing a positively charged carbon.
condensed structure	a structure that does not show some (or all) of the covalent bonds.

conjugate acid	the species formed when a base accepts a proton.
conjugate base	the species formed when an acid loses a proton.
core electrons	electrons in filled shells.
covalent bond	a bond created as a result of sharing electrons.
degenerate orbitals	orbitals that have the same energy.
delocalized electrons	electrons that do not belong to a single atom nor are they shared in a bond between two atoms.
dipole	a positive end and a negative end.
dipole moment (μ)	a measure of the separation of charge in a bond or in a molecule.
double bond	composed of a sigma bond and a pi bond.
electron affinity	the energy given off when an atom acquires an electron.
electronegative	describes an element that readily acquires an electron.
electronegativity	the tendency of an atom to pull electrons toward itself.
electropositive	describes an element that readily loses an electron.
electrostatic attraction	an attractive force between opposite charges.
electrostatic potential map	a map that allows you to see how electrons are distributed in a molecule.
equilibrium constant	the ratio of products to reactants at equilibrium (or the ratio of the rate constants for the forward and reverse reactions).
excited-state electronic configuration	the electronic configuration that results when an electron in the ground state has been moved to a higher energy orbital.
formal charge	the number of valence electrons – (the number of nonbonding electrons + 1/2 the number of bonding electrons).
free radical (radical)	a species with an unpaired electron.
ground-state electronic configuration	a description of which orbitals the electrons of an atom occupy when they are all in their lowest energy orbitals.
Heisenberg uncertainty principle	states that both the precise location and the momentum of an atomic particle cannot be simultaneously determined.

Henderson-Hasselbalch equation $pK_a = pH + \log[HA]/[A^-]$

Hund's rule states that when there are degenerate orbitals, an electron will occupy an empty orbital before it will pair up with another electron.

hybrid orbital an orbital formed by hybridizing (mixing) atomic orbitals.

hydride ion a negatively charged hydrogen (a hydrogen atom without its electron).

hydrogen ion (proton) a positively charged hydrogen (a hydrogen atom with an extra electron).

inductive electron withdrawal the pull of electrons through sigma bonds by an atom or a group of atoms.

ionic bond a bond formed through the attraction of opposite charges.

ionic compound a compound composed of a positive ion and negative ion.

ionization energy the energy required to remove an electron from an atom.

isotopes atoms with the same number of protons but a different number of neutrons.

Kekulé structure a model that represents the bonds between atoms as lines.

Lewis acid a substance that accepts an electron pair.

Lewis base a substance that donates an electron pair.

Lewis structure a model that represents the bonds between atoms as lines or dots and the lone-pair electrons as dots.

lone-pair electrons valence electrons not used in bonding.

mass number the number of protons plus the number of neutrons in an atom.

molecular orbital an orbital associated with a molecule.

molecular orbital (MO) theory describes a model in which the electrons occupy orbitals as they do in atoms but the orbitals extend over the entire molecule.

node a region within an orbital where there is zero probability of finding an electron.

nonbonding electrons valence electrons not used in bonding.

nonpolar covalent bond a bond formed between two atoms that share the bonding electrons equally.

octet rule	states that an atom will give up, accept, or share electrons in order to achieve a filled outer shell (or an outer shell that contains eight electrons) and no electrons of higher energy. Because a filled second shell contains eight electrons, this is known as the octet rule.
orbital	the volume of space around the nucleus where an electron is most likely to be found.
orbital hybridization	mixing of atomic orbitals.
organic compound	a compound that contains carbon.
Pauli exclusion principle	states that no more than two electrons can occupy an orbital and that the two electrons must have opposite spin.
pH	the pH scale is used to describe the acidity of a solution ($pH = -\log [H^+]$).
pi (π) bond	a bond formed as a result of side-to-side overlap of *p* orbitals.
pK_a	describes the tendency of a compound to lose a proton ($pK_a = -\log K_a$, where K_a is the acid dissociation constant).
polar covalent bond	a bond formed between two atoms that do not share the bonding electrons equally.
potential map (electrostatic potential map)	a map that allows you to see how electrons are distributed in a molecule.
proton	a positively charged hydrogen; a positively charged atomic particle.
proton transfer reaction	a reaction in which a proton is transferred from an acid to a base.
quantum mechanics	the use of mathematical equations to describe the behavior of electrons in atoms or molecules.
radical (free radical)	a species with an unpaired electron.
resonance	having delocalized electrons.
resonance contributors	structures with localized electrons that approximate the true structure of a compound with delocalized electrons.
resonance hybrid	the actual structure of a compound with delocalized electrons.
sigma (σ) bond	a bond with a symmetrical distribution of electrons about the internuclear axis.
single bond	a single pair of electrons shared between two atoms.

tetrahedral bond angle the bond angle (109.5°) formed by an sp^3 hybridized central atom.

tetrahedral carbon an sp^3 hybridized carbon; a carbon that forms covalent bonds using four sp^3 hybrid orbitals.

trigonal planar carbon an sp^2 hybridized carbon.

triple bond composed of a sigma bond and two pi bonds.

valence electrons an electron in an outermost shell.

valence shell electron pair repulsion (VSEPR) model a model that combines the concepts of atomic orbitals and shared electron pairs with minimization of electron repulsions.

wave equation an equation that describes the behavior of each electron in an atom or a molecule.

wave functions a series of solutions of a wave equation.

Solutions to Problems

1. The mass number of an atom = the number of protons + the number of neutrons
The atomic number = the number of protons.
All isotopes have the same atomic number; in the case of oxygen it is 8.
Therefore:

The isotope of oxygen with a mass number of 16 has 8 protons and 8 neutrons.
The isotope of oxygen with a mass number of 17 has 8 protons and 9 neutrons.
The isotope of oxygen with a mass number of 18 has 8 protons and 10 neutrons.

2. All four atoms have two core electrons (in their filled shell); the valence electrons are in the outer shell.
a. 3 **b.** 5 **c.** 6 **d.** 7

3. **a.** Potassium is in the first column of the periodic table; therefore, like lithium and sodium, potassium has one valence electron.
b. It occupies a 4 *s* orbital.

4. **a.** Using the aufbau principle (electrons go into available orbitals with the lowest energy) and the Pauli exclusion principle (no more than 2 electrons are in each atomic orbital), and remembering that the relative energies of the atomic orbitals are:

$$1s < 2s < 2p < 3s < 3p < 4s < 3d < 4p < 5s < 4d < 5p$$

The question can be answered if you remember that there is one *s* atomic orbital and 3 degenerate *p* atomic orbitals. To write the electronic configurations for Br and I, you need to remember that there are 5 degenerate *d* atomic orbitals.

Cl $1s^2\ 2s^2\ 2p^6\ 3s^2\ 3p^5$
Br $1s^2\ 2s^2\ 2p^6\ 3s^2\ 3p^6\ 4s^2\ 3d^{10}\ 4p^5$
I $1s^2\ 2s^2\ 2p^6\ 3s^2\ 3p^6\ 4s^2\ 3d^{10}\ 4p^6\ 5s^2\ 4d^{10}\ 5p^5$

b. Notice that because the three atoms are in the same column of the periodic table, they all have the same number of valence electrons (7), and the valence electrons are in similar orbitals (2 are in an *s* orbital and 5 are in *p* orbitals).

5. The atomic numbers can be found in the periodic table on the last page of the text. Notice that elements in the same column of the periodic table have the same number of valence electrons and their valence electrons are in similar orbitals.

a. carbon (atomic number = 6): $1s^2\ 2s^2\ 2p^2$
silicon (atomic number = 14): $1s^2\ 2s^2\ 2p^6\ 3s^2\ 3p^2$

b. oxygen (atomic number = 8): $1s^2\ 2s^2\ 2p^4$
sulfur (atomic number = 16): $1s^2\ 2s^2\ 2p^6\ 3s^2\ 3p^4$

c. fluorine (atomic number = 9): $1s^2\ 2s^2\ 2p^5$
bromine (atomic number = 35): $1s^2\ 2s^2\ 2p^6\ 3s^2\ 3p^6\ 4s^2\ 3d^{10}\ 4p^5$

d. magnesium (atomic number = 12): $1s^2\ 2s^2\ 2p^6\ 3s^2$
calcium (atomic number = 20): $1s^2\ 2s^2\ 2p^6\ 3s^2\ 3p^6\ 4s^2$

6. The polarity of a bond can be determined by the difference in the electronegativities (given in Table 1.3) of the atoms sharing the bonding electrons. The greater the difference in electronegativity, the more polar the bond.

a. Cl—CH_3 **b.** H—OH **c.** H—F **d.** Cl—CH_3

7. **a.** KCl has the most polar bond $(3.0 - 0.8 = 2.2)$, whereas it is 1.8 for LiBr, 1.6 for NaI, and 0 for Cl_2.
b. Cl_2 has the least polar bond because the two chlorine atoms share the bonding electrons equally.

8. Solved in the text.

9. To answer this question, compare the electronegativities of the two atoms sharing the bonding electrons using Table 1.3 on page 11 of the text. (Note that if the atoms being compared are in the same row of the periodic chart, the atom on the right is the more electronegative; if the atoms being compared are in the same column, the one closer to the top of the column is the more electronegative.)

a. $\overset{\delta-}{HO}—\overset{\delta+}{H}$ **c.** $\overset{\delta+}{H_3C}—\overset{\delta-}{NH_2}$ **e.** $\overset{\delta-}{HO}—\overset{\delta+}{Br}$ **g.** $\overset{\delta+}{I}—\overset{\delta-}{Cl}$

b. $\overset{\delta-}{F}—\overset{\delta+}{Br}$ **d.** $\overset{\delta+}{H_3C}—\overset{\delta-}{Cl}$ **f.** $\overset{\delta-}{H_3C}—\overset{\delta+}{MgBr}$ **h.** $\overset{\delta+}{H_2N}—\overset{\delta-}{OH}$

10. **a.** LiH and HF are polar (they have a red end and a blue end).
b. A potential map marks the edges of the molecule's electron cloud. The electron cloud is largest around the H in LiH because that H has more electrons around it than the H's in the other molecules.
c. Because the hydrogen of HF is blue, we know that that compound has the most positively charged hydrogen and, therefore, will be most apt to attract a negatively charged molecule.

11. By answering this question you will see that a formal charge is a book-keeping device. It does *not necessarily* tell you which atom has the greatest electron density or is the most electron deficient.
a. oxygen
b. oxygen (it is the more red)
c. oxygen
d. hydrogen (it is the most blue)

Notice that in hydroxide ion, the atom with the formal negative charge **is** the atom with the greater electron density. In the hydronium ion, however, the atom with the formal positive charge **is not** the most electron deficient atom.

12. formal charge = number of valence electrons
– (number of lone-pair electrons + ½ the number of bonding electrons)

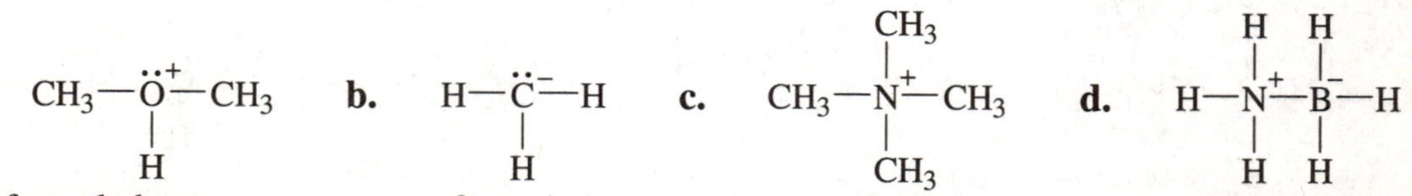

a. formal charge 6–(2+3) = +1
b. formal charge 4–(2+3) = –1
c. formal charge 5–(0+4) = +1
d. formal charge N 5–(0+4) = +1; B 3–(0+4) = –1

13.

a. 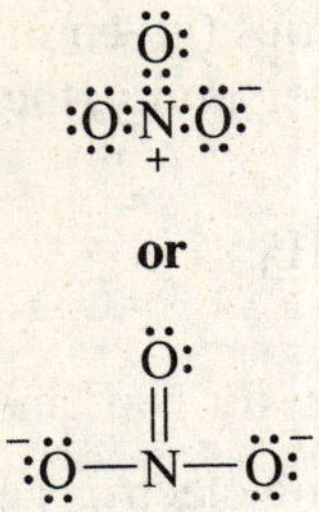

c. 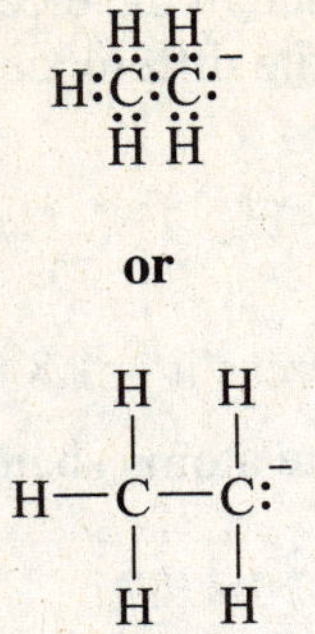

e. 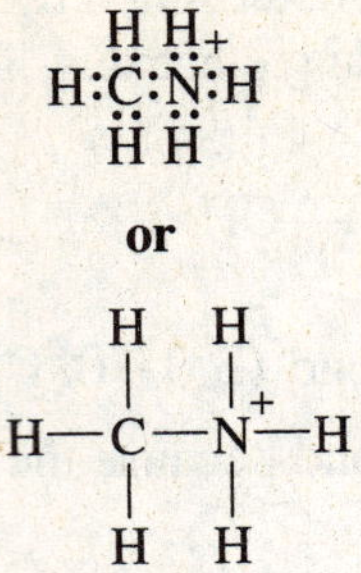

g. 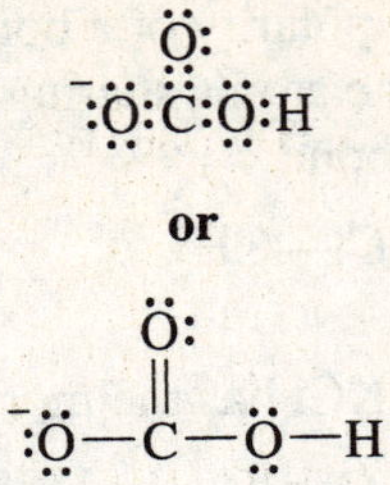

b. 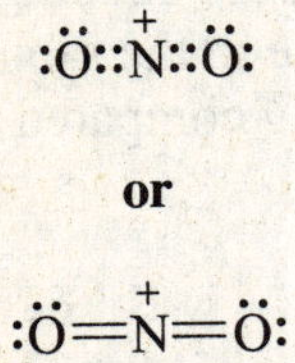

d. 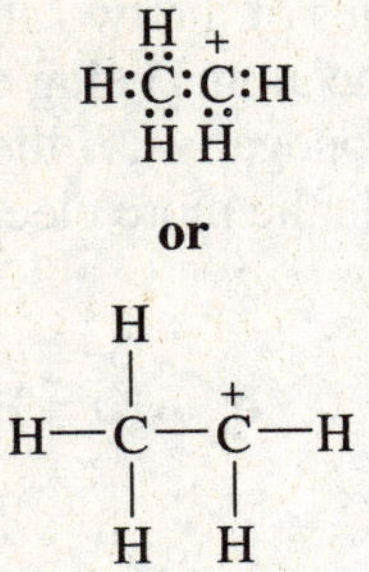

f. 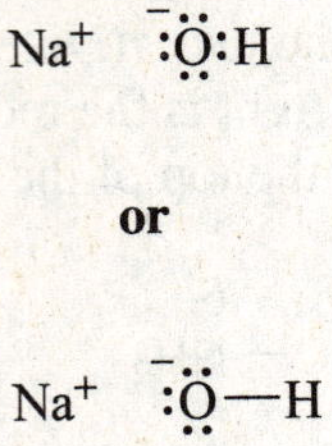

h. 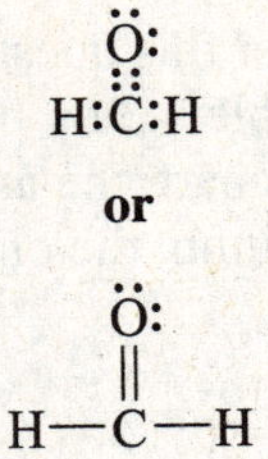

14.

a. 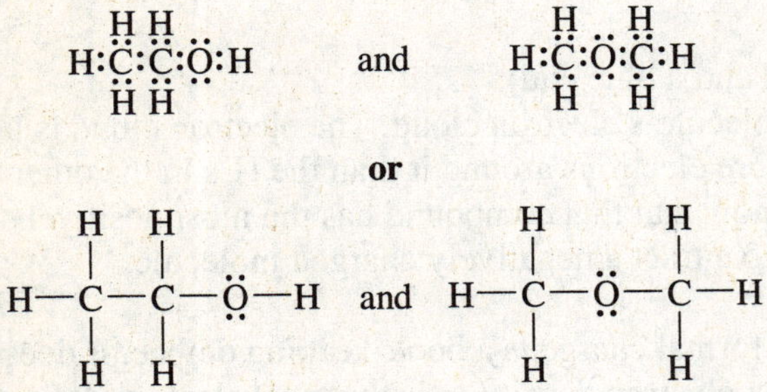

b. 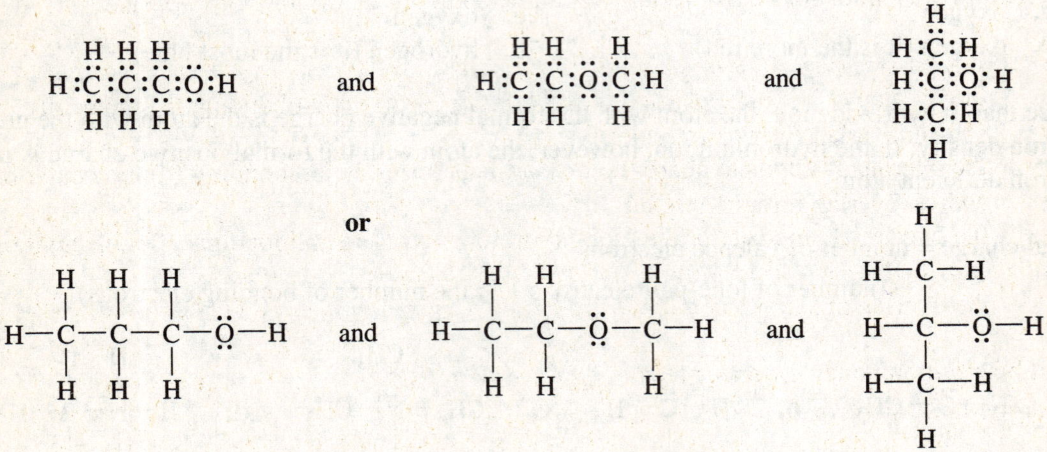

15.

a. $CH_3CH_2\ddot{N}H_2$ **c.** $CH_3CH_2\ddot{\underset{\cdot\cdot}{O}}H$ **e.** $CH_3CH_2\ddot{\underset{\cdot\cdot}{Cl}}:$

b. $CH_3\ddot{N}HCH_3$ **d.** $CH_3\ddot{\underset{\cdot\cdot}{O}}CH_3$ **f.** $H\ddot{\underset{\cdot\cdot}{O}}\ddot{N}H_2$

16. **a.** $CH_3CH_2CH_2Cl$ **c.** 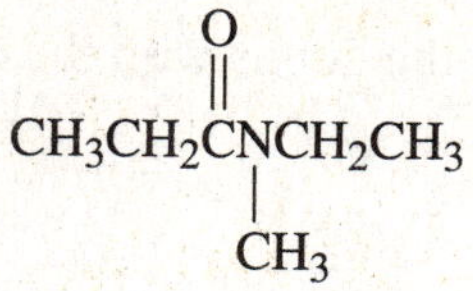

b. 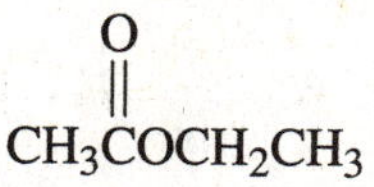 **d.** $CH_3CH_2C{\equiv}N$

17. **a.** the (green) chlorine atom
b. the (red) oxygen atoms
c. the (blue) nitrogen atoms
d. the (black) carbon atoms and (white) hydrogen atoms

18. **a.** 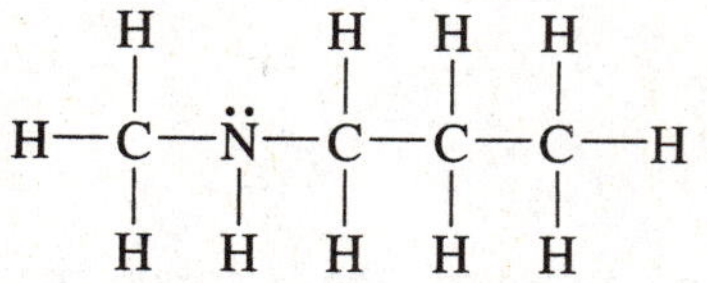 **b.** 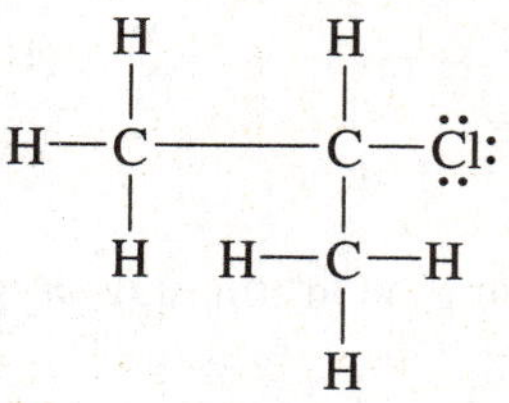 **c.** 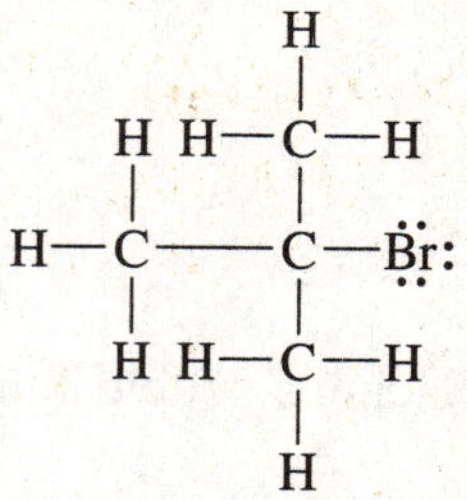

d. 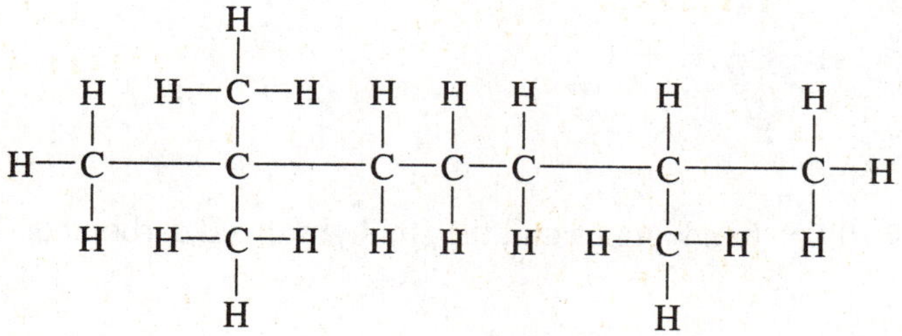

19. He_2^+ has three electrons. Two are in a bonding orbital and one is in an antibonding orbital. Because there are more electrons in the bonding orbital than in the antibonding orbital, He_2^+ exists.

20. **a.** π^* side-to-side overlap forms a π bond;
the overlap of opposite phase (colors) orbitals forms an antibonding (*) molecular orbital.
b. σ^* end-on overlap forms a σ bond;
the overlap of opposite phase (colors) orbitals forms an antibonding (*) molecular orbital.
c. σ^* end-on overlap forms a σ bond;
the overlap of opposite phase (colors) orbitals forms an antibonding (*) molecular orbital.
d. σ end-on overlap forms a σ bond;
the overlap of same phase (colors) orbitals forms a bonding molecular orbital.

21. The carbon-carbon bonds form as a result of sp^3—sp^3 overlap.
The carbon-hydrogen bonds form as a result of sp^3—s overlap.

22. **a(2).** The carbon forms fours bonds and each chlorine forms one bond.

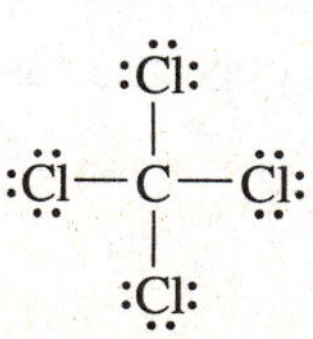

b(2). The carbon uses sp^3 orbitals to form the bonds with the chlorine atoms so the bond angles are all 109.5°

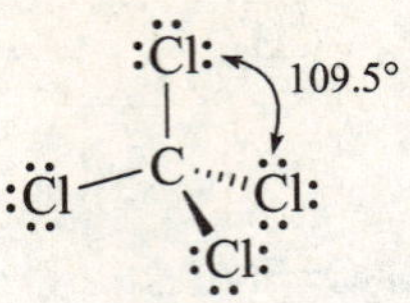

a(3). The first attempt at Lewis structure shows that carbon does not have a complete octet and does not form the needed number of bonds.

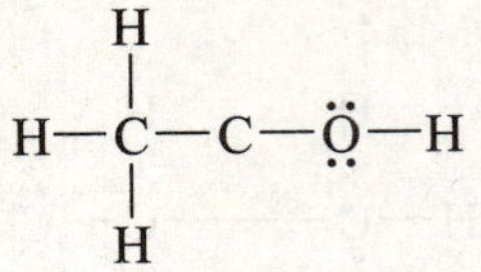

Placing a double bond between the carbon and oxygen and removing the H solves the problem.

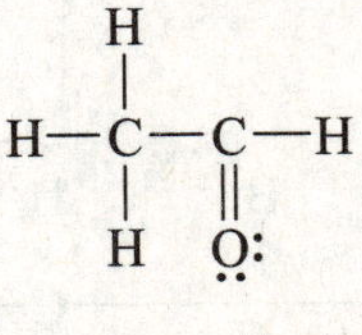

b(3). The sp^3 hybridized carbon has 109.5° bond angles and the sp^2 hybridized carbon has 120° bond angles.

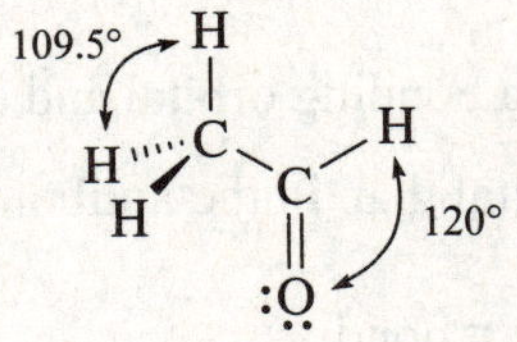

a(4). In order to fill their octets and from the required number of bonds, carbon and nitrogen must form a triple bond.

H—C≡N:

b(4). Because the carbon is *sp* hybridized, the bond angle is 180°.

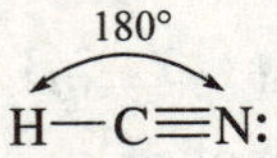

23. CH_4 with no lone pairs has bond angles of 109.5°
H_2O with 2 lone pairs has bond angles of 104.5°
From the above we see that the bond angle decreases as the number of lone pairs increases, because a lone pair is more diffuse than a bonding pair.

Therefore, H_3O^+ with 1 lone pair has bond angles in between those two; its bond angles will be greater than 104.5° (two lone pairs) and less than 109.5° (no lone pairs).

24. The hydrogens of the ammonium ion are the bluest atoms. Therefore, they have the least electron density. In other words, they have the most positive (least negative) electrostatic potential.

25. Water is the most polar—it has a deep red area and the most intense blue area.
Methane is the least polar—it is all the same color with no red or blue areas.

26. The methyl carbanion has one lone pair so we expect it to have bond angles of ~107.3°, about the same bond angles that ammonia has, another molecule with one lone pair.

27. Bonding electrons in shells farther from the nucleus form **longer** and **weaker** bonds due to poorer overlap of the bonding orbitals. Therefore:

a. **relative lengths** of the bonds in the halogens are: $Br_2 > Cl_2$

relative strengths of the bonds are: $Cl_2 > Br_2$

b. **relative lengths**: CH_3—Br > CH_3—Cl > CH_3—F

relative strengths: CH_3—F > CH_3—Cl > CH_3—Br

28. a. **longer:** 1. C—I 2. C—Cl 3. H—Cl
b. **stronger:** 1. C—Cl 2. C—C 3. H—H

29. We know the σ bond is stronger than a π bond because the σ bond in ethane has a bond dissociation energy of 90 kcal/mol, whereas the bond dissociation energy of the double bond ($\sigma + \pi$) in ethene is 174 kcal/mole, which is less than twice as strong.
Because the σ bond is stronger, we know that it has more effective orbital-orbital overlap.

30. The carbon-carbon sigma bond formed by sp^2—sp^2 overlap is stronger because an sp^2 orbital has 33.3% *s* character, whereas an sp^3 orbital has 25% *s* character.
Because electrons in an *s* orbital are closer on average to the nucleus than those in a *p* orbital, the greater the *s* character in the interacting orbitals, the stronger (and shorter) bond.

31. a.

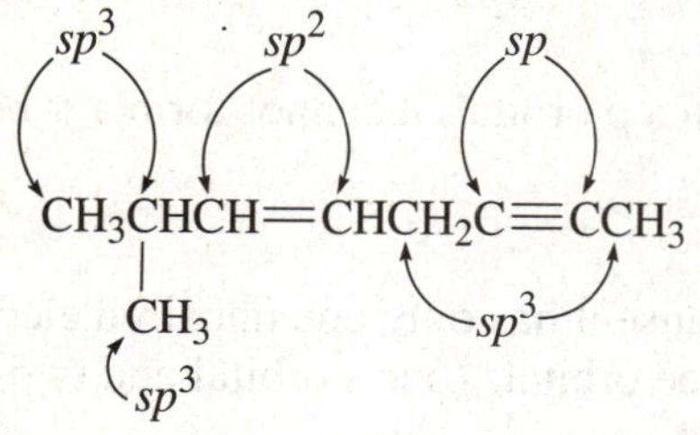

b.

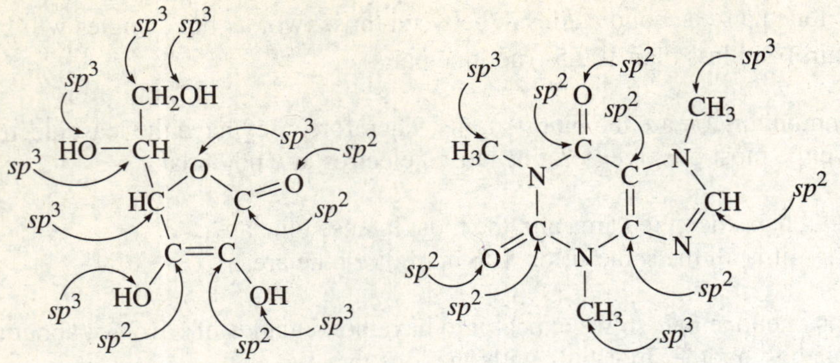

32. The bond angle depends on the central atom.

a. sp^3 nitrogen with no lone pair: 109.5°

b. sp^3 carbon: 109.5°

c. sp^3 carbon: 109.5°

d. sp^3 carbon: 109.5°

33. **a.** $\mathbf{BeH_2}$

Because beryllium does not have any unpaired electrons in its ground state, it cannot form any bonds unless it promotes an electron. After promotion, hybridization of the two orbitals (an *s* orbital and a *p* orbital) that contain unpaired electrons results in two *sp* hybrid orbitals.

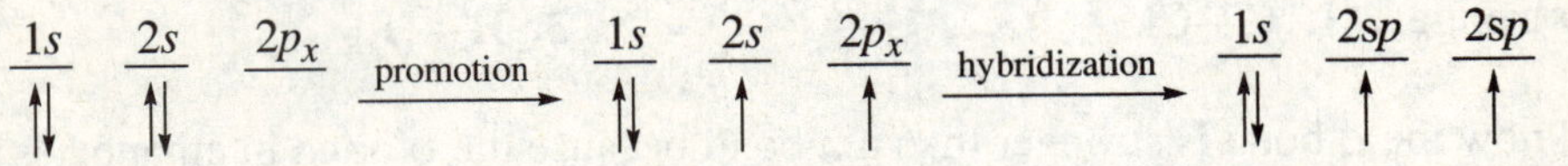

Each *sp* orbital of beryllium overlaps with the *s* orbital of a hydrogen. The two *sp* orbitals orient themselves to get as far away from each other as possible, resulting in a bond angle of **180°**.

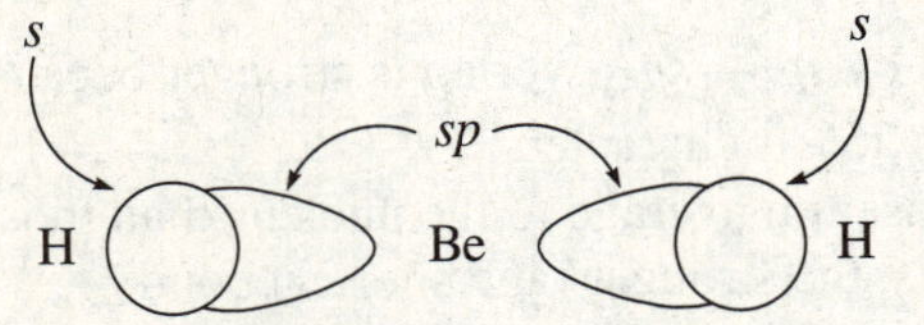

bond angle = 180°

H—Be—H

H_2Be

Notice that because beryllium does not have an electron in a *p* orbital, it cannot form a *π* bond.

b. $\mathbf{BH_3}$

Without promotion, boron could form only one bond because it has only one unpaired electron. Promotion gives it three unpaired electrons. When the three orbitals (one *s* orbital and two *p* orbitals) containing the unpaired electrons are hybridized, three sp^2 orbitals result.

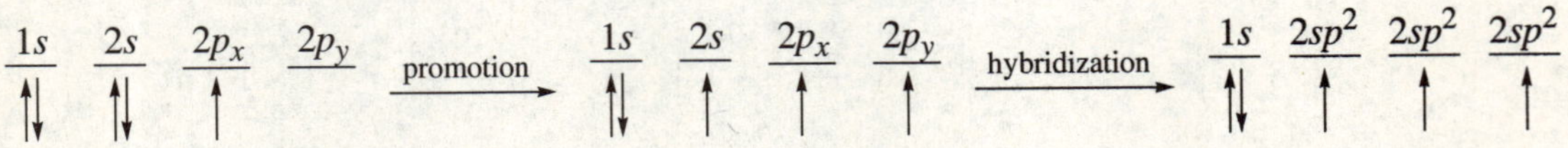

Each sp^2 hybrid orbital overlaps with the *s* orbital of hydrogen. The three sp^2 orbitals orient themselves to get as far away from each other as possible, resulting in bond angles in of **120°**.

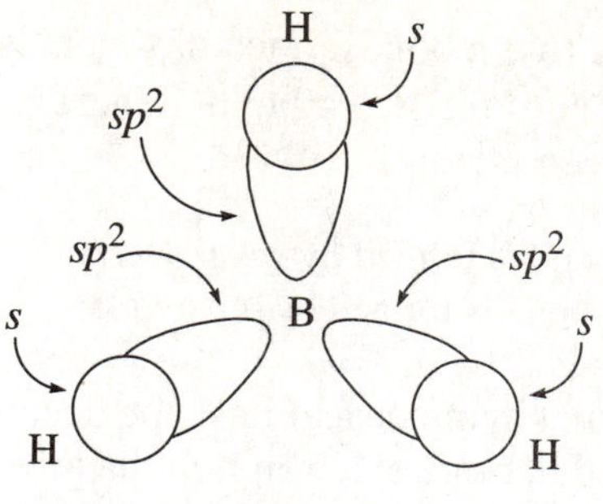

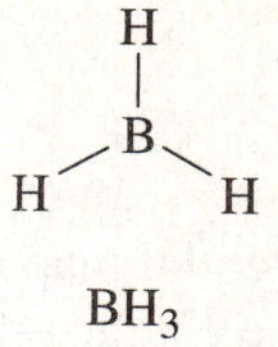

c. $\mathbf{CCl_4}$

The carbon in CCl_4 is bonded to four atoms, so it uses four sp^3 hybrid orbitals.
Each carbon-chlorine bond is formed by the overlap of an sp^3 orbital of carbon with an sp^3 orbital of chlorine. Because the four sp^3 orbitals of carbon orient themselves to get as far away from each other as possible, the bond angles are all **109.5°**.

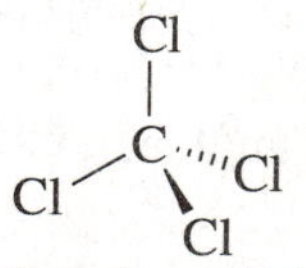

bond angles = 109.5°

d. $\mathbf{CO_2}$

The carbon in CO_2 is bonded to two atoms, so it uses two *sp* hybrid orbitals. Each carbon-oxygen bond is a double bond. One of the bonds of each double bond is formed by the overlap of an *sp* orbital of carbon with an sp^2 orbital of oxygen. The second bond of the double bond is formed as a result of side-to-side overlap of a *p* orbital of carbon with a *p* orbital of oxygen. Because the two *sp* orbitals orient themselves to get as far away from each other as possible, the bond angle in CO_2 is **180°**.

O=C=O
bond angle = 180°

e. The double-bonded carbon and the double-bonded oxygen in HCOOH use sp^2 hybrid orbitals; thus, the bonds around the double-bonded carbon are all 120°. The single-bonded oxygen uses sp^3 hybrid orbitals and each hydrogen uses an *s* orbital; thus, the C—O—H bond angle is 104.5°.

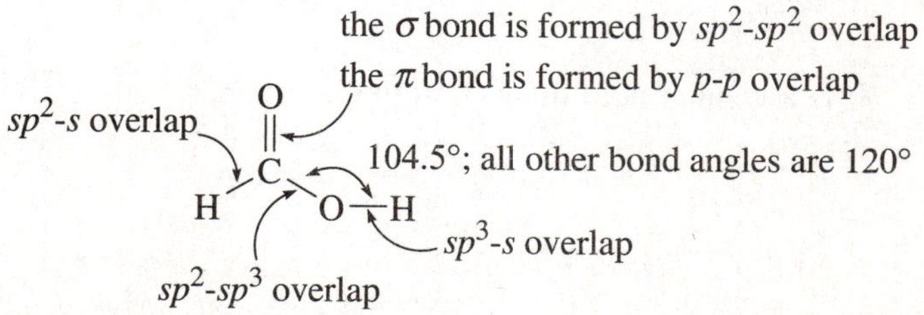

f. The triple bond consists of one σ bond and two π bonds. Each nitrogen uses an *sp* hybrid orbital to form the σ bond and a *p* orbital to form each of the two π bonds. There are no bond angles in this compound; a bond angle is the angle between 3 atoms.

N≡N the σ bond is formed by *sp-sp* overlap
each π bond is formed by *p-p* overlap

34. The electrostatic potential map of ammonia is not symmetrical in shape because it is not symmetrical in the distribution of the charge—the nitrogen end is more electron rich than the three hydrogens. The electrostatic potential map of the ammonium ion is symmetrical in shape because it is symmetrical in the distribution of the charge. Its symmetry results from the fact that the nitrogen atom forms a bond with each of the four hydrogens and the four bond angles are all the same.

35. **a, e, g, h**

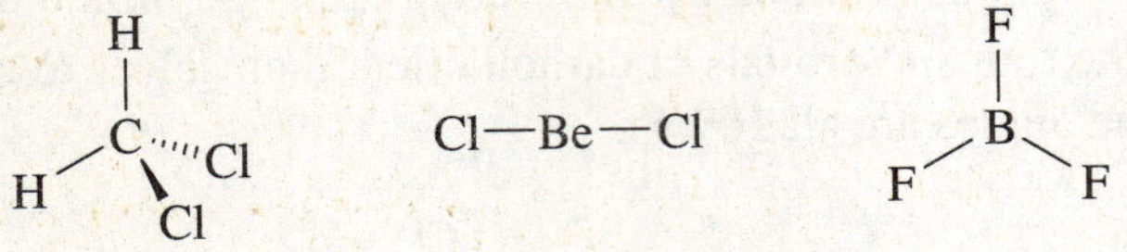

Because C is sp^3 hyridized, the bond angles in CH_2Cl_2 are 109.5° and the compound has a dipole moment.
Because Be is *sp* hybridized, the bond angle in $BeCl_2$ is 180° and the compound does not have a dipole moment.
Because B is sp^2 hybridized, the bond angles in BF_3 are 120° and the compound does not have a dipole moment.

36. **a.** (1) $^+NH_4$ (2) HCl (3) H_2O (4) H_3O^+

b. (1) $^-NH_2$ (2) Br^- (3) NO_3^- (4) HO^-

37. **a.** The lower the pK_a, the stronger the acid, so the compound with $pK_a = 5.2$ is the stronger acid.
b. The greater the dissociation constant, the stronger the acid, so the compound with dissociation constant $= 3.4 \times 10^{-3}$ is the stronger acid.

38. $K_a = K_{eq}[H_2O]$

$$K_{eq} = \frac{K_a}{[H_2O]} = \frac{4.53 \times 10^{-6}}{55.5} = 8.16 \times 10^{-8}$$

39. $K_a = 1.51 \times 10^{-5}$ It is a weaker acid than vitamin C.

40. a. $HO^- + H^+ \longrightarrow H_2O$

b. $HCO_3^- + H^+ \rightleftharpoons H_2CO_3 \rightleftharpoons H_2O + CO_2$

c. $CO_3^{2-} + 2\,H^+ \rightleftharpoons H_2CO_3 \rightleftharpoons H_2O + CO_2$

41. a. basic b. acidic c. basic

42. a. CH_3COO^- is the stronger base.
Because HCOOH is the stronger acid, it has the weaker conjugate base.

b. $^-NH_2$ is the stronger base.
Because H_2O is the stronger acid, it has the weaker conjugate base.

c. H_2O is the stronger base.
Because $CH_3OH_2^+$ is the stronger acid, it has the weaker conjugate base.

43. The conjugate acids have the following relative strengths:

$$CH_3\overset{+}{O}H_2 > CH_3\overset{O}{\overset{\|}{C}}OH > CH_3\overset{+}{N}H_3 > CH_3OH > CH_3NH_2$$

The bases, therefore, have the following relative strengths:

$$CH_3\overset{-}{N}H > CH_3O^- > CH_3NH_2 > CH_3\overset{O}{\overset{\|}{C}}O^- > CH_3OH$$

44. <u>if the lone pairs are not shown:</u>

a. CH_3OH as an acid $CH_3OH + NH_3 \rightleftharpoons CH_3O^- + \overset{+}{N}H_4$

CH_3OH as a base $CH_3OH + HCl \rightleftharpoons CH_3\overset{+}{O}H + Cl^-$ (with H on O: $CH_3\overset{+}{O}H_2$)

b. NH_3 as an acid $NH_3 + CH_3O^- \rightleftharpoons {}^-NH_2 + CH_3OH$

NH_3 as a base $NH_3 + HBr \rightleftharpoons \overset{+}{N}H_4 + Br^-$

<u>if the lone pairs are shown:</u>

a. CH_3OH as an acid $CH_3\ddot{\underset{..}{O}}H + \ddot{N}H_3 \rightleftharpoons CH_3\ddot{\underset{..}{O}}{:}^- + \overset{+}{N}H_4$

CH_3OH as a base $CH_3\ddot{\underset{..}{O}}H + H\ddot{\underset{..}{Cl}}{:} \rightleftharpoons CH_3\overset{+}{\ddot{O}}H + {:}\ddot{\underset{..}{Cl}}{:}^-$ (with H on O)

b. NH_3 as an acid $\quad \ddot{N}H_3 + CH_3\ddot{\underset{..}{O}}\!:^- \rightleftharpoons {}^-\!:\ddot{N}H_2 + CH_3\ddot{\underset{..}{O}}H$

NH_3 as a base $\quad \ddot{N}H_3 + H\ddot{\underset{..}{Br}}\!: \rightleftharpoons \overset{+}{N}H_4 + :\ddot{\underset{..}{Br}}\!:^-$

45. Notice that in each case, the equilibrium goes away from the strong acid and toward the weak acid.

a. $CH_3OH + HO^- \rightleftharpoons CH_3O^- + H_2O$

$pK_a = 15.5 \qquad pK_a = 15.7$

$CH_3OH + H_3O^+ \rightleftharpoons CH_3\overset{+}{O}H_2 + H_2O$

$pK_a = -1.7 \qquad pK_a = -2.5$

$CH_3\overset{O}{\overset{\|}{C}}OH + H_2O \rightleftharpoons CH_3\overset{O}{\overset{\|}{C}}O^- + H_3O^+$

$pK_a = 4.8 \qquad pK_a = -1.7$

$CH_3NH_2 + HO^- \rightleftharpoons CH_3\bar{N}H + H_2O$

$pK_a = 4.8 \qquad pK_a = 15.7$

$CH_3NH_2 + H_3O^+ \rightleftharpoons CH_3\overset{+}{N}H_3 + H_2O$

$pK_a = -1.7 \qquad pK_a = 10.7$

b. $HCl + H_2O \rightleftharpoons H_3O^+ + Cl^-$

$pK_a = -7 \qquad pK_a = -1.7$

$NH_3 + H_2O \rightleftharpoons {}^+NH_4 + HO^-$

$pK_a = 15.7 \qquad pK_a = 9.4$

46. **a.** $HC\equiv CH + HO^- \rightleftharpoons HC\equiv C^- + H_2O$

$pK_a = 25 \qquad pK_a = 15.7$

b. $HC\equiv CH + {}^-NH_2 \rightleftharpoons HC\equiv C^- + NH_3$

$pK_a = 25 \qquad pK_a = 36$

c. ${}^-NH_2$ would be a better base because the equilibrium favors the products. When HO^- removes a proton, the equilibrium favors the reactants.

47.

$$K_{eq} = \frac{K_a \text{ of the reactant acid}}{K_a \text{ of the product acid}}$$

For **a** the reactant acid is HCl and the product acid is H_3O^+.

For **b** the reactant acid is CH_3COOH and the product acid is H_3O^+.

For **c** the reactant acid is H_2O and the product acid is $CH_3\overset{+}{N}H_3$.

For **d** the reactant acid is $CH_3\overset{+}{N}H_3$ and the product acid is H_3O^+.

a. $K_{eq} = \dfrac{10^{7}}{10^{1.7}} = 10^{7-1.7} = 10^{5.3} = 2.0 \times 10^{5}$

b. $K_{eq} = \dfrac{10^{-4.8}}{10^{1.7}} = 10^{-4.8-1.7} = 10^{-6.5} = 3.2 \times 10^{-7}$

c. $K_{eq} = \dfrac{10^{-15.7}}{10^{-10.7}} = 10^{-15.7+10.7} = 10^{-5} = 1.0 \times 10^{-5}$

d. $K_{eq} = \dfrac{10^{-10.7}}{10^{1.7}} = 10^{-10.7-1.7} = 10^{-12.4} = 4.0 \times 10^{-13}$

48. **a.** HBr **b.** $CH_3CH_2CH_2\overset{+}{O}H_2$ **c.** $CH_3CH_2CH_2OH$ **d.** 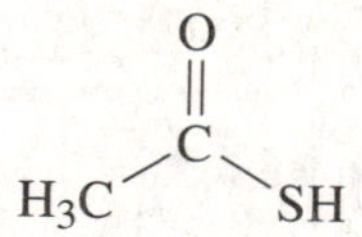

49. **a.** Because HF is the weakest acid, F^- is the strongest base.

b. Because HI is the strongest acid, I^- is the weakest base.

50. **a.** oxygen **b.** H_2S **c.** CH_3SH

The size of an atom is more important than its electronegativity in determining stability. Therefore, even through oxygen is more electronegative than sulfur, H_2S is a stronger acid than H_2O, and CH_3SH is a stronger acid than CH_3OH because the sulfur atom is larger—therefore, the electrons in its conjugate base are spread out over a greater volume, which stabilizes it. The more stable the base, the stronger its conjugate acid.

51. Compare the acid strengths of the conjugate acids, recalling that the weaker acid has the stronger conjugate base.

a. HO^- because H_2O is a weaker acid than H_3O^+.

b. NH_3 because $^+NH_4$ is a weaker acid than H_3O^+.

c. CH_3O^- because CH_3OH is a weaker acid than CH_3COOH.

d. CH_3O^- because CH_3OH is a weaker acid than CH_3SH.

52. **a.** $CH_3OCH_2CH_2OH$ because of the CH_3O group that withdraws electrons inductively.

b. $CH_3CH_2CH_2\overset{+}{O}H_2$ because oxygen is more electronegative than nitrogen and, therefore is better at withdrawing electrons inductively.

c. $CH_3CH_2OCH_2CH_2OH$ because the electron-withdrawing oxygen is closer to the OH group.

d. $CH_3CH_2\overset{O}{\overset{\|}{C}}OH$ because the electron-withdrawing C═O is closer to the OH group.

53. 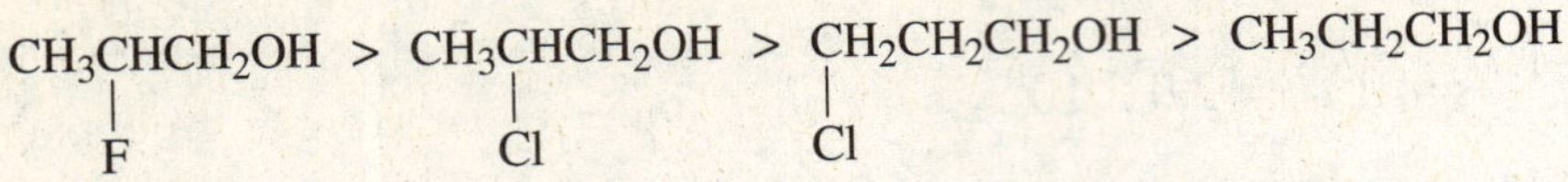

The first two compounds are the most acidic because they have the electron-withdrawing substituent closest to the O—H bond. The first compound is more acidic than the second because fluorine is more electronegative than chlorine.
The second listed compound is a stronger acid than the third listed compound because the chlorine in the third compound is farther away from the O—H bond.
The compound on the far right does not have a substituent that withdraws electrons inductively, so it is the least acidic of the compounds.

54. The weaker acid has the stronger conjugate base.

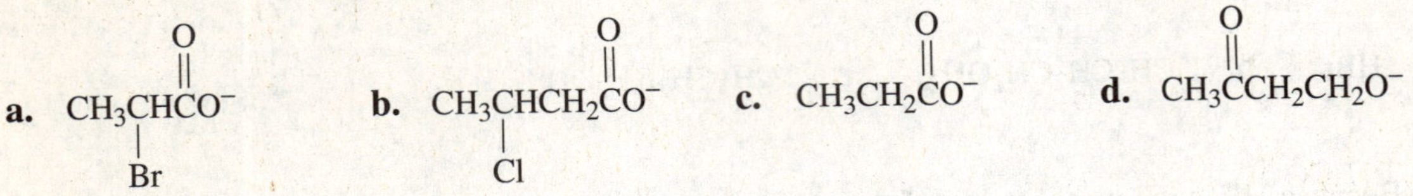

55. Solved in the text.

56. When a sulfonic acid loses a proton, the electrons left behind are shared by three oxygen atoms. In contrast, when a carboxylic acid loses a proton, the electrons left behind are shared by only two oxygen atoms. The sulfonate ion, therefore, is more stable than the carboxylate ion.
The more stable the base, the stronger is its conjugate acid. Thus the sulfonic acid is a stronger acid than the carboxylic acid.

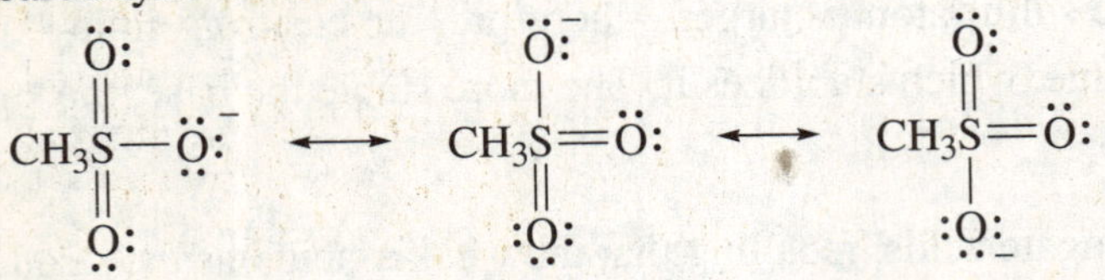

a sulfonate ion

a carboxylate ion

57. **a.** 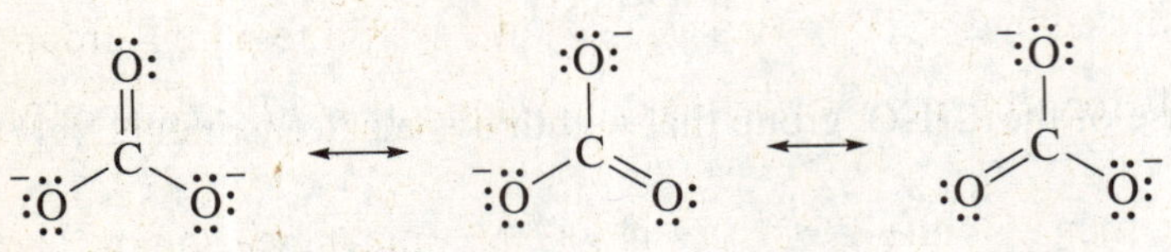

b. 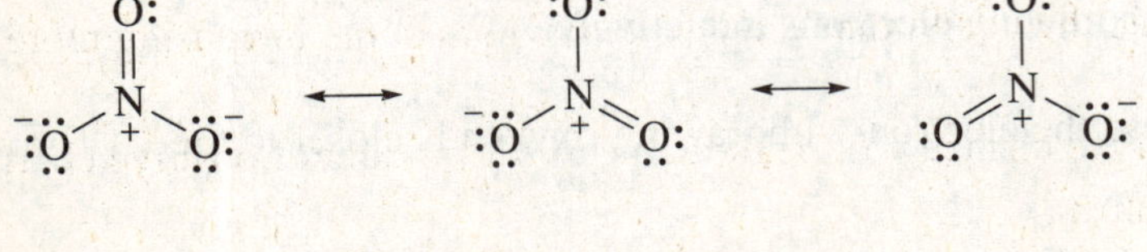

58. **a.** $CH_3C{\equiv}\overset{+}{N}H$

b. CH_3CH_3

c. $F_3C\overset{O}{\overset{\|}{C}}OH$

d. an sp^2 hybridized oxygen.

e. $CH_3C{\equiv}\overset{+}{N}H$ > $CH_3\underset{CH_3}{\underset{|}{C}}{=}\overset{+}{N}HCH_3$ > $CH_3CH_2\overset{+}{N}H_3$

f. HNO_3 is more acidic than HNO_2
When the structure of the conjugate base of each acid is drawn out, you can see that NO_3 has a positively charged nitrogen and NO_2 does not. The positive charge decreases the electron density on the oxygen by inductive electron withdraw thereby stabilizing it. A stable base is a weak base. Therefore, it has the stronger conjugate acid.

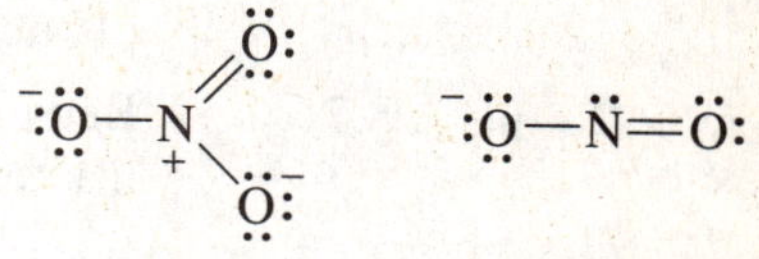

59. **a.** CH_3COO^- **b.** $CH_3CH_2\overset{+}{N}H_3$ **c.** H_2O **d.** Br^- **e.** $\overset{+}{N}H_4$ **f.** $HC{\equiv}N$ **g.** NO_2^- **h.** NO_3^-

60. **a.** **1.** neutral
2. neutral
3. charged
4. charged
5. charged
6. charged

b. **1.** charged
2. charged
3. charged
4. charged
5. neutral
6. neutral

c. **1.** neutral
2. neutral
3. neutral
4. neutral
5. neutral
6. neutral

61. **a.** The COOH group will be in its basic form because the solution is more basic than its pK_a value.
The $^+NH_3$ group will be in its acidic form because the solution is more acidic than its pK_a value.

$$CH_3\underset{^+NH_3}{\underset{|}{C}}H\overset{O}{\overset{\|}{C}}O^-$$

b. No, alanine can never be without a charge. To be without a charge would require a group with a pK_a value of 9.69 to lose a proton before (be more acidic than) a group with a pK_a value of 2.34.

c. As the pH becomes more basic than 2.34 the COOH group will become increasing more negatively charged.
As the pH becomes more acidic than 9.69, the $^+NH_3$ group will become increasing more positively charged.
Therefore, the amount of negative change will be the same as the amount of positive charge at the intersection of the two pK_a values.

$$\frac{2.34 + 9.69}{2} = 6.02$$

62. **a.** 10.4 (two log units more basic than the pK_a)
b 2.7 (one log unit more acidic than the pK_a)
c. 6.4 (two log units more acidic than the pK_a)
d. 7.3 (when the pH is equal to the pK_a)
e. 5.6 (one log unit more basic than the pK_a)

63. **a.** **1.** pH = 4.9
2. pH = 10.7

b. **1.** pH > 6.9 — Because the basic form is the form in which the compound is charged, the pH needs to be more than two units more basic than the pK_a.
2. pH < 8.7 — Because the acidic form is the form in which the compound is charged, the pH needs to be more than two units more acidic than the pK_a.

64. greater than pH 10.4 (As long as the pH is greater than the pK_a of the compound, the majority of the compound will be in its basic form.)

65. **a.** $CH_3COO^- + H^+ \rightleftharpoons CH_3COOH$

b. $CH_3COOH + HO^- \rightleftharpoons CH_3COO^-$

66.

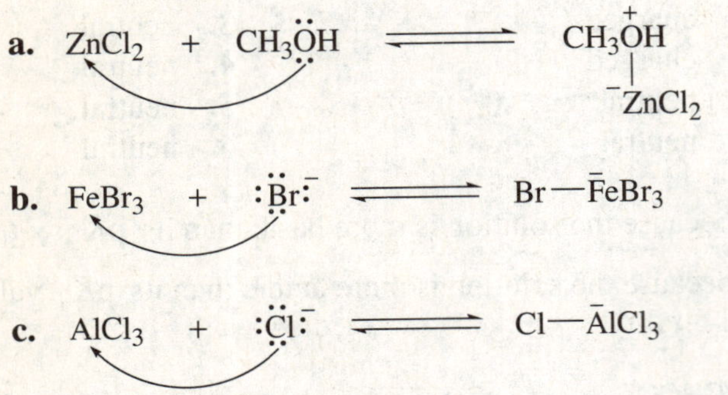

67. **a, b, c,** and **h** are Brønsted acids (protonating-donating acids). Therefore, they react with HO^- by donating a proton to it.
d, e, f, and **g** are Lewis acids. They react with HO^- by accepting a pair of electrons from it.

a. $CH_3OH + HO^- \rightleftharpoons CH_3O^- + H_2O$

b. $\overset{+}{N}H_4 + HO^- \rightleftharpoons NH_3 + H_2O$

c. $CH_3\overset{+}{N}H_3 + HO^- \rightleftharpoons CH_3NH_2 + H_2O$

d. $BF_3 + HO^- \rightleftharpoons HO-\overset{-}{B}F_3$

e. $\overset{+}{C}H_3 + HO^- \rightleftharpoons CH_3OH$

f. $FeBr_3 + HO^- \rightleftharpoons HO-\bar{F}eBr_3$

g. $AlCl_3 + HO^- \rightleftharpoons HO-\bar{A}lCl_3$

h. $CH_3COOH + HO^- \rightleftharpoons CH_3COO^- + H_2O$

68.

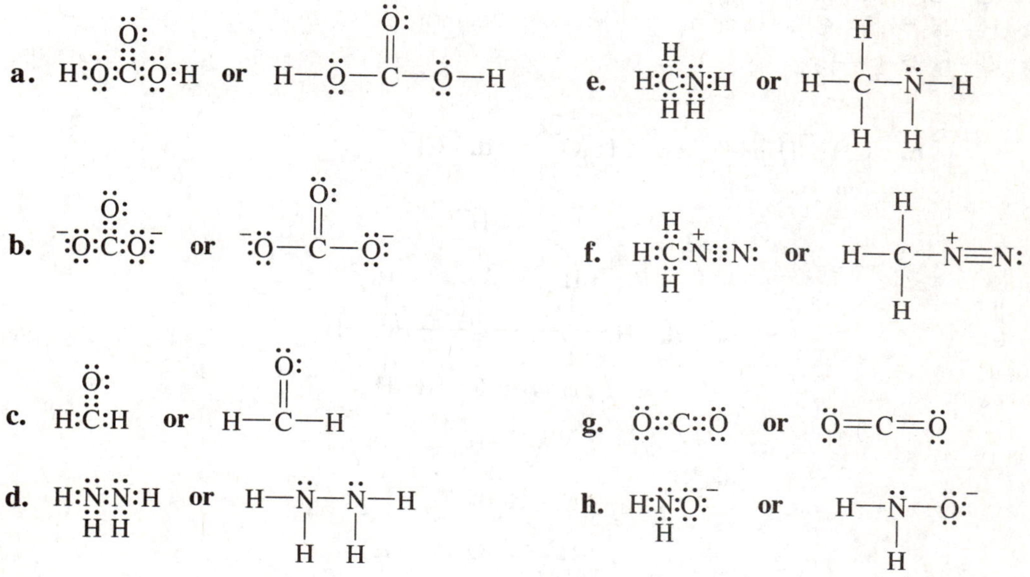

69.

a. sp^3, tetrahedral
b. sp^2, trigonal planar
c. sp^3, tetrahedral
d. sp^2, trigonal planar
e. sp^3, tetrahedral
f. sp^2, trigonal planar
g. sp, linear
h. sp^3, tetrahedral
i. sp^3, tetrahedral

70. a. $CH_3CH_2CH_3$ b. $CH_3CH{=}CH_2$ c. $CH_3C{\equiv}CCH_3$ or $CH_3CH_2C{\equiv}CH$

71. The hybridization of the central atom determines the bond angle. If it is sp^3, the number of lone pairs on the central atom determines the bond angle.

a. 107.3°
b. 107.3°
c. 109.5°
d. 104.5°*
e. 104.5°
f. 120°(sp^2)
g. 109.5°
h. 180°(sp)
i. 109.5°

*104.5° is the correct prediction based on the bond angle in water.
However, the bond angle is actually somewhat larger (111.7°) because the bond opens up to minimize the interaction between the electron clouds of the relatively bulky CH_3 groups.

72.

		1*s*	2*s*	2p_x	2p_y	2p_z	3*s*	3p_x	3p_y	3p_z	4*s*
a.	Ca	⇅	⇅	⇅	⇅	⇅	⇅	⇅	⇅	⇅	⇅
b.	Ca^{2+}	⇅	⇅	⇅	⇅	⇅	⇅	⇅	⇅	⇅	
c.	Ar	⇅	⇅	⇅	⇅	⇅	⇅	⇅	⇅	⇅	
d.	Mg^{2+}	⇅	⇅	⇅	⇅	⇅					

73.
a. C—F > C—O > C—N
b. C—Cl > C—Br > C—I
c. H—O > H—N > H—C
d. C—N > C—H > C—C

74. **a.** HO^- **b.** $CH_3\bar{N}H$ **c.** CH_3O^- **d.** Cl^-

75. **a.** 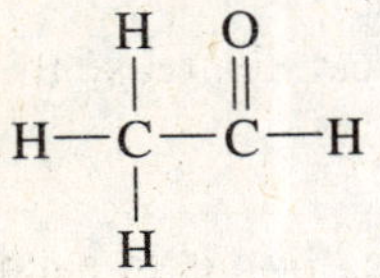

d. 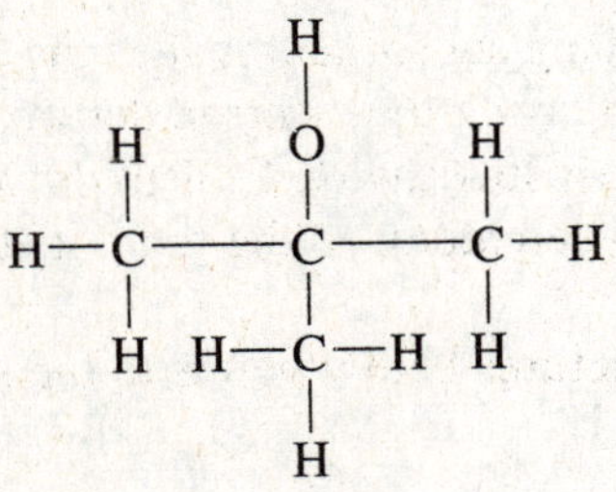

b. 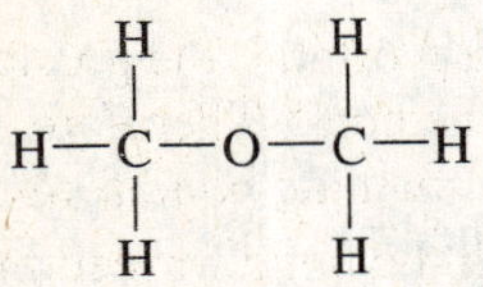

e. 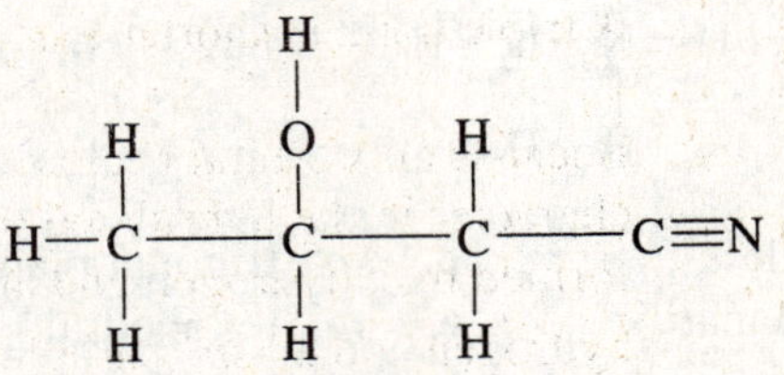

c. 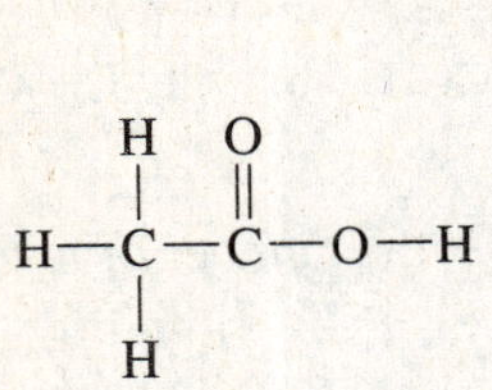

f. 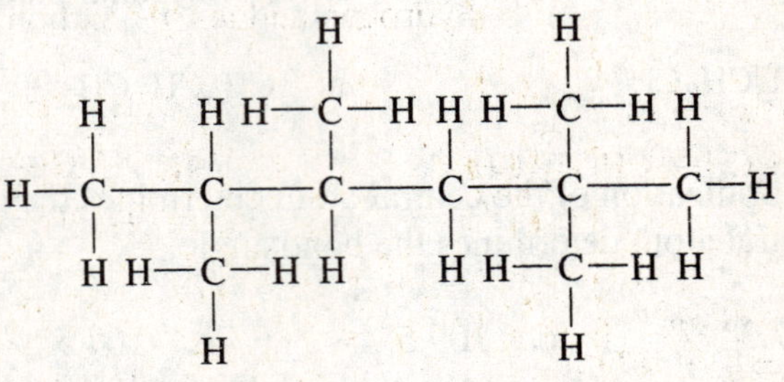

76. **a.** CH_3-Br (→) **c.** $HO-NH_2$ (←) **e.** CH_3-OH (→)

b. CH_3-Li (←) **d.** $I-Br$ (→) **f.** $(CH_3)_2N-H$ (←)

77. **a.** $CH_3CH{=}CH_2$ (sp^2 ⇓ at CH) **c.** CH_3CH_2OH (sp^3 ⇓ at CH₂) **e.** $CH_3CH{=}NCH_3$ (sp^2 ⇓ at N)

b. 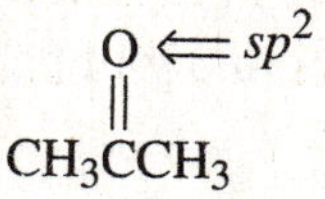

d. 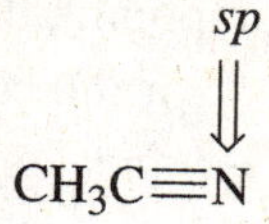

f. 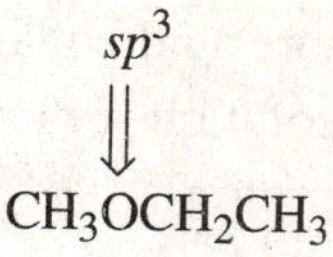

78. formal charge = number of valence electrons
– (number of lone-pair electrons + ½ the number of bonding electrons)

a. 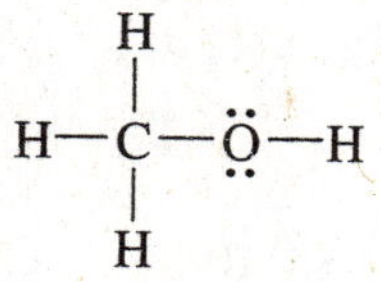

formal charge
6 – (4 + 2) = 0

b. 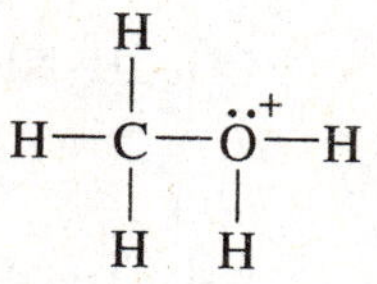

formal charge
6 – (2 + 3) = +1

c. 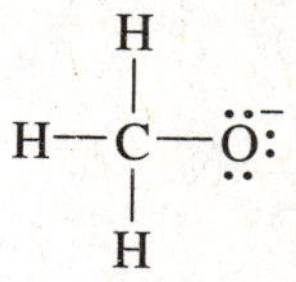

formal charge
6 – (6 + 1) = –1

d. 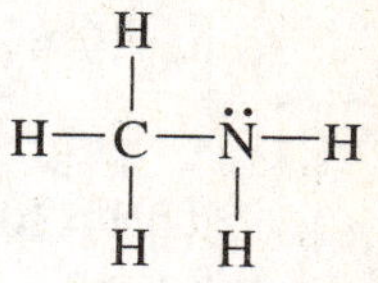

formal charge
5 – (2 + 3) = 0

79. **a.** $CH_3CH_2\underset{Cl}{\underset{|}{C}H}COOH > CH_3\underset{Cl}{\underset{|}{C}H}CH_2COOH > ClCH_2CH_2CH_2COOH > CH_3CH_2CH_2COOH$

b. The electron-withdrawing substituent makes the carboxylic acid more acidic, because it stabilizes its conjugate base by decreasing the electron density around the oxygen atom.

c. The closer the electron-withdrawing Cl is to the acidic proton, the more it can decrease the electron density around the oxygen atom, so the more it stabilizes the conjugate base.

80. The open arrow in the structures below points to the shorter of the two indicated bonds in each compound.

For **1**, **2**, and **3**: A triple bond is shorter than a double bond, which is shorter than a single bond.

For **4** and **5**: Because an *s* orbital is closer to the nucleus than a *p* orbital, the greater the *s* character in the hybridized orbital, the shorter is the bond. Therefore, the bond formed by a hydrogen and an *sp* carbon is shorter than the bond formed by a hydrogen and an sp^2 carbon, which is shorter than the bond formed by a hydrogen and an sp^3 carbon. (See Table 1.7 on page 41 of the text.)

1. 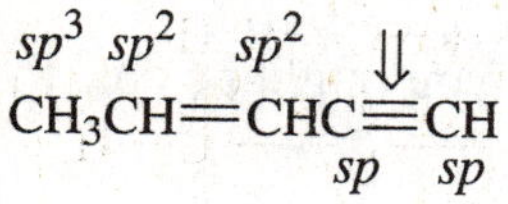

4. 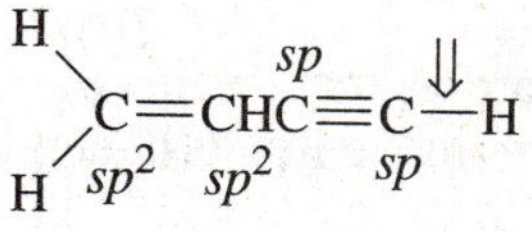

2. 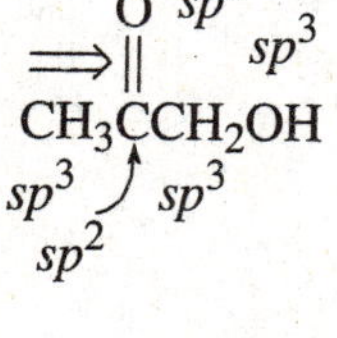

5. 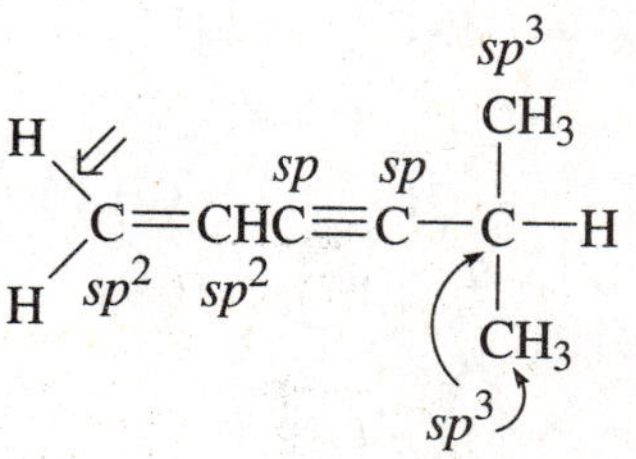

3. 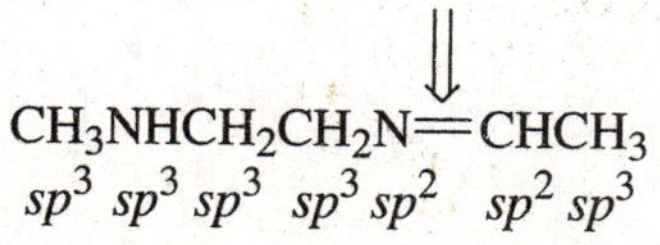

81. If the solution is more acidic than the pK_a of the compound, the compound will be in its acidic form (with the proton). If the solution is more basic than the pK_a of the compound, the compound will be in its basic form (without the proton).

a. at pH = 3	CH_3COH (C=O)	**b.** at pH = 3	$CH_3CH_2\overset{+}{N}H_3$	**c.** at pH = 3	CF_3CH_2OH
at pH = 6	CH_3CO^- (C=O)	at pH = 6	$CH_3CH_2\overset{+}{N}H_3$	at pH = 6	CF_3CH_2OH
at pH = 10	CH_3CO^- (C=O)	at pH = 10	$CH_3CH_2\overset{+}{N}H_3$	at pH = 10	CF_3CH_2OH
at pH = 14	CH_3CO^- (C=O)	at pH = 14	$CH_3CH_2NH_2$	at pH = 14	$CF_3CH_2O^-$

82. **a.** $CCl_3CH_2OH > CHCl_2CH_2OH > CH_2ClCH_2OH$

b. The greater the number of electron-withdrawing chlorine atoms equidistant from the OH group, the stronger the acid.

83. In an alkene, six atoms are in the same plane: the two sp^2 hybridized carbons and the two atoms that are bonded to each of the two sp^2 hybridized carbons. The other atoms in the molecule will not be in the same plane with these six atoms.

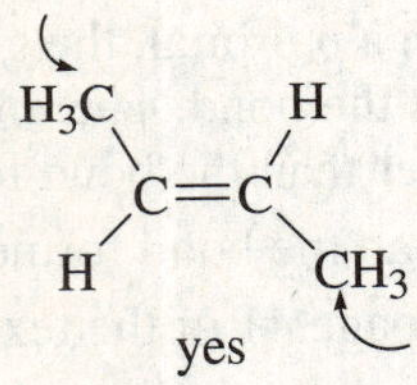

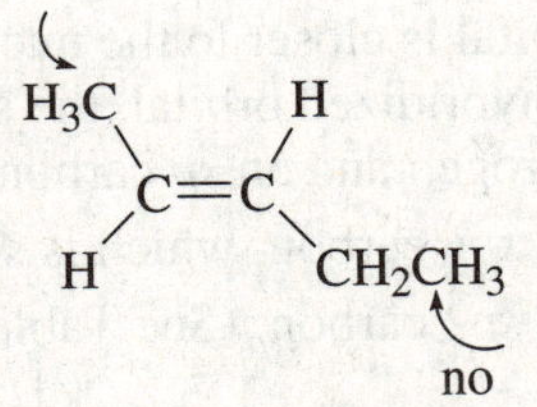

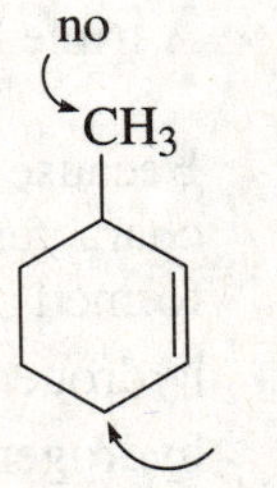

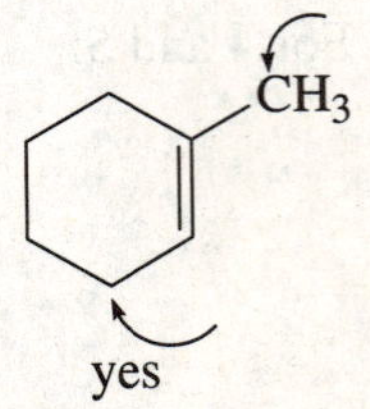

If you put stars next to the six atoms that lie in a plane in each molecule, you might be able to see more clearly whether the indicated atoms lie in the same plane.

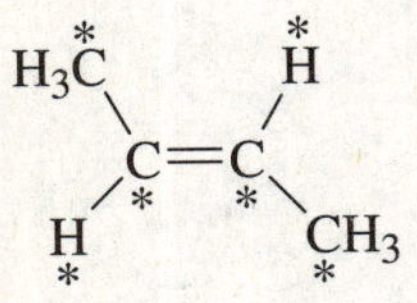

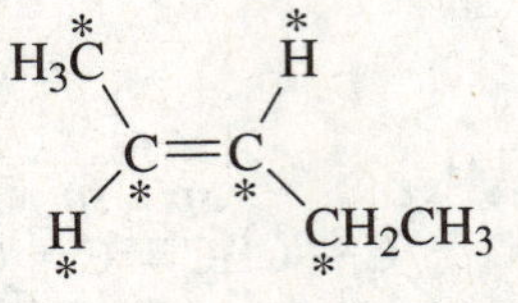

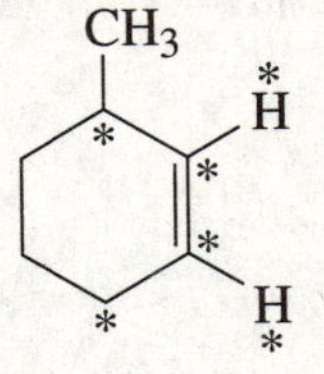

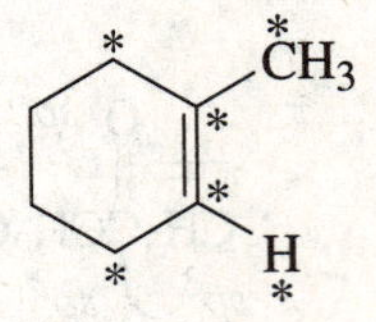

84. **a.** $CH_3COOH + CH_3O^- \rightleftharpoons CH_3COO^- + CH_3OH$

b. $CH_3CH_2OH + {}^-NH_2 \rightleftharpoons CH_3CH_2O^- + NH_3$

c. $CH_3COOH + CH_3NH_2 \rightleftharpoons CH_3COO^- + CH_3\overset{+}{N}H_3$

d. $CH_3CH_2OH + HCl \rightleftharpoons CH_3CH_2\overset{+}{O}H_2 + Cl^-$

85. If the central atom is *sp*^{3} hybridized and it does not have a lone pair, the molecule will have tetrahedral bond angles (109.5°). Therefore, only $\overset{+}{N}H_4$ has tetrahedral bond angles. The following species are close to being perfectly tetrahedral: H_2O, H_3O^+, NH_3, $^-CH_3$.

86.

a.

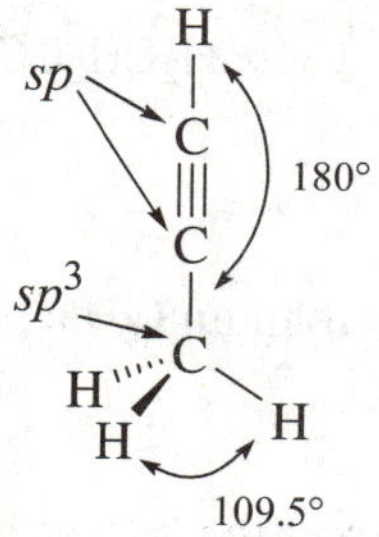

b.

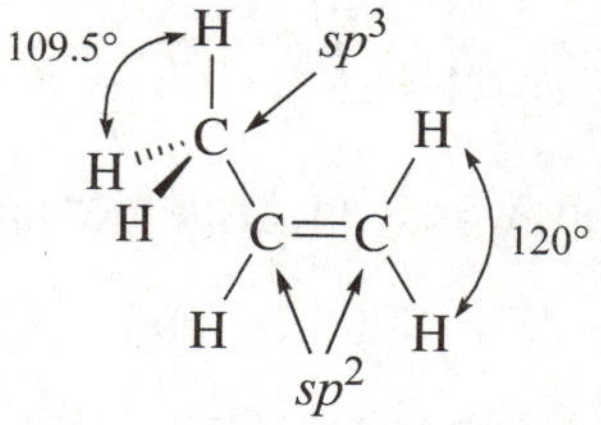

c.

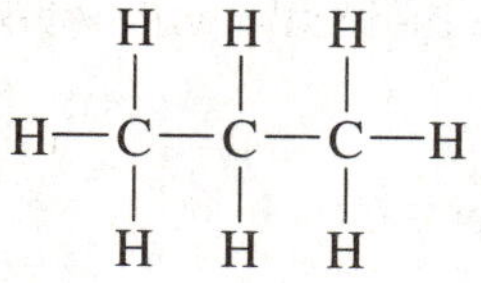

d.

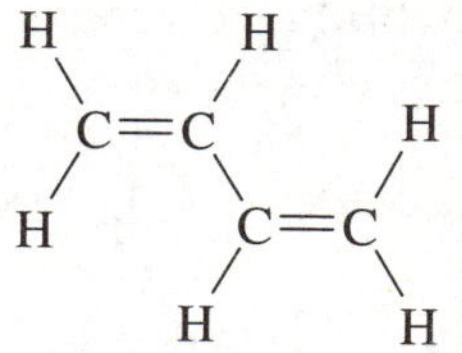

The 3 carbons are all *sp*^{3} hybridized.
All the bond angles are 109.5°.

The 4 carbons are all *sp*^{2} hybridized.
All the bond angles are 120°.

87. The log of $10^{-4} = -4$, the log of $10^{-5} = -5$ the log of $10^{-6} = -6$ etc.
Because the $pK_a = -\log K_a$ the pK_a of an acid with a K_a of 10^{-4} = is $-(-4) = 4$.
An acid with a K_a of 4.0×10^{-4} is a stronger acid than one with a K_a of 1.0×10^{-4}.
Therefore, the pK_a can be estimated as being between 3 and 4.

a.
1. between 3 and 4
2. between –2 and –1
3. between 10 and 11
4. between 9 and 10
5. between 3 and 4

b.
1. $pKa = 3.4$
2. $pKa = -1.3$
3. $pKa = 10.2$
4. $pKa = 9.1$
5. $pKa = 3.7$

c. Because the lower the pK_a the stronger the acid, nitric acid (HNO_3) is the strongest acid.

88. The nitrogen in the upper left-hand corner is slightly red, indicating that it is more basic than the other ring nitrogen. Delocalization of the lone-pair electrons of the bottom nitrogen causes the left-hand nitrogen to have a greater concentration of negative charge (more basic) and the bottom nitrogen to have less negative charge (less basic). The nitrogen that is not in the ring is the most basic of the three nitrogens; that why it is the one that is most apt to be protonated. Protonating it puts a positive formal charge on the nitrogen, as indicated by the blue area.

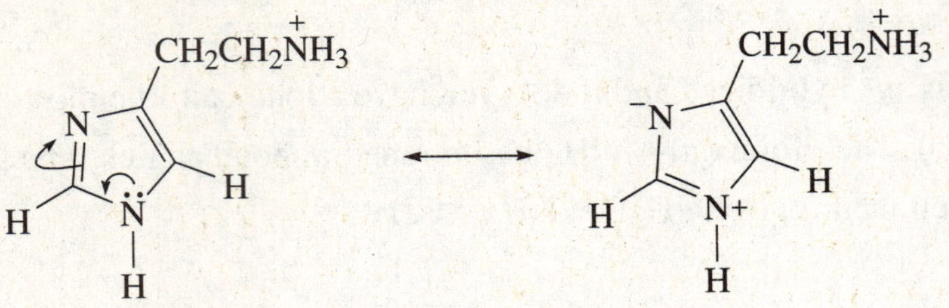

89. **a.** $CH_3CH_2CH(Cl)COOH$ > $CH_3CH_2CH(Br)COOH$ > $CH_3CH_2CH(OH)COOH$ > $CH_3CH_2CH_2COOH$

b. The relative acidities show that a Cl is a more electronegative substituent than an OH.

90.

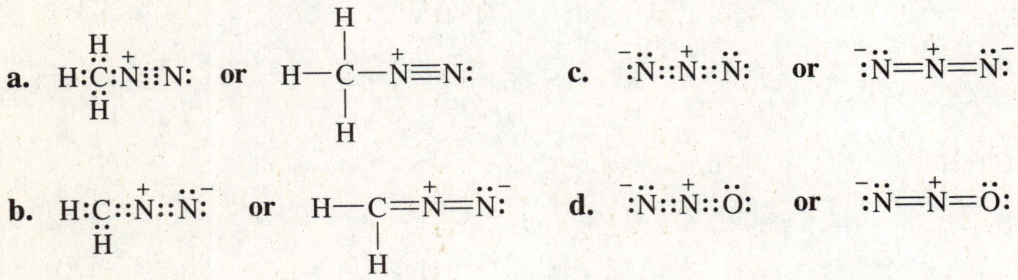

91. The reaction with the most favorable equilibrium constant is the one with the greatest difference between the strengths of the reactant acid and the product acid. (that is, the one that has the strongest reactant acid and the weakest product acid).

a. **1.** CH_3OH is a stronger reactant acid ($pK_a = 15.5$) than CH_3CH_2OH ($pK_a = 15.9$), and both reactions form the same product acid ($^+NH_4$). Therefore, the reaction of CH_3OH with NH_3 has the more favorable equilibrium constant.

2. Both reactions have the same reactant acid (CH_3CH_2OH). The product acids are different: $^+NH_4$ is a stronger product acid ($pK_a = 9.4$) than $CH_3NH_3^+$ ($pK_a = 10.7$). Therefore, the reaction of CH_3CH_2OH with CH_3NH_2 has the more favorable equilibrium constant.

b. Now we have to compare "apples" and "oranges" because the reaction with the most favorable equilibrium constant in **1** and the reaction with the most favorable quilibrium constant in **2** do not have any species in common: **1** has the stronger reactant acid, whereas **2** has the weaker product acid. Therefore, the equilibrium constants have to be calculated.

The reaction of CH_3OH with NH_3 has an equilibrium constant $= 7.9 \times 10^{-7}$

$$(K_{eq} = 10^{-15.5} / 10^{-9.4} = 10^{-6.1} = 7.9 \times 10^{-7}).$$

The reaction of CH_3CH_2OH with CH_3NH_2 has an equilibrium constant $= 6.3 \times 10^{-6}$

$$(K_{eq} = 10^{-15.9} / 10^{-10.7} = 10^{-5.2} = 6.3 \times 10^{-6}).$$

Therefore, the reaction of CH_3CH_2OH with CH_3NH_2 has the greatest equilibrium constant of the four reactions.

92.

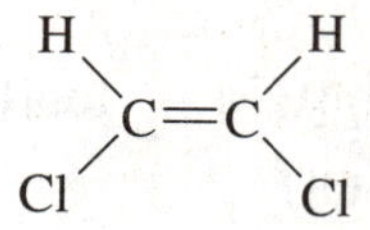

The dipole moment is 2.95 D because the two Cl's are withdrawing electrons in the same direction.

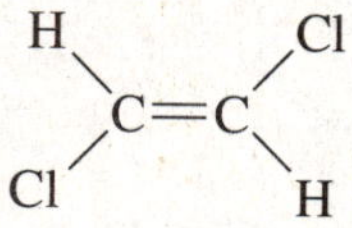

The dipole moment is 0 D because the two Cl's are withdrawing electrons in opposite direction.

93. The bond angle of the triple bonded carbons and the atoms adjacent to them is 180°. A 180° angle cannot fit into the ring structure. Therefore, the overlap between the *sp* orbital and the adjacent sp^3 orbital is poor (see Figure 2.6 on page 105 of the text), which causes the compound to be unstable.

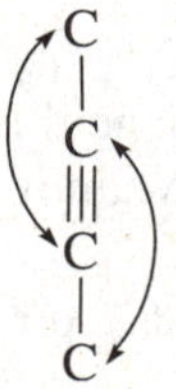

94.

$$K_a = \frac{[H^+][HO^-]}{[H_2O]}$$

$$K_a = \frac{(1 \times 10^{-7})(1 \times 10^{-7})}{55.5}$$

$$K_a = 1.80 \times 10^{-16}$$

$$pK_a = -\log 1.80 \times 10^{-16}$$

$$pK_a = 15.7$$

The answer can also be obtained in the following way:

$$K_a = \frac{[H^+][HO^-]}{[H_2O]}$$

$$K_a[H_2O] = [H^+][HO^-]$$

take the log of both sides

$$\log K_a + \log[H_2O] = \log[H^+] + \log[HO^-]$$

multiply both sides by -1

$$-\log K_a - \log[H_2O] = -\log[H^+] - \log[HO^-]$$

$$pK_a - \log[H_2O] = pH + pOH$$

$$pK_a - \log[H_2O] = 14$$

$$pK_a = 14 + \log[H_2O]$$

$$pK_a = 14 + \log 55.5$$

$$pK_a = 14 + 1.7$$

$$pK_a = 15.7$$

95. From the following equilibria you can see that a carboxylic acid is neutral when it is in its acidic form and charged when it is in its basic form. An amine is charged when it is in its acidic form and neutral when it is in its basic form.

$$RCOOH \rightleftharpoons RCOO^- + H^+$$

$$R\overset{+}{N}H_3 \rightleftharpoons RNH_2 + H^+$$

Charged species will dissolve in water and neutral species will dissolve in ether.
In separating compounds you want essentially all (100:1) of each compound in either its acidic form or its basic form.

From the Henderson-Hasselbalch equation it can be calculated that in order to obtain a 100:1 ratio of acidic form:basic form, the pH must be two pH units lower than the pK_a of the compound; and in order to obtain a 100:1 ratio of basic form:acidic form, the pH must be two pH units greater than the pK_a of the compound.

a. If both compounds are to dissolve in water, they both must be charged. Therefore, the carboxylic acid must be in its basic form, and the amine must be in its acidic form. To accomplish this, the pH will have to be at least two pH units greater than the pK_a of the carboxylic acid and at least two pH units less than the pK_a of the ammonium ion. In other words, it must be between pH 6.8 and pH 8.7.

b. For the carboxylic acid to dissolve in water, it must be charged (in its basic form), so the pH will have to be greater than 6.8. For the amine to dissolve in ether, it will have to be neutral (in its basic form), so the pH will have to be greater than 12.7 to have essentially all of it in the neutral form. Therefore, the pH of the water layer must be greater than 12.7.

c. To dissolve in ether, the carboxylic acid will have to be neutral, so the pH will have to be less than 2.8 to have essentially all the carboxylic acid in the acidic (neutral) form. To dissolve in water, the amine will have to be charged, so the pH will have to be less than 8.7 to have essentially all the amine in the acidic form. Therefore, the pH of the water layer must be less than 2.8.

96. Charged compounds will dissolve in water and uncharged compounds will dissolve in ether. The acidic forms of carboxylic acids and alcohols are neutral and the basic forms are charged. The acidic forms of amines are charged and the basic forms are neutral.

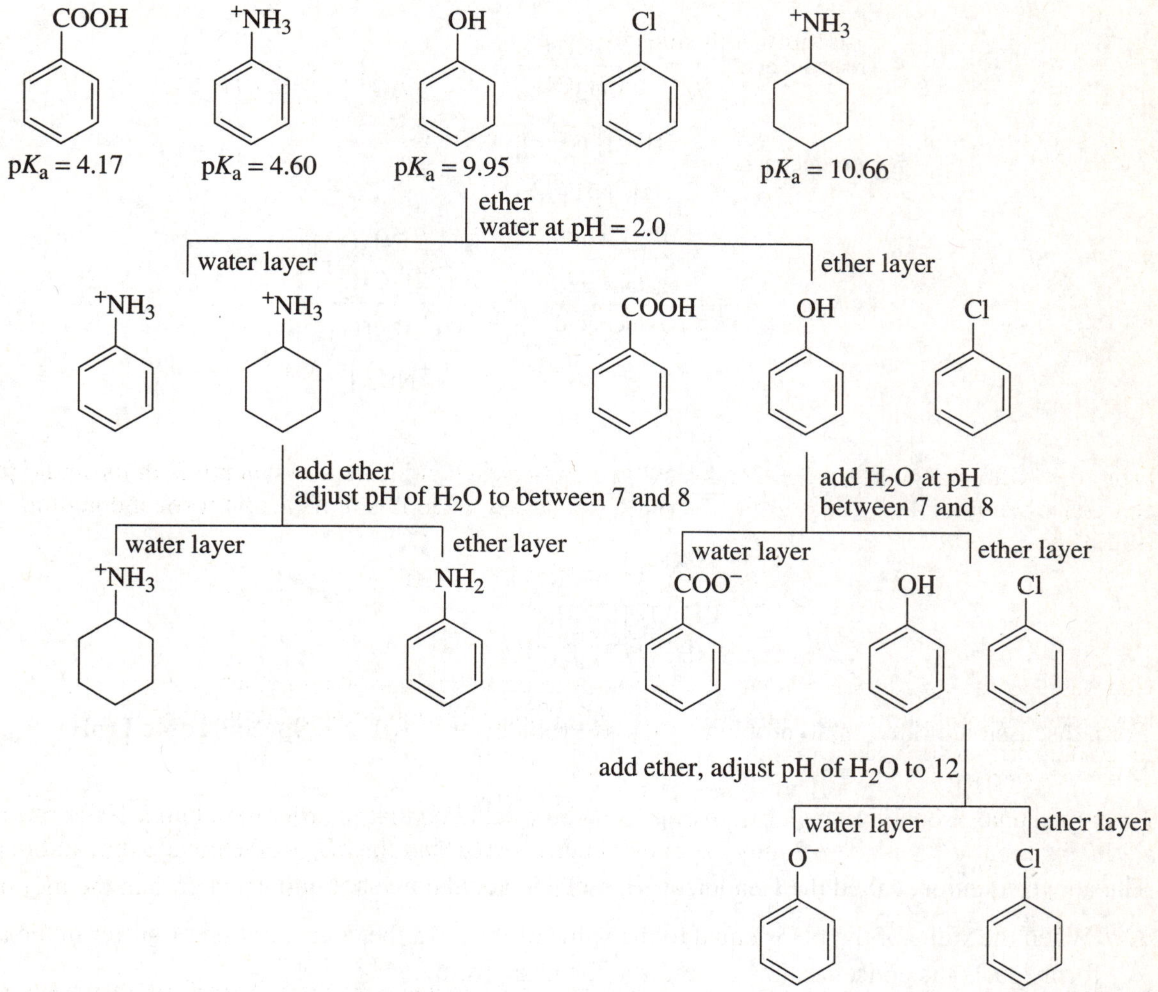

97. If light is shone on a molecule of H_2 or Br_2, one of the two electrons in a bonding MO can be promoted to an empty antibonding MO. The electron in the antibonding MO will cancel out the electron that is left in the bonding MO, so overall there will be no bonding. In other words, H_2 or Br_2 will have been broken up into hydrogen atoms or bromine atoms. The energy difference between the bonding and antibonding MO's is greater in the case of H_2 than in the case of Br_2. Therefore, less energy is required to break Br_2 into bromine radicals.

98. Using the following reaction as an example:

$$CH_3OH + NH_3 \rightleftharpoons CH_3O^- + \overset{+}{N}H_4$$

$$K_{eq} = \frac{[\text{products}]}{[\text{reactants}]}$$

$$K_{eq} = \frac{[CH_3O^-][^+NH_4]}{[CH_3OH][NH_3]}$$

$$K_a \text{ reactant acid} = \frac{[H^+][CH_3O^-]}{[CH_3OH]}$$

$$K_a \text{ product acid} = \frac{[H^+][NH_3]}{[^+NH_4]}$$

$$K_{eq} = \frac{K_a \text{ reactant acid}}{K_a \text{ product acid}} = \frac{\dfrac{[H^+][CH_3O^-]}{[CH_3OH]}}{\dfrac{[H^+][NH_3]}{[^+NH_4]}}$$

$$= \frac{\dfrac{[CH_3O^-]}{[CH_3OH]}}{\dfrac{[NH_3]}{[^+NH_4]}}$$

$$= \frac{[CH_3O^-][^+NH_4]}{[NH_3][CH_3OH]}$$

99. For a discussion of how to do problems such as Problems 99 - 101, see **Special Topic I** (pH, pK_a, and Buffers).

$$pK_a = pH + \log\frac{[HA]}{[A^-]}$$

The above equation, called the Henderson-Hasselbalch equation, shows that:

1. When the value of the pH is equal to the value of the pK_a, the concentration of buffer in the acidic form equals the concentration of buffer in the basic form.
2. When the solution is more acidic than the pK_a, more buffer species is in the acidic form than in the basic form
3. When the solution is more basic than the pK_a, more buffer species is in the basic form than in the acidic form.

$$pK_a = pH + \log\frac{[HA]}{[A^-]}$$

Because the pH of the blood (~7.4) is greater than the pK_a of the buffer (6.1), more buffer species is in the basic form than in the acidic form. Therefore, the buffer is better at neutralizing excess acid.

100. a.

$$\text{fraction present in the acidic from} = \frac{\text{amount in the acidic form}}{\text{amount in the acidic form} + \text{amount in the basic form}}$$

$$= \frac{[HA]}{[HA]+[A^-]}$$

Because there are two unknowns, we must define one in terms of the other.
By using the definition of the acid dissociation constant, we can determine [A⁻] in terms of

$$K_a = \frac{[H^+][A^-]}{[HA]}$$

$$[A^-] = \frac{K_a[HA]}{[H^+]}$$

Substituting the value of [A⁻] into the equation gives the fraction that is present in the acidic form

$$\frac{[HA]}{[HA]+[A^-]} = \frac{[HA]}{[HA]+\frac{K_a[HA]}{[H^+]}} = \frac{1}{1+\frac{K_a}{[H^+]}} = \frac{[H^+]}{[H^+]+K_a}$$

Therefore, the percent that is present in the acidic form is given by:

$$\frac{[H^+]}{[H^+]+K_a} \times 100$$

Because the pK_a of the acid is given as 5.3, we know that K_a is 5.0×10^{-6} (p$K_a = -\log K_a$).

Because the pH of the solution is given as 5.7, we know that $[H^+]$ is 2.0×10^{-6} ($\text{pH} = -\log[H^+]$).
Substituting into the equation that gives the percent that is present in the acidic form:

$$\frac{2.0 \times 10^{-6}}{2.0 \times 10^{-6} + 5.0 \times 10^{-6}} \times 100$$

$$\frac{2.0 \times 10^{-6}}{7.0 \times 10^{-6}} \times 100 = 29\%$$

b. percent present in the acidic form $= \frac{[H^+]}{[H^+]+K_a} = .80$

$$[H^+] = .80\left([H^+]+K_a\right)$$
$$[H^+] = .80\,[H^+]+.80\,K_a$$
$$.20\,[H^+] = .80\,K_a$$
$$[H^+] = 4\,K_a$$
$$[H^+] = 4\times 5.0\times 10^{-6}$$
$$[H^+] = 20\times 10^{-6}$$
$$pH = 4.7$$

101. **a.**

$$K_a = \frac{[H^+][A^-]}{[HA]}$$
$$1.74\times 10^{-5} = \frac{x^2}{1.0-x}$$
$$1.74\times 10^{-5} = x^2$$
$$x = 4.16\times 10^{-3}$$
$$pH = 2.38$$

b.

$$K_a = \frac{[H^+][A^-]}{[HA]}$$
$$2.00\times 10^{-11} = \frac{x^2}{0.1-x}$$
$$2.00\times 10^{-12} = x^2$$
$$x = 1.41\times 10^{-6}$$
$$pH = 5.85$$

c. This question can be answered by plugging the numbers into the Henderson-Hasselbalch equation.

$$pK_a = pH + \log\frac{[acid]}{[base]}$$
$$3.76 = pH + \frac{0.3}{0.1}$$
$$3.76 = pH + \log 3$$
$$3.76 = pH + 0.48$$
$$pH = 3.76 - 0.48 = 3.28$$

Chapter 1 Practice Test

1. Answer the following:

a. Which is a stronger acid, HCl or HBr?

b. Which is a stronger base, NH_3 or H_2O?

c. Which bond has a greater dipole moment, a carbon-oxygen bond or a carbon-fluorine bond?

d. Which has a dipole moment of zero, $CHCl_3$ or CCl_4?

2. What is the hybridization of the carbon atom in each of the following compounds?

$^{+}CH_3$ $\quad$ $^{-}CH_3$ $\quad$ $\cdot CH_3$

3. Draw the Lewis structure for HCO_3^-.

4. The following compounds are drawn in their acidic forms, and their pK_a's are given. Draw the form in which each compound would predominantly exist at pH = 8.

CH_3COOH	CH_3CH_2OH	$CH_3\overset{+}{O}H_2$	$CH_3CH_2\overset{+}{N}H_3$
pK_a = 4.8	pK_a = 15.9	pK_a = −2.5	pK_a = 11.2

5. Which compound has greater bond angles, H_3O^+ or $^{+}NH_4$?

6. What is the conjugate base of NH_3?

7. Give the structure of a compound that contains five carbons, two of which are sp^2 hybridized and three of which are sp^3 hybridized.

8. **a.** What products would be formed from the following reaction?

$$CH_3OH + \overset{+}{N}H_4 \rightleftharpoons$$

b. Does the reaction favor reactants or products?

9. **a.** What orbitals do carbon's electrons occupy before promotion?

b. What orbitals do carbon's electrons occupy after promotion?

10. Which of the following compounds is a stronger acid?

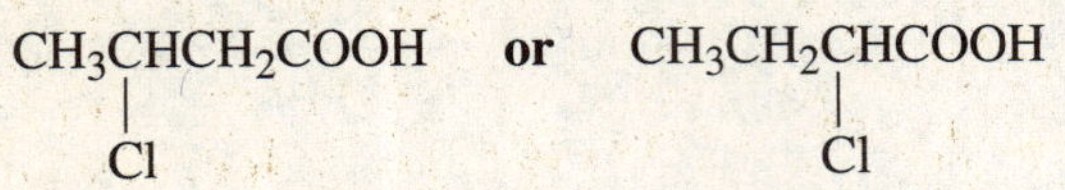

11. For each of the following compounds indicate the hybridization of the atom to which the arrow is pointing:

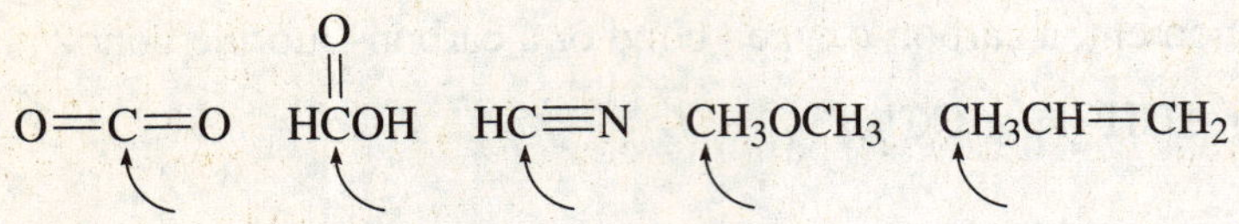

12. Indicate whether each of the following statements is true or false:

a. A pi bond is stronger than a sigma bond. T F

b. A triple bond is shorter than a double bond. T F

c. The oxygen-hydrogen bonds in water are formed by the overlap of an sp^2 orbital of oxygen with an *s* orbital of hydrogen. T F

d. HO^- is a stronger base than $^-NH_2$. T F

e. A double bond is stronger than a single bond. T F

f. A tetrahedral carbon has bond angles of 107.5°. T F

g. A Lewis acid is a compound that accepts a share in a pair of electrons. T F

h. CH_3CH_3 is more acidic than $H_2C{=}CH_2$. T F

ANSWERS TO ALL THE PRACTICE TESTS CAN BE FOUND AT THE END OF THE SOLUTIONS MANUAL

SPECIAL TOPIC I

pH, pK_a, and Buffers

This is a continuation of the discussion on acids and bases found in Sections 1.16 - 1.26 on pages 44 – 65 of the text. Now we will see how the pH of solutions of acids and bases can be calculated. We will look at three different kinds of solutions.

1. A solution made by dissolving a strong acid or a strong base in water.
2. A solution made by dissolving a weak acid or a weak base in water.
3. A solution made by dissolving a weak acid and its conjugate base in water. Such a solution is known as a **buffer solution**.

Before we start, we need to review a few terms.

An acid is a compound that donates a proton, and a base is a compound that accepts a proton.

The degree to which an acid (HA) dissociates is described by its acid dissociation constant (K_a).

$$HA \rightleftharpoons H^+ + A^-$$

$$K_a = \frac{[H^+][A^-]}{[HA]}$$

The strength of an acid can be indicated by its acid dissociation constant or by its pK_a value.

$$\mathbf{p}K_a = -\mathbf{\log} K_a$$

The stronger the acid, the **larger** its dissociation constant and the **smaller** its pK_a value. For example, an acid with a dissociation constant of 1×10^{-2} (pK_a = 2) is stronger than an acid with a dissociation constant of 1×10^{-4} (pK_a = 4).

While the pK_a scale is used to describe the strength of an acid, the pH scale is used to describe the acidity of a solution. In other words, the pH scale describes the concentration of hydrogen ions in a solution.

$$pH = -\log [H^+]$$

The smaller the pH, the more acidic the solution:
acidic solutions have pH values < 7; a neutral solution has a pH = 7; basic solutions have pH values > 7.

A solution with a pH = 2 is more acidic than a solution with a pH = 4.
A solution with a pH = 12 is more basic than a solution with a pH = 8.

Determining the pH of a Solution

To determine the pH of a solution, the concentration of hydrogen ion [H^+] in the solution must be determined.

Strong Acids

A strong acid is one that dissociates completely in solution. Strong acids have pK_a values < 1.

Because a strong acid dissociates completely, the concentration of hydrogen ions is the same as the concentration of the acid: a 1.0 M HCl solution contains 1.0 M [H^+]; a 1.5 M HCl solution contains 1.5 M [H^+]. Therefore, to determine the pH of a strong acid, the [H^+] value does not have to be calculated; it is the same as the molarity of the strong acid.

solution	[H^+]	pH
1.0 M HCl	1.0 M	0
1.0×10^{-2} M HCl	1.0×10^{-2} M	2.0
6.4×10^{-4} M HCl	6.4×10^{-4} M	3.2

Strong Bases

Strong bases are compounds such as NaOH or KOH that dissociate completely in water.
Because they dissociate completely, the [HO^-] is the same as the molarity of the strong base.

The pOH scale describes the basicity of a solution. The smaller the pOH, the more basic the solution; just like the smaller the pH, the more acidic the solution.

$$\text{pOH} = -\log\,[\text{HO}^-]$$

[HO^-] and [H^+] are related by the ionization constant for water (K_w).

$$K_w = [\text{H}^+][\text{HO}^-] = 10^{-14}$$

$$\text{pH} + \text{pOH} = 14$$

solution	[HO^-]	pOH	pH
1.0 M NaOH	1.0 M	0	14.0 – 0 = 14.0
1.0×10^{-4} M NaOH	1.0×10^{-4} M	4.0	14.0 – 4.0 = 10.0
7.8×10^{-2} M NaOH	7.8×10^{-2} M	1.1	14.0 – 1.1 = 12.9

Weak Acids

A weak acid does not dissociate completely in solution. This means that $[H^+]$ must be calculated before the pH can be determined.

Acetic acid (CH_3COOH) is an example of a weak acid. It has an acid dissociation constant of 1.74×10^{-5} ($pK_a = 4.76$). The pH of a 1.00 M solution of acetic acid can be calculated as follows:

$$CH_3COOH \rightleftharpoons H^+ + CH_3COO^-$$

$$K_a = \frac{[H^+][CH_3COO^-]}{[CH_3COOH]}$$

Each molecule of acetic acid that dissociates forms one proton and one molecule of acetate ion. Thus the concentration of protons in solution equals the concentration of acetate ions. Each has a concentration that can be represented by x. The concentration of acetic acid, therefore, is whatever we started with minus x.

$$1.74 \times 10^{-5} = \frac{(x)(x)}{1.00 - x}$$

The denominator $(1.00 - x)$ can be simplified to 1.00 because 1.00 is much greater than x. (When we actually calculate the value of x, we see that it is 0.004. And $1.00 - 0.004 = 1.00$.)

$$1.74 \times 10^{-5} = \frac{x^2}{1.00}$$

$$x = 4.17 \times 10^{-3}$$

$$pH = -\log 4.17 \times 10^{-3}$$

$$pH = 2.38$$

Formic acid (HCOOH) has a pK_a value of 3.75. The pH of a 1.50 M solution of formic acid can be calculated as follows:

$$HCOOH \rightleftharpoons H^+ + HCOO^-$$

$$K_a = \frac{[H^+][HCOO^-]}{[HCOOH]}$$

A compound with a $pK_a = 3.75$ has an acid dissociation constant of 1.78×10^{-4}.

$$1.78 \times 10^{-4} = \frac{(x)(x)}{1.50 - x} = \frac{x^2}{1.50}$$
$$x^2 = 2.67 \times 10^{-4}$$
$$x = 1.63 \times 10^{-2}$$
$$\text{pH} = -\log 1.63 \times 10^{-2}$$
$$\text{pH} = 1.79$$

Weak Bases

When a weak base is dissolved in water, it accepts a proton from water, creating hydroxide ion.

Determining the concentration of hydroxide allows the pOH to be determined, and this in turn allows the pH to be determined.

The pH of a 1.20 M solution of sodium acetate can be calculated as follows:

$$CH_3COO^- + H_2O \rightleftharpoons CH_3COOH + HO^-$$

$$\frac{K_w}{K_a} = \frac{[HO^-][CH_3COOH]}{[CH_3COO^-]}$$
$$\frac{1.00 \times 10^{-14}}{1.74 \times 10^{-5}} = \frac{(x)(x)}{1.20 - x}$$
$$5.75 \times 10^{-10} = \frac{x^2}{1.20}$$
$$x^2 = 6.86 \times 10^{-10}$$
$$x = 2.62 \times 10^{-5}$$
$$\text{pOH} = -\log 2.62 \times 10^{-5}$$
$$\text{pOH} = 4.58$$
$$\text{pH} = 14.00 - 4.58$$
$$\text{pH} = 9.42$$

Notice that by setting up the equation equal to K_w/K_a, we can avoid the introduction of a new term (K_b).

Buffer Solutions

A buffer solution is a solution that maintains nearly constant pH in spite of the addition of small amounts of H^+ or HO^-. That is because a buffer solution contains both a weak acid and its conjugate base. The weak acid can donate a proton to any HO^- added to the solution, and the conjugate base can accept any H^+ that is added to the solution, so the addition of HO^- or H^+ does not significantly change the pH of the solution.
(In order to maintain approximately constant pH, the amount of H^+ or HO^- added to the solution cannot exceed the concentration of the conjugate acid or base in the solution.)

A buffer can maintain nearly constant pH in a range of one pH unit on either side of the pK_a of the conjugate acid. For example, an acetic acid/sodium acetate mixture can be used as a buffer in the pH range 3.76 – 5.76 because acetic acid has a $pK_a = 4.76$; methylammonium ion/methylamine can be used as a buffer in the pH range 9.7 – 11.7 because the methylammonium ion has a $pK_a = 10.7$.

The pH of a buffer solution can be determined from the Henderson-Hasselbalch equation. This equation comes directly from the expression defining the acid dissociation constant. Its derivation is found on page 60 of the text.

Henderson-Hasselbalch equation

$$pK_a = pH + \log\frac{[HA]}{[A^-]}$$

The pH of an acetic acid/sodium acetate buffer solution (pK_a of acetic acid $= 4.76$) that is 1.00 M in acetic acid and 0.50 M in sodium acetate is calculated as follows:

$$\begin{aligned} pK_a &= pH + \log\frac{[HA]}{[A^-]} \\ 4.76 &= pH + \log\frac{1.00}{0.50} \\ 4.76 &= pH + \log 2 \\ 4.76 &= pH + 0.30 \\ pH &= 4.46 \end{aligned}$$

Remember from Section 1.24 that compounds exist primarily in their acidic forms in solutions that are more acidic than their pK_a values and primarily in their basic forms in solutions that are more basic than their pK_a values. Therefore, it could have been predicted that the above solution will have a pH less than the pK_a of acetic acid because there is more conjugate acid than conjugate base present in the solution.

There are three ways a buffer solution can be prepared:

1. Weak Acid and Weak Base

A buffer solution can be prepared by mixing a solution of a weak acid with a solution of its conjugate base.

The pH of a formic acid/sodium formate buffer (pK_a of formic acid $= 3.75$) solution prepared by mixing 25 mL of 0.10 M formic acid and 15 mL of 0.20 M sodium formate is calculated as follows:

$$\text{molarity} = \frac{\text{moles}}{\text{liters}} = \frac{\text{millimoles}}{\text{milliliters}}$$

The number of millimoles (mmol) of each of the buffer components can be determine by multiplying the number of milliliters (mL) by the molarity (M).

$$25\text{ mL} \times 0.10\text{ M} = 2.5\text{ mmol formic acid}$$
$$15\text{ mL} \times 0.20\text{ M} = 3.0\text{ mmol sodium formate}$$

$$\begin{aligned} pK_a &= \text{pH} + \log\frac{[\text{HA}]}{[\text{A}^-]} \\ 3.75 &= \text{pH} + \log\frac{2.5}{3.0} \\ 3.75 &= \text{pH} + \log 0.83 \\ 3.75 &= \text{pH} - 0.08 \\ \text{pH} &= 3.83 \end{aligned}$$

It could have been predicted that the above solution would have a pH greater than the pK_a of formic acid because there is more conjugate base than conjugate acid present in the solution.

Notice that we use mmol rather than molarity (mmol/mL) because both the acid and the conjugate base are in the same solution so they have the same volume. Therefore, volumes will cancel in the equation.

2. Weak Acid and Strong Base

A buffer solution can be prepared by mixing a solution of a weak acid with a strong base such as NaOH. The NaOH reacts completely with the weak acid, thereby creating the conjugate base. For example, if 20 mmol of a weak acid and 5 mmol of a strong base are added to a solution, the 5 mmol of strong base will react with 5 mmol of weak acid, creating 5 mmol weak base and leaving behind 15 mmol of weak acid.

The pH of a solution prepared by mixing 10 mL of a 2.0 M solution of a weak acid with a pK_a of 5.86 with 5.0 mL of a 1.0 M solution of sodium hydroxide can be calculated as follows:

When the 20 mmol of HA and the 5.0 mmol of HO^- are mixed, the 5.0 mmol of strong base will react with 5.0 mmol of HA, with the result that 5.0 mmol of A^- will be formed and 15 mmol (20 mmol – 5.0 mmol) of HA will be left unreacted.

$$10\text{ mL} \times 2.0\text{ M} = 20\text{ mmol HA} \longrightarrow 15\text{ mmol HA}$$
$$5.0\text{ mL} \times 1.0\text{ M} = 5.0\text{ mmol HO}^- \longrightarrow 5.0\text{ mmol A}^-$$

$$\begin{aligned} pK_a &= \text{pH} + \log\frac{[\text{HA}]}{[\text{A}^-]} \\ 5.86 &= \text{pH} + \log\frac{15}{5} \\ 5.86 &= \text{pH} + \log 3 \\ 5.86 &= \text{pH} + 0.48 \\ \text{pH} &= 5.38 \end{aligned}$$

3. Weak Base and Strong Acid

A buffer solution can be prepared by mixing a solution of a weak base with a strong acid such as HCl. The strong acid will react completely with the weak base, thereby forming the conjugate acid. The pH of an ethylammonium ion/ethylamine buffer (pK_a of $CH_3CH_2NH_3^+ = 11.0$) prepared by mixing 30 mL of 0.20 M ethylamine with 40 mL of 0.10 M HCl can be calculated as follows:

$$30 \text{ mL} \times 0.20 \text{ M} = 6.0 \text{ mmol } RNH_2 \longrightarrow 2.0 \text{ mmol } RNH_2$$

$$40 \text{ mL} \times 0.10\text{M} = 4.0 \text{ mmol } H^+ \longrightarrow 4.0 \text{ mmol } R\overset{+}{N}H_3$$

$$pK_a = pH + \log \frac{[HA]}{[A^-]}$$

$$11.0 = pH + \log \frac{4.0}{2.0}$$

$$11.0 = pH + \log 2.0$$

$$11.0 = pH + \log 0.30$$

$$pH = 10.7$$

Fraction Present in Acidic or Basic Form

A common question to ask is what fraction of a buffer will be in a particular form; either what fraction will be in the acidic form or what fraction will be in the basic form. This is an easy question to answer if you remember the following formulas, which are derived after the following calculations:

$$\textbf{fraction present in the acidic form} = \frac{[H^+]}{K_a + [H^+]}$$

$$\textbf{fraction present in the basic form} = \frac{K_a}{K_a + [H^+]}$$

What fraction of an acetic acid/sodium acetate buffer (pK_a of acetic acid $= 4.76$) is present in the acidic form at pH = 5.20?

$$\frac{[H^+]}{K_a + [H^+]} = \frac{6.31 \times 10^{-6}}{1.74 \times 10^{-5} + 6.31 \times 10^{-6}}$$

$$= \frac{6.31 \times 10^{-6}}{17.4 \times 10^{-6} + 6.31 \times 10^{-6}}$$

$$= \frac{6.31 \times 10^{-6}}{23.7 \times 10^{-6}} = \frac{6.31}{23.7}$$

$$= 0.26$$

What fraction of a formic acid/sodium formate buffer (pK_a formic acid = 3.75) is present in the basic form at pH = 3.90?

$$\frac{K_a}{K_a + [H^+]} = \frac{1.78 \times 10^{-4}}{1.78 \times 10^{-4} + 1.26 \times 10^{-4}}$$

$$= \frac{1.78 \times 10^{-4}}{3.04 \times 10^{-4}} = \frac{1.78}{3.04}$$

$$= 0.586$$

$$= 0.59$$

The formulas describing the fraction present in the acidic or basic form are obtained from the definition of the acid dissociation constant.

$$\text{fraction present in the acidic form} = \frac{[HA]}{[HA]+[A^-]}$$

First we need to define [A⁻] in terms of [HA], so we can get rid of [A⁻].

$$K_a = \frac{[H^+][A^-]}{[HA]}$$

$$[A^-] = \frac{K_a[HA]}{[H^+]}$$

$$\text{fraction present in the acidic form} = \frac{[HA]}{[HA]+[A^-]} = \frac{[HA]}{[HA]+\frac{K_a[HA]}{[H^+]}} = \frac{1}{1+\frac{K_a}{[H^+]}}$$

$$= \frac{[H^+]}{K_a+[H^+]}$$

$$\text{fraction present in the basic form} = \frac{[A^-]}{[HA]+[A^-]}$$

First we need to define [HA] in terms of [A⁻], so we can get rid of [HA].

$$K_a = \frac{[H^+][A^-]}{[HA]}$$

$$[HA] = \frac{[H^+][A^-]}{K_a}$$

$$\text{fraction present in the basic form} = \frac{[A^-]}{[HA]+[A^-]} = \frac{[A^-]}{[A^-]+\dfrac{[H^+][A^-]}{K_a}}$$

$$= \frac{1}{1+\dfrac{[H^+]}{K_a}}$$

$$= \frac{K_a}{K_a+[H^+]}$$

Preparing Buffer Solutions

The type of calculations just shown can be used to determine how to make a buffer solution. For example, how can 100 mL of a 1.00 M buffer solution of pH = 4.24 be prepared if you have available to you 1.50 M solutions of acetic acid, sodium acetate, HCl, and NaOH?

acetic acid has a $pK_a = 4.76$

$$\text{fraction present in the acidic form at pH} = 4.24 = \frac{[H^+]}{K_a+[H^+]}$$

$$\frac{[H^+]}{K_a+[H^+]} = \frac{5.75\times 10^{-5}}{1.74\times10^{-5}+5.75\times10^{-5}}$$

$$= \frac{5.75\times10^{-5}}{7.49\times10^{-5}}$$

$$= 0.77$$

If a 1.00 M buffer solution is desired, the buffer must be 0.77 M in acetic acid and 0.23 M in sodium acetate.

Recalling that:

$$M = \frac{\text{moles}}{\text{liter}}$$

$$M = \frac{\text{millimoles}}{\text{milliliters}} = \frac{\text{mmol}}{\text{mL}}$$

There are three ways such a buffer solution can be prepared:

1. **By mixing the appropriate amounts of acetic acid and sodium acetate in water, and adding water to obtain a final volume of 100 mL.**

The amount of acetic acid needed:

$$[CH_3COOH] = 0.77\ \text{M}$$

$$\frac{x\ \text{mmol}}{100\ \text{mL}} = 0.77\ \text{M}$$

$$x = 77\ \text{mmol}$$

Therefore, we need to have 77 mmol of acetic acid in the final solution.

To obtain 77 mmol of acetic acid from a 1.50 M solution of acetic acid:

$$\frac{77\ \text{mmol}}{y\ \text{mL}} = 1.50\ \text{M}$$

$$y = 51.3\ \text{mL}$$

Notice that the formula M = mmol/mL was used twice. The first time it was used to determine the number of mmol of acetic acid that was needed in the final solution. The second time it was used to determine how that number of mmol could be obtained from an acetic acid solution of a known concentration.

The amount of sodium acetate needed:

$$[CH_3COO^-] = 0.23\ \text{M}$$

$$\frac{x\ \text{mmol}}{100\ \text{mL}} = 0.23$$

$$x = 23\ \text{mmol}$$

To obtain 23 mmol of sodium acetate from a 1.50 M solution of sodium acetate:

$$\frac{23\ \text{mmol}}{y\ \text{mL}} = 1.50\ \text{M}$$

$$y = 15.3\ \text{mL}$$

The desired buffer solution can be prepared using: 51.3 mL 1.50 M acetic acid

15.3 mL 1.50 M sodium acetate

33.4 mL H_20

2. By mixing the appropriate amounts of acetic acid and sodium hydroxide, and adding water to obtain a final volume of 100 mL.

Sodium hydroxide is used to convert some of the acetic acid into sodium acetate. This means that acetic acid will be the source of both acetic acid and sodium acetate.

The concentrations needed are: $[CH_3COOH] = 1.00\ M$

$$[NaOH] = 0.23\ M$$

The amount of acetic acid needed: $[CH_3COOH] = 1.00\ M$

$$\frac{x\ \text{mmol}}{100\ \text{mL}} = 1.00\ M$$

$$x = 100\ \text{mmol}$$

To obtain 100 mmol of acetic acid from a 1.50 M solution of acetic acid:

$$\frac{100\ \text{mmol}}{y\ \text{mL}} = 1.50\ M$$

$$y = 66.7\ \text{mL}$$

The amount of sodium hydroxide needed: $[NaOH] = 0.23\ M$

$$\frac{x\ \text{mmol}}{100\ \text{mL}} = 0.23\ M$$

$$x = 23\ \text{mmol}$$

To obtain 23 mmol of sodium hydroxide from a 1.50 M solution of NaOH:

$$\frac{23\ \text{mmol}}{y\ \text{mL}} = 1.50\ M$$

$$y = 15.3\ \text{mL}$$

The desired buffer solution can be prepared using: 66.7 mL 1.50 M acetic acid

15.3 mL 1.50 M NaOH

18.0 mL H_2O

3. By mixing the appropriate amounts of sodium acetate and hydrochloric acid, and adding water to obtain a final volume of 100 mL.

Hydrochloric acid is used to convert some of the sodium acetate into acetic acid.

This means that sodium acetate will be the source of both acetic acid and sodium acetate.

The concentrations needed are: $[CH_3COONa] = 1.00\ M$

$[HCl] = 0.77\ M$

The amount of sodium acetate needed: $[CH_3COONa] = 1.00\ M$

$$\frac{x\ \text{mmol}}{100\ \text{mL}} = 1.00\ M$$

$$x = 100\ \text{mmol}$$

To obtain 100 mmol of sodium acetate from a 1.50 M solution of sodium acetate:

$$\frac{100\ \text{mmol}}{y\ \text{mL}} = 1.5\ M$$

$$y = 66.7\ \text{mL}$$

The amount of hydrochloric acid needed:

$$[HCl] = 0.77\ M$$

$$\frac{x\ \text{mmol}}{100\ \text{mL}} = 0.77\ M$$

$$x = 77\ \text{mmol}$$

To obtain 77 mmol of hydrochloric acid from a 1.50 M solution of HCl:

$$\frac{77\ \text{mmol}}{y\ \text{mL}} = 1.50\ M$$

$$y = 51.3\ \text{mL}$$

100 mL of a 1.00 M acetic acid/acetate buffer cannot be made from these reagents, because the volumes needed (66.7 mL + 51.3 mL) add up to more than 100 mL. To make this buffer using sodium acetate and hydrochloric acid, you would need to use a more concentrated solution (>1.50 M) of sodium acetate or a more concentrated solution (>1.50 M) of HCl.

Problems on pH, pK_a, and Buffers

1. Calculate the pH of each of the following solutions.

a. 1×10^{-3} M HCl

b. 0.60 M HCl

c. 1.40×10^{-2} M HCl

d. 1×10^{-3} M KOH

e. 3.70×10^{-4} M NaOH

f. 1.20 M solution of an acid with a $pK_a = 4.23$

g. 1.60×10^{-2} M sodium acetate (pK_a of acetic acid $= 4.76$)

2. Calculate the pH of each of the following buffer solutions:

a. A buffer prepared by mixing 20 mL of 0.10 M formic acid and 15 mL of 0.50 M sodium formate (pK_a of formic acid $=$ 3.75).

b. A buffer prepared by mixing 10 mL of 0.50 M aniline and 15 mL of 0.10 M HCl (pK_a of the anilinium ion $=$ 4.60).

c. A buffer prepared by mixing 15 mL of 1.00 M acetic acid and 10 mL of 0.50 M NaOH (pK_a of acetic acid $=$ 4.76).

3. What fraction of a carboxylic acid with $pK_a = 5.23$ would be ionized at pH = 4.98?

4. What would be the concentration of formic acid and sodium formate in a 1.00 M buffer solution with a pH = 3.12?

5. You have found a bottle labeled 1.00 M RCOOH. You want to determine what carboxylic acid it is, so you decide to determine its pK_a value. How would you do this?

6. **a.** How would you prepare 100 mL of a buffer solution that is 0.30 M in acetic acid and 0.20 M in sodium acetate using a 1.00 M acetic acid solution and a 2.00 M sodium acetate solution?

b. The pK_a of acetic acid is 4.76. Would the pH of the above solution be greater or less than 4.76?

7. You have 100 mL of a 1.50 M acetic acid/sodium acetate buffer solution that has a pH = 4.90. How could you change the pH of the solution to 4.50?

8. You have 100 mL of a 1.00 M solution of an acid with a $pK_a = 5.62$ to which you add 10 mL of 1.00 M sodium hydroxide. What fraction of the acid will be in the acidic form? How much more sodium hydroxide will you need to add in order to have 40% of the acid in the acidic form?

9. Describe three ways to make a 1.00 M acetic acid/sodium acetate buffer solution with a pH = 4.00.

10. You have available to you 1.50 M solutions of acetic acid, sodium acetate, sodium hydroxide, and hydrochloric acid. How would you make 50 mL of each of the buffers described in the preceding problem?

11. How would you make a 1.0 M buffer solution with a pH = 3.30?

12. You are planning to carry out a reaction that will produce protons. In order for the reaction to take place at constant pH, it will be carried in a solution buffered at pH = 4.2. Would it be better to use a formic acid/formate buffer or an acetic acid/acetate buffer?

Answers to Problems on pH, p*Ka*, and Buffers

1. **a.** $pH = -\log 1 \times 10^{-3}$

$pH = 3$

d. $pOH = -\log 1 \times 10^{-3}$

$pOH = 3$

$pH = 14 - 3 = 11$

b. $pH = -\log 0.60$

$pH = 0.22$

e. $pOH = -\log 3.70 \times 10^{-4}$

$pOH = 3.43$

$pH = 10.57$

c. $pH = -\log 1.40 \times 10^{-2}$

$pH = 1.85$

f. $pK_a = 4.23, \quad K_a = 5.89 \times 10^{-5}$

$$K_a = \frac{[H^+][A^-]}{[HA]}$$

$$5.89 \times 10^{-5} = \frac{x^2}{1.20}$$

$$x^2 = 7.07 \times 10^{-5}$$

$$x = 8.41 \times 10^{-3}$$

$$pH = 2.08$$

g. $$\frac{K_w}{K_a} = \frac{[HO^-][HA]}{[A^-]}$$

$$\frac{1.0 \times 10^{-14}}{1.74 \times 10^{-5}} = \frac{x^2}{1.60 \times 10^{-2}} = \quad (K_a = 10^{-4.76} = 1.74 \times 10^{-5})$$

$$5.75 \times 10^{-10} = \frac{x^2}{1.60 \times 10^{-2}}$$

$$x^2 = 9.20 \times 10^{-12}$$

$$x = 3.03 \times 10^{-6}$$

$$pOH = 5.52$$

$$pH = 14.00 - 5.52 = 8.48$$

2. **a.** formic acid: 20 mL × 0.10 M = 2.0 mmol
sodium formate: 15 mL × 0.50 M = 7.5 mmol

$$pK_a = pH + \log\frac{[HA]}{[A^-]}$$

$$3.75 = pH + \log\frac{2.0}{7.5}$$

$$3.75 = pH + \log 0.27$$

$$3.75 = pH + (-0.57)$$

$$pH = 4.32$$

b. aniline: 10 mL × 0.50 = 5.0 mmol ⟶ 3.5 mmol anline (RNH_2)

HCl: 15 mL × 0.10 = 1.5 mmol ⟶ 1.5 mmol anilinium hydrochloride ($R\overset{+}{N}H_3$)

$$pK_a = pH + \log\frac{[HA]}{[A^-]}$$

$$4.60 = pH + \log\frac{1.5}{3.5}$$

$$4.60 = pH + \log 0.43$$

$$4.60 = pH + (-0.37)$$

$$pH = 4.97$$

c. acetic acid: 15 mL × 1.00 = 15 mmol ⟶ 10 mmol acetic acid

NaOH: 10 mL × 0.50 = 5.0 mmol ⟶ 5.0 mmol sodium acetate

$$pK_a = pH + \log\frac{[HA]}{[A^-]}$$

$$4.76 = pH + \log\frac{10}{5.0}$$

$$4.76 = pH + \log 2$$

$$4.76 = pH + 0.30$$

$$pH = 4.46$$

3. The ionized form is the basic form.

$$\frac{K_a}{K_a + [H^+]} = \frac{5.89 \times 10^{-6}}{5.89 \times 10^{-6} + 10.47 \times 10^{-6}} = \frac{5.89 \times 10^{-6}}{16.36 \times 10^{-6}}$$

$$= 0.36$$

4.

$$pK_a = pH + \log\frac{[HA]}{[A^-]}$$

$$3.75 = 3.12 + \log\frac{[HA]}{[A^-]}$$

$$0.63 = \log\frac{[HA]}{[A^-]}$$

$$4.27 = \frac{[HA]}{[A^-]}$$

$$[HA] = 4.27[A^-]$$

$$[HA] + [A^-] = 1.0\ M$$

$$4.27[A^-] + [A^-] = 1.0\ M$$

$$5.27[A^-] = 1.0\ M$$

$$[A^-] = 0.19\ M$$

$$[\text{sodium formate}] = 0.19\ M$$

$$[\text{formic acid}] = 0.81\ M$$

5.

$$pK_a = pH + \log\frac{[HA]}{[A^-]}$$

when $[HA] = [A^-]$,

$$pK_a = pH$$

Thus, in order to have a solution in which the pH will be the same as the pK_a, the number of mm of acid must equal the number of mm of conjugate base.

Preparing a solution of x mmol of RCOOH and 1/2 x mmol NaOH will give a solution in which $[RCOOH] = [RCOO^-]$.

For example: 20 mL of 1.00 M RCOOH = 20 mmol
10 mL of 1.00 M NaOH = 10 mmol

This will give a solution that contains 10 mmol RCOOH and 10 mmol $RCOO^-$.

The pH of this solution is the pK_a of RCOOH.

6. **a.** $\frac{x \text{ mmol}}{100 \text{ mL}} = 0.30 \text{ M}$

$x = 30$ mmol of acetic acid

$\frac{x \text{ mmol}}{100 \text{ mL}} = 0.20 \text{ M}$

$x = 20$ mmol of sodium acetate

$\frac{30 \text{ mmol}}{y \text{ mL}} = 1.00 \text{ M}$

$y = 30$ mL of 1.00M acetic acid

$\frac{20 \text{ mmol}}{y \text{ ml}} = 2.00 \text{ M}$

$y = 10$ mL of 2.00 M acetic acid

The buffer solution could be prepared by mixing:
30 mL of 1.00 M acetic acid
10 mL of 2.00 M sodium acetate
60 mL of water

b. Because the concentration of buffer in the acidic form (0.30 M) is greater than the concentration of buffer in the basic form (0.20 M), the pH of the solution will be less than 4.76.

7. $$pK_a = \text{pH} + \log\frac{[\text{HA}]}{[\text{A}^-]}$$

original solution

$$\begin{aligned}
4.76 &= 4.90 + \log\frac{[\text{HA}]}{[\text{A}^-]} \\
-0.14 &= \log\frac{[\text{HA}]}{[\text{A}^-]} \\
0.72 &= \frac{[\text{HA}]}{[\text{A}^-]} \\
[\text{HA}] &= 0.72[\text{A}^-] \\
[\text{HA}]+[\text{A}^-] &= 1.50 \text{ M} \\
0.72[\text{A}^-]+[\text{A}^-] &= 1.50 \text{ M} \\
1.72[\text{A}^-] &= 1.50 \text{ M} \\
[\text{A}^-] &= 0.87 \text{ M} \\
[\text{HA}] &= 0.63 \text{ M}
\end{aligned}$$

desired solution

$$4.76 = 4.50 + \log \frac{[HA]}{[A^-]}$$

$$0.26 = \log \frac{[HA]}{[A^-]}$$

$$1.82 = \frac{[HA]}{[A^-]}$$

$$[HA] = 1.82\,[A^-]$$

$$[HA]+[A^-] = 1.50\text{ M}$$

$$1.82\,[A^-]+[A^-] = 1.50\text{ M}$$

$$2.82\,[A^-] = 1.50\text{ M}$$

$$[A^-] = 0.53\text{ M}$$

$$[HA] = 0.97\text{ M}$$

The original solution contains 87 mmol of A^- (100 mL x 0.87 M).

The desired solution with a pH = 4.50 must contain 53 mmol of A^-.

Therefore, 34 mmol of A^- $(87 - 53 = 34)$ must be converted to HA.

This can be done by adding 34 mmol of HCl to the original solution.

If you have a 1.00 M HCl solution, you will need to add 34 mL to the original solution in order to change its pH from 4.90 to 4.50.

$$\frac{34\text{ mmol}}{x\text{ mL}} = 1.00\text{ M}$$

$$x = 34\text{ mL}$$

Note that after adding HCl to the original solution, it will no longer be a 1.50 M buffer; it will be more dilute (150 mm/134 mL = 1.12 M).

The change in the concentration of the buffer solution will be less if a more concentrated solution of HCl is used to change the pH. For example, if you have a 2.00 M HCl solution:

$$\frac{34\text{ mmol}}{x\text{ mL}} = 2.00\text{ M}$$

$$x = 17\text{ mL}$$

You will need to add 17 mL to the original solution, and the concentration of the buffer will be 1.28 M (150 mm/117 mL = 1.28 M).

8. acid: $100 \text{ mL} \times 1.00 = 100 \text{ mmol} \longrightarrow 90 \text{ mmol HA}$

NaOH: $10 \text{ mL} \times 1.00 = 10 \text{ mmol} \longrightarrow 10 \text{ mmol A}^-$

Therefore, 90% is in the acidic form.

For 40% to be in the acidic form you need:

40 mmol HA

60 mmol A^-

You need to have 60 mmol rather than 10 mmol in the basic form. To get the additional 50 mmol in the basic form, you would need to add 50 mL of 1.0 M NaOH.

9.

$$pK_a = pH + \log\frac{[HA]}{[A^-]}$$

$$4.76 = 4.00 + \log\frac{[HA]}{[A^-]}$$

$$0.76 = \log\frac{[HA]}{[A^-]}$$

$$5.75 = \frac{[HA]}{[A^-]}$$

$$[HA] = 5.75[A^-]$$

$$[HA] + [A^-] = 1.0 \text{ M}$$

$$5.75[A^-] + [A^-] = 1.0 \text{ M}$$

$$6.75[A^-] = 1.0 \text{ M}$$

$$[A^-] = 0.15 \text{ M}$$

$$[HA] = 0.85 \text{ M}$$

a. [acetic acid] = 0.85 M

[sodium acetate] = 0.15 M

b. [acetic acid] = 1.00 M

[NaOH] = 0.15 M

c. [sodium acetate] = 1.00 M

[HCl] = 0.85 M

10. **a.** $\frac{x \text{ mmol}}{50 \text{ mL}} = 0.85 \text{ M}$

$x = 42.5$ mmol of acetic acid

$\frac{42.4 \text{ mmol}}{y \text{ mL}} = 1.50 \text{ M}$

$y = 28.3$ mL of 1.50 M acetic acid

$\frac{x \text{ mmol}}{50 \text{ mL}} = 0.15 \text{ M}$

$x = 7.5$ mmol of sodium acetate

$\frac{7.5 \text{ mmol}}{y \text{ mL}} = 1.50 \text{ M}$

$y = 5.0$ mL of 1.50 M sodium acetate

28.3 mL of 1.50 M acetic acid

5.0 mL of 1.50 M sodium acetate

16.7 mL of H_2O

b. $\frac{x \text{ mmol}}{50 \text{ mL}} = 1.00 \text{ M}$

$x = 50$ mmol of acetic acid

$\frac{50 \text{ mmol}}{y \text{ mL}} = 1.50 \text{ M}$

$y = 33.3$ mL of 1.50 M acetic acid

$\frac{x \text{ mmol}}{50 \text{ mL}} = 0.15 \text{ M}$

$x = 7.5$ mmol of NaOH

$\frac{7.5 \text{ mmol}}{y \text{ mL}} = 1.50 \text{ M}$

$y = 5.0$ mL of 1.50 M NaOH

33.3 mL of 1.50 M acetic acid

5.0 mL of 1.50 M NaOH

11.7 mL of H_2O

c. $$\frac{x \text{ mmol}}{50 \text{ mL}} = 1.00 \text{ M}$$

$$x = 50 \text{ mmol of sodium acetate}$$

$$\frac{50 \text{ mmol}}{y \text{ mL}} = 1.50 \text{ M}$$

$$y = 33.3 \text{ mL of } 1.50 \text{ M sodium acetate}$$

$$\frac{x \text{ mmol}}{50 \text{ mL}} = 0.85 \text{M}$$

$$x = 42.5 \text{mm}$$

$$\frac{42.5 \text{ mmol}}{y \text{mL}} = 1.5 \text{M}$$

$$y = 28.3 \text{mL of } 1.5 \text{M HCl}$$

We cannot make the required buffer with these solutions, because 33.3 mL + 28.3 mL > 50 mL.

11. Because formic acid has a $\text{p}K_a = 3.75$, a formic acid/formate buffer can be a buffer at pH = 3.30.

$$\text{p}K_a = \text{pH} + \log \frac{[\text{HA}]}{[\text{A}^-]}$$

$$3.75 = 3.30 + \log \frac{[\text{HA}]}{[\text{A}^-]}$$

$$0.45 = \log \frac{[\text{HA}]}{[\text{A}^-]}$$

$$2.82 = \frac{[\text{HA}]}{[\text{A}^-]}$$

$$[\text{HA}] = 2.82[\text{A}^-]$$

$$[\text{HA}] + [\text{A}^-] = 1.0 \text{ M}$$

$$2.82[\text{A}^-] + [\text{A}^-] = 1.0 \text{ M}$$

$$3.82[\text{A}^-] = 1.0 \text{ M}$$

$$[\text{A}^-] = 0.26 \text{ M}$$

$$[\text{HA}] = 0.74 \text{ M}$$

The solution must have [formic acid] = 0.74 M and [sodium formate] = 0.26 M.

12. At pH = 4.20, 74% of the formate buffer will be in the basic form.

$\text{pH} = 4.20, \ [\text{H}^+] = 6.31 \times 10^{-5}$ Formic acid has a $\text{p}K_a = 3.75$; $K_a = 1.78 \times 10^{-4}$.

$$\frac{K_a}{K_a + [\text{H}^+]} = \frac{1.78\times10^{-4}}{1.78\times10^{-4} + 6.31\times10^{-5}} = \frac{1.78\times10^{-4}}{1.78\times10^{-4} + 0.63\times10^{-4}}$$

$$= \frac{1.78\times10^{-4}}{2.41\times10^{-4}}$$

$$= 0.74$$

At pH = 4.20, 22% of the acetate buffer will be in the basic form.

$\text{pH} = 4.20, \ [\text{H}^+] = 6.31 \times 10^{-5}$ Acetic acid has a $\text{p}K_a = 4.76$; $K_a = 1.74 \times 10^{-5}$.

$$\frac{K_a}{K_a + [\text{H}^+]} = \frac{1.74\times10^{-5}}{1.74\times10^{-5} + 6.31\times10^{-5}}$$

$$= \frac{1.78\times10^{-5}}{8.05\times10^{-5}}$$

$$= 0.22$$

The reaction to be carried out will generate protons that will react with the basic form of the buffer in order to keep the pH constant. Therefore, the formate buffer is preferred because it has a greater percentage of the buffer in the basic form.

CHAPTER 2

An Introduction to Organic Compounds: Nomenclature, Physical Properties, and Representation of Structure

Important Terms

alcohol	a compound with an OH group in place of one of the hydrogens of an alkane (ROH).
alkane	a hydrocarbon that contains only single bonds.
alkyl halide	a compound with a halogen in place of one of the hydrogens of an alkane.
alkyl substituent	a substituent formed by removing a hydrogen from an alkane.
amine	a compound in which one or more of the hydrogens of NH_3 is replaced by an alkyl substituent (RNH_2, R_2NH, R_3N).
angle strain	the strain introduced into a molecule as a result of its bond angles being distorted from their ideal values.
anti conformer	the staggered conformer in which the largest substituents bonded to the two carbons are opposite each other. It is the most stable of the staggered conformers.
axial bond	a bond of the chair form of cyclohexane that is perpendicular to the plane in which the chair is drawn (an up-down bond).
banana bonds	the bonds in small rings that are slightly bent as a result of orbitals overlapping at an angle rather than overlapping head-on.
boat conformation	the conformation of cyclohexane that roughly resembles a boat.
boiling point	the temperature at which the vapor pressure of a liquid equals the atmospheric pressure.
chair conformation	the conformation of cyclohexane that roughly resembles a chair. It is the most stable conformation of cyclohexane.
cis fused	two rings fused together in such a way that if the second ring were considered to be two substituents of the first ring, the two substituents would be on the same side of the first ring.
cis isomer	the isomer with both hydrogens on the same side of the double bond.
cis-trans stereoisomers (geometric isomers)	geometric (or *E*,*Z*) isomers.
common name	non-systematic nomenclature.
conformation	the three-dimensional shape of a molecule at a given instant.

conformational analysis — the investigation of various conformations of a compound and their relative stabilities.

conformers — different conformations of a molecule.

constitutional isomers (structural isomers) — molecules that have the same molecular formula but differ in the way the atoms are connected.

cycloalkane — an alkane with its carbon chain arranged in a closed ring.

1,3-diaxial interaction — the interaction between an axial substituent and the other two axial substituents on the same side of the cyclohexane ring.

dipole-dipole interaction — an interaction between the dipole of one molecule and the dipole of another.

eclipsed conformation — a conformation in which the bonds on adjacent carbons are parallel to each other as viewed looking down the carbon-carbon bond.

equatorial bond — a bond of the chair form of cyclohexane that juts out from the ring in approximately the same plane that contains the chair.

ether — a compound in which an oxygen is bonded to two alkyl groups (ROR).

flagpole hydrogens — the two hydrogens in the boat conformation of cyclohexane that are closest to each other.

functional group — the center of reactivity of a molecule.

gauche conformer — a staggered conformer in which the largest substituents bonded to the two carbons are gauche to each other; that is, they have a dihedral angle of approximately 60°.

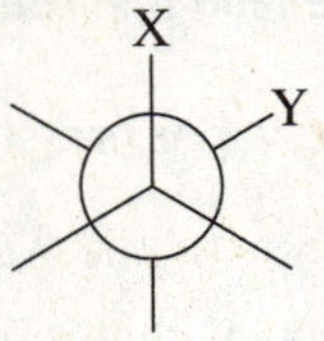

The substituents are gauche to each other

gauche interaction — the interaction between two atoms or groups that are gauche to each other.

geometric isomers (cis-trans stereoisomers) — cis-trans (or *E,Z*) isomers.

half-chair conformer — the least stable conformation of cyclohexane.

homolog — a member of a homologous series.

homologous series — a family of compounds in which each member differs from the next by one methylene group.

hydrocarbon	a compound that contains only carbon and hydrogen.
hydrogen bond	an unusually strong dipole-dipole attraction (5 kcal/mol) between a hydrogen bonded to O, N, or F and the lone pair of a different O, N, or F.
hyperconjucation	delocalization of electrons by the overlap of a σ orbital with an empty orbital.
induced dipole-induced dipole interaction	an interaction between a temporary dipole in one molecule and the dipole that the temporary dipole induces in another molecule.
IUPAC nomenclature	systematic nomenclature.
melting point	the temperature at which a solid becomes a liquid.
methylene group	a CH_2 group.
Newman projection	a way to represent the three-dimensional spatial relationships of atoms by looking down the length of a particular carbon-carbon bond.
packing	the property that determines how well individual molecules fit into a crystal lattice.
parent hydrocarbon	the longest continuous carbon chain in a molecule.
perspective formula	a way to represent the three-dimensional spatial relationships of atoms using two adjacent solid lines, one solid wedge and one hatched wedge.
polarizability	the ease with which an electron cloud of an atom can be distorted.
primary alcohol	an alcohol in which the OH group is bonded to a primary carbon.
primary alkyl halide	an alkyl halide in which the halogen is bonded to a primary carbon.
primary amine	an amine with one alkyl group bonded to the nitrogen.
primary carbon	a carbon bonded to only one other carbon.
primary hydrogen	a hydrogen bonded to a primary carbon.
quaternary ammonium salt	a nitrogen compound with four alkyl groups bonded to the nitrogen.
ring-flip (chair-chair interconversion)	the conversion of a chair conformer of cyclohexane into the other chair conformer. Bonds that are axial in one chair onformer are equatorial in the other chair conformer.
sawhorse projection	a way to represent the three-dimensional spatial relationships of atoms by looking at the carbon-carbon bond from an oblique angle.

secondary alcohol an alcohol in which the OH group is bonded to a secondary carbon.

secondary alkyl halide an alkyl halide in which the halogen is bonded to a secondary carbon.

secondary amine an amine with two alkyl groups bonded to the nitrogen.

secondary carbon a carbon bonded to two other carbons.

secondary hydrogen a hydrogen bonded to a secondary carbon.

skeletal structure a structure that shows the carbon-carbon bonds as lines and does not show the carbon-hydrogen bonds.

skew-boat conformer one of the conformations of a cyclohexane ring.

solubility the extent to which a compound dissolves in a solvent.

solvation the interaction between a solvent and another molecule (or ion).

staggered conformation a conformation in which the bonds on one carbon bisect the bond angle on the adjacent carbon when viewed looking down the carbon-carbon bond.

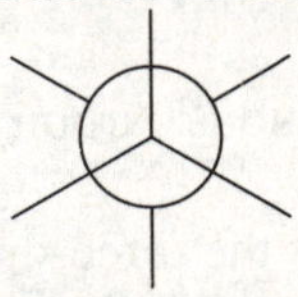

steric hindrance hindrance due to groups occupying a volume of space.

steric strain the repulsion between the electron cloud of an atom or group of atoms and the electron cloud of another atom or group of atoms.

straight-chain alkane an alkane in which the carbons form a continuous chain with no branches.

structural isomers (constitutional isomers) molecules that have the same molecular formula but differ in the way the atoms are connected.

symmetrical ether an ether with two identical substituents bonded to the oxygen.

systematic nomenclature IUPAC nomenclature.

tertiary alcohol an alcohol in which the OH group is bonded to a tertiary carbon.

tertiary alkyl halide an alkyl halide in which the halogen is bonded to a tertiary carbon.

tertiary amine an amine with three alkyl groups bonded to the nitrogen.

tertiary carbon a carbon bonded to three other carbons.

tertiary hydrogen a hydrogen bonded to a tertiary carbon.

trans-fused two rings fused together in such a way that if the second ring were considered to be two substituents of the first ring, the two substituents would be on opposite sides of the first ring.

trans isomer the isomer that has the hydrogens on opposite sides of the double bond.

twist-boat conformer one of the conformations of a cyclohexane ring.

unsymmetrical ether an ether with two different substituents bonded to the oxygen.

van der Waals forces induced dipole-induced dipole interactions.

Solutions to Problems

1. a. propyl alcohol b. dimethyl ether c. propylamine

2. a. 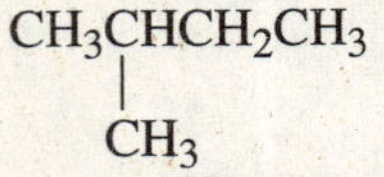
2-methylbutane

b. 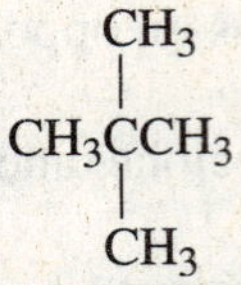
2,2-dimethylpropane

3. Notice that each carbon forms fours bonds and each hydrogen and bromine forms one bond.

$CH_3CH_2CH_2CH_2Br$
n-butyl bromide
or
butyl bromide

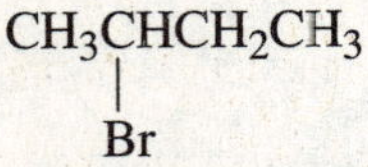
sec-butyl bromide

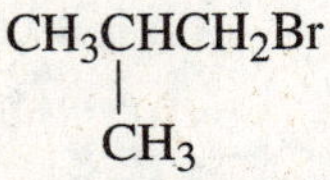
isobutyl bromide

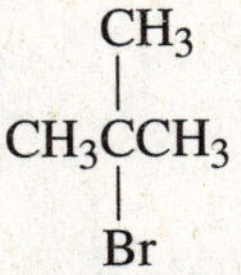
tert-butyl bromide

4. "Dibromomethane does not have constitutional isomers" proves that carbon is tetrahedral.

If carbon were not tetrahedral but flat, dibromomethane would have constitutional isomers because the two structures shown below would be different since the Br's would be 90° apart in one compound and 180° apart in the other compound. Only because carbon is tetrahedral are the two structures are identical, giving dibromomethane no constitutional isomers.

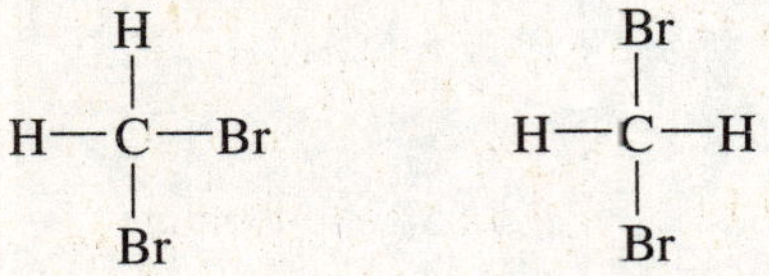

5. a. 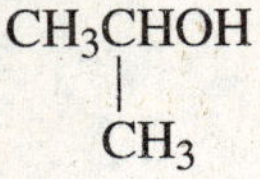

c. 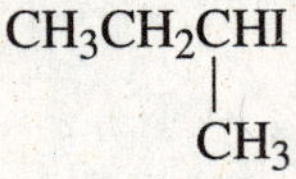

e. 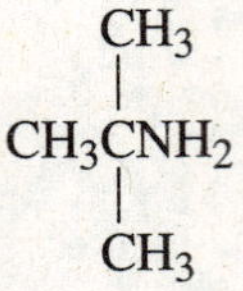

b. 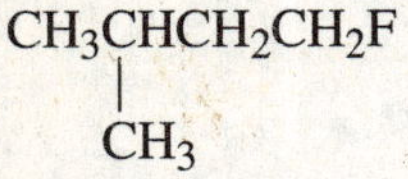

d. 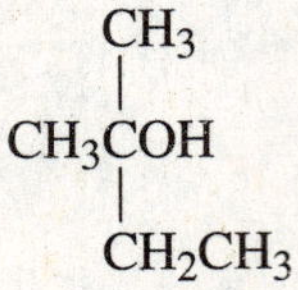

f. $CH_3CH_2CH_2CH_2CH_2CH_2CH_2CH_2Br$

6. a. ethyl methyl ether
b. methyl propyl ether
c. *sec*-butylamine
d. *n*-butyl alcohol
e. isobutyl bromide
f. *sec*-butyl chloride

7.

a. 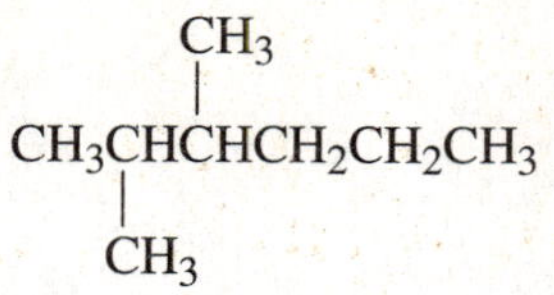

b. 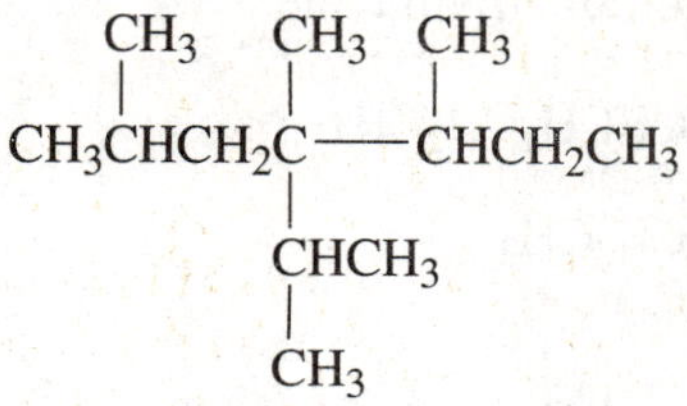

c. 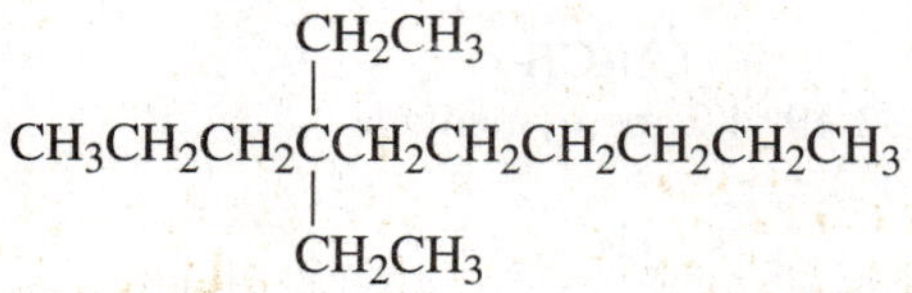

d. 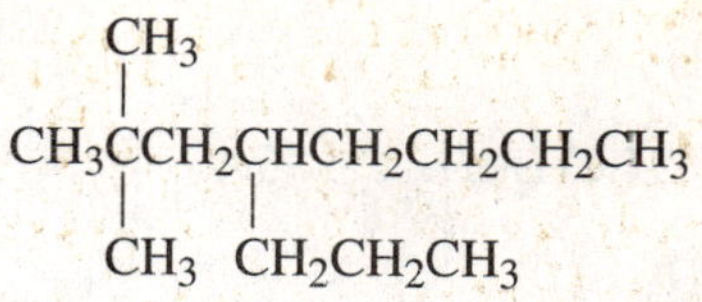

e. 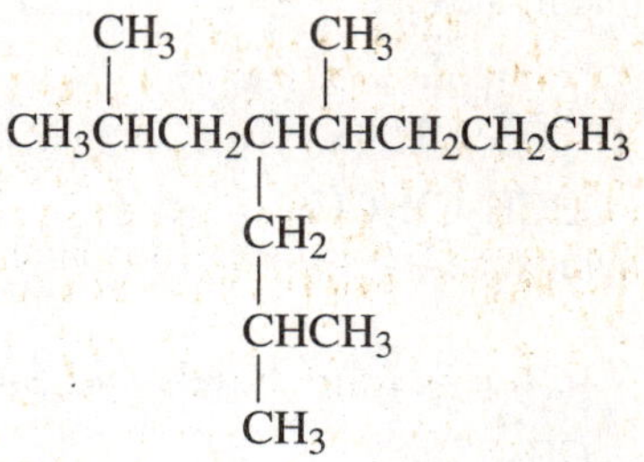

f. 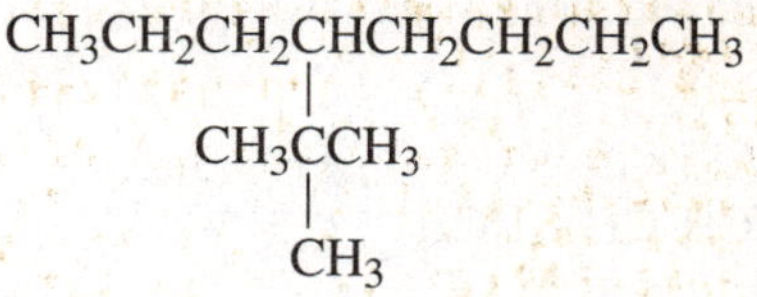

8.

a. #1 $CH_3CH_2CH_2CH_2CH_2CH_2CH_2CH_3$
octane

#2 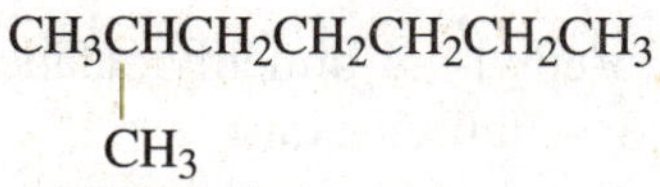
2-methylheptane

#3 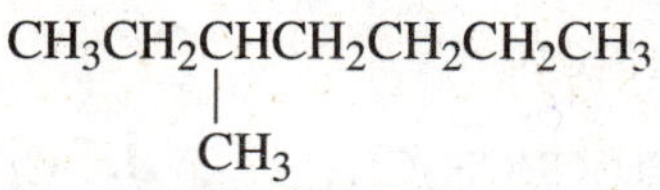
3-methylheptane

#4 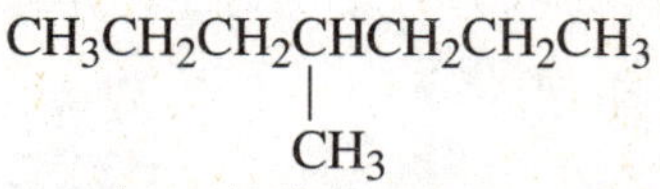
4-methylheptane

#5 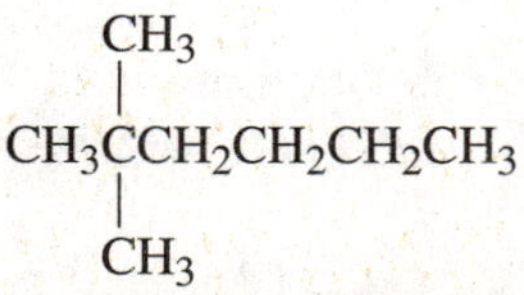
2,2-dimethylhexane

#10 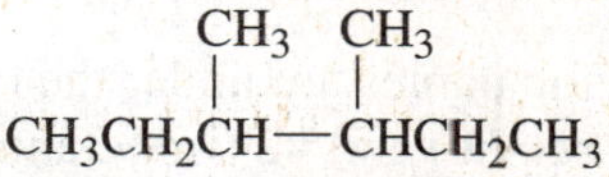
3,4-dimethylhexane

#11 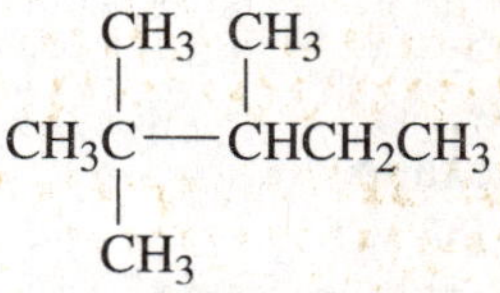
2,2,3-trimethylpentane

#12 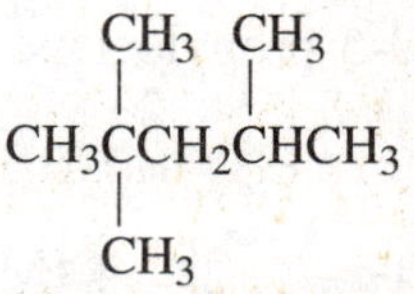
2,2,4-trimethylpentane

#13 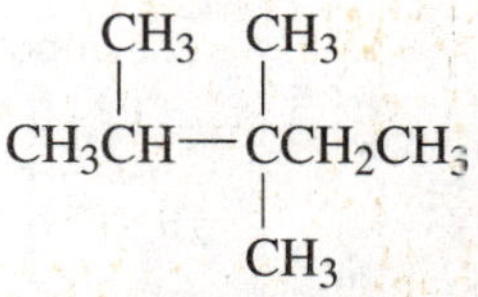
2,3,3-trimethylpentane

#14 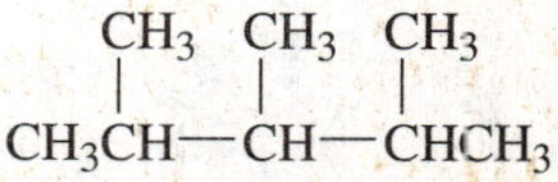
2,3,4-trimethylpentane

#6

CH_3
$CH_3CH_2CCH_2CH_2CH_3$
CH_3

3,3-dimethylhexane

#7

CH_3 CH_3
$CH_3CH{-}CHCH_2CH_2CH_3$

2,3-dimethylhexane

#8

CH_3 CH_3
$CH_3CHCH_2CHCH_2CH_3$

2,4-dimethylhexane

#9

CH_3 CH_3
$CH_3CHCH_2CH_2CHCH_3$

2,5-dimethylhexane

#15

CH_3 CH_3
$CH_3C{-}CCH_3$
CH_3 CH_3

2,2,3,3-tetramethylbutane

#16

$CH_3CH_2CHCH_2CH_2CH_3$
CH_2CH_3

3-ethylhexane

#17

CH_3
$CH_3CH_2CHCHCH_3$
CH_2CH_3

3-ethyl-2-methylpentane

#18

CH_3
$CH_3CH_2CCH_2CH_3$
CH_2CH_3

3-ethyl-3-methylpentane

b. The systematic names are under the compound.
c. Only #1 (*n*-octane) and #2 (isooctane) have common names.
d. #2, #7, #8, #9, #12, #13, #14, #17
e. #3, #8, #10, #11
f. #5, #11, #12, #15

9.
a. 2,2,4-trimethylhexane
b. 2,2-dimethylbutane
c. 3-methyl-4-propylheptane
d. 2,2,5-trimethylhexane
e. 3,3-diethyl-4-methyl-5-propyloctane
f. 5-ethyl-4,4-dimethyloctane
g. 3,3-diethylhexane
h. 4-isopropyloctane
i. 2,5-dimethylheptane

10.
a. $CH_3CH_2CH_2CH_2CH_3$
pentane

b.
CH_3
CH_3CCH_3
CH_3
2,2-dimethylpropane

c.
CH_3
$CH_3CHCH_2CH_3$
2-methylbutane

d.
CH_3
$CH_3CHCH_2CH_3$
2-methylbutane

11.

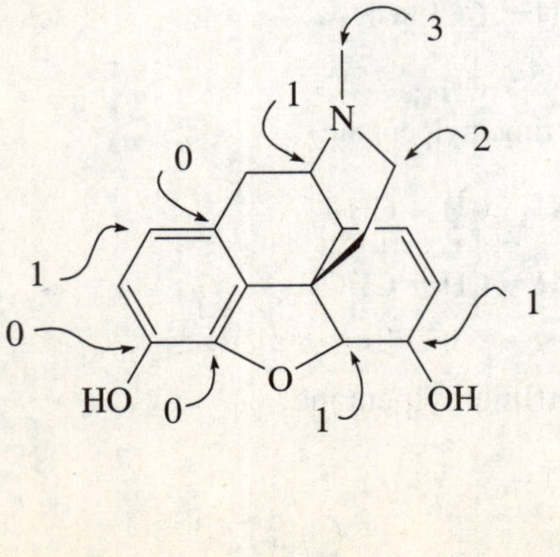

12. **a.** 

d. 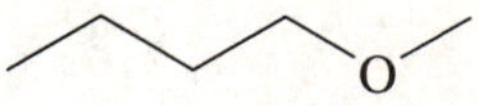

b. 

e. 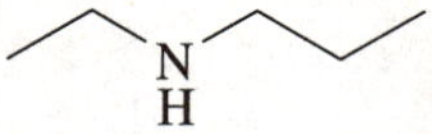

c. 

f. 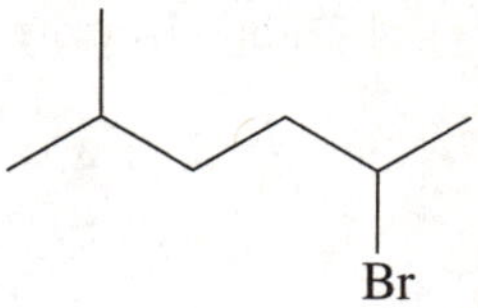

13. **a.** 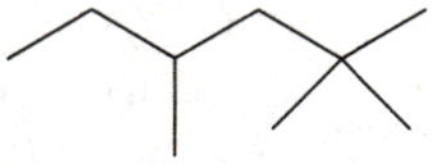

d. 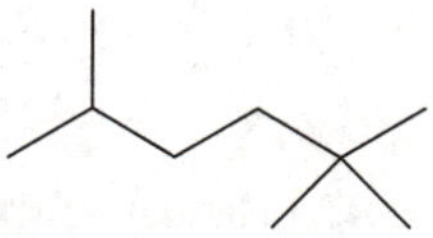

g. 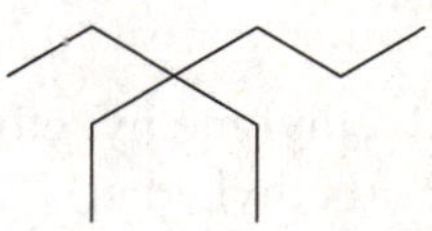

b. 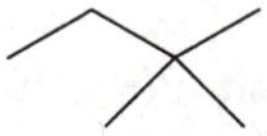

e. 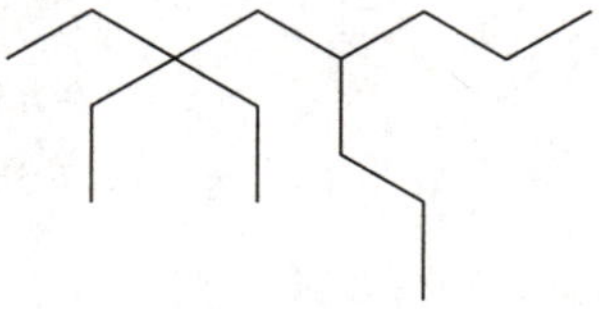

h. 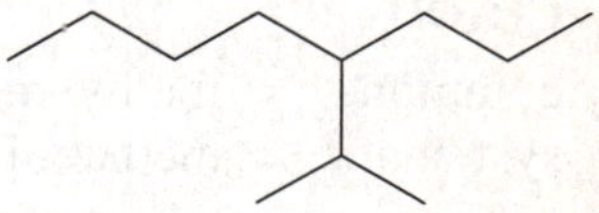

c. 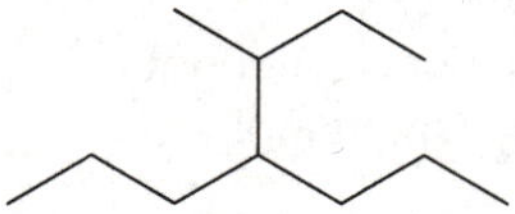

f. 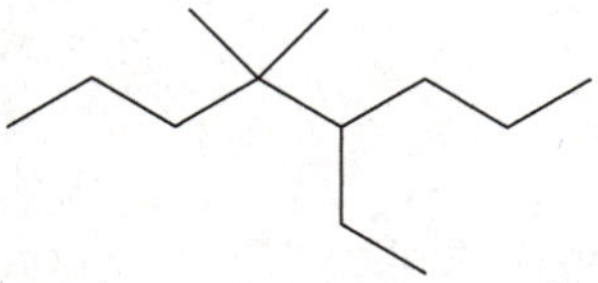

i. 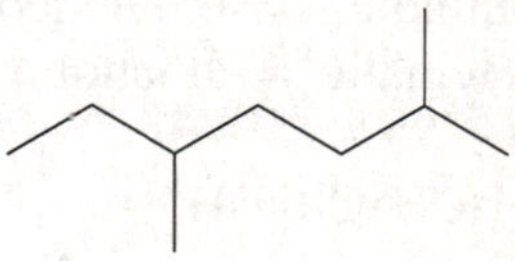

14. **a.** 1-ethyl-2-methylcyclopentane
b. ethylcyclobutane
c. 4-ethyl-1,2-dimethylcyclohexane
d. 3,6-dimethyldecane
e. 2-cyclopropylpentane
f. 1-ethyl-3-isobutylcyclohexane
g. 5-isopropylnonane
h. 1-*sec*-butyl-4-isopropylcyclohexane

15. **a.** *sec*-butyl chloride
2-chlorobutane
secondary

b. isohexyl chloride
1-chloro-4-methylpentane
primary

c. cyclohexyl bromide
bromocyclohexane
secondary

d. isopropyl fluoride
2-fluoropropane
secondary

16. **a.** 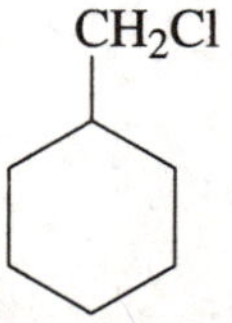

chloromethylcyclohexane

Note that the name of a CH_2Cl sustituent is "chloromethyl", because a Cl is in place of one of the H's of a methyl substituent

b. 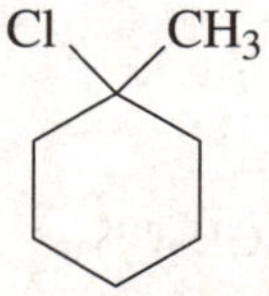

1-chloro-1-methylcyclohexane

c.

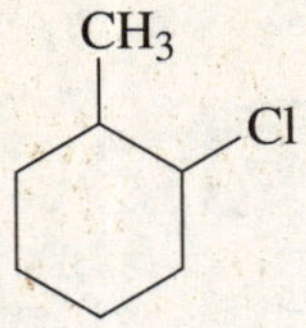
1-chloro-2-methylcyclohexane

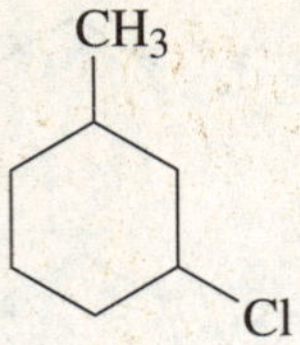
1-chloro-3-methylcylclohexane

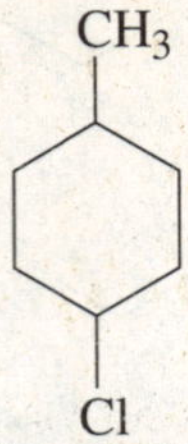
1-chloro-4-methylcyclohexane

17. **a.** **1.** methoxyethane
2. ethoxyethane
3. 4-methoxyoctane
4. 1-isopropoxy-3-methylbutane
5. 1-propoxybutane
6. 2-isopropoxypentane

b. No.

c. **1.** ethyl methyl ether
2. diethyl ether
3. no common name
4. isopentyl isopropyl ether
5. butyl propyl ether
6. no common name

18. CH_3OH
common = methyl alcohol
systematic = methanol

CH_3CH_2OH
common = ethyl alcohol
systematic = ethanol

$CH_3CH_2CH_2OH$
common = propyl alcohol or *n*-propyl alcohol
systematic = 1-propanol

$CH_3CH_2CH_2CH_2OH$
common = butyl alcohol or *n*-butyl alcohol
systematic = 1-butanol

$CH_3CH_2CH_2CH_2CH_2OH$
common = pentyl alcohol or *n*-pentyl alcohol
systematic = 1-pentanol

$CH_3CH_2CH_2CH_2CH_2CH_2OH$
common = hexyl alcohol or *n*-hexyl alcohol
systematic = 1-hexanol

19. **a.** 1-pentanol
primary

b. 4-methylcyclohexanol
secondary

c. 5-chloro-2-methyl-2-pentanol
tertiary

d. 5-methyl-3-hexanol
secondary

e. 4-chloro-3-ethylcyclohexanol
secondary

f. 2,6-dimethyl-4-octanol
secondary

20.

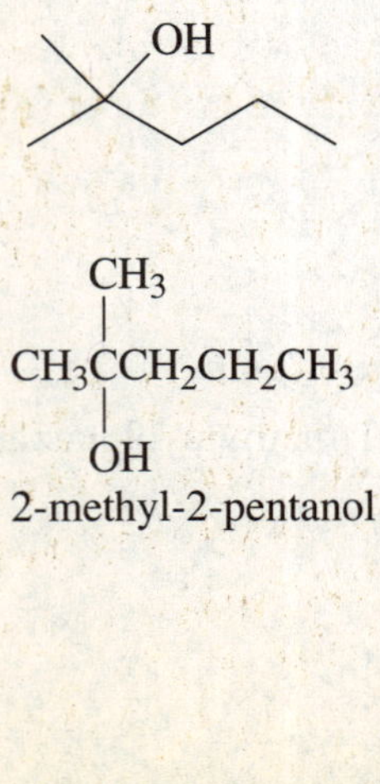
2-methyl-2-pentanol

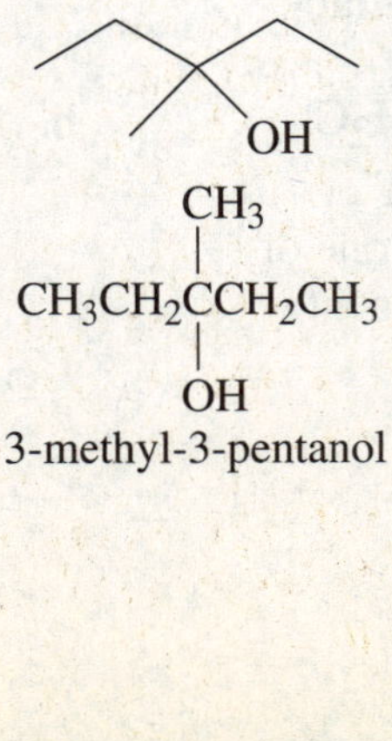
3-methyl-3-pentanol

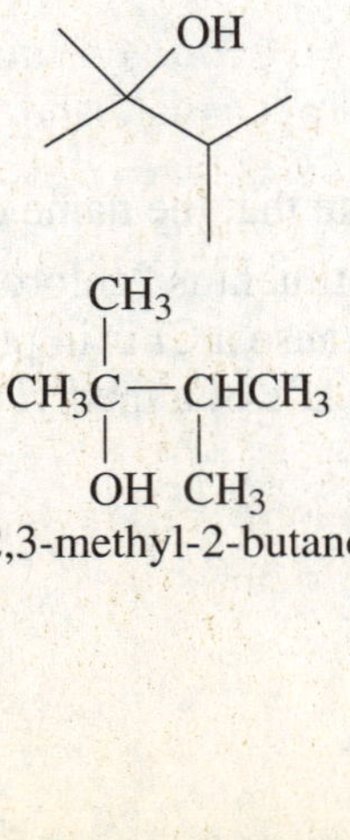
2,3-methyl-2-butanol

21. **a.** tertiary **b.** tertiary **c.** primary

22.

a. hexylamine
1-hexanamine
primary

b. *sec*-butylisobutylamine
N-isobutyl-2-butanamine
secondary

c. cyclohexylamine
cyclohexanamine
primary

d. butylpropylamine
N-propyl-1-butanamine
secondary

e. diethylpropylamine
N,N-diethyl-1-propanamine
tertiary

f. *N*-ethyl-3-methylcyclohexanamine
it does not have a common name
secondary

23.

a. 

b. $CH_3CH_2NHCH_2CH_3$

c. 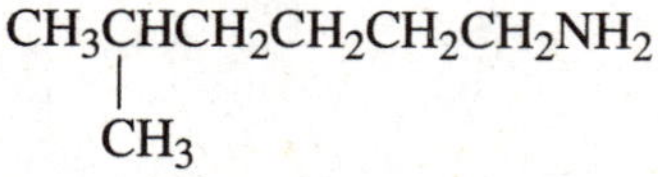

d. 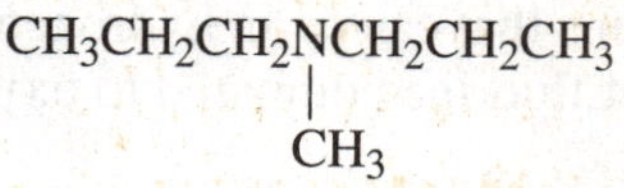

e. 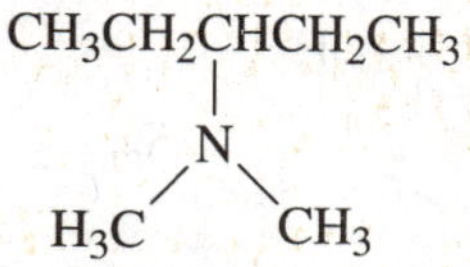

f. 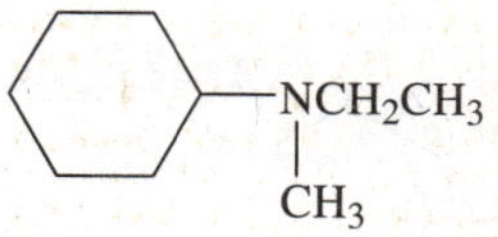

24.

a. 6-methyl-1-heptanamine
isooctylamine
primary

b. 3-methyl-*N*-propyl-1-butanamine
isopentylpropylamine
secondary

c. *N*-ethyl-*N*-methylethanamine
diethylmethylamine
tertiary

d. 2,5-dimethylcyclopentanamine
no common name
primary

25.

a. The bond angle is predicted to be similar to the bond angle in water (104.5°)
b. The bond angle is predicted to be similar to the bond angle in ammonia (107.3°)
c. The bond angle is predicted to be similar to the bond angle in water (104.5°)
d. The bond angle is predicted to be similar to the bond angle in the ammonium ion (109.5°)

26.

a. 1, 4, and 5
b. 1, 2, 4, 5, and 6

27.

a. Each water molecule has two hydrogens that can form hydrogen bonds, whereas each alcohol molecule has only one hydrogen that can form a hydrogen bond. Therefore, there are more hydrogen bonds between water molecules than between alcohol molecules.

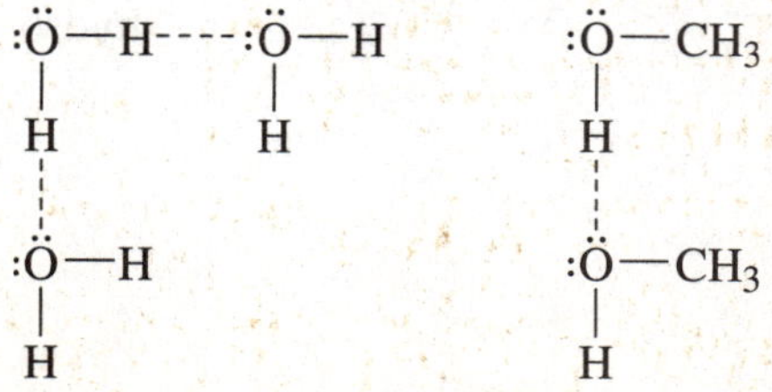

b. Each water molecule has two hydrogens that can form hydrogen bonds, whereas each ammonia has three hydrogens that can form hydrogen bonds. However, oxygen is more electronegative than nitrogen, so the hydrogen bonds between water molecules are stronger than the hydrogen bonds between ammonia molecules. Because the number of hydrogen bonds supports ammonia as having the higher boiling point but the strength of the hydrogen bonds supports water, we could not have predicted which would have the higher boiling point. However, being told that water has the higher boiling point we can conclude that the greater electronegativity of oxygen compared to nitrogen is more important than the number of hydrogens that can form hydrogen bonds.

c. Each water molecule has two hydrogens that can form hydrogen bonds, whereas each molecule of hydrogen fluoride has only one hydrogen that can form a hydrogen bond. However, fluorine is more electronegative than oxygen. Again we cannot predict which will have the higher boiling point, but we can conclude from the fact that water has the higher boiling point that *in this case* the greater number of hydrogens that can form hydrogen bonds is more important than the greater electronegativity of fluorine compared to oxygen.

28.

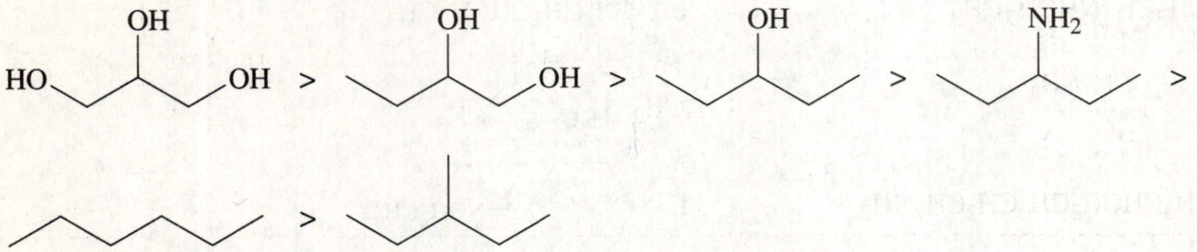

29. **a.** $CH_3CH_2CH_2CH_2CH_2CH_2Br$ > $CH_3CH_2CH_2CH_2CH_2Br$ > $CH_3CH_2CH_2CH_2Br$

b. $CH_3CH_2CH_2CH_2CH_2CH_2CH_2CH_2CH_3$ > $CH_3CH_2CH_2CH_2CH_2CH_2CH_2CH_3$ >

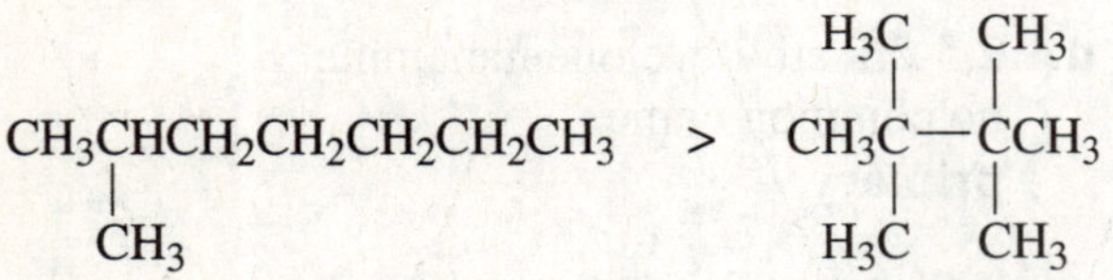

c. $CH_3CH_2CH_2CH_2CH_2OH$ > $CH_3CH_2CH_2CH_2OH$ > $CH_3CH_2CH_2CH_2Cl$ > $CH_3CH_2CH_2CH_2CH_3$

30. **a.** $HOCH_2CH_2CH_2OH$ > $CH_3CH_2CH_2OH$ > $CH_3CH_2CH_2CH_2OH$ > $CH_3CH_2CH_2CH_2Cl$

b.

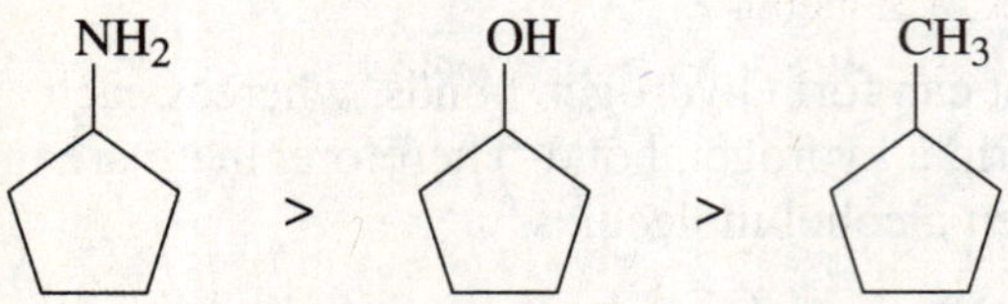

The amine is more soluble in water than the alcohol, because the amine has two hydrogens that can form hydrogen bonds with water.

31. Because cyclohexane is a nonpolar compound it will have the lowest solubility in the most polar solvent, which, of the solvents given, is ethanol.

$CH_3CH_2CH_2CH_2CH_2OH$	$CH_3CH_2OCH_2CH_3$	CH_3CH_2OH	$CH_3CH_2CH_2CH_2CH_2CH_3$
1-pentanol	diethyl ether	ethanol	hexane

32. **a.**

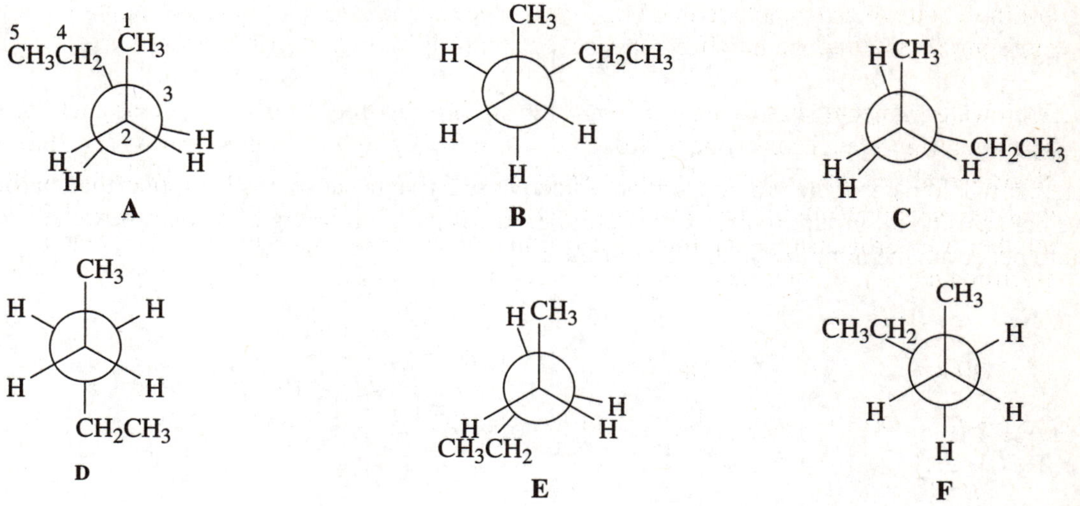

b.

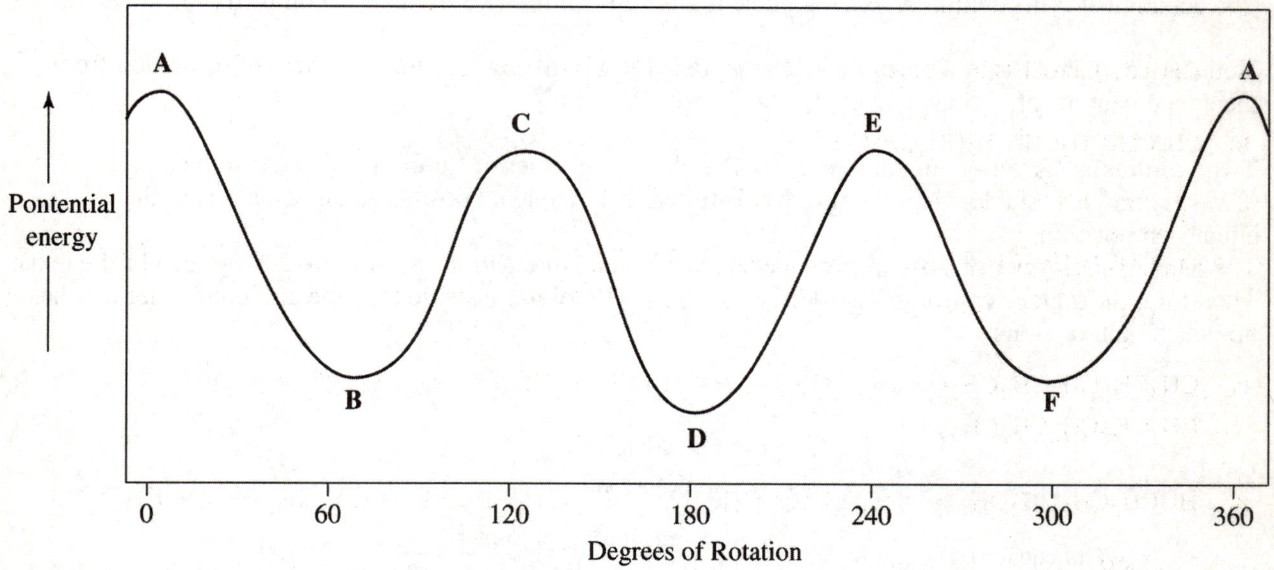

33. **a.**

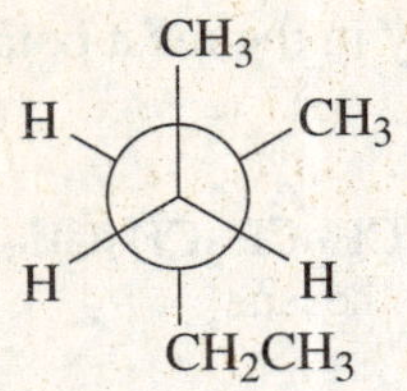

b.

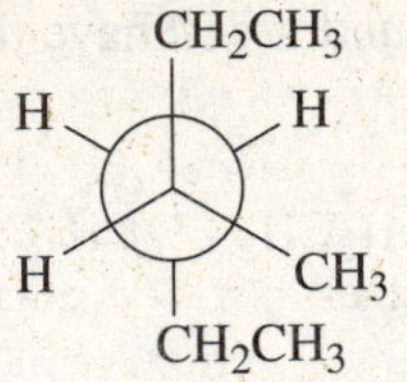

c.

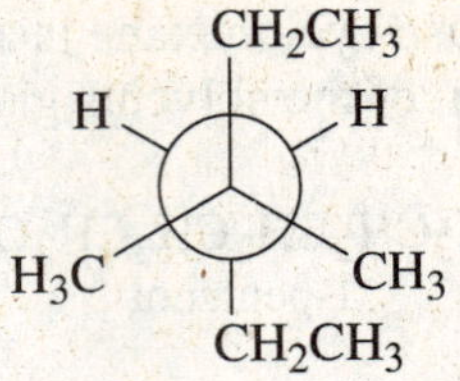

34. **a.** $180° - \frac{360°}{8}$

$180° - 45° = 135°$

b. $180° - \frac{360°}{9}$

$180° - 40° = 140°$

35. Hexethal would be expected to be the more effective sedative because it is less polar than barbital since hexethal has a hexyl group in place of the ethyl group of barbital. Being less polar, hexethal will be better able to penetrate the nonpolar membrane of the cell.

36. **a.**

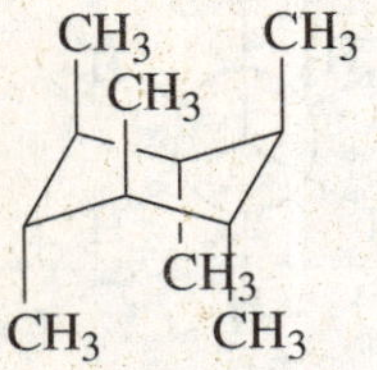

b.

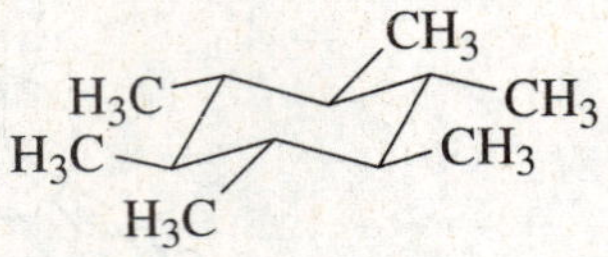

37. The "strainless" heat of formation of cycloheptane is 7 (– 4.92) = – 34.4 kcal/mol.
The actual heat of formation of cycloheptane is – 28.2 kcal/mol (from Table 2.9 on p. 109 of the text).

You can get the total strain energy of cycloheptane by subtracting the strainless heat of formation from the actual heat of formation: – 28.2 – (– 34.4) = 6.2 kcal/mol.

38. Two 1,3-diaxial (gauche) interactions cause the chair conformer of fluorocyclohexane to be 0.25 kcal/mol less stable when the fluoro substituent is in the axial position than when it is in the equatorial position.
The gauche conformer of 1-fluoropropane has one gauche interaction (See Figure 2.14 on p. 111 the text). Therefore, the gauche conformer is (0.25/2) = 0.13 kcal/mol more stable than the anti conformer that has no gauche interactions.

39.

$$K_{eq} = \frac{[\text{equatorial conformer}]}{[\text{axial conformer}]} = \frac{5.4}{1}$$

$$\%\text{ of equatorial conformer} = \frac{[\text{equatorial conformer}]}{[\text{equatorial conformer}] + [\text{axial conformer}]} \times 100$$

$$= \frac{5.4}{5.4 + 1} \times 100 = \frac{5.4}{6.4} \times 100 = 84\%$$

40. **a.** cis **b.** cis **c.** cis **d.** trans **e.** trans **f.** trans

41. Both *trans*-1,4-dimethylcyclohexane and *cis*-1-*tert*-butyl-3-methylcyclohexane have a conformer with two substituents in the equatorial position and a conformer with two substituents in the axial position. *cis*-1-*tert*-Butyl-3-methylcyclohexane will have a higher percentage of the diequatorial-substituted conformer because the bulky *tert*-butyl substituent will have a greater preference for the equatorial position.

42. **a.** 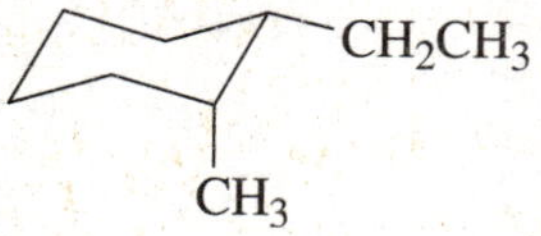

b. 

c. *trans*-1-Ethyl-2-methylcyclohexane is more stable because both substituents are in equatorial positions.

43.
a. one equatorial and one axial in each
b. both equatorial in one and both axial in the other
c. both equatorial in one and both axial in the other
d. one equatorial and one axial in each
e. one equatorial and one axial in each
f. both equatorial in one and both axial in the other

44. **a.** One chair conformer of *trans*-1,4-dimethylcyclohexane has both substituents in equatorial positions, so it does not have any 1,3-diaxial interactions. The other chair conformer has both substituents in axial positions. When a substituent is in an axial position, it experiences two 1,3-diaxial interactions, so this chair conformer has a total of four 1,3-diaxial interactions. Since the 1,3-diaxial interaction between a methyl group and a hydrogen causes a strain energy of 0.9 kcal/mol, the chair conformer with both substituents in axial positions is $4 \times 0.9 = 3.6$ kcal/mol less stable than the chair conformer with both substituents in equatorial positions.

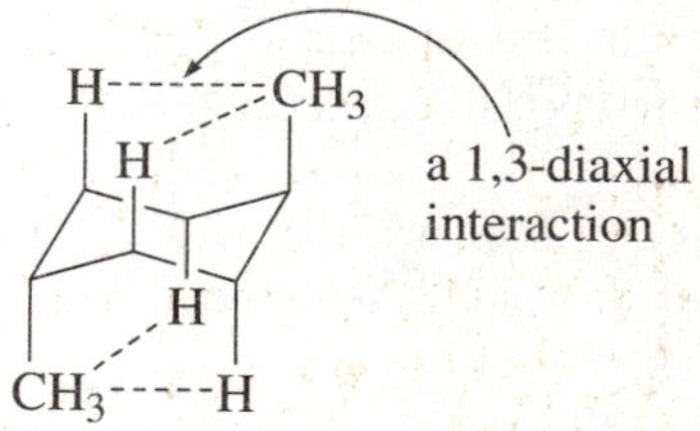

b. Each of the chair conformers of *cis*-1,4-dimethylcyclohexane has one substituent in an equatorial position and one in an axial position. Therefore, the two conformers have the same energy. (Thus, the difference in energy between them is 0 kcal/mol).

45. Both Kekulé and skeletal structures are shown.

a. 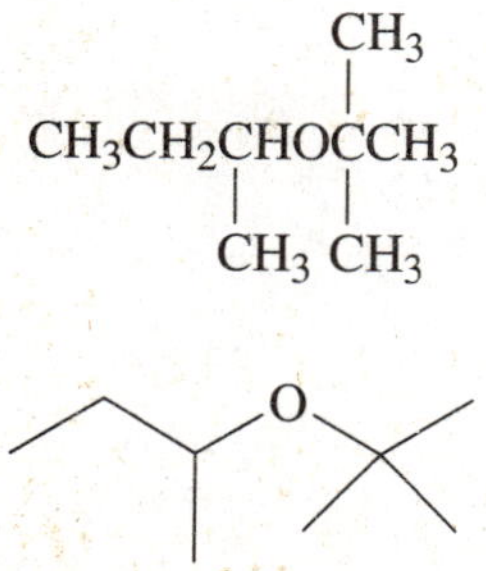

b. 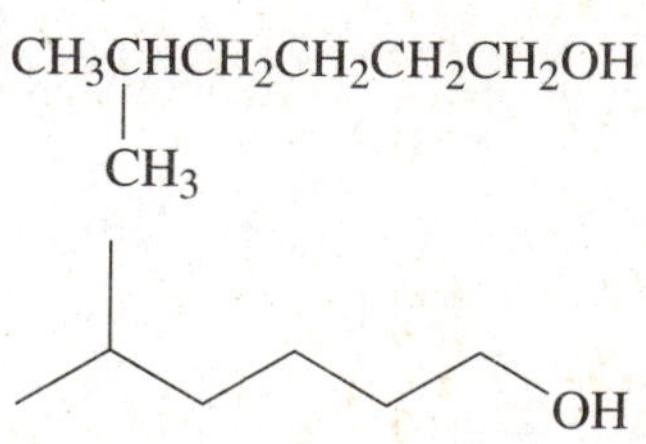

c.

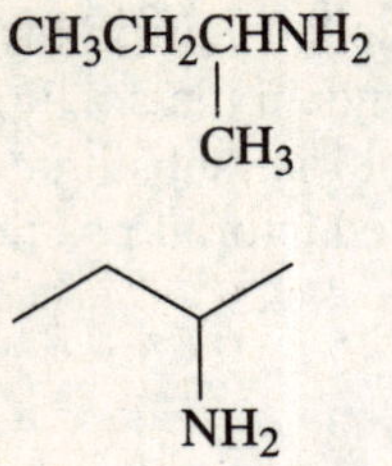

d.

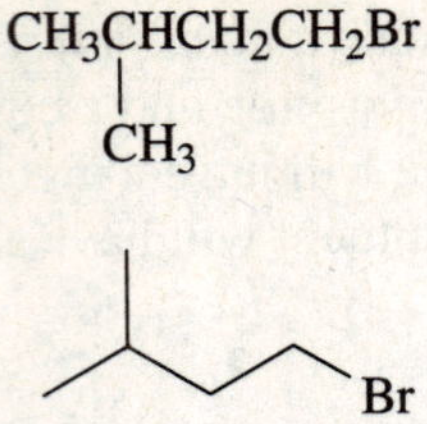

e.

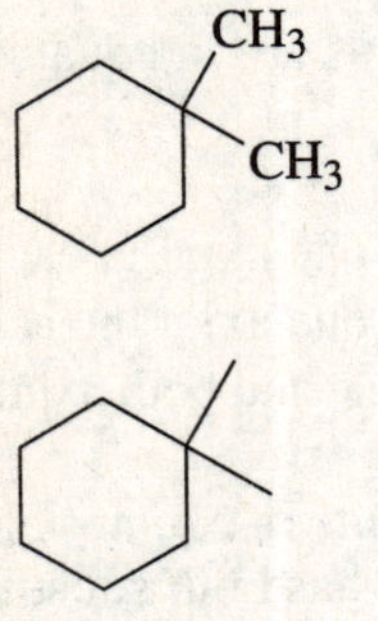

f.

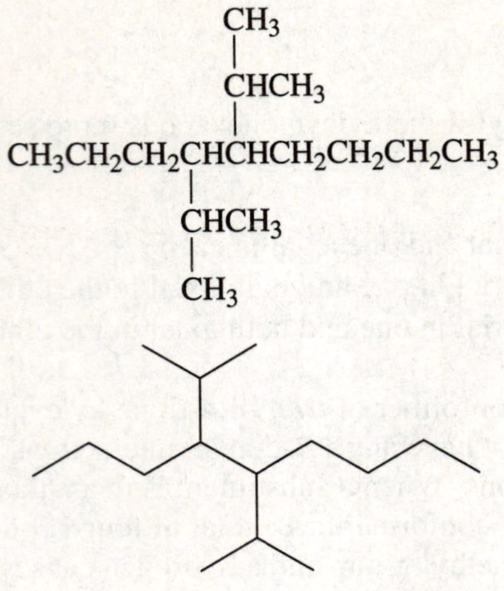

g.

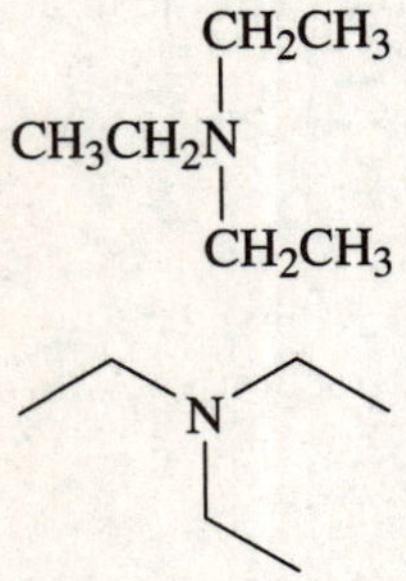

h.

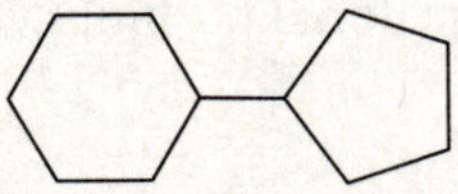

i.

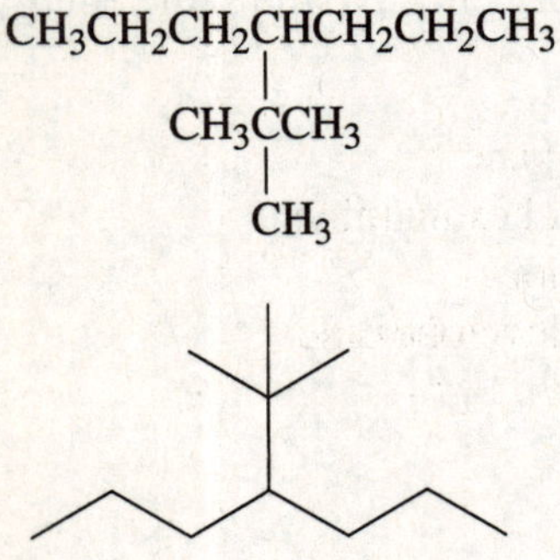

j.

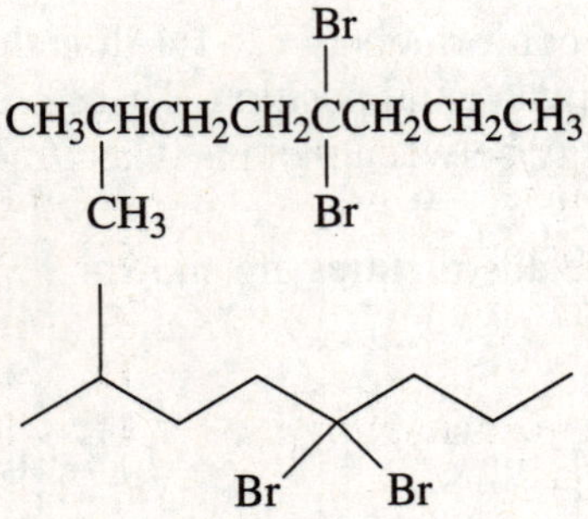

k. 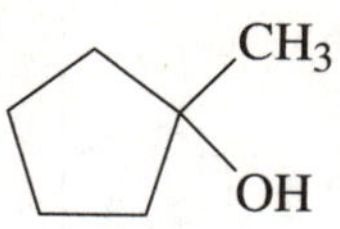

l. 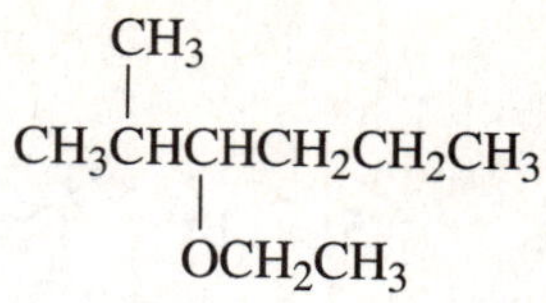

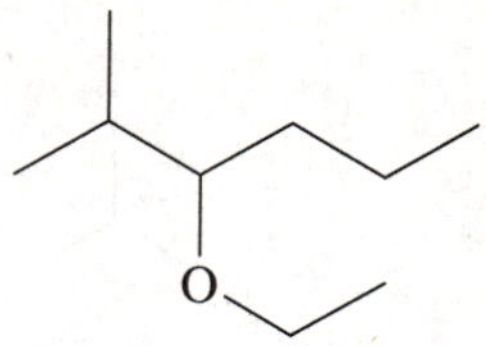

m. 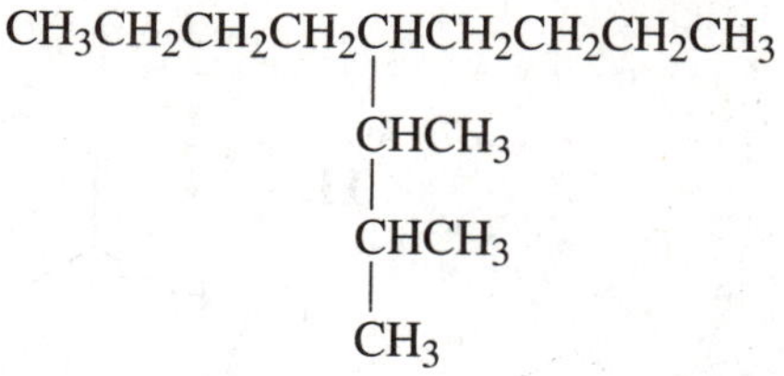

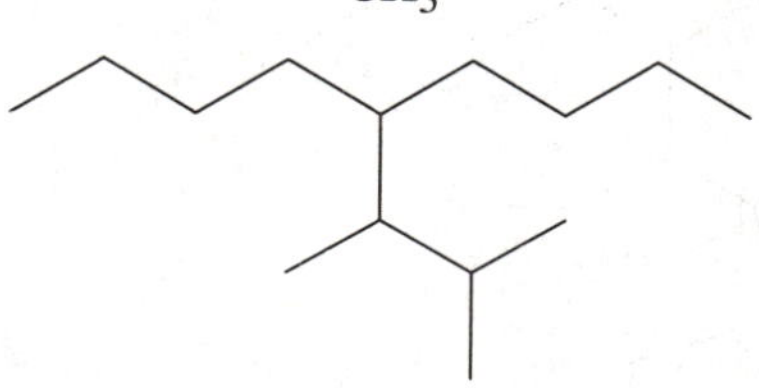

n. 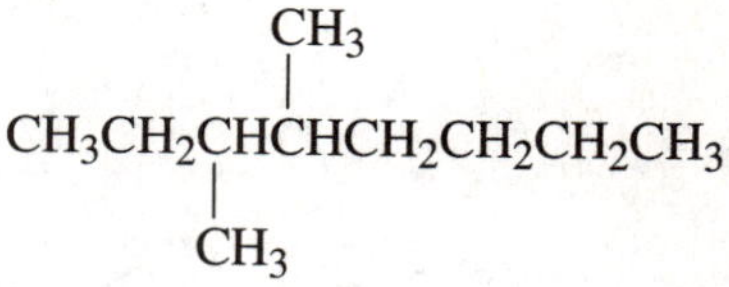

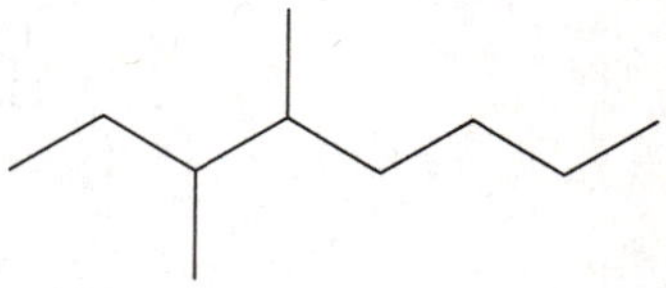

o. 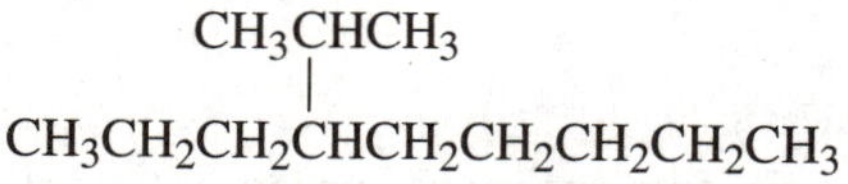

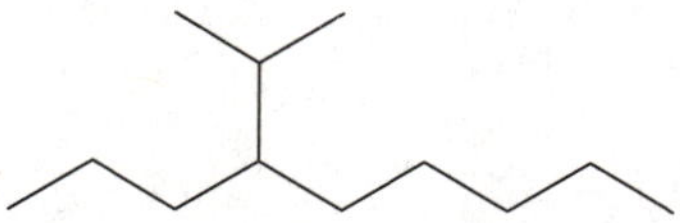

46. **a.**

1. 2,2,6-trimethylheptane
2. 5-bromo-2-methyloctane
3. 5-methyl-3-hexanol
4. 3,3-diethylpentane
5. 5-bromo-*N*-ethyl-1-pentanamine
6. 2,3,5-trimethylhexane
7. 3-ethoxyheptane
8. 1,3-dimethoxypropane
9. *N*,*N*-dimethylcyclohexanamine
10. 3-ethylcyclohexanol
11. 1-bromo-4-methylcyclohexane

b.

1. 

5. 

9. 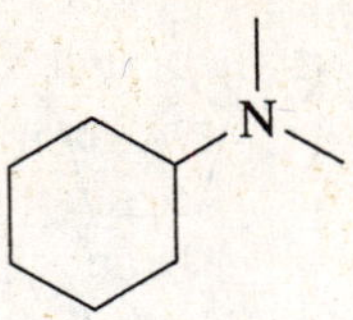

2. 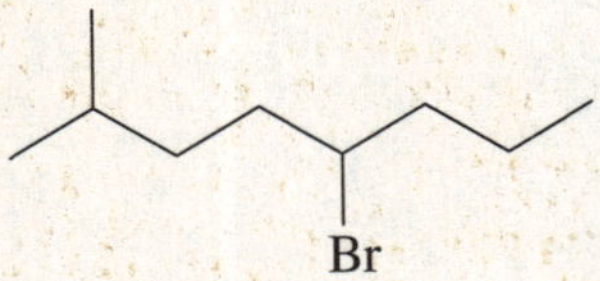

6. 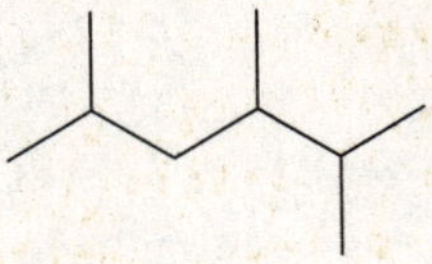

10. 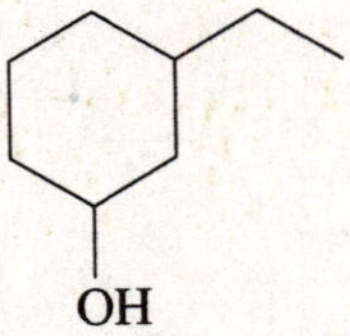

3. 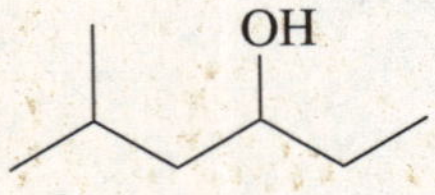

7. 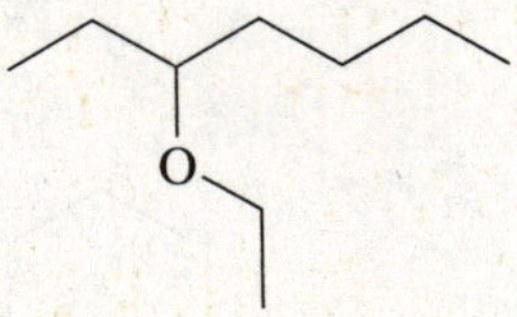

11. 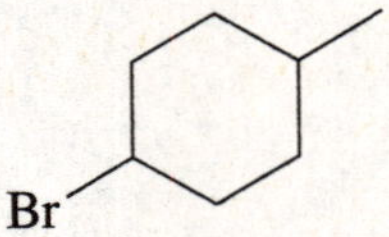

4. 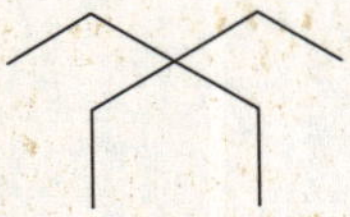

8. 

47. **a.** 3 **b.** 6 **c.** 3

48. The first conformer is the most stable because the three substituents are more spread out, so its gauche interactions will not be as large—the Cl in the first conformer is between a CH_3 and an H, whereas the Cl in the other two conformers is between two CH_3 groups.

49. **a.** 

b.
```
   CH3 CH3
    |   |
CH3C — CCH3
    |   |
   CH3 CH3
```

c.
```
CH3CHCH2CHCH3
    |     |
   CH3   CH3
```

50.
a. 2-butanamine
sec-butylamine

b. 2-chlorobutane
sec-butyl chloride

c. *N*-ethyl-2-butanamine
sec-butylethylamine

d. 1-ethoxypropane
ethyl propyl ether

e. 2-methylpentane
isohexane

f. 2-propanamine
isopropylamine

g. 2-bromo-2-methylbutane
tert-pentyl bromide

h. 4-methyl-1-pentanol
isohexyl alcohol

i. bromocyclopentane
cyclopentyl bromide

j. cyclohexanol
cyclohexyl alcohol

51.
a. 1-bromohexane
b. pentyl chloride
c. 1-butanol
d. 1-hexanol
e. hexane
f. 1-pentanol
g. 1-bromopentane
h. butyl alcohol
i. octane
j. isopentyl alcohol
(The alcohol has the higher boiling point because it forms stronger hydrogen bonds. If you were asked to compare their solubilities in water, the amine is more soluble because it forms more hydrogen bonds.)
k. hexylamine (it can form more hydrogen bonds)

52.

#1 $CH_3CH_2CH_2CH_2CH_2CH_2CH_3$
heptane

#2
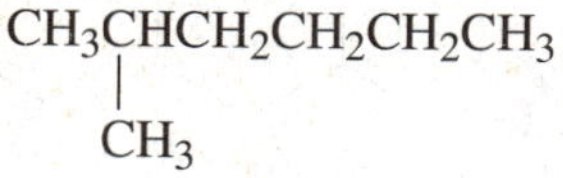
2-methylhexane

#3
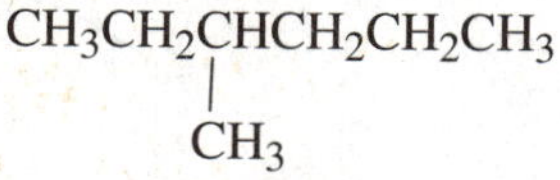
3-methylhexane

#4
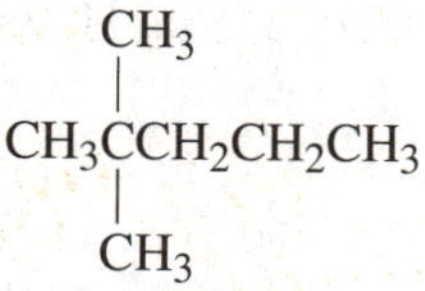
2,2-dimethylpentane

#5
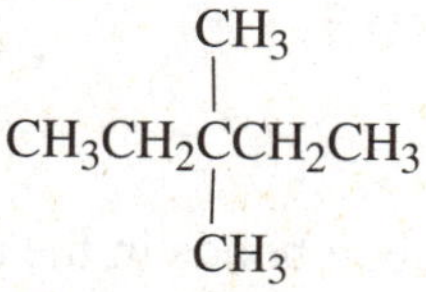
3,3-dimethylpentane

#6
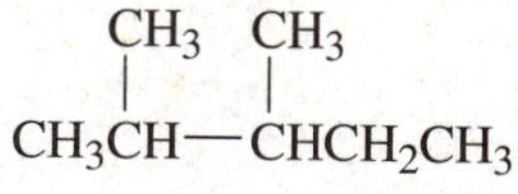
2,3-dimethylpentane

#7
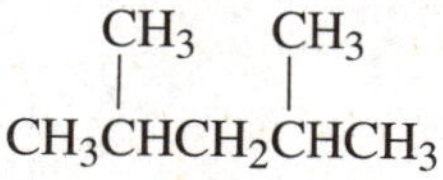
2,4-dimethylpentane

#8
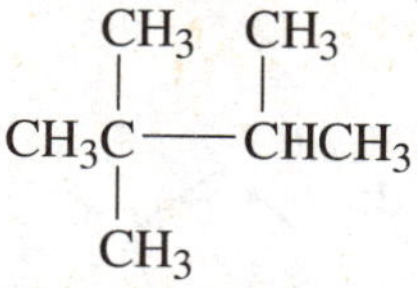
2,2,3-trimethylbutane

#9
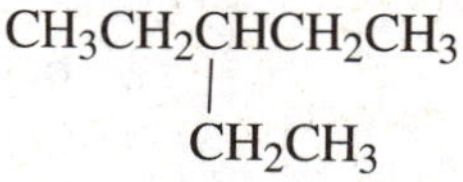
3-ethylpentane

53. Ansaid is more soluble in water. It has a fluoro substituent that can hydrogen bond to water. Hydrogen bonding increases its solubility in water.
Motrin has a nonpolar isobutyl substituent in place of Ansaid's fluoro substituent.

54. Al Kane named only one compound correctly.

a. 2-bromo-3-pentanol
b. 4-ethyl-2,2-dimethylheptane
c. 3-methylcyclohexanol
g. correct
h. 2,5-dimethylheptane
i. 5-bromo-2-pentanol

d. 2,2-dimethylcyclohexanol
e. 5-(2-methylpropyl)nonane
f. 1-bromo-3-methylbutane
j. 3-ethyl-2-methyloctane
k. 2,3,3-trimethyloctane
l. 5-methyl-*N*,*N*-dimethyl-3-hexanamine

55. B has the highest energy. All three compounds are diaxial-substituted cyclohexanes, so each one has four 1,3-diaxial interactions. Only B has a 1,3-diaxial interaction between CH_3 and Cl, which will be greater than a 1,3-diaxial interaction between CH_3 and H or a 1,3-diaxial interaction between Cl and H.

56. The only one is 2,2,3-trimethylbutane.

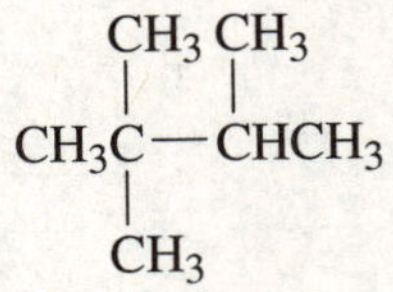

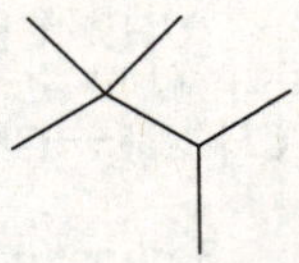

57. **a.**

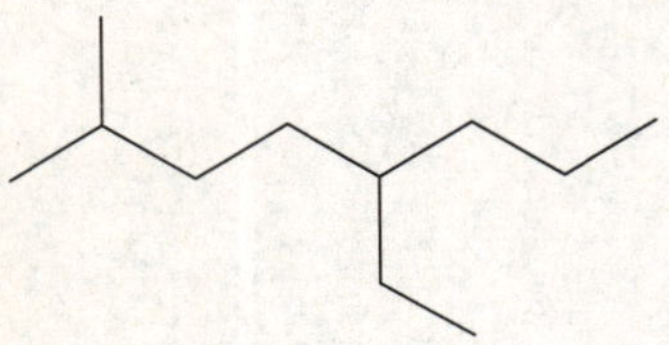

c.

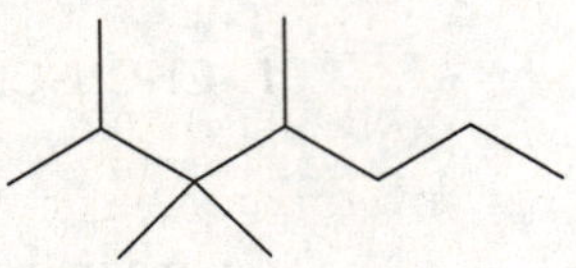

e.

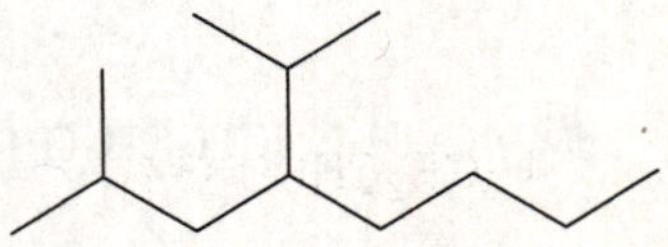

b.

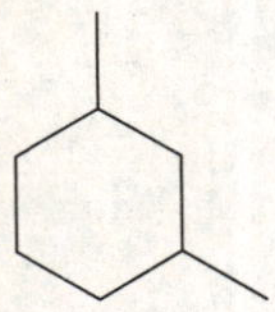

d.

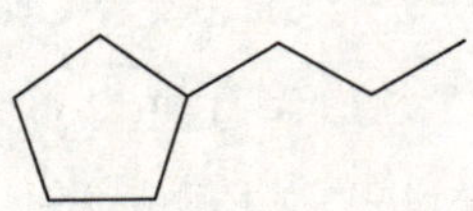

f.

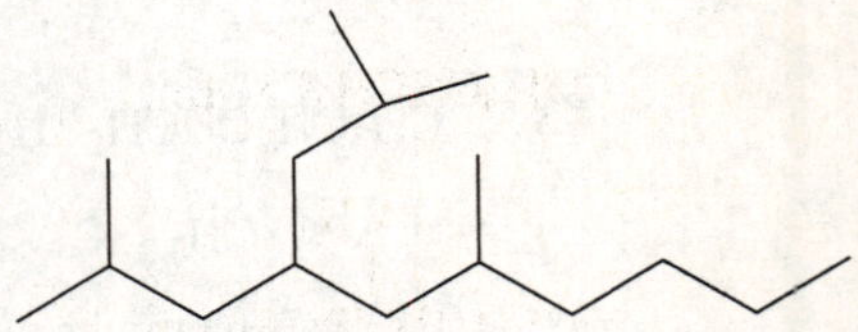

58. **a.**

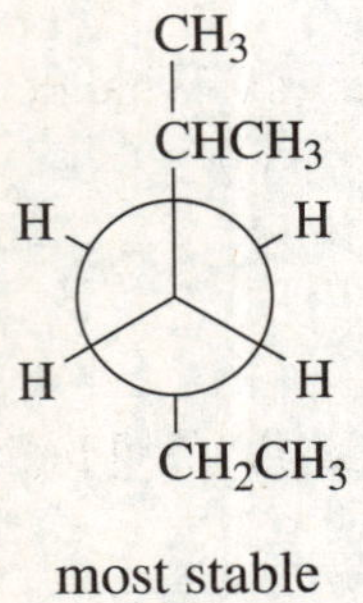

most stable

b.

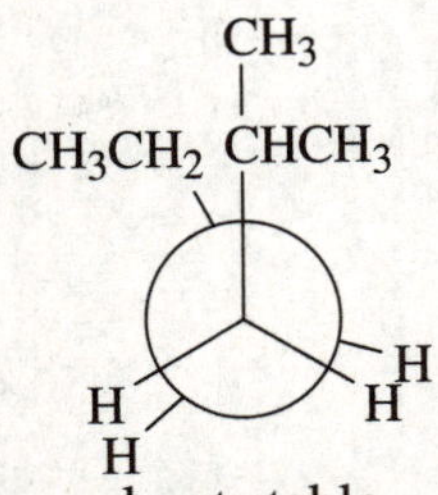

least stable

c. Rotation can occur about all the C—C bonds. There are six carbon-carbon bonds in the compound, so there are five other carbon-carbon bonds, in addition to the C_3-C_4 bond, about which rotation can occur.

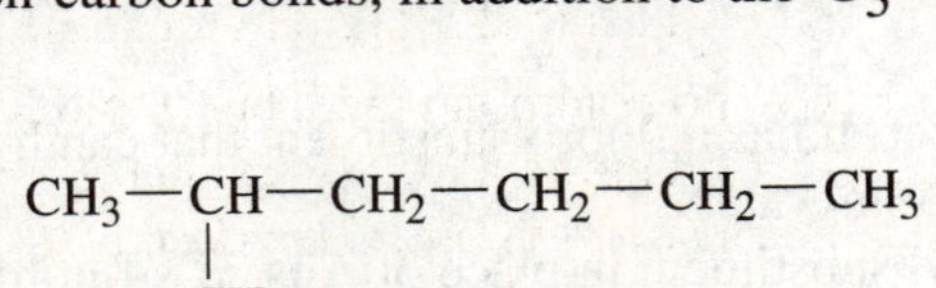

d. Three of the carbon-carbon bonds have staggered conformers that are equally stable because each is bonded to a carbon with three identical substituents.

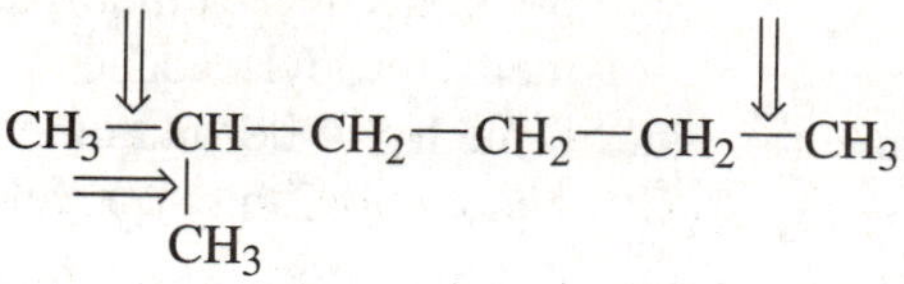

59. **C** and **D** are cis isomers. (In **C** both substituents are downward pointing; in **D** both substituents are upward pointing.)

60. $CH_3CH_2CH_2CH_2CH_2Br$

a. 1-bromopentane — primary alkyl halide
b. pentyl bromide

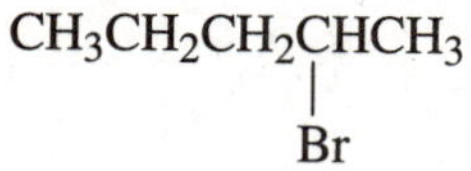

a. 2-bromopentane — secondary alkyl halide
b. no common name

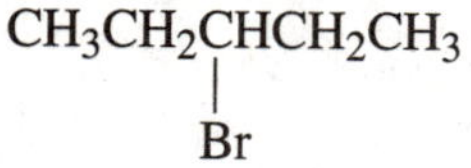

a. 3-bromopentane — secondary alkyl halide
b. no common name

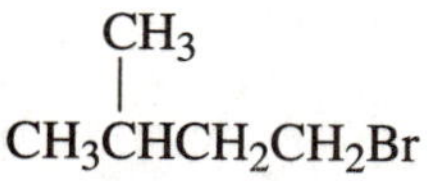

a. 1-bromo-3-methylbutane — primary alkyl halide
b. isopentyl bromide

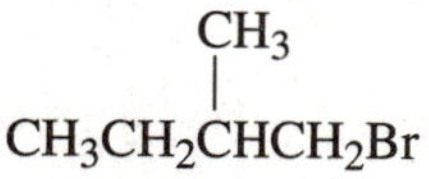

a. 1-bromo-2-methylbutane — primary alkyl halide
b. no common name

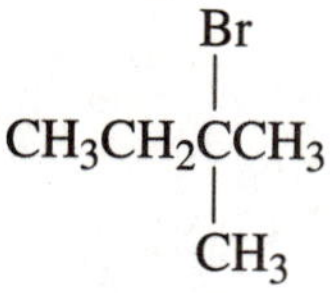

a. 2-bromo-2-methylbutane — tertiary alkyl halide
b. *tert*-pentyl bromide

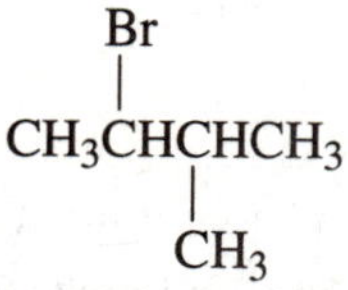

a. 2-bromo-3-methylbutane — secondary alkyl halide
b. no common name

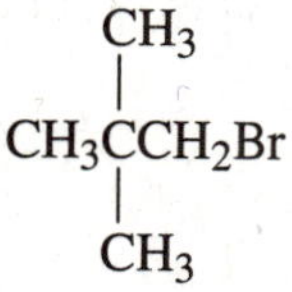

a. 1-bromo-2,2-dimethylpropane — primary alkyl halide
b. no common name, but in older literature the common name neopentyl bromide is used.

c. Four isomers do not have common names.
d. Four isomers are primary alkyl halides.
e. Three isomers are secondary alkyl halides.
f. One isomer is a tertiary alkyl halide.

61.
a. butane
b. 1-propanol
c. 4-propyl-1-nonanol
d. 5-isopropyl-2-methyloctane or 2-methyl-5-(1-methylethyl)octane
e. 6-chloro-4-ethyl-3-methyloctane
f. 1-methoxy-5-methyl-3-propylhexane
g. 6-isobutyl-2,3-dimethyldecane or 2,3-dimethyl-6-(2-methylpropyl)decane
h. 8-methyl-4-decanamine
i. 1-isobutyl-2-methylcyclohexane

62. a.

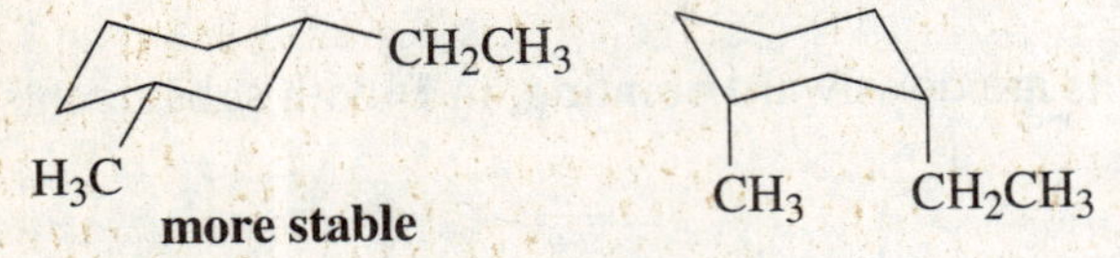

b.

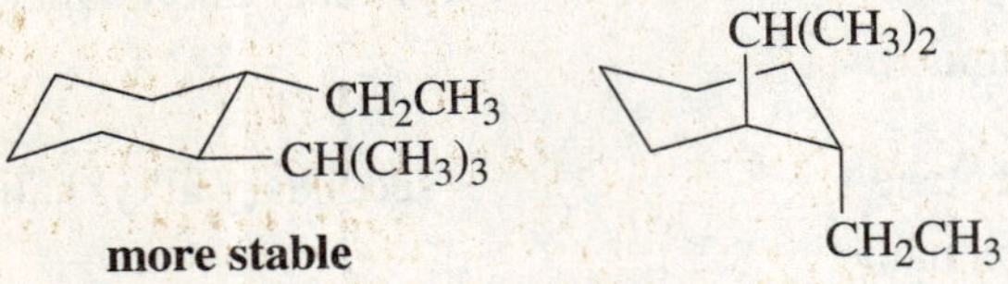

c.

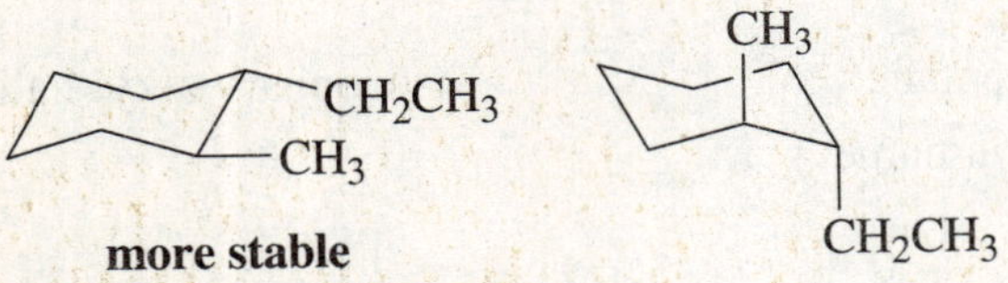

d.

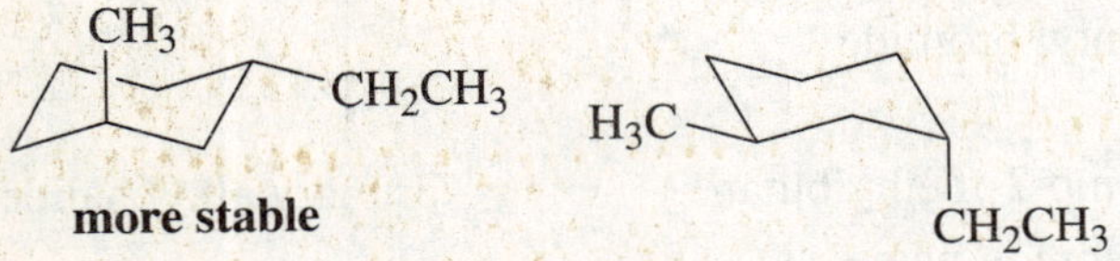

e.

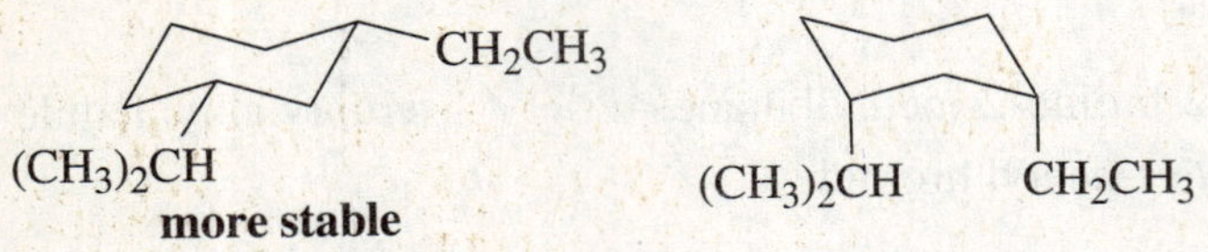

f.

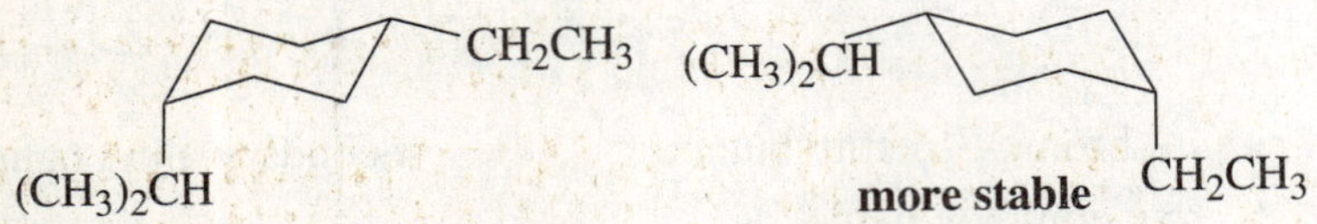

63. Alcohols with low molecular weights are more water soluble than alcohols with high molecular weights because, as a result of having fewer carbons, they have a smaller nonpolar component that has to be dragged into water.

64. Six ethers have molecular formula = $C_5H_{12}O$.

$CH_3OCH_2CH_2CH_2CH_3$

1-methoxybutane
butyl methyl ether

$CH_3CHCH_2CH_3$ with OCH_3 on C-2

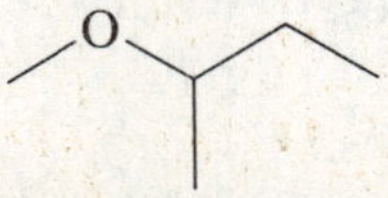

2-methoxybutane
sec-butyl methyl ether

$CH_3CH_2OCH_2CH_2CH_3$

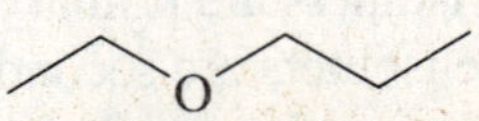

1-ethoxypropane
ethyl propyl ether

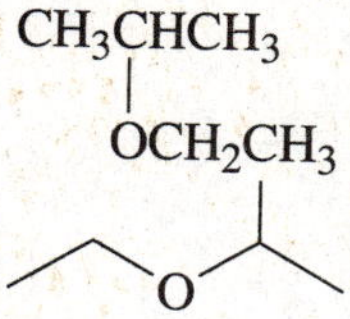

2-ethoxypropane
ethyl isopropyl ether

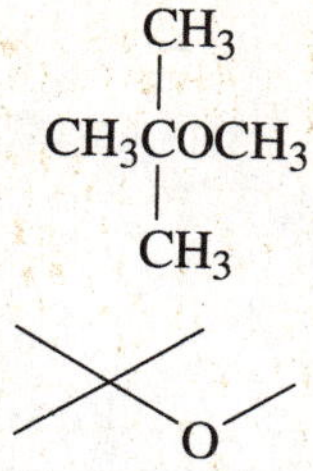

2-methoxy-2-methylpropane
tert-butyl methyl ether

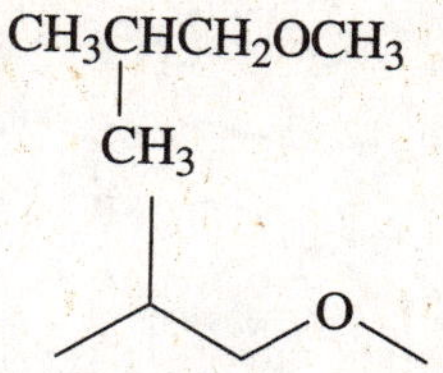

1-methoxy-2-methylpropane
isobutyl methyl ether

65. The most stable conformer has two CH3 groups in equatorial positions and one in an axial position. (The other conformer would have two CH3 groups in axial positions and one in an equatorial position.)

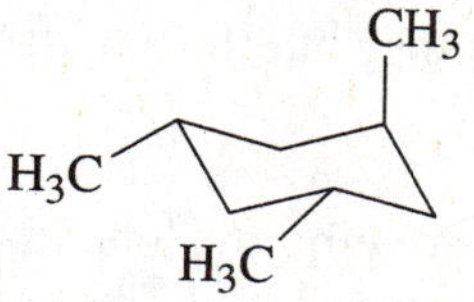

66.
a. 6-methyl-*N*-methyl-3-heptanamine
b. 3-ethyl-2,5-dimethylheptane
c. 1,2-dichloro-3-methylpentane
d. 2,3-dimethylpentane
e. 5-(1,1-dimethylpropyl)nonane
f. 5-(2-ethylbutyl)-3,3-dimethyldecane

67. One chair conformer of *trans*-1,2-dimethylcyclohexane has both substituents in equatorial positions, so it does not have any 1,3-diaxial interactions. However, Figure 2.12 on p. 111 of the textbook shows that the two methyl substituents are gauche to one another (as they would be in gauche butane), giving it a strain energy of 0.9 kcal/mol.

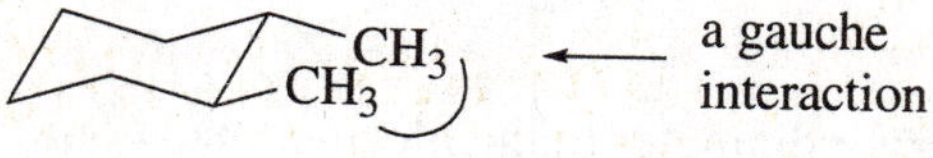

The other chair conformer of *trans*-1,2-dimethylcyclohexane has both substituents in axial positions. When a substituent is in an axial position, it experiences two 1,3-diaxial interactions. This chair conformer, therefore, has a total of four 1,3-diaxial interactions. Each diaxial interaction is between a CH_3 and an H, so each results in a strain energy of 0.9 kcal/mol. Therefore this chair conformer has a strain energy of 3.6 kcal/mol ($4 \times 0.9 = 3.6$).

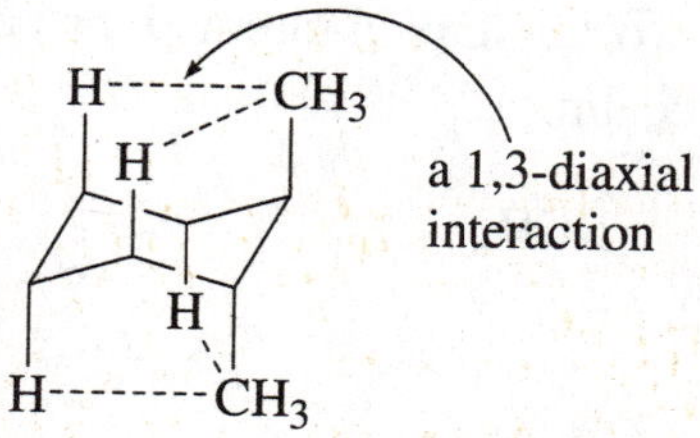

Thus, one conformer is 2.7 kcal/mole more stable than the other.

68.

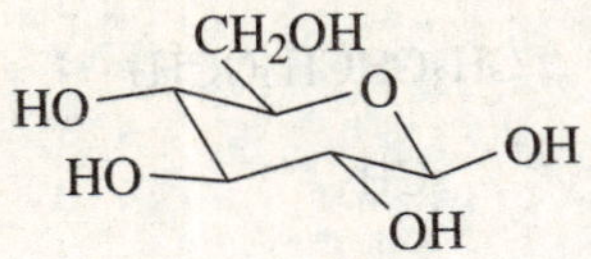

69.
a. 5-methyl-3-hexanol
b. bromocylopentane
c. 2-methyl-3-pentanol
d. 5-bromo-2-methyloctane
e. 5-ethyl-6-propyldecane
f. 6-bromo-2-hexanol
g. 4-ethyl-3-methylcyclohexanol
h. 4-bromo-1-ethyl-2-methylcyclohexane

70. **a.** 1-Hexanol has a higher boiling point than 3-hexanol because the alkyl group in 1-hexanol can better engage in van der Waals interactions. The OH group of 3-hexanol makes it more difficult for is 6 carbons to lie close to the 6 carbons of another molecule of 3-hexanol.

b. The floppy ethyl groups in diethyl ether interfere with the ability of the oxygen atom to engage in hydrogen bonding with water.

71. One of the chair conformers of *cis*-1,3-dimethylcyclohexane has both substituents in equatorial positions, so there are no unfavorable 1,3-diaxial interactions. The other chair conformer has three 1,3-diaxial interactions, two between a CH_3 and an H and one between two CH_3 groups. We know that a 1,3-diaxial interaction between a CH_3 and an H is 0.9 kcal/mol. Subtracting 1.8 from 5.4 (the energy difference between the two conformers) results in a value of 3.6 kcal/mol for the 1,3-diaxial interaction between the two CH_3 groups.

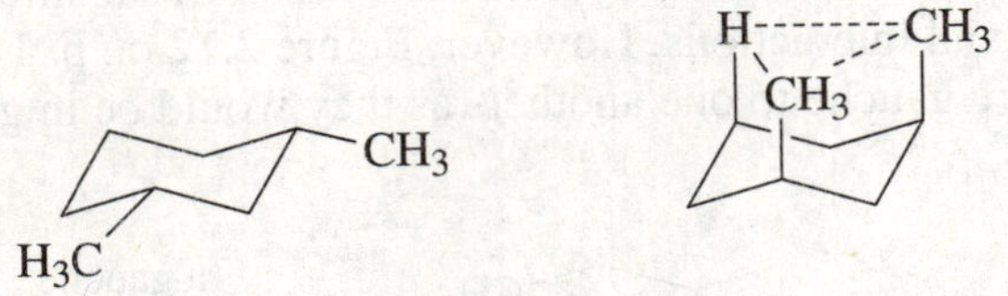

72. Because bromine has a larger diameter than chlorine, one would expect bromine to have a greater preference for the equatorial position as would be indicated by a larger ΔG°. However, Table 2.10 indicates that it has a smaller ΔG°, indicating that it has less preference for the equatorial position than chlorine has. The C—Br bond is longer than the C—Cl bond, which causes bromine to be farther away than chlorine from the other axial substituents. Apparently, the longer bond more than offsets the larger diameter.

73. The conformer on the left has two 1,3-diaxial interactions between a CH_3 and an H (2×0.9 kcal/mole) for a total strain energy of 1.8 kcal/mol.

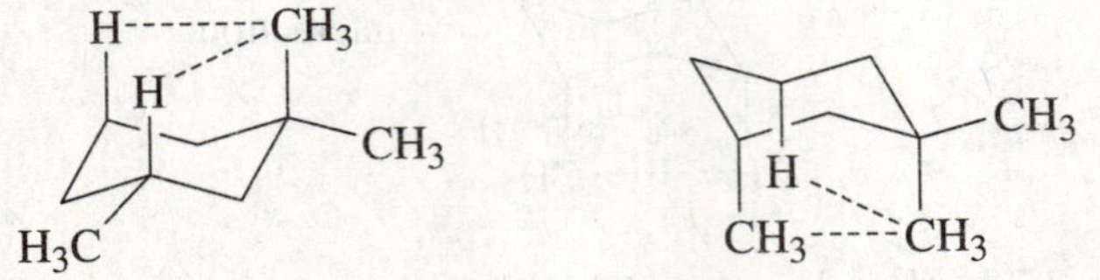

The conformer on the right has three 1,3-diaxial interactions, two between a CH_3 and an H (1.8 kcal/mole) and one between two CH_3 groups (3.6 kcal/mole; see Problem 71) for a total strain energy of 5.4 kcal/mol. Therefore, the conformer on the left will predominate at equilibrium.

Chapter 2 Practice Test

1. Name the following compound:

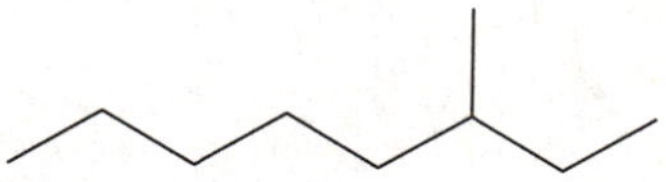

2. Draw the following conformers of hexane considering rotation about the C3—C4 bond:

   a. the most stable of all the conformers

   b. the least stable of all the conformers

   c. a gauche conformer

3. Give two names for each of the following compounds:

   a. $CH_3CH_2CHCH_3$ (with Cl on the CH)

   b. $CH_3CHCH_2CH_2CH_2OH$ (with CH_3 on the CH)

   c. 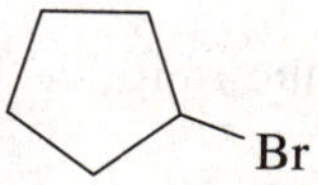

4. Label the three compounds in each set in order of decreasing boiling point. (Label the highest boiling compound #1, the next #2, and the lowest boiling #3.)

   a. $CH_3CH_2CH_2CH_2CH_2Br$ $\quad$ $CH_3CH_2CH_2Br$ $\quad$ $CH_3CH_2CH_2CH_2Br$

   b. $CH_3CH_2CH_2CH_2CH_3$ $\quad$ $CH_3CH_2CH_2CH_2OH$ $\quad$ $CH_3CH_2CH_2CH_2Cl$

   c. $CH_3C(CH_3)_2—C(CH_3)_2CH_3$ $\quad$ $CH_3CH_2CH_2CH_2CH_2CH_2CH_2CH_3$ $\quad$ $CH_3CHCH_2CH_2CH_2CH_2CH_3$ (with CH_3 on the CH)

5. Give the systematic name for each of the following compounds:

   a. $CH_3CHCH_2CH_2CHCH_2CH_3$ (with CH_3 on the first CH and OH on the second CH)

   b. $CH_3CH_2CHOCH_2CH_3$ (with $CH_2CH_2CH_2CH_3$ on the CH)

   c. 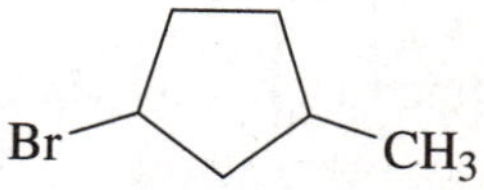

   d. $CH_3CHCHCH_2CH_2CH_2Cl$ (with CH_2CH_3 on the first CH and Cl on the second CH)

6. Draw the other chair conformer.

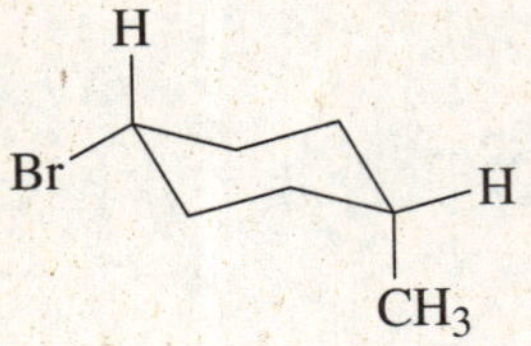

7. Draw the most stable conformer of *trans*-1-isopropyl-3-methylcyclohexane.

8. Which of the following has:

a. the higher boiling point: diethyl ether or butyl alcohol?
b. the greater solubility in water: 1-butanol or 1-pentanol?
c. the higher boiling point: hexane or isohexane?
d. the higher boiling point: pentylamine or ethylmethylamine?
e. the greater solubility in water: ethyl alcohol or ethyl chloride?

9. Give two names for each of the following compounds:

a. $CH_3CHCH_2CH_2Br$ (with CH_3 on the second carbon: $\overset{}{CH_3}\underset{\displaystyle |\atop \displaystyle CH_3}{CH}CH_2CH_2Br$)

b. $CH_3\underset{\displaystyle |\atop \displaystyle CH_3}{CH}CH_2CH_2OH$

c. $CH_3\underset{\displaystyle |\atop \displaystyle CH_3}{CH}CH_2CH_2NH_2$

10. Which is more stable:

a. A staggered conformer or an eclipsed conformer?

b. The chair conformer of methylcyclohexane with the methyl group in the axial position or the chair conformer of methylcyclohexane with the methyl group in the equatorial position?

c. Cyclohexane or cyclobutane?

11. Give the structure of the following:

a. a secondary alkyl bromide that has three carbons

b. a secondary amine that has three carbons

c. an alkane with no secondary hydrogens

d. a constitutional isomer of butane

e. three compounds with molecular formula C_3H_8O

CHAPTER 3

Alkenes: Structure, Nomenclature, and an Introduction to Reactivity
Thermodynamics and Kinetics

Important Terms

acyclic — noncyclic.

addition reaction — a reaction in which atoms or groups are added to the reactant.

alkene — a hydrocarbon that contains a double bond.

allyl group — $CH_2{=}CHCH_2{-}$

allylic carbon — an sp^3 carbon adjacent to a vinyl carbon.

Arrhenius equation — relates the rate constant of a reaction to the energy of activation and to the temperature at which the reaction is carried out ($k = Ae^{-E_a/RT}$).

cis isomer — the isomer with the hydrogens on the same side of the double bond.

degree of unsaturation — the number of π bonds and/or rings in a hydrocarbon.

***E* isomer** — the isomer with the high-priority groups on opposite sides of the double bond.

electrophile — an electron deficient atom or molecule.

electrophilic addition reaction — an addition reaction in which the first species that adds to the reactant is an electrophile.

endergonic reaction — a reaction with a positive $\Delta G°$.

endothermic reaction — a reaction with a positive $\Delta H°$.

enthalpy — the heat given off ($-\Delta H°$) or the heat absorbed ($+\Delta H°$) during the course of a reaction.

entropy — a measure of the freedom of motion in a system.

exergonic reaction — a reaction with a negative $\Delta G°$.

exothermic reaction — a reaction with a negative $\Delta H°$.

experimental energy of activation ($E_a = \Delta H^\dagger - RT$) — a measure of the approximate energy barrier to a reaction. (It is approximate because it does not contain an entropy component.)

first-order rate constant — the rate constant of a first-order reaction.

first-order reaction (unimolecular reaction)	a reaction whose rate is dependent on the concentration of one reactant.
free energy of activation ($\Delta G^{\dagger}$)	the true energy barrier to a reaction.
functional group	the center of reactivity of a molecule.
geometric isomers (cis-trans stereoisomers)	cis-trans (or *E,Z*) isomers.
Gibbs free energy change	the difference between the free energy content of the products and the free energy content of the reactants at equilibrium under standard conditions (1M, 25 °C, 1 atm).
intermediate	a species formed during a reaction that is not the final product of the reaction.
kinetics	the field of chemistry that deals with the rates of chemical reactions.
kinetic stability	is indicated by $\Delta G^{\dagger}$. If $\Delta G^{\dagger}$ is large, the compound is kinetically stable (not very reactive). If $\Delta G^{\dagger}$ is small, the compound is kinetically unstable (is very reactive).
Le Châtelier's principle	states that if an equilibrium is disturbed, the components of the equilibrium will adjust in a way that will offset the disturbance.
mechanism of the reaction	a description of the step-by-step process by which reactants are changed into products.
nucleophile	an electron-rich atom or molecule.
rate constant	a measure of how easy it is to reach the transition state of a reaction (to get over the energy barrier of the reaction).
rate-determining step or rate-limiting step	the step in a reaction that has the transition state with the highest energy.
reaction coordinate diagram	describes the energy changes that take place during the course of a reaction.
saturated hydrocarbon	a hydrocarbon that is completely saturated with hydrogen (contains no double or triple bonds).
second-order rate constant	the rate constant of a second-order reaction.
second-order reaction (bimolecular reaction)	a reaction whose rate is dependent on the concentration of two reactants, or on the square of the concentration of a single reactant.
solvation	the interaction between a solvent and another molecule (or ion).

thermodynamics the field of chemistry that describes the properties of a system at equilibrium.

thermodynamic stability is indicated by $\Delta G°$. If $\Delta G°$ is negative, the products are more stable than the reactants. If $\Delta G°$ is positive, the reactants are more stable than the products.

trans isomer the isomer with the hydrogens on opposite sides of the double bond.

transition state the highest point on a hill in a reaction coordinate diagram. In the transition state, bonds in the reactant that will break are partially broken and bonds in the product that will form are partially formed.

unsaturated hydrocarbon a hydrocarbon that contains one or more double or triple bonds.

vinyl group $CH_2{=}CH{-}$

vinylic carbon a carbon that is doubly bonded to another carbon.

***Z* isomer** the isomer with the high-priority groups on the same side of the double bond.

Solutions to Problems

1. a. C_5H_8 b. C_4H_6 c. $C_{10}H_{16}$ d. C_8H_{10}

2. a. 3 b. 4 c. 1 d. 3 e. 13

3. a. $CH_3CH{=}CH_2$

b. $CH_3C{\equiv}CH$

$CH_2{=}C{=}CH_2$

c. $HC{\equiv}CCH_2CH_3$

$CH_3C{\equiv}CCH_3$

$CH_2{=}CHCH{=}CH_2$

$CH_2{=}C{=}CHCH_3$

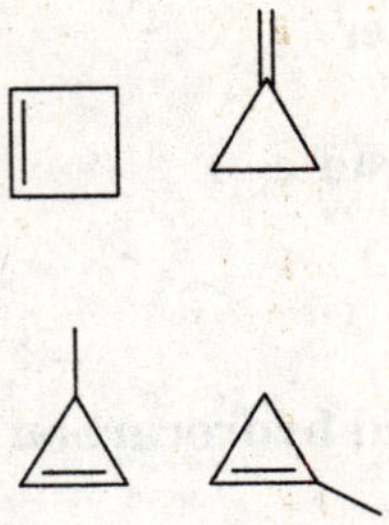

4. a.

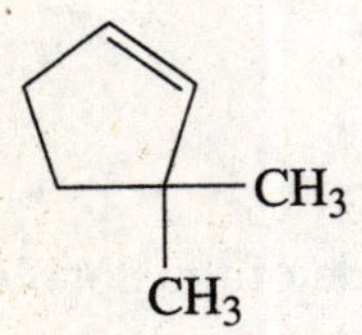

b.

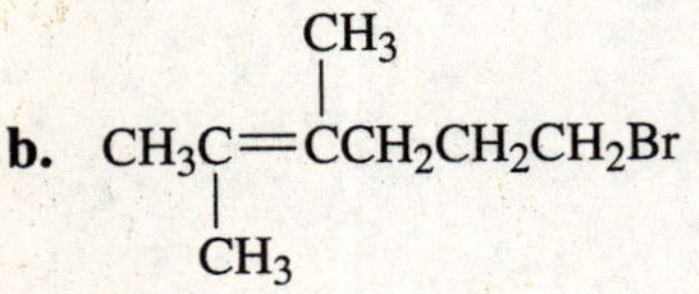

c. $CH_3CH_2OCH{=}CH_2$

d. $CH_2{=}CHCH_2OH$

5. a. 4-methyl-2-pentene
b. 2-chloro-3,4-dimethyl-3-hexene
c. 1-bromocyclopentene
d. 1-bromo-4-methyl-3-hexene
e. 1,5-dimethylcyclohexene
f. 1-butoxy-1-propene

6. a. 5 b. 4 c. 4 d. 6

7. a. 1 and 3

b. 1.

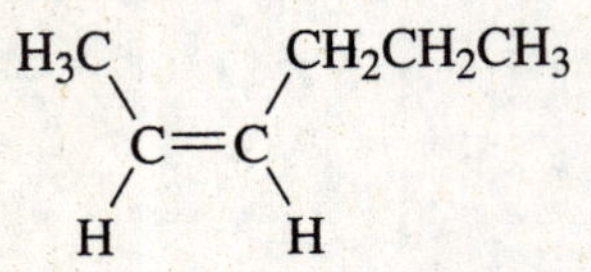

cis

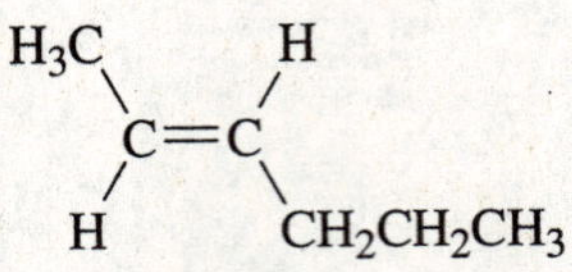

trans

3.

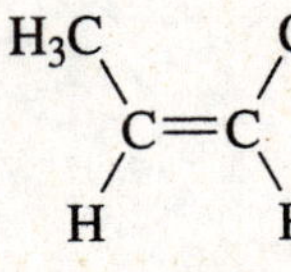

cis

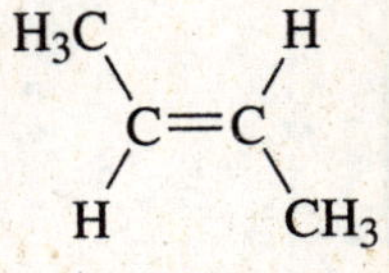

trans

8. $CH_3CH_2CH_2CH{=}CH_2$

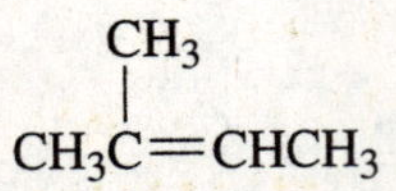

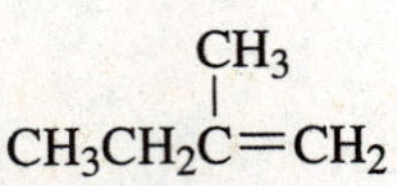

9. Only **C** has a dipole moment of zero because the bond dipoles cancel since they are in opposite directions.

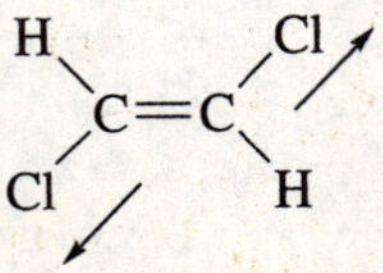

10. **a.** –I > –Br > –OH > –CH_3

b. –OH > –CH_2Cl > –CH=CH_2 > –CH_2CH_2OH

11. **a.**

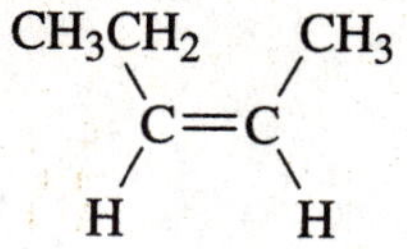

Z

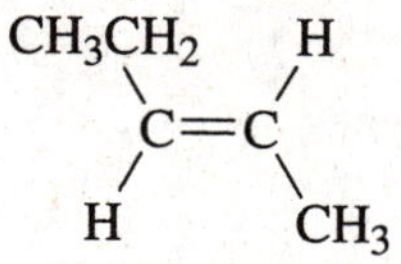

E

b.

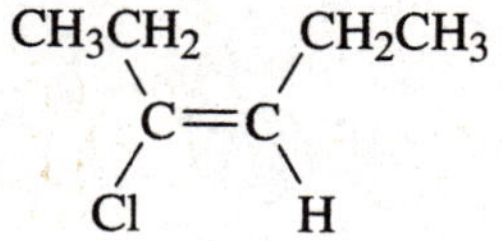

E

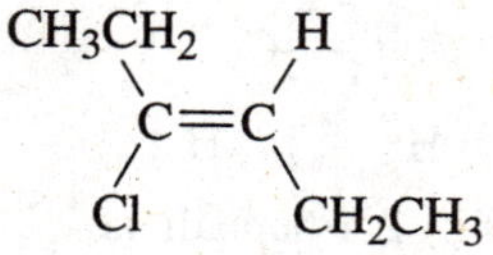

Z

c.

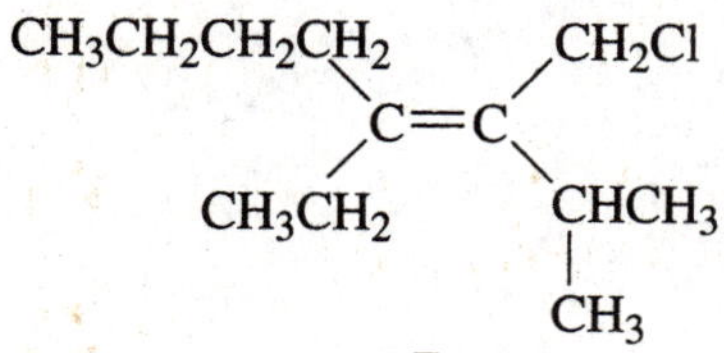

Z

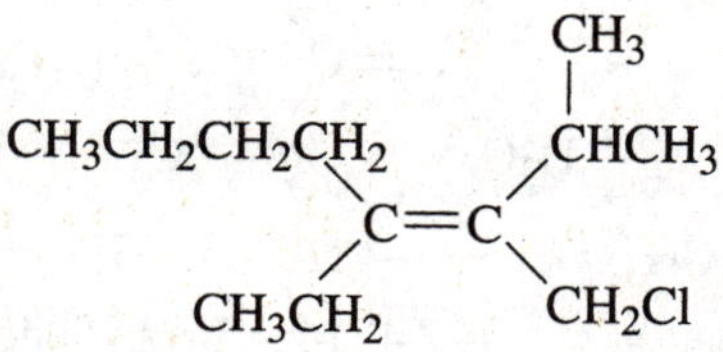

E

d.

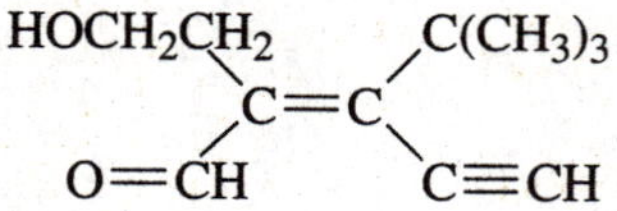

Z

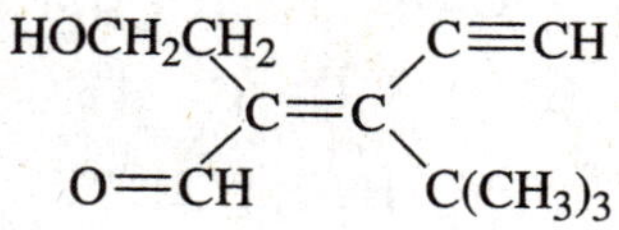

E

12. **a.** (*E*)-2-heptene **b.** (*Z*)-3,4-dimethyl-2-penene **c.** (*Z*)-1-chloro-3-ethyl-4-methyl-3-hexene

13.

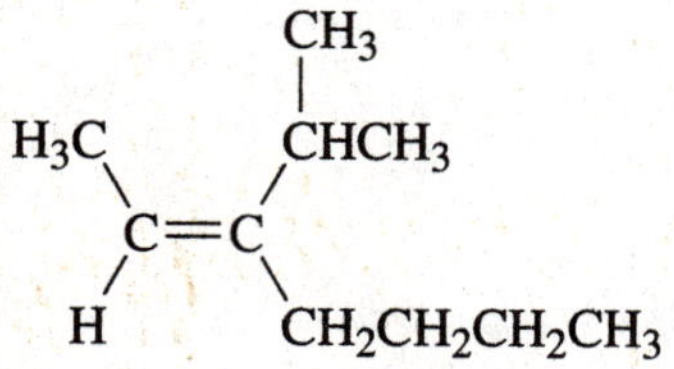

14. **a.**

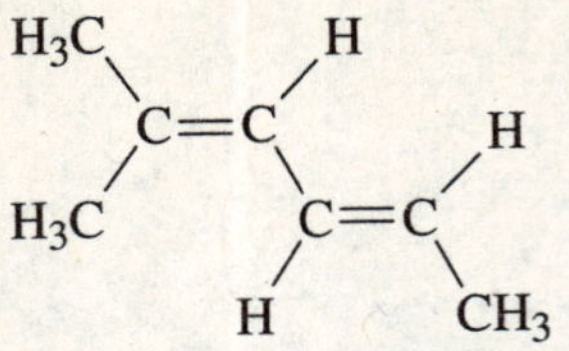

(*E*)-2-methyl-2,4-hexadiene

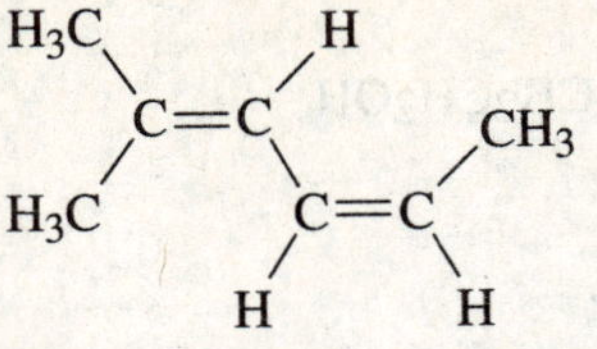

(*Z*)-2-methyl-2,4-hexadiene

b.

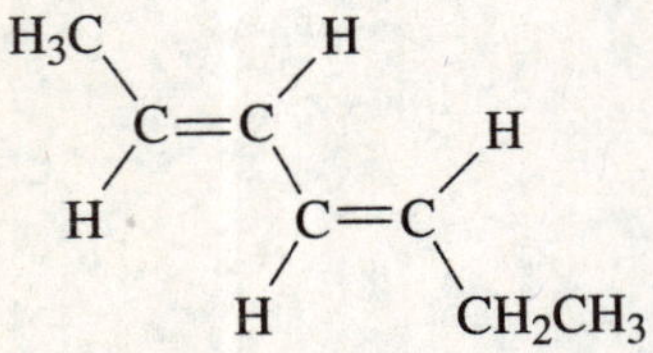

(2*E*,4*E*)-2,4-heptadiene

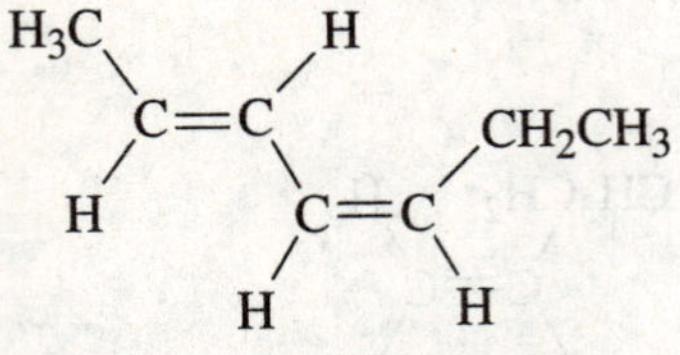

(2*E*,4*Z*)-2,4-heptadiene

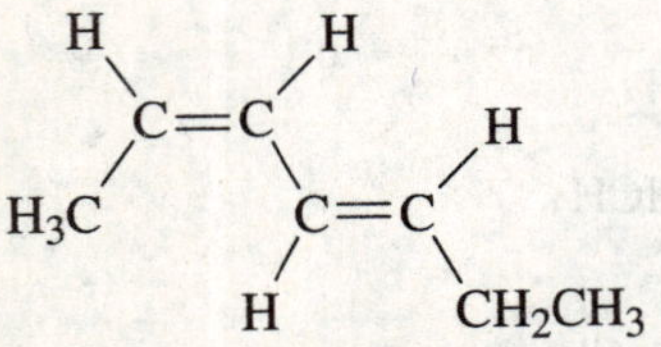

(2*Z*,4*E*)-2,4-heptadiene

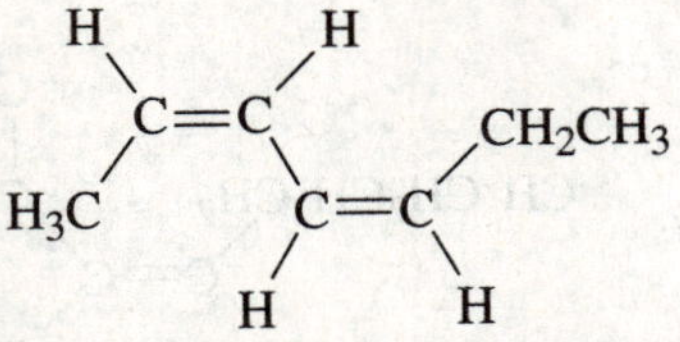

(2*Z*,4*Z*)-2,4-heptadiene

c.

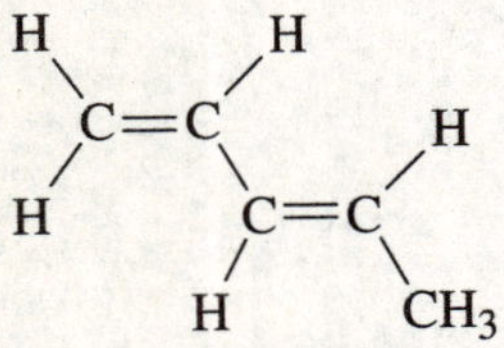

(*E*)-1,3-pentadiene

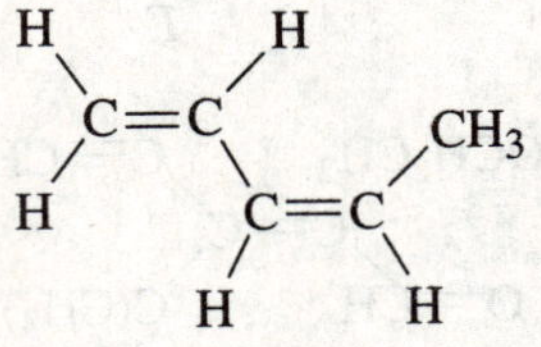

(*Z*)-1,3-pentadiene

b has four stereoisomers because each double bond can have either the *E* or the *Z* configuration.

a and **c** have only two stereoisomers because in each case there are two identical substituents bonded to one of the sp^2 carbons, so only one of the double bonds can have either the *E* or the *Z* configuration.

15. **a.** $AlCl_3$ is the electrophile and NH_3 is the nucleophile.

b. The H^+ of HBr is the electrophile and HO^- is the nucleophile.

16. nucleophiles: H^- CH_3O^- $CH_3C{\equiv}CH$ NH_3

electrophiles: $CH_3\overset{+}{C}HCH_3$

17. **1.**

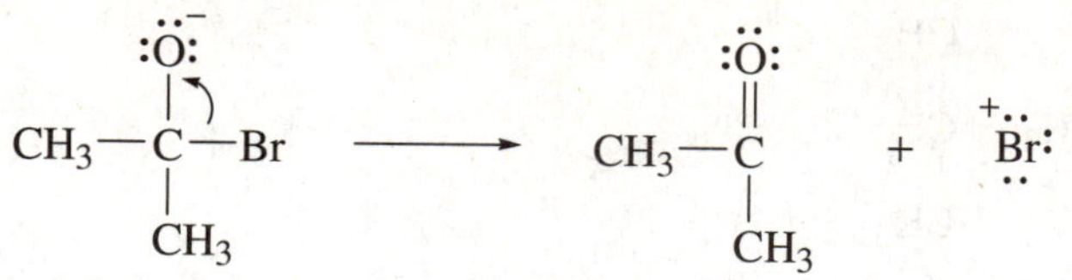

Notice that drawing the arrows incorrectly leads to a bromine with an incomplete octet and a positive charge.

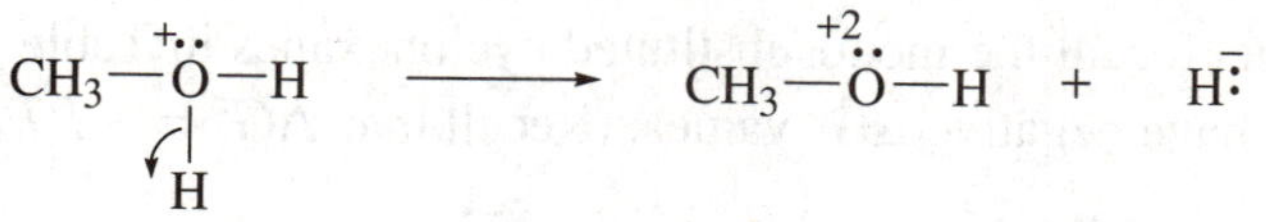

Notice that drawing the arrows incorrectly leads to an oxygen with an incomplete octet and two positive charges.

2. This one can't be drawn because the arrow is supposed to show where the electrons move to but there are no electrons on the H to go anywhere.

3.

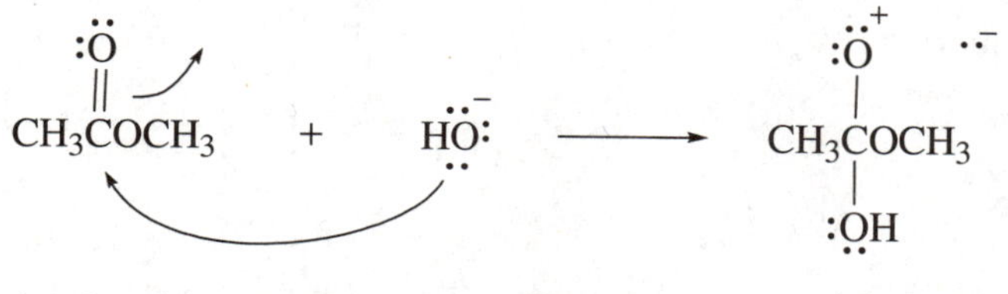

Notice that drawing the arrows incorrectly leads to an oxygen with an incomplete octet and a positive charge and a pair of electrons in outer space.

4. This one can't be drawn because the arrow is supposed to show where the electrons move to but there are no electron on the C to go anywhere.

18. **a.**

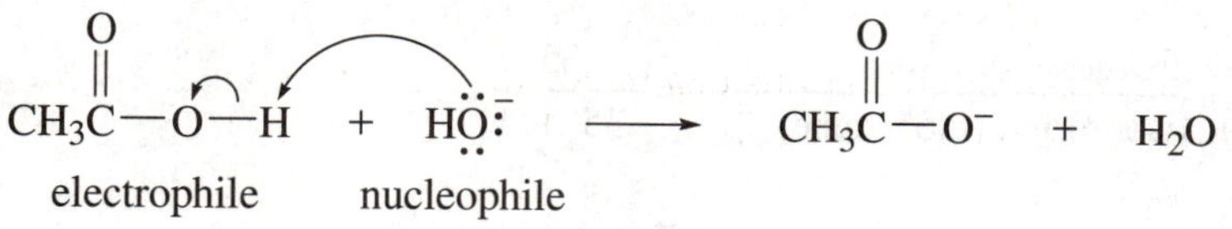

b.

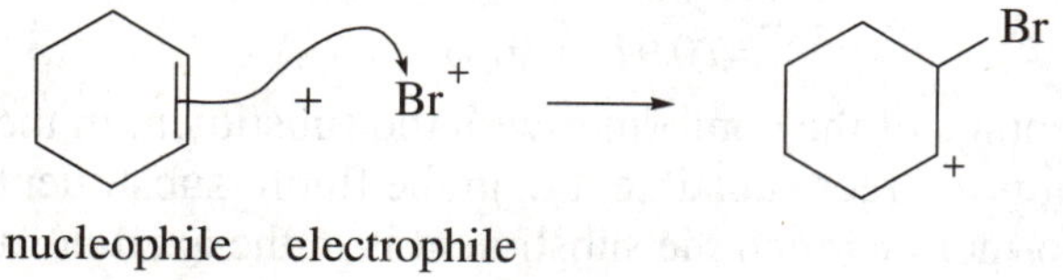

c.

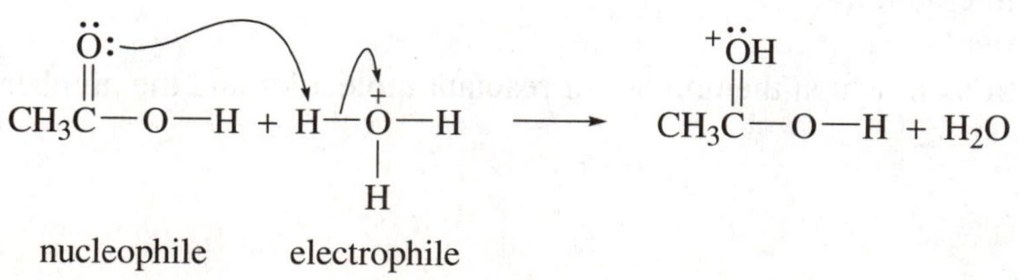

d. 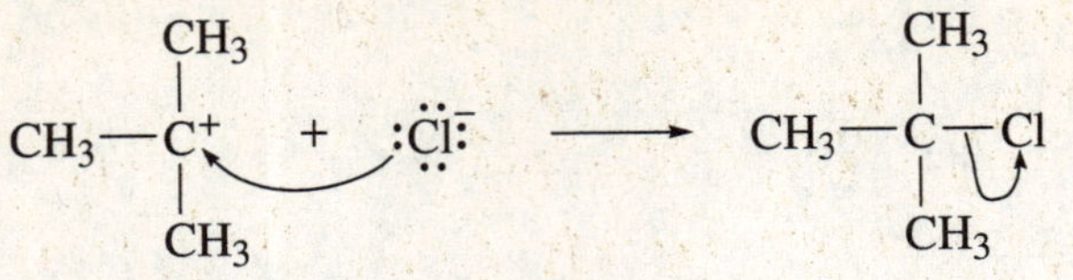

electrophile nucleophile

19. See above.

20. **a.** Because all the equilibrium constants for all the monosubstituted cyclohexanes in Table 2.10 on page 112 of the text are positive, they all have negative $\Delta G°$ values. (Recall that $\Delta G° = -RT \ln K_{eq}$.)

b. the *tert*-butyl substituent

c. the *tert*-butyl substituent

d.

$$\Delta G° = -RT \times \ln K_{eq}$$
$$\Delta G° = -1.986 \times 10^{-3} \times 298 \times \ln 18$$
$$\Delta G° = -0.59 \times \ln 18$$
$$\Delta G° = -0.59 \times 2.89$$
$$\Delta G° = -1.7 \text{ kcal/mol (or } -7.1 \text{ kJ/mol)}$$

21. **a.** Solved in the text.

b.

$$\Delta G° = -RT \ln K_{eq}$$
$$-2.1 = -1.986 \times 10^{-3} \times 298 \times \ln K_{eq}$$
$$\ln K_{eq} = 3.56$$
$$K_{eq} = 35$$

$$K_{eq} = \frac{[\text{isopropylcyclohexane}]_{\text{equatorial}}}{[\text{isopropylcyclohexane}]_{\text{axial}}} = \frac{35}{1}$$

$$\frac{[\text{isopropylcyclohexane}]_{\text{equatorial}}}{[\text{isopropylcyclohexane}]_{\text{equatorial}} + [\text{isopropylcyclohexane}]_{\text{axial}}} = \frac{35}{35 + 1}$$
$$= \frac{35}{36}$$
$$= 0.97 = 97\%$$

c. Isopropylcyclohexane has a greater percentage of the conformer with the substituent in the equatorial position because the isopropyl substituent is a larger substituent than the fluoro substituent. The larger the substituent, the less stable is the conformer in which the substituent is in the axial position because of the 1,3-diaxial interactions.

22. $\Delta S°$ is more significant in reactions in which the number of reactant molecules and the number of product molecules are not the same.

a. **1.** $A + B \rightleftharpoons C$
2. $A + B \rightleftharpoons C$

b. None of the four reactions has a positive $\Delta S°$.
In order to have a positive $\Delta S°$, the products must have greater freedom of motion than the reactants (there should be more molecules of products than molecules of reactant).

23. **a.** **(1)**

$$\Delta G° = \Delta H° - T\Delta S°$$
(recall that $T = °C + 273$)
$$\Delta G° = -12 - (273 + 30)(.01)$$
$$\Delta G° = -12 - 3 = -15 \text{ kcal/mol}$$

$$\Delta G° = -RT \ln K_{eq}$$
$$\Delta G° = -(1.986 \times 10^{-3})(303) \ln K_{eq}$$
$$-15 = -0.60 \ln K_{eq}$$
$$\ln K_{eq} = 25$$
$$K_{eq} = 7.2 \times 10^{10}$$

(2)
$$\Delta G° = \Delta H° - T\Delta S°$$
$$\Delta G° = -12 - (273 + 150)(.01)$$
$$\Delta G° = -12 - 4 = -16 \text{ kcal/mol}$$

$$\Delta G° = -(1.986 \times 10^{-3})(423) \ln K_{eq}$$
$$-16 = -0.84 \ln K_{eq}$$
$$\ln K_{eq} = 19$$
$$K_{eq} = 1.8 \times 10^{8}$$

b. For this reaction: the calculations show that increasing the temperature causes $\Delta G°$ to be more negative.
c. For this reaction: the calculations show that increasing the temperature causes K_{eq}, to be smaller. (Remember that $\ln K_{eq} = -\Delta G°/RT$).

24. **a.**

bonds broken		bonds formed		
π bond of ethene	62	C–H	101	
H–Cl	103	C–Cl	85	
	165		186	$\Delta H° = 165 - 186 = -21$ kcal/mol

b.

bonds broken		bonds formed		
π bond of ethene	62	C–H	101	
H–H	104	C–H	101	
	166		202	$\Delta H° = 166 - 202 = -36$ kcal/mol

c. Both are exothermic.
d. Both reactions have sufficiently negative $\Delta H°$ values to expect that they would be exergonic as well.

25. **a.** **a** and **b** because the product is more stable than the reactant
b. **b** because it has the smallest rate constant (highest hill) leading from the product to the transition state
c. **c** because it has the largest rate constant (smallest hill) leading from the product to the transition state

26. **a.** A thermodynamically **unstable** product is one that is less stable than the reactant.
A kinetically **unstable** product is one that has a large rate constant (reforms reactant rapidly).

b. A kinetically **stable** product is one that has a small rate constant (reforms reactant slowly).

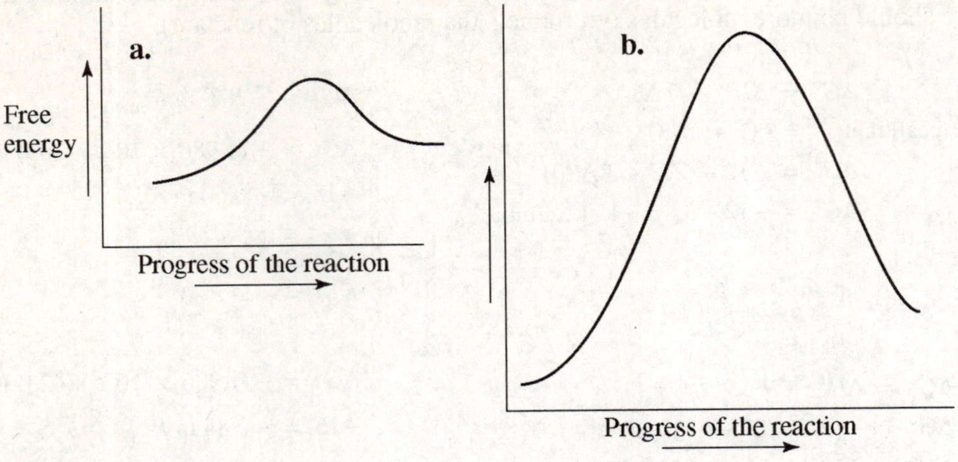

27. **a.** Solved in the text.
b. Decreasing the concentration of methyl chloride (by a factor of 10) will decrease the rate of the reaction (by a factor of 10) to 1×10^{-8} Ms^{-1}.
c. Changing the concentration will not affect the rate constant (k) of the reaction, because rate constants do not depend on the concentration of the reactants.

28. The rate constant for a reaction can be increased by **decreasing** the stability of the reactant (increasing its energy) or by **increasing** the stability of the transition state (decreasing is energy).

29. Taking the logarithm of both sides of the Arrhenius equation gives the following equation, which we can use to answer the questions:

$$\ln K = \ln A - \frac{E_a}{RT}$$

a. Increasing the experimental activation energy (E_a) will decrease the rate constant of a reaction (will cause the reaction to be slower).
b. Increasing the temperature (T) will increase the rate constant of a reaction (will cause the reaction to go faster).

30. **a.** The first reaction has the greater equilibrium constant.

$$K_{eq} = \frac{1 \times 10^{-3}}{1 \times 10^{-5}} = 1 \times 10^{2} \qquad K_{eq} = \frac{1 \times 10^{-2}}{1 \times 10^{-3}} = 1 \times 10$$

b. Because both reactions start with the same concentration, the first reaction will form the most product because it has the greater equilibrium constant.

31.

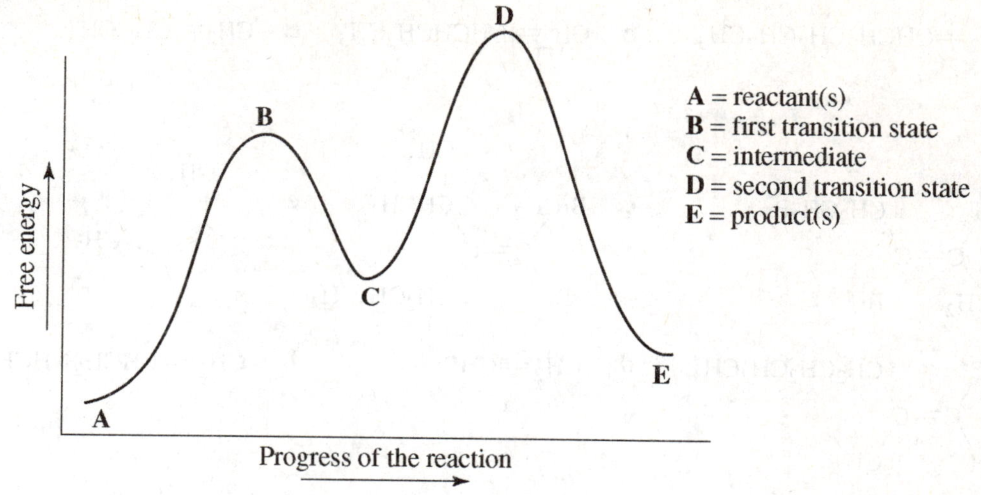

32. **a.** The first step (in the forward direction) has the greatest free energy of activation.
b. The first-formed intermediate is more apt to revert to reactants.
c. The second step is the rate-determining step because it has the transition state with the highest energy.

Notice that the second step is rate-determining even though the first step has the greatest energy of activation (steepest hill to climb). That is because it is easier for the intermediate that is formed in the first step to go back to starting material than to undergo the second step of the reaction. So, the second step is the rate-limiting step.

33.

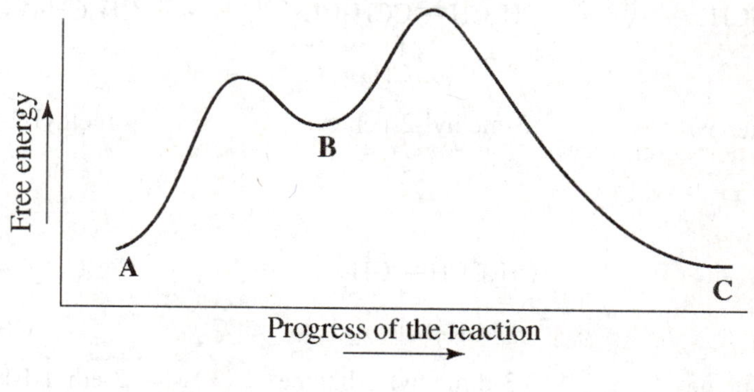

a. one
b. two
c. the second step (k_2) (In this particular diagram, $k_2 > k_1$: if you had made the transition state for the second step a lot higher, you could have had a diagram in which $k_1 > k_2$.)
d. the second step (k_{-1})
e. the second step in the reverse direction (k_{-1})
f. B to C
g. C to B

34. **a.** 3,8-dibromo-4-nonene
b. (*Z*)-4-ethyl-3,7-dimethyl-3-octene
c. 1,5-dimethylcyclopentene
d. 3-ethyl-2-methyl-2-heptene
e. 4-methylcyclohexene
f. 4-ethyl-5-methylcyclohexene

35. **a.** $CH_2{=}CHCH_2CH_2CH_2CH_3$ **b.** 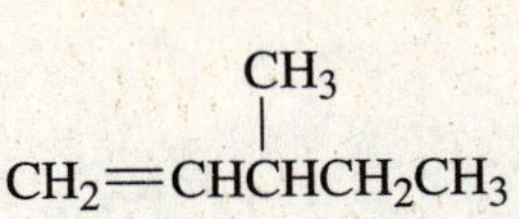 **c.** 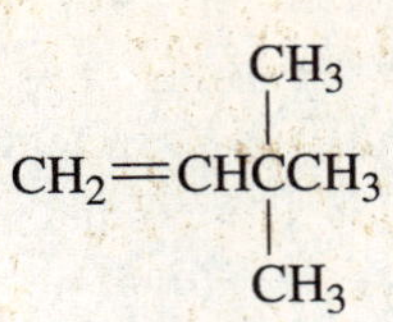

36. **a.** 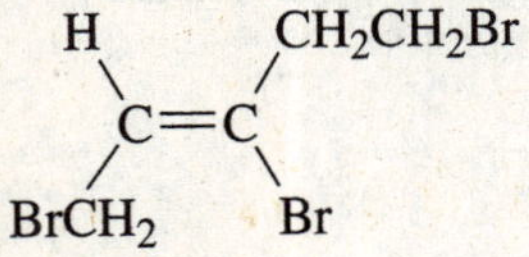 **c.** 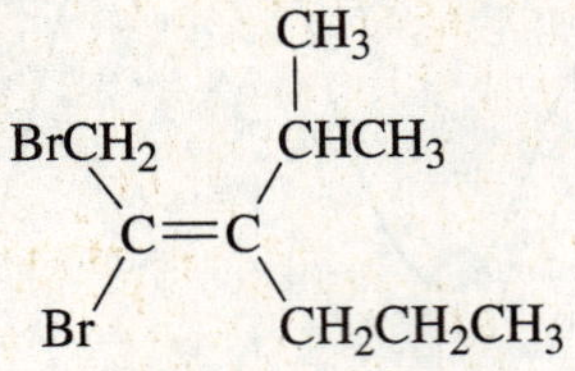 **e.** 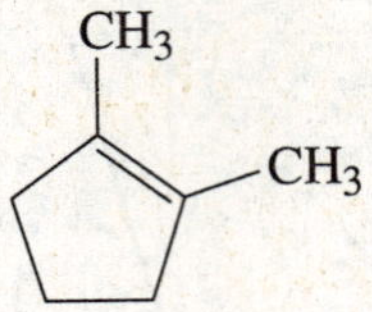

b. 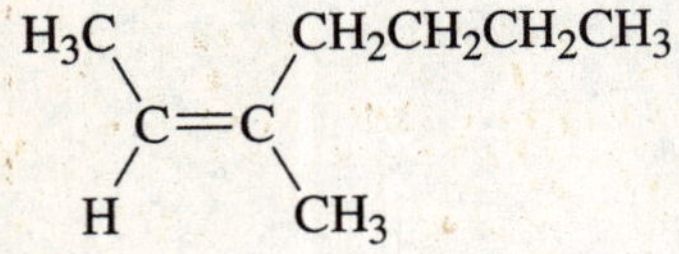 **d.** $CH_2{=}CHBr$ **f.** $CH_2{=}CHCH_2NHCH_2CH{=}CH_2$

37. **a.** $CH_2{=}CHCH_2CH_2CH_2CH_3$

1-hexene

$CH_3CH{=}CHCH_2CH_2CH_3$

2-hexene

$CH_3CH_2CH{=}CHCH_2CH_3$

3-hexene

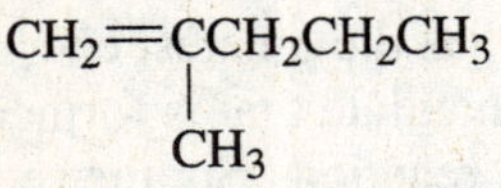

2-methyl-1-pentene

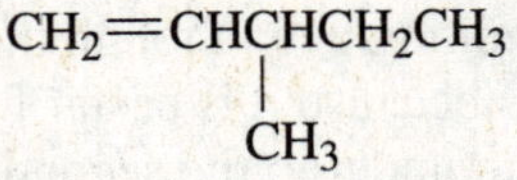

3-methyl-1-pentene

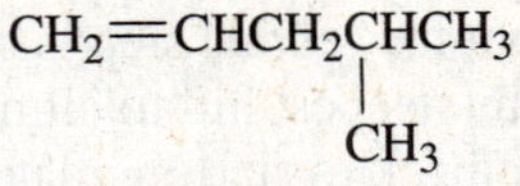

4-methyl-1-pentene

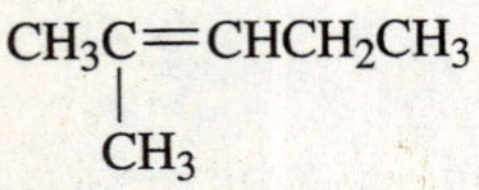

2-methyl-2-pentene

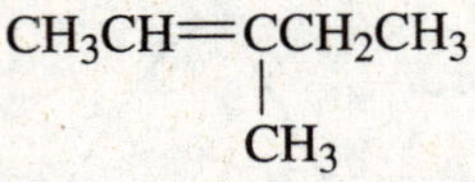

3-methyl-2-pentene

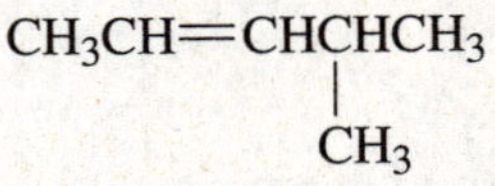

4-methyl-2-pentene

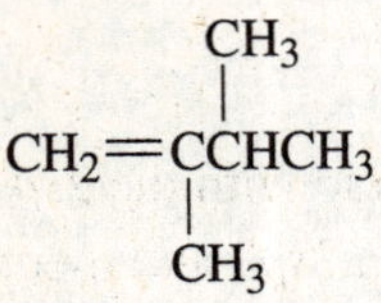

2,3-dimethyl-1-butene

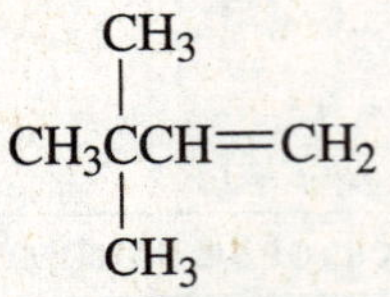

3,3-dimethyl-1-butene

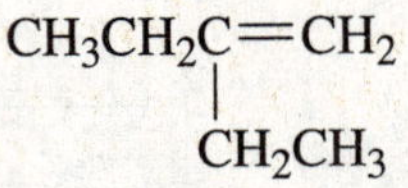

2-ethyl-1-butene

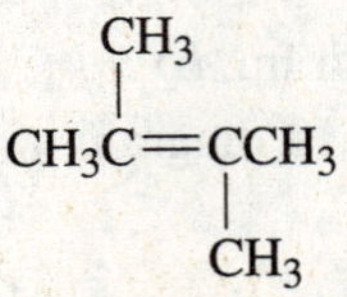

2,3-dimethyl-2-butene

b. Of the compounds shown above, the following have *E* and *Z* isomers:
2-hexene, 3-hexene, 3-methyl-2-pentene, 4-methyl-2-pentene

38. **a.** (*E*)-3-methyl-3-hexenes
b. *trans*-8-methyl-4-nonene or (*E*)-8-methyl-4-nonene
c. *trans*-9-bromo-2-nonene or (*E*)-9-bromo-2-nonene
d. 2,4-dimethyl-1-pentene
e. 2-ethyl-1-pentene
f. *cis*-2-pentene or (*Z*)-2-pentene

39.

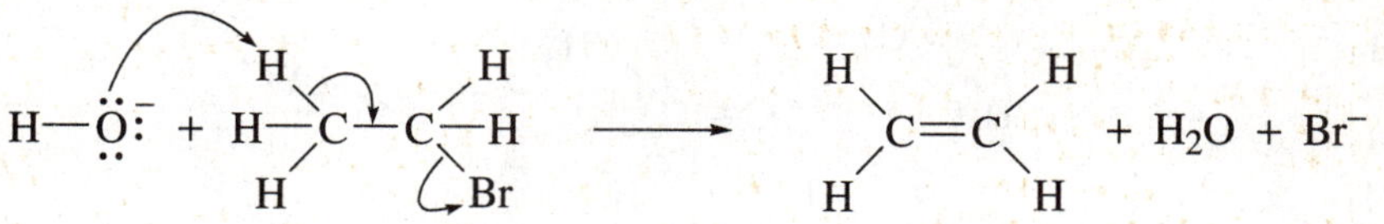

40. Tamoxifen is an *E* isomer.

41.
$\Delta G^\circ = -RT \ln K_{eq}$
$\ln K_{eq} = -\Delta G^\circ/RT$
$\ln K_{eq} = -\Delta G^\circ/0.59$ kcal/mol

a. $\ln K_{eq} = -2.72/0.59$ kcal/mol
$\ln K_{eq} = -4.6$
$K_{eq} = [B]/[A] = 0.01$

b. $\ln K_{eq} = -0.65/0.59$ kcal/mol
$\ln K_{eq} = -1.10$
$K_{eq} = [B]/[A] = 0.33$

c. $\ln K_{eq} = 2.72/0.59$ kcal/mol
$\ln K_{eq} = 4.6$
$K_{eq} = [B]/[A] = 100$

d. $\ln K_{eq} = 0.65/0.59$ kcal/mol
$\ln K_{eq} = 1.10$
$K_{eq} = [B]/[A] = 3.0$

42. If the number of carbons is 40, $C_nH_{2n+2} = C_{40}H_{82}$. Therefore, a compound with molecular formula $C_{40}H_{56}$ is missing (82-56) 26 hydrogens. This means that it has a total of 13 rings and π bonds, since each ring and π bond causes the compound to have two fewer hydrogens. Knowing that it has two rings and no triple bonds, we can conclude that it has 11 double bonds.

43. **a.** The C—Cl bond is a stronger bond because Cl uses a $3sp^3$ orbital to overlap with the $2sp^3$ orbital of carbon, whereas Br uses a $4sp^3$ orbital. A $4sp^3$ has a greater volume than a $3sp^3$ and therefore has less electron density and forms a weaker bond.
b. The Br—Br bond is a stronger bond because Br uses a $3sp^3$ orbital to overlap with the $3sp^3$ orbital of the other bromine, whereas I uses a $5sp^3$ orbital. A $5sp^3$ has a greater volume than a $4sp^3$ and therefore has less electron density and forms a weaker bond.

44. **a.** *Z* **b.** *E* **c.** *E* **d.** *Z*

45. If the number of carbons is 30, $C_nH_{2n+2} = C_{30}H_{62}$. A compound with molecular formula $C_{30}H_{50}$ is missing 12 hydrogens. Because it has no rings, squalene has 6 π bonds (12/2 = 6).

46. **a.** —CH=CH$_2$ > —CH(CH$_3$)$_2$ > —CH$_2$CH$_2$CH$_3$ > —CH$_3$

b. —OH > —NH$_2$ > —CH$_2$OH > —CH$_2$NH$_2$

c. —Cl > —COCH$_3$ > —C≡N > —CH=CH$_2$

47. **a.**

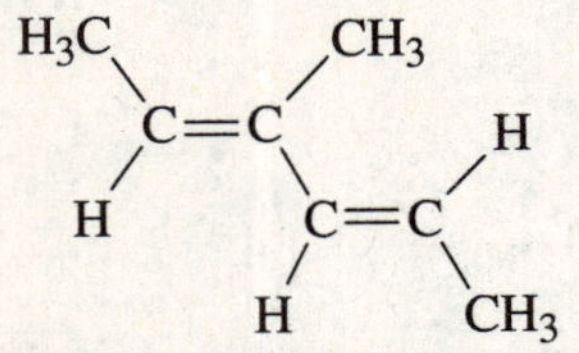

(2*E*,4*E*)-3-methyl-2,4-hexadiene

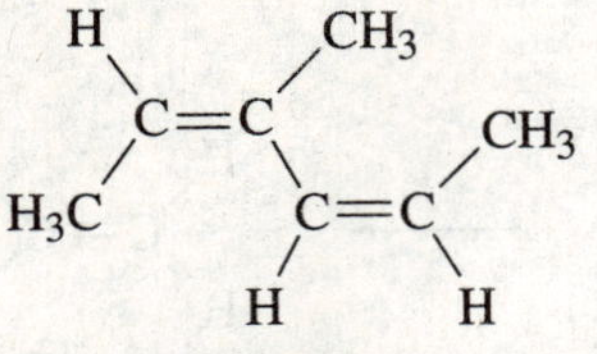

(2*Z*,4*Z*)-3-methyl-2,4-hexadiene

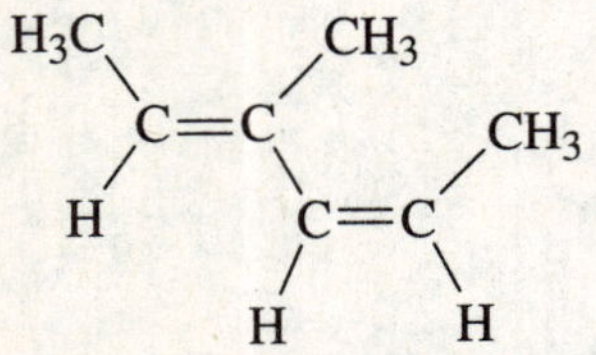

(2*E*,4*Z*)-3-methyl-2,4-hexadiene

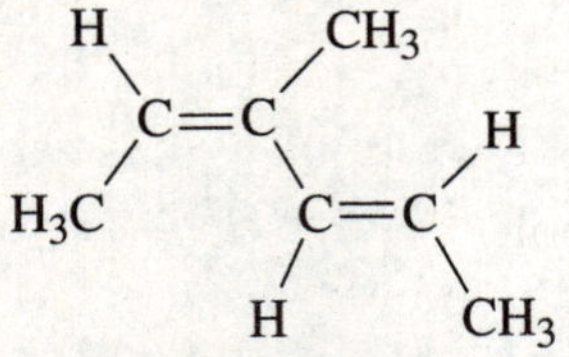

(2*Z*,4*E*)-3-methyl-2,4-hexadiene

b.

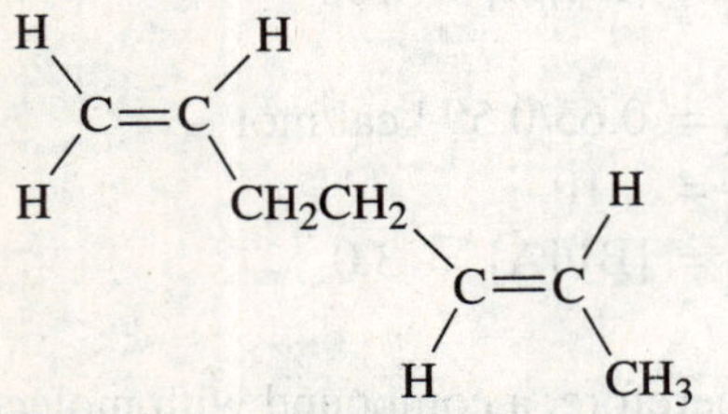

(*E*)-1,5-heptadiene

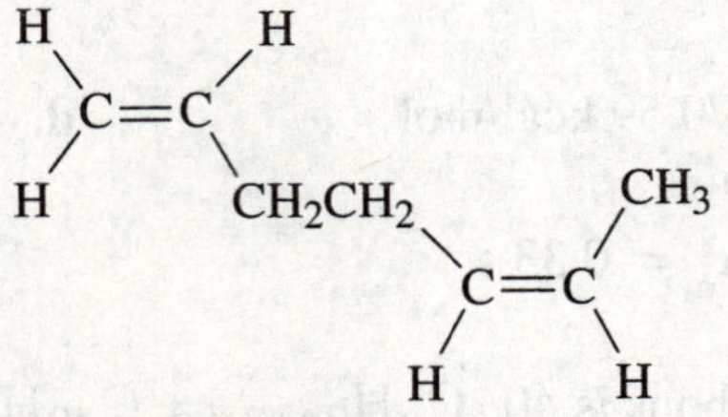

(*Z*)-1,5-heptadiene

c.

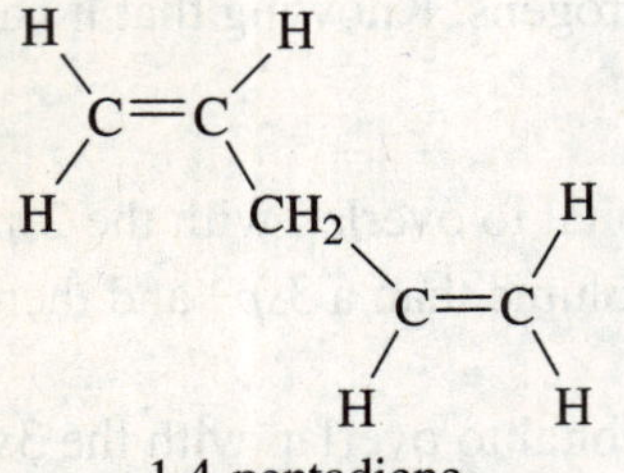

1,4-pentadiene

48. Only one of Molly's names is correct.

a. 2-pentene
b. correct
c. 3-methyl-1-hexene
d. 3-heptene
e. 4-ethylcyclohexene
f. 2-chloro-3-hexene
g. 3-methyl-2-pentene
h. 2-methyl-1-hexene
i. 1-methylcyclopentene or methylcyclopentene

49. a. 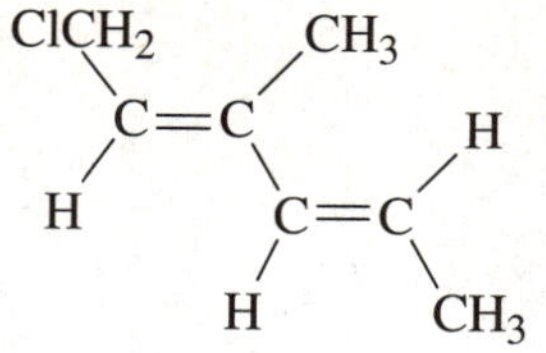

c. 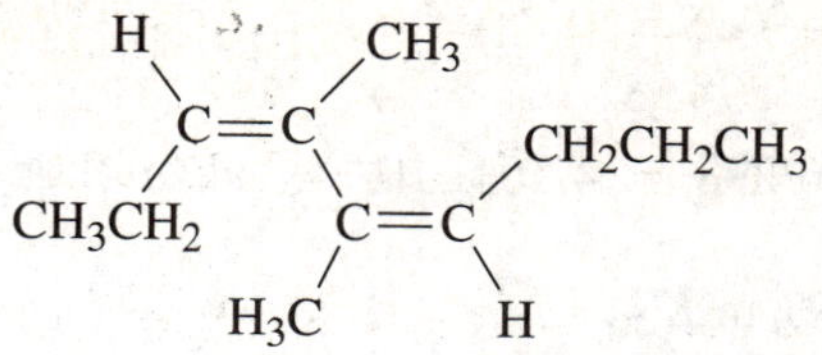

b. 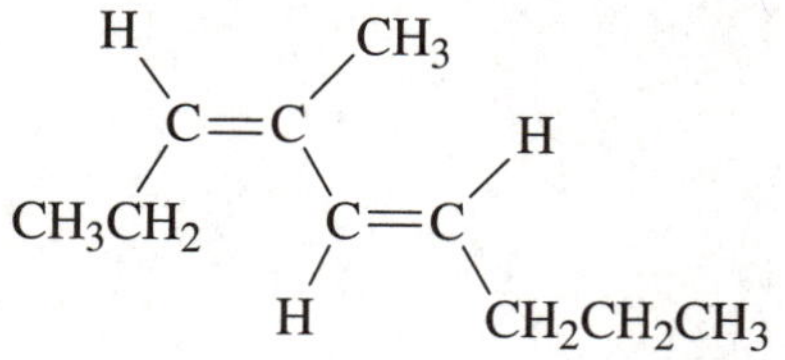

d. 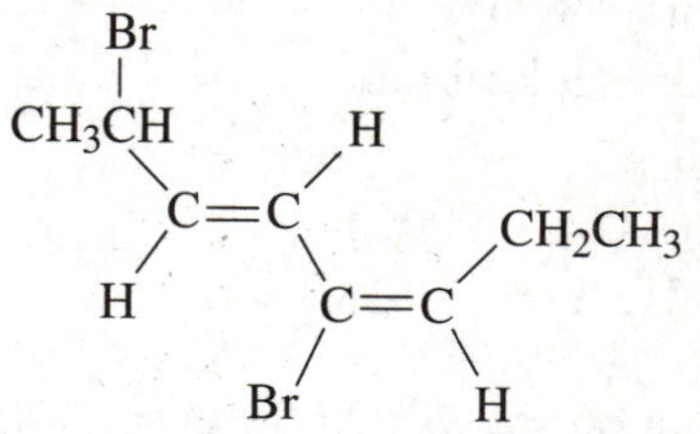

50.

a. 2
b. 3
c. C to D
d. the products (D) are more stable than the reactants (A)
e. B
f. endergonic
g. exergonic
h. B

51. $\Delta G^\circ = \Delta H^\circ - T\Delta S^\circ$

a. $\Delta G^\circ = 20 - (298)(25 \times 10^{-3}) = 20 - 7.5$
$\Delta G^\circ = 12.5$ kcal/mol

$\Delta G^\circ = -RT \ln K_{eq}$
$12.5 = -1.986 \times 10^{-3} \times 298 \ln K_{eq}$
$12.5 = -0.59 \ln K_{eq}$
$-21.2 = \ln K_{eq}$
$6.2 \times 10^{-10} = K_{eq}$

b. $\Delta G^\circ = 20 - (398)(25 \times 10^{-3}) = 20 - 10$
$\Delta G^\circ = 10.0$ kcal/mol

$10.0 = -1.986 \times 10^{-3} \times 398 \ln K_{eq}$
$10.0 = -0.79 \ln K_{eq}$
$-12.7 = \ln K_{eq}$
$3.1 \times 10^{-6} = K_{eq}$

52. **a.** $\Delta G^\circ = -RT \ln K_{eq}$
$\Delta G^\circ = -RT \ln 10$
$\Delta G^\circ = -1.986 \times 10^{-3} \times 298 \times 2.3$
$\Delta G^\circ = -1.36$ kcal/mol

b. $\Delta G^\circ = \Delta H^\circ - 0$
$-1.36 = \Delta H^\circ - 0$
$\Delta H^\circ = -1.36$ kcal/mol

c. $\Delta G^\circ = 0 - T\Delta S^\circ$

$-1.36 = 0 - 298\Delta S^\circ$

$\Delta S^\circ = 1.36/298 = 4.56 \times 10^{-3}$ kcal/(mol deg) = 4.56 cal/(mol deg)

53. $\Delta G^\circ = -RT \ln K_{eq}$

$\ln K_{eq} = -\Delta G^\circ/RT$

$\ln K_{eq} = -\Delta G^\circ/0.59$ kcal/mol

$\ln K_{eq} = -5.3/0.59$ kcal/mol

$\ln K_{eq} = -9.0$

$K_{eq} = [B]/[A] = 0.00013$

$[B]/[A] = 0.00013/1 = 0.13/1000$

The previous calculation shows that for every 1000 molecules in the chair conformation there is 0.13 molecule in a twist-boat conformation.
This agrees with the statement in Section 2.12 that for every thousand molecules of cyclohexane in a chair conformation, there is no more than one molecule in a twist-boat conformation.

54. A step-by-step description of how to solve this problem is given in the box entitled "Calculating Kinetic Parameters".

E_a can be determined from the Arrhenius equation ($\ln k = -E_a/RT$), because a plot of $\ln k$ versus $1/T$ gives a slope $= -E_a/R$

$\ln 2.11 \times 10^{-5} = -10.77$	$T = 304$	$1/T = 3.29 \times 10^{-3}$
$\ln 4.44 \times 10^{-5} = -10.02$	$T = 313$	$1/T = 3.19 \times 10^{-3}$
$\ln 1.16 \times 10^{-4} = -9.06$	$T = 324.5$	$1/T = 3.08 \times 10^{-3}$
$\ln 2.10 \times 10^{-4} = -8.47$	$T = 332.8$	$1/T = 3.00 \times 10^{-3}$
$\ln 4.34 \times 10^{-4} = -7.74$	$T = 342.2$	$1/T = 2.92 \times 10^{-3}$

slope $= -8290$

$E_a = -(\text{slope})\, R$

$E_a = -(-8290) \times 1.98 \times 10^{-3}$ kcal/mol

$E_a = 16.4$ kcal/mol

<u>To find ΔG°:</u>

$-\Delta G^\circ = RT \ln kh/Tk_b$

From the graph used to determine E_a, one can find the rate constant (k) at 30°.
(It is $1.84 \times 10^{-5}\ s^{-1}$.)

$-\Delta G^\circ = 1.98 \times 10^{-3} \times 303 \ln (1.84 \times 10^{-5} \times 6.625 \times 10^{-27})/(303 \times 1.3805 \times 10^{-10})$

$-\Delta G^\circ = 1.98 \times 10^{-3} \times 303 \ln (1.22 \times 10^{-31})/(4.18 \times 10^{-8})$

$-\Delta G^\circ = 1.98 \times 10^{-3} \times 303 \ln 2.92 \times 10^{-24}$

$-\Delta G^\circ = 1.98 \times 10^{-3} \times 303 \times (-54.2)$

$\Delta G^\circ = 32.5$ kcal/mol

To find ΔH°:

$$\Delta H^\circ = E_a - RT$$

$$\Delta H^\circ = 16.4 - 1.98 \times 10^{-3} \times 303$$

$$\Delta H^\circ = 16.4 - 0.6$$

$$\Delta H^\circ = 15.8 \text{ kcal/mol}$$

To find ΔS°:

$$\Delta S^\circ = (\Delta H^\circ - \Delta G^\circ)/T$$

$$\Delta S^\circ = (15.8 - 32.5)/303$$

$$\Delta S^\circ = (-16.7)/303$$

$$\Delta S^\circ = 0.055 \text{ kcal/(mol deg)} = 55 \text{ cal/(mol deg)}$$

Chapter 3 Practice Test

1. Name the following compounds:

a. $CH_3CH_2CHCH_2CH{=}CH_2$
$\quad\quad\quad\quad |$
$\quad\quad\quad\; CH_3$

c. $CH_3CH_2CH{=}CHCH_2CH_2CHCH_3$
$\quad\quad\quad\quad\quad\quad\quad\quad\quad\quad |$
$\quad\quad\quad\quad\quad\quad\quad\quad\; CH_2CH_3$

b. 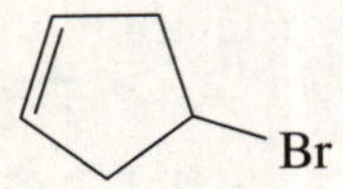

d. 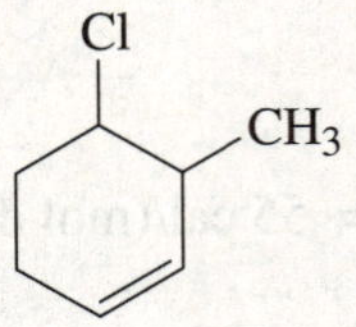

2. Label the following substituents in order of decreasing priority in the *E, Z* system of nomenclature. Label the highest priority #1.

$-\overset{\displaystyle O}{\overset{\|}{C}}CH_3 \qquad -CH{=}CH_2 \qquad -Cl \qquad -C{\equiv}N$

3. Correct the incorrect names.

a. 3-pentene

b. 2-vinylpentane

c. 2-ethyl-2-butene

d. 2-methylcyclohexene

4. Indicate whether each of the following statements is true or false:

a. Increasing the energy of activation, increases the rate of the reaction. T F

b. Decreasing the entropy of the products compared to the entropy of the reactants makes the equilibrium constant more favorable. T F

c. An exergonic reaction is one with a $-\Delta G^\circ$. T F

d. An alkene is an electrophile. T F

e. The higher the energy of activation, the more slowly the reaction will take place. T F

f. Another name for *trans*-2-butene is (*Z*)-2-butene. T F

g. *trans*-2-Butene has a dipole moment of zero. T F

h. A reaction with a negative ΔG° has an equilibrium constant greater than one. T F

i. Increasing the energy of the reactants, increases the rate of the reaction. T F

j. Increasing the energy of the products, increases the rate of the reaction. T F

5. Do the following compounds have the *E* or the *Z* configuration?

a. 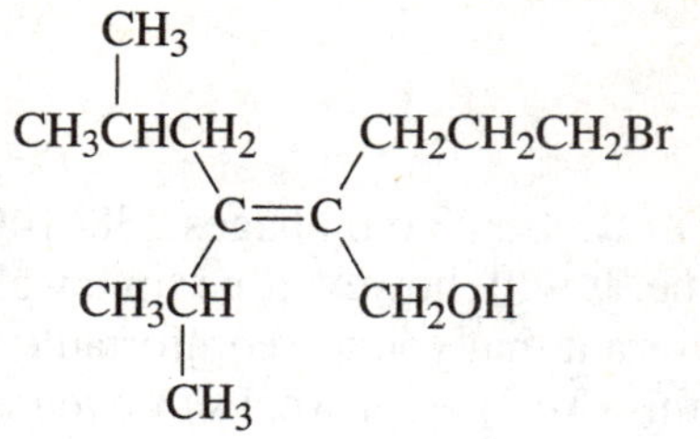

b. 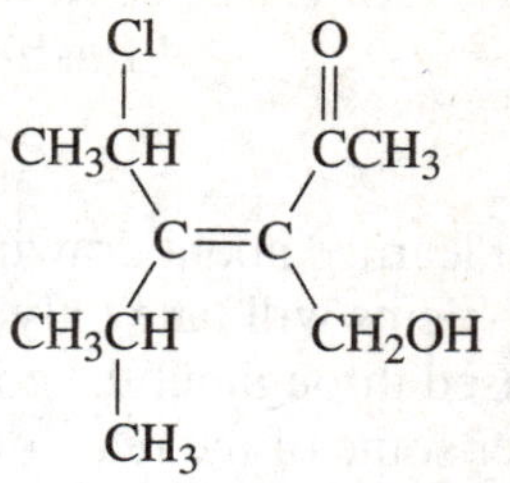

6. Draw structures for the following:

a. allyl alcohol

b. 3-methylcyclohexene

c. *cis*-3-heptene

d. vinyl bromide

7. How many π bonds and/or rings does a hydrocarbon have if it has a molecular formula of C_8H_8?

8. Using curved arrows show the movement of electrons in the following reaction:

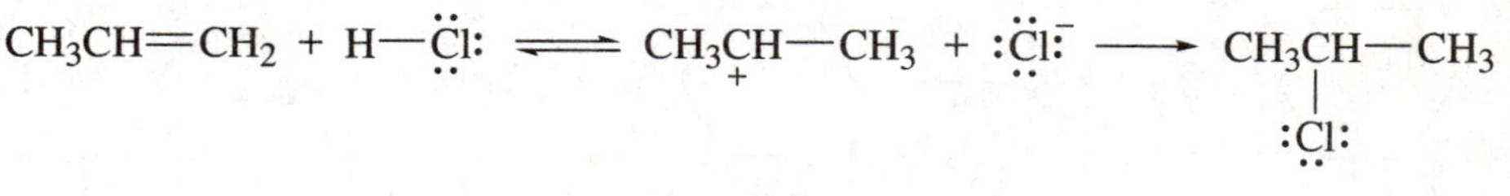

9. Which of the following have cis-trans isomers?

a. 1-pentene

b. 4-methyl-2-hexene

c. 2-bromo-3-hexene

d. 2-methyl-2-hexene

10. Which of the following has a more favorable equilibrium constant (that is, which reaction favors products more)?

a. A reaction with a $\Delta H°$ of 4 kcal/mol or a reaction with a $\Delta H°$ of 7 kcal/mol?

b. A reaction that takes place at 25 °C or the same reaction that takes place at 35 °C?

c. A reaction in which two reactants form one product or a reaction in which one reactant forms two products?

SPECIAL TOPIC II

An Exercise in Drawing Curved Arrows "Pushing Electrons"

This is an extension of what you learned about drawing curved arrows in Section 3.6 on pages 138–140 of the text. Working through these problems will take only a little of your time. It will, however, be time well spent, because curved arrows will be used throughout the course and it is important that you are comfortable with this notation. (You will not encounter some of the reaction steps shown in this exercise for weeks or even months, so don't worry about why the chemical changes take place.)

Chemists use curved arrows to show how electrons move as covalent bonds break and/or new covalent bonds form. The tail of the arrow is positioned at the point where the electrons are in the reactant, and the head of the arrow points to where these same electrons end up in the product.

In the following reaction step, the bond between bromine and a carbon of the cyclohexane ring breaks and both electrons in the bond end up with bromine in the product. Thus, **the arrow starts at the electrons that carbon and bromine share in the reactant,** and **the head of the arrow points at bromine** because this is where the two electrons end up in the product.

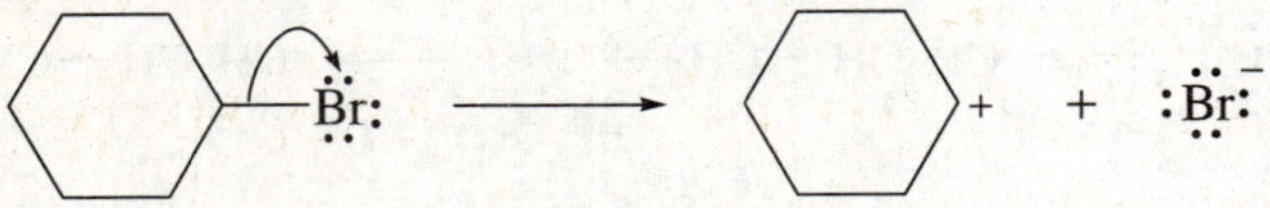

Notice that the carbon of the cyclohexane ring is positively charged in the product because it has lost the two electrons it was sharing with bromine. The bromine is negatively charged in the product because it has gained the electrons that it shared with carbon in the reactant. The fact that two electrons move in this example is indicated by the two barbs on the arrowhead.

Notice that the arrow always starts at a bond or at a lone pair. It does not start at a negative charge.

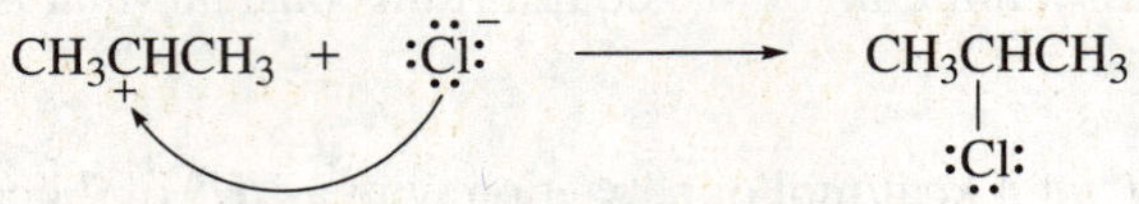

In the following reaction step a bond is being formed between the oxygen of water and a carbon of the other reactant. The arrow starts at one of the lone pairs of the oxygen and points at the atom (the carbon) that will share the electrons in the product. The oxygen in the product is positively charged because the electrons that oxygen had to itself in the reactant are now being shared with carbon. The carbon that was positively charged in the reactant is not charged in the product, because it has gained a share in a pair of electrons.

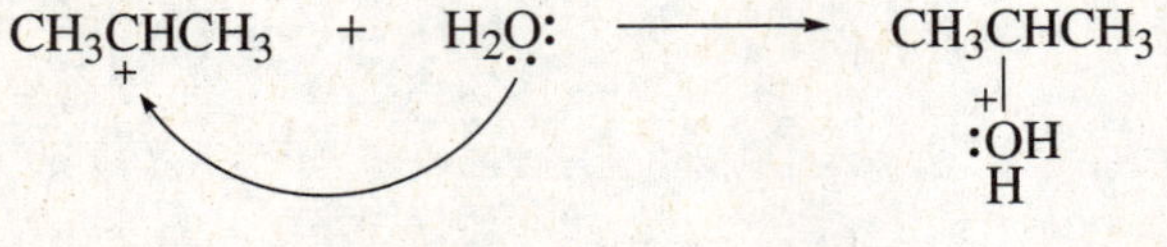

Problem 1. Draw curved arrows to show the movement of the electrons in the following reaction steps. (You will find the answers immediately after Exercise 10.)

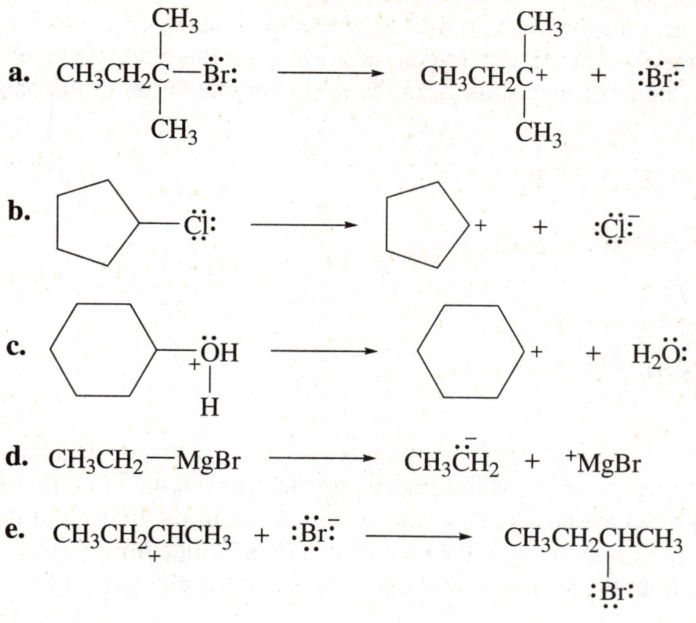

Frequently chemists do not show the lone-pair electrons when they write reactions. The following are the same reaction steps you just saw except that the lone pairs are not shown.

Problem 2. Draw curved arrows to show the movement of the electrons.

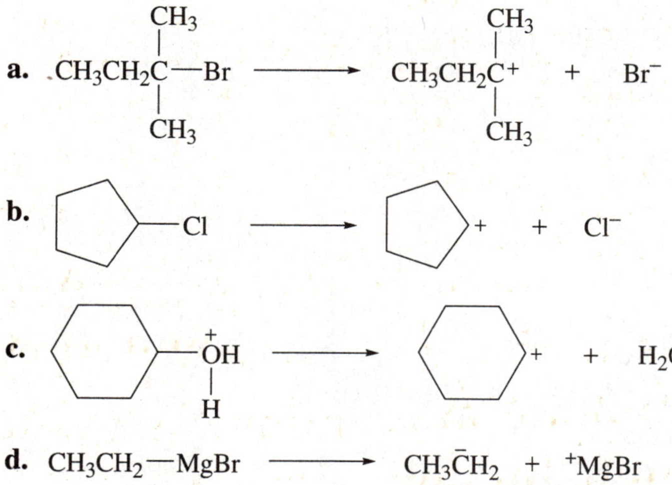

The lone-pair electrons on Br^- in example **e** have to be shown in the reactant because an arrow can start only at a bond or at a lone pair. Bromine's lone pairs do not have to be shown in the product.

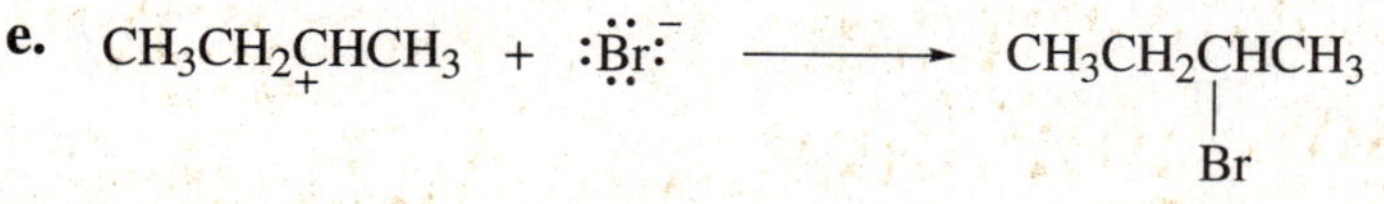

Many reaction steps involve both bond breaking and bond formation. In the following examples, the electrons in the bond that breaks are the same as the electrons in the bond that forms, so only one arrow is needed to show how the electrons move. As in the previous examples, the arrow starts at the point where the electrons are in the reactant, and the head of the arrow points to where these same electrons end up in the product (at the carbon atom in the first example, and between the carbons in the next example). Notice that the atom that loses a share in a pair of electrons (C in the first example, H in the second) ends up with a positive charge.

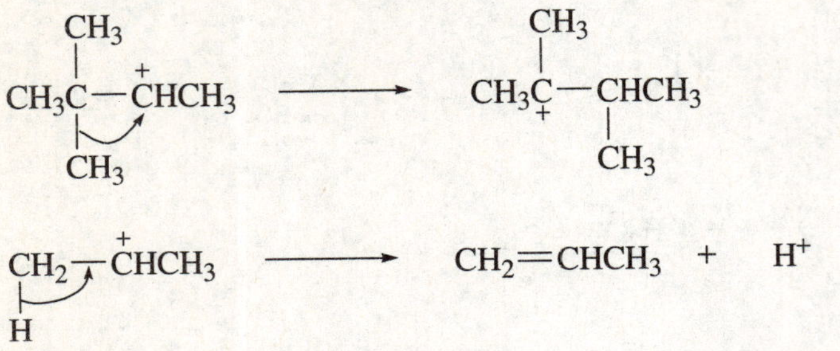

Frequently, the electrons in the bond that breaks are not the same as the electrons in the bond that forms. In such cases, two or more arrows are needed to show the movement of the electrons. In each of the following examples, look at the arrows that illustrate how the electrons move. Notice how the movement of the electrons allows you to determine both the structure of the products and the charges on the products.

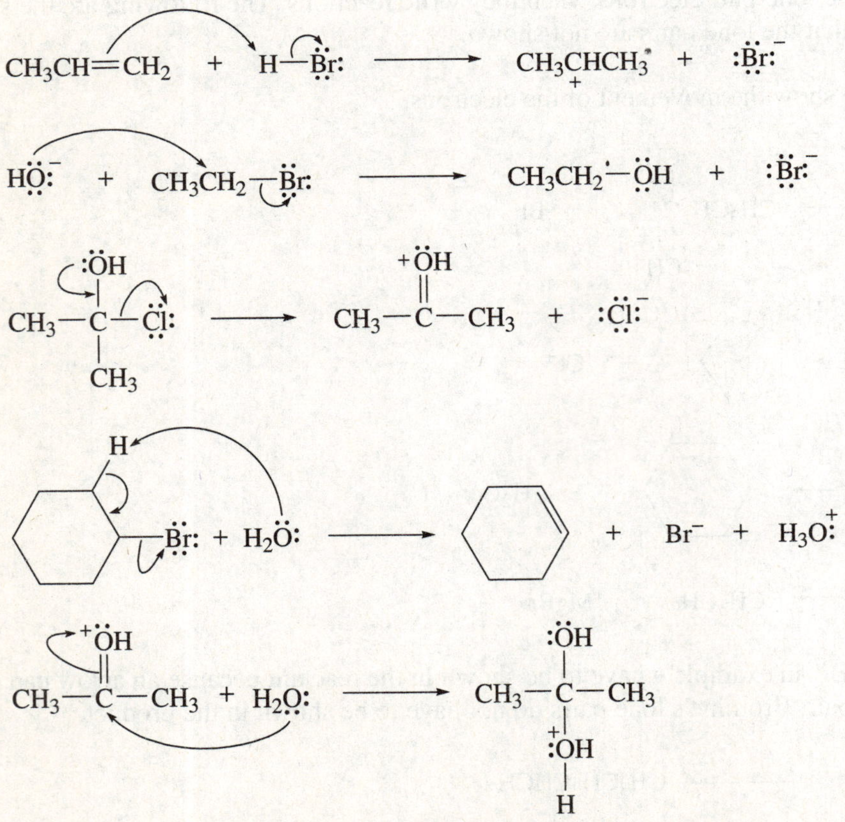

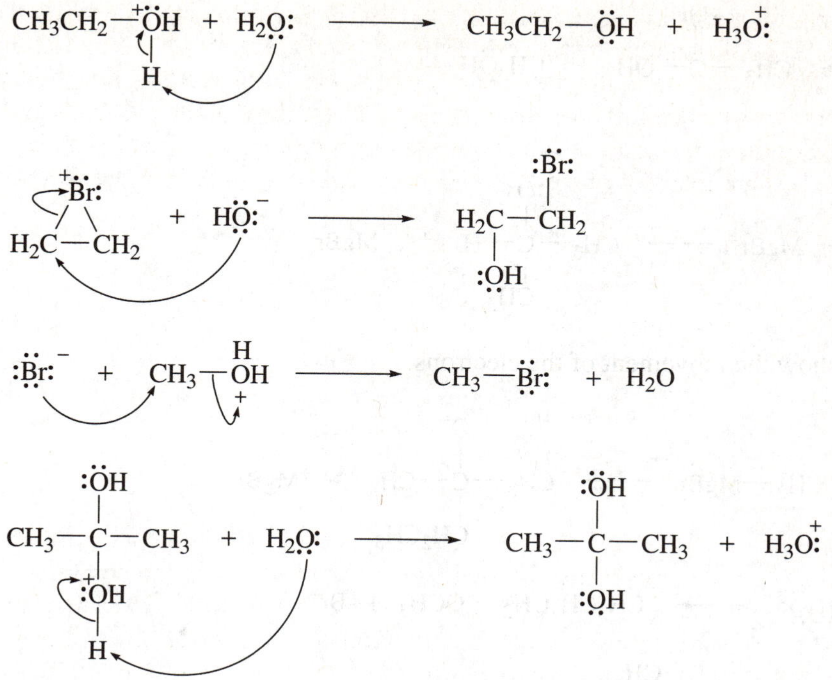

Problem 3. Draw curved arrows to show the movement of the electrons that result in the formation of the given product.

a. 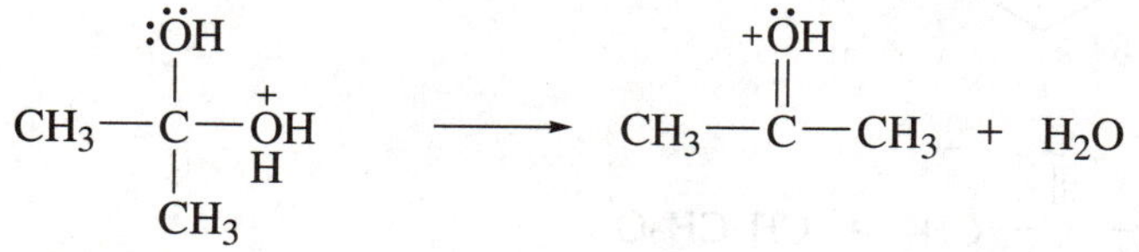

b. $CH_3CH_2CH{=}CH_2 + H{-}Cl \longrightarrow CH_3CH_2\overset{+}{C}H{-}CH_3 + Cl^-$

c. $CH_3CH_2{-}Br + \ddot{N}H_3 \longrightarrow CH_3CH_2{-}\overset{+}{N}H_3 + Br^-$

d. 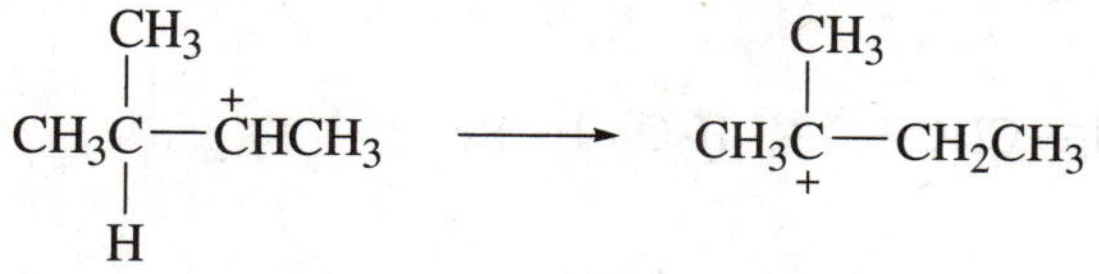

Problem 4. Draw curved arrows to show the movement of the electrons.

a. $CH_3CH{=}CHCH_3 + H{-}\overset{+}{\ddot{O}}(H){-}H \longrightarrow CH_3\overset{+}{C}H{-}CH_2CH_3 + H_2\ddot{O}:$

b. $CH_3CH_2CH_2CH_2{-}\ddot{Cl}: + {}^{-}\ddot{C}{\equiv}N \longrightarrow CH_3CH_2CH_2CH_2{-}C{\equiv}N + :\ddot{Cl}:^-$

c. 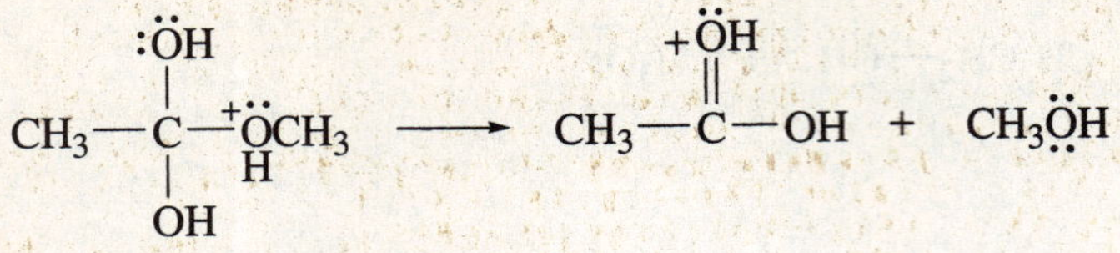

d. 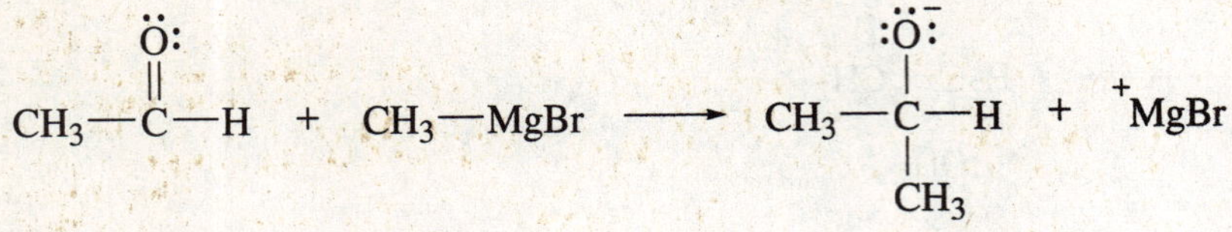

<u>Problem 5.</u> Draw curved arrows to show the movement of the electrons.

a. 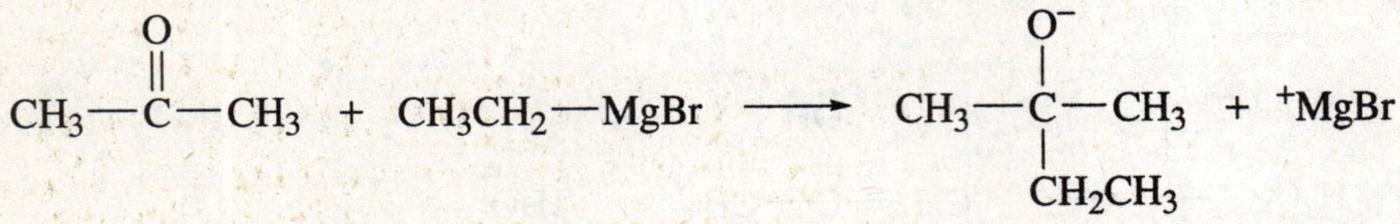

b. 

c. 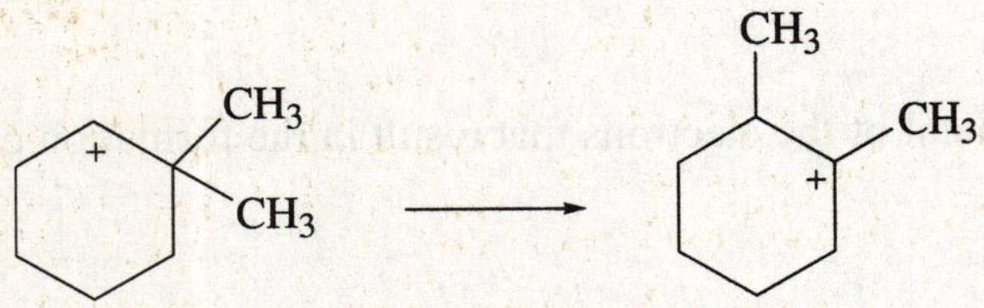

d. 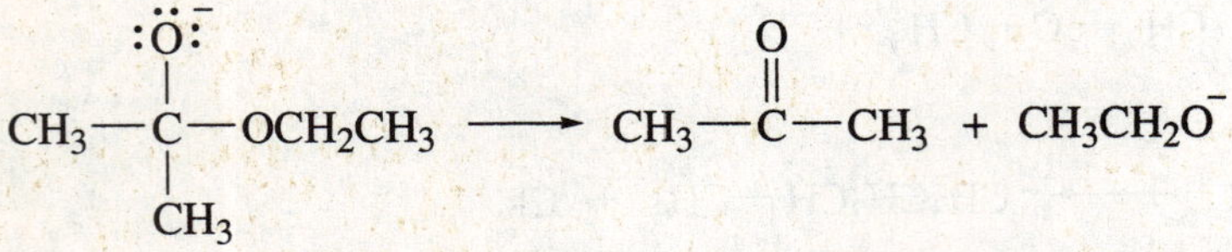

<u>Problem 6.</u> Draw curved arrows to show the movement of the electrons.

a. 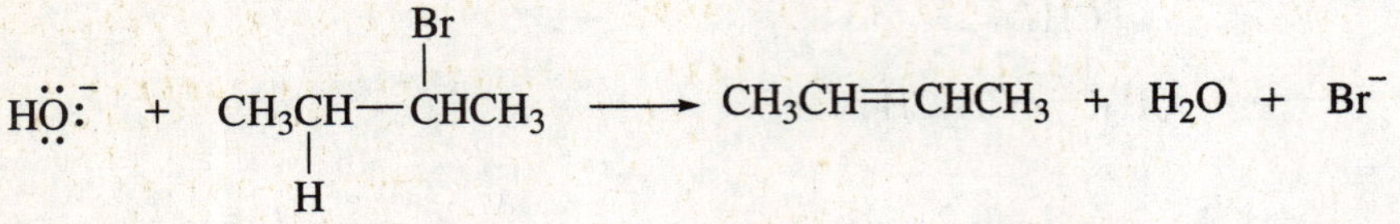

b. 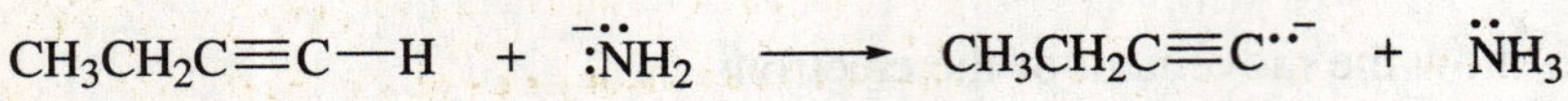

c. 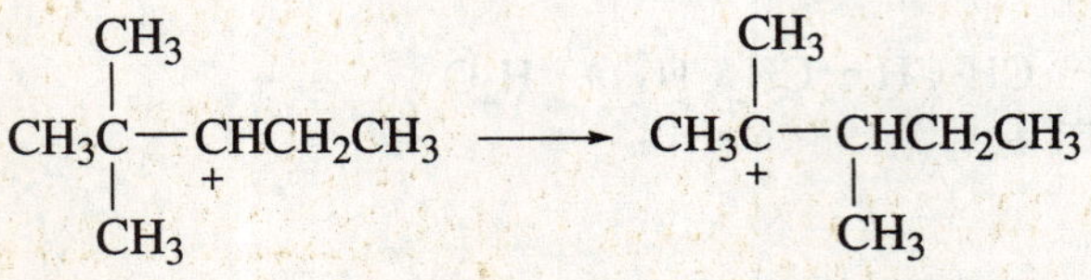

d. 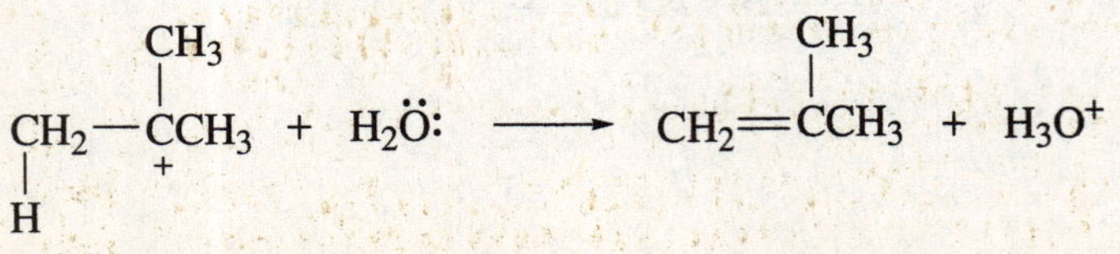

Problem 7. Draw curved arrows to show the movement of the electrons.

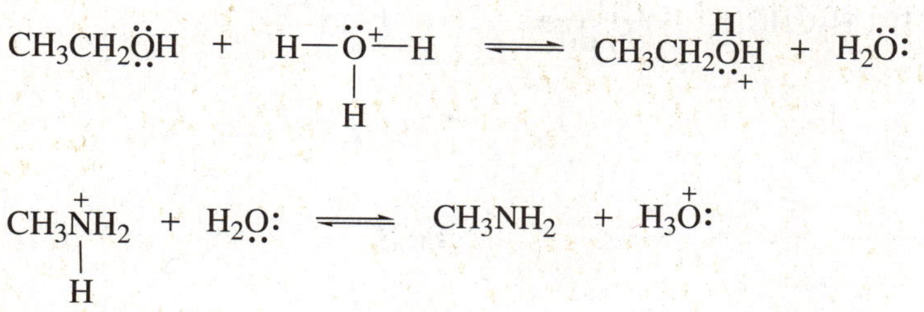

Curved arrows are used to show the movement of electrons in each step of a reaction.

Problem 8. Draw curved arrows to show the movement of the electrons in each step of the following reaction sequences.

a.

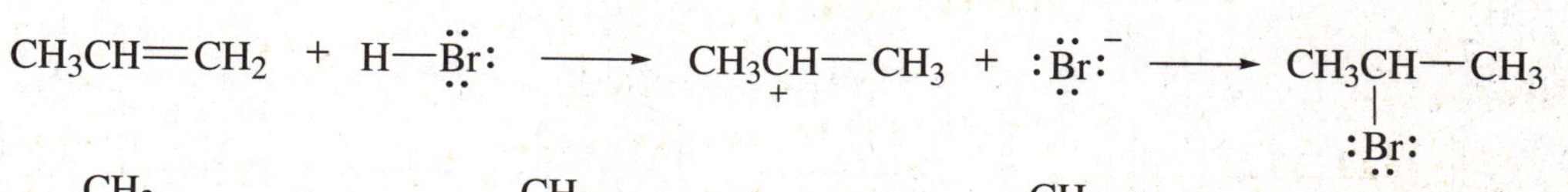

b.

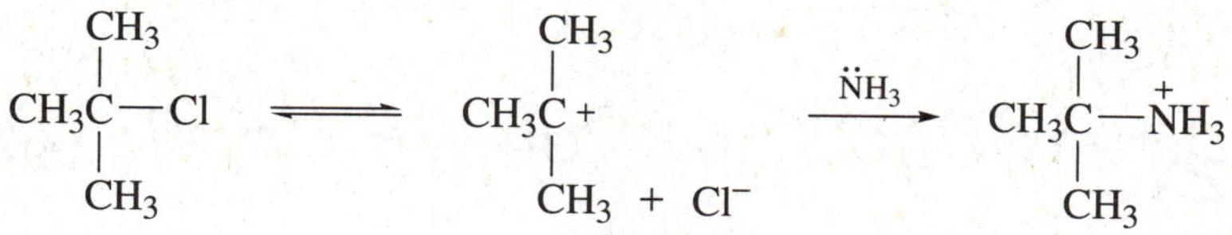

c.

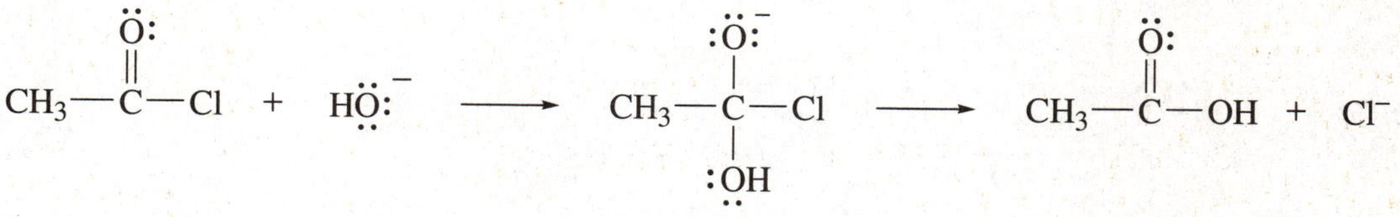

Problem 9. Draw curved arrows to show the movement of the electrons in each step of the following reactions.

a.

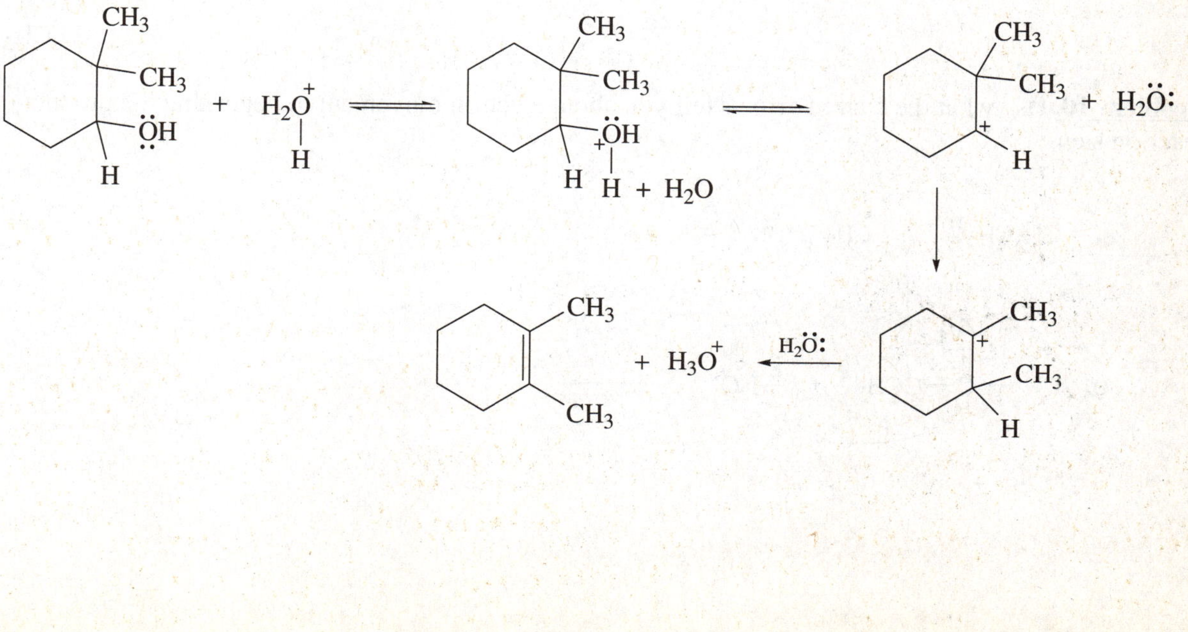

b. 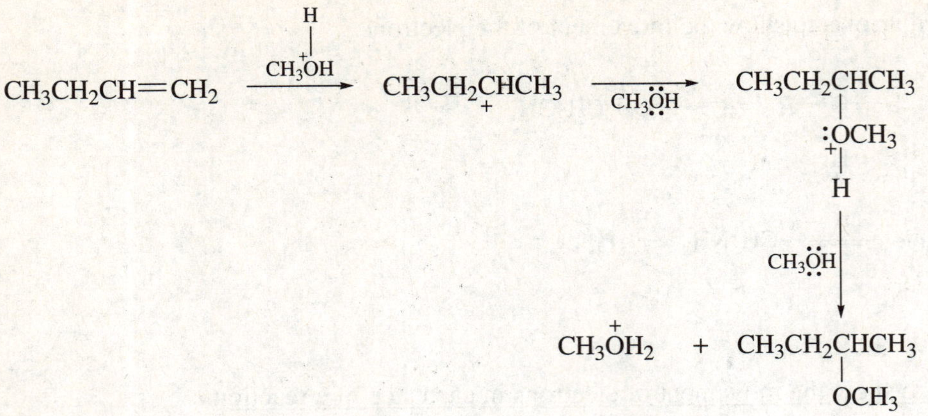

c. In the following example, the acid and base are not specified: HB^+ represents any acid present in the reaction mixture and B: can be any base.

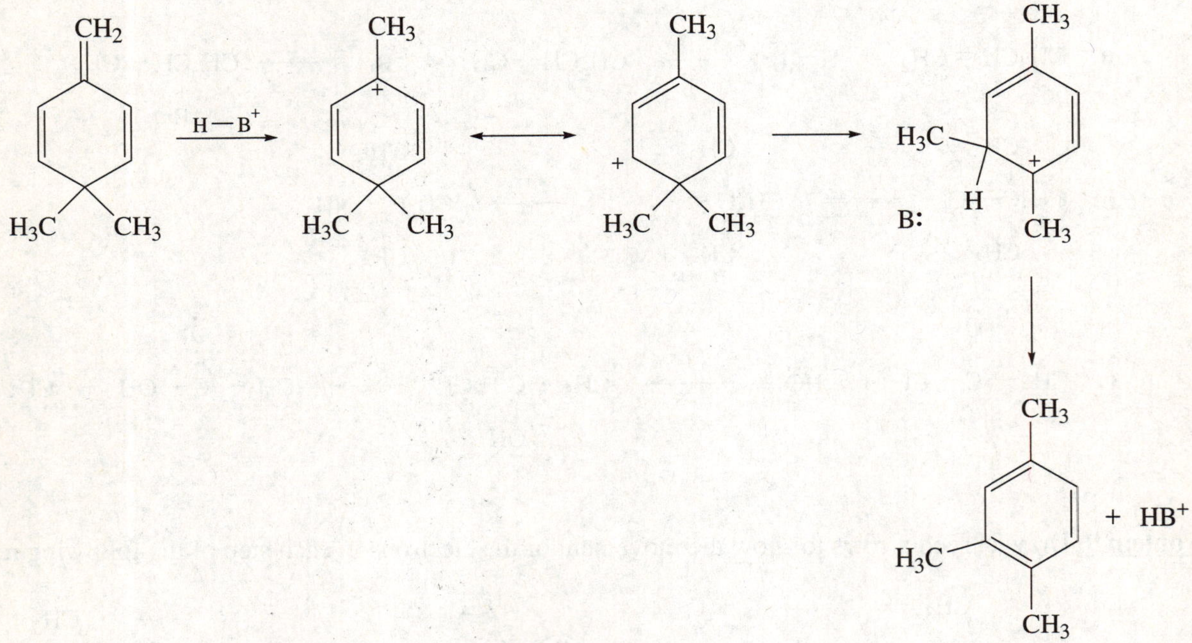

<u>Problem 10.</u> Use what the curved arrows tell you about electron movement to determine the product of each reaction step.

a. 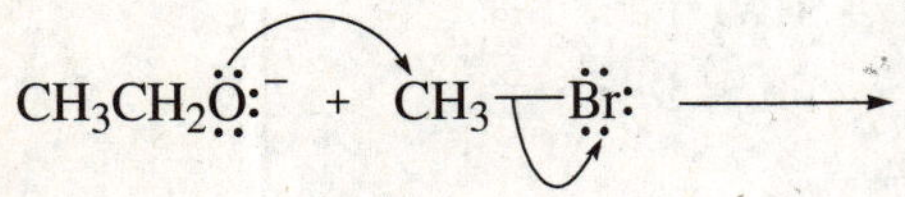

b. 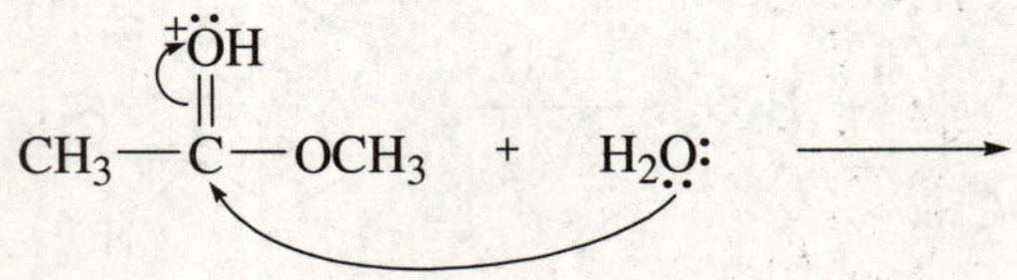

c. 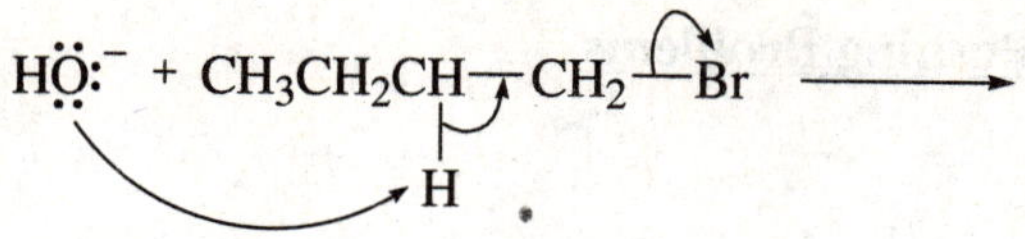

d. 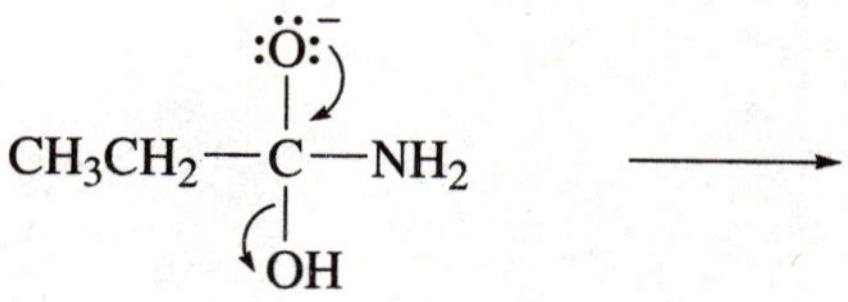

e. 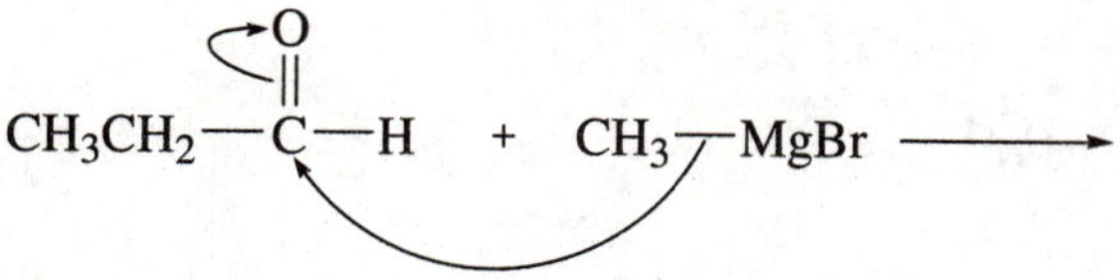

f. 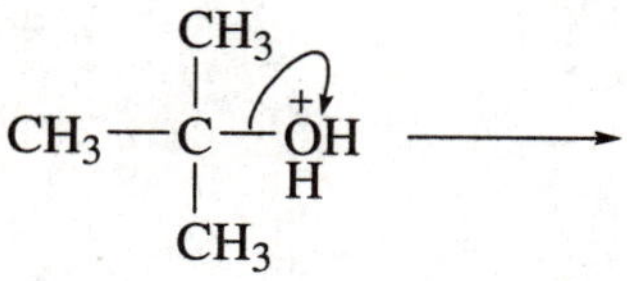

g. 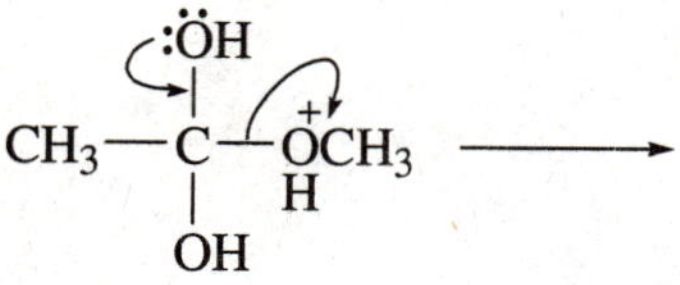

Answers to Electron Pushing Problems

Problem 1

a. 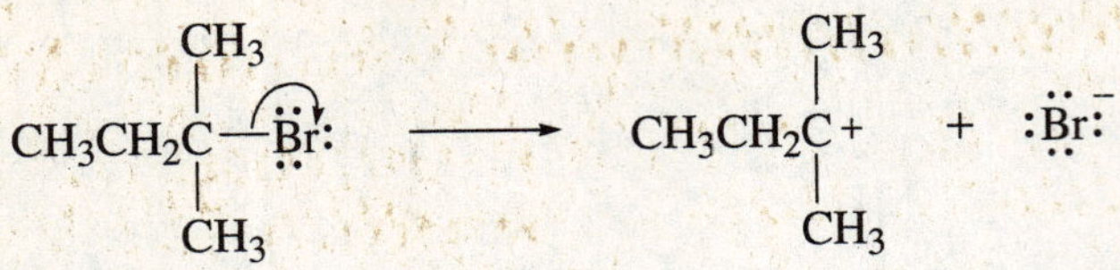

b. 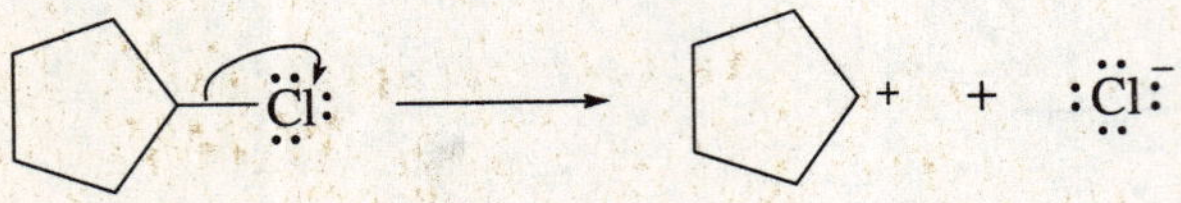

c. 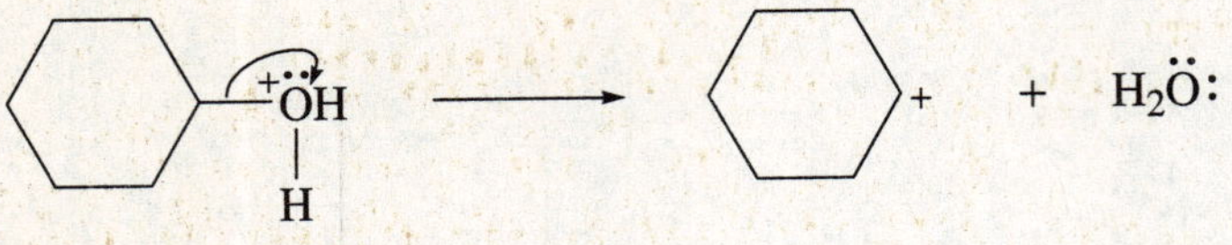

d. 

e. 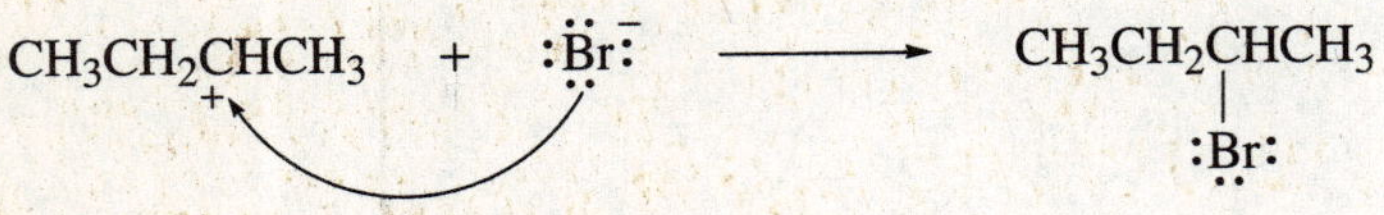

Problem 2

a. 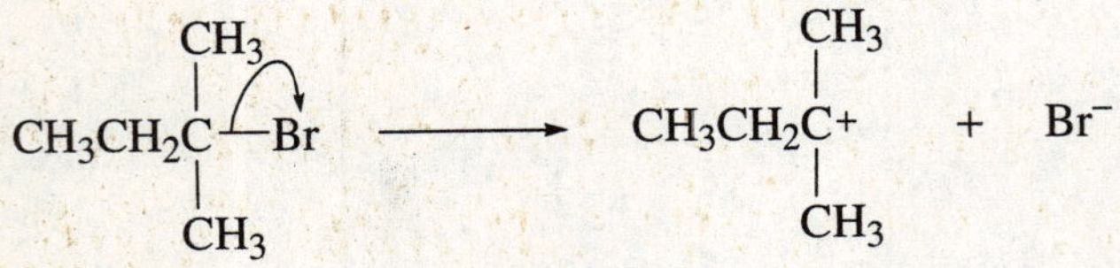

b. 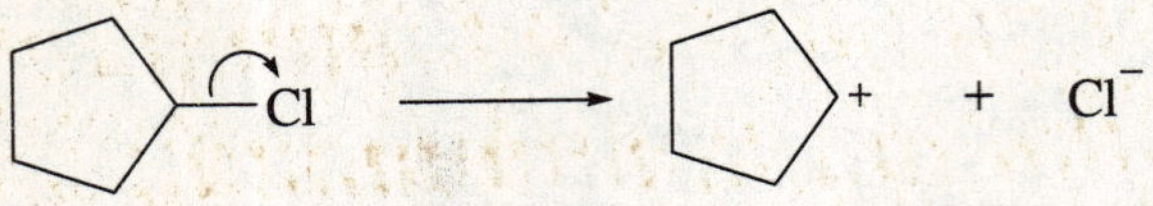

c. 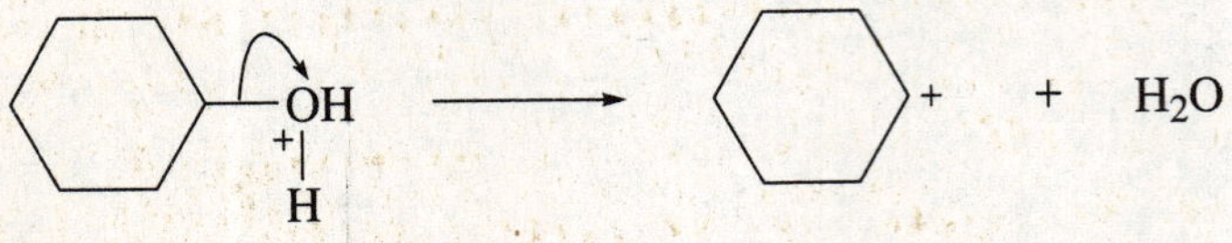

d. 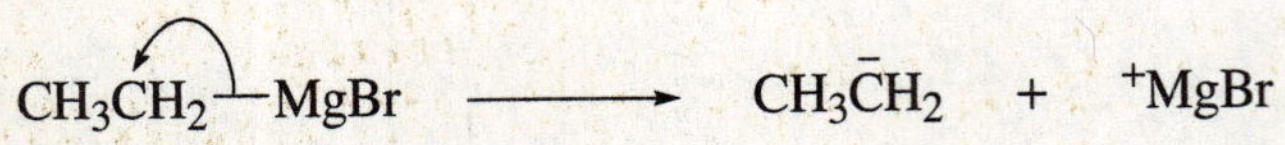

Problem 3

a. 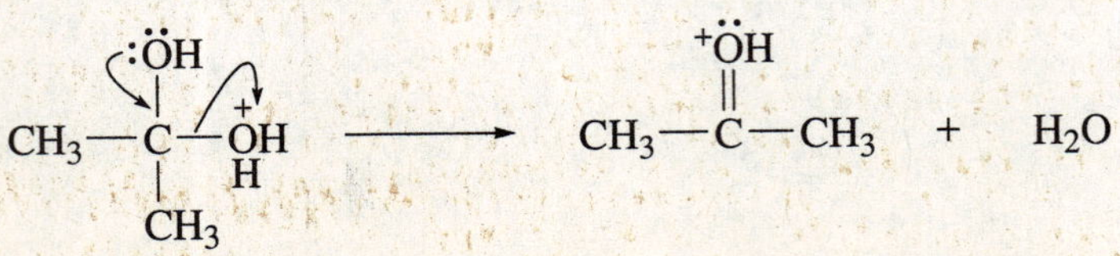

b. 

c. 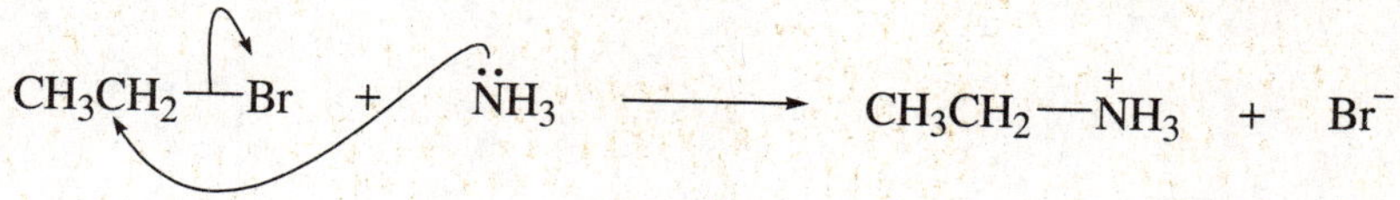

d. 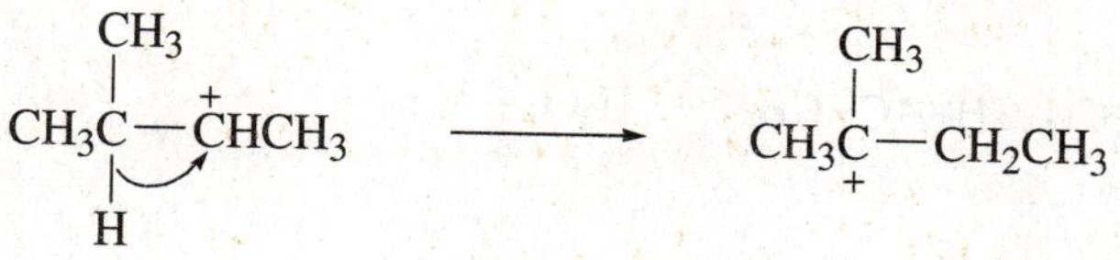

Problem 4

a. 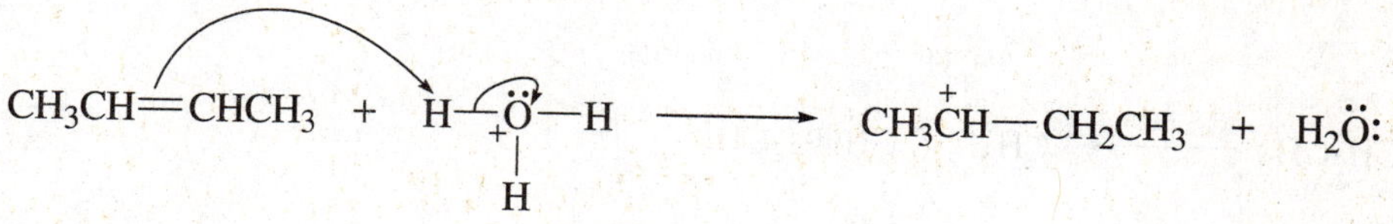

b. 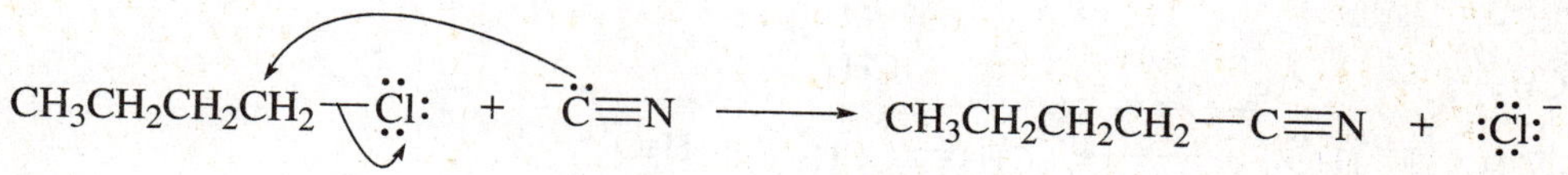

c. 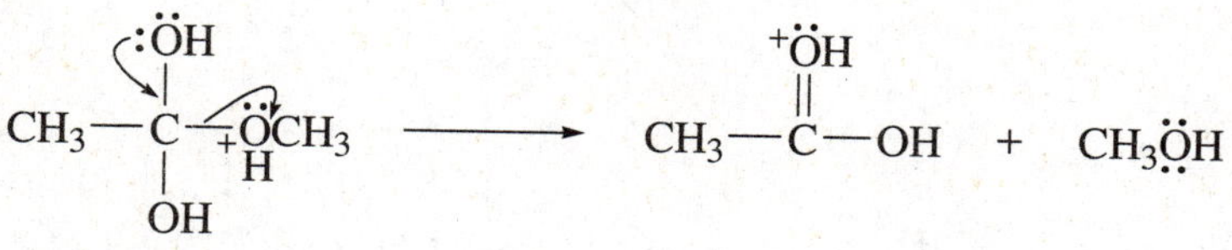

d. 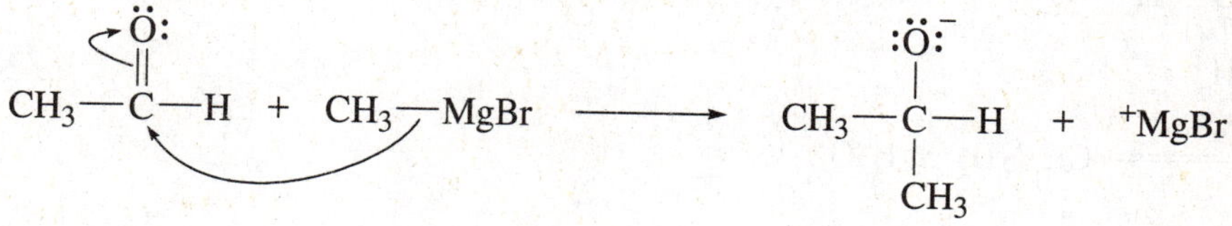

Problem 5

a. 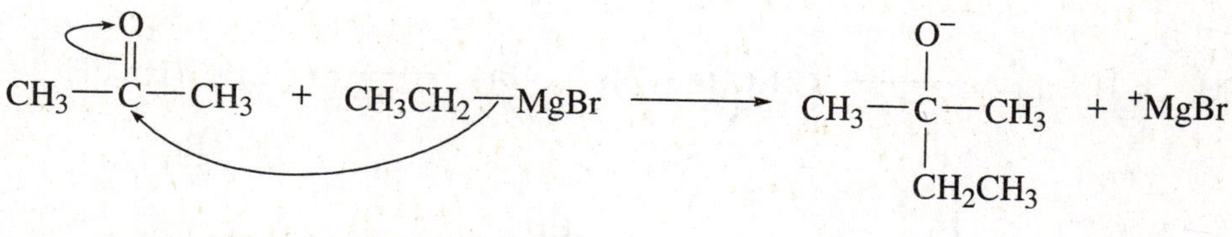

b. 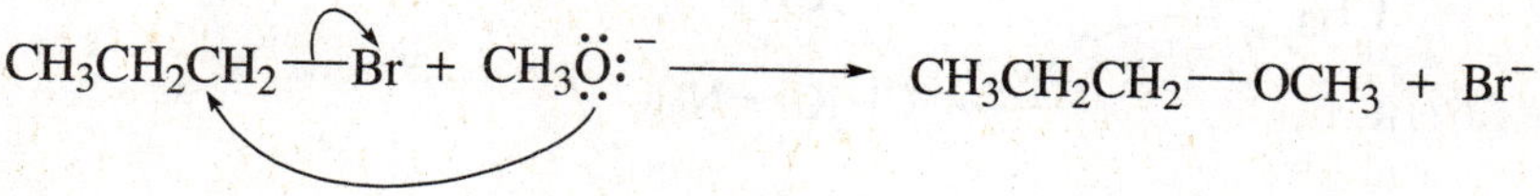

c. 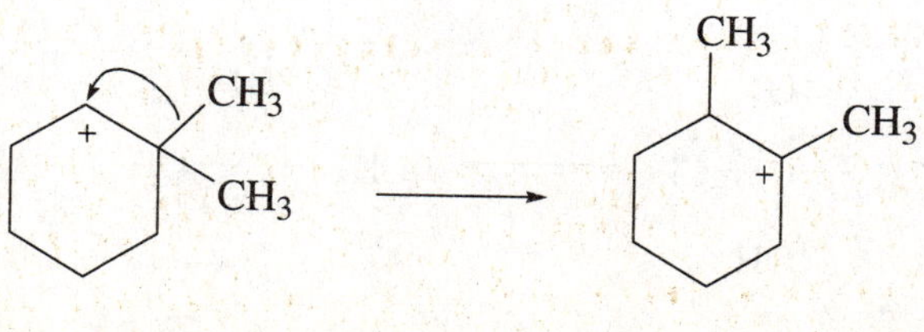

d. 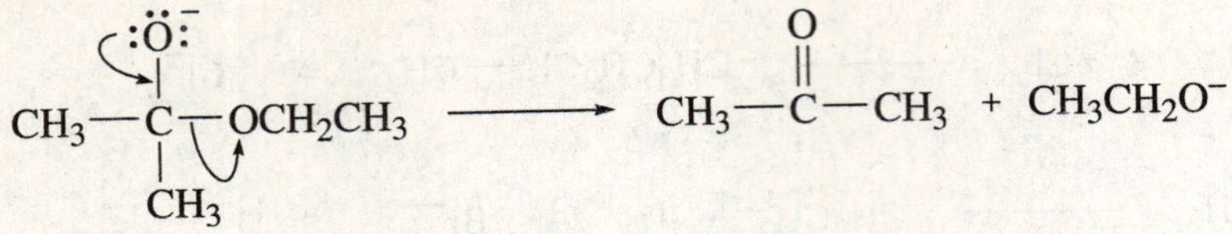

Problem 6

a. 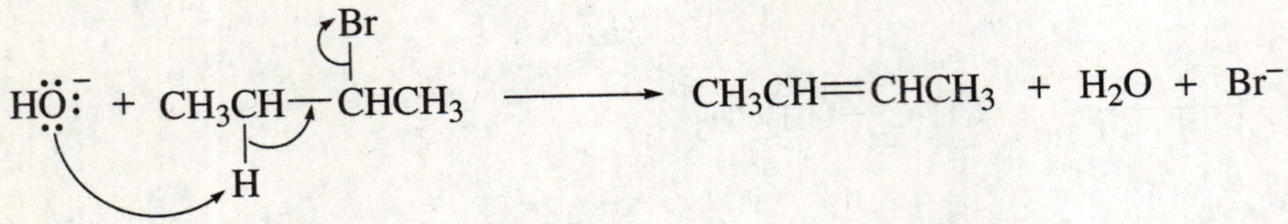

b. 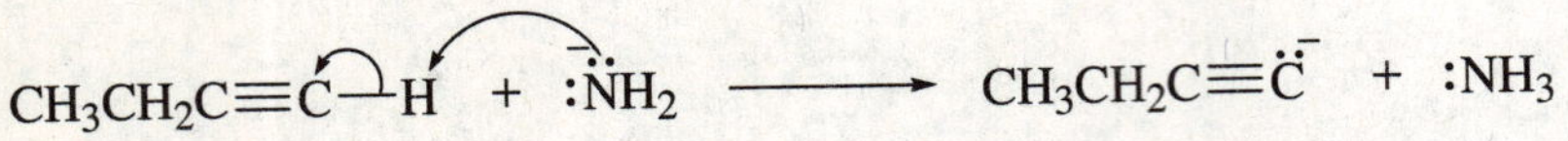

c. 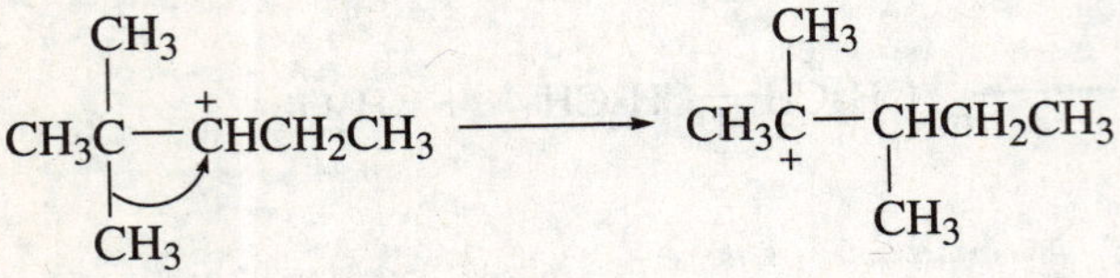

d. 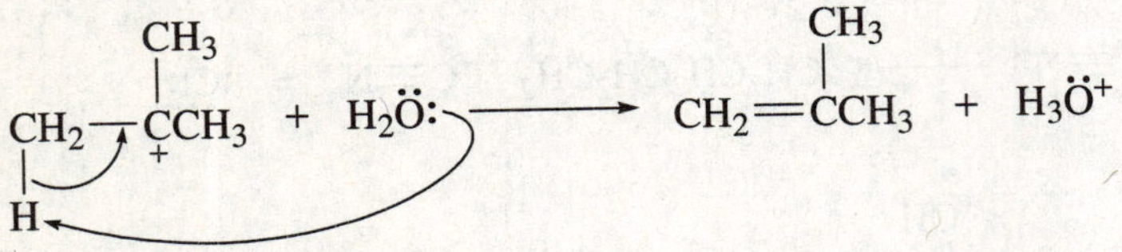

Problem 7

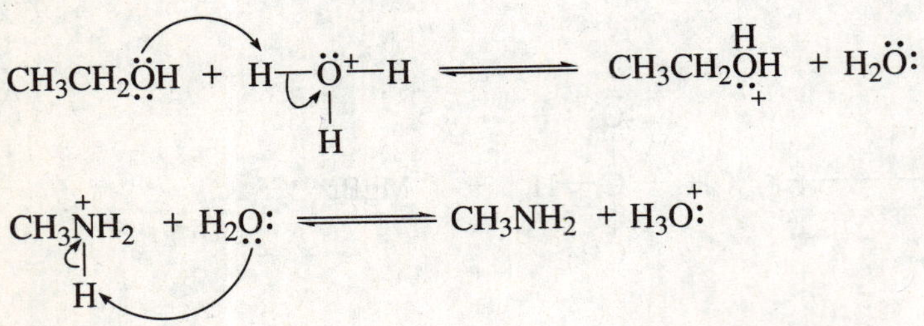

Problem 8

a. 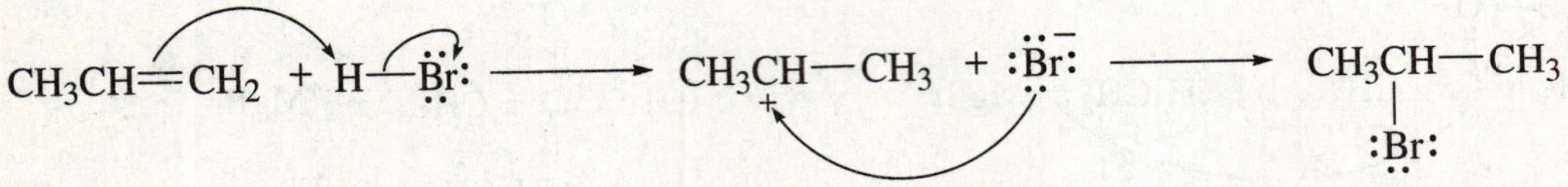

b. 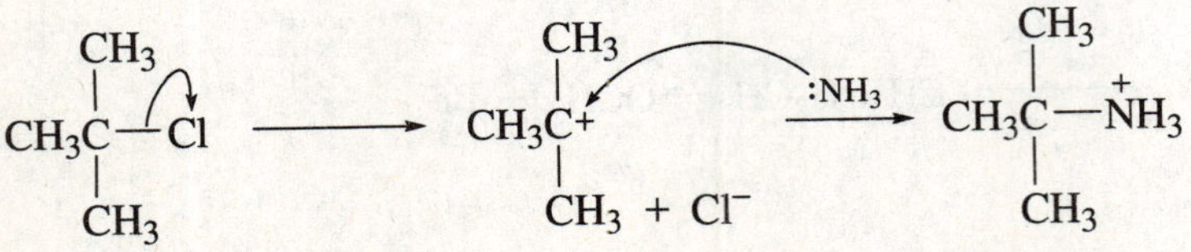

c. 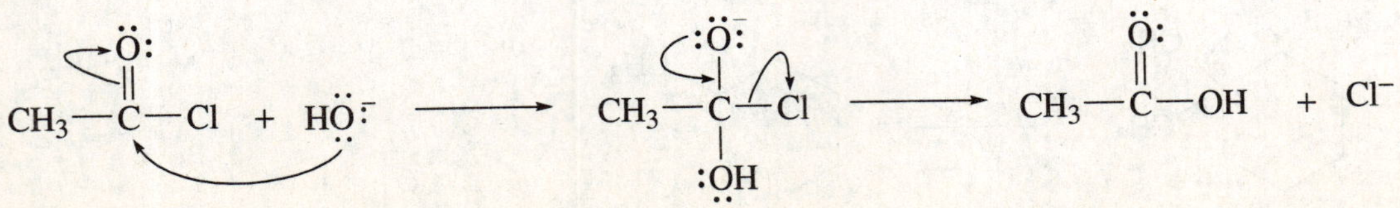

Problem 9

a.

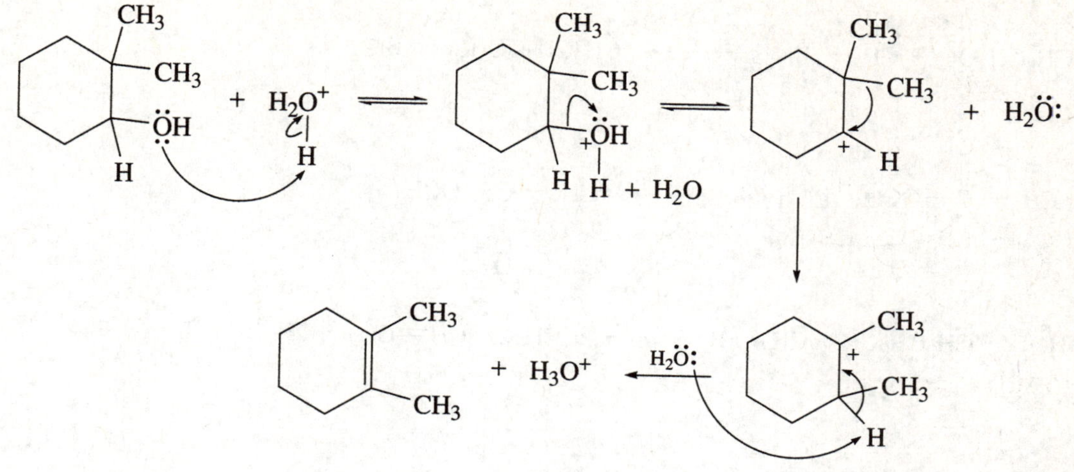

b.

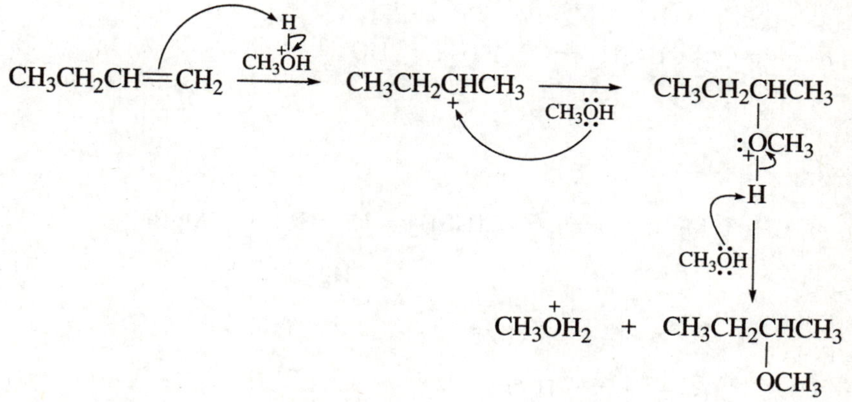

c.

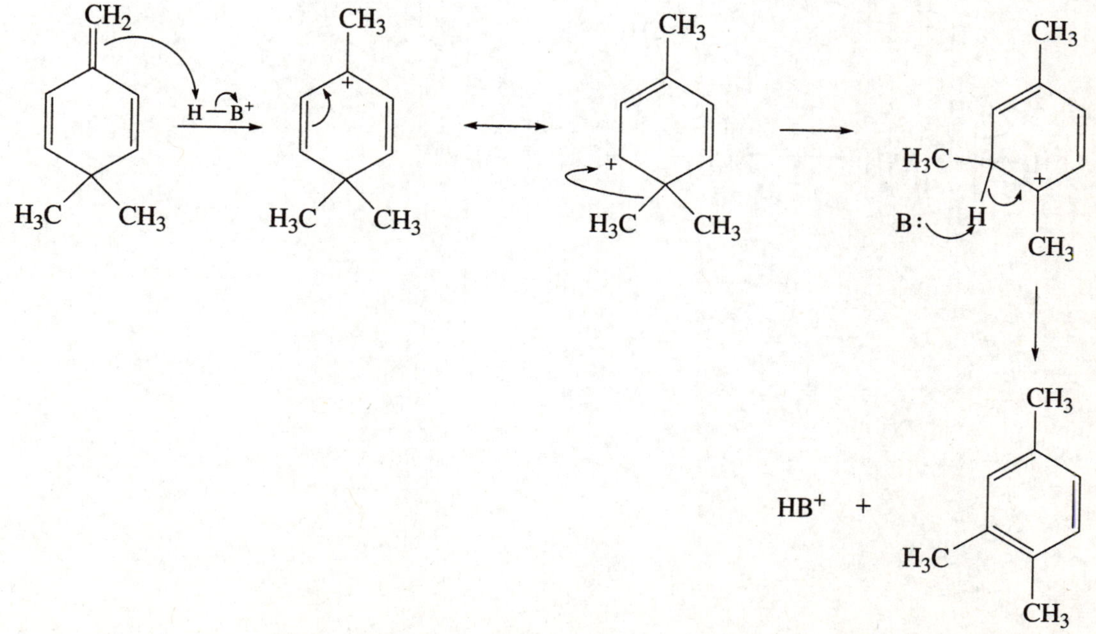

Problem 10

a. 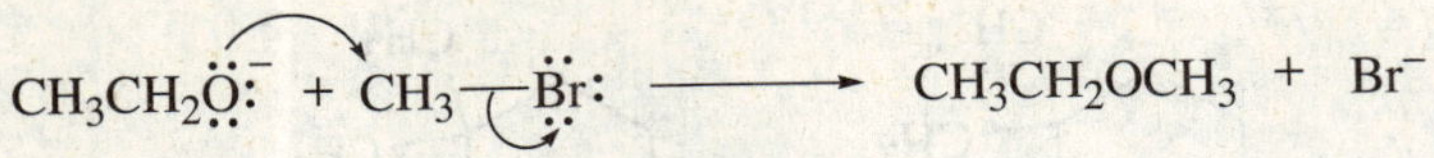

b. 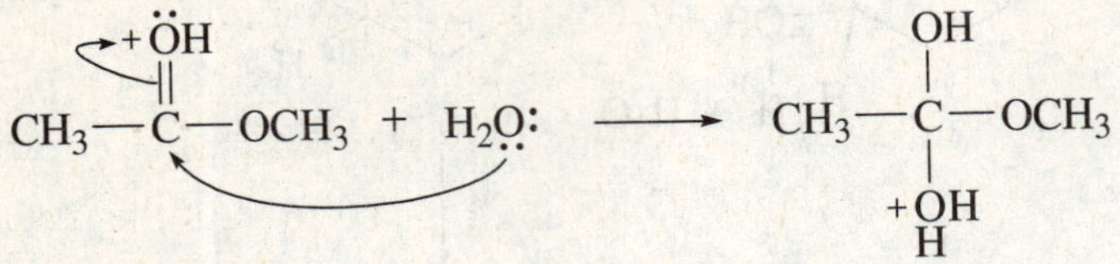

c. 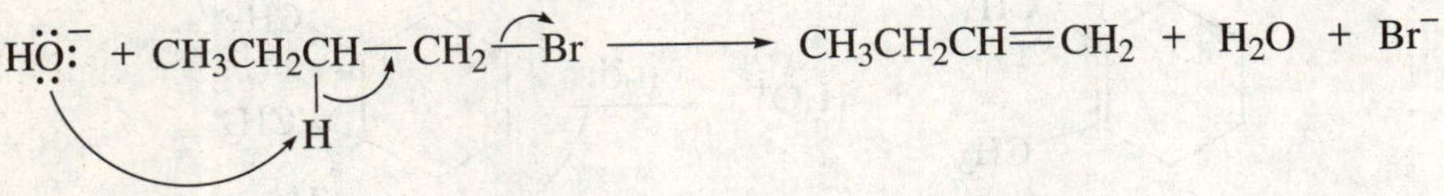

d. 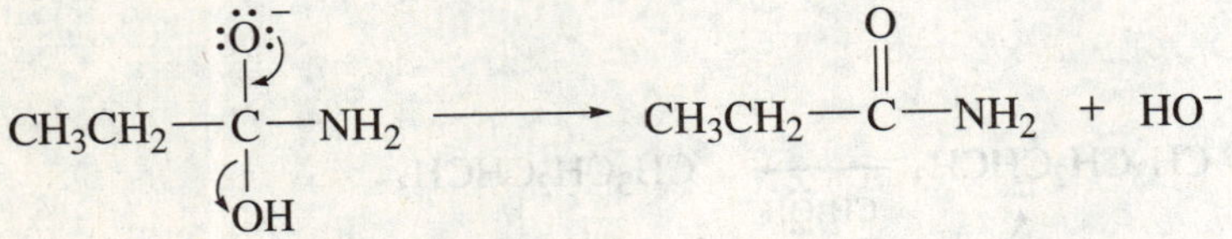

e. 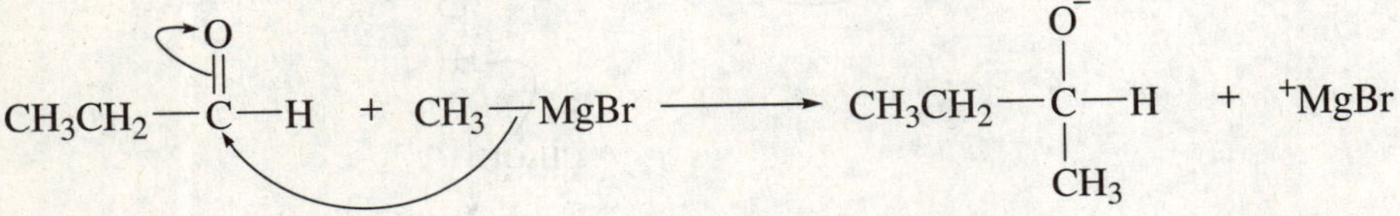

f. 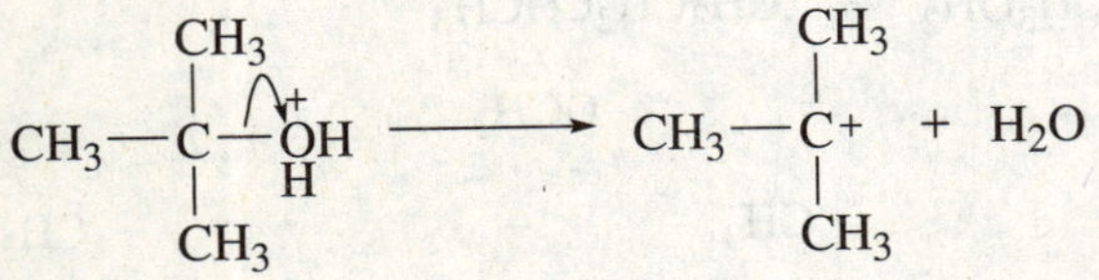

g. 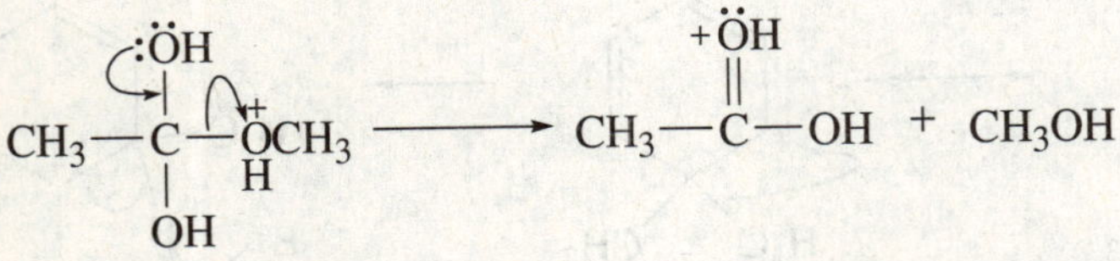

SPECIAL TOPIC III

Kinetics

This is a continuation of the discussion on kinetics found in Section 3.7 of the text. The rate laws are derived in Appendix III.

1. How long would it take for the reactant of a first-order reaction to decrease to one-half its initial concentration if the rate constant is $4.5 \times 10^{-3}\ s^{-1}$ and the initial concentration of the reactant is:
a. 1.0 M? **b.** 0.50 M?

2. How long would it take for the reactants of a second-order reaction to decrease to one-half their initial concentration if the rate constant is $2.3 \times 10^{-2}\ M^{-1}s^{-1}$ and the initial concentration of both reactants is:
a. 1.0 M? **b.** 0.50 M?

3. How many half-lives are required for a first-order reaction to reach > 99% completion?

4. The initial concentration of a reactant undergoing a first-order reaction is 0.40 M. After 5 minutes, the concentration of the reactant is 0.27 M; after an additional 5 minutes, the concentration of the reactant is 0.18 M; and after an additional 5 minutes, the concentration of the reactant is 0.12 M.

a. What is the average rate of the reaction during each five-minute interval?
b. What is the rate constant of the reaction?

5. What percentage of a compound, undergoing a first-order reaction with a rate constant of $2.70 \times 10^{-5}\ s^{-1}$, would have reacted at the end of two hours?

6. How long would it take for a first-order reaction with a rate constant of $5.3 \times 10^{-4}\ s^{-1}$ to reach 70% completion?

7. The following data were obtained in a study of the rate of inversion of sucrose at 25°C. The initial concentration of sucrose was 1.00 M.

Time (minutes)	0	30	60	90	130	180
Sucrose inverted (M)	0	0.100	0.195	0.277	0.373	0.468

a. What is the order of the reaction?
b. What is the rate constant of the reaction?

8. Calculate the activation energy of a first-order reaction that is 20% complete in 15 minutes at 40 °C and is 20% complete in 3 minutes at 60 °C.

9. Analysis of an aqueous solution of sucrose shows that 80 grams of the original 100 grams of sucrose remain after 10 hours. At this rate, how much sucrose would be left after 24 hours?

Solutions to Problems in Special Topic III

1. **a.** half-life of a first-order reaction $= t_{1/2} = \dfrac{\ln 2}{k_1}$

$$= \frac{0.693}{k_1}$$

$$= \frac{0.693}{4.5\times10^{-3}\ \text{s}^{-1}}$$

$$= 154 \text{ seconds}$$

$$= 2 \text{ minutes, } 34 \text{ seconds}$$

b. The half-life of a first-order reaction is independent of concentration, so the answer is the same as the answer for **a** (15.4 seconds).

2. **a.** half-life of a second-order reaction $= t_{1/2} = \dfrac{1}{k_2 a}$

$$= \frac{1}{2.3\times10^{-2}\ \text{M}^{-1}\text{s}^{-1}\ (1.00\ \text{M})}$$

$$= 43 \text{ seconds}$$

b. $t_{1/2} = \dfrac{1}{2.3\times10^{-2}\ \text{M}^{-1}\text{s}^{-1}\ (0.50\ \text{M})}$

$$= \frac{1}{1.5\times10^{-2}\ \text{s}^{-1}}$$

$$= 87 \text{ seconds}$$

3. One-half of the compound reacts during the first half-life. One half of what is left after the first half-life reacts during the second half-life. One half of what is left after the second half-life reacts during the third half-life, etc. The following Table shows that seven half-lives are required to reach > 99% completion.

number of half-lives	percentage completion	
1	0.50 (100) = 50	50
2	0.50 (50) = 25	75
3	0.50 (25) = 12.5	87.5
4	0.50 (12.5) = 6.25	93.8
5	0.50 (6.25) = 3.125	96.9
6	0.50 (3.125) = 1.5625	98.4
7	0.50 (1.5625) = 0.78125	99.2

4. **a.**

$$\text{rate} = \frac{\text{change in concentration}}{\text{change in time}}$$

$$\mathbf{1^{st}\ interval:}\ \text{rate} = \frac{0.40\ \text{M} - 0.27\ \text{M}}{5\ \text{min}} = \frac{0.13\ \text{M}}{5\ \text{min}} = 2.60\times10^{-2}\ \text{M min}^{-1}$$

$$\mathbf{2^{nd}\ interval:}\ \text{rate} = \frac{0.27\ \text{M} - 0.18\ \text{M}}{5\ \text{min}} = \frac{0.09\ \text{M}}{5\ \text{min}} = 1.80\times10^{-2}\ \text{M min}^{-1}$$

$$\mathbf{3^{rd}\ interval:}\ \text{rate} = \frac{0.18\ \text{M} - 0.12\ \text{M}}{5\ \text{min}} = \frac{0.06\ \text{M}}{5\ \text{min}} = 1.20\times10^{-2}\ \text{M min}^{-1}$$

Thus, we see that the **rate** of the reaction decreases with decreasing concentration.

b. rate = k [reactant], using the average concentration of the reactant during the 5 minute interval.

1st interval: $2.60 \times 10^{-2}\ \text{M min}^{-1} = \mathbf{k}\ (0.335\ \text{M})$

$\mathbf{k} = 7.8 \times 10^{-2}\ \text{min}^{-1}$

2nd interval: $1.80 \times 10^{-2}\ \text{M min}^{-1} = \text{k}\ (0.225\ \text{M})$

$\mathbf{k} = 8.0 \times 10^{-2}\ \text{min}^{-1}$

3rd interval: $1.20 \times 10^{-2}\ \text{M min}^{-1} = \mathbf{k}\ (0.15\ \text{M})$

$\mathbf{k} = 8.0 \times 10^{-2}\ \text{min}^{-1}$

The **rate constant**, as its name indicates, is constant during the course of the reaction.

5. $\boldsymbol{a}$ = the initial concentration
$\boldsymbol{x}$ = the concentration that has reacted at time = $\boldsymbol{t}$

$$\ln\frac{a}{a-x} = k_1 t$$

$$\ln\frac{100}{100-x} = 2.70\times10^{-5}\ \text{s}^{-1}\ (2\ \text{hours})$$

$$\ln\frac{100}{100-x} = 2.70\times10^{-5}\ \text{s}^{-1}\ (7200\ \text{seconds})$$

$$\ln\frac{100}{100-x} = 0.194$$

$$\frac{100}{100-x} = 1.21$$

$$100 = 121 - 1.21\,x$$

$$1.21\,x = 21$$

$$x = 17.4\,\%$$

6.

$$\ln\frac{a}{a-x} = k_1 t$$

$$\ln\frac{100}{100-70} = 5.30\times10^{-4}\ \text{s}^{-1}\ t$$

$$t = \frac{\ln 3.33}{5.30\times10^{-4}\ \text{s}^{-1}}$$

$$t = \frac{1.20}{5.30\times10^{-4}\ \text{s}^{-1}}$$

$$t = 2260 \text{ seconds}$$

$$t = 37.7 \text{ minutes}$$

7.

<u>1st order</u>

$$k_1 = \frac{-\ln\dfrac{a-x}{a}}{t}$$

$$k_1 = \frac{-\ln\dfrac{1.000-0.100}{1.000}}{30} = \mathbf{3.51\times10^{-3}}$$

$$k_2 = \frac{-\ln\dfrac{1.000-0.195}{1.000}}{60} = \mathbf{3.62\times10^{-3}}$$

$$k_1 = \frac{-\ln\dfrac{1.000-0.277}{1.000}}{90} = \mathbf{3.60\times10^{-3}}$$

$$k_1 = \frac{-\ln\dfrac{1.000-0.373}{1.000}}{130} = \mathbf{3.59\times10^{-3}}$$

$$k_1 = \frac{-\ln\dfrac{1.000-0.468}{1.000}}{180} = \mathbf{3.51\times10^{-3}}$$

<u>2nd order</u>

$$k_2 = \frac{\dfrac{1}{a-x}-\dfrac{1}{a}}{t}$$

$$k_2 = \frac{\dfrac{1}{1.000-0.100}-\dfrac{1}{1.000}}{30} = \frac{1.111-1.000}{30} = \mathbf{3.70\times10^{-3}}$$

$$k_2 = \frac{\dfrac{1}{1.000-0.195}-1.000}{60} = \mathbf{4.03\times10^{-3}}$$

$$k_2 = \frac{\dfrac{1}{1.000-0.277}-1.000}{90} = \mathbf{4.26\times10^{-3}}$$

$$k_2 = \frac{\dfrac{1}{1.000-0.373}-1.000}{130} = \mathbf{4.57\times10^{-3}}$$

$$k_2 = \frac{\dfrac{1}{1.000-0.468}-1.000}{180} = \mathbf{4.89\times10^{-3}}$$

a. Because the calculated rate constants are relatively constant when the data are plugged into a first-order equation but vary considerably when the data are plugged into a second-order equation, one can conclude that the reaction is first-order.

b. $3.59 \times 10^{-3}\ \text{min}^{-1}$

8.

$$k \text{ at } 40^\circ\text{C} = \frac{\ln \dfrac{1.00}{1.00 - 0.20}}{15 \text{ min}} = 1.49 \times 10^{-2}\ \text{min}^{-1}$$

$$k \text{ at } 60^\circ\text{C} = \frac{\ln \dfrac{1.00}{1.00 - 0.20}}{3 \text{ min}} = 7.44 \times 10^{-2}\ \text{min}^{-1}$$

$$\ln k_2 - \ln k_1 = \frac{-E_a}{R}\left(\frac{1}{T_2} - \frac{1}{T_1}\right)$$

$$\ln 7.44 \times 10^{-2} - \ln 1.49 \times 10^{-2} = \frac{-E_a}{1.986 \times 10^{-3}\ \text{kcal}}\left(\frac{1}{333} - \frac{1}{313}\right)$$

$$-2.60 - (-4.21) = \frac{-E_a}{1.986 \times 10^{-3}}(0.00300 - 0.00319)$$

$$1.61 = \frac{-E_a}{1.986 \times 10^{-3}}(-0.00019)$$

$$E_a = \frac{1.61 \times 1.986 \times 10^{-3}}{0.00019}$$

$$E_a = 16.8\ \text{kcal/mol}$$

9. Sucrose is hydrolyzed to form a mixture of glucose and fructose. Because there is excess water (it is the solvent), the reaction is a pseudo first-order reaction.

First, the rate constant of the reaction must be determined:

$$\ln \frac{a}{a - x} = k_1 t$$

$$\ln \frac{100}{80} = k_1 \times 10 \text{ hours}$$

$$k_1 = 2.23 \times 10^{-2}\ \text{hr}^{-1}$$

Now we can calculate the amount of sucrose that has reacted, and therefore, the amount that would be left.

$$\ln \frac{a}{a-x} = k_1 t$$

$$\ln \frac{100}{100-x} = 2.23\times10^{-2}\ (24\ \text{hr})$$

$$\ln \frac{100}{100-x} = 0.535$$

$$\frac{100}{100-x} = 1.71$$

$$100 = 171 - 1.71x$$

$$1.71x = 71$$

$$x = 41.5\ \text{g have reacted}$$

Therefore, 58.5 g would be left.

CHAPTER 4
The Reactions of Alkenes

Important Terms

acid-catalyzed reaction — a reaction catalyzed by an acid.

alkoxymercuration-reduction — addition of an alcohol to an alkene using mercuric acetate followed by sodium borohydride and sodium hydroxide.

carbene — a neutral carbon with a lone-pair of electrons and an empty orbital.

carbocation rearrangement — the rearrangement of a carbocation to a more stable carbocation.

catalyst — a compound that increases the rate at which a reaction occurs without being consumed in the reaction. Because it does not change the equilibrium constant of the reaction, it does not change the amount of product that is formed.

catalytic hydrogenation — the addition of hydrogen to a double or a triple bond with the aid of a metal catalyst.

concerted reaction — a reaction in which all the bond-making and bond-breaking processes take place in a single step.

constitutional isomers (structural isomers) — molecules that have the same molecular formula but differ in the way the atoms are connected.

dimer — a molecule formed by joining together two identical molecules.

electrophilic addition reaction — an addition reaction in which the first species that adds to the reactant is an electrophile.

epoxide (oxirane) — an ether in which the oxygen is incorporated into a three-membered ring.

halohydrin — an organic molecule that contains a halogen atom and an OH group.

Hammond postulate — states that the transition state will be more similar in structure to the species (reactants or products) that it is closer to energetically.

heat of hydrogenation — the heat ($\Delta H°$) released in a hydrogenation reaction.

heterogeneous catalyst — a catalyst that is insoluble in the reaction mixture.

hormone — a compound that controls growth and other changes in tissues.

hydration — addition of water to a compound.

1,2-hydride shift — the movement of a hydride ion from one carbon to an adjacent carbon.

hydroboration-oxidation — the addition of borane to an alkene (or to an alkyne) followed by reaction with hydrogen peroxide and hydroxide ion.

hydrogenation	addition of hydrogen.
hyperconjugation	delocalization of electrons by overlap of carbon-hydrogen or carbon-carbon σ bonds with an empty *p* orbital.
Markovnikov's rule	the actual rule is: "When a hydrogen halide adds to an asymmetrical alkene, the addition occurs such that the halogen attaches itself to the carbon atom of the alkene bearing the least number of hydrogen atoms." Chemists use the rule as follows: The hydrogen adds to the sp^2 carbon that is bonded to the greater number of hydrogens. A more general rule is: The electrophile adds to the sp^2 carbon that is bonded to the greater number of hydrogens.
mechanism of the reaction	a description of the step-by-step process by which reactants are changed into products.
1,2-methyl shift	the movement of a methyl group with its bonding electrons from one carbon to an adjacent carbon.
oxidation reaction	a reaction that decreases the number of C—H bonds in the reactant or increases the number of C—O, C—N, or C—X bonds bonds (X denotes a halogen).
oxymercuration-reduction	addition of water using a mercuric salt of a carboxylic acid as a catalyst, followed by reaction with sodium borohydride.
pheromone	a chemical substance used for the purpose of communication.
primary carbocation	a carbocation with the positive charge on a primary carbon.
reduction reaction	a reaction that increases the number of C—H bonds in the reactant or deccreases the number of C—O, C—N, or C—X bonds bonds (X denotes a halogen).
regioselective reaction	a reaction that leads to the preferential formation of one constitutional isomer over another.
secondary carbocation	a carbocation with the positive charge on a secondary carbon.
steric effect	an effect due to the space occupied by a substituent.
steric hindrance	refers to bulky groups at the site of a reaction that make it difficult for the reactants to approach one another.
tertiary carbocation	a carbocation with the positive charge on a tertiary carbon.
vicinal	vicinal substituents are substituents on adjacent carbon.

Solutions to Problems

1. **a.** The bond orbitals of the carbon that is adjacent to the positively charged carbon are the σ bond orbitals that are available for overlap with the vacant *p* orbital. Because the methyl cation does not have a carbon adjacent to the positively charged carbon, there are no σ bond orbitals available for overlap with the vacant *p* orbital.

b. An ethyl cation is more stable because the carbon adjacent to the positively charged carbon has three σ bond orbitals available for overlap with the vacant *p* orbital, whereas a methyl cation does not have any σ bond orbitals available for overlap with the vacant *p* orbital.

2. **a.** **1.** 3 **2.** 3 **3.** 6

b. *sec*-butyl cation

3. **a.** $CH_3CH_2\overset{+}{C}(CH_3)CH_3 > CH_3CH_2\overset{+}{C}HCH_3 > CH_3CH_2CH_2\overset{+}{C}H_2$

b. The halogen atoms decrease the stability of the carbocation because, since they are more electronegative than a hydrogen, they are more effective than a hydrogen at withdrawing electrons away from the positively charged carbon. This increases the concentration of positive charge on the carbocation which makes it less stable.

$CH_3CH(CH_3)CH_2\overset{+}{C}H_2 > CH_3CH(Cl)CH_2\overset{+}{C}H_2 > CH_3CH(F)CH_2\overset{+}{C}H_2$

Because fluorine is more electronegative than chlorine, the fluorine-substituted carbocation is less stable than the chlorine-substituted carbocation.

4. The transition state will resemble the one that it is closer to on the reaction coordinate diagram; that is, the one that it is closer to in energy.

a. products **b.** reactants **c.** reactants **d.** products

5. **a.** $CH_3CH_2CH(Br)CH_3$

c. 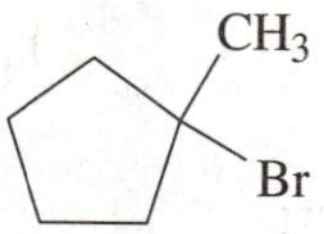

e. 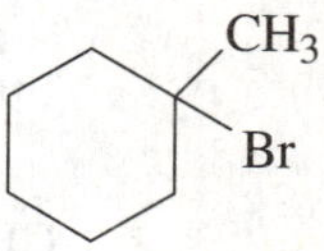

b. $CH_3CH_2C(CH_3)(Br)CH_3$

d. $CH_3C(CH_3)(Br)CH_2CH_2CH_3$

f. $CH_3CH_2CH(Br)CH_3$

6. **a.** $CH_2{=}C(CH_3)CH_3$

c. 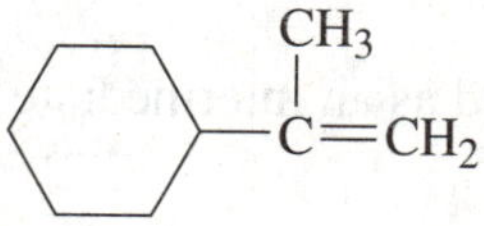

b. 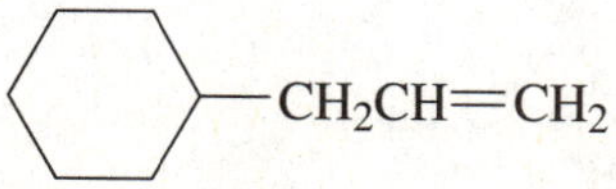

d. 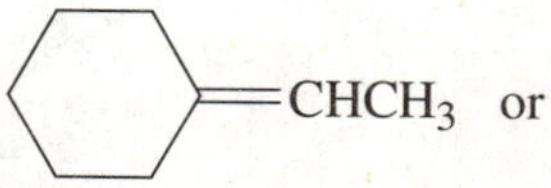 or 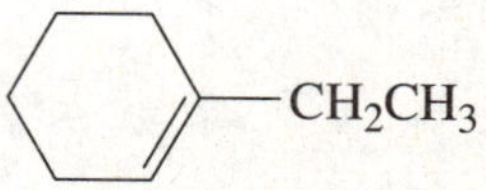

7. **a.**

$$\begin{array}{c} CH_3 \\ | \\ CH_3CH_2C{=}CH_2 \end{array}$$

b.

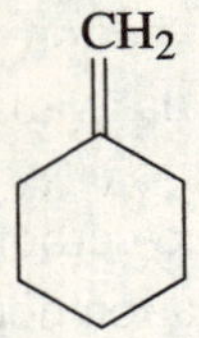

In both **a** and **b**, the compound that is more highly regioselective is the one where the choice is between forming a tertiary carbocation or a primary carbocation.
In the less regioselective compound, the choice is between forming a tertiary carbocation or a secondary carbocation, so the difference in the stability of the two possible carbocations is not as great as it is when the choice is between a tertiary and a primary carbocation.

8. As long as the pH is greater than about –2.5 and less than about 15, more than 50% of 2-propanol would be in its neutral, nonprotonated form.

$$\overset{+}{R}OH_2 \overset{pk_a = -2.5}{\rightleftharpoons} ROH \overset{pk_a = \sim 15}{\rightleftharpoons} RO^-$$

Recall that when the pH = pK_a, half the compound is in its acid form and half is in its basic form. Therefore at a pH less than – 2.5, more than half of the compound will be in its positively charged protonated form. At a pH greater than ~ 15, more than half of the compound will exist as the negatively charged anion.
Thus at a pH between –2.5 and ~ 15, more than half of the compound will exist in the neutral nonprotonated form.

9. **a.** 3 transition states **b.** 2 intermediates
c. the first step is the slowest step, so it has the smallest rate constant; the second step is fast because no bonds are being broken; the third step is fast because transfer of a proton from or to an O or an N is always a fast reaction.

10. **a.** $\begin{array}{c} CH_3CH_2CH_2CHCH_3 \\ \quad\quad\quad | \\ \quad\quad\quad OH \end{array}$

c. $\begin{array}{c} CH_3CH_2CH_2CH_2CHCH_3 \\ \quad\quad\quad\quad | \\ \quad\quad\quad\quad OH \end{array}$ and $\begin{array}{c} CH_3CH_2CH_2CHCH_2CH_3 \\ \quad\quad\quad | \\ \quad\quad\quad OH \end{array}$

b. 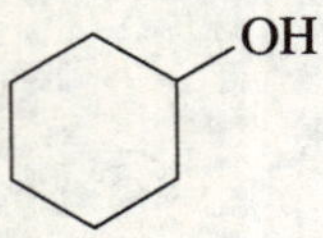

d. 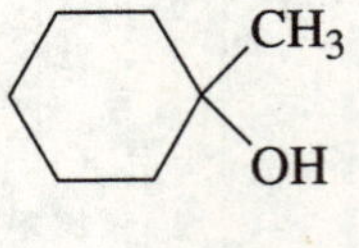

11. **a.** 1. $\begin{array}{c} CH_3 \\ | \\ CH_3CCH_3 \\ | \\ Cl \end{array}$ 2. $\begin{array}{c} CH_3 \\ | \\ CH_3CCH_3 \\ | \\ Br \end{array}$ 3. $\begin{array}{c} CH_3 \\ | \\ CH_3CCH_3 \\ | \\ OH \end{array}$ 4. $\begin{array}{c} CH_3 \\ | \\ CH_3CCH_3 \\ | \\ OCH_3 \end{array}$

b. The first step in all the reactions is addition of an electrophilic proton (H^+) to the carbon of the CH_2 group.
A *tert*-butyl carbocation is formed as an intermediate in each of the reactions.

c. The nucleophile that adds to the *tert*-butyl carbocation is different in each reaction. In reactions #3 and #4, there is a third step—a proton is lost from the group that was the nucleophile in the second step of the reaction.

3. 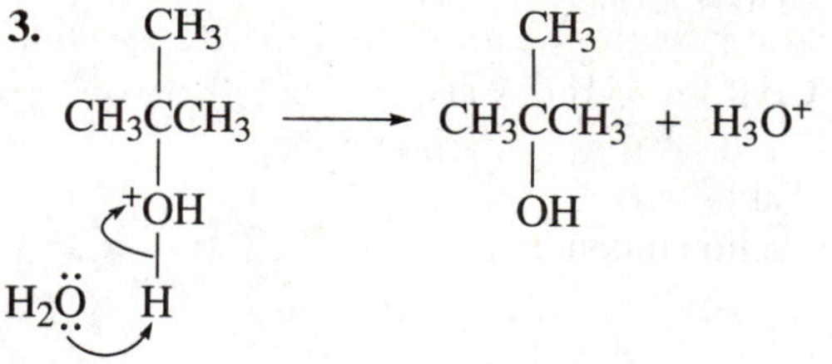

4. 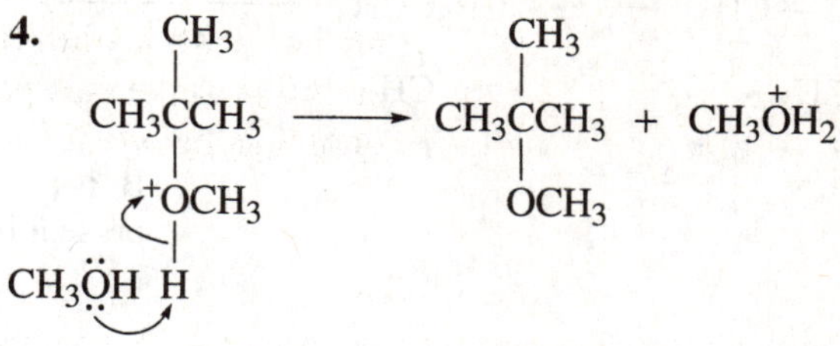

12. a. 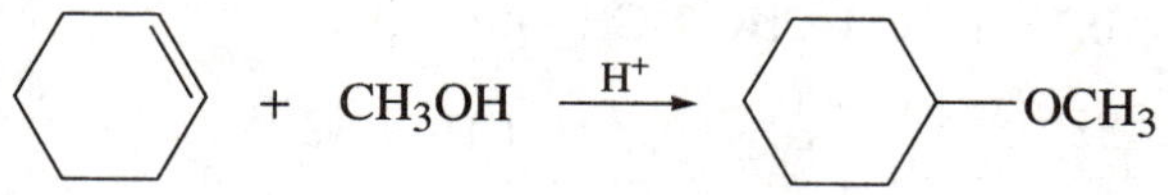

b. 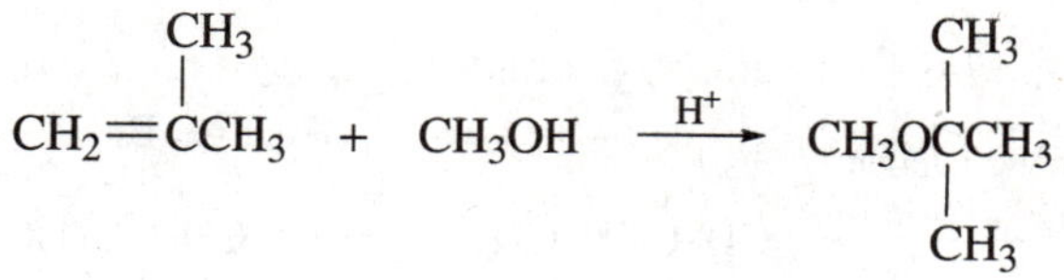

c. 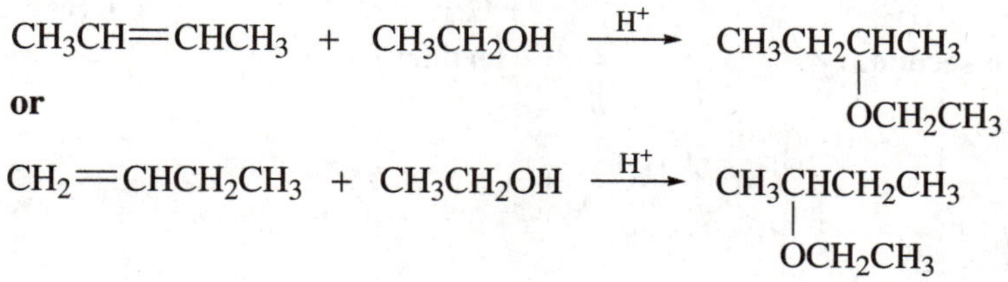

d. 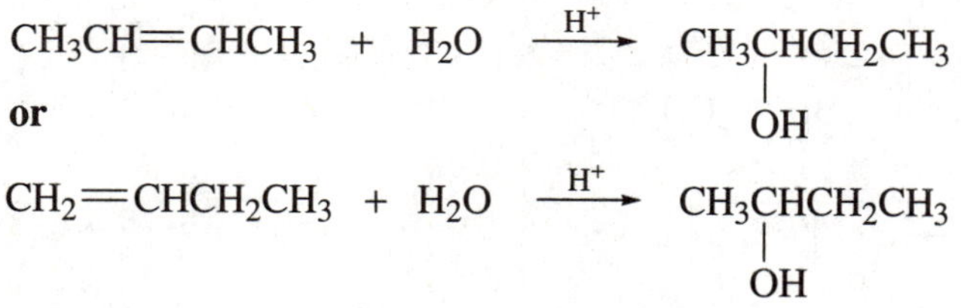

e. 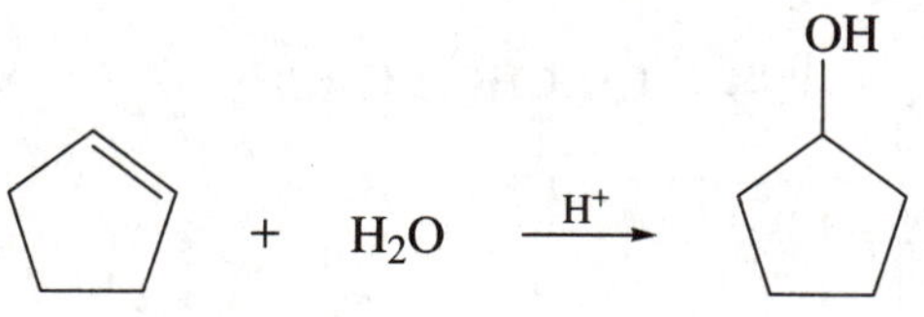

f. 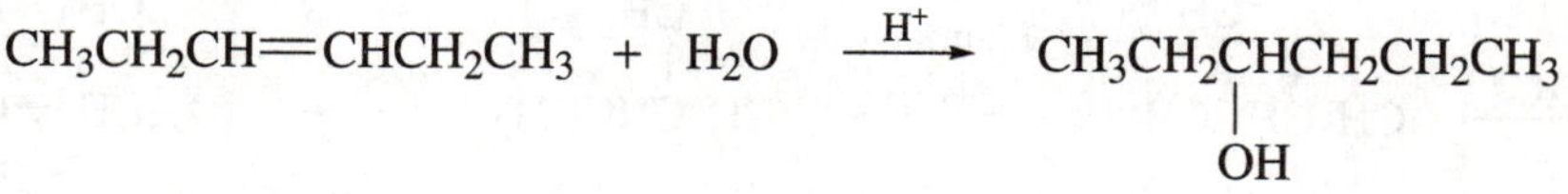

13. 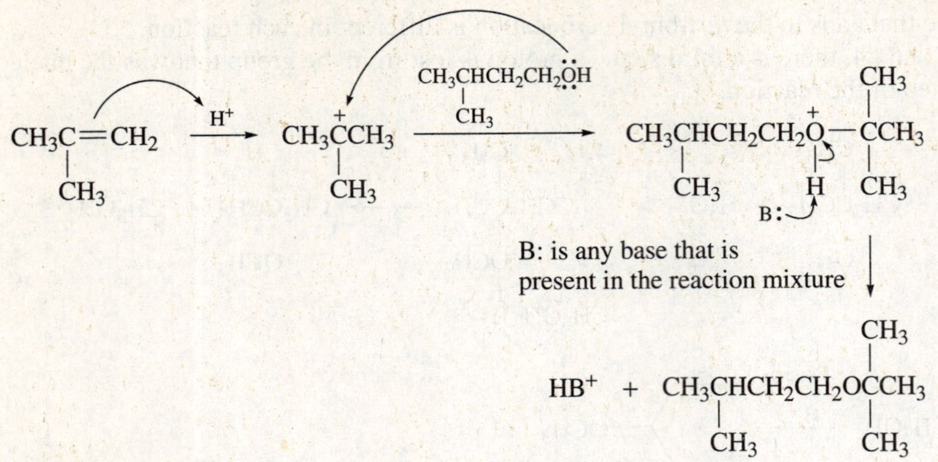

14. Solved in the text.

15. **a.** 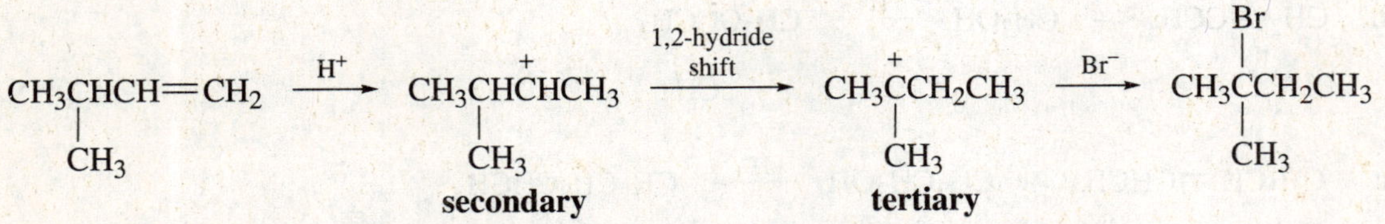

b. 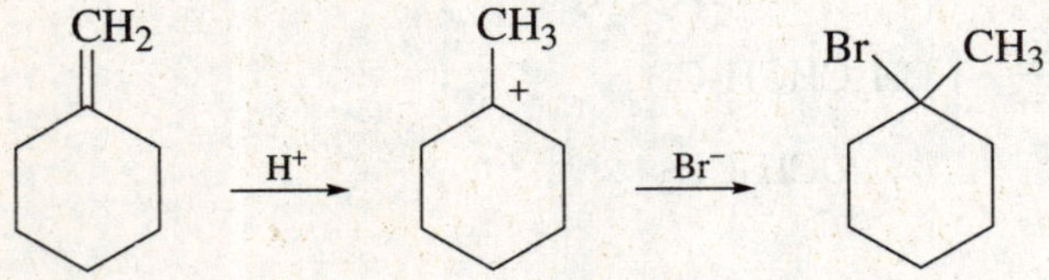

c. 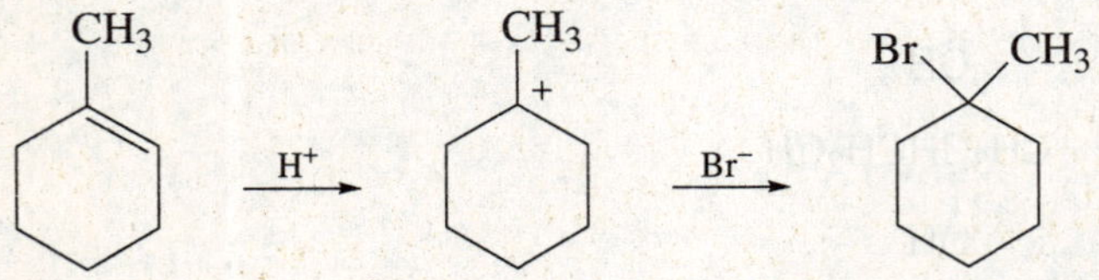

d. 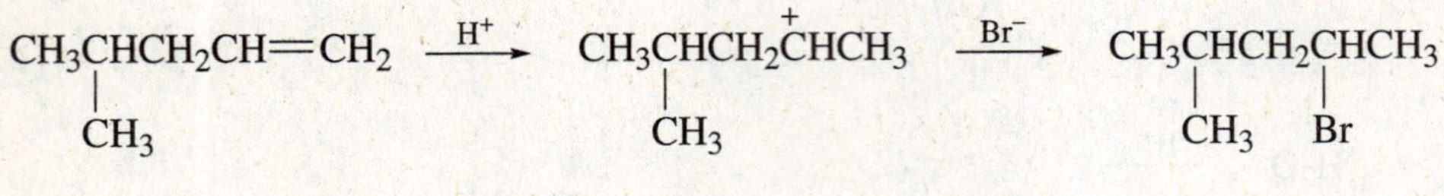

e. 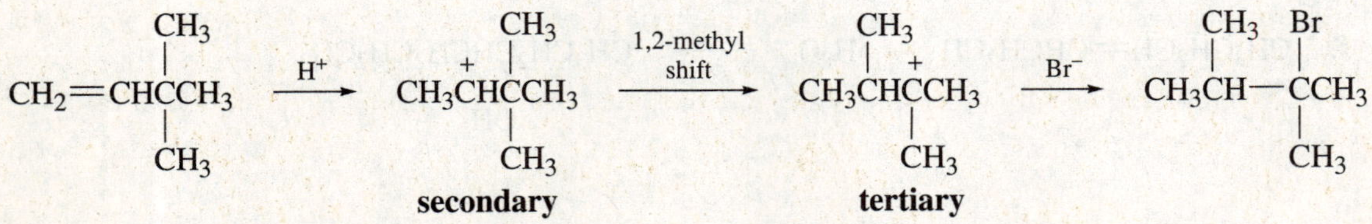

f.

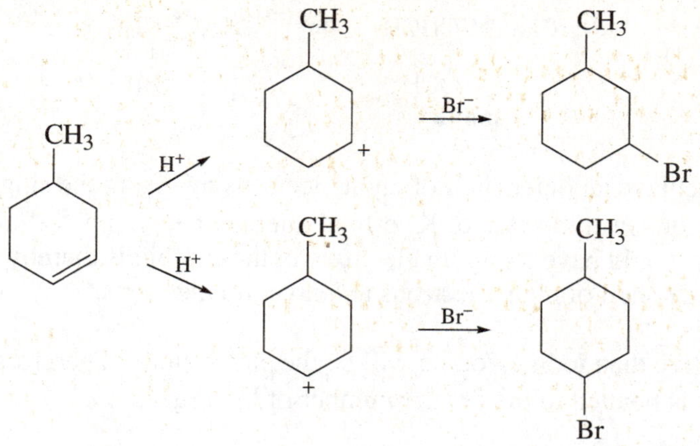

1-Bromo-3-methylcyclhexane and 1-bromo-4-methylcyclohexane will be obtained in approximately equal amounts because in each case the initially formed carbocation is secondary.

16.

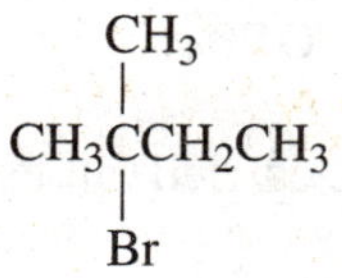

Addition of H^+ would from a carbocation that could rearrange.

Addition of Br^+ forms cyclic bromonium ion rather than a carbocation so there is no rearrangement.

17. **a.** The first step in the reaction with Br_2 forms a cyclic bromonium ion, whereas the first step in the reaction with HBr forms a carbocation.

b. If the bromide ion were to attack the positively charged bromine atom, a highly unstable compound (with a negative charge on carbon and a positive charge on bromine) would be formed.

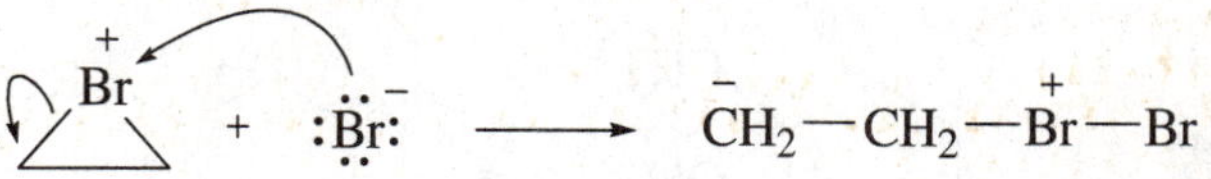

Notice that the electrostatic potential maps of the cyclic bromonium ions on p. 176 of the text show that the ring carbons are the least electron dense (most blue) atoms in the intermediate.

18. The nucleophile that is present in greater concentration is more apt to collide with the carbocation intermediate. Therefore, if the solvent is a nucleophile, the major product will come from reaction of the solvent with the carbocation or cyclic bromonium (or chloronium) ion intermediate, because the concentration of the solvent is much greater than the concentration of the other nucleophile. (For example, in "a" the concentration of CH_3OH is much greater than the concentration of Cl^-.)

a. 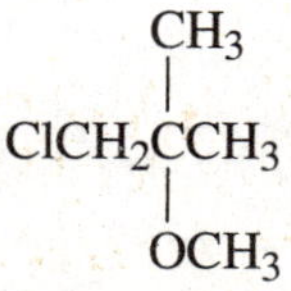 and 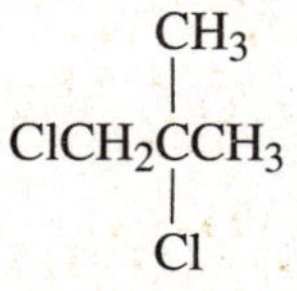

major

c. 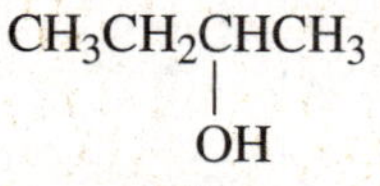 and 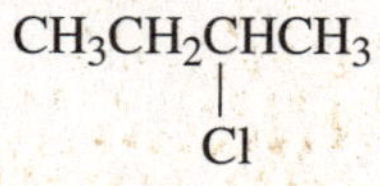

major

b. 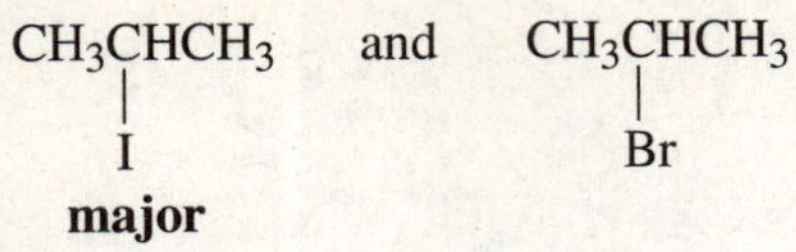

d. 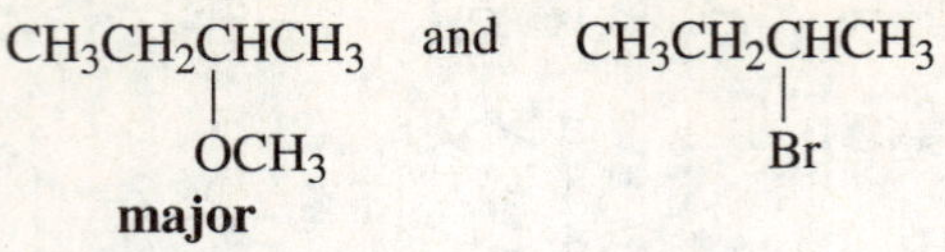

19. As elements, sodium and potassium achieve an outer shell of eight electrons by losing the single electron they have in the 3*s* (in the case of Na) or 4*s* (in the case of K) orbital, thereby becoming Na^+ and K^+. In order to form a covalent bond, they would have to regain electrons in these orbitals,thereby losing the stability associated with having an outer shell of eight electrons and no extra electrons.

20. Because chlorine is more electronegative than iodine, iodine will be the electrophile. Therefore, it will end up attached to the sp^2 carbon that is bonded to the greater number of hydrogens.

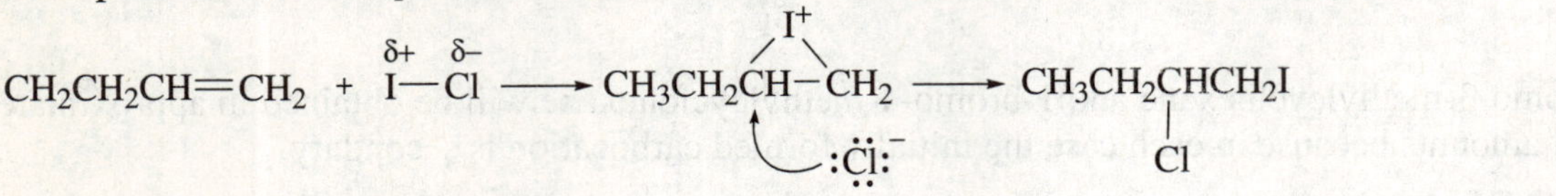

21. **a.** 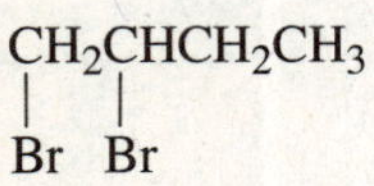 **b.** 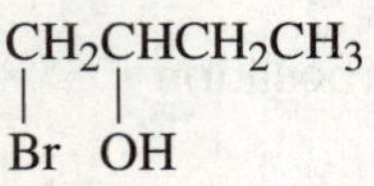 **c.** 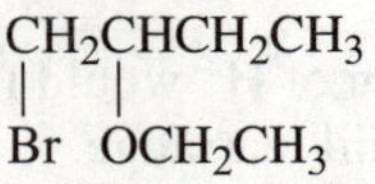 **d.** 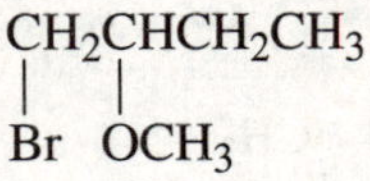

22. Notice that the addition of water or the addition of an alcohol to an alkene can be carried out using an acid catalyst (Section 4.5) or by mercuration/demercuration (Section 4.8).

a. 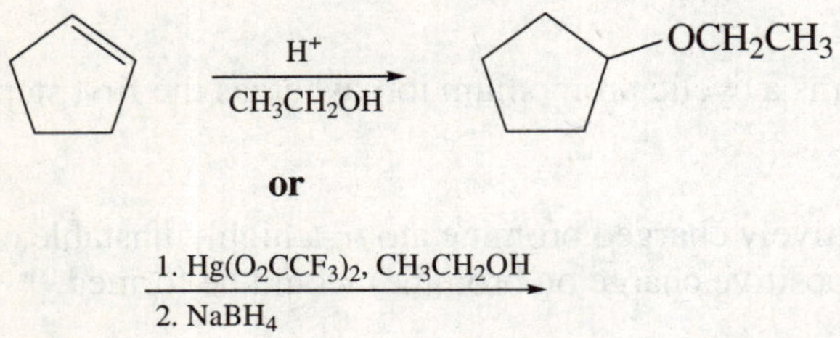

b. 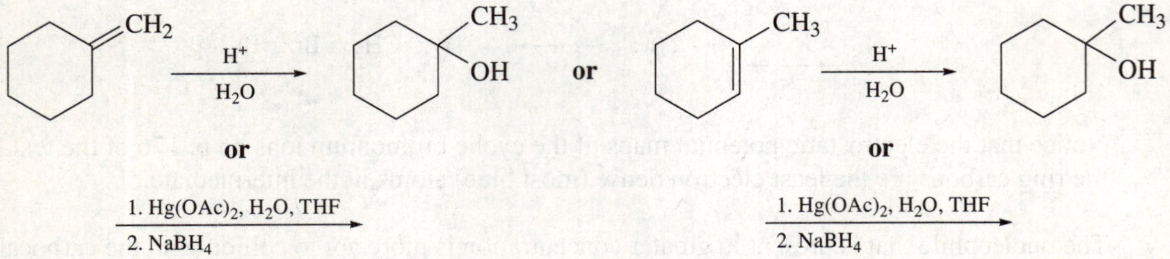

c. 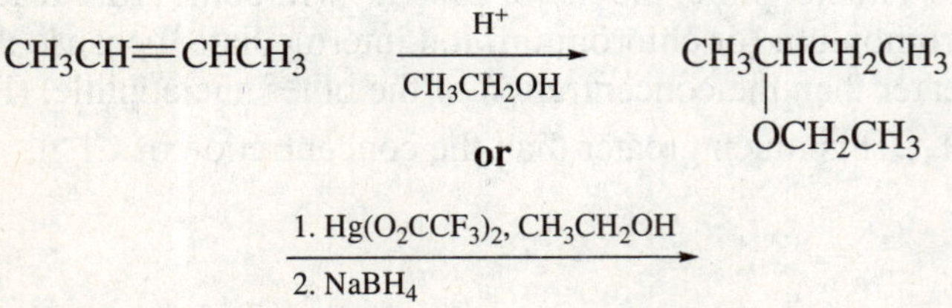

d. 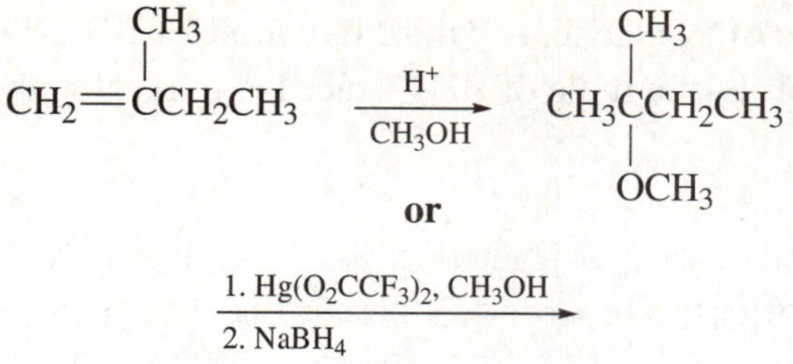

23. **a.** 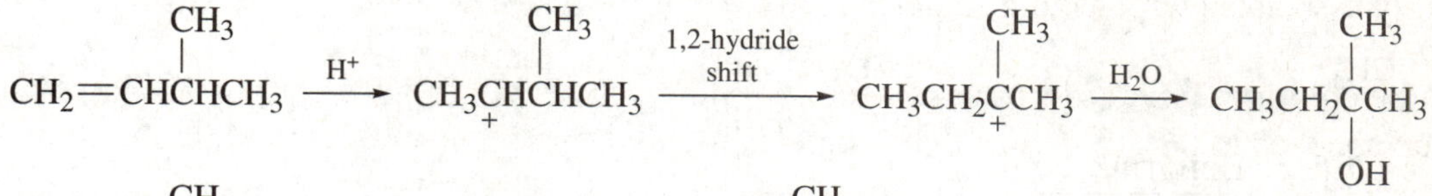

b. 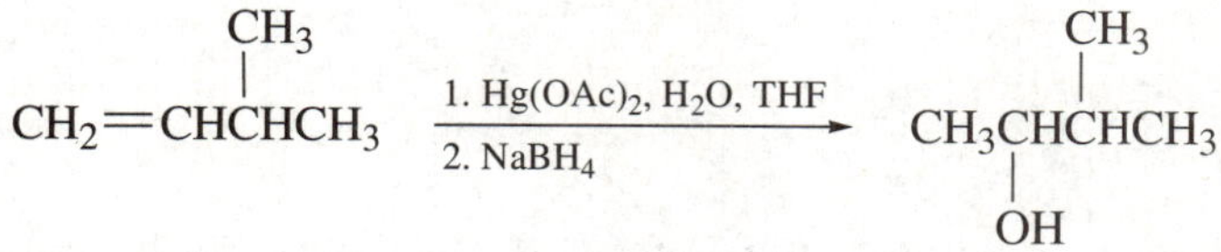

In **a**, a carbocation rearrangement is required to get the desired product from the given starting materials

In **b,** the desired product must be obtained from a reaction that will not form a carbocation intermediate.

24. **a.** 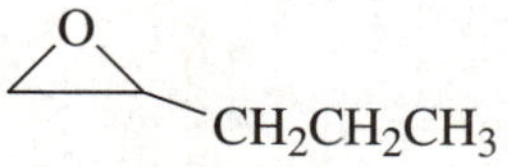

c. 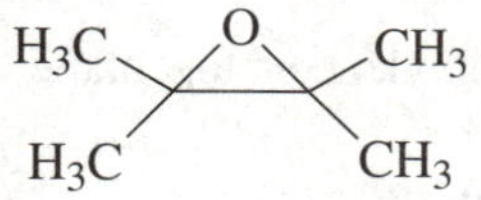

b. 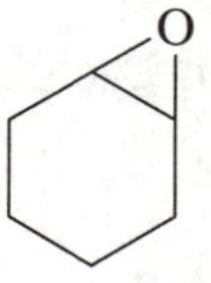

d. 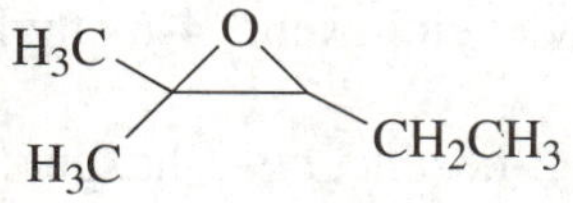

25. **a.** cyclohexene **b.** 1-butene

26. Addition of a proton to the carbon that is bonded to the greater number of hydrogens forms a carbocation intermediate. The alcohol group in the same molecule is the nucleophile that reacts with the carbocation.

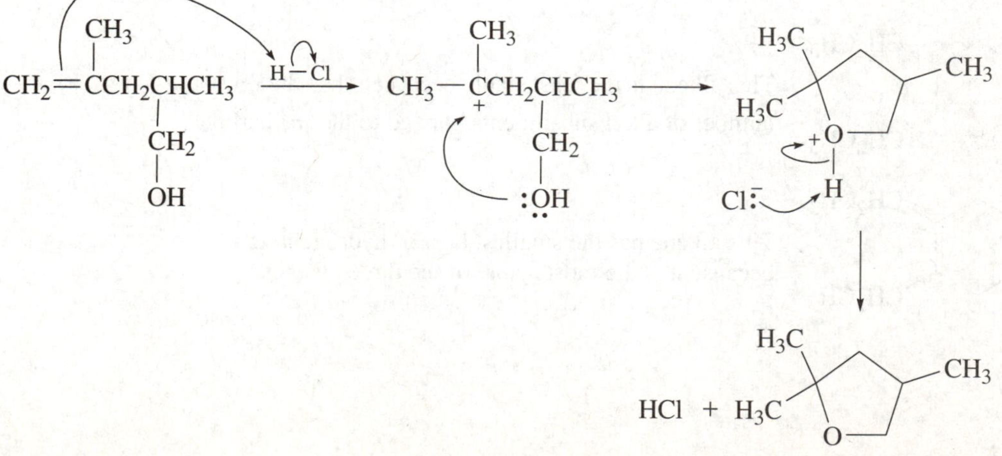

27. Because one mole of BH_3 reacts with three moles of an alkene, one third of a mole of BH_3 is needed to react with each mole of alkene. Therefore, two thirds of a mole of BH_3 is needed to react with two moles of an alkene (in this case, 1-pentene).

28. **a.**

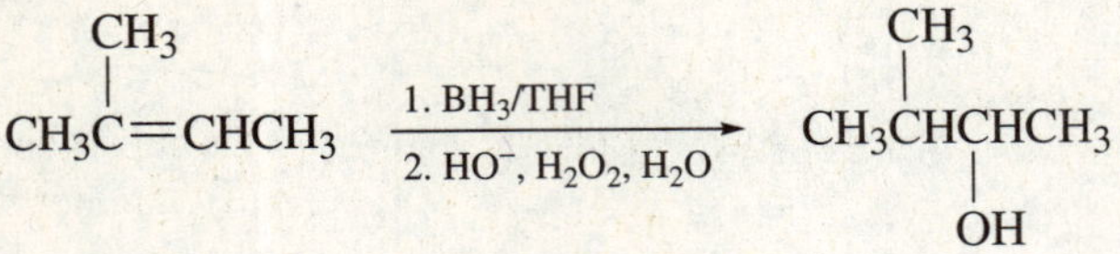

b.

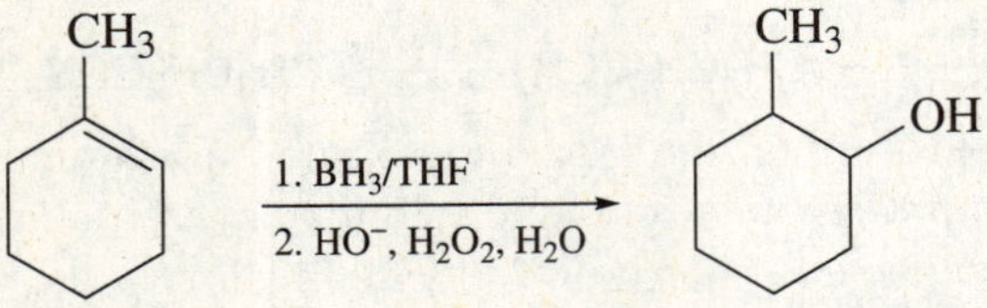

29. **a.** 1-pentene or 2-pentene
b. 1-methylcyclopentene, 3-methylcyclopentene, 4-methylcyclopentene, or methylenecyclopentane

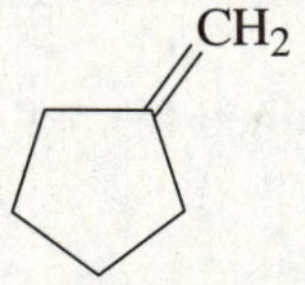

methylenecyclopentane

30. **a.** 3 (1-butene, *cis*-2-butene, *trans*-2-butene)

b. 4 (1-methylcyclohexene, 3-methylcyclohexene, 4-methylcyclohexene, methylenecyclohexane)

c. 5 (1-hexene, *cis*-2-hexene, *trans*-2-hexene, *cis*-3-hexene, *trans*-3-hexene)

31. Because alkene A has the smaller heat of hydrogenation, it is more stable.

32. **a.**

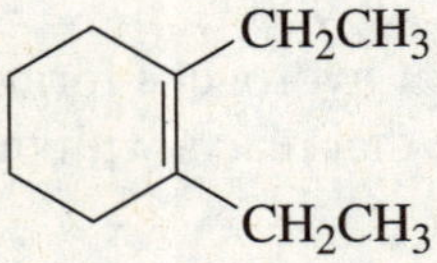

This alkene is the most stable because it has the greatest number of alkyl substituents bonded to the sp^2 carbons.

b.

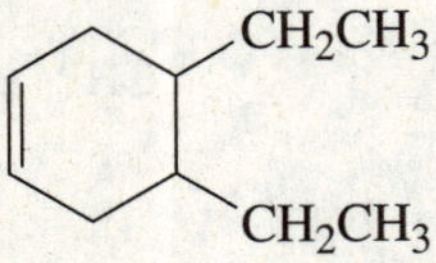

This alkene is the least stable because it has the fewest number of alkyl substituents bonded to the sp^2 carbons.

c.

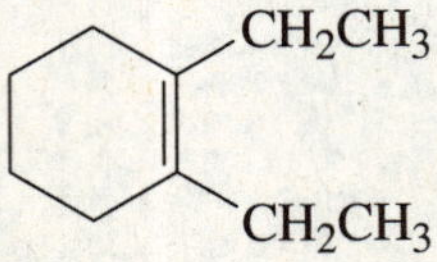

This alkene has the smallest heat of hydrogenation because it is the most stable of the three alkenes.

33. 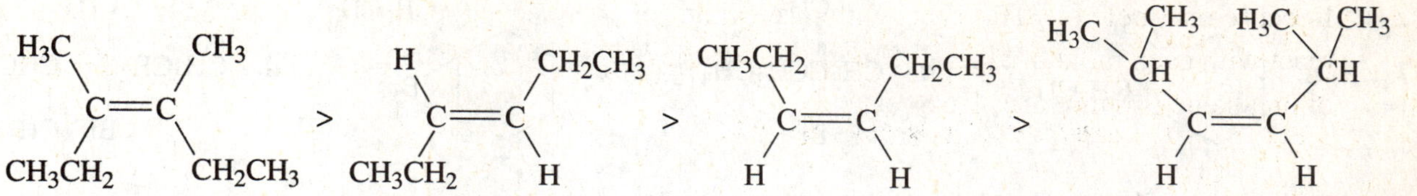

4 alkyl substituents > 2 trans substituents > 2 cis substituents > 2 cis substituents that cause greater steric strain

34. Solved in the text.

35. Hydroboation of of 3-methylcyclohexene would form both of 2-methylcyclohexanol (the desired product) and 3-methylcyclohexanol because borane could add to either the 1-position or the 2-position of the alkene since the two transition states would have approximately the same stability. Therefore, less of the desired product would be formed from 3-methylcyclohexene than from 1-methylcyclohexene.

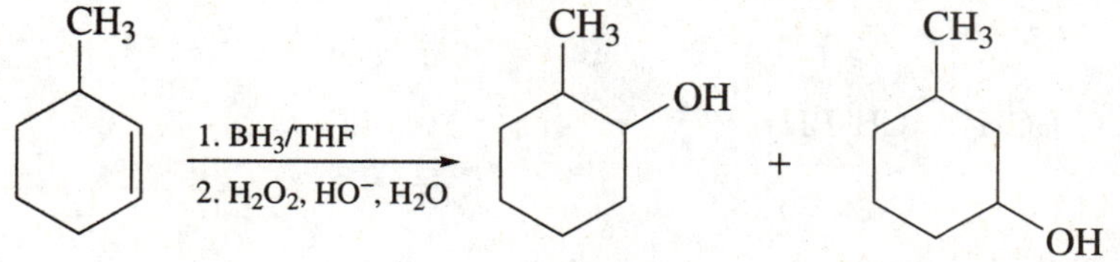

36. **a.** $CH_3CH{=}CH_2 \xrightarrow[CH_3OH]{H^+} CH_3CH(OCH_3)CH_3$

b. $CH_3CH_2CH{=}CHCH_3 + Br_2 \longrightarrow CH_3CH_2CHBrCHBrCH_3$

c. 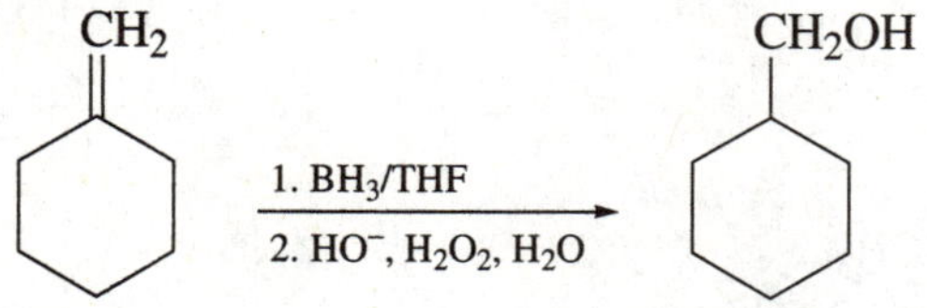

d. 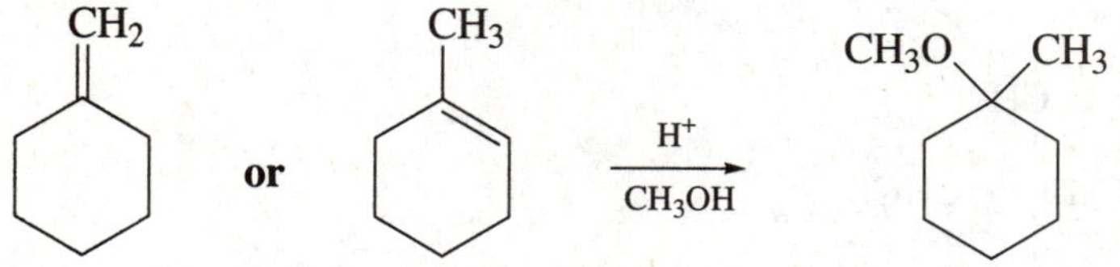

e. 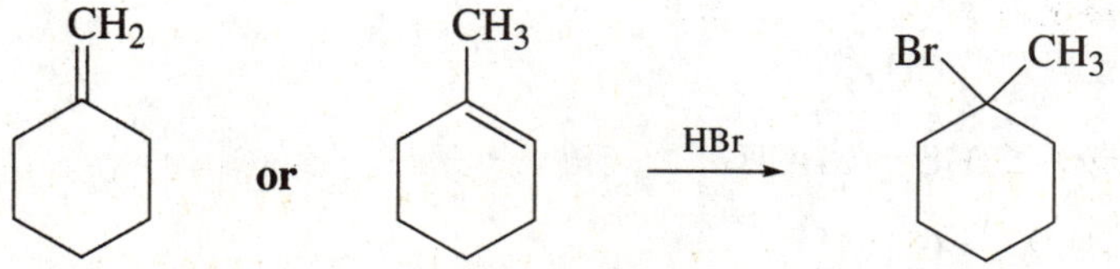

f. 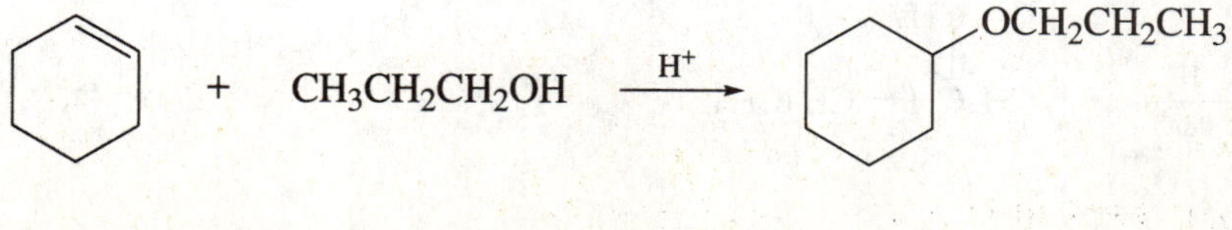

37. **a.** 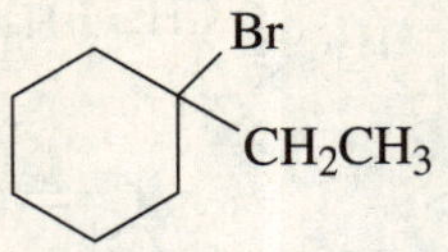

b. 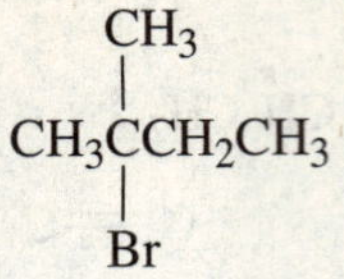

c. 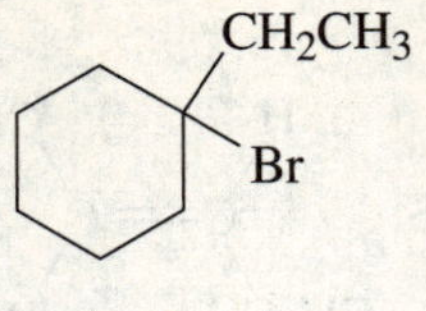

d. 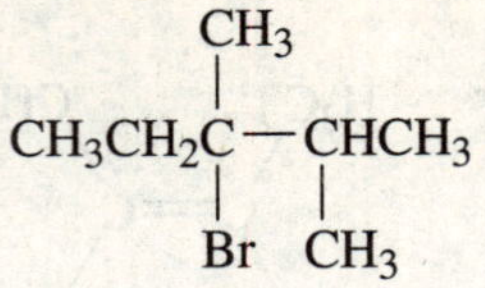

38. **a.** 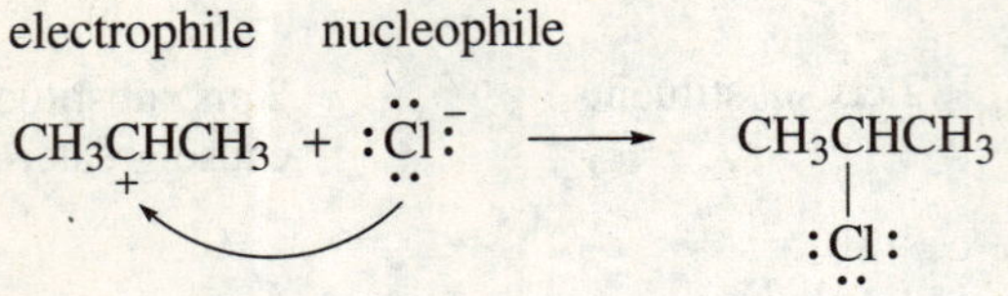

b. 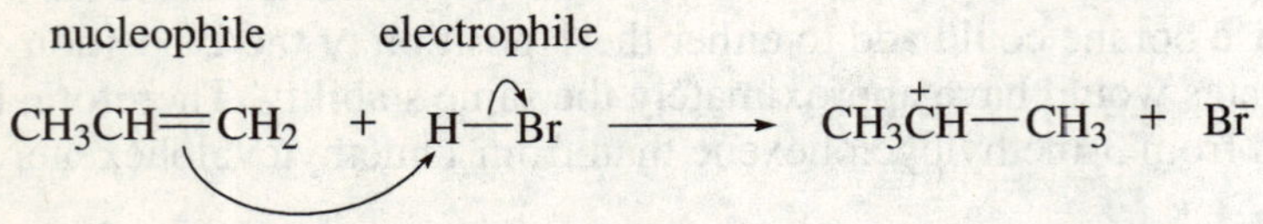

c. 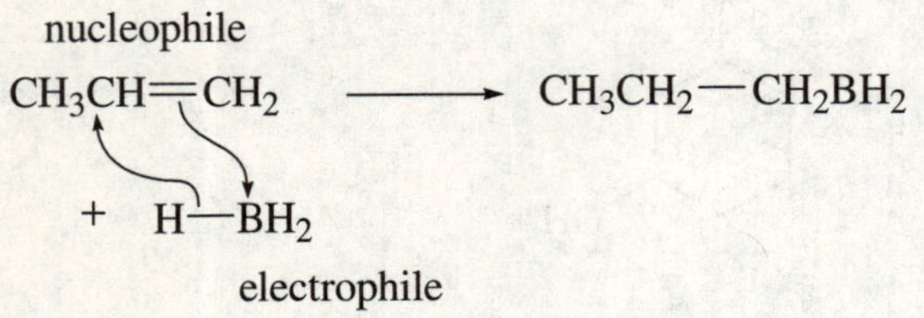

39. **a.** 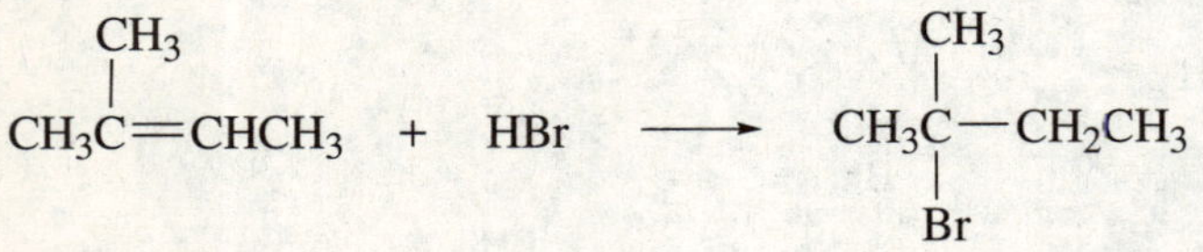

b. 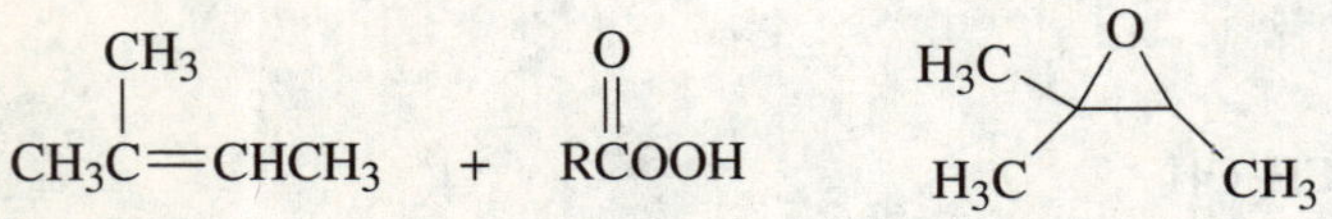

c. 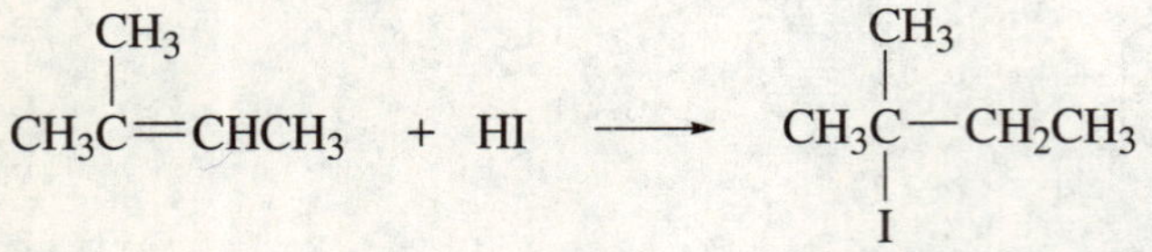

d. 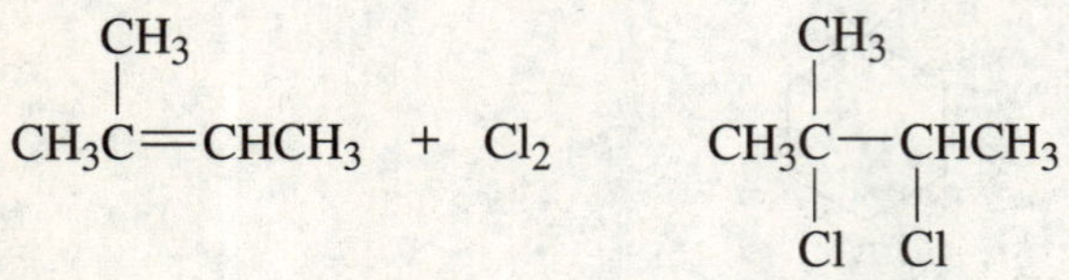

e. 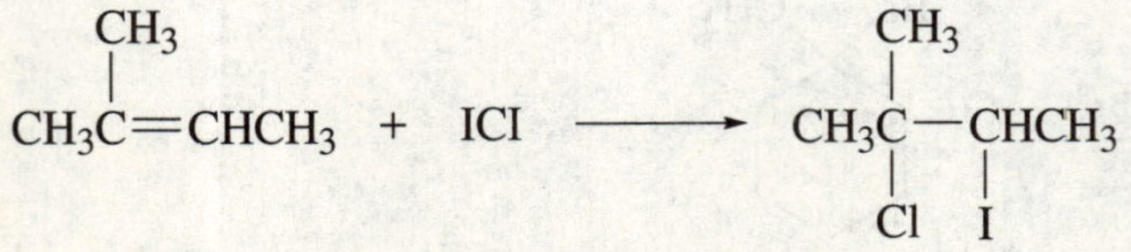

f. 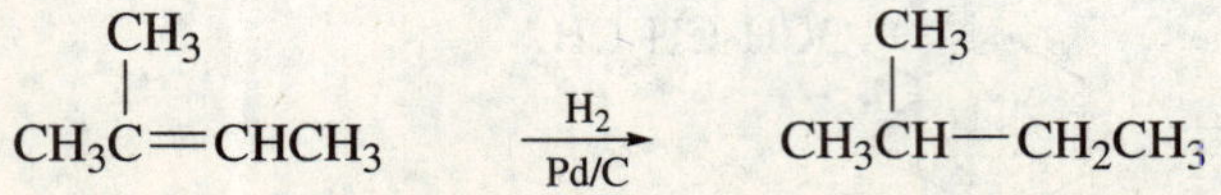

g. 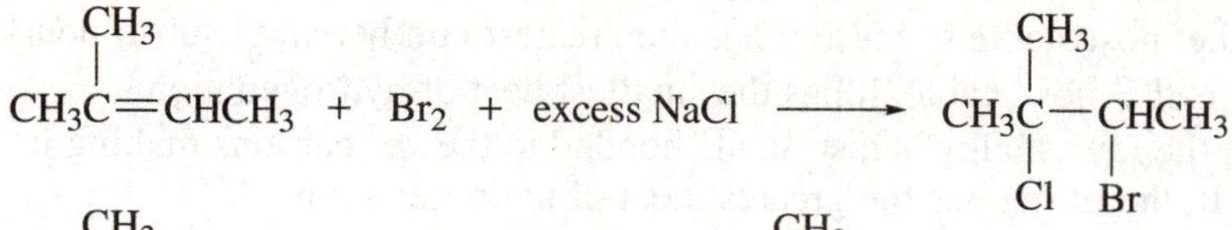

h. 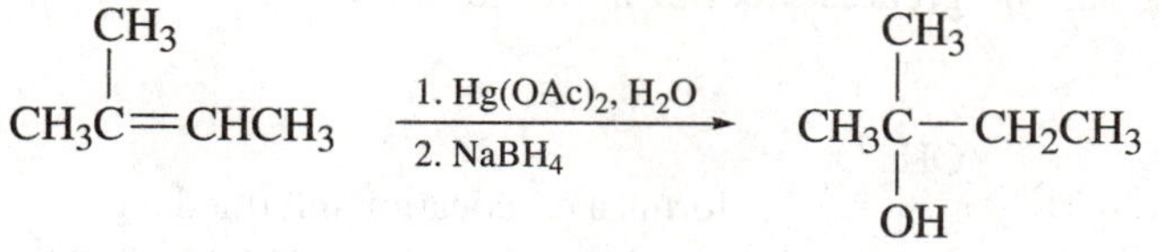

i. 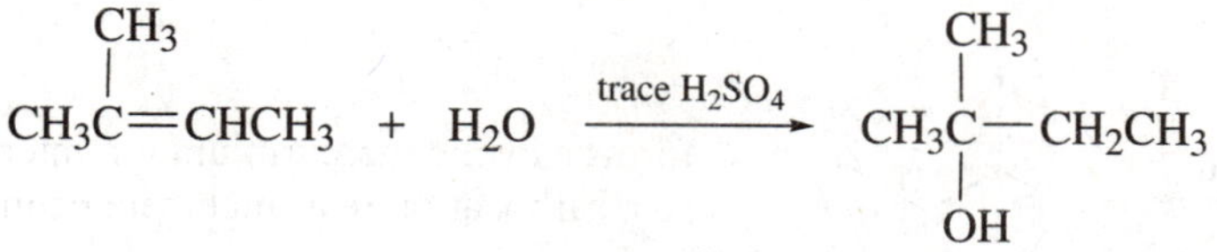

j. 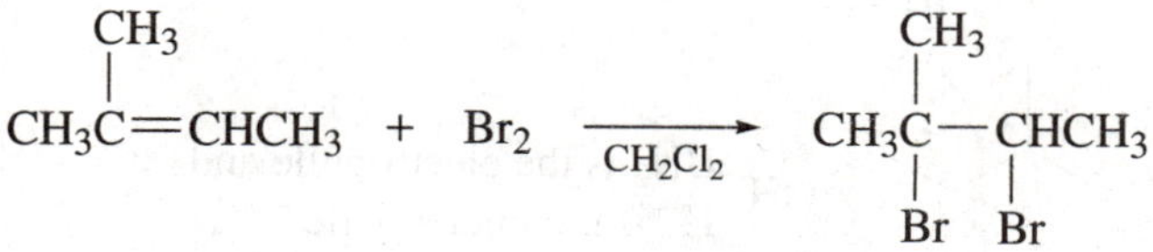

k. 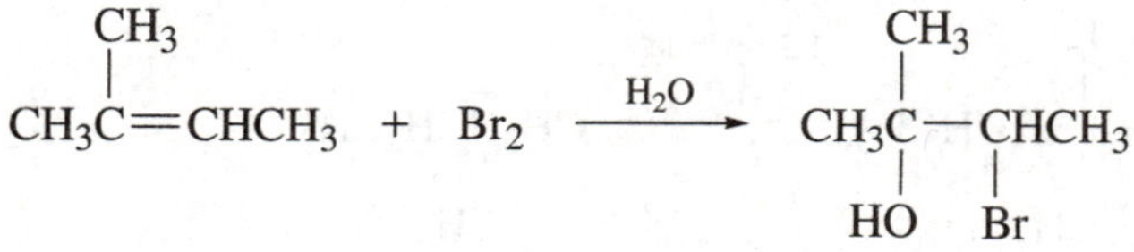

l. 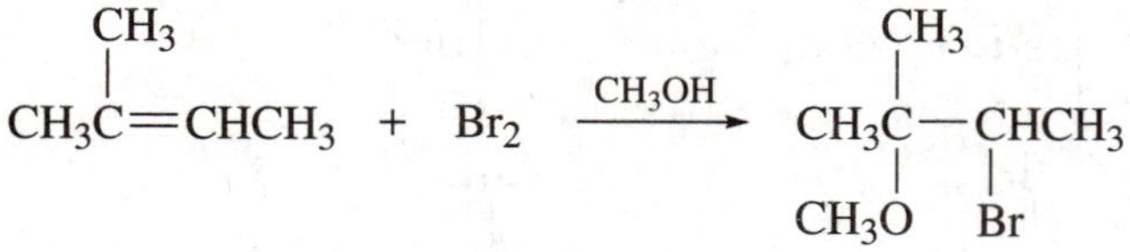

m. 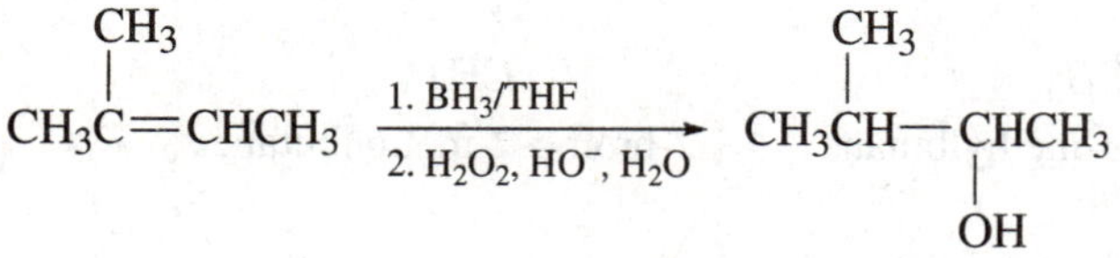

n. 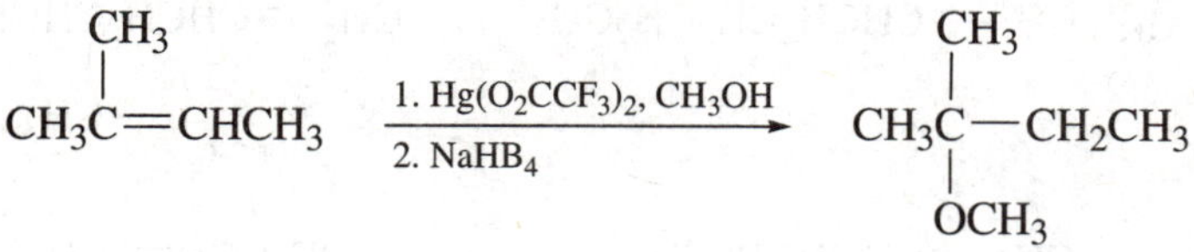

40. **a.** 2,2-diethyl-3-isopropyloxirane or 2,3-epoxy-4-ethyl-2-methylhexane
b. 3-ethyl-2,2-dimethyloxirane or 2,3-epoxy-2-methylpentane

41. 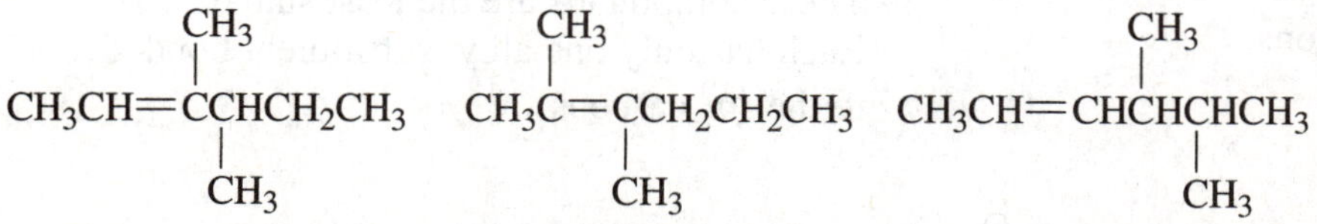

3,4-dimethyl-2-hexene 2,3-dimethyl-2-hexene 4,5-dimethyl-2-hexene

2,3-Dimethyl-2-hexene is the most stable because it has the greatest number alkyl substituents bonded to the sp^2 carbons. Because it is the most stable, it has the smallest heat of hydrogenation. 4,5-Dimethyl-2-hexene has the fewest alkyl substituents bonded to the sp^2 carbons making it the least stable of the three alkenes. It, therefore, has the greatest heat of hydrogenation.

42.

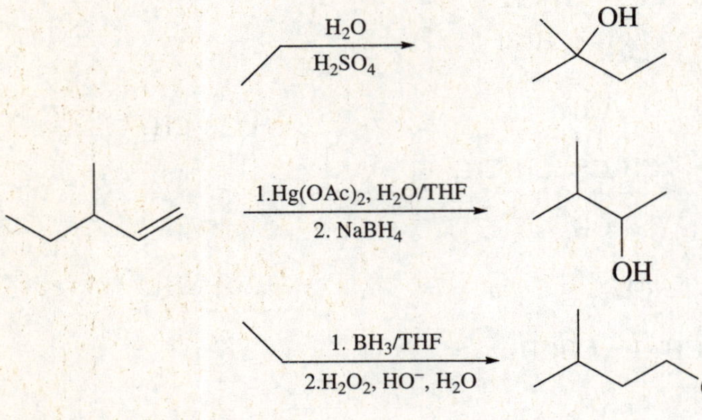

forms a carbocation intermediate, so a carbocation rearrrangement can occur

forms a cyclic mercurinium ion intermediate, so a carbocation rearrangement cannot occur

BH_3 is the electrophile and H^- is the nucleophile

43.

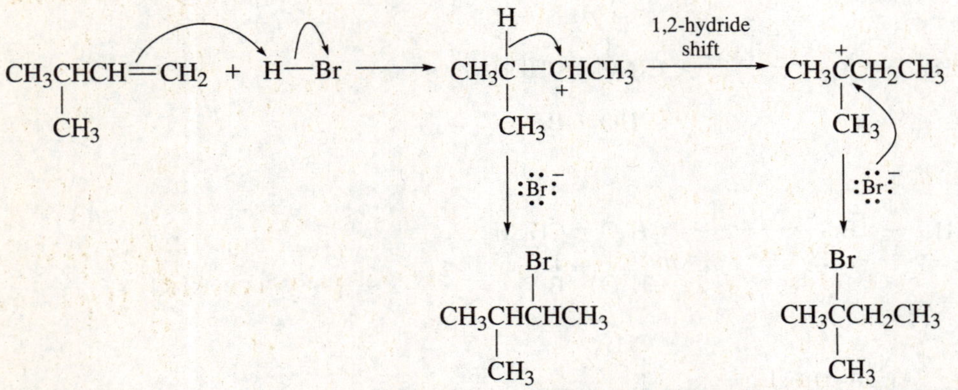

2-bromo-3-methylbutane   2-bromo-2-methylbutane

44.

a.

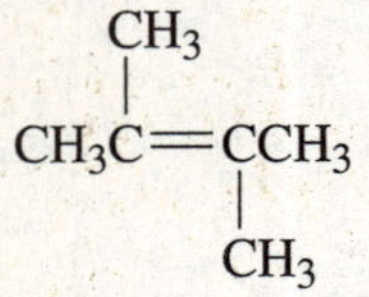

This compound is the most stable.
It has 4 alkyl substituents bonded to the sp^2 carbons.

b.

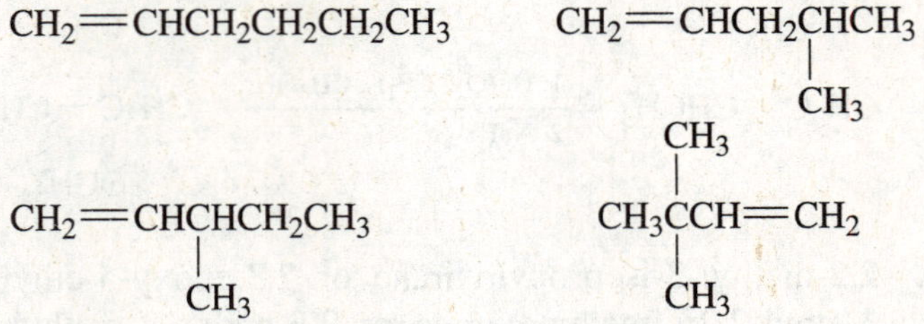

These compounds are the least stable.
Each has only one alkyl substituent bonded to the sp^2 carbons.

45. **a.**

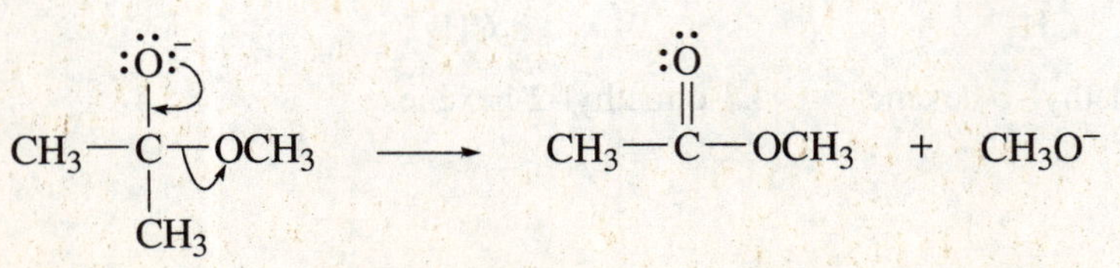

b. $CH_3C{\equiv}C{-}H + {:}\ddot{N}H_2^- \longrightarrow CH_3C{\equiv}C^- + {:}NH_3$

c. $CH_3CH_2{-}Br + CH_3\ddot{O}{:}^- \longrightarrow CH_3CH_2{-}\ddot{O}CH_3 + Br^-$

46.

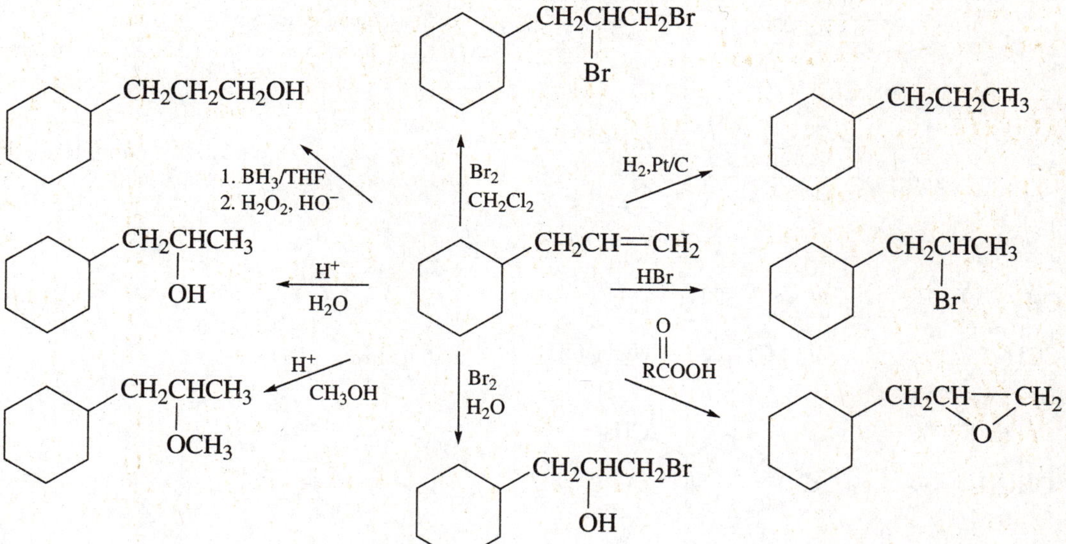

47.

a. 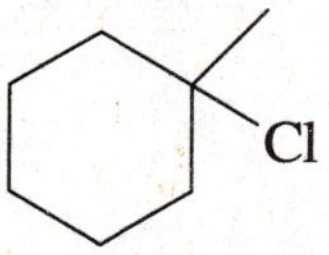

b. 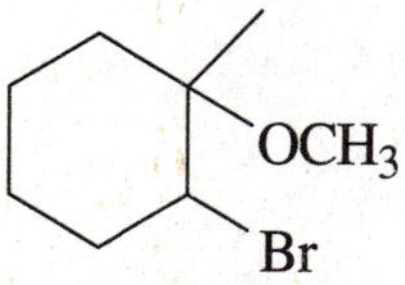

c. 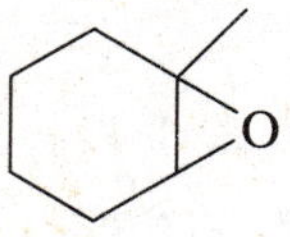

d. 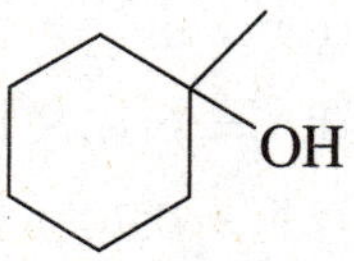

e. 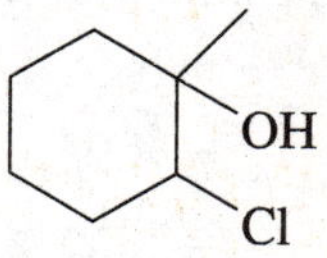

f. 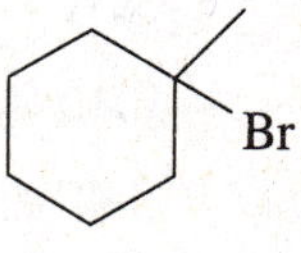

g. 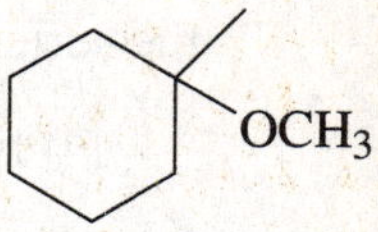

h. 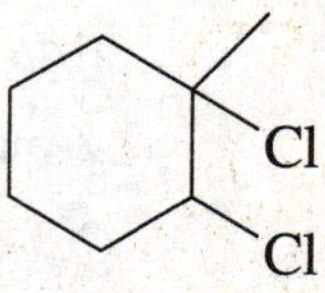

48.

a. 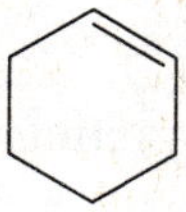 $\xrightarrow[Pd/C]{H_2}$

b. $CH_3CH_2CH_2CH{=}CH_2 \xrightarrow{HCl}$

d. $CH_3CH_2CH{=}CHCH_2CH_3 \xrightarrow[H_2O]{Br_2}$

e. 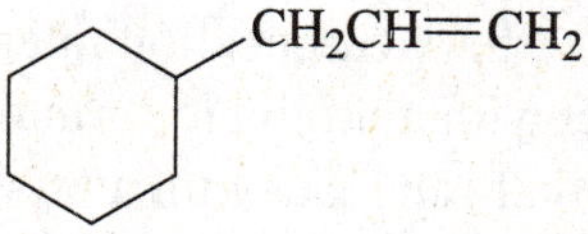

$\xrightarrow[H_2O]{H^+}$

c. 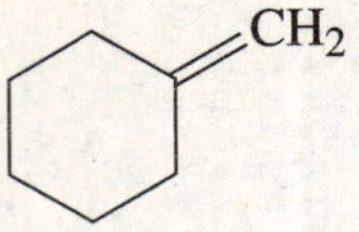 $\xrightarrow[\text{2. } H_2O_2\text{, } HO^-\text{, } H_2O]{\text{1. } BH_3/THF}$

f. $CH_3CH_2CH{=}CHCH_2CH_3 \xrightarrow[\text{NaBr excess}]{Cl_2}$

or

$CH_3CH_2CH{=}CHCH_2CH_3 \xrightarrow[\text{NaCl excess}]{Br_2}$

49. 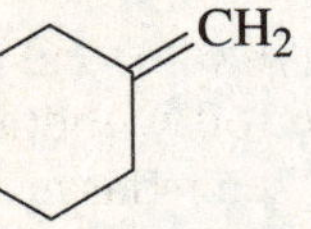 and 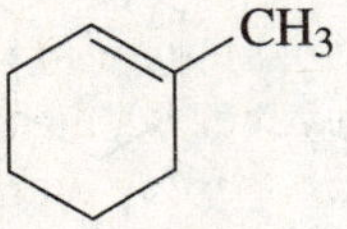

50. **a.** $CH_3\overset{CH_3}{\underset{+}{C}}CH_3$ **c.** $CH_3\overset{CH_3}{C}{=}CHCH_2CH_3$

b. $CH_3\underset{+}{C}HCH_3$ **d.** 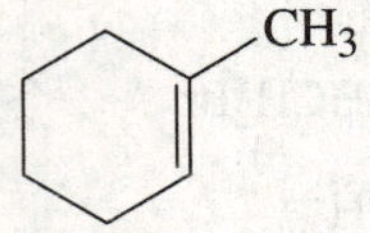

51. **a.** To determine relative rates, the rate constant of each alkene is divided by the smallest rate constant of the series (3.51×10^{-8}).

propene	$= 4.95 \times 10^{-8}/3.51 \times 10^{-8}$	$= 1.41$
cis-2-butene	$= 8.32 \times 10^{-8}/3.51 \times 10^{-8}$	$= 2.37$
trans-2-butene	$= 3.51 \times 10^{-8}/3.51 \times 10^{-8}$	$= 1$
2-methyl-2-butene	$= 2.15 \times 10^{-4}/3.51 \times 10^{-8}$	$= 6.12 \times 10^3$
2,3-dimethyl-2-butene	$= 3.42 \times 10^{-4}/3.51 \times 10^{-8}$	$= 9.74 \times 10^3$

b. Both compounds form the same carbocation, but since (*Z*)-2-butene is less stable than (*E*)-2-butene, (*Z*)-2-butene has a smaller free energy of activation.

c. 2-Methyl-2-butene reacts faster because it forms a tertiary carbocation in the rate-limiting step, whereas *cis*-2-butene forms a less stable secondary carbocation.

d. Both compounds form a tertiary carbocation intermediate. However, 2,3-dimethyl-2-butene has two sp^2 carbons that can react with a proton to form the tertiary carbocation, whereas 2-methyl-2-butene has only one sp^2 carbon that can react with a proton to form the tertiary carbocation. Therefore there will be more collisions with the proper orientation that lead to a productive reaction in the case of 2,3-dimethyl-2-butene.

52. **a.** 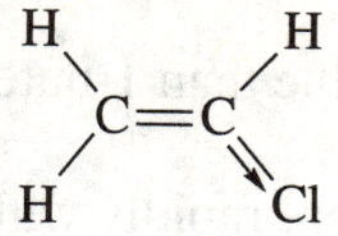 has a greater dipole moment than 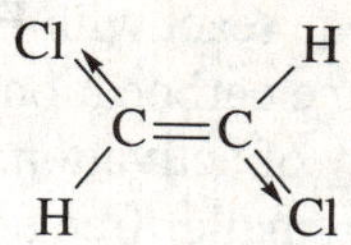

b. 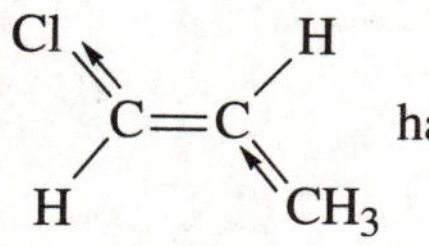 has a greater dipole moment than 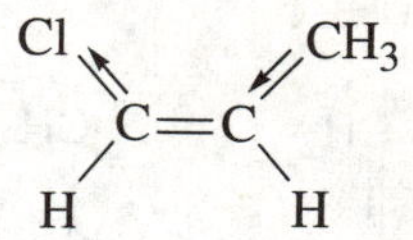

53. No, he should not follow the student's advice. Markovnikov's rule would indicate that the secondary carbocation is more stable than the primary carbocation. However because of the electron-withdrawing fluoro substituents in this compound, the primary carbocation is more stable than the secondary carbocation, since in the latter the positive charge is closer to the fluoro substituents. So the major product will be 1,1,1-trifluoro-3-iodopropane, not 1,1,1-trifluoro-2-iodopropane, the compound that would be predicted to be the major product by Markovnikov's rule.

$F_3CCH_2—\overset{+}{C}H_2$	$F_3C\overset{+}{C}H—CH_3$
more stable	less stable because of the nearby electron-withdrawing fluorines

54. **a.** 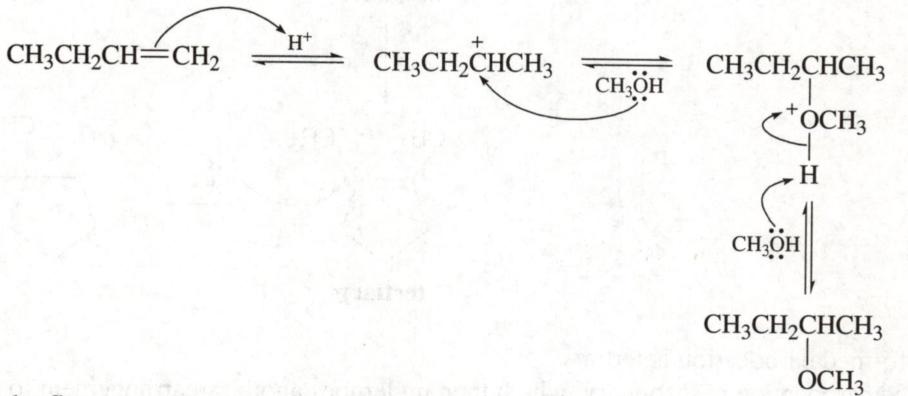

b. the first step (See the answer to Problem 9.)

c. H^+

d. 1-butene

e. the *sec*-butyl cation

f. methanol

55. **a.** 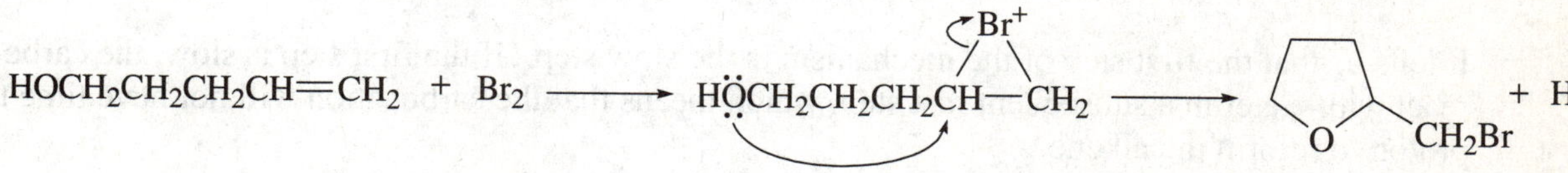

b. 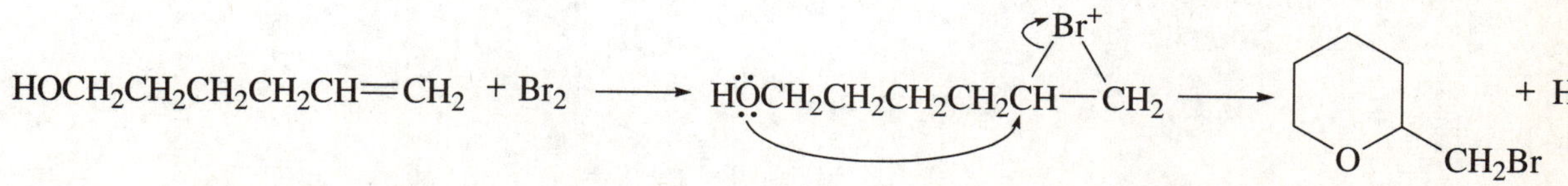

56. **a.** Both 1-butene and 2-butene react with HCl to form 2-chlorobutane.
b. Both alkenes form the same carbocation, but because 2-butene is more stable than 1-butene, 2-butene has the greater free energy of activation.
c. Because 1-butene has the smaller free energy of activation, it will react more rapidly with HCl.
d. Both compounds form the same carbocation, but since (*Z*)-2-butene is less stable, it will react more rapidly with HCl.

57. **a.** 4 alkenes

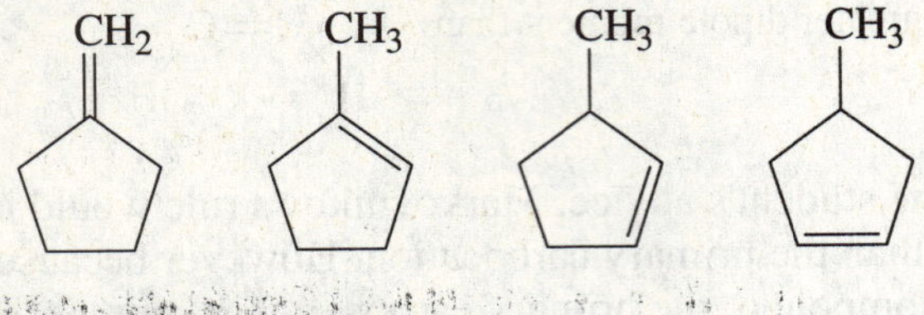

b. 1-Methylcyclopentene is the most stable.
c. Because 1-methylcyclopentene is the most stable, it has the smallest heat of hydrogenation.

58. **a.**

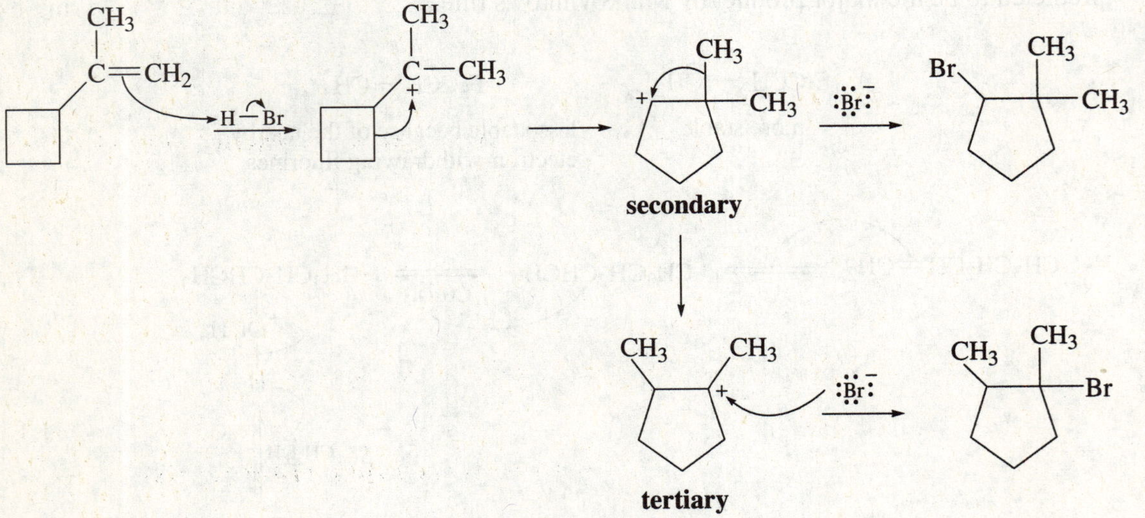

b. The initially formed carbocation is tertiary.
c. The rearranged carbocation is secondary, which then undergoes another rearrangement to a more stable tertiary carbocation.
d. The initially formed carbocation rearranges in order to release the strain in the four-membered ring. (A tertiary carbocation with a strained four-membered ring is less stable than a secondary carbocation with an unstrained five-membered ring.)

59. It tells us that the first step of the mechanism is the slow step. If the first step is slow, the carbocation will react with water in a subsequent fast step, which means that the carbocation will not have time to lose a proton to reform the alkene.

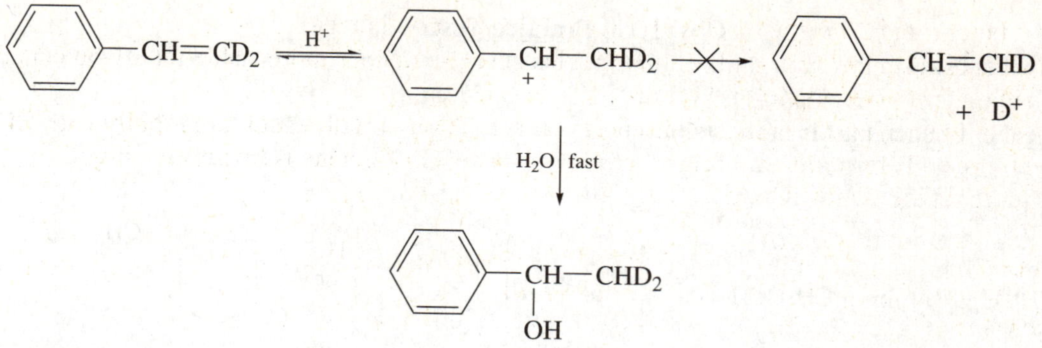

If the first step were not the slow step, an equilibrium would be set up between the alkene and the carbocation, and because the carbocation could lose either H^+ or D^+ when it reformed the alkene, all the deuterium (D) would not be retained in the alkene.

60. **a.** A proton adds to the alkene, forming a secondary carbocation, which undergoes a ring-expansion rearrangement to form a more stable tertiary carbocation.

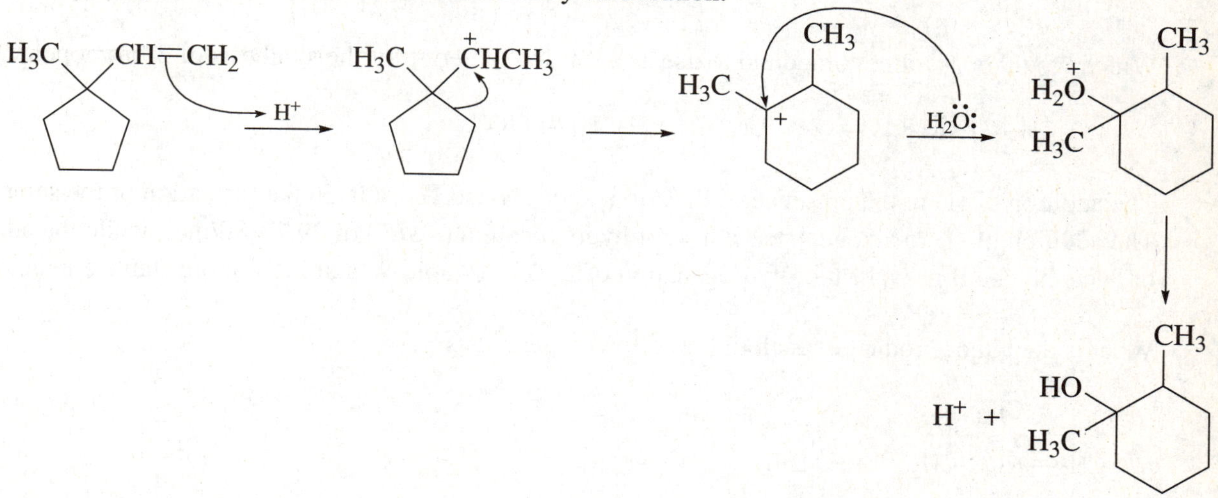

61. **a.**

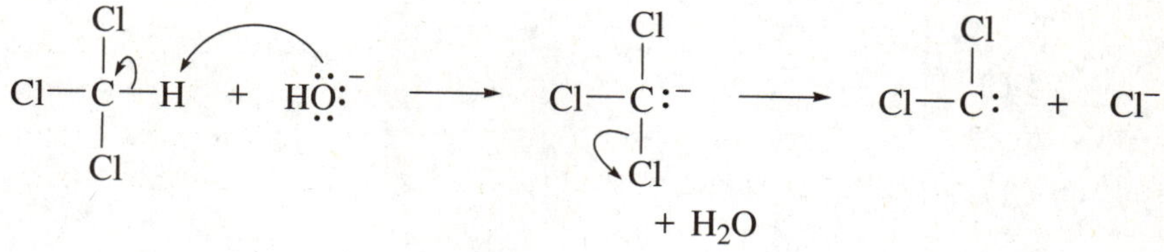

b.

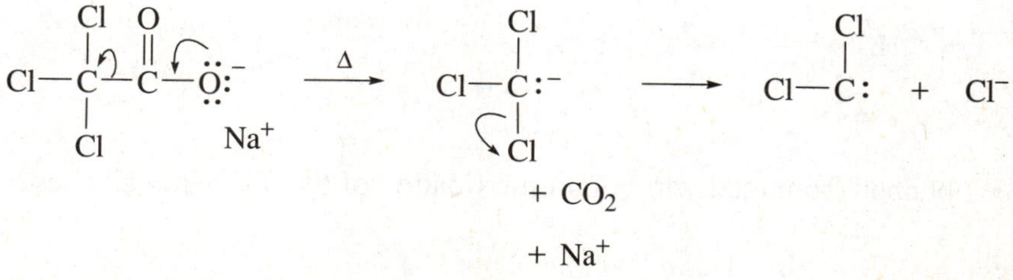

Chapter 4 Practice Test

1. Which member of each pair is more stable?

a. $CH_3\underset{+}{C}HCH_2CH_3$ or $CH_3\overset{CH_3}{\underset{+}{C}}CH_3$

b. $CH_3CH_2CH_2\overset{+}{C}H_2$ or $CH_3CH_2\overset{+}{C}HCH_3$

c. $CH_3CH_2\overset{+}{C}H_2$ or $ClCH_2CH_2\overset{+}{C}H_2$

d. 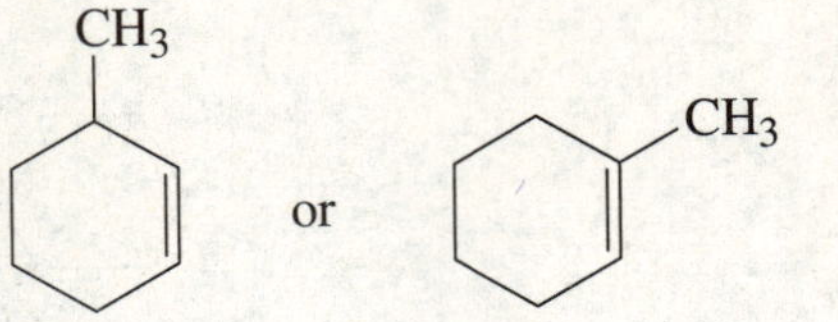

e. 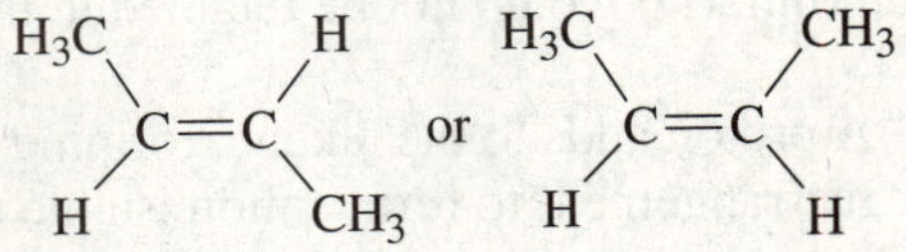

2. Which would be a better compound to use as a starting material for the synthesis of 2-bromopentane?

$CH_3CH_2CH_2CH{=}CH_2$ or $CH_3CH_2CH{=}CHCH_3$

3. The addition of H_2 in the presence of Pd/C to alkenes **A** and **B** results in the formation of the same alkane. The addition of H_2 to alkene **A** has a heat of hydrogenation ($-\Delta H°$) of 29.7 kcal/mol, while the addition of H_2 to alkene **B** has a heat of hydrogenation of 27.3 kcal/mol. Which is the more stable alkene, **A** or **B**?

4. What is the major product of each of the following reactions?

a. $CH_2{=}\overset{CH_3}{C}CH_2CH_3 \; + \; HBr \longrightarrow$

b. $CH_3\overset{CH_3}{C}HCH{=}CH_2 \; + \; HCl \longrightarrow$

c. $CH_3CH_2CH{=}CH_2 \; + \; Cl_2 \xrightarrow{H_2O}$

d. $CH_3\overset{CH_3}{\underset{CH_3}{C}}CH{=}CH_2 \; + \; HBr \longrightarrow$

5. What product would be obtained from the hydroboration-oxidation of the following alkenes?

a. 2-methyl-2-butene   b. 1-ethylcyclopentene

6. Indicate how each of the following compounds could be synthesized using an alkene as one of the starting materials:

a. → 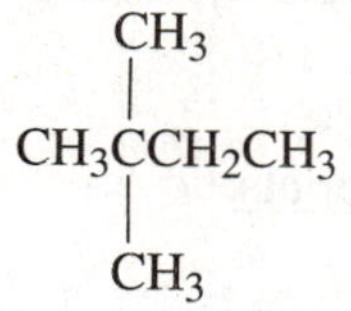

b. → 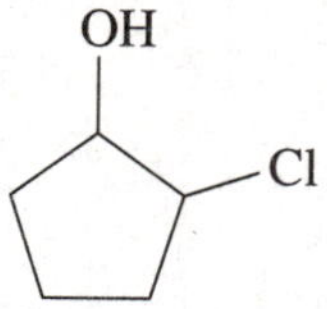

c. → 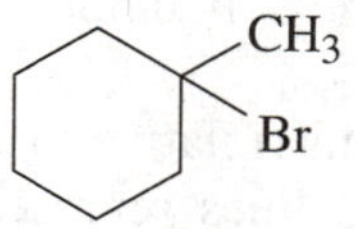

d. → 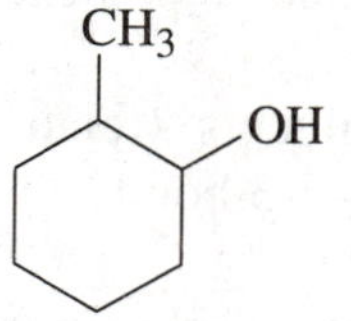

7. Indicate the carbocations that you would expect to rearrange to give a more stable carbocation.

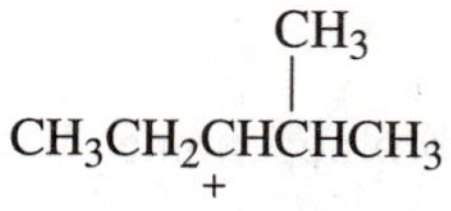 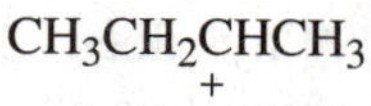 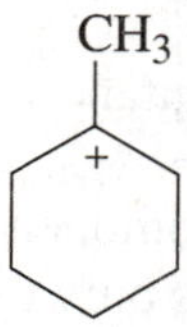 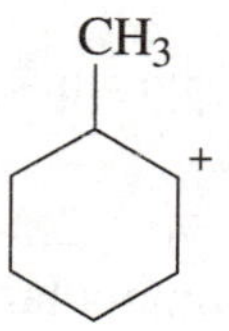

8. Indicate whether each of the following statements is true or false:

a. The addition of Br_2 to 1-butene to form 1,2-dibromobutane is a concerted reaction. T F

b. The reaction of 1-butene with HCl will form 1-chlorobutane as the major product. T F

c. 2,3-Dimethyl-2-pentene is more stable than 3,4-dimethyl-2-pentene. T F

d. The reaction of HBr with 3-methylcyclohexene is more highly regioselective than the reaction of HBr with 1-methylcyclohexene. T F

e. The reaction of an alkene with a carboxylic acid forms an epoxide. T F

f. A catalyst increases the equilibrium constant of a reaction. T F

Special Topic IV

Exercise in Model Building

Do the following exercises using molecular models.

1. Build the enantiomers of 2-bromobutane.

- **a.** Try to superimpose them.
- **b.** Show that they are mirror images.
- **c.** Which one is (*R*)-2-bromobutane?

2. Build the erythro enantiomers of 3-bromo-2-butanol.

- **a.** Where are the Br and OH substituents (relative to each other) in the Fischer projection? (Recall that in a Fischer projection, the horizontal lines represent bonds that point out of the plane of the paper toward the viewer, and the dashed lines represent bonds that point back from the plane of the paper away from the viewer.)
- **b.** Where are the Br and OH substituents (relative to each other) in the most stable conformer considering rotation about the C-2—C-3 bond?

3.
- **a.** Build the compounds labeled "1" and "2" shown on the top of page 221 of the text.
- **b.** Build their mirror images.
- **c.** Show that "1" is superimposable on its mirror image but '2" is not.

4. Build the three stereoisomers of 2,3-dibromobutane.

5. Build the four stereoisomers of 2,3-dibromopentane. Why does 2,3-dibromopentane have four stereoisomers, whereas 2,3-dibromobutane has only three stereoisomers?

6. Build (*S*)-2-pentanol.

7. Build (2*S*,3*R*)-3-bromo-2-butanol. Rotate the model so it is in a Fischer projection. Compare its structure with the structure of (2*S*,3*R*)-3-bromo-2-butanol shown on page 226 of the text.

8. Rotate the model of (2*S*,3*R*)-3-bromo-2-butanol so it is in a perspective formula as a staggered conformer. Compare its structure with the structure of (2*S*,3*R*)-3-bromo-2-butanol shown on page 226 of the text.

9. Build the compounds shown in Problem 37 on page 229 of the text. Name the compounds.

10. Build (*S*)-1-bromo-2-methylbutane. Substitute the Br with an HO to form 2-methyl-1-butanol. What is the configuration of your model of 2-methyl-1-butanol?

11. Build two models of *trans*-2-pentene. Add Br_2 to opposite sides of the double bond, forming the two enantiomers shown on page 242 of the text. Rotate them so they are Fischer projections. Are they erythro or threo enantiomers? Compare your answer with that given in the text.

12. Build models of the molecules shown in Problem 94 on page 257 of the text. What is the configuration of the asymmetric center in each of the molecules?

13. **a.** Build *cis*-1-bromo-4-chlorocyclohexane. Build its mirror image. Are they superimposable?

b. Build *cis*-1-bromo-2-chlorocyclohexane. Build its mirror image. Are they superimposable?

14. **a.** Build *cis*-cyclooctene.

b. Build *trans*-cyclooctene.

c. Build *cis*-cyclohexene.

d. Can you build *trans*-cyclohexene?

CHAPTER 5
Stereochemistry: The Arrangement of Atoms in Space;
The Stereochemistry of Addition Reactions

Important Terms

absolute configuration	the three-dimensional structure of a chiral compound. The configuration is designated by *R* or *S*. The configuration is called absolute to distinguish it from a relative configuration.
achiral (optically inactive)	a molecule or object that contains an element (a plane or a point) of symmetry.
amine inversion	results when a compound containing an sp^3 hybridized nitrogen with a nonbonding pair of electrons rapidly turns inside out.
anti addition	an addition reaction in which the two added substituents add to opposite sides of the molecule.
asymmetric center	a atom that is bonded to four different substituents.
biochemistry	the chemistry associated with living organisms.
chiral (optically active)	a chiral molecule has a nonsuperimposable mirror image.
chiral catalyst	a catalyst that will cause the synthesis of one enantiomer in great excess over the other.
chiral probe	something capable of distinguishing between enantiomers.
cis isomer	the isomer with substituents on the same side of a cyclic structure, or the isomer with the hydrogens on the same side of a double bond.
cis-trans isomers	geometric (or *E,Z*) isomers.
chromatography	a separation technique in which the mixture to be separated is dissolved in a solvent and the solution is passed through a column packed with an adsorbent stationary phase.
configuration	the three-dimensional structure of a chiral compound. The configuration is designated by *R* or *S*.
configurational isomers	stereoisomers that cannot interconvert unless a covalent bond is broken. Cis-trans isomers and optical isomers are configurational isomers.
constitutional isomers (structural isomers)	molecules that have the same molecular formula but differ in the way the atoms are connected.
dextrorotatory	the enantiomer that rotates polarized light in a clockwise direction.

diastereomer	a configurational isomer that is not an enantiomer.
enantiomers	nonsuperimposable mirror-image molecules.
enantiomerically pure	only one enantiomer is present in an enantiomerically pure sample.
enantiomeric excess	how much excess of one enantiomer is present in a mixture of a pair of enantiomers.
enzyme	a protein that catalyzes a biological reaction.
erythro enantiomers	the pair of enantiomers with similar groups on the same side (in the case of stereoisomers with adjacent asymmetric centers) when drawn in a Fischer projection.
Fischer projection	a method of representing the spatial arrangement of groups bonded to an asymmetric center. The asymmetric center is the point of intersection of two perpendicular lines; the horizontal lines represent bonds that project out of the plane of the paper toward the viewer, and the vertical lines represent bonds that project back from the plane of the paper away from the viewer.
isomers	nonidentical compounds with the same molecular formula.
levorotatory	the enantiomer that rotates polarized light in a counterclockwise direction.
meso compound	a compound that possesses asymmetric centers and a plane of symmetry.
observed rotation	the amount of rotation observed in a polarimeter.
optical purity	how much excess of one enantiomer is present in a mixture of a pair of enantiomers.
optically active (chiral)	rotates the plane of polarized light.
optically inactive (achiral)	does not rotate the plane of polarized light.
perspective formula	a method of representing the spatial arrangement of groups bonded to an asymmetric center. Two bonds are drawn in the plane of the paper; a solid wedge is used to depict a bond that projects out of the plane of the paper toward the viewer, and a hatched wedge is used to represent a bond that projects back from the paper away from the viewer.
plane of symmetry	an imaginary plane that bisects a molecule into a pair of mirror images.
plane-polarized light	light that oscillates in a single plane passing through the direction the light travels.
polarimeter	an instrument that measures the rotation of polarized light.

polarized light light that oscillates in only one plane.

racemic mixture (racemic modification, racemate) a mixture of equal amounts of a pair of enantiomers.

***R* configuration** after assigning relative priorities to the four groups bonded to an asymmetric center, if the lowest-priority group is on a vertical axis in a Fischer projection (or pointing away from the viewer in a perspective formula), an arrow drawn from the highest-priority group to the next highest-priority group goes in a clockwise direction.

receptor a protein or other compound that binds a particular molecule.

regioselective describes a reaction that leads to the preferential formation of one constitutional isomer over another.

relative configuration the configuration of a compound relative to the configuration of another compound.

resolution of a racemic mixture separation of a racemic mixture into the individual enantiomers.

***S* configuration** after assigning relative priorities to the four groups bonded to an asymmetric center, if the lowest-priority group is on a vertical axis in a Fischer projection (or pointing away from the viewer in a perspective formula), an arrow drawn from the highest-priority group to the next highest-priority group goes in a counterclockwise direction.

specific rotation the amount of rotation that will be caused by a compound with a concentration of 1.0 g/mL in a sample tube 1.0 decimeter long.

stereocenter (stereogenic center) an atom at which the interchange of two groups produces a stereoisomer.

stereochemistry the field of chemistry that deals with the structure of molecules in three dimensions.

stereoisomers isomers that differ in the way the atoms are arranged in space.

stereoselective describes a reaction that leads to the preferential formation of one stereoisomer over another.

stereospecific describes a reaction in which the reactant can exist as stereoisomers and each stereoisomeric reactant leads to a different stereoisomeric product.

syn addition an addition reaction in which the two added substituents add to the same side of the molecule.

threo enantiomers — the pair of enantiomers with similar groups on opposite sides (in the case of stereoisomers with adjacent asymmetric centers) when drawn in a Fischer projection.

trans isomer — the isomer with substituents on the opposite sides of a cyclic structure, or the isomer with the hydrogens on the opposite sides of a double bond.

Do not confuse the terms **conformation** and **configuration.**
(To help you understand the difference, see the dogs on the next two pages.)

Conformations are different spatial arrangments of the same compound (see page 103 of the text).
They result from rotation about single bonds. They cannot be separated.
Some conformations are more stable than others.

Configurational isomers are different compounds.
They do not readily interconvert; bonds have to be broken to convert one configurational isomer to another.

Different Conformations

Stable

Unstable

Different Configurations

Solutions to Problems

1. **a.** 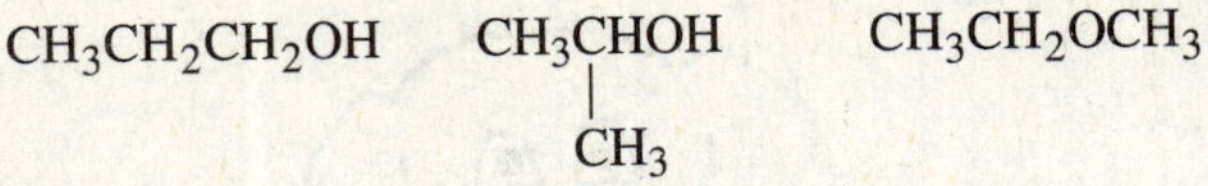

b. There are seven constitutional isomers with molecular formula $C_4H_{10}O$.

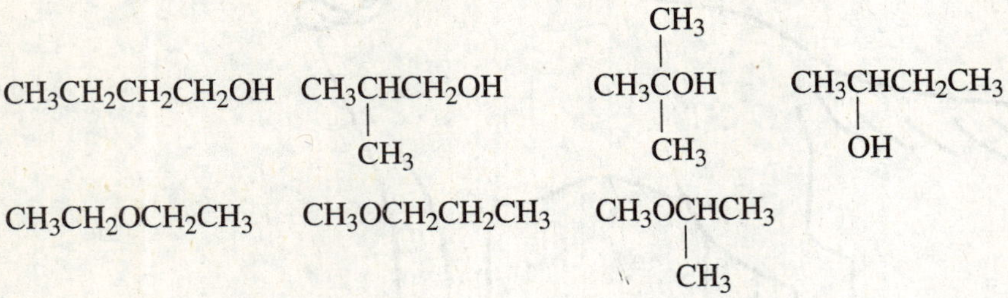

2. **a.** 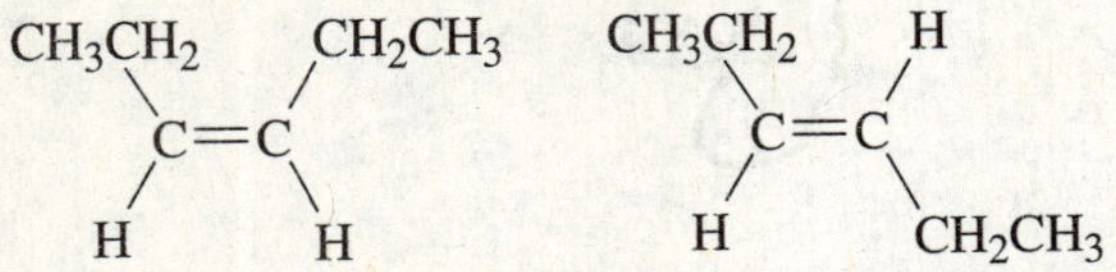

b. 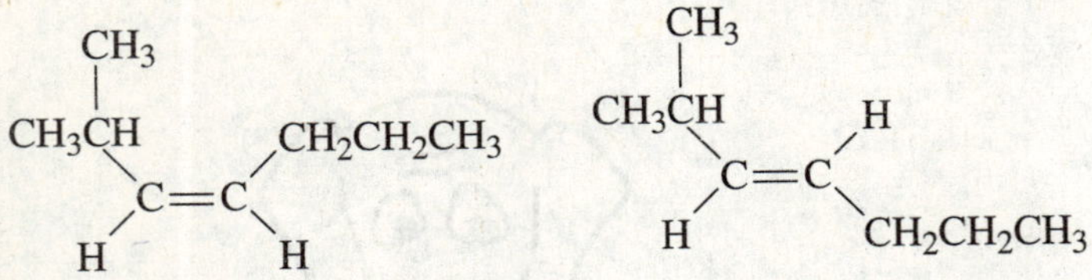

c. 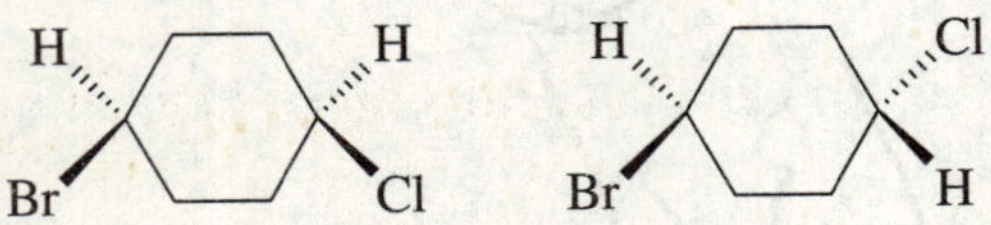

d. 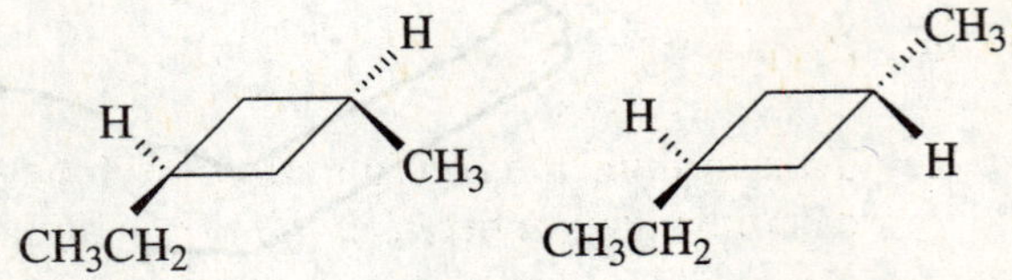

3. **a**, **b**, **c**, **f**, and **h** are chiral.
d, **e**, and **g** are each superimposable on its mirror image. These, therefore, are achiral.

4. **a.** F, G, J, L, N, P, Q, R, S, Z

b. A, H, I, M, O, T, U, V, W, X, Y

5. **a**, **c**, and **f** have asymmetric centers.

6. Solved in the text.

7. **a**, **c**, and **f**, because in order to be able to exist as a pair of enantiomers, the compound must have an asymmetric center (except in the case of certain compounds with unusual structures. See Problem 96.)

8. Draw the first enantiomer with the groups in any order you want. Then draw the second enantiomer by drawing the mirror image of the first enantiomer. Your answer might not look exactly like the ones shown below because the first enantiomer can be drawn with the 4 groups on any of the 4 bonds. The next one is the mirror image of the first one.

a. **1.**

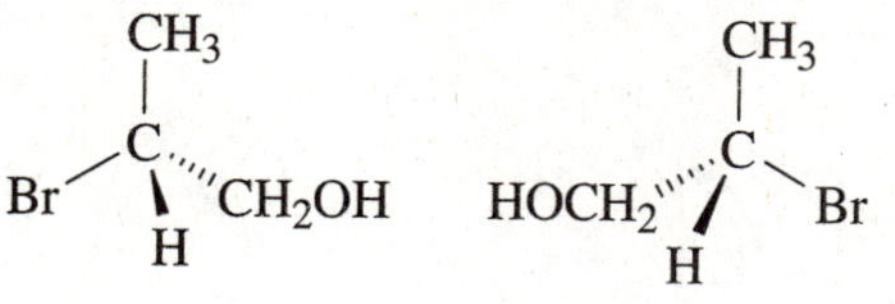

2.

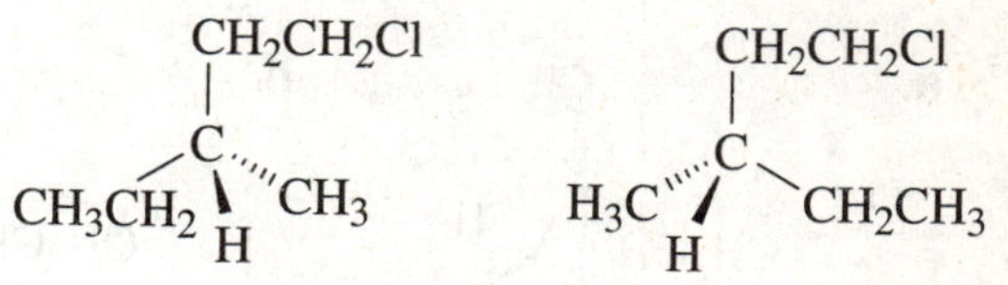

3.

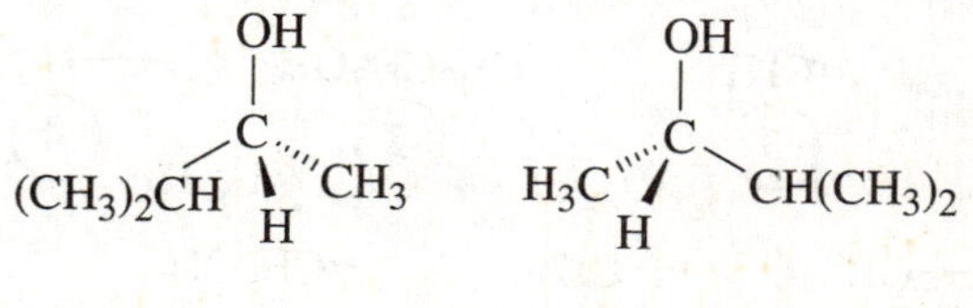

b. **1.**

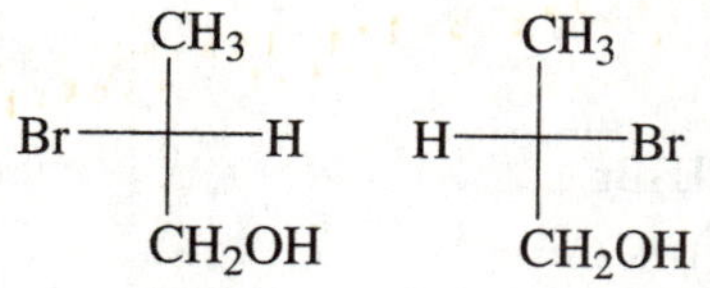

2.

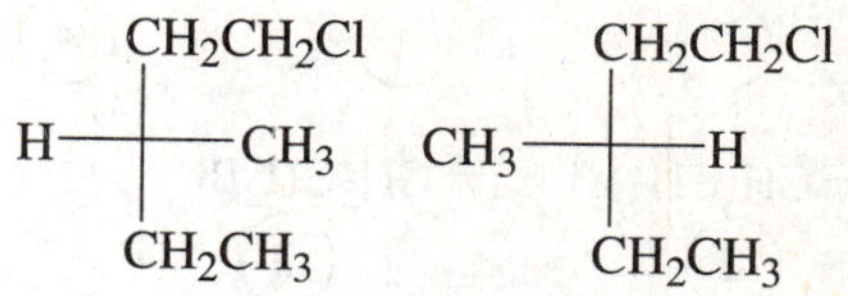

3.

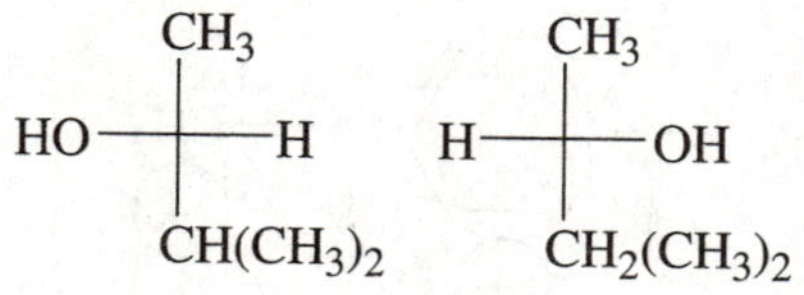

9. **A**, **B**, and **C** are identical.

10. **a.** *R* **b.** *R*

c. To determine the configuration. Add the fourth bond to the asymmetric center. Remember that it cannot be drawn between the two solid bonds.

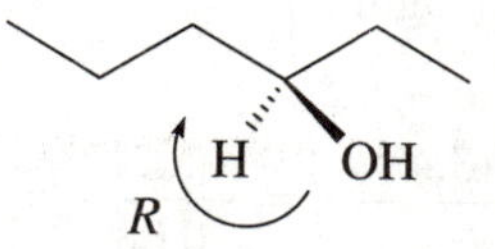

d. Draw the structure with the fourth bond to the asymmetric center. Switch a pair so that the H is on a dotted bond. The configuration of the compound with the switched pair is *S*. Therefore, the configuration of the compound given in the question is *R*.

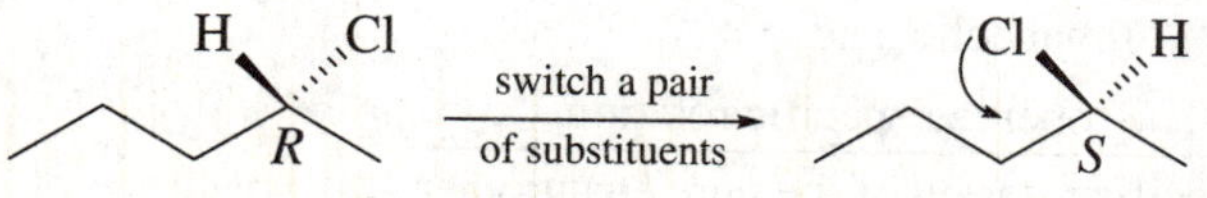

11. **a.** *S* **b.** *R* **c.** *S* **d.** *S*

12. The easiest way to determine whether two compounds are identical or enantiomers is to determine their configurations: If both are *R* (or both are *S*), they are identical. If one is *R* and the other is *S*, they are enantiomers.

a. identical **b.** enantiomers **c.** enantiomers **d.** enantiomers

13. **a.**

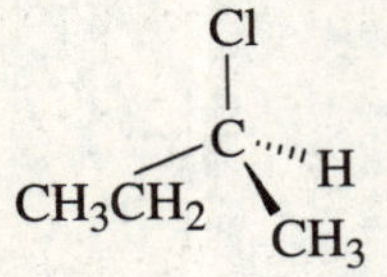

b.

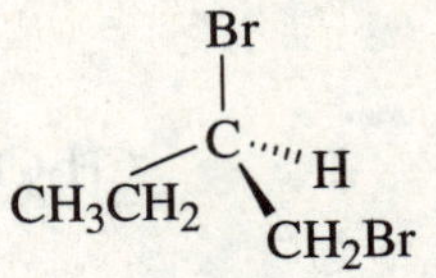

14. **a.** —CH_2OH (1) —CH_3 (3) —CH_2CH_2OH (2) —H (4)

b. —CH=O (2) —OH (1) —CH_3 (4) —CH_2OH (3)

c. —$CH(CH_3)_2$ (2) —CH_2CH_2Br (3) —Cl (1) —$CH_2CH_2CH_2Br$ (4)

d. —CH=CH_2 (2) —CH_2CH_3 (3)

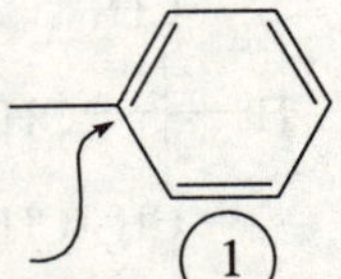

(1) —CH_3 (4)

C attached to 2 C's

C attached to 3 C's

15. **a.** levorotatory **b.** dextrorotatory

16.

$$\text{specific rotation} = \frac{\text{observed rotation}}{\text{concentration} \times \text{length}}$$

$$[\alpha] = \frac{+13.4°}{\frac{2}{50} \times 2 \text{ dm}} = \frac{+13.4°}{0.08} = +168$$

17. **a.** – 24 **b.** 0

18. **a.**

50% of the mixture is excess (+)-mandelic acid

$$\text{optical purity} = 0.50 = \frac{\text{oberved specific rotation}}{\text{specific rotation of the pure enantiomer}}$$

$$0.50 = \frac{\text{observed specific rotation}}{+158}$$

observed specific rotation = +79

b. 0 (It is a racemic mixture.)

c. 50% of the mixture is excess (–)-mandelic acid.
observed specific rotation = –79

19. **a.** From the data given, you cannot determine what the configuration of naproxen is.

b. 97% of the commercial preparation is (+)-naproxen; 3% is a racemic mixture
Therefore, the commercial preparation forms 98.5% (+)-naproxen and 1.5% (–)-naproxen.

20. Solved in the text.

21. As a result of the double bond, the compound has a cis isomer and a trans isomer. Because the compound also has an asymmetric center, the cis isomer can exist as a pair of enantiomers and the trans isomer can exist as a pair of enantiomers.

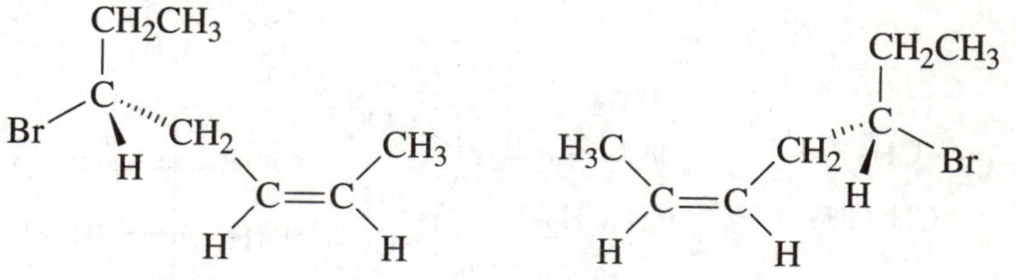

cis enantiomers

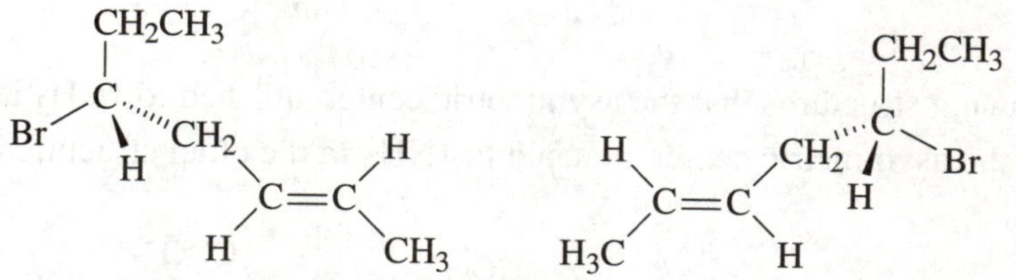

trans enantiomers

22. The statement is not correct. For example, the sp^2 carbons of cis and trans isomers are stereocenters. Therefore, *trans*-2-butene has two stereocenters but only two stereoisomers. Another example is the compound shown in Problem 21; it has three stereocenters but only four stereoisomers.

23. **a.** enantiomers
b. identical compounds (Therefore, they are not isomers.)
c. diastereomers

24. **a.** First find the sp^3 carbons that are bonded to four different substituents; these are the asymmetric centers. Cholesterol has eight asymmetric centers. They are indicated by arrows.

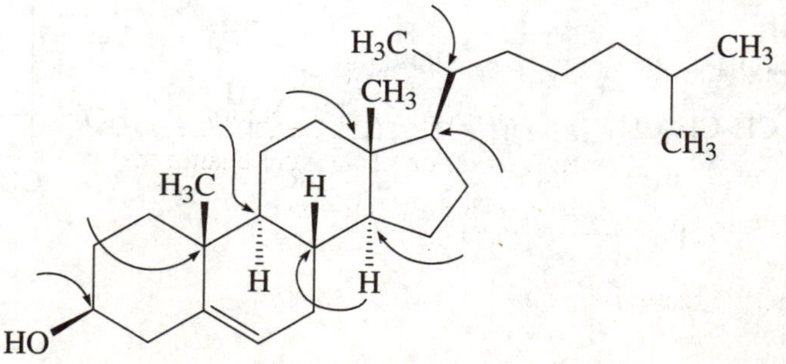

b. $2^8 = 256$ Only the stereoisomer shown above is found in nature.

25. Your perspective formulas may not look exactly like the ones drawn here because you can draw the first one with the groups attached to any bonds you want.
Just make certain that the second one is a mirror image of the first one.

a.

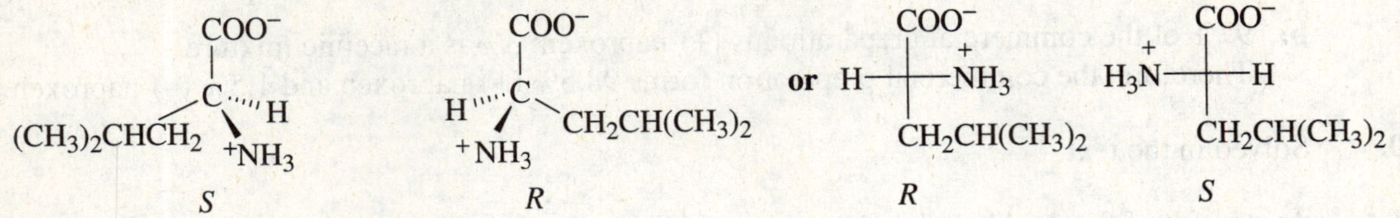

Again your perspective formulas may mot look exactly like the ones drawn here. To make sure you have all four, determine the configuration of each of the asymmetric centers. You should have *R,R*, *S,S*, *R,S*, and *S,R*.

b.

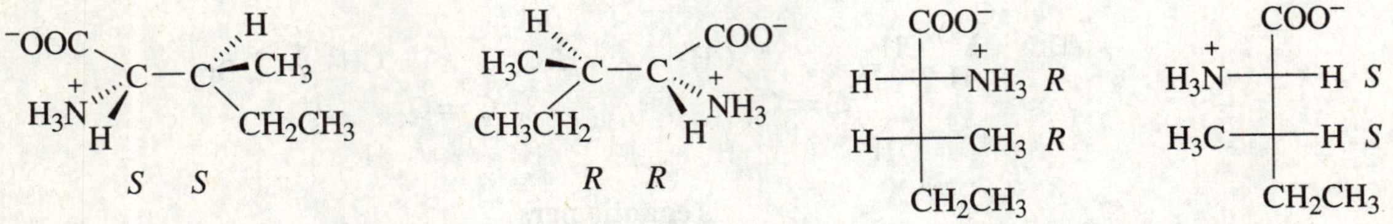

or

Notice in the following structures that the asymmetric center attached to $+NH_3$ in one structure has the *R* configuration and the asymmetric center attached to $+NH_3$ in the other structure has the *S* configuration.

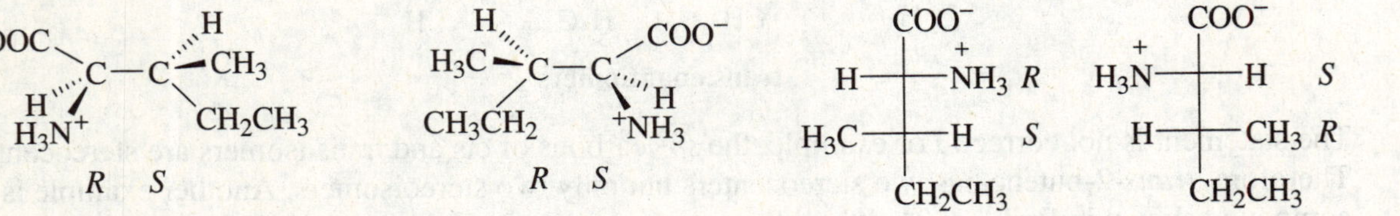

26. **a.**

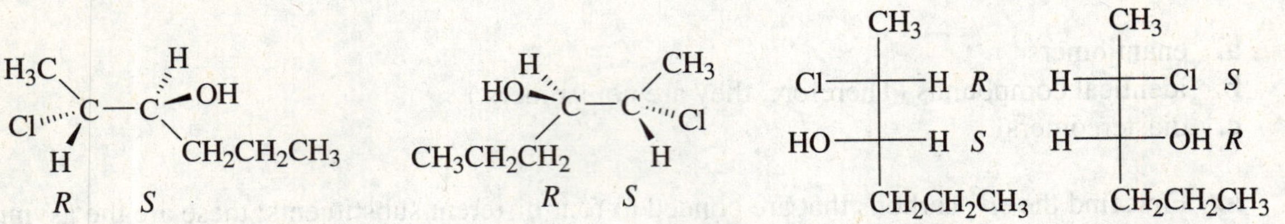

or

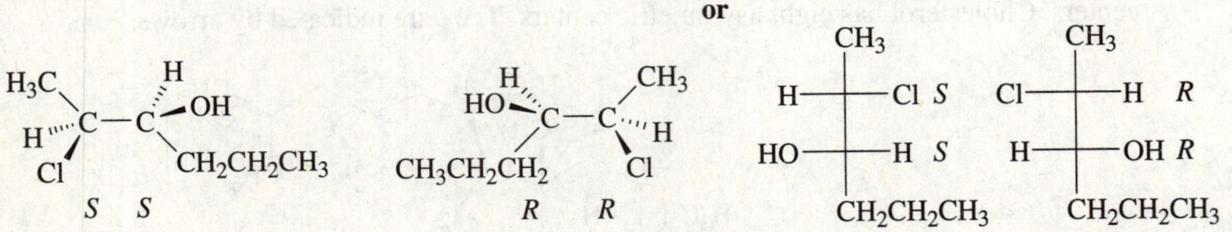

b.

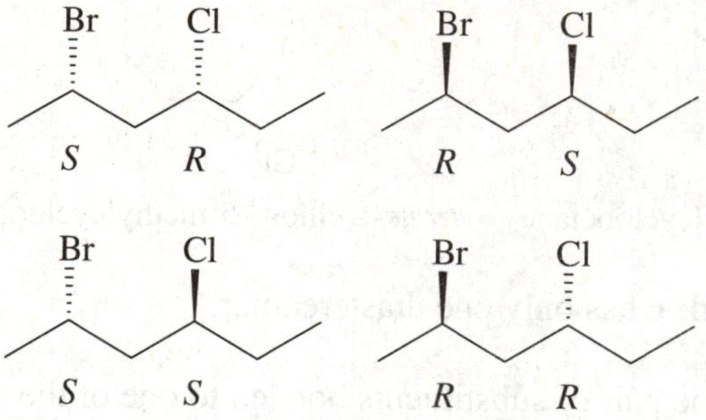

or

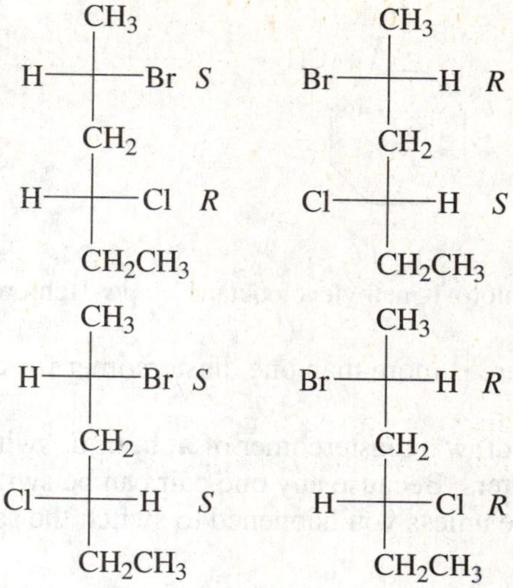

c.

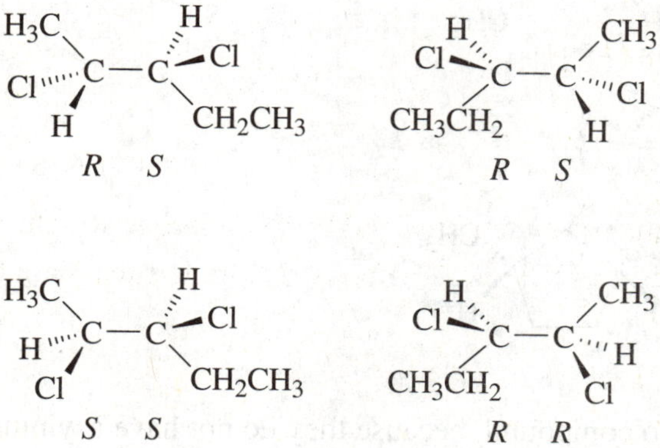

or

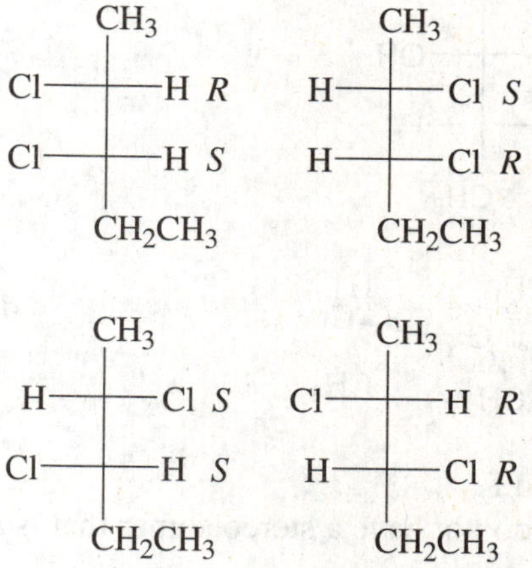

d.

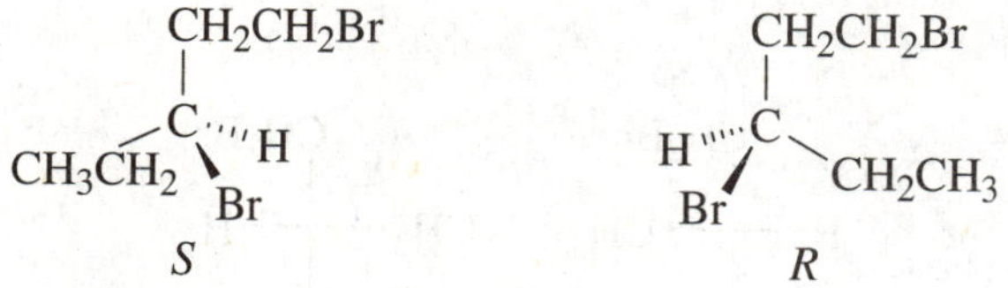

or

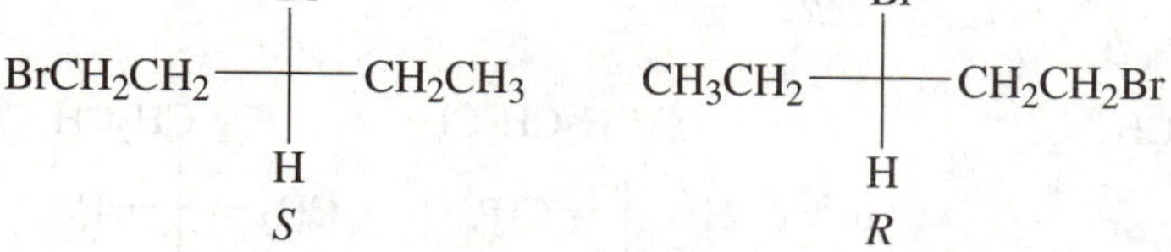

27.

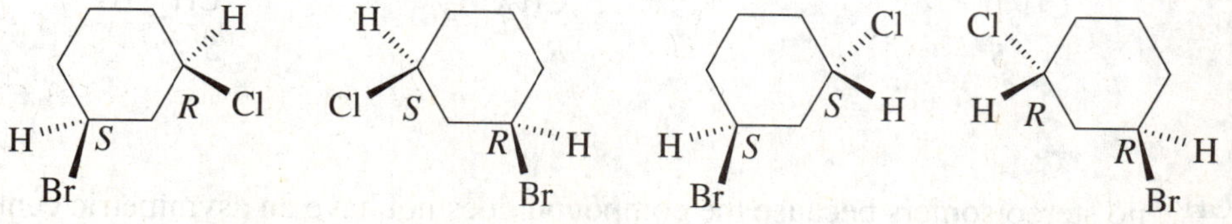

28.

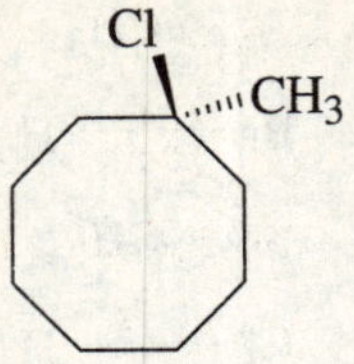

1-chloro-1-methylcyclooctane

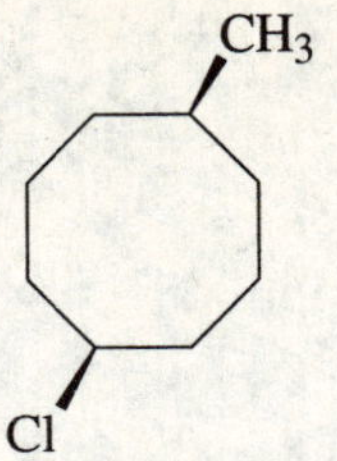

cis-1-chloro-5-methylcyclooctane

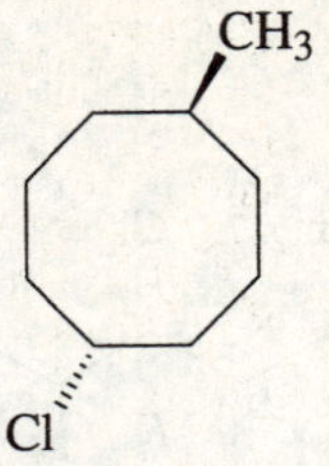

trans-1-chloro-5-methylcyclooctane

29. There is more than one diastereomer for **a, b,** and **d; c** has only one diastereomer.

To draw a diastereomer of **a, b,** or **d**, switch any one pair of substituents bonded to one of the asymmetric centers. Because any one pair can be switched, your diastereomer won't be the same as the one drawn here unless you happened to switch the same pair.

a.

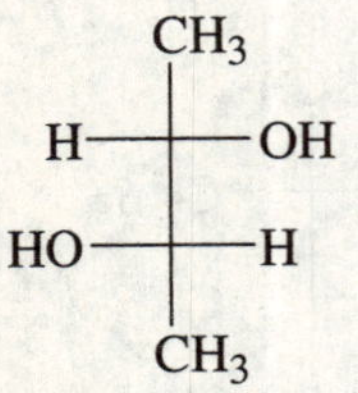

c.

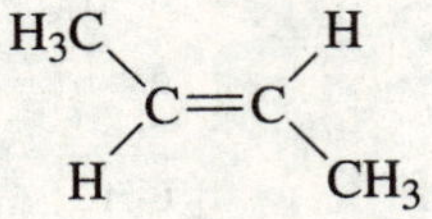

b.

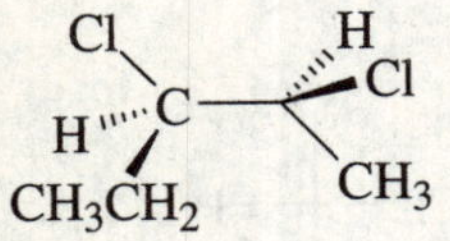

d.

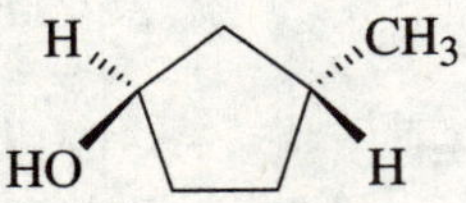

30. **b, d,** and **f**
c and **e** do not have a stereoisomer that is a meso compound, because they do not have asymmetric centers

31. Solved in the text.

32. **a.**

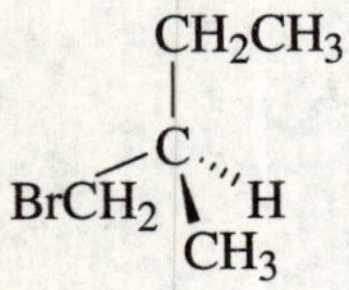

R

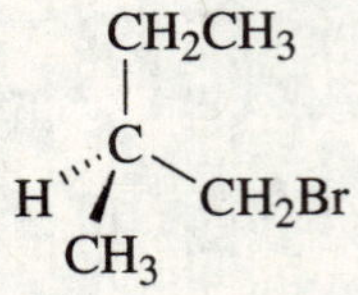

S

or

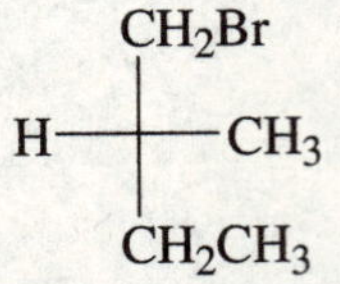

R

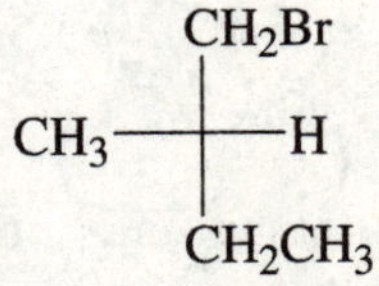

S

b.

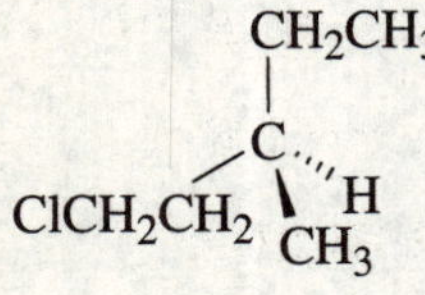

R

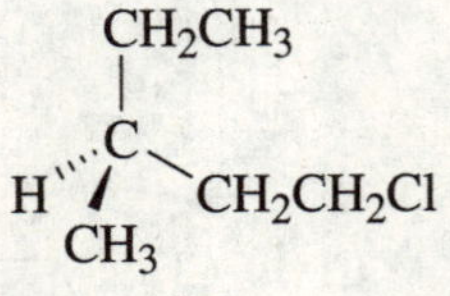

S

or

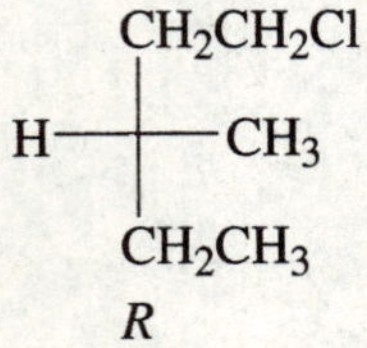

R

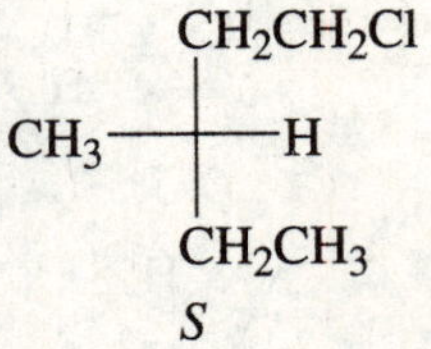

S

c.

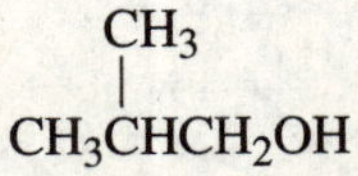

no stereoisomers because the compound does not have an asymmetric center

d.

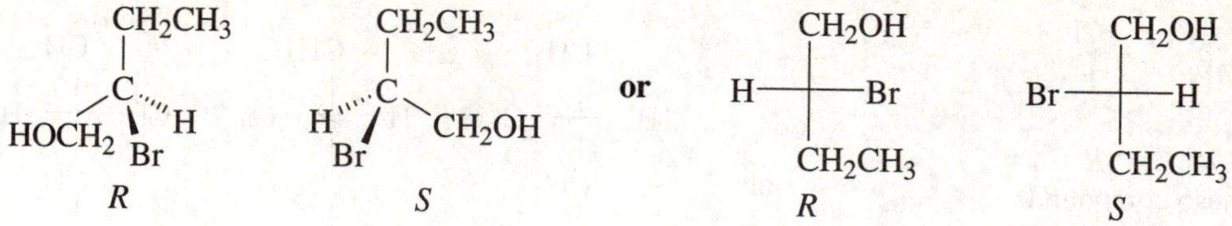

e.

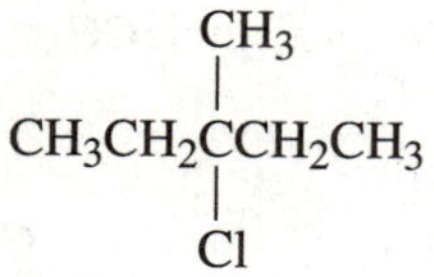

no stereoisomers because the compound does not have an asymmetric center

f.

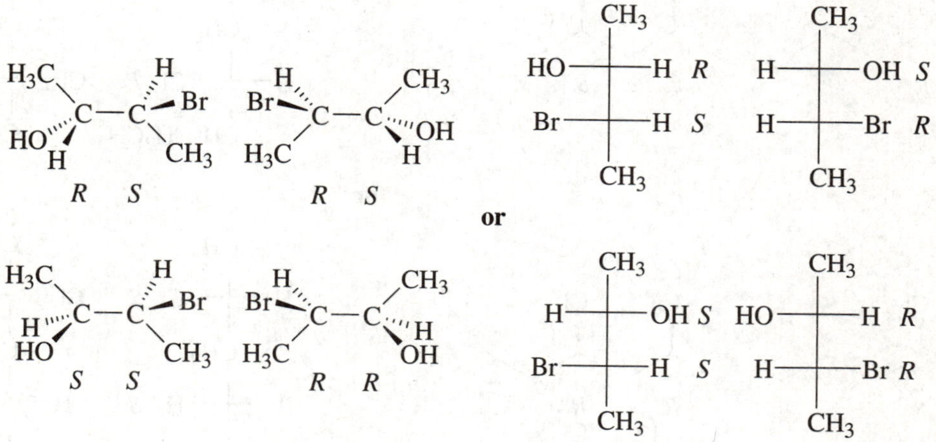

g.

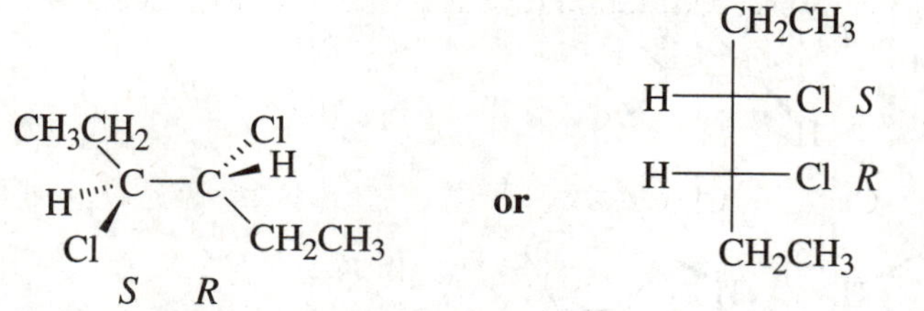

a meso compound   a meso compound

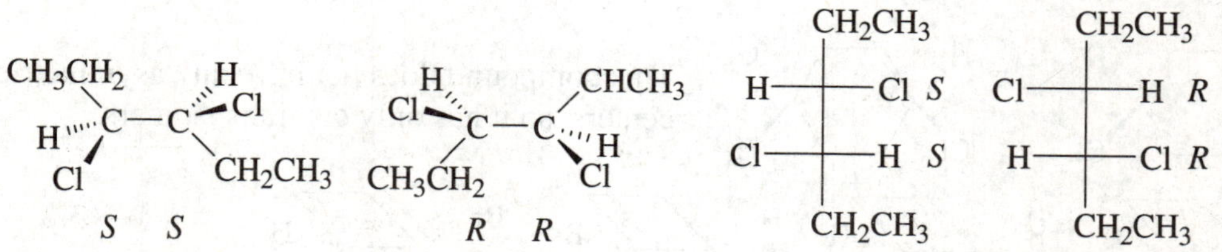

h.

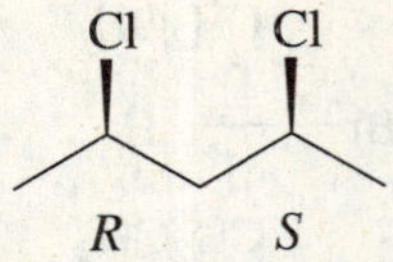

a meso compound

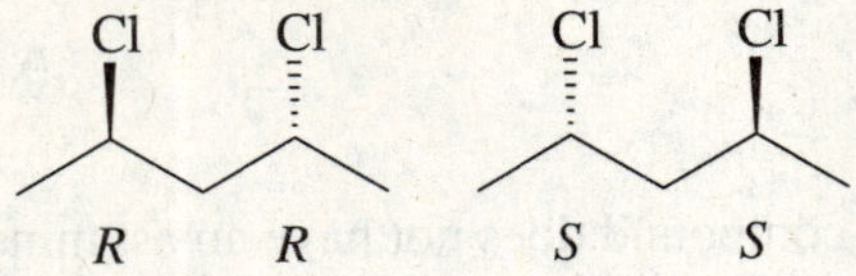

or

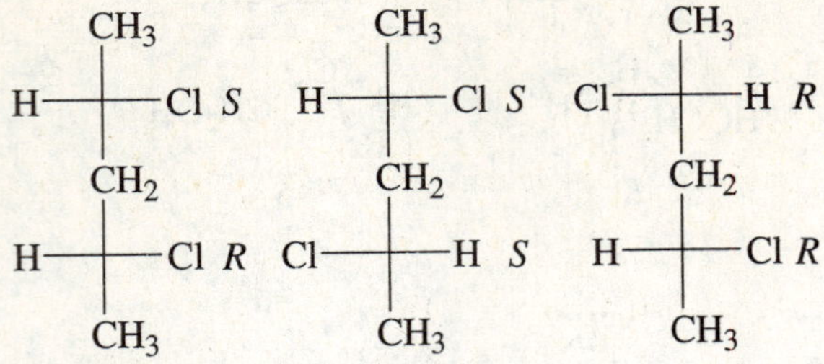

a meso compound

i.

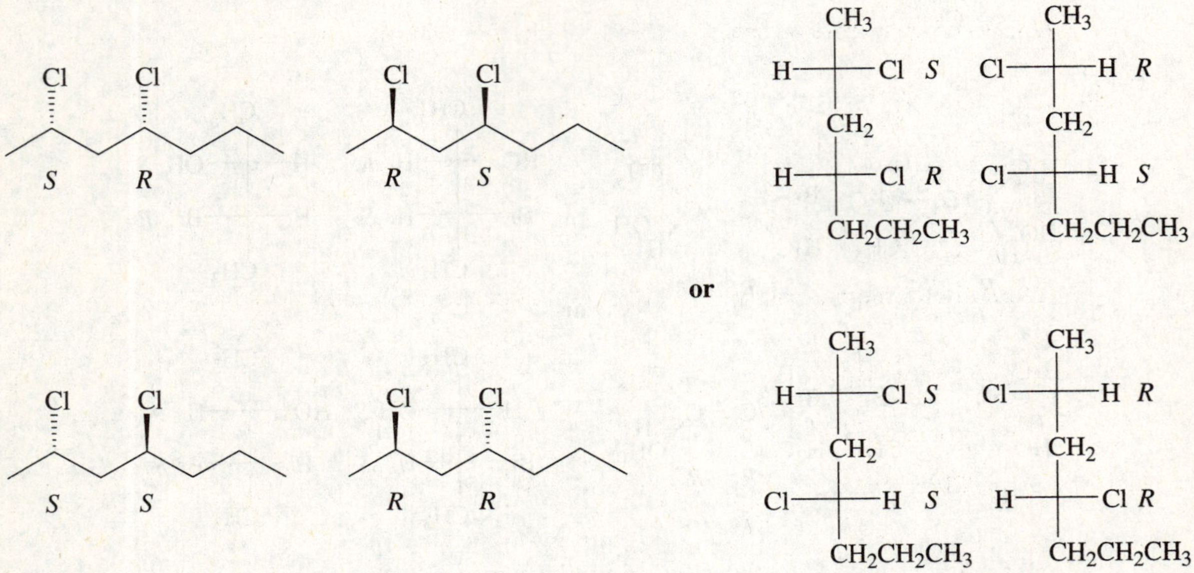

j.

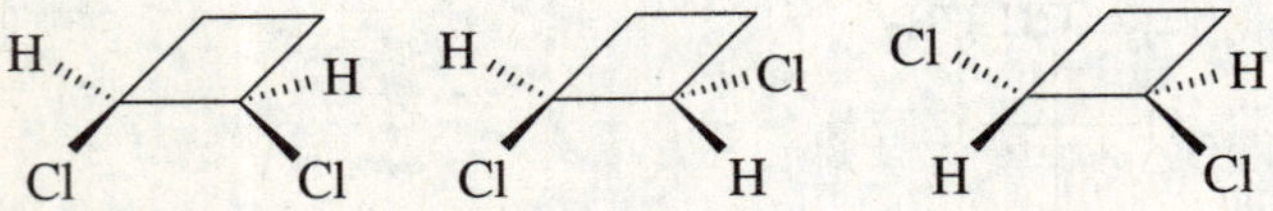

k.

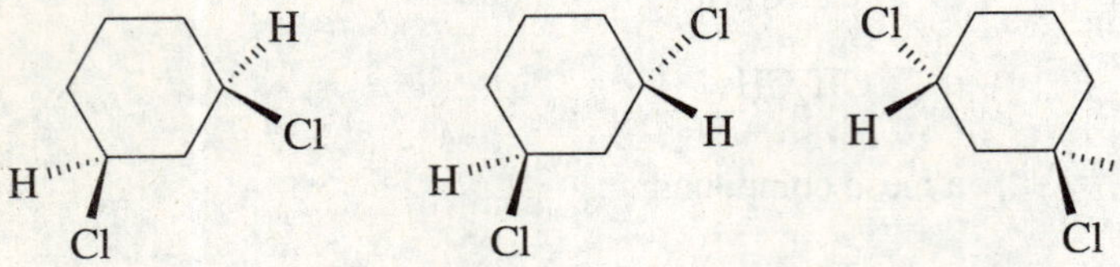

l.

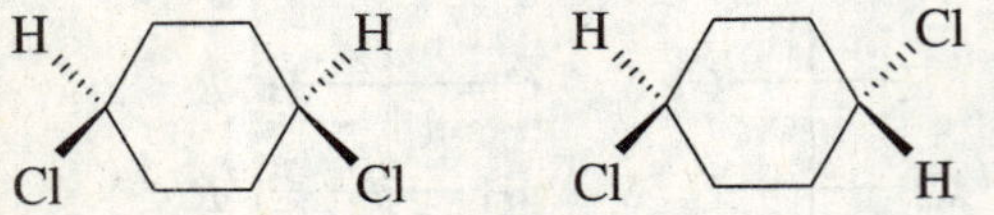

This compound does not have any asymmetric centers, so it has only cis-trans isomers.

m.

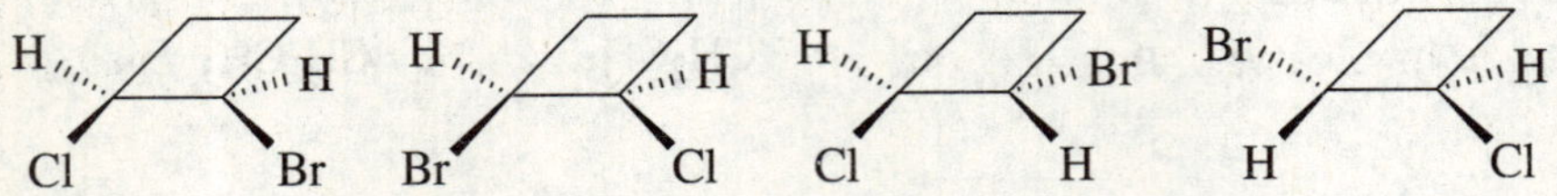

n.

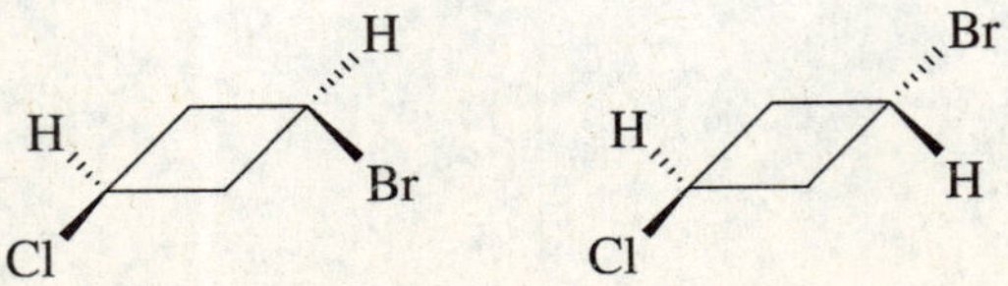

This compound dos not have any asymmetric centers, so it has only cis-trans isomers.

33. **a.**

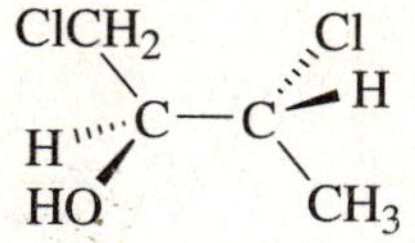

(2*S*,3*R*)-1,3-dichloro-2-butanol

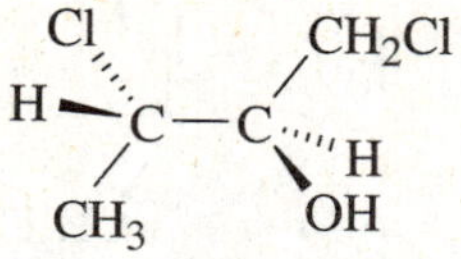

(2*R*,3*S*)-1,3-dichloro-2-butanol

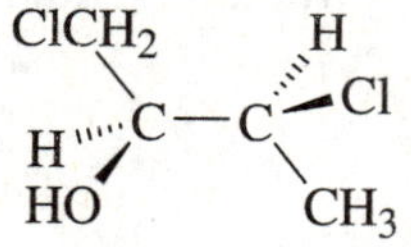

(2*S*,3*S*)-1,3-dichloro-2-butanol

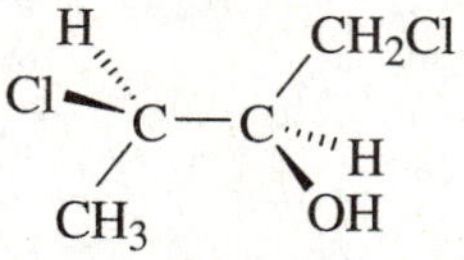

(2*R*,3*R*)-1,3-dichloro-2-butanol

b.

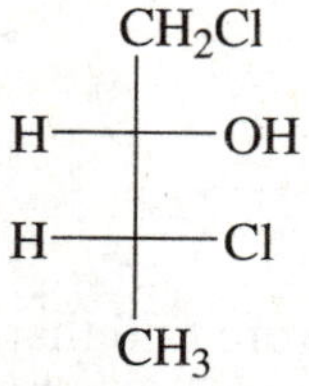

(2*S*,3*R*)-1,3-dichloro-2-butanol

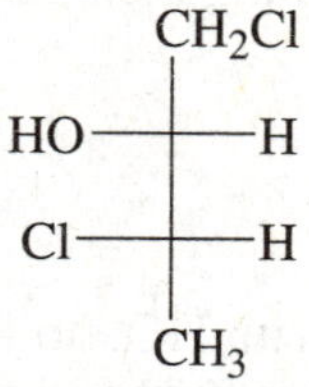

(2*R*,3*S*)-1,3-dichloro-2-butanol

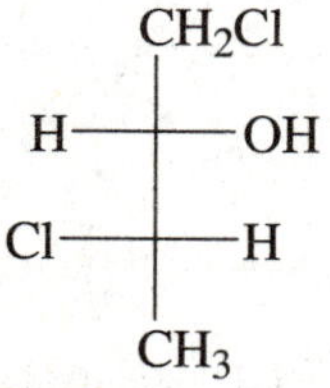

(2*S*,3*S*)-1,3-dichloro-2-butanol

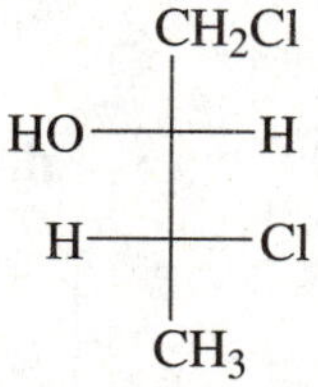

(2*R*,3*R*)-1,3-dichloro-2-butanol

34.

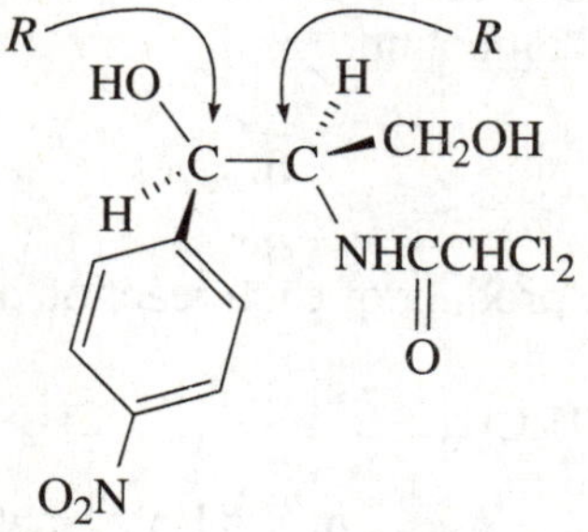

35. Your answer might be correct yet not look like the answers shown here. If you can get the answer shown here by interchanging **two** pairs of groups bonded to an asymmetric center, then your answer is correct. If you get the answer shown here by interchanging **one** pair of groups bonded to an asymmetric center, then your answer is not correct.

a.

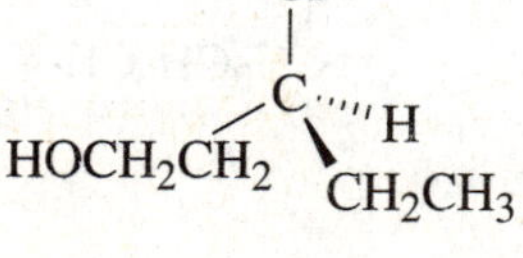

c.

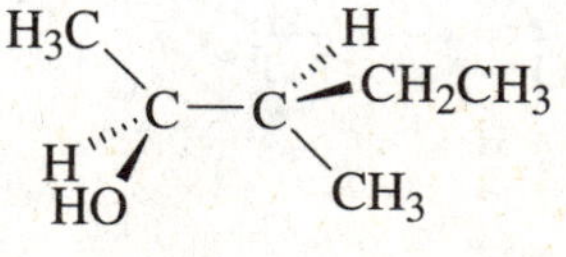

b. 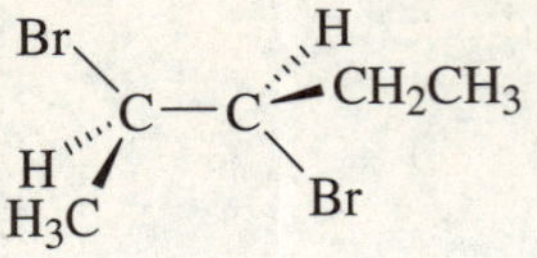

d. 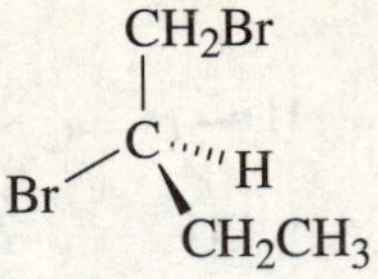

36. The carbon of the COO^- group is C-1. The first structure is 2*R*,3*S*. Therefore, naturally occurring threonine, with a configuration of 2*S*,3*R*, is the mirror image of the first structure. Thus, the second structure is naturally occurring threonine.

37.
a. (2*R*,3*R*)-2,3-dichloropentane
b. (2*R*,3*R*)-2-bromo-3-chloropentane
c. (1*R*,3*S*)-1,3-cyclopentanediol (naming the compound clockwise) or (1*S*,3*R*)-1,3-cyclopentanediol (naming the compound counterclockwise)
d. (3*R*,4*S*)-3-chloro-4-methylhexane

38. Solved in the text.

39. We see that the *R*-alkyl halide reacts with HO^- to form the *S*-alcohol. We are told that the product (the *R*-alcohol) is (–). We can, therefore, conclude that the (+) alcohol has the *S* configuration.

40. From the structures given on page 231 of the text, you can determine the configuration of the asymmetric center in each compound.

a. *R* **b.** *R* **c.** *S* **d.** *S*

41. Only **b** is true.

42. Compound A has two stereoisomers because it has an asymmetric center.
Compound B does not have stereoisomers because it does not have an asymmetric center.
Compound C has an asymmetric center but, because of the lone pair, the two enantiomers rapidly interconvert, so it exists as a single compound.

43. **a.** no **b.** no **c.** no **d.** yes **e.** no **f.** no

44. Only the stereoisomers of the major product of each reaction are shown.

a. 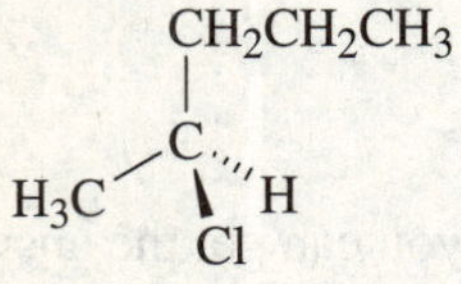
S
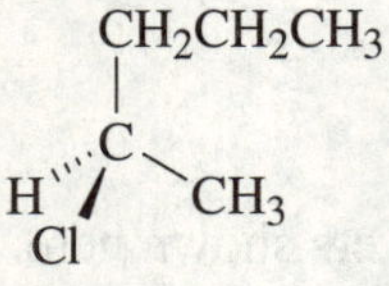
R
or 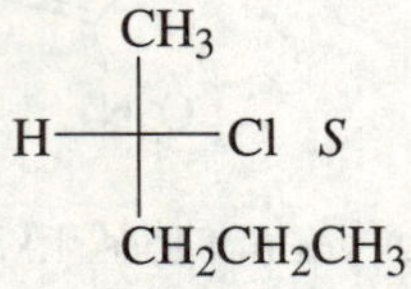 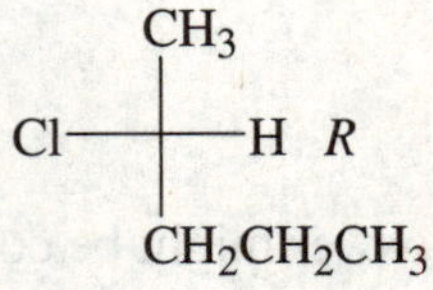

b. 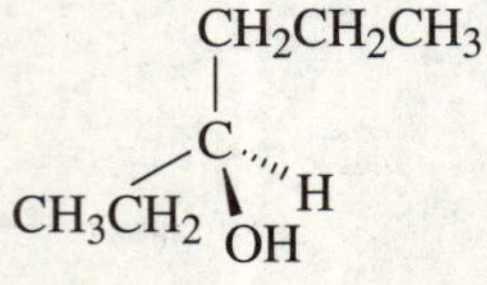
S
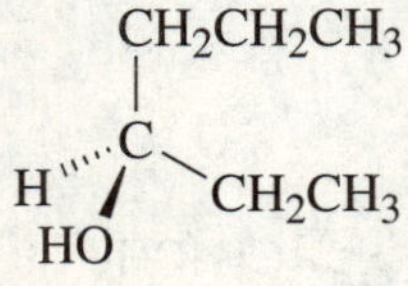
R
or 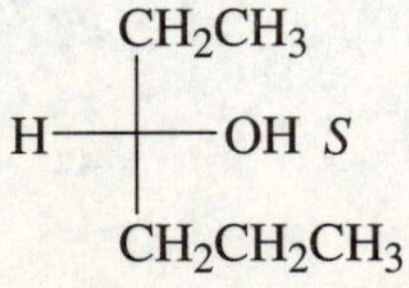 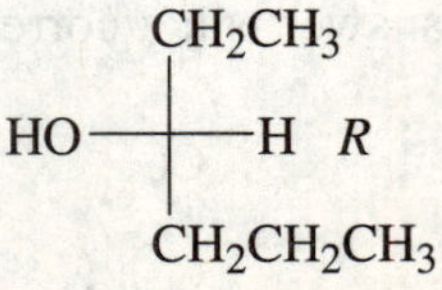

c. 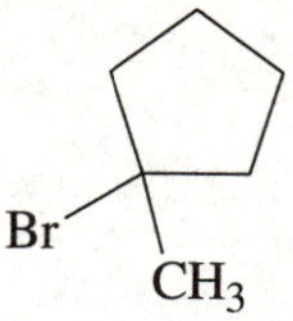

This compound does not have any stereoisomers because it does not have an asymmetric center.

d. 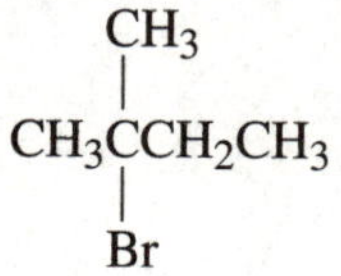

This compound does not have any stereoisomers because it does not have an asymmetric center.

45. **a.** 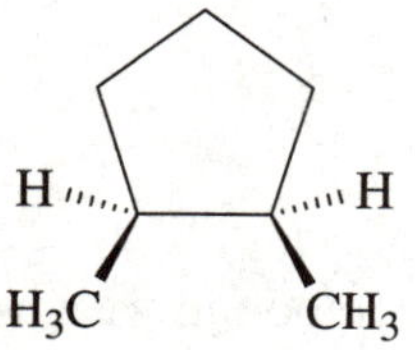 and 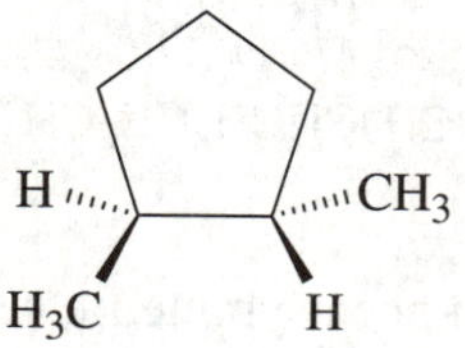

b. 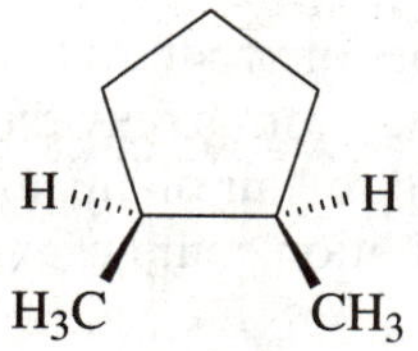

Hydrogen will be more likely to add to the side of the ring where the hydrogen substituent is than to side of the ring where the methyl substituent is because hydrogen provides less steric hindrance than a methyl group does.

46. **a.** **1.** *trans*-3-heptene **2.** *cis*-3-heptene

b. **1.** 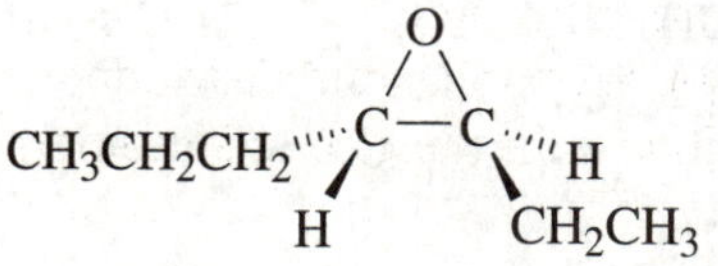 **2.** 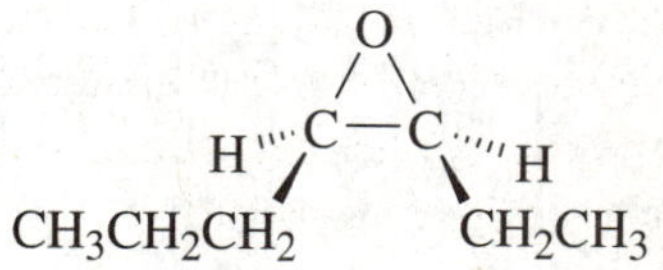

47. **a.** 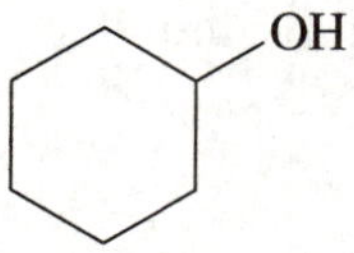

b. 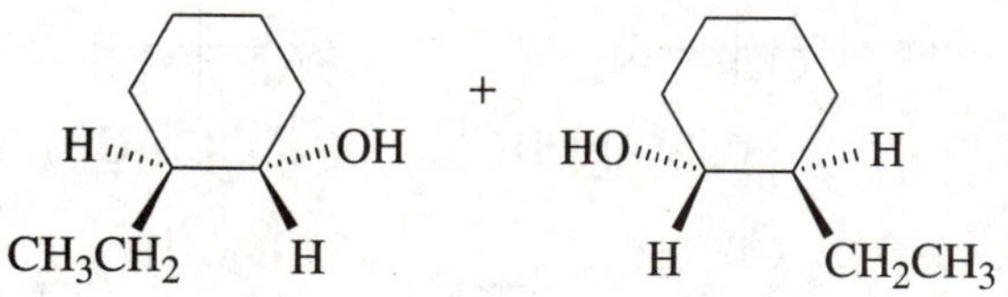

c. 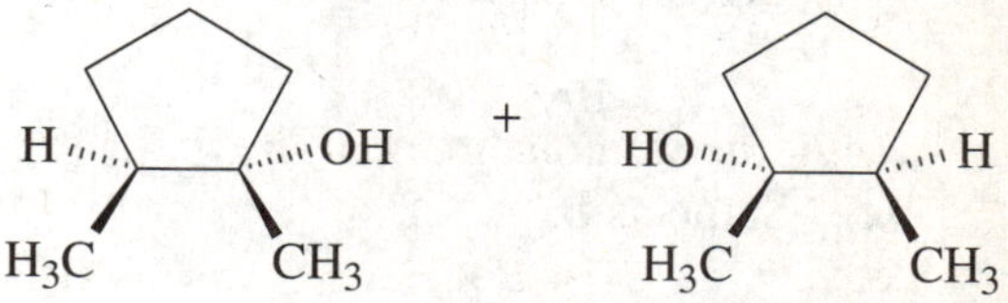

d. 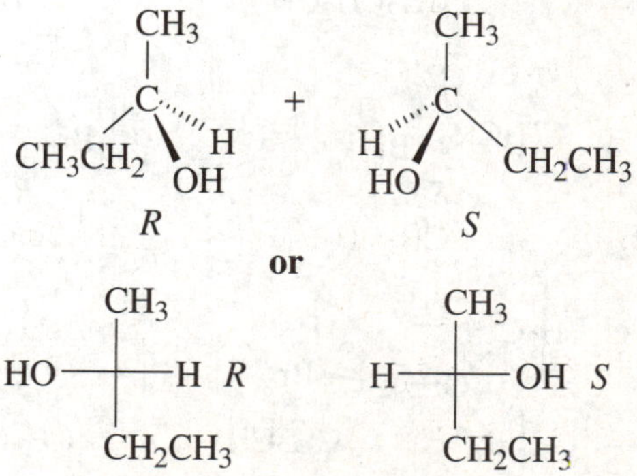

48. 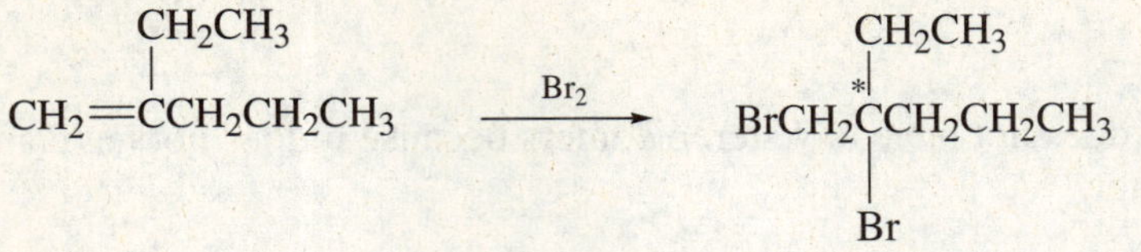

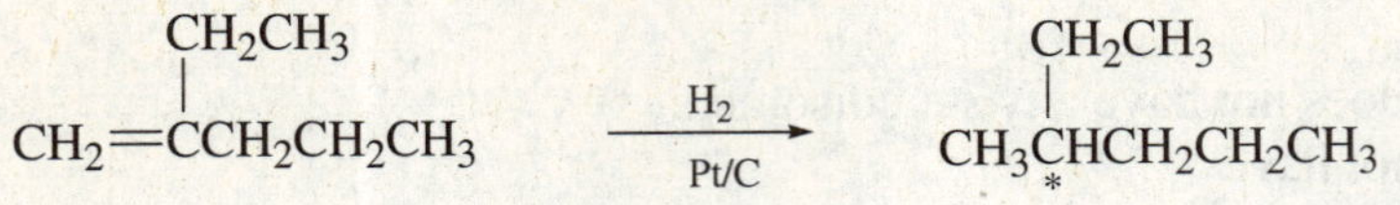

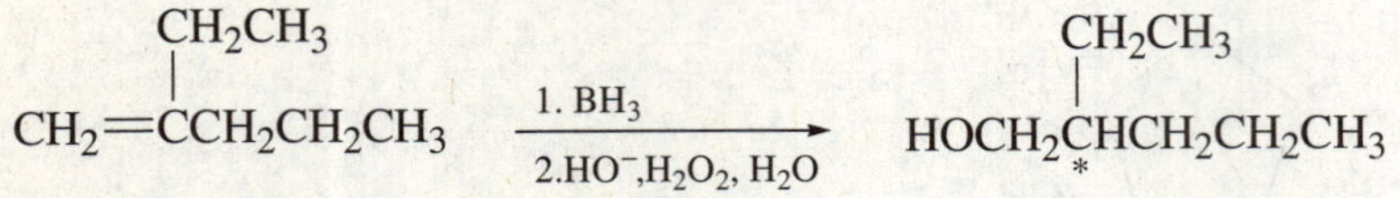

(* indicates an asymmetric center)
Each of the reactions forms a compound with one asymmetric center from a compound with no asymmetric centers. Therefore, each of the products is a racemic mixture.

49. You could identify the intermediate that is formed by determining the number of products that are formed. Because *trans*-2-butene forms a cyclic bromonium ion intermediate, it forms only the erythro enantiomers. If the reaction formed a carbocation intermediate, both the erythro pair of enantiomers and the threo pair of enantiomers would be formed because both syn and anti addition could occur.

50. **a.** 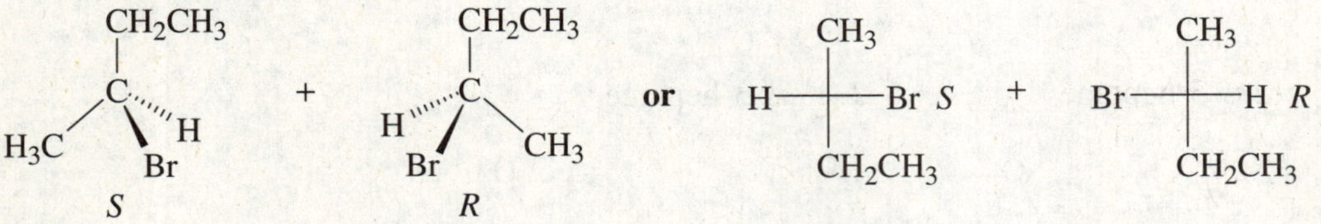

b. 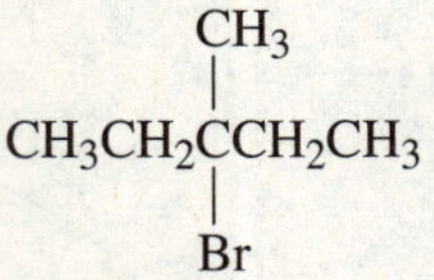 This compound does not have an asymmetric center

c. Same as **b**.

d. 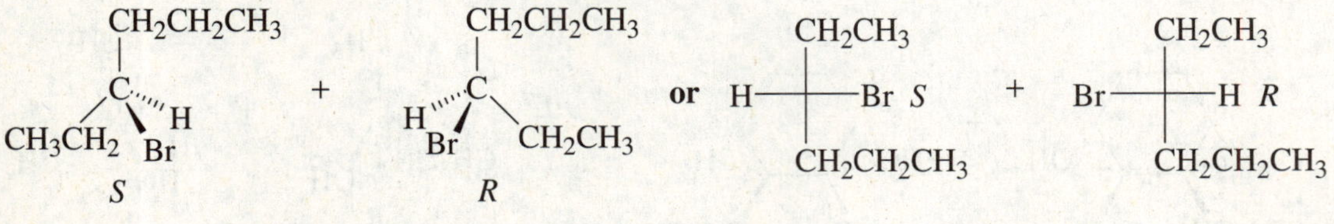

e. 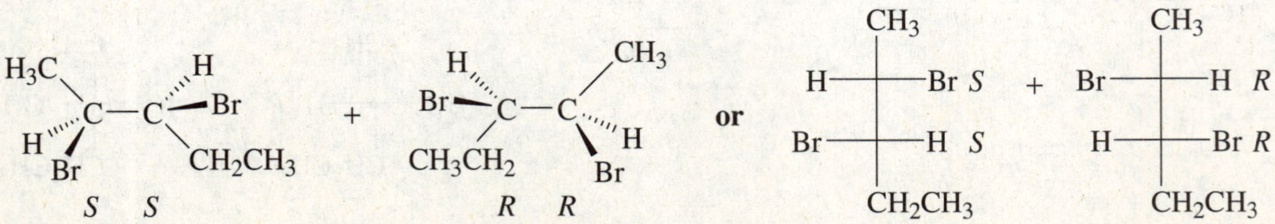

f.

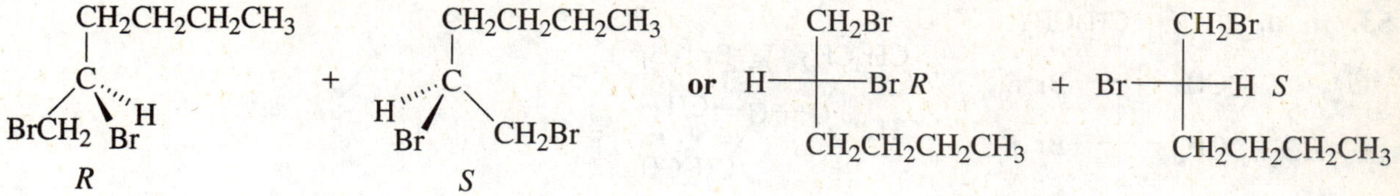

51. Two different bromonium ions are formed because Br^+ can add to the double bond either from the top of the plane or from the bottom of the plane defined by the alkene, and the two bromonium ions are formed in equal amounts. Attacking the less hindered carbon of one bromonium ion forms one stereoisomer, while attacking the less hindered carbon of the other bromonium ion forms the other stereoisomer.

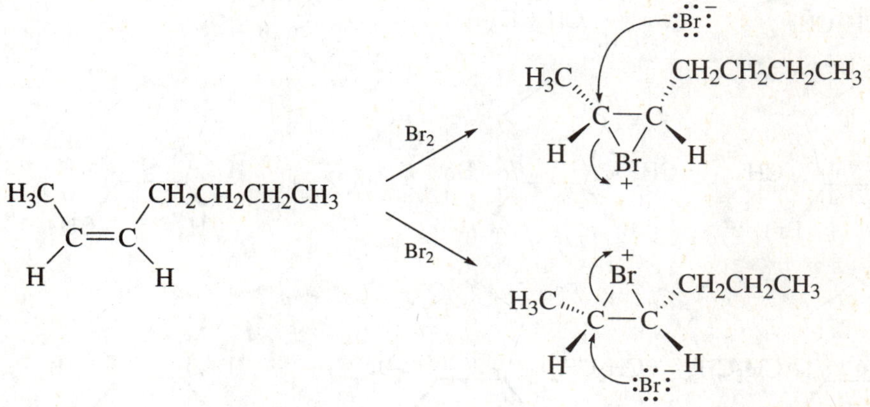

52. a.

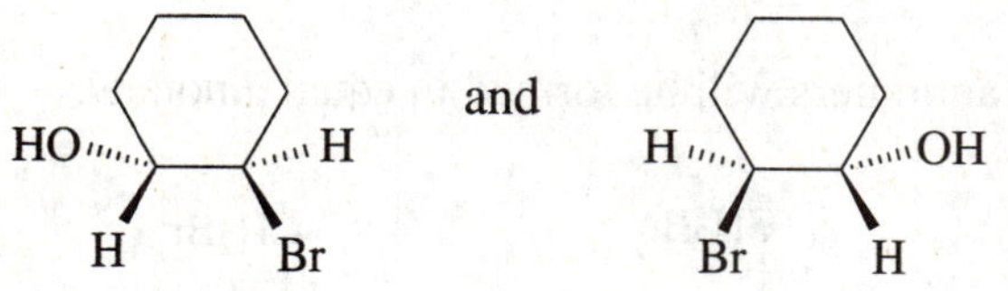

The addition of Br and OH are anti, so in a cyclic compound these two substituents are trans to one another.

b. mechanism for the reaction

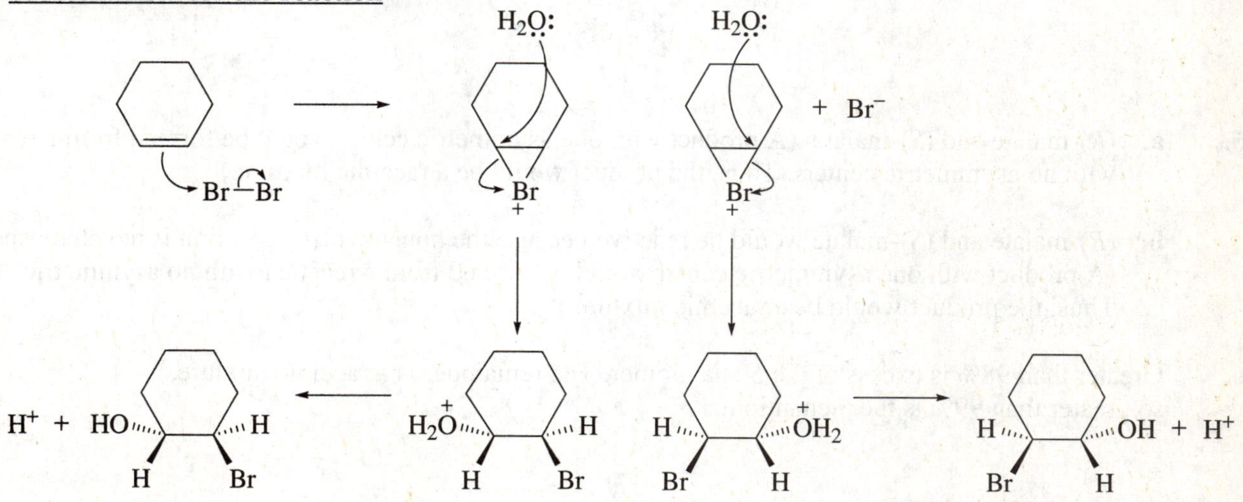

53. **a.**

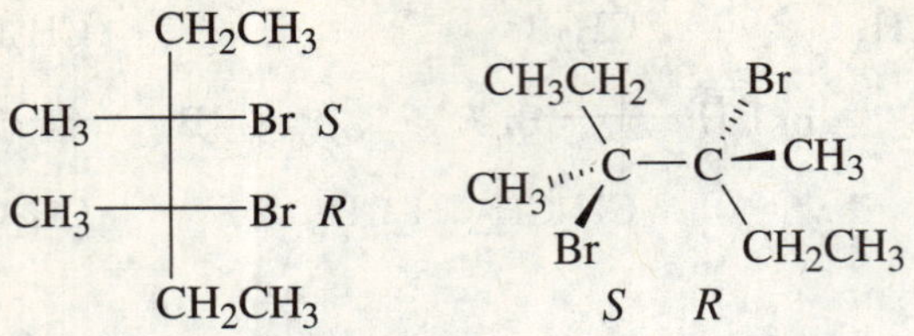

b.

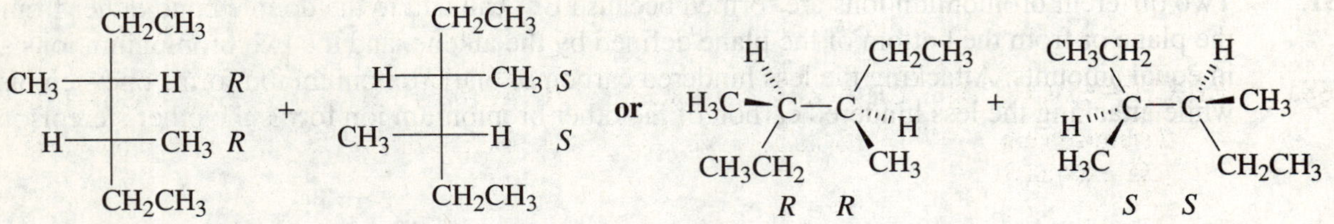

c.

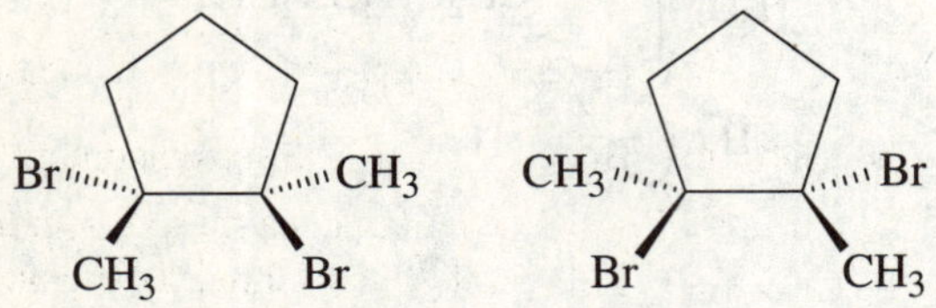

e.

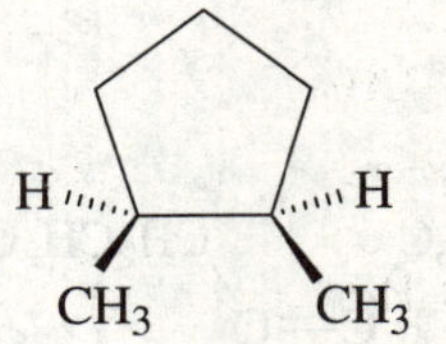

d.

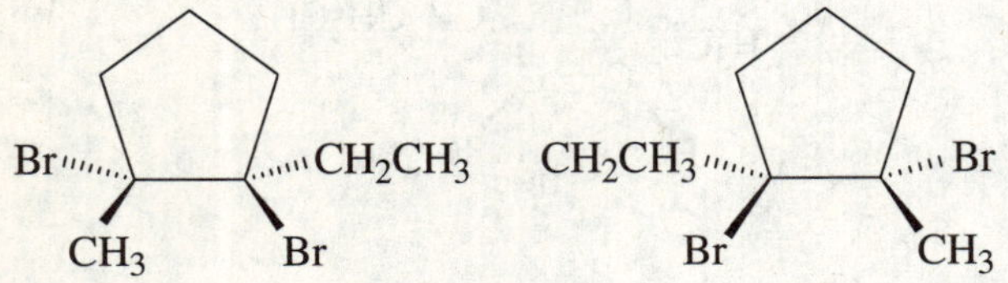

f.

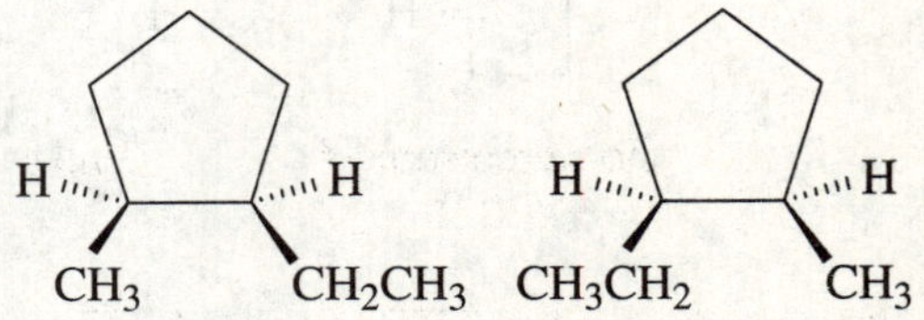

54. **a.** $CH_3\underset{Cl}{\underset{|}{C}}HCH_2Br$

b. The *R* and *S* enantiomers will be formed in equal amounts.

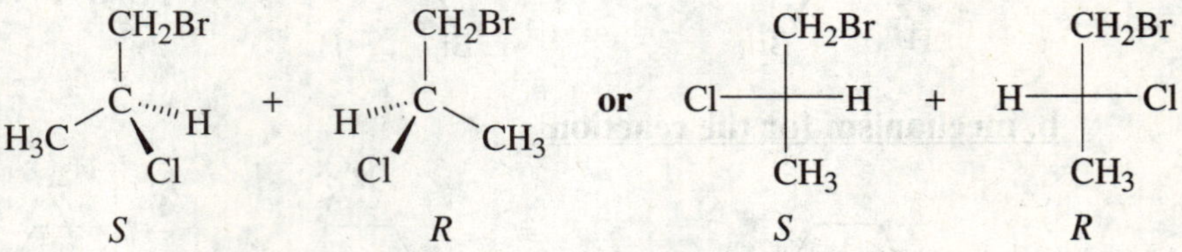

55. **a.** (*R*)-malate and (*S*)-malate (A product with one asymmetric center would be formed from a reactant with no asymmetric centers. Thus, the product would be a racemic mixture.)

b. (*R*)-malate and (*S*)-malate would be reactive because the nonenzymatic reaction is not stereospecific. (A product with one asymmetric center would be formed from a reactant with no asymmetric centers. Thus, the product would be a racemic mixture.)

56. Greater than 98% is excess of the *S* enantiomer. The remainder is a racemic mixture, so greater than 99% is the *S* enantiomer.

57.

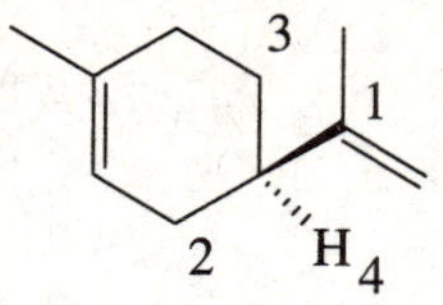

(+)-limonene is the *R* isomer

58.

$CH_3CH{=}CHCH_2CH_3$
2 steroisomers
[cis and trans]

$CH_2{=}CHCH_2CH_2CH_3$
no stereoisomers

CH_3
$CH_3C{=}CHCH_3$
no stereoisomers

CH_3
$CH_3CHCH{=}CH_2$
no stereoisomers

CH_3
$CH_3CH_2C{=}CH_2$
no stereoisomers

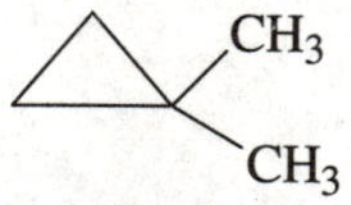

no stereoisomers

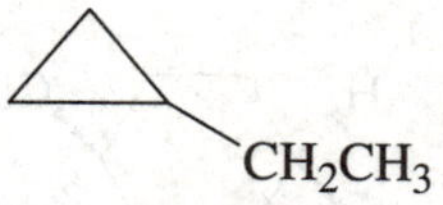

no stereoisomers

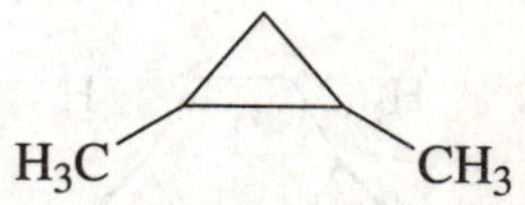

3 stereoisomers
[Cis is a meso compound.]
[Trans is a pair of enantiomers.]

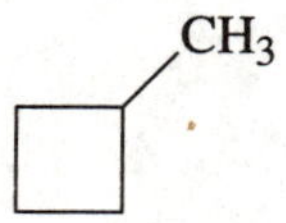

no stereoisomers

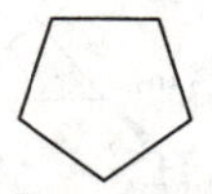

no stereoisomers

59. **a.**

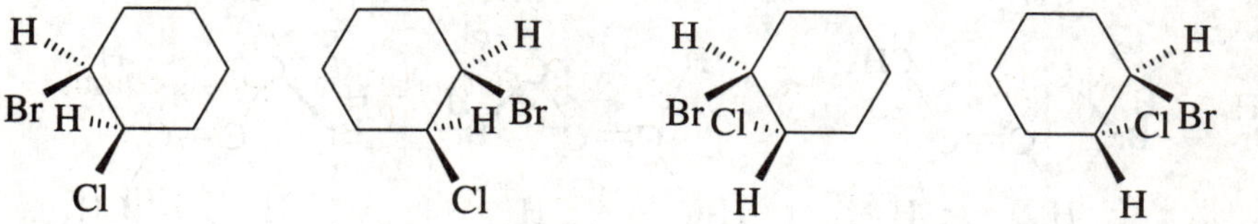

cis trans

b.

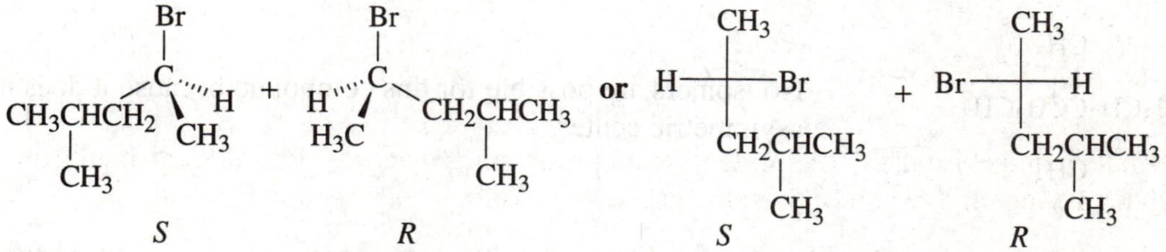

S *R* *S* *R*

c.

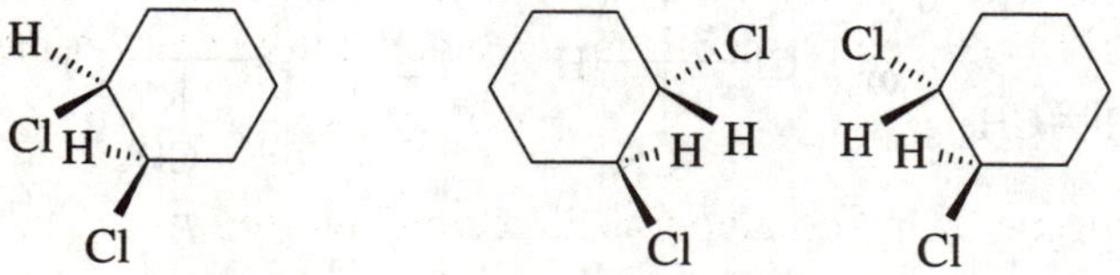

cis trans

d.

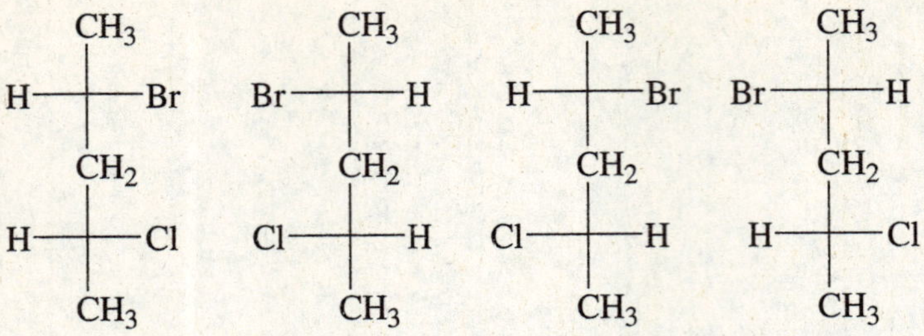

e.

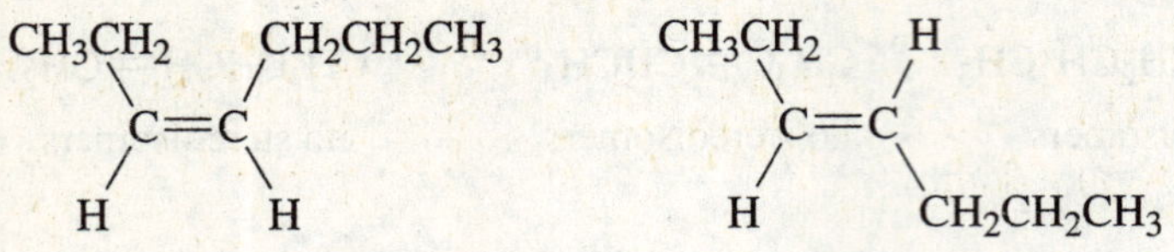

f.

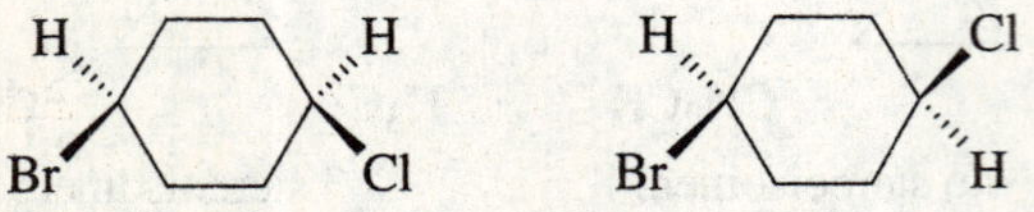

cis trans

g.

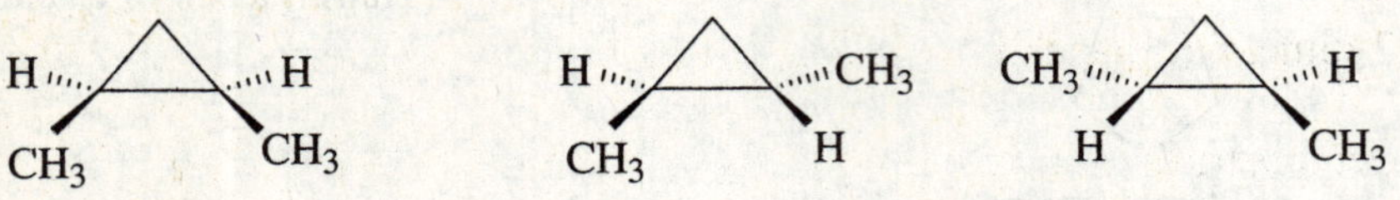

cis trans

h.

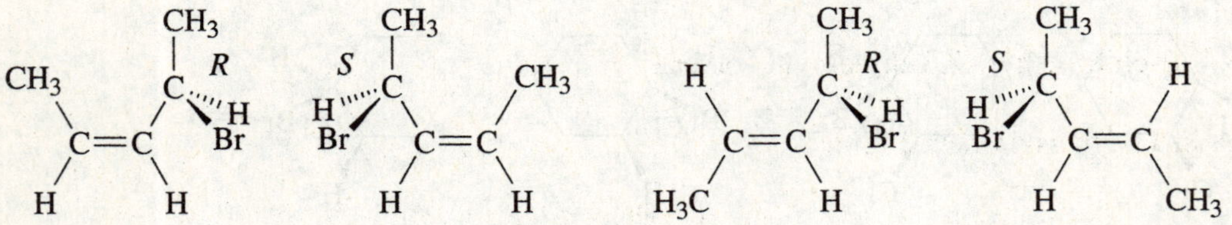

cis trans

i.

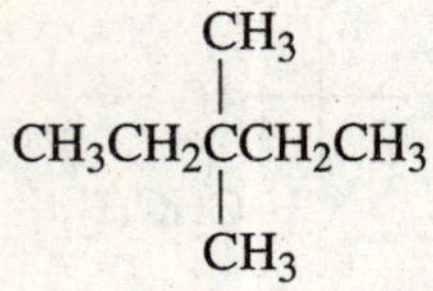

No isomers are possible for this compound because it does not have an asymmetric center.

j.

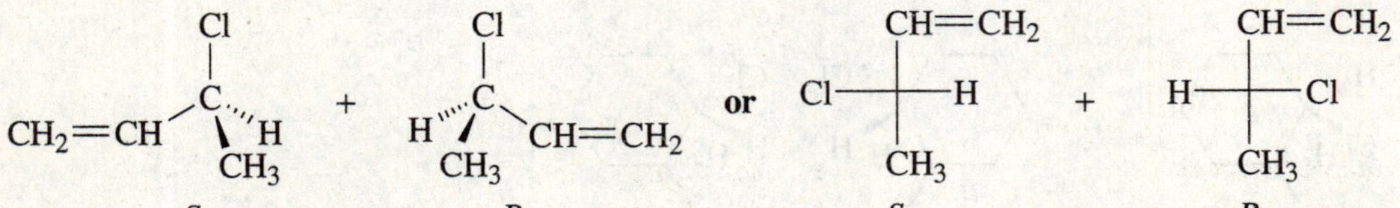

S *R* *S* *R*

k.

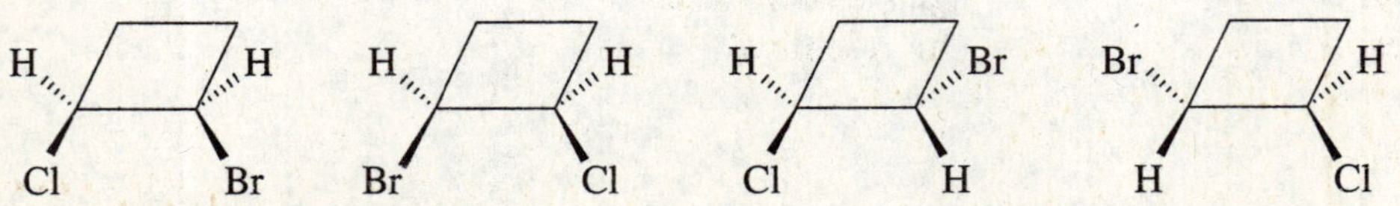

cis trans

l. 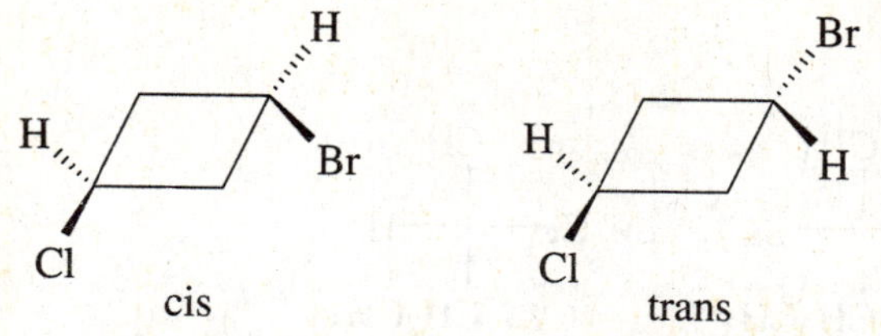

60. Only the fourth one (CHFBrCl) is optically active.

61. **a.** (*R*)-3-methyl-1-pentene

b. (*E*)-1-bromo-2-chloro-2-fluoro-1-iodoethene

c. (2*R*,3*R*)-3-chloro-2-pentanol

d. (*Z*)-2-bromo-1-chloro-1-fluoroethene

e. 8-bromo-2-ethyl-1-octene

f. (*E*)-1,3-dibromo-4,7-dimethyl-3-octene

g. (*S*)-2-methyl-1,2,5-pentanetriol

h. (2*S*,3*S*)-1,2-dibromo-2-methyl-3-pentanol

i. (*E*)-4-(2-chloroethyl)-2,3-dimethyl-3-octene

62. Mevacor has eight asymmetric centers.

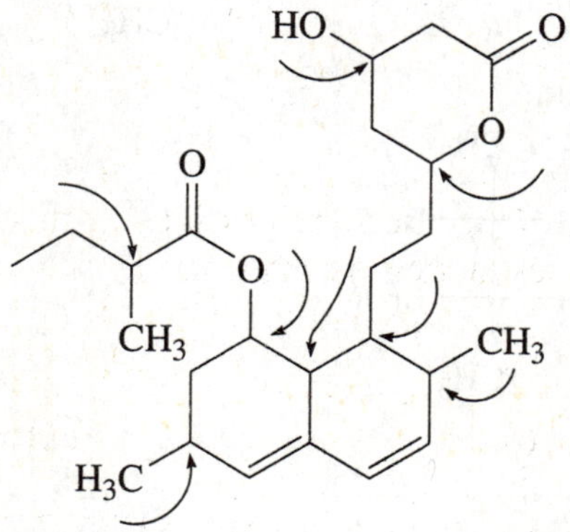

63.

a. diastereomers
b. enantiomers
c. constitutional isomers
d. diastereomers
e. diastereomers
f. identical
g. diastereomers
h. identical

64. 1. Both cis and trans give these products.

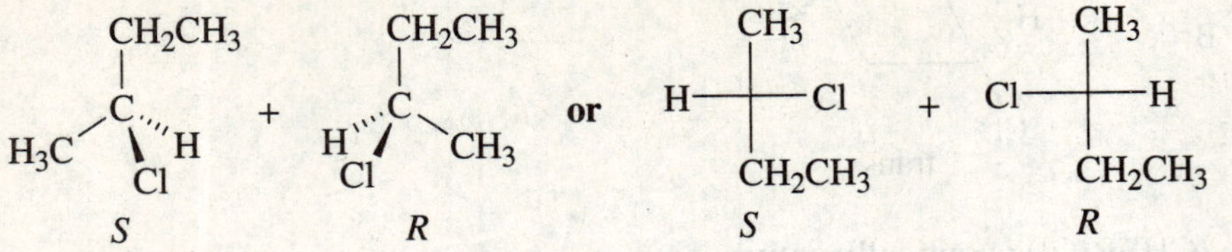

2. Both cis and trans give these products.

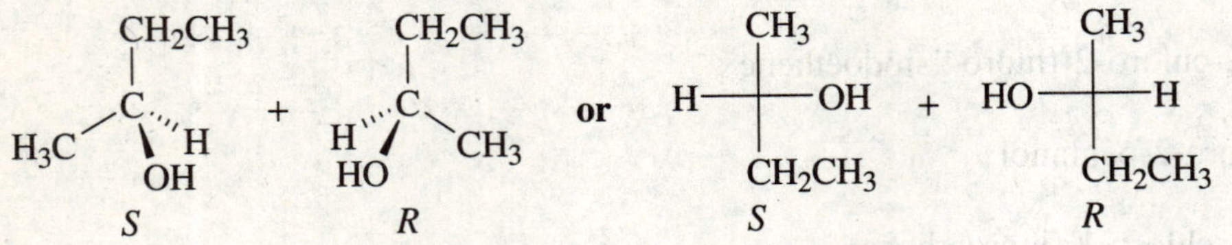

3. Cis gives a meso compound.

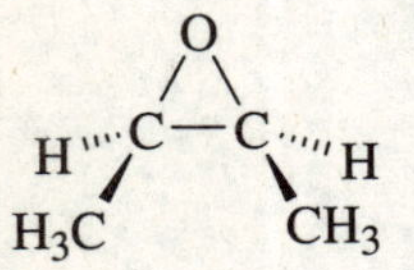

Trans gives a pair of enantiomers.

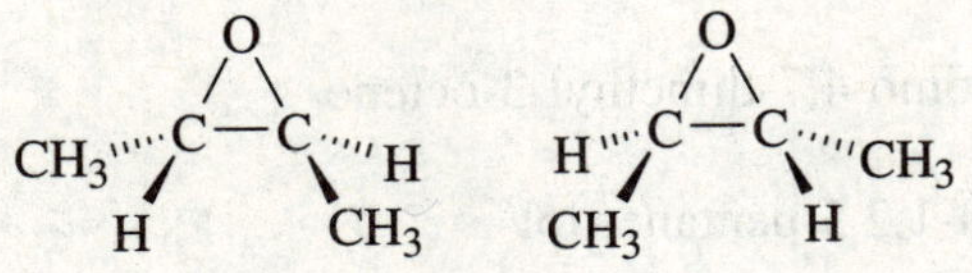

4. Cis gives the threo pair.

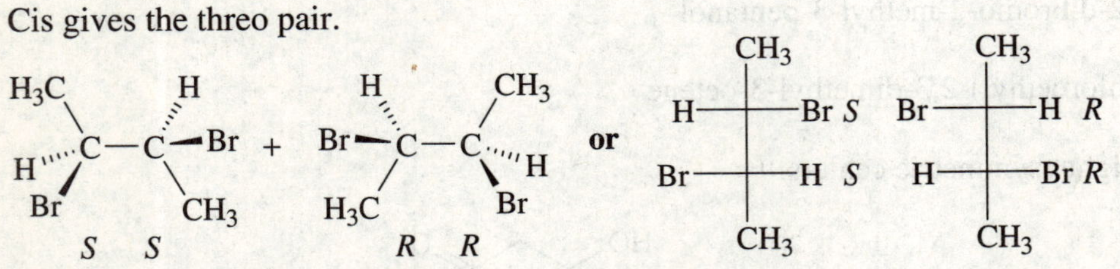

Trans gives a meso compound.

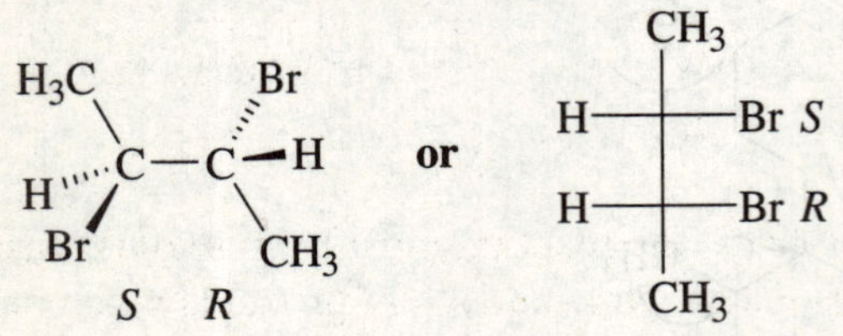

5. Cis gives the threo pair.

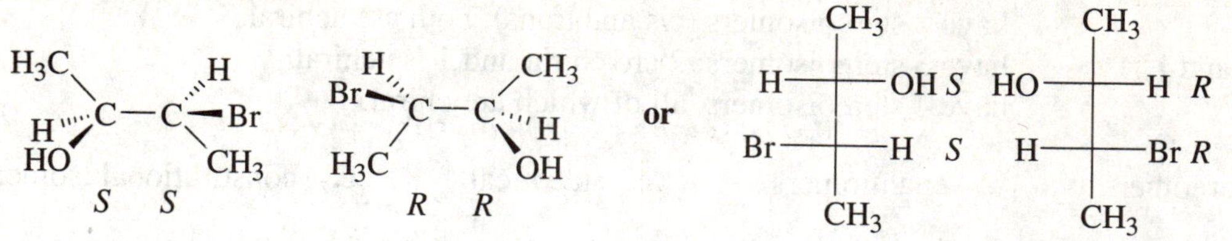

Trans gives the erythro pair.

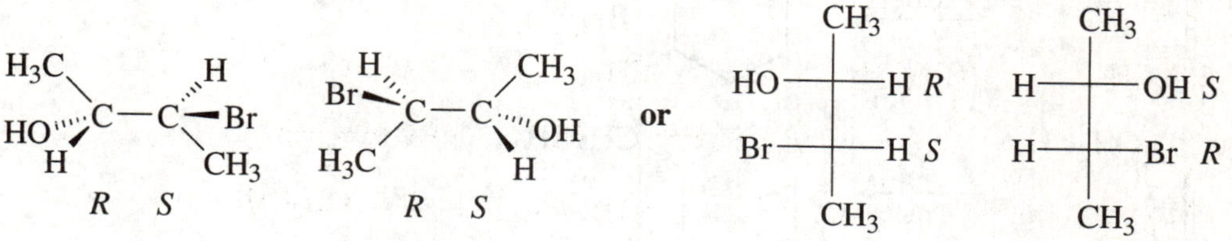

6. Both cis and trans give this product.
$CH_3CH_2CH_2CH_3$

7. Both cis and trans give these products.

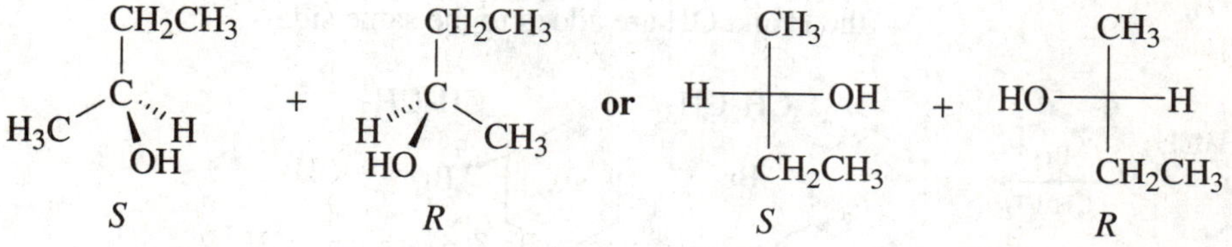

8. Both cis and trans give these products.

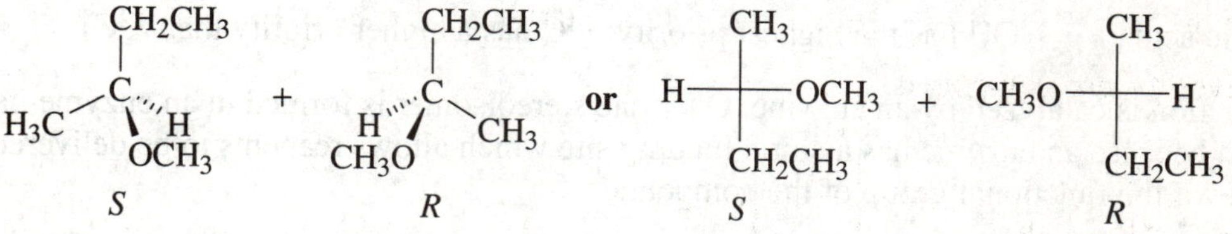

b. For the cis and trans isomers to form different products, the reaction must form two new asymmetric centers in the product. Therefore, the cis and trans alkenes form different products when they react with Br_2 in CH_2Cl_2 and when they react with Br_2 in H_2O.

65. **a.** Because there are two asymmetric centers, there are four possible stereoisomers.

b.

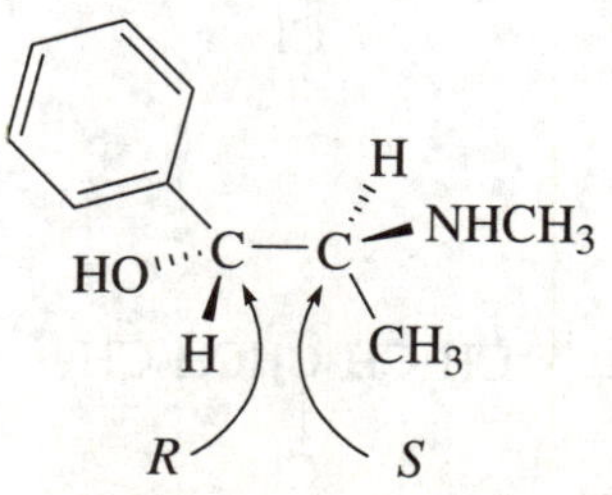

66. **c** does not have stereoisomers; it is, therefore, achiral.
e and **h** have 2 stereoisomers (cis and trans); both are achiral.
a, **d**, **f**, **i**, and **j** have 3 stereoisomers; 2 are chiral and 1 is achiral.
b and **g** have 4 stereoisomers, all of which are chiral.

67. **a.** diastereomers **b.** enantiomers **c.** identical **e.** constitutional isomers

68. **a.**

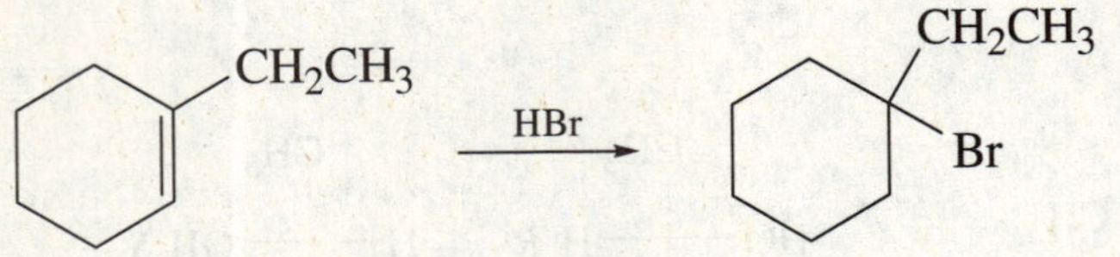

b.

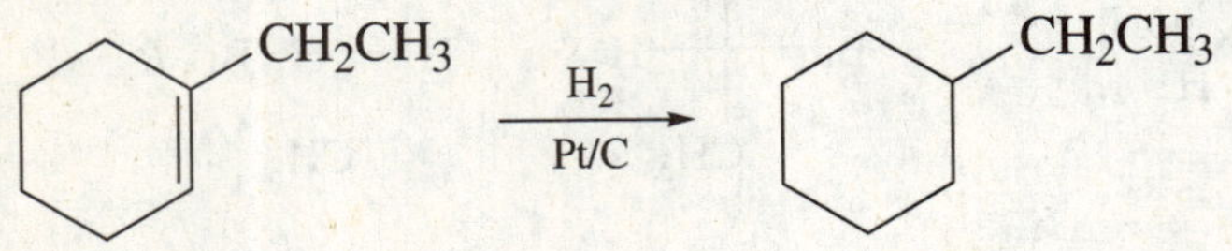

c.

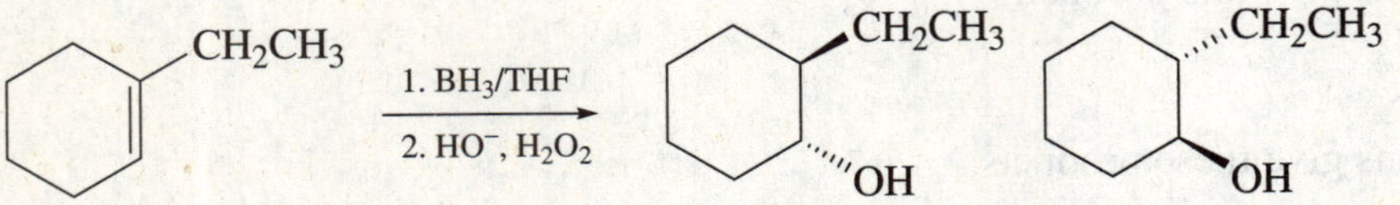

(the H and OH are added to the same side)

d.

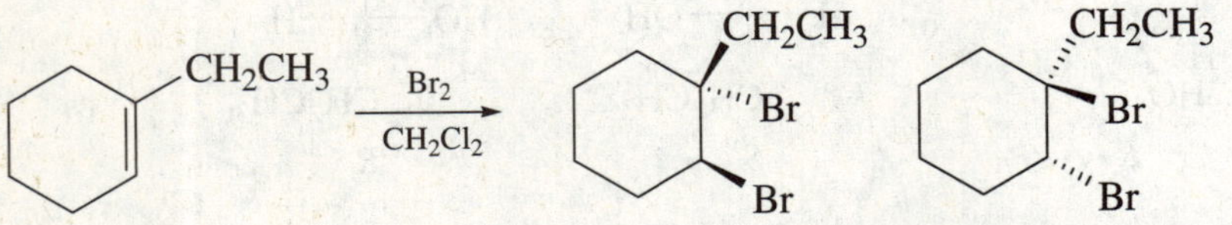

(the two Br's are added to opposite sides)

69. **a.** (*S*)-citric acid (OH has the highest priority; ^{14}C has a higher priority than ^{12}C)

b. The reaction is catalyzed by an enzyme. Only one stereoisomer is formed in an enzyme-catalyzed reaction because an enzyme has a chiral binding site which allows reagents to be delivered to only one side of the functional group of the compound.

c. The product of the reaction will be achiral because if it doesn't have a ^{14}C, the two CH_2COOH groups will be identical so it will not have an asymmetric center.

70. **a.**

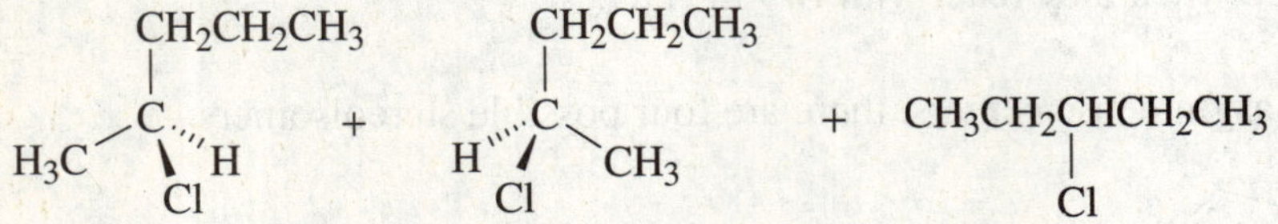

S *R*

or

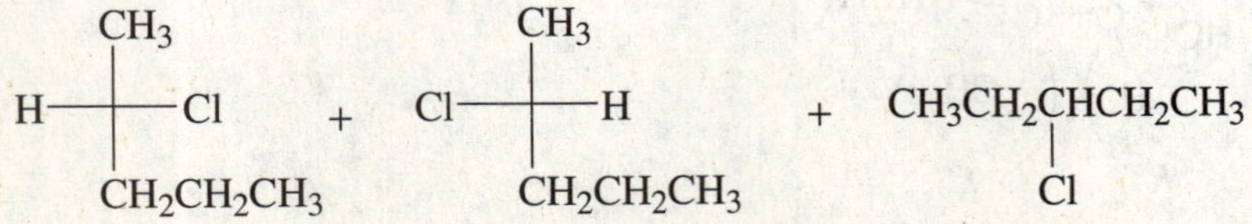

S *R*

b.

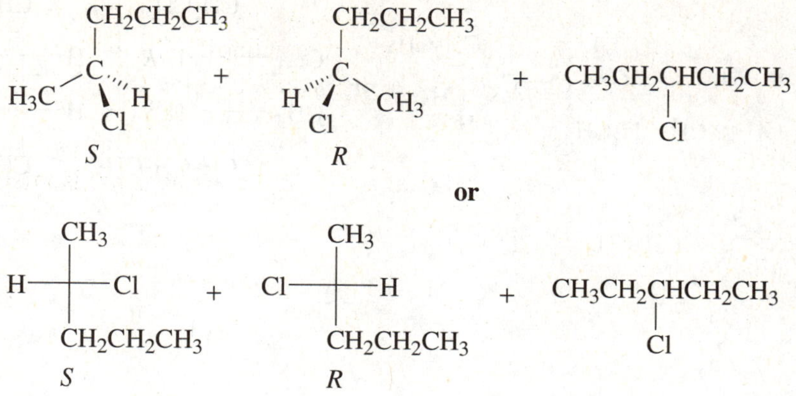

c.

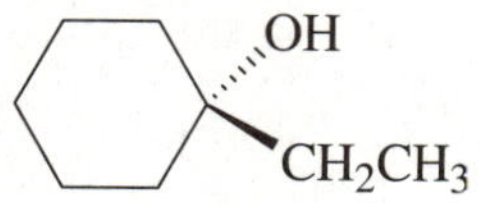

d.

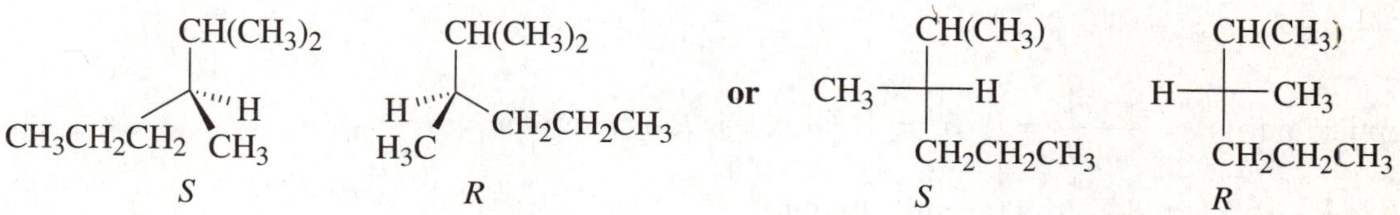

e.

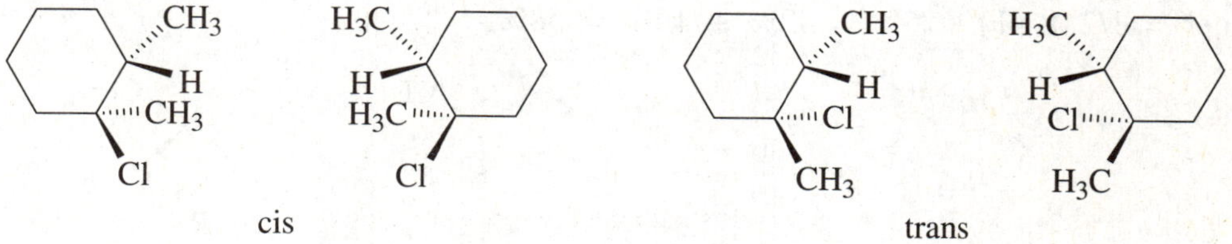

f.

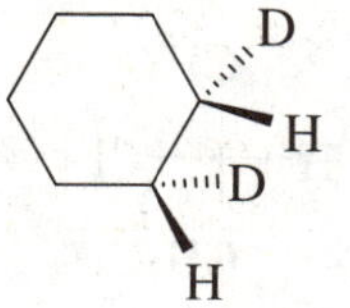

g.

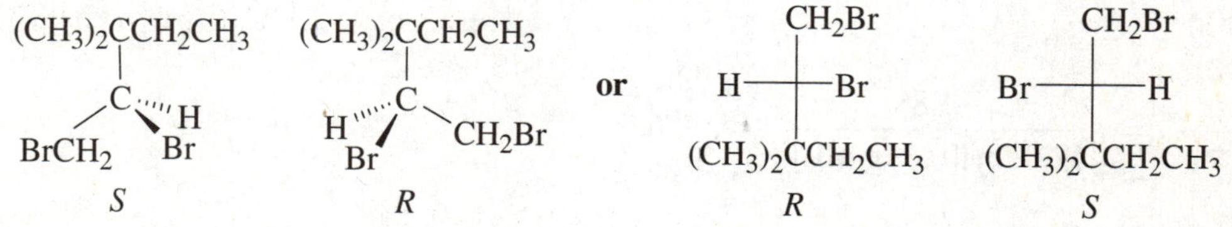

h.

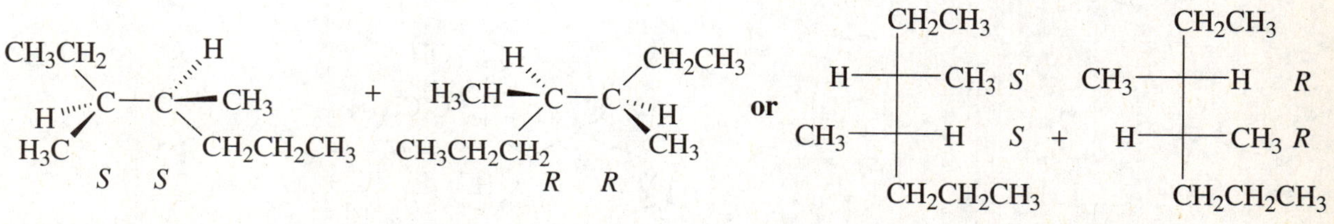

i. 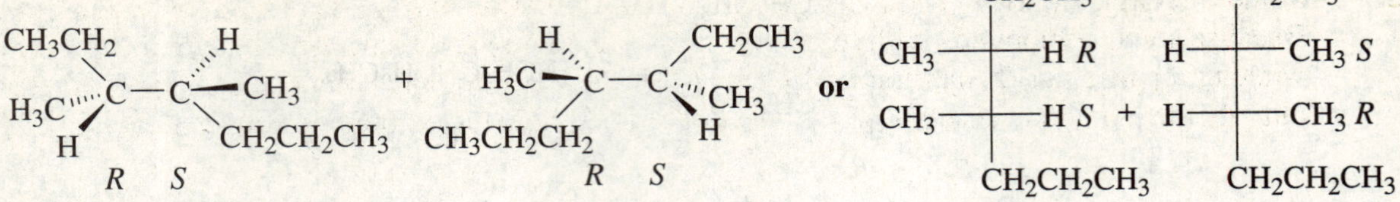

j. 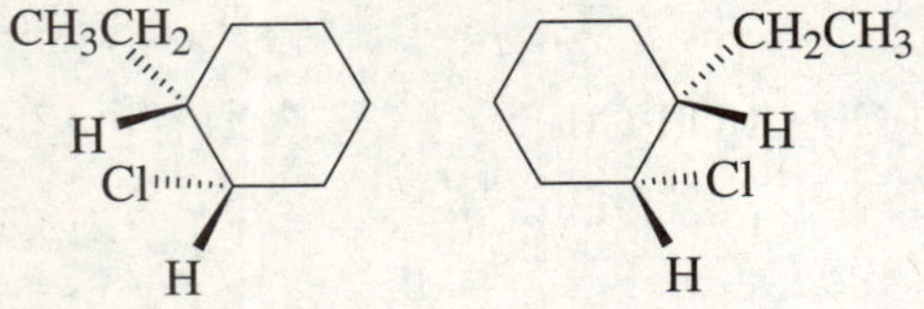

71.

a. identical
b. identical
c. enantiomers
d. constitutional isomers
e. constitutional isomers
f. diasereomers
g. constitutional isomers
h. enantiomers

72.

$$\text{optical purity} = \frac{+1.4}{+8.7} = .16 = 16\% \text{ excess } R \text{ enantiomer}$$

$$100\% - 16\% = 84\% \text{ is a racemic mixture}$$

$$R \text{ enantiomer} = 1/2\,(84\%) + 16\% = 42\% + 16\% = 58\%$$

73. **a.** *R* **b.** *S* **c.** 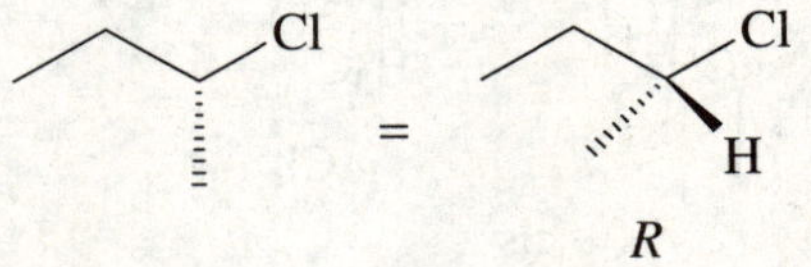

d. 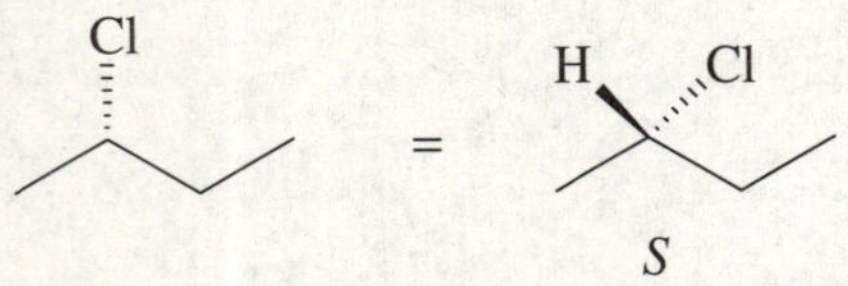

e. 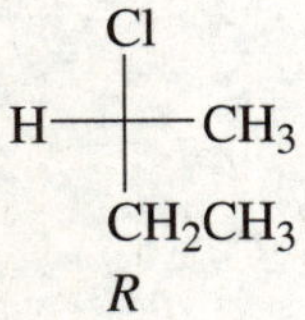

f. 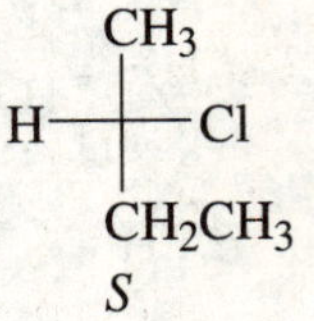

74.

$$[\alpha] = \frac{\alpha}{l \times c} = \frac{-1.8°}{[2.0\text{dm}][0.15\text{g/mL}]} = -6.0$$

75. Butaclamol has four asymmetric centers; three of them are carbons and one is a nitrogen.

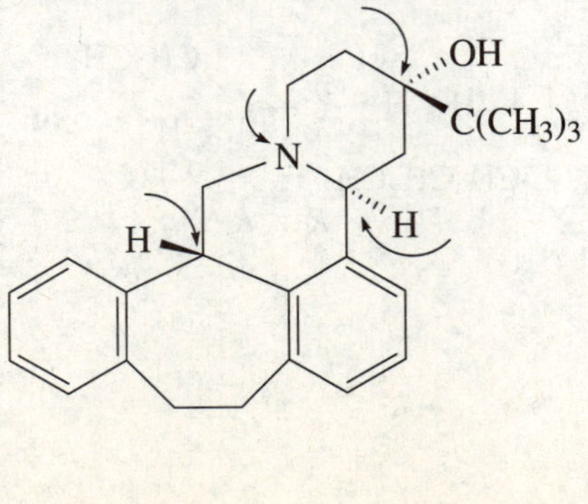

76. *R* and *S* are related to (+) and (–) in that if one configuration (say, *R*) is (+), the other one is (–). Because some compounds with the *R* configuration are (+) and some are (–), there is no way to determine whether a particular *R* enantiomer is (+) or (–) without putting the compound in a polarimeter or finding out whether someone else has previously determined how the compound rotates light.

77. **a.**

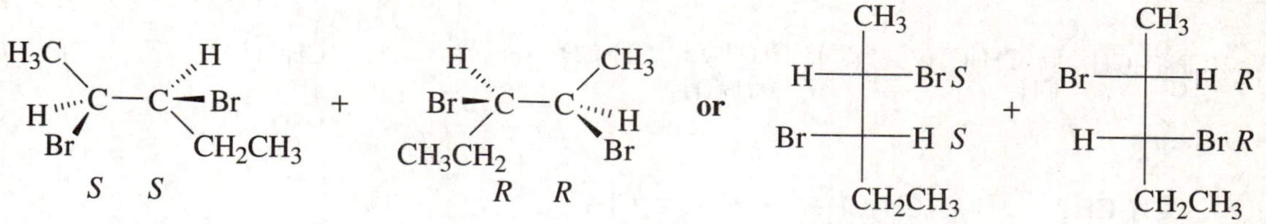

b.

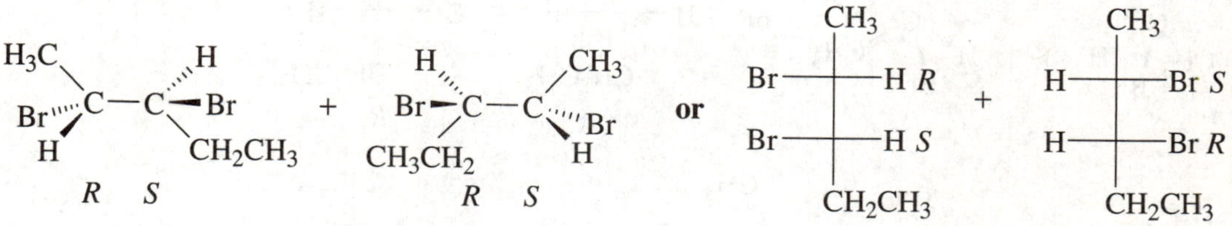

c.

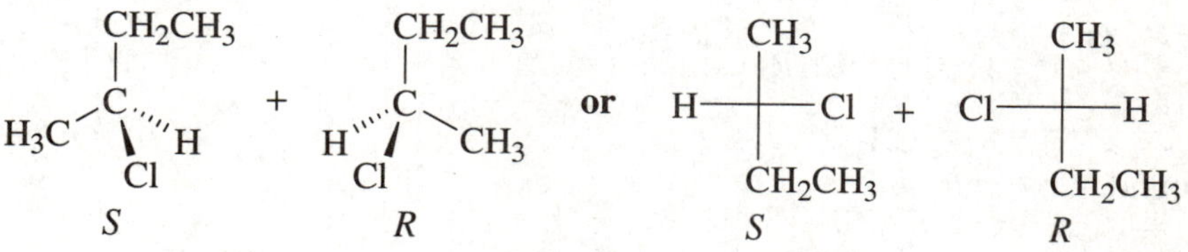

d.

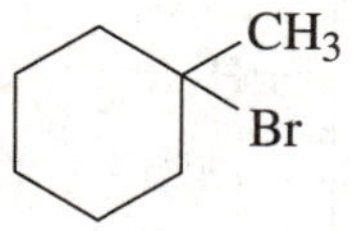

e.

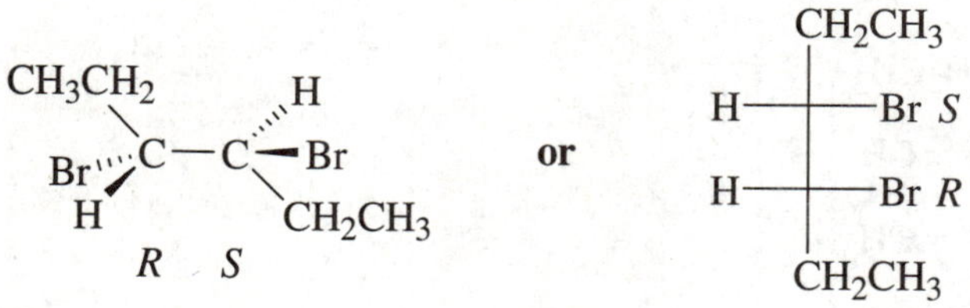

f.

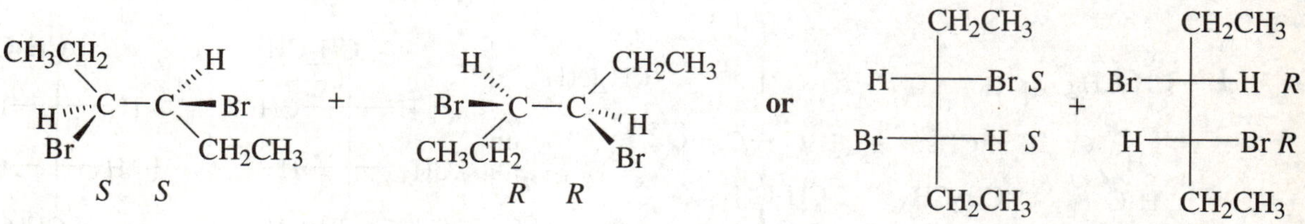

g. The initially formed carbocation is secondary. It undergoes a 1,2-methyl shift to form a tertiary carbocation that gives the products shown below.

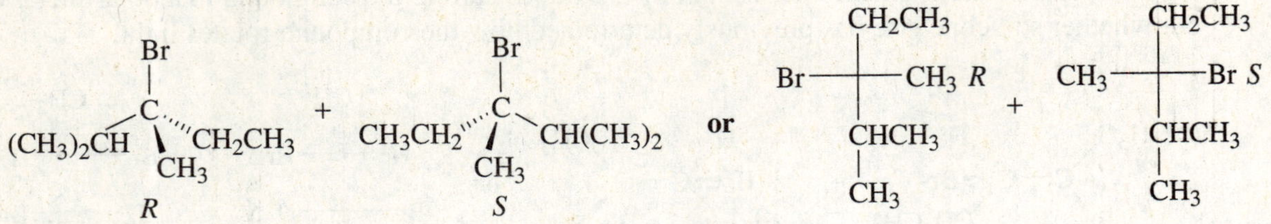

h.

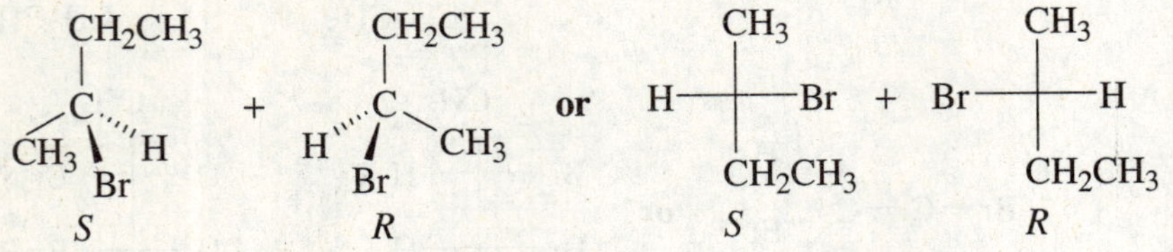

i.

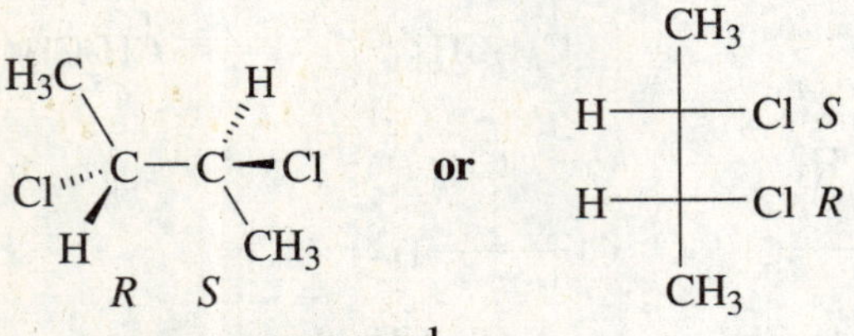

a meso compound

j.

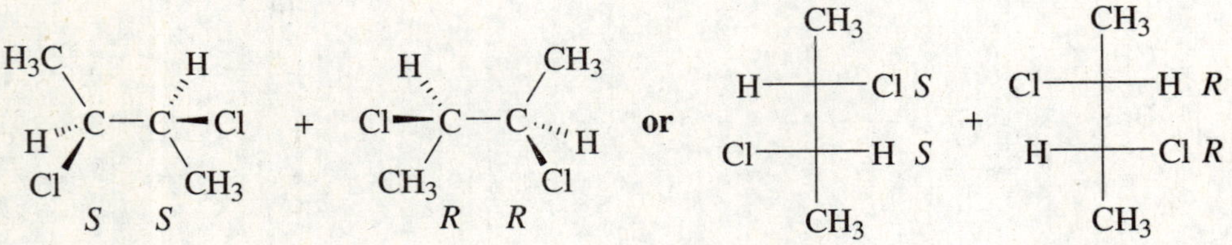

k.

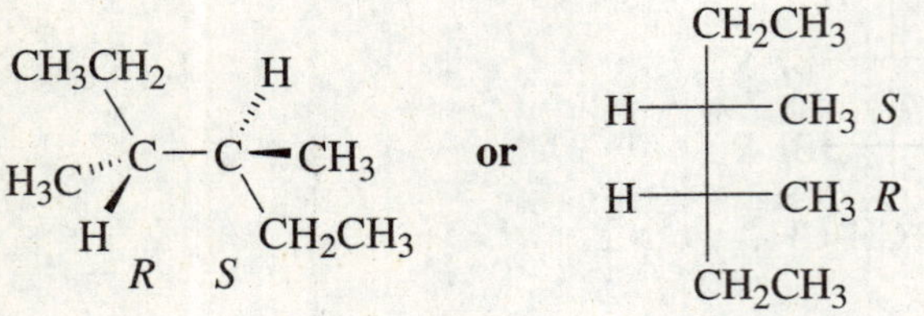

l.

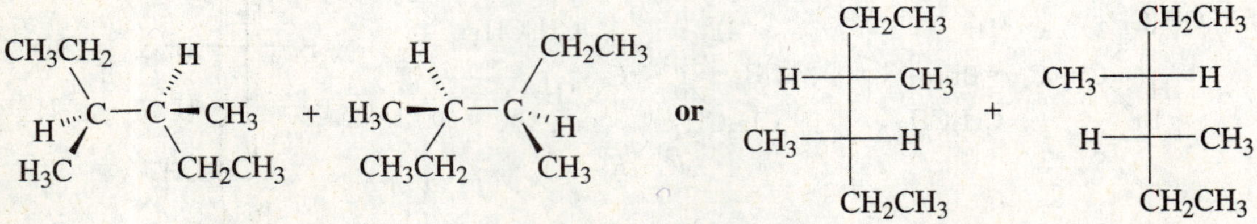

78. **a.** The compound has four stereoisomers.

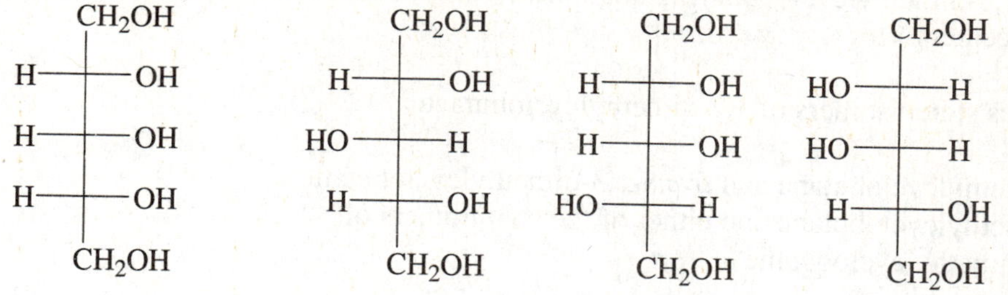

b. The first two stereoisomers are optically inactive because they are meso compounds. (They have a plane of symmetry.)

79. **a.**

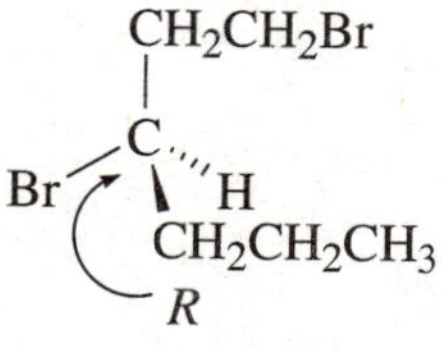

b.

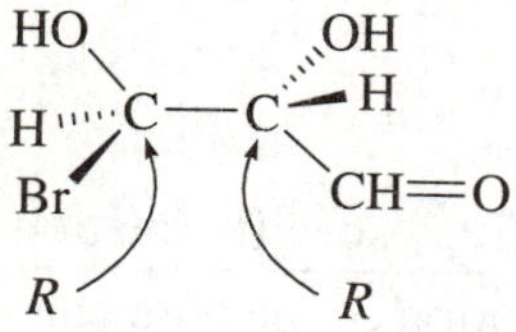

c.

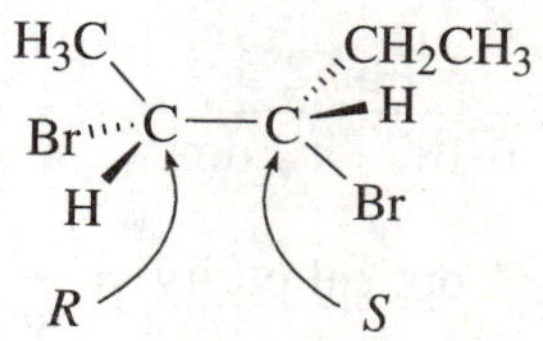

80. **a. and b.**

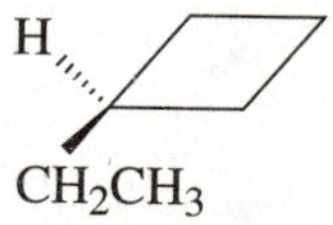

ethylcyclobutane

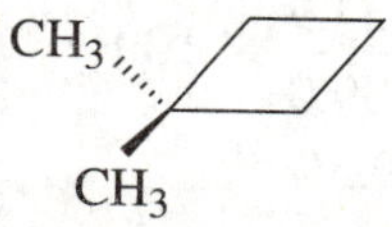

1,1-dimethylcyclobutane

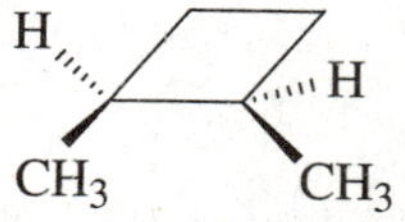

cis-1,2-dimethylcyclobutane

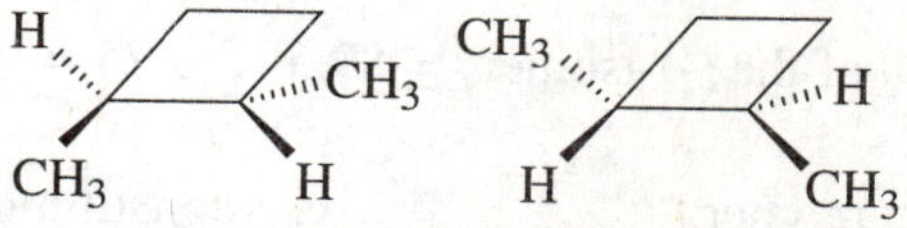

trans-1,2-dimethylcyclobutane

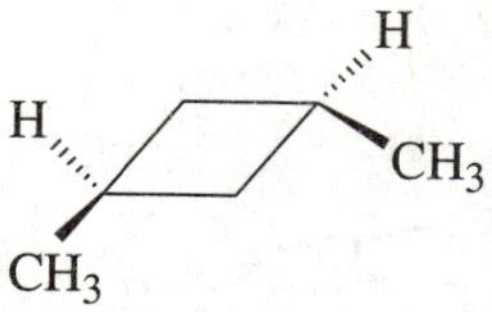

cis-1,3-dimethylcyclobutane

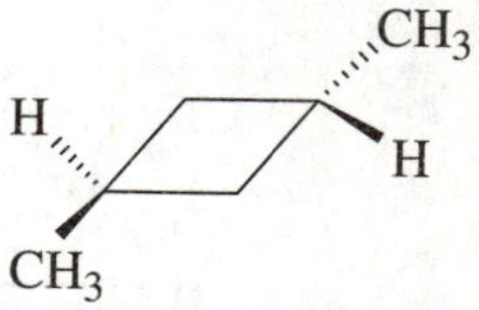

trans-1,3-dimethylcyclobutane

c. **1.** ethylcyclobutane
1,1-dimethylcyclobutane
1,2-dimethylcyclobutane
1,3-dimethylcyclobutane

2. the three isomers of 1,2-dimethylcyclobutane
the two isomers of 1,3-dimethylcyclobutane

3. *cis*-and *trans*-1,2-dimethylcyclobutane
cis-and *trans*-1,3-dimethylcyclobutane

4. the two trans stereoisomers of 1,2-dimethylcyclobutane

5. all the isomers except the two trans stereoisomers of 1,2-dimethylcyclobutane

6. *cis*-1,2-dimethylcyclobutane
(Note: *cis*-1,3-dimethylcyclobutane is not a meso compound, because it does not have any symmetric centers.)

7. the two trans stereoisomers of 1,2-dimethylcyclobutane

8. *cis*-1,3-dimethylcyclobutane and *trans*-1,3-dimethylcyclobutane
cis-1,2-dimethylcyclobutane and either of the enantiomers of *trans*-1,2-dimethylcyclobutane

81.

$$\text{observed specific rotation} = \frac{\text{observed rotation}}{\text{concentration} \times \text{length}}$$

$$\frac{-6.52°}{0.187 \times 1\text{ dm}} = -34.9$$

$$\%\text{ optical purity} = \frac{\text{observed specific rotation}}{\text{specific rotation of the pure enantiomer}} \times 100$$

$$= \frac{-34.9}{-39.0} \times 100$$

$$= 89.5\%$$

$$\%\text{ of the (+)-isomer} = \frac{100 - 89.5}{2} = 5.25\%$$

$$\%\text{ of the (-)-isomer} = 89.5 + 5.25 = 94.75\%$$

82. **a.** diastereomers
b. identical
c. constitutional isomers
d. diastereomers

83. **a.**

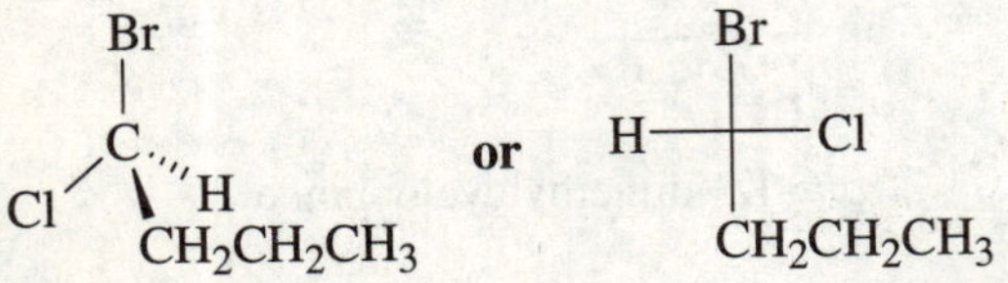

b.

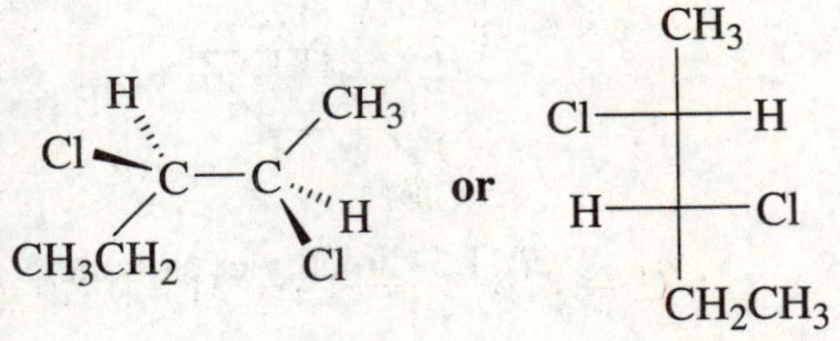

c.

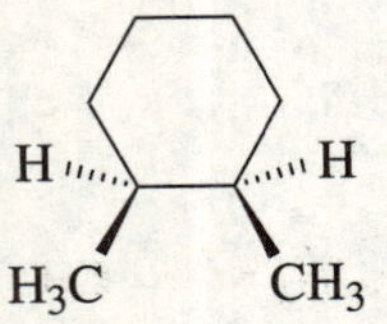

d.

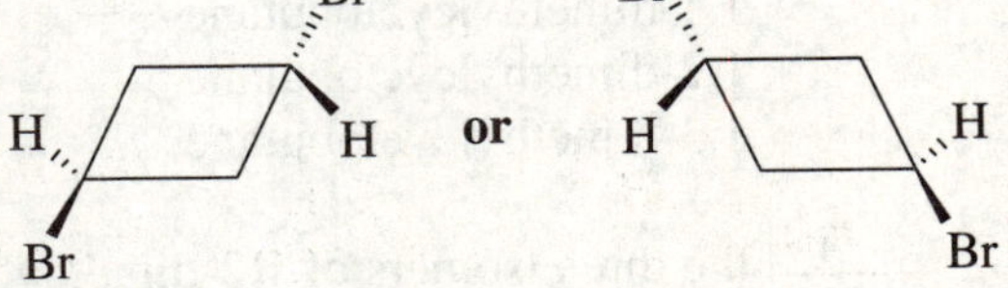

e.

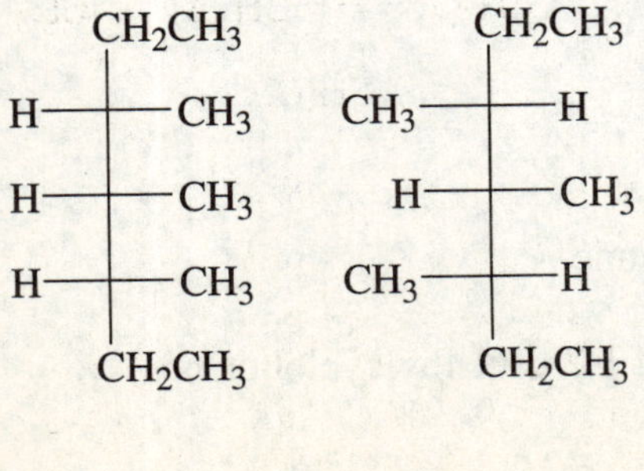

84. In the transition state for amine inversion, the nitrogen atom is sp^2 hybridized, which means it has bond angles of 120°. A nitrogen atom in a three-membered ring cannot achieve a 120° bond angle, so the amine inversion that would interconvert the enantiomers cannot occur. Therefore, the enantiomers can be separated.

85. The first product would not be formed, because none of the bonds attached to C-2 were broken during the reaction. Therefore, the configuration at C-2 cannot change.

86. The fact that the optical purity is 72% means that there is 72% enantiomeric excess of the *S* isomer and 28% racemic mixture. Therefore, the actual amount of the *S* isomer in the sample is 72% + 1/2(28%) = 86%. So the amount of the *R* isomer in the sample is 100% - 86% = 14%.

87. The compound is not optically active because it has a center of symmetry.
A center of symmetry is a point, and if a line is drawn to this point from an atom or group and the line then extended an equal distance beyond the point, the line would touch an identical atom or group.

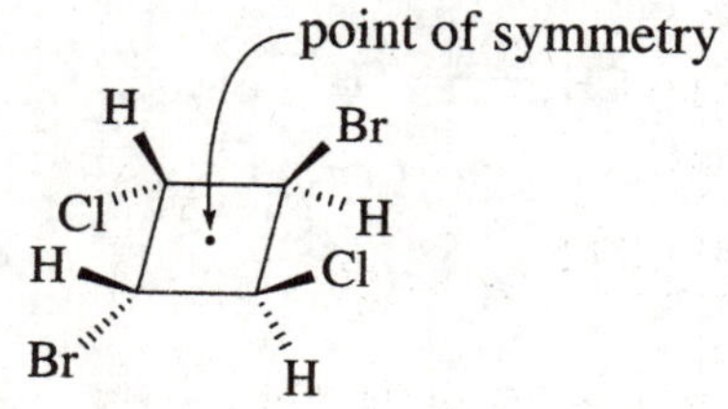

88.

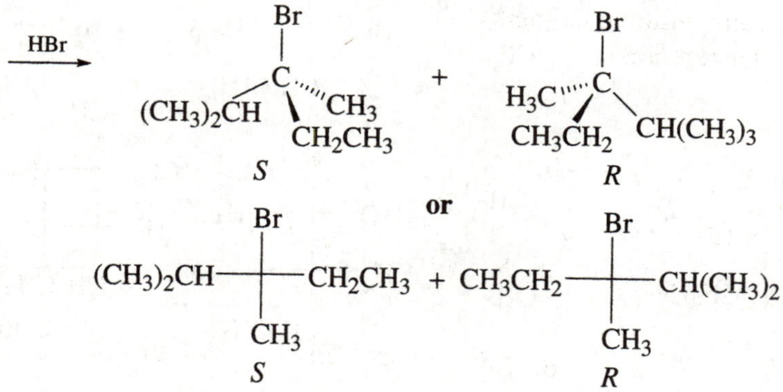

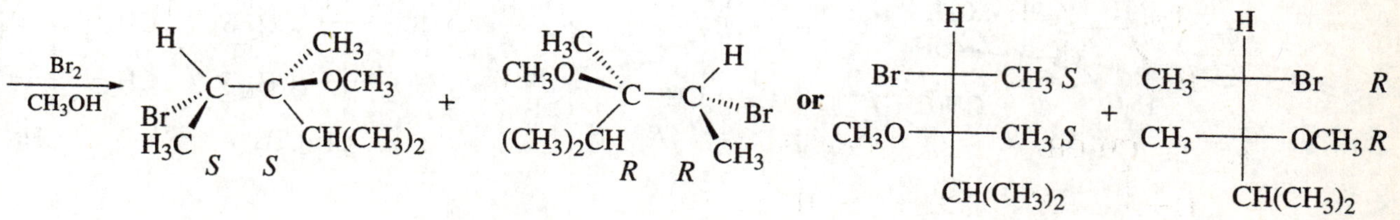

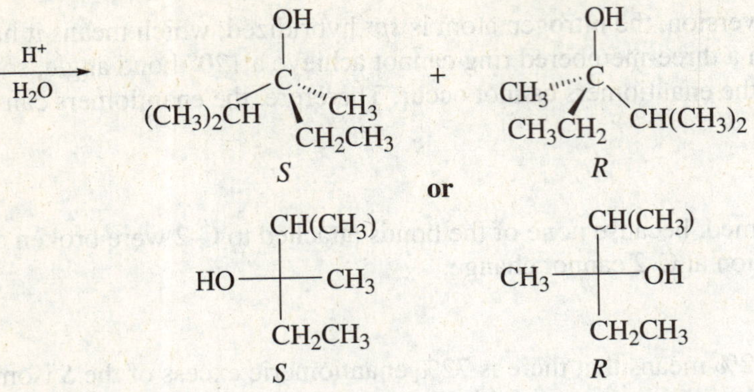
H+
H2O
or

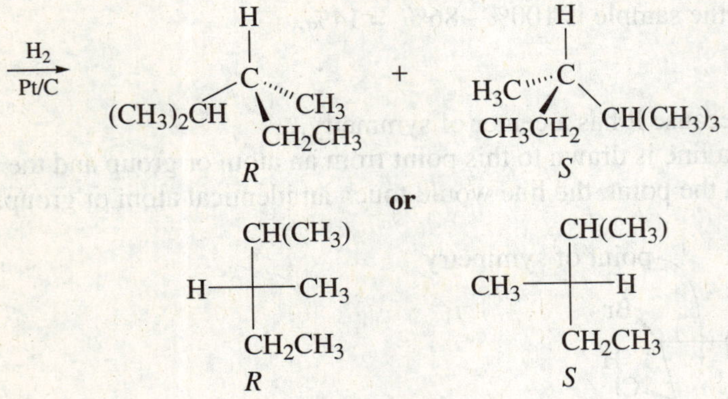
H2
Pt/C
or

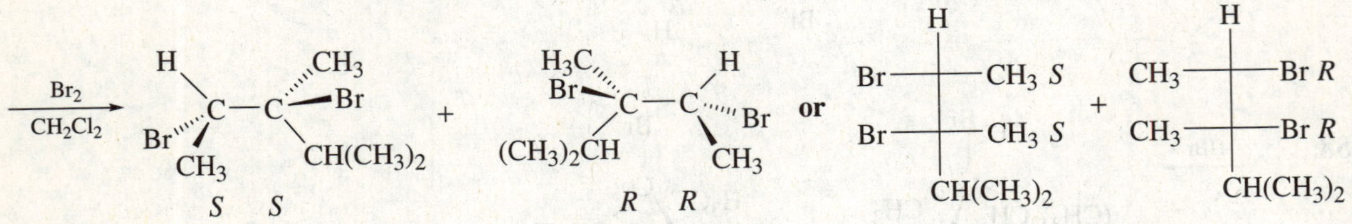
Br2
CH2Cl2
or

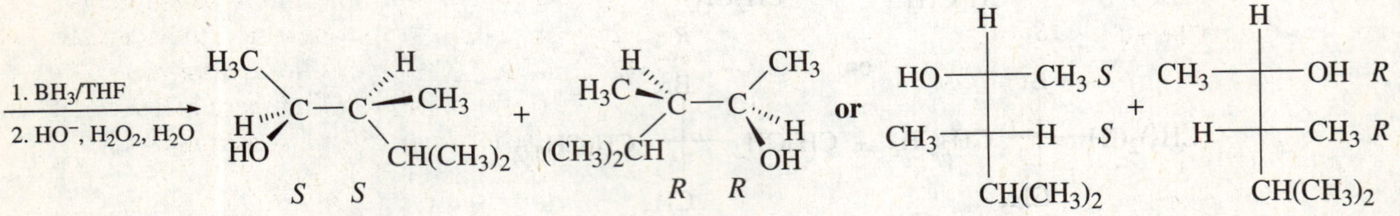
1. BH3/THF
2. HO−, H2O2, H2O
or

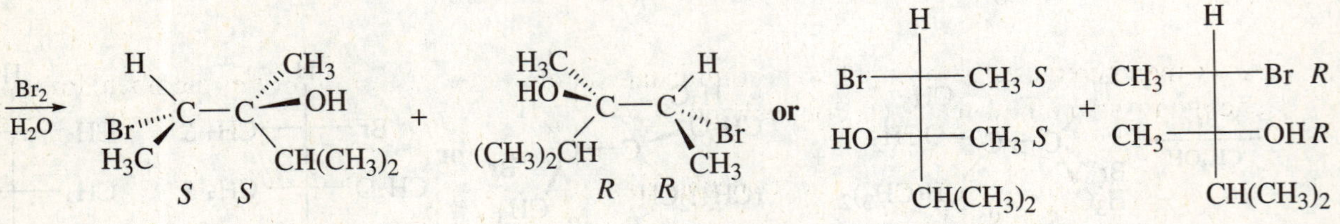
Br2
H2O
or

89. **a.**

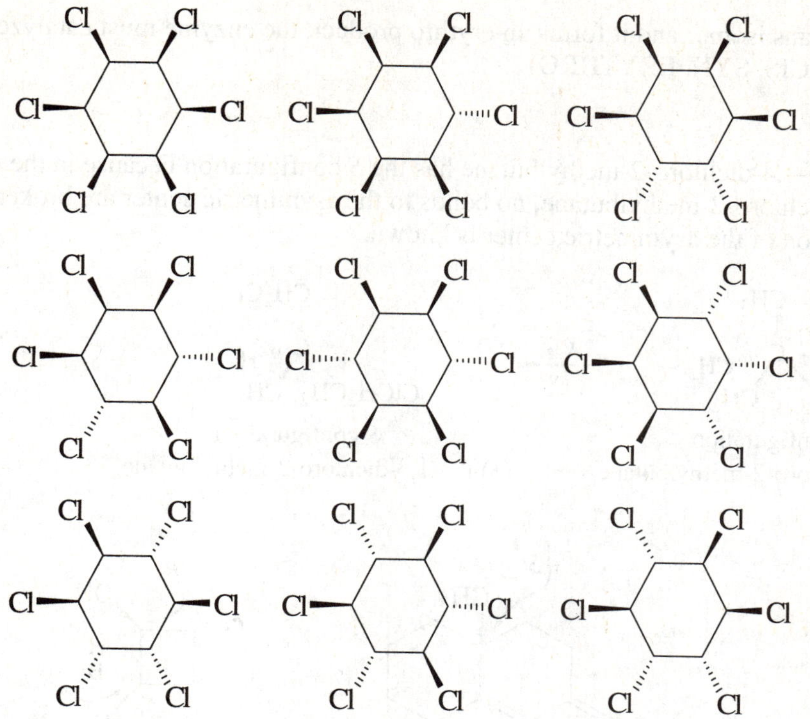

b.

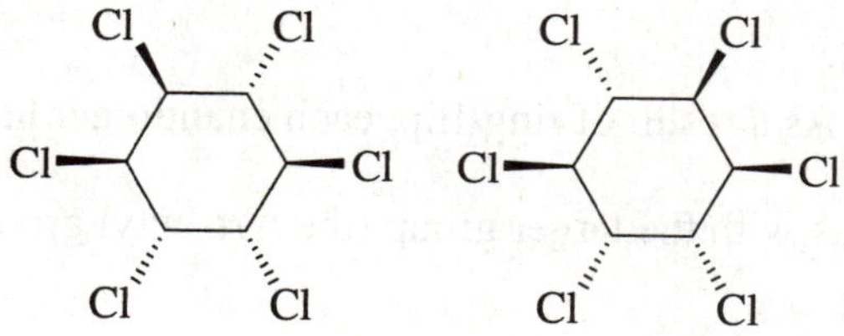

This is a pair of enantiomers because they are nonsuperimposable mirror images.

c.

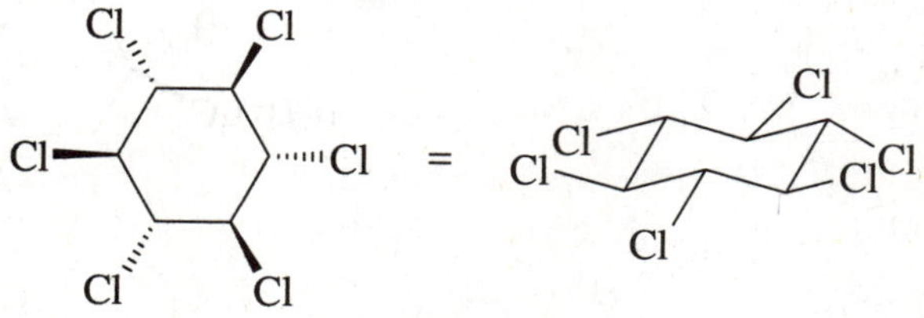

This is the most stable isomer because, since the chloro substituents are all trans to each other, they can all be in the more stable equatorial position. (Recall that there is less steric strain when a substituent is in the equatorial position.)

90. Yes.

91. A six-membered ring is too small to accommodate a trans double bond, but a ten-membered ring is large enough to accommodate a trans double bond.

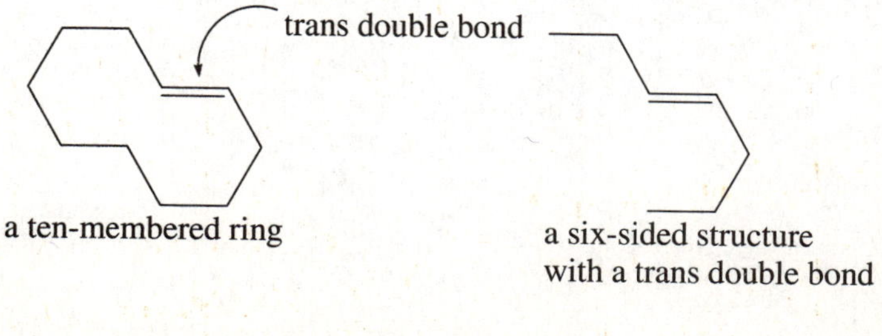

92. Because fumarate is the trans isomer and it forms an erythro product, the enzyme must catalyze an anti addition of D_2O. (Recall: CIS, SYN, ERYTHRO)

93. We can determine that (–)-1,4-dichloro-2-methylbutane has the *S* configuration because in the process of forming it from (*S*)-(+)-1-chloro-2-methylbutane, no bonds to the asymmetric center are broken. Therefore, the configuration of the asymmetric center is known.

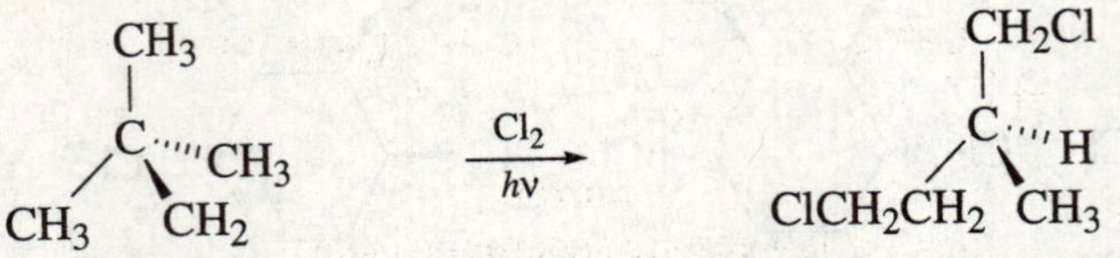

S-configuration
(*S*)-(+)-1-chloro-2-methylbutane

S-configuration
(*S*)-(–)-1,4-dichloro-2-methylbutane

94. **a.**

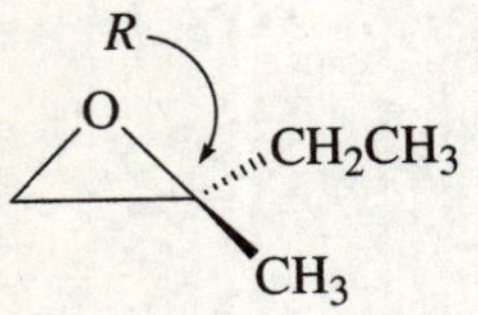

b.

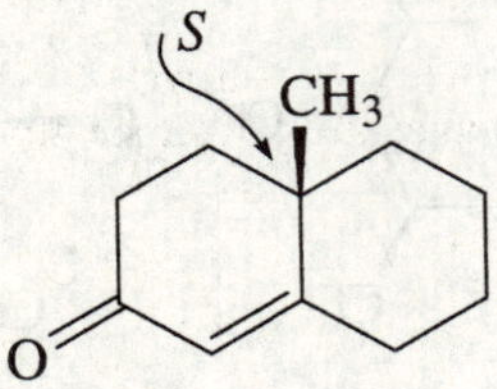

c.

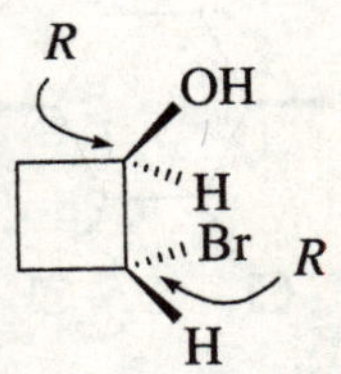

95. The trans compound exits as a pair of enantiomers. As a result of ring-flip, each enantiomer has two chair conformations.
In each case, the more stable conformation is the one with the larger group (the *tert*-butyl group) in the equatorial position.

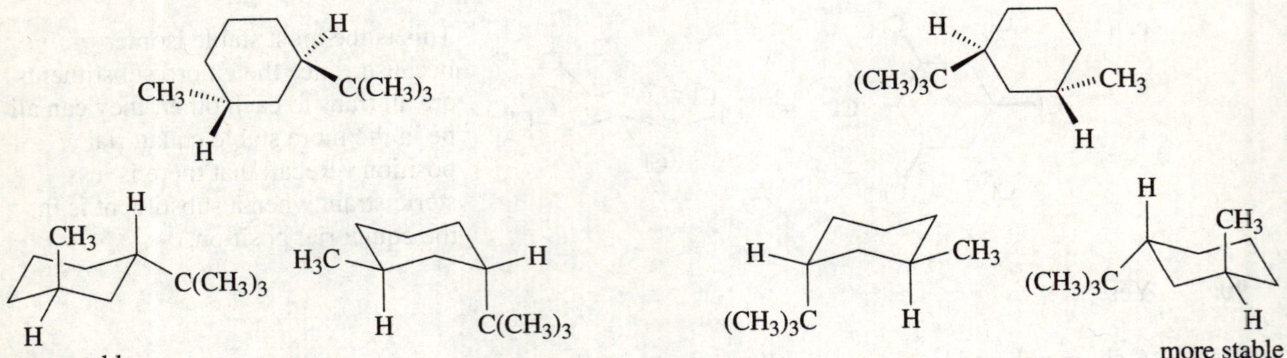

more stable

more stable

96. **a.** The compounds do not have any asymmetric centers.

b. **1.** is not chiral.

2. is chiral. Because of its unusual geometry, it is a chiral molecule, even though it does not have any asymmetric centers, because it cannot be superimposed on its mirror image. This will be easier to understand if you build models.

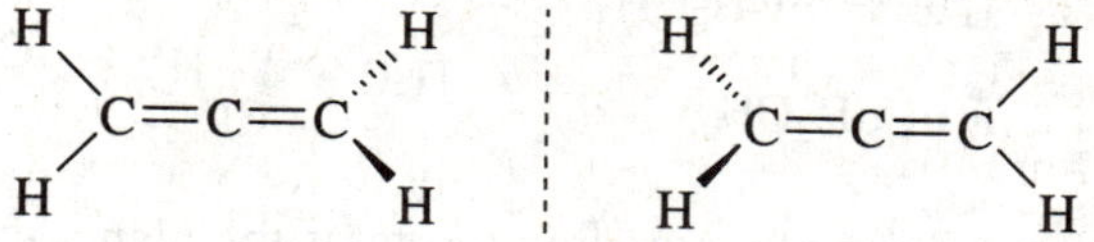

mirror
images are superimposable

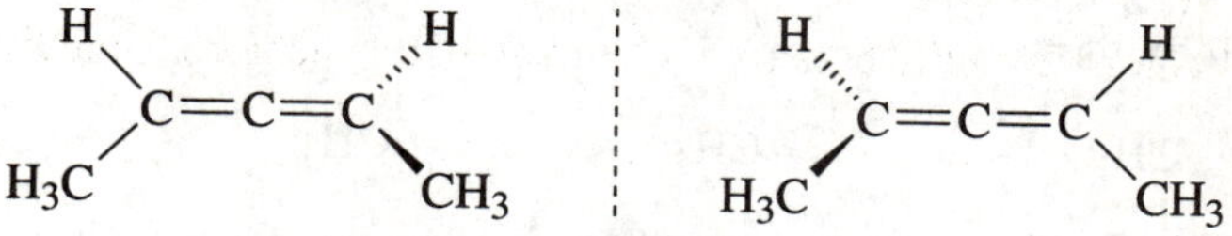

mirror
images are not superimposable

Chapter 5 Practice Test

1. Are the following compounds identical or a pair of enantiomers?

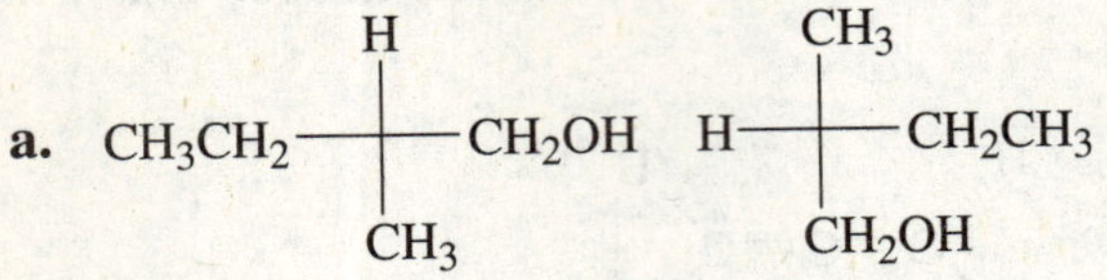

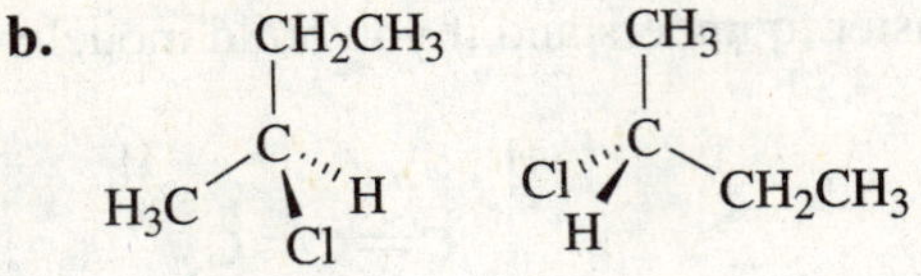

2. 30 mL of a solution containing 2.4 g of a compound rotates the plane of polarized light –0.48° in a polarimeter with a 2 decimeter sample tube. What is the specific rotation of the compound?

3. Which are meso compounds?

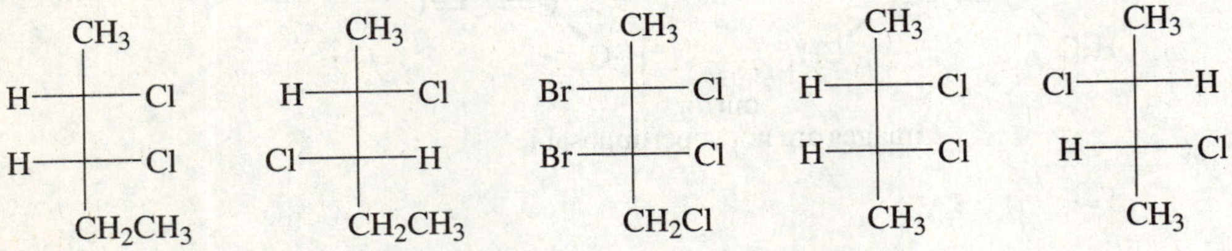

4. Draw the constitutional isomers with molecular formula C_4H_9Cl.

5. Draw the stereoisomers of:

a. 2,3-dibromobutane **b.** 2,3-dibromopentane

6. Give the stereochemistry of the products that would be obtained from each of the following reactions:

a. 1-butene + HCl **c.** *trans*-3-hexene + Br_2

b. 2-pentene + HBr **d.** *trans*-3-heptene + Br_2

7. *R*-(–)-2-methyl-1-butanol can be oxidized to (+)-2-methylbutanoic acid without breaking any of the bonds to the asymmetric center. What is the configuration of (–)-2-methylbutanoic acid?

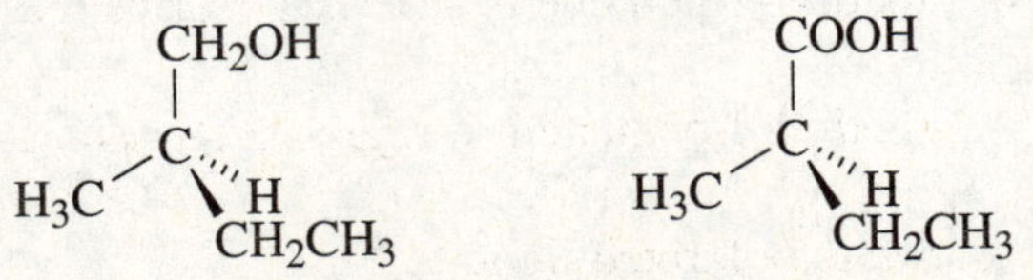

(*R*)-(–)-2-methyl-1-butanol (+)-2-methylbutanoic acid

8. What stereoisomers would be obtained from each of the following reactions?

a. 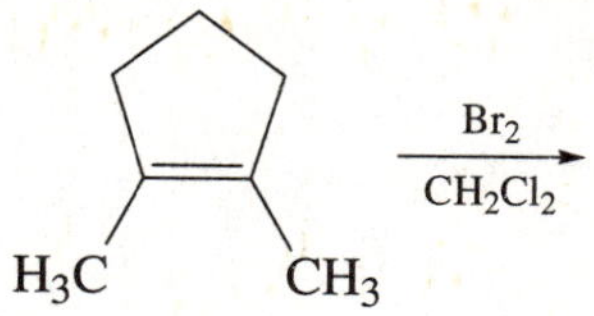

c. 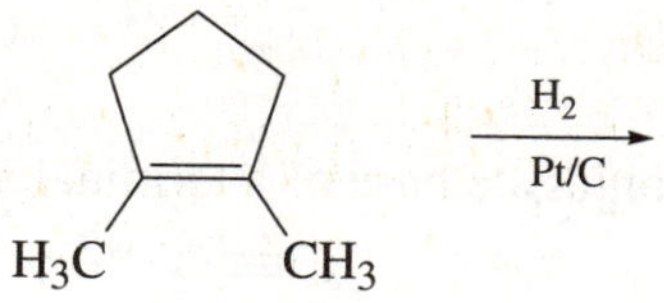

b. 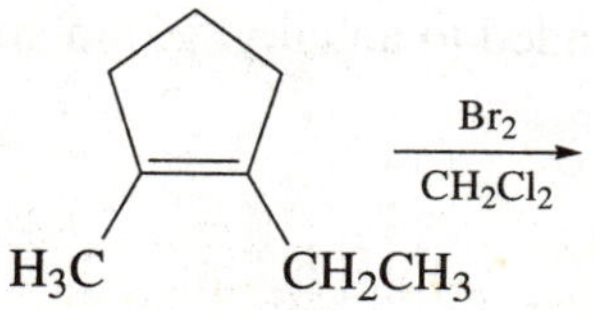

d. 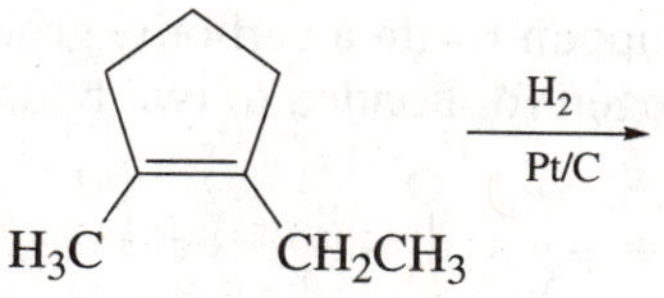

9. Which of the following have the *R* configuration?

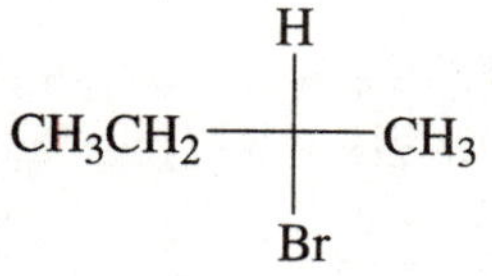
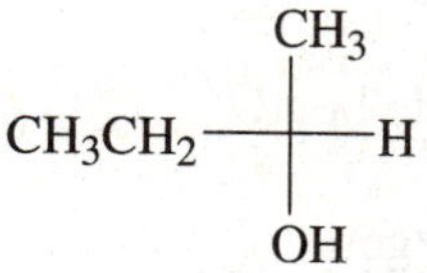
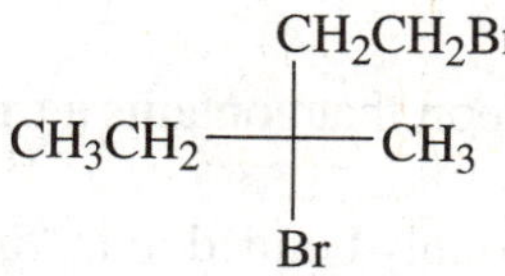
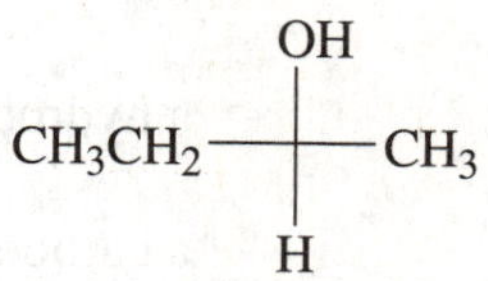

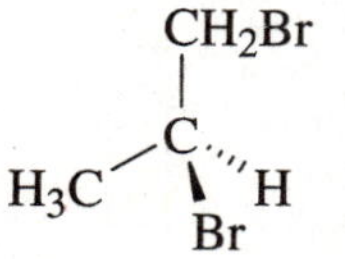
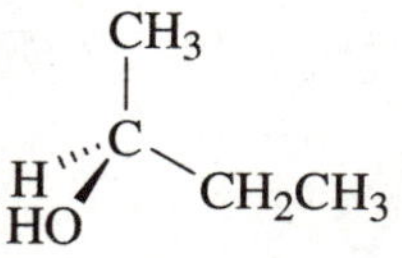
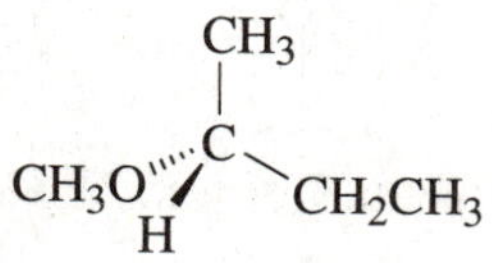
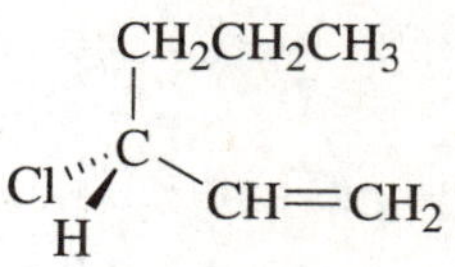

10. Draw a diastereomer of the following compound:

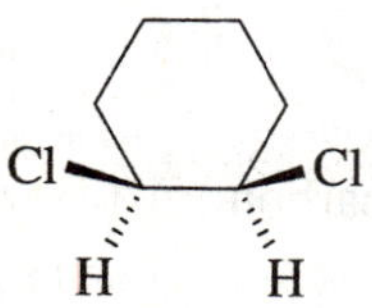

11. Indicate whether each of the following statements is true or false:

a. Diastereomers have the same melting points. T F

b. The addition of HBr to 3-methyl-2-pentene is a stereospecific reaction. T F

c. The addition of HBr to 3-methyl-2-pentene is a stereoselective reaction. T F

d. The addition of HBr to 3-methyl-2-pentene is a regioselective reaction. T F

e. Meso compounds do not rotate polarized light. T F

f. 2,3-Dichloropentane has a stereoisomer that is a meso compound. T F

g. All chiral compounds with the *R* configuration are dextrorotatory. T F

h. A compound with three asymmetric centers can have a maximum of nine stereoisomers. T F

CHAPTER 6
The Reactions of Alkynes • An Introduction to Multistep Synthesis

Important Terms

acetylide ion — the conjugate base of a terminal alkyne

$$RC{\equiv}C^-$$

aldehyde — a compound with a carbonyl group that is bonded to an alkyl group and to a hydrogen (or bonded to two hydrogens).

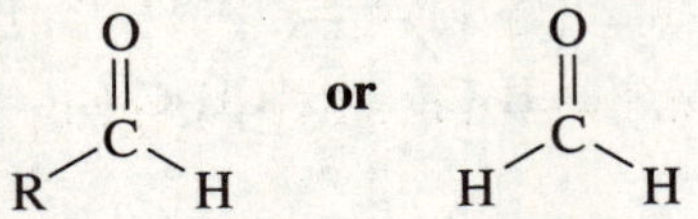

alkylation reaction — a reaction that adds an alkyl group to a reactant.

alkyne — a hydrocarbon that contains a triple bond.

carbonyl group — a carbon doubly bonded to an oxygen.

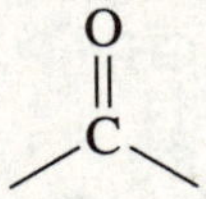

enol — an α,β-unsaturated alcohol.

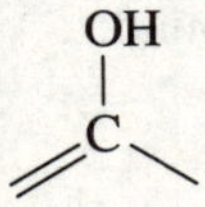

geminal dihalide — a compound with two halogen atoms bonded to the same carbon.

internal alkyne — an alkyne with its triple bond not at the end of the carbon chain.

keto-enol tautomers — a ketone and its isomeric α,β-unsaturated alcohol.

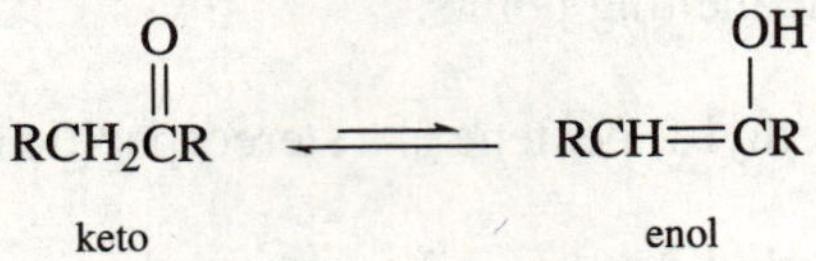

ketone — a compound with a carbonyl group that is bonded to two alkyl groups.

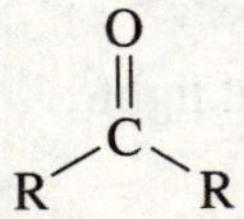

π-complex — a complex formed between an electrophile and a triple bond.

polymer — a large molecule made by linking up many small molecules called monomers.

radical anion — a species with a negative charge and an unpaired electron.

retrosynthetic analysis or retrosynthesis	working backward (on paper) from a target molecule to available starting materials.
tautomerization	interconversion of tautomers.
tautomers	isomers that are in rapid equilibrium. The keto and enol tautomers differ in the location of a double bond and a hydrogen.
terminal alkyne	an alkyne with its triple bond at the end of the carbon chain.
vinylic cation	a compound with a positive charge on a vinylic carbon.
vinylic radical	a compound with an unpaired electron on a vinylic carbon.

Solutions to Problems

1. The molecular formula of a noncyclic hydrocarbon with 14 carbons and no π bonds is $C_{14}H_{30}$ (C_nH_{2n+2}).
Because a compound has two fewer hydrogens for every ring and π bond, a compound with one ring and 4 π bonds (2 triple bonds) would have 10 fewer hydrogens than the C_nH_{2n+2} formula. Thus, the molecular formula is $\mathbf{C_{14}H_{20}}$.

2. **a.** $ClCH_2CH_2C{\equiv}CCH_2CH_3$

b. 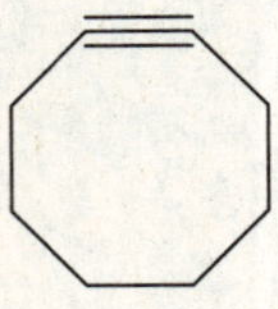

c. $CH_3CHC{\equiv}CH$ (with CH_3 on the CH)

d. $HC{\equiv}CCH_2Cl$

e. $HC{\equiv}CCH_2CCH_3$ (with two CH_3 groups on the C)

f. $CH_3C{\equiv}CCH_3$

3. To answer this question, you need to have the information in Section 6.2.
a. 1-bromo-1,3-pentadiene **b.** 1-hepten-5-yne **c.** 4-hepten-1-yne

4. $HC{\equiv}CCH_2CH_2CH_2CH_3$
1-hexyne
butylacetylene

$CH_3C{\equiv}CCH_2CH_2CH_3$
2-hexyne
methylpropylacetylene

$CH_3CH_2C{\equiv}CCH_2CH_3$
3-hexyne
diethylacetylene

$CH_3CH_2CHC{\equiv}CH$ (with CH_3 on the CH)
3-methyl-1-pentyne
sec-butylacetylene

$CH_3CHCH_2C{\equiv}CH$ (with CH_3 on the CH)
4-methyl-1-pentyne
isobutylacetylene

$CH_3CHC{\equiv}CCH_3$ (with CH_3 on the CH)
4-methyl-2-pentyne
isopropylmethylacetylene

$CH_3CC{\equiv}CH$ (with two CH_3 groups on the C)
3,3-dimethyl-1-butyne
tert-butylacetylene

5. **a.** 5-bromo-2-pentyne
b. 6-bromo-2-chloro-4-octyne
c. 1-methoxy-2-pentyne
d. 3-ethyl-1-hexyne

6. **a.** 1-hepten-4-yne
b. 4-methyl-1,4-hexadiene
c. 5-vinyl-5-octen-1-yne
d. 3-butyn-1-ol
e. 1,3,5-heptatriene
f. 2,4-dimethyl-4-hexen-1-ol

7. pentane 1-pentene 1-pentyne

8. **a.** *sp*²–*sp*²　**b.** *sp*²–*sp*³　**c.** *sp*–*sp*²　**d.** *sp*–*sp*³　**e.** *sp*–*sp*　**f.** *sp*²–*sp*²　**g.** *sp*²–*sp*³　**h.** *sp*–*sp*³　**i.** *sp*²–*sp*

9. The less stable reactant will be the more reactive reactant, if the less stable reactant has the more stable transition state, **or** if the less stable reactant has the less stable transition state **and** the difference in the stabilities of the reactants is greater than the difference in the stabilities of the transition states.

10. **a.** 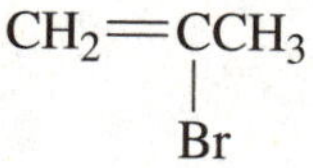　**d.** 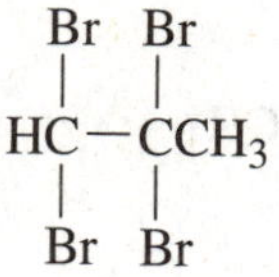

b. 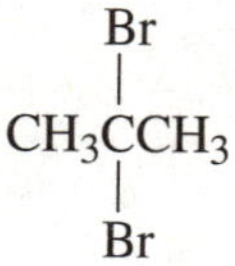　**e.** 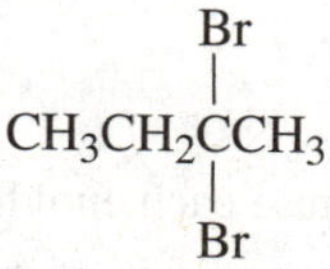

c. 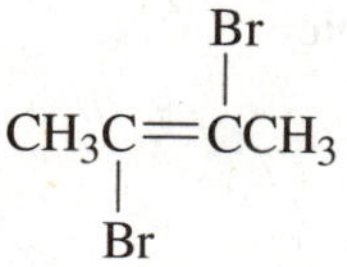　**f.** 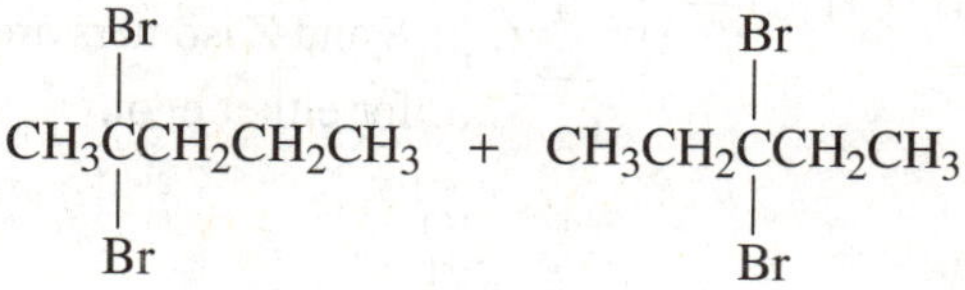

11. **a.** 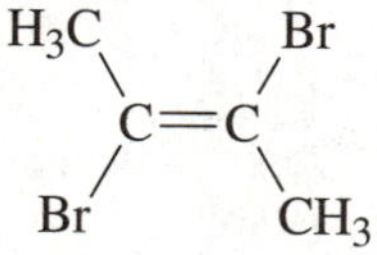

Only anti addition occurs, because the intermediate is a cyclic bromonium ion. Therefore the product has the trans configuration.

12. Because the alkyne is not symmetrical, two ketones will be obtained.

 and 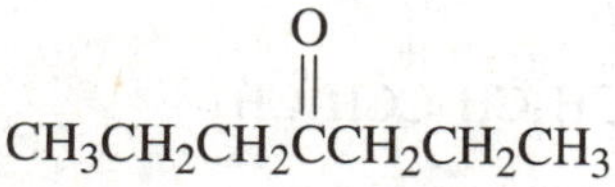

13. **a.** $CH_3C{\equiv}CH$　**b.** $CH_3CH_2C{\equiv}CCH_2CH_3$　**c.** 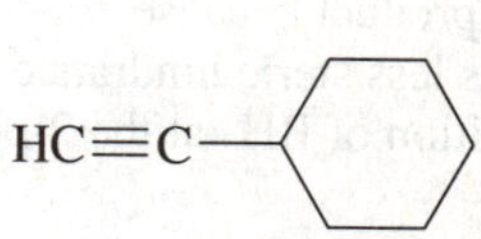

The best answer for "**b**" is 3-hexyne, because it would form only the desired ketone. 2-Hexyne would form two different ketones, so only half of the product would be the desired ketone.

14. **a.** 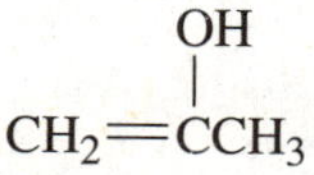　Because the ketone has identical substituents bonded to the carbonyl carbon, it has only one enol tautomer.

b. $CH_3CH{=}\overset{OH}{\overset{|}{C}}CH_2CH_2CH_3$ and $CH_3CH_2\overset{OH}{\overset{|}{C}}{=}CHCH_2CH_3$ *E* and *Z* isomers are possible for each of these enols.

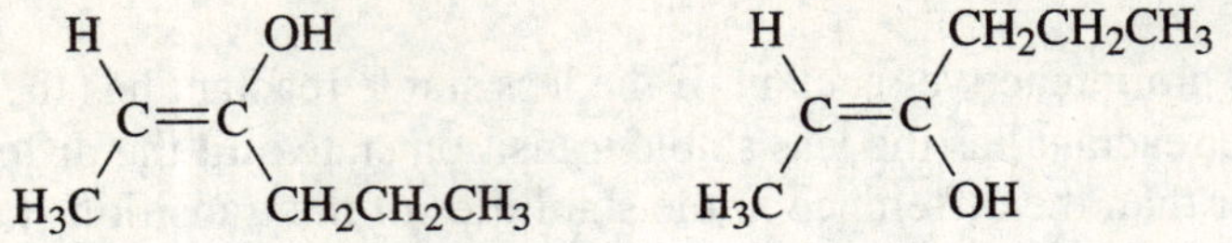

E *Z*

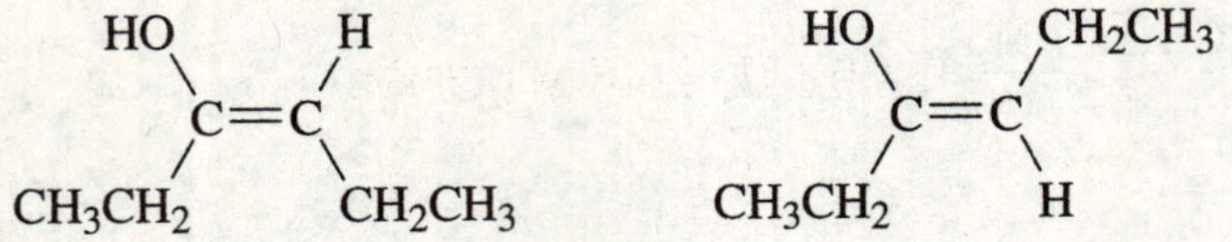

E *Z*

c. 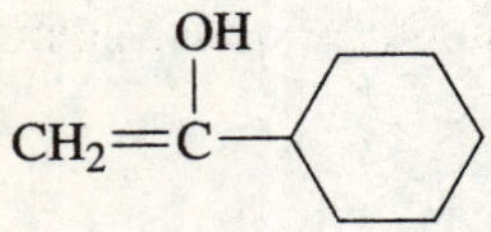 and 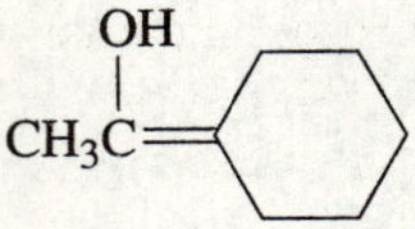 Because each enol has identical groups on one of its sp^2 carbons, *E* and *Z* isomers are not possible for either enol.

15. a. (1) $CH_3CH_2\overset{O}{\overset{\|}{C}}CH_3$ b. (1) $CH_3CH_2\overset{O}{\overset{\|}{C}}CH_3$

(2) $CH_3CH_2CH_2\overset{O}{\overset{\|}{C}}H$ (2) $CH_3CH_2\overset{O}{\overset{\|}{C}}CH_3$

c. (1) $CH_3CH_2CH_2\overset{O}{\overset{\|}{C}}CH_3$ and $CH_3CH_2\overset{O}{\overset{\|}{C}}CH_2CH_3$

(2) $CH_3CH_2CH_2\overset{O}{\overset{\|}{C}}CH_3$ and $CH_3CH_2\overset{O}{\overset{\|}{C}}CH_2CH_3$

major product because there is less steric hindrance to addition of BH_3 at the 2-position

16. Ethyne (acetylene)
An alkyne can form an aldehyde only if the OH group adds to a terminal *sp* carbon. When water adds to a terminal alkyne, the proton adds to the terminal *sp* carbon. Therefore, the only way the OH group can add to a terminal *sp* carbon is if there are two terminal *sp* carbons in the alkyne. In other words, the alkyne must be ethyne.

17. a. $CH_3CH_2CH_2C{\equiv}CH$ or $CH_3CH_2C{\equiv}CCH_3$ $\xrightarrow[Pt/C]{H_2}$ $CH_3CH_2CH_2CH_2CH_3$

1-pentyne 2-pentyne

b. $CH_3C{\equiv}CCH_3 \xrightarrow[\text{Lindlar catalyst}]{H_2}$ 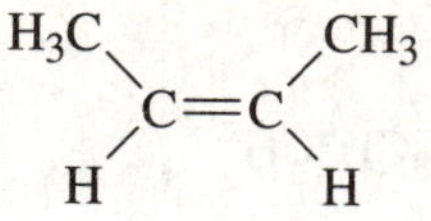

2-butyne

c. $CH_3CH_2C{\equiv}CCH_3 \xrightarrow[NH_3\ \text{liq}]{Na}$ 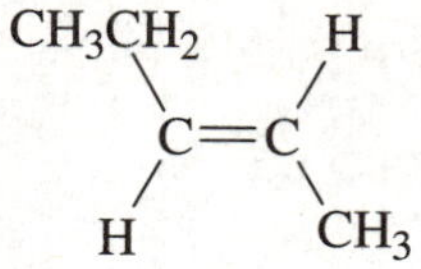

2-pentyne

d. $CH_3CH_2CH_2CH_2C{\equiv}CH \xrightarrow[\text{Lindlar catalyst}]{H_2}$ **or** $\xrightarrow[NH_3]{Na}$ $CH_3CH_2CH_2CH_2CH{=}CH_2$

1-hexyne

18. The base used to remove a proton must be stronger than the base that is formed as a result of proton removal. A terminal alkyne has a pK_a ~ 25. Therefore, the base used to remove a proton from a terminal alkyne must be a stronger base than the terminal alkyne. In other words, any base whose conjugate acid has a pK_a greater than 25 can be used.

19. The electronegativities of carbon atoms decrease in the order: $sp > sp^2 > sp^3$.
The more electronegative the carbon atom, the less stable it will be with a positive charge.

a. $CH_3\overset{+}{C}H_2$ **b.** $H_2C{=}\overset{+}{C}H$

20. The reaction will not favor products because the carbanion that would be formed is a stronger base than the amide ion. (Recall that the equilibrium favors reaction of the strong and formation of the weak; Section 1.19)

$$CH_3CH_3 + {}^-NH_2 \rightleftharpoons CH_3\bar{C}H_2 + NH_3$$

CH_3CH_3	$^-NH_2$	$CH_3\bar{C}H_2$	NH_3
pK_a > 60			pK_a = 36
weaker acid	weaker base	stronger base	stronger acid

21. **a.** $CH_3CH_2CH_2\bar{C}H_2 > CH_3CH_2CH{=}\bar{C}H > CH_3CH_2C{\equiv}C^-$

b. $^-NH_2 > CH_3C{\equiv}C^- > CH_3CH_2O^- > F^-$

22. Solved in the text.

23. **a.** $HC{\equiv}CH \xrightarrow[2.\ CH_3CH_2CH_2Br]{1.\ ^-NH_2} CH_3CH_2CH_2C{\equiv}CH$

b. $HC{\equiv}CH \xrightarrow[2.\ CH_3Br]{1.\ ^-NH_2} CH_3C{\equiv}CH \xrightarrow[\text{Na/NH}_3\ \text{liq}]{H_2\ /\text{Lindlar catalyst or}} CH_3CH{=}CH_2$

c. $HC{\equiv}CH \xrightarrow[2.\ CH_3Br]{1.\ ^-NH_2} CH_3C{\equiv}CH \xrightarrow[2.\ CH_3Br]{1.\ ^-NH_2} CH_3C{\equiv}CCH_3 \xrightarrow[\text{Lindlar catalyst}]{H_2}$ 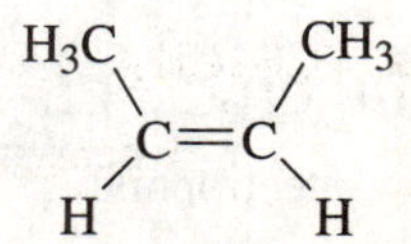

d. product of **a** 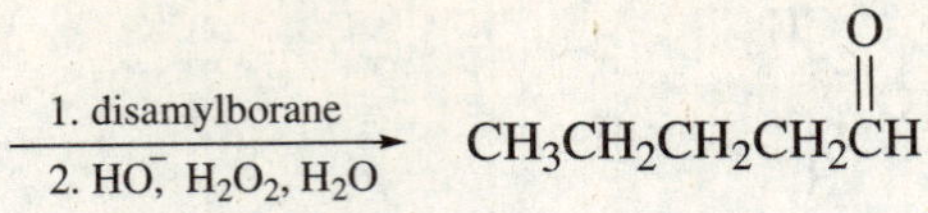

e. product of **b** 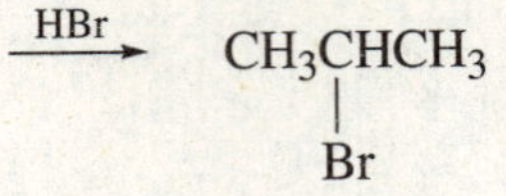

f. 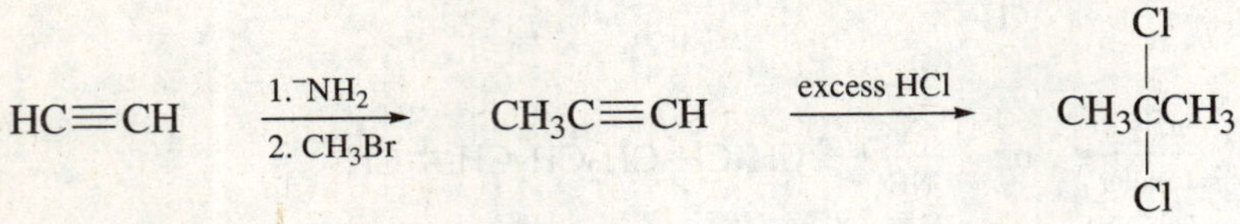

24. **a.** 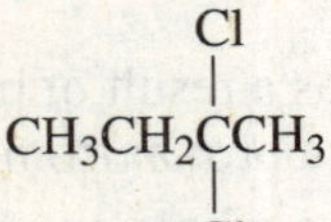

b. 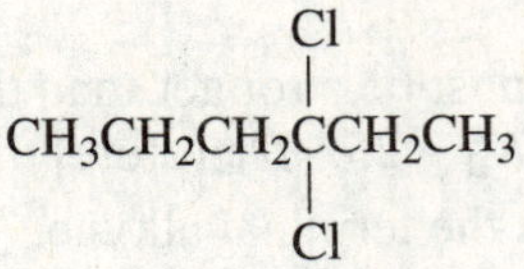

c. 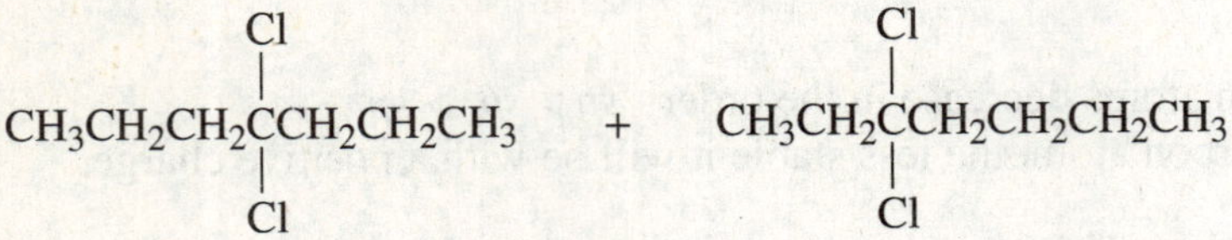

25. **a.** $CH_3C{\equiv}CCH_2CH_2CH_3$

b. 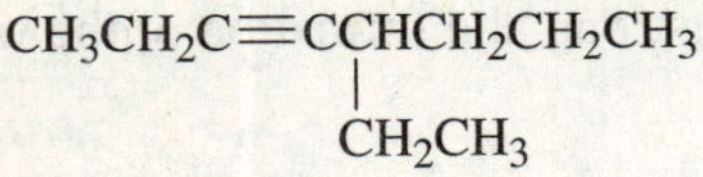

c. $CH_3C{\equiv}CH$

d. $CH_2{=}CHC{\equiv}CH$

e. $CH_3OC{\equiv}CH$

f. 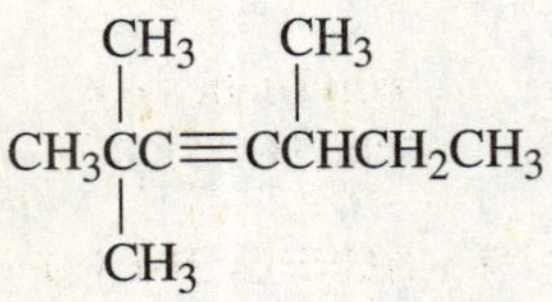

g. $BrC{\equiv}CCH_2CH_2CH_3$

h. $HC{\equiv}CCH_2Br$

i. $CH_3CH_2C{\equiv}CCH_2CH_3$

j. 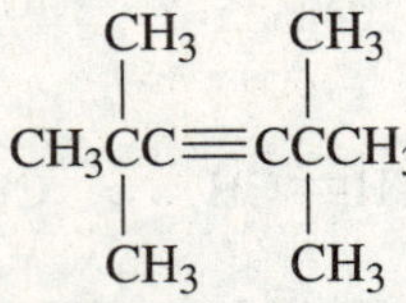

k. 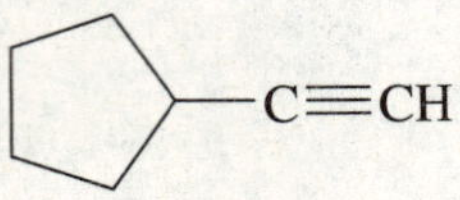

l. 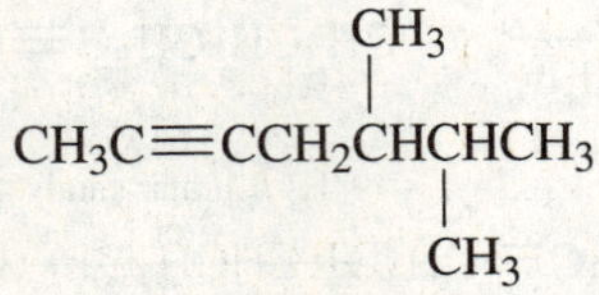

26. 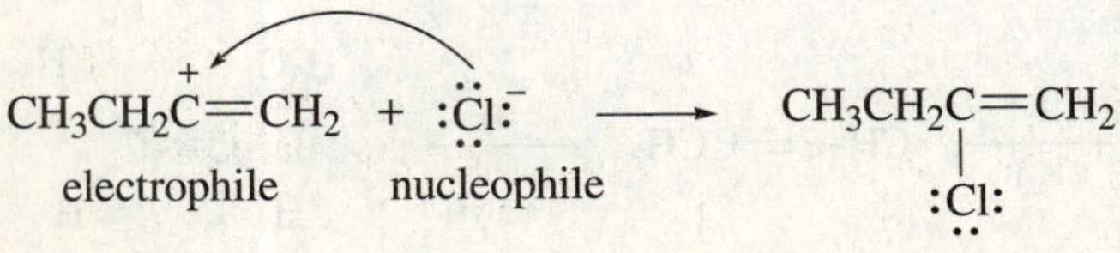

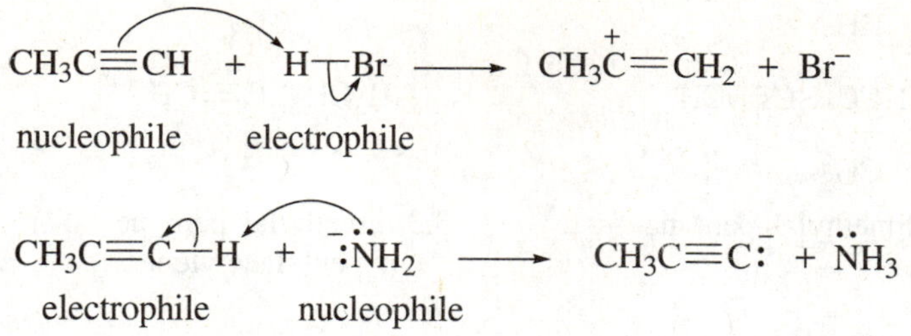

27. **a.** 5-bromo-2-hexyne
b. 5-methyl-2-octyne
c. 5,5-dimethyl-2-hexyne
d. 6-chloro-2-methyl-3-heptyne
e. 1,6-cyclooctadiene
f. 1,6-dimethyl-1,3-cylohexadiene

28.

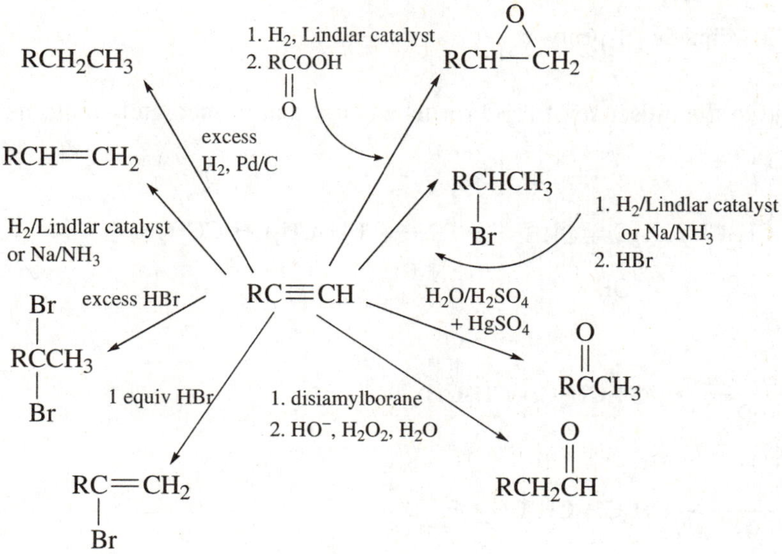

29. Al named only one correctly.
a. 4-methyl-2-hexyne
b. 7-bromo-3-heptyne
c. correct
d. 2-pentyne

30.

$HC{\equiv}CCH_2CH_2CH_2CH_2CH_3$
1-heptyne
pentylacetylene

$CH_3C{\equiv}CCH_2CH_2CH_2CH_3$
2-heptyne
butylmethylacetylene

$CH_3CH_2C{\equiv}CCH_2CH_2CH_3$
3-heptyne
ethylpropylacetylene

$CH_3CH_2CH_2CH(CH_3)C{\equiv}CH$
3-methyl-1-hexyne

$CH_3CH_2CH(CH_3)CH_2C{\equiv}CH$
4-methyl-1-hexyne

$CH_3CH(CH_3)CH_2CH_2C{\equiv}CH$
5-methyl-1-hexyne
isopentylacetylene

$CH_3CH_2CH(CH_3)C{\equiv}CCH_3$
4-methyl-2-hexyne
sec-butylmethylacetylene

$CH_3CH(CH_3)CH_2C{\equiv}CCH_3$
5-methyl-2-hexyne
isobutylmethylacetylene

$CH_3CH(CH_3)C{\equiv}CCH_2CH_3$
2-methyl-3-hexyne
ethylisopropylacetylene

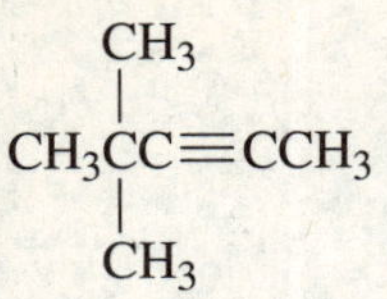

4,4-dimethyl-2-pentyne
tert-butylmethylacetylene

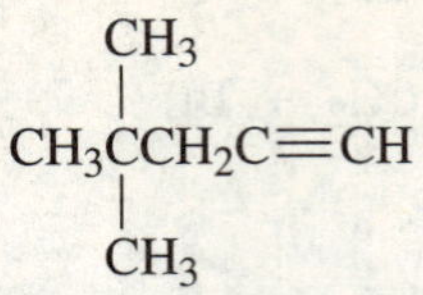

4,4-dimethyl-1-pentyne

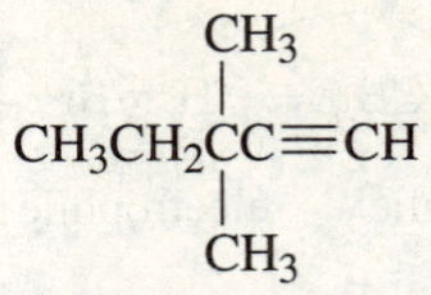

3,3-dimethyl-1-pentyne
tert-pentylacetylene

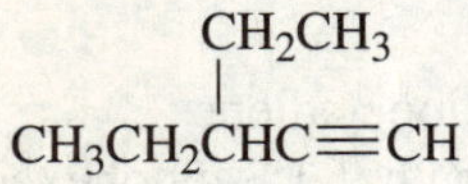

3-ethyl-1-pentyne

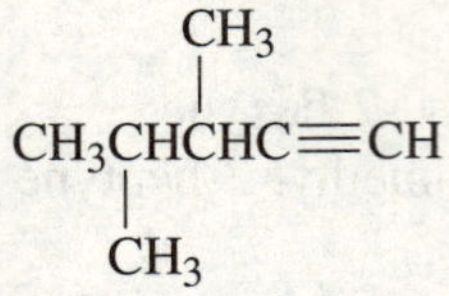

3,4-dimethyl-1-pentyne

31. Addition of Br_2 to the triple bond in the presence of H_2O forms an enol that immediately tautomerizes to a ketone.

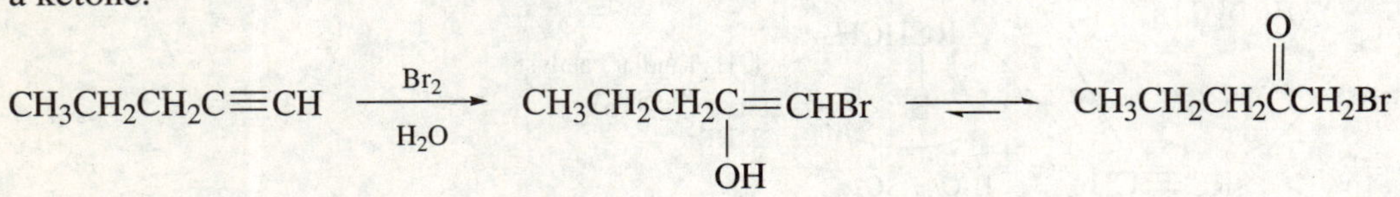

32. **a.** 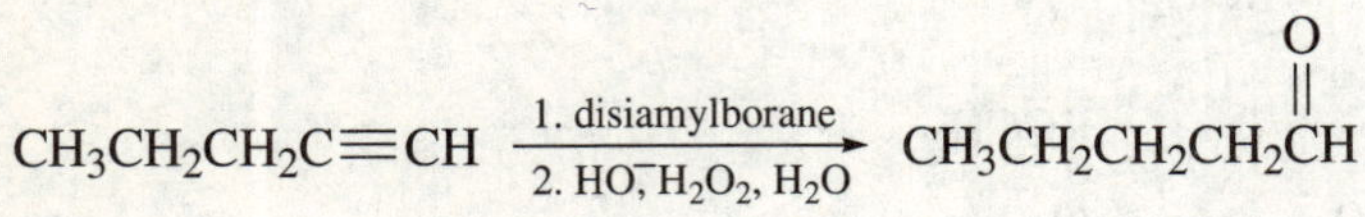

b. $CH_3CH_2CH{=}CH_2 \xrightarrow[\text{2. HO}^-\text{, } H_2O_2\text{, } H_2O]{\text{1. } BH_3} CH_3CH_2CH_2CH_2OH$

c. 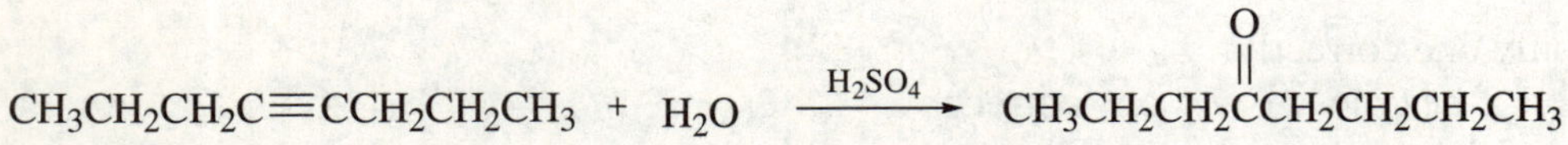

This symmetrical alkyne will give the greatest yield of the desired ketone. Because the reactant is not a terminal alkyne, the reaction can take place without the mercuric ion catalyst.

33. **a.** H_2, Lindlar catalyst **b.** Na, NH_3 (liq) **c.** excess H_2, Pt/C

34.
a. 1-octen-6-yne
b. *cis*-3-hexen-1-ol
c. 1,5-octadiyne
d. 5-chloro-1,3-cyclohexadiene
e. 1-methyl-1,3,5-heptatriene

35. The molecular formula of the hydrocarbon is $C_{32}H_{56}$.

$$C_nH_{2n+2} = C_{32}H_{66}$$

With one triple bond, two double bonds, and one ring, the degree of unsaturation is 5. Therefore, the compound is missing 10 hydrogens from C_nH_{2n+2}.

36. **a.** 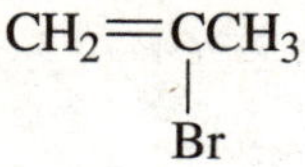

b. 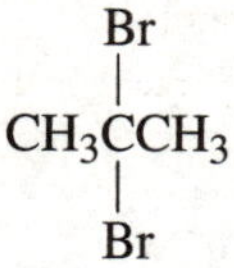

c. 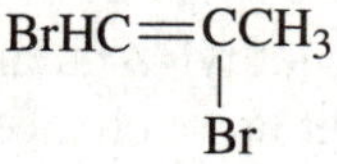

d. 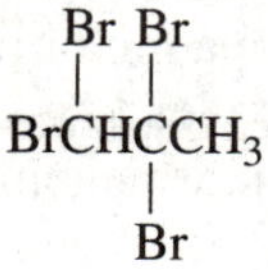

e. 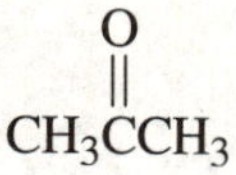

f. 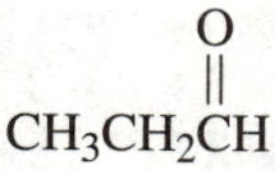

g. $CH_3CH_2CH_3$

h. $CH_3CH{=}CH_2$

i. $CH_3CH{=}CH_2$

j. $CH_3C{\equiv}C^-$

k. $CH_3C{\equiv}CCH_2CH_2CH_2CH_2CH_3$

37. **a.** 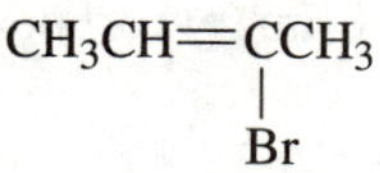

b. 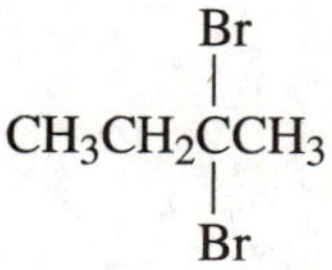

c. 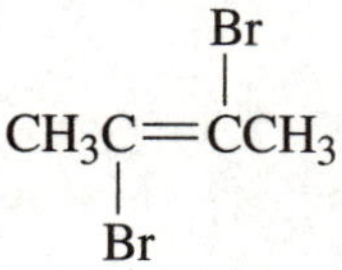

d. 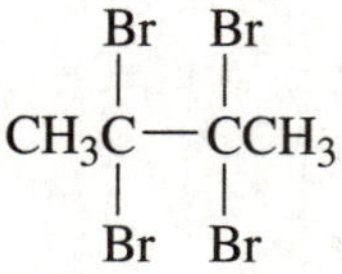

e. 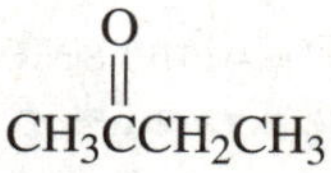

f. 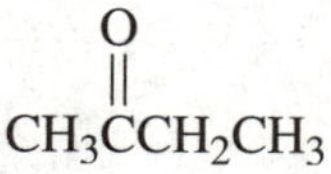

g. $CH_3CH_2CH_2CH_3$

h. 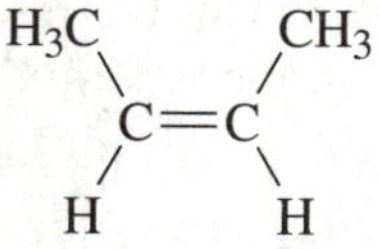

i. 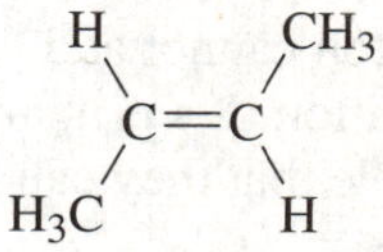

j. no reaction

k. no reaction

38. **a.** **1.** 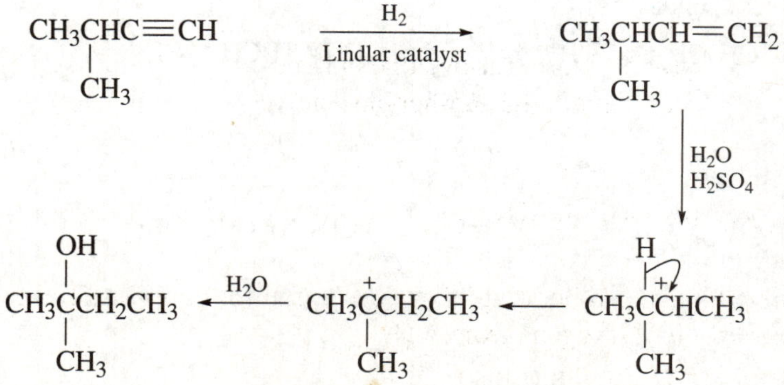

2. 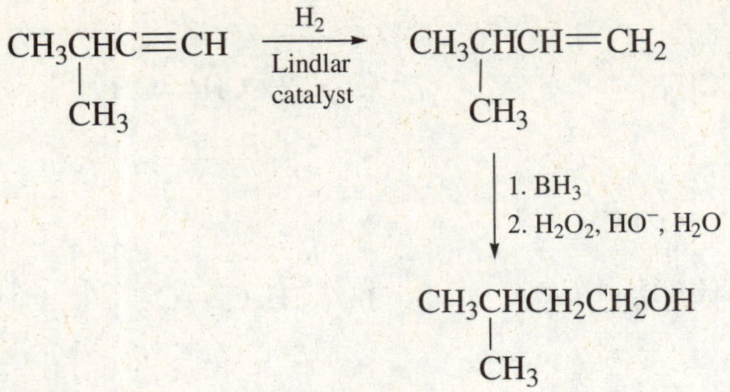

b. 3-Methyl-2-butanol will be a minor product obtained from both **1.** and **2.** 3-Methyl-2-butanol will be obtained from **1.** because water can attack the secondary carbocation before it has a chance to rearrange to the tertiary carbocation. 3-Methyl-2-butanol will be obtained from **2.** because in the second step of the synthesis boron can also add to the other sp^2 carbon; it will be a minor product because the transition state for its formation is less stable than the transition state leading to the major product. Because a carbocation is not formed as an intermediate, a carbocation rearrangement cannot occur.

(The proton cannot add to the other sp^2 carbon in the second step of part **1.** because that would require the formation of a primary carbocation in the second step of the synthesis. Primary carbocations are so unstable that they can never be formed.)

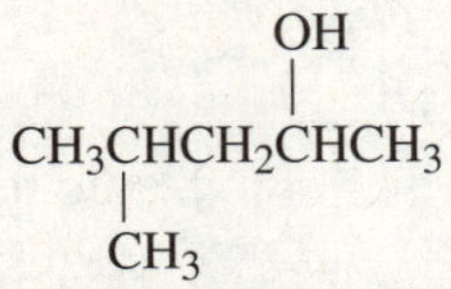

3-methyl-2-butanol

39. Three of the names are correct.

a. 3-heptyne
b. 5-methyl-3-heptyne
c. correct
d. 6,7-dimethyl-3-octyne
e. correct
f. correct

40. only **c** and **e** are keto-enol tautomers.

41. **a.** 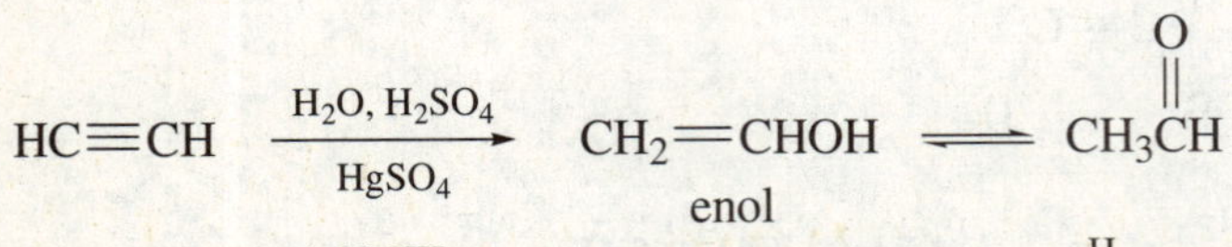

b. 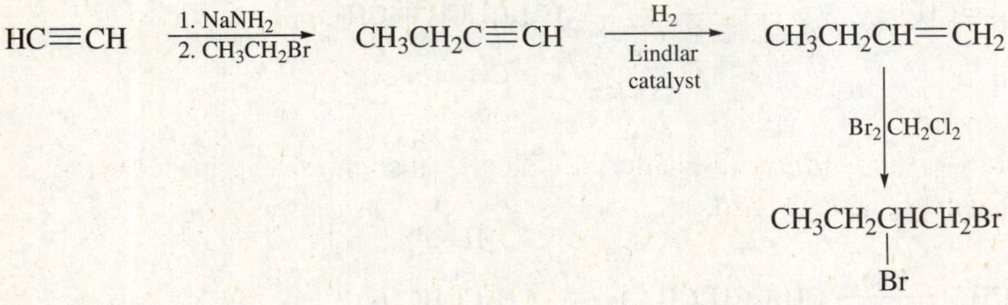

c. 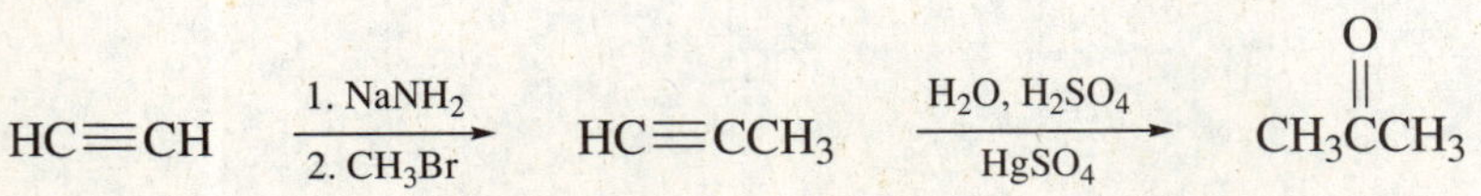

d. 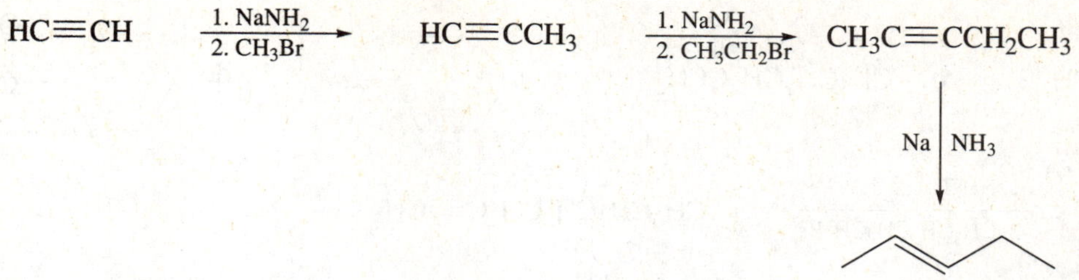

e. 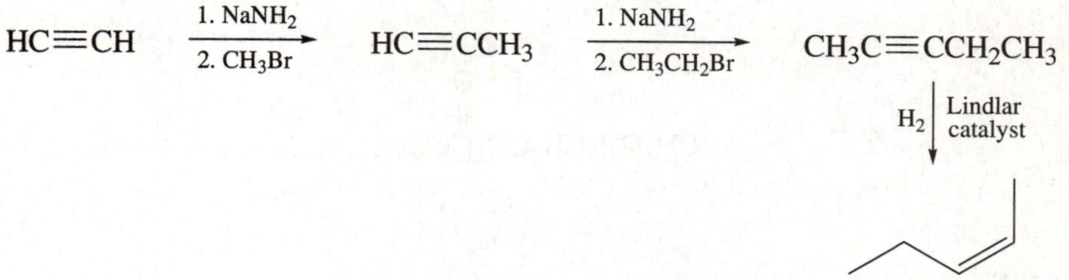

f. 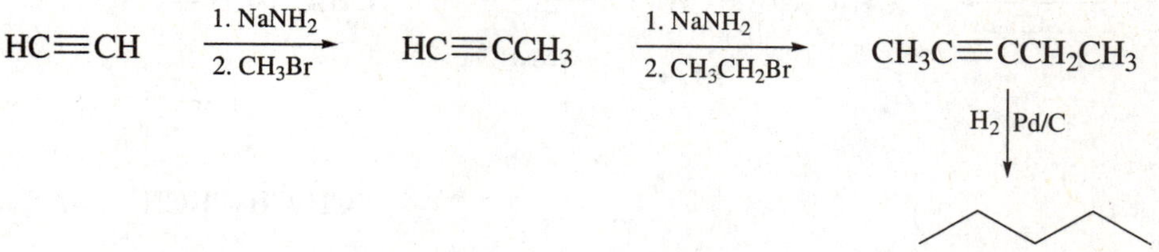

42. **a.** Syn addition of H_2 forms *cis*-2-butene. When Br_2 adds to *cis*-2-butene, the threo pair of enantiomers is formed.

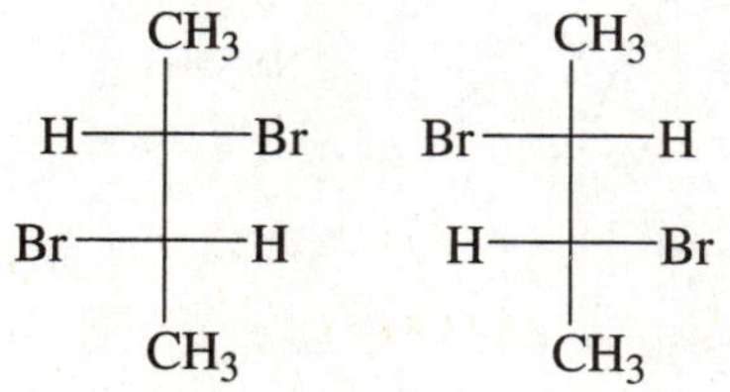

b. Reaction with sodium and liquid ammonia forms *trans*-2-butene. And when Br_2 adds to *trans*-2-butene, a meso compound is formed.

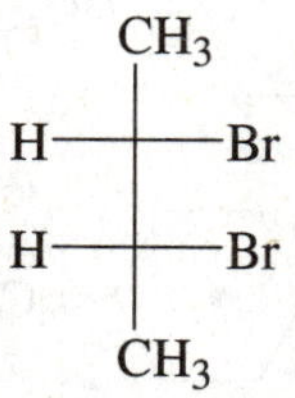

c. Anti addition of Cl_2 forms *trans*-2,3-dichloro-2-butene. And when Br_2 adds to *trans*-2,3-dichloro-2-butene, a meso compound is formed.

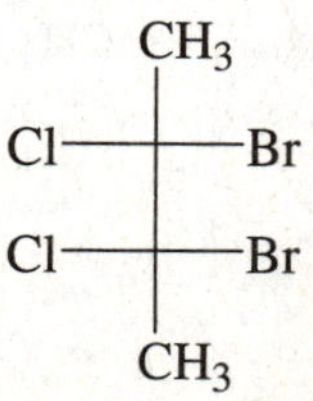

43. **a.** $CH_3CH_2\overset{\overset{O}{\|}}{C}CH_3$ **b.** $CH_3CH_2CH_2\overset{\overset{O}{\|}}{C}CH_3$ **c.** 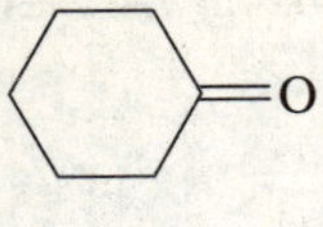 **d.** 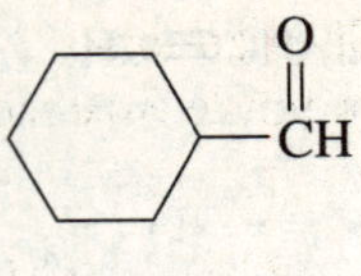

44. **a.** $HC{\equiv}CH \xrightarrow[\text{2. } CH_3CH_2CH_2CH_2Br]{\text{1. } NaNH_2} CH_3CH_2CH_2CH_2C{\equiv}CH$

$$CH_3CH_2CH_2CH_2C{\equiv}CH \xrightarrow[H_2SO_4]{H_2O,\ HgSO_4} CH_3CH_2CH_2CH_2\overset{\overset{O}{\|}}{C}CH_3$$

b. $HC{\equiv}CH \xrightarrow[\text{2. } CH_3CH_2Br]{\text{1. } NaNH_2} CH_3CH_2C{\equiv}CH \xrightarrow[\text{or } Na/NH_3 \text{ liq}]{H_2,\ \text{Lindlar catalyst}} CH_3CH_2CH{=}CH_2$

$$CH_3CH_2CH{=}CH_2 \xrightarrow{HBr} CH_3CH_2\underset{\underset{Br}{|}}{C}HCH_3$$

c. $HC{\equiv}CH \xrightarrow[\text{2. } CH_3CH_2CH_2Br]{\text{1. } NaNH_2} CH_3CH_2CH_2C{\equiv}CH \xrightarrow[\text{or } Na/NH_3 \text{ liq}]{H_2,\ \text{Lindlar catalyst}} CH_3CH_2CH_2CH{=}CH_2$

$$CH_3CH_2CH_2CH{=}CH_2 \xrightarrow[H_2SO_4]{H_2O} CH_3CH_2CH_2\underset{\underset{OH}{|}}{C}HCH_3$$

d. 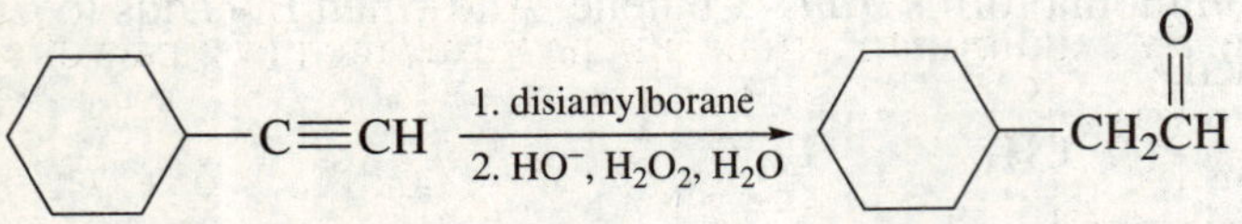

e. 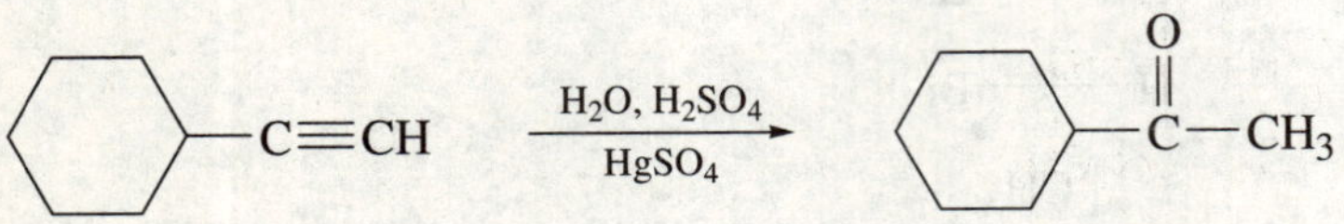

f. 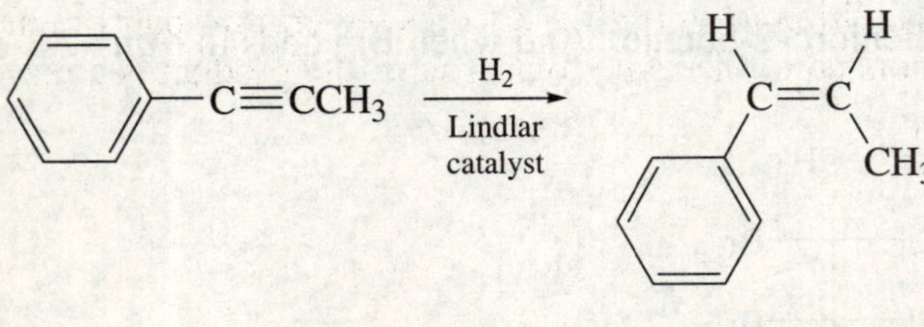

45. She can make 3-octyne by using 1-hexyne instead of 1-butyne. She would then need to use ethyl bromide (instead of butyl bromide) for the alkylation step:

$$CH_3CH_2CH_2CH_2C{\equiv}CH \xrightarrow[\text{2. } CH_3CH_2Br]{\text{1. } NaNH_2} CH_3CH_2CH_2CH_2C{\equiv}CCH_2CH_3$$

Or she could make the 1-butyne she needed by alkylating ethyne:

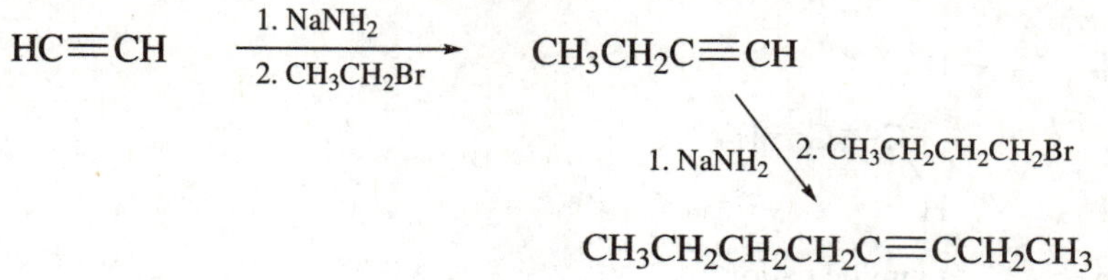

46. **a.** Only one product is obtained from the hydroboration-oxidation of 2-butyne because the alkyne is symmetrical. Two different products can be obtained from hydroboration-oxidation of unsymmetrical 2-pentyne.

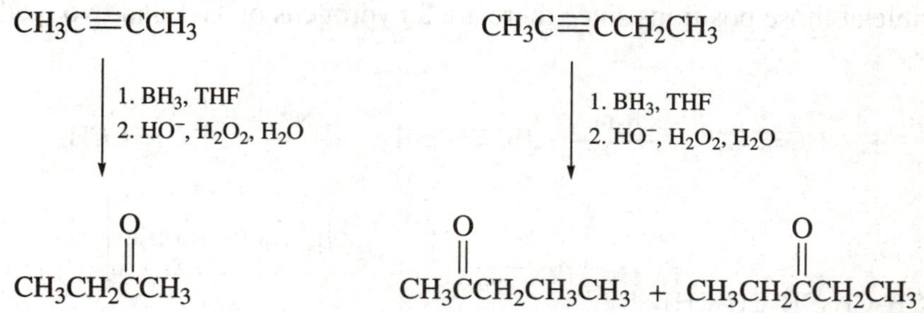

b. Only one product will be obtained from hydroboration-oxidation of any symmetrical alkyne, such as 3-hexyne or 4-octyne.

$CH_3CH_2C{\equiv}CCH_2CH_3$ (3-hexyne) $\qquad$ $CH_3CH_2CH_2C{\equiv}CCH_2CH_2CH_3$ (4-octyne)

47. **a.** The first step forms a trans alkene. Syn addition to a trans alkene forms the threo pair of enantiomers.

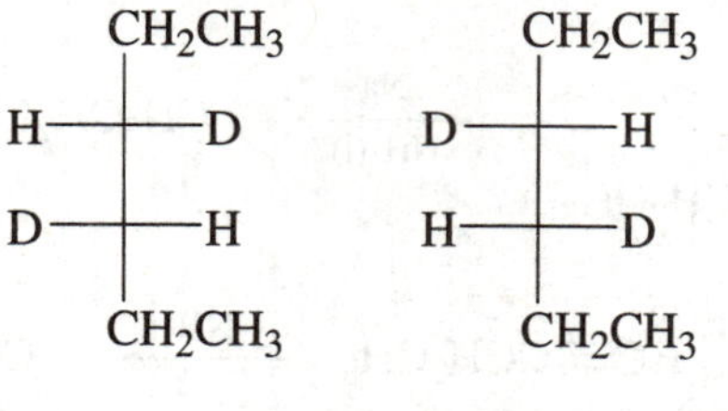

b. The first step forms a cis alkene. Syn addition to a cis alkene forms the erythro pair of enantiomers, but since each asymmetric carbon is bonded to the same four groups, the product is a meso compound.

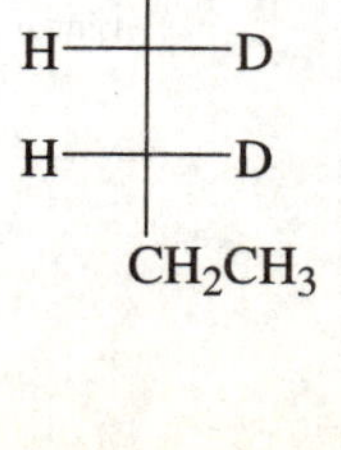

48. There are two reasons why hyperconjugation is less effective in stabilizing a positive charge on a vinylic cation than a positive charge on an alkyl cation.
First, an sp^2–s bond is stronger than an sp^3—s bond, so the sp^2 orbitals are less able to donate electron density to the adjacent positively charge carbon.
Second, an sp^2 carbon has 120° bond angles compared to the 109.5° bond angles of an sp^3 carbon, so the orbitals of an sp^2 carbon are farther away, which makes them less able to interact with the positively charged carbon.

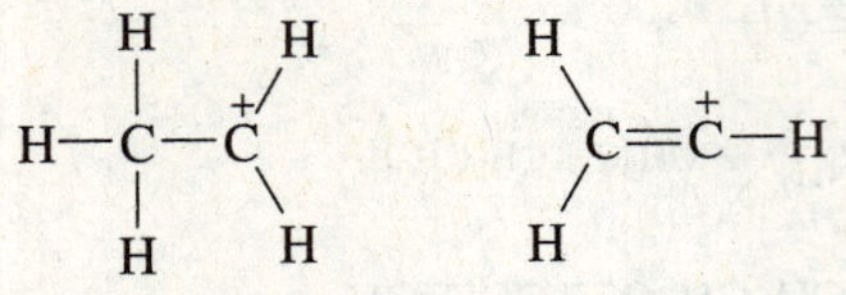

an alkyl cation a vinylic cation

49. (3*E*,6*E*)-3,7,11-trimethyl-1,3,6,10-dodecatetraene

The configuration of the double bond at the 1-position and at the 10-position is not specified because isomers are not possible at those positions since there are 2 hydrogens on C-1 and two methyl groups on C-11.

50. **a.**

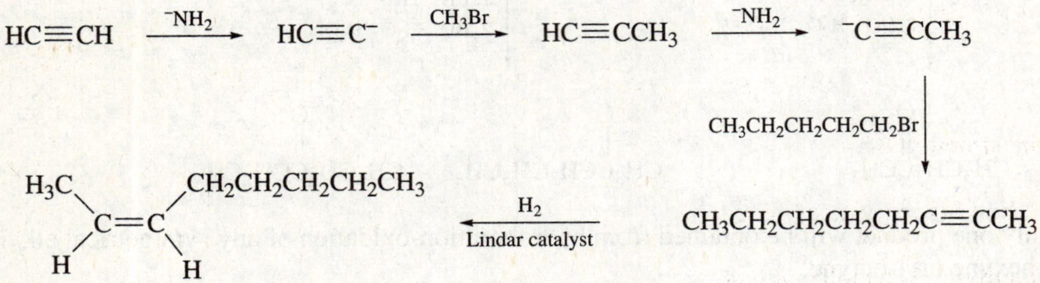

b.

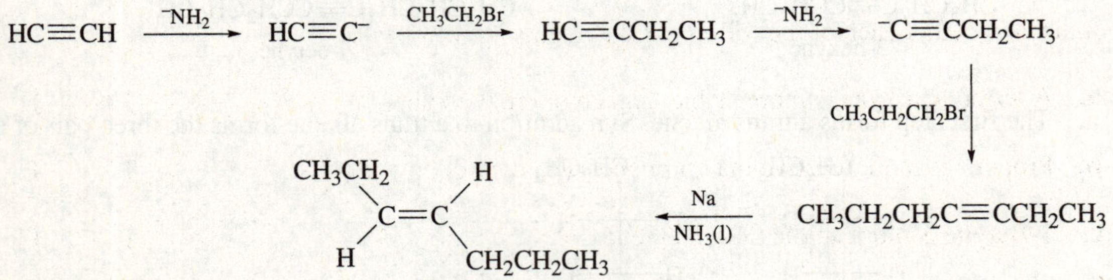

c.

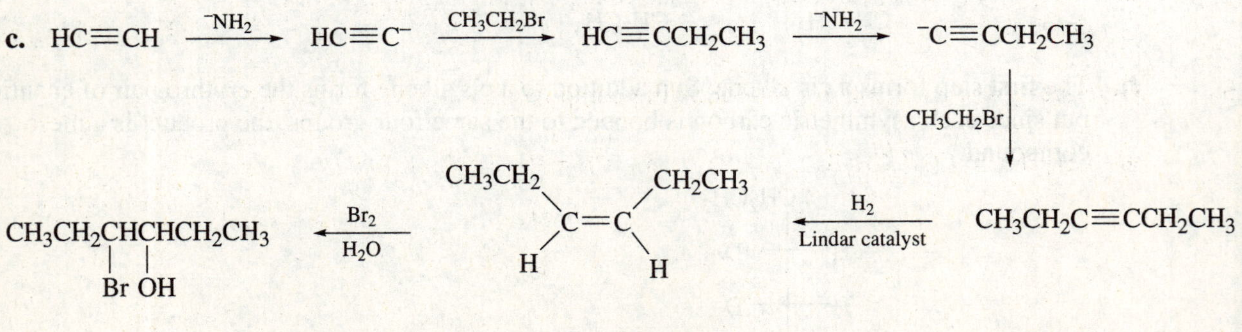

Chapter 6 Practice Test

1. What reagents could be used to convert the given starting material into the desired product?

a. 

b. 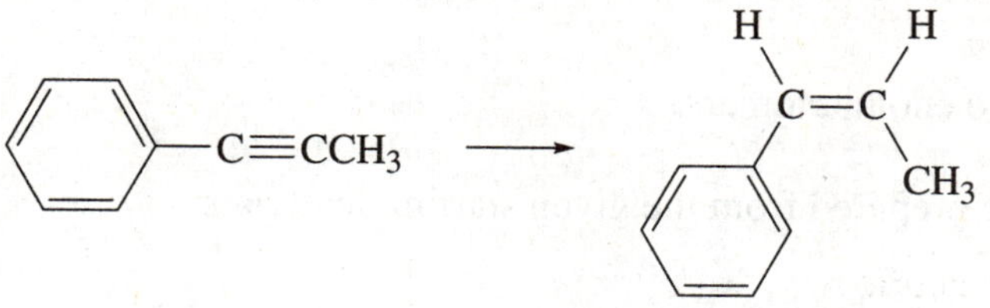

2. Draw the enol tautomer of the following compound:

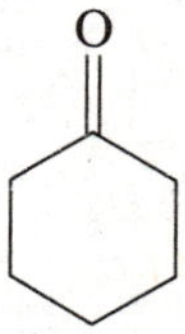

3. Give the structure of

a. *sec*-butyl isobutyl acetylene

b. 2-methyl-1,3-cyclohexadiene

4. Indicate whether each of the following statements is true or false:

a. A terminal alkyne is more stable than an internal alkyne. T F

b. Propyne is more reactive than propene toward reaction with HBr. T F

c. 1-Butyne is more acidic than 1-butene. T F

d. An sp^2 hybridized carbon is more electronegative than an sp^3 hybridized carbon. T F

5. Give the systematic name for the following compounds:

a. $CH_3CHC\equiv CCH_2CH_2Br$ (with CH_3 on the second carbon)

b. $CH_3CHC\equiv CCH_2CH_2OH$ (with CH_3 on the second carbon)

6. What alkyne would be the best reagent to use for the synthesis of the following ketone?

$$CH_3CH_2CH_2\overset{\overset{O}{\|}}{C}CH_3$$

7. Rank the following compounds in order of decreasing acidity:
(Label the most acidic compound #1.)

NH_3 $\quad$ $CH_3C{\equiv}CH$ $\quad$ CH_3CH_3 $\quad$ H_2O $\quad$ $CH_3CH{=}CH_2$

8. Give an example of a ketone that has two enol tautomers.

9. Show how the target molecules could be prepared from the given starting materials.

a. $CH_3CH_2C{\equiv}CH \longrightarrow CH_3CH_2CH_2CH_2CH_2CH_3$

b. $CH_3CH_2C{\equiv}CH \longrightarrow CH_3CH_2\underset{\underset{Br}{|}}{CH}CH_2CH_2CH_3$

c. $CH_3CH_2C{\equiv}CH \longrightarrow CH_3CH_2\overset{\overset{O}{\|}}{C}CH_2CH_2CH_3$

CHAPTER 7

Delocalized Electrons and Their Effect on Stability, Reactivity, and pK_a • More About Molecular Orbital Theory

Important Terms

1,2-addition (direct addition) addition to the 1- and 2-positions of a conjugated system.

1,4-addition (conjugate addition) addition to the 1- and 4-positions of a conjugated system.

allene a compound with two adjacent double bonds.

allylic carbon a carbon adjacent to an sp^2 carbon of a carbon-carbon double bond.

allylic cation a compound with a positive charge on an allylic carbon.

antibonding molecular the molecular orbital formed when out-of-phase orbitals overlap.

antiymmetric molecular orbital a molecular orbital that does not have a plane of symmetry but would have if one half of the MO were turned upside down.

aromatic compound compounds that are unusually stable because of large delocalization energies (see Section 14.1).

benzylic carbon a carbon, joined to other atoms by single bonds, that is bonded to a benzene ring.

benzylic cation a compound with a positive charge on a benzylic carbon.

bonding molecular orbital the molecular orbital formed when in-phase orbitals overlap.

bridged bicyclic compound a bicyclic compound in which the rings share two nonadjacent carbons.

common intermediate an intermediate that two compounds have in common.

concerted reaction a reaction in which all the bond making and breaking processes occur in a single step.

conjugate addition addition to the 1- and 4-positions of a conjugated system.

conjugated diene a compound with two conjugated double bonds.

conjugated double bonds double bonds separated by one single bond.

contributing resonance structure a structure with localized electrons that approximates the true structure of a compound with delocalized electrons.

cumulated double bonds double bonds that are adjacent to one another.

cycloaddition reaction a reaction in which two π-electron-containing reactants combine to form a cyclic product.

[4 + 2] cycloaddition reaction a cycloaddition reaction in which four π electrons come from one reactant and two π electrons come from the other reactant.

delocalization energy (resonance energy) the extra stability associated with a compound as a result of having delocalized electrons.

delocalized electrons electrons that are not localized on a single atom or between two atoms.

Diels-Alder reaction a [4 + 2] cycloaddition reaction.

diene a hydrocarbon with two double bonds.

dienophile an alkene that reacts with a diene in a Diels-Alder reaction.

direct addition (1,2-addition) addition to the 1- and 2-positions of a conjugated system.

donation of electrons by resonance donation of electrons through π bonds.

electron delocalization the sharing of electrons by more than two atoms.

endo a substituent is endo if it and the bridge are on opposite sides of the bicyclic compound.

equilibrium control thermodynamic control.

exo a substituent is exo if it and the bridge are on the same side of the bicyclic compound.

highest occupied molecular orbital (HOMO) the highest energy molecular orbital that contains electrons.

isolated double bonds double bonds separated by more than one single bond.

kinetic control when a reaction is under kinetic control, the relative amounts of the products depend on the rates at which they are formed.

kinetic product the product that is formed the fastest.

linear combination of molecular orbitals (LCAO) the combination of atomic orbitals to produce molecular orbitals.

localized electrons electrons that are restricted to a particular locality.

lowest unoccupied molecular orbital (LUMO) the lowest energy molecular orbital that does not contain an electron.

nonbonding molecular orbital	because the *p* orbitals are too far apart to overlap significantly, the molecular orbital that results neither favors nor disfavors bonding.
pericyclic reaction	a reaction that takes place in one step as a result of a cyclic reorganization of electrons.
polyene	a compound that has several double bonds.
proximity effect	an effect caused by one species being close to another.
resonance	electron delocalization.
resonance contributor (resonance structure)	a structure with localized electrons that approximates the true structure of a compound with delocalized electrons.
resonance electron donation	donation of electrons through π bonds.
resonance electron withdrwal	withdrawal of electrons through π bonds.
resonance energy (delocalization energy) (resonance stabilization energy)	the extra stability a compound has as a result of having delocalized electrons.
resonance hybrid	the actual structure of a compound with delocalized electrons; it is represented by two or more structures with localized electrons.
***s*-cis-conformation**	the conformation in which two double bonds of a conjugated diene are on the same side of a connecting single bond.
separated charges	a positive and a negative charge that can be neutralized by the movement of electrons.
***s*-trans-conformation**	the conformation in which two double bonds of a conjugated diene are on opposite sides of a connecting single bond.
symmetric molecular orbital	an orbital with a plane of symmetry so that one half is the mirror image of the other half.
thermodynamic control	when a reaction is under thermodynamic control, the relative amounts of the products depend on their stabilities.
thermodynamic product	the most stable product.
withdrawal of electrons by resonance	withdrawal of electrons through π bonds.

Solutions to Problems

1. **a.** If stereoisomers are not included, 3 different monosubstituted compounds can be formed. If stereoisomers are included, 4 different monosubstituted compounds can be formed because the second listed compound has an asymmetric center.

$BrC{\equiv}CC{\equiv}CCH_2CH_3$ $HC{\equiv}CC{\equiv}CCHCH_3$ $HC{\equiv}CC{\equiv}CCH_2CH_2Br$

Br

b. If stereoisomers are not included, 2 different monosubstituted compounds can be formed.

$BrCH{=}CHC{\equiv}CCH{=}CH_2$ $CH_2{=}CC{\equiv}CCH{=}CH_2$

Br

c. If stereoisomers are included, 3 different monosubstituted compounds can be formed because the first compound has a double bond that can have cis–trans isomers.

2. Ladenburg benzene is a better proposal. It would form 1 monosubstituted compound, 3 disubstituted compounds, and would not add Br_2, all in accordance with what early chemists knew about the structure of benzene.

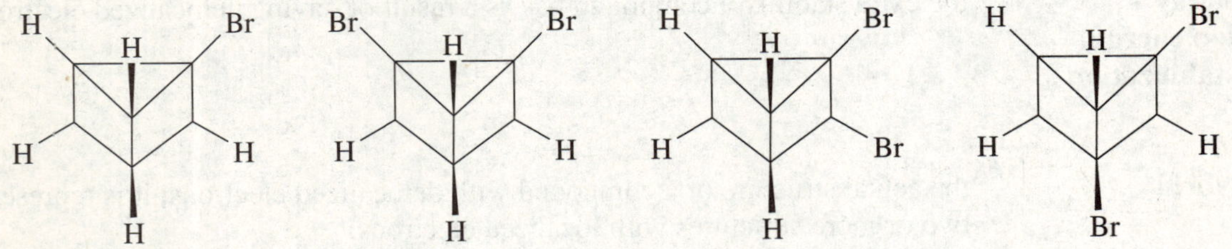

Dewar benzene is not in accordance with what early chemists knew about the structure of benzene, because it would form 2 monosubstituted compounds, 6 disubstituted compounds, and it would add Br_2.

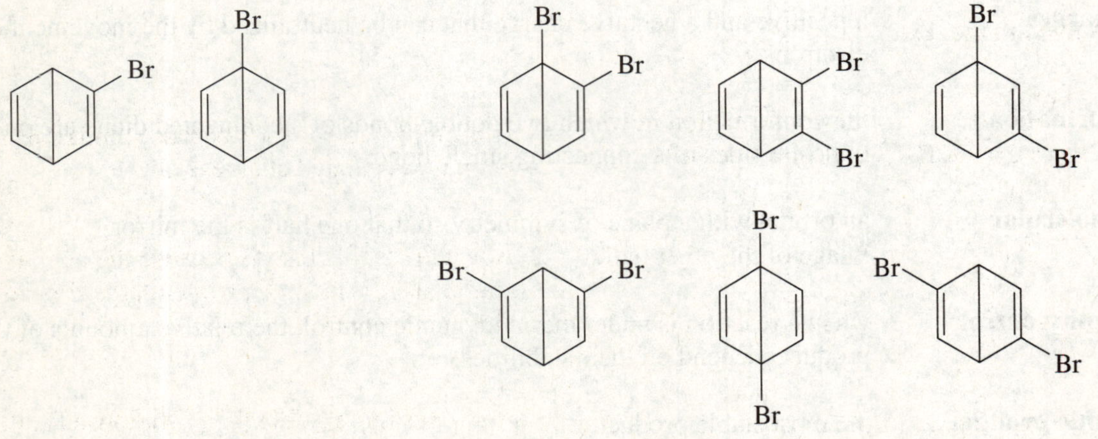

3. **a.** All the carbon-oxygen bonds in the carbonate ion should be the same length because each carbon-oxygen bond is represented in one resonance contributor with a double bond and in two resonance contributors with single bonds.

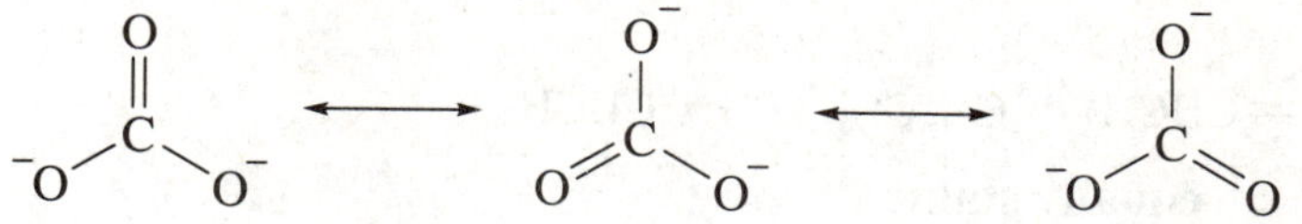

b. Because the two negative charges are shared equally by three oxygens, each oxygen should have two thirds of a negative charge.

4. **a.** **2, 4,** and **7**

2.

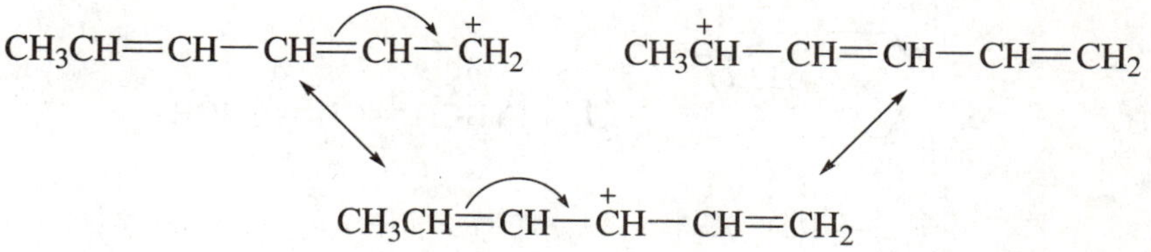

4.

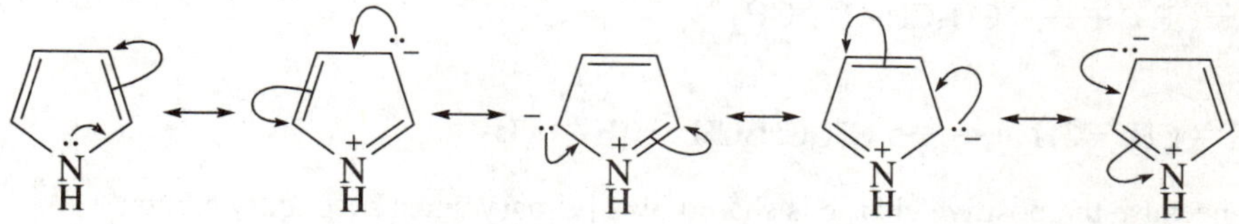

7.

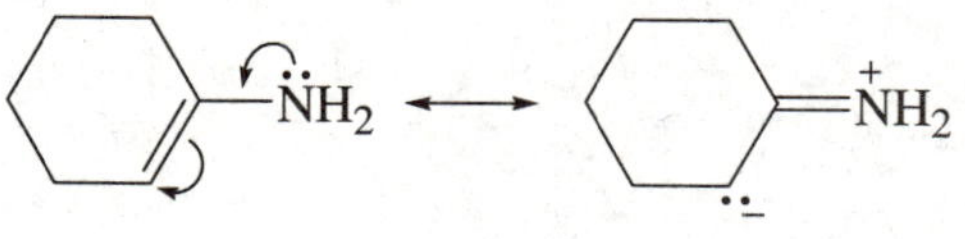

5. The resonance contributor that makes the greatest contribution to the hybrid is labeled "**A**". "**B**" contributes less to the hybrid than "**A**", and "**C**" contributes less to the hybrid than "**B**".

a. Solved in the text.

b.

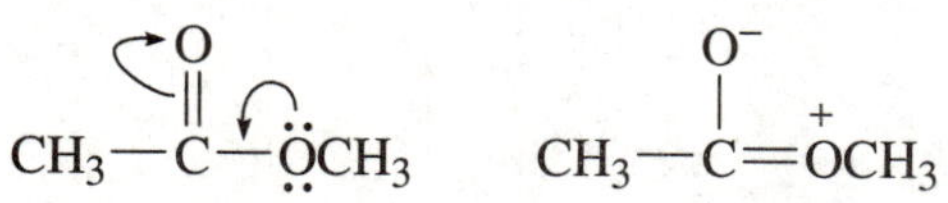

A **B**

A is more stable than B because, unlike B, A does not have separated charges.

c.

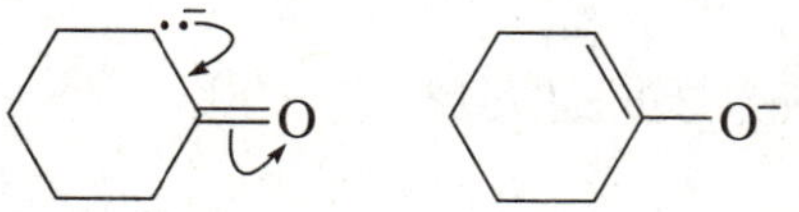

B **A**

A is more stable than B because the negative charge in A is on an O, while that is B is on a less electronegative C.

d.

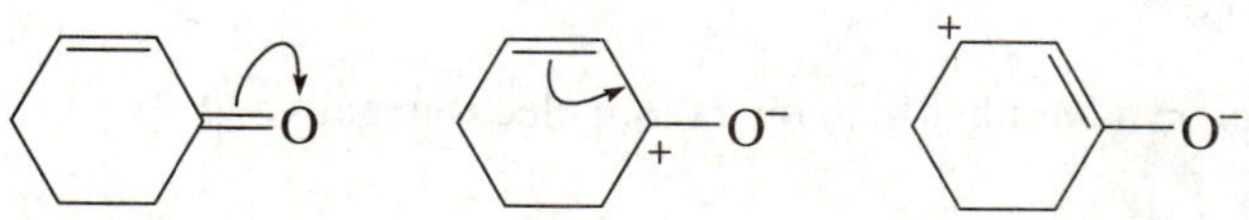

A **C** **B**

B is more stable than C because the electronegative oxygen atom is closer to the postive charge in C.
C is so unstable that it can be neglected.

e. 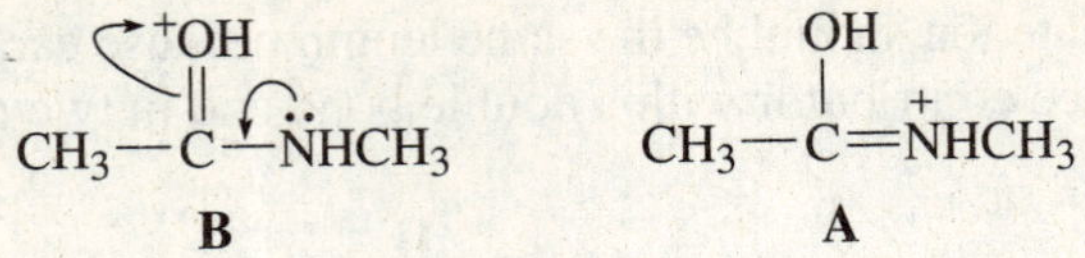

A is more stable than B because the positive charge in A is on a less electronegative atom.

f. 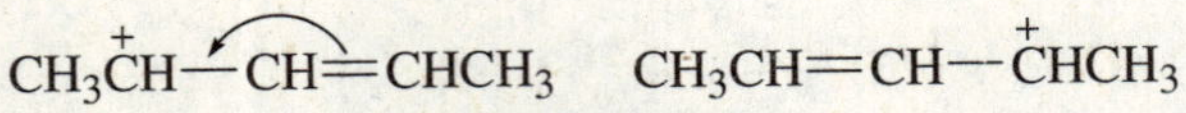

equally stable

6. a. 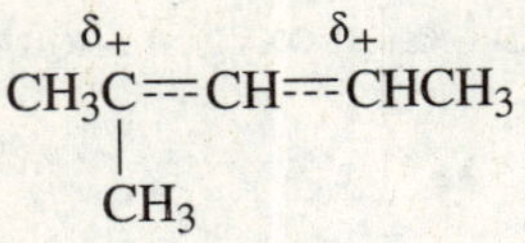

c. 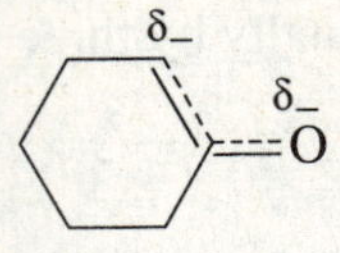

e. 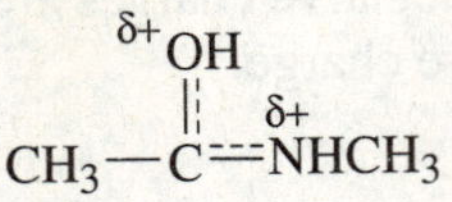

b. 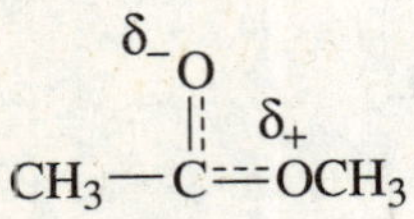

d. 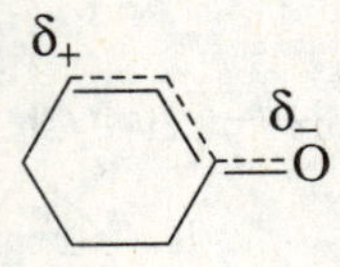

f. 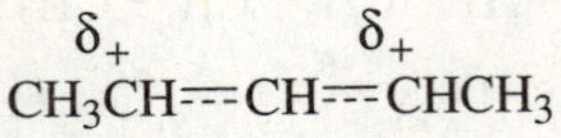

7. a. 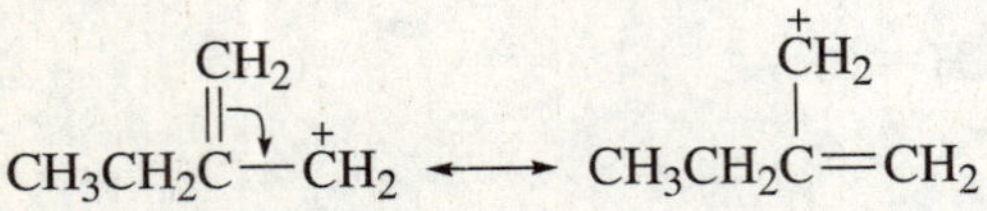

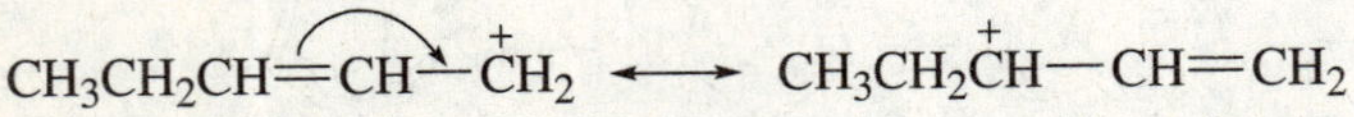

more stable because the positive charge is shared by a primary and a secondary carbon

b. 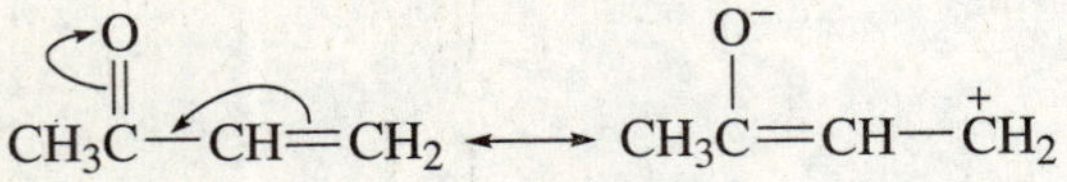

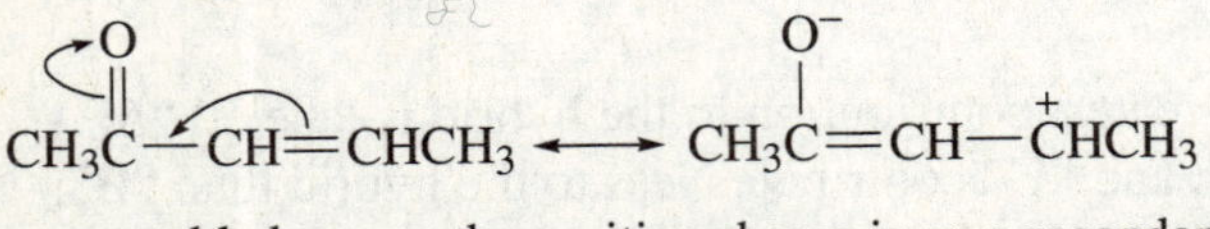

more stable because the positive charge is on a secondary carbon

c. 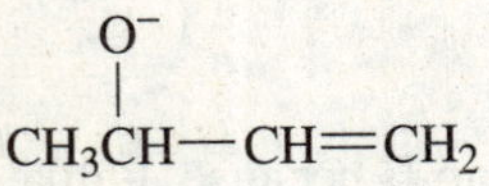

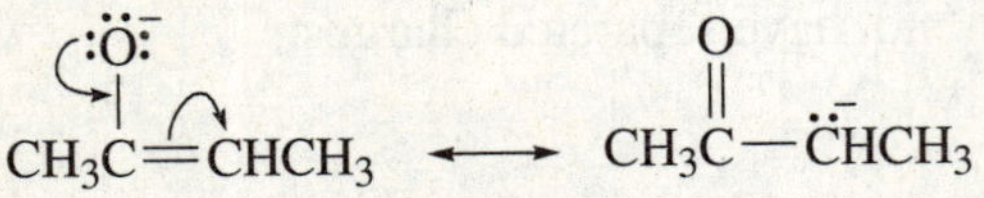

more stable because only in this compound is the negative charge delocalized

d. 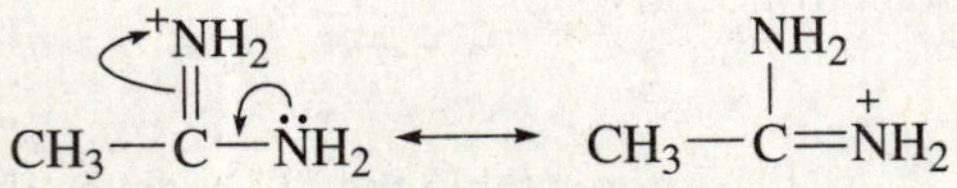

more stable because the positive charge is on a N rather than on a more electronegative O

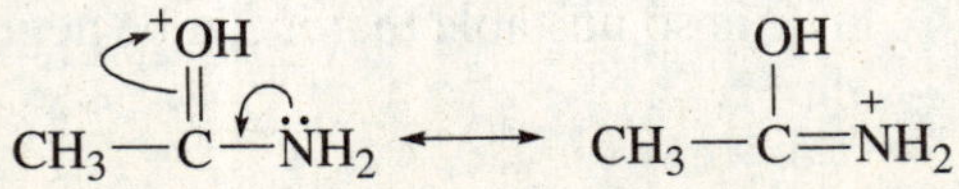

8. The second species has the greatest delocalization energy; it has three resonance contributors and none of them have separated charges. (See the answer to Problem 3.)
The first species has two resonance contributors and none of them have separated charges.
The third species has two resonance contributors, one of which has separated charges.

9. The smaller the heat of hydrogenation (the absolute value of $\Delta H°$), the more stable the compound. Therefore, the relative stabilities of the dienes are:

conjugated diene > isolated diene > cumulated diene

10. $CH_2{=}CHCH_2CH{=}CH_2 < CH_3CH{=}CHCH{=}CH_2 < CH_3CH{=}CHCH{=}CHCH_3 < CH_3\overset{\overset{CH_3}{|}}{C}{=}CHCH{=}\overset{\overset{CH_3}{|}}{C}CH_3$

1,4-pentadiene < 1,3-pentadiene < 2,4-hexadiene < 2,5-dimethyl-2,4-hexadiene

11. **a.** The compound with delocalized electrons is more stable than the compound in which all the electrons are localized.

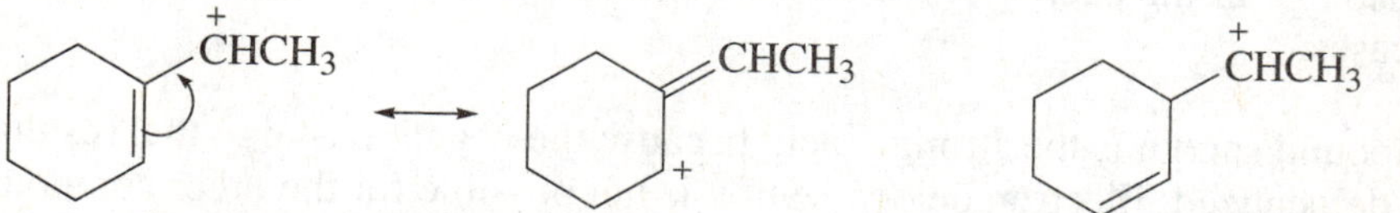

electrons are delocalized — electrons are localized

b. Because nitrogen is less electronegative than oxygen, it is better able to share the positive charge.

$CH_3\ddot{O}{-}\overset{+}{C}H_2 \longleftrightarrow CH_3\overset{+}{\ddot{O}}{=}CH_2$ $\quad$ $CH_3\ddot{N}H{-}\overset{+}{C}H_2 \longleftrightarrow CH_3\overset{+}{N}H{=}CH_2$

more stable

c. In order for electron delocalization to occur, the atoms that share the electrons must be in the same plane. The two *tert*-butyl groups prevent the positively charged carbon and the benzene ring from being in the same plane. Therefore, the carbocation cannot be stabilized by electron delocalization.

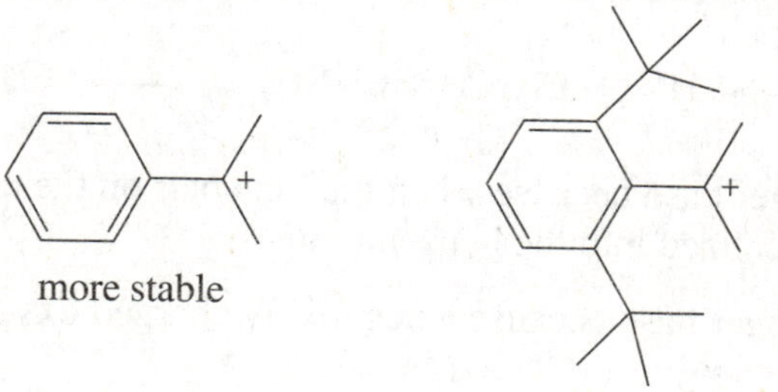

more stable

12. The ψ_3 molecular orbital of 1,3-butadiene has 3 nodes (two vertical and one horizontal).
The ψ_4 molecular orbital of 1,3-butadiene has 4 nodes (three vertical and one horizontal).

13. **a.** ψ_1 and ψ_2 are bonding molecular orbitals, and ψ_3 and ψ_4 are antibonding molecular orbitals.
b. ψ_1 and ψ_3 are symmetric molecular orbitals, and ψ_2 and ψ_4 are antisymmetric molecular orbitals.
c. ψ_2 is the HOMO and ψ_3 is the LUMO in the ground state.
d. ψ_3 is the HOMO and ψ_4 is the LUMO in the excited state.
e. If the HOMO is symmetric, the LUMO is antisymmetric and vice versa.

14. **a.** ψ_1, ψ_2, and ψ_3 are bonding molecular orbitals, and ψ_4, ψ_5, and ψ_6 are antibonding molecular orbitals.

b. ψ_1, ψ_3, and ψ_5 are symmetric molecular orbitals, and ψ_2, ψ_4, and ψ_6 are antisymmetric molecular orbitals.

c. ψ_3 is the HOMO and ψ_4 is the LUMO in the ground state.

d. ψ_4 is the HOMO and ψ_5 is the LUMO in the excited state.

e. If the HOMO is symmetric, the LUMO is antisymmetric and vice versa.

15. **a.** The ψ_1 molecular orbital of 1,3-butadiene has 3 bonding interactions and the ψ_2 molecular orbital has 2 bonding interactions.

b. The ψ_1 molecular orbital of 1,3,5,7-octatetraene has 7 bonding interactions and the ψ_2 molecular orbital has 6 bonding interactions.
Notice that the ψ_1 molecular orbital has one bonding interaction between each of the overlapping *p* orbitals. Notice also that as the energy of the molecular orbital increases, the number of bonding interactions decreases.

16. In each case, the compound shown is the stronger acid because the negative charge that results when it loses a proton can be delocalized. Electron delocalization is not possible for the other compound in each pair.

a. $CH_3CH{=}CHOH \longrightarrow H^+ + CH_3CH{=}CHO^- \longleftrightarrow CH_3\bar{C}HCH{=}O$

b. $CH_3C({=}O){-}OH \longrightarrow H^+ + CH_3C({=}O){-}O^- \longleftrightarrow CH_3C(O^-){=}O$

c. $CH_3CH{=}CHOH \longrightarrow H^+ + CH_3CH{=}CHO^- \longleftrightarrow CH_3\bar{C}HCH{=}O$

d. $CH_3CH{=}CH\overset{+}{N}H_3 \longrightarrow H^+ + CH_3CH{=}CHNH_2 \longleftrightarrow CH_3\bar{C}HCH{=}\overset{+}{N}H_2$

17. **a.** Ethylamine is a stronger base because when the lone pair on the nitrogen in aniline is protonated, it can no longer be delocalized into the benzene ring.

b. Ethoxide ion is a stronger base because a negatively charged oxygen is a stronger base than a neutral nitrogen.

c. Ethoxide ion is a stronger base because when the phenolate ion is protonated, the pair of electrons that is protonated can no longer be delocalized into the benzene ring.

18. The carboxylic acid is the most acidic because its conjugate base has greater resonance stabilization than does the conjugate base of phenol. The alcohol is the least stable because, unlike the negative change on the conjugate base of phenol, the negative change on its oxygen atom cannot be delocalized.

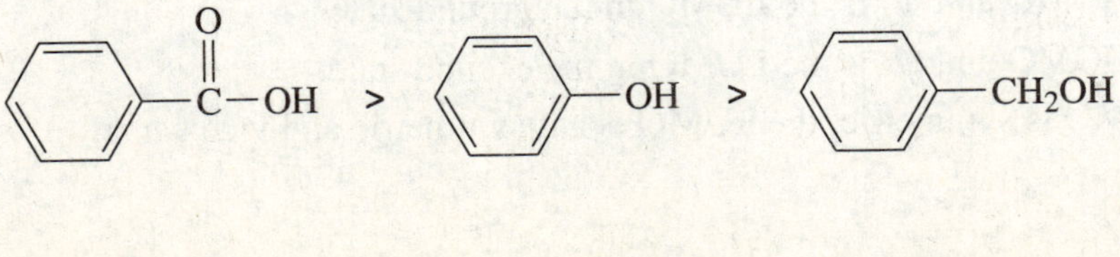

19. Solved in the text.

20. Electron withdrawal will make it a stronger acid (because it will make the conjugate base weaker). Electron donation will make it a weaker acid (because it will make the conjugate base stronger).

The pK_a values show that the methoxy substituted acid is a weaker acid (it has a larger pK_a value). Therefore, we know that resonance electron donation is a more important effect than inductive electron withdrawal.

21.

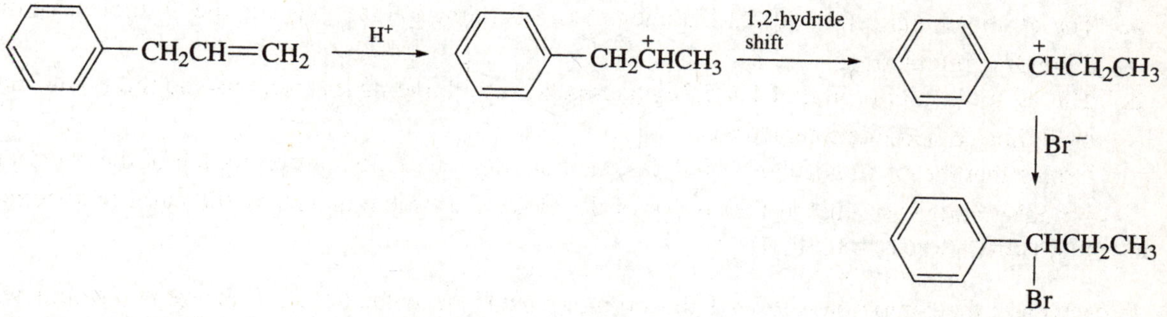

22. **a.** Solved in the text.

b. The contributing resonance structures show that there are two sites that could be protonated.

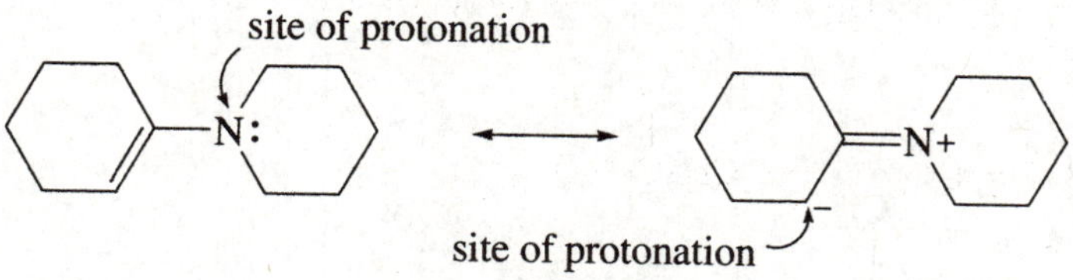

23. **a.**

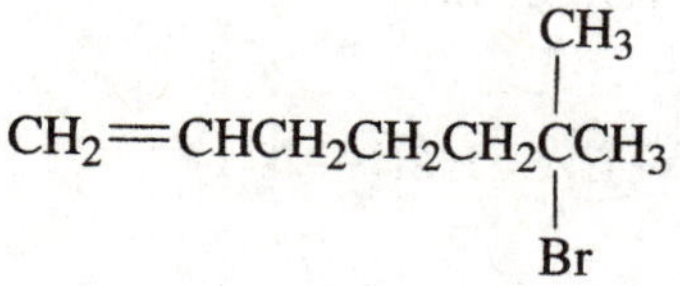

The more reactive double bond is the one that forms a tertiary carbocation.

b.

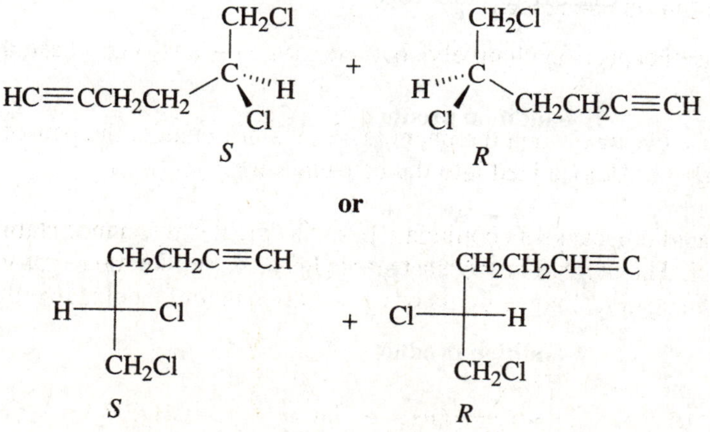

The double bond is more reactive than the triple bond. The reaction forms a new asymmetric center so a pair of enantiomers is formed.

c. 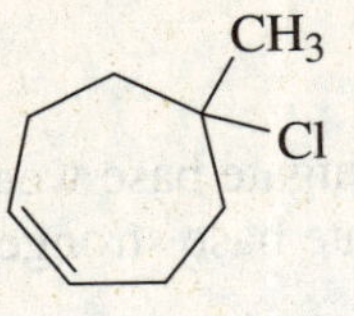 The more reactive double bond is the one that forms a tertiary carbocation.

24. The indicated double bond is the most reactive in an electrophilic addition reaction because addition of an electrophile to this double bond forms the most stable carbocation (a tertiary allylic carbocation).

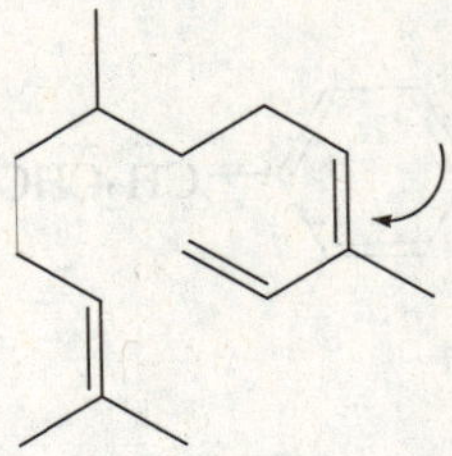

25. a. $CH_3CH{=}CH{-}CH{=}CHCH_3$

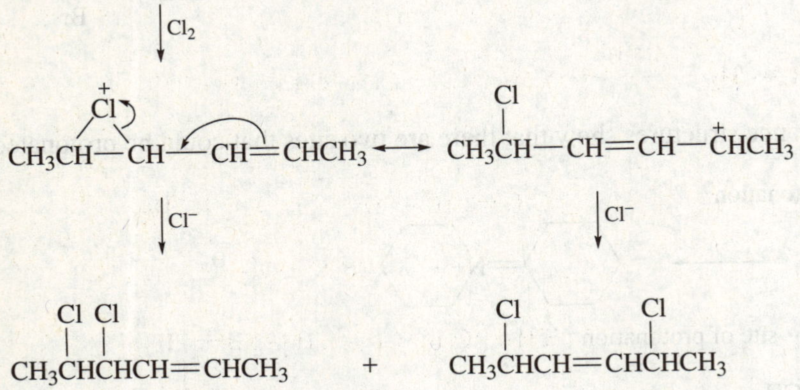

1,2- addition product + 1,4- addition product

b. 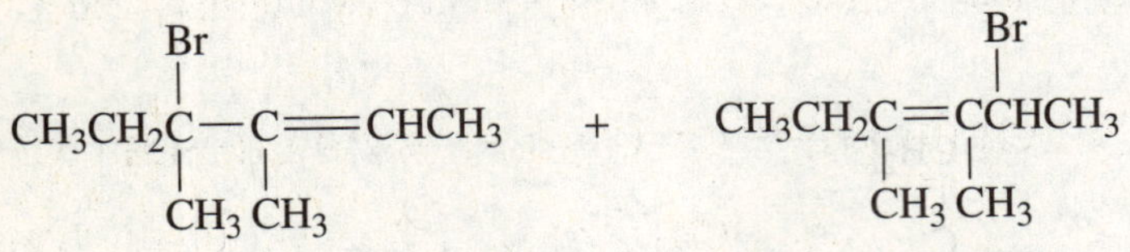

1, 2-addition product + 1, 4-addition product

c. 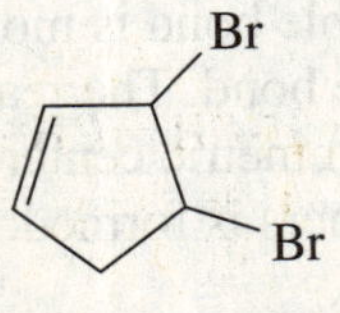 + 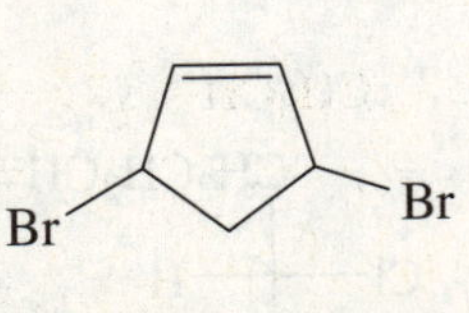

1, 2-addition product 1, 4-addition product

26.

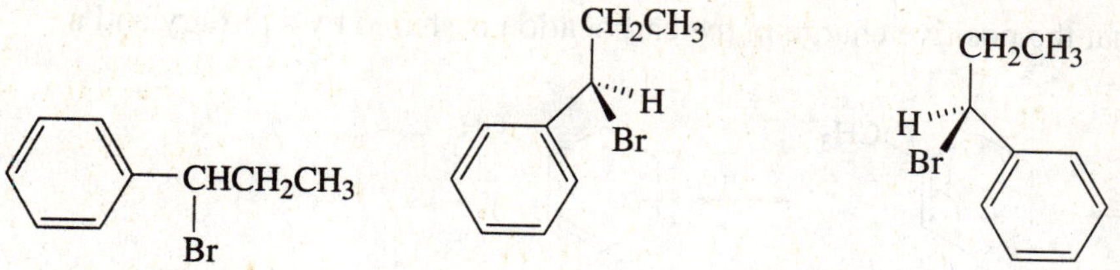

product

stereoisomers of the product

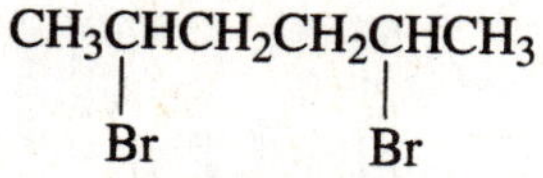

product

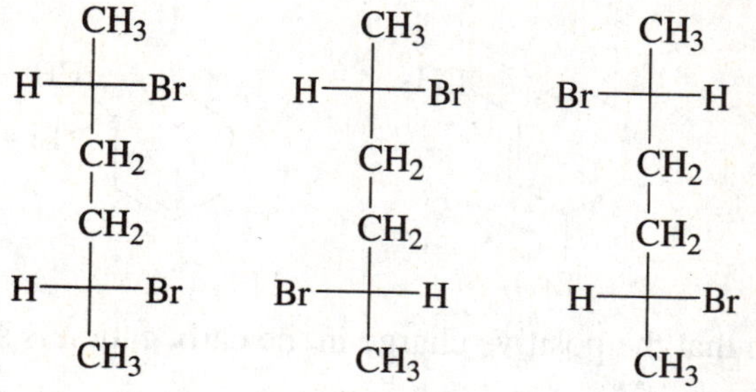

27.

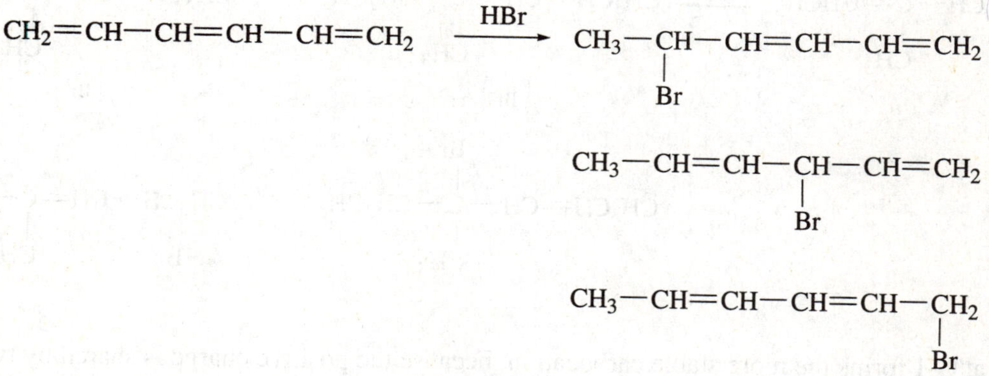

28. **a.** The proton adds so that the positive charge in the carbocation is shared by a tertiary and a secondary carbon.

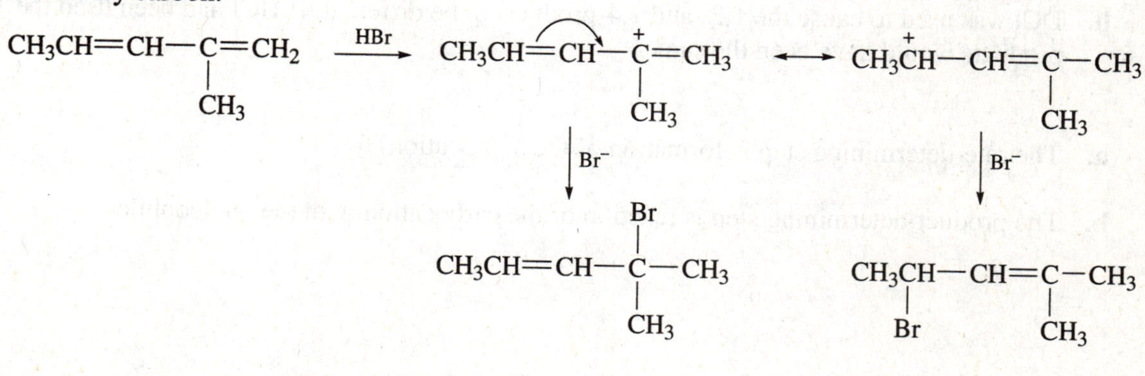

b. The proton adds so that the positive charge in the carbocation is shared by a tertiary and a secondary carbon.

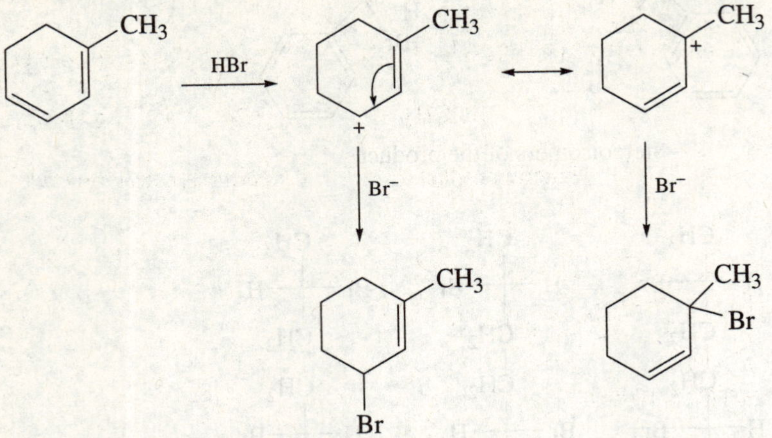

c. The proton adds so that the positive charge in the carbocation is shared by a tertiary and a secondary carbon.

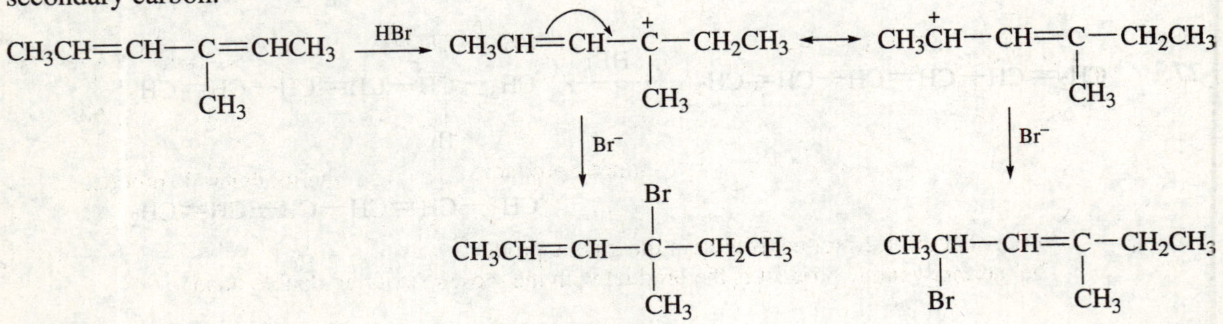

29. **a.** Addition at C-1 forms the more stable carbocation, because the positive charge is shared by two secondary allylic carbons. If it had added to C-4, the positive charge would have been shared by a secondary and a primary carbocation.

b. DCl was used to cause the 1,2- and 1,4-products to be different. If HCl had been used the 1,2-and 1,4-products would have been the same.

30. **a.** The rate-determining step is formation of the carbocation

b. The product-determining step is reaction of the carbocation with the nucleophile.

31. **a.** Solved in the text.

b.

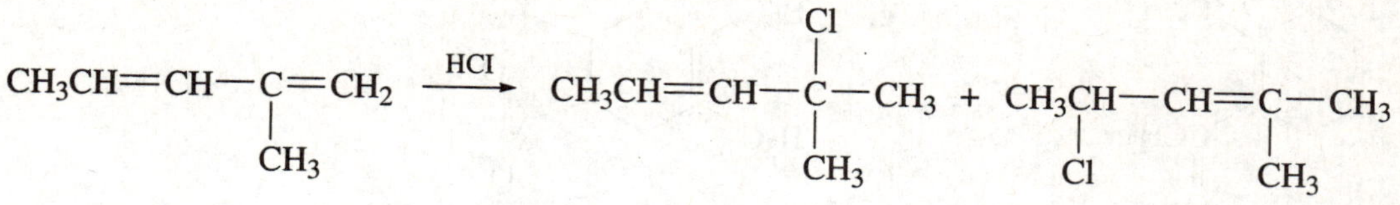

c.

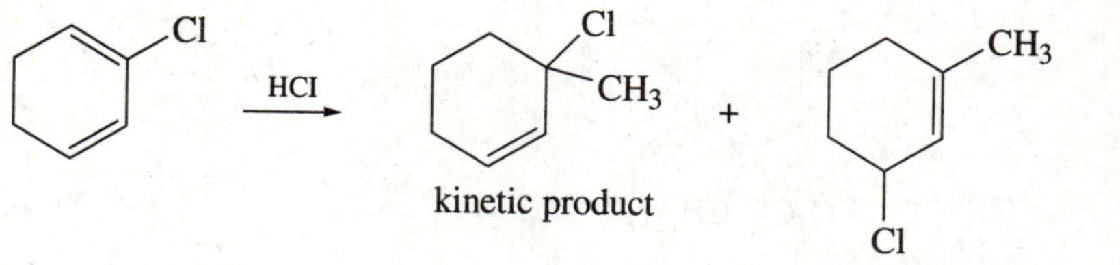

d.

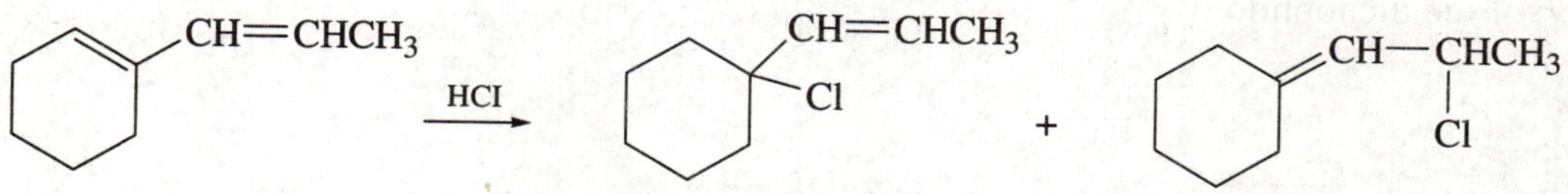

Notice that the 1,2-product is always the kinetic product.
The thermodynamic product is the product with the most substituted double bond

32. In order for a Diels-Alder reaction to occur, the overlapping orbitals of the reactants must have the same color (the same symmetry). In other words, they must both be symmetric or both be antisymmetric. In a [2+2] cycloaddition reaction at room temperature (in the ground state electronic configuration), the HOMO of one of the reactants will be symmetric and the LUMO of the other will be antiymmetric (see Figure 7.11 on p. 328 of the text). Thus, they will not have the same symmetry and the reaction will not occur.

In contrast, a [2+2] cycloaddition reaction does occur under photochemical conditions. Under photochemical conditions one of the alkenes will be in an excited state. Therefore, its HOMO will be antiymmetric and will be able to overlap with the antiymmetric LUMO of the other alkene.

33. **a.**

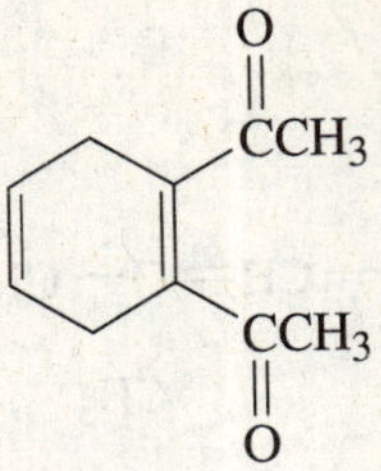

c.

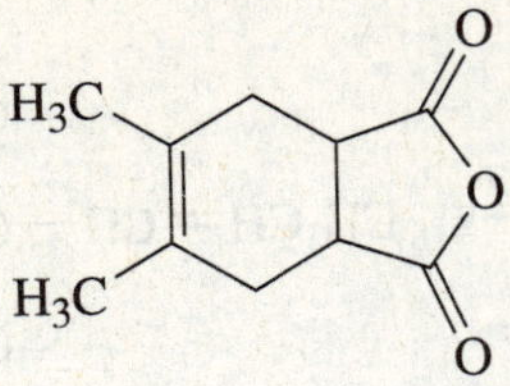

b.

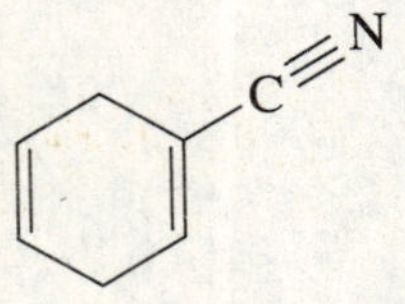

d.

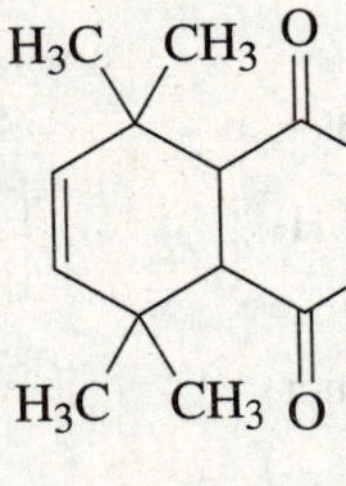

34. First draw the resonance contributors to determine where the charges are on the reactants. The major product is obtained by joining the negatively charged carbon of the diene with the positively charged carbon of the dienophile.

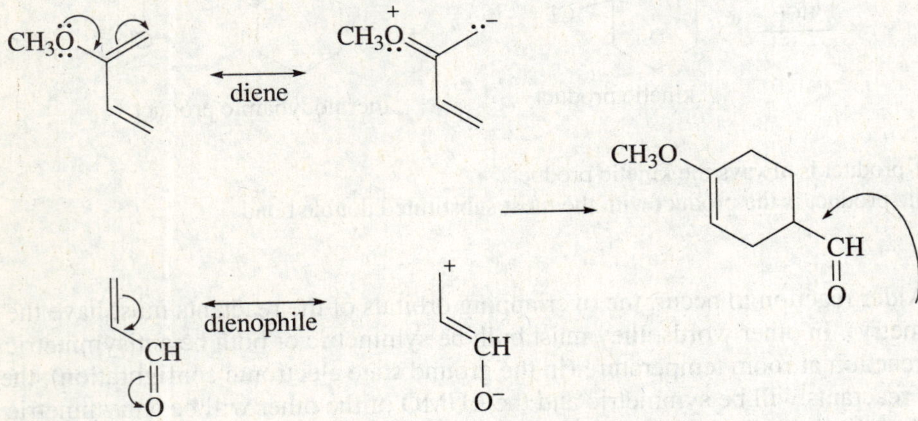

Because the reaction creates an asymmetric center, the product will be a racemic mixture.

35. **a.**

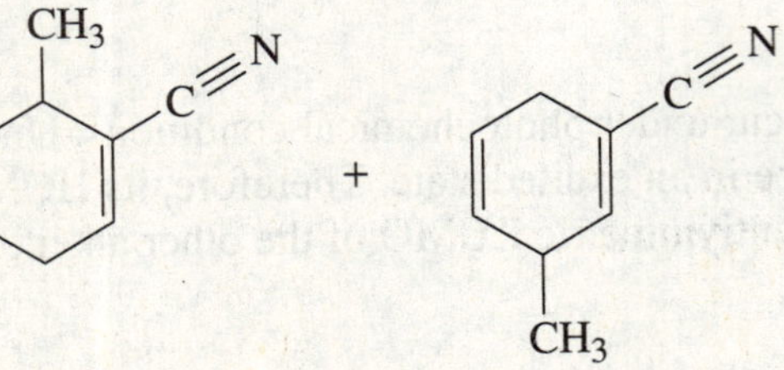

b.

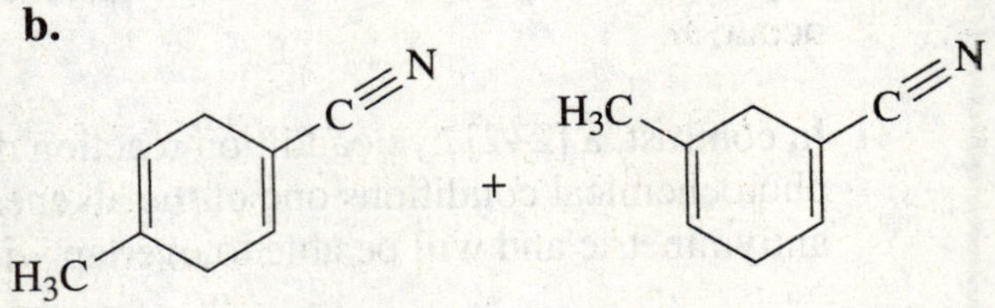

36. **a** and **d** will not react, because they are both locked in an *s*-trans conformation.
c and **e** will react, because they are both locked in an *s*-cis conformation.
b and **f** will react, because they can rotate into an *s*-cis conformation.

37. Solved in the text.

38. **a.** It is not optically active because it is a meso compound.
(It has 2 asymmetric centers and a plane of symmetry.)

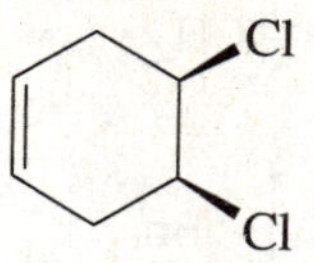

b. It is not optically active because it is a racemic mixture. (Identical amounts of the enantiomers will be obtained.)

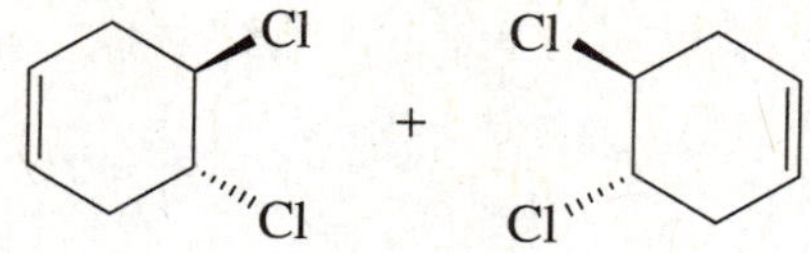

39. **a.** 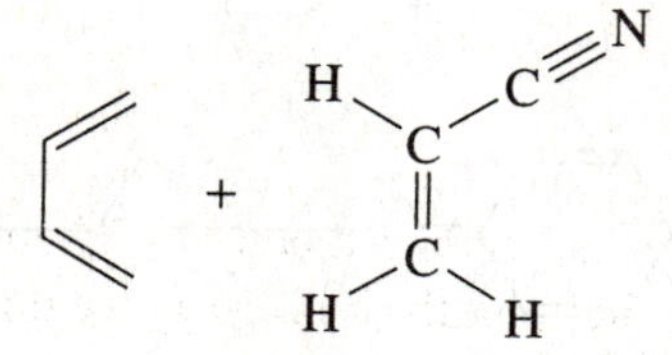 **c.** 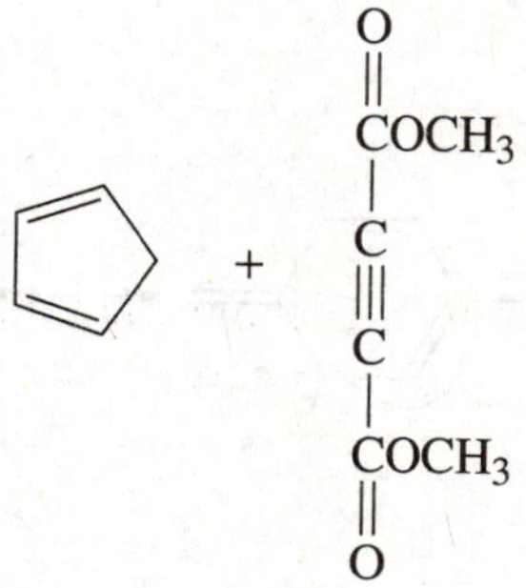 **e.** 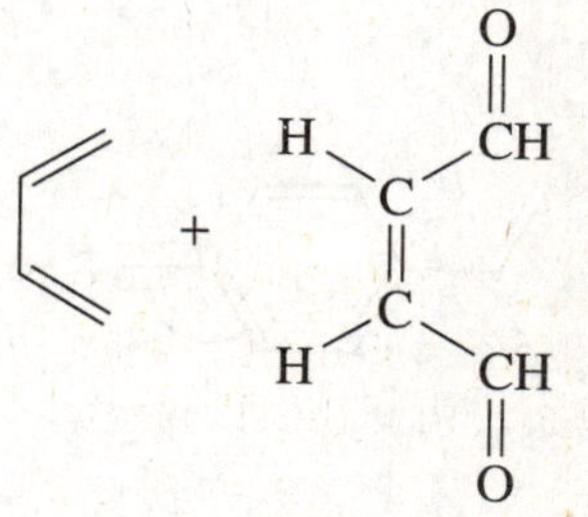

b. 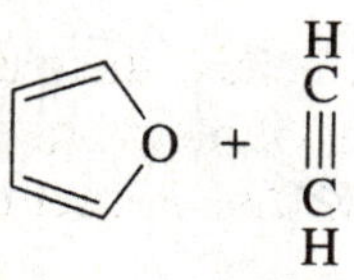 **d.** 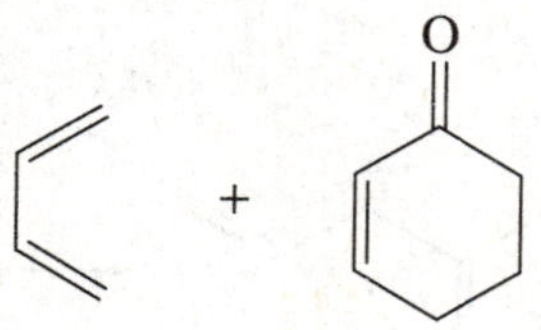 **f.** 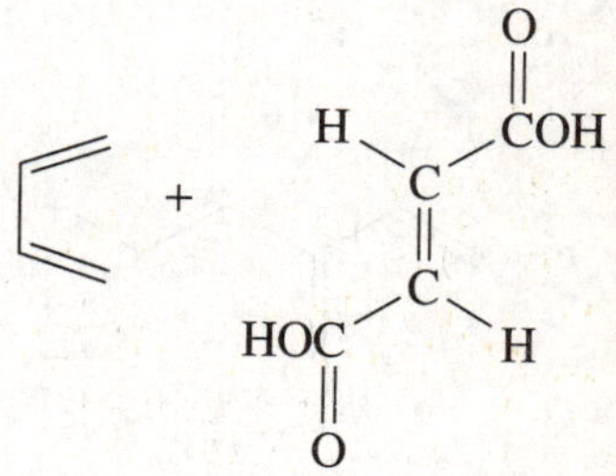

40. **a, b, d, e, i, l, m, n**

41. **a.** and **b.**

1. $\ddot{\text{C}}\text{H}_2^{-}\text{—}\overset{+}{\text{N}}\text{≡N:}$ ⟷ $\text{CH}_2\text{=}\overset{+}{\text{N}}\text{=}\ddot{\text{N}}\text{:}^{-}$

More stable, because the negative charge is on nitrogen rather than on carbon.

2. $\text{:N≡}\overset{+}{\text{N}}\text{—}\ddot{\text{O}}\text{:}^{-}$ ⟷ $^{-}\text{:}\ddot{\text{N}}\text{=}\overset{+}{\text{N}}\text{=}\ddot{\text{O}}\text{:}$

More stable, because the negative charge is on oxygen rather than on nitrogen.

3. $^{-}\text{:}\ddot{\text{O}}\text{—}\ddot{\text{N}}\text{=}\ddot{\text{O}}\text{:}$ ⟷ $\text{:O=}\ddot{\text{N}}\text{—}\ddot{\text{O}}\text{:}^{-}$

Both are equally stable.

Additional resonance structures could be drawn for each of these three species, but they are relatively unstable because each has an incomplete octet.

42. **a.** 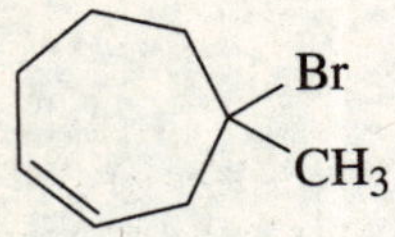

b. 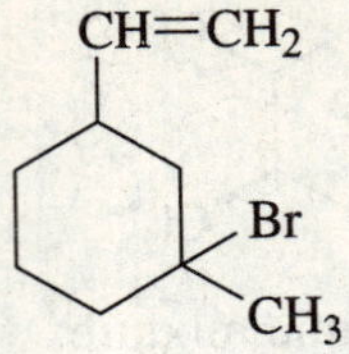

43. **a.** 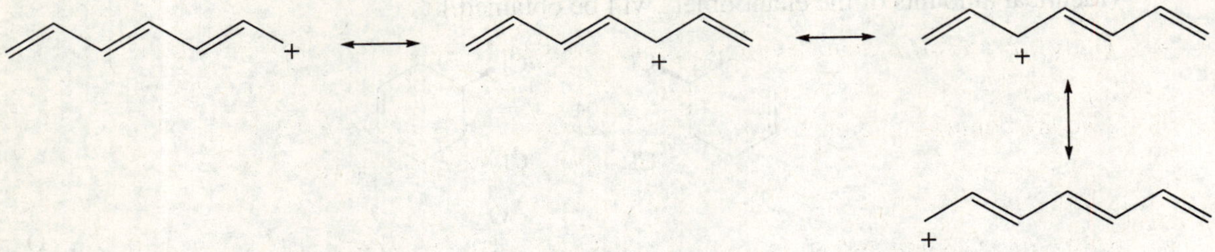

b. 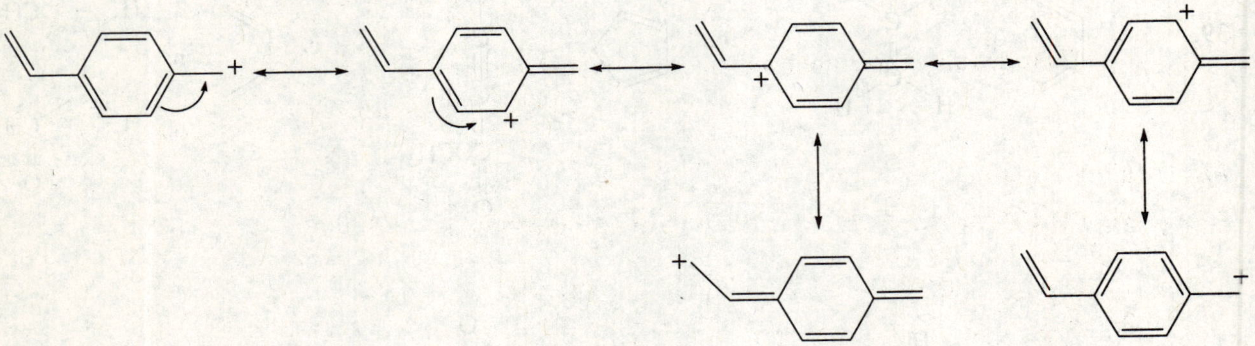

c. 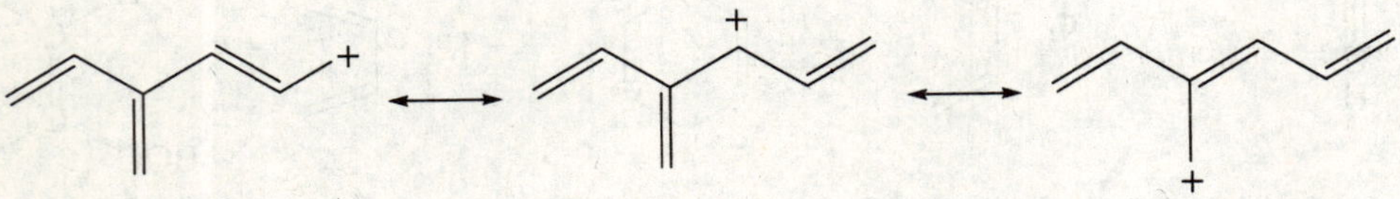

d. 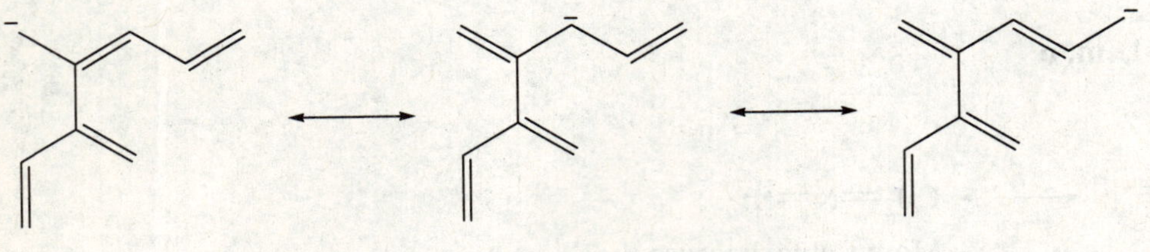

44. 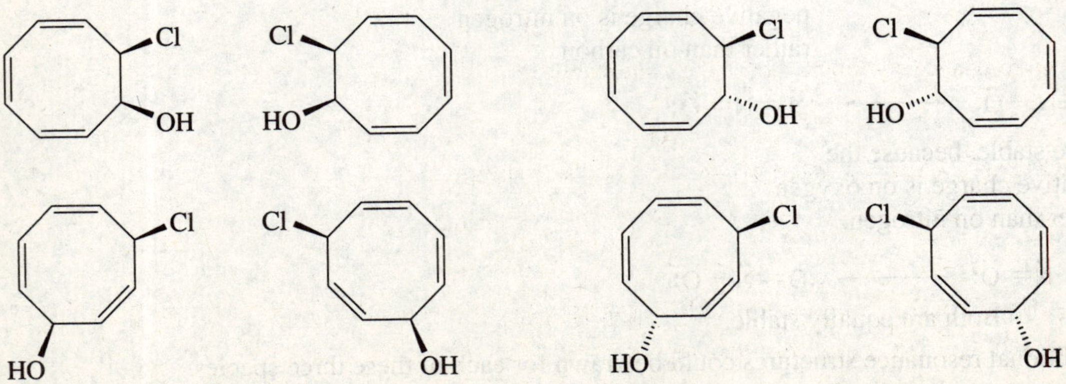

45.
a. different compounds
b. different compounds
c. resonance contributors
d. resonance contributors
e. different compounds

Notice that in the structures that are different compounds, both atoms and electrons have changed their locations. In contrast, in structures that are resonance contributors, only the electrons have moved.

46.
a. There are six linear dienes with molecular formula C_6H_{10} .

b. Two are conjugated dienes.
$CH_2{=}CHCH{=}CHCH_2CH_3$
$CH_3CH{=}CHCH{=}CHCH_3$

c. Two are isolated dienes.
$CH_2{=}CHCH_2CH{=}CHCH_3$
$CH_2{=}CHCH_2CH_2CH{=}CH_2$

There are also two cumulated dienes.
$CH_2{=}C{=}CHCH_2CH_2CH_3$
$CH_3CH{=}C{=}CHCH_2CH_3$

47.
a.

1. $CH_3CH{=}CH{-}\ddot{O}CH_3 \longleftrightarrow CH_3\overset{-}{\ddot{C}}H{-}CH{=}\overset{+}{O}CH_3$

2. 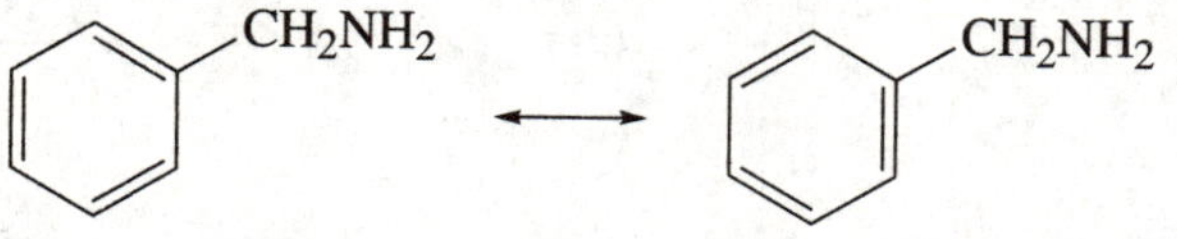

the two resonance contributors have the same stability

3. 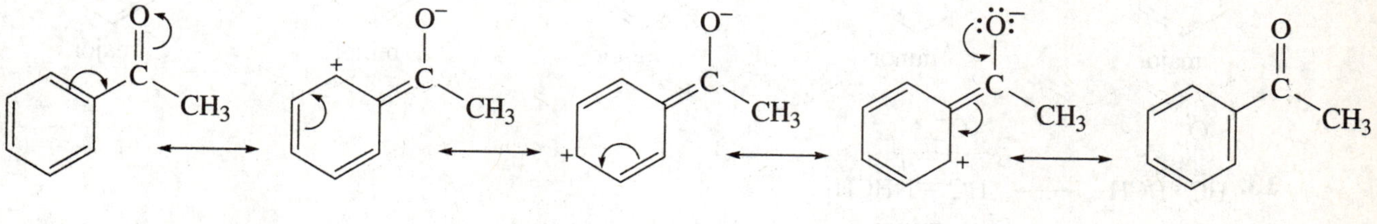

4. 

the two resonance contributors have the same stability

5. 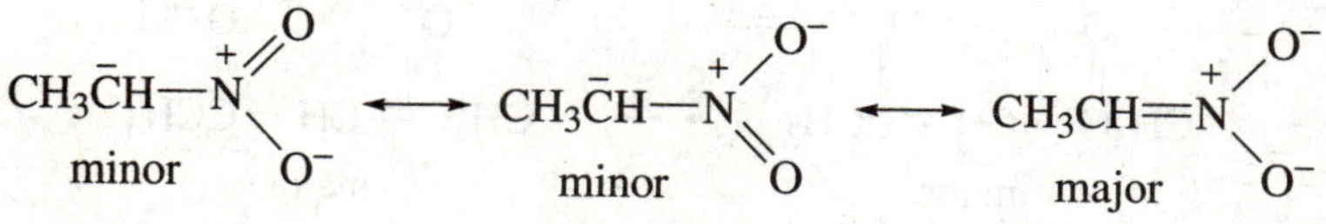

6. $CH_3CH{=}CH\overset{+}{C}H_2 \rightleftharpoons CH_3\overset{+}{C}HCH{=}CH_2$
minor major

7. 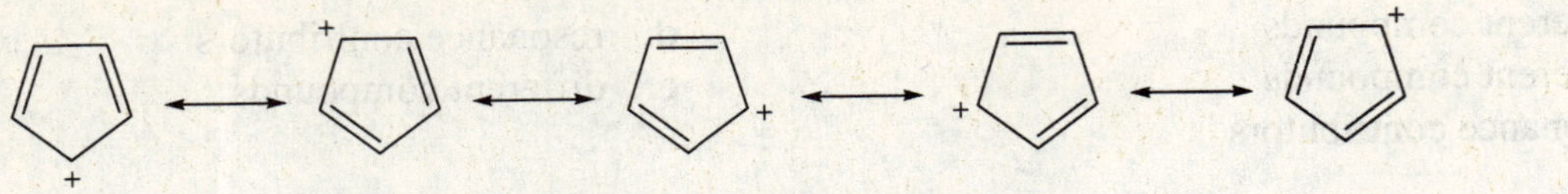

The five contributors are equally stable

8. $CH_3CH_2\overset{O}{\overset{\|}{C}}-OCH_2CH_3$ (major) $\longleftrightarrow$ $CH_3CH_2\overset{O^-}{\overset{|}{C}}=\overset{+}{O}CH_2CH_3$ (minor)

9. $CH_3CH=CHCH=CH\overset{+}{C}H_2$ (minor) $\longleftrightarrow$ $CH_3\overset{+}{C}HCH=CHCH=CH_2$ (major) $\longleftrightarrow$ $CH_3CH=CH\overset{+}{C}HCH=CH_2$ (minor (isolated double bonds))

10. $^-CH_2\overset{O}{\overset{\|}{C}}OCH_2CH_3$ (minor) $\longleftrightarrow$ $CH_2=\overset{O^-}{\overset{|}{C}}OCH_2CH_3$ (major)

11. $CH_3\bar{C}HC\equiv N$ (minor) $\longleftrightarrow$ $CH_3CH=C=N^-$ (major)

12. 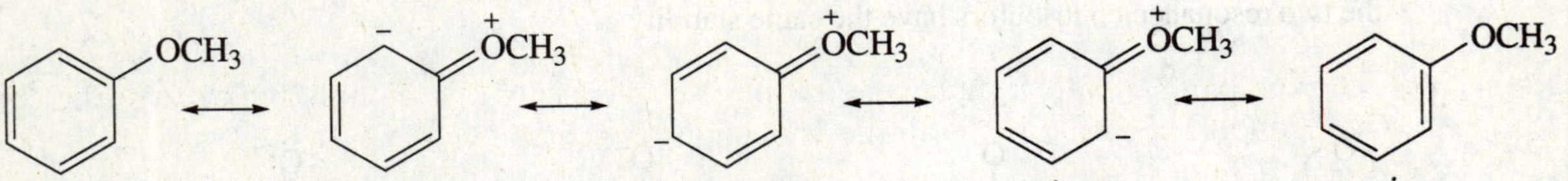

major minor minor minor major

13. $H\overset{O}{\overset{\|}{C}}NHCH_3$ (major) $\longleftrightarrow$ $H\overset{O^-}{\overset{|}{C}}=\overset{+}{N}HCH_3$ (minor)

14. $H\overset{O}{\overset{\|}{C}}CH=CH\bar{C}H_2$ (minor) $\longleftrightarrow$ $H\overset{O}{\overset{\|}{C}}\bar{C}HCH=CH_2$ (minor) $\longleftrightarrow$ $H\overset{O^-}{\overset{|}{C}}=CHCH=CH_2$ (major)

15. $CH_3\overset{O}{\overset{\|}{C}}-\bar{C}H-\overset{O}{\overset{\|}{C}}CH_3$ (minor) $\longleftrightarrow$ $CH_3\overset{O}{\overset{\|}{C}}-CH=\overset{O^-}{\overset{|}{C}}CH_3$ (major) $\longleftrightarrow$ $CH_3\overset{O^-}{\overset{|}{C}}=CH-\overset{O}{\overset{\|}{C}}CH_3$ (major)

b. 2, 4 and 7

48. Both compounds form the same product when they are hydrogenated, so the difference in heats of hydrogenation will depend only on the difference in the stabilities of the reactants. Because 1,2-pentadiene has cumulated double bonds and 1,4-pentadiene has isolated double bonds, 1,2-pentadiene is less stable and, therefore, will have a greater heat of hydrogenation (a more negative $\Delta H°$).

$$CH_2{=}C{=}CHCH_2CH_3 \xrightarrow[\text{Pd/C}]{H_2} CH_3CH_2CH_2CH_2CH_3$$

1,2-pentadiene

$$CH_2{=}CHCH_2CH{=}CH_2 \xrightarrow[\text{Pd/C}]{H_2} CH_3CH_2CH_2CH_2CH_3$$

1,4-pentadiene

49. **a.** $CH_3\overset{+}{C}HCH{=}CH_2$

This makes the greater contribution because the positive charge is on a secondary carbon.

b. 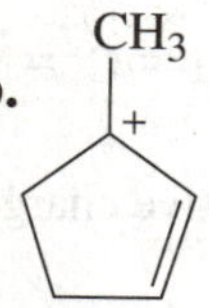

This makes the greater contribution because the positive charge is on a tertiary carbon.

c. 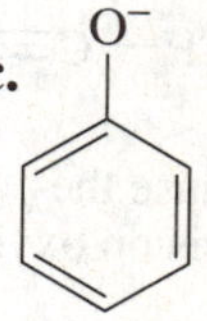

This makes the greater contribution because the negative charge is on an oxygen.

50. **a.** The resonance contributors show that the carbonyl oxygen has the greater electron density.

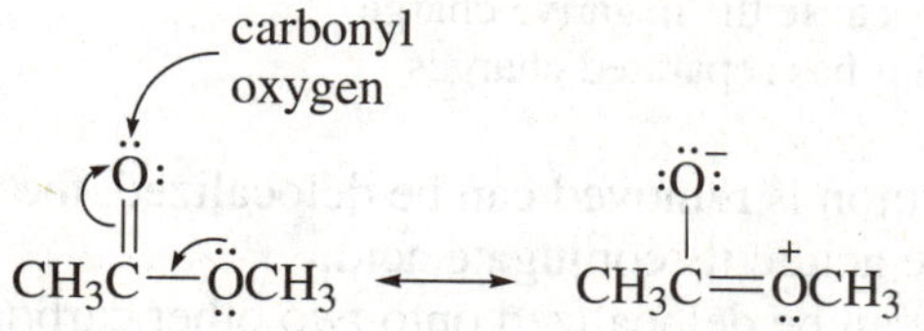

b. The compound on the right has the greater electron density on its nitrogen because the compound on the left has a resonance contributor with a positive charge on the nitrogen.

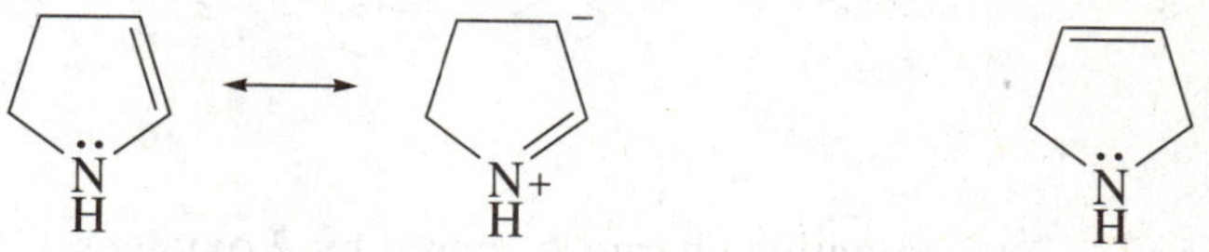

c. The compound with the cyclohexane ring has the greater electron density on its oxygen because the lone pair on the nitrogen can be delocalized onto the oxygen.
There is less delocalization onto oxygen by the lone pair in the compound with the benzene ring because the lone pair can also be delocalized away from the oxygen into the benzene ring.

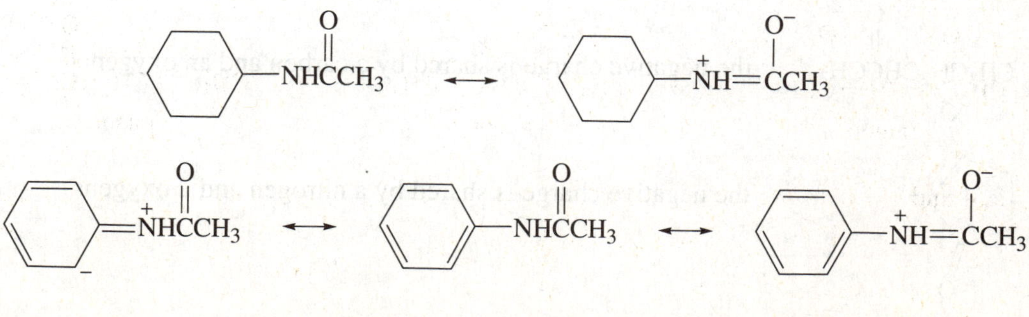

51. The methyl group on benzene can lose a proton easier than the methyl group on cyclohexane because the electrons left behind on the carbon in the former can be delocalized into the benzene ring. In contrast, the electrons left behind in the other compound are localized on the carbon.

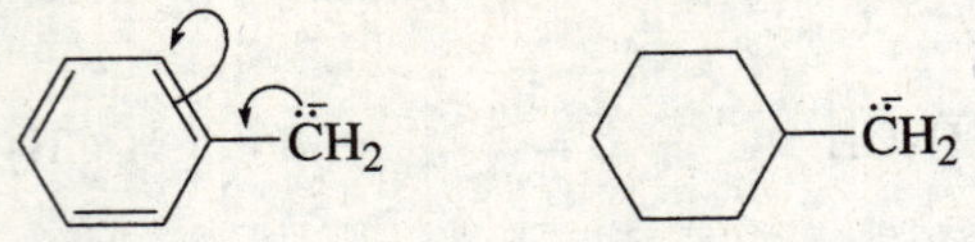

52. The carbocation is stable because the positive charge is shared by 10 carbon atoms (the central carbon and 3 carbons of each of the 3 benzene rings).

53.

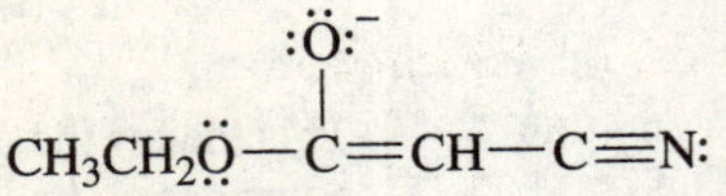

#1 most stable because the negative charge is on oxygen

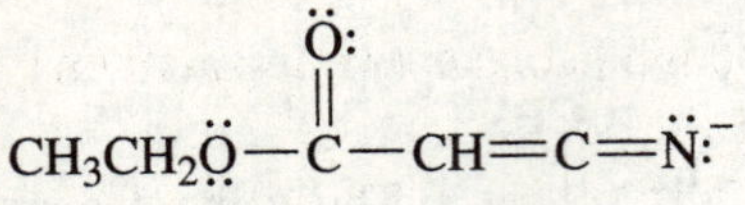

#2 because the negative charge is on nitrogen

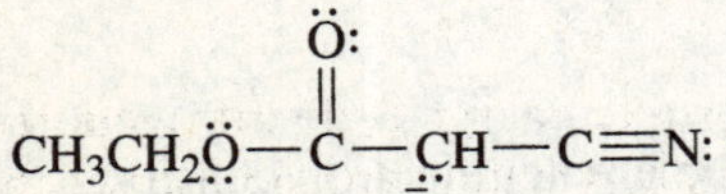

#3 because the negative charge is on carbon

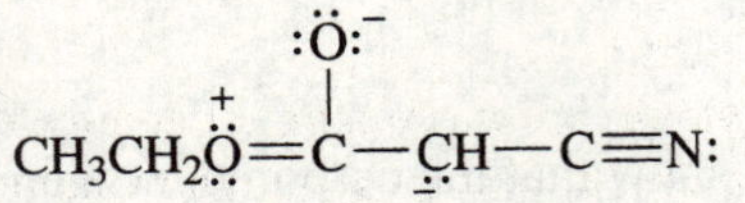

#4 the least stable because the negative charge is on carbon and it has separated charges

54. The more the electrons that are left behind when the proton is removed can be delocalized, the greater the stability of the base. The more stable the base, the more acidic its conjugate acid. The negative charge on the base in the first compound can be delocalized onto two other carbons; the negative charge on the base in the second compound can be delocalized onto one other carbon; the negative charge on the base in the last compound cannot be delocalized.

55. **a.** $\mathrm{CH_3\overset{\overset{O}{\|}}{C}O^-}$ the negative charge is shared by 2 oxygens

b. $\mathrm{CH_3\overset{\overset{O}{\|}}{C}\underline{C}H\overset{\overset{O}{\|}}{C}CH_3}$ the negative charge is shared by a carbon and 2 oxygens

c. $\mathrm{CH_3CH_2\underline{C}H\overset{\overset{O}{\|}}{C}CH_3}$ the negative charge is shared by a carbon and an oxygen

d. 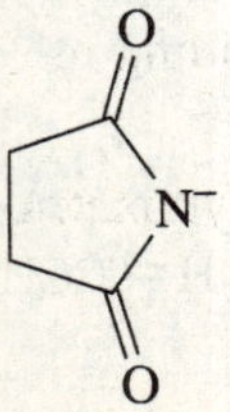 the negative charge is shared by a nitrogen and 2 oxygens

56. The stronger base is the less stable base of each pair in Problem 55.

a. $CH_3CH_2O^-$ Less stable because the negative charge cannot be delocalized.

b. $CH_3\overset{O}{\overset{\|}{C}}\underline{C}HCH_2\overset{O}{\overset{\|}{C}}CH_3$ Less stable because the negative charge can be delocalized onto only one carbonyl oxygen.

c. $CH_3\bar{C}HCH_2\overset{O}{\overset{\|}{C}}CH_3$ Less stable because the negative charge cannot be delocalized.

d. 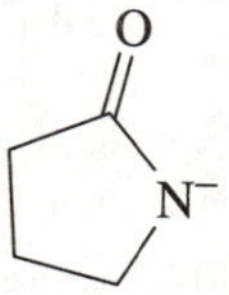

Less stable because the negative charge can be delocalized onto only one carbonyl oxygen.

57. The resonance contributors of pyrrole are more stable because the positive charge is on nitrogen. In furan, the positive charge is on oxygen which, being more electronegative, is less stable with a positive charge.

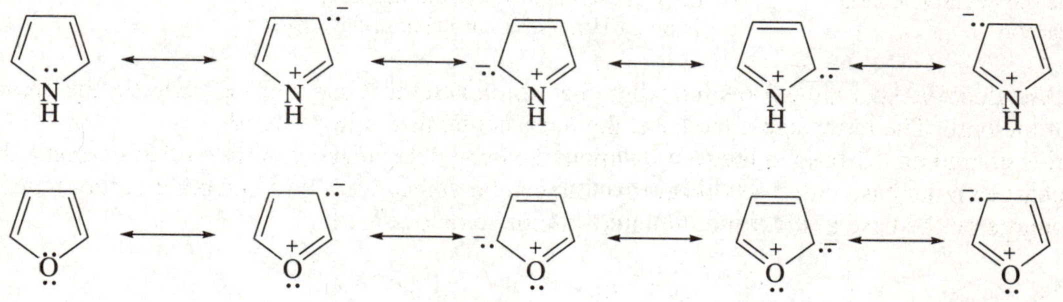

58. **A** is the most acidic because the electrons left behind when the proton is removed can be delocalized onto two oxygen atoms. **B** is the next most acidic because the electrons left behind when the proton is removed can be delocalized onto one oxygen atom. **C** is the least acidic because the electrons left behind when the proton is removed cannot be delocalized.

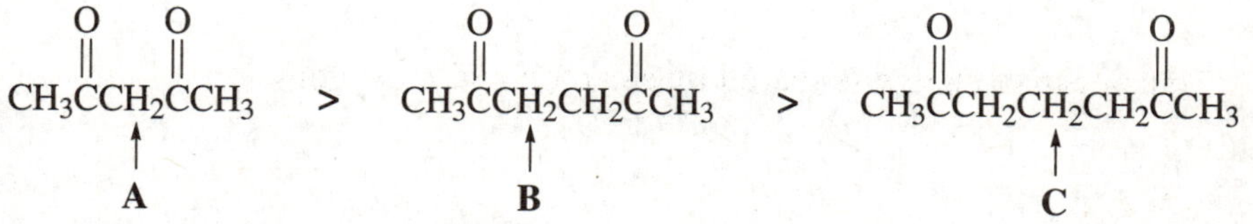

59. **a.** It has 8 molecular orbitals.

b. ψ_1, ψ_2, ψ_3, and ψ_4 are bonding molecular orbitals; ψ_5, ψ_6, ψ_7, and ψ_8 are antibonding molecular orbitals.

c. ψ_1, ψ_3, ψ_5 and ψ_7 are symmetric molecular orbitals; ψ_2, ψ_4, ψ_6, and ψ_8 are antisymmetric molecular orbitals.

d. ψ_4 is the HOMO and ψ_5 is the LUMO in the ground state.

e. ψ_5 is the HOMO and ψ_6 is the LUMO in the excited state.

f. The HOMO is symmetric, the LUMO is antisymmetric and vice versa.

g. It has 7 nodes between the nuclei. It also has one node that passes through the nuclei.

60. The reaction of 1,3-cyclohexadiene with Br_2 forms 3,4-dibromocyclohexene as the 1,2-addition product and 3,6-dibromocyclohexene as the 1,4-addition product. (Recall that in naming the compounds, the double bond is at the 1,2-position.) The reaction of 1,3-cyclohexadiene with HBr forms 3-bromocyclohexene as both the 1,2-addition product and the 1,4-addition product.

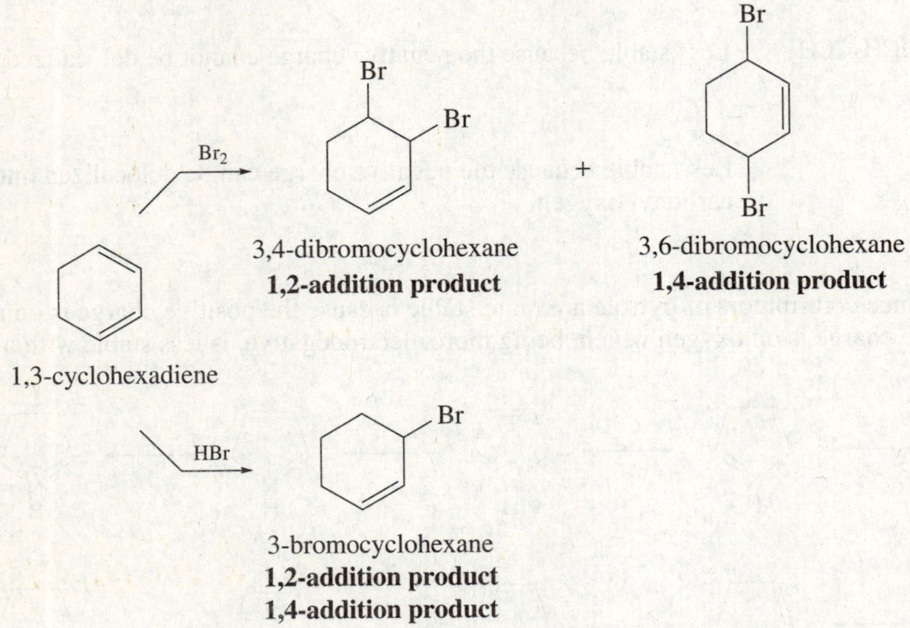

61. a.

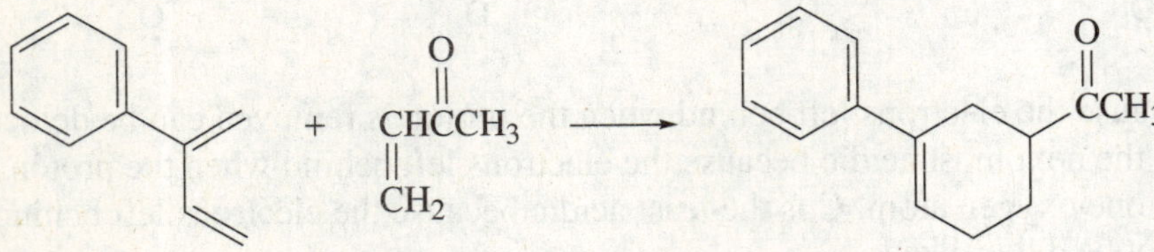

b.

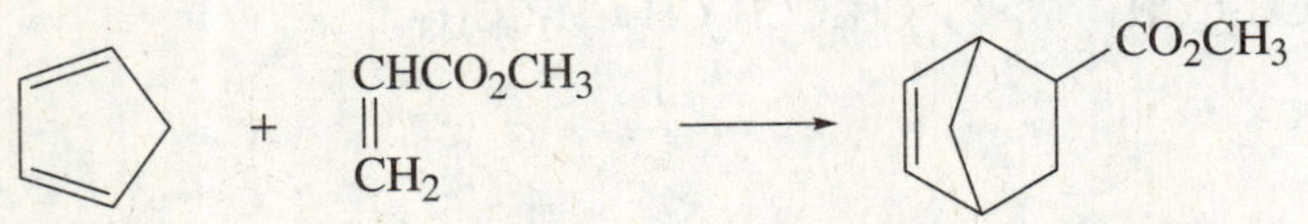

c.

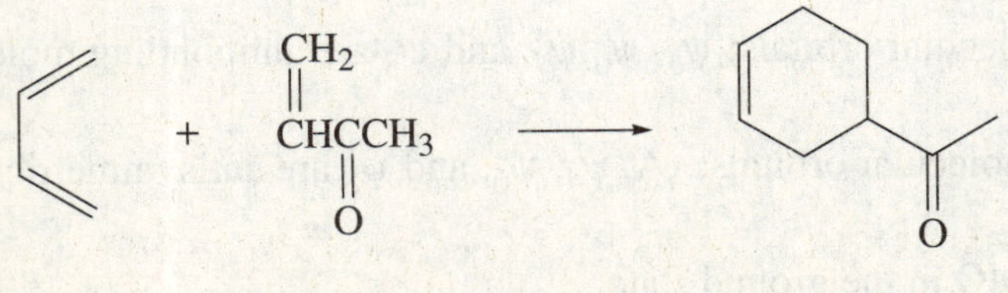

d. 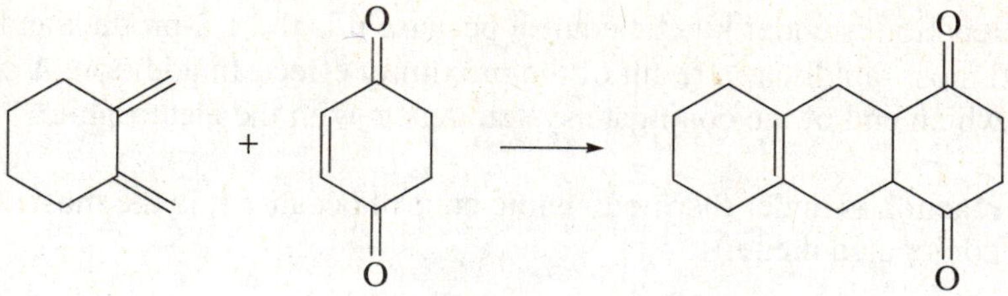

62. a. 1. 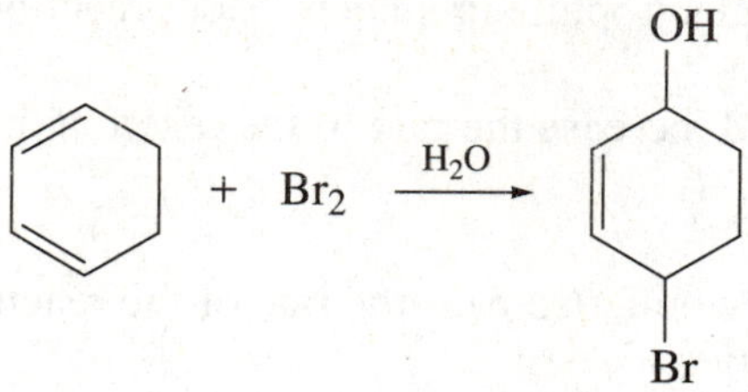

2. 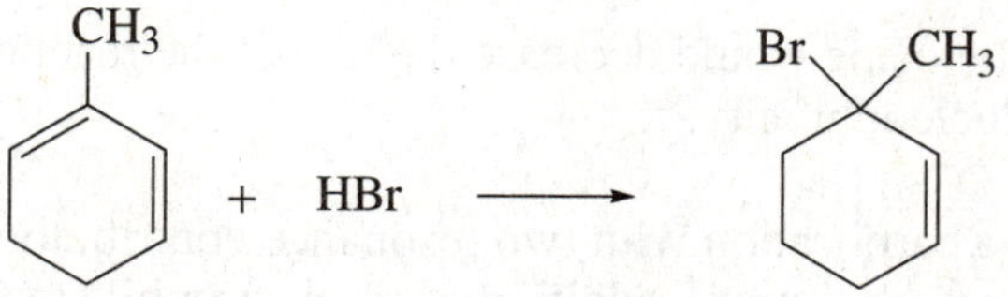 or

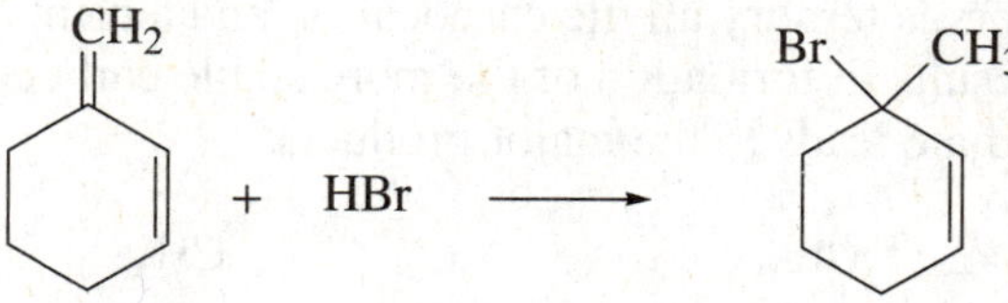

3. 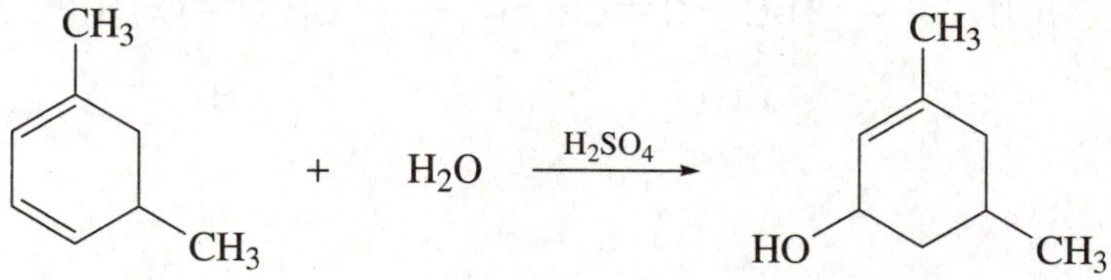 or

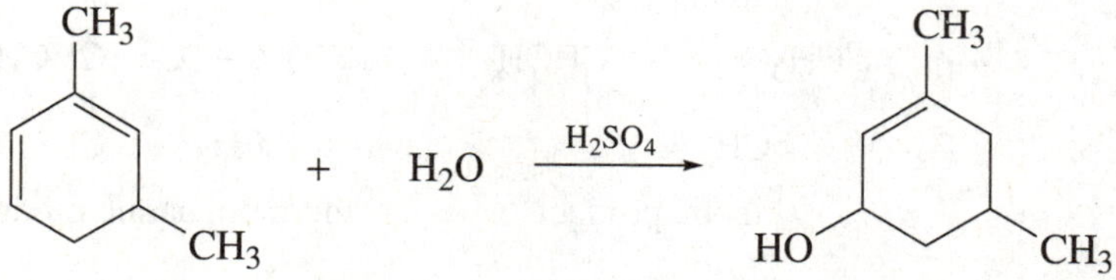

b. 1. 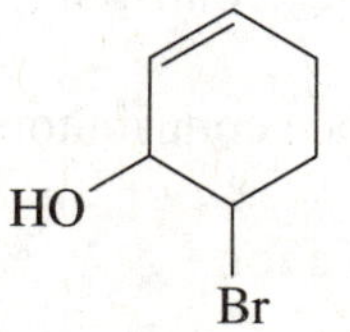 2. 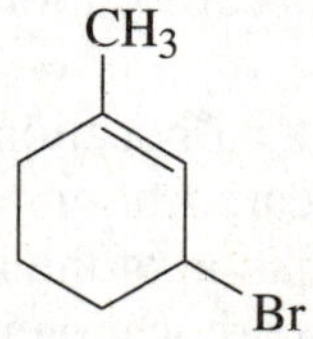 3. 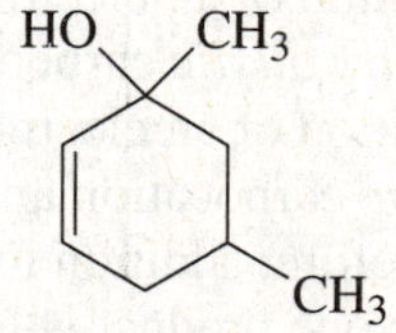

63. a. $CH_2{=}CHCH{=}CHCH{=}CH_2 \xrightarrow{HBr} CH_3\underset{|}{C}HCH{=}CHCH{=}CH_2$ +

1,3,5-hexatriene — Br **A**

$CH_3CH{=}CHCHCH{=}CH_2$ + $CH_3CH{=}CHCH{=}CHCH_2Br$

B Br **C**

b. **A** will predominate if the reaction is under kinetic control because it is the 1,2-product and therefore will be the product formed most rapidly as a result of the proximity effect. In addition, **A** will be the 1,2-product regardless of which end of the conjugate system reacts with the electrophile.

c. **C** will predominate if the reaction is under thermodynamic control because it is the most stable diene. (It is the most substituted conjugated diene.)

64. The diene is the nucleophile, and the dienophile is the electrophile in a Diels-Alder reaction.

a. An electron-donating substituent in the diene would increase the rate of the reaction, because electron donation would increase its nucleophilicity.

b. An electron-donating substituent in the dienophile would decrease the rate of the reaction, because electron donation would decrease its electrophilicity.

c. An electron-withdrawing substituent in the diene would decrease the rate of the reaction, because electron withdrawal would decrease its nucleophilicity.

65. **a.** Addition of an electrophile to C-1 forms a carbocation with two resonance contributors, a tertiary allylic carbocation and a secondary allylic carbocation. Addition of an electrophile to C-4 forms a carbocation with two resonance contributors, a tertiary allylic carbocation and a primary allylic carbocation. Therefore, addition to C-1 results in formation of the more stable carbocation intermediate, and the more stable intermediate leads to the major products.

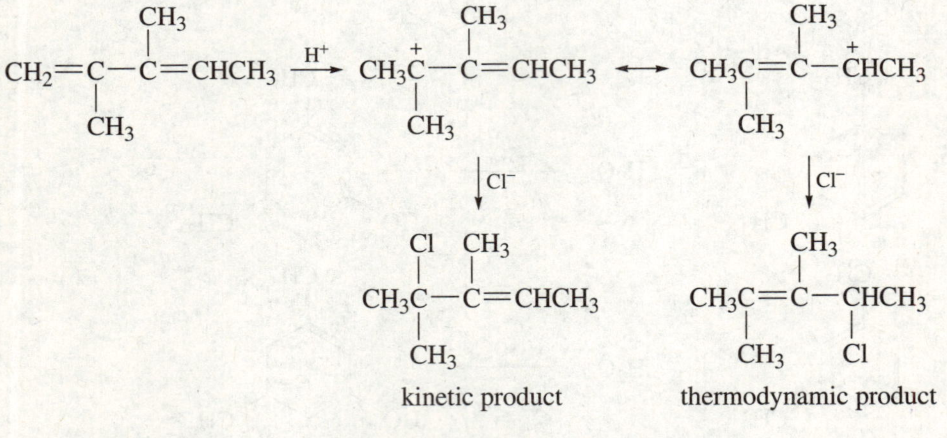

b. Addition of an electrophile to C-1 forms a carbocation with two resonance contributors; both are *tertiary allylic* carbocations.
Addition of an electrophile to C-4 forms a carbocation with two resonance contributors, a *secondary allylic* carbocation and a *primary allylic* carbocation.
Therefore, addition to C-1 results in formation of the more stable carbocation.
Only one product is formed, because the carbocation is symmetrical.

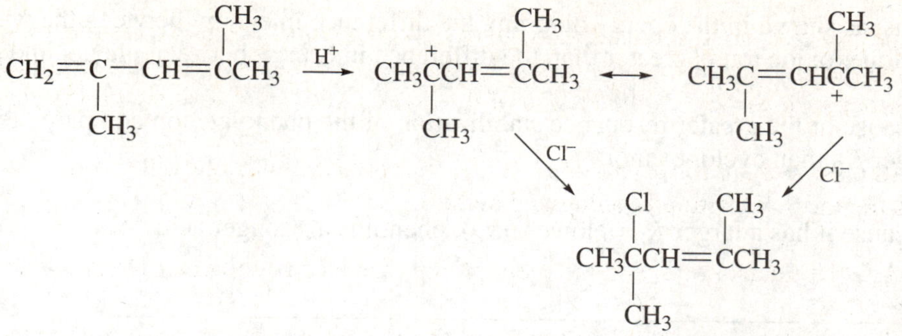

This is the only product because the carbocation is symmetrical.

66. **a.** and **d.**

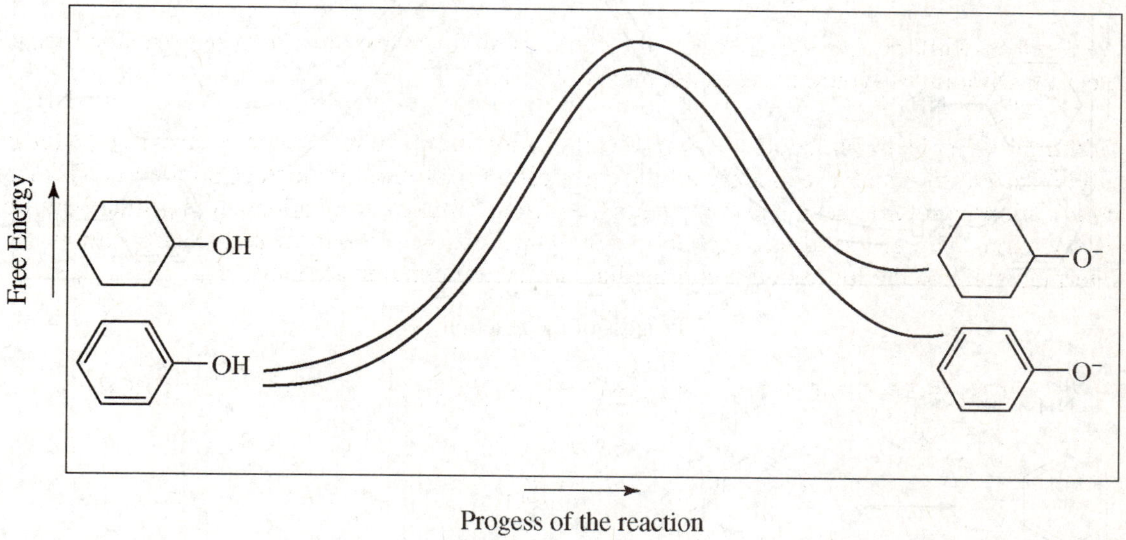

b.

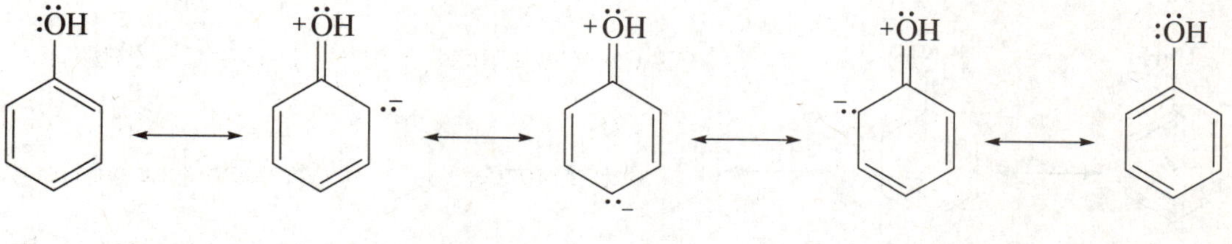

c.

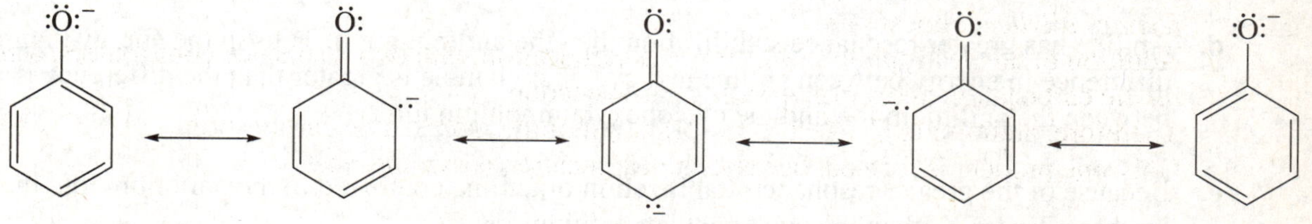

d. The resonance contributors in "**c**" are more stable than the resonance contributors in "**b**" because in "**b**" a positive charge is on the most electronegative atom (the oxygen).
Therefore, the phenolate ion has greater resonance stabilization than phenol.

Thus, as shown in the energy diagram, the difference in energy between the phenolate ion and the cyclohexoxide ion is greater than the difference in energy between phenol and cyclohexanol.

e. Because of the greater resonance stabilization of the phenolate ion compared to phenol, phenol has a larger K_a than cyclohexanol.

f. Because it has a larger K_a (a lower pK_a), phenol is a stronger acid.

67.

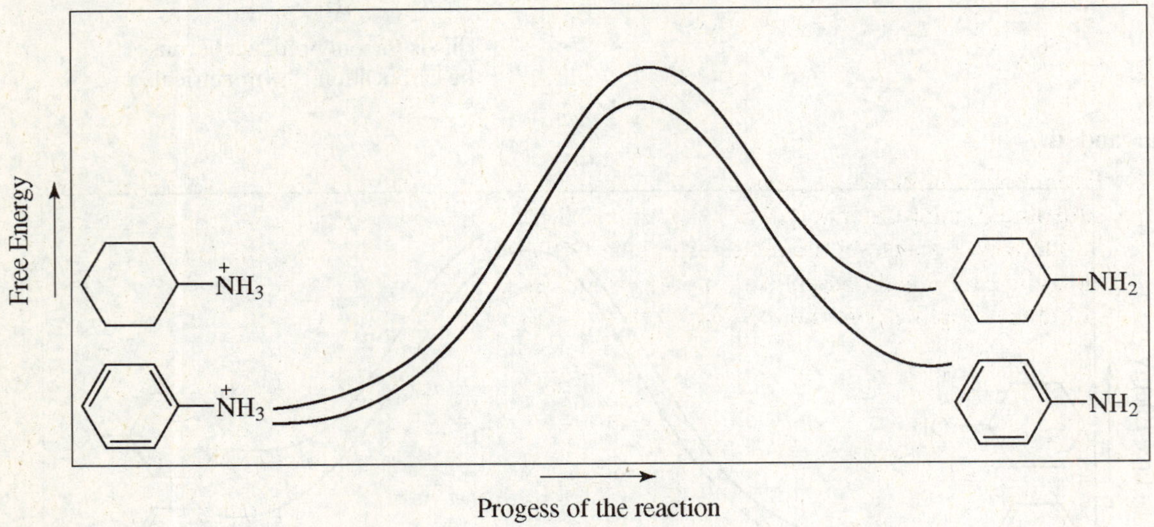

b.

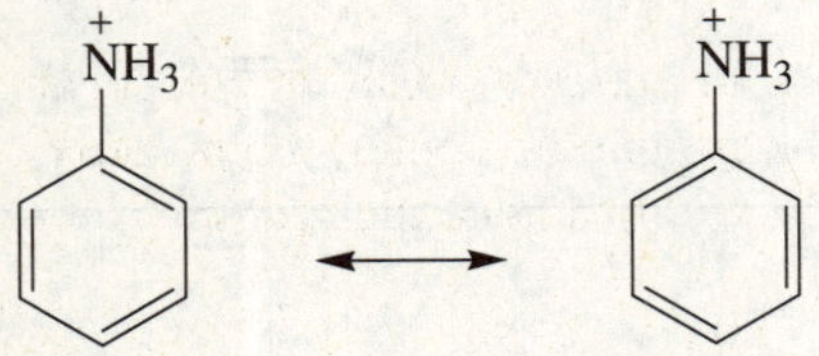

c.

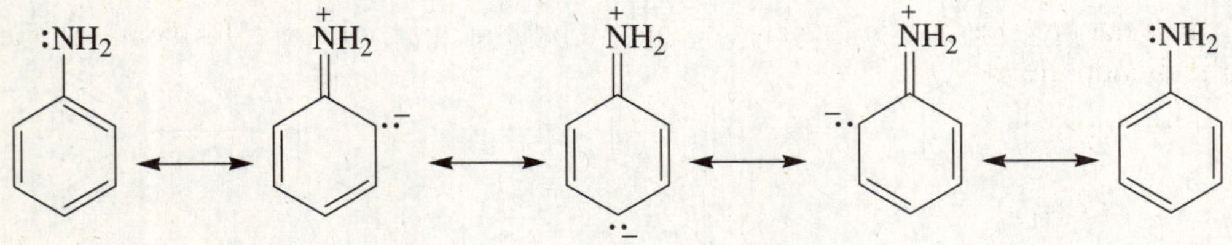

d. Aniline has greater resonance stabilization than the anilinium ion. Thus in the energy diagram, the difference in energy between aniline and cyclohexylamine is greater than the difference in energy between the anilinium ion and the cyclohexylammonium ion.

e. Because of the greater resonance stabilization of aniline compared to the anilinium ion, the anilinium ion has a larger K_a than the cyclohexylammonium ion.

f. Because it has a larger K_a (a lower pK_a), the anilinium ion is a stronger acid than the cyclohexylammonium ion. Therefore, cyclohexylamine is a stronger base than aniline. (The stronger the acid, the weaker its conjugate base.)

68. **a.** 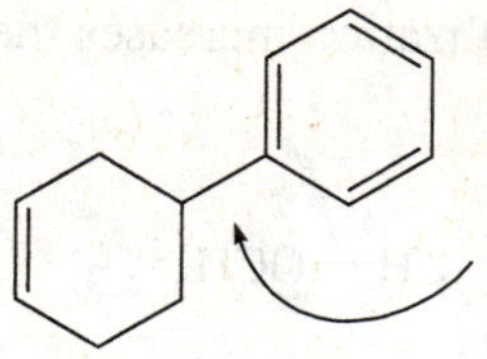

Because the reaction creates an aymmetric center in the product, the product will be a racemic mixture

b. 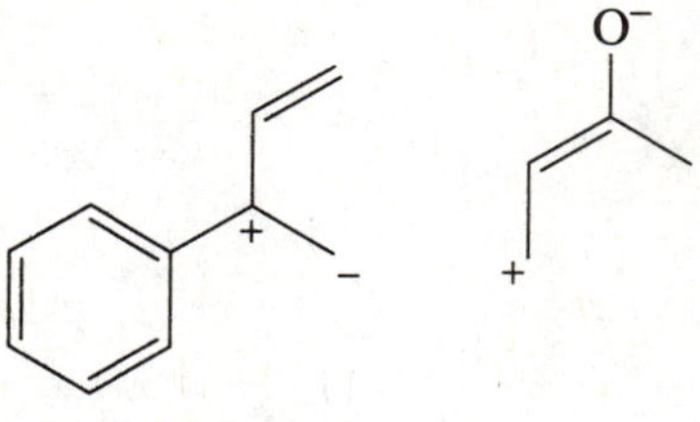

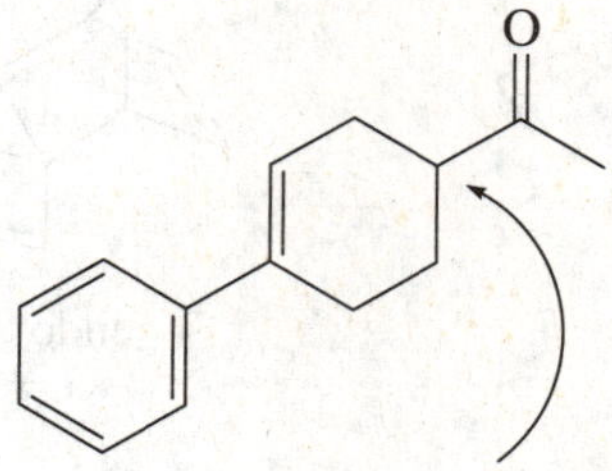

Even though both reacts are unsymmetrically substituted, the reactants will be aligned primarily in one way, because of the relatively stable tertiary benzylic cation and delocalization of the π electrons of the dieneophile onto the oxygen.

Because the reaction creates an aymmetric center, the product will be a racemic mixture.

c. 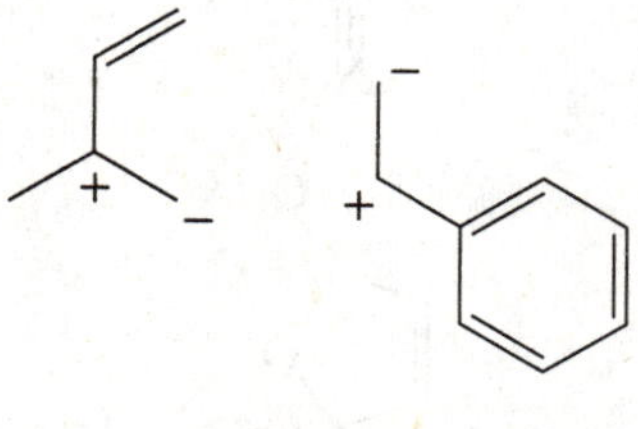

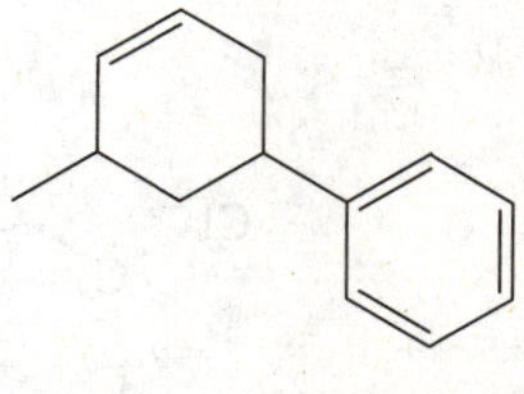

The products will be aligned primarily as shown because in the diene, a tertiary allylic carbocation is more stable than a secondary allylic carbocation. In the dienophile, a primary carbanion is more stable than a primary carbocation would be.

Because the reaction creates an aymmetric center, the product will be a racemic mixture.

69. The first pair is the preferred set of reagents because it has the more nucleophilic diene and the more electrophilic dienophile.

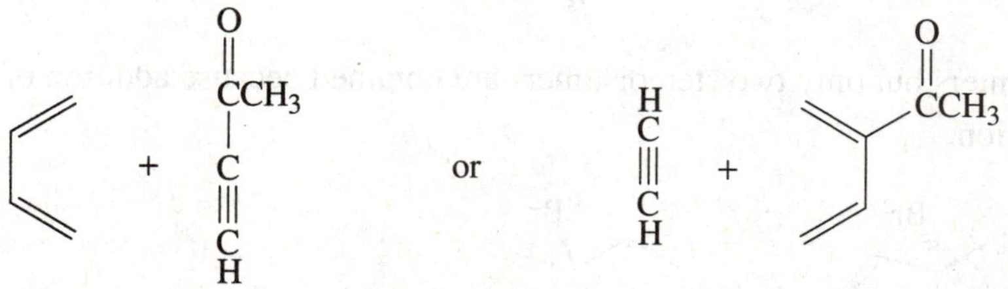

70. A Diels-Alder reaction is a reaction between a nucleophilic diene and an electrophilic dienophile.

a. The compound shown below is more reactive in both **1** and **2**, because electron delocalization increases the electrophilicity of the dienophile.

b. The compound shown below is more reactive, because electron delocalization increases the nucleophilicity of the diene.

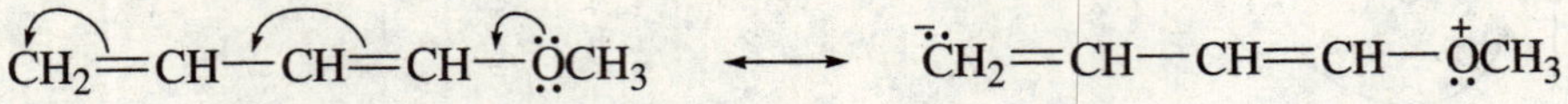

71.

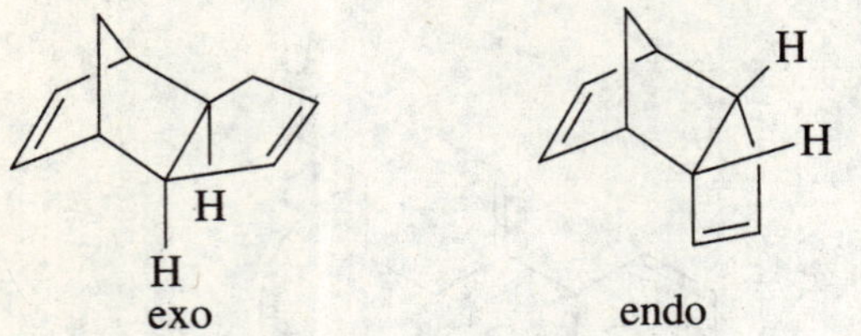

exo endo

72. **a.**

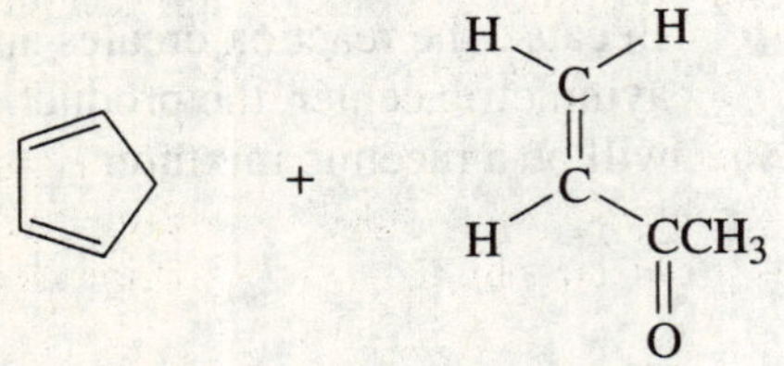

c.

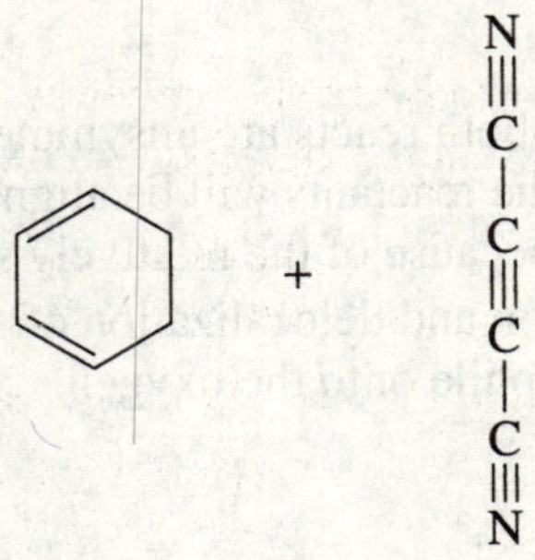

b.

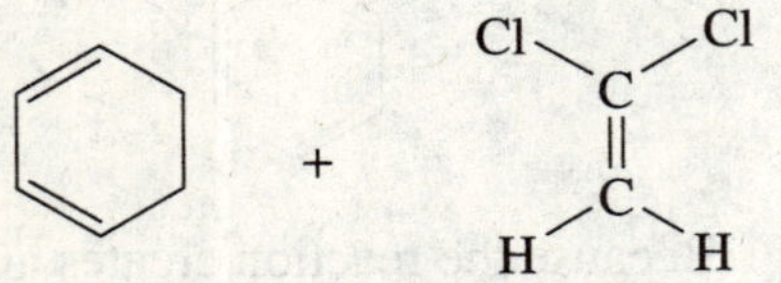

d.

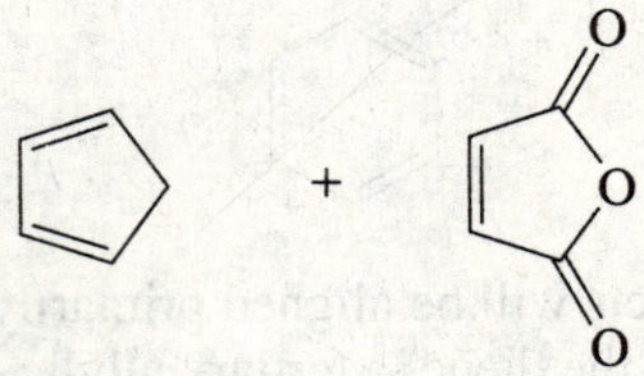

73. **a.**

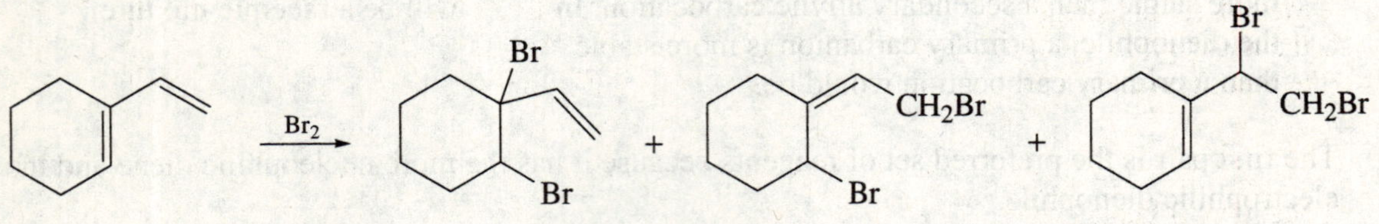

A B C

b. A has two asymmetric centers but only two stereoisomers are obtained because addition of Br_2 can occur only in an anti fashion.

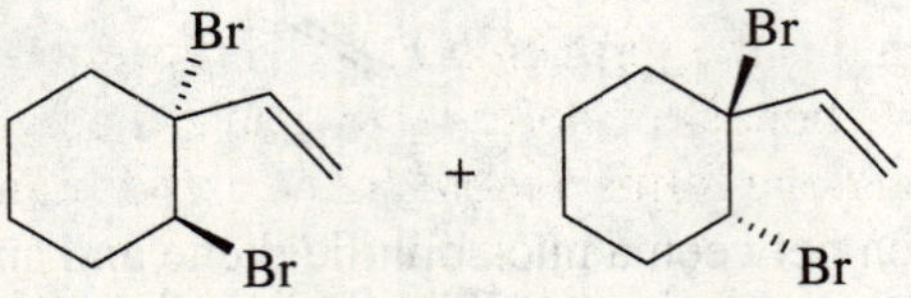

B has four stereoisomers because it has an asymmetric center and a double bond that can be in either the *E* or *Z* configuration.

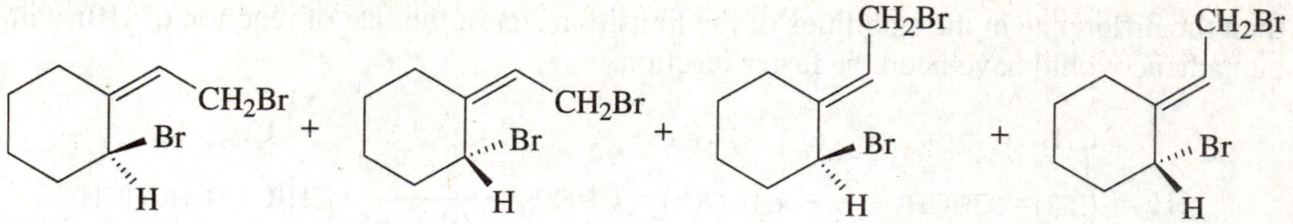

C has two stereoisomers because it has one asymmetric center

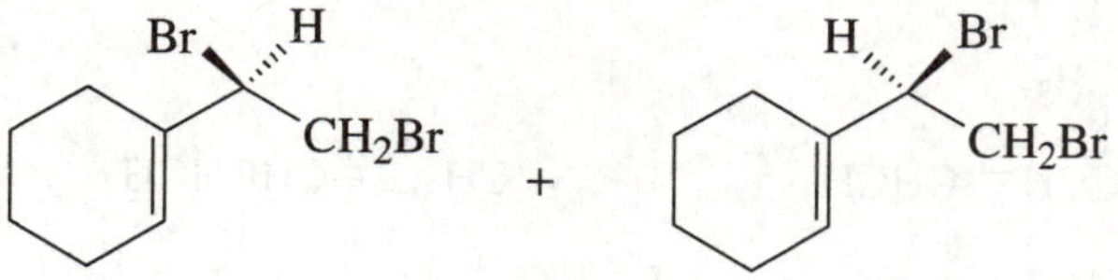

74. Nine of the compounds are shown below. Since each has one asymmetric center, each can have either the *R* or the *S* configuration. Therefore, 18 different products can be obtained.

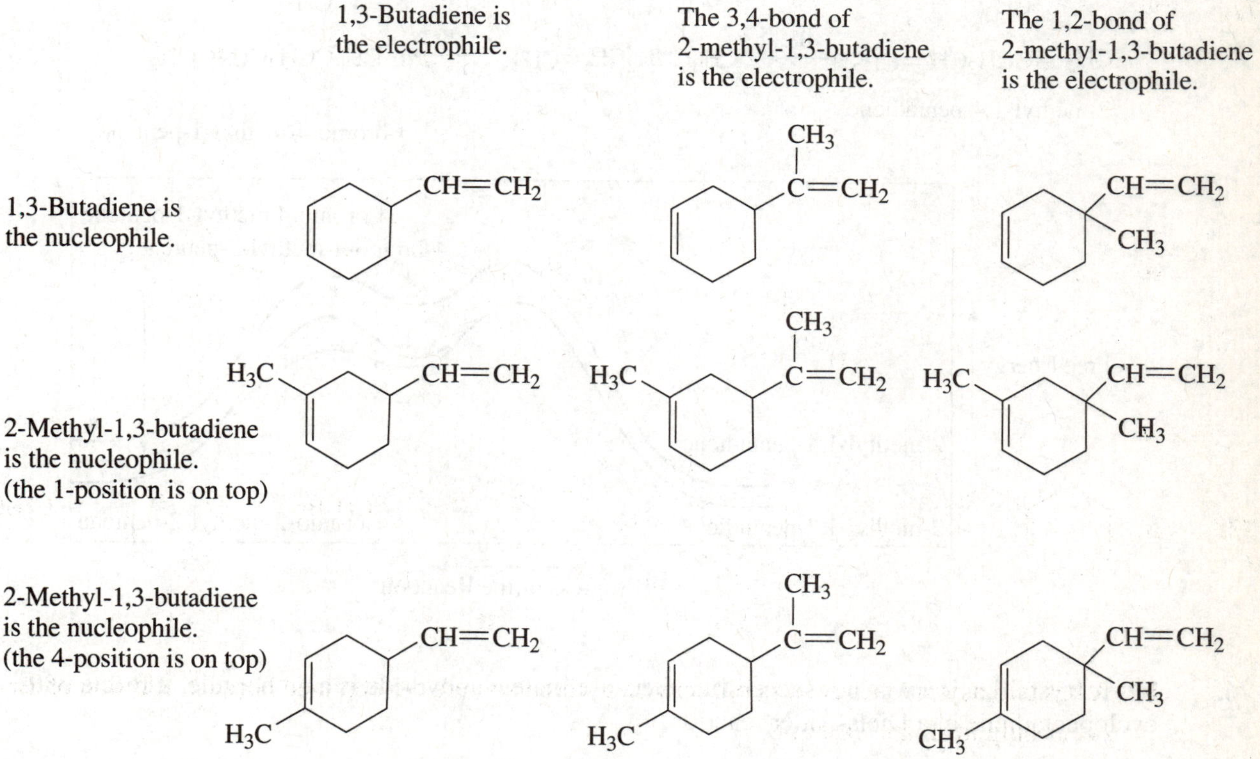

75. The rate-limiting step of the reaction is the formation of the carbocation intermediate. 2-Methyl-1,3-pentadiene (with conjugated double bonds) is more stable than 2-methyl-1,4-pentadiene (with isolated double bonds). 2-Methyl-1,3-pentadiene forms a more stable carbocation than does 2-methyl-1,4-pentadiene.

Since the more stable reactant forms the more stable carbocation, the relative free energies of activation of the rate-limiting steps of the two reactions depend on whether the difference in the stabilities of the reactants is greater or less than the difference in the stabilities of the transition states (which depend on the difference in stabilities of the carbocations). Because the difference in the stabilities of the reactants is less than the difference in the stabilities of the transition states, the rate of reaction of HBr with 2-methyl-1,3-pentadiene is the faster reaction. (If the difference in the stabilities of the reactants had been greater than the difference in the stabilities of the transition states, the rate of reaction of HBr with 2-methyl-1,4-pentadiene would have been the faster reaction.)

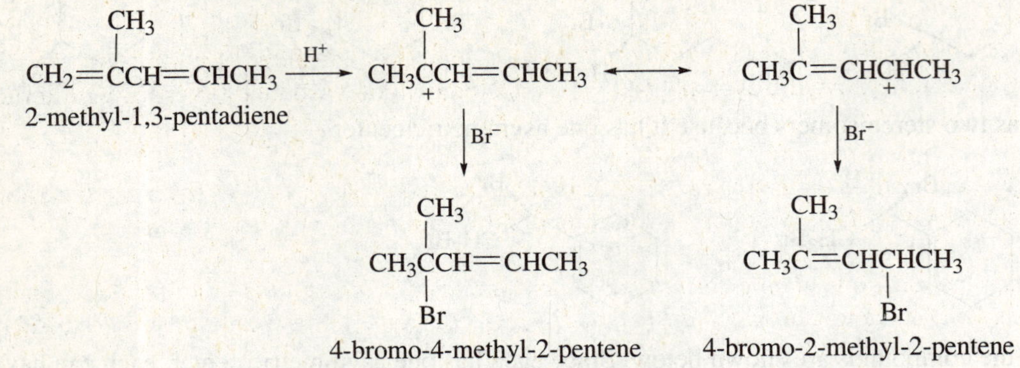

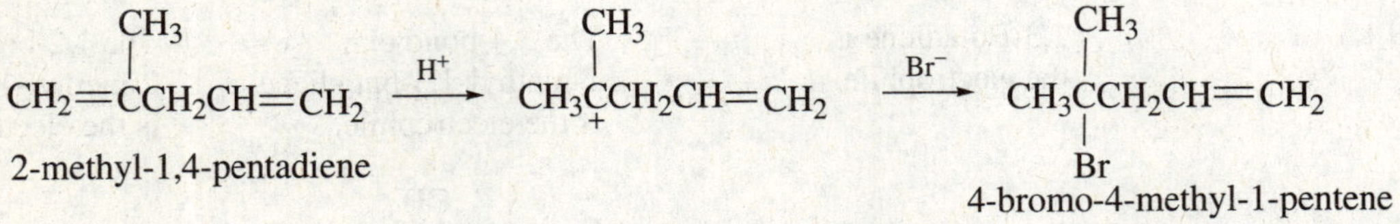

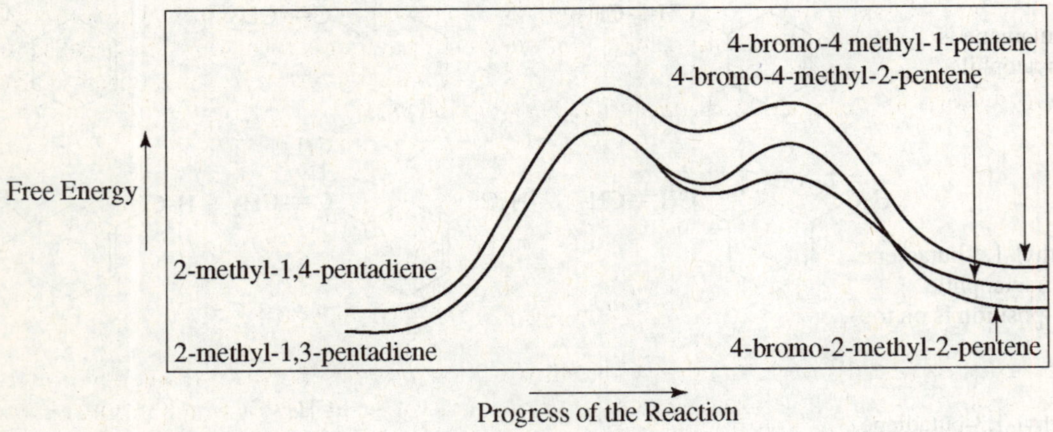

76. His recrystallization was not successful. Because maleic anhydride is a dienophile, it reacts with cyclopentadiene in a Diels-Alder reaction.

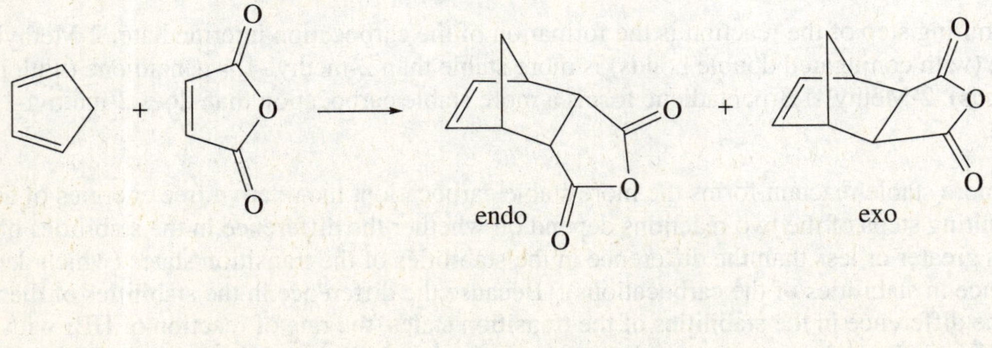

77. High temperatures are required in order to break the bonds formed by the overlapping in-phase orbitals. The Diels-Alder product will not reform at high temperatures, because a [4+2] cycloaddition will not occur unless both of the reactants are in their ground states.

78. We saw in Problem 76 that maleic anhydride reacts with cyclopentadiene. The function of maleic acid in this reaction is to remove the cyclopentadiene, since removal of a product drives the equilibrium toward products. (See Le Châtelier's principle on page 144 of the text.)

79. The bridgehead carbon cannot have the 120° bond angle required for the *sp*^{2} hybridized carbon of a double bond. With a 120° bond angle, the compound would be too strained to exist.

80. **a.** Unless the reaction is being carried out under kinetic control, the amount of product obtained is not dependent on the rate at which the product is formed, so the relative amounts of products obtained will not tell you which product was formed faster.

b. In a thermodynamically controlled reaction, the product distribution depends on the realative stabilites of the products since the products come to equilibrium. Thus if the distribution of products that is obtained does not reflect the relative stabilities of the products, the reaction must have been kinetically controlled.

81. He should follow his friend's advice. If he uses 2-methyl-1-3,cyclohexadiene, the product that is formed faster will be 3-chloro-3-methylcyclohexene both if the proximity effect controls which product is formed faster and if the more stable transition state controls which product is formed faster, because this product is formed through a transition state in which the positive charge on primarily on a tertiary carbon. Thus the experiment will not be able to differentiate between the two.

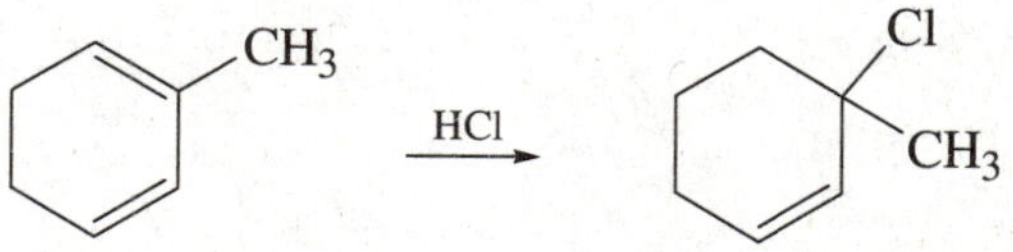

3-chloro-3-methylcyclohexene

If he follows his friend's advice and uses 1-methyl-1-3,cyclohexadiene, the product that is formed faster will be 3-chloro-1-methylcyclohexene only if the proximity effect controls which product is formed faster. The product will be 3-chloro-3-methylcyclohexene if the more stable transition state controls which product is formed faster, because this is the product that is formed through a transition state in which the positive charge on primarily on a tertiary carbon.

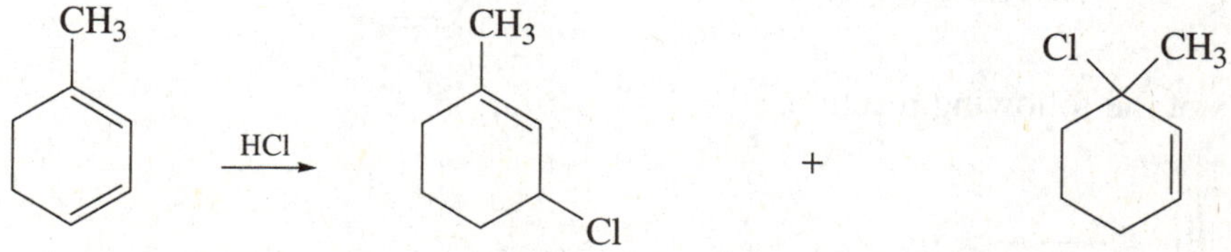

3-chloro-1-methylcyclohexene 3-chloro-3-methylcyclohexene

Chapter 7 Practice Test

1. For each of the following pairs of compounds indicate the one that is the more stable:

a. 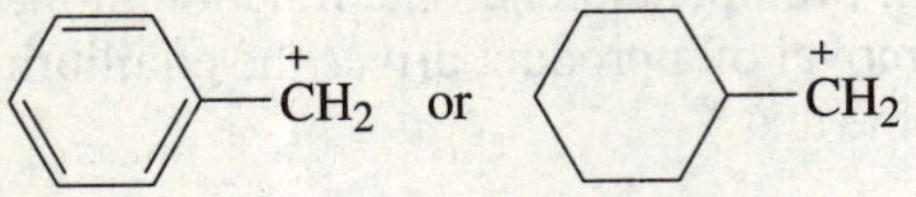

b. $CH_3\overset{-}{C}HCH_3$ or $CH_3CHC{\equiv}\overset{-}{C}H$

c. $CH_3\overset{-}{C}HCH_2\overset{O}{\overset{\|}{C}}CH_3$ or $CH_3\overset{-}{C}H\overset{O}{\overset{\|}{C}}CH_3$

d. $CH_2{=}CH\overset{+}{C}H_2$ or $CH_2{=}CHCH_2\overset{+}{C}H_2$

e. 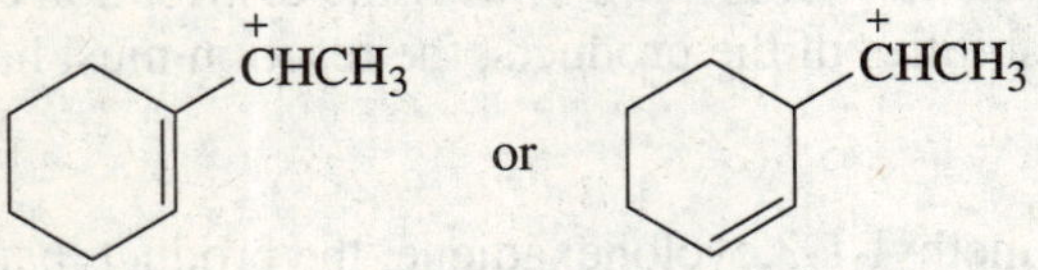

2. Draw resonance contributors for the following:

a. $CH_3CH{=}CH{-}\ddot{\underset{\cdot\cdot}{O}}CH_3$

b. $CH_3CH{=}CH{-}CH{=}CH{-}\overset{+}{C}H_2$

c. $\overset{-\cdot\cdot}{C}H_2{-}CH{=}CH{-}\overset{O}{\overset{\|}{C}}H$

3. Which compounds do not have delocalized electrons?

$CH_3CH_2NHCH{=}CHCH_3$ $\quad CH_3\underset{+}{\overset{CH_3}{\overset{|}{C}}}CH_2CH{=}CH_2$ $\quad CH_2{=}CHCH_2CH{=}CH_2$

$CH_2{=}CH\overset{O}{\overset{\|}{C}}CH_3$ $\quad CH_3CH_2NHCH_2CH{=}CHCH_3$ 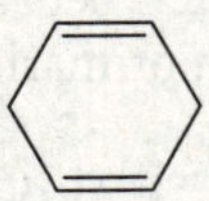

4. Give the products of the following reactions:

a. 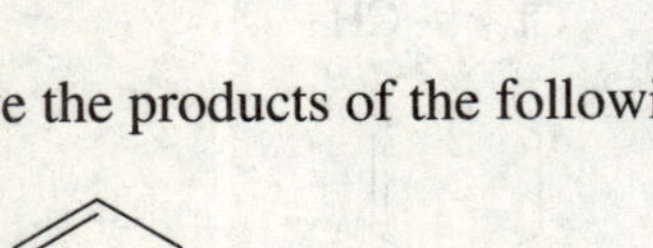

b. 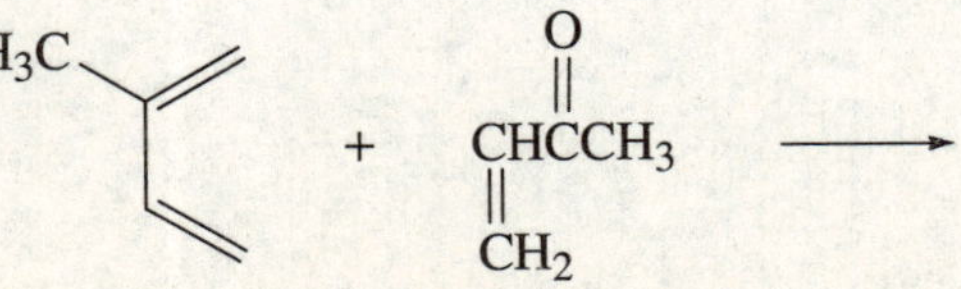

5. Which of the following pairs are resonance contributors?

a. CH_3CH_2OH and CH_3OCH_3

b. $CH_3\overset{O}{\overset{\|}{C}}OH$ and $CH_3\overset{O}{\overset{\|}{C}}O^-$

c. $CH_3\overset{O}{\overset{\|}{C}}OH$ and $CH_3\overset{O^-}{\overset{|}{C}}{=}\overset{+}{O}H$

d. $CH_3CH_2\overset{O}{\overset{\|}{C}}H$ and $CH_3CH{=}\overset{OH}{\overset{|}{C}}H$

6. Which of the following dienes can be used in a Diels-Alder reaction?

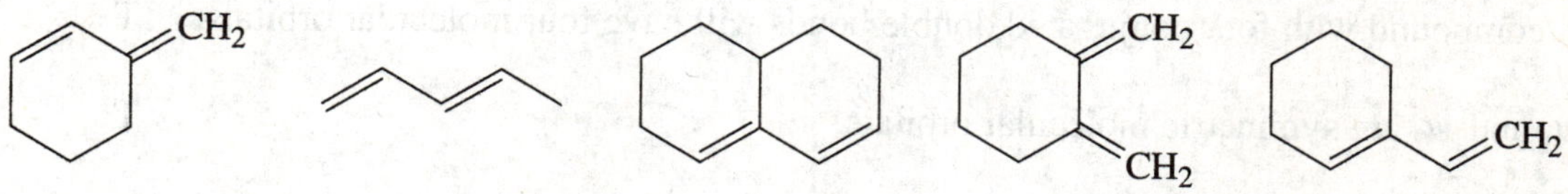

7. Which is a stronger base?

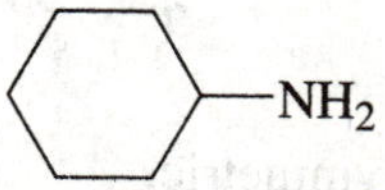 or 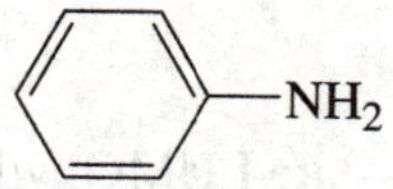

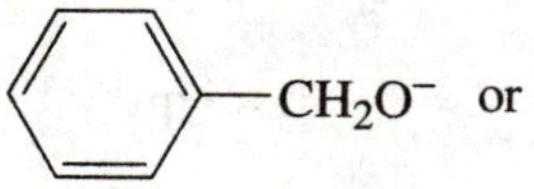 or 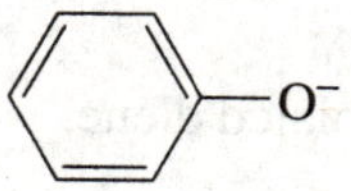

8. Draw resonance contributors for the following:

a. 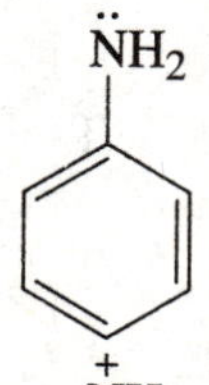

b. 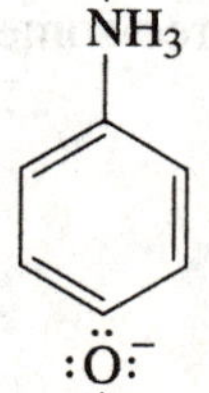

c. 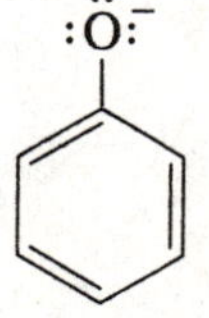

9. Which resonance contributor makes a greater contribution to the resonance hybrid?

a. 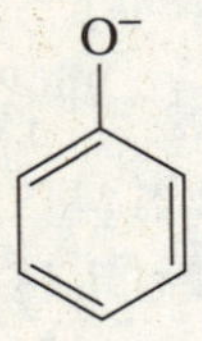 or 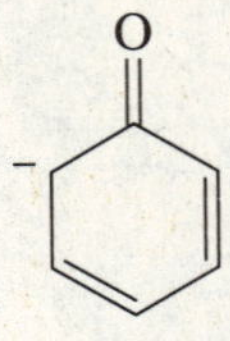

b. 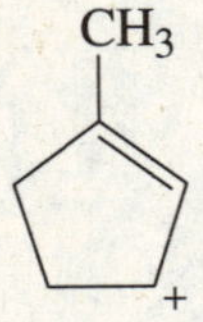 or 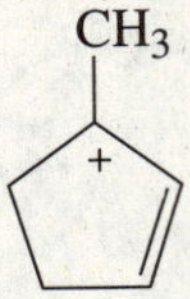

10. Indicate whether the following statements is true or false:

a. A compound with four conjugated double bonds will have four molecular orbitals. T F

b. ψ_1 and ψ_2 are symmetric molecular orbitals. T F

c. If ψ_3 is the HOMO in the ground state, ψ_4 will be the HOMO in the excited state. T F

d. If ψ_3 is the LUMO, ψ_4 will be the HOMO. T F

e. If the ground-state HOMO is symmetric, the ground-state LUMO will be antiymmetric. T F

f. A conjugated diene is more stable than an isomeric isolated diene. T F

g. A single bond formed by an sp^2—sp^2 overlap is longer than a single bond formed by an sp^2—sp^3 overlap. T F

h. The thermodynamically controlled product is the major product obtained when the reaction is carried out under mild conditions. T F

i. 1,3-Hexadiene is more stable than 1,4-hexadiene. T F

11. Give the four products that would be obtained from the following reaction. Ignore stereoisomers.

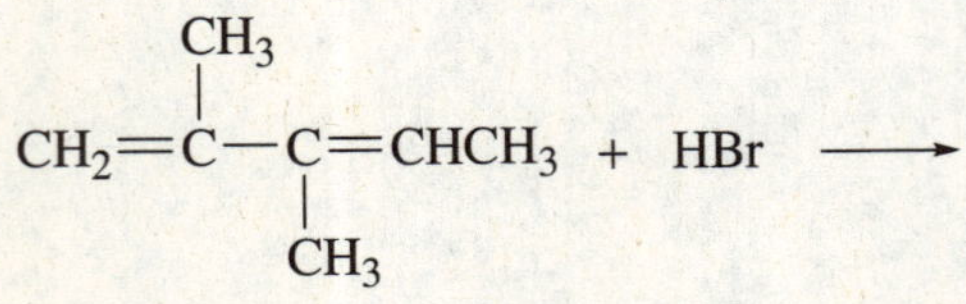

12. What reactants are necessary for the synthesis of the following compound via a Diels-Alder reaction?

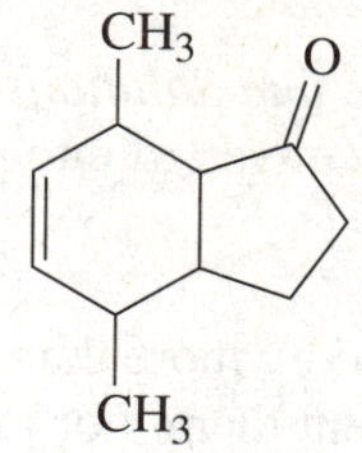

13. Rank the following carbocations in order of decreasing stability:

$CH_3CH{=}CH\overset{+}{C}H_2$ $CH_3CH{=}CH\overset{+}{C}HCH_3$ $CH_3CH{=}CHCH_2\overset{+}{C}H_2$ $CH_3CH{=}CH\overset{+}{C}(CH_3)CH_3$

14. Two 1,2-products and two 1,4-products can be obtained from the following reaction:

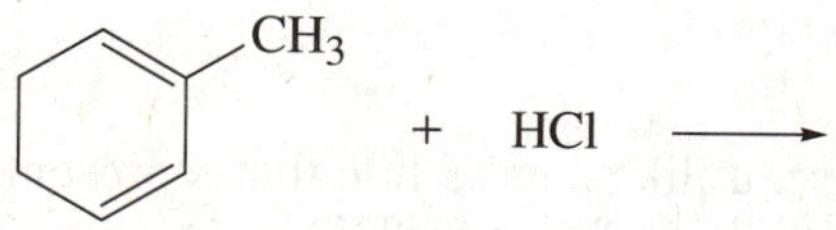

a. Which are the predominant 1,2- and 1,4-products?
b. Which of the above two products is the product of thermodynamic control?

15. What reagents could be used to convert the given starting material into the given product?

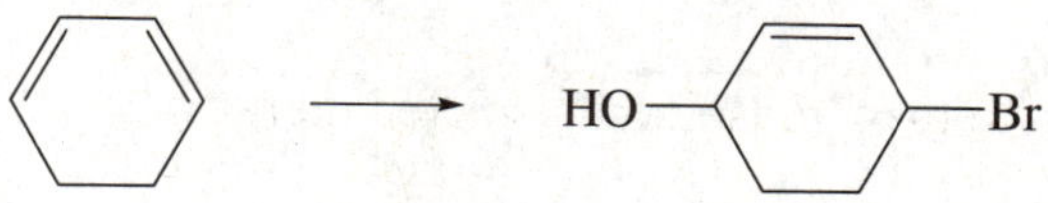

16. Give the product of the following reaction, showing its configuration:

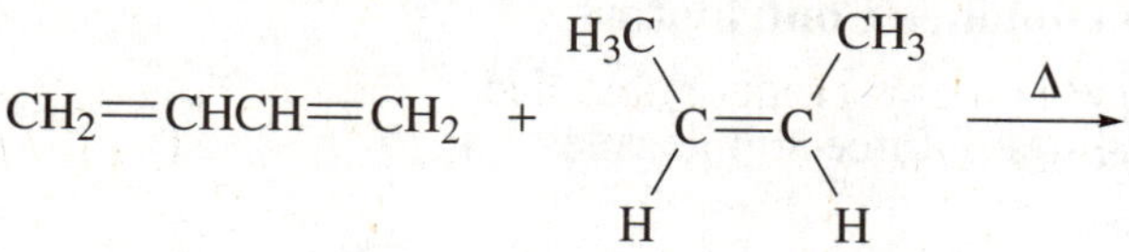

17. For each of the following reactions give the major 1,2- and 1,4-products. Label the product of kinetic control and the product of thermodynamic control.

a. $CH_2{=}CH{-}C(CH_3){=}CH_2$ + HCl →

b.

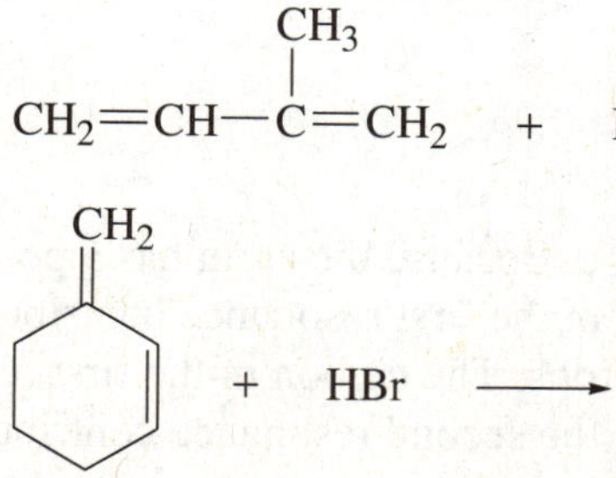

SPECIAL TOPIC V

Drawing Resonance Contributors

On p. 130-140 of the text and in Special Topic II of the *Study Guide and Solutions Manual*, we saw that chemists use curved arrows to show how electrons move when reactants are converted into products. Chemists also use curved arrows when they draw resonance contributors.

We have seen that delocalized electrons are electrons that are shared by more than two atoms. When electrons are shared by more than two atoms, we cannot use solid lines to represent them. For example in the carboxylate ion, a pair of electrons are shared by a carbon and two oxygens. Thus the two oxygen atoms share the negative charge. We show the pair of delocalized electrons by a dotted line spread over the three atoms. We have seen that this structure is called a resonance hybrid.

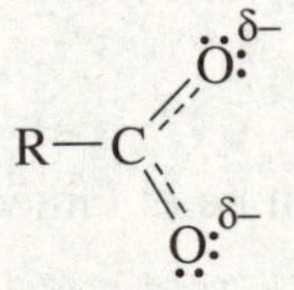

carboxylate ion

Chemists do not like to use dotted lines when drawing structures because, unlike a solid line that represents two electrons, the dotted lines do not specify the number of electrons they represent. Therefore, chemists use structures with localized electrons (solid lines) to approximate the resonance hybrid that has delocalized electrons (dotted lines). These approximate structures are called resonance contributors. Curved arrows are used to show the movement of electrons in going from one resonance contributor to the next.

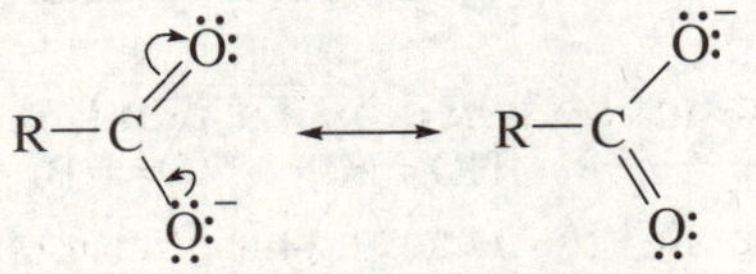

Rules for Drawing Resonance Contributors

Resonance contributors are easy to draw if you remember these three simple rules.

1. Only electrons move; atoms NEVER move.
2. The only electrons that can move are π electrons (electrons in π bonds) and lone pair electrons.
3. Electrons always move toward a positive charge or toward a pair of atoms attached by a π bond. (In Special Topic VI we will see that electrons can also move toward a radical.)

Moving π electrons toward a positive charge.

In the following example, π electrons are moved toward a positive charge. Because the atom has a positive charge, it can accept the electrons. The carbon that is positively charged in the first resonance contributor is neutral in the second resonance contributor because it has received electrons. The carbon in the first resonance contributor that loses its share of the π electrons is positively charged in the second resonance contributor.

$$CH_3CH{=}CH{-}\overset{+}{C}HCH_3 \longleftrightarrow CH_3\overset{+}{C}H{-}CH{=}CHCH_3$$

We see that the following carbocation has three resonance contributors.

$$CH_2{=}CH{-}CH{=}CH{-}\overset{+}{C}HCH_3 \longleftrightarrow CH_2{=}CH{-}\overset{+}{C}H{-}CH{=}CHCH_3$$

$$\updownarrow$$

$$\overset{+}{C}H_2{-}CH{=}CH{-}CH{=}CHCH_3$$

Notice that in going from one resonance contributor to the next, the total number of electrons in the structure does not change. Therefore, each of the resonance contributors must have the same charge.

Problem 1. Draw the resonance contributors for the following carbocation:
(The answers can be found immediately after Problem 12.)

$$CH_3CH{=}CH{-}CH{=}CH{-}CH{=}CH{-}\overset{+}{C}H_2 \longleftrightarrow$$

$$\updownarrow$$

$$\longleftrightarrow$$

Problem 2. Draw the resonance contributors for the following carbocation:

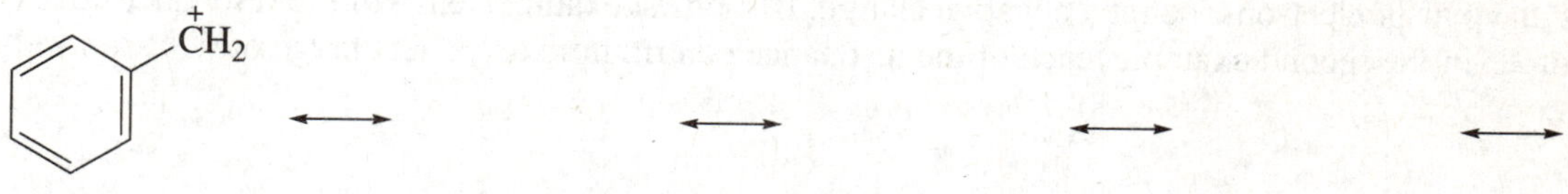

Moving π electrons toward a pair of atoms attached by a π bond.

In the following example, π electrons are moved toward a pair of atoms attached by a π bond. The atom to which the electrons are moved can accept them because the π bond can break.

Problem 3. Draw the resonance contributor for the following compound:

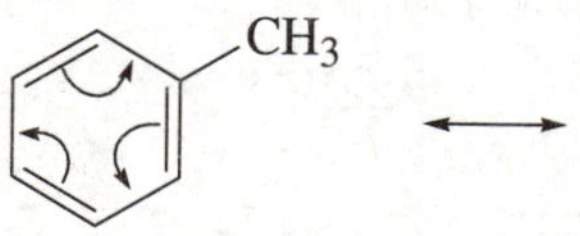

In the next example, the electrons can be moved equally easily to the left or to the right. The charges on the end carbons cancel, so there is no charge on any of the carbons in the resonance hybrid.

$$\overset{-}{\ddot{C}}H_2{-}CH{=}CH{-}\overset{+}{C}H_2 \longleftrightarrow CH_2{=}CH{-}CH{=}CH_2 \longleftrightarrow \overset{+}{C}H_2{-}CH{=}CH{-}\overset{-}{\ddot{C}}H_2$$

$$CH_2{\overset{...}{-}}CH{\overset{...}{-}}CH{\overset{...}{-}}CH_2$$

resonance hybrid

When electrons can be moved in either of two directions and there is a difference in the electronegativity of the atoms to which they can be moved, always move the electrons toward the more electronegative atom. For example, in the following example the electrons are moved toward oxygen, not toward carbon.

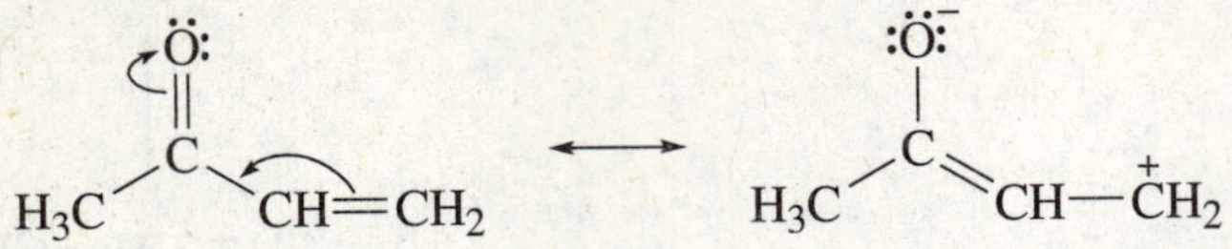

Notice that the first resonance contributor has a net charge of 0. Since the number of electrons in the molecule does not change, the other resonance contributor must have the same net charge. (A net charge of 0 does not mean that there is no charge on any of the atoms; a resonance contributor with a positive charge on one atom and a negative charge on another has a net charge of 0.)

Problem 4. Draw the resonance contributor for the following:

$$CH_2{=}CH{-}C{\equiv}N \longleftrightarrow$$

Moving a lone pair toward a pair of atoms attached by a π bond.

In the following examples, lone pair electrons are moved toward a pair of atoms attached by a π bond. Notice that the arrow starts at a pair of electrons, not at a negative charge. In the first example, each of the resonance contributors has a net charge of -1; in the second example, each of the resonance contributors has a net charge of 0.

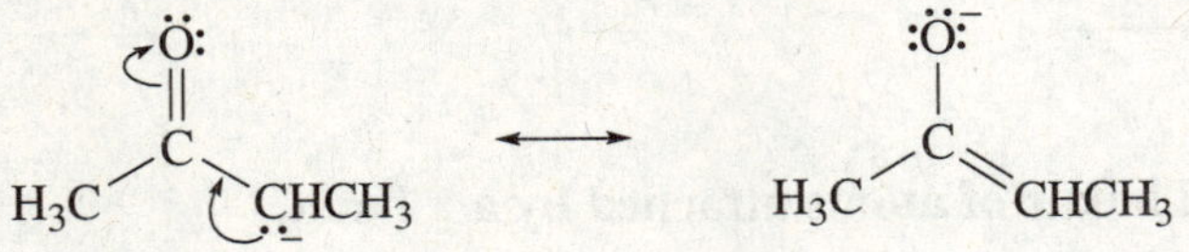

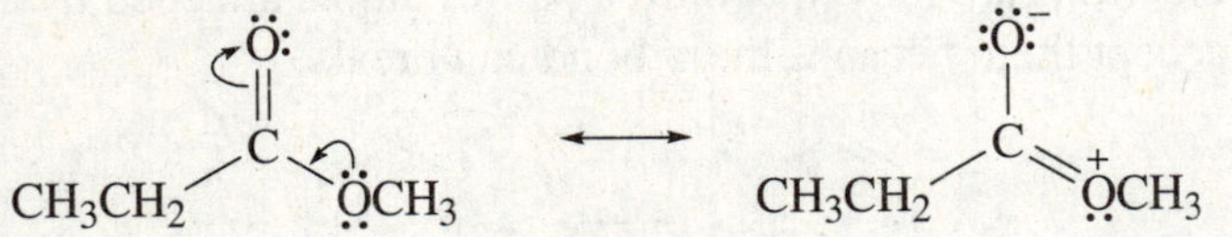

The following species has three resonance contributors. Notice again that the arrow starts at a lone pair, not at a negative charge. The three oxygen atoms share the two negative charges. Therefore, each oxygen atom in the hybrid has 2/3 of a negative charge.

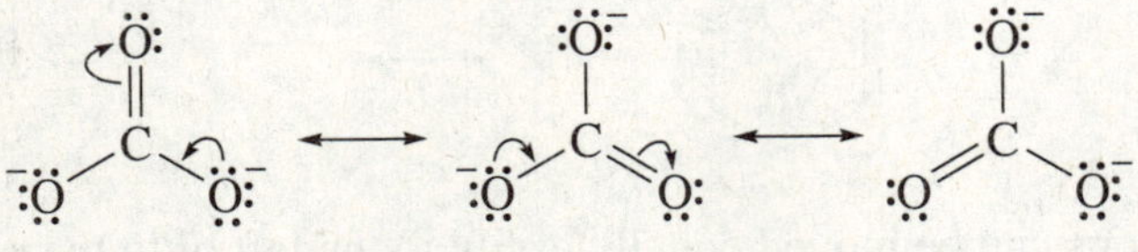

Problem 5. Draw the resonance contributor for the following:

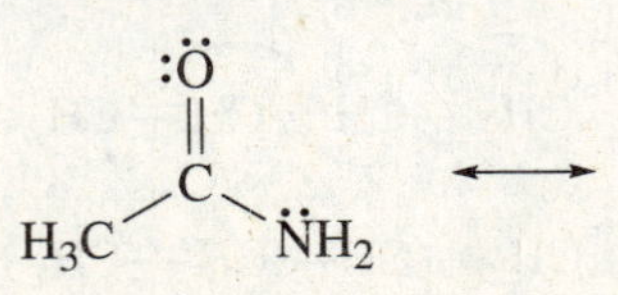

Notice that in the next example the lone pair electron moves away from the most electronegative atom in the molecule. They do so because that is the only way electron delocalization can occur. The π electrons cannot move toward the oxygen because the oxygen atom has a complete octet. Recall that electrons can move only toward a positive charge or toward a pair of atoms attached by a π bond.

$$CH_3CH{=}CH{-}\ddot{\underset{..}{O}}CH_3 \longleftrightarrow CH_3\overset{-}{\ddot{C}}H{-}CH{=}\overset{+}{\underset{..}{O}}CH_3$$

The next compound has five resonance contributors. To get to the second resonance contributor, a lone pair on nitrogen is moved toward a pair of atoms attached by a π bond. Notice that the first and fifth resonance contributors are not the same; they are similar to the two resonance contributors of benzene.

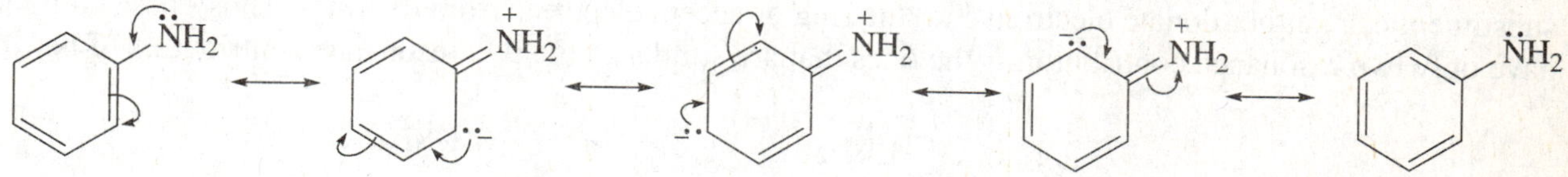

Problem 6. Draw the resonance contributors for the following:

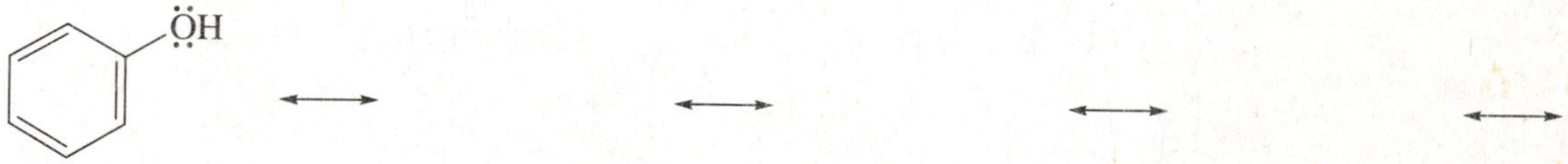

The following species do not have delocalized electrons. Electrons cannot be moved toward an sp^3 hybridized atom because an sp^3 hybridized atom has a complete octet and, therefore, cannot accept any more electrons.

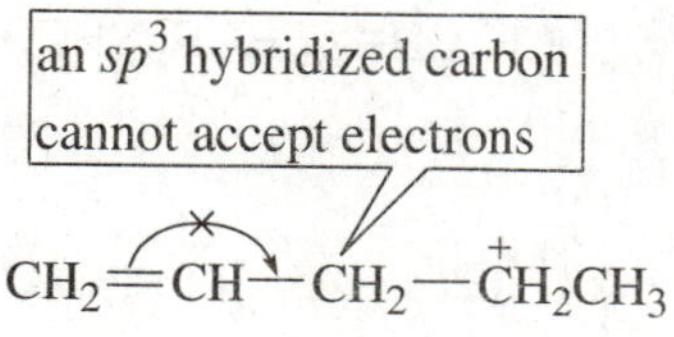

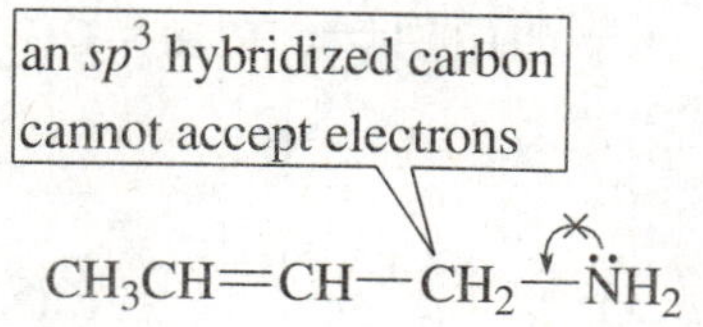

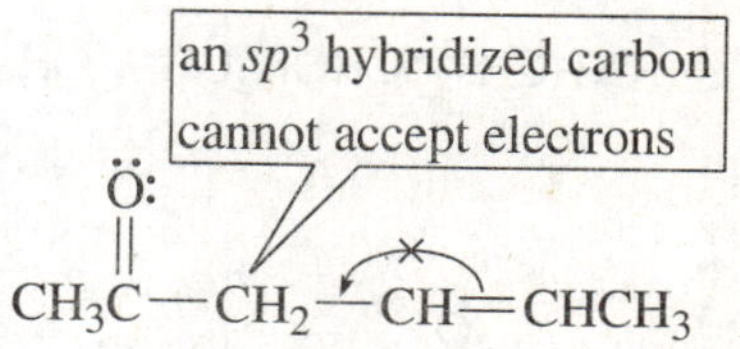

Compare the above structures that do not have delocalized electrons with those that do. Notice that:

1. there can be only one single bond between the atom with the π bond and the atom with the positive charge.
2. there can be only one single bond between the atom with the π bond and the atom with the lone pair.
3. there can be only one single bond between the atoms with the π bonds (that is, the double bonds must be conjugated).

Notice the difference in the resonance contributors for the four structures shown below. In the following compound, a lone pair on the atom attached to the ring moves toward a pair of atoms attached by a π bond.

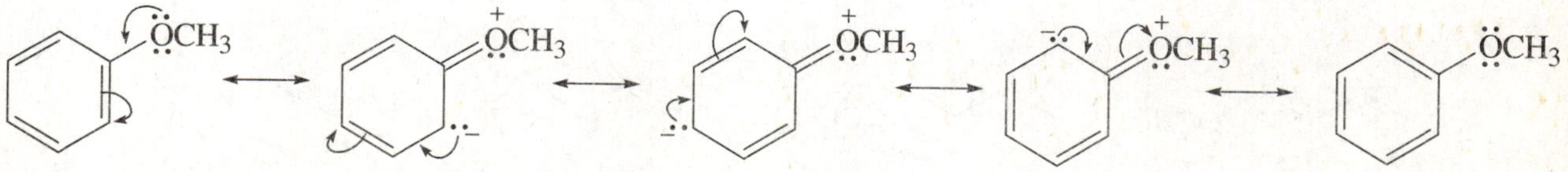

In the next compound, first a π bond moves toward a pair of atoms attached by a π bond; the electron movement is toward the oxygen since it is a more electronegative atom than carbon. Then a π bond moves to a positive charge.

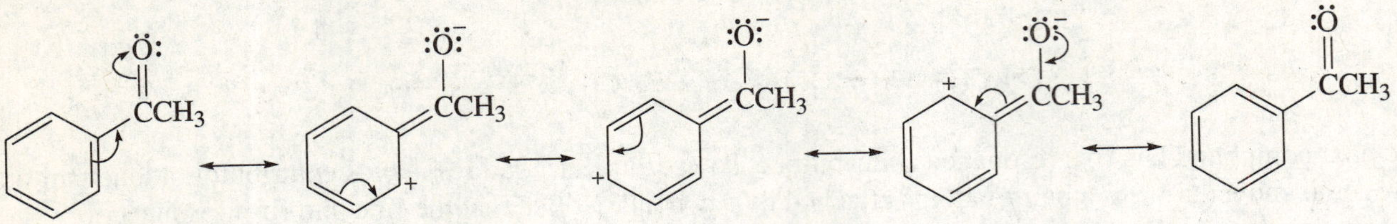

In the next two compounds, the atom attached to the ring has neither a lone pair or a π bond. Therefore, the substituent can neither donate electrons into the ring or accept electrons from the ring. Thus, these compounds have only two resonance contributors—the ones that are similar to the two resonance contributors of benzene.

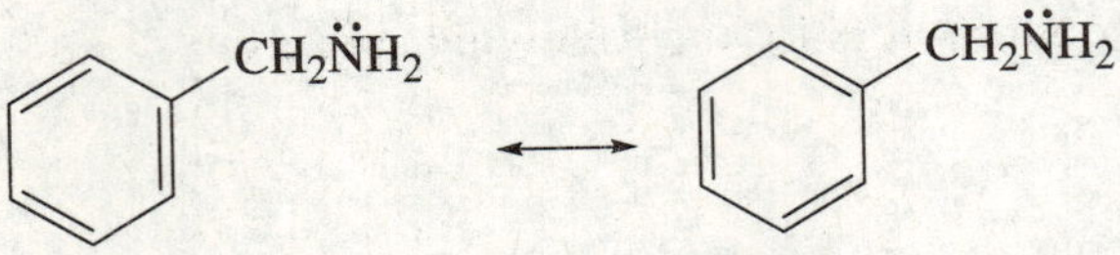

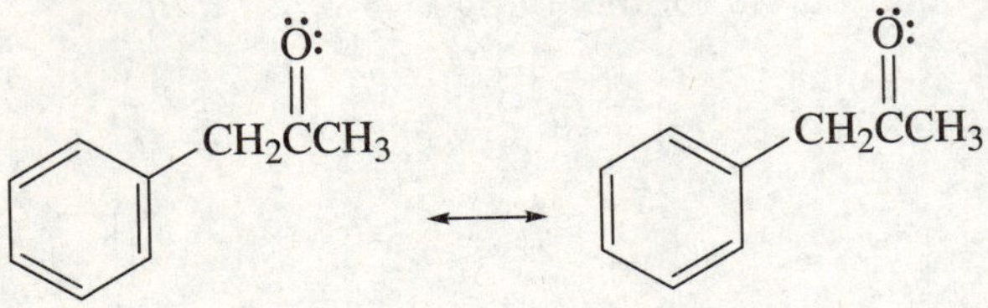

Problem 7. Which of the following species have delocalized electrons?

$CH_3CH{=}CHCH_2CH{=}CH_2$

A

$CH_3CH{=}CHCH{=}CH_2$

B

$CH_3CH{=}CHCH_2\ddot{N}HCH_3$

C

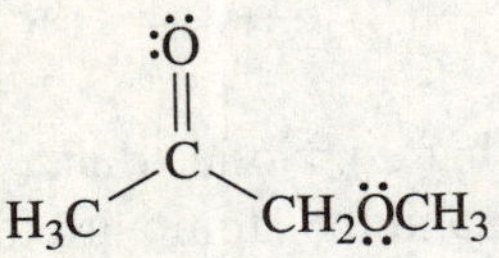

D

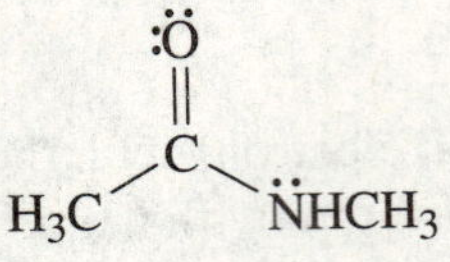

E

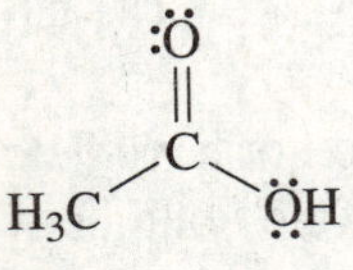

F

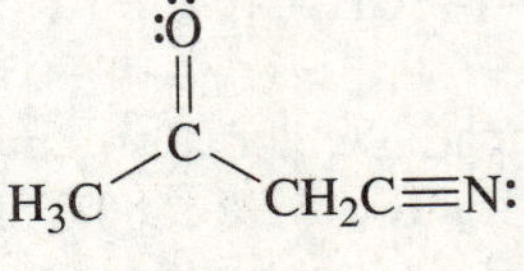

G

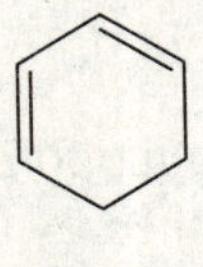

H

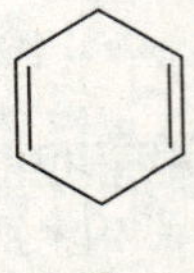

I

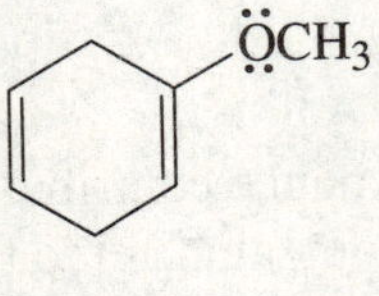

J

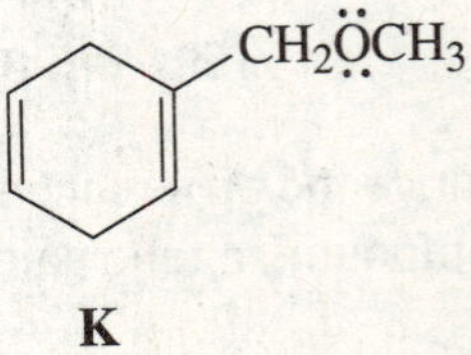

K

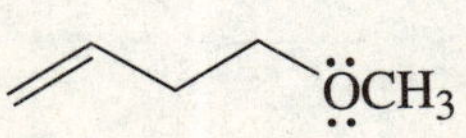

L

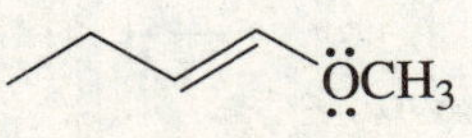

M

N

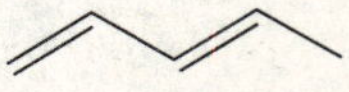

O

Problem 8. Draw the resonance contributors for those compounds in Problem 7 that have delocalized electrons.

Problem 9. Draw the arrows to show how one resonance contributor leads to the next one.

a. 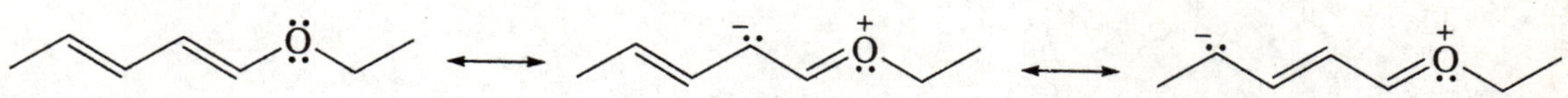

b. 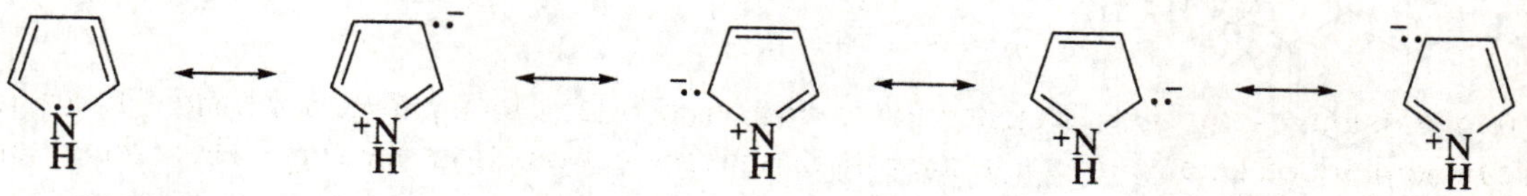

Problem 10. Draw the resonance contributors for the following species:

a. 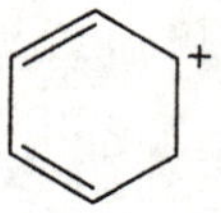

b. 

c. 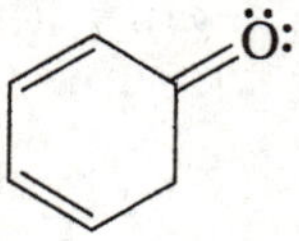

d. $CH_2{=}CH{-}CH{=}CH{-}\ddot{N}H_2$

e. 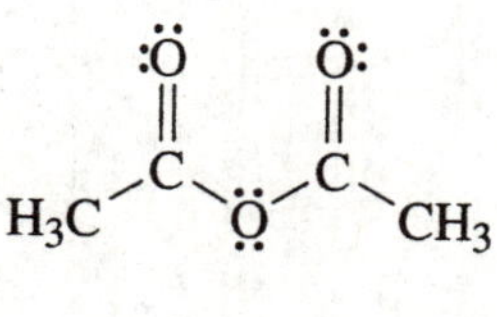

f. 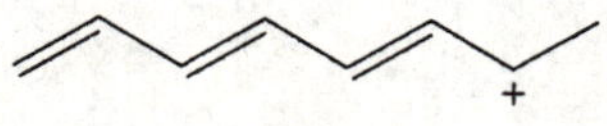

Problem 11. Draw the resonance contributors for the following species:

a. 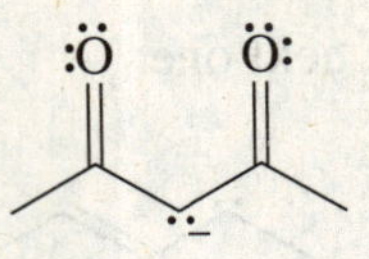

b. 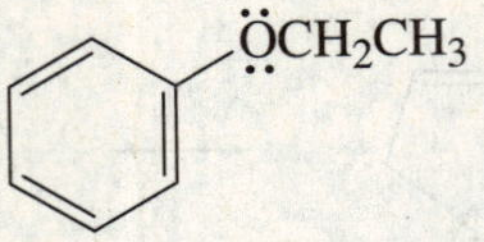

c. 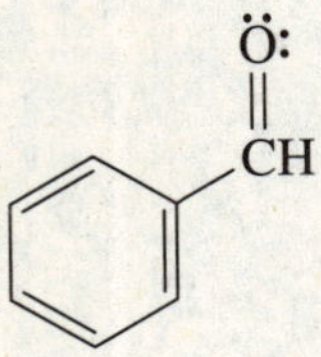

d. 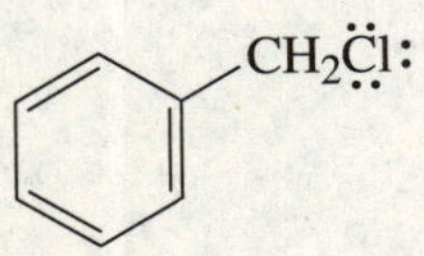

e. 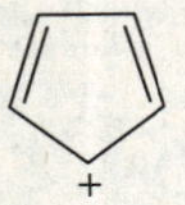

f. 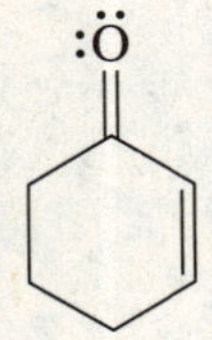

Problem 12. Draw the resonance contributors for the following species:

a.

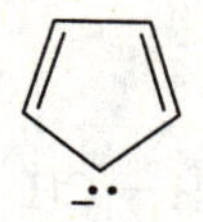

b.

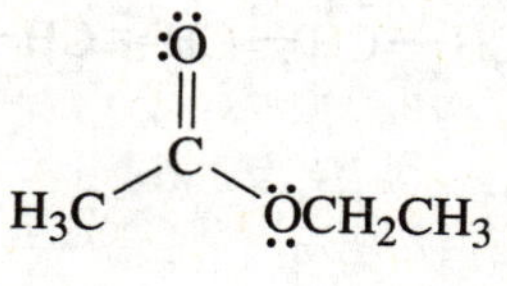

c.

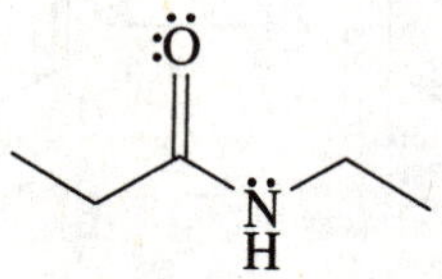

d.

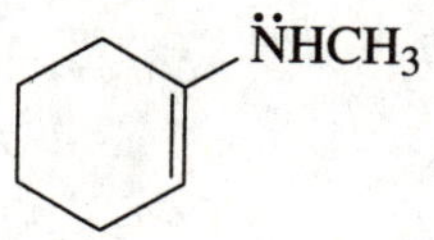

e.

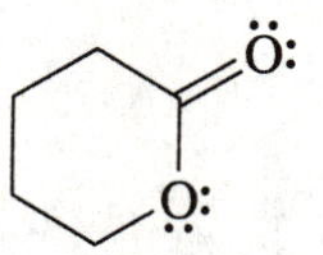

f.

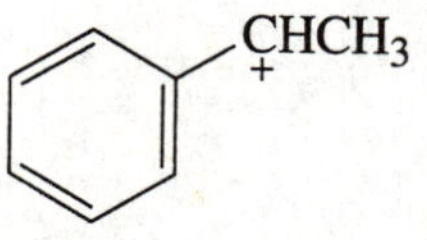

Answers to the Problems

Problem 1.

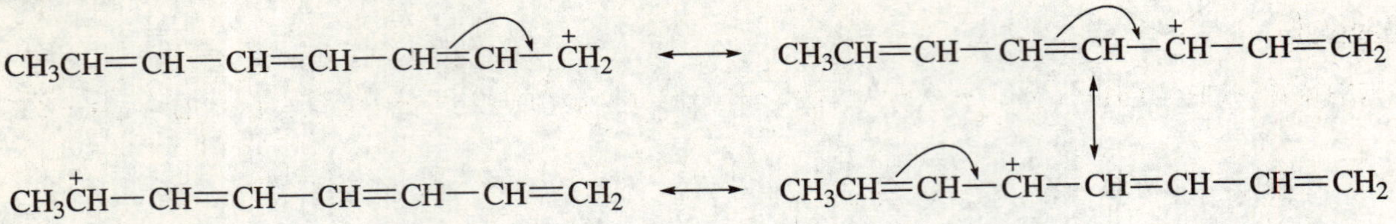

Problem 2.

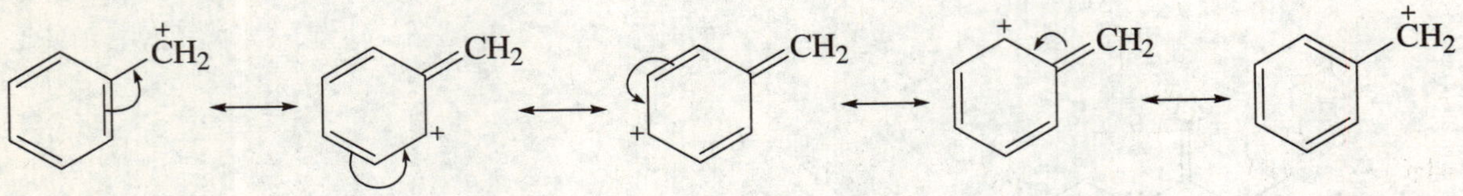

Problem 3.

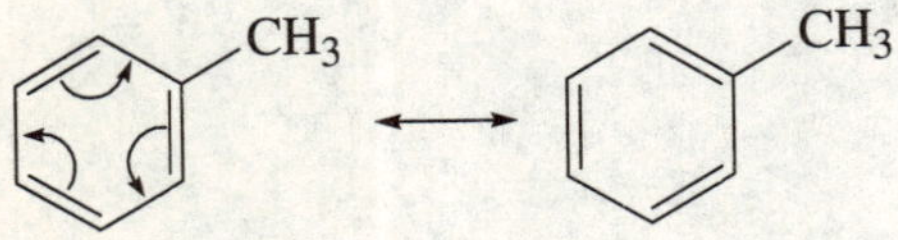

Problem 4.

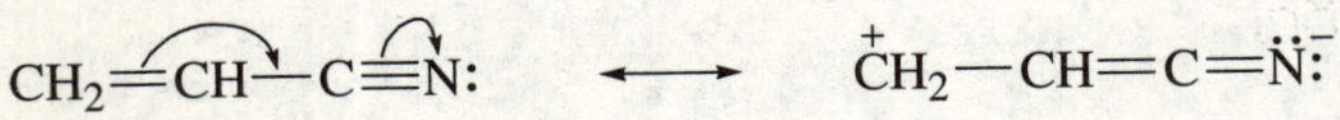

Problem 5.

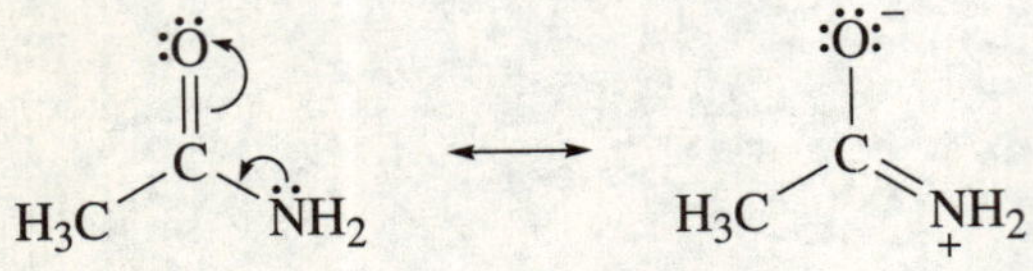

Problem 6.

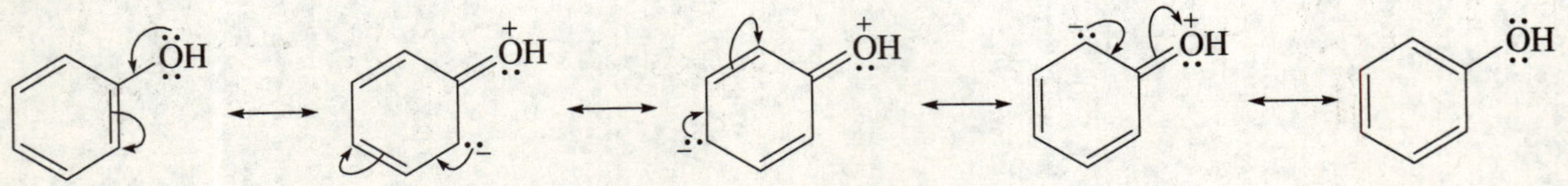

Problem 7.

B, E, F, H, J, M, O

Problem 8.

B
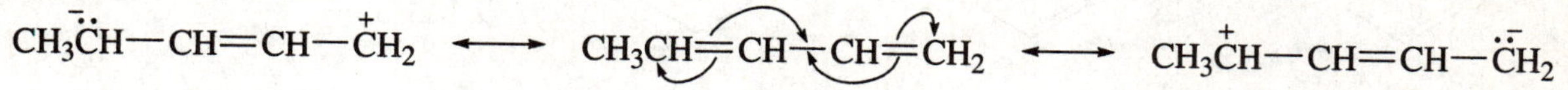

E
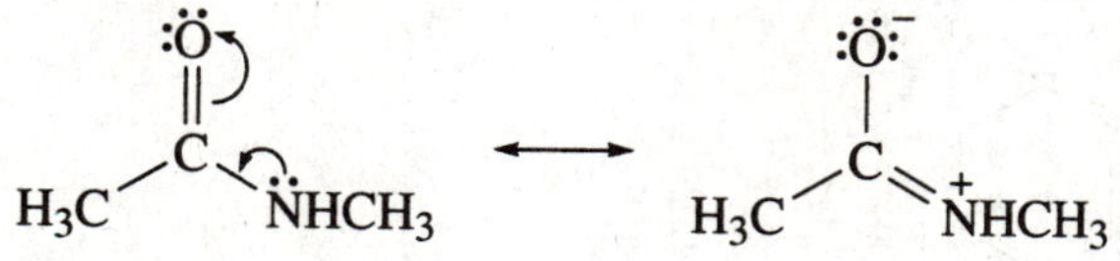

F
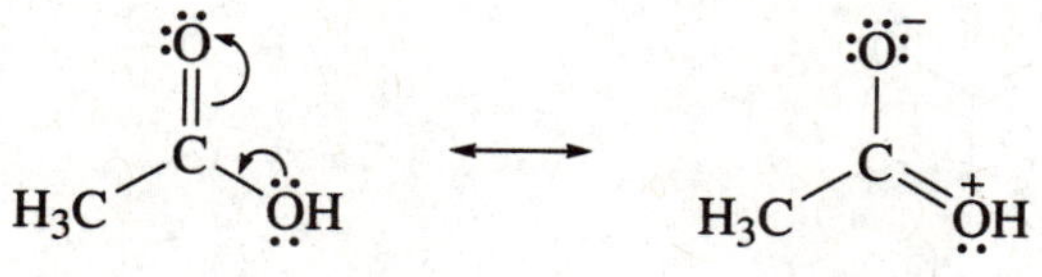

H

J
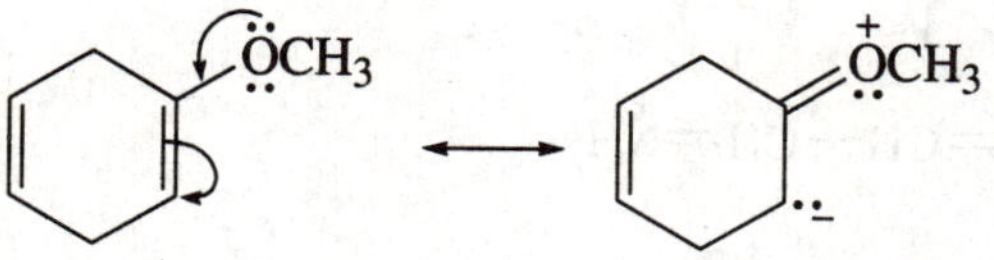

M

O
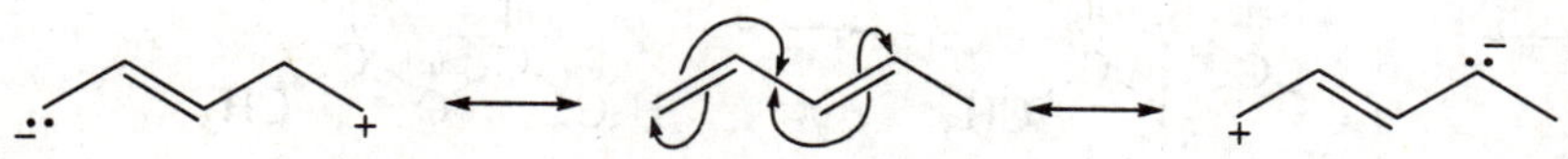

Problem 9.

1.
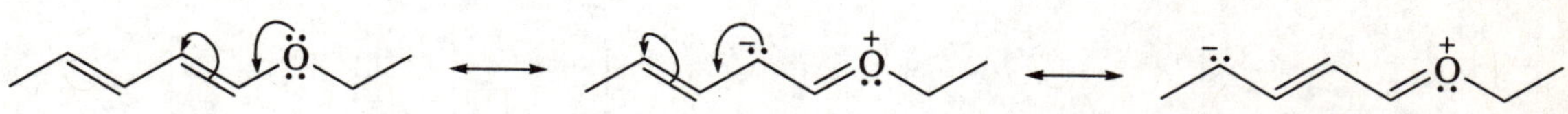

2.
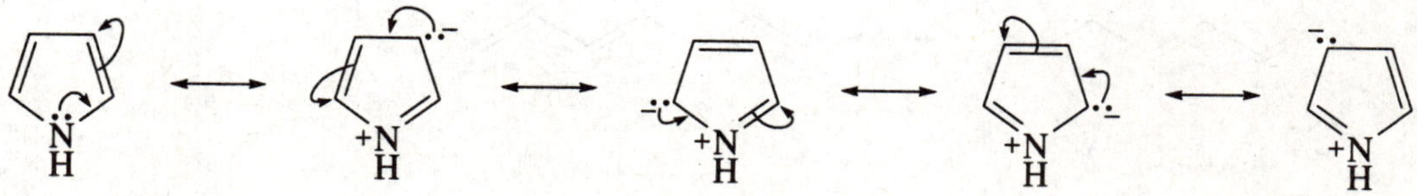

Problem 10.

a.

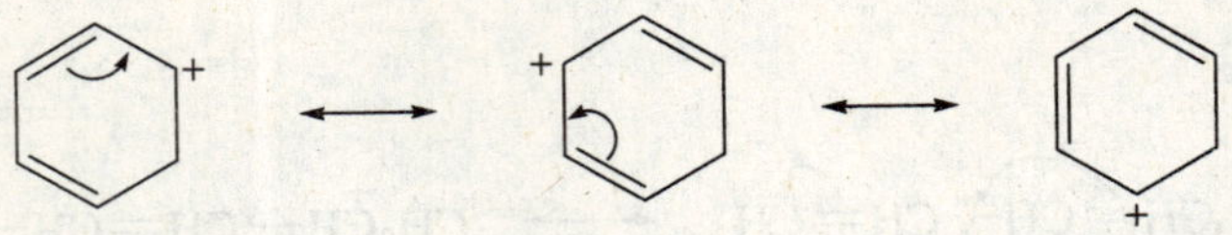

b. Notice in the following example that the electrons can move either clockwise or counterclockwise.

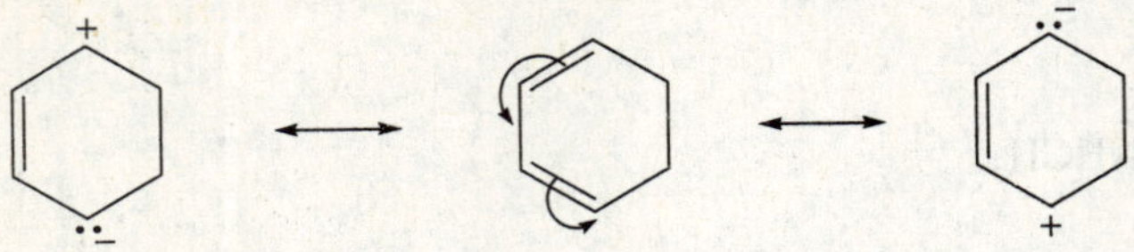

c.

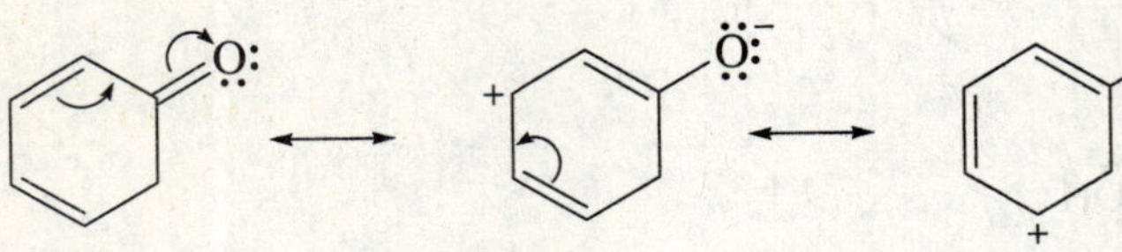

d.

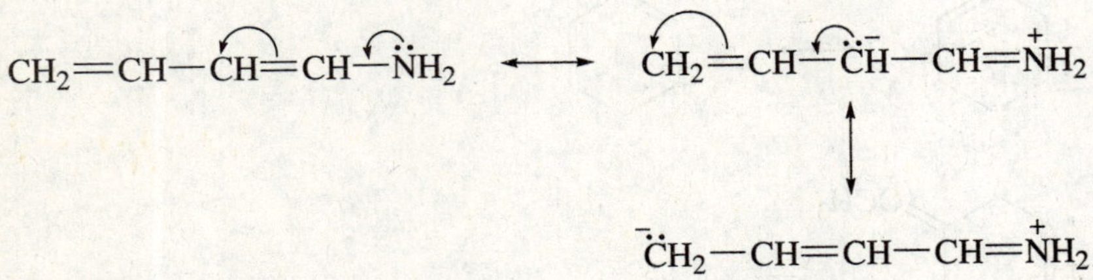

e. Notice in the following example that the electrons can be delocalized onto either of the two sp^2 oxygens.

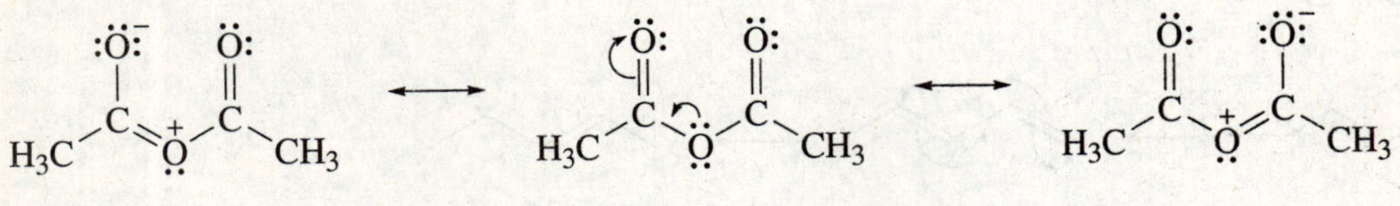

f.

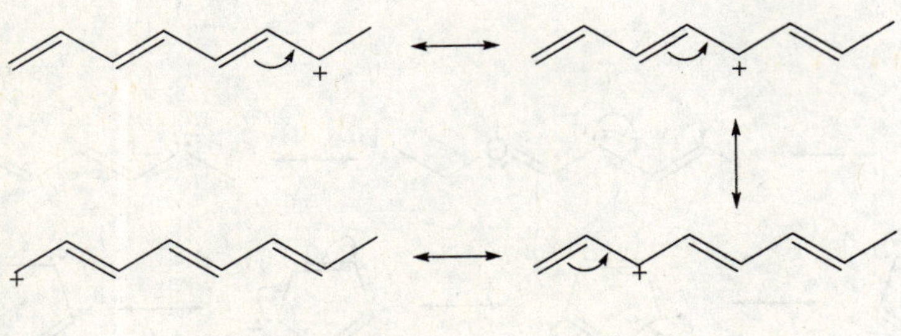

Problem 11.

a.

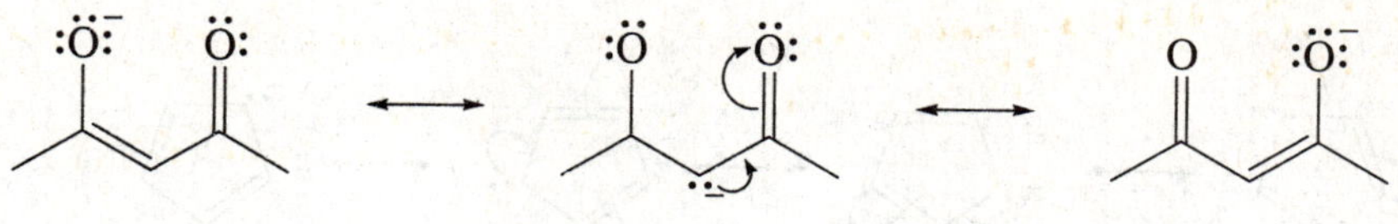

b.

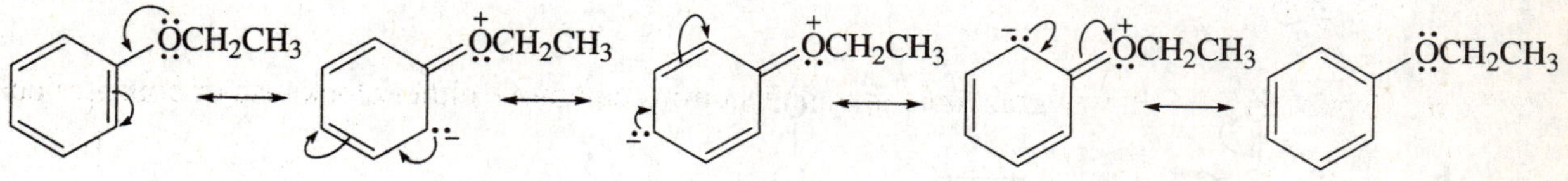

c.

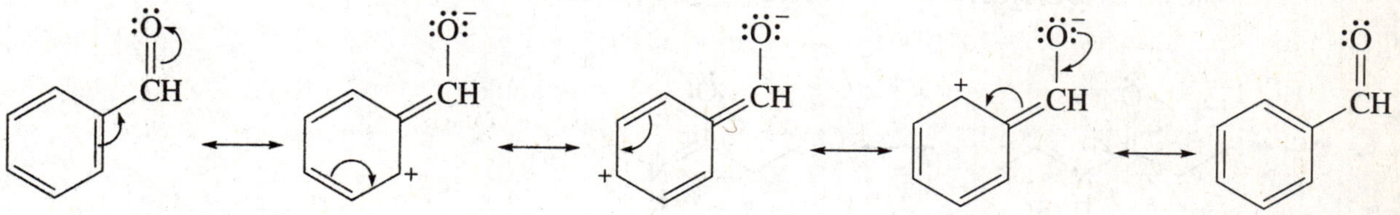

d.

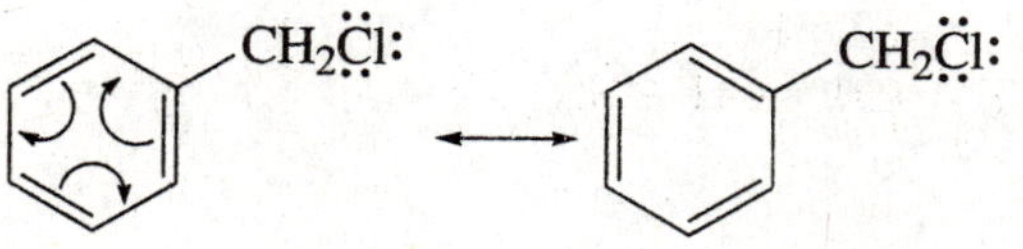

e.

f.

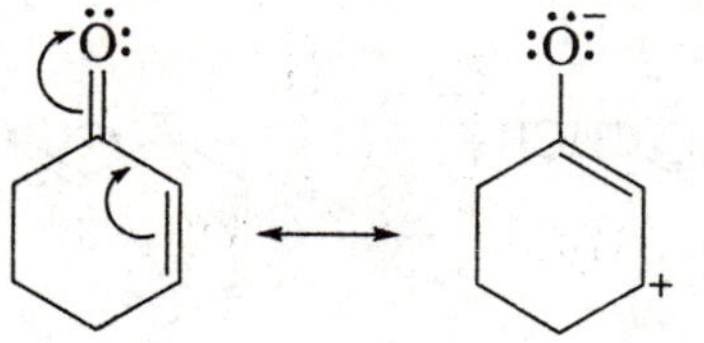

Problem 12.

a.

b.

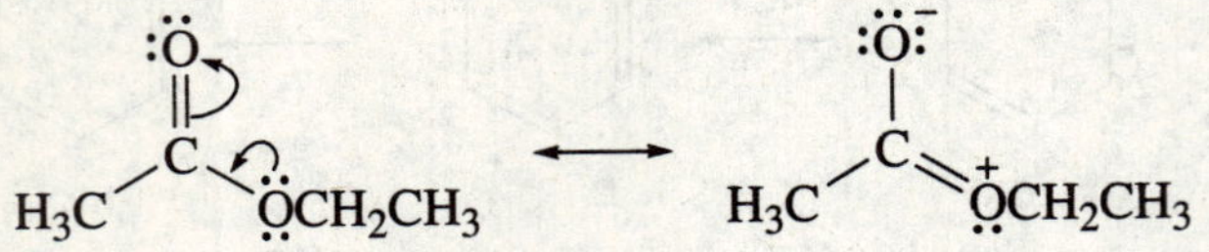

c.

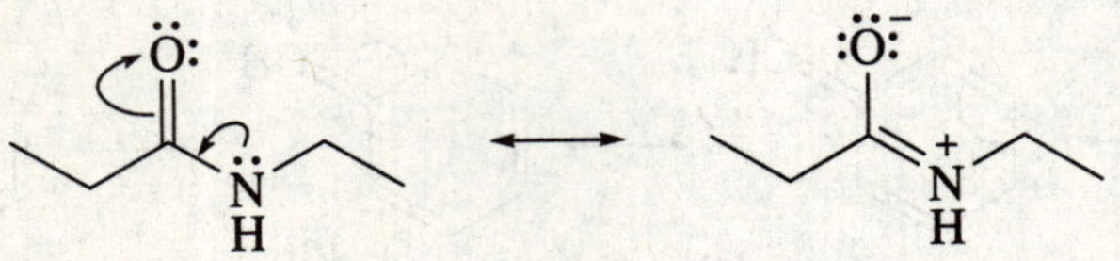

d.

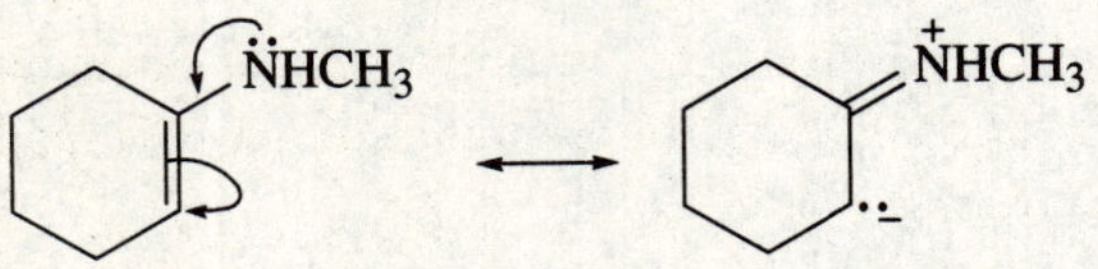

e.

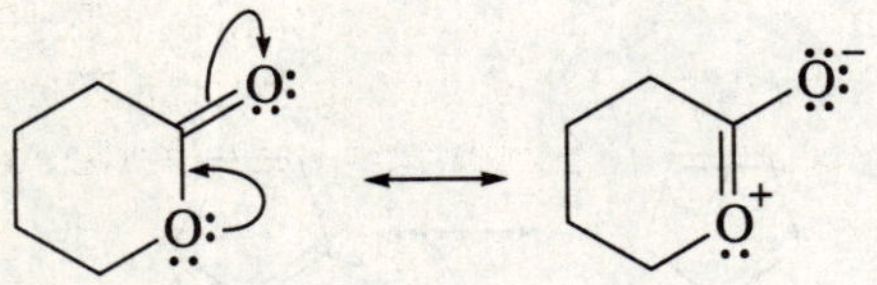

f.

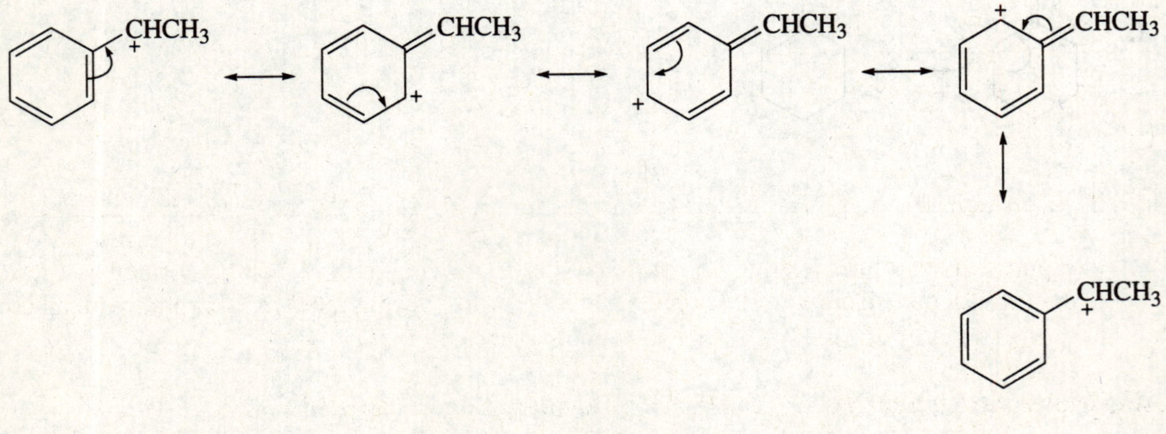

Special Topic VI

Molecular Orbital Theory

Chemists use models to describe such things as the bonding in molecules, the stability of molecules, and the reactions between molecules. The model used is generally the one that provides the best description of the molecule under consideration. One very powerful model is molecular orbital theory.

You were introduced to molecular orbital (MO) theory in Section 1.6 where you saw that electrons are assigned to a volume of space called an orbital. According to MO theory, covalent bonds are formed when atomic orbitals combine to form molecular orbitals. Let's review some important principles:

1. Orbitals are conserved. In other words, the number of molecular orbitals formed must equal the number of orbitals combined. For example, when two orbitals overlap, two molecular orbitals are formed—one lower in energy (a bonding MO) and one higher in energy (an antibonding MO, indicated by a *) than the overlapping orbitals.
2. Electrons always occupy the available *atomic orbitals* with the lowest energy, and no more than two electrons can occupy an *atomic orbital.* Similarly, electrons always occupy the available *molecular orbitals* with the lowest energy, and no more than two electrons can occupy a *molecular orbital.*
3. The relative energies of the molecular orbitals are: $\sigma < \pi < \pi^* < \sigma^*$.
4. The strongest covalent bonds are formed by electrons that occupy the molecular orbitals with the lowest energy. For example, the energy of a σ MO is lower than that of a π MO, and we have seen that a σ bond is stronger than a π bond (Section 1.14).

There are two ways orbitals can overlap to form molecular orbitals; these are shown in Figures 1 and 2. In Figure 1, each of the overlapping orbitals contributes one electron to the bond. In Figure 2, a filled orbital (it has two electrons) overlaps an empty orbital. In each case, we see that the electrons are stabilized (are lower in energy) as a result of orbital overlap.

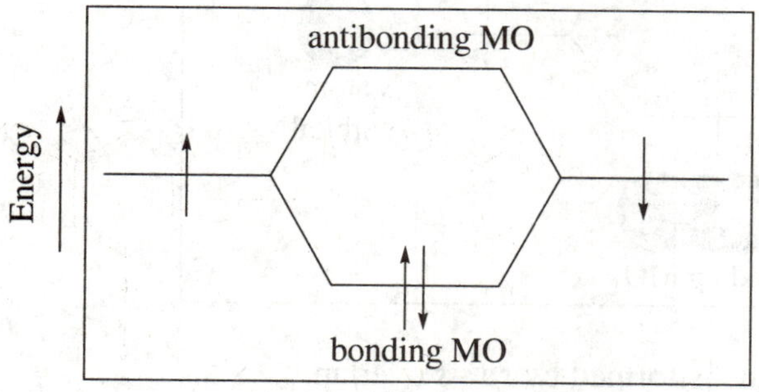

Figure 1. Each of the overlapping orbitals contributes one electron to the bond.

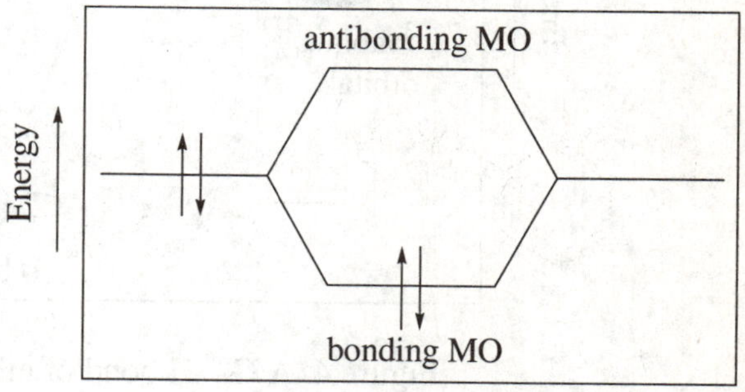

Figure 2. A bond is formed as a result of the overlap of a filled orbital with an empty orbital.

Using Molecular Orbital Theory to Describe Covalent Bond Formation

Take a look at Figure 1.4 on page 24 of the text. There you can see that an H—H bond is formed by the overlap of a 1*s* orbital of a hydrogen atom with the 1*s* orbital of another hydrogen atom; each of the atomic orbitals contributes one electron to the molecular orbital.

Figure 3 shows that the C—C bond in ethane is formed by the overlap of an sp^3 orbital of carbon with an sp^3 orbital of another carbon; again, each of the overlapping orbitals contributes one electron to the bond. Figure 4 shows that the C—H bond of ethane is formed by the overlap of an sp^3 orbital of carbon with the *s* orbital of hydrogen. Because an *s* orbital is more stable than an sp^3 orbital (page 29 of the text), the MO formed by sp^3—*s* overlap is more stable than the MO formed by sp^3—sp^3 overlap. As a result, the C—H bond is stronger (and shorter) than the C—C bond.

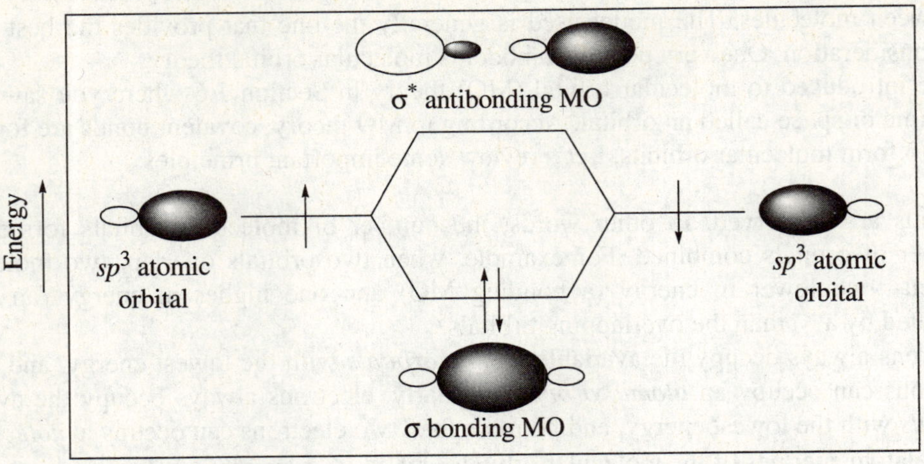

Figure 3. The C—C bond of ethane is formed by sp^3–sp^3 overlap.

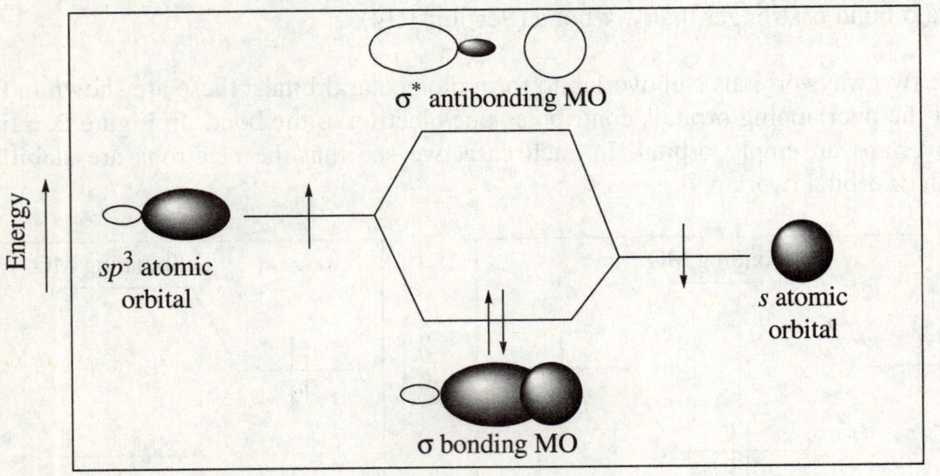

Figure 4. A C—H bond of ethane is formed by sp^3–*s* overlap.

Figure 5 shows that the two sp^3 orbitals that overlap to form the C—O bond of an alcohol or of an ether do not have the same energy. An electron is more stable in the orbital of the more electronegative atom. Thus, the C—O bond is a little stronger and shorter than the C—C bond.

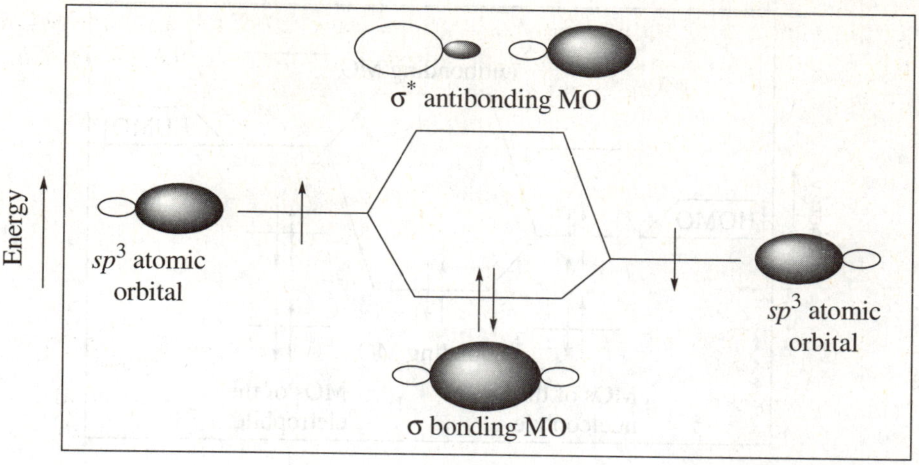

Figure 5. The C—O bond of an ether or of an alcohol is formed by sp^3–sp^3 overlap.

Figure 6 shows that the C—C σ bond of ethene is formed by the overlap of an sp^2 orbital of carbon with an sp^2 orbital of another carbon. The π bond of ethene is formed by the side-to-side overlap of two *p* orbitals (see Figure 1.6 on page 26 of the text.) A π molecular orbital is less stable than a σ molecular orbital. The π bond, therefore, is weaker than the σ bond. Examine Figure 1.8 on page 27 of the text to see the molecular orbitals that are formed when the overlapping *p* orbitals do not each belong to the same kind of atom.

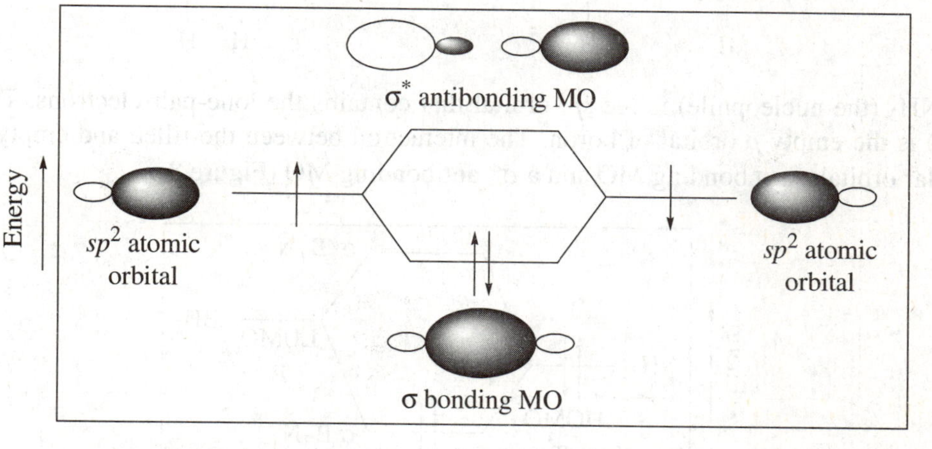

Figure 6. The C—C σ bond of ethene is formed by sp^2–sp^2 overlap.

Using Molecular Orbital Theory to Describe Chemical Reactions

We have seen that most organic reactions involve the reaction of a nucleophile with an electrophile. Molecular orbital theory describes a reaction between a nucleophile and an electrophile as occurring as the result of the interaction of the HOMO (highest occupied MO) of the nucleophile with the LUMO (lowest unoccupied MO) of the electrophile, because the most stabilizing interaction is between orbitals closest in energy. Notice that in these examples, a filled orbital overlaps an empty orbital. The interaction of a filled orbital with an empty orbital is stabilizing, because the two electrons involved in bond formation end up in the lower-energy bonding MO and no electrons have to be placed in an antibonding MO (Figure 7).

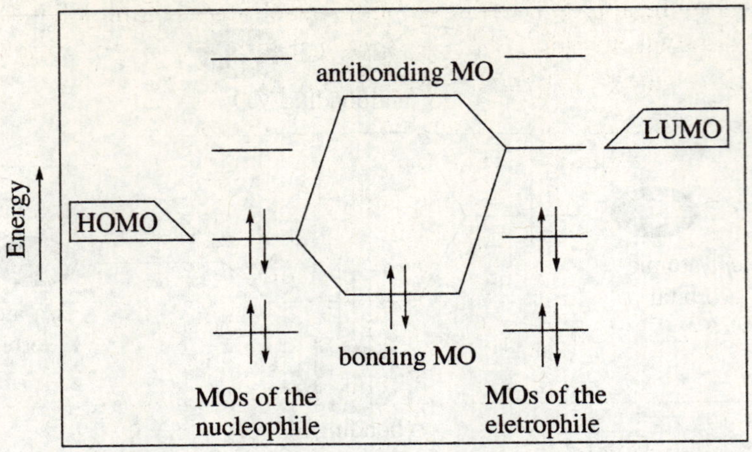

Figure 7. The interaction of the HOMO of the nucleophile with the LUMO of the electrophile.

We will start by looking at the reaction of NH_3 (a Lewis base) with BH_3 (a Lewis acid), a reaction discussed in Section 1.26 of the text. The bond formed between a Lewis base (such as NH_3) and a Lewis acid (such as BH_3) results from the base sharing a pair of electrons with the acid.

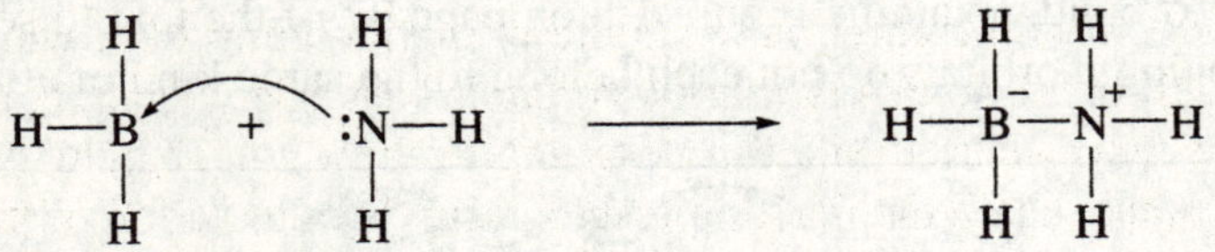

The HOMO of NH_3 (the nucleophile) is the sp^3 orbital that contains the lone-pair electrons. The LUMO of BH_3 (the electrophile) is the empty *p* orbital of boron. The interaction between the filled and empty orbitals produces two new molecular orbitals, a σ bonding MO and a σ^* antibonding MO (Figure 8).

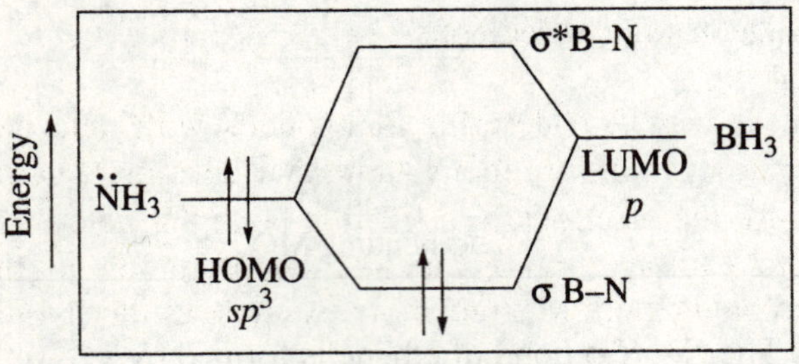

Figure 8. The interaction of the HOMO of NH_3 with the LUMO of BH_3.

Now we will look at the reaction of 2-butene with HBr, the first electrophilic addition reaction we examined in Section 3.6.

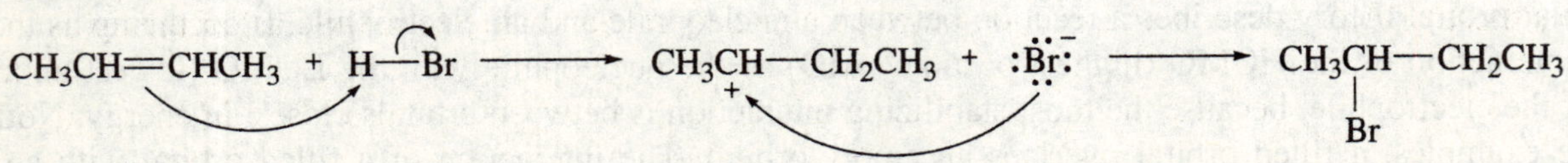

In the first step of the reaction, the alkene is the nucleophile; the electrons of the π bond are in the π bonding MO; this is the HOMO. HBr is the electrophile. The electrons that form the H—Br bond are in a σ bonding MO. Therefore, the LUMO of HBr is the σ^* antibonding MO (Figure 9).

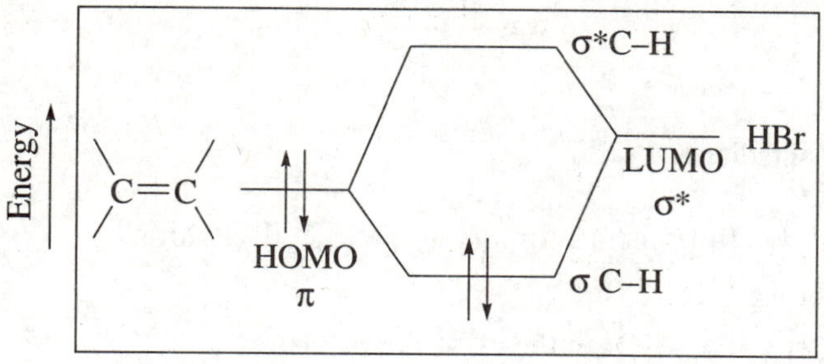

Figure 9. Interaction of the HOMO of the alkene with the LUMO of HBr.

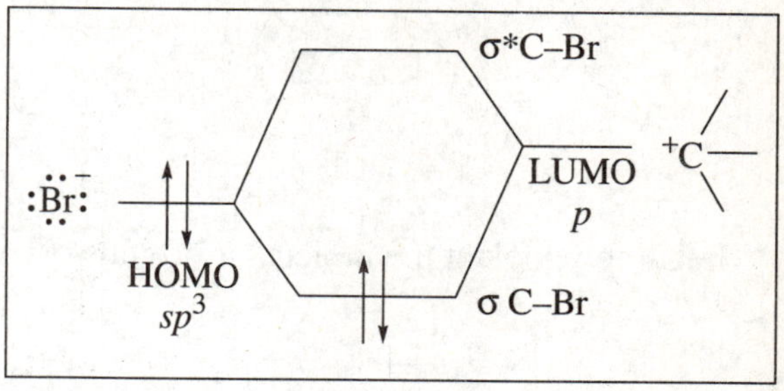

Figure 10. Interaction of the HOMO of Br^- with the LUMO of the carbocation.

In the second step of the reaction, Br^- is the nucleophile; the HOMO is occupied by one of its lone pairs in an sp^3 orbital. The carbocation is the electrophile; the LUMO is carbon's empty *p* orbital (Figure 10).

In Section 8.2 of the text, we used MO theory to explain why the alkyl halide in an S_N2 reaction undergoes back-side attack, resulting in inversion of configuration. A lone pair of the nucleophile is in the HOMO. The LUMO of the electrophilic alkyl halide is the σ^* antibonding MO, which has its largest lobe at the back of the carbon in the C—Br bond, so this is where overlap is best. (See Figure 8.1 on page 349 of the text).

In Section 16.5 of the text, we used MO theory to explain how a nucleophile reacts with a carbonyl group to form a tetrahedral species. Once again the reaction results from the interaction of the HOMO of the nucleophile and the LUMO of the electrophile. In this case, a lone pair of the nucleophile is in the HOMO and the LUMO is the π^* antibonding orbital of the carbonyl group. The π^* antibonding orbital is largest at the carbon atom, so this is where the nucleophile overlaps (See Figure 16.3 on page 736 of the text.)

Molecular Orbital Theory and Delocalized Electrons

Molecular orbital theory is useful to describe compounds with delocalized electrons. For example, in Section 4.2 we saw that tertiary carbocations are more stable than secondary carbocations, which are more stable than primary carbocations, because the electrons in a filled σ bond are delocalized by overlapping the empty *p* orbital of the positively charged carbon. (See Figure 4.2 on page 163 of the text). This kind of electron delocalization is known as hyperconjugation; it is one more example of the stabilization that results when a filled orbital overlaps an empty orbital.

Molecular orbital theory avoids having to use contributing resonance structures, because the electrons in the most stable MO are delocalized over the entire molecule. For example, in Section 7.8, we saw that 1,3-butadiene is more stable than 1,4-pentadiene, because the π electrons in 1,3-butadiene are delocalized over four sp^2 carbons, whereas the intervening methylene group in 1,4-pentadiene prevents the *p* orbitals of C2 and C4 from overlapping. Thus the π electrons of 1,4-pentadiene are localized, causing its molecular orbitals to have the same energy as those of ethene—another compound with localized electrons. (See pages 306–310 of the text.)

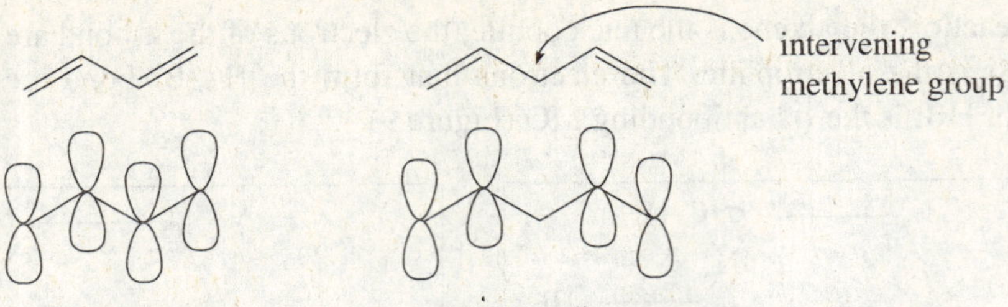

1,3-butadiene 1,4-pentadiene

Let's now look at the molecular orbitals of the allyl cation, the allyl radical, and the allyl anion.

$CH_2{=}CH{-}\overset{+}{C}H_2$ the allyl cation

$CH_2{=}CH{-}\dot{C}H_2$ the allyl radical

$CH_2{=}CH{-}\overset{..-}{C}H_2$ the allyl anion

The three *p* atomic orbitals of the three allyl carbons combine to produce three π molecular orbitals (Figure 11).

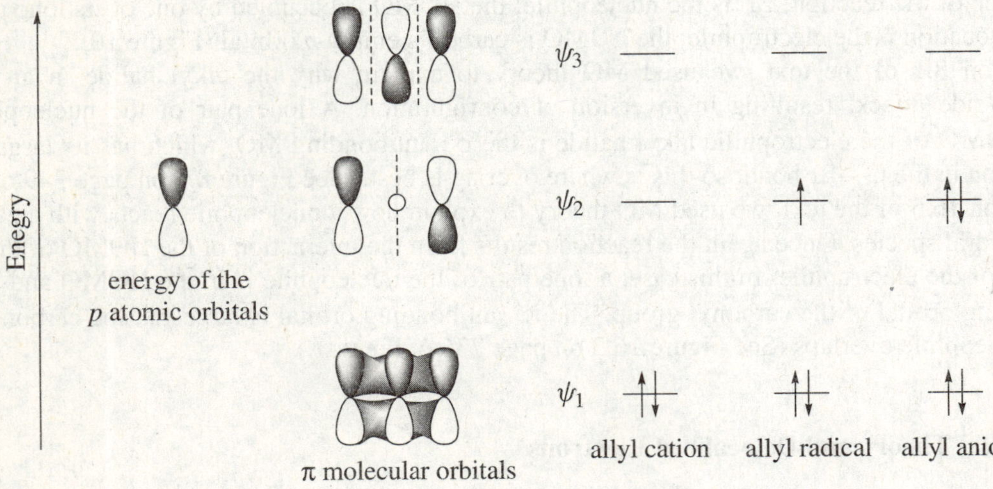

Figure 11. The distribution of the electrons in the molecular orbitals of the allyl cation, the allyl radical, and the allyl anion.

The bonding MO (ψ_1) encompasses the three carbons. In an acyclic system, the number of bonding MOs always equals the number of antibonding MOs. Therefore, when there is an odd number of MOs, one of them must be a nonbonding molecular orbital; ψ_2 is the nonbonding MO. We have seen (Section 7.8) that as the energy of the MO increases, the number of nodes increases. Consequently ψ_2 must have a node—in addition to the one that ψ_1 has—that bisects the *p* orbitals. The only symmetrical position for a node to pass through in ψ_2 is through the middle carbon. You also know that it needs to pass through the middle carbon because that is the only way ψ_2 can be fully antisymmetric, which it must be since ψ_1 and ψ_3 are symmetric. (Recall that MOs alternate between being symmetric and antisymmetric; Section 7.8)

You can see why ψ_2 is called a nonbonding molecular orbital—there is no overlap between the *p* orbital on the middle carbon and the *p* orbital on either of the end carbons. Notice that a nonbonding MO has the same energy as the isolated *p* atomic orbitals. The third MO (ψ_3) is an antibonding MO.

The two π electrons of the allyl cation are in the bonding MO, which means they are spread out over all three carbons. Consequently, the two carbon-carbon bonds are identical, with each having some double-bond character. The resonance contributors show that the positive charge is shared equally by the end carbon atoms, which is another way of showing that the stability of the allyl cation is due to electron delocalization.

$$CH_2{=}CH{-}\overset{+}{C}H_2 \longleftrightarrow \overset{+}{C}H_2{-}CH{=}CH_2$$

The contributing resonance structures show that when a nucleophile such as Br^- reacts with an allyl cation, the Br^- can bond to either of the end carbons—because they share the positive charge—but cannot bond to the middle carbon. Likewise, MO theory shows that only the end carbons have an empty orbital with which the filled orbital of Br^- can overlap. The central carbon has a node so there can be no interaction with this carbon.

The allyl radical has two electrons in the bonding MO, so these electrons are spread over all three carbon atoms. The third electron is in the nonbonding MO. The MO diagram shows that the third electron is shared equally by the end carbons with none of the electron density residing on the middle carbon. This agrees with what the resonance contributors show.

$$CH_2{=}CH{-}\dot{C}H_2 \longleftrightarrow \dot{C}H_2{-}CH{=}CH_2$$

Finally, the allyl anion has two electrons in the nonbonding MO. These two electrons are shared equally by the end carbons. This too agrees with what the resonance contributors show.

$$CH_2{=}CH{-}\overset{\cdot\cdot\,-}{C}H_2 \longleftrightarrow \overset{\cdot\cdot\,-}{C}H_2{-}CH{=}CH_2$$

We have seen that both molecular orbital theory and contributing resonance structures can be used to explain electron delocalization. The choice is a matter of preference. Hyperconjugation is probably best shown by molecular orbital theory, because contributing resonance structures require breaking sigma bonds. Contributing resonance structures are sometimes preferred when one needs to see on which atoms charges reside.

CHAPTER 8
Substitution Reactions of Alkyl Halides

Important Terms

aprotic solvent — a solvent that does not have a hydrogen bonded to an oxygen or to a nitrogen; some aprotic solvents are polar, others are polar.

back-side attack — nucleophilic attack on the side of the carbon opposite to the side bonded to the leaving group.

base — a substance that accepts a proton.

basicity — describes the tendency of a compound to share its electrons with a proton.

bifunctional molecule — a molecule with two functional groups.

bimolecular — describes a process involving two molecules.

complete racemization — formation of a pair of enantiomers in equal amounts.

dielectric constant — a measure of how well a solvent can insulate opposite charges from one another.

elimination reaction — a reaction that removes atoms or groups from the reactant to form a π bond.

first-order reaction — a reaction whose rate is dependent on the concentration of one reactant.

intermolecular reaction — a reaction that takes place between two molecules.

intimate ion pair — results when the covalent bond that joined the cation and anion has broken, but the cation and anion are still next to each other.

intramolecular reaction — a reaction that takes place within a molecule.

inversion of configuration — turning the carbon inside out like an umbrella so that the resulting product has a configuration opposite to that of the reactant.

ion-dipole interaction — the interaction between an ion and the dipole of a molecule.

kinetics — the field of chemistry that deals with the rates of chemical reactions.

leaving group — the group that is displaced in a substitution reaction.

nucleophile — an electron-rich atom or molecule.

nucleophilicity — a measure of how readily an atom or molecule with a lone pair attacks an atom.

nucleophilic substitution reaction — a reaction in which a nucleophile substitutes for an atom or group.

partial racemization — formation of a pair of enantiomers in unequal amounts.

protic solvent	a solvent that has a hydrogen bonded to an oxygen or to a nitrogen.
rate constant	describes how difficult it is to overcome the energy barrier of a reaction.
rate law	the equation that shows the relationship between the rate of a reaction and the concentration of the reactants.
second-order reaction	a reaction whose rate is dependent on the concentration of two reactants.
S_N1 reaction	a unimolecular nucleophilic substitution reaction.
S_N2 reaction	a bimolecular nucleophilic substitution reaction.
solvent-separated ion pair	results when the cation and anion are separated by one or more solvent molecules.
solvolysis	reaction with the solvent.
steric effects	effects due to the fact that groups occupy a certain volume of space.
steric hindrance	refers to bulky groups at the site of a reaction that make it difficult for the reactants to approach each other.
substitution reaction	a reaction that changes one substituent of a reactant for another.
unimolecular	describes a process involving one molecule.

Solutions to Problems

1. The rate of the reaction will be 0.05 of the original rate, since the concentration is 0.05 of the original concentration.

2. It decreases the magnitude of the rate constant, which causes the reaction to be slower.

3. The closer the methyl group is to the site of nucleophilic attack, the greater the steric hindrance to nucleophilic attack and the slower the rate of the reaction.

$$CH_3CH_2CH_2CH_2CH_2Br > CH_3\overset{CH_3}{\overset{|}{C}}HCH_2CH_2Br > CH_3CH_2\overset{CH_3}{\overset{|}{C}}HCH_2Br > CH_3CH_2\underset{\underset{CH_3}{|}}{\overset{\overset{CH_3}{|}}{C}}Br$$

4. **a.** Solved in the text.

b. The S_N2 reaction of (*R*)-2-bromobutane with hydroxide ion will form (*S*)-2-butanol.

c. The S_N2 reaction of (*S*)-3-chlorohexane and with hydroxide ion will form (*R*)-3-hexanol.

d. The S_N2 reaction of 3-iodopentane (it does not have an asymmetric center) with hydroxide ion will form 3-pentanol (it does not have an asymmetric center).

5. **a.** RO^- because ROH is weaker acid than RSH .

b. RS^- because it is less well solvated by water and sulfur is more polarizable than oxygen.

6. **a.** aprotic **b.** aprotic **c.** protic **d.** aprotic

7. Solved in the text.

8. **a.** $CH_3CH_2Br + HO^-$ — HO^- is a stronger nucelophile than H_2O.

b. $CH_3\underset{\underset{CH_3}{|}}{C}HCH_2Br + HO^-$ — The alkyl halide has less steric hindrance.

c. $CH_3CH_2Cl + CH_3S^-$ — CH_3S^- is a stronger nucelophile than CH_3O^- in a solvent that can form hydrogen bonds.

d. $CH_3CH_2Br + I^-$ — Br^- is a weaker base than $Cl,^-$ so Br^- is better leaving group.

9. Solved in the text.

10. In each reaction, the conjugate acid of the leaving group is stronger than the conjugate acid of the nucleophile. That means the leaving group is the weaker base (better leaving group).

H_2O $pK_a = 15.7$	HCl $pK_a = -7.0$	NH_3 $pK_a = 36$	HCl $pK_a = -7.0$
H_2S $pK_a = 7.0$	HBr $pK_a = -9.0$	$RC{\equiv}CH$ $pK_a = 25$	HBr $pK_a = -9.0$
ROH $pK_a = 15.5$	HI $pK_a = -10.0$	$HC{\equiv}N$ $pK_a = 9.1$	HI $pK_a = -10.0$
RSH $pK_a = 10.5$	HBr $pK_a = -9.0$		

11. **a.** $CH_3CH_2OCH_2CH_2CH_3$ **c.** $CH_3CH_2\overset{+}{N}(CH_3)_3\ Br^-$
b. $CH_3CH_2C{\equiv}CCH_3$ **d.** $CH_3CH_2SCH_2CH_3$

12. Solved in the text.

13. **a.** Reaction of an alkyl halide with ammonia gives a low yield of primary amine because as soon as the primary amine is formed, it can react with another molecule of alkyl halide to form a secondary amine; the secondary amine can react with the alkyl halide to form a tertiary amine which can then react with an alkyl halide to form a quaternary ammonium salt.
b. The alkyl azide is not treated with hydrogen until after all the alkyl halide has reacted with azide ion. Therefore, when the primary amine is formed, there is no alkyl halide for it to react with to form a secondary amine.

14.

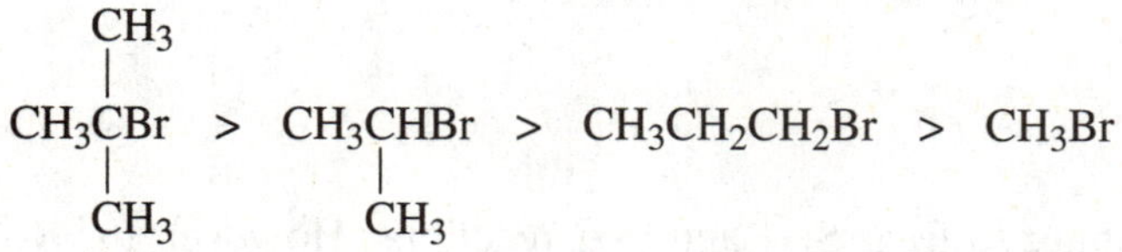

15. Because ethyl bromide and methyl bromide cannot dissociate to form a carbocation (the carbocation is too unstable to be formed), the alkyl halides have to react by an S_N2 pathway. The nucleophile (H_2O) is very weak, so the S_N2 reactions are very slow.

16.

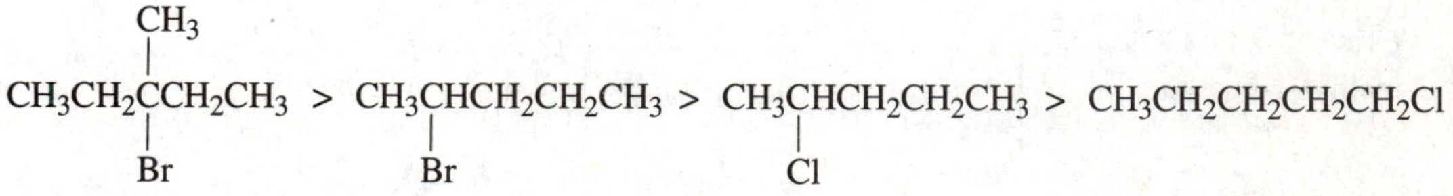

17. **a.** The S_N2 and S_N1 reactions will form different products because secondary carbocation that is formed initially in the S_N1 reaction will undergo a 1,2-hydride shift to form a tertiary carbocation.

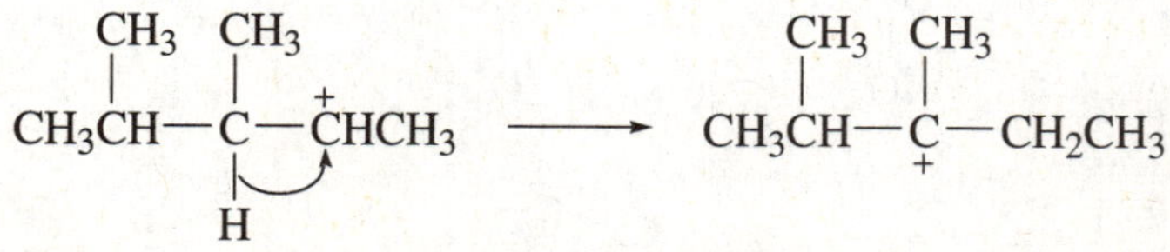

b. The S_N2 and S_N1 reactions will form different products because secondary carbocation that is formed initially in the S_N1 reaction will undergo a 1,2-hydride shift to form a tertiary carbocation.

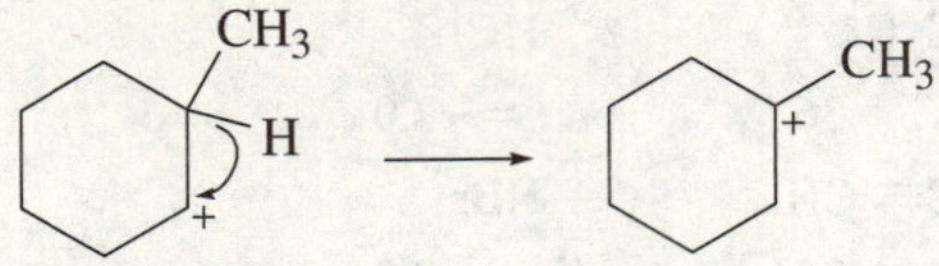

c. The S_N2 and S_N1 reactions will form different products because secondary carbocation that is formed initially in the S_N1 reaction will undergo a 1,2-methyl shift to form a tertiary carbocation.

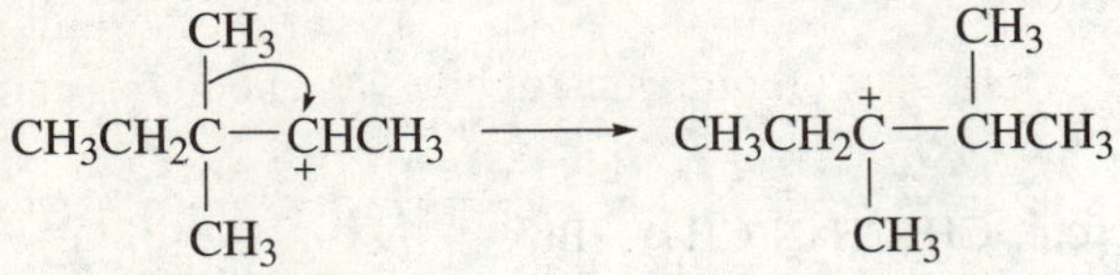

d. The configuration of the asymmetric center is not specified; therefore, the configuration of the product cannot be specified. If the configuration at C-4 were specified, one stereoisomer would be obtained from an S_N2 reaction (the product with the inverted configuration compared to the configuration of the reactant), and two products would be obtained from an S_N1 reaction (one with the inverted configuration and one with the retained configuration).

e. The S_N1 and S_N2 reactions will form different products because the secondary carbocation formed initially in the S_N1 reaction will undergo a 1,2-hydride shift to form a tertiary carbocation.

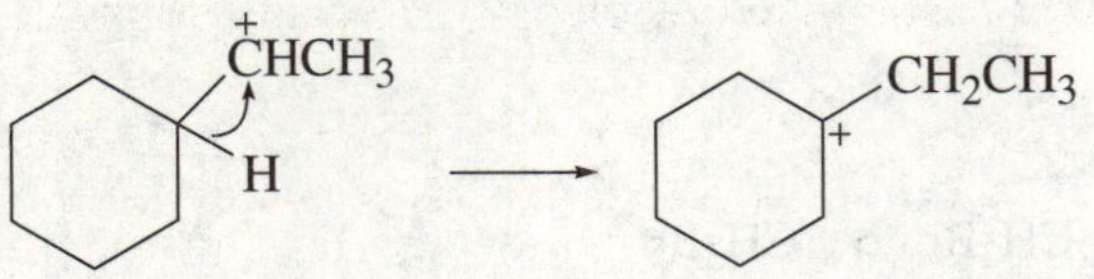

f. The same substitution product will be obtained from S_N1 and S_N2 reactions. However, the S_N2 reaction would be very slow because of steric hindrance.

18.

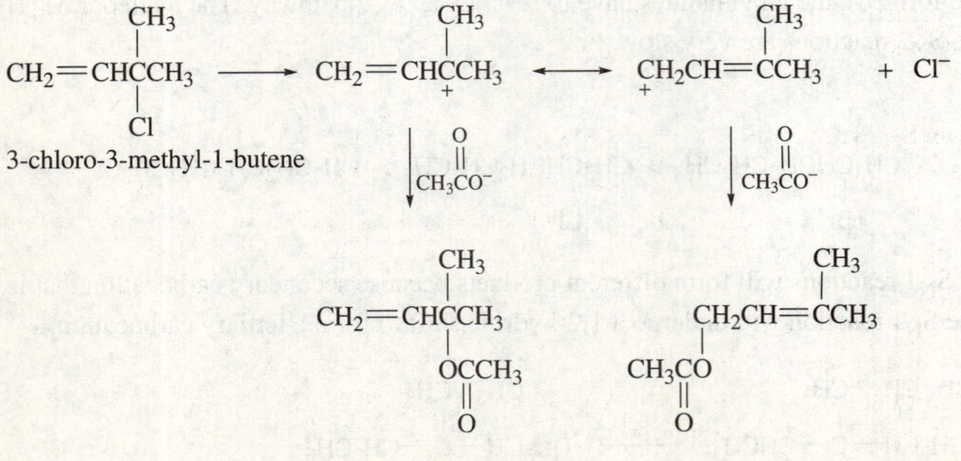

19. **a.** **1.** The S_N2 reaction proceeds with back-side attack.

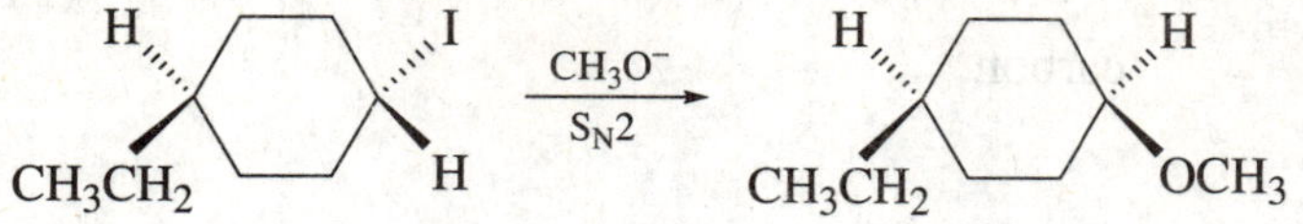

2.

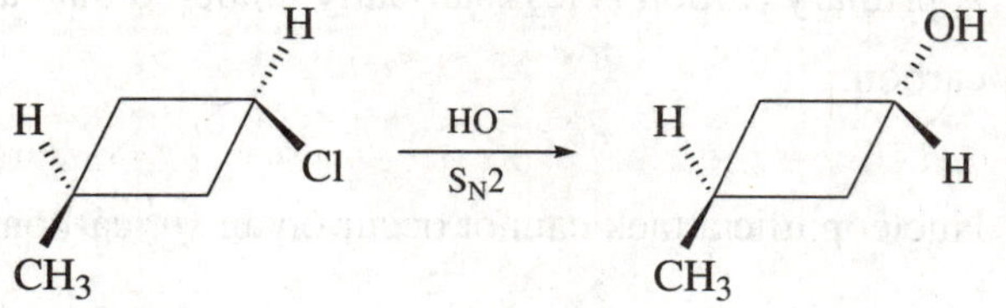

b. **1.** The S_N1 reaction forms a carbocation; the nucleophile can attack either side of the carbocation.

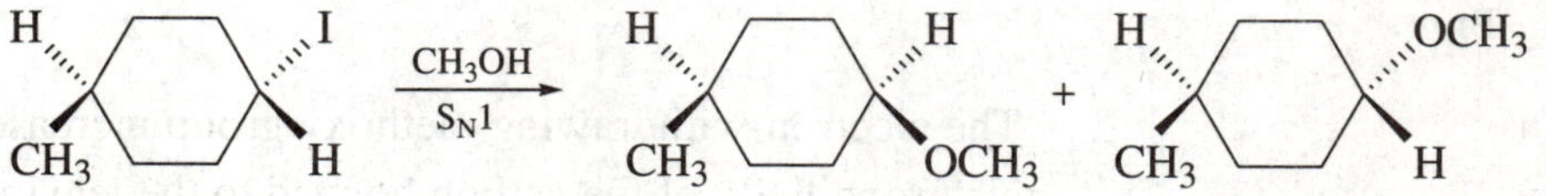

2.

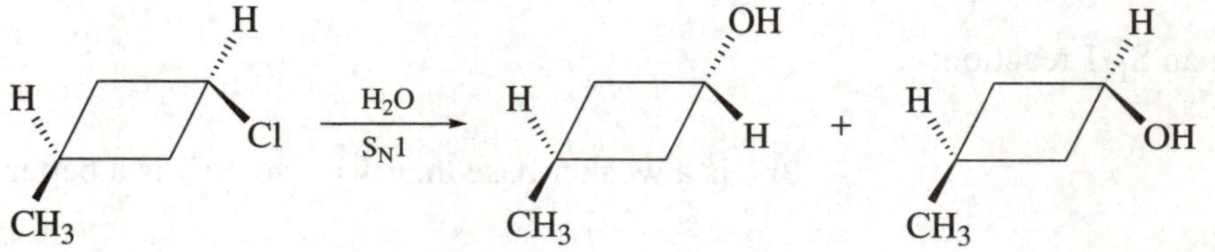

20. *trans*-4-Bromo-2-hexene (the compound on the right) is more reactive in an S_N1 solvolysis reaction, because the carbocation that is formed is stabilized by electron delocalization. (It is a secondary allylic carbocation.)

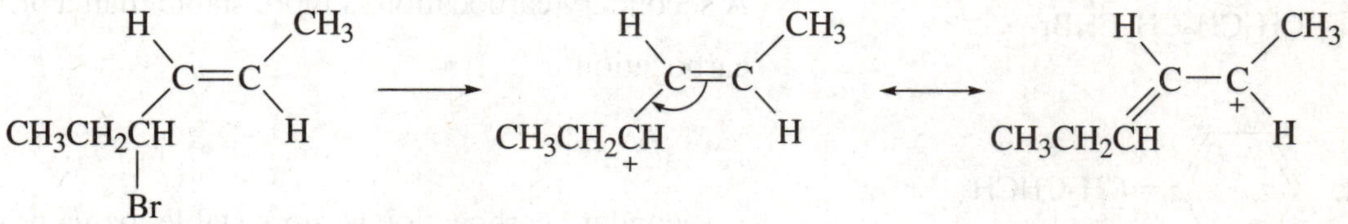

The carbocation formed by the other alkyl halide is less stable because the secondary carbocation cannot be stabilized by resonance.

21. Most reactive in an S_N2 reaction:

a.

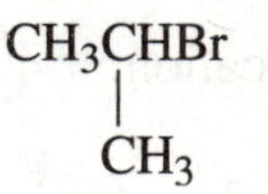

Br^- is a weaker base than Cl^-, so Br^- is a better leaving group.

b.

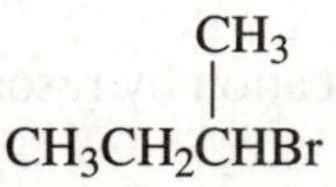

This compound has less steric hindrance.

c. $CH_3CH_2CH(CH_3)CH_2Br$ — A primary carbon is less sterically hindered than a secondary carbon.

d. 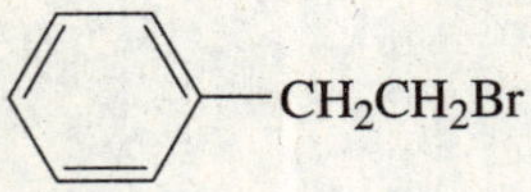 — A primary carbon is less sterically hindered than a secondary carbon.

e. 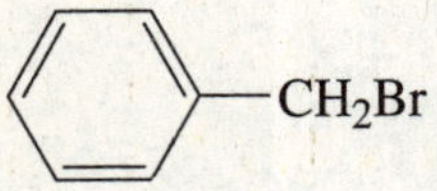 — Nucleophilic attack cannot occur on an *sp*^{2} carbon.

f. $CH_3CH{=}CHCH(Br)CH_3$ — Nucleophilic attack cannot occur on an *sp*^{2} carbon.

g. CH_3OCH_2Cl — The electron-withdrawing methoxy group increases the electrophilicity of the carbon bonded to the leaving group.

22. Most reactive in an S_N1 reaction:

a. $CH_3CH(CH_3)Br$ — Br^- is a weaker base than Cl^-, so Br^- is a better leaving group.

b. The two compounds are equally reactive.

c. $CH_3CH_2CH_2CH(CH_3)Br$ — A secondary carbocation is more stable than a primary carbocation.

d. 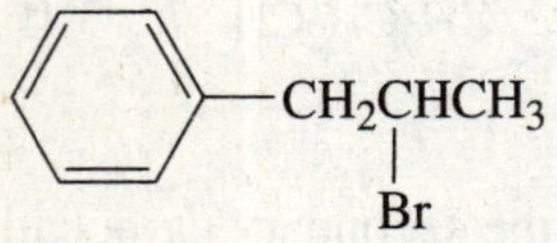 — A secondary carbocation is more stable than a primary carbocation.

e. 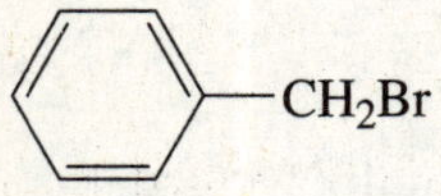 — A benzyl cation is more stable than an aryl cation.

f. $CH_3CH{=}CHCH(Br)CH_3$ — An allyl cation is more stable than a vinyl cation.

g. CH_3OCH_2Cl — The methoxy group stabilizes the carbocation by resonance electron donation.

23. **a.** $CH_3CH{=}CHCH_2Br \xrightarrow{CH_3O^-} CH_3CH{=}CHCH_2OCH_3$

b. $CH_3CH{=}CHCH_2Br \longrightarrow CH_3CH{=}CH\overset{+}{C}H_2 \longleftrightarrow CH_3\overset{+}{C}HCH{=}CH_2$

$\downarrow CH_3O^-$ (from each resonance form)

$CH_3CH{=}CHCH_2OCH_3 + CH_3CH(OCH_3)CH{=}CH_2$

24. **a.** An S_N2 reaction, because of the high concentration of a good mucleophile.

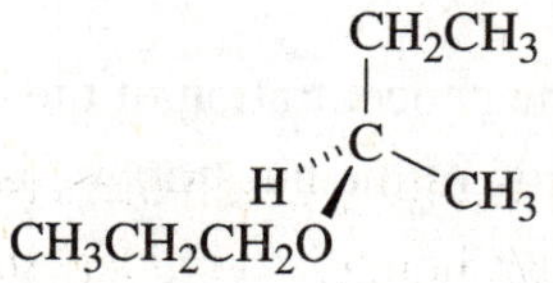

b. An S_N2 reaction, because a good nucleophile is used.

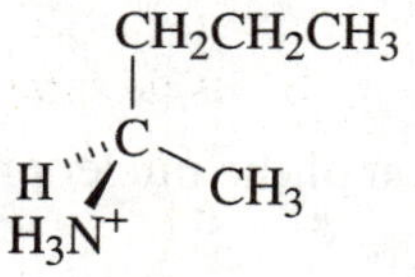

Even though ammonia is not negatively charged, it is a good nucleophile. Recall that an amine is a much stronger base than an alcohol.
(The pK_a of a protonated alcohol is ~ – 2.5. The pK_a of the ammonium ion is 9.4: see Appendix II in the text.)

c. An S_N2 reaction, because of the high concentration of a good nucleophile.

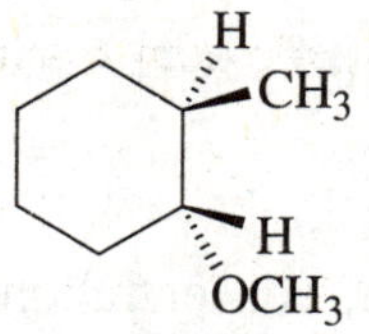

d. An S_N1 reaction, because a poor nucleophile is used. Notice that the initially formed secondary carbocation rearranges to a more stable tertiary carbocation.

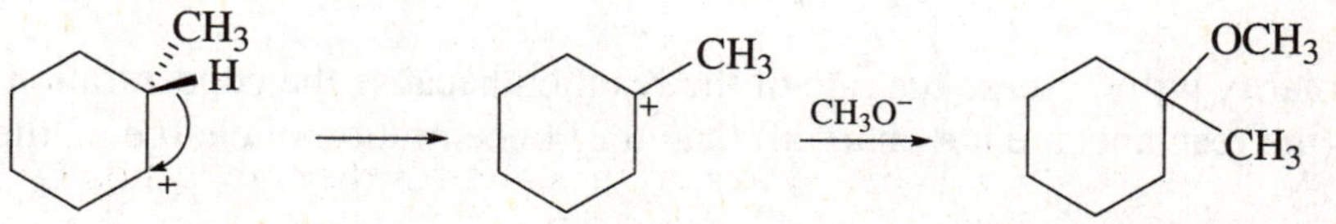

The product does not
have an asymmetric center.

e. An S_N2 reaction, because of the high concentration of a good nucleophile.

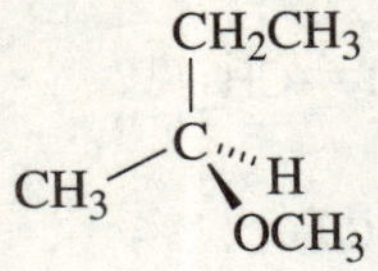

f. An S_N1 reaction, because a poor nucleophile is used.

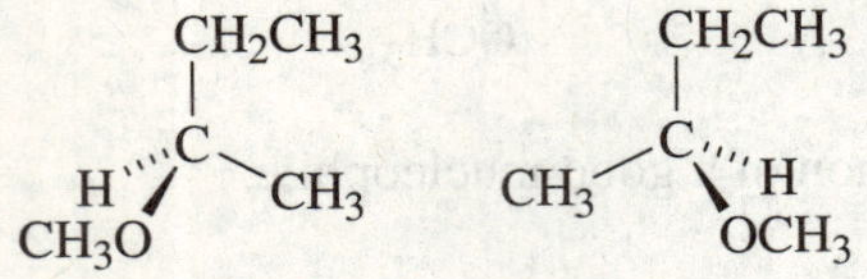

25. Because the rate of an S_N1 reaction is not affected by increasing the concentration of the nucleophile, while the rate of an S_N2 reaction is increased when the concentration of the nucleophile is increased, you first have to determine whether the reactions are S_N1 or S_N2 reactions.

- **a.** is an S_N2 reaction because the configuration of the product is inverted compared with that of the reactant.
- **b.** is an S_N2 reaction because the reactant is a primary alkyl halide.
- **c.** is an S_N1 reaction because the reactant is a tertiary alkyl halide.

Because they are S_N2 reactions, **a** and **b** will go faster if the concentration of the nucleophile is increased.
Because it is an S_N1 reaction, the rate of **c** will not change if the concentration of the nucleophile is increased.

26. **a.** Solved in the text

b.

$$\frac{S_N2}{S_N2 + S_N1} = \frac{k_2[\text{2-bromobutane}]\,[\text{HO}^-]}{k_2[\text{2-bromobutane}]\,[\text{HO}^-] + k_1[\text{2-bromobutane}]}$$

$$\frac{S_N2}{S_N2 + S_N1} = \frac{3.20 \times 10^{-5} \times 1 \times 10^{-3}}{3.20 \times 10^{-8} + 1.50 \times 10^{-6}} = \frac{3.20 \times 10^{-8}}{3.20 \times 10^{-8} + 150 \times 10^{-8}} = \frac{3.20}{153} = .02$$

$$= 2\%$$

27. **a.** Increasing the polarity will decrease the rate of the reaction because the concentration of charge on the reactants is greater (the reactants are charged) than the concentration of charge on the transition state.

b. Increasing the polarity will decrease the rate of the reaction because the concentration of charge on the reactants is greater (the reactants are charged) than the concentration of charge on the transition state.

c. Increasing the polarity will increase the rate of the reaction because the concentration of charge on the reactants is less (the reactants are not charged) than the concentration of charge on the transition state.

28. **a.** $CH_3Br + HO^- \longrightarrow CH_3Br + Br^-$

HO^- is better nucleophile than H_2O.

b. $CH_3I + HO^- \longrightarrow CH_3Br + I^-$

I^- is better leaving group than Cl^-.

c. $CH_3Br + NH_3 \longrightarrow CH_3\overset{+}{N}H_3 + Br^-$

NH_3 is a better nucleophile than H_2O.

d. $CH_3Br + HO^- \xrightarrow{\text{DMSO}} CH_3OH + Br^-$

DMSO will not stabilize the nucleophile by hydrogen bonding.

e. $CH_3Br + NH_3 \xrightarrow{\text{EtOH}} CH_3NH_3 + Br^-$

The more polar solvent will be more able to stabilize the transition state. (EtOH is ethanol.)

29. Solved in the text.

30. Acetate ion will be a more reactive nucleophile in dimethyl sulfoxide because dimethyl sulfoxide will not stabilize the nucleophile by hydrogen bonding, whereas methanol will stabilize it by hydrogen bonding.

31. Only an S_N1 reaction will give the product with retention of configuration. Since the S_N1 reaction is favored by a polar solvent, a greater percentage of the reaction will take place by an S_N1 pathway in 50% water/50% ethanol, the more polar of the two solvents.

32. **a.** 

because it forms a six-membered ring, whereas the other compound would form a seven-membered ring. A seven-membered ring is more strained than a six-membered ring so the six-membered ring is formed more easily (see Table 2.9 on p. 109 of the text).

b. 

because it forms a five-membered ring, whereas the other compound would form a four-membered ring. A four-membered ring is more strained than a five-membered ring so the five-membered ring is formed more easily.

c. 

because it forms a seven-membered ring, whereas the other compound would form an eight-membered ring. An eight-membered ring is more strained than a seven-membered ring so the seven-membered ring is formed more easily; also, the Br and OH are less likely to be in the proper position relative to one another for reaction because there are more bonds around which rotation to an unfavorable conformation can occur in the compound that leads to the eight-membered ring (Section 23.6).

33. **a.** CH_3OH **b.** CH_3NH_2 **c.** CH_3SH **d.** CH_3SH **e.** CH_3OCH_3 **f.** $CH_3\overset{+}{N}H_2CH_3$

(Notice that the product in "c" is not protonated because its pK_a is ~ –7; the product in "f" is protonated because its pK_a is ~11).

34. **a.** **1.** 3° > 2° > 1°

2. An S_N1 reaction is not affected by the reactivity of the nucleophile, but a poor (weakly reactive) nucleophile favors an S_N1 by disfavoring an S_N2 reaction.

3. An S_N1 reaction is not affected by the concentration of the nucleophile, but a low concentration of a nucleophile favors an S_N1 by disfavoring an S_N2 reaction.

4. If the reactant is charged, an S_N1 reaction will be favored by the least polar solvent that will dissolve the reactant.
If the reactant is not charged, an S_N1 reaction will be favored by a protic polar solvent.

b. **1.** 1° > 2° > 3°

2. A highly reactive nucleophile favors an S_N2 reaction.

3. A high concentration of a nucleophile favors an S_N2 reaction.

4. If the reactant is charged, an S_N2 reaction will be favored by the least polar solvent that will dissolve the reactant (generally an aprotic polar solvent).
If the reactant is not charged, an S_N2 reaction will be favored by a protic polar solvent.

35. If the atoms are in the same row, the stronger base is the better nucleophile. If the atoms are in the same column, the larger atom is the better nucleophile because the solvent will form stronger hydrogen bonds with the smaller atom.

a. HO^- **c.** H_2S **e.** I^-

b. NH_3 **d.** HS^- **f.** Br^-

36. The weaker base is the better leaving group.

a. H_2O **c.** H_2S **e.** I^-

b. H_2O **d.** HS^- **f.** Br^-

37. **a.** HO^- **c.** HS^- **e.** CH_3NH_2 **g.** $CH_3\overset{\overset{O}{\|}}{C}O^-$

b. CH_3O^- **d.** $CH_3CH_2S^-$ **f.** $^-C{\equiv}N$ **h.** $CH_3C{\equiv}C^-$

38. **a.** 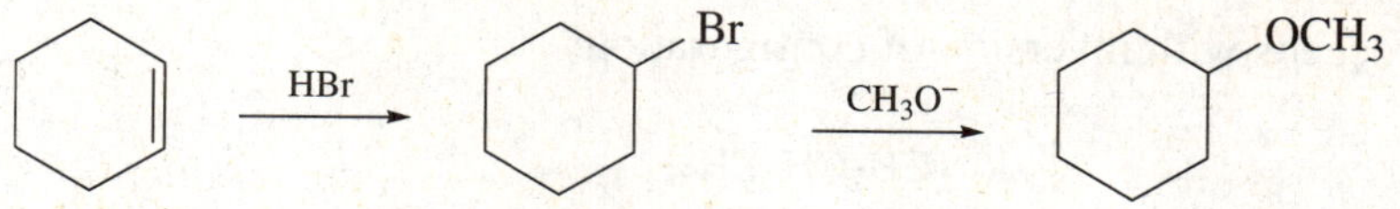

b. 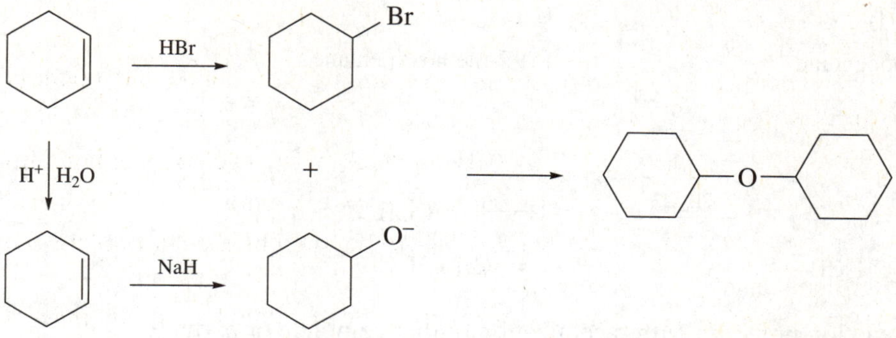

c. 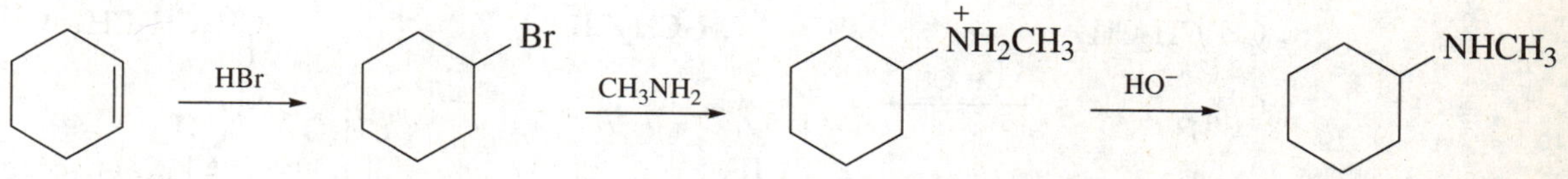

39. **a.** $CH_3CH_2S^- > CH_3CH_2O^- > CH_3\overset{O}{\overset{\|}{C}}O^-$ **c.** $NH_3 > H_2O$

b. 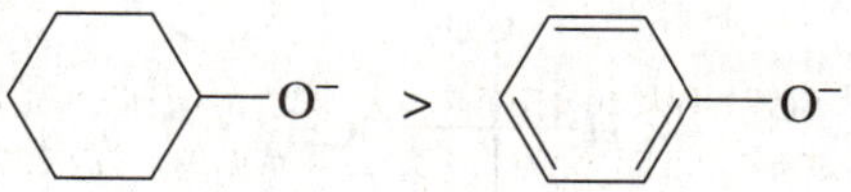 **d.** $I^- > Br^- > Cl^-$

40. The pK_a would increase (it would be a weaker acid) because of a decreased tendency to form a charged species in a less polar solvent. (See Problem 29.)

41. **a.** 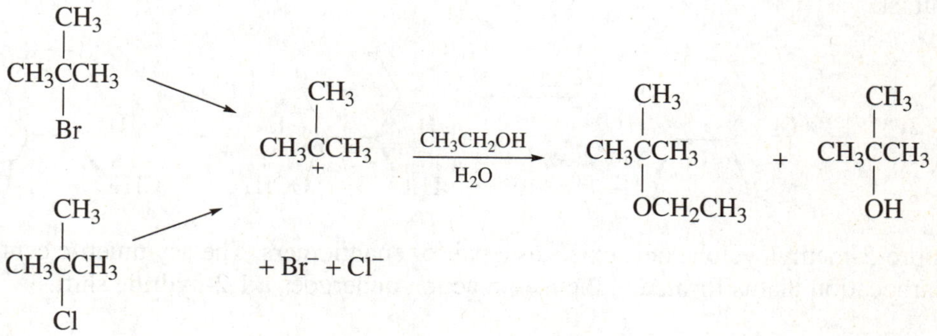

b. The products are obtained as a result of reacting with the carbocation. 2-Bromo-2-methylpropane and 2-chloro-2-methylpropane form the same carbocation, so both alkyl halides form the same products.

42. **a.** The S_N2 reaction proceeds with inversion of configuration.

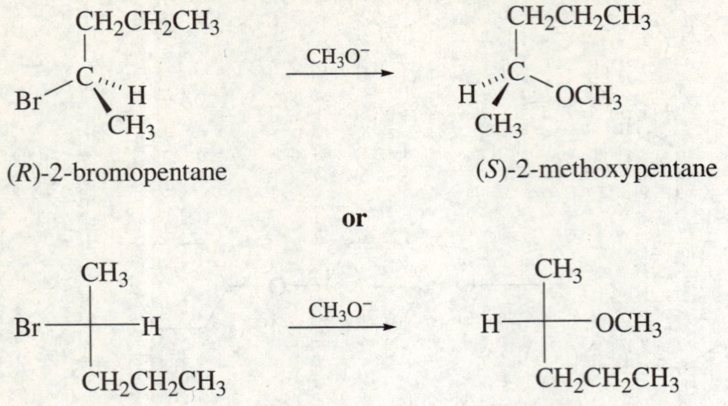

b. The S_N1 reaction proceeds with racemization (either complete or partial).

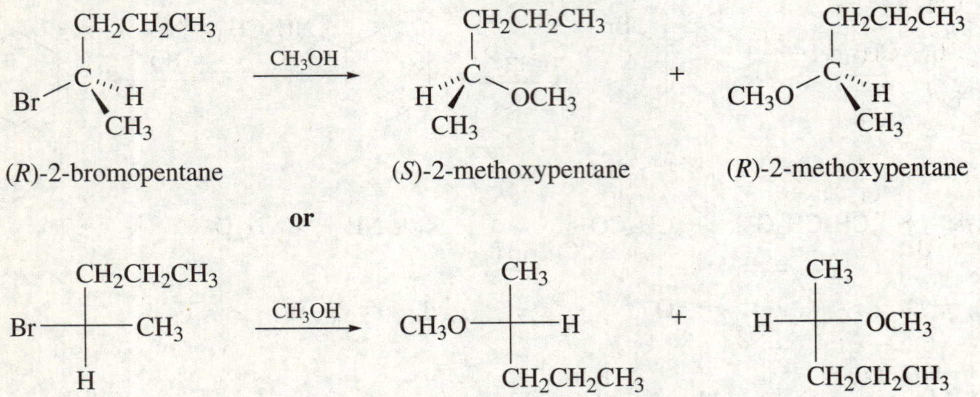

c. *trans*-1-Chloro-2-methylcyclohexane exists as a pair of enantiomers. The S_N2 reaction proceeds with back-side attack.

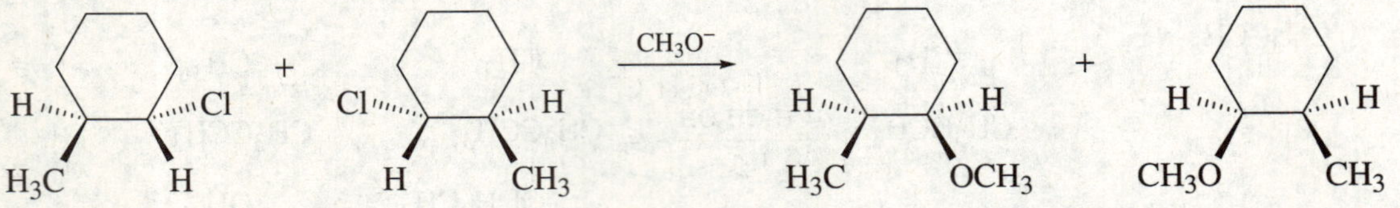

d. *trans*-1-Chloro-2-methylcyclohexane exists as a pair of enantiomers. The asymmetric center is lost when the carbocation that is formed in the S_N1 reaction undergoes a 1,2-hydride shift.

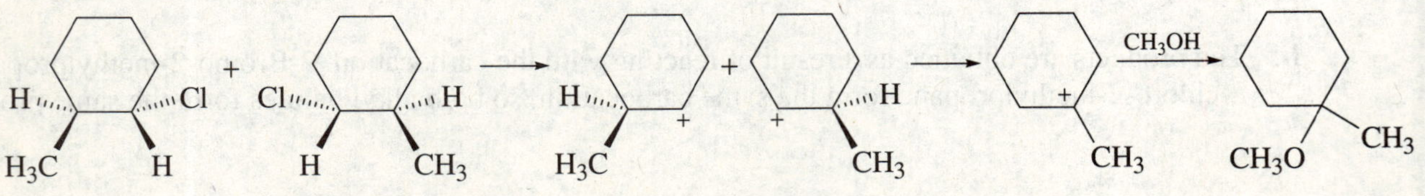

e. 3-Bromo-2-methylpentane exists as a pair of enantiomers. The asymmetric center is lost when the carbocation that is formed in the S_N1 reaction undergoes a 1,2-hydride shift.

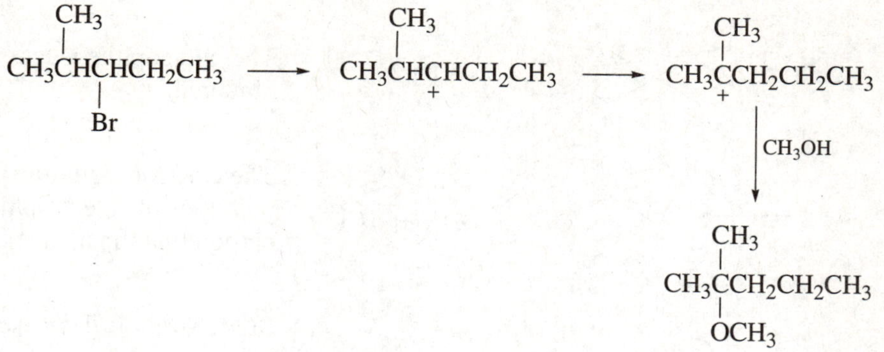

f. 3-Bromo-3-methylpentane does not have an asymmetric center.

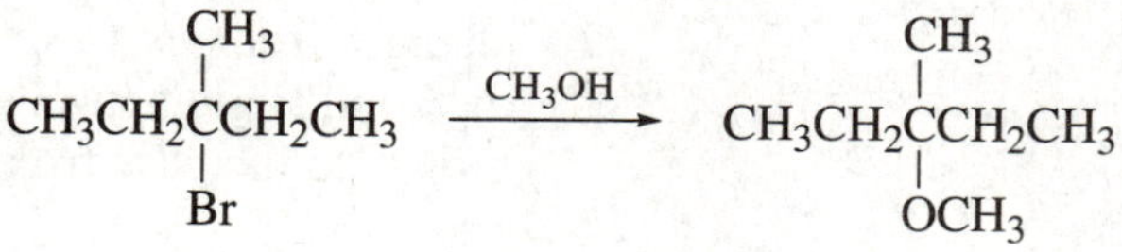

43. **a.**

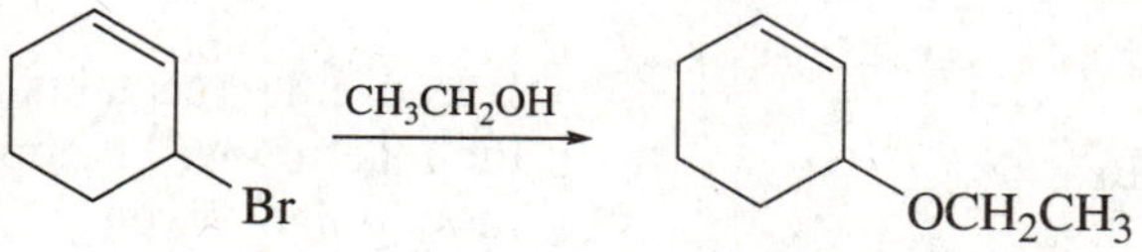

b.

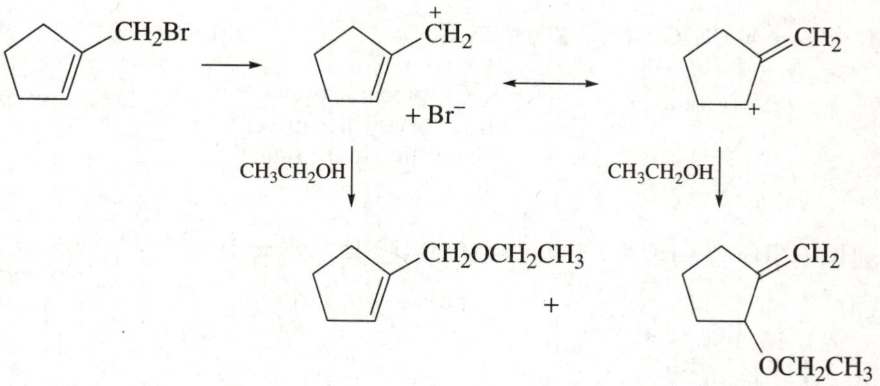

c.

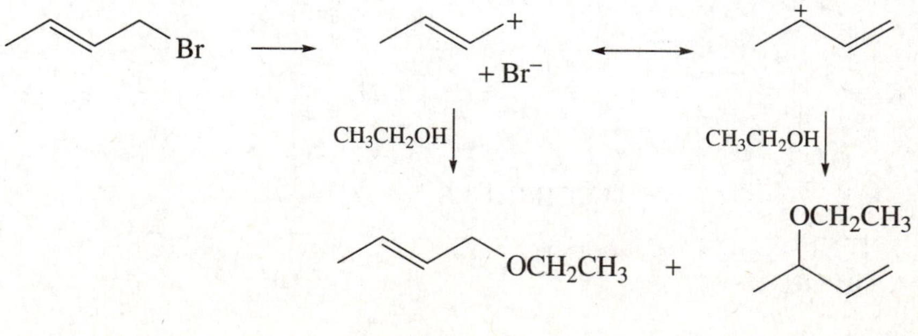

44. Methoxide ion will be a stronger nucleophile in DMSO because DMSO cannot stabilize the anion by hydrogen bonding.

45.

a. 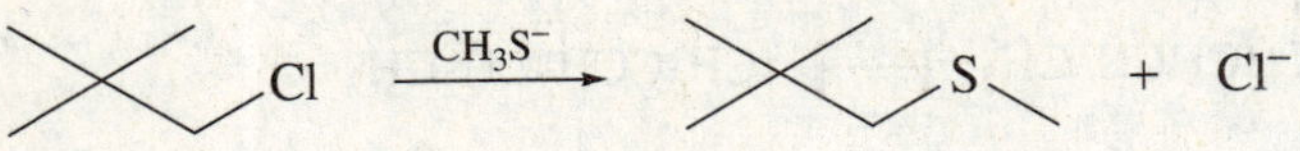

The nucleophile is less sterically hindered.

b. 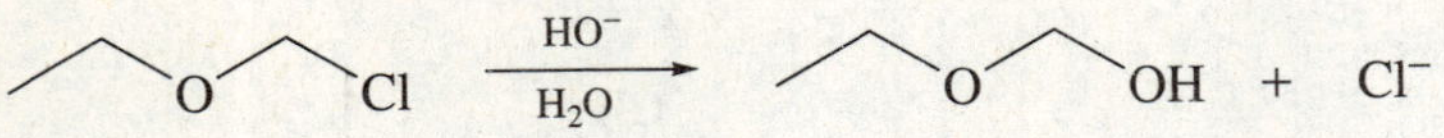

The electron-withdrawing oxygen increases the electrophilicity of the carbon that the nucleophile attacks.

c. 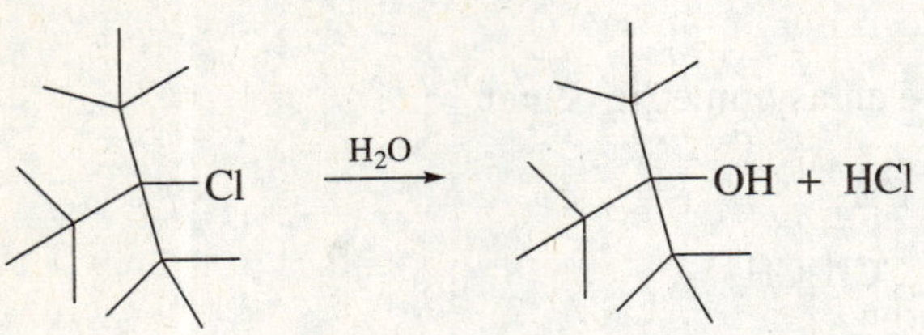

Steric strain is decreased when Cl^- dissociates to form the carbocation in the rate-limiting step since the hybridization of the carbon atom changes from sp^2 to sp^3, allowing the bond angle between the bulky groups to increase from 109.5° to 120°.

d. $(CH_3)_3CBr \xrightarrow{H_2O} (CH_3)_3COH + HBr$

Because the reactants are neutral the reaction will be faster in the more polar solvent.

46. $CH_3CH_2Cl + K^+I^- \underset{}{\overset{\text{acetone}}{\rightleftharpoons}} CH_3CH_2I + K^+Cl^-$

K^+Cl^- precipitates out in acetone, which drives the reaction to the right

47. 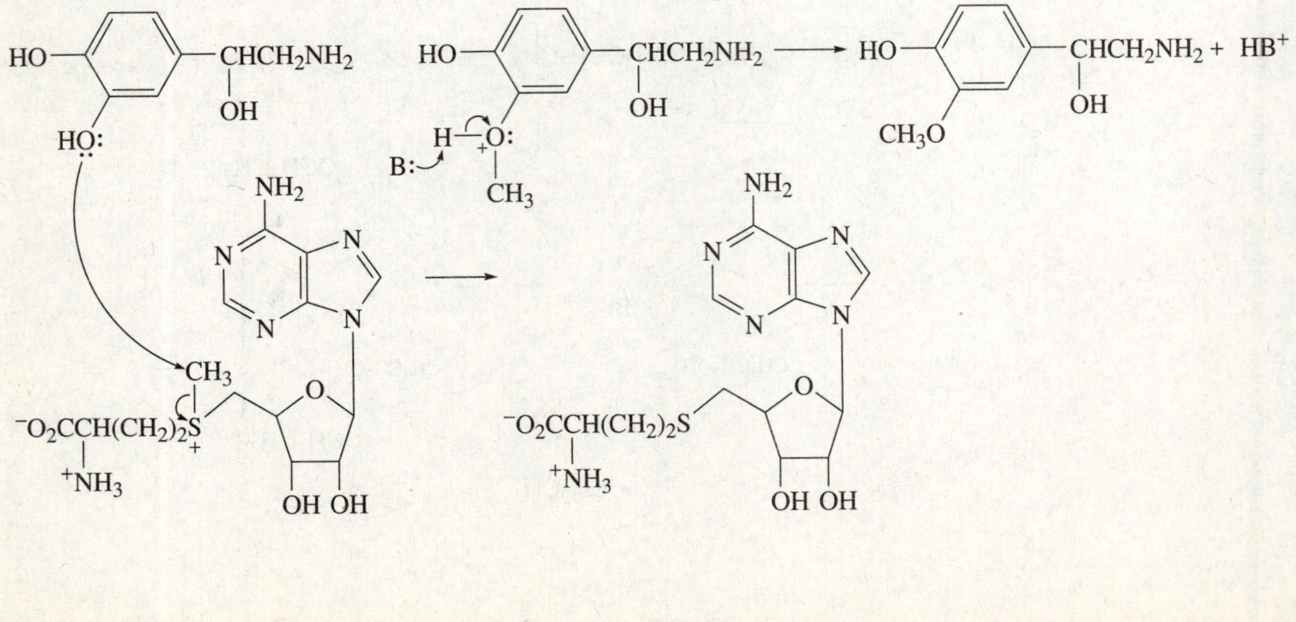

48. The slow step of the solvolysis reaction is formation of the carbocation. Since the products of the slow step are charged, they will be better stabilized by formic acid than by acetic acid because formic acid has a greater dielectric constant. Because the products are more stabilized, the transition state leading to them will also be more stabilized so the reaction will be faster in the solvent with the greater dielectric constant.

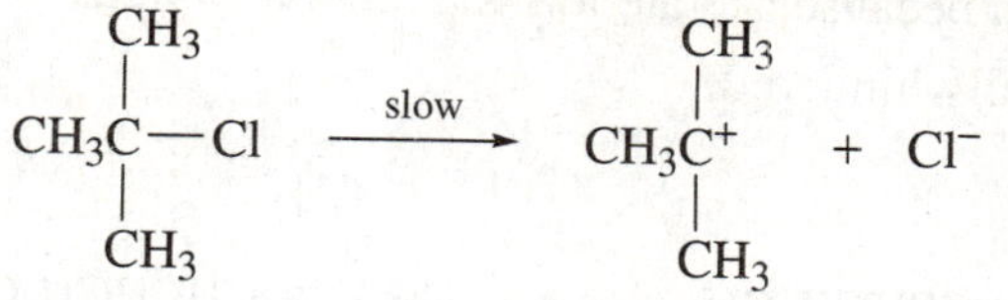

49. In parts a-d, the configuration of the asymmetric center attached to the Cl in the reactant will be inverted in the product.

a.

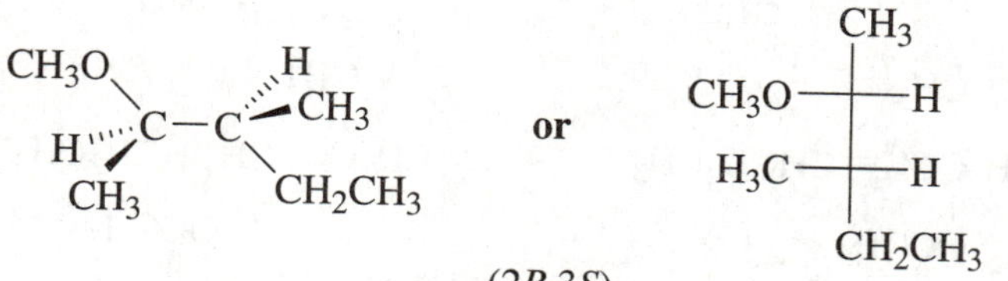

(2*R*,3*S*)

b.

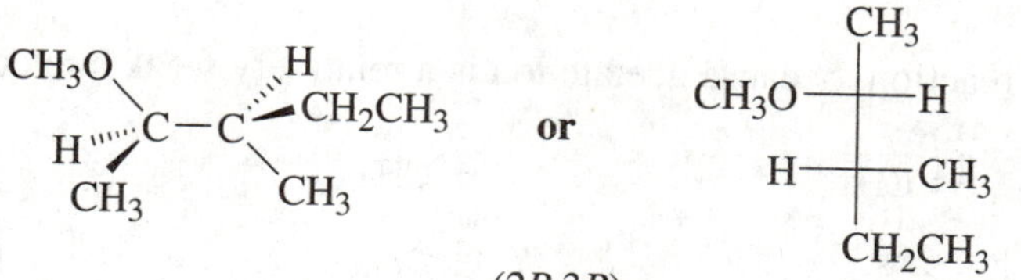

(2*R*,3*R*)

c.

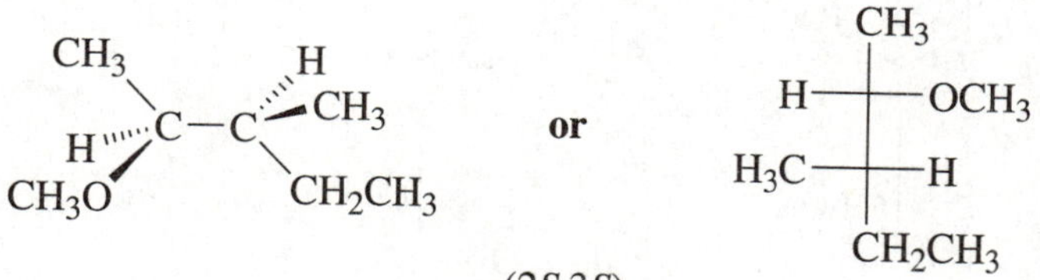

(2*S*,3*S*)

d.

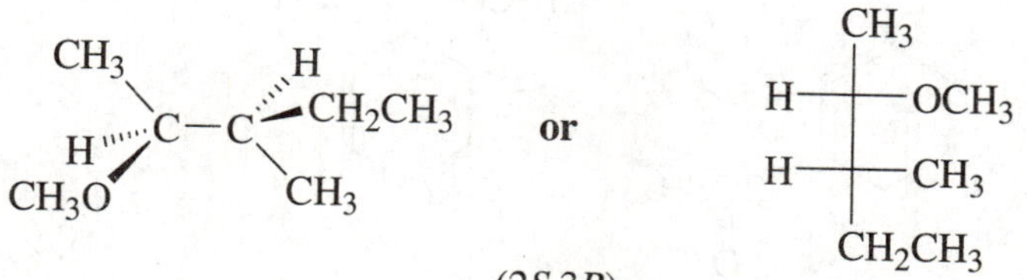

(2*S*,3*R*)

e.

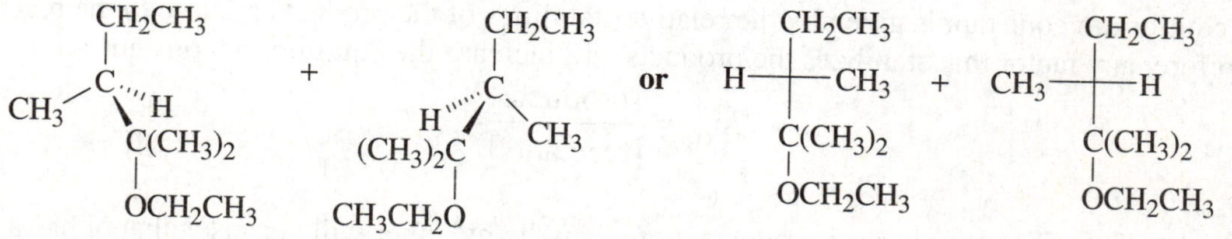

The alkyl halide dissociates to form a carbocation that undergoes a 1,2-methyl shift. Because the carbocation is planar, the methyl groups with its pair of electrons can add either to the top or the bottom of the carbocation. Therefore, the asymmetric center in the product can have either the *R* or the *S* configuration.

f. 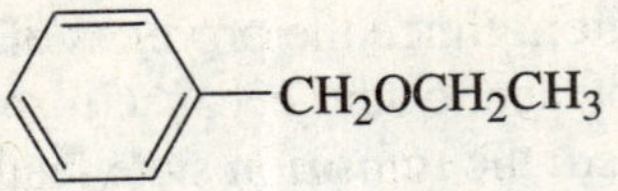

50. **a.** We can predict that this is an S_N1 reaction, because acetate ion is a relatively weak nucleophile and the alkyl halide are sterically hindered.

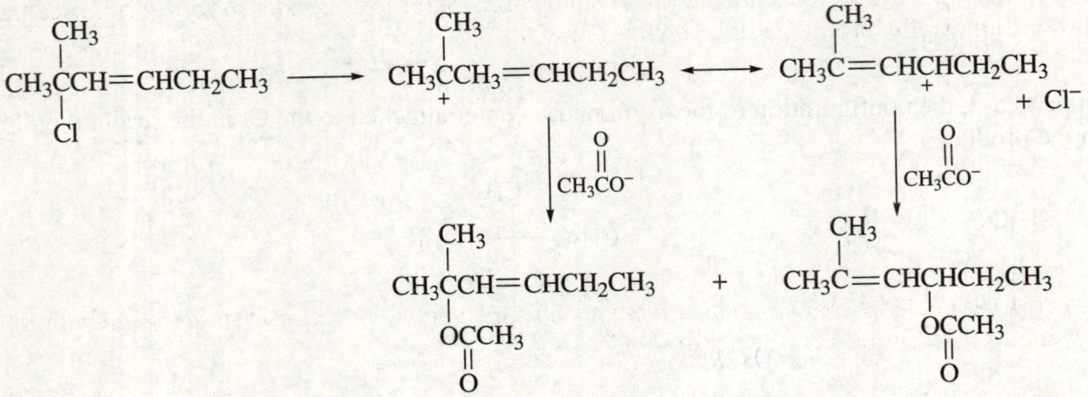

b. We can predict that this is an S_N1 reaction, because acetate ion is a relatively weak nucleophile.

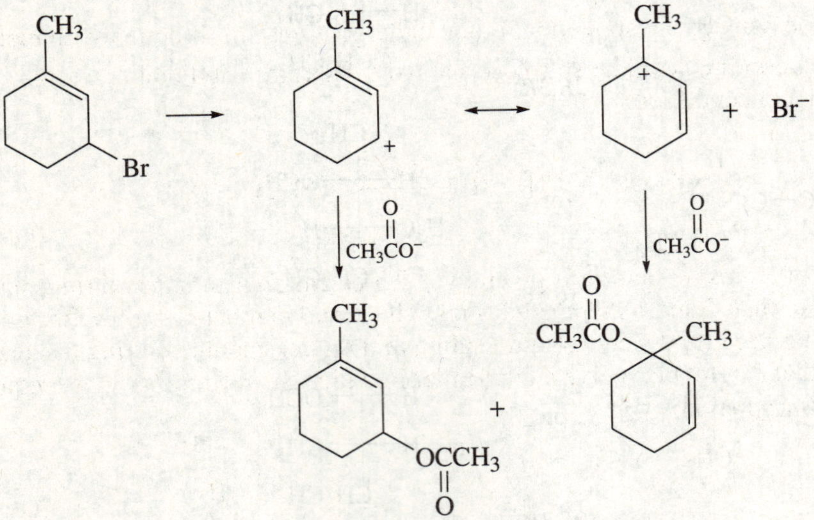

51. The equlibrium contstant is given by the relative stabilities of the products relative to the reactants. Therefore, any factor that stabilizes the products will increase the equilibrium constant.

$$K_{eq} = \frac{[\text{products}]}{[\text{reactants}]}$$

Ethanol will stabilize the charged products more than diethyl ether will because ethanol has a greater dielectric constant. Therefore the equilibrium will lie farther to the right (towards products) in ethanol.

52. **a.** The reaction with quinuclidine had the larger rate constant because quinuclidine is less sterically hindered as a result of the substituents on the nitrogen being pulled back into a ring structure.

b. The reaction with quinuclidine had the larger rate constant for the same reason given in **a**.

c. Isopropyl iodide had the larger ratio because, since it is more sterically hindered than methyl iodide, it is more affected by differences in the amount of steric hindrance in the nucleophile.

53. Because methanol is a poor nucleophile, the reaction will take place predominately via an S_N1 pathway. A secondary benzylic carbocation is more stable and, therefore, is easier to form than a secondary carbocation, so the product will be the one shown.

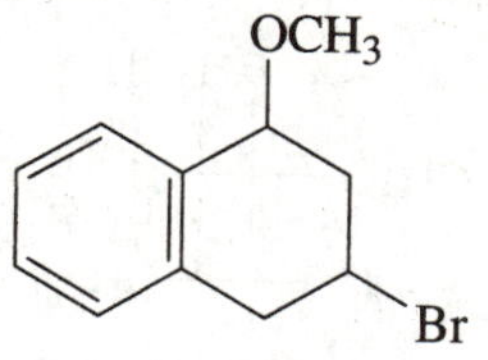

54. I^- is a good nucleophlie in a polar solvent such as methanol, so it reacts rapidly with methyl bromide, causing its concentration to decrease rapidly.

I^- is a good leaving group so methyl iodide undergoes an S_N2 reaction with methanol. Methanol is a poor nucleophile so it reacts slowly. Therefore, iodide ion returns slowly to its original concentration.

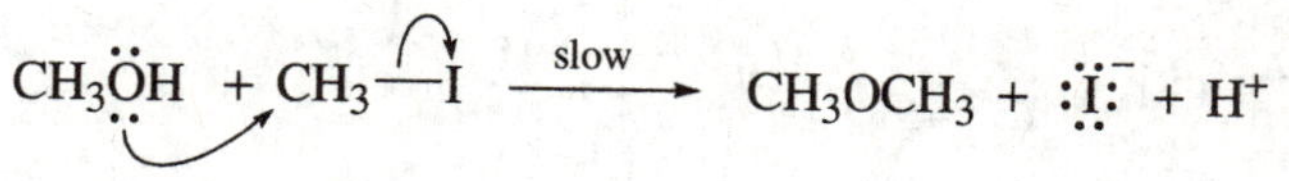

55. The number of atoms in the ring is given by *n*. Three and four membered rings have strain, so they are harder to make than 5 and 6 membered rings. The three-membered ring is formed faster than the four-membered ring because the compound leading to the three-membered ring has one less carbon-carbon single bond that can rotate to give a conformer in which the reacting groups are positioned too far from one another for reaction.

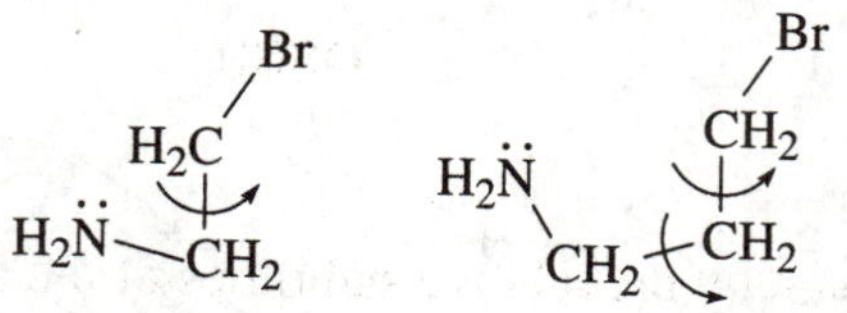

Even though the compound that forms the five-membered ring has one more bond that can rotate to give a conformer in which the reacting groups are positioned too far from one another for reaction than the compound that forms the four-membered ring, the five-membered ring is formed faster because it is relatively strain free. So lack of strain more than makes up for the lower probability of having the reacting groups in the proper position for reaction.

The rate of the ring-forming reaction gets slower as the size of the ring being formed gets larger, because the reactant has more bonds that can rotate to give conformers in which the reacting groups are positioned too far from one another for reaction.

56. The configuration of the asymmetric center attached to the Br in the reactant will be inverted in the product.

a.

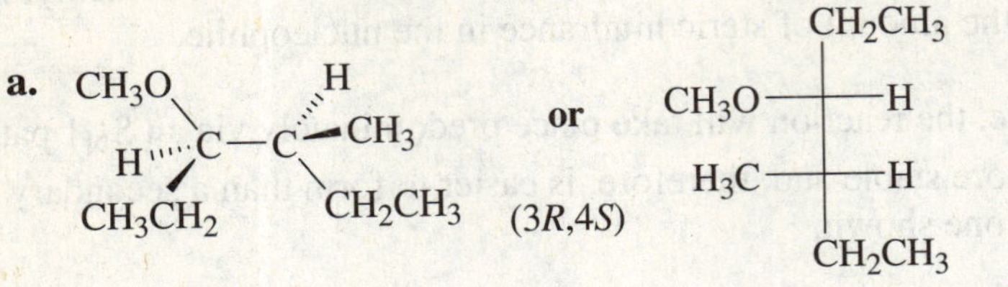

b.

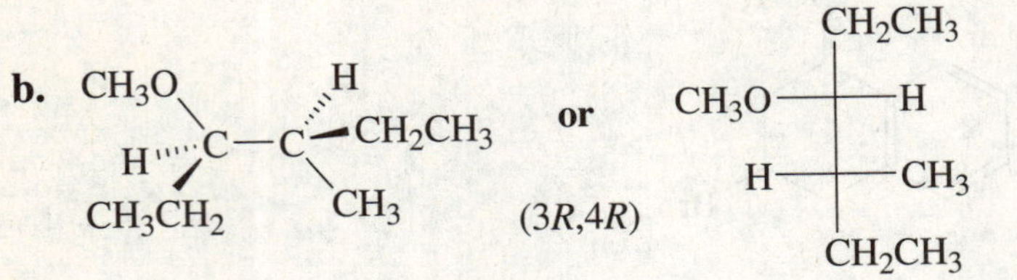

c.

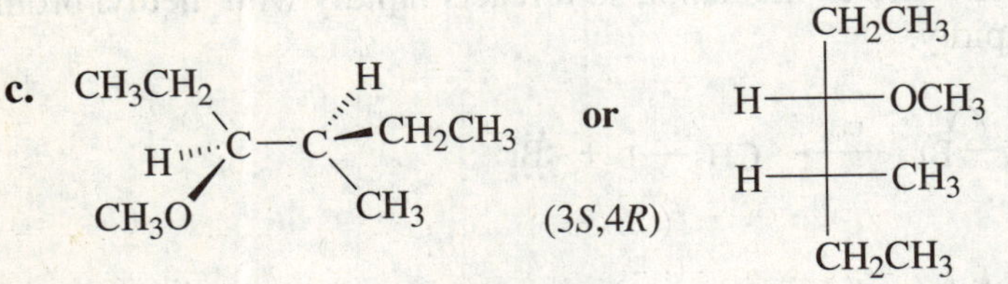

d.

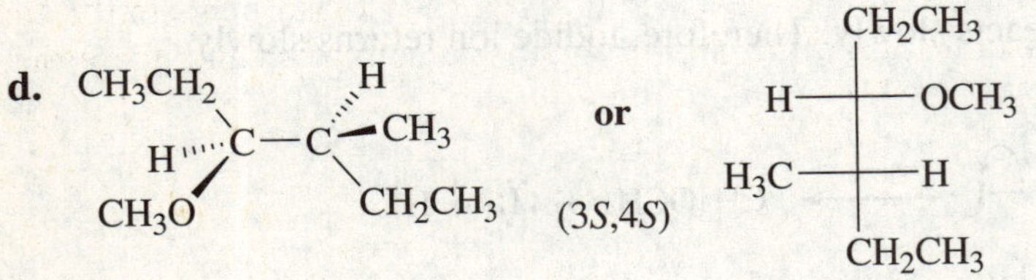

57. Tetrahydrofuran can solvate a charge better because the floppy ethyl substituents of diethyl ether provide steric hindrance, making it difficult for the nonbonding electrons of the oxygen to approach the positive charge that is to be solvated.

58. **a.**

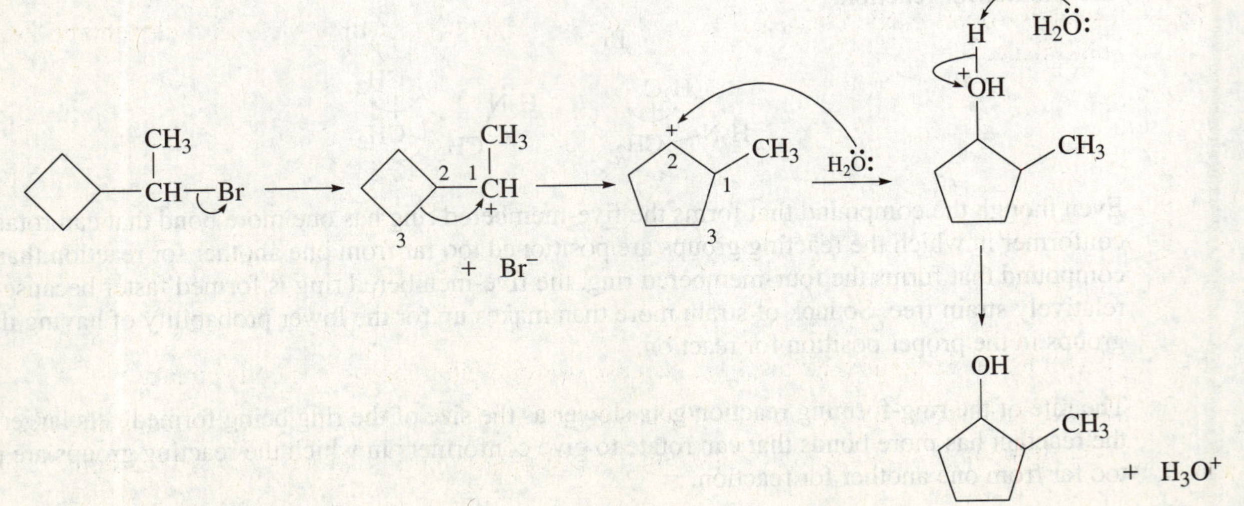

b.

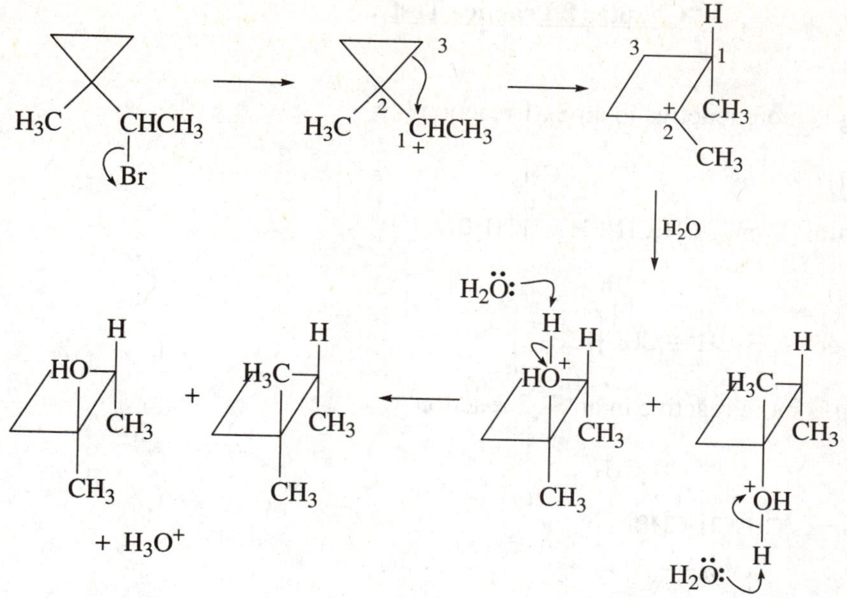

59. The cis isomer would be expected to react faster in an S_N1 reaction.
The rate-limiting step in an S_N1 reaction is formation of the carbocation intermediate.
Both isomers form the same carbocation. The cis isomer (with one substituent in the equatorial position and one in the axial position) is less stable than the trans isomer (with both substituents in the equatorial position). Since the cis isomer is less stable, it has the smaller energy of activation and, therefore, the faster reaction rate.

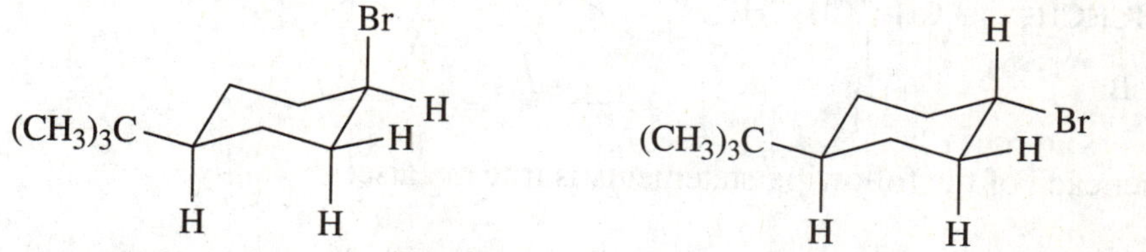

cis-1-bromo-4-*tert*-butylcyclohexane *trans*-1-bromo-4-*tert*-butylcyclohexane

60. A Diels-Alder reaction between hexachlorocyclopentadiene and 3,4-dichlorocyclopentene forms chlordane.

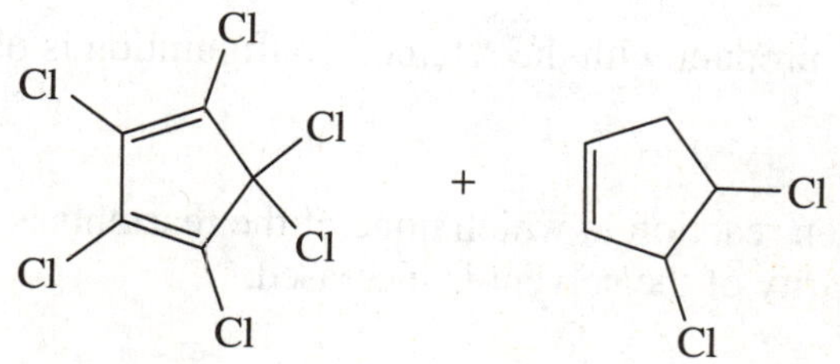

61. It will not undergo an S_N2 reaction, because of steric hindrance to backside attack.
It will not undergo an S_N1 reaction, because the carbocation that would be formed is unstable, since the ring structure prevents it from achieving the 120° bond angles required for an sp^2 carbon.

Chapter 8 Practice Test

1. Which of the following is more reactive in an S_N1 reaction?

a. 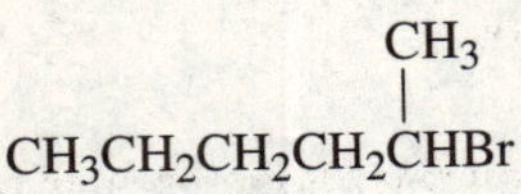 **or** 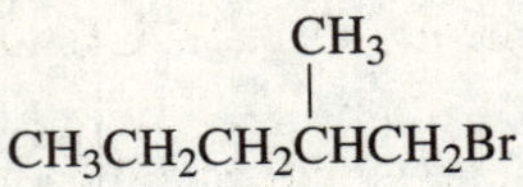

b. 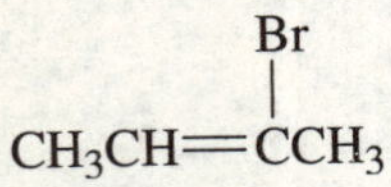 **or** 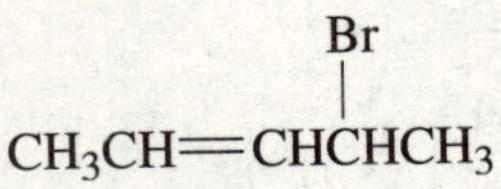

2. Which of the following is more reactive in an S_N2 reaction?

a. 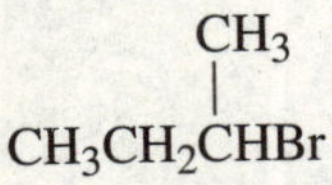 **or** 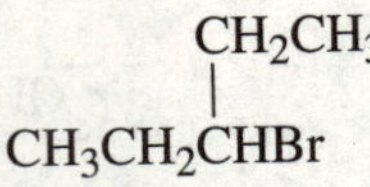

b. 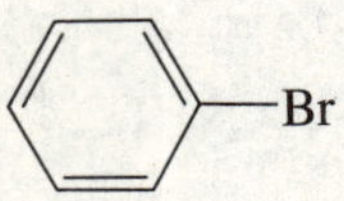 **or** 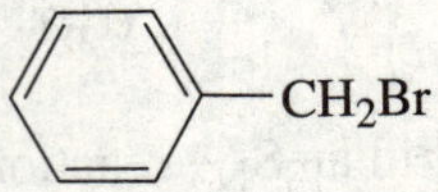

3. Which of the following alkyl halides forms a constitutional isomer as a result of an S_N1 reaction that is different from the isomer formed as a result of an S_N2 reaction?

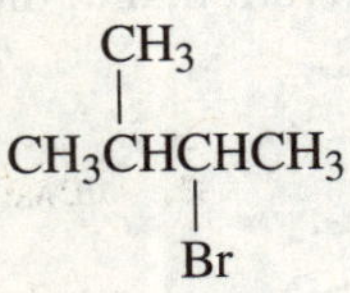 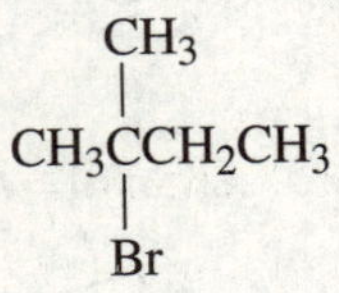 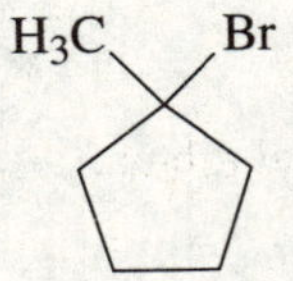 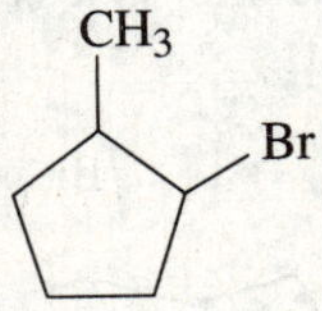

4. Indicate whether each of the following statements is true or false:

a. Increasing the concentration of the nucleophile favors an S_N1 reaction over an S_N2 reaction. T F

b. Ethyl iodide is more reactive than ethyl chloride in an S_N2 reaction. T F

c. In an S_N1 reaction, the product with the retained configuration is obtained in greater yield. T F

d. The rate of a substitution reaction in which none of the reactants is charged will increase if the polarity of the solvent is increased. T F

e. An S_N2 reaction is a two-step reaction. T F

f. The pK_a of a carboxylic acid is greater in water than it is in a mixture of dioxane and water. T F

g. 4-Bromo-1-butanol will form a cyclic ether faster than will 3-bromo-1-propanol. T F

5. Answer the following:

a. Which is a stronger base, CH_3O^- or CH_3S^-?

b. Which is a better nucleophile in an aqueous solution, CH_3O^- or CH_3S^-?

6. For each of the following pairs of S_N2 reactions, indicate the one that occurs with the greater rate constant:

a. $CH_3CH_2CH_2Cl + HO^-$ **or** $CH_3\underset{\underset{\displaystyle Cl}{|}}{C}HCH_3 + HO^-$

b. $CH_3CH_2CH_2Cl + HO^-$ or $CH_3CH_2CH_2I + HO^-$

c. $CH_3CH_2CH_2Br + HO^-$ or $CH_3CH_2CH_2Br + H_2O$

d. $CH_3\underset{\underset{\displaystyle Br}{|}}{C}HCH_3 \xrightarrow[H_2O/CH_3OH]{CH_3O^-}$ or $CH_3\underset{\underset{\displaystyle Br}{|}}{C}HCH_3 \xrightarrow[CH_3OH]{CH_3O^-}$

e. $BrCH_2CH_2CH_2CH_2NHCH_3$ or $BrCH_2CH_2CH_2NHCH_3$

7. Circle the aprotic solvents:

a. dimethyl sulfoxide **b.** diethyl ether **c.** ethanol **d.** hexane

8. How would increasing the polarity of the solvent affect the following?

a. the rate of the S_N2 reaction of methyl amine with 2-bromobutane

b. the rate of the S_N1 reaction of methyl amine with 2-bromobutane

c. the rate of the S_N2 reaction of methoxide ion with 2-bromobutane

d. the pK_a of acetic acid

e. the pK_a of phenol

CHAPTER 9

Elimination Reactions of Alkyl Halides • Competition Between Substitution and Elimination

Important Terms

anti elimination — an elimination reaction in which the substituents being eliminated are removed from opposite sides of the molecule.

anti-periplanar — substituents are parallel on opposite sides of a molecule.

dehydrohalogenation — elimination of a proton and a halide ion.

deuterium kinetic isotope effect — ratio of the rate constant obtained for a compound containing hydrogen and the rate constant obtained for an identical compound in which one or more of the hydrogens have been replaced by deuterium.

elimination reaction — a reaction that removes atoms or groups from the reactant to form a π bond.

β-elimination reaction or 1,2-elimination reaction — an elimination reaction where the groups being eliminated are bonded to adjacent carbons.

E1 reaction — a unimolecular elimination reaction.

E2 reaction — a bimolecualr elimination reaction.

regioselectivity — the preferential formation of a constitutional isomer.

sawhorse projection — a way to represent the three-dimensional spatial relationships of atoms by looking at the carbon-carbon bond from an oblique angle.

syn elimination — an elimination reaction in which substituents being eliminated are removed from the same side of the molecule.

syn-periplanar — substituents are parallel on the same side of a molecule.

target molecule — the desired end product of a synthesis.

Williamson ether synthesis — formation of an ether from the reaction of an alkoxide ion with an alkyl halide.

Zaitsev's rule — the rule that states that the more stable alkene product is obtained by removing a proton from the β-carbon that is bonded to the fewest hydrogens.

Solutions to Problems

1. a. 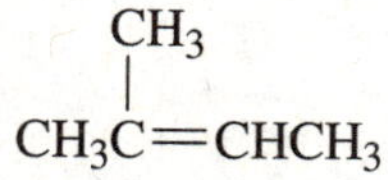 b. 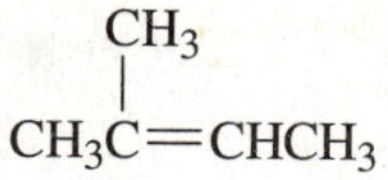 c. 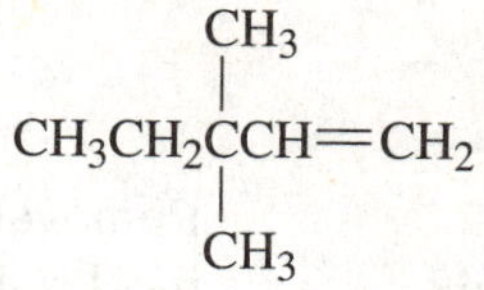

2. a. 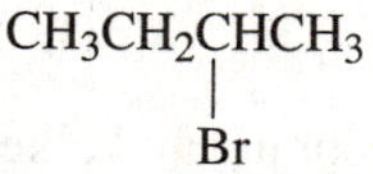 The alkene that is formed is more stable than the alkene that is formed from the other alkyl halide.

b. 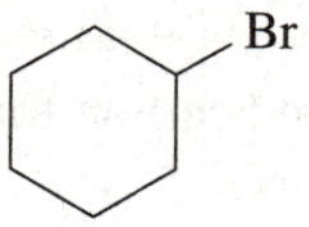 Br is a better leaving group (weaker base) than Cl.

c. 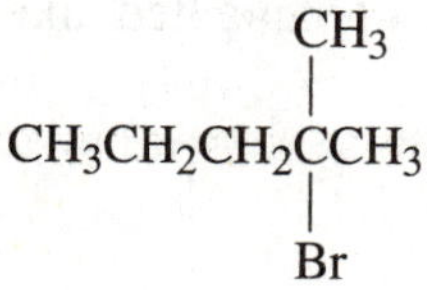 The alkene that is formed is more stable than the alkene that is formed from the other alkyl halide.

d. 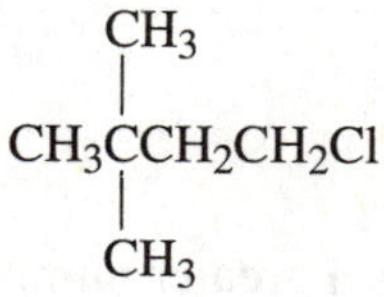 The other alkyl halide cannot undergo an E2 reaction because it doesn't have any β-hydrogens.

3. The reaction of 2-bromo-2,3-dimethylbutane with sodium *tert*-butoxide forms two alkene elimination products. Because of steric hindrance, the less stable alkene (2,3-dimethyl-1-butene) is the one that is easier to make.

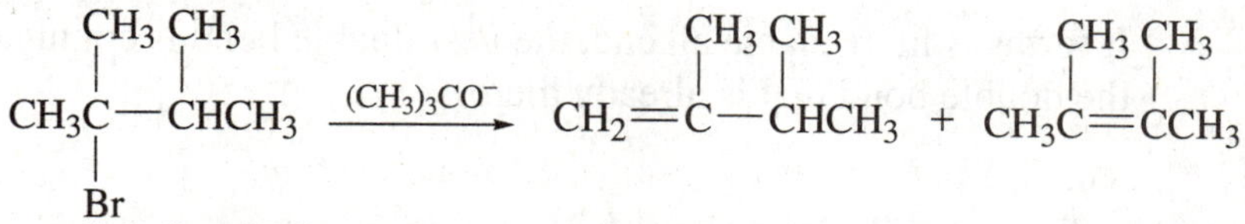

2-bromo-2,3-dimethylbutane 2,3-dimethyl-1-butene 2,3-dimethyl-1-butene

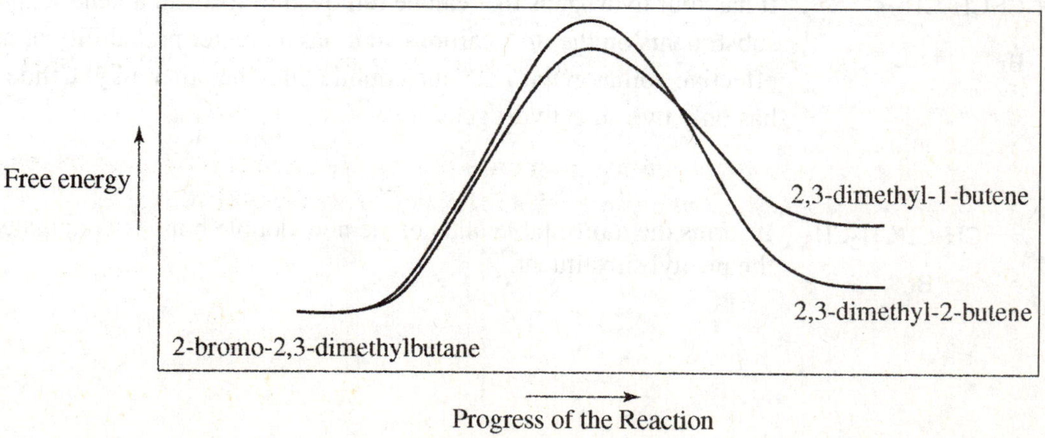

4. **a.** $CH_3CH{=}CHCH_3$ Removal of a β-hydrogen from the most substituted carbon forms the most stable "alkene-like" transition state.

b. 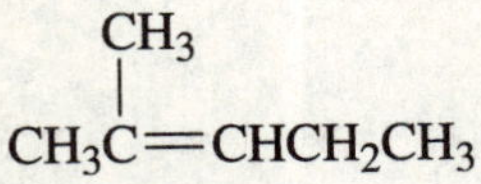 Removal of a β-hydrogen from the most substituted carbon forms the most stable "alkene-like" transition state.

c. $CH_3CH{=}CHCH{=}CH_2$ The β-hydrogen is removed that will lead to a conjugated alkene.

d. $CH_2{=}CHCH_2CH_3$ Removal of a β-hydrogen from the least substituted carbon forms the most stable "carbanion-like" transition state.

e.  The β-hydrogen is removed that will lead to a conjugated alkene.

f. 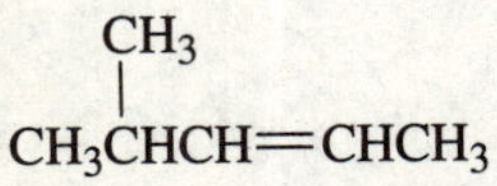 Removal of a β-hydrogen from the least substituted carbon forms the most stable "carbanion-like" transition state.

5. **a.** 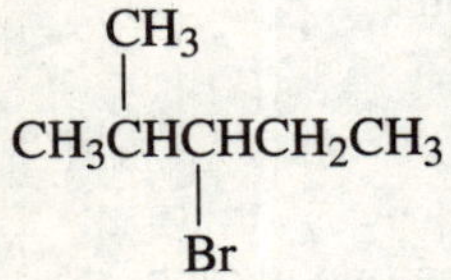 It forms the more stable alkene; the alkene has a greater number of substituents bonded to the sp^2 carbons.

b. 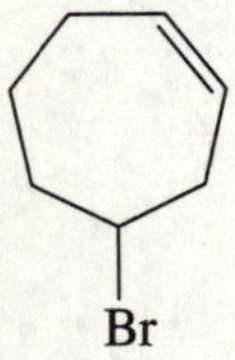 It forms is more stable alkene; the new double bond is conjugated with the double bond that is already there.

c. 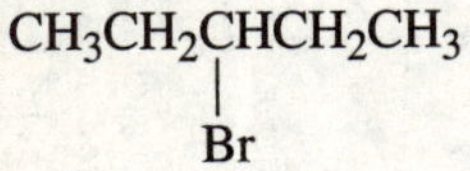 It has four hydrogens that can be removed to form an alkene with two substituents on the sp^2 carbons so it has a greater probability of an effective collision with the nucleophile than the other alkyl halide that has only two such hydrogens.

d. 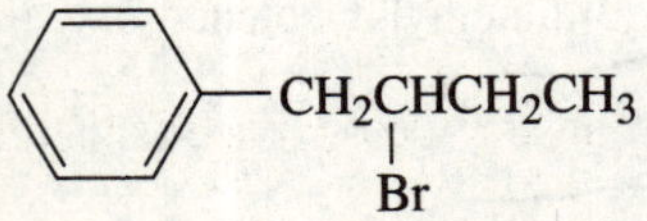 It forms the more stable alkene; the new double bond is conjugated with the phenyl substituent.

6. 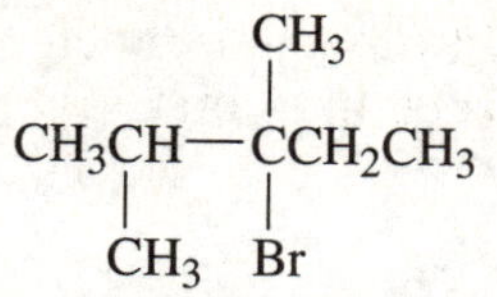

3-bromo-2,3-dimethylpentane

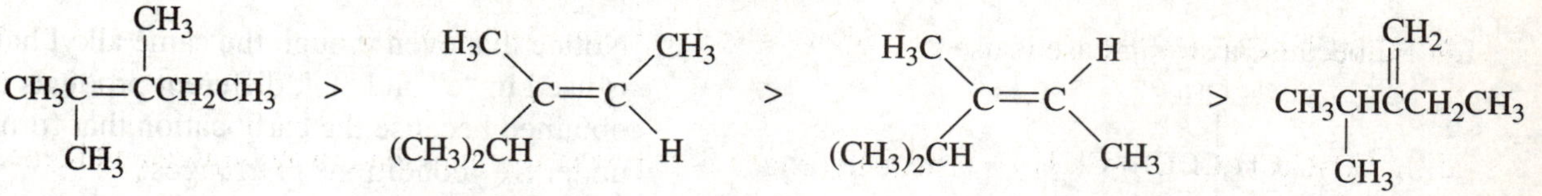

Four alkyl substituents are bonded to the sp^2 carbons. > Two alkyl substituents are bonded to the sp^2 carbons; the largest groups are on opposite sides of the double bond. > Two alkyl substituents are bonded to the sp^2 carbons; the largest groups are on the same side of the double bond. > Two alkyl substituents are bonded to the sp^2 carbons.

7. The major product is the one predicted by Zaitsev's rule, because the fluoride ion dissociates in the first step, forming a carbocation. Loss of a proton from the carbocation will follow Zaitsev's rule, as it does in all E1 reactions.

8. **a.** B because it forms the more stable carbocation.
b. B because it forms the more stable alkene.
c. B because it forms the more stable carbocation.
d. A because it is less sterically hindered.

9. 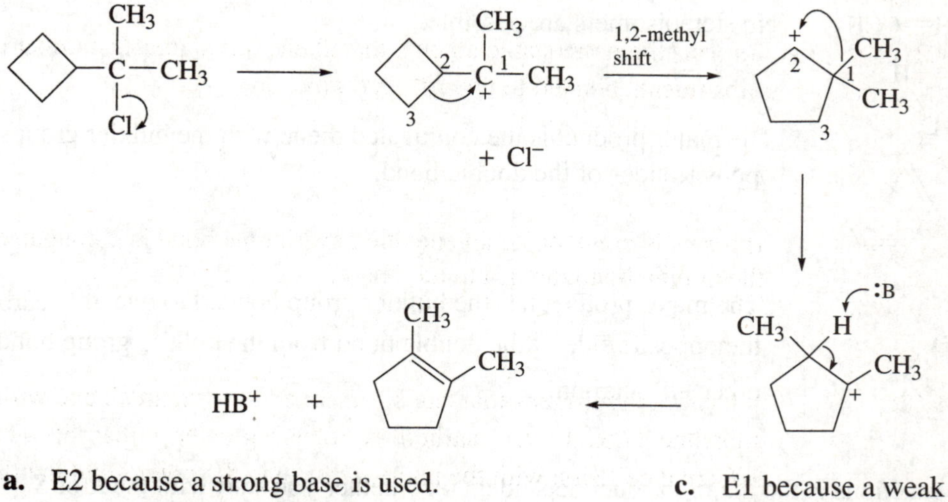

10. **a.** E2 because a strong base is used.
$CH_3CH{=}CHCH_3$

b. E1 because a weak base is used.
$CH_3CH{=}CHCH_3$

c. E1 because a weak base is used.
$CH_3C(CH_3){=}CH_2$

d. E2 because a strong base is used.
$CH_3C(CH_3){=}CH_2$

e. E1 because a weak base is used.

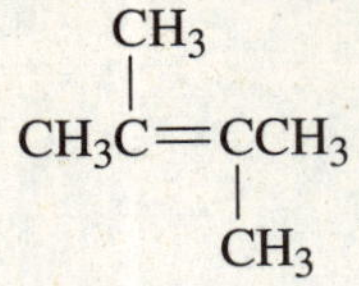

f. E2 because a strong base is used.

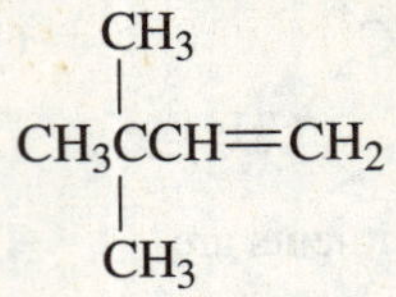

Notice that even though the same alkyl halide is used in "e" and "f", different products are obtained because the carbocation that forms under S_N1 conditions rearranges.

11. **a.**

$$\frac{E2}{E2 + E1} = \frac{k_2[\textit{tert-butyl bromide}][HO^-]}{k_2[\textit{tert-butyl bromide}][HO^-] + k_1[\textit{tert-butyl bromide}]}$$

$$\frac{E2}{E2 + E1} = \frac{7.1 \times 10^{-5} \times 5.0}{7.1 \times 10^{-5} \times 5.0 + 1.5 \times 10^{-5}} = \frac{35.5 \times 10^{-5}}{35.5 \times 10^{-5} + 1.5 \times 10^{-5}} = \frac{35.5}{37} = .96$$

$$= 96\%$$

b.

$$\frac{E2}{E2 + E1} = \frac{7.1 \times 10^{-5} \times 2.5 \times 10^{-3}}{7.1 \times 10^{-5} \times 2.5 \times 10^{-3} + 150 \times 10^{-7}} = \frac{1.78 \times 10^{-7}}{1.78 \times 10^{-7} + 150 \times 10^{-7}} = \frac{1.78}{152} = 1.2\%$$

12. **a.** **1.** 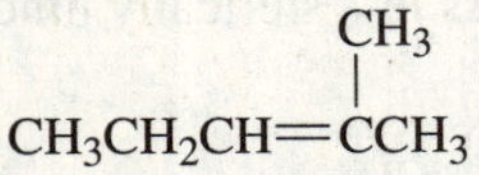 No stereoisomers are possible.

2. 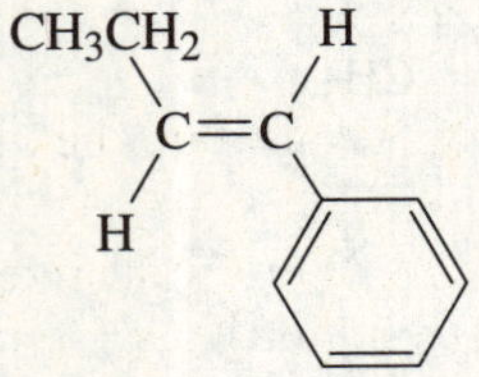

The major product is the conjugated diene with the bulkier groups on the opposite sides of the double bond.

3. 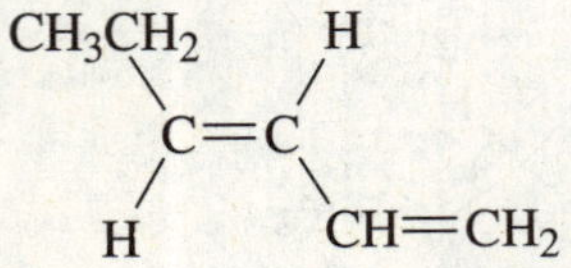

The major product has the bulkier group bonded to one sp^2 carbon on the opposite side of the double bond from the bulkier group bonded the other sp^2 carbon.

b. In none of the reactions is the major product dependent on whether you start with the *R* or *S* enantiomer of the reactant.

13. **a.** Solved in the text.

c. 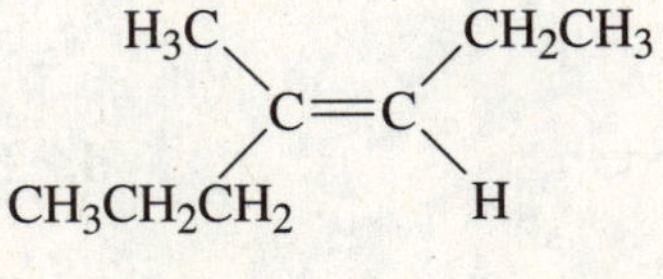

d. 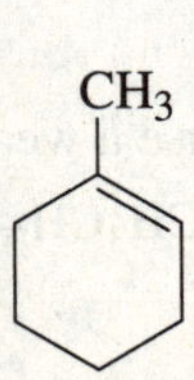

b. 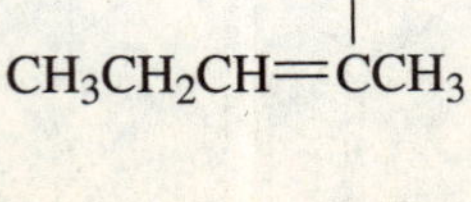

14. Elimination occurs only when the H and Br to be eliminated are in axial positions.

When Br is in an axial position in the cis isomer, it has an axial hydrogen on each of the adjacent carbons. The one bonded to the same carbon as the ethyl group will be more apt to be the one eliminated with Br because the product formed is more stable and, therefore, more easily formed than the product formed when the other H is eliminated with Br.
When Br is in an axial position in the trans isomer, it has an axial hydrogen on only one adjacent carbon, and it is not the carbon that is bonded to the ethyl group. Therefore, a different product is formed.

cis-1-bromo-2-ethylcyclohexane *trans*-1-bromo-2-ethylcyclohexane

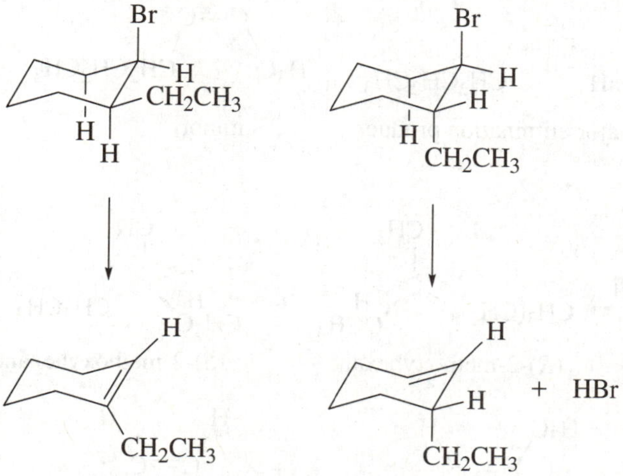

1-ethylcyclohexene 3-ethylcyclohexene

15. In order to undergo an E2 reaction, the substituents that are to be eliminated must both be in axial positions.
When bromine and an adjacent hydrogen are in axial positions, the large *tert*-butyl substituent is in an equatorial position in the cis isomer and in an axial position in the trans isomer. The rate constant for the reaction is $k'K_{eq}$.
Because a large substituent is more stable in an equatorial position than in an axial position, elimination of the cis isomer occurs through its more stable chair conformer (K_{eq} is large; see p. 408 of the text), while elimination of the trans isomer has to occur through its less stable chair conformer (K_{eq} is small). The cis isomer, therefore, reacts more rapidly in an E2 reaction.

cis-1-bromo-4-*tert*-butylcyclohexane *trans*-1-bromo-4-*tert*-butylcyclohexane

16. **a.**

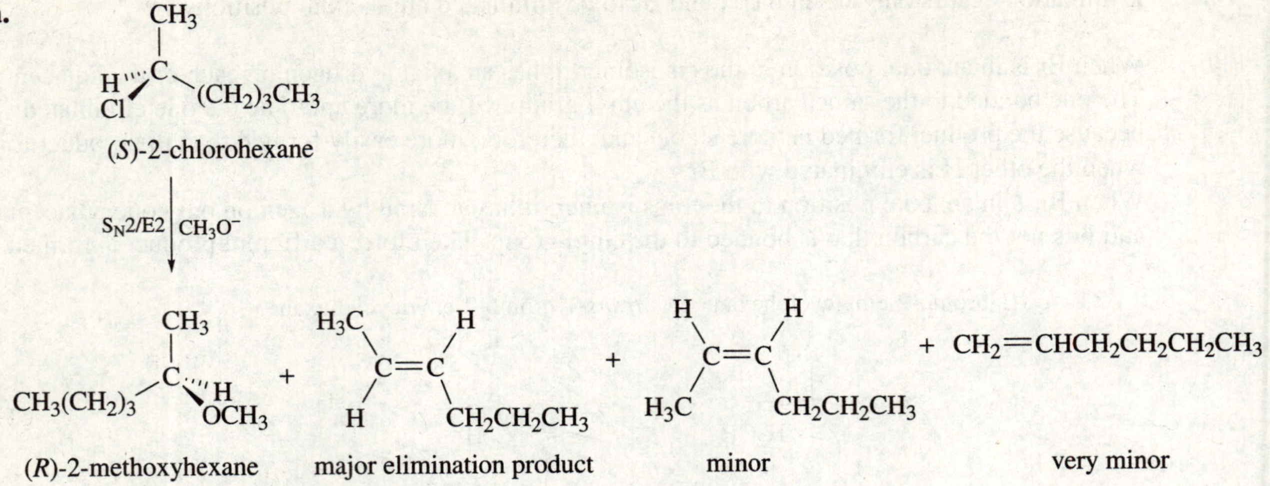
(S)-2-chlorohexane
SN2/E2 CH3O−
(R)-2-methoxyhexane
major elimination product
minor
very minor

b.

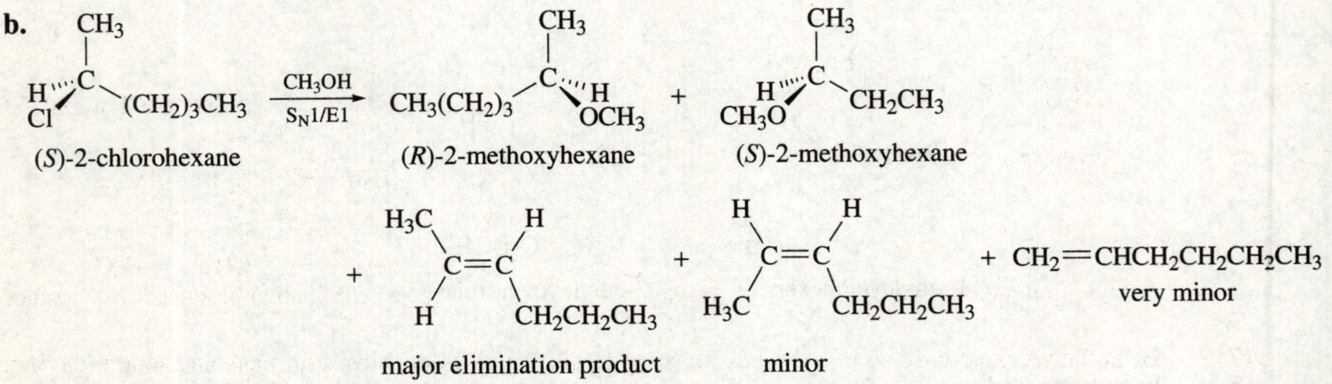
(S)-2-chlorohexane
CH3OH
SN1/E1
(R)-2-methoxyhexane
(S)-2-methoxyhexane
major elimination product
minor
very minor

c.

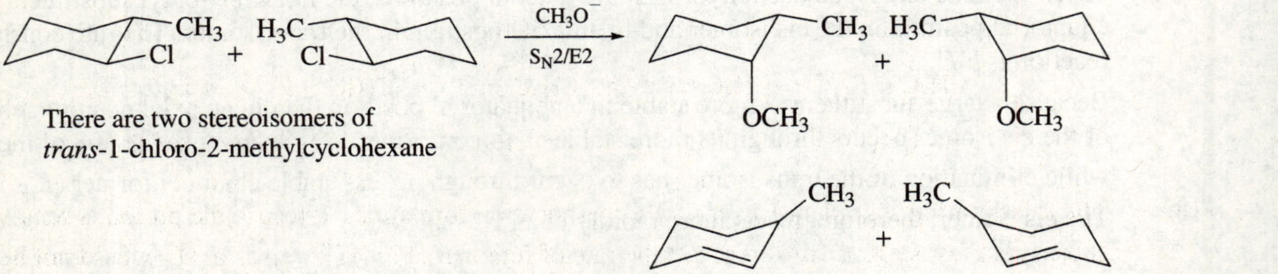
CH3O−
SN2/E2
There are two stereoisomers of
trans-1-chloro-2-methylcyclohexane

d. 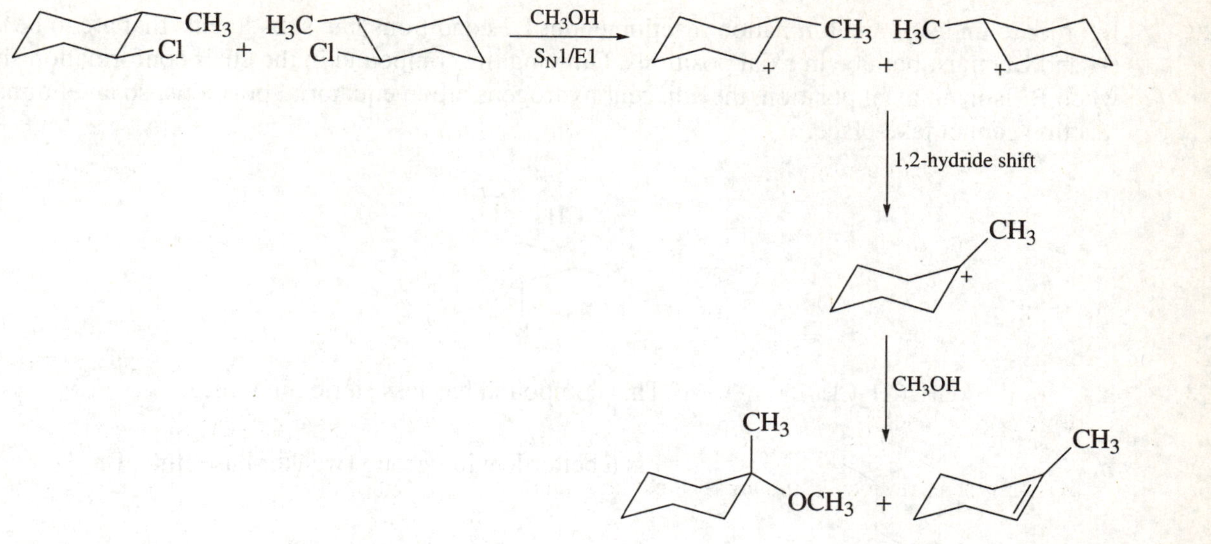

e. 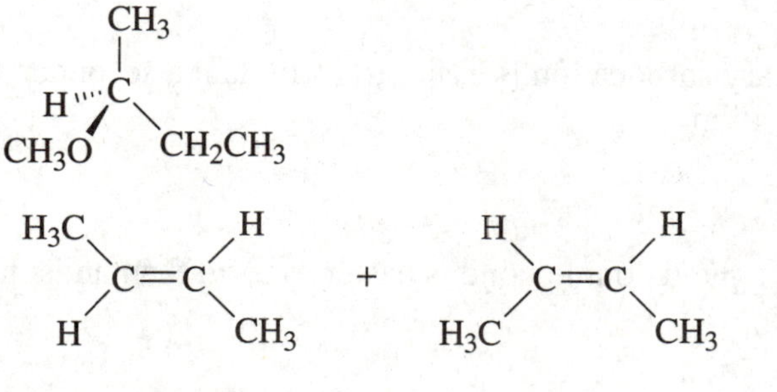

major elimination product minor

f. 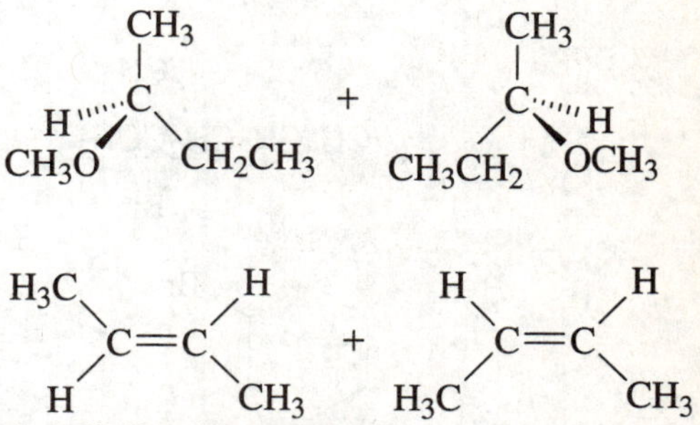

major elimination product minor

17. Br is the weakest base so it is the best leaving group, and F is the strongest base and worst leaving group. It is easier to break a C—H bond than a C—D bond, so the more β-hydrogens that are replaced by a deuterium, the slower the rate of the E2 reaction will be.

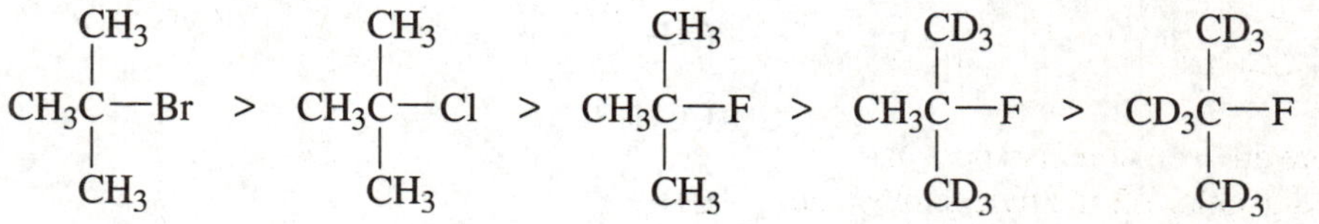

18. The rate-limiting step in an E1 reaction is carbocation formation. Because the proton is removed in a subsequent fast step, the difference in the rate of removal of an H^+ versus a D^+ would not be reflected in the rate constant. Therefore, the deuterium kinetic isotope effect would be close to 1.

19. Because CH_3S^- is a stronger nucleophile and weaker base than CH_3O^-, the ratio of substitution (where CH_3S^- reacts as a nucleophile) to elimination (where CH_3S^- reacts as a base) will increase when the nucleophile is changed from CH_3O^- to CH_3S^-.

20. In order to undergo an elimination reaction under E2 conditions, the substituents that are to be eliminated (H and Br) must both be in axial positions. Drawing the compound in the chair conformation shows that when Br is in an axial position, the adjacent hydrogens are in equatorial positions, so an elimination reaction cannot take place.

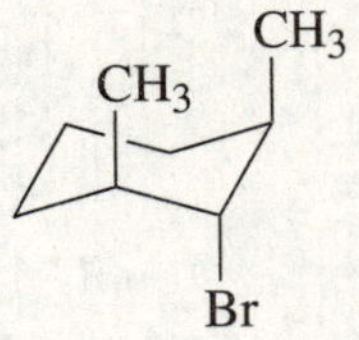

21. **a.** $CH_3CH_2CH_2Br$ — This compound has less steric hindrance.

b. 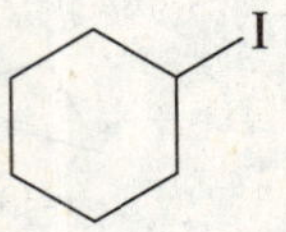 — I is a better leaving group (weaker base) than Br.

c. $CH_3CH_2CH_2C(CH_3)(Br)CH_3$ — A tertiary carbocation is easier to form than a secondary carbocation.

d. 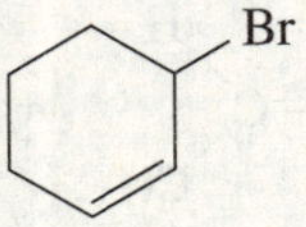 — A conjugated double bond is easier to form than an isolated double bond.

22. **a.**
1. no reaction
2. no reaction
3. substitution and elimination
4. substitution and elimination

b.
1. primarily substitution
2. substitution and elimination
3. substitution and elimination
4. elimination

23. $CH_3C(CH_3)_2CH_2Br$

1-bromo-2,2-dimethylpropane

a. The bulky *tert*-butyl substituent blocks the backside of the carbon bonded to the bromine to nucleophilic attack, making an S_N2 reaction difficult. An S_N1 reaction is difficult because the carbocation formed when the bromide ion departs is an unstable primary carbocation.

b. It cannot undergo an E2 reaction, because the β-carbon is not bonded to a hydrogen. It cannot undergo an E1 reaction, because that would require formation of a primary carbocation.

24. Because a strong base is used in the Williamson ether synthesis, the reaction is an S_N2 reaction, so a competing E2 reaction can also occur.

a. $CH_3CH_2CH_2Br + CH_3CH_2CH_2CH_2O^- \longrightarrow CH_3CH_2CH_2CH_2OCH_2CH_2CH_3$ (butyl propyl ether) $+ CH_3CH{=}CH_2$ (propene)

b. $CH_3CH_2CH_2CH_2Br + CH_3CH_2CH_2O^- \longrightarrow CH_3CH_2CH_2CH_2OCH_2CH_2CH_3$ (butyl propyl ether) $+ CH_3CH_2CH{=}CH_2$ (1-butene)

25. **a.**

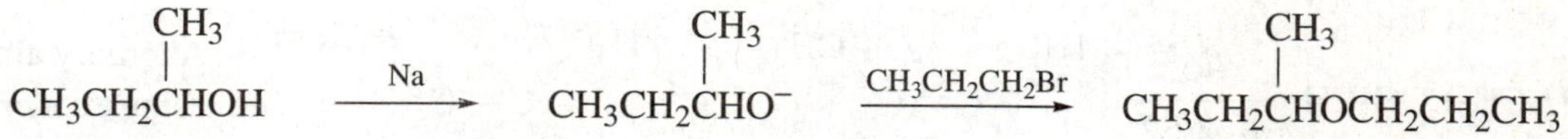

b.

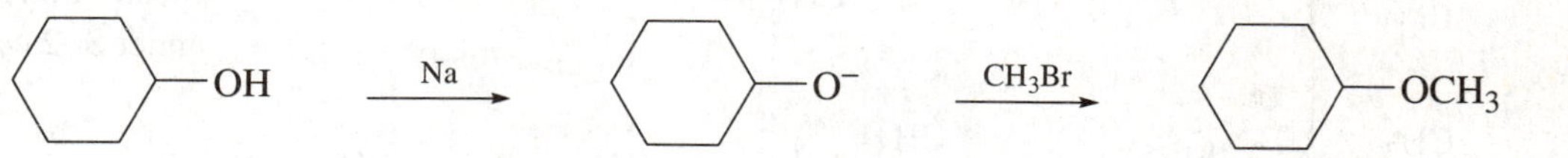

c.

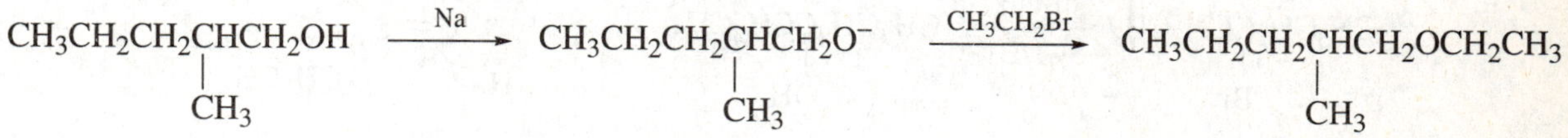

or

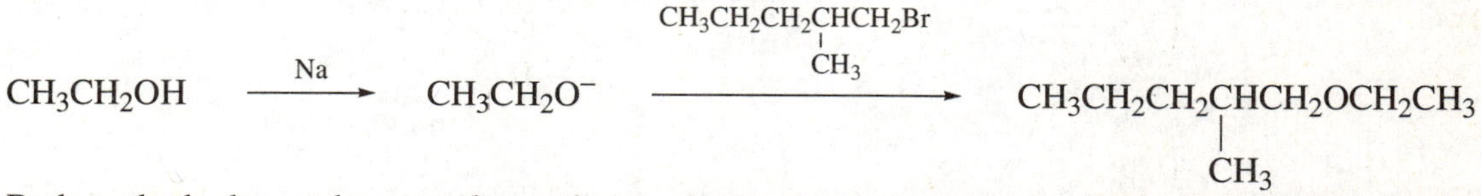

Both methods shown above can be used to synthesize the target molecule. The first method is preferred because ethyl bromide has less steric hindrance toward nucleophilic attack than does1-bromo-2-methylpentane.

d.

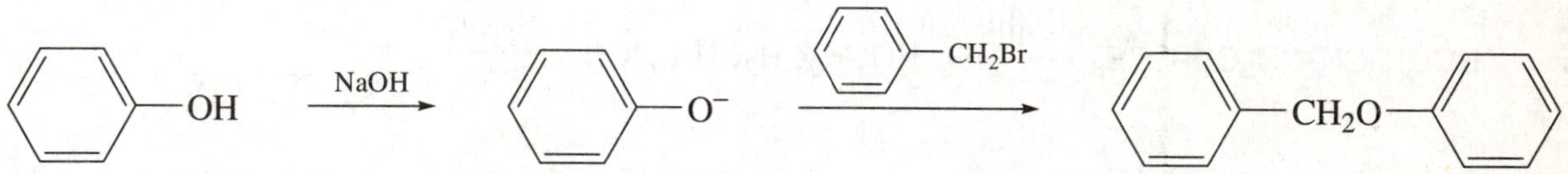

26.

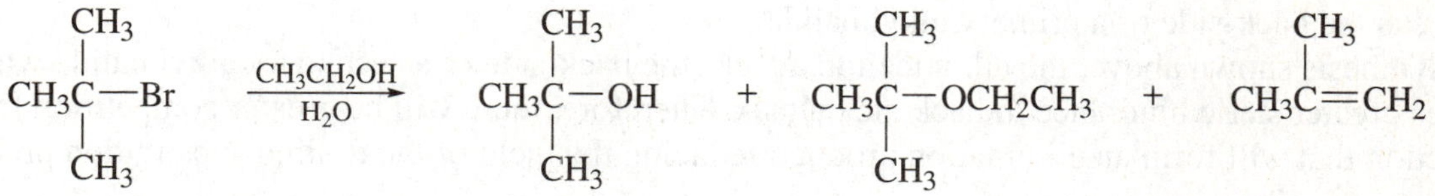

27. **a.**

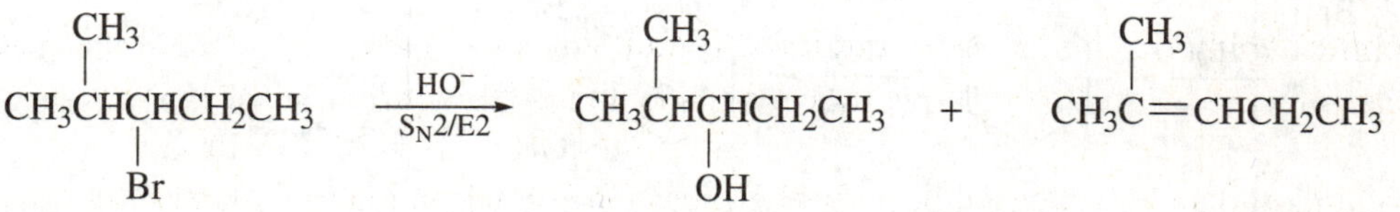

The S_N2 reaction will form the substitution product with the inverted configuration compared to the configuration of the starting material. But, because the configuration of the starting material is not specified (we do not know if it is the *R* isomer, the *S* isomer, or a mixture of *R* and *S*), we cannot specify the configuration of the S_N2 product.

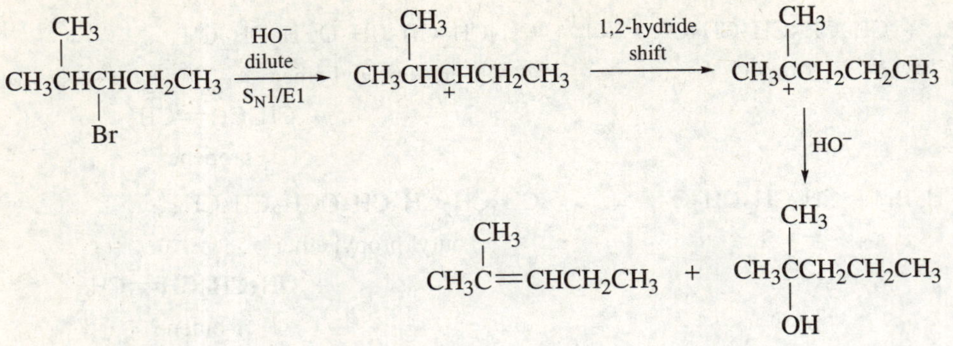

b. 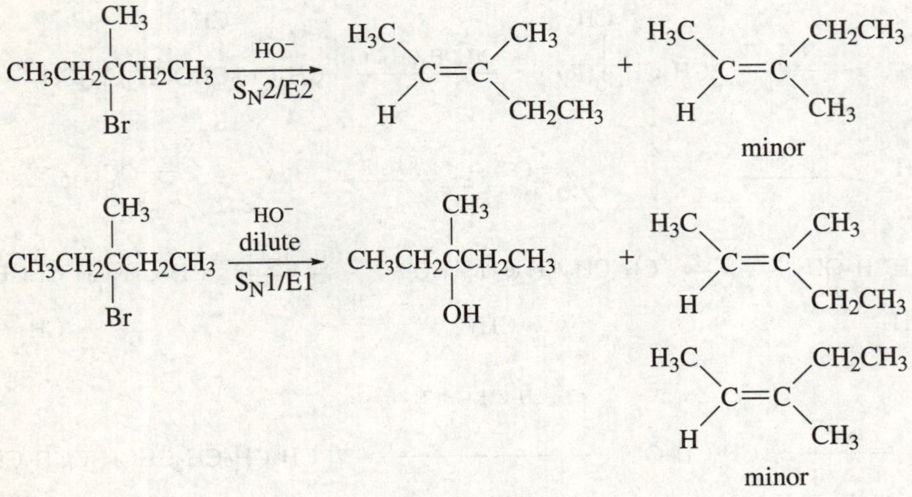

A tertiary alkyl halide cannot undergo a substitution reaction under S_N2 conditions.

28. Because a cumulated diene is less stable than an alkyne, the transition state for its formation is less stable so it is harder to make.

29. 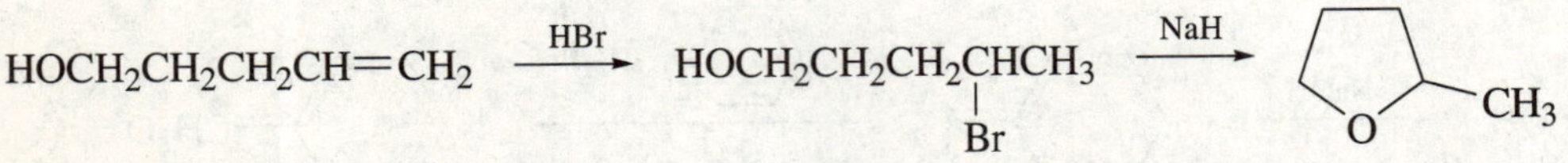

The synthesis shown in the text will give a higher yield of the target molecule because the alkoxide ion attacks the back-side of a primary alkyl halide.
In the synthesis shown above, the alkoxide ion attacks the back-side of a secondary alkyl halide which provides greater steric hindrance to back-side attack. Therefore, there will be greater competition from the E2 reaction that will form an elimination product, reducing the yield of the desired substitution product.

30. **a.** 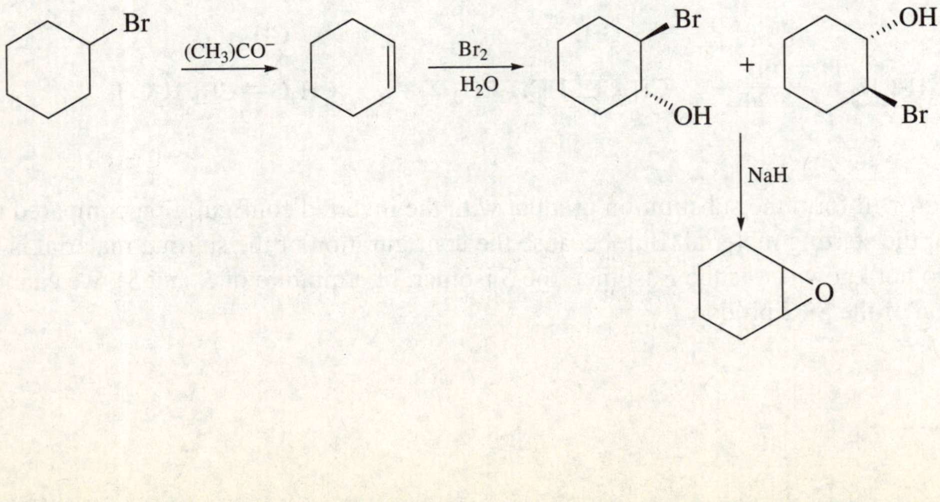

b. 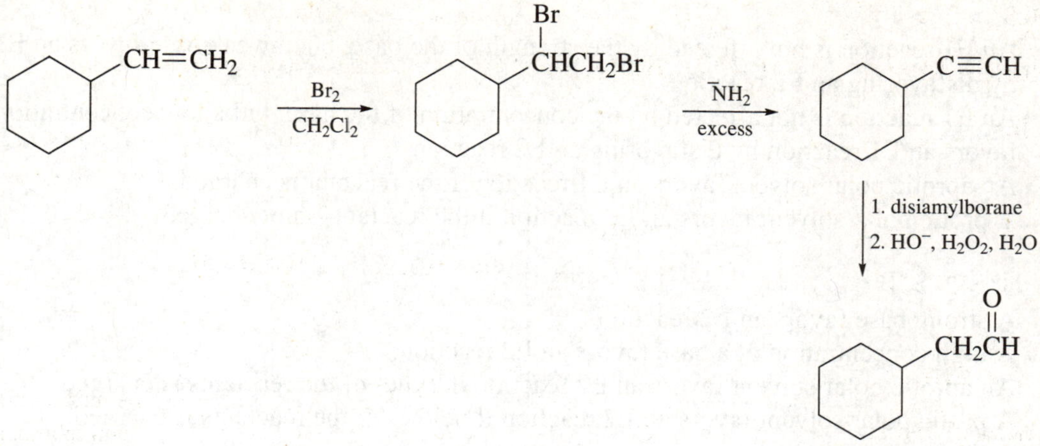

c. 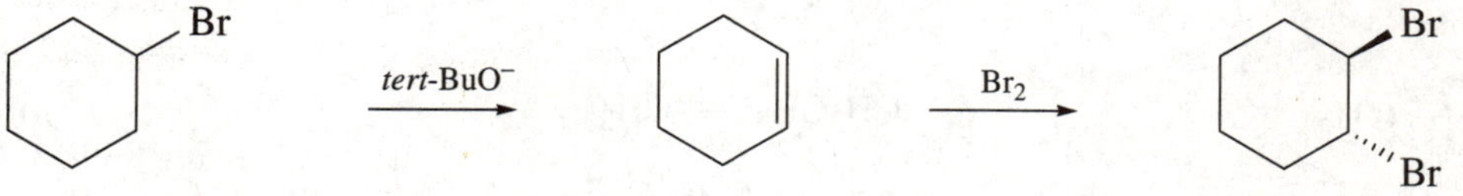

d. 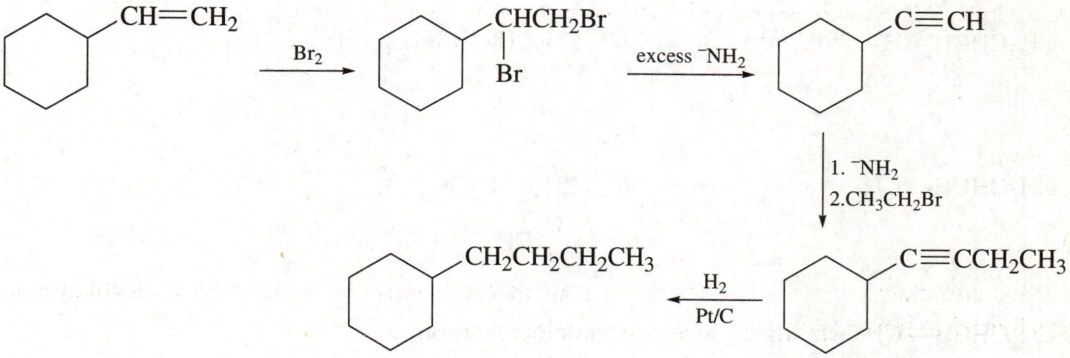

31. a. 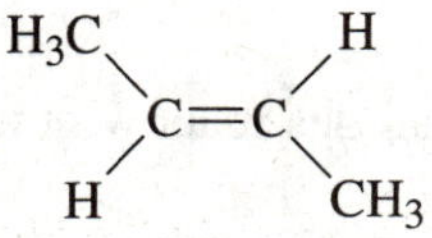

c. 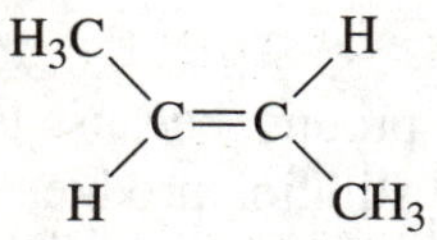

e. 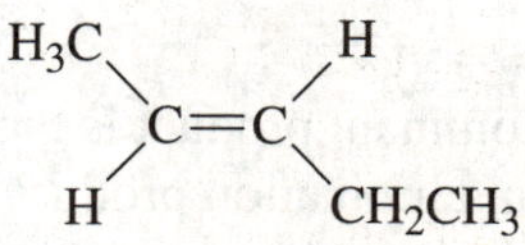

b. 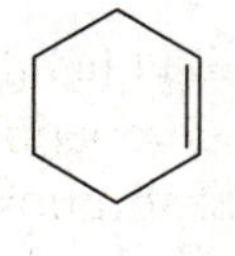

d. 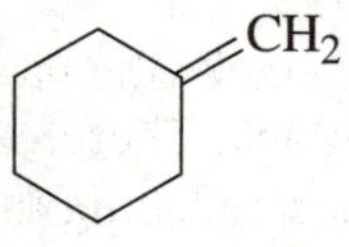

f. 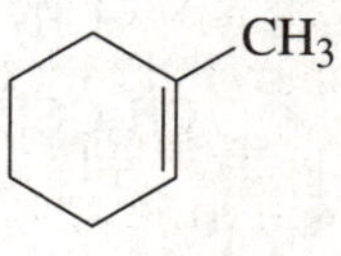

32. a. 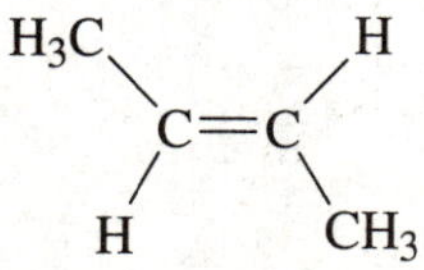

c. 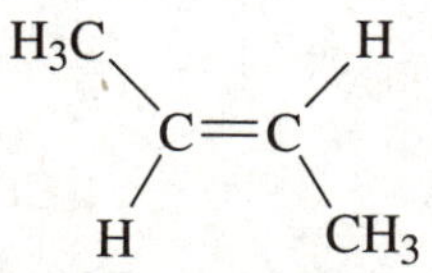

e. 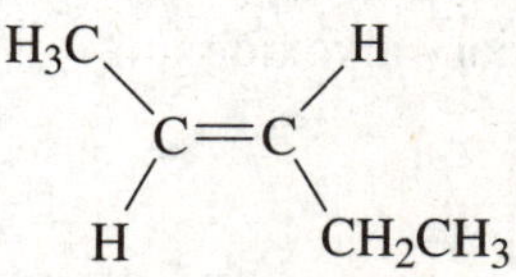

b. 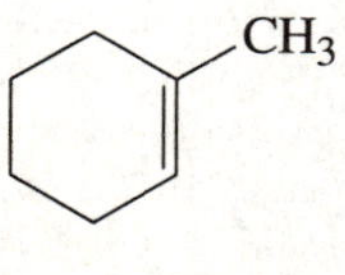

d. No reaction because a primary carbocation cannot be formed.

f. 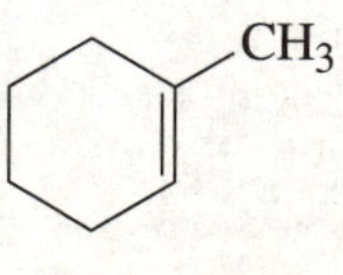

33. **a.** **1.** 3° > 2° > 1°

2. An E1 reaction is not affected by the strength of the base, but a weak base favors an E1 reaction by disfavoring an E2 reaction.

3. An E1 reaction is not affected by the concentration of the base, but a low concentration of a base favors an E1 reaction by disfavoring an E2 reaction.

4. An aprotic polar solvent favors an E1 reaction if the reactant is charged.
A protic polar solvent favors an E1 reaction if the reactant is not charged.

b. **1.** 3° > 2° > 1°

2. A strong base favors an E2 reaction.

3. A high concentration of a base favors an E2 reaction.

4. An aprotic polar solvent favors an E2 reaction if either of the reactants is charged.
A protic polar solvent favors an E2 reaction if neither of the reactants is charged.

34. **a.** $CH_3CH_2CH_2\overset{-}{C}H_2$

b. $CH_3\overset{+}{C}HCH_2CH_3$

c. $CH_2\overset{+}{C}HCH{=}CH_2$

d. $CH_2\overset{-}{C}HCH{=}CH_2$ stabilized by electron delocalization

e. 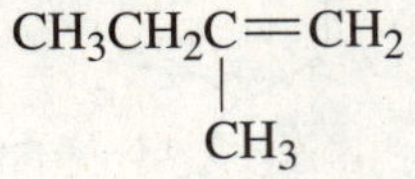

f. 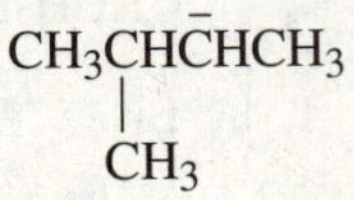

g. 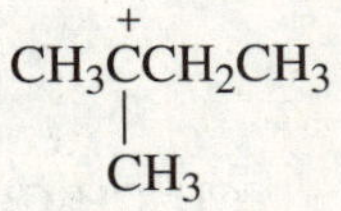

35. The predominant product is the elimination product because tertiary alkyl halides react with nucleophiles to form an elimination product and little substitution product.

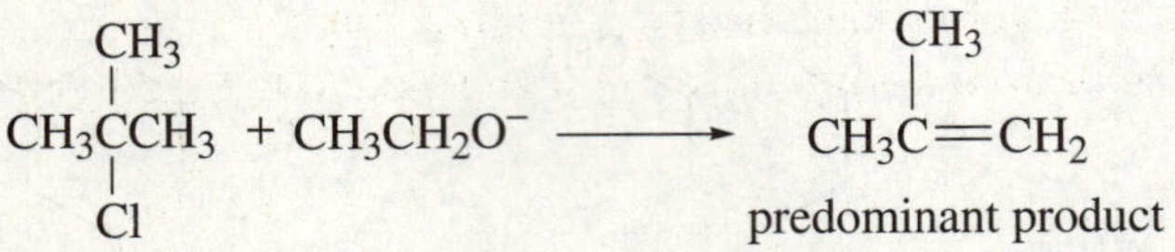

Rather than a tertiary alkyl halide and a primary alkoxide ion, he should have used a primary alkyl halide and a tertiary alkoxide ion.

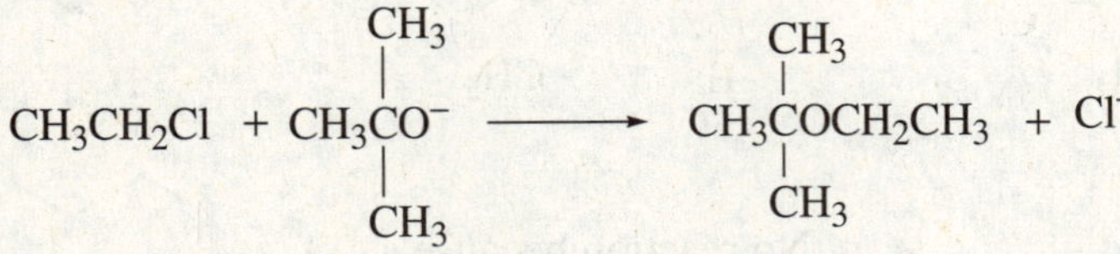

36. **a.** 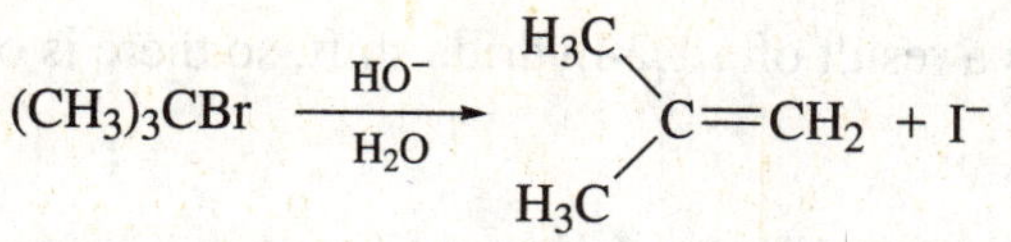

Because Br^- is better leaving group than Cl^-

b. 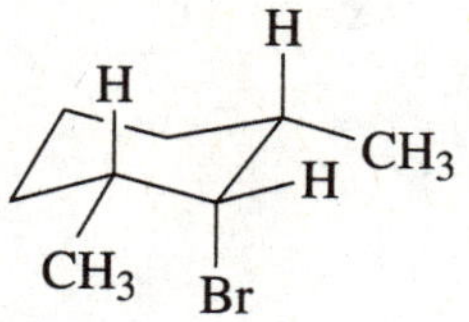

This compound is the only one that can undergo an E2 elimination reaction because the other compound does not have a hydrogen trans to the Br.

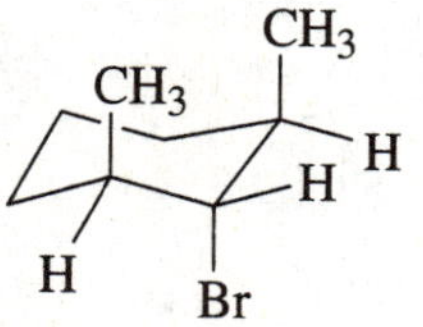

This compound will undergo an elimination reaction more slowly because it can react only by an E1 pathway.

37. The very minor products that would be obtained from "anti-Zaitsev" elimination (that is, the less substituted alkenes) are not shown.

a. 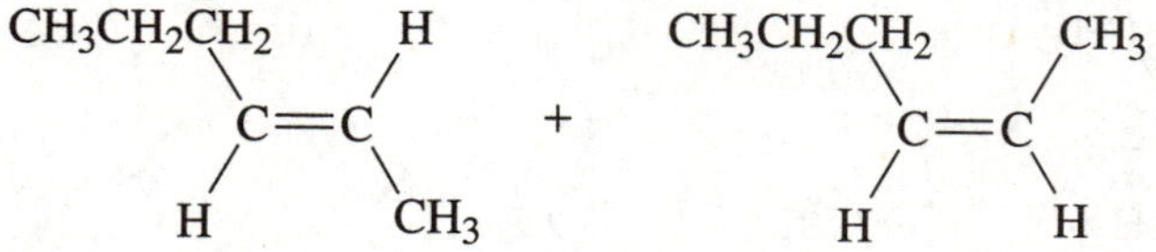

minor

b. 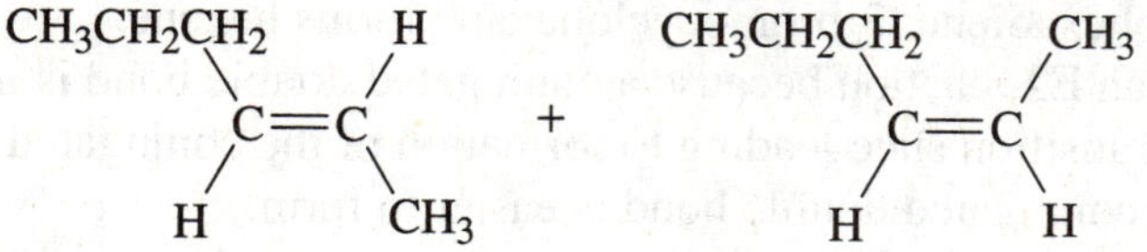

major minor

c. *trans*-1-Chloro-2-methylcyclohexane has two stereosiomers.
A hydrogen cannot be removed from the carbon attached to the methyl group because it is not axial to the Cl.

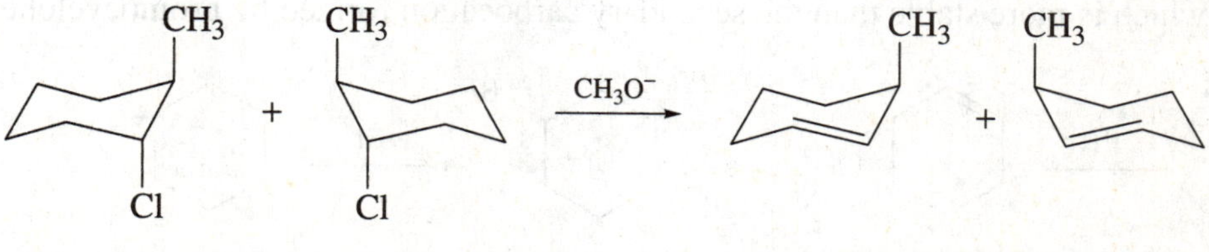

d. Both carbocations from the same carbocation as a result of a 1,2-hydride shift, so there is only one elimination product.

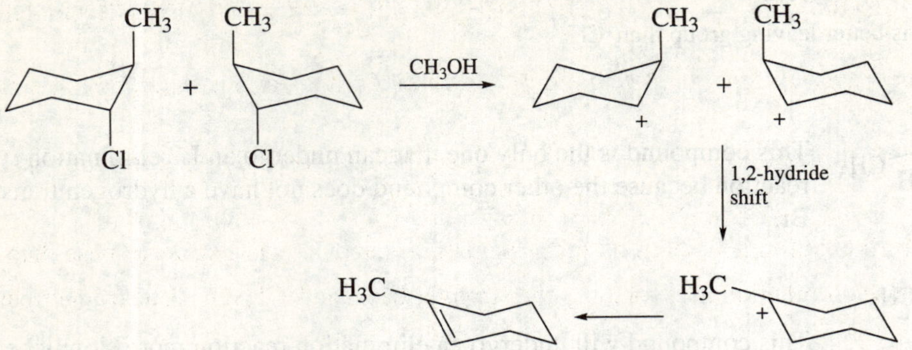

e. 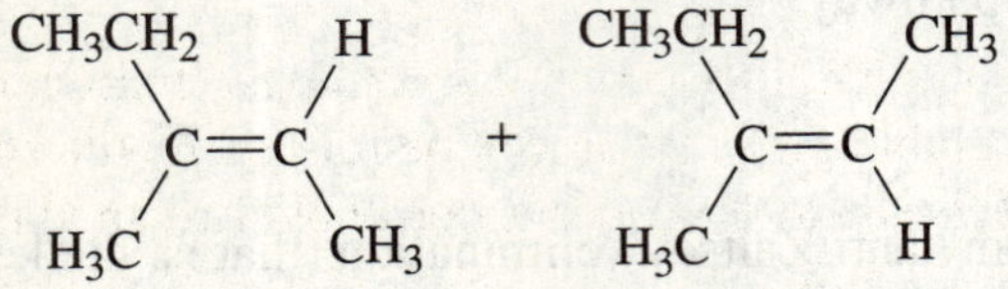

minor

f. 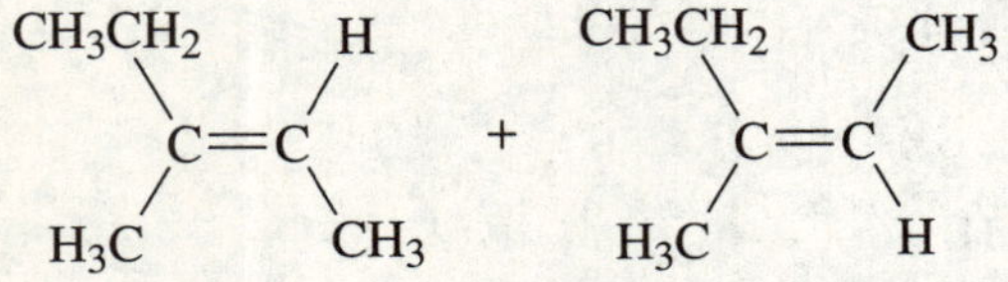

major minor

38. **a.** 3-Bromocyclohexene forms 1,3-cyclohexadiene; 3-bromocyclohexane forms hexene. 3-Bromocyclohexene reacts faster in an E2 reaction because a conjugated double bond is more stable than an isolated double bond, so the transition state leading to formation of the conjugated double bond is more stable and therefore the conjugated double bond is easier to form.

b. 3-Bromocyclohexene also reacts faster in an E1 reaction because it forms an allylic secondary carbocation which is more stable than the secondary carbocation formed by bromocyclohexane.

39. **a.** 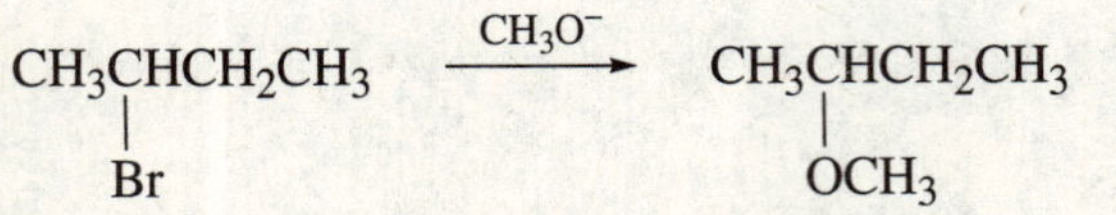

b. $CH_3CH_2CH_2CH_2Br \xrightarrow{CH_3O^-} CH_3CH_2CH_2CH_2OCH_3$

c. $CH_3CH_2CH_2CH_2Br \xrightarrow{CH_3NH_2} CH_3CH_2CH_2CH_2\overset{+}{N}H_2CH_3 \xrightarrow{HO^-} CH_3CH_2CH_2CH_2NHCH_3$

40. **a.** ethoxide ion because elimination is favored by the bulkier base, and *tert*-butoxide ion is bulkier than ethoxide ion

b. ^-SCN because elimination is favored by the stronger base, and ^-OCN is a stronger base than ^-SCN

c. Br^- because elimination is favored by the stronger base, and Cl^- is a stronger base than Br^-

d. CH_3S^- because elimination is favored by the stronger base, and CH_3O^- is a stronger base than CH_3S^-

41. The first compound has two axial hydrogens adjacent to Br, the second has one axial hydrogen adjacent to Br but it cannot form the more substituted (more stable) alkene that can be formed by the first compound. The last compound cannot undergo an E2 reaction because it does not have an axial hydrogen adjacent to Br.

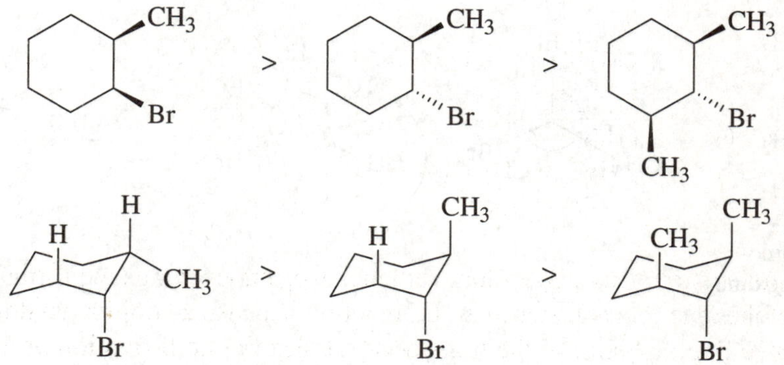

42. These are all E2 reactions.

a. The stereoisomer formed in greaest yield is the one with the largest groups on opposite sides of the double bond.

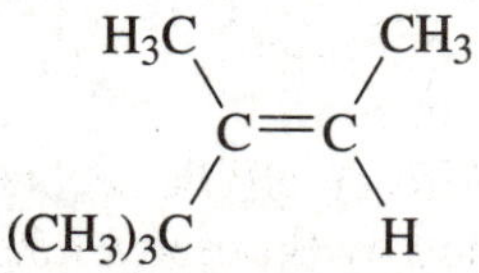

b. No isomers are possible for this compound because one of the sp^2 carbons is bonded to two H's.

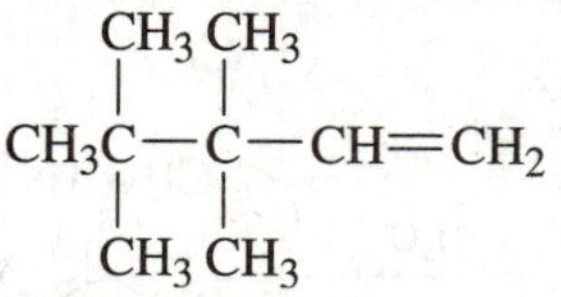

c. No isomers are possible for this compound because one of the sp^2 carbons is bonded to two CH_3's.

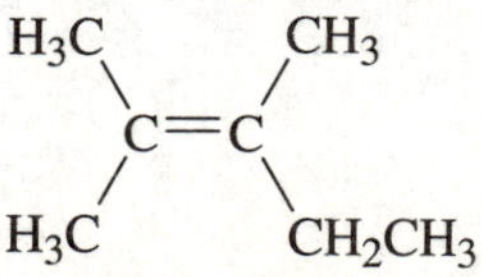

d. Because it is an E2 reaction and there is only one hydrogen on the β–carbon, the stereoisomer formed in greater yield depends on the configuratuion of the reactant. The reactant can have four different configurations: S,S; S,R; R,R; and R,S. To determine the product of the reaction:

1. Draw the reactant in a Fischer projection.
2. From the Fischer projection, draw the Newman projection, remembering that the Fischer projection shows the compound in the eclipsed conformation.
3. Rotate the Newman projection so the H and Br are anti.
4. Put a double bond between where H and Br; that allow you to determine the configuration of the alkene.

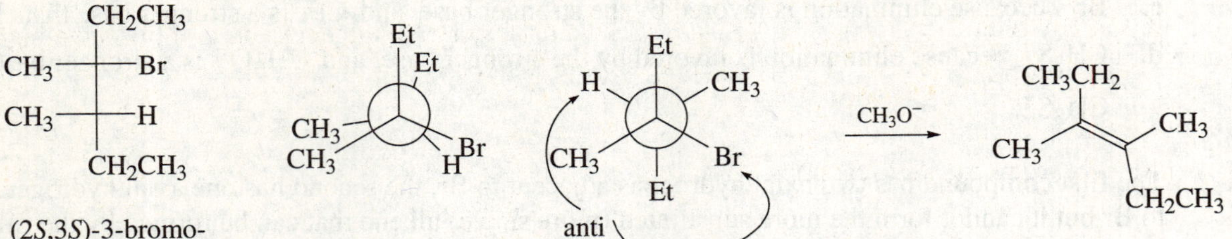

(2*S*,3*S*)-3-bromo-
3,4-dimethylhexane

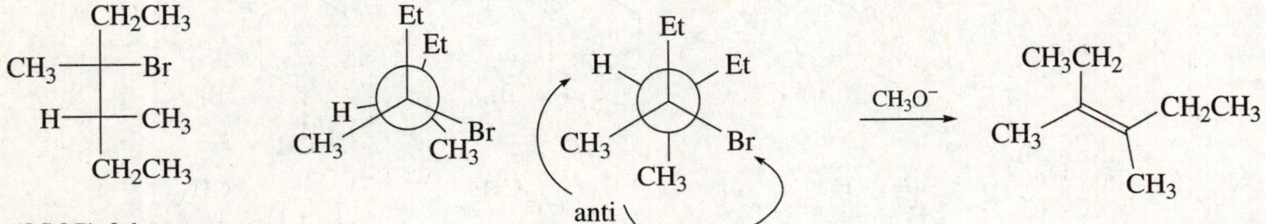

(2*S*,3*R*)-3-bromo-
3,4-dimethylhexane

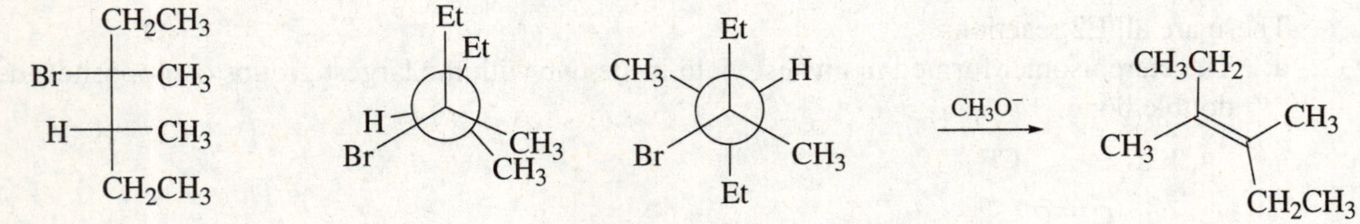

(2*R*,3*R*)-3-bromo-
3,4-dimethylhexane

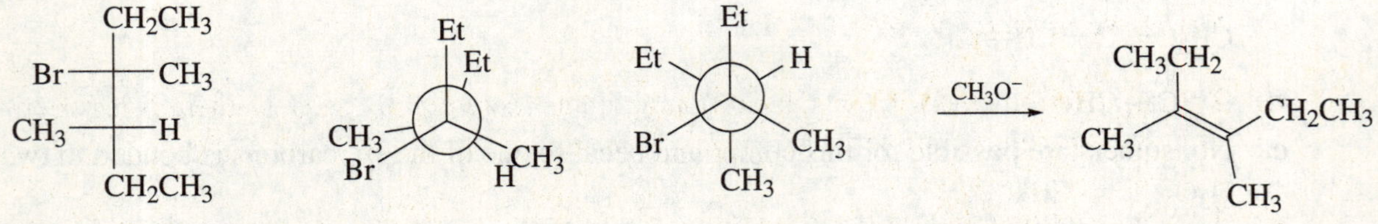

(2*R*,3*S*)-3-bromo-
3,4-dimethylhexane

43. 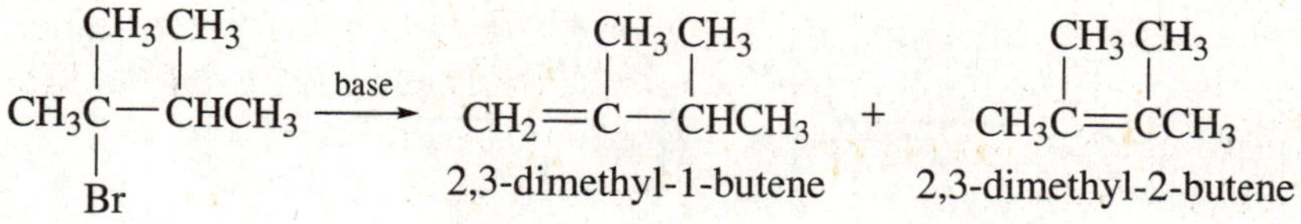

a. 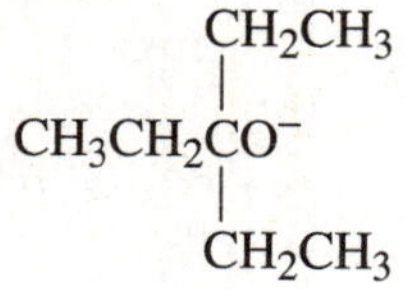

Because it is the most sterically hindered base, it gives the highest percentage of the 1-alkene because the H that is removed is the one that is most accessible to the base.

b. $CH_3CH_2O^-$

Because it is the least sterically hindered base, it gives the highest percentage of the 2-alkene.

44. **a.** These are S_N2/E2 reactions.

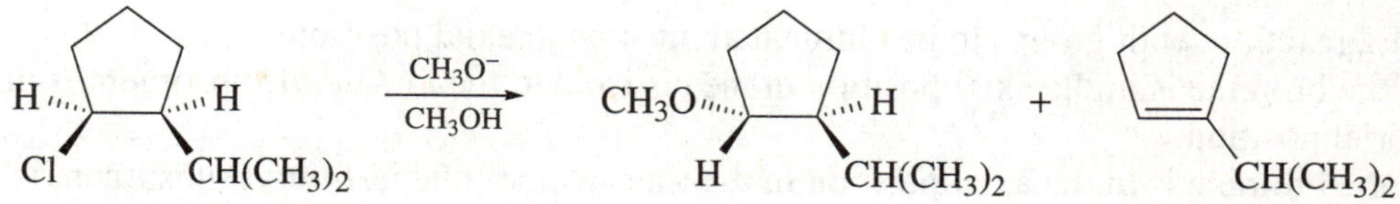

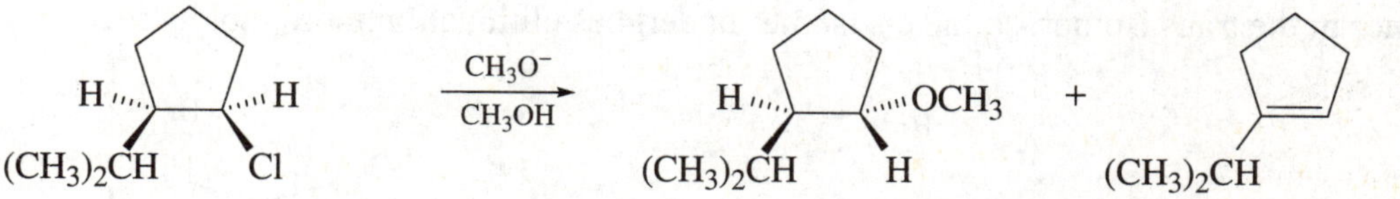

b. Only the substitution products are optically active.

c. 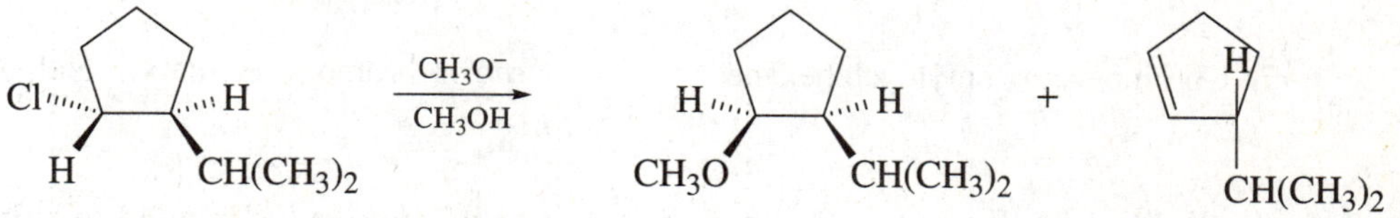

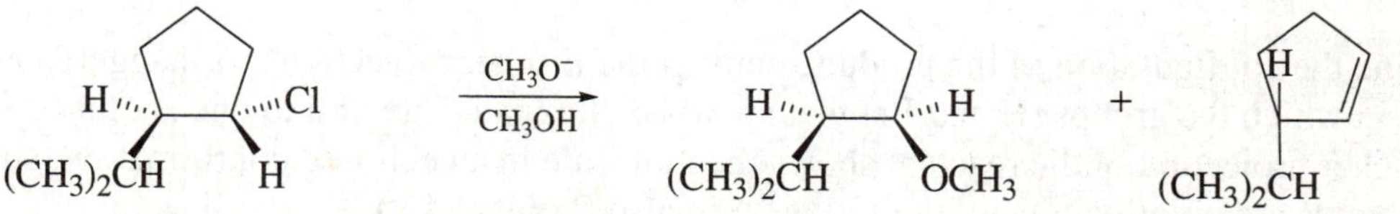

Yes, all the products are optically active.

d. The cis enantiomers form the substitution products more rapidly because there is less steric hindrance from the adjacent substituent to removal of the H.

e. The cis enantiomers form the elimination products more rapidly because the alkenes formed from the cis enantiomers are more substituted and, therefore, more stable. The more stable the alkene, the lower the energy of the transition state leading to its formation.

45. 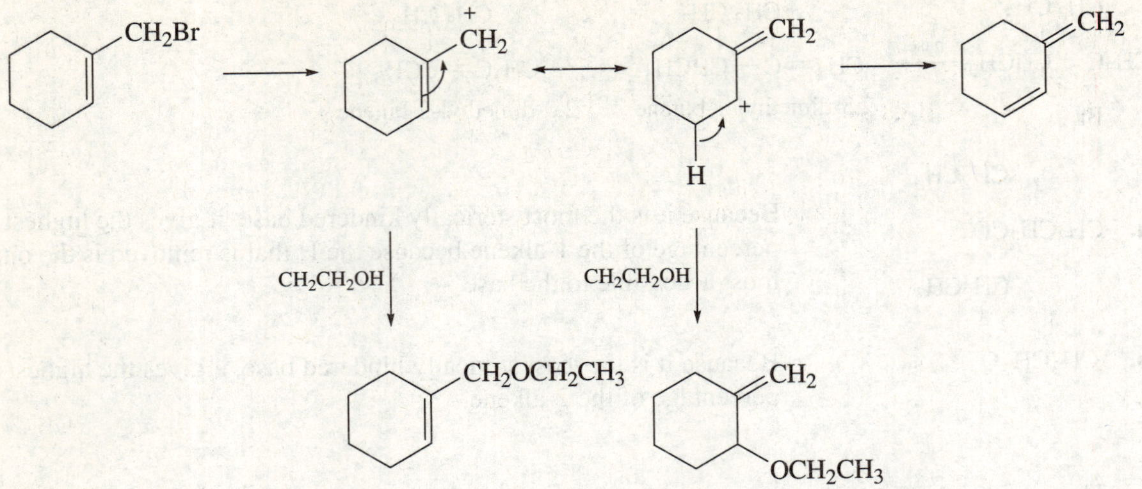

46. In an E2 reaction, both groups to be eliminated must be in axial positions.
When the bromine is in the axial position in the cis isomer, the *tert*-butyl substituent is in the more stable equatorial position.
When the bromine is in the axial position in the trans isomer, the *tert*-butyl substituent is in the less stable axial position.
Thus, elimination takes place via the most stable conformer in the cis isomer and via a less stable chair conformer in the trans isomer, so the cis isomer undergoes elimination more rapidly.

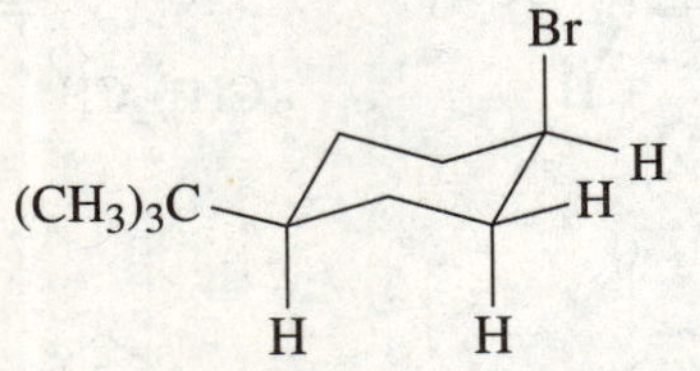

cis-1-bromo-4-*tert*-butylcyclohexane

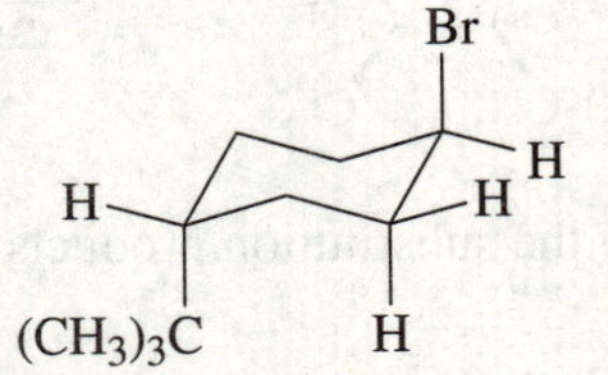

trans-1-bromo-4-*tert*-butylcyclohexane

47. In reactions a-d, the β-carbon from which the hydrogen is to be removed is bonded to only one hydrogen. Therefore, the configuration of the reactant determines the configuration of the E2 elimination product.

To determine the configuration of the product, convert the Fischer projection to a staggered Newman projection, in which the groups (H and Br) that are to be eliminated are anti to one another. (Remember that the Fischer projection of the reactant shows the molecule in an eclipsed conformation, since both horizontal bonds are pointing toward the viewer. See also Problem 42.)
Thus the elimination product formed from (2*S*,3*S*)-2-chloro-3-methypentane is the *E* isomer.

a.

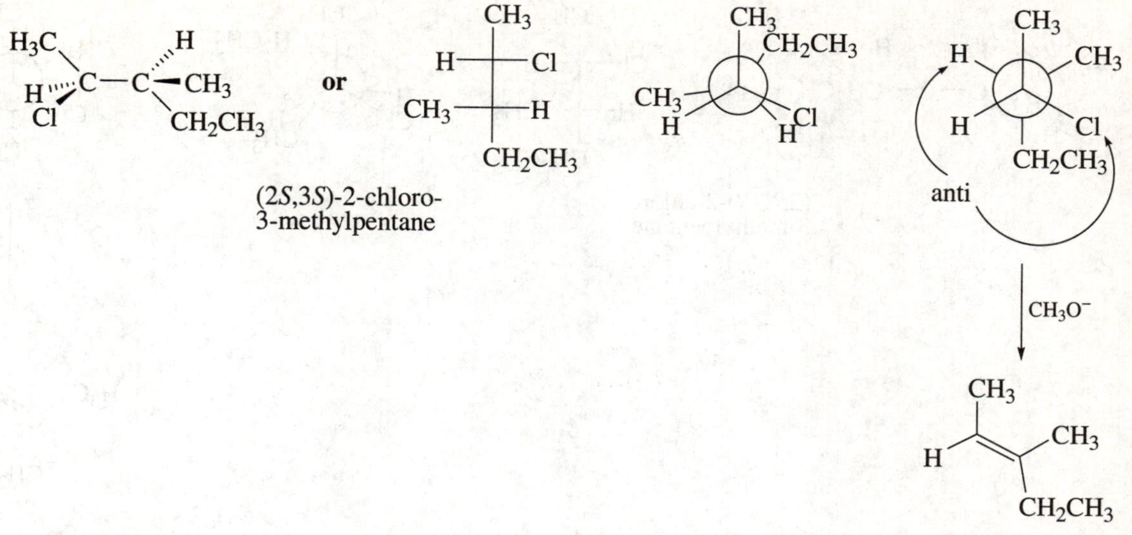

H3C
H
H
C
C
CH3
Cl
CH2CH3
or
CH3
H
Cl
CH3
H
CH2CH3
(2S,3S)-2-chloro-
3-methylpentane
CH3
CH2CH3
CH3
H
H
Cl
CH3
H
CH3
H
Cl
CH2CH3
anti
CH3O−
CH3
CH3
H
CH2CH3

b.

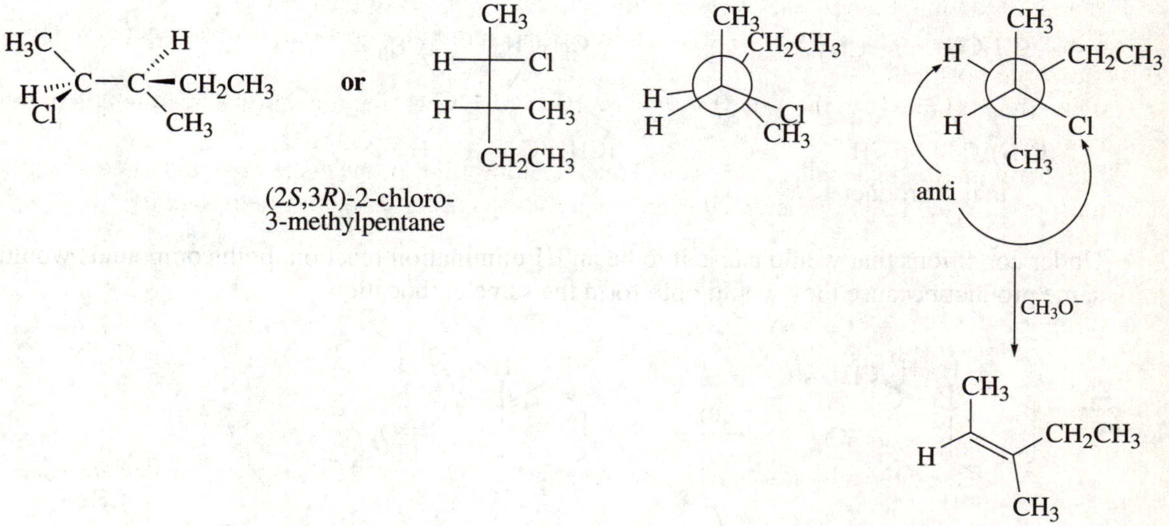

H3C
H
H
C
C
CH2CH3
Cl
CH3
or
CH3
H
Cl
H
CH3
CH2CH3
(2S,3R)-2-chloro-
3-methylpentane
CH3
CH2CH3
H
H
Cl
CH3
CH3
H
CH2CH3
H
Cl
CH3
anti
CH3O−
CH3
CH2CH3
H
CH3

c.

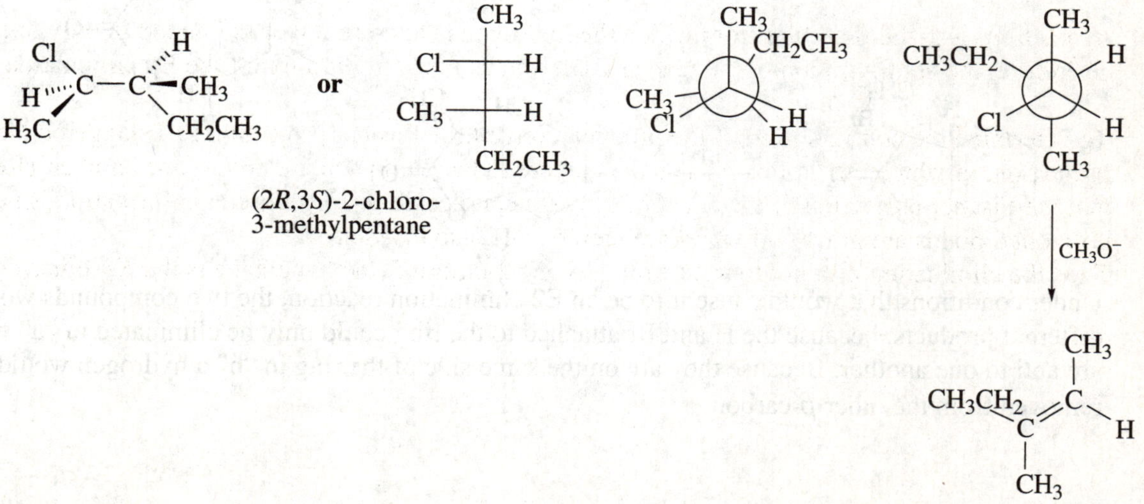

Cl
H
H
C
C
CH3
H3C
CH2CH3
or
CH3
Cl
H
CH3
H
CH2CH3
(2R,3S)-2-chloro-
3-methylpentane
CH3
CH2CH3
CH3
Cl
H
H
CH3
CH3CH2
H
Cl
H
CH3
CH3O−
CH3
CH3CH2
C
C
H
CH3

d.

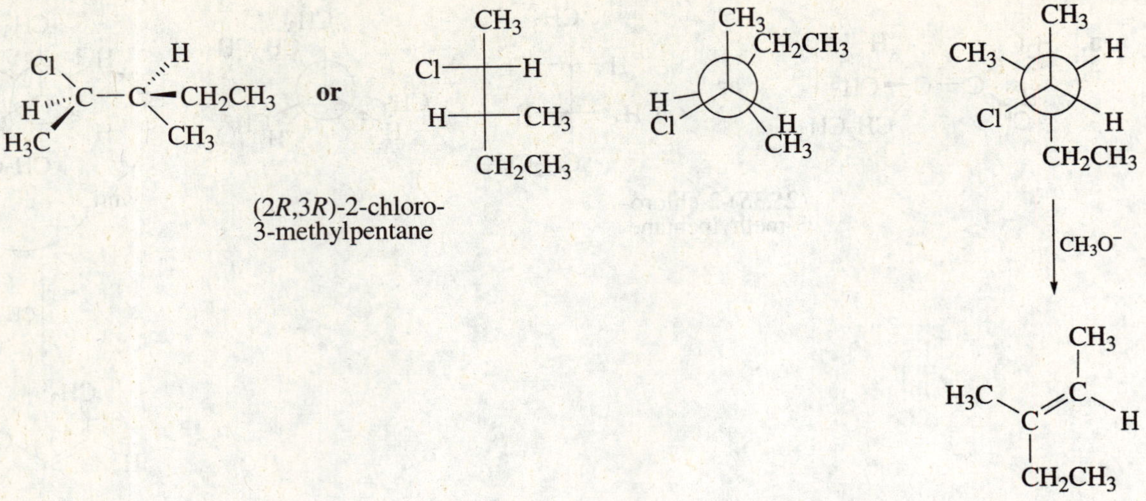

(2*R*,3*R*)-2-chloro-3-methylpentane

e. This reactant has two hydrogens on the β-carbon

major product

48. Under conditions that would cause it to be an E1 elimination reaction, both compounds would form the same product because they would both form the same carbocation.

a.

b.

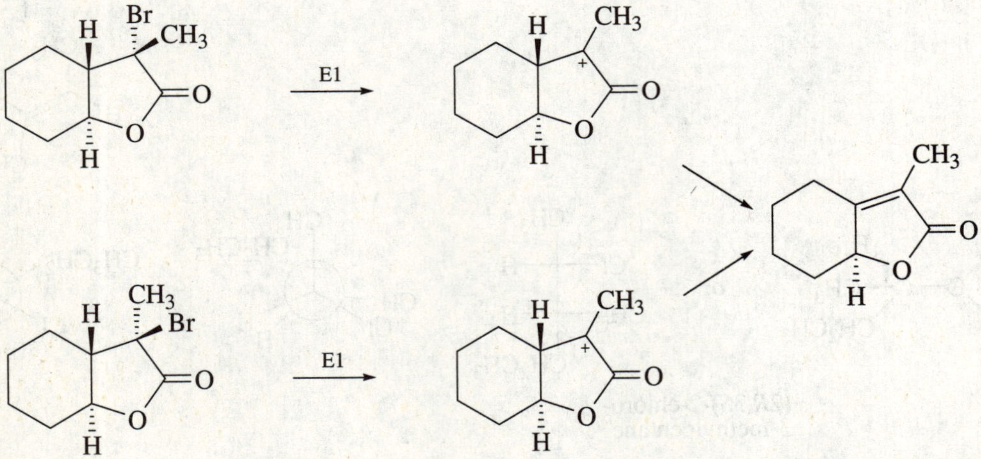

Under conditions that would cause it to be an E2 elimination reaction, the two compounds would from different products. because the H and Br attached to the ring could only be eliminated in "a" where they are anti to one another. Because they are on the same side of the ring in "b" a hydrogen would have to be removed from the other β-carbon.

a.

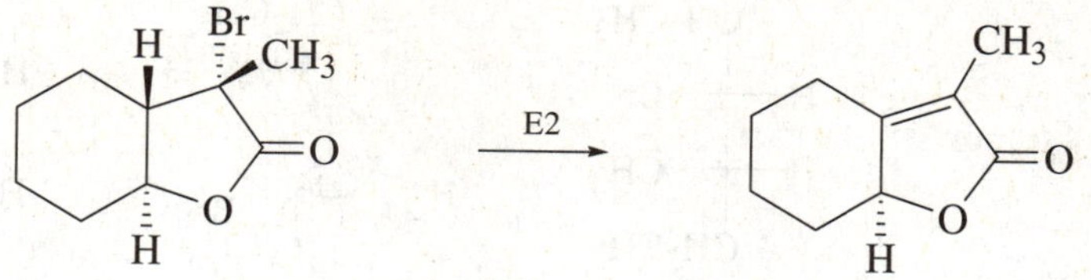

b.

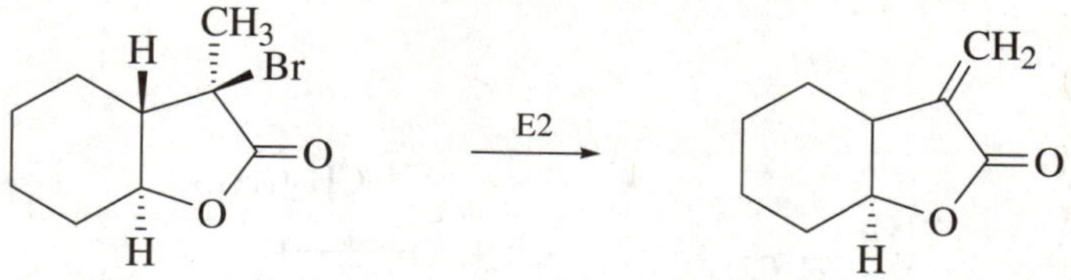

49. In order to undergo an E2 reaction, a chlorine and a hydrogen on an adjacent carbon must be trans to one another so they can both be in the required axial positions. Every Cl in the following compound has a Cl trans to it so no Cl has a hydrogen trans to it. Thus, it is the least reactive of the isomers.

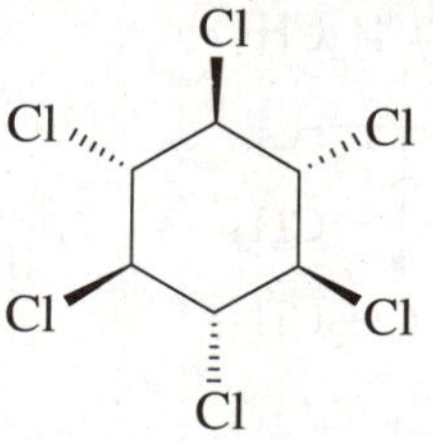

50. The silver ion increases the ease of departure of the halogen atom.

$$CH_3CH{=}CHCH_2{-}Br{:} + Ag^+ \longrightarrow CH_3CH{=}CHCH_2{-}Br^+{:}Ag \longrightarrow CH_3CH{=}CH\overset{+}{C}H_2 + AgBr$$

51. For a description of how to do this problem, see Problem 42.

a.

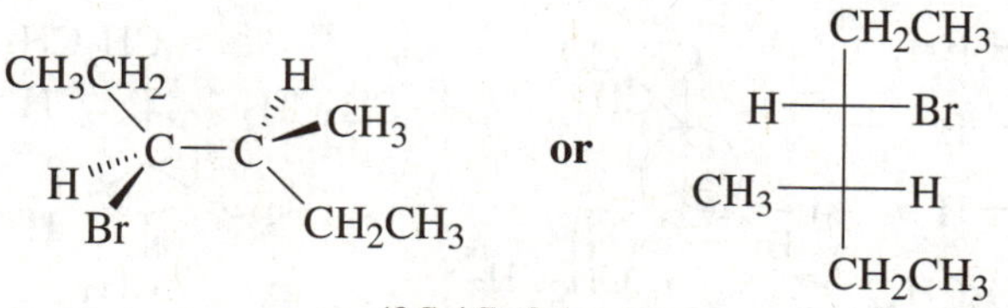

(3*S*,4*S*)-3-bromo-4-methylhexane

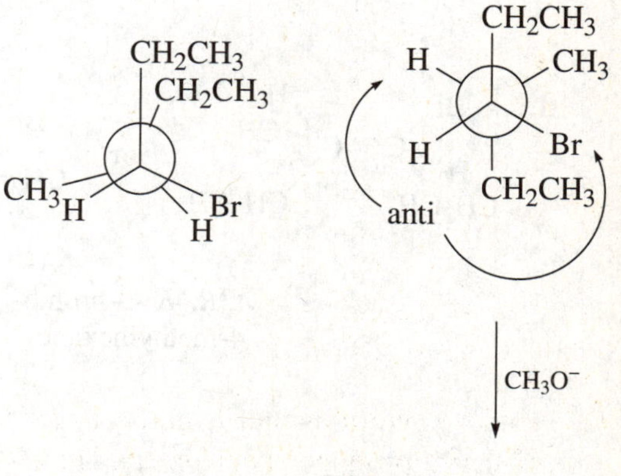

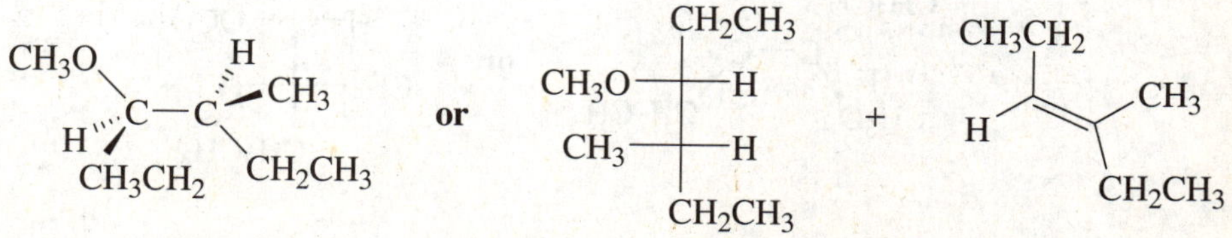

b.

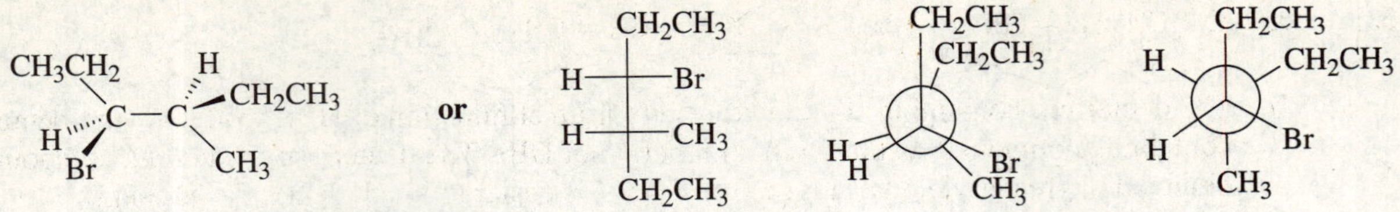

(3*S*,4*R*)-3-bromo-
4-methylhexane

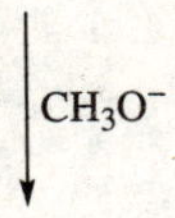

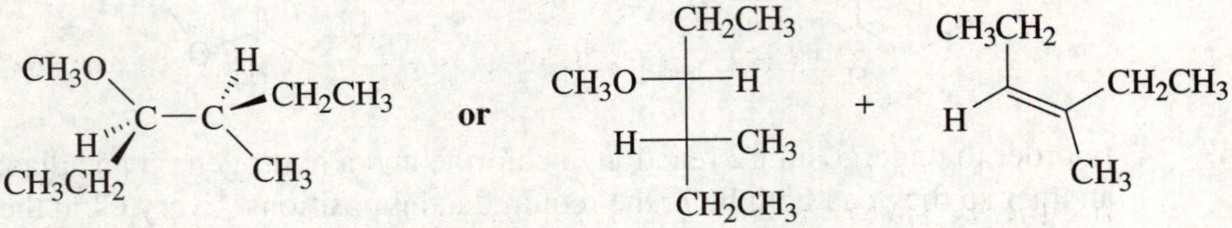

c.

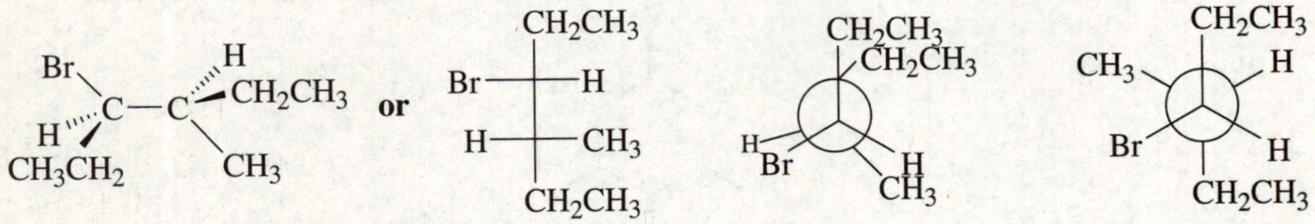

(3*R*,4*R*)-3-bromo-
4-methylhexane

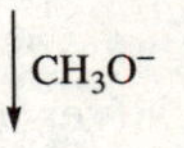

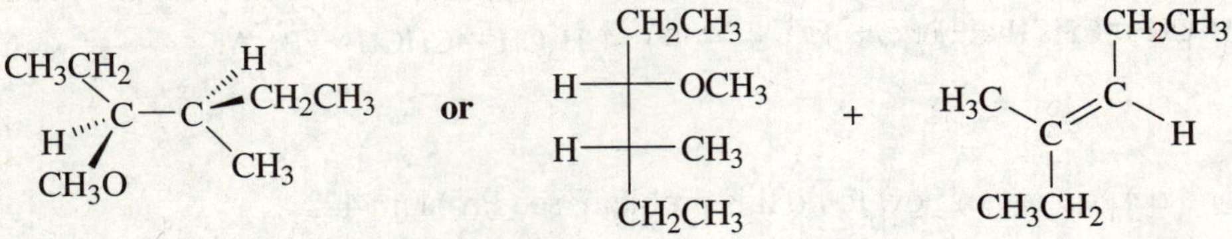

d.

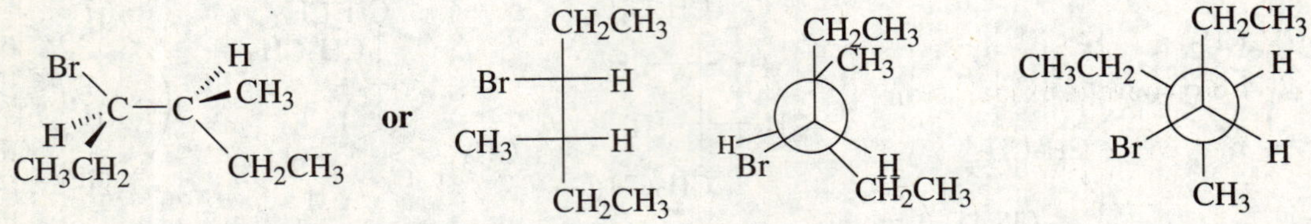

(3*R*,4*R*)-3-bromo-
4-methylhexane

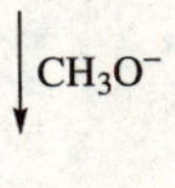

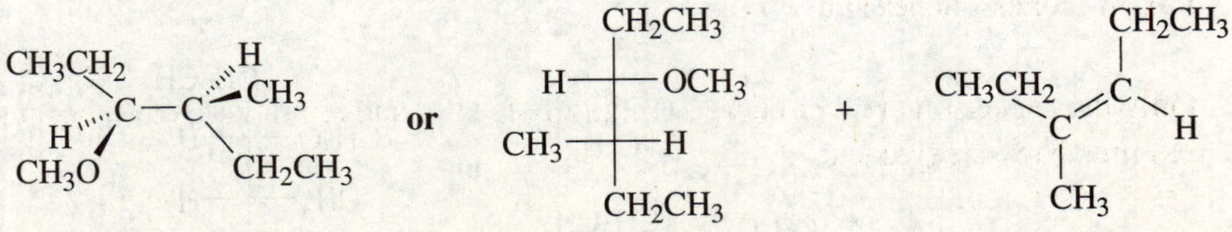

52. **a.** $CH_3CH_2CD{=}CH_2$ and $CH_3CH_2CH{=}CH_2$

b. The deuterium-containing compound results from elimination of HBr, while the non-deuterium-containing compound results from elimination of DBr. The deuterium-containing compound will be obtained in greater yield because a C—H bond is easier to break than a C—D bond.

53. **a.**

$$CH_3CH_2CH_2CH_2Br \xrightarrow{tert\text{-}BuO^-} CH_3CH_2CH{=}CH_2 \xrightarrow{Br_2} CH_3CH_2CH(Br)CH_2Br \xrightarrow[\ ^-NH_2]{\text{excess}} CH_3CH_2C{\equiv}CH$$

$$CH_3CH_2C{\equiv}CH \xrightarrow{^-NH_2} CH_3CH_2C{\equiv}C^- \xrightarrow{CH_3CH_2Br} CH_3CH_2C{\equiv}CCH_2CH_3 \xrightarrow[H_2SO_4]{H_2O} CH_3CH_2C({=}O)CH_2CH_2CH_3$$

b. 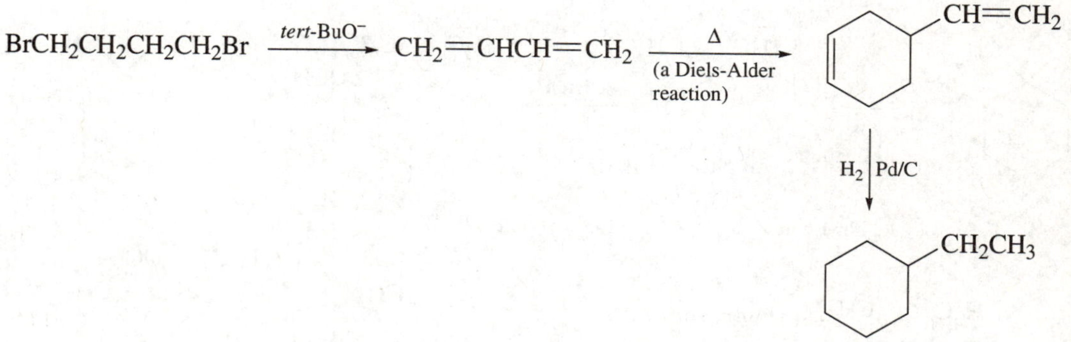

54. **a.** Both compounds form the same elimination product because they both have hydrogens on the same β-carbons that are anti to the bromine.

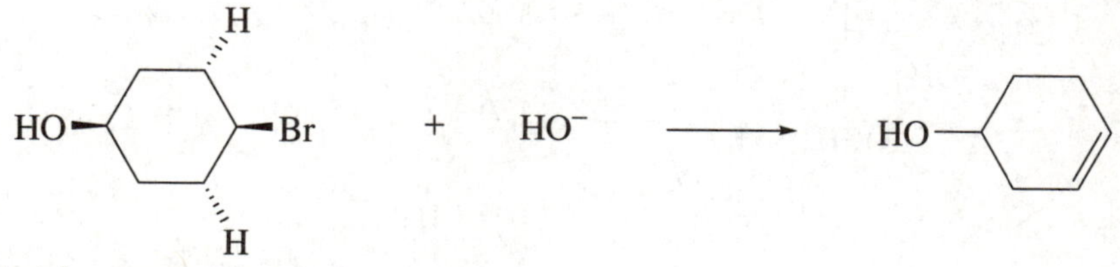

cis-4-bromocyclohexanol

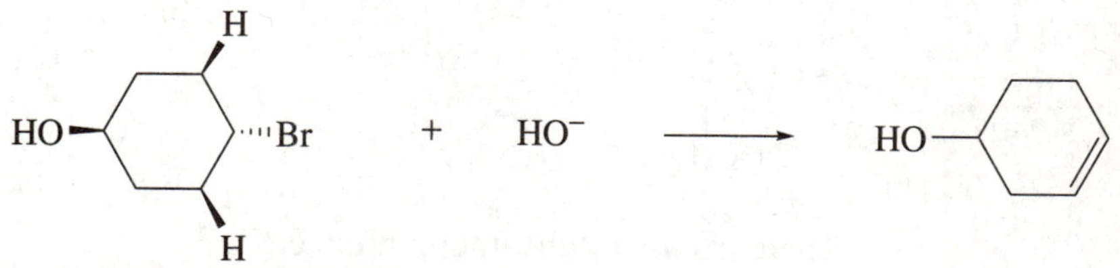

trans-4-bromocyclohexanol

Only the trans isomer can undergo an intramolecular substitution reaction because the S_N2 reaction requires back-side attack.

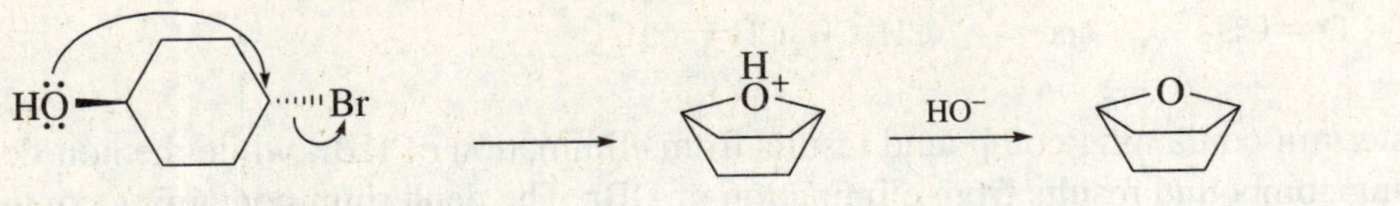

trans-4-bromocyclohexanol

The cis isomer can undergo only an intermolecular substitution reaction.

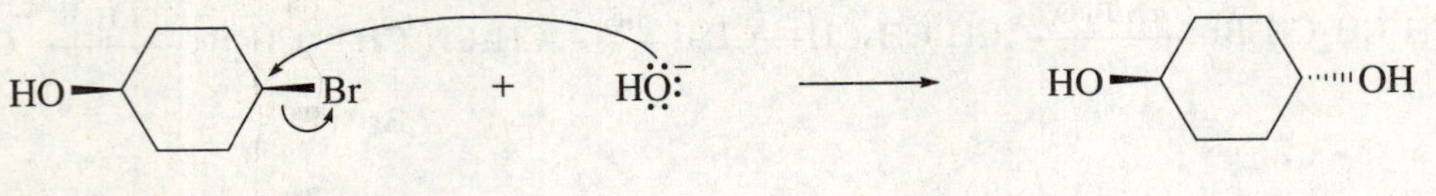

cis-4-bromocyclohexanol

b. The elimination reaction forms two stereoisomers because the reaction forms an asymmetric center in the product. Both substitution reactions form a single stereoisomer.

55.

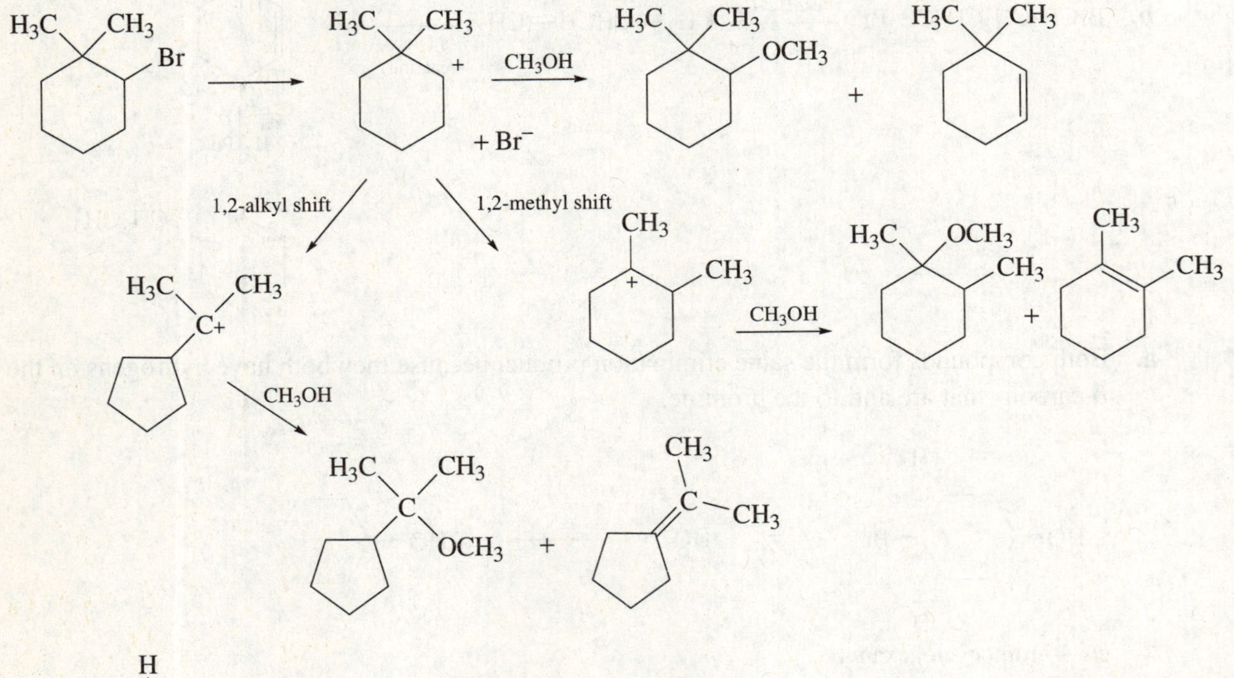

56.

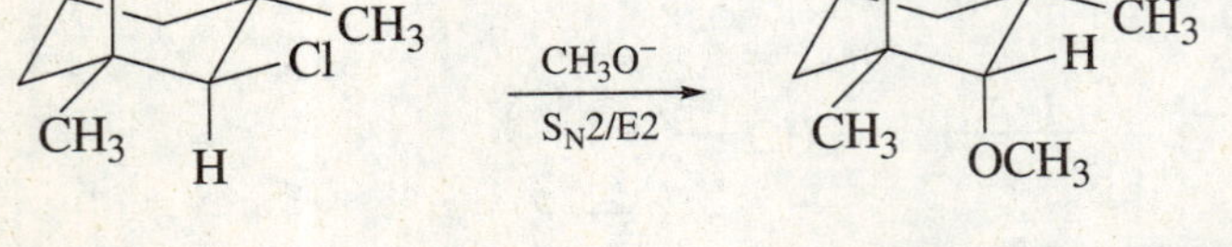

There is only a substitution product.
The reactant does not undergo elimination because when Cl is in an axial position, there is not an adjacent hydrogen in an axial position.

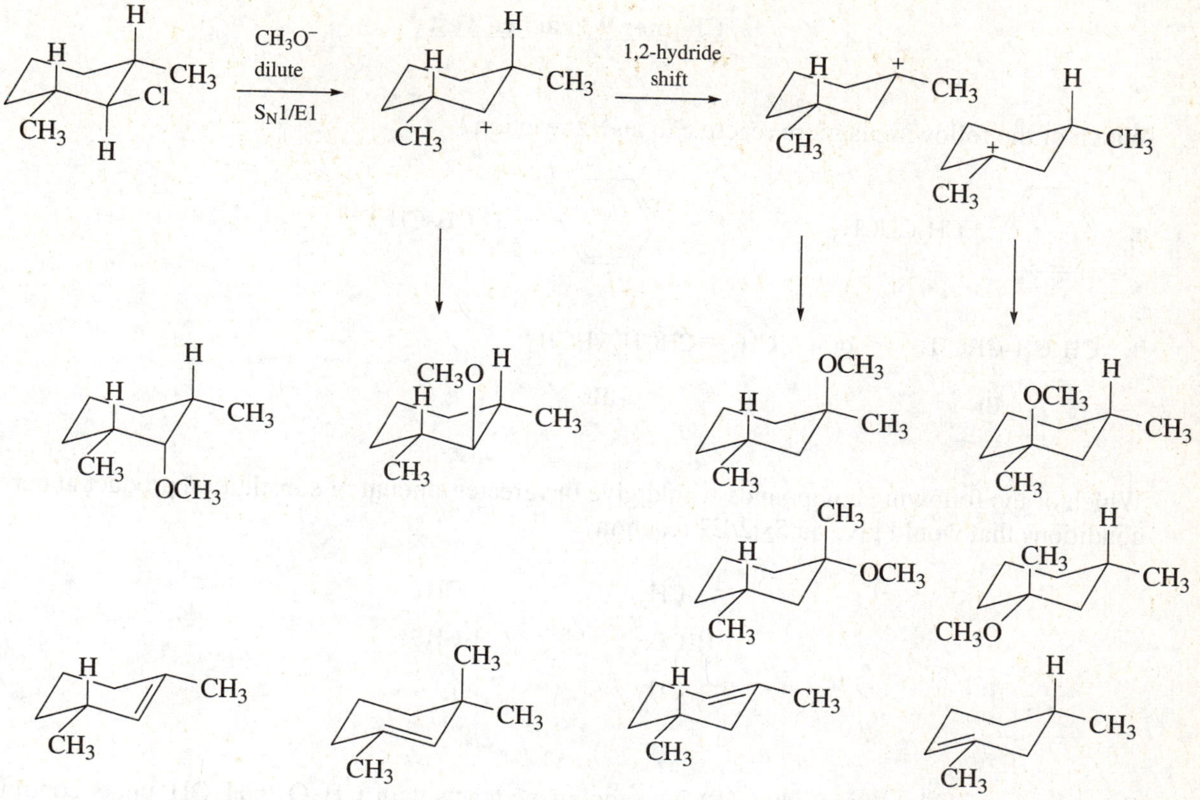

57. **a.**

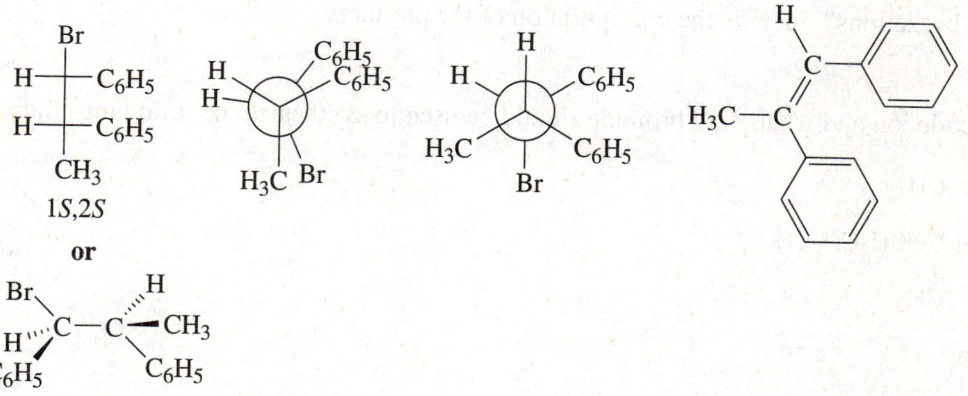

b.

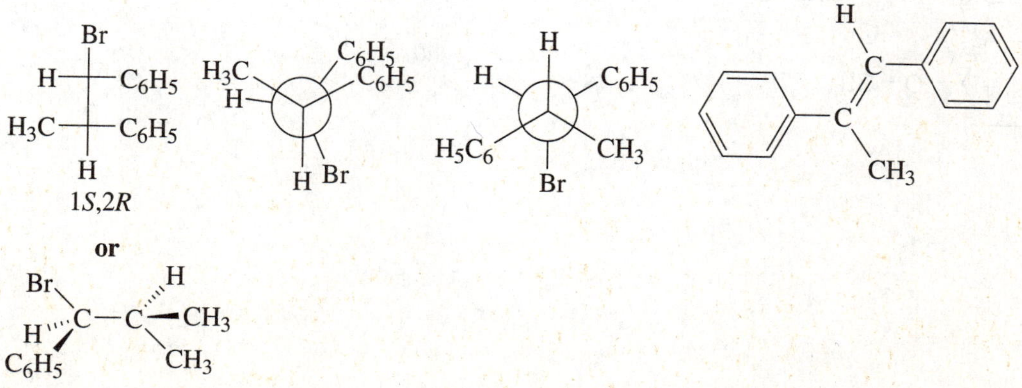

Chapter 9 Practice Test

1. Which of the following is more reactive in an E2 reaction?

a. 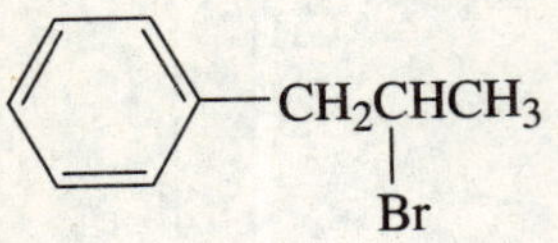 or 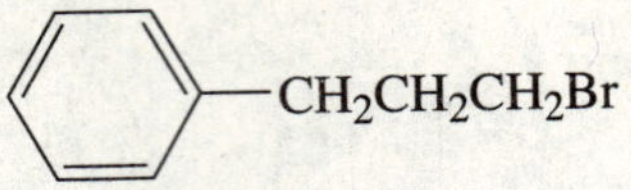

b. $CH_3CH_2CHCH_3$ (Br on C-2) or $CH_2{=}CHCH_2CHCH_3$ (Br on C-4)

$$\begin{array}{c} CH_3CH_2\underset{\displaystyle |\atop \displaystyle Br}{C}HCH_3 \end{array} \quad \text{or} \quad \begin{array}{c} CH_2{=}CHCH_2\underset{\displaystyle |\atop \displaystyle Br}{C}HCH_3 \end{array}$$

2. Which of the following compounds would give the greater amount of substitution product under conditions that would give an $S_N2/E2$ reaction?

$$\begin{array}{c} CH_3 \\ | \\ CH_3CBr \\ | \\ CH_3 \end{array} \quad \text{or} \quad \begin{array}{c} CH_3 \\ | \\ CH_3CHBr \end{array}$$

3. What products are obtained when (*R*)-2-bromobutane reacts with CH_3O^-/CH_3OH under conditions that favor S1/E1 reactions? Include the configuration of the products.

4. What alkoxide ion and what alkyl bromide should be used to synthesize the following ethers?

a.
$$\begin{array}{c} CH_3 \\ | \\ CH_3CH_2COCH_2CH_2CH_3 \\ | \\ CH_3 \end{array}$$

b. 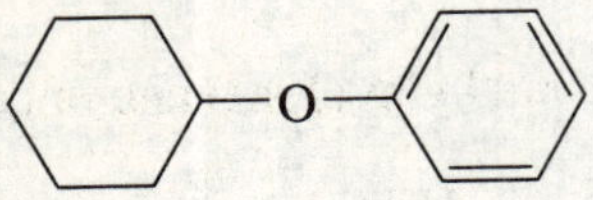

c. 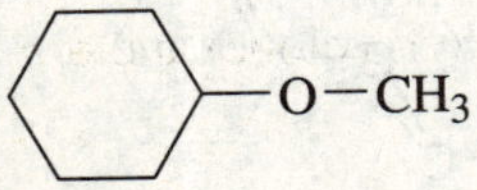

5. For each of the following pairs of E2 reactions, indicate the one that occurs with the greater rate constant:

a. $CH_3CH_2CH_2Cl + HO^-$ or $CH_3CH(Cl)CH_3 + HO^-$

b. $CH_3CH_2CH_2Cl + HO^-$ or $CH_3CH_2CH_2I + HO^-$

c. $CH_3CH_2CH_2Br + HO^-$ or $CH_3CH_2CH_2Br + H_2O$

d. $CH_3CH(Br)CH_3 \xrightarrow[H_2O/CH_3OH]{CH_3O^-}$ or $CH_3CH(Br)CH_3 \xrightarrow[CH_3OH]{CH_3O^-}$

e. $CH_3CH(Br)CH_3 + HO^-$ or $CH_3C(CH_3)(Br)CH_3 + HO^-$

6. Give the major product of an E2 reaction of each of the following compounds with hydroxide ion:

a. 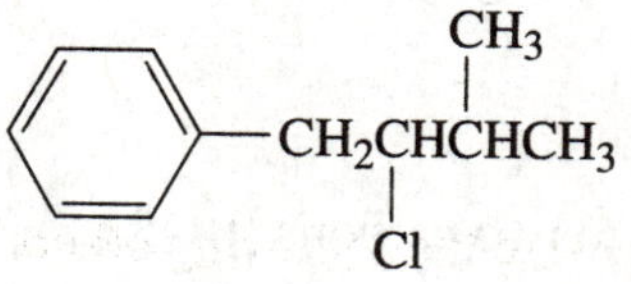

b. 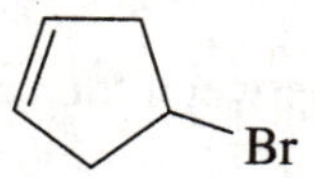

c. $CH_3CH(Br)CH(CH_3)CH_3$

d. $CH_3CH(F)CH(CH_3)CH_3$

7. a. Which would be more reactive in an E2 reaction, *cis*-1-bromo-2-methylcyclohexane or *trans*-1-bromo-2-methylcyclohexane?

b. Which would be more reactive in an E1 reaction, *cis*-1-bromo-2-methylcyclohexane or *trans*-1-bromo-2-methylcyclohexane?

CHAPTER 10
Reactions of Alcohols, Amines, Ethers, Epoxides, and Sulfur-Containing Compounds • Organometallic Compounds

Important Terms

alcohol	an organic compound with an OH functional group (ROH).
alkaloid	a natural product with a nitrogen heteroatom found in the leaves, bark, or seeds of plants.
alkyl tosylate	an ester of *para*-toluenesulfonic acid.
alkyl triflate	an ester of trifluoromethanesulfonic acid.
antibiotic	a compound that interferes with the growth of a microorganism.
arene oxide	an aromatic compound that has had one of its double bonds converted to an epoxide.
coupling reaction	a reaction that joins two carbon-hydrogen containing groups.
crown ether	a cyclic molecule that possesses several ether linkages.
crown-guest complex	the complex formed when a crown ether binds a substrate.
dehydration	loss of water.
ether	a compound containing an oxygen bonded to two carbons (ROR).
Gilman reagent	a dialkylcopper reagent used to replace a halogen with an alkyl group.
Grignard reagent	the compound that results when magnesium is inserted between the carbon and halogen of an alkyl halide (RMgBr, RMgCl).
Heck reaction	couples an aryl, benzyl, or vinyl halide or triflate with an alkene.
inclusion compound	a compound that specifically binds a metal ion or an organic molecule.
mercapto group	an SH group.
molecular recognition	the recognition of one molecule by another as a result of specific interactions.
organocuprate	$(R)_2CuLi$; prepared by treating an organolithium reagent with cuprous iodide.
organolithium compound	RLi; prepared by adding lithium to an alkyl halide.
organomagnesium compound	RMgBr; prepared by adding an alkyl halide to magnesium shavings.

organometallic compound	a compound with a carbon-metal bond.
phase-transfer catalyst	a compound that carries a polar reagent into a nonpolar phase.
Stille reaction	couples an aryl, benzyl, or vinyl halide or triflate with a stannane.
sulfide (thioether)	the sulfur analog of an ether (RSR).
sulfonate ester	the ester of a sulfonic acid (RSO_2OR).
sulfonium salt	$(R)_3S^+X^-$
Suzuki reaction	couples an aryl, benzyl, or vinyl halide or triflate with an organoborane.
thioether (sulfide)	the sulfur analog of an ether (RSR).
thiol (mercaptan)	the sulfur analog of an alcohol (RSH).
transmetallation	metal exchange.
vicinal diol (vicinal glycol)	a compound with OH groups bonded to adjacent carbons.

Solutions to Problems

1. They no longer have a lone pair of electrons.

2.
a. Solved in the text.
b. The conjugate acid of the leaving group of $CH_3\overset{+}{O}H_2$ is H_3O^+; its pK_a is = −1.7.
The conjugate acid of the leaving group of CH_3OH is H_2O; its pK_a is = 15.5.
Because H_3O^+ is a much stronger acid than H_2O, H_3O^+ is the weaker base and the better leaving group. Thus, $CH_3\overset{+}{O}H_2$ is more reactive than CH_3OH.

3. All the syntheses below are shown to be accomplished by converting the alcohol into a sulfonate ester and then treating the sulfonate ester with the desired nucleophile:
They also could have been carried out by converting the alcohol into an alkyl halide and then adding the desired nucleophile.

a. $CH_3CH_2CH_2CH_2OH \xrightarrow[\text{2. } CH_3O^-]{\text{1. TsCl}} CH_3CH_2CH_2CH_2OCH_3$

b. $CH_3CH_2CH_2CH_2OH \xrightarrow[\text{2. } CH_3CH_2CO^- \text{ (C=O)}]{\text{1. TsCl}} CH_3CH_2CH_2CH_2OCCH_2CH_3$ (C=O)

c. $CH_3CH_2CH_2CH_2OH \xrightarrow[\text{2. } CH_3CH_2NH_2]{\text{1. TsCl}} CH_3CH_2CH_2CH_2\overset{+}{N}H_2CH_2CH_3 \xrightarrow{HO^-} CH_3CH_2CH_2CH_2NHCH_2CH_3$

Notice that in "c" a neutral amine ($CH_3CH_2NH_2$) is used instead of $CH_3CH_2NH^-$, because the weaker base ($CH_3CH_2NH_2$) favors the desired substitution reaction, whereas the stronger base ($CH_3CH_2NH^-$) would favor elimination at the expense of substitution (Section 9.8).

d. $CH_3CH_2CH_2CH_2OH \xrightarrow[\text{2. } ^-C{\equiv}N]{\text{1. TsCl}} CH_3CH_2CH_2CH_2C{\equiv}N$

4. The relative reactivity would be: tertiary > primary > secondary.
If secondary alcohols underwent an S_N2 reaction, they would be less reactive than primary alcohols because they are more sterically hindered than primary alcohols.

5. All four alcohols undergo an S_N1 reaction because they are either secondary or tertiary alcohols.

a. $CH_3CH_2CH(OH)CH_3 \overset{H^+}{\rightleftharpoons} CH_3CH_2CH(\overset{+}{O}H_2)CH_3 \longrightarrow CH_3CH_2\overset{+}{C}HCH_3 + H_2O \xrightarrow{Br^-} CH_3CH_2CH(Br)CH_3$

b.

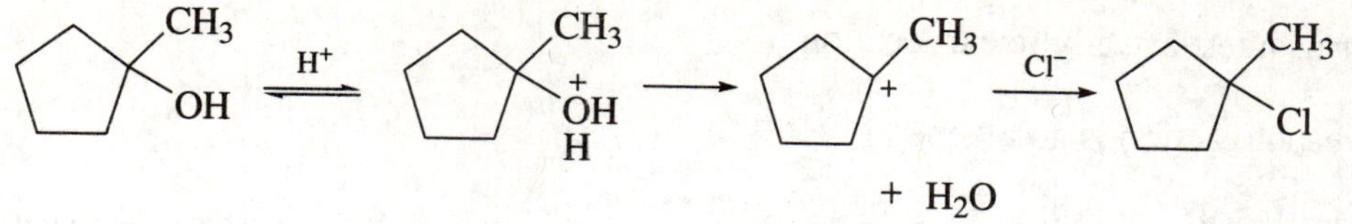

The carbocation that is initially formed in **c** and the carbocation that is initially formed in **d** rearrange in order to form more stable carbocations.

c.

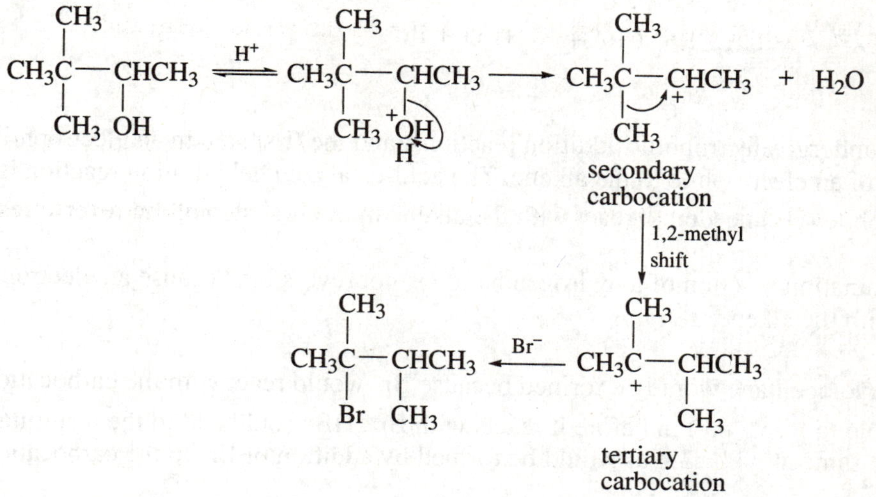

d.

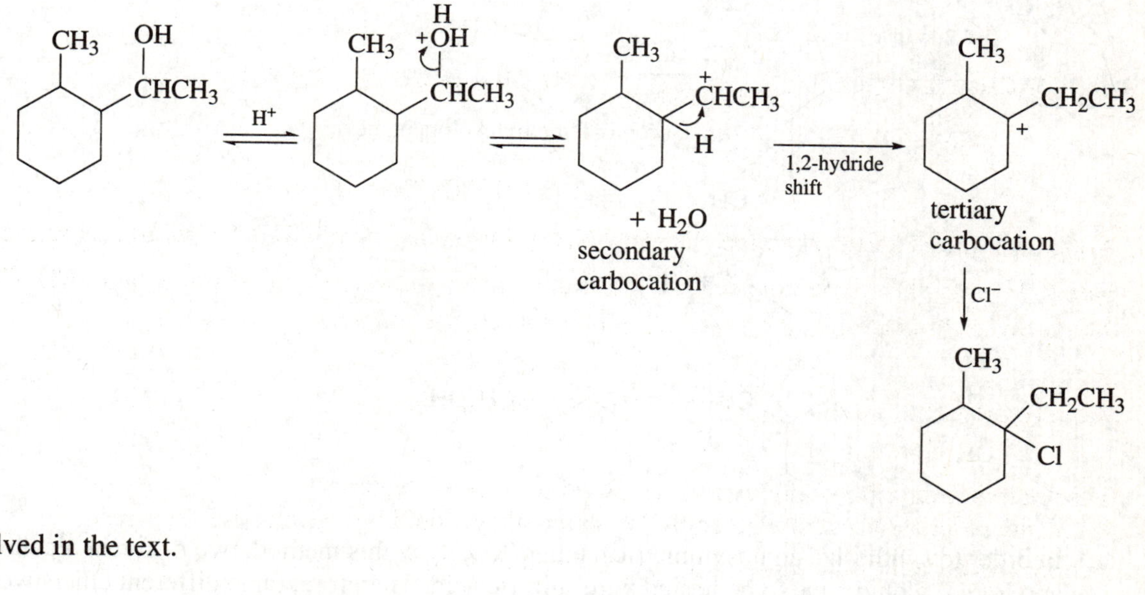

6. Solved in the text.

7. **a.** $CH_3CH_2CH_2OH \xrightarrow[\text{2. } CH_3CH_2S^-]{\text{1. TsCl/pyridine}} CH_3CH_2CH_2SCH_2CH_3$

b. $CH_3CH_2CH_2OH \xrightarrow[\text{2. } CH_3CH(CH_3)CH_2O^-]{\text{1. TsCl/pyridine}} CH_3CH_2CH_2OCH_2CH(CH_3)CH_3$

8. **D** because it forms a tertiary allylic carbocation.

9. The product of each reaction is an alkene.

$$\underset{\substack{|\\ \text{OH}}}{CH_3CH_2CHCH_3} \xrightleftharpoons[\Delta]{H_2SO_4} \underset{\substack{|\\ +\text{OH}\\ \text{H}}}{CH_3CH_2CHCH_3} \rightleftharpoons \overset{}{CH_3CH_2\overset{+}{C}HCH_3} + H_2O \rightleftharpoons CH_3CH{=}CHCH_3 + H_3O^+$$

$$\underset{\substack{|\\ \text{Br}}}{CH_3CH_2CHCH_3} \xrightarrow{HO^-} CH_3CH{=}CHCH_3 + H_2O + Br^-$$

Recall that alkenes undergo electrophilic addition reactions, and the first step in an electrophilic addition reaction is addition of an electrophile to the alkene. The acid-catalyzed dehydration reaction is reversible because the electrophile (H^+) needed to react with the alkene in the first step of the reverse reaction is available.
Base-promoted elimination reaction of a hydrogen halide is not reversible because an electrophile is not available to react with the alkene.

10. 1-Bromo-1-methylcyclopentane would be formed because Br^- would react with the carbocation. If the carbocation were able to lose a proton before it reacts with Br^-, HBr could add to the resulting alkene, thereby forming the same alkyl halide as would be formed by addition of Br^- to the carbocation.

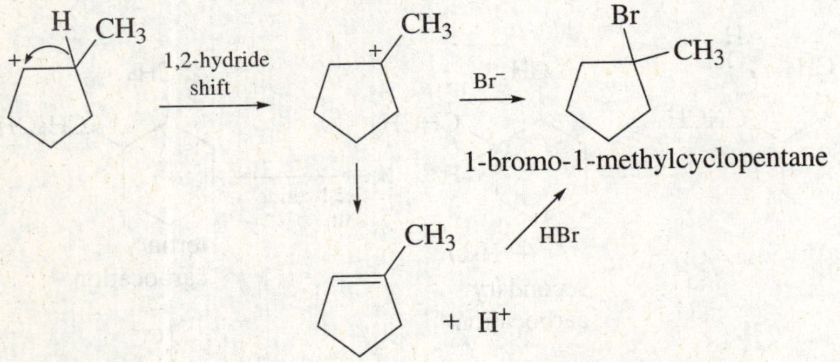

11.

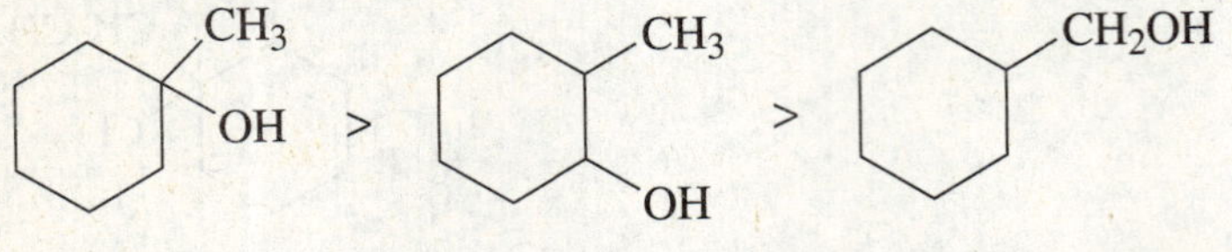

12. **a.** In order to synthesize an unsymmetrical ether (ROR') by this method, two different alcohols (ROH and R'OH) would have to be heated with sulfuric acid. Therefore, three different ethers would be obtained as products. Consequently, the desired ether would account for considerably less than half of the total amount of ether that is synthesized.

$$ROH + R'OH \xrightarrow[\Delta]{H_2SO_4} ROH + ROH' + R'OR'$$

b. It could be synthesized by a Williamson ether synthesis. (See Section 9.9 on page 418 of the text.)

$$CH_3CH_2CH_2OH \xrightarrow{Na} CH_3CH_2CH_2O^- \xrightarrow{CH_3CH_2Br} CH_3CH_2CH_2OCH_2CH_3 + Br^-$$

13. a. 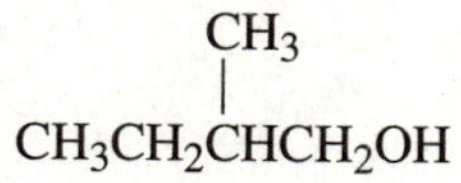 b. 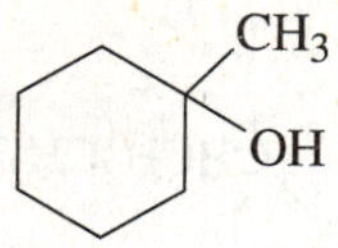 c. 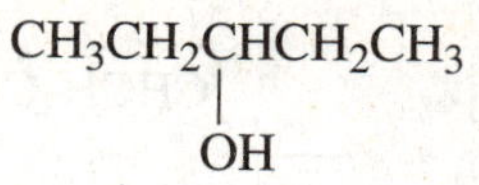 d. 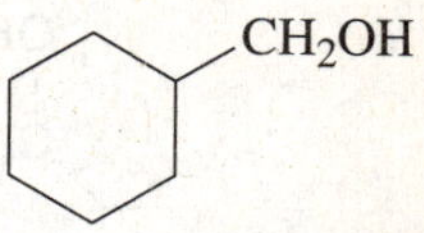

14. a. 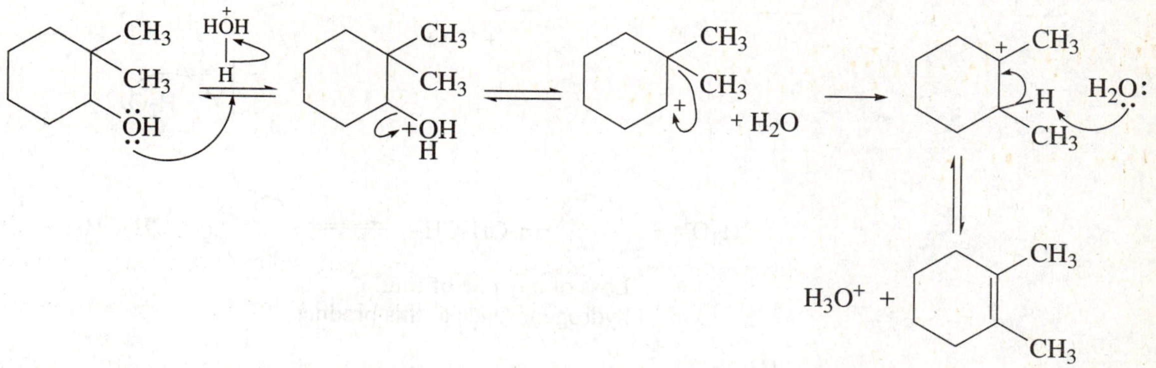

b. Because of the difficulty of forming a primary carbocation, dehydration is an E2 reaction. The alkene that results is protonated and the proton that is removed is the one that results in formation of the most stable alkene.

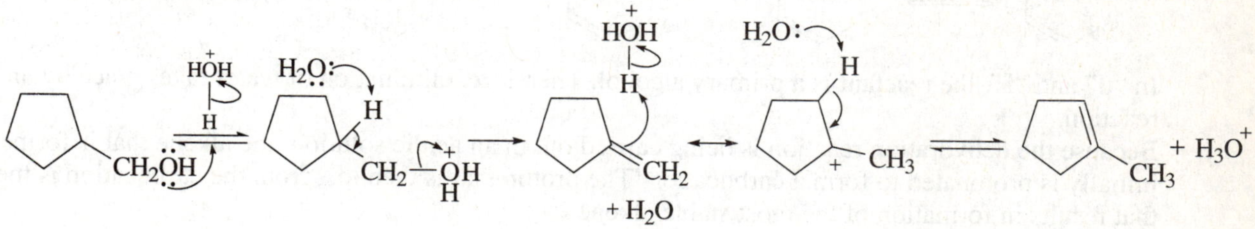

c. 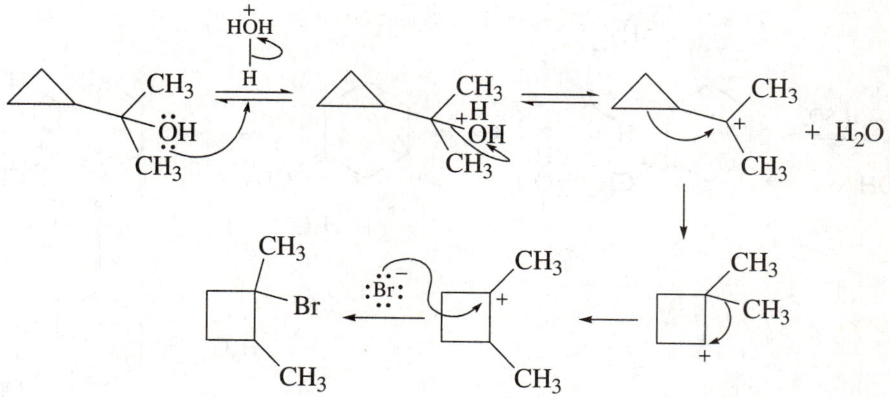

15. a. 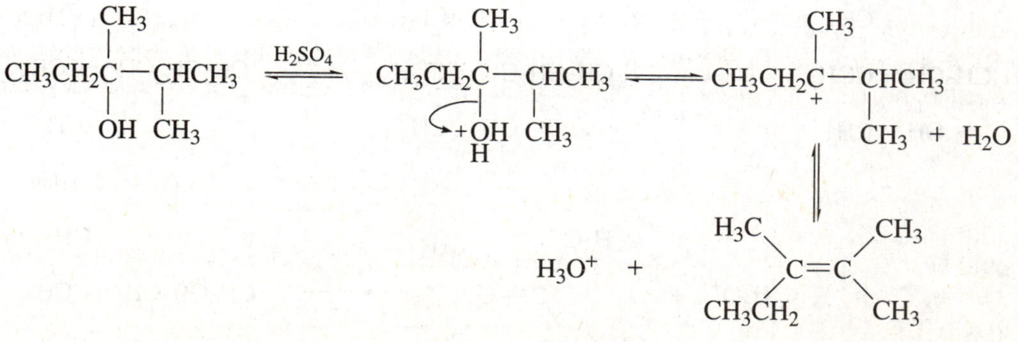

b.

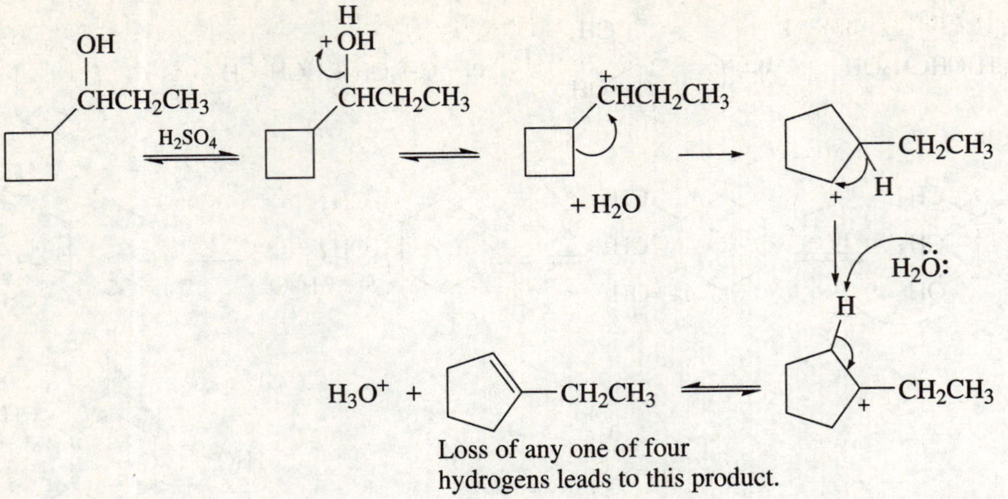

c.

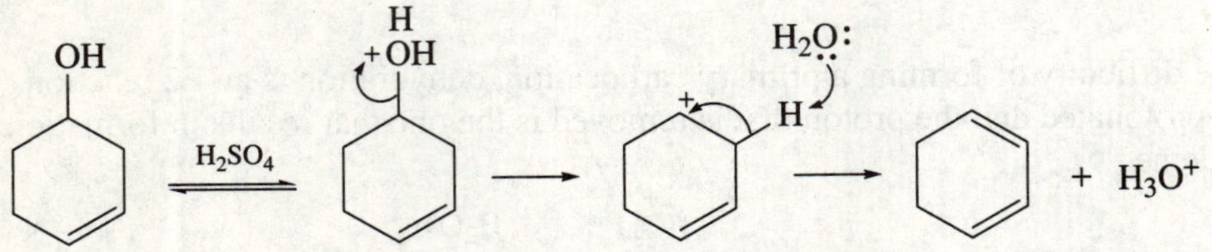

In "**d**" and "**f**", the reactant is a primary alcohol. Therefore, elimination of water takes place by an E2 reaction.
Because the dehydration reaction is being carried out in an acidic solution, the alkene that is formed initially is protonated to form a carbocation. The proton that is then lost from the carbocation is the one that results in formation of the most stable alkene.

d.

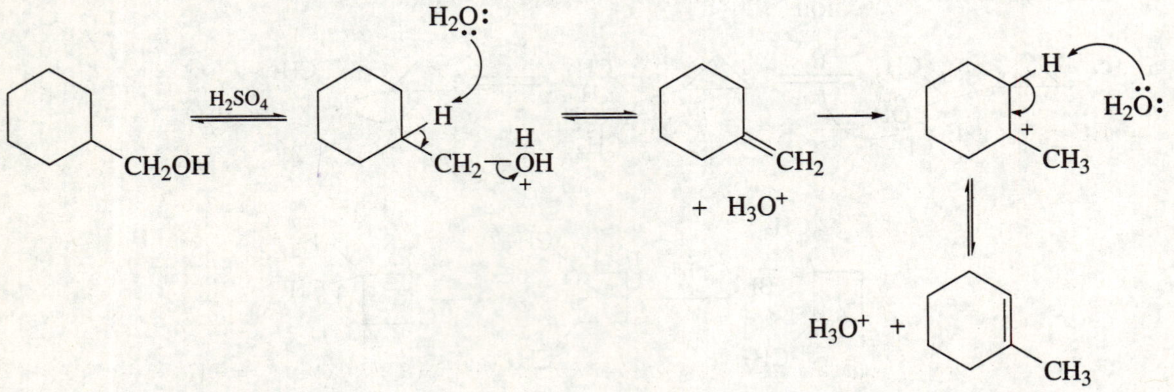

e.

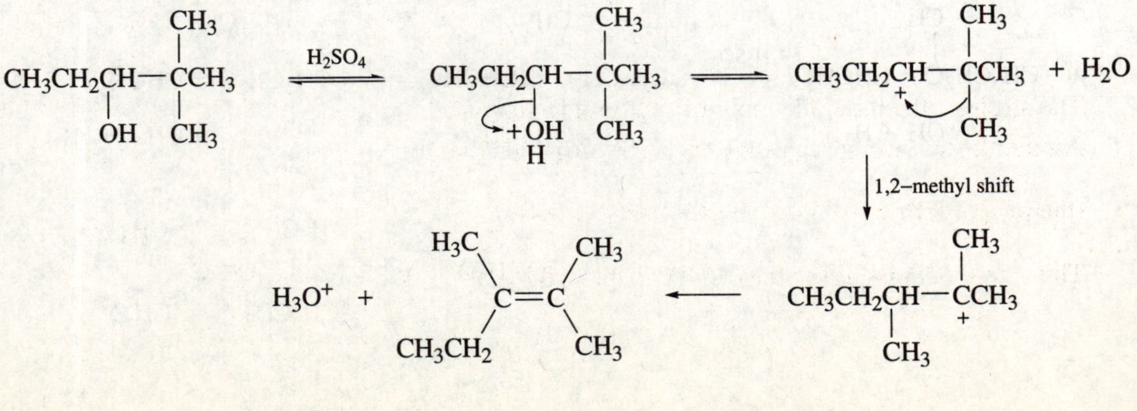

f. 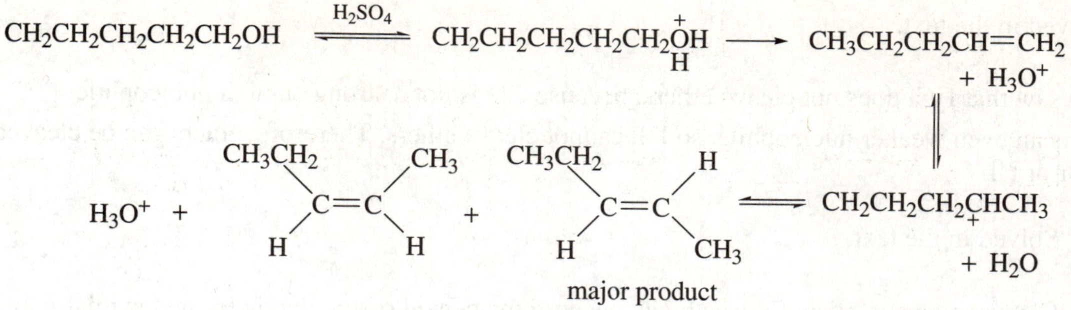

16. 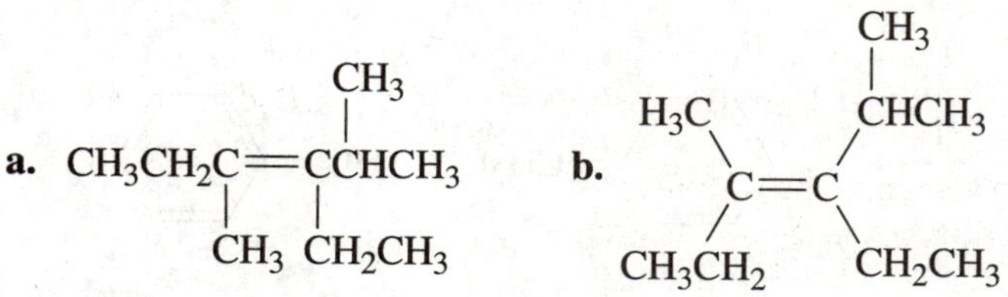

17. **a.** 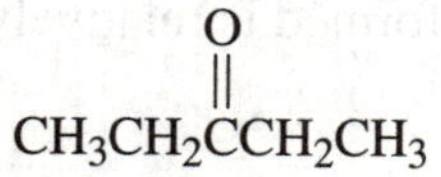

c. A tertiary alcohol is not oxidized to a carbonyl compound.

e. 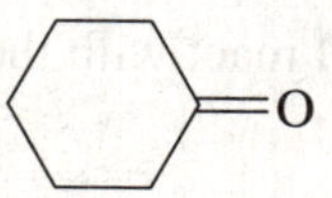

b. 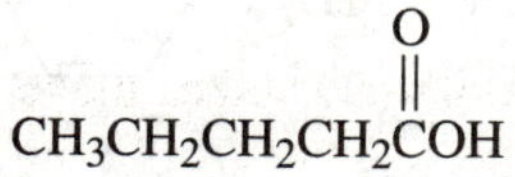

d. 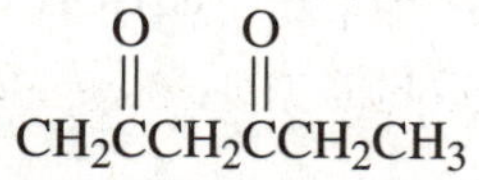

f. 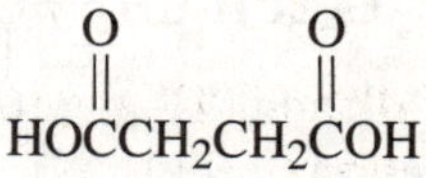

18. Chromic acid needs to be protonated to convert a poor leaving group (^{-}OH) into a good leaving group (H_2O). The alcohol displaces water to form a protonated chromate ester, which loses a proton. The aldehyde is then formed in an elimination reaction.

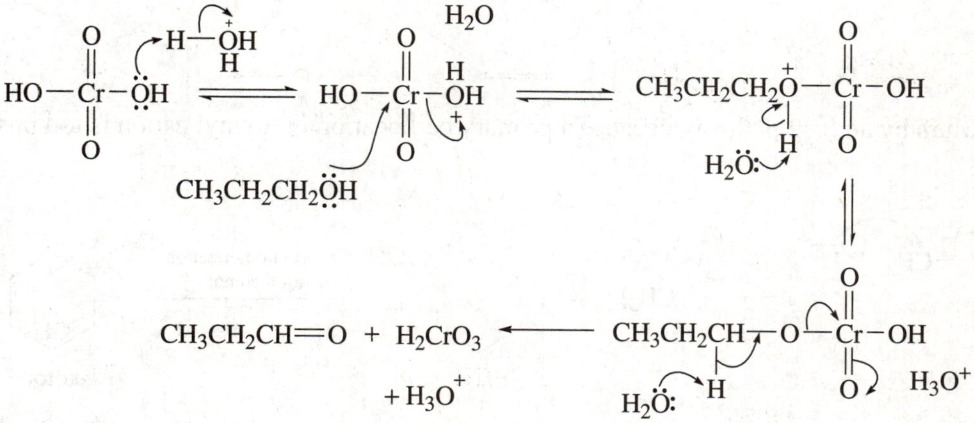

19. The stronger the base, the more difficult it is to displace.
Ammonia is a stronger base by about 18 pK_a units than Br^-;

(the pK_a of HBr = – 9; the pK_a $CH_3\overset{+}{N}H_3$ = 9.4).

Thus, ammonia is far too basic to be displaced by Br^-.

$$Br^- + CH_3\overset{+}{N}H_3 \xrightarrow{\Delta} \text{no reaction}$$

20. Solved in the text.

21. We saw that HCl does not cleave ethers, because Cl^- is not a strong enough nucleophile.
F^- is an even weaker nucleophile, so HF cannot cleave ethers. Therefore, ethers can be cleaved only with HBr or HI.

22. **a.** Solved in the text.

b. Cleavage occurs by an SN1 pathway because the benzyl cation that is formed is relatively stable; I^- will react with the benzyl cation.

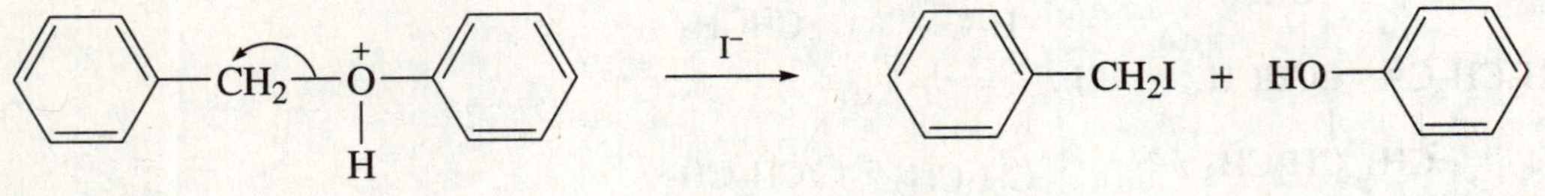

c. Cleavage occurs by an SN1 pathway because the benzyl cation that is formed is relatively stable; I^- will react with the benzyl cation.

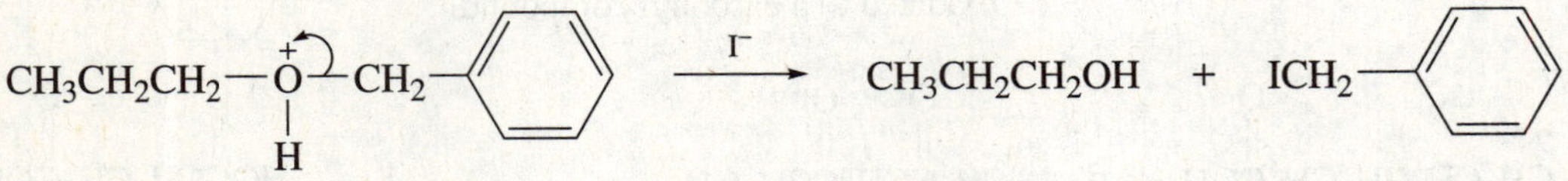

d. Cleavage occurs by an SN2 pathway because a primary carbocation is too unstable to be formed.

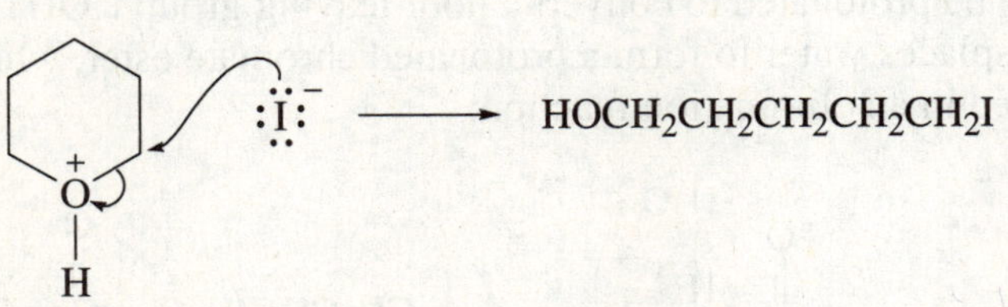

e. Cleavage occurs by an SN2 pathway because a primary carbocation or a vinyl cation is too unstable to be formed.

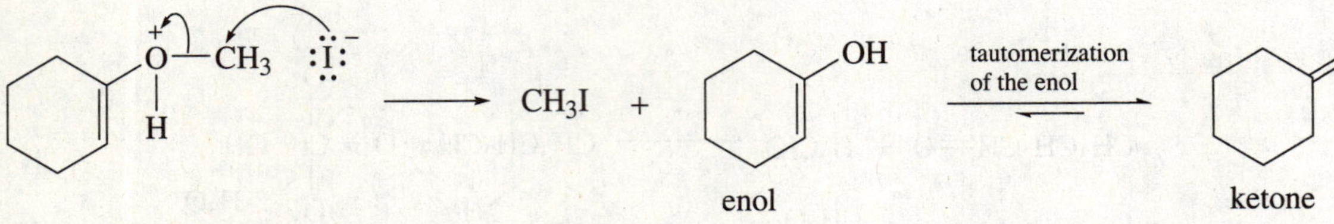

f. Cleavage will occur by an SN1 pathway because the tertiary carbocation that is formed is relatively stable; I^- will react with the tertiary carbocation.

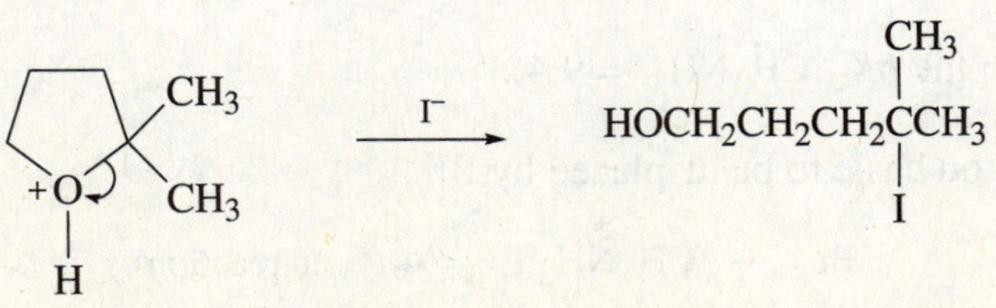

23.

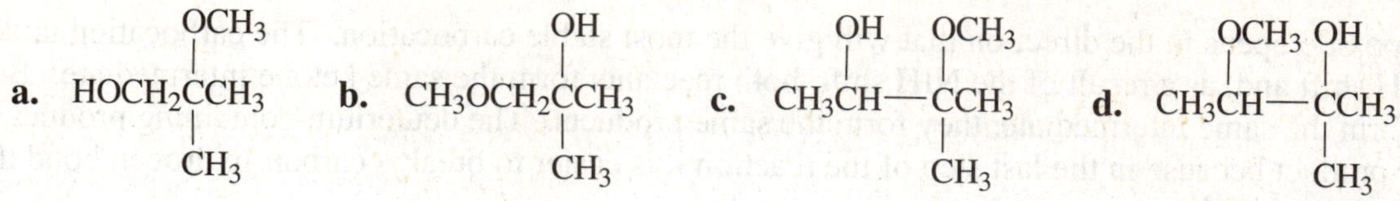

24. The reactivity of tetrahydrofuran is more similar to a noncyclic ether because the five-membered ring does not have the strain that makes the epoxide reactive.

25. The carbocation leading to 1-naphthol can be stabilized without destroying the aromaticity of the intact benzene ring. The carbocation leading to 2-naphthol can be stabilized by electron delocalization only by destroying the aromaticity of the intact benzene ring. Therefore, the carbocation leading to 1-naphthol is more stable.

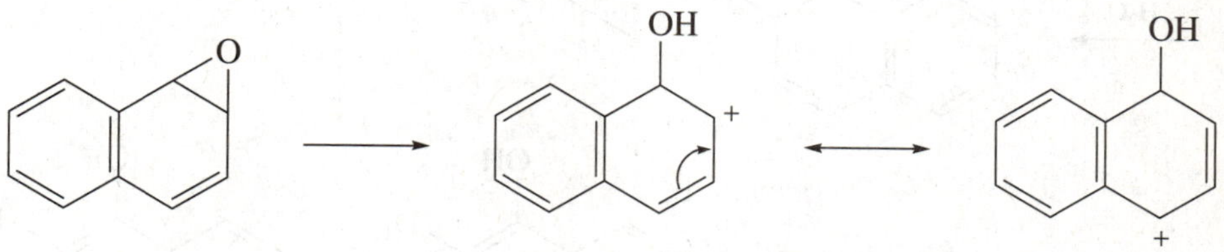

carbocation that leads to 1-naphthol

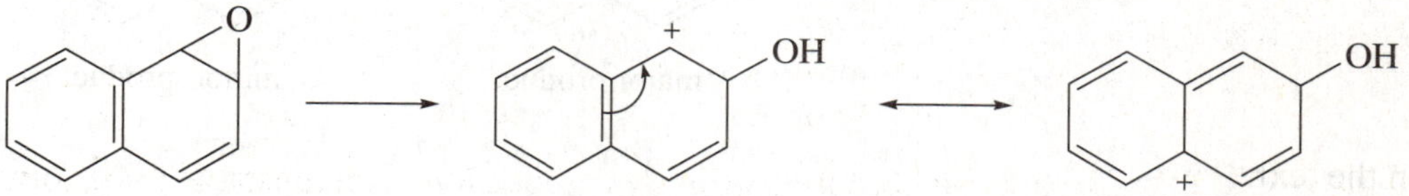

carbocation that leads to 2-naphthol

26. <u>**without an NIH shift**</u>

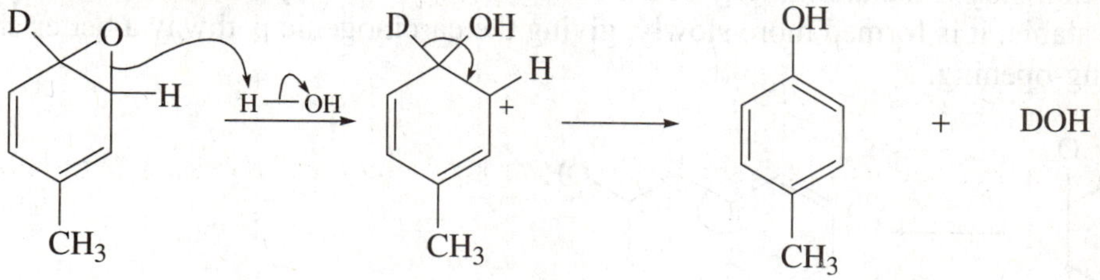

<u>**with an NIH shift**</u>

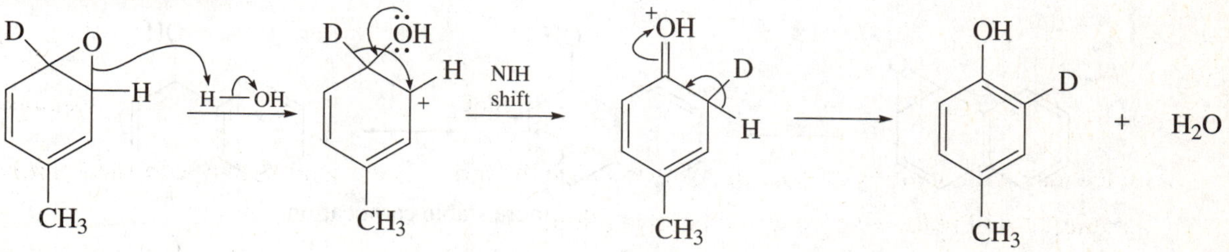

27. The epoxide opens in the direction that will give the most stable carbocation. The carbocation undergoes an NIH shift and, as a result of the NIH shift, both reactants form the same ketone intermediate. Because they form the same intermediate, they form the same products. The deuterium-containing product is the major product because in the last step of the reaction it is easier to break a carbon-hydrogen bond than a carbon-deuterium bond.

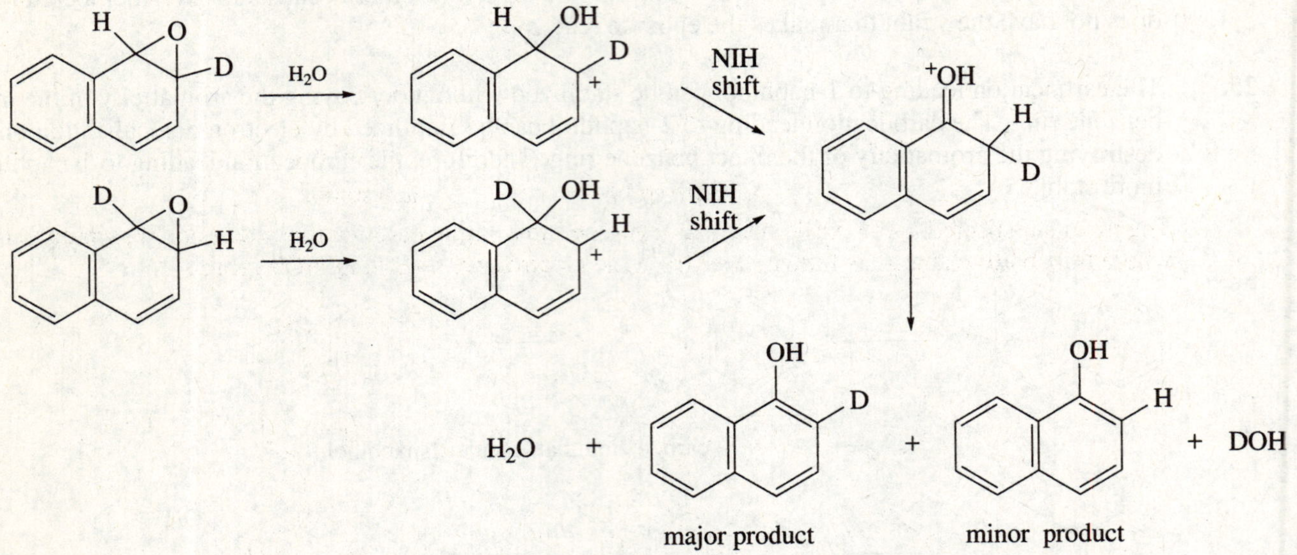

28. **a.** Solved in the text.

b. The compound without the double bond in the second ring is more apt to be carcinogenic. It opens to form a less stable carbocation than the other compound because it can be stabilized by electron delocalization only if the aromaticity of the benzene ring is destroyed. Because the carbocation is less stable, it is formed more slowly, giving the carcinogenic pathway a better chance to compete with ring-opening.

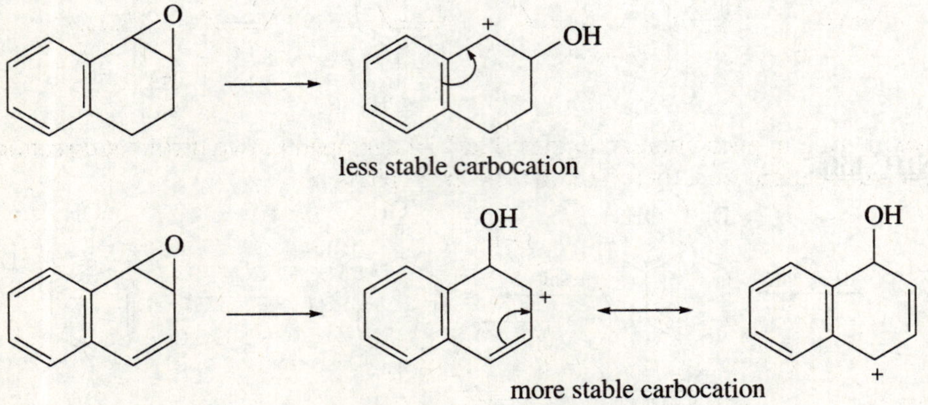

29. Each arene oxide will open in the direction that forms the most stable carbocation. Thus, the methoxy-substituted arene oxide opens so the positive charge can be stabilized by electron donation from the methoxy group.

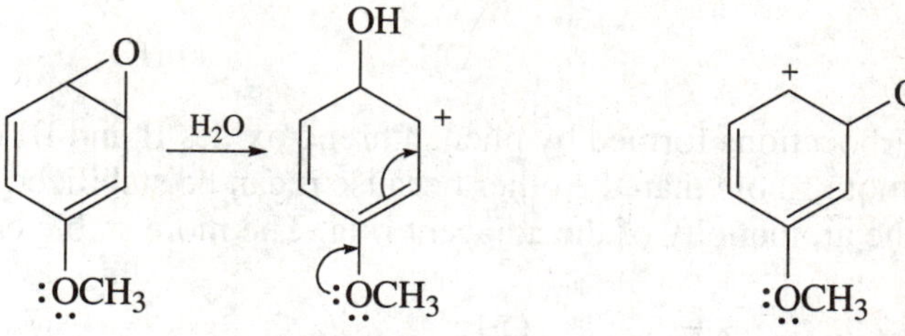

more stable **less stable**

The nitro-substituted arene oxide opens to form the most stable carbocation intermediate, which is the one where the positive charge is father away from the electron-withdrawing NO_2 group.

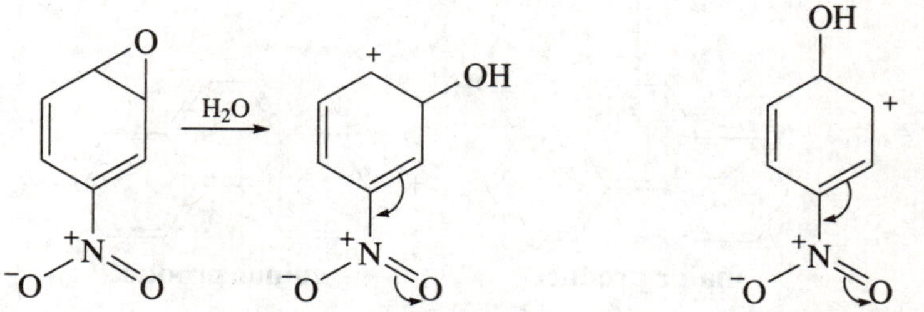

more stable **less stable**

30. **a.** Note that a bond joining two rings cannot be epoxidized.

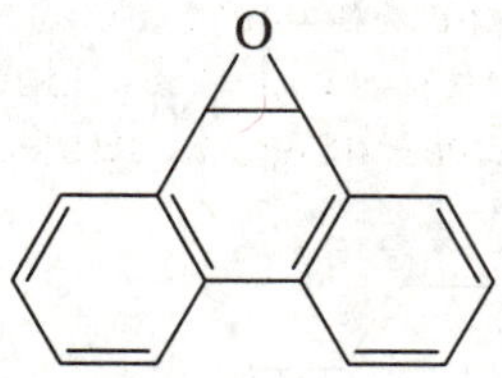

I

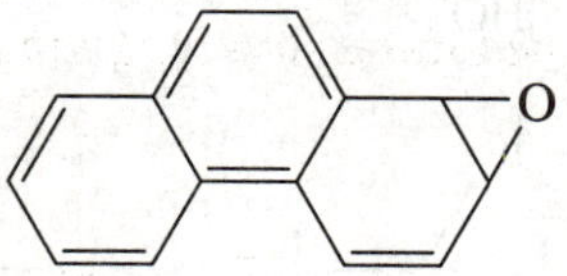

II

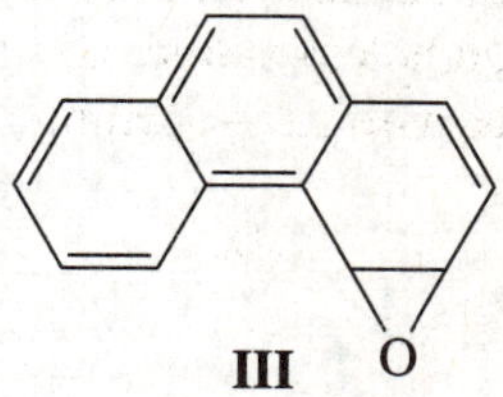

III

b. The epoxide ring in phenanthrene oxides II and III can open in two different directions to give two different carbocations and, therefore, two different phenols.

I →

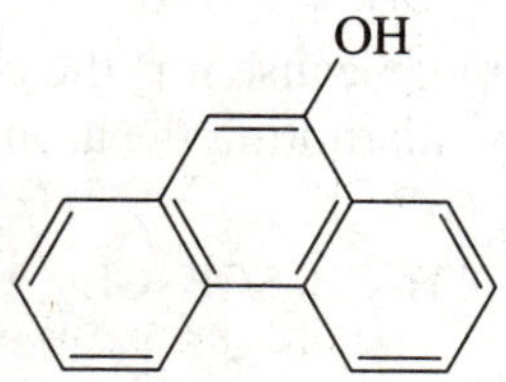

II →

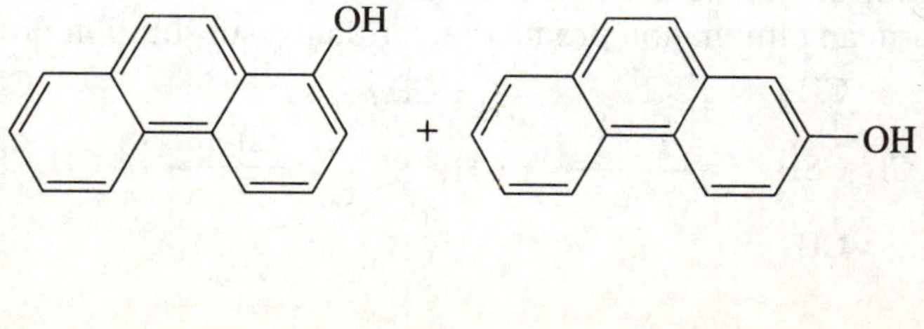

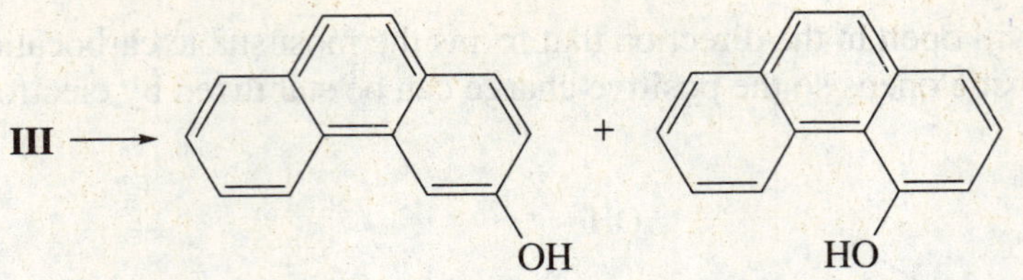

c. The two different carbocations formed by phenanthrene oxides II and III differ in stability. One carbocation is more stable than the other because it can be stabilized by electron delocalization without disrupting the aromaticity of the adjacent ring. The more stable carbocation leads to the major product.

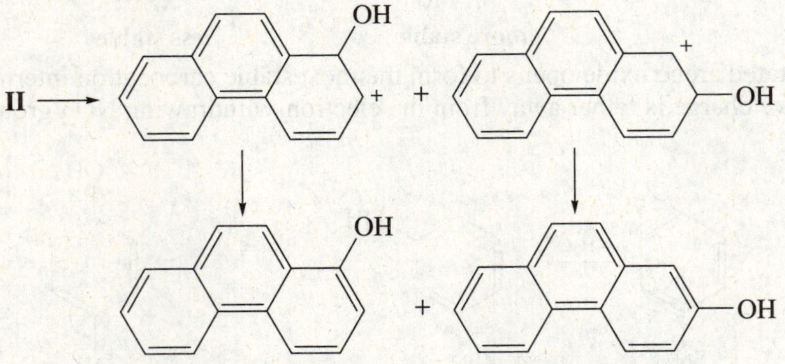

major product **minor product**

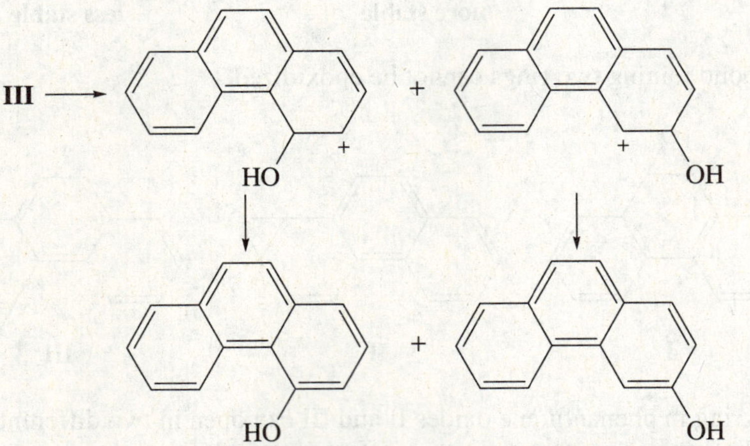

major product **minor product**

d. Phenanthrene oxide I is the most carcinogenic because it is the only one that opens to form a carbocation that cannot be stabilized without disrupting the aromaticity of the other ring(s).

31. a. $CH_3CH_2SH \xrightarrow{HO^-} CH_3CH_2S^- \xrightarrow{CH_3CH_2Br} CH_3CH_2SCH_2CH_3 + Br^-$

b. The reaction cannot be done with methanethiol and *tert*-butyl bromide, because a tertiary alkyl halide would form an elimination product rather than a substitution product.

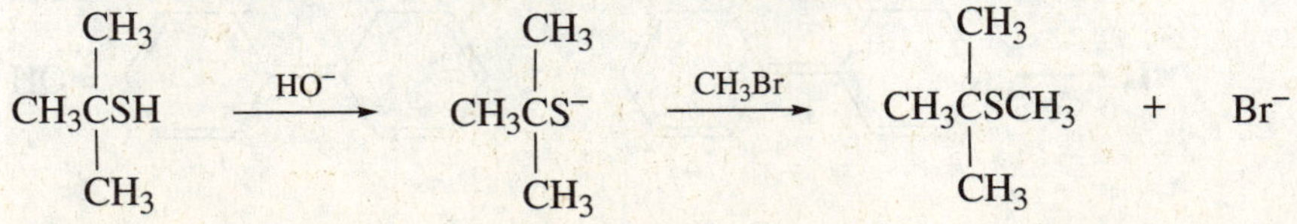

c. The highest yield is obtained by having the less substituted of the two R groups of the thioether be the alkyl halide and the more substituted be the thiol.

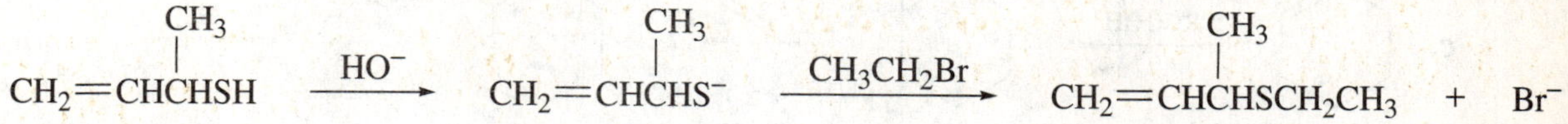

d. The synthesis must be done this way because the sp^2 carbon of the benzene ring cannot undergo back-side attack.

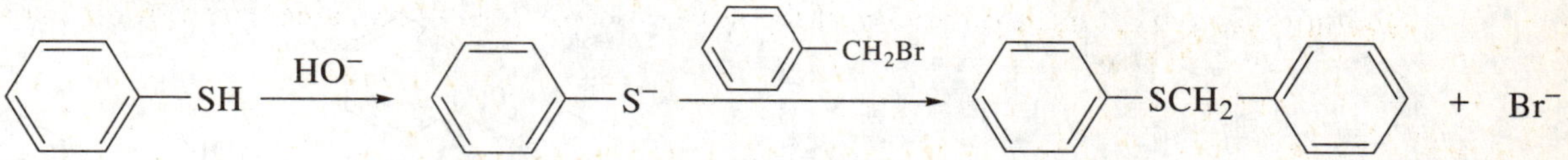

32. The first compound is too insoluble.
The second compound is too reactive.
The third compound is more soluble than the first because of the oxygen-containing substituent.
The third compound is less reactive than the second because the lone pair can be delocalized into the benzene ring, so the lone pair is less apt to displace a chloride ion.

33. **a.** $CH_3CH_2CH_2CH_2CH_2OH$ **b.** 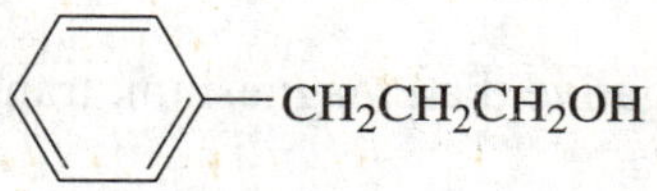 **c.** 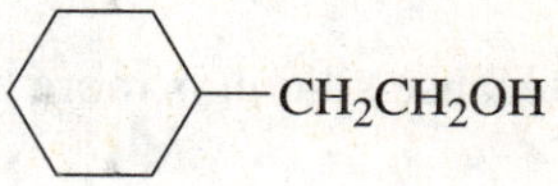

34. **a.**

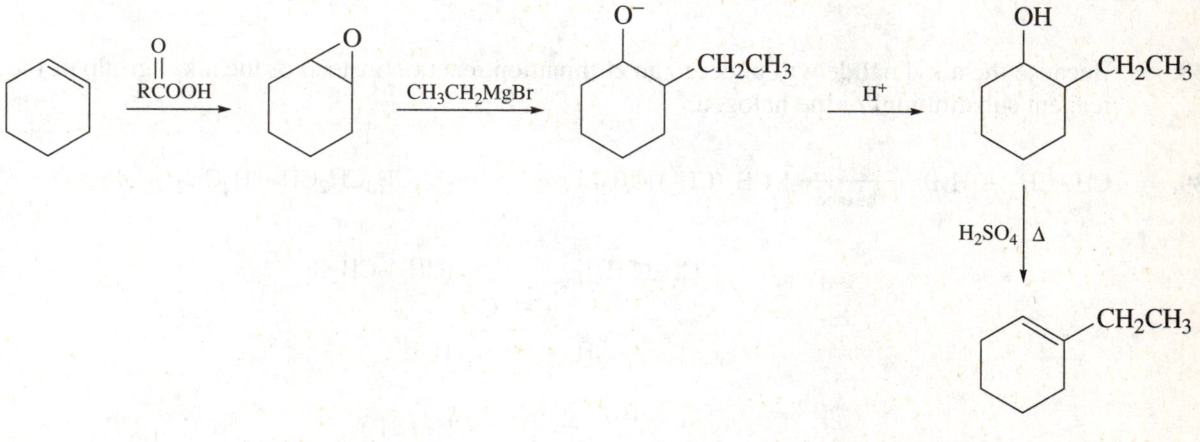

b.

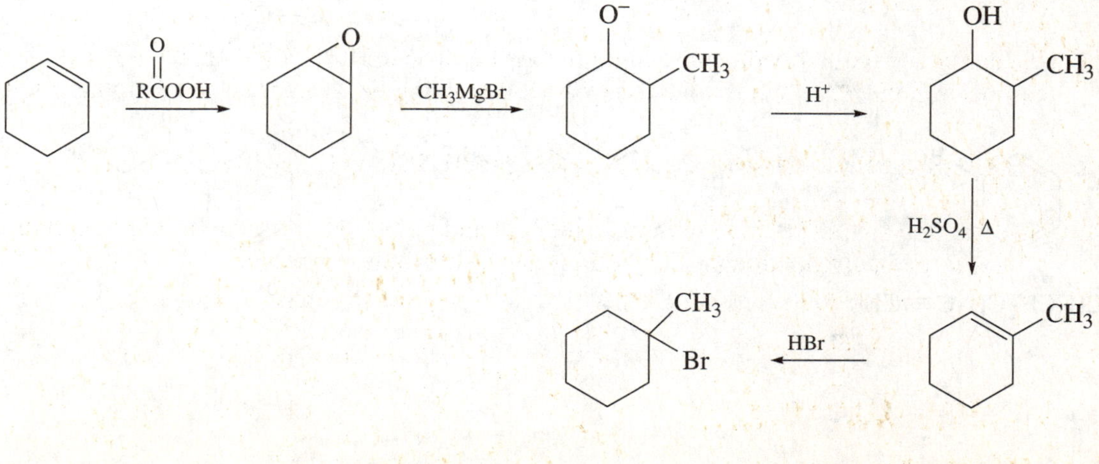

c.

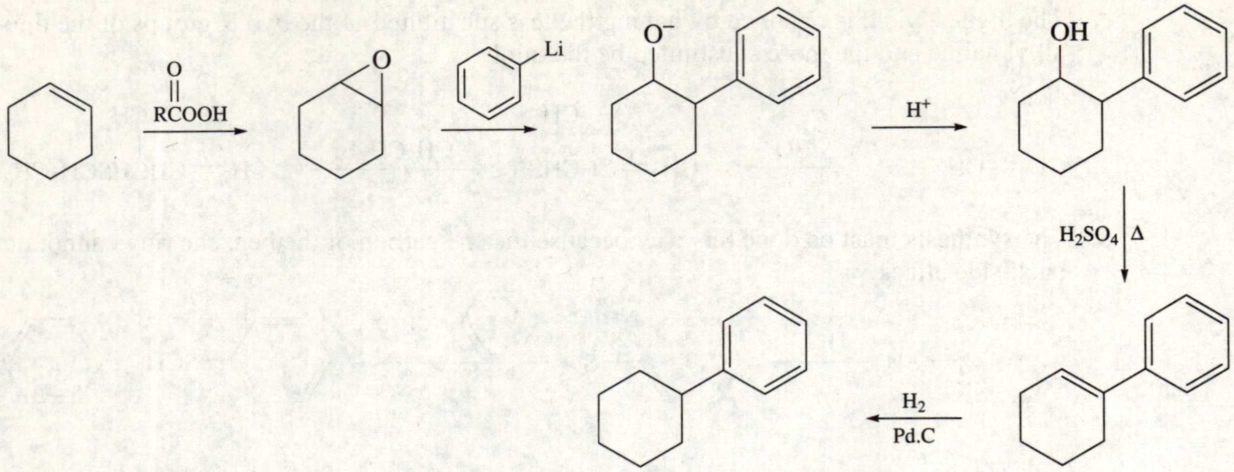

35. Solved in the text.

36. All the reactions occur because, in each case, the reactant acid is a stronger acid than the product acid (methane, $pK_a \sim 60$).

37. Because silicon is more electronegative than magnesium, transmetallation will occur.

$$4\ CH_3MgCl + SiCl_4 \longrightarrow (CH_3)_4Si + 4\ MgCl_2$$

38. Because the alkyl halide will undergo an elimination reaction instead of the alkyl group of the Gilman reagent substituting for the halogen.

39. $CH_3(CH_2)_3CH_2Br \xrightarrow[\text{hexane}]{\text{Li}} CH_3(CH_2)_3CH_2Li \xrightarrow{\text{CuI}} (CH_3CH_2CH_2CH_2CH_2)_2CuLi$

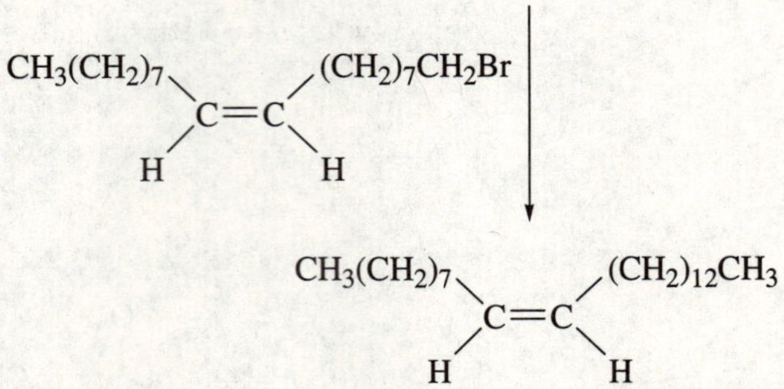

40. **a.**

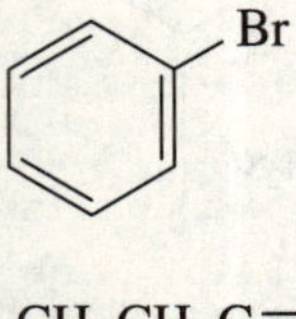

b.

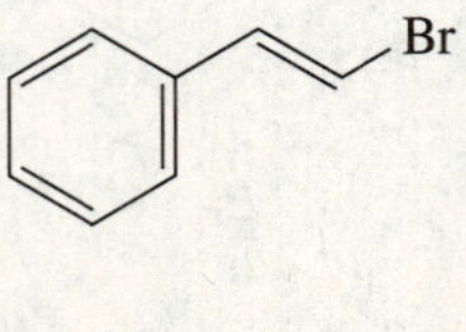

c.

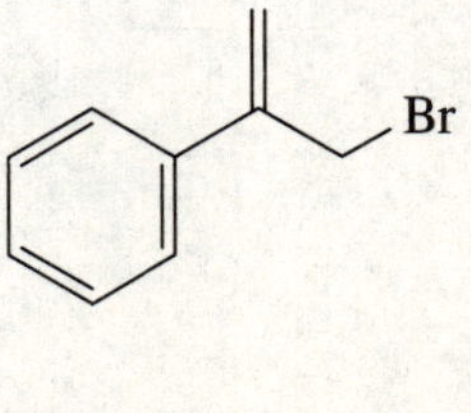

41. $CH_3CH_2CH_2C{\equiv}CH$

42. 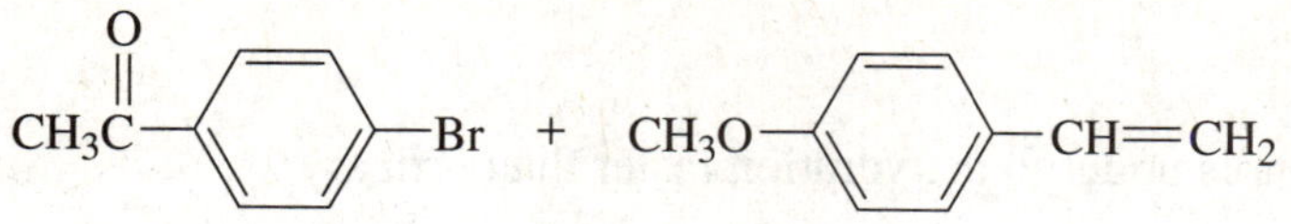

or

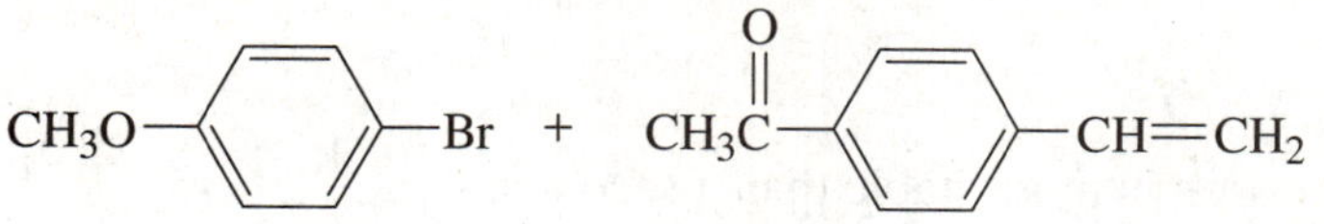

43. **a.** $CH_3CH_2CH_2O\overset{O}{\overset{\|}{C}}CH_3$

b. $CH_3CH_2CH_2CH_2Br$

c. 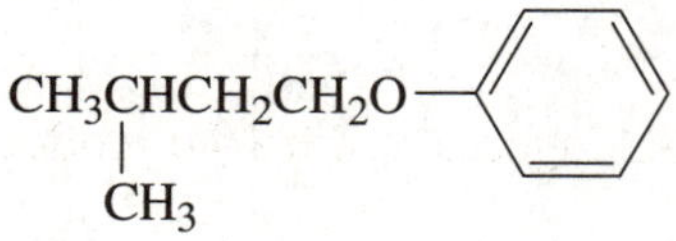

d. 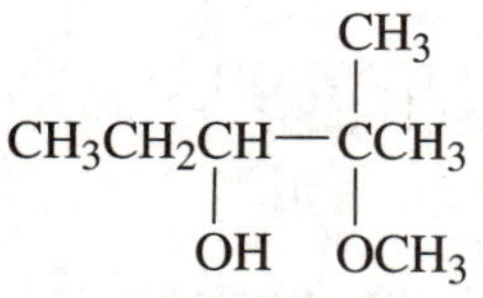

e. 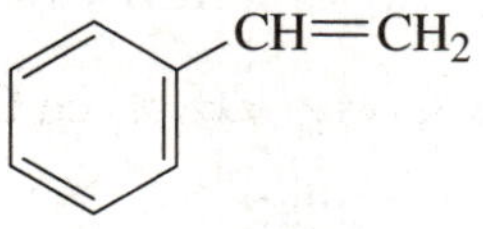

f. 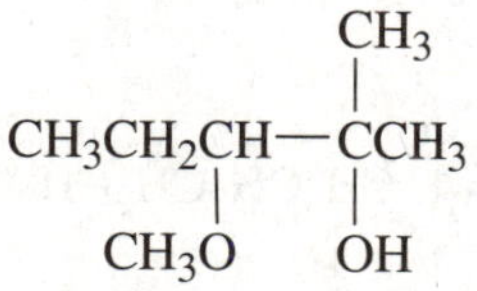

g. $CH_3CH{=}CHCH_3$

h. $CH_3\underset{CH_3}{\underset{|}{C}}HCH_2CH_2Cl$

i. 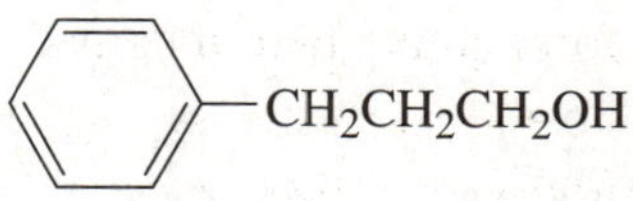

j. 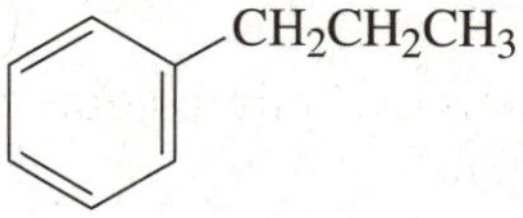

44. **a.** 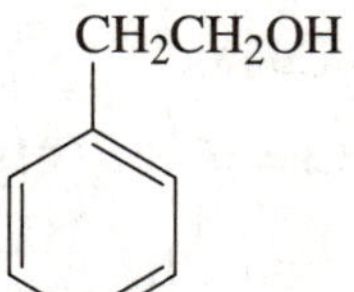

The other alcohol cannot undergo dehydration because its β-carbon is not bonded to a hydrogen.

b. 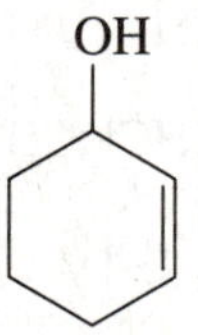

A secondary allylic cation is more stable than a secondary carbocation.

c. 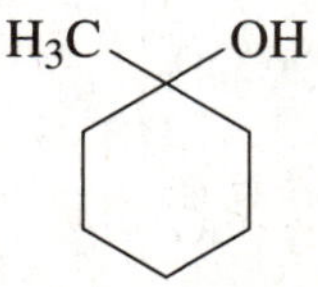

A tertiary carbocation is more stable than a secondary carbocation.

d. 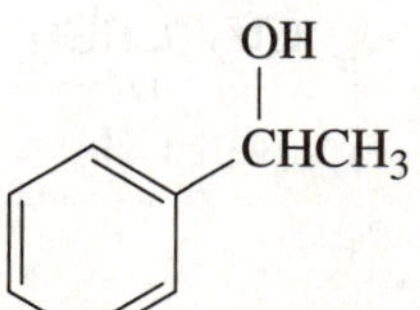

The rate-limiting step in both dehydrations is carbocation formation: a secondary benzylic cation is more stable than a secondary carbocation and, therefore, easier to form.

e. 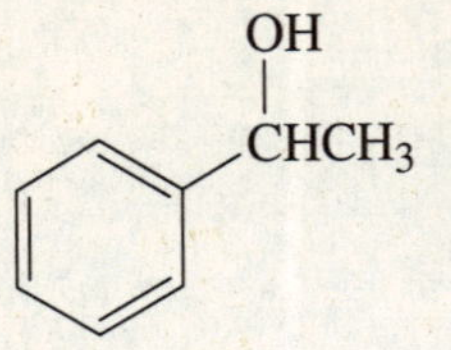

Secondary alcohols undergo dehydration faster than primary alcohols.

f. 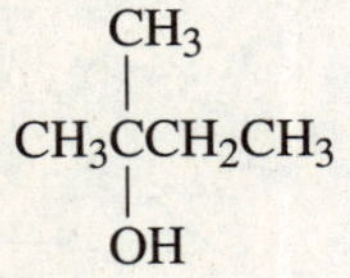

A tertiary carbocation is more stable than a secondary carbocation.

45. **a.** 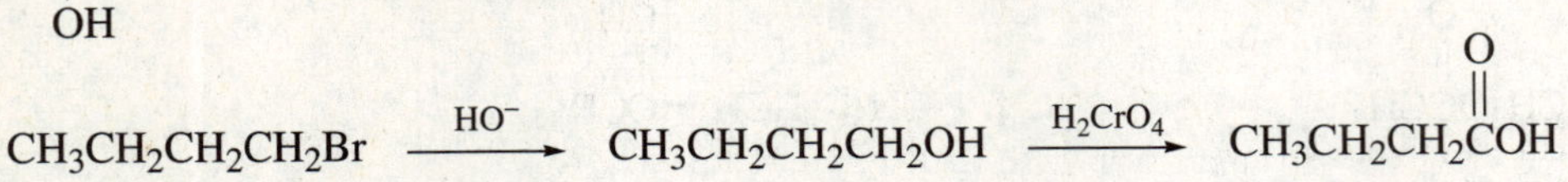

b. 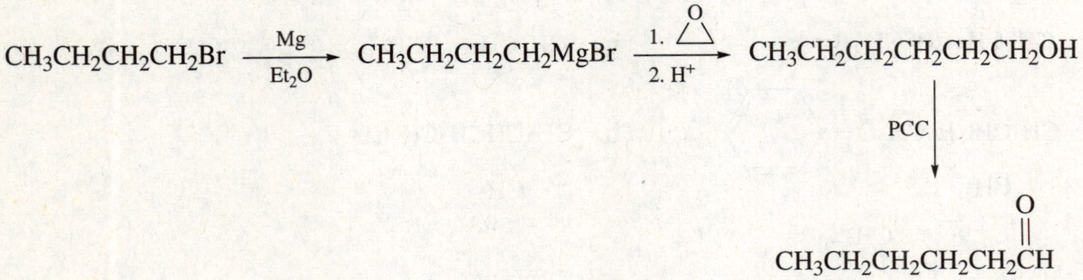

c. 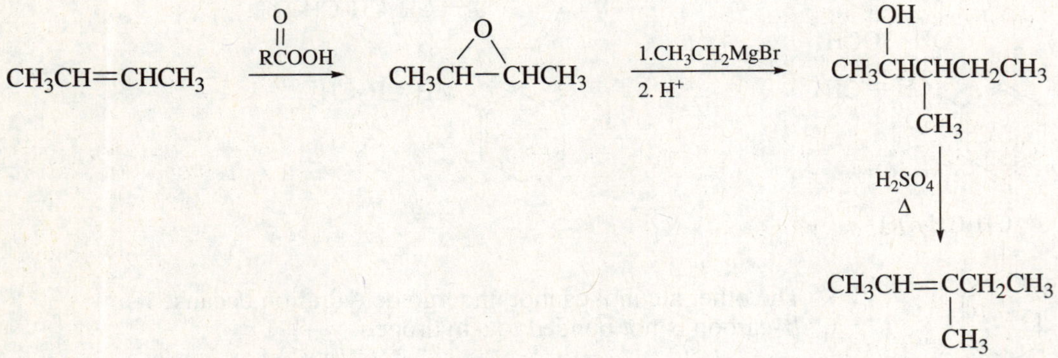

d. 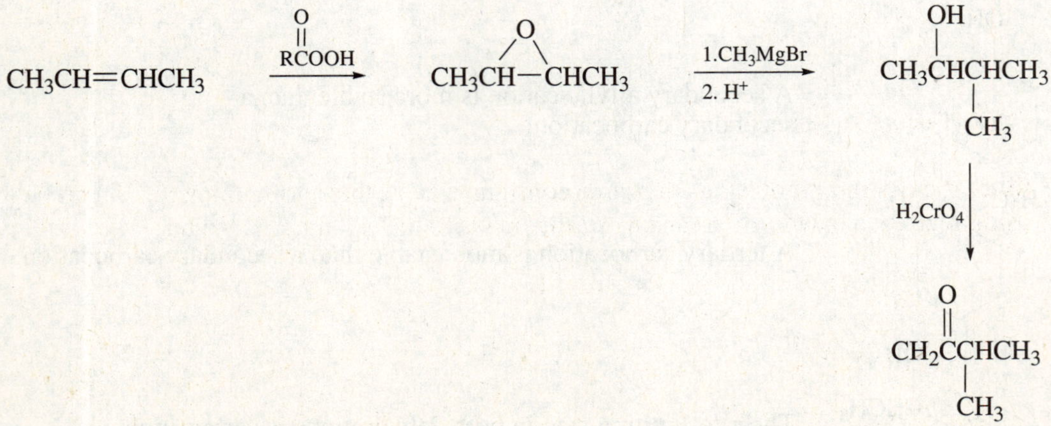

46. **c** is the only one that can be used to form a Grignard reagent.
The Grignard reagent formed from **a** will be destroyed by reacting with the proton of the alcohol group.
The Grignard reagent formed from **b** will be destroyed by reacting with the proton of the carboxylic acid group.
The Grignard reagent formed from **d** will be destroyed by reacting with a proton of the amino group.

47. Only one S_N2 reaction takes place in reactions "**a**" and "**b**", so the product has the inverted configuration compared to the configuration of the reactant.

a.

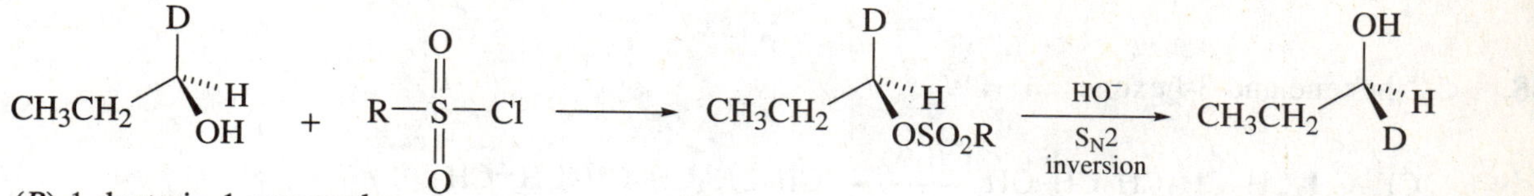

(*R*)-1-deuterio-1-propanol (*S*)-1-deuterio-1-propanol

or

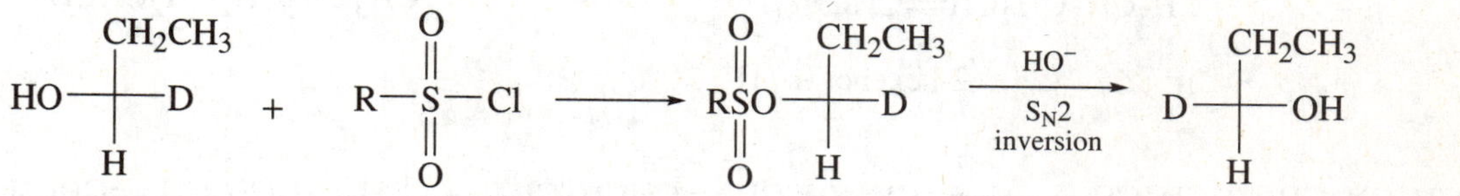

(*R*)-1-deuterio-1-propanol (*S*)-1-deuterio-1-propanol

b.

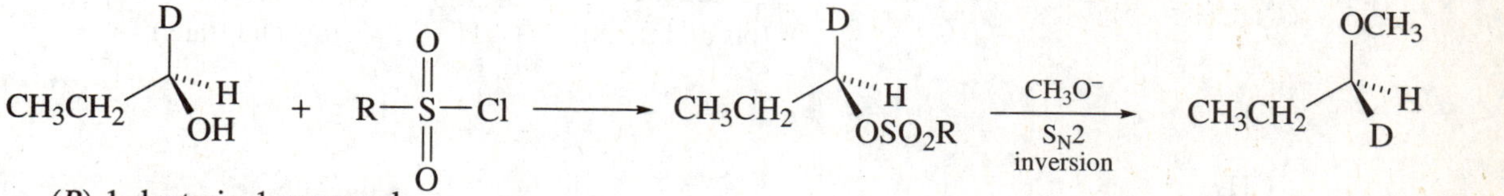

(*R*)-1-deuterio-1-propanol (*S*)-1-deuterio-1-methoxypropane

or

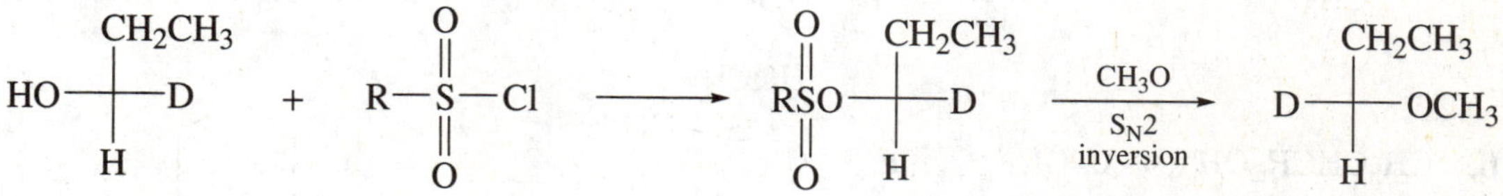

(*R*)-1-deuterio-1-propanol (*S*)-1-deuterio-1-methoxypropane

c. Since the desired product has the same configuration as the starting material, it can be synthesized using reactions that don't break any of the bonds to the asymmetric center.

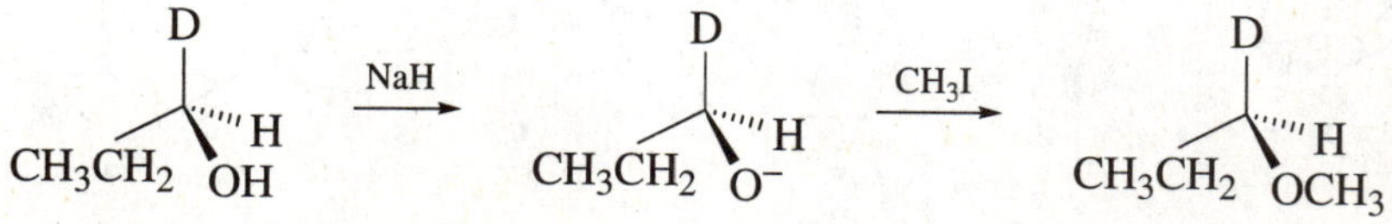

(*R*)-1-deuterio-1-propanol (*R*)-1-deuterio-1-methoxypropane

It could also be synthesized using two consecutive S_N2 reactions reactions. Since each one involves inversion of configuration, the final product will have the same configuration as the starting material.

For example, treating the starting material with PBr_3 forms (*S*)-1-bromo-1-deuteriopropane. Treating (*S*)-1-bromo-1-deuteriopropane with methoxide ion forms (*R*)-1-bromo-1-deutero-1-methoxypropane.

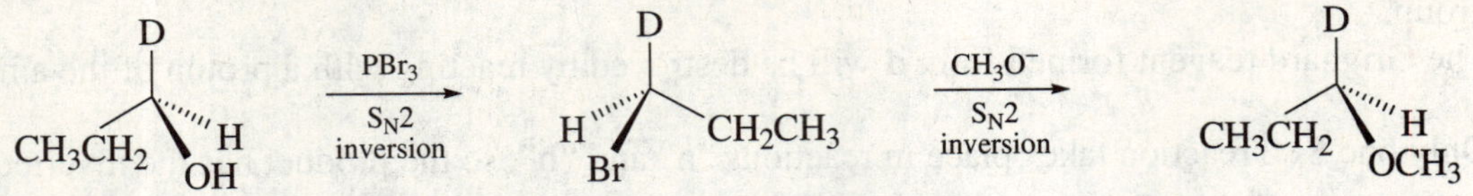

(*R*)-1-deuterio-1-propanol (*S*)-1-bromo-1-deuteriopropane (*R*)-1-deuterio-1-methoxypropane

48. 2-hexene and 3-hexene

$CH_3CH_2CH_2CH_2CH_2CH_2OH \xrightarrow[\Delta]{H_2SO_4} CH_3CH_2CH_2CH_2CH{=}CH_2$

$\xrightarrow{H^+} CH_3CH_2CH_2CH_2\overset{+}{C}HCH_3 \longrightarrow CH_3CH_2CH_2CH{=}CHCH_3$ (2-hexene) $+ H^+$

$\xrightarrow{H^+} CH_3CH_2CH_2\overset{+}{C}HCH_2CH_3 \longrightarrow CH_3CH_2CH{=}CHCH_2CH_3 + CH_3CH_2CH_2CH{=}CHCH_3$

3-hexene (cis and trans) + H^+ 2-hexene (cis and trans)

49. **a.** $CH_3CBr(CH_3)_2$ + CH_3CH_2OH

b. $CH_3CH(CH_3)CH_2OH$ + CH_3I

c.

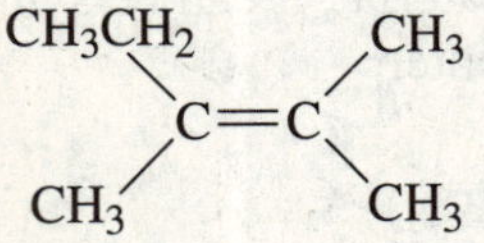

d.

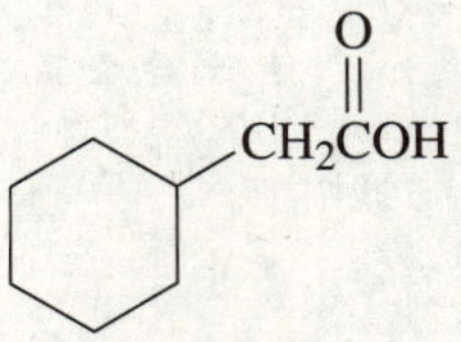

e.

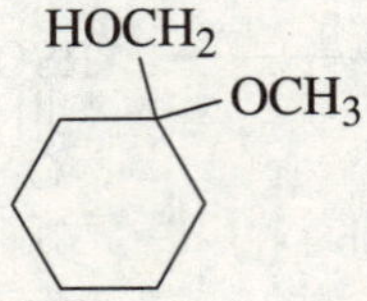

f.

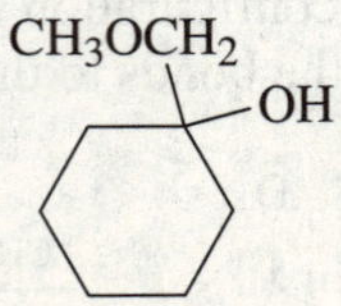

g.

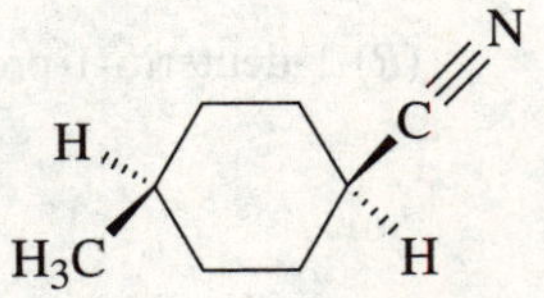

h.

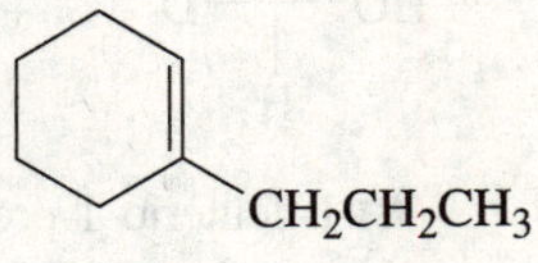

i.

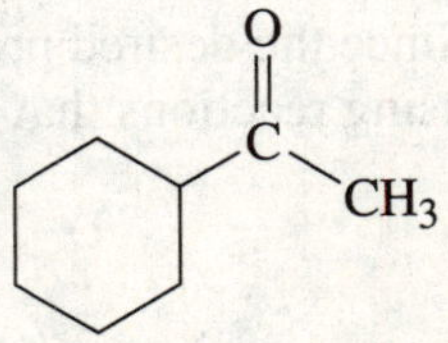

50. 2,3-Dimethyl-2-butanol will dehydrate faster because it is a tertiary alcohol, whereas 3,3-dimethyl-2-butanol is a secondary alcohol.

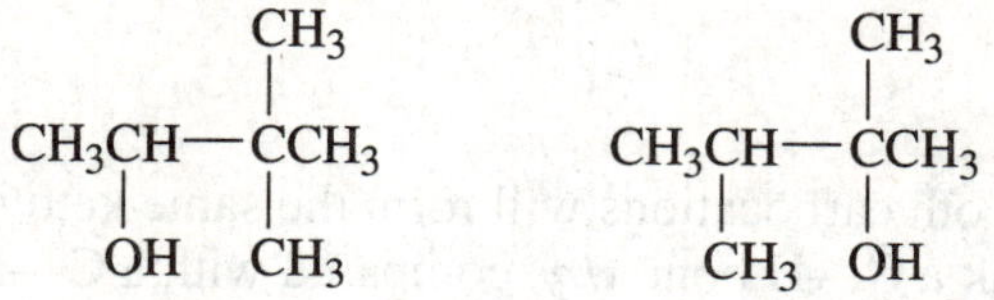

3,3-dimethyl-2-butanol **a secondary alcohol**

2,3-dimethyl-2-butanol **a tertiary alcohol**

51. a.

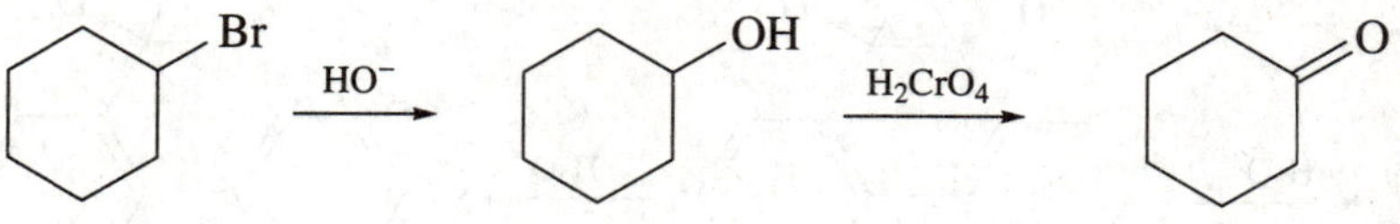

b.

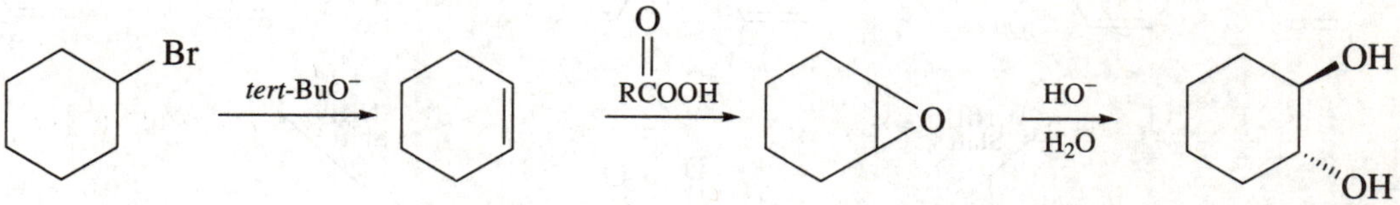

c.

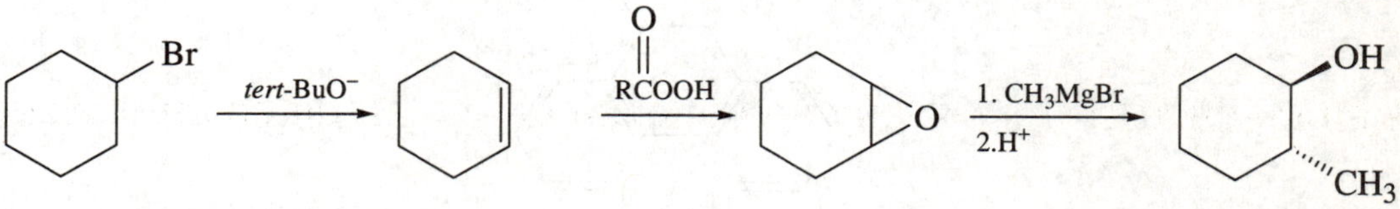

d.

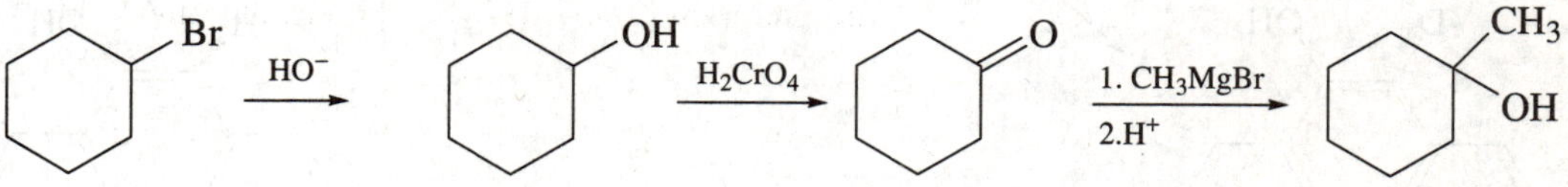

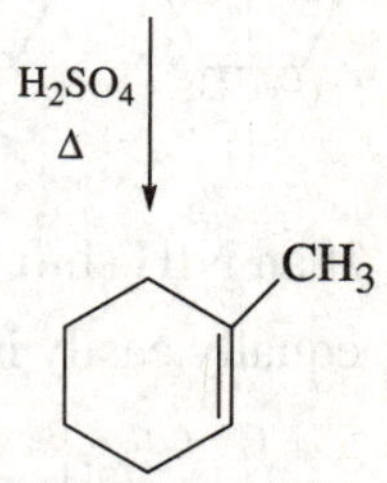

or

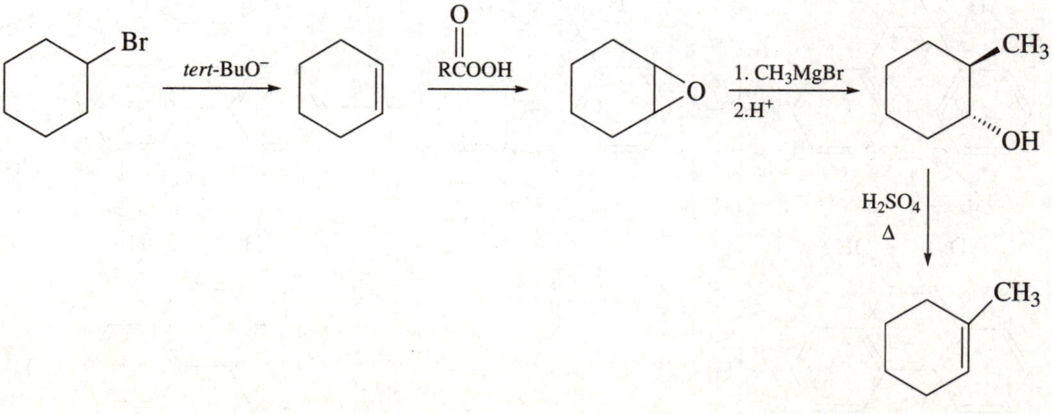

52. 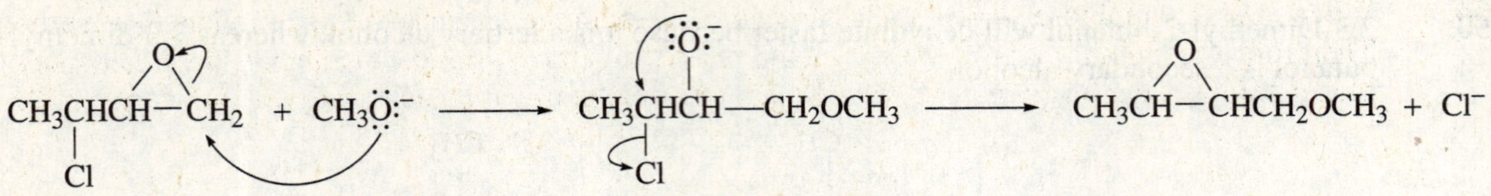

53. **a.** If an NIH shift occurs, both carbocations will form the same keto intermediate. Because it is about four times easier to break a C—H bond (k_3) compared with a C—D bond (k_4), about 80% of the deuterium will be retained.

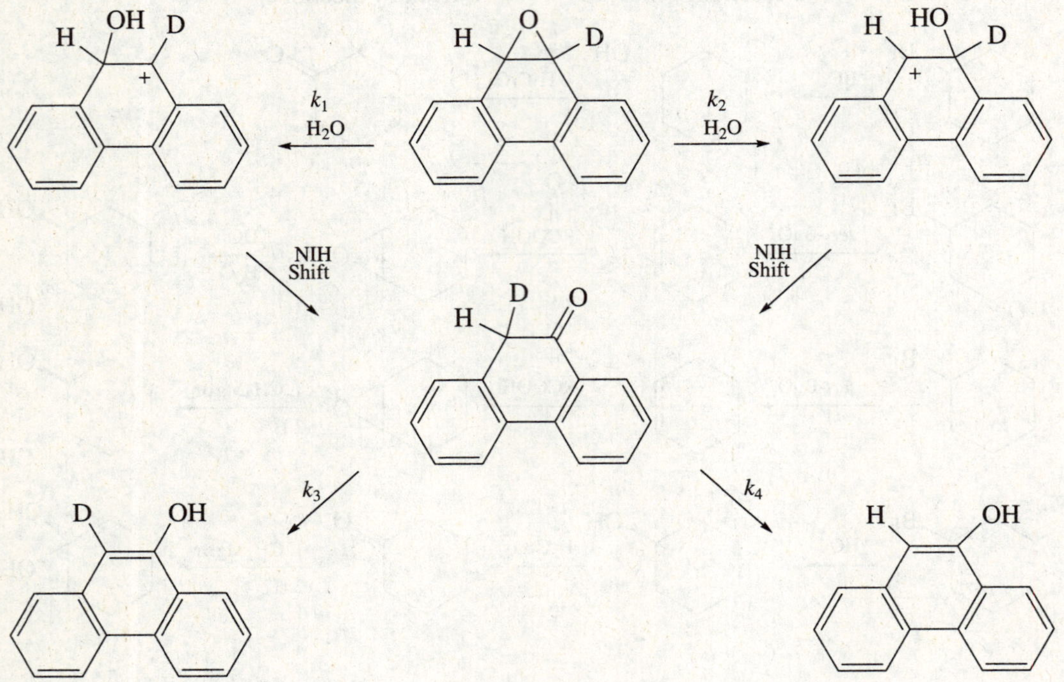

b. If an NIH shift does not occur, 50% of the deuterium will be retained because the epoxide can open equally easily in either direction (k_1 is equal to k_2) and subsequent loss of H^+ or D^+ is fast.

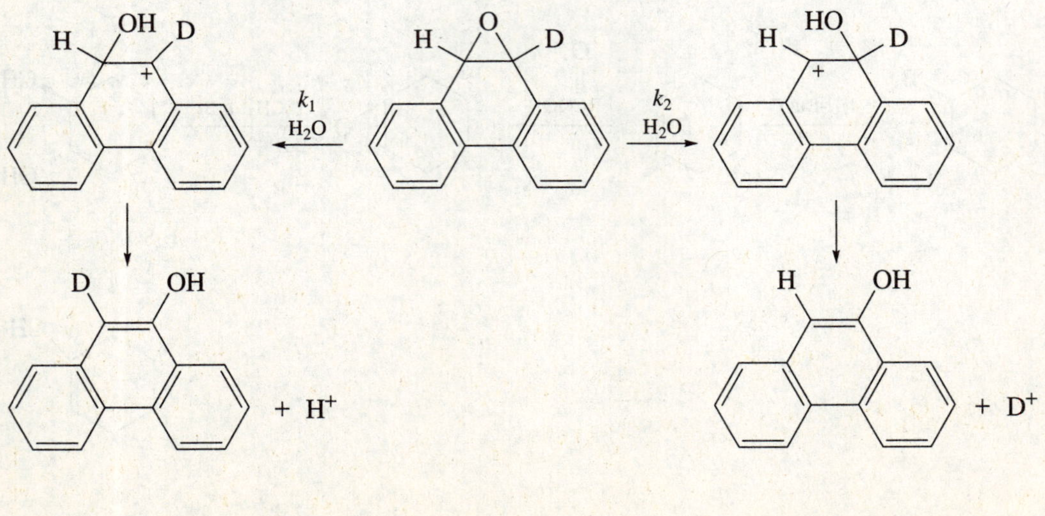

54. When (*S*)-2-butanol loses water, the asymmetric center in the reactant becomes a planar sp^2 carbon. Therefore, the chirality is lost. When water attacks the carbocation, it can attack from either side of the planar carbocation, forming (*S*)-2-butanol and (*R*)-2-butanol with equal ease.

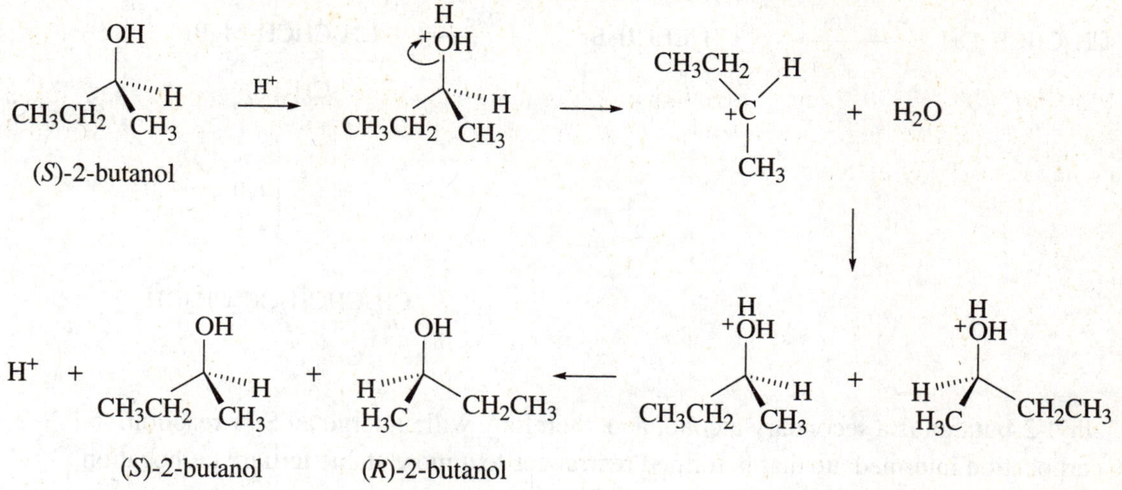

55. **a.**

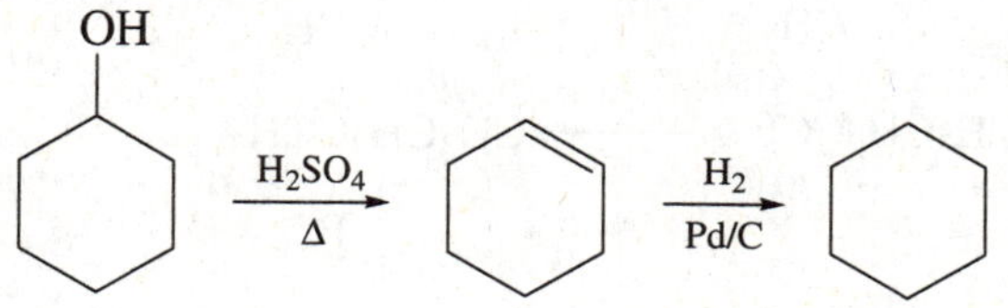

b.

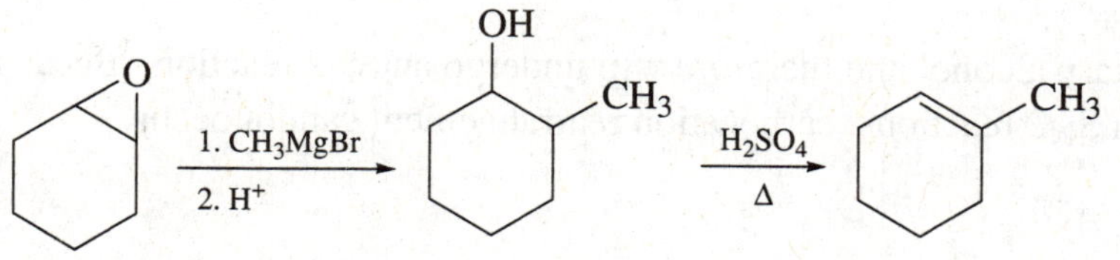

c.

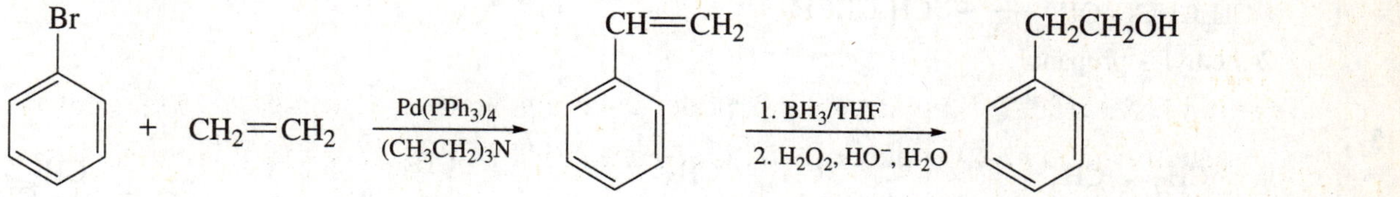

d.

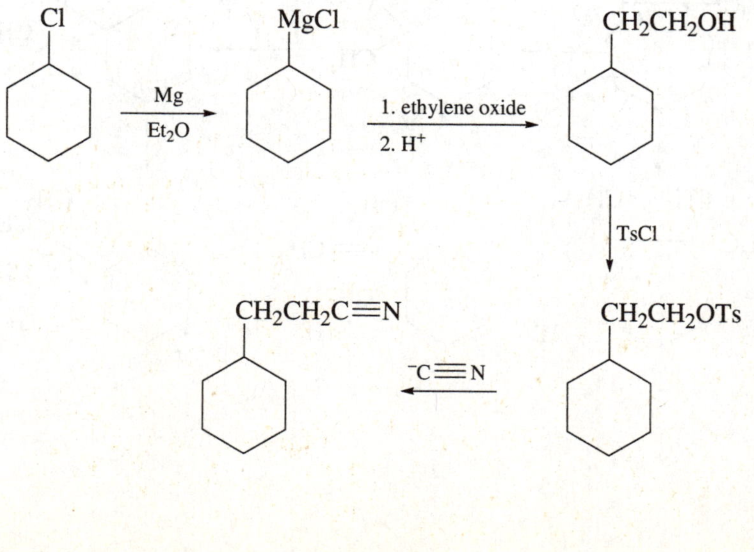

e. 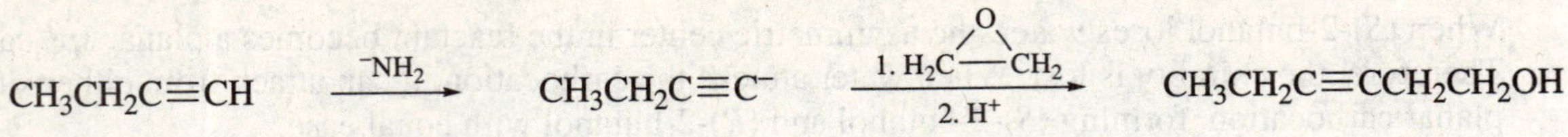

f. 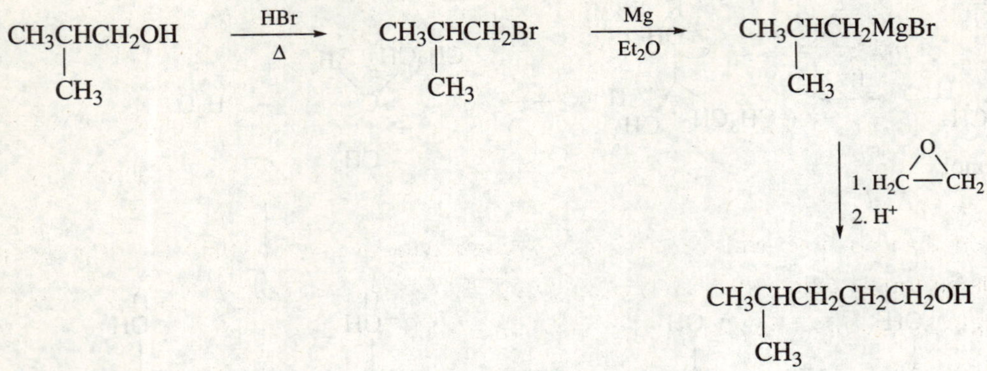

56. 3-Methyl-2-butanol is a secondary alcohol and, therefore, will undergo an S_N1 reaction. The carbocation intermediate that is formed rearranges to a more stable tertiary carbocation.

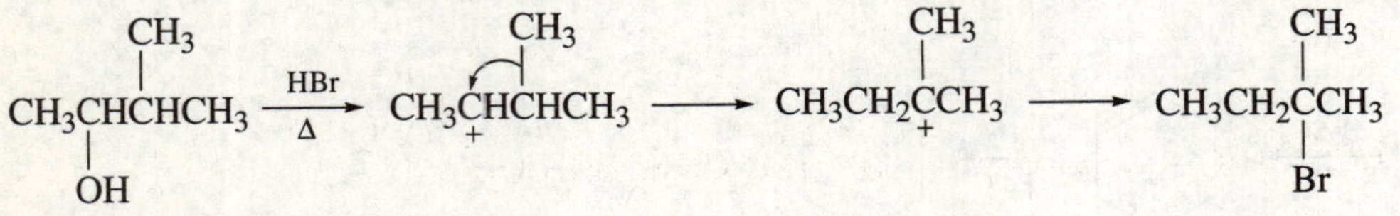

3-methyl-2-butanol

2-Methyl-1-propanol is a primary alcohol and therefore will undergo an S_N2 reaction. Because carbocations are not formed in S_N2 reactions, carbocation rearrangement cannot occur.

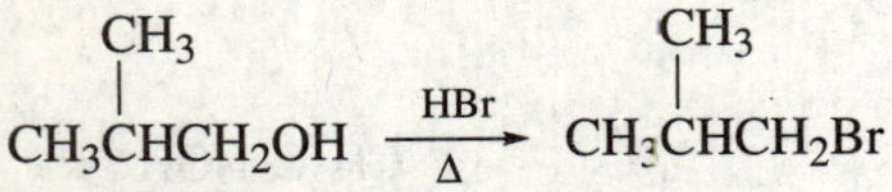

2-methyl-1-propanol

57.

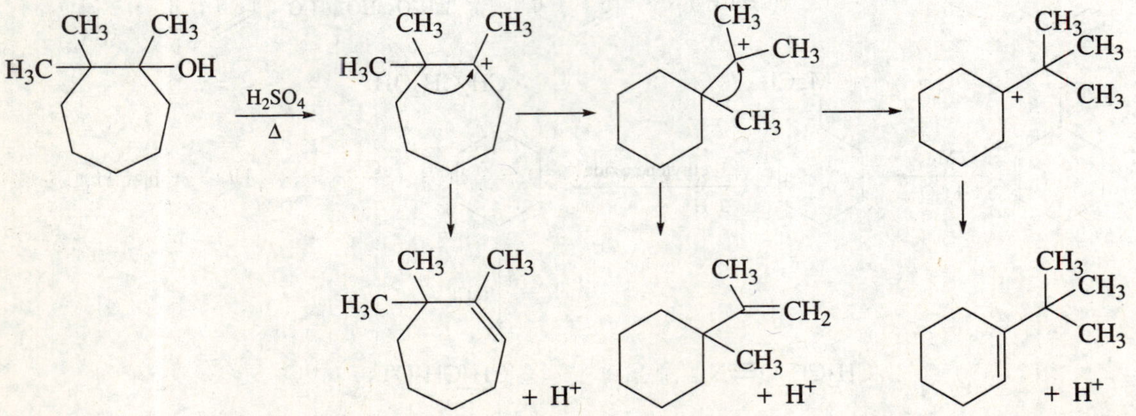

58.

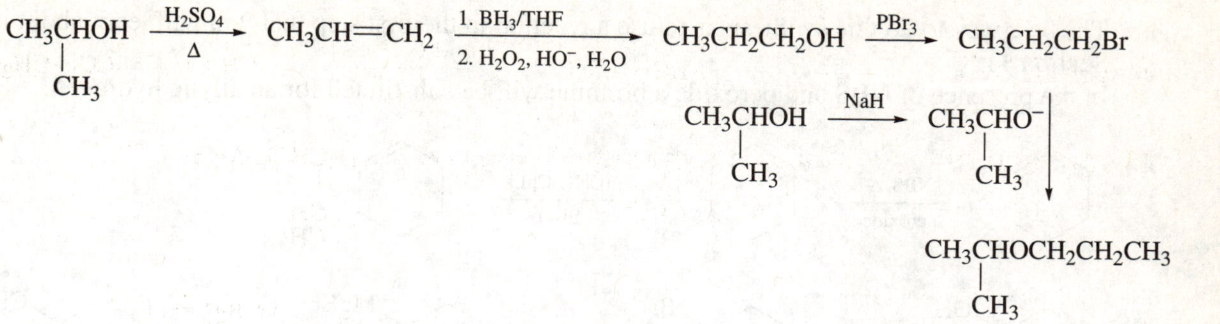

59. Cyclopropane does not react with HO^-, because cyclopropane does not contain a leaving group; a carbanion is far too basic to serve as a leaving group. Ethylene oxide reacts with HO^- because ethylene oxide contains an RO^- leaving group.

60. **Diethyl ether** is the ether that would be obtained in greatest yield, because it is a symmetrical ether. Since it is symmetrical, only one alcohol is used in its synthesis. Therefore, it is the only ether that would be formed. In contrast, the synthesis of an unsymmetrical ether requires two different alcohols. Therefore, the unsymmetrical ether is one of three different ethers that can be formed.

61. **a.**

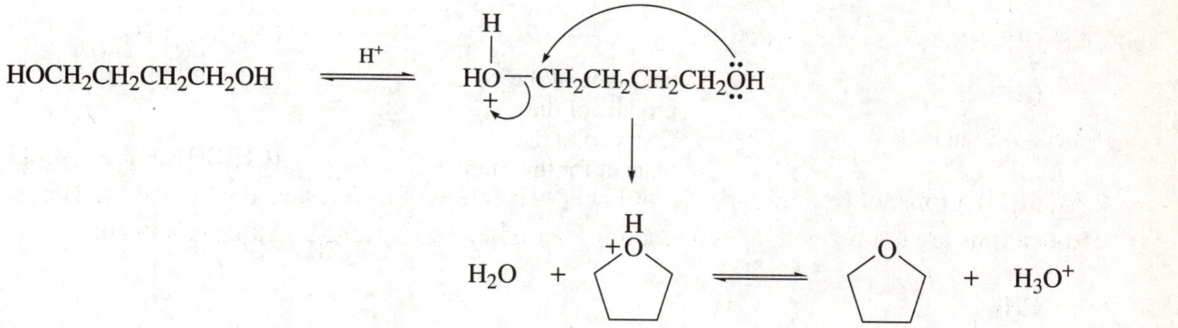

b.

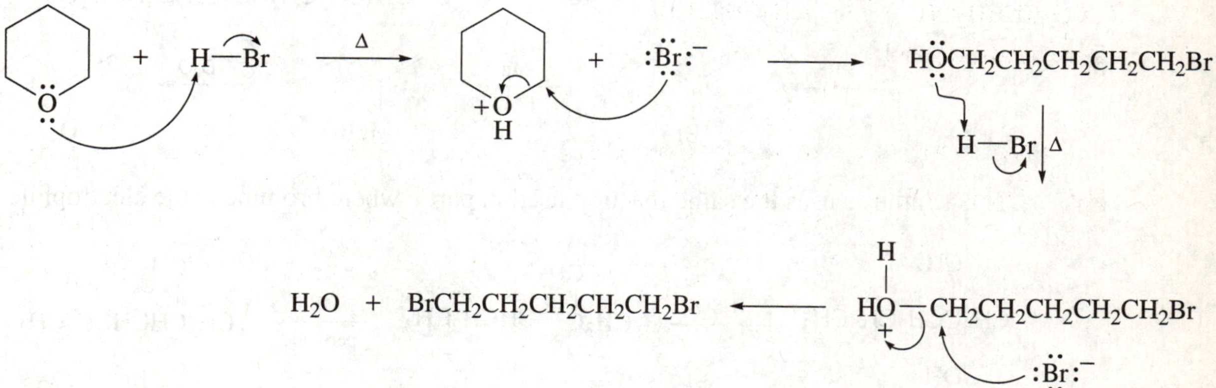

62. **a.** The best way to do this synthesis is to use a reaction in the first step that you will learn about in Section 11.8.
In the presence of NBS and peroxide a bromine will be substituted for an allylic hydrogen.

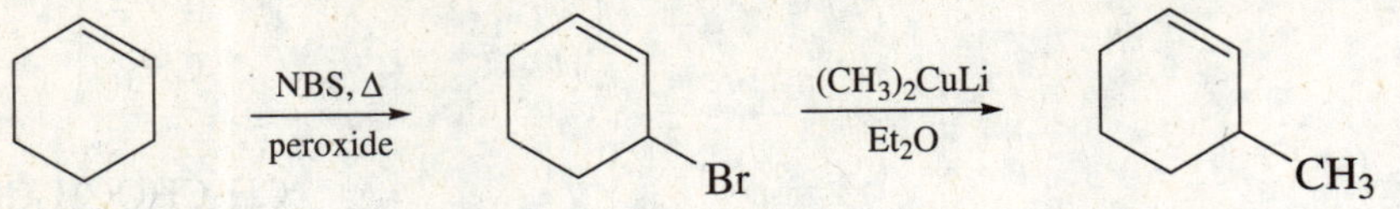

b.

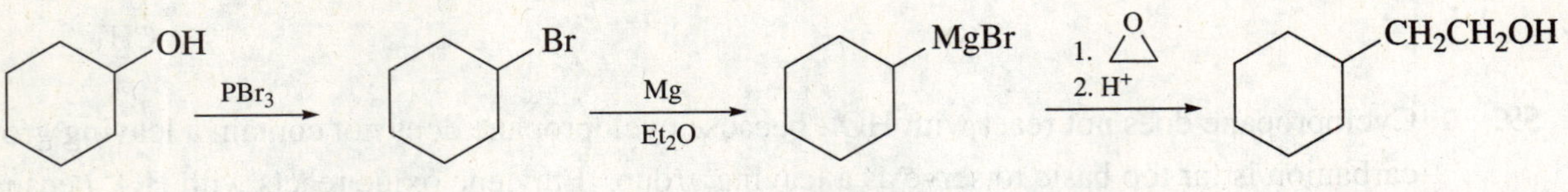

c. This synthesis requires a reaction you will learn about in Section 11.6. In the presence of peroxide, HBr will add to the double bond in a way that causes Br to be the electrophile. Therefore, Br is the first to add to the double bond and, like other electrophiles, it adds to the *sp*[2] carbon that is bonded to the greater number of hydrogens.

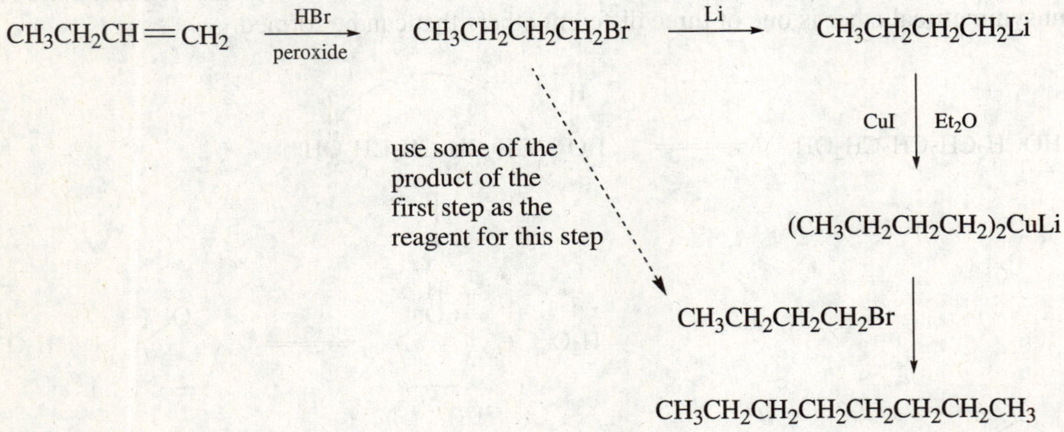

d.

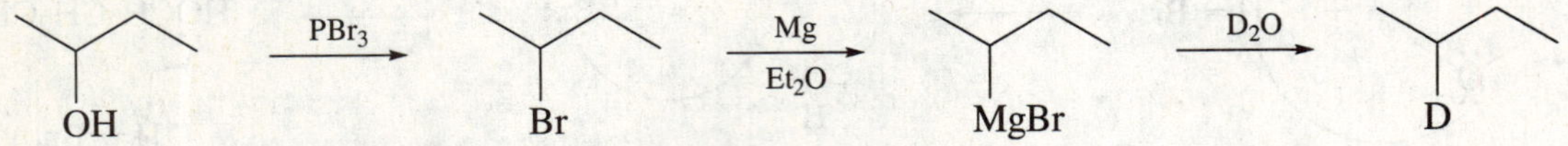

e. This synthesis uses the same reaction used in part **c** where bromine is the electrophile.

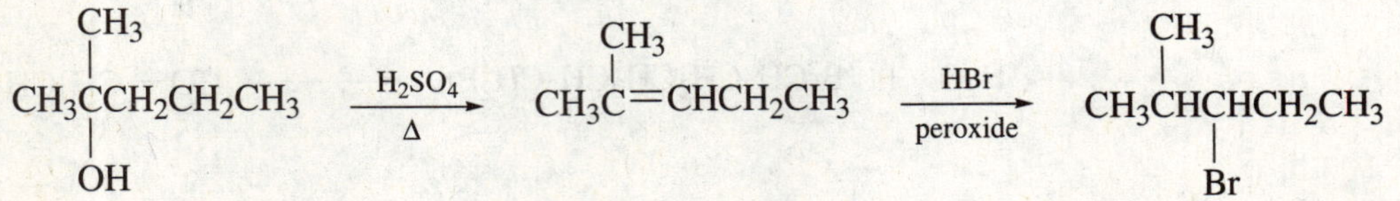

63.

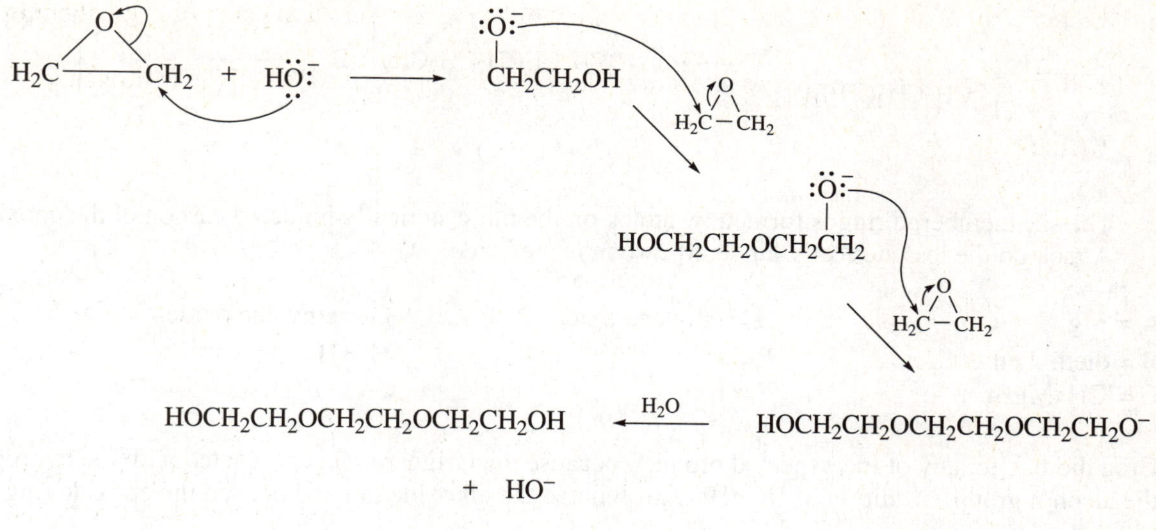

64.

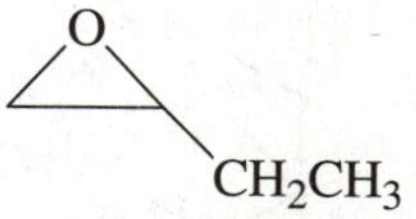

2-ethyloxirane

a. $CH_3CH_2CHCH_2OH$
 $|$
 OH

0.1 M HCl is a dilute solution of HCl in water

b. $CH_3CH_2CHCH_2OH$
 $|$
 OCH_3

c. $CH_3CH_2CHCH_2CH_2CH_3$
 $|$
 OH

d. $CH_3CH_2CHCH_2OH$
 $|$
 OH

e. $CH_3CH_2CHCH_2OCH_3$
 $|$
 OH

65. Ethyl alcohol is not obtained as a product, because it reacts with the excess HI and forms ethyl iodide.

$$CH_3CH_2OCH_2CH_3 \xrightarrow[\Delta]{HI} CH_3CH_2I + CH_3CH_2OH \xrightarrow[\Delta]{HI} CH_3CH_2I + H_2O$$

66. **a.**

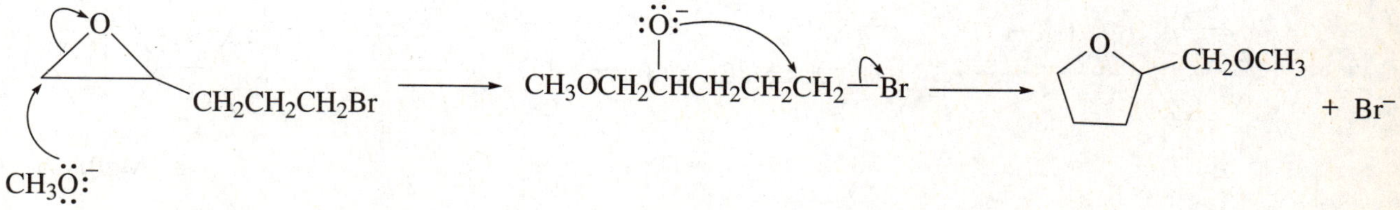

b.

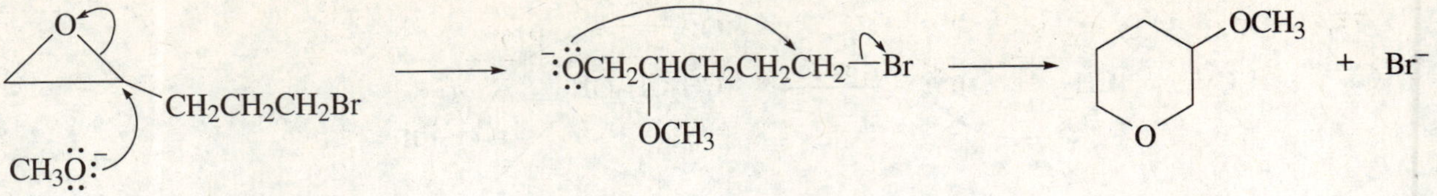

c. The six-membered ring is formed by attack on the more sterically hindered carbon of the epoxide. Attack on the less sterically hindered carbon is preferred.

67.

A = Mg	**D** = ethylene oxide	**G** = ethylene oxide
B = diethyl ether	**E** = H^+	**H** = H^+
C = CH_3MgBr	**F** = Na	

68. Greg did not get any of the expected product, because the Grignard reagent reacted with the hydrogen of the alcohol group. Addition of HCl/H_2O protonated the alkoxide ion and opened the epoxide ring.

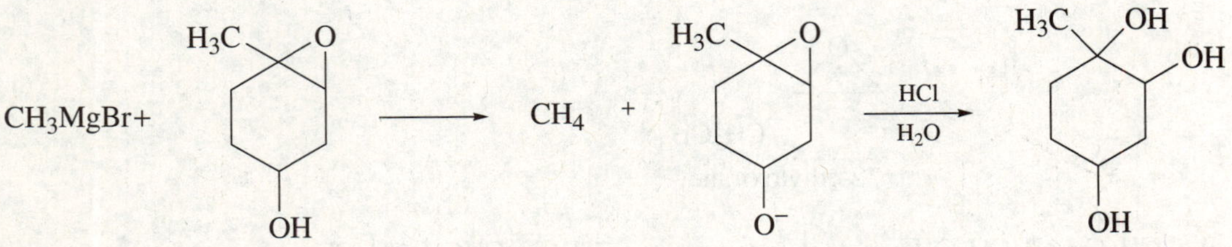

69. **A** is the substitution product that is formed by methoxide ion attacking a carbon of the three-membered ring and eliminating the amino group, thereby opening the ring.
B is the product of a Hofmann elimination reaction: methoxide ion removes a proton from a methyl group bonded to a carbon, eliminating the amino group. The red color disappears when Br_2 is added to **B**, because Br_2 adds to the double bond.
When the aziridinium ion reacts with methanol, only the substitution product is formed.

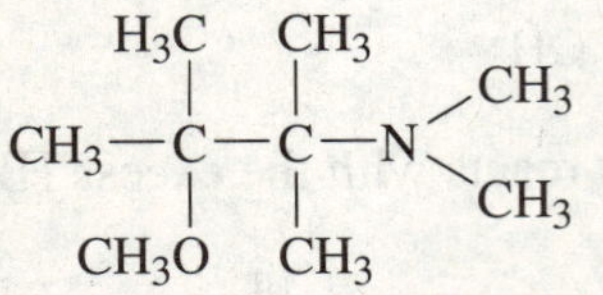

A

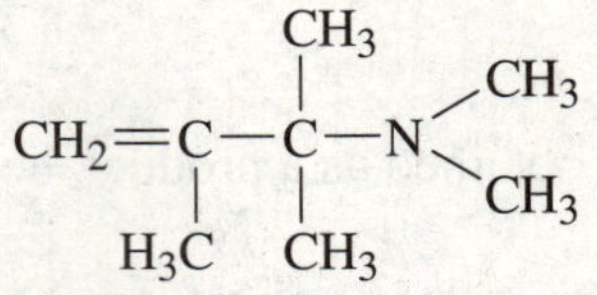

B

70.

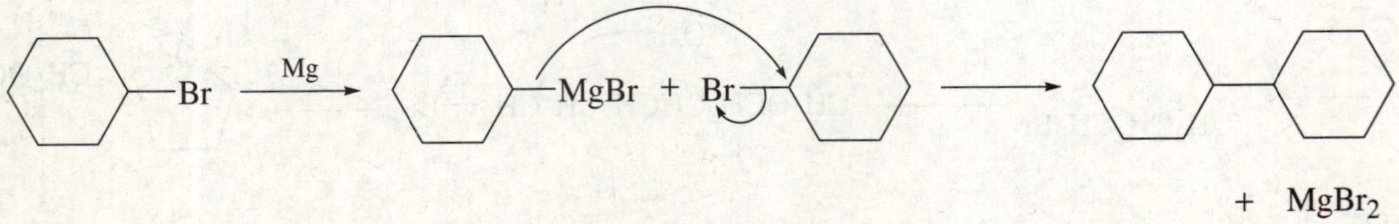

71. **a.**

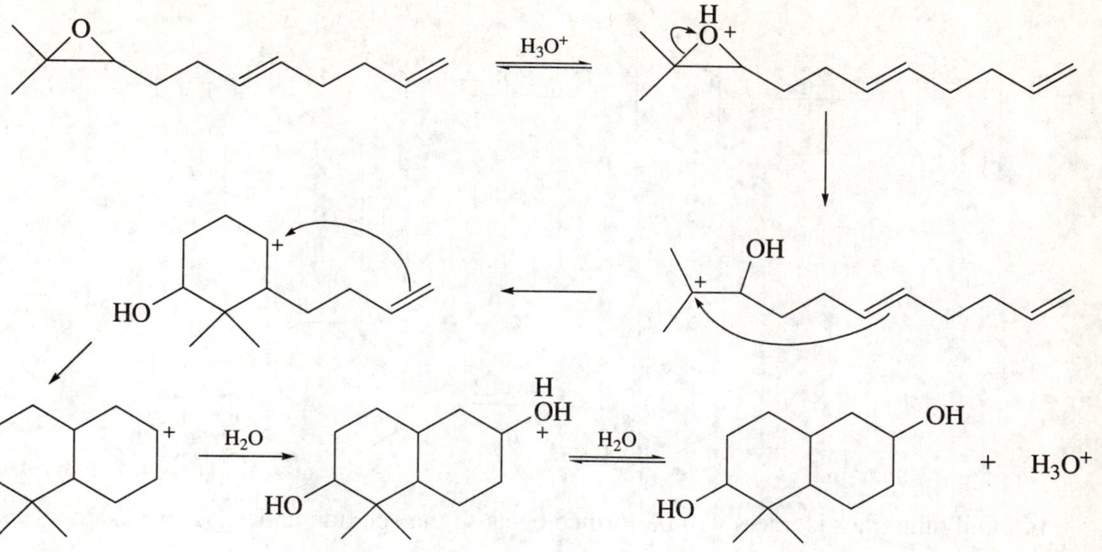

O
H₃O⁺
H
O+
OH
+
HO
H₂O
H
OH
H₂O
OH
+ H₃O⁺
HO

b.

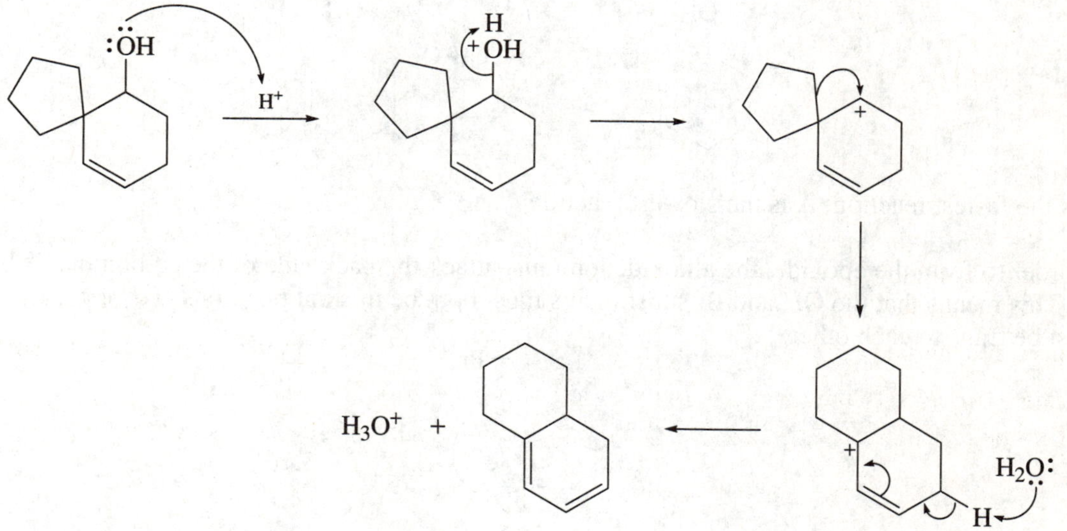

:ÖH
H⁺
H
+OH
+
H₃O⁺ +
H₂O:
H

c.

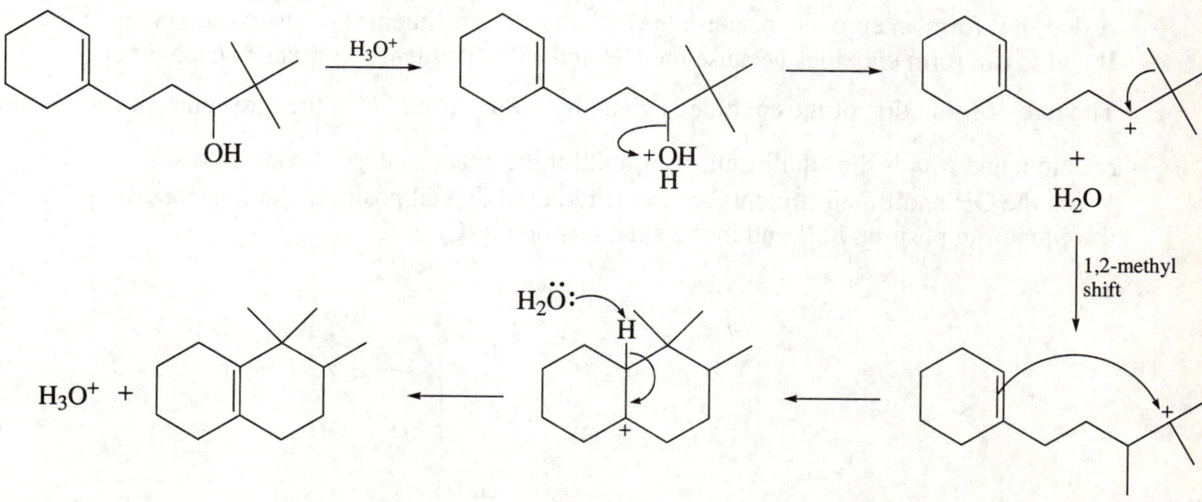

H₃O⁺
OH
+OH
H
+
+
H₂O
1,2-methyl
shift
H₂Ö:
H
H₃O⁺ +

72. **a.**

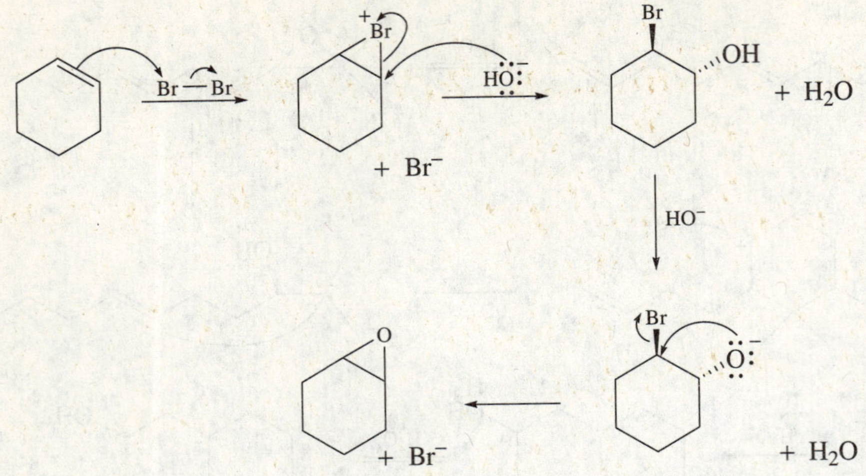

b. Only the trans isomers will be formed because the epoxide undergoes backside attack by methoxide ion.

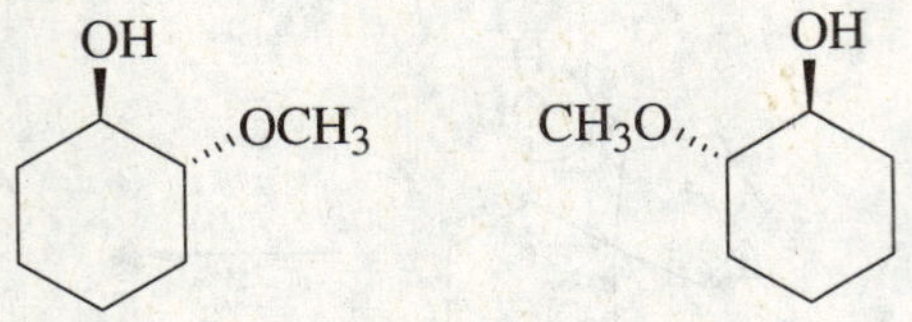

73. **B** is the fastest reaction; **A** is the slowest reaction

In order to form the epoxide, the alkoxide ion must attack the back-side of the carbon that is bonded to Br. This means that the OH and Br substituents must both be in axial positions. To be 1,2-diaxial, they must be trans to each other.

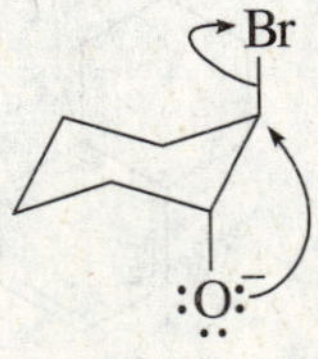

A does not form an epoxide, because the OH and Br substituents are cis to each other.
B and **C** can form epoxides because the OH and Br substituents are trans to each other.

The rate of formation of the epxoide is given by $k'K_{eq}$, where k' is the rate constant for the substitution reaction, and K_{eq} is the equilibrium constant for the equatorial/axial conformers.
When the OH and Br substituents are in the required diaxial position, the large *tert*-butyl substituent is in the equatorial position in **B** and in the axial position in **C**.

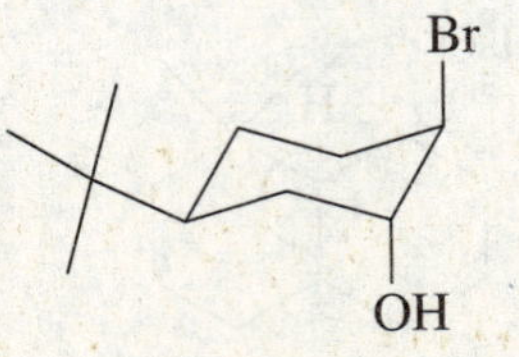

B

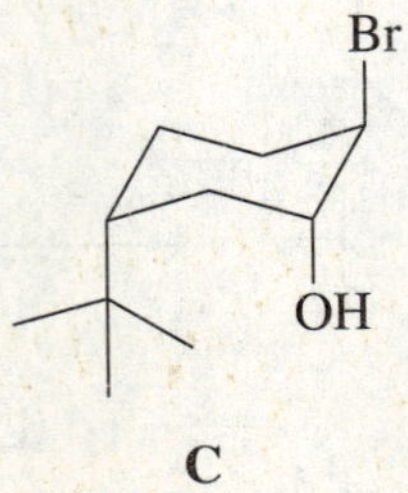

C

Because the more stable conformer is the one with the large *tert*-butyl group in the equatorial position, the more stable conformer of **B** has the OH and Br substituents in the required diaxial position (K_{eq} is large), whereas the OH and Br substituents in **C** are in the required diaxial position in its less stable conformer (K_{eq} is small). Therefore, **B** reacts faster than **C**.

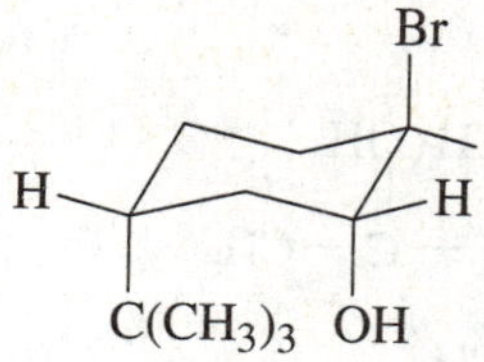

B

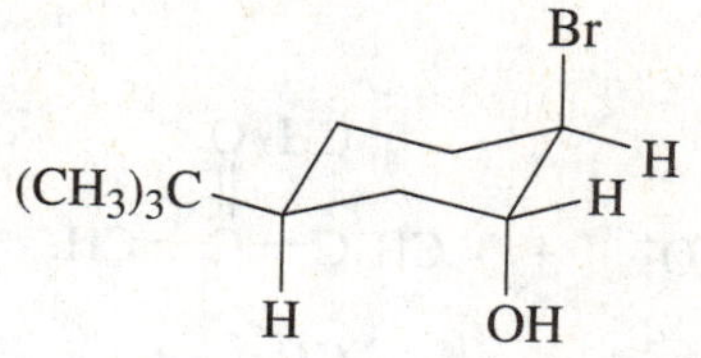

C

74. The benzene ring can become attached to either of the sp^2 carbons of the alkene, resulting in the formation of a pair of constitutional isomers.

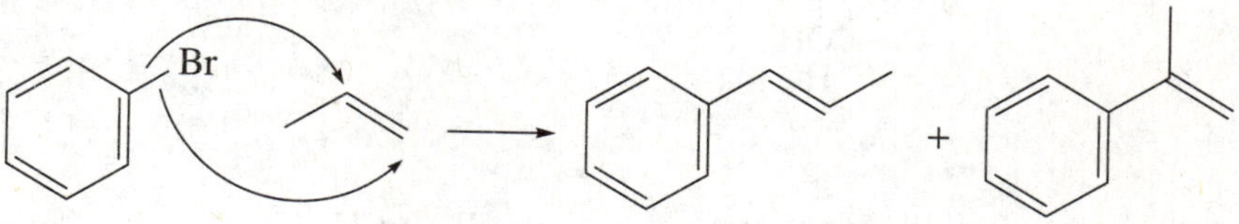

75. **a.** The reaction of 2-chlorobutane with HO^- is an intermolecular reaction, so the two compounds have to find one another in the solution.

$$CH_3CH(Cl)CH_2CH_3 + HO^- \longrightarrow CH_3CH(OH)CH_2CH_3 + Cl^-$$

The following reaction takes place in two steps. The first is an intramolecular S_N2 reaction, so the reaction is much faster than the above reaction since the two reactants are in the same molecule. The second S_N2 reaction is fast because the strain of the three-membered ring and the positive charge on the nitrogen make it a very good leaving group.

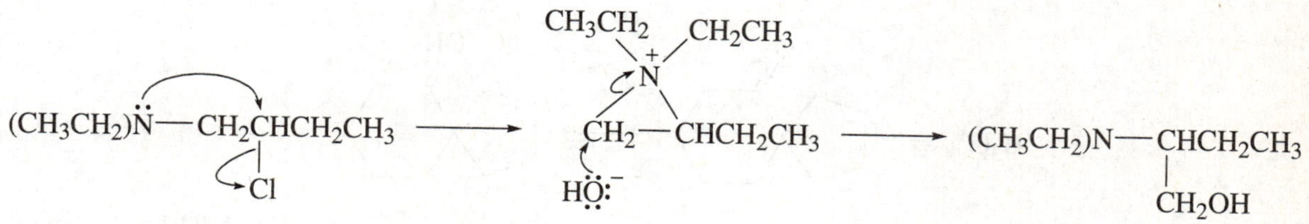

b. The HO group is bonded to a different carbon because HO^- will attack the least sterically hindered carbon of the three-membered ring.

76. 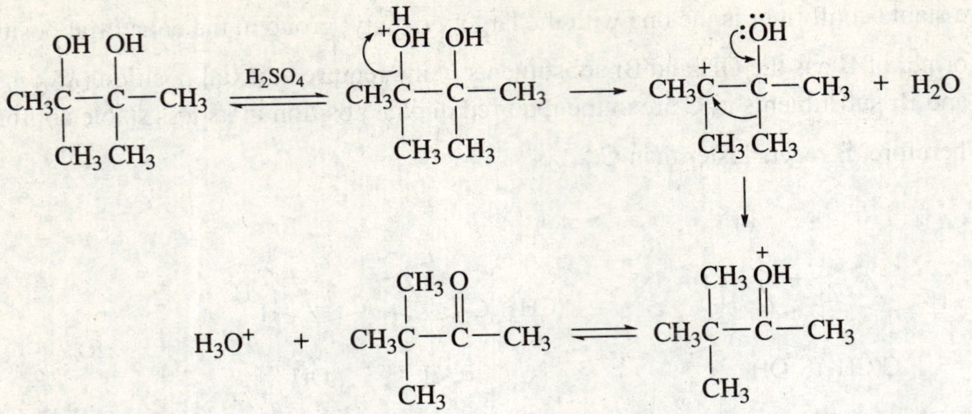

77. **a.** 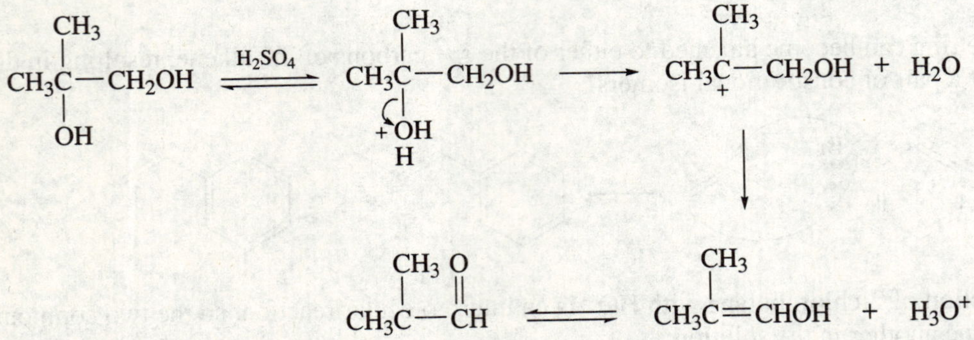

b. Dehydration of the primary alcohol group cannot occur because it cannot lose water via an E1 pathway, because a primary carbocation cannot be formed. It cannot lose water via an E2 pathway, because the β-carbon is not bonded to a hydrogen.

78. 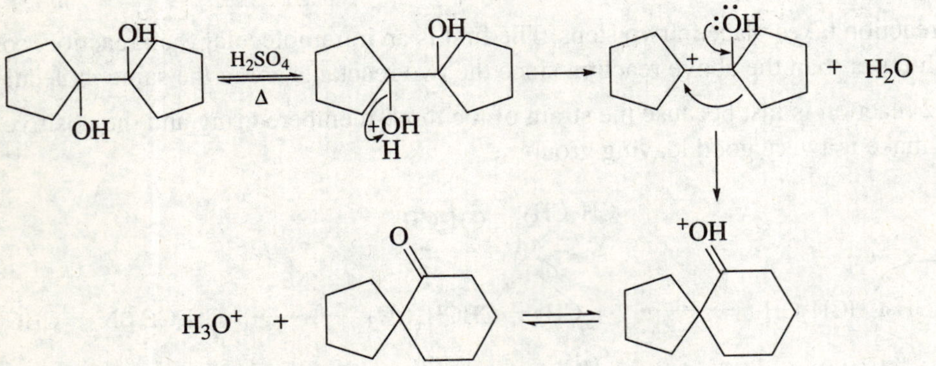

79. 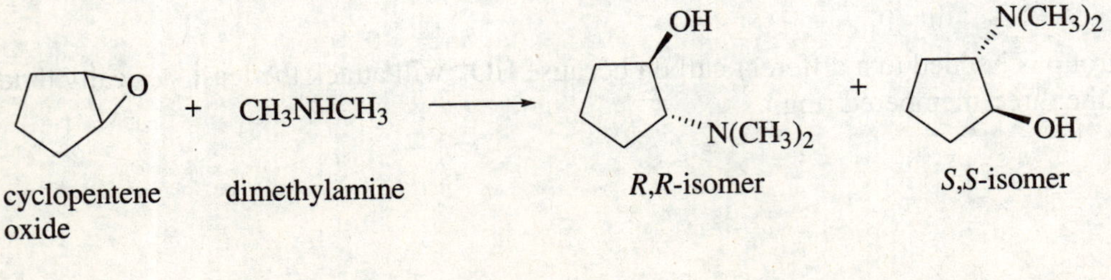

80. **a.**

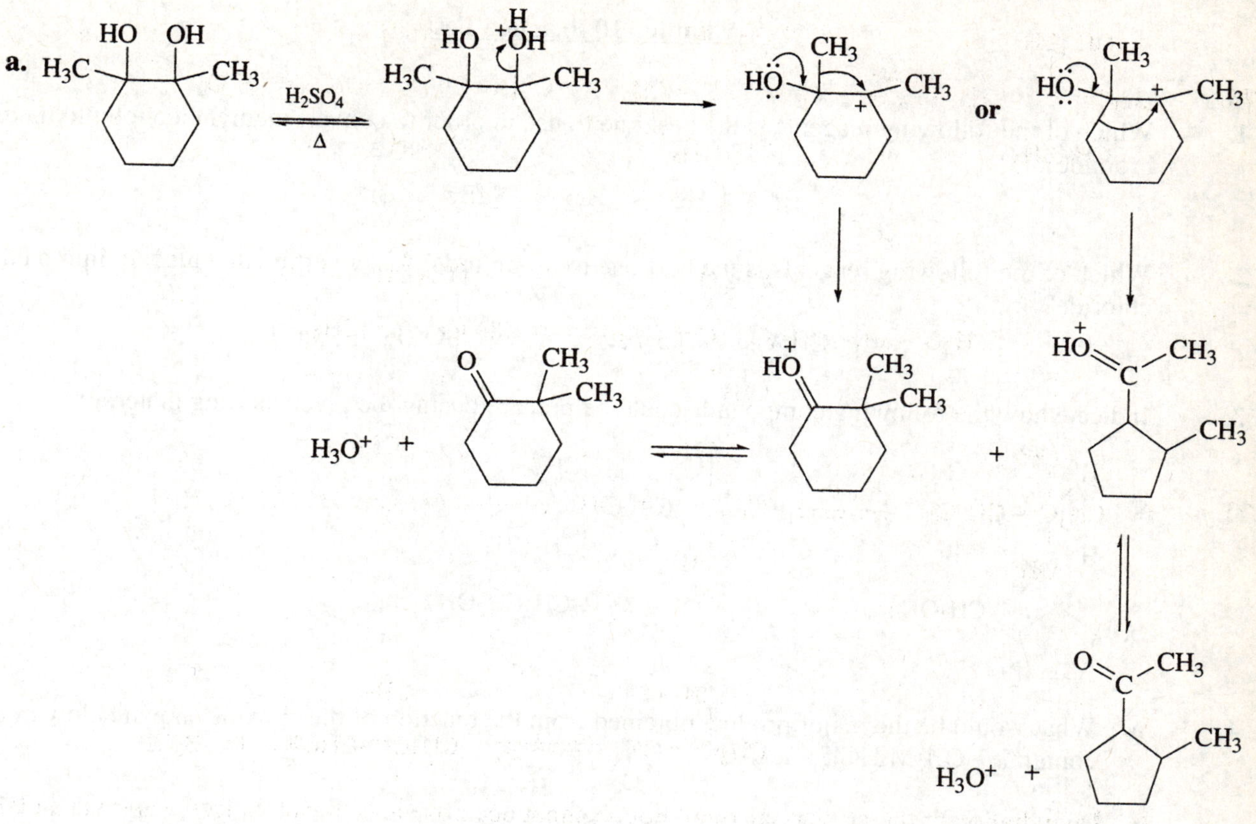
HO
OH
H3C
CH3
H2SO4
Δ
H
+OH
CH3
HÖ
+
or
H3O+
+
O
C

b.

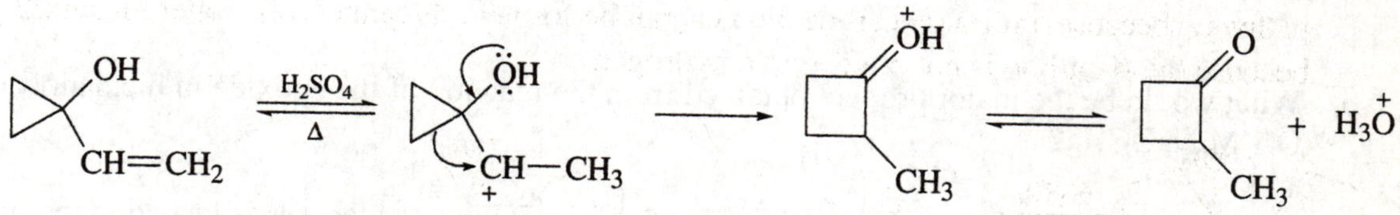
OH
CH=CH2
H2SO4
Δ
ÖH
CH—CH3
+
+OH
CH3
O
+ H3O+

Chapter 10 Practice Test

1. Which of the following reagents is the best one to use in order to convert methyl alcohol into methyl bromide?

Br^- HBr Br_2 NaBr Br^+

2. Which of the following reagents is the best one to use in order to convert methyl alcohol into methyl chloride?

Cl_2/CH_2Cl_2 $Cl_2/h\nu$ Cl^- $SOCl_2$ NaCl

3. Indicate how the following compounds could be prepared using the given starting material:

a. 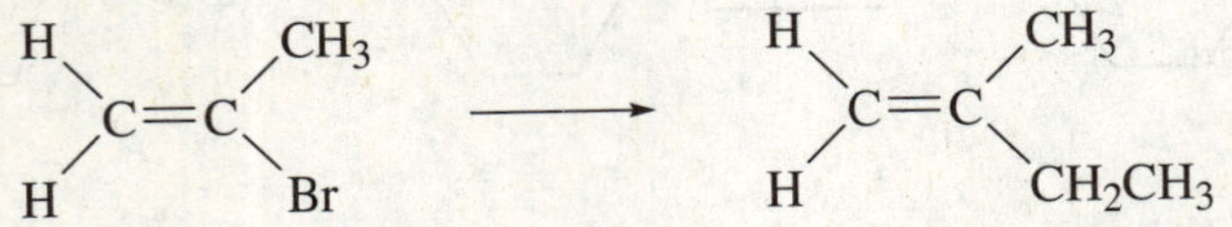

b. 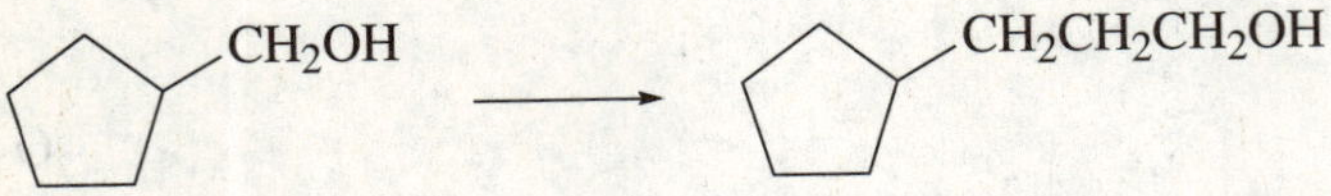

4. a. What would be the major product obtained from the reaction of the epoxide shown below in methanol containing O.1 M HCl?

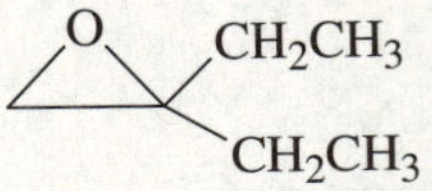

b. What would be the major product obtained from the reaction of the epoxide in methanol containing O.1 M $NaOCH_3$?

5. Give the major product that is obtained when each of the following alcohols is heated in the presence of H_2SO_4.

a. 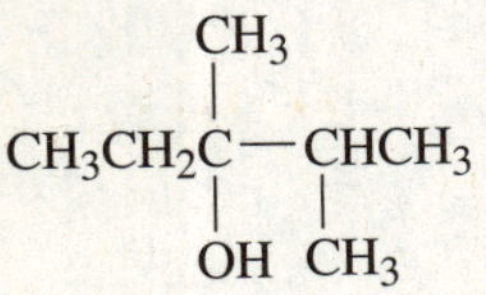

c. $CH_3CH_2CH_2CH_2CH_2OH$

b. 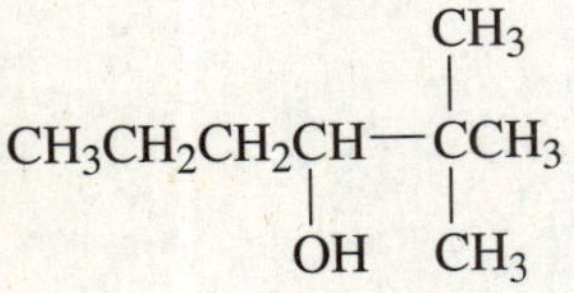

d. 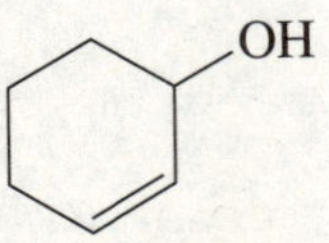

6. Indicate whether the following statements are true or false:

a. Tertiary alcohols are easier to dehydrate than secondary alcohols. T F

b. Alcohols are more acidic than thiols. T F

c. Alcohols have higher boiling points than thiols. T F

7. What products would be obtained from heating the following ethers with one equivalent of HI?

a. 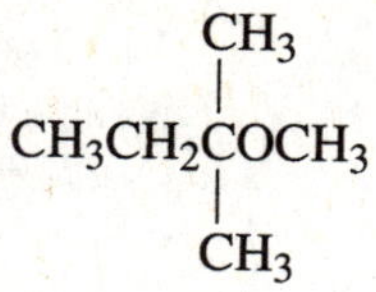

b. 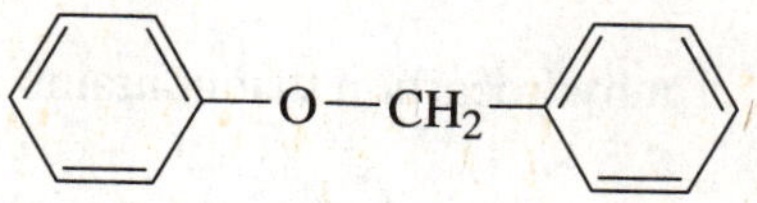

8. What alcohols would be formed from the reaction of ethylene oxide with the following Grignard reagents?

a. $CH_3CH_2CH_2MgBr$

b. 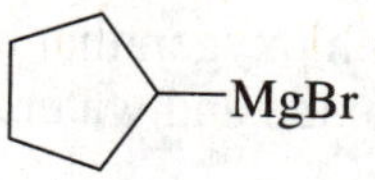

9. Give the major product of each of the following reactions:

a. 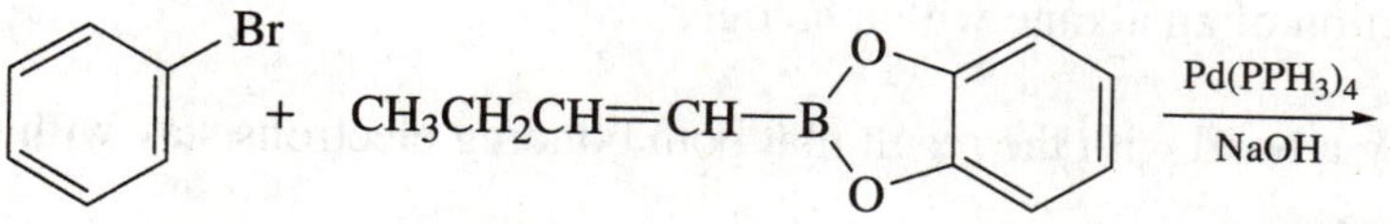

b. 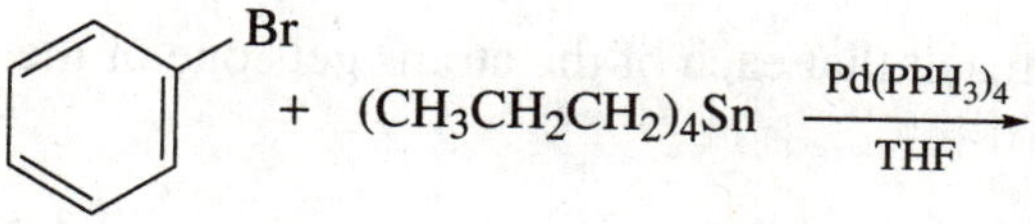

c. 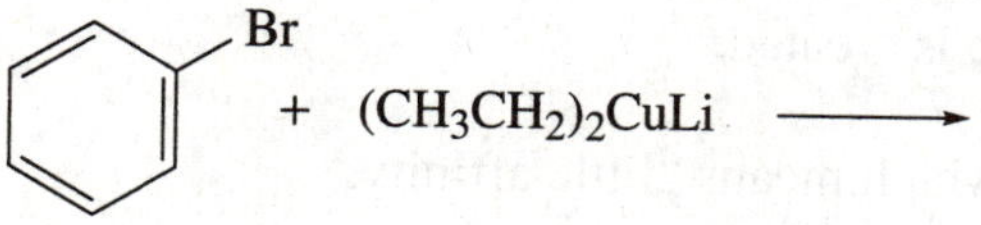

d. 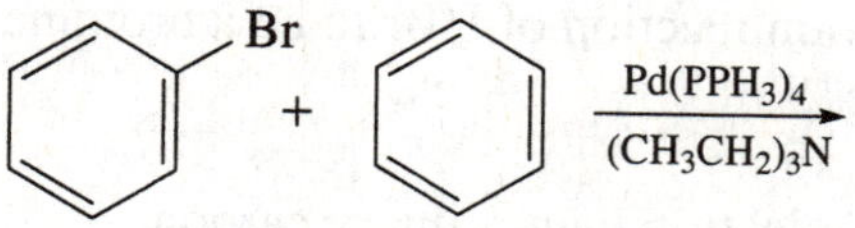

CHAPTER 11
Radicals • Reactions of Alkanes

Important Terms

alkane — a hydrocarbon that contains only single bonds.

allylic radical — a compound with an unpaired electron on an allylic carbon.

benzylic radical — a compound with an unpaired electron on a benzylic carbon.

combustion — a reaction with oxygen that takes place at high temperatures and converts alkanes to carbon dioxide and water.

free radical — an atom or a molecule with an unpaired electron.

halogenation — the reaction of an alkane with a halogen.

heterolytic bond cleavage (heterolysis) — breaking a bond with the result that both bonding electrons stay with one the atoms.

homolytic bond cleavage (homolysis) — breaking a bond with the result that each of the atoms gets one of the bonding electrons.

initiation step — the step in which radicals are created and or the step in which the radical needed for the first propagating step is created.

paraffin — an older word for alkanes, which means "little affinity."

peroxide effect — causes the electrophile in an addition reaction of HBr to be a bromine radical instead of a proton.

primary alkyl radical — an alkyl radical with the unpaired electron on a primary carbon.

propagation step — in the first of a pair of propagation steps, a radical (or an electrophile or a nucleophile) reacts to produce another radical (or an electrophile or a nucleophile) that reacts in the second propagation step to produce the radical (or the electrophile or the nucleophile) that was the reactant in the first propagation step.

radical (often called a free radical) — an atom or a molecule with an unpaired electron.

radical addition reaction — an addition reaction in which the first species that adds is a radical.

radical chain reaction — a reaction in which radicals are formed and react in repeating propagating steps.

radical inhibitor — a compound that traps radicals.

radical initiator — a compound that creates radicals.

radical substitution reaction a substitution reaction that has a radical intermediate.

reactivity-selectivity principle the greater the reactivity of a species, the less selective it will be.

saturated hydrocarbon a hydrocarbon that contains only single bonds (is saturated with hydrogen).

secondary alkyl radical an alkyl radical with the unpaired electron on a secondary carbon.

termination step two radicals combine to produce a molecule in which all the electrons are paired.

tertiary alkyl radical an alkyl radical with the unpaired electron on a tertiary carbon.

Solutions to Problems

1.

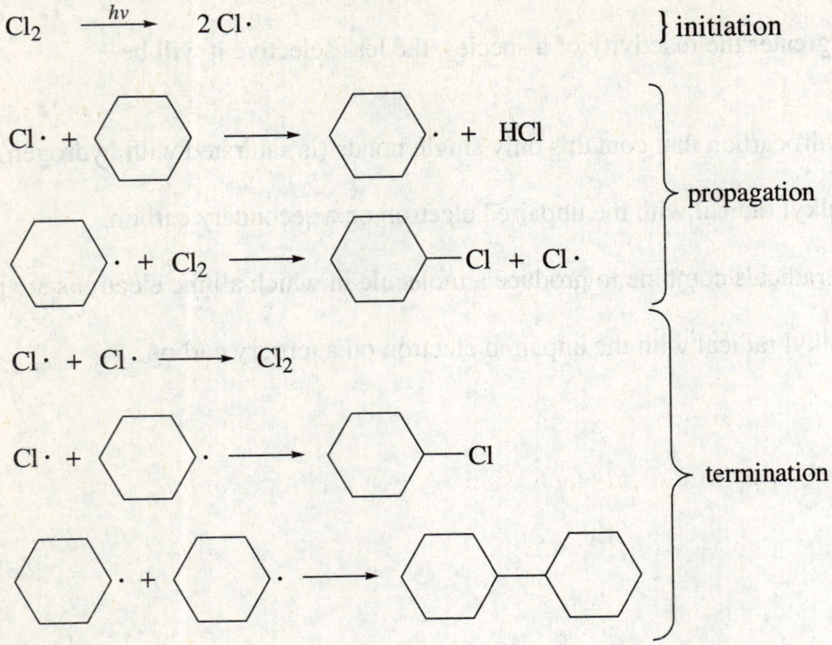

2.

$Cl_2 \xrightarrow{h\nu} 2\,Cl\cdot$

$Cl\cdot + CH_4 \longrightarrow \cdot CH_3 + HCl$

$\cdot CH_3 + Cl_2 \longrightarrow CH_3Cl + Cl\cdot$

$Cl\cdot + CH_3Cl \longrightarrow \cdot CH_2Cl + HCl$

$\cdot CH_2Cl + Cl_2 \longrightarrow CH_2Cl_2 + Cl\cdot$

$Cl\cdot + CH_2Cl_2 \longrightarrow \cdot CHCl_2 + HCl$

$\cdot CHCl_2 + Cl_2 \longrightarrow CHCl_3 + Cl\cdot$

$Cl\cdot + CHCl_3 \longrightarrow \cdot CCl_3 + HCl$

$\cdot CCl_3 + Cl_2 \longrightarrow CCl_4 + Cl\cdot$

$Cl\cdot + \cdot CCl_3 \longrightarrow CCl_4$

3. **a.**

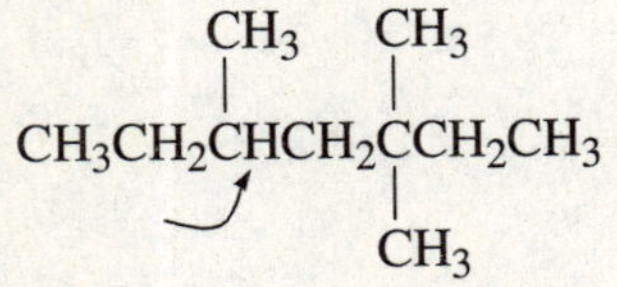

b. 6

4.

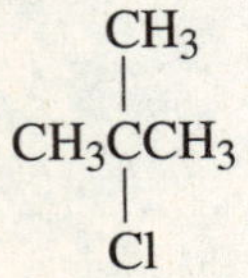

$1 \times x = x$

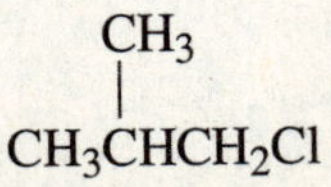

$9 \times 1 = 9$

$$\text{fraction of the total that is 1-chloro-2-methylpropane} = \frac{\text{1-chloro-2-methylpropane}}{\text{1-chloro-2-methylpropane + 2-chloro-2-methylpropane}} = 0.64$$

$$\frac{9}{9+x} = 0.64$$

$$9 = 0.64\,(9+x)$$

$$9 = 5.76 + 0.64\,x$$

$$3.24 = 0.64\,x$$

$$x = 5$$

It is 5 times easier for a chloride radical to remove a hydrogen atom from a tertiary carbon than from a primary carbon.

5. **a.** 3 **b.** 3 **c.** 5 **d.** 1 **e.** 5 **f.** 4 **g.** 2 **h.** 1 **i.** 4

6. Note: the denominator used in each of these problems is obtained by adding the numerators.

a. $CH_3CH_2CH_2CH_2CH_2Cl$ $\quad 6 \times 1 = 6 \quad \frac{6}{28.8} = 21\%$

$CH_3CH_2CH_2CH(Cl)CH_3$ $\quad 4 \times 3.8 = 15.2 \quad \frac{15.2}{28.8} = 53\%$

$CH_3CH_2CH(Cl)CH_2CH_3$ $\quad 2 \times 3.8 = 7.6 \quad \frac{7.6}{28.8} = 26\%$

b. $ClCH_2CH(CH_3)CH_2CH_2CH(CH_3)CH_3$ $\quad 12 \times 1 = 12 \quad \frac{12}{37.2} = 32\%$

$CH_3C(CH_3)(Cl)CH_2CH_2CH(CH_3)CH_3$ $\quad 2 \times 5 = 10 \quad \frac{10}{37.2} = 27\%$

$CH_3CH(CH_3)CH(Cl)CH_2CH(CH_3)CH_3$ $\quad 4 \times 3.8 = 15.2 \quad \frac{15.2}{37.2} = 41\%$

c. $ClCH_2CH(CH_3)CH_2CH_2CH_3$ $\quad 6 \times 1 = 6 \quad \frac{6}{29.2} = 21\%$

$CH_3C(CH_3)(Cl)CH_2CH_2CH_3$ $\quad 1 \times 5 = 5 \quad \frac{5}{29.2} = 17\%$

$CH_3CH(CH_3)CH(Cl)CH_2CH_3$ $\quad 2 \times 3.8 = 7.6 \quad \frac{7.6}{29.2} = 26\%$

$CH_3CH(CH_3)CH_2CH(Cl)CH_3$ $\quad 2 \times 3.8 = 7.6 \quad \frac{7.6}{29.2} = 26\%$

$CH_3CH(CH_3)CH_2CH_2CH_2Cl$ $\quad 3 \times 1 = 3 \quad \frac{3}{29.2} = 10\%$

7. Solved in the text.

8. **a.** 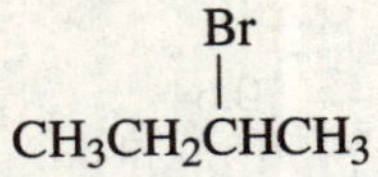 $CH_3CH_2CH_2CH_2Br$

$4 \times 82 = 328$ $\qquad$ $6 \times 1 = 6$

$$\frac{328}{328+6} = \frac{328}{334} = 98\%$$

b. 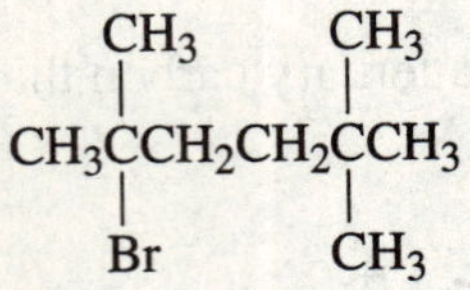

$1 \times 1600 = 1600$

$$\frac{1600}{1943} = 82\% \qquad \text{other products} \begin{cases} 1600 \\ 9 \times 1 = 9 \\ 2 \times 82 = 164 \\ 2 \times 82 = 164 \\ 6 \times 1 = 6 \end{cases}$$

$$\overline{1943}$$

9.
a. Chlorination, because the halogen is substituting for a primary hydrogen
b. Bromination, because the halogen is substituting for a tertiary hydrogen.
c. Because the molecule has only one kind of hydrogen, only one monohalogenated product will be obtained by both bromination and chlorination.

10. Solved in the text.

11. 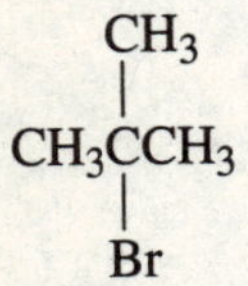 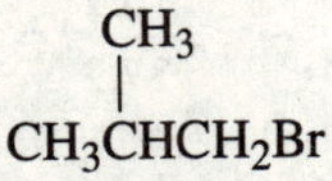

$1 \times 1600 = 1600$ $\qquad$ $9 \times 1 = 9$

$$\frac{1600}{1609} = 99.4\% \text{ for bromination compared with}$$

36% for chlorination

12. **a.** $CH_3CH_2CH_2CH_2CH_2Br$ 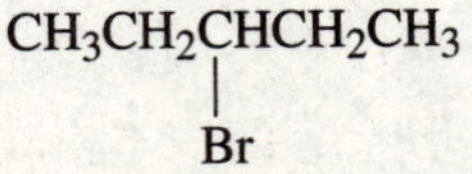 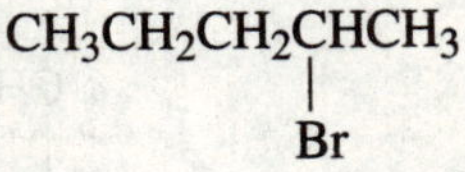

$$\frac{6}{498} = 1\% \qquad \frac{164}{498} = 33\% \qquad \frac{328}{498} = 66\%$$

b. 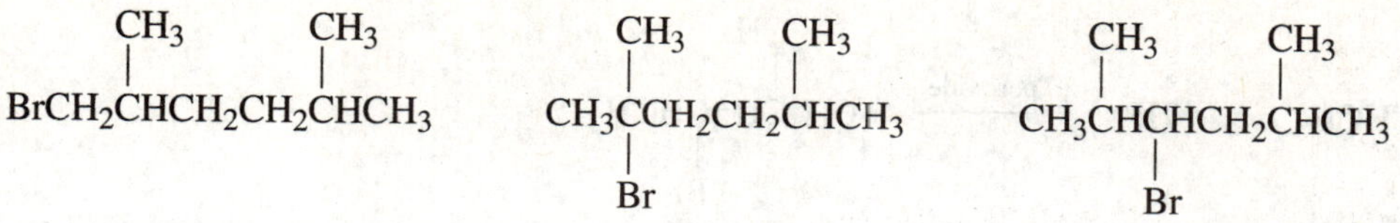

$$\frac{12}{3540} = 0.3\% \qquad \frac{3200}{3540} = 90.4\% \qquad \frac{328}{3540} = 9.3\%$$

c. 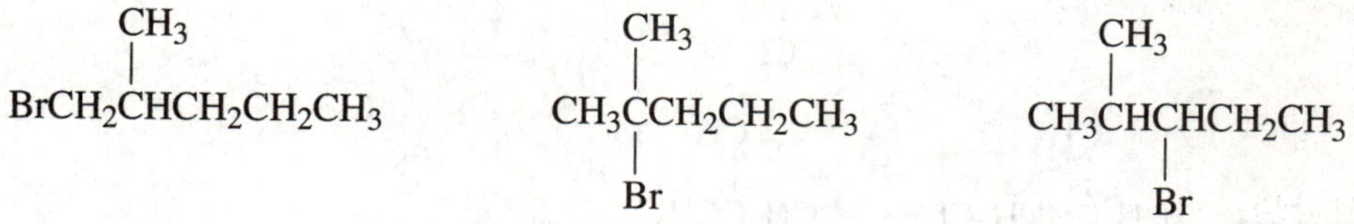

$$\frac{6}{1937} = 0.3\% \qquad \frac{1600}{1937} = 82.6\% \qquad \frac{164}{1937} = 8.5\%$$

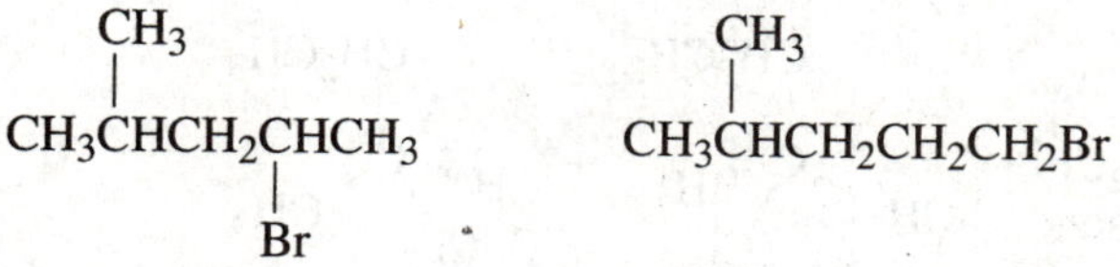

$$\frac{164}{1937} = 8.5\% \qquad \frac{3}{1937} = 0.2\%$$

13. 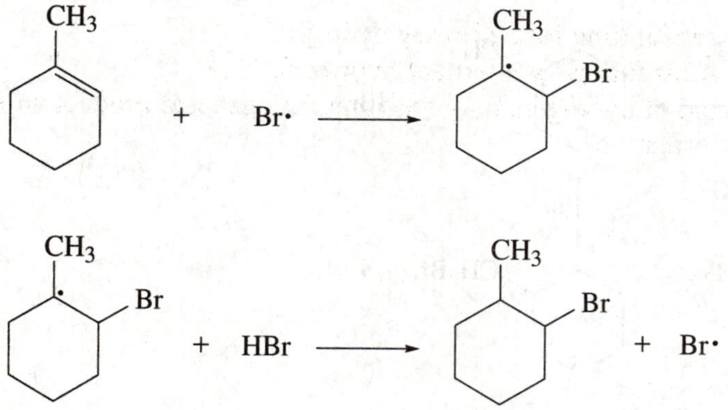

14. **a.** 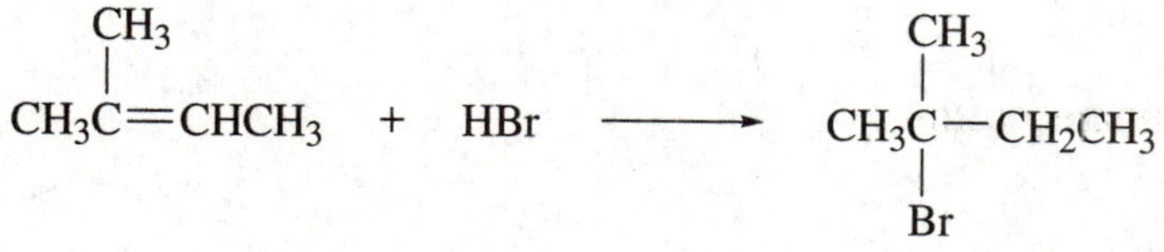

b. 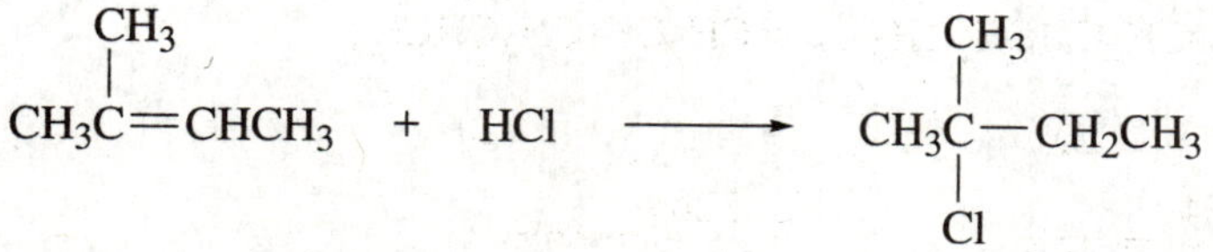

c. 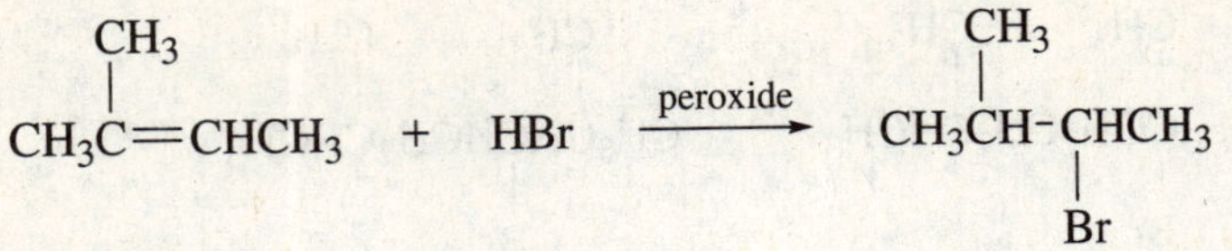

d. 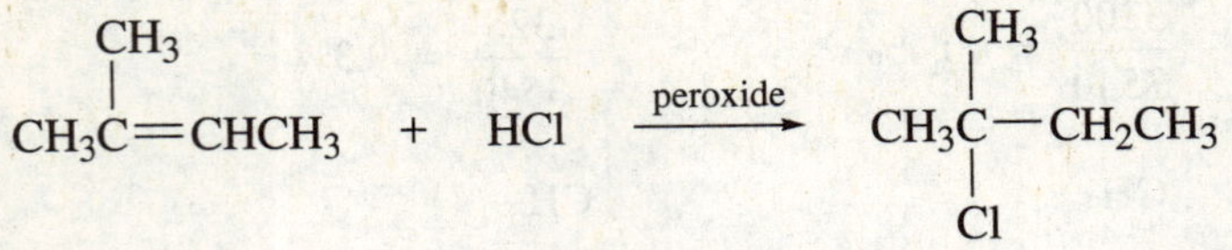

15. **a.** 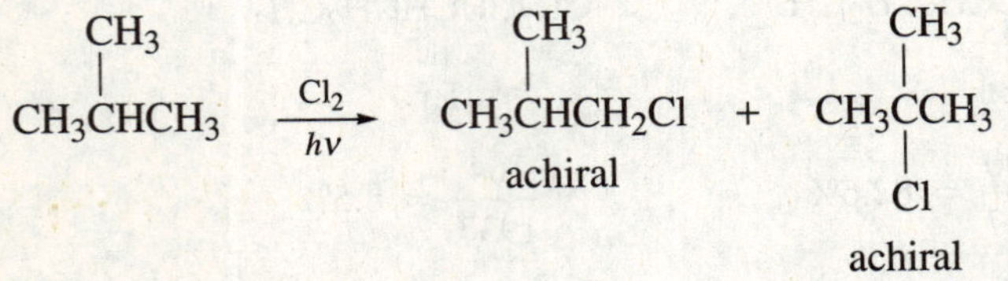

b. 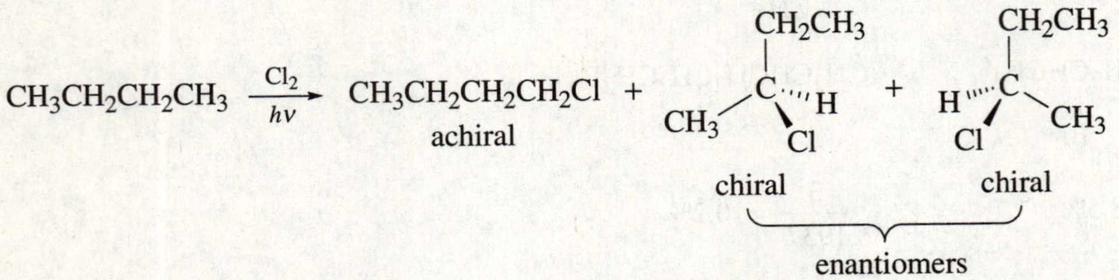

16. 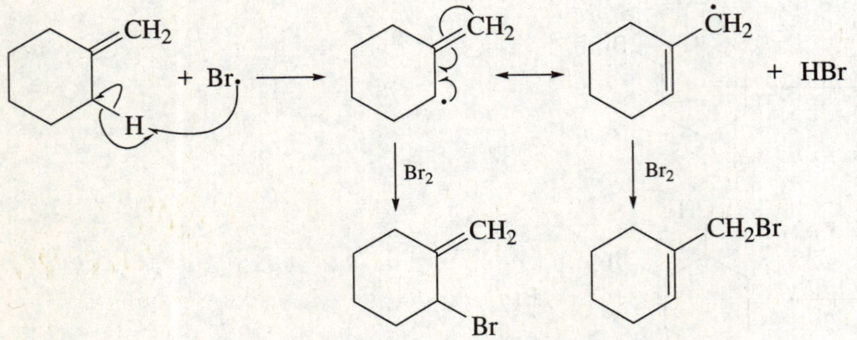

17. Solved in the text.

18. 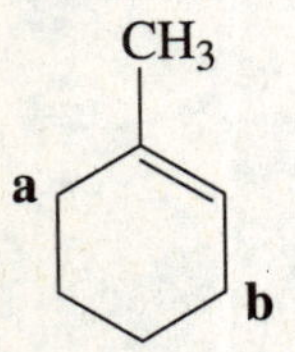

1-methylcyclohexene

a. **1.** There are two sets of allylic hydrogens, **a** and **b**. Removal of one of the **a** allylic hydrogens forms an intermediate in which the unpaired electron is shared by two secondary carbons. Removal of one of the **b** allylic hydrogens forms an intermediate in which the unpaired electron is shared by a tertiary carbon and a secondary carbon.

Therefore the major products are obtained by removing a **b** allylic hydrogen.

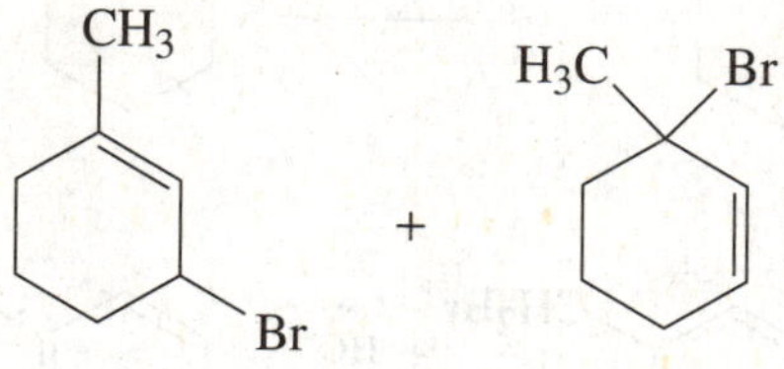

2. 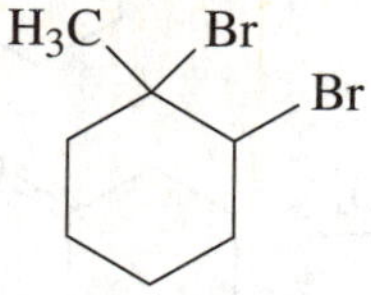

3. 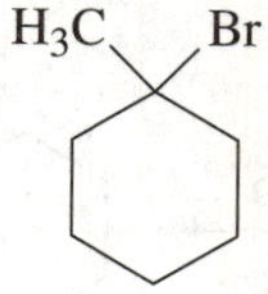

4. 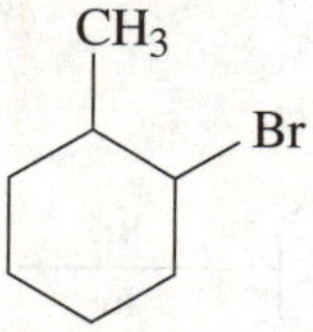

b. 1. Each of the products has one asymmetric center.
The *R* and the *S* isomer will be obtained for each product.

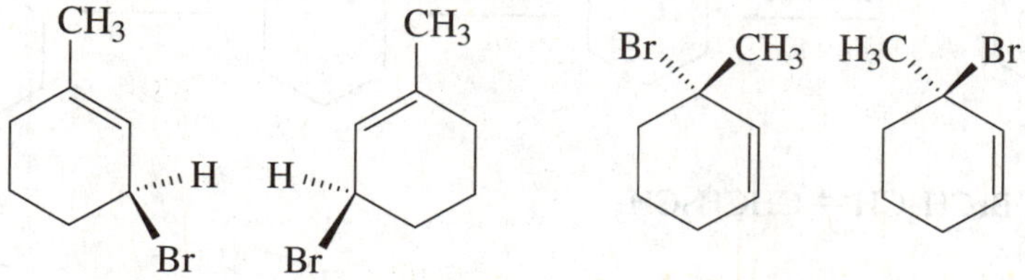

2. The product has two asymmetric centers.
Because addition of Br_2 is anti, only the products with the bromine atoms on opposite sides of the ring are obtained.

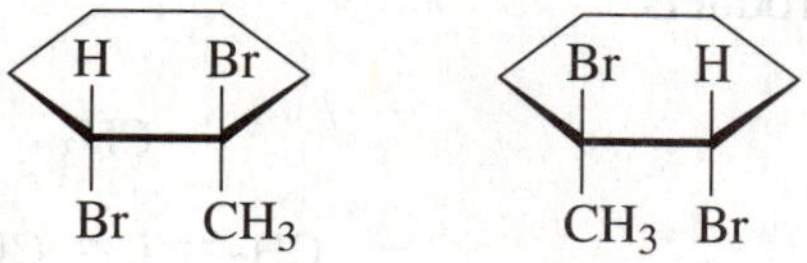

3. The products does not have an asymmetric center, so it does not have stereoisomers.

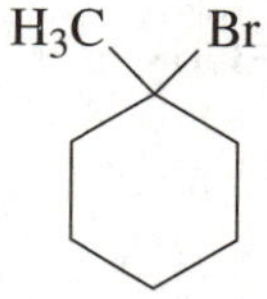

4. The product has two asymmetric centers.
Because radical addition of HBr can be either syn or anti, four products are obtained.

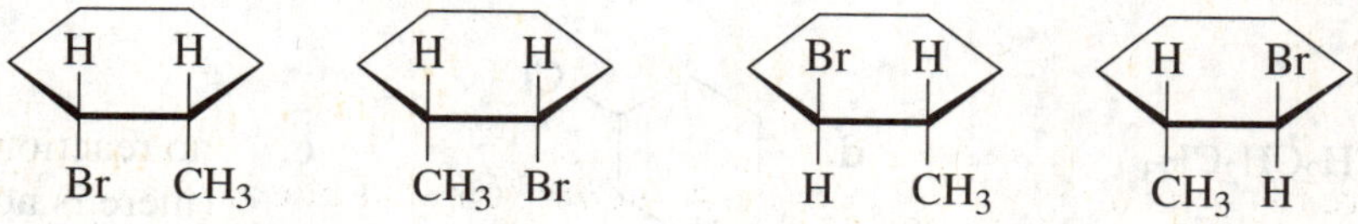

19. a.

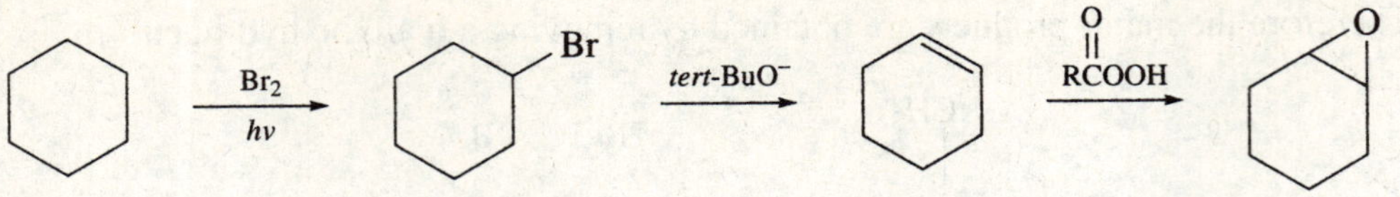

b.

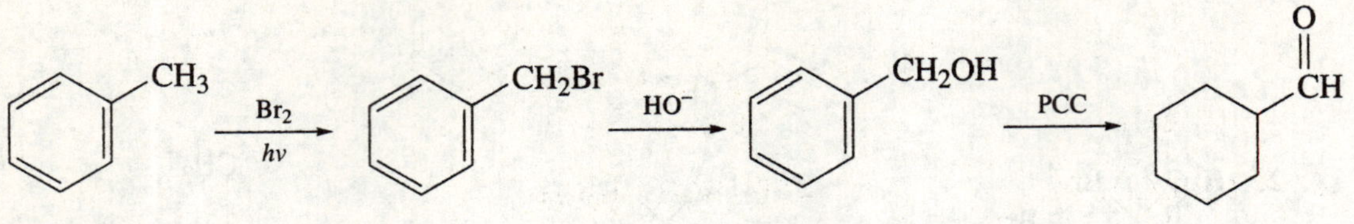

c.

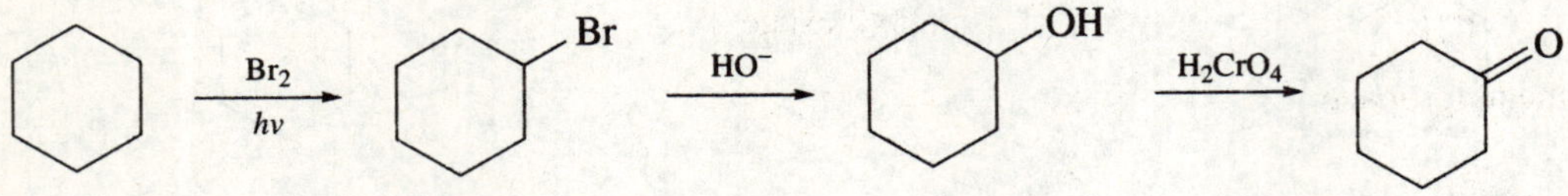

d.

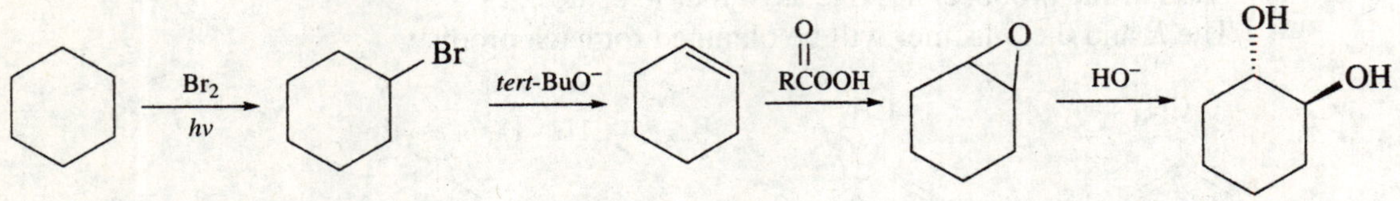

20. a. $CH_2{=}CHCHCH_2CH_3$ (Br on C-3) + $BrCH_2CH{=}CHCH_2CH_3$

b. In the radical intermediate that leads to the major products, the unpaired electron is shared by a primary carbon and a tertiary carbocation. In the radical intermediate that leads to the minor products, the unpaired electron is shared by a primary and a secondary carbocation. It, therefore, is not as stable as the intermediate that leads to the major products.

$CH_3\overset{\delta\cdot}{C}(CH_3){\cdots}CH{\cdots}\overset{\delta\cdot}{C}H_2$ $\qquad$ $\overset{\delta\cdot}{C}H_2{\cdots}C(CH_3){\cdots}\overset{\delta\cdot}{C}HCH_3$

$CH_3C(CH_3){=}CHCH_2Br$ + $CH_3C(CH_3)(Br)CH{=}CH_2$

major products

$BrCH_2C(CH_3){=}CHCH_3$ + $CH_2{=}C(CH_3)CH(Br)CH_3$

minor products

In bromination, selectivity is much more important than probability, so even though twice as many hydrogens are available for removal by a bromine radical that leads to the minor products, they will be minor products because the easier to remove hydrogens lead to the major products.

c. $CH_3CH_2C(CH_3)(Br)CH_2CH_2CH_3$ major product

d.

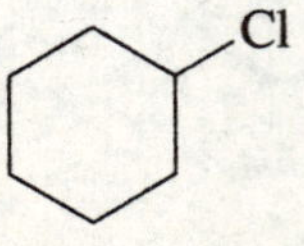

e. no reaction (there is no light or heat that is required to form radicals)

f. 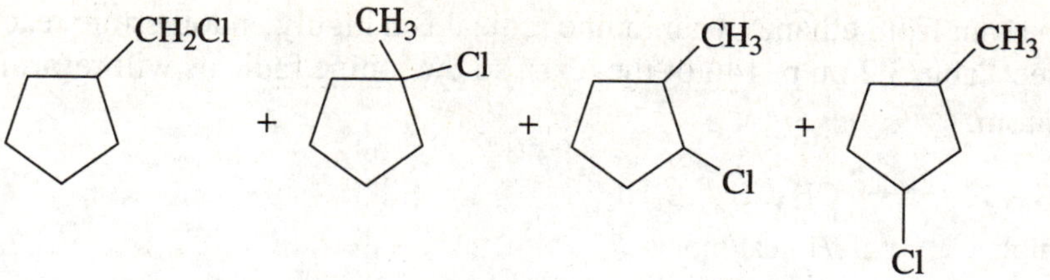

21. **a.**

$$\begin{array}{c} CH_3 \\ | \\ CH_3CCH_3 \\ | \\ CH_3 \end{array}$$

dimethylpropane

$$\begin{array}{l} \quad\quad\;\; CH_3 \\ \quad\quad\;\; | \\ CH_3CH_2CHCH_2CH_2CH_3 \end{array}$$

3-methylhexane

22. **a.** 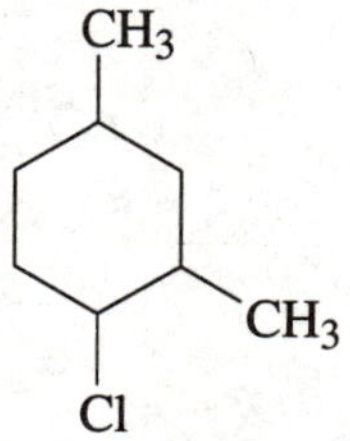

b. 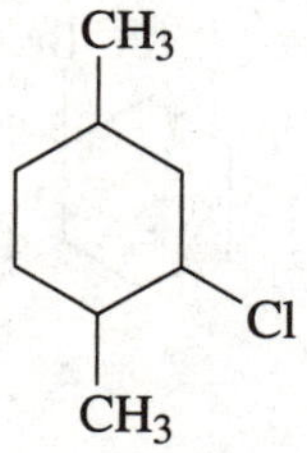

c. 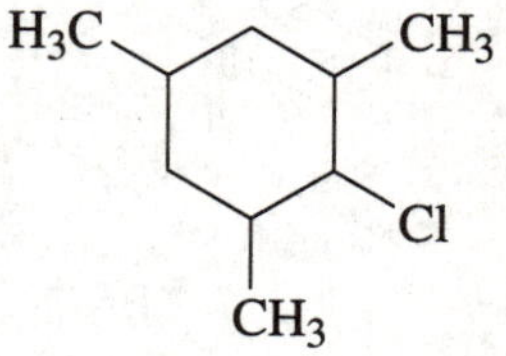

23. **a.** 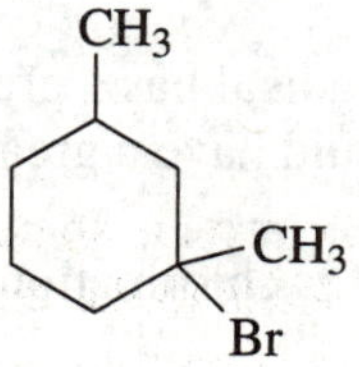

b. 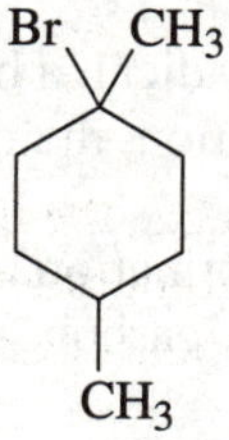

c. 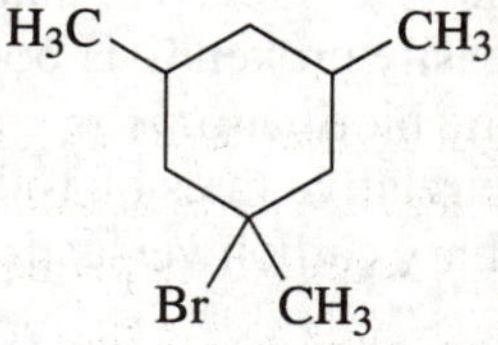

24. **a.** 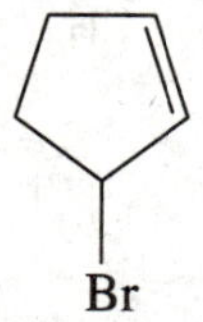

d. 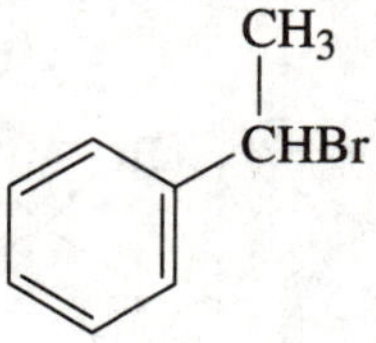

b. 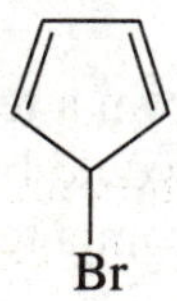

e. 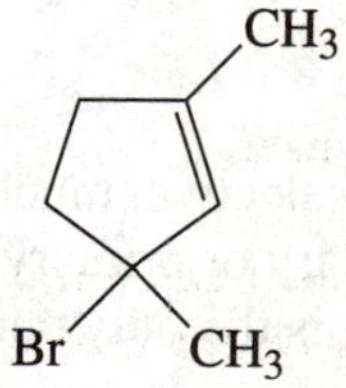

c. 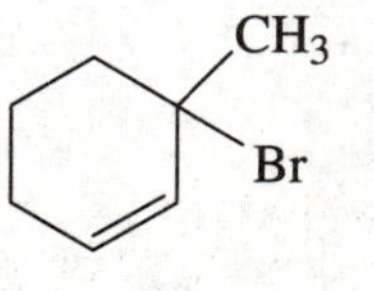

or 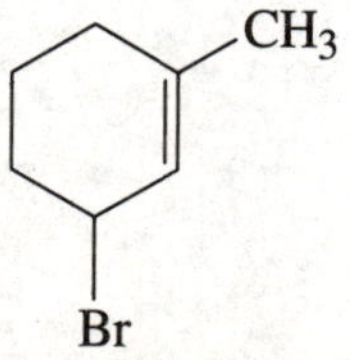

f. 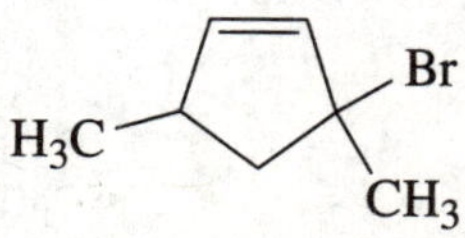

In **c**, because one product is under kinetic control and the other is under thermodynamic control, the major product will depend on the conditions under which the reaction is carried out.

25. Abstraction of a hydrogen atom from ethane by an iodine radical is a highly endothermic reaction ($\Delta H° = 101 - 71 = 30$ kcal/mol; see Table 3.2 on p. 146 of the text), so the iodine radicals will reform I_2 rather than abstract a hydrogen atom.

$$\underset{101\text{ kcal/mol}}{CH_3CH_3} + I\cdot \longrightarrow \underset{71\text{ kcal/mol}}{CH_3\dot{C}H_2} + HI$$

26.

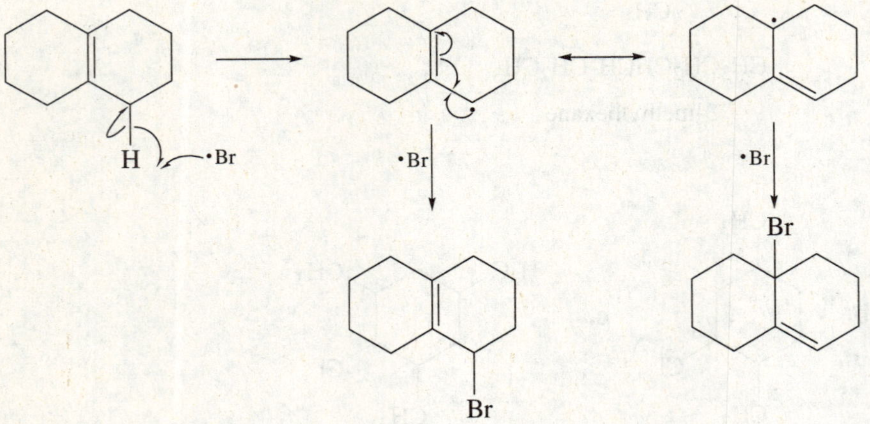

27. It is easier to break a C-H bond than a C-D bond.
Because a bromine radical is less reactive than a chlorine radical, a bromine radical has a greater preference for the more easily broken C-H bond. Bromination, therefore, would have a greater deuterium kinetic isotope effect than chlorination.
This is reminiscent of the relative rates of 1600 : 1 for bromination and 5 : 1 for chlorination for removal of a hydrogen from a tertiary carbon versus from a primary carbon.

28. **a.** Theoretically five monobromination products are possible, but some will be obtained in very small amounts.

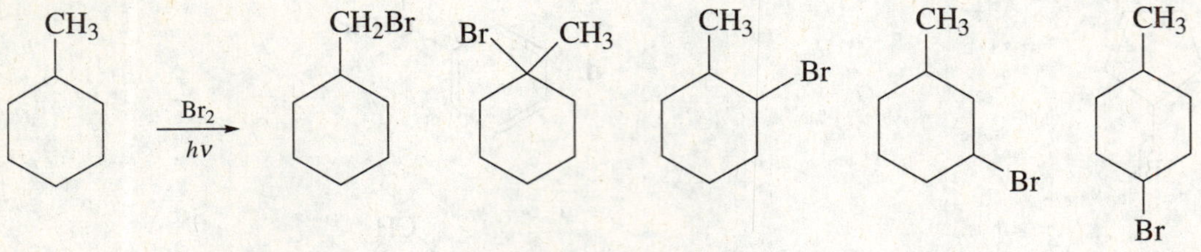

b. A bromine radical is very selective, much preferring to remove a hydrogen atom from a tertiary carbon than from a secondary or primary carbon. 1-Bromo-1-methylcyclohexane would be obtained in greatest yield because it is the only one formed by removing a hydrogen atom from a tertiary carbon.

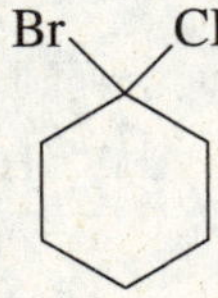

Therefore the major products are obtained by removing a **b** allylic hydrogen.

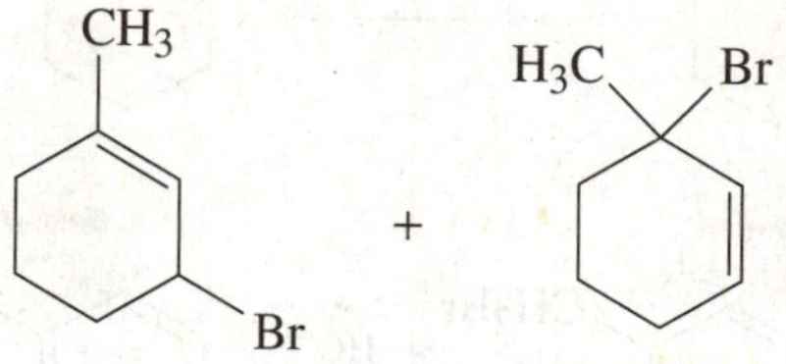

2. 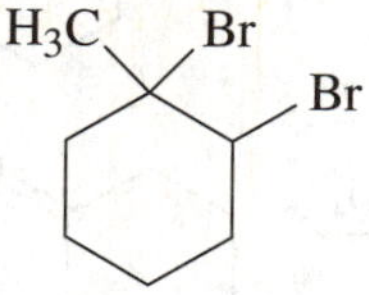

3. 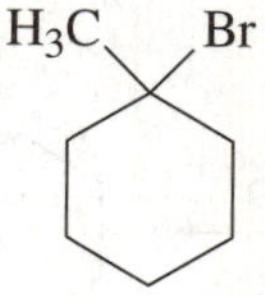

4. 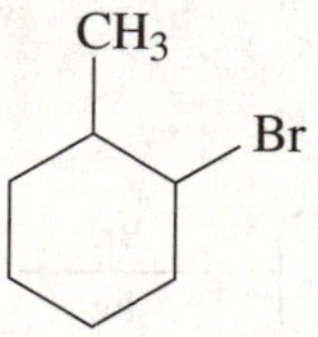

b. **1.** Each of the products has one asymmetric center.
The *R* and the *S* isomer will be obtained for each product.

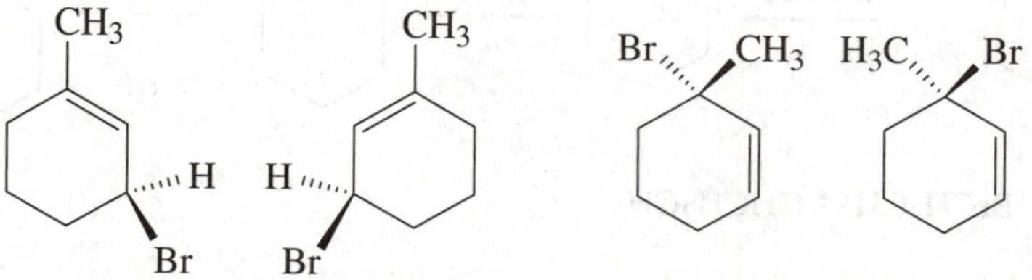

2. The product has two asymmetric centers.
Because addition of Br_2 is anti, only the products with the bromine atoms on opposite sides of the ring are obtained.

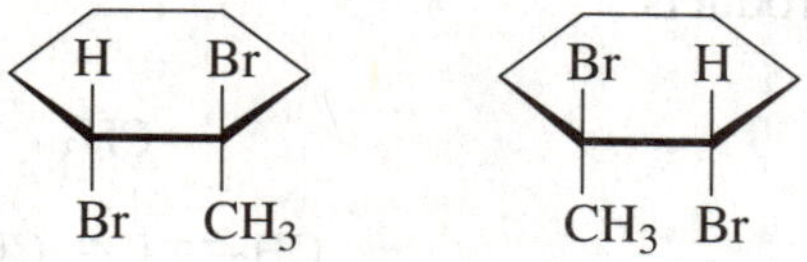

3. The products does not have an asymmetric center, so it does not have stereoisomers.

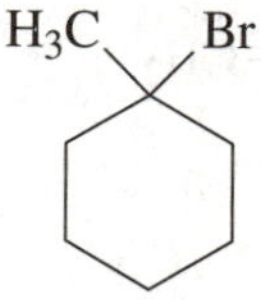

4. The product has two asymmetric centers.
Because radical addition of HBr can be either syn or anti, four products are obtained.

19.

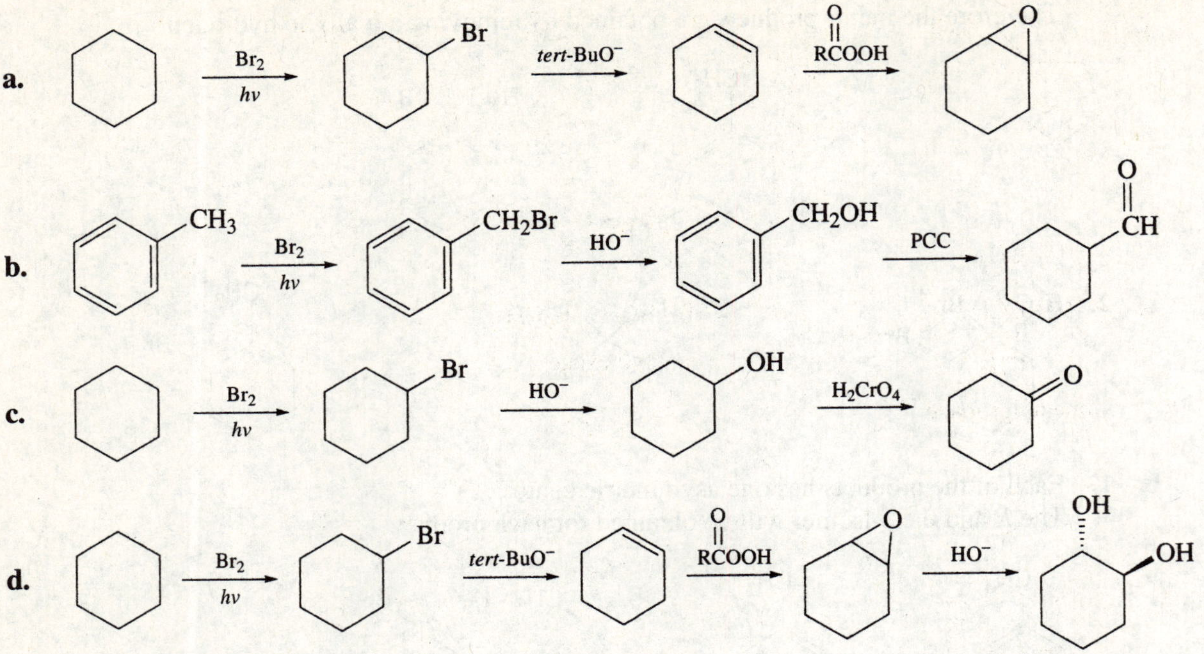

20. **a.** $CH_2{=}CHCH(Br)CH_2CH_3$ + $BrCH_2CH{=}CHCH_2CH_3$

b. In the radical intermediate that leads to the major products, the unpaired electron is shared by a primary carbon and a tertiary carbocation. In the radical intermediate that leads to the minor products, the unpaired electron is shared by a primary and a secondary carbocation. It, therefore, is not as stable as the intermediate that leads to the major products.

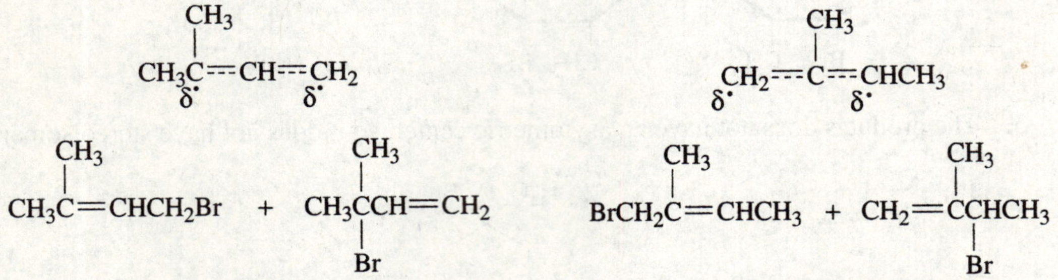

major products minor products

In bromination, selectivity is much more important than probability, so even though twice as many hydrogens are available for removal by a bromine radical that leads to the minor products, they will be minor products because the easier to remove hydrogens lead to the major products.

c. $CH_3CH_2C(CH_3)(Br)CH_2CH_2CH_3$ major product

d.

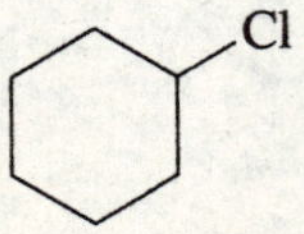

e. no reaction (there is no light or heat that is required to form radicals)

f. 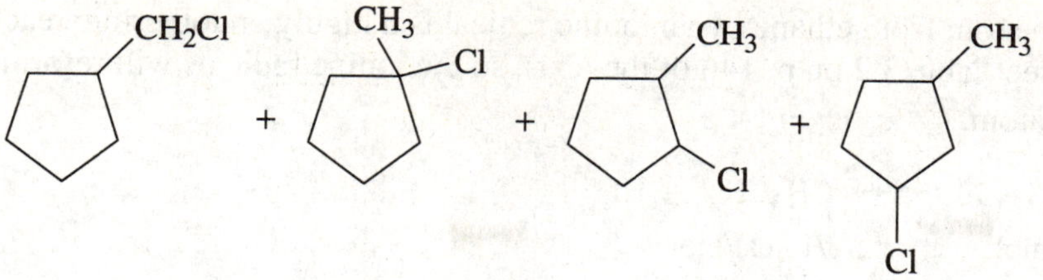

21. **a.** 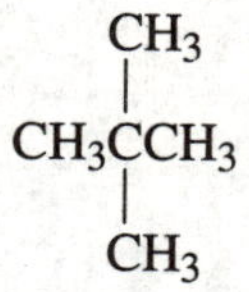

dimethylpropane

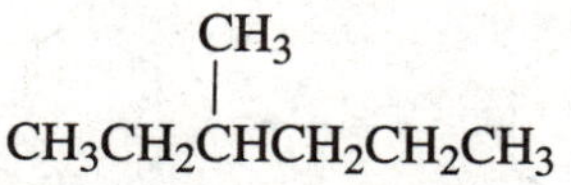

3-methylhexane

22. **a.** 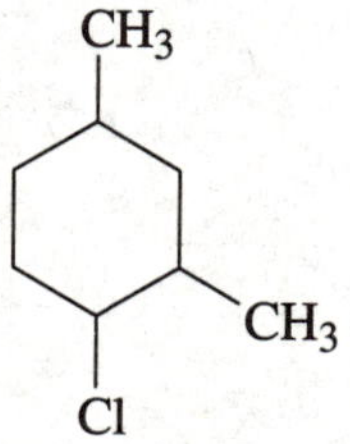 **b.** 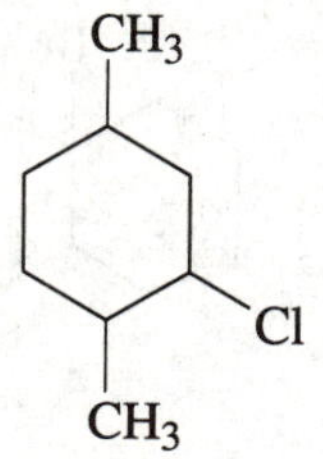 **c.** 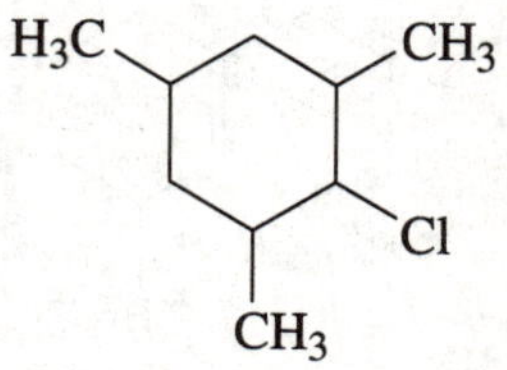

23. **a.** 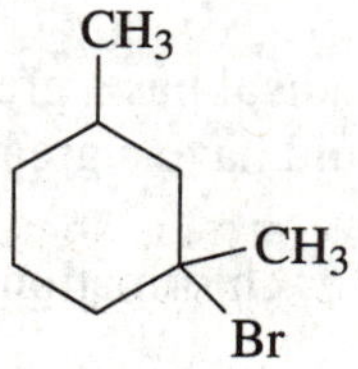 **b.** 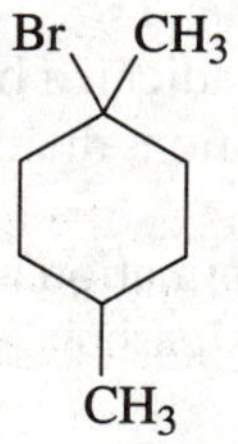 **c.** 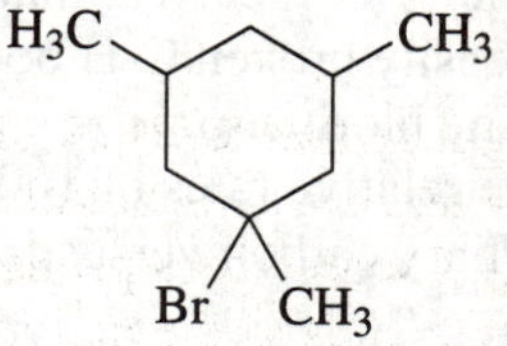

24. **a.** 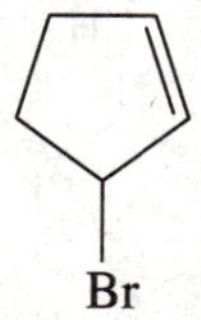 **d.** 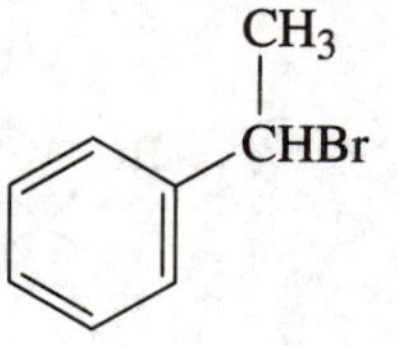

b. 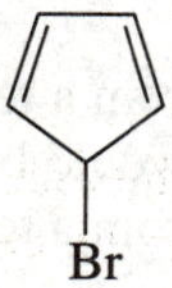 **e.** 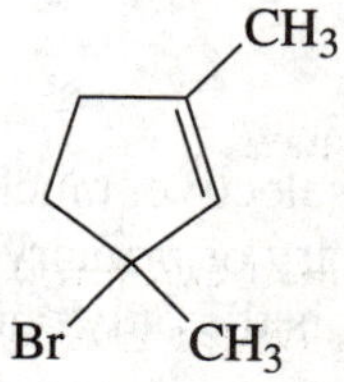

c. 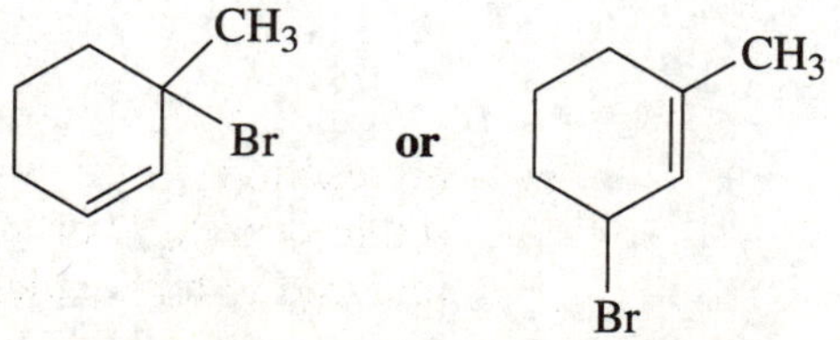 **f.** 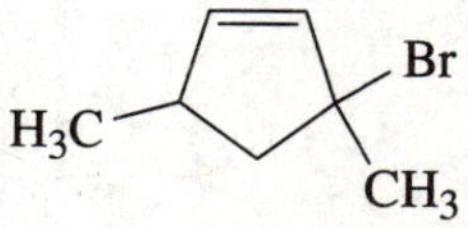

In **c**, because one product is under kinetic control and the other is under thermodynamic control, the major product will depend on the conditions under which the reaction is carried out.

25. Abstraction of a hydrogen atom from ethane by an iodine radical is a highly endothermic reaction ($\Delta H° = 101 - 71 = 30$ kcal/mol; see Table 3.2 on p. 146 of the text), so the iodine radicals will reform I_2 rather than abstract a hydrogen atom.

$$\underset{101\text{ kcal/mol}}{CH_3CH_3} + I\cdot \longrightarrow \underset{71\text{ kcal/mol}}{CH_3\dot{C}H_2} + HI$$

26.

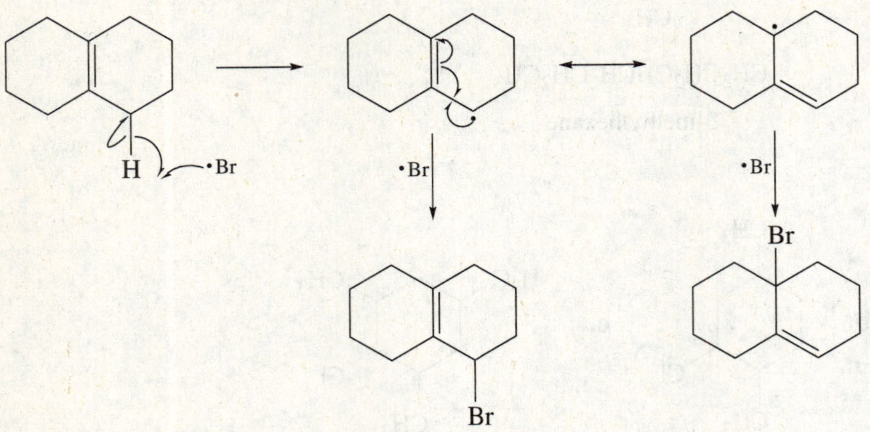

27. It is easier to break a C-H bond than a C-D bond.
Because a bromine radical is less reactive than a chlorine radical, a bromine radical has a greater preference for the more easily broken C-H bond. Bromination, therefore, would have a greater deuterium kinetic isotope effect than chlorination.
This is reminiscent of the relative rates of 1600 : 1 for bromination and 5 : 1 for chlorination for removal of a hydrogen from a tertiary carbon versus from a primary carbon.

28. **a.** Theoretically five monobromination products are possible, but some will be obtained in very small amounts.

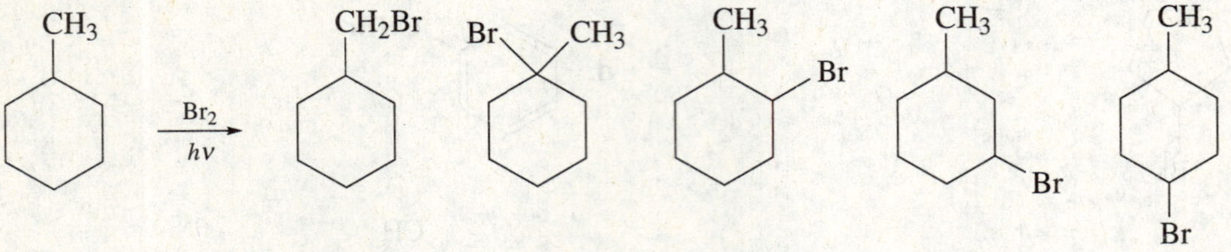

b. A bromine radical is very selective, much preferring to remove a hydrogen atom from a tertiary carbon than from a secondary or primary carbon. 1-Bromo-1-methylcyclohexane would be obtained in greatest yield because it is the only one formed by removing a hydrogen atom from a tertiary carbon.

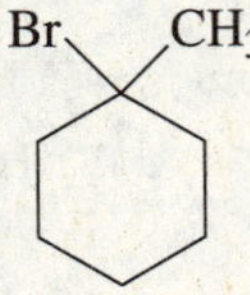

c. The number of possible stereoisomers for each compound is indicated below each structure.

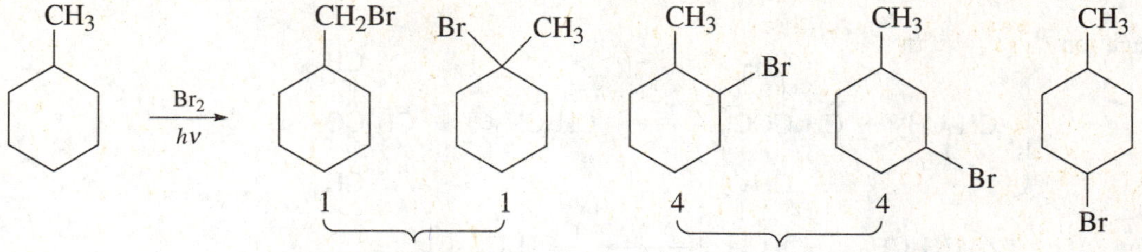

2

These have 0 asymmetric centers.

These have 2 asymmetric centers.

This does not have an asymmetric center.

A total of 12 stereoisomers can be obtained.

29. **a.** 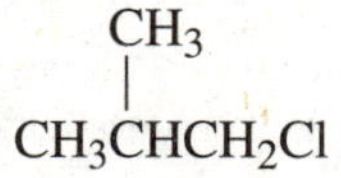

c. 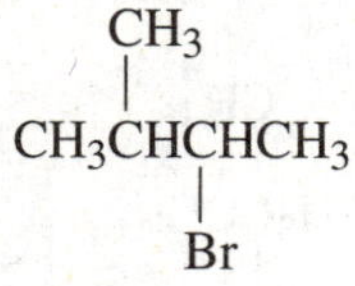

e. 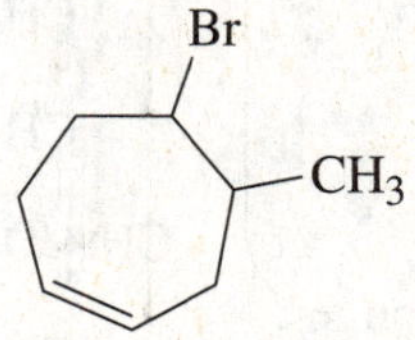

b. 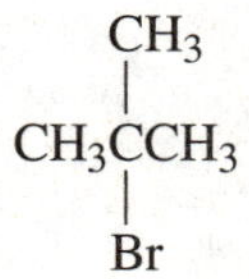

d. 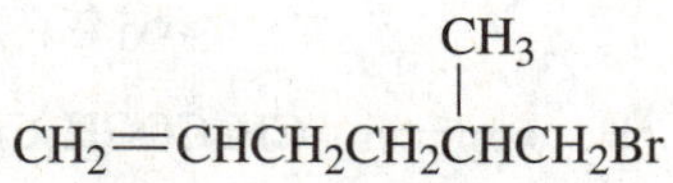

f. 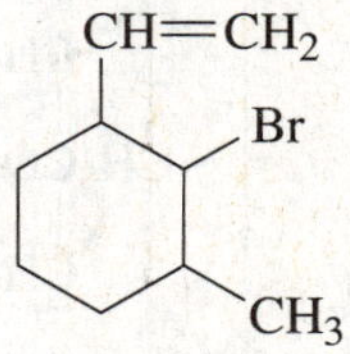

30. **a.** One difference between this reaction and the monochlorination of ethane is the source of the chlorine radical. This reaction has two sets of propagation steps because two different radicals are generated in the initiation step. This reaction also has several more termination steps because of the two different radicals generated in the initiation step.

initiation 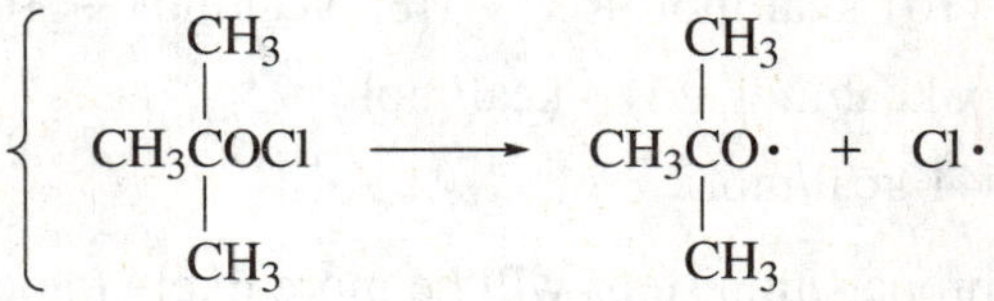

propagation 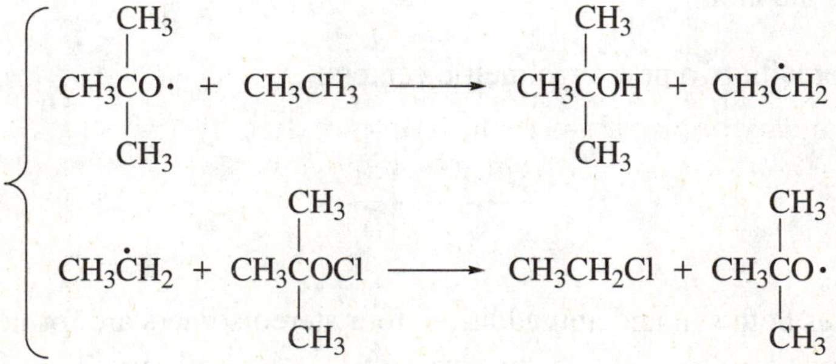

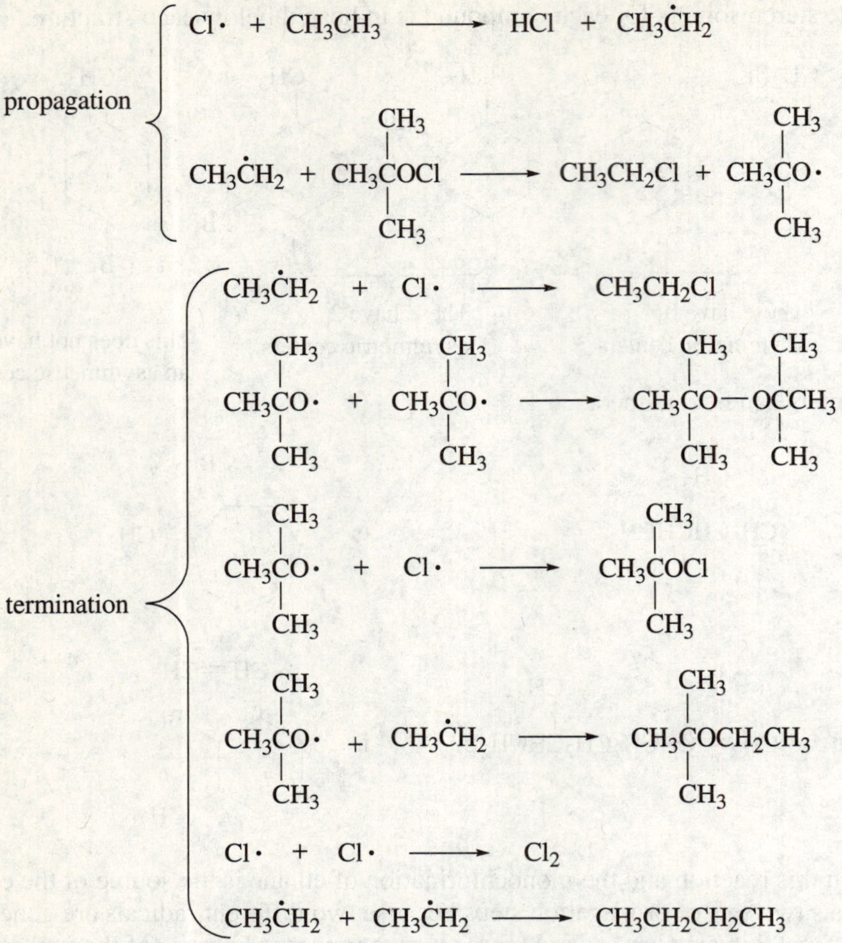

b. $$\Delta H^\circ = \text{bonds broken} - \text{bonds formed}$$

$$-42 \text{ kcal/mol} = [101 \text{ kcal/mol} + x] - [82 \text{ kcal/mol} + 102 \text{ kcal/mol}]$$

$$-143 \text{ kcal/mol} = x \text{ kcal/mol} - 184 \text{ kcal/mol}$$

$$x = 41 \text{ kcal/mol}$$

c. The second set of propagation steps will be more likely to occur because the HCl bond is stronger than the OH bond of the alcohol.

31. The reaction forms a product with two new asymmetric centers.

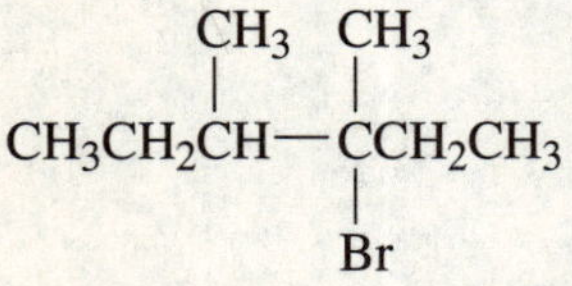

Because the reaction involves both syn and anti addition, four stereoisomers are formed.

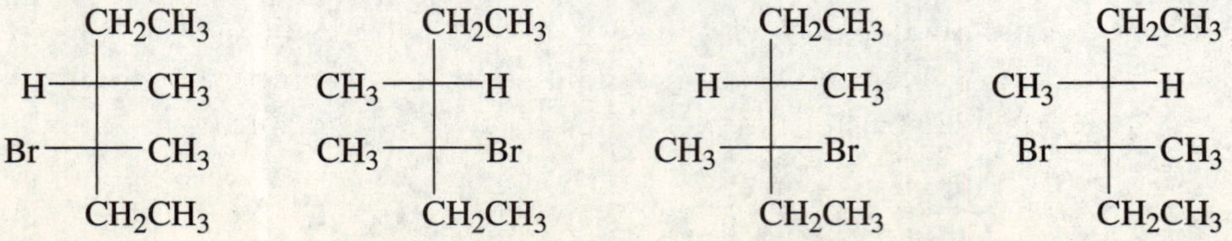

32. **a.** $HC{\equiv}CH \xrightarrow[\text{2. } CH_3CH_2CH_2Br]{\text{1. } ^-NH_2} CH_3CH_2CH_2C{\equiv}CH \xrightarrow[\text{Lindlar catalyst}]{H_2} CH_3CH_2CH_2CH{=}CH_2 \xrightarrow{\text{HBr, peroxide}} CH_3CH_2CH_2CH_2CH_2Br$

b. $HOCH_2CH_2CH{=}CH_2 \xrightarrow[\text{peroxide}]{HBr} HOCH_2CH_2CH_2CH_2Br \xrightarrow{NaH} {}^-OCH_2CH_2CH_2CH_2Br \longrightarrow$

c. $CH_3CH_2CH{=}CH_2 \xrightarrow[\text{peroxide}]{\text{NBS, }\Delta} CH_3CH(Br)CH{=}CH_2 \xrightarrow{tert\text{-BuO}^-} CH_2{=}CHCH{=}CH_2$

d.

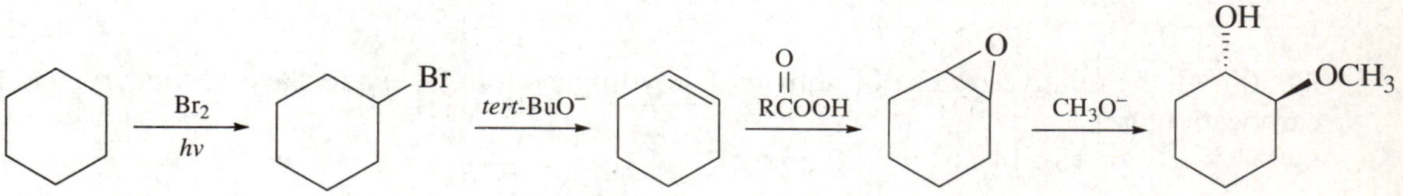

e.

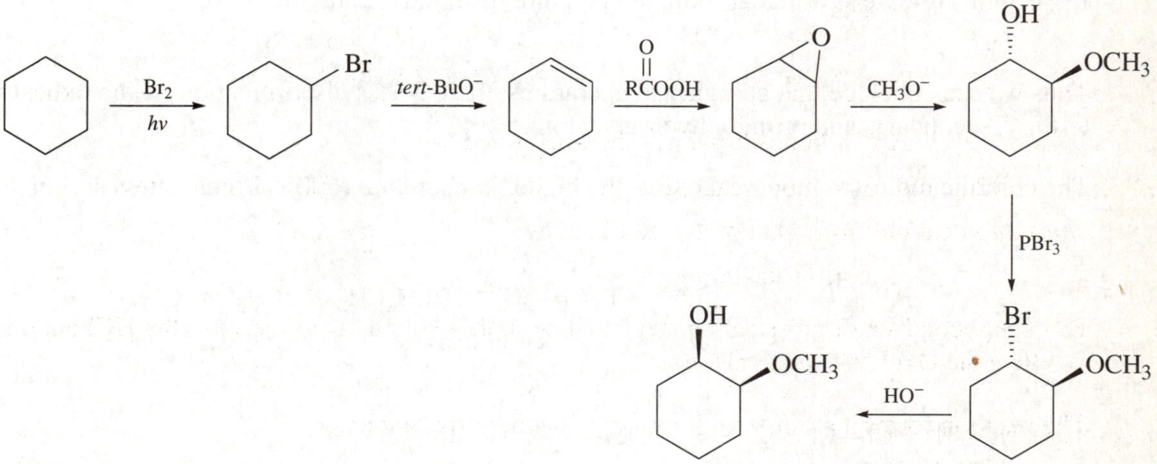

33.

$CH_3CHCH_2CH_3$ (with CH_3 on C-2)

2-methylbutane

2-Methylbutane forms two primary alkyl halides, one secondary alkyl halide, and one tertiary alkyl halide. First we need to calculate how much alkyl halide is formed from each primary hydrogen available, from each secondary hydrogen available, and from each tertiary hydrogen available.

a primary alkyl halide

$ClCH_2CHCH_2CH_3$ (with CH_3 on C2)	Substitution of any one of 6 hydrogens leads to this product. percentage of this product that is formed = 36 percentage formed per hydrogen available = 36/6 = 6
$CH_3CHCH_2CH_2Cl$ (with CH_3 on C2)	Substitution of any one of 3 hydrogens leads to this product. percentage of this product that is formed = 18 percentage formed per hydrogen available = 18/3 = 6

a secondary alkyl halide

$CH_3CHCHCH_3$ (Cl on C3, CH_3 on C2)	Substitution of any one of 2 hydrogens leads to this product. percentage of this product that is formed = 28 percentage formed per hydrogen available = 28/2 = 14

a tertiary alkyl halide

$CH_3CCH_2CH_3$ (Cl and CH_3 on C2)	Substitution of any one of 1 hydrogens leads to this product. percentage of this product that is formed = 18 percentage formed per hydrogen available = 18/1 = 18

At 300 °C, the relatives rates of removal of a hydrogen atom from a tertiary, secondary, and primary carbocation are:

$$18 : 14 : 6 \;=\; 3 : 2.3 : 1$$

In Section 11.4, we saw that at room temperature the relative rates are:

$$5 : 3.8 : 1$$

Thus we can conclude that at higher temperatures, there is less discrimination by the radical between a tertiary, secondary, and primary hydrogen atom.

34. The chlorine radical is more reactive at the higher temperature (600°C), so it is less selective.

35. **a.** 

b. 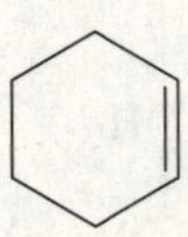 $CH_3CH_2CH{=}CHCH_2CH_3$ (cis or trans) $CH_3C{=}CCH_3$ (with CH_3 on each alkene carbon)

36. **a.** $CH_3{-}H \quad Cl{-}Cl \longrightarrow CH_3{-}Cl \quad H{-}Cl$

105 58 84 103

$\Delta H°$ = bonds broken – bonds formed

$\Delta H°$ = [105 kcal/mol + 58 kcal/mol] – [84 kcal/mol + 103 kcal/mol]

$\Delta H°$ = 163 – 187 = –24 kcal/mol

b. $CH_3{-}H + \dot{Cl} \longrightarrow \dot{C}H_3 + H{-}Cl \qquad 105 - 103 = \dfrac{\Delta H^{o}}{2\ \text{kcal/mol}}$

$\dot{C}H_3 + Cl{-}Cl \longrightarrow CH_3{-}Cl + \dot{Cl} \qquad 58 - 84 = \dfrac{-26\ \text{kcal/mol}}{-24\ \text{kcal/mol}}$

c. If you cancel the things that are the same on opposite sides of the equations in part b an then add the two equations, you are left with the equation in part a.

$$\begin{array}{lcl} CH_3{-}H + \dot{\not{C}l} & \longrightarrow & \dot{C}\not{H}_3 + H{-}Cl \\ \not{C}\not{H}_3 + Cl{-}Cl & \longrightarrow & CH_3{-}Cl + \dot{\not{C}l} \\ \hline CH_4 + Cl_2 & \longrightarrow & CH_3Cl + HCl \end{array}$$

37. The first propagation step is very endothermic, so it would not be able to compete with the first propagation step shown in Problem 36b.

$$\begin{array}{lcll} & & & \Delta H^\circ \\ CH_3{-}H + \cdot Cl & \longrightarrow & CH_3{-}Cl + \cdot H & 105-84 = 21\,kcal/mol \\ H\cdot + Cl{-}Cl & \longrightarrow & H{-}Cl + \cdot Cl & 58-103 = \underline{-45\,kcal/mol} \\ & & & \qquad\qquad -24\,kcal/mol \end{array}$$

38. $RO{-}OR \longrightarrow 2RO\cdot$

$RO\cdot + HCCl_3 \longrightarrow ROH + \cdot CCl_3$

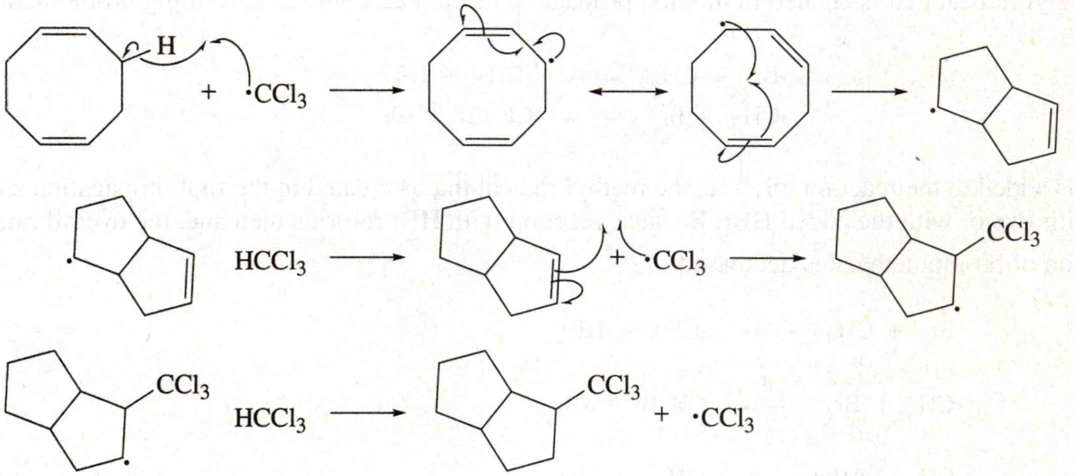

39. **a.**

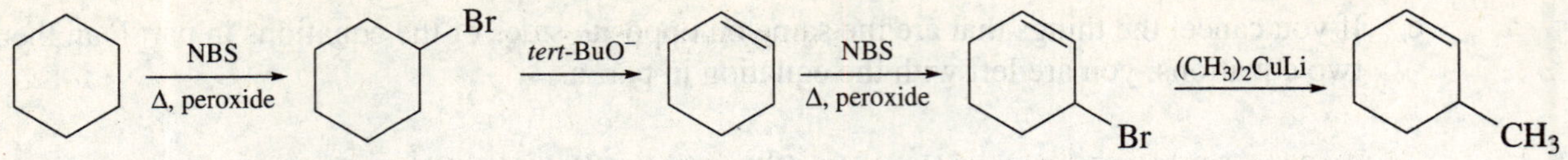

b.
$$CH_3CH_2CH{=}CH_2 \xrightarrow{Br_2} CH_3CH_2\underset{\displaystyle Br}{\underset{|}{C}}H{-}CH_2Br \xrightarrow{^-NH_2} CH_3CH_2C{\equiv}CH \xrightarrow{^-NH_2} CH_3CH_2C{\equiv}C^-$$

$$CH_3CH_2CH{=}CH_2 \xrightarrow[\text{peroxide}]{HBr} CH_3CH_2CH_2CH_2Br$$

$$CH_3CH_2C{\equiv}C^- \xrightarrow{CH_3CH_2CH_2CH_2Br} CH_3CH_2C{\equiv}CCH_2CH_2CH_2CH_3 \xrightarrow[\text{Pd/C}]{H_2} CH_3CH_2CH_2CH_2CH_2CH_2CH_2CH_3$$

c.
$$CH_3\overset{\displaystyle CH_3}{\overset{|}{\underset{\underset{\displaystyle OH}{|}}{C}}}CH_2CH_2CH_3 \xrightarrow[\Delta]{H_2SO_4} CH_3\overset{\displaystyle CH_3}{\overset{|}{C}}{=}CHCH_2CH_3 \xrightarrow[\text{peroxide}]{HBr} CH_3\overset{\displaystyle CH_3}{\overset{|}{C}}H\underset{\displaystyle Br}{\underset{|}{C}}HCH_2CH_3$$

40. The methyl radical that is created in the first propagation step reacts with Br_2, forming bromomethane.

$$\cdot Br + CH_4 \longrightarrow \cdot CH_3 + HBr$$
$$\cdot CH_3 + Br_2 \longrightarrow CH_3Br + \cdot Br$$

If HBr is added to the reaction mixture, the methyl radical that is created in the first propagation step can react with Br_2 or with the added HBr. Because reaction with HBr reforms methane, the overall rate of formation of bromomethane is decreased.

$$\cdot Br + CH_4 \longrightarrow \cdot CH_3 + HBr$$

$$\cdot CH_3 + Br_2 \longrightarrow CH_3Br + \cdot Br$$

$$\cdot CH_3 + HBr \longrightarrow CH_4 + \cdot Br$$

Chapter 11 Practice Test

1. How many monochlorinated products would be obtained from the following reaction? (Ignore stereoisomers.)

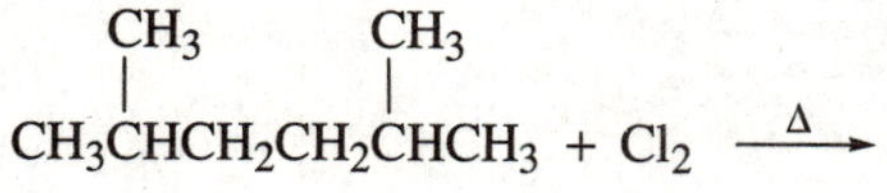

2. Give the first propagation step in the monochlorination of ethane.

3. When (*S*)-2-bromopentane is brominated, 2,3-dibromopentanes are formed. Which of the following compounds are **not** formed?

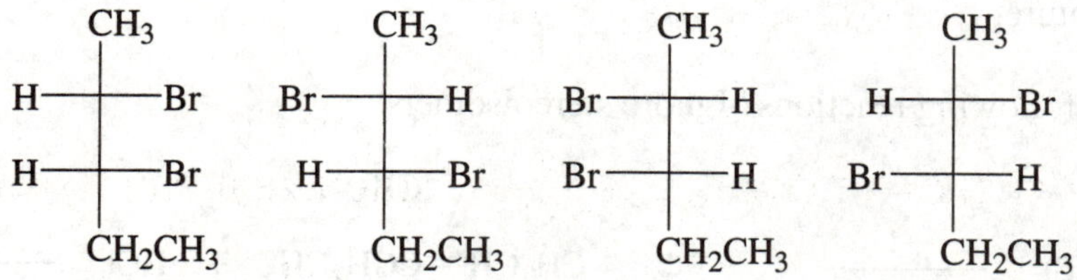

4. Determine the $\Delta H°$ of the two propagation steps in the monochlorination of ethane, using Table 3.2 on page 146 of the text.

5. Label the radicals in order of decreasing stability. (Label the most stable #1.)

$\dot{C}H_2CH_2CH_2CH{=}CH_2$ $\qquad$ $CH_3CH_2\dot{C}HCH{=}CH_2$

$CH_3CH_2CH_2CH{=}\dot{C}H$ $\qquad$ $CH_3\dot{C}HCH_2CH{=}CH_2$

6. Give the product(s) of the following reactions, ignoring stereoisomers:

a. 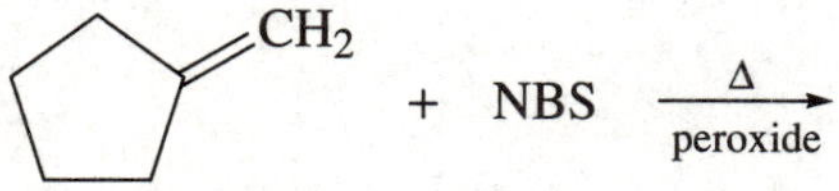

b. $CH_3CH_2CH{=}CH_2 + NBS \xrightarrow[\text{peroxide}]{\Delta}$

7. a. Give the products that will be obtained from monochlorination of the following alkane at room temperature: (Ignore stereoisomers.)

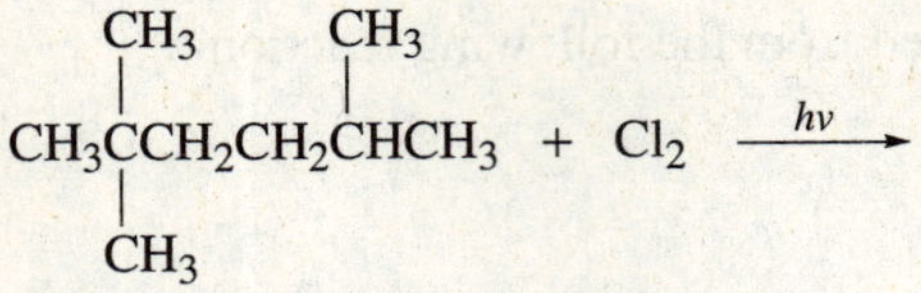

b. What product will be obtained in greatest yield?

c. What product would be obtained in greatest yield if the alkane were brominated instead of chlorinated?

8. Calculate the yield of 2-chloro-2-methylbutane formed when 2-methylbutane is chlorinated in the presence of light at room temperature.

9. What is the major product of the following reactions? Ignore stereoisomers.

a. 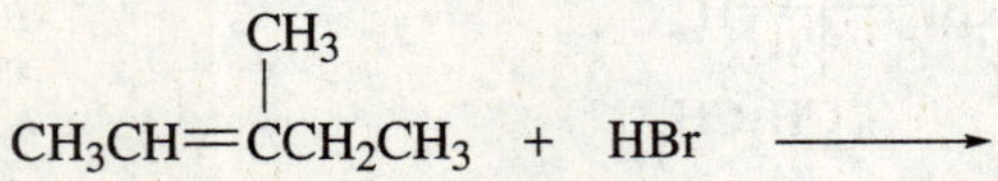

c. 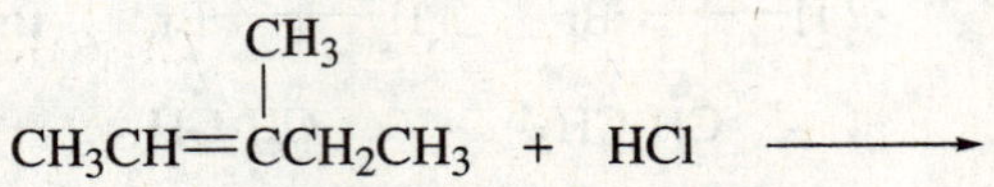

b. 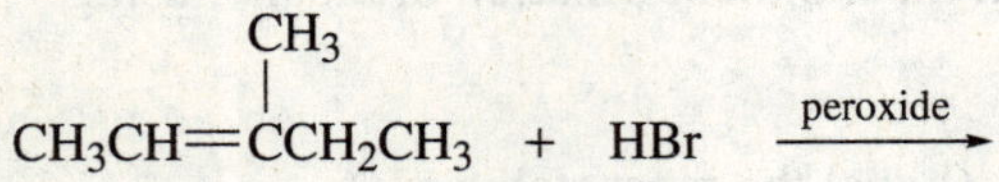

d. 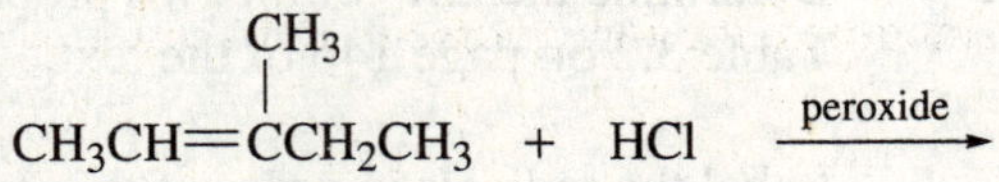

SPECIAL TOPIC VII

Drawing Curved Arrows in Radical Systems

We have seen that chemists use curved arrows:

1. to show how electrons move in going from a reactant to a product.
2. to show how electrons move in going from one resonance contributor to the next.

In Special Topic II (using curved arrows in reactions) and Special Topic V (using curved arrows in drawing resonance contributors), the electron movement involved the simultaneous movement of two electrons. Here we will show electron movement that involves the movement of only one electron.

Drawing Curved Arrows in Radical Reactions

When a bond breaks in a way that allows each of the bonded atoms to retain one of the bonding electrons, an arrowhead with one barb is used to represent the movement of the single electron.

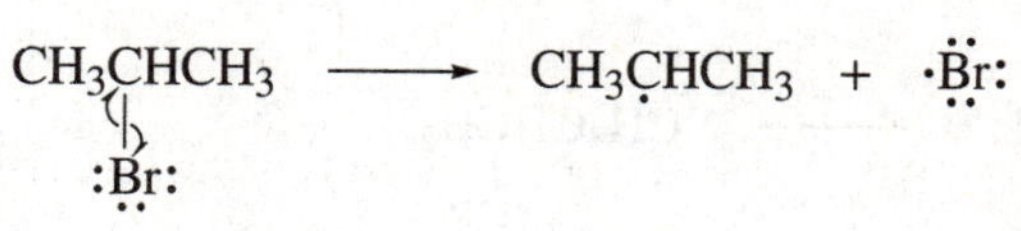

Sometimes, the lone pairs are not shown.

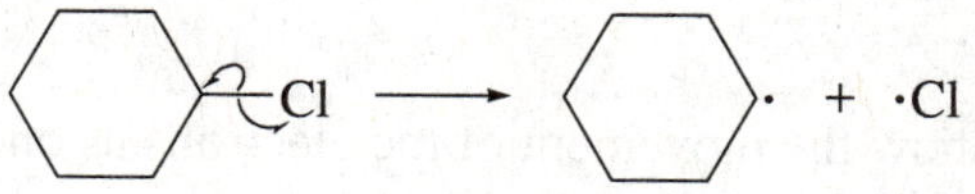

When a bond is formed using one electron from one atom and one electron from another, an arrowhead with one barb is used to represent the movement of the single electron.

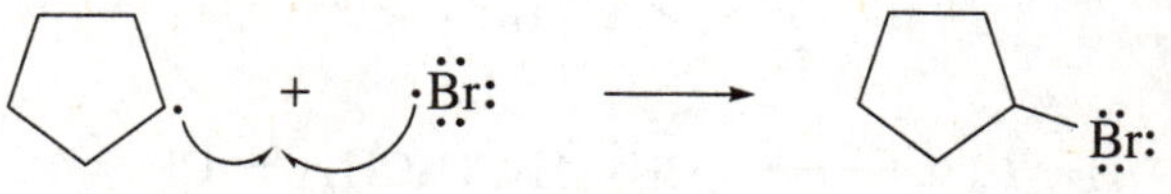

Sometimes, the lone pairs are not shown.

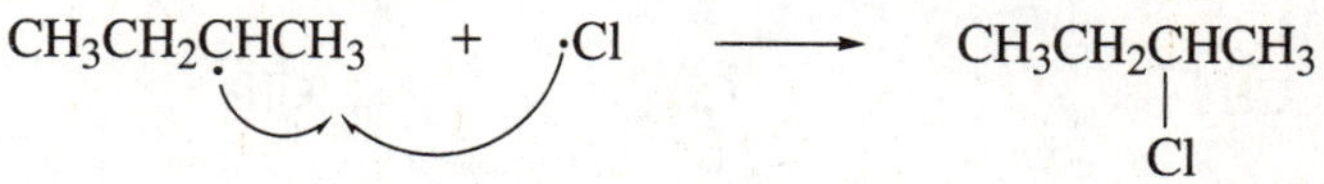

Problem 1. Draw curved arrows to show the movement of the electrons as the bond breaks.

a. $CH_3CH_2—\ddot{\underset{..}{Cl}}: \longrightarrow CH_3\dot{C}H_2 + \cdot\ddot{\underset{..}{Cl}}:$

b.

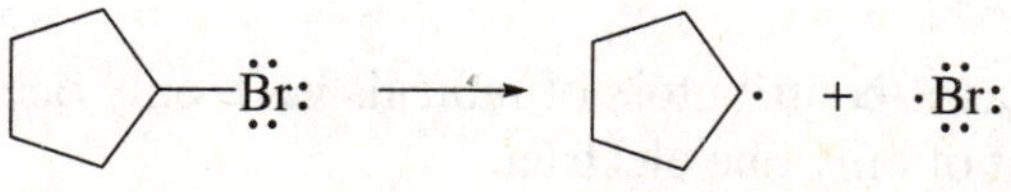

c. $CH_3\ddot{\underset{..}{O}}—\ddot{\underset{..}{O}}CH_3 \longrightarrow CH_3\ddot{\underset{..}{O}}\cdot + \cdot\ddot{\underset{..}{O}}CH_3$

Problem 2. Draw curved arrows to show the movement of the electrons as the bond forms.

a. 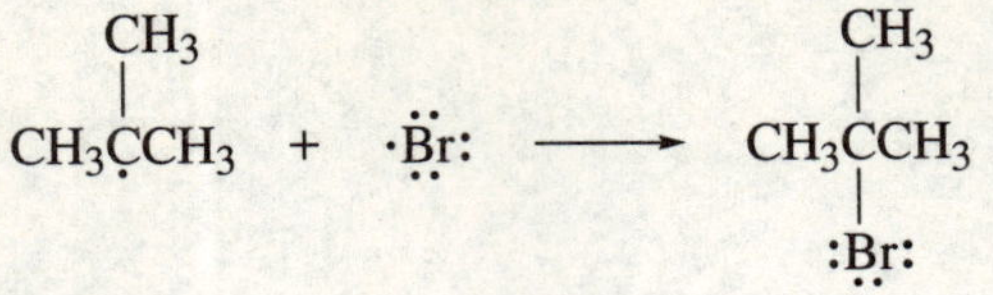

b. $\cdot Br + \cdot Br \longrightarrow Br_2$

c. 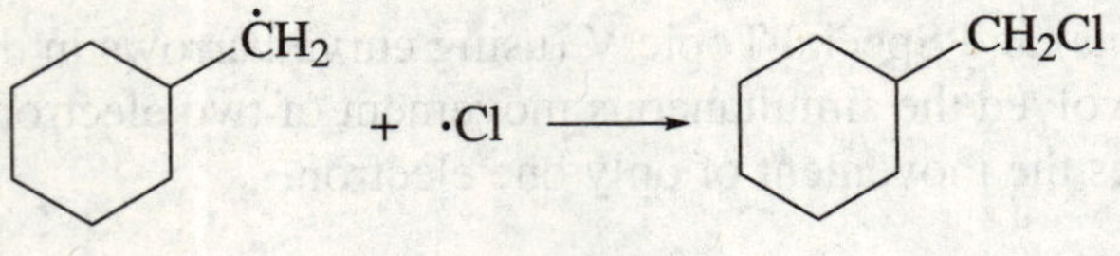

Often a bond breaks and a bond forms in the same step.

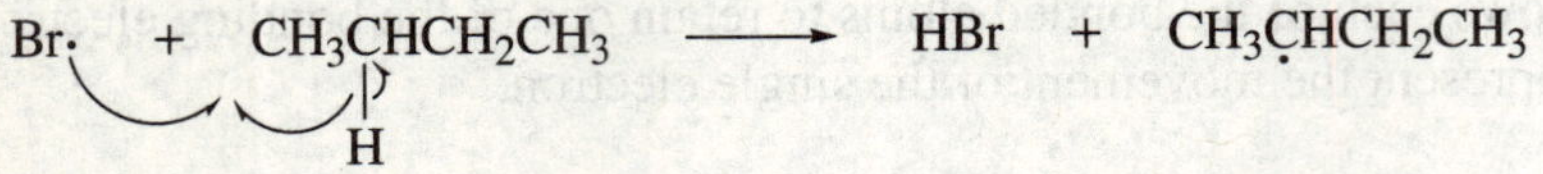

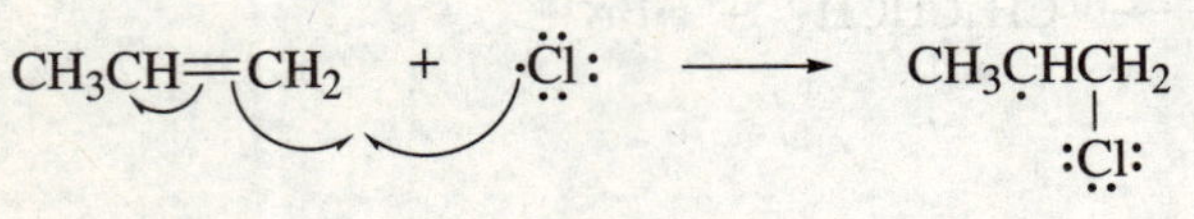

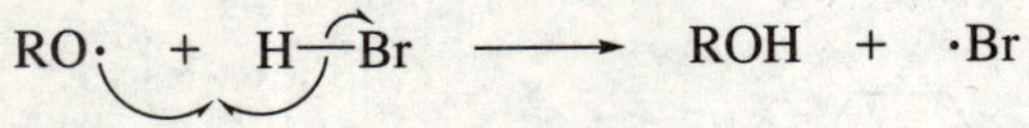

Problem 3. Draw curved arrows to show the movement of the electrons as one bond breaks and another bond forms.

a. 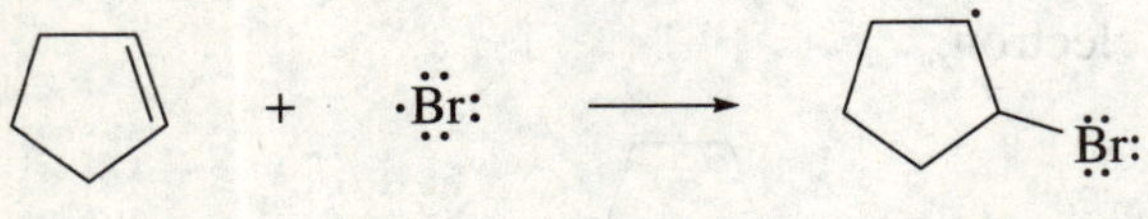

b. $CH_3\dot{C}HCH_2CH_3 + H—Cl \longrightarrow CH_3CH_2CH_2CH_3 + \cdot Cl$

c. $CH_3\dot{C}H_2 + Br—Br \longrightarrow CH_3CH_2Br + \cdot Br$

d. 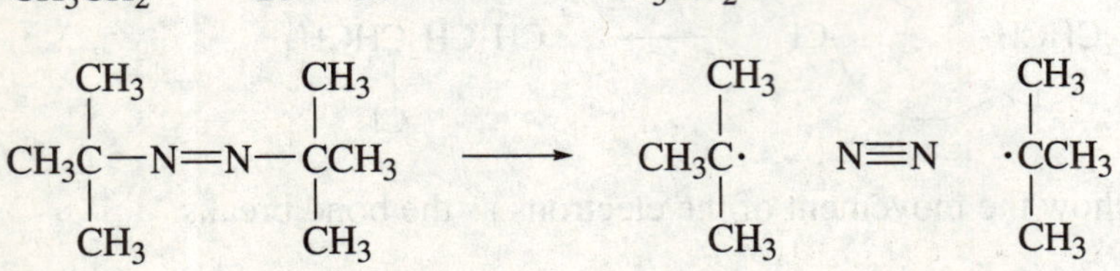

Drawing Curved Arrows in Contributing Resonance Structures that are Radicals

The arrows that represent electron movement in resonance contributors of radicals have only one barb on the arrowhead because the arrow represents the movement of only one electron.

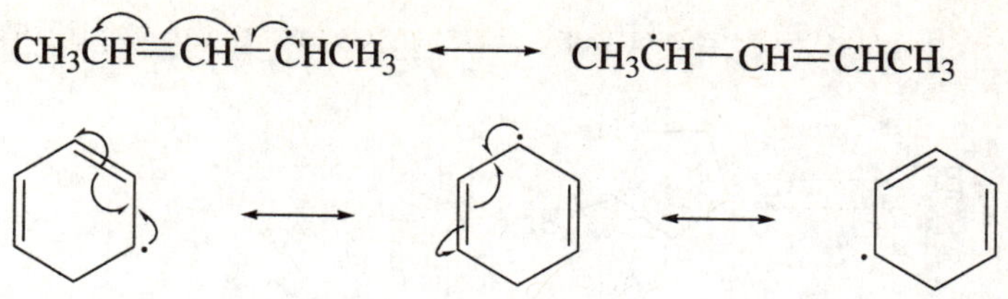

Problem 4. Draw curved arrows to show the movement of the electrons as one resonance contributor is converted to the next.

a. 

b. 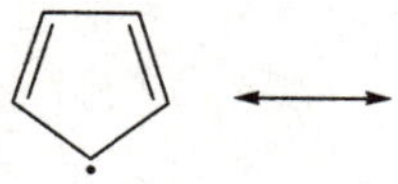

c. 

d. 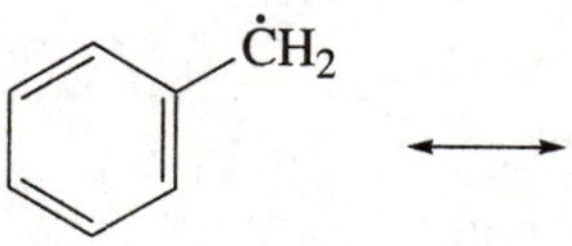

Answers to Problems

Problem 1.

a. 

b. 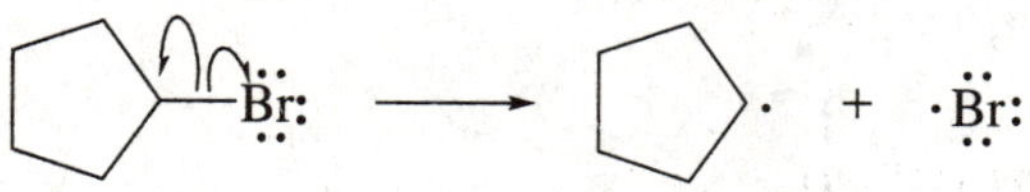

c. 

Problem 2.

a. 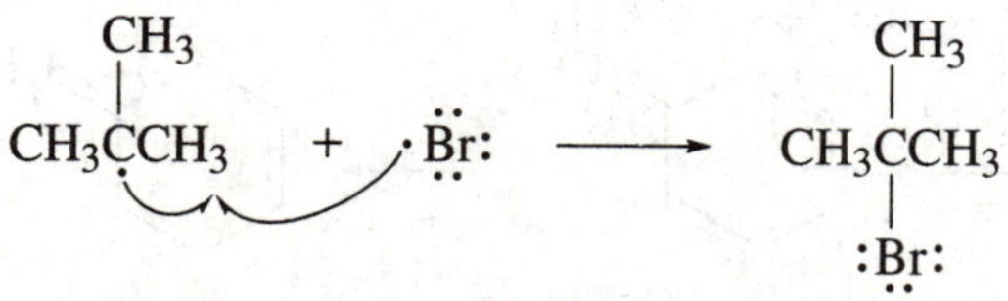

b. 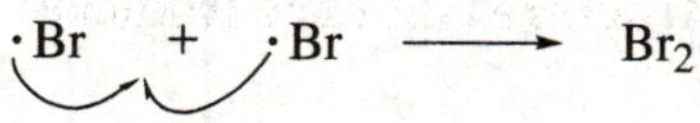

c. 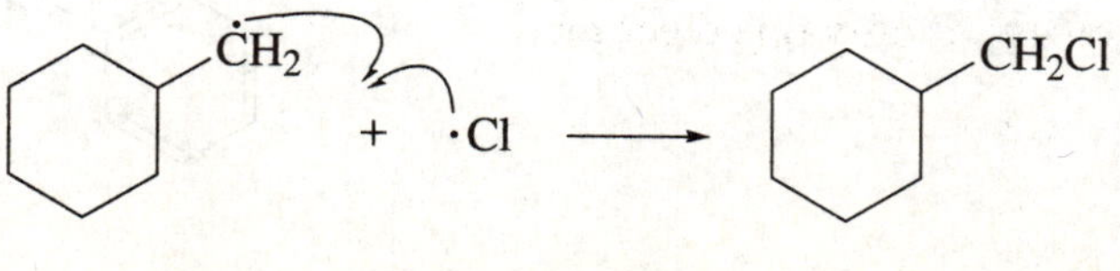

Problem 3.

a.

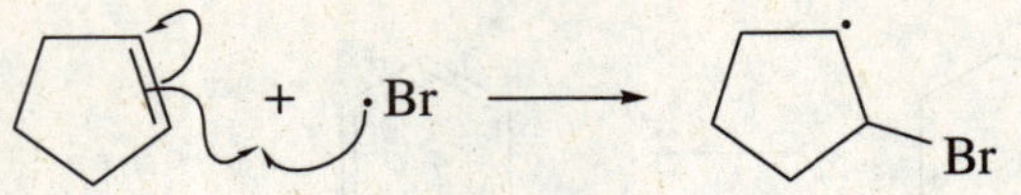

b.

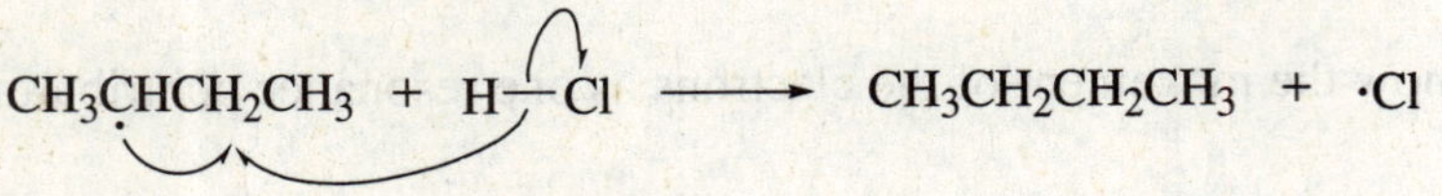

c.

d.

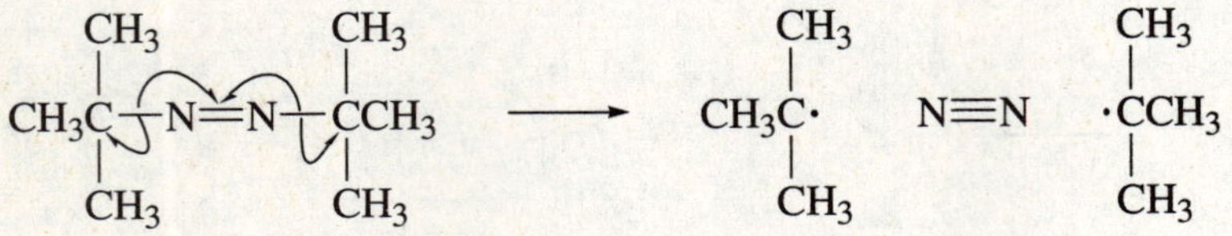

Problem 4.

a.

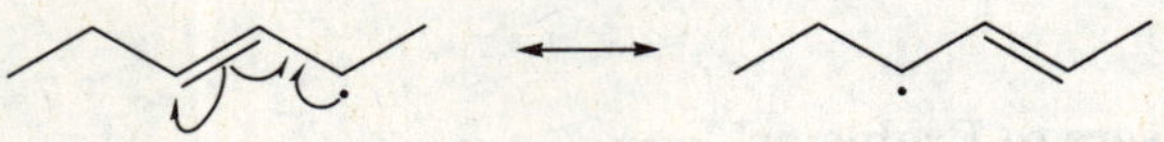

b.

c.

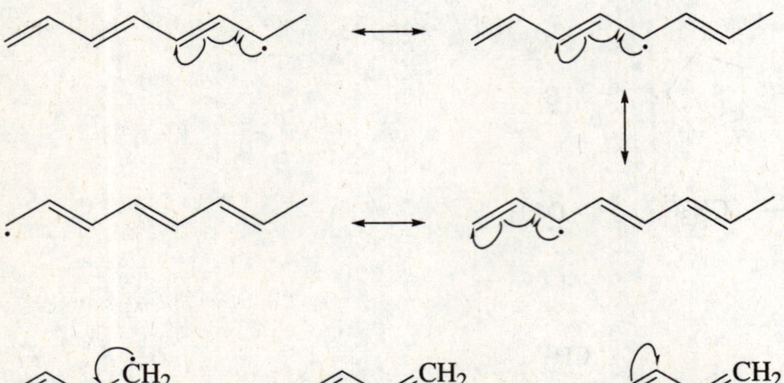

d.

CHAPTER 12

Mass Spectrometry, Infrared Spectroscopy, and Ultraviolet/Visible Spectroscopy

Important Terms

absorption band	a peak in a spectrum that occurs as a result of absorption of energy.
auxochrome	a substituent that when attached to a chromophore alters the λ_{max} and intensity of absorption of UV/Vis radiation.
base peak	the peak in a mass spectrum with the greatest intensity.
Beer-Lambert Law	relationship between the absorbance of UV/Vis light, the concentration of the sample, the length of the light path, and the molar absorbtivity.
bending vibration	a vibration that does not occur along the line of the bond.
α cleavage	homolytic cleavage of an alpha substituent.
chromophore	the part of a molecule responsible for a UV or visible spectrum.
electromagnetic radiation	radiant energy that displays wave properties.
electronic transition	promotion of an electron from its HOMO to its LUMO.
fingerprint region	the right-hand third of an IR spectrum where the absorption bands are characteristic of the compound as a whole.
frequency	the velocity of a wave divided by its wavelength. (It has units of cycles/second.)
functional group region	the left-hand two thirds of an IR spectrum, where most functional groups show absorption bands.
highest occupied molecular orbital(HOMO)	the highest energy molecular orbital that contains electrons.
Hooke's law	an equation that describes the motion of a vibrating spring.
λ_{max}	the wavelength at which there is maximum UV/Vis absorbance.
infrared radiation	electromagnetic radiation familiar to us as heat.
infrared spectroscopy	spectroscopy that uses infrared energy to provide a knowledge of the functional groups in a compound.
infrared spectrum (IR spectrum)	a plot of relative absorption versus wavenumber (or wavelength) of absorbed infrared radiation.
lowest unoccupied molecular orbital(LUMO)	the lowest energy molecular orbital that does not contain electrons.

mass spectrometry provides a knowledge of the molecular weight and certain structural features of a compound.

mass spectrum a plot of the relative abundance of the positively charged fragments produced in a mass spectrometer versus their *m/z* values.

McLafferty rearrangement rearrangement of the molecular ion of a ketone that contains a γ-hydrogen; the bond between the α- and β-carbons breaks, and a γ-hydrogen migrates to the oxygen.

molar absorptivity the absorbance obtained from a 1.00 M solution in a cell with a 1.00 cm light path.

molecular ion (M) the radical cation formed by removing one electron from a molecule.

nominal molecular mass mass to the nearest whole number.

radical cation a species with a positive charge and an unpaired electron.

spectroscopy study of the interaction of matter and electromagnetic radiation.

stretching frequency the frequency at which a stretching vibration occurs.

stretching vibration a vibration occurring along the line of the bond.

ultraviolet light electromagnetic radiation with wavelengths ranging from 180 to 400 nm.

UV/Vis spectroscopy the absorption of electromagnetic radiation to determine information about conjugated systems.

visible light electromagnetic radiation with wavelengths ranging from 400 to 780 nm.

wavelength distance from any point on one wave to the corresponding point on the next wave.

wavenumber the number of waves in one centimeter.

Solutions to Problems

1. Only positively charged fragments are accelerated through the analyzer tube.

$CH_3CH_2\overset{+}{C}H_2$ $\qquad$ $CH_3CH_2\overset{\bullet+}{C}H_2$ $\qquad$ $\overset{+}{C}H_2CH{=}CH_2$

2. The peak at $m/z = 57$ will be more intense for 2,2-dimethylpropane than for 2-methylbutane or for pentane because the peak at $m/z = 57$ is due to loss of a methyl group: loss of a methyl group from 2,2-dimethylpropane results in the formation of a tertiary carbocation, whereas loss of a methyl group from 2-methylbutane and pentane results in the formation of a secondary and primary carbocation, respectively.

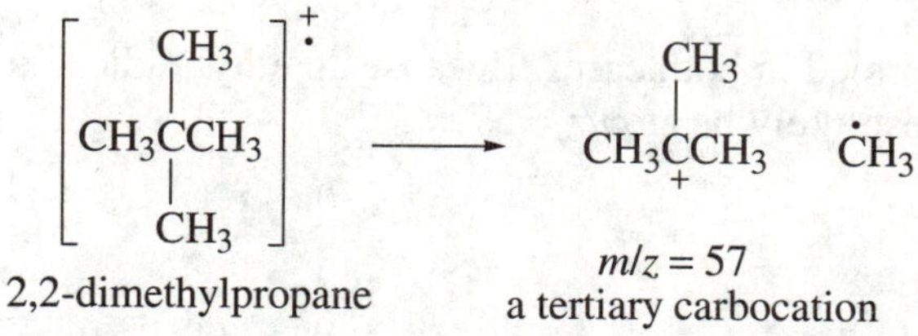

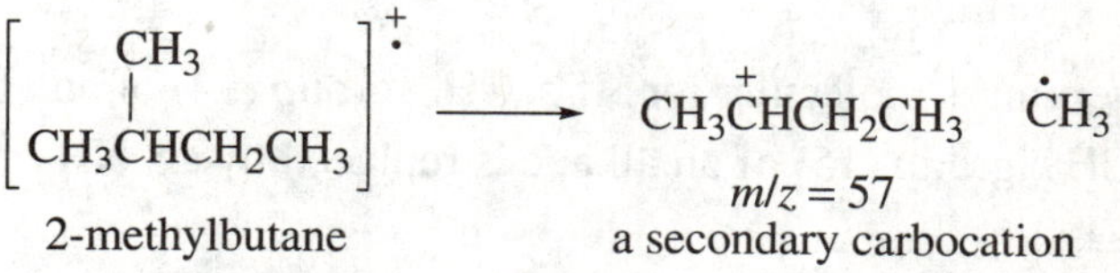

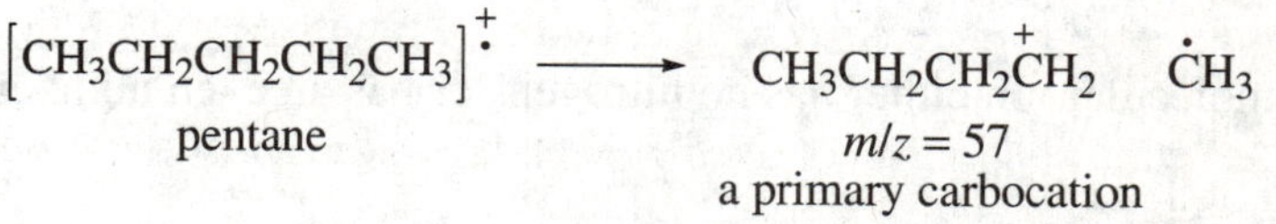

Notice that the mass spectrum of 2-methylbutane can be distinguished from those of the other isomers by the peak at $m/z = 43$. The peak at $m/z = 43$ will be most intense for 2-methylbutane because such a peak is due to loss of an ethyl group, which results in the formation of a secondary carbocation. Pentane gives a less intense peak at $m/z = 43$ because loss of an ethyl group from pentane forms a primary carbocation. 2,2-Dimethylpropane cannot form a peak at $m/z = 43$, because it does not have an ethyl group.

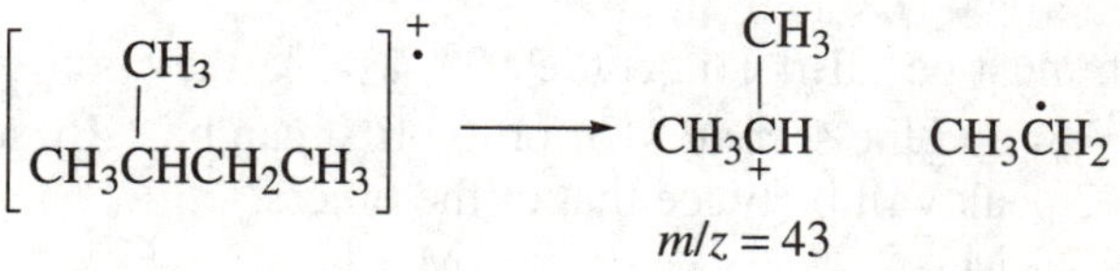

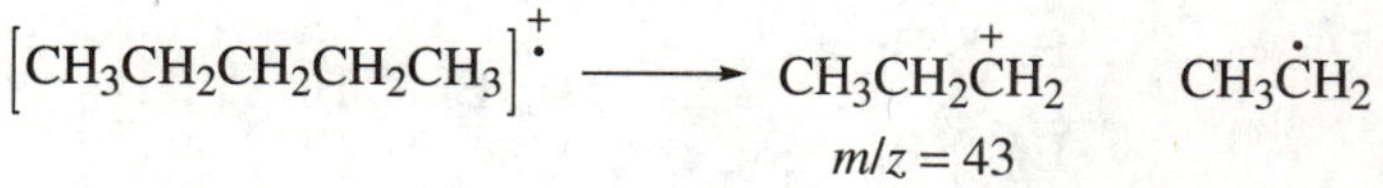

3. Peaks should occur at $m/z = 57$ for loss of an ethyl group (86-29), and at $m/z = 71$ for loss of a methyl group (86-15).

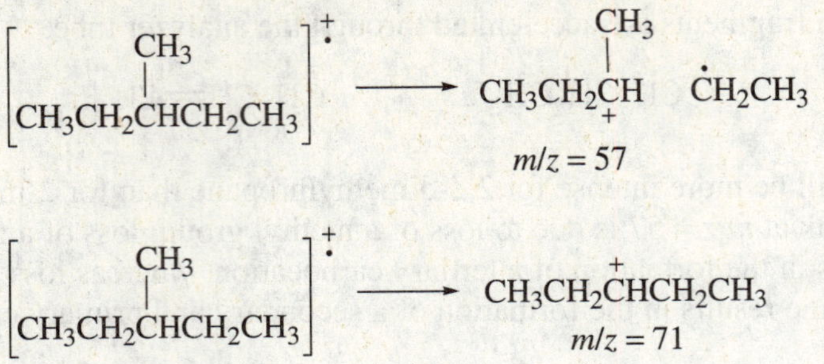

A secondary carbocation is formed in both cases. Because an ethyl radical is more stable than a methyl radical, the base peak will most likely be at $m/z = 57$.

4. Solved in the text.

5. **a.** **1.** $15 + (3 \times 14) + 16 = 73$ **2.** $16 + (3 \times 14) + 16 = 74$

b. An alkane has an even-mass molecular ion. If a CH_2 group (14) of an alkane is replaced by an NH group (15), or if a CH_3 group (15) of an alkane is replaced by an NH_2 group (16), the molecular ion will have an odd mass.
A second nitrogen in the molecular ion will cause it to have an even mass.
Thus, for a molecular ion to have an odd mass, an odd number of nitrogens must be present.

c. An even-mass molecular ion either has no nitrogens or has an even number of nitrogens.

6.

$$\text{number of carbon atoms} = \frac{\text{relative intensity of M} + 1 \text{ peak}}{0.011 \times \text{relative intensity of M peak}}$$

$$= \frac{3.81}{0.011 \times 43.27} = \frac{3.81}{0.467} = 8 \text{ carbon atoms}$$

7. The ratio is 1:2:1.
To get the M peak, they both must be ^{35}Br. To get the M + 4 peak, they both must be ^{37}Br. To get the M + 2 peak the first can be ^{35}Br and the second ^{37}Br or the first can be ^{37}Br and the second ^{37}Br. So the relative intensity of the M + 2 peak will be twice that of the others.

M	M + 2	M + 4
^{35}Br ^{35}Br	^{35}Br ^{37}Br ^{37}Br ^{35}Br	^{37}Br ^{37}Br

8. The calculated exact masses show that only C_6H_{14} has an exact mass of 86.10955.

C_6H_{14}	6(12.00000) =	72.00000
	14(1.007825) =	14.10955
		86.10955

$C_4H_6O_2$	4(12.00000) =	48.00000
	6(1.007825) =	6.04695
	2(15.9949) =	31.9898
		86.03675

$C_4H_{10}N_2$	4(12.00000) =	48.00000
	10(1.007825) =	10.07825
	2(14.0031) =	28.0064
		86.08465

9. Because the compound contains chlorine, the M + 2 peak is one third the size of the M peak. Loss of a chlorine atom from either the M + 2 peak (80 – 37) or the M peak (78 – 35) gives a peak with *m/z* = 43.

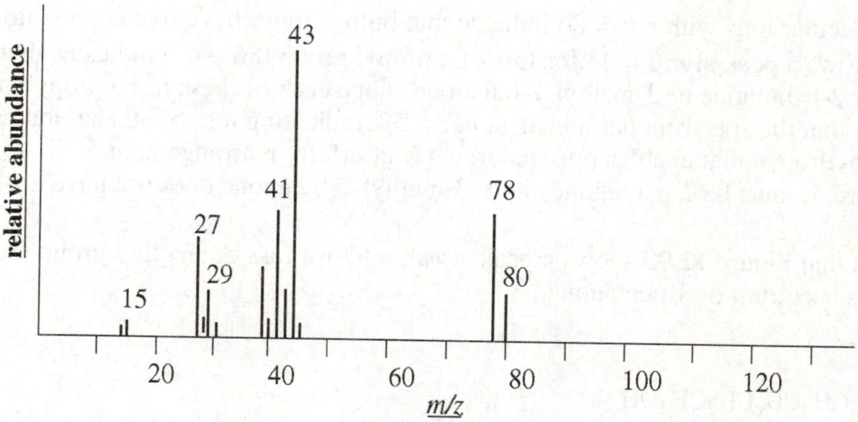

Mass spectrum of 1-chloropropane

10. The base peak at *m/z* = 73, due to loss of a methyl group (88 – 15), indicates that Figure 12.7a is the mass spectrum of **2-methoxy-2-methylpropane**.

The base peak at *m/z* = 59 (88 – 29), due to loss of an ethyl group, indicates that Figure 12.7b is the mass spectrum of **2-methoxybutane**.

The base peak at *m/z* = 45 (88 – 43), due to loss of a propyl group, indicates that Figure 12.7c is the mass spectrum of **1-methoxybutane**.

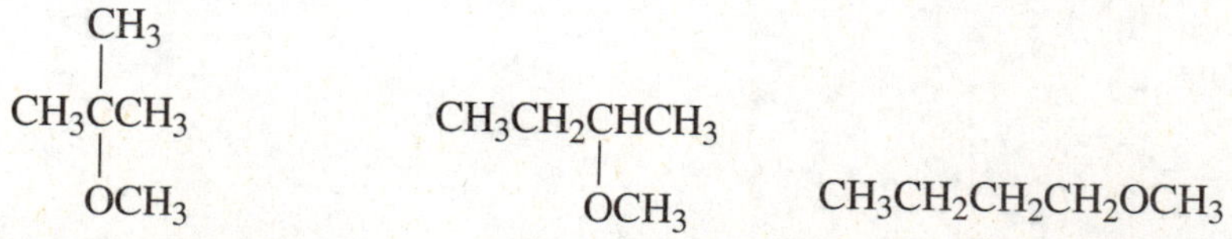

2-methoxy-2-methylpropane   2-methoxybutane   1-methoxybutane

11.

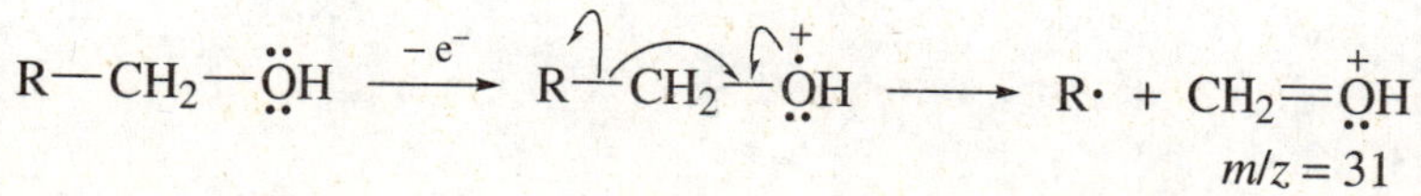

12. All three ketones will have a molecular ion with $m/z = 86$.

3-pentanone

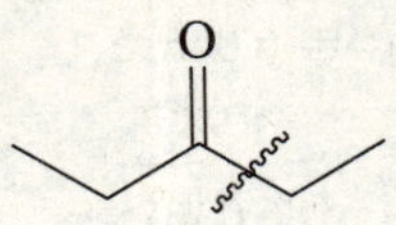

$86 - 29 = 57$

2-pentanone

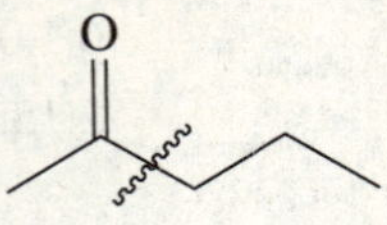

$86 - 41 = 43$

3-methyl-2-butanone

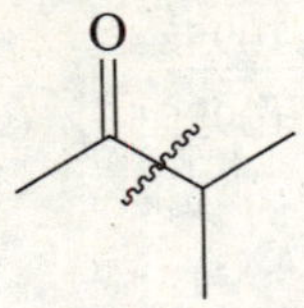

$86 - 41 = 43$

3-Pentanone will have a base peak at $m/z = 57$, whereas the other two ketones will have base peaks at $m/z = 43$.
2-Pentanone will have a peak at at $m/z = 58$ due to a McLafferty rearrangement.
3-Methyl-2-butanone does not have any γ-hydrogens. Therefore, it can't undergo a McLafferty rearrangement so it will not have a peak at at $m/z = 58$.

13. The molecular ions with $m/z = 86$ indicate that both ketones have five carbon atoms. The fact that Figure 12.9a shows a peak at $m/z = 43$ for loss of a propyl group $(86 - 43)$ indicates that it is the mass spectrum of either 2-pentanone or 3-methyl-2-butanone, since each of these has a propyl group.
The fact that the spectrum has a peak at $m/z = 58$, indicating loss of ethene, indicates that the compound has a γ-hydrogen that enables it to undergo a McLafferty rearrangement.
Therefore, it must be 2-pentanone, since 3-methyl-2-butanone does not have a γ-hydrogen.

The fact that Figure 12.9b shows a peak at $m/z = 57$ for loss of an ethyl group $(86 - 29)$, indicates that it is the mass spectrum of 3-pentanone.

14. **a.** $CH_3CH_2CH_2CH_2CH_2\ddot{O}H$

$$\downarrow -e^-$$

$$CH_3CH_2CH_2CH_2-CH_2-\dot{\overset{+}{O}}H \xrightarrow{\alpha\text{-cleavage}} CH_3CH_2CH_2\dot{C}H_2 + CH_2{=}\overset{+}{\ddot{O}}H$$

$m/z = 88$ $\qquad m/z = 31$

$$CH_3CH_2\overset{H}{C}HCH_2CH_2-\dot{\overset{+}{O}}H \longrightarrow CH_3CH_2\dot{\overset{+}{C}}HCH_2CH_2 + H_2O$$

$m/z = 70$

b.

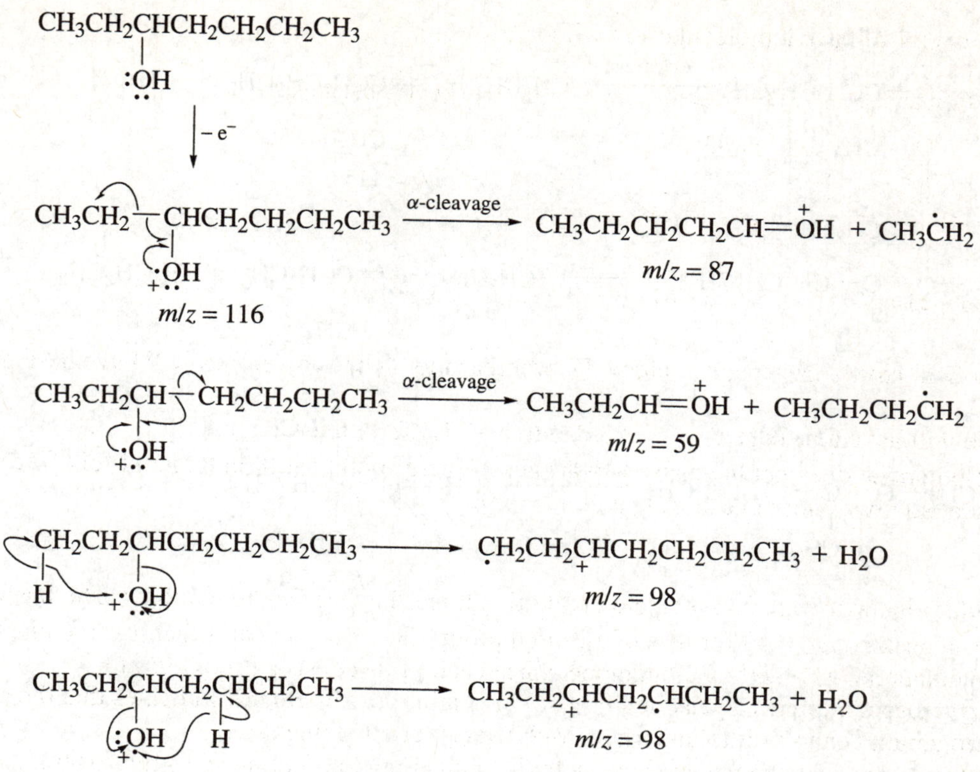
CH3CH2CHCH2CH2CH2CH3
:OH
– e⁻
CH3CH2CHCH2CH2CH2CH3
m/z = 116
α-cleavage
CH3CH2CH2CH2CH=OH + CH3CH2
m/z = 87
CH3CH2CH=OH + CH3CH2CH2CH2
m/z = 59
CH2CH2CHCH2CH2CH2CH3 + H2O
m/z = 98
CH3CH2CHCH2CHCH2CH3 + H2O
m/z = 98

c.

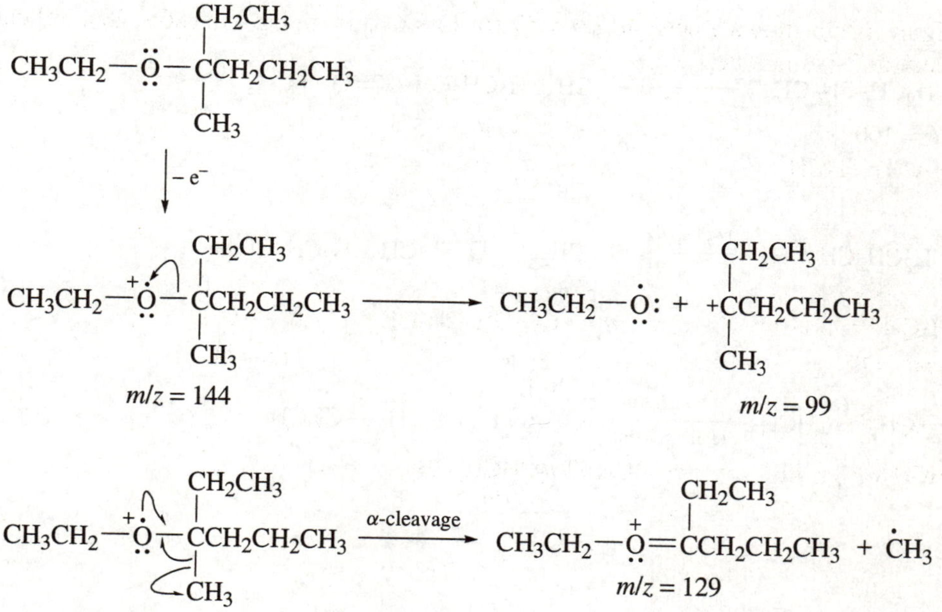
CH2CH3
CH3CH2—O—CCH2CH2CH3
CH3
– e⁻
m/z = 144
CH3CH2—O: + +CCH2CH2CH3
m/z = 99
α-cleavage
CH3CH2—O=CCH2CH2CH3 + CH3
m/z = 129

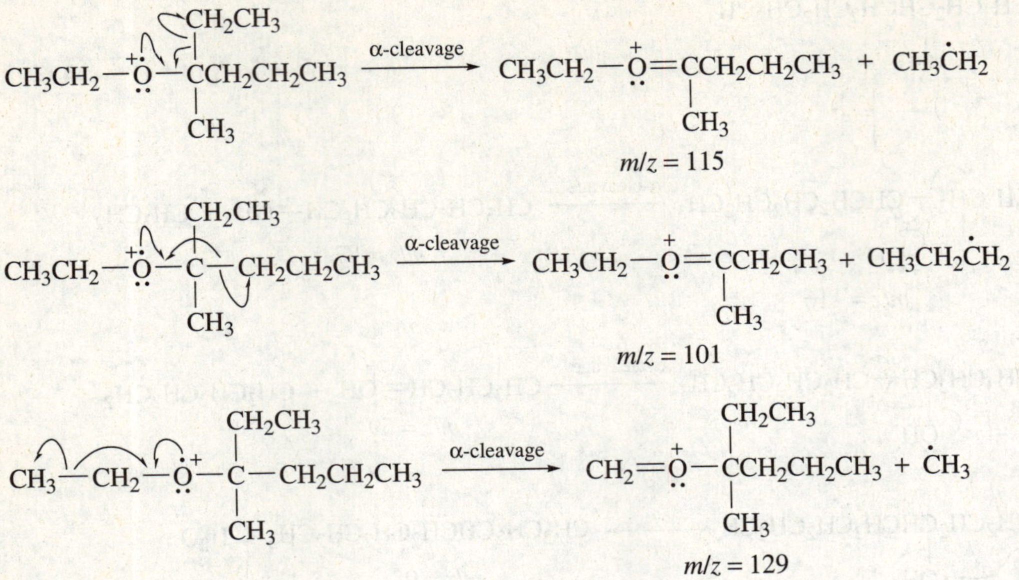
α-cleavage
m/z = 115
α-cleavage
m/z = 101
α-cleavage
m/z = 129

d.

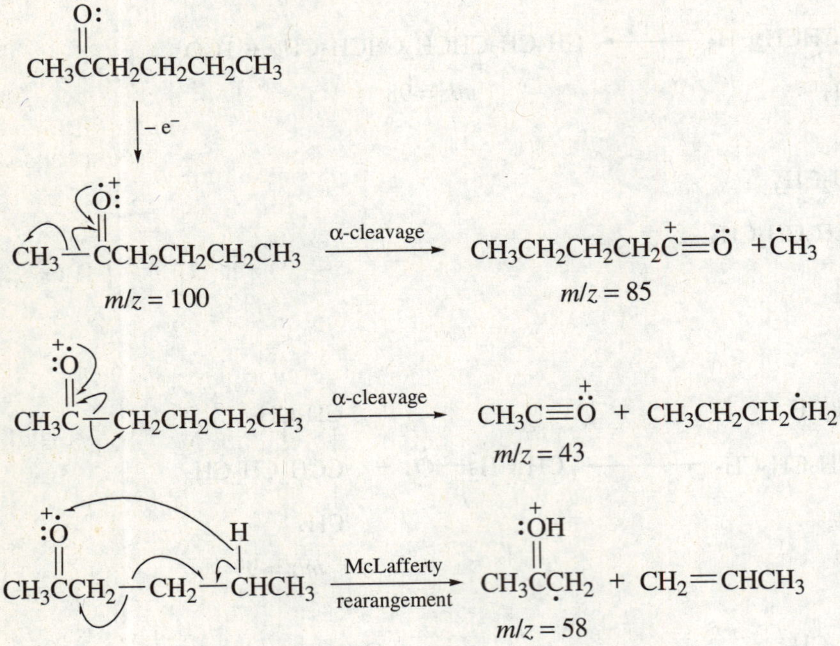
– e⁻
m/z = 100
α-cleavage
m/z = 85
α-cleavage
m/z = 43
McLafferty
rearangement
m/z = 58

e.

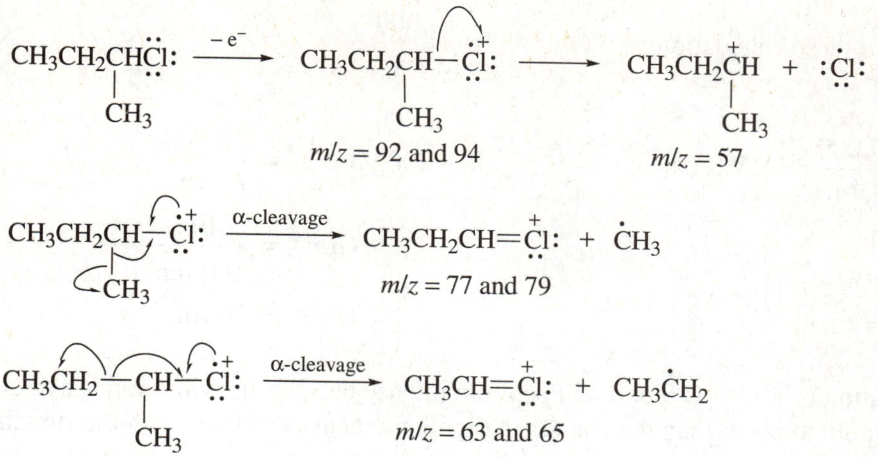

f.

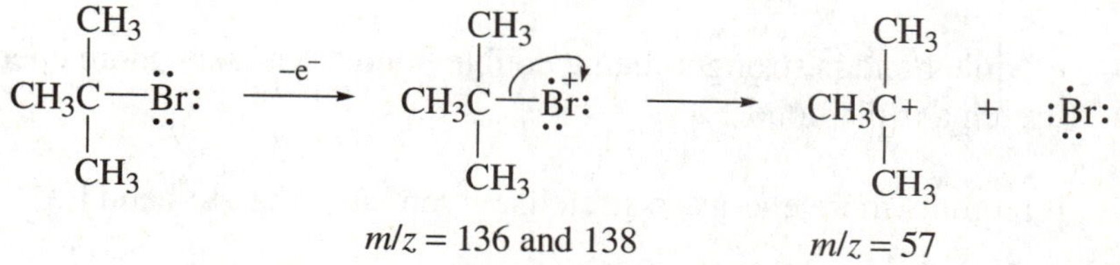

15. When (*Z*)-2-pentene reacts with water and an acid catalyst, 3-pentanol and 2-pentanol are formed. Both alcohols have a molecular weight of 88. (Notice that the first step in solving this problem is to use chemical knowledge to identify the products.) The absence of a molecular ion peak is consistent with the fact that the compounds are alcohols.

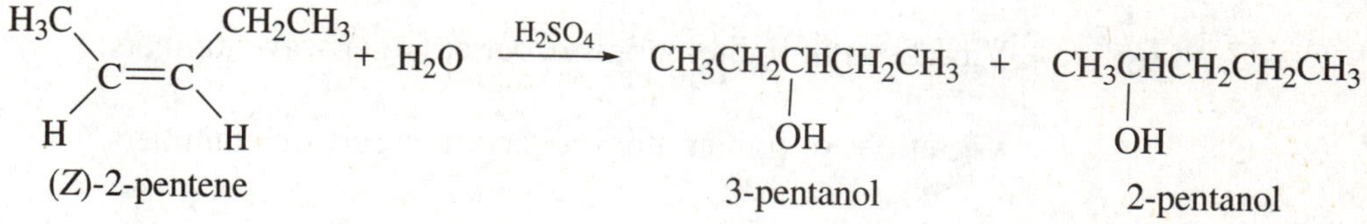

Figure 12.10a shows a base peak at $m/z = 59$ due to loss of an ethyl group (88 – 29), indicating that it is the spectrum of 3-pentanol.

Figure 12.10b shows a base peak at $m/z = 45$ due to loss of a propyl group (88 – 43), indicating that it is the spectrum of 2-pentanol.

16. The wavelength is the distance from the top of one wave to the top of the next wave. We see that "a" has a longer wavelength than "b".
Infrared radiation has longer wavelengths than ultraviolet light because infrared radiation is lower in energy. Therefore, "a" depicts infrared radiation and "b" depicts ultraviolet light.

17. **a.** 2000 cm^{-1} (The larger the wavenumber, the higher the energy.)

b. 8 μm (The shorter the wavelength, the higher the energy.)

c. 2 μm, because 2 μm = 5000 cm^{-1} (5000 cm^{-1} is a larger wavenumber than 3000 cm^{-1})

18. the equation used for these calculations: $\tilde{\nu}\ (\text{cm}^{-1}) = \dfrac{10{,}000}{\lambda\ (\mu\text{m})}$

a. $\tilde{\nu}\ (\text{cm}^{-1}) = \dfrac{10{,}000}{4\ (\mu\text{m})} = 2500\ \text{cm}^{-1}$

b. $200\ (\text{cm}^{-1}) = \dfrac{10{,}000}{\lambda\ (\mu\text{m})}$

$\lambda\ (\mu\text{m}) = \dfrac{10{,}000}{200\ (\text{cm}^{-1})}$

$\lambda\ (\mu\text{m}) = 50\ \mu\text{m}$

19. The stretching vibration of a C═O bond will be more intense because the vibration is associated with a greater change in dipole moment than the change in dipole moment associated with the stretching vibration of a C═C bond.

20. **a.** **1.** C≡C stretch — A triple bond is stronger than a double bond, so it takes more energy to stretch a triple bond.

2. C—H stretch — It requires more energy to stretch a given bond than to bend it.

3. C═N stretch — A double bond is stronger than a single bond, so it takes more energy to stretch a double bond.

4. C═O stretch — A double bond is stronger than a single bond, so it takes more energy to stretch a double bond.

b. **1.** C—O — Vibrations of lighter atoms occur at larger wavenumbers.

2. C—C — Vibrations of lighter atoms occur at larger wavenumbers.

21. **a.** The carbon-oxygen stretch of phenol because it has partial double-bond character as a result of electron delocalization.

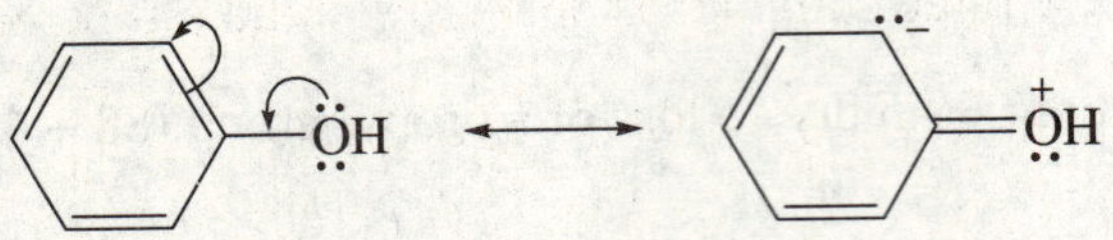

b. The carbon-oxygen double-bond stretch of a ketone because it has more double-bond character. The double bond character of the carbonyl group of the amide is reduced by electron delocalization.

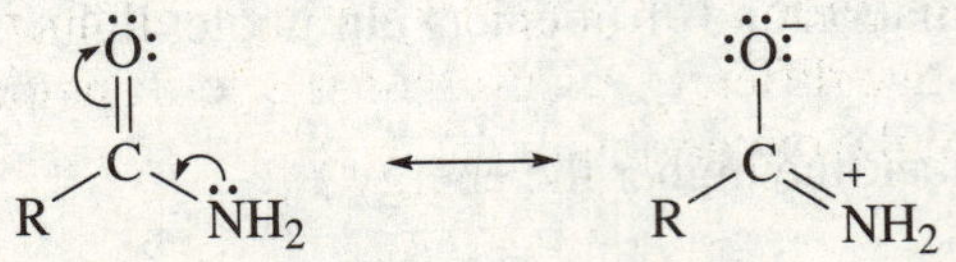

c. The C—O stretch, because a bond stretches at a higher frequency than it bends.

22. A carbonyl group bonded to an sp^3 carbon will exhibit an absorption band at a larger wavenumber because a carbonyl group bonded to an sp^2 carbon will have greater single-bond character as a result of electron delocalization.

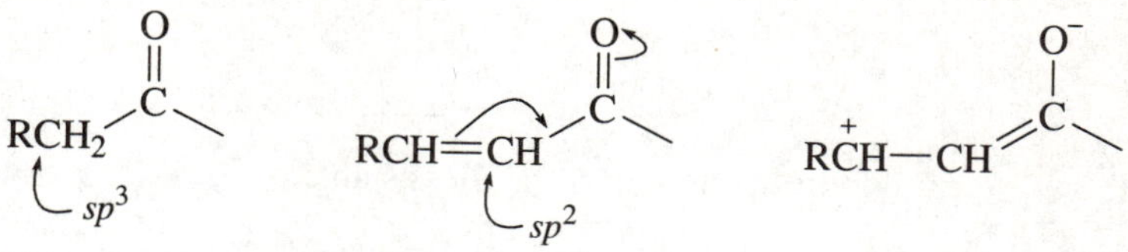

23. **a.** Because alkyl groups are more electron-donating than a hydrogen because of hyperconjugation, the absorption band for the carbonyl group bonded to two relatively electron-withdrawing hydrogens has the largest wavenumber, while the absorption band for the carbonyl group bonded to two alkyl groups has the largest wavenumber.

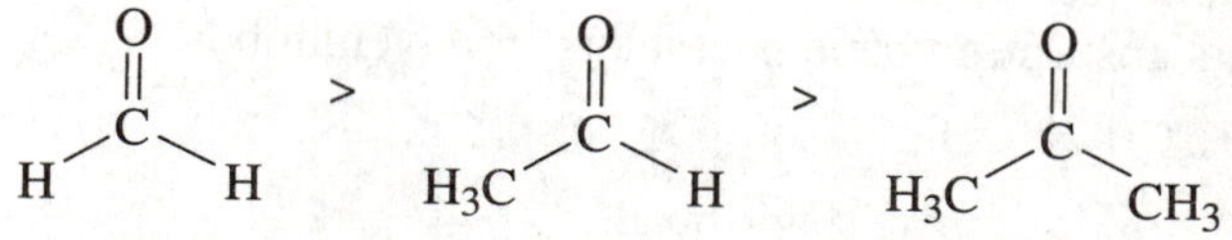

b. The C=O absorption band of an ester occurs at the largest wavenumber because the carbonyl group of an ester has the most double bond character since the predominant effect of the ester oxygen atom is inductive electron withdrawal.
The C=O absorption band of an amide occurs at the smallest wavenumber because the carbonyl group of an amide has the least double bond character since the predominant effect of the amide nitrogen atom is electron donation by resonance.

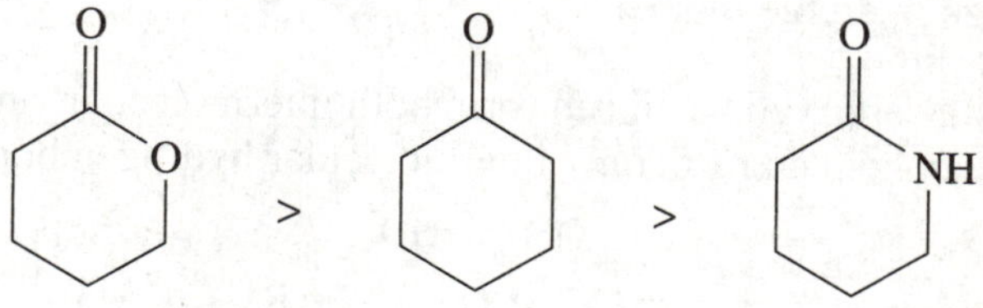

c. The C=O absorption band of the three compounds decreases in the following order.

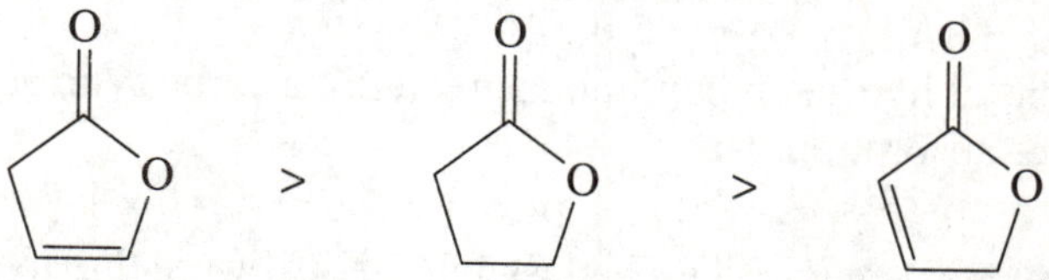

The first compound has the C=O absorption band at the largest wavenumber because the lone pair on the oxygen can be delocalized onto two different atoms; thus it is less apt than the lone pair in the other compounds to be delocalized onto the oxygen atom.

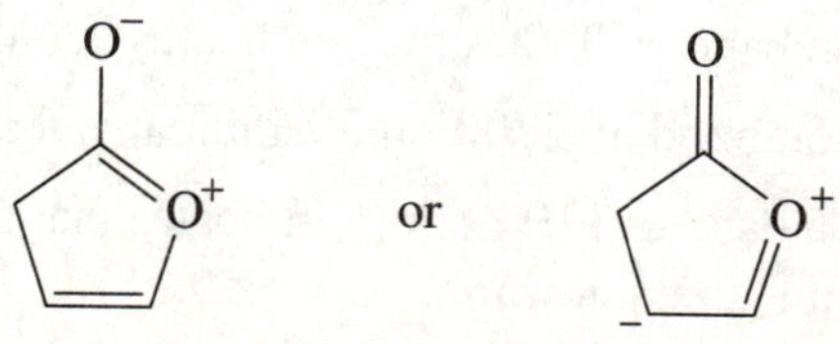

The third compound has the C=O absorption band at the smallest wavenumber because its resonance contributor with the C—O single bond is the more stable than that of the second compound.

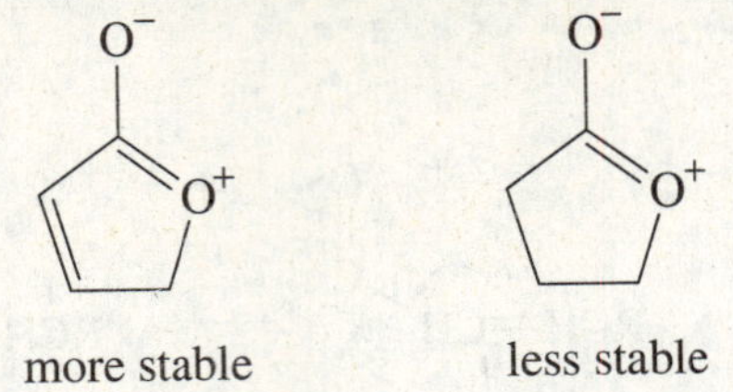

24. Ethanol dissolved in carbon disulfide will show the oxygen-hydrogen stretch at a larger wavenumber because there is extensive hydrogen bonding in the undiluted alcohol, and an oxygen-hydrogen bond is easier to stretch if it is hydrogen bonded.

25. The C—O bond in 1-hexanol will occur at a smaller wavenumber; it is a pure single bond whereas the C—O bond in pentanoic acid has some double bond character which makes it harder to stretch. Because it is harder to stretch, its absorption band is at a larger wavenumber.

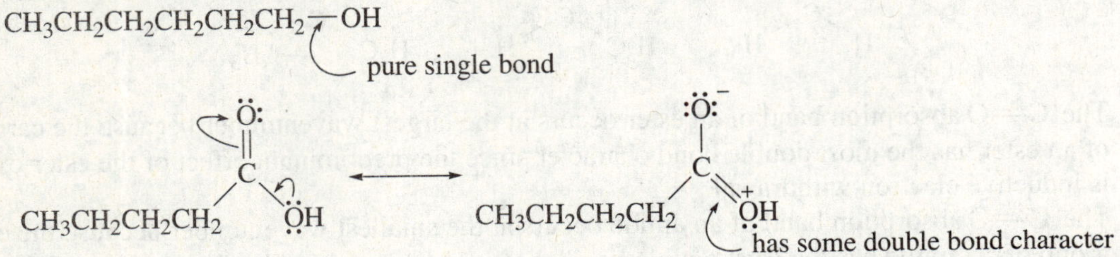

26. **a.** Because oxygen is more electronegative than nitrogen, the O—H stretching vibration is associated with a greater change in dipole moment.

b. The O—H group of a carboxylic acid can form both intermolecular and intramolecular hydrogen bonds, whereas an alcohol can form only intermolecular hydrogen bonds.

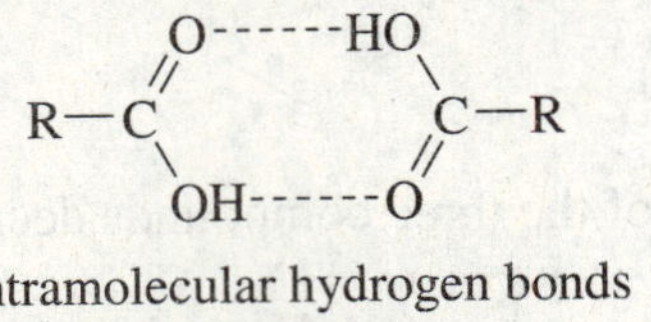

Therefore, the extent of hydrogen bonding is greater in a carboxylic acid, and hydrogen bonded OH groups have broader absorption bands.

27. The absorption band at 1100 cm^{-1} would be less intense if it were due to a C—N bond because a smaller change in dipole moment is associated with a C—N bond compared to the change in dipole moment associated with the stretch of a C—O bond.

28. **a.** The absorption band at 1700 cm^{-1} indicates that the compound has a carbonyl group.
The absence of an absorption band at 3300 cm^{-1} indicates that the compound is not a carboxylic acid.
The absence of an absorption band at 2700 cm^{-1} indicates that the compound is not an aldehyde.
The absence of an absorption band at 1100 cm^{-1} indicates that the compound is not an ester or an amide.
The compound, therefore, must be a **ketone**.

b. The absence of an absorption band at 3400 cm^{-1} indicates that the compound does not have an N—H bond.

The absence of an absorption band between 1700 cm^{-1} and 1600 cm^{-1} indicates that the compound is not an amide.
The compound, therefore, must be a **tertiary amine**.

29. **a.** An aldehyde would show absorption bands at 2820 and 2720 cm^{-1}. A ketone would not have these absorption bands.

b. A open-chain ketone would have a methyl substituent and therefore an absorption band at 1385-1365 cm^{-1} that a cyclic ketone would not have.

c. Cyclohexene would show an *sp*^{3} C—H stretch slightly to the right of 3000 cm^{-1}. Benzene would not show an absorption band in this region.

d. The cis isomer would show a carbon-hydrogen bending vibration at 730-675 cm^{-1}, whereas the trans isomer would show a carbon-hydrogen bending vibration at 960-980 cm^{-1}.

e. Cyclohexene would show a carbon-carbon double-bond stretching vibration at 1680-1600 cm^{-1} and a carbon-hydrogen stretching vibration at 3100-3020 cm^{-1}. Cyclohexane would not show these absorption bands.

f. A primary amine would show a nitrogen-hydrogen stretch at 3500-3300 cm^{-1}, and a tertiary amine would not have this absorption band.

30. **a.** An absorption band at 1150-1050 cm^{-1} would be present for the ether and absent for the alkane.

b. An absorption band at 3300-2500 cm^{-1} would be present for the carboxylic acid and absent for the ester.

c. An absorption band at 1385-1365 cm^{-1} would be present for methylcyclohexane and absent for cyclohexane.

d. Only the terminal alkyne would show an absorption band at 3300 cm^{-1}.

e. An absorption band at 1780-1650 cm^{-1} would be present for the carboxylic acid and absent for the alcohol.

f. An absorption band 2960-2850 cm^{-1} would be present for the compound with the methyl group and absent for benzene.

31. 2-butyne, H_2, Cl_2, ethene

32. The absorption bands in the vicinity of 3000 cm^{-1} indicate that the compound has hydrogens attached to both *sp*^{2} and *sp*^{3} carbons. The lack of absorption at 1600 cm^{-1} and 1500 cm^{-1} indicates the compound does not have a benzene ring. The *sp*^{2} hydrogens, therefore, must be those of an **alkene**.

The lack of significant absorption at 1600 cm^{-1} indicates that the compound must be an alkene with a relatively small (if any) dipole moment change when the vibration occurs. The absorption band at 965 cm^{-1} indicates the compound is a ***trans*-alkene**.

The molecular ion with $m/z = 84$ suggests the compound has a molecular formula of C_6H_{12}. The base peak with $m/z = 55$ indicates the compound most easily loses an ethyl group ($84 - 29 = 55$), but it can also lose a methyl group ($84–15 = 69$) or a propyl group ($84–43 = 41$).
Because it most easily loses an ethyl group, the ethyl group must be attached to an allylic carbon.
The compound, therefore, is ***trans*-2-hexene**.

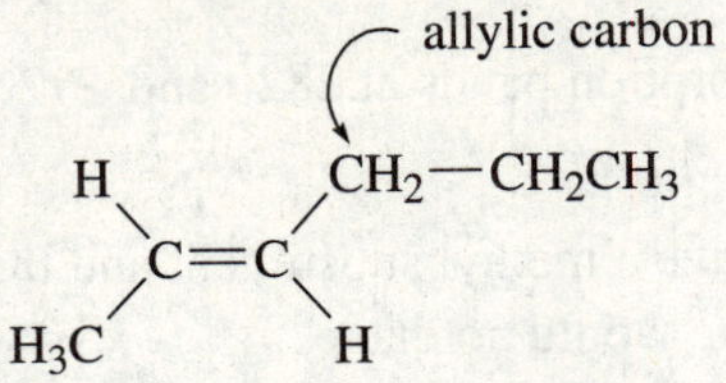

trans-2-hexene

33. The absorption band at ~1700 cm^{-1} indicates the compound has a carbonyl group, and the absorption band at ~1600 cm^{-1} indicates the compound has a carbon-carbon double bond. The absorption bands in the vicinity of 3000 cm^{-1} indicate that the compound has hydrogens attached to both sp^2 and sp^3 carbons. The absorption band at ~1380 cm^{-1} indicates the compound has a methyl group.
Because the compound has only four carbons and one oxygen, it must be **methyl vinyl ketone**. Notice that the carbonyl stretch is at a lower frequency (1700 cm^{-1}) than expected for a ketone (1720 cm^{-1}), because the carbonyl group has partial single bond character because of electron delocalization.

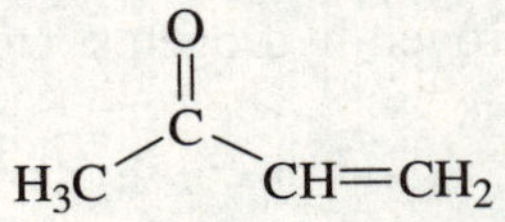

34. Diethyl ether does not have a π bond, so it does not have a π* antibonding molecualr orbital. Therefore, it cannot have an $n \rightarrow \pi^*$ transition.

35. $A = c\,l\,\varepsilon$

$$c = \frac{A}{l\,\varepsilon}$$

$$c = \frac{0.52}{12{,}600} = 4.1 \times 10^{-5}\ \text{M}$$

36. $A = c\,l\,\varepsilon$

$$c = \frac{A}{l\,\varepsilon}$$

$$\varepsilon = \frac{0.40}{4.0 \times 10^{-5}} = 10{,}000\ \text{M}^{-1}\text{cm}^{-1}$$

37. **a.**

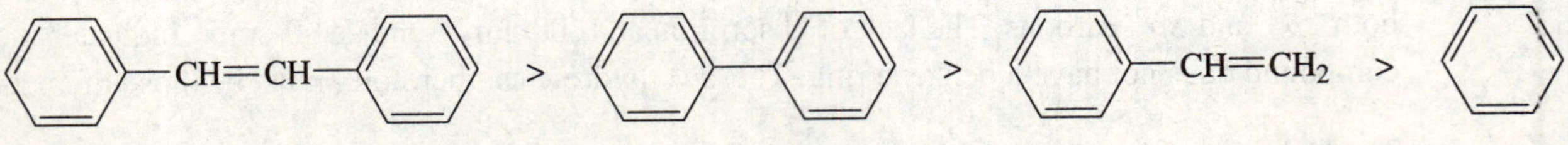

b.

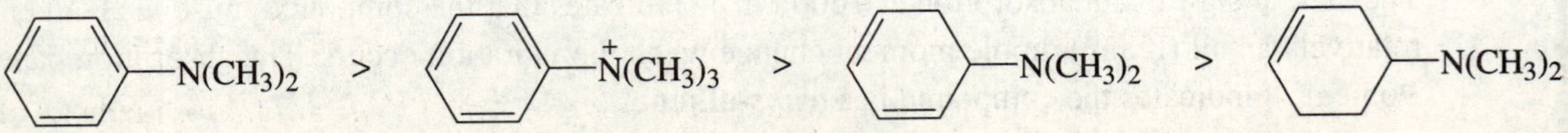

38. **a.** Blue is the result of absorption of light of a longer wavelength than light that produces purple. The compound on the right has two ($N(CH_3)_2$) auxochromes that will cause it to absorb at a longer wavelength than the compound on the left that has only one $N(CH_3)_2$ auxochrome. Therefore, it is the blue compound.

b. They will be the same color at pH = 3 because the $N(CH_3)_2$ groups will be protonated and, therefore, will not possess the lone pair that causes the compound to absorb light of a longer wavelength.

39. NADH is formed as a product; it absorbs light at 340 nm. Therefore, the rate of the oxidation reaction can be determined by monitoring the increase in absorbance at 340 nm as a function of time.

$$CH_3CH_2OH + NAD^+ \xrightarrow{\text{alcohol dehydrogenase}} CH_3\overset{\overset{\displaystyle O}{\|}}{C}H + NADH$$

40. The Henderson-Hassselbalch equation (Section 1.24) shows us that when the pH equals the pK_a, the concentration of the species in the acidic form is the same as the concentration of the species in the basic form. From the data given, we see that the absorbance of the acid is 0. We also see that the absorbance ceases to change with increasing pH after the absorbance reaches 1.60. That means that all of the compound is in the basic form when the absorbance is 1.60. Thus when the absorbance is half of 1.60 (or 0.80), half of the compound is in the basic form and half is in the acidic form; in other words, the concentration in the acid form is the same as the concentration in the basic form. We see that the absorbance is 0.80 at pH = 5. Therefore, the pK_a of the compound is 5.0.

41. The molecular ion peak for these compounds is $m/z = 86$; the peak at $m/z = 57$ is due to loss of an ethyl group (86 – 29), and the peak at $m/z = 71$ is due to loss of a methyl group (86 – 15).

a. 3-Methylpentane will be more apt to lose an ethyl group (forming a secondary carbocation and a primary radical) than a methyl group (forming a secondary carbocation and a methyl radical), so the peak at $m/z = 57$ would be more intense than the peak at $m/z = 71$.

$$CH_3CH_2\underset{\underset{\displaystyle CH_3}{|}}{C}HCH_2CH_3$$

3-methylpentane

b. 2-Methylpentane has two pathways to lose a methyl group (forming a secondary carbocation and a methyl radical in each pathway), and it cannot form a secondary carbocation by losing an ethyl group. (Loss of an ethyl group would form a primary carbocation and a primary radical.) Therefore, it will be more apt to lose a methyl group than an ethyl group, so the peak at $m/z = 71$ would be more intense than the peak at $m/z = 57$.

$$CH_3\underset{\underset{\displaystyle CH_3}{|}}{C}HCH_2CH_2CH_3$$

2-methylpentane

42.
1. the change in the dipole moment when the bond stretches or bends
2. the number of bonds that give rise to the absorption band
3. the concentration of the sample

43. **a.** An absorption band at ~1250 cm^{-1} would be present for the ester and absent for the ketone.

b. An absorption band at 720 cm^{-1} would be present for heptane and absent for methylcyclohexane.

c. An absorption band at 3650-3200 cm^{-1} would be present for the alcohol and absent for the ether.

d. An absorption band at 3500-3300 cm^{-1} would be present for the amide and absent for the ester.

e. The secondary alcohol would have an absorption band at 1385-1365 cm^{-1} for the methyl group. The primary alcohol does not have a methyl group, so it would not have this absorption band.

f. The trans isomer would have an absorption band at 980-960 cm^{-1}, while the cis isomer would have an absorption band at 730-675 cm^{-1}. In addition, a weak absorption band at 1680-1600 cm^{-1} would be present for the cis isomer and absent for the trans isomer.

g. The C=O absorption band will be at a larger wavenumber for the ester (1740 cm^{-1}) than for the ketone (1720 cm^{-1}).

h. The C=O absorption band will be at a larger wavenumber for the β,γ-unsaturated ketone (1720 cm^{-1}) than for the α,β-unsaturated ketone since the double bonds in the latter are conjugated (1680 cm^{-1}).

i. The alkene would have an absorption band at 1680-1600 cm^{-1} and 3100-3020 cm^{-1} that the alkyne would not have. The alkyne would have an absorption band at 2260-2100 cm^{-1} that the alkene would not have.

j. An absorption band at ~2820 and ~2720 cm^{-1} would be present for the aldehyde and absent for the ketone.

k. Absorption bands at 1600 cm^{-1} and 1500 cm^{-1} and at 3100-3020 cm^{-1} would be present for the compound with the benzene ring and absent for the compound with the cyclohexane ring. An absorption band at 2960-2850 cm^{-1} would be present for the compound with the cyclohexane ring and absent for the compound with the benzene ring.

l. Absorption bands at 990 cm^{-1} and 910 cm^{-1} would be present for the terminal alkene and absent for the internal alkene.

44. **a.** The absorption at 236 nm is due to a $\pi \rightarrow \pi^*$ transition, and the absorption at 314 nm is due to an $n \rightarrow \pi^*$ transition.

b. The one at 236 nm (the $\pi \rightarrow \pi^*$ transition) shows the greater absorbance.

45. **a.** If the reaction had occurred, the intensity of the absorption bands at ~1700 cm^{-1} (due to the carbonyl group) and ~2700 cm^{-1} (due to the aldehyde hydrogen) of the reactant would have decreased. If all the aldehyde had reacted, these absorption bands would have disappeared.

b. If all the NH_2NH_2 had been removed, there would be no N—H absorption at ~3400 cm^{-1}.

46. **a.** 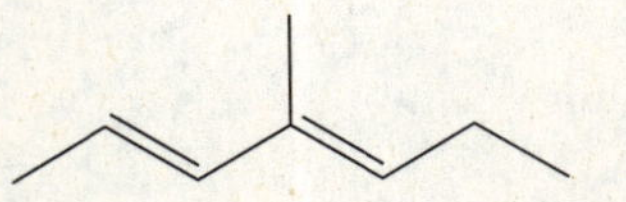 and 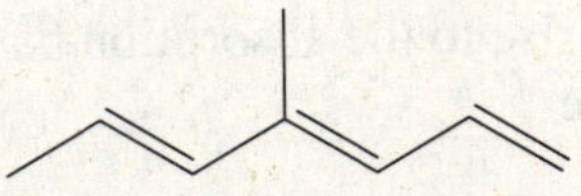

The λ_{max} will be at a longer wavelength.

b. $CH_2{=}CHCH{=}CHCH{=}CH_2$ and $CH_2{=}CHCH{=}CH\overset{O}{\overset{\|}{C}}CH_3$

It will show 1 absorption band. a $\pi \rightarrow \pi^*$ transition

It will show 2 absorption bands. a $\pi \rightarrow \pi^*$ transition and an $n \rightarrow \pi^*$ transition

c. 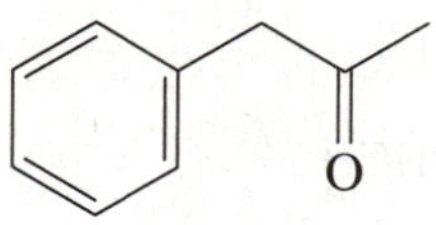 and 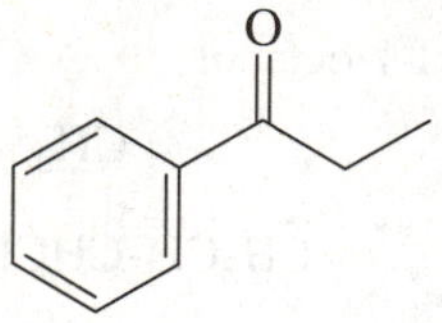

The λ_{max} will be at a longer wavelength because the carbonyl group is conjugated with the benzene ring.

d. 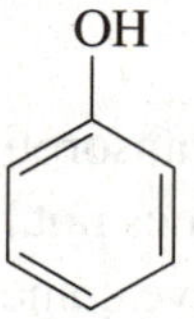 and 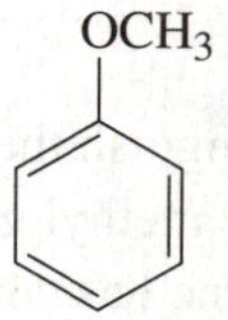

The λ_{max} of the phenolate ion is at a longer wavelength than the λ_{max} of phenol. Since the pK_a of phenol is ~10, the λ_{max} will be at a longer wavelength at pH = 11 than at pH = 7.

The λ_{max} is pH independent from pH 0-14.

47. The compound would have an M and an M + 4 peak of equal intensity, and there would be an M + 2 peak of twice the intensity of the M peak.

48. If the force constants are approximately the same, the lighter atoms absorb at higher frequencies.

C—C > C—N > C—O

49.

3600 cm^{-1} — 3000	3000 — 1800	1800 — 1400	1400 — 1000
OH 3300-3000	sp^3 CH 2950	C=O 1700	C—O 1250-1050
NH 3600-3200	$R\overset{O}{\overset{\|}{C}}H$ 2700	C=C 1600	C—N 1230-1030
sp CH 3300	C≡C 2100	 1500	
sp^2 CH 3050	C≡N 2250		

50. The molecular weight of each of the alcohols is 158. The peak at *m/z* = 140 is due to loss of water (158 – 18); each of the alcohols could show such a peak. The peaks at *m/z* = 87, 115, and 143 are due to loss of a group with five carbons, a group with three carbons, and a methyl group, respectively. Only 2,2,4-trimethyl-4-heptanol would lose all three groups.

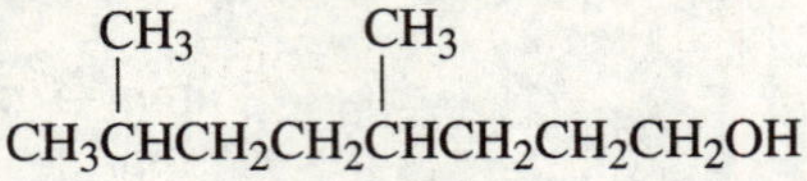

4,7-dimethyl-1-octanol

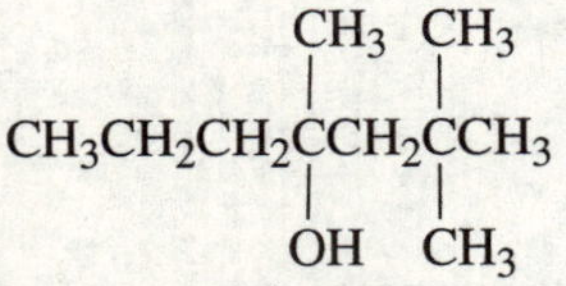

2,2,4-trimethyl-4-heptanol

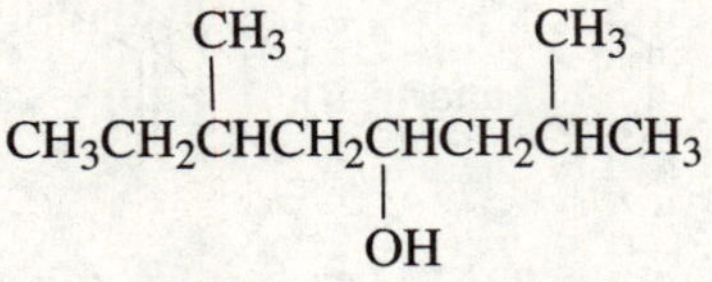

2,6-dimethyl-4-octanol

51. $CH_2{=}CHCH_2CH_2CH{=}CH_2$
1,5-hexadiene

$CH_3CH{=}CHCH{=}CHCH_3$
2,4-hexadiene

The easiest way to distinguish the two compounds is by the presence or absence of an absorption band at ~1370 cm^{-1} due to the methyl group that 2,4-hexadiene has but that 1,5-hexadiene does not have. In addition, 2,4-hexadiene has conjugated double bonds; therefore its double bonds have some single-bond character due to electron delocalization. Consequently, they are easier to stretch than the isolated double bonds of 1,5-hexadiene. Thus, the carbon-carbon double bond stretch of 2,4-hexadiene will be at a smaller wavenumber than the carbon-carbon double-bond stretch of 1,5-hexadiene.

52. The fact that the abundance of the M + 2 peak is 30% of the abundance of the M peak indicates that the compound has one chlorine atom. The peak at *m/z* = 77 is due to loss of the chlorine atom (112 – 35 = 77). The peak at *m/z* = 77, and fact that the this peak does not fragment, indicates it is a benzene ring. The compound is chlorobenzene.

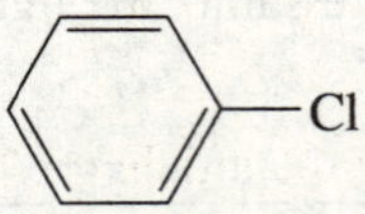

53. A hydrocarbon with eight carbons (C_nH_{2n+2}) has a molecular formula of C_8H_{18} and a molecular weight of 114. Since the molecular ion is 112, the compound must be a hydrocarbon with eight carbons and one double bond or one ring. Possible compounds are cyclooctane, methylcycloheptane, all the dimethylcyclohexanes, ethylcyclohexane, propylcyclopentane, isopropylcyclopentane, all the ethylmethyl cyclopentanes, all the trimethylcyclopentanes, all octenes, all methylheptenes, all ethylhexenes, etc.

54. **a.** The absorption band at ~2100 cm^{-1} indicates a carbon-carbon triple bond, and the absorption band at ~3300 cm^{-1} indicates a hydrogen bonded to an *sp* carbon.

$$CH_3CH_2CH_2CH_2C{\equiv}CH$$

b. The absence of an absorption band at ~2700 cm^{-1} indicates that the compound is not an aldehyde, and the absence of a broad absorption band in the vicinity of 3000 cm^{-1} indicates that the compound is not a carboxylic acid. The ester and the ketone can be distinguished by the absorption band that is present at ~1200 cm^{-1} that indicates the carbon-oxygen single bond of an ester.

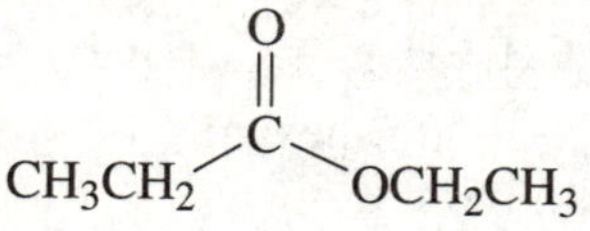

c. The absorption band at ~1360 cm^{-1} indicates the presence of a methyl group.

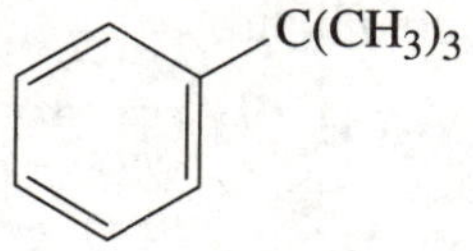

55. 4-Methyl-2-pentanone would show peaks at $m/z = 85$ (loss of a methyl group) and at $m/z = 43$ (loss of an isobutyl group) and a peak at $m/z = 58$ due to a McLafferty rearrangement.

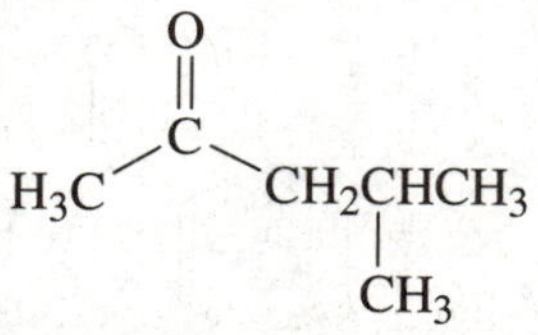

4-methyl-2-pentanone

2-Methyl-3-pentanone would show peaks at $m/z = 71$ (loss of an ethyl group) and at $m/z = 57$ (loss of an isopropyl group). Because it doesn't have a γ-hydrogen, it cannot undergo a McLafferty rearrangement.

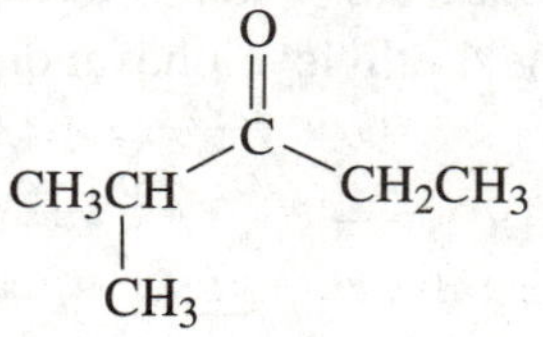

2-methyl-3-pentanone

56. The absorption bands at ~2700 cm^{-1} for the aldehydic hydrogen and at ~1380 cm^{-1} for the methyl group would distinguish the compounds.

A would have the band at ~2700 cm^{-1} but not the one at ~1380 cm^{-1}.
B would have the band at ~1380 cm^{-1} but not the one at ~2700 cm^{-1}.
C would have the both the band at ~2700 cm^{-1} and the one at ~1380 cm^{-1}.

57. 1-Hexyne will show absorption bands at ~3300 cm^{-1} for a hydrogen bonded to an *sp* carbon and at ~2100 cm^{-1} for the triple bond.

$$CH_3CH_2CH_2CH_2C{\equiv}CH_3$$

1-hexyne

2-Hexyne will show the absorption band at ~2100 cm^{-1} but not the one at ~3300 cm^{-1}.

$$CH_3CH_2CH_2C{\equiv}CCH_3$$

2-hexyne

3-Hexyne will show neither the absorption band at ~ 3300 cm^{-1} nor the one at ~ 2100 cm^{-1} (there is no change in dipole moment).

$CH_2CH_2C{\equiv}CCH_2CH_3$

3-hexyne

58. **a.** The broad absorption band at ~ 3300 cm^{-1} is characteristic of the oxygen-hydrogen stretch of an alcohol, and the absence of absorption bands at ~ 1600 cm^{-1} and ~ 3100 cm^{-1} indicates that it is not the alcohol with a carbon-carbon double bond.

$CH_3CH_2CH_2CH_2OH$

b. The absorption band at ~ 1685 cm^{-1} indicates a carbon-oxygen double bond. The absence of a strong and broad absorption band at ~ 3000 cm^{-1} rules out the carboxylic acid, and the absence of an absorption band at ~ 2700 cm^{-1} rules out the aldehyde. Thus, it must be the ketone.

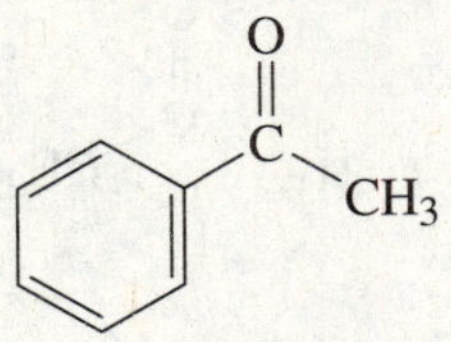

c. The absorption band at ~ 1700 cm^{-1} indicates a carbon-oxygen double bond. The absence of an absorption band at ~ 1600 cm^{-1} rules out the ketones with the benzene or cyclohexene rings. The absence of absorption bands at ~ 2100 cm^{-1} and ~ 3300 cm^{-1} rules out the ketone with the carbon-carbon triple bond. Thus, it must be 4-ethylcyclohexanone.

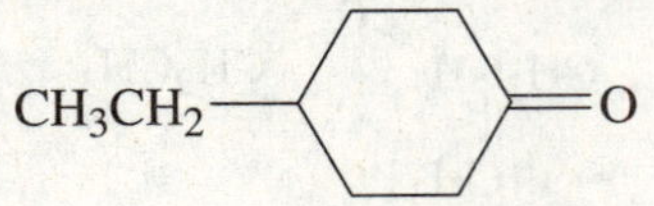

59. Since only benzene absorbs light of 260 nm, the concentration of benzene can be determined by measuring the amount of absorbance at 260 nm (A), using the Beer-Lambert law, since the length of the light path of the cell (l) is known and the molar absorptivity of benzene at 260 nm (ε) can be looked up (or can be determined from the absorbance at 260 nm given by a solution of benzene with a known concentration).

the Beer-Lambert law: $A = c\, l\, \varepsilon$

60. **a.** The absorption band at ~ 2700 cm^{-1} indicates that the compound is an aldehyde (carbon-hydrogen stretch of an aldehyde hydrogen). The absence of an absorption band at ~ 1600 cm^{-1} rules out the aldehyde with the benzene ring. Thus, it must be the other aldehyde.

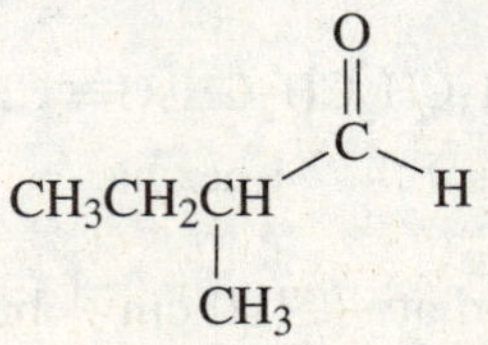

b. The absorption bands at ~ 3350 cm^{-1} and ~ 3200 cm^{-1} indicate that the compound is an amide (nitrogen-hydrogen stretch). The absence of an absorption band at ~ 3050 cm^{-1} indicates that the compound does not have hydrogens bonded to *sp*^{2} carbons. Therefore, it is not the amide that has a benzene ring.

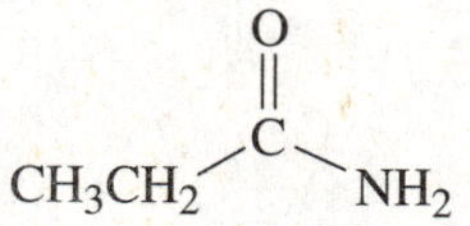

c. The absence of absorption bands at ~1600 cm^{-1} and ~1500 cm^{-1} indicates that the compound does not have a benzene ring. Thus, it must be the ketone. This is confirmed by the absence of an absorption band at ~1380 cm^{-1}, indicating that the compound does not have a methyl group. (Notice the first fundamental of the 1700 cm^{-1} absorption band at 3400 cm^{-1}.)

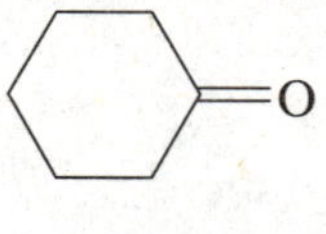

61. **a.** RC(=O)—H 2700
C=O 1700
C=C 1600

b. C=O 1700
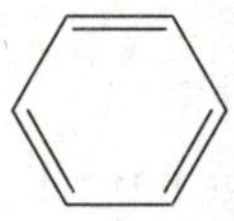 1500, 1600
C—O ~1250, ~1050

c. C—O ~1050
O—H 3600-3200 (broad)

d. C=O 1700
C—N ~1030
N—H 3500-3300

e. *sp* CH 3300 (narrow)
C≡C 2100

f. C=O 1700
C(=O)OH ~3000 (broad)
C—O ~1250

62. Calculating the term in Hooke's law that depends on the masses of the atoms joined by a bond for a C—H bond and for a C—C bond shows why smaller atoms give rise to greater wavenumbers.

for a carbon-hydrogen bond

$$\frac{m_1 + m_2}{m_1 \times m_2} = \frac{12 + 1}{12 \times 1} = \frac{13}{12} = 1.08$$

for a carbon-carbon bond

$$\frac{m_1 + m_2}{m_1 \times m_2} = \frac{12 + 12}{12 \times 12} = \frac{24}{144} = 0.17$$

63. The absorption band at ~1700 cm^{-1} indicates that the compound has a carbonyl group, and the absence of an absorption band at ~1380 cm^{-1} indicates that it has no methyl groups. The absence of an absorption band at ~1600 cm^{-1} indicates the compound does not have a carbon-carbon double bond, and the absence of an absorption band at ~3050 cm^{-1} indicates the compound does not have hydrogens bonded to *sp*^{2} carbons.

From the molecular formula you can deduce that the compound is **cyclopentanone**.

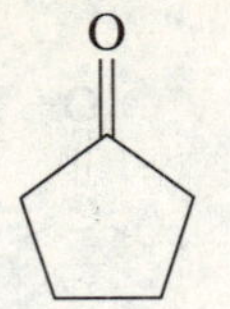

64. Before the addition of acid, the compound is colorless because the benzene rings are not conjugated with each other, since they are separated from each other by two single bonds. In the presence of acid, a carbocation is formed in which the three benzene rings are conjugated with each other. The conjugated carbocation is highly colored.

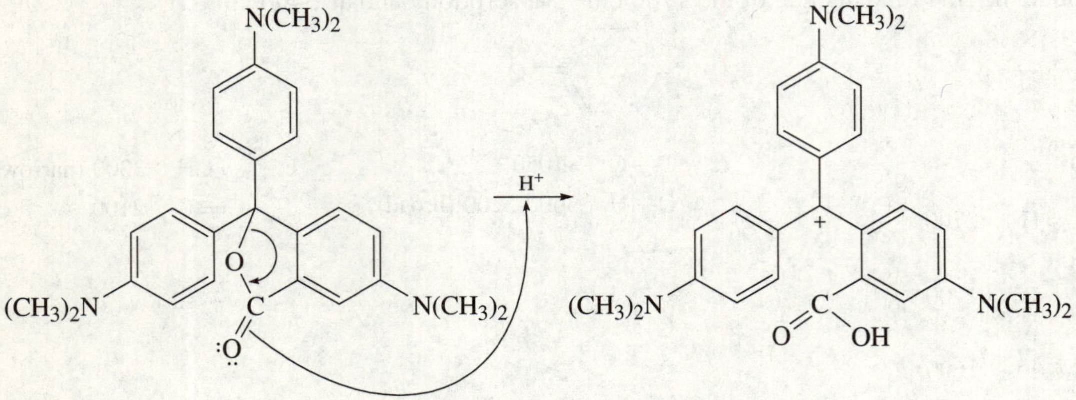

65. The broad absorption band at ~ 3300 cm^{-1} indicates that the compound has an OH group. The absorption bands at ~ 2900 cm^{-1} indicate that the compound has hydrogens attached to an sp^3 carbon. The compound, therefore, is **benzyl alcohol**.

66. The broad absorption band at ~ 3300 cm^{-1} indicates that the compound has an OH group. The absence of absorption at ~ 2950 cm^{-1} indicates the compound does not have any hydrogens bonded to sp^3 carbons. Therefore, the compound is phenol.

67. The M peak at 100 indicates that the compound has a molecular weight of 100, and the relative intensities of the M and M + 1 peaks indicate that the compound has seven carbons.

$$\text{number of carbon atoms} = \frac{M + 1}{0.011 \times M} = \frac{2.10}{0.011 \times 27.32} = \frac{2.10}{0.301} = 7$$

When $n = 7$, $(C_nH_{2n+2}) = C_7H_{16}$. A compound with that molecular formula has a molecular weight of 100 in agreement with the M peak.

68. $$\tilde{\nu} = \frac{1}{2 \times 3.1416 \times 3 \times 10^{10} \text{ cm s}^{-1}} \sqrt{\frac{10 \times 10^{5} \text{ g s}^{-2}\left(\frac{12}{6.02} + \frac{12}{6.02}\right) \times 10^{-23} \text{ g}}{\frac{12}{6.02} \times 10^{-23} \text{ g} \times \frac{12}{6.02} \times 10^{-23} \text{ g}}}$$

$$\tilde{\nu} = \frac{1}{18.85 \times 10^{10}} \sqrt{10.0 \times 10^{28}}$$

$$\tilde{\nu} = \frac{1}{18.85 \times 10^{10}} \times 3.16 \times 10^{14}$$

$$\tilde{\nu} = 1676 \text{ cm}^{-1}$$

69. **a.** The IR spectrum indicates that the compound is an aliphatic ketone with at least one methyl group. The M peak at *m/z* = 100 indicates that the ketone is a hexanone. The peak at 43(100 – 57) for loss of a butyl group and the peak at 85 for loss of a methyl group (100 – 15) suggest that the compound is **2-hexanone**.

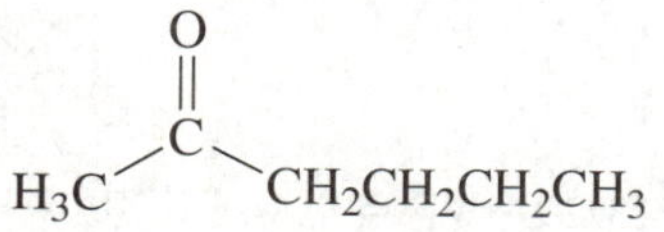

This is confirmed by the peak at 58 for loss of propene (100 – 42) as a result of a McLafferty rearrangement.

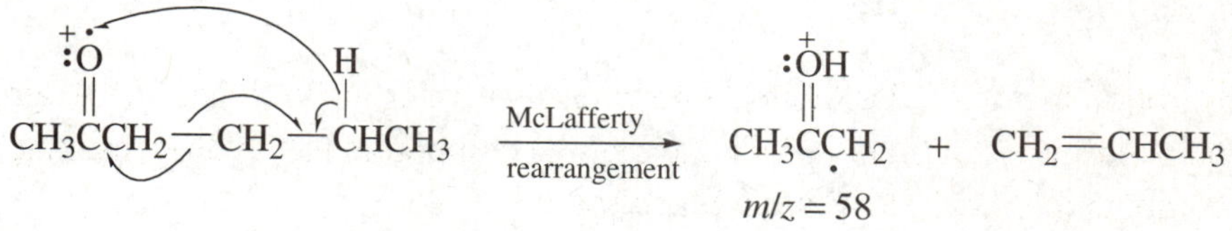

b. The equal heights of the M and M + 2 peaks at 162 and 164 indicate that the compound contains bromine. The peak at *m/z* = 83 (162 – 79) is for the carbocation that is formed when the bromine atom is eliminated. The IR spectrum does not indicate the presence of any functional groups, and it shows that there are no methyl groups present. The *m/z* peak = 83 indicates a carbocation with a molecular formula of C_6H_{11}. The fact that the compound does not contain a methyl group indicates that the compound is **bromocyclohexane**.

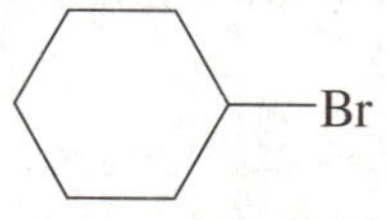

bromocyclohexane

c. The absorption bands at ~1700 cm^{-1} and ~2700 cm^{-1} indicate that the compound is an aldehyde. The molecular ion peak at *m/z* = 72 indicates that the aldehyde contains four carbons. The peak at *m/z* = 44 for loss of a group with molecular weight 28 indicates that ethene has been lost as a result of a McLafferty rearrangement.

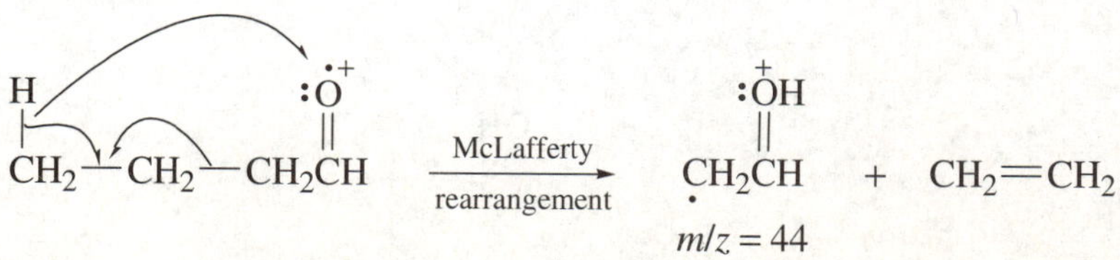

A McLafferty rearrangement can occur only if the aldehyde has a γ-hydrogen.
The only four-carbon aldehyde that has a γ-hydrogen is **butanal.**

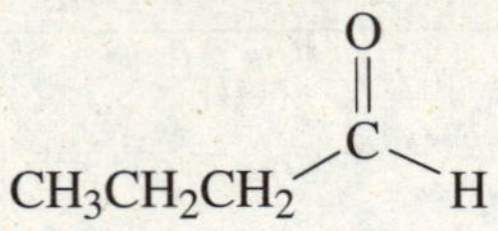

70. **a.** At pH < 4, the lone pair of the nitrogen of the $N(CH_3)_2$ group is protonated, so it cannot interfere with the fully conjugated system.

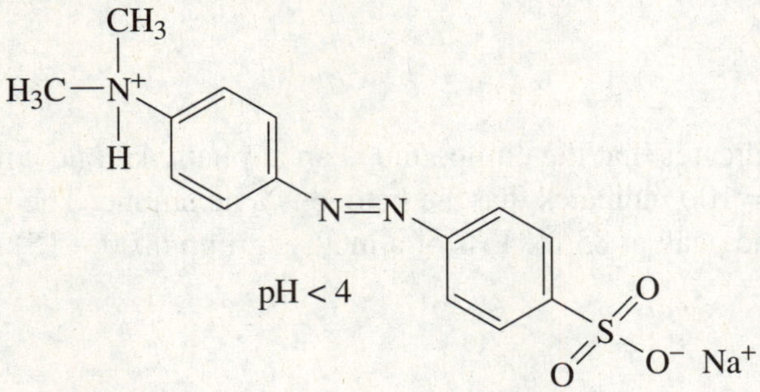

At pH > 4, the nitrogen of the $N(CH_3)_2$ group is not protonated and the lone pair can be delocalized into the benzene ring. This decreases the conjugation and, therefore, light of shorter wavelengths will be absorbed; hence, the change in color from red to yellow.

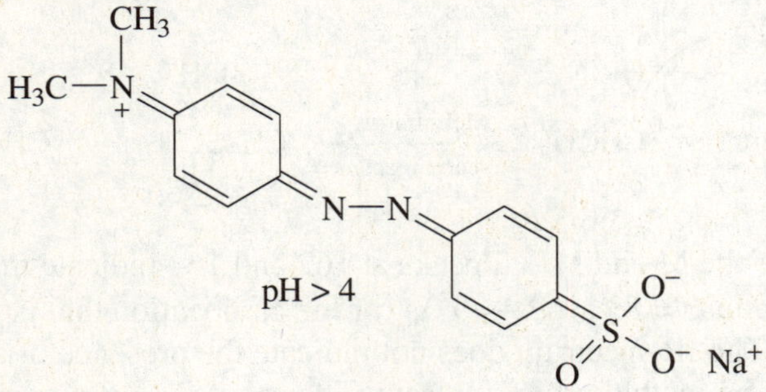

b. In acidic solutions, the three benzene rings are isolated from one another.
In basic solutions, as a result of loss of the proton from one of the OH groups, there is a greater degree of conjugation.

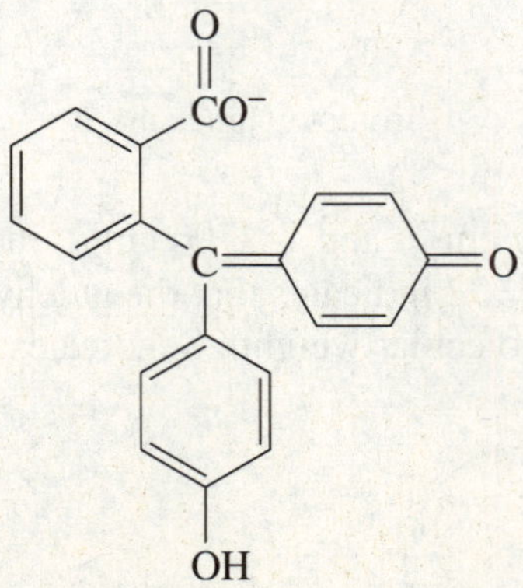

Chapter 12 Practice Test

1. Give one IR absorption band that could be used to distinguish each of the following pairs of compounds. Indicate the compound for which the band would be present.

a. $CH_3CH_2CH_2\overset{\overset{O}{\|}}{C}H$ and $CH_3CH_2CH_2\overset{\overset{O}{\|}}{C}CH_3$

b. $CH_3CH_2\overset{\overset{O}{\|}}{C}NH_2$ and $CH_3CH_2\overset{\overset{O}{\|}}{C}OCH_3$

c. $CH_3CH_2CH_2CH_2OH$ and $CH_3CH_2CH_2OCH_3$

d. 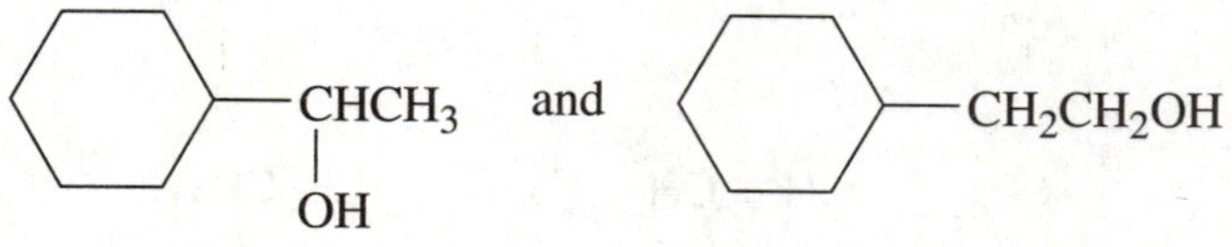

e. $CH_3CH_2CH_2\overset{\overset{O}{\|}}{C}OCH_3$ and $CH_3CH_2CH_2\overset{\overset{O}{\|}}{C}CH_3$

f. $CH_3CH_2CH{=}CHCH_3$ and $CH_3CH_2C{\equiv}CCH_3$

g. $CH_3CH_2C{\equiv}CH$ and $CH_3CH_2C{\equiv}CCH_3$

2. Indicate whether each of the following is true or false.

a. The O—H stretch of a concentrated solution of an alcohol occurs at a higher frequency than the O—H stretch of a dilute solution. T F

b. Light of 2 μm is of higher energy than light of 3 μm. T F

c. It takes more energy for a bending vibration than for a stretching vibration. T F

d. Propyne will not have an absorption band at 3100 cm^{-1} because there is no change in the dipole moment. T F

e. Light of 8 μm has the same energy as light of 1250 cm^{-1}. T F

f. The M + 2 peak of an alkyl chloride is half the height of the M peak. T F

g. A chromophore that exhibits both $n \rightarrow \pi^*$ and $\pi \rightarrow \pi^*$ transitions will have the $n \rightarrow \pi^*$ transition at a longer wavelength. T F

3. The major peaks shown in the mass spectrum of a tertiary alcohol are at $m/z = 73$, 87, 98, and 101. Identify the alcohol.

4. A 3.8×10^{-4} M solution of cyclohexanone shows an absorbance of 0.75 at 280 nm in a 1.00 cm cell. What is the molar absorptivity of cyclohexanone at 280 nm?

5. How could you distinguish between the IR spectra of the following compounds?

a.
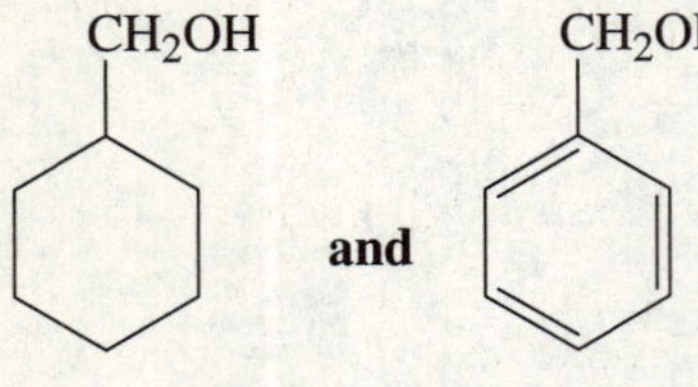

d.
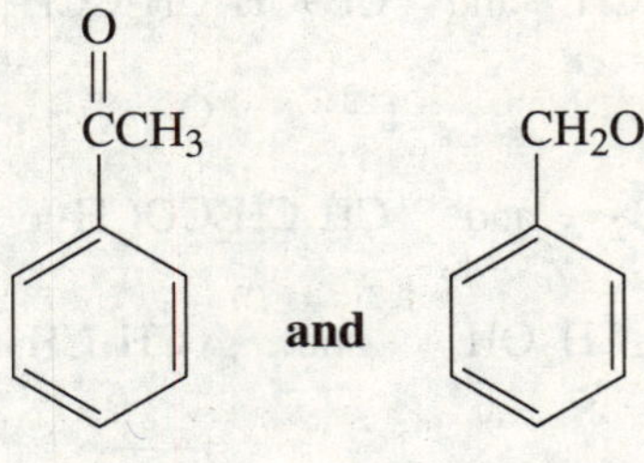

b.
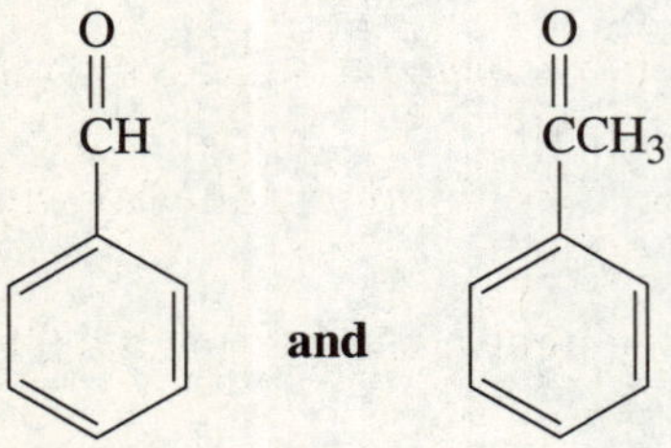

e.
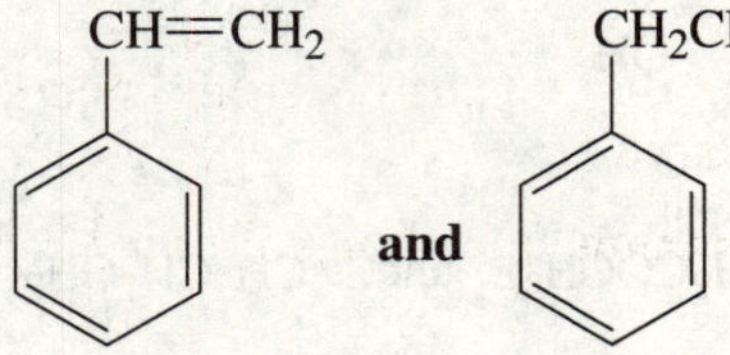

c.
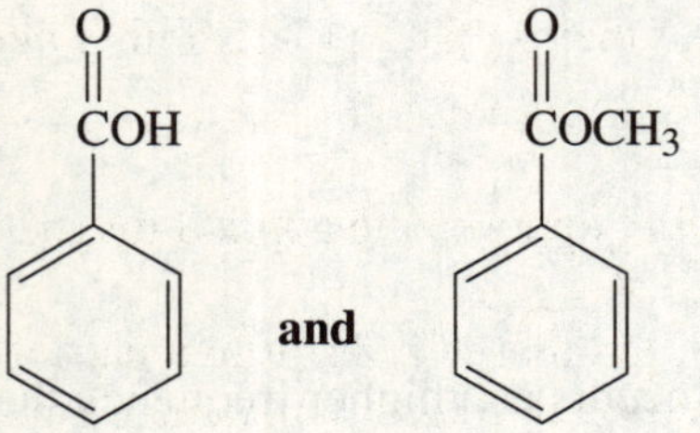

6. Which compound has the greater λ_{max}? (300 nm is a greater λ_{max} than 250 nm.)

a.
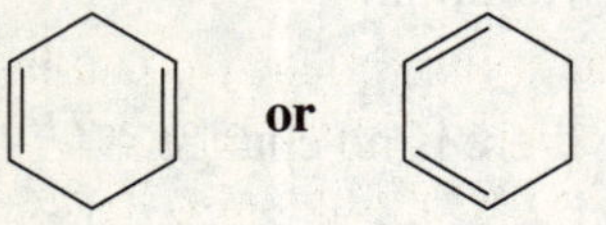

c.
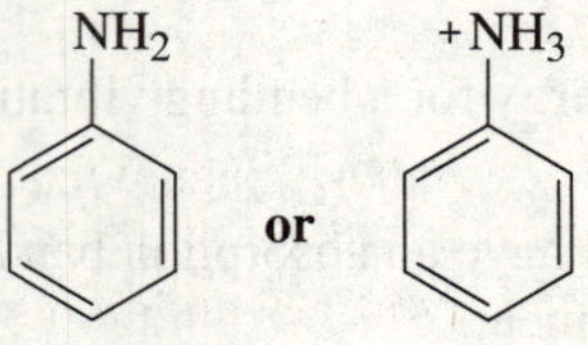

b.
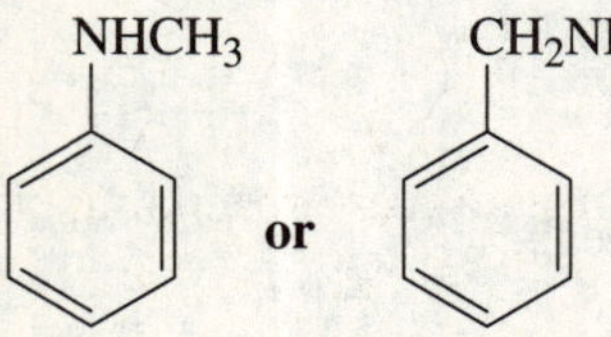

7. A solution of a compound with a molar absorbtivity of 1,200 $M^{-1}cm^{-1}$ at 297 nm gives an absorbance of 0.76 at that wavelength in a 1 cm quartz cell. What is the concentration of the solution?

CHAPTER 13
NMR Spectroscopy

Important Terms

applied magnetic field — the externally applied magnetic field.

chemically equivalent protons — protons with the same connectivity relationship to the rest of the molecule.

chemical shift — location of a signal occurring in an NMR spectrum. It is measured downfield from a reference compound (most often TMS).

^{13}C NMR — nuclear magnetic resonance that shows carbon (^{13}C) nuclei.

COSY spectrum — a 2-D NMR spectrum showing ^1H-^1H correlations.

coupled protons — protons that split each other. Coupled protons have the same coupling constant.

coupling constant — the distance (in hertz) between two adjacent peaks of a split NMR signal.

DEPT ^{13}C NMR spectrum — a group of four ^{13}C NMR spectra that distinguish CH_3, CH_2, and CH groups.

diamagnetic anisotropy — the term used to describe the greater freedom of π electrons to move in response to a magnetic field as a consequence of their greater polarizability compared with σ electrons.

diamagnetic shielding — shielding by the local magnetic field that opposes the applied magnetic field.

diastereotopic hydrogens — two hydrogens bonded to the same carbon that will result in a pair of diastereomers when each of them is replaced in turn with deuterium.

2-D NMR — two-dimensional nuclear magnetic resonance.

doublet — an NMR signal that is split into two peaks.

doublet of doublets — an NMR signal that is split into four peaks of approximately equal height. A doublet of doublets is caused by splitting a signal into a doublet by one hydrogen and into another doublet by another (nonequivalent) hydrogen.

downfield — farther to the left-hand side of the spectrum.

effective magnetic field — the magnetic field that a nucleus "senses" through the surrounding cloud of electrons.

enantiotopic hydrogens — two hydrogens bonded to a carbon that is bonded to two other groups that are nonidentical.

Fourier transform NMR (FT-NMR) — a technique in which all the nuclei are excited simultaneously by an rf pulse, their relaxation monitored, and the data mathematically converted to a frequency spectrum.

geminal coupling — the mutual splitting of two nonidentical protons bonded to the same carbon.

gyromagnetic ratio — the ratio of the magnetic moment of a rotating charged particle to its angular momentum.

HETCOR spectrum — a 2-D NMR spectrum showing ^{13}C-^{1}H correlations.

^{1}H NMR — nuclear magnetic resonance that shows hydrogen nuclei.

high-resolution NMR spectroscopy — NMR spectroscopy that uses a spectrometer with a high operating frequency.

long-range coupling — splitting of a proton by a proton more than 3 σ bonds away.

magnetic resonance imaging (MRI) — NMR used in medicine. The difference in the way water is bound in different tissues produces the signal variation between organs as well as between healthy and diseased states.

methine hydrogen — a tertiary hydrogen.

MRI scanner — an NMR spectrometer used in medicine for whole-body NMR.

multiplet — an NMR signal split by two non-equivalent sets of protons.

multiplicity — the number of peaks in an NMR signal.

***N* + 1 rule** — an ^{1}H NMR signal for a hydrogen with N equivalent hydrogens bonded to an adjacent carbon is split into $N + 1$ peaks; a ^{13}C NMR signal for a carbon bonded to N hydrogens is split into $N + 1$ peaks.

NMR spectroscopy — the absorption of electromagnetic radiation to determine the structural features of an organic compound. In the case of ^{1}H NMR spectroscopy, it determines the carbon-hydrogen framework.

operating frequency — the frequency at which an NMR spectrometer operates.

prochiral carbon — a carbon (bonded to two hydrogens) that will become an asymmetric center if one of the hydrogens is replaced by deuterium.

pro-*R*-hydrogen — replacing this hydrogen with deuterium creates an asymmetric center with the *R* configuration.

pro-*S*-hydrogen — replacing this hydrogen with deuterium creates an asymmetric center with the *S* configuration.

proton exchange	the transfer of a proton from one molecule to another.
quartet	an NMR signal that is split into four peaks.
reference compound	a compound added to the sample whose NMR spectrum is to be taken. The position of the signals in the NMR spectrum are measured from the position of the signal given by the reference compound.
rf radiation	radiation in the radiofrequency region of the electromagnetic spectrum.
shielding	caused by electron donation to the environment of a proton. The electrons shield the proton from the full effect of the applied magnetic field. The more a proton is shielded, the farther to the right its signal appears in an NMR spectrum.
singlet	an unsplit NMR signal.
spin-coupled ^{13}C NMR spectrum	a ^{13}C NMR spectrum in which each signal for a carbon is split by the hydrogens bonded to that carbon.
spin-coupling	coupling between the carbon nucleus and certain nearby atoms.
spin-spin coupling	the splitting of a signal in an NMR spectrum described by the $N + 1$ rule.
α-spin state	nuclei in this spin state have their magnetic moments oriented in the same direction as the applied magnetic field.
β-spin state	nuclei in this spin state have their magnetic moments oriented opposite to the direction of the applied magnetic field.
splitting diagram (splitting tree)	a diagram that describes the splitting of a set of protons.
triplet	an NMR signal that is split into three peaks.
upfield	farther to the right-hand side of the spectrum.

Solutions to Problems

1. $$\nu = \frac{\gamma}{2\pi}$$

$$= \frac{26.75 \times 10^7 \text{ rad T}^{-1} \text{ sec}^{-1} \times 1.0\text{ T}}{2(3.1416) \text{ rad}}$$

$$= 43 \times 10^6 \text{ sec}^{-1}$$

$$= 43 \times 10^6 \text{ Hz} = 43 \text{ MHz}$$

2. **a.** $$\nu = \frac{\gamma}{2\pi} B_0$$

$$B_0 = \frac{\nu \times 2\pi}{\gamma}$$

$$B_0 = \frac{360 \times 10^6 \times 2(3.1416)}{26.75 \times 10^7}$$

$$B_0 = \frac{226.2}{26.75}$$

$$B_0 = 8.46 \text{ T}$$

b. $$B_0 = \frac{\nu \times 2\pi}{\gamma}$$

$$B_0 = \frac{500 \times 10^6 \times 2(3.1416)}{26.75 \times 10^7}$$

$$B_0 = \frac{314.2}{26.75}$$

$$B_0 = 11.75 \text{ T}$$

From these calculations, you can see that the greater the operating frequency of the instrument (360 MHz versus 500 MHz), the more powerful is the magnet (8.46 T versus 11.75 T) required to operate it.

3. **a.** 2 **b.** 1 **c.** 3 **d.** 3 **e.** 1 **f.** 4 **g.** 3 **h.** 3 **i.** 5 **j.** 4 **k.** 2 **l.** 3 **m.** 3 **n.** 4 **o.** 3

4. **a** would give two signals, **b** would give one signal, and **c** would give three signals.

5. #1

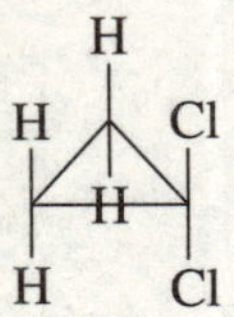

All the H's are equivalent.

#2

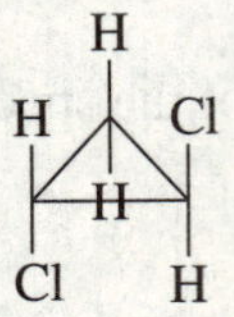

The H's attached to the front of the molecule are equivalent, and the methylene H's are equivalent.

#3

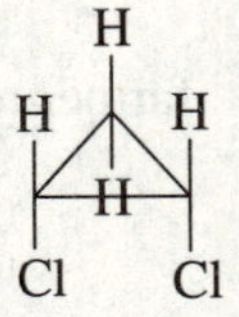

The H's attached to the front of the molecule are equivalent and the methylene H's are not equivalent.

6. $$\text{chemical shift in ppm} = \frac{\text{downfiled from TMS in Hertz}}{\text{operating frequency in Megahertz}}$$

a. $$1 \text{ ppm} = \frac{300 \text{ Hz}}{300 \text{ MHz}}$$

300 hertz

b. $$1 \text{ ppm} = \frac{500 \text{ Hz}}{500 \text{ MHz}}$$

500 hertz

7. a. $\dfrac{600\,\text{Hz}}{300\,\text{MHz}} = 2.0\ \text{ppm}$

b. The answer would still be 2.0 ppm because the chemical shift is independent of the operating frequency of the spectrometer.

c. $\dfrac{x\,\text{Hz}}{100\,\text{MHz}} = 2.0\ \text{ppm}$

$x = 200$

200 Hz downfield from TMS

8. a. The chemical shift is independent of the operating frequency. Therefore, if the two signals differ by **1.5 ppm** in a 300-MHz spectrometer, they will still differ by **1.5 ppm** in a 100-MHz spectrometer.

b. $\dfrac{\text{Hz}}{\text{MHz}} = \text{ppm}$

$\dfrac{90\ \text{Hz}}{300\ \text{MHz}} = 0.3\ \text{ppm}$

$\dfrac{x\ \text{Hz}}{100\ \text{MHz}} = 0.3\ \text{ppm}$

$x = 30\ \text{Hz}$

9. Magnesium is less electronegative than silicon. (See Table 10.3 on page 466 of the text.) Therefore, the peak for $(CH_3)_2Mg$ would be upfield from the TMS peak.

10. **a. and b.**

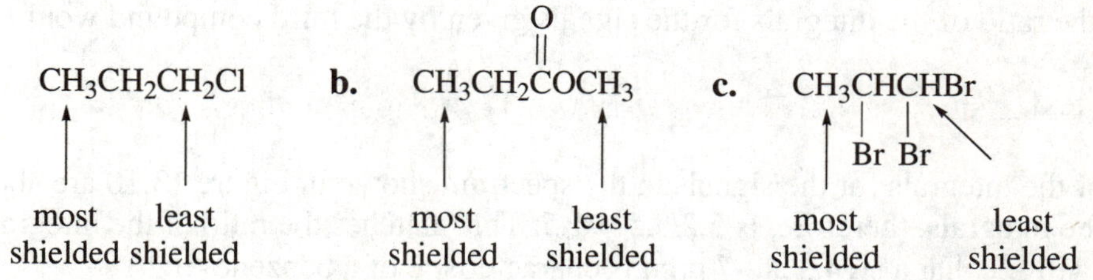

11.

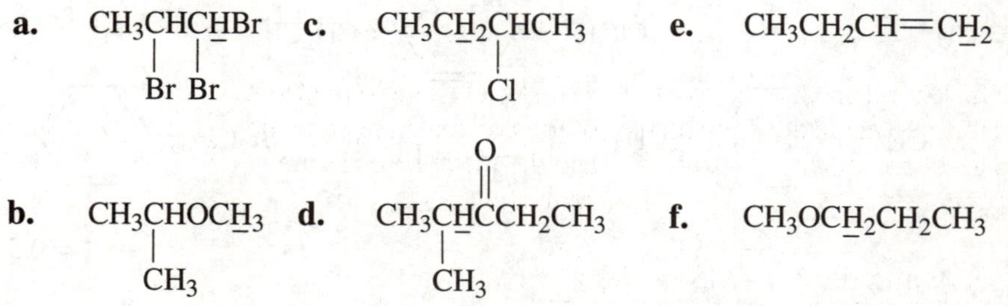

12.

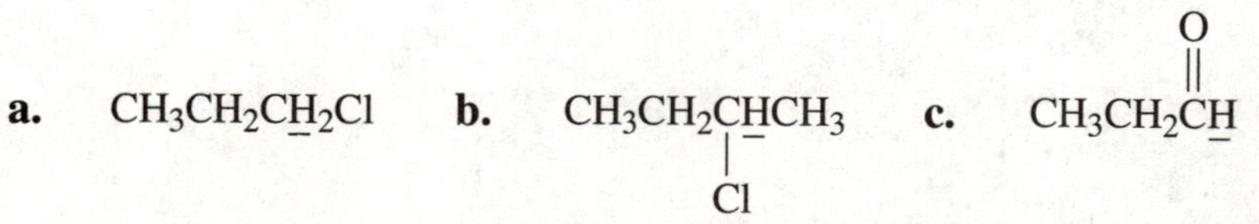

13.

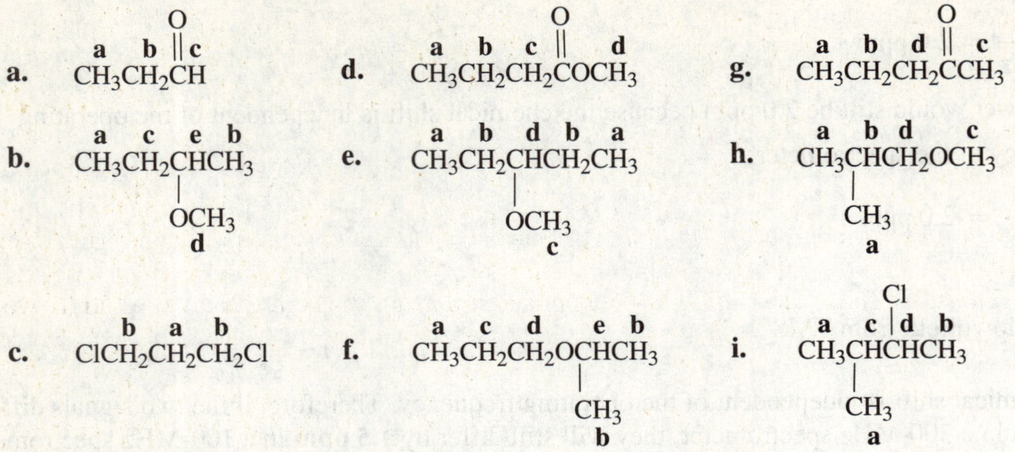

14. From the direction of the electron flow around the benzene ring pictured in Figure 13.6 on page 582 of the text, you can see that the magnetic field induced in the region of the hydrogens that protrude out from the compound shown in this problem is in the same direction as the applied magnetic field, whereas the magnetic field induced in the region of the hydrogens that protrude into the center of the compound is in the opposite direction of the applied magnetic field.

Thus, the signal at 9.25 ppm is for the hydrogens that protrude out because they need a smaller applied magnetic field to come into resonance due to the fact that the induced and applied magnetic fields are in the same direction. The signal at -2.88 ppm is for the hydrogens that protrude inward because a greater applied field is necessary to make up for the magnetic field that is induced in the opposite direction.

15. Each of the compounds would show two signals, but the ratio of the integrals for the two signals would be different for each of the compounds. The ratio of the integrals for the signals given by the first compound would be 2 : 9 (or 1 : 4.5), the ratio of the integrals for the signals given by the second compound would be 1 : 3, and the ratio of the integrals for the signals given by the third compound would be 1 : 2.

16. Solved in the text.

17. The heights of the integrals for the signals in the spectrum shown in Figure 13.10 are about 3.5 and 5.2. The ratio of the integrals, therefore, is 5.2/3.5 = 1.5. This matches the ratio of the integrals calculated for **B**. (Later we will see that a signal at ~7 ppm is characteristic of a benzene ring.)

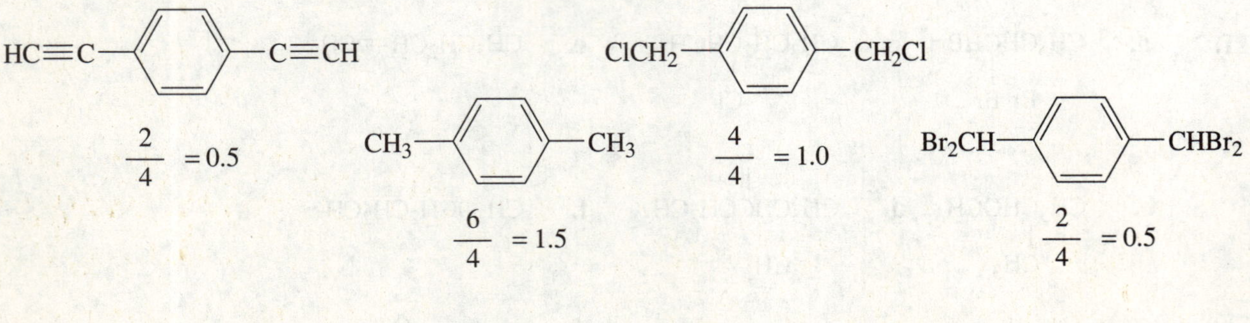

18. The signal farthest downfield in both spectra is the signal for the hydrogens bonded to the carbon that is also bonded to the halogen. Because chlorine is more electronegative than iodine, the farthest downfield signal should be farther downfield in the ^{1}H NMR spectrum for 1-chloropropane than in the ^{1}H NMR spectrum for 1-iodopropane. Therefore, the **first spectrum** in Figure 13.5 is the ^{1}H NMR spectrum for **1-iodopropane**, and the **second spectrum** in Figure 13.5 is ^{1}H the NMR spectrum for **1-chloropropane**.

19. **a.** A triplet is caused by coupling to two equivalent protons on an adjacent carbon. The two protons can be aligned in three different ways: both with the field, one with the field and one is against the field, or both against the field. That is why the signal is a triplet. There is only one way to align two protons that are with the field or two protons that are against the field. However, there are two ways to align two protons if one is with and one against the field: with and against or against and with. Consequently, the peaks in a triplet have relative intensities of 1 : 2 : 1.

↑↑	↑↓	↓↓
	↓↑	

b. A quintet is caused by coupling to four equivalent protons on adjacent carbons. The four protons can be aligned in five different ways. The following possible arrangements for the alignment of four protons explain why the relative intensities of a quintet are 1: 4 : 6 : 4 : 1.

↑↑↑↑	↑↑↑↓	↑↑↓↓	↑↓↓↓	↓↓↓↓
	↑↑↓↑	↓↓↑↑	↓↑↓↓	
	↑↓↑↑	↑↓↑↓	↓↓↑↓	
	↓↑↑↑	↓↑↓↑	↓↓↓↑	
		↑↓↓↑		
		↓↑↑↓		

20. The signal at 12.2 ppm indicates that the compound is a carboxylic acid. From the molecular formula and the splitting patterns of the signals, the spectra can be identified as the ^{1}H NMR spectrum of:

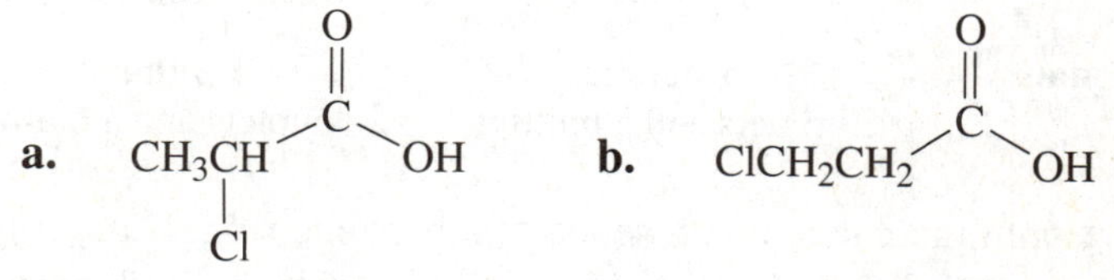

21. **a.** 3 signals **b.** 2 signals **c.** 3 signals

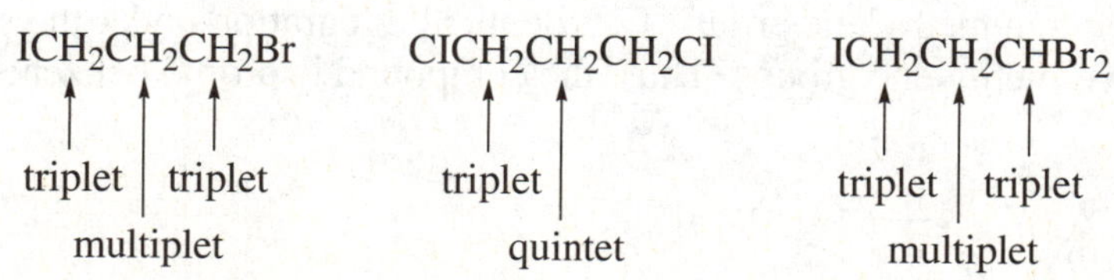

22. The contributing resonance structures show that the nitro group's ability to withdraw electrons by electron delocalization causes the C-2 and C-4 positions of the resonance hybrid (the actual structure of the compound) to have a partial positive charge. The nitro group's ability to withdraw electrons inductively (through the sigma bonds) causes the partial positive charge to be greater at C-2 than at C-4. Therefore, the H_c proton has the least electron dense (less shielded) environment, so its signal appears at the highest frequency. The C-3 position does not have a partial positive charge. Therefore, the H_a proton has the most electron dense (most shielded) environment, so its signal appears at the lowest frequency.

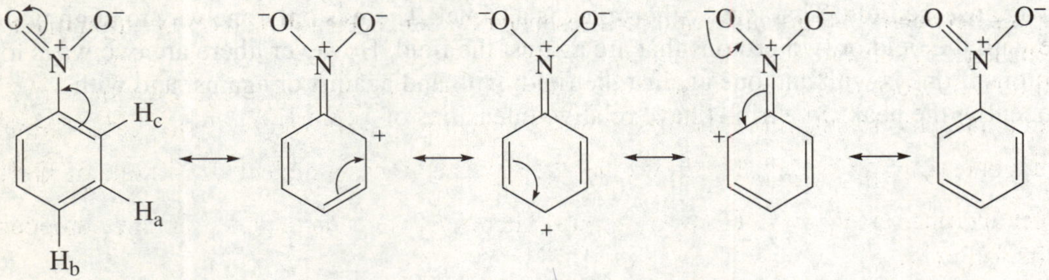

23. Each compound will have two doublets. In addition, **A** will have 2 doublet of doublets, **B** will have one doublet of doublets and a singlet, and **C** will have no other signals.

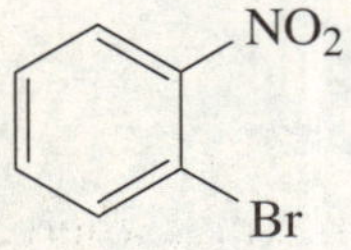

A

4 signals

(2 doublets and
2 doublet of doublets)

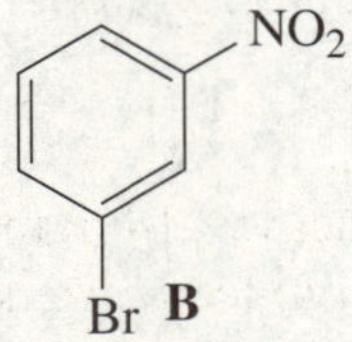

B

4 signals

(1 singlet,
2 doublets and,
1 doublet of doublets)

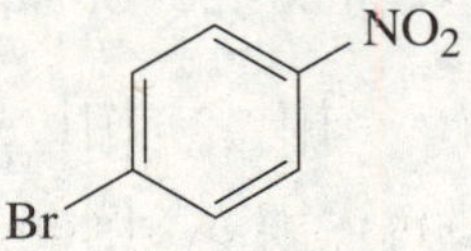

C

2 signals

(2 doublets)

24.

$CH_3C(Br)_2CH_3$	$BrCH_2CH_2CH_2Br$	$CH_3CH_2CH(Br)Br$	$CH_3CH(Br)CH_2Br$
1 signal	**2 signals**	**3 signals** 2 triplets and a multiplet	**3 signals** 2 doublets and a multiplet

25. **a.** The signal at ~7.2 ppm indicates the presence of a benzene ring. If the ring has a single substituent, it has a molecular formula of C_6H_5. Subtracting C_6H_5 from the molecular formula of the compound $(C_9H_{12} - C_6H_5)$ gives a substituent with a molecular formula of C_3H_7. The triplet at ~0.9 ppm indicates a methyl group adjacent to a methylene group. The identical integration under the signals at 1.6 ppm and 2.6 ppm indicate two methylene groups. Thus, the compound is propyl benzene.

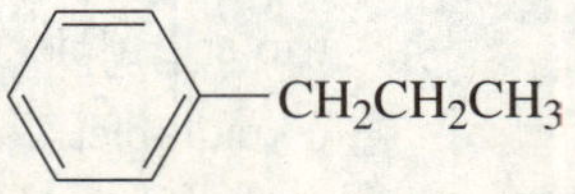

b. The triplet and quartet indicate a CH_3CH_2 group bonded to an atom that is not bonded to any hydrogens. The molecular formula of $C_5H_{10}O$ indicates that the compound is diethyl ketone.

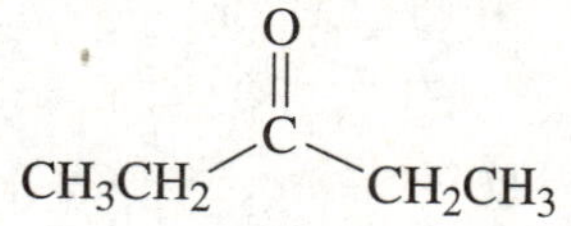

c. The signals between ~7-8 ppm indicate the presence of a benzene ring. The observation that there are 3 benzene ring signals indicates the compound has a mono-substituted benzene ring or a 1,3-disubstituted benzene ring. Since none of the 3 benzene ring signals is a singlet, we know that the compound has a mono-substituted benzene ring. Subtracting C_6H_5 from the molecular formula of the compound $(C_9H_{10}O_2 - C_6H_5)$ shows that the substituent has a molecular formula of $C_3H_5O_2$.

The triplet and quartet suggest a CH_3CH_2 group. Therefore, we can conclude that the compound is one of the following.

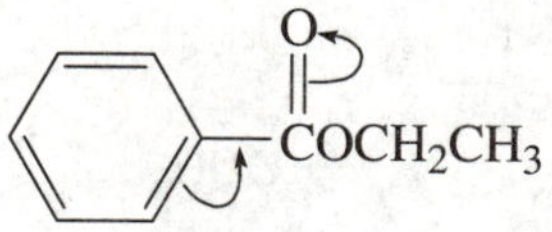

A

or

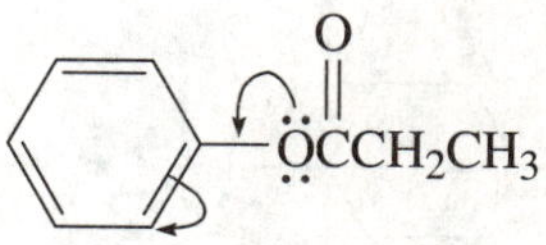

B

The substituent attached to Compound **A** withdraws electrons from the benzene ring, whereas the substituent attached to compound **B** donates electrons into the benzene ring. Because one of the peaks of the benzene ring signal is at a higher frequency (farther downfield) than usual (> 8 ppm) suggests that an electron-withdrawing substituent is present on the benzene ring. This is confirmed by the signal for the CH_2 group at 3.8 ppm, indicating that the methylene group is adjacent to the electron-withdrawing oxygen. Thus, the compound is compound **A**.

26. **a.** 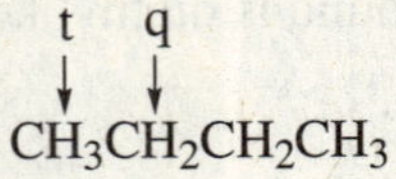

b. 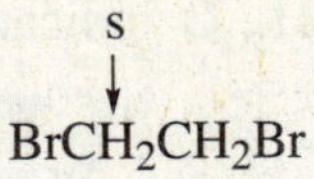

c. 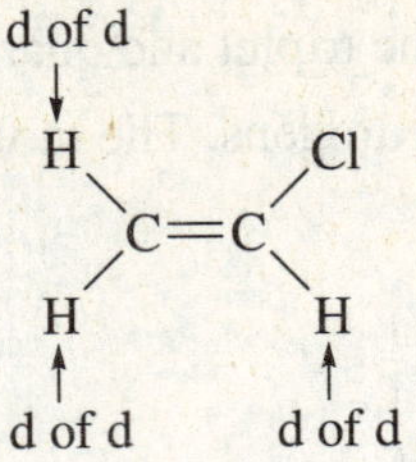

d. 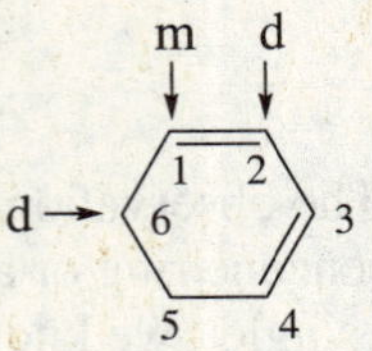

The H on C-2 will not be split by the H on C-3, since they are equivalent. The H's on C-6 will not be split by the H's on C-5, since the H's on C-5 and C-6 are equivalent.

e. 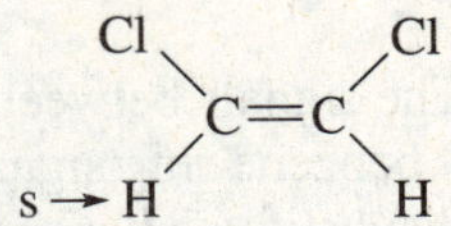

f. 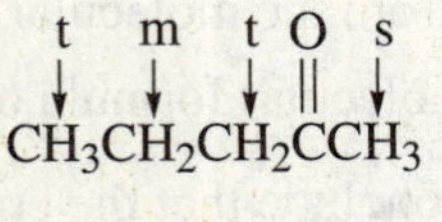

g. 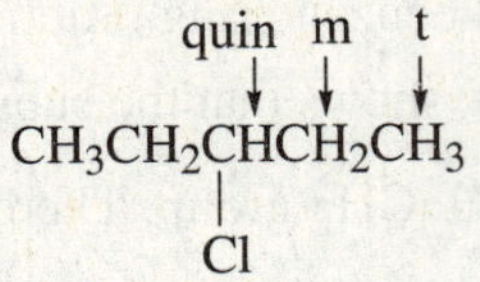

h. 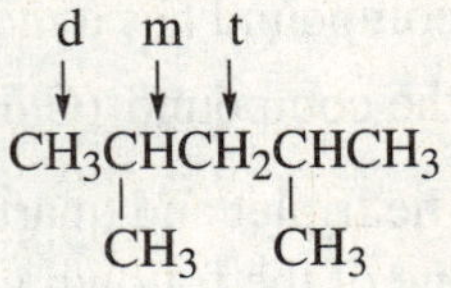

i. 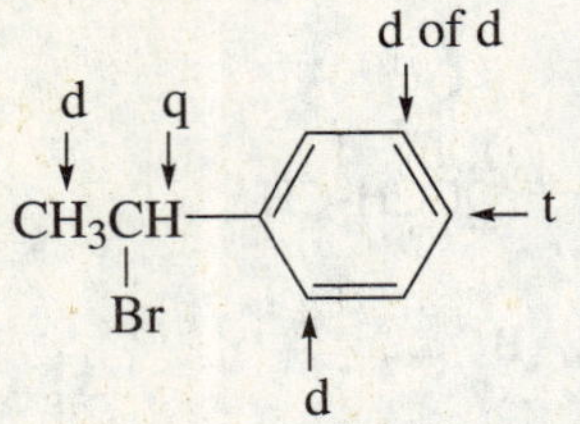

j. 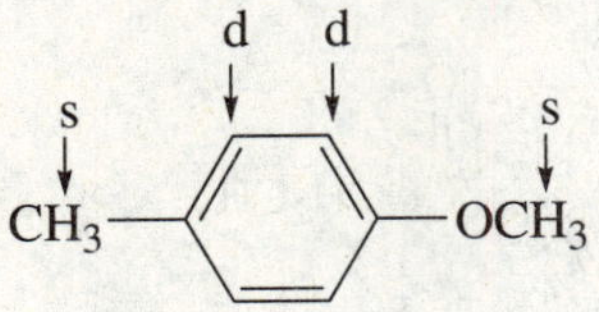

k. 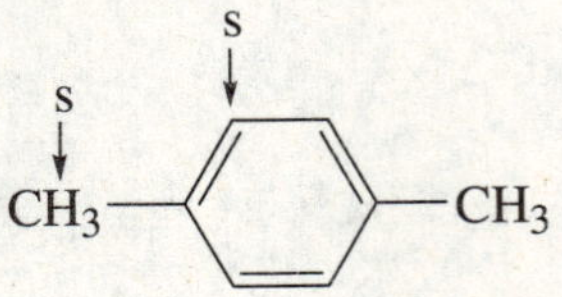

l. 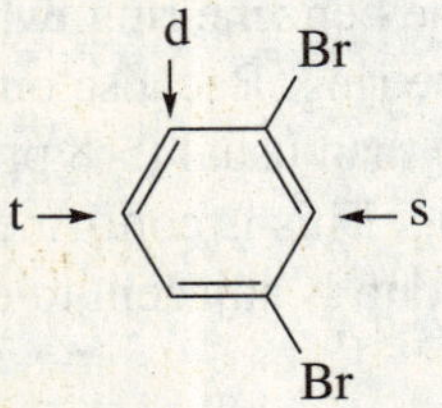

m. 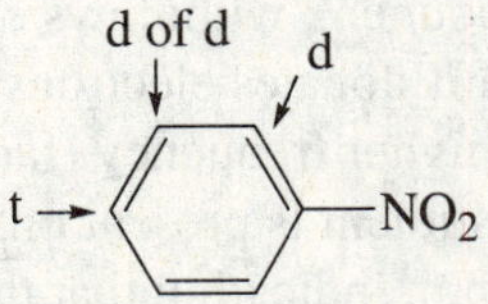

n. 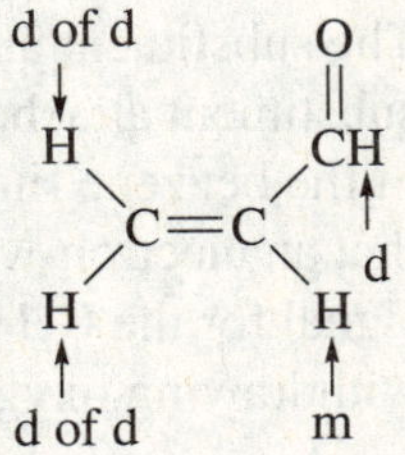

o. 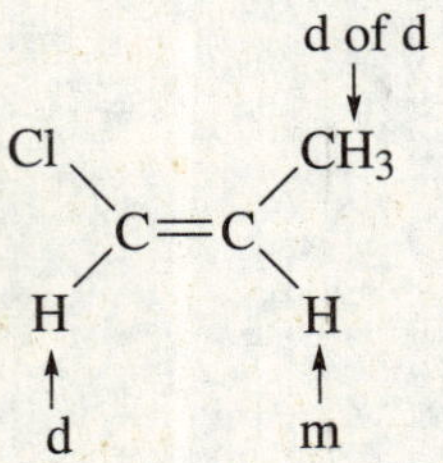

27. **a.** $CH_3OCH_2CH_2OCH_3$

b. There are 3 possibilities.

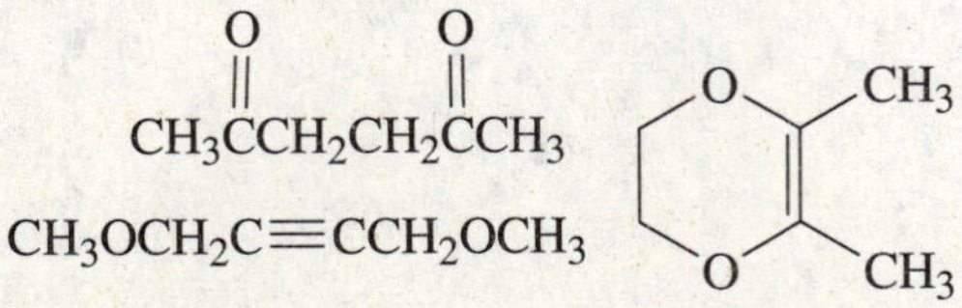

c. 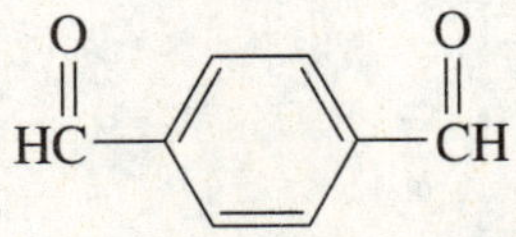

28. Each spectrum is described going from left to right.

a. $BrCH_2CH_2CH_2CH_2Br$
triplet triplet

b. two triplets (close to each other) singlet multiplet
(Appendix VI on page A-20 of the text indicates that a methylene adjacent to an RO and a methylene adjacent to a Br appear at about the same place.)

c. singlet (Equivalent H's do not split each other.)

d. quartet singlet triplet

e. three singlets

f. three doublets of doublets

g. quartet triplet

h. singlet quartet triplet

i. doublet multiplet doublet

j. triplet quintet

k. triplet doublet of doublets multiplet doublet

l. triplet singlet quintet

m. singlet (Equivalent H's do not split each other.)

n. singlet (Equivalent H's do not split each other.)

o. singlet

29. There is no coupling between H_a and H_c because they are separated by four σ bonds. (We will see in Section 13.16 that unless the sample is pure and dry, a hydrogen on an oxygen will not undergo coupling.)
There is no coupling between H_b and H_c because they are separated by four σ bonds and one π bond.

30. The IR spectrum indicates a benzene ring (1600 cm^{-1} and 1500^{-1}) with hydrogens on sp^2 carbons, and no carbonyl group. The absorption bands in the 1250^{-1}–1000^{-1} region suggests there are two C-O single bonds, one with no double bond character and one with some double bond character. The two singlets in the ^1H NMR spectrum with about the same integration suggest two methyl groups, one of which is adjacent to an oxygen. That the benzene ring protons (6.7-7.1 ppm) consist of two doublets indicates a 1,4-disubstituted benzene ring.

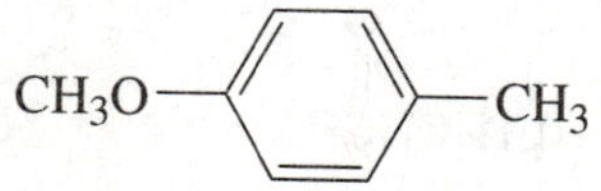

31. a.

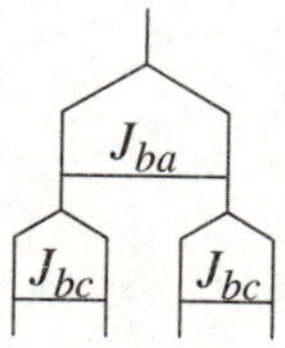

b.

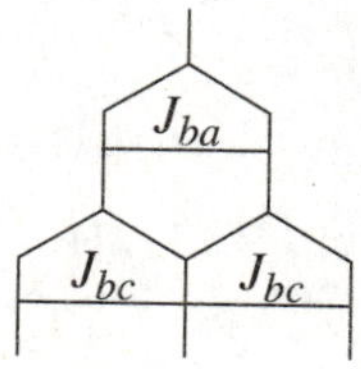

32. a. A b. B and C

33. Solved in the text.

34. The H of a carboxylic acid is more deshielded than the H of an alcohol as a result of electron delocalization.

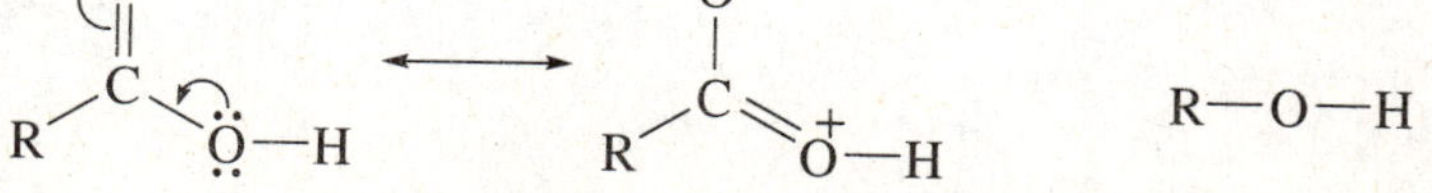

35. The greater the extent of hydrogen bonding, the greater the chemical shift. Therefore, the ^{1}H NMR spectrum of pure ethanol would have the signal for the OH proton at a greater chemical shift because it would be hydrogen bonded to a greater extent.

36. 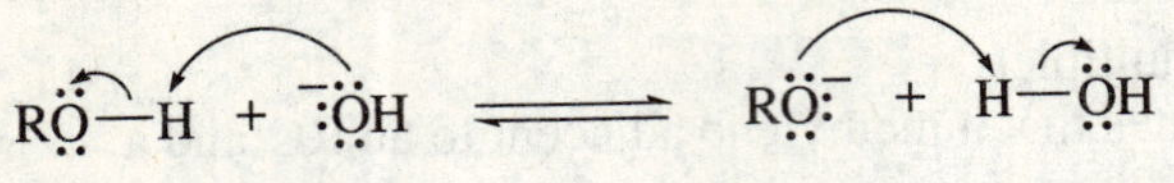

37. Figure 13.31 is the ^{1}H NMR spectrum of propanamide. Notice that the signals for the N-H protons are unusually broad. Because of the partial double bond character of the C-N bond, the two N-H protons are not chemically equivalent. The quartet and triplet are characteristic of an ethyl group.

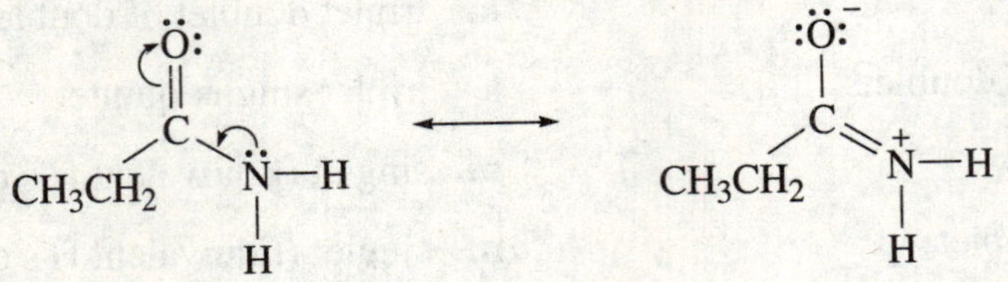

38. **a.** **1.** 3 **4.** 4 **7.** 3
2. 3 **5.** 3 **8.** 3
3. 2 **6.** 4 **9.** 2

b. An arrow is drawn to the carbon that gives the signal at the lowest frequency.

1. 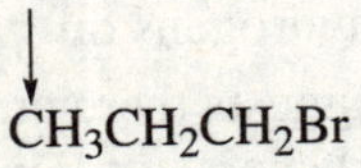
4. 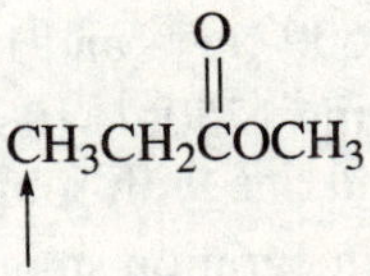
7. 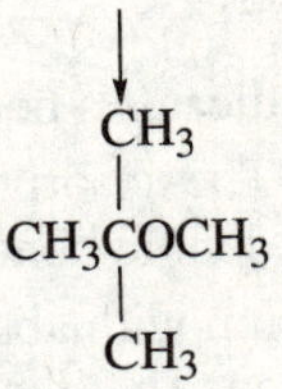

2. 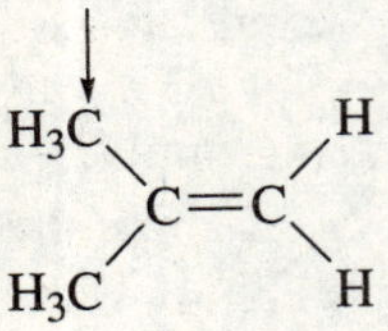
5. 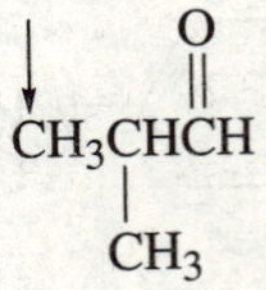
8. 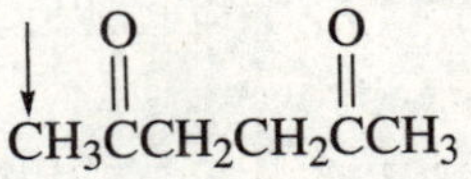

3. 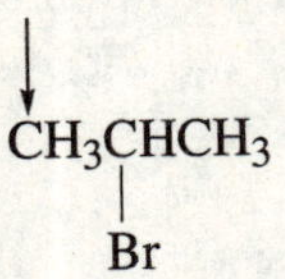
6. 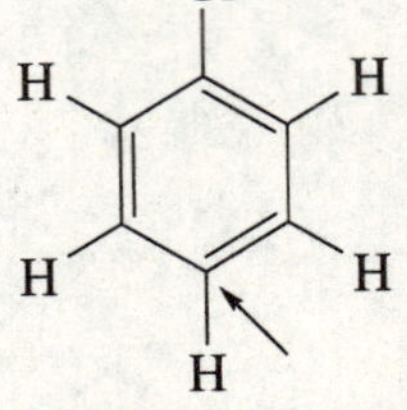
9. 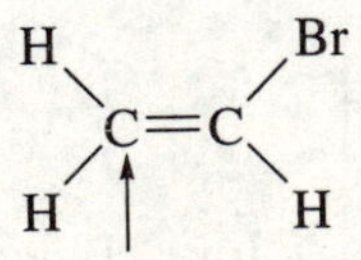

39. Each spectrum is described going from left to right.

1. triplet triplet quartet
3. triplet quartet quartet
5. singlet quartet triplet quartet

40.

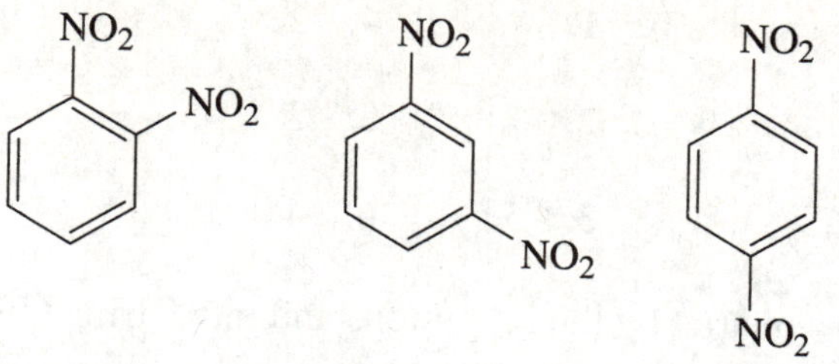

a.	^1H NMR	2 signals	3 signals	1 signal
b.	^{13}C NMR	3 signals	4 signals	2 signals

41. **a.** The signal at 210 ppm is for a carbonyl carbon. There are ten other carbons in the compound and five other signals. That suggests the compound is a ketone with identical five-carbon alkyl groups.

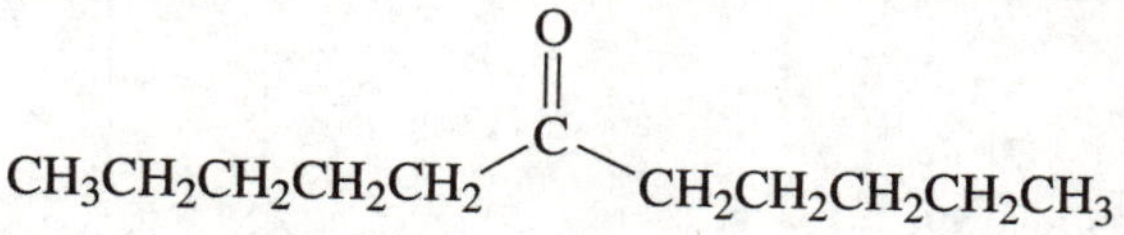

b. Because there are only four signals for the six carbons of the benzene ring (the signals between 110–117), the compound must be a 1,4-disubstituted benzene.

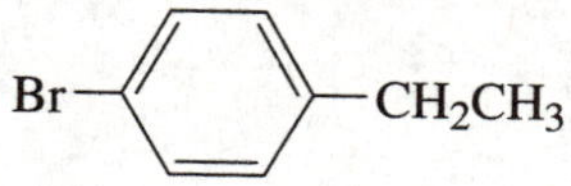

c. The signal at 212 ppm is for a carbonyl carbon. There are five other carbons in the compound and three other signals. That suggests the compound is a cyclic ketone.

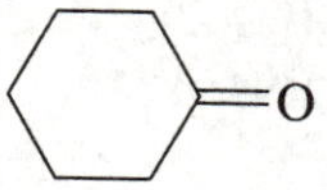

d. The molecular formula indicates the compound has one double bond or one ring. Each of the two sp^2 carbons must be bonded to the same groups because the six carbons only exhibit three signals. Whether the compound is cyclohexene, *cis*-3-hexene, or *trans*-3-hexene cannot be determined from the spectrum. An ^1H NMR spectrum could distinguish the compounds because, unlike the other two, cyclohexene would not have a quartet. The *cis* and *trans* compounds could be distinguished by their coupling constants because the coupling constant for trans protons is greater than the coupling constant for cis protons.

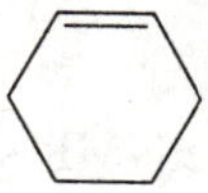 or $CH_3CH_2CH{=}CHCH_2CH_3$

42.

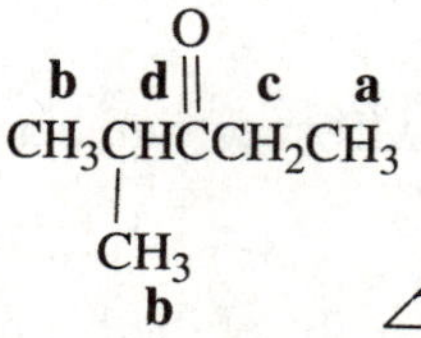

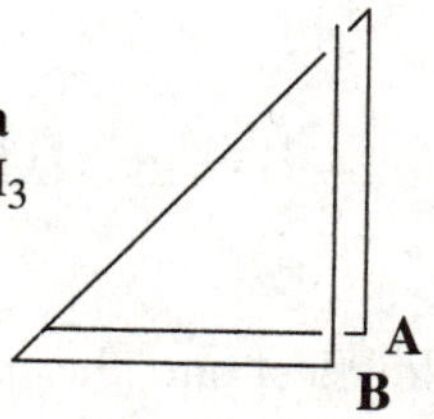

Point **A** shows that the "**a**" protons are split by the "**c**" protons.
Point **B** shows that the "**b**" protons are split by the "**d**" protons.

43. **a.** **1.** 5 **2.** 5 **3.** 4 **4.** 2 **5.** 3 **6.** 3

b. **1.** 7 **2.** 7 **3.** 5 **4.** 2 **5.** 2 **6.** 4

44. The H_b proton will be split into a quartet by the H_a protons and into a triplet by the H_c protons.

a.

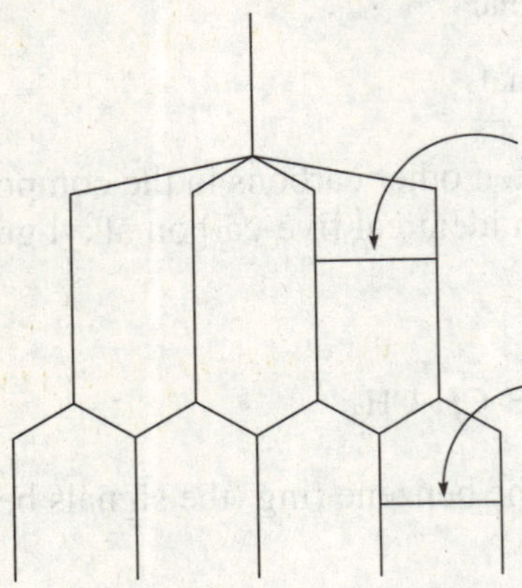

5 peaks

b.

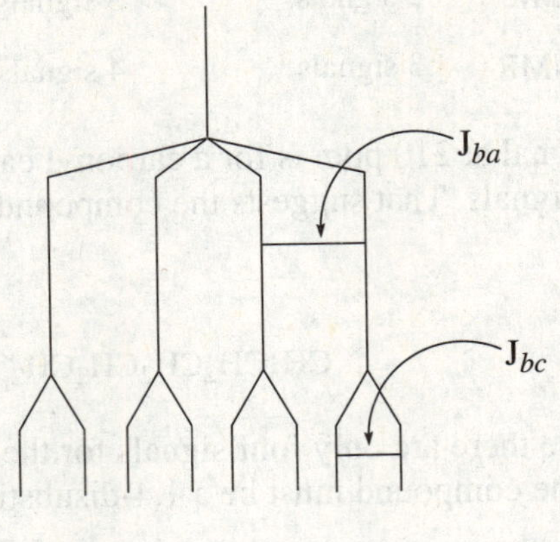

8 peaks

45. **a.**

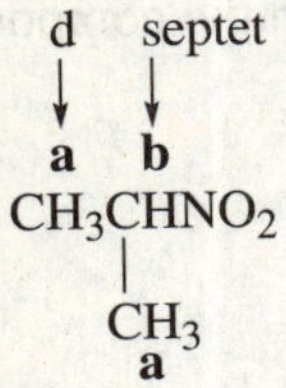

c.

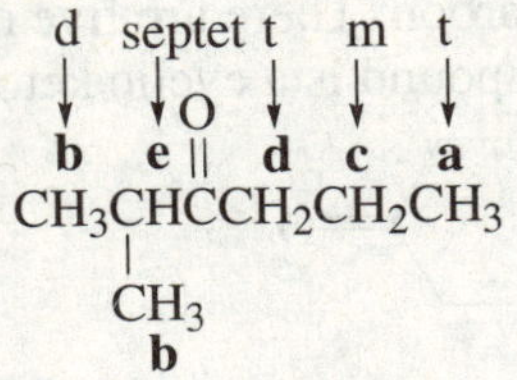

e.

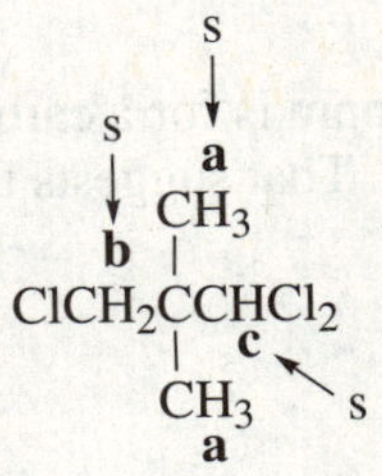

b.

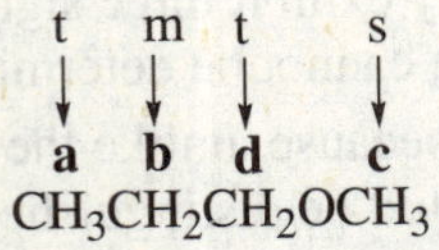

d.

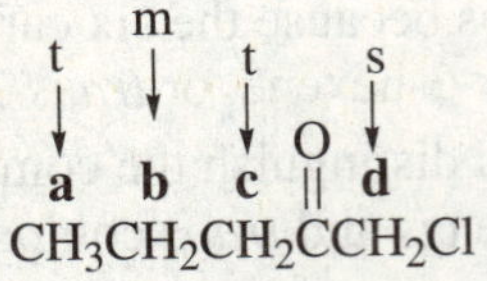

f.

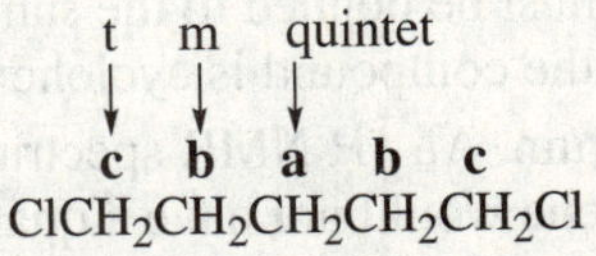

46. **a.** The spectrum must be that of **2-bromopropane** because the NMR spectrum has two signals and the lowest frequency (farthest upfield) signal is a doublet.

b. The spectrum must be that of **1-nitropropane** because the NMR spectrum has three signals and the both the lowest frequency (farthest upfield) and highest frequency (farthest downfield) signals are triplets.

c. The spectrum must be that of **ethyl methyl ketone** because the NMR spectrum has three signals and the signals are a triplet, a singlet, and a quartet.

47. By dividing the value of the integration steps by the smallest one, the ratios of the hydrogens are found to be 3 : 2 : 1 : 9.

$$\frac{40.5}{13} = 3.1 \qquad \frac{27}{13} = 2.1 \qquad \frac{13}{13} = 1 \qquad \frac{118}{13} = 9.1$$

Assuming that the ratios are given in the highest frequency to lowest frequency direction, a possible compound could be the following ester.

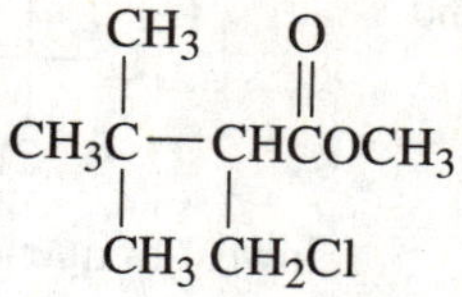

48. **a.** $CH_3CH_2CH_2OCH_3$ (4 signals) and $CH_3CH_2OCH_2CH_3$ (2 signals)

b. $BrCH_2CH_2CH_2Br$ (2 signals) and $BrCH_2CH_2CH_2NO_2$ (3 signals)

c. 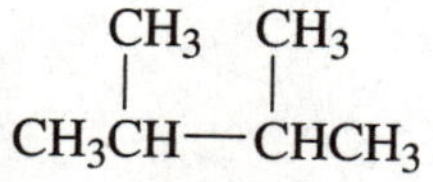 and 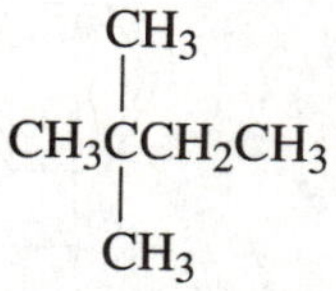

2 signals 3 signals

d. 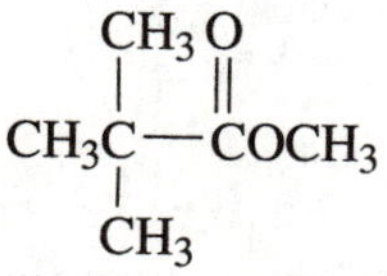 and 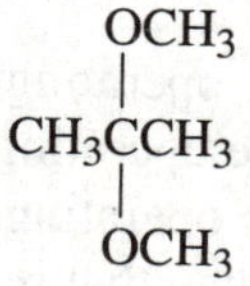

2 signals with integration 3:1 2 signals with integration 1:1

e. 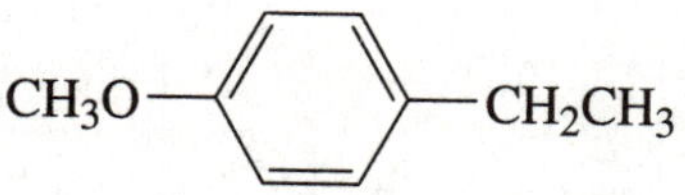 and 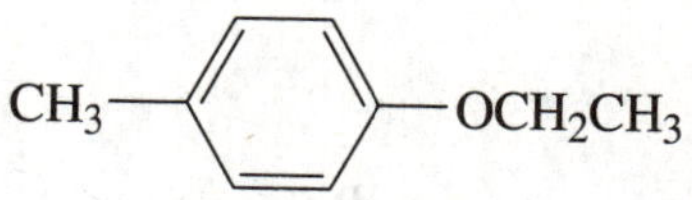

the singlet would be at a higher frequency in this compound the triplet would be at a higher frequency in this compound

f. 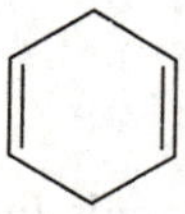 and 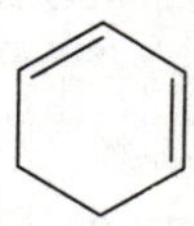

2 signals 3 signals

g. 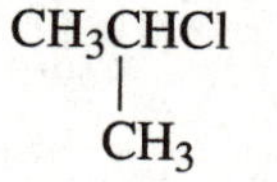 and 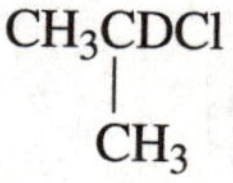

2 signals 1 signal

h. 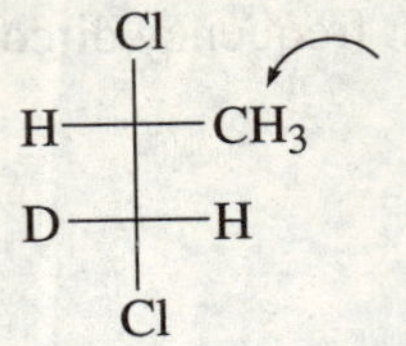 and 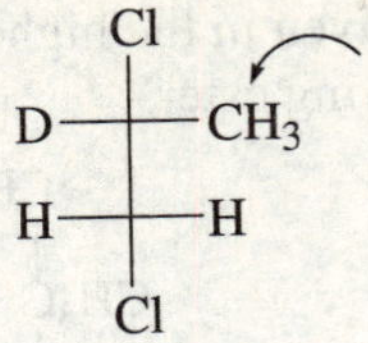

lowest frequency signal = doublet

lowest frequency signal = singlet

i. 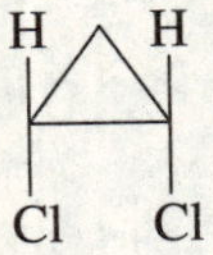 and 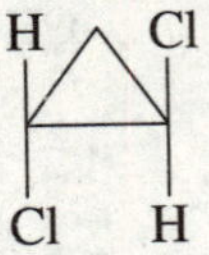

3 signals

2 signals

j. 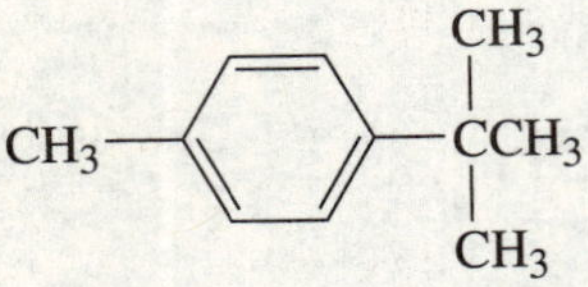 and 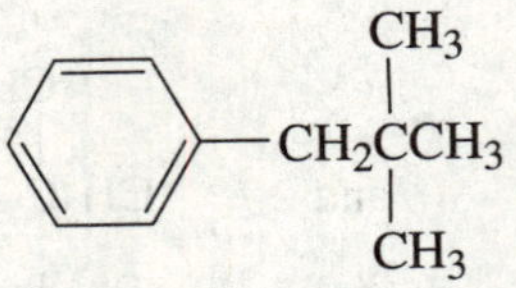

the signals for the benzene ring protons plus two signals with integration 3:1

the signals for the benzene ring protons plus two signals with integration 4.5:1

49. **a.** Chemical shift in ppm is independent of the operating frequency.
b. Chemical shift in hertz is proportional to the operating frequency.
c. The coupling constant is independent of the operating frequency.
d. The frequency required for NMR is lower than that required for IR, which is lower than that required for UV/Vis.

50. **a.** 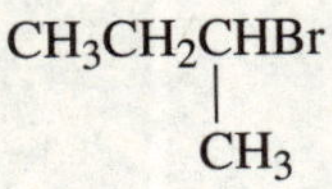 **b.** $CH_3CH_2CH_2CH_2Br$ **c.** 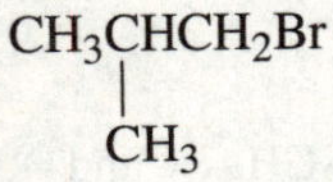

51. **a.** 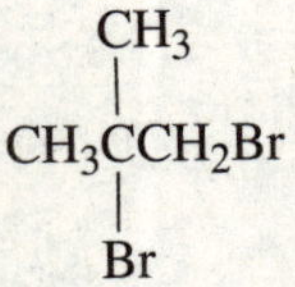 **b.** 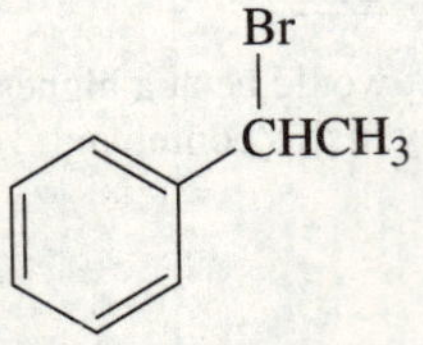 **c.** 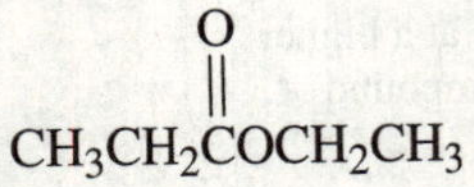

52. The singlet at 210 ppm indicates a carbonyl group. The splitting of the other two signals indicates an isopropyl group. The molecular formula indicates that it must have two isopropyl groups.

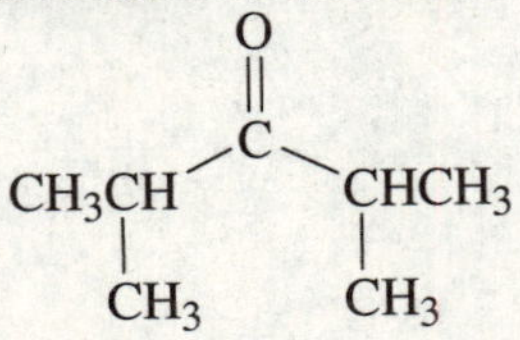

53. 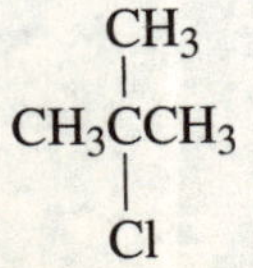

tert-butyl chloride
Compound A

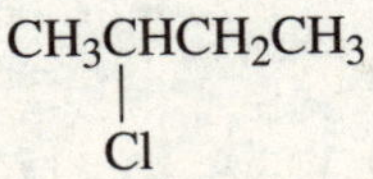

sec-butyl chloride
Compound B

54. **a.** 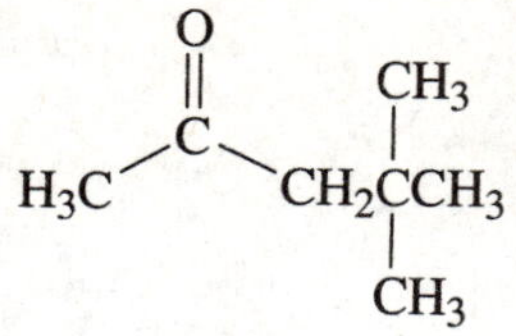 **b.** 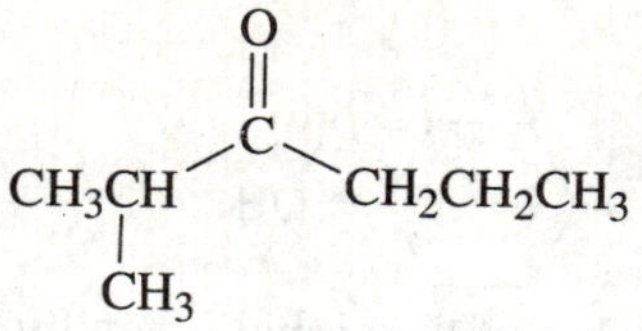 **c.** 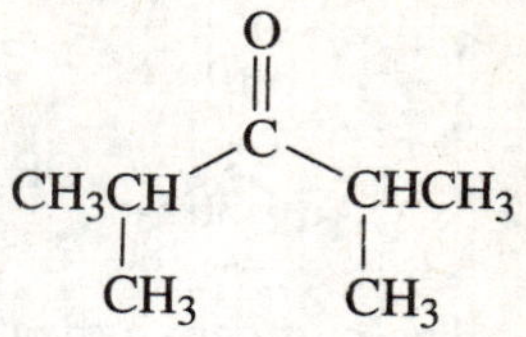

55. $CH_3CH_2CH{=}CH_2$ 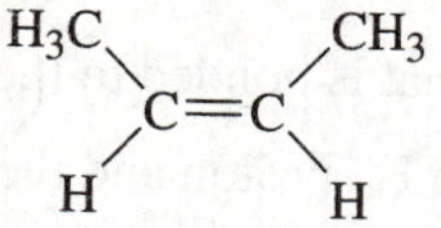 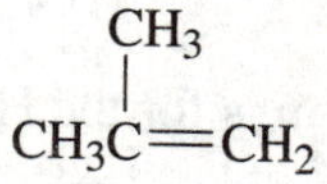

1-butene *cis*-2-butene 2-methylpropene

It would be better to use ^{13}C NMR because you would only have to look at the number of signals in each spectrum: 1-butene will show four signals, *cis*-2-butene will show two signals, and 2-methylpropene will show three signals. (In the ^{1}H NMR spectrum, 1-butene will show four signals, and *cis*-2-butene and 2-methylpropene will both show two signals.)

56. **a.** $CH_3CH_2CH_2CH_2OCH_3$

c. 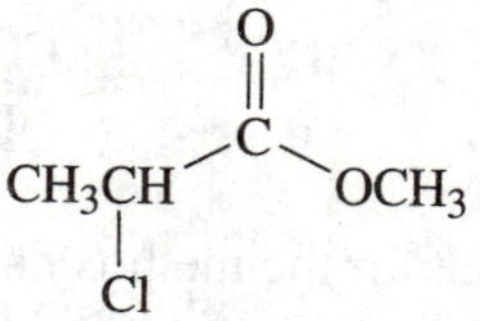

b. 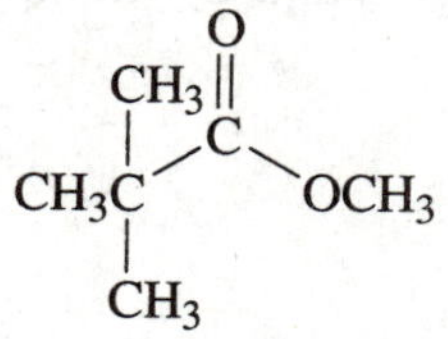

d. 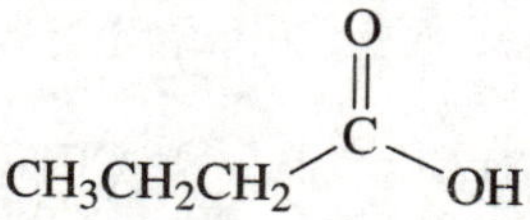

57.

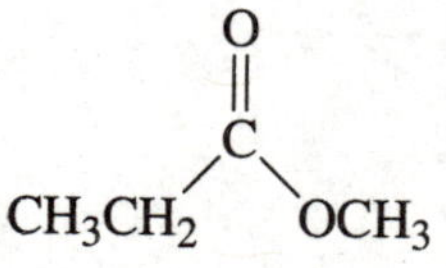

A

3 signals

singlet, quartet, triplet (singlet farthest downfield)

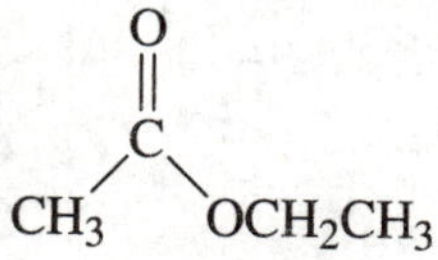

B

3 signals

singlet, quartet, triplet (quartet farthest downfield)

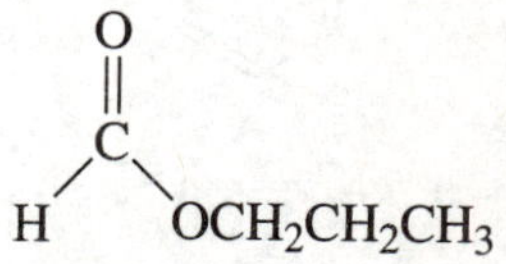

C

4 signals

singlet, triplet multiplet, triplet

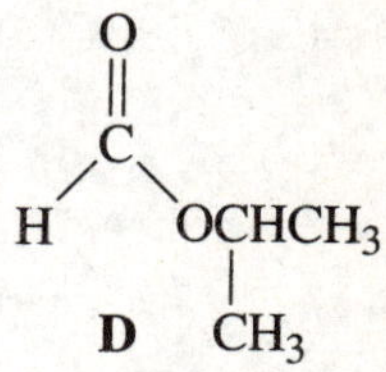

D

3 signals

singlet, septet doublet

C can be distinguished from **A**, **B**, and **D** because **C** has 4 signals and the others have 3 signals.

D can be distinguished from **A** and **B** because the 3 signals of **D** are a singlet, a septet, and a doublet, whereas the 3 signals of **A** and **B** are a singlet, a quartet, and a triplet.

A and **B** can be distinguished because the highest frequency signal in **A** is a singlet, whereas the lowest frequency signal in **B** is a quartet.

58. It is the ^{1}H NMR spectrum of *tert*-butyl methyl ether.

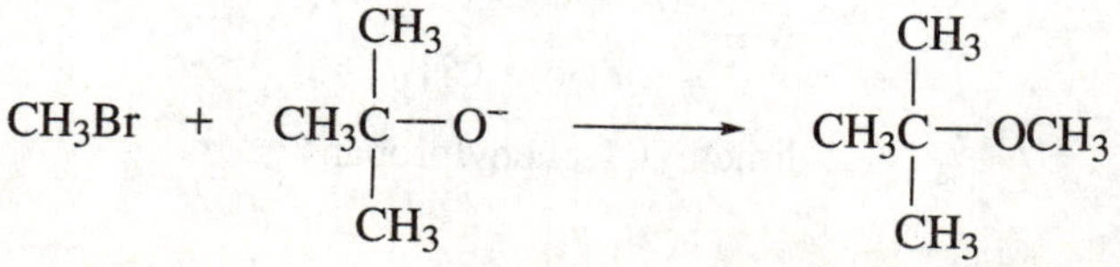

tert-butyl methyl ether

59. **a.** **b.**

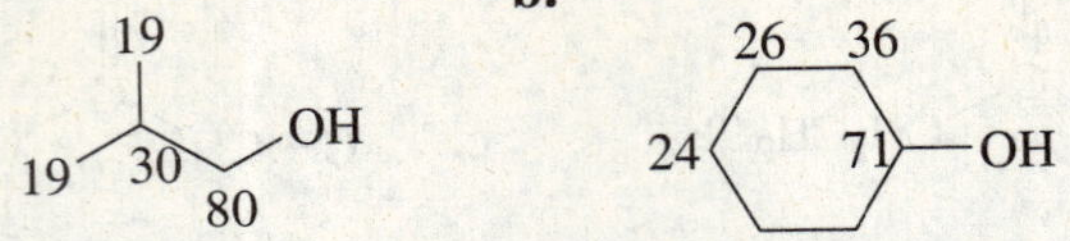

The numbers indicate the ppm value of the signal given by that carbon.

60. The broad signal at ~2.9 ppm is for the H_a proton that is bonded to the oxygen. The signal for the H_b protons is split into a doublet by the H_e proton. The H_c proton and the H_d are each split by the H_e proton; the coupling constant is greater for the trans protons. The H_e proton is split by the H_b protons and by the H_c proton and the H_d protons.

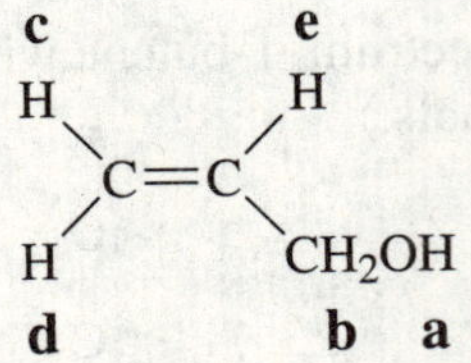

61. Each compound will have a singlet for the methoxy substituent and two doublets. In addition, **A** will have 2 doublet of doublets, **B** will have one doublet of doublets and a second singlet, and **C** will have no other signals.

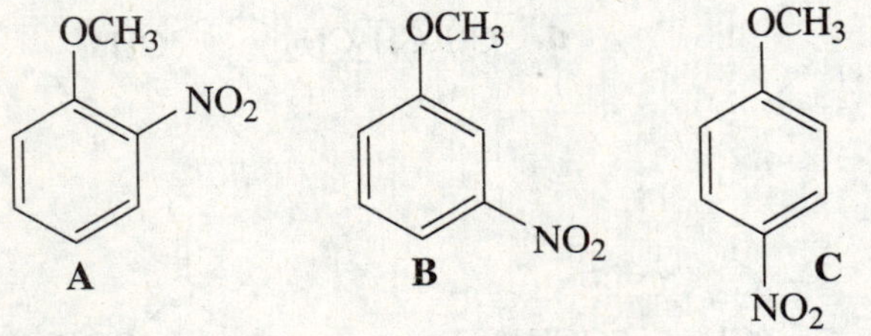

5 signals	5 signals	3 signals
(1 singlet, 2 doublets and 2 doublet of doublets)	(2 singlets, 2 doublets and, 1 doublet of doublets)	(1 singlet and 2 doublets)

62. The signals at ~7.2 ppm indicate that the compounds whose spectra are shown in **a** and **b** both contain a benzene ring. From the molecular formula, you know that there are five additional carbons in both **a** and **b**.

The hydrogens on the five carbons in **a** must all be accounted for by two singlets with integral ratios of ~ 1 : 4.5, indicating that the compound is **2,2-dimethyl-1-phenylpropane**.

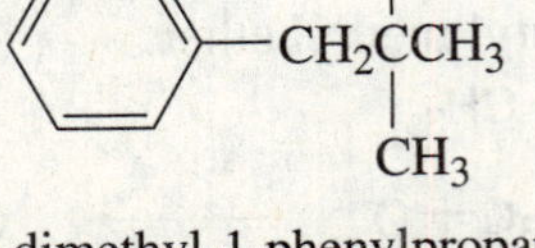

2,2-dimethyl-1-phenylpropane

b. The two singlets in **b** have integral ratios of ~ 1 : 3, indicating a methyl substituent and a *tert*-butyl substituent. The signals at ~7.1 and ~7.3 ppm indicate that the benzene ring has two kinds of hydrogens. Thus, the compound is **4-methyl-1-*tert*-butylbenzene**.

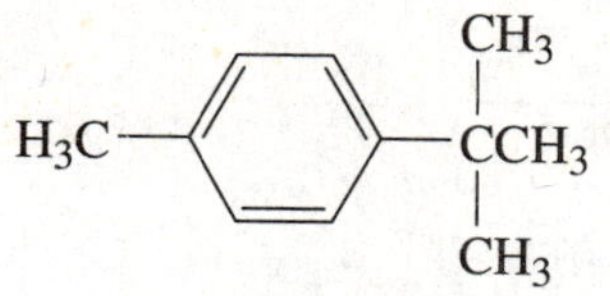

4-*tert*-butyltoluene

63.

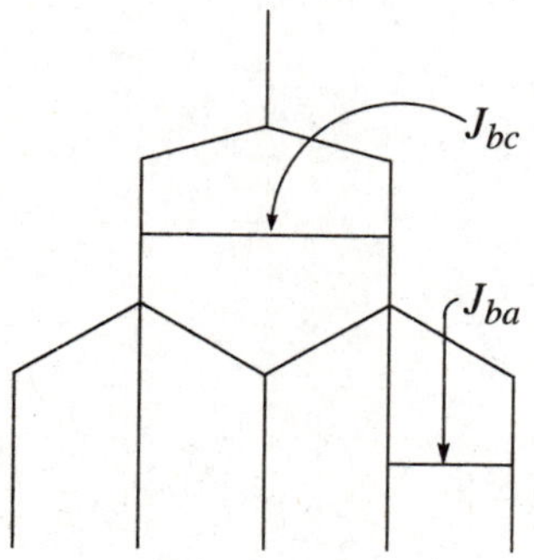

5 signals

64. Oxygen is more electronegative than chlorine. (See Table 1.3 on p. 11 of the text.)

a.

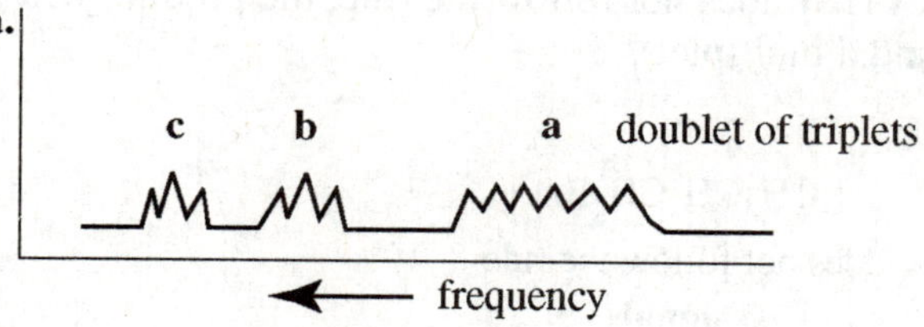

b a c
$ClCH_2CH_2OH$

b.

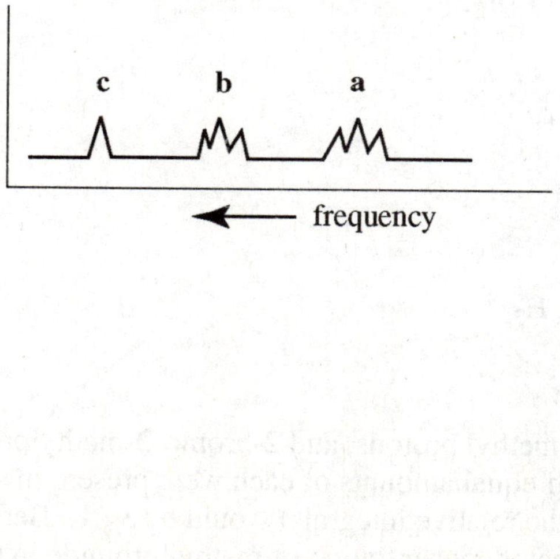

c. $\overset{a}{ClCH_2}\overset{b}{CH_2OH}$

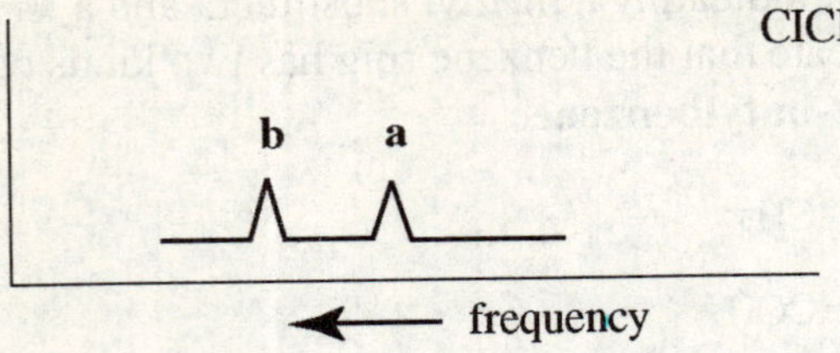

d.

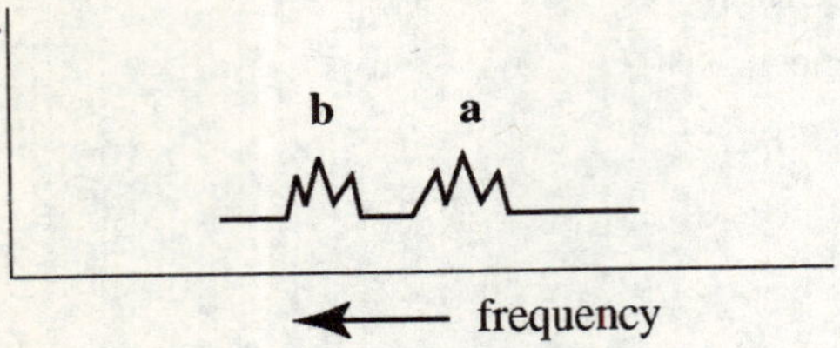

e.

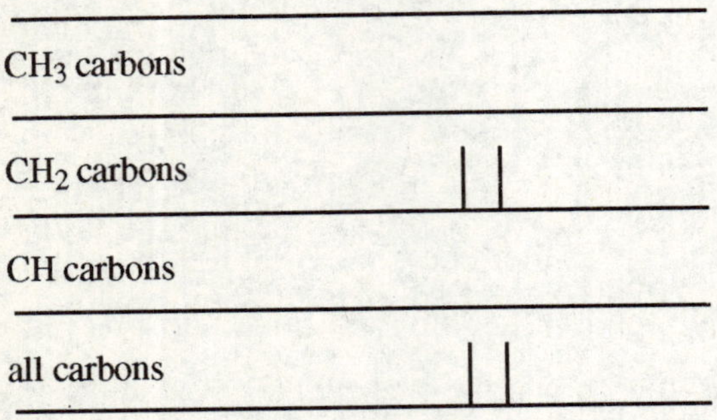

65. If addition of HBr to propene follows the rule that says the electrophile adds to the sp^2 carbon that is bonded to the greater number of hydrogens, the product of the reaction will give an NMR spectrum with two signals (a doublet and a septet). If addition of HBr does not follow the rule, the product will give an NMR spectrum with three signals (two triplets and a multiplet).

$$CH_3CH{=}CH_2 + HBr \longrightarrow CH_3CHBrCH_3 \qquad CH_3CH_2CH_2Br$$

follows the rule — **2 signals**; does not follow the rule — **3 signals**

66. **a.** 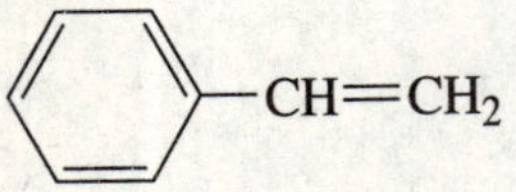

c. 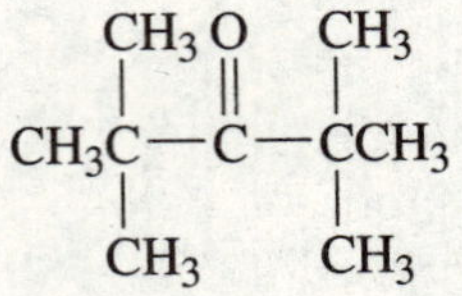

b. 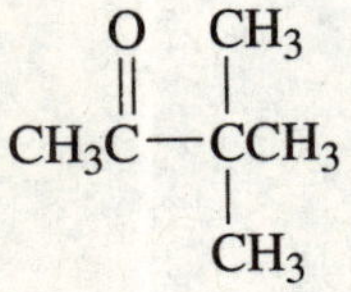

d. $CH_2{=}CHOCH_2CH_3$

67. Methyl bromide gives a signal at 2.7 ppm for its three methyl protons, and 2-bromo-2-methylpropane gives a signal at 1.8 ppm for its nine methyl protons. If equal amounts of each were present in solution, the ratio of the hydrogens (and, therefore, the ratio of the relative integrals) would be 3 : 1. Because the relative integrals are 1 : 6, there must be an 18-fold greater concentration of methyl bromide in the mixture.

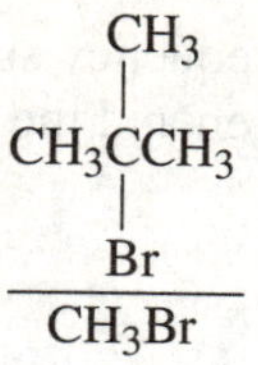

$$= \frac{3}{1} \times \frac{1}{18} = \frac{1}{6}$$

68. Using the formula given on page 571 of the text:

$$\Delta E = h\nu = \frac{h\gamma}{2\pi} B_o$$

Given: Planck's constant $= h = 6.626 \times 10^{-34}$ Js (page 157 in the text)

1 cal = 4.184 J (page 143 in the text)

60 MHz = 1.4092 T (page 571 in the text)

γ for ^{1}H $= 2.675 \times 10^8\ T^{-1}s^{-1}$

$$\Delta E = \frac{h\gamma}{2\pi} B_o$$

$$= \frac{6.626 \times 10^{-34}\ \text{Js} \times 2.675 \times 10^8\ T^{-1}s^{-1}}{2(3.1416)} \times 1.4092\ T \times \frac{1\ \text{cal}}{4.184\ \text{J}}$$

$$= 9.50 \times 10^{-27}\ \text{cal}$$

69. All four spectra show a singlet at ~2.0 ppm suggesting they are all esters with a methyl group attached to the carbonyl group. So now the problem becomes determining the nature of the group attached to the oxygen.

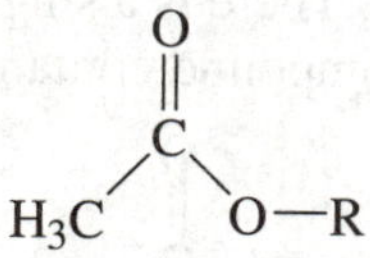

a. The highest frequency signal in the first spectrum is a triplet, indicating that the carbon attached to the oxygen is bonded to a CH_2 group. The lowest frequency signal is also a triplet, indicating that it too is attached to a CH_2 group. The presence of two multiplets confirms the structure.

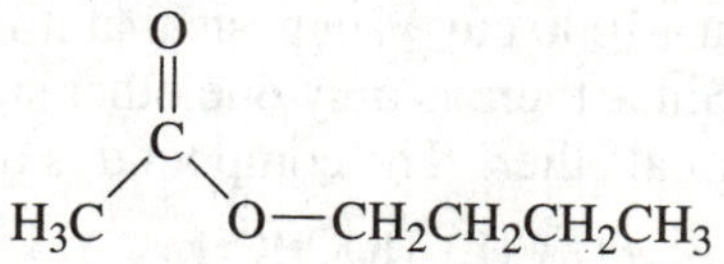

b. The highest frequency signal in the second spectrum is a multiplet, indicating that the carbon attached to the oxygen is attached to two non-equivalent carbons bonded to hydrogens.The lowest frequency signal is a triplet, indicating that it is attached to a CH_2 group.The doublet at ~1.2 ppm is a methyl group attached to a carbon bonded to a one hydrogen.

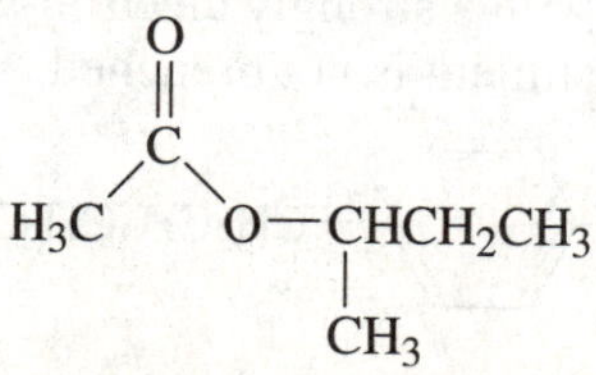

c. The highest frequency signal and the lowest frequency signal in the third spectrum are doublets, indicating that the carbon attached to the oxygen and the terminal carbon of the group are both attached to a CH group.

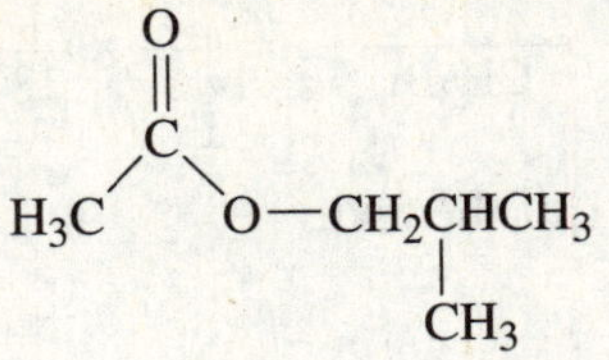

d. The group attached to the oxygen in the fourth spectrum has only one kind of hydrogen.

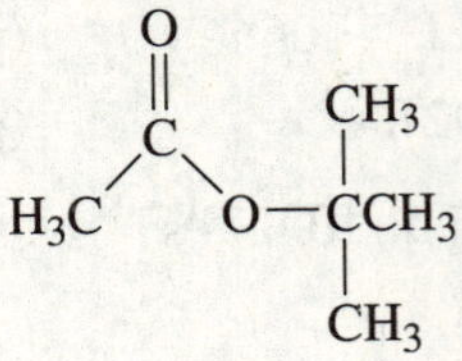

70. 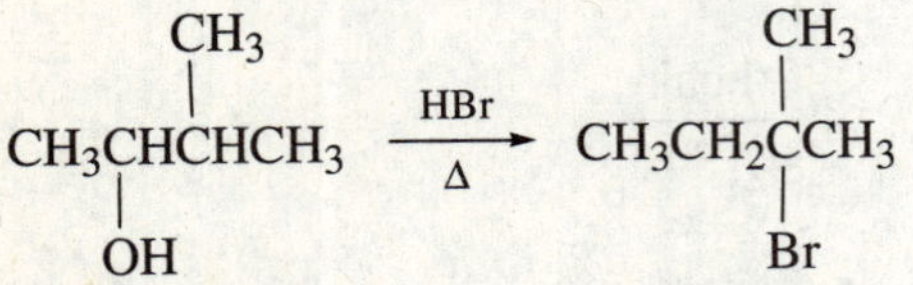

compound A compound B

71. **a.** The IR spectrum indicates that the compound is a ketone. The doublet in the NMR spectrum at ~0.9 ppm suggests an isopropyl group. There is a singlet at ~2.1 ppm on top of a multiplet, and a doublet at ~2.2 ppm. Knowing that the compound contains 6 carbons helps in the identification.

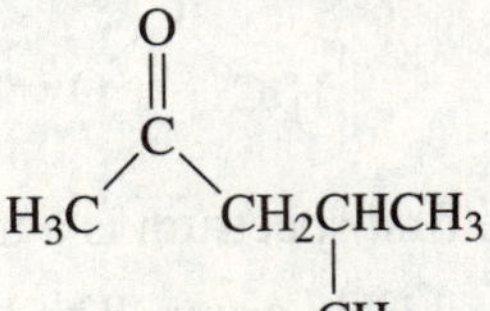

b. The IR spectrum indicates that this oxygen-containing compound is not a carbonyl compound or an alcohol; the absorption band at ~1000 cm^{-1} suggests that it is an ether. The doublet at ~1.1 ppm suggests an isopropyl group. Since there is only one other signal in the NMR spectrum, the compound must be a symmetrical ether. The compound is diisopropyl ether.

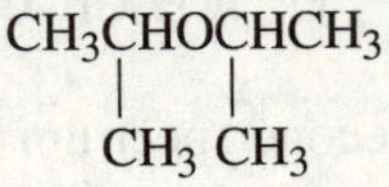

c. The absorption band at ~3400 cm^{-1} indicates an alcohol. The two signals at 7.3 and 8.1 ppm indicate a 1,4-disubstituted benzene ring with a strongly electron-withdrawing substituent. The two triplets indicate that the four-carbon substituent is not branched.

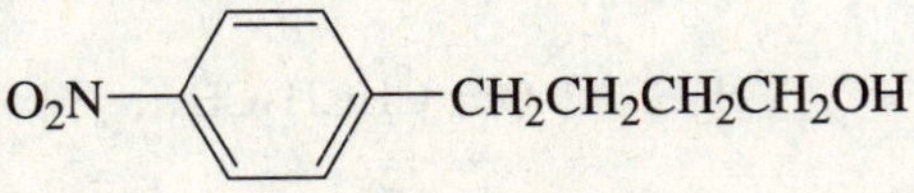

d. The absorption bands at ~1700 cm^{-1} and 2700 cm^{-1} indicate the compound is an aldehyde. The two doublets at ~7.0 and 7.8 ppm indicate a 1,4-disubstituted benzene ring. That none of the NMR signals is a doublet suggests that the aldehyde group is attached directly to the benzene ring. The two triplets and two multiplets indicate an unbranched substituent. The triplet at ~ 4.0 ppm indicates that the group giving this signal is next to an electron-withdrawing group.

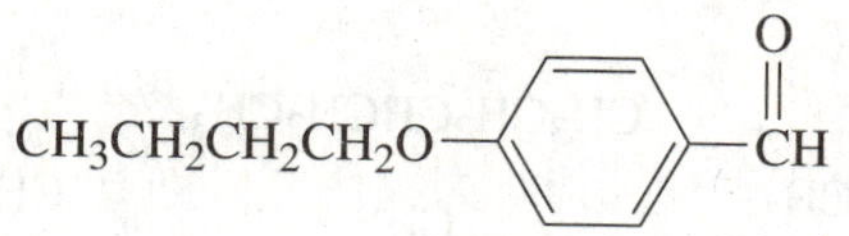

72. **a.** The IR spectrum indicates it is a ketone; the mass spectrum tells you it is a ketone with 7 carbons. The fact that there are only 3 signals in the NMR spectrum suggests it is a symmetrical ketone. The splitting pattern confirms the structure.

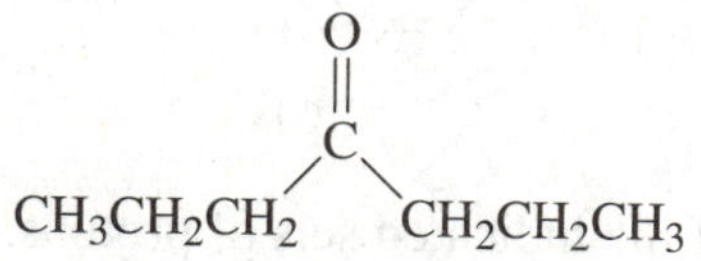

b. The M + 2 peak tells you the compound contains chlorine; the IR spectrum indicates it is a ketone; the NMR spectrum indicates it has a monosubstituted benzene ring. The singlet at ~4.7 indicates that the group giving this signal is in a strongly electron-withdrawing environment. The integration tells you that the two kinds of hydrogens are in a 2.5 : 1 ratio.

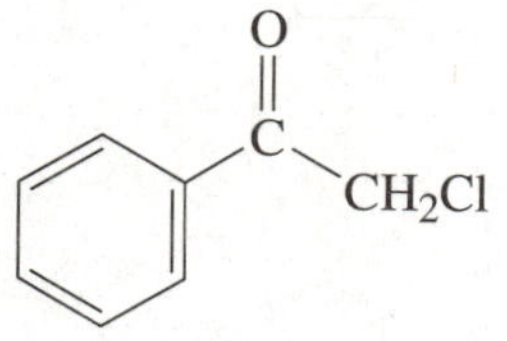

73. The DEPT ^{13}C NMR spectrum indicates that the compound has six carbons. The molecular formula tells us that the compound has six carbons and an oxygen. The signal at 220 ppm suggests a ketone. The following ketone has the single CH_3 group, the single CH group, and the three CH_2 groups indicated by the DEPTspectrum.

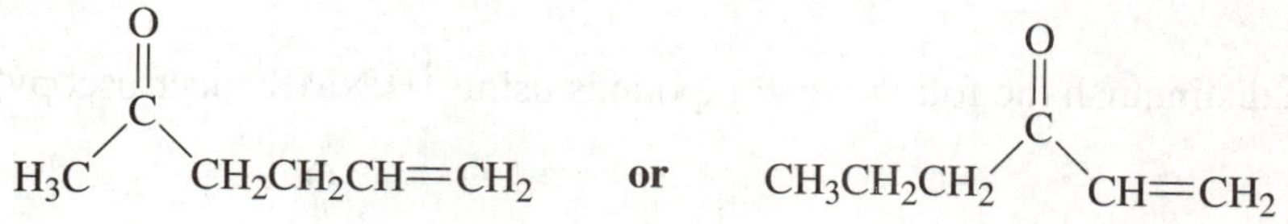

74. There is a singlet for the three methyl groups that are attached to the quaternary carbon. The singlet is on top of the triplet for the other methyl group. The methylene group shows the expected quartet.

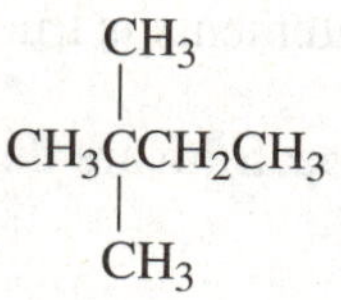

Chapter 13 Practice Test

1. How many signals would you expect to see in the ^{1}H NMR spectrum of each of the following compounds?

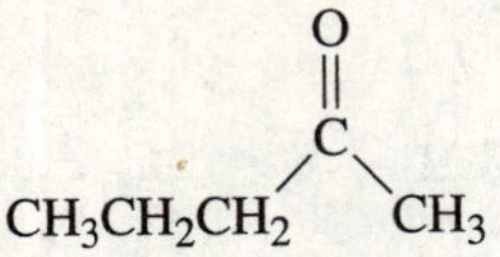
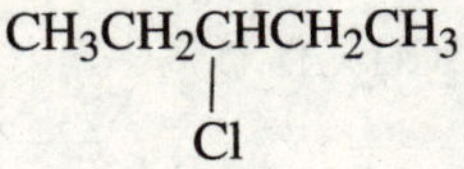
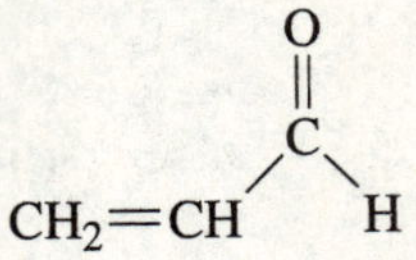
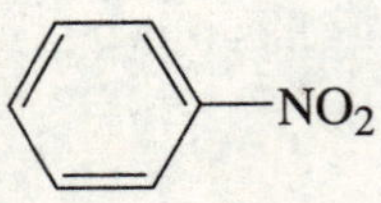
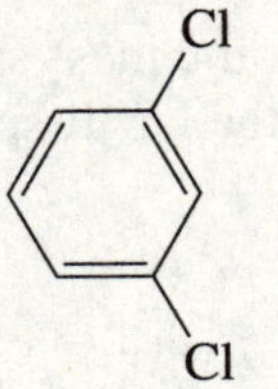
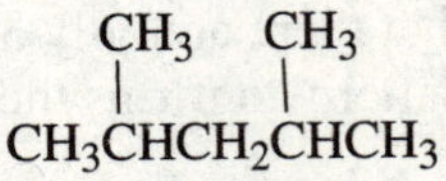

2. Indicate the multiplicity of each of the indicated sets of protons. (that is, indicate whether it is a singlet, doublet, triplet, quartet, quintet, multiplet, or doublet of doublets.)

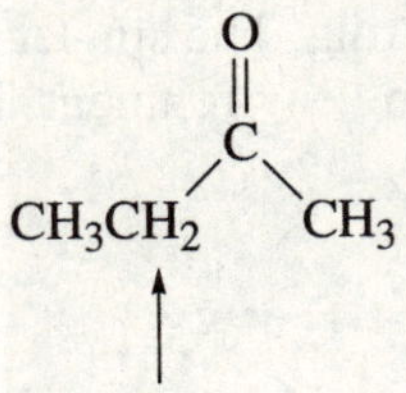
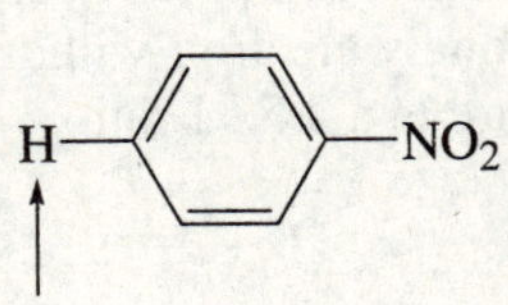
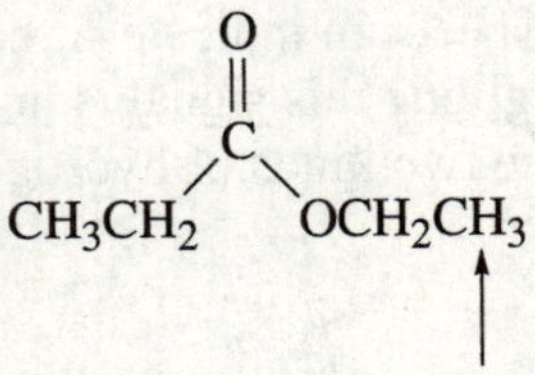
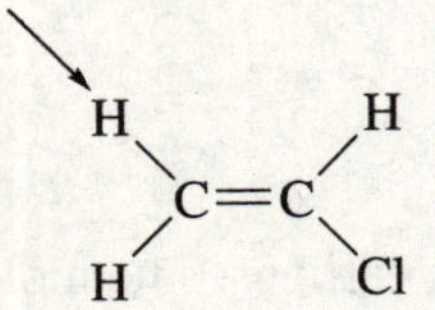
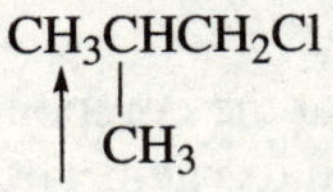
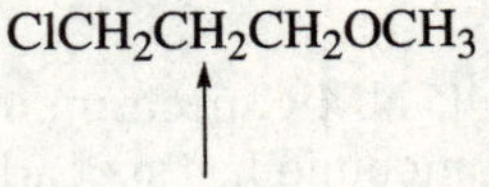
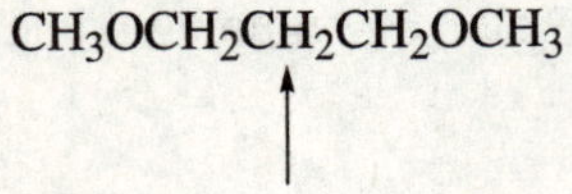
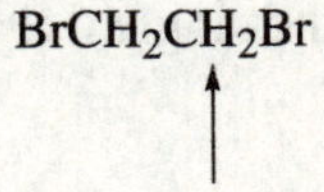

3. How could you distinguish the following compounds using ^{1}H NMR spectroscopy?

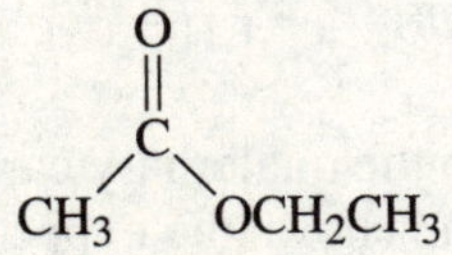
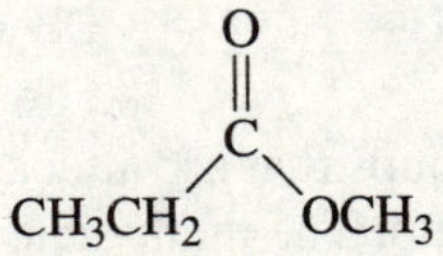
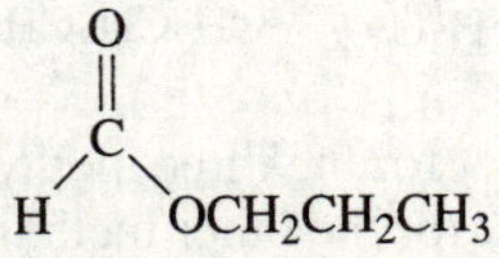

4. Indicate whether each of the following statements is true or false:

a. The signals on the right of an NMR spectrum are deshielded compared to the signals on the left. T F

b. Dimethyl ketone has the same number of signals in its ^{1}H NMR spectrum as in its ^{13}C NMR spectrum. T F

c. In the ^1H NMR spectrum of the compound shown below, the lowest frequency signal (the one farthest upfield is a singlet and the highest frequency signal (the one farthest downfield) is a doublet. T F

d. The greater the frequency of the signal, the greater its chemical shift in ppm. T F

5. For each compound:

a. Indicate the number of signals you would expect to see in ^1H NMR spectrum.

b. Indicate the hydrogen or set of hydrogens that would give the highest frequency (farthest downfield) signal.

c. Indicate the multiplicity of that signal.

1. $CH_3CH_2CH_2Cl$ 2. $CH_3CH_2COCH_3$ (with C=O) 3. CH_3CHCH_3 (with Br on the middle C)

6. For each compound of Problem 5:

a. Indicate the number of signals you would expect to see in its proton-decoupled ^{13}C NMR spectrum.

b. Indicate the carbon that would give the highest frequency (farthest downfield) signal.

c. Indicate the multiplicity of that signal in a spin-coupled ^{13}C NMR spectrum.

CHAPTER 14
Aromaticity • Reactions of Benzene

Important Terms

aliphatic compound — an organic compound with a higher hydrogen-to-carbon ratio than an aromatic compound.

annulene — a monocyclic hydrocarbon with alternating double and single bonds.

antiaromatic compound — a cyclic and planar compound with an uninterrupted cloud of electrons containing an even number of pairs of π electrons.

aromatic compound — a cyclic and planar compound with an uninterrupted cloud of electrons containing an odd number of pairs of π electrons.

benzyl group —

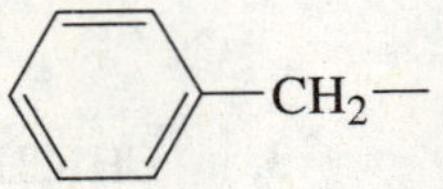

Clemmensen reduction — a reaction that reduces the carbonyl group of a ketone to a methylene group using Zn(Hg)/HCl, Δ.

electrophilic aromatic substitution reaction — a reaction in which an electrophile substitutes for a hydrogen of an aromatic ring.

Friedel-Crafts acylation — an electrophilic substitution reaction that puts an acyl group on an aromatic ring.

Friedel-Crafts alkylation — an electrophilic substitution reaction that puts an alkyl group on an aromatic ring.

Gatterman-Koch reaction — a reaction that uses a high-pressure mixture of carbon monoxide and HCl to form benzaldehyde.

halogenation — reaction with a halogen (Br_2, Cl_2).

heteroatom — an atom other than a carbon atom or a hydrogen atom.

heterocyclic compound (heterocycle) — a cyclic compound in which one or more of the atoms of the ring are heteroatoms.

Hückel's rule or the 4*n* + 2 rule — the number of π electrons in a cyclic uninterrupted π cloud that cyclic and planar compound must have in order to be aromatic.

nitration — substitution of a nitro group (NO_2) for a hydrogen of an aromatic ring.

phenyl group —

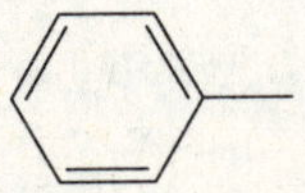

C_6H_5—

principle of microscopic reversibility	states that the mechanism for a reaction in the forward direction has the same intermediates and the same transition states as the mechanism for the reaction in the reverse direction.
Stille reaction	a reaction that will replace a bromine substituent on an aromatic ring with an alkyl substituent.
sulfonation	substitution of a hydrogen of an aromatic ring with a sulfonic acid group (SO_3H).
Suzuki reaction	a reaction that will replace a bromine substituent on an aromatic ring with an alkyl substituent.
Wolff-Kishner reduction	a reaction that reduces the carbonyl group of a ketone to a methylene group using NH_2NH_2/HO^-, Δ.

Solutions to Problems

1. **a.** In the case of 9 pairs of π electrons, there are 18 electrons.
Therefore, $4n + 2 = 18$ and $n = 4$.

b. Because it has an odd number of pairs of π electrons, it will be aromatic if it is cyclic, planar, and if every atom in the ring has a *p* orbital.

2. **a.** Notice that each resonance contributor has a charge of –1.

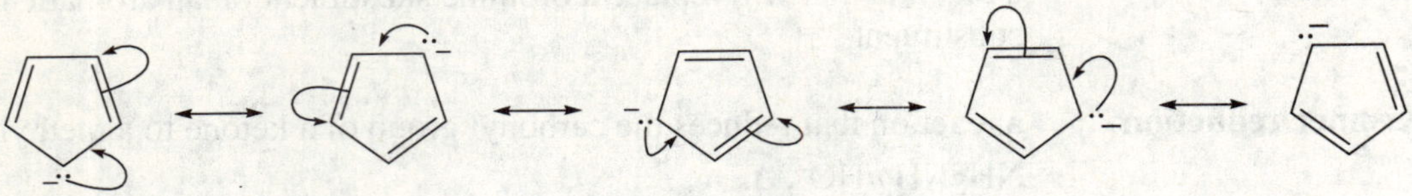

b. All five ring atoms share the negative charge.

3. **a.**

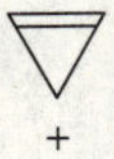

This is the only one that is aromatic; it is cyclic, planar every ring atom has a *p* orbital and it has one pair of π electrons.

The first compound is not aromatic because one of the atoms is sp^3 hybridized and, therefore, does not have a *p* orbital.

The third compound is not aromatic because it has two pairs of π electrons.

b.

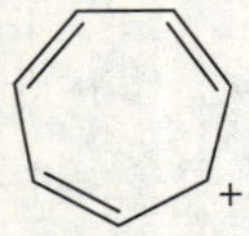

This is the only one that is aromatic; it is cyclic, planar every ring atom has a *p* orbital and it has three pairs of π electrons.

The first compound is not aromatic because one of the atoms is sp^3 hybridized and, therefore, does not have a *p* orbital.

The third compound is not aromatic because it has four pairs of π electrons.

4. **d** and **e** are aromatic.

d is aromatic, because it is cyclic, planar, every atom in the ring has a *p* orbital, and it has seven pairs of π electrons.

e is aromatic, because it is cyclic, planar, every atom in the ring has a *p* orbital, and it has three pairs of π electrons.

c is not aromatic, because it has two pairs of π electrons.

a and **b** are not aromatic, because each compound has two pairs of π electrons and every atom in the ring does not have a *p* orbital

f is not aromatic, because it is not cyclic.

5. **a.** Solved in the text.

b. There are five monobromophenanthrenes.

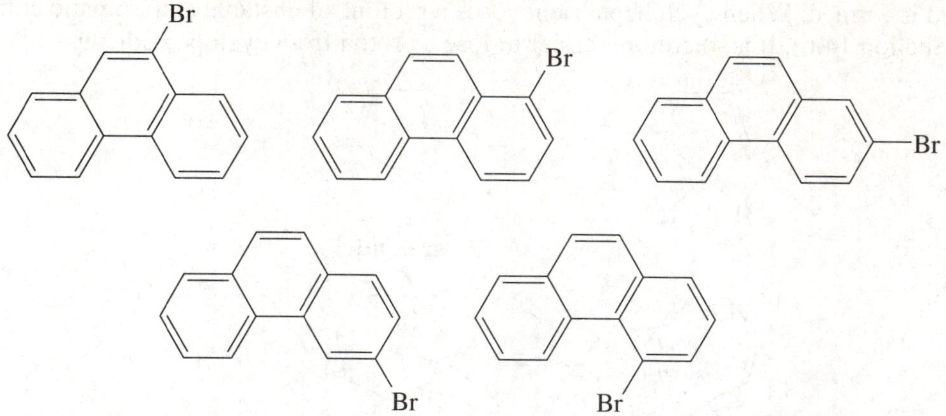

6. To be aromatic, a compound must be cyclic and planar, every atom in the ring must have a *p* orbital, and the π cloud must contain an odd number of pairs of π electrons.

[10]-Annulene is cyclic, every atom in the ring has a *p* orbital, and it has the correct number of π electrons to be aromatic (five pairs). Knowing it is not aromatic, we can conclude that it is not planar.

[12]-Annulene has an even number of pairs of π electrons (six pairs) in its π cloud, so it cannot be aromatic.

7. **a.** Notice that each resonance contributor has a net charge of 0.

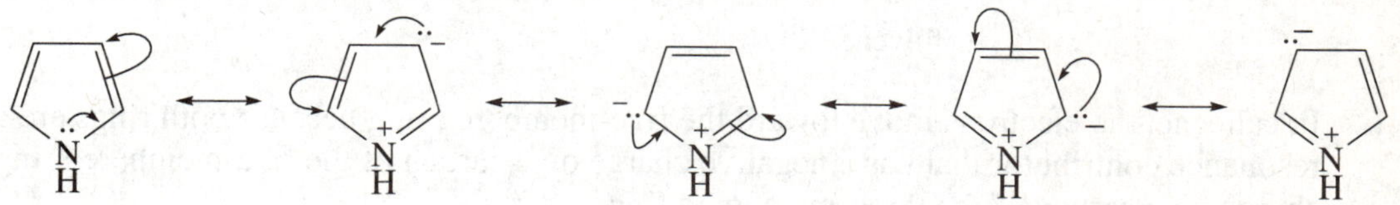

b. Four ring atoms share the negative charge.

8.

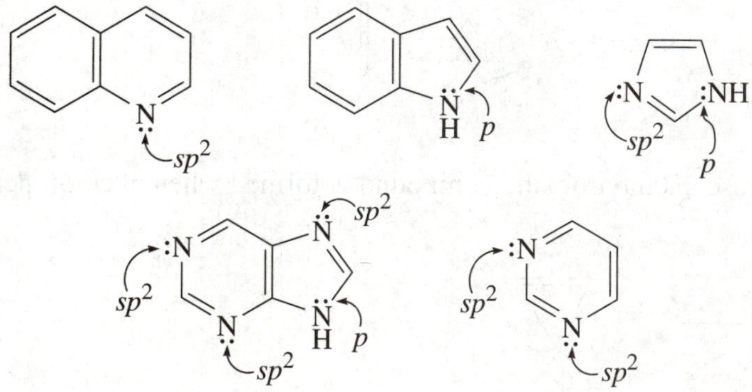

9. **a.** The nitrogen atom (the atom at the bottom of the epm) in pyrrole has a partial positive charge because it donates electrons by resonance into the ring.

b. The nitrogen atom (the atom at the bottom of the epm) in pyridine cannot donate electrons by resonance; it withdraws electrons from the ring inductively because it is the most electronegative atom in the molecule.

c. The relatively electronegative nitrogen atom in pyridine withdraws electrons from the ring.

10. Cyclopentadiene has a lower pK_a. When cyclopentadiene loses a proton, a relatively stable aromatic compound is formed. When cycloheptatriene loses a proton, an unstable antiaromatic compound is formed (Section 14.6). It is, therefore, easier to lose a proton from cyclopentadiene.

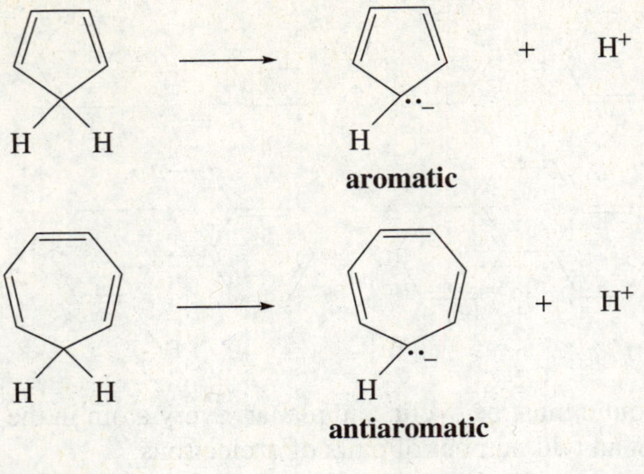

11. Solved in the text.

12. **a.** In fulvene, the electrons move toward the five-membered ring because the resonance contributor that has a negative charge on a ring carbon is aromatic.

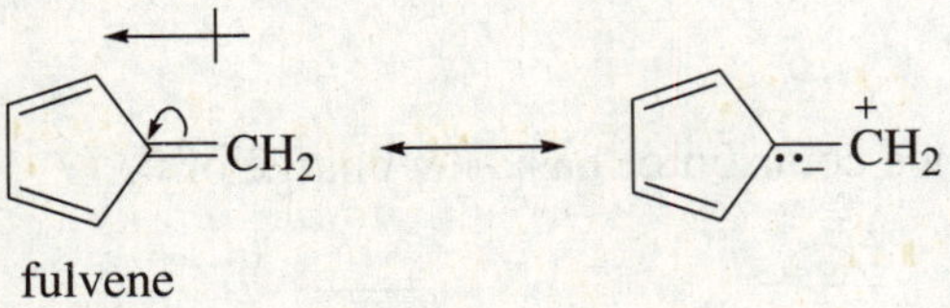

b. In calicene, the electrons move toward the five-membered ring because both rings are aromatic in the resonance contributor that has a negative charge on a carbon of the five-membered ring and a positive charge on a carbon of the three-membered ring.

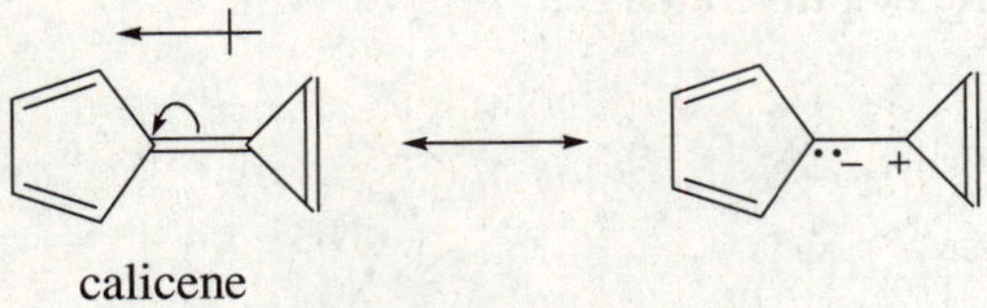

13. **a.** Cyclopropane has a lower pK_a because an antiaromatic compound is formed when cyclopropene loses a proton.

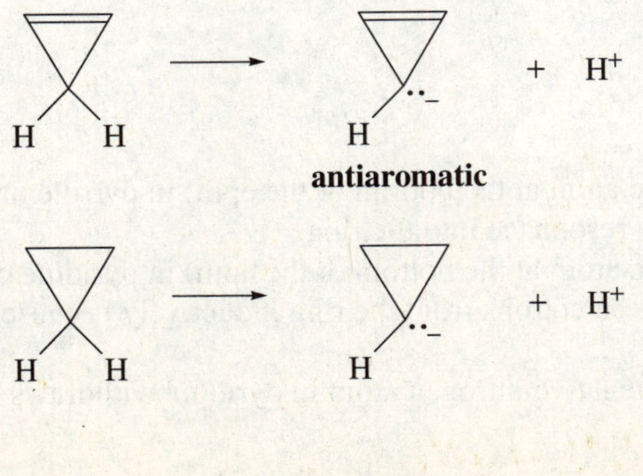

b. 3-Bromocyclopropene is more soluble in water, because it is more apt to ionize since heterolytic cleavage of its carbon-bromine bond forms an aromatic compound.

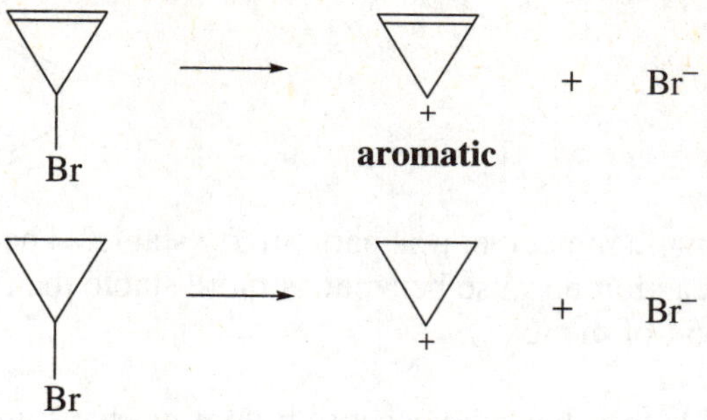

14. **c** is antiaromatic because it is cyclic, planar, every atom in the ring has a *p* orbital, and it has two pairs of π electrons.

When you did Problem 4, you found that **d** and **e** are aromatic.

a, b, and **f** are neither aromatic nor antiaromatic. **a** and **b** have ring atoms that do not have *p* orbitals, and **f** is not cyclic.

15. Cyclobutadiene has 1 bonding molecular orbital, 2 nonbonding molecular orbitals, and 1 antibonding molecular orbital.

Because each of its four atoms has a *p* atomic orbital, it has 4 π electrons; 2 π electrons are in the bonding π molecular orbital, and each of the two nonbonding molecular orbitals contains one π electron.

16. No. To be aromatic, all the bonding molecular orbitals must be filled and there must be no partially filled orbitals. Each orbital contains two electrons. Since a radical has an unpaired electron, there will be an orbital with a single electron (a partially filled molecular orbital). A radical, therefore, cannot be aromatic. (A radical can, however, be attached to an aromatic compound. For example, the phenyl radical has a methylene radical attached to an aromatic benzene ring.)

17. The Frost circles show that compounds with completely filled bonding orbitals and electrons in no other orbitals are aromatic (the cycloheptatrienyl cation and the cylopropenyl cation).

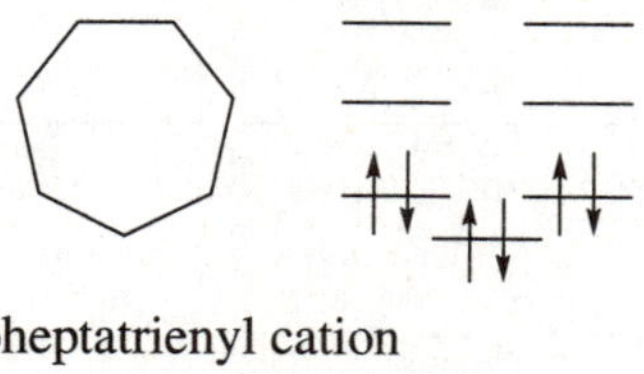

cycloheptatrienyl cation

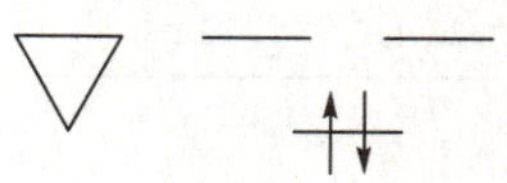

cyclopropenyl cation

The compound with unpaired electrons in degenerate orbitals is antiaromatic (the cycloheptatrienyl anion).

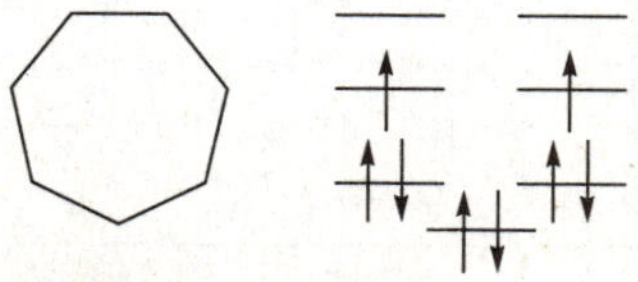

cycloheptatrienyl anion

18. **a.** 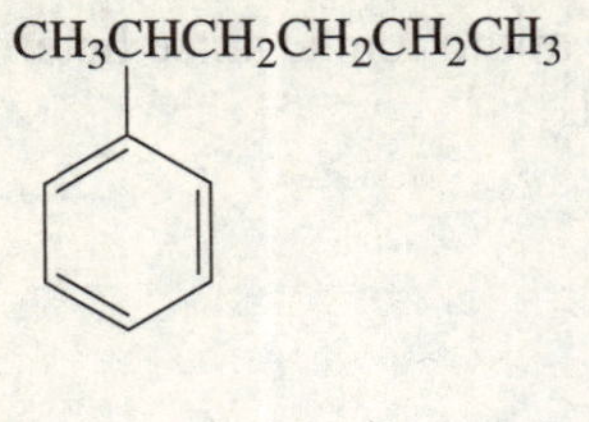 **b.** 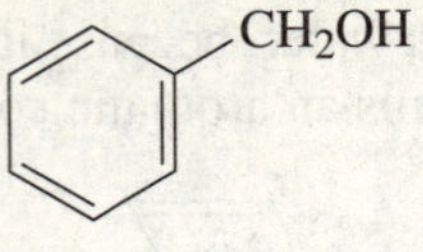 **c.** 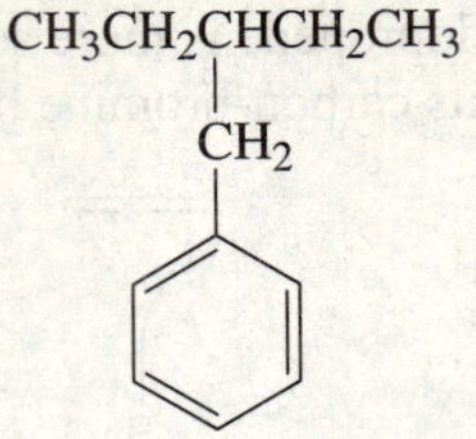 **d.** 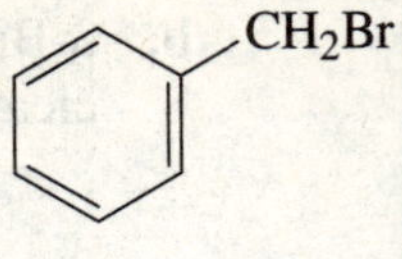

19. Because benzene is an aromatic molecule, it is particularly stable. Therefore, electrophilic addition to benzene is an endergonic reaction because benzene is more stable than the non-aromatic addition product. (See Figure 14.5 on page 654 of the text.)

An alkene is much less stable than benzene because it does not have any delocalized electrons. Electrophilic addition to an alkene is an exergonic reaction because an alkene is less stable than the addition product. (The overall reaction converts a σ bond and a π bond into two σ bonds; a σ bond is stronger and, therefore, of lower energy than a π bond.)

20. Ferric bromide activates Br_2 for nucleophilic attack by accepting a pair of electrons from it. Hydrated ferric bromide cannot do this because it has already accepted a pair of electrons from water.

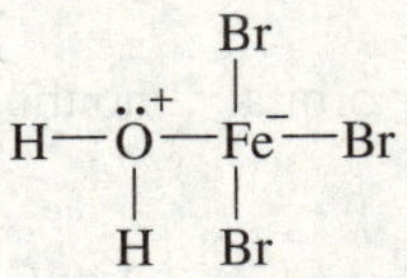

21. Solved in the text.

22. In sulfonation, **A** to **B** has a smaller rate constant (a higher energy barrier and, therefore, a slower reaction) than **B** to **C**; **A** to **B** is the rate-determining step.

In desulfonation (the reverse reaction), **C** to **B** has a smaller rate constant than **B** to **A**, but **B** to **A** is the rate-determining step. That is because once **B** is formed, it is easier for it to go back to **C** than to proceed to **A**, so **B** to **A** is the bottleneck (or rate-determining) step.
This example shows that the rate-determining step of a reaction is not necessarily the step with the smallest rate constant; it is the step that has the transition state with the highest energy on the reaction coordinate.

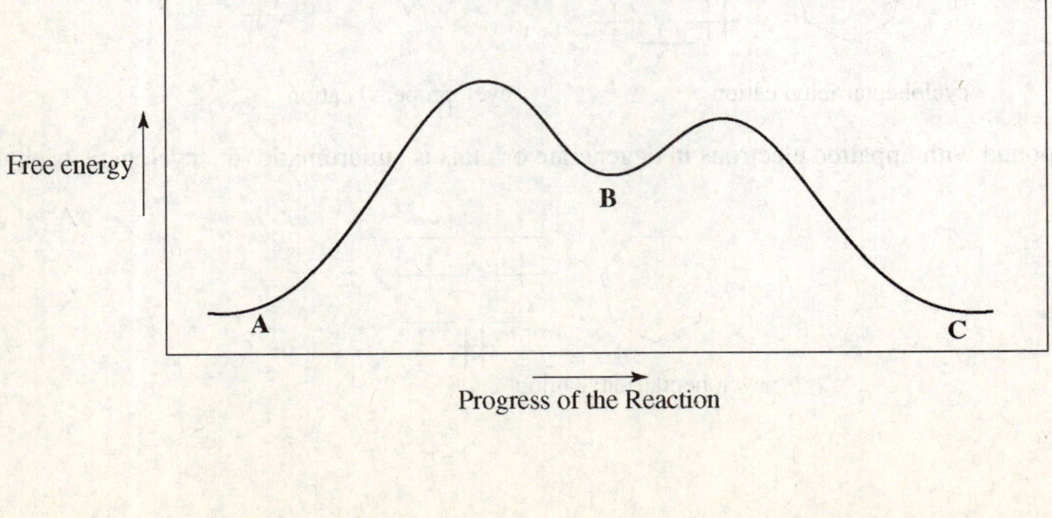

23.

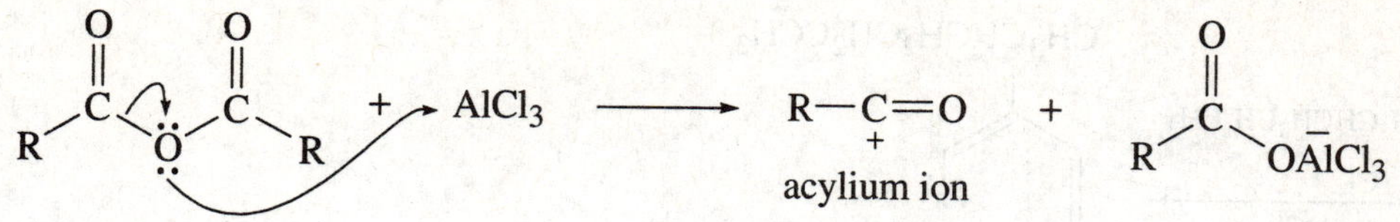

24.

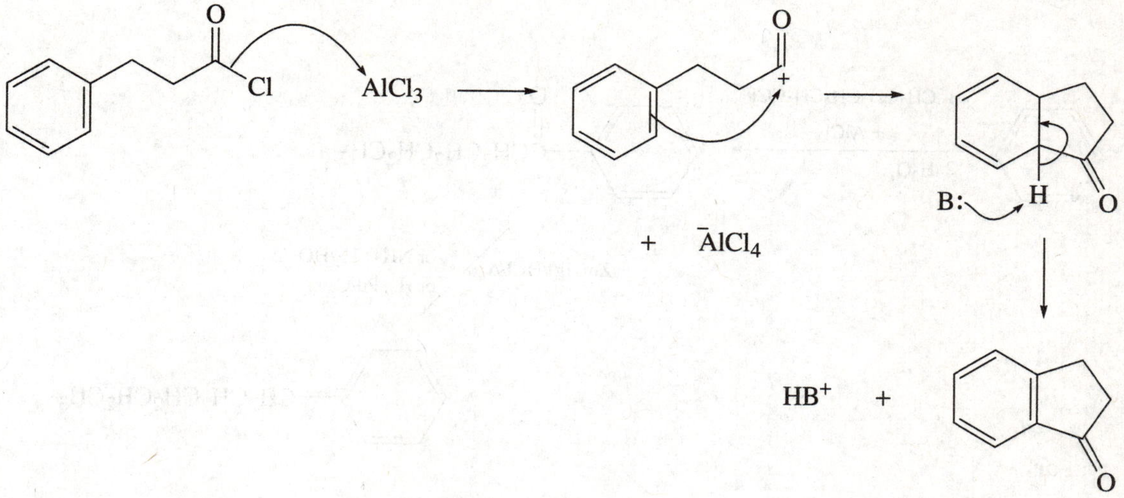

25.

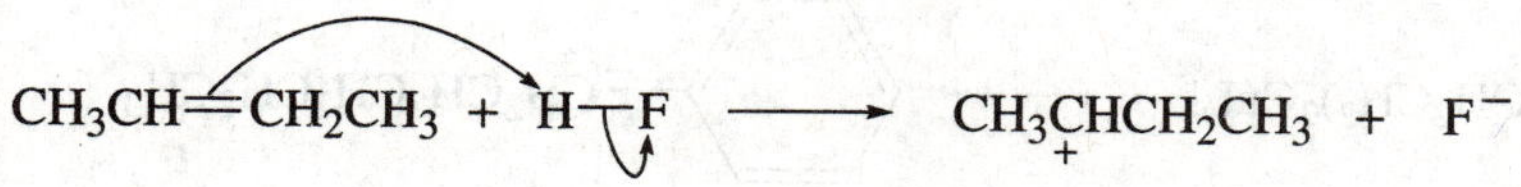

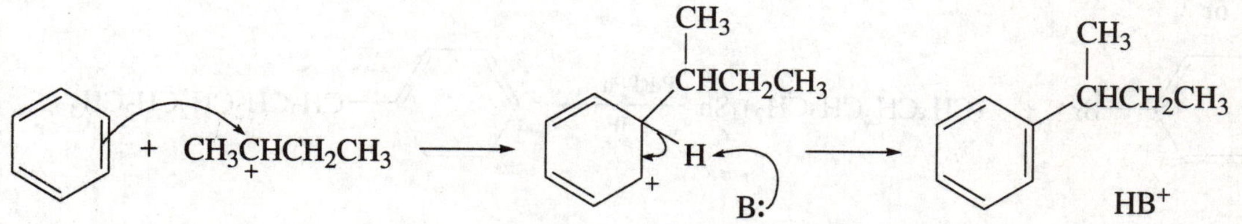

26.

a.

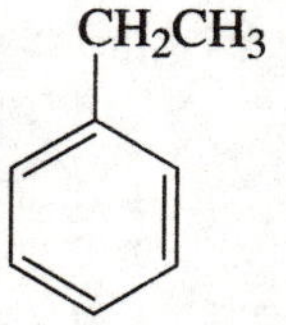

c.

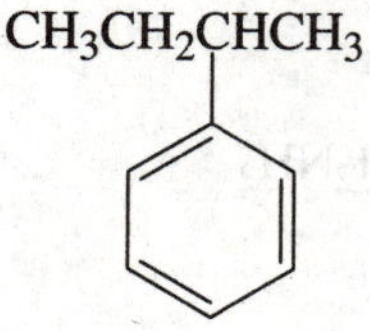

e.

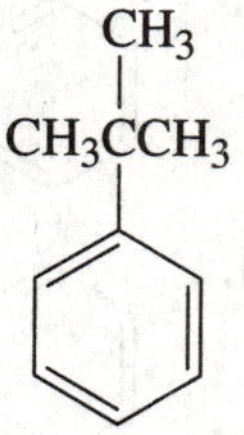

b.

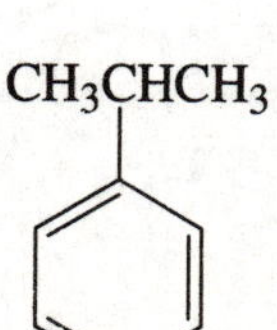

d.

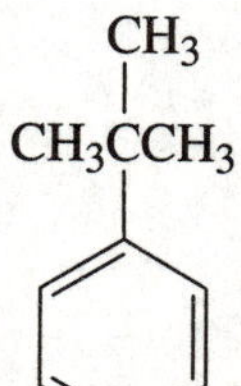

f.

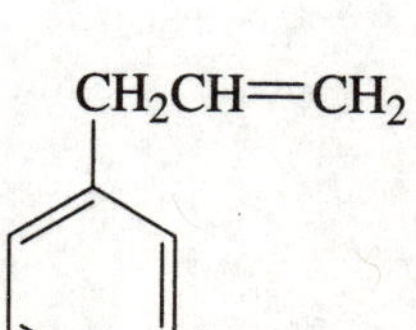

27. **a.**

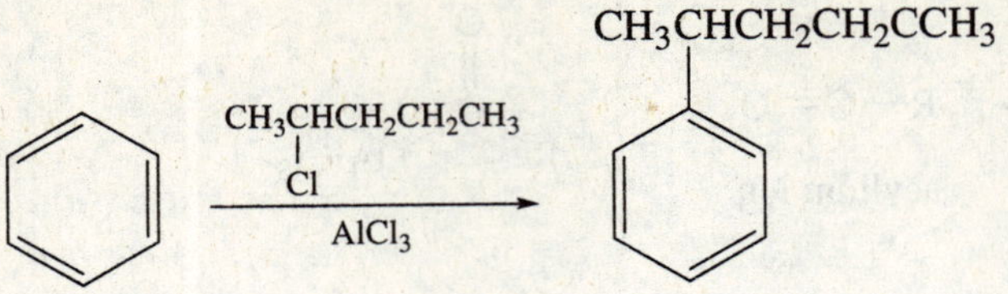

b.

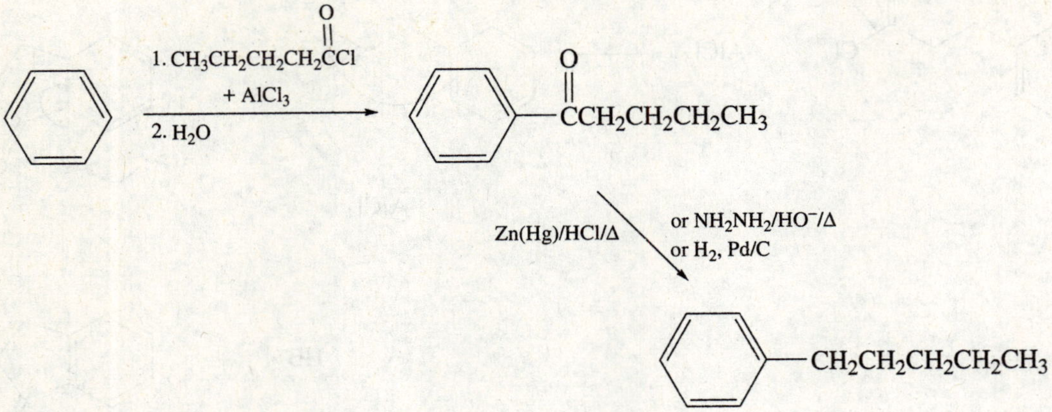

or

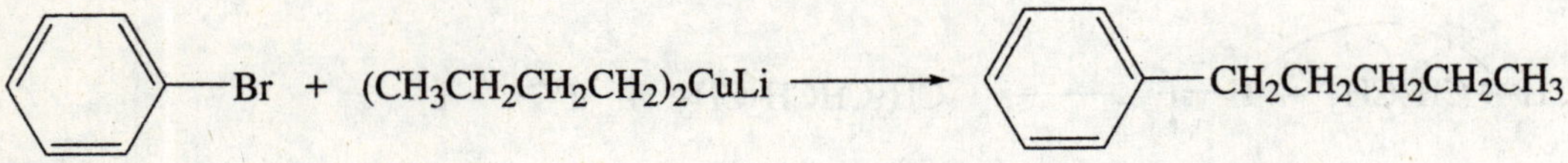

or

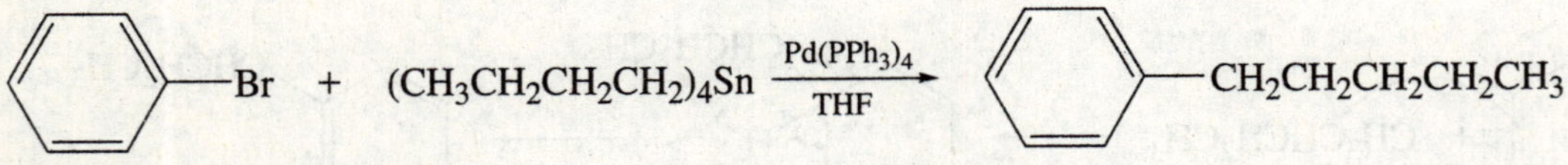

28. **a.**

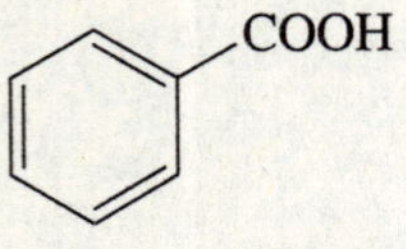

c.

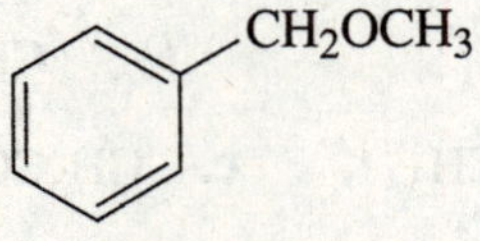

b.

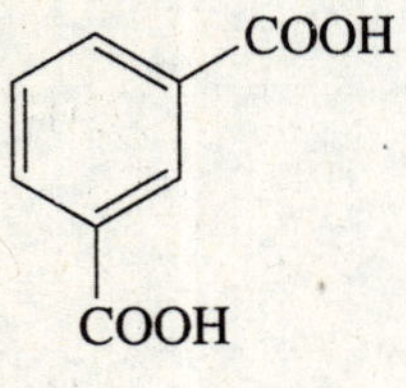

d.

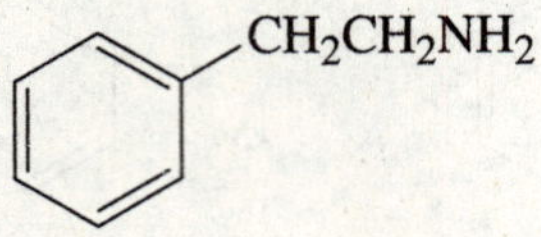

29. **a.** Solved in the text.

b.

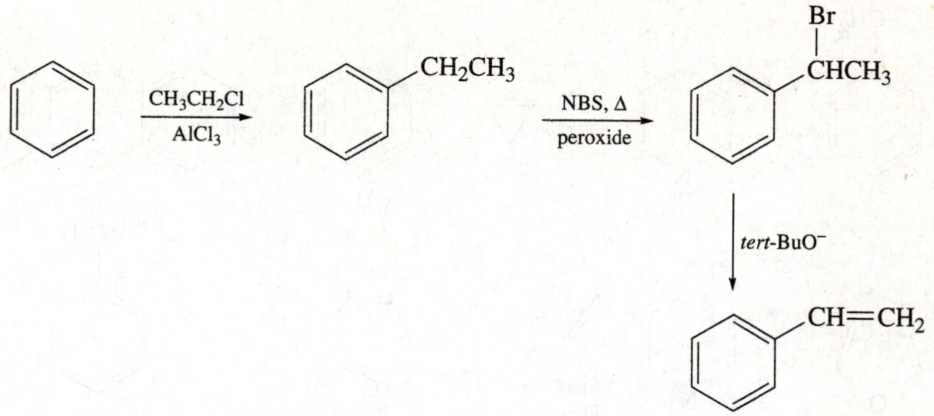

c.

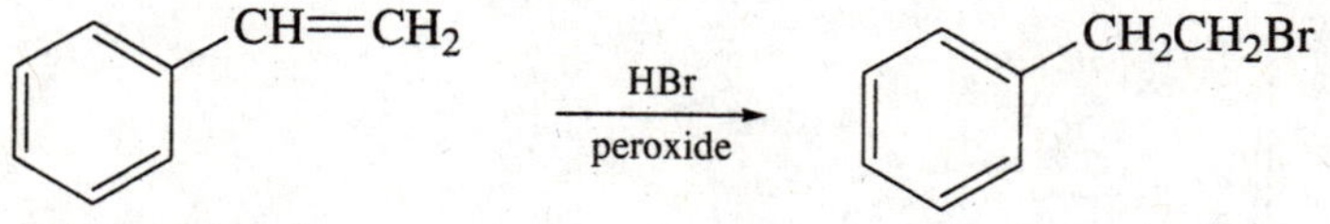

product of **b**

d.

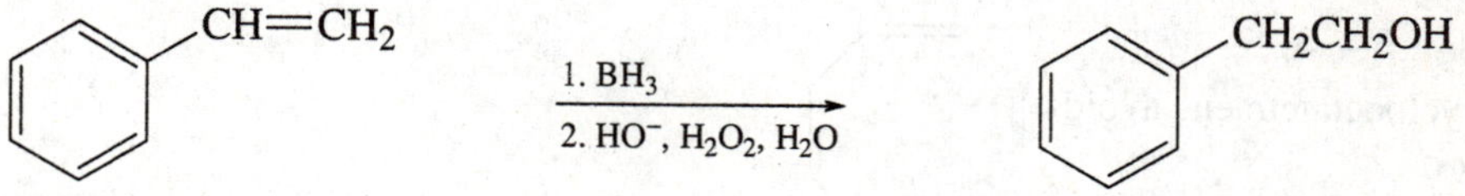

product of **b**

e.

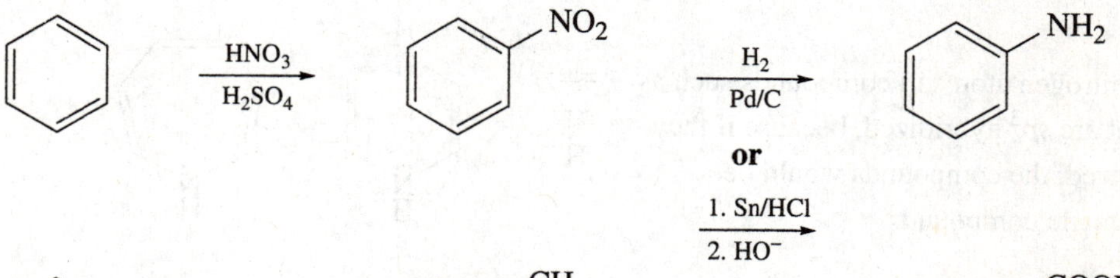

f.

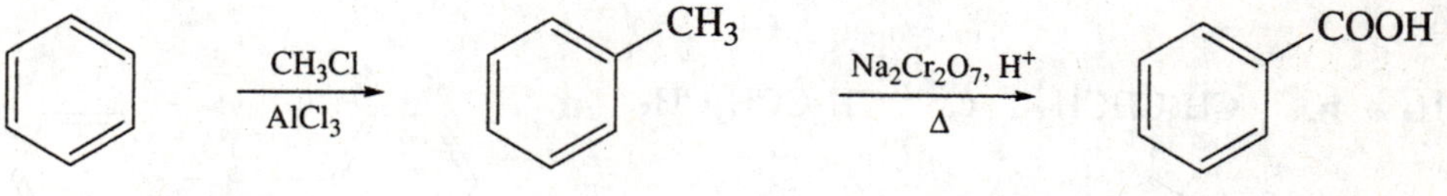

30. a.

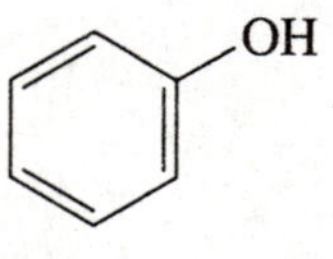

d.

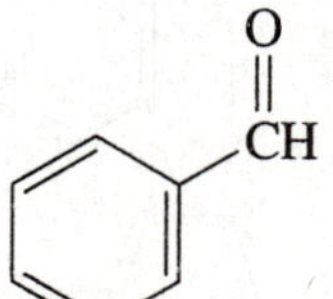

g.

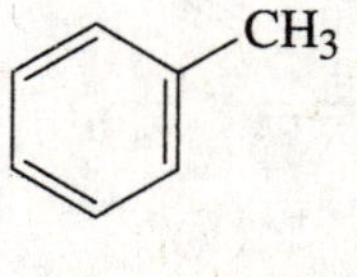

b.

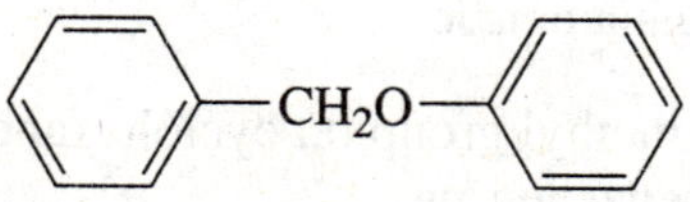

e.

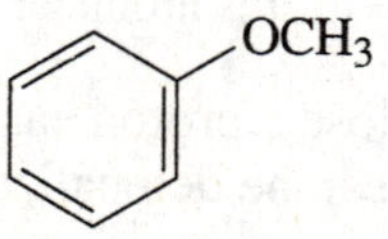

h.

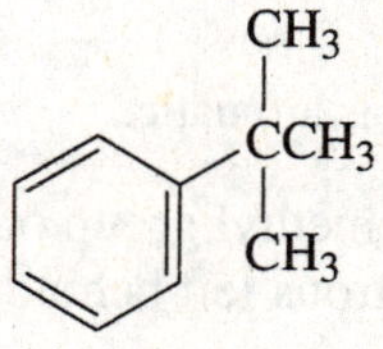

c.

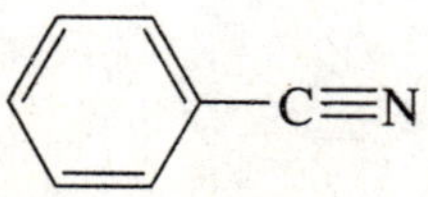

f.

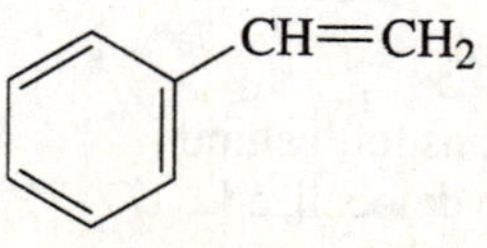

i.

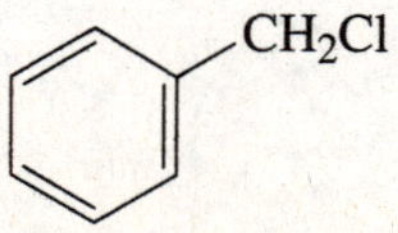

31. **aromatic**

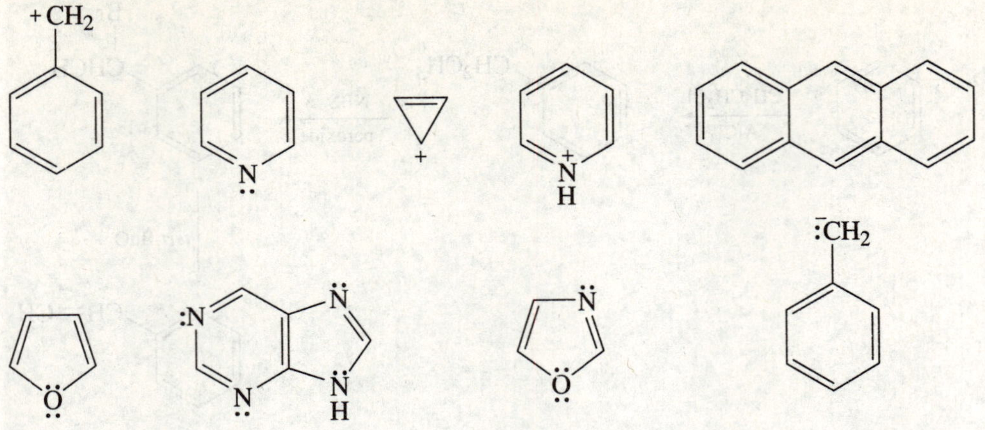

antiaromatic

By not being flat, cyclooctatetraene avoids being antiaromatic.

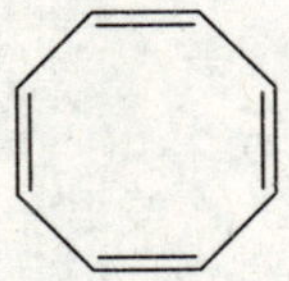

Notice that the nitrogen atoms in compounds such as those to the right are sp^3 hybridized, because if they were sp^2 hybridized, the compounds would be unstable antiaromatic compounds.

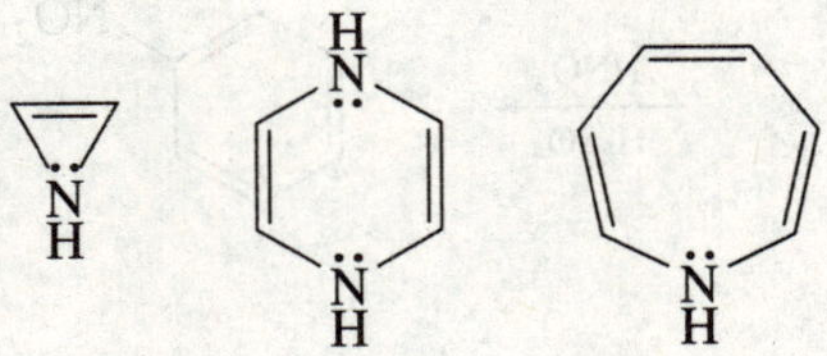

32. **a.**

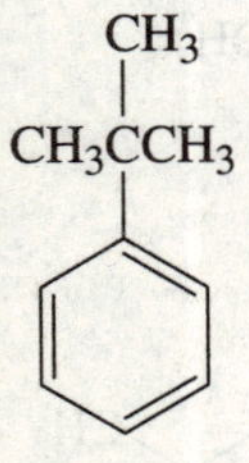

b.

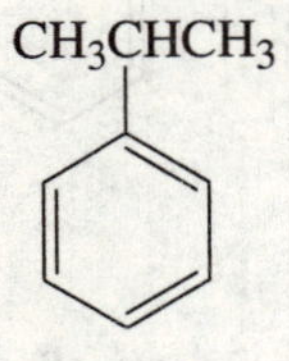

c.

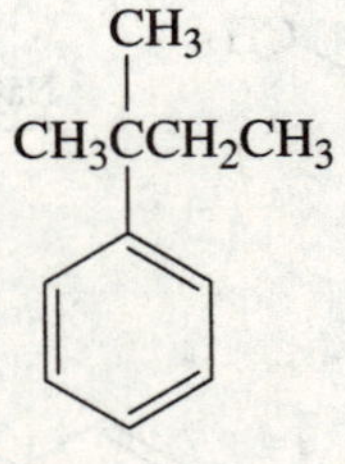

d.

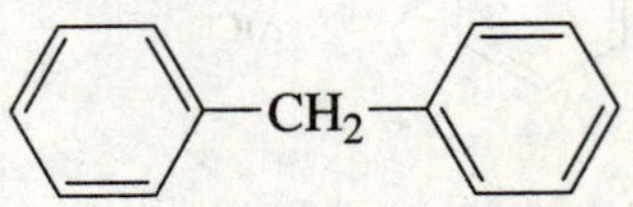

33. **a.**

It is aromatic.

b.

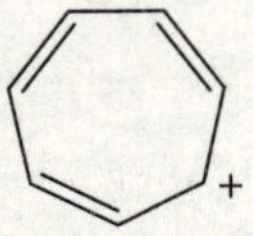

It is aromatic.

c.

It is aromatic.

d.

It is not antiaromatic.

34. The methyl group on benzene can lose a proton easier than the methyl group on cyclohexane because the electrons left behind in the former can be delocalized into the benzene ring.

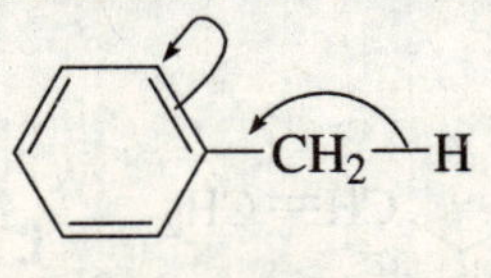

electrons left behind can be delocalized

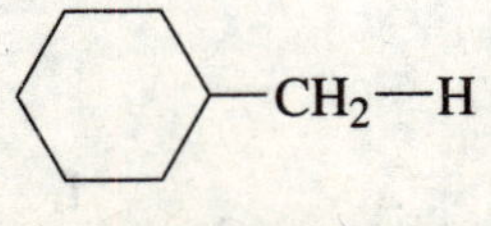

electrons left behind cannot be delocalized

35. **a.**

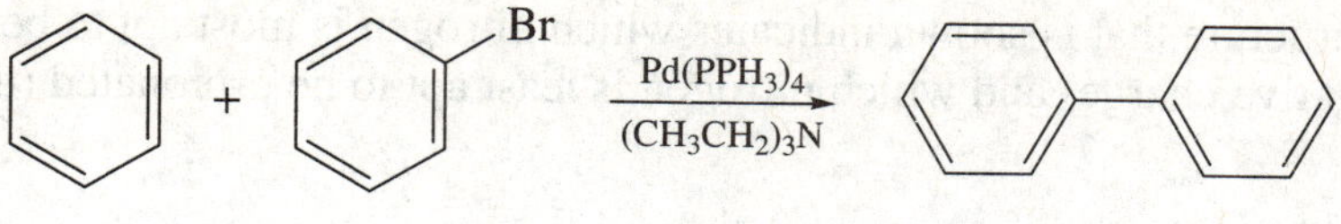

b.

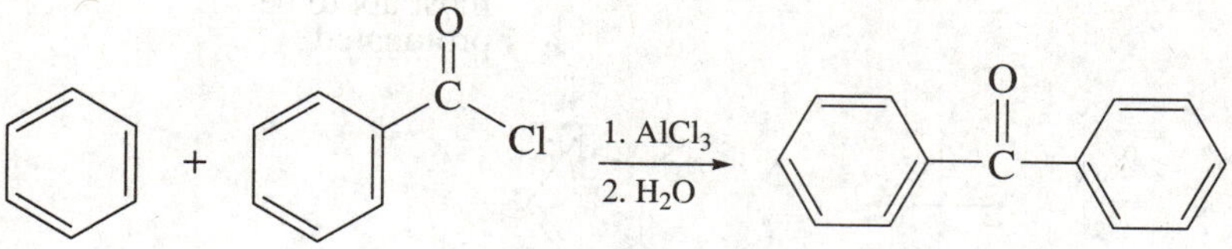

36. The ^{1}H NMR spectrum is the spectrum of 1-phenylbutane: the benzene ring protons show a signal at ~ 7.2 ppm. The two triplets and two multiplets indicate a butyl substituent.

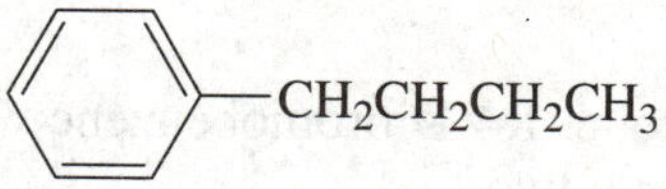

Therefore, the acyl chloride has a straight chain propyl group and a carbonyl group that will be reduced to a methylene group.

$$CH_3CH_2CH_2\overset{\overset{\displaystyle O}{\|}}{C}Cl$$

37. **a.**

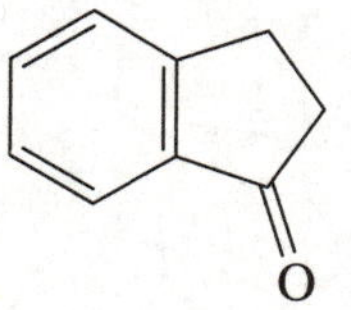

b.

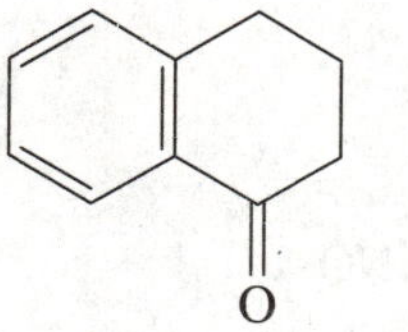

38.

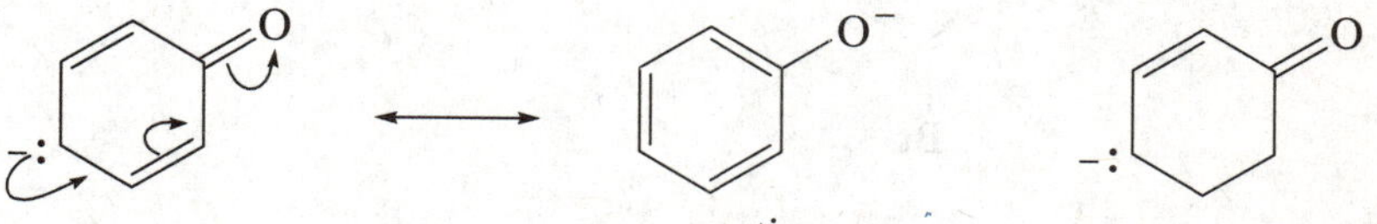

aromatic

This is the most stable (weakest) base because it has a resonance contributor that is aromatic. Therefore the other base (the one on the right) is the stronger base.

39. The following alkyl halide reacts more rapidly in an S_N1 reaction because it forms the more stable carbocation; the carbocation is more stable because it has a resonance contributor that is aromatic.

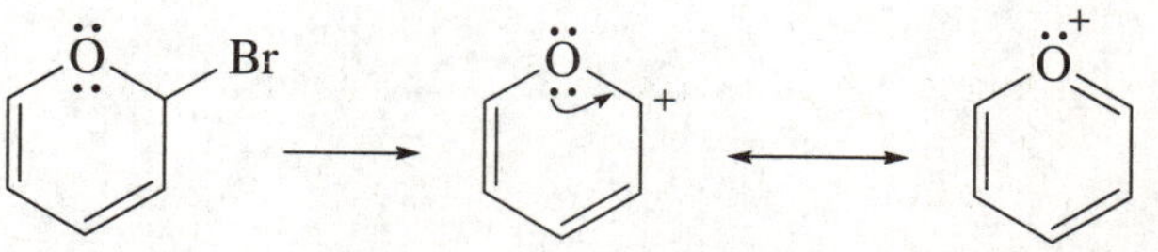

+ Br^- aromatic

40. The contributing resonance structure that is shown indicates which nitrogen is most apt to be protonated (the one with the greatest negative charge) and which nitrogen is least apt to be protonated (the one with the greatest positive charge).

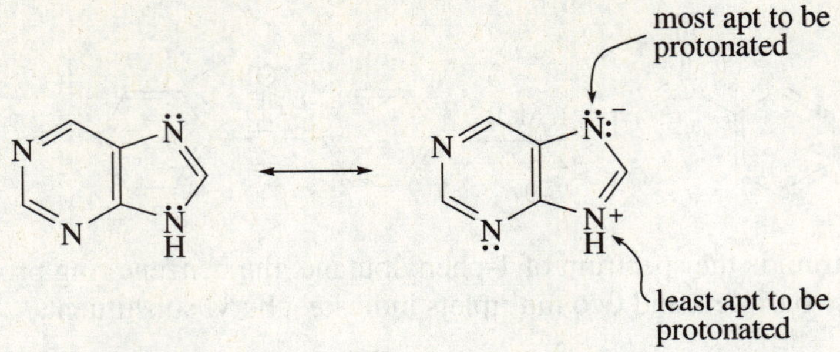

41. The compound is 1,3-dibromobenzene. 1,4-Dibromobenzene would form only one product and 1,2-dibromobenzene would form only two.

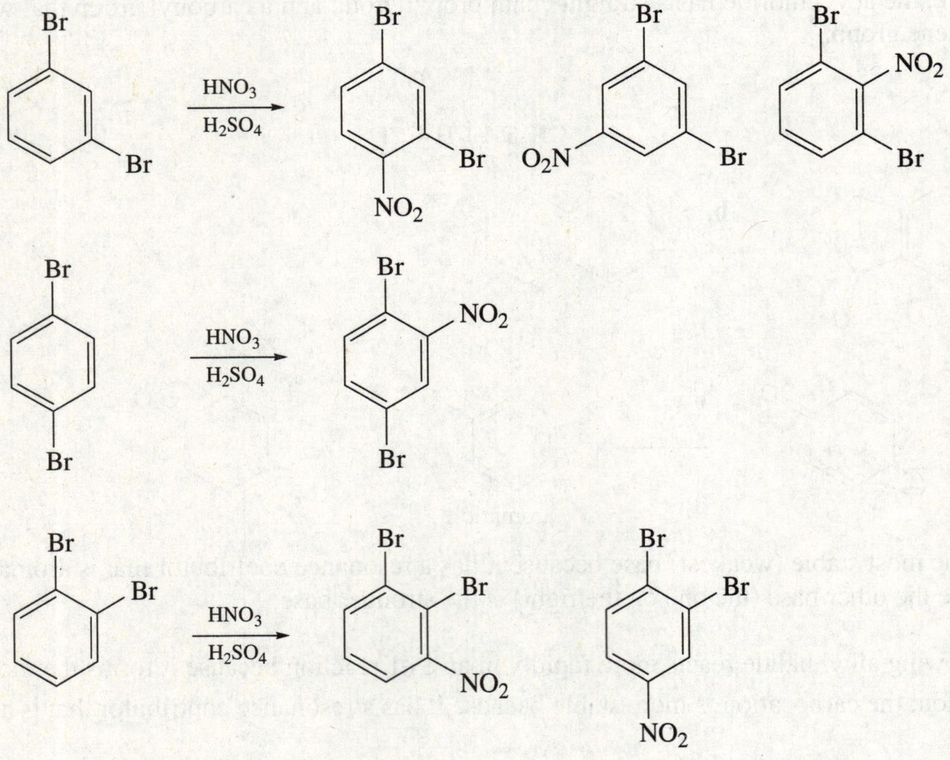

42. **a.**

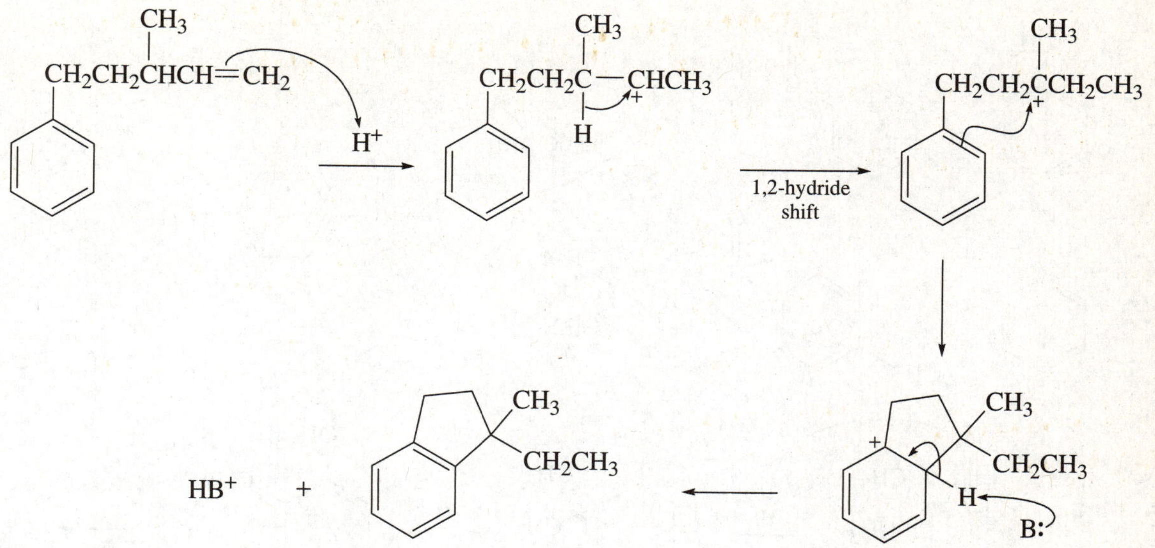

b.

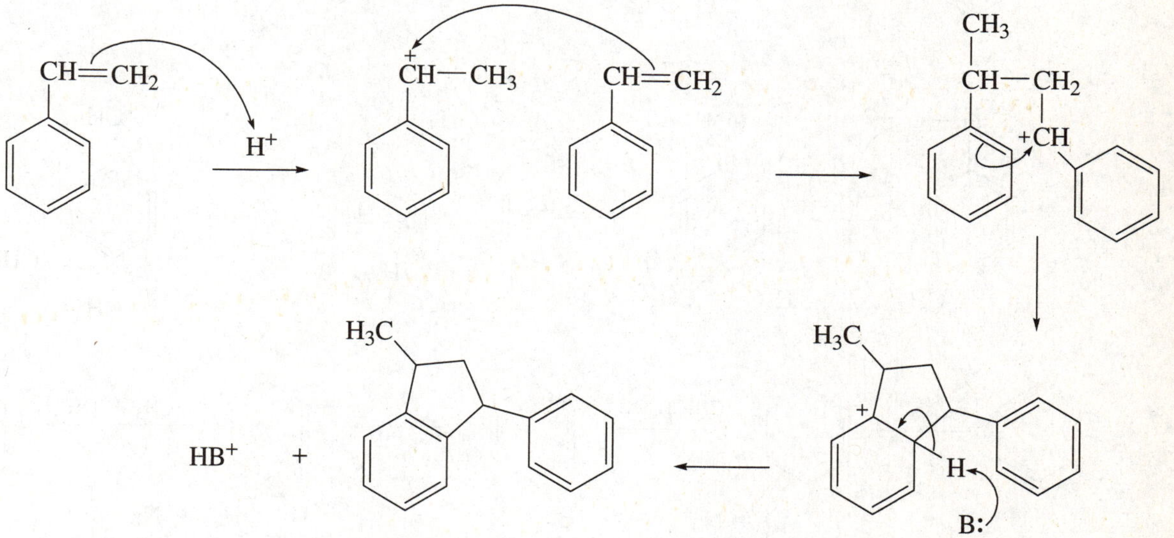

43. **a.**

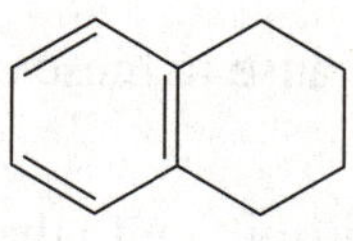

b.

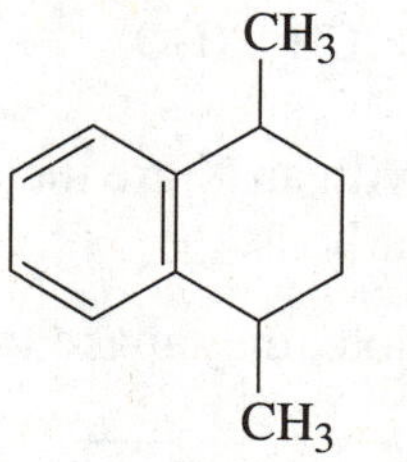

44. The following compound is the strongest acid because it is the only one that forms a base that is aromatic. Recall that the more stable (weaker) the base, the stronger is its conjugate acid.

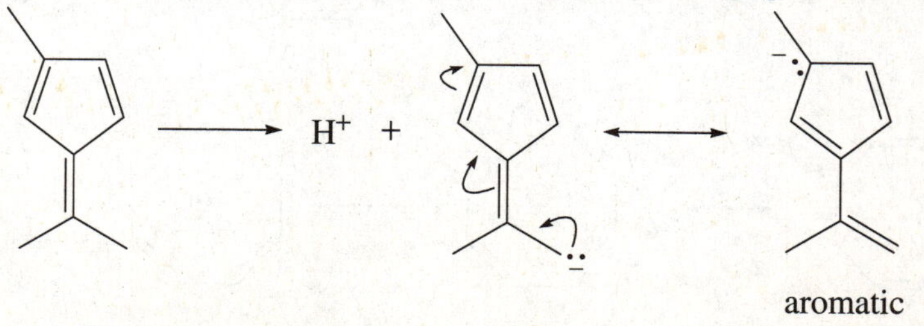

45.

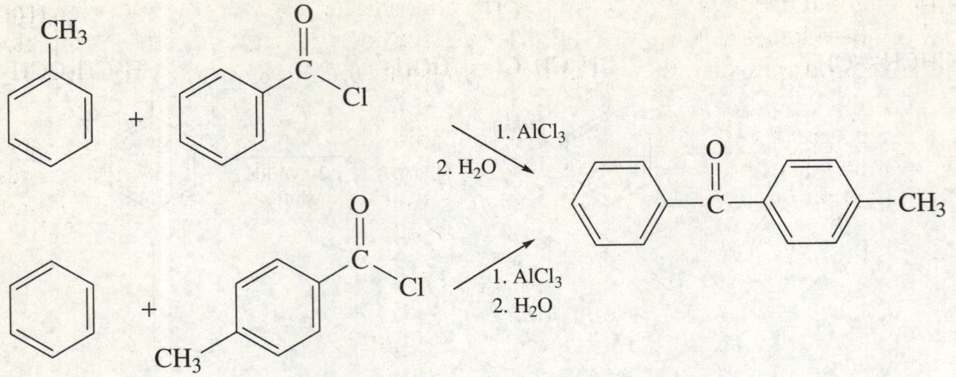

46. **a.**

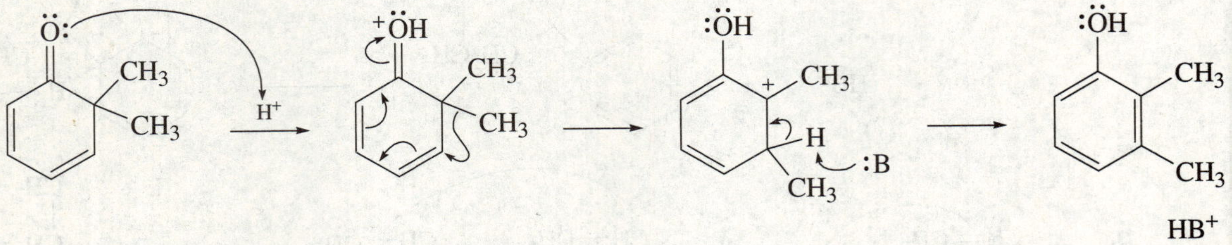

b.

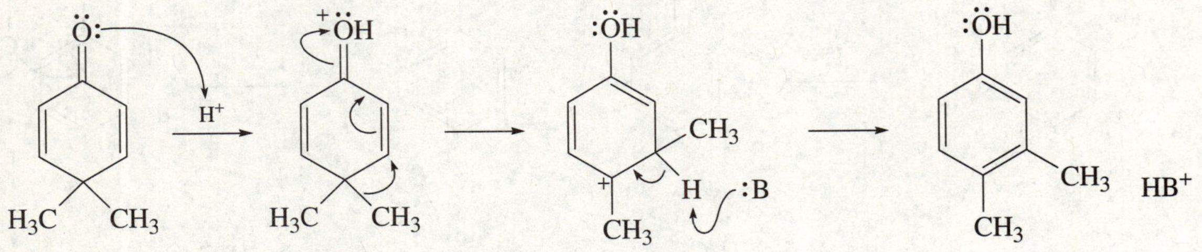

47.

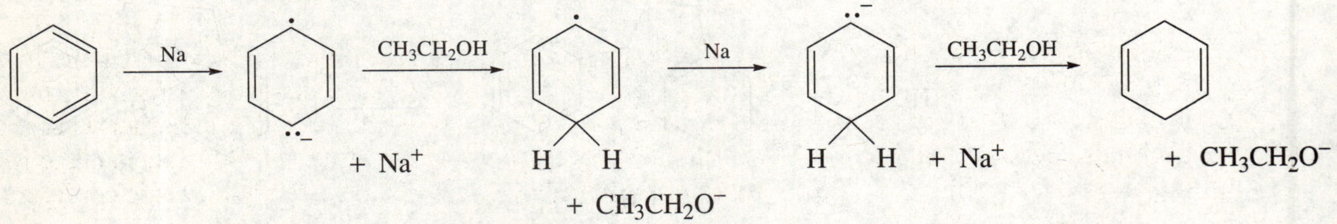

48. **a.** The three resonance contributors marked with an X are the least stable because in these contributors the two negative charges are on adjacent carbons.

b. Because these contributors are the least stable, they make the smallest contribution to the hybrid.

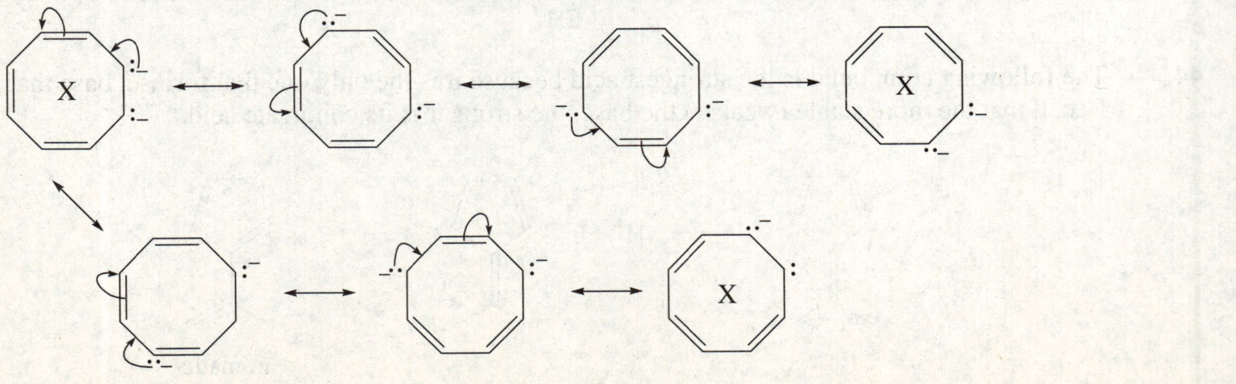

49. The observation that cyclobutadiene is rectangular and the observation that there are two different 1,2-dideuterio-1,3-cyclobutadienes both indicate that the π electrons are localized rather than delocalized. Localization of the π electrons prevents the compound from being antiaromatic.

50. From the three resonance structures for the cyclohexadienyl anion, the bond orders in the resonance hybrid can be calculated. (For example, the C1-C2 bond is represented by a single bond in one resonance structure and by a double bond in two resonance structures, which gives it a bond order 1+ 2/3 in the hybrid.)

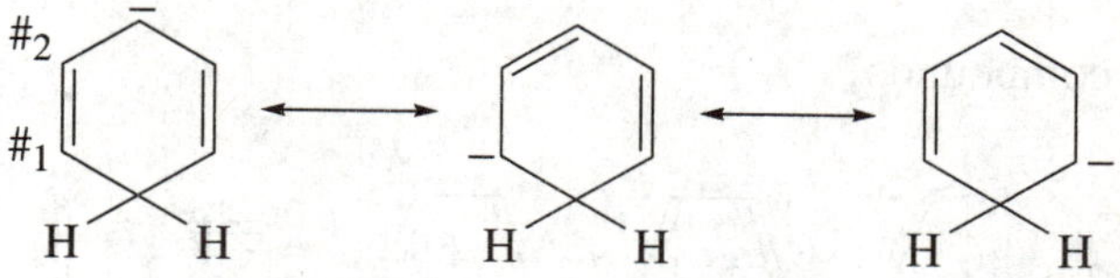

contributing resonance structures

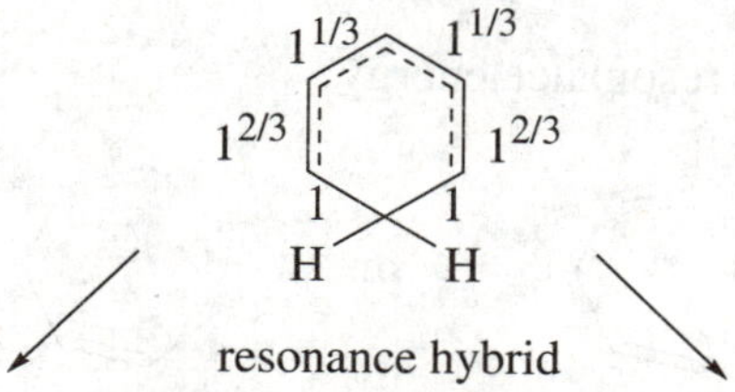

resonance hybrid

1,4-cyclohexadiene 1,3-cyclohexadiene

Comparing the bond order difference in each bond in the anion and in the corresponding bond in the possible products, there is a (1/3 + 1/3 + 1/3 + 1/3 = 4/3) 4/3 difference in bond order when 1,4-cyclohexadiene is formed and a (1/3 + 1/3 + 2/3 + 2/3 = 6/3) 6/3 difference in bond order when 1,3-cyclohexadiene is formed. Thus, the principle of least motion predicts that 1,4-cyclohexadiene will be the major product, since its formation from the anion involves less of a change in electronic configuration.

Chapter 14 Practice Test

1. Which are aromatic compounds?

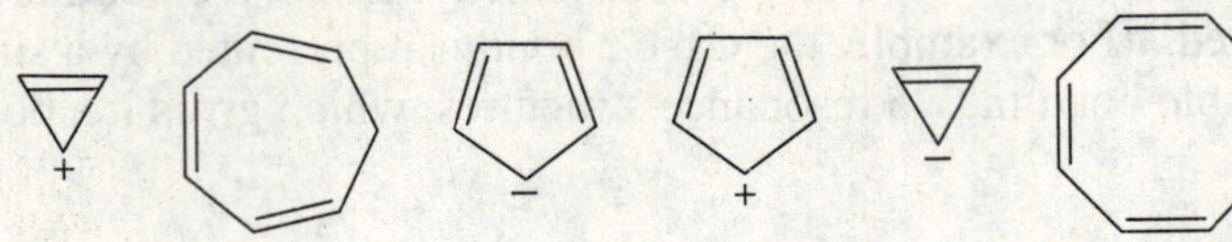

2. Which are antiaromatic compounds?

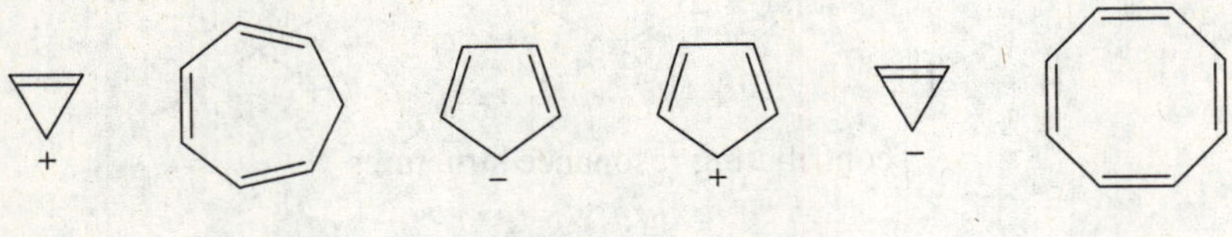

3. Which compound has the greater resonance energy?

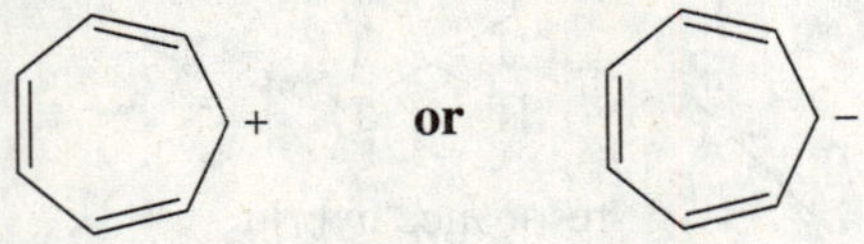

4. Draw the resonance contributors for the carbocation intermediate that is formed when benzene reacts with an electrophile ($\overset{+}{Y}$).

5. Which of the following is a stronger acid?

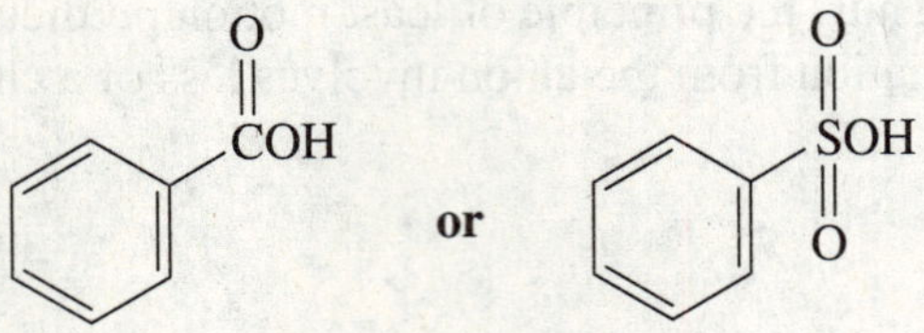

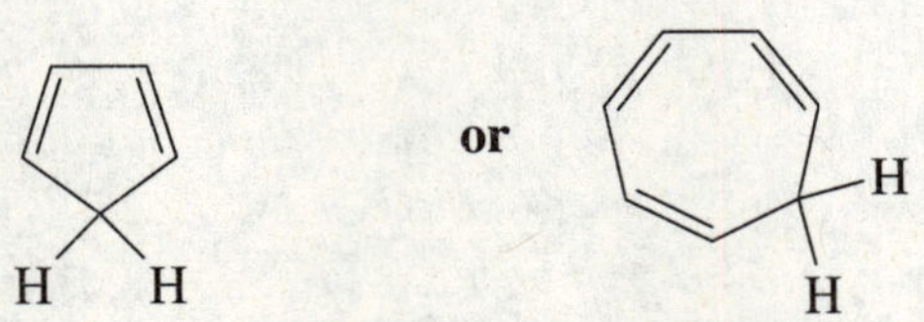

6. Give two alkyl halides, two alkenes, and two alcohols that could be used in a reaction with benzene to form 2-phenylbutane.

7. What acid anhydride would you use in a synthesis of 1-phenylpropane?

8. Which in each of the following pairs is more stable?

a. 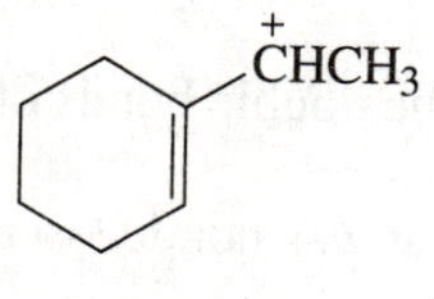 or 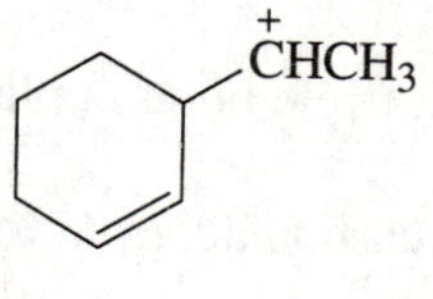

b. $CH_3\overset{+}{C}HCH_2CH_3$ or $CH_3\overset{+}{C}HCH{=}CH_2$

c. $CH_3\overset{-}{C}HCH_2\overset{\overset{O}{\|}}{C}CH_3$ or $CH_3\overset{-}{C}H\overset{\overset{O}{\|}}{C}CH_3$

9. Give the mechanism for formation of the nitronium ion from nitric acid and sulfuric acid.

CHAPTER 15
Reactions of Substituted Benzenes

Important Terms

activating substituent a substituent that increases the reactivity of an aromatic ring. Electron-donating substituents activate aromatic rings toward electrophilic attack, and electron-withdrawing substituents activate aromatic rings toward nucleophilic attack.

arenediazonium salt $Ar\overset{+}{N}{\equiv}N\ X^-$

azo linkage an $-N{=}N-$ bond.

benzyne intermediate a compound with a triple bond in place of one of the double bonds of benzene.

cine substitution substitution at the carbon adjacent to the carbon that was bonded to the leaving group.

deactivating substituent a substituent that decreases the reactivity of an aromatic ring. Electron-withdrawing substituents deactivate aromatic rings toward electrophilic attack, and electron-donating substituents deactivate aromatic rings toward nucleophilic attack.

direct substitution substitution at the carbon that was bonded to the leaving group.

donate electrons by resonance donation of electrons through *p* orbital overlap with neighboring π bonds.

fused rings rings that share two adjacent carbons.

inductive electron donation donation of electrons through a σ bond.

inductive electron withdrawal withdrawal of electrons through a σ bond.

meta director a substituent that directs an incoming substituent meta to an existing substituent.

nitrosamine (*N*-nitroso compound) an amine with a nitroso $(N{=}O)$ substituent bonded to its nitrogen atom.

nucleophilic aromatic substitution a reaction in which a nucleophile substitutes for an atom or group bonded to an aromatic ring.

ortho-para director a substituent that directs an incoming substituent ortho and para to an existing substituent.

resonance electron donation donation of electrons through *p* orbital overlap with neighboring π bonds.

resonance electron withdrawal	withdrawal of electrons through *p* orbital overlap with neighboring π bonds.
Sandmeyer reaction	the reaction of an arenediazonium ion with a cuprous salt.
Schiemann reaction	the reaction of an arenediazonium ion with HBF_4.
S_NAr reaction	a nucleophilic aromatic substitution reaction.
withdraw electrons by resonance	withdrawal of electrons through *p* orbital overlap with neighboring π bonds.

Solutions to Problems

1. **a.** *ortho*-ethylphenol or 2-ethylphenol

b. *meta*-bromochlorobenzene or 3-bromochlorobenzene

c. *meta*-bromobenzaldehyde or 3-bromobenzaldehyde

d. *ortho*-ethyltoluene or 2-ethyltoluene

2. **a.**

b.

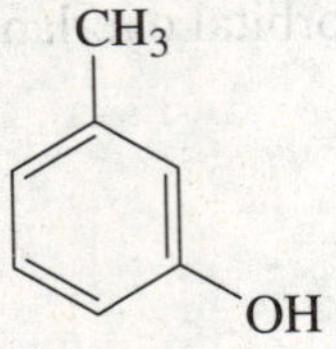

c.

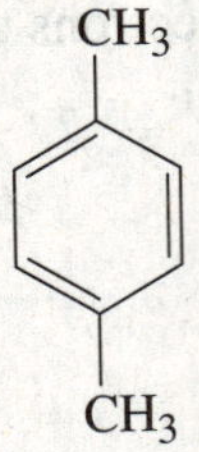

d.

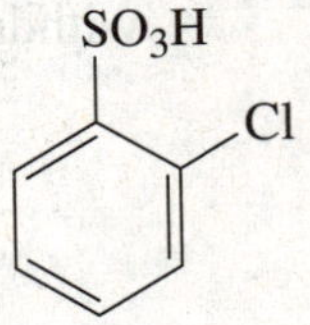

3. **a.**

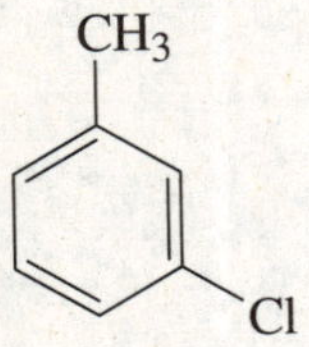

b.

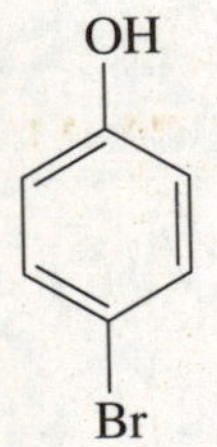

c.

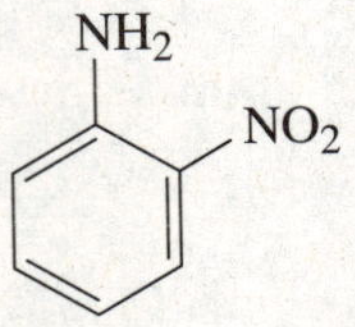

d.

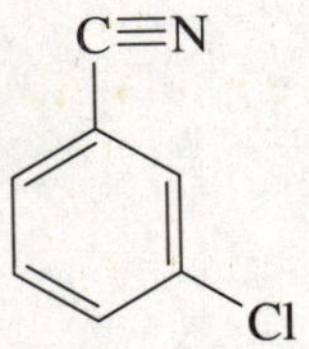

e.

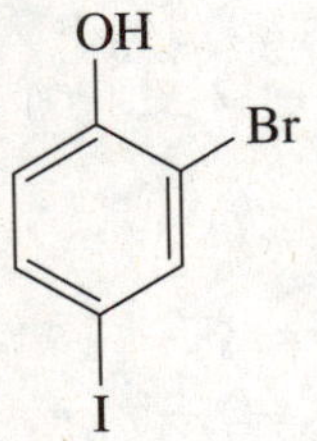

f.

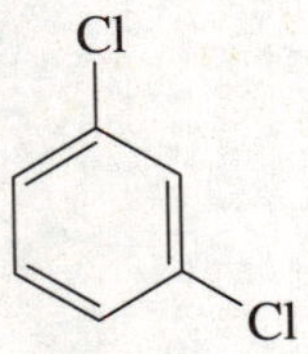

g.

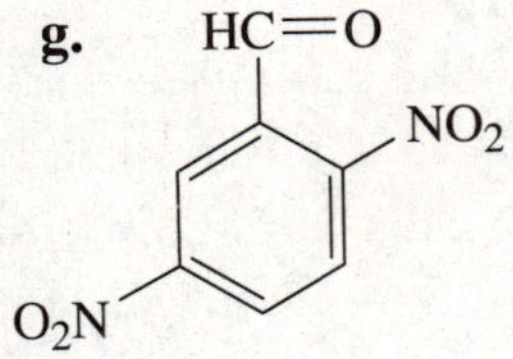

h.

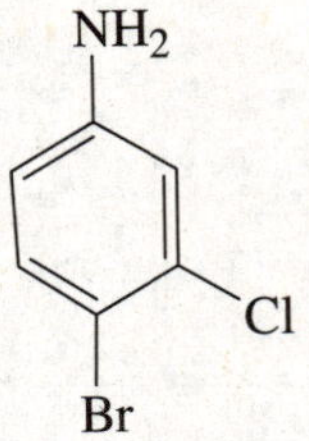

4. **a.** 1,3,5-tribromobenzene

b. *meta*-nitrophenol or 3-nitrophenol

c. *para*-bromotoluene or 4-bromotoluene

d. *ortho*-dichlorobenzene or 1,2-dichlorobenzene

5. **a.** donates electrons by resonance and withdraws electrons inductively

b. donates electrons by hyperconjugation

c. withdraws electrons by resonance and withdraws electrons inductively

d. donates electrons by resonance and withdraws electrons inductively

e. donates electrons by resonance and withdraws electrons inductively

f. withdraws electrons inductively

6. **a.** phenol > toluene > benzene > bromobenzene > nitrobenzene

b. toluene > chloromethylbenzene > dichloromethylbenzene > difluoromethylbenzene

7. Solved in the text.

8. **a.**

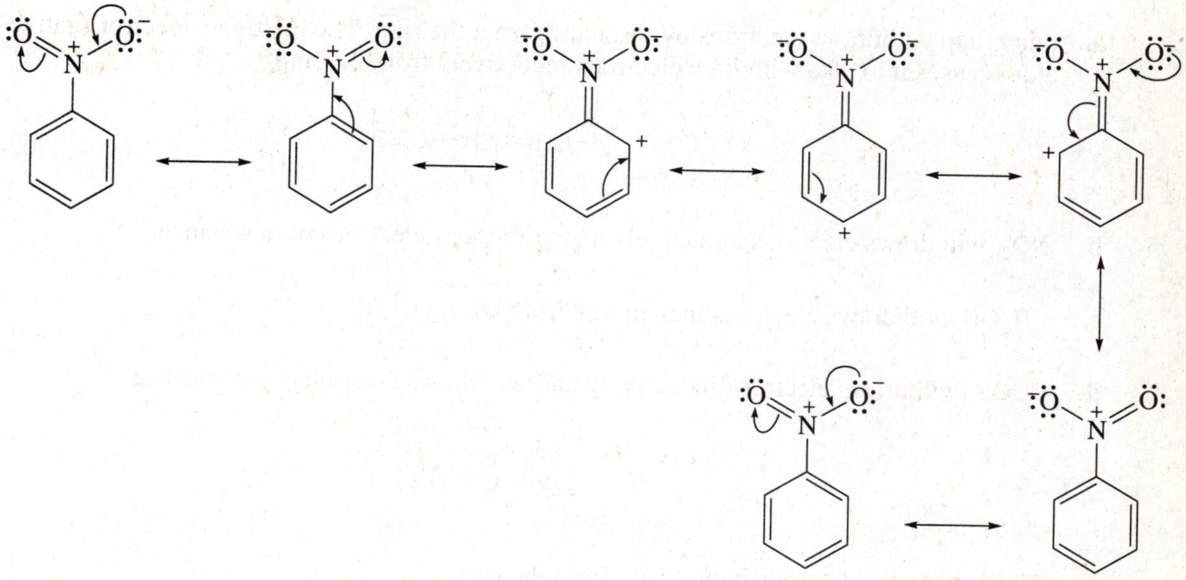

b.

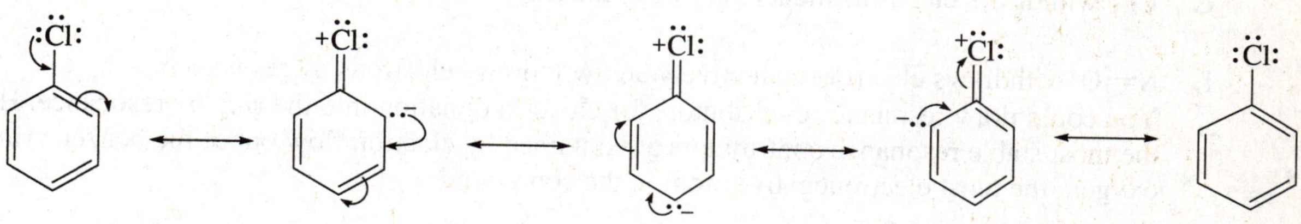

9. **a.**

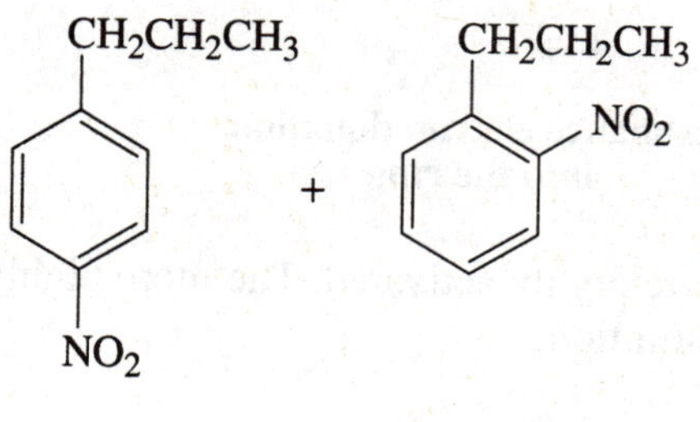

b.

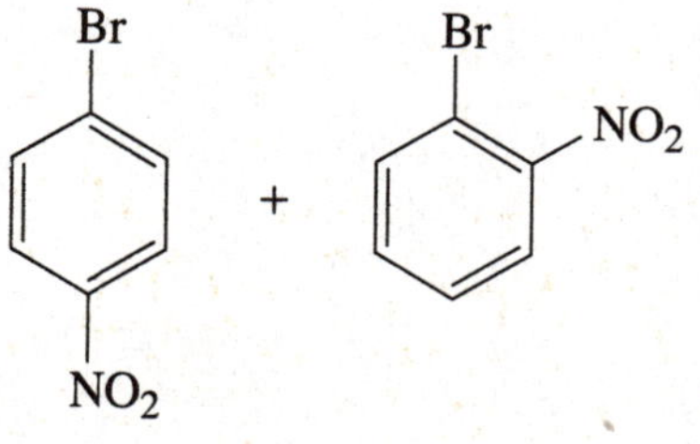

c.

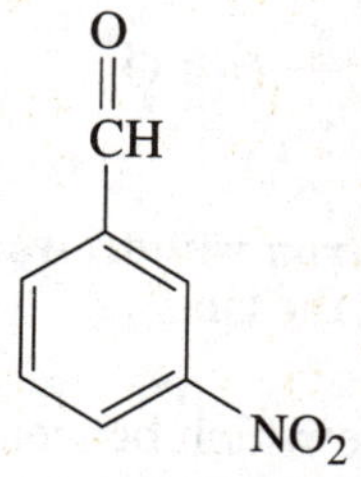

d.

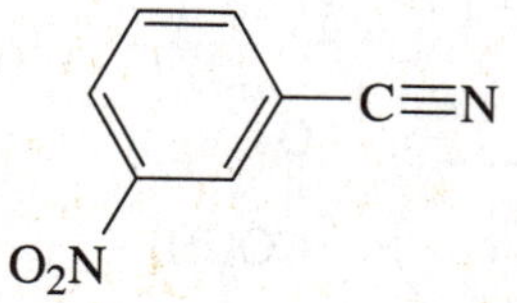

e.

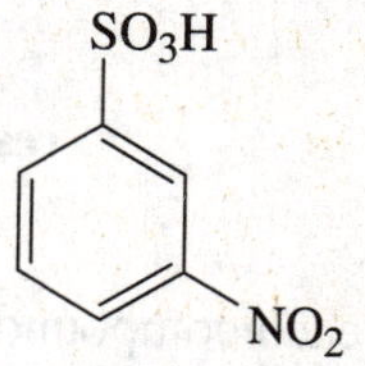

f.

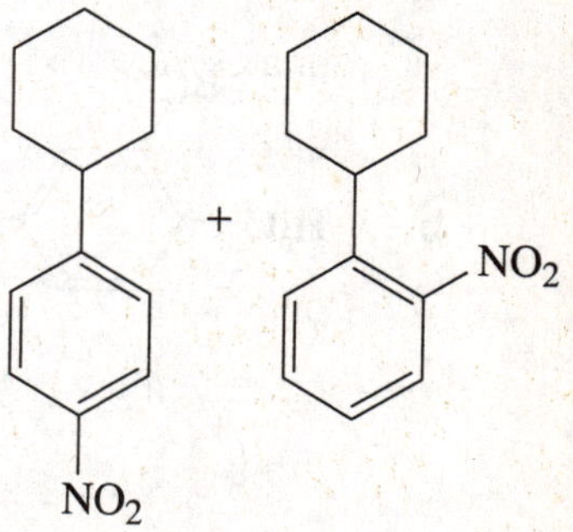

10. They are all meta directors:

a. this group withdraws electrons by resonance from the ring. The relatively electronegative nitrogen atom causes it to also withdraw electrons inductively from the ring.

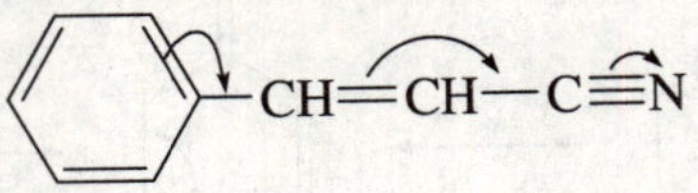

b. NO_2 withdraws electrons inductively and withdraws electrons by resonance.

c. CH_2OH withdraws electrons inductively from the ring.

d. COOH withdraws electrons inductively and withdraws electrons by resonance.

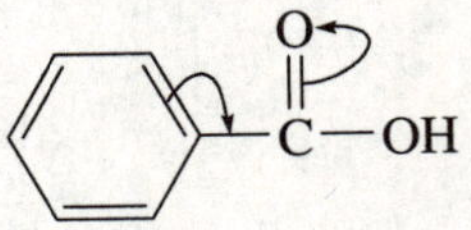

e. CF_3 withdraws electrons inductively from the ring.

f. N=O withdraws electrons inductively and withdraws electrons by resonance.
You could draw resonance contributors for electron donation into the ring by resonance. However, the most stable resonance contributors are obtained by electron flow out of the benzene ring toward oxygen, the most electronegative atom in the compound.

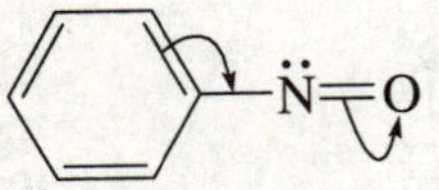

resonance eletron withdrawal out of the ring

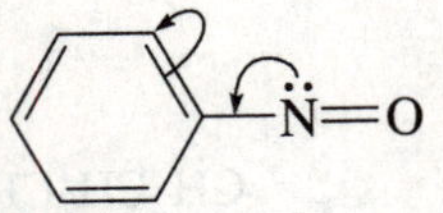

resonance eletron donation into the ring

11. For each compound, determine which benzene ring is more highly activated. The more highly activated ring is the one that undergoes aromatic electrophilic substitution.

a. Solved in the text.

b. 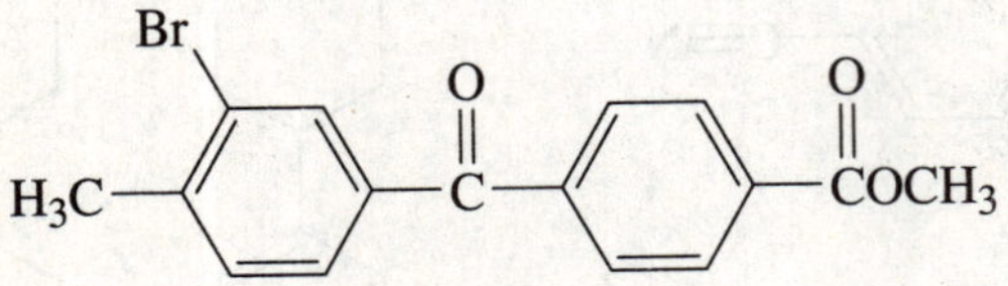

c. Because the right hand ring is highly activated and a catalyst is employed, monobrominated, dibrominated, and tribrominated compounds can be obtained depending on how much Br_2 is available.

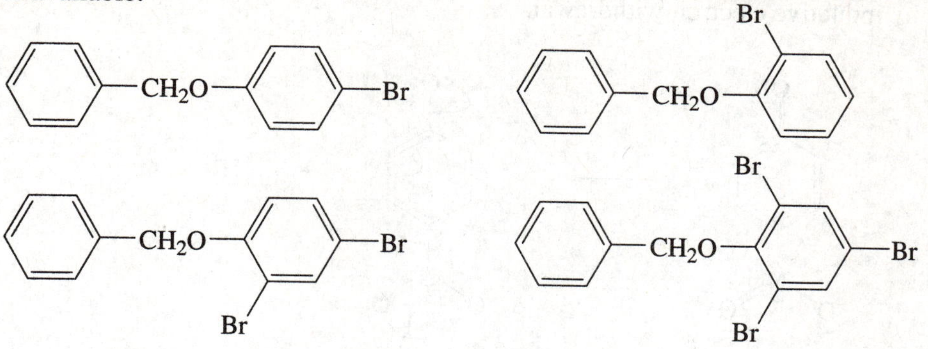

d. 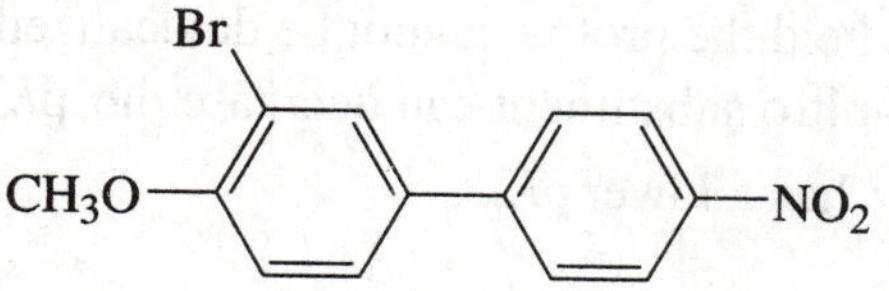

12. **a.** $ClCH_2\overset{O}{\overset{\|}{C}}OH$

b. $O_2NCH_2\overset{O}{\overset{\|}{C}}OH$

c. $H_3\overset{+}{N}CH_2\overset{O}{\overset{\|}{C}}OH$

d. 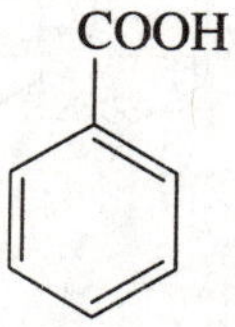

e. $HO\overset{O}{\overset{\|}{C}}CH_2\overset{O}{\overset{\|}{C}}OH$

The negatively charged compound has a greater amount of electron donation by resonance.

f. $H\overset{O}{\overset{\|}{C}}OH$

A hydrogen is electron-withdrawing compared to a methyl group.

g. $FCH_2\overset{O}{\overset{\|}{C}}OH$

Fluorine is more eletronegative than chlorine.

h. 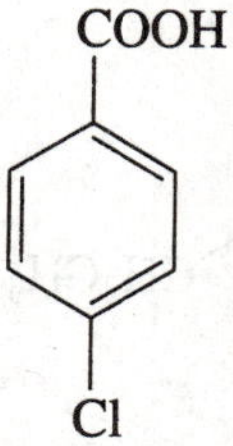

We know that F withdraws electrons inductively more than Cl does, because F is more electronegative than Cl.
We know that F donates electrons by resonance better than Cl does, because F donation involves 2*p*-2*p* overlap, whereas Cl donation involves 3*p*-2*p* overlap.

Because Table 15.1 on page 684 of the text shows that F is more activating than Cl, we know that overall F donates electrons better than Cl. Thus resonance electron donation is more important than inductive electron withdrawal.

13. When *para*-nitrophenol loses a proton, the electrons that held the proton can be delocalized by resonance onto the nitro substituent. Therefore, the *para*-nitro substituent decreases the pK_a by resonance electron withdrawal and by inductive electron withdrawal.

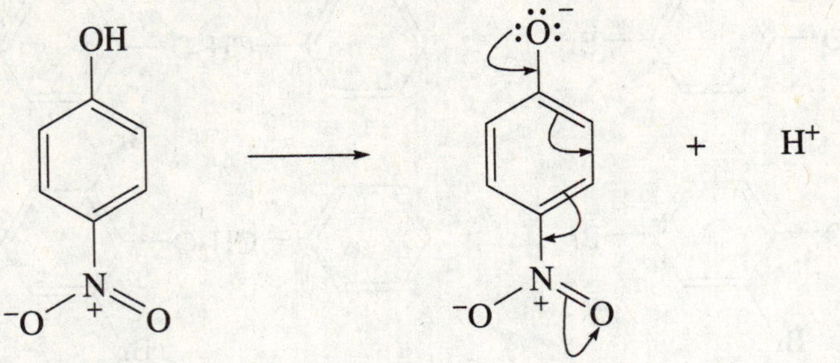

When *meta*-nitrophenol loses a proton, the electrons that held the proton cannot be delocalized by resonance onto the nitro substituent. Therefore, the *meta*-nitro substituent can decrease the pK_a only by inductive electron withdrawal. Therefore, the para isomer has a lower pK_a .

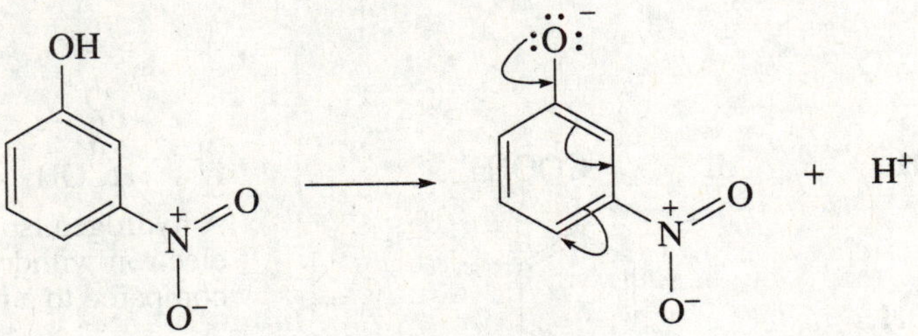

14. Notice that in all three syntheses, the Friedel-Crafts reaction has to be done first. Both NO_2 and SO_3H are meta directors and a Friedel-Crafts reaction cannot be carried out if there is a meta director on the ring.

a.

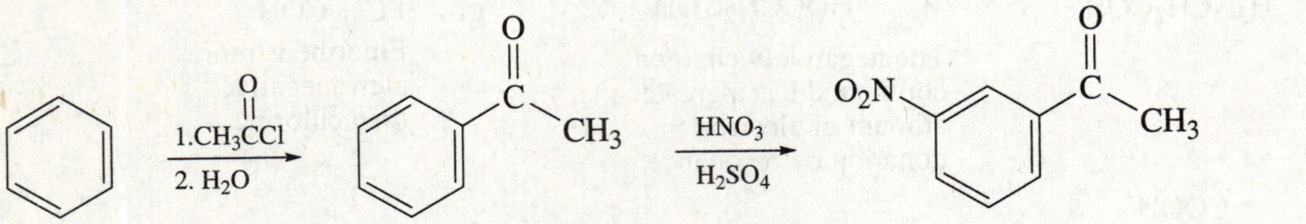

b.

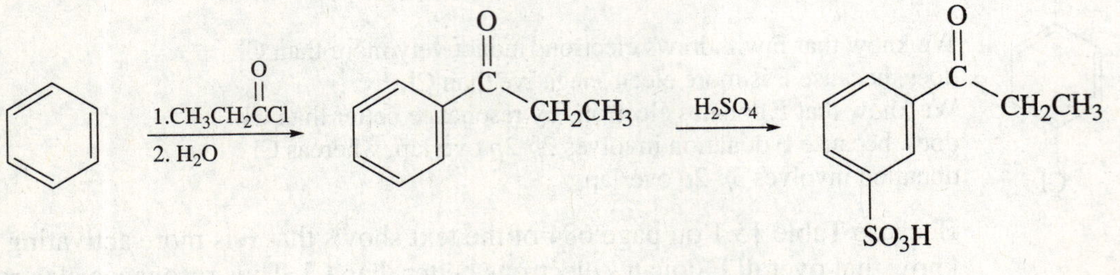

c.

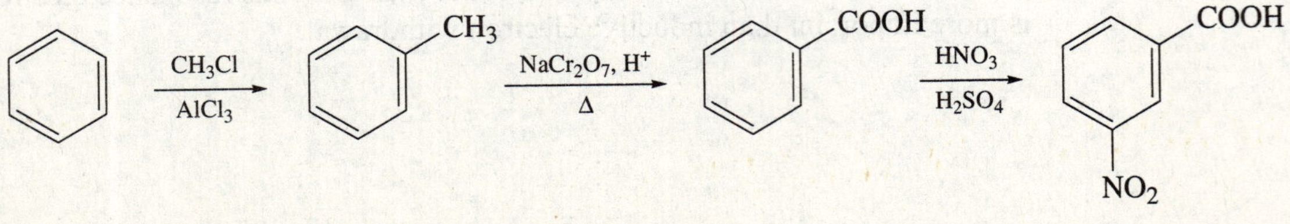

15. No reaction will occur in **a** and **c**, because a Friedel-Crafts reaction cannot be carried out on a ring that possesses a meta director.

a. no reaction **c.** no reaction

b.

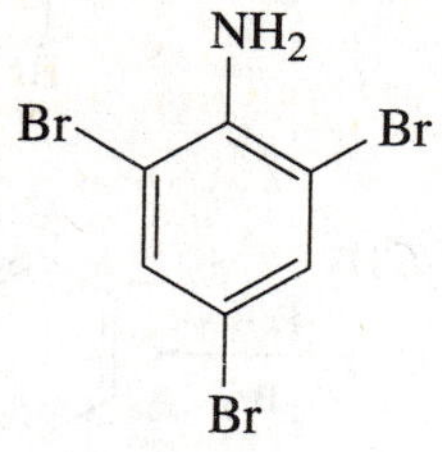

d.

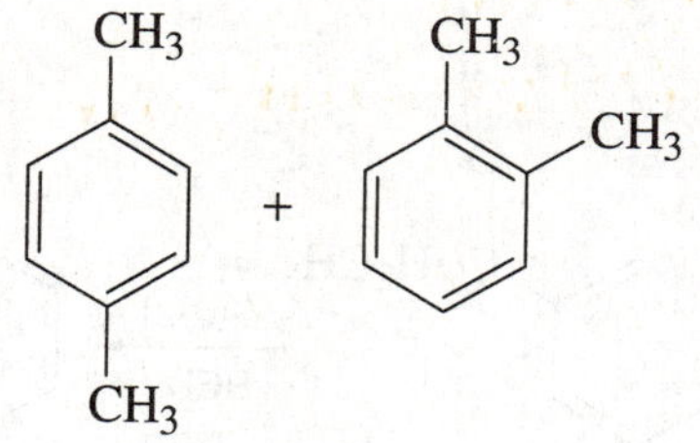

16. a.

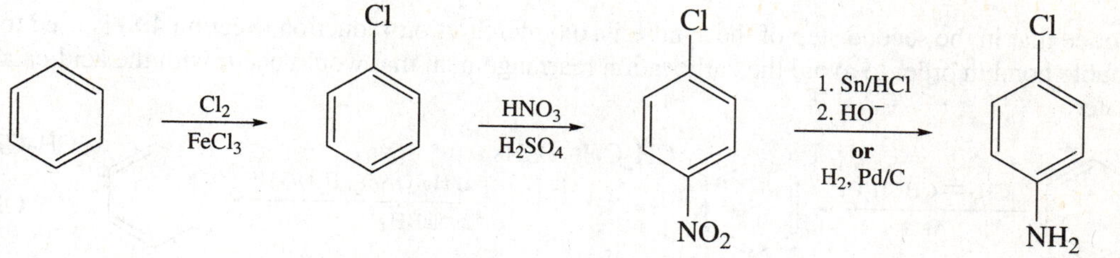

b.

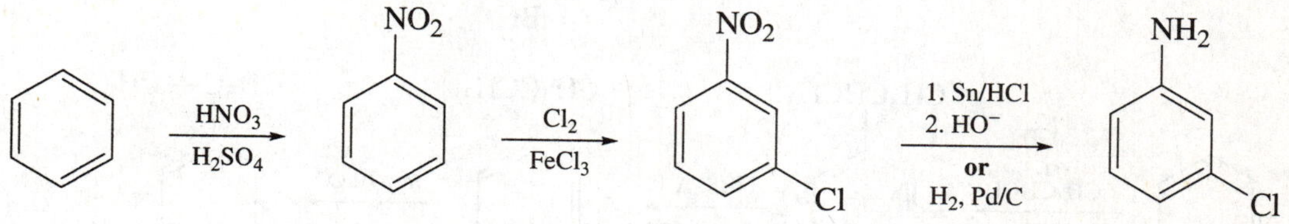

c.

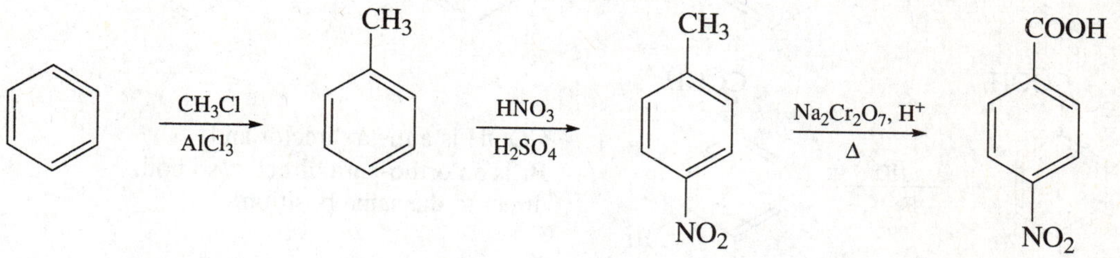

d.

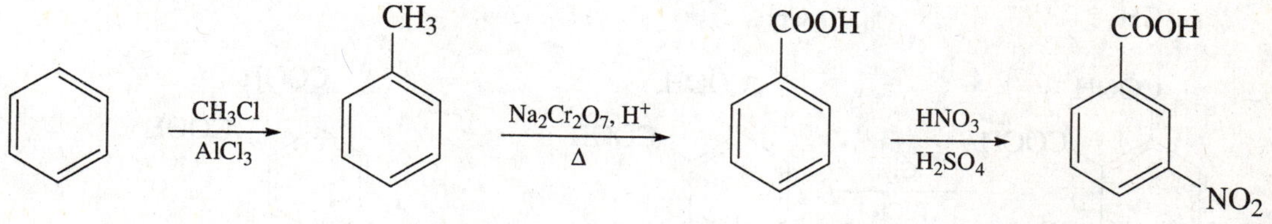

In the next two problems ("e" and "f"), the carbonyl group can be converted to a methylene group by H_2 Pd/C, or by a Wolff-Kishner reduction (as is used in "e") or by a Clemmensen reduction (as is used in "f").

e.

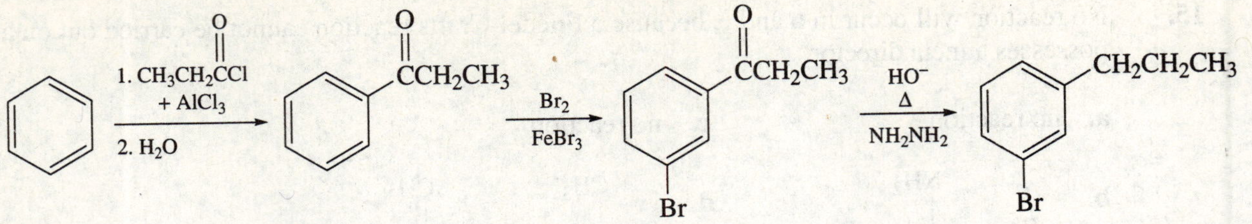

f.

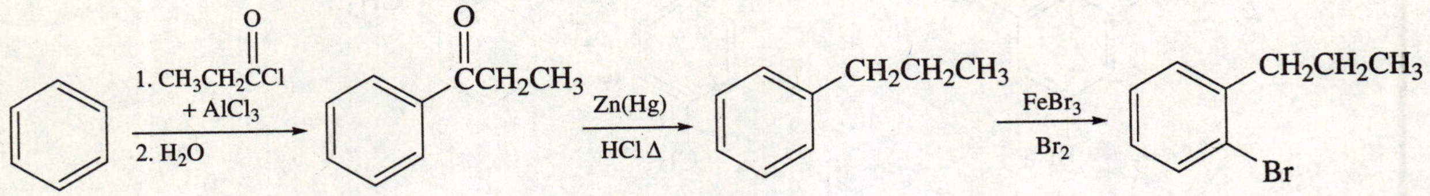

g. Notice that in the second step of the synthesis, oxymercuration-reduction (Section 4.8) is used to add water to the double bond in order to avoid the carbocation rearrangement that would occur with the acid-catalyzed addition of water.

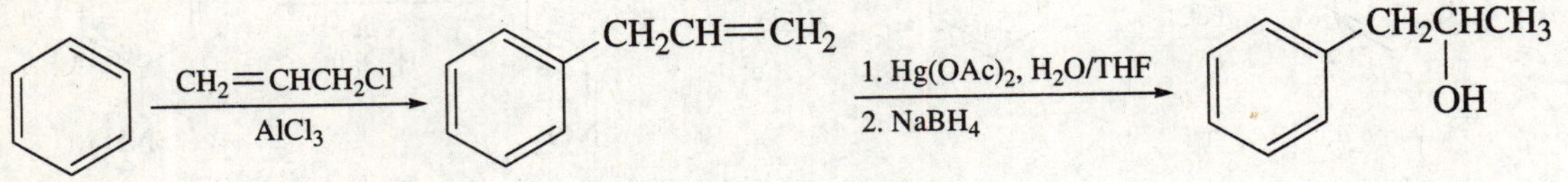

h.

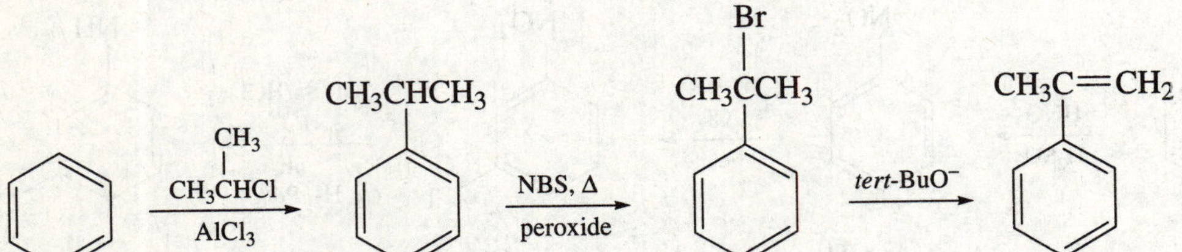

17. **a.**

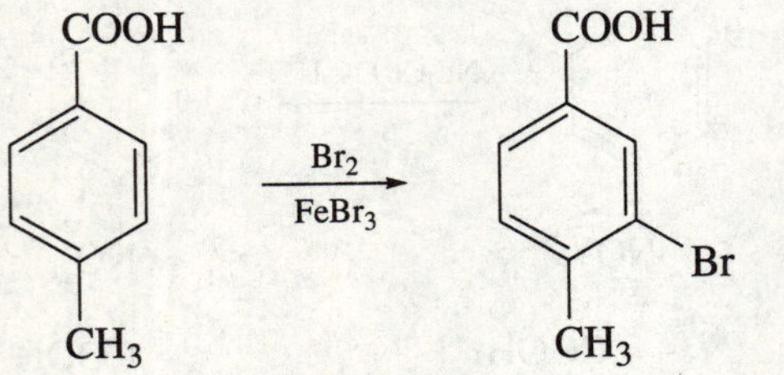

COOH is a meta director and Br is an ortho-para director, so both direct to the same position.

b.

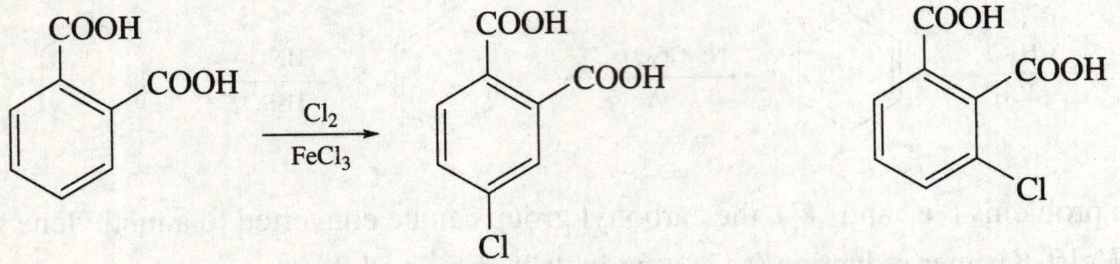

COOH directs to the meta position. The same product will be obtained regardless of which COOH is the director.

Less of this compound will be obtained because of steric hindrance.

c.

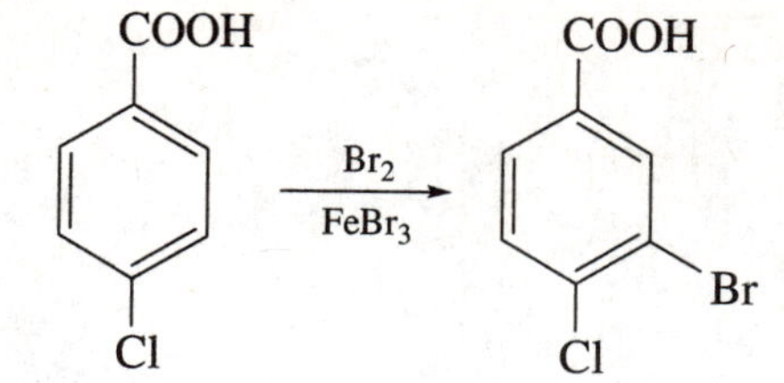

COOH directs to its meta position and Cl directs to its ortho position, so they both direct to the same position on the ring

d.

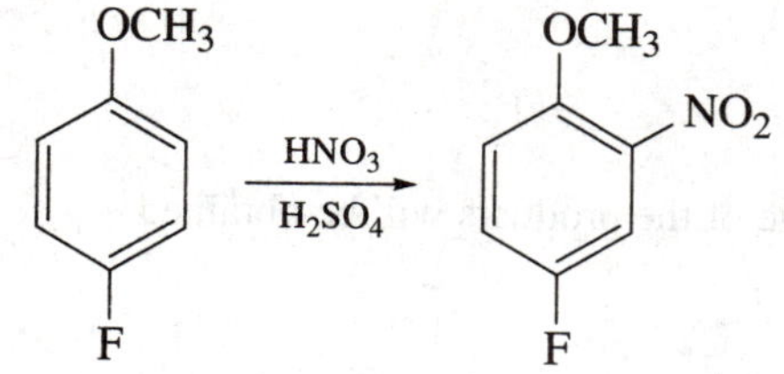

A methoxy substituent is strongly activating whereas a fluorine substituent is deactivating, so the methoxy substituent will do the directing.

e.

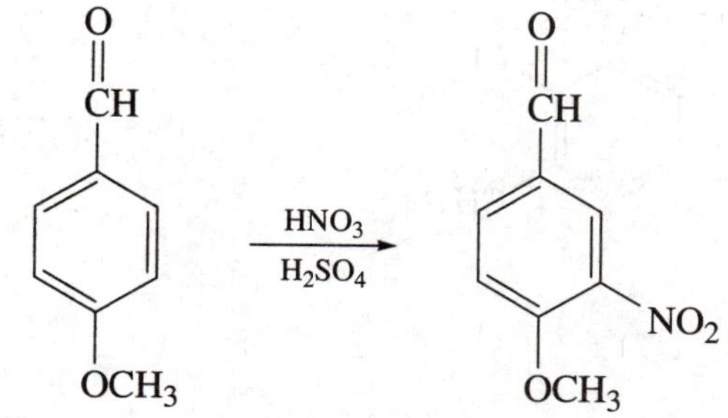

The aldehyde group is a meta director and the methoxy group is an ortho-para director, so both direct to the same position.

f.

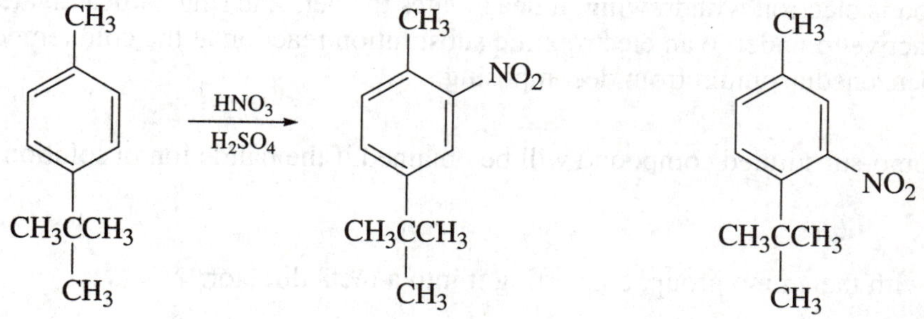

Less of this product will be obtained because of steric hindrance.

18. **a.** Two products will be obtained.

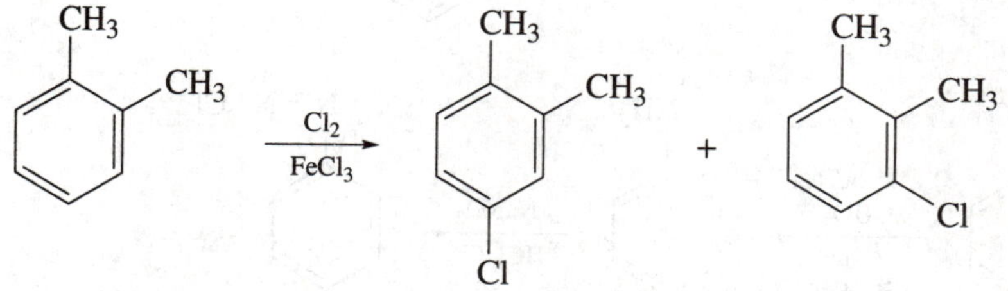

b. One product will be obtained.

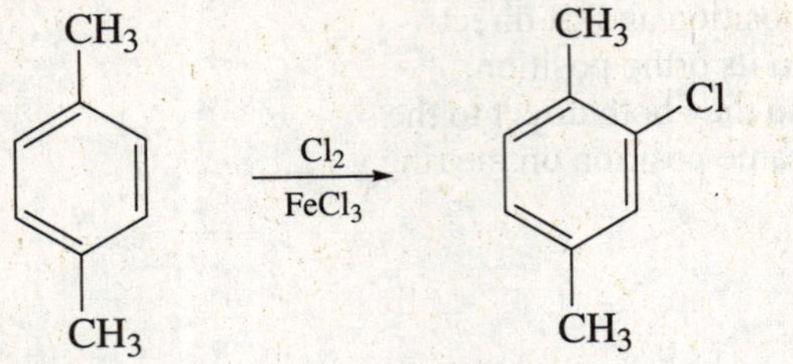

c. Two products could be obtained, but little, if any, of one of the products will be obtained because of steric hindrance.

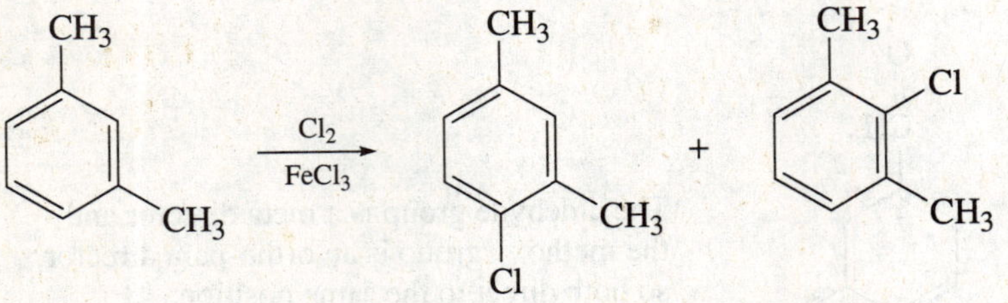

19. Solved in the text.

20. Because a diazonium ion is electron withdrawing, it deactivates the benzene ring. A deactivated benzene ring would be too unreactive to undergo an electrophilic substitution reaction at the cold temperature necessary to keep the benzenediazonium from decomposing.

21. More of the desired bromo-substituted compound will be obtained if the halide ion in solution is Br^- than if it is Br^- and Cl^-.

22. $FeBr_3$ would complex with the amino group, converting it into a meta director.

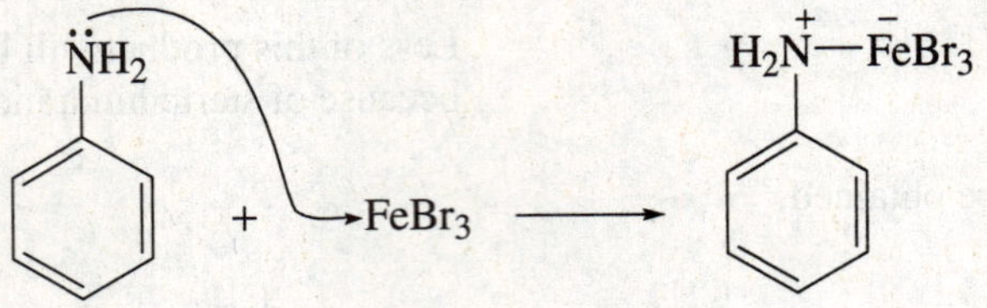

23. 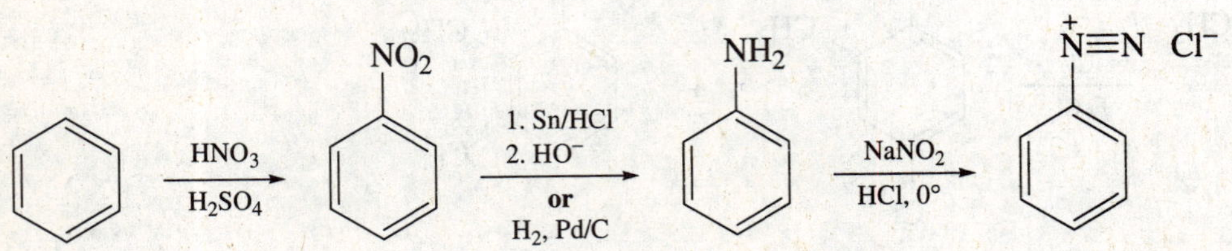

24. **a.** Solved in the text.

b. The reaction steps are the same as those in **a** (answered on page 702 of the text) except for the last step. In the last step the meta-bromo-substituted ion should undergo reaction with water rather than with CuBr .

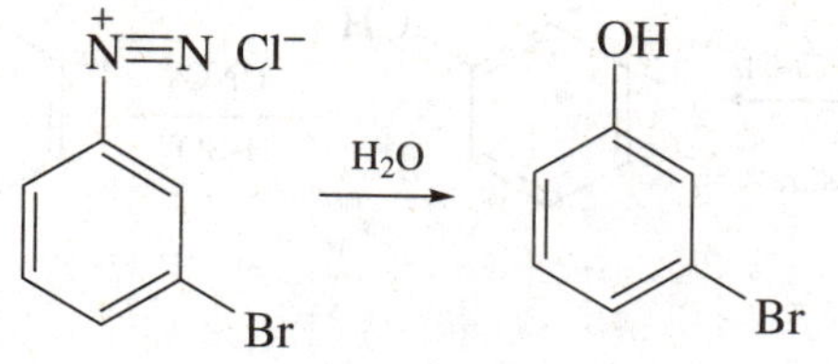

c.

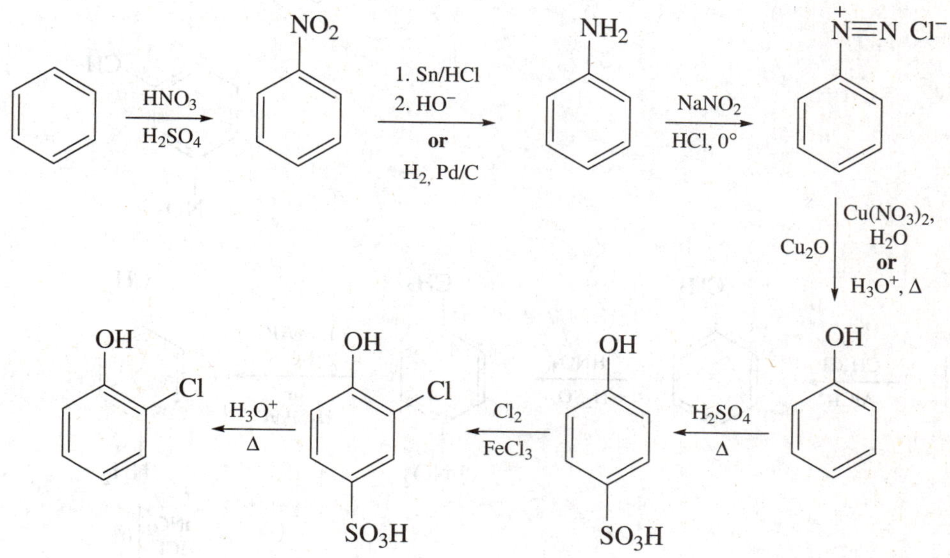

or

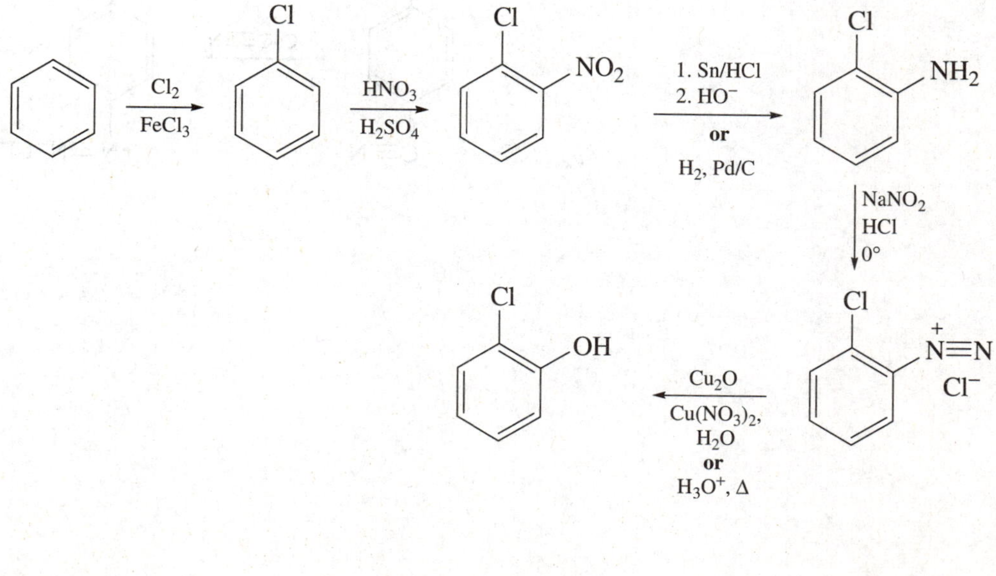

d. The nitro group cannot be placed on the benzene ring first, because a Friedel-Crafts reaction cannot be carried out on a ring with a meta director. Because formyl chloride is too unstable to be purchased, benzaldehyde is prepared by the Gatterman-Koch reaction p. 613 in the text).

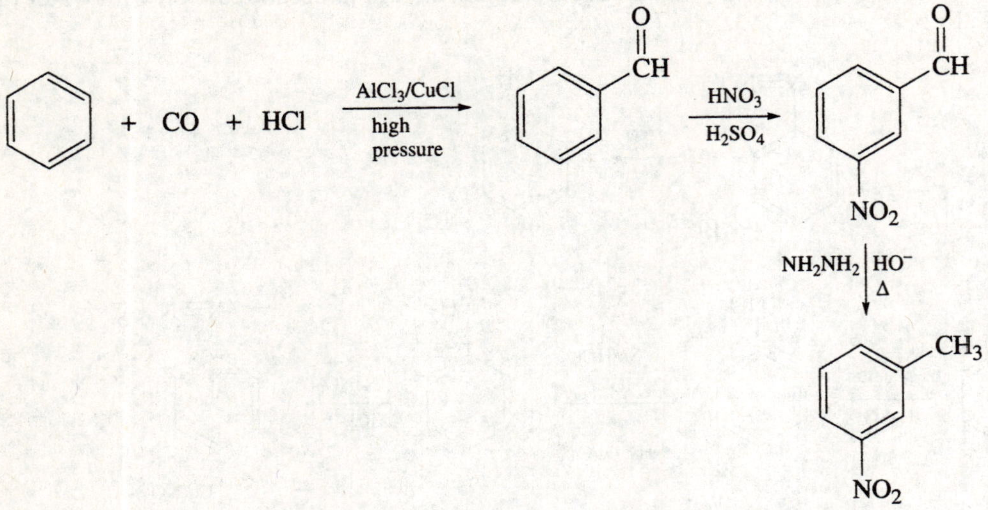

e.

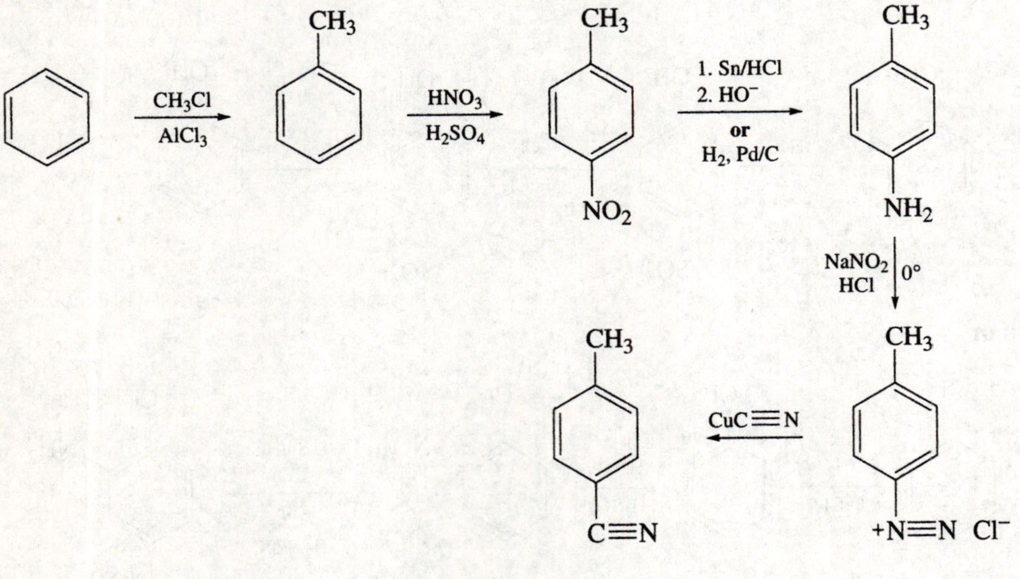

f.

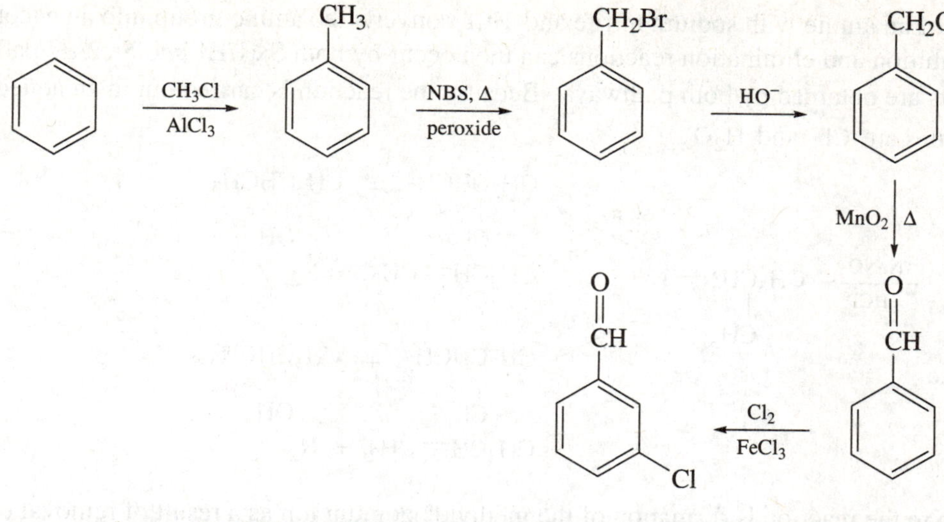

25. You can see why nucleophilic attack occurs on the neutral nitrogen if you compare the products of nucleophilic attack on the two nitrogens. Nucleophilic attack on the neutral nitrogen forms a stable product, whereas nucleophilic attack on the positively charged nitrogen would form an unstable compound with two charged nitrogen atoms.

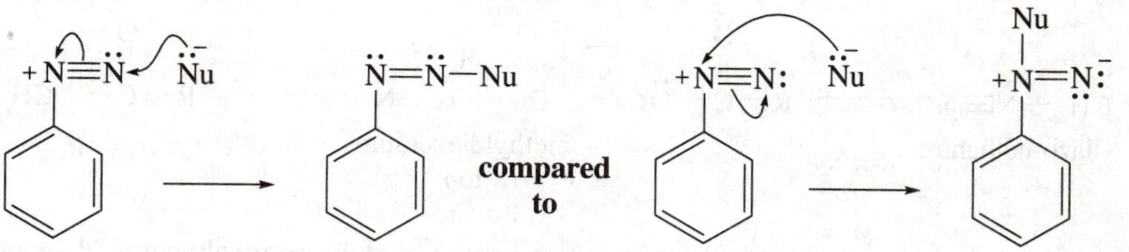

The terminal nitrogen is electrophilic because of electron withdrawal by the positively charged nitrogen. If you draw the resonance contributors, you can see that the "neutral" nitrogen is electron deficient.

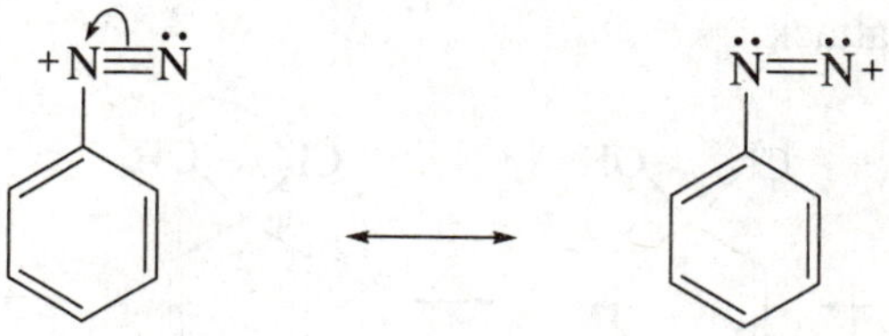

26. **a.**

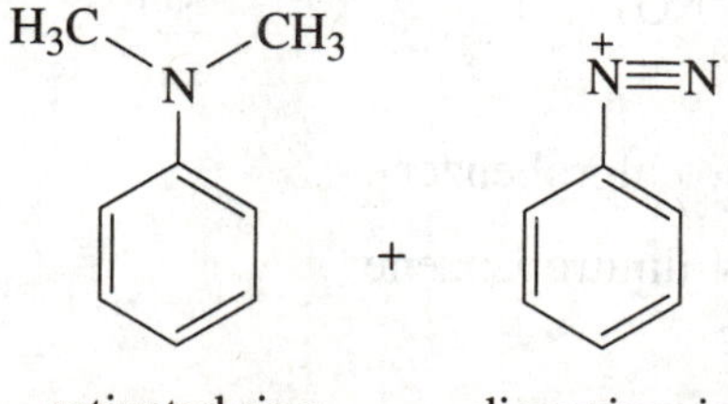

activated ring diazonium ion

b.

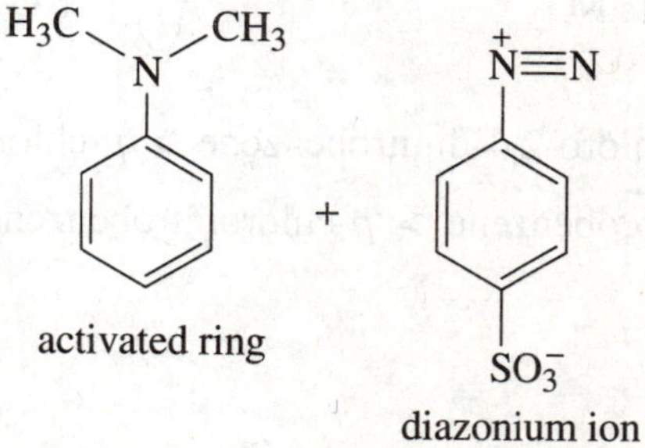

activated ring

diazonium ion

27. The reaction of an amine with sodium nitrite and HCl converts the amino group into an excellent leaving group. Substitution and elimination reactions can then occur by both S_N1/E1 and S_N2/E2 pathways. The same products are obtained by both pathways. Because the reaction is carried out in an aqueous solution, the nucleophiles are Cl^- and H_2O.

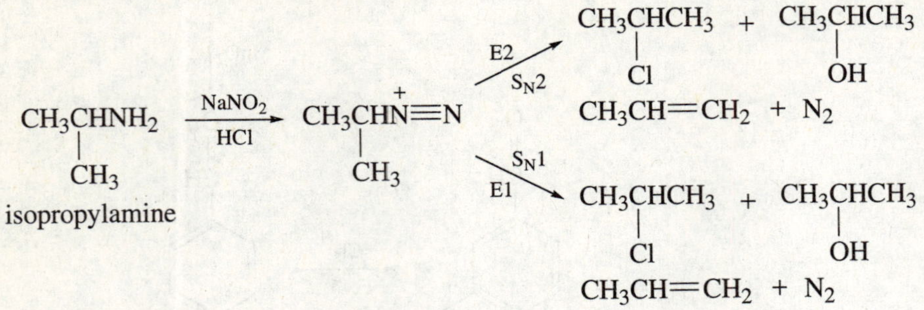

28. The first step in the reaction is formation of the methyldiazonium ion as a result of removal of a proton from the carboxylic acid by diazomethane. Diazomethane is both explosive and toxic, so it should be synthesized only in small amounts by experienced laboratory workers.
In the second step of the reaction, the carboxylate ion displaces nitrogen from the methyldiazonium ion. High yields are obtained, since the only side product is N_2 gas.

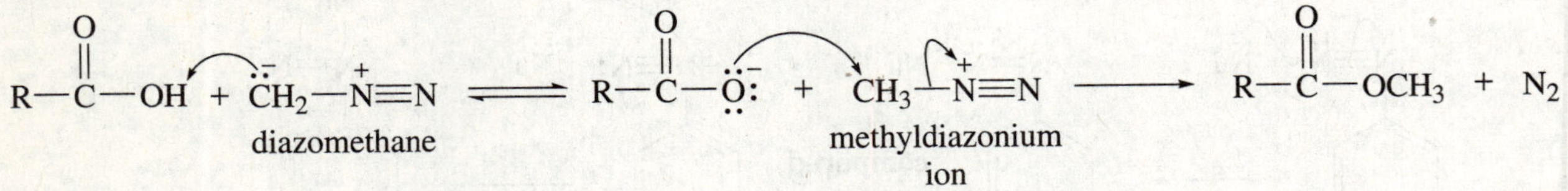

29. From the resonance contributors, you can see that the reason that *meta*-chloronitrobenzene does not react with hydroxide ion is because the negative charge that is generated on the benzene ring cannot be delocalized onto the nitro substituent.
Electron delocalization onto the nitro substituent can occur only if the nitro substituent is ortho or para to the site of nucleophilic attack.

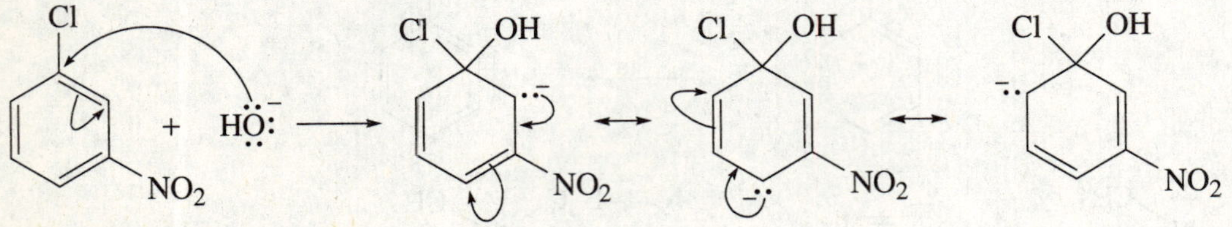

30. **a.** 1-chloro-2,4-dinitrobenzene > *p*-chloronitrobenzene > chlorobenzene

b. chlorobenzene > *p*-chloronitrobenzene > 1-chloro-2,4-dinitrobenzene

31. **a.**

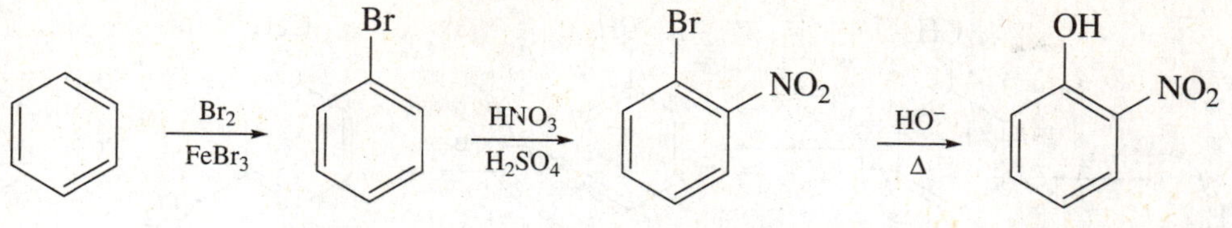

b.

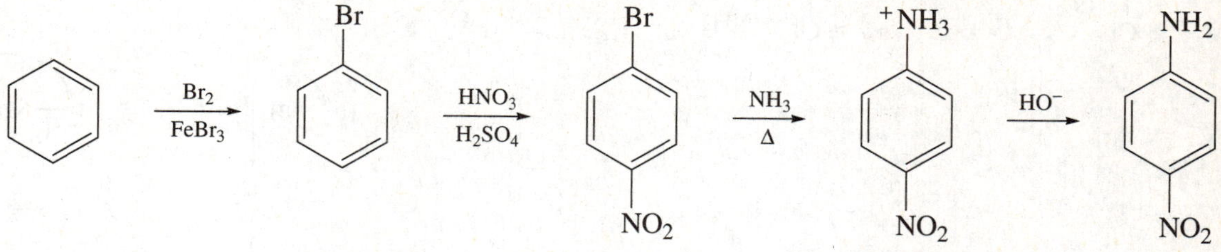

c.

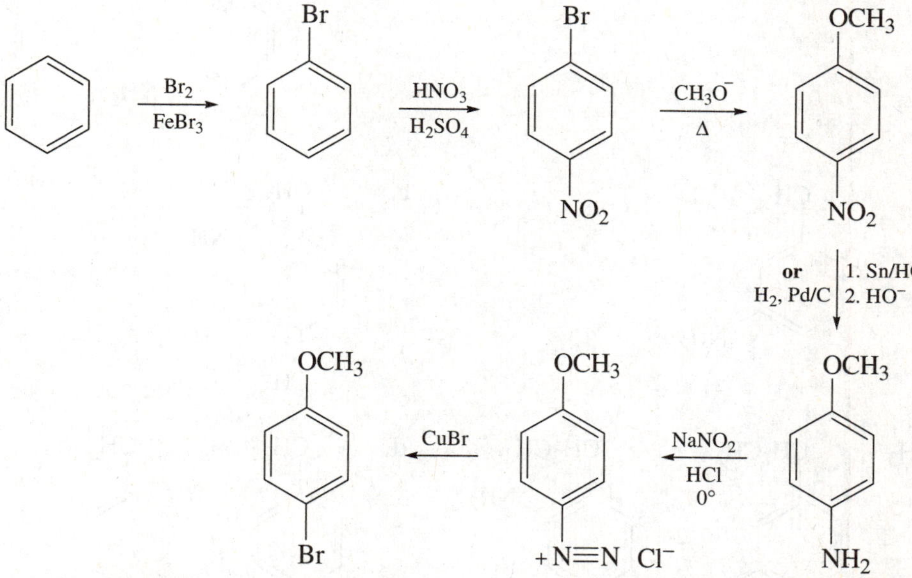

The target compound can also be prepared by the method shown in Problem 66a, omitting the last step.

d. This synthesis would be carried out exactly as "**c**" (above) except for the last step, which would be:

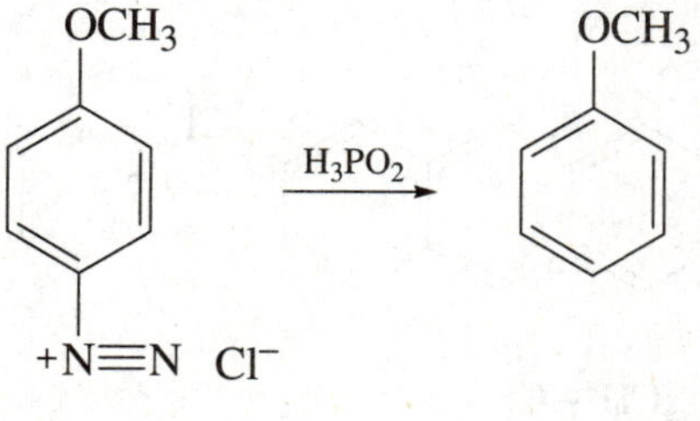

32.

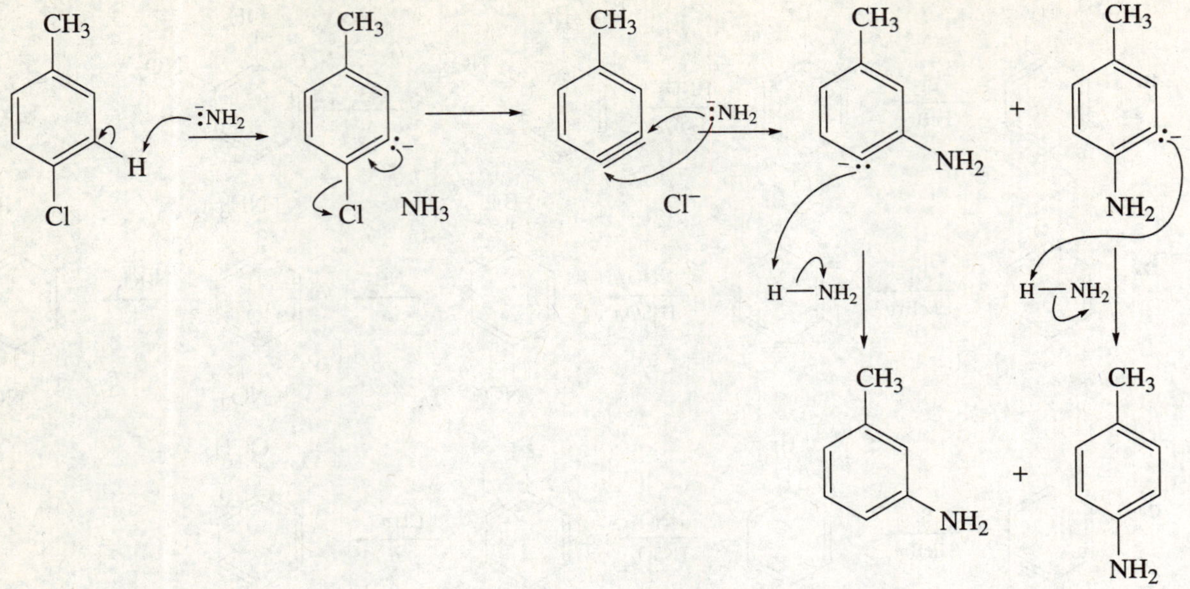

33. **a.**

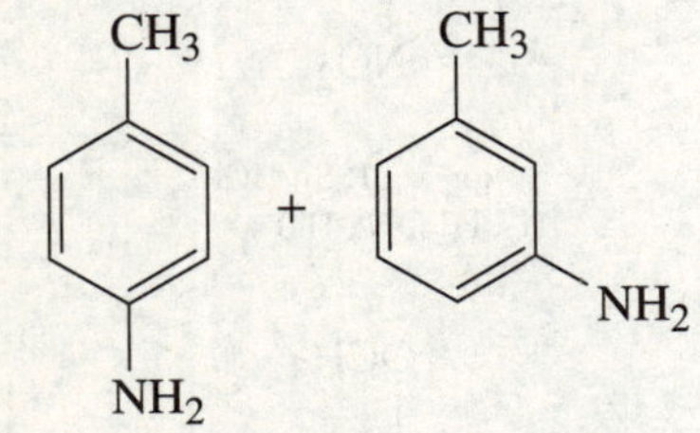

c.

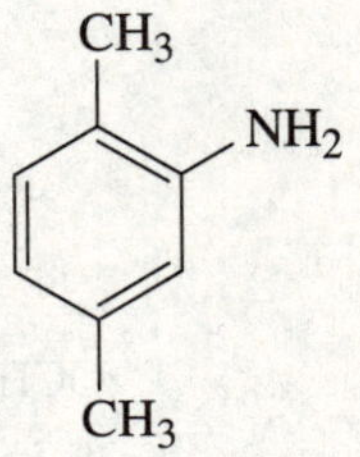

b.

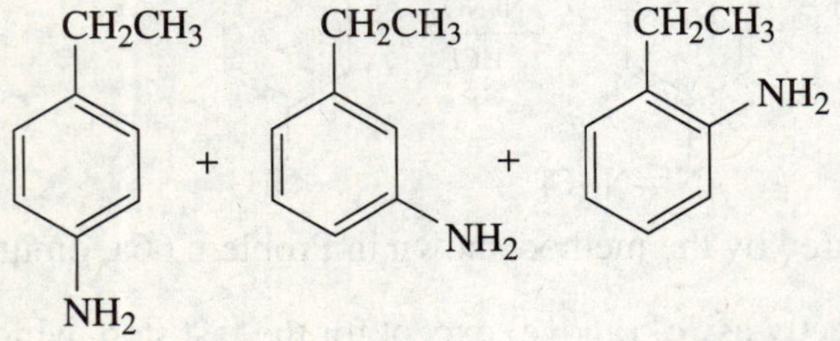

d.

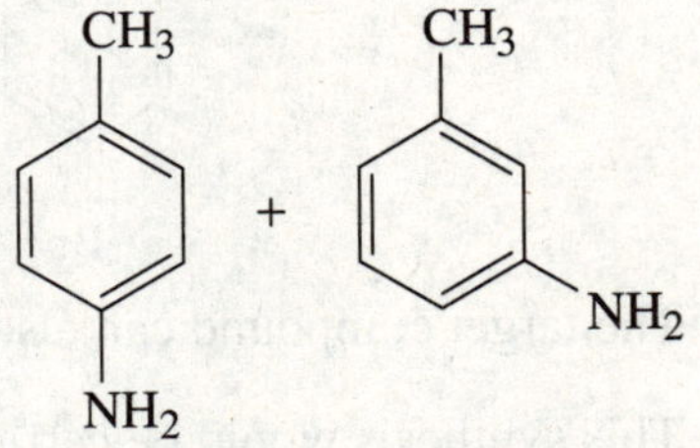

34. **a.**

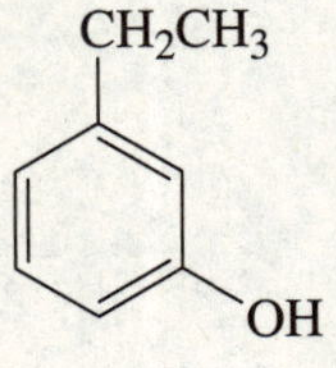

c.

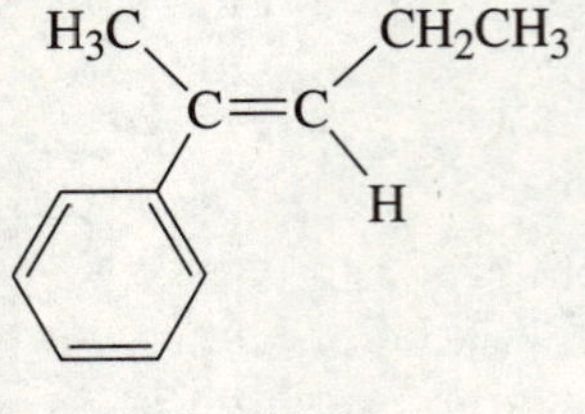

e.

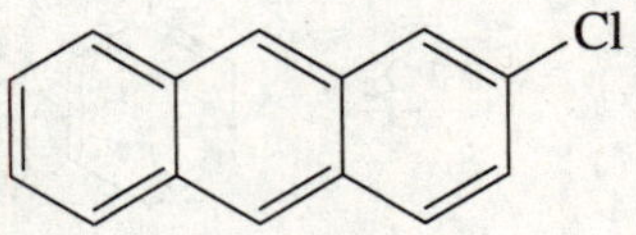

g.

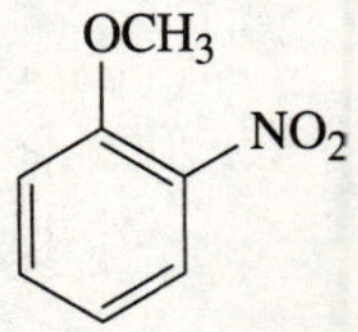

b.

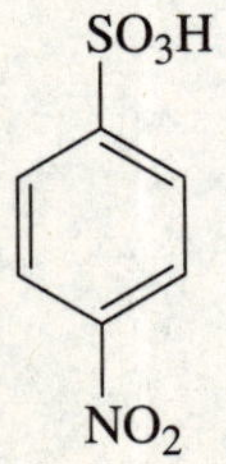

d.

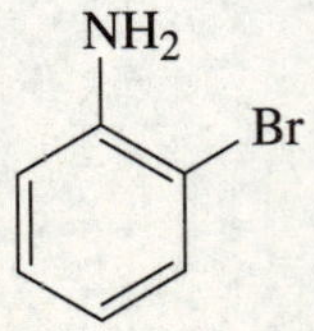

f.

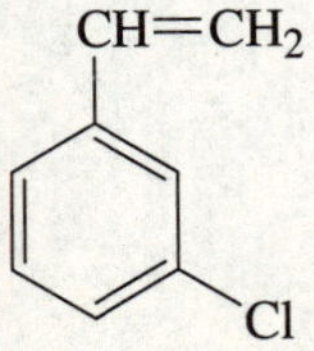

h.

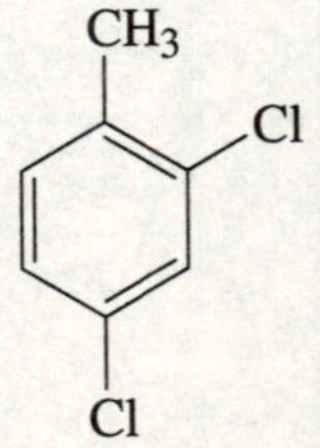

35.

a. *m*-bromobenzoic acid
b. 1,2,4-tribromobenzene
c. 2,6-dimethylphenol
f. 3,5-dichlorobenzenesulfonic acid
g. *o*-bromotoluene
h. *p*-cyclohexyltoluene

d. *p*-nitrostyrene
e. *m*-ethylanisole

i. 2-chloro-4-ethylazobenzene

36.

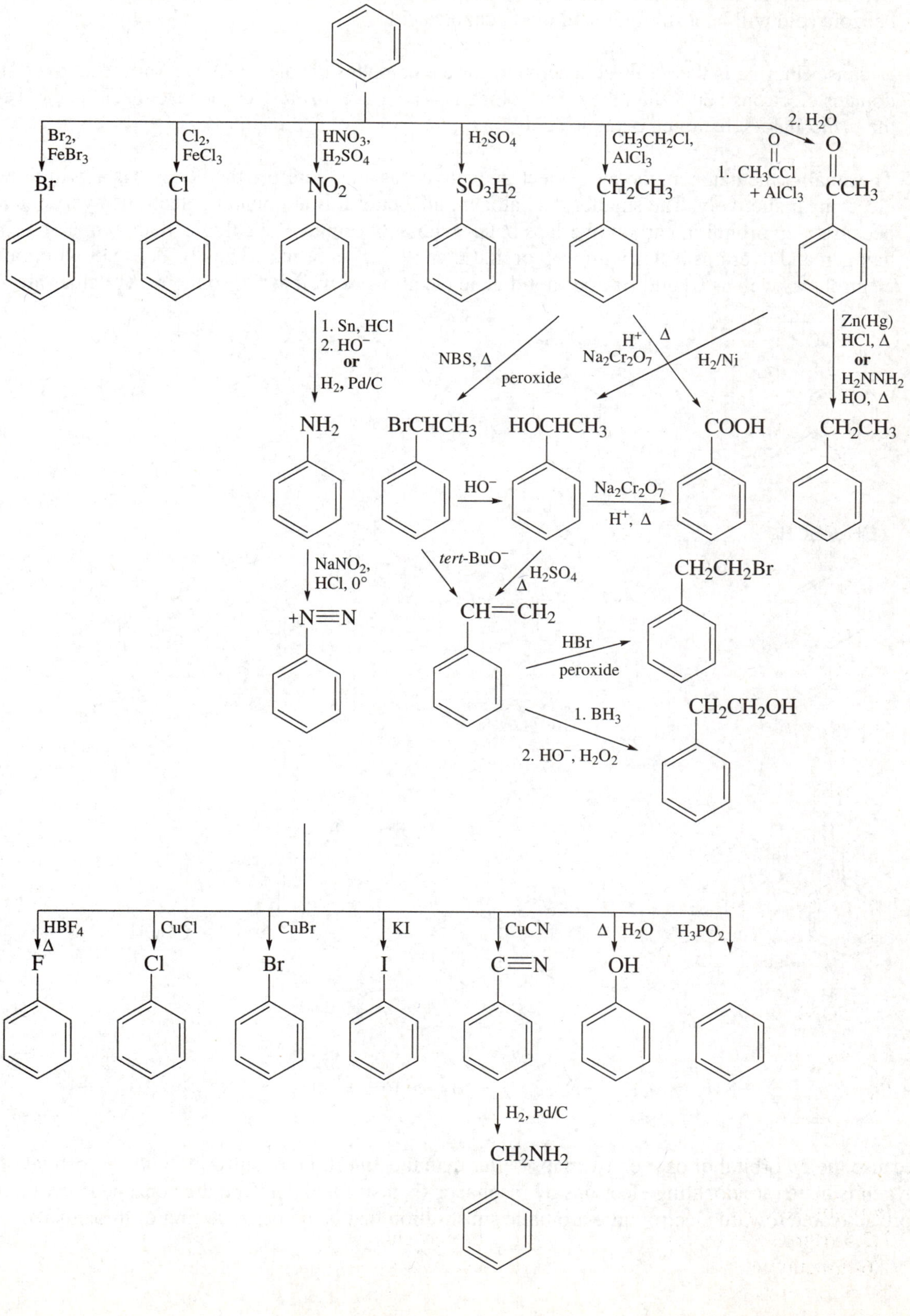

37. **a.** The halogens withdraw electrons inductively and donate electrons by resonance. Because they all deactivate the benzene ring toward electrophilic aromatic substitution, we know that their electron withdrawing effect is stronger than their electron donating effect. Therefore, an *ortho*-halo-substituted benzoic acid will be a stronger acid than benzoic acid.

b. Because fluorine is the weakest deactivator of the halogens (Table 15.1), we know that overall it donates electrons better than the other halogens. Therefore, *ortho*-fluorobenzoic acid is the weakest of the *ortho*-halo-substituted benzoic acids.

c. The smaller the halogen, the more electronegative it is and therefore the better it is at withdrawing electrons inductively. The smaller the halogen, the better it is at donating electrons by resonance because a 2*p* orbital of carbon overlaps better with a 2*p* orbital of a halogen than with a 3*p* orbital of a halogen and overlaps better with a 3*p* orbital than with a 4*p* orbital. Thus Br does not withdraw electrons as well as Cl and Br does not donate electrons as well as Cl, so their pK_a values are similar.

38. **a.**

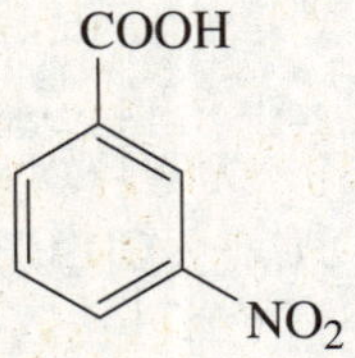

d.

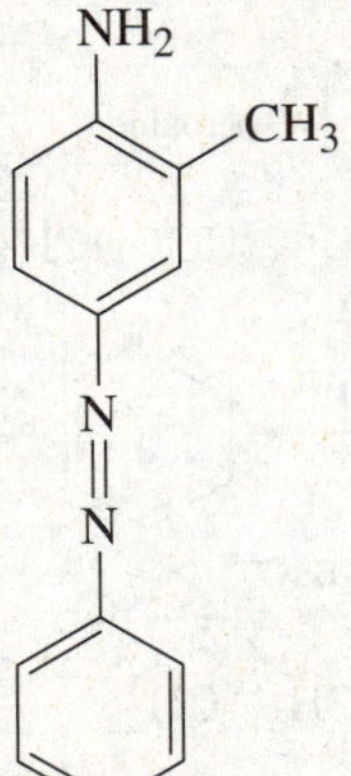

f.

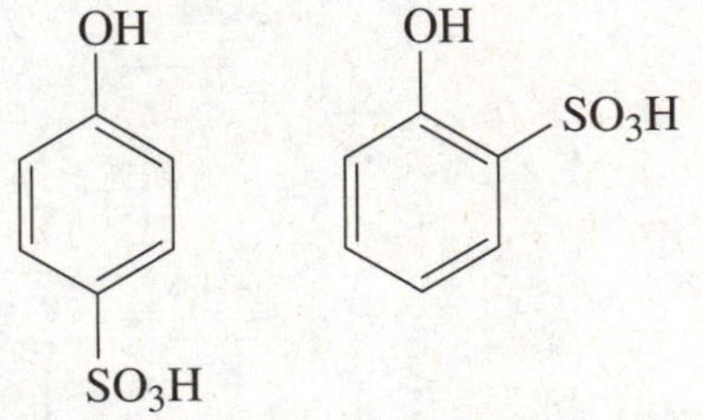

b.

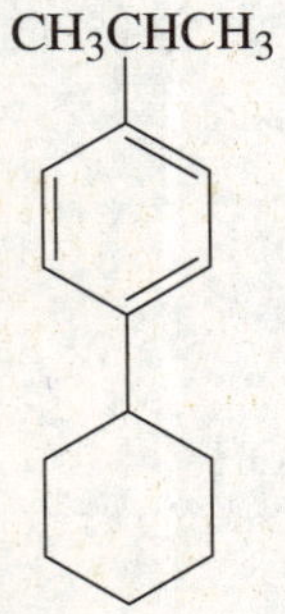

g.

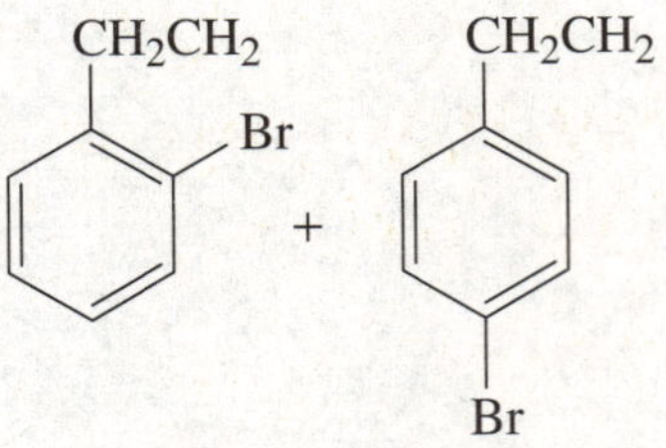

e.

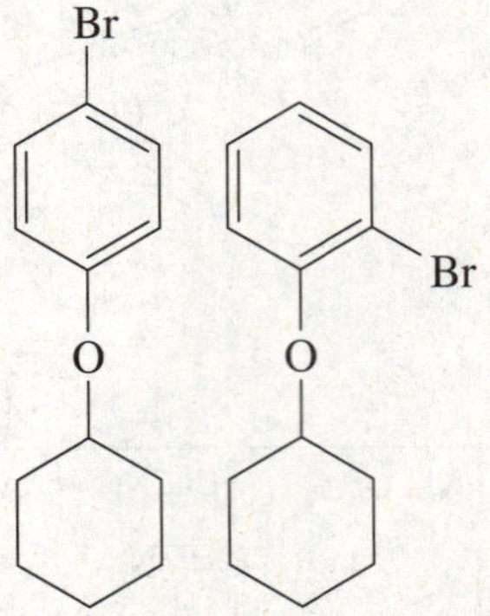

h.

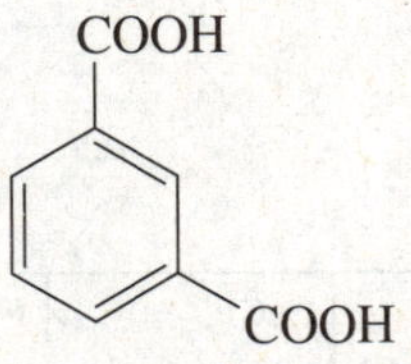

c.

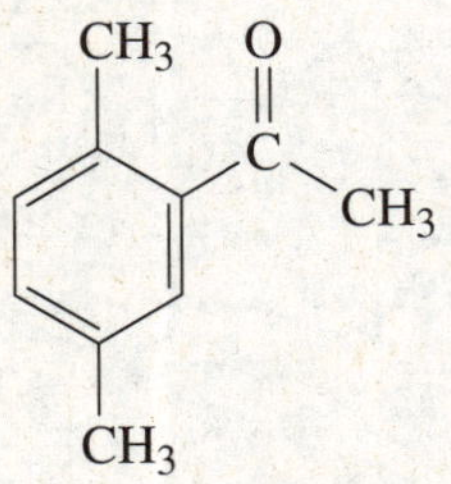

39.

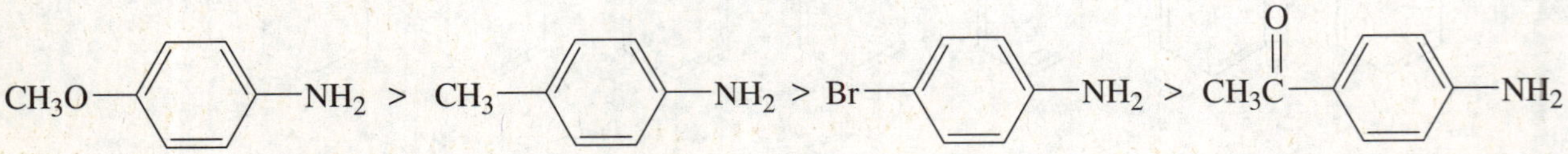

40. Because the 2*p* orbital of oxygen overlaps better than the 3*p* orbital of sulfur with the 2*p* orbital of carbon, oxygen is better are donating electrons by resonance than sulfur. Therefore the benzene ring of anisole is more activated toward electrophilic aromatic substitution that is the benzene ring of thioanisole.

41.

CH3O
OCH3

42. **a.**

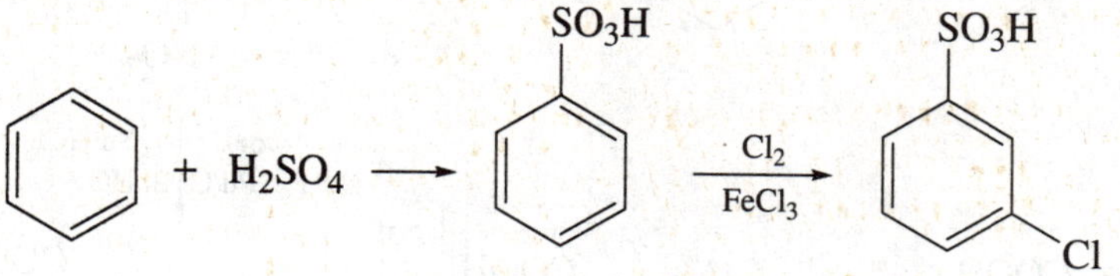
SO3H
+ H2SO4
Cl2
FeCl3
SO3H
Cl

b.

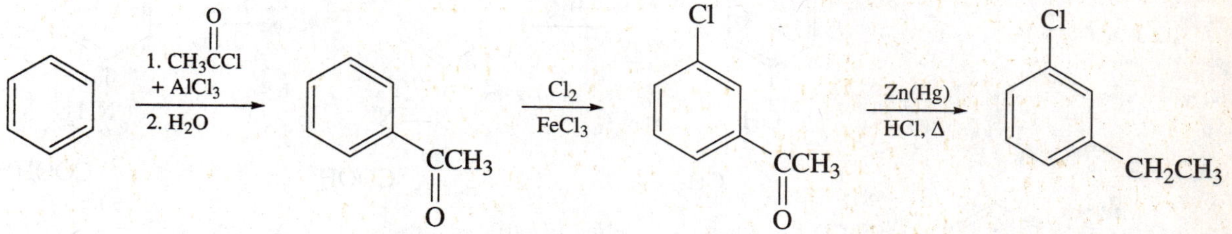
O
1. CH3CCl
+ AlCl3
2. H2O
CCH3
O
Cl2
FeCl3
Cl
CCH3
O
Zn(Hg)
HCl, Δ
Cl
CH2CH3

c.

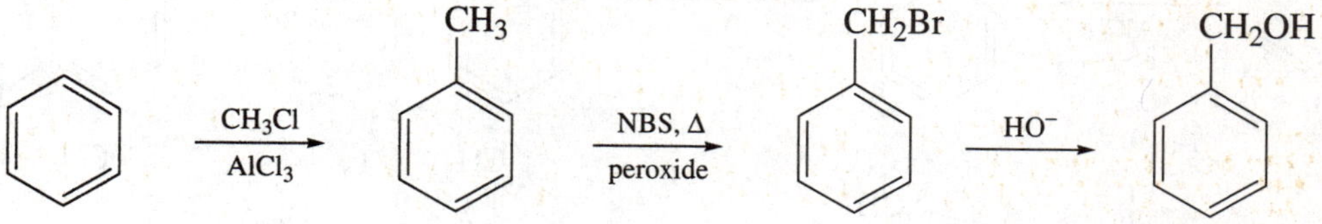
CH3Cl
AlCl3
CH3
NBS, Δ
peroxide
CH2Br
HO⁻
CH2OH

d.

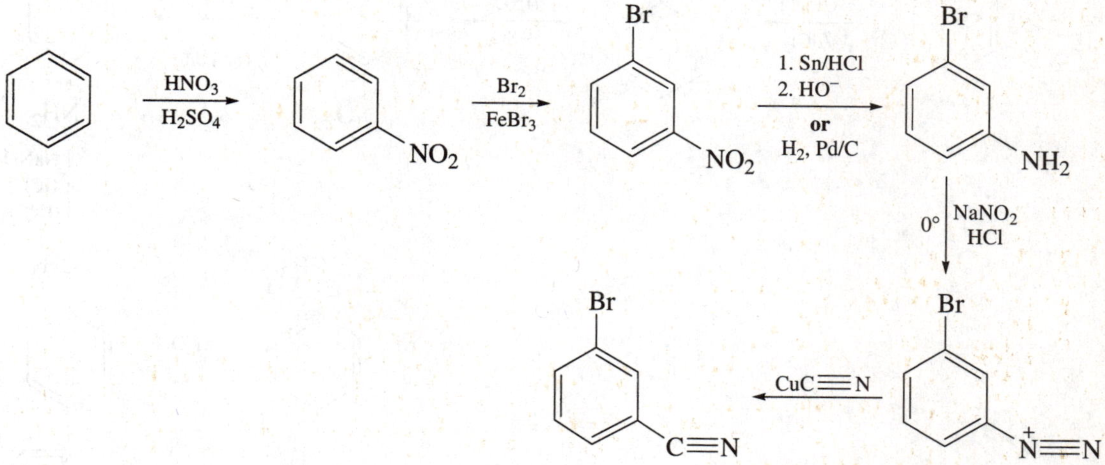
HNO3
H2SO4
NO2
Br2
FeBr3
Br
NO2
1. Sn/HCl
2. HO⁻
or
H2, Pd/C
Br
NH2
0°
NaNO2
HCl
Br
N≡N
CuC≡N
Br
C≡N

e.

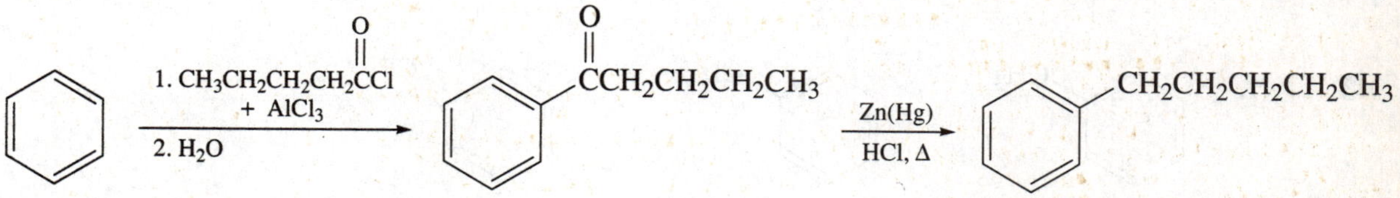
O
1. CH3CH2CH2CH2CCl
+ AlCl3
2. H2O
O
CCH2CH2CH2CH3
Zn(Hg)
HCl, Δ
CH2CH2CH2CH2CH3

f.

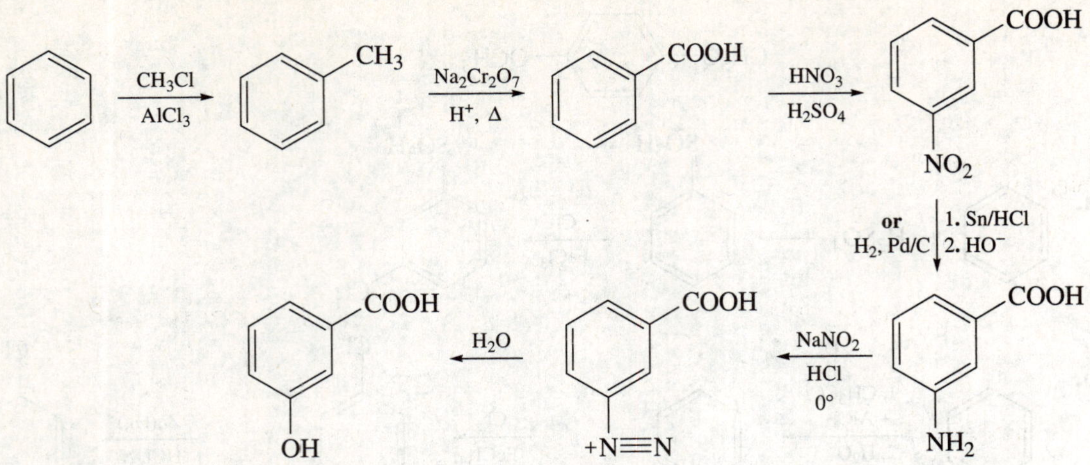

g.

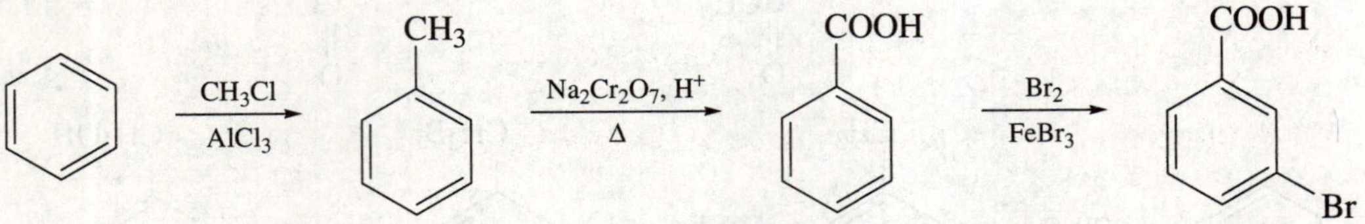

h.

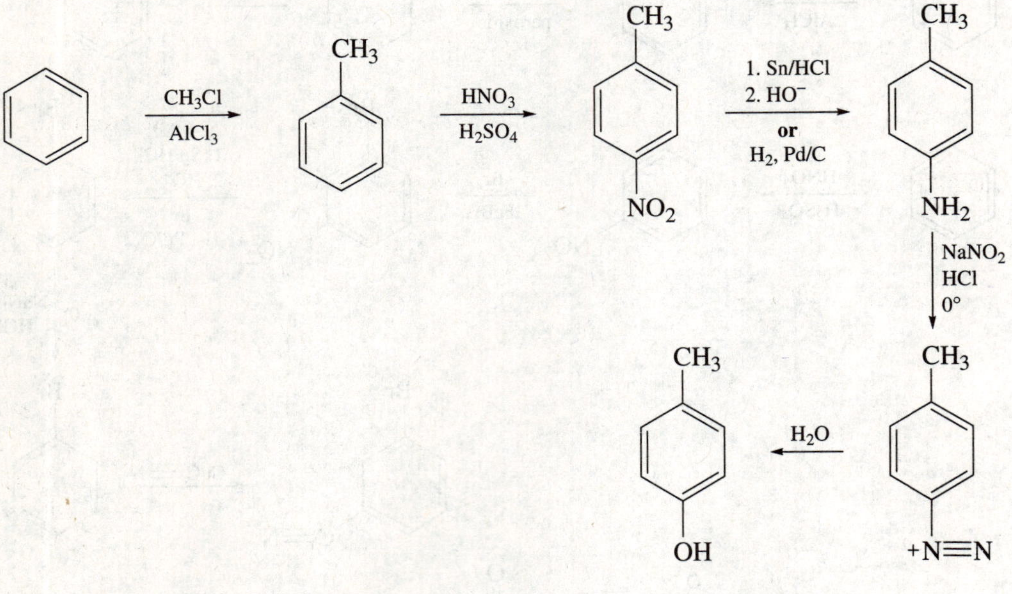

43. 1.

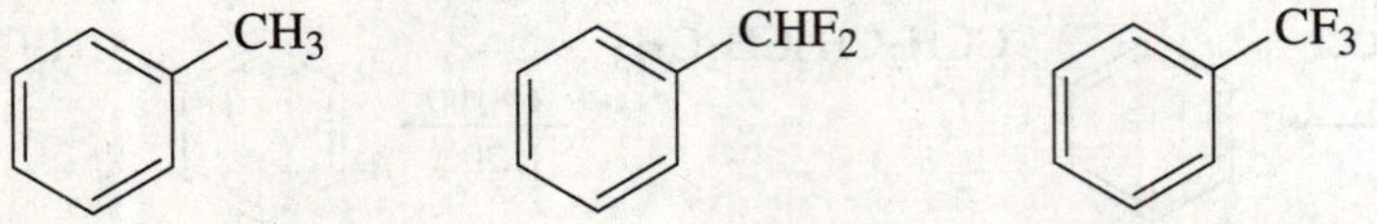

most reactive

least reactive
highest % meta product

2. 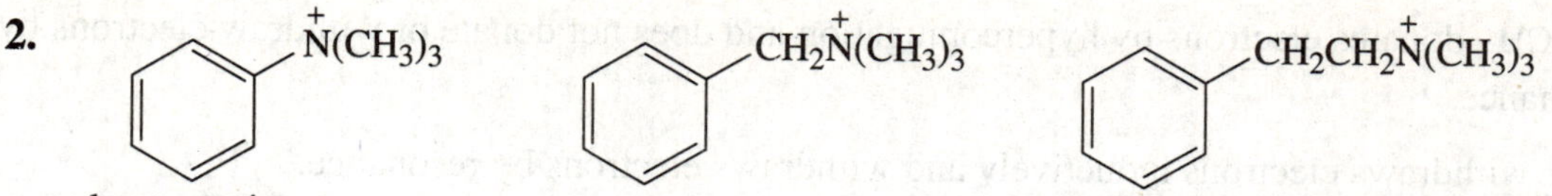

least reactive
highest % meta product

most reactive

3. 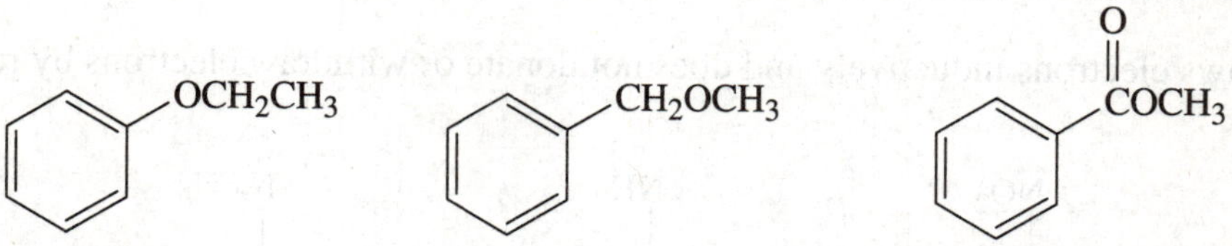

most reactive

least reactive
highest % meta product

44. a. anisole > ethylbenzene > benzene > chlorobenzene > nitrobenzene

b. 2,4-dinitrophenol > 2,4-dinitrotoluene > 1-chloro-2,4-dinitrobenzene

c. *p*-cresol > *p*-xylene > toluene > benzene

d. phenol > propylbenzene > benzene > benzoic acid

e. *p*-chlorotoluene > *p*-nitrotoluene > 2-chloro-4-nitrotoluene > 2,4-dinitrotoluene

f. fluorobenzene > chlorobenzene > bromobenzene > iodobenzene

45. a. 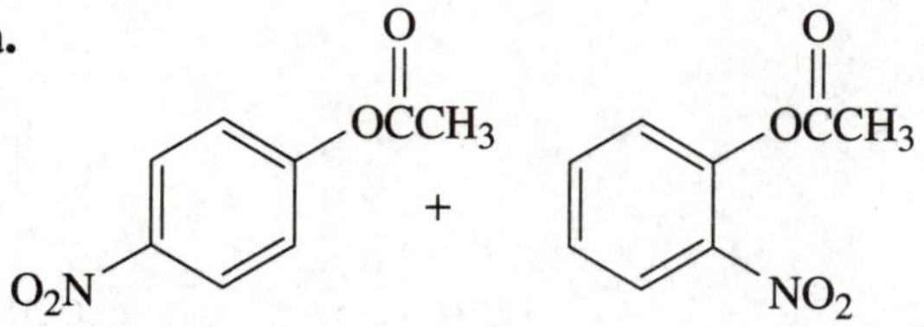

c. 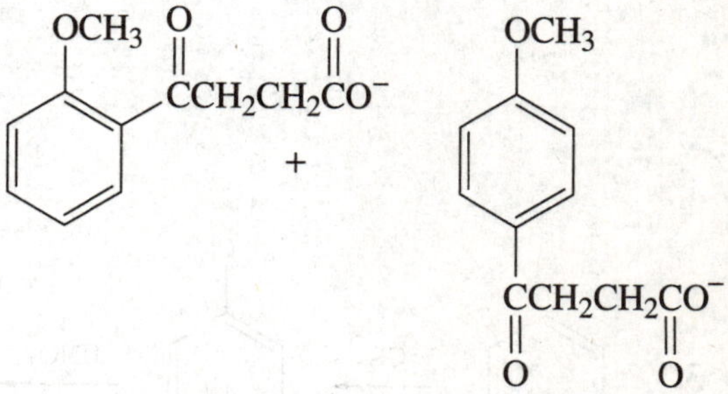

b. 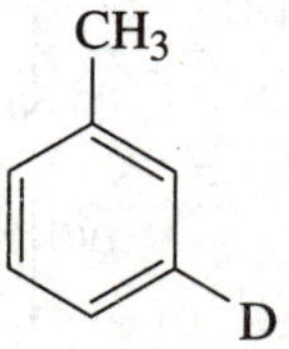

d. 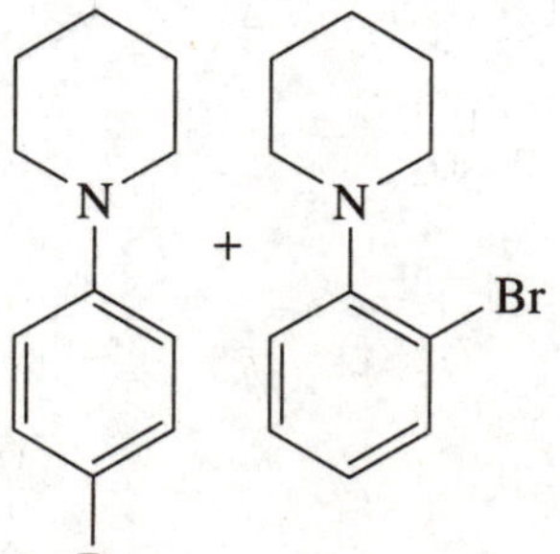

e. 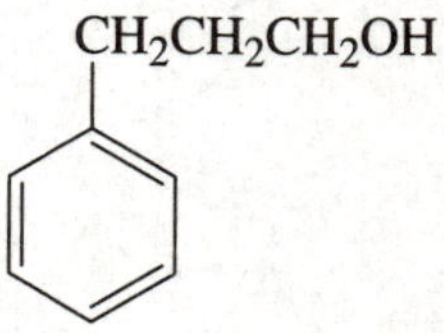

f. 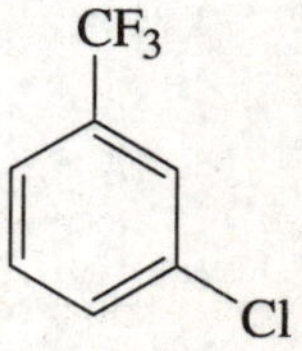

46. **a.** CH_2CH_3 donates electrons by hyperconjugation and does not donate or withdraw electrons by resonance.

b. NO_2 withdraws electrons inductively and withdraws electrons by resonance.

c. Br deactivates the ring and directs ortho/para.

d. OH withdraws electrons inductively, donates electrons by resonance, and activates the ring.

e. $^+NH_3$ withdraws electrons inductively and does not donate or withdraw electrons by resonance.

47. **1.**

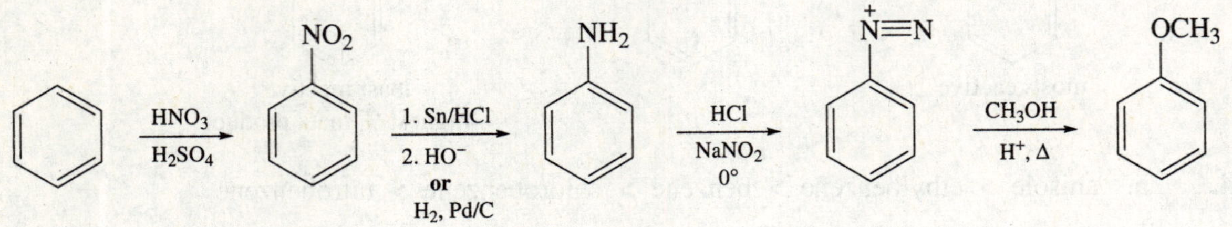

2.

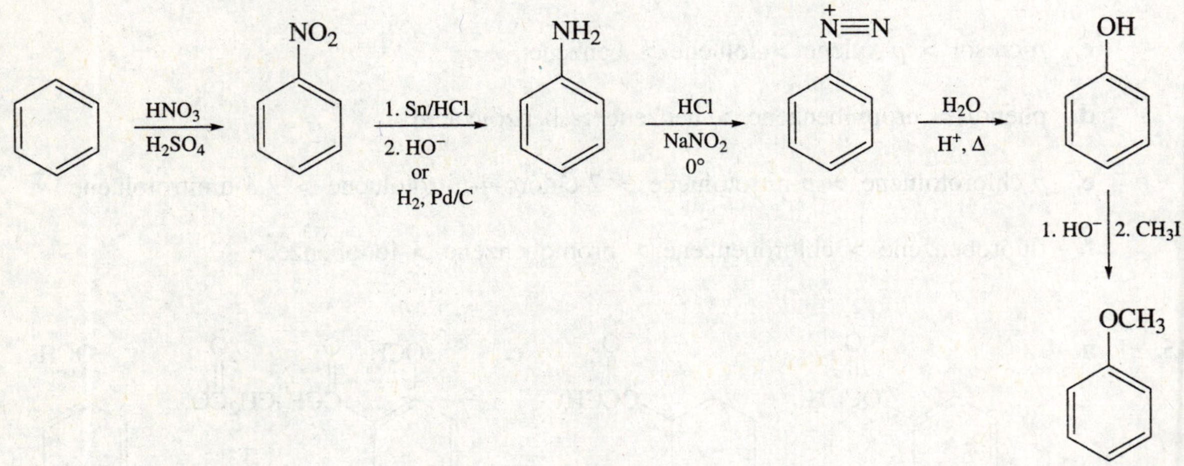

3.

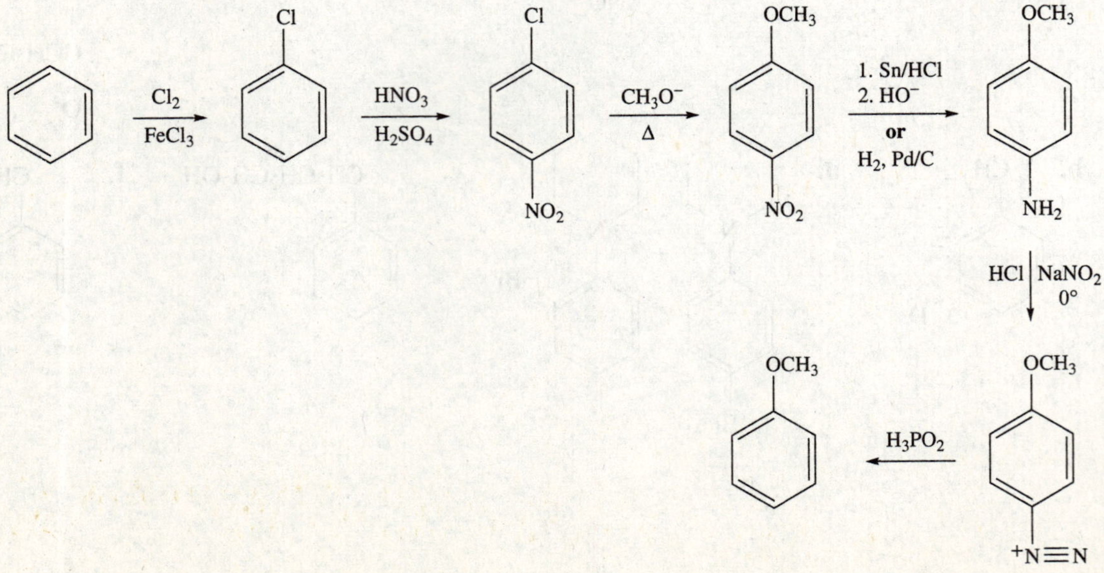

48.

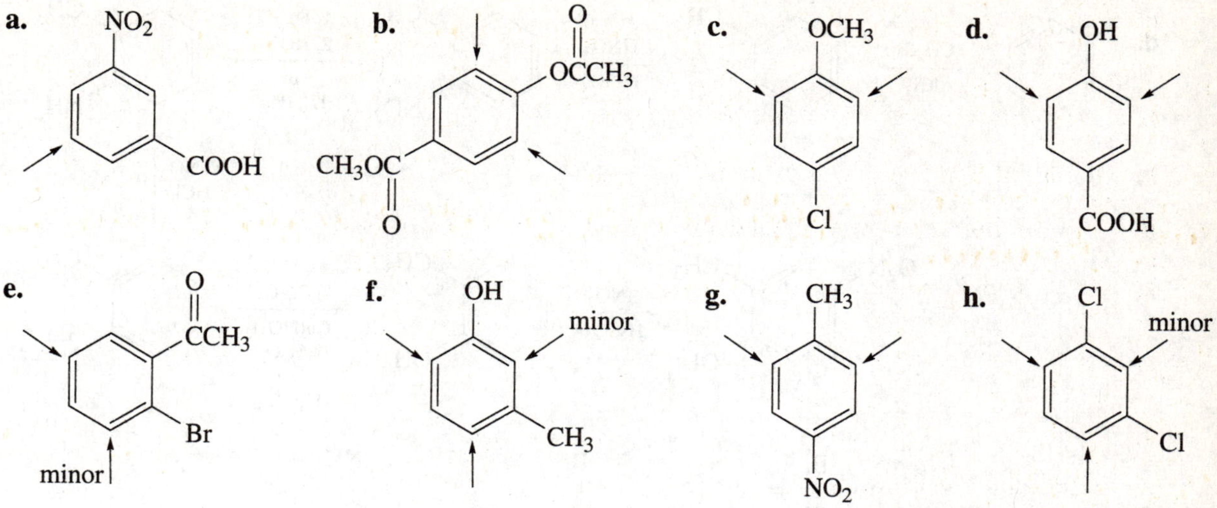
a.
NO2
COOH
b.
O
OCCH3
CH3OC
O
c.
OCH3
Cl
d.
OH
COOH
e.
O
CCH3
Br
minor
f.
OH
minor
CH3
g.
CH3
NO2
h.
Cl
minor
Cl

49.

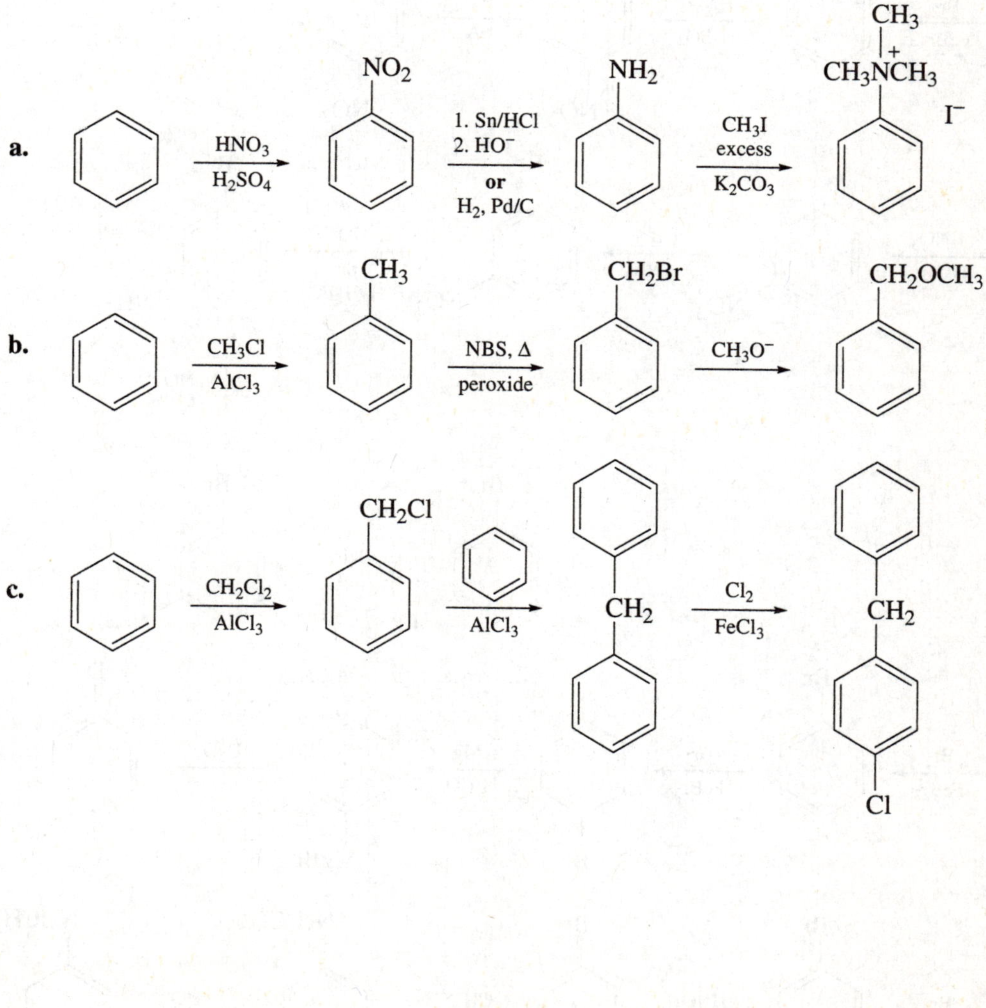
a.
HNO3
H2SO4
NO2
1. Sn/HCl
2. HO−
or
H2, Pd/C
NH2
CH3I
excess
K2CO3
CH3
CH3NCH3
I−
b.
CH3Cl
AlCl3
CH3
NBS, Δ
peroxide
CH2Br
CH3O−
CH2OCH3
c.
CH2Cl2
AlCl3
CH2Cl
AlCl3
CH2
Cl2
FeCl3
CH2
Cl

d.

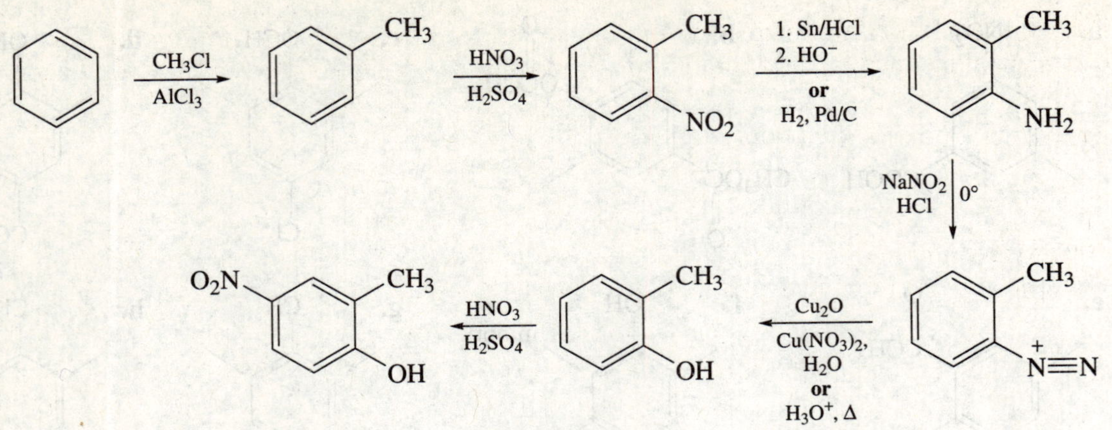
CH3Cl
AlCl3
CH3
HNO3
H2SO4
CH3
NO2
1. Sn/HCl
2. HO−
or
H2, Pd/C
CH3
NH2
NaNO2
HCl
0°
CH3
+
N≡N
Cu2O
Cu(NO3)2,
H2O
or
H3O+, Δ
CH3
OH
HNO3
H2SO4
O2N
CH3
OH

e.

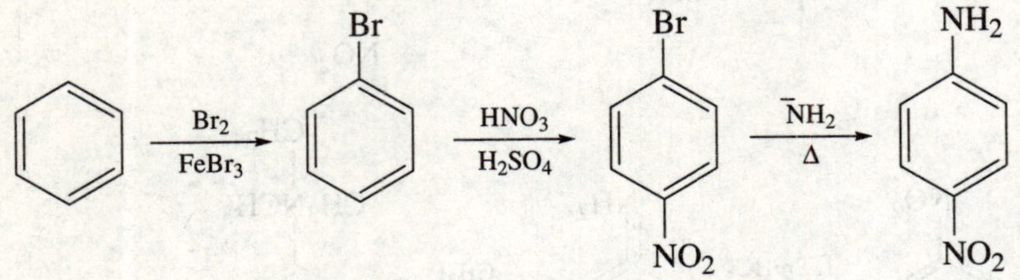
Br2
FeBr3
Br
HNO3
H2SO4
Br
NO2
−NH2
Δ
NH2
NO2

f.

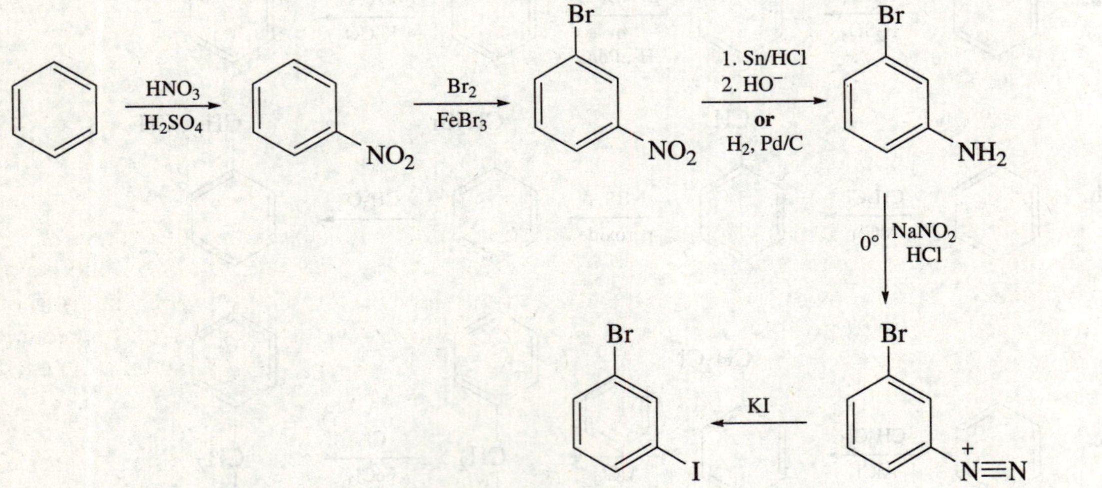
HNO3
H2SO4
NO2
Br2
FeBr3
Br
NO2
1. Sn/HCl
2. HO−
or
H2, Pd/C
Br
NH2
NaNO2
HCl
0°
Br
+
N≡N
KI
Br
I

g.

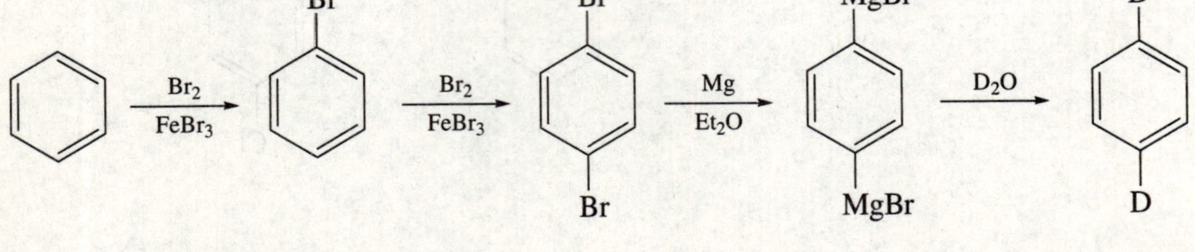
Br2
FeBr3
Br
Br2
FeBr3
Br
Br
Mg
Et2O
MgBr
MgBr
D2O
D
D

h.

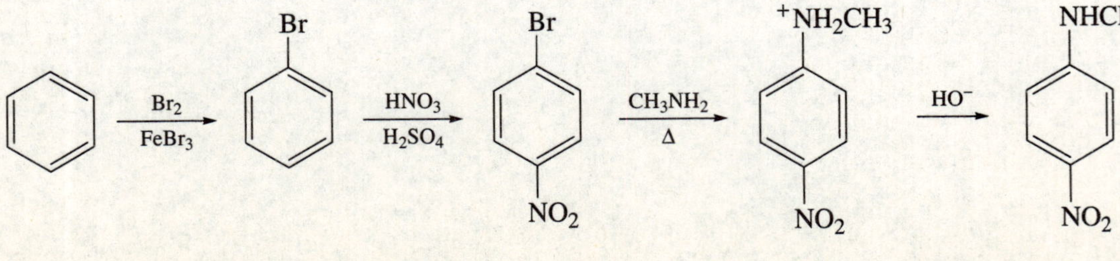
Br2
FeBr3
Br
HNO3
H2SO4
Br
NO2
CH3NH2
Δ
+NH2CH3
NO2
HO−
NHCH3
NO2

50. The compound with the methoxy substituent is the more reactive because it forms the more stable carbocation intermediate. The carbocation intermediate is stabilized by resonance electron donation.

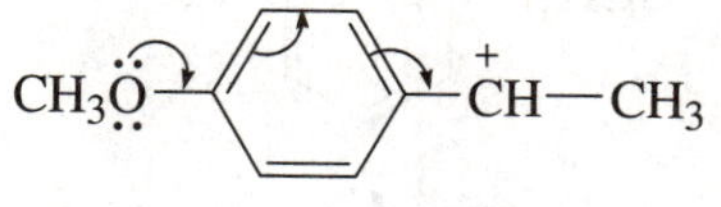

51. *Meta*-xylene will react more rapidly. In *meta*-xylene both methyl groups activate the same position, whereas in *para*-xylene each methyl group activates a different position.

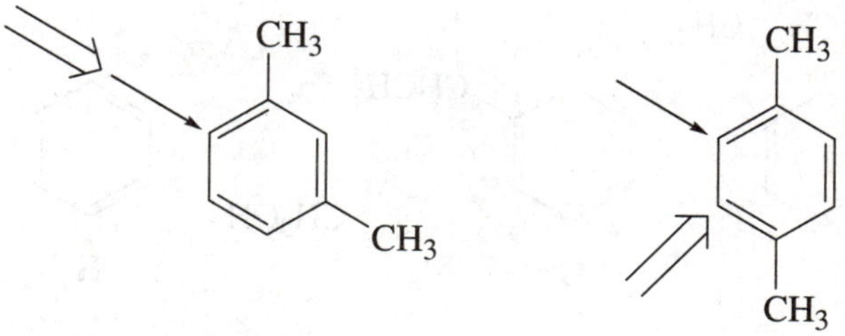

52. **a.**

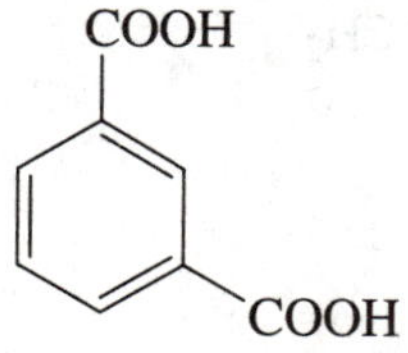

b.

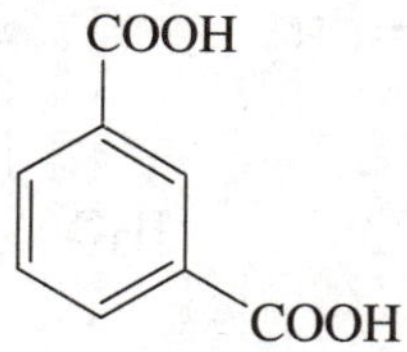

c.

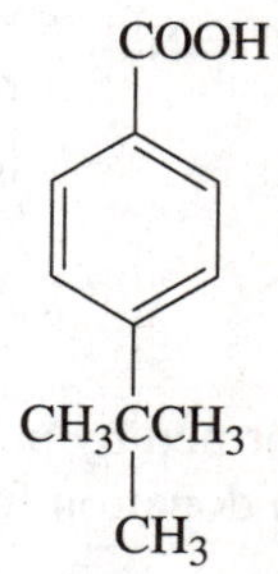

53. Yes, the advice is sound. The para isomer will form one product, because the formyl and ethyl groups both direct to the same positions and both positions result in the same product. The ortho isomer will form two products, because the formyl and ethyl groups both direct to the same positions but different products are obtained from each position. The meta isomer will form as many as four products, because the formyl and ethyl groups direct to four different positions and a different product is obtained from each position.

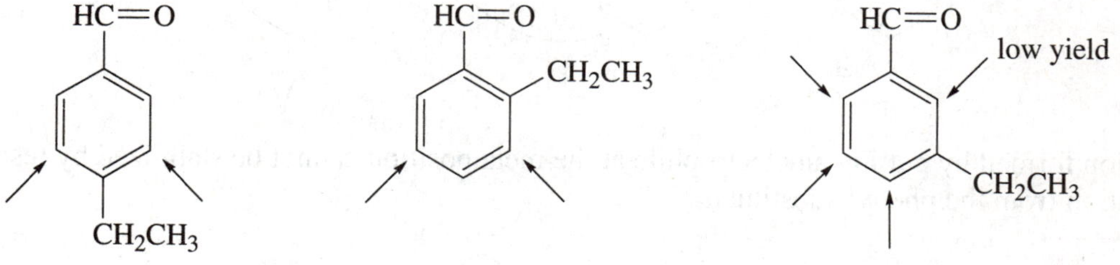

54.

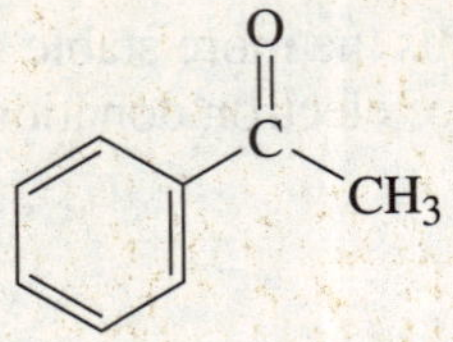

A

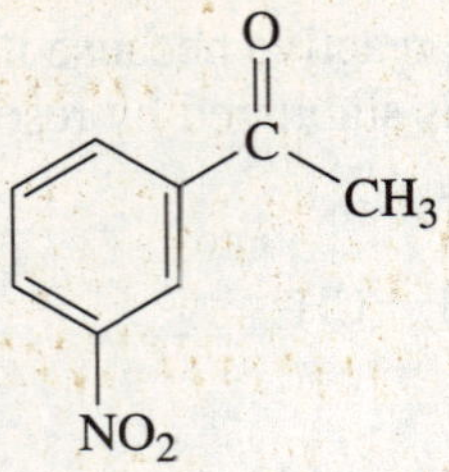

B

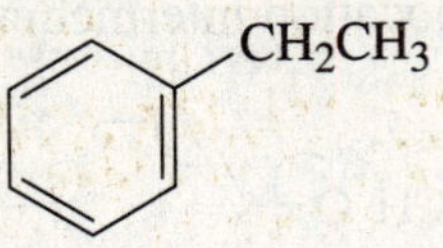

C

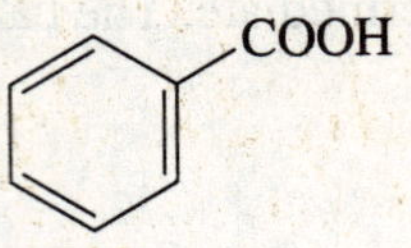

D

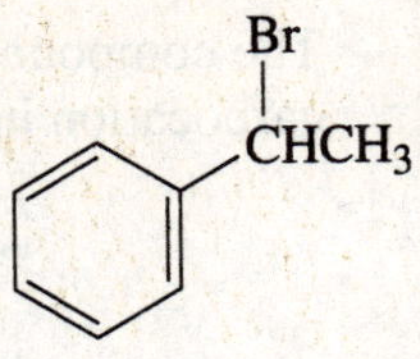

E

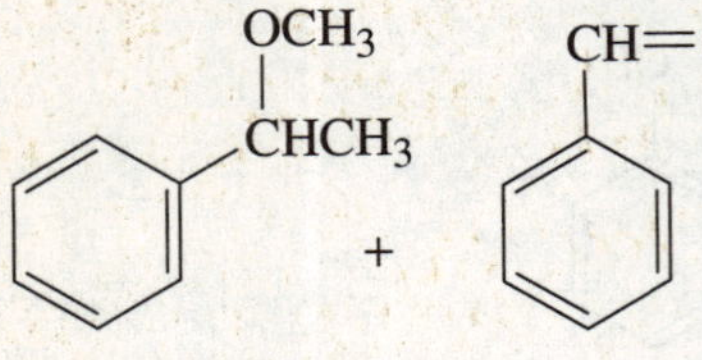

F

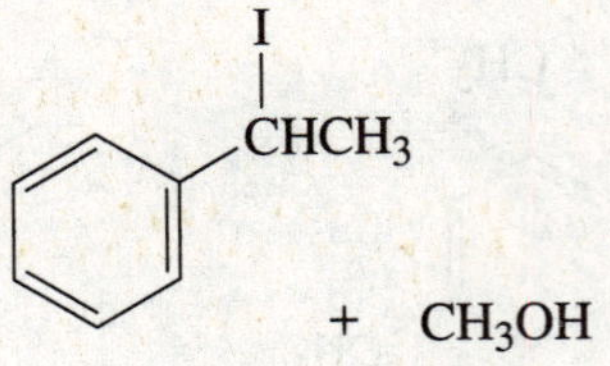

G

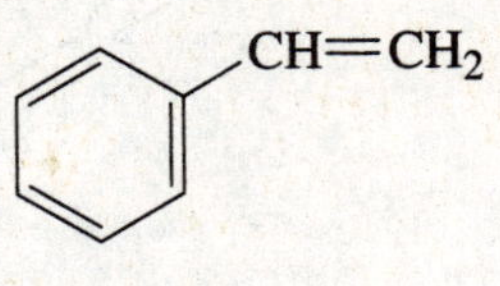

H

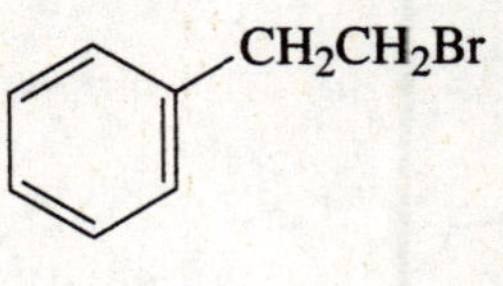

I

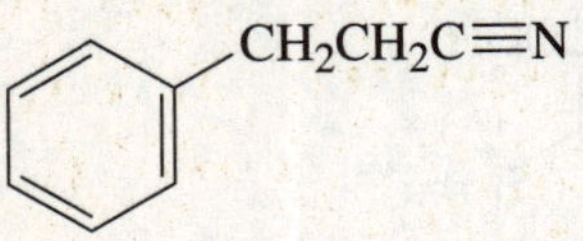

J

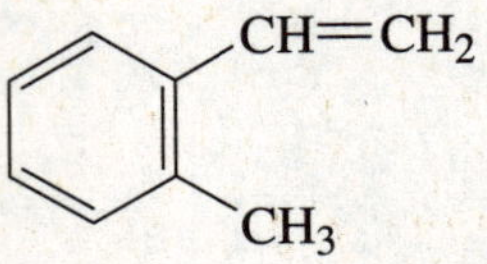

K

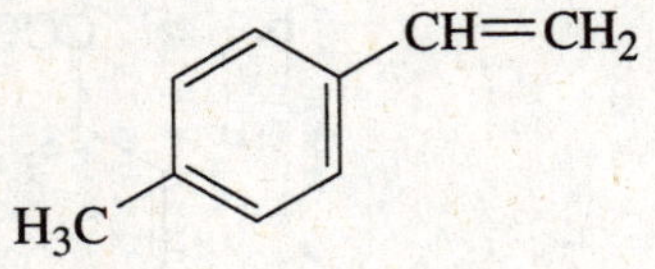

L

55. The carbocation formed by putting an electrophile at the ortho or para positions can be stabilized by resonance electron donation from the phenyl substituent.

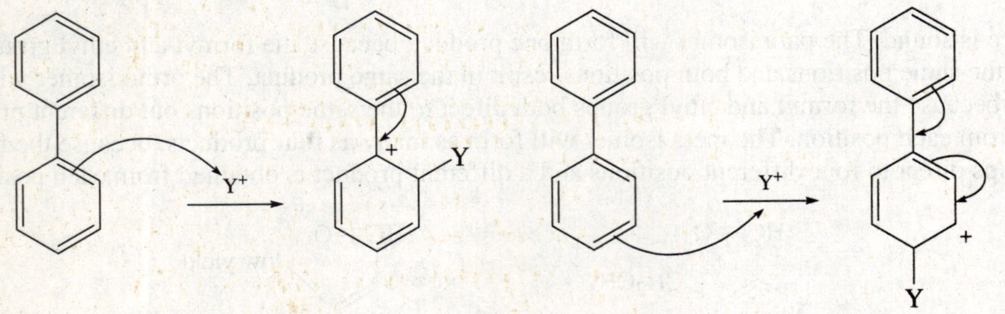

The carbocation formed by putting an electrophile at the meta position cannot be stabilized by resonance electron donation from the phenyl substituent.

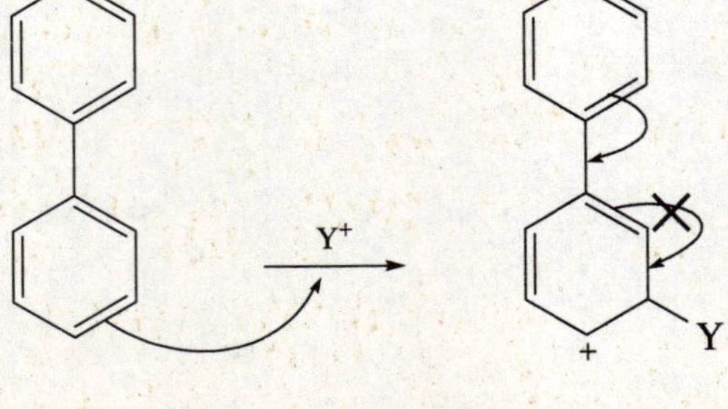

56. The chloro substituent primarily withdraws electrons inductively. (It only minimally donates electrons by resonance.) The closer it is to the COOH group, the more it can withdraw electrons from the OH bond and the stronger the acid.

The nitro substituent withdraws electrons inductively. It also withdraws electrons by resonance if it is ortho or para to the COOH group. Therefore, the ortho and para isomers are the strongest acids, and the ortho isomer is a stronger acid than the para isomer because of the greater inductive electron withdrawal from the closer position.

The amino substituent primarily donates electrons by resonance, but it can donate electrons by resonance to the COOH group only if it is ortho or para to it. From the pK_a values it is apparent that resonance electron donation to the COOH group is more efficient from the ortho position.

57. The spectrum indicates that the benzene ring has a substituent with two different kinds of hydrogens. The doublet and multiplet indicate that the substituent is an isopropyl group. Therefore, Compound A is isopropylbenzene.

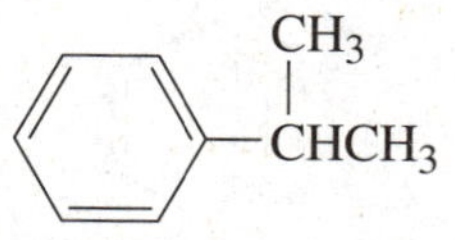

isopropylbenzene

58. 1.

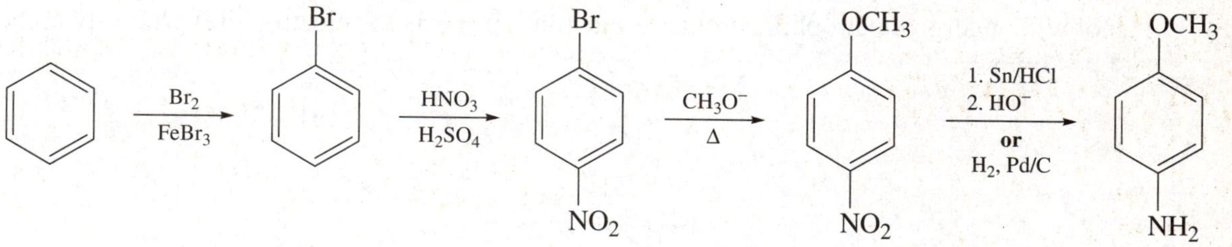

2.

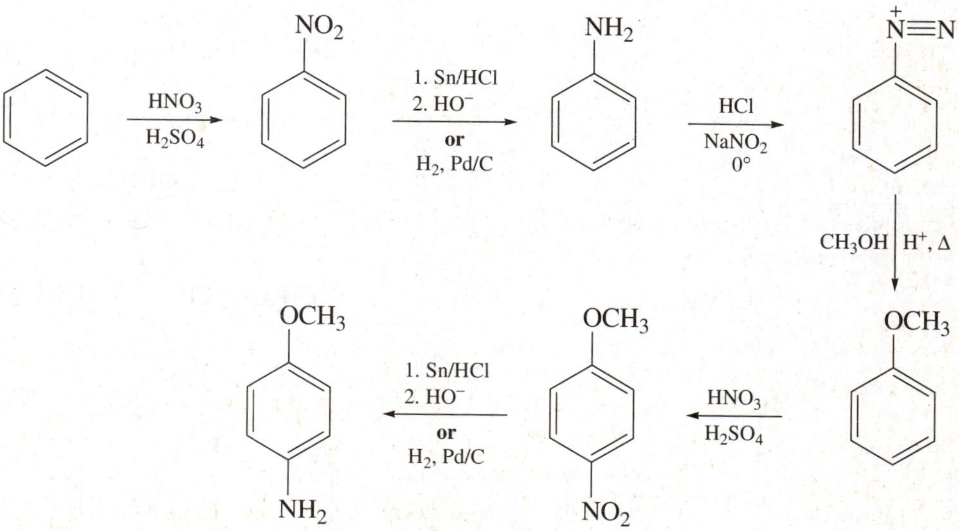

59. **a.** The hydroxy-substituted carbocation intermediate is more stable because the positive charge can be stabilized by resonance electron donation from the OH group.

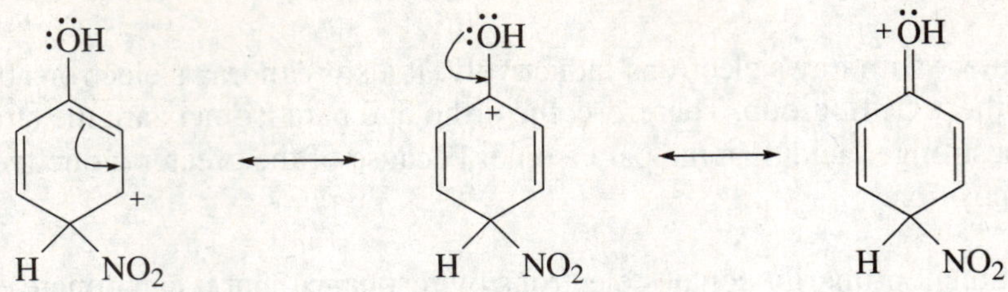

b. The carbanion with the negative charge meta to the nitro group is more stable because a negative charge in the meta position can be delocalized onto the nitro group but a negative charge in the ortho position cannot.

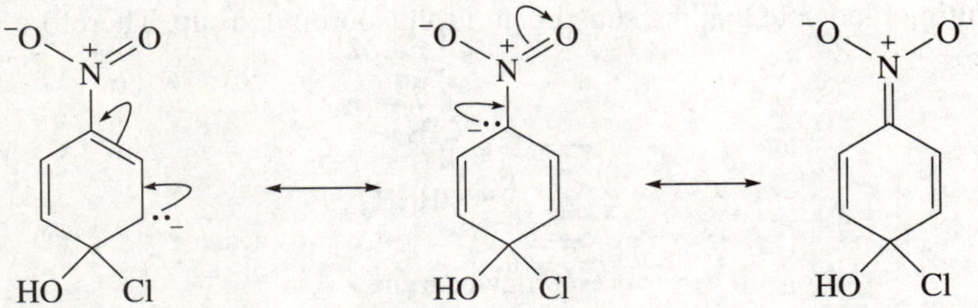

60. Phenol will undergo electrophilic aromatic substitution (D+ is the electrophile) primarily at the ortho and para positions.

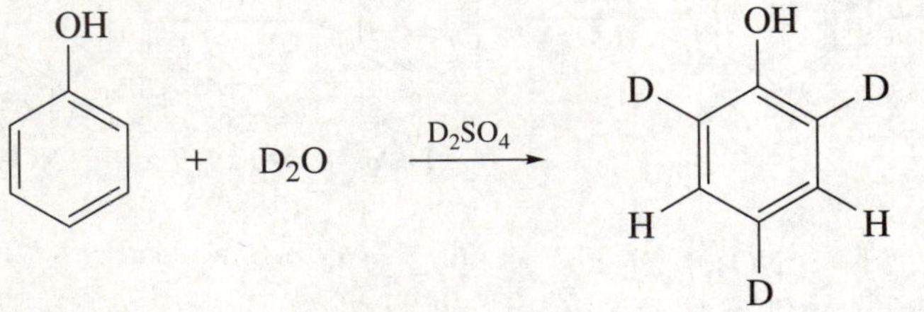

61. **a.**

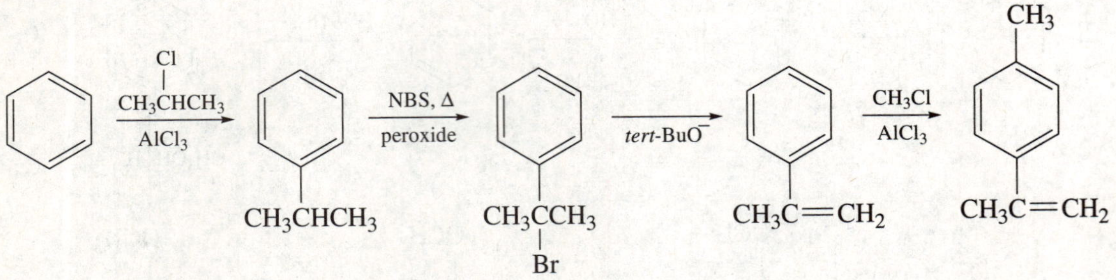

b.

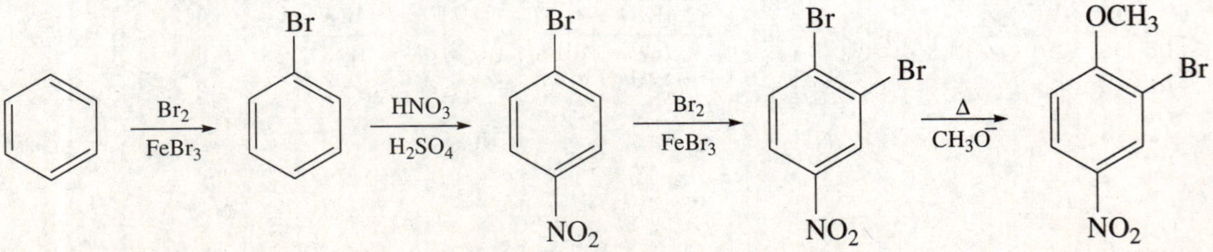

c.

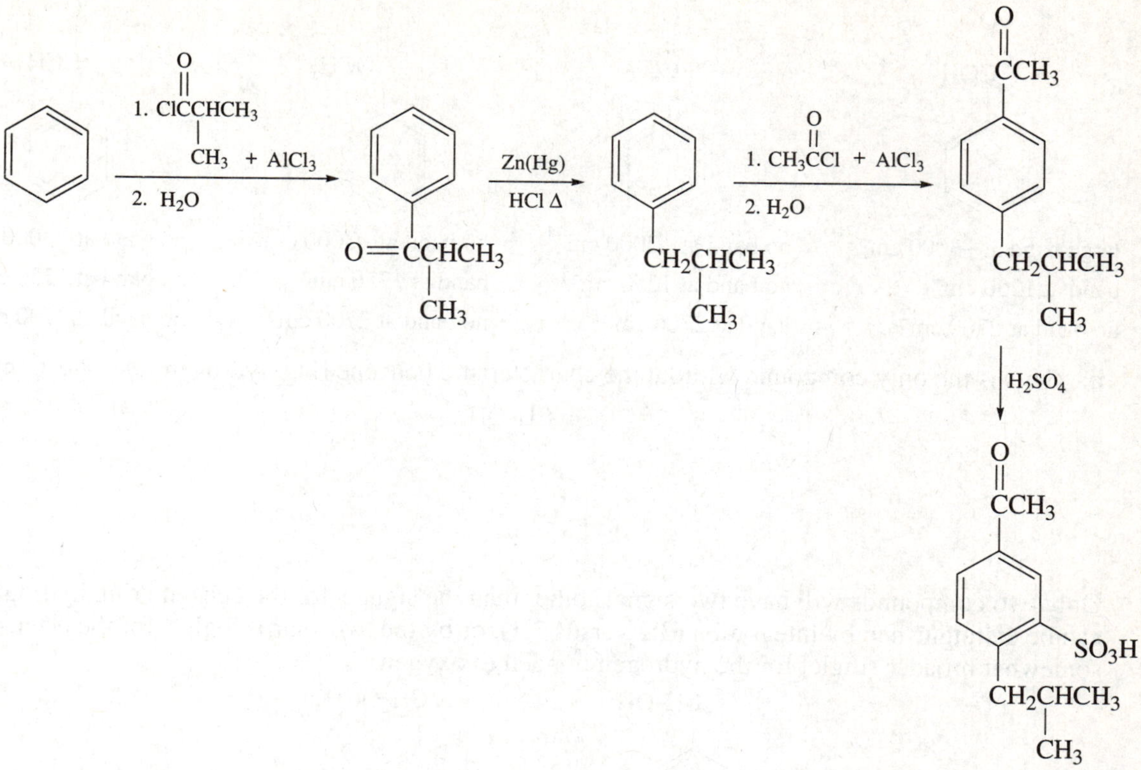

62. The unsplit signal at ~ 7.1 ppm suggests that all the hydrogens of the benzene ring in the product are chemically equivalent.

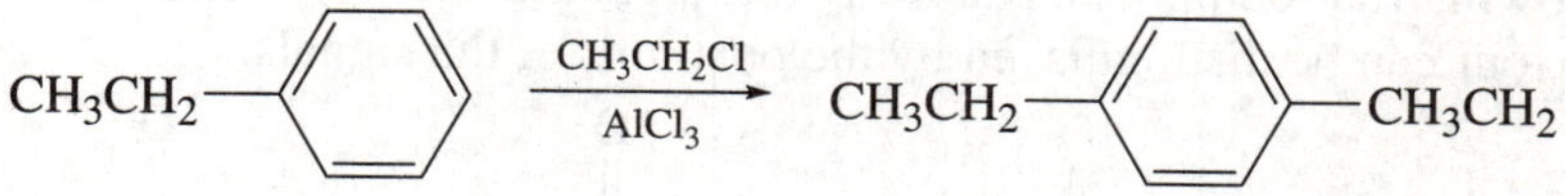

63. **a.** The first three compounds will not show a carbonyl stretch at 1700 cm^{-1} and the bottom four will show this absorption band. The first three can be distinguished from one another by the presence or absence of the indicated absorption bands.

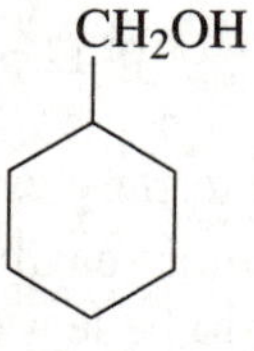

band at 3300 cm^{-1}
no band at 1600 cm^{-1}

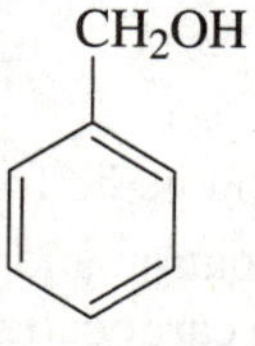

band at 3300 cm^{-1}
band at 1600 cm^{-1}

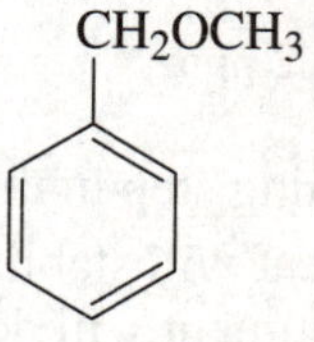

no band at 3300 cm^{-1}
band at 1600 cm^{-1}

The last four compounds all have an absorption band at 1700 cm^{-1}.
They can be distinguished by the presence or absence of the indicated absorption bands.

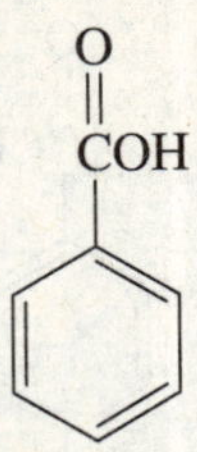

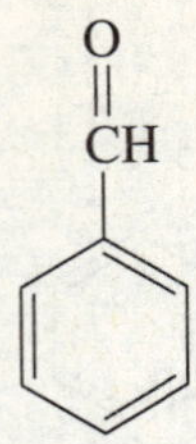

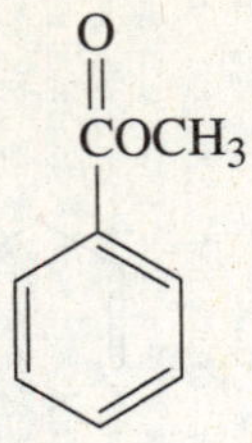

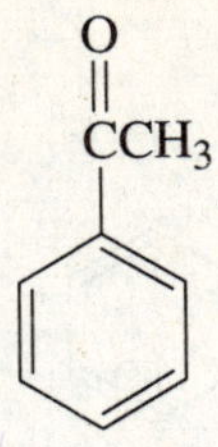

large band at ~3000 cm^{-1}	no band at ~3000 cm^{-1}	no band at ~3000 cm^{-1}	no band at ~3000 cm^{-1}
band at 1250 cm^{-1}	no band at 1250 cm^{-1}	band at 1250 cm^{-1}	no band at 1250 cm^{-1}
no band at 2700 cm^{-1}	band at 2700 cm^{-1}	no band at 2700 cm^{-1}	no band at 2700 cm^{-1}

b. This is the only compound without the characteristic benzene ring hydrogens at ~ 7-8 ppm

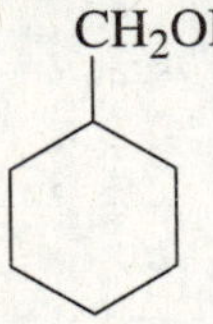

Only two compounds will have two signals other than the signals for the benzene ring hydrogens. They can be distinguished by integration (3:2 versus 2:1), or by the two sharp singlets for the ester, versus the somewhat broader singlet for the hydrogen bonded to oxygen.

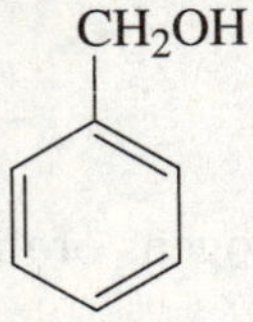

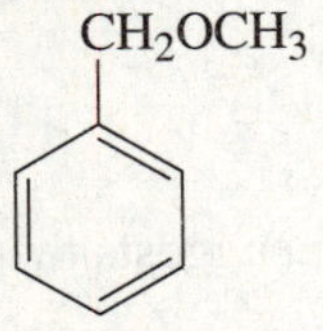

Each of the following four compounds has only one signal (a singlet) in addition to the benzene ring hydrogens. The four can be distinguished by the positions of the signals.

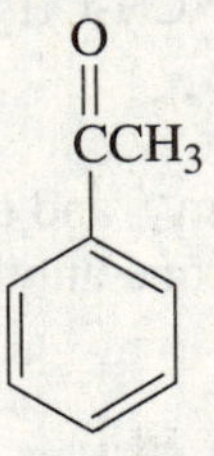

~ 2 ppm

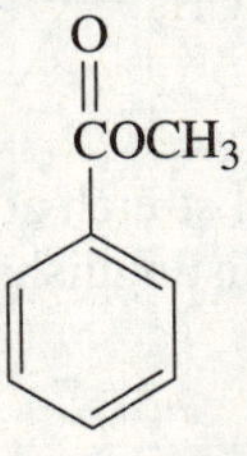

~ 4 ppm

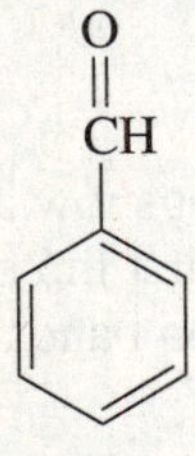

~ 9–10 ppm

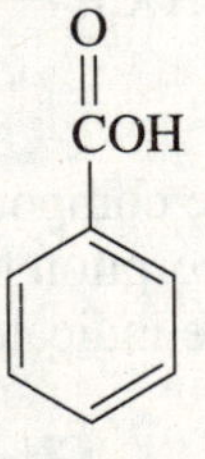

~ 10–12 ppm

64. The rate-determining step in the S_N1 reaction is the formation of the tertiary carbocation. An electron-donating substituent will stabilize the carbocation and cause it to be more easily formed. An electron-withdrawing substituent will destabilize the carbocation and cause it to be less easily formed.

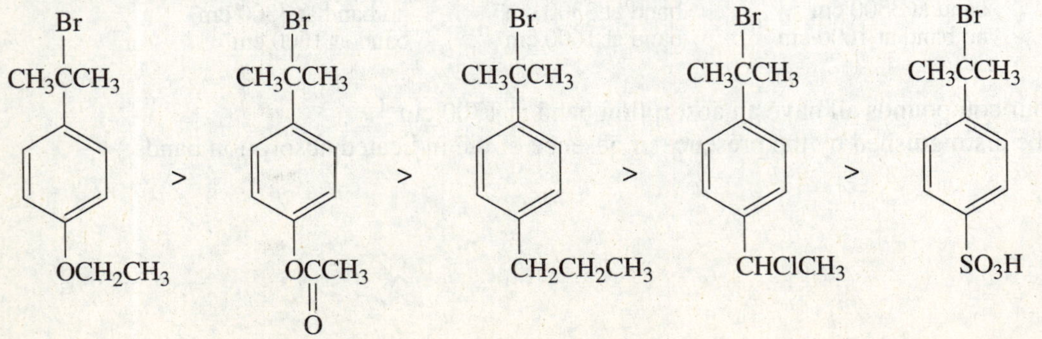

65. A fluoro substituent is more electronegative than a chloro substituent. Therefore, nucleophilic attack on the carbon bearing the fluoro substituent will be easier than nucleophilic attack on the carbon bearing the chloro substituent.

A fluoro substituent is a stronger base than a chloro substituent, so elimination of the halogen in the second step of the reaction will be harder for a fluoro-substituted benzene than for a chloro-substituted benzene.

The fact that the fluoro-substituted compound is more reactive tells you that attack of the nucleophile on the aromatic ring is the rate-determining step of the reaction.

66. **a.**

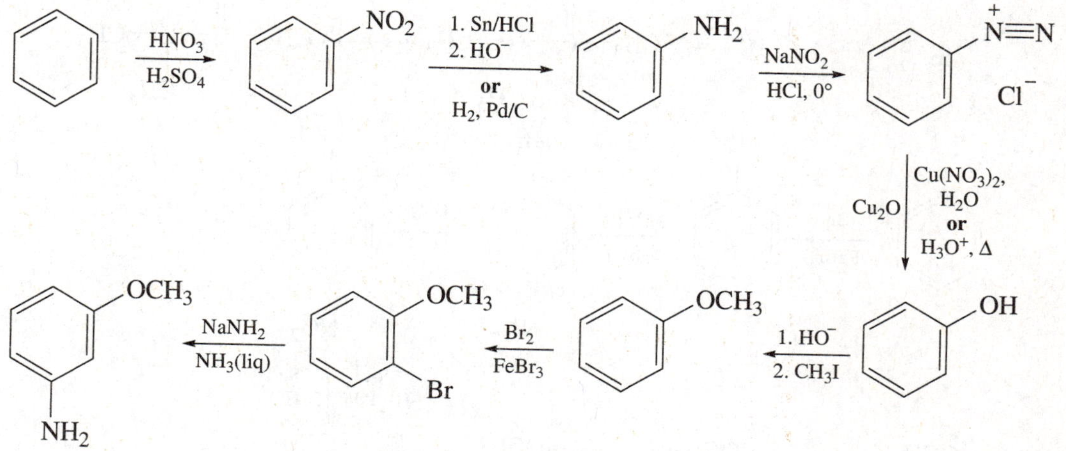

b.

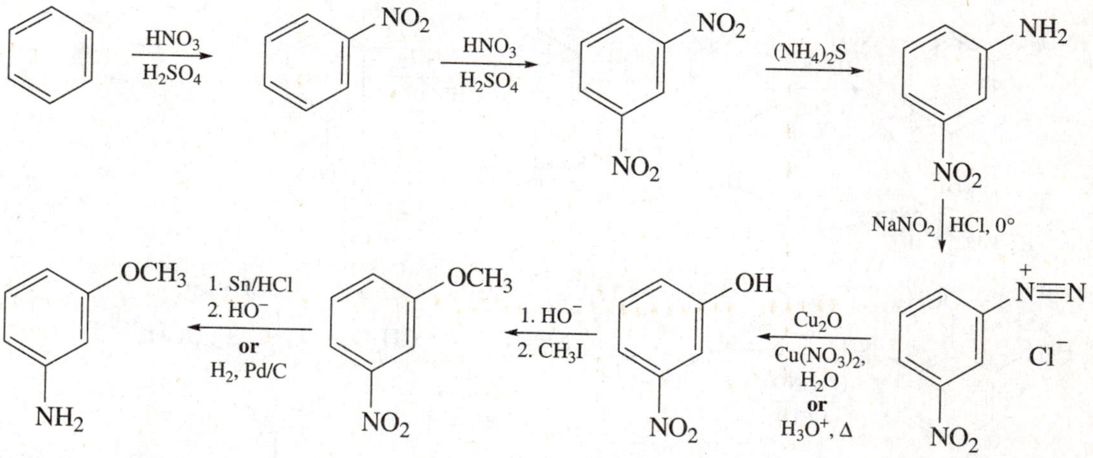

67. **a.** The alkyl diazonium ion is very unstable. Loss of N_2 and a 1,2-hydride shift forms a *tert*-butyl carbocation, which can undergo either substitution or elimination.

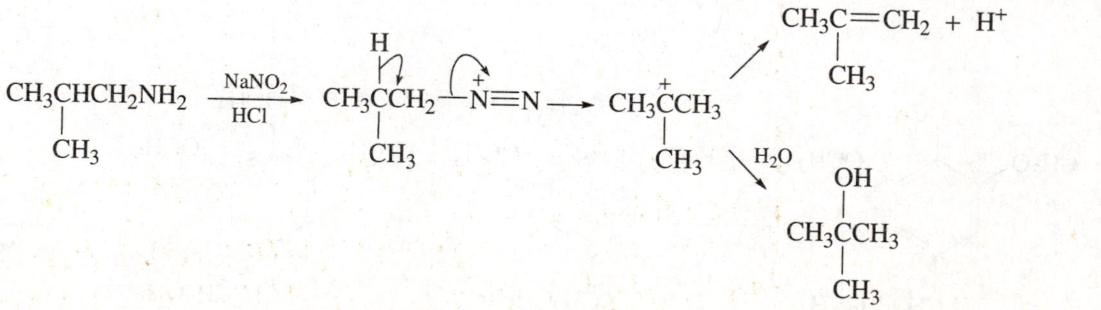

b. The cation formed from the diazonium ion will undergo a pinacol rearrangement. (See Chapter 10, Problem 76.)

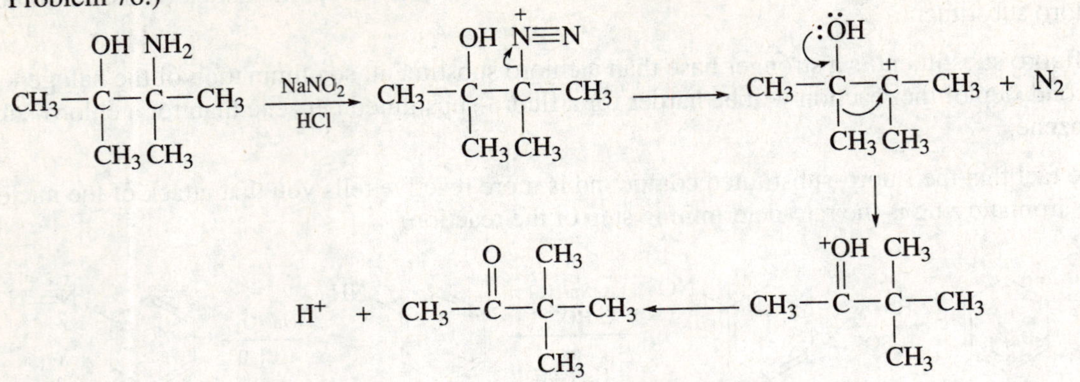

68.

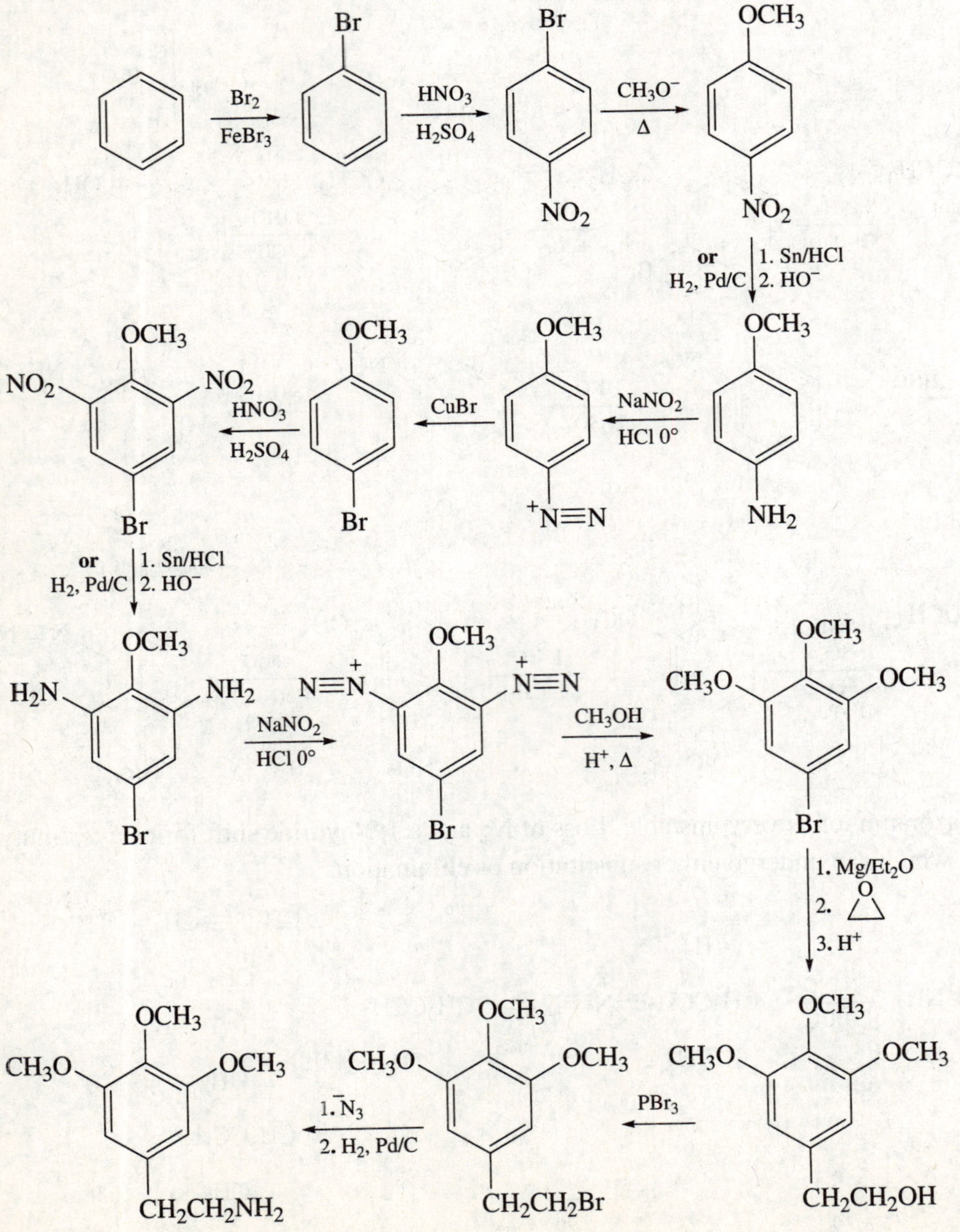

69. The configuration of the asymmetric center in the reactant will be retained only if the asymmetric center undergoes two successive S_N2 reactions.

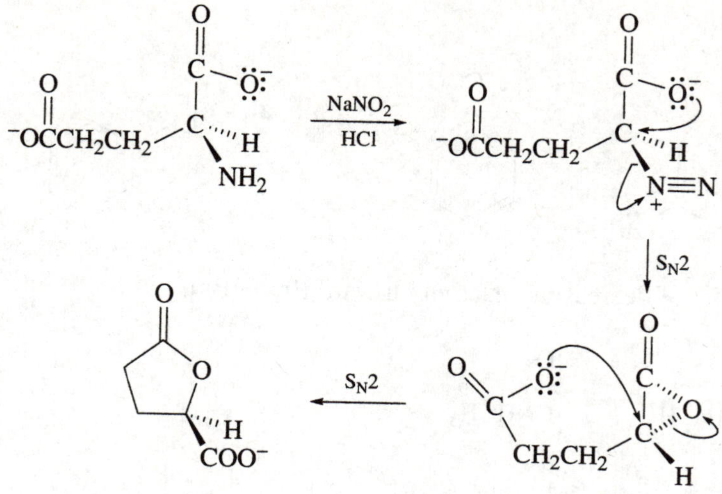

70. A chloro group is a better leaving group than the ammonium group, so the product is formed without hydroxide ion catalysis.

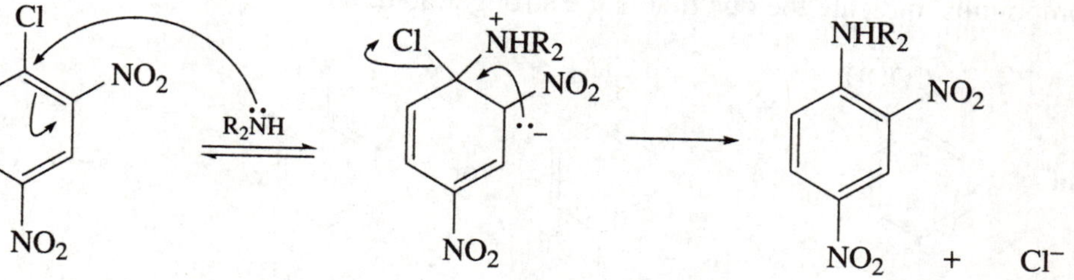

A methoxy group is a poorer leaving group than the ammonium group, so the ammonium group is eliminated, reforming starting materials. If hydroxide is added to the solution, hydroxide ion converts the ammonium group into an amino group. Since the amino group is a poorer leaving group than the methoxy group, the methoxy group is eliminated.

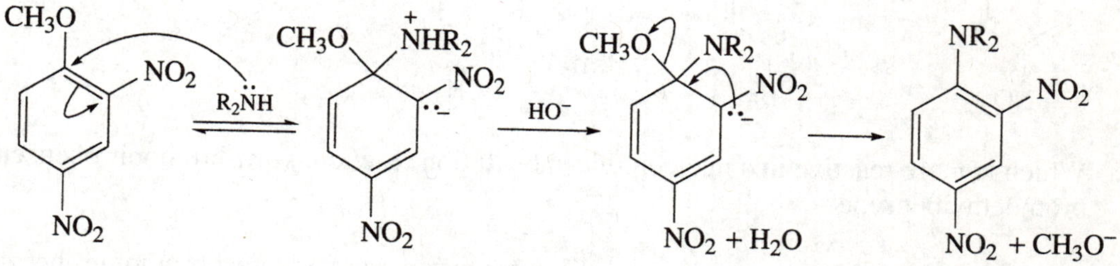

Chapter 15 Practice Test

1. Give one name for each of the following.

a. 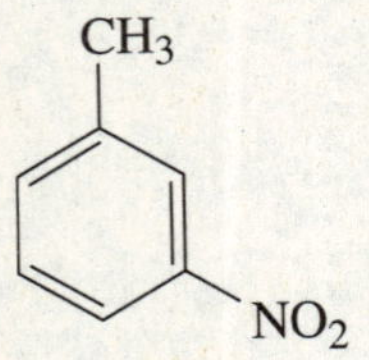
b. 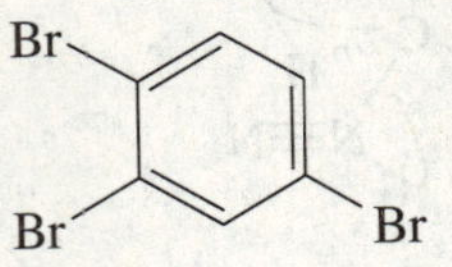
c. 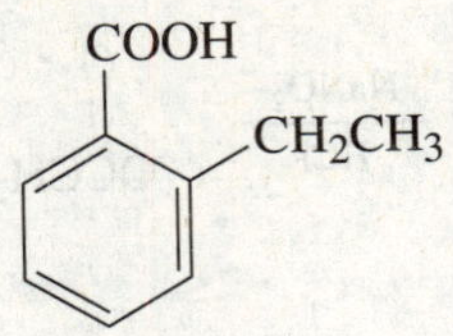
d. 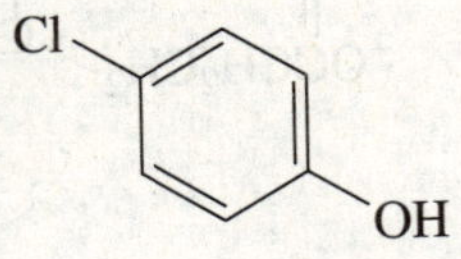

2. Rank the following compounds in order of decreasing reactivity toward $Br_2/FeBr_3$.

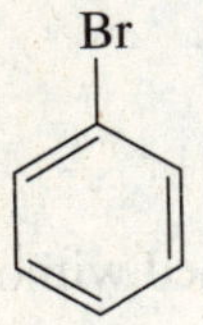 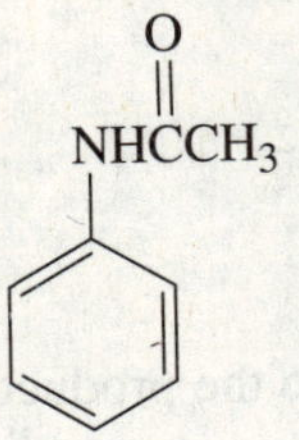 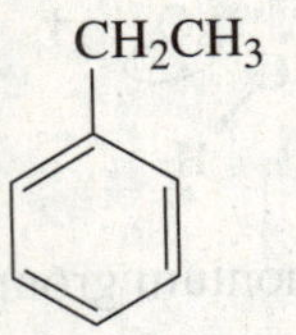 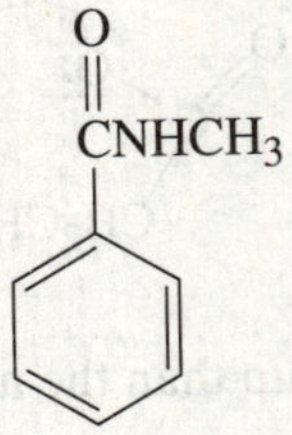 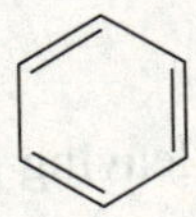

3. For each pair of compounds, indicate the one that is the stronger acid.

a. 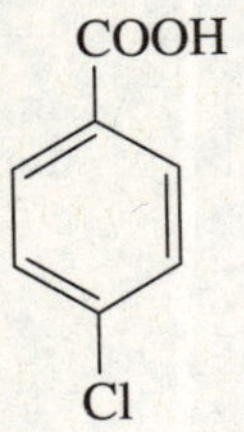 **or** 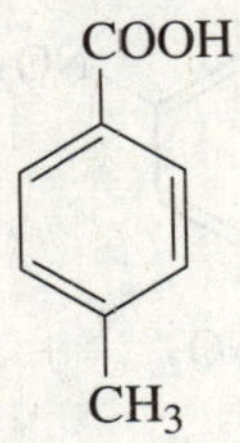

c. 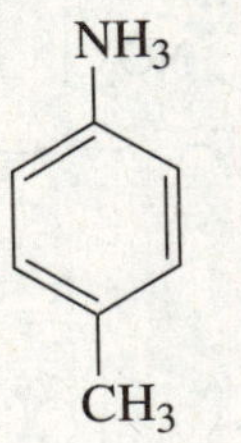 **or** 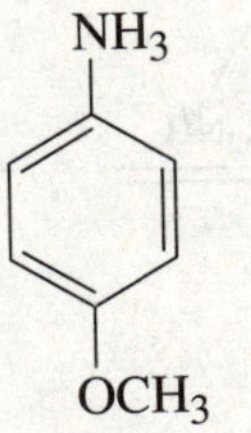

b. 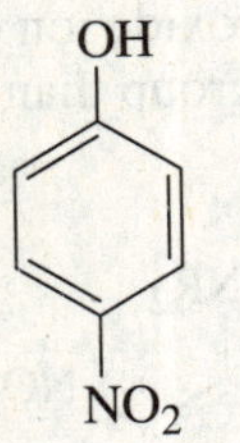 **or** 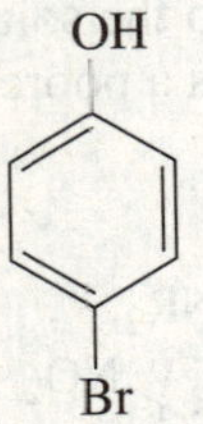

d. 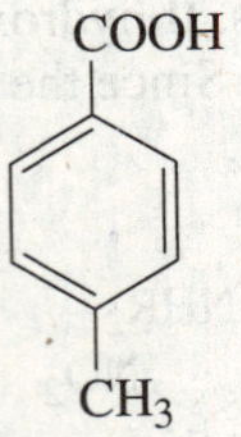 **or** 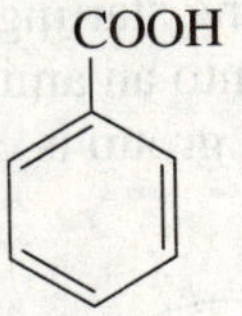

4. **a.** Which is more reactive in a nucleophilic substitution reaction, *para*-bromonitrobenzene or *para*-bromoethylbenzene?

b. Which is more reactive in an electrophilic substitution reaction, *para*-bromonitrobenzene or *para*-bromoethylbenzene?

5. Give the major product(s) of each of the following reactions.

a.

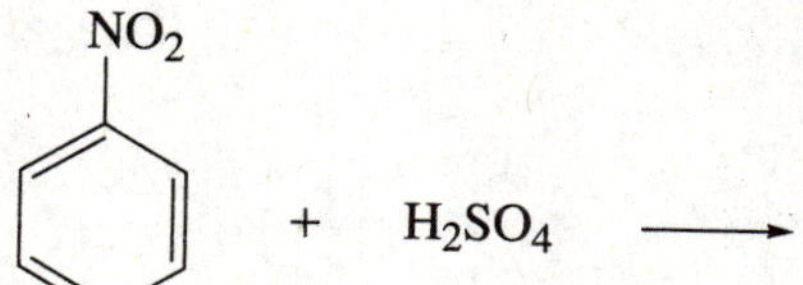

b.

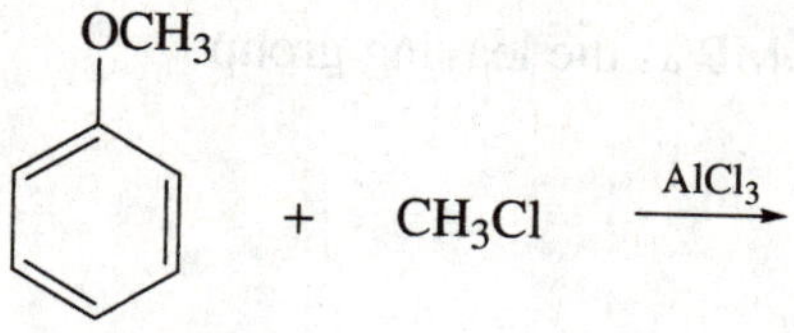

c.

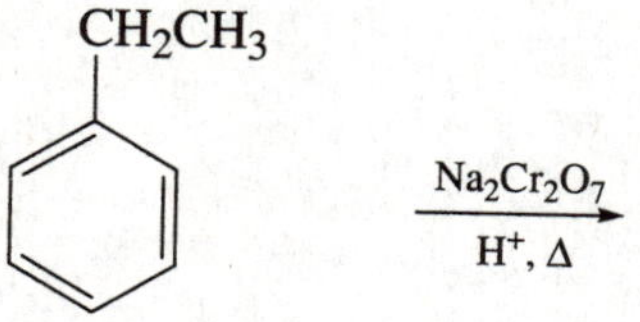

d.

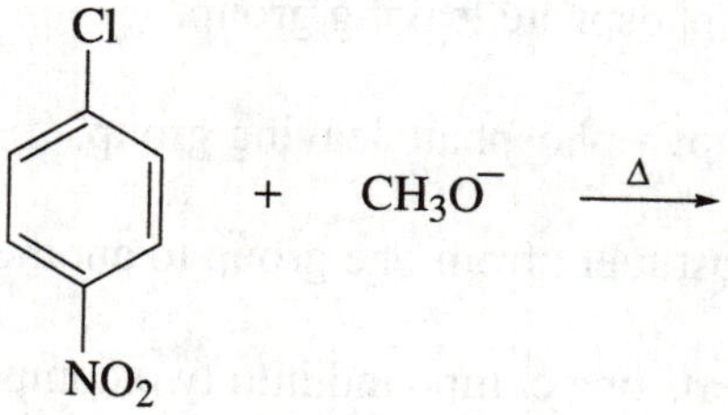

e.

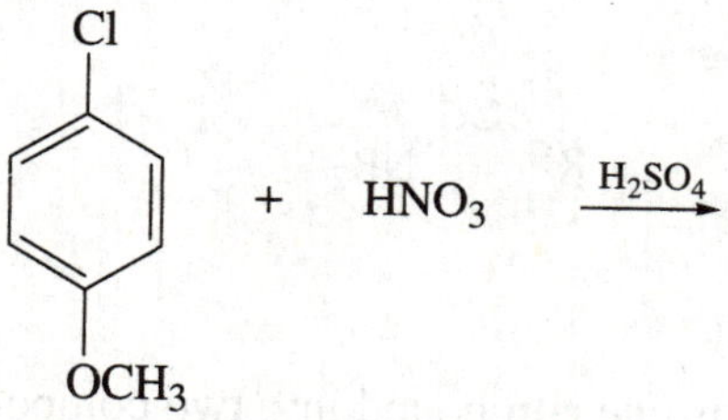

f.

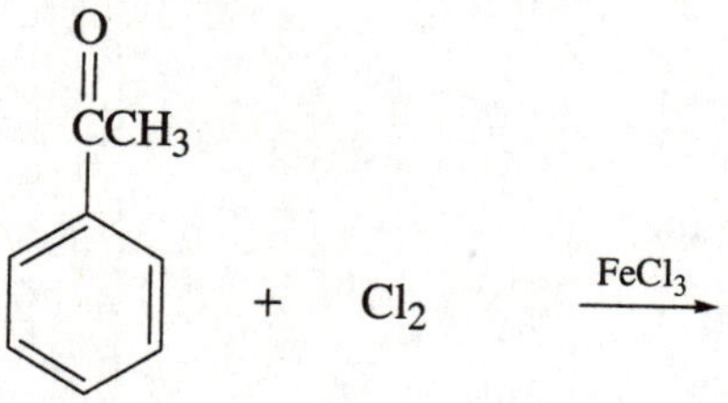

6. Indicate whether each of the following statements is true or false.

a. Benzoic acid is more reactive than benzene towards electrophilic substitution. T F

b. *para*-Chlorobenzoic acid is more acidic than *para*-methoxybenzoic acid. T F

c. A CH=CH_2 group is a meta director. T F

d. *para*-Nitroaniline is more basic than *para*-chloroaniline. T F

CHAPTER 16
Carbonyl Compounds I: Nucleophilic Acyl Substitution

Important Terms

acid anhydride 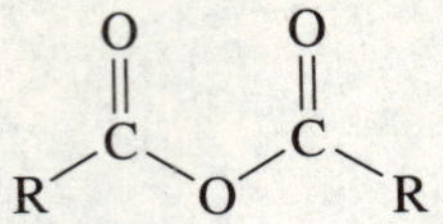

acyl adenylate a carboxylic acid derivative with AMP as the leaving group.

acyl group 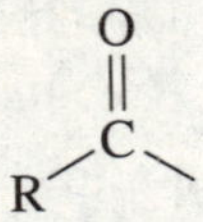

acyl halide 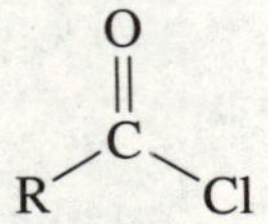 or 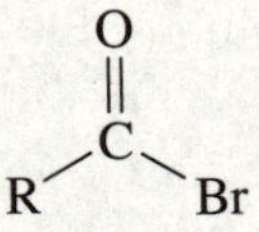

acyl phosphate a carboxylic acid derivative with a phosphate leaving group.

acyl pyrophosphate a carboxylic acid derivative with a pyrophosphate leaving group.

acyl transfer reaction a reaction that transfers an acyl substituent from one group to another.

alcoholysis reaction with an alcohol that converts one compound into two compounds.

amide 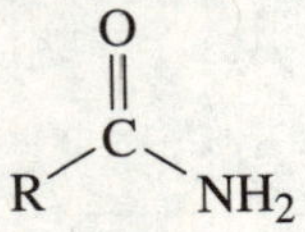 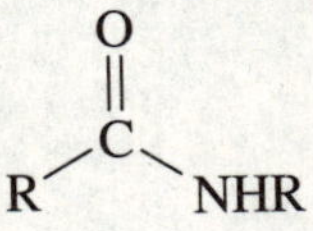 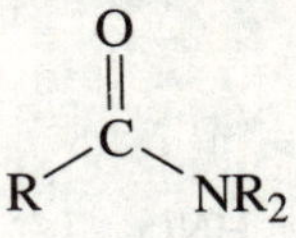

amino acid an α-amino carboxylic acid.

aminolysis reaction with an amine that converts one compound into two compounds.

biosynthesis synthesis in a biological system.

α-carbon a carbon adjacent to a carbonyl group.

carbonyl carbon the carbon of a carbonyl group.

carbonyl compound a compound that contains a carbonyl group.

carbonyl group a carbon doubly bonded to an oxygen.

carbonyl oxygen the oxygen of a carbonyl group.

carboxyl group 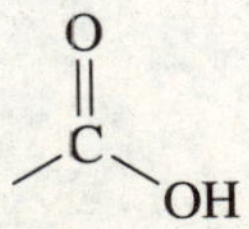 or $—COOH$ or $—CO_2H$

carboxylic acid

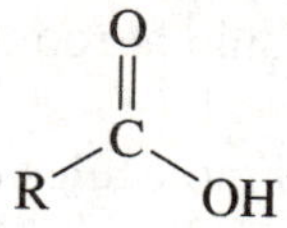

carboxylic acid derivative a compound that is hydrolyzed to a carboxylic acid.

carboxyl oxygen the single-bonded oxygen of a carboxylic acid or ester.

catalyst a species that increases the rate of a reaction without being consumed in the reaction.

detergent a salt of a sulfonic acid.

ester

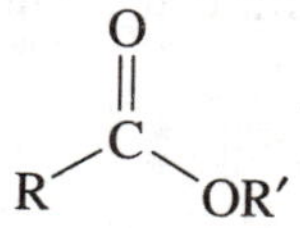

fat a triester of glycerol that exists as a solid at room temperature.

fatty acid a long-chain carboxylic acid.

Fischer esterification reaction reaction of a carboxylic acid with excess alcohol and an acid catalyst.

Gabriel synthesis a method used to convert an alkyl halide into a primary amine.

hydrolysis reaction with water that converts one compound into two compounds.

hydrophobic interactions the attractive forces of hydrocarbon chains in water.

imide a compound with two acyl groups bonded to a nitrogen.

lactam a cyclic amide.

lactone a cyclic ester.

micelle a spherical aggregation of molecules, each with a long hydrophobic tail and a polar head, arranged so the polar head points to the outside of the sphere.

mixed anhydride an acid anhydride with two different R groups.

neurotransmitter a compound that transmits nerve impulses across the synapses between nerve cells.

nitrile a compound that contains a carbon-nitrogen triple bond.

$$R{-}C{\equiv}N$$

nucleophilic acyl substitution reaction a reaction in which a group bonded to an acyl group is substituted by another group.

oil a triester of glycerol that exists as a liquid at room temperature.

phosphoanhydride bond the bond holding two phosphoric acid molecules together.

Ritter reaction the reaction of a nitrile with a secondary or tertiary alcohol to form a secondary amine.

saponification hydrolysis of a fat under basic conditions.

soap a sodium or potassium salt of a fatty acid.

symmetrical anhydride an acid anhydride with identical R groups.

tetrahedral intermediate the intermediate formed in a nucleophilic acyl substitution reaction.

thioester the sulfur analog of an ester.

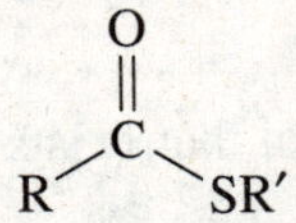

transesterification reaction the reaction of an ester with an alcohol to form a different ester.

Solutions to Problems

1. The lactone would be a three-membered ring, which is too strained to form.

2. **a.** butanenitrile
propyl cyanide

b. ethanoic propanoic anhydride
acetic propionic anhydride

c. potassium butanoate
potassium butyrate

d. pentanoyl chloride
valeryl chloride

e. isobutyl butanoate
isobutyl butyrate

f. *N,N*-dimethylhexanamide

g. γ-butyrolactam or
2-azacyclopentanone

h. cyclopentanecarboxylic acid

i. β-methyl-δ-valerolactone or
5-methyl-2-oxacyclohexanone

3. **a.** 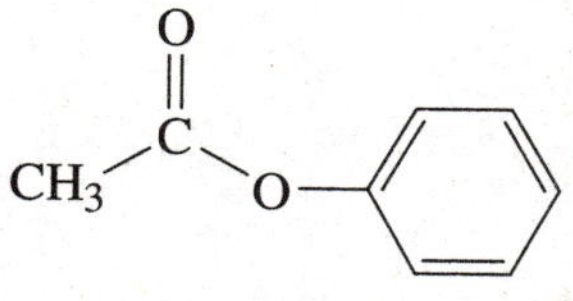

b. 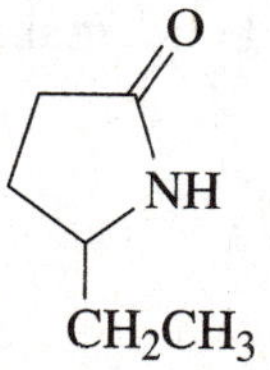

c. $CH_3CH_2CH_2C{\equiv}N$

d. 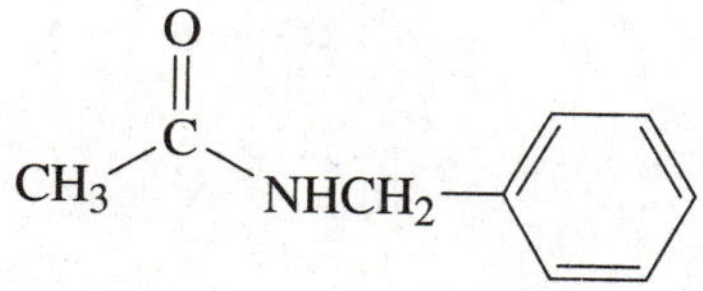

e. 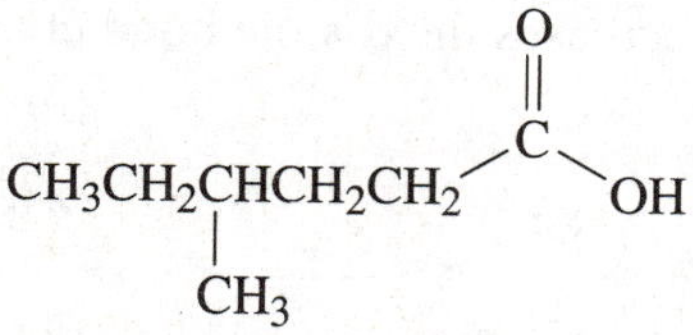
γ-methylcaproic acid

f. 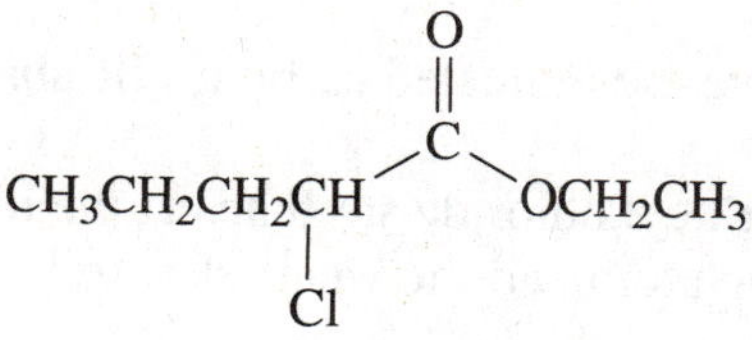

g. 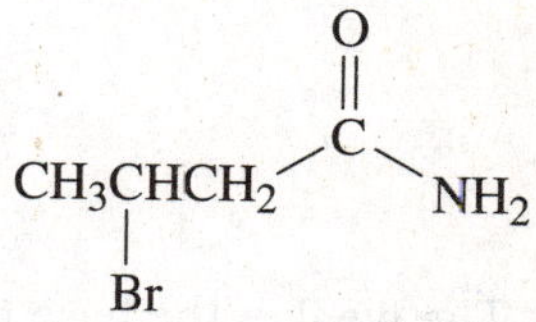

h. 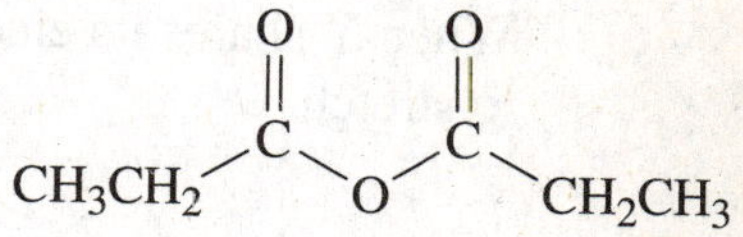

i. 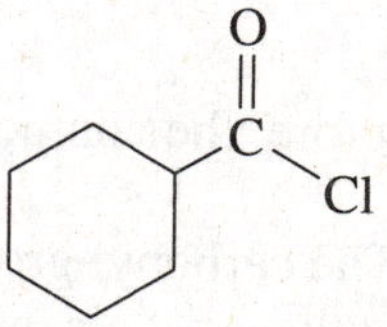

4. The carbon-oxygen single bond in an alcohol is longer because, as a result of electron delocalization, the carbon-oxygen single bond in a carboxylic acid has some double-bond character.

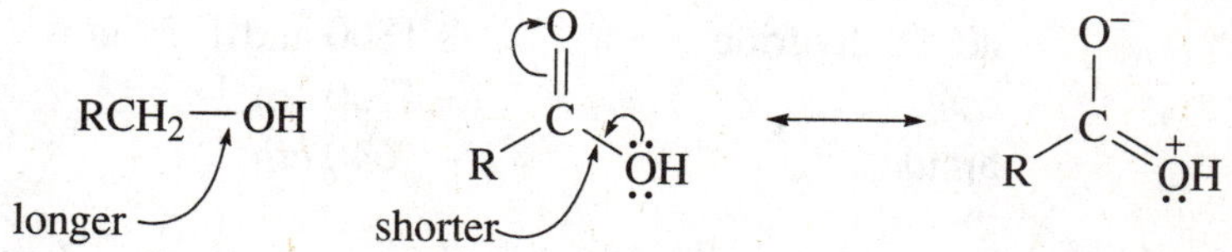

5. **a.** The bond between oxygen and the methyl group is the longest because it is a pure single bond, whereas the other two carbon-oxygen bonds have some double-bond character.

The bond between carbon and the carbonyl oxygen is the shortest because it has the most double bond character.

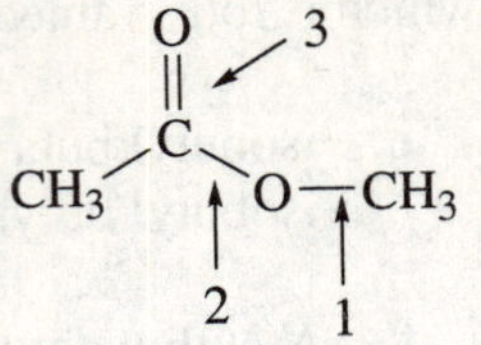

1 = longest
3 = shortest

b. Notice that the longer the bond, the lower its IR stretching frequency.

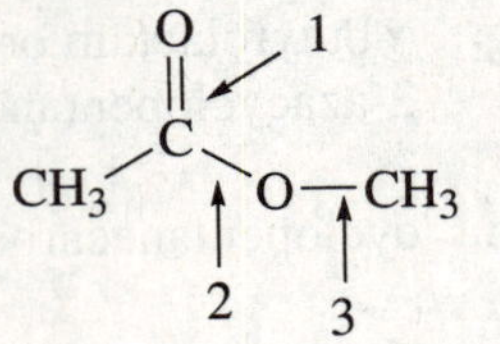

1 = highest frequency
3 = lowest frequency

6. The more electronegative the base (Y) attached to the carbonyl carbon, the greater the double bond character of the C=O bond.

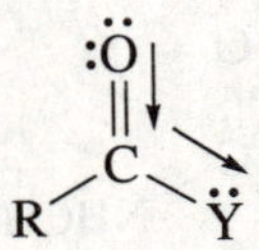

The weaker the base (Y) attached to the carbonyl carbon the less well it shares it electrons. When Y shares its electrons, it decreases the double bond character of the C=O bond, making it easier to stretch.

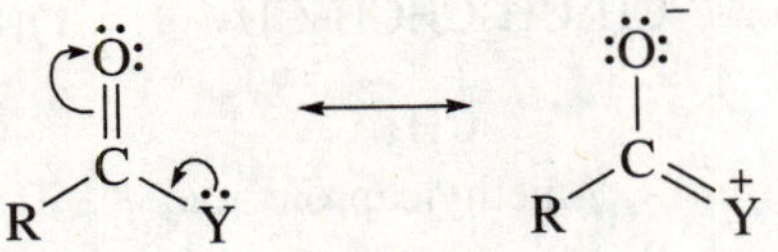

Therefore, the compounds have the indicated carbonyl IR absorption bands.

The carbonyl group of the acyl chloride stretches at the highest frequency because the chlorine atom is the most electronegative atom and the weakest base.

The carbonyl group of the amide stretches at the lowest frequency because the nitrogen atom is the least electronegative atom and the strongest base.

acyl chloride	~ 1800 cm^{-1}
acid anhydride	~ 1800 and 1750 cm^{-1}
ester	~ 1730 cm^{-1}
amide	~ 1640 cm^{-1}

The resonance contributors show why the acid anhydride has two carbonyl IR absorption bands.

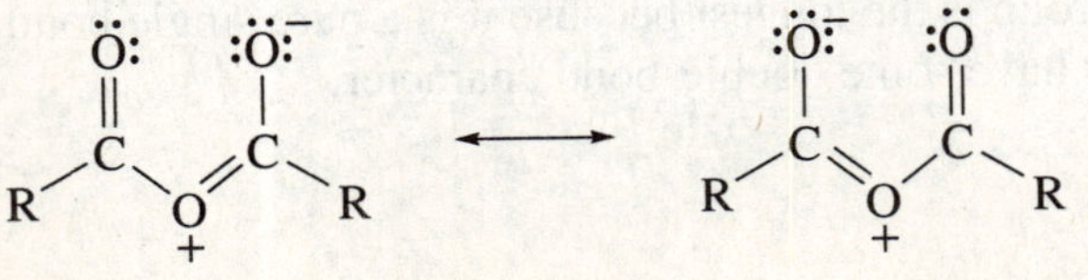

7. **a.** Because HCl is a stronger acid than H_2O, Cl^- is a weaker base than HO^-. Therefore, Cl^- will be eliminated from the tetrahedral intermediate, so the product of the reaction will be acetic acid. However, since the solution is basic, acetic acid will be in its basic form as a result of losing a proton.

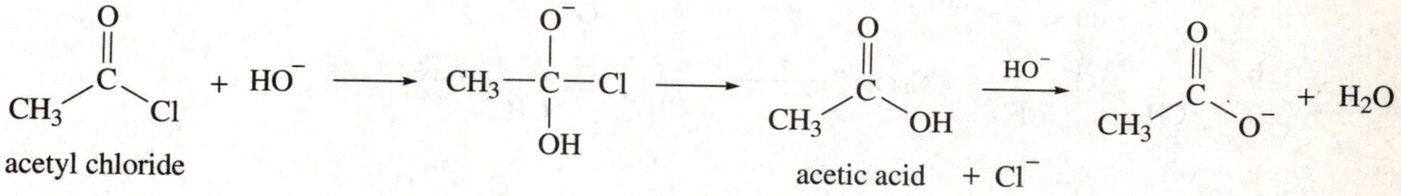

b. Because H_2O is a stronger acid than NH_3, HO^- is a weaker base than $^-NH_2$. Therefore, HO^- will be eliminated from the tetrahedral intermediate, so the product of the reaction will be acetamide.

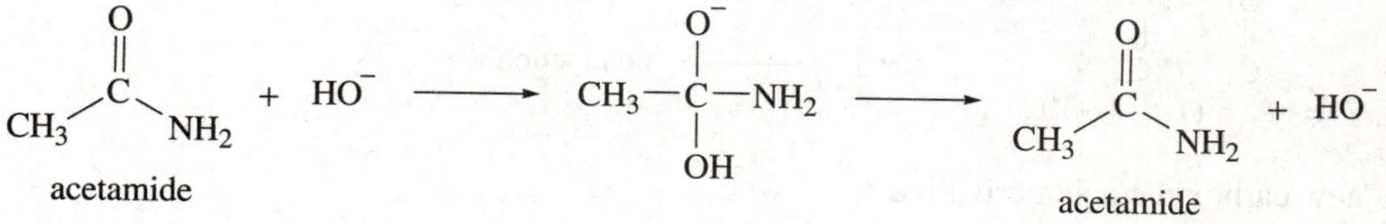

8. It is a true statement.

If the nucleophile is the stronger base, it will be harder to eliminate the nucleophile from the tetrahedral intermediate (**B**) than to eliminate the group attached to the acyl group in the reactant. In other words, the hill that has to be climbed from the intermediate (**B** to **A**) back to the reactants is higher than the hill that has to be climbed from the intermediate (**B** to **C**) to the products. Since the transition state with the highest energy is the transition state of the rate-limiting step, the first step is the rate-limiting step.

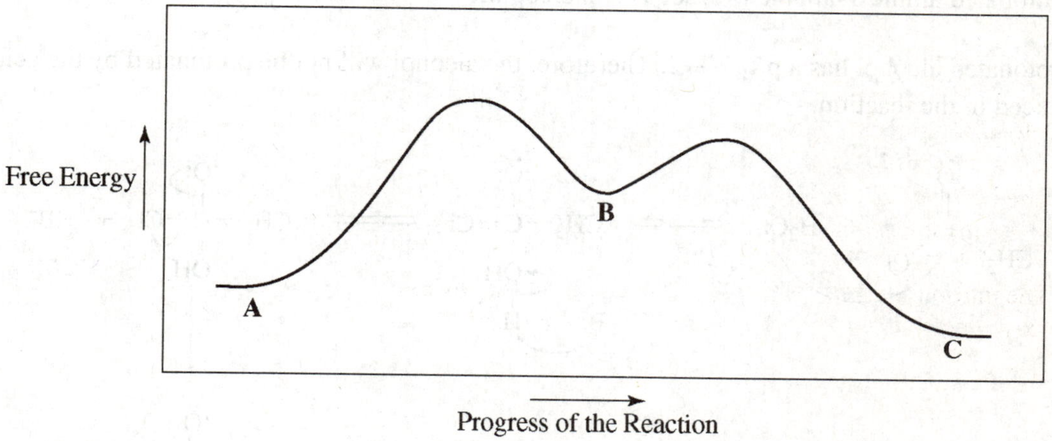

9. **a.** 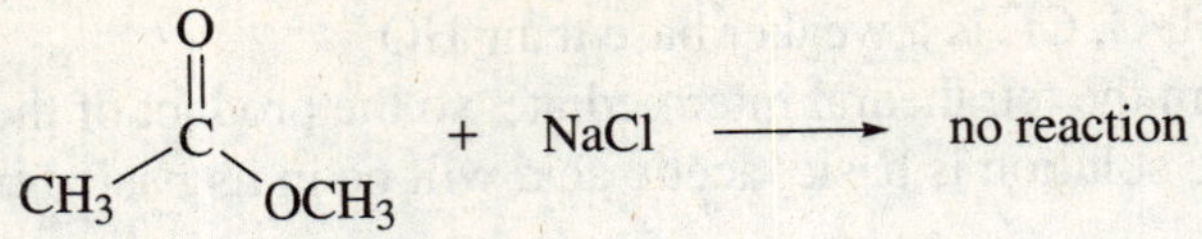

b. 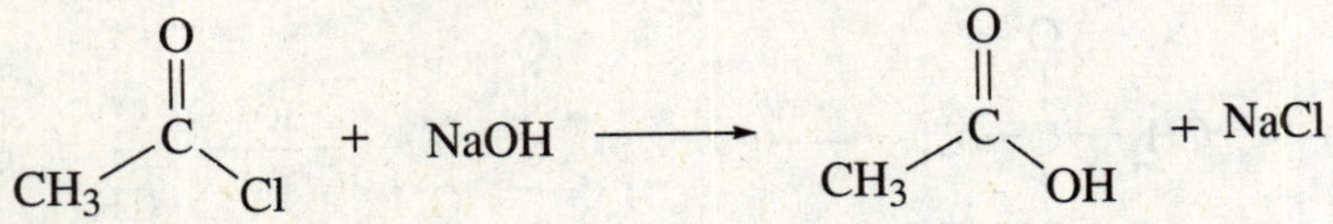

c. 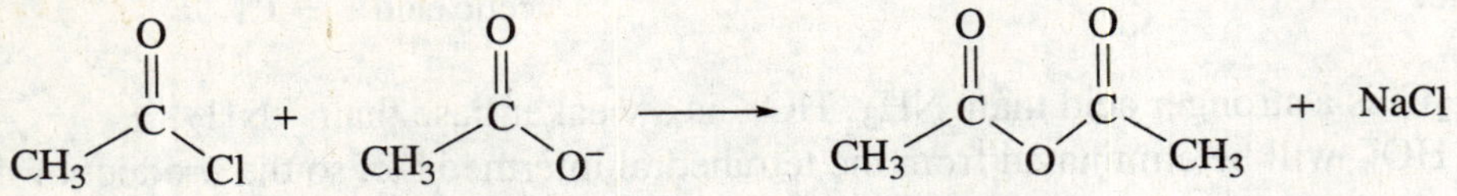

d. 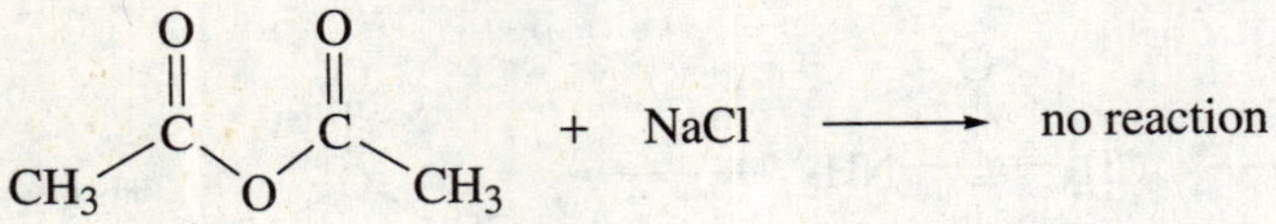

10. **a.** a new carboxylic acid derivative

b. no reaction

c. a mixture of two carboxylic acid derivatives

11. Solved in the text.

12. A protonated amine has a pK_a ~ 11. Therefore, the amine will be protonated by the acid that is produced in the reaction, and a protonated amine is not a nucleophile. Excess amine is used in order to have some unprotonated amine available to react as a nucleophile.

A protonated alcohol has a pK_a ~ – 2. Therefore, the alcohol will not be protonated by the acid that is produced in the reaction.

13. **a.** 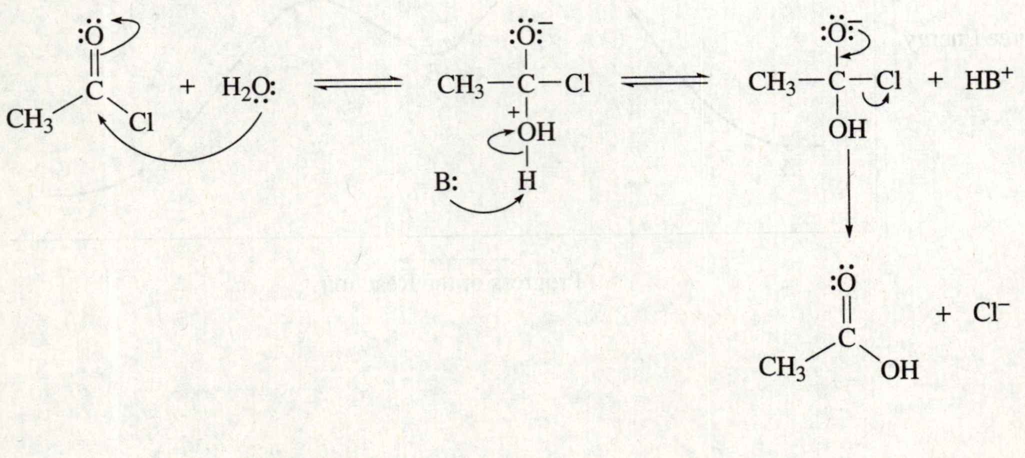

b.

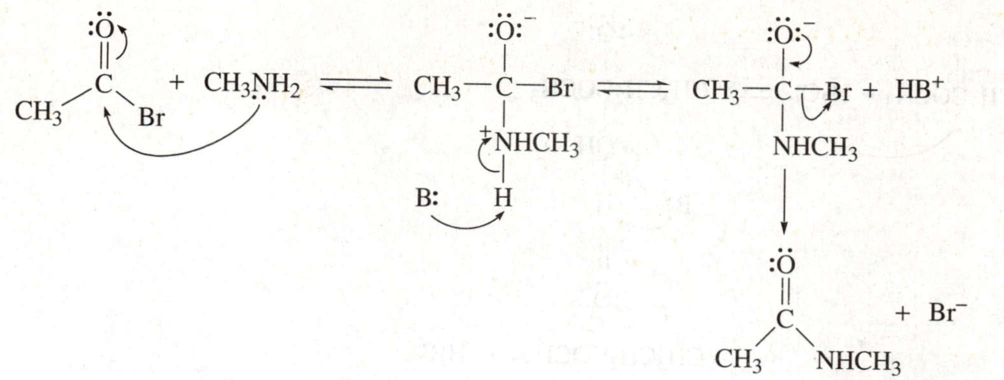

14. **a.** $CH_3CH_2CH_2OH$ **c.** $(CH_3)_2NH$ **e.** $CH_3C(=O)O^-$

b. $CH_3CH_2NH_2$ **d.** H_2O **f.**

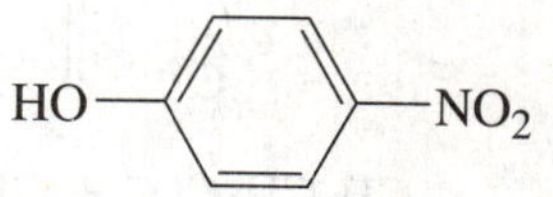

15. **a.**

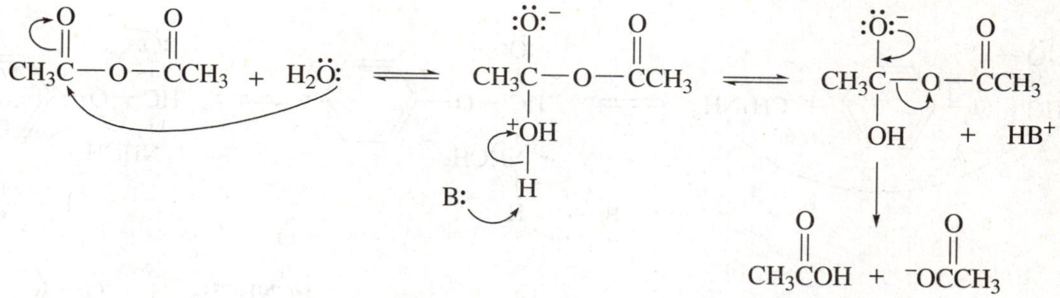

b. The mechanisms are exactly the same.

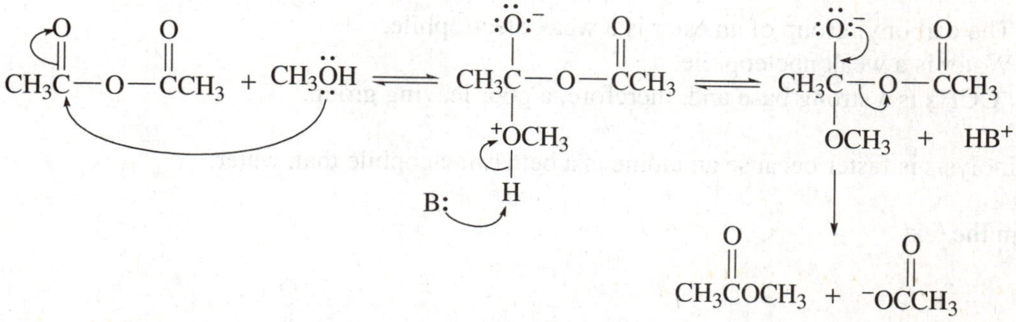

16. When an acid anhydride reacts with an amine, the tetrahedral intermediate does not have to lose a proton before it eliminates the carboxylate ion, because the carboxylate ion is a weaker base (pK_a of its conjugate acid is ~ 5) than an amine (pK_a of its conjugate acid is ~ 10).

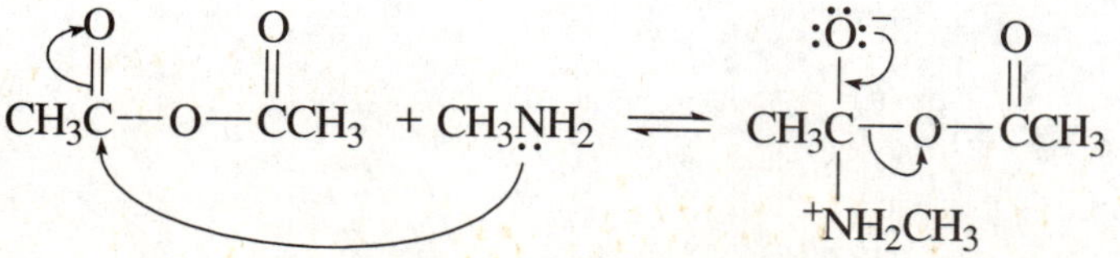

17. **a.**

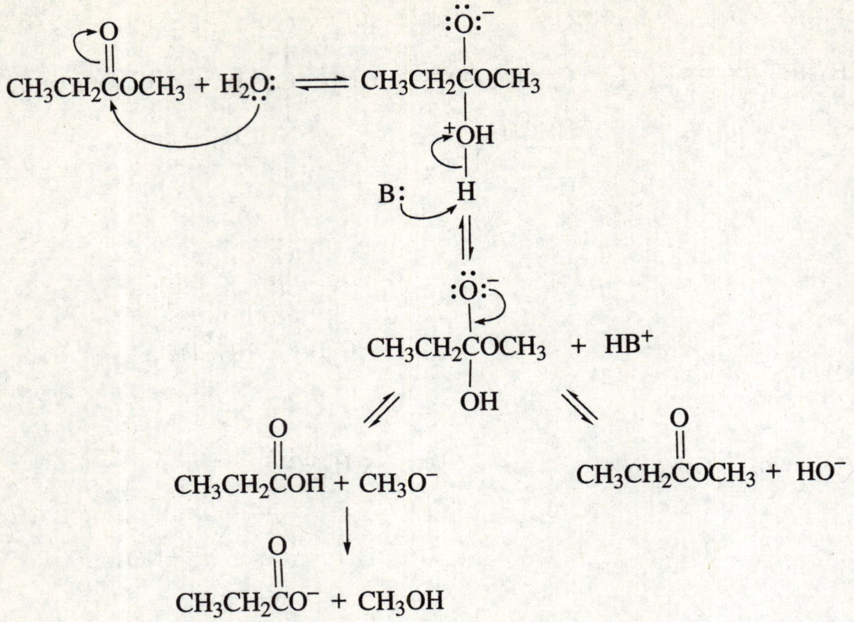

b.

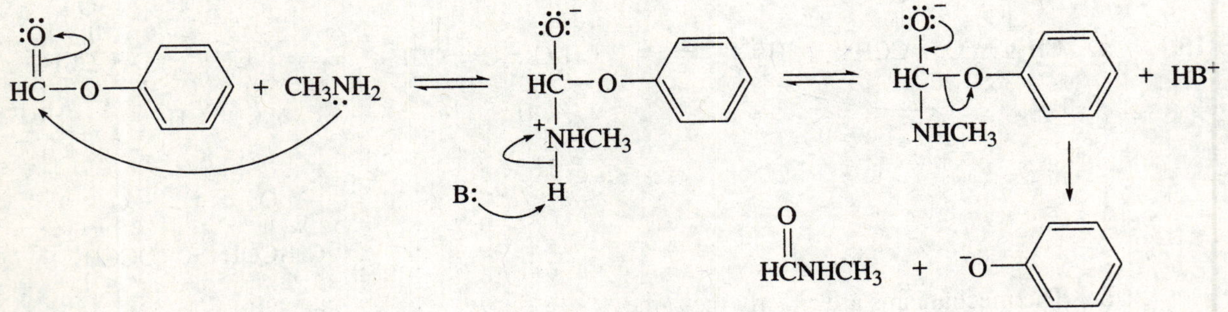

18. **a.** **1.** The carbonyl group of an ester is a weak electrophile.

2. Water is a weak nucleophile.

3. $^{-}OCH_3$ is a strong base and, therefore, a poor leaving group.

b. Aminolysis is faster because an amine is a better nucleophile than water.

19. Solved in the text.

20. **a.**

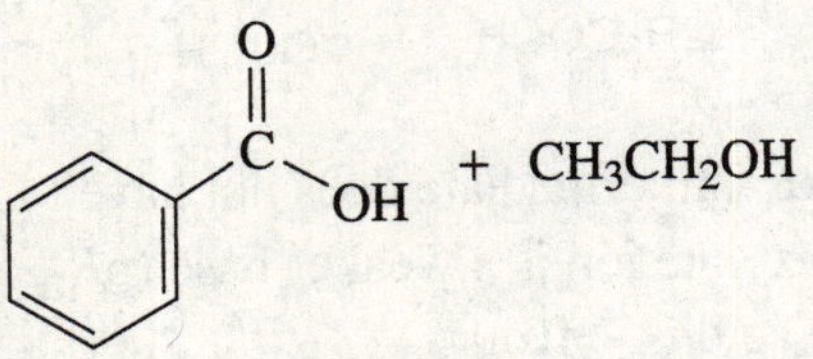

b.

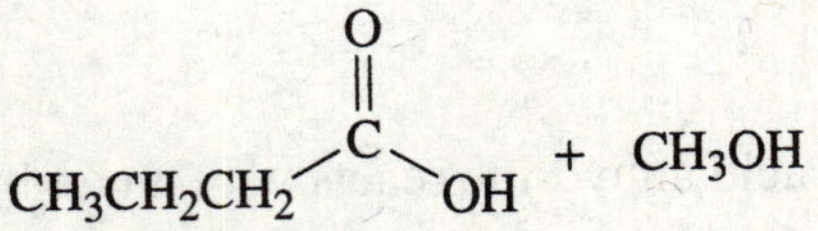

c.

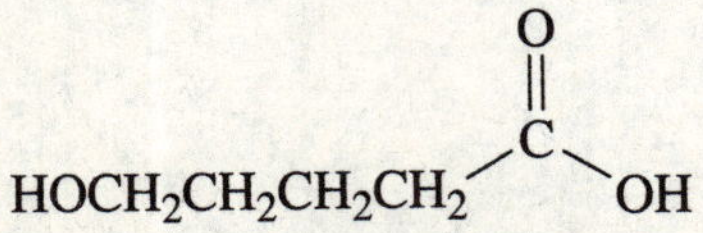

21. The mechanism for the acid-catalyzed reaction of acetic acid and methanol is the exact reverse of the mechanism for the acid-catalyzed hydrolysis of methyl acetate.

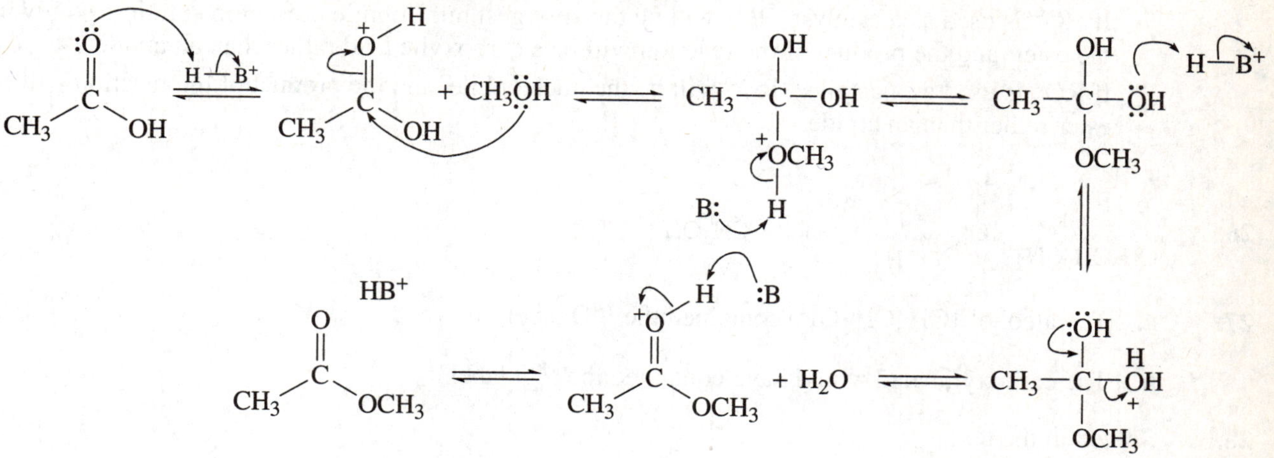

22. **a.** 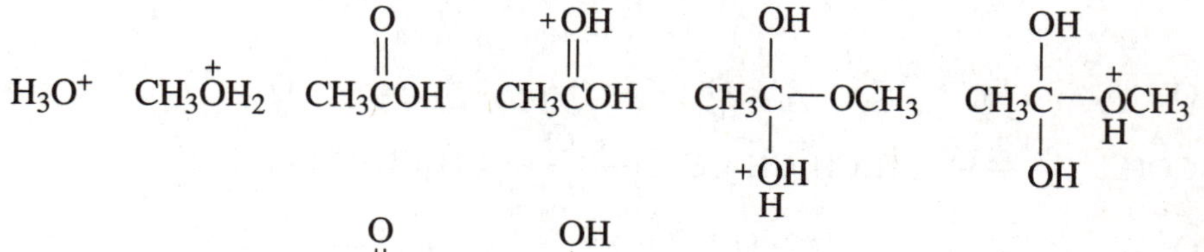

b. 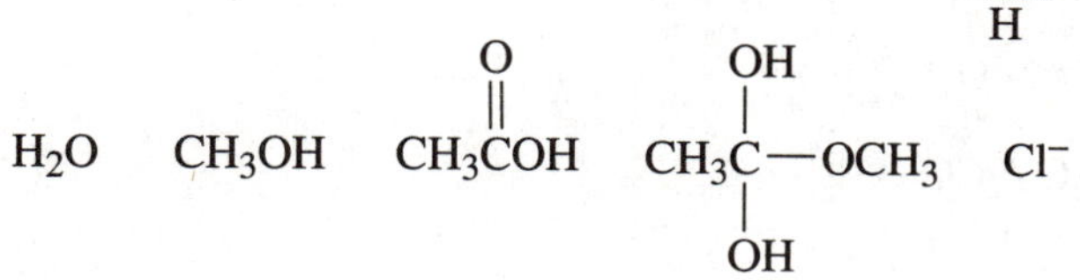

c. H_3O^+

d. H_3O^+ if excess water is used, $CH_3\overset{+}{O}H_2$ if not

23. Solved in the text.

24. 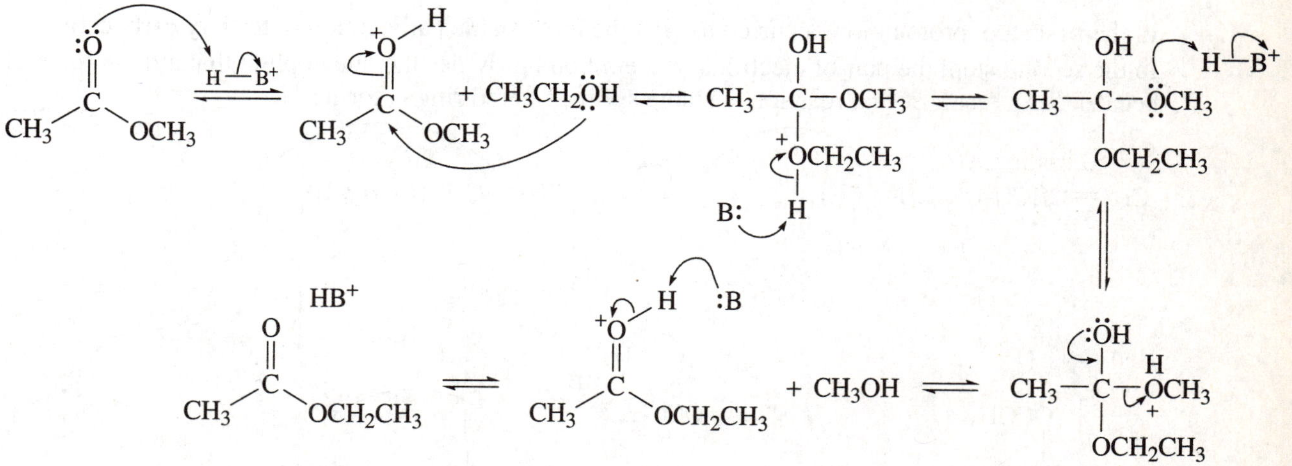

25. **a.** The conjugate base ($CH_3CH_2CH_2O^-$) of the reactant alcohol ($CH_3CH_2CH_2OH$) can be used to catalyze the reaction.

b. If H^+ is used as a catalyst, the amine will be protonated in the acidic solution and, therefore, will not be able to react as a nucleophile.

If HO^- is used as a catalyst, HO^- will be the strongest nucleophile present in solution, so it will attack the ester, and the product of the reaction will be a carboxylic acid rather than an amide.

If RO^- is used as a catalyst, RO^- will be the nucleophile, and the product of the reaction will be an ester rather than an amide.

26. $CH_3CH_2CH_2C(=O)\overset{18}{O}CH_3$ and $CH_3\overset{18}{O}H$

27. **a.** The alcohol (CH_3CH_2OH) contained the ^{18}O label.

b. The carboxylic acid would have contained the ^{18}O label.

28. Solved in the text.

29. Solved in the text.

30. **a.** $CH_3CH_2CH_2\overset{O}{\overset{\|}{C}}OH \xrightarrow{SOCl_2} CH_3CH_2CH_2\overset{O}{\overset{\|}{C}}Cl \xrightarrow{CH_3OH} CH_3CH_2CH_2\overset{O}{\overset{\|}{C}}OCH_3$

or

$CH_3CH_2CH_2\overset{O}{\overset{\|}{C}}O^- \xrightarrow{CH_3Br} CH_3CH_2CH_2\overset{O}{\overset{\|}{C}}OCH_3$

b. $CH_3\overset{O}{\overset{\|}{C}}OH \xrightarrow{SOCl_2} CH_3\overset{O}{\overset{\|}{C}}Cl \xrightarrow{CH_3(CH_2)_7OH} CH_3\overset{O}{\overset{\|}{C}}O(CH_2)_7CH_3$

or

$CH_3\overset{O}{\overset{\|}{C}}O^- \xrightarrow{CH_3(CH_2)_7Br} CH_3\overset{O}{\overset{\|}{C}}O(CH_2)_7CH_3$

31. In the first step, protonation occurred to give the most stable carbocation (a tertiary carbocation). In the second step, the pair of electrons of the π bond provides the nucleophile that attacks the carbocation because, as a result of that reaction, a stable six-membered ring is formed.

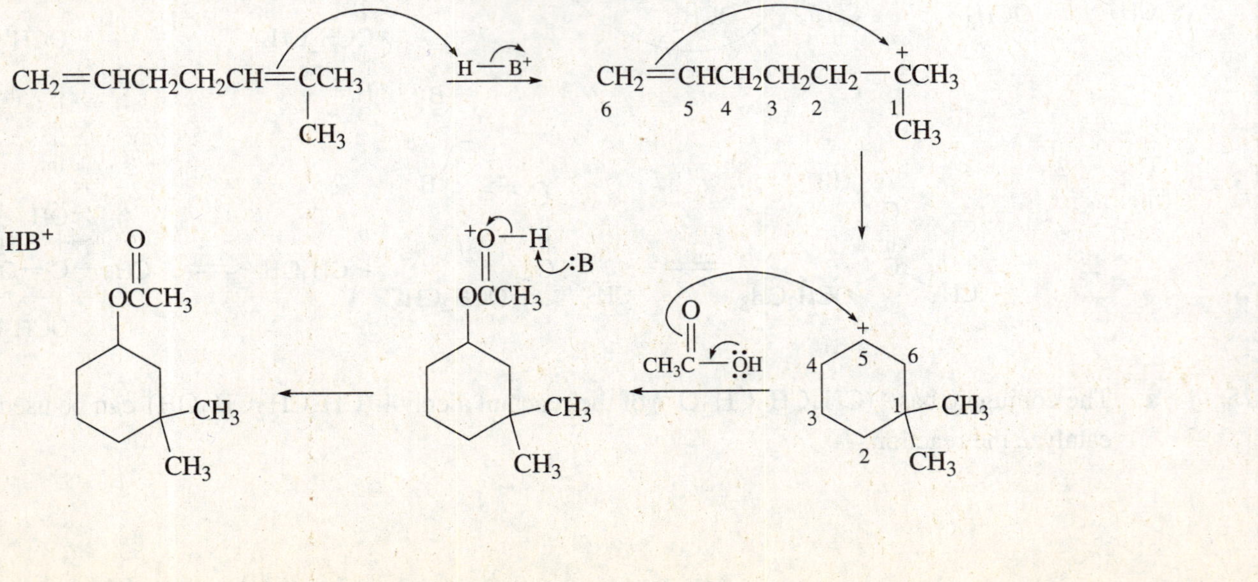

32. **a.** 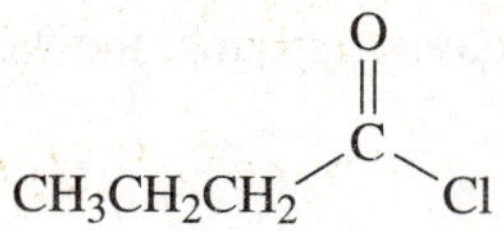 + 2 $CH_3CH_2NH_2$ **b.** 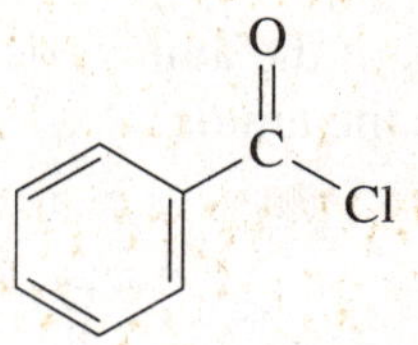 + 2 CH_3NHCH_3

33. **a.** #2 and #5 will form amides

For #2: if the nucleophile were $CH_3\bar{N}H$ instead of CH_3NH_2, the better nucleophile will increase the rate of amide formation.

For #5: use a large excess of methylamine to force the equilibrium to the right.

34. We know that the electrophile is thionyl chloride and that its electrophilic site is the sulfur atom. We know that the amide is the nucleophile and its nucleophilic site is the oxygen atom.

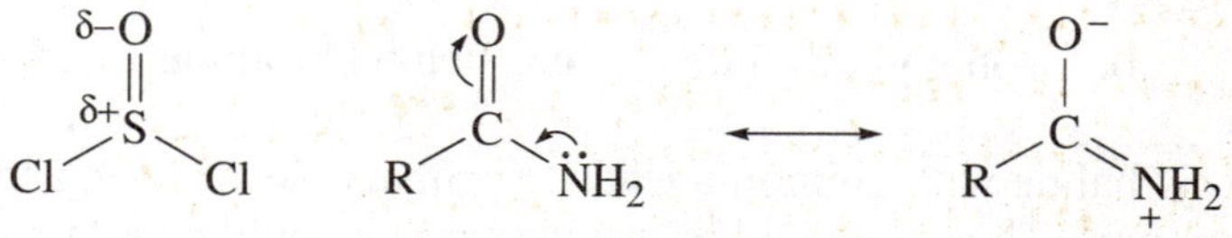

Therefore, we should let the nucleophile attack the electrophile and then see if that product helps us figure out how we can arrive at the given product (the nitrile).

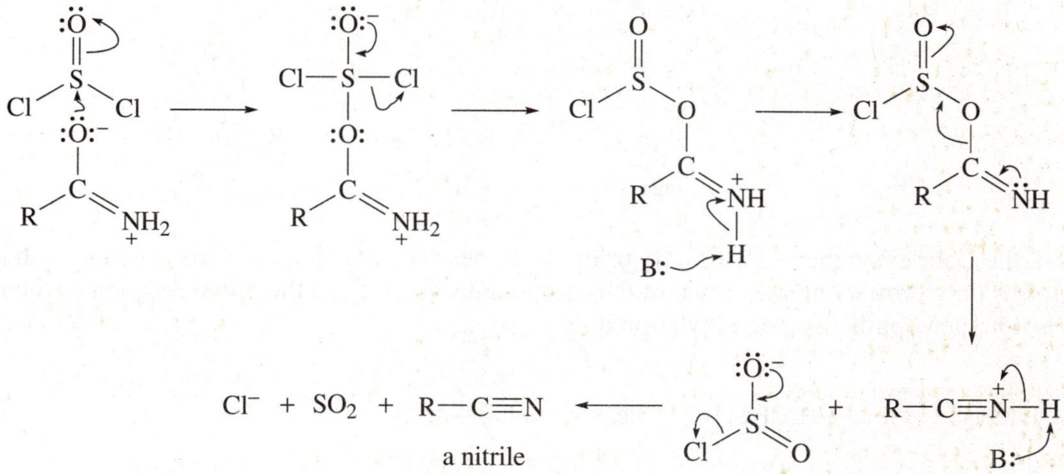

35. Because methicillin is missing the methylene group, its benzene ring is closer to the four-membered ring. This plus the bulky methoxy groups may prevent penicillinase from being able to bind it at its active site; therefore, it would not be able to be used as a substrate. Alternatively, the extra steric hindrance associated with methicillin may prevent water from attacking the carbonyl group in order to hydrolyze it.

36. The relative reactivities of the amides depend on the basicities of their leaving groups: the weaker the base, the more reactive the amide.

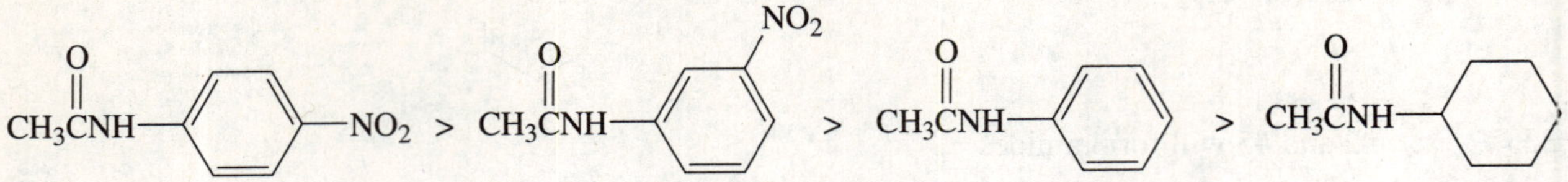

C B D A

A *para*-nitro-substituted aniline is less basic than a *meta*-substituted aniline because when the nitro group is in the para position, electrons can be delocalized onto the nitro group.

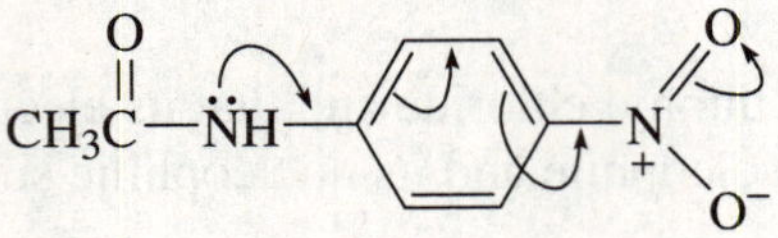

37. **a.** pentyl bromide **b.** isohexyl bromide **c.** benzyl bromide **d.** cyclohexyl bromide

38. The reaction of an alkyl halide with ammonia leads to primary, secondary, and tertiary amines, and even quaternary ammonium ions. Thus, the yield of primary amine could be relatively low.

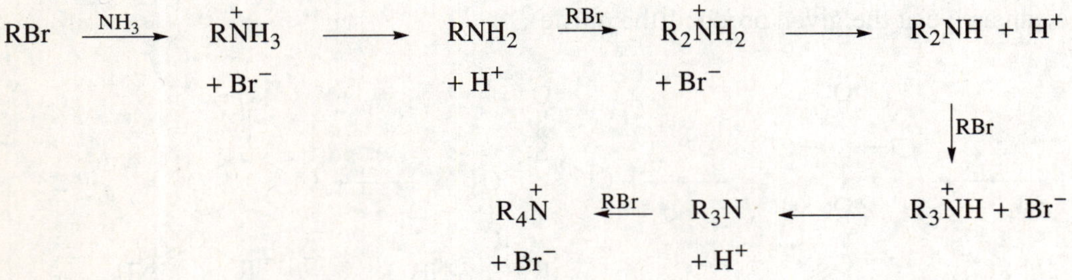

In contrast, the Gabriel synthesis forms only primary amines. The reaction of an alkyl halide with azide ion also forms only primary amines because the compound formed from the initial reaction of the two reagents is not nucleophilic, so polyalkylation does not occur.

39. **a.** $CH_3CH_2CH_2Br$ **b.** 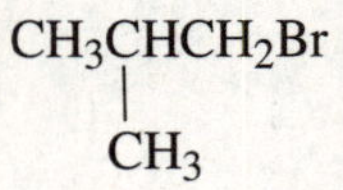 **c.** 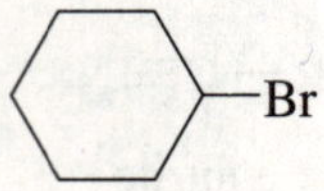

40. **a.** 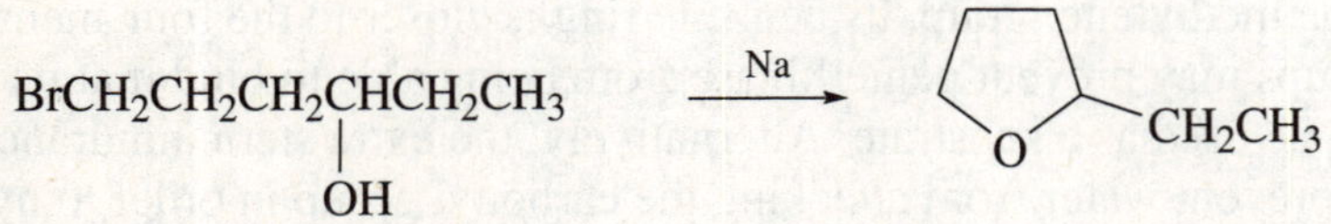

b. 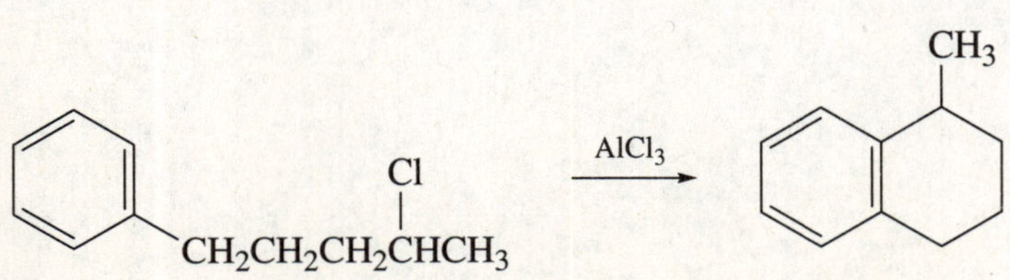

c. In this synthesis and in the synthesis of "**d**", a primary alkyl halide cannot be used in the first step because the carbocation will rearrange to a secondary carbocation and will result in the formation of a methyl-subsititued five-membered ring.

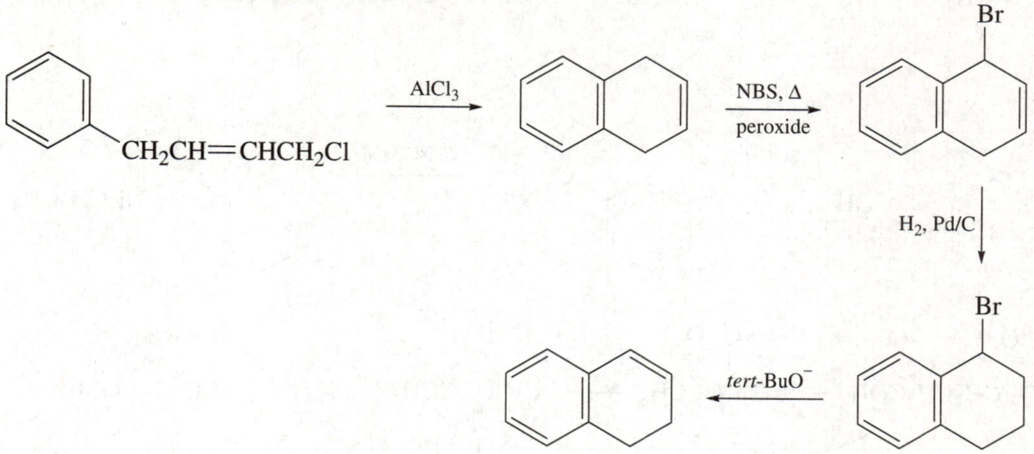

d.

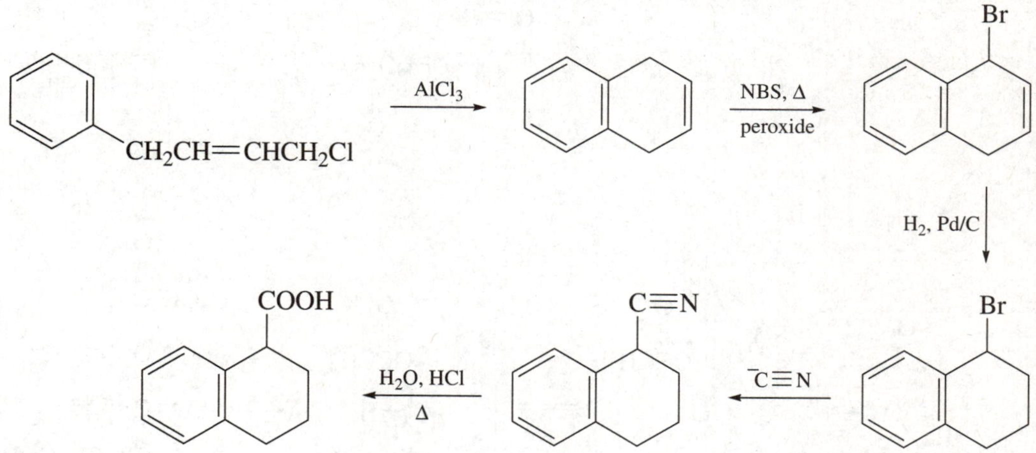

e.

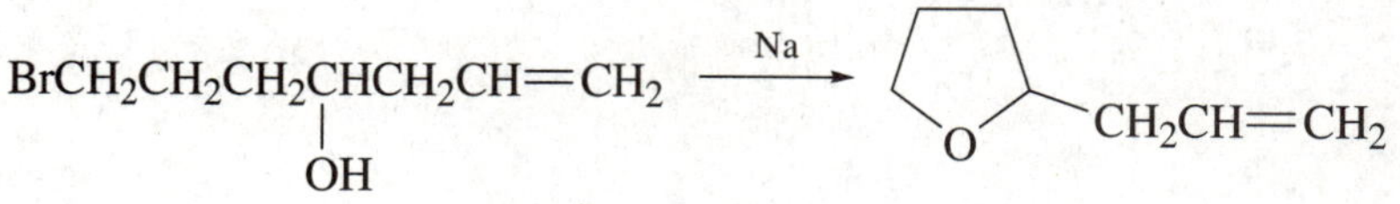

f.

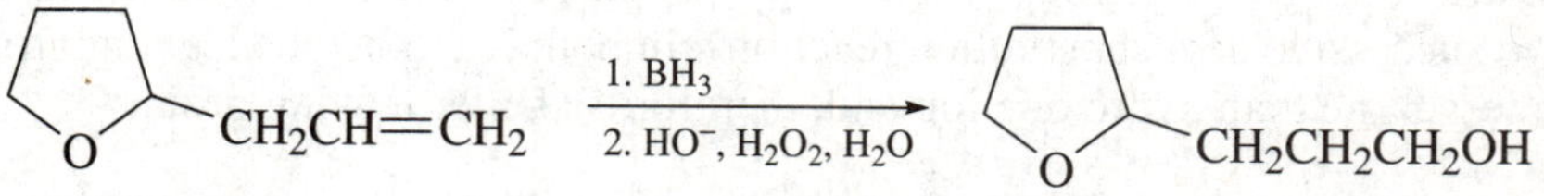

product of **e**

41. **a.**

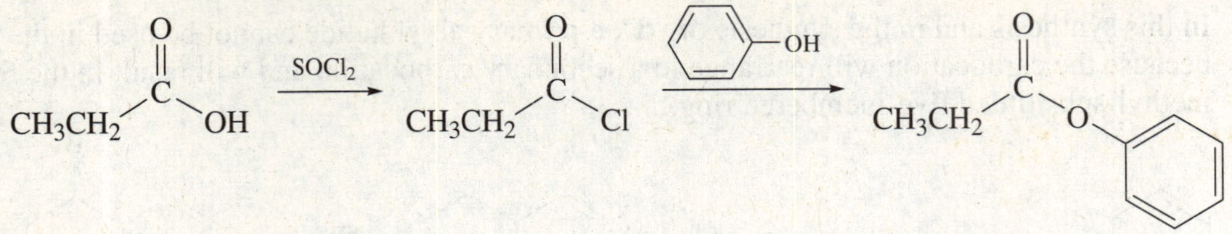

b.

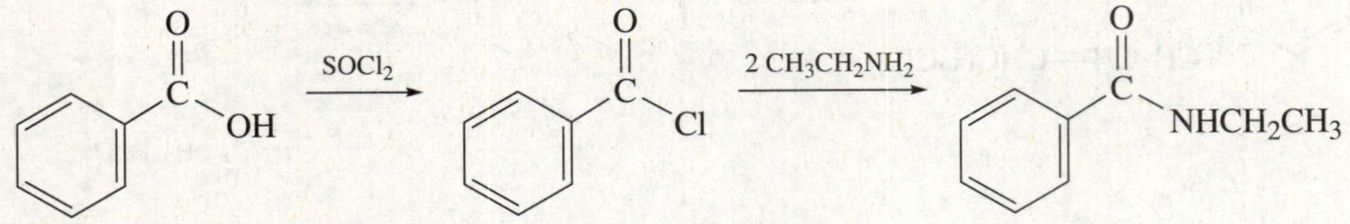

42. **a.**

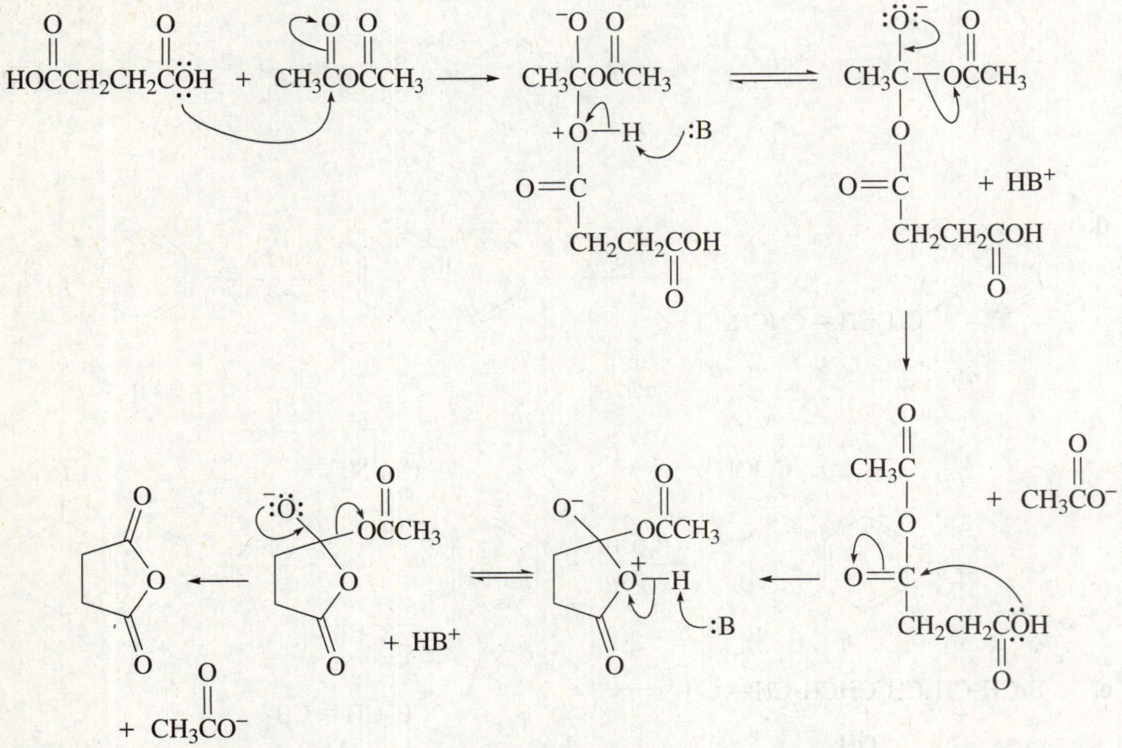

b. Without acetic anhydride, the leaving group would be hydroxide ion. Acetic anhydride causes the reaction to take place via two successive acyl substitution reactions. In both reactions the leaving group is acetate ion, which is much less basic than hydroxide ion and, therefore, a better leaving group.

43. PDS lasts longer because its ester linkage will hydrolyze more slowly because its alcohol leaving group is a stronger base. You can determine which of the alcohols is a stronger base by comparing how close the negatively-charged oxygen atom is to an electron-withdrawing group. An electron-withdrawing group stabilizes the base, making it a weaker base and, therefore, a better leaving group. In Dexon, there is a double-bonded oxygen attached to the CH_2 group that is next to the negatively charged oxygen. In PDS, there are two hydrogen attached to the CH_2 group that is next to the negatively charged oxygen. Thus, the alcohol is Dexon is a weaker base and a better leaving group.

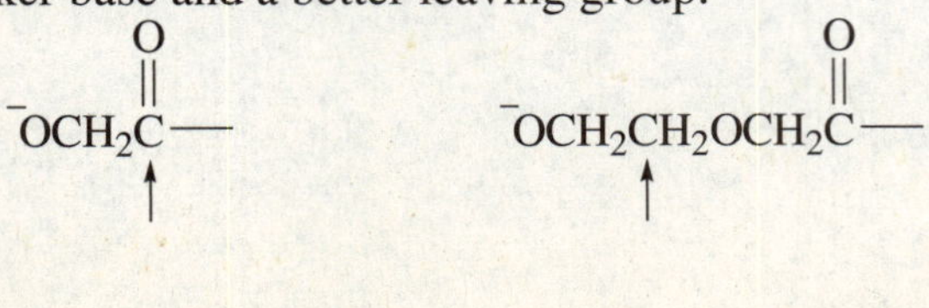

44. **a.** 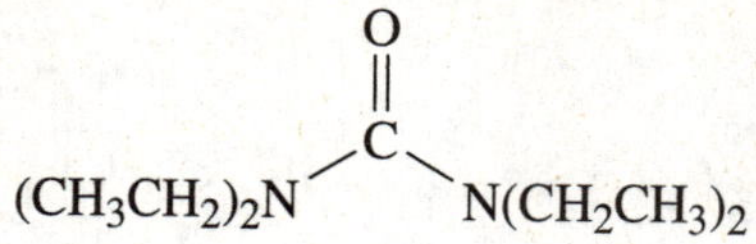

e. 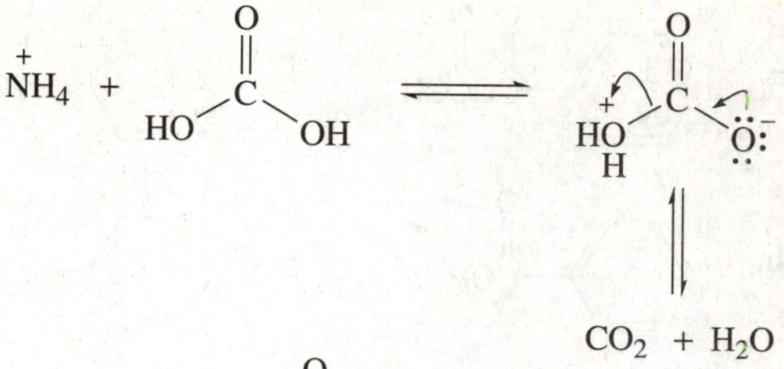

b. 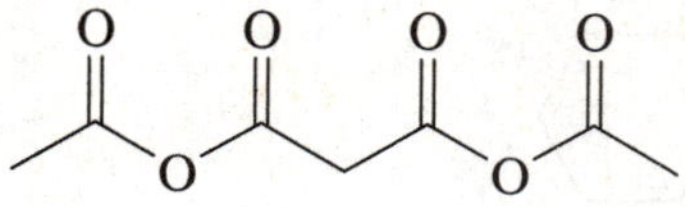

c. 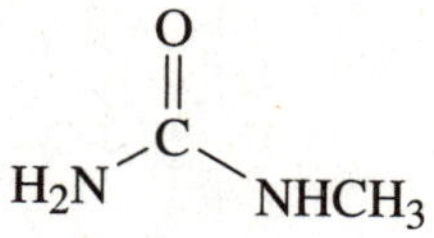

f. 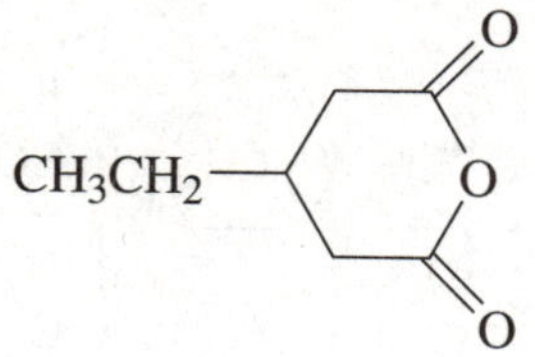

d. no reaction

45. **a.** 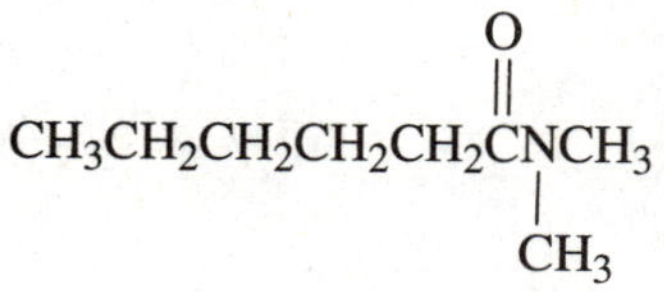

d. $CH_3CH_2C{\equiv}N$

h. 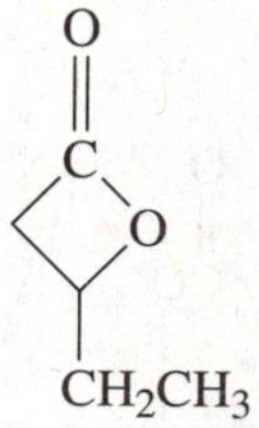

b. 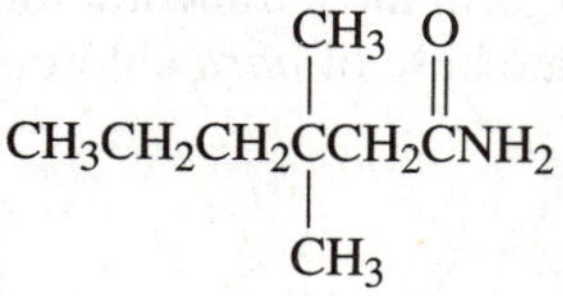

e. 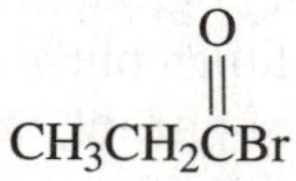

i. 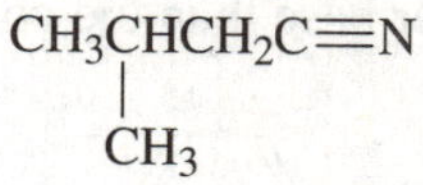

c. 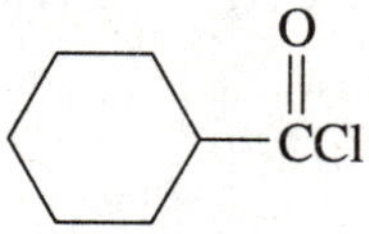

f. 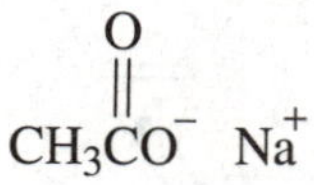

j. 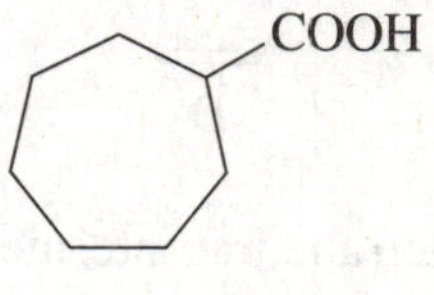

g. 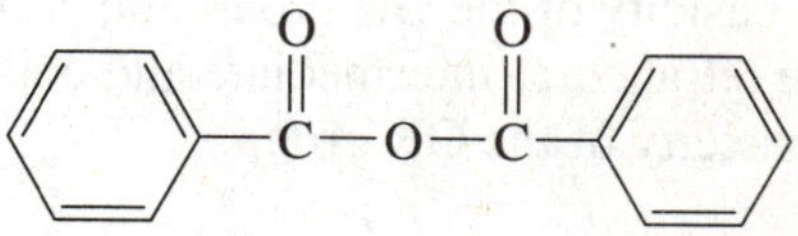

46. **a.** 5-ethylheptanoic acid

b. propyl propanoate
propyl propionate

c. (*S*)-3-methylpentanoic acid
(*S*)-β-methylvaleric acid

d. *N,N*-dimethylbutanamide
N,N-dimethylbutyramide

e. pentanoyl chloride
valeryl chloride

f. acetic benzoic anhydride

g. *N*-methyl-3-butenamide

h. propanoic anhydride
propionic anhydride

i. (*R*)-3-methylhexanenitrile

47.

a. 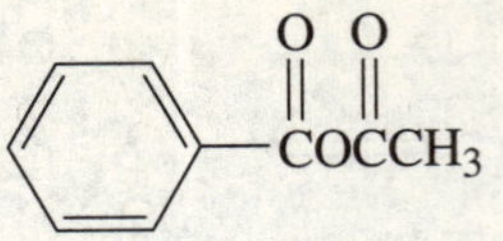

f. 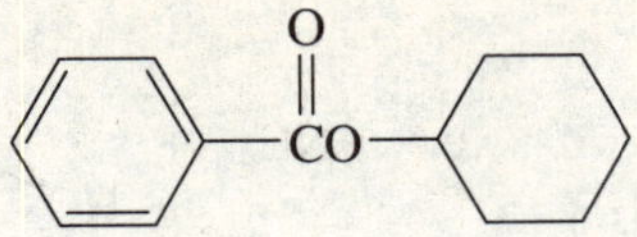

b. 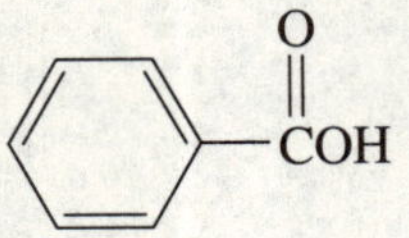

g. 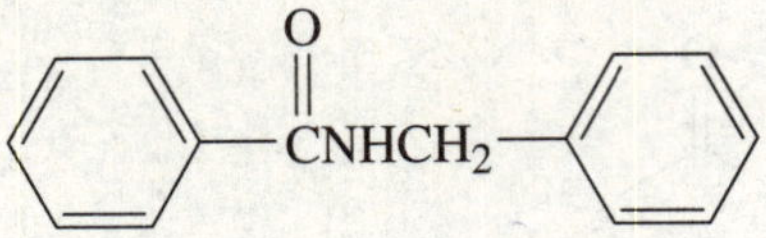

c. 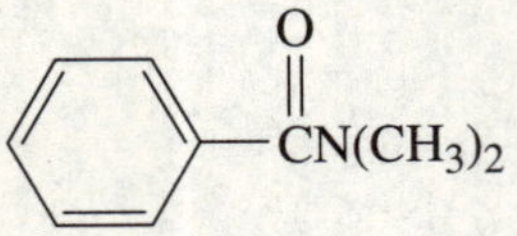

h. 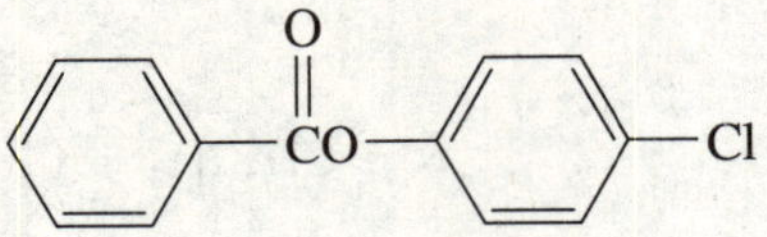

d. 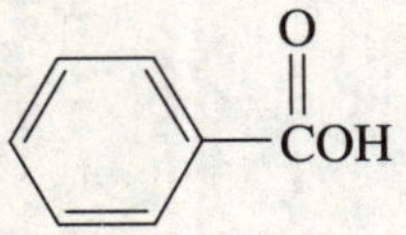

i. 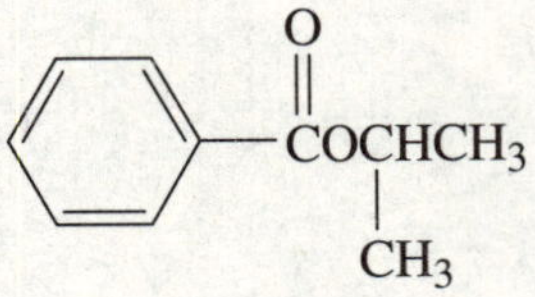

e. 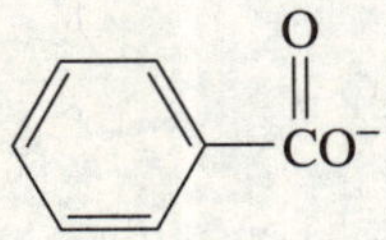

j. 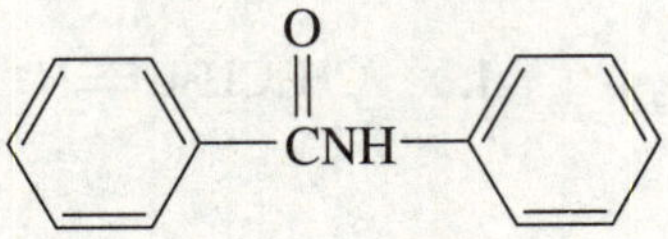

48. **a.** The weaker the base attached to the acyl group, the easier it is to form the tetrahedral intermediate. (*para*-Chlorophenol is a stronger acid than phenol so the conjugate base of *para* chlorophenol is a weaker base than the conjugate base of phenol, etc.)

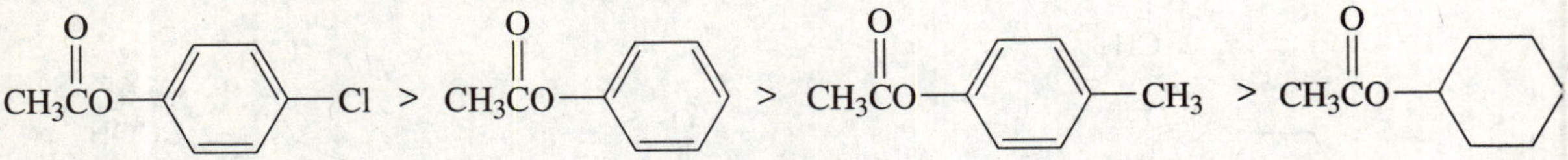

D **A** **C** **B**

b. The tetrahedral intermediate collapses by eliminating the OR group of the tetrahedral intermediate. The weaker the basicity of the OR group, the easier it is to eliminate it. Thus, the rate of both formation of the tetrahedral intermediate and collapse of the tetrahedral intermediate is decreased by increasing the basicity of the OR group.

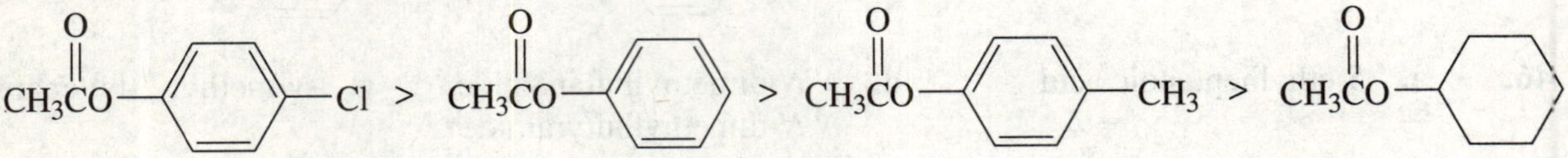

49. Because acetate ion is a weak base, the S_N2 reaction will from only a substitution product (see page 415 of the text). The product of the S_N2 reaction is an ester which, when hydrolyzed, forms cyclohexanol (the target compound) and acetic acid.

50. **a.** Methyl acetate has a resonance contributor that butanone does not have, and this resonance contributor causes methyl acetate to be more polar than butanone. Because butanone is less polar, it has the lower dipole moment.

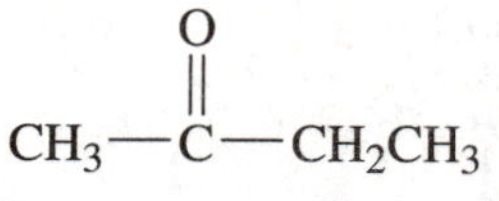

b. Because it is more polar, the intermolecular forces holding methyl acetate molecules together are stronger, so we would expect methyl acetate to have a higher boiling point.

51. Propyl formate is easy to distinguish because it is the only ester that will show four signals. The other three esters show three signals. Isopropyl formate can be distinguished by its unique splitting pattern: a singlet, a doublet, and a septet. The splitting patterns of the other two esters are the same: a singlet, a triplet, and a quartet. They can be distinguished because the highest frequency signal in ethyl acetate is a quartet, whereas the highest frequency signal in methyl propionate is a singlet.

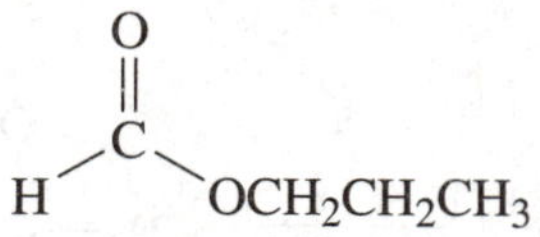

4 signals

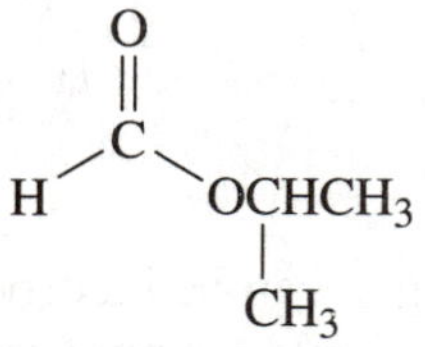

3 signals

singlet, doublet, septet

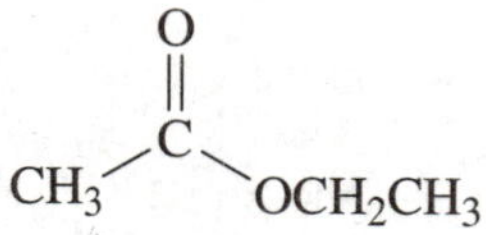

3 signals

singlet, triplet, quartet

The signal farthest downfield is a quartet.

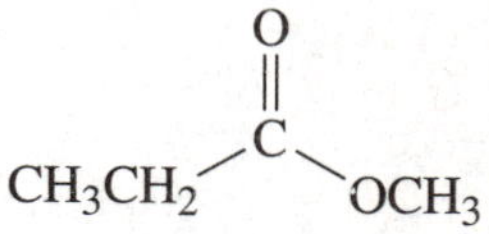

3 signals

singlet, triplet, quartet

The signal farthest downfield is a singlet.

52. The reaction of methylamine with propionyl chloride generates a proton that will protonate unreacted amine, thereby destroying its nucleophilicity. If two equivalents of CH_3NH_2 are used, one equivalent will remain unprotonated and be able to react with propionyl chloride to form *N*-methylpropanamide.

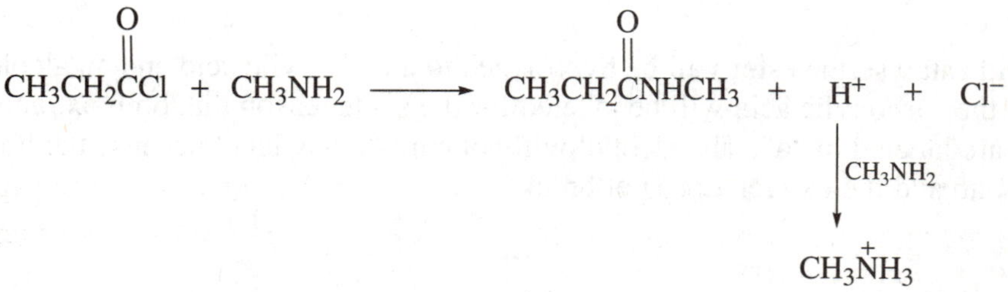

53. **a.**

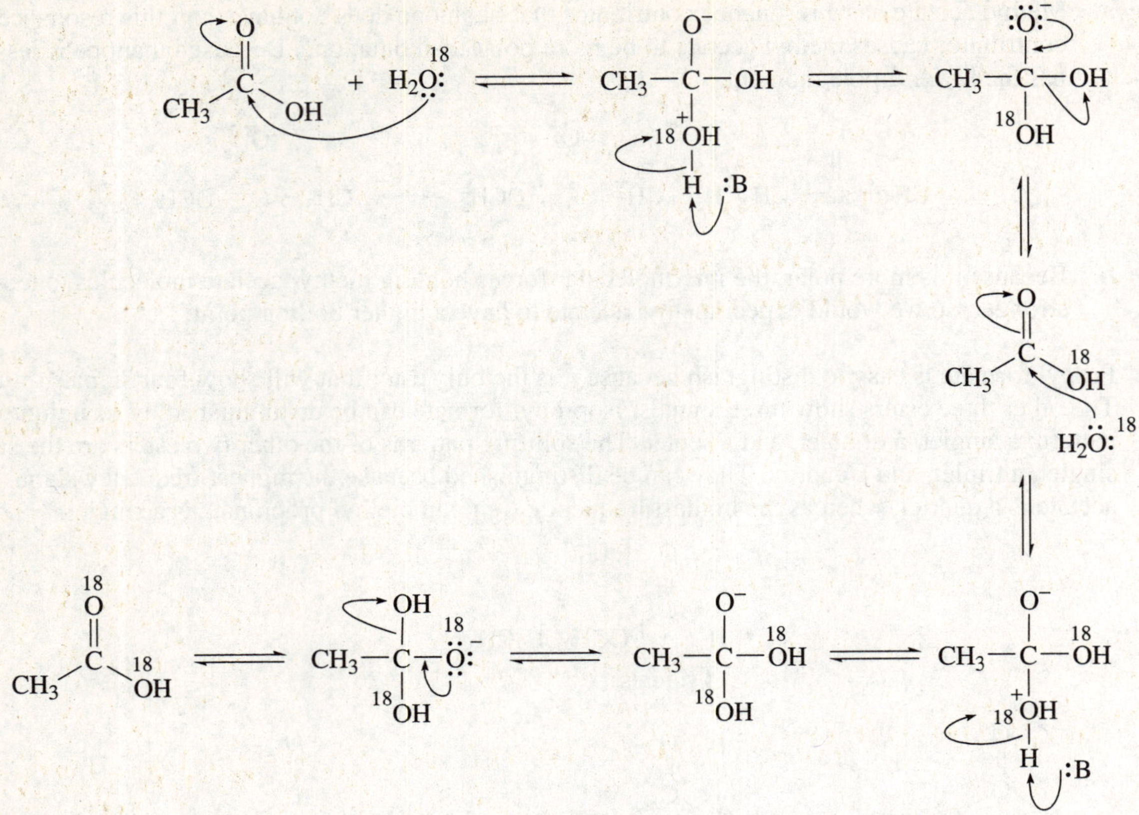

b. The carboxyl oxygen will be labeled. Only one isotopically labeled oxygen can be incorporated into the ester because the bond between the methyl group and the labeled oxygen does not break, so there is no way for the carbonyl oxygen to become labeled.

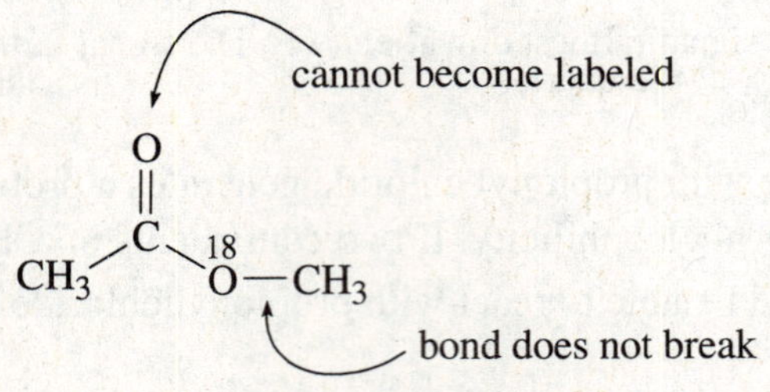

c. In the present of an acid catalyst, the ester will be hydrolyzed to a carboxylic acid and an alcohol. Both oxygen atoms of the carboxylic acid will be labeled for the same reason that both oxygen atoms of the carboxylic acid are labeled in "**a**". The alcohol will not contain any label because the bond between the methyl group and the oxygen does not break.

54. **a.** $CH_3\overset{O}{\overset{\|}{C}}OH \xrightarrow{SOCl_2} CH_3\overset{O}{\overset{\|}{C}}Cl \xrightarrow{CH_3CH_2CH_2OH} CH_3\overset{O}{\overset{\|}{C}}OCH_2CH_2CH_3$

$CH_3\overset{O}{\overset{\|}{C}}O^- \xrightarrow{CH_3CH_2CH_2Br} CH_3\overset{O}{\overset{\|}{C}}OCH_2CH_2CH_3$

b. $CH_3\overset{O}{\overset{\|}{C}}OH \xrightarrow{SOCl_2} CH_3\overset{O}{\overset{\|}{C}}Cl \xrightarrow{(CH_3)CH(CH_2)_2OH} CH_3\overset{O}{\overset{\|}{C}}O(CH_2)_2CH(CH_3)_2$

$CH_3\overset{O}{\overset{\|}{C}}O^- \xrightarrow{(CH_3)CH(CH_2)_2Br} CH_3\overset{O}{\overset{\|}{C}}O(CH_2)_2CH(CH_3)_2$

c. $CH_3CH_2CH_2\overset{O}{\overset{\|}{C}}OH \xrightarrow{SOCl_2} CH_3CH_2CH_2\overset{O}{\overset{\|}{C}}Cl \xrightarrow{CH_3CH_2OH} CH_3CH_2CH_2\overset{O}{\overset{\|}{C}}OCH_2CH_3$

$CH_3CH_2CH_2\overset{O}{\overset{\|}{C}}O^- \xrightarrow{CH_3CH_2Br} CH_3CH_2CH_2\overset{O}{\overset{\|}{C}}OCH_2CH_3$

d.

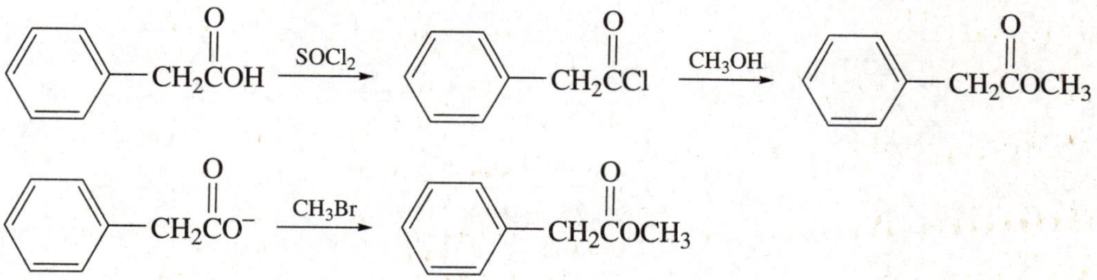

55. **a.** isopropyl alcohol and HCl
b. aqueous sodium hydroxide
c. ethylamine
d. water and HCl

56. The offset in the NMR spectrum shows that there is a signal at ~ 10 ppm, which is where the proton of a COOH group shows a signal. The two triplets and the multiplet are characteristic of a propyl group. The compound is butanoic acid.

$C_5H_{10}O_2$

The molecular formula shows that the unknown compound has one more carbon atom than butanoic acid; the IR spectrum shows it is an ester. Since butanoic acid is formed from acid hydrolysis of the ester, the compound must be the methyl ester of butanoic acid.

57. Aspartame has an amide group and an ester group that will be hydrolyzed in an aqueous solution of HCl. Because the hydrolysis is carried out in an acidic solution, the carboxylic acid groups and the amino groups will be in their acidic forms.

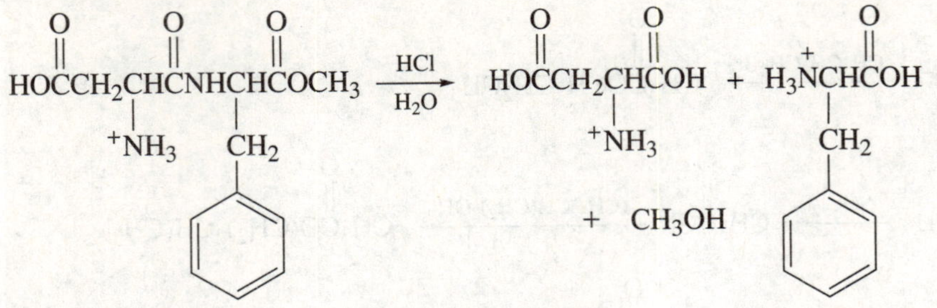

58. **a.** 1, 3, 4, 6, 7, 9 will not form the indicated products under the given conditions.

b. 9 will form the product shown in the presence of an acid catalyst.

59. The tertiary amine is a stronger nucleophile than the alcohol, so formation of the charged amide will be faster than formation of the new ester would have been. The charged amide is more reactive than an ester, so formation of the new ester by reaction of the alcohol with the charged amide will be faster than formation of the ester by reaction of the alcohol with the starting ester would have been. In other words, both reactions that occur in the presence of the tertiary amine are faster than the single reaction that occurs in the absence of the tertiary amine.

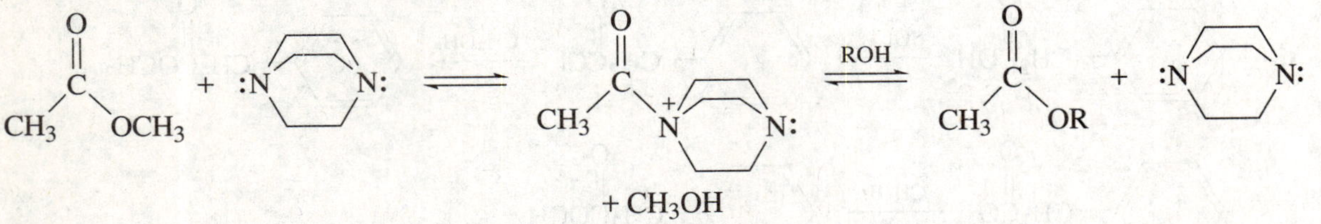

60. 1-Bromobutane undergoes an S_N2 reaction with NH_3 to form butylamine (**A**). Butylamine can then react with 1-bromobutane to form dibutylamine (**B**). The amines each form an amide upon reaction with acetyl chloride. The IR spectrum of **C** exhibits an NH stretch at about 3300 cm^{-1}, whereas the IR spectrum of **D** does not exhibit this absorption band, because the nitrogen in **D** is not bonded to a hydrogen.

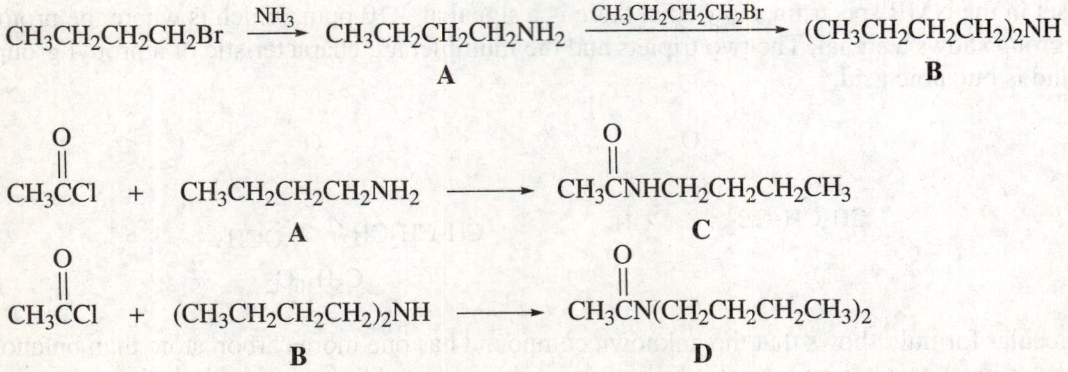

61. **1.** $Cl\overset{O}{\overset{\|}{C}}OCH_3$ **2.** $CH_3O\overset{O}{\overset{\|}{C}}OCH_3$ **3.** $CH_3CH_2CH_2NH\overset{O}{\overset{\|}{C}}NHCH_2CH_2CH_3$ **4.** $CH_3CH_2O\overset{O}{\overset{\|}{C}}NHCH_3$

62. 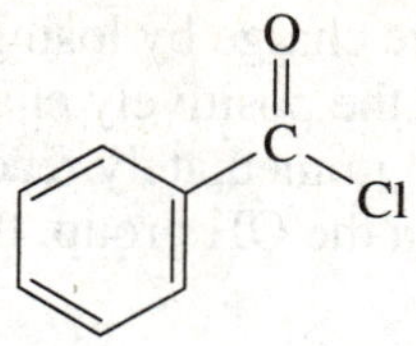

63. 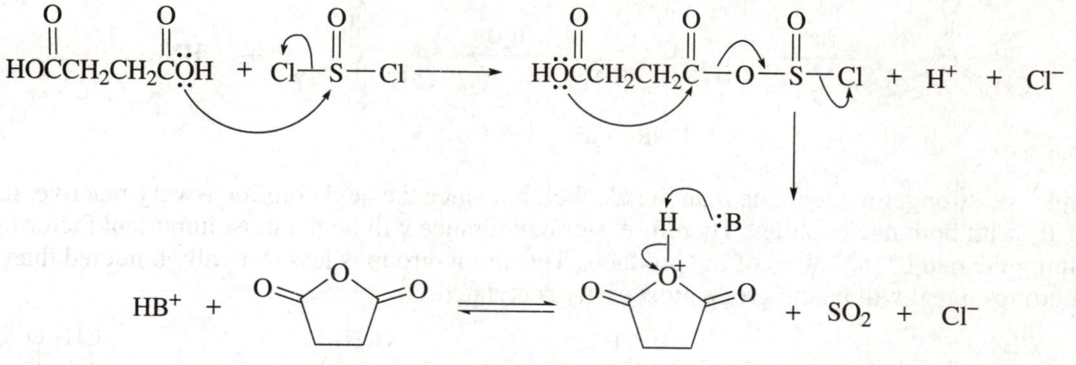

64. a. $CH_3\overset{O}{\overset{\|}{C}}F$

b. $HO\overset{O}{\overset{\|}{C}}CH_2CH_2CH_2\overset{+}{N}H_3\ Cl^-$

c. 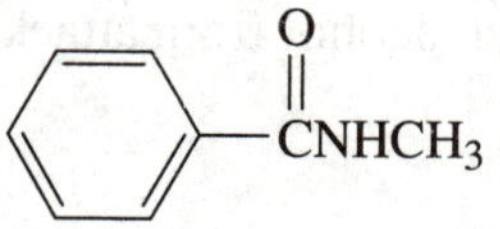

d. $HO\overset{O}{\overset{\|}{C}}CH_2CH_2\overset{O}{\overset{\|}{C}}OH$

e. 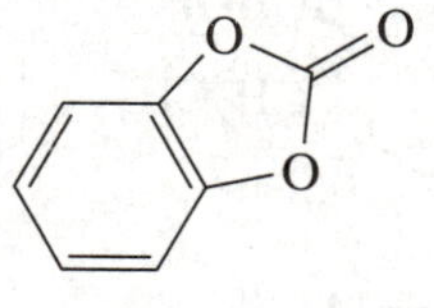

f. $HO\overset{O}{\overset{\|}{C}}CH_2CH_2CH_2OH$

g. $CH_3\overset{O}{\overset{\|}{C}}CH_2OH + CH_3O\overset{O}{\overset{\|}{C}}CH_3$

h. 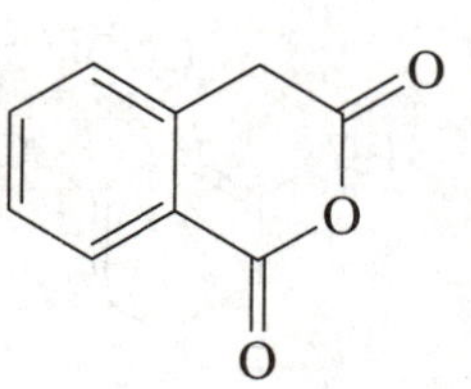

i. 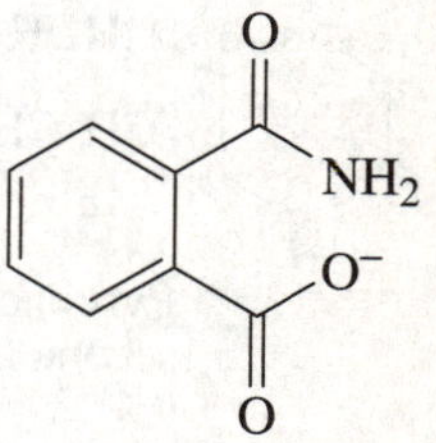

j. 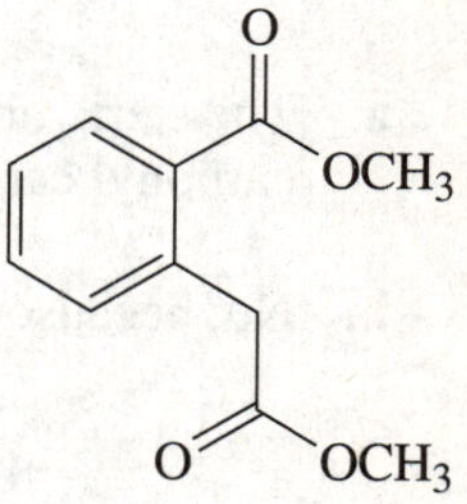

65. $ClCH_2CH_2CH_2\overset{O}{\overset{\|}{C}}Cl + CH_3OH \longrightarrow ClCH_2CH_2CH_2\overset{O}{\overset{\|}{C}}OCH_3$

Compound A

66. a. $CH_3\overset{O}{\overset{\|}{C}}NH\underset{CH_3}{\underset{|}{C}}HCH_2CH_2O\overset{O}{\overset{\|}{C}}CH_3 + CH_3\overset{O}{\overset{\|}{C}}OH$ **b.** $CH_3\overset{O}{\overset{\|}{C}}NH\underset{CH_3}{\underset{|}{C}}HCH_2CH_2OH$

Because the amino group is a stronger nucleophile than the OH group, the predominant product will be the amide if the reaction is stopped prematurely.

67. If the amine is tertiary, the nitrogen atom in the amide cannot get rid of its positive charge by losing a proton. An amide with a positively charged amino group is very reactive because the positively charged amino group is a weak base and, therefore, an excellent leaving group. Water will immediately react with the amide, and because the positively charged amine is a better leaving group than the OH group, the amine will be expelled and the product will be a carboxylic acid.

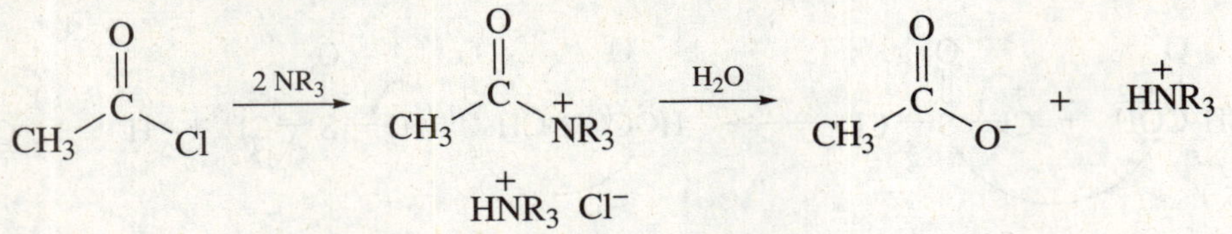

68. The amine is a stronger nucleophile than the alcohol, but since the acyl chloride is very reactive, it can react easily with both nucleophiles. Therefore, steric hindrance will be the most important factor in determining the rate of formation of the products. The amino group is less sterically hindered than the alcohol group so that will be the group most easily acetylated.

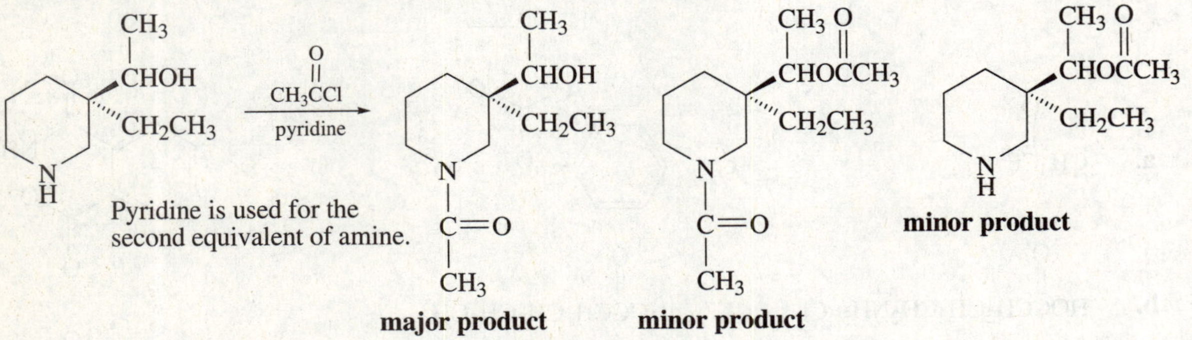

69. **a.** The steric hindrance provided by the methyl groups prevents methyl alcohol from attacking the carbonyl carbon.

b. No, because there would be no steric hindrance to nucleophilic attack.

c.

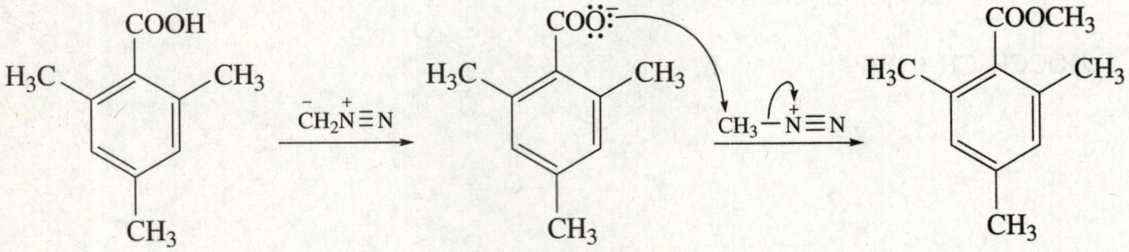

70. The spectrum shows that the compound has four different kinds of hydrogens with relative ratios 2 : 1 : 1 : 6. The doublet at 0.9 ppm and the ratio of 6 hydrogens suggests that the compound has an iso group. The signal at 3.4 ppm indicates hydrogens on a carbon that is attached to an oxygen atom. The shape of the signal at 2.4 ppm suggests an OH group. We can conclude that the spectrum is a spectrum of isobutyl alcohol. The molecular formula indicates the compound that undergoes hydrolysis is an ester. Subtracting the atoms due to the isobutyl group lets us identify the ester as isobutyl benzoate.

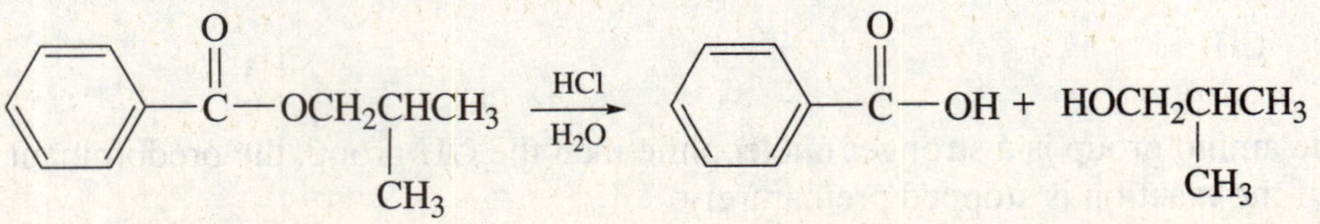

71. The more electronegative the base (Y) attached to the carbonyl carbon, the greater the double bond character of the C═O bond.

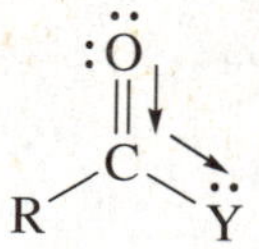

The weaker the base (Y) attached to the carbonyl carbon the less well it shares it electrons. When Y shares its electrons, it decreases the double bond character of the C═O bond, making it easier to stretch.

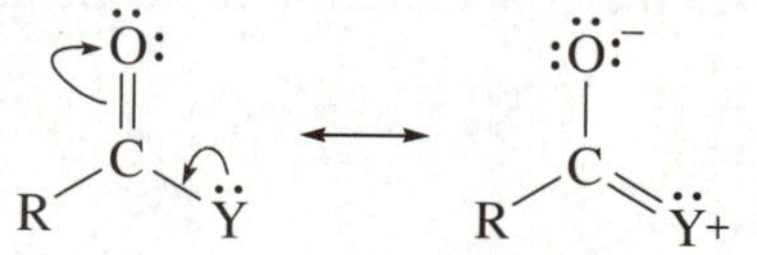

Therefore, the carbonyl IR absorption band decreases in the order:

$$CH_3\overset{\overset{O}{\|}}{C}Cl \;>\; CH_3\overset{\overset{O}{\|}}{C}OCH_3 \;>\; CH_3\overset{\overset{O}{\|}}{C}H \;>\; CH_3\overset{\overset{O}{\|}}{C}NH_2$$

The carbonyl group of the acyl chloride stretches at the highest frequency because the chlorine atom is the most electronegative atom and the weakest base.

The predominant effect of the oxygen of an ester is inductive electron withdrawal, so its carbonyl group stretches at a higher frequency than the carbonyl group of the aldehyde.

The carbonyl group of the amide stretches at the lowest frequency because the nitrogen atom is the strongest base, so it is best at sharing its electrons.

72. **a.**

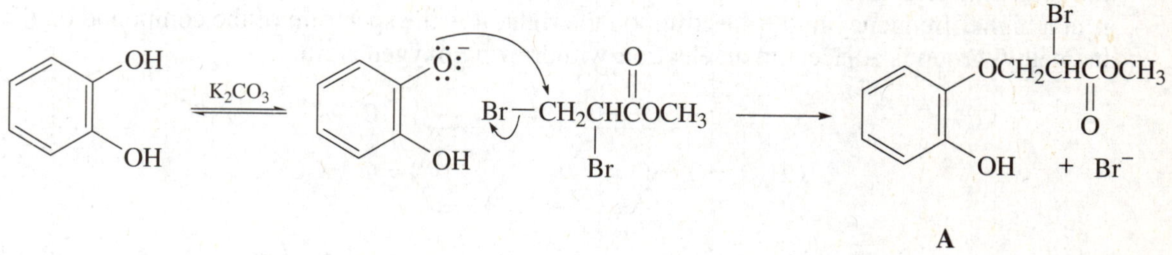

b.

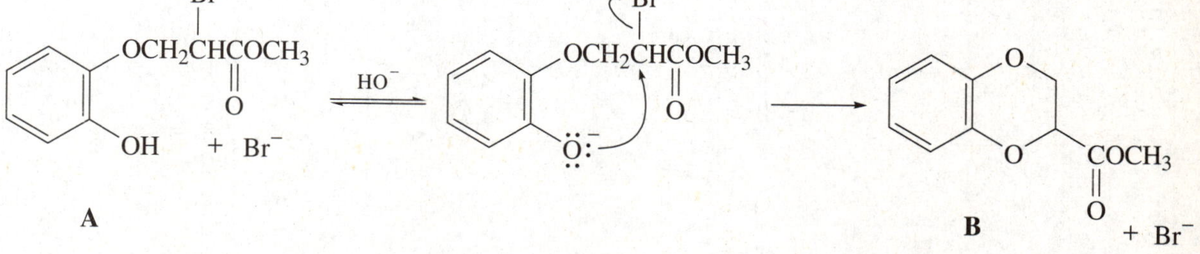

The intramolecular formation of **B** will occur more rapidly than the intermolecular formation of **A** because in the former reaction the reagents don't have to wander through the solution to find one another.

73. **a.** $4.02 = \dfrac{x^2}{(1-x)^2}$

$2.00 = \dfrac{x}{(1-x)}$

$2 - 2x = x$

$2 = 3x$

$x = 0.667$

[ethyl acetate] = 0.667 times the concentration of acetic acid used

b. $4.02 = \dfrac{x^2}{(10-x)(1-x)}$

$x = 0.974$

[ethyl acetate] = 0.974 times the concentration of acetic acid used

c. $4.02 = \dfrac{x^2}{(100-x)(1-x)}$

$x = 0.997$

[ethyl acetate] = 0.997 times the concentration of acetic acid used

74. Each of the NMR spectra has signals between about 7-8 ppm, indicating that the compound has a benzene ring with one substituent (the benzene ring has hydrogens in three different environments) and one additional signal that is a singlet. From the molecular formulas it can be determined that the esters have the following structures. The singlet in each spectrum is due to a methyl group. Because the methyl group is at a higher frequency in the spectrum on the right, it is the spectrum of the compound on the right, since its methyl group is adjacent to an electron-withdrawing oxygen atom.

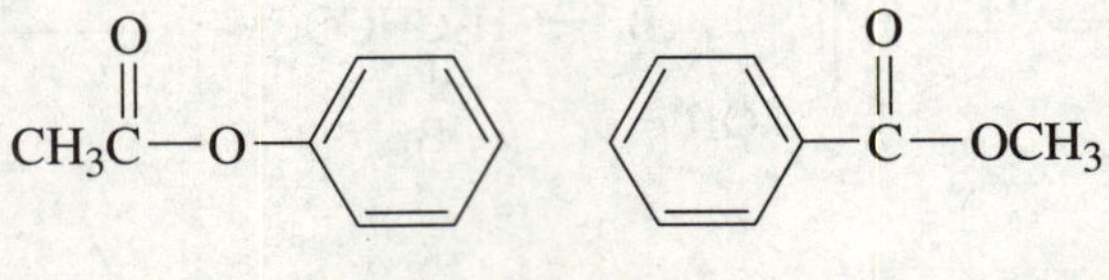

The ester on the left will hydrolyze faster because phenol is a much weaker base than methyl alcohol.

75. **a.**

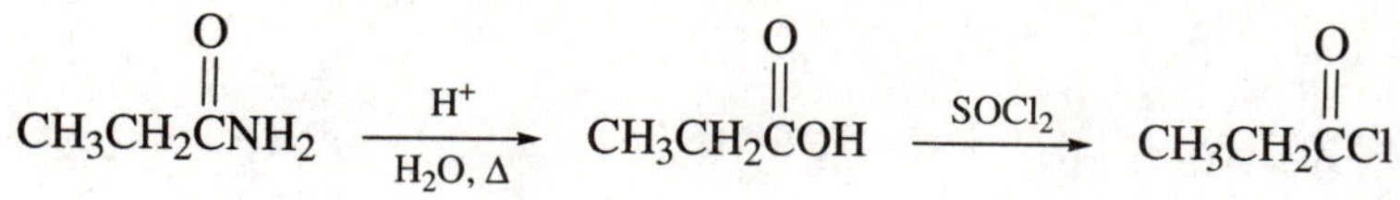

b.

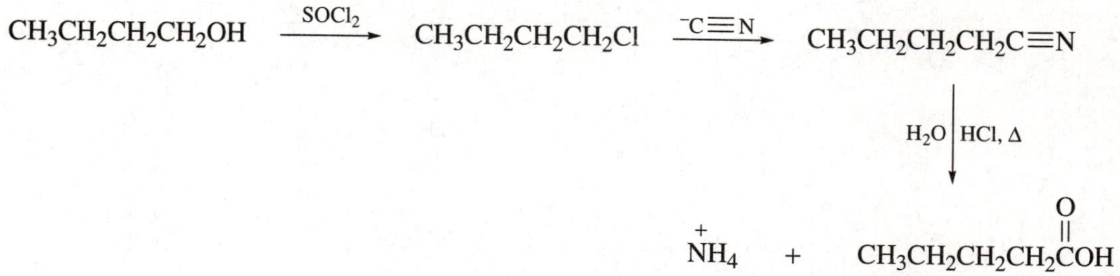

c.

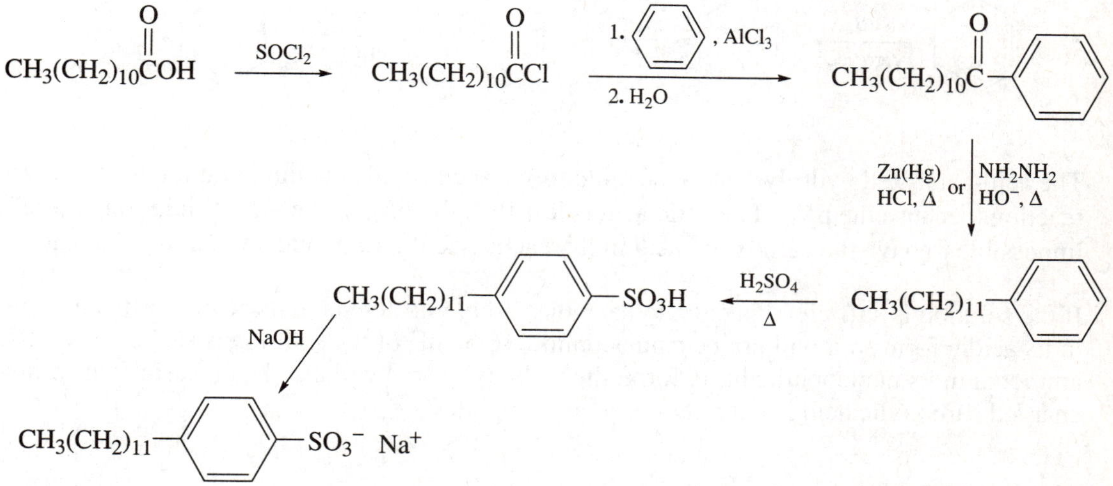

d.

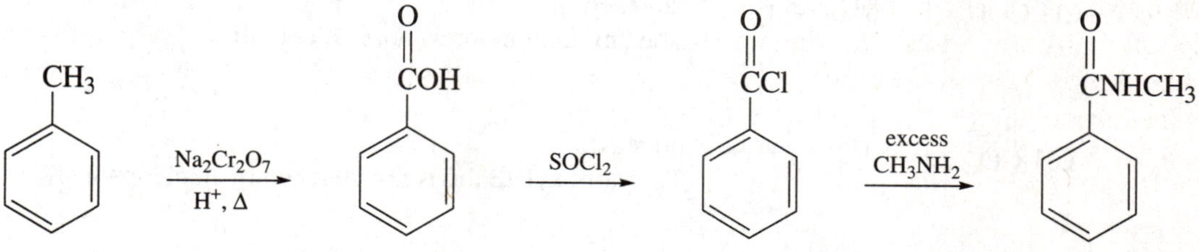

e. 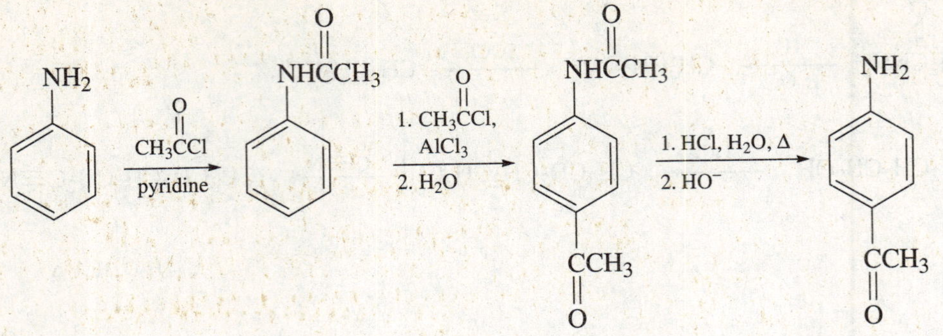

f. 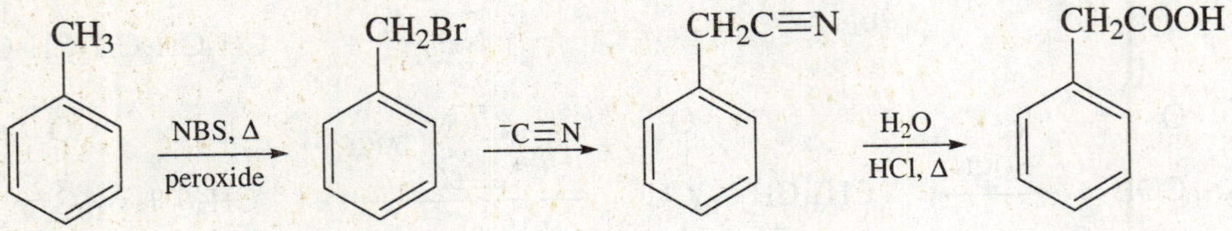

76. The acid-catalyzed hydrolysis of acetamide forms acetic acid and ammonium ion. It is an irreversible reaction, because the pK_a of a acetic acid is less than the pK_a of the ammonium ion. Therefore, it is impossible to have the carboxylic acid in its reactive acidic form and ammonia in its reactive basic form.

If the solution is sufficiently acidic to have the carboxylic acid is in its acidic form, ammonia will also be in its acidic form so it will not be a nucleophile. If the pH of the solution is sufficiently basic to have ammonia in its nucleophilic basic form, the carboxylic acid will also be its basic form; a negatively charged carboxylic acid is not attacked by nucleophiles

$CH_3\overset{O}{\overset{\|}{C}}OH + \overset{+}{N}H_4 \longrightarrow$ no reaction
The ammonium ion is not nucleophilic.

$CH_3\overset{O}{\overset{\|}{C}}O^- + NH_3 \longrightarrow$ no reaction
The carboxylate ion is not attacked by nucleophiles.

77. **a.** 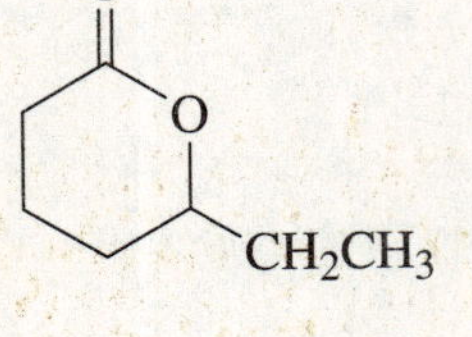

b. 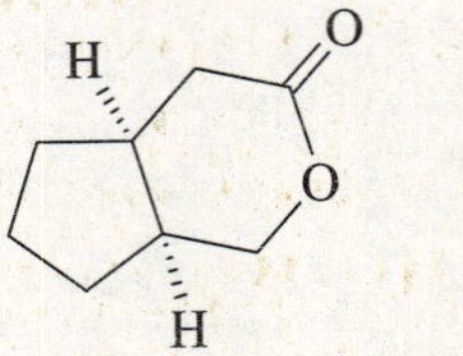

$+ CH_3CH_2OH$

c. 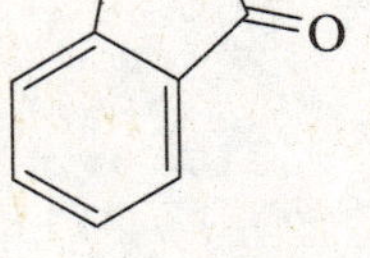

78.

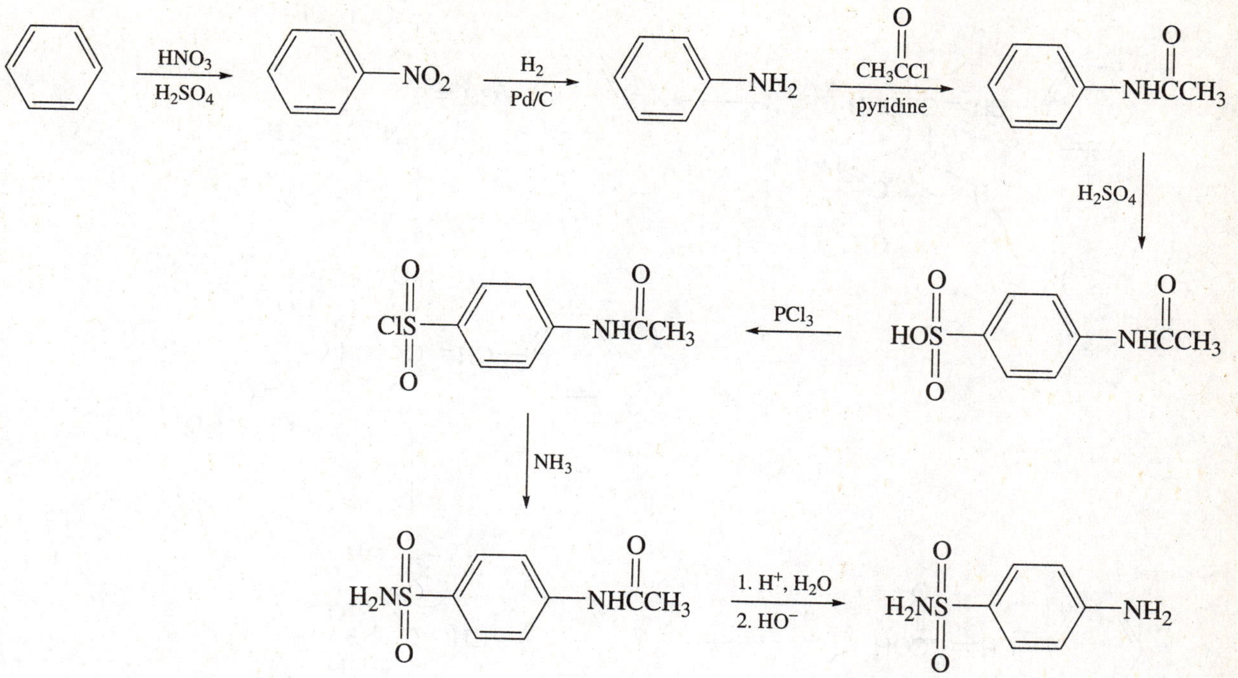

79. **a.**

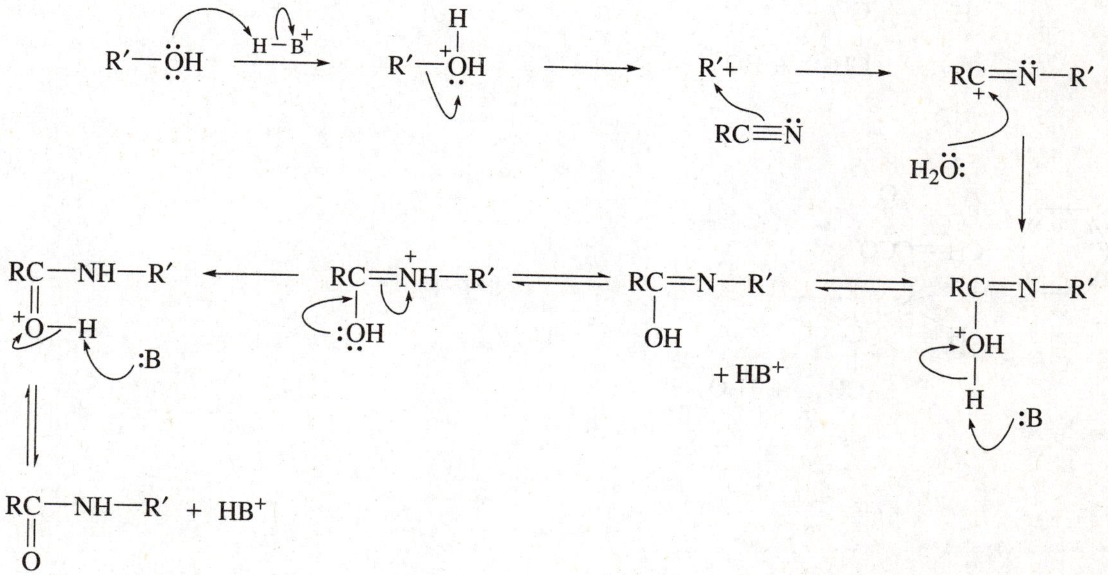

b. The Ritter reaction does not work with primary alcohols, because primary alcohols do not form carbocations.

c. The only difference in the two reactions is the electrophile that attaches to the nitrogen of the nitrile: it is a carbocation in the Ritter reaction and a proton in the acid-catalyzed hydrolysis of a nitrile.

80.

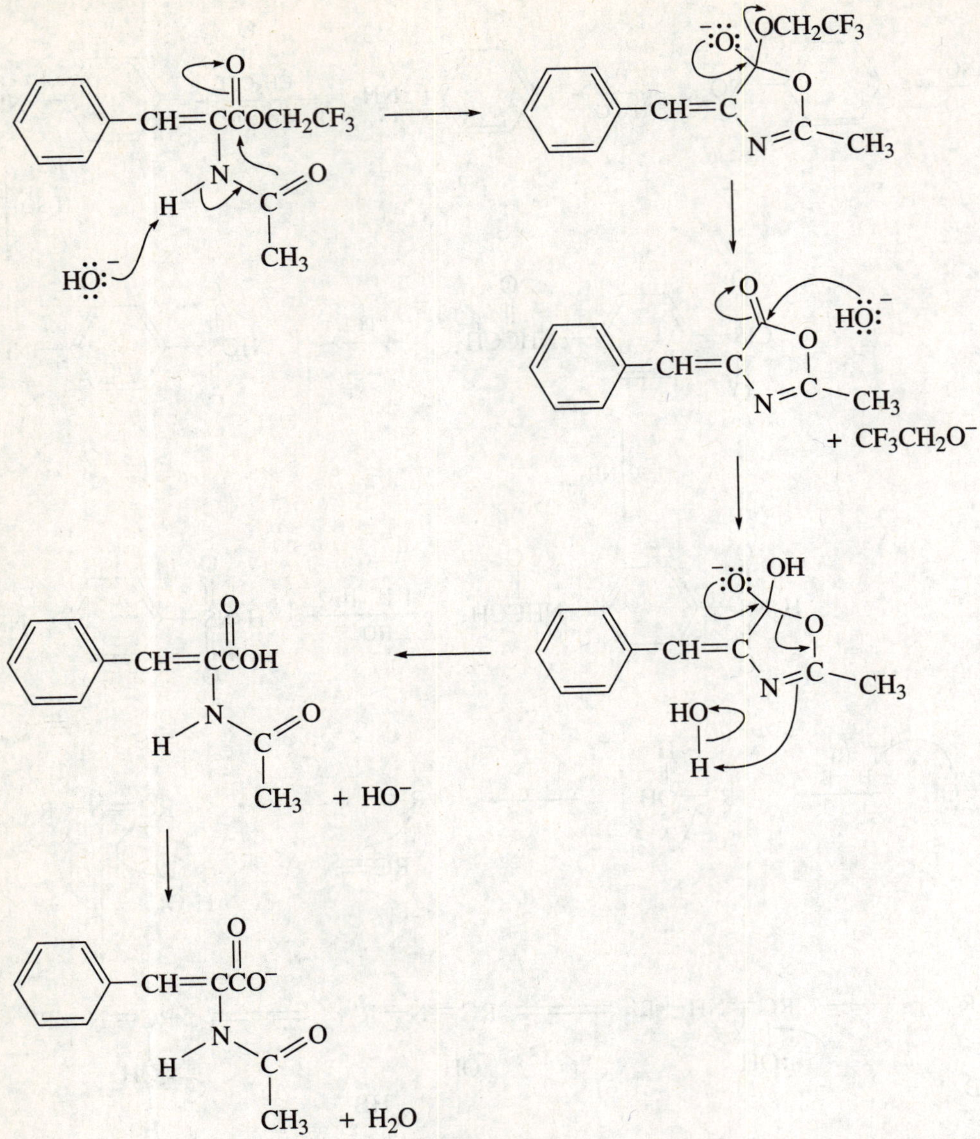

81. **a.**

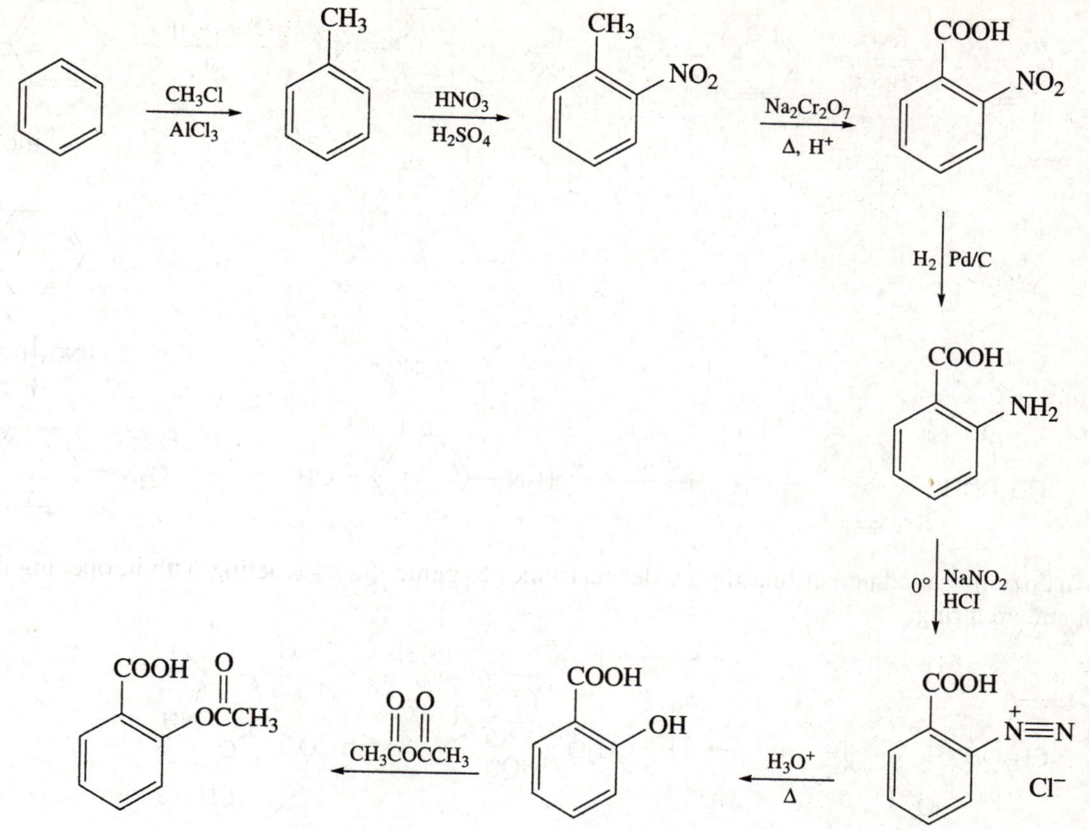
CH3Cl
AlCl3
CH3
HNO3
H2SO4
NO2
Na2Cr2O7
Δ, H+
COOH
H2 Pd/C
NH2
0° NaNO2
HCl
+
N≡N
Cl−
H3O+
Δ
OH
O O
CH3COCCH3
O
OCCH3

b.

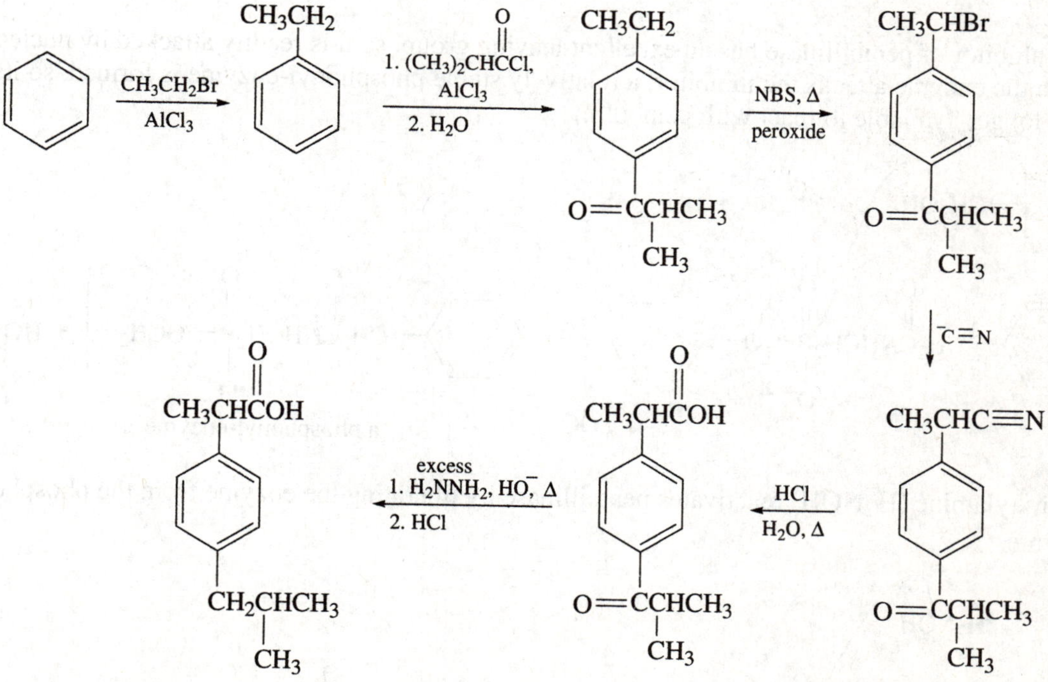
CH3CH2Br
AlCl3
CH3CH2
O
1. (CH3)2CHCCl,
AlCl3
2. H2O
O═CCHCH3
CH3
NBS, Δ
peroxide
CH3CHBr
−C≡N
CH3CHC≡N
HCl
H2O, Δ
CH3CHCOH
excess
1. H2NNH2, HO−, Δ
2. HCl
CH2CHCH3

c.

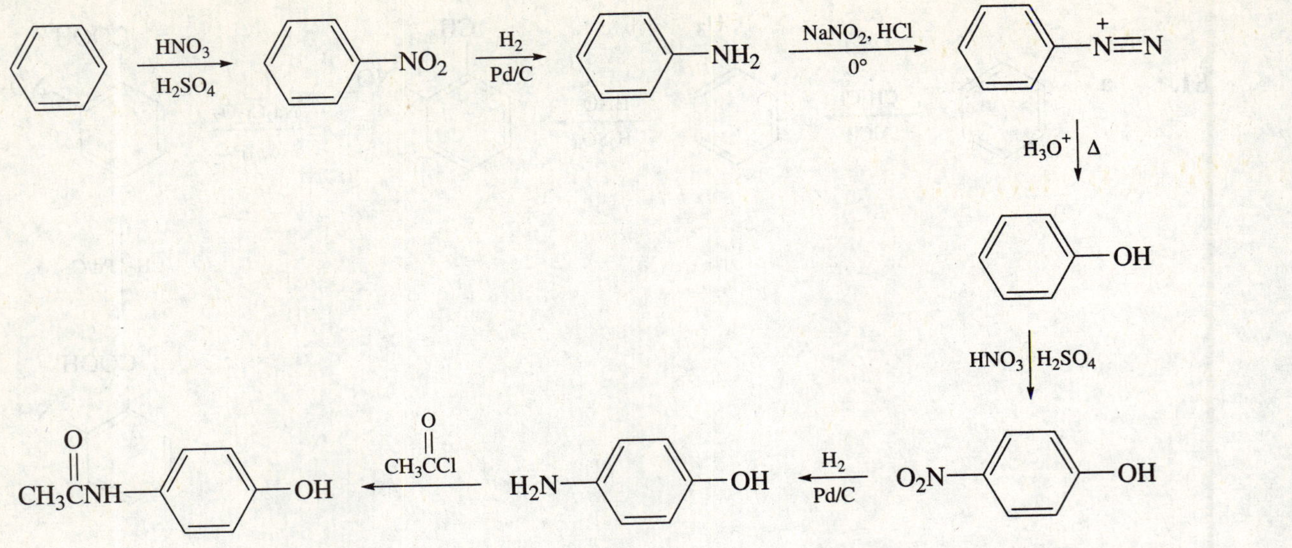

82. An enzyme called penicillinase provides resistance to penicillin by reacting with it, opening the four-membered ring.

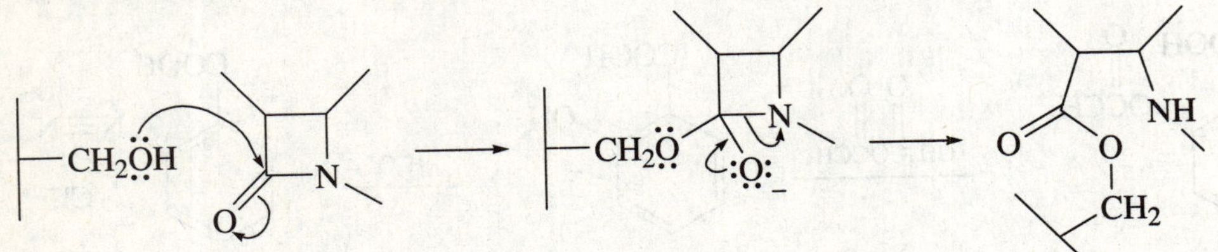

The inhibitor of penicillinase has an excellent leaving group, so it is readily attacked by nucleophiles. When the enzyme attacks the inhibitor, a relatively stable phosphonyl-enzyme is formed, so its OH group is no longer available to react with penicillin.

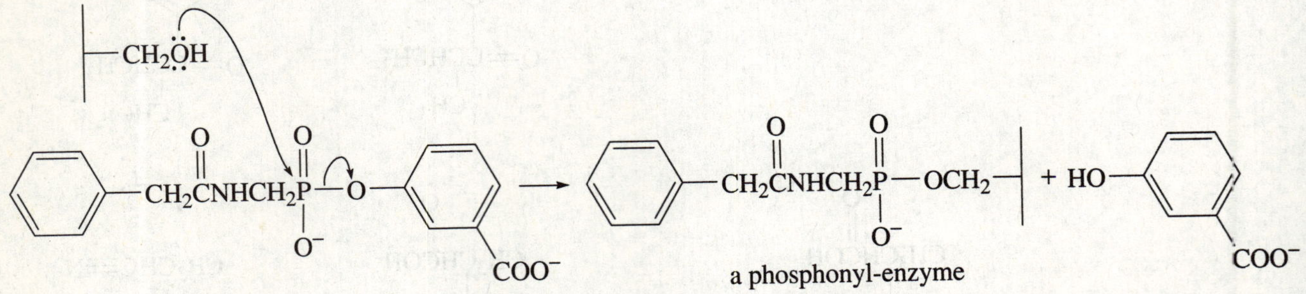

Hydroxylamine (H_2NOH) reactivates penicillinase by liberating the enzyme from the phosphonyl-enzyme.

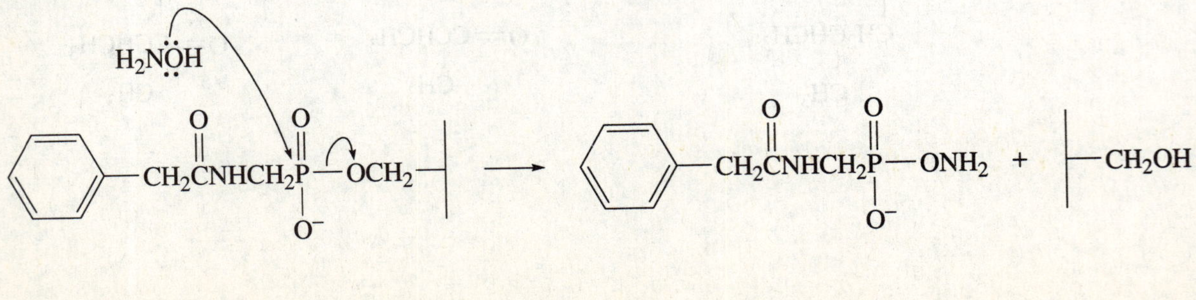

83. **a.**

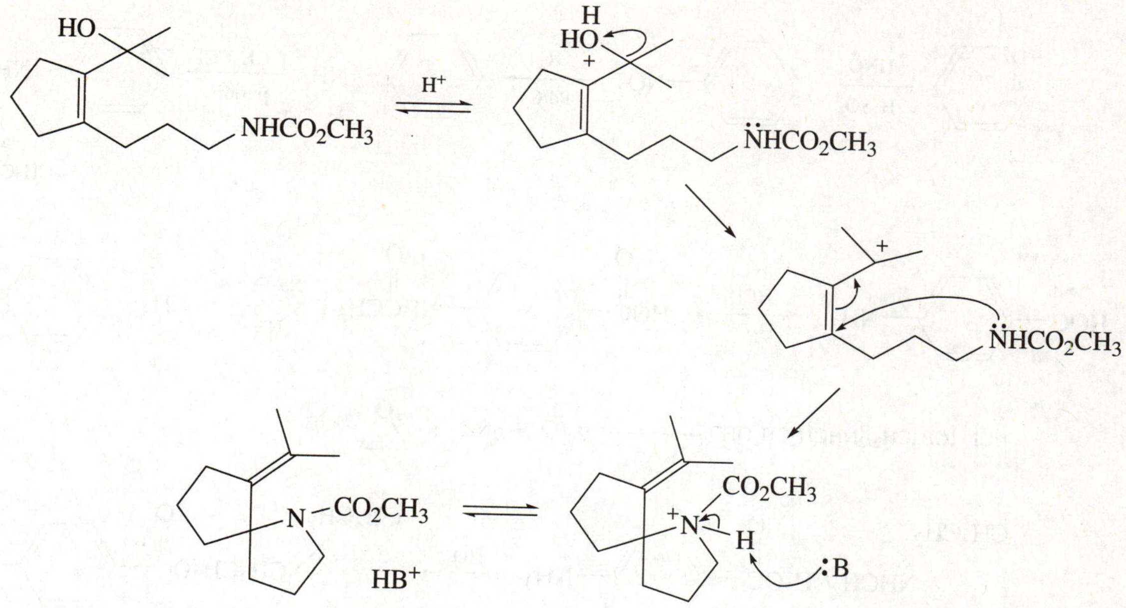
HO
H+
H
HO
+
NHCO2CH3
N̈HCO2CH3
+
N̈HCO2CH3
N–CO2CH3
HB+
CO2CH3
+N
H
:B

b.

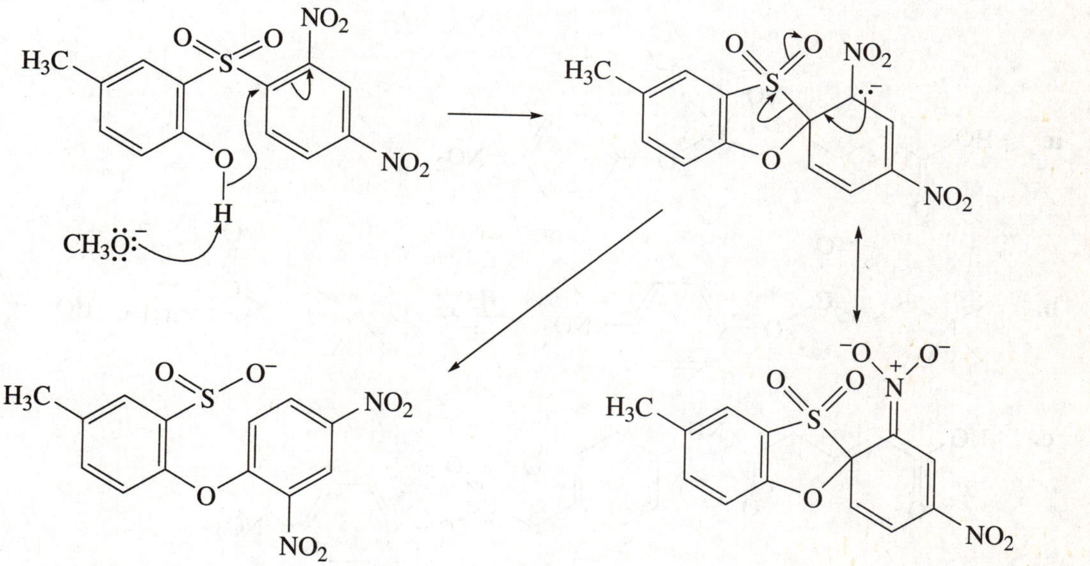
NO2
H3C
O
S
O
O
H
CH3Ö:−
NO2
H3C
NO2
NO2
−O
+
N
O−
H3C
S
O
NO2
O−
H3C
S
O
NO2
NO2

84.

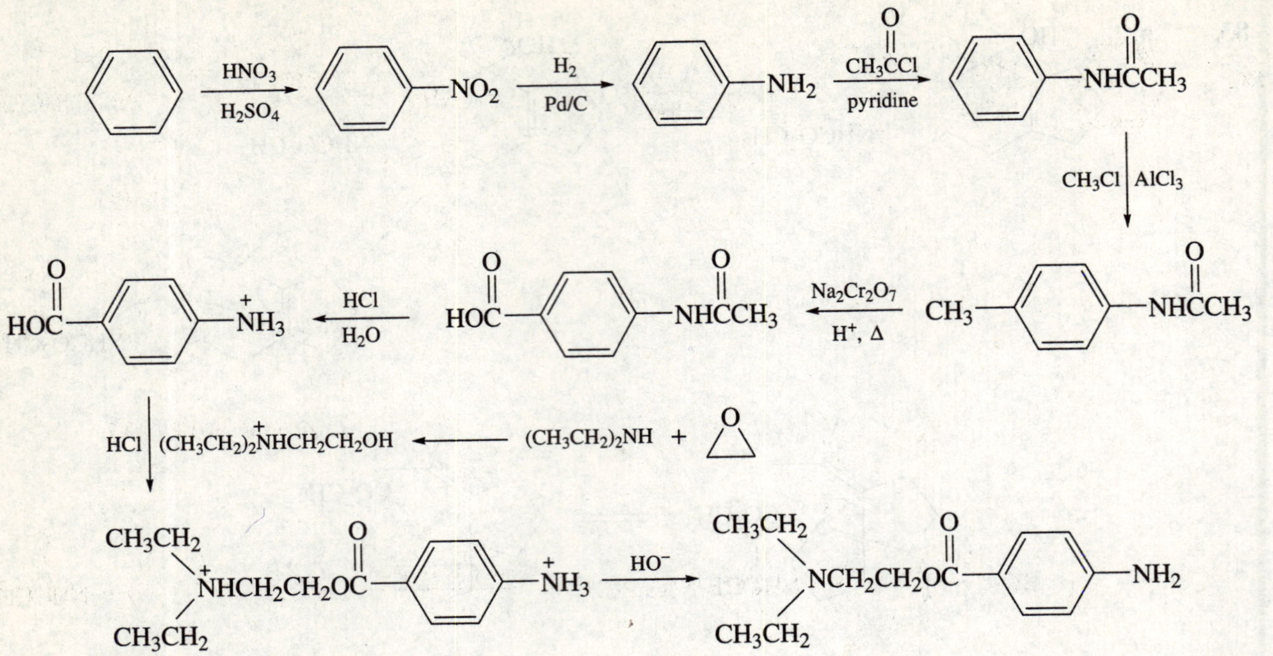
HNO3
H2SO4
NO2
H2
Pd/C
NH2
O
CH3CCl
pyridine
O
NHCCH3
CH3Cl AlCl3
O
CH3
NHCCH3
Na2Cr2O7
H+, Δ
O
HOC
O
NHCCH3
HCl
H2O
O
HOC
NH3
HCl (CH3CH2)2NHCH2CH2OH
(CH3CH2)2NH +
O
CH3CH2
NHCH2CH2OC
CH3CH2
NH3
HO−
CH3CH2
NCH2CH2OC
CH3CH2
NH2

85. **a.**

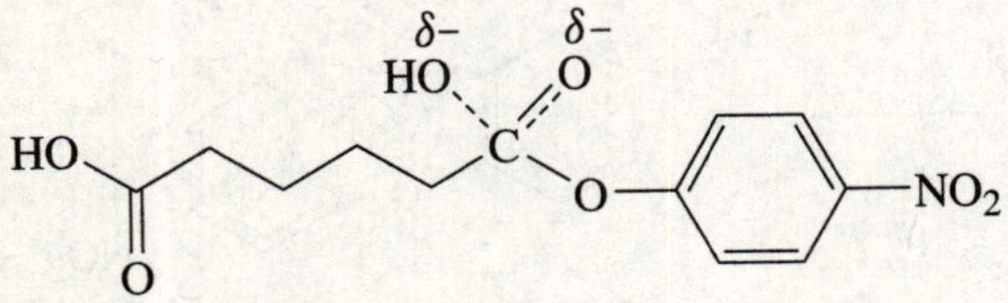
δ−
δ−
HO
O
HO
C
O
O
NO2

b.

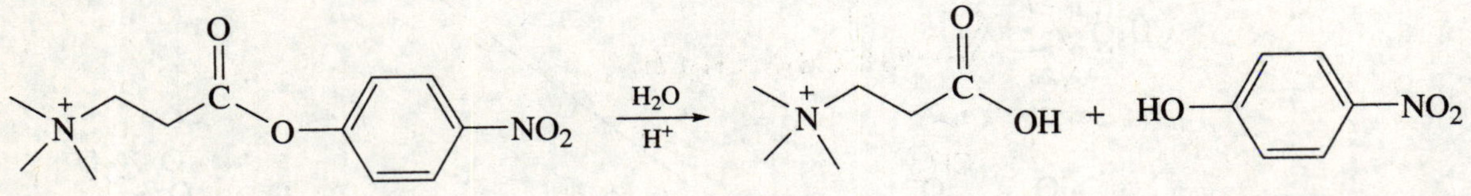
O
N
O
NO2
H2O
H+
O
N
C
OH
+
HO
NO2

c.

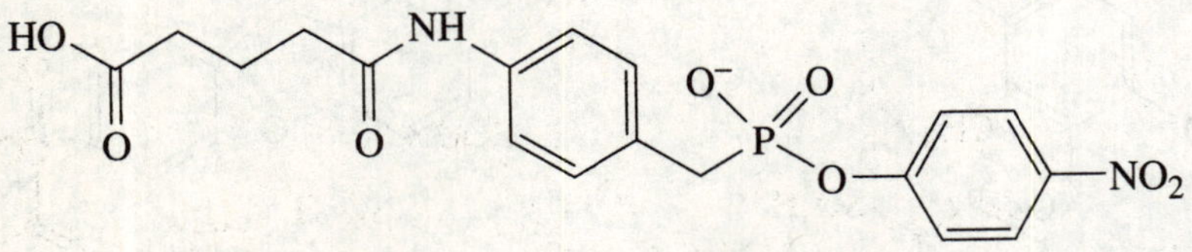
HO
O
NH
O
O−
O
P
O
NO2

86.

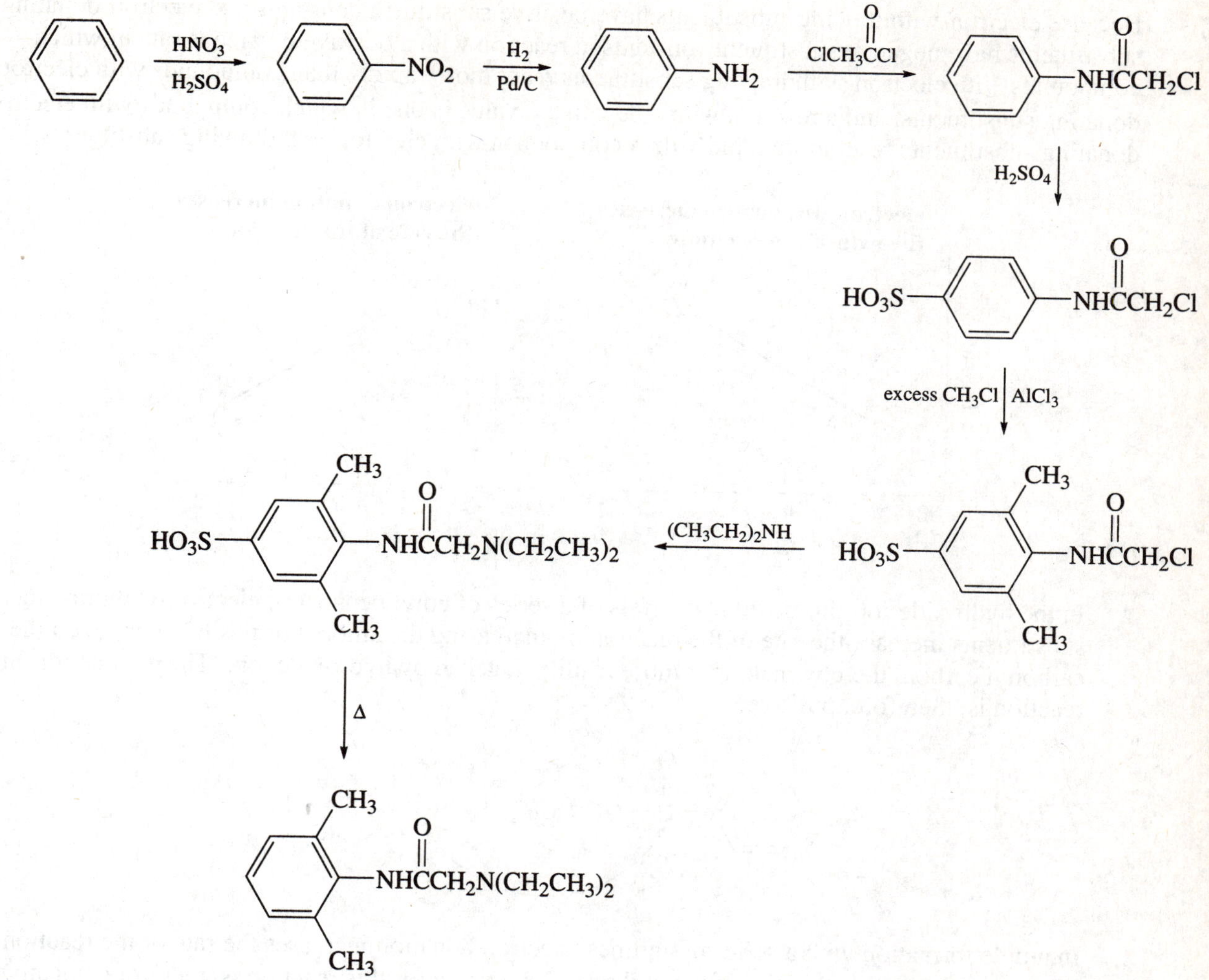

87.

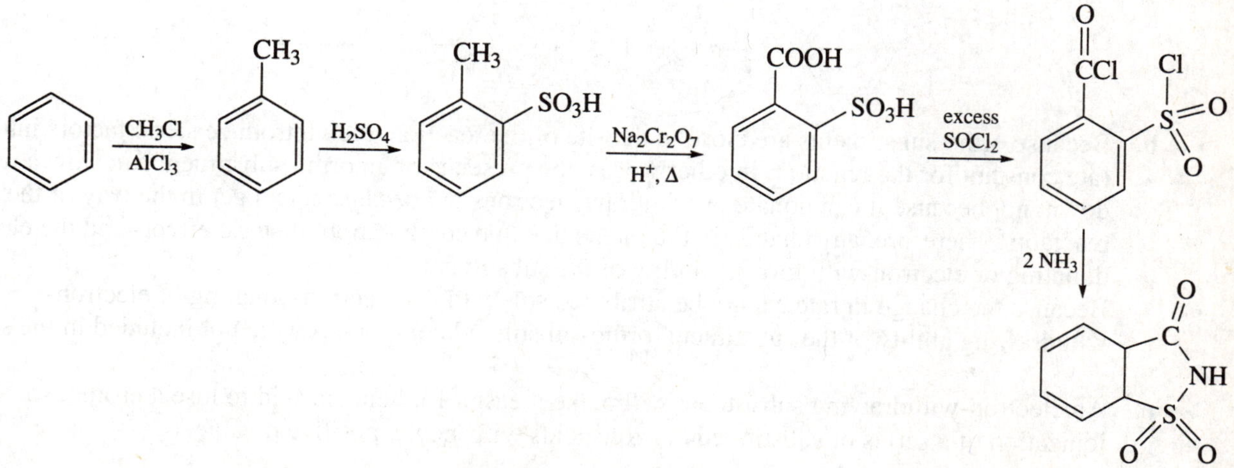

88. Because electron-withdrawing substituents have positive substituent constants and electron-donating substituents have negative substituent constants, a reaction with a positive ρ value is one in which compounds with electron-withdrawing substituents react more rapidly than compounds with electron-donating substituents, and a reaction with a negative ρ value is one in which compounds with electron-donating substituents react more rapidly than compounds with electron-withdrawing substituents.

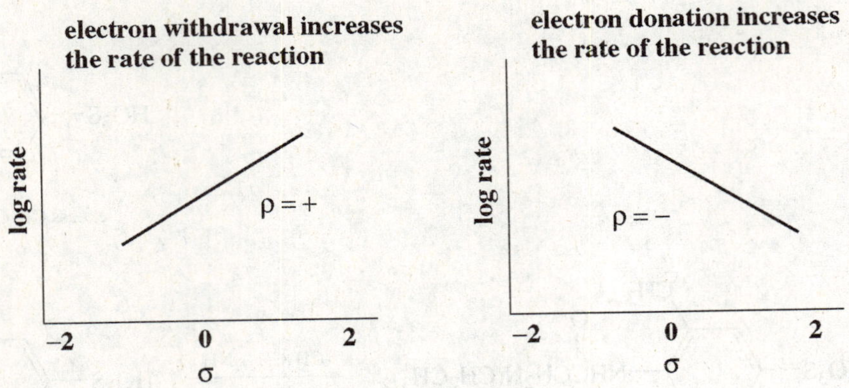

a. In the hydroxide-ion-promoted hydrolysis of a series of ethyl benzoates, electron-withdrawing substituents increase the rate of the reaction by increasing the amount of positive charge on the carbonyl carbon, thereby making it more readily attacked by hydroxide ion. The ρ value for this reaction is, therefore, positive.

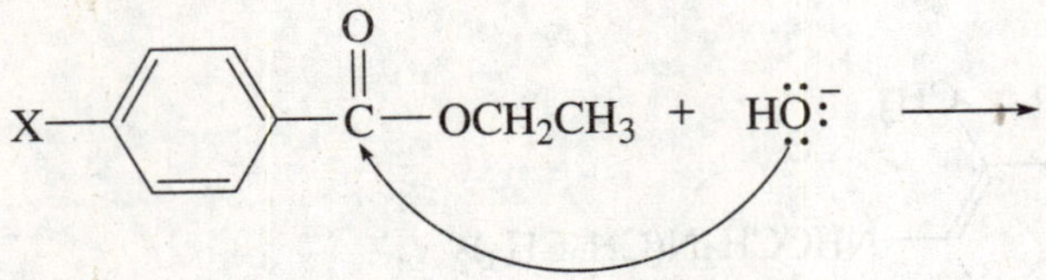

In amide formation with a series of anilines, electron donation increases the rate of the reaction by increasing the nucleophilicity of the aniline. The ρ value for this reaction is, therefore, negative.

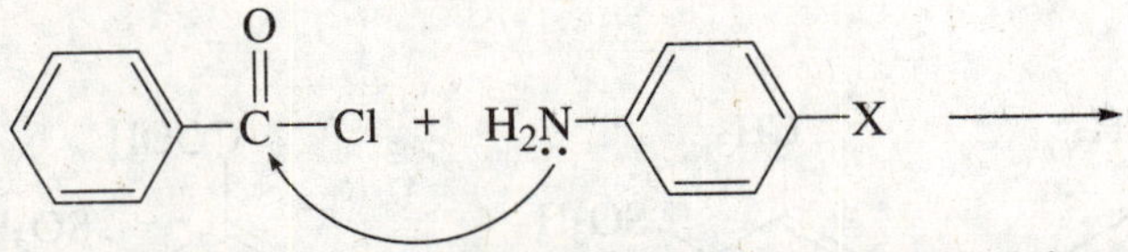

b. Because ortho substituents are close to the site of the reaction, they introduce steric factors into the rate constant for the reaction. In other words, the presence of an ortho substituent can slow a reaction down, not because it can donate or withdraw electrons but because it can get in the way of the reactants. Therefore, any change in the rate is due to a combination of steric effects and the electron-donating or electron-withdrawing ability of the substituent.
Because the change in rate cannot be attributed solely to the electron-donating or electron-withdrawing ability of the substituent, ortho-substituted compounds were not included in the study.

c. An electron-withdrawing substituent will make it easier for benzoic acid to lose a proton, so ionization of a series of substituted benzoic acids will show a positive ρ value.

Chapter 16 Practice Test

1. Circle the compound in each pair that is more reactive toward nucleophilic acyl substitution.

a. $CH_3\overset{\overset{O}{\|}}{C}OCH_3$ or $CH_3\overset{\overset{O}{\|}}{C}NHCH_3$

b. $CH_3\overset{\overset{O}{\|}}{C}OCH_3$ or 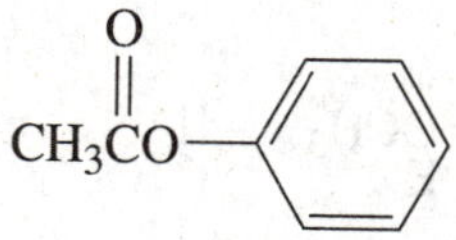

c. 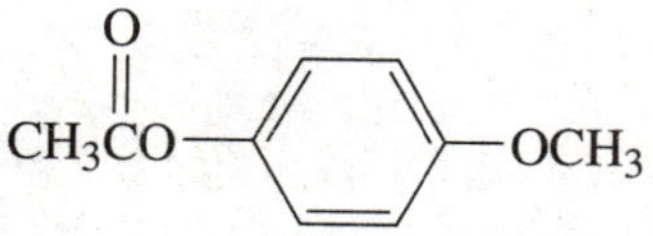 or 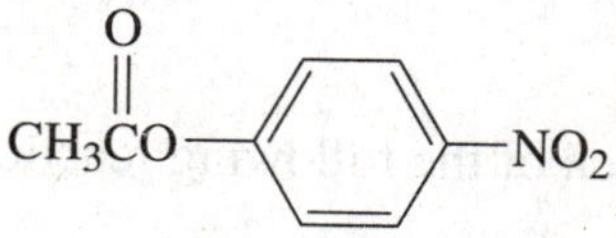

d. $CH_3\overset{\overset{O}{\|}}{C}O\overset{\overset{O}{\|}}{C}CH_3$ or $CH_3\overset{\overset{O}{\|}}{C}Cl$

2. Give the systematic name for each of the following:

a. $CH_3CH_2CH_2CH_2\overset{\overset{O}{\|}}{C}NHCH_2CH_3$

b. $CH_3CH_2\underset{\underset{CH_3}{|}}{C}HCH_2\overset{\overset{O}{\|}}{C}OH$

c. 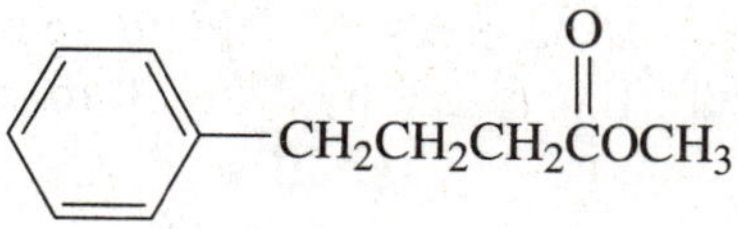

d. $CH_3CH_2\overset{\overset{O}{\|}}{C}O\overset{\overset{O}{\|}}{C}CH_3$

3. Give an example of each of the following:

a. a symmetrical anhydride

b. a hydrolysis reaction

c. a transesterification reaction

4. What carbonyl compound would be obtained from collapse of each of the following tetrahedral intermediates?

a. 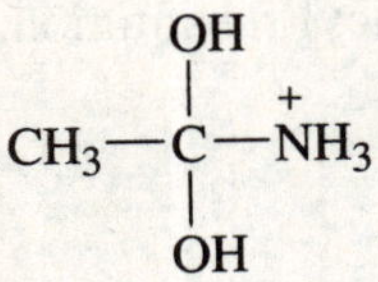

c. 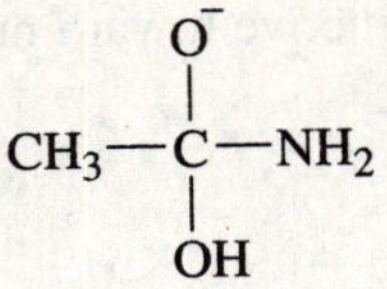

b. 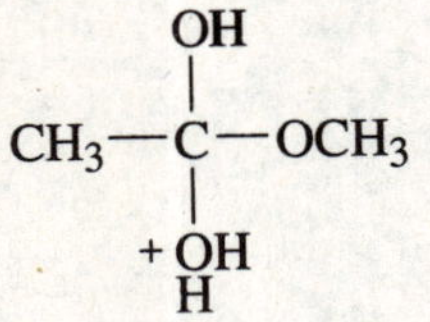

d. 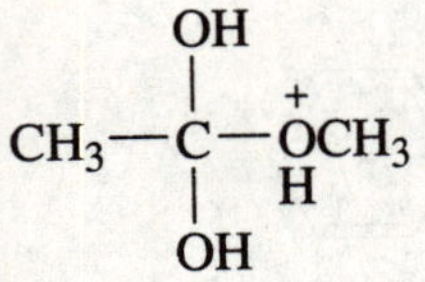

5. Give the product of each of the following reactions.

a. $CH_3CH_2C{\equiv}N + H_2O \xrightarrow[\Delta]{HCl}$

b. $CH_3CH_2\overset{O}{\overset{\|}{C}}OCH_2CH_2CH_3 + H_2O \xrightarrow[\Delta]{HO^-}$

c. $CH_3CH_2CH_2\overset{O}{\overset{\|}{C}}O\overset{O}{\overset{\|}{C}}CH_2CH_3 + H_2O \longrightarrow$

d. $CH_3CH_2\overset{O}{\overset{\|}{C}}Cl + 2\ CH_3CH_2NH_2 \longrightarrow$

e. 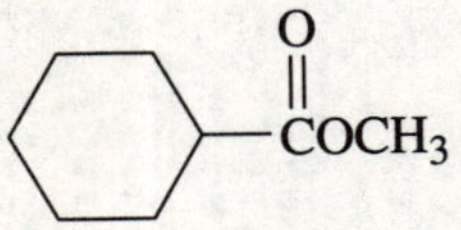 $+ \underset{\text{excess}}{CH_3CH_2OH} \xrightarrow{HCl}$

f. $CH_3CH_2CH_2\overset{O}{\overset{\|}{C}}OH \xrightarrow[\text{2. } CH_3CH_2CH_2OH]{\text{1. } SOCl_2}$

g. $CH_3CH_2\overset{O}{\overset{\|}{C}}NHCH_3 + H_2O \xrightarrow[\Delta]{HCl}$

h. 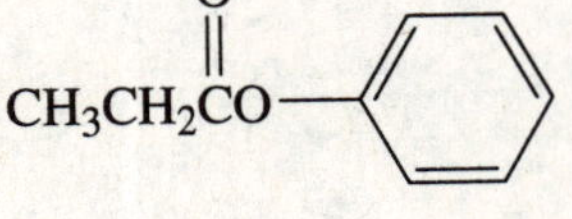 $+ \underset{\text{excess}}{H_2O} \xrightarrow{HCl}$

CHAPTER 17

Carbonyl Compounds II: Reactions of Aldehydes and Ketones • More Reactions of Carboxylic Acid Derivatives • Reactions of α,β-Unsaturated Carbonyl Compounds

Important Terms

acetal

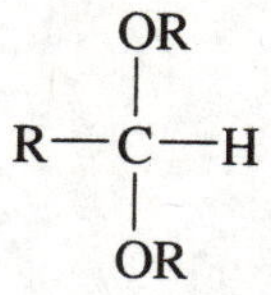

aldehyde

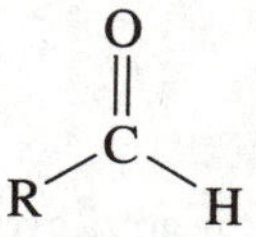

conjugate addition — nucleophilic addition to the β-carbon of an α,β-unsaturated carbonyl compound.

cyanohydrin

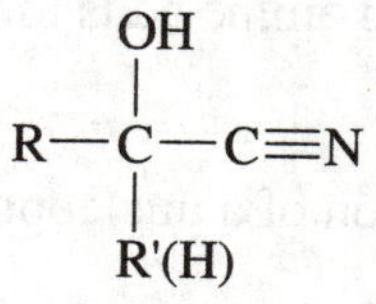

deoxygenation — removal of an oxygen from a reactant.

***gem*-diol (hydrate)** — a molecule with two OH groups on the same carbon.

direct addition — nucleophilic addition to the carbonyl carbon.

disconnection — breaking a bond to carbon, in a retrosynthetic analysis, to give a simpler molecule.

enamine — an α,β-unsaturated tertiary amine.

hemiacetal

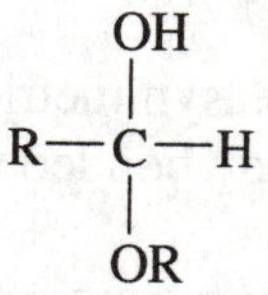

hemiketal

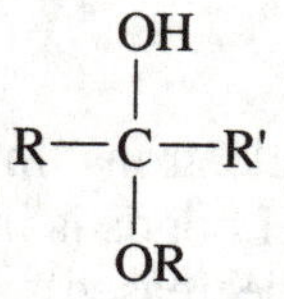

hydrate — a molecule with two OH groups on the same carbon.

hydrazone

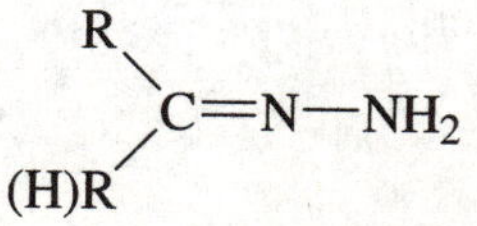

imine (Schiff base) 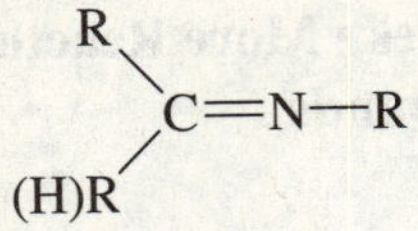

ketal 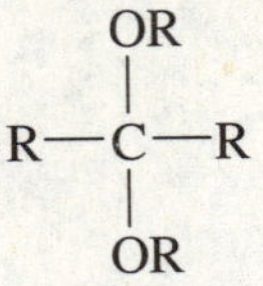

ketone 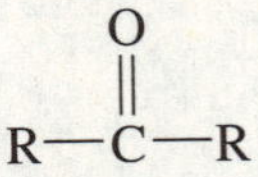

nucleophilic acyl substitution reaction a reaction in which a group bonded to an acyl group is substituted by another group.

nucleophilic addition-elimination reaction a nucleophilic addition reaction that is followed by an elimination reaction. Imine formation is an example: an amine adds to the carbonyl carbon, and water is eliminated.

nucleophilic addition reaction a reaction that involves the addition of a nucleophile to a reagent.

oxime 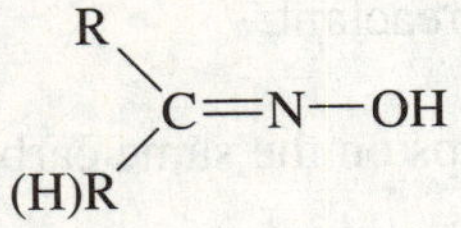

phenylhydrazone 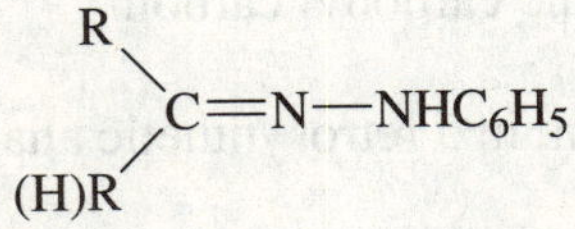

pH-rate profile a plot of the rate constant of a reaction versus the pH of the reaction mixture.

prochiral carbonyl carbon a carbonyl carbon that will become an asymmetric center if it is attacked by a group unlike any of the groups already bonded to it.

protecting group a reagent that protects a functional group from a synthetic operation that it would otherwise not survive.

reduction reaction a reaction in which a molecule gains electrons. In the case of an organic molecule, a reaction in which the number of C-H bonds is increased or the number of C-O, C-N, or C-X (X = halogen) bonds in decreased

reductive amination the reaction of an aldehyde or a ketone with ammonia or with a primary amine in the presence of a reducing agent.

semicarbazone

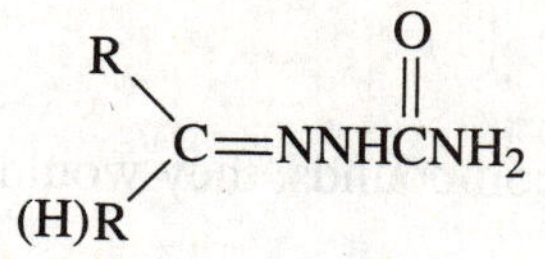

synthon a fragment of a disconnection

Wittig reaction the reaction of an aldehyde or a ketone with a phosphonium ylide, resulting in formation of an alkene.

ylide a compound with opposite charges on adjacent, covalently bonded atoms with complete octets.

Solutions to Problems

1. If the carbonyl group were anywhere else in these compounds, they would not be ketones and, therefore, would not have the "one" suffix.

2. a. 3-methylpentanal, β-methylvaleraldehyde
   b. 4-heptanone, dipropyl ketone
   c. 2-methyl-4-heptanone, isobutyl propyl ketone
   d. 4-phenylbutanal, γ-phenylbutyraldehyde
   e. 4-ethylhexanal, γ-ethylcaproaldehyde
   f. 1-hepten-3-one, butyl vinyl ketone

3. **a.** 6-hydroxy-3-heptanone **b.** 2-oxocyclohexylmethanenitrile **c.** 3-formylpentanamide

4. **a.** 2-Heptanone is more reactive because it has less steric hindrance. There is little difference in the amount of steric hindrance provided at the carbonyl carbon (the site of nucleophilic attack) by a pentyl and a propyl group because they differ at a point somewhat removed from the site of nucleophilic attack. The difference in size between a methyl group and a propyl group is significant at the site of nucleophilic attack.

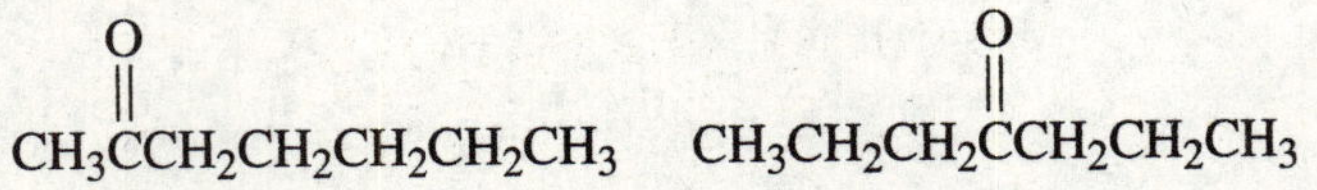

2-heptanone 4-heptanone

b. *para*-Nitroacetophenone is more reactive because the electron-withdrawing nitro group makes the carbonyl carbon more susceptible to nucleophilic attack compared with an electron-donating methoxy group.

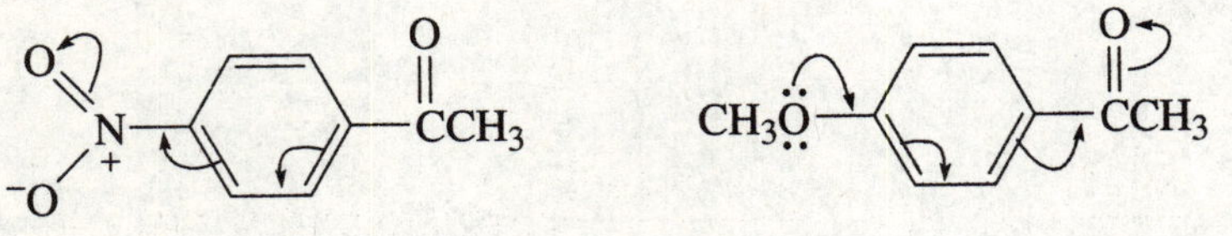

5. 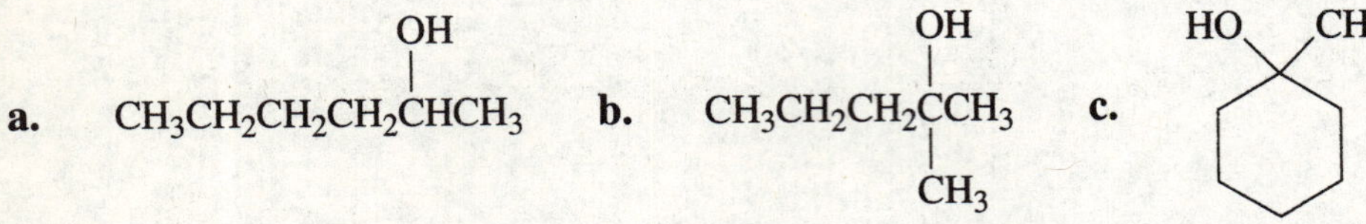

6. 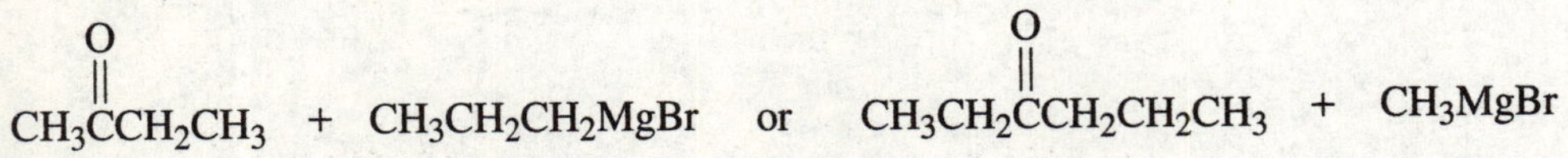

7. **a.** Two isomers are obtained because the reaction creates an asymmetric center in the product.

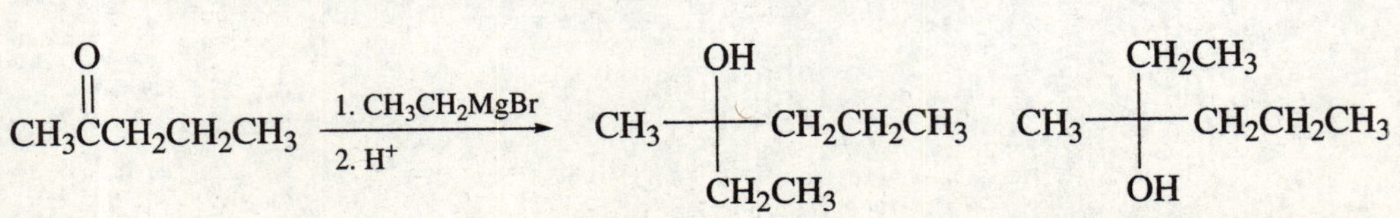

(*S*)-3-methyl-3-hexanol (*R*)-3-methyl-3-hexanol

b. Only one compound is obtained because the product does not have an asymmetric center.

$$\mathrm{CH_3\overset{O}{\overset{\|}{C}}CH_2CH_2CH_3 \xrightarrow[2.\ H^+]{1.\ CH_3MgBr} CH_3\underset{CH_3}{\overset{OH}{C}}CH_2CH_2CH_3}$$

8. **a.** Solved in the text.

b. **1.** $\mathrm{CH_3\overset{O}{\overset{\|}{C}}OR + 2\ CH_3MgBr}$

2. Solved in the text.

4. $\mathrm{CH_3\overset{O}{\overset{\|}{C}}OR + 2\ CH_3CH_2MgBr}$

6. 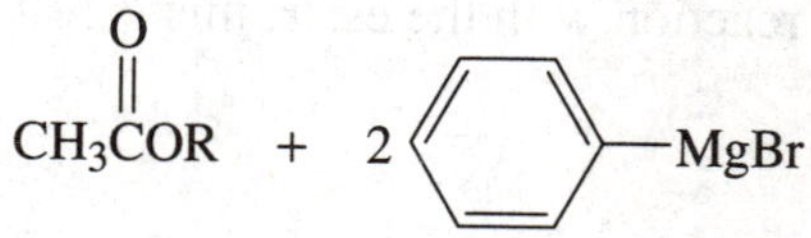

9. If a secondary alcohol is formed from the reaction of a formate ester with excess Grignard reagent, the two alkyl substituents of the alcohol will be identical because they both come from the Grignard reagent. Therefore, only the following two alcohols (B and D) can be prepared that way.

$$\mathrm{CH_3\underset{OH}{CH}CH_3 \qquad CH_3CH_2\underset{OH}{CH}CH_2CH_3}$$

10. **A** and **C** will not undergo nucleophilic addition with a Grignard reagent. **A** has a H on a nitrogen and **C** has an H on an oxygen; these acidic hydrogens will react with the Grignard reagent, converting it into an alkane, before it has a chance to react as a nucleophile.

11. If the compound is alkylated after nucleophilic attack, the alkylation on oxygen will compete with alkylation on carbon.

a. $\mathrm{HC{\equiv}CH \xrightarrow{^-NH_2} HC{\equiv}C^- \xrightarrow{CH_3CH_2\overset{O}{\overset{\|}{C}}H} HC{\equiv}C\overset{O^-}{CH}CH_2CH_3 \xrightarrow{H^+} HC{\equiv}C\overset{OH}{CH}CH_2CH_3}$

b. 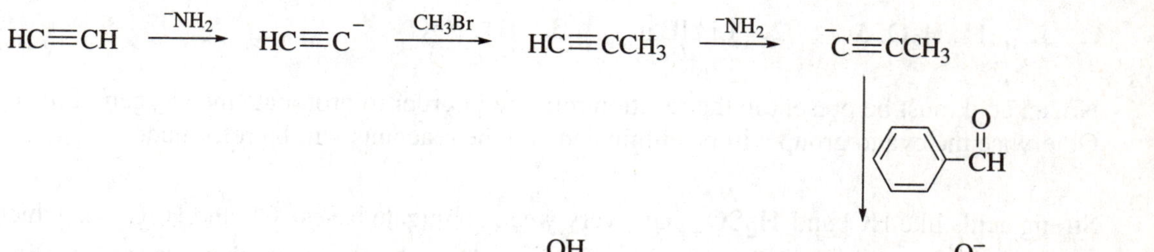

c. 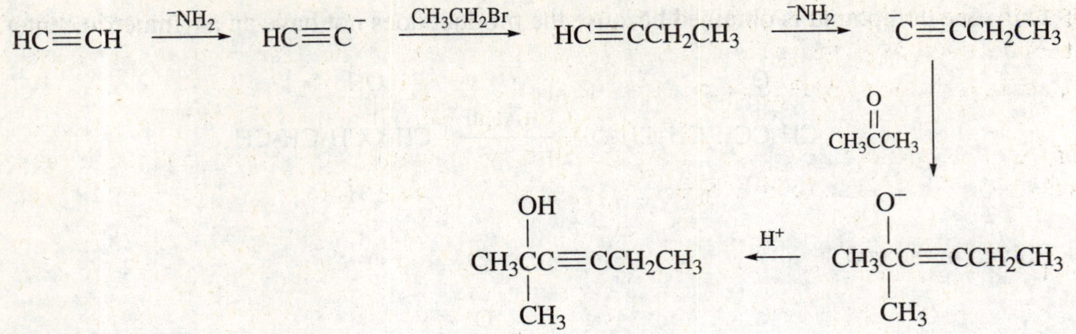

12. Similar to the way a Grignard reagent reacts with an ester, an acetylide ion first undergoes a nucleophilic acyl substitution reaction with the ester; this is followed by a nucleophilic addition reaction.

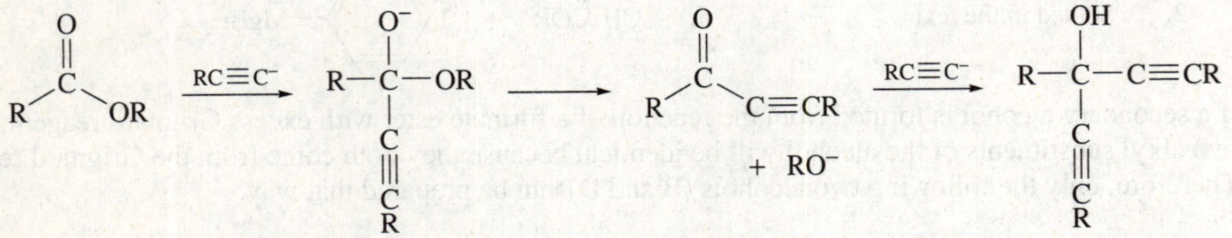

13. 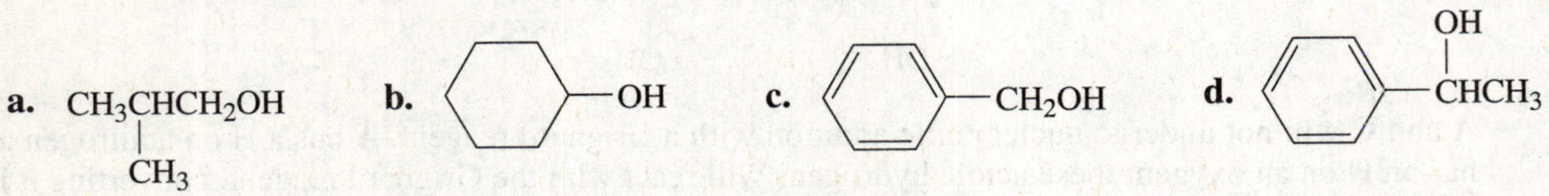

14. 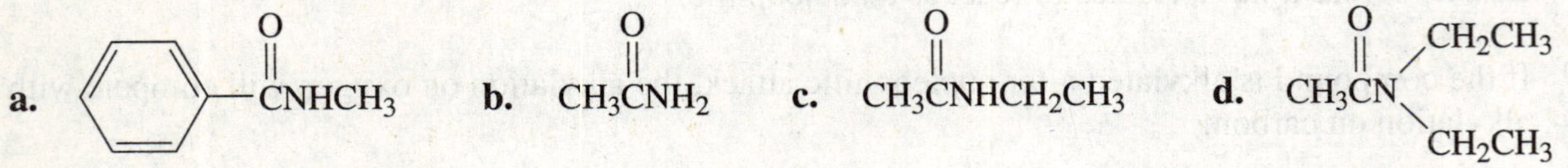

15. a. 1. $LiAlH_4$ 2. H_2O

b. H^+, H_2O, Δ

c. 1. H^+, H_2O, Δ 2. $LiAlH_4$ 3. H^+, H_2O

16. No, an acid must be present in the reaction mixture in order to protonate the oxygen of the cyanohydrin. Otherwise the cyano group will be eliminated and the reactants will be reformed.

17. Strong acids like HCl and H_2SO_4 have very weak conjugate bases (Cl^- and HSO_4^-), which are excellent leaving groups. When these bases add to the carbonyl group, they are readily eliminated, reforming the starting materials. Cyanide ion is a strong enough base, so it is not eliminated unless the oxygen in the product is negatively charged.

18. Solved in the text.

19. Figure 17.2 for imine formation when the amine is hydroxylamine ($pK_a = 6.0$) shows that the maximum rate is obtained when the pH is 1.5 units lower than the pK_a of the amine. Thus if the amine has a pK_a of 10, imine formation should be carried out at pH = 8.5.

20. For the derivations of the equations used to calculate the amount of a compound that is present in either its acidic or basic form, refer to Special Topic I in this study guide.

a. fraction present in the acidic form = $\dfrac{[H^+]}{K_a + [H^+]}$

$$\frac{[H^+]}{K_a + [H^+]} = \frac{3.2 \times 10^{-5}}{3.2 \times 10^{7} + 3.2 \times 10^{-5}}$$

$$= \frac{3.2 \times 10^{-5}}{3.2 \times 10^{7}}$$

$$= 1 \times 10^{-12}$$

b. fraction present in the acidic form = $\dfrac{[H^+]}{K_a + [H^+]}$

$$\frac{[H^+]}{K_a + [H^+]} = \frac{3.2 \times 10^{-2}}{3.2 \times 10^{7} + 3.2 \times 10^{-2}}$$

$$= \frac{3.2 \times 10^{-2}}{3.2 \times 10^{7}}$$

$$= 1 \times 10^{-9}$$

c. fraction present in the basic form = $\dfrac{K_a}{K_a + [H^+]}$

$$\frac{K_a}{K_a + [H^+]} = \frac{1.0 \times 10^{-6}}{1.0 \times 10^{-6} + 3.2 \times 10^{-2}}$$

$$= \frac{1.0 \times 10^{-6}}{3.2 \times 10^{-2}}$$

$$= 3.1 \times 10^{-3}$$

21. The nitrile reacts with the Grignard reagent to form an imine which could then be hydrolyzed to a ketone.

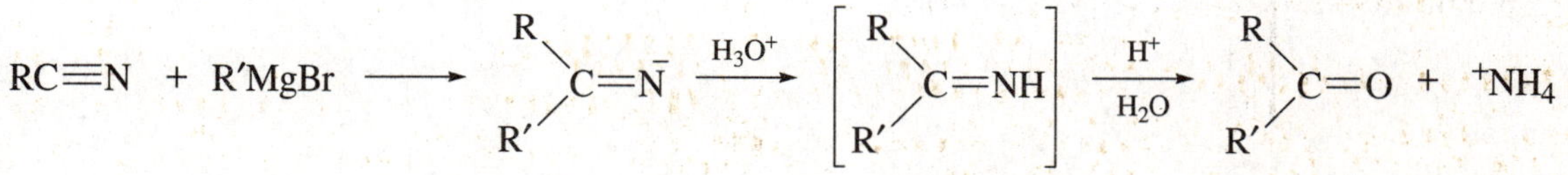

22. **a.**

1.

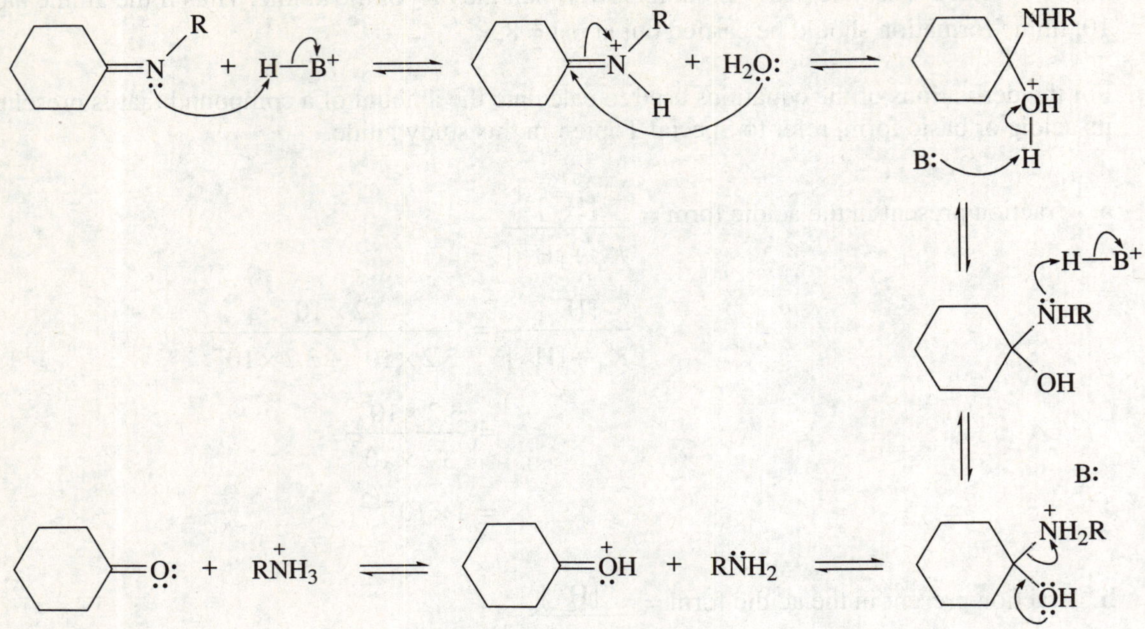

2.

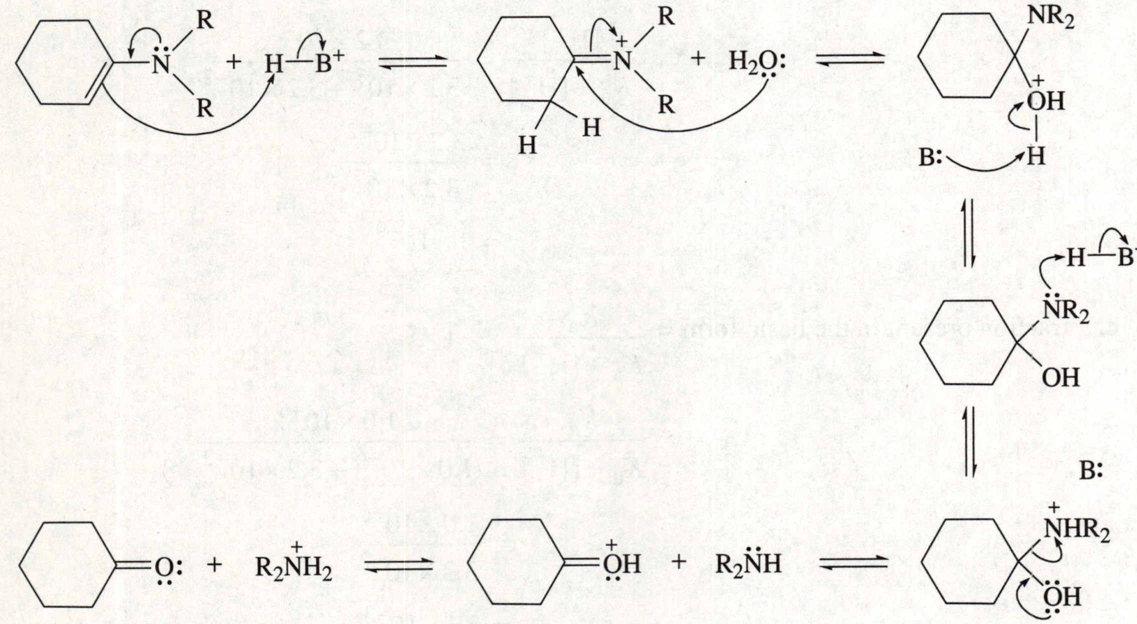

b. The only difference is in the first step of the mechanism: in imine hydrolysis the acid protonates the nitrogen; in enamine hydrolysis, the acid protonates the β-carbon.

23. **a.**

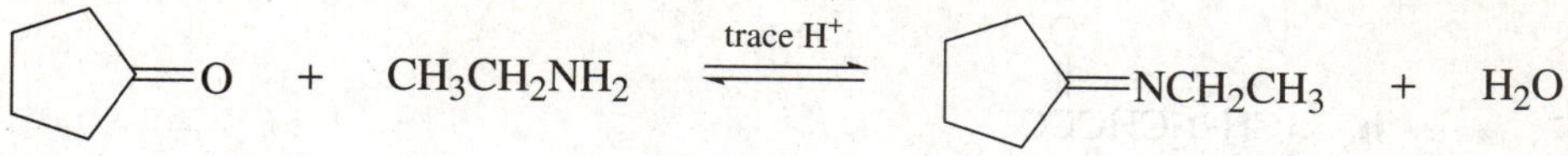

b.

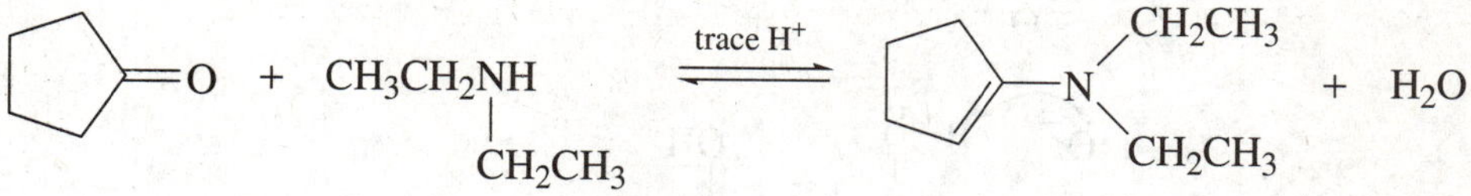

c.

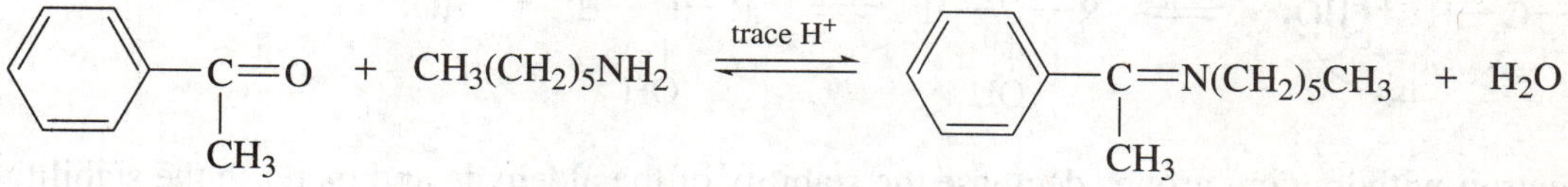

d.

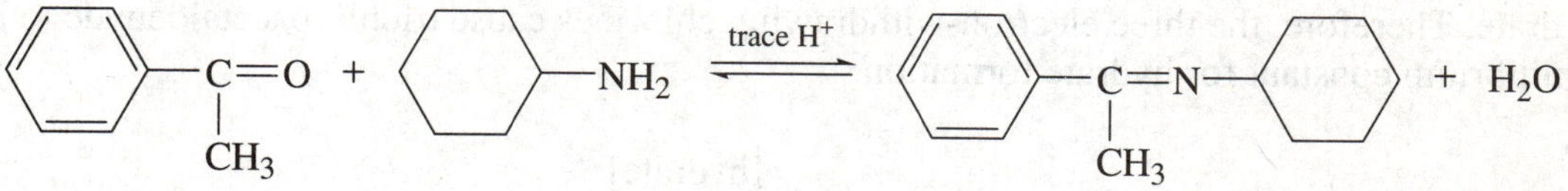

24. **a.**

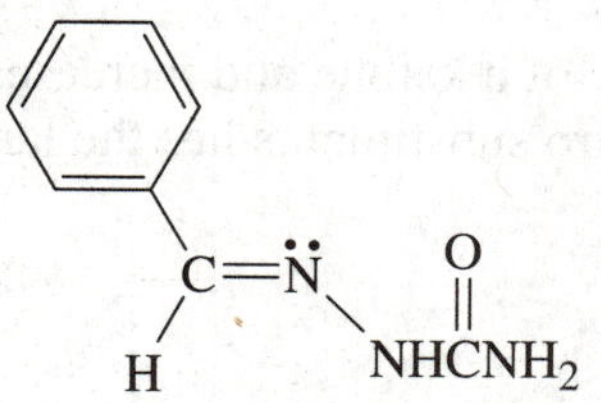

b.

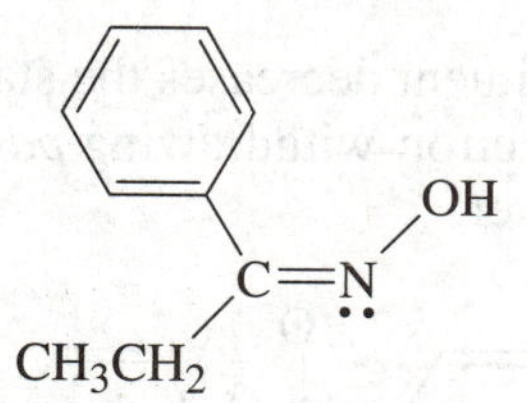

c.

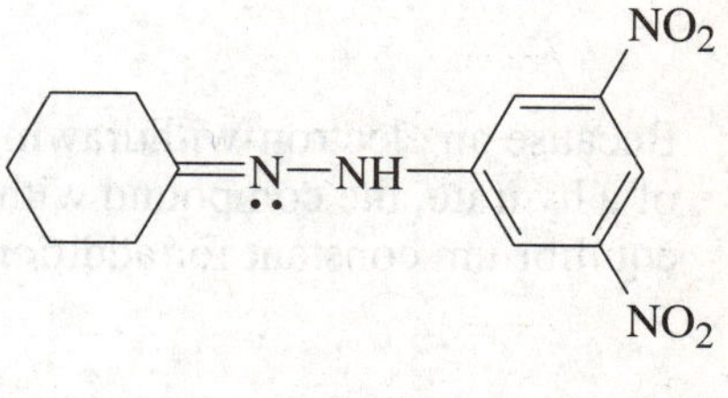

25. The lone pair electrons that are on each of the nitrogens attached to the carbonyl group are delocalized onto the carbonyl group. These nitrogens, therefore, cannot act as nucleophiles since their lone pair electrons are not available for nucleophilic attack.

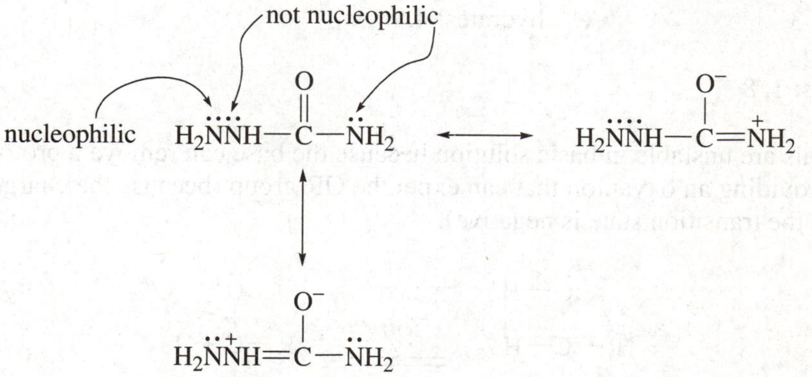

26. A tertiary amine will be obtained because the primary amine synthesized in the first part of the reaction will react with the excess carbonyl compound, forming an imine that will be reduced to a secondary amine. The secondary amine will then react with the carbonyl compound, forming an enamine that will be reduced to a tertiary amine.

$$R_2C{=}O + NH_3 \xrightarrow[\text{Pd/C}]{H_2} R_2CHNH_2 \xrightarrow[\text{H}_2,\ \text{Pd/C}]{R_2C{=}O} R_2CHNHCHR_2 \xrightarrow[\text{H}_2,\ \text{Pd/C}]{R_2C{=}O} (R_2CH)_3N$$

27. **a.** 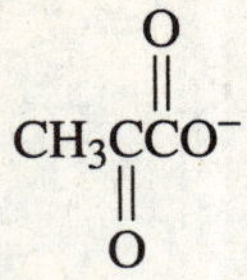 **b.** 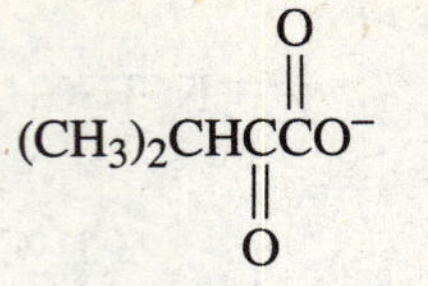

28. 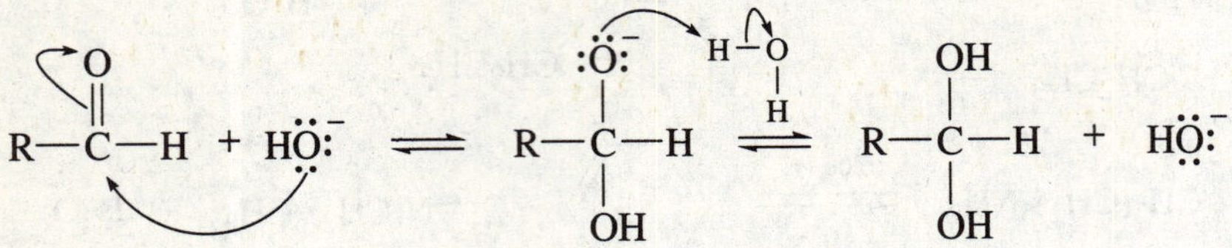

29. Electron-withdrawing groups decrease the stability of the aldehyde and increase the stability of the hydrate. Therefore, the three electron-withdrawing chlorines cause trichloroacetaldehyde to have a large equilibrium constant for hydrate formation.

$$K_{eq} = \frac{[\text{hydrate}]}{[\text{aldehyde}][H_2O]}$$

30. Because an electron-withdrawing substituent decreases the stability of a ketone and increases the stability of a hydrate, the compound with the electron-withdrawing *para*-nitro substituents has the largest equilibrium constant for addition of water.

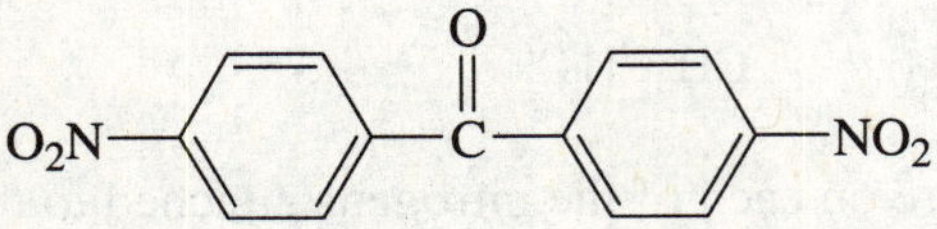

31. **a.** hemiacetals: 7

b. acetals: 2, 3

c. hemiketals: 1, 8

d. ketals: 5

e. hydrates: 4, 6

32. **a.** Hemiacetals are unstable in basic solution because the base can remove a proton from an OH group, thereby providing an oxyanion that can expel the OR group (because the charge that develops on the oxygen in the transition state is negative).

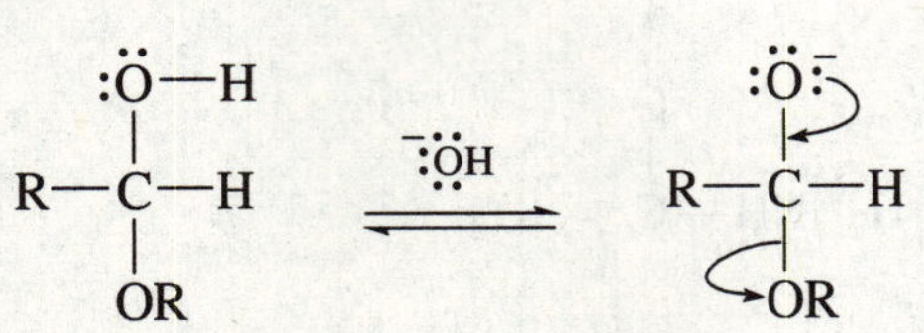

b. In order for an acetal to form, the CH_3O group in the hemiacetal must eliminate an OH group. Hydroxide ion is too basic to be eliminated by a CH_3O group (because the charge that develops on the oxygen in the transition state is positive) but water, a much weaker base, can be eliminated by a CH_3O group. In other words, the OH group must be protonated before it can be eliminated by a CH_3O group. Therefore, acetal formation must be carried out in an acidic solution.

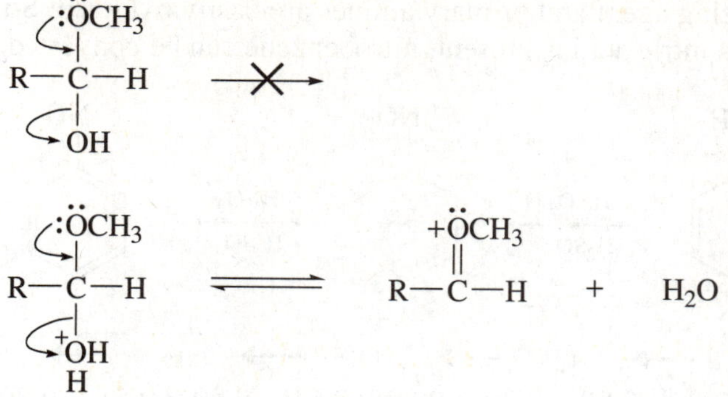

c. Hydrate formation can be catalyzed by hydroxide ion because a group does not have to be eliminated after hydroxide ion attacks the aldehyde or ketone.

33. When a tetrahedral intermediate collapses, the intermediate that is formed is very unstable because of the positive charge on the sp^2 oxygen atom. In the case of an acetal or ketal, the only way to form a neutral species is to reform the acetal or ketal. In the case of a hydrate, a neutral species can be formed by loss of a proton.

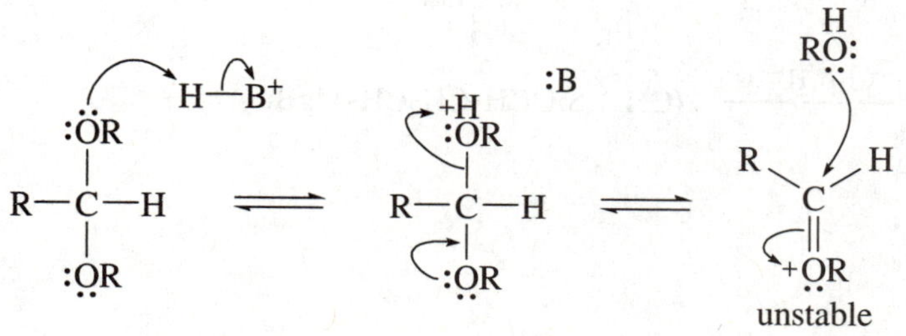

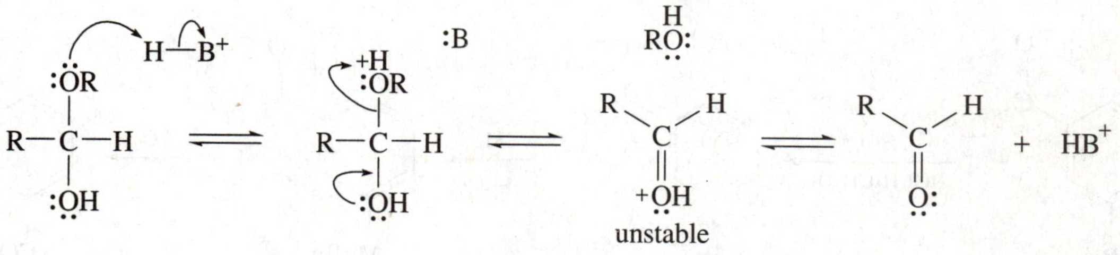

34. a. 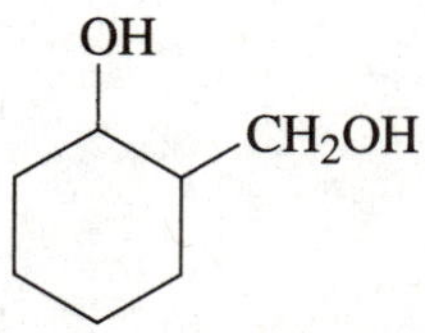 b. $NaBH_4$

35. An acetal has a very poor leaving group (CH_3O^-).

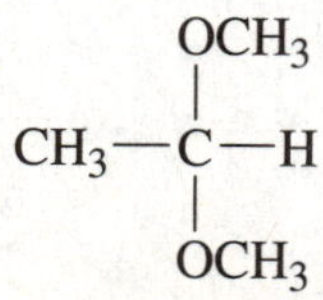

36. Nitric acid is an oxidizing agent and primary amines are easily oxidized. So one product would be nitrobenzene. If excess nitric acid is present, nitrobenzene can be converted to *meta*-dinitrobenzene.

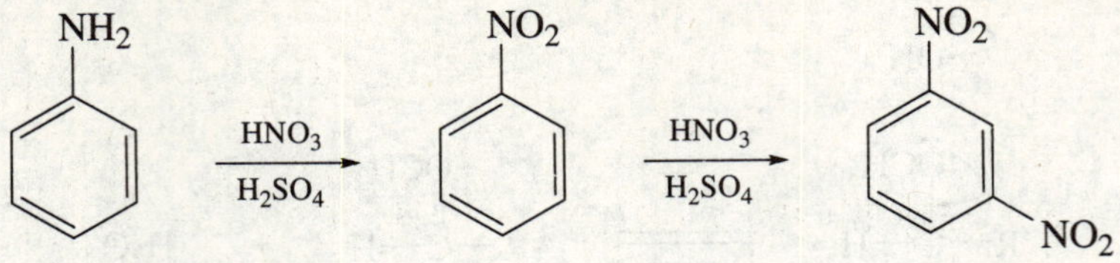

37. **a.**

A	→ B	→ C	→ D	→ E	→ F	→ G
	80%	0.80 × 0.80	0.80 × 0.64	0.80 × 0.51	0.80 × 0.41	0.80 × 0.33
		64%	51%	41%	33%	26%

b.

G	→ H	→ I
26%	0.80 × 0.26	0.80 × 0.21
	21%	17%

38. **a.** $HOCH_2CH_2CH_2Br$ + $(CH_3)_3SiCl$ $\xrightarrow{(CH_3CH_2)_3N}$ $(CH_3)_3SiOCH_2CH_2CH_2Br$

$\downarrow$ Mg | Et_2O

$(CH_3)_3SiOCH_2CH_2CH_2CHCH_3$ (with $O^- Mg^+Br$ on CH) $\xleftarrow{CH_3CH{=}O}$ $(CH_3)_3SiOCH_2CH_2CH_2MgBr$

$\downarrow$ H_3O^+

$HOCH_2CH_2CH_2CHCH_3$ (with OH on CH)

b.

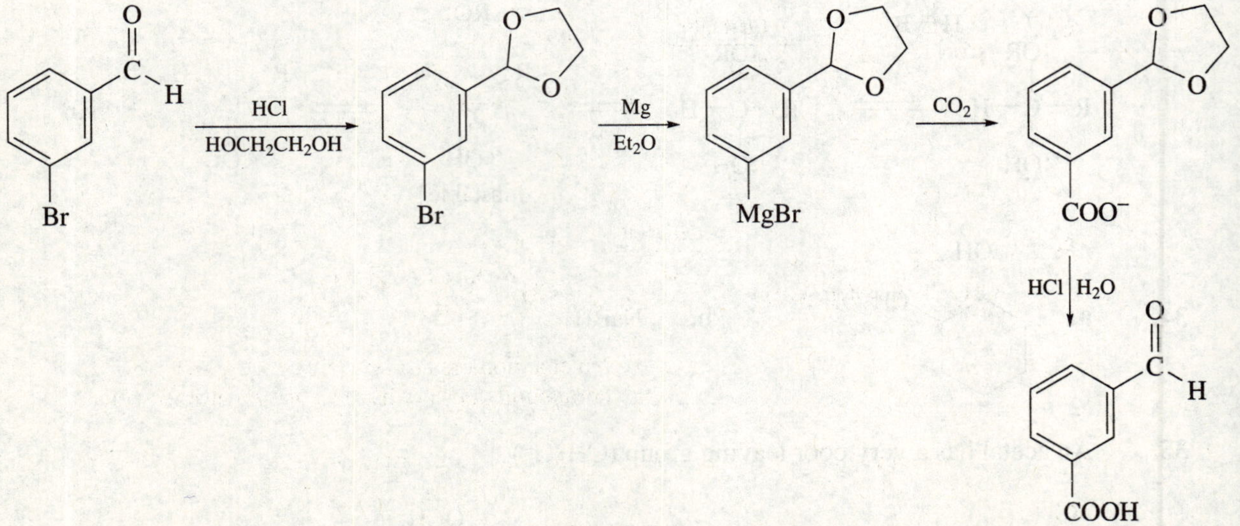

39. **a.**

1. Solved in the text

3. $(C_6H_5)_2C{=}O + CH_3CH{=}P(C_6H_5)_3$

or $(C_6H_5)_2C{=}P(C_6H_5)_3 + CH_3CH{=}O$

2. 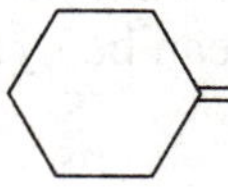 $=O + CH_3CH_2CH{=}P(C_6H_5)_3$

or 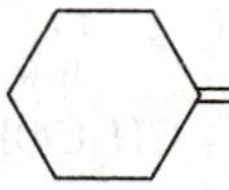 $=P(C_6H_5)_3 + CH_3CH_2CH{=}O$

4. 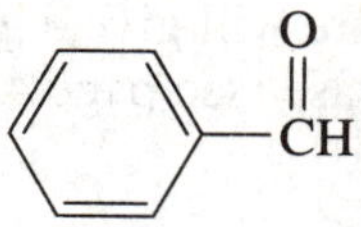 $CH_2{=}P(C_6H_5)_3$

or 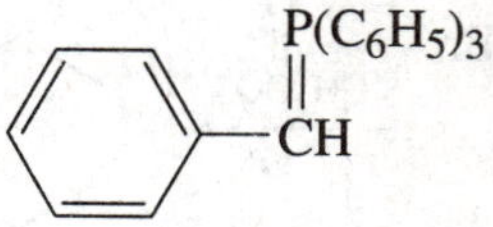 $+ CH_2{=}O$

b.

1. Solved in the text
2. $CH_3CH_2CH_2Br$ or 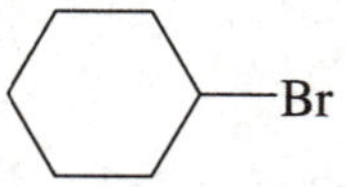
3. CH_3Br or 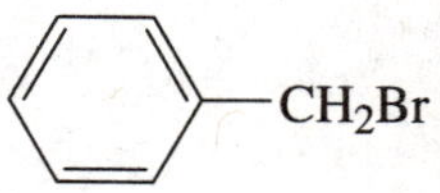
4. CH_3CH_2Br or $(C_6H_5)_2CHBr$

40. To review how to determine whether an asymmetric center has the *R* or *S* configuration, see Section 5.7.

a.

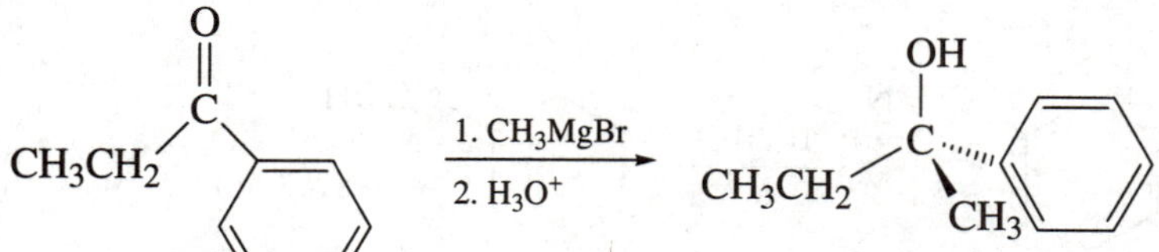

S

b.

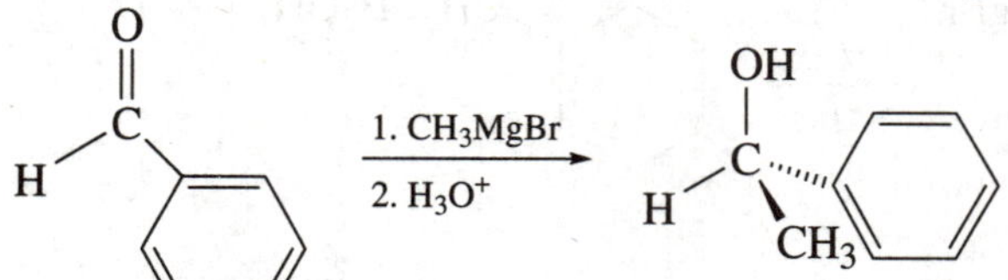

R

c.

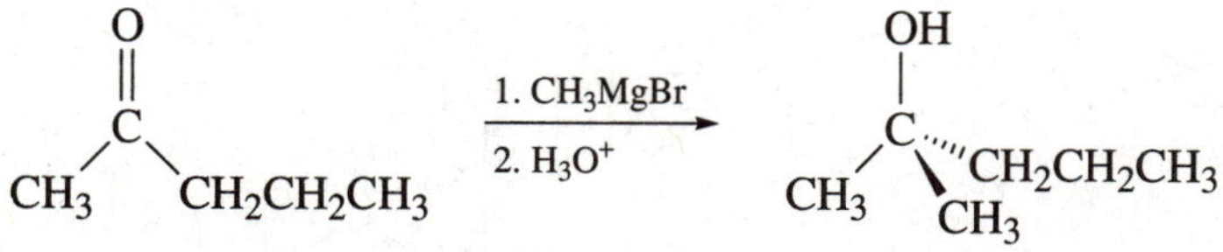

No enantiomers are possible because the compound does not have an asymmetric center.

d.

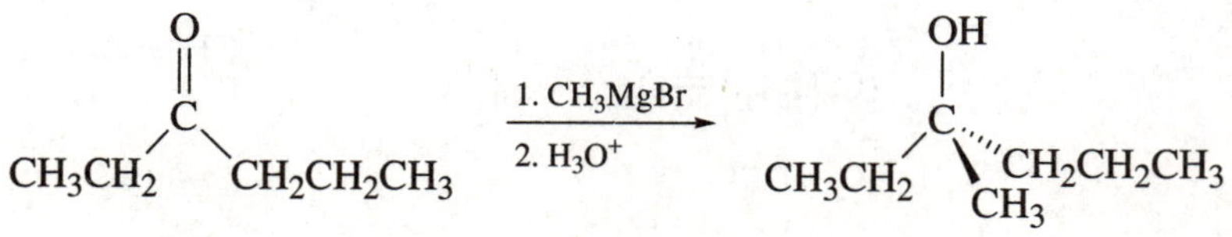

S

41. **a.**

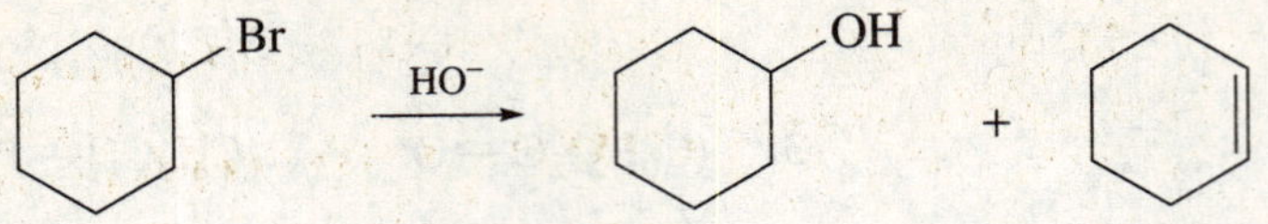

The above synthesis will also form an elimination product (cyclohexene).
The following procedure will give a greater yield of the target molecule because the substitution product is favored by a weak base (see page 415 of the text), and the substitution product can be hydrolyzed to the target compound.

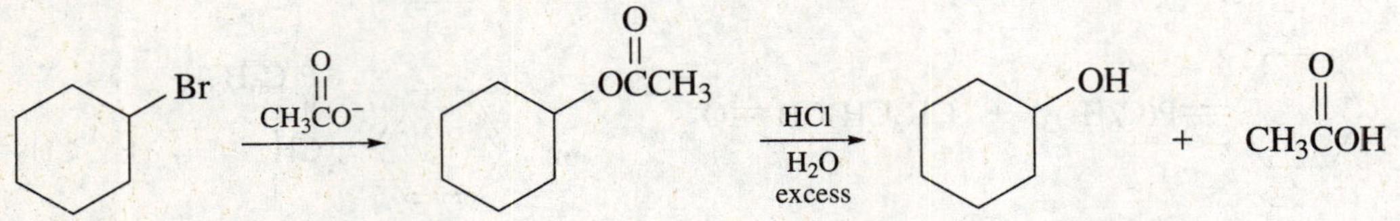

b.

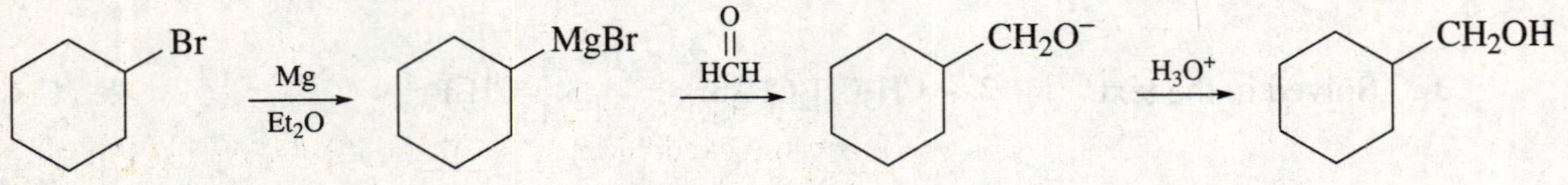

c.

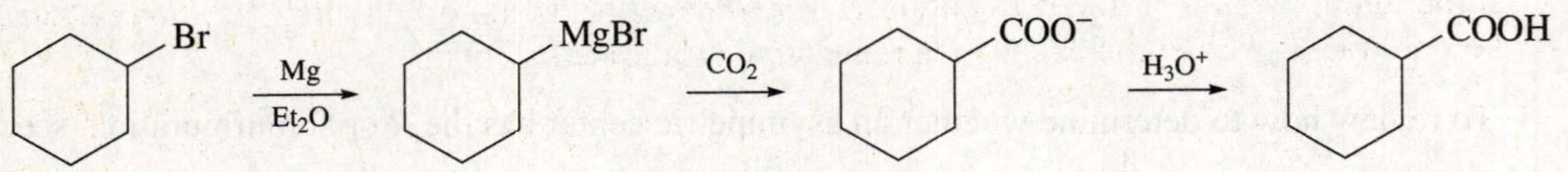

or

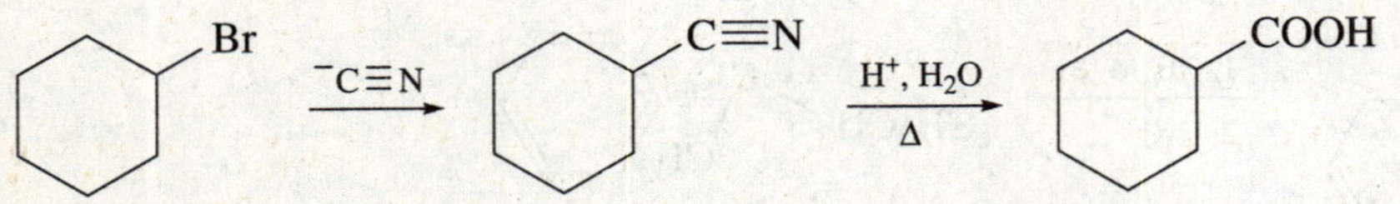

d.

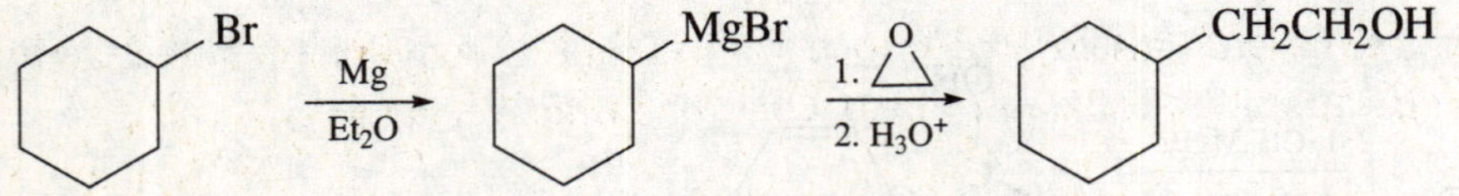

e.

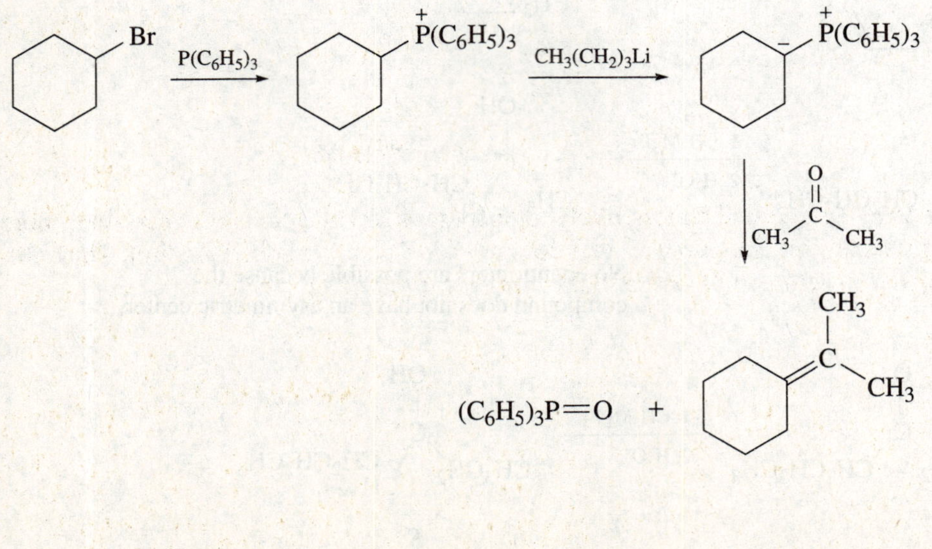

f. 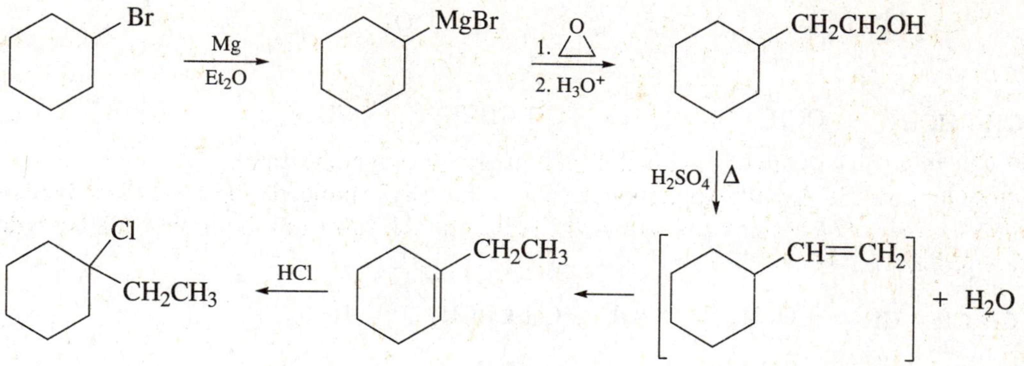

42. a. 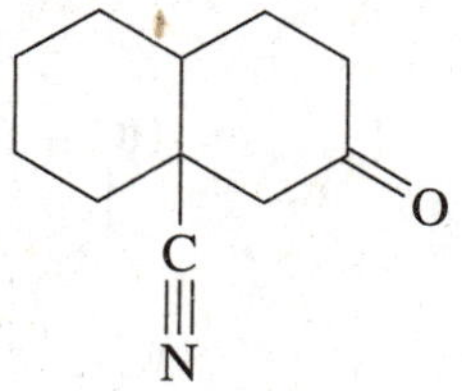

b. 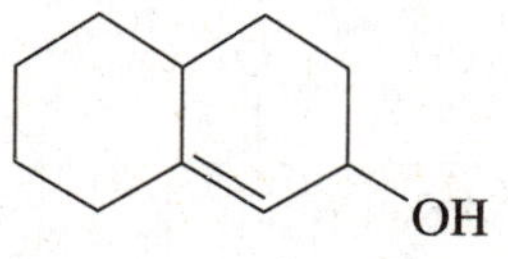

Steric hindrance
favors direct addition.

c. 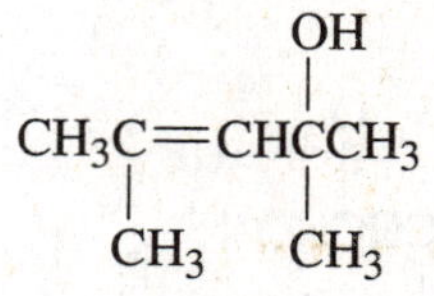

d. $CH_3CH{=}CHCH_2OH$

43. If the initial reaction of the α,β–unsaturated ketone with sodium borohydrate is conjugate addition, a ketone will be formed, which is then reduced to a saturated alcohol.

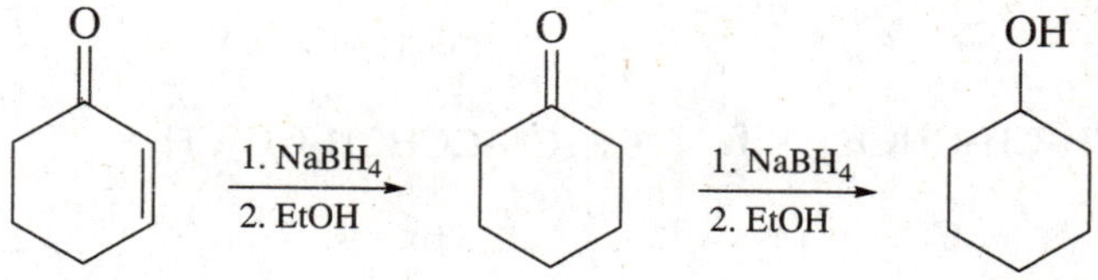

a saturated alcohol

If the initial reaction of the α,β–unsaturated ketone with sodium borohydrate is direct addition, the final product will be an α,β–unsaturated alcohol, which won't be further reduced.

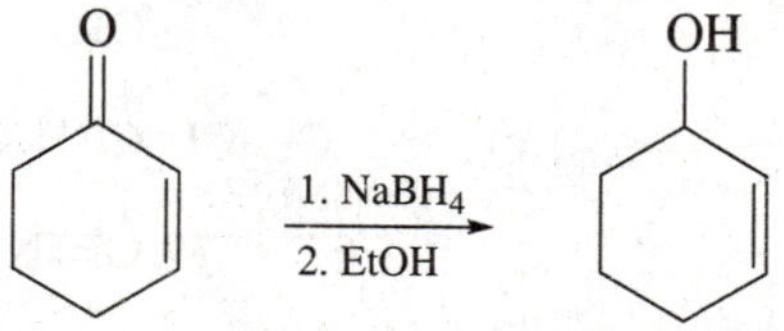

unsaturated alcohol

A sterically hindered ketone will be less likely to undergo direct addition and, therefore, more likely to undergo conjugate addition, the pathway that does not form an unsaturated alcohol. Thus, the nonsterically hindered ketone will give a higher yield of unsaturated alcohol.

44. **a.** 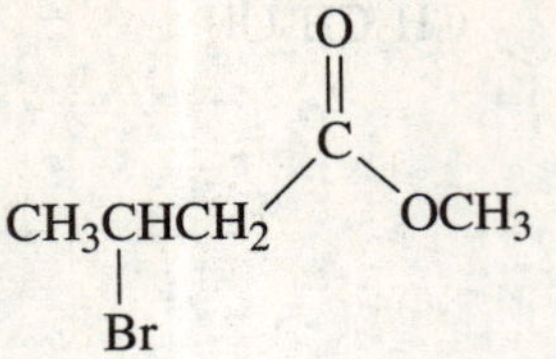 **c.** 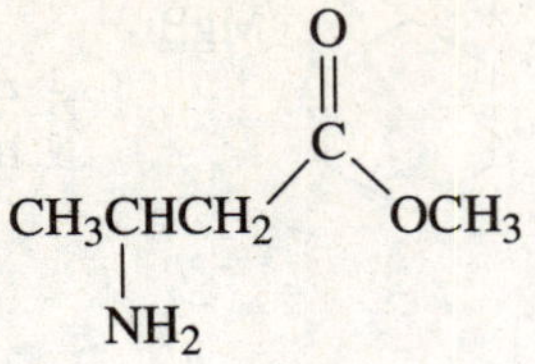

b. 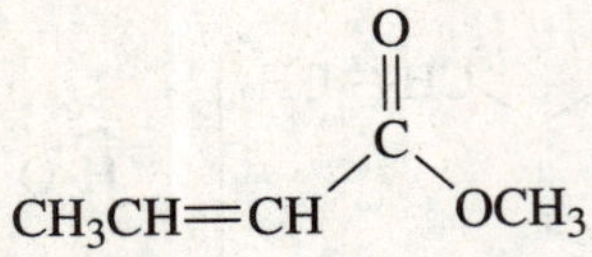 **d.** 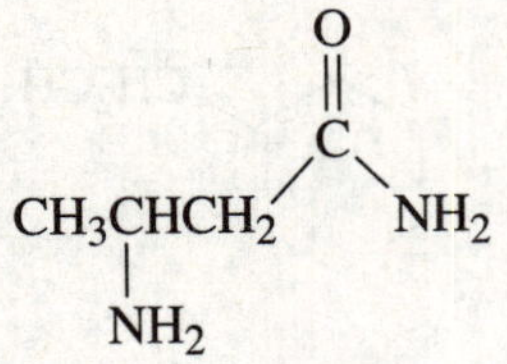

45. **a.** 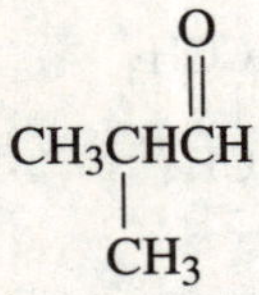 **d.** 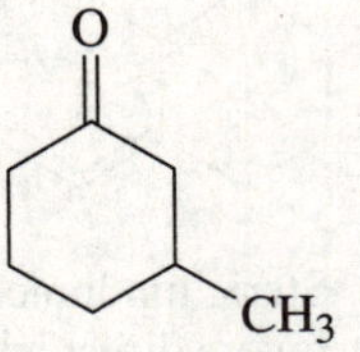 **h.** 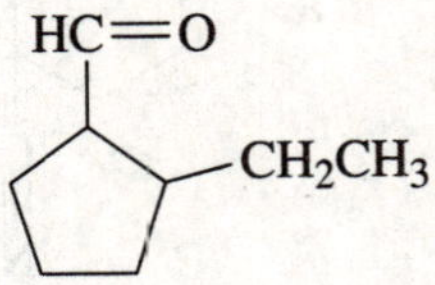

b. 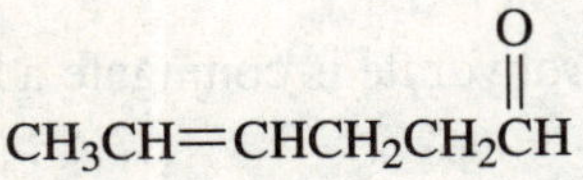 **e.** 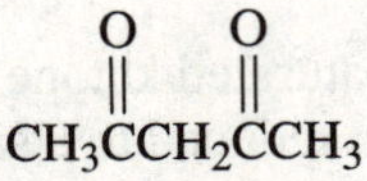 **i.** 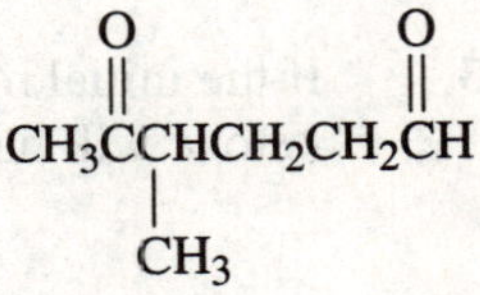

c. 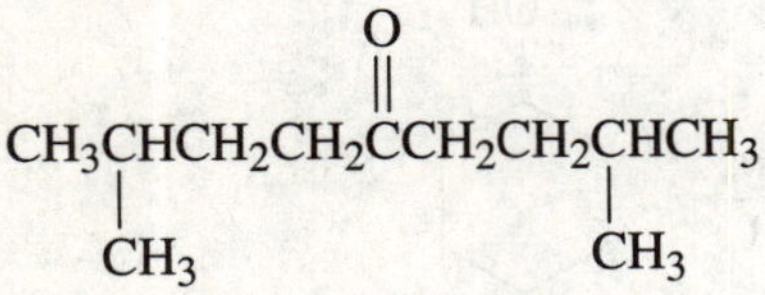 **f.** 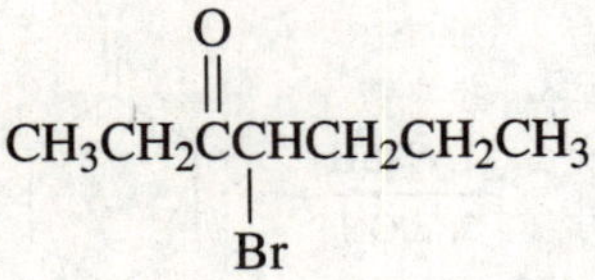 **j.** 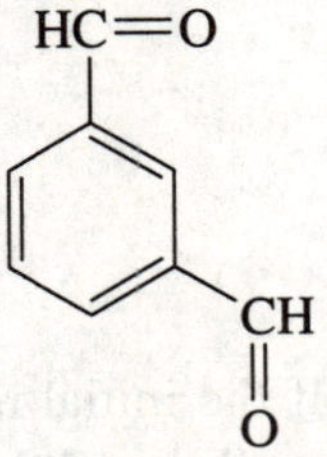

g. 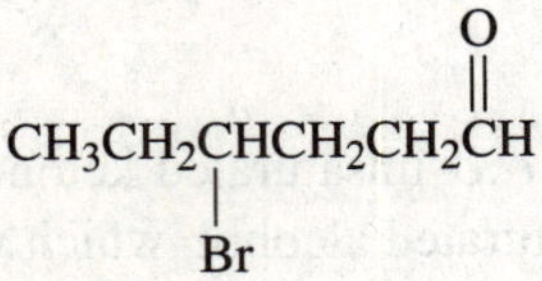

46. **a.** 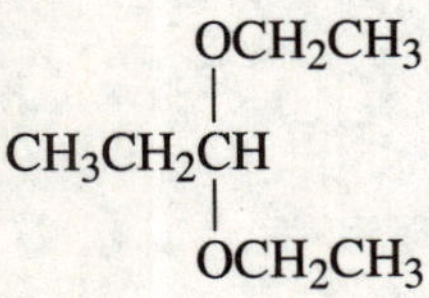 **e.** 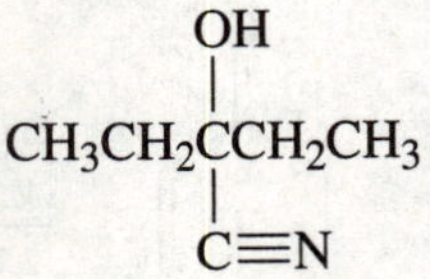

b. 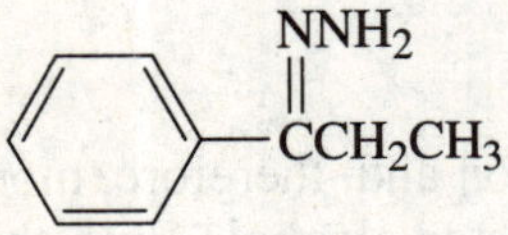 **f.** $CH_3CH_2CH_2CH_2OH$ + CH_3CH_2OH

c. 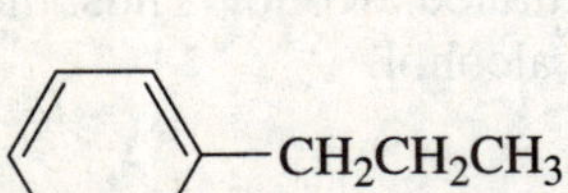 **g.** 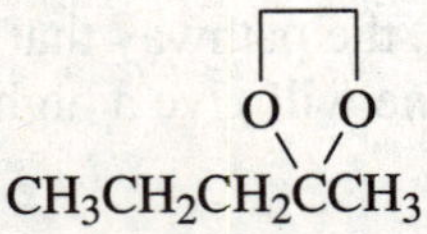

d. 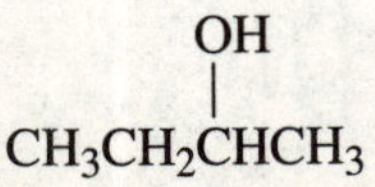 **h.** 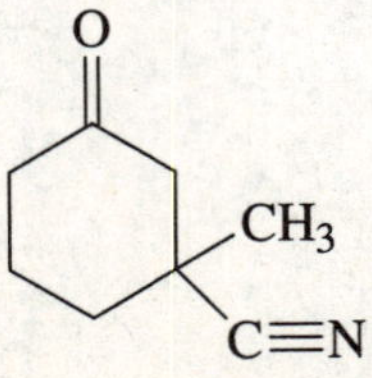

47. 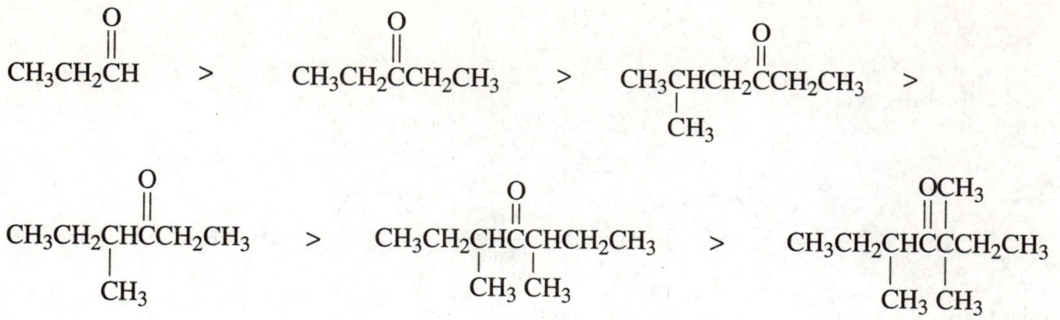

48. a. 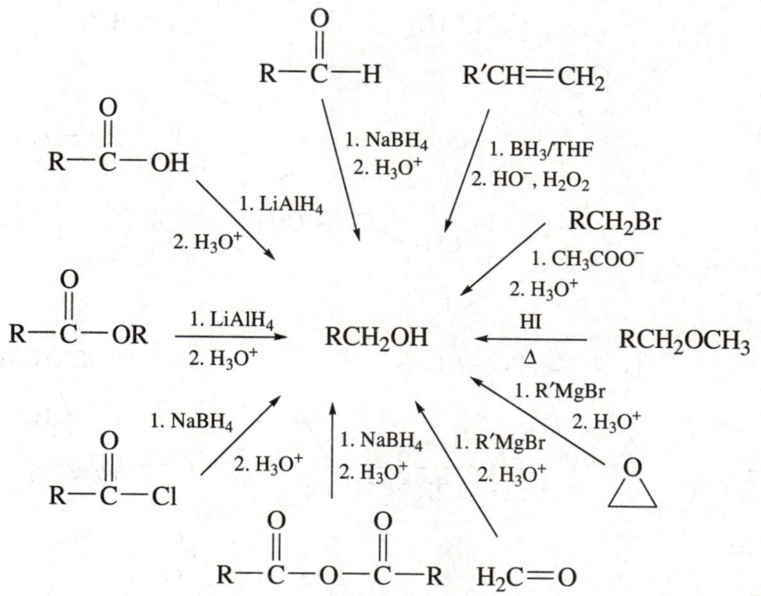

b. The reaction of a Grignard reagent with an epoxide.

c. The reaction of a Grignard reagent with an epoxide and the reaction of a Grignard reagent with formaldehyde. In each of these reactions, a new carbon-carbon bond is formed.

49. a. 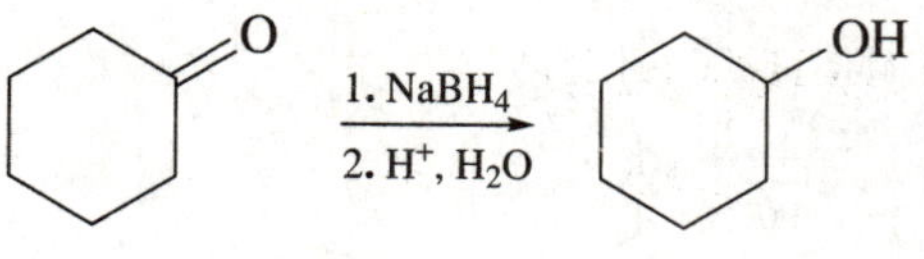

b. 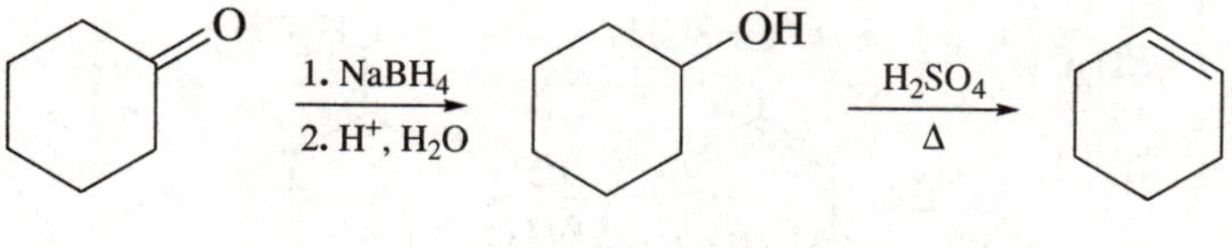

c. 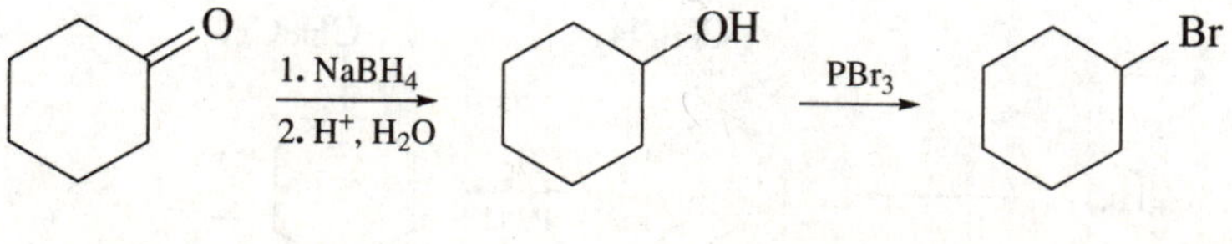

d.

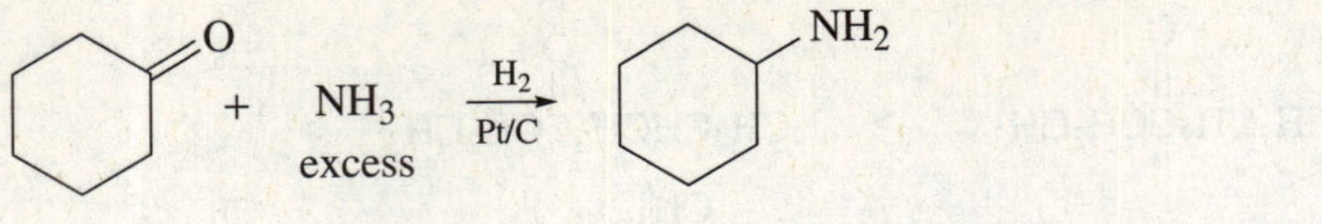

e.

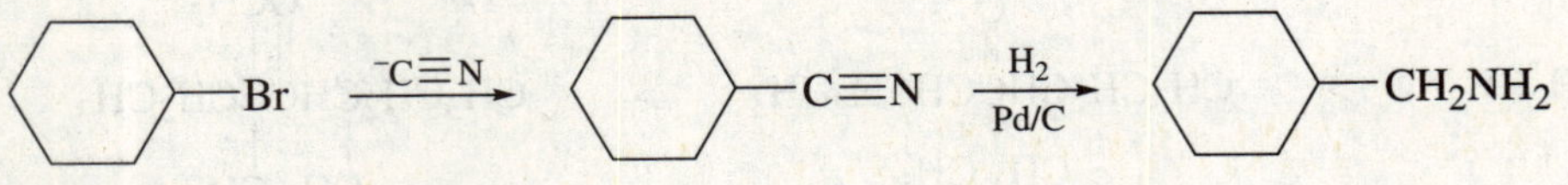

product of **c**

f.

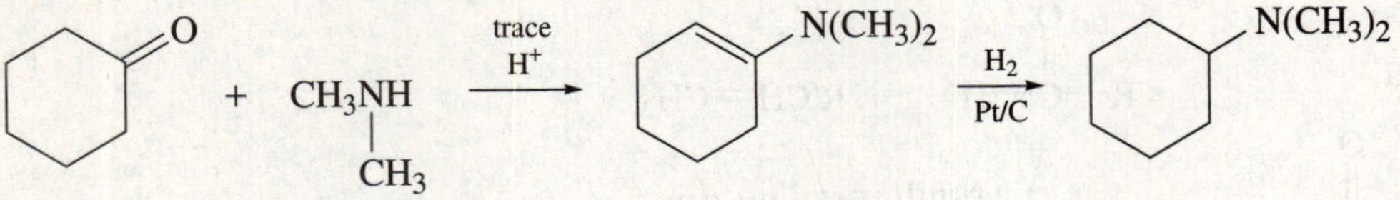

g.

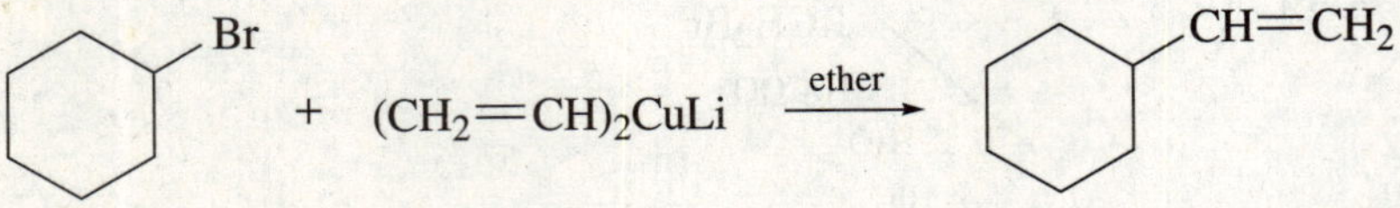

product of **c**

or

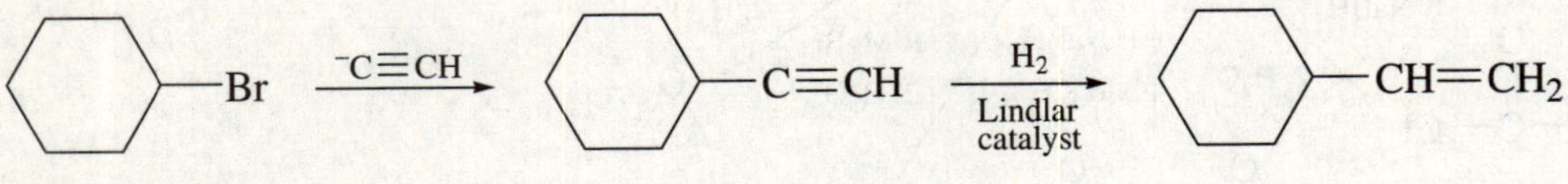

product of **c**

h.

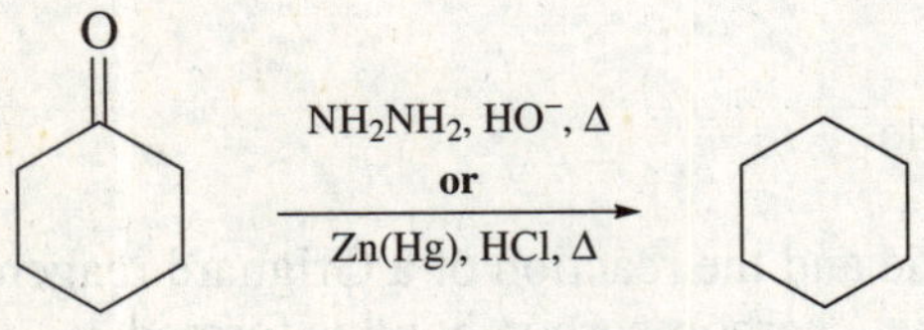

or

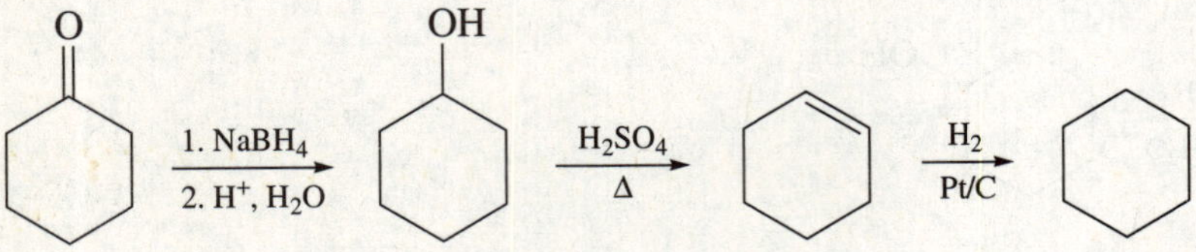

i.

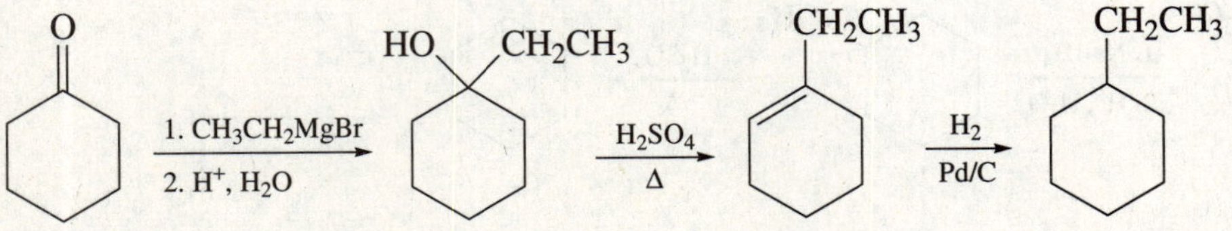

or

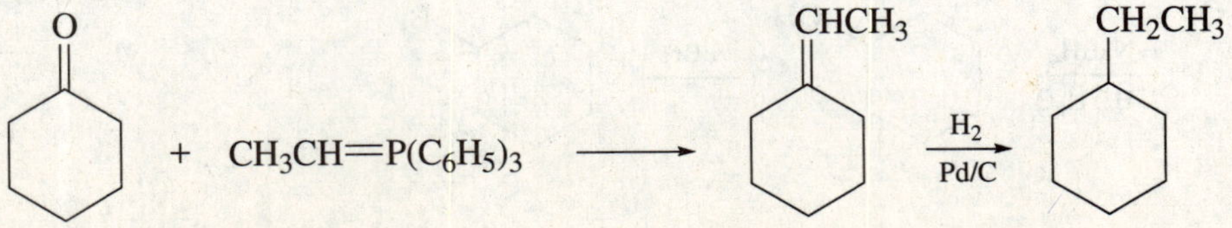

50.

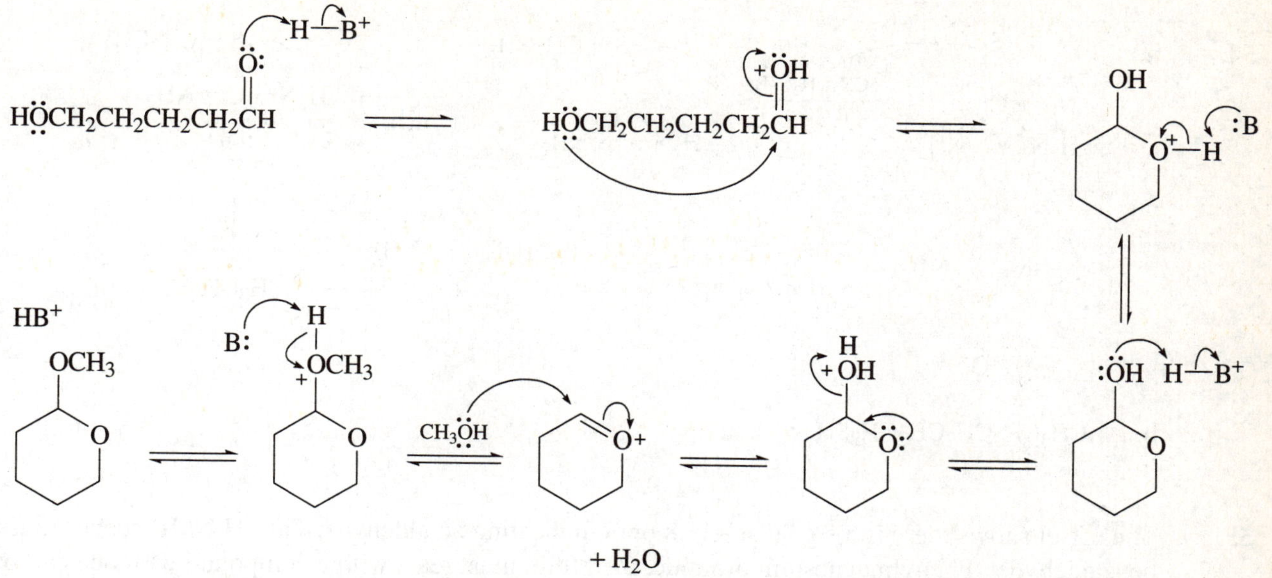

51. The greater the electron-withdrawing ability of the para substituent, the greater the K_{eq} for hydrate formation.

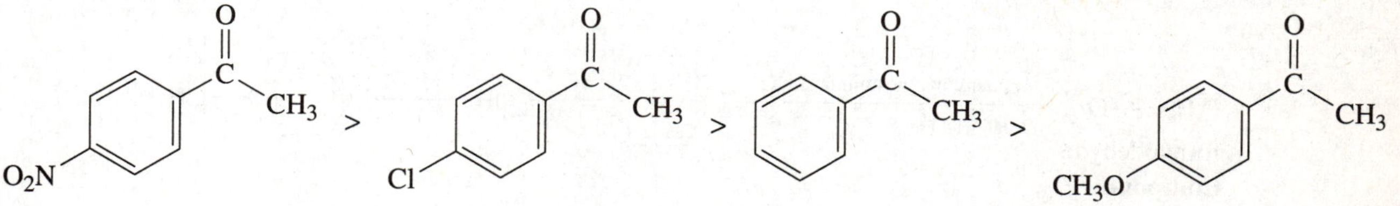

52. **a.** $CH_3OH \xrightarrow{PBr_3} CH_3Br \xrightarrow[Et_2O]{Mg} CH_3MgBr \xrightarrow[2.\ H^+,\ H_2O]{1.\ H_2C{=}O} CH_3CH_2OH$

b. $CH_4 \xrightarrow[h\nu]{Br_2} CH_3Br \xrightarrow[Et_2O]{Mg} CH_3MgBr \xrightarrow[2.\ H^+,\ H_2O]{1.\ \text{ethylene oxide}} CH_3CH_2CH_2OH$

53. **a.**

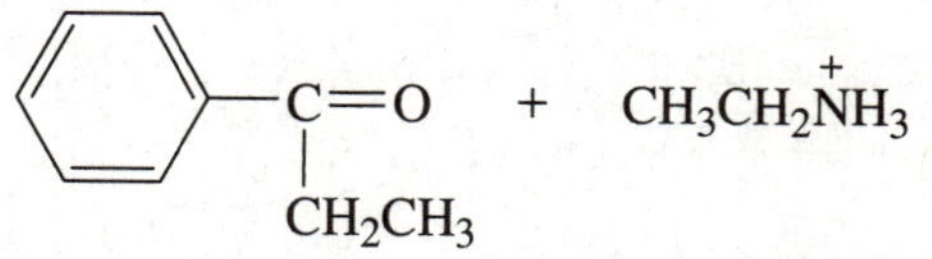

b.

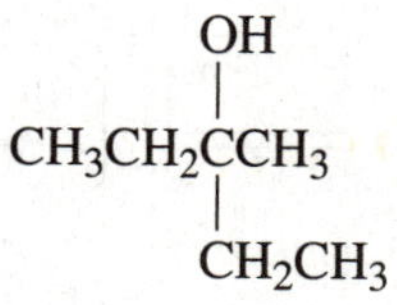

c.

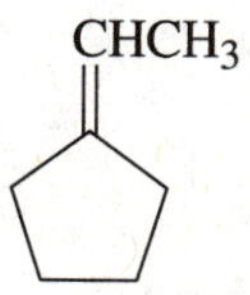

d.

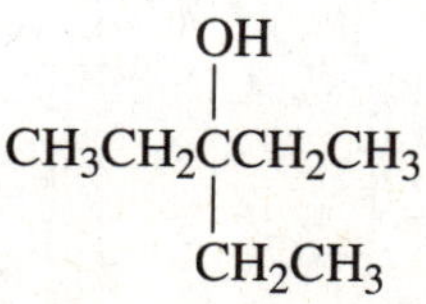

e.

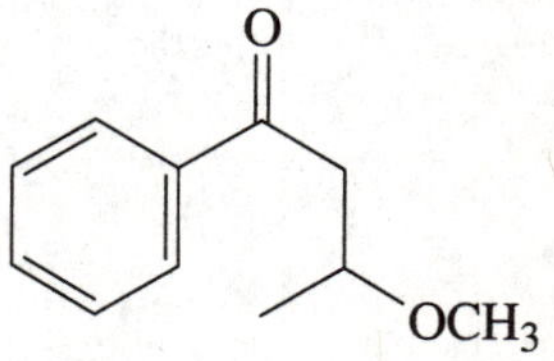

f.

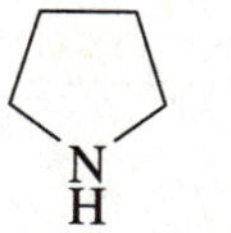

g.

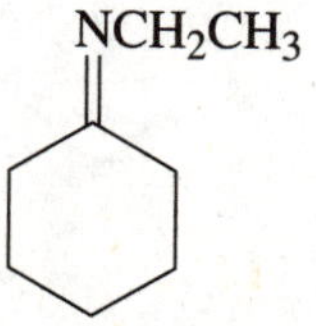

h.

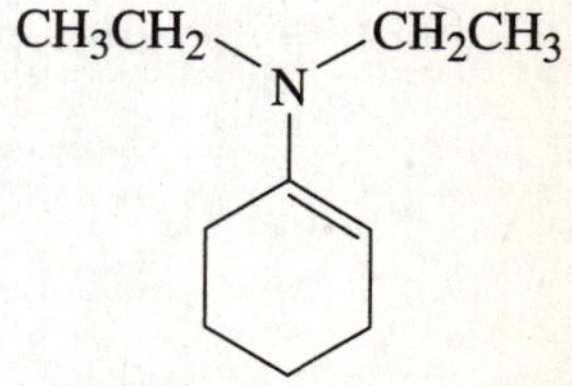

i.

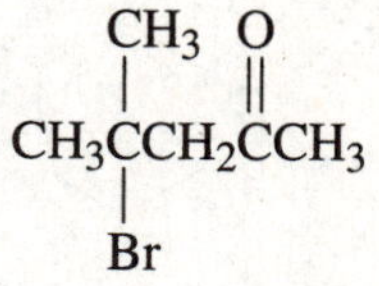

j.

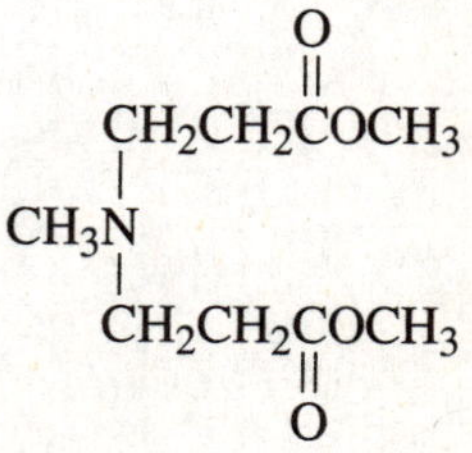

54. **a.**

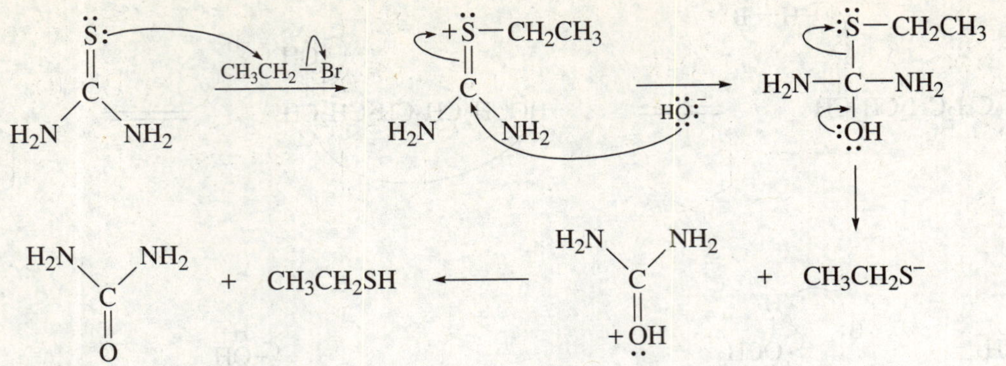

b. $CH_3CH_2CH_2CH_2CH_2SH$

55. The offset shows there is a signal at ~11.8 ppm, indicating an aldehyde. The ^1H NMR spectrum is that of benzaldehyde. Phenylmagnesium bromide, therefore, must react with a compound with one carbon atom to form an alcohol that can be oxidized by the mild oxidizing agent (MnO_2) to benzaldehyde. Therefore, Compound Z must be formaldehyde.

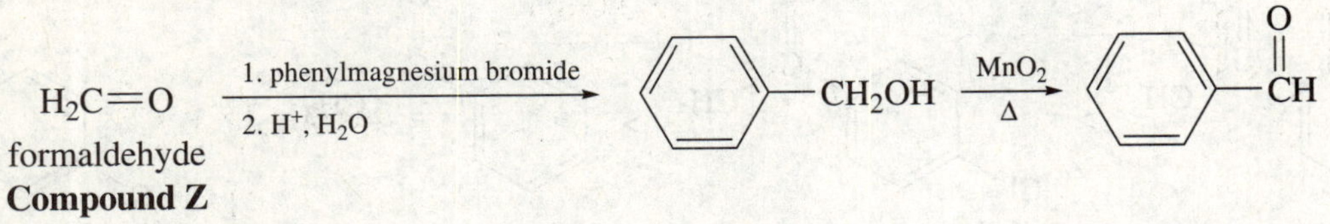

56. **a.**

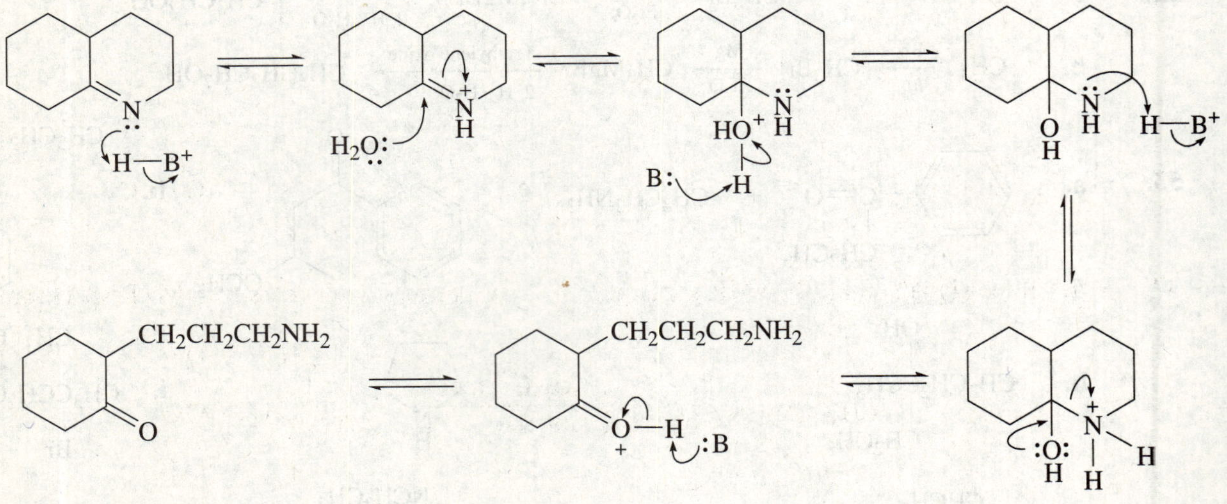

b.

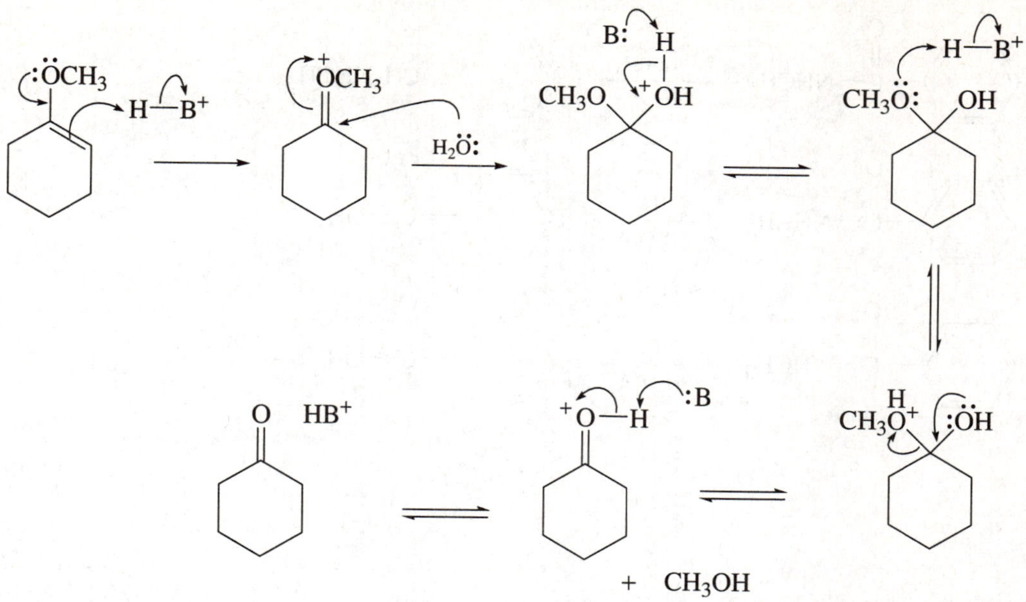

c.

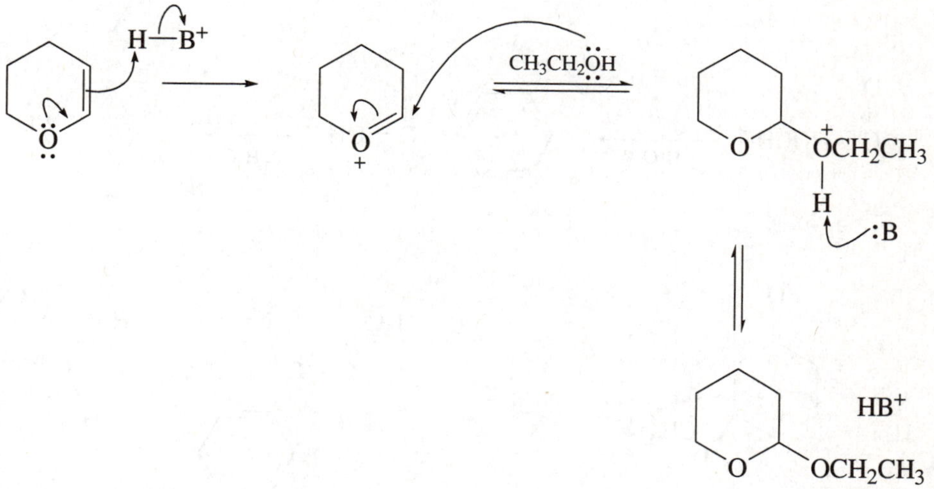

57. **a.** three signals in its ^{1}H NMR spectrum **b.** three signals in its ^{13}C NMR spectrum

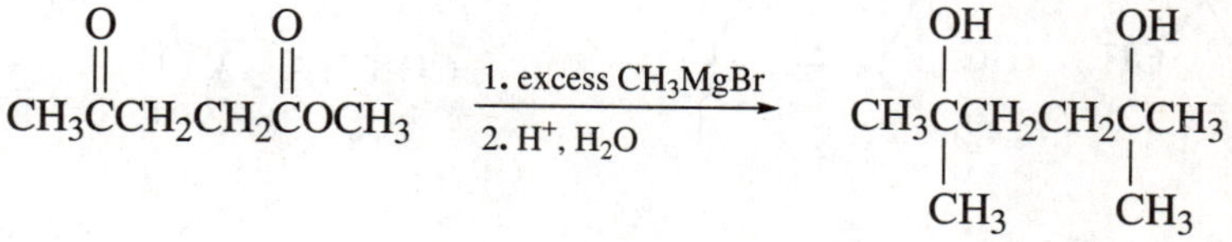

58.

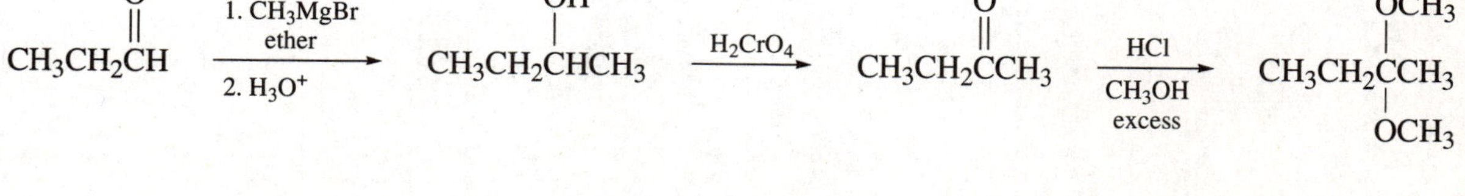

59. **a.**

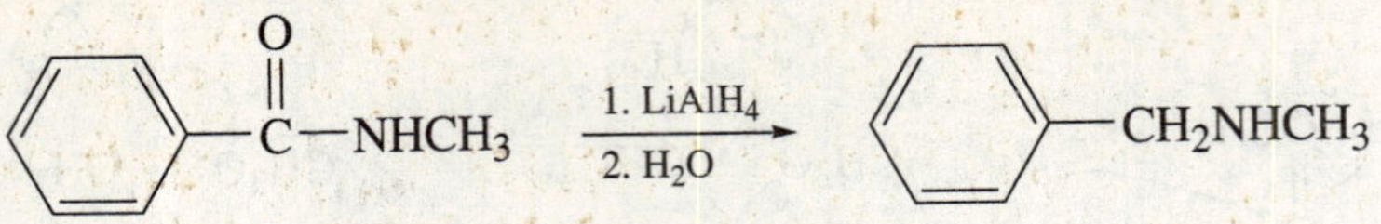

b.

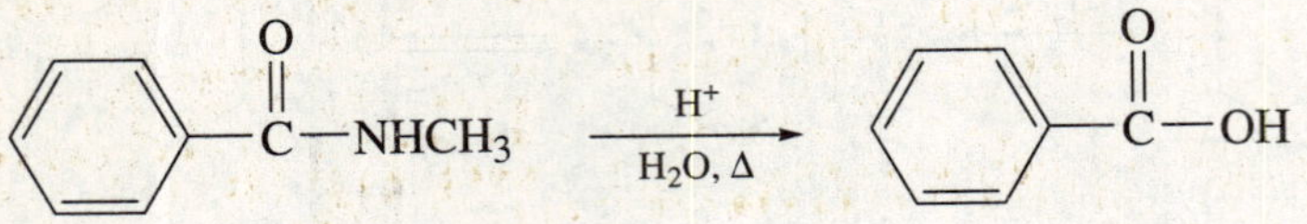

c.

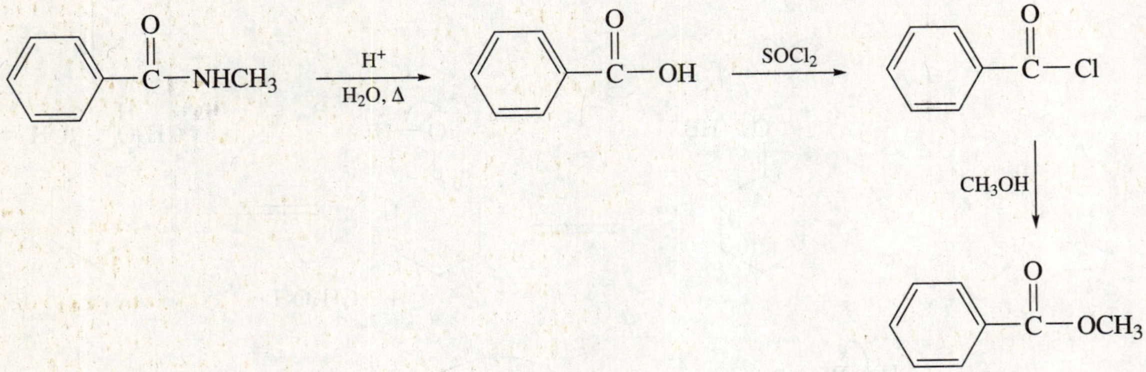

d.

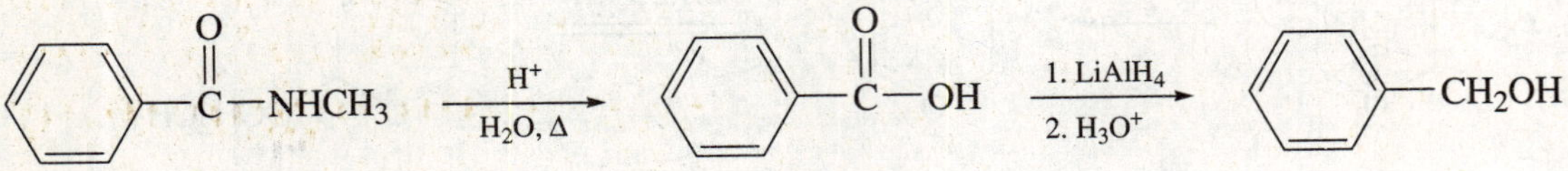

60. **a.**

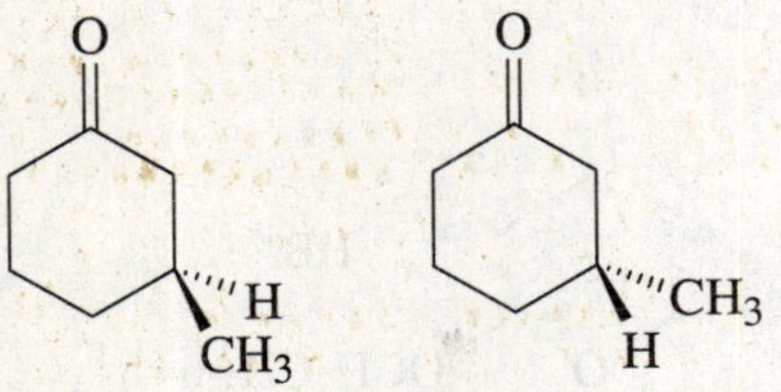

b.

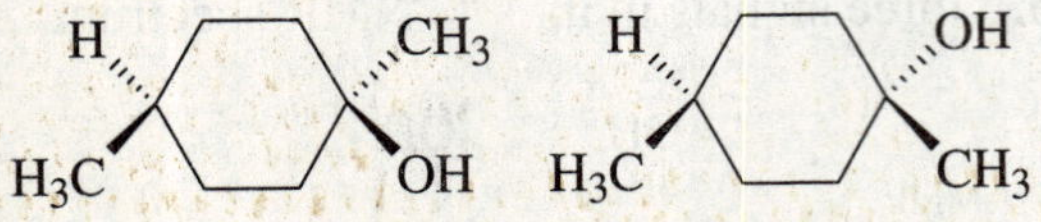

c.

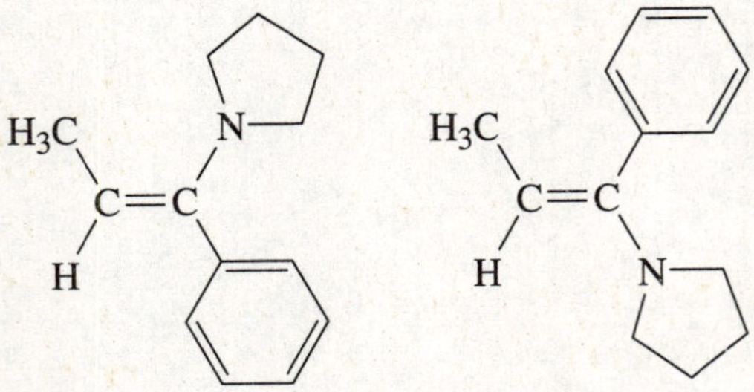

d.

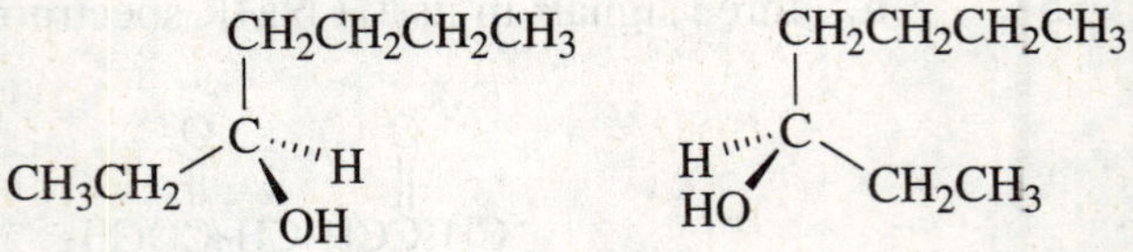

61. **a.**

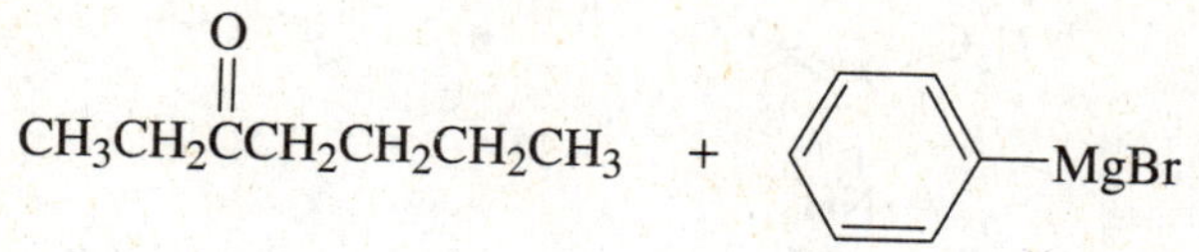

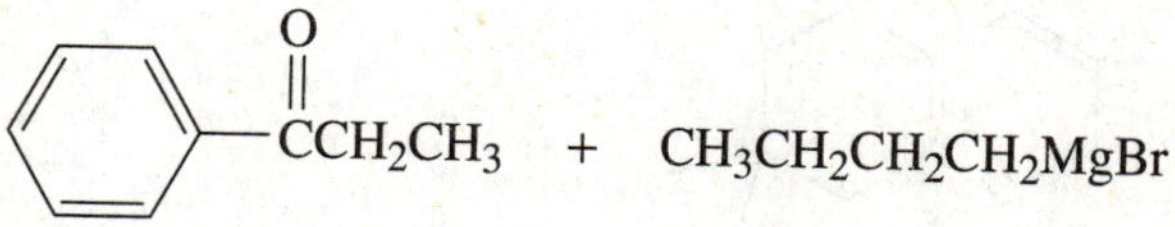

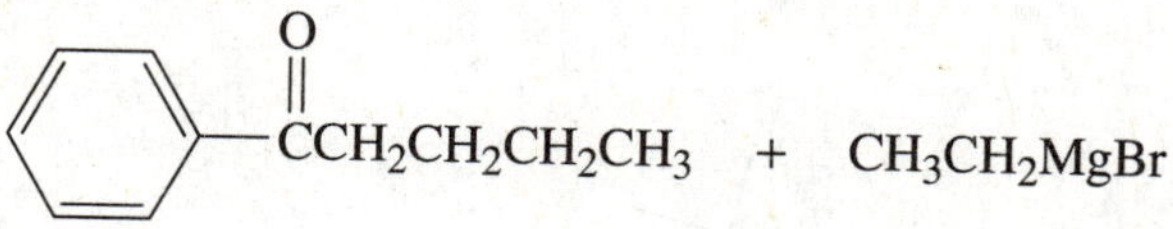

b. $CH_3CH_2C(=O)CH_2CH_2CH_3$ + CH_3CH_2MgBr

$CH_3CH_2C(=O)CH_2CH_3$ + $CH_3CH_2CH_2MgBr$

$CH_3CH_2CH_2C(=O)OCH_2CH_3$ + 2 CH_3CH_2MgBr

62. **a.**

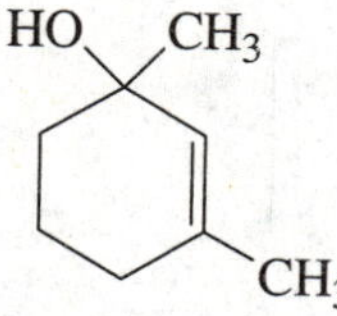

b.

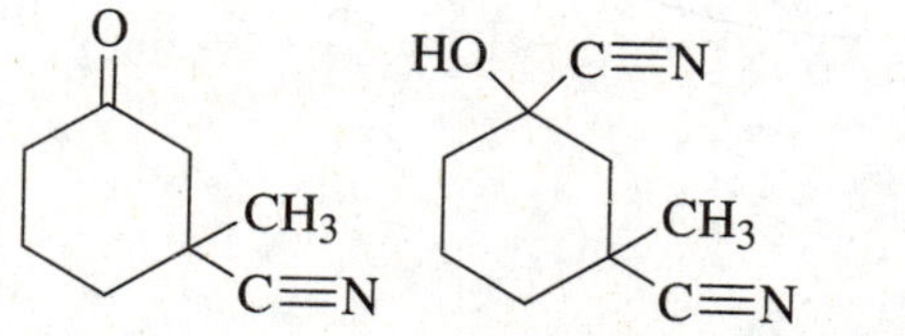

Because there is excess cyanide ion.

c.

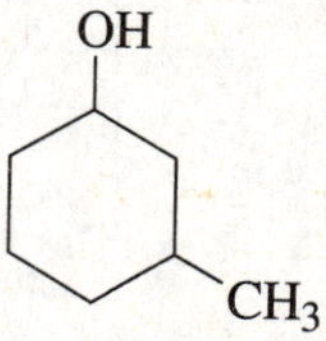

d.

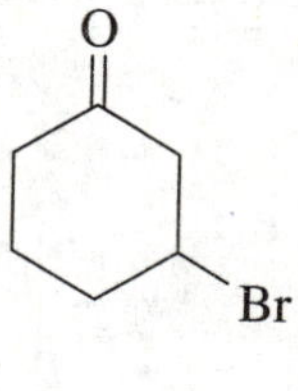

e.

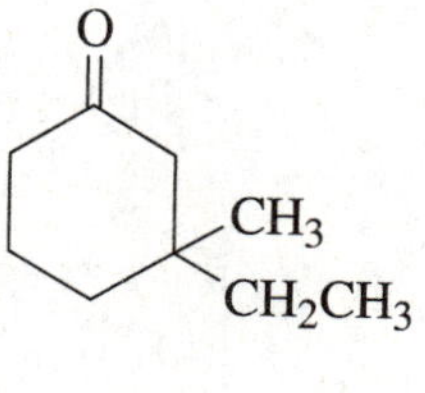

f.

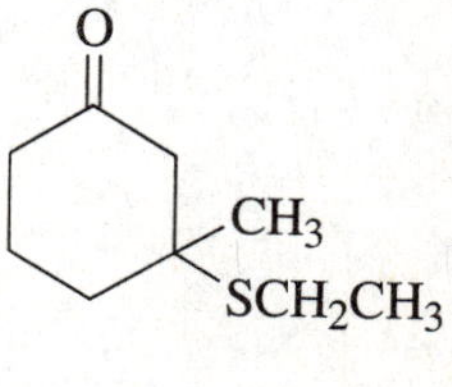

63. Enovid would have its carbonyl stretch at a higher frequency. The carbonyl group in Norlutin has some single-bond character because of the conjugated double bonds. This causes the carbon-oxygen bond to be easier to stretch than the carbon-oxygen bond in Enovid, which has isolated double bonds.

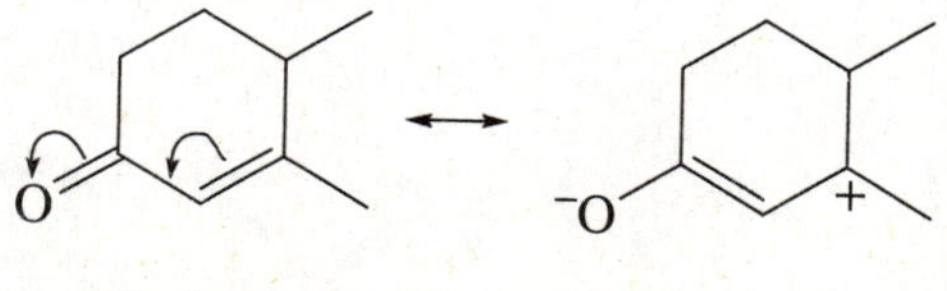

Norlutin

64. **a.**

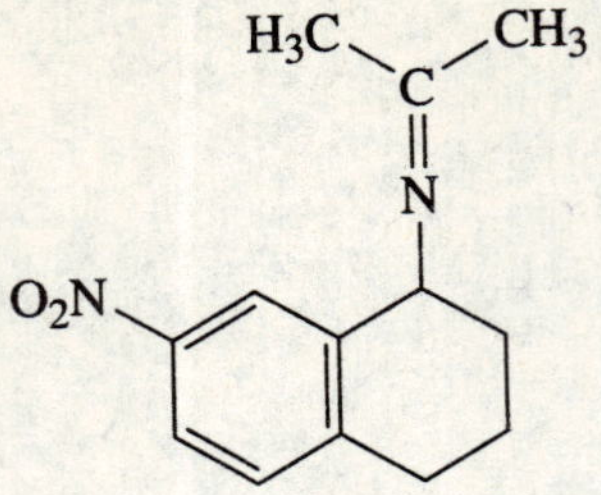

c.

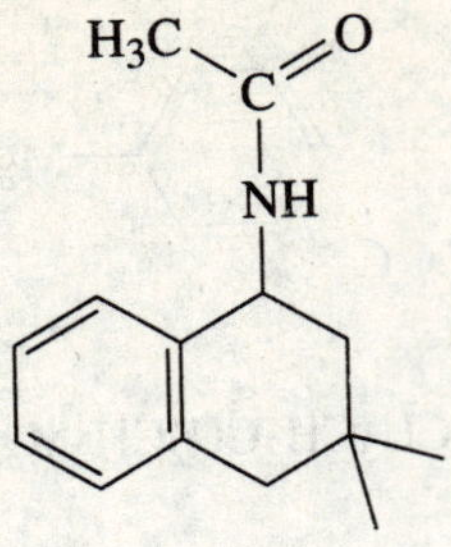

b.

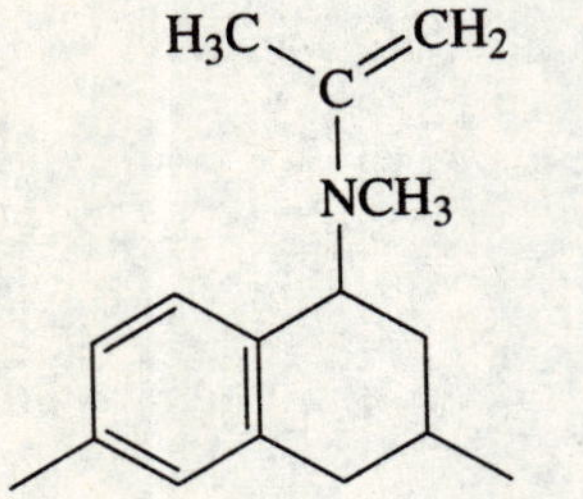

d.

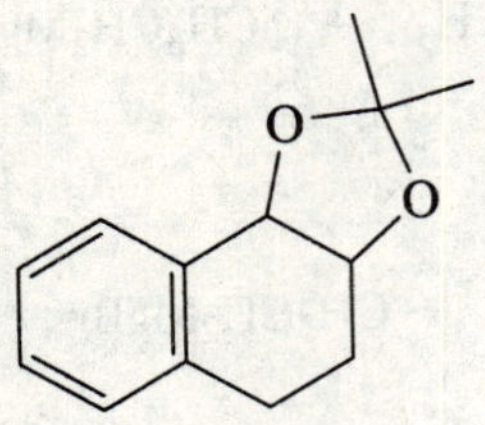

65. **a.**

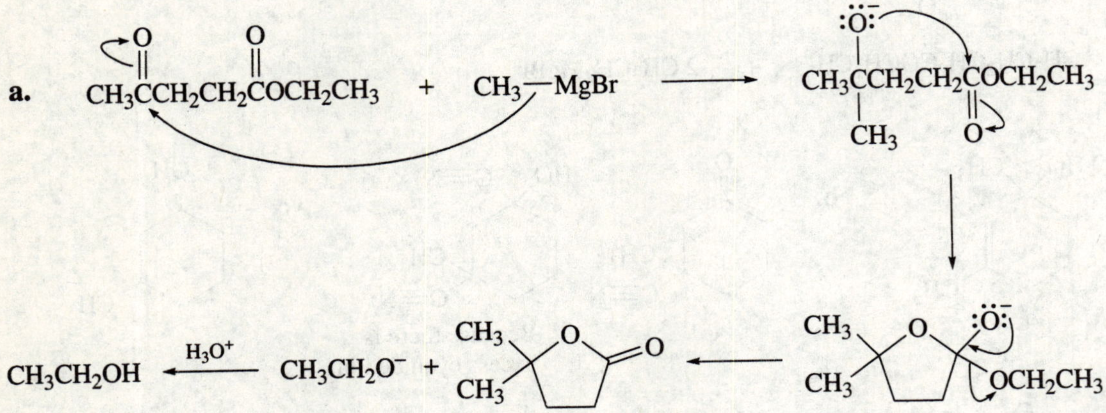

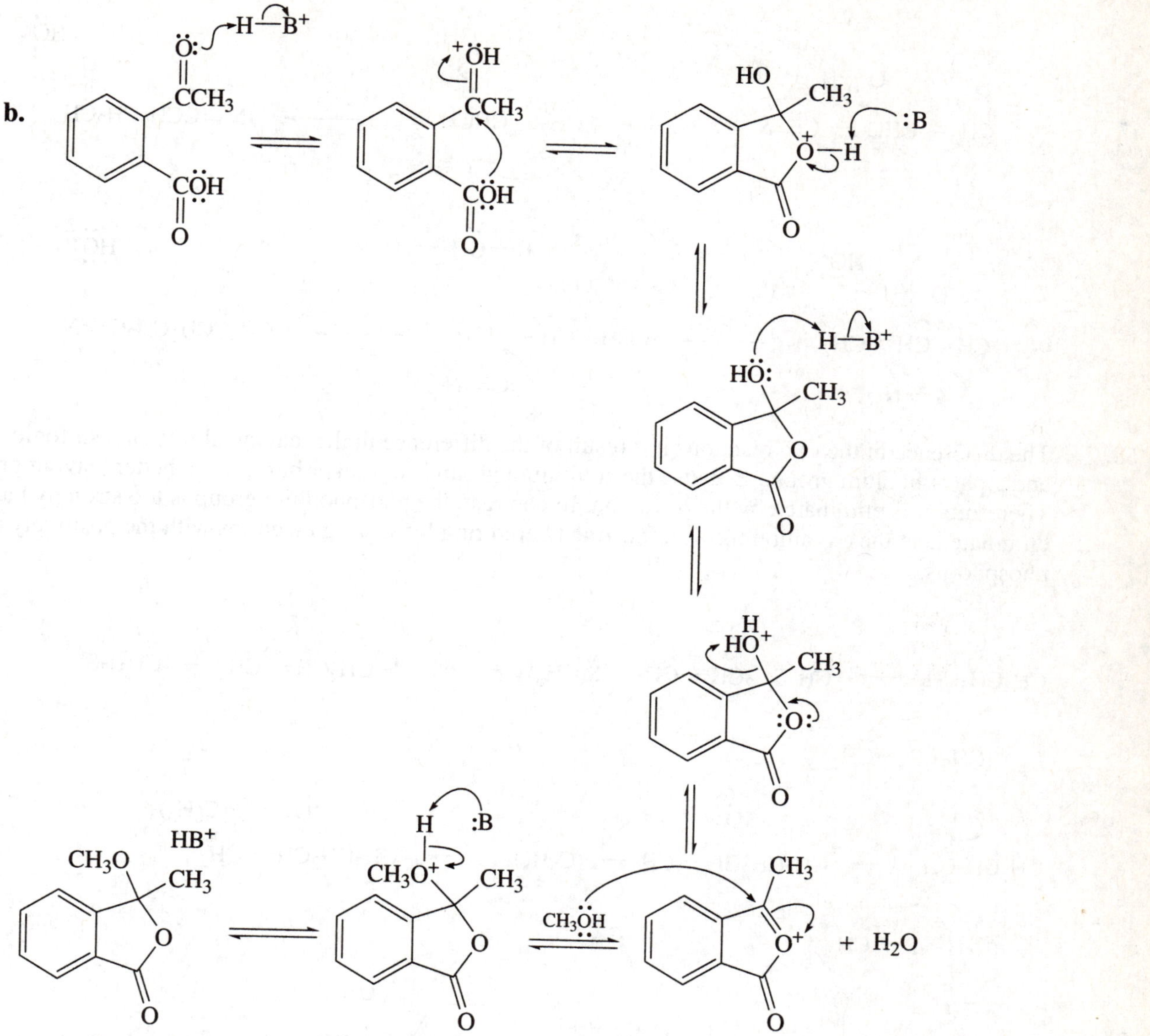

66. The absorption bands at 1600 cm^{-1}, 1500 cm^{-1}, and > 3000 cm^{-1} in the IR spectrum indicate that the compound has a benzene ring. The absorption band 1720 cm^{-1} suggests it is a ketone with the carbonyl group not conjugated with the benzene ring.

The broad signal at 1.8 ppm in the ^1H NMR spectrum indicates an OH group. We also see there are three different kinds of hydrogens (other than the OH group and the benzene ring hydrogens). The doublet at ~ 1.2 ppm indicates a methyl group adjacent to a carbon bonded to one hydrogen, and the doublet at 3.6 ppm is due to protons on a carbon that has the carbon bonded to one hydrogen on one side of it and the electron-withdrawing OH group on the other side.

The IR spectrum is the spectrum of 2-phenylpropanal and the ^1H NMR spectrum is the spectrum of 2-phenyl-1-propanol.

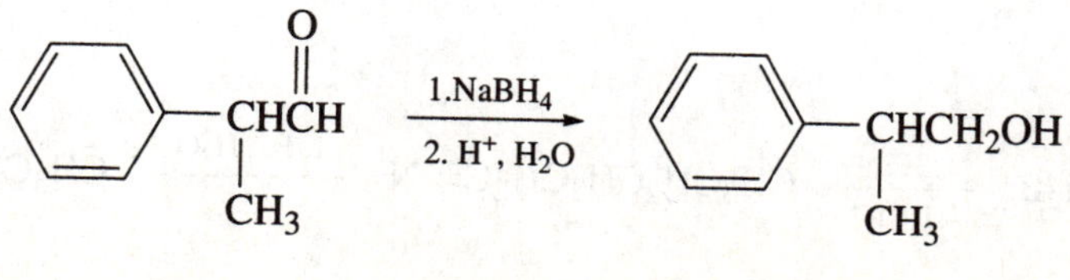

67. **a.**

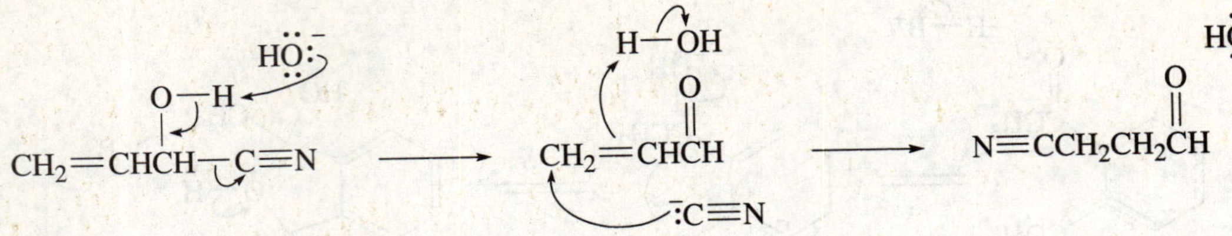

b.

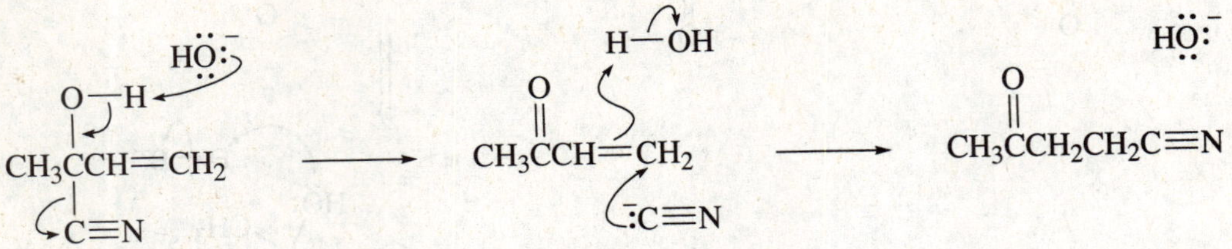

68. The difference in the two reactions is a result of the difference in the leaving ability of a sulfonium group and a phosphonium group. Because the sulfonium group is a weaker base, it is a better leaving group. Therefore, it is eliminated by the oxyanion. In contrast, the phosphonium group is too strongly basic to be eliminated, so the oxyanion forms a four-membered ring by sharing electrons with the positively charged phosphorus.

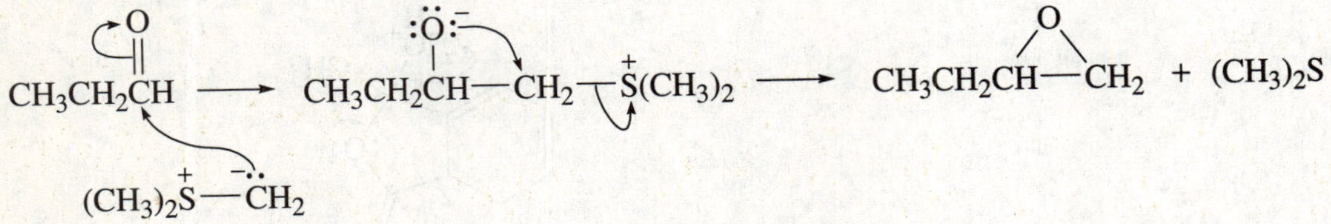

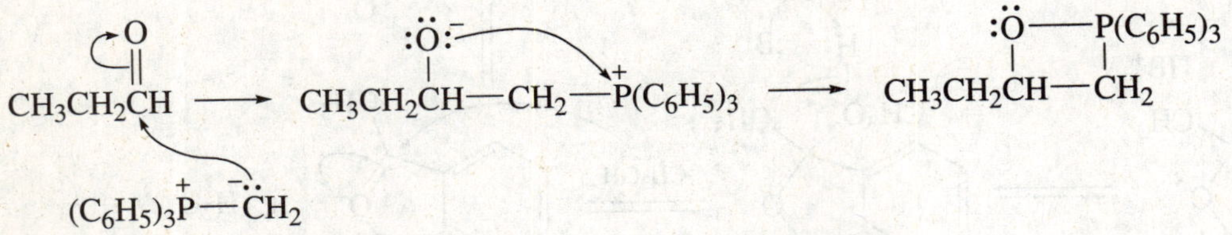

69. **a.**

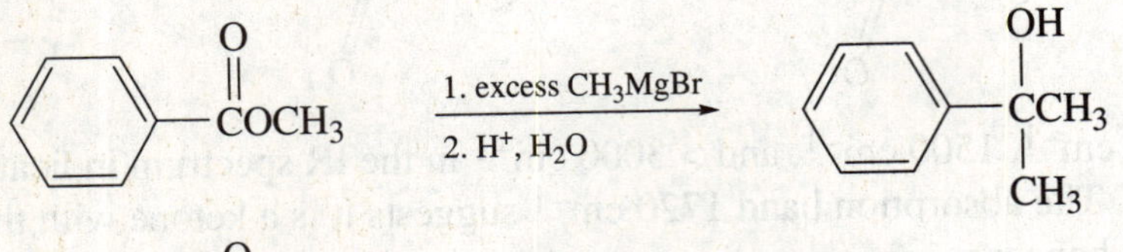

b.

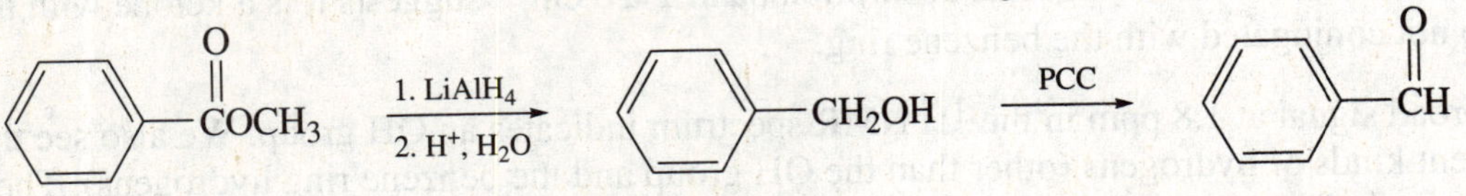

After you study Chapter 19, return to this problem and show how it can be done in one step.

c.

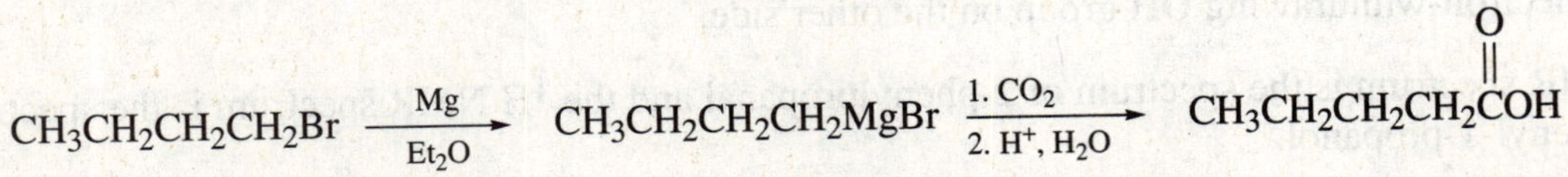

or

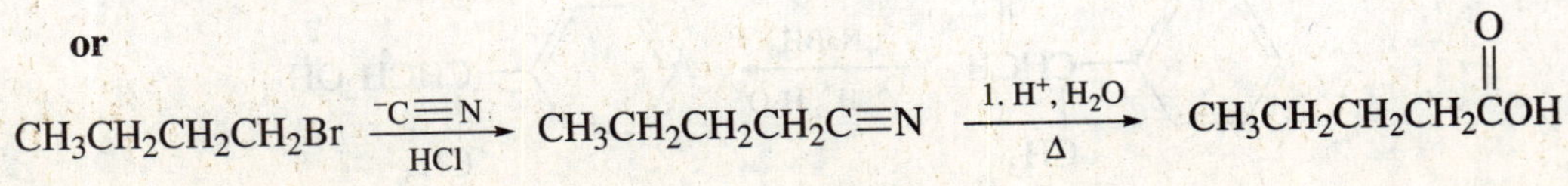

d.

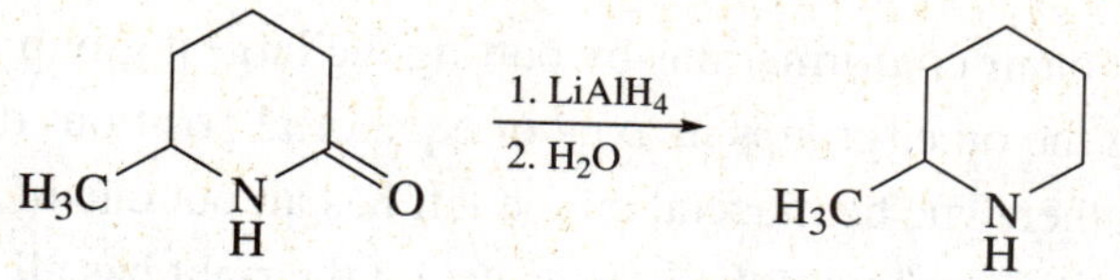

e.

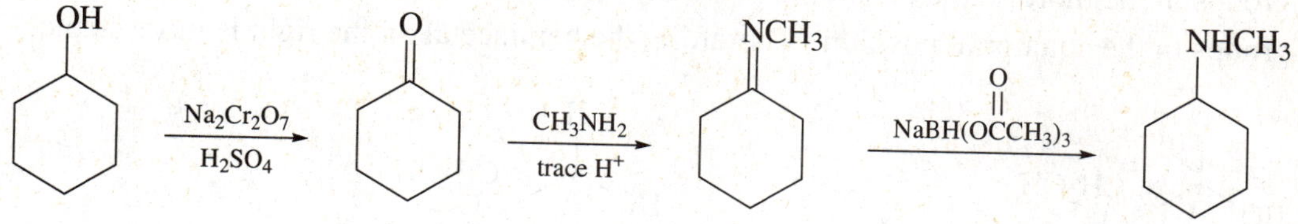

70. **a.**

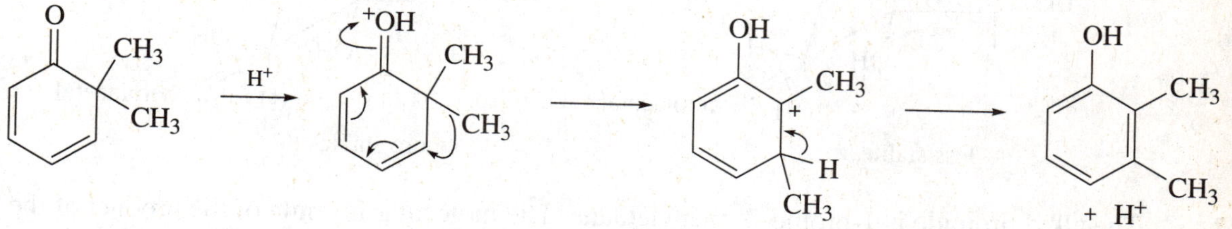

b.

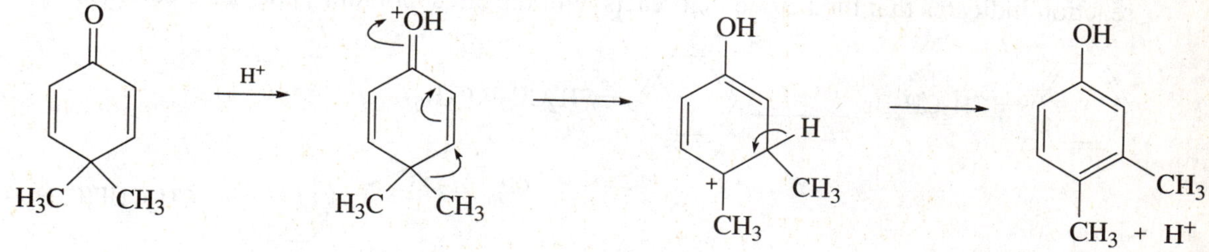

71. **a.** The OH group on C-5 of glucose reacts with the aldehyde group in an intramolecular reaction, forming a cyclic hemiacetal. Because the reaction creates a new asymmetric center, two cyclic hemiacetals can form, one with the *R* configuration at the new asymmetric center and one with the *S* configuration.

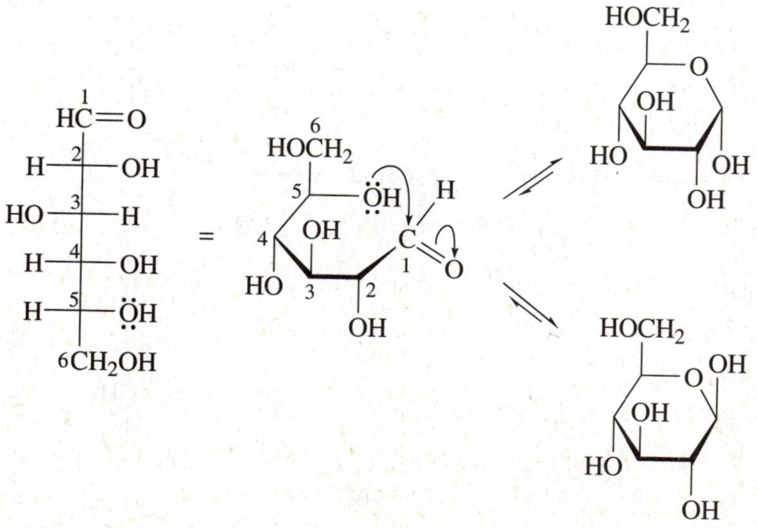

b. The two products can be drawn in their chair conformations by putting the largest group (CH_2OH) in the equatorial position and then putting the other groups in axial or equatorial positions depending on whether they are cis or trans to one another. The hemiacetal on the left has all but one of its OH groups in the more stable equatorial position, whereas the hemiacetal on the right has all its OH groups in the equatorial position. Therefore, the hemiacetal on the right is more stable.

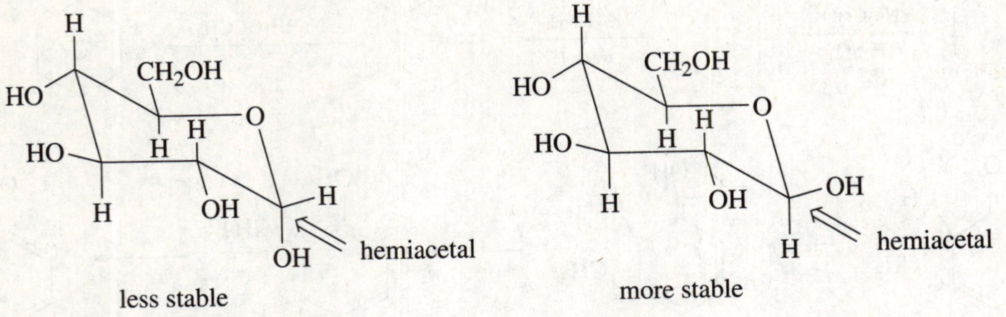

72. The alkyl bromide is 1-bromo-2-phenylethane. The molecular formula of the product of the Witting reaction indicates that the ketone that reacts with the phosphonium ylide has 3 carbons.

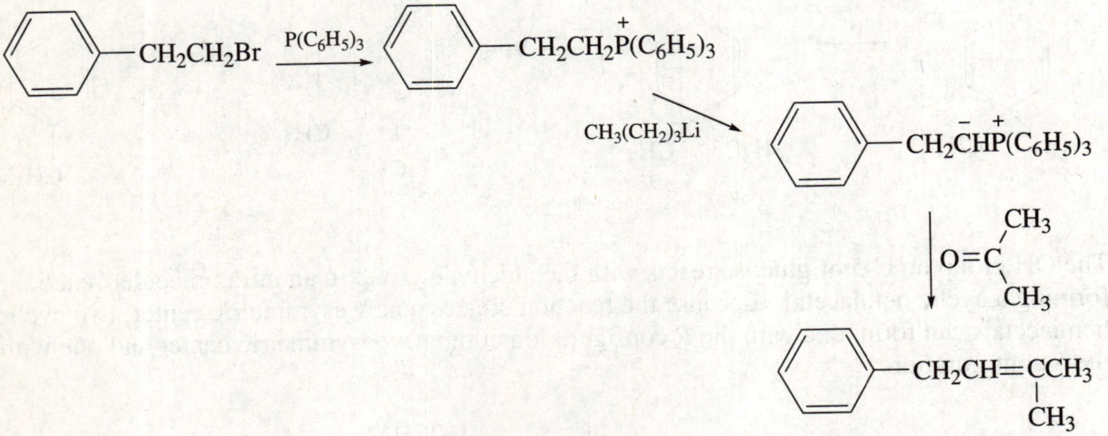

73.

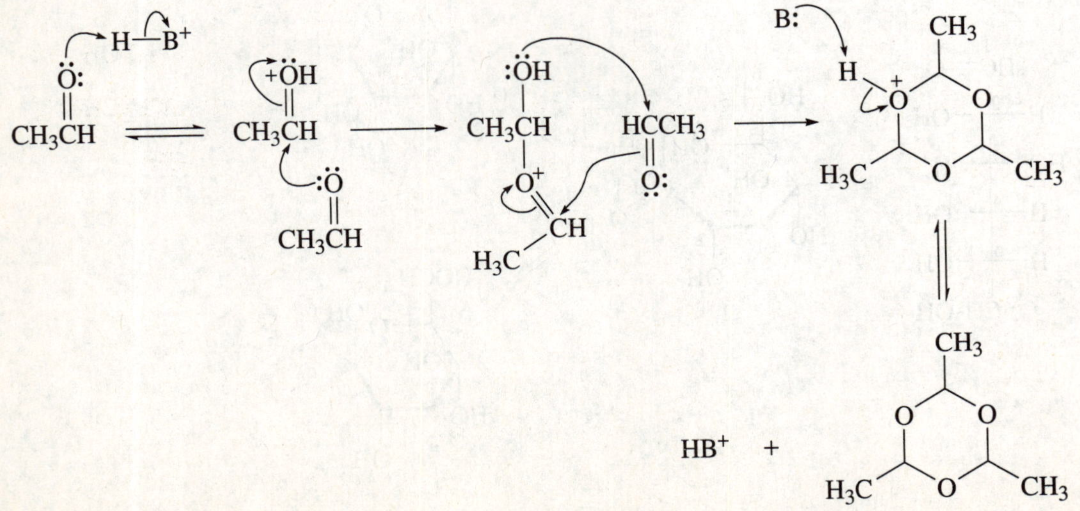

74. Converting the Fischer projection of (*R*)-mandelonitrile into a wedge-and-dash structure allows you to see that attack of cyanide ion occurred on the *Si* face (decreasing priorities on the face closest to the observer are in a counterclockwise direction).

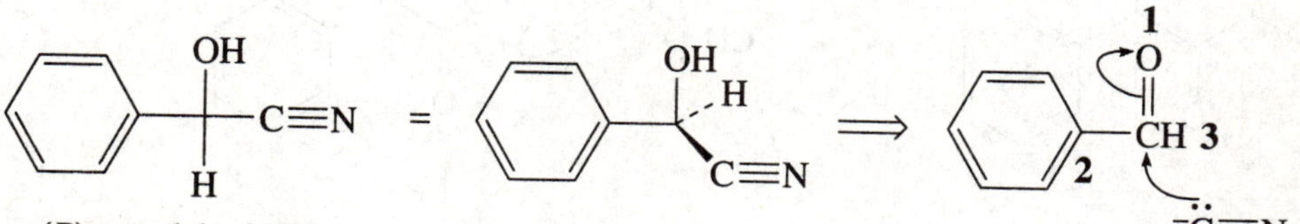

(*R*)-mandelonitrile

75. **a.**

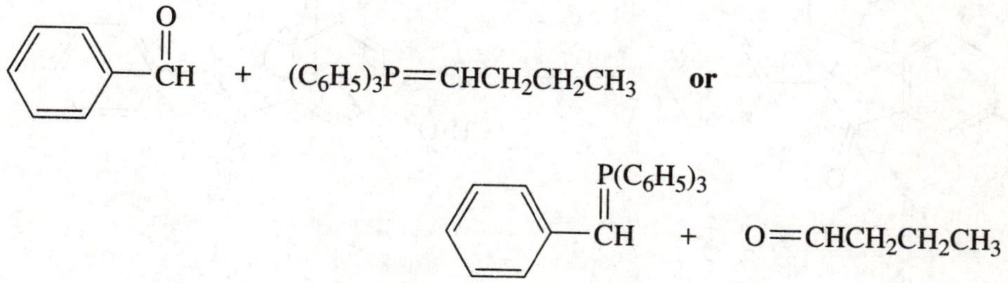

b.

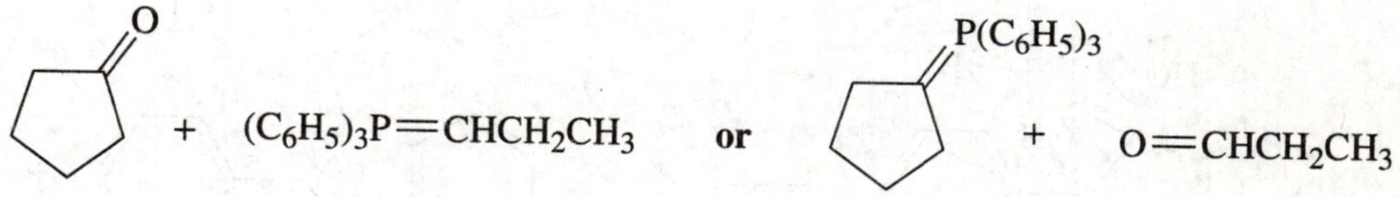

c.

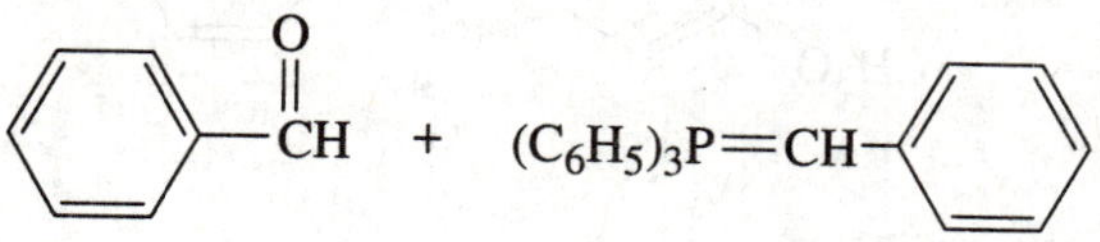

d.

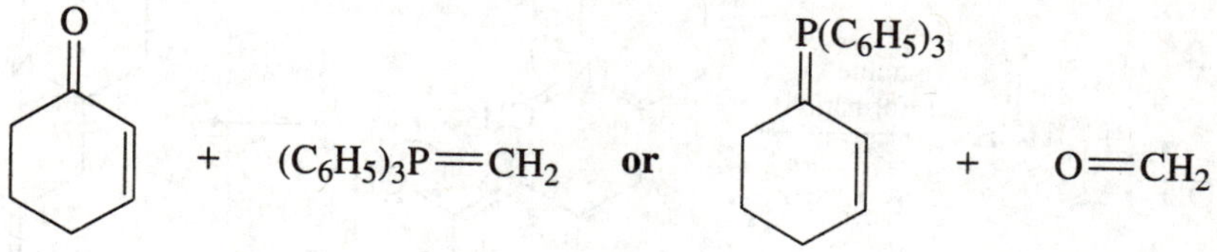

76.

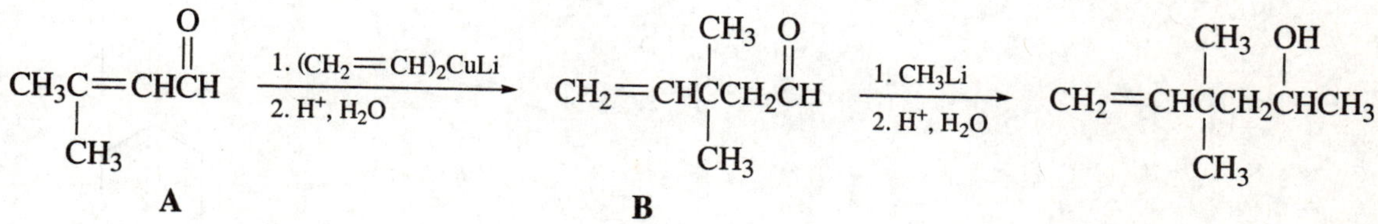

77. **a.**

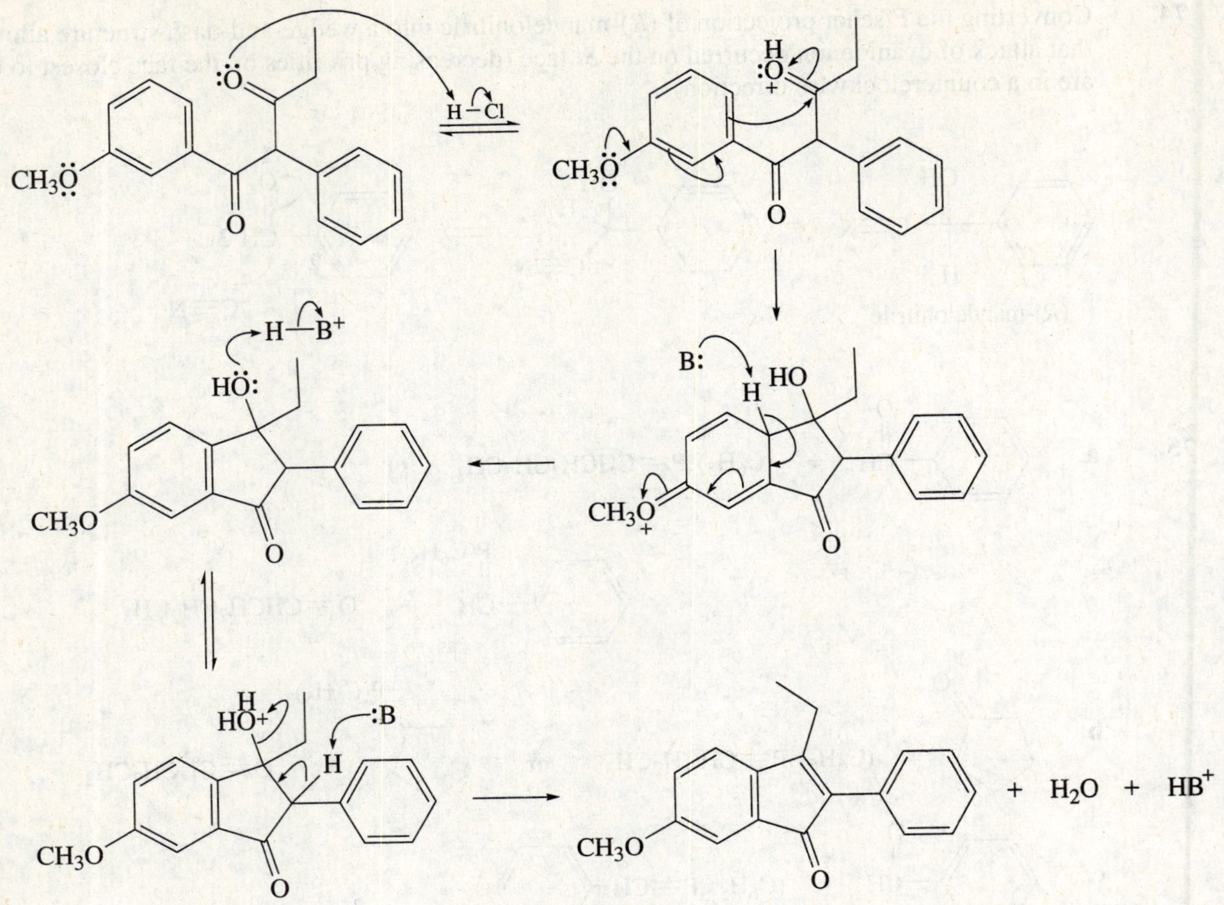
CH3Ö
H—Cl
CH3Ö
H—B+
HÖ:
CH3O
B:
HO
CH3O
HO+
:B
CH3O
CH3O
+ H2O + HB+

b.

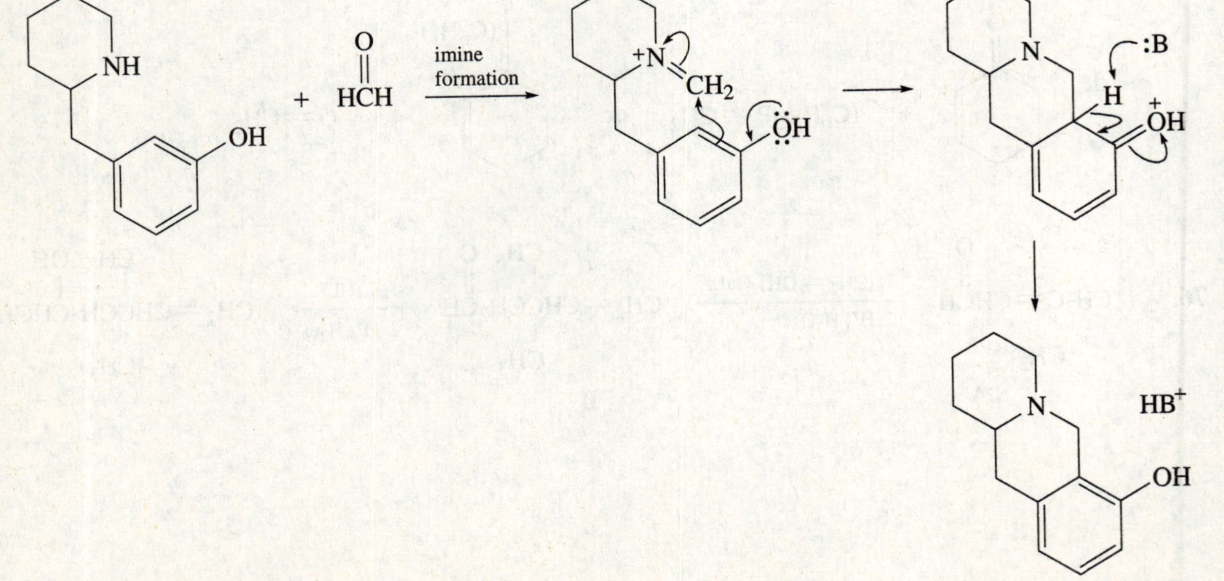
NH
OH
HCH
imine
formation
+N
CH2
ÖH
N
:B
H
ÖH
N
HB+
OH

78. The compound that gives the ^1H NMR spectrum is 2-phenyl-2-butanol.

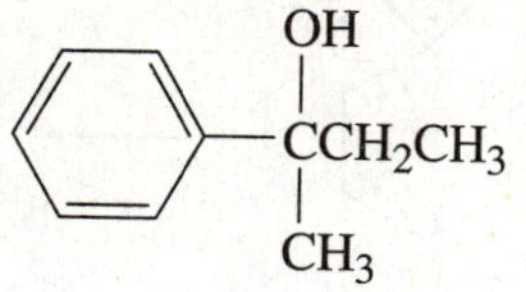

Therefore, the compound that reacts with methylmagnesium bromide is 1-phenyl-1-propanone.

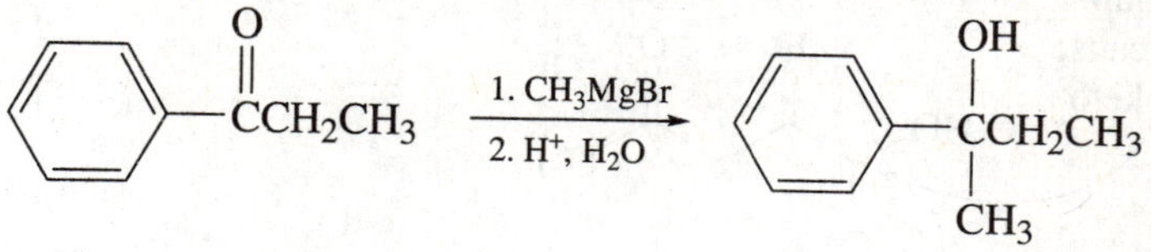

79. **a.**

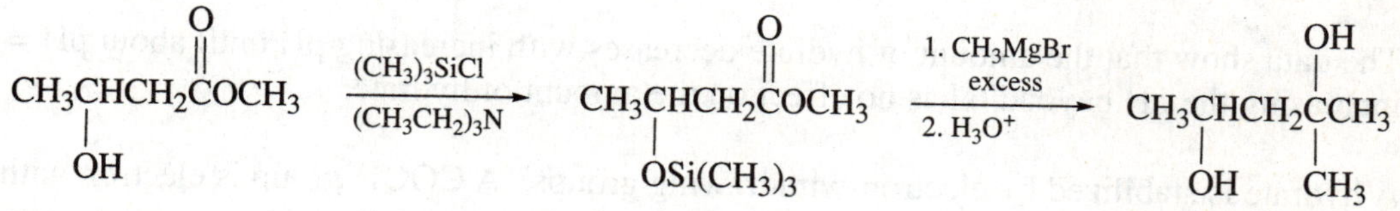

b.

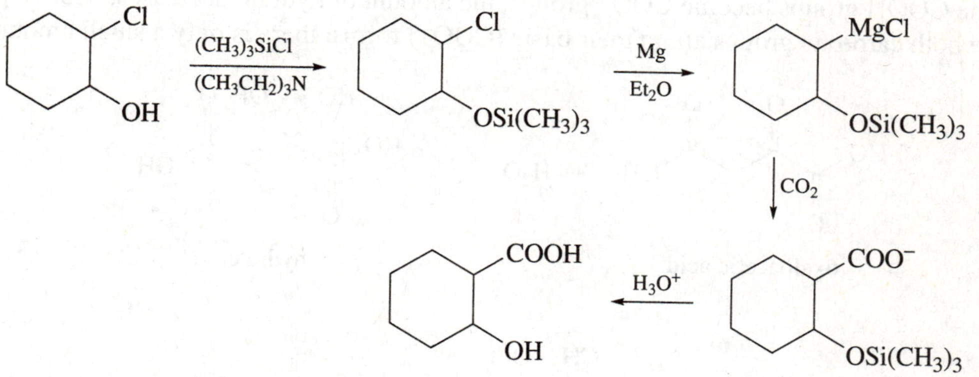

c.

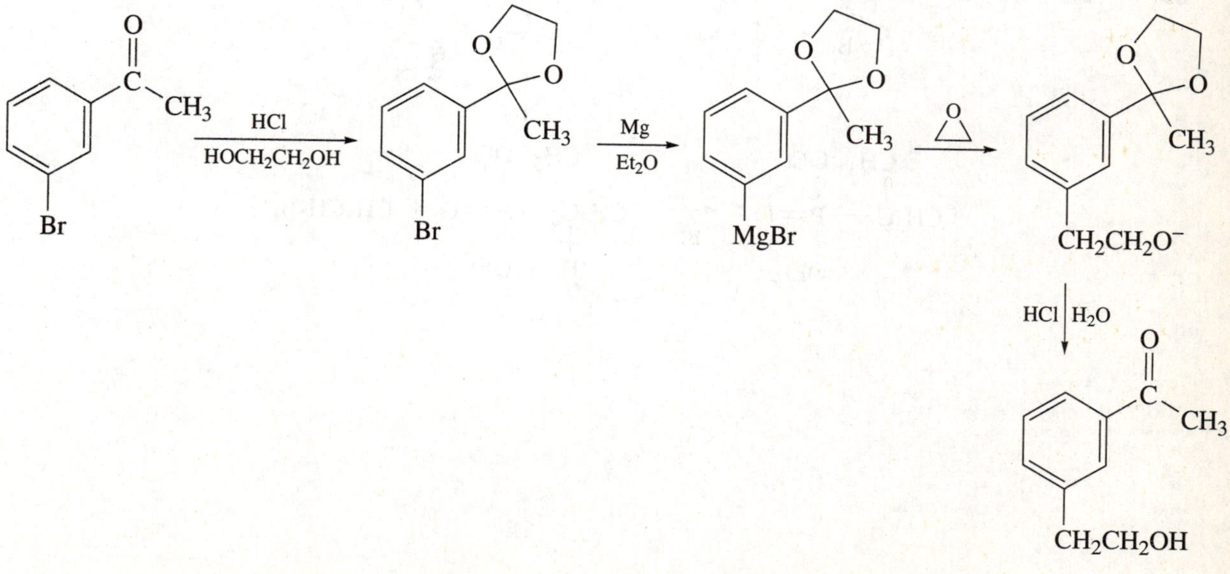

80. 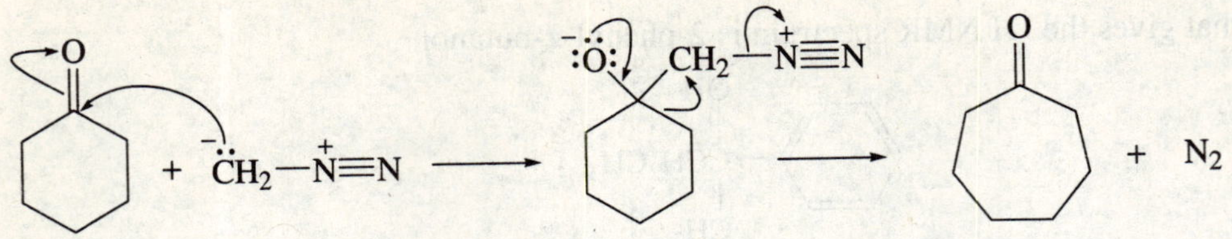

81. **a.** This carboxyl group is more acidic because it is closer to the keto group.

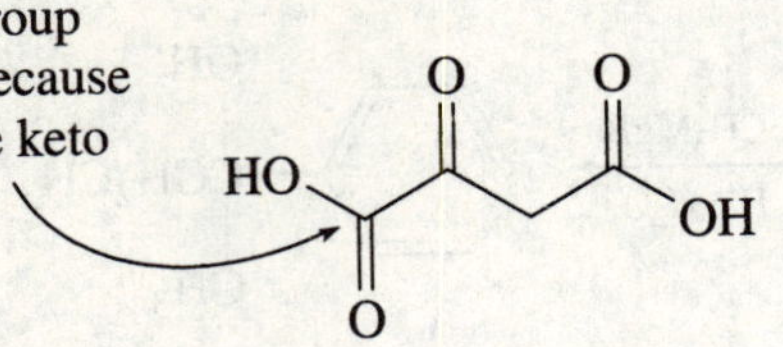

b. The data show that the amount of hydrate decreases with increasing pH until about pH = 6, and that increasing the pH beyond 6 has no effect on the amount of hydrate.

A hydrate is stabilized by electron-withdrawing groups. A COOH group is electron withdrawing, but a COO⁻ group is less so. In acidic solutions, where both carboxylic acid groups are in their acidic (COOH) forms, the compound exists as essentially all hydrate. As the pH of the solution increases and the COOH groups become COO⁻ groups, the amount of hydrate decreases. Above pH = 6, where both carboxyl groups are in their basic (COO⁻) forms, there is only a small amount of hydrate.

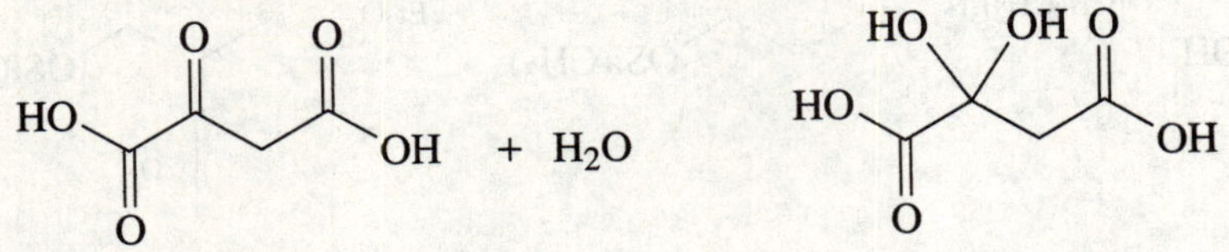

oxaloacetic acid   hydrate

82. **a.** 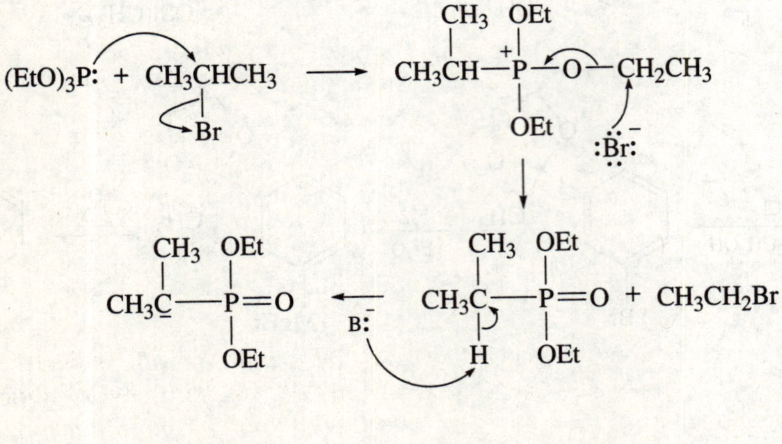

b.

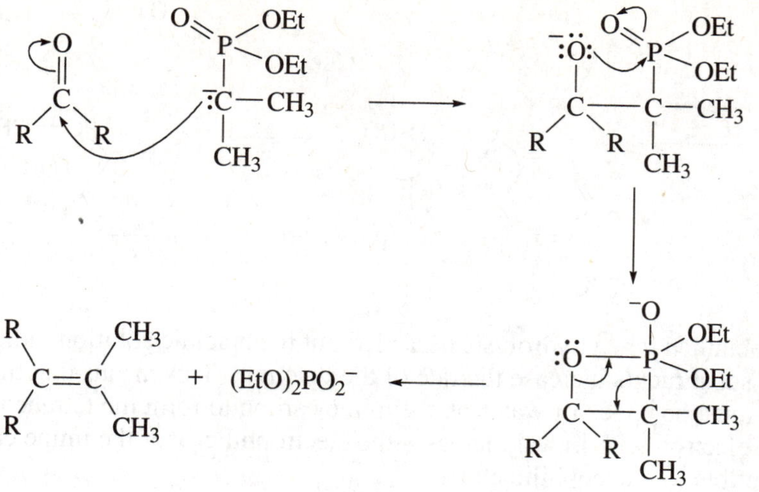

c. 1.

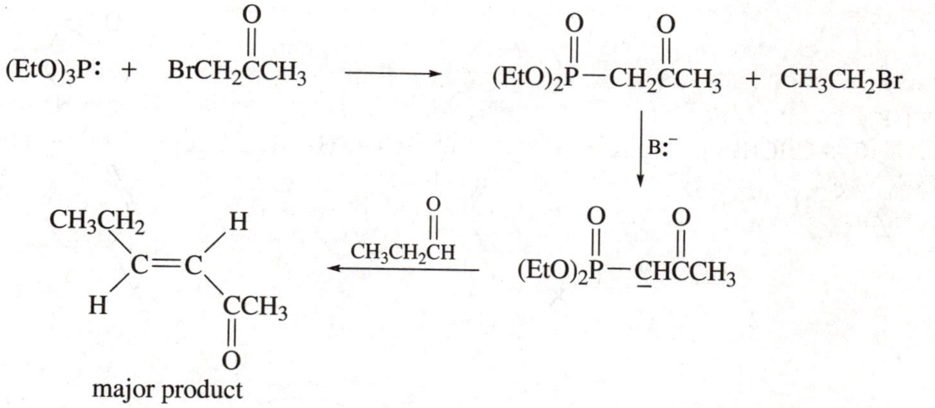

2.

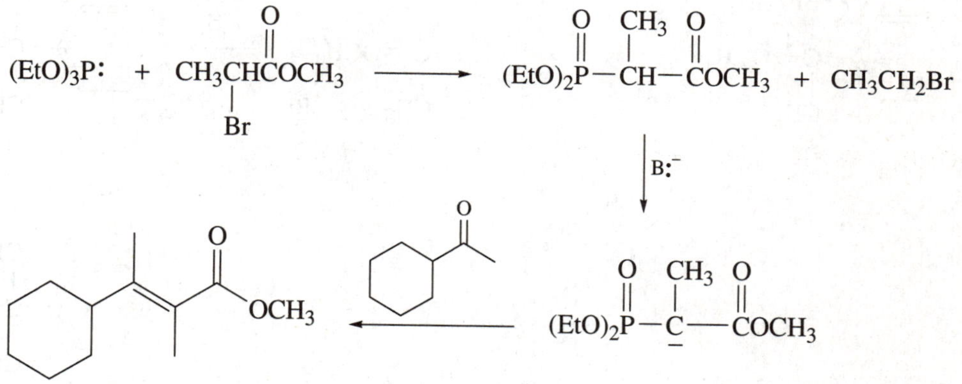

83. **a.** The negative ρ value obtained when hydrolysis is carried out in a basic solution indicates that electron-donating substituents increase the rate of the reaction. This means that the rate-determining step must be protonation of the carbon (the first step), because the more electron-donating the substituent, the greater the negative charge on the carbon and the easier it will be to protonate.

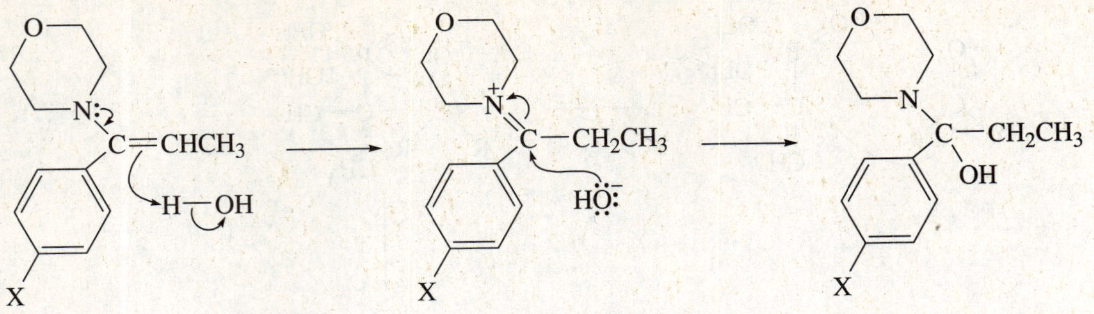

b. The positive ρ value obtained when hydrolysis is carried out in an acidic solution indicates that electron-withdrawing substituents increase the rate of the reaction. This means that the rate-determining step must be attack of water on the imine carbon to form the tetrahedral intermediate, because electron withdrawal increases the electrophilicity of the imine carbon, making it more susceptible to nucleophilic attack.

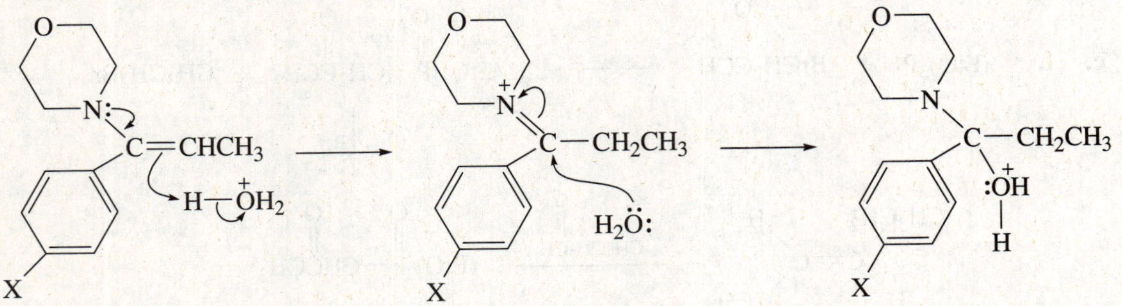

84. **a.**

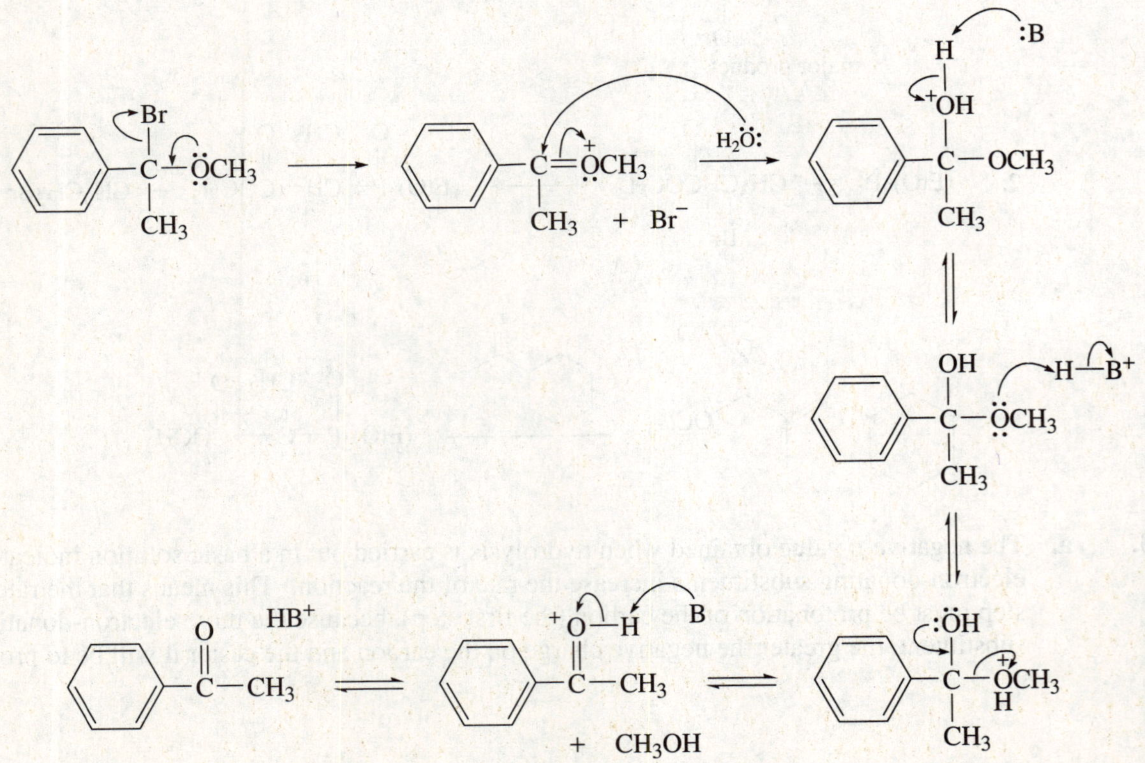

b.

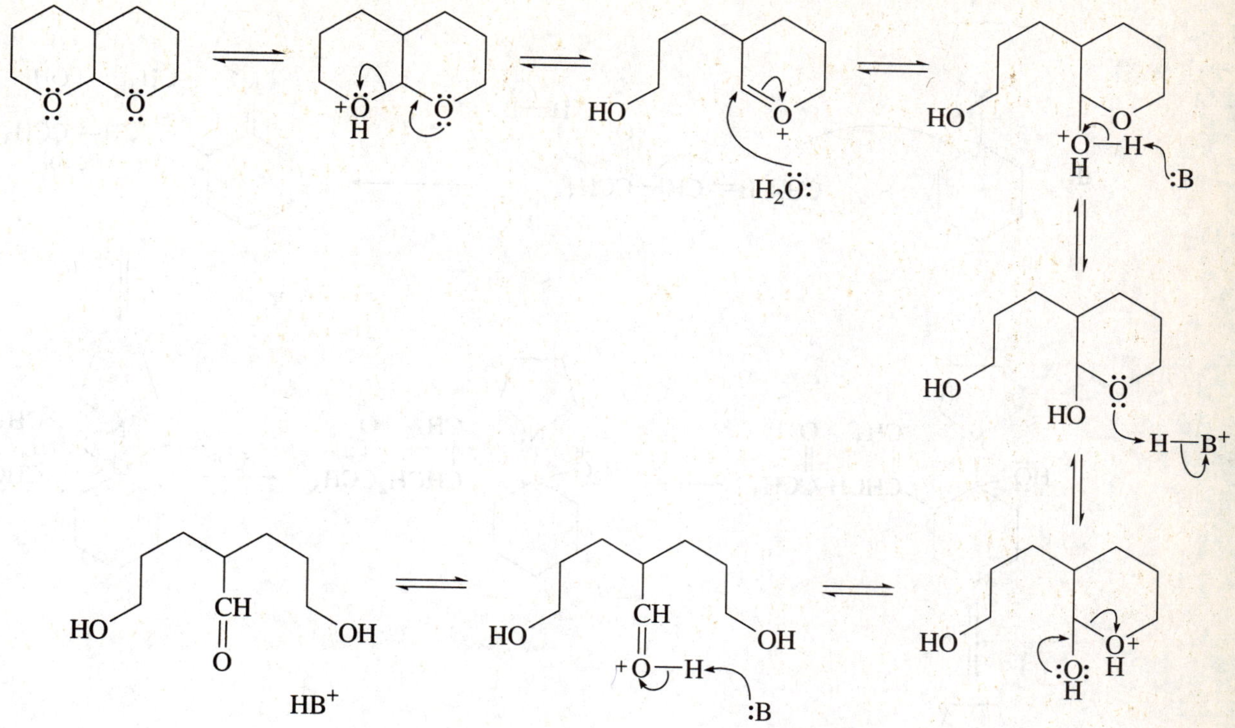

c.

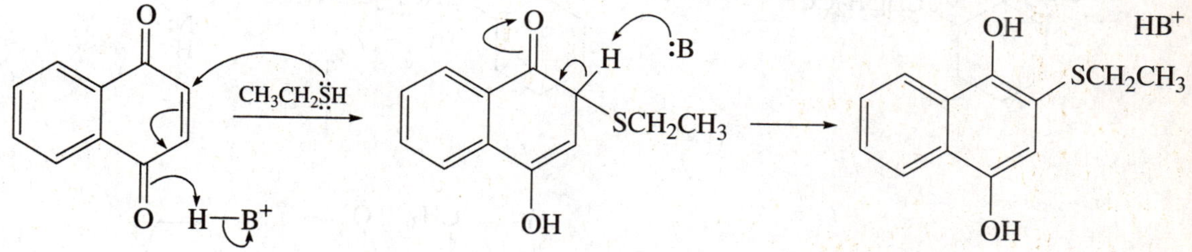

d.

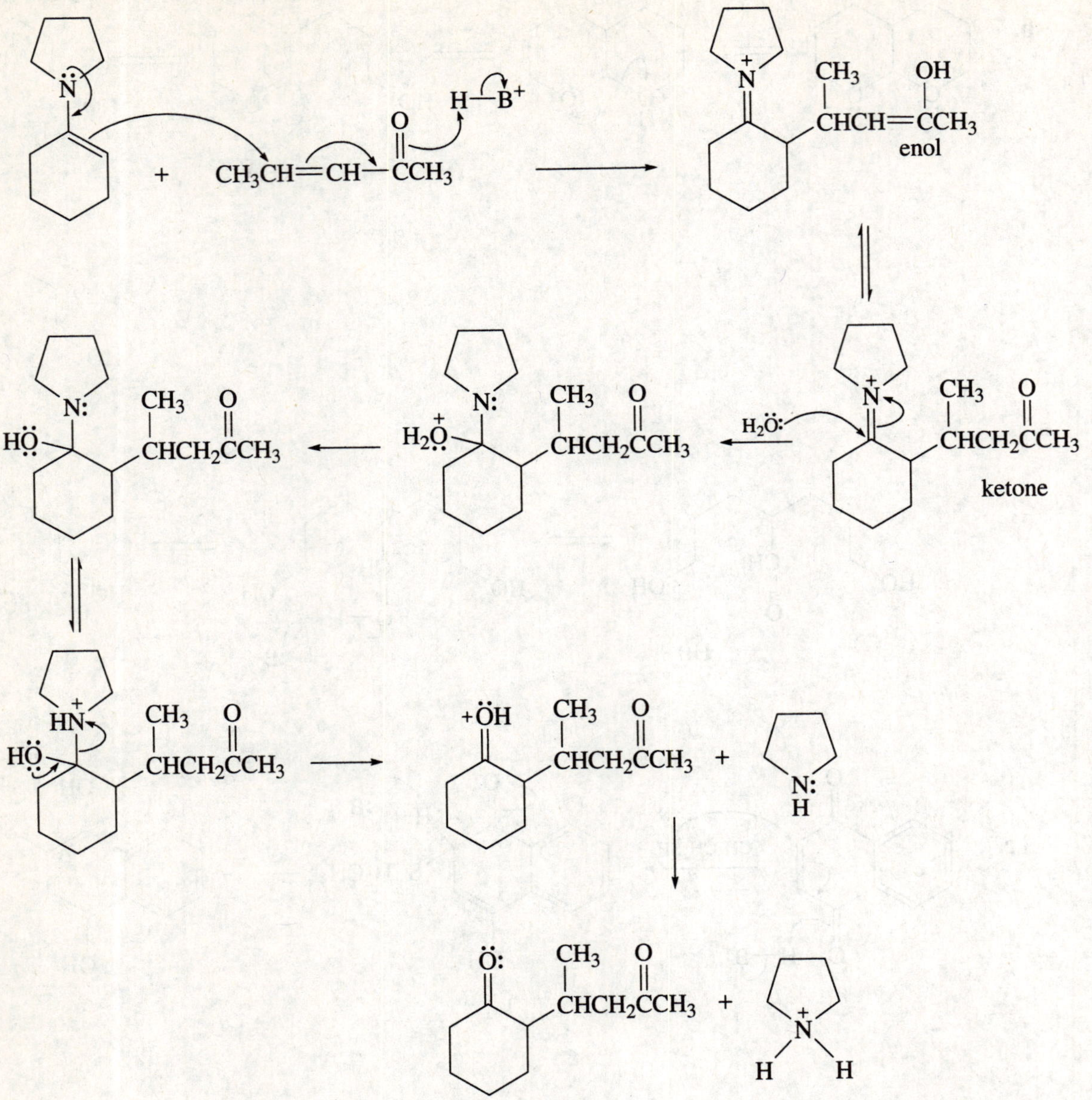

Chapter 17 Practice Test

1. Give the product of each of the following reactions.

a. 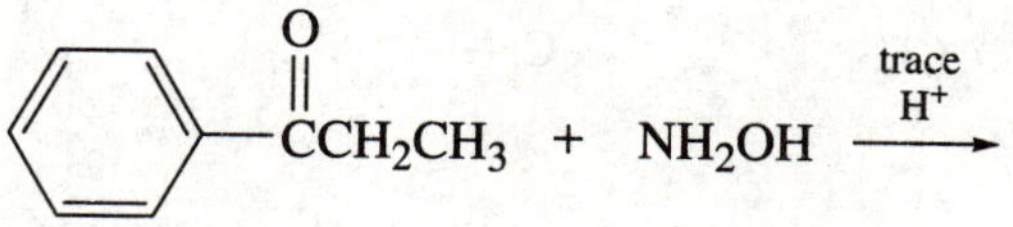

b. 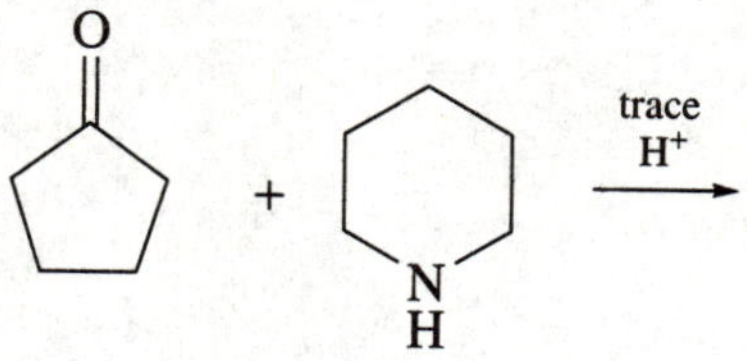

c. $CO_2 \xrightarrow[2.\ H^+,\ H_2O]{1.\ CH_3CH_2CH_2MgBr}$

d. 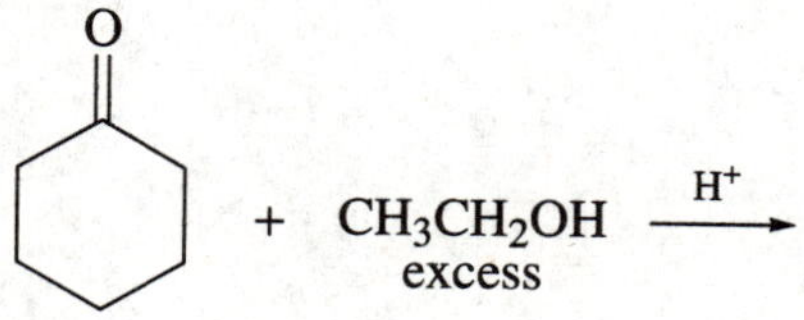

e. $CH_3CH_2\overset{O}{\overset{\|}{C}}CH_2CH_3 \xrightarrow[2.\ H^+,\ H_2O]{1.\ CH_3MgBr}$

f. 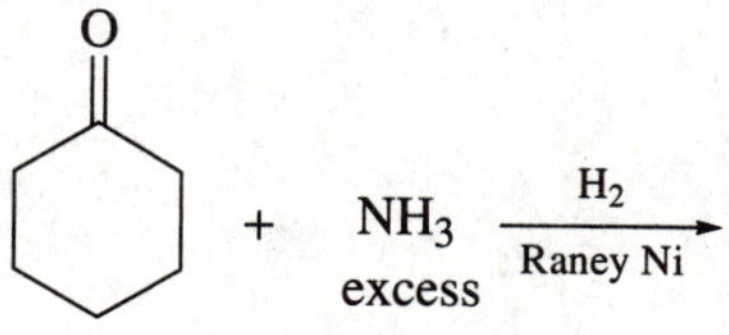

g. $CH_3CH_2\overset{O}{\overset{\|}{C}}CH_2CH_3 \xrightarrow[HCl]{^-C{\equiv}N}$

h. $CH_3CH{=}CH\overset{O}{\overset{\|}{C}}CH_3 \xrightarrow[2.\ H^+,\ H_2O]{1.\ LiAlH_4}$

i. 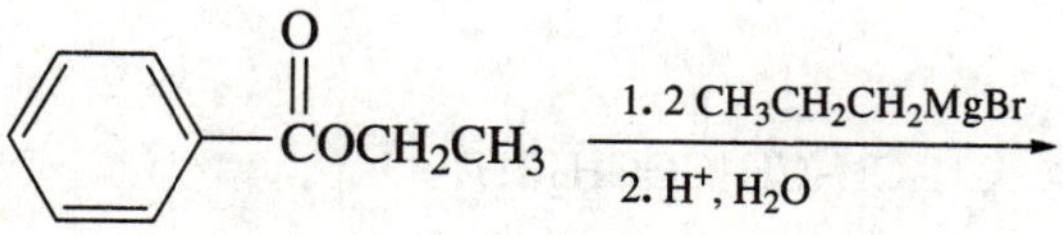

2. Which of the following alcohols cannot be prepared by the reaction of an ester with excess Grignard reagent?

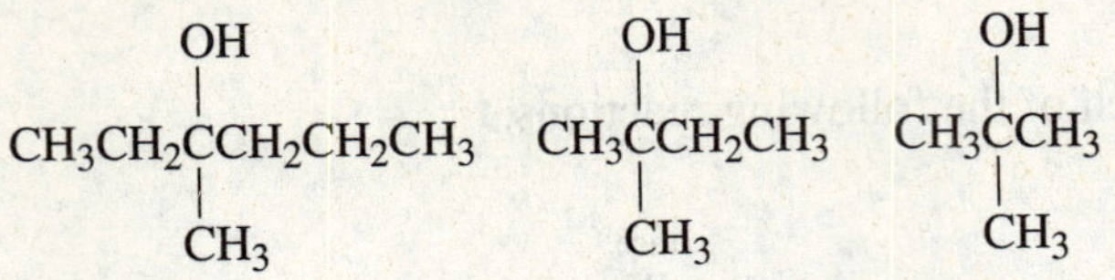

3. Which of the following ketones would form the greatest amount of hydrate in an aqueous solution?

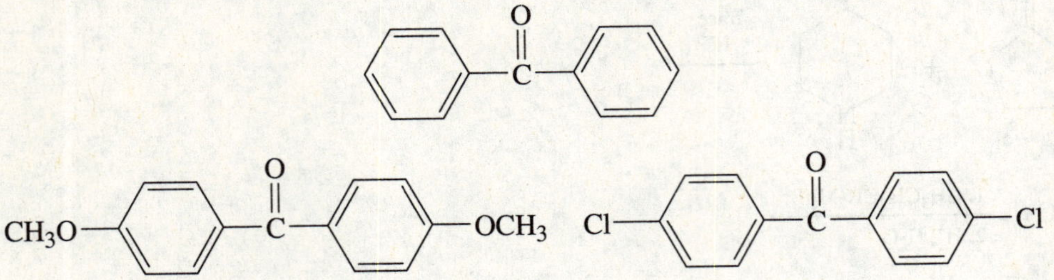

4. Give an example of each the following:

a. an enamine

b. an acetal

c. an imine

d. a hemiacetal

e. a phenylhydrazone

5. Which is more reactive toward nucleophilic addition?

a. butanal or methyl propyl ketone

b. 4-heptanone or 3-pentanone

6. Indicate how the following compound could be prepared using the given starting material.

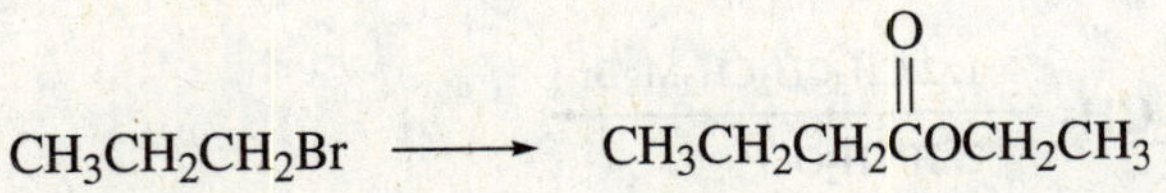

CHAPTER 18

Carbonyl Compounds III: Reactions at the α-Carbon

Important Terms

acetoacetic ester synthesis	synthesis of a methyl ketone using ethyl acetoacetate as the starting material.
aldol addition	a reaction between two molecules of an aldehyde (or two molecules of a ketone) that connects the α-carbon of one with the carbonyl carbon of the other.
aldol condensation	an aldol addition followed by elimination of water.
ambident nucleophile	a nucleophile with two nucleophilic sites.
annulation reaction	a ring-forming reaction.
α-carbon	a carbon adjacent to a carbonyl carbon.
carbon acid	a compound that contains a carbon bonded to a relatively acidic hydrogen.
Claisen condensation	a reaction between two molecules of an ester that connects the α-carbon of one with the carbonyl carbon of the other and eliminates an alkoxide ion.
condensation reaction	a reaction combining two molecules while removing a small molecule (usually water or an alcohol).
decarboxylation	loss of carbon dioxide.
Dieckmann condensation	an intramolecular Claisen condensation.
β-diketone	a ketone with a second ketone carbonyl group at the β-position.
enolization	keto-enol interconversion.
gluconeogenesis	the synthesis of D-glucose from pyruvate.
glycolysis	the breakdown of D-glucose into two molecules of pyruvate.
haloform reaction	the conversion of a methyl ketone to a carboxylic acid and haloform.
Hell-Volhard-Zelinski (HVZ) reaction	conversion of a carboxylic acid into an α-bromocarboxylic acid using PBr_3 + Br_2.
α-hydrogen	a hydrogen bonded to the carbon adjacent to a carbonyl carbon.
keto-enol tautomerism (keto-enol interconversion)	interconversion of keto and enol tautomers.
β-keto ester	an ester with a ketone carbonyl group at the β-position.

Kolbe-Schmidt carboxylation reaction	reaction of a phenolate ion with carbon dioxide under pressure.
malonic ester synthesis	synthesis of a carboxylic acid using diethyl malonate as the starting material.
Michael reaction	the addition of an α-carbanion to the β-carbon of an α,β-unsaturated carbonyl compound.
mixed aldol addition (crossed aldol addition)	an aldol addition in which two different carbonyl compounds are used.
mixed Claisen condensation	a Claisen condensation in which two different esters are used.
Robinson annulation	a Michael reaction followed by an intramolecular aldol condensation.
Stork enamine reaction	uses an enamine as a nucleophile in a Michael reaction.
α-substitution reaction	a reaction that puts a substituent on an α-carbon in place of an α-hydrogen.
tautomers	isomers that differ in the location of a double bond and a hydrogen.

Solutions to Problems

1. The electrons left behind when a proton is removed from propene are delocalized—they are shared by three carbon atoms. In contrast, the electrons left behind when a proton is removed from an alkane are localized—they belong to a single (carbon) atom. Because electron delocalization allows charge to be distributed over more than one atom, it stabilizes the base. Thus, base with delocalized electrons is more stable so it has the stronger conjugate acid. Thus, propene is a stronger acid than an alkane.

$$CH_2{=}CHCH_3 \rightleftharpoons CH_2{=}CH\ddot{C}H_2^{-} \longleftrightarrow \ddot{C}H_2^{-}CH{=}CH_2 \quad + H^+$$

$$CH_3CH_2CH_3 \rightleftharpoons CH_3\ddot{C}^{-}HCH_3 + H^+$$

Propene, however, is not as acidic as the carbon acids in Table 18.1, because the electrons left behind when a proton is removed from these carbon acids are delocalized onto an oxygen or a nitrogen, which are more electronegative atoms than carbon and, therefore, better able to accommodate the electrons.

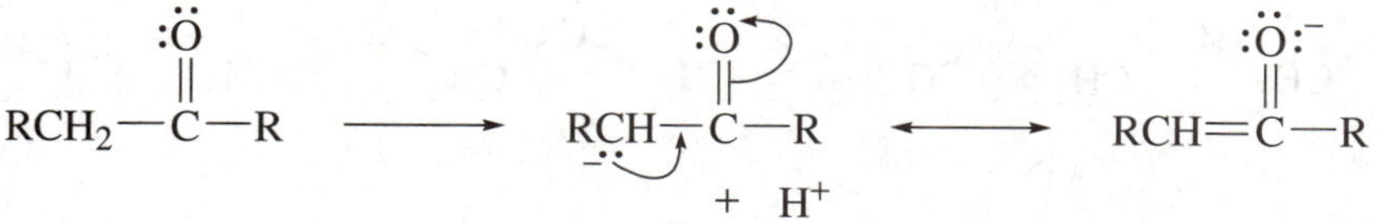

2. a. 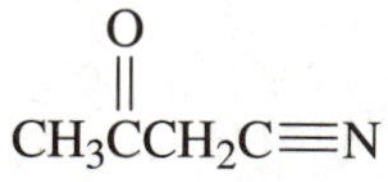

a β-keto nitrile

b. 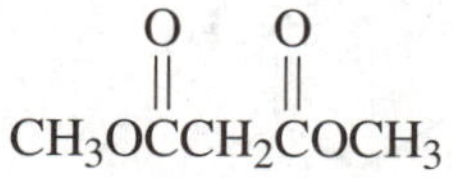

a β-diester

3. A proton cannot be removed from the α-carbon of *N*-methylethanamide or ethanamide because these compounds have a hydrogen bonded to the nitrogen and this hydrogen is more acidic that the one attached to the α-carbon.

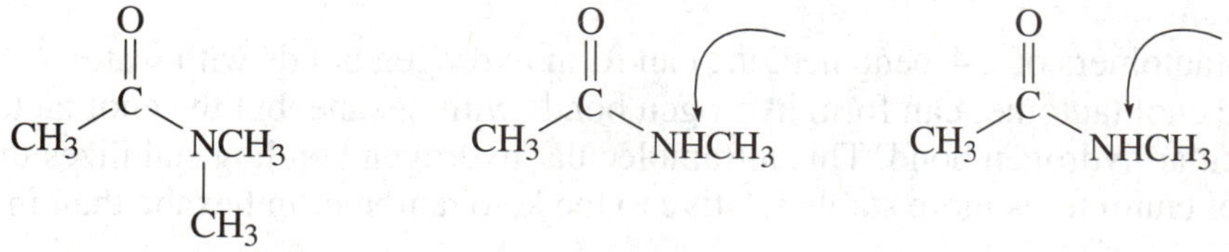

N,N-dimethylethanamide *N*-methylethanamide ethanamide

The following contributing resonance structures show why the hydrogen attached to the nitrogen is more acidic (the nitrogen has a partial positive charge) than the hydrogen attached to the α-carbon.

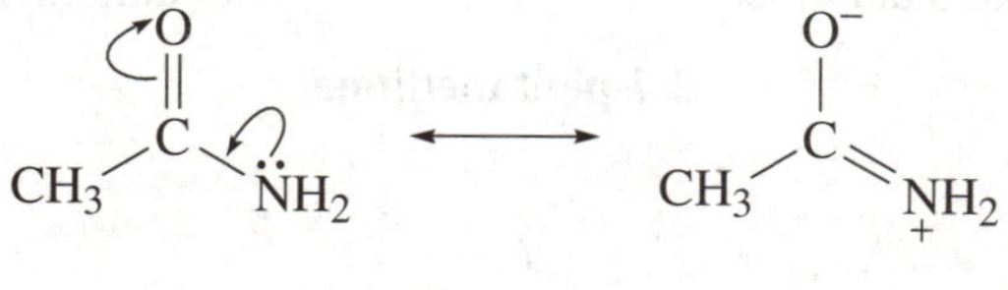

4. Electron delocalization of the lone pair on nitrogen or oxygen competes with electron delocalization of the electrons left behind on the α-carbon when it loses a proton. The lone pair on nitrogen is more delocalized than the lone pair on oxygen, because nitrogen is better able to accommodate a positive charge. Therefore the amide competes better with the carbanion for electron delocalization, so the α-carbon is less acidic.

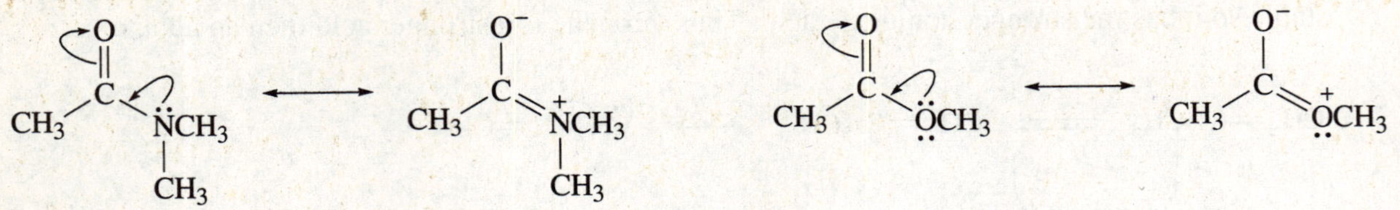

5. **a.** $CH_3\overset{\overset{O}{\|}}{C}H > HC{\equiv}CH > CH_2{=}CH_2 > CH_3CH_3$

b.

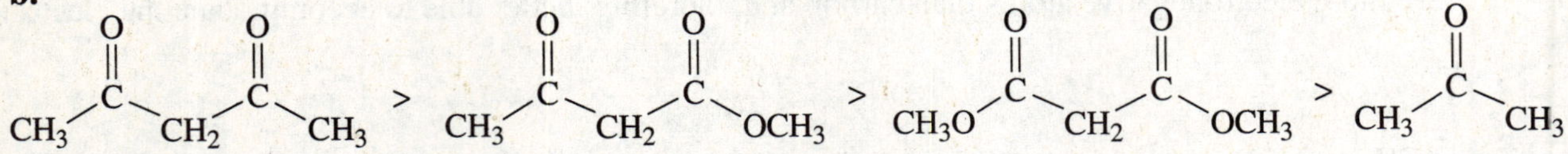

c.

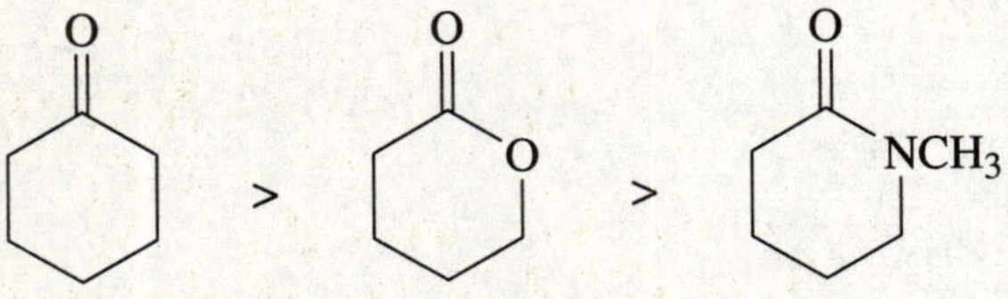

The ketone is the strongest acid because there is no competition for delocalization of the electrons that are left behind when the α-hydrogen is removed. The lactam is the weakest acid because nitrogen, being less electronegative, can better accommodate a positive charge and, therefore, is better than oxygen at delocalizing its lone pair onto the carbonyl oxygen. Therefore, nitrogen is better at competing with the electrons left behind when an α-hydrogen is removed for delocalization onto the carbonyl oxygen.

6. Both the keto and enol tautomers of 2,4-pentanedione can form hydrogen bonds with water. Neither the keto nor the enol tautomer can form hydrogen bonds with hexane, but the enol tautomer can still form an intramolecular hydrogen bond. This intramolecular hydrogen bonding stabilizes the enol tautomer. Thus, the enol tautomer is more stable relative to the keto tautomer in hexane than in water.

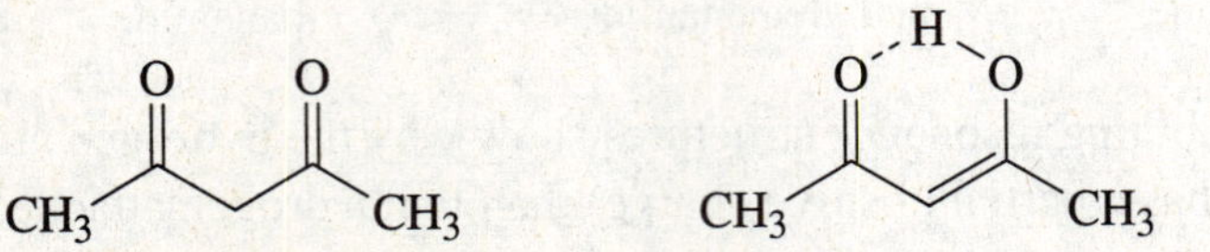

keto tautomer enol tautomer

2,4-pentanedione

7. $pK_a = pH + \log\frac{[\text{enol}]}{[\text{enolate}]}$

a. $20 = 8 + \log\frac{[\text{enol}]}{[\text{enolate}]}$

$12 = \log\frac{[\text{enol}]}{[\text{enolate}]}$

$10^{12} = \frac{[\text{enol}]}{[\text{enolate}]}$

b. $16 = 8 + \log\frac{[\text{enol}]}{[\text{enolate}]}$

$8 = \log\frac{[\text{enol}]}{[\text{enolate}]}$

$10^{8} = \frac{[\text{enol}]}{[\text{enolate}]}$

c. $10 = 8 + \log\frac{[\text{enol}]}{[\text{enolate}]}$

$2 = \log\frac{[\text{enol}]}{[\text{enolate}]}$

$10^{2} = \frac{[\text{enol}]}{[\text{enolate}]}$

8. **a.** 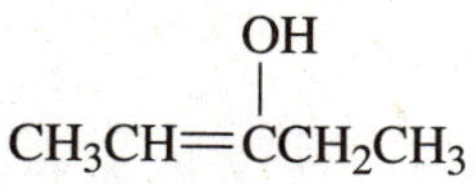

b. 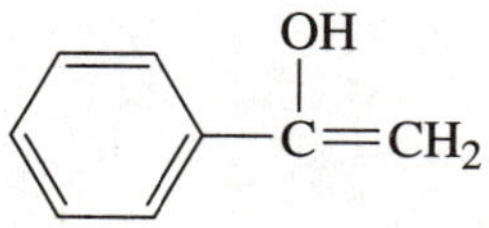

c. 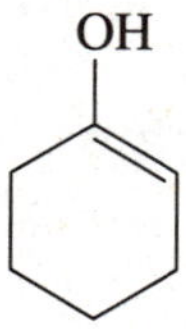

d. 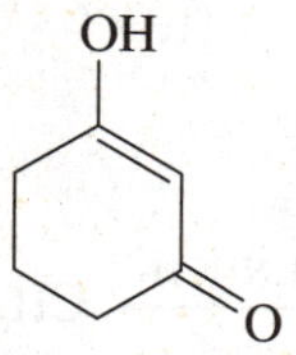 and 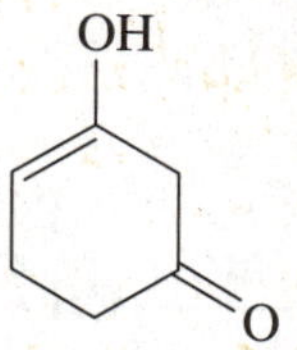

more stable
because the double bonds
are conjugated

e. 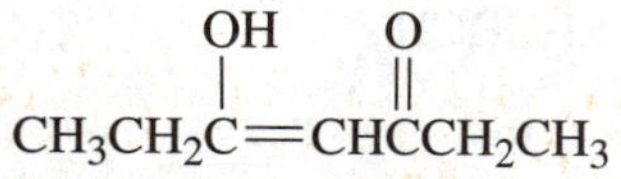 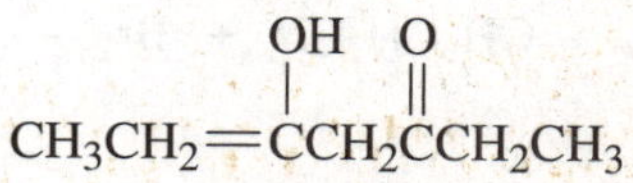

more stable
because the double bonds are conjugated

f. 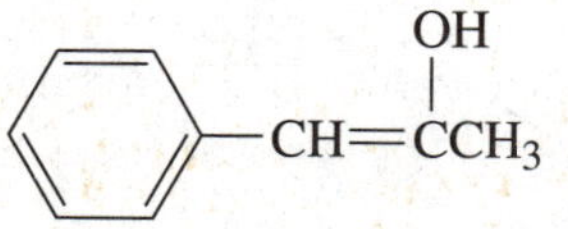 and 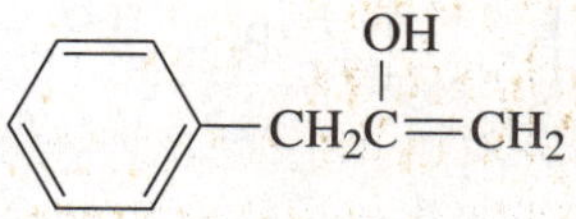

more stable
because the double bond is conjugated with the benzene ring

9. The aldehyde hydrogen cannot be removed by a base. The aldehyde hydrogen is not acidic, because the electrons left behind if it were to be removed cannot be delocalized.

10. The haloform reaction requires that a group be created that is a weaker base than hydroxide ion so that hydroxide ion is not the group eliminated from the tetrahedral intermediate. For an alkyl group to be the weaker base, it must be bonded to three halogen atoms. The only alkyl group that can become bonded to three halogen atoms (because it is bonded to three hydrogens) is a methyl group.

11. A Br—Br bond is weaker and easier to break than a Cl—Cl bond, which in turn is weaker and easier to break than a D—O bond. Because the rates of bromination, chlorination, and deuterium exchange are about the same, you know that breaking the Br—Br, Cl—Cl, or D—O bond takes place after the rate-determining step. Therefore, the rate-determining step must be removal of the proton from the α-carbon of the ketone.

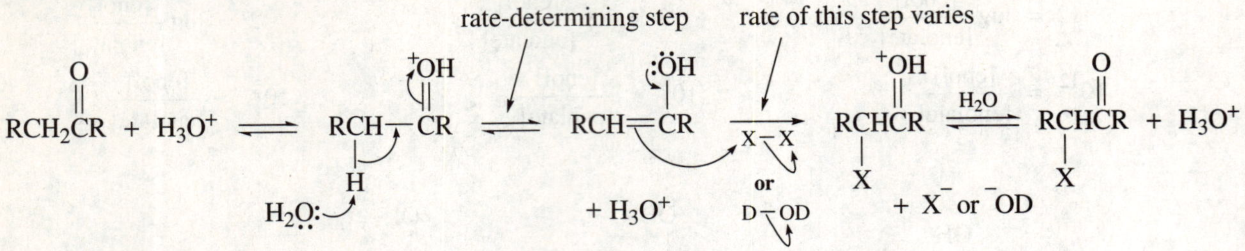

12. a. 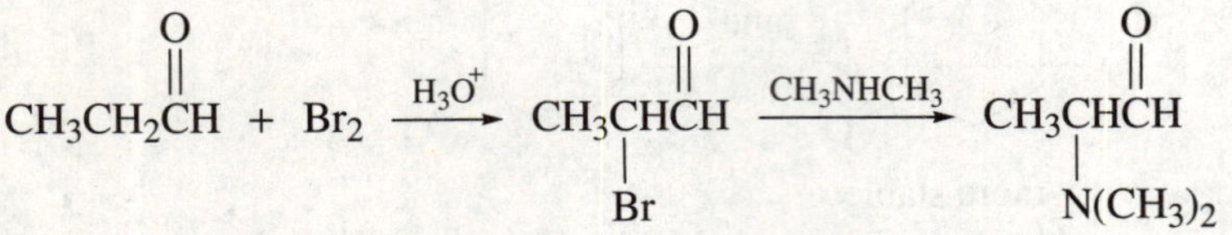

b. 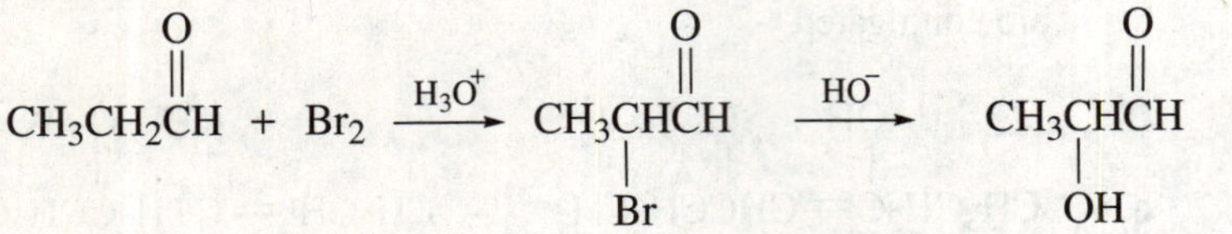

c. 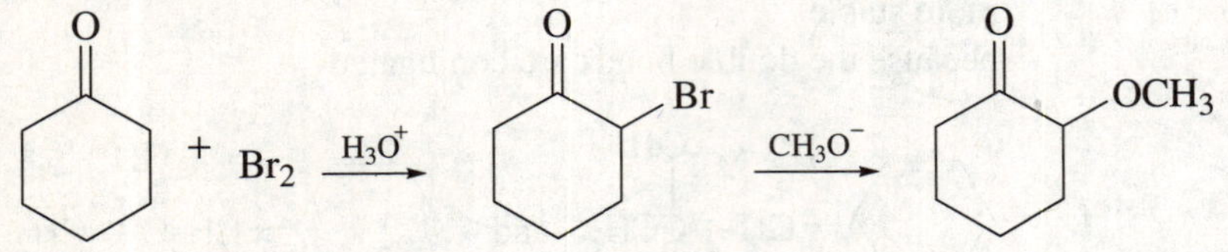

d. 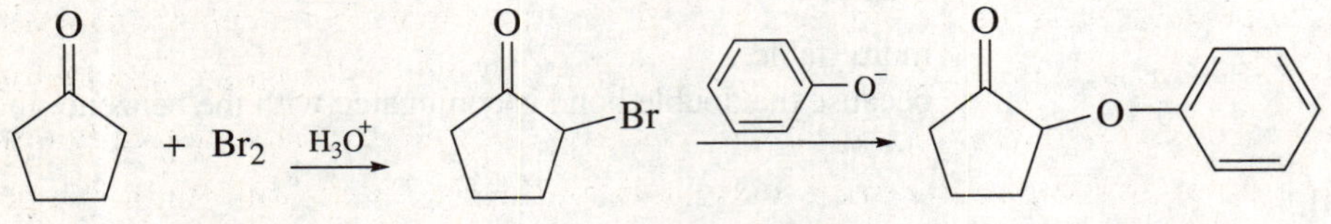

13. a. 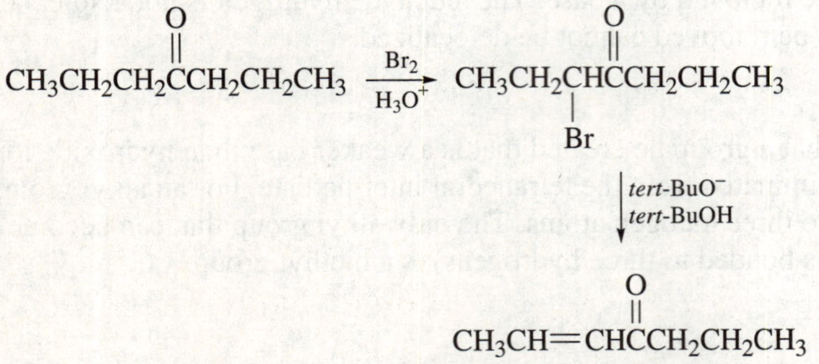

b. 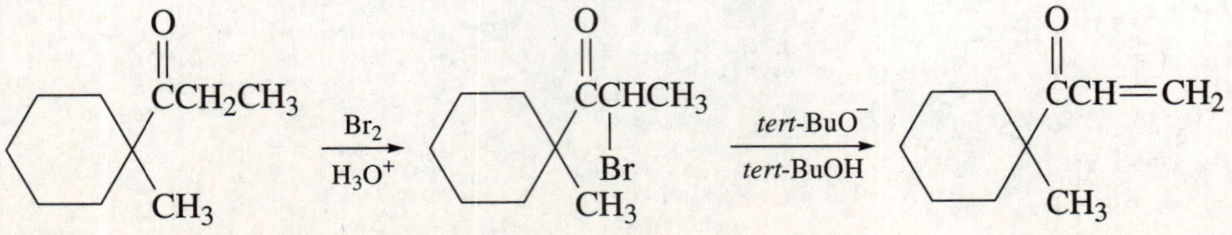

14. a.

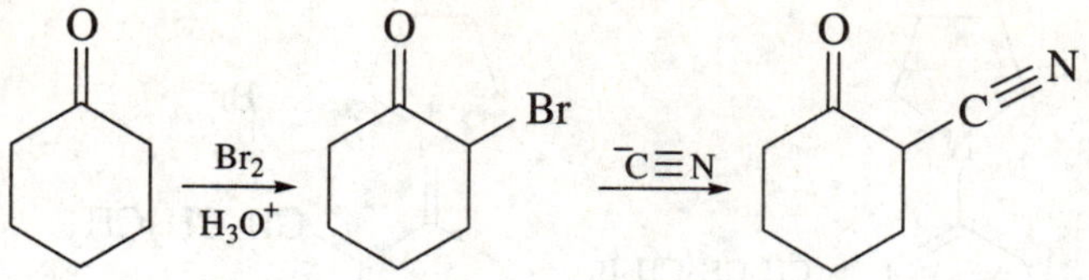

b.

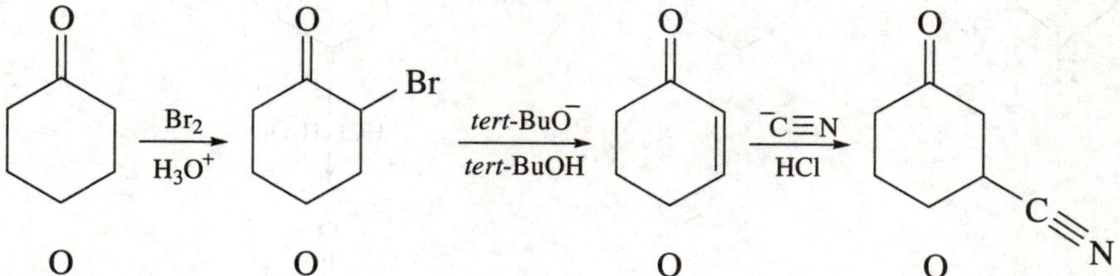

c.

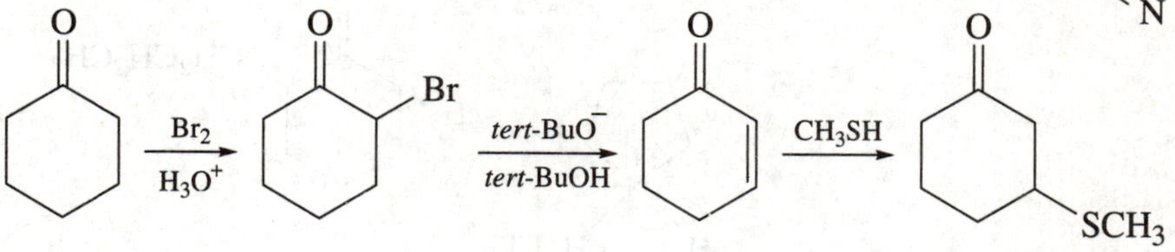

d.

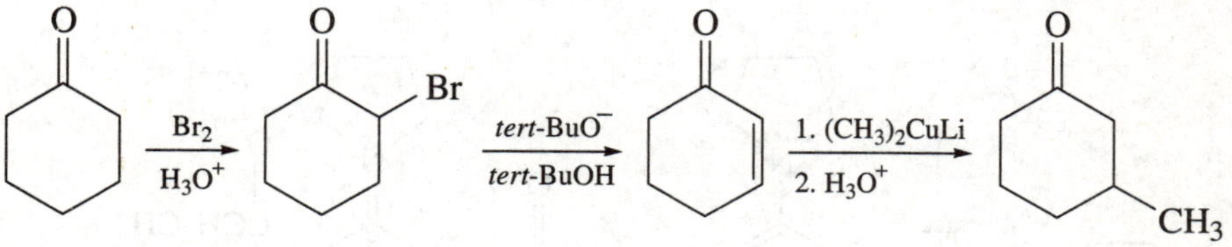

15.

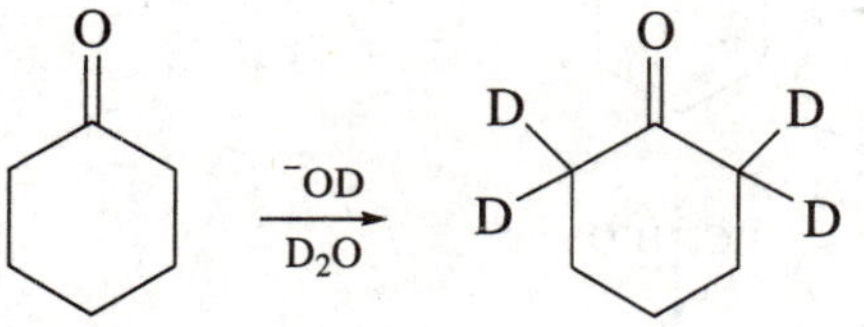

16. Alkylation of an alpha carbon is an S_N2 reaction. S_N2 reactions work best with primary alkyl halides because there is less steric hindrance in a primary alkyl halide than in a secondary alkyl halide. S_N2 reactions don't work at all with tertiary alkyl halides because they are the most sterically hindered of the alkyl halides. Therefore, in the case of tertiary alkyl halides, the S_N2 reaction cannot compete with the E2 elimination reaction.

17. a. $$CH_3\overset{\overset{O}{\|}}{C}CH_3 \xrightarrow[2.\ ICH_2CH=CH_2]{1.\ LDA/THF} CH_3\overset{\overset{O}{\|}}{C}CH_2CH_2CH=CH_2$$

b.

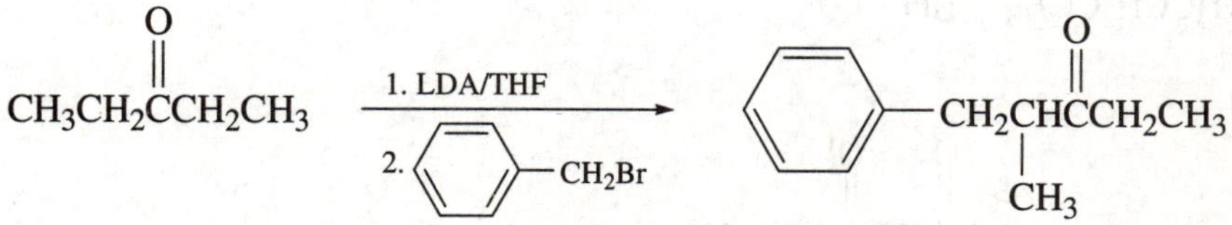

18. **a.**

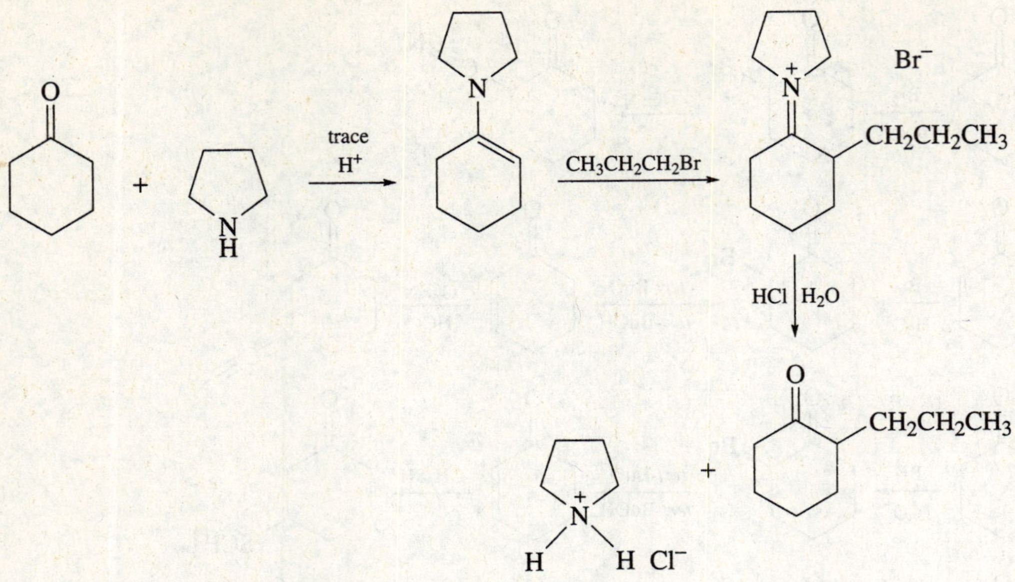

b.

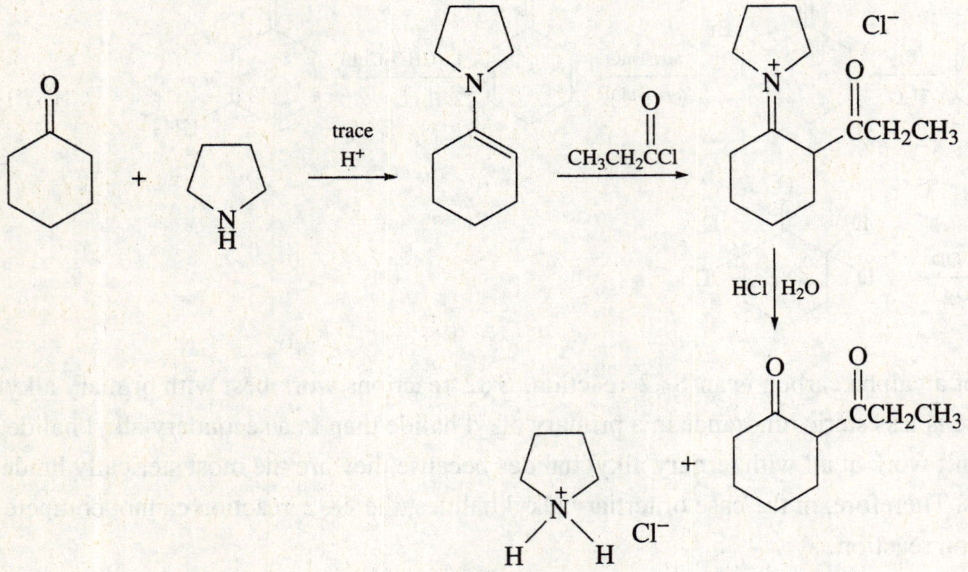

19. **a.**

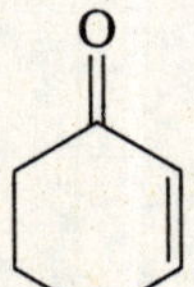

$CH_3\overset{O}{\overset{\|}{C}}CH_2\overset{O}{\overset{\|}{C}}CH_3$ HO^-

b. $CH_3\overset{O}{\overset{\|}{C}}CH{=}CH_2$ $CH_3CH_2O\overset{O}{\overset{\|}{C}}CH_2\overset{O}{\overset{\|}{C}}OCH_2CH_3$ $CH_3CH_2O^-$

20. a.

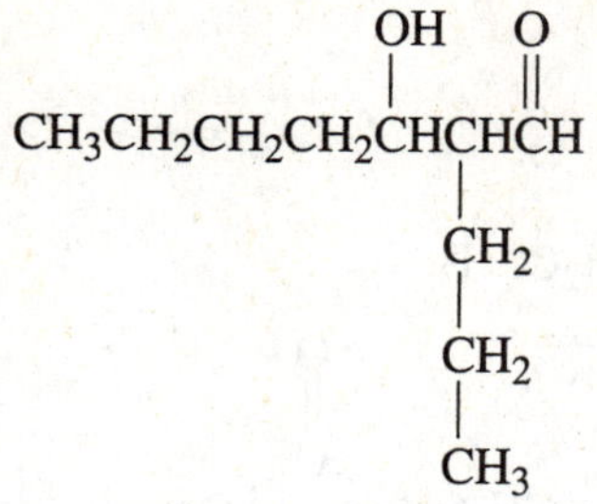

c.

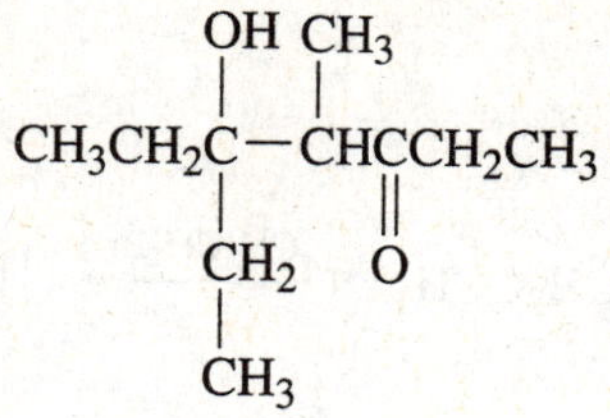

b.

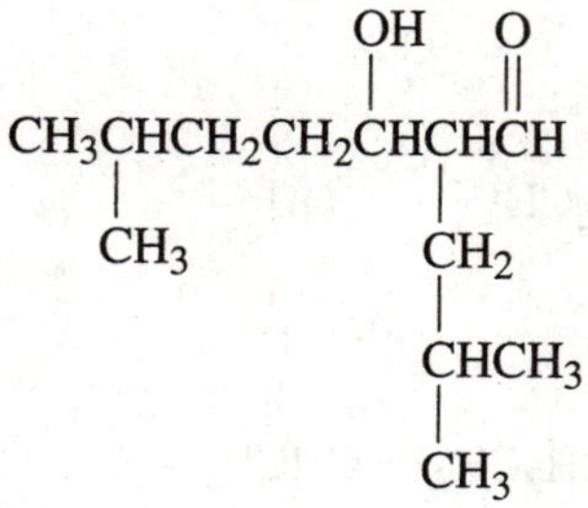

d.

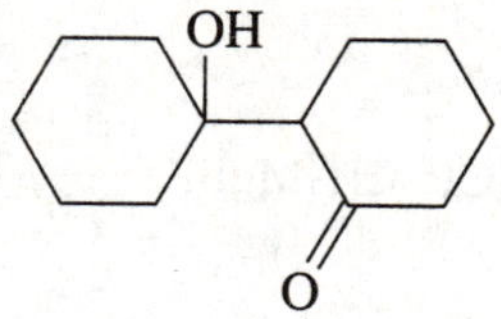

21. a.

b.

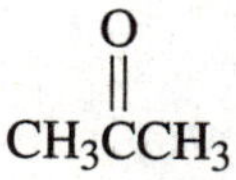

c.

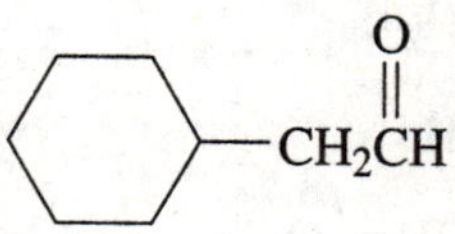

d.

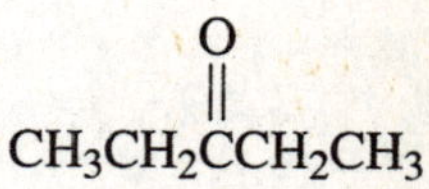

22.

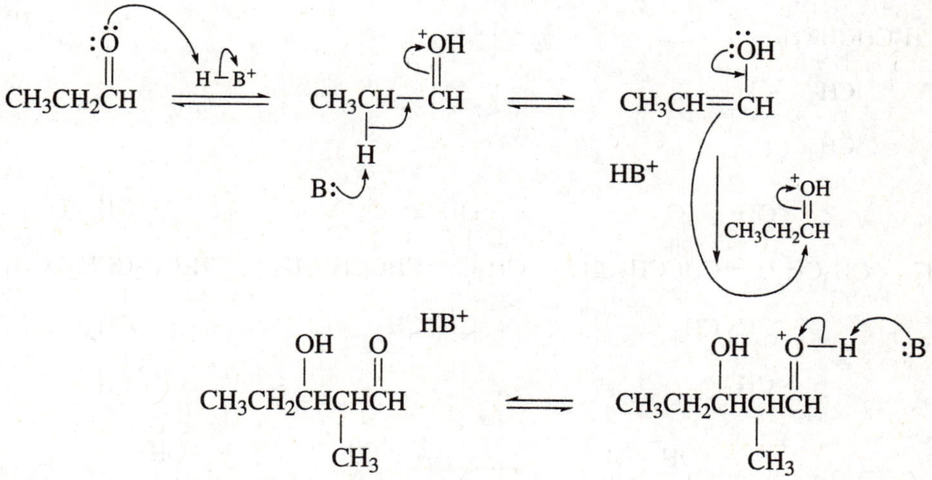

23.

aldol addition

aldol condensation

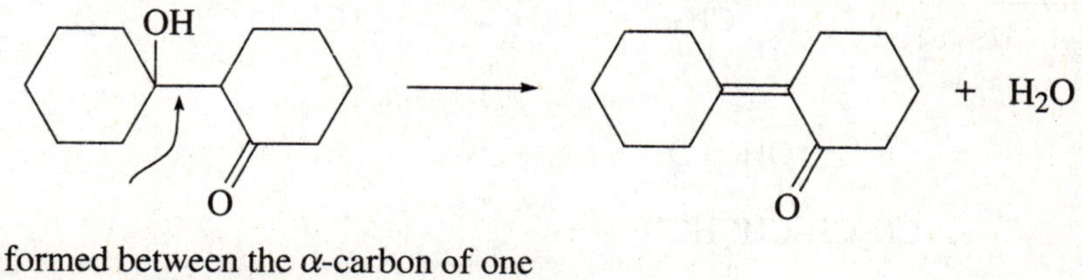

a bond has formed between the α-carbon of one molecule and the carbonyl carbon of another

24. **a.** Solved in the text.

b. 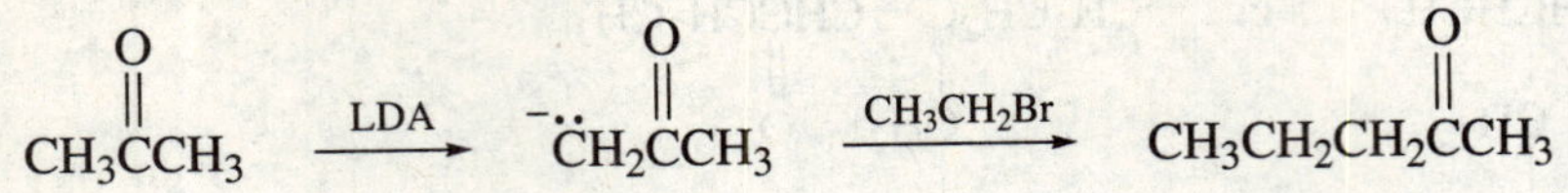

c. 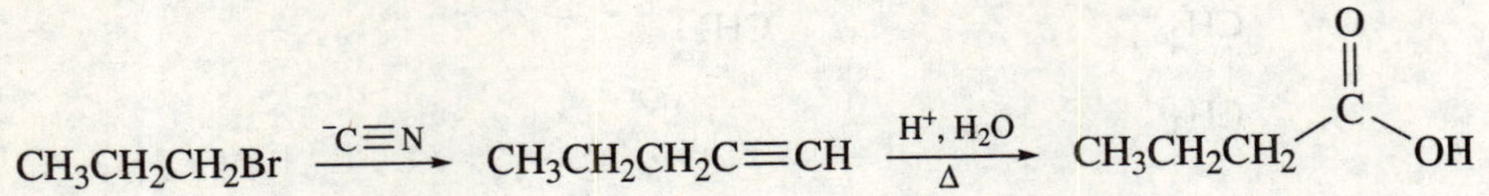

or

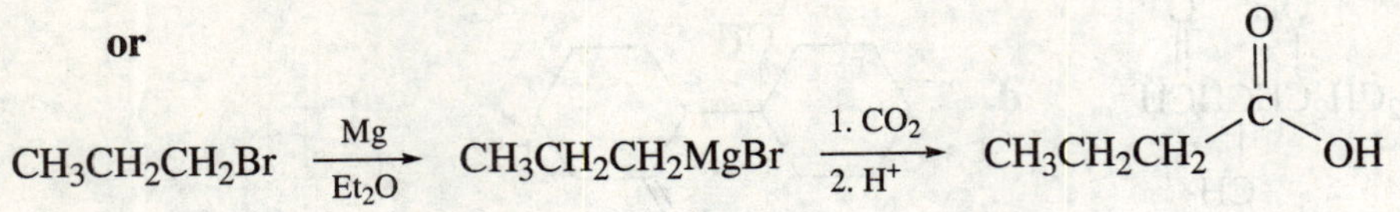

25. **a.** 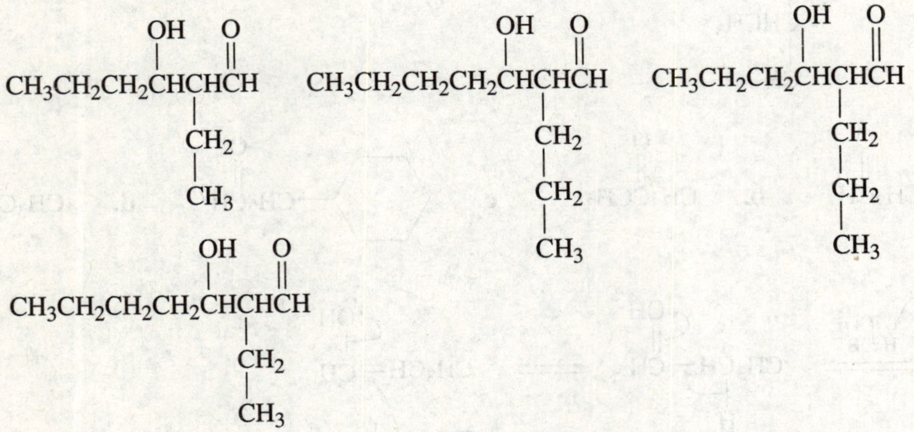

b. 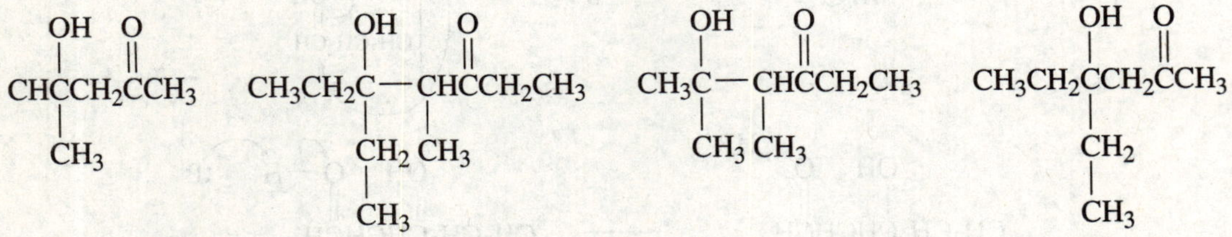

c. 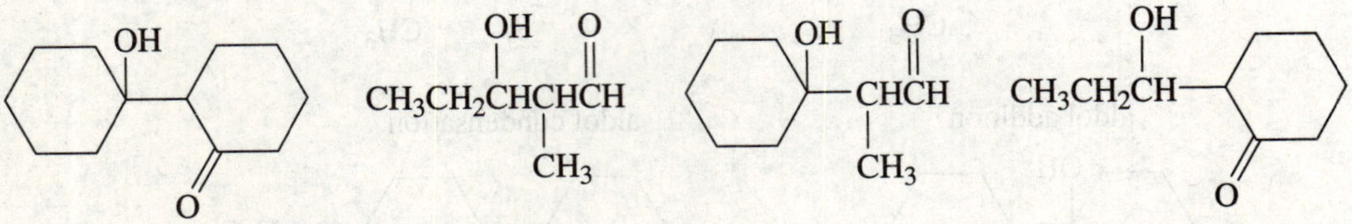

d. 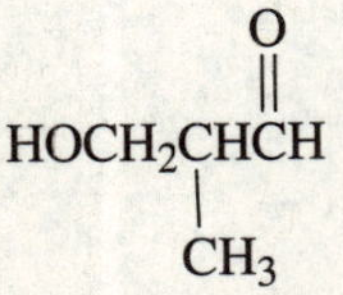

major product

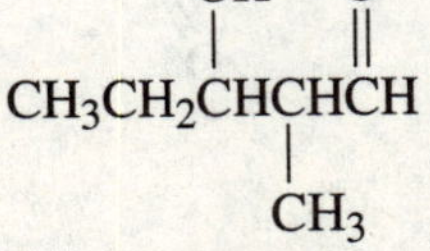

minor product

26. **a.**

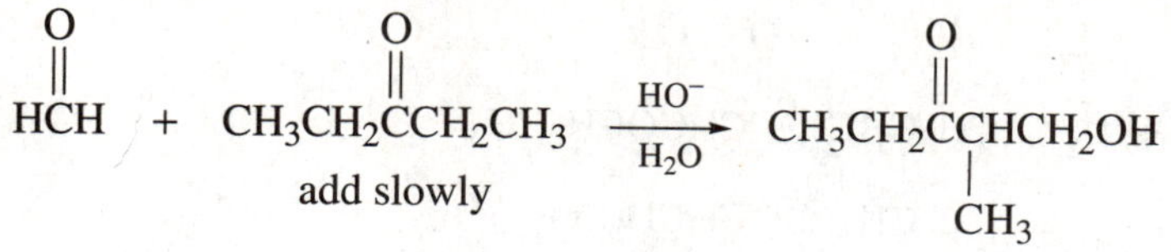
HCH
CH3CH2CCH2CH3
add slowly
HO−
H2O
CH3CH2CCHCH2OH
CH3

b.

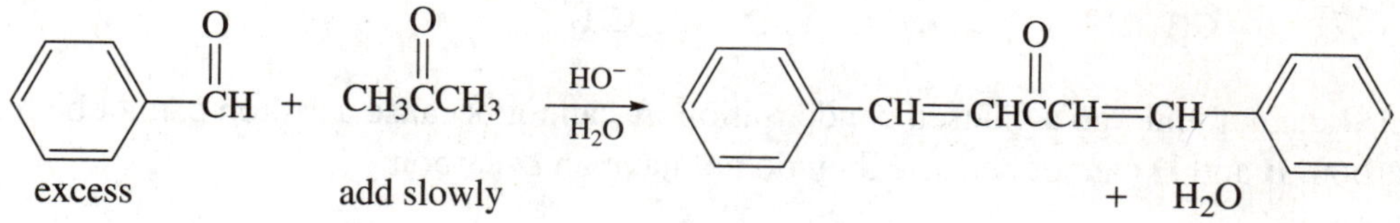
CH
excess
CH3CCH3
add slowly
HO−
H2O
CH=CHCCH=CH
+ H2O

c.

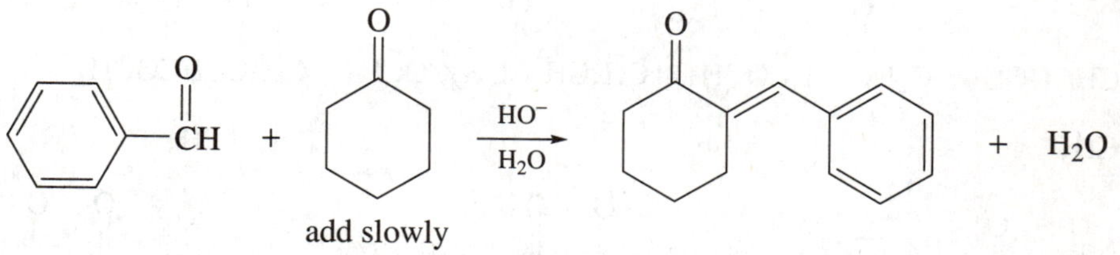
CH
add slowly
HO−
H2O
+ H2O

27.

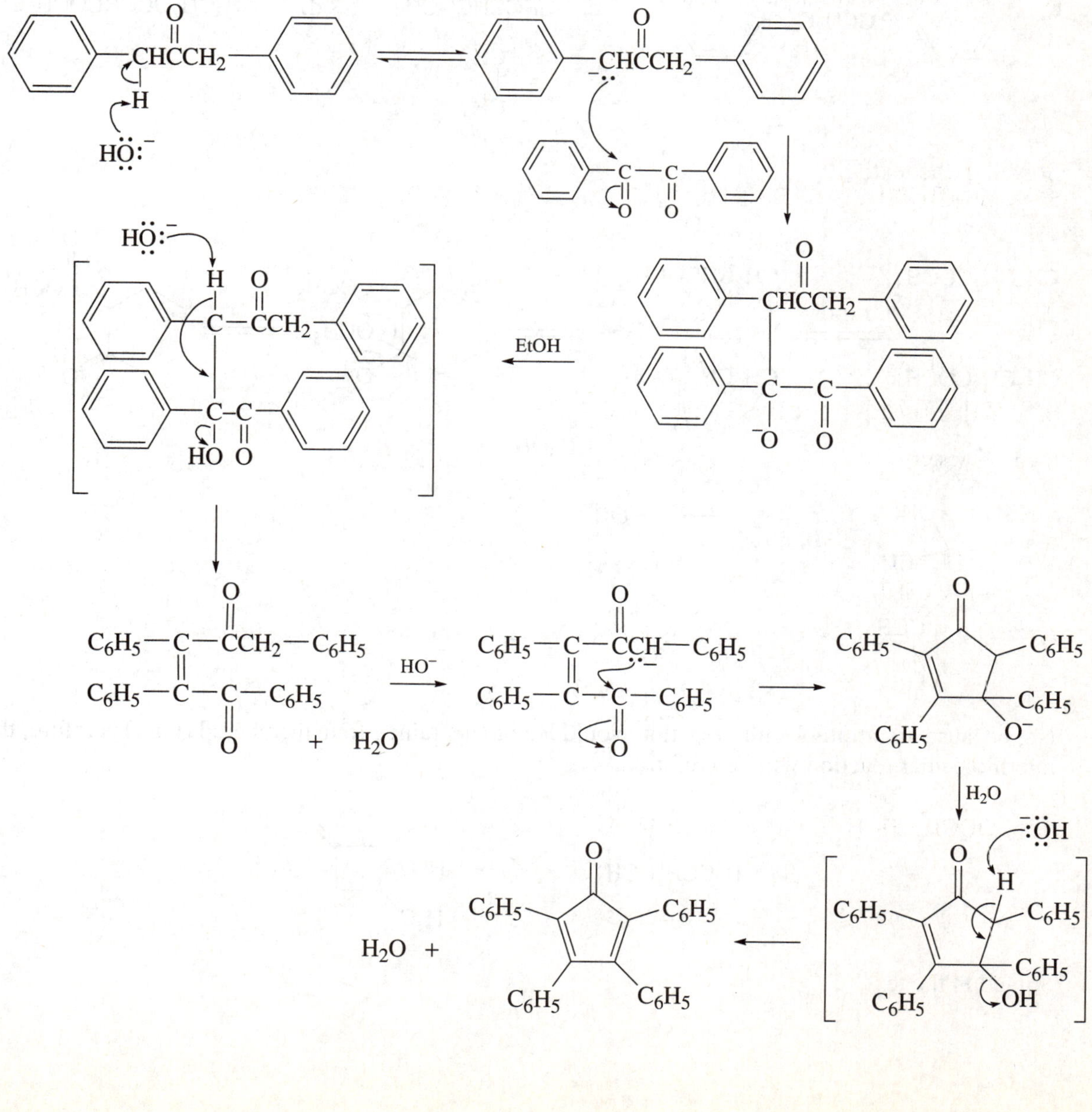
EtOH
HO−
H2O

28. **a.** 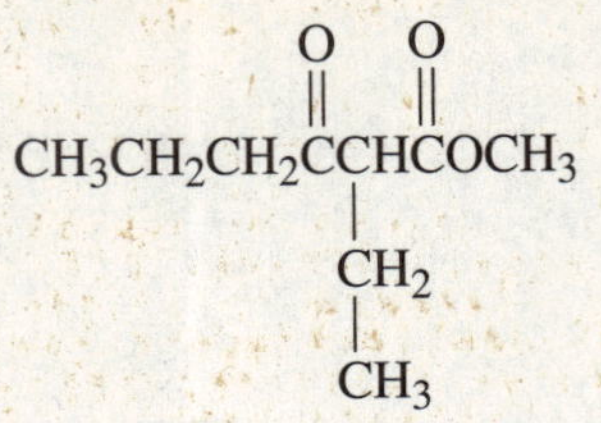 **b.** 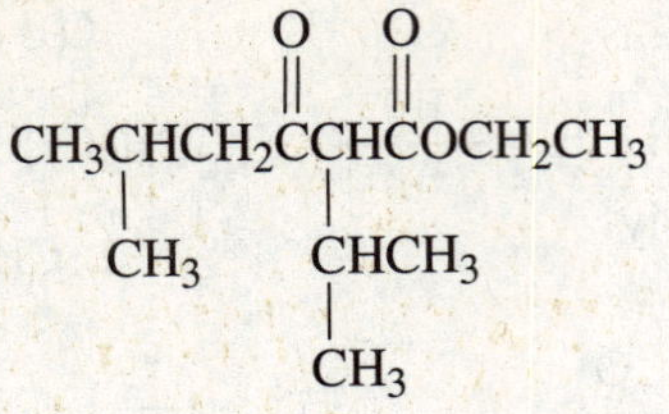

29. **A**, **B**, and **D** cannot undergo a Claisen condensation. **A** cannot because a proton cannot be removed from an *sp*^{2} carbon. **B** and **D** cannot because they do not have an α-carbon.

30. **a.** 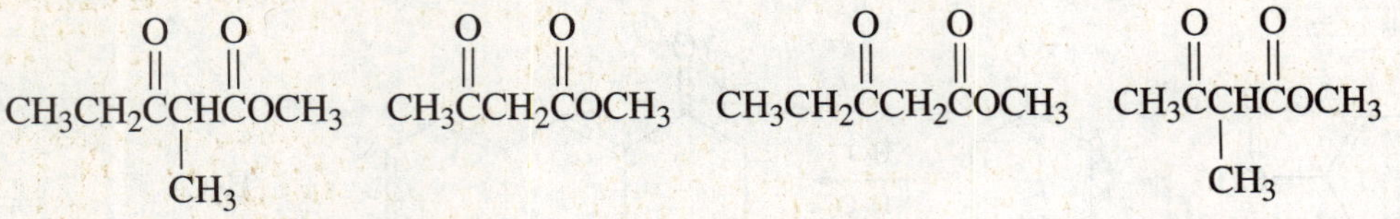

b. 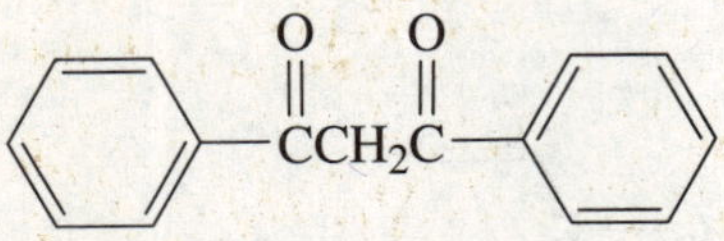 **c.** 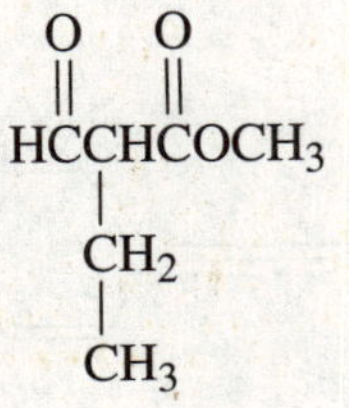 **d.** 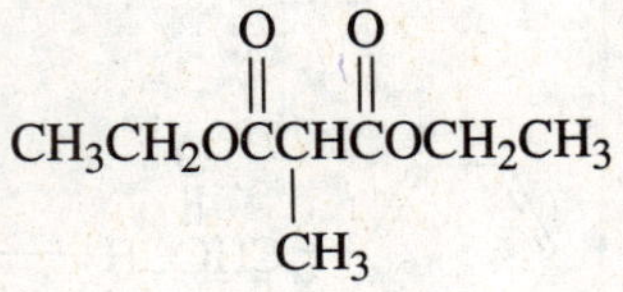

31. Solved in the text.

32. 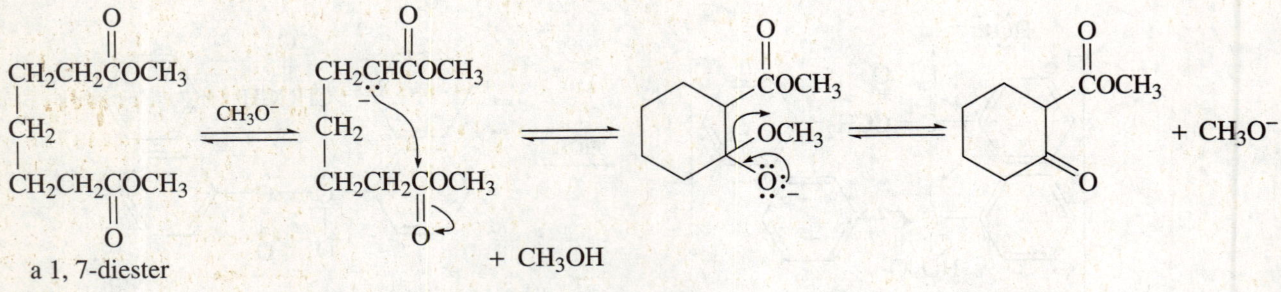

33. **a.** 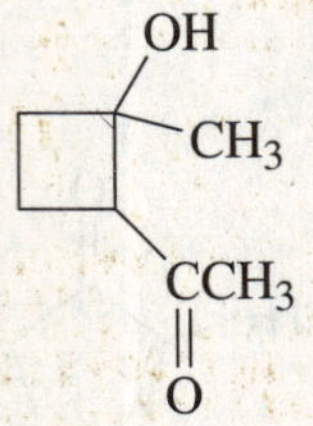 **b.** 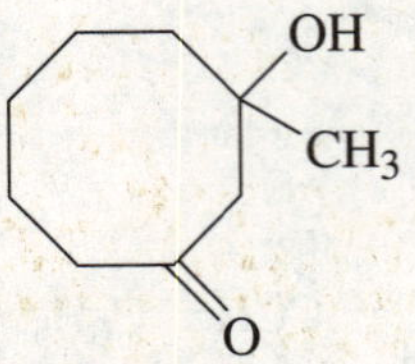

34. No, because an intramolecular reaction would lead to a strained four-membered ring. Therefore, the intermolecular reaction will be preferred.

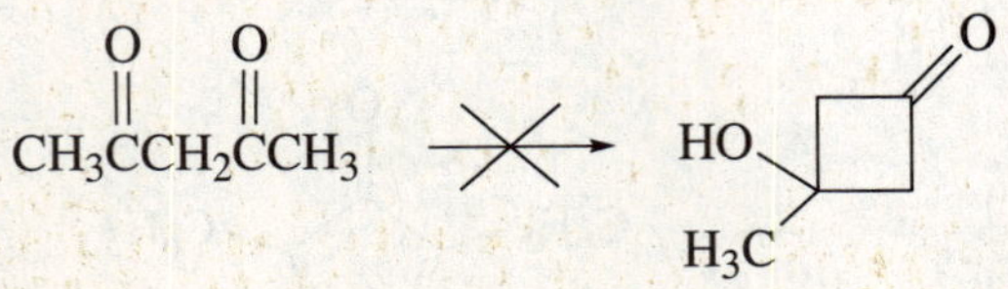

35. Solved in the text.

36. **a.** 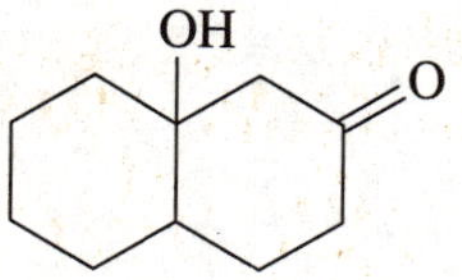 **b.** 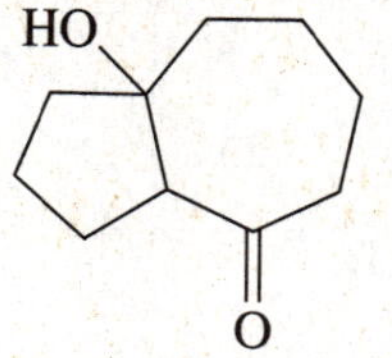 **c.** 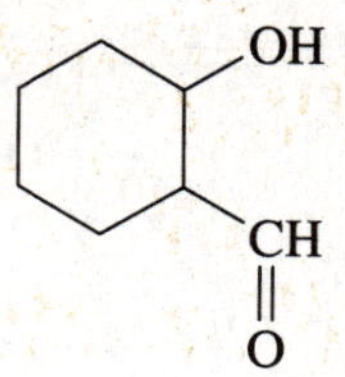 **d.** 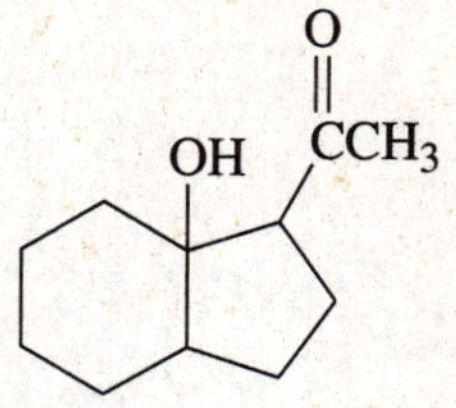

37. **a.**

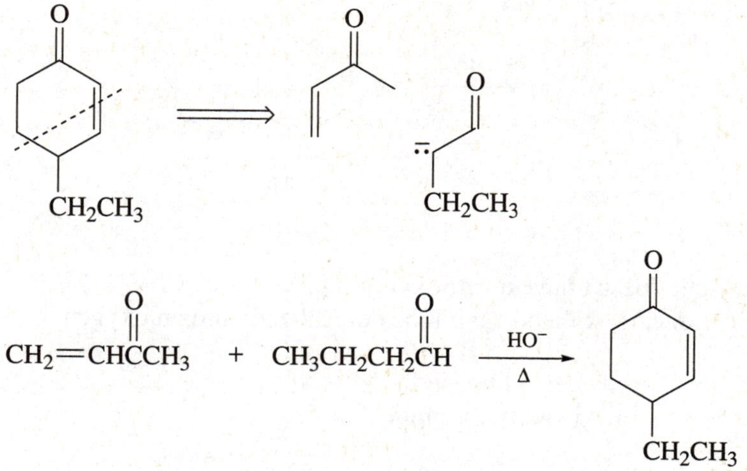

b.

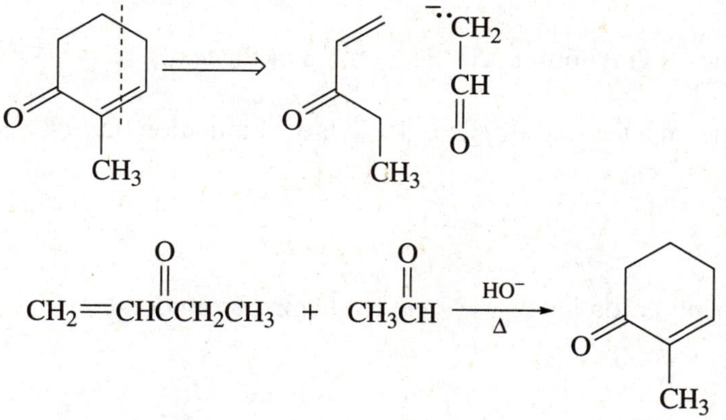

c.

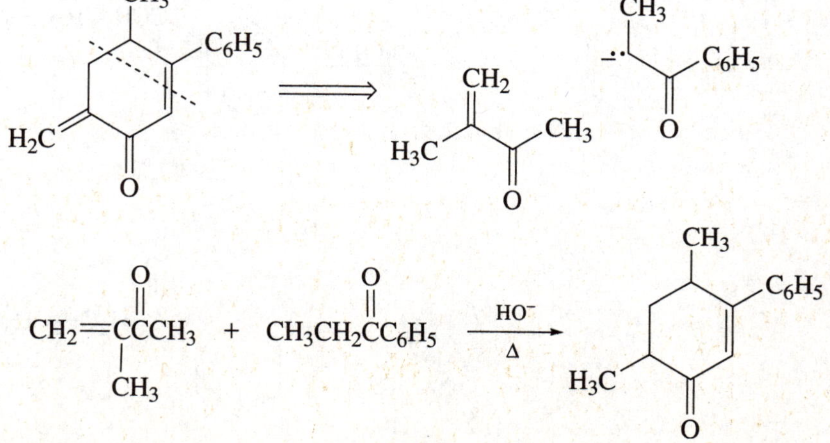

d. 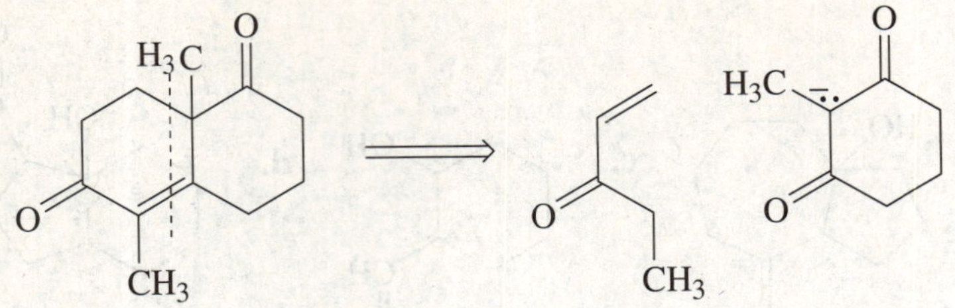

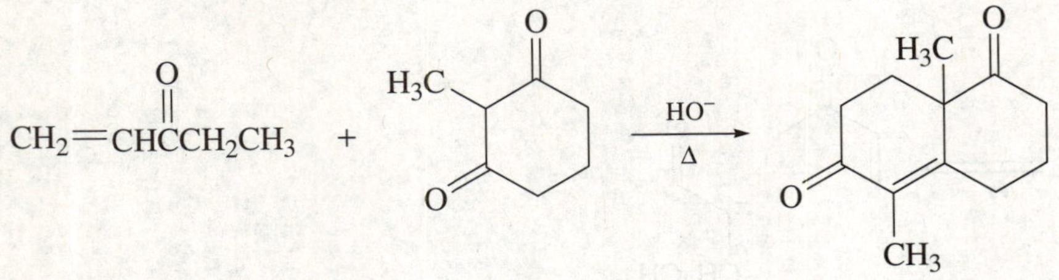

38. **A** and **D** can be decarboxylated.
B can't be decarboxylated, because it doesn't have a carboxyl group.
The electrons left behind if **C** were decarboxylated cannot be delocalized onto an oxygen.

39. **a.** methyl bromide **c.** benzyl bromide
b. methyl bromide (twice) **d.** isobutyl bromide

40. **a.** An S_N2 reaction cannot be done on bromobenzene. (Section 8.8 of the text.)

b. An S_N2 reaction cannot be done on vinyl bromide. (Section 8.8 of the text.)

c. An S_N2 reaction cannot be done on a tertiary alkyl halide. (Only elimination occurs; Section 9.8 of the text.)

41. Solved in the text.

42. **a.** ethyl bromide **b.** pentyl bromide **c.** benzyl bromide

43. **a.** 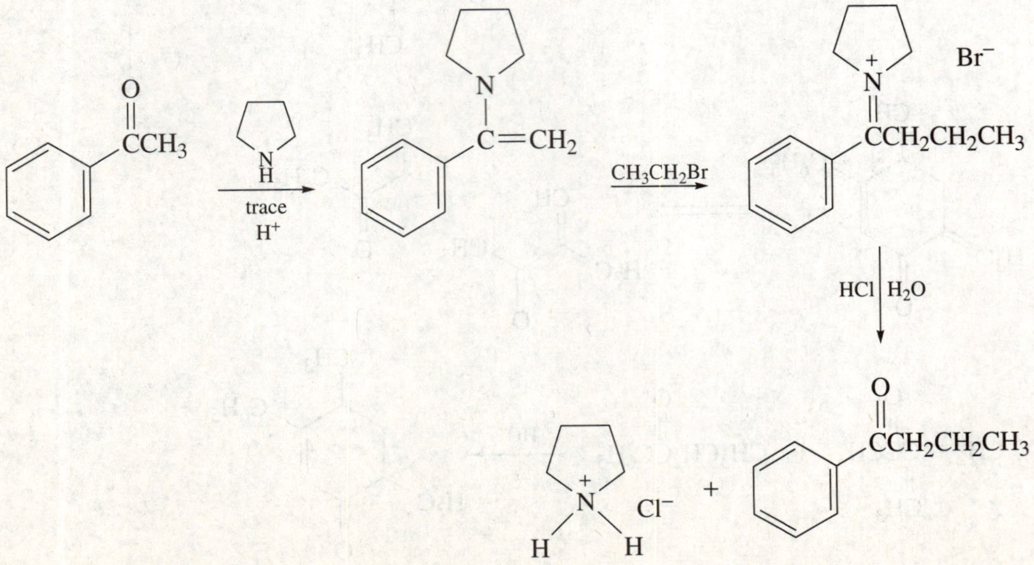

b.

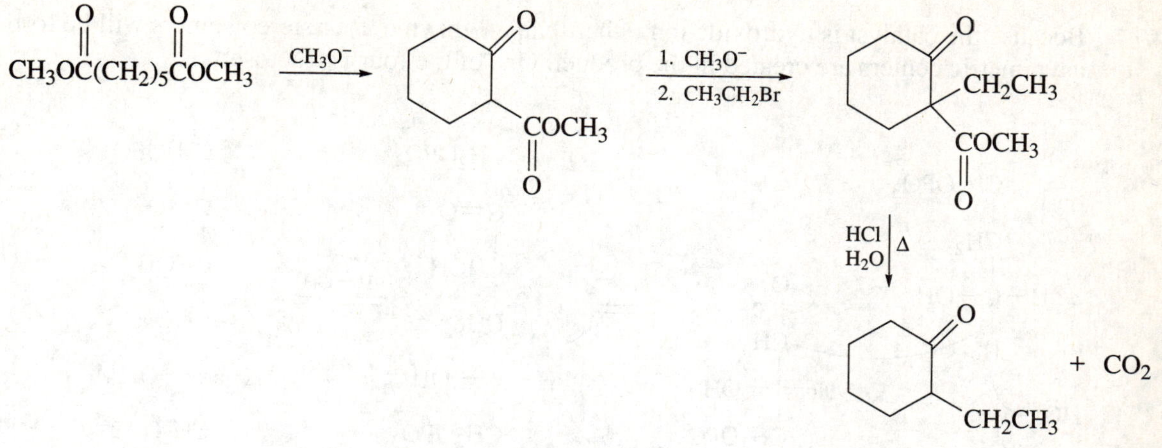
CH3OC(CH2)5COCH3
CH3O−
COCH3
1. CH3O−
2. CH3CH2Br
CH2CH3
COCH3
HCl
H2O
Δ
CH2CH3
+ CO2

c.

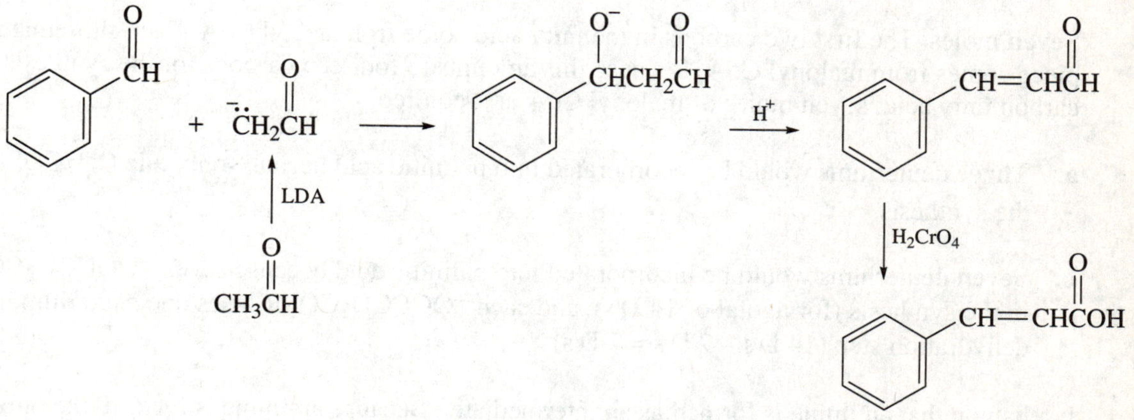
CH
+ CH2CH
LDA
CH3CH
CHCH2CH
H+
CH=CHCH
H2CrO4
CH=CHCOH

d.

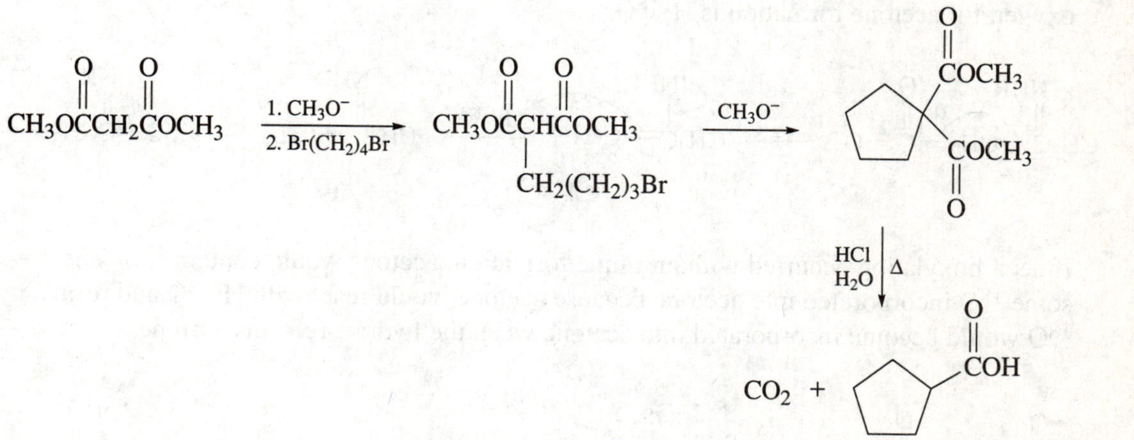
CH3OCCH2COCH3
1. CH3O−
2. Br(CH2)4Br
CH3OCCHCOCH3
CH2(CH2)3Br
CH3O−
COCH3
COCH3
HCl
H2O
Δ
CO2 +
COH

44. Because the catalyst is hydroxide ion rather than an enzyme, four stereoisomers will be formed since two asymmetric centers are created in the product. One of the four is fructose.

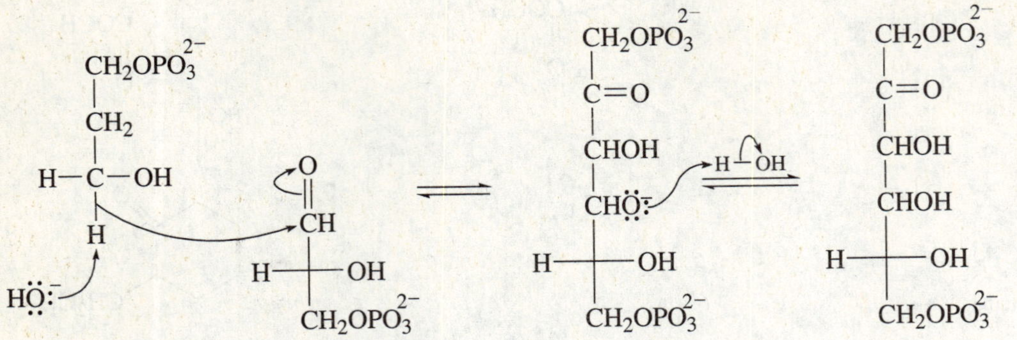

45. **Seven** moles. The first two carbons in the fatty acid come from acetyl CoA. Each subsequent two-unit piece comes from malonyl CoA. Because this amounts to fourteen carbons for the synthesis of the 16-carbon fatty acid, seven moles of malonyl CoA are required.

46. **a.** **Three** deuteriums would be incorporated into palmitic acid because only one CD_3COSR is used in the synthesis.

b. **Seven** deuteriums would be incorporated into palmitic acid because seven $^{-}OOCCD_2COSR$ are used in the synthesis (for a total of 14 D's), and each $^{-}OOCCD_2COSR$ loses one deuterium in the dehydration step (14 D's – 7 D's = 7 D's).

47. It tells you that an imine is formed as an intermediate. Because an imine is formed, the only source of oxygen for acetone formation is $H_2{}^{18}O$.

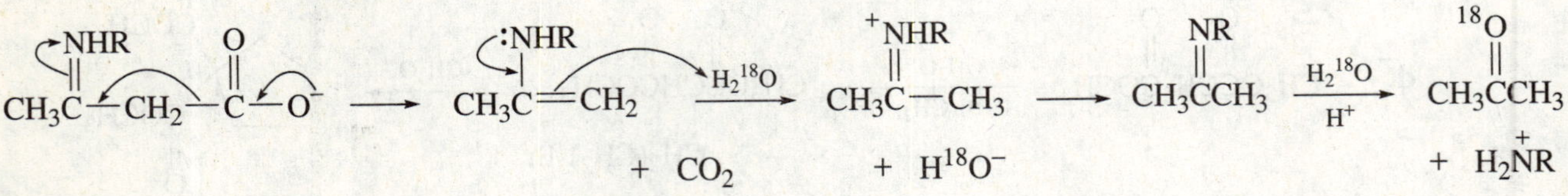

If decarboxylation occurred without imine formation, acetone would contain ^{16}O. There would also be some ^{18}O incorporated into acetone because acetone would react with $H_2{}^{18}O$ and form a hydrate. The ^{18}O would become incorporated into acetone when the hydrate reforms acetone.

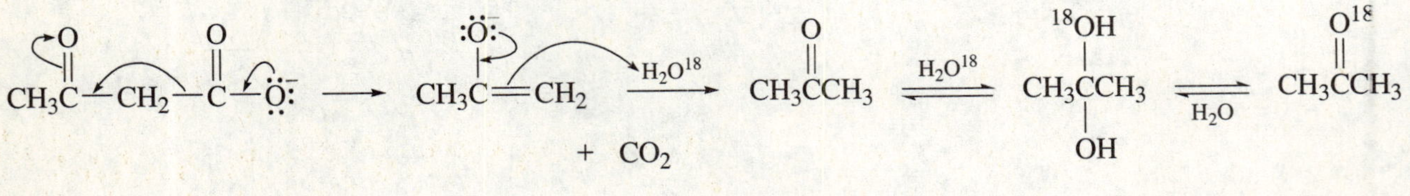

48. a. $CH_3\overset{O}{\overset{\|}{C}}CH_2\overset{O}{\overset{\|}{C}}OCH_2CH_3$ c. $R\overset{O}{\overset{\|}{C}}CH_2\overset{O}{\overset{\|}{C}}R'$ e. $CH_3CH_2CH_2CH_2\overset{O}{\overset{\|}{C}}OH$

b. $HO\overset{O}{\overset{\|}{C}}\underset{CH_3}{\underset{|}{C}}H\overset{O}{\overset{\|}{C}}OH$ d. 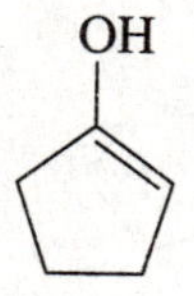

49. a. 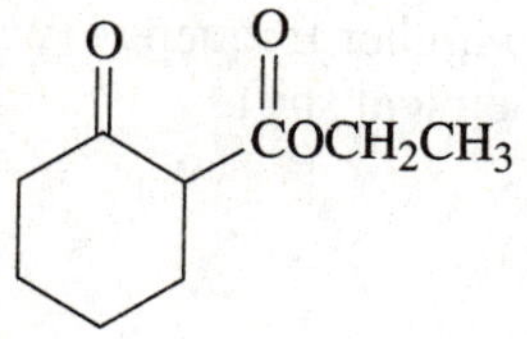

g. 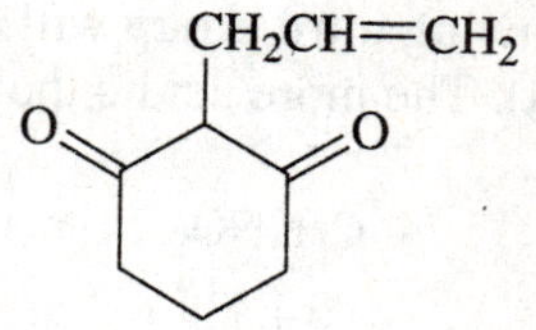

b. $CH_3CH_2CH_2\underset{Br}{\underset{|}{C}}H\overset{O}{\overset{\|}{C}}OH$

h. 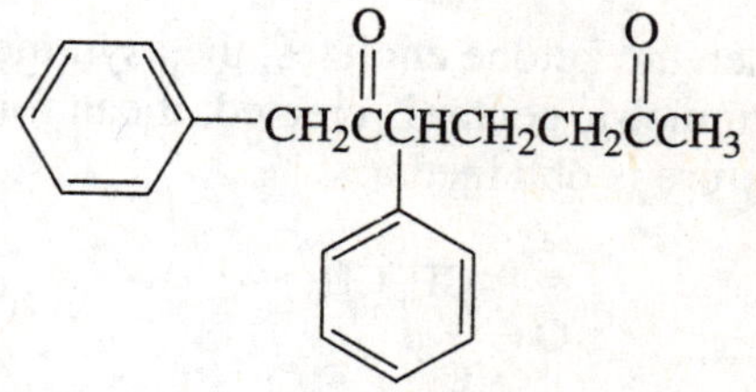

c. $CH_3\overset{O}{\overset{\|}{C}}CH_2\overset{O}{\overset{\|}{C}}CH_3$

i. 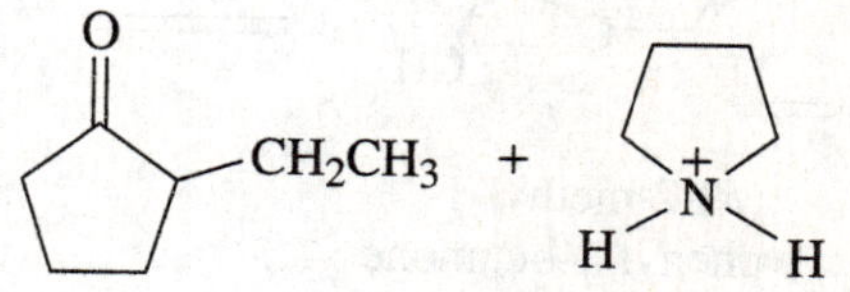

d. 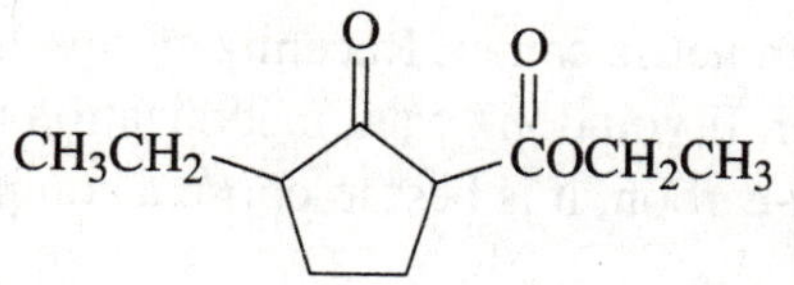

j. 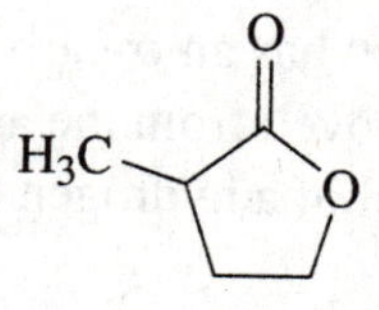

e. $CH_3\underset{CH_3}{\underset{|}{C}}HCH_2CH_2\overset{O}{\overset{\|}{C}}OH + CO_2$

k. 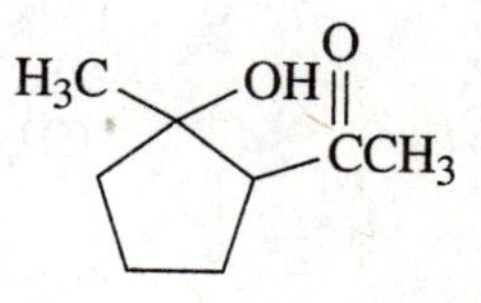

f. 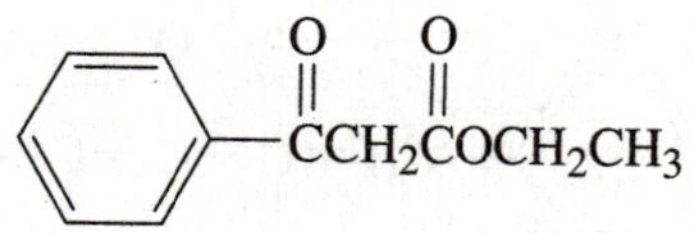

l. 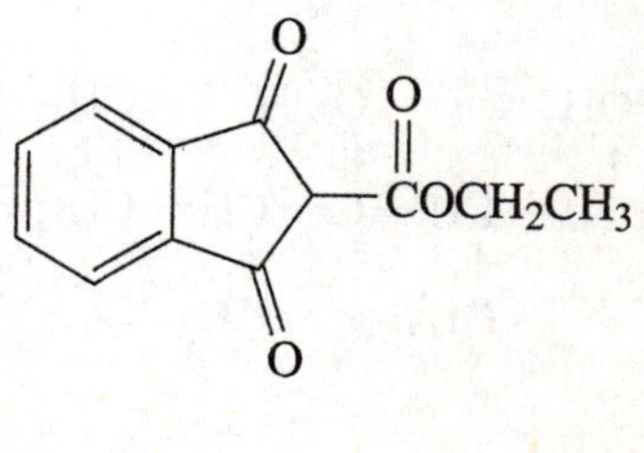

50. The compound on the far right loses CO_2 at the lowest temperature because the electrons left behind when CO_2 is removed can be delocalized.

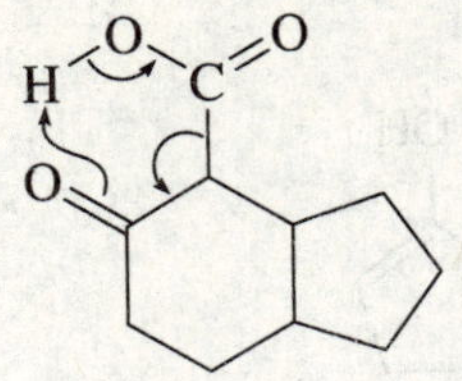

51. The electron withdrawing nitro group will cause the signal to occur at a higher frequency (will have a larger chemical shift). The more acidic the hydrogen, the greater the chemical shift.

CH_3NO_2	$CH_2(NO_2)_2$	$CH(NO_2)_3$
δ 4.33	δ 6.10	δ 7.52

52. **a.** When the ketone enolizes, the asymmetric center is lost. When the enol reforms the ketone, an asymmetric center is created; it can form the *R* and *S* enantiomer equally as easily, so a racemic mixture is obtained.

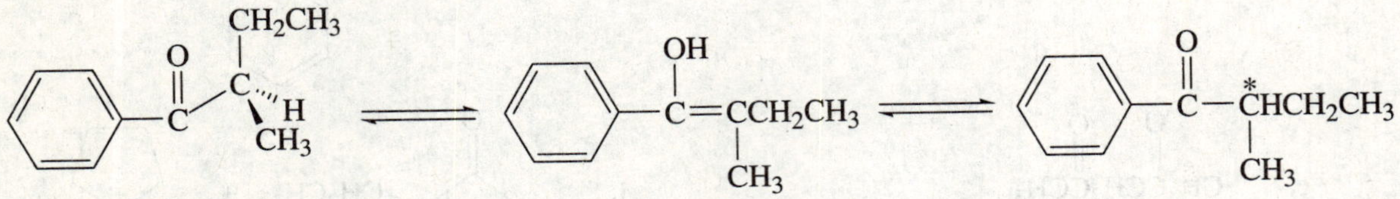

(*R*)-2-methyl-1-phenyl-1-butanone

b. You need a ketone that has an α-carbon that is an asymmetric center. Racemization will occur when an α-hydrogen is removed from the asymmetric center. If you don't want racemization to have to compete with removal of a hydrogen from a second α-carbon, it is best to chose a compound that can form only one enol.

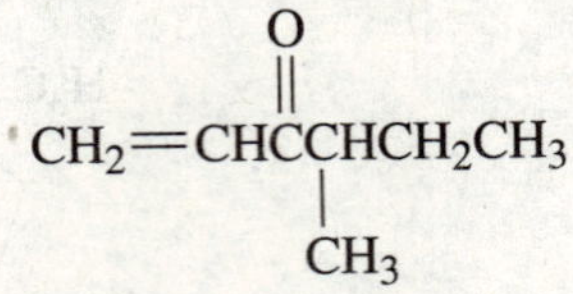

53.

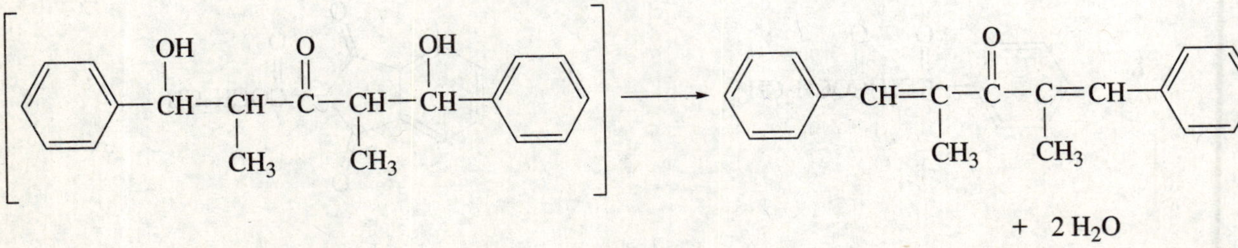

54.

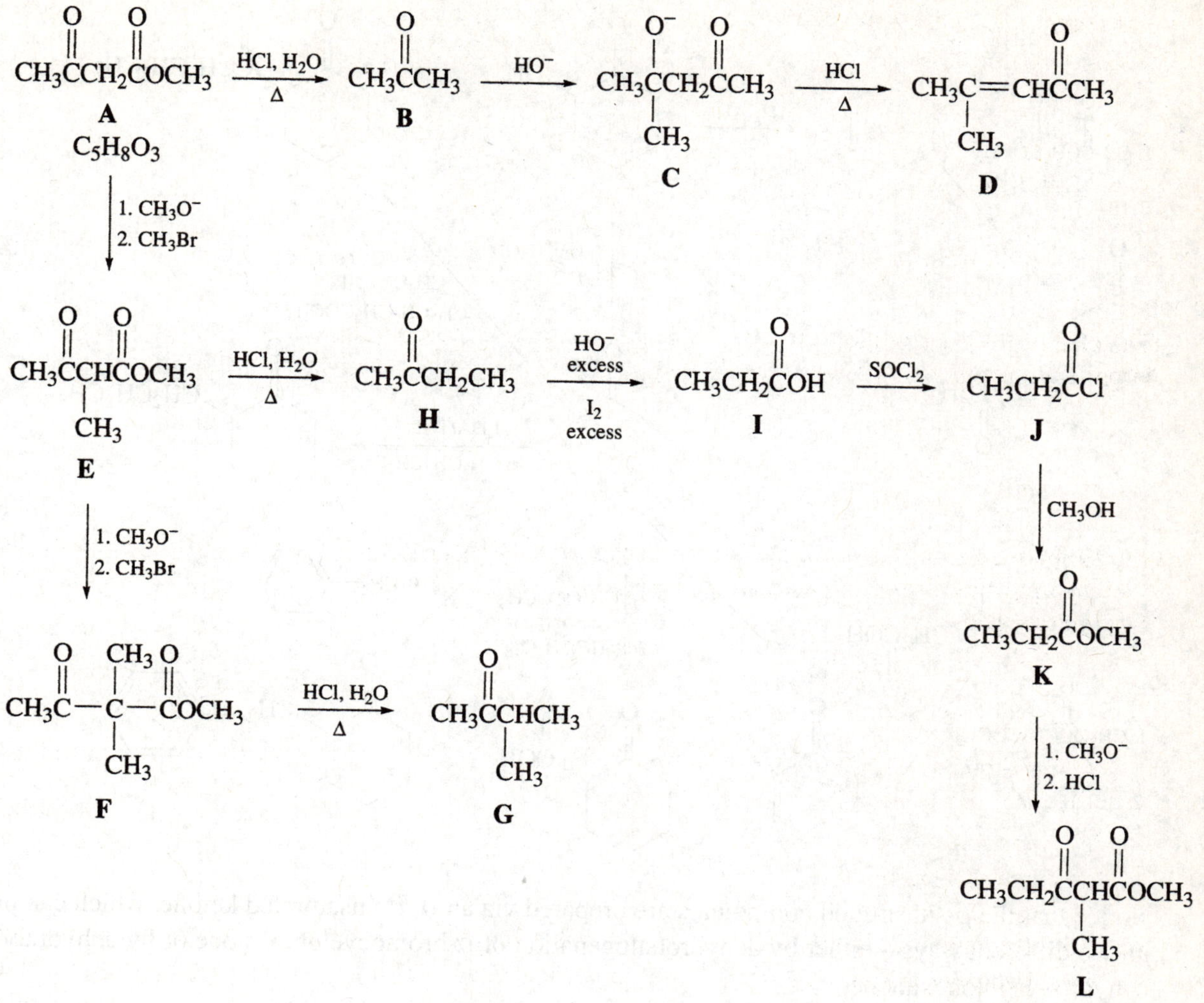

55.

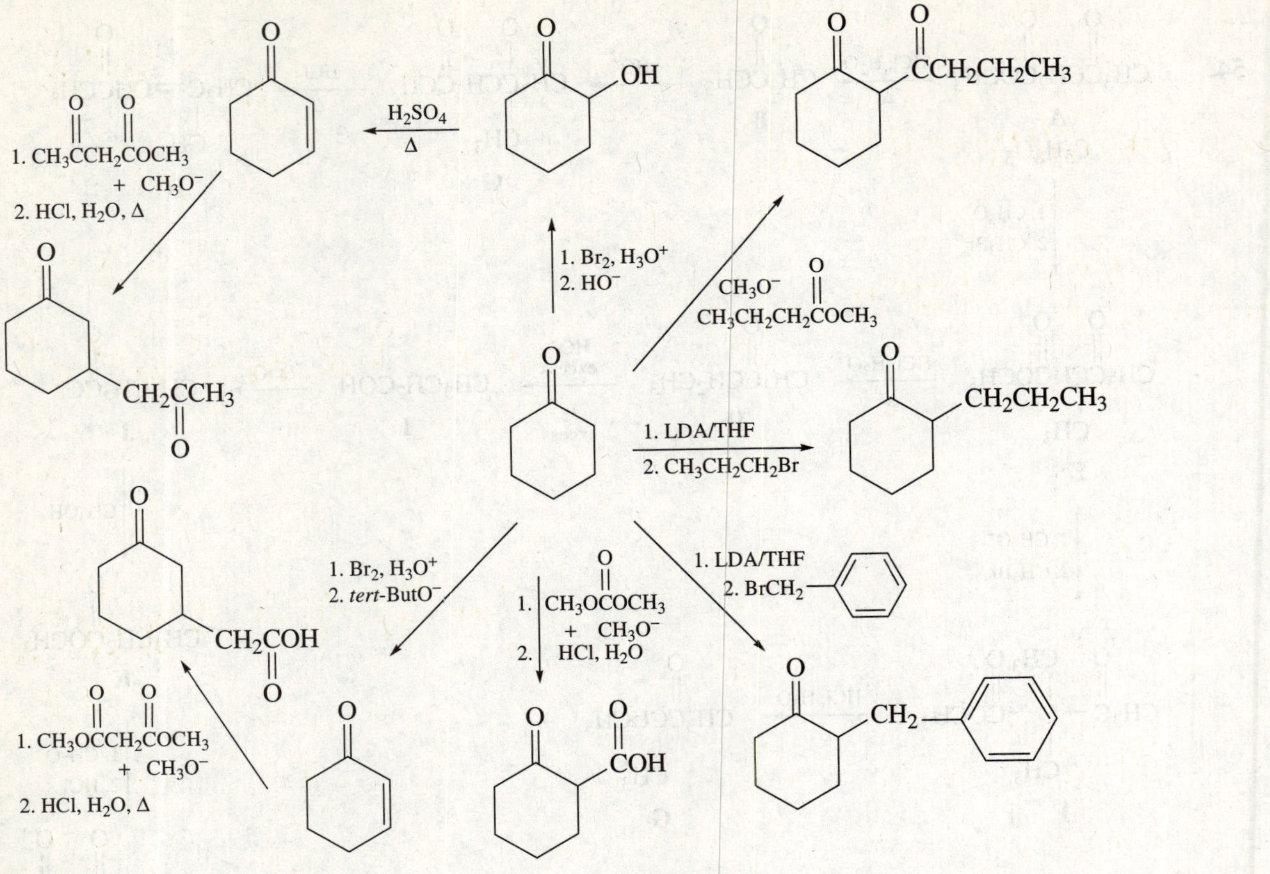

Notice that the β-substituted compounds are prepared via an α, β-unsaturated ketone, which can prepared in two different ways—either by dehydrohalogenation of α-bromocyclohexanone or by dehydration of α-hydroxycyclohexanone.

56. Remember that there are no positively charged organic reactants, intermediates, or products in a basic solution, and no negatively charged organic reactants, intermediates, or products in an acidic solution.

a.

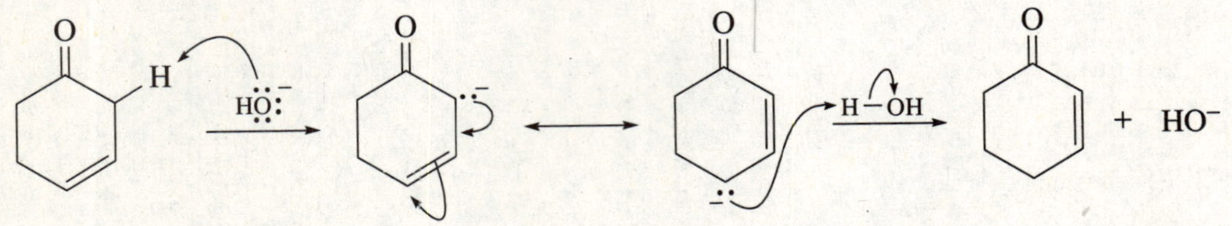

b.

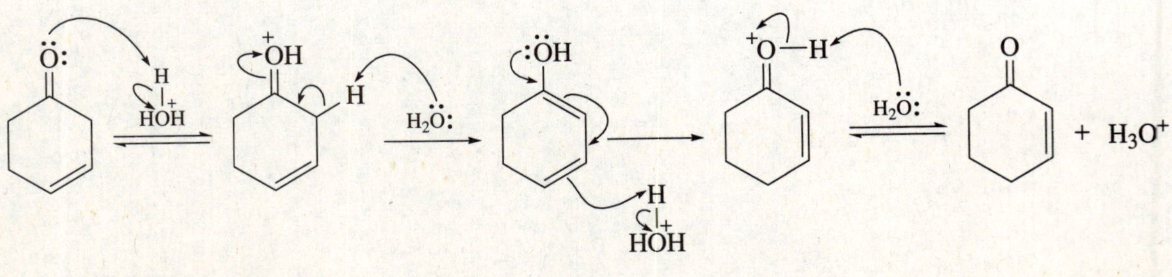

57. **a.**

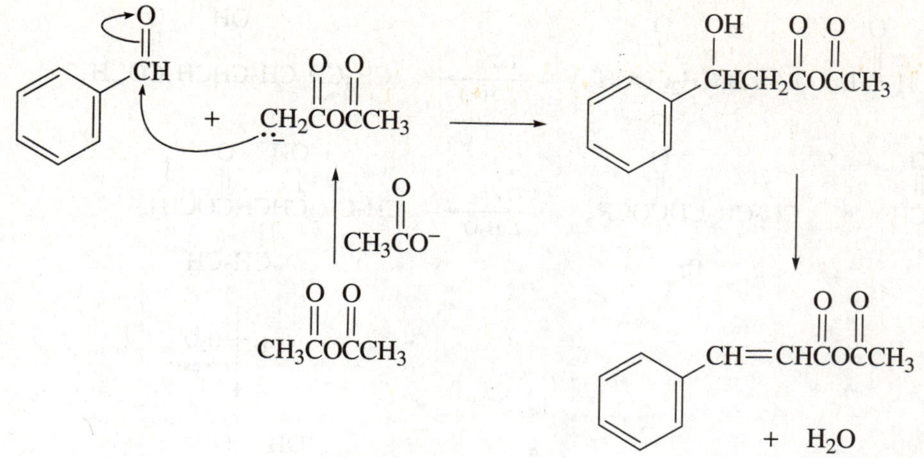

O
CH
O O
+ CH2COCCH3
OH O O
CHCH2COCCH3
O
CH3CO−
O O
CH3COCCH3
O O
CH=CHCOCCH3
+ H2O

b.

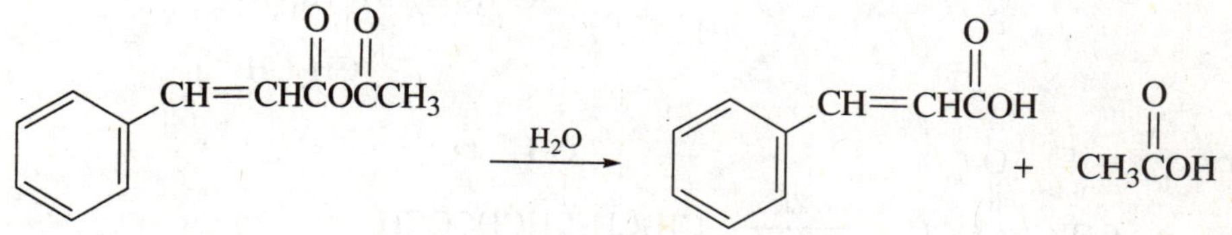

O O
CH=CHCOCCH3
H2O
O
CH=CHCOH
O
+ CH3COH

c.

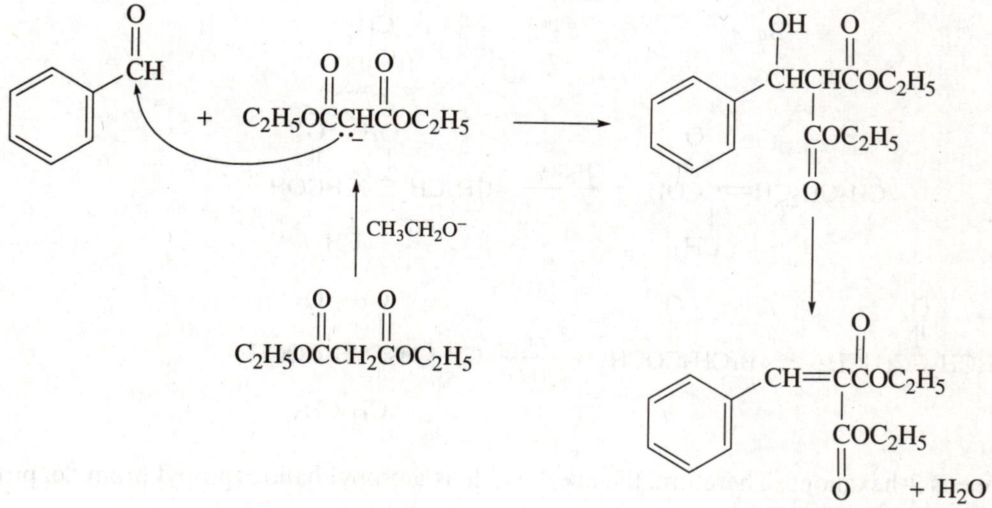

O
CH
O O
+ C2H5OCCHCOC2H5
OH O
CHCHCOC2H5
COC2H5
O
CH3CH2O−
O O
C2H5OCCH2COC2H5
O
CH=CCOC2H5
COC2H5
O
+ H2O

d.

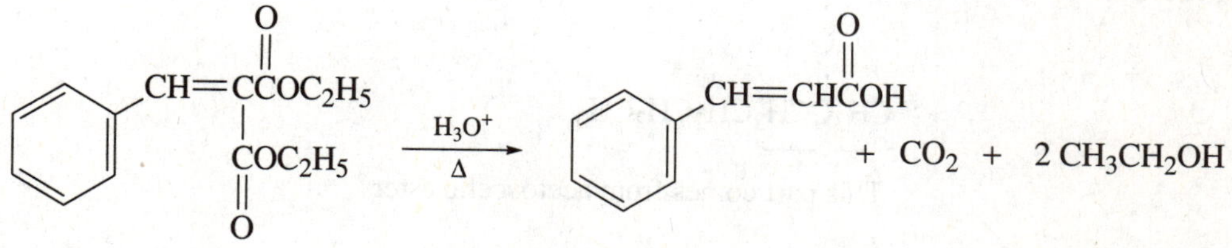

O
CH=CCOC2H5
COC2H5
O
H3O+
Δ
O
CH=CHCOH
+ CO2 + 2 CH3CH2OH

58. **a.** 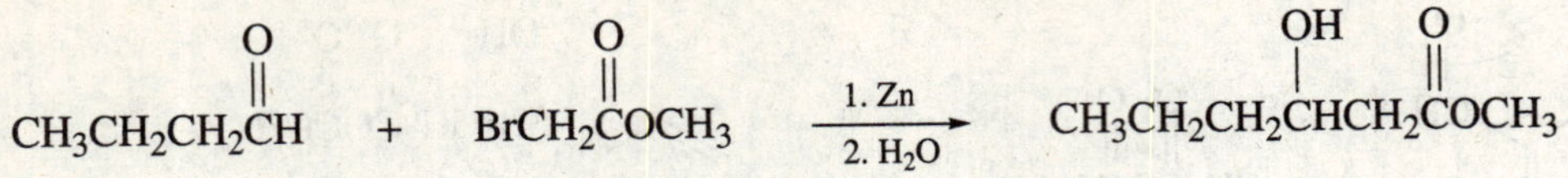

b. 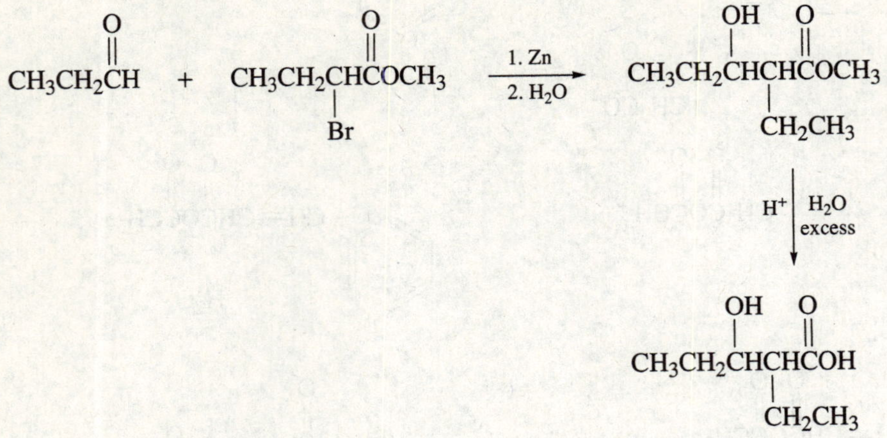

c. 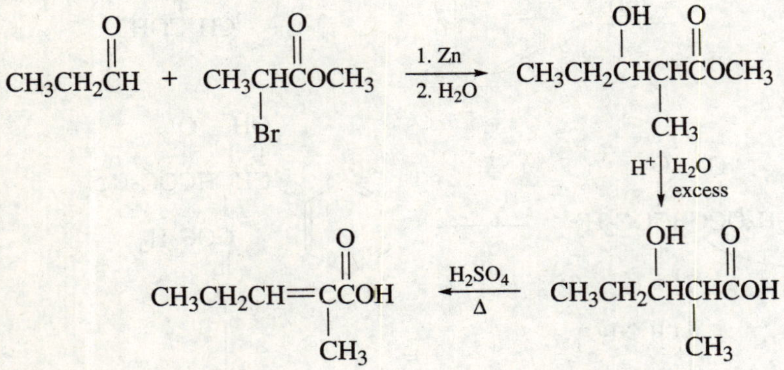

d. 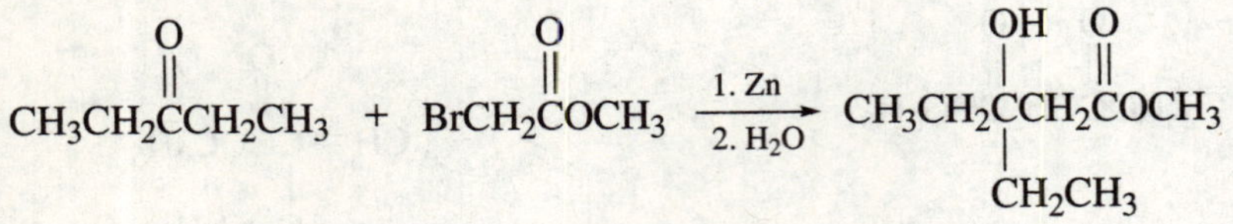

59. The ketone is 2-hexanone. Therefore, the alkyl halide is a propyl halide (propyl bromide, propyl chloride, or propyl iodide).

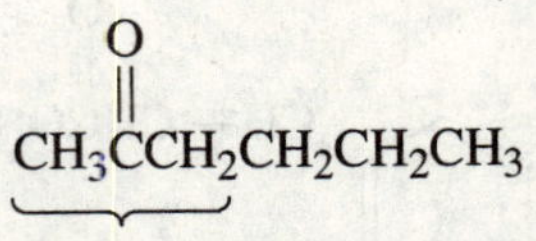

This part comes from acetoacetic ester

60. **a.** 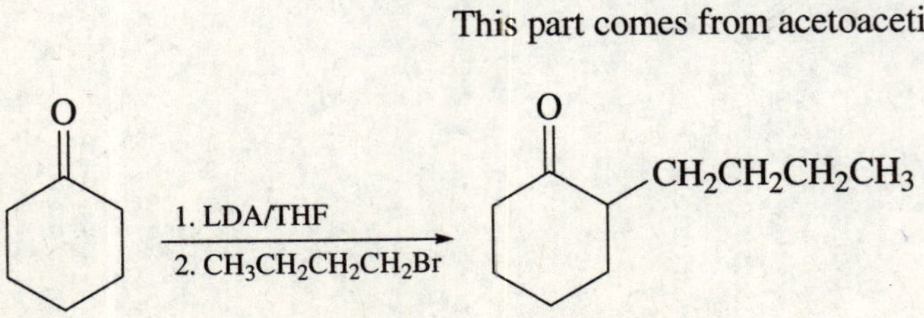

b. 1.

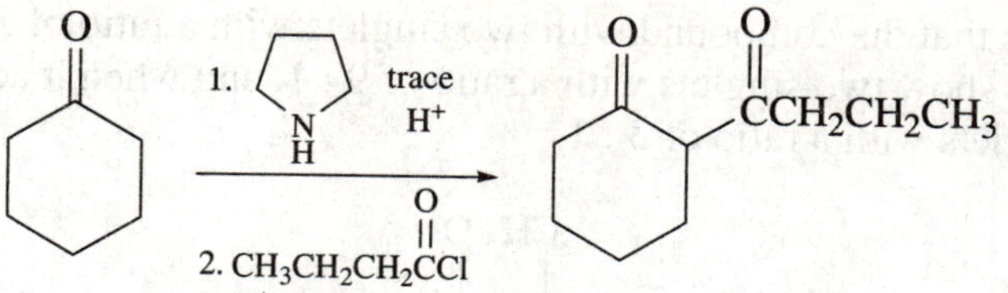
1.
trace
H⁺
2. CH₃CH₂CH₂CCl
3. H⁺, H₂O
CCH₂CH₂CH₃

2.

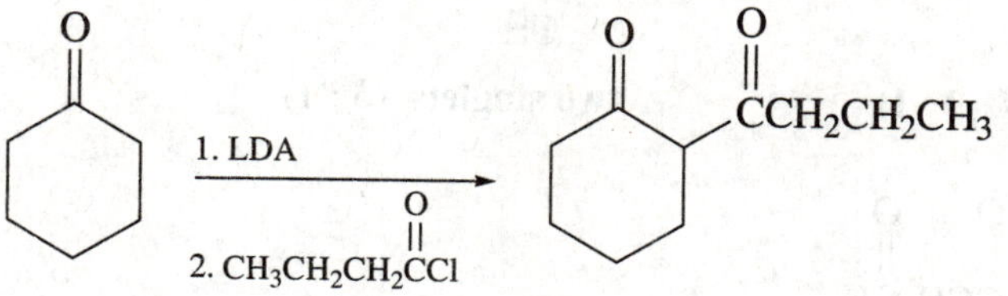
1. LDA
2. CH₃CH₂CH₂CCl
CCH₂CH₂CH₃

c.

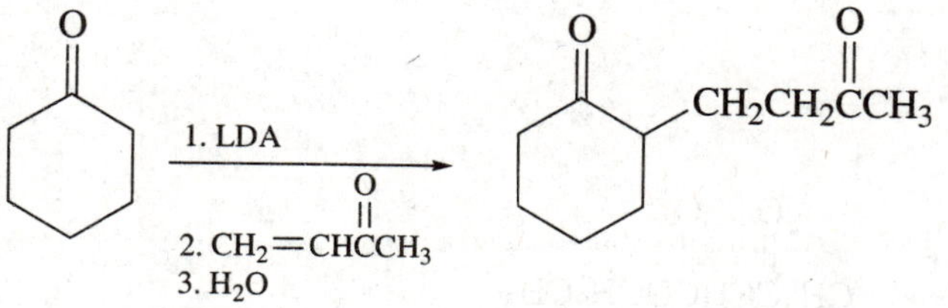
1. LDA
2. CH₂=CHCCH₃
3. H₂O
CH₂CH₂CCH₃

d.

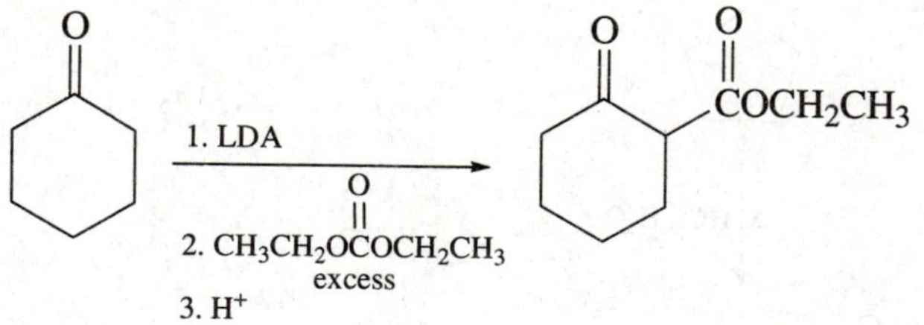
1. LDA
2. CH₃CH₂OCOCH₂CH₃
excess
3. H⁺
COCH₂CH₃

e.

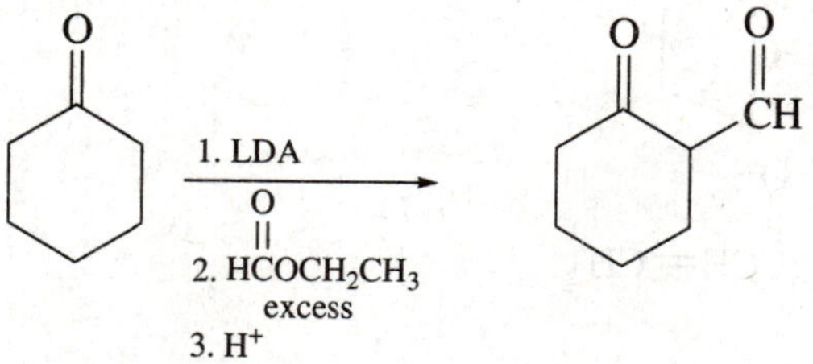
1. LDA
2. HCOCH₂CH₃
excess
3. H⁺
CH

f.

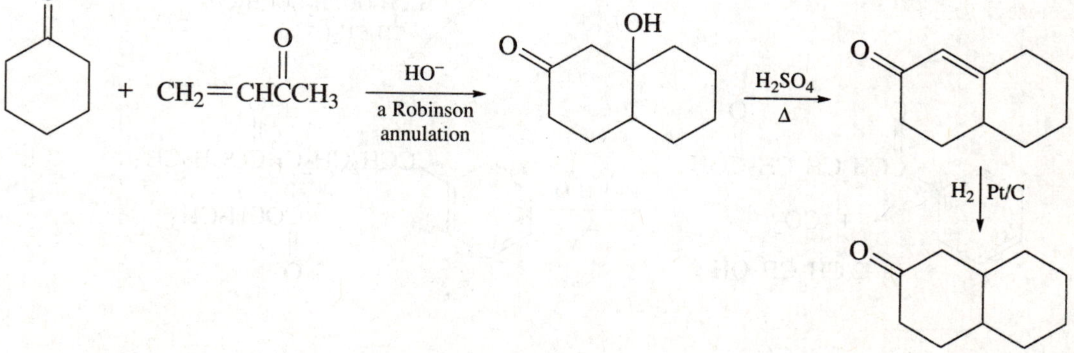
CH₂=CHCCH₃
HO⁻
a Robinson
annulation
OH
H₂SO₄
Δ
H₂
Pt/C

61. The positive iodoform test indicates that the compound with two singlets with a ratio of 3 : 1 is a methyl ketone. 3,3-Dimethyl-1-butyne will show two singlets with a ratio of 9 : 1, and when it adds water, it will form a methyl ketone with two singlets with a ratio of 3 : 1.

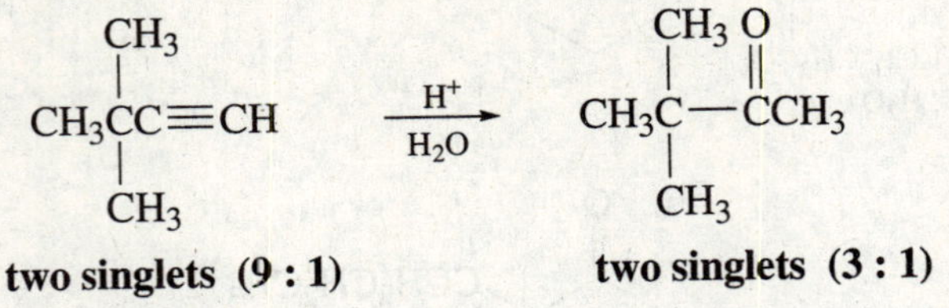

62. **a.**

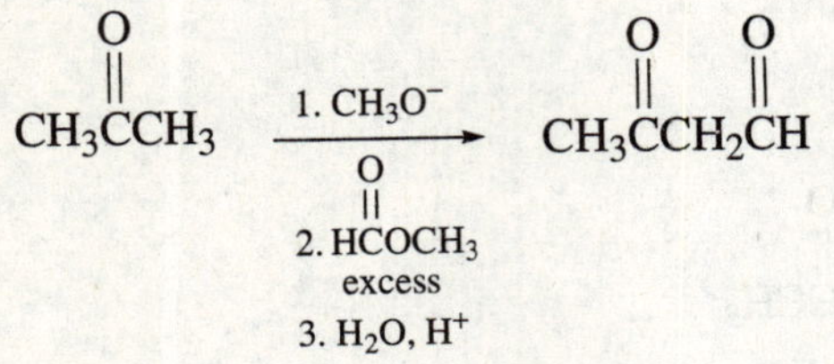

b.

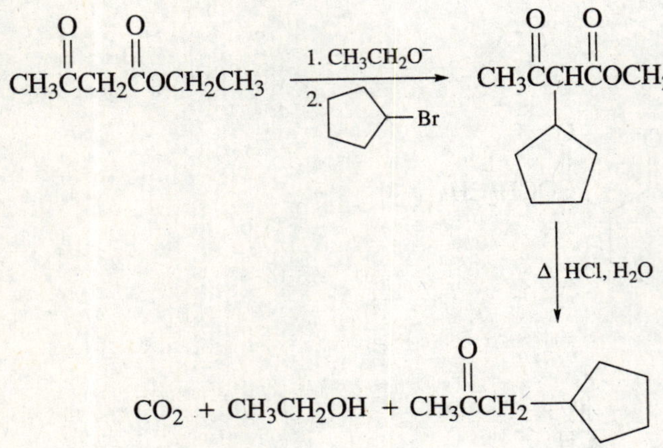

c.

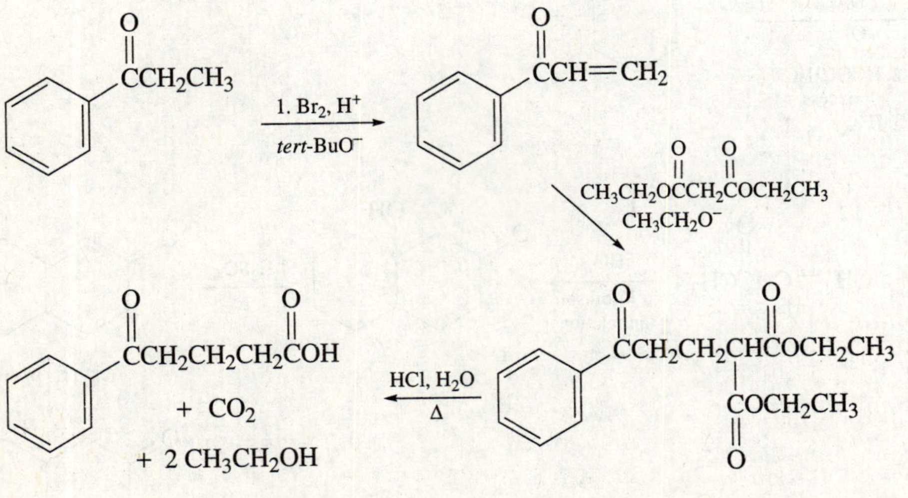

d. 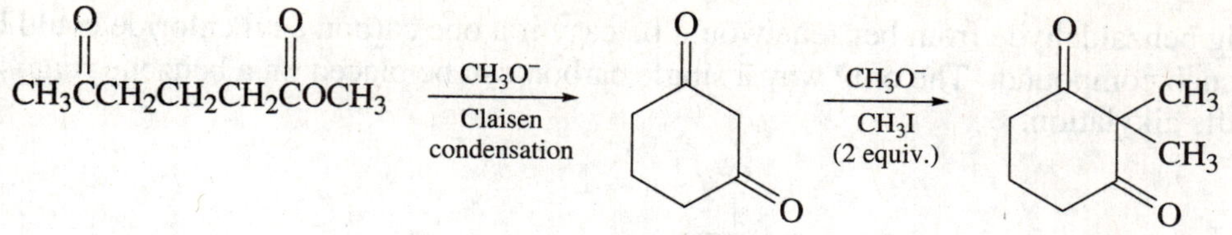
CH3CCH2CH2CH2COCH3
CH3O−
Claisen
condensation
CH3O−
CH3I
(2 equiv.)
CH3
CH3

e. 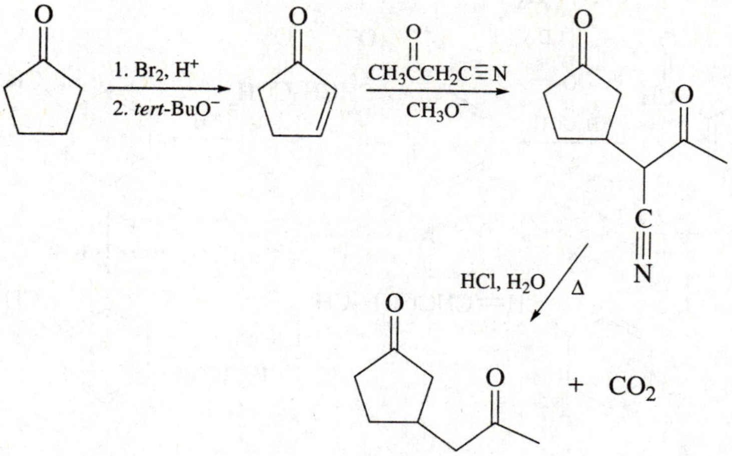
1. Br2, H+
2. tert-BuO−
CH3CCH2C≡N
CH3O−
HCl, H2O
Δ
+ CO2

f. 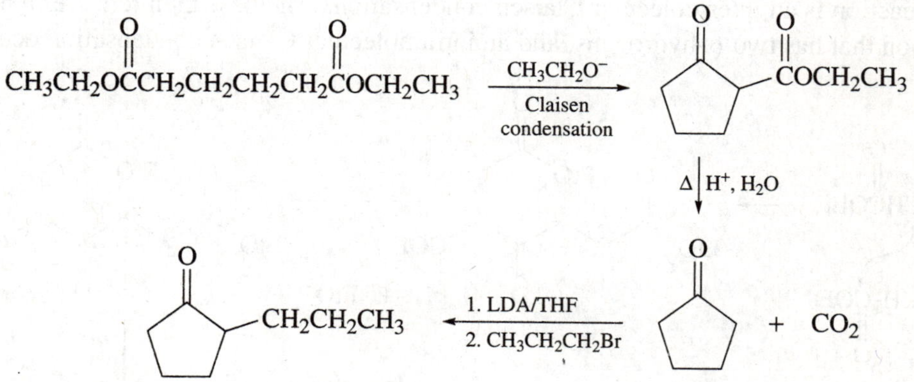
CH3CH2OCCH2CH2CH2CH2COCH2CH3
CH3CH2O−
Claisen
condensation
COCH2CH3
Δ
H+, H2O
+ CO2
1. LDA/THF
2. CH3CH2CH2Br
CH2CH2CH3

63. 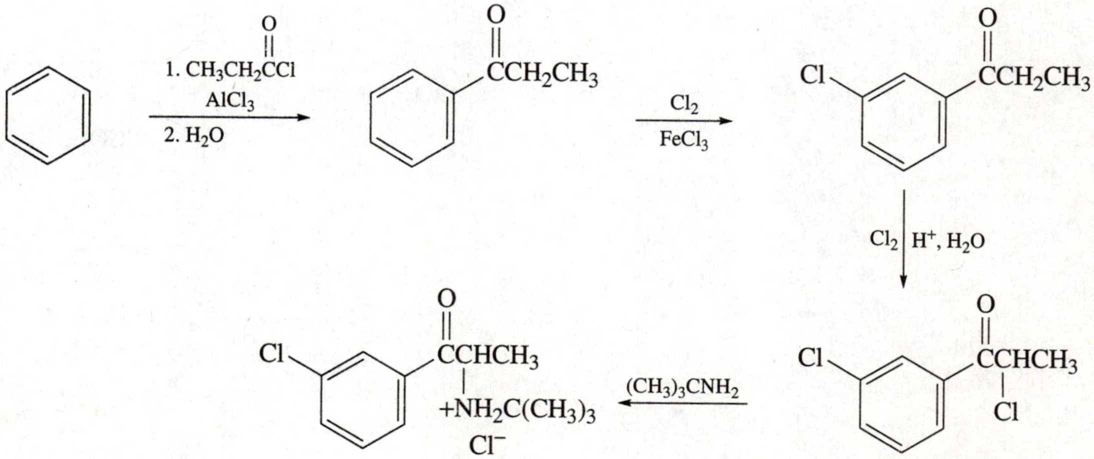
1. CH3CH2CCl
AlCl3
2. H2O
CCH2CH3
Cl2
FeCl3
Cl
Cl2
H+, H2O
CCHCH3
(CH3)3CNH2
+NH2C(CH3)3
Cl−

64. Synthesizing benzaldehyde from benzene would be easy if a one carbon acyl chloride could be used, but there is no such compound. The only way a single carbon can be placed on a benzene ring is via a Friedel-Crafts alkylation.

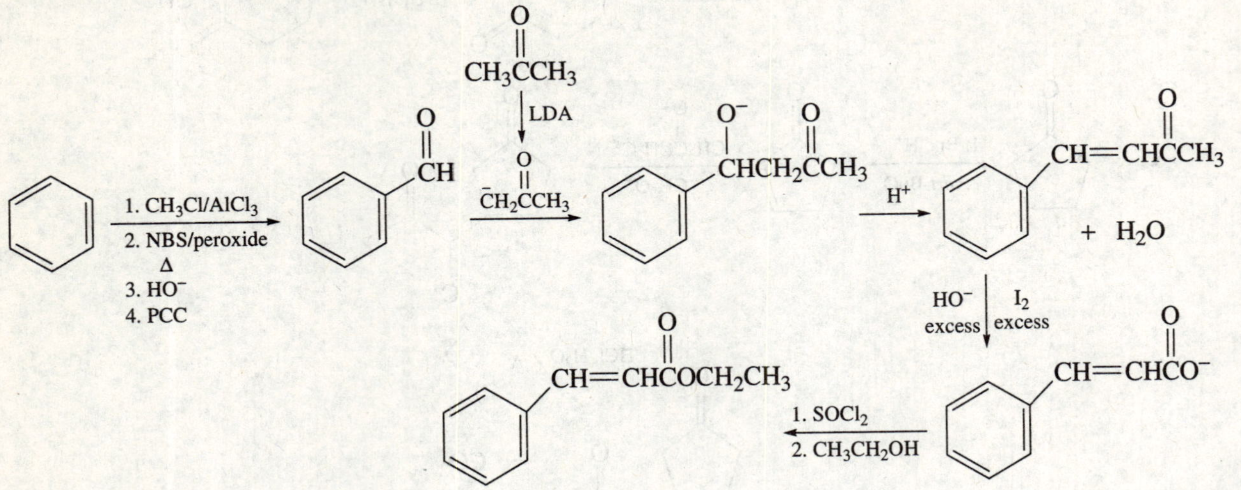

65. **a.** The first reaction is an intermolecular Claisen condensation. The base then removes a hydrogen from the α-carbon that has two α-hydrogens, and an intramolecular Claisen condensation occurs.

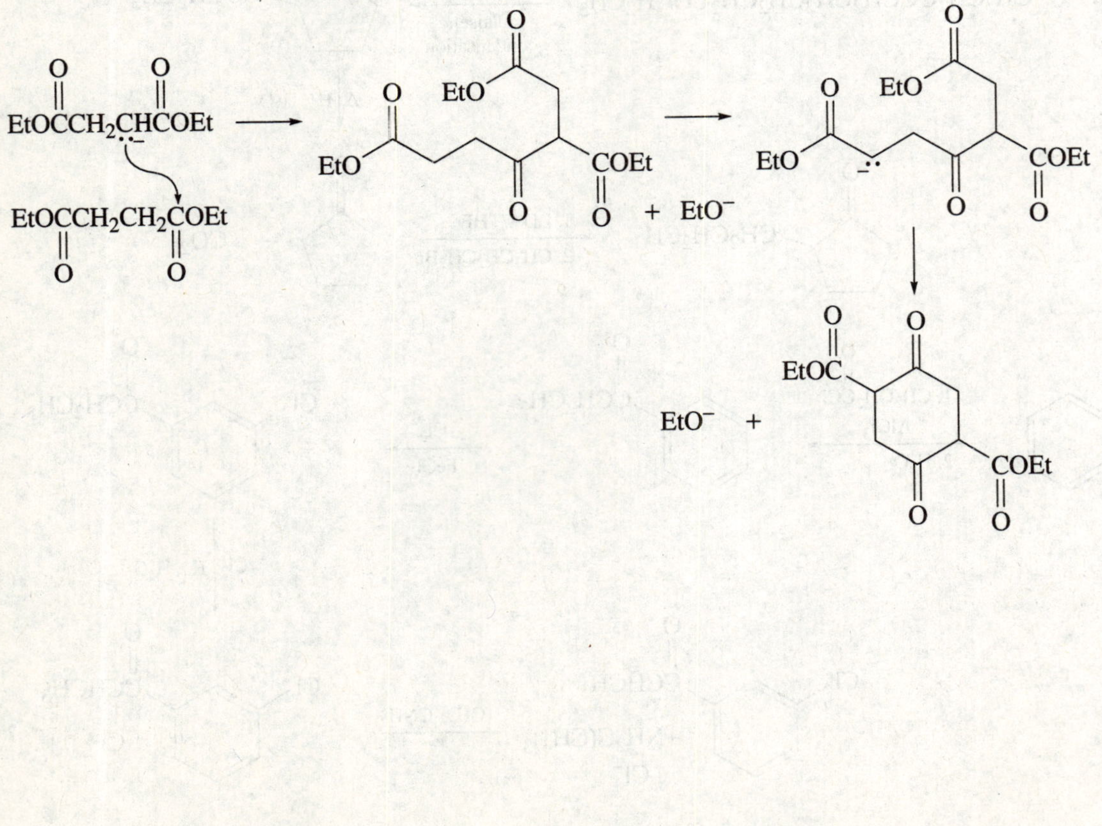

b. The reaction involves two successive aldol condensations.

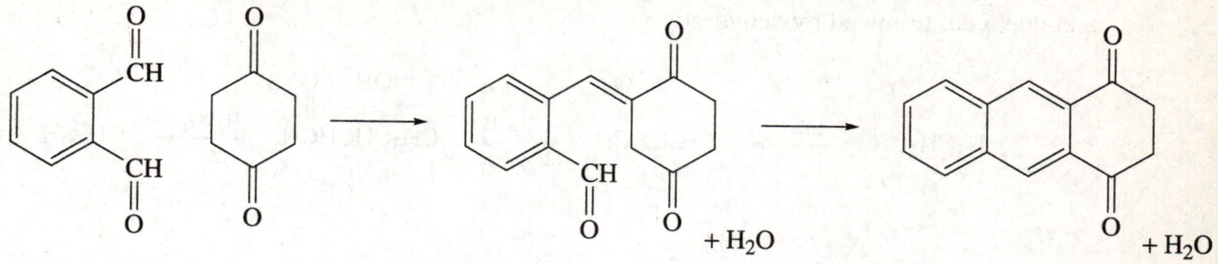

c. The base removes a proton from the carbon that is flanked by two carbonyl groups.

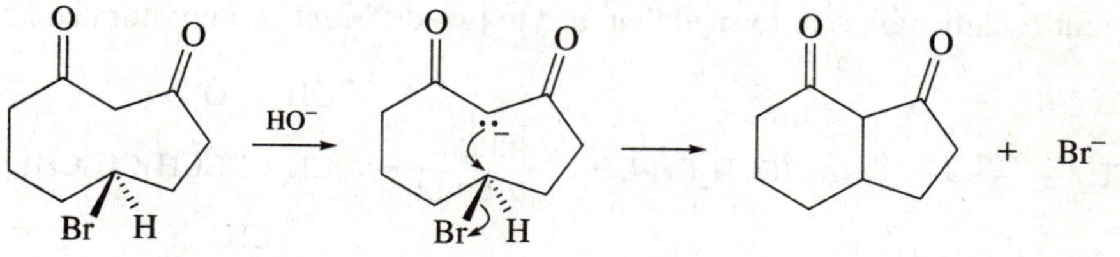

66. **a.** CH_3CH_2COH (C=O) $\xrightarrow[2.\ H_3O^+]{1.\ Br_2,\ PBr_3}$ CH_3CHCOH (C=O; Br on CH) $\xrightarrow[2.\ H_3O^+]{\text{excess } 1.\ NH_3}$ CH_3CHCOH (C=O; $^+NH_3$ on CH)

b.

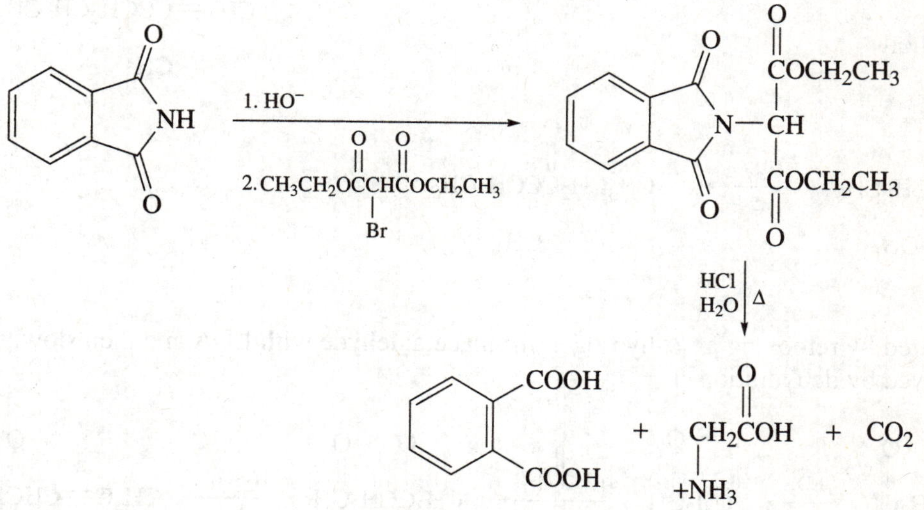

67. **a** can be prepared by removing an α-hydrogen from acetone with LDA and then adding formaldehyde, followed by dehydration.

CH_3CCH_3 (C=O) $\xrightarrow{\text{LDA}}$ $\ddot{C}H_2CCH_3$ (C=O, carbanion) $\xrightarrow{HCH \text{ (C=O)}}$ $CH_2CH_2CCH_3$ (OH on first C; C=O) $\xrightarrow[\Delta]{H_2SO_4}$ $CH_2{=}CHCCH_3$ (C=O)

b can be prepared by removing an α-hydrogen from propionaldehyde with LDA and then slowly adding acetaldehyde, followed by dehydration.

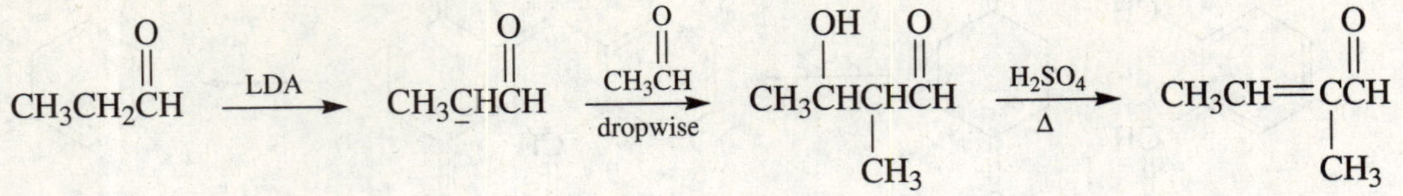

c can be prepared by removing an α-hydrogen from 3-hexanone with LDA and then adding formaldehyde, followed by dehydration. The yield is poor because 3-hexanone is an asymmetrical ketone; therefore, two different α-carbanions are formed that lead to two different α,β-unsaturated ketones.

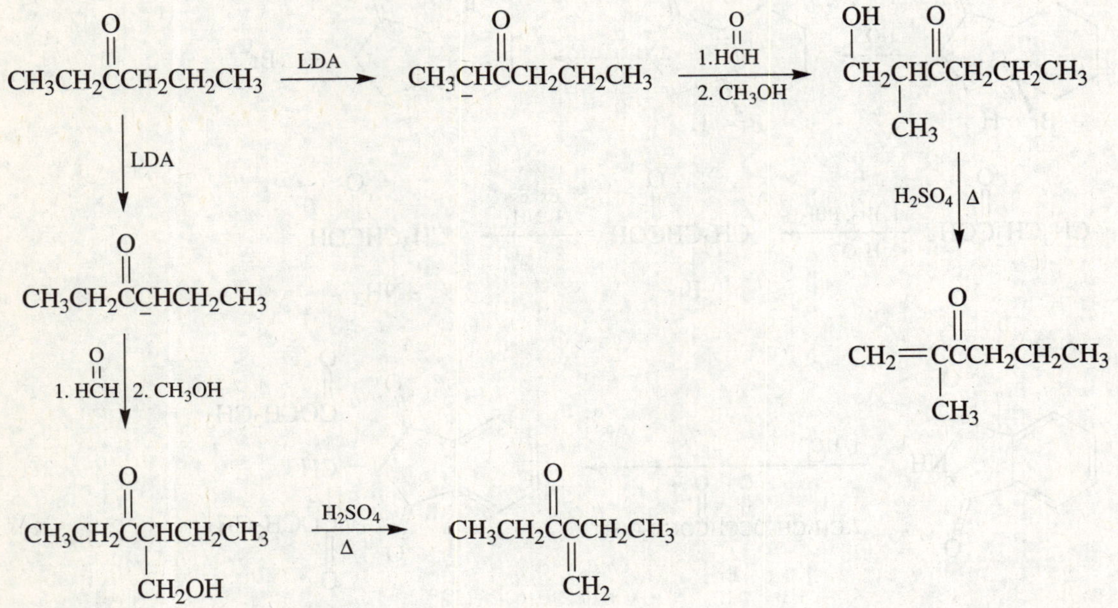

d can be prepared by removing an α-hydrogen from acetaldehyde with LDA and then slowly adding acetone, followed by dehydration.

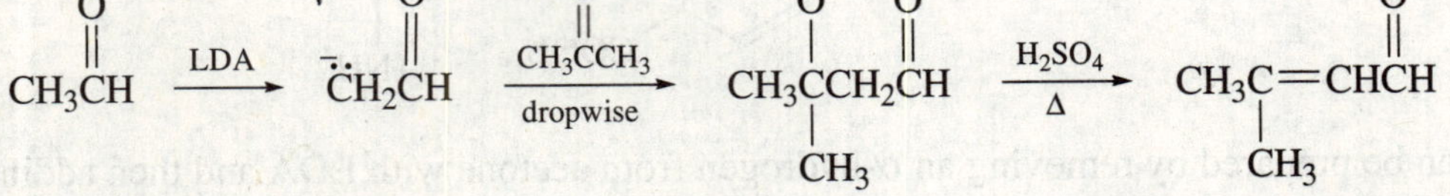

e cannot be prepared by a mixed aldol condensation. It can be prepared via an intramolecular aldol condensation using 5-oxooctanal, followed by dehydration.

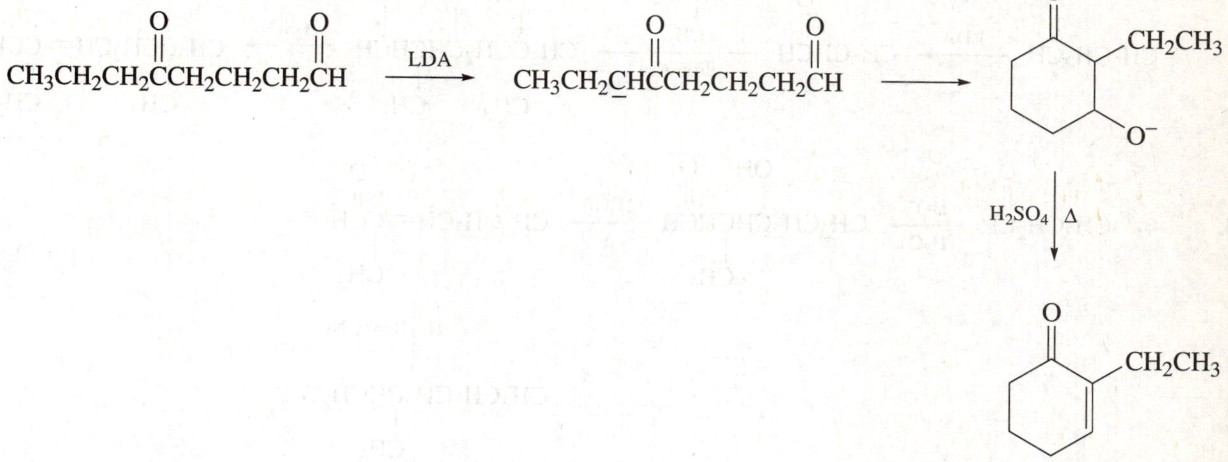

f cannot be prepared by a mixed aldol condensation. It can be prepared via an intramolecular aldol condensation using 1,7-heptanedial, followed by dehydration.

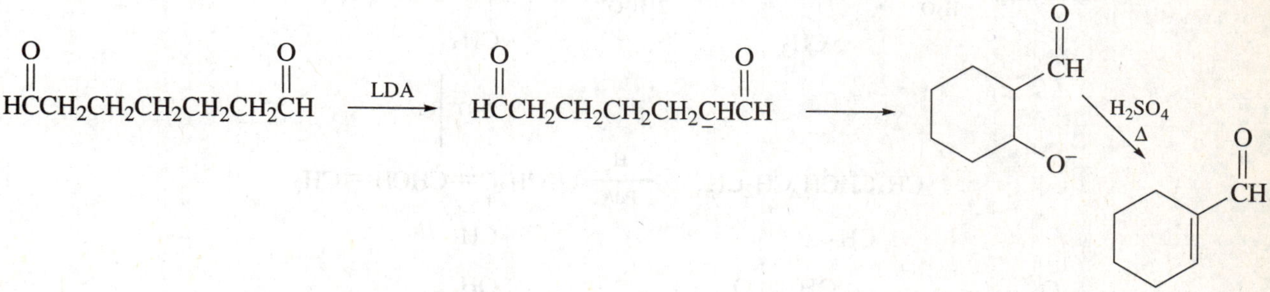

g cannot be prepared by a mixed aldol condensation. A small amount of g can be formed via an intramolecular aldol condensation using 7-oxooctanal, but this compound can form two different α-carbanions that can react with a carbonyl group to form a six-membered ring. Because an aldehyde is more reactive than a ketone, the desired compound will be formed only as a minor product.

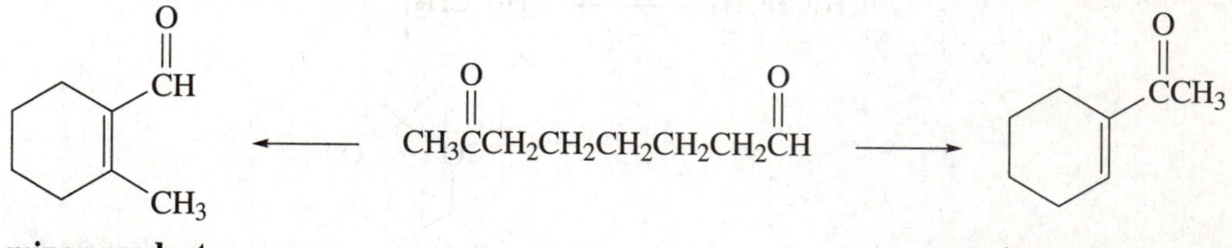

minor product **major product**

Theoretically **h** can be prepared by a mixed aldol condensation. However, because of steric hindrance, the reaction of propanal with dimethylpropanal to form the desired product would be very slow.

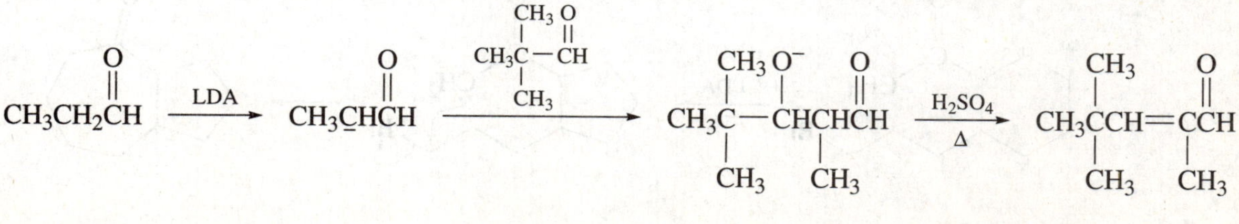

i can be prepared was using propanal and 3,3-dimethylbutanal.

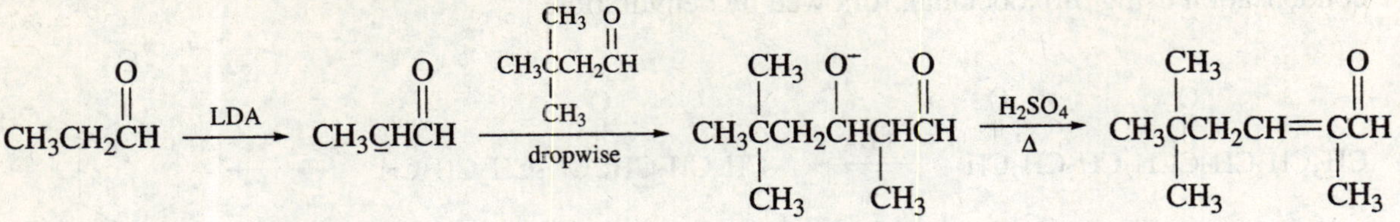

68.

a.

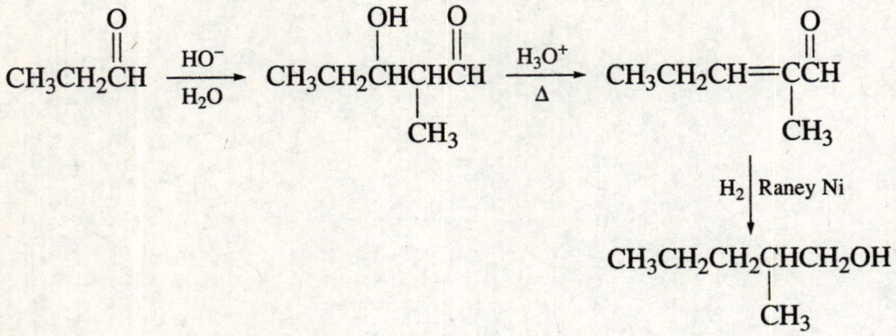

b.

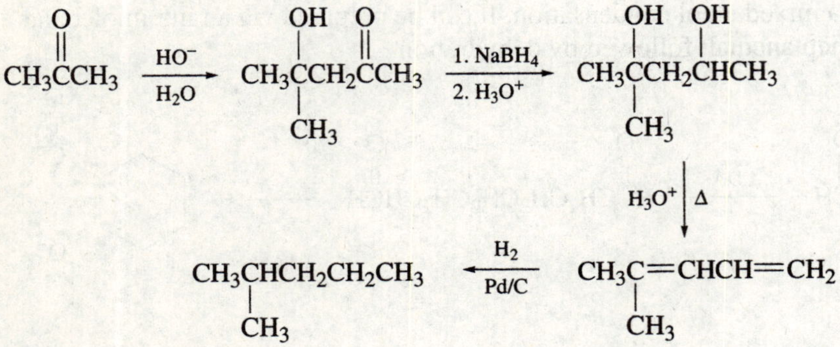

c.

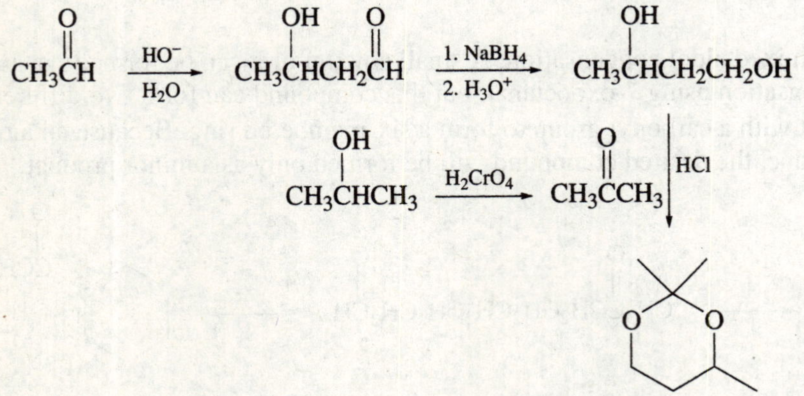

69. At low temperatures (– 78°), the proton will be more apt to be removed from the methyl group because its hydrogens are the most accessible and are slightly more acidic.

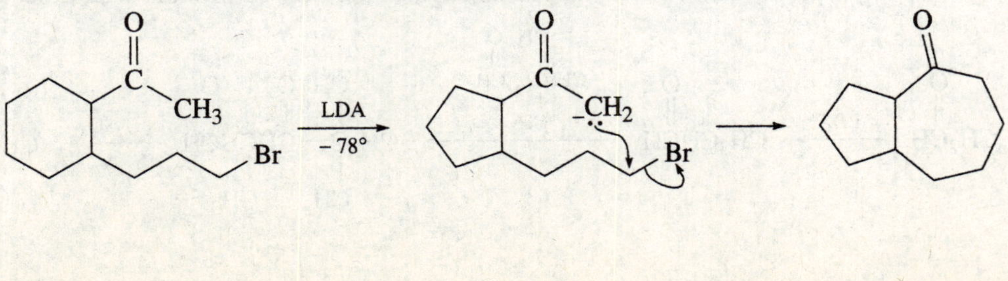

At higher temperatures (25°), the proton will be more apt to be removed from the more substituted α-carbon because in that way the more stable enolate is formed (the one with the more stable double bond).

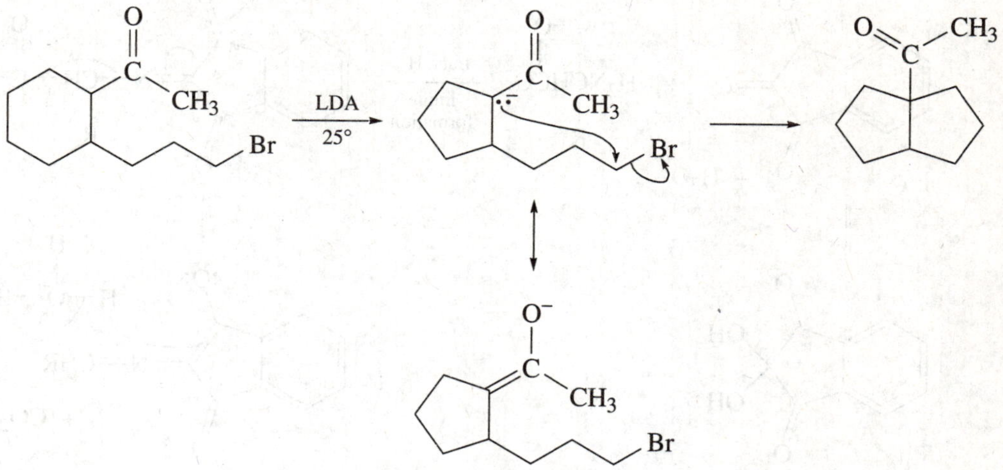

70.

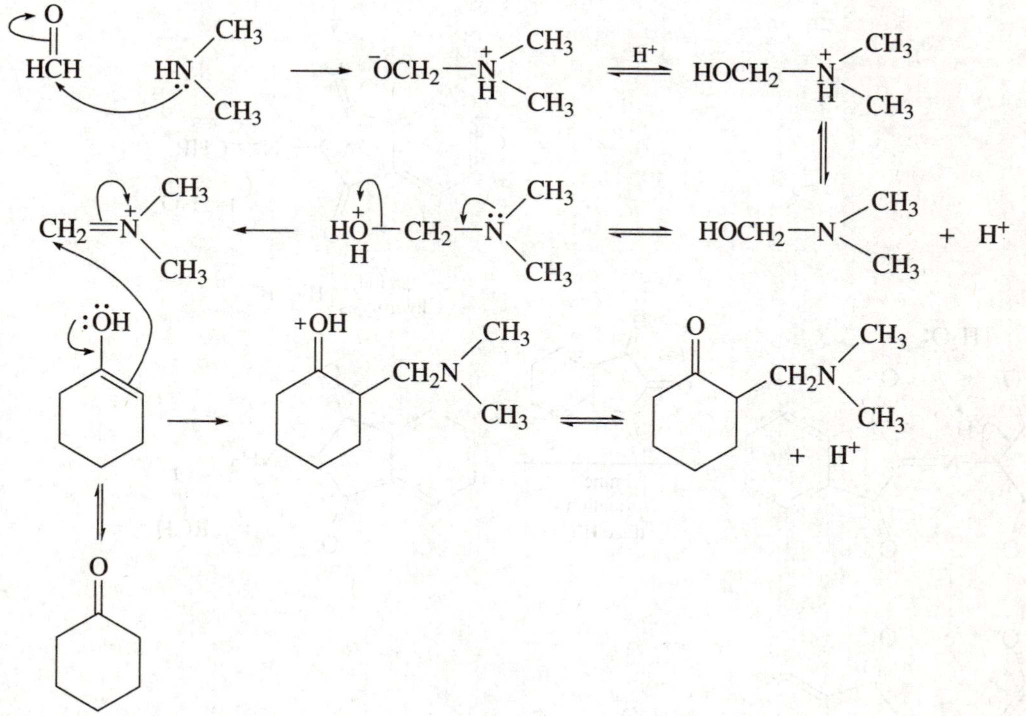

71. The compound that gives the ^{1}H NMR spectrum is 4-phenyl-3-propen-2-one. The singlet at 2.3 ppm is the methyl group, the doublet at ~ 6.7 ppm and the doublet buried under the benzene ring protons are the hydrogens bonded to the sp^2 carbons. The compounds that would form this compound (via an aldol condensation) are benzaldehyde and acetone.

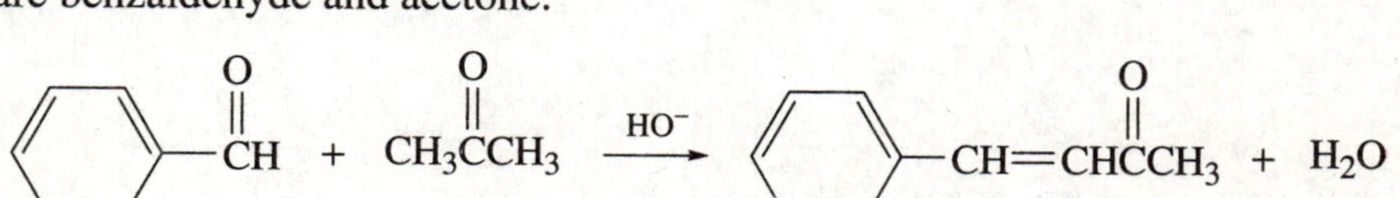

4-phenyl-3-buten-2-one

72. The middle carbonyl group is hydrated because it is stabilized by the electron-withdrawing carbonyl groups on either side of it.

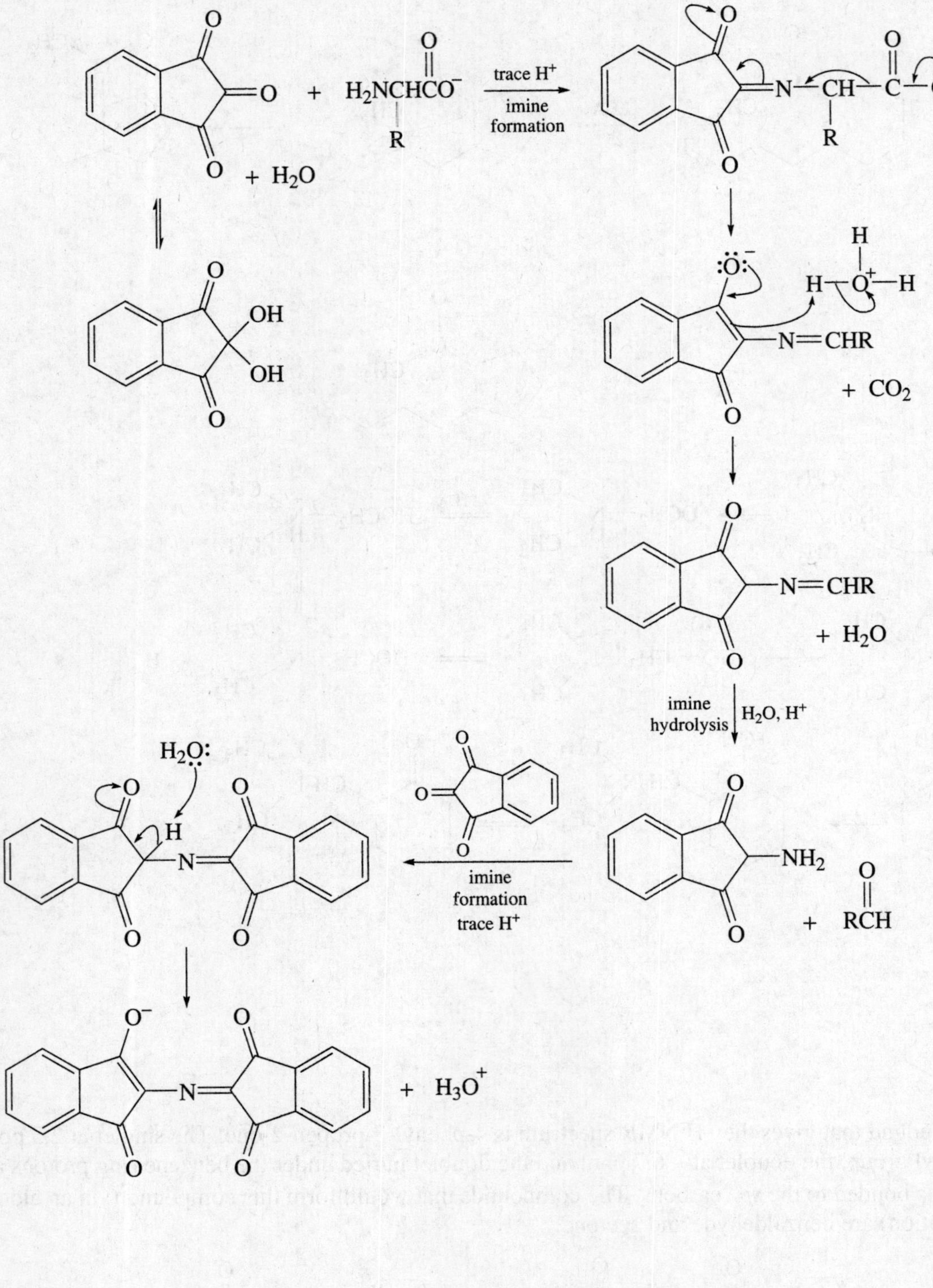

73.

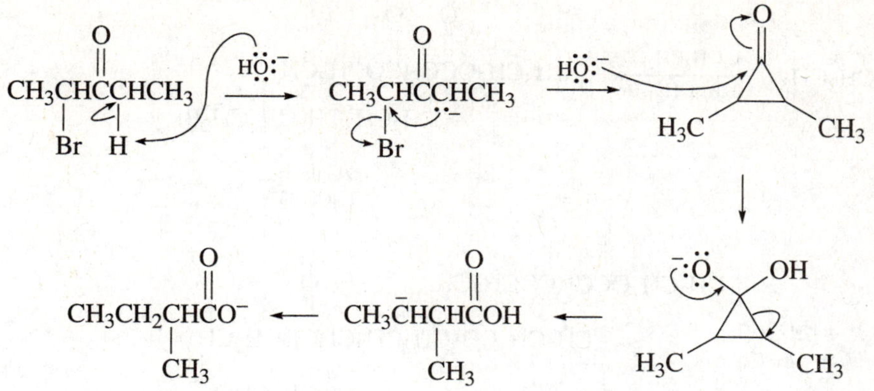

74.

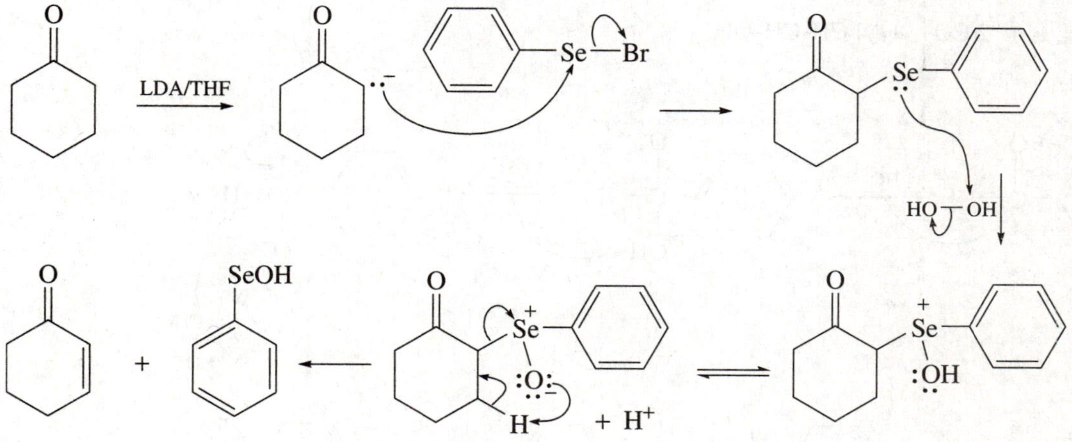

75. **a.**

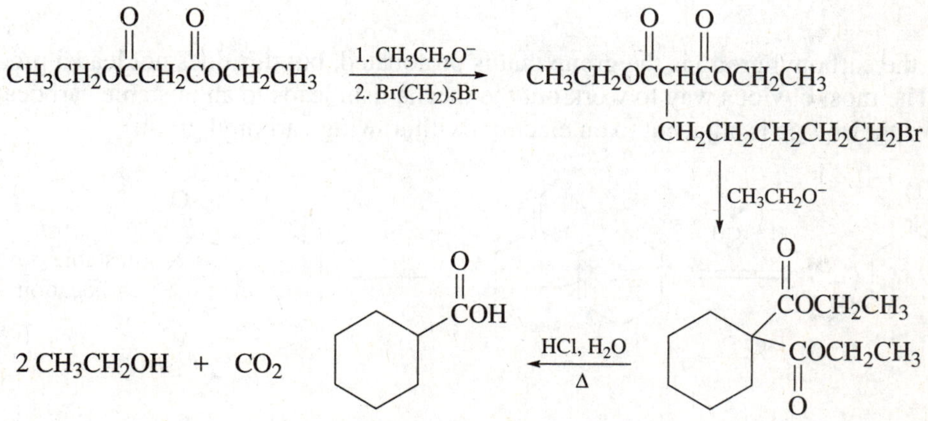

b.

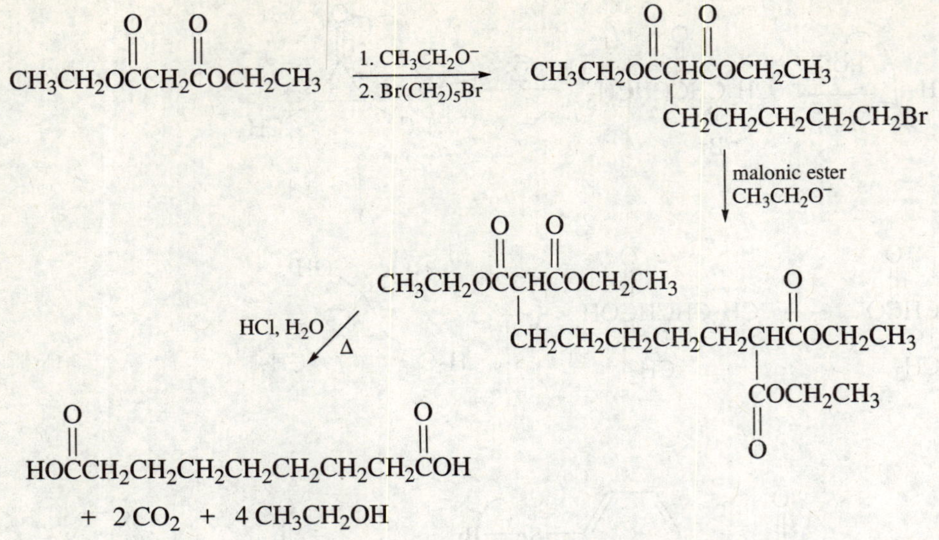

76.

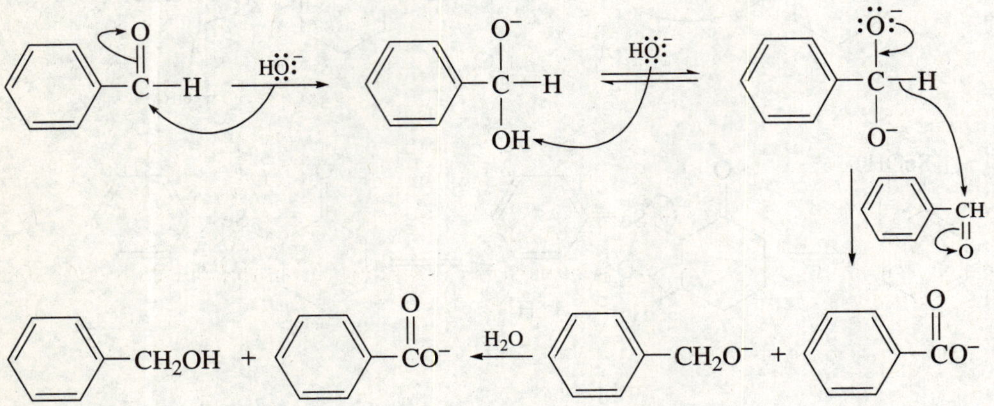

77. **a.** Generally the carbonyl group is the group that is protonated, but this does not lead a productive reaction. The most obvious way to work out the mechanism leads to an unstable carbocation, since the positive the charge is adjacent to an electron-withdrawing carbonyl group.

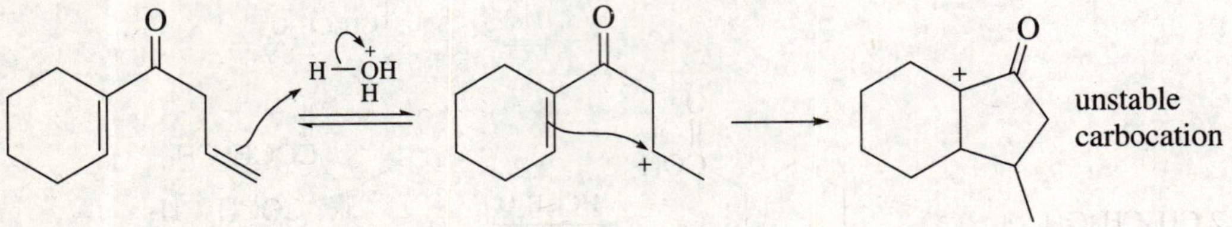

The following mechanism avoids the formation of the unstable carbocation

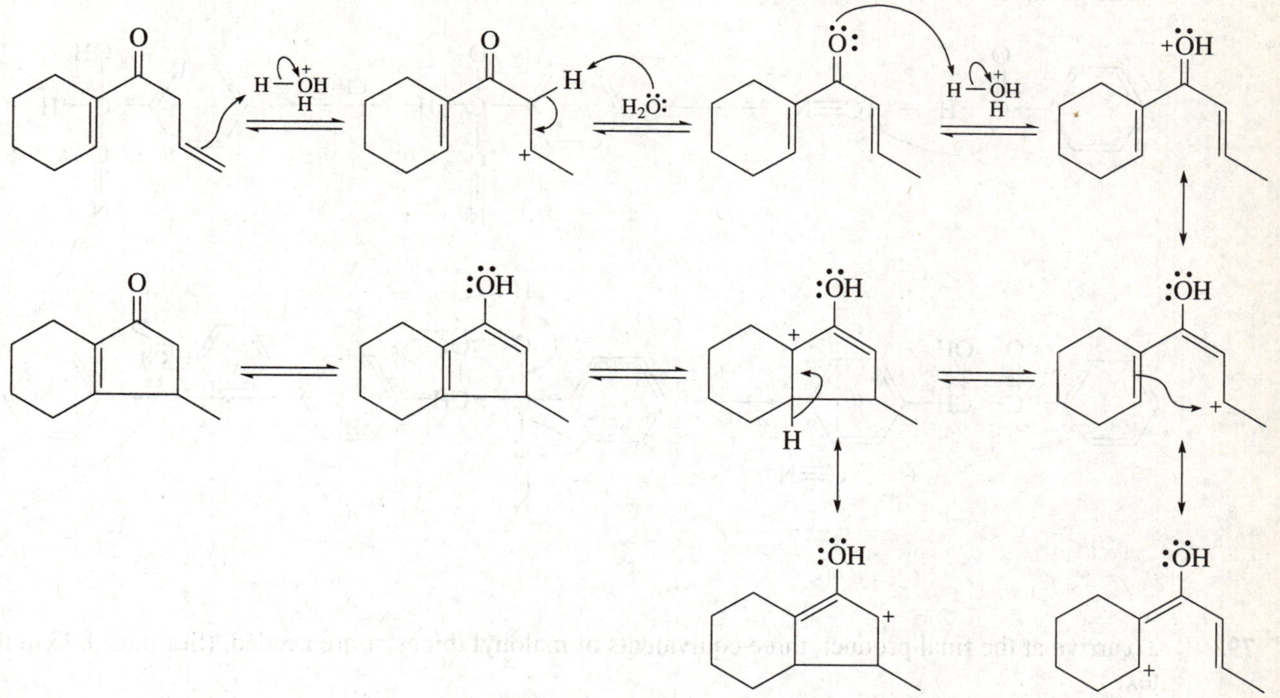

b. The carbonyl group on the left is protonated because electron delocalization causes it to be more basic than the other carbonyl group.

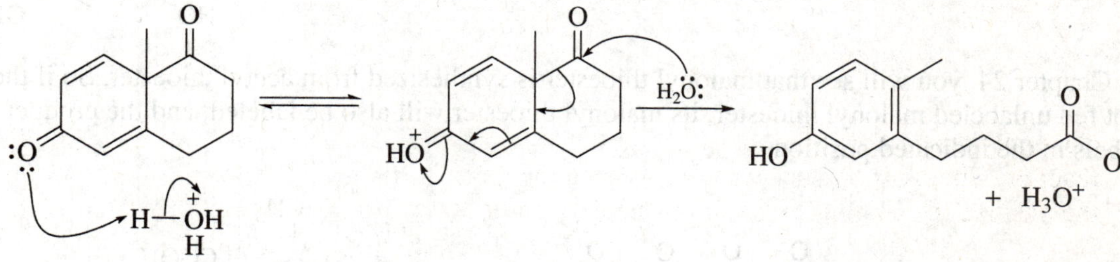

78. Hydroxide ion cannot be used in place of cyanide ion because the aldehydic hydrogen cannot be removed unless the electrons left behind can be delocalized onto an electronegative atom. The nitrogen of the cyano group serves that purpose.

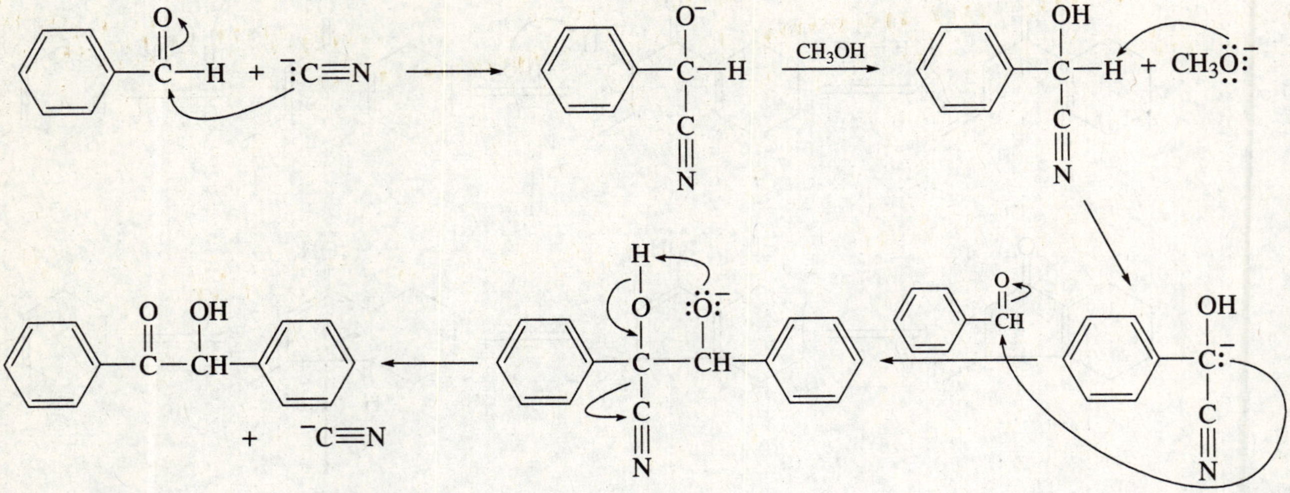

79. To arrive at the final product, three equivalents of malonyl thioester are needed. (See page 893 in the text.)

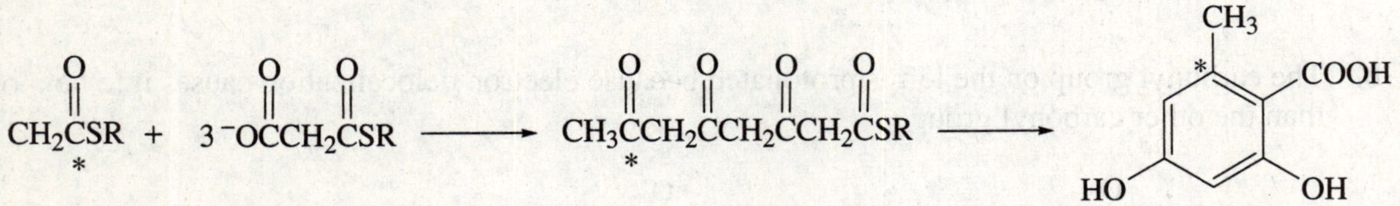

In Chapter 24, you will see that malonyl thioester is synthesized from acetyl thioester. So if the lichen is not fed unlabeled malonyl thioester, its malonyl thioester will also be labeled, and the product will have labels in the indicated positions.

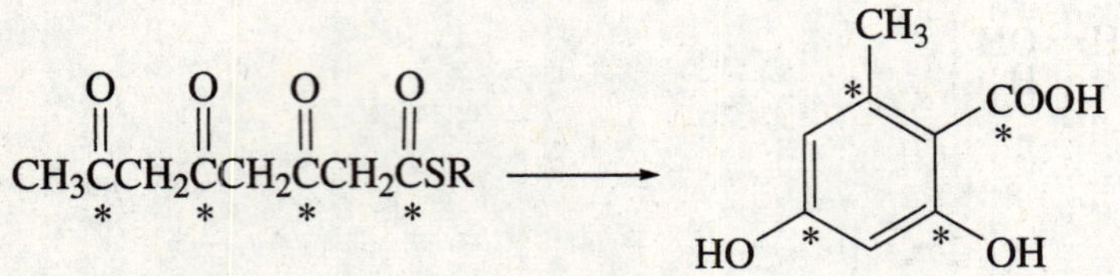

80.

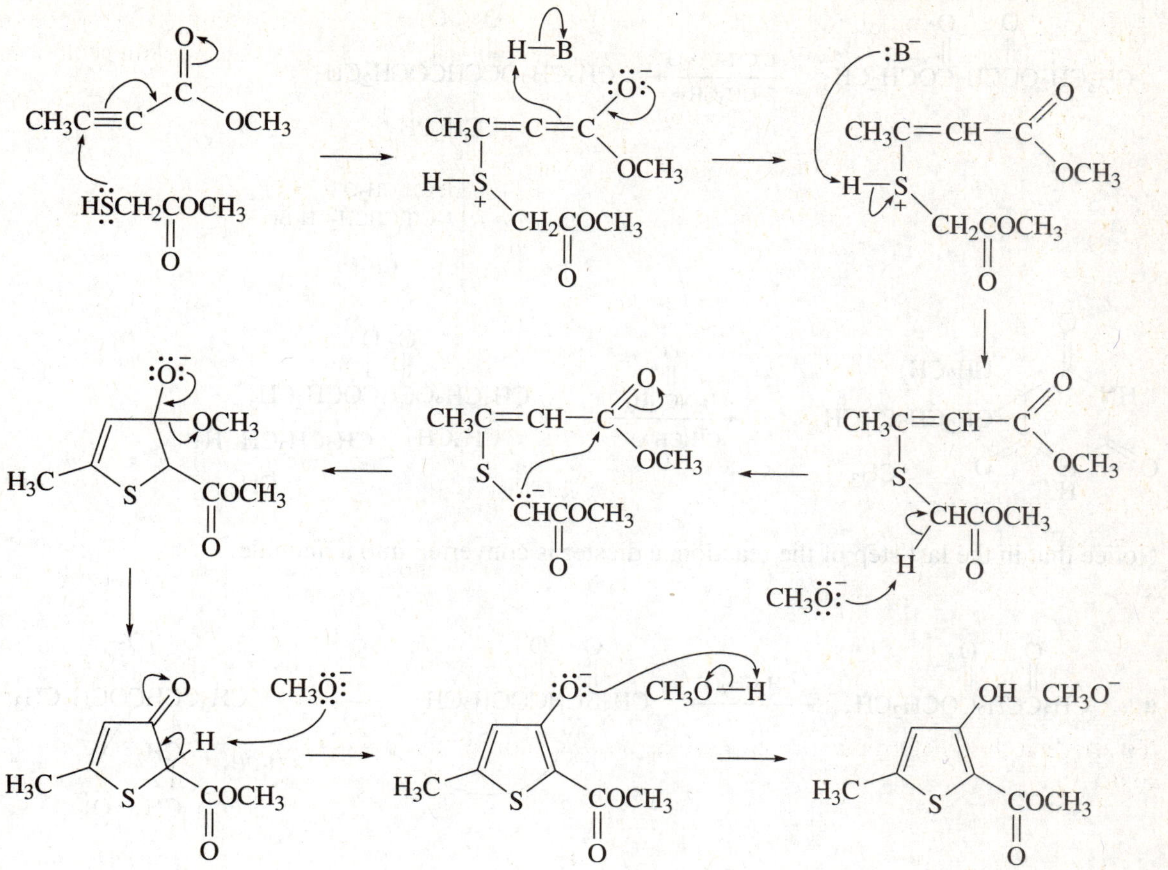

81.

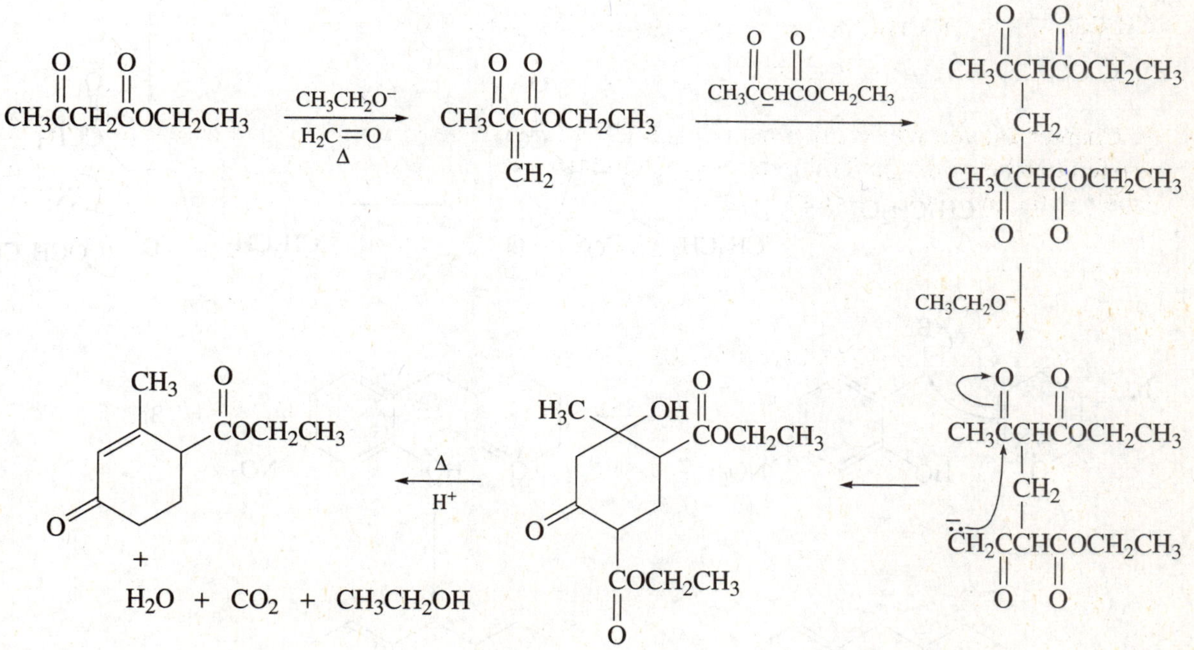

82.

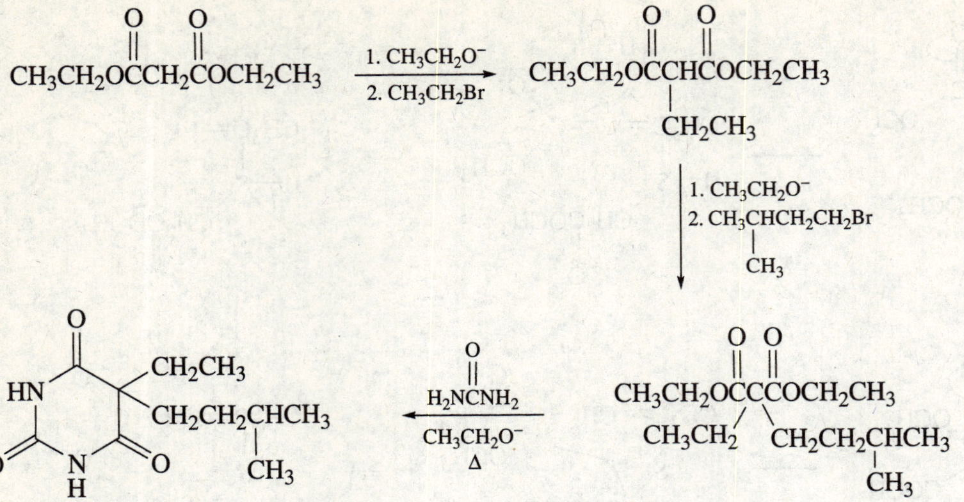

Notice that in the last step of the reaction, a diester is converted into a diamide.

83. **a.**

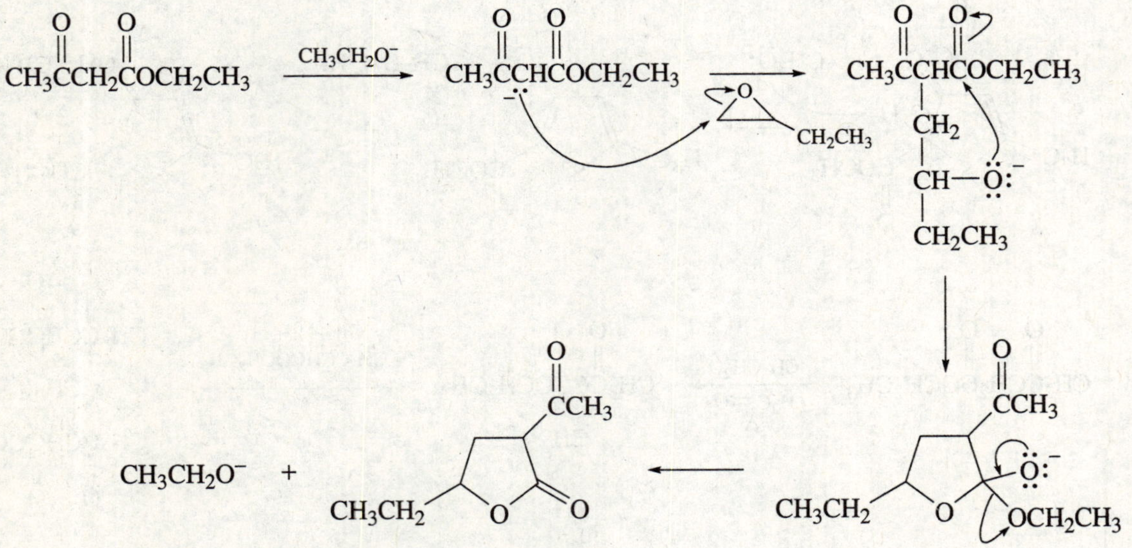

b.

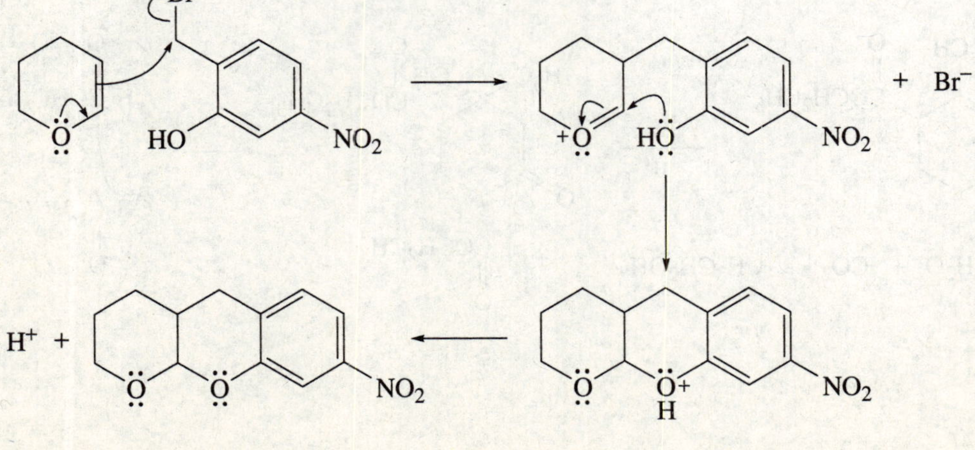

84. **a.**

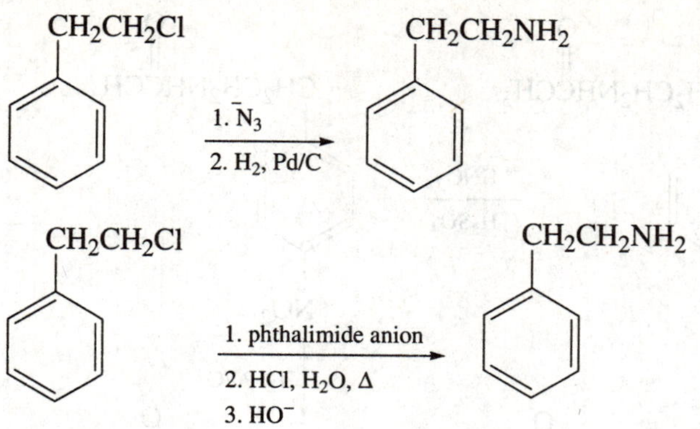

b.

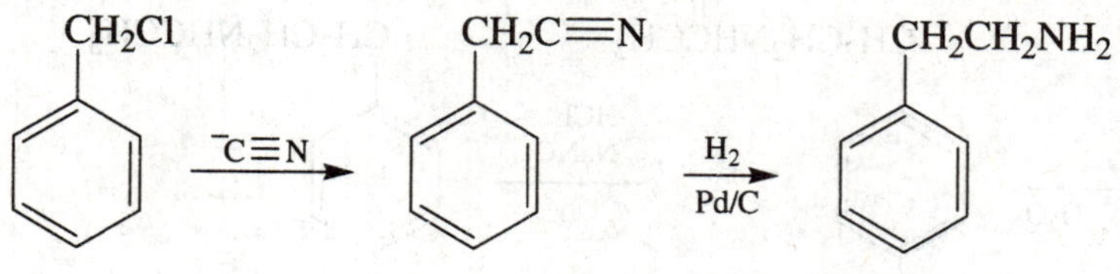

c.

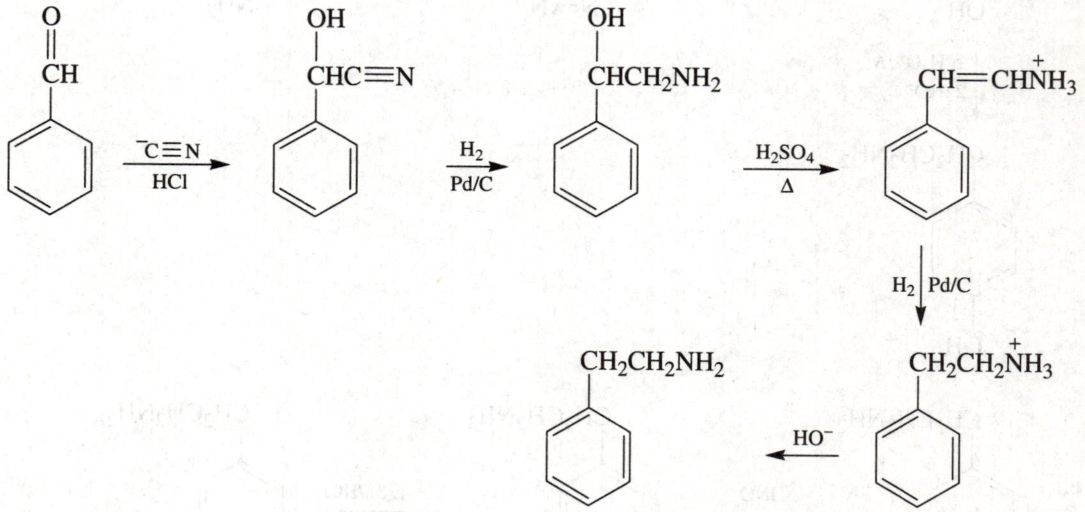

or

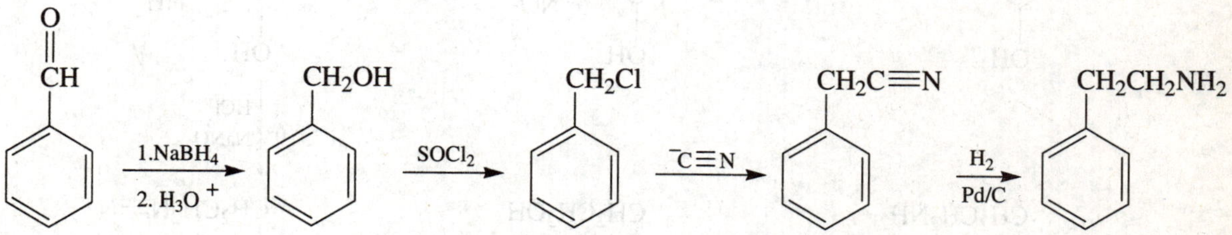

d.

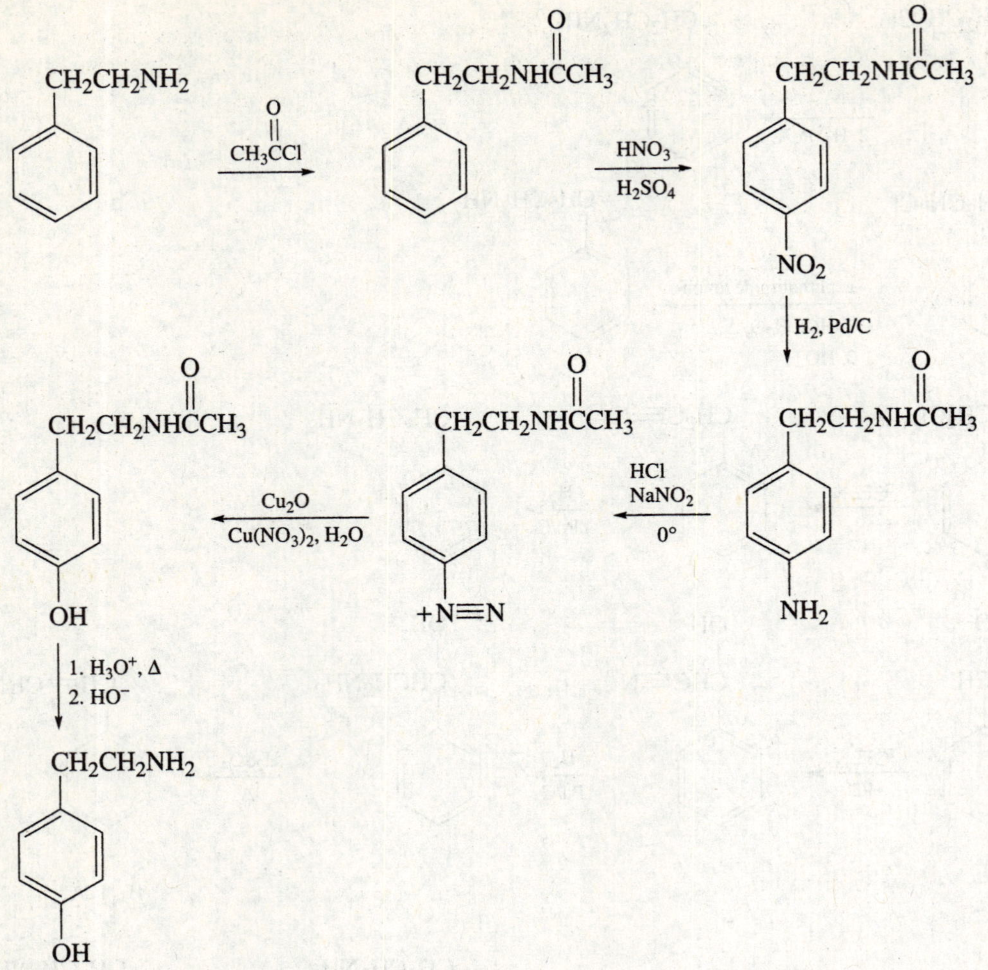
CH2CH2NH2
O
CH3CCl
O
CH2CH2NHCCH3
HNO3
H2SO4
O
CH2CH2NHCCH3
NO2
H2, Pd/C
O
CH2CH2NHCCH3
NH2
HCl
NaNO2
0°
O
CH2CH2NHCCH3
+N≡N
Cu2O
Cu(NO3)2, H2O
O
CH2CH2NHCCH3
OH
1. H3O+, Δ
2. HO−
CH2CH2NH2
OH

e.

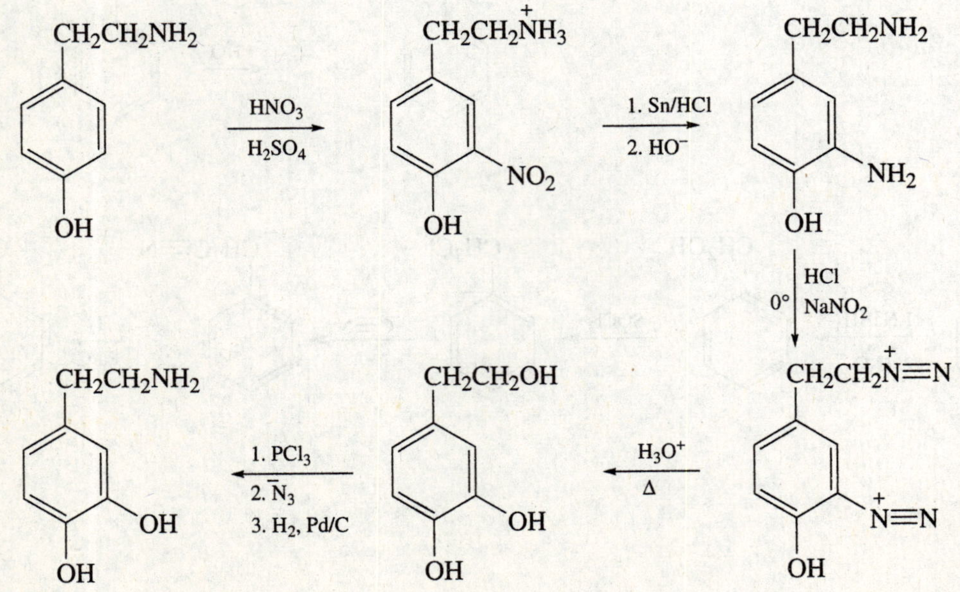
CH2CH2NH2
OH
HNO3
H2SO4
+
CH2CH2NH3
NO2
OH
1. Sn/HCl
2. HO−
CH2CH2NH2
NH2
OH
HCl
0° NaNO2
+
CH2CH2N≡N
+
N≡N
OH
H3O+
Δ
CH2CH2OH
OH
OH
1. PCl3
2. −N3
3. H2, Pd/C
CH2CH2NH2
OH
OH

85. The substituent in ibuprofen is placed on the ring by a Friedel-Crafts acylation reaction. A Friedel-Crafts reaction cannot be done in the synthesis of ketoprofen because the benzene rings are deactivated and deactivated rings cannot undergo Friedel-Crafts reactions.

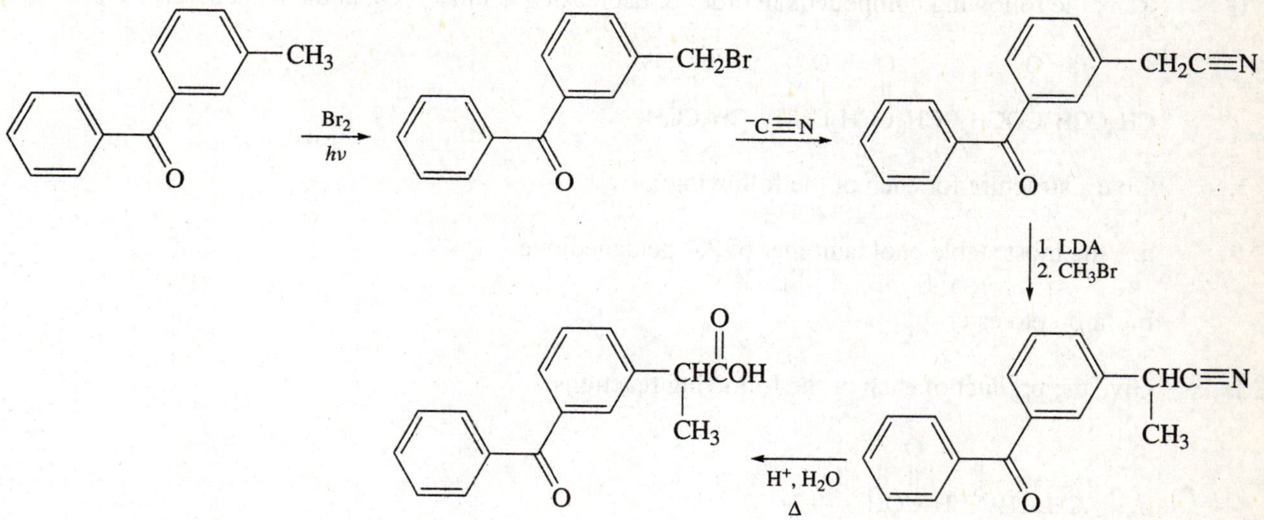

Chapter 18 Practice Test

1. Rank the following compounds in order of decreasing acidity. (Label the most acidic #1.)

$CH_3C(=O)CH_2C(=O)OCH_3$ $CH_3C(=O)CH_2C(=O)CH_3$ $CH_3C(=O)CH_3$

2. Give a structure for each of the following:

a. the most stable enol tautomer of 2,4-pentanedione

b. a *β*-keto ester

3. Give the product of each of the following reactions.

a. $CH_3CH_2C(=O)CH_2C(=O)OH \xrightarrow{\Delta}$

b. $CH_3CH_2CH_2C(=O)H \xrightarrow{Br_2,\ HO^-}$

c. $2\ CH_3CH_2CH_2C(=O)OCH_3 \xrightarrow[2.\ H_3O^+]{1.\ CH_3O^-}$

d. $CH_2{=}CHC(=O)CH_3 + CH_3C(=O)CH_2C(=O)CH_3 \xrightarrow{HO^-}$

e.

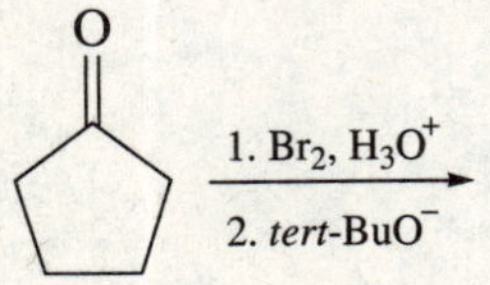

f.

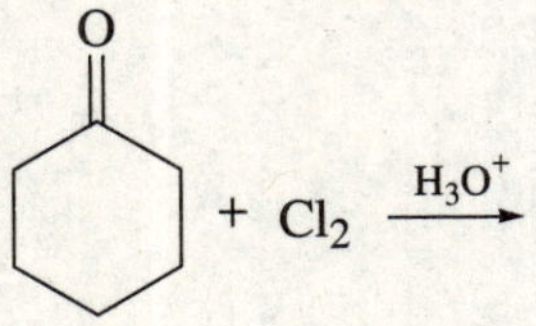

g. $CH_3CH_2OC(=O)CH_2C(=O)OCH_2CH_3 \xrightarrow{1.\ CH_3CH_2O^-;\ 2.\ CH_3CH_2CH_2Br;\ 3.\ HCl,\ H_2O,\ \Delta}$

4. Give an example of each of the following:

a. an aldol addition

b. an aldol condensation

c. a Claisen condensation

d. a Dieckmann condensation

e. a malonic ester synthesis

f. an acetoacetic ester synthesis

5. Give the products of the following crossed aldol addition:

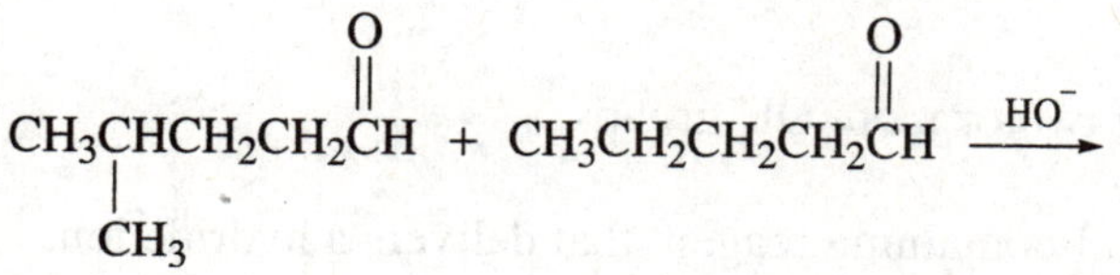

6. What ester would be required to prepare each β-ketoester?

a.

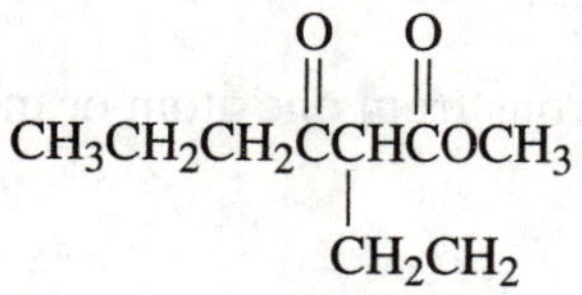

b.

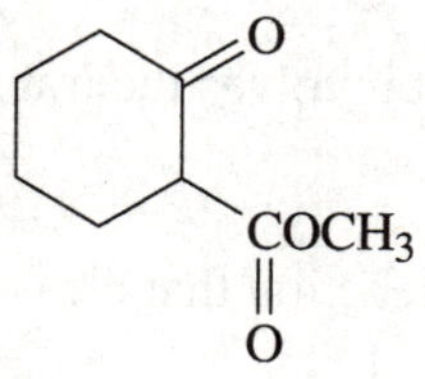

CHAPTER 19
More About Oxidation-Reduction Reactions

Important Terms

Baeyer-Villiger oxidation	oxidation of aldehydes or ketones with H_2O_2 to form carboxylic acids or esters, respectively.
catalytic hydrogenation	addition of hydrogen to a double or triple bond in the presence of a metal catalyst.
chemoselective reaction	a reaction in which a reagent reacts with one functional group in preference to another.
dissolving metal reduction	a reduction using sodium or lithium metal dissolved in liquid ammonia.
enantioselective reaction	a reaction that forms an excess of one enantiomer.
functional group interconversion	a reaction that converts one functional group into another.
glycol	a compound containing two or more OH groups.
metal hydride reduction	a reduction using a metal-containing reagent that delivers a hydride ion.
molozonide	an unstable intermediate containing a five-membered ring with three oxygens in a row that is formed from the reaction of an alkene with ozone.
oxidation	loss of electrons by an atom or molecule.
oxidation-reduction reaction	a reaction that involves the transfer of electrons from one atom or molecule to another.
oxidative cleavage	an oxidation reaction that cleaves the reactant into two or more compounds.
oxidizing agent	the compound that is reduced in a redox reaction as it oxidizes the other compound.
ozonide	a five-membered ring compound formed as a result of rearrangement of a molozonide.
ozonolysis	reaction of a carbon-carbon double or triple bond with ozone.
peroxyacid	a carboxylic acid with an OOH group instead of an OH group.
redox reaction	an oxidation-reduction reaction.
reducing agent	the compound that is oxdized in a redox reaction as it reduces the other compound.
reduction	gain of electrons by an atom or molecule.

Rosenmund reduction the reduction of an acyl chloride to an aldehyde using H_2/Lindlar catalyst.

Swern oxidation uses dimethyl sulfoxide, oxalyl chloride, and triethylamime to oxidize primary alcohols to aldehydes and secondary alcohols to ketones.

Tollens test a test to determine the presence of an aldehyde. A positive result is indicated by the formation of a silver mirror.

vicinal diol (vicinal glycol) a 1,2-diol.

Solutions to Problems

1. **a.** reduction **b.** neither **c.** oxidation **d.** oxidation **e.** reduction **f.** neither

2. **a.** $CH_3CH_2CH_2CH_2CH_2OH$

b. $CH_3CH_2CH_2CH_2NH_2$

c. 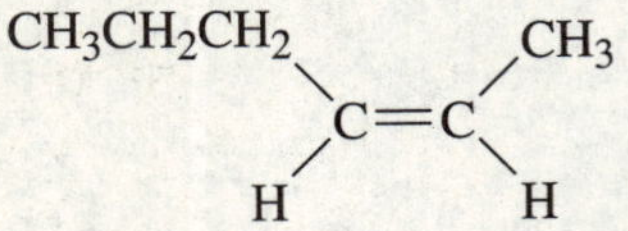

d. 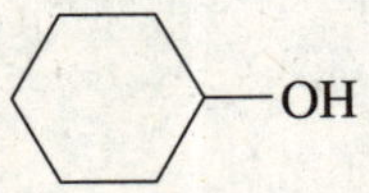

e. no reaction

f. CH_3CH_2OH

g. 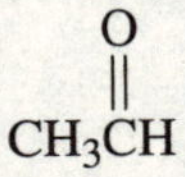

h. 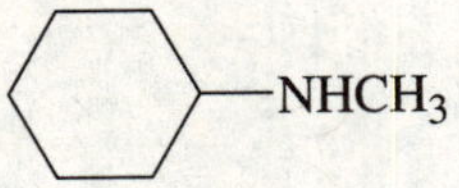

3. A terminal alkyne cannot be reduced by Na, because Na reacts with a terminal alkyne to form an acetylide ion rather than a radical anion (Section 6.9).

$$2\,CH_3C{\equiv}CH + 2\,Na \longrightarrow 2\,CH_3C{\equiv}C^- + 2\,Na^+ + H_2$$

4. **a.** 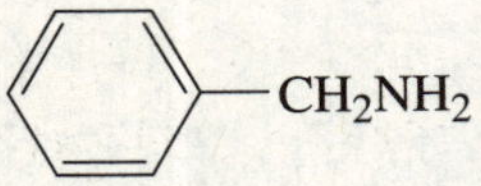

b. 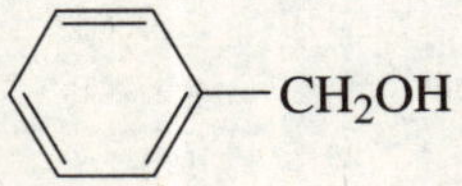

c. 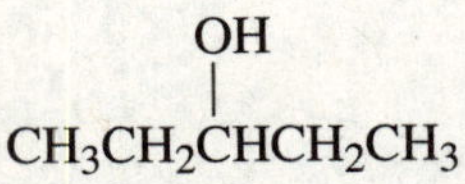

d. 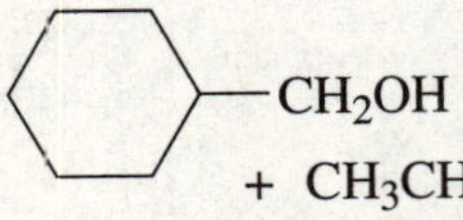 + CH_3CH_2OH

e. $CH_2CH_2CH_2NHCH_2CH_3$

f. $CH_3CH_2CH_2CH_2OH$

5. Carbon-nitrogen double and triple bonds can be reduced by $LiAlH_4$ because the bonds are polar. The hydride ion will be attracted to the partially positively charged carbon.

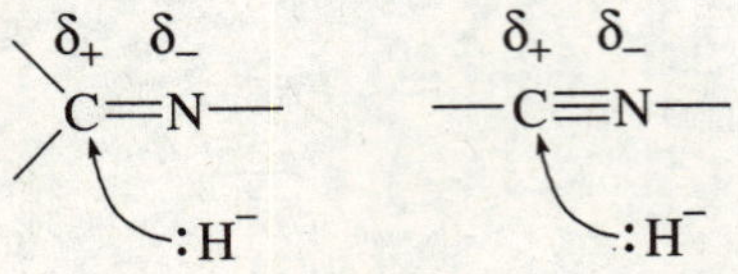

6. **a.** 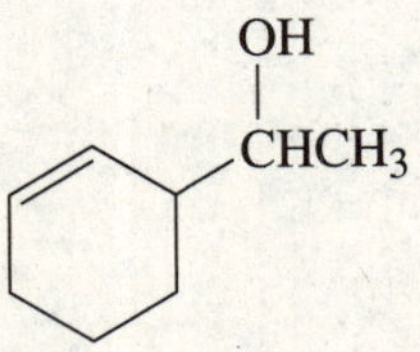

b. 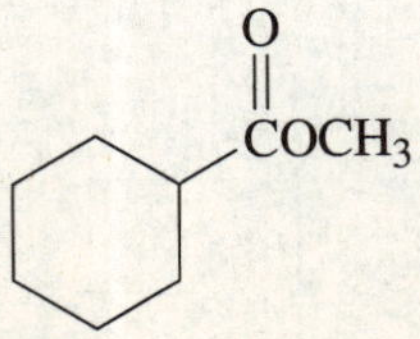

c. 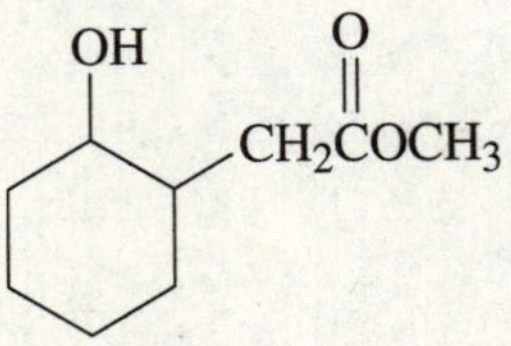

d. 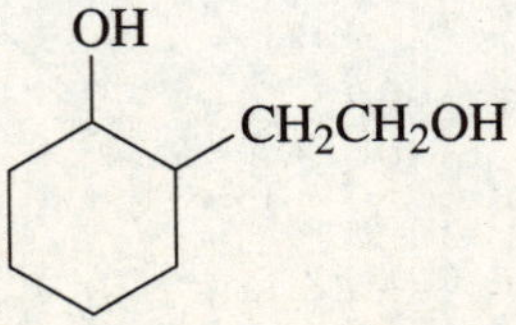 + CH_3OH

7. **a.** solved in the text.

b.

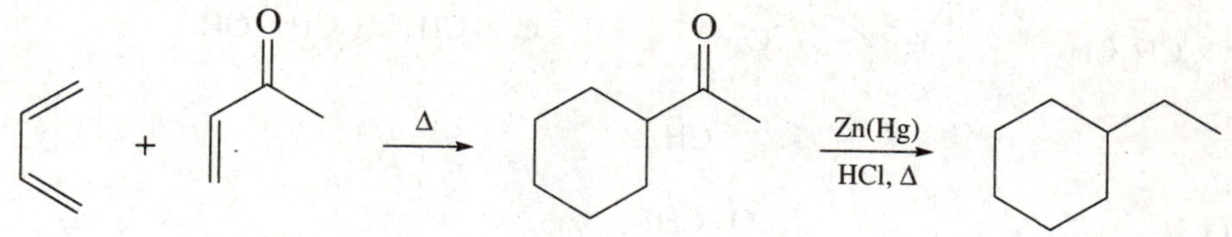

8. **a.**

1. $CH_3CH_2\overset{O}{\overset{\|}{C}}CH_2CH_3$

2. $CH_3CH_2CH_2CH_2\overset{O}{\overset{\|}{C}}OH$

3. A tertiary alcohol is not oxdized to a carbonyl compound.

4. $CH_3\overset{O}{\overset{\|}{C}}CH_2\overset{O}{\overset{\|}{C}}CH_2CH_3$

5. 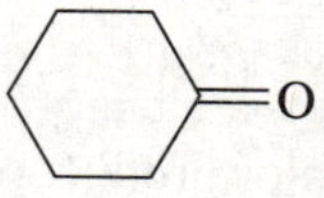

6. $HO\overset{O}{\overset{\|}{C}}CH_2CH_2\overset{O}{\overset{\|}{C}}OH$

b.

1. $CH_3CH_2\overset{O}{\overset{\|}{C}}CH_2CH_3$

2. $CH_3CH_2CH_2CH_2\overset{O}{\overset{\|}{C}}H$

3. A tertiary alcohol is not oxdized to a carbonyl compound.

4. $CH_3\overset{O}{\overset{\|}{C}}CH_2\overset{O}{\overset{\|}{C}}CH_2CH_3$

5. 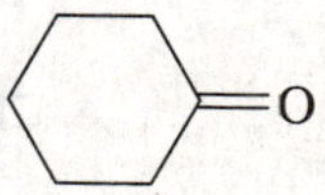

6. $H\overset{O}{\overset{\|}{C}}CH_2CH_2\overset{O}{\overset{\|}{C}}H$

9. Chromic acid needs to be protonated to convert a poor leaving group (HO^-) into a good leaving group (H_2O). The alcohol displaces water to form a protonated chromate ester, which loses a proton. The aldehyde is then formed in an elimination reaction.

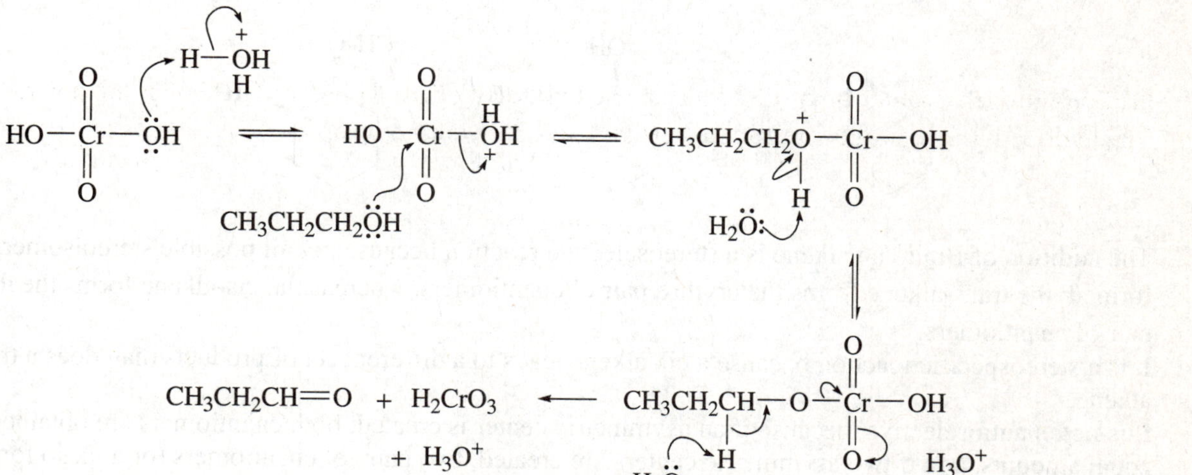

10. **a.** 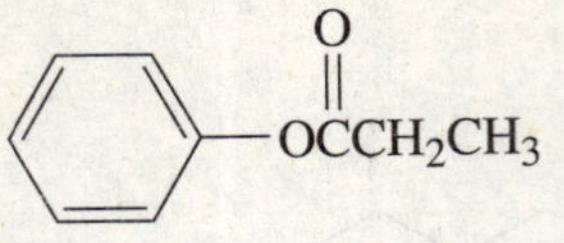 **c.** 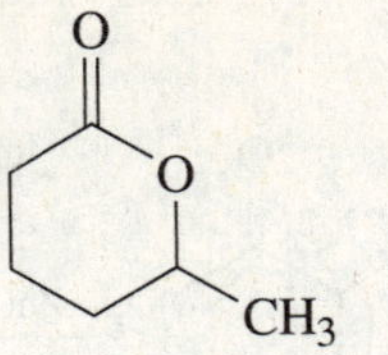 **e.** 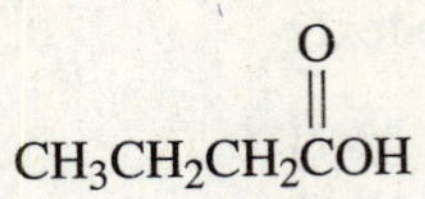

b. 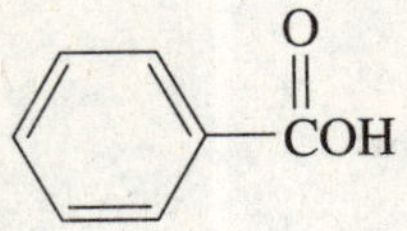 **d.** 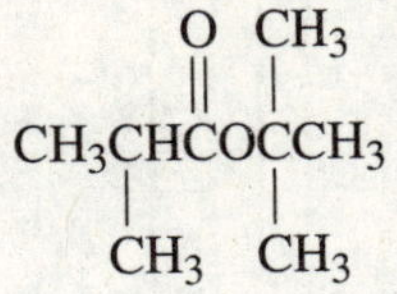 **f.** no reaction

11. The major product results from attack of the Grignard reagent on the less sterically hindered carbon of the epoxide (the one bonded to a hydrogen and to a methyl group). It will be easier to answer this question if you use molecular models.

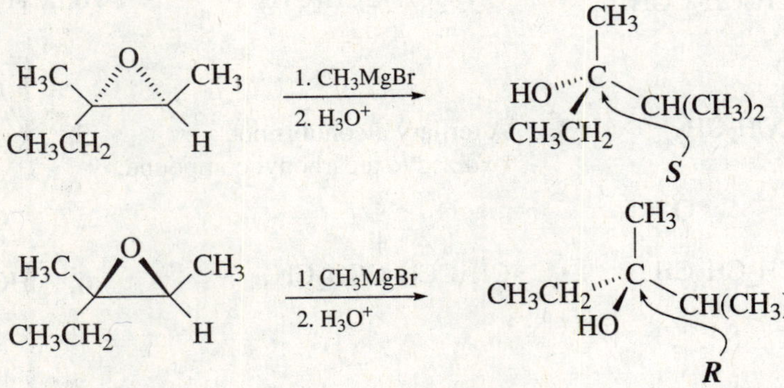

12. Methoxide ion attacks a carbon of the three-membered ring from the backside. (It is an S_N2 reaction.) Either of the two ring carbons can be attacked. Therefore, the products are the trans stereoisomers.

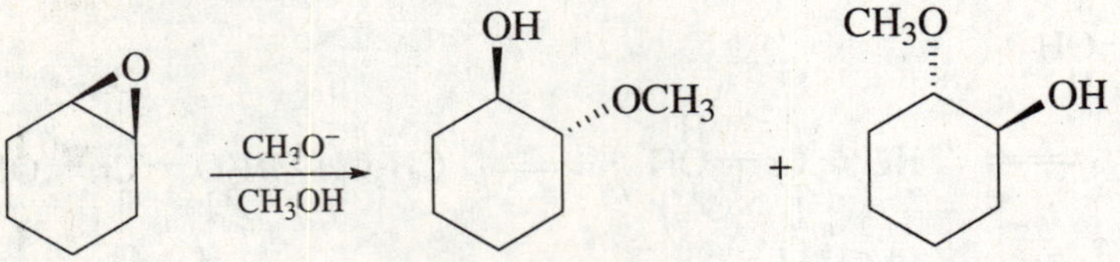

13. The addition of Br_2 to an alkene is a stereoselective reaction because not all possible stereoisomers are formed: the trans-alkene forms the erythro pair of enantiomers, whereas the cis-alkene forms the threo pair of enantiomers.
It is a stereospecific reaction because a cis-alkene leads to a different set of products than does a trans-alkene.
It is not enantioselective, because if an asymmetric center is created, both enantiomers are obtained in equal amounts, and if two asymmetric centers are created, two pairs of enantiomers (or a meso form and a pair of enantiomers) are obtained.

14. **a.** 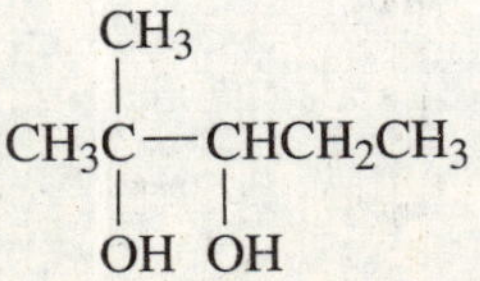 **b.** 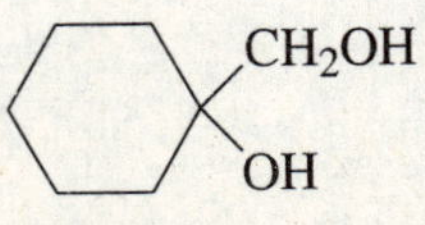

15. **a.** Syn addition to the trans isomer forms the threo pair of enantiomers. (See Section 5.19)

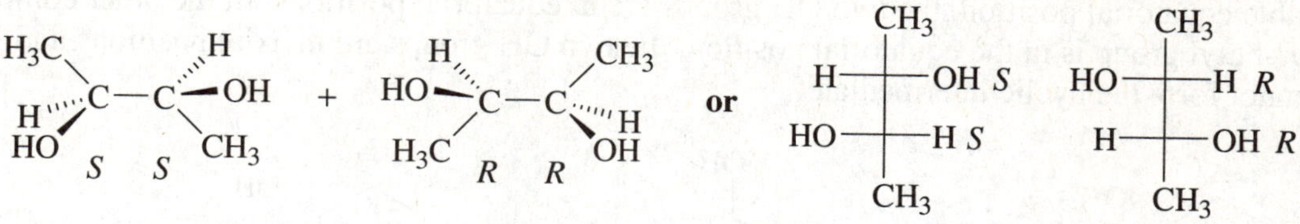

b. Syn addition to the cis isomer forms the erythro pair of enantiomers. In this case, the product is a meso compound because each asymmetric center is bonded to the same four substituents, so only one stereoisomer is formed.

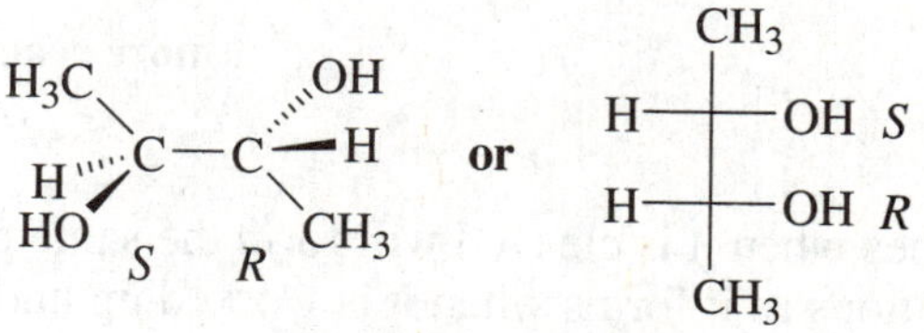

c. Syn addition to the cis isomer forms the erythro pair of enantiomers.

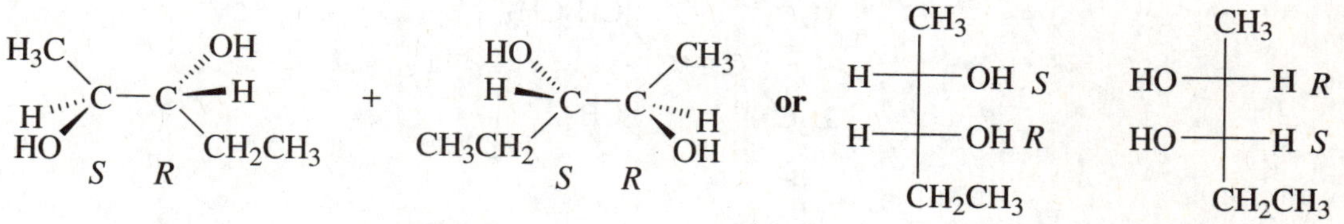

d. Syn addition to the trans isomer forms the threo pair of enantiomers.

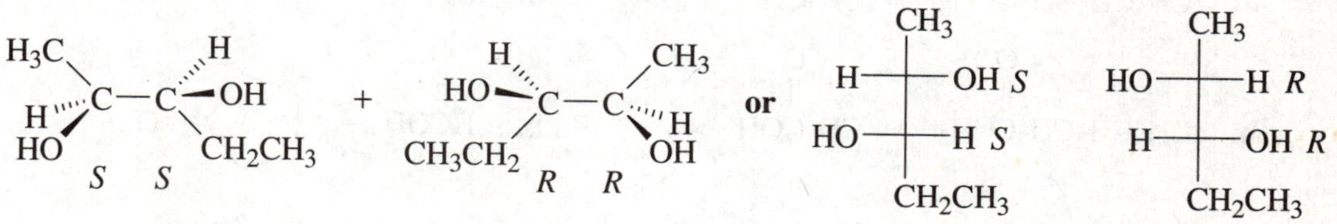

16. The fact that only one ketone is obtained means that the alkene must be symmetrical. The cyclic ketone that is obtained is cyclohexanone. Therefore, the alkene is

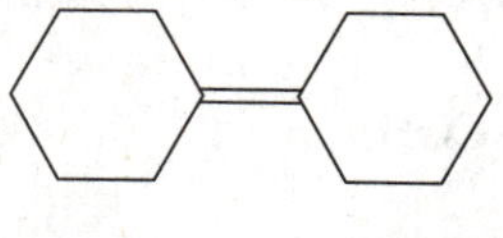

17. **a.** The compound on the left is more reactive because when the bulky *tert*-butyl group is in the more stable equatorial position, the two OH groups are in equatorial positions. In the other compound, the *tert*-butyl group is in the equatorial position, the two OH groups are in axial positions and, therefore, are too far away one another to form the cyclic intermediate

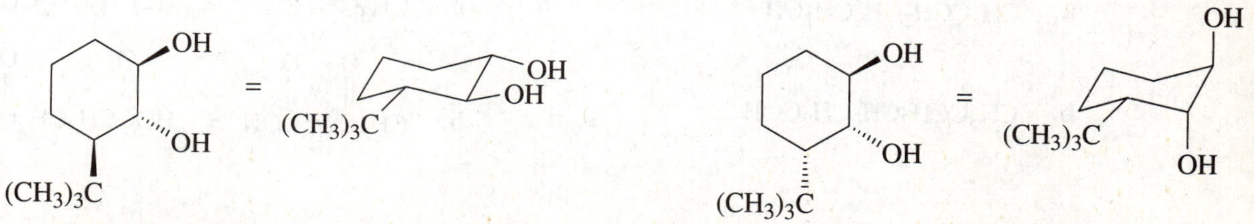

more reactive **less reactive**

b. The compound on the right is more reactive because when the bulky *tert*-butyl group is in the more stable equatorial position, the two OH groups are in equatorial positions. In the other compound, the *tert*-butyl group is in the equatorial position, the two OH groups are in axial positions and, therefore, cannot form the cyclic intermediate

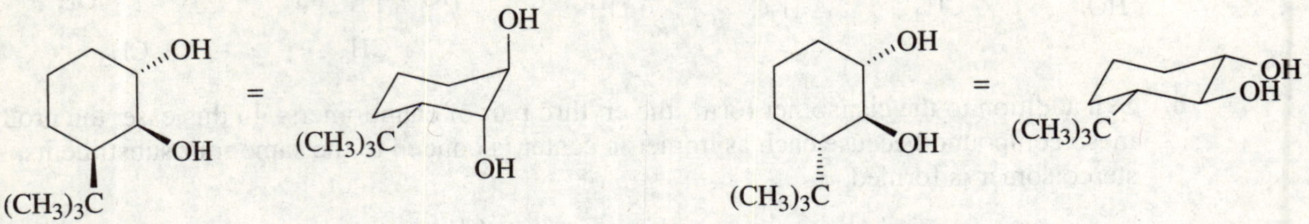

less reactive **more reactive**

18. Any alkene that forms two ketones when it is cleaved will form the same products when the ozonide is worked up under reducing conditions as it forms when it is worked up under oxidizing conditions. In other words, each sp^2 carbon must be bonded to two alkyl groups.

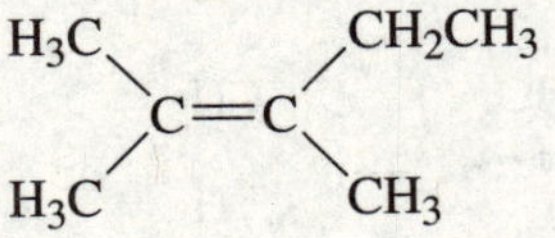

19. **1.** **a.** $CH_3CH_2CH_2C(=O)CH_3$ + $CH_3C(=O)H$

b. $CH_3CH_2CH_2C(=O)CH_3$ + $CH_3C(=O)OH$

2. **a.** $HC(=O)H$ + $CH_3CH_2CH_2CH_2C(=O)H$

b. $HC(=O)OH$ + $CH_3CH_2CH_2CH_2C(=O)OH$

3. **a.** $CH_3C(=O)CH_2CH_2CH_2C(=O)H$

b. $CH_3C(=O)CH_2CH_2CH_2C(=O)OH$

4. **a.** $HC(=O)H$ + 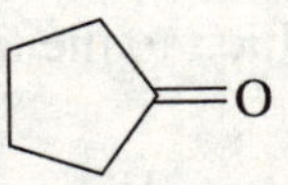

b. $HC(=O)OH$ + 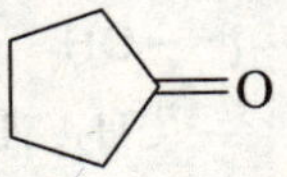

5. **a.** $CH_3CH_2CH_2C(=O)H$ (2 equivalents)

b. $CH_3CH_2CH_2C(=O)OH$ (2 equivalents)

6. **a.** $CH_3C(=O)-C(=O)H$ + $HC(=O)CH_2CH_2C(=O)H$

b. $CH_3C(=O)-C(=O)OH$ + $HOC(=O)CH_2CH_2C(=O)OH$

20. **a.**

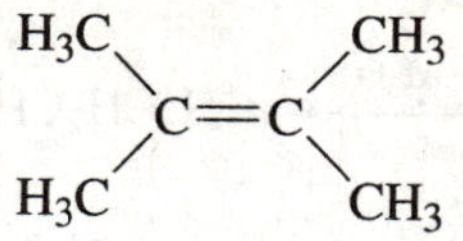

2,3-dimethyl-2-butene

b.

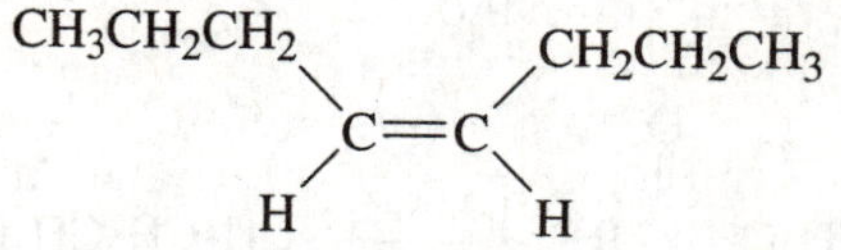

cis-4-octene

or

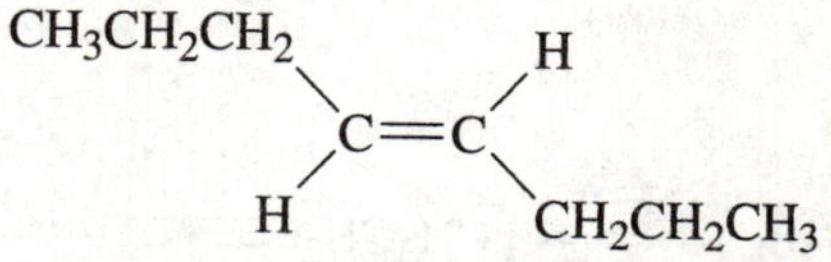

trans-4-octene

21. Solved in the text.

22. It does not tell you whether the double bond has the cis or the trans configuration.

23. Solved in the text.

24. **a.**

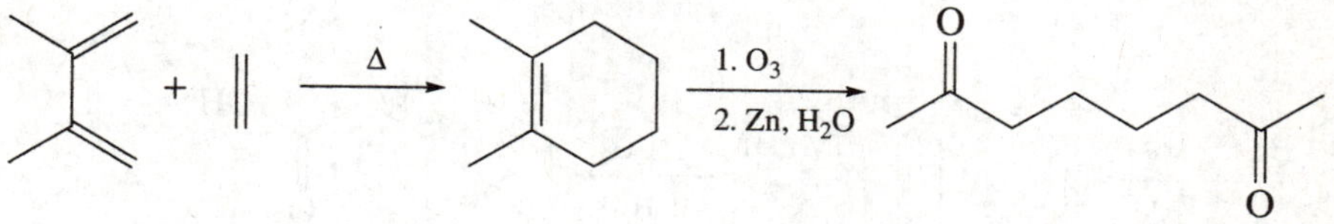

b.

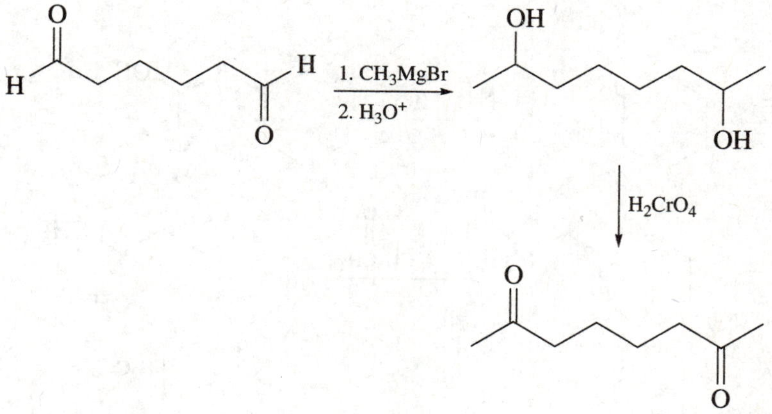

25. **a.**

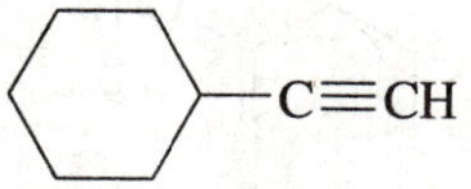

b. $CH_3CH_2C{\equiv}CCH_2CH_2CH_2C{\equiv}CCH_2CH_3$

26. **a.**

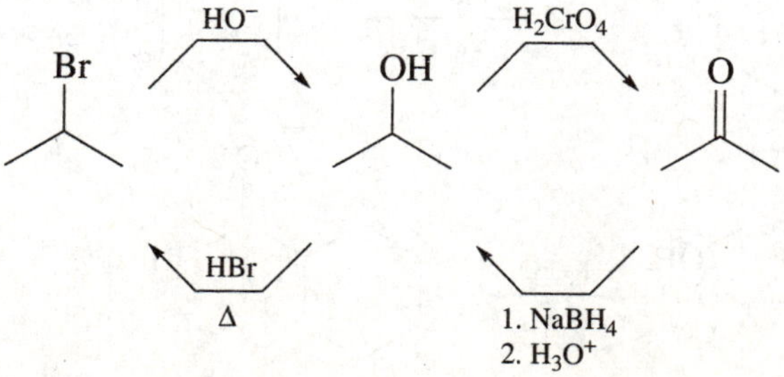

b.

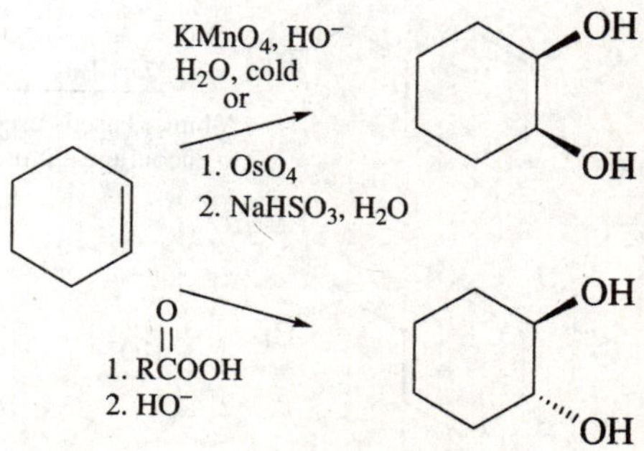

27. **a.**

1. $CH_3CH_2Br \xrightarrow[Et_2O]{Mg} CH_3CH_2MgBr \xrightarrow{H_2C=O} CH_3CH_2CH_2O^- \xrightarrow{H_3O^+} CH_3CH_2CH_2OH$

2. $CH_3CH_2Br \xrightarrow[Et_2O]{Mg} CH_3CH_2MgBr \xrightarrow{CO_2} CH_3CH_2\overset{O}{\overset{\|}{C}}O^- \xrightarrow[2.\ H_3O^+]{1.\ LiAlH_4} CH_3CH_2CH_2OH$

3. $CH_3CH_2Br \xrightarrow[HCl]{^-C\equiv N} CH_3CH_2C\equiv N \xrightarrow[\Delta]{H_3O^+} CH_3CH_2\overset{O}{\overset{\|}{C}}OH \xrightarrow[2.\ H_3O^+]{1.\ LiAlH_4} CH_3CH_2CH_2OH$

b. $CH_3CH_2Br \xrightarrow{^-C\equiv N} CH_3CH_2C\equiv N \xrightarrow[Pd/C]{H_2} CH_3CH_2CH_2NH_2$

c. $CH_3CH_2CH_2Br \xrightarrow{tert\text{-}BuO^-} CH_3CH{=}CH_2 \xrightarrow[2.\ Zn,\ H_2O]{1.\ O_3} CH_3CH{=}O \xrightarrow[2.\ H_2\ \text{Raney Nickel}]{1.\ NH_3} CH_3CH_2NH_2$

28. **a.**

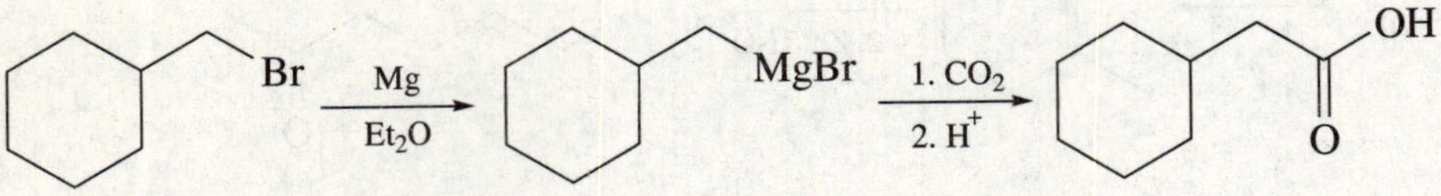

or

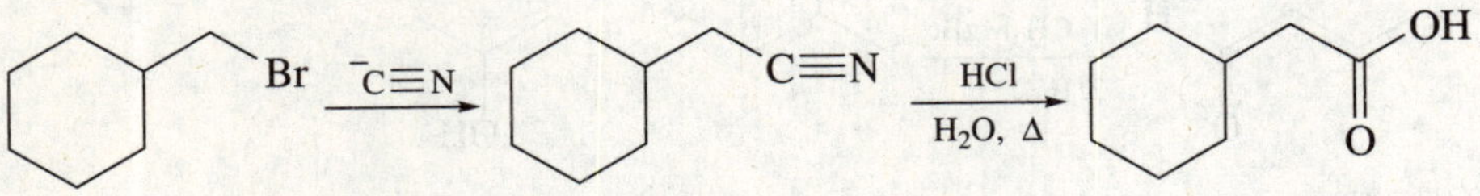

b.

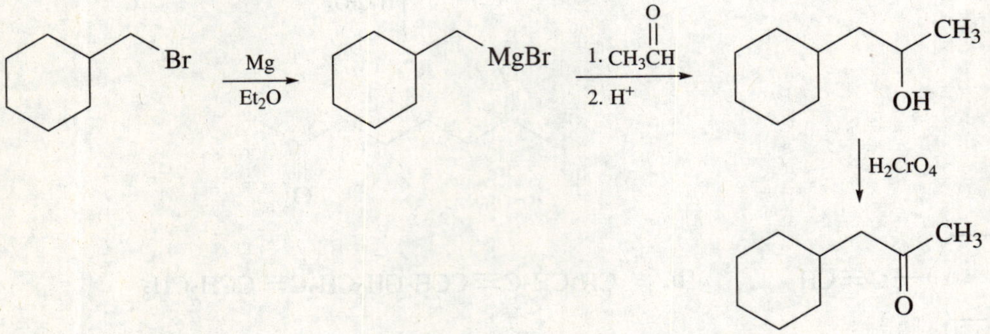

c.

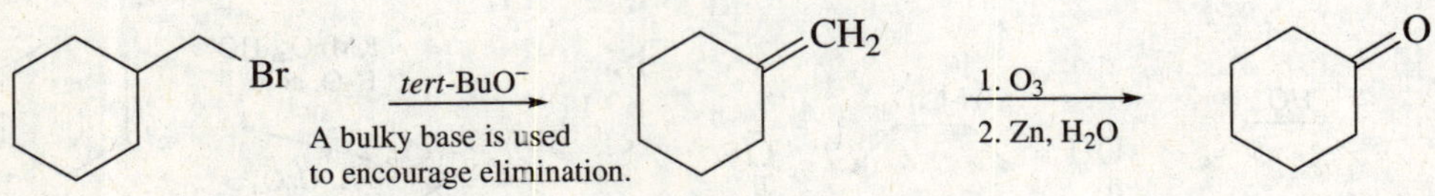

d.

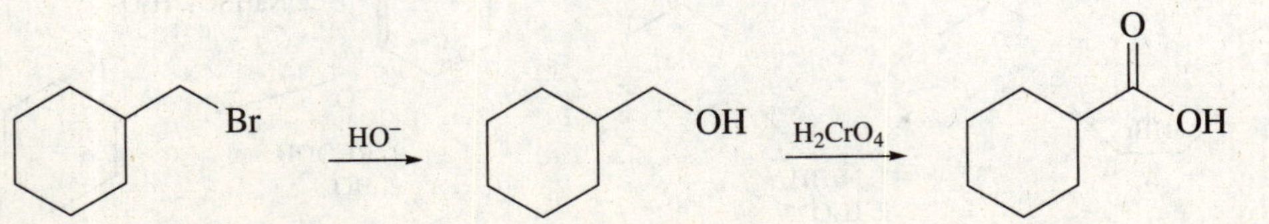

29. aldehyde
carboxylic acid
ester
acyl halide
acid anhydride
alkene
alkyl halide
ether
epoxide

30. **a.** oxidized **b.** reduced **c.** reduced **d.** oxidized **e.** oxidized **f.** oxidized **g.** reduced

31. **a.** $CH_3CH_2CH_2CH_2\overset{\overset{\textstyle O}{\|}}{C}OH$
oxidation

b. 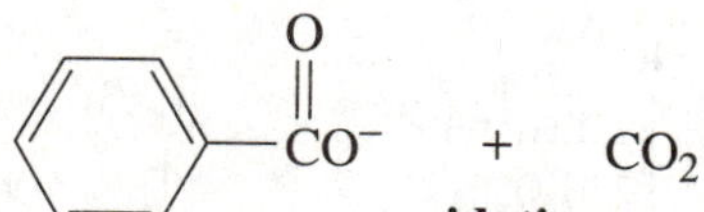
oxidation

c. $CH_3CH_2CH_2\overset{\overset{\textstyle O}{\|}}{C}H$ **reduction**

d. $CH_3CH_2CH_2CH_2OH$
Formation of the aldehyde after steps 1and 2 is neither oxidation or reduction.
Formation of the alcohol after steps 3 and 4 is a **reduction.**

e. $CH_3CH_2\overset{\overset{\textstyle O}{\|}}{C}H$ **oxidation**

f. $CH_3CH_2CH_2CH_2NHCH_3$ **reduction**

g. 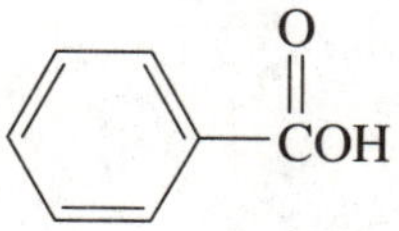 **oxidation**

h. 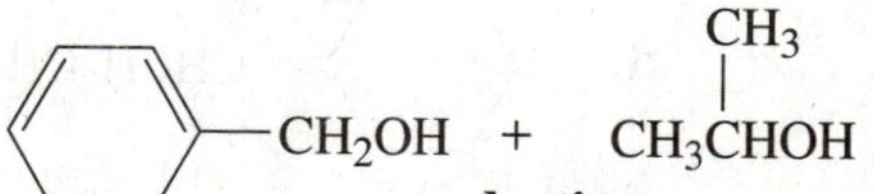
reduction

i. 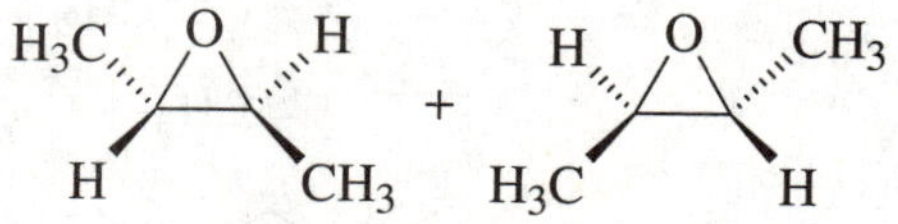
oxidation

j. 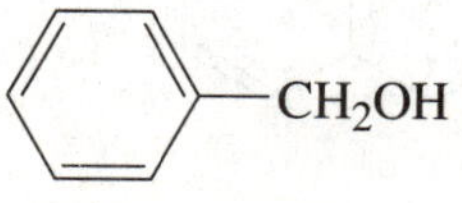 **reduction**

k. 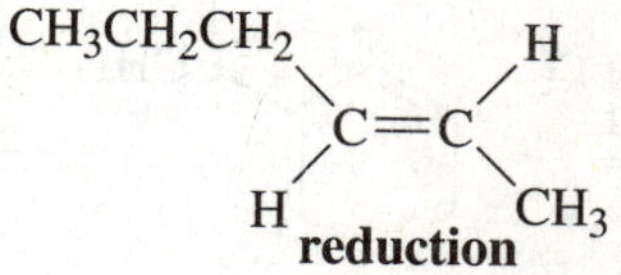
reduction

l. $CH_3CH_2CH_2\overset{\overset{\textstyle O}{\|}}{C}OH$ + $CH_3\overset{\overset{\textstyle O}{\|}}{C}OH$
oxidation

m. 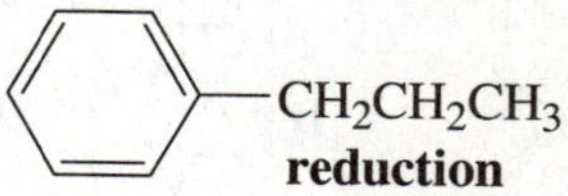
reduction

n. 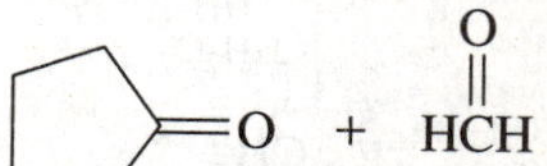
oxidation

o. 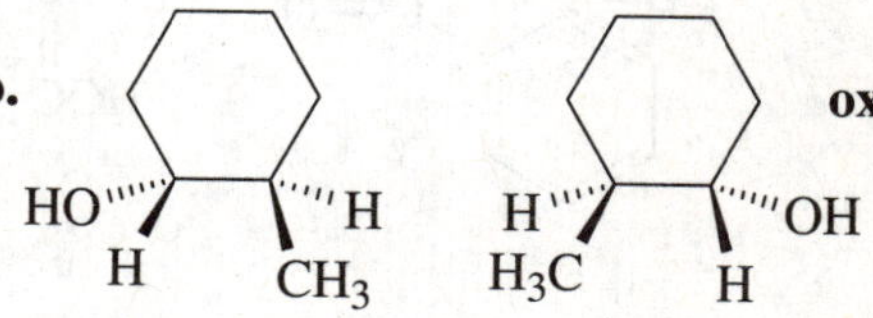 **oxidation**

p. 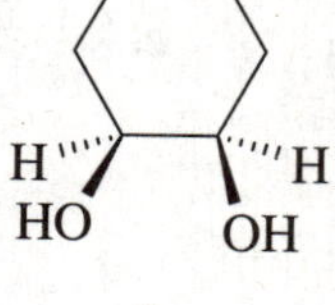 **oxidation**

q. $^{-}O\overset{\overset{\textstyle O}{\|}}{C}CH_2CH_2CH_2CH_2\overset{\overset{\textstyle O}{\|}}{C}O^{-}$
oxidation

r. $HO\overset{\overset{\textstyle O}{\|}}{C}CH_2\overset{\overset{\textstyle O}{\|}}{C}OH$ **oxidation**

32. **a.** $CH_3CH_2CH_2\overset{\overset{\textstyle O}{\|}}{C}H \xrightarrow{H_2CrO_4} CH_3CH_2CH_2\overset{\overset{\textstyle O}{\|}}{C}OH$

b. $CH_3CH_2CH_2CH_2OH \xrightarrow{H_2CrO_4} CH_3CH_2CH_2\overset{\overset{\textstyle O}{\|}}{C}OH$

c. $CH_3CH_2CH_2CH_2Br \xrightarrow{HO^-} CH_3CH_2CH_2CH_2OH \xrightarrow{H_2CrO_4} CH_3CH_2CH_2\overset{O}{\overset{\|}{C}}OH$

d. $CH_3CH_2CH{=}CH_2 \xrightarrow[2.\ HO^-,\ H_2O_2]{1.\ BH_3} CH_3CH_2CH_2CH_2OH \xrightarrow{H_2CrO_4} CH_3CH_2CH_2\overset{O}{\overset{\|}{C}}OH$

33.

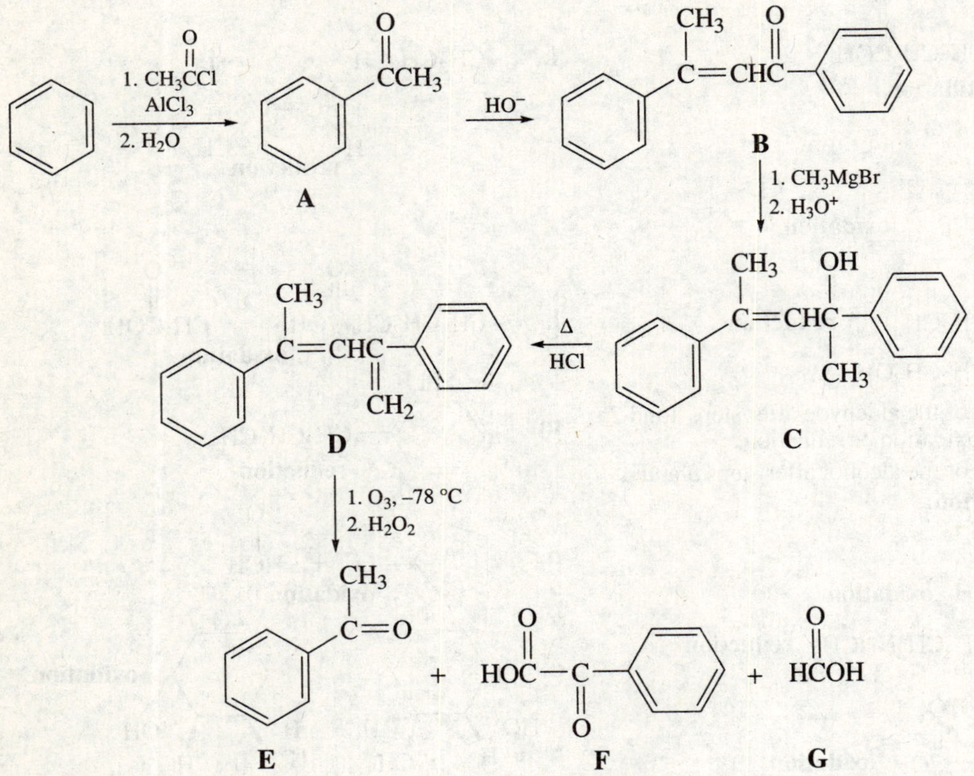

34.

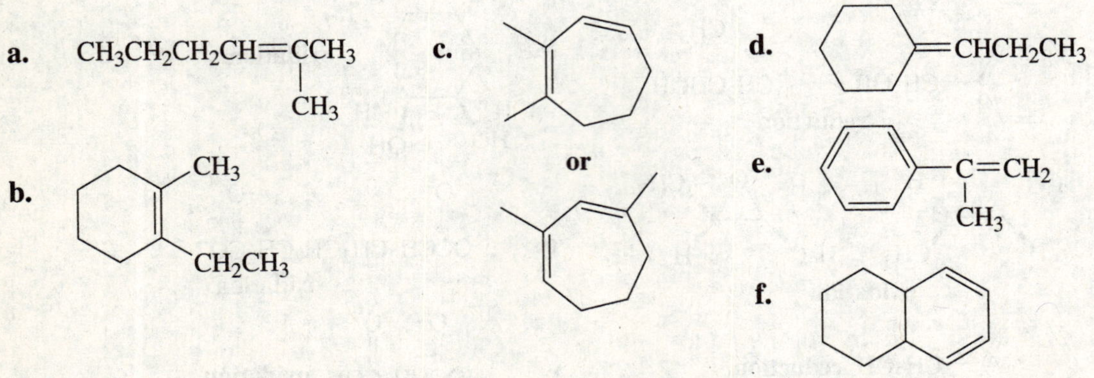

35. a. $CH_3CH_2CH{=}CH_2 \xrightarrow[2.\ H_2O_2,\ HO^-]{1.\ BH_3} CH_3CH_2CH_2CH_2OH$

$CH_3CH_2CH_2CH_2OH \underset{\Delta}{\overset{H_2SO_4}{\rightleftharpoons}} CH_3CH{=}CHCH_3 + H_2O$

$CH_3CH{=}CHCH_3 \xrightarrow[H_2SO_4]{KMnO_4} 2\ CH_3\overset{O}{\overset{\|}{C}}OH$

b.

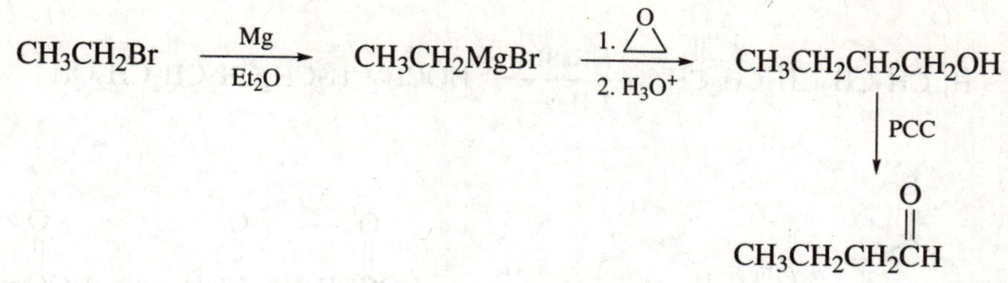
CH3CH2Br
Mg
Et2O
CH3CH2MgBr
1.
O
2. H3O+
CH3CH2CH2CH2OH
PCC
O
CH3CH2CH2CH

c.

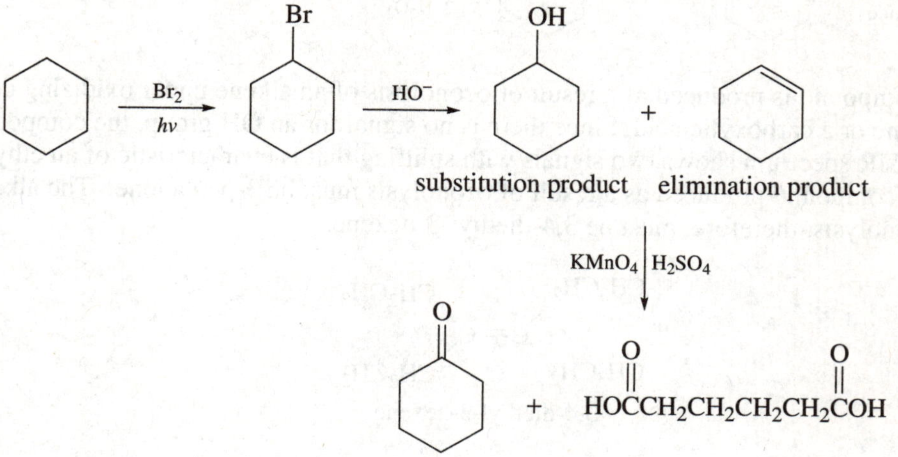
Br
OH
Br2
hv
HO−
+
substitution product
elimination product
KMnO4
H2SO4
O
O
O
+
HOCCH2CH2CH2CH2COH

36. **a.**

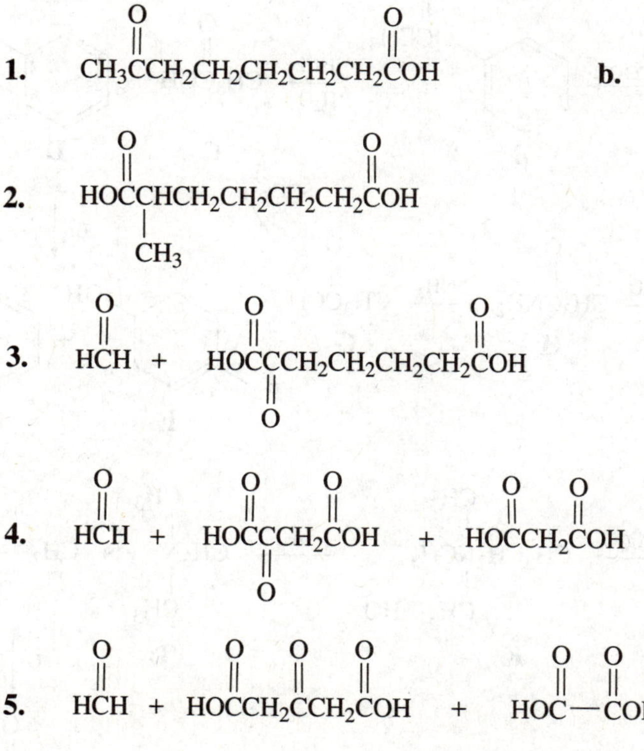
1.
CH3CCH2CH2CH2CH2CH2COH
2.
HOCCHCH2CH2CH2CH2COH
CH3
3.
HCH +
HOCCCH2CH2CH2CH2COH
4.
HCH +
HOCCCH2COH
+
HOCCH2COH
5.
HCH +
HOCCH2CCH2COH
+
HOC—COH

b.

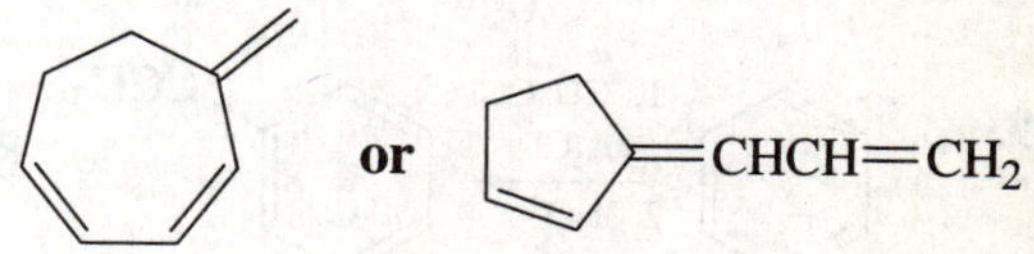
or
CHCH=CH2

37. **a.**

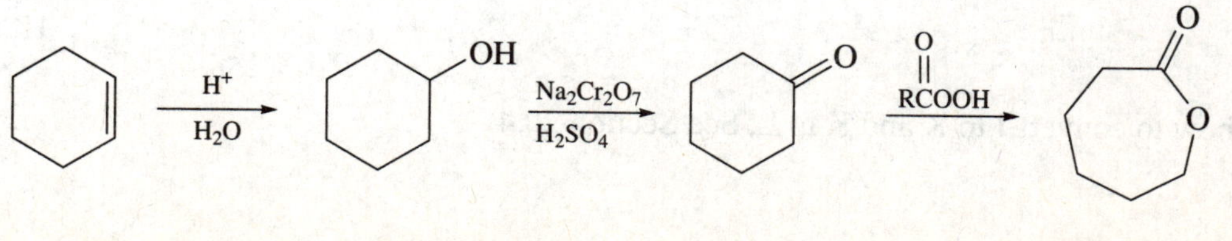
H+
H2O
OH
Na2Cr2O7
H2SO4
O
O
RCOOH
O
O

b.

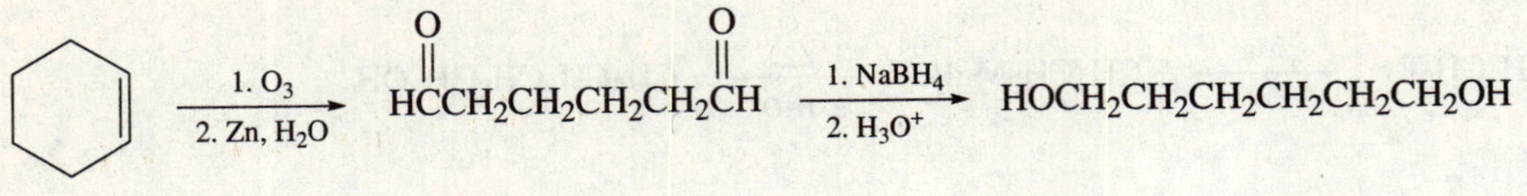

c.

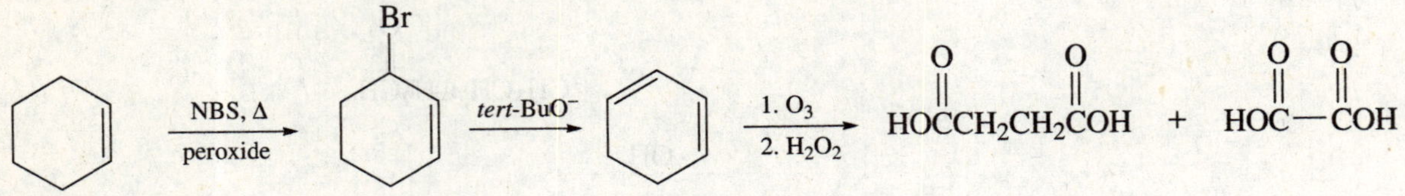

38. Because the compound is produced as a result of ozonolysis of an alkene under oxidizing conditions, it must be a ketone or a carboxylic acid. Since there is no signal for an OH group, the compound must be a ketone. The NMR spectrum shows two signals with splitting that is characteristic of an ethyl group. Therefore, the compound produced as a result of ozonolysis must be 3-pentanone. The alkene that underwent ozonolysis, therefore, must be 3,4-diethyl-3-hexene.

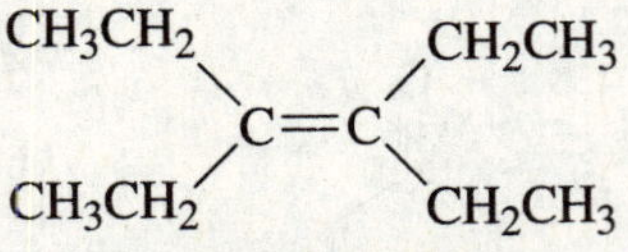

3,4-diethyl-3-hexene

39.

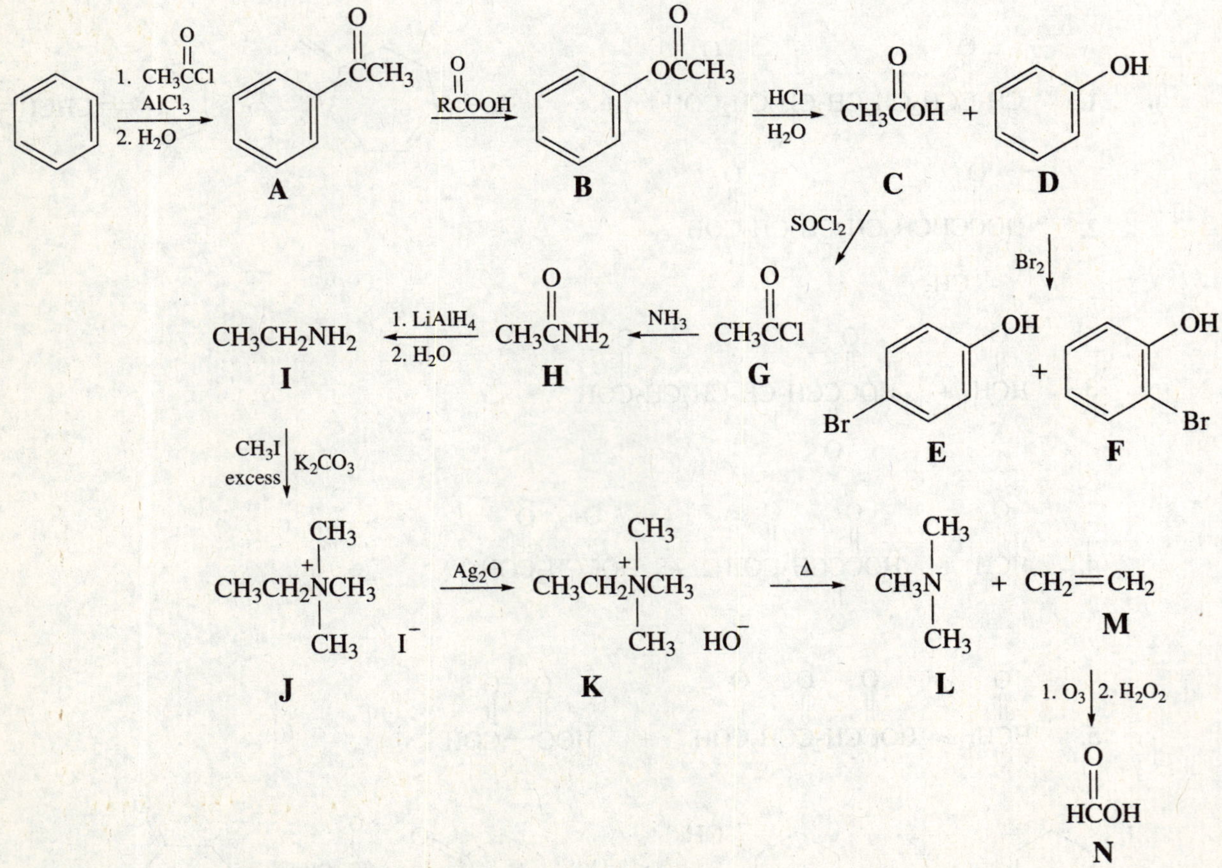

To see how to convert J to K and K to L, See Section 20.4.

40. The rate-determining step in the chromic acid oxidation of an alcohol is the E2 elimination reaction of the chromate ester. 2-Propanol is oxidized more rapidly than 2-deutero-2-propanol because it is easier to break the carbon-hydrogen bond than the carbon-deuterium bond.

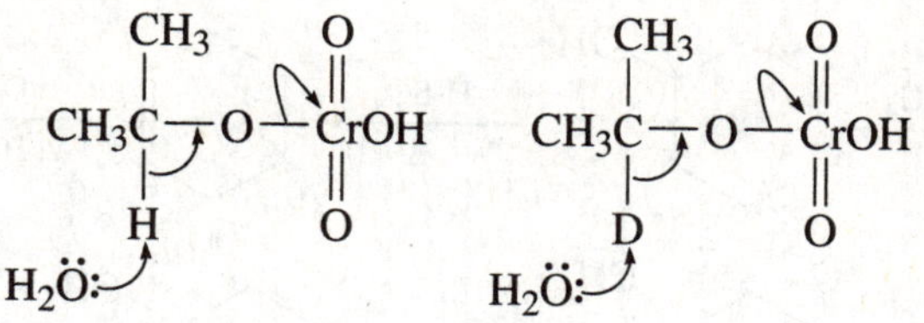

41.

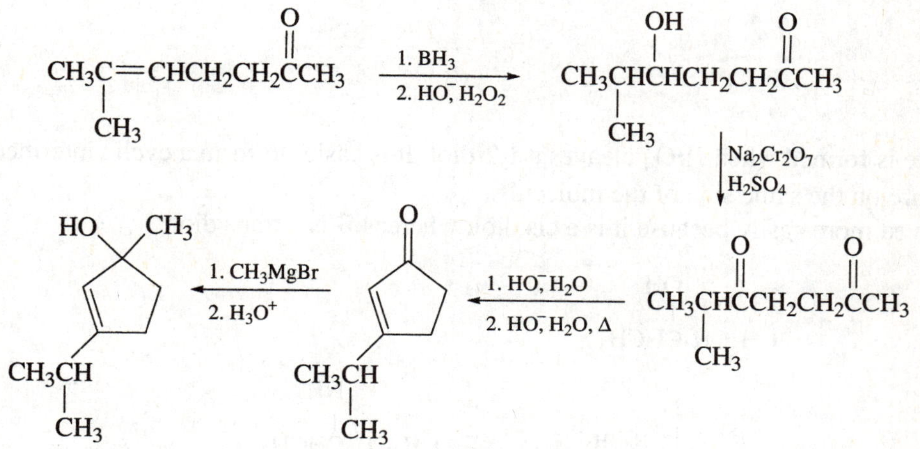

42. **a.**

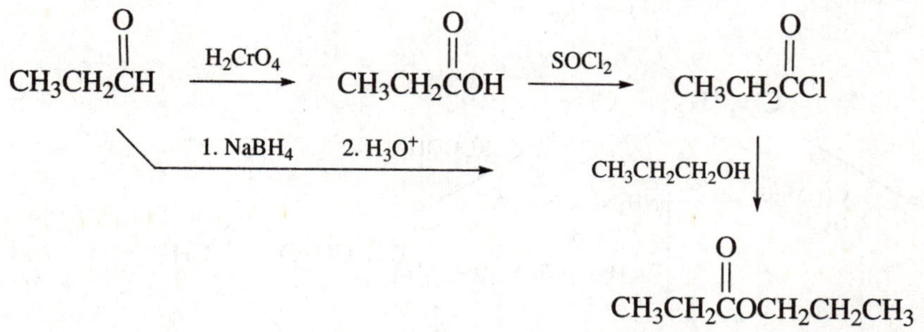

b.

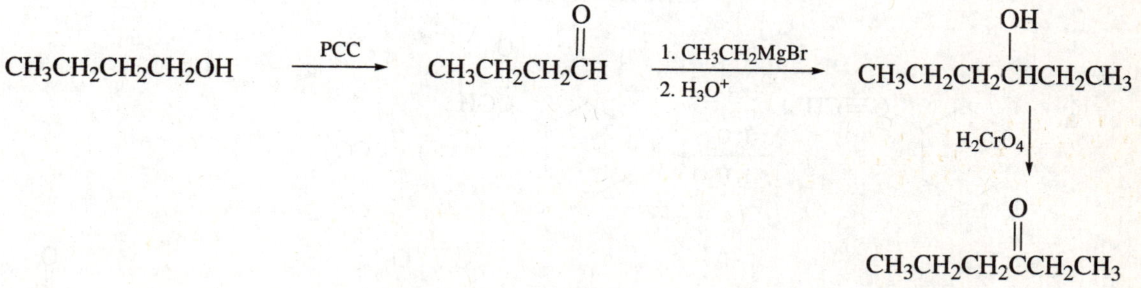

c.

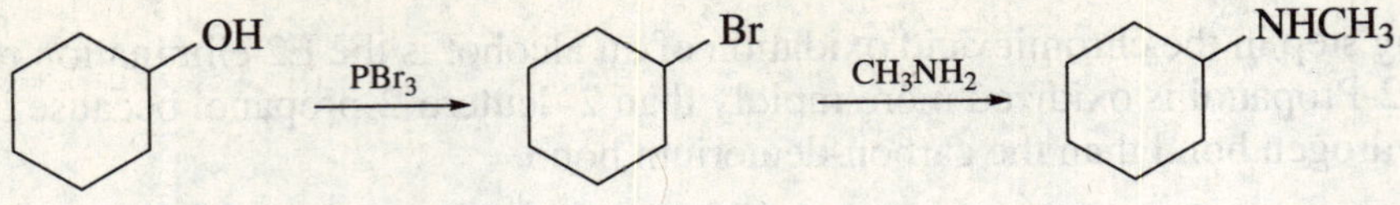

d.

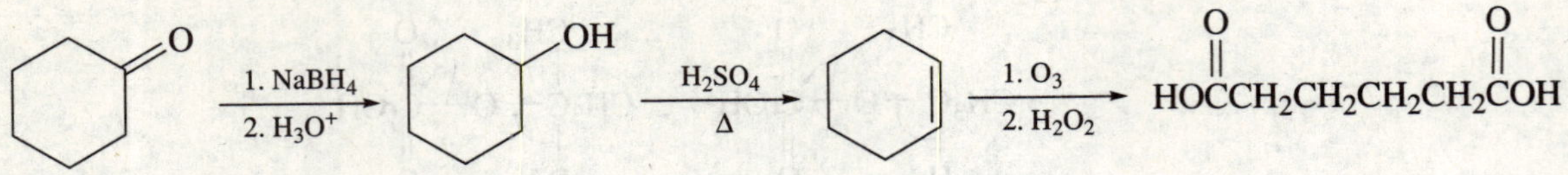

e.

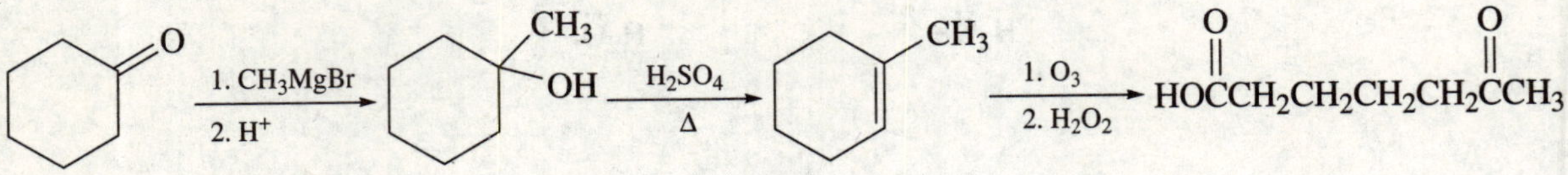

43. $CH_3CH_2C(CH_3){=}CH_2$

44. A cyclic intermediate is formed when HIO_4 cleaves a 1,2-diol. It is easier to form a cyclic intermediate if the two OH groups are on the same side of the molecule.
Therefore, **A** is cleaved more easily because it is a cis diol, whereas **B** is a trans diol.

45.

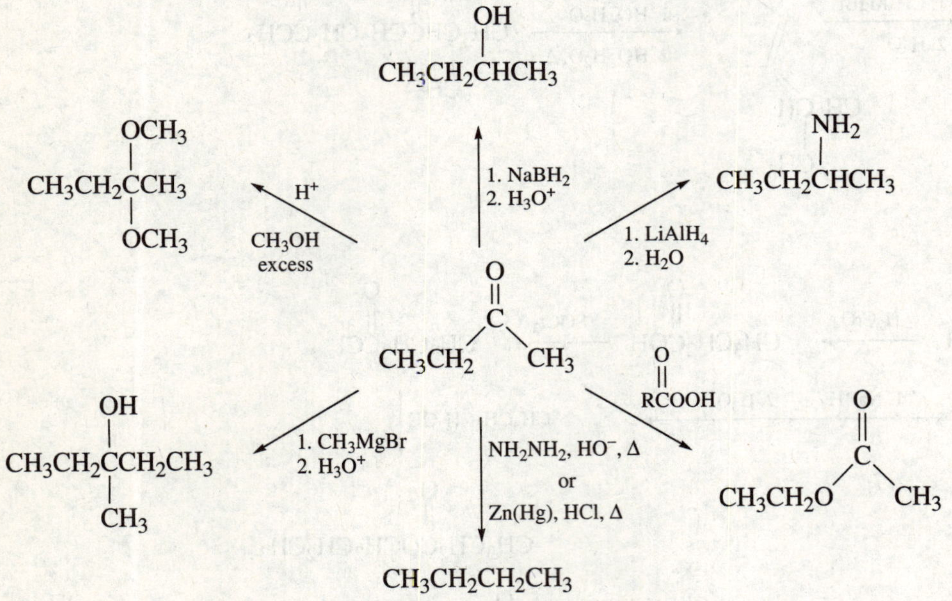

46. a.

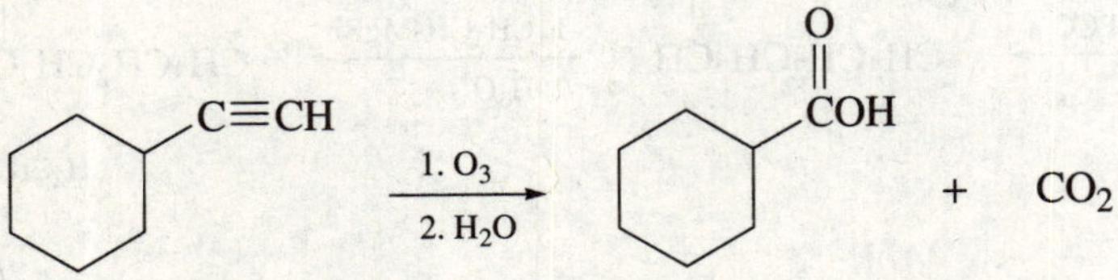

b.

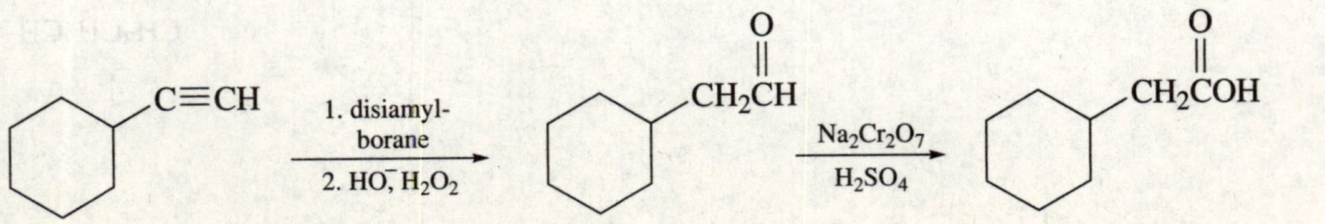

47. **a.**

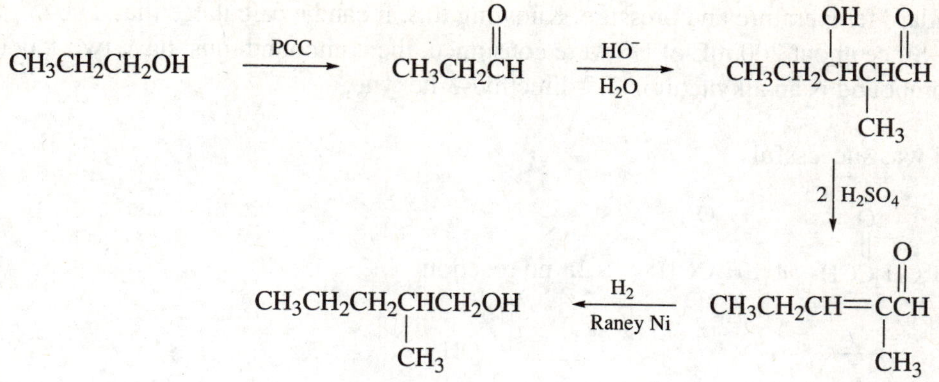

b.

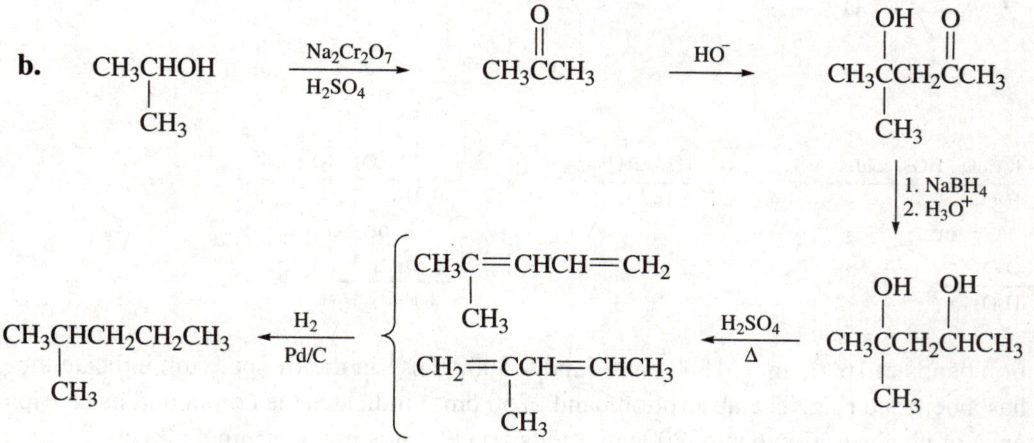

c.

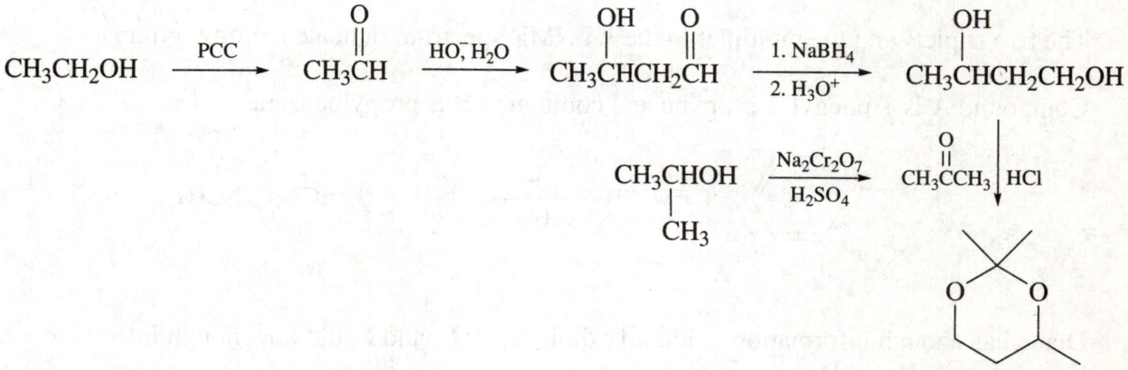

48. Because oxidative cleavage formed only one four-carbon carboxylic acid, you know the compound is either a symmetrical eight-carbon alkene or a symmetrical eight-carbon alkyne. Therefore, the compound has a molecular formula of C_8H_{14} or C_8H_{16} and, consequently, a molecular weight of 110 or 112.

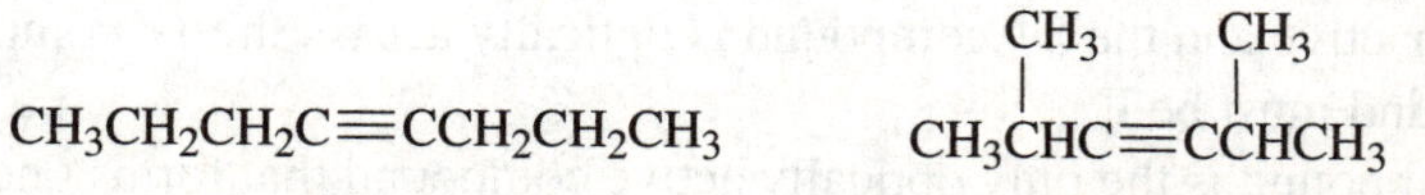

4-octyne 2,5-dimethyl-3-hexyne

0.5 g of the hydrocarbon was hydrogenated; this corresponds to 0.0045 mol (0.5/112 = 0.0045, 0.5/110 = 0.0045). The key here is to remember from general chemistry that 1 mol of a gas has a volume of 22.4

liters at standard temperature and pressure. Knowing this, it can be calculated that 0.0045 mol of H_2 = 100 mL H_2. Since about 200 mL of H_2 were consumed, the compound must have two π bonds. Thus, the unknown compound is an alkyne; it is 2,5-dimethyl-3-hexyne.

49. **a.** Only #4 was successful.

b. **1.** $CH_3CH_2\overset{O}{\overset{\|}{C}}CH_3 + CH_3\overset{O}{\overset{\|}{C}}OH$ **2.** no reaction

3.

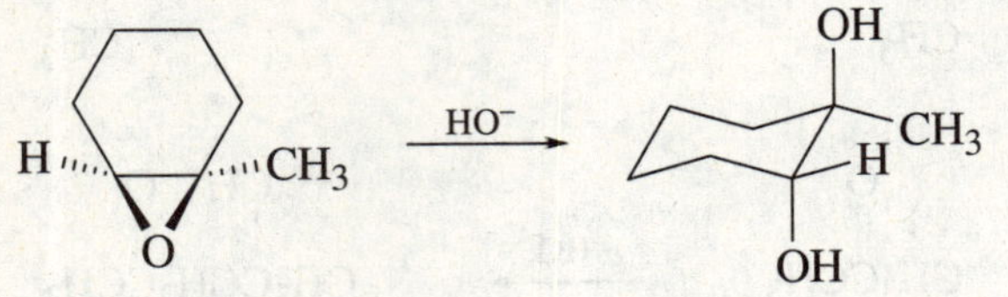

c. **1.** $\xrightarrow[\text{2. } H_2O]{\text{1. } KMnO_4,\ HO^-,\ \text{cold}}$ or $\xrightarrow[\text{2. } H_2O_2]{\text{1. } OsO_4}$

2. $\xrightarrow[\text{2. } H_3O^+]{\text{1. } LiAlH_4}$

3. $\xrightarrow[\text{2. } H_2O]{\text{1. } KMnO_4,\ HO^-,\ \text{cold}}$ or $\xrightarrow[\text{2. } H_2O_2]{\text{1. } OsO_4}$

50. The absorption bands at 1600 cm^{-1}, 1500 cm^{-1}, and > 3000 cm^{-1} in the IR spectrum indicate the compound has a benzene ring. The absorption band 2250 cm^{-1} indicates the compound has a triple bond, and the absence of an absorption band 3200 cm^{-1} tells you that it is not a terminal alkyne.

The two triplets and the multiplet in the ^{1}H NMR spectrum indicate a propyl group.

Compound **A** is 1-phenyl-1-propyne and compound **B** is propylbenzene.

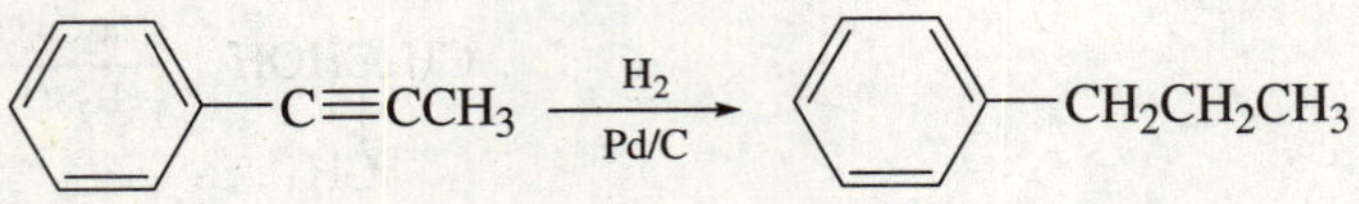

A **B**

51. Diane has enough information to identify diols **A**, **C**, **E**, and **F**, but not enough information to distinguish between diols **B** and **D**.

Only the 4th compound will form two products upon cleavage with periodic acid. So it must be **F**.
The 1st and 5th compounds will not react with periodic acid. Because the 1st compound is optically inactive and the 5th compound is optically active, the 1st compound must be **C** and the 5th compound must be **E**.
The 3rd compound is the only optically active compound that forms one compound with periodic acid. Therefore, it must be **A**.

Diane could distinguish between **B** and **D** if she analyzed the products obtained from the reaction of **B** and **D** with periodic acid. **B** will form acetaldehyde (a 2-carbon aldehyde), whereas **D** will form propanal (a 3-carbon aldehyde).

1 **2** **3** **4** **5** **6**

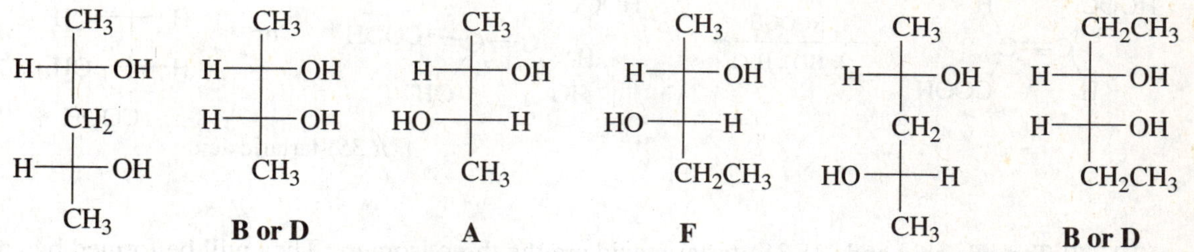

C — **B or D** — **A** — **F** — **E** — **B or D**

52.

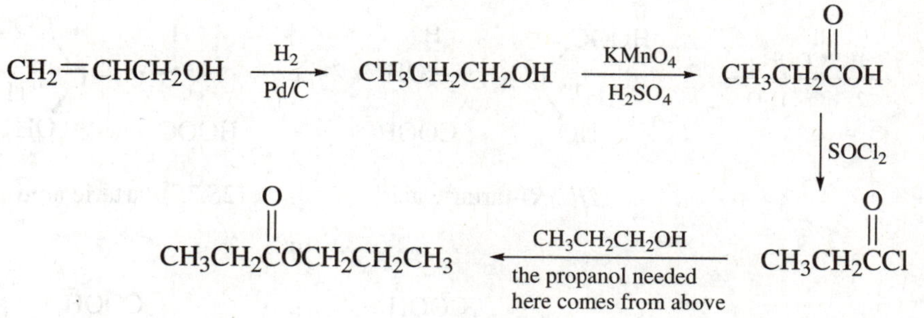

53.

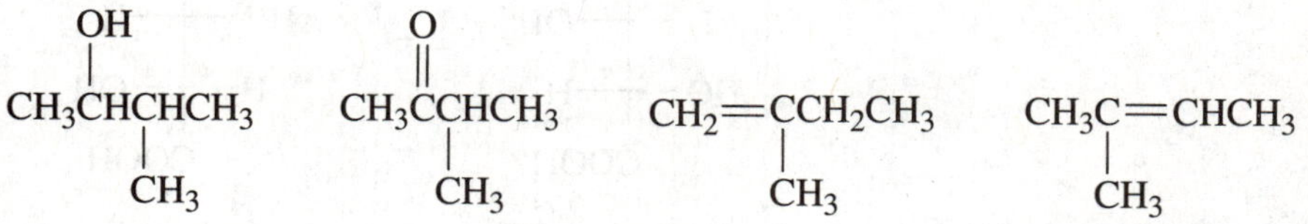

A **B** **C** **D**

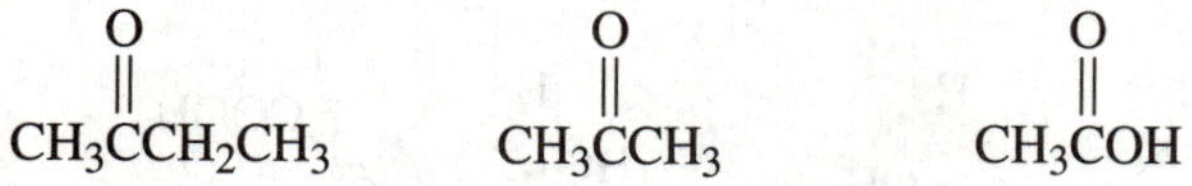

E **F** **G**

54.

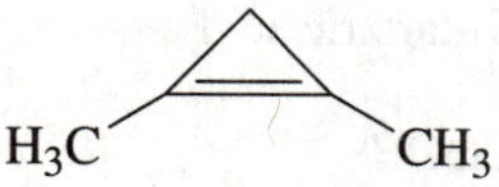

55. **a.** (2*R*,3*S*)-Tartaric acid is a meso compound. It will be formed by syn addition to the cis isomer. Syn addition of two OH groups can be carried out using a cold, basic solution of potassium permanganate or using osmium tetroxide followed by an aqueous solution of sodium bisulfite.

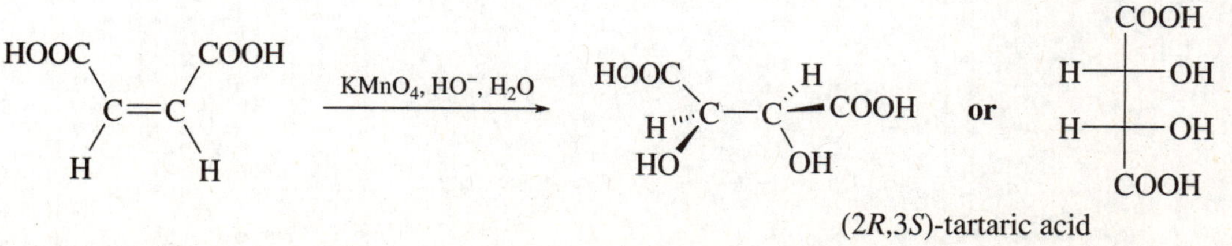

(2*R*,3*S*)-tartaric acid

b. (2*R*,3*S*)-Tartaric acid can also be formed by anti addition to the trans isomer. Anti addition of two OH groups can be carried out by first forming an epoxide and then treating the epoxide with an aqueous solution of hydroxide.

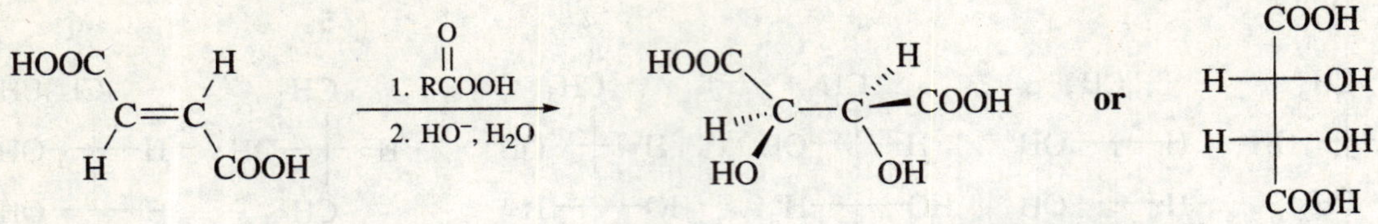

(2*R*,3*S*)-tartaric acid

c. (2*R*,3*R*)-Tartaric acid and (2*S*,3*S*)-tartaric acid are the threo isomers. They will be formed by anti addition to the cis isomer.

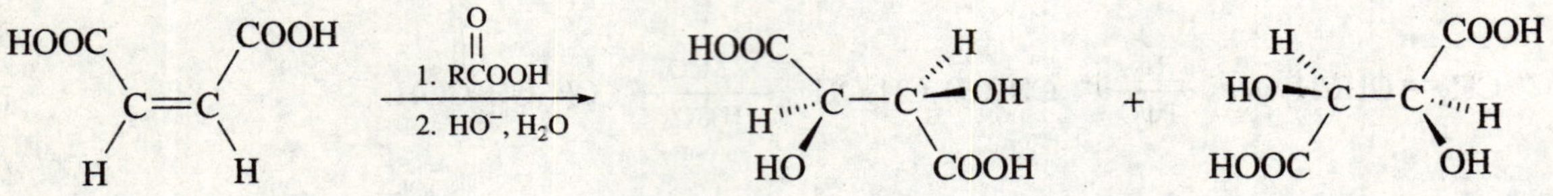

(2*R*,3*R*)-tartaric acid (2*S*,3*S*)-tartaric acid

or

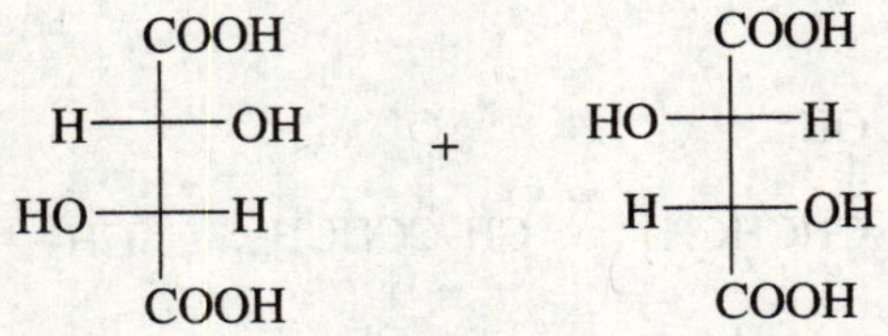

d. The threo isomers can also be formed by syn addition to the trans isomer.

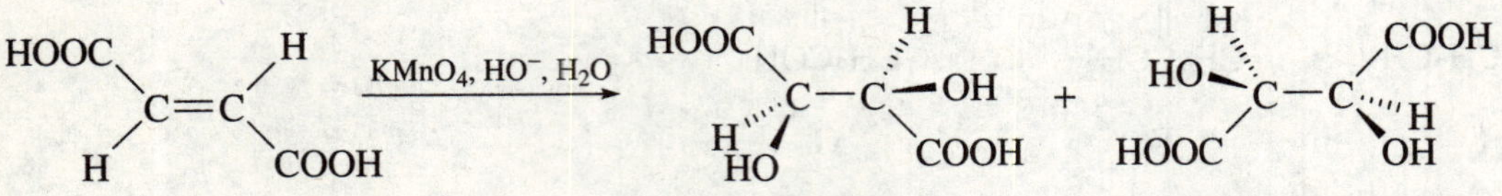

(2*R*,3*R*)-tartaric acid (2*S*,3*S*)-tartaric acid

or

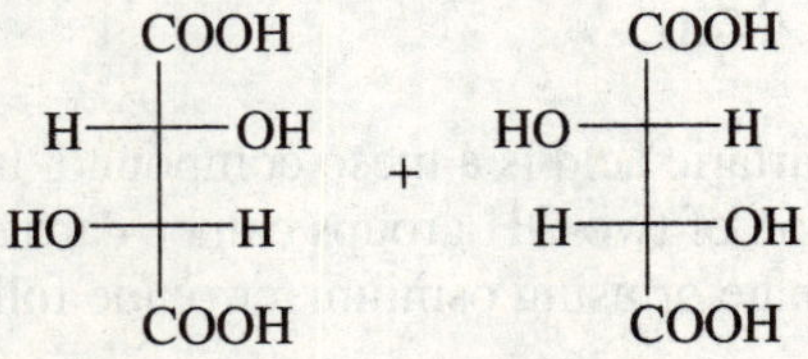

56.

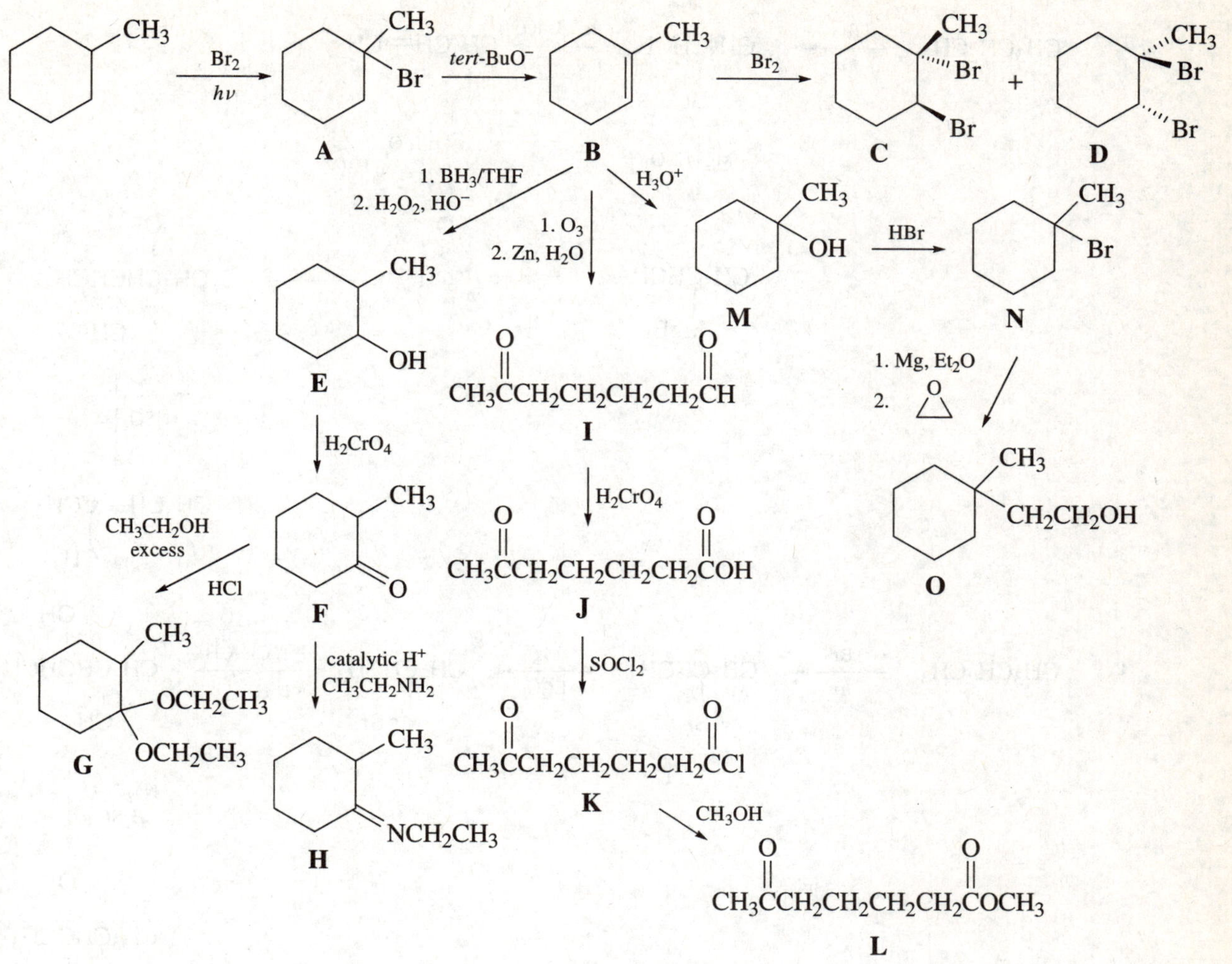

57. **a.**

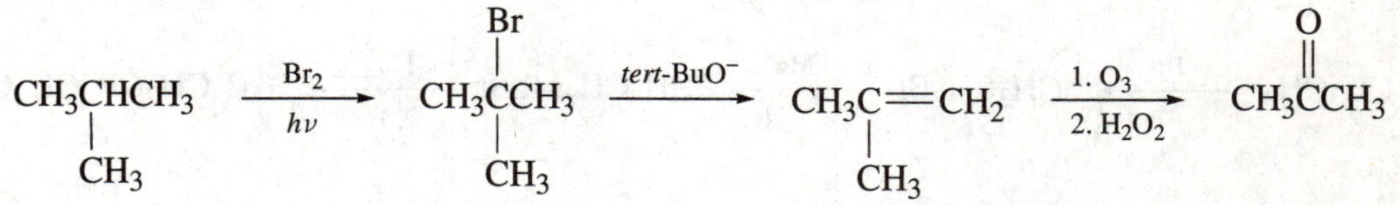

b. Note that 2-bromopropane (obtained from the first step) is divided into two portions; one portion is used to form propene, and the other is used to form the Grignard reagent.

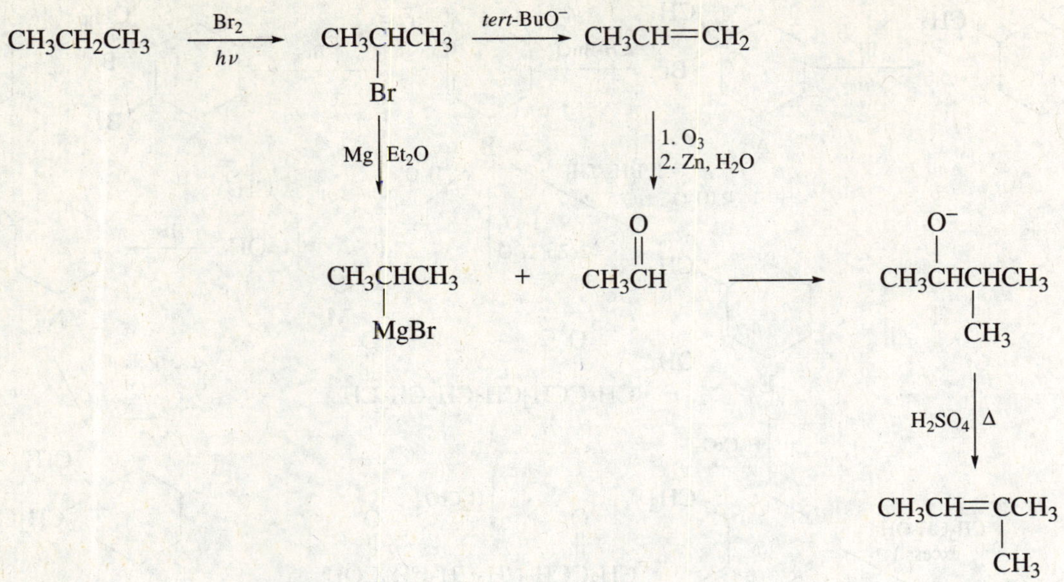

c.

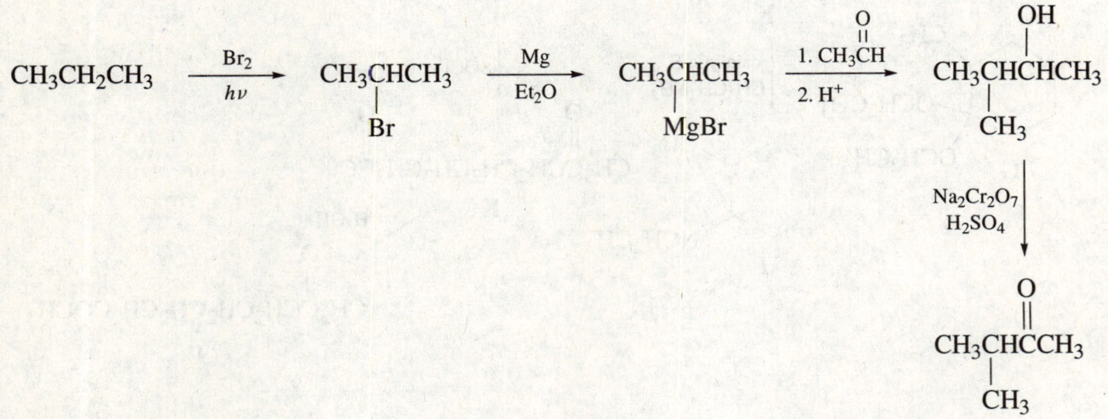

d.

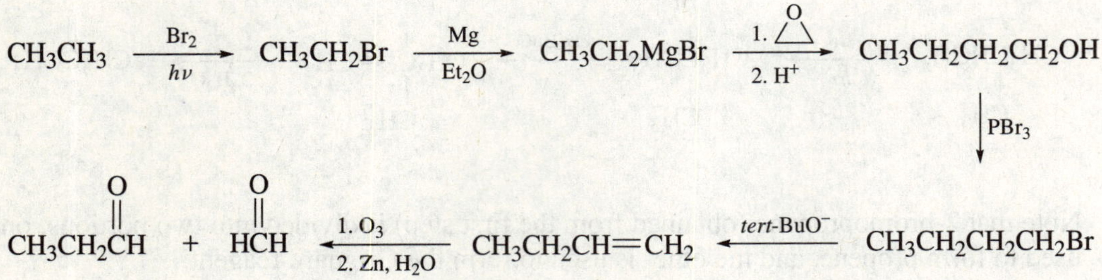

The ethylene oxide needed for the reaction can be obtained from some of the bromoethane prepared in the first step.

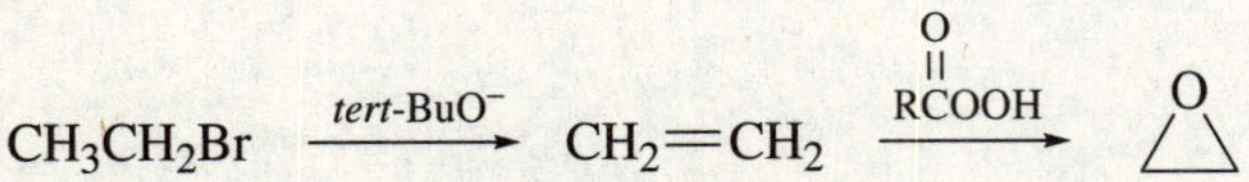

58. 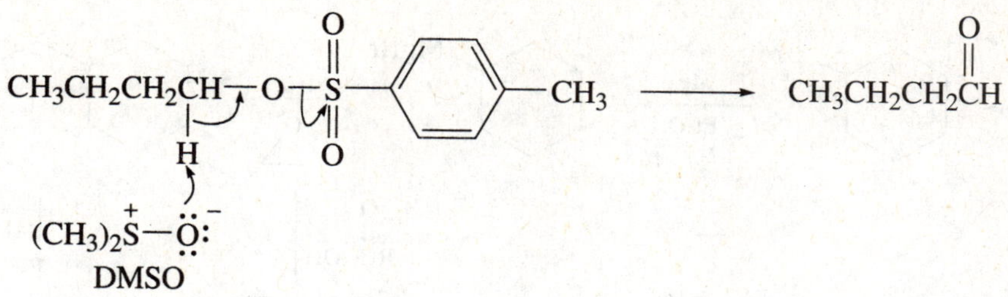

59. **a.** 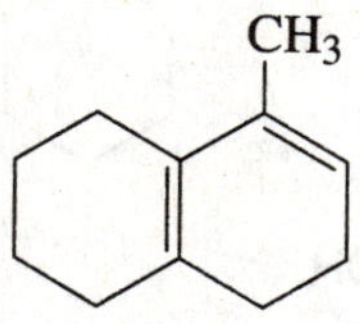 **b.** 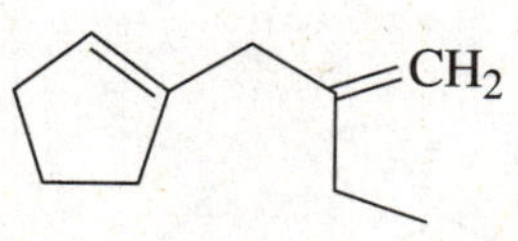

60. 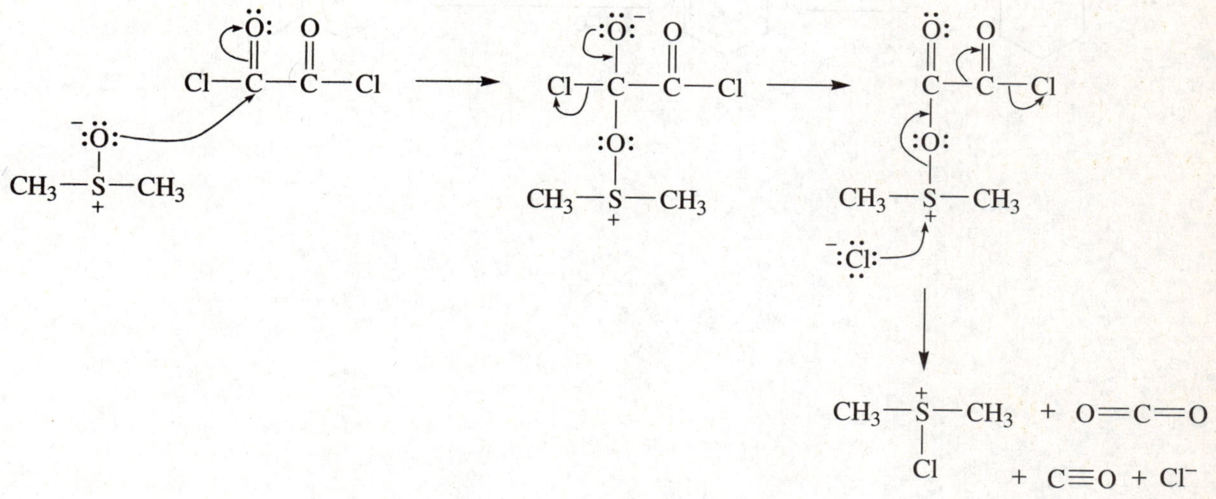

61. **a.** 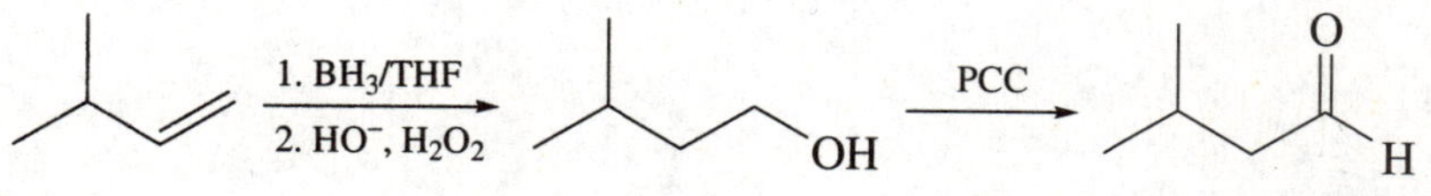

b. 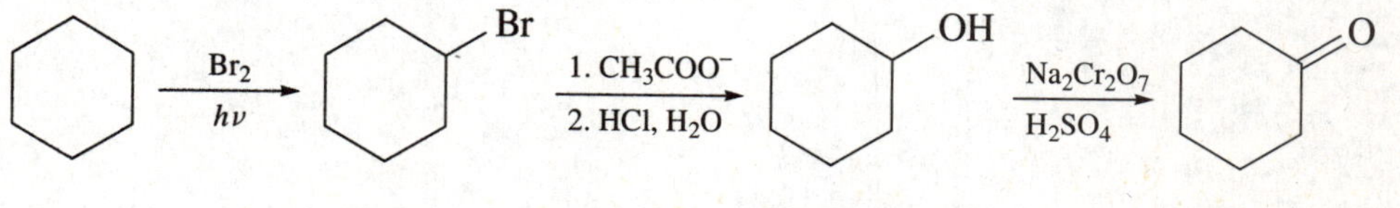

c. 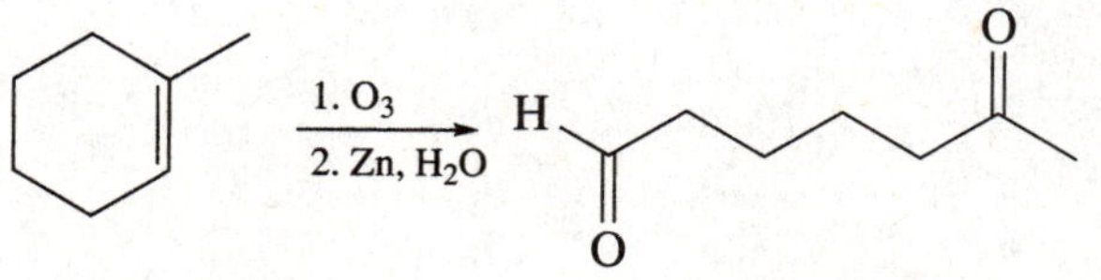

d. 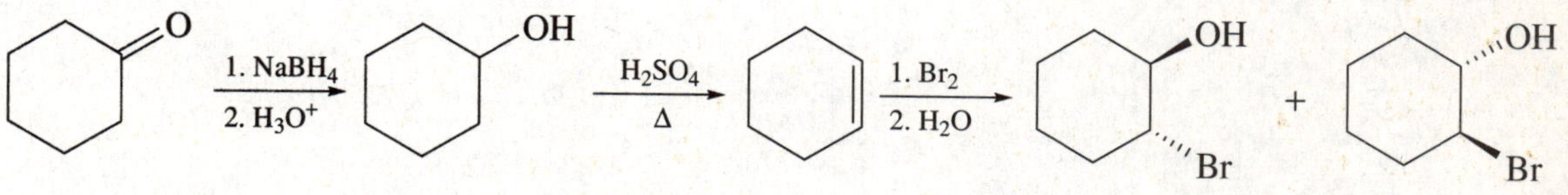

e.

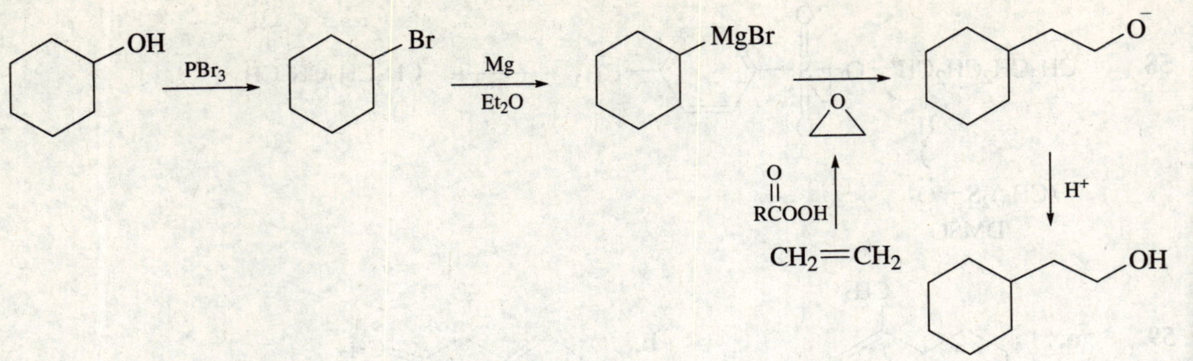
OH
PBr3
Br
Mg
Et2O
MgBr
O
O
RCOOH
CH2=CH2
O⁻
H⁺
OH

f.

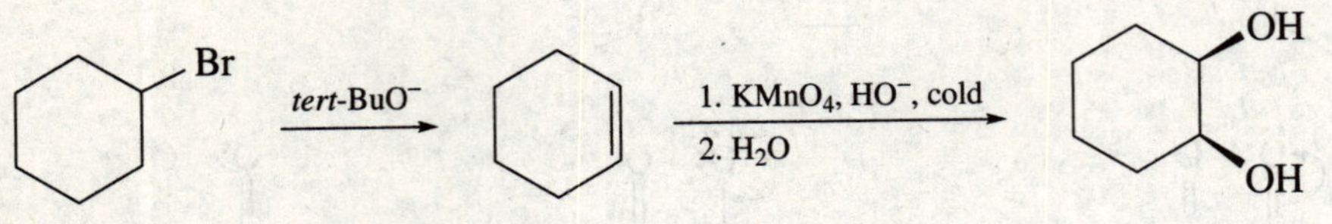
Br
tert-BuO⁻
1. KMnO4, HO⁻, cold
2. H2O
OH
OH

Chapter 19 Practice Test

1. Indicate whether each of the following reactions is an oxsidation, a reduction, or neither:

a. $CH_3CH_2\overset{O}{\overset{\|}{C}}CH_3 \longrightarrow CH_3CH_2\overset{OH}{\overset{|}{C}}HCH_3$

b. $CH_3CH_2\overset{O}{\overset{\|}{C}}H \longrightarrow CH_3CH_2\overset{O}{\overset{\|}{C}}OH$

2. Give the product of each of the following reactions, or if no reaction will occur, so state:

a. $CH_3CH_2CH_2CH_2OH \xrightarrow[H_2SO_4]{Na_2Cr_2O_7}$

b. $CH_3CH_2CH_2\overset{O}{\overset{\|}{C}}OCH_3 \xrightarrow[Pd/C]{H_2}$

c. $CH_3CH_2\overset{O}{\overset{\|}{C}}NHCH_2CH_3 \xrightarrow[2.\ H_2O]{1.\ LiAlH_4}$

d. 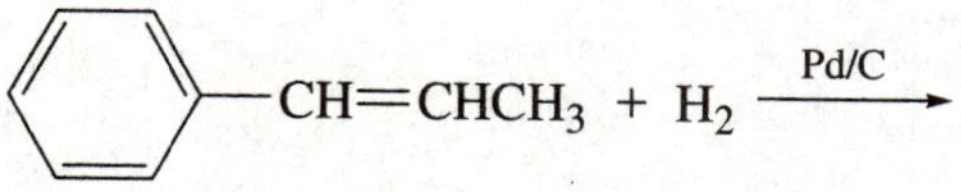

e. $CH_3CH_2CH_2CH_2\overset{O}{\overset{\|}{C}}H + R\overset{O}{\overset{\|}{C}}OOH \longrightarrow$

f. 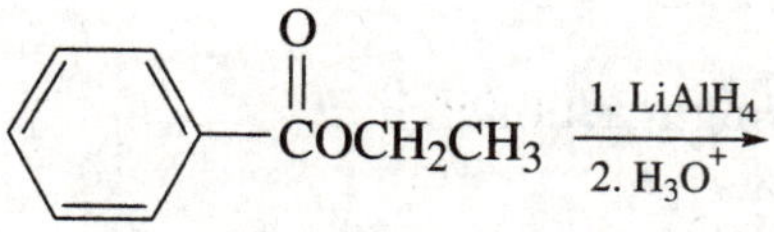

g. $CH_3CH_2CH_2\overset{O}{\overset{\|}{C}}CH_2CH_3 \xrightarrow[2.\ H_3O^+]{1.\ NaBH_4}$

h. $CH_3CH_2CH_2\overset{CH_3}{\overset{|}{C}}{=}CHCH_2CH_3 \xrightarrow[2.\ H_2O_2]{1.\ O_3}$

3. Which of the following reagents **cannot** be used to convert an aldehyde into an alcohol?

H_2, Pd/C	1. $NaBH_4$ 2. H_3O^+	1. $LiAlH_4$ 2. H_3O^+	1. Ag_2O, NH_3 2. H_3O^+	H_2, Raney Nickel

4. Two alkenes, when treated with ozone and then with $(CH_3)_2S$, both form only the ketone shown below. Identify the alkenes.

$$CH_3CH_2\overset{\overset{O}{\|}}{C}CH_3$$

5. Indicate whether each of the following statements is true or false:

a. $NaBH_4$ is a weaker reducing agent than $LiAlH_4$. T F

b. Esters are easier to reduce than ketones. T F

c. In an oxidation-reduction reaction the oxidizing agent is oxidized. T F

d. Ketones are reduced to primary alcohols. T F

e. Aldehydes are oxidized to carboxylic acids. T F

f. Acyl halides are oxidized to aldehydes. T F

g. Alkenes cannot be reduced with $NaBH_4$. T F

6. Describe how the following compound can be prepared from the given starting material:

$$CH_3CH_2CH_2CH_3 \longrightarrow CH_3CH_2\overset{\overset{O}{\|}}{C}CH_3$$

7. Which pair of reagents could be used to carry out the following reaction?

$$CH_3\overset{\overset{CH_3}{|}}{C}{=}CHCH_3 \longrightarrow CH_3\overset{\overset{O}{\|}}{C}CH_3 + CH_3\overset{\overset{O}{\|}}{C}H$$

a. 1. O_3 2. Zn, H_2O and 1. O_3 2. H_2O_2

b. 1. OsO_4 2. H_2O_2 and $KMnO_4$, H^+

c. 1. O_3 2. Zn, H_2O and 1. O_3 2. $(CH_3)_2S$

CHAPTER 20
More About Amines • Heterocyclic Compounds

Important Terms

Cope elimination reaction	elimination from an amine oxide.
exhaustive methylation	reaction of an amine with excess methyl iodide, resulting in the formation of a quaternary ammonium iodide.
furan	a five-membered ring aromatic compound containing an oxygen heteroatom.
heteroatom	an atom other than a carbon atom or a hydrogen atom.
heterocyclic compound (heterocycle)	a cyclic compound in which one or more of the atoms of the ring are heteroatoms.
Hofmann elimination (anti-Zaitsev elimination)	a hydrogen is removed from the β-carbon bonded to the most hydrogens.
Hofmann elimination reaction	elimination of a proton and a tertiary amine from a quaternary ammonium hydroxide.
imidazole	a five-membered ring compound with two nitrogen heteroatoms and two double bonds.
ligation	sharing of nonbonded electrons with a metal.
phase-transfer catalysis	catalysis of a reaction by providing a way to bring a polar reagent into a nonpolar phase so that the reaction between a polar and a nonpolar compound can occur.
phase-transfer catalyst	a compound that carries a polar reagent into a nonpolar phase.
porphyrin ring system	consists of four pyrrole rings joined by one carbon bridge.
protoporphyrin IX	the porphyrin ring system of heme.
purine	a pyrimidine ring fused to an imidazole ring.
pyrimidine	a benzene ring with nitrogens at the 1- and 3-positions.
pyrrole	a five-membered ring aromatic compound containing a nitrogen heteroatom.
quaternary ammonium ion	a cation containing a nitrogen bonded to four alkyl groups (R_4N^+).
thiophene	a five-membered ring aromatic compound containing a sulfur heteroatom.

Solutions to Problems

1. a. 2, 2-dimethylaziridine
   b. 4-ethylpiperidine
   c. 3-methylazacyclobutane
   d. 2-methylthiacyclopropane
   e. 2,3-dimethyltetrahydrofuran
   f. 2-ethyloxacyclobutane

2. The electron-withdrawing oxygen atom of morpholine stabilizes the conjugate base, making protonated morpholine the stronger acid.

$pK_a = 9.28$ morpholine $pK_a = 11.12$ piperidine

3. a.

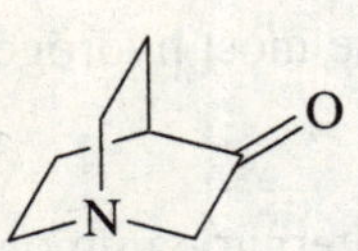

b. The conjugate acid of 3-quinuclidinone has a lower pK_a than the conjugate acid of morpholine (pK_a = 9.28) because the electron-withdrawing oxygen atom in 3-quinuclidinone is closer to the nitrogen atom. The pK_a of the conjugate acid of 3-quinucidinone is 7.46.

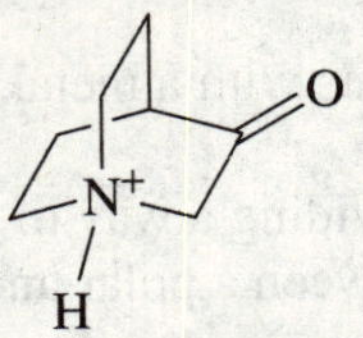

conjugate acid of 3-quinuclidinone

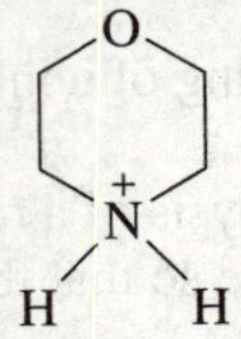

conjugate acid of morpholine

c. The conjugate acid of 3-chloroquinuclidine has a lower pK_a than the conjugate acid of 3-bromoquinuclidine because chlorine is more electronegative than bromine. This means that 3-bromoquinuclidine is a stronger base than 3-chloroquinuclidine.

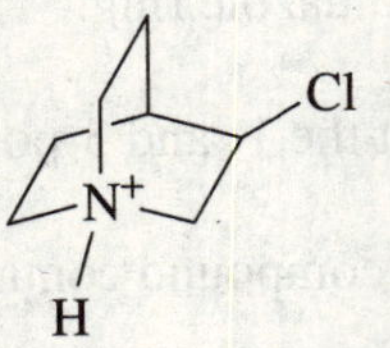

conjugate acid of 3-chloroquinuclidine

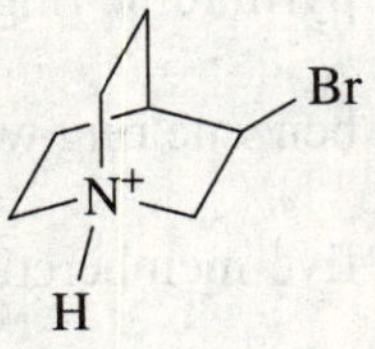

conjugate acid of 3-bromoquinuclidine

4. **a.** 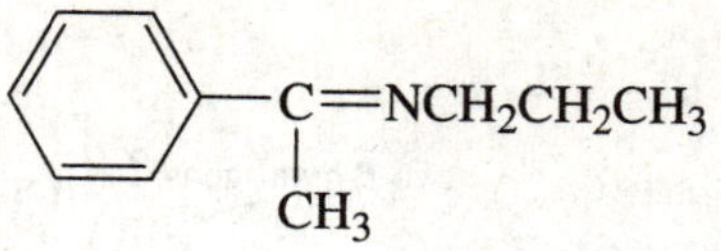 **c.** 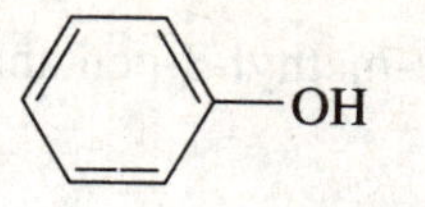

b. 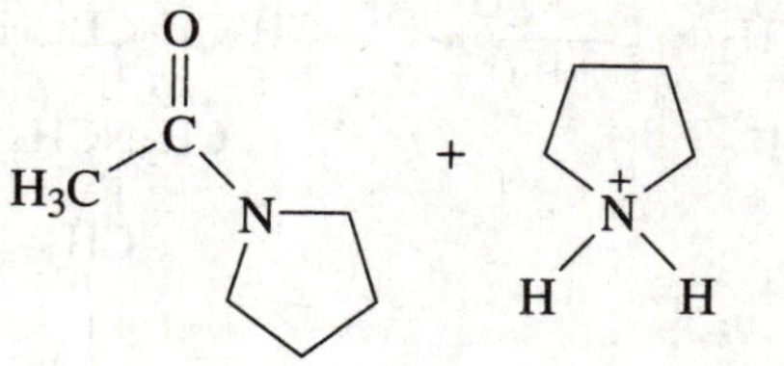 **d.** 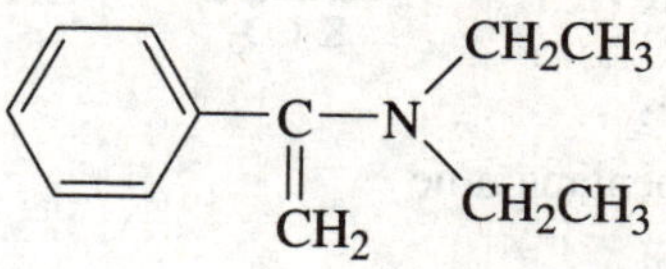

5. The only difference is the leaving group.

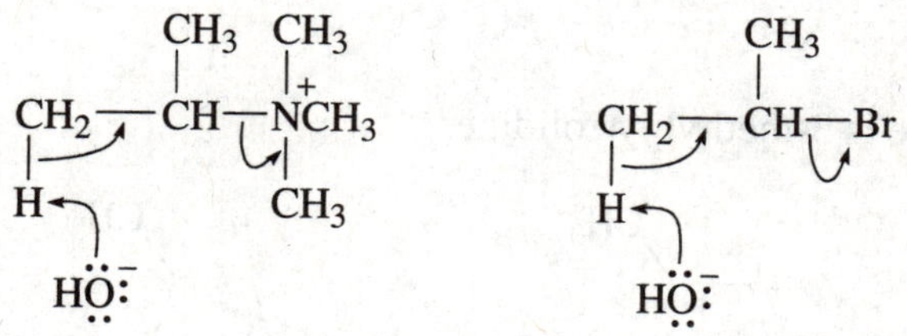

6. 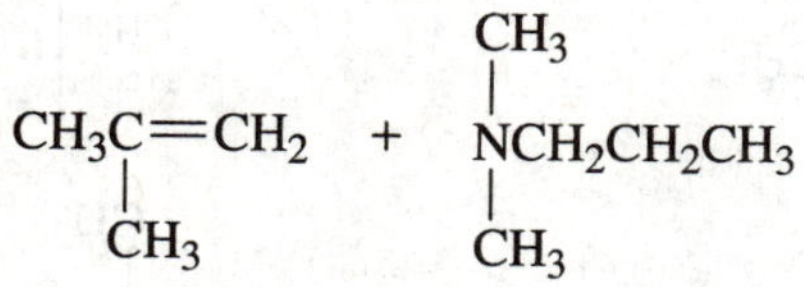

7. In each case, a proton is removed from the β-carbon that is bonded to the greater number of hydrogens.

a. 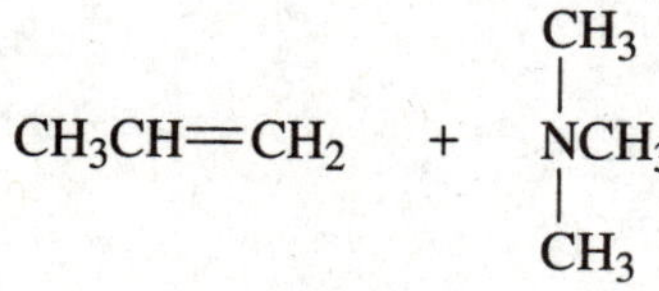 **c.** 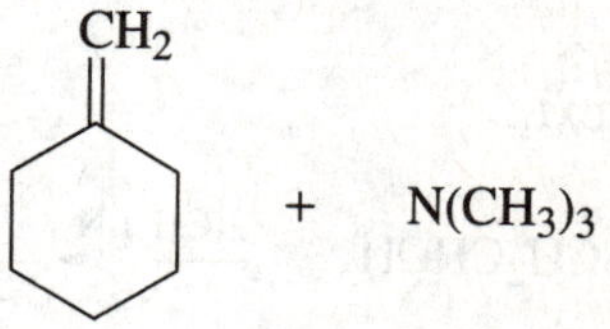

b. 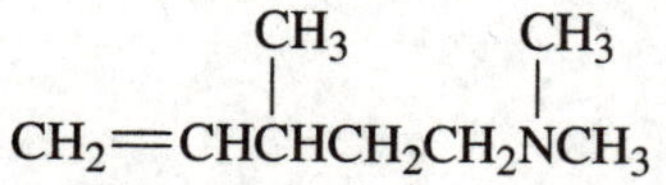 **d.** 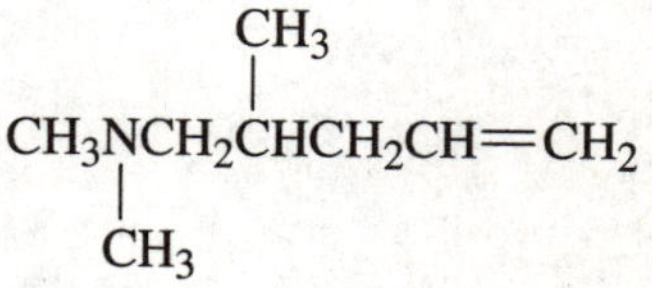

8. **a.** The primary amine is 2-methyl-3-pentanamine.

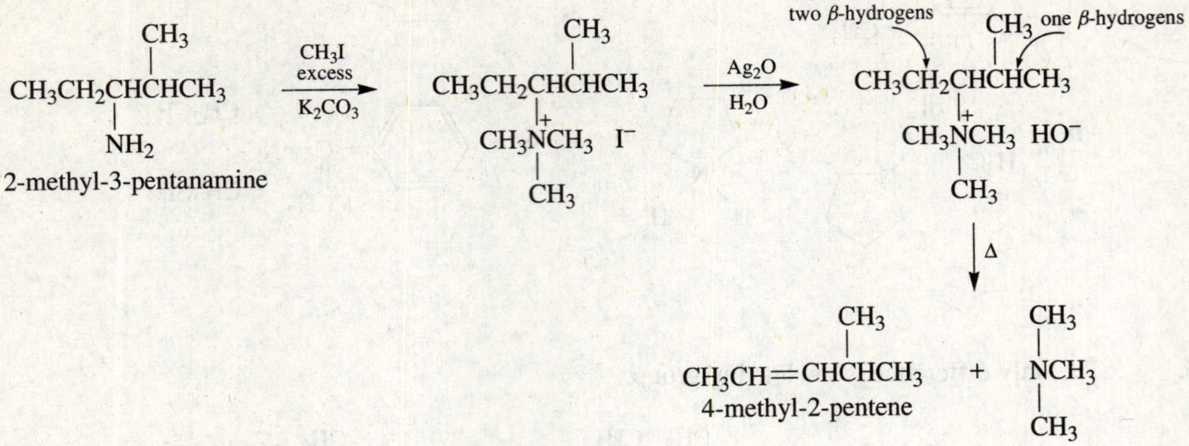

b. The secondary amine is 3-methylpyrrolidine.

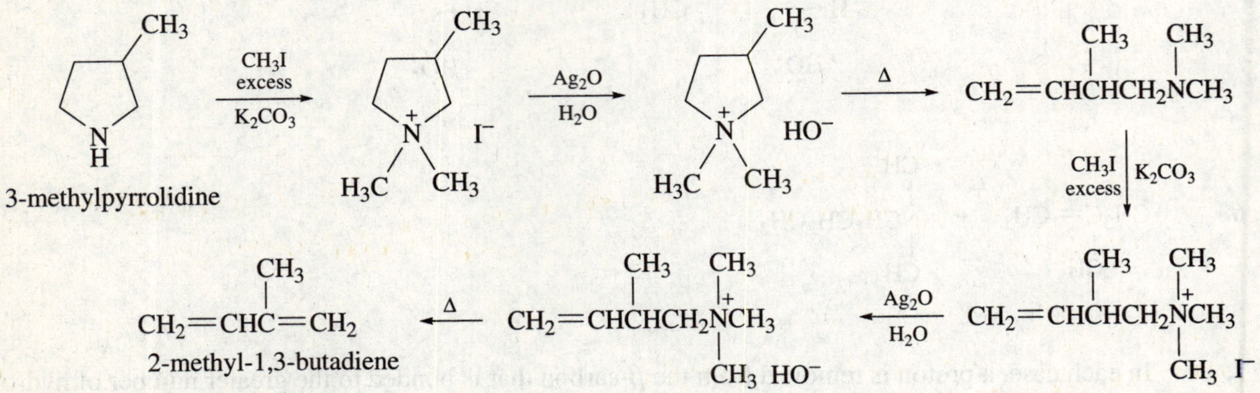

9. **a.** Solved in the text.

b.

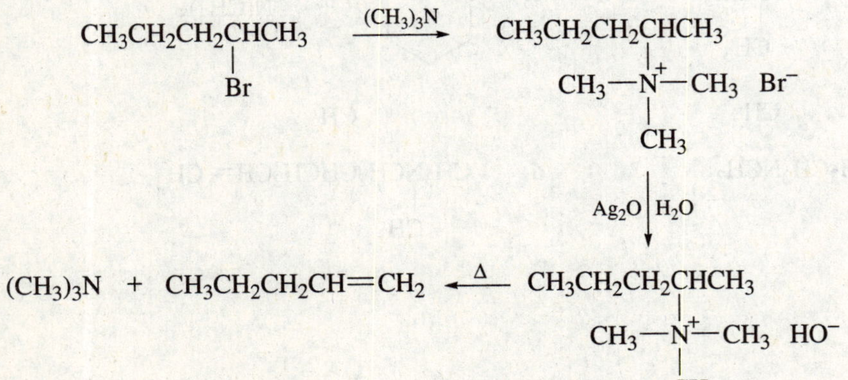

c.

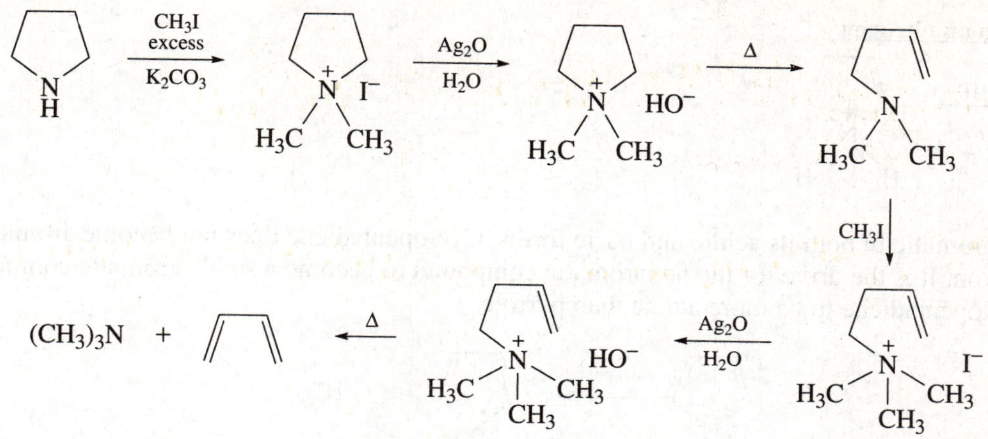

10. Because the major product is the one obtained by removing a proton from the β-carbon bonded to the most hydrogens, we can conclude that the Cope elimination has a "carbanion-like" transition state.

11. a.

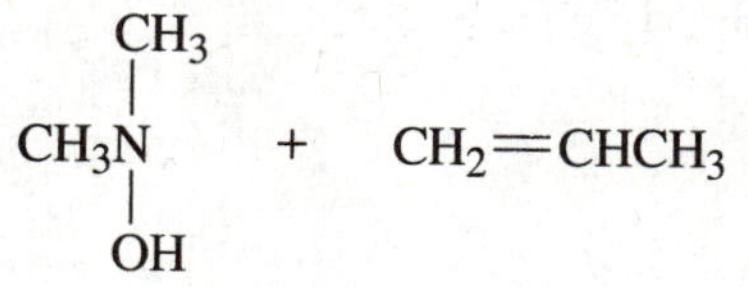

c.

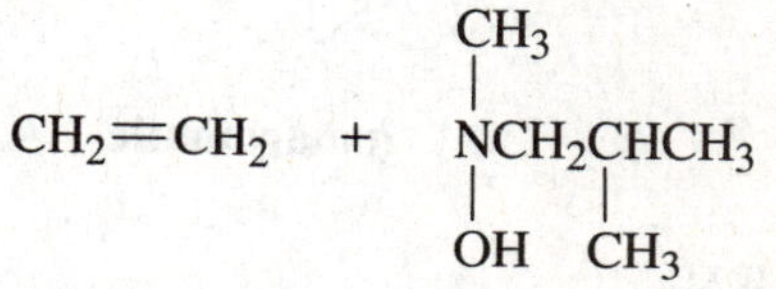

b.

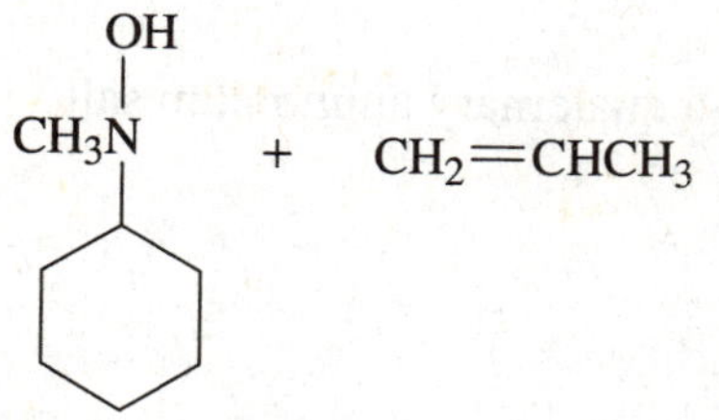

d.

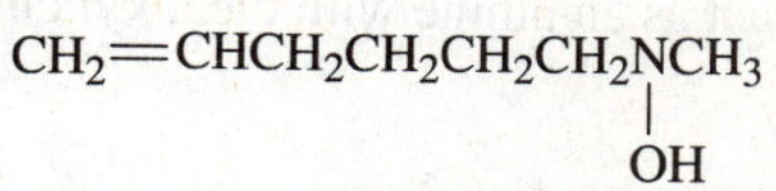

Notice that the methyl group is the β-carbon with the greatest number of hydrogens.

12.

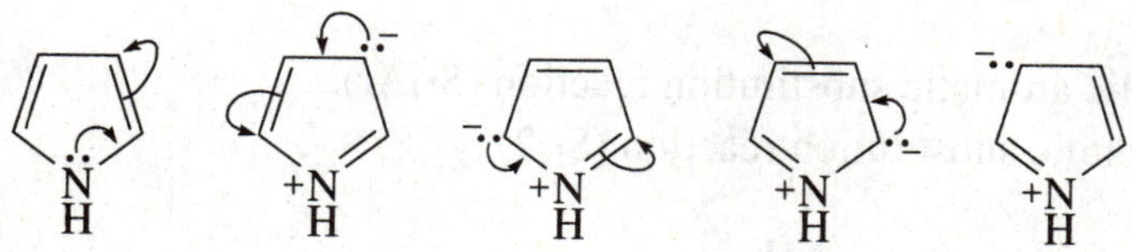

13.

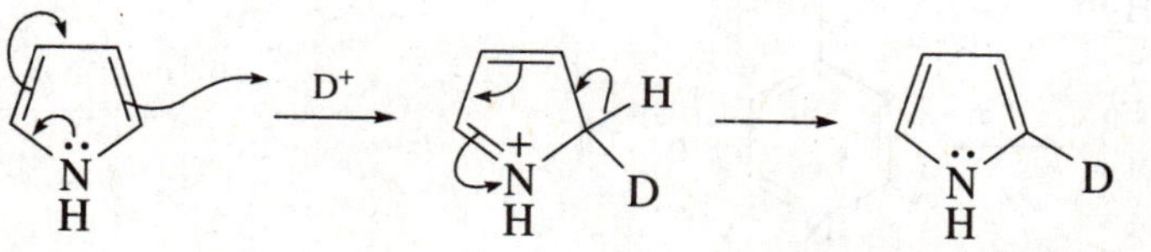

2-deuteriopyrrole

14. Protonation on C-2 leads to a cation with three resonance contributors. Protonation on nitrogen leads to a cation with no resonance contributors. (The positively charged nitrogen cannot accept electrons, because that would put ten electrons around the nitrogen.)

protonation on C-2

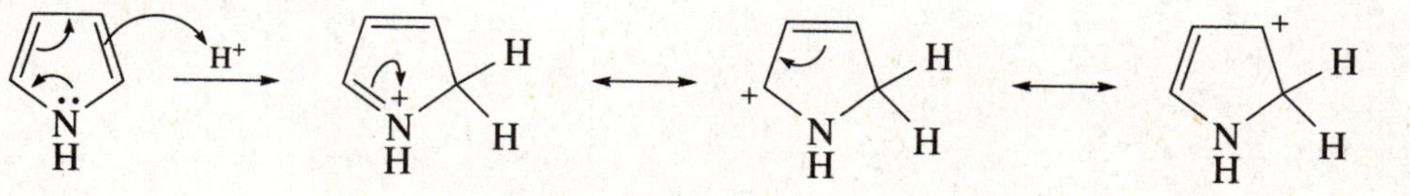

protonation on nitrogen

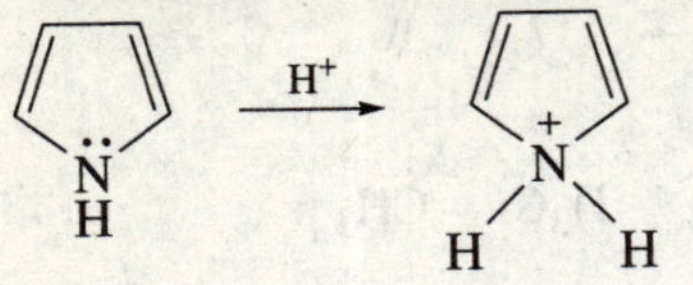

15. Pyrrole is aromatic in both its acidic and basic forms. Cyclopentadiene does not become aromatic until it loses a proton. It is the drive for the nonaromatic compound to become a stable aromatic compound that causes cyclopentadiene to be more acidic than pyrrole.

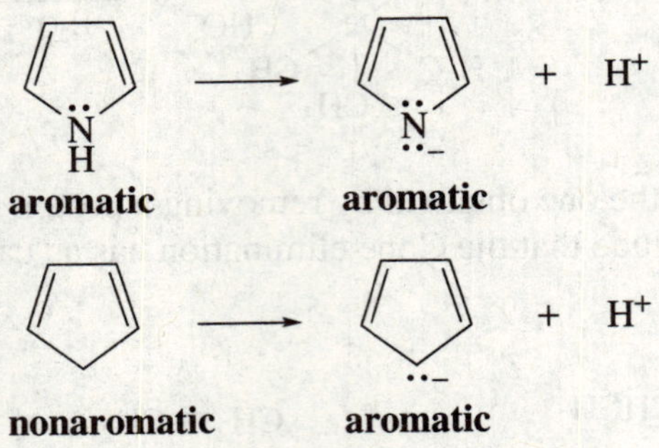

16. Solved in the text.

17. Pyridine will act as an amine with the alkyl chloride, forming a quaternary ammonium salt.

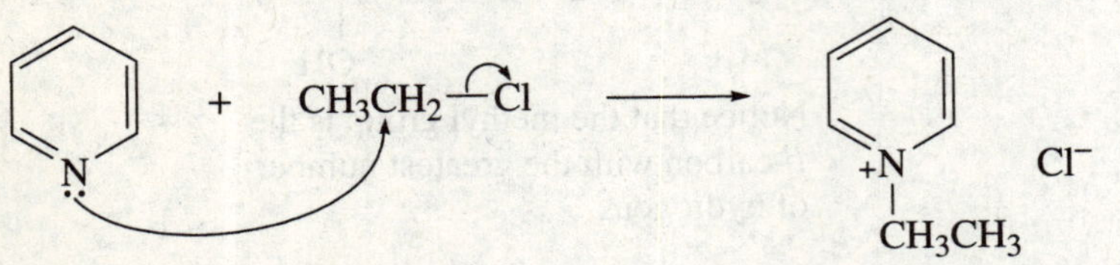

18. The first reaction is a two-step nucleophilic aromatic substitution reaction (S_NAr). The second reaction is a one-step nucleophilic substitution reaction (S_N2).

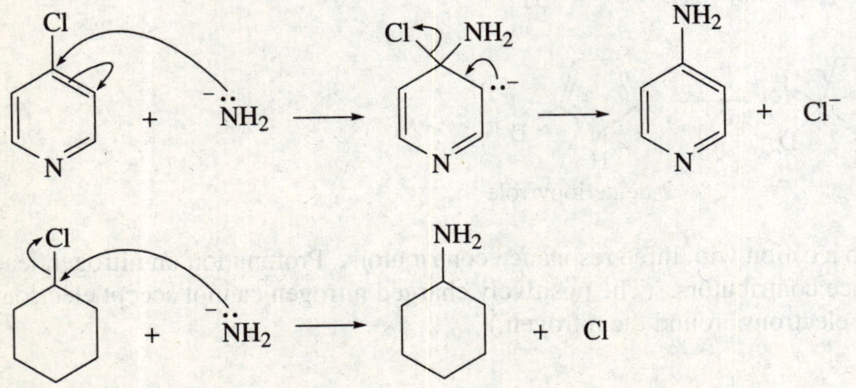

19. **a.**

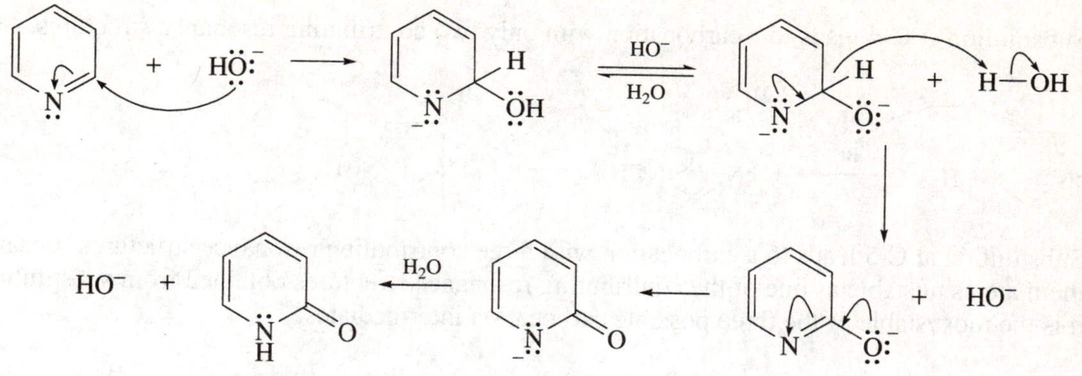

2-pyridinone

b. 4-Pyridinone is also formed because nucleophilic attack by hydroxide ion can take place at the 4-position as well as at the 2-position.

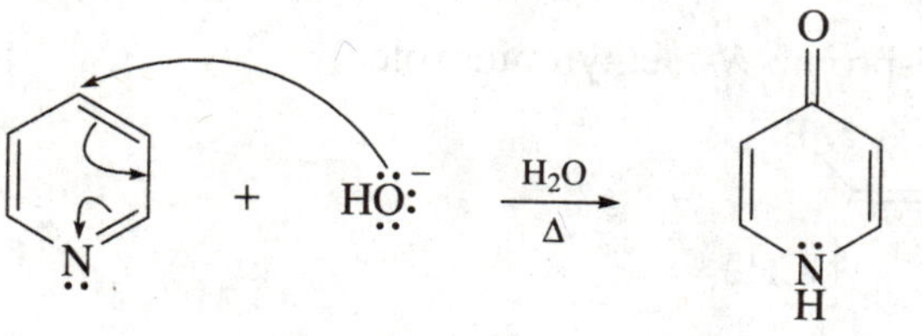

4-pyridinone

20. It is easiest to remove a proton from the methyl group of the *N*-alkylated pyridine because the electrons left behind when the proton is removed can be delocalized onto the positively charged nitrogen atom. It is easier to remove a proton from 4-methylpyridine than from 3-methylpyridine because in the former, the electrons left behind when the proton is removed can be delocalized onto the electronegative nitrogen atom.
In 3-methylpyridine, the electrons can be delocalized only onto carbon atoms.

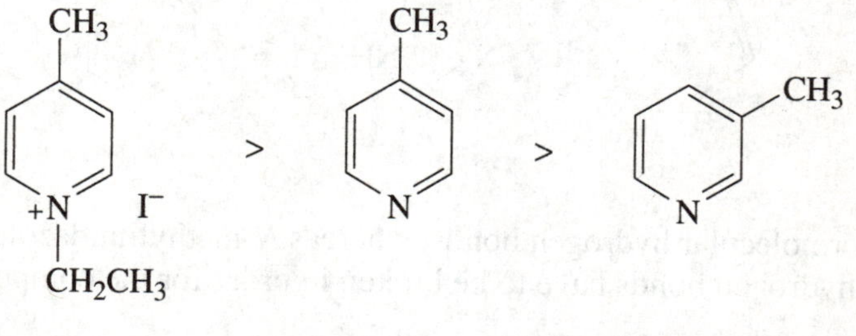

21. There are three possible sites for electrophilic substitution: C-2, C-4, and C-5. To determine the major product, compare the relative stabilities of the carbocations formed in the first step of the reaction.

Substitution at C-2 leads to a carbocation with three contributing resonance structures.
One of them is particularly unstable, because the positive charge is on a nitrogen atom that does not have a complete octet.

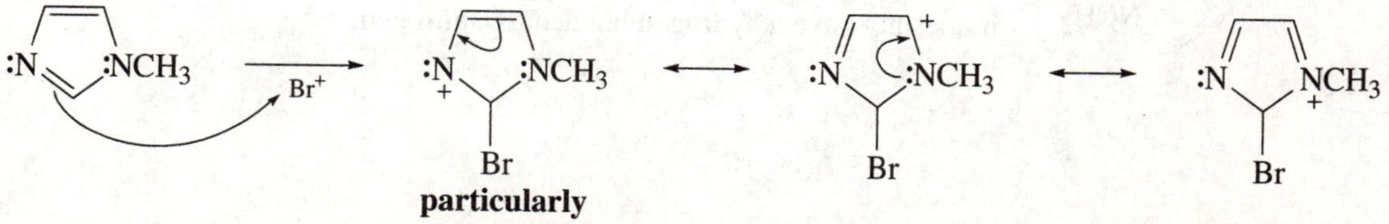

particularly unstable

Substitution at C-4 leads to a carbocation with only two contributing resonance structures.

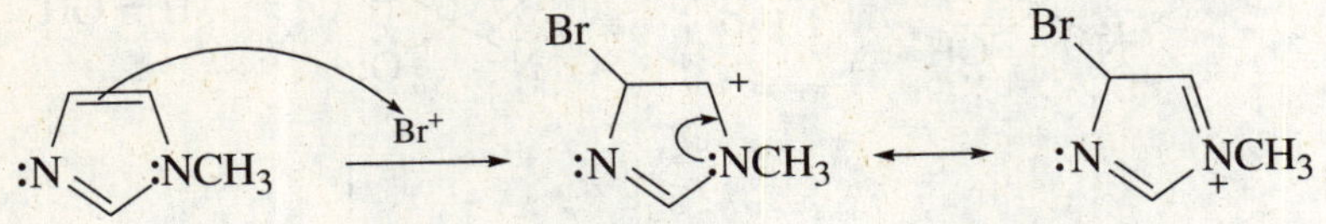

Substitution at C-5 leads to a carbocation with three contributing resonance structures. Because none of them are as unstable as one of the contributing resonance structures obtained from substitution at C-2, it is the most stable of the three possible carbocation intermediates.

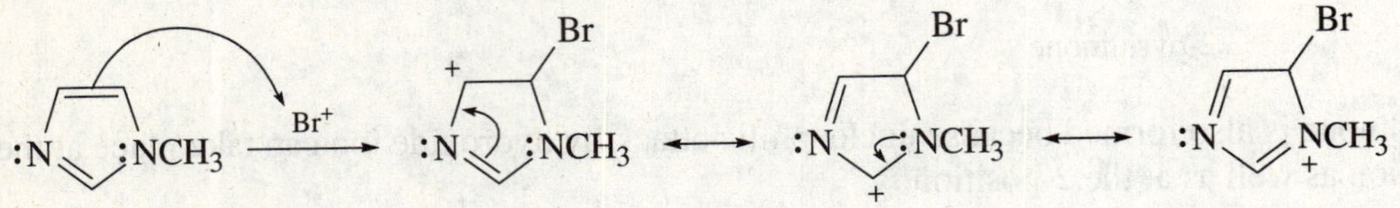

Therefore, the major product of the reaction is 5-bromo-*N*-methylimidazole.

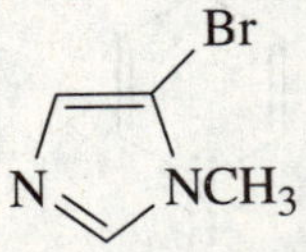

5-bromo-*N*-methylimidazole

22. Pyrrole and imidazole are more reactive than benzene because each has a carbocation intermediate that can be stabilized by resonance electron donation into the ring by a nitrogen atom.
Pyrrole is more reactive than imidazole because the second nitrogen atom of imidazole cannot donate electrons into the ring by resonance but can only withdraw electrons from the ring inductively.

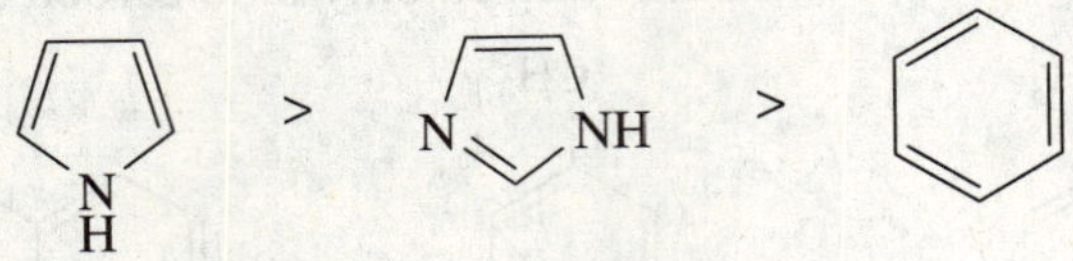

23. Imidazole forms intermolecular hydrogen bonds, whereas *N*-methylimidazole cannot form hydrogen bonds. Because the hydrogen bonds have to be broken in order for the compound to boil, imidazole has a higher boiling point.

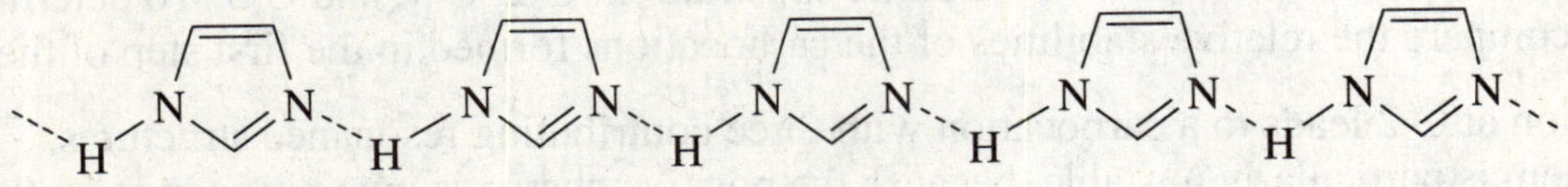

This compound cannot form hydrogen bonds, because it does not have a hydrogen bonded to a nitrogen.

24. fraction of imidazole in the acidic form $= \dfrac{[H^+]}{K_a + [H^+]}$

$$pH = 7.3; [H^+] = 5.0 \times 10^{-8}$$

$$pK_a = 6.8; K_a = 1.58 \times 10^{-7}$$

fraction of imidazole in the acidic form $= \dfrac{5.0 \times 10^{-8}}{1.58 \times 10^{-7} + 5.0 \times 10^{-8}}$

$$= \frac{5.0 \times 10^{-8}}{2.08 \times 10^{-7}}$$

$$= 0.24$$

percent of imidazole in the acidic form = 24%

25. Yes, porphyrin is aromatic, because it fulfills the two requirements for aromaticity. It has an uninterrupted ring of *p* orbital-bearing atoms (it is cyclic and planar), and the π cloud contains an odd number of pairs (thirteen pairs) of π electrons.

26. **a.** 2-methylazacyclobutane
b. 2, 3-dimethylpiperidine
c. 3-chloropyrrole
d. 2-ethyl-5-methylpiperidine

27. **a.**

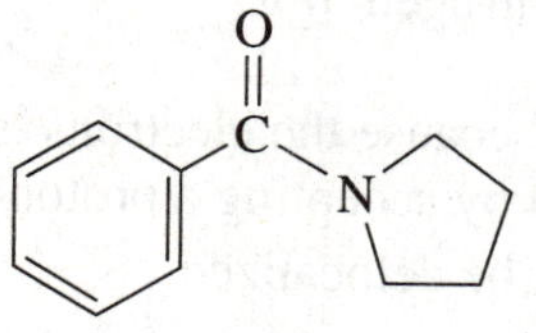

b.

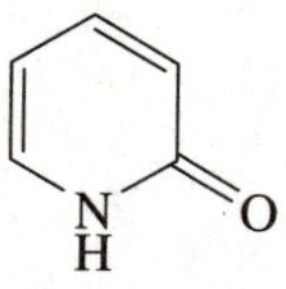

c. $CH_3CH_2CH_2CH_2CH_2NH_2$

d.

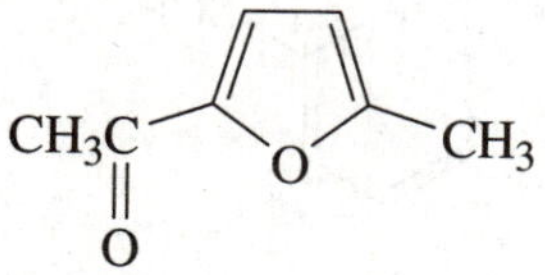

e.

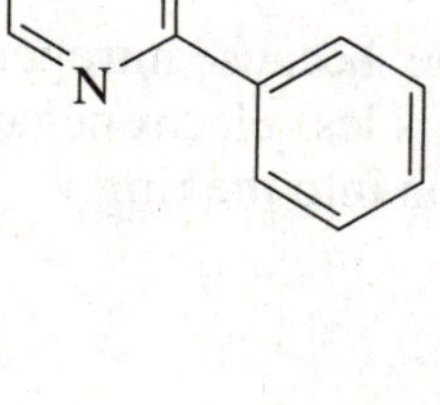

f.

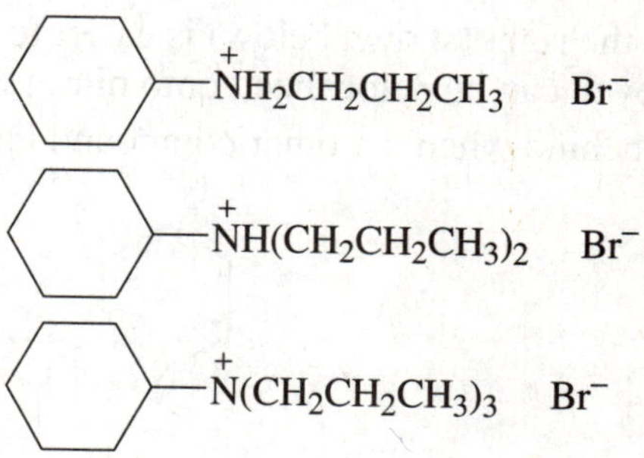

g.

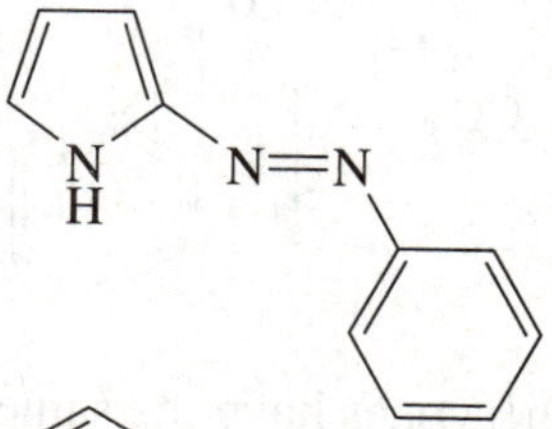

h.

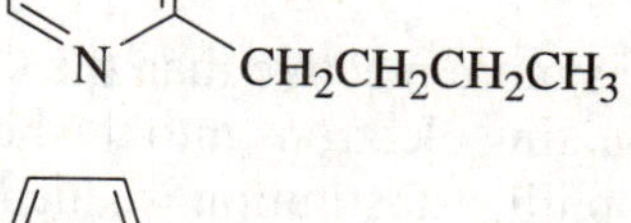

i.

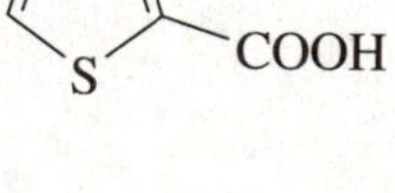

28.

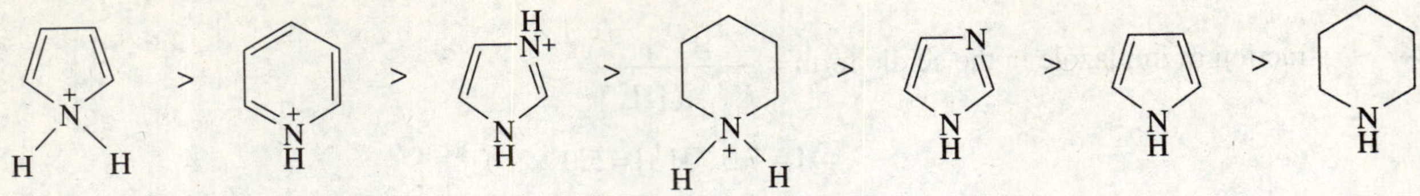

A **B** **C** **D** **E** **F** **G**

A is the most acidic because it becomes aromatic when it loses a proton.

B, **C**, and **D** are the next most acidic because in all three, the proton is bonded to a positively charged nitrogen atom.

B and **C** are more acidic than **D** because in **B** and **C** the proton to be lost is bonded to an *sp*^{2} nitrogen atom, which is more electronegative than the *sp*^{3} nitrogen atom in **D**.

B is more acidic than **C** because the uncharged nitrogen atom in **C** can donate electrons by resonance to the positively charged nitrogen atom.

Neutral compounds **E**, **F**, and **G** are the least acidic, with **E** and **F** more acidic than **G**, because **E** and **F** lose a proton from an *sp*^{2} nitrogen atom, whereas a proton is lost from a less electronegative *sp*^{3} nitrogen atom in **G**.

E is more acidic than **F** because of the electron-withdrawing second nitrogen in **E**.

29. The compound on the right (shown below) is easier to decarboxylate because the electrons left behind when CO_2 is removed can be delocalized onto nitrogen and stabilized by accepting a proton from oxygen. The electrons left behind when the other compound loses CO_2 cannot be delocalized.

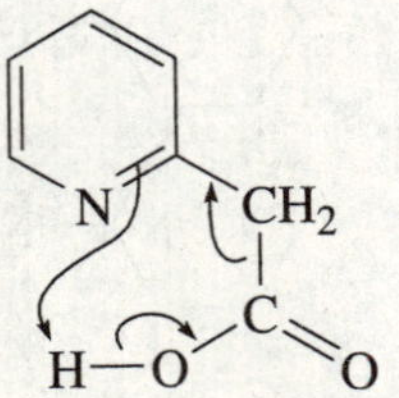

30.

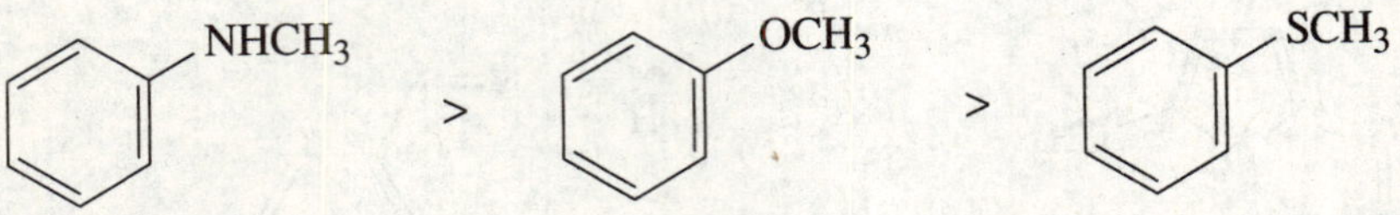

The N-, O-, and S-substituted benzenes have the same relative reactivity toward electrophilic aromatic substitution as the N-, O-, and S-containing five-membered heterocyclic rings and for the same reason.

The N-substituted benzene is more reactive than the O-substituted benzene, because nitrogen is more effective than oxygen at donating electrons into the benzene ring, since it is less electronegative than oxygen. (Recall that electrophilic substitution is aided by electron donation into the ring.)

The S-substituted benzene is the least reactive because the lone-pair electrons of sulfur are in a 3*p* orbital, whereas the lone-pair electrons of nitrogen and oxygen that are delocalized are in a 2*p* orbital. The overlap of a 3*p* orbital with the 2*p* orbital of carbon is less effective than the overlap of a 2*p* orbital with the 2*p* orbital of carbon.

31. By comparing the carbocation intermediates formed in the rate-determining step when the amino-substituted compound undergoes electrophilic substitution at C-3 and C-4, you can see that the carbocation formed in the case of substitution at C-3 is more stable; one of its resonance contributors is particularly stable because it is the only contributor that does not have an atom with an incomplete octet. As a result the amino- substituted compound undergoes electrophilic substitution predominantly at C-3. Therefore, the keto-substituted compound is the one that undergoes electrophilic substitution predominantly at C-4.

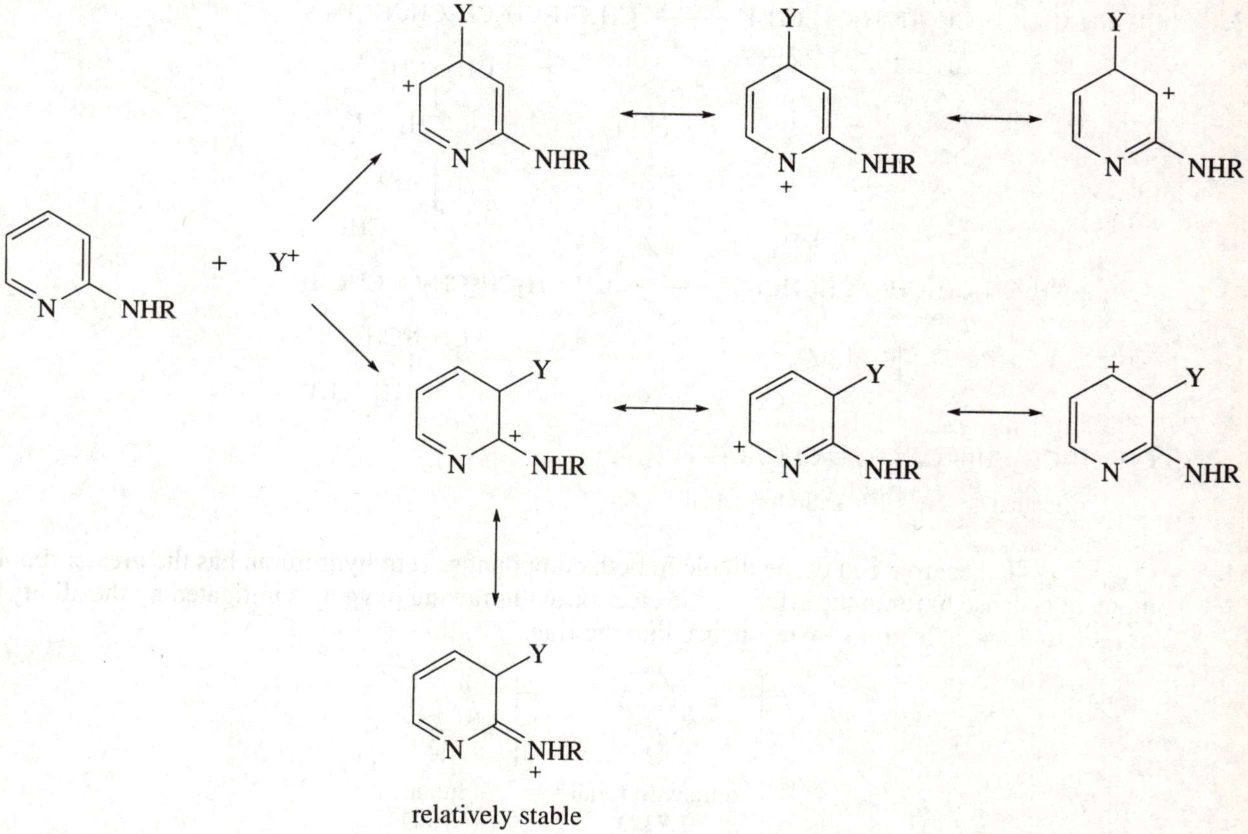

32. **a.** The Lewis acid, $AlCl_3$, complexes with nitrogen, causing the aziridine ring to open when it is attacked by the nucleophilic benzene ring. The ring will open in the direction that puts the partial positive charge on the more substituted carbon, because that will from the more stable carbocation.

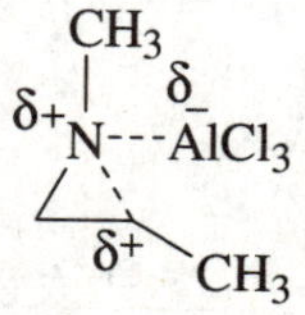

Therefore, the major and minor products are those shown.

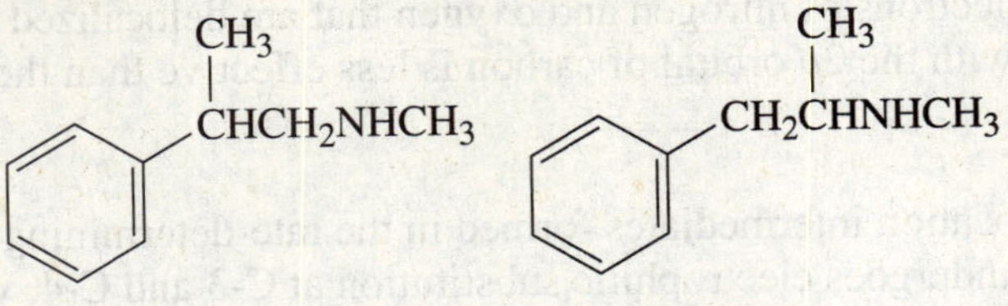

major minor

b. Yes, epoxides can undergo similar reactions.

33.

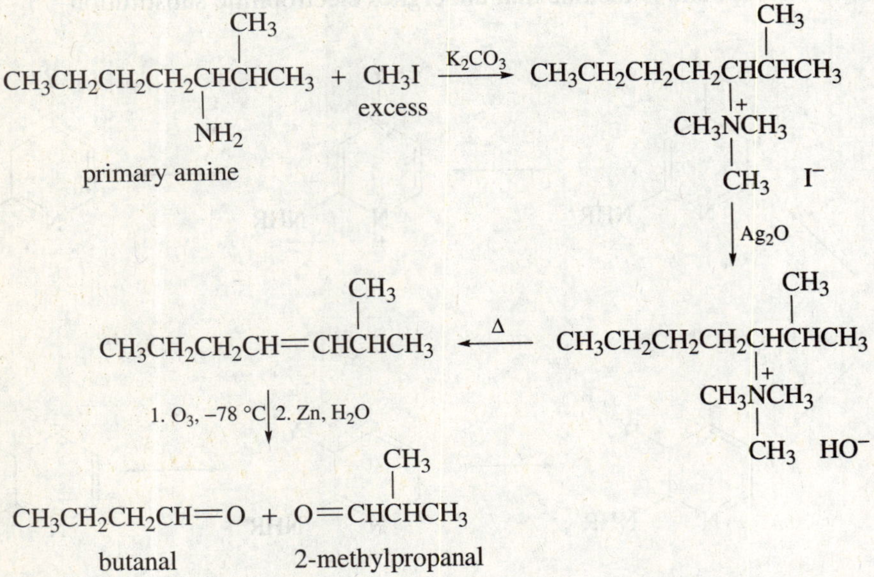

34. Oxygen is the negative end of the dipole in both compounds. Tetrahydrofuran has the greater dipole moment because in furan the effect of the electron-withdrawing oxygen is mitigated by the ability of oxygen to donate electrons by resonance into the ring.

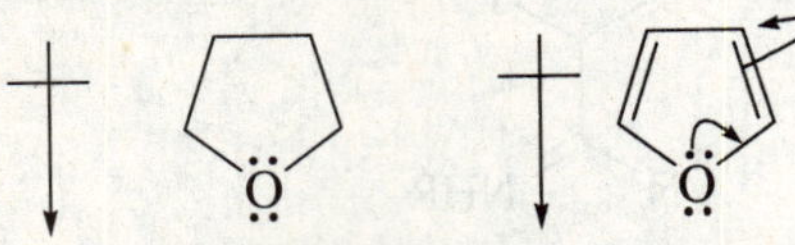

tetrahydrofuran 1.73 D

furan 0.70 D

35. We saw in Section 14.19 of the text that an alkyl substituent bonded to a benzene ring can be oxidized to a carboxylic acid substituent. Similarly, an alkyl substituent bonded to a pyridine ring can be oxidized to a carboxylic acid substituent. Nicotine is an alkaloid found in tobacco leaves. It is used in agriculture as an insecticide.

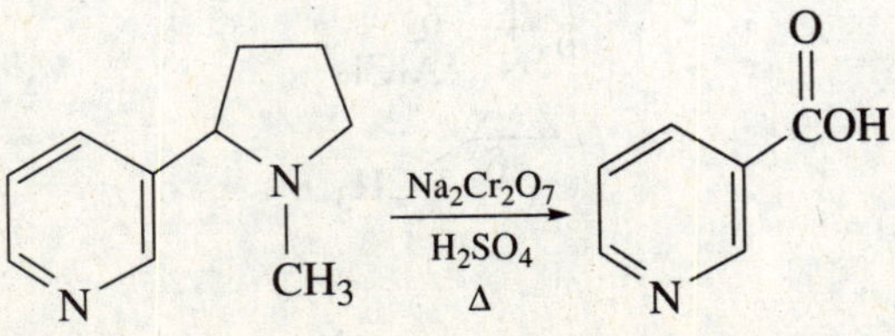

nicotine niacin

36. Pyrrolidine is a saturated nonaromatic compound, whereas pyrrole and pyridine are unsaturated aromatic compounds. The C-2 hydrogens of pyrrolidine are at δ 2.82, about where one would expect the signal for hydrogens bonded to an sp^3 carbon adjacent to an electron-withdrawing amino group.

The C-2 hydrogens of pyrrole and pyridine are expected to be at a higher frequency because of magnetic anisotropy. Because the nitrogen of pyrrole donates electrons into the ring and the nitrogen of pyridine withdraws electrons from the ring, the C-2 hydrogens of pyrrole are in an environment with a greater electron density, so they should show a signal at a lower frequency relative to the C-2 hydrogens of pyridine. Thus, the C-2 hydrogens of pyrrole are at δ 6.42, and the C-2 hydrogens of pyridine are at δ 8.50.

37. A UV spectrum results from the π electron system. The lone-pair electrons on the nitrogen atom in aniline are delocalized into the benzene ring and, thus, are part of the π system. Protonation of aniline removes two electrons from the π system, which has a significant effect on its UV spectrum.

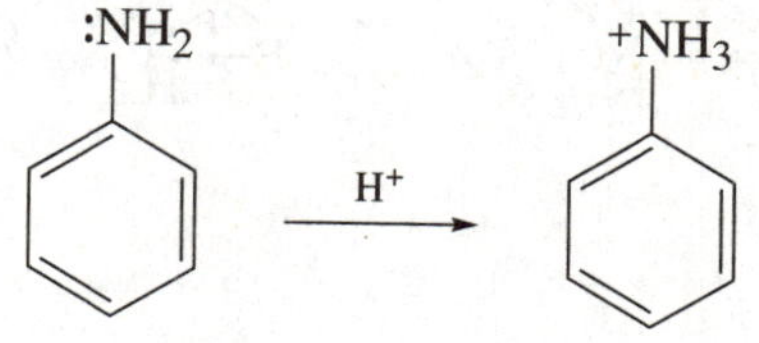

The lone-pair electrons on the nitrogen atom in pyridine are sp^2 electrons and thus are not part of the π system. Protonation of pyridine, therefore, does not remove any electrons from the π system and has only a minor effect on the UV spectrum.

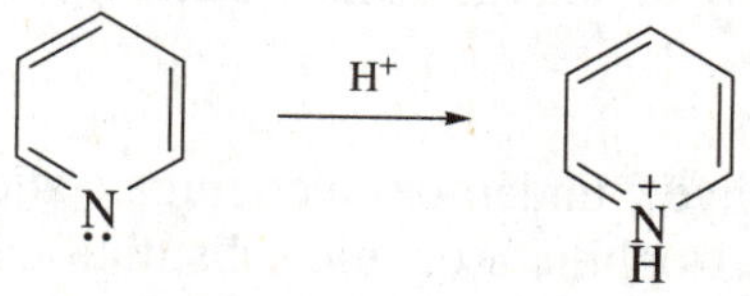

38. When ammonia loses a proton, the electrons left behind remain on nitrogen.

$$\ddot{N}H_3 \longrightarrow {}^{-}\!:\!\ddot{N}H_2 + H^+$$

When pyrrole loses a proton, the electrons left behind on the nitrogen can be delocalized onto the four ring carbons. Electron delocalization stabilizes the anion and makes it easier to form.

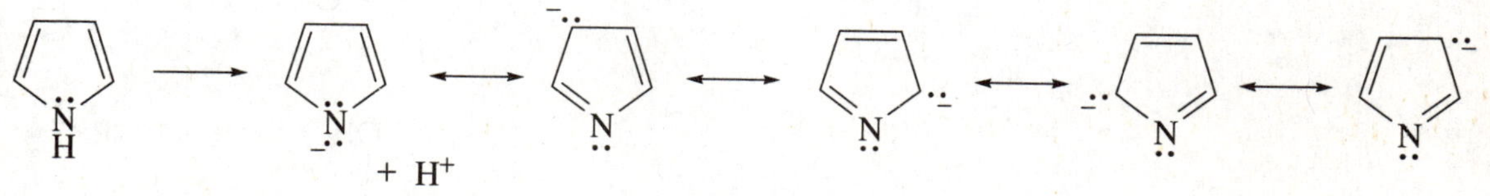

39.

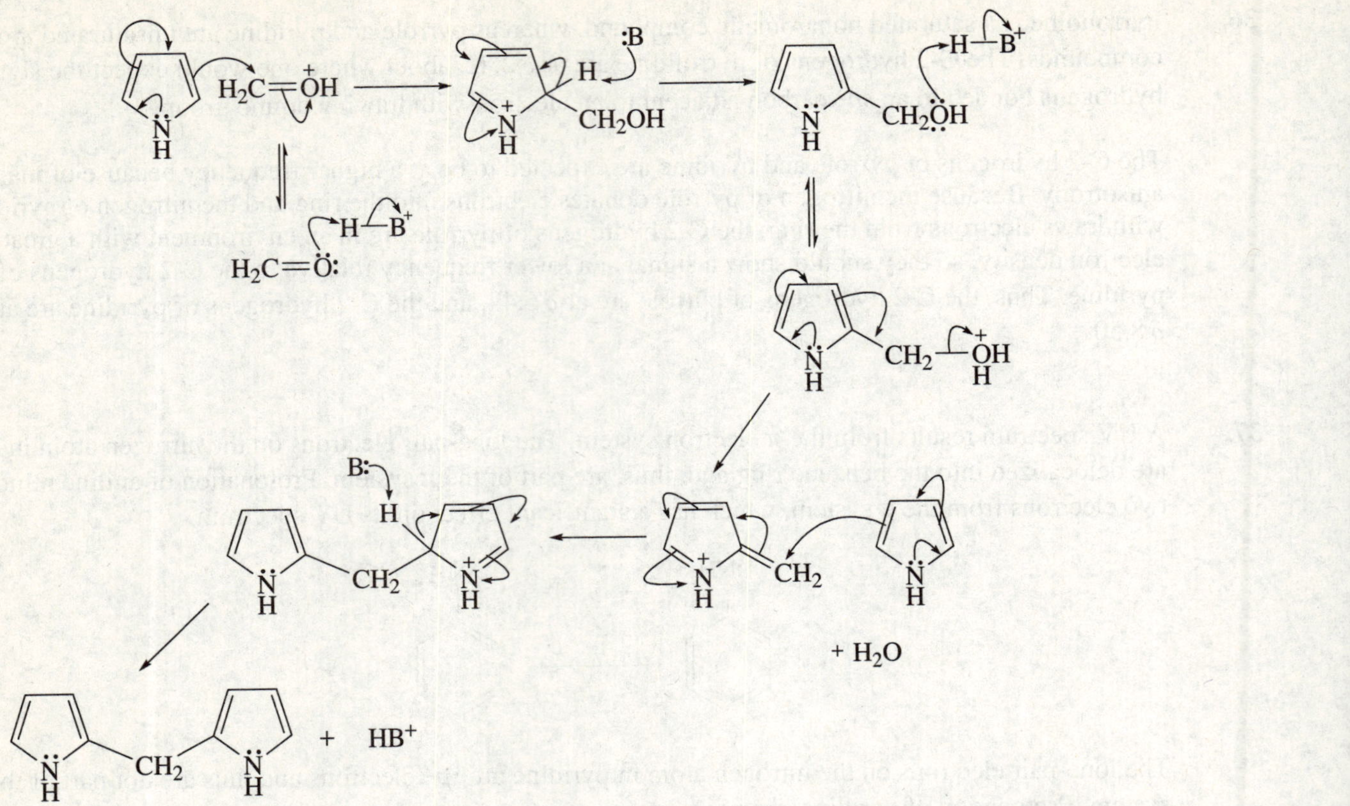

40. Before we can answer the questions, we must figure out the mechanism of the reaction. Once the mechanism is known, it will be easy to determine how a change in a reactant will affect the product. The mechanism is shown below:

Propenal, an *α,β*-unsaturated aldehyde, undergoes a conjugate addition reaction with aniline. This is followed by an intramolecular electrophilic aromatic substitution reaction. Dehydration of the alcohol results in 1,2-dihydroquinoline, which is oxidized to quinoline by nitrobenzene.

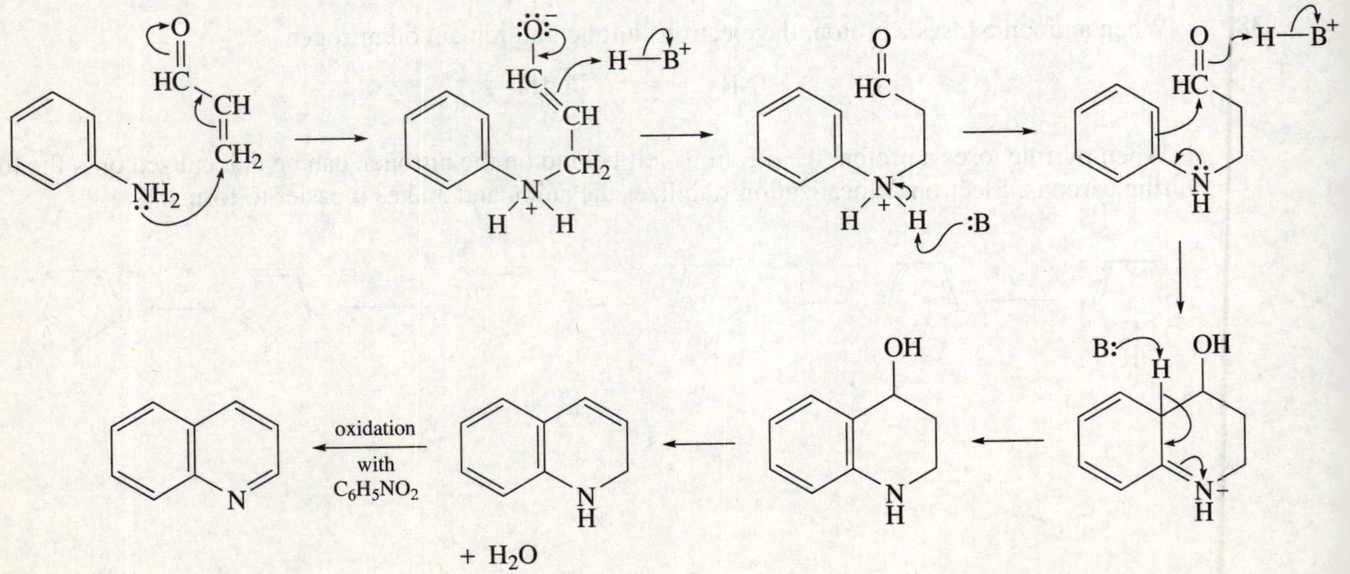

a.

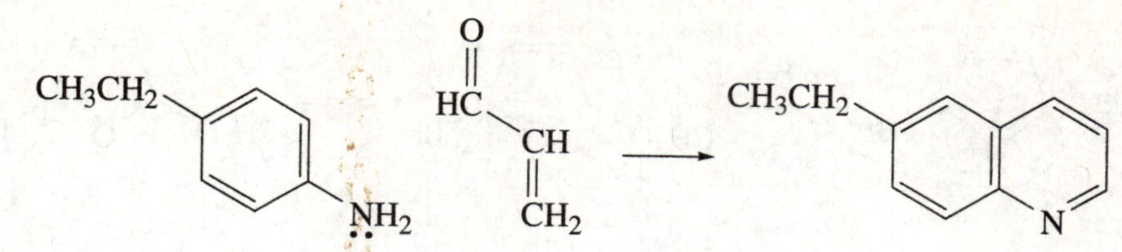

b.

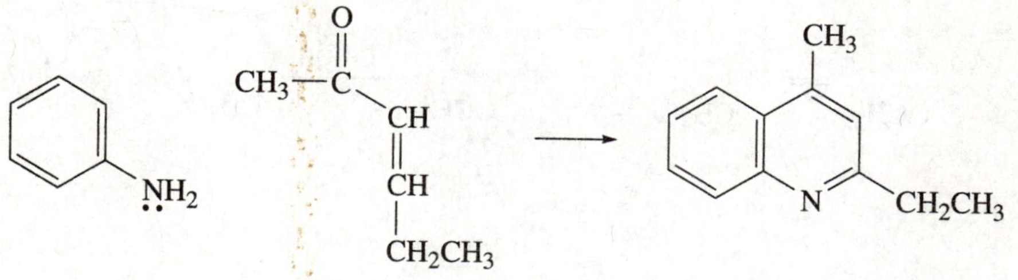

c.

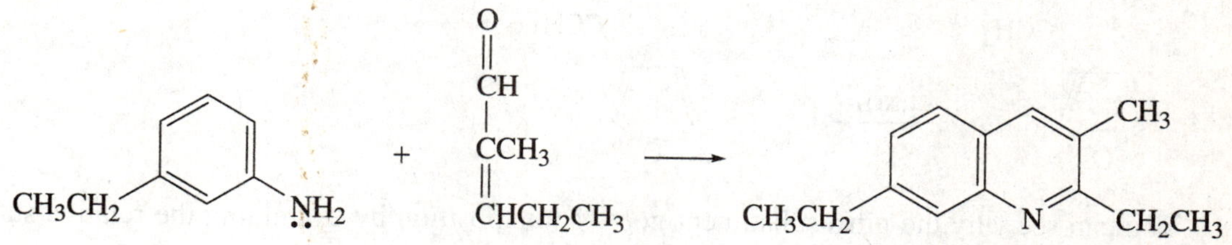

meta-ethylaniline 2- methyl-2-pentenal 2,7-diethyl-3-methyquinoline

41. **a.**

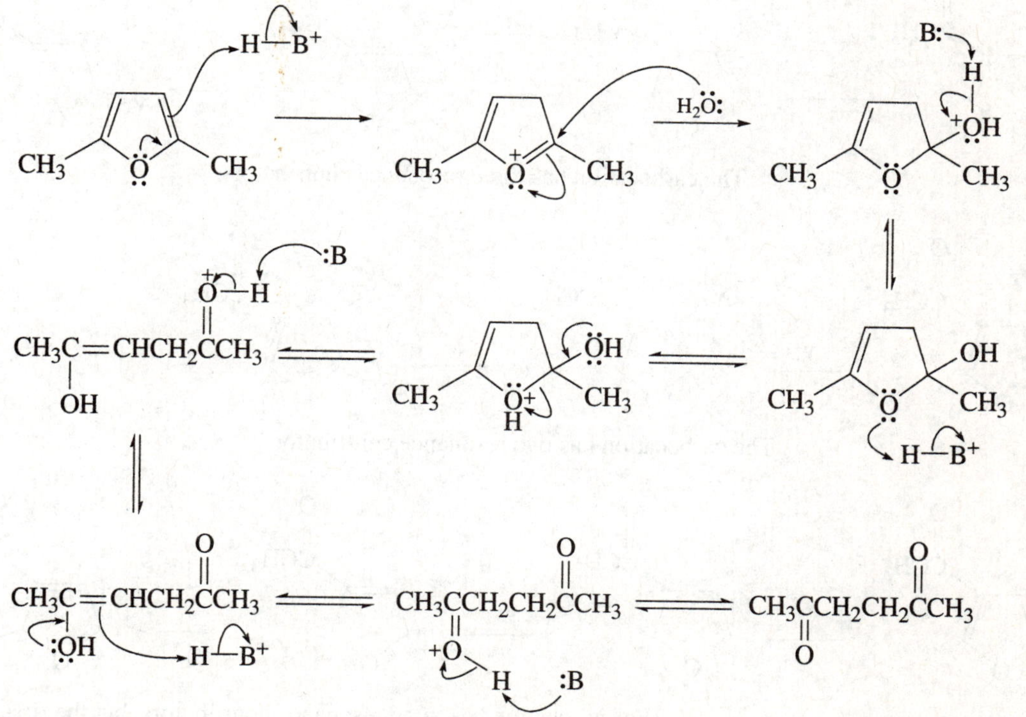

b.

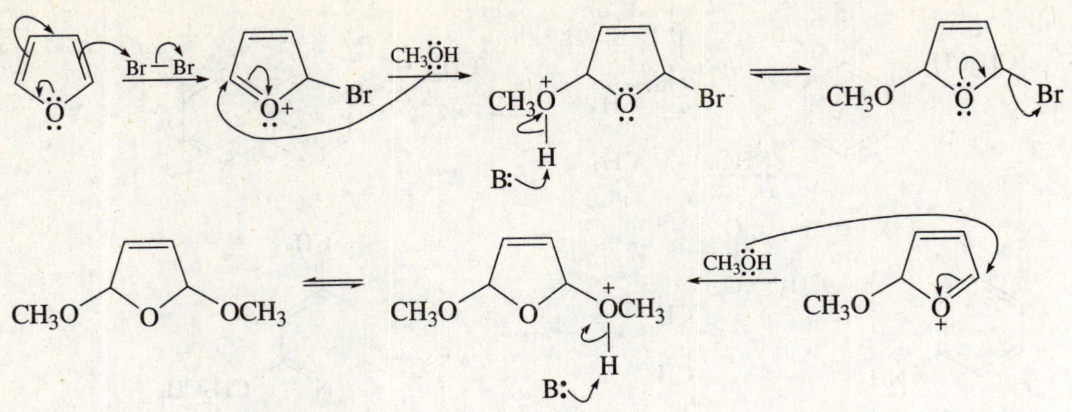

42. **a.**

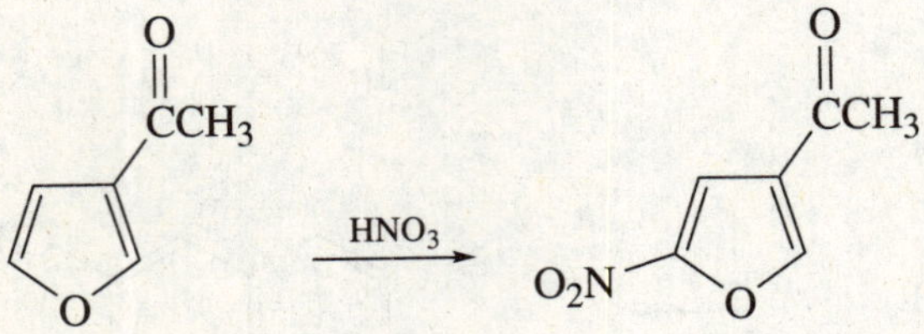

You can see why the nitro substituent goes to this position by examining the relative stabilities of the possible carbocation intermediates.

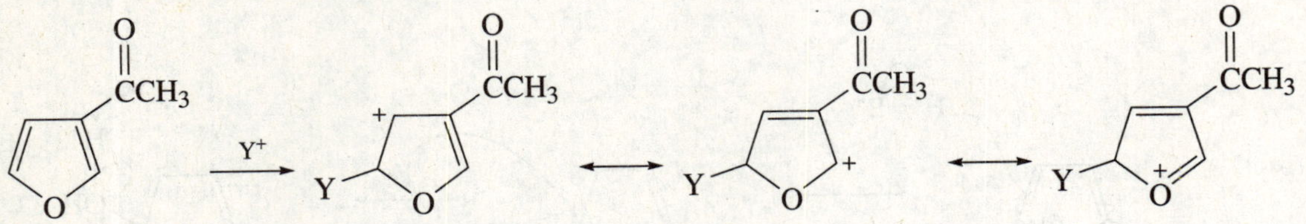

The carbocation has three resonance contributors.

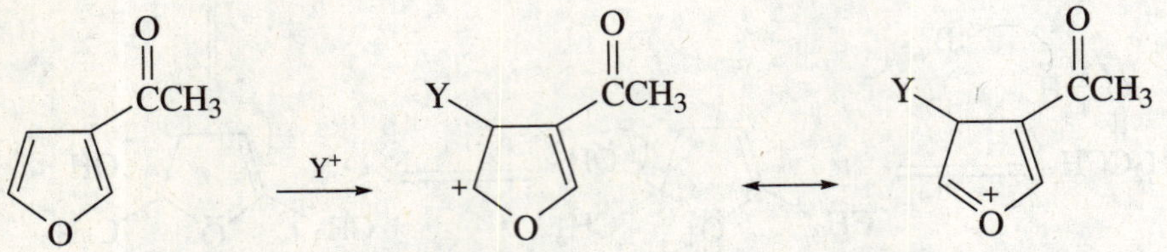

The carbocation has two resonance contributors.

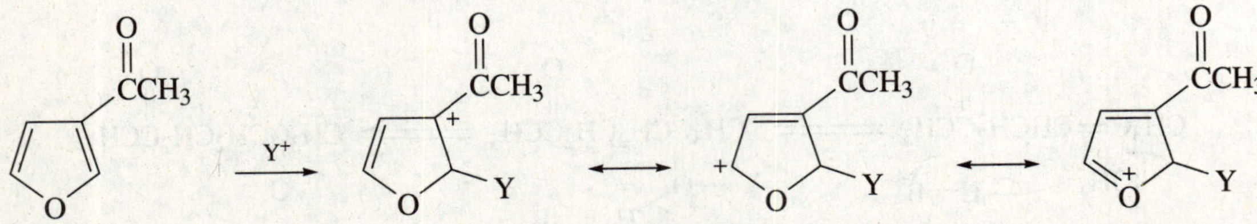

The carbocation has three resonance contributors, but the first one is relatively unstable becasuse the positive charge is on the carbon attached to the electron-withdrawing substituent.

b.

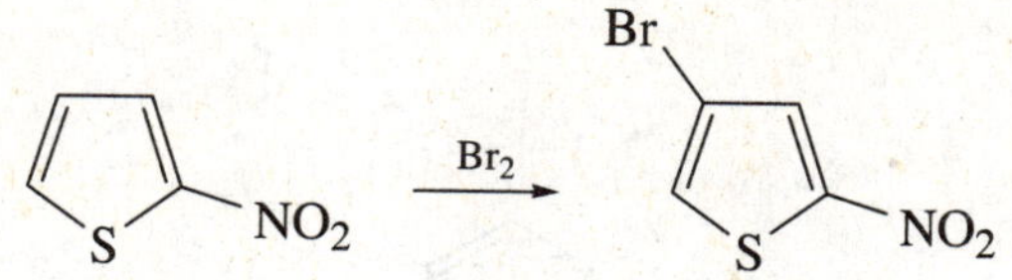

You can understand why the bromo substituent goes to this position by examining the relative stabilities of the possible carbocation intermediates.

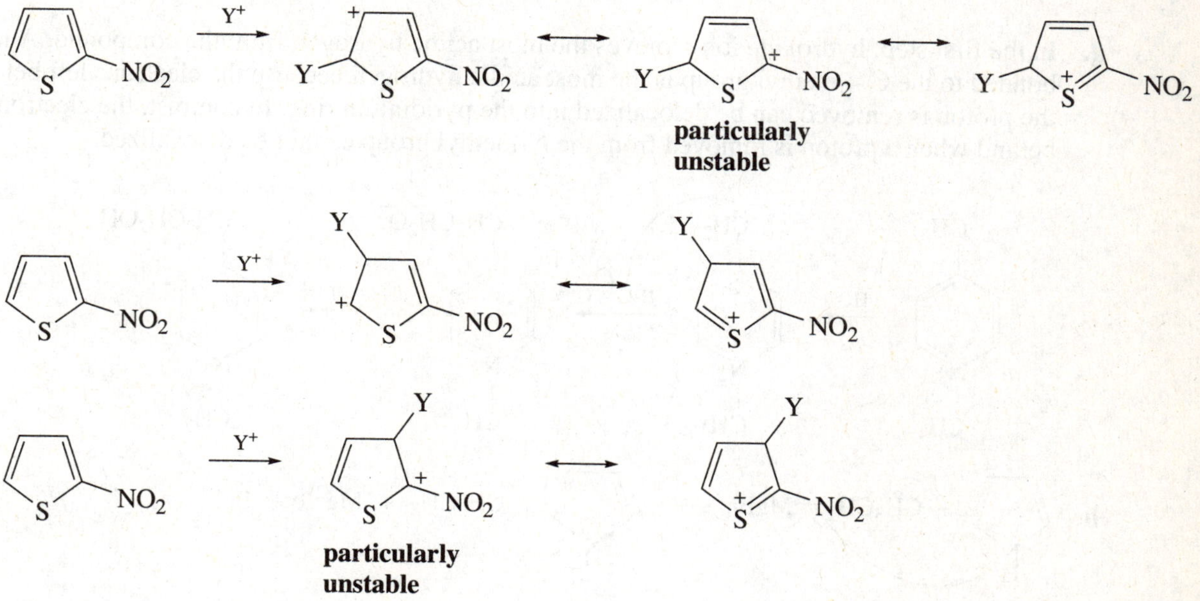

c.

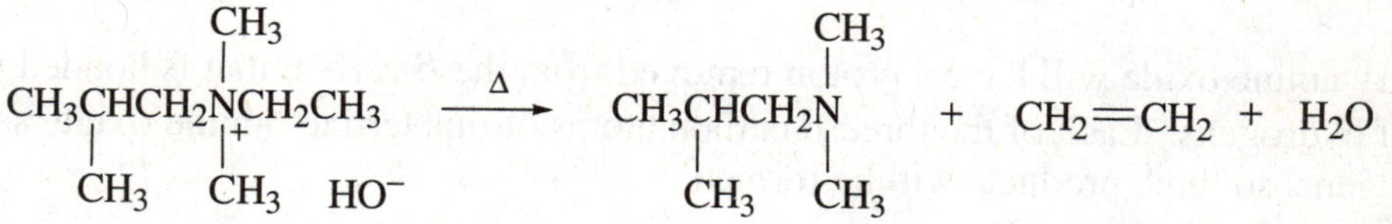

d.

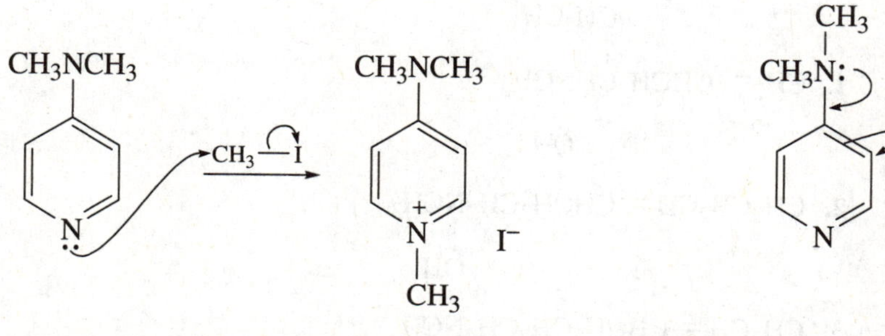

The other nitrogen is not alkylated, because its lone pair is delocalized into the pyridine ring, so it is not available to react with the alkyl halide.

e.

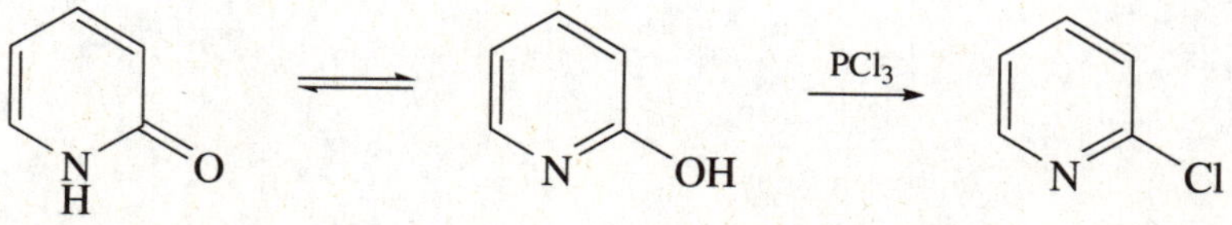

PCl_3 substitutes a Cl for an OH, as it does in alcohols and carboxylic acids.

f. 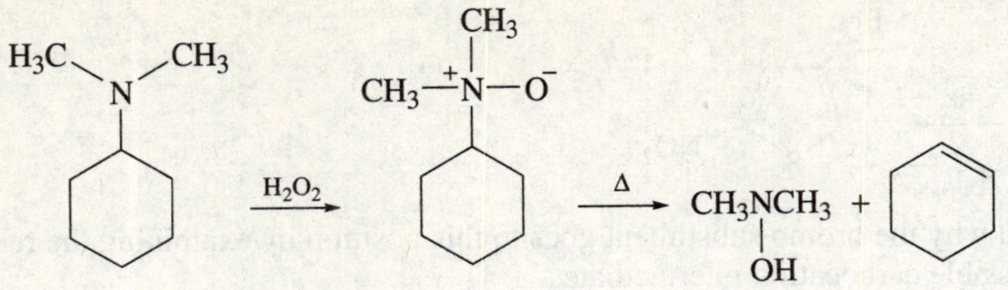

g. In the first step, hydroxide ion removes the most acidic hydrogen from the compound. A hydrogen bonded to the C-4 methyl group is the most acidic hydrogen because the electrons left behind when the proton is removed can be delocalized into the pyridinium ring. In contrast, the electrons left behind when a proton is removed from the N-methyl group cannot be delocalized.

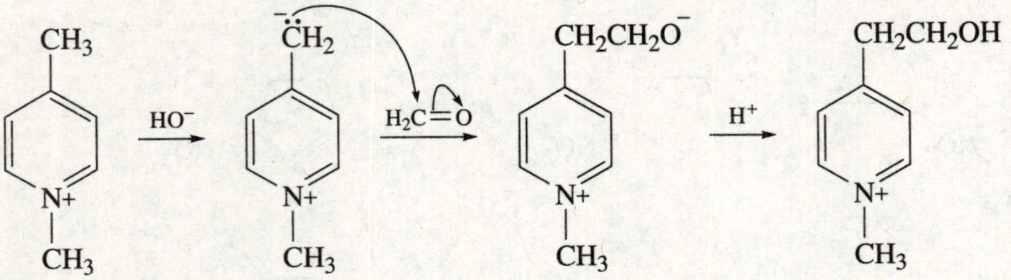

h. 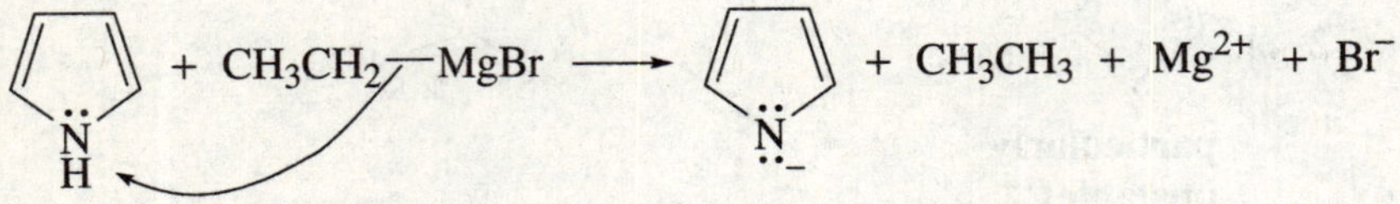

i. The tertiary amine oxide will have a proton removed from the β-carbon that is bonded to the greater number of hydrogens. Each of the three β-carbon atoms of this tertiary amine oxide is attached to two hydrogens, so three products will be formed.

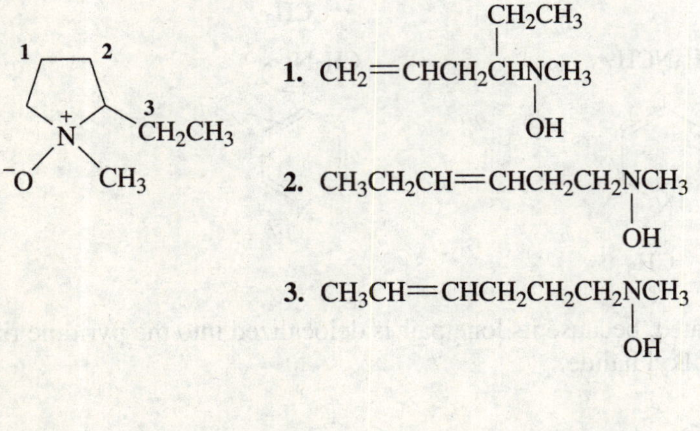

43. *N*-Methylpiperidine forms 1,4-pentadiene.

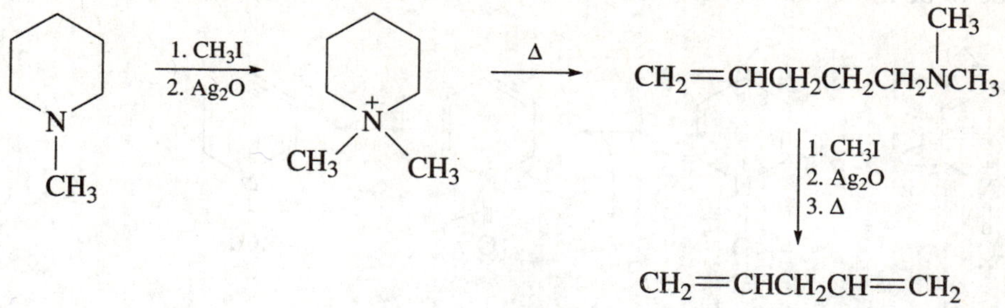

2-Methylpiperidine forms 1,5-hexadiene.

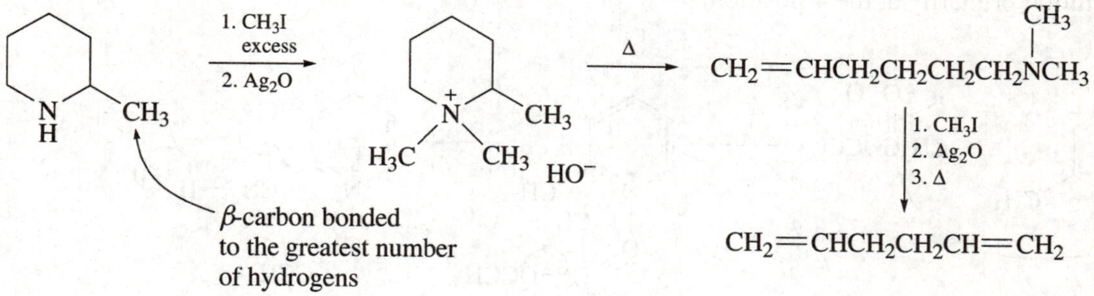

3-Methylpiperidine forms 2-methyl-1,4-pentadiene.

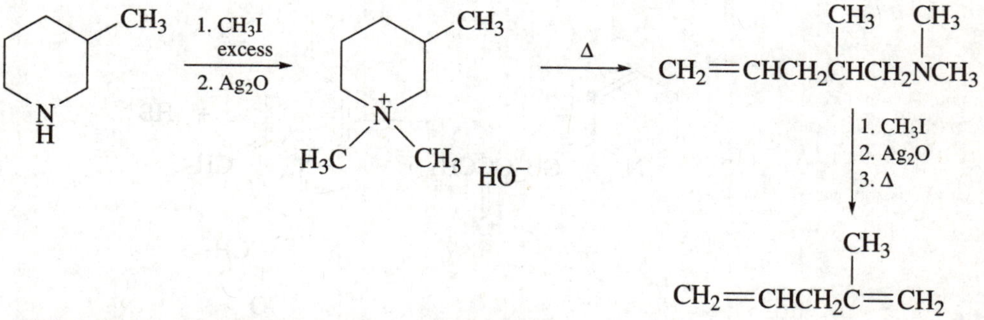

4-Methylpiperidine forms 3-methyl-1,4-pentadiene.

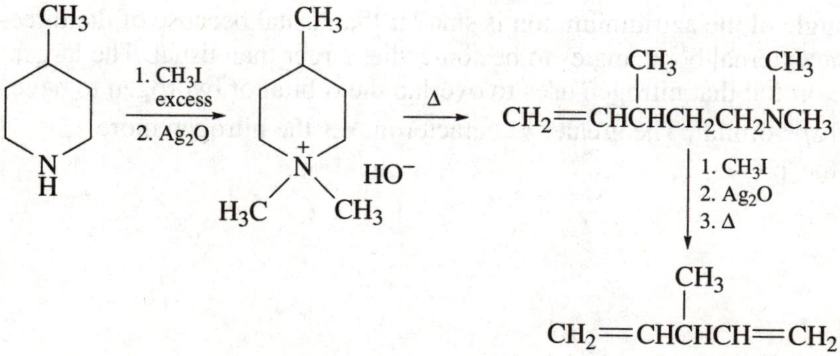

44. The increase in the electron density of the ring as a result of resonance donation of electrons by oxygen causes pyridine-*N*-oxide to be more reactive toward electrophilic substitution than pyridine.

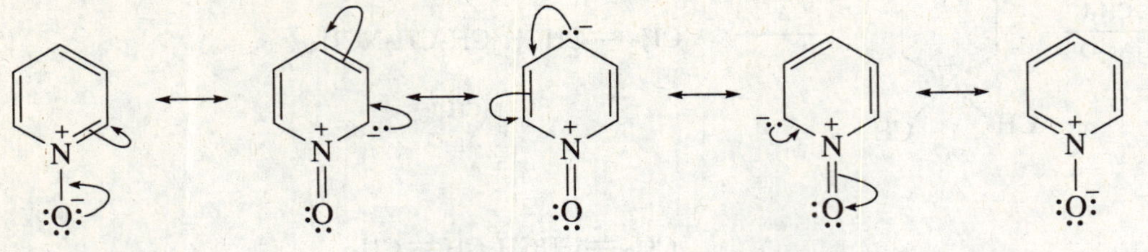

From the resonance structures you can see that the increased electron density is at the 2- and 4-positions. Because the 2-position is somewhat sterically hindered, pyridine-*N*-oxide undergoes electrophilic substitution primarily at the 4-position.

45.

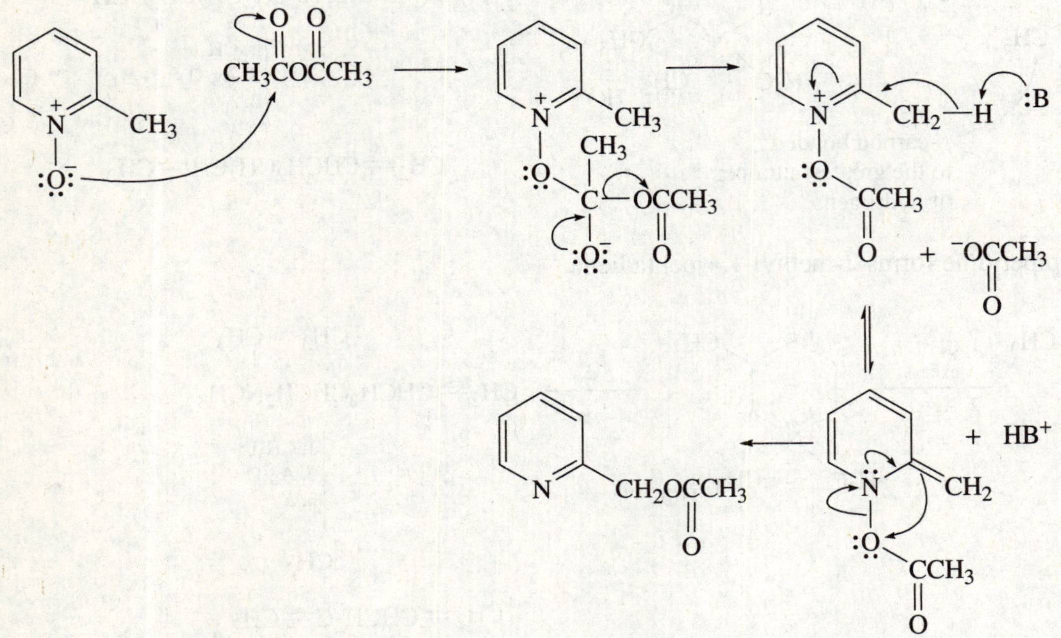

46. The C—N—C internal bond angle of the aziridinium ion is smaller than usual because of the three-membered ring. This causes the external bond angles to be somewhat larger than usual. The larger external bond angles cause the orbital that nitrogen uses to overlap the orbital of hydrogen to have more *s* character than a typical sp^3 orbital. The greater *s* character makes the nitrogen more electronegative which lowers the pK_a.

47.

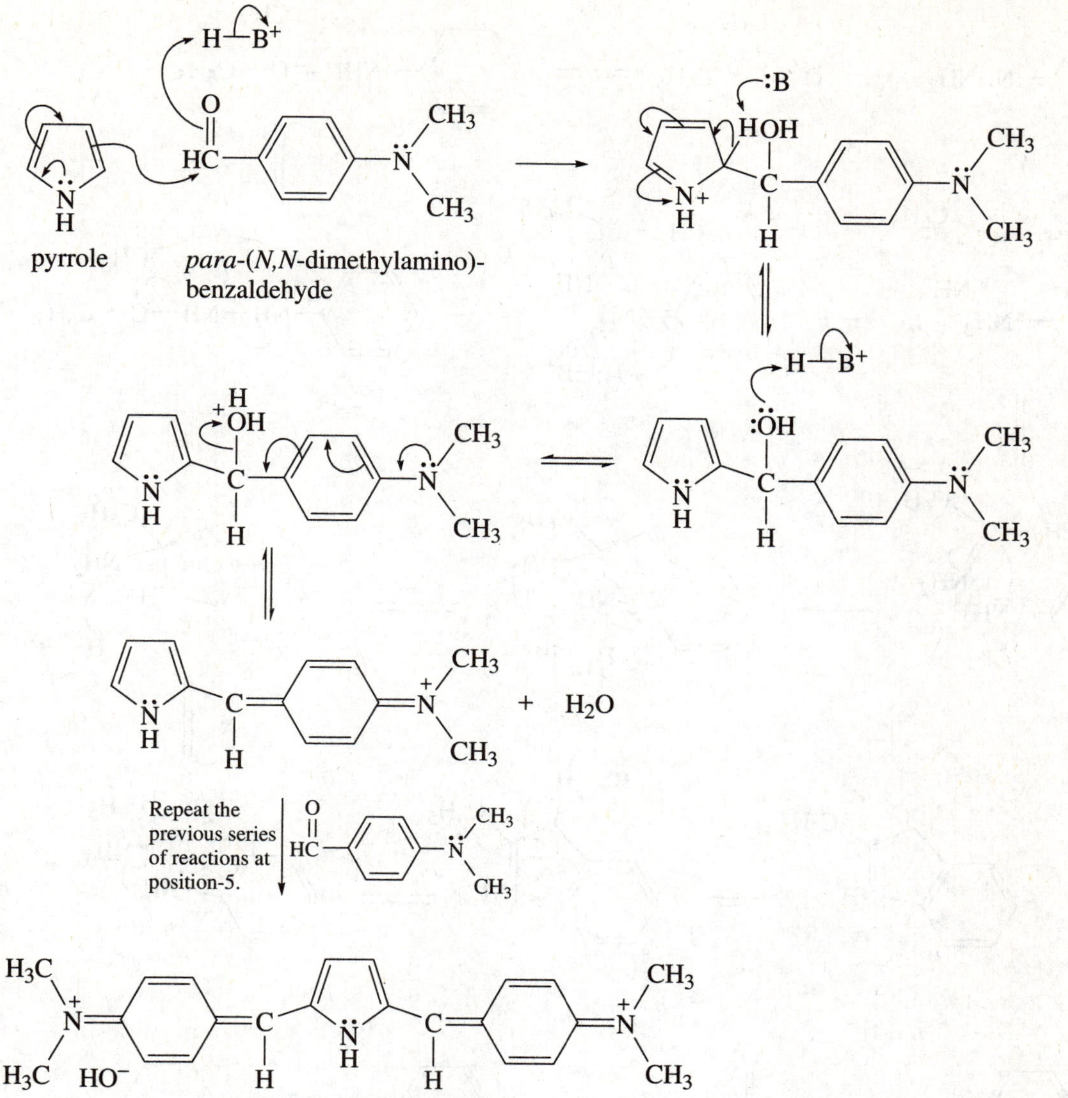

colored compound

48.

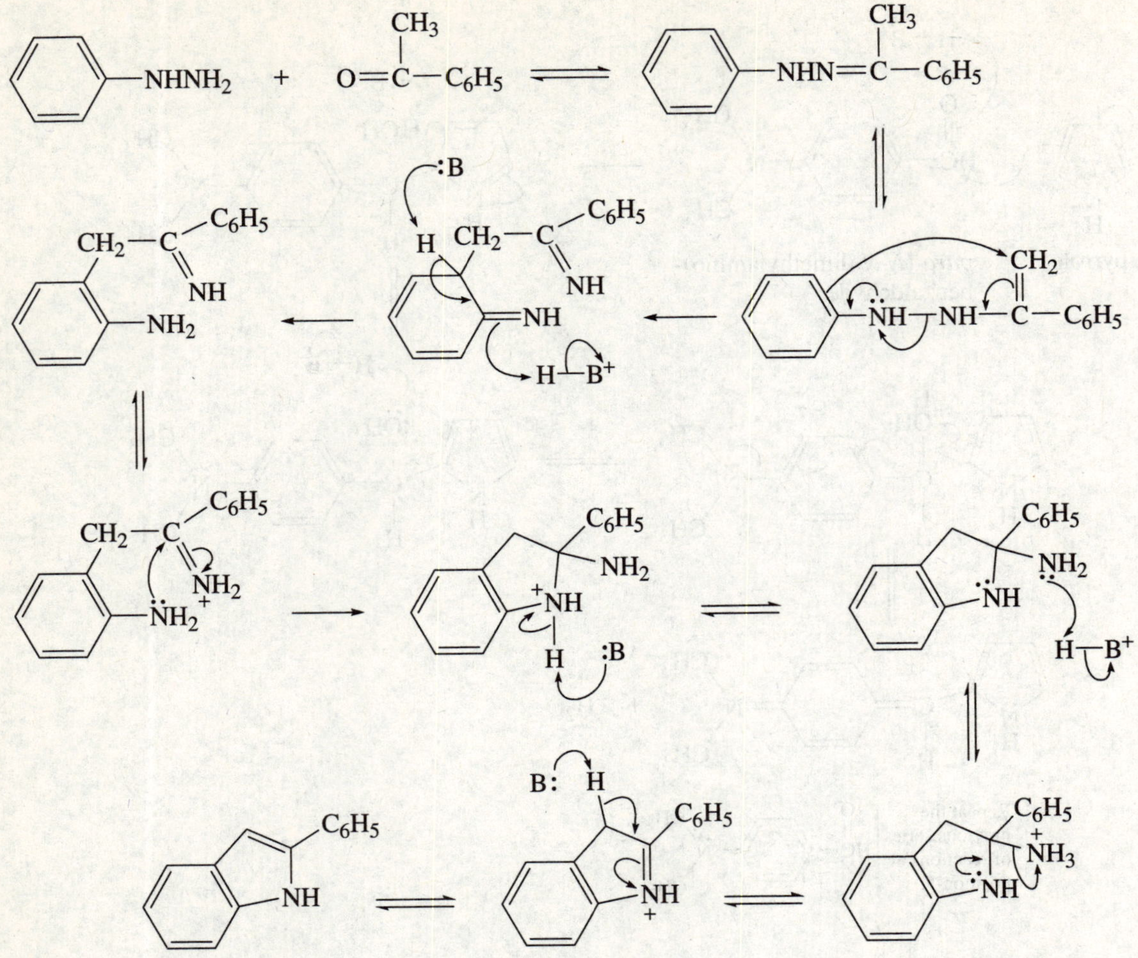

49. **a.**

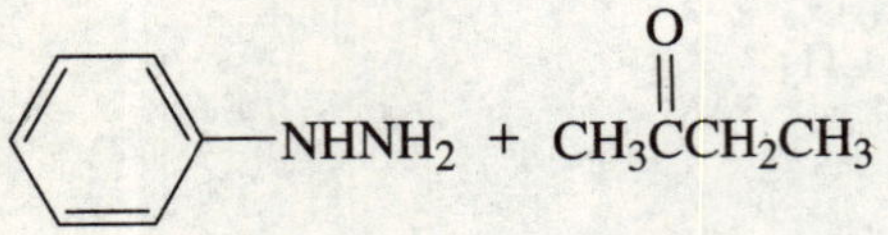

c.

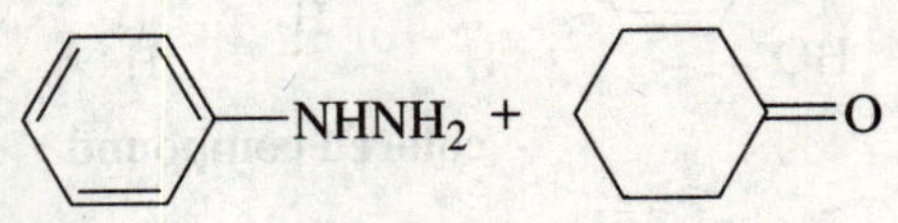

b.

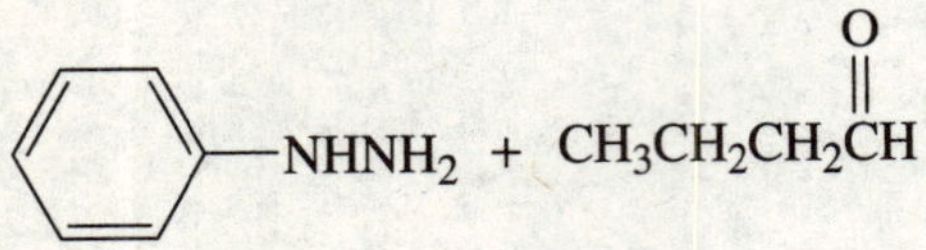

50.

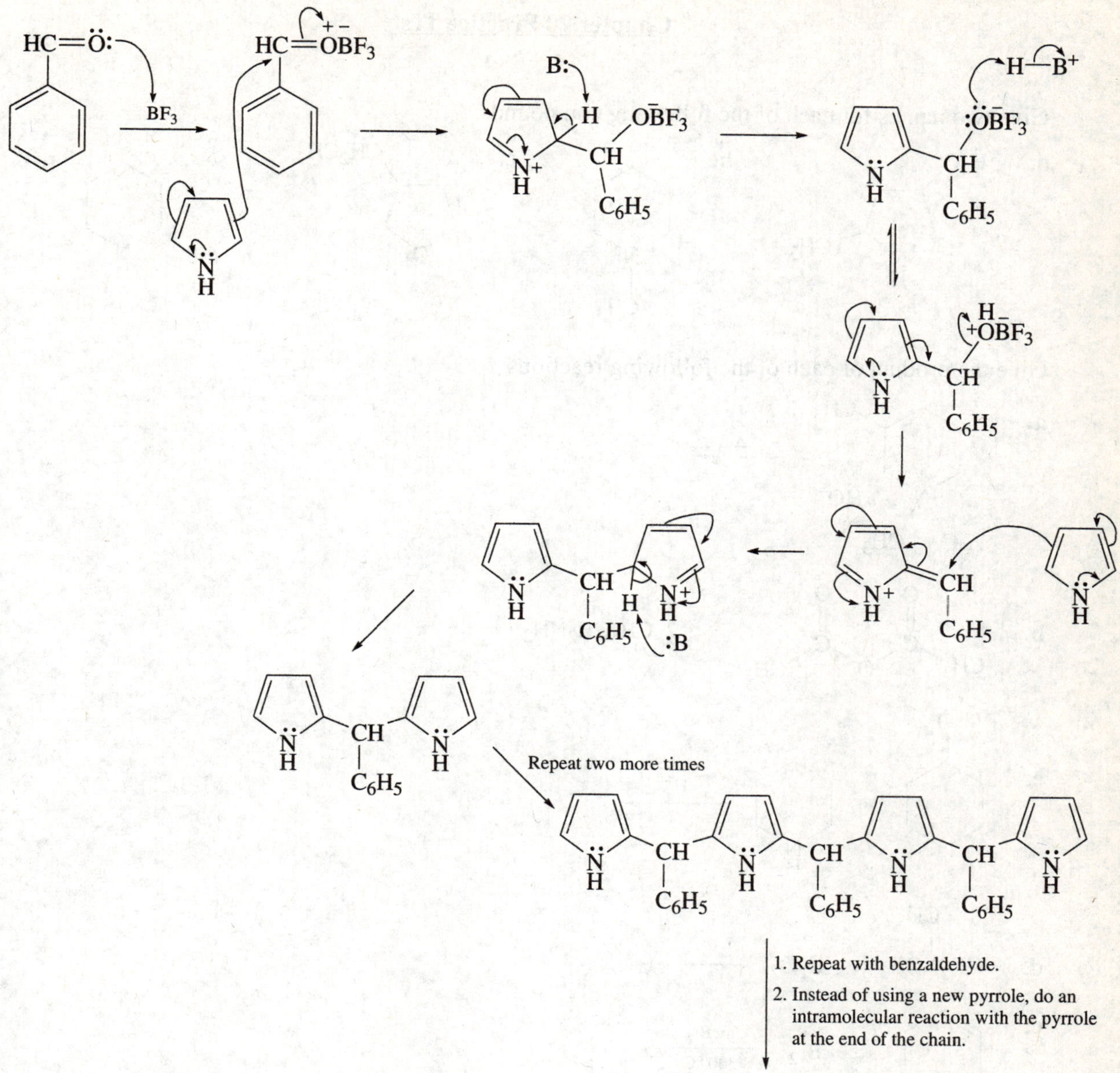
HC=Ö:
BF3
HC=OBF3
B:
H
OBF3
CH
N+
H
C6H5
H—B+
:ÖBF3
N
H
CH
C6H5
H
+OBF3
CH
C6H5
CH
N+
H
C6H5
N
H
CH
H
N+
H
C6H5
:B
CH
C6H5
Repeat two more times
CH
C6H5
1. Repeat with benzaldehyde.
2. Instead of using a new pyrrole, do an intramolecular reaction with the pyrrole at the end of the chain.

Chapter 20 Practice Test

1. Give two names for each of the following compounds:

a. 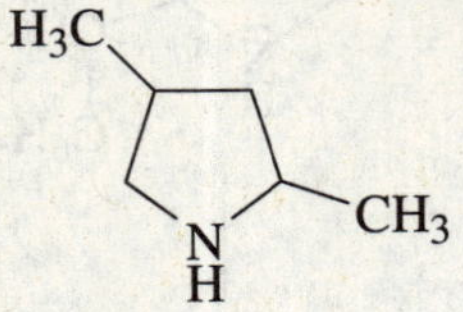

b. 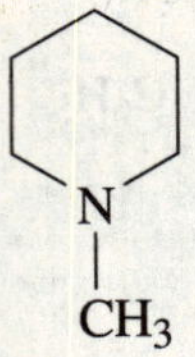

c. 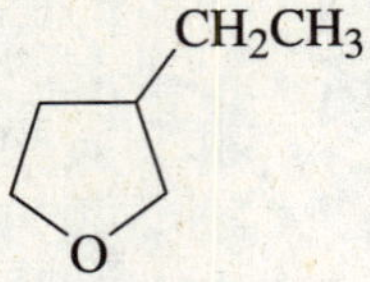

d. 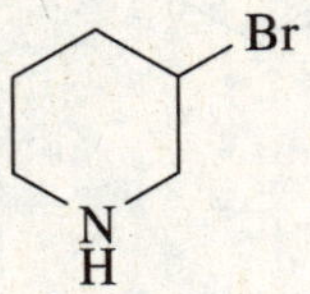

2. Give the product of each of the following reactions:

a. 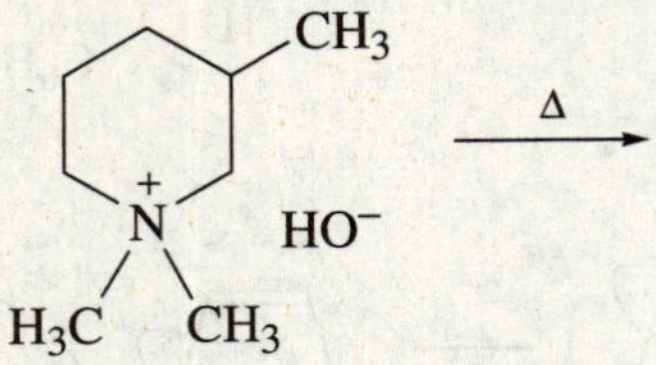

b. 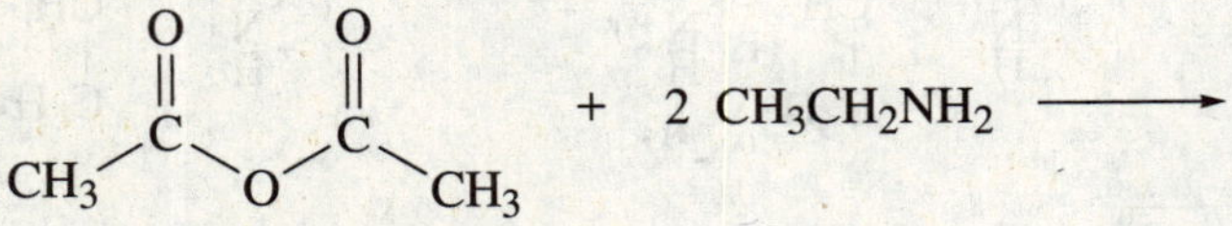

c. 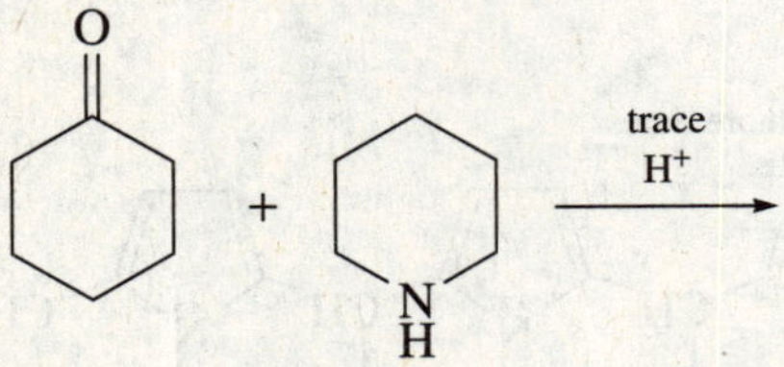

d. $CH_3N(CH_3)CH_3 + H_2O_2 \longrightarrow$

e. 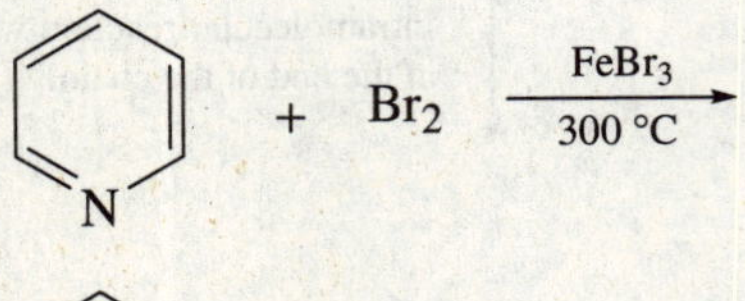

f. 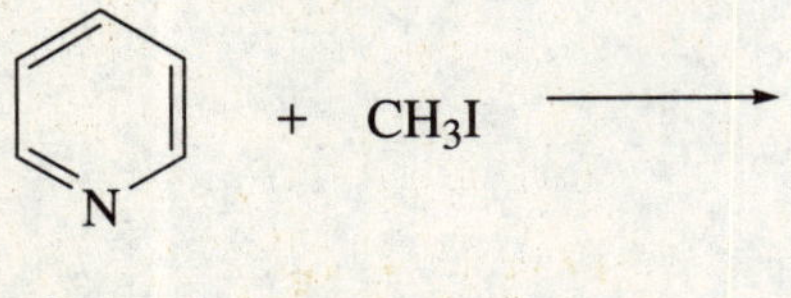

g. 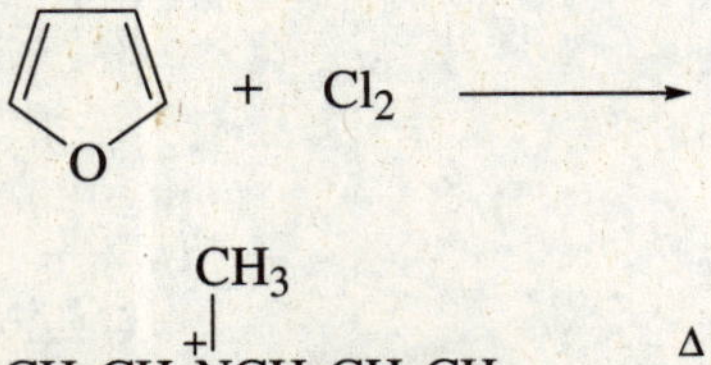

h. $CH_3CH_2\overset{+}{N}(CH_3)(O^-)CH_2CH_2CH_3 \xrightarrow{\Delta}$

3. Which is a stronger acid?

a. 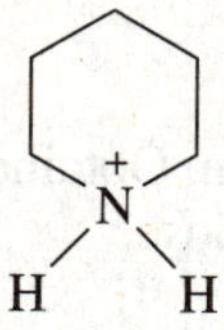 or 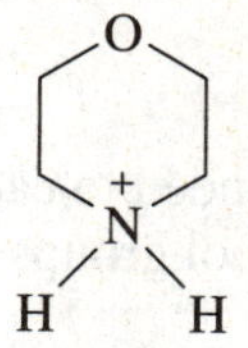

b. 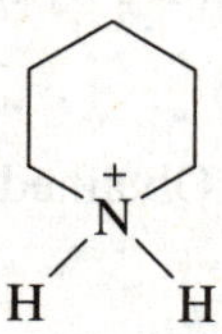 or 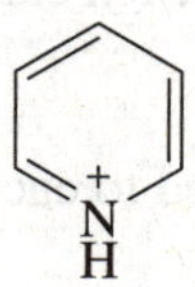

c. 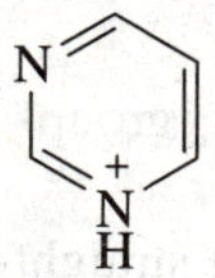 or 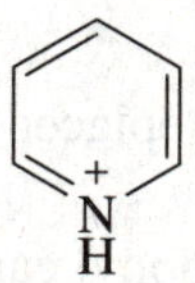

d. 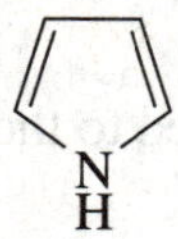 or 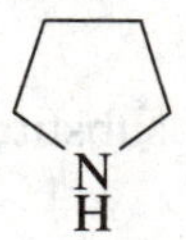

4. Indicate whether each of the following is true or false:

a. Pyridine is more reactive toward nucleophilic aromatic substitution than is pyrrole. T F

b. Pyrrole is more reactive toward electrophilic aromatic substitution than is furan. T F

c. Pyrrole is more reactive toward electrophilic aromatic substitution than is benzene. T F

d. Pyridine is more reactive toward electrophilic aromatic substitution than is benzene. T F

CHAPTER 21
Carbohydrates

Important Terms

aldaric acid — a dicarboxylic acid with an OH group bonded to each carbon. Obtained by oxidizing the aldehyde and primary alcohol groups of an aldose.

alditol — a compound with an OH group bonded to each carbon. Obtained by reducing an aldose or a ketose.

aldonic acid — a carboxylic acid with an OH group bonded to each carbon. Obtained by oxidizing the aldehyde group of an aldose.

aldose — a polyhydroxyaldehyde.

amino sugar — a sugar in which one of the OH groups is replaced by an NH_2 group.

anomeric carbon — the carbon in a cyclic sugar that is the carbonyl carbon in the straight-chain form.

anomeric effect — preference for the axial position by certain substituents bonded to the anomeric carbon.

anomers — two cyclic sugars that differ in configuration only at the carbon that is thecarbonyl carbon in the straight-chain form.

antibody — a compound that recognizes foreign particles in the body.

antigen — a compound that can generate a response from the immune system.

bioorganic compound — an organic compound that is found in a biological system.

carbohydrate — a sugar, a saccharide. Naturally occurring carbohydrates have the D-configuration.

complex carbohydrate — contains two or more sugar molecules linked together; it can be hydrolyzed to simple sugars.

deoxy sugar — a sugar in which one of the OH groups has been replaced by a hydrogen.

disaccharide — a compound containing two sugar molecules linked together.

epimers — monosaccharides that differ in configuration at only one carbon.

furanose — a five-membered ring sugar.

furanoside — a five-membered ring glycoside.

glycoprotein — a protein that is covalently bonded to oligosaccharide.

glycoside the acetal of a sugar.

***N*-glycoside** a glycoside with a nitrogen instead of an oxygen at the glycosidic linkage.

glycosidic bond the bond between the anomeric carbon and the alcohol residue in a glycoside.

α-1,4'-glycosidic linkage a glycosidic linkage between the C-1 of one sugar and the C-4 of a second sugar with the oxygen atom at C-1 in the axial position.

α-1,6'-glycosidic linkage a glycosidic linkage between the C-1 of one sugar and the C-6 of a second sugar with the oxygen atom at C-1 in the axial position.

β-1,4'-glycosidic linkage a glycosidic linkage between the C-1 of one sugar and the C-4 of a second sugar with the oxygen atom at C-1 in the equatorial position.

Haworth projection a way to show the structure of a sugar in which the five- and six-membered rings are represented as being flat.

heptose a monosaccharide with seven carbons.

hexose a monosaccharide with six carbons.

ketose a polyhydroxyketone.

Kiliani-Fischer synthesis a method used to increase the number of carbons in an aldose by one, resulting in the formation of a pair of C-2 epimers.

molecular recognition the ability of molecules to recognize one another.

monosaccharide a single sugar molecule.

mutarotation a slow change in optical rotation to an equilibrium value.

nonreducing sugar a sugar that cannot be oxidized by reagents such as Ag^+ and Cu^+. Nonreducing sugars are not in equilibrium with the open-chain aldose or ketose.

oligosaccharide three to ten sugar molecules linked by glycosidic bonds.

osazone the product obtained by reacting an aldose or a ketose with excess phenylhydrazine. An osazone contains two imine bonds.

oxocarbenium ion an ion in which the positive charge is shared by a carbon and an oxygen.

pentose a monosaccharide with five carbons.

photosynthesis the synthesis of glucose and O_2 from CO_2 and H_2O.

polysaccharide a compound containing ten or more sugar molecules linked together.

pyranose	a six-membered ring sugar.
pyranoside	a six-membered ring glycoside.
reducing sugar	a sugar that can be oxidized by reagents such as Ag^+ and Cu^+. Reducing sugars are in equilibrium with the open-chain aldose or ketose.
simple carbohydrate	a single sugar molecule.
tetrose	a monosaccharide with four carbons.
triose	a monosaccharide with three carbons.
Wohl degradation	a method used to shorten an aldose by one carbon.

Solutions to Problems

1. D-Ribose is an aldopentose.
D-Sedoheptulose is a ketoheptose.
D-Mannose is an aldohexose.

2. Notice that an L-sugar is the mirror image of a D-sugar.

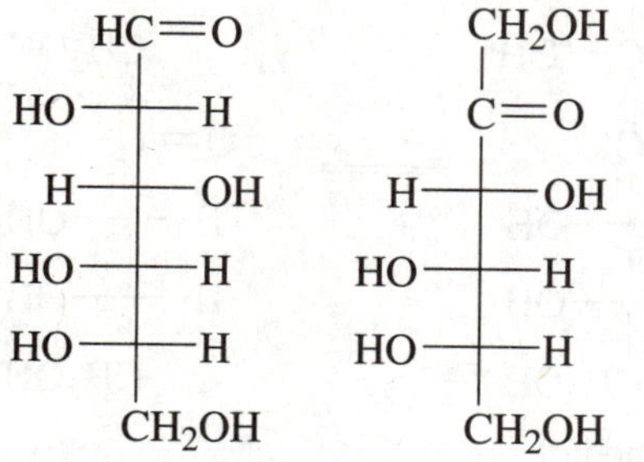

L-glucose L-fructose

3. The easiest way to answer this question is to determine whether the structure represents *R*-glyceraldehyde or *S*-glyceraldehyde.
R-glyceraldehyde = D-glyceraldehyde;
S-glyceraldehyde = L-glyceraldehyde.

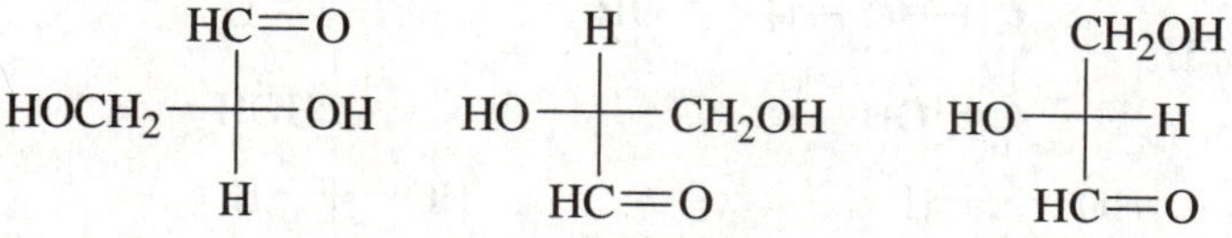

L-glyceraldehyde L-glyceraldehyde D-glyceraldehyde

4. **a.** enantiomers **b.** diastereomers

5. **a.** D-ribose **b.** L-talose **c.** L-allose

6. **a.** D-glucose = (2*R*,3*S*,4*R*,5*R*)-2,3,4,5,6-pentahydroxyhexanal

b. Because D-mannose is a C-2 epimer of D-glucose, the systematic name of D-mannose can be obtained just by changing the configuration of the C-2 carbon in the systematic name of D-glucose:
D-mannose = (2*S*,3*S*,4*R*,5*R*)-2,3,4,5,6-pentahydroxyhexanal.

c. D-Galactose is the C-4 epimer of D-glucose. Therefore, each of its carbon atoms, except C-4, has the same configuration as it has in D-glucose.
D-galactose = (2*R*,3*S*,4*S*,5*R*)-2,3,4,5,6-pentahydroxyhexanal

d. L-Glucose is the mirror image of D-glucose, so each carbon in D-glucose has the opposite configuration in L-glucose.
L-glucose = (2*S*,3*R*,4*S*,5*S*)-2,3,4,5,6-pentahydroxyhexanal

7. D-psicose

8. **a.** A ketoheptose has four asymmetric centers ($2^4 = 16$ stereoisomers).

b. An aldoheptose has five asymmetric centers ($2^5 = 32$ stereoisomers).

c. A ketotriose does not have an asymmetric center; therefore, it has no stereoisomers.

9.

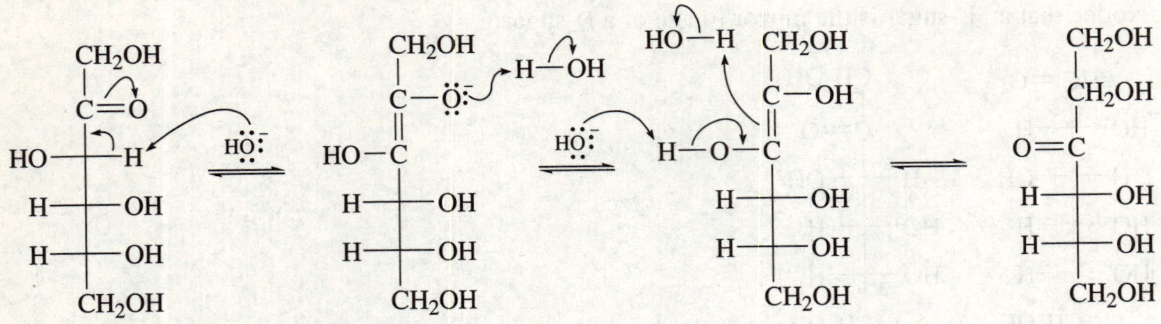

D-fructose an enolate ion an enediol

10. Removal of an α-hydrogen creates an enol that can enolize back to the ketone (using the OH at C-2) or can enolize to an aldehyde (using the OH at C-1). The aldehyde has a new asymmetric center (indicated by an *); one of the epimers is D-glucose, and the other is D-mannose.

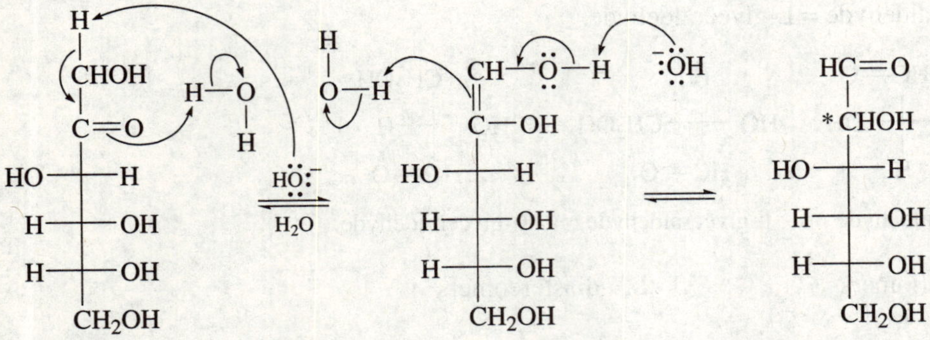

D-fructose D-glucose D-mannose

11. D-tagatose, D-galactose, and D-talose

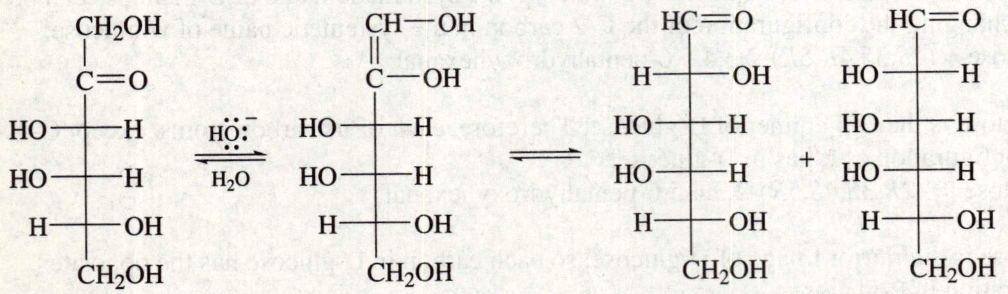

D-tagatose D-galactose D-talose

12. **a.** When D-idose is reduced, D-iditol is formed.

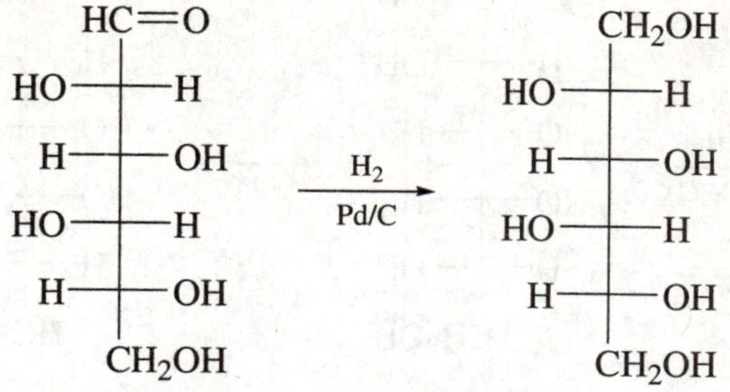

D-idose D-iditol

b. When D-sorbose is reduced, C-2 becomes an asymmetric center, so both D-iditol and the C-2 epimer of D-iditol (D-gulitol) are formed.

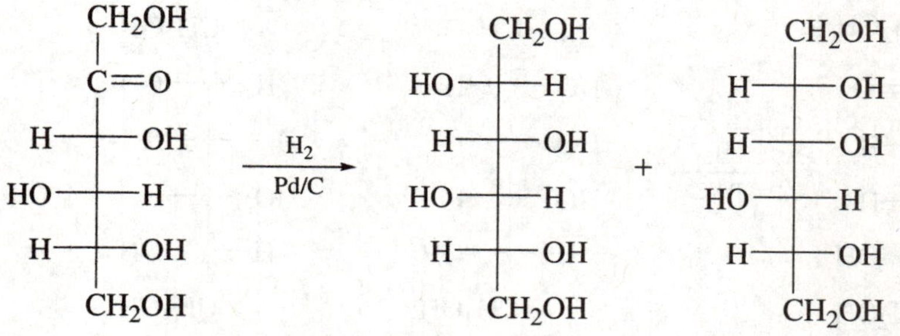

D-sorbose D-iditol D-gulitol

13. **a.** **1.** D-Altrose is reduced to the same alditol as D-talose.
The easiest way to answer this question is to draw D-talose and its alditol. Then draw the monosaccharide with the same configuration at C-2, C-3, C-4, and C-5 as D-talose, reversing the functional groups at C-1 and C-6. (Put the alcohol group at the top and the aldehyde group at the bottom.) The resulting Fischer projection can be rotated 180° in the plane of the paper, and the monosaccharide can then be identified.

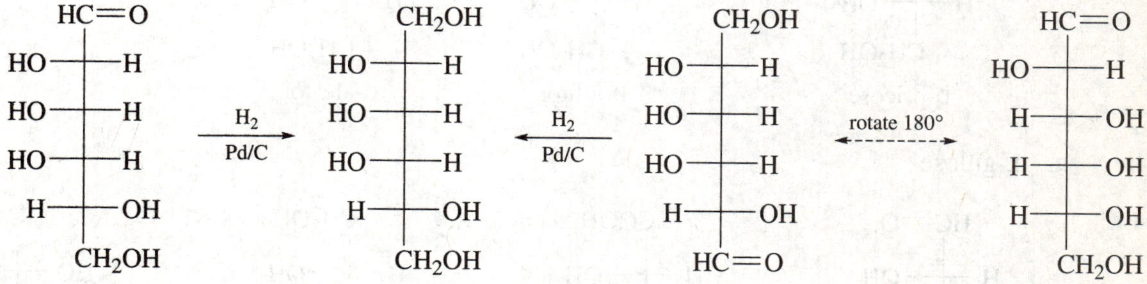

D-talose D-altrose

2. L-Galactose is reduced to the same alditol as D-galactose.

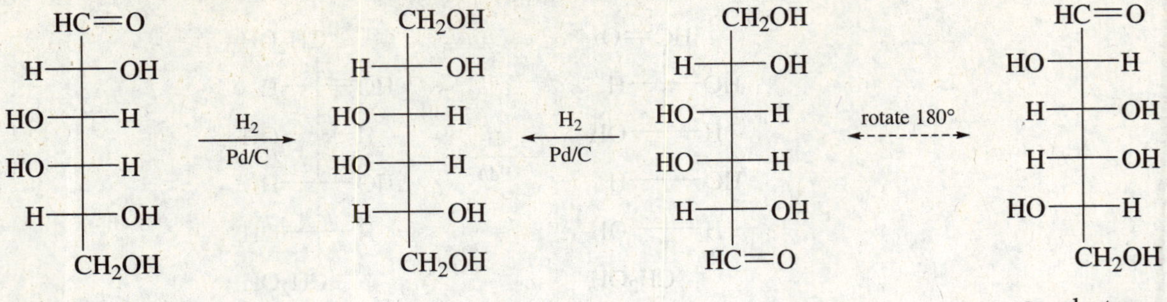

D-galactose L-galactose

b. 1. The ketohexose (D-tagatose) with the same configuration at C-3, C-4, and C-5 as D-talose will give the same alditol as D-talose. The other alditol is the one with the opposite configuration at C-2.

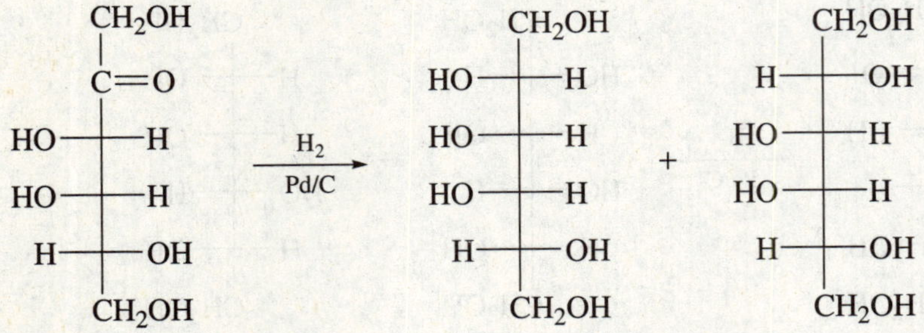

D-tagatose D-talitol D-galactitol

2. D-psicose

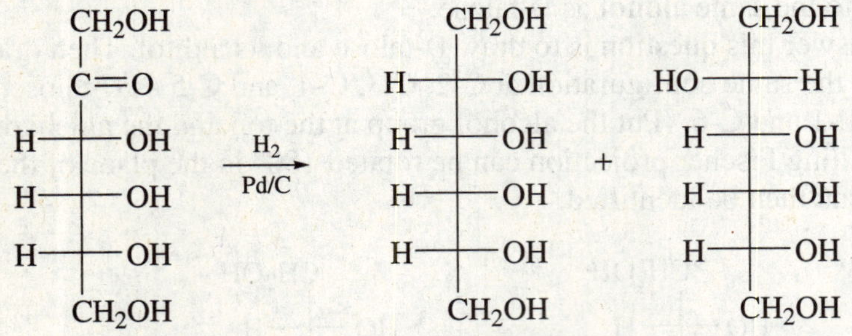

D-psicose D-allitol D-altritol

14. a. L-gulose

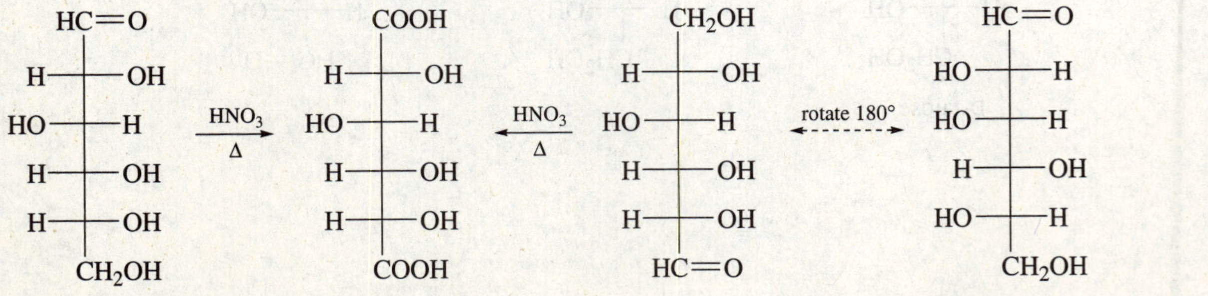

D-glucose D-glucaric acid L-gulose

b. L-Gularic acid, because it is also the oxidation product of L-gulose.

c. D-allose and L-allose, D-altrose and D-talose, L-altrose and L-talose, D-galactose and L-galactose

15. **a.** D-arabinose and D-ribulose **c.** L-gulose and L-sorbose

b. D-allose and D-psicose **d.** D-talose and D-tagatose

16. D-gulose and D-idose

17. **a.** D-gulose and D-idose **b.** L-xylose and L-lyxose

18. **a.** D-allose and D-altrose **c.** L-allose and L-altrose

b. D-glucose and D-mannose

19. Solved in the text.

20.

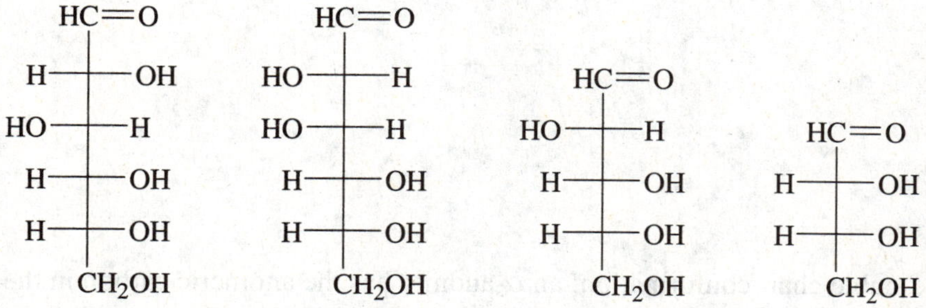

D-glucose D-mannose D-arabinose D-erythrose

A **B** **C** **D**

21. The hemiacetals in **a** and **b** have one asymmetric center; therefore, each has two stereoisomers. The hemiacetals in **c** and **d** have two asymmetric centers; therefore, each has four stereoisomers.

a. Solved in the text. **b.**

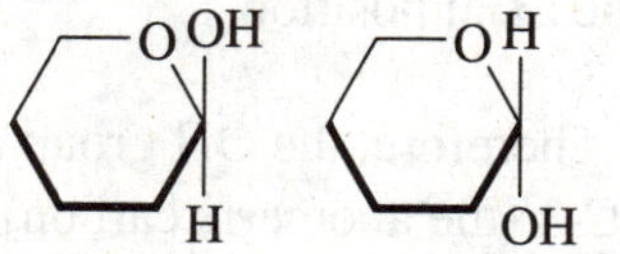

c.

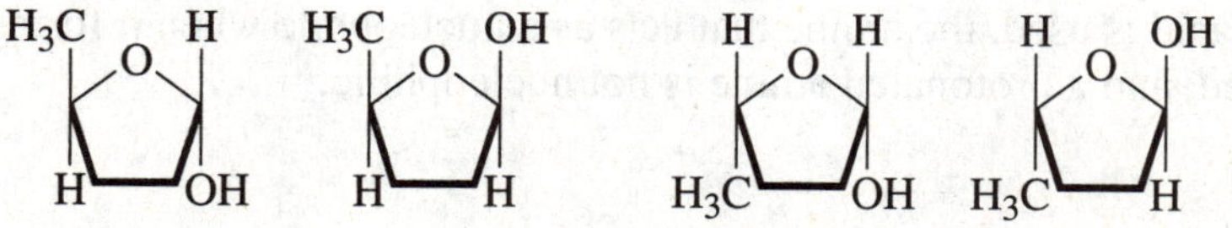

d.

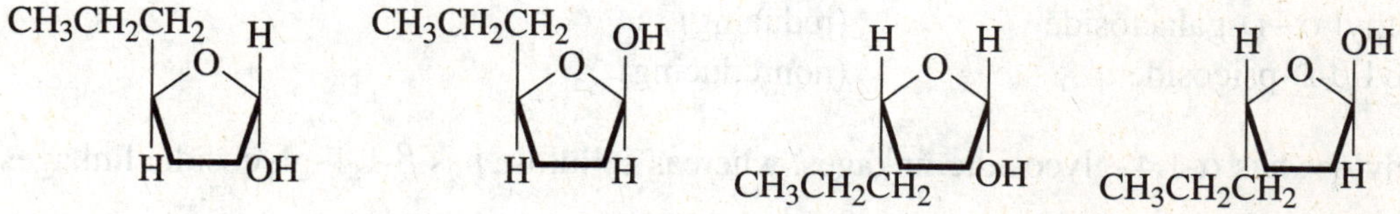

22. **a.**

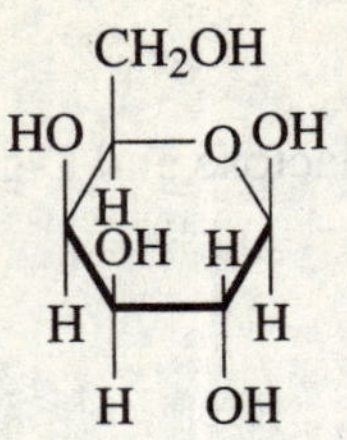

b.

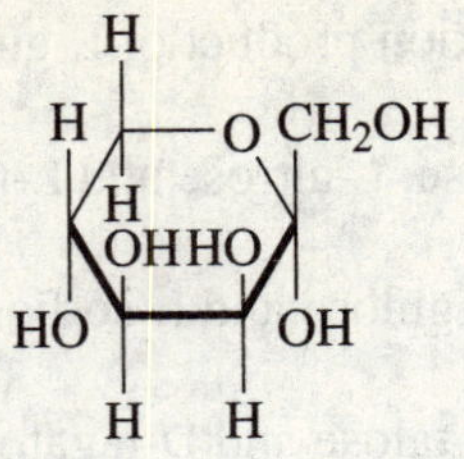

c.

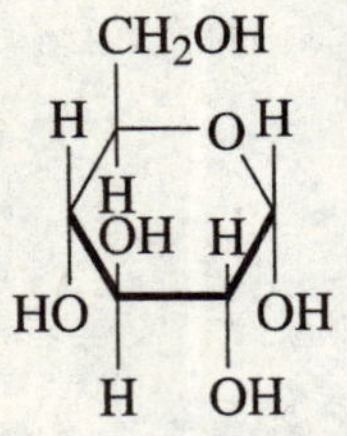

α-D-glucose

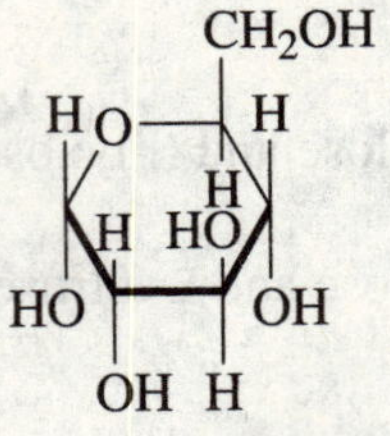

α-L-glucose
the mirror image of α-D-glucose

23.

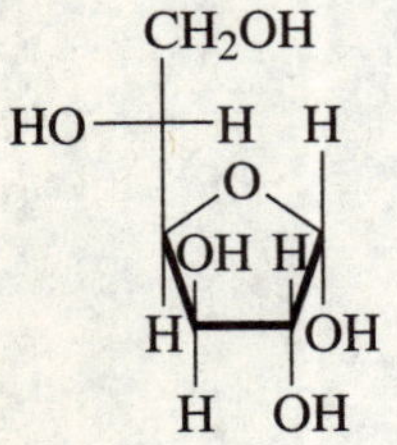

24. First recall that in the more stable chair conformation: an α-anomer has the anomeric carbon in the axial position, whereas a β-anomer has the anomeric carbon in the equatorial position. Then recall that glucose has all its OH groups in equatorial positions. Now this question can be answered easily.

a. Solved in the text.

b. Idose differs in configuration from glucose at C-2, C-3, and C-4. Therefore, the OH groups at C-2, C-3, and C-4 in β-D-idose are in the axial position.

c. Allose is a C-3 epimer of glucose. Therefore, the OH group at C-3 is in the axial position and, since it is the α-anomer, the OH group at C-1 (the anomeric carbon) is in the axial position.

25. If more than a trace amount of acid is used, the amine that acts as a nucleophile when it forms the *N*-glycoside becomes protonated, and a protonated amine is not nucleophilic.

26. **a.** Solved in the text.
b. α– D-talose (nonreducing)
c. methyl α– D-galactoside (reducing)
d. ethyl β-D-psicoside (nonreducing)

27. **a.** Amylose has α-1,4'-glycosidic linkages, whereas cellulose has β-1,4'-glycosidic linkages.

b. Amylose has α-1,4'-glycosidic linkages, whereas amylopectin has both α-1,4'-glycosidic linkages and α-1,6'-glycosidic linkages.

c. Glycogen and amylopectin have the same kind of linkages, but glycogen has a higher frequency of α-1,6'-glycosidic linkages.

d. Cellulose has a hydroxyl at C-2, whereas chitin has an *N*-acetylamino group at that position.

28. A proton is more easily lost from the C-3 OH group because the anion that is formed when the proton is removed is more stable than the anion that is formed when a proton is removed from the C-2 OH group. When a proton is removed from the C-3 OH group, the electrons that are left behind can be delocalized onto another oxygen atom. When a proton is removed from the C-2 OH group, the electrons that are left behind are delocalized onto a carbon atom. Because oxygen is more electronegative than carbon, a negatively charged oxygen is more stable than a negatively charged carbon.

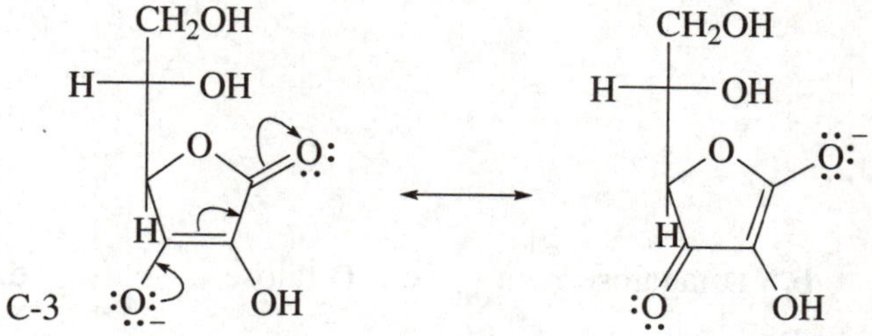

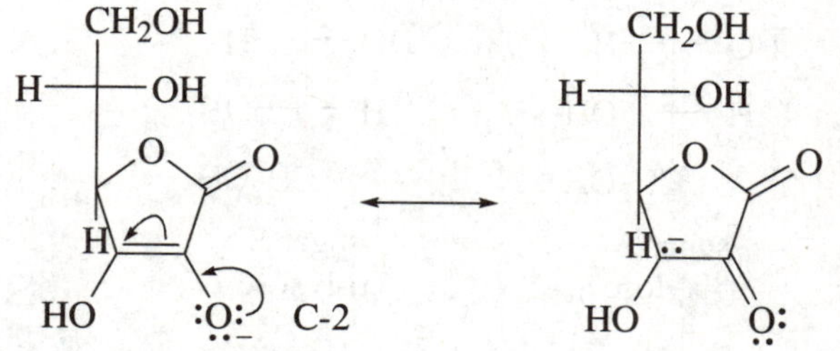

29. **a.** People with Type O blood can receive blood only from other people with Type O blood, because Type A, B, and AB blood have sugar components that Type O blood does not have.

b. People with Type AB blood can give blood only to other people with Type AB blood, because Type AB blood has sugar components that Type A, B, or O blood does not have.

30. **a.**

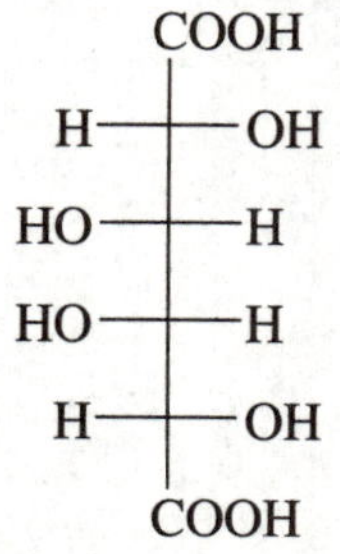

b.

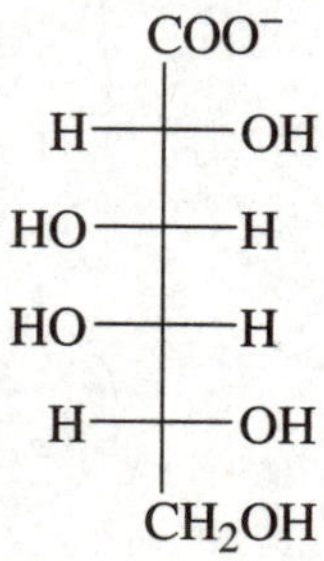

c.

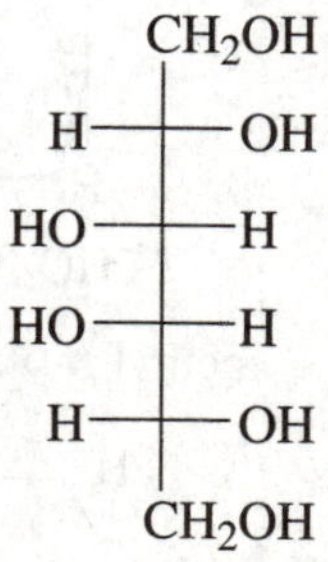

d.

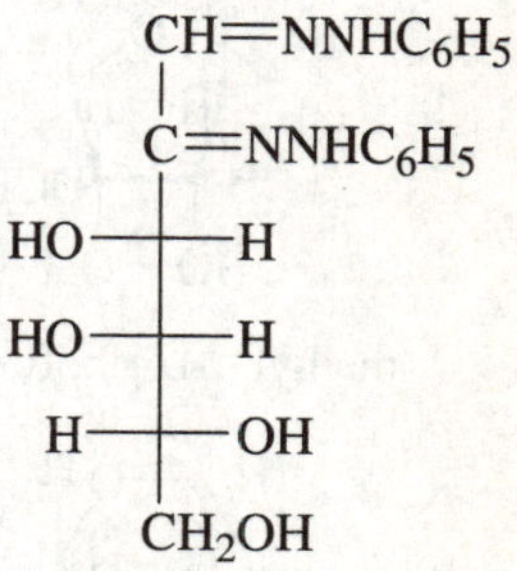

e. 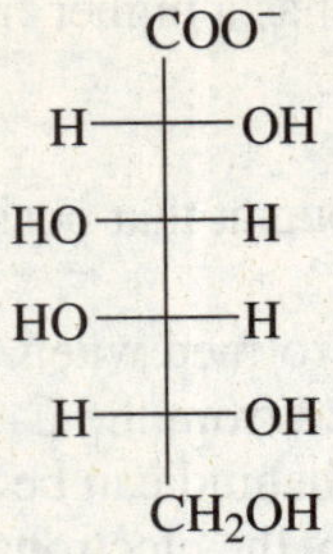

f. 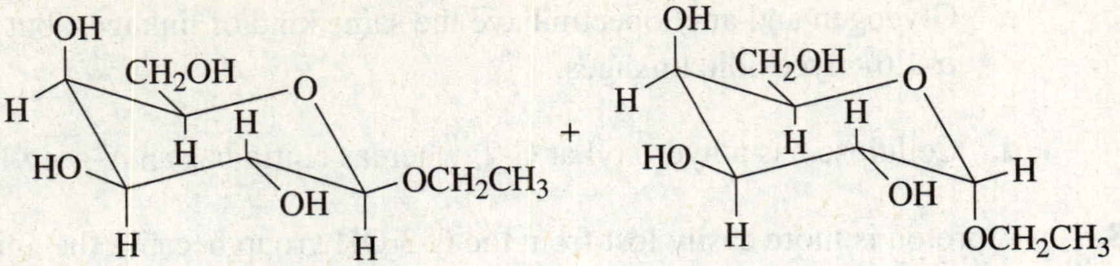

g. 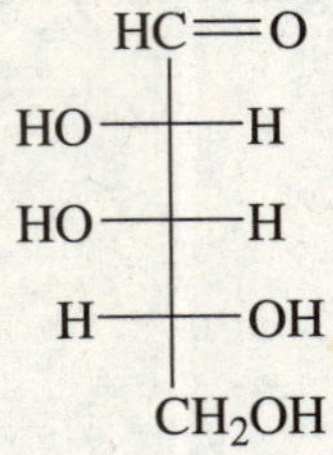

31. **a.** D-lyxose **b.** D-tagatose **c.** D-talose **d.** D-psicose

32. 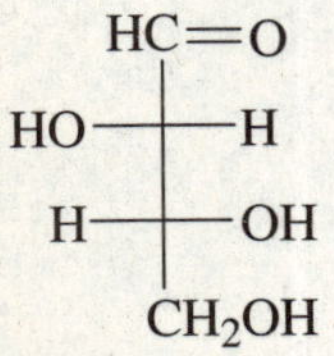

D-threose

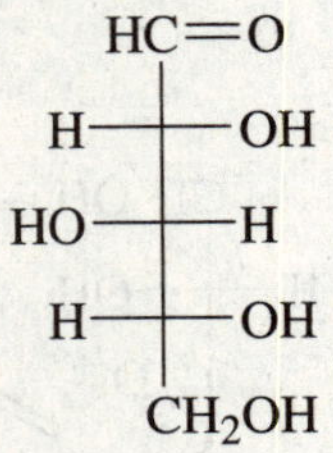

sugar A
D-xylose

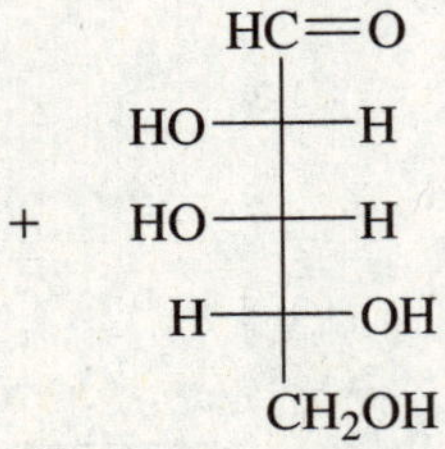

sugar B
D-lyxose

33. **a.** D-ribose and L-ribose, D-arabinose and L-arabinose, D-xylose and L-xylose, D-lyxose and L-lyxose

b. D-ribose and D-arabinose, L-ribose and L-arabinose, D-xylose and D-lyxose, L-xylose and L-lyxose

c. D-arabinose, L-arabinose, D-lyxose, and L-lyxose

34. 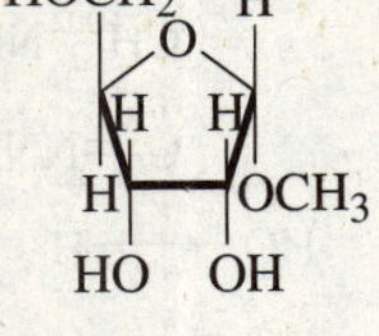

methyl α-D-ribofuranose

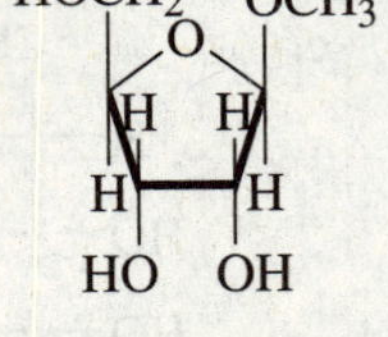

methyl β-D-ribofuranose

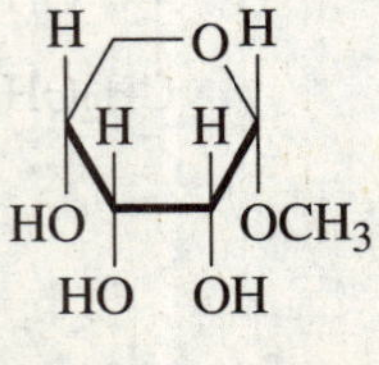

methyl α-D-ribopyranose

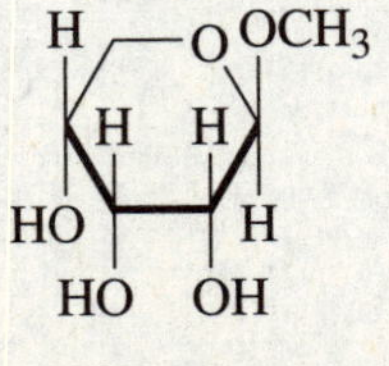

methyl β-D-ribopyranose

35. D-Altrose and D-talose are reduced to the same alditol.

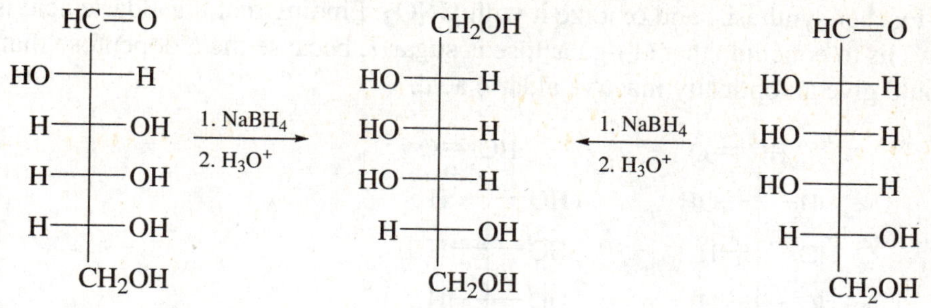

D-altrose alditol D-talose

If D-altrose is **A** and D-talose is **B**, then **C** must be D-galactose, the C-2 epimer of **B**.

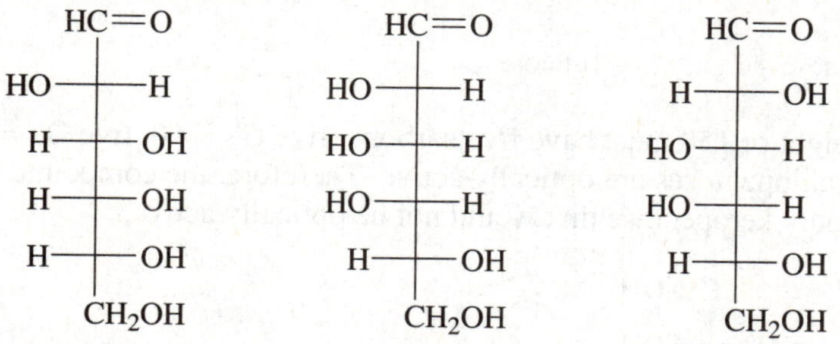

D-altrose **A** D-talose **B** D-galactose **C**

If D-altrose is **B** and D-talose is **A**, then **C** must be D-allose, the C-2 epimer of **B**.

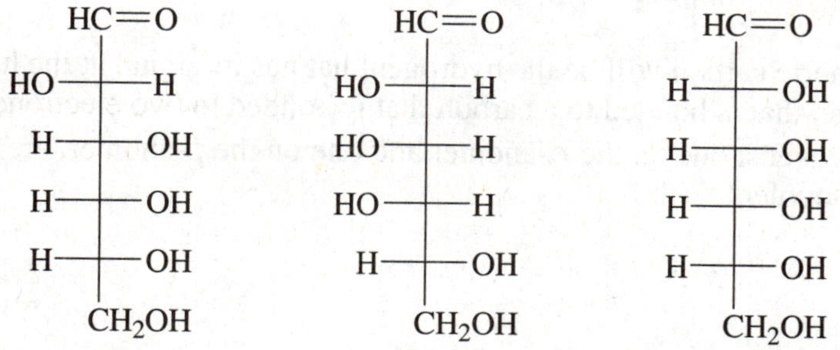

D-altrose **B** D-talose **A** D-allose **C**

36. **1.** As Fischer did, we can narrow our search to eight 6-carbon sugars, since there are eight pairs of enantiomers. First we need to find an aldopentose that forms (+)-galactose as a product of a Kiliani-Fischer synthesis. That sugar is the one known as (–)-lyxose. The Kiliani-Fischer synthesis on (–)-lyxose yields two sugars with melting points that show them to be the sugars known as (+)-galactose and (+)-talose. Now we know that (+)-galactose and (+)-talose are C- 2 epimers. They are sugars 1 and 2, 3 and 4, 5 and 6, or 7 and 8. (See page 990 of the text.)

2. When (+)-galactose and (+)-talose react with HNO_3, (+)-galactose forms an optically inactive product and (+)-talose forms an optically active product. Thus, (+)-galactose and (+)-talose are sugars 1 and 2 or 7 and 8. Since (+)-galactose is the one that forms the optically inactive oxidation product, it is either sugar 1 or 7.

3. To determine the structure of (+)-galactose, we can go back to (−)-lyxose, the sugar that forms sugars 7 and 8 by a Kiliani-Fischer synthesis, and oxidize it with HNO_3. Finding that the aldaric acid is optically active allows us to conclude that (+)-galactose is sugar 7, because the aldopentose that leads to sugars 1 and 2 would give an optically inactive aldaric acid.

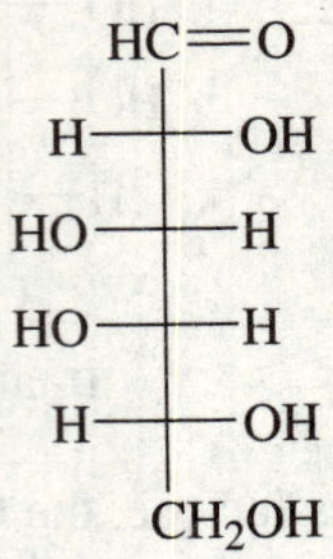

7
D-galactose

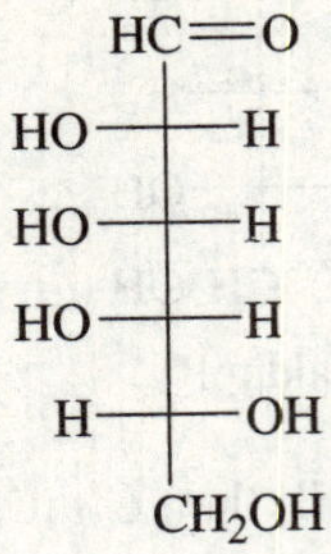

8
D-talose

37. A monosaccharide with a molecular weight of 150 must have five carbons (five C's = 60, five O's = 80, and 10 H's = 10 for a total of 150). All aldopentoses are optically active. Therefore, the compound must be a ketopentose. The following is the only ketopentose that would not be optically active.

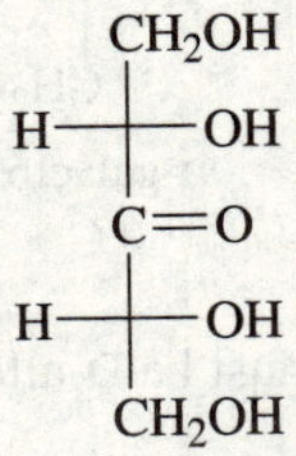

38. The hydrogen that is bonded to the anomeric carbon will be the hydrogen that has its signal at the highest frequency, because it is the only hydrogen that is bonded to a carbon that is bonded to two electronegative oxygen atoms. So the two anomeric hydrogens, one on the α-anomer and one on the β-anomer, are responsible for the two high-frequency doublets.

39.

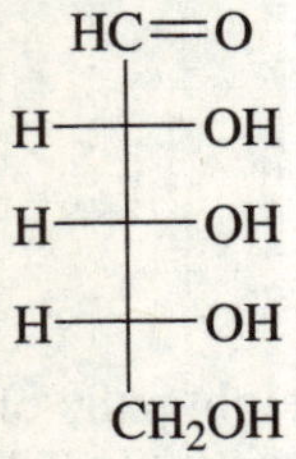

D-ribose
A

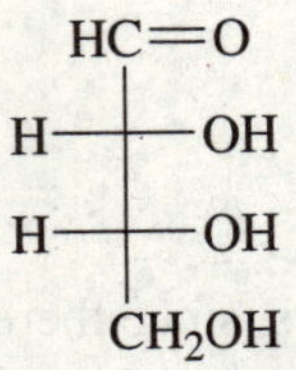

D-erythrose
B

40. The hexose is D-altrose.

1. Knowing that (+)-glyceraldehyde is D-glyceraldehyde gives the configuration at C-5.
2. Knowing that the second Wohl degradation gives D-erythrose gives the configuration at C-4.

3. Knowing that the first Wohl degradation followed by oxidation gives an optically inactive aldaric acid, gives the configuration at C-3

4. Knowing that the original hexose forms an optically active aldaric acid gives the configuration at C-2.

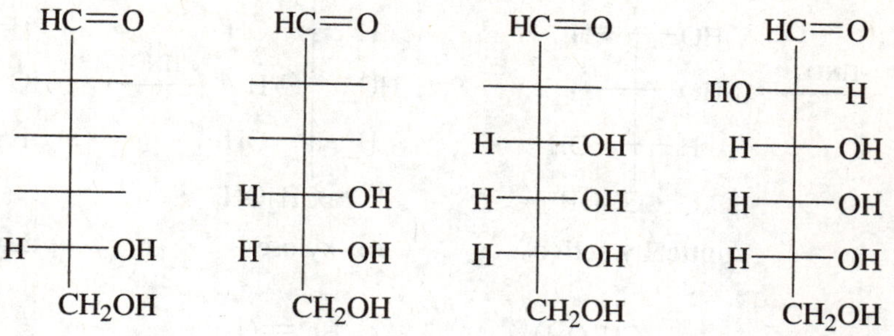

D-altrose

41.

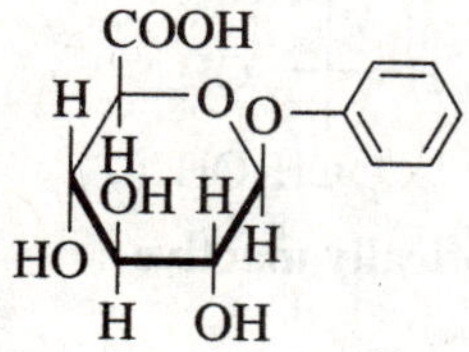

the β-D-glucuronide

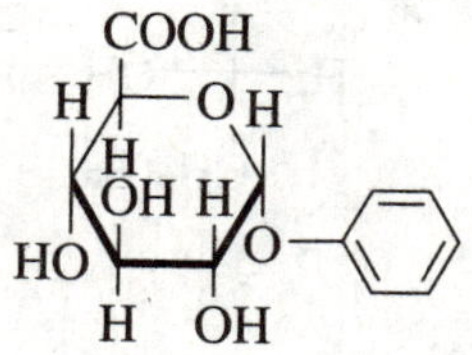

the α-D-glucuronide

42. **D-arabinose**. The only D-aldopentoses that are oxidized to optically active aldaric acids are D-arabinose and D-lyxose. A Ruff degradation of D-arabinose forms D-erythrose, whereas a Ruff degradation of D-lyxose forms D-threose. Since D-erythrose forms an optically inactive aldaric acid but D-threose does not, the D-aldopentose is D-arabinose.

43.

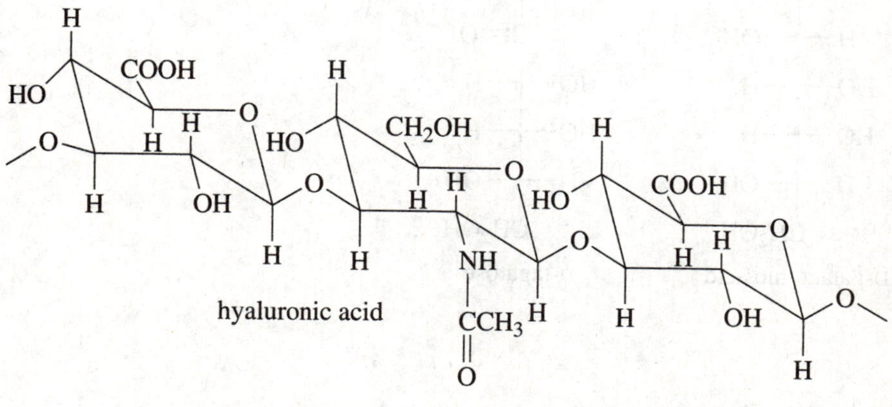

hyaluronic acid

44. She can take a sample of one of the sugars and oxidize it with nitric acid to an aldaric acid or reduce it with sodium borohydride to an alditol. If the product is optically active, the sugar was D-lyxose; if the product is not optically active, the sugar was D-xylose.

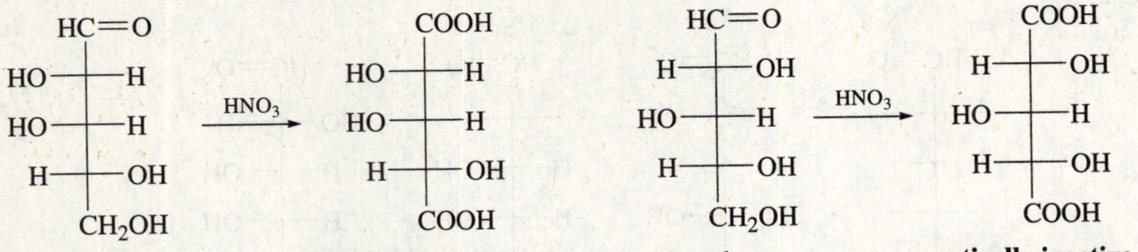

D-lyxose — **optically active** — D-xylose — **optically inactive**

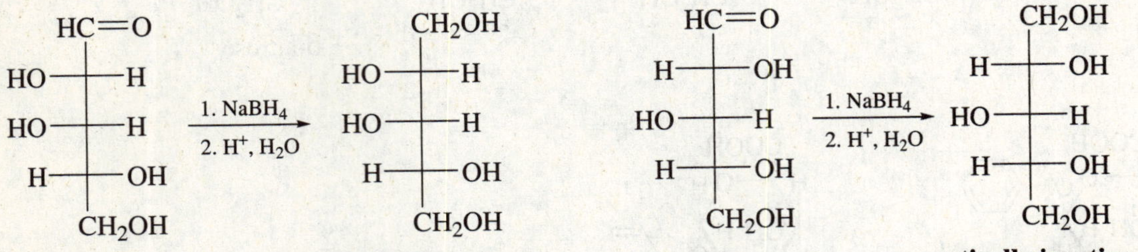

D-lyxose — **optically active** — D-xylose — **optically inactive**

45. We know the hexose is a ketohexose since it does not react with Br_2; we know it is a 2-ketohexose because if it were a 3-ketohexose it wouldn't be able to form a hemiacetal (because the hemiacetal would have an unstable four-membered ring) so it wouldn't under mutarotation. Knowing that it is oxidized by Tollens reagent to the aldonic acids D-talonic acid and D-galactonic acid tells us that the aldonic acids and the ketohexose have the same configuration at C-3, C-4, and C-5. Therefore, the hexose is D-tagatose.

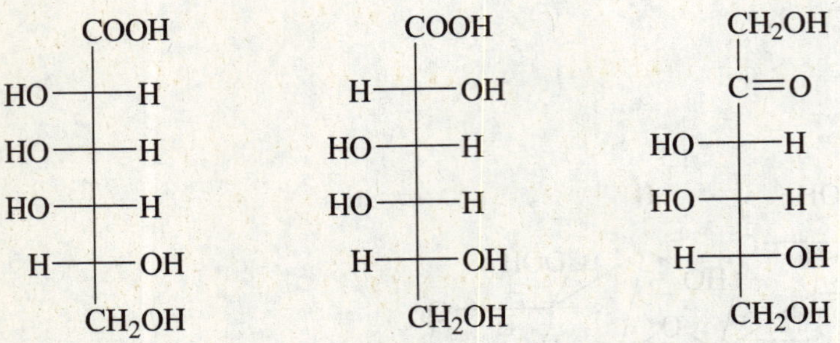

D-talonic acid — D-galactonic acid — D-tagatose

46.

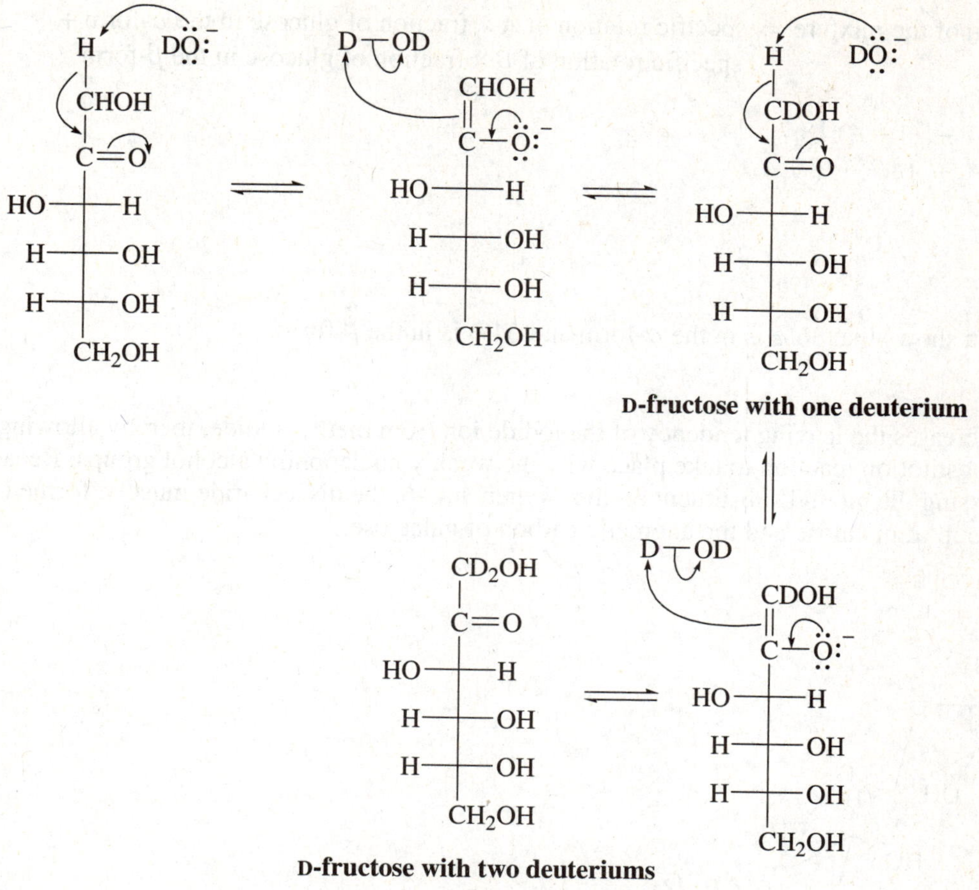

47. **10** aldaric acids

Each of the following pairs forms the same aldaric acid:

D-allose and L-allose	L-altrose and L-talose
D-galactose and L-galactose	D-glucose and L-gulose
D-altrose and D-talose	L-glucose and D-gulose

Thus twelve aldohexoses form six aldaric acids. The other four aldohexoses each form a distinctive aldaric acid, and (6 + 4 = 10).

48. Let A = the fraction of glucose in the α-form and B = the fraction of glucose in the β-form.

$$A + B = 1$$
$$B = 1 - A$$

specific rotation of A = 112.2
specific rotation of B = 18.7
specific rotation of the equilibrium mixture = 52.7

specific rotation of the mixture = specific rotation of $A \times$ fraction of glucose in the α-form + specific rotation of $B \times$ fraction of glucose in the β-form

$$52.7 = 112.2\,A + (1 - A)\,18.7$$
$$52.7 = 112.2\,A + 18.7 - 18.7\,A$$
$$34.0 = 93.5\,A$$
$$A = 0.36$$
$$B = 0.64$$

This calculation shows that 36% is in the α-form and 64% is in the β-form.

49. Silver oxide increases the leaving tendency of the iodide ion from methyl iodide, thereby allowing the nucleophilic substitution reaction to take place with the weakly nucleophilic alcohol groups. Because mannose is missing the methyl substituent on the oxygen at C-6, the disaccharide must be formed using the C-6 OH group of mannose and the anomeric carbon of galactose.

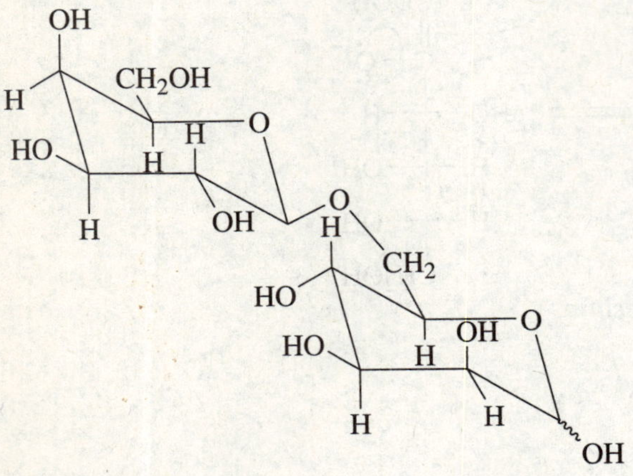

50. D-Altrose will most likely exist as a furanose because
(1) the furanose is particularly stable, because all the large substituents are trans to each other, and
(2) the pyranose has two of its OH groups in the unstable axial position.

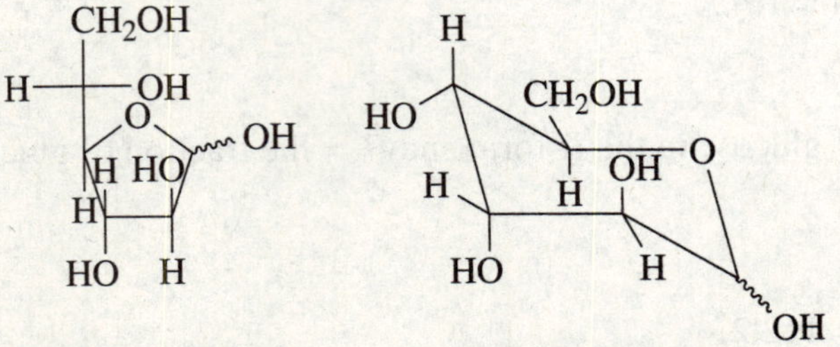

51. a. From its molecular formula and the fact that only glucose is formed when trehalose is hydrolyzed, we know that it is a disaccahride. Trehalose can be a nonreducing sugar only if the anomeric carbon of one glucose is connected to the anomeric carbon of the other glucose. (Notice that the second glucose is drawn upside down and is reversed horizontally.)

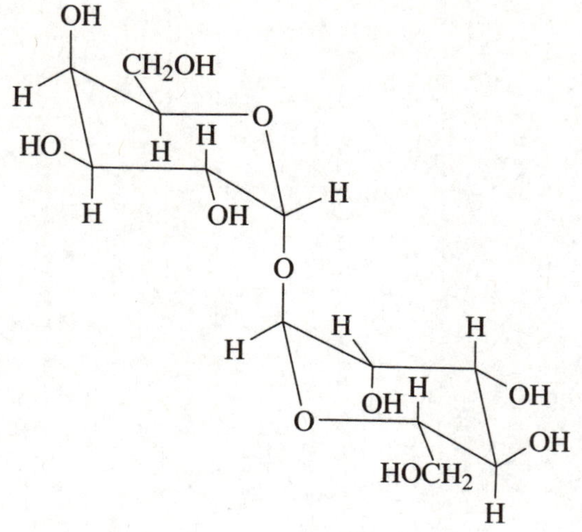

52.

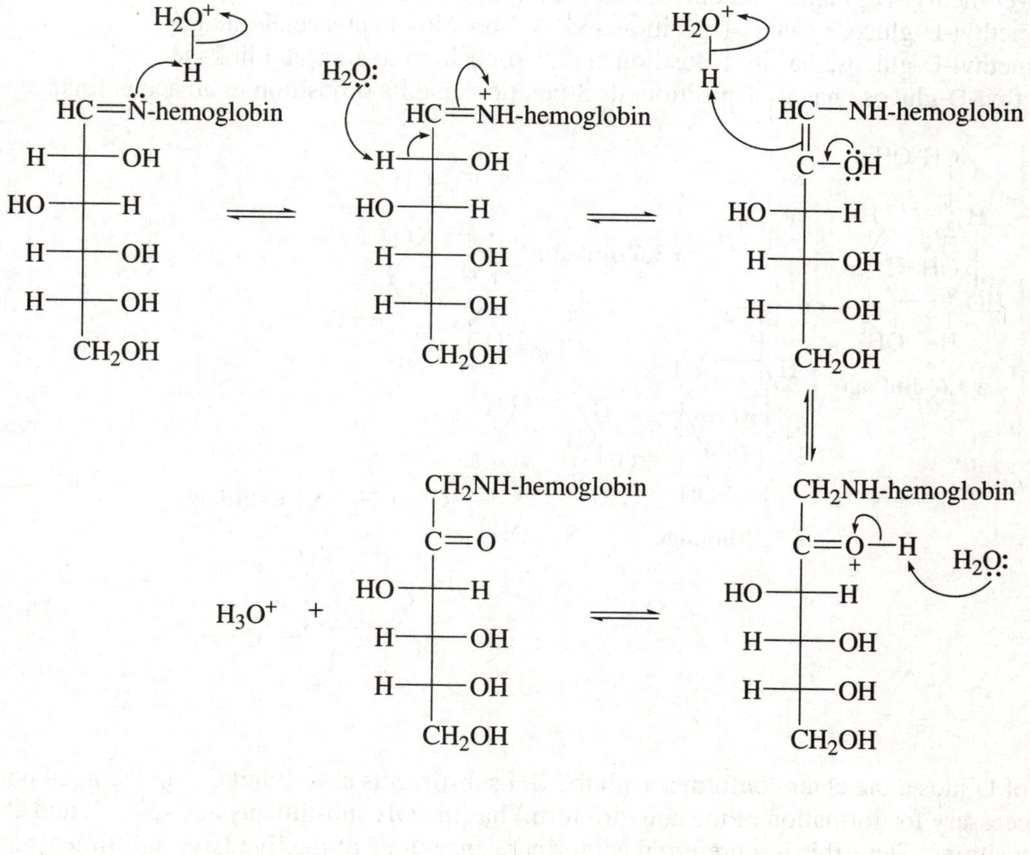

53. Because mannose is missing the methyl substituent on the oxygen at C-6, the disaccharide must be formed using the C-6 OH group of mannose and the anomeric carbon of galactose. Silver oxide increases the leaving tendency of the iodide ion from methyl iodide, thereby allowing the nucleophilic substitution reaction to take place with the weakly nucleophilic alcohol groups.

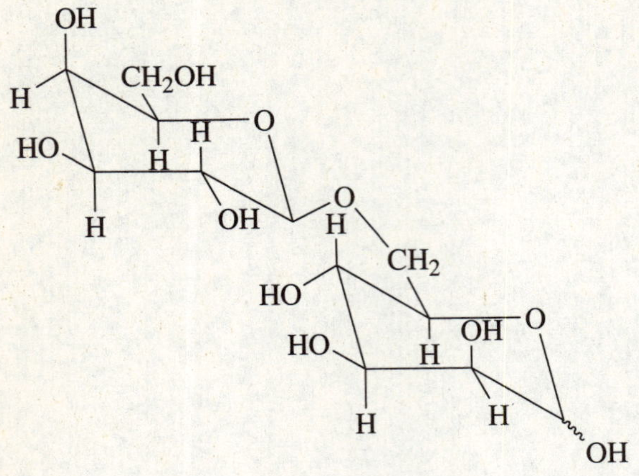

54. Because all the glucose units have six-membered rings, the 5-position is never methylated.
2,3,4,6-tetra-*O*-methyl-D-glucose has only its 1-position in an acetal linkage.
2,3,6-tri-*O*-methyl-D-glucose has its 1-position and its 3-position in an acetal linkage.
2,3,4-tri-*O*-methyl-D-glucose has its 1-position and its 6-position in an acetal linkage.
2,4-di-*O*-methyl-D-glucose has its 1-position, its 3-position, and its 6-position in an acetal linkage.

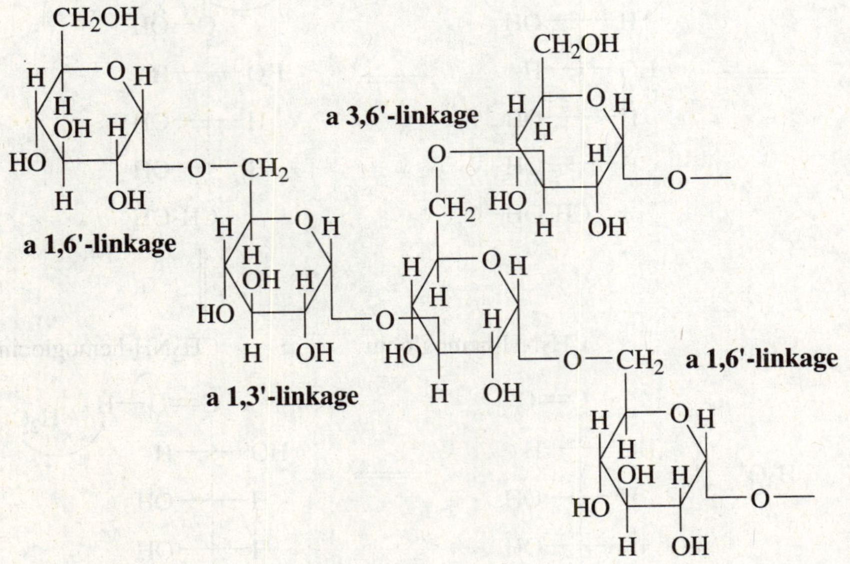

55. In the case of D-idose, the chair conformer with the OH substituents at C-1 and C-5 in the axial position (which is necessary for formation of the anhydro form) has the OH substituents at C-2, C-3, and C-4 all in equatorial positions. Thus, this is a preferred conformer, since three of the five large substituents are in the more stable equatorial position.

In the case of D-glucose, the chair conformer with the OH substituents at C-1 and C-5 in the axial position has the OH substituents at C-2, C-3, and C-4 all in axial positions. This is a relatively unstable conformer, since all the large substituents are in less stable axial positions.

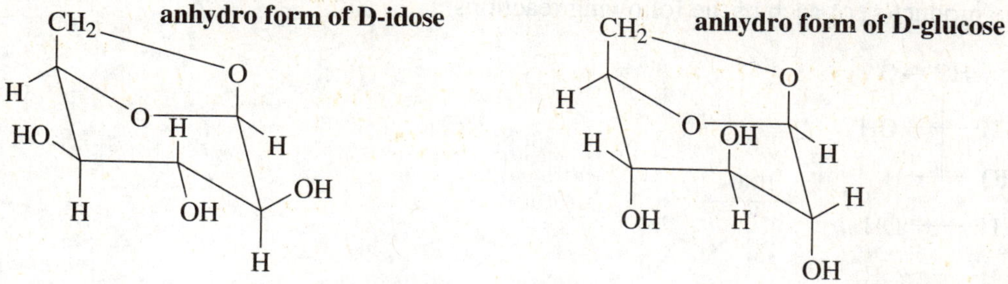

56.

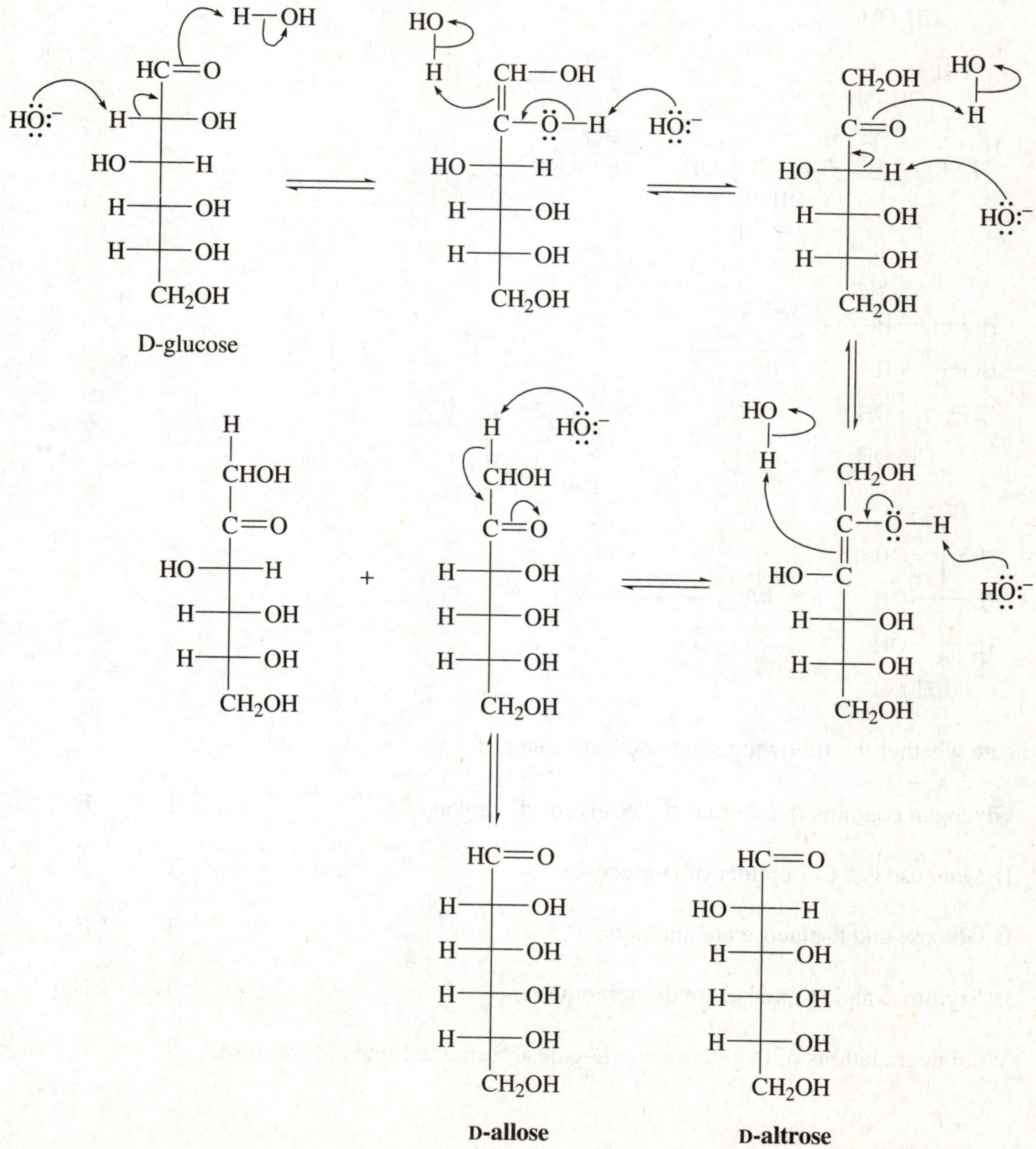

Chapter 21 Practice Test

1. Give the product(s) of each of the following reactions:

a.

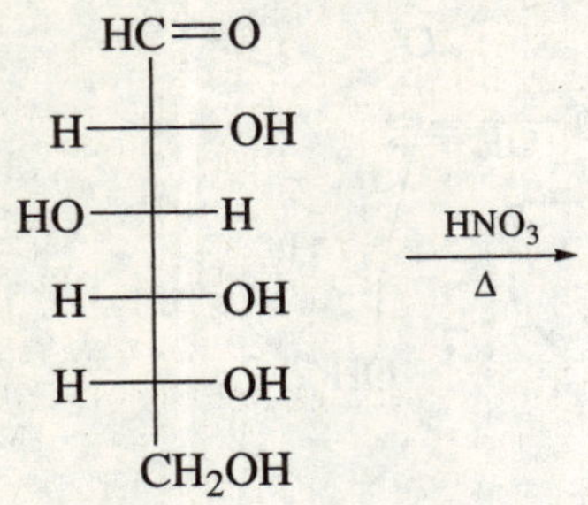

b.

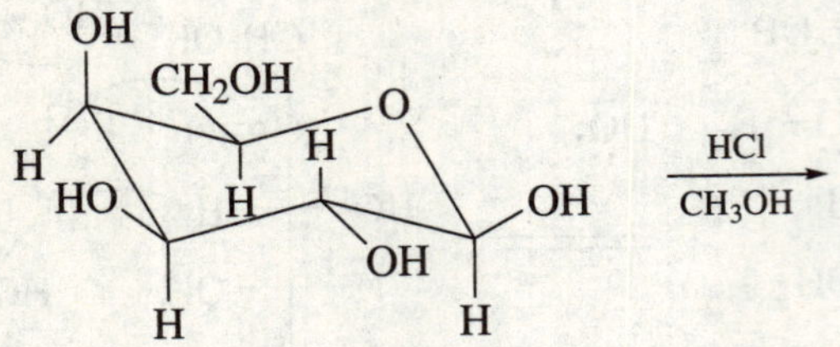

c.

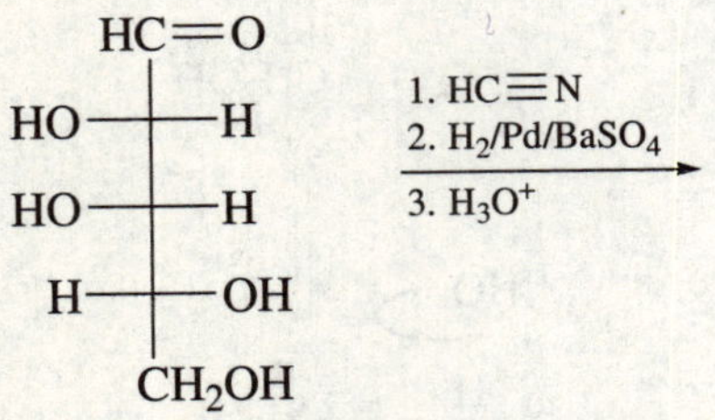

d.

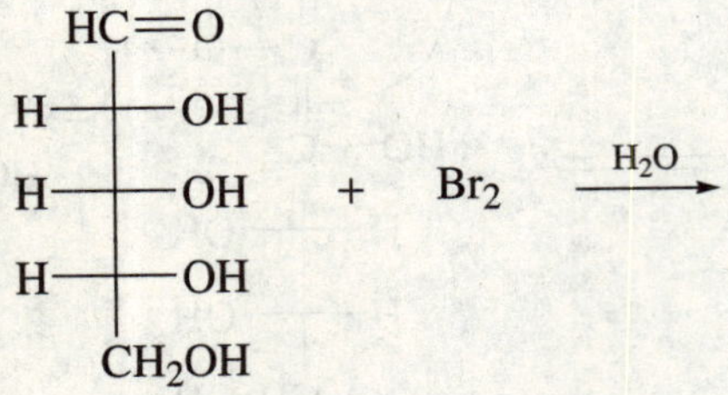

2. Indicate whether the following statements are true or false:

a. Glycogen contains α-1,4'- and β-1,6'-glycosidic linkages. T F

b. D-Mannose is a C-1 epimer of D-glucose. T F

c. D-Glucose and L-glucose are anomers. T F

d. D-Erythrose and D-threose are diastereomers. T F

e. Wohl degradations of D-glucose and D-gulose form the same aldopentose. T F

3. Which of the following sugars will form an optically active aldaric acid?

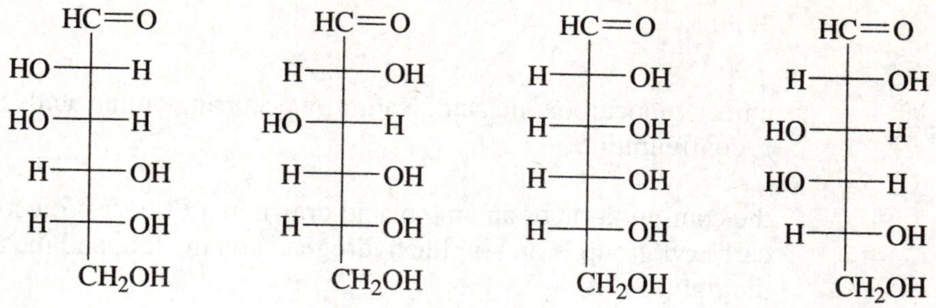

4. When crystals of D-fructose are dissolved in a basic aqueous solution, two aldohexoses are obtained. Identify the aldohexoses.

5. A sugar forms the same osazone as D-galactose but is not oxidized by an aqueous solution of Br_2. Identify the sugar.

6. D-Talose and ______________ are reduced to the same alditol.

7. What is the main structural difference between amylose and cellulose?

8. What aldohexoses are formed from a Kiliani-Fischer synthesis starting with D-xylose?

9. What aldohexose is the C-3 epimer of D-glucose?

10. Draw the most stable chair conformer of β-D-allose, a C-3 epimer of β-D-glucose.

CHAPTER 22
Amino Acids, Peptides, and Proteins

Important Terms

amino acid — an α-aminocarboxylic acid. Naturally occurring amino acids have the L-configuration.

D-amino acid — the configuration of an amino acid drawn in a Fischer projection such that the carboxyl group is on top, the hydrogen is on the left, and the amino group is on the right.

L-amino acid — the configuration of an amino acid drawn in a Fischer projection such that the carboxyl group is on top, the hydrogen is on the right, the amino group is on the left.

amino acid analyzer — an instrument that automates the ion-exchange separation of amino acids.

amino acid residue — a monomeric unit of a peptide or protein.

anion-exchange resin — a resin that binds anions.

antiparallel β-pleated sheet — the adjacent hydrogen-bonded peptide chains in a β-pleated sheet run in opposite directions.

automated solid-phase peptide synthesis — an automated technique that synthesizes a peptide (in the C-terminal to N-terminal direction) while its C-terminal amino acid is attached to a solid support.

cation-exchange resin — a resin that binds cations.

coil conformation (loop conformation) — that part of a protein that is highly ordered but not in an α-helix or a β-sheet.

C-terminal amino acid — the terminal amino acid of a peptide (or protein) that has a free carboxyl group.

N-terminal amino acid — the terminal amino acid of a peptide (or protein) that has a free amino group.

denaturation — destruction of the highly organized tertiary structure of a protein.

dipeptide — two amino acids linked together by an amide bond.

disulfide — a compound with an -S—S- bond formed by two cysteine residues.

disulfide bridge — a disulfide (-S—S-) bond in a peptide or protein.

Edman's reagent — phenyl isothiocyanate. A reagent used to determine the N-terminal amino acid of a polypeptide.

electrophoresis — a technique that separates amino acids on the basis of their pI values.

endopeptidase	an enzyme that hydrolyzes a peptide bond that is not at the end of a peptide chain.
enkephalin	a pentapeptide synthesized by the body to control pain.
enzyme	a protein that is a catalyst.
essential amino acid	an amino acid that humans must obtain from their diet because they either cannot synthesize it at all or they cannot synthesize it in adequate amounts.
exopeptidase	an enzyme that hydrolyzes a peptide bond at the end of a peptide chain.
fibrous protein	a water insoluble protein that has its polypeptide chains arranged in bundles.
globular protein	a water soluble protein that tends to have a roughly spherical shape.
α-helix	the backbone of a polypeptide coiled in a right-handed spiral with hydrogen bonding occurring within the helix.
hydrophobic interactions	interactions between nonpolar groups. They increase stability by decreasing the amount of structured water (increasing entropy).
interchain disulfide bridge	a disulfide bridge between two cysteine residues in different peptide chains.
intrachain disulfide bridge	a disulfide bridge between two cysteine residues in the same peptide chain.
ion-exchange chromatography	a technique that uses a column packed with an insoluble resin to separate compounds on the basis of their charge and polarity.
isoelectric point (pI)	the pH at which there is no net charge on an amino acid.
kinetic resolution	separating enantiomers based on the difference in their rate of reaction with an enzyme.
oligomer	a protein with more than one peptide chain.
oligopeptide	three to ten amino acids linked by amide bonds.
paper chromatography	a technique that separates amino acids based on polarity.
parallel β-pleated sheet	the adjacent hydrogen-bonded peptide chains in a β-pleated sheet run in the same direction.
partial hydrolysis	a technique that hydrolyzes only some of the peptide bonds in a polypeptide.
peptidase	a peptidase is an enzyme that catalyzes the hydrolysis of a peptide bond.
peptide	polymer of amino acids linked together by amide bonds.

peptide bond the amide bond that links the amino acids in a peptide or protein.

β-pleated sheet the backbone of a polypeptide extended in a zigzag structure with hydrogen bonding between neighboring chains.

polypeptide many amino acids linked by amide bonds.

primary structure the sequence of amino acids in a protein.

protein a polymer of amino acids linked together by amide bonds.

quaternary structure a description of the way the individual polypeptide chains of a protein are arranged with respect to one other.

random coil the conformation of a totally denatured protein.

secondary structure a description of the conformation of the backbone of a protein.

structural protein a protein that gives strength to a biological structure.

subunit an individual chain of an oligomer.

tertiary structure a description of the three-dimensional arrangement of all the atoms in a protein.

thin-layer chromatography a technique that separates compounds on the basis of their polarity.

tripeptide three amino acids linked by amide bonds.

zwitterion a compound with a negative charge and a positive charge on nonadjacent atoms.

Solutions to Problems

1. **a.** Protonation of the doubled-bonded nitrogen forms a conjugate acid that is stabilized by electron delocalization. Protonation of the other nitrogen forms a conjugate acid that is not stabilized by electron delocalization. In other words, protonation of the double-bonded nitrogen allows the aromatic compound to retain its aromaticity. The aromaticity of the basic form is lost when protonation occurs on the other nitrogen. The resonance-stabilized conjugate acid is the one that is more readily formed.

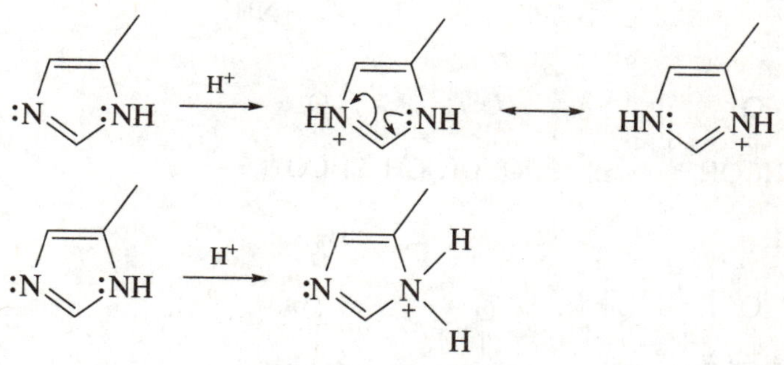

b. Protonation of the doubled-bonded nitrogen forms a conjugate acid that is stabilized by electron delocalization. Protonation of either of the other nitrogen atoms forms a conjugate acid that is not stabilized by electron delocalization. The resonance-stabilized conjugate acid is the one that is more readily formed.

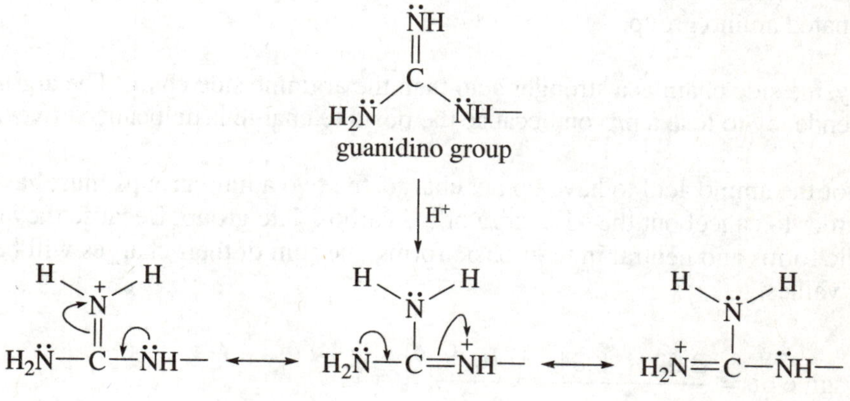

2. **a.** (*R*)-alanine **b.** (*R*)-aspartate

c. The α-carbons of all the D-amino acids except cysteine have the *R*-configuration. Similarly, the α-carbon of all the L-amino acids except cysteine has the *S*-configuration. (In all the amino acids except cysteine, the amino group has the highest priority and the carboxyl group has the second-highest priority. In cysteine, the thiomethyl group has a higher priority than the carboxyl group because sulfur has a greater atomic number than oxygen.)

3. Solved in the text.

4. Isoleucine is the only other amino acid that has more than one asymmetric center. (Like threonine, it has two asymmetric centers.)

5. The electron-withdrawing $^+NH_3$ substituent on the α-carbon increases the acidity of the carboxyl group.

6. **a.** Solved in the text.

c. 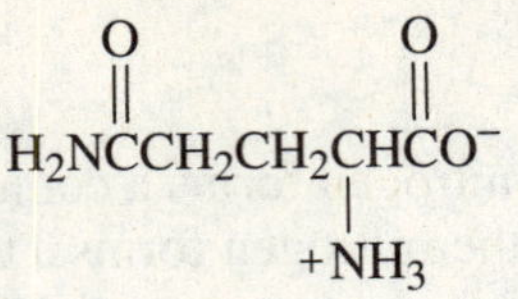

e. 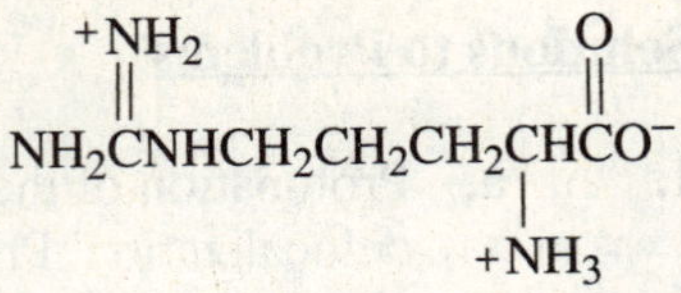

b. 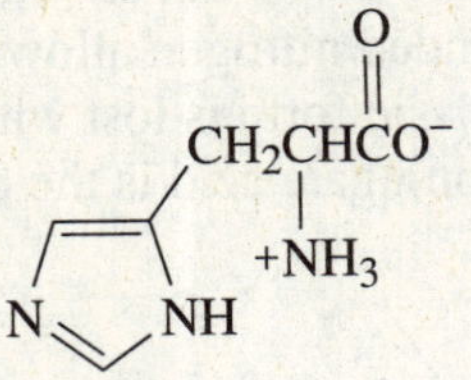

d. 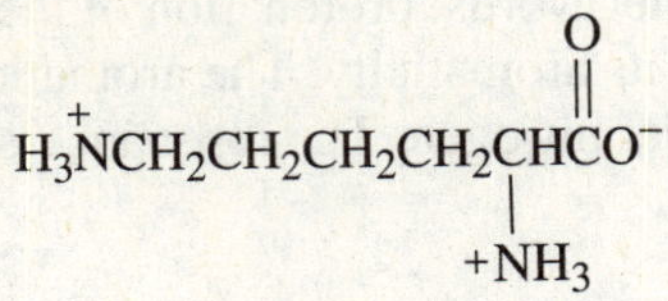

f. 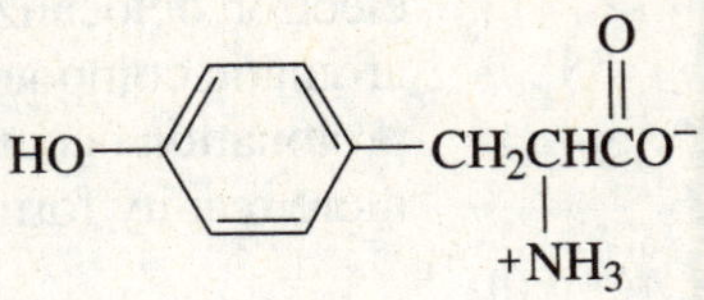

7. **a.** 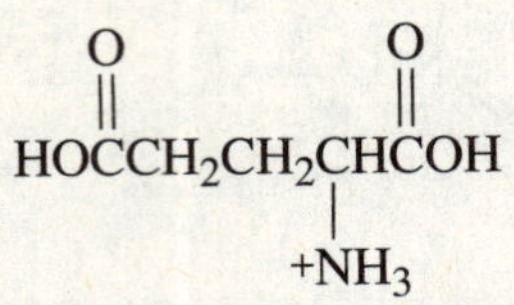

c. 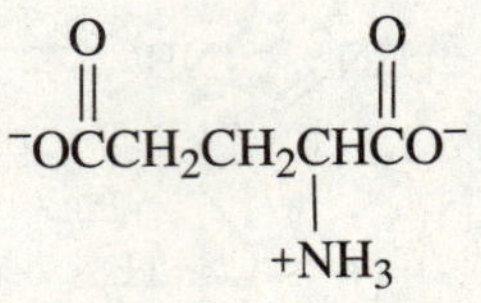

b. 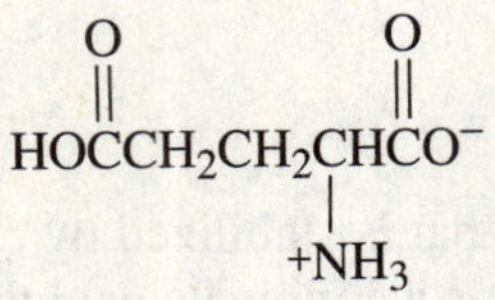

d. 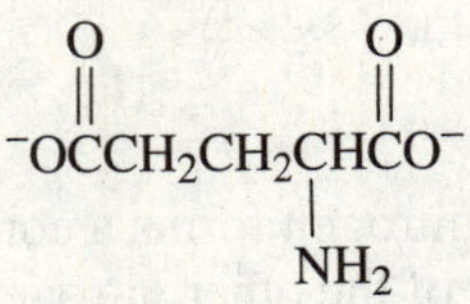

8. **a.** The carboxyl group of the aspartic acid side chain is a stronger acid than the carboxyl group of the glutamic acid side chain because the aspartic acid side chain is closer to the electron-withdrawing protonated amino group.

b. The lysine side chain is a stronger acid than the arginine side chain. The arginine side chain has less of a tendency to lose a proton because the positive charge is delocalized over three nitrogen atoms.

9. In order for the amino acid to have no net charge, the two amino groups must have a +1 charge between them in order to cancel out the –1 charge of the carboxylate group. Because they are positively charged in their acidic forms and neutral in their basic forms, the sum of their charges will be +1 at the midpoint of their pK_a values.

10. **a.** asparagine $pI = \dfrac{2.02 + 8.84}{2} = \dfrac{10.86}{2} = 5.43$

b. arginine $pI = \dfrac{9.04 + 12.48}{2} = \dfrac{21.52}{2} = 10.76$

c. serine $pI = \dfrac{2.21 + 9.15}{2} = \dfrac{11.36}{2} = 5.68$

d. aspartate $pI = \dfrac{2.09 + 3.86}{2} = \dfrac{5.95}{2} = 2.98$

11. **a.** Asp (pI = 2.98)

b. Arg (pI = 10.76)

c. Asp

d. Met, because at pH = 6.20 Met is farther away from the pH at which it has no net charge (pI of Met = 5.75, pI of Gly = 5.97).

12. Tyrosine and cysteine each have two groups that are neutral in their acidic forms and negatively charged in their basic forms. Unlike other amino acids that have similarly ionizing groups, the pK_a values of the two similarly ionizing groups in tyrosine and cysteine are not close in value and one of them is relatively close in value to the pK_a of the group that ionizes differently. Therefore, the third group cannot be ignored in calculating the pI.

13. 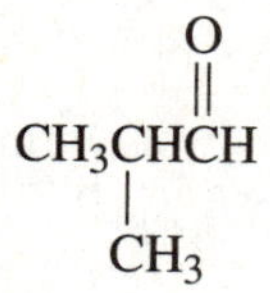

14. Leucine and isoleucine both have butyl side chains and, therefore, have the same polarity. Consequently, the spots for both amino acids appear at the same place on the chromatographic plate.

15. Because the amino acid analyzer contains a cation-exchange resin (it binds cations), the less positively charged the amino acid, the less tightly it is bound to the column. Using buffer solutions of increasingly higher pH to elute the column causes the amino acids bound to the column to become increasingly less positively charged, so they can be released from the column.

16. Cation-exchange chromatography releases amino acids in order of their pI values. The amino acid with the lowest pI is released first because at a given pH it will be the amino acid with the highest concentration of negative charge, and negatively charged molecules are not bound by the negatively charged resin. The relatively nonpolar resin will release polar amino acids before nonpolar amino acids.

a. Asp (pI = 2.98) is more negative than Ser (pI = 5.68).
b. Gly is more polar than Ala.
c. Val is more polar than Leu.
d. Tyr is more polar than Phe.

17. A column containing an anion-exchange resin releases amino acids in reverse order of their pI values. The amino acid with the highest pI is released first, because at a given pH it will be the amino acid with the highest concentration of positive charge.

$$\text{His} > \text{Val} > \text{Ser} > \text{Asp}$$

18. The first equivalent of ammonia will react with the acidic proton of the carboxylic acid to form ammonium ion, which is not nucleophilic and, therefore, cannot substitute for Br. A second equivalent of ammonia therefore is needed for the desired nucleophilic substitution reaction.

19. **a.** The reactions below show that pyruvic acid forms alanine, oxaloacetic acid forms aspartate, and α-ketoglutarate forms glutamate. If reductive amination (p. 812) is carried out in the cell, only the L-isomer of each amino acid will be formed.

b. If reductive amination is carried out in the laboratory, both the D- and L-isomer of each amino acid will be formed.

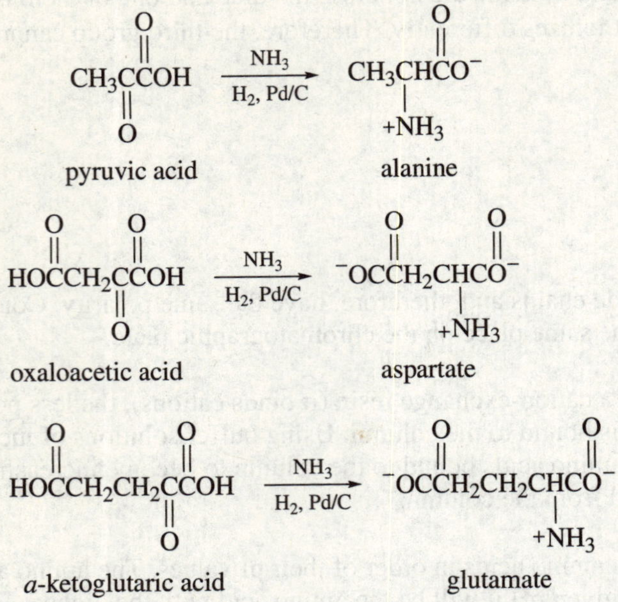

20. **a.** leucine **b.** methionine

21. **a.** 4-bromo-1-butanamine **b.** benzyl bromide

22. **a.** alanine **b.** isoleucine **c.** leucine

23. Convert the amino acids into esters using $SOCl_2$ followed by ethanol. The products obtained after treatment with pig liver esterase will be an L-amino acid and unreacted ester of the D-amino acid. These compounds can be readily separated, and the D-amino acid can be obtained by acid-catalyzed hydrolysis of the ester. This technique is called a *kinetic resolution* because the enantiomers are separated (resolved) as a result of reacting at different rates in the enzyme-catalyzed reaction.

24.

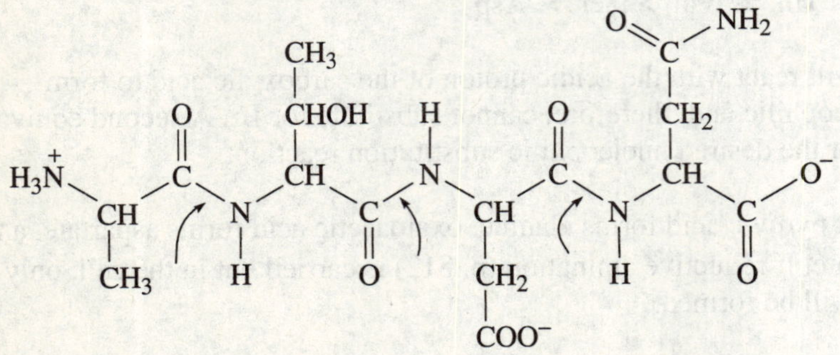

25. **A-G-M A-M-G M-G-A M-A-G G-A-M G-M-A**

26. 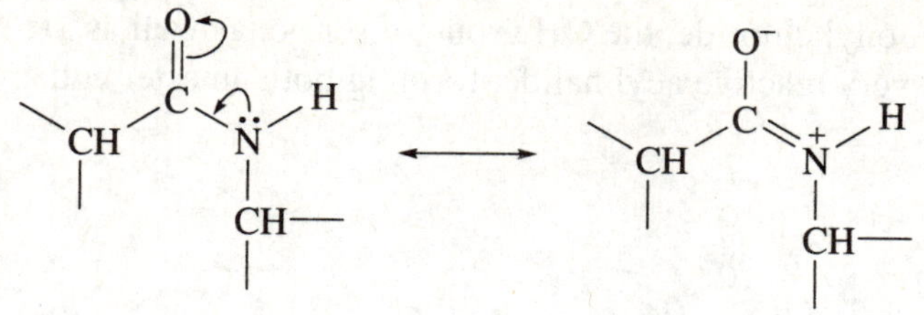

27. The bonds on either side of the α-carbon can freely rotate. In other words, the bond between the α-carbon and the carbonyl carbon and the bond between the α-carbon and the nitrogen (the bonds indicated by arrows) can freely rotate.

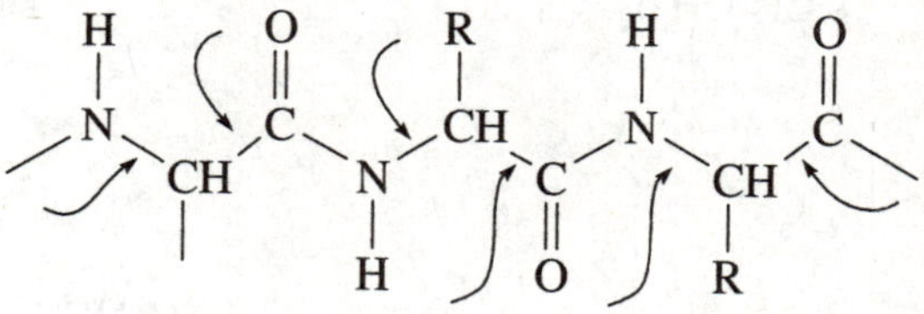

28. In forming the amide linkage, the amino group of cysteine reacts with the γ-carboxyl group rather than with the α-carboxyl group of glutamate.

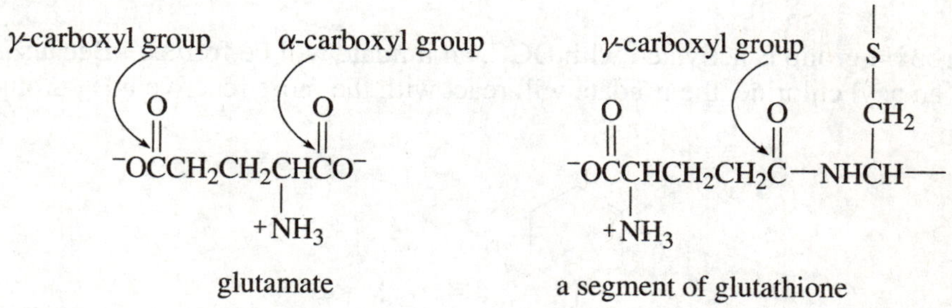

29. **Leu-Val** and **Val-Val** will be formed because the amino group of leucine is not reactive (so leucine could not be the C-terminal amino acid) but the amino group of valine would react equally easily with the carboxyl group of leucine and the carboxyl group of valine.

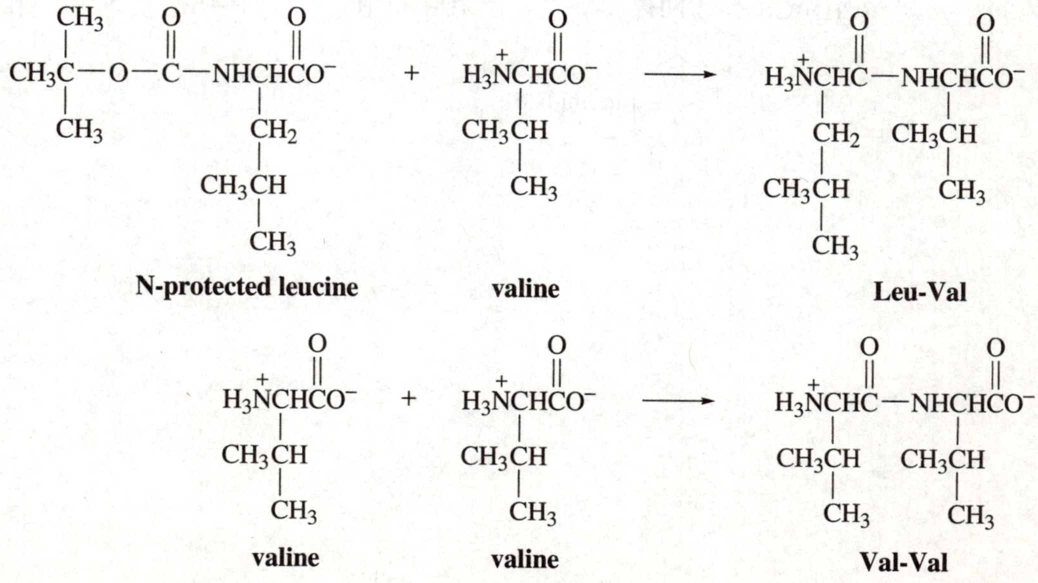

30. If valine's carboxyl group is activated with thionyl chloride, the OH group of serine, as well as the NH_2 group of serine, would react readily with the very reactive acyl halide, forming both an ester and an amide.

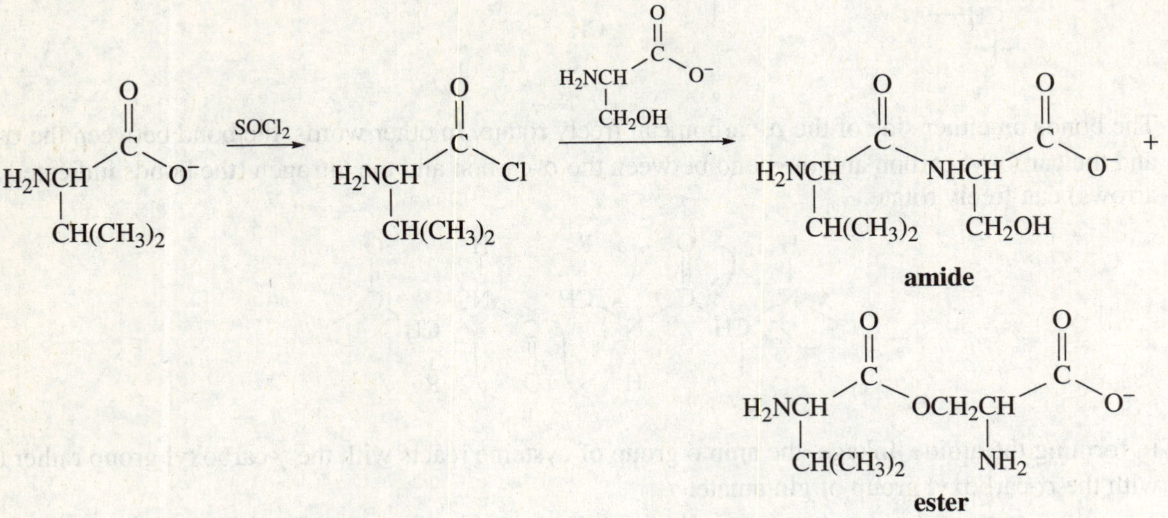

If valine's carboxyl group is activated with DCC, an imidate will be formed. Because an imidate is less reactive than an acyl chloride, the imidate will react with the more reactive NH_2 group in preference to the OH group.

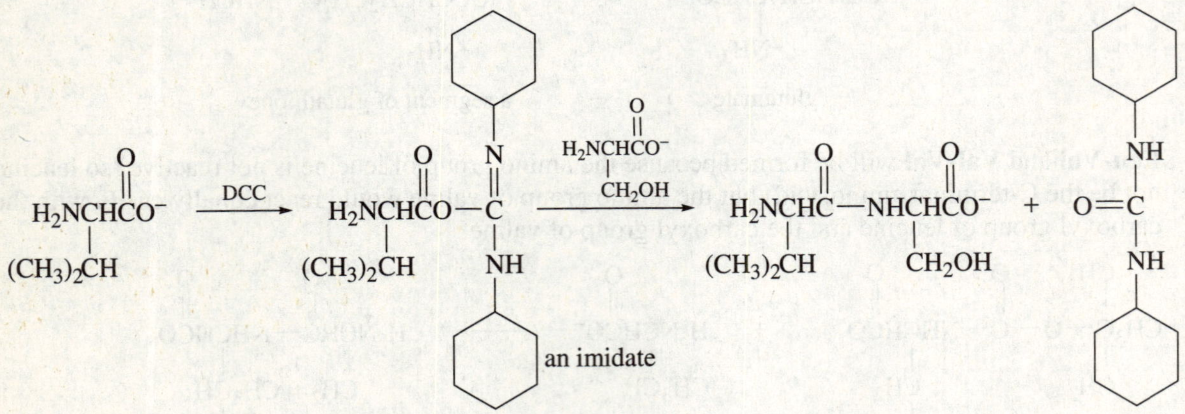

31.

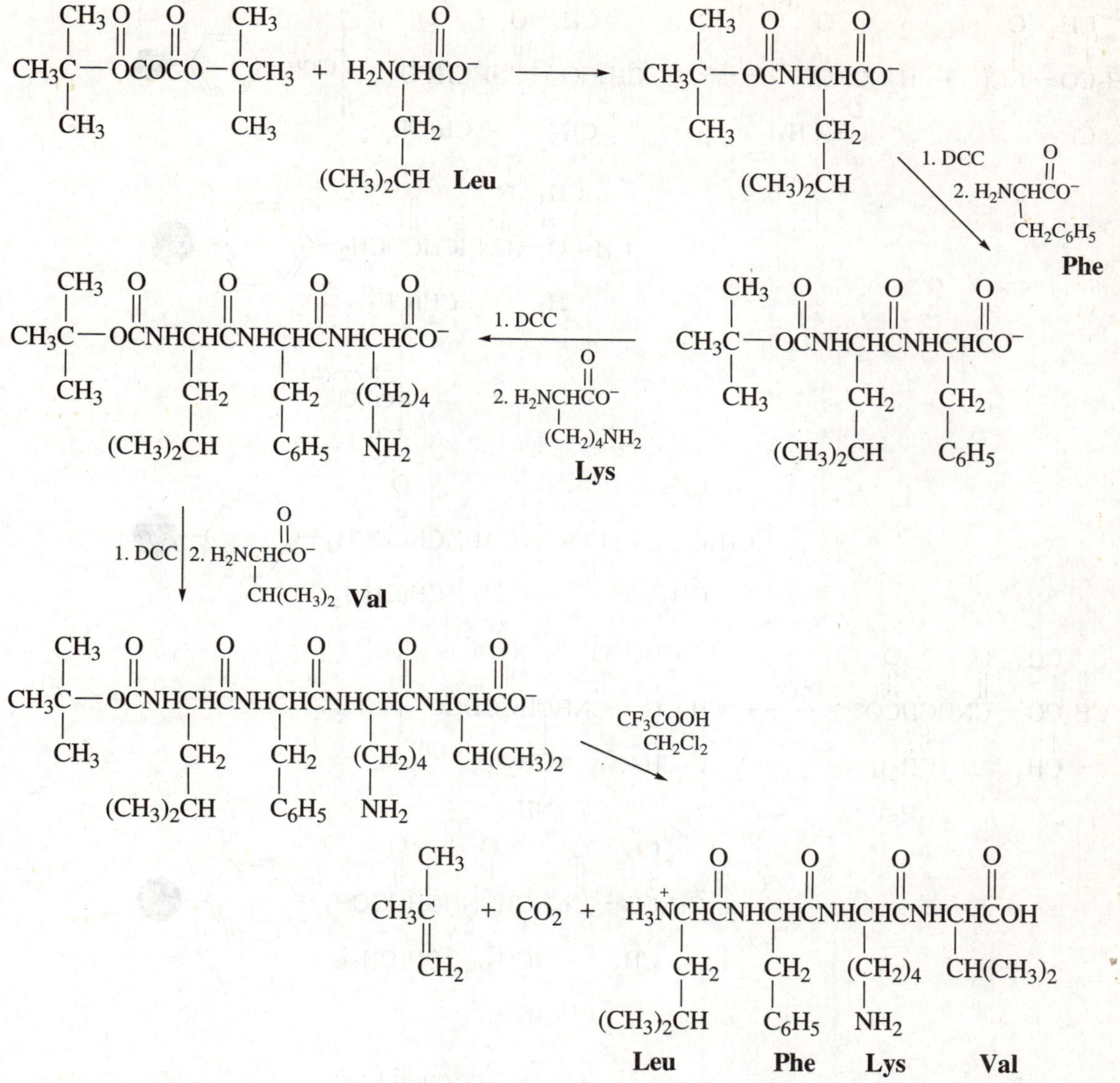

32. **a.** 5.8%

2	**3**	**4**	**5**	**6**	**7**	**8**	**9**
70%	49%	34%	24%	17%	12%	8.2%	5.8%

b. 4.4%

2	**3**	**4**	**5**	**6**	**7**	**8**	**9**	**10**	**11**	**12**	**13**	**14**	**15**
80%	64%	51%	41%	33%	26%	21%	17%	13%	11%	8.6%	6.9%	5.5%	4.4%

33.

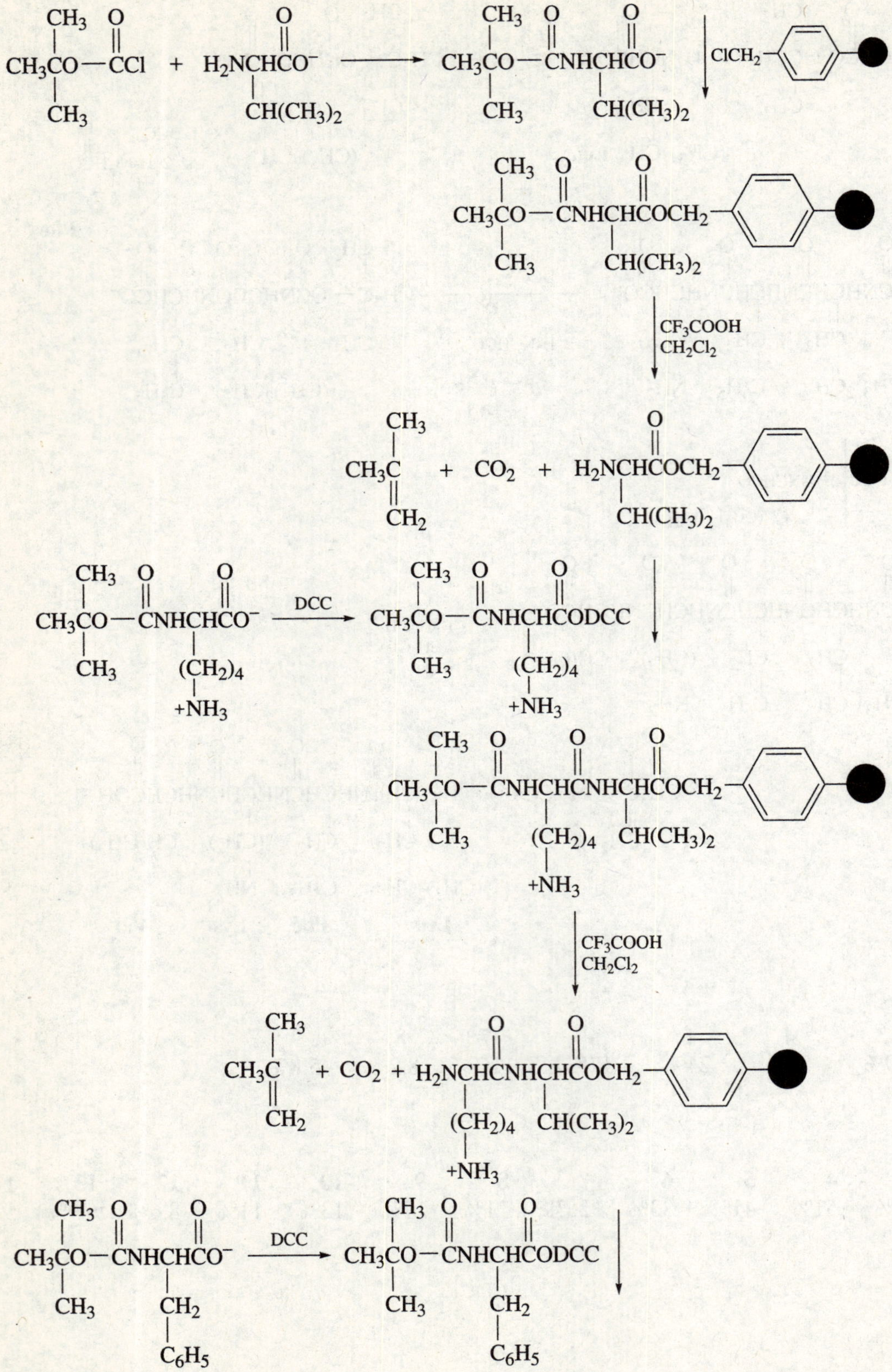

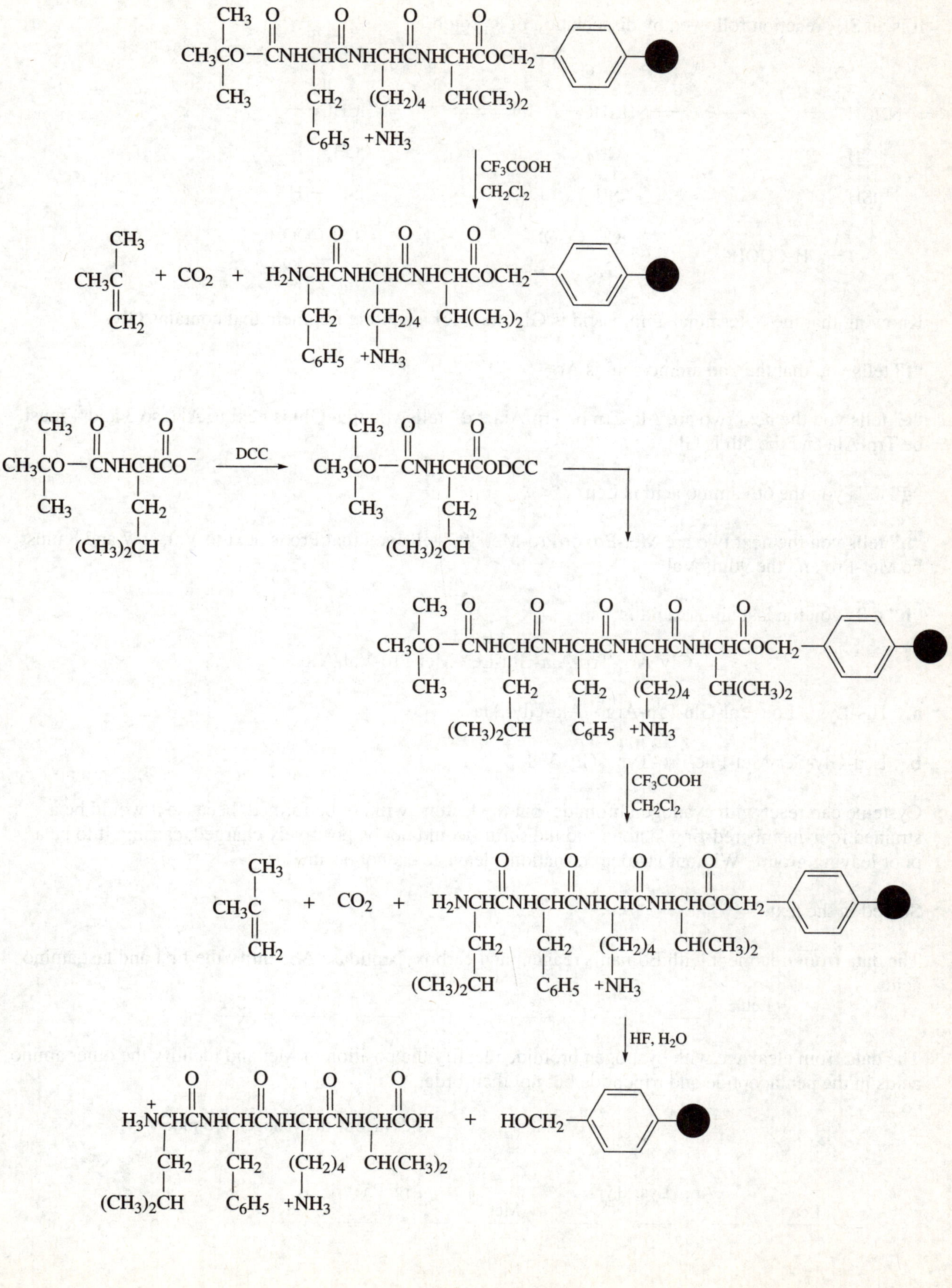
CF3COOH
CH2Cl2
DCC
CF3COOH
CH2Cl2
HF, H2O

34. It is an S_N2 reaction followed by dissociation of a proton.

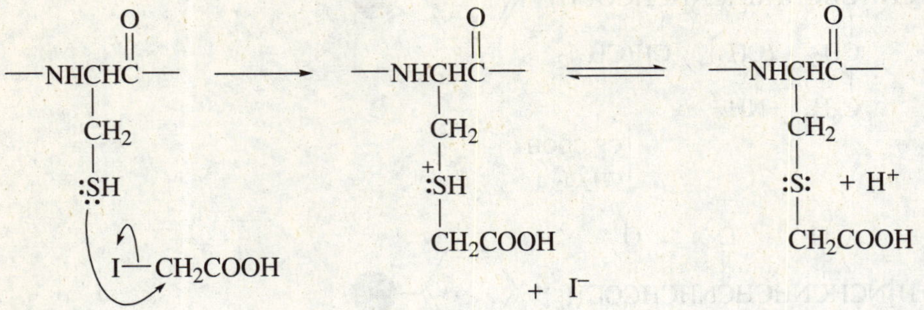

35. Knowing that the N-terminal amino acid is Gly, look for a peptide fragment that contains Gly.

"f" tells you that the 2nd amino acid is Arg

"e" tells you the next two are Ala-Trp or Trp-Ala. **"d"** tells you that Glu is next to Ala, so 3 and 4 must be Trp-Ala and the 5th is Glu

"g" tells you the 6th amino acid is Leu

"h" tells you the next two are Met-Pro or Pro-Met. **"c"** tells you that Pro is next to Val, so 7 and 8 must be Met-Pro and the 9th is Val

"b" tells you the last amino acid is Asp

Gly-Arg-Trp-Ala-Glu-Leu-Met-Pro-Val-Asp

36. **a.** His-Lys Leu-Val-Glu-Pro-Arg Ala-Gly-Ala

b. Leu-Gly-Ser-Met-Phe-Pro-Tyr Gly-Val

37. Cysteine can react with cyanogen bromide, but the lactone will not be formed, because it would be a strained four-membered-ring lactone and the sulfur would not be positively charged, causing it to be a poor leaving group. Without lactone formation, cleavage cannot occur.

38. Solved in the text.

39. The data from treatment with Edman's reagent and carboxypeptidase A identify the first and last amino acids.

Leu ___ ___ ___ ___ ___ ___ Ser

The data from cleavage with cyanogen bromide identify the position of Met and identify the other amino acids in the pentapeptide and tripeptide but not their order.

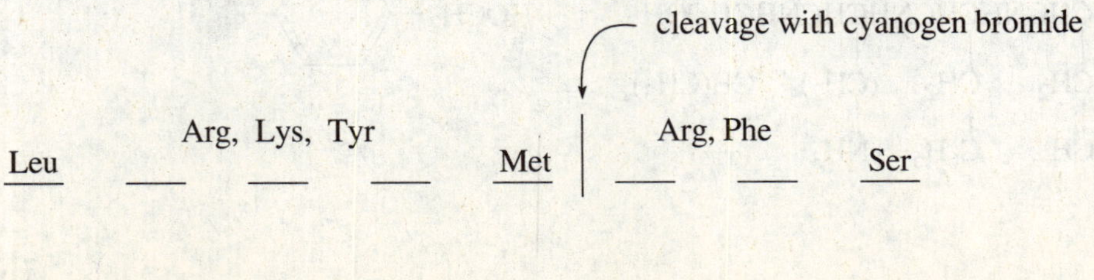

The data from treatment with trypsin put the remaining amino acids in the correct position.

Leu Tyr Lys | Arg | Met Phe Arg | Ser

40. Treatment with Edman's reagent would release two PTH-amino acids in approximately equal amounts.

41. 74 amino acids/3.6 amino acids per turn of the helix = 20.6 turns of the helix
20.6×5.4 Å = **110 Å** in an α-helix

74 amino acids $\times$ 3.5 Å = **260 Å** in a straight chain

42. It would fold so that its nonpolar residues are on the outside of the protein in contact with the nonpolar membrane and its polar residues are on the inside of the protein.

43. **a.** A cigar-shaped protein has the greatest surface area to volume ratio, so it has the highest percentage of polar amino acids.

b. A subunit of a hexamer would have the smallest percentage of polar amino acids because part of the surface of the subunit can be on the inside of the hexamer and, therefore, have nonpolar amino acids on its surface.

44. An amino acid is insoluble in diethyl ether (a relatively nonpolar solvent) because an amino acid exists as a zwitterion at neutral pH. In contrast, carboxylic acids and amines have a single charge at neutral pH and, therefore, are less polar.

45. **a.** Val-Arg-Gly-Met-Arg-Ala Ser

b. Ser-Phe-Lys-Met Pro-Ser-Ala-Asp

c. Arg Ser-Pro-Lys Lys Ser-Glu-Gly

46.

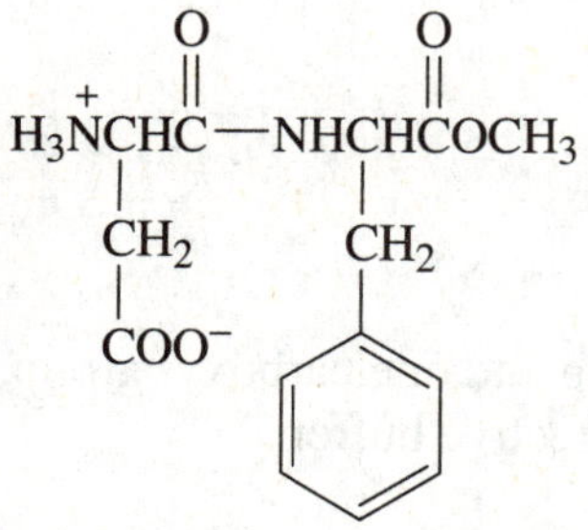

47. **a.**

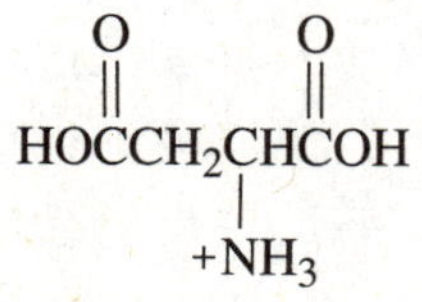

b.

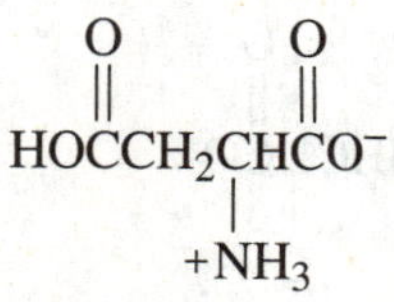

c.

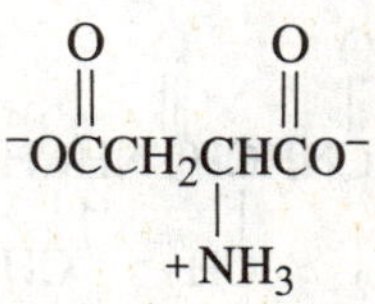

d.

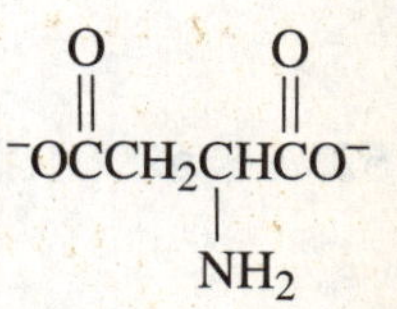

48. The student is correct. At the pI the total of the positive charges on the tripeptide's amino groups must be 1 to balance the one negative charge of the carboxyl group. When the pH of the solution is equal to the pK_a of a lysine residue, the three lysine groups each have one-half a positive charge for a total of one and one-half positive charges. Thus, the solution must be more basic than this in order to have just one positive charge.

49. Since the mixture of amino acids is in a solution of pH = 5, **His** will have an overall positive charge and **Glu** will have an overall negative charge. **His**, therefore, will migrate to the cathode, and **Glu** will migrate to the anode. **Ser** is more polar than **Thr** (both have OH groups, but **Thr** has an additional carbon); **Thr** is more polar than **Met** (**Met** has S instead of O and an additional carbon); and **Met** is more polar than **Leu**.

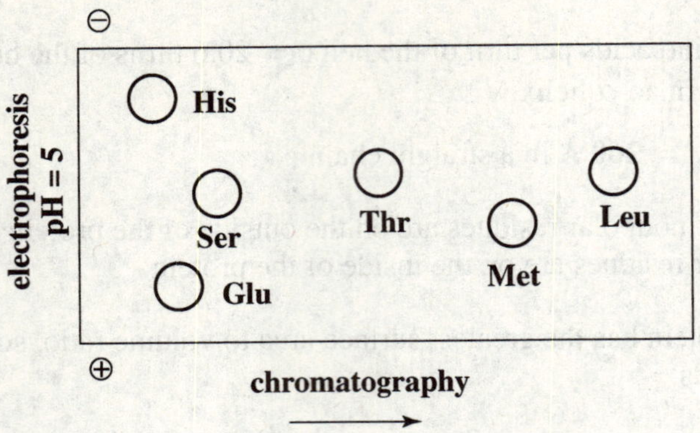

50. You would (correctly) expect serine and cysteine to have lower pK_a value than alanine since a hydroxymethyl and a thiomethyl group are more electron withdrawing than a methyl group. Because oxygen is more electronegative than sulfur, one would probably expect serine to have a lower pK_a than cysteine. The fact that cysteine has a lower pK_a than serine can be explained by stabilization of serine's carboxyl proton by hydrogen bonding to the β-OH group of serine, which causes it to have less of a tendency to be removed by base.

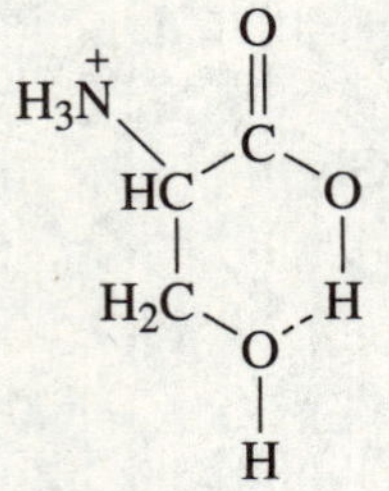

51. Each compound has two groups that can act as a buffer, one amino group and one carboxyl group. Thus, the compound in higher concentration (0.2 M glycine) will be a more effective buffer.

52. **a.**

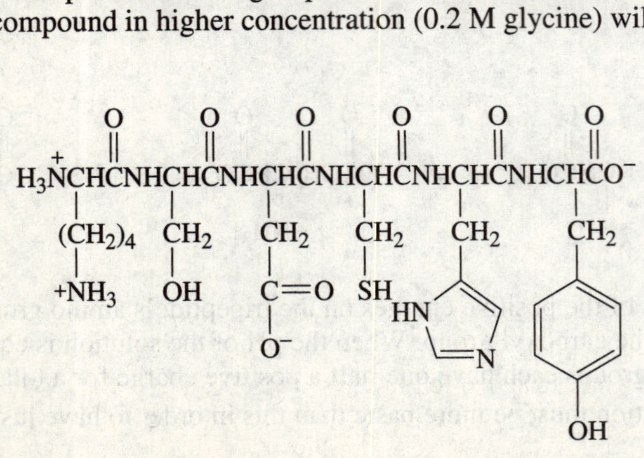

b.

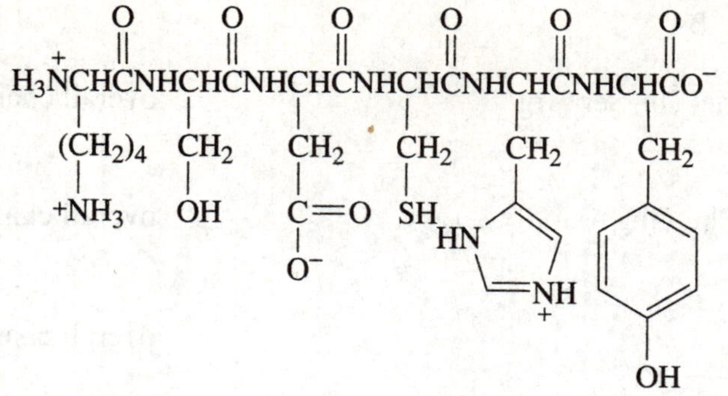

c.

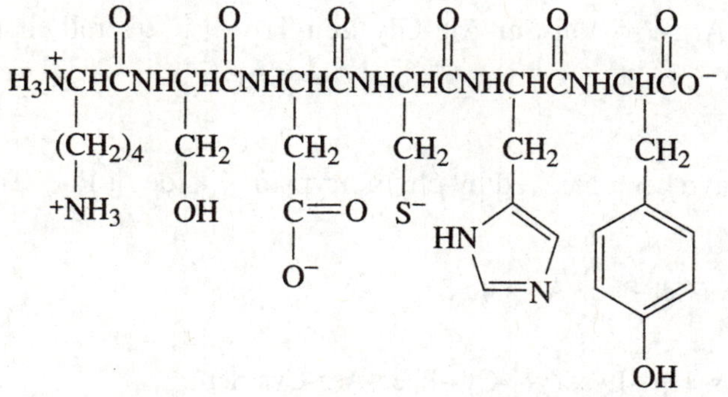

53.

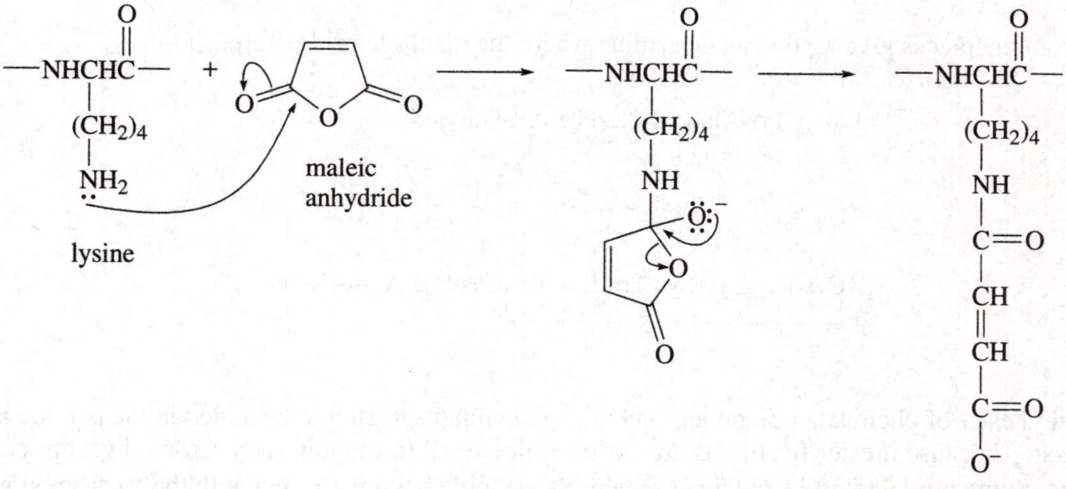

54. **a.** When the polypeptide is treated with maleic anhydride, lysine reacts with maleic anhydride (See Problem 53), but the amino group of arginine is not sufficiently nucleophilic to react. Therefore, trypsin will cleave only at arginine residues, since it will no longer recognize lysine residues.

b. Four fragments will be obtained from the polypetide. Remember that trypsin will not cleave the Arg-Pro bond.

c. The N-terminal end of each fragment will be positively charged because of the $^{+}NH_3$ group, the C-terminal end will be negatively charged because of the COO^- group; arginine residues will be positively charged; aspartate and glutamate residues will be negatively charged, and lysine residues will be negatively charged because they are attached to the maleic acid group.

Elution order: **A > D > C > B**

A Gly-Ser-Asp-Ala-Leu-Pro-Gly-Ile-Thr-Ser-Arg — overall charge = 0
+ −   + −

B Asp-Val-Ser-Lys-Val-Glu-Tyr-Phe-Glu-Ala-Gly-Arg — overall charge = −3
+ −   −   −   −   + −

C Ser-Glu-Phe-Lys-Glu-Pro-Arg — overall charge = −2
+ −   −   −   + −

D Leu-Tyr-Met-Lys-Val-Glu-Gly-Arg-Pro-Val-Ser-Ala-Gly-Leu-Trp — overall charge = −1
+   −   −   +   −

55. First mark off where the chains would have been cleaved by chymotrypsin (C-side of Phe, Trp, Tyr).

Val- Met-Tyr | -Ala-Cys-Ser-Phe | -Ala-Glu-Ser

Ser-Cys-Phe | -Lys-Cys-Trp | -Lys-Tyr | -Cys-Phe | -Arg-Cys-Ser

Then, from the pieces given, you can determine where the disulfide bridges are.

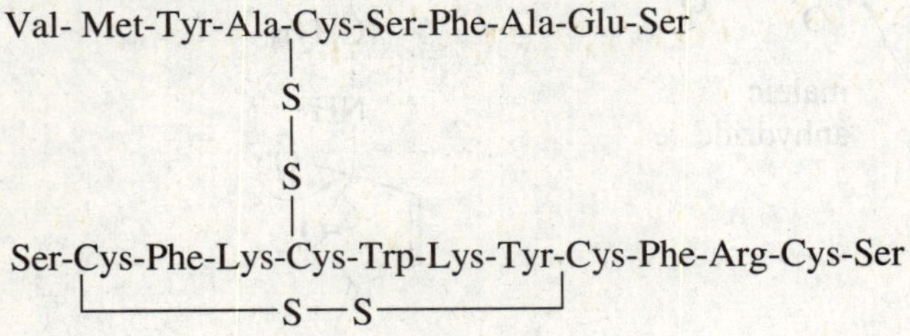

56. The methyl ester of phenylalanine rather than phenylalanine itself should be added in the peptide bond-forming step because if esterification is done after amide bond formation, both carboxyl groups could be esterified. Some product will be obtained in which the amide bond is formed with the γ-carboxyl group of aspartate rather than with the α-carboxyl group.

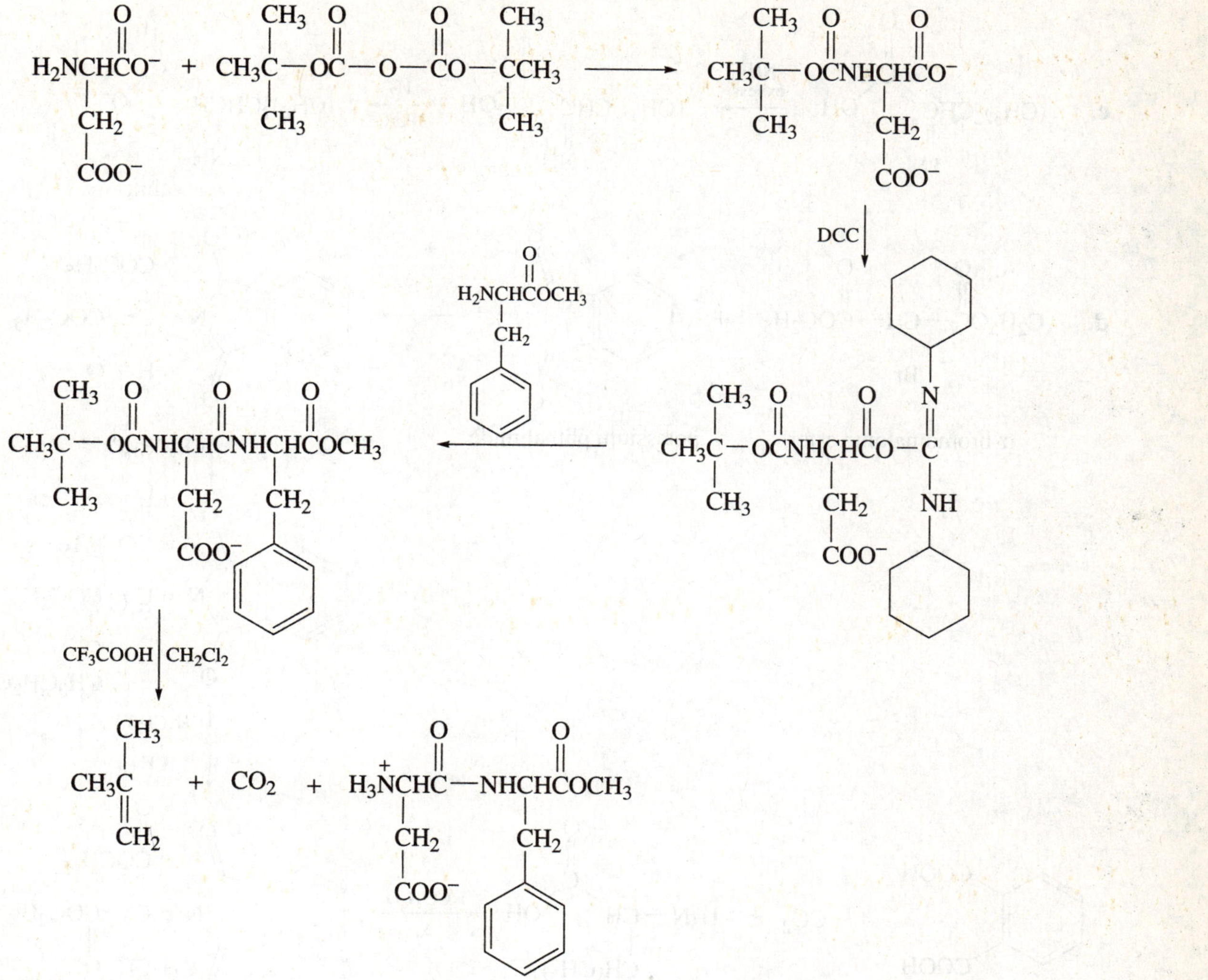
DCC
CF_3COOH CH_2Cl_2

57. **a.**

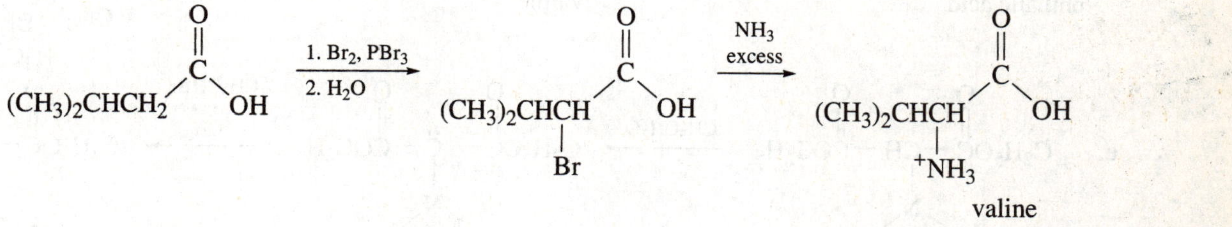
1. Br_2, PBr_3
2. H_2O
NH_3 excess
valine

b.

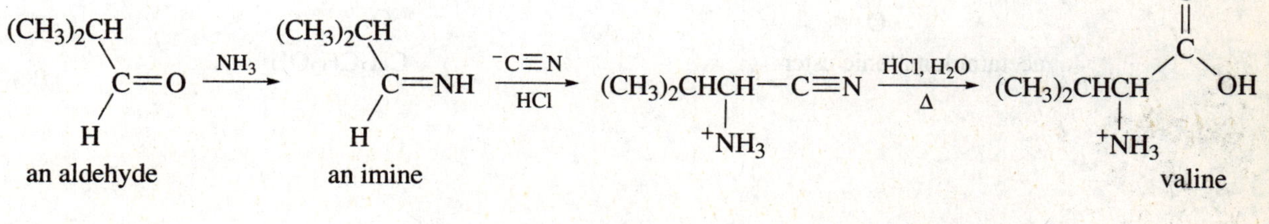
NH_3
HCl
HCl, H_2O
Δ
an aldehyde
an imine
valine

c.

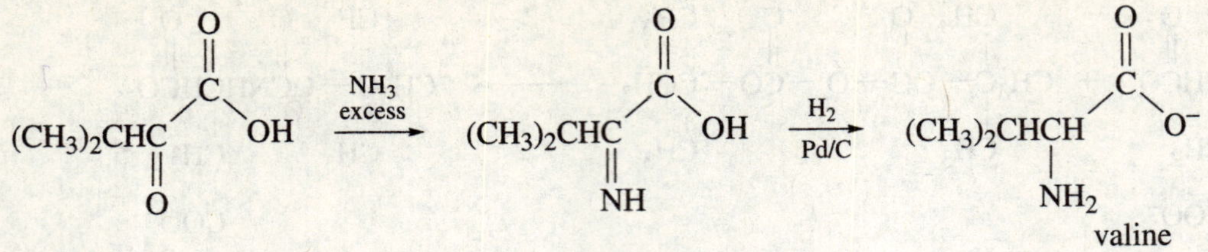
NH3
excess
H2
Pd/C
valine

d.

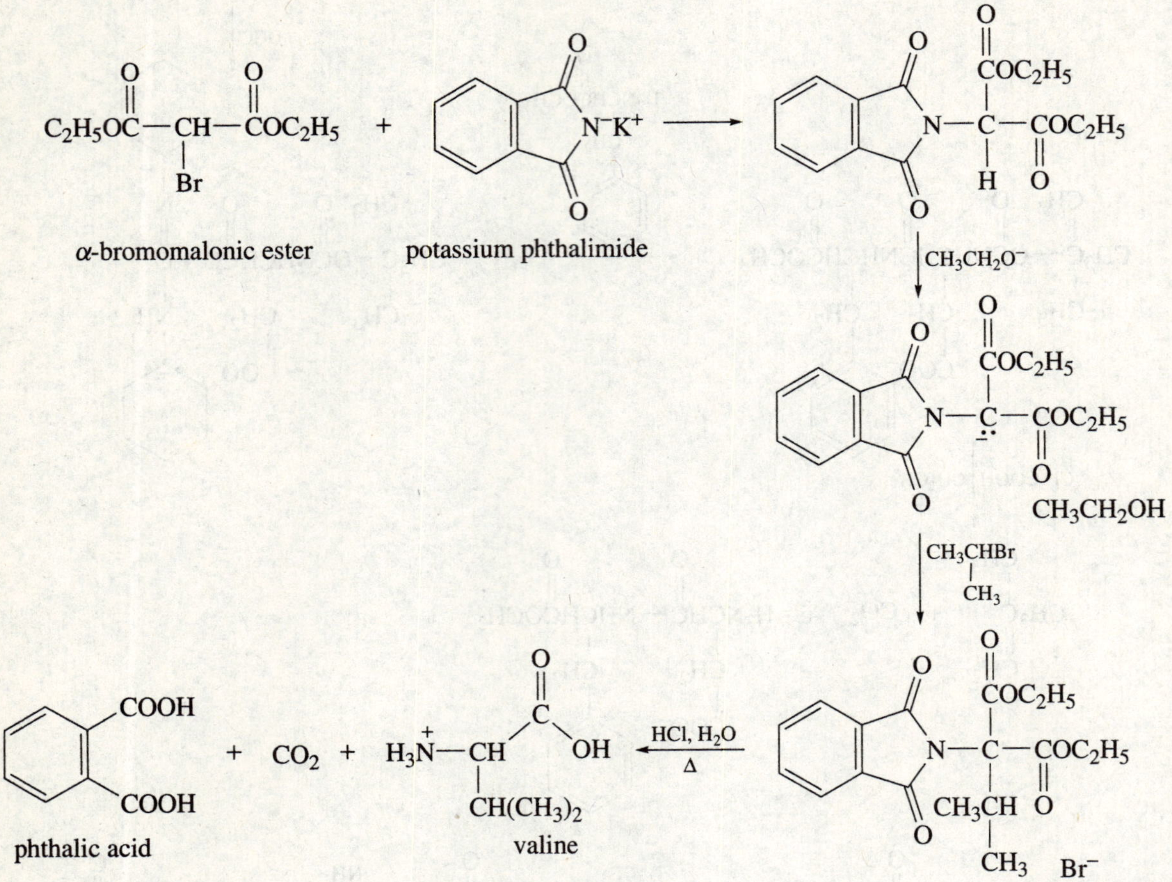
α-bromomalonic ester
potassium phthalimide
CH3CH2O−
CH3CH2OH
CH3CHBr
CH3
HCl, H2O
Δ
phthalic acid
valine

e.

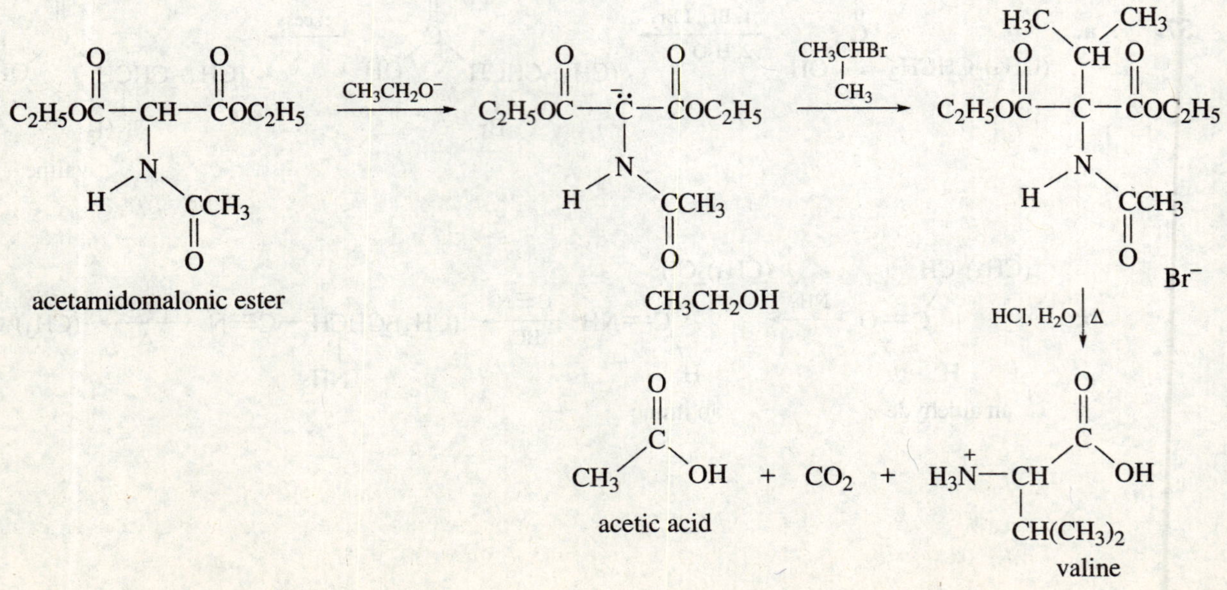
CH3CH2O−
CH3CHBr
CH3
acetamidomalonic ester
CH3CH2OH
Br−
HCl, H2O Δ
acetic acid
valine

58. Ser-Glu-Leu-Trp-Lys-Ser-Val-Glu-His-Gly-Ala-Met

From the experiment with carboxypeptidase, we know the C-terminal amino acid is Met.

“l” tells us the amino acid adjacent to Met is Ala

“e” tells us the next amino acid is Gly

“b” tells us the next amino acid is His

“g” tells us the next amino acid is Glu

“j” tells us the next amino acid is Val

“c” tells us the next amino acid is Ser

“i” tells us the next amino acid is Lys

“a” tells us the next amino acid is Trp

“h” tells us the next amino acid is Leu

“k” tells us the next amino acid is Glu

“k” tells us the next (first) amino acid is Ser

59. **a.** $20^8 = 25{,}600{,}000{,}000$ **b.** 20^{100}

60. The pK_a of the carboxylic acid of the dipeptide is higher than the pK_a of the carboxylic acid of the amino acid because the positively charged amino group of the amino acid is more strongly electron withdrawing than the amide group of the peptide. This causes the amino acid to be a stronger acid and have a lower pK_a.

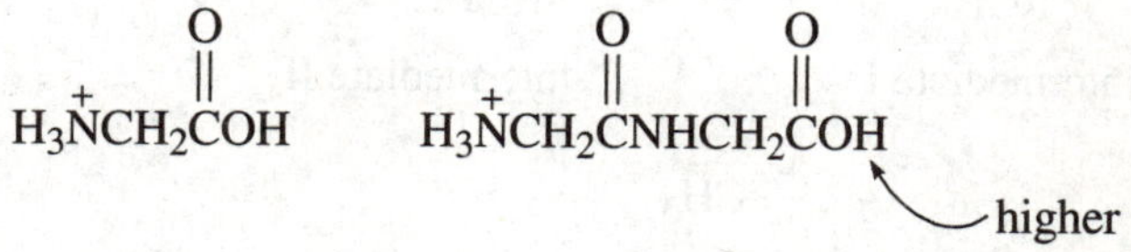

The pK_a of the amino group of the peptide is lower than the pK_a of the amino group of the amino acid because the amide group of the peptide is more strongly electron withdrawing than the carboxylate group of the amino acid.

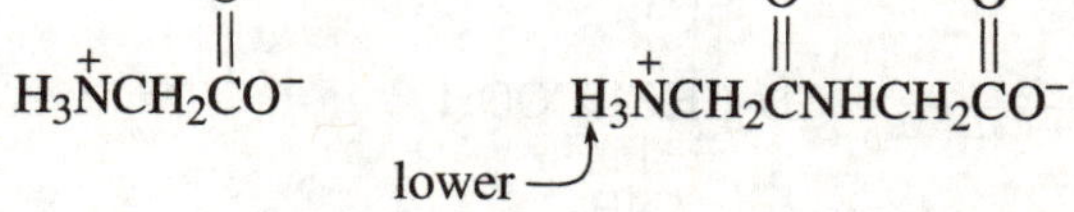

61. Finding that there is one less spot than the number of amino acids tells you that the spots for two of the amino acids superimpose. Since leucine and isoleucine have identical polarities, they are good candidates for being the amino acids that migrate to the same location.

Val (pI = 5.97), Trp (pI = 5.89), and Met (pI = 5.75) can be ordered based on their pI values because the one with the greatest pI will be the one with the greatest amount of positive charge at pH = 5.

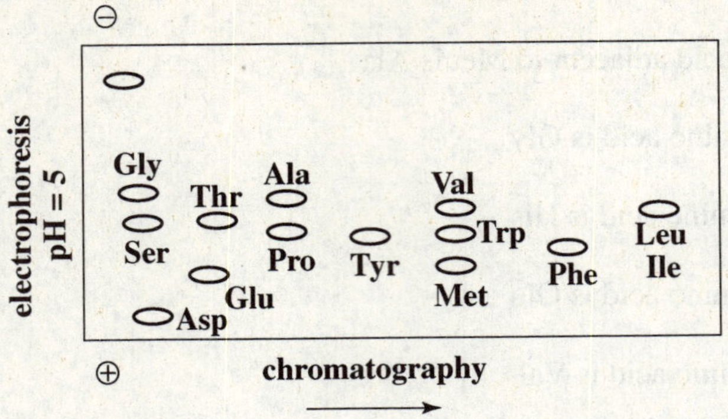

62. Oxidation of dithiothreitol is an intramolecular reaction so it occurs with a larger rate constant than the oxidation of 2-mercaptoethanol, which is an intermolecular reaction. The reverse reduction reaction should occur with about the same rate constant in both cases. Increasing the rate of the oxidation reaction while keeping the rate of the reduction reaction constant is responsible for the greater equilibrium constant, since $K_{eq} = k_1/k_{-1}$.

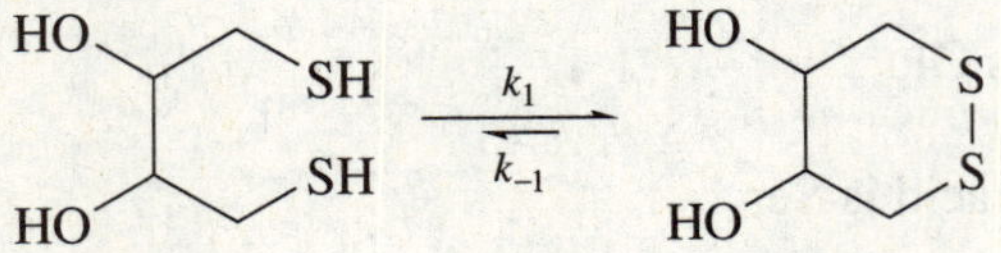

$$2\ HOCH_2CH_2SH \underset{k_{-1}}{\overset{k_1}{\rightleftharpoons}} HOCH_2CH_2S{-}SCH_2CH_2OH$$

63. **a.**

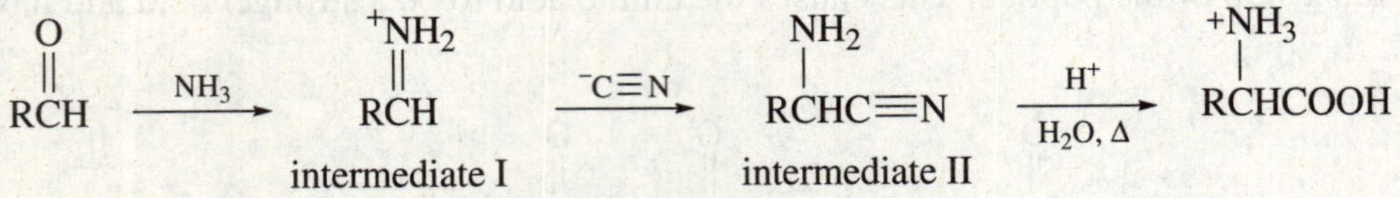

b.

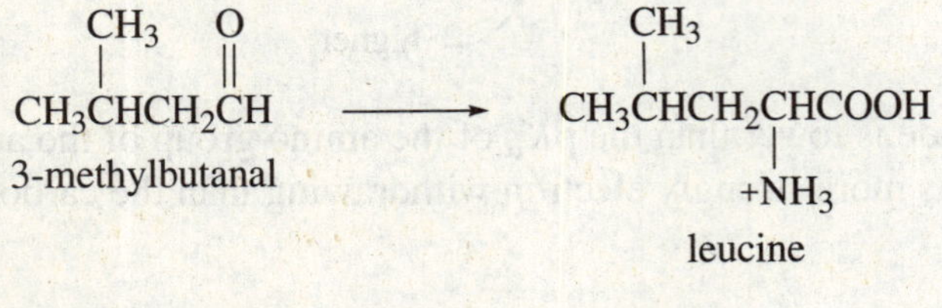

c.

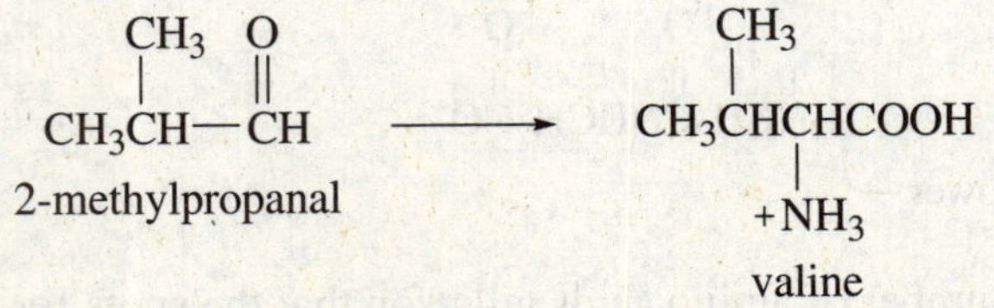

64.

amino acid	**Absorbance at 280 nm**	**molar absorptivity = absorbance/concentration**
tryptophan	0.61	610
tyrosine	0.15	150
phenylalanine	0.01	1

65. The spot marked with an **X** is the peptide that is different in the normal and mutant polypeptide. The spot is closer to the cathode and farther to the right, indicating that the substituted amino acid in the mutant has a greater pI and is less polar.

The fingerprints are those of hemoglobin (normal) and sickle-cell hemoglobin (mutant). In sickle-cell hemoglobin, a glutamate in the normal polypeptide is substituted with a valine. This agrees with our observation that the substituted amino acid is less negative and more nonpolar. (See the discussion of sickle cell anemia on page 1216 of the text.)

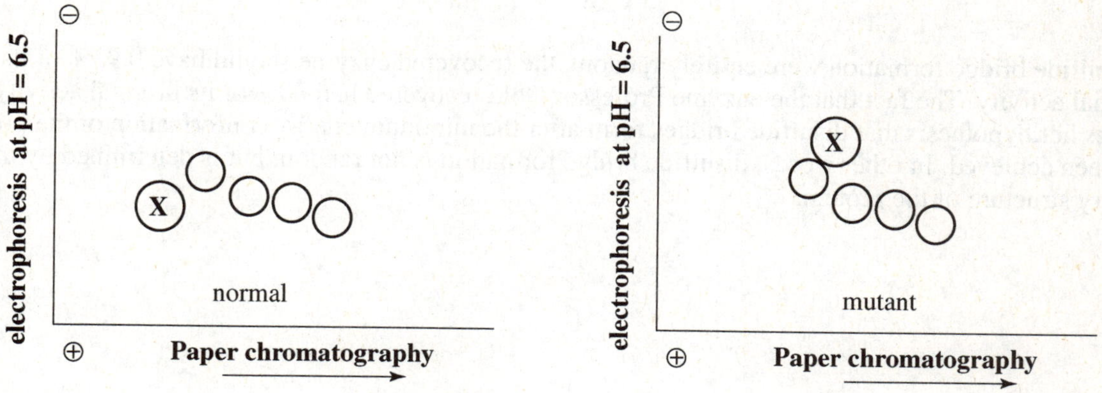

66. **a.** Acid-catalyzed hydrolysis indicates the peptide contains 12 amino acids.

__ __ __ __ __ __ __ __ __ __ __ __

b. Treatment with Edman's reagent indicates that Val is the N-terminal amino acid.

Val __ __ __ __ __ __ __ __ __ __ __

c. Treatment with carboxypeptidase A indicates that Ala is the C-terminal amino acid.

Val __ __ __ __ __ __ __ __ __ __ Ala

d. Treatment with cyanogen bromide indicates that Met is the 5th amino acid with Arg, Gly, Ser in an unknown order in positions 2, 3, and 4.

Val __ __ __ Met __ __ __ __ __ __ Ala

Arg, Gly, Ser

e. Treatment with trypsin indicates that Arg is the 3rd amino acid, and Ser is 2nd, Gly is 4th, Tyr is 6th, and Lys is 7th. Since Lys is in the terminal fragment means that cleavage did not occur at Lys, so Pro must be at lysine's cleavage sit.

Val Ser Arg Gly Met Tyr Lys ___ ___ ___ ___ ___ Ala

Lys- Pro, Phe, Ser

f. Treatment with chymotrypsin indicates that Phe is the 10th amino acid and Ser is the 11th .

Val Ser Arg Gly Met Tyr Lys Lys Pro Phe Ser Ala

67. Because the native enzyme has four disulfide bridges, the denatured enzyme has eight cysteine residues. The first cysteine has a 1 in 7 chance of forming a disulfide bridge with the correct cysteine. The first cysteine of the next pair has a 1 in 5 chance, and the first cysteine of the third pair has a 1 in 3 chance.

$$\frac{1}{7} \times \frac{1}{5} \times \frac{1}{3} = 0.0095$$

If disulfide bridge formation were entirely random, the recovered enzyme should have 0.95% of its original activity. The fact that the enzyme Professor Gold recovered had 80% of its original activity proves her hypothesis that disulfide bridges form after the minimum energy conformation of the protein has been achieved. In other words, disulfide bridge formation is not random but is determined by the tertiary structure of the protein.

Chapter 22 Practice Test

1. Give the structure of the following amino acids at pH = 7:

   a. glutamic acid **b.** lysine **c.** isoleucine **d.** arginine **e.** asparagine

2. Draw the form of histidine that predominates at:

   a. pH = 1 **b.** pH = 4 **c.** pH = 8 **d.** pH = 11

3. Answer the following:

   a. Alanine has a pI = 6.02 and serine has a pI = 5.68. Which would have the highest concentration of positive charge at pH = 5.50?

   b. Which amino acid is the only one that does not have an asymmetric center?

   c. Which are the two most nonpolar amino acids?

   d. Which amino acid has the lowest pI?

4. Why does the carboxyl group of alanine have a lower pK_a than the carboxyl group of propanoic acid?

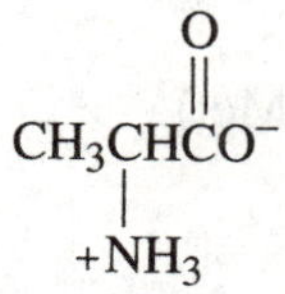

$pK_a = 2.2$

alanine

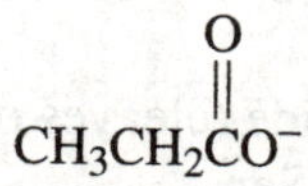

$pK_a = 4.7$

propanoic acid

5. Indicate whether each of the following is true or false:

   a. A cigar-shaped protein has a greater percentage of polar residues than a spherical protein. T F

   b. Naturally occurring amino acids have the L-configuration. T F

   c. There is free rotation about a peptide bond. T F

6. Give the compound obtained from mild oxidation of cysteine.

7. Define the following:

   a. the primary structure of a protein

   b. the tertiary structure of a protein

   c. the quaternary structure of a protein

8. Identify the spots.

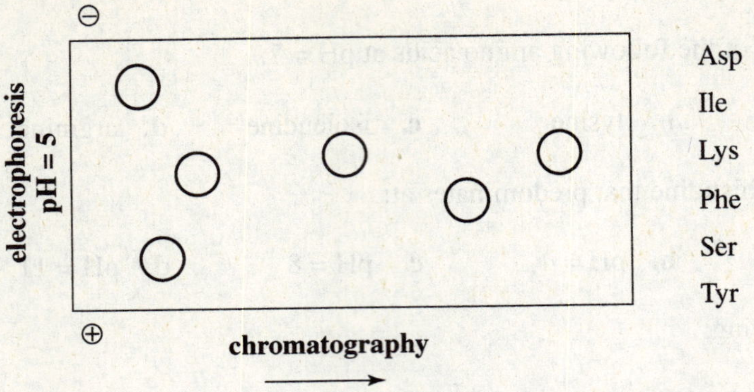

9. Calculate the pI of each of the following amino acids:

a. phenylalanine (pK_a's = 2.16, 9.18) **b.** arginine (pK_a's = 2.17, 9.04, 12.48)

10. From the following information, determine the primary sequence of the decapeptide:

a. Acid hydrolysis gives: Ala, 2 Arg, Gly, His, Ile, Lys, Met, Phe, Ser

b. Reaction with Edman's reagent liberated Ala

c. Reaction with carboxypeptidase A liberated Ile

d. Reaction with cyanogen bromide (cleaves on the C-side of Met)
1. Gly, 2 Arg, Ala, Met, Ser
2. Lys, Phe, Ile, His

e. Reaction with trypsin (cleaves on the C-side of Arg and Lys)
1. Arg, Gly
2. Ile
3. Phe, Lys, Met, His
4. Arg, Ser, Ala

f. Reaction with thermolysin (cleaves on the N-side of Leu, Ile, Phe, Trp, Tyr)
1. Lys, Phe
2. 2 Arg, Ser, His, Gly, Ala, Met
3. Ile

Chapter 22 Practice Test

1. Give the structure of the following amino acids at pH = 7:

a. glutamic acid **b.** lysine **c.** isoleucine **d.** arginine **e.** asparagine

2. Draw the form of histidine that predominates at:

a. pH = 1 **b.** pH = 4 **c.** pH = 8 **d.** pH = 11

3. Answer the following:

a. Alanine has a pI = 6.02 and serine has a pI = 5.68. Which would have the highest concentration of positive charge at pH = 5.50?

b. Which amino acid is the only one that does not have an asymmetric center?

c. Which are the two most nonpolar amino acids?

d. Which amino acid has the lowest pI?

4. Why does the carboxyl group of alanine have a lower pK_a than the carboxyl group of propanoic acid?

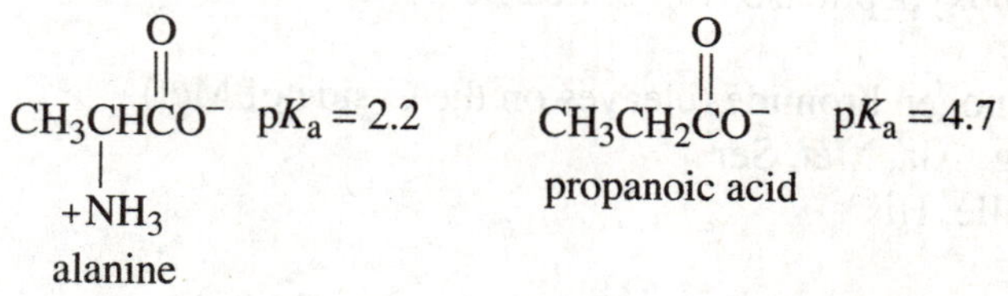

5. Indicate whether each of the following is true or false:

a. A cigar-shaped protein has a greater percentage of polar residues than a spherical protein. T F

b. Naturally occurring amino acids have the L-configuration. T F

c. There is free rotation about a peptide bond. T F

6. Give the compound obtained from mild oxidation of cysteine.

7. Define the following:

a. the primary structure of a protein

b. the tertiary structure of a protein

c. the quaternary structure of a protein

8. Identify the spots.

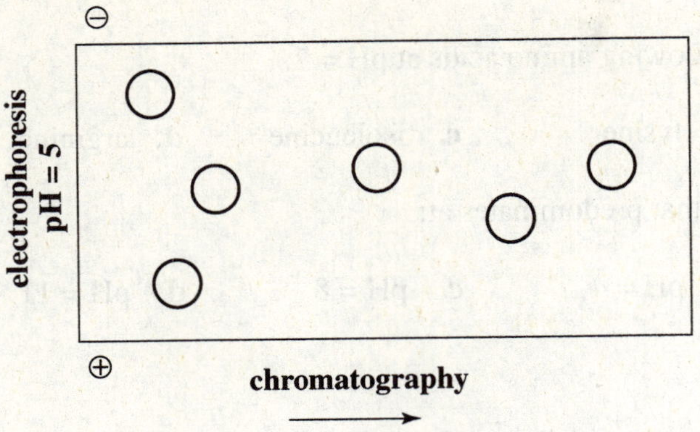

Asp
Ile
Lys
Phe
Ser
Tyr

9. Calculate the pI of each of the following amino acids:

a. phenylalanine (pK_a's = 2.16, 9.18) **b.** arginine (pK_a's = 2.17, 9.04, 12.48)

10. From the following information, determine the primary sequence of the decapeptide:

a. Acid hydrolysis gives: Ala, 2 Arg, Gly, His, Ile, Lys, Met, Phe, Ser

b. Reaction with Edman's reagent liberated Ala

c. Reaction with carboxypeptidase A liberated Ile

d. Reaction with cyanogen bromide (cleaves on the C-side of Met)
1. Gly, 2 Arg, Ala, Met, Ser
2. Lys, Phe, Ile, His

e. Reaction with trypsin (cleaves on the C-side of Arg and Lys)
1. Arg, Gly
2. Ile
3. Phe, Lys, Met, His
4. Arg, Ser, Ala

f. Reaction with thermolysin (cleaves on the N-side of Leu, Ile, Phe, Trp, Tyr)
1. Lys, Phe
2. 2 Arg, Ser, His, Gly, Ala, Met
3. Ile

<u>CHAPTER 23</u>
Catalysis

<u>Important Terms</u>

acid catalyst — a catalyst that increases the rate of a reaction by donating a proton.

active site — a pocket or cleft in an enzyme where the substrate is bound.

acyl-enzyme intermediate — an amino acid residue of an enzyme that has been acylated while catalyzing a reaction.

base catalyst — a catalyst that increases the rate of a reaction by removing a proton.

catalyst — a substance that increases the rate of a reaction without itself being consumed in the overall reaction.

covalent catalysis (nucleophilic catalysis) — catalysis that occurs as a result of a nucleophile forming a covalent bond with one of the reactants.

effective molarity — the concentration of the reagent that would be required in an intermolecular reaction for it to have the same rate as an intramolecular reaction.

electrophilic catalyst — an electrophile that facilitates a reaction.

electrostatic catalysis — stabilization of a charge by an opposite charge.

enzyme — a protein that is a catalyst.

***gem*-dialkyl effect** — two alkyl groups on a carbon whose effect is to increase the probability that the molecule will be in the proper conformation for ring closure.

general-acid catalysis — catalysis in which a proton is transferred to the reactant during the slow step of the reaction.

general-base catalysis — catalysis in which a proton is removed from the reactant during the slow step of the reaction.

induced fit model — a model that describes the specificity of an enzyme for its substrate: the shape of the active site does not become completely complementary to the shape of the substrate until after the enzyme has bound the substrate.

intramolecular catalysis (anchimeric assistance) — catalysis in which the catalyst that facilitates the reaction is part of the molecule undergoing reaction.

lock-and-key model — a model that describes the specificity of an enzyme for its substrate: the substrate fits the enzyme like a key fits into a lock.

metal-ion catalysis — catalysis in which the species that facilitates the reaction is a metal ion.

molecular recognition	the recognition of one molecule by another as a result of specific interactions; for example, the specificity of an enzyme for its substrate.
nucleophilic catalysis (covalent catalysis)	catalysis that occurs as a result of a nucleophile forming a covalent bond with one of the reactants.
nucleophilic catalyst	a catalyst that increases the rate of a reaction by acting as a nucleophile.
pH-activity profile or pH-rate profile	a plot of the activity of an enzyme as a function of the pH of the reaction mixture.
relative rate	obtained by dividing the actual rate constant by the rate constant of the slowest reaction in the group being compared.
site-specific mutagenesis	a technique that substitutes one amino acid of a protein for another.
specific-acid catalysis	catalysis in which the proton is fully transferred to the reactant before the slow step of the reaction.
specific-base catalysis	catalysis in which the proton is completely removed from the reactant before the slow step of the reaction.
substrate	the reactant of an enzyme-catalyzed reaction.

Solutions to Problems

1. $\Delta H^{\ddagger}$, E_a, $\Delta S^{\ddagger}$, $\Delta G^{\ddagger}$, k_{rate} (A catalyst increases the rate of a reaction by decreasing the distance between the reactant and the transition state. These are the parameters that measure the difference in energy between the reactant and the transition state.)

2. Notice that **(1)** and **(2)** have only the first part (because the final product of the reaction is a tetrahedral intermediate), **(3)** has only the second part (because the initial reactant is a tetrahedral intermediate) and **(4)** has both the first and second parts.

First part

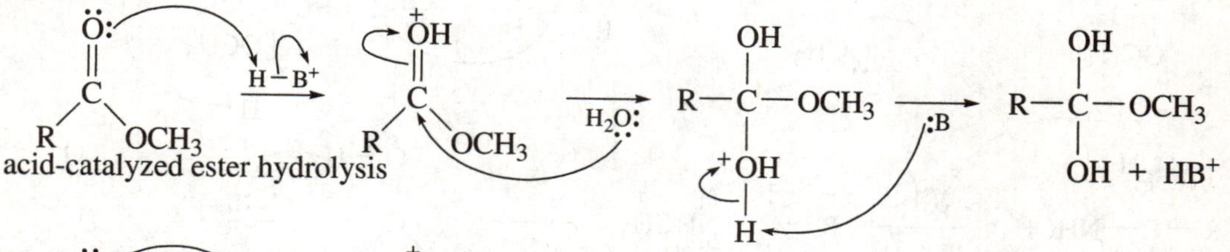

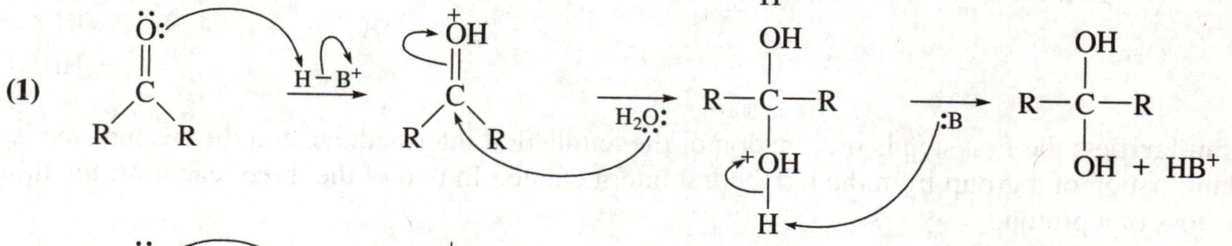

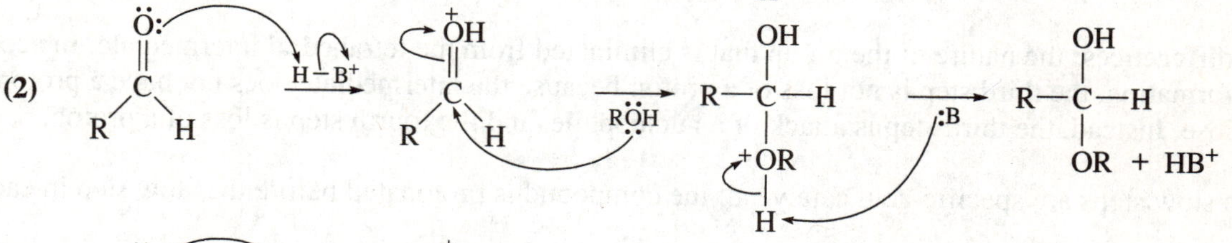

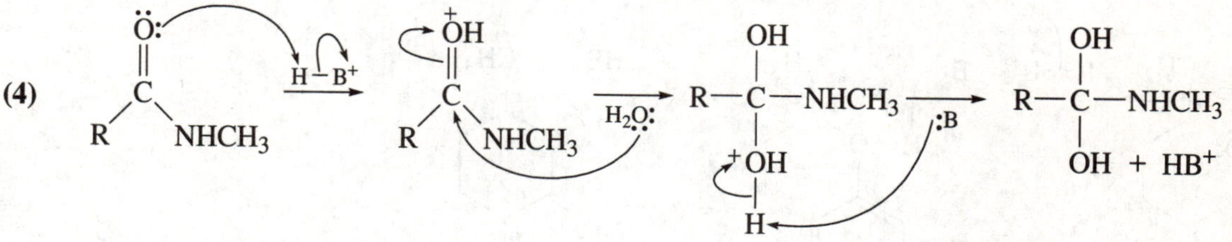

a. similarities: the first step is protonation of the carbonyl compound, the second step is attack of a nucleophile on the protonated carbonyl compound, and the third step is loss of a proton.

b. differences: the carbonyl compound that is used as the starting material; the nucleophile used in **(2)** is an alcohol rather than water.

Second part

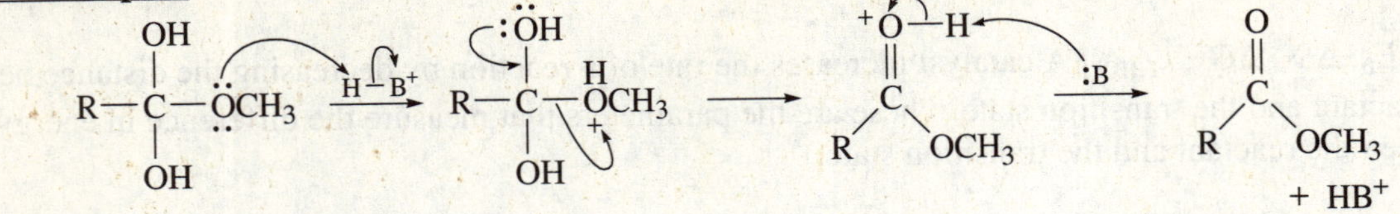

acid-catalyzed ester hydrolysis

(3) 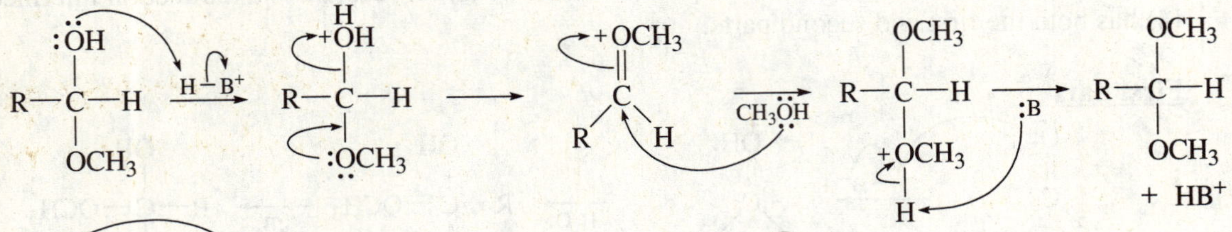

(4) 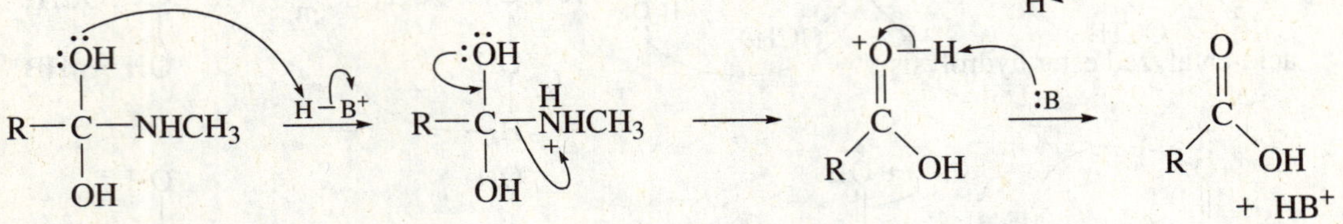

a. **similarities:** the first step is protonation of the tetrahedral intermediate, and the second step is elimination of a group from the tetrahedral intermediate. In two of the three reactions, the third step is loss of a proton.

b. **differences:** the nature of the group that is eliminated from the tetrahedral intermediate. In acetal formation, the third step is not loss of a proton because the intermediate does not have a proton to lose. Instead, the third step is attack of a nucleophile, and the fourth step is loss of a proton.

3. Both slow steps are specific-acid catalyzed; the compound is protonated before the slow step in each case.

4. 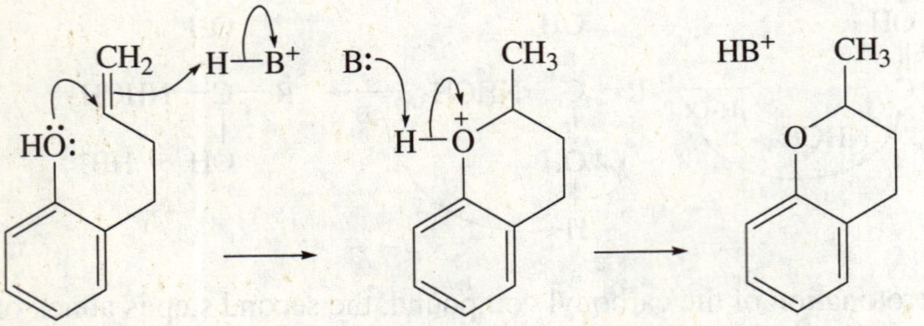

5. Solved in the text.

6. 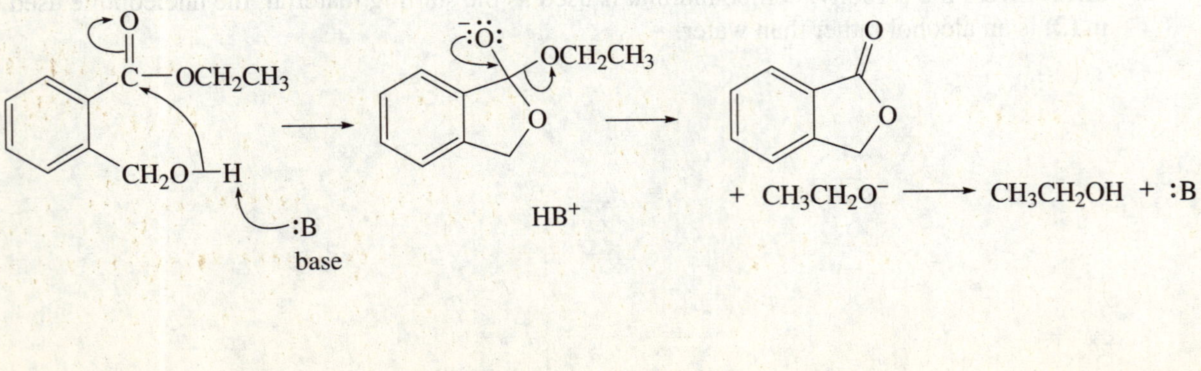

7. The metal ion catalyzes the decarboxylation reaction by complexing with the negatively charged oxygen of the carboxyl group and carbonyl oxygen of the β-keto group, thereby making it easier for the carbonyl oxygen to accept the electrons that are left behind when CO_2 is eliminated.

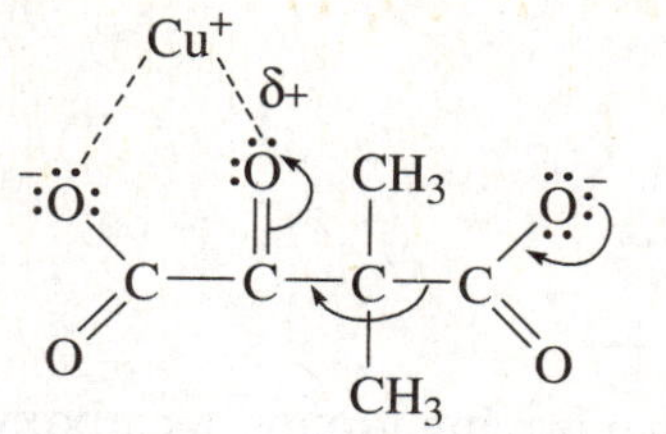

Because acetoacetate and the monoethyl ester of dimethyloxaloacetate do not have a negatively charged oxygen on one carbon and a carbonyl group on an adjacent carbon with which to form a complex, a metal ion does not catalyze decarboxylation of these compounds.

8. Co^{2+} can catalyze the reaction in three different ways. It can complex with the reactant, increasing the susceptibility of the carbonyl group to nucleophilic attack. It can also complex with water, increasing the tendency of water to lose a proton, resulting in a stronger nucleophile for hydrolysis. And it can complex with the leaving group, decreasing its basicity and thereby making it a better leaving group.

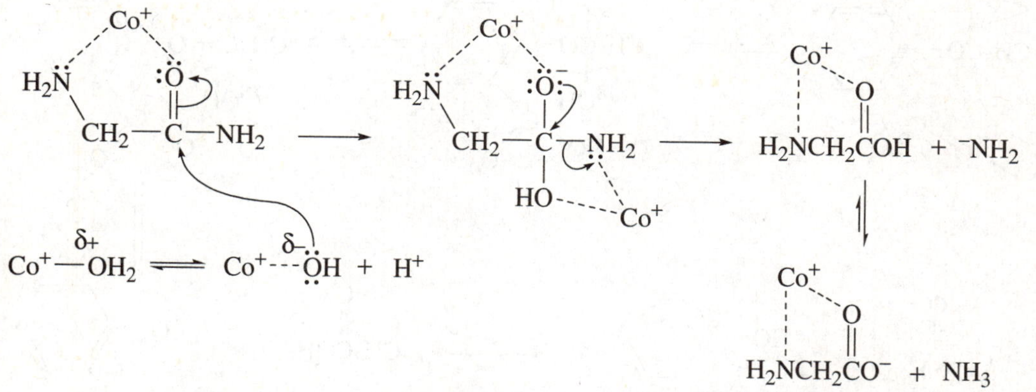

9. Because the reacting groups in the trans isomer are pointed in opposite directions, they cannot react in an intramolecular reaction. Because they can only react via an intermolecular pathway, they will have approximately the same rate of reaction as they would have if the reacting groups were in separate molecules. Consequently, the relative rate would be expected to be close to one.

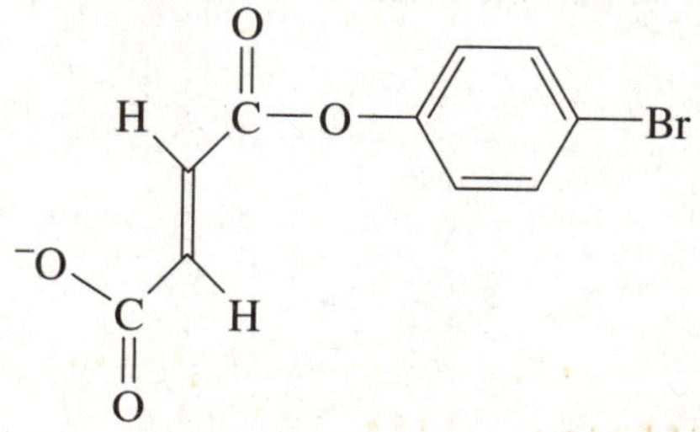

10. **a.** The nucleophile can attack the back side of either of the two ring carbons to which the sulfur is bonded, thereby forming two trans products. There are two nucleophiles (water and ethanol), so a total of four products will be formed.

b.

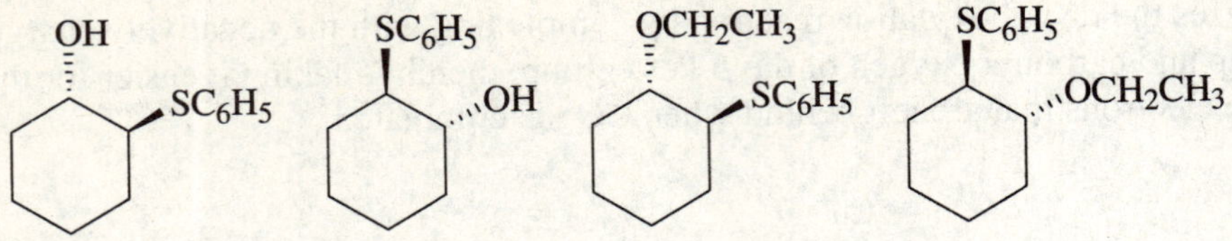

11. Solved in the text.

12. The tetrahedral intermediate has two leaving groups, a carboxylate ion and a phenolate ion. The carboxylate ion is a weaker base (a better leaving group) than the phenolate ion, so the tetrahedral intermediate reforms **A**. The 2,4-dinitrophenoxide ion is a weaker base (better leaving group) than the carboxylate ion, so the tetrahedral intermediate forms **B**.

13. If the *ortho*-carboxyl substituent acts as an intramolecular base catalyst, ^{18}O would not be incorporated into salicylic acid.

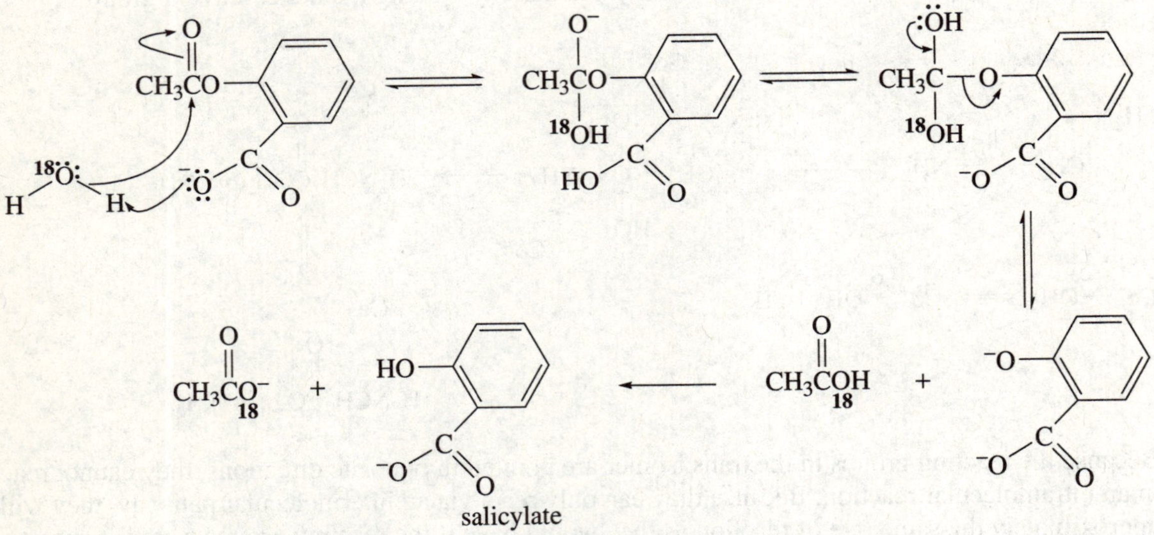

If the *ortho*-carboxyl substituent acts as an intramolecular nucleophilic catalyst, ^{18}O would be incorporated into salicylic acid.

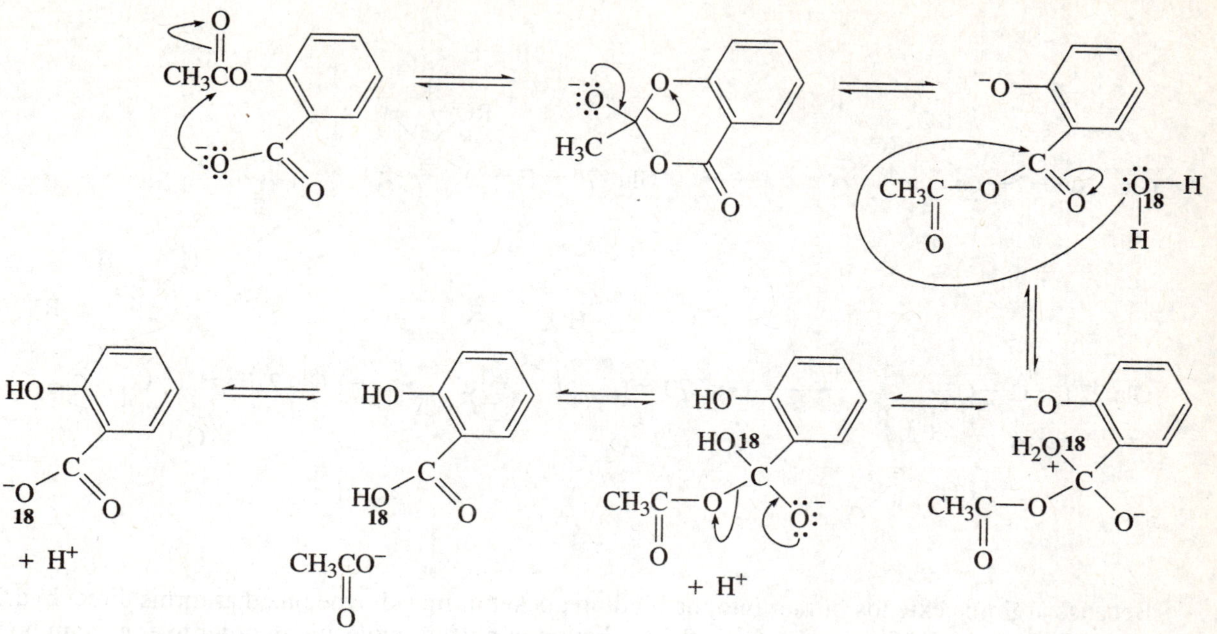

Not all the salicylic acid would contain ^{18}O, because the anhydride intermediate can be hydrolyzed in two different ways. If water attacks the benzyl carbonyl group (above), salicylic acid will contain ^{18}O. If, however, water attacks the acetyl carbonyl group (below), acetic acid will contain ^{18}O.

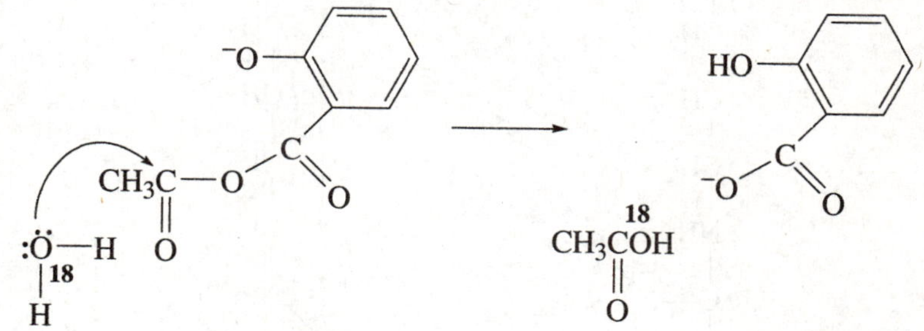

14. Solved in the text.

15. 2, 3, and 4 are bases, so they can help remove a proton.

16. Ser-Ala-Phe would be more readily cleaved by carboxypeptidase A because the phenyl substituent of phenylalanine would be more attracted to the hydrophobic pocket of the enzyme than would the negatively charged substituent of aspartate.

17. Glu 270 attacks the carbonyl group of the ester, forming a tetrahedral intermediate. Collapse of the tetrahedral intermediate is most likely catalyzed by an acid group of the enzyme in order to increase the leaving ability of the RO group. (Perhaps the HO substituent of tyrosine is close enough in the esterase to act as the catalyst.) The group that donates the proton can then act as a base catalyst to remove a proton from water as it hydrolyzes the anhydride.

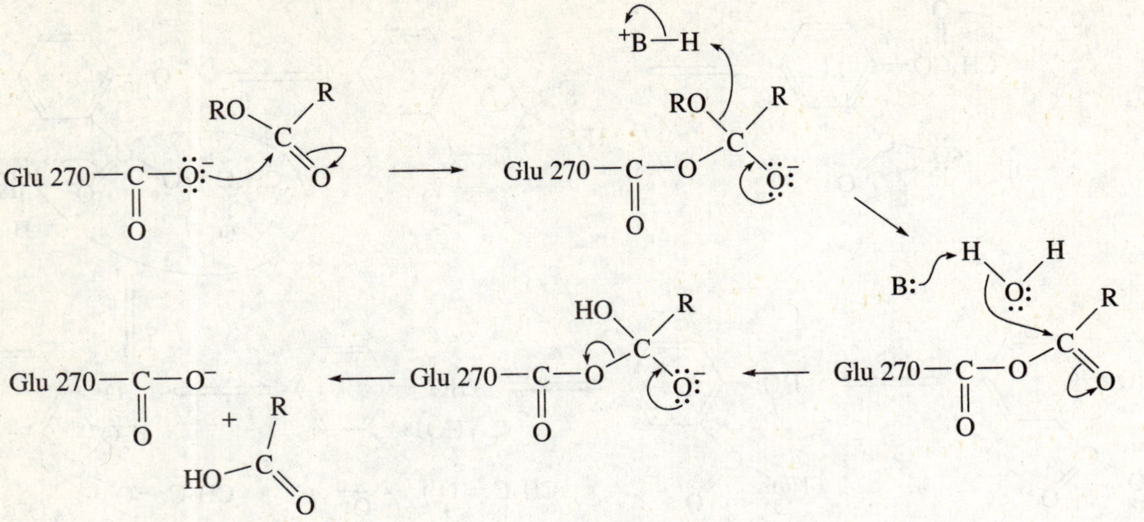

18. Because arginine extends farther into the binding pocket, it must be the one that forms direct hydrogen bonds. Lysine, which is shorter, needs the mediation of a water molecule in order to engage in bond formation with aspartate.

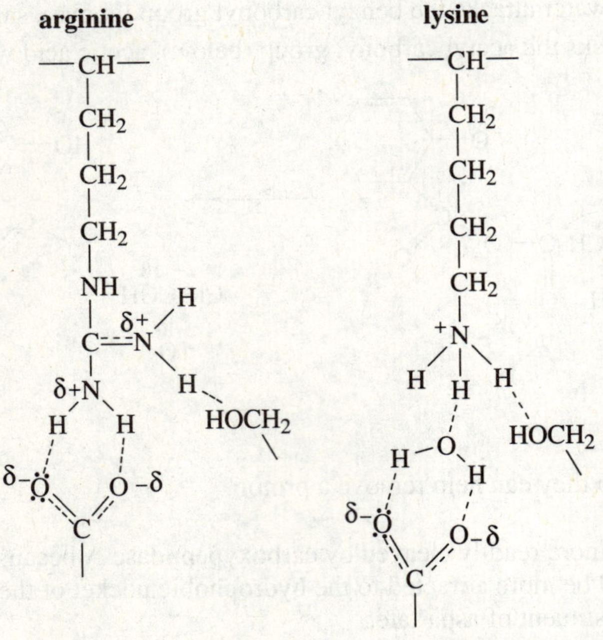

19. The side chains of D-Arg and D-Lys are not positioned to bind correctly at the active site.

20. NAM would contain ^{18}O because it is the ring that would undergo nucleophilic attack by $H_2{}^{18}O$.

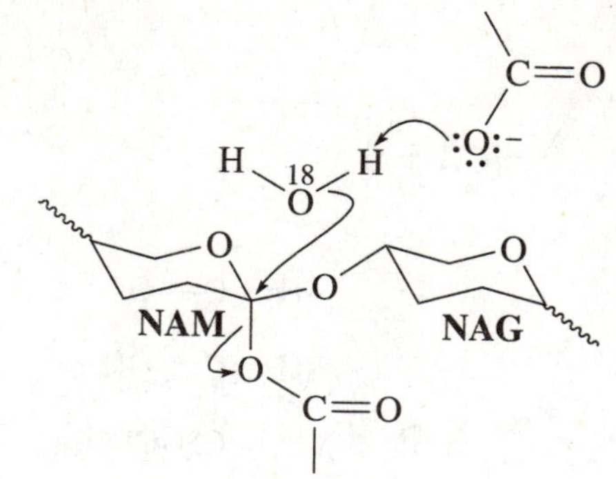

21. Lemon juice contains citric acid. Some of the side chains of an enzyme will become protonated in an acidic solution. This will change the charge of the group (for example a negatively charged aspartate, when protonated, becomes neutral), and because the shape of an enzyme is determined by the interaction of the side chains, changing the charges of the side chains will cause the enzyme to undergo a conformational change that leads to denaturation. When the enzyme is denatured, it loses its ability to catalyze the reaction that causes apples to turn brown.

22. In the absence of imine formation, D-fructose isomerizes to D-glucose and D-mannose (an equilibrium is set up between D-fructose, D-glucose, and D-mannose) because a new asymmetric center is formed at C-2 and it can have either the *R* or the *S* configuration. Enzyme catalyzed reactions are enantioselective—the enzyme catalyzes the formation of a single enantiomer of a pair. Thus D-fructose isomerizes only to D-glucose.

23. From the pK_a value it could be lysine (pK_a = 10.79), tyrosine (pK_a = 10.07), or cysteine (pK_a = 8.33).

The fact that the pK_a is given by a descending leg means that the side chain of the amino acid is functioning as an acid catalyst. From what we are told about the mechanism on page 1091, we know it is a lysine.

24. Only a primary amine can form an imine, so only **2** can form an imine with the substrate. Notice that **1** cannot form an imine because the lone pair on the NH_2 group is delocalized onto the oxygen atom, so this NH_2 group is not a nucleophile.

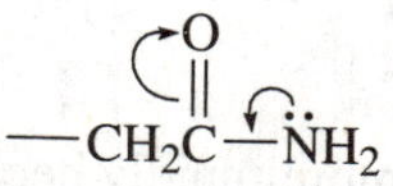

25. The positively charged nitrogen atom of the imine can accept the electrons that are left behind when the C3-C4 bond breaks.

In the absence of the imine, the electrons would be delocalized onto a neutral oxygen. The neutral oxygen is not as electron withdrawing as the positively charged nitrogen. In other words, imine formation makes it easier to break the C—C bond.

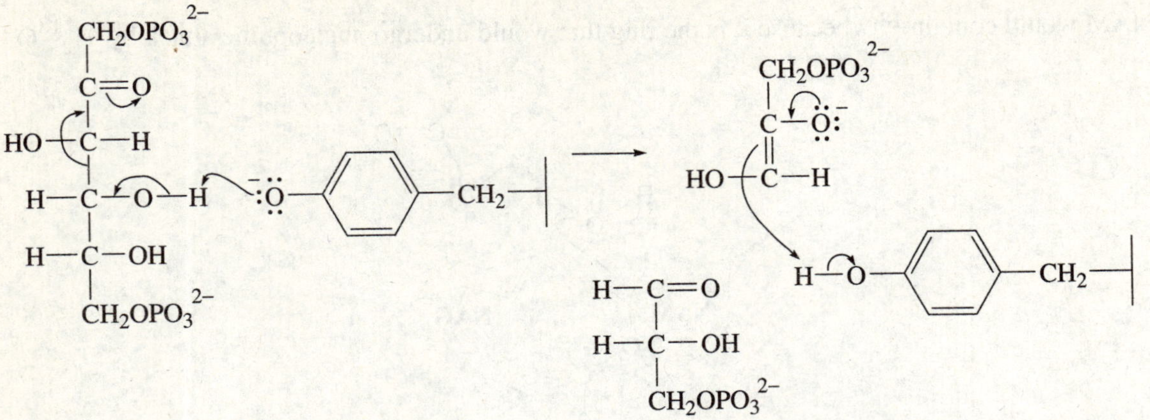

26. In order to break the C3-C4 bond, the carbonyl group has to be at the 2-position as it is in fructose, so it can accept the electrons; the carbonyl group at the 1-position in glucose cannot accept electrons. Therefore, glucose must isomerize to a ketose, so the carbonyl group will be at the 2-position.

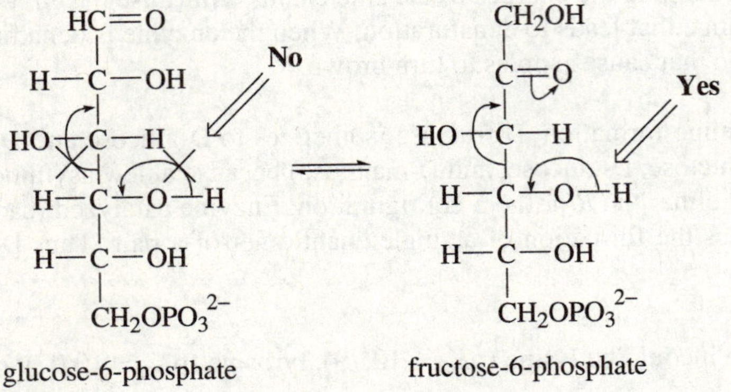

27. Cysteine residues are known to react with iodoacetic acid. If a cysteine residue is at the active site of an enzyme, adding a substituent to the sulfur atom could interfere with the enzyme's being able to bind the substrate or it could interfere with positioning the tyrosine residue that is involved in catalyzing the reaction. Adding a substituent to cysteine might also cause a conformational change in the enzyme that could destroy its activity

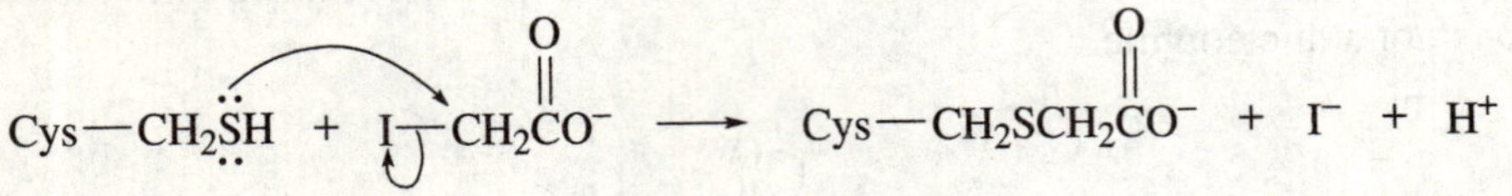

28. The following compound will lose HBr more rapidly because the negatively charged oxygen is in position to act as an intramolecular base catalyst.

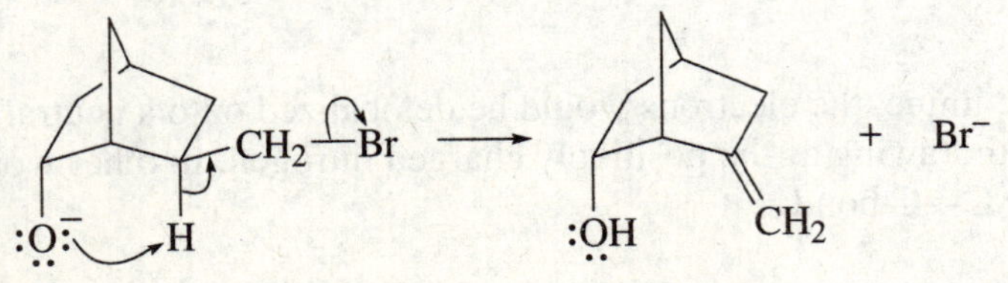

29. The compounds shown below will be the more reactive because the methyl substituent causes the reacting groups (COOH and OH) to stay in a more favorable conformation for reaction.

30. The compound shown below will form a lactone more rapidly because it forms a five-membered-ring lactone, which is less strained and than the seven-membered-ring lactone formed by the other compound. The greater stability of the five-membered-ring product will cause the transition state leading to its formation to be more stable than the transition state leading to the seven-membered-ring product.

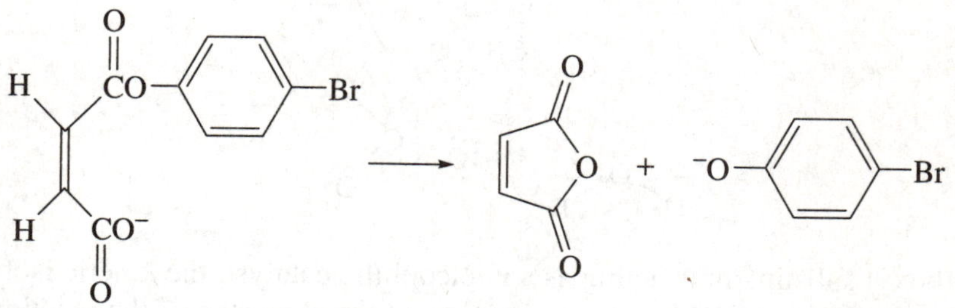

31. In order to hydrolyze an amide, the NH_2 group in the tetrahedral intermediate has to leave in preference to the less basic OH group. This can happen if the NH_2 group is protonated because $^+NH_3$ is a weaker base and, therefore, easier to eliminate than OH. Of the four compounds, two have substituents that can protonate the NH_2 by acting as acid catalysts, *ortho*-carboxy-benzamide and *ortho*-hydroxybenzamide.

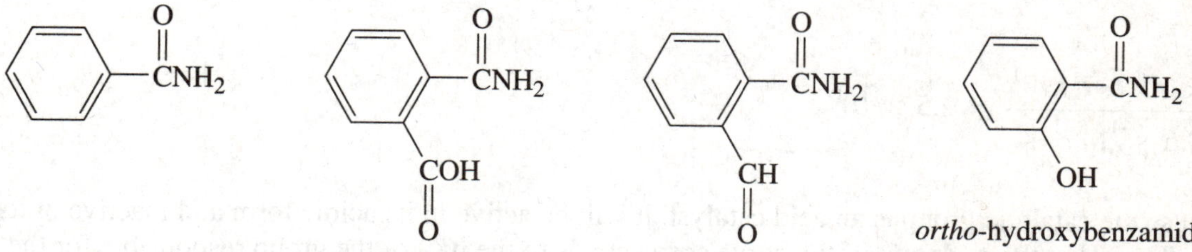

ortho-carboxybenzamide *ortho*-hydroxybenzamide

Because the carboxy group is electron-withdrawing and the phenolic OH group is electron-donating, formation of the tetrahedral intermediate will be faster for *o*-carboxybenzamide.
Because the carboxy group is a stronger acid than the phenolic OH group, the tetrahedral intermediate of *o*-carboxybenzamide will collapse to products faster. Therefore, the *o*-carboxybenzamide has the faster rate of hydrolysis.

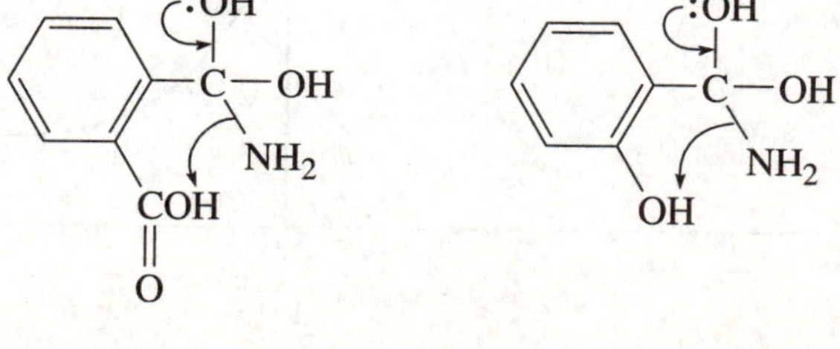

32. a. 

intramolecular nucleophilic catalysis

b. 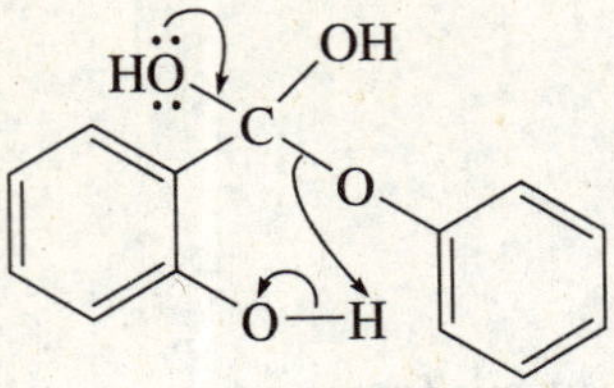

intramolecular general-acid catalysis

The OH substituent is protonating the leaving group as it departs, causing it to be a weaker base and, thus, a better leaving group.

33. If the *ortho*-carboxy substituent is acting as a base catalyst, the kinetic isotope effect will be greater than 1.0 because an OH (or OD) bond is broken in the slow step of the reaction.

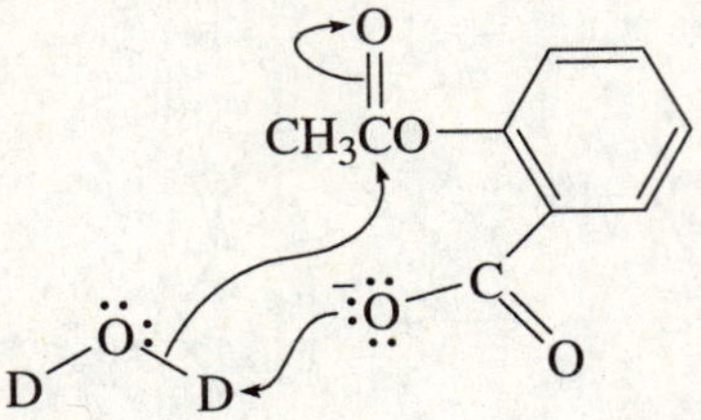

If the *ortho*-carboxyl substituent is acting as a nucleophilic catalyst, the kinetic isotope effect will be about 1.0 because an OH (or OD) bond is not broken in the slow step of the reaction.

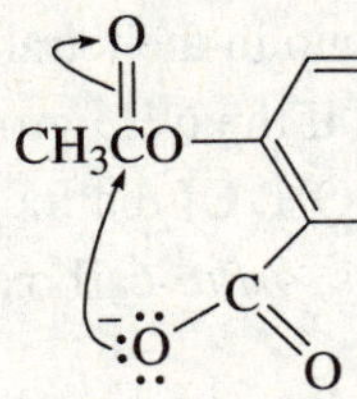

34. $$\frac{15 \times 10^6\ s^{-1}M^{-1}}{0.6\ s^{-1}M^{-1}} = 2.5 \times 10^6$$

35. Because the catalytic group is an acid catalyst, it will be active in its acidic form and inactive in its basic form. The pH at the midpoint of the curve corresponds to the pK_a of the group responsible for the catalysis.

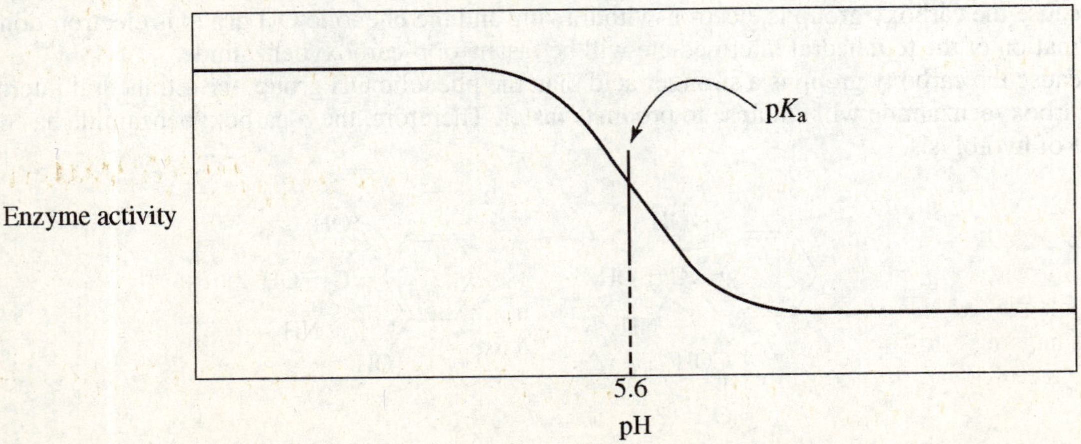

36. Co^{2+} can catalyze the hydrolysis reaction by complexing with three nitrogen atoms in the substrate as well as with water. Complexation increases the acidity of water, thereby providing a stronger nucleophile for the hydrolysis reaction. Complexation with the substrate locks the nucleophile into the correct position for attack on the carbonyl carbon.

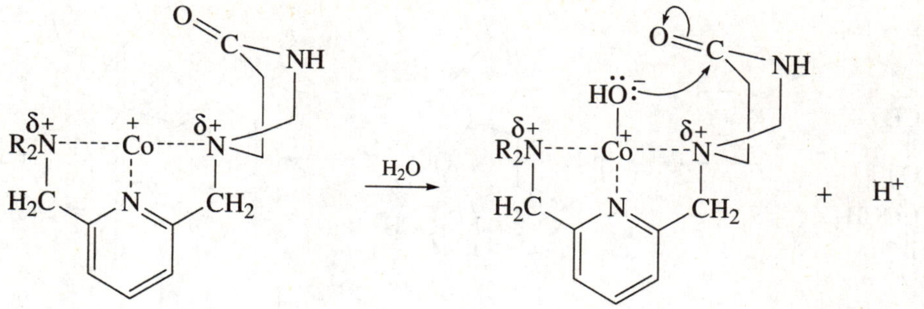

37.

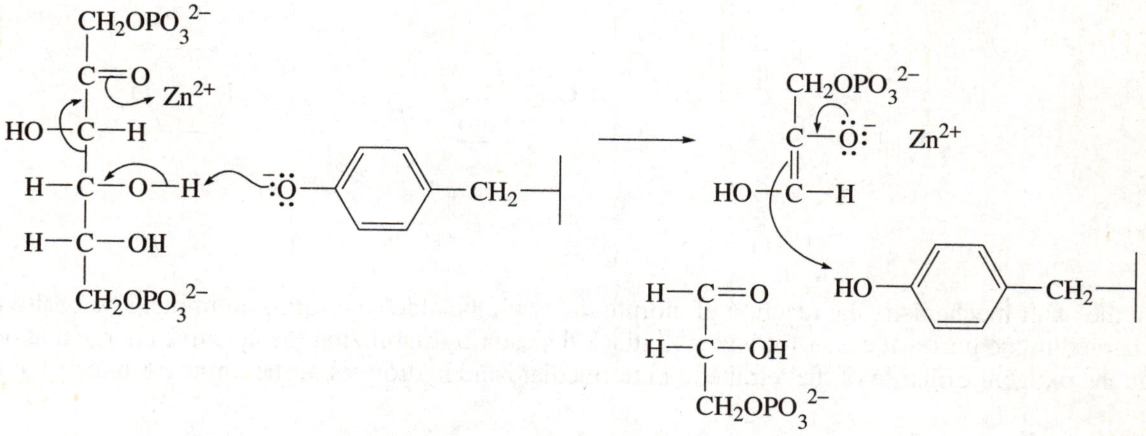

38.

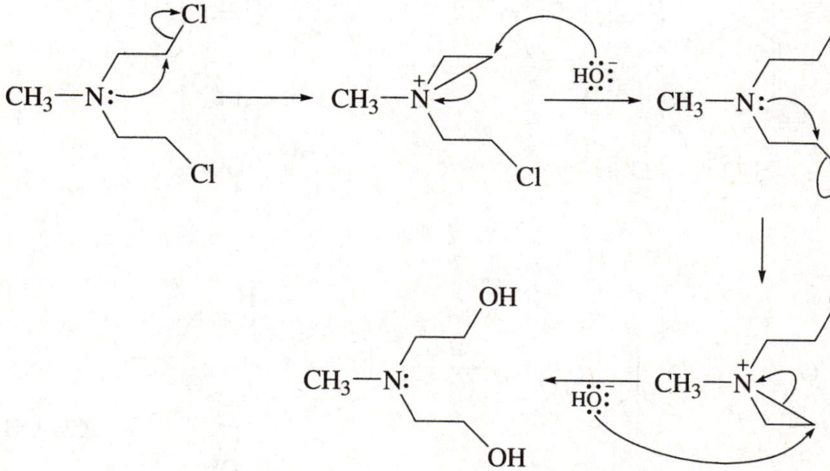

39. There are two possible mechanisms that involve morpholine as a nucleophilic catalyst that can account for the observed catalysis by morpholine. In the absence of morpholine, the first step in the reaction is attack of water on the **ester**. In the presence of morpholine, the first step in the reaction (in both mechanisms) is attack of morpholine on the **aldehyde**, which is a faster reaction because morpholine is a stronger nucleophile than water and an aldehyde is more susceptible to nucleophilic attack than an ester.

In one mechanism, the negatively charged aldehyde oxygen (which is a much stronger nucleophile than water) is then the nucleophile that attacks the ester. The tetrahedral intermediate collapses to form a lactone. Imine formation followed by imine hydrolysis gives the final product.

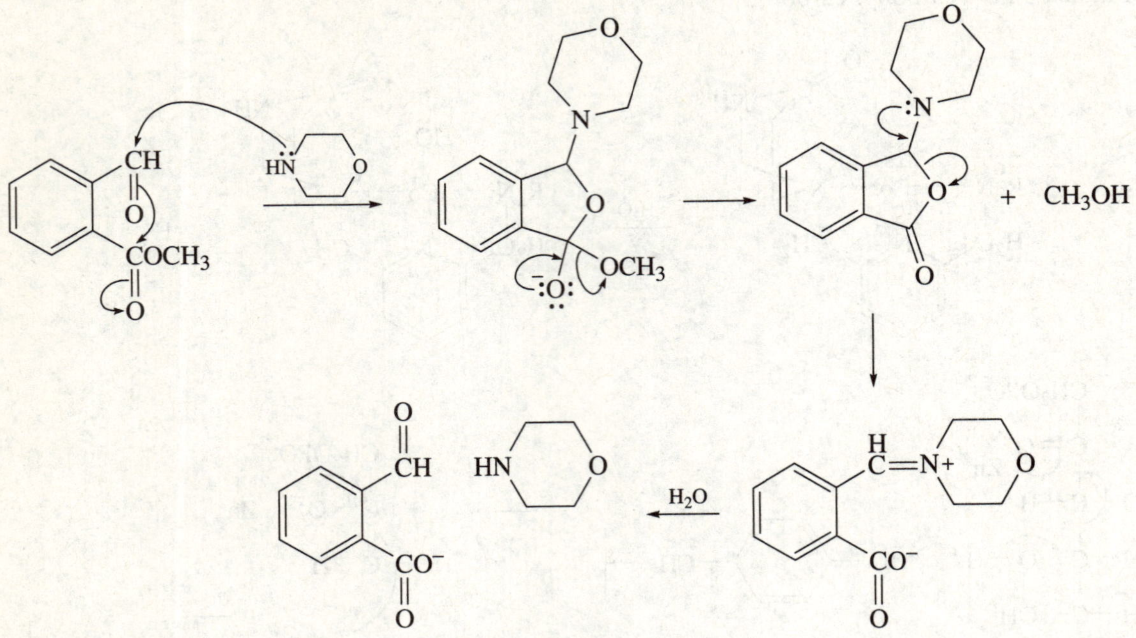

In the other mechanism, the reaction of morpholine with the aldehyde forms an imine. The positively charged imine makes it easier for water to attack the ester by stabilizing the negative charge that develops on the oxygen. Collapse of the tetrahedral intermediate and hydrolysis of the imine form the final product.

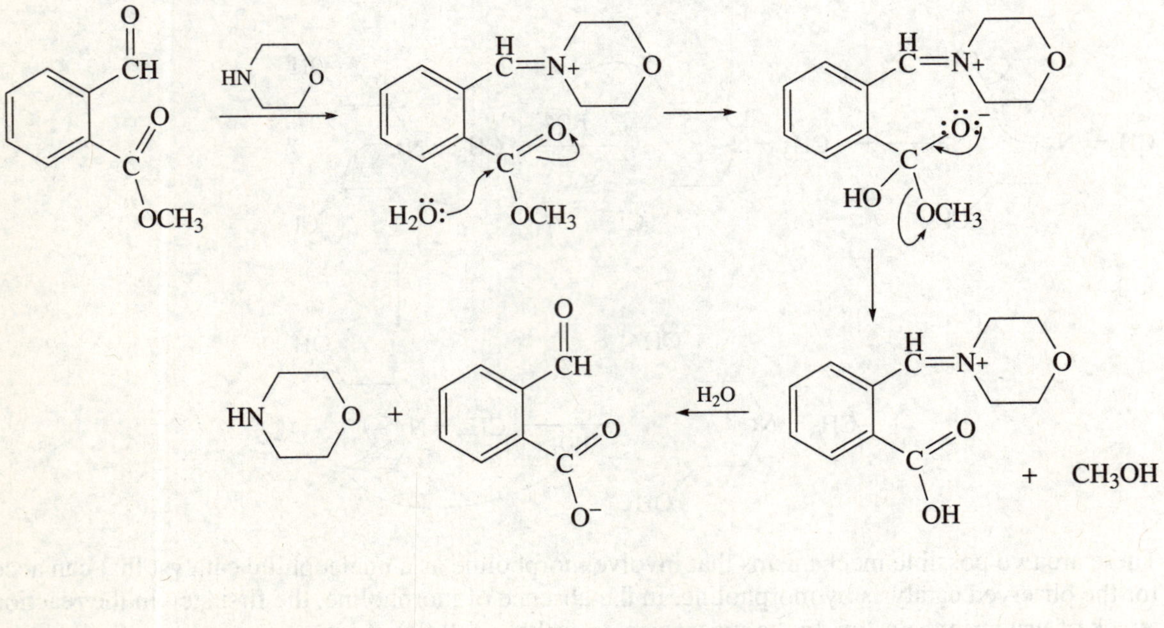

40.

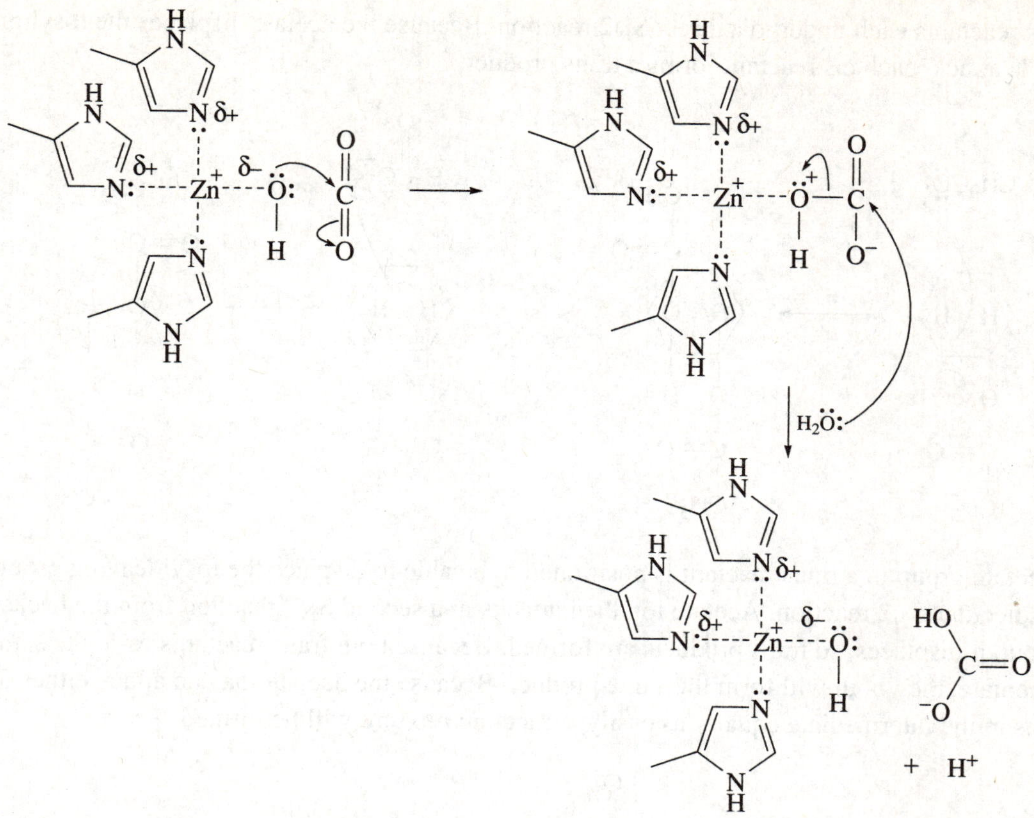

41. At pH = 12, the nucleophile is hydroxide ion. Attack of hydroxide ion on the carbonyl group is faster in **A** because the negative charge that is created in the tetrahedral intermediate is stabilized by the positively charged nitrogen.

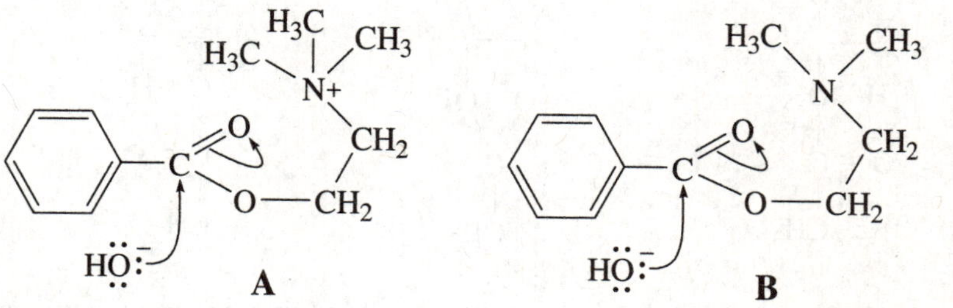

At pH = 8, the nucleophile is water. Attack of water on the carbonyl group is faster in **B** because the amino group can act as a base catalyst to make water a better nucleophile.

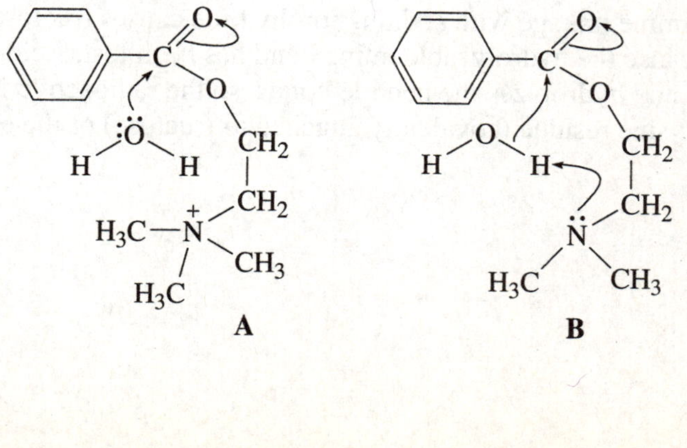

42. **a.** The cis reactants each undergo a direct S_N2 reaction. Because the acetate displaces the tosyl group by backside attack, each cis reactant forms a trans product.

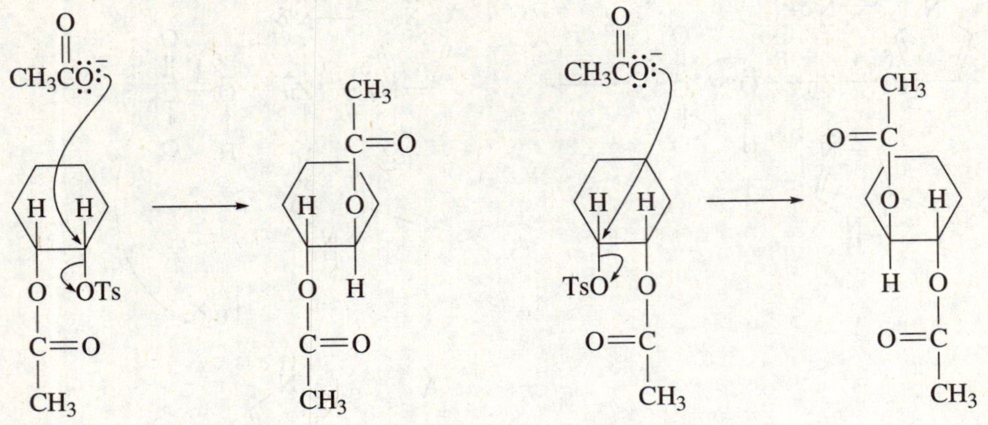

b. The acetate group in a trans reactant is positioned to be able to displace the tosyl leaving group by an intramolecular S_N2 reaction. Acetate ion then attacks in a second S_N2 reaction from the backside of the group it displaces, so trans products are formed. Because both trans reactants form the same intermediate, they both will form the same product. Because the acetate ion can attack either of the carbons in the intermediate equally as easily, a racemic mixture will be formed.

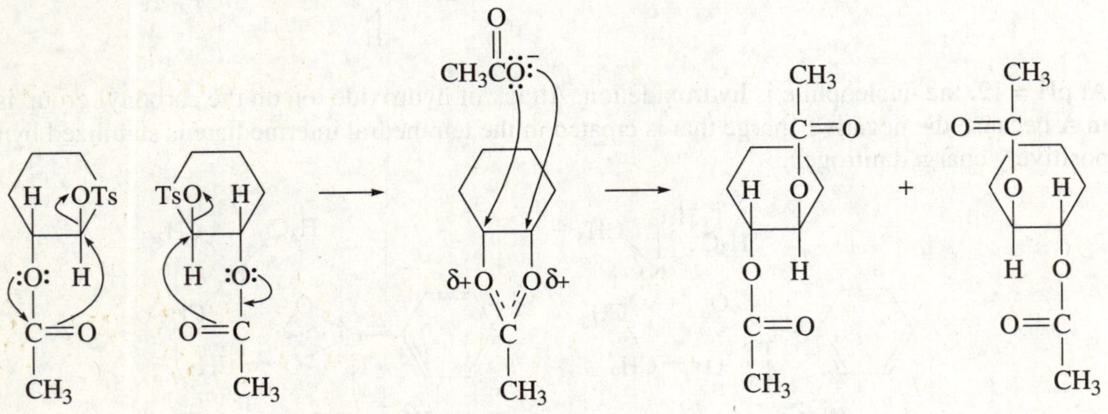

c. The trans reactant is more reactive because the tosyl leaving group is displaced in an intramolecular reaction, forming a positively charged cis intermediate that is considerably more reactive than the neutral cis isomer.

43. Reduction of the imine linkage with sodium borohydride causes fructose to become permanently attached to the enzyme because the hydrolyzable imine bond has been lost. Acid-catalyzed hydrolysis removes the phosphate groups and hydrolyzes the peptide bonds, so the radioactive fragment that is isolated after hydrolysis is the lysine residue (covalently attached to fructose) of the enzyme that originally formed the imine.

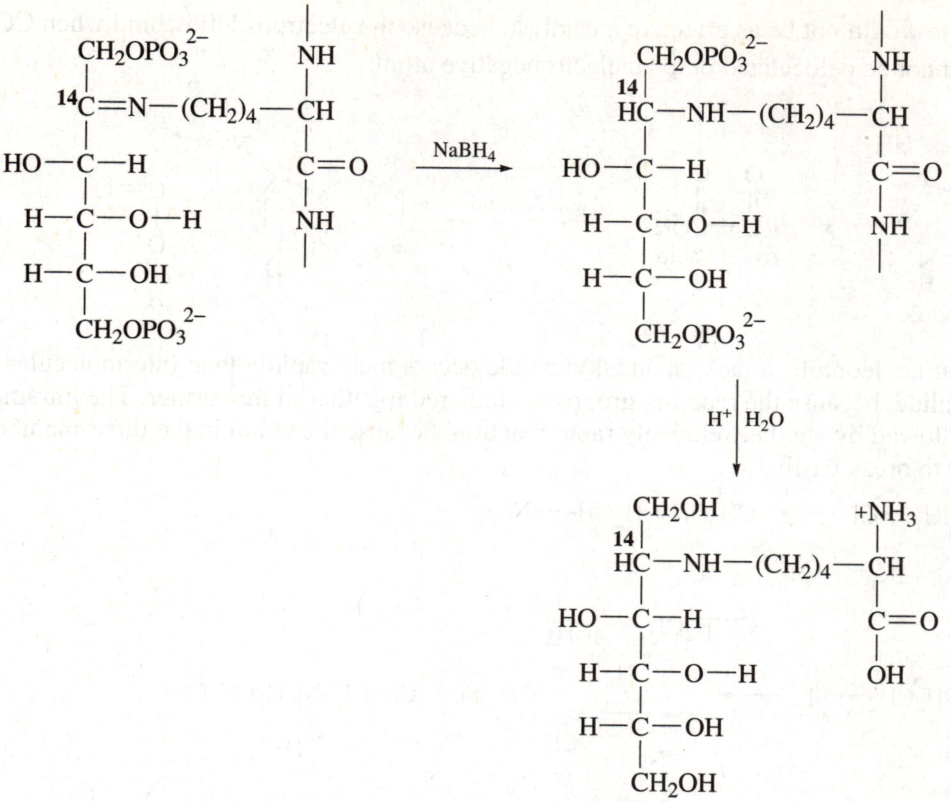

44. **a.** 3-Amino-2-oxindole catalyzes the decarboxylation of an α-keto acid by first forming an imine that is followed by a prototropic shift.

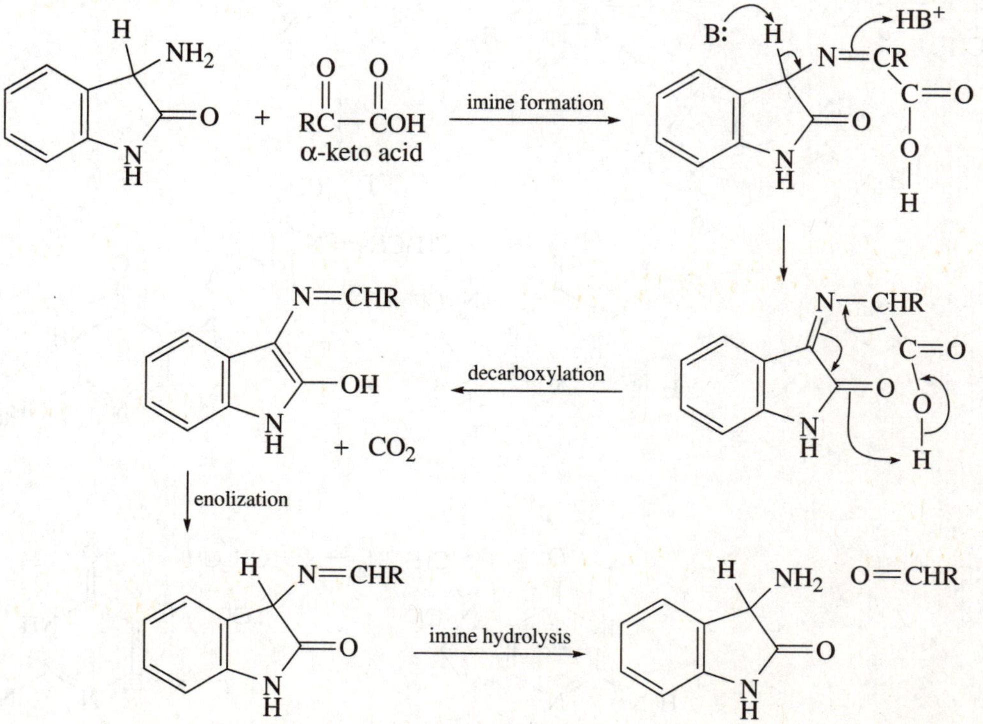

b. 3-Aminoindole would not be as effective a catalyst, because the electrons left behind when CO_2 is eliminated cannot be delocalized onto an electronegative atom.

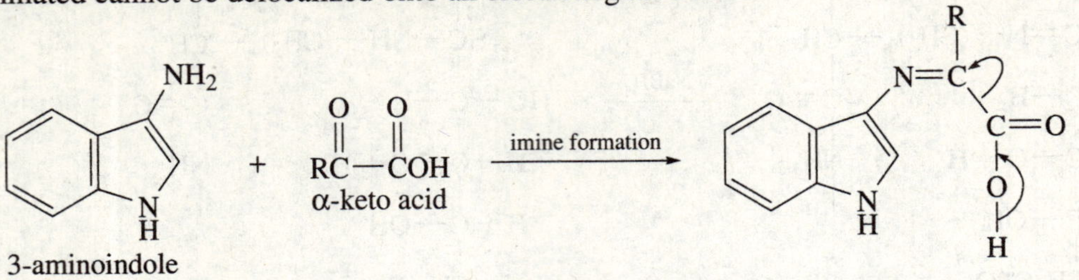

45. **a.** Intramolecular nucleophilic attack on an alkyl halide occurs more rapidly than Intermolecular attack on an alkyl halide, because the reacting groups are tethered together in the former. The intramolecular reaction is followed by another relatively rapid reaction, because the strain in the three-membered ring causes it to break easily.

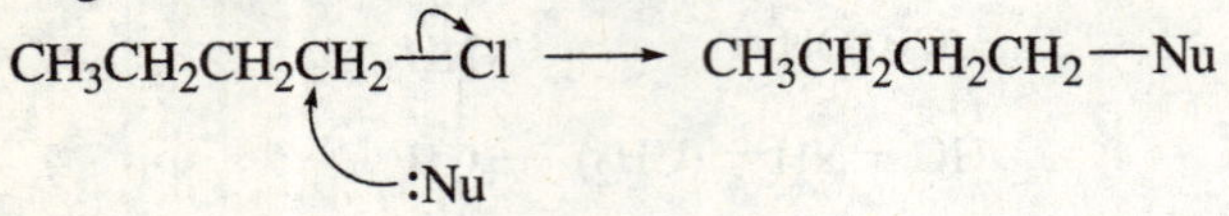

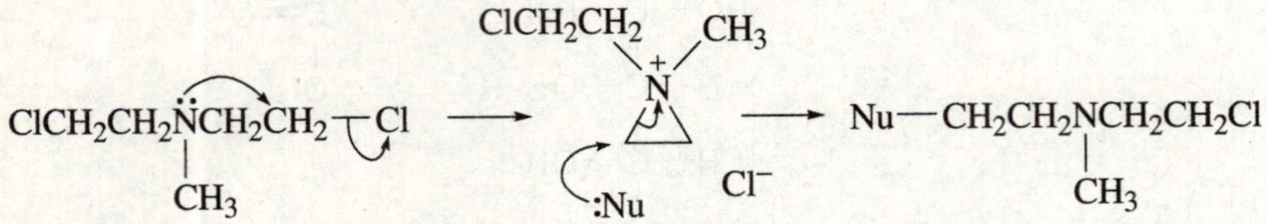

b.

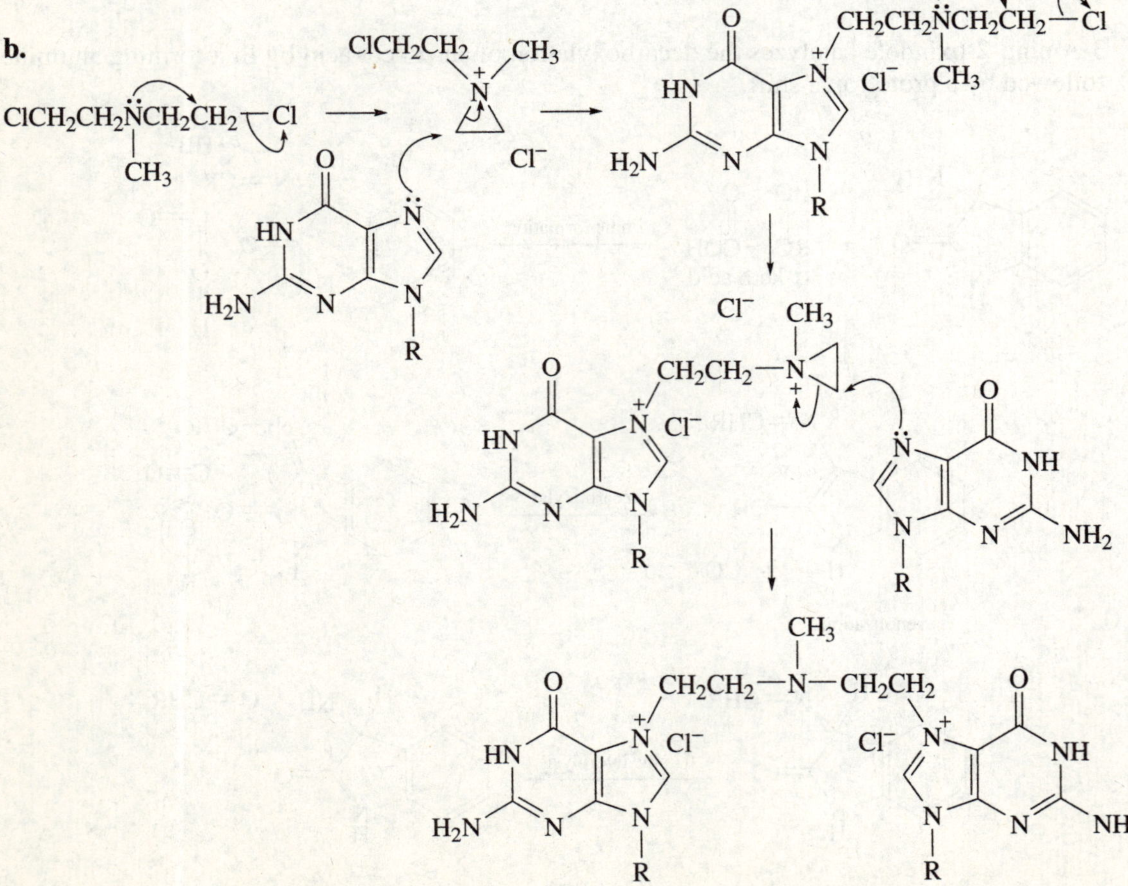

46.

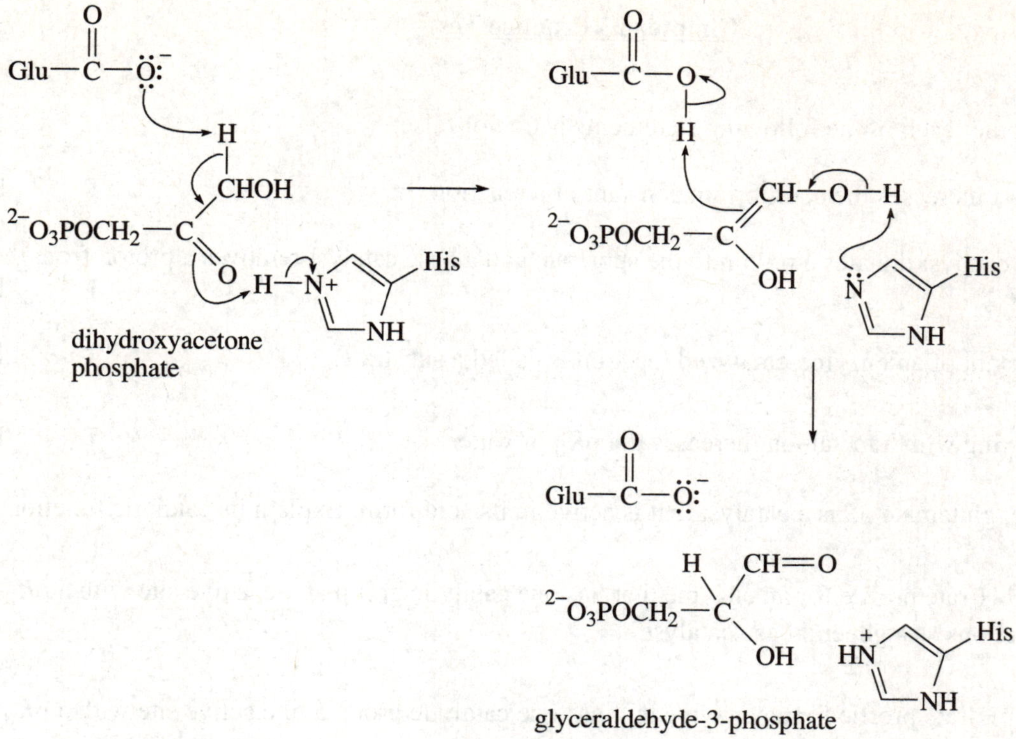
Glu
His
NH
dihydroxyacetone phosphate
OH
glyceraldehyde-3-phosphate

Chapter 23 Practice Test

1. Indicate whether each of the following statements is true or false:

a. A catalyst increases the equilibrium constant of a reaction T F

b. An acid catalyst donates a proton to the substrate and a base catalyst removes a proton from the substrate. T F

c. The reactant of an enzyme-catalyzed reaction is called a substrate T F

d. Complexing with a metal ion increases the pK_a of water T F

2. In lysozyme, glutamate 35 is a catalyst that is active in its acid form. Explain its catalytic function.

3. **a.** Draw a pH-rate profile for an enzyme that has one catalytic group at the active site with a pK_a = 5.7 that functions as a general base catalyst.

b. Draw a pH-rate profile for an enzyme that has one catalytic group at the active site with a pK_a = 5.7 that functions as a general acid catalyst.

4. Using arrows, show the mechanism arrows for the first step in the mechanism for chymotrypsin.

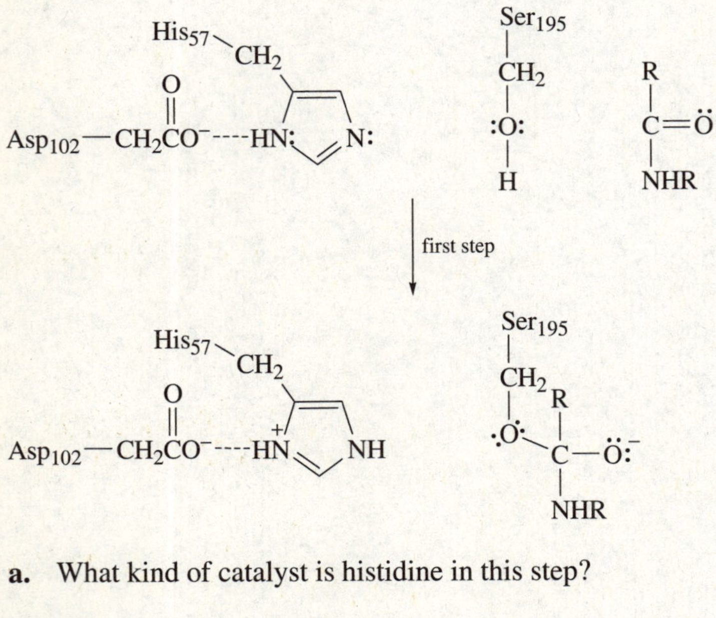

a. What kind of catalyst is histidine in this step?

b. What kind of catalyst is serine in this step?

c. How does aspartate catalyze the reaction?

CHAPTER 24
The Organic Mechanisms of the Coenzymes

Important Terms

anabolic reaction — the reactions living organisms carry out that result in the synthesis of complex biomolecules from simple precursor molecules.

apoenzyme — an enzyme (that needs a cofactor to catalyze a reaction) without its cofactor.

biotin — the coenzyme required by enzymes that catalyze carboxylation of a carbon adjacent to an ester or a keto group.

catabolism — the reactions living organisms carry out to provide energy and simple precursor molecules for synthesis.

coenzyme — a cofactor that is an organic molecule.

coenzyme A — a thiol used by biological organisms to form thioesters.

coenzyme B_{12} — the coenzyme required by enzymes that catalyze certain rearrangement reactions.

cofactor — an organic molecule or a metal ion that an enzyme needs in order to catalyze a reaction.

competitive inhibitor — a compound that inhibits an enzyme by competing with the substrate for binding at the active site.

dehydrogenase — an enzyme that carries out an oxidation reaction by removing hydrogen from the substrate.

electron sink — a site to which electrons can be delocalized.

flavin adenine dinucleotide (FAD) — a coenzyme required in certain oxidation reactions. It is reduced to $FADH_2$, which is a coenzyme required in certain reduction reactions.

flavin mononucleotide (FMN) — a coenzyme required in certain oxidation reactions. It is reduced to $FMNH_2$, which is a coenzyme required in certain reduction reactions.

heterocycle (heterocyclic compound) — a cyclic compound in which one or more of the ring atoms is an atom other than carbon.

holoenzyme — an enzyme plus its cofactor.

lipoate — a coenzyme required in certain oxidation reactions.

mechanism-based inhibitor — an inhibitor that inactivates an enzyme by undergoing part of the normal catalytic mechanism.

metabolism	reactions living organisms carry out in order to obtain the energy they need and to synthesize the compounds they require.
metalloenzyme	an enzyme that has a tightly bound metal ion.
nicotinamide adenine dinucleotide (NAD^+)	a coenzyme required in certain oxidation reactions. It is reduced to NADH, which is a coenzyme required in certain reduction reactions.
nicotinamide adenine dinucleotide phosphate ($NADP^+$)	a coenzyme required in certain oxidation reactions. It is reduced to NADPH, which is a coenzyme required in certain reduction reactions.
nucleotide	a heterocycle attached in the β-position to a phosphorylated ribose.
pyridoxal phosphate	the coenzyme required by enzymes that catalyze certain transformations of amino acids.
suicide inhibitor (mechanism-based inhibitor)	a compound that inactivates an enzyme by undergoing part of its normal catalytic mechanism.
tetrahydrofolate (THF)	the coenzyme required by enzymes that catalyze a reaction that donates a group containing a single carbon to its substrate.
thiamine pyrophosphate (TPP)	the coenzyme required by enzymes that catalyze a reaction that transfers a two-carbon fragment to its substrate.
transamination	a reaction in which an amino group is transferred from one compound to another.
transimination	the reaction of a primary amine with an imine to form a new imine and a primary amine derived from the original imine.
vitamin	a substance needed in small amounts for normal body function that the body cannot synthesize or cannot synthesize in adequate amounts.
vitamin KH_2	the coenzyme required by the enzyme that catalyzes the carboxylation of glutamate side chains.

Solutions to Problems

1. The metal ion (Zn^{2+}) makes the carbonyl carbon more susceptible to nucleophilic attack, increases the nucleophilicity of water by making it more like a hydroxide ion, and stabilizes the negative charge on the transition state.

2. The alcohol is oxidized to a ketone.

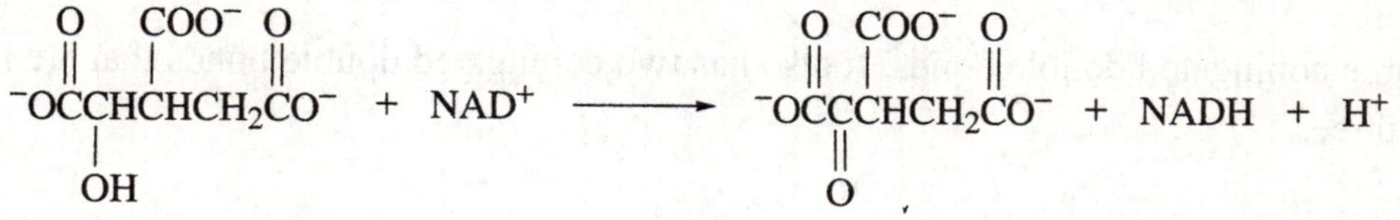

3. The ketone is reduced to an alcohol.

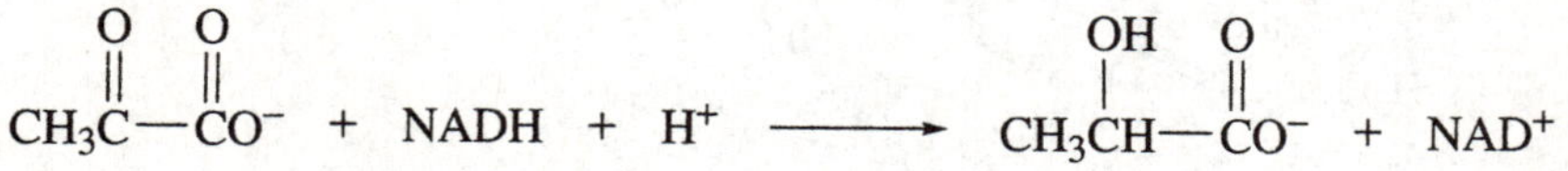

4. FAD contains a diphosphate linkage. One of the phosphate groups comes from FMN and the other comes from ATP. Since ATP uses only one of its three phosphate groups in forming FAD, the other product of the reaction must be pyrophosphate to account for loss of the two phosphate groups. In other words, the phosphate group of FMN must attack the α-phosphorus of ATP, eliminating pyrophosphate. (See pages 770 and 1143 of the text.)

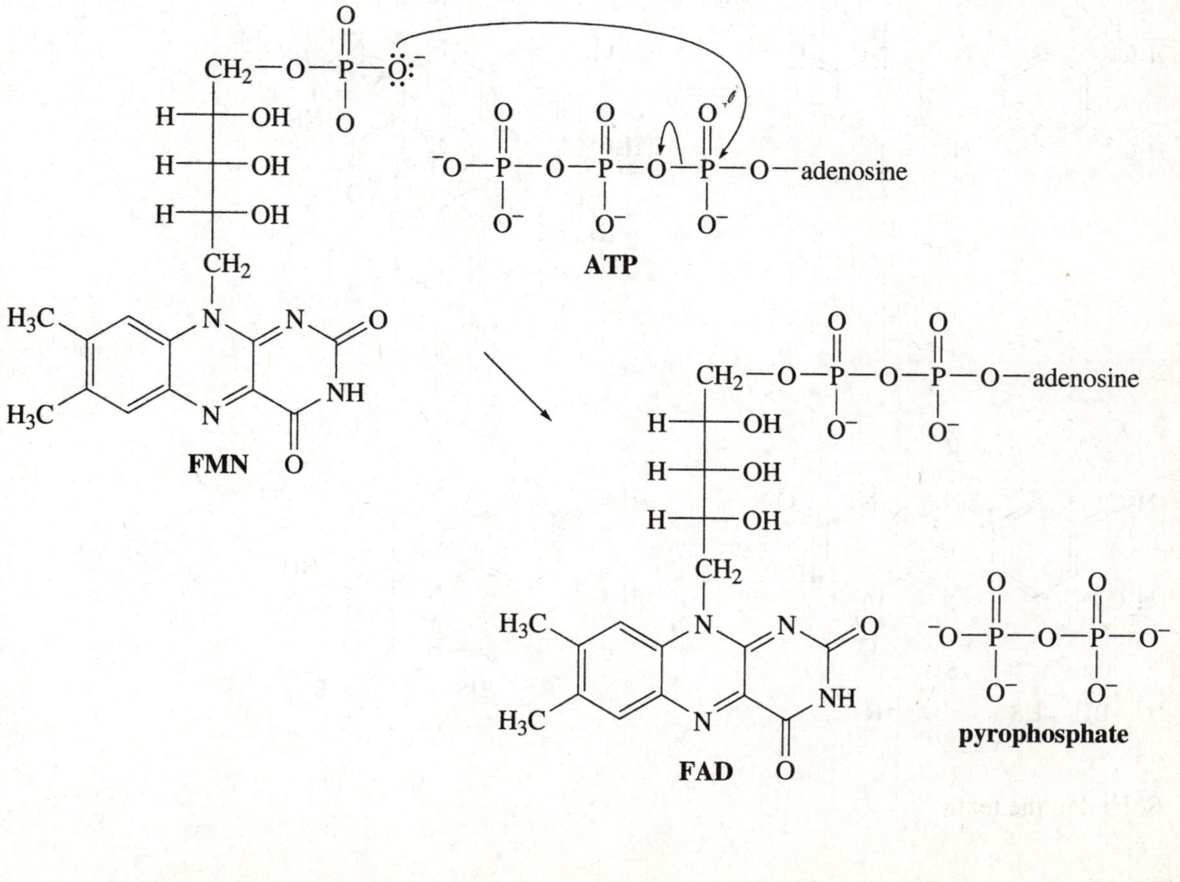

5. **a.** FAD has seven conjugated double bonds. (The conjugated double bonds are indicated by *.)

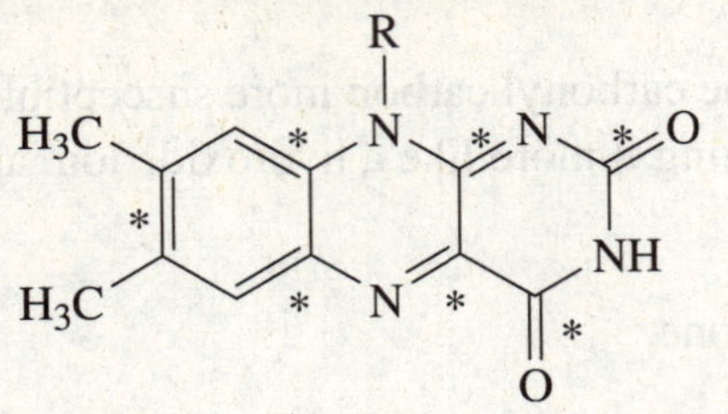

b. $FADH_2$ has three conjugated double bonds. It also has two conjugated double bonds that are isolated from the other three.

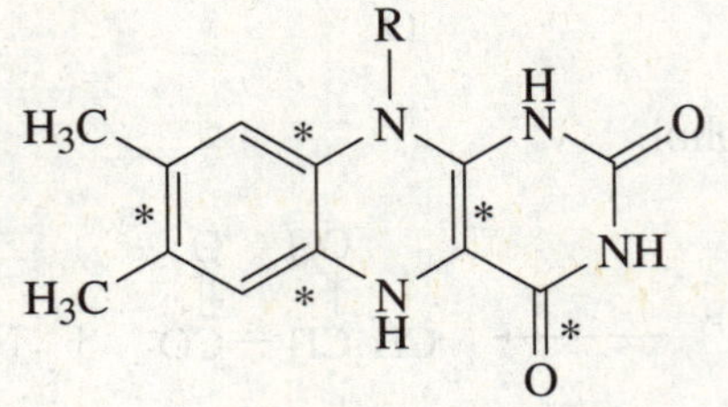

6. The mechanism for reduction of lipoate by $FADH_2$ is the reverse of the mechanism for oxidation of dihydrolipoate by FAD. Because the pK_a of the N-1 hydrogen is 7, $FADH_2$ is present in both the acidic and basic forms at physiological pH (7.3).

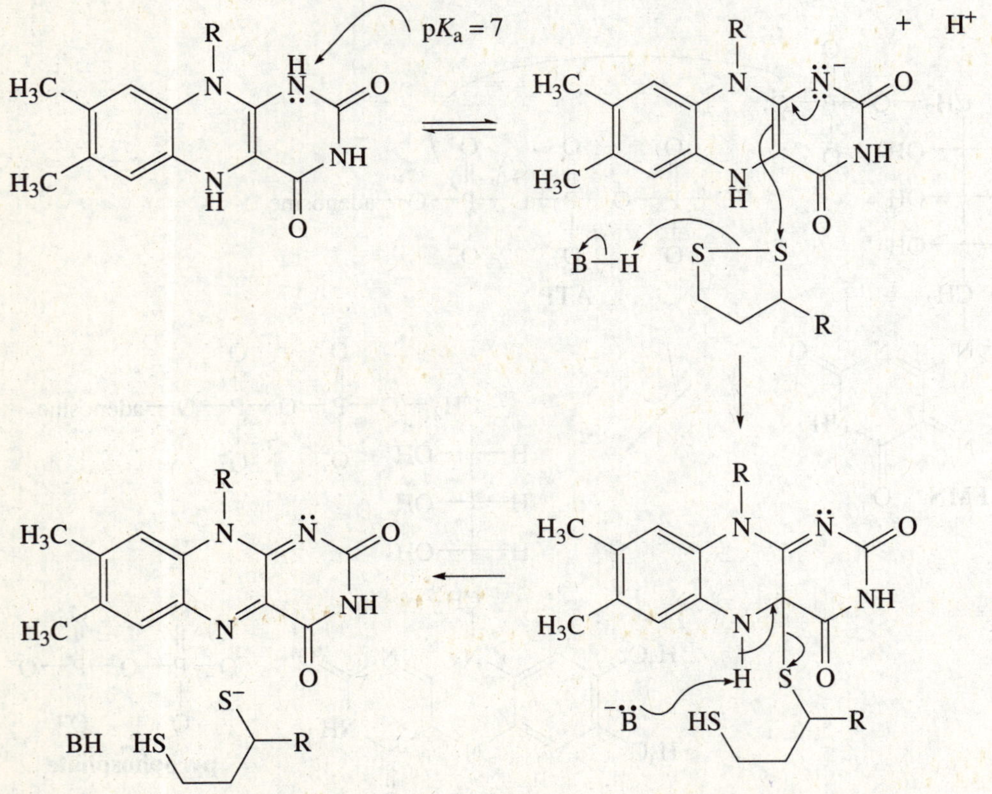

7. Solved in the text.

8. When a proton is removed from the methyl group at C-8, the electrons that are left behind can be delocalized onto the oxygen at the 2-position or onto the oxygen at the 4-position.

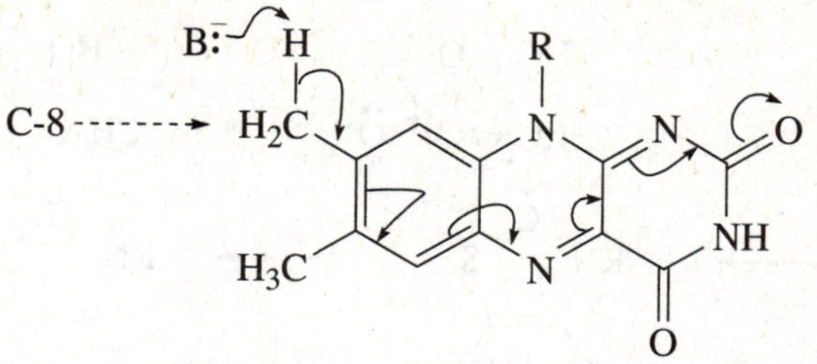

When a proton is removed from the methyl group at C-7, the electrons that are left behind can be delocalized only onto carbon atoms that, being less electronegative than oxygen atoms, are less able to accommodate the electrons. (Try pushing electrons to prove to yourself that this is so.)

9.

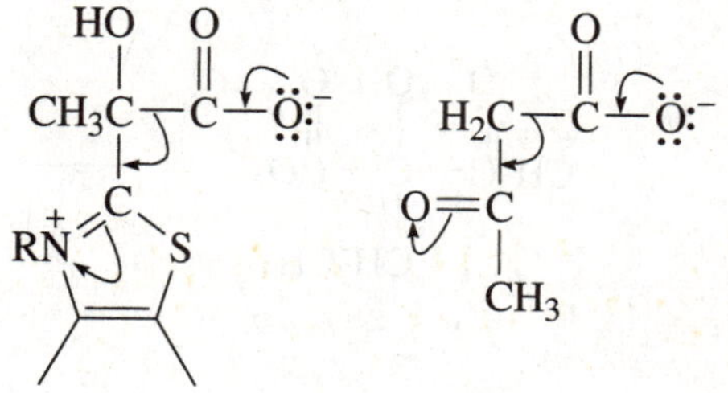

10. The only difference in the mechanisms of pyruvate decarboxylase and acetolactate synthase is the species the two-carbon fragment (the carbanion) is transferred to: a proton in the case of pyruvate carboxylase and pyruvate in the case of acetolactate synthase.

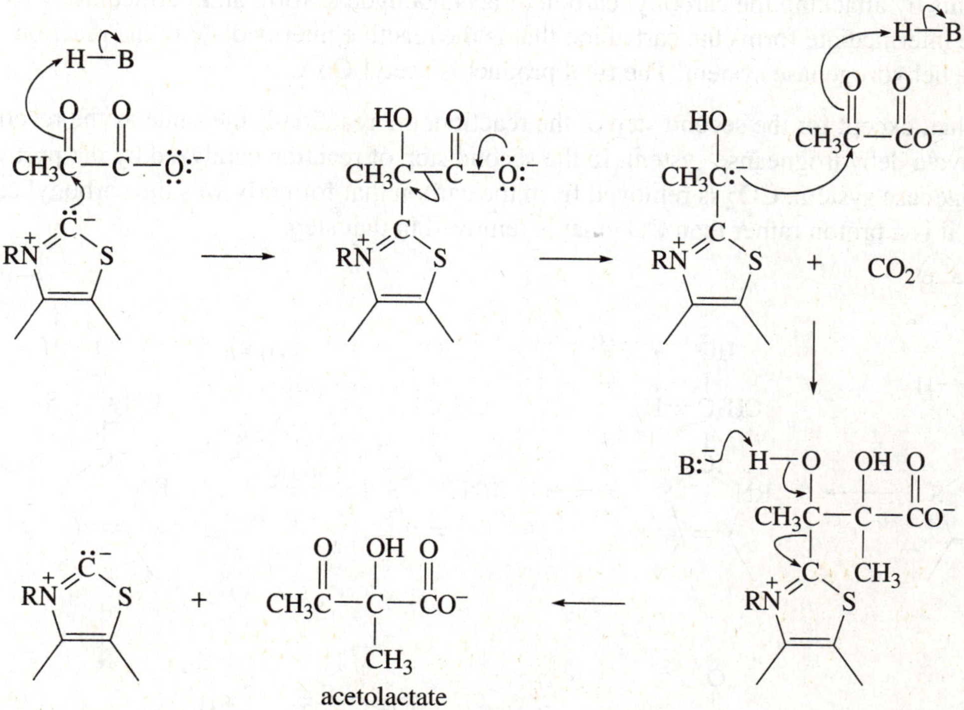

acetolactate

11. Notice that the only difference in this reaction and that in Problem 8 is that the species to which the two-carbon fragment is transferred has an ethyl group in place of the methyl group.

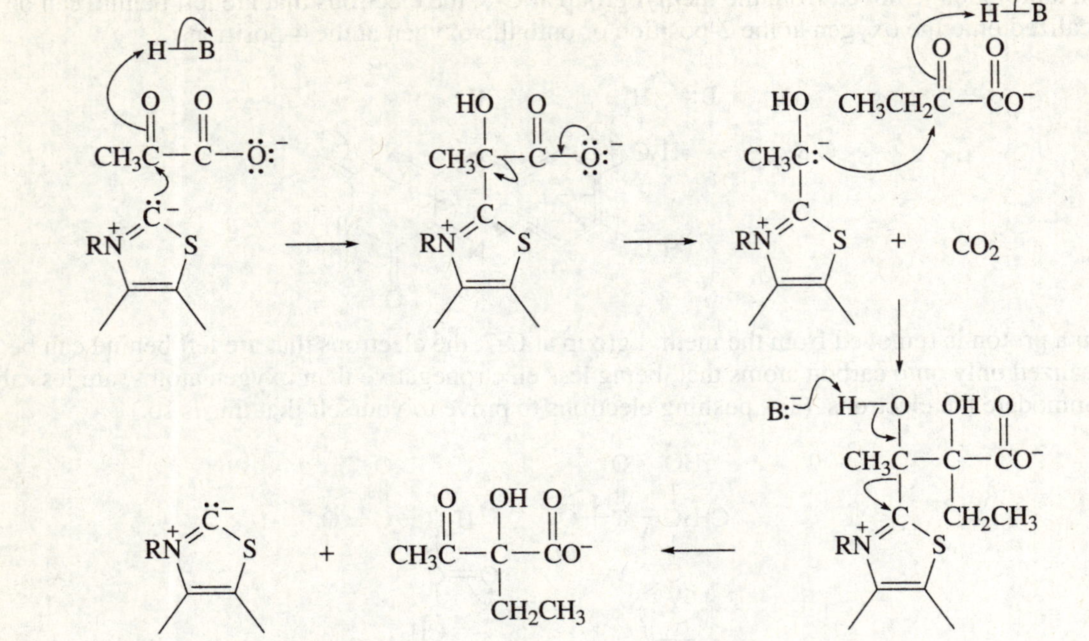

12. Solved in the text.

13. Acetaldehyde is responsible for the physiological effects known as a hangover. A way to "cure" a hangover, therefore, is to get rid of the acetaldehyde that is formed from the oxidation of ethanol. Thiamine pyrophosphate (the coenzyme form of vitamin B_1) can convert acetaldehyde into acetyl-CoA. It does this by attacking the carbonyl carbon of acetaldehyde to form an intermediate. Loss of a proton from the intermediate forms the carbanion that is the reactive intermediate in the reaction catalyzed by the pyruvate dehydrogenase system. The final product is acetyl-CoA.

Notice that, except for the second step of the reaction, the reaction is the same as the reaction catalyzed by the pyruvate dehydrogneanse system. In the second step of reaction catalyzed by the pyruvate dehydrogenase system, CO_2 is removed from the carbon that formerly was the carbonyl carbon; in this reaction it is a proton rather than CO_2 that is removed in that step.

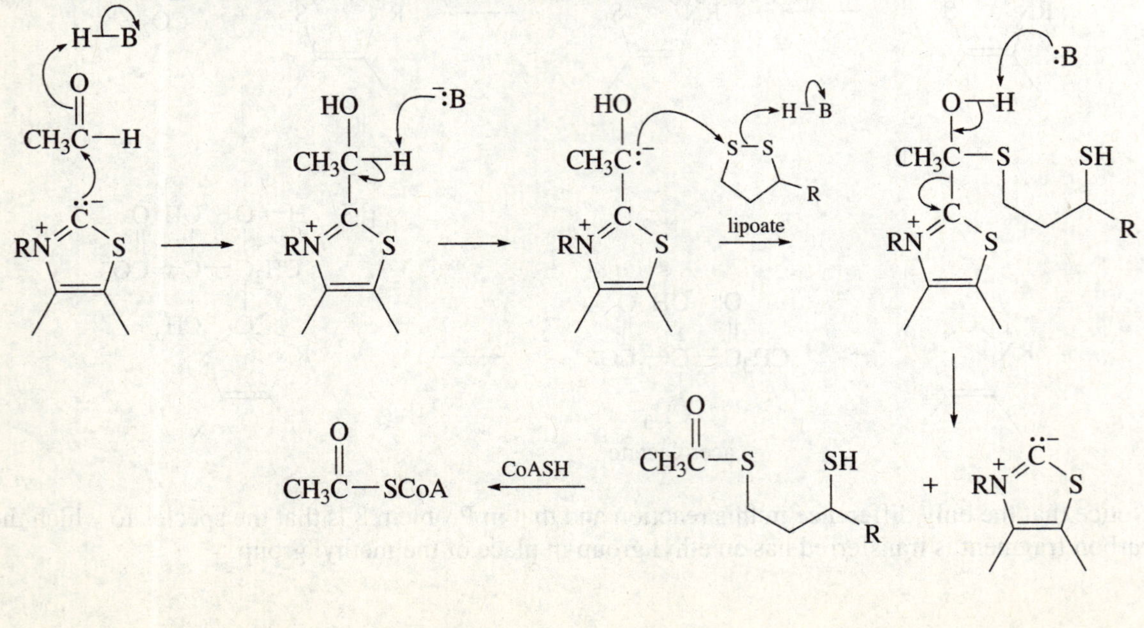

14. The α-keto group that accepts the amino group from pyridoxamine is converted into an amino group.

a.

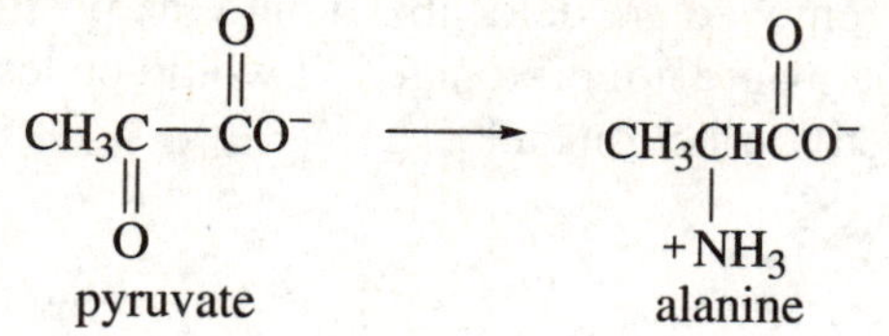

b.

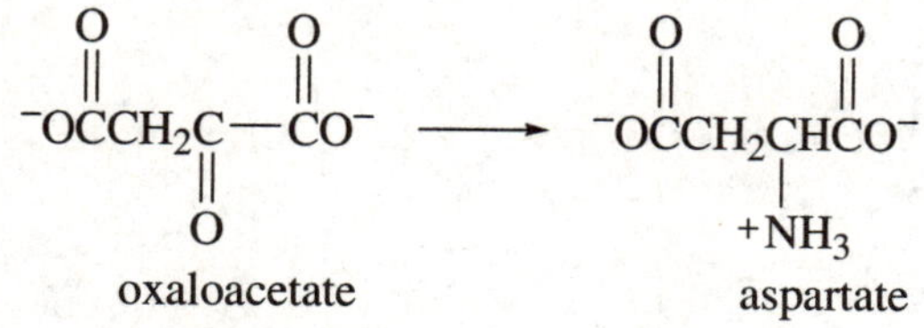

15. In the first step, a proton is removed from the α-carbon, and the electrons left behind expel the leaving group from the β-carbon. Transimination produces an enamine. The enamine tautomerizes to an imine that is hydrolyzed to the final product.

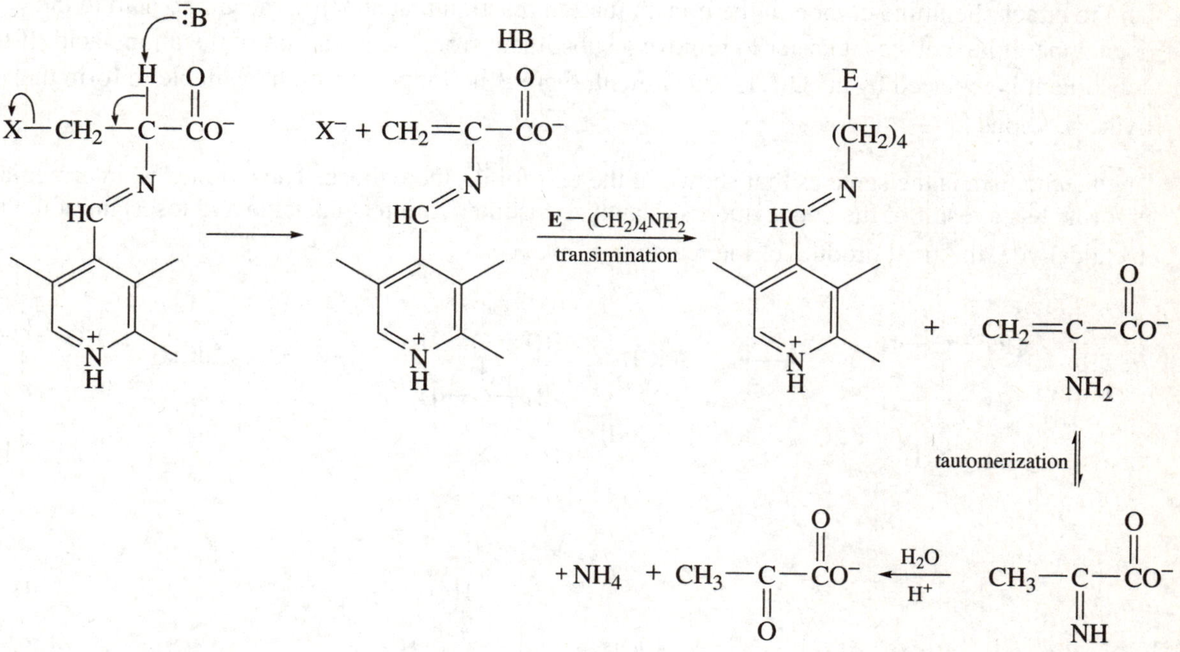

16. The compound on the right is more easily decarboxylated because the electrons left behind when CO_2 is eliminated are delocalized onto the positively charged nitrogen of the pyridine ring. The electrons left behind if the other compound is decarboxylated cannot be delocalized.

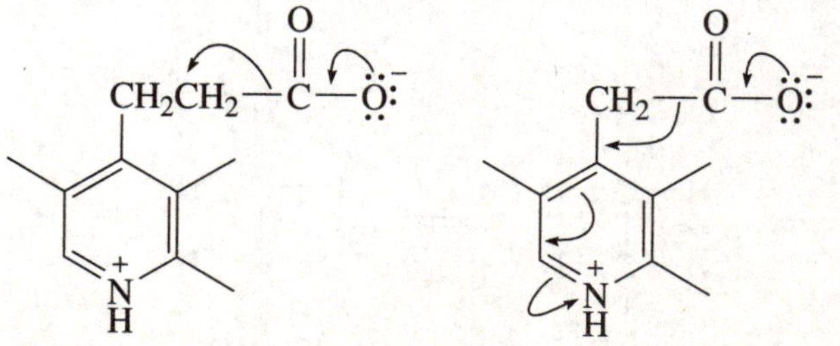

17. The first step in all amino acid transformations is removal of a substituent from the α-carbon of the amino acid. The electrons left behind when the substituent is removed are delocalized onto the positively charged nitrogen of the pyridine ring. If the ring nitrogen were not protonated, it would be less attractive to the electrons. In other words, it would be a less effective electron sink.

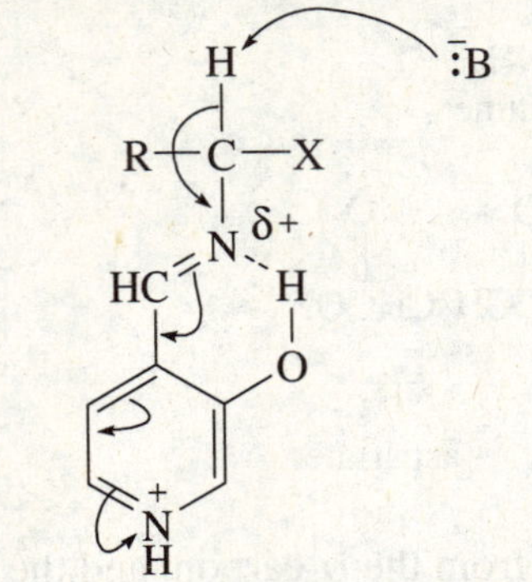

18. The hydrogen of the OH substituent forms a hydrogen bond with the nitrogen of the imine linkage (see above structure). This puts a partial positive charge on the nitrogen, which makes it easier for the amino acid to attack the imine carbon in the transimination reaction that attaches the amino acid to the coenzyme. It also makes it easier to remove a substituent from the α-carbon of the amino acid. If the OH substituent is replaced by an OCH_3 substituent, there is no longer a proton available to form the hydrogen bond.

19. The mechanism is the same as that shown in the text for dioldehydrase. The tetrahedral intermediate that is formed as a result of the coenzyme B_{12}-catalyzed isomerization is unstable and loses ammonia to give acetaldehyde, the final product of the reaction.

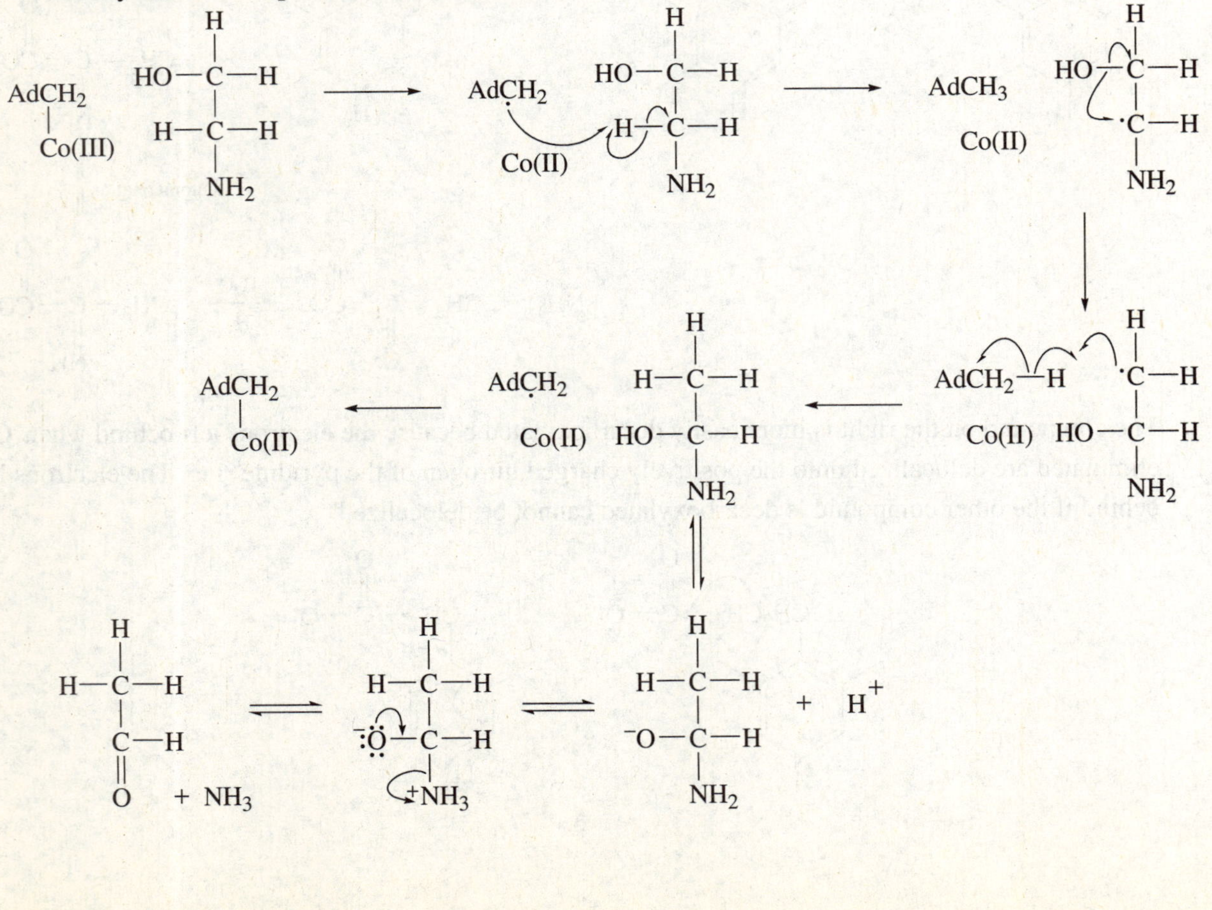

20. 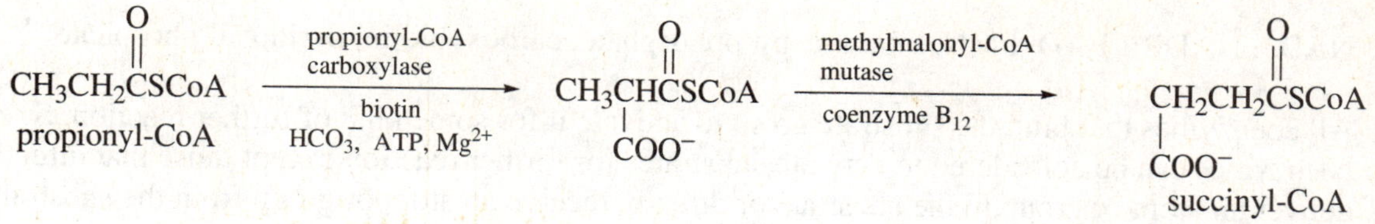

21. The methyl group in thymidine comes from the methylene group of N^5,N^{10}-methylene-THF. The methylene group of N^5,N^{10}-methylene-THF comes from the CH_2OH group of serine by means of a PLP-catalyzed C_α-C_β cleavage.

22. Two thiol groups are necessary for each of the two oxidations in the conversion of vitamin K epoxide to vitamin KH_2. Since lipoate has two thiol groups, each oxidation involves an intramolecular reaction. When thiols such as ethanethiol or propanethiol are used, the oxidation involves an intermolecular reaction. These thiols react more slowly than lipoate because the two thiol groups are not in the same molecule. Therefore, they have to find each other in order to react.

23. **a.** NAD^+, $NADP^+$, FAD, FMN, lipoate, vitamin K epoxide

b. N^5-methyl-THF, N^5,N^{10}-methylene-THF, N^5,N^{10}-methenyl-THF, N^5-formyl-THF, N^{10}-formyl-THF, N^5-formimino-THF, *S*-adenosylmethionine

One could include biotin; it accepts a one carbon group (from HCO_3^-) and transfers it to the substrate.

c. 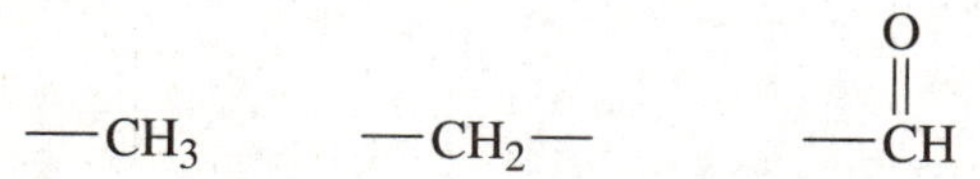

d. FAD oxidizes dihydrolipoate back to lipoate.

e. NAD^+ oxidizes $FADH_2$ back to FAD.

f. 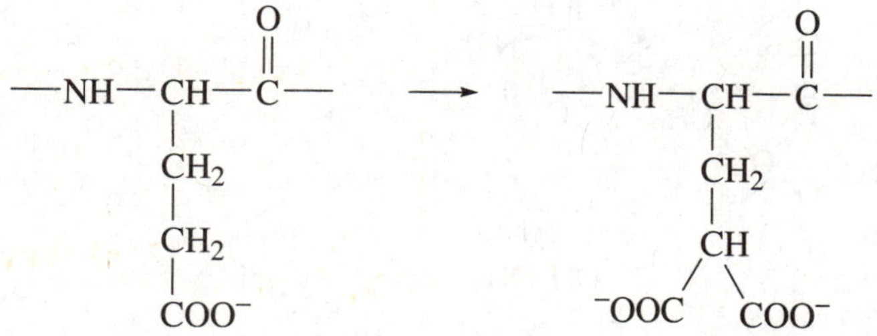

g. thiamine pyrophosphate and pyridoxal phosphate

h. Thiamine pyrophosphate is used for the decarboxylation of α-keto acids. Pyridoxal phosphate is used for the decarboxylation of amino acids.

i. biotin and vitamin KH_2

j. Biotin carboxylates a carbon adjacent to a carbonyl group. Vitamin KH_2 carboxylates the γ-carbon of a glutamate.

24. **a.** NAD^+, $NADP^+$, FAD, FMN, thiamine pyrophosphate, carboxybiotin, pyridoxal phosphate

b. All coenzymes that bind the substrate do so to activate it for some kind of further reaction. So all the coenzymes can be considered to activate substrates for further reaction except those that directly deliver an atom or group to the substrate or directly receive an atom or group from the substrate without binding it, such as NAD^+, $NADP^+$, NADH, NADPH, and the tetrahydrofolate coenzymes that deliver a one-carbon fragment in one step such as N^5-methyl-THF. In other words, the answers are: FAD, FMN, $FADH_2$, $FMNH_2$, thiamine pyrophosphate, biotin, pyridoxal phosphate, coenzyme B_{12}, some tetrahydrofolate coenzymes, and vitamin KH_2.

c. thiamine pyrophosphate

d. vitamin KH_2

25. **a.** acetyl-CoA carboxylase, biotin

b. dihydrolipoyl dehydrogenase, FAD

c. methylmalonyl-CoA mutase, coenzyme B_{12}

d. lactate dehydrogenase, NADH (This enzyme is named according to the reaction it catalyzes in the reverse direction.)

e. aspartate transaminase, pyridoxal phosphate

f. propionyl-CoA carboxylase, biotin

26.

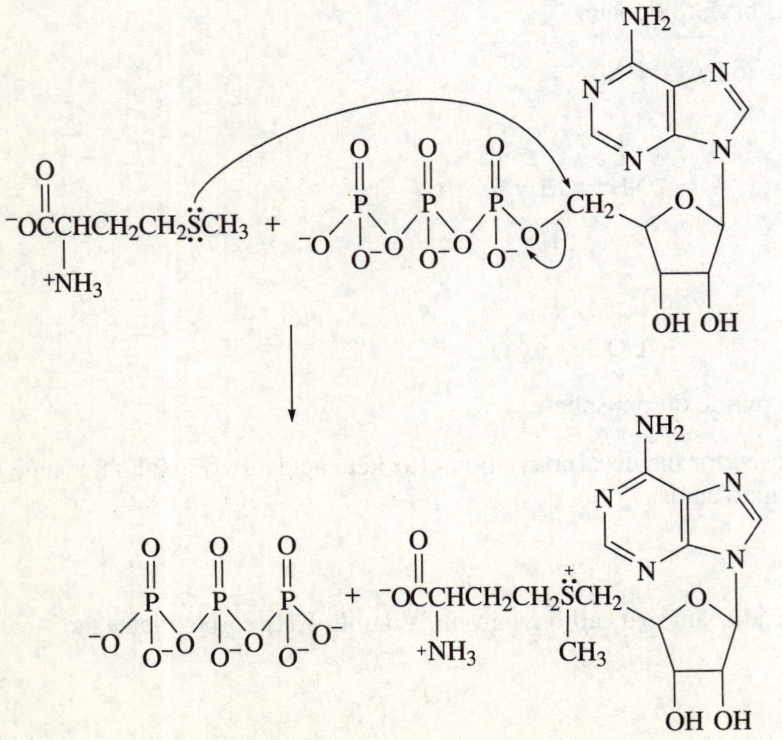

27. **a.** thiamine pyrophosphate, lipoate, coenzyme A, FAD, NAD^+, indicated by numbers 1–5 in the mechanism below

b.

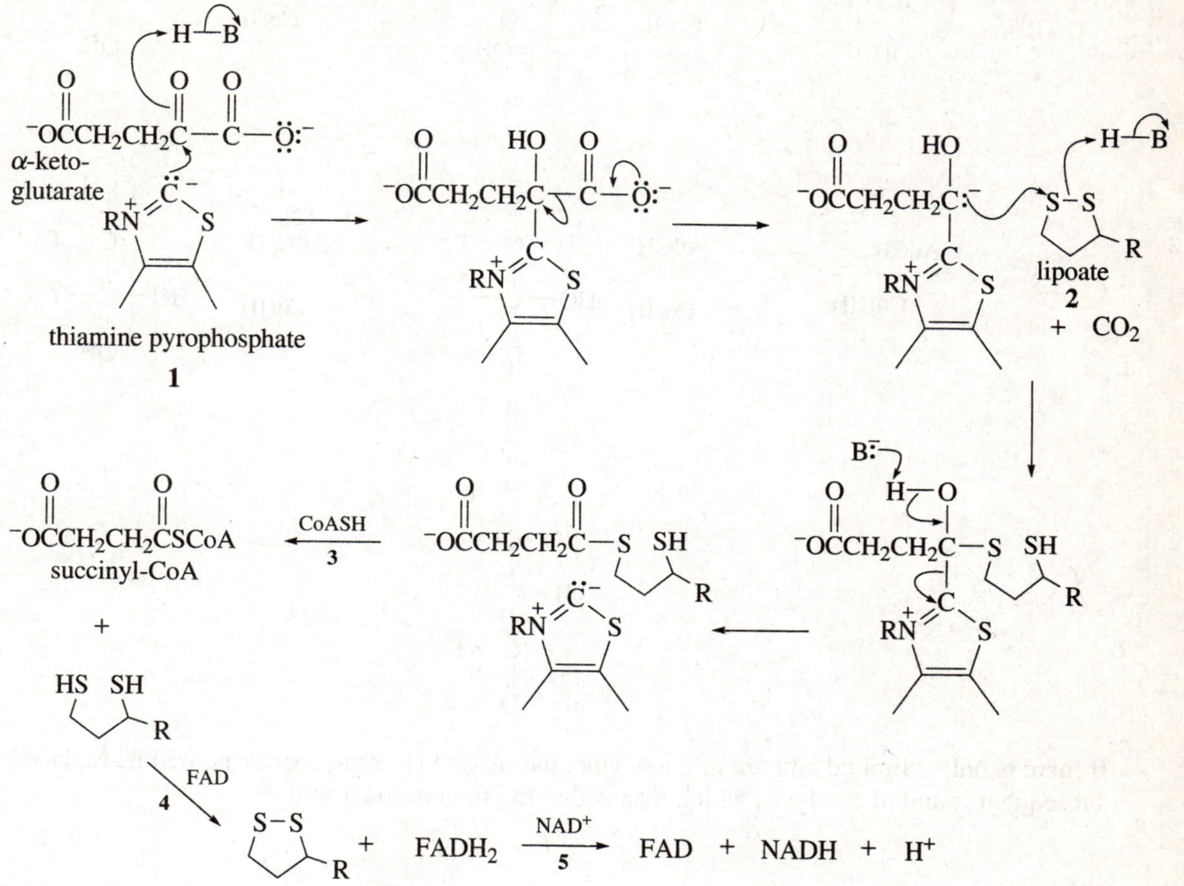

28. The products of the reaction are shown below.

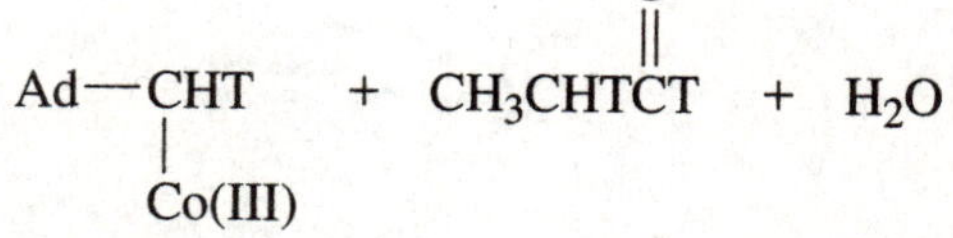

The following mechanism explains the products.

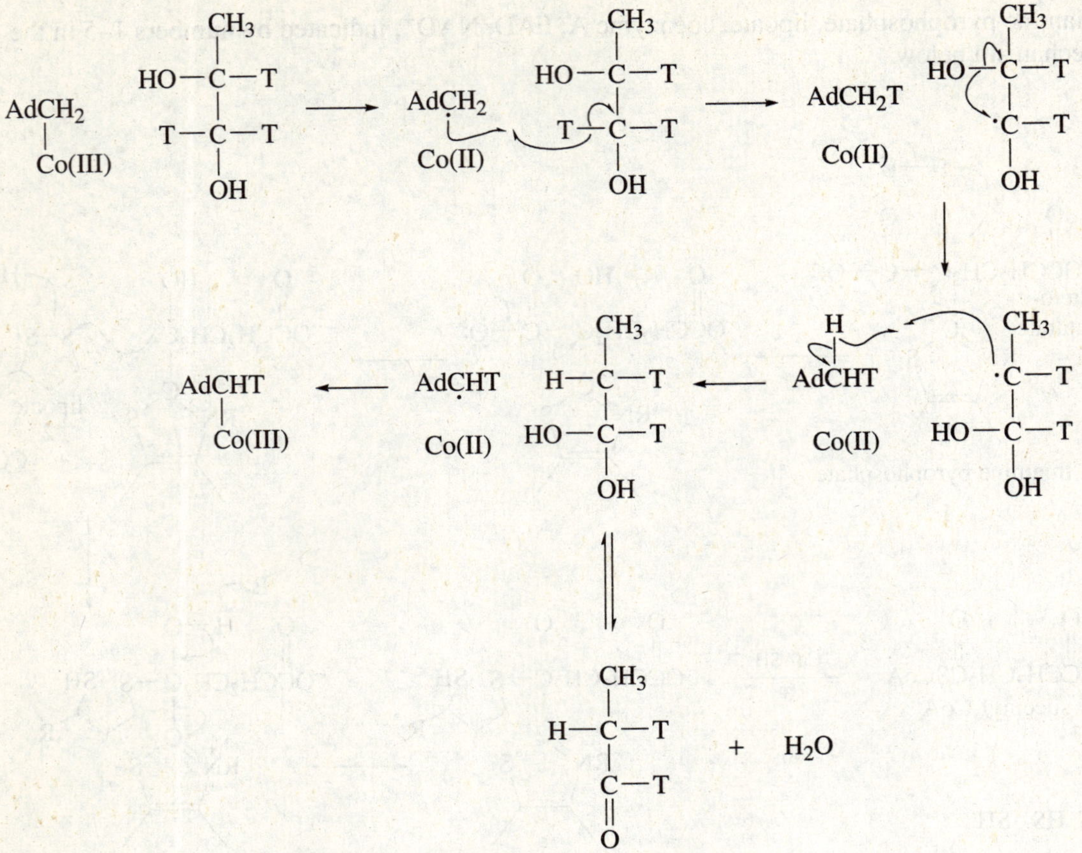

If there is only a limited amount of coenzyme, the second H of the coenzyme will be replaced by T in a subsequent round of catalysis, which means that the final product will be:

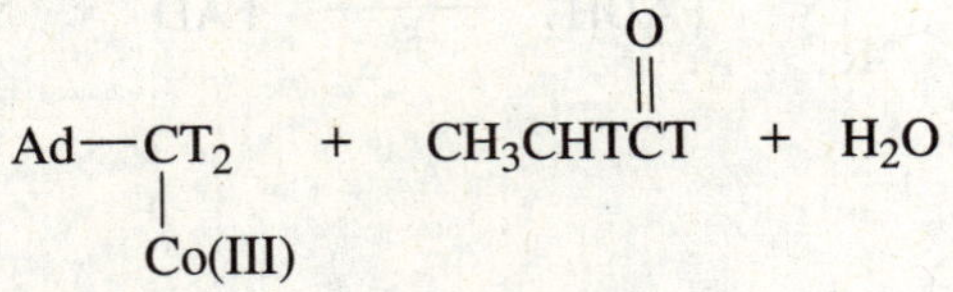

29.

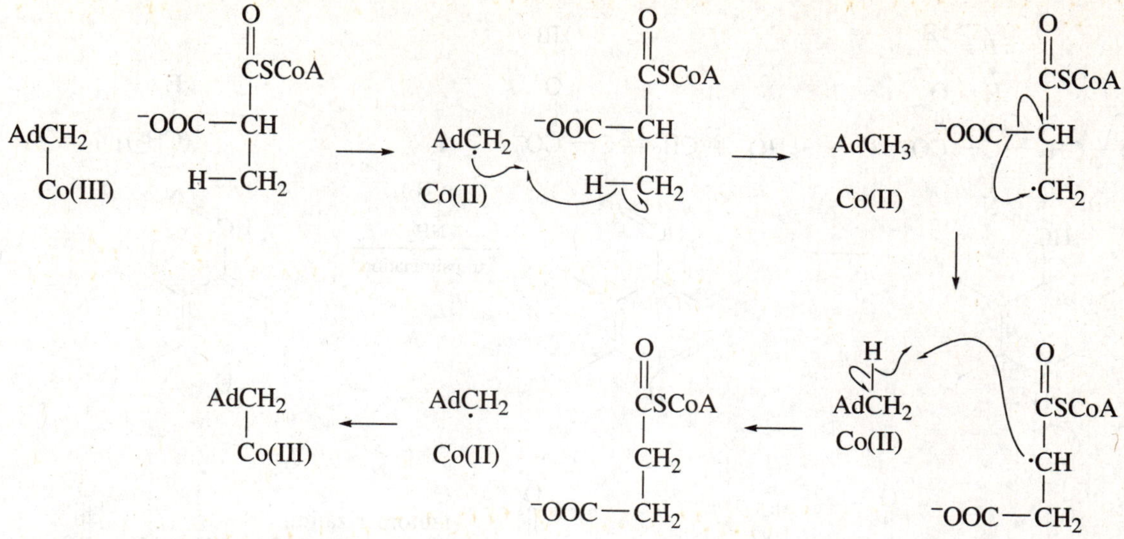

30. **a.**

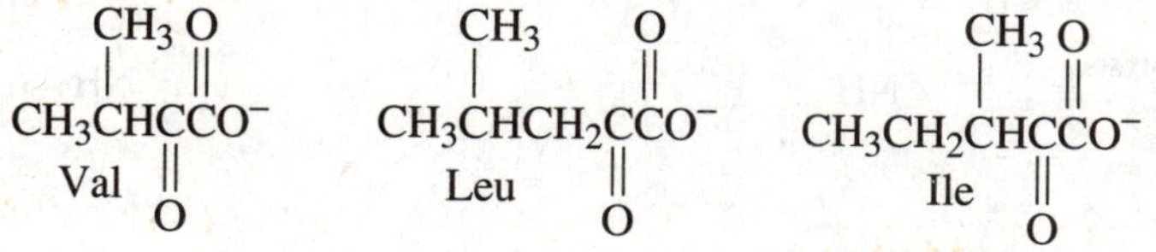

b.

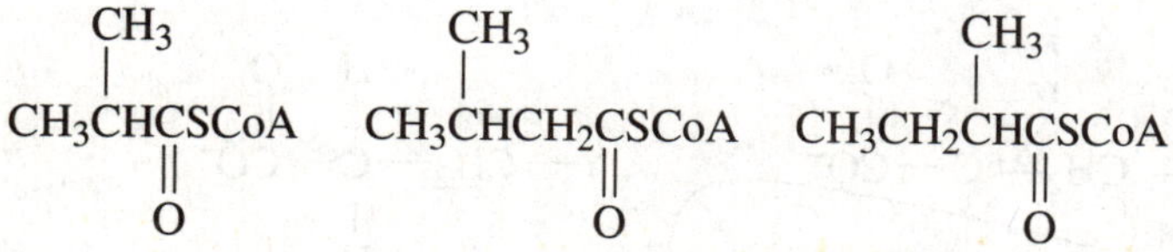

c. thiamine pyrophosphate, lipoate, coenzyme A, FAD, NAD^+

d. The disease can be treated by a diet low in branched-chain amino acids.

31.

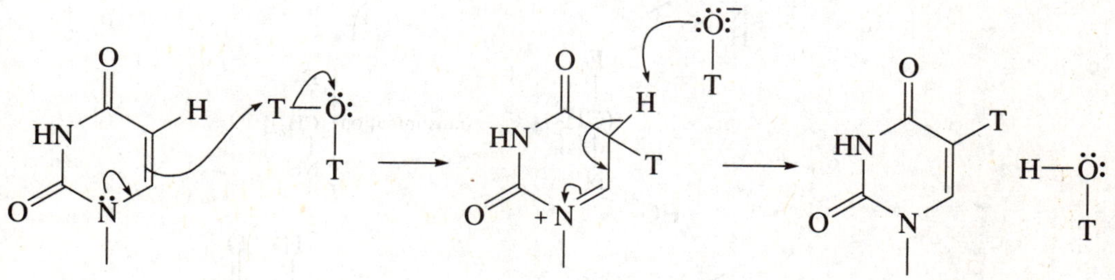

32.

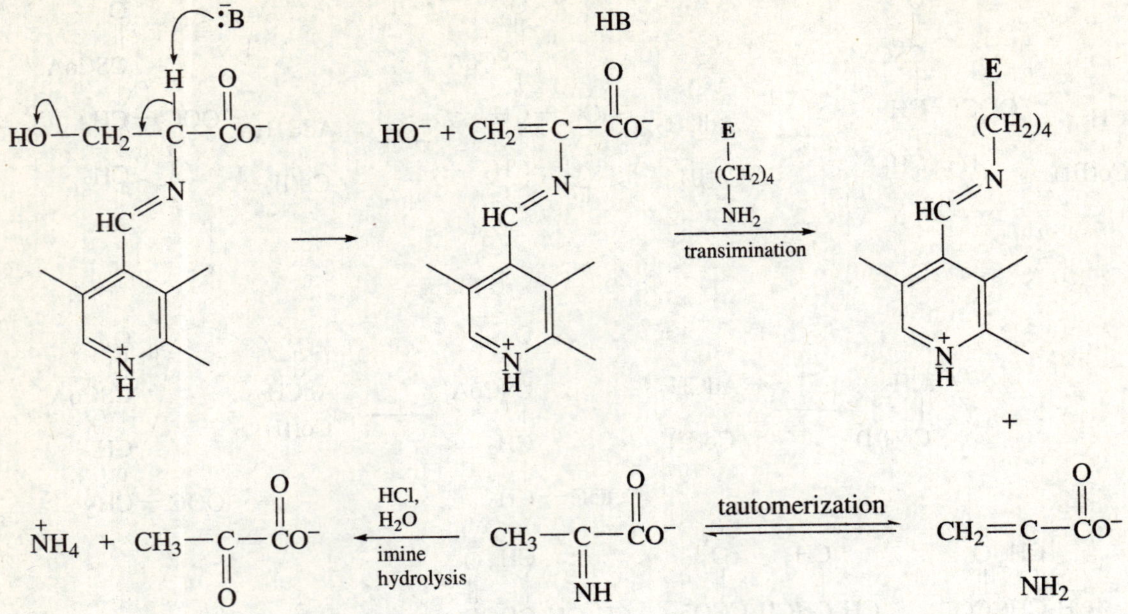
:B̄
H
O
HO
CH₂
C
CO⁻
N
HC
N
H
HB
HO⁻ + CH₂
C
CO⁻
E
(CH₂)₄
NH₂
transimination
+
CH₃
C
CO⁻
NH
tautomerization
CH₂
C
CO⁻
NH₂
HCl,
H₂O
imine
hydrolysis
N̄H₄ + CH₃
C
CO⁻
O

33.

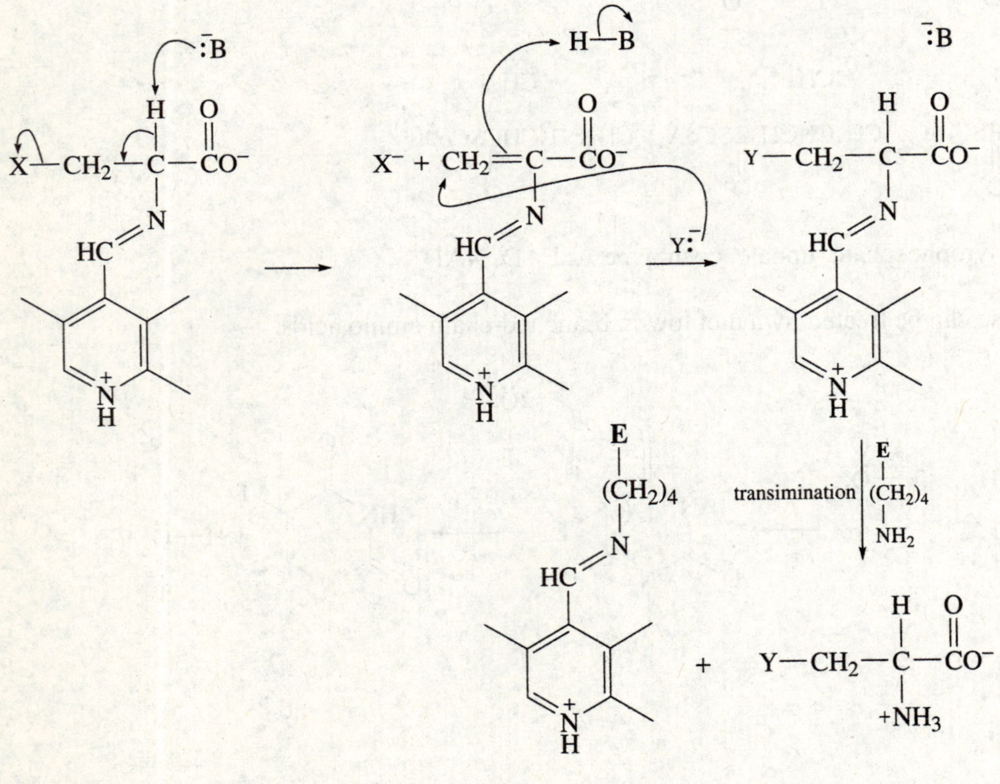
:B̄
H
O
X
CH₂
C
CO⁻
N
HC
H
X⁻ + CH₂
C
CO⁻
H—B
Y:⁻
:B̄
Y—CH₂
C
CO⁻
E
(CH₂)₄
NH₂
transimination
E
(CH₂)₄
N
+
Y—CH₂
C
CO⁻
+NH₃

34.

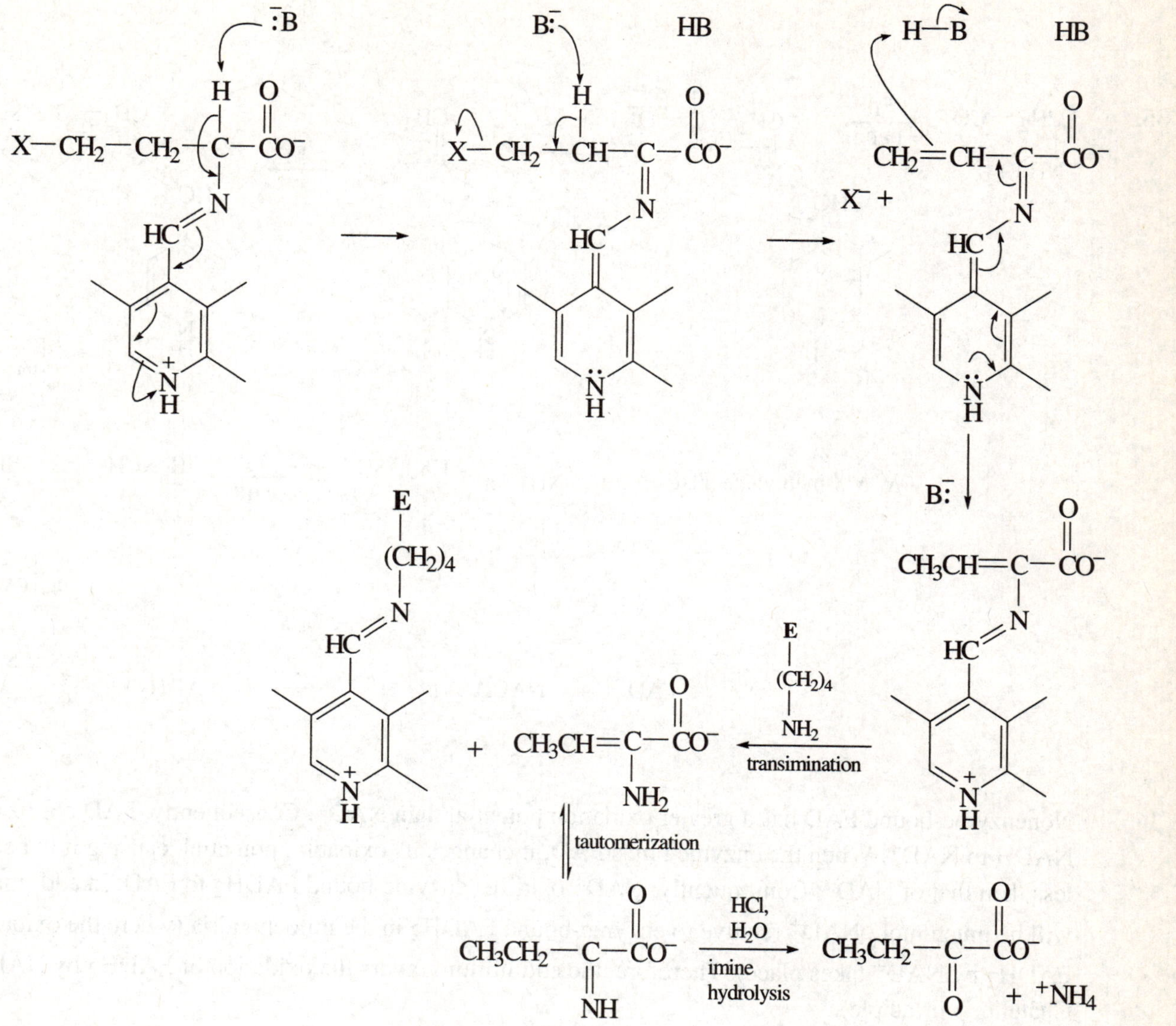

:B
B:
HB
H—B
X—CH2—CH2—C—CO−
X− +
CH2=CH—C—CO−
CH3CH=C—CO−
transimination
tautomerization
CH3CH2—C—CO−
HCl,
H2O
imine
hydrolysis
+ +NH4

35.

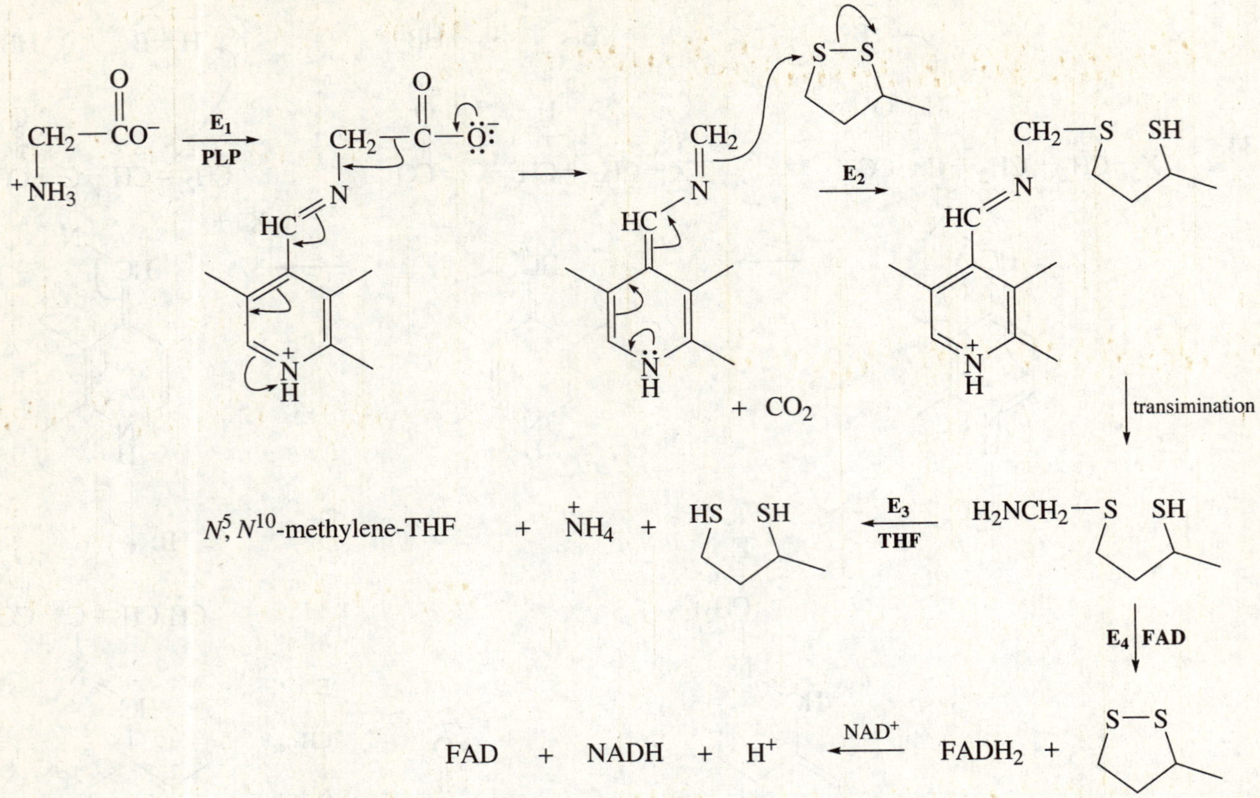

36. Nonenzyme-bound FAD has a greater oxidation potential than NAD^+. Consequently, FAD oxidizes NADH to NAD^+. When the enzyme binds FAD, it changes its oxidation potential, causing it to become less than that of NAD^+. Consequently, NAD^+ oxidizes enzyme-bound $FADH_2$ to FAD. In addition, there will be much more NAD^+ relative to enzyme-bound $FADH_2$ in the mitochondria (where the oxidation of $FADH_2$ by NAD^+ takes place). Therefore, the equilibrium favors the oxidation of $FADH_2$ by NAD^+ (Le Châtelier's principle).

$$\text{E-FADH}_2 + \underset{\text{excess}}{\text{NAD}^+} \rightleftharpoons \text{E-FAD} + \text{NADH} + \text{H}^+$$

37.

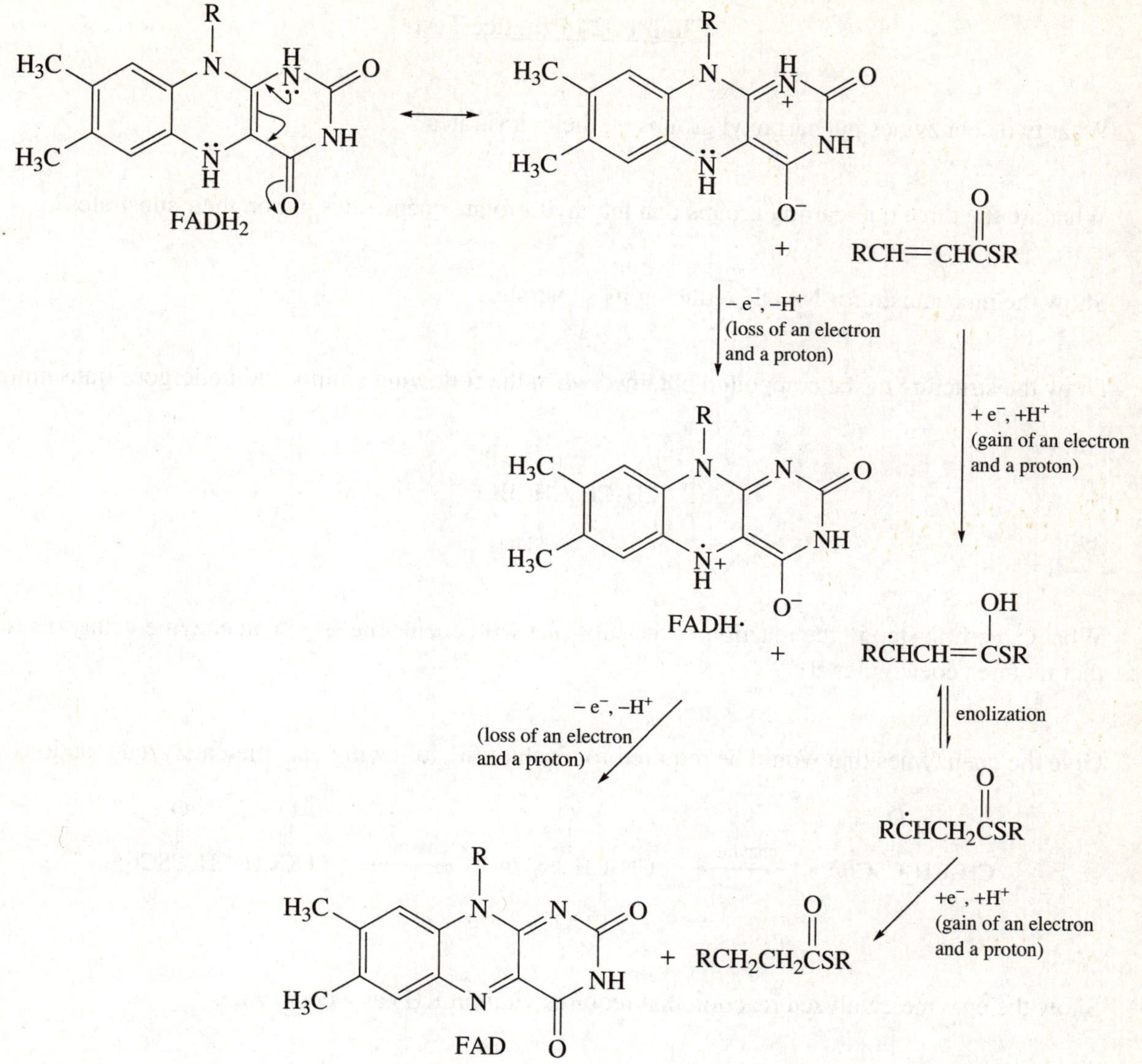
FADH2
+
RCH=CHCSR
− e⁻, −H⁺
(loss of an electron
and a proton)
+ e⁻, +H⁺
(gain of an electron
and a proton)
FADH·
+
RCHCH=CSR
OH
enolization
RCHCH2CSR
− e⁻, −H⁺
(loss of an electron
and a proton)
+e⁻, +H⁺
(gain of an electron
and a proton)
FAD
+ RCH2CH2CSR

Chapter 24 Practice Test

1. What two coenzymes put carboxyl groups on their substrates?

2. What are the three one-carbon groups that tetrahydrofolate coenzymes put on their substrates?

3. Show the mechanism for NADH reducing its substrate.

4. Draw the structure of the compound obtained when the following amino acid undergoes transamination.

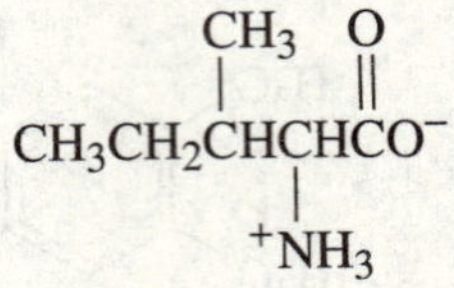

5. What is the first step in the reaction of the substrate with coenzyme B_{12} in an enzyme-catalyzed reaction that requires coenzymes B_{12}?

6. Give the coenzymes that would be required for each of the following enzyme-catalyzed reactions.

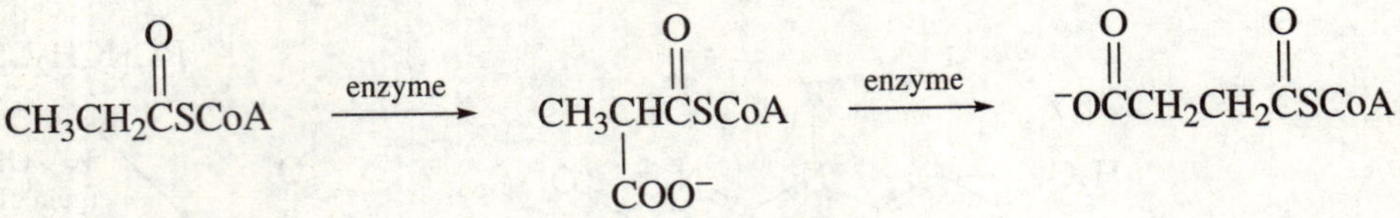

7. Show the enzyme-catalyzed reaction that requires vitamin KH_2 as a coenzyme.

8. Give the product of the enzyme-catalyzed reaction that requires biotin and whose substrate is acetyl-CoA.

9. Other than the substrate, enzyme, and coenzyme, what 3 additional reagents are needed by a reaction that requires biotin as a coenzyme?

10. What is the function of FAD in the pyruvate dehydrogenase system?

11. Give the structures of the two products obtained from the following transamination reaction.

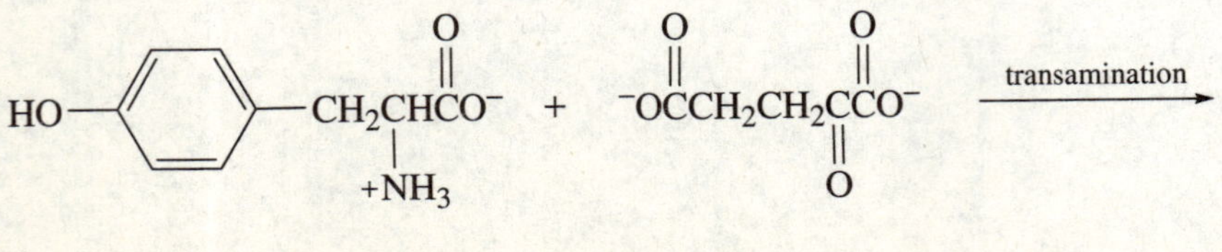

12. Indicate whether each the following statements is true or false:

a. Vitamin B_1 is the only water-insoluble vitamin that has a coenzyme function. T F

b. $FADH_2$ is reducing agent. T F

c. Thiamine pyrophospate is vitamin B_6. T F

d. Cofactors that are organic molecules are called coenzymes. T F

e. Vitamin K is a water-soluble vitamin. T F

f. Lipoic acid is covalently bound to its enzyme by an amide linkage. T F

CHAPTER 25
The Chemistry of Metabolism

Important Terms

acyl adenylate

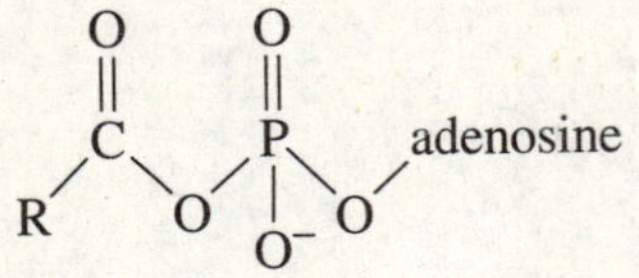

acyl phosphate

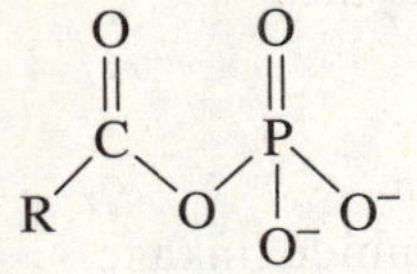

acyl pyrophosphate

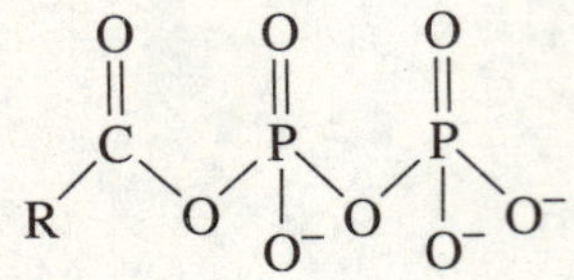

anabolism — the reactions living organisms carry out that result in the synthesis of complex biomolecules from simple precursor molecules.

catabolism — the reactions living organisms carry out to provide energy and simple precursor molecules for synthesis.

citric acid cycle — a series of reactions that converts the acetyl group of acetyl-CoA into two molecules of CO_2 and a molecule of CoASH.

dehydrogenase — an enzyme that catalyzes an oxidation reaction.

Glycolysis — the series of reactions that converts glucose into two molecules of pyruvate.

high-energy bond — a bond that releases a great deal of energy when it is broken.

kinase — an enzyme that puts a phosphate group on its substrate.

metabolism — reactions living organisms carry out in order to obtain the energy they need and to synthesize the compounds they require.

***β*-oxidation** — a repeating series of four reactions that converts a fatty acyl-CoA molecule into molecules of acetyl-CoA.

oxidative phosphorylation — the fourth stage of catabolism in which NADH and $FADH_2$ are oxidized back to NAD^+ and FAD: for each NADH that is oxidized, 3 ATPs are formed; for each $FADH_2$ that is oxidized, two ATPs are formed.

phosphoanyhdride bond — the bond that holds two phosphoric acid molecules together.

phosphoryl transfer reaction — the transfer of a phosphate group from one compound to another.

Solutions to Problems

1. Solved in the text.

2. Pyruvate is a good leaving group because the electrons left behind when the P—O bond breaks can be delocalized, resulting in the formation of a relatively stable keto group.

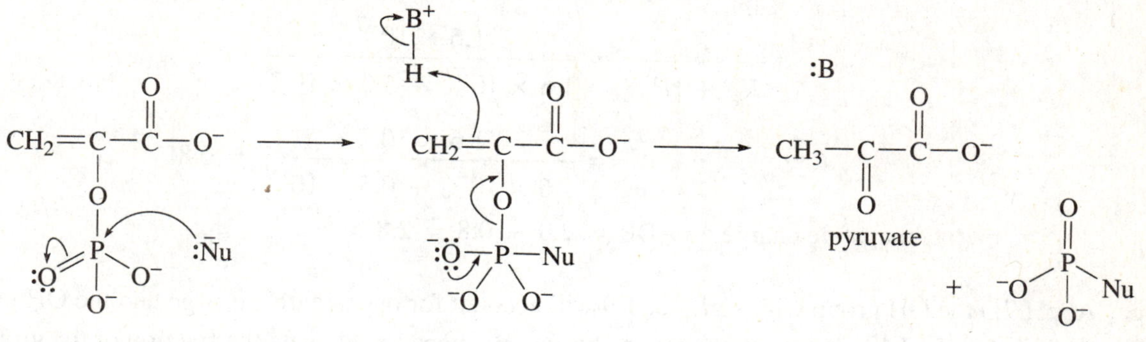

3. The $\Delta G°$ for formation of ATP is +7.3 kcal/mol. This means that for a compound to "drive the formation" of ATP, its $\Delta G°$ for hydrolysis must be more negative than −7.3 kcal/mol. Phosphocreatine is the only one of the four compounds that hydrolyzes with sufficient energy.

4. Phosphoric acid has three pK_a's: the first OH group to lose a proton has a pK_a of 1.9, the second OH group has a pK_a of 6.7, and the third OH group a pK_a of 12.4.

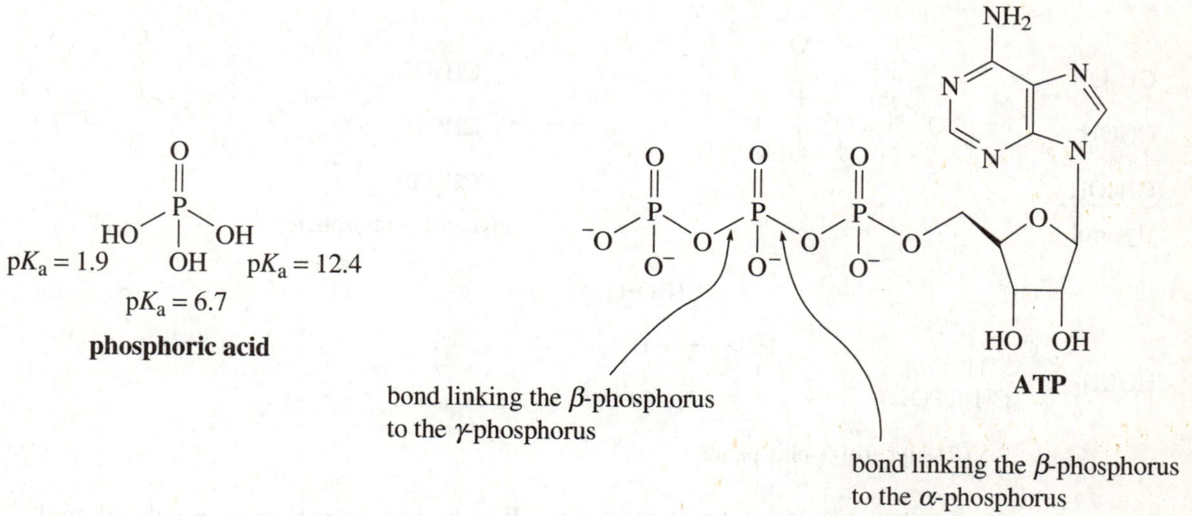

When the bond linking the β-phosphorus to the α-phosphorus breaks, the phosphorus leaving group attains its second negative charge which means that the pK_a of the conjugate acid of the leaving group is ~6.7. When the bond linking the β-phosphorus to the γ-phosphorus breaks, the phosphorus leaving group attains its third negative charge, which means that the pK_a of the conjugate acid of the leaving group is ~12.4. Consequently, the bond to the γ-phosphorus never breaks because the leaving group is a much stronger base than the leaving group that leaves when the bond to the α-phosphorus breaks.

5. **a.** Solved in the text.

b. Because pH 7.3 is much more basic than the pK_a values of the first two ionizations of ADP, these two groups will be in their basic forms at that pH, giving ADP two negative charges. We can determine the fraction of the group with pK_a of 6.8 that will be in its basic form at pH 7.3 using the method shown in part **a**.

$$\frac{K_a}{K_a + [H^+]} = \frac{1.6 \times 10^{-7}}{1.6 \times 10^{-7} + 5.0 \times 10^{-8}}$$

$$= \frac{1.6 \times 10^{-7}}{1.6 \times 10^{-7} + 0.5 \times 10^{-7}} = 0.8$$

total negative charge on ADP = 2.0 + 0.8 = 2.8

c. At pH 7.3 the OH group with a pK_a of 1.9 will account for one negative charge and the OH group with a pK_a of 12.4 will have no negative charge. We need to calculate the fraction of the group with a pK_a value of 6.7 that will be negatively charged at pH 7.3.

$$\frac{K_a}{K_a + [H^+]} = \frac{2.0 \times 10^{-7}}{2.0 \times 10^{-7} + 0.5 \times 10^{-7}} = 0.8$$

total negative charge on phosphate = 1.0 + 0.8 = 1.8

6.

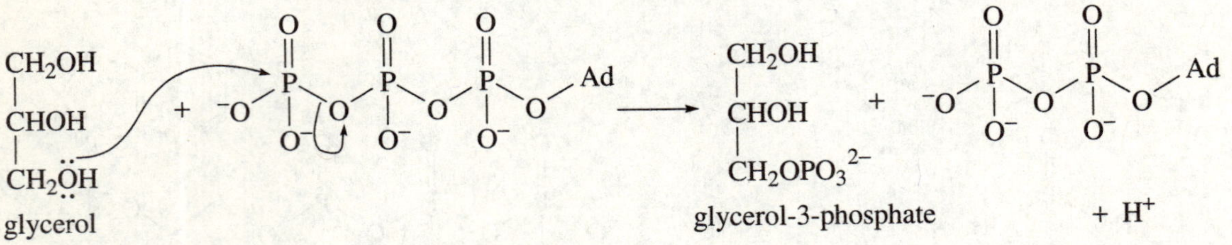

7.

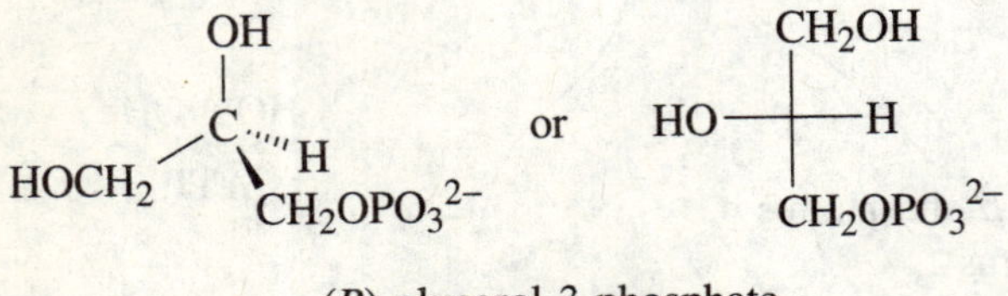

(*R*)-glycerol-3-phosphate

8. The resonance contributor on the right shows that the *β*-carbon of the *α*,*β*-unsaturated carbonyl compound has a partial positive charge and the *α*-carbon has no charge. The nucleophilic OH group, therefore, is attracted to the positively charged *β*-carbon.

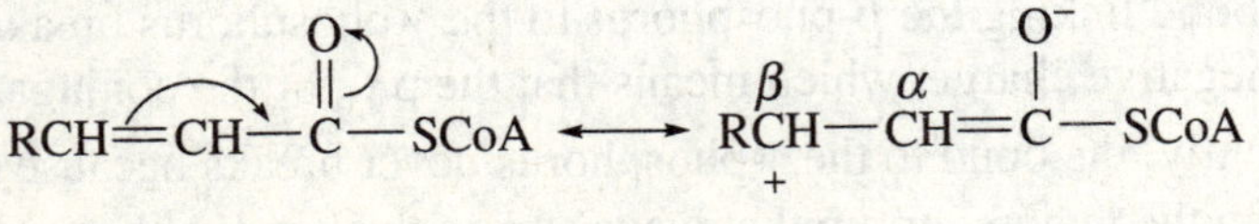

9. eight

10. seven: one mole of NADH will be obtained from each of the following rounds of *β*–oxidation.

$$16 \xrightarrow{1} 14 + 2 \xrightarrow{2} 12 + 2 \xrightarrow{3} 10 + 2 \xrightarrow{4} 8 + 2 \xrightarrow{5} 6 + 2 \xrightarrow{6} 4 + 2 \xrightarrow{7} 2 + 2$$

11.

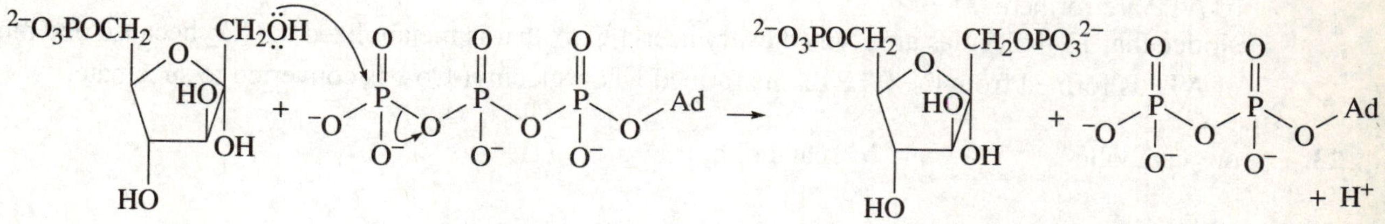

fructose-6-phosphate ATP fructose-1,6-diphosphate ADP

12. **a.** conversion of glucose to glucose-6-phosphate
conversion of glucose-6-phosphate to glucose-1,6-diphosphate

b. conversion of 1,3-diphosphoglycerate to 3-phosphoglycerate
conversion of 2-phosphoenolpyruvate to pyruvate

13. two; each mole of D-glucose is converted to two moles of glyceraldehyde-3-phosphate and each mole of glyceraldehyde-3-phosphate requires one mole of NAD^+ for it to be converted to a mole of pyruvate.

14. a ketone

15. thiamine pyrophosphate

16. an aldehyde

17.

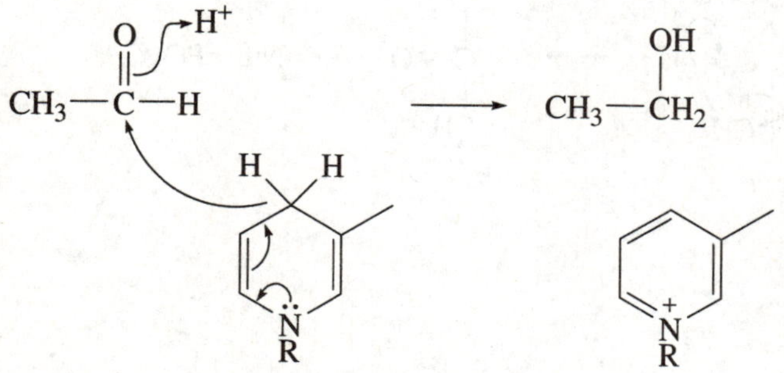

18. pyridoxal phosphate

19.

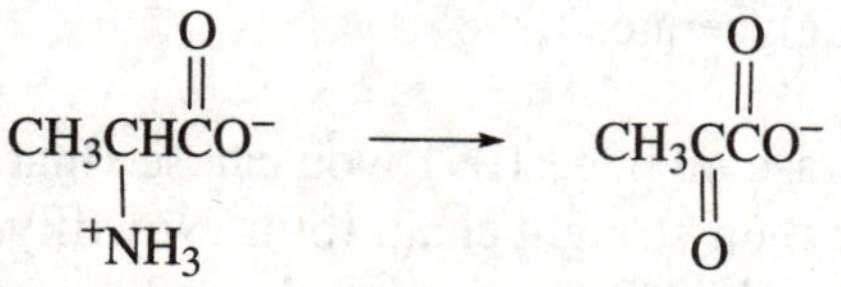

alanine **pyruvate**

20. a secondary alcohol

21. citrate and isocitrate

22. When one molecule of acetyl-CoA is converted to CO_2, 3 molecules of NADH and one molecule of $FADH_2$ are formed.
Since each NADH forms 3 molecules of ATP and each $FADH_2$ forms 2 molecules of ATP, 11 molecules of ATP are formed.
Notice that 12 molecules are formed every acetyl-CoA that is metabolized to CO_2 because one molecule of ATP is formed from the GTP that is formed when succinyl-CoA is converted to succinate.

23. **a.** catabolic **b.** catabolic

24.

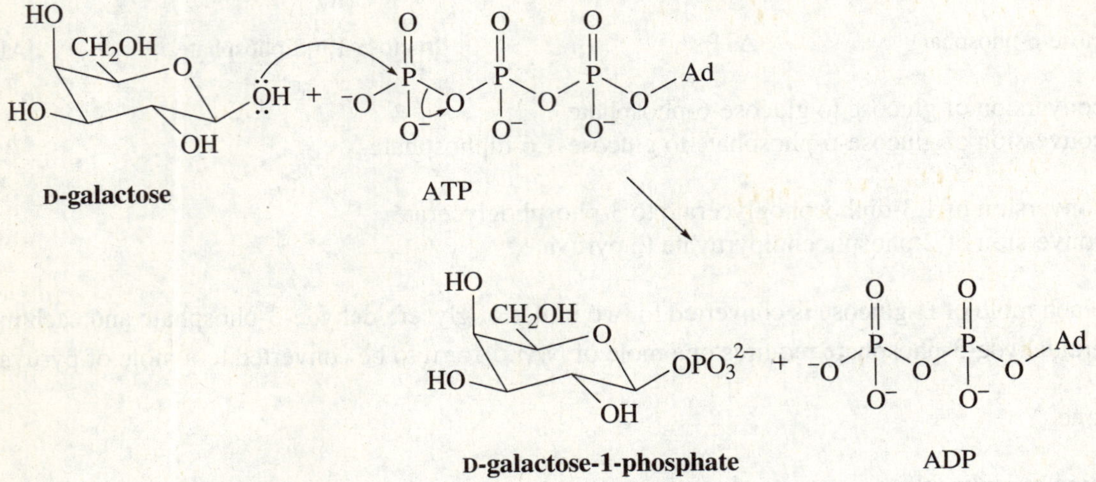

25. Pyruvate is a good leaving group because the electrons left behind when pyruvate leaves are delocalized.

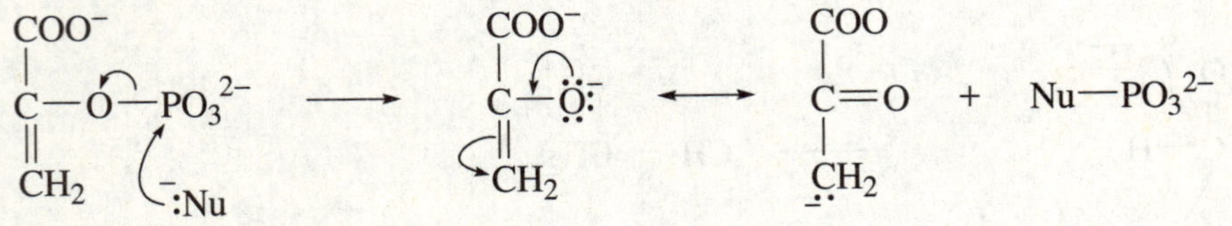

26. The hydrogen on the α-carbon.

27. The conversion of citrate to isocitrate.
The conversion of fumarate to (*S*)-malate.

28. The label will be on the phosphate group that is attached to the enzyme (phosphoglycerate mutase) that catalyzes the isomerization of 3-phosphoglycerate to 2-phosphoglycerate.

29. If you examine the mechanism for glucose-6-phosphate isomerase on page 1091, you can see that C-1 in D-glucose is also C-1 D-fructose, when the sugars are drawn in their straight chain form. Now if you examine the mechanism for aldolase on page 1093, you can see which carbons in D-glucose correspond to the carbons in dihydroxyacetone phosphate and D-glyceraldehyde-3-phosphate. Then, if you look at page 1149, you will see how the carbons in dihydroxyacetone phosphate correspond to the carbons in D-glyceraldehyde-3-phosphate.

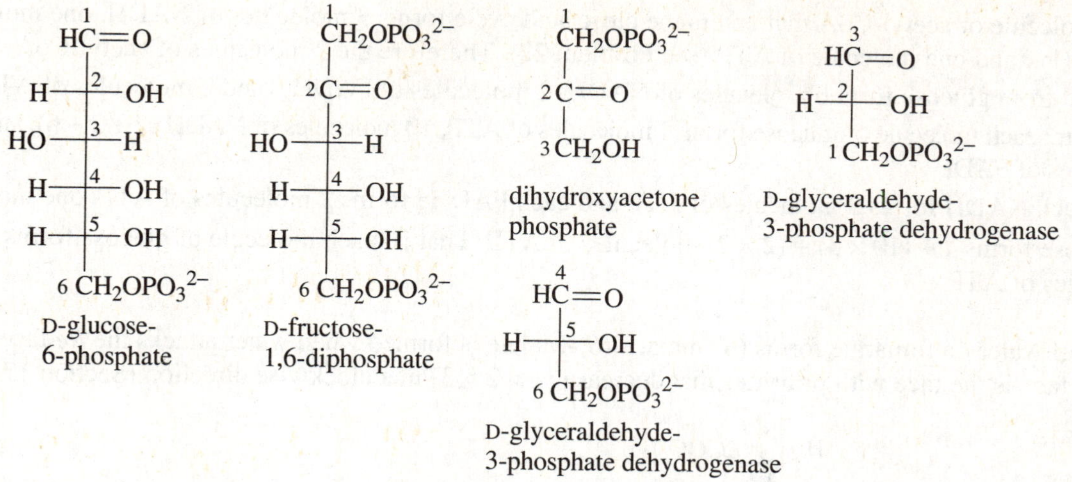

Finally we see how the carbons in D-glyceraldehyde-3-phosphate correspond to the carbons in pyruvate. Therefore, C-3 and C-4 of glucose each becomes a carboxyl group in pyruvate.

30. Pyruvate loses its carboxyl group when it is converted to ethanol. Since the carboxyl group is C-3 or C-4 of glucose, half of the ethanol molecules contain C-1 and C-2 of glucose and the other half contain C-5 and C-6 of glucose.

31. The β-oxidation of a molecule of a 16-carbon fatty acyl-CoA will form 8 molecules of acetyl-CoA.

32. Each molecule of acetyl-CoA forms two molecules of CO_2. Therefore, the 8 molecules of acetyl-CoA obtained from a molecule of a 16-carbon fatty acyl-CoA will form 16 molecules of CO_2.

33. No ATP is formed from β-oxidation.

34. Each molecule of acetyl-CoA that is cleaved from the 16-carbon fatty acyl-CoA forms one molecule of $FADH_2$ and one molecule of NADH. Since a 16-carbon fatty acyl-CoA undergoes 7 cleavages, 7 molecules of $FADH_2$ and 7 molecules of NADH are formed from the 16-carbon fatty acyl-CoA.

35. Since each NADH forms 3 molecules of ATP and each $FADH_2$ forms 2 molecules of ATP, the 7 molecules of NADH form 21 molecules of ATP and the 7 molecules of $FADH_2$ form 14 molecules of ATP. Therefore, 35 molecules of ATP are formed.

36. We have seen that each molecule acetyl-CoA that enters the citric acid cycle forms 12 molecules of ATP (Section 25.11). A molecule of a 16-carbon fatty acid will form 8 molecules of acetyl-CoA. These will form 96 molecules of ATP. When these are added to the number of ATP molecules formed from the NAD and $FADH_2$ generated in β-oxidation, we see that 131 molecules of ATP are formed.

37. Each molecule of glucose, while being converted to 2 molecules of pyruvate, forms 2 molecules of ATP and 2 molecules of NADH.
The 2 molecules of pyruvate form 2 molecules of NADH while being converted to 2 molecules of acetyl-CoA.

Each molecule of acetyl-CoA that enters the citric acid cycle forms 3 molecules of NADH, one molecule of $FADH_2$, and one molecule of ATP (see Problem 22). Therefore the 2 molecules of acetyl-CoA obtained from glucose form 6 molecules of NADH, 2 molecules of $FADH_2$, and 2 molecules of ATP. Therefore, each molecule of glucose forms 4 molecules of ATP, 10 molecules of NADH (2 + 2 + 6), and 2 molecules of $FADH_2$.
Since each NADH forms 3 molecules of ATP and each $FADH_2$ forms 2 molecules of ATP, one molecule of glucose forms $4 + (10 \times 3) + (2 \times 2)$ molecules of ATP. That is, each molecule of glucose forms 38 molecules of ATP.

38. Attack of water on fumarate forms (*S*)-malate. (*S*)-Malate is formed when water attacks the *Re* face. The *Re* face is the face with priorities that decrease ($1 > 2 > 3$) in a clockwise direction (Section 17.14).

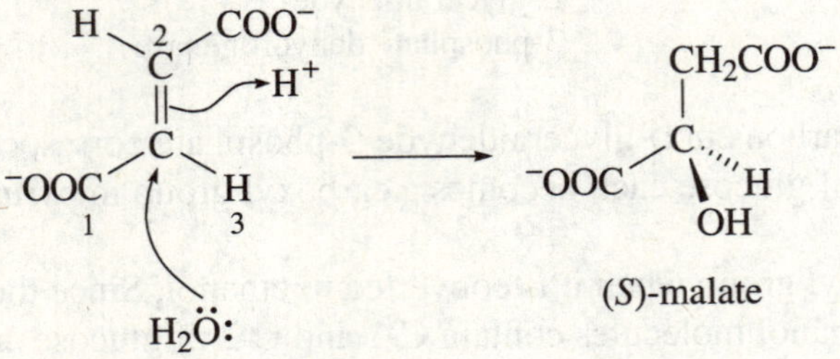

(*S*)-malate

39. The conversion of propionyl-CoA to methylmalonyl-CoA requires biotin.
The conversion of methylmalonyl-CoA to succinyl-CoA requires coenzyme B_{12}.

40. In Problem 29 we saw how the carbons in D-glyceraldehyde-3-phosphate correspond to the carbons in pyruvate.

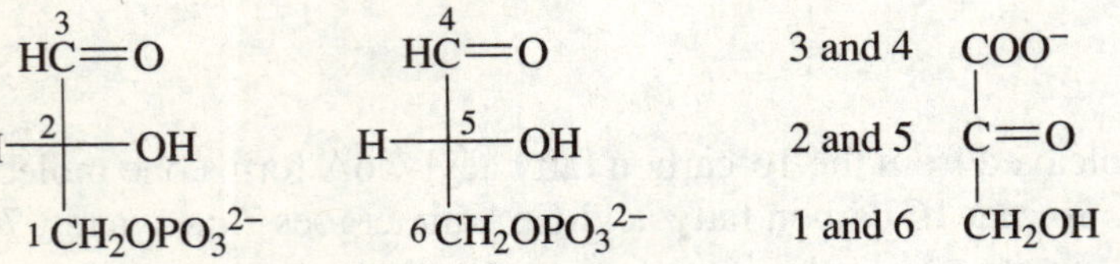

D-glyceraldehyde-3-phosphate dehydrogenase

pyruvate

Now we can answer the questions. The label in pyruvate is indicated by a *.

a. 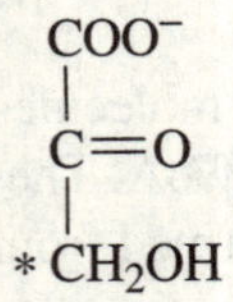

c. 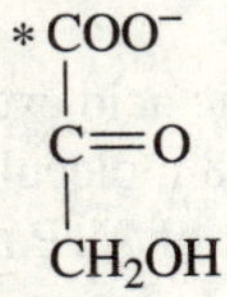

e. 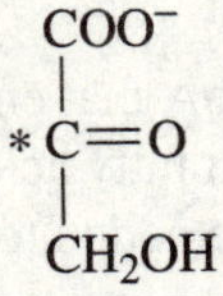

b. 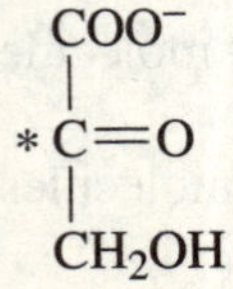

d. 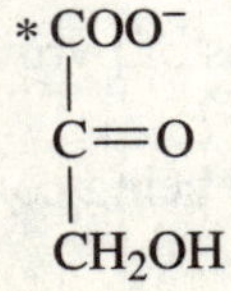

f. 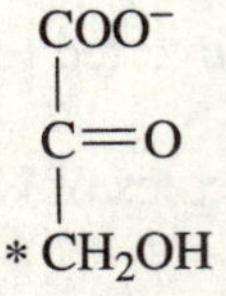

41. A Claisen condensation between two molecules of acetyl-coA forms acetoacetyl-CoA, which when hydrolyzed forms acetoacetate.

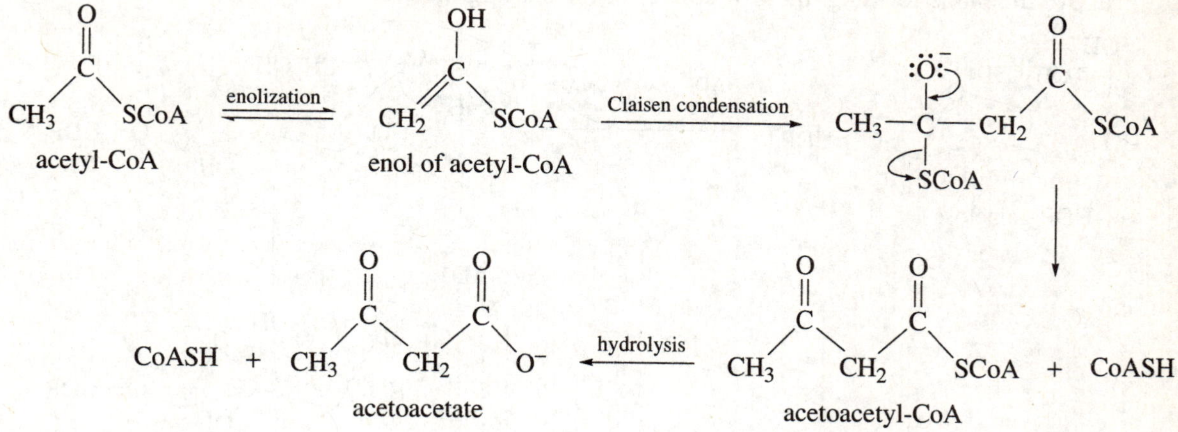

Acetoacetate can undergo decarboxylation to form acetone or it can be reduced to 3-hydroxybutyrate.

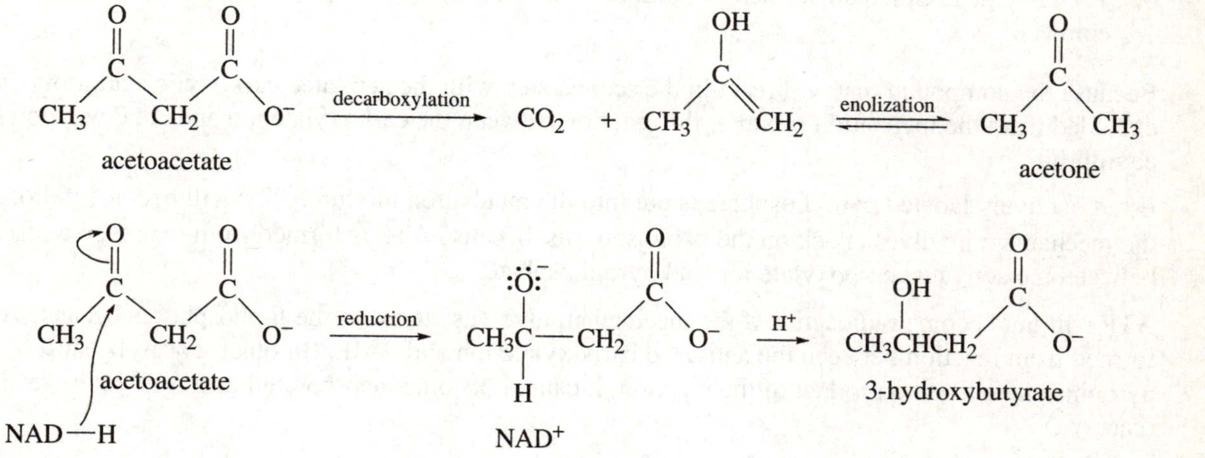

42.

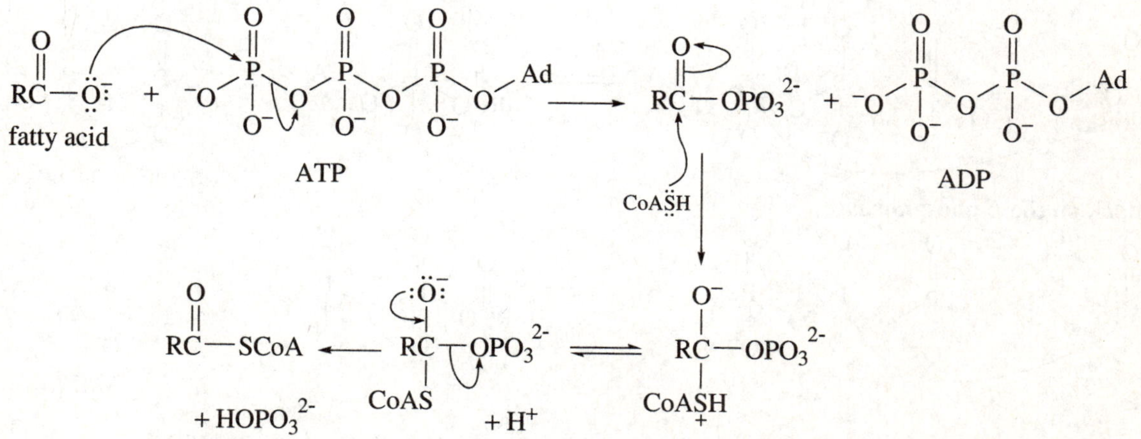

43. **a.** UDP-Galactose and UDP-glucose are C-4 epimers. NAD^+ oxidizes the C-4 OH group of UDP-galactose to a ketone. When NADH reduces the ketone back to an OH, it attacks the sp^2 carbon from above the plane forming, the C-4 epimer of the starting material.

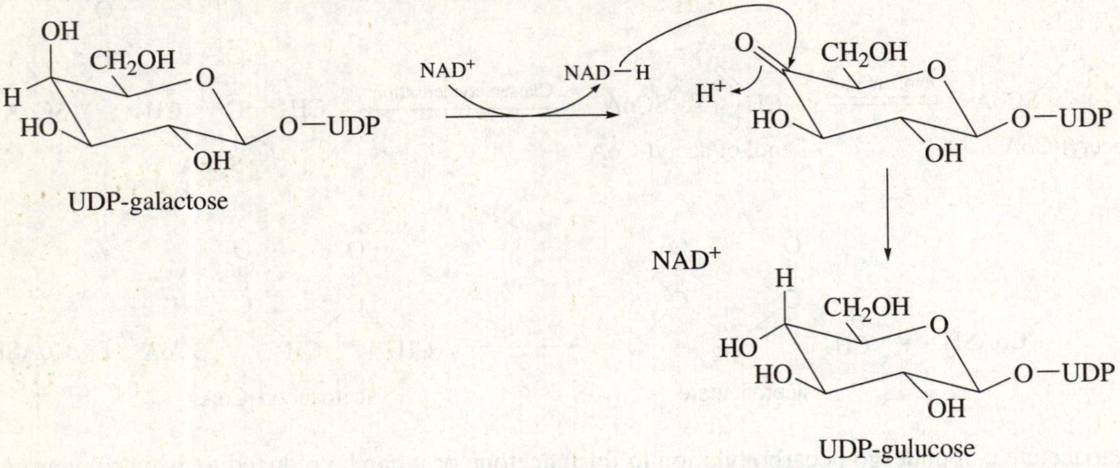

b. The enzyme is called an epimerase because it converts a compound into an epimer (in this case a C-4 epimer).

44. Because the compound that will react in the second step with the activated carboxylic acid group is excluded from the incubation mixture, the reaction between the carboxylate ion and ATP will come to equilibrium.

If radioactively labeled pyrophosphate is put into the incubation mixture, ATP will become radioactive if the mechanism involves attack on the α-phosphorus, because ATP is formed from the reverse reaction between the activated carboxylate ion and pyrophosphate.

ATP will not become radioactive if the mechanism involves attack on the β-phosphorus because ATP is formed from reaction between the activated carboxylate ion and AMP. (In other words, because pyrophosphate is not a product of the reaction, it cannot become incorporated into ATP in the reverse reaction.)

attack on the α-phosphorus

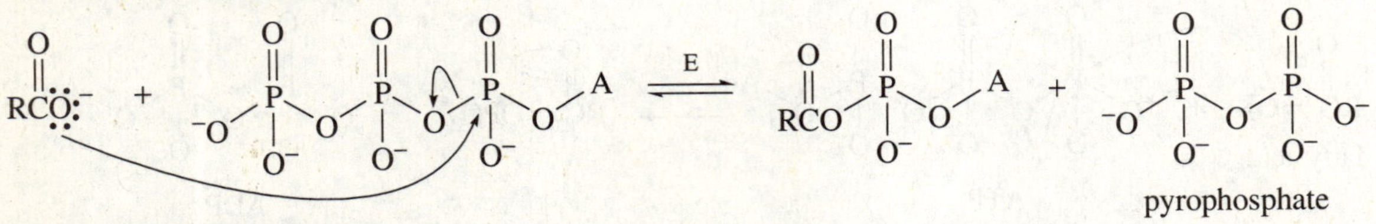

attack on the β-phosphorus

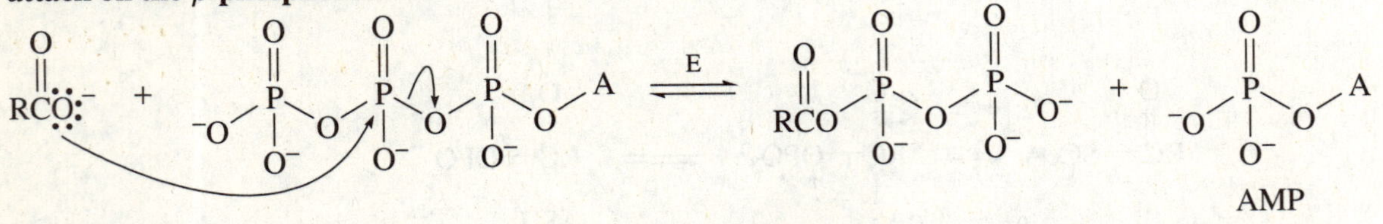

45. If radioactive AMP is added to the reaction mixture, the results will be opposite. If the mechanism involves attack on the α-phosphorus, ATP will not become radioactive, because AMP is not a product when attack occurs on the α–phosphorus. If the mechanism involves attack on the β-phosphorus, ATP will become radioactive, because AMP is a product when attack occurs on the β–phosphorus.

Chapter 25 Practice Test

1. Give a structure for each of the following:

   a. the intermediate formed when a nucleophile (RO^-) attacks the β–phosphorus of ATP

   b. an acyl pyrophosphate

   c. pyrophosphate

2. Fill in the six blanks in the following scheme.

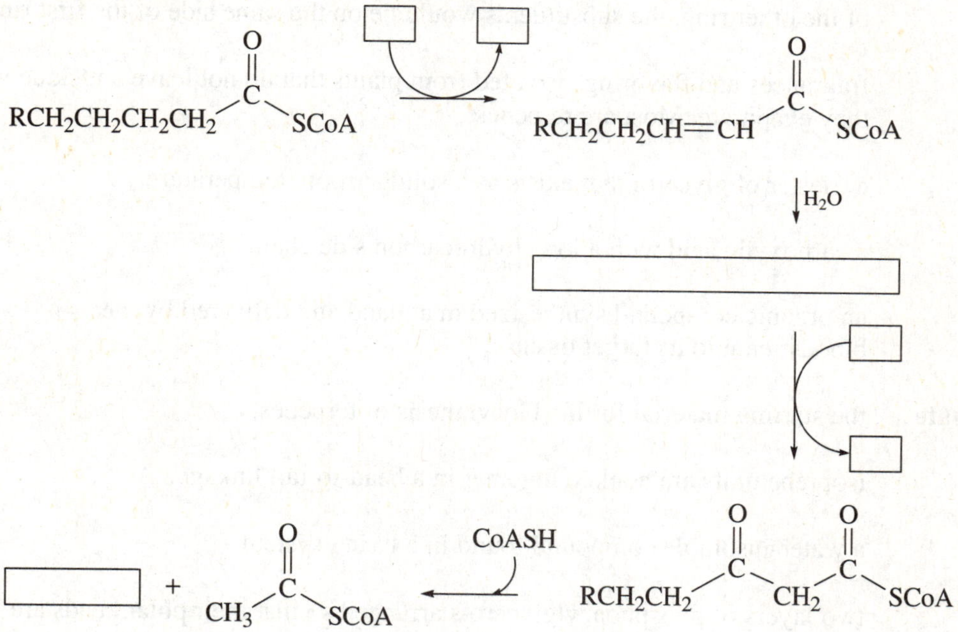

3. Which of the following are not citric acid cycle intermediates: fumarate, acetate, citrate?

4. Which provide energy to the cell, anabolic reactions or catabolic reactions?

5. What compounds are formed when proteins undergo the first stage of catabolism?

6. What compound is formed when fatty acids undergo the second stage of catabolism?

7. Indicate whether each of the following statements is true or false:

a. Each molecule of $FADH_2$ forms 3 molecules of ATP in the fourth stage of catabolism.	T	F
b. $FADH_2$ is oxidized to FAD.	T	F
c. NAD^+ is oxidized to NADH.	T	F
d. Acetyl-CoA is a citric acid cycle intermediate.	T	F

CHAPTER 26
Lipids

Important Terms

angular methyl group	a methyl substituent at the 10- or 13-position of a steroid ring system.
carotenoid	a class of compound (a tetraterpene) responsible for the red and orange colors of fruits, vegetables, and fall leaves.
cholesterol	a steroid that is the precursor of all other steroids.
cis fused	two rings fused together such that if one ring is considered to be two substituents of the other ring, the substituents would be on the same side of the first ring.
essential oil	fragrances and flavorings isolated from plants that do not leave a residue when they evaporate. Most are terpenes.
fat	a triester of glycerol that exists as a solid at room temperature.
fatty acid	a carboxylic acid with a long hydrocarbon side chain.
hormone	an organic compound synthesized in a gland and delivered by the bloodstream to its target tissue.
isopentenyl pyrophosphate	the starting material for the biosynthesis of terpenes.
isoprene rule	isoprene units are hooked together in a head-to-tail linkage.
lipid	a water-insoluble compound found in a living system.
lipid bilayer	two layers of phosphoacylglycerols arranged so that their polar heads are on the outside and their nonpolar fatty acid chains are on the inside.
membrane	the material the surrounds the cell in order to isolate its contents.
mixed triacylglycerol	a triacylglycerol in which the fatty acid components are different.
monoterpene	a terpene that contains 10 carbons.
oil	a triester of glycerol that exists as a liquid at room temperature.
phosphatidic acid	a phosphoacylglycerol in which only one of the OH groups of phosphate is in an ester linkage.
phosphoacylglycerol (phosphoglyceride)	formed when two OH groups of glycerol form esters with fatty acids and the terminal OH group is part of a phosphodiester.
phospholipid	a lipid that contains a phosphate group.
polyunsaturated fatty acid	a fatty acid with more than one double bond.

progestins a class of steroid hormones.

prostaglandin a carboxylic acid, derived from arachidonic acid, that is responsible for a variety of physiological functions.

sesquiterpene a terpene that contains 15 carbons.

simple triacylglycerol a triacylglycerol in which the fatty acid components are the same.

sphingolipid a lipid that contains sphingosine.

sphingomyelin a sphingolipid in which the primary OH group of sphingosine is bonded to a phosphocholine or to a phosphoethanolamine.

squalene a triterpene that is a precursor of steroid molecules.

steroid a class of compounds that contains a steroid ring system.

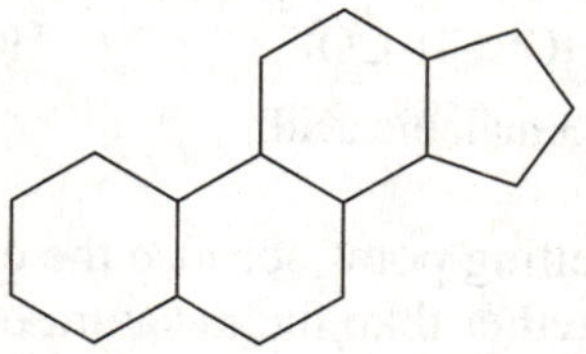

steroid ring system

α-substituent a substituent on the opposite side of a steroid ring system as the angular methyl groups.

β-substituent a substituent on the same side of a steroid ring system as the angular methyl groups.

terpene a lipid isolated from a plant that contains carbon atoms in multiples of five.

terpenoid a terpene that contains oxygen.

tetraterpene a terpene that contains 40 carbons.

trans fused two rings fused together such that if one ring is considered to be two substituents of the other ring, the substituents would be on opposite sides of the first ring.

triacylglycerol the compound formed when the three OH groups of glycerol are esterified with fatty acids.

triterpene a terpene that contains 30 carbons.

wax an ester formed from a long straight-chain carboxylic acid and a long straight-chain alcohol.

Solutions to Problems

1. **a.** Stearic acid has the higher melting point because it has two more methylene groups (giving it a greater surface area) than palmitic acid.

   b. Palmitic acid has the higher melting point because it does not have any carbon-carbon double bonds, whereas palmitoleic acid has a cis double bond that prevents the molecules from packing closely together.

   c. Oleic acid has the higher melting point because it has one double bond, while linoleic acid has two double bonds, which give greater interference to close packing of the molecules.

2. $CH_3CH_2CH_2CH_2CH_2CH{=}CHCH_2CH{=}CHCH_2CH{=}CHCH_2CH{=}CHCH_2CH_2CH_2COOH$

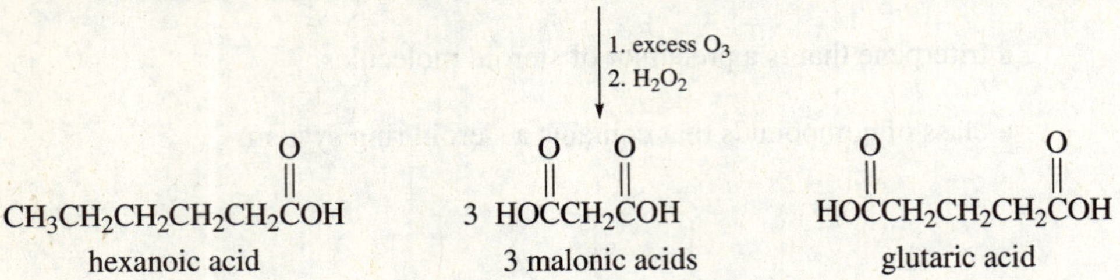

3. Glyceryl tripalmitate has a higher melting point because the carboxylic acid components are saturated and can, therefore, pack more closely together than the unsaturated carboxylic acid components of glyceryl tripalmitoleate.

4. To be optically inactive, the fat must have a plane of symmetry. In other words, the fatty acids at C-1 and C-3 must be identical. Therefore, stearic acid must be at C-1 and C-3.

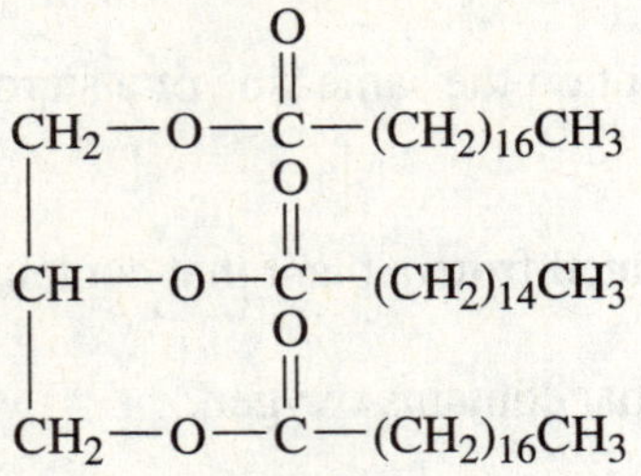

5. To be optically active, the fat must not have a plane of symmetry. Therefore, stearic acid must be at C-1 and C-2 or at C-2 and C-3.

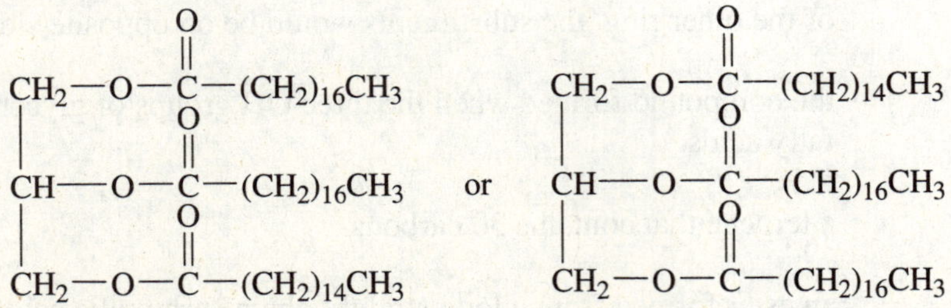

6. Solved in the text.

7. 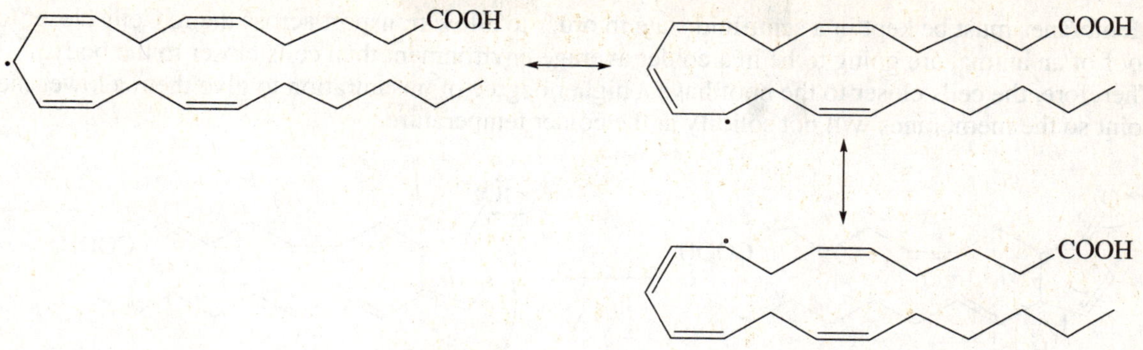

8. No, the identities of R^1 and R^2 have no affect on the configuration of the asymmetric center at C-2 because:

   the group with priority #1 will always be the carboxyl group on C-2.
   the group with priority #2 will always be the group that contains the phosphorus.
   the group with priority #3 will always be the group that contains the other carboxyl group.
   the group with priority #4 will always be the H on C-2.

9. Because the interior of a membrane is nonpolar and the surface of a membrane is polar, integral proteins will have a higher percentage of nonpolar amino acids.

10. The bacteria could synthesize phosphoacylglycerols with more saturated fatty acids because these triacylglycerols would pack more tightly in the lipid bilayer and, therefore, would have higher melting points and be less fluid.

11. a. The sphingomyelins can differ in the fatty acid component of the amide and have either choline or ethanolamine attached to the phosphate group.

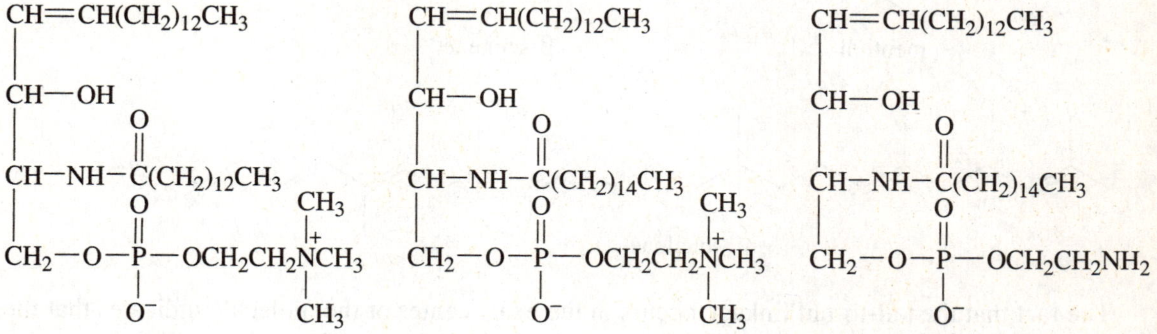

b. 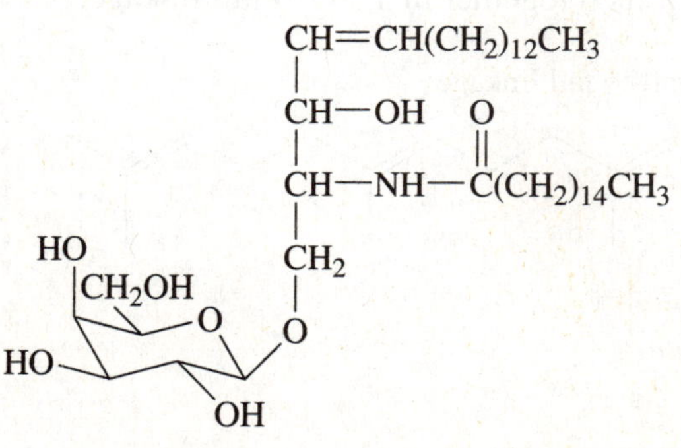

12. Membranes must be kept in a semifluid state in order to allow transport across them. Cells closer to the hoof of an animal are going to be in a colder average environment than cells closer to the body. Therefore, the cells closer to the hoof have a higher degree of unsaturation to give them a lower melting point so the membranes will not solidify at the colder temperature.

13.

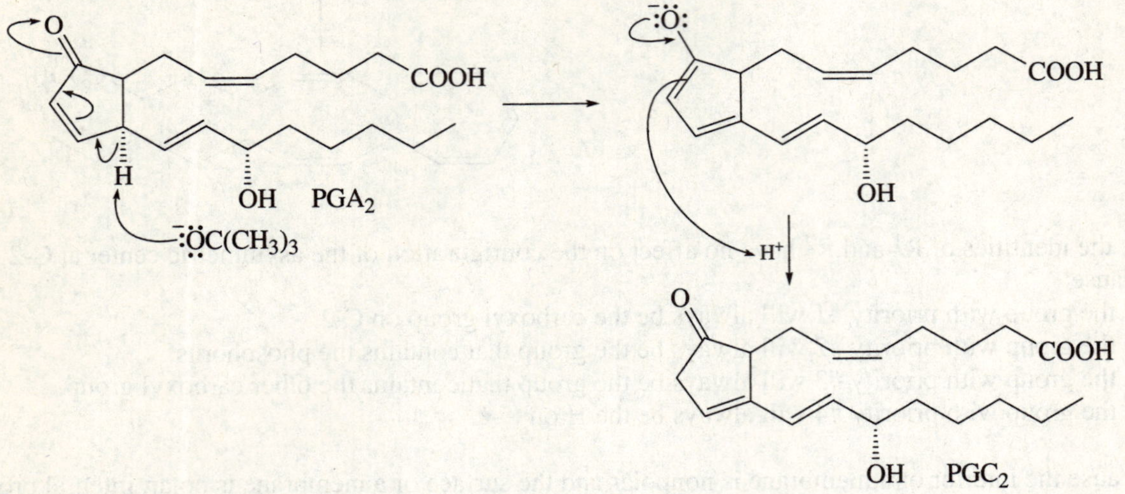

14. (+)-Limonene is found in oranges.

15.

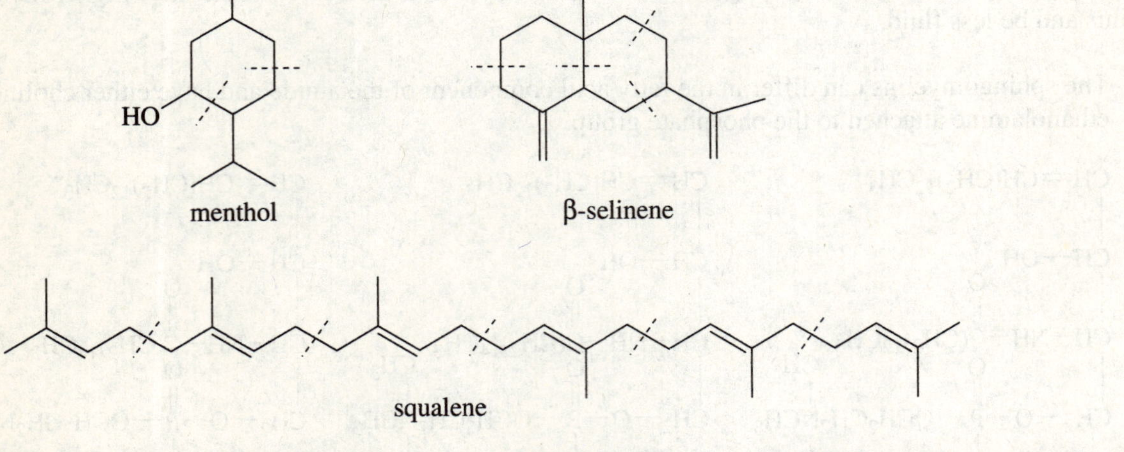

16. The fact that the tail-to-tail linkage occurs in the exact center of the molecule indicates that the two halves are synthesized (in a head-to-tail fashion) and then joined together in a tail-to-tail linkage.

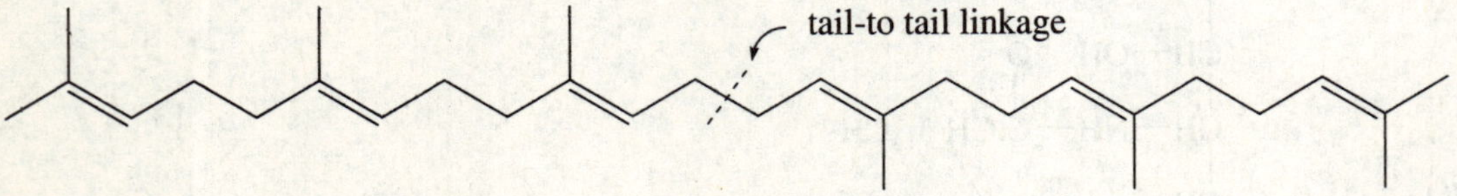

17. Squalene, lycopene, and β-carotene are all synthesized in the same way. In each case, two halves are synthesized (in a head-to-tail fashion) and then joined together in a tail-to-tail linkage.

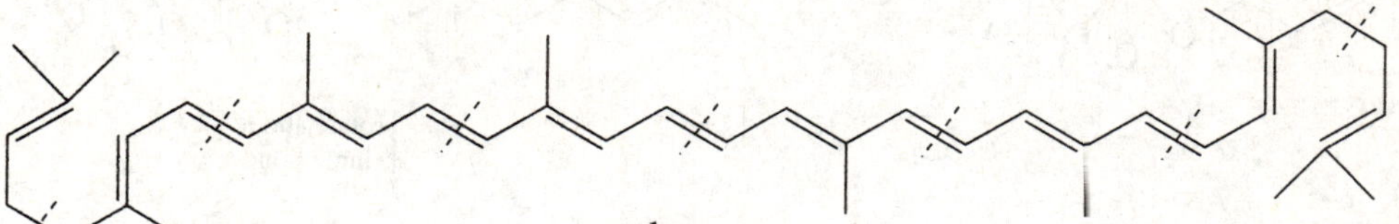

lycopene

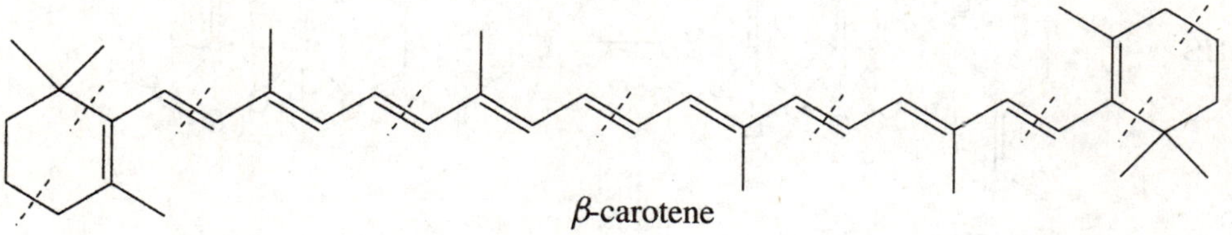

β-carotene

18. Solved in the text.

19.

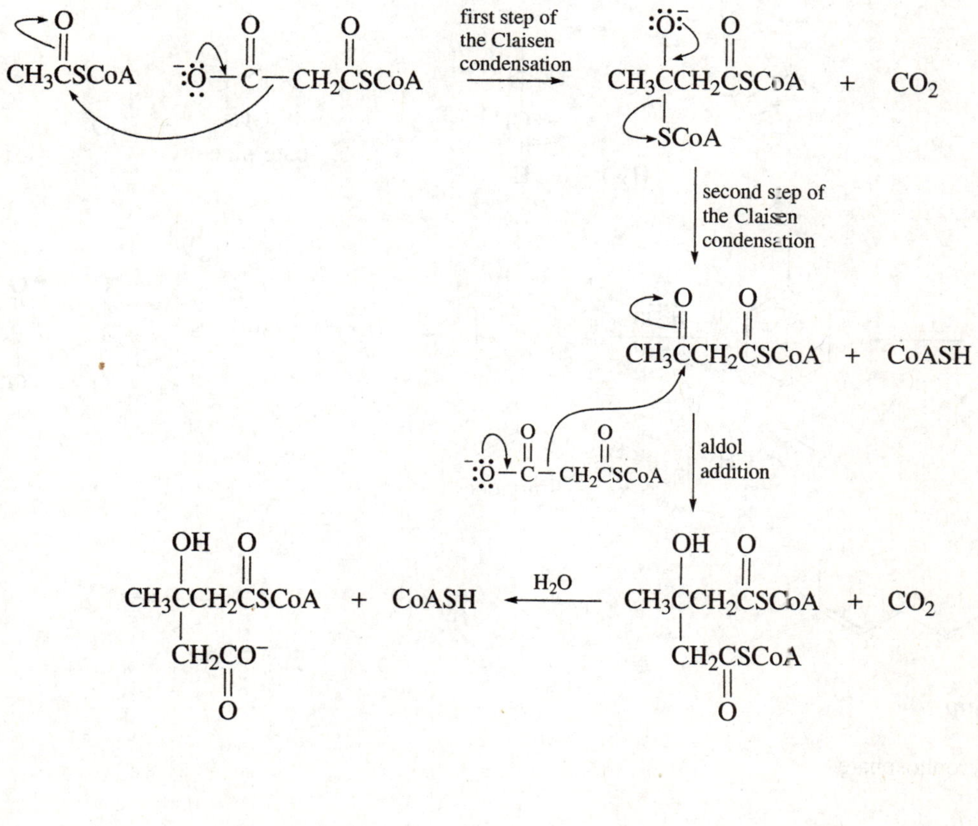

20.

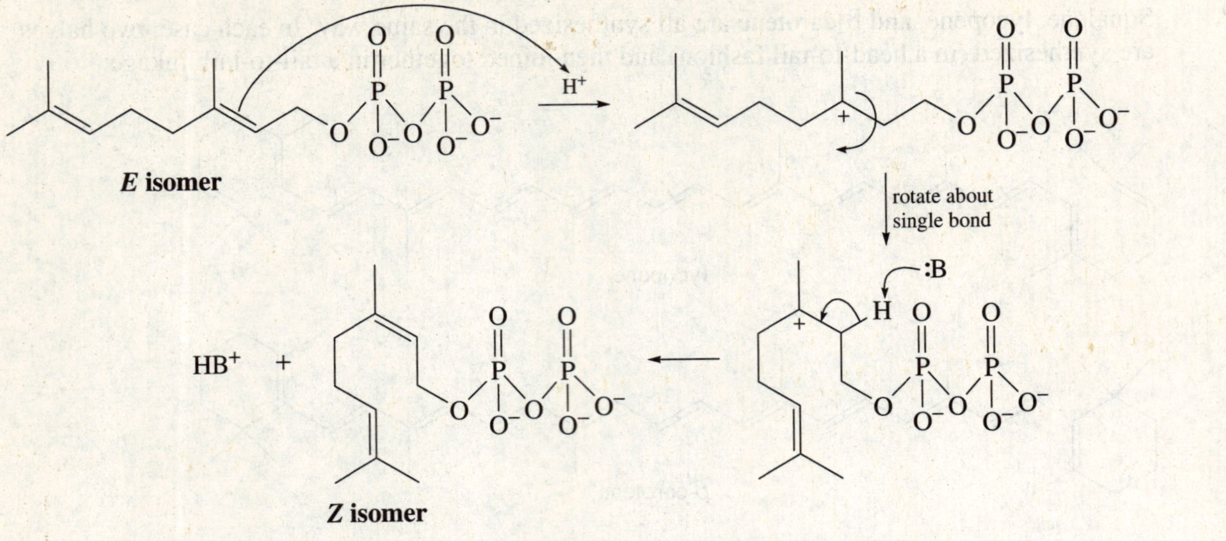

E isomer **Z isomer**

21.

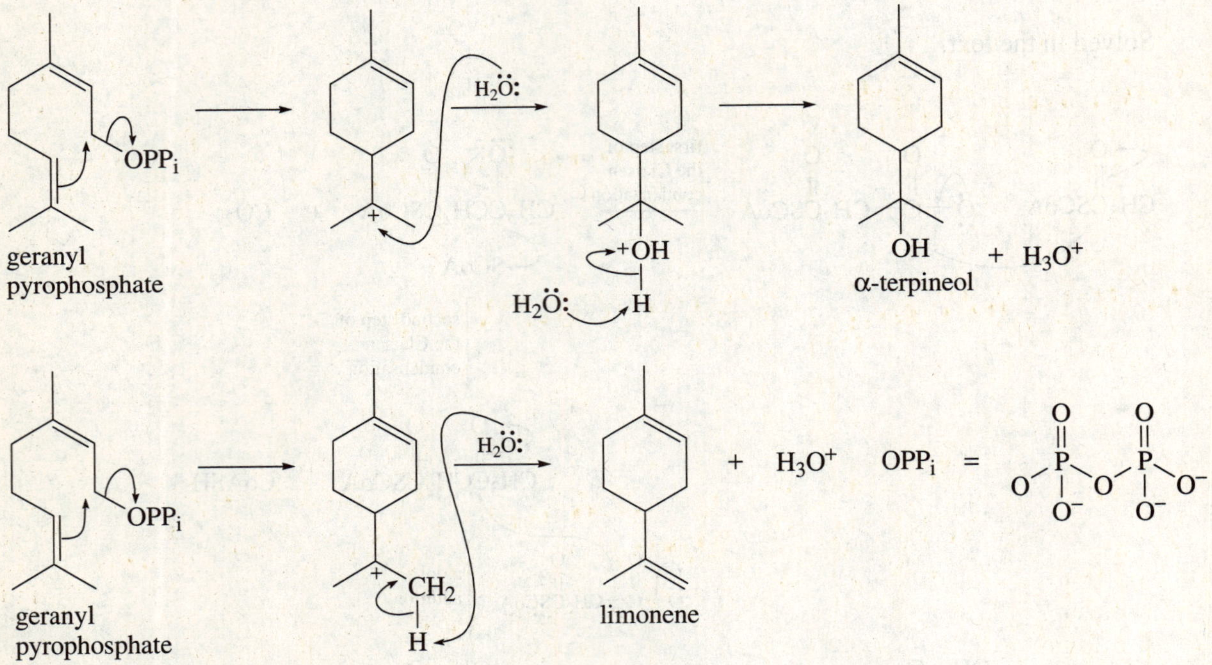

22.

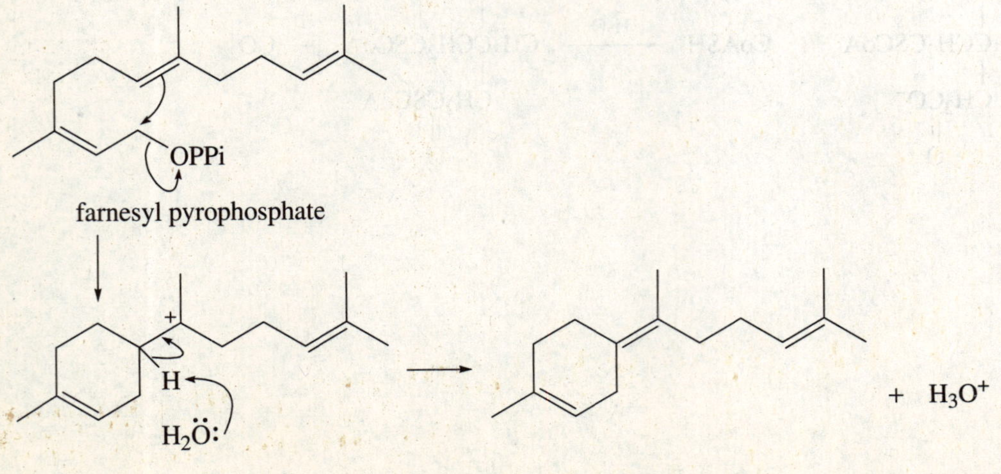

23. Solved in the text.

24. A **β-hydrogen** at C-5 means that the A and B rings are **cis** fused, whereas an **α-hydrogen** at C-5 means that the A and B rings are **trans** fused.

25. Because the OH substituent is on the **same side** of the steroid ring system as the angular methyl groups, it is a **β-substituent**.

26. The hemiacetal is formed by reaction of the primary alcohol with the aldehyde.

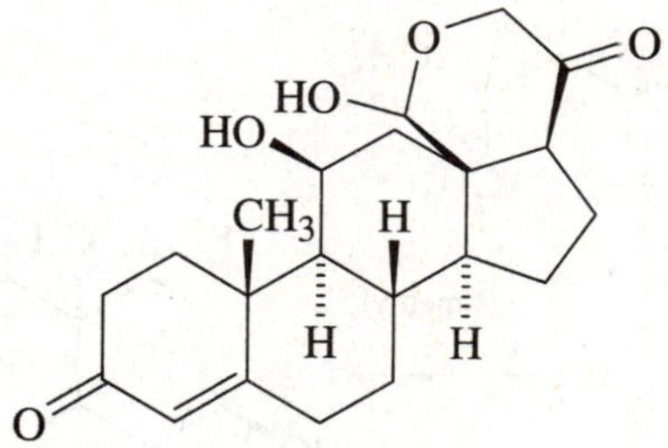

27.

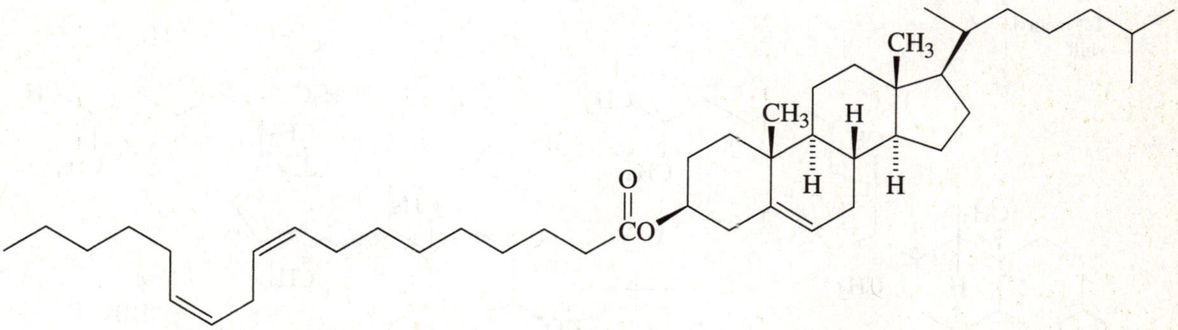

28. Notice that since the methyl and hydrogen at the juncture of the A and B rings are on the same side, we know that the A and B rings are cis fused. The three OH groups are all in axial positions.

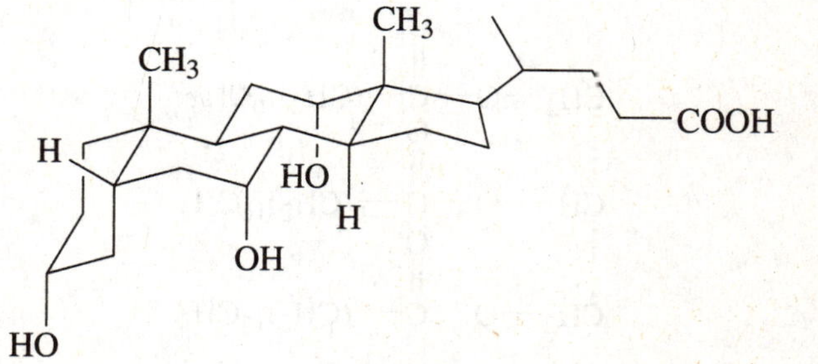

29. There are two hydride shifts and two methyl shifts. The last step is elimination of a proton.

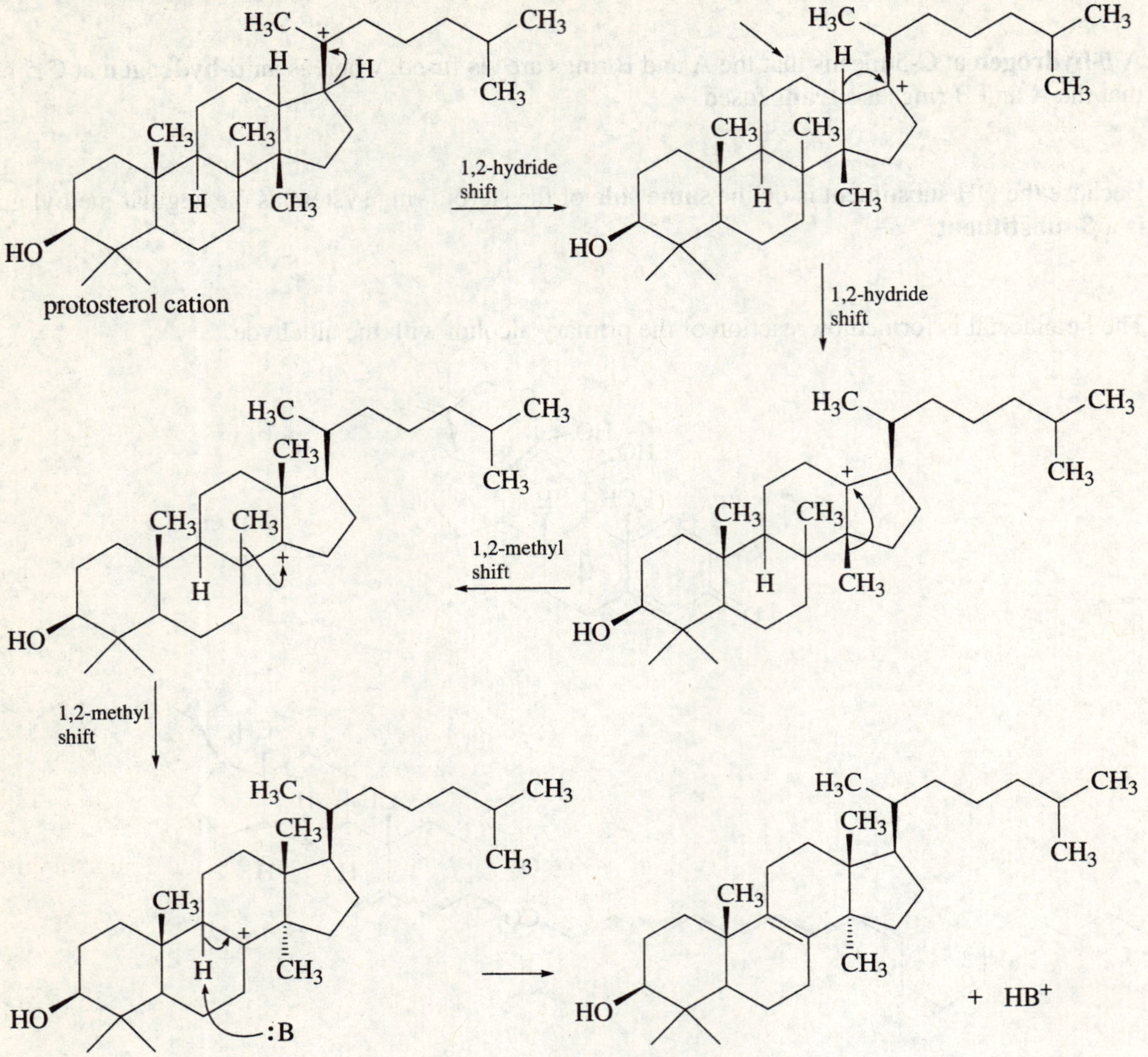

30. If stearic acid were at C-1 and C-3, the fat would not be optically active. Therefore, stearic acid must be at C-1 and C-2 (or C-2 and C-3).

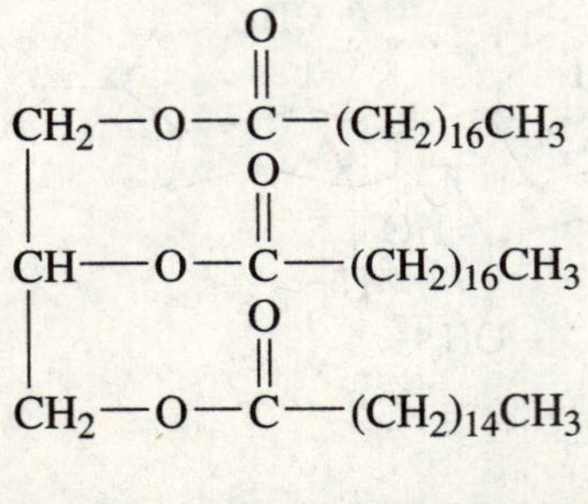

31. All triacylglycerols do not have the same number of asymmetric centers. If the carboxylic acid components at C-1 and C-3 of glycerol are not identical, the triacylglycerol has one asymmetric center (C-2). If the carboxylic acid components at C-1 and C-3 of glycerol are identical, the triacylglycerol has no asymmetric centers.

32. **a.** There are three triacylglycerols in which one of the fatty acid components is lauric acid and two are myristic acid. Myristic acid can be at C-1 and C-3 of glycerol, in which case the triacylglycerol does not have any asymmetric centers. If myristic acid is at C-1 and C-2 of glycerol, C-2 is an asymmetric center, and consequently, the compound can have a pair of enantiomers.

b. There are six triacylglycerols in which one of the fatty acid components is lauric acid, one is myristic acid, and one is palmitic acid. The three possible arrangements are shown below (with the fatty acid components abbreviated as L, M, and P). Since each has an asymmetric center, each can exist as a pair of enantiomers for a total of six triacylglycerols.

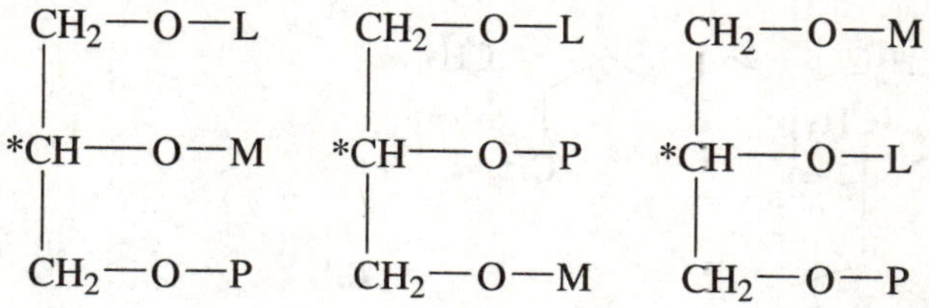

33. 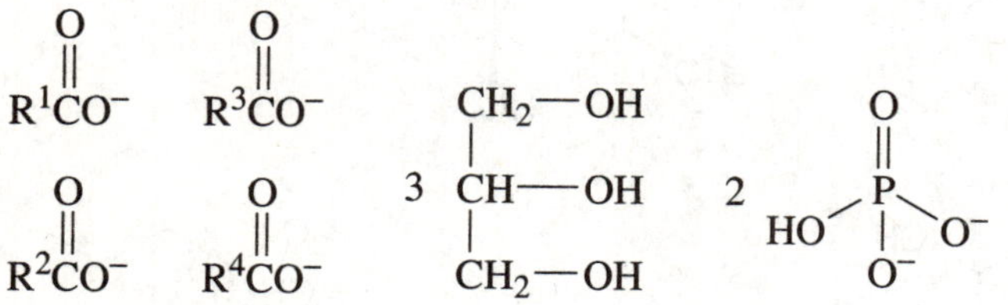

34. 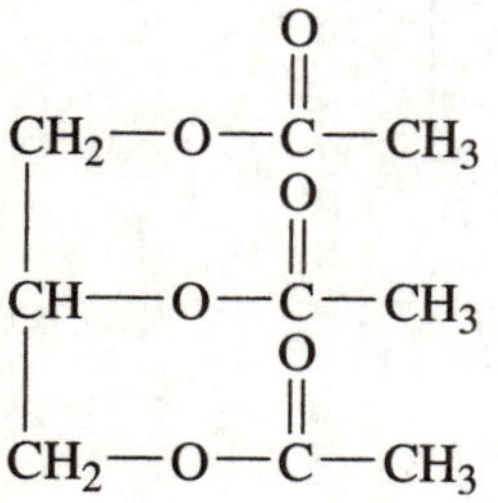

The structure at left has a molecular formula = $C_9H_{14}O_6$ and a molecular weight = 218.

Subtracting 218 from the total molecular weight gives the molecular weight of the methylene (CH_2) groups in the triacylglycerol.

$$722 - 218 = 504$$

Dividing 504 by the molecular weight of a methylene group (14) will give the number of methylene groups.

$$\frac{504}{14} = 36$$

Since there are 36 methylene groups, each fatty acid in the triacylglycerol has 12 methylene groups.

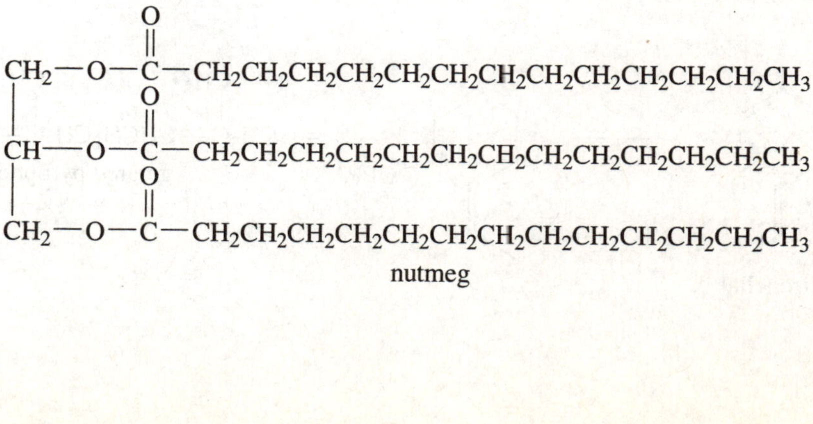

nutmeg

35.

a. 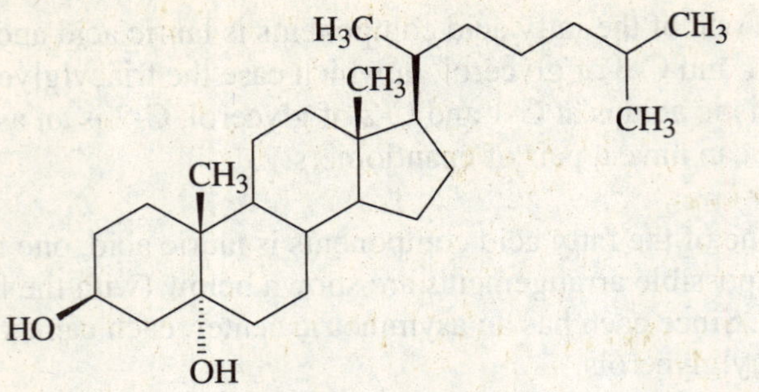

b. 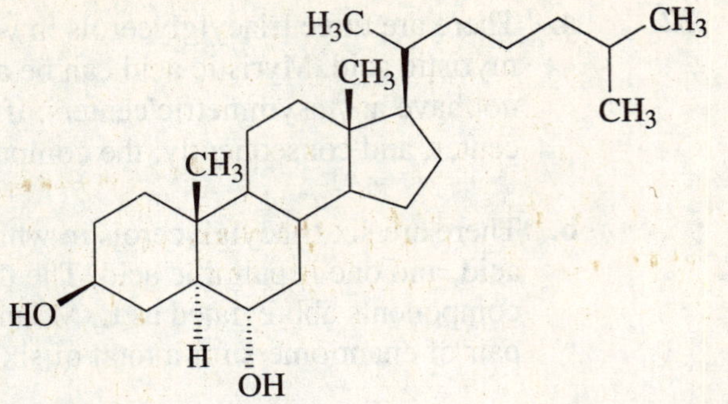

c. 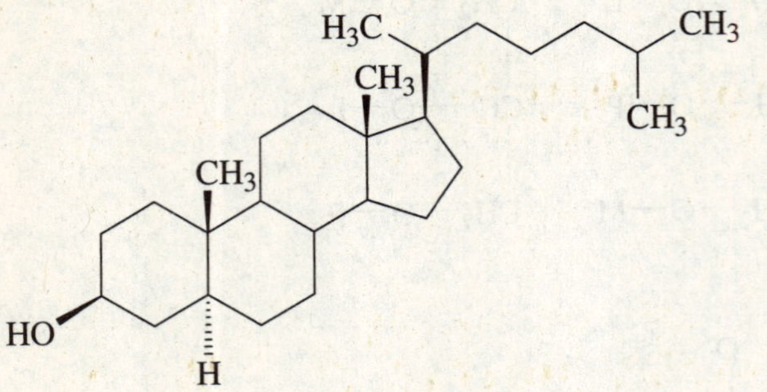

d. 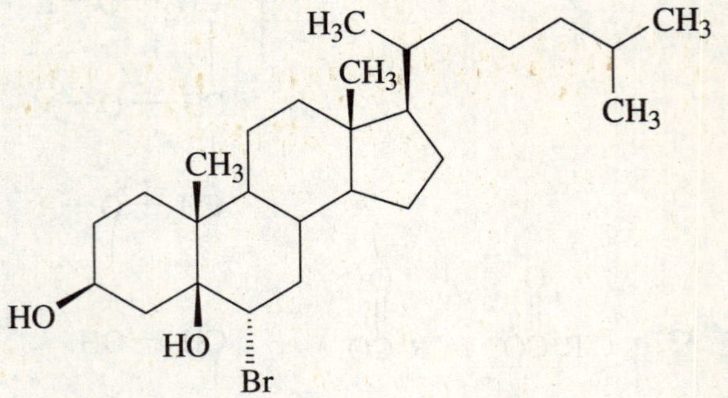

e. 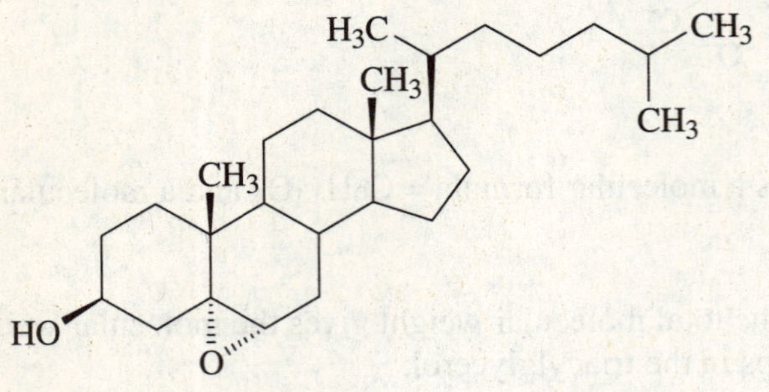

f. 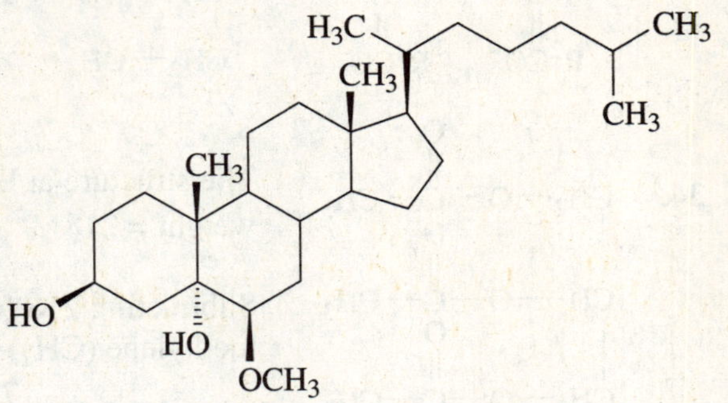

36. **a.** **Starting with mevalonyl pyrophosphate, you can trace the location of the label in the compounds that lead to the formation of geranyl pyrophosphate, and geranyl pyrophosphate is converted to citronellal.**

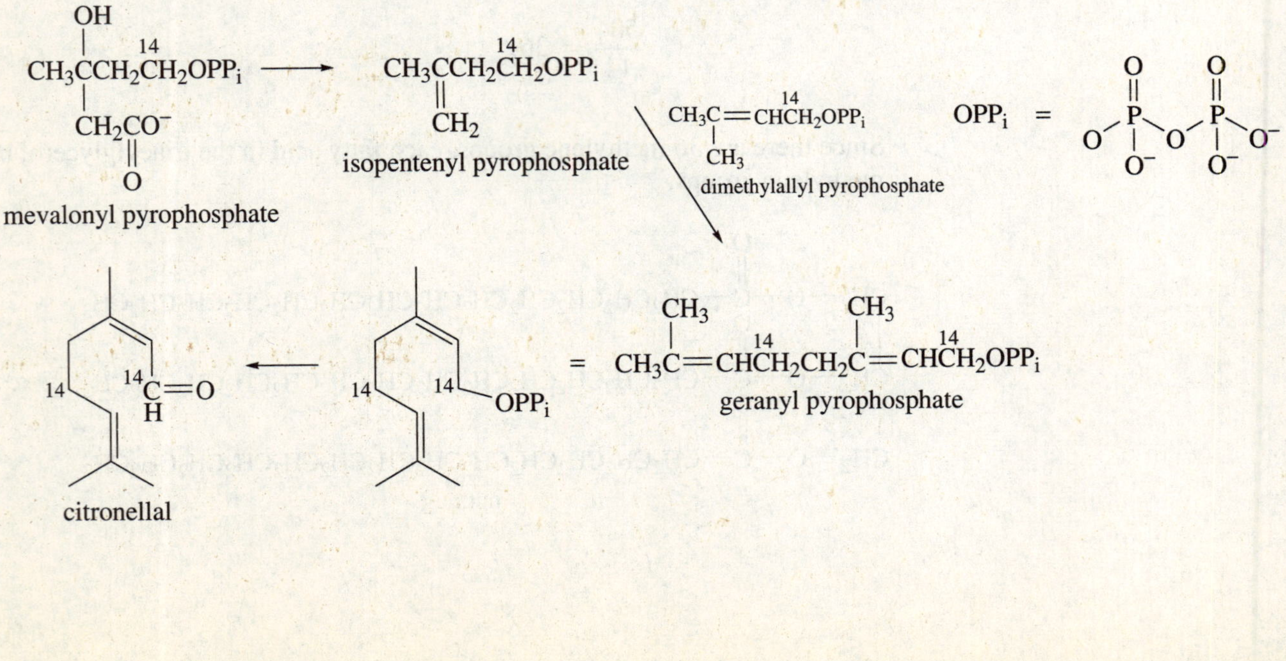

b. The label is lost from sample B when mevalonic pyrophosphate loses CO_2 to form isopentenyl pyrophosphate, so none of the carbons will be labeled in citronellal.

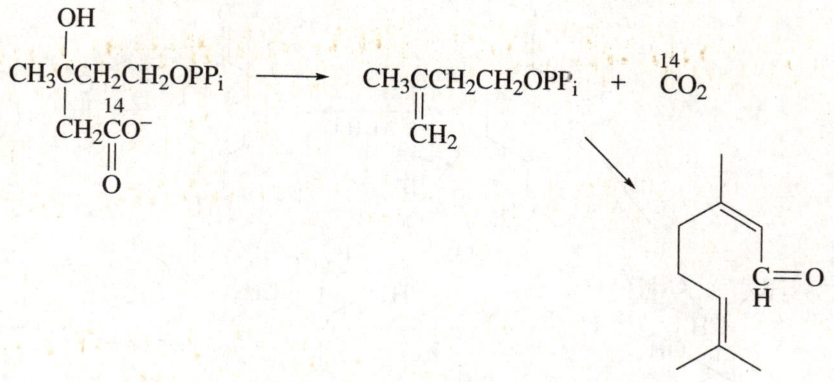

c. Because the methyl groups are equivalent in the carbocation that is formed as an intermediate when isopentenyl pyrophosphate is converted to dimethylallyl pyrophosphate, either of the methyl groups can be labeled in dimethylallyl pyrophosphate. This means that either of the two methyl groups can be labeled in geranyl pyrophosphate and in citronellal.

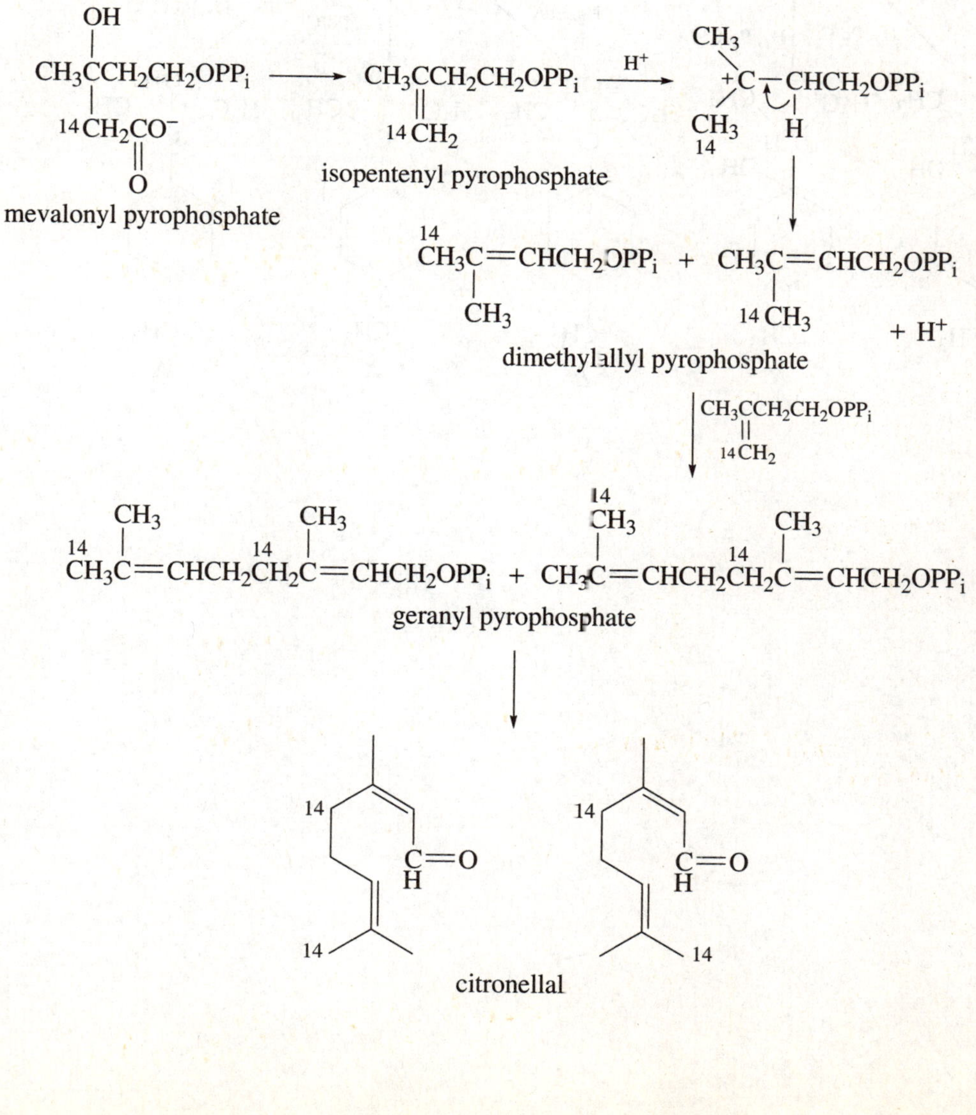

37. There are five possible structures for compound A, and two possible structures for compound B.

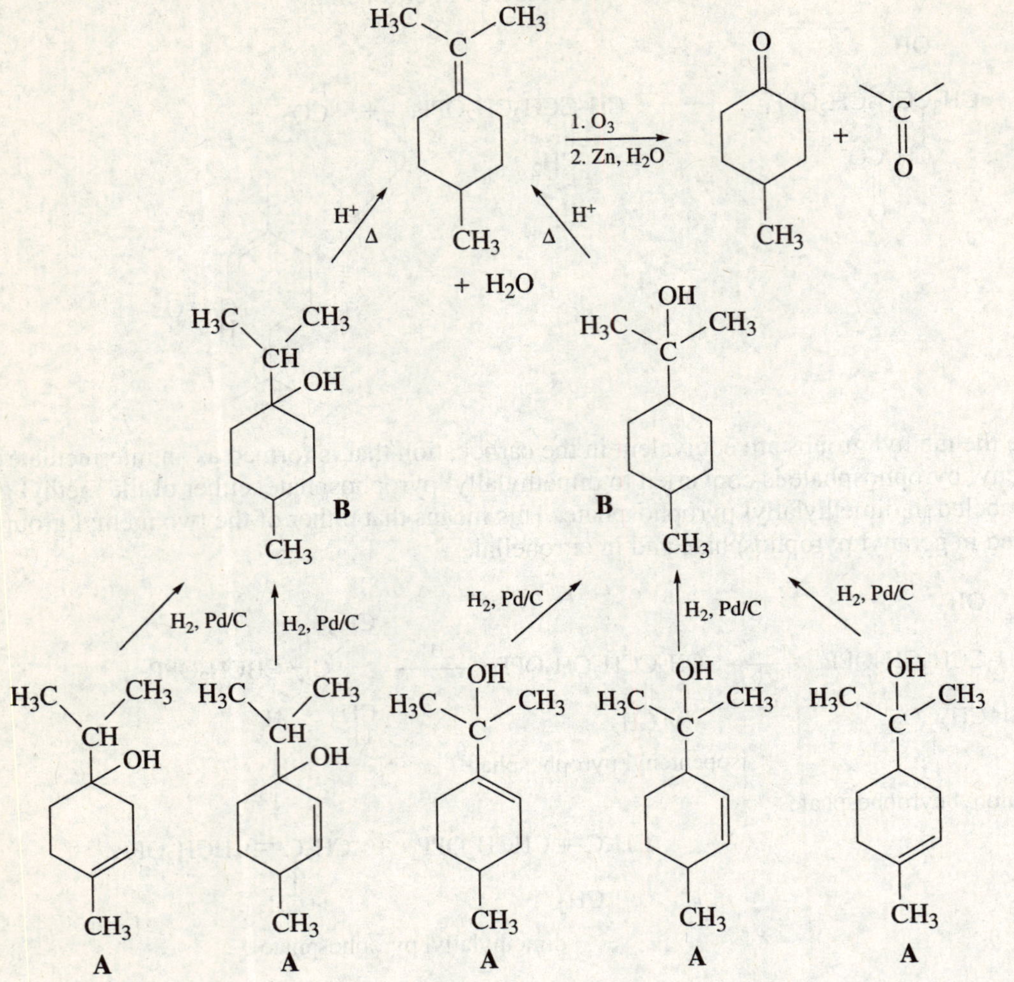

38. The mechanism for the conversion of isopentenyl pyrophosphate to geranyl pyrophospate to farnesyl pyrophosphate is shown in the text on pages 1182-1184. Now we need to show how farnesyl pyrophosphate is converted to eudesmol.

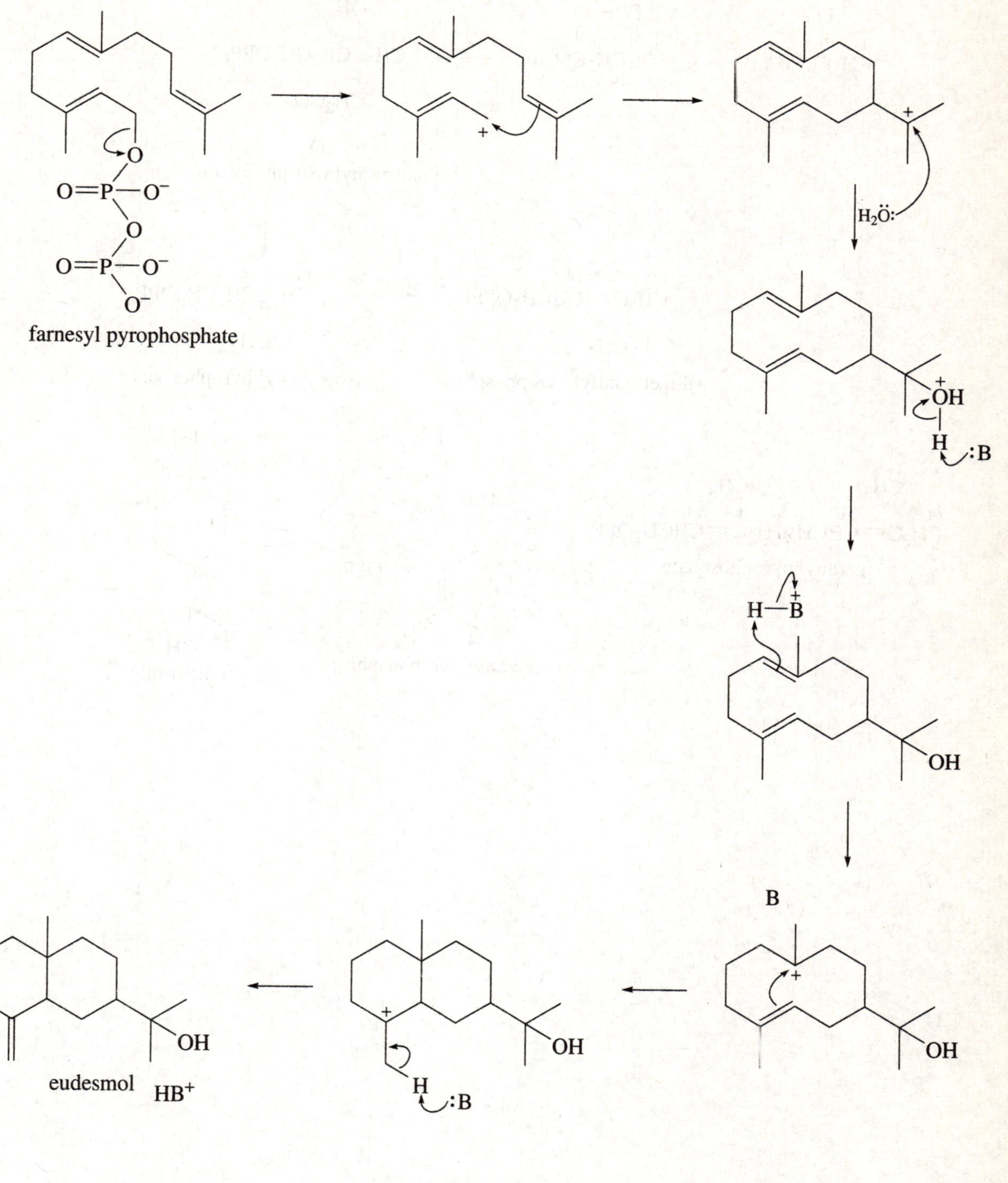

39. Because acetyl-CoA is converted into malonyl-CoA (see Section 26.8), mevalonyl pyrophosphate will contain three labeled carbons, which means that juniper oil will contain six labeled carbons.

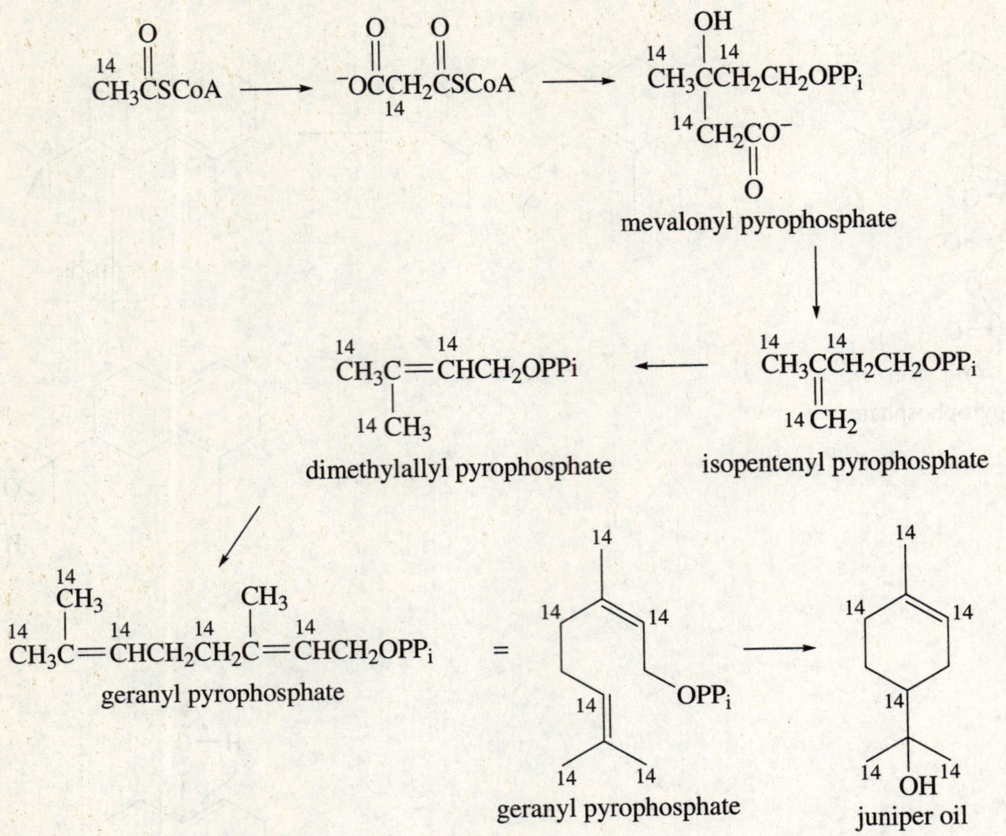

40. **a.**

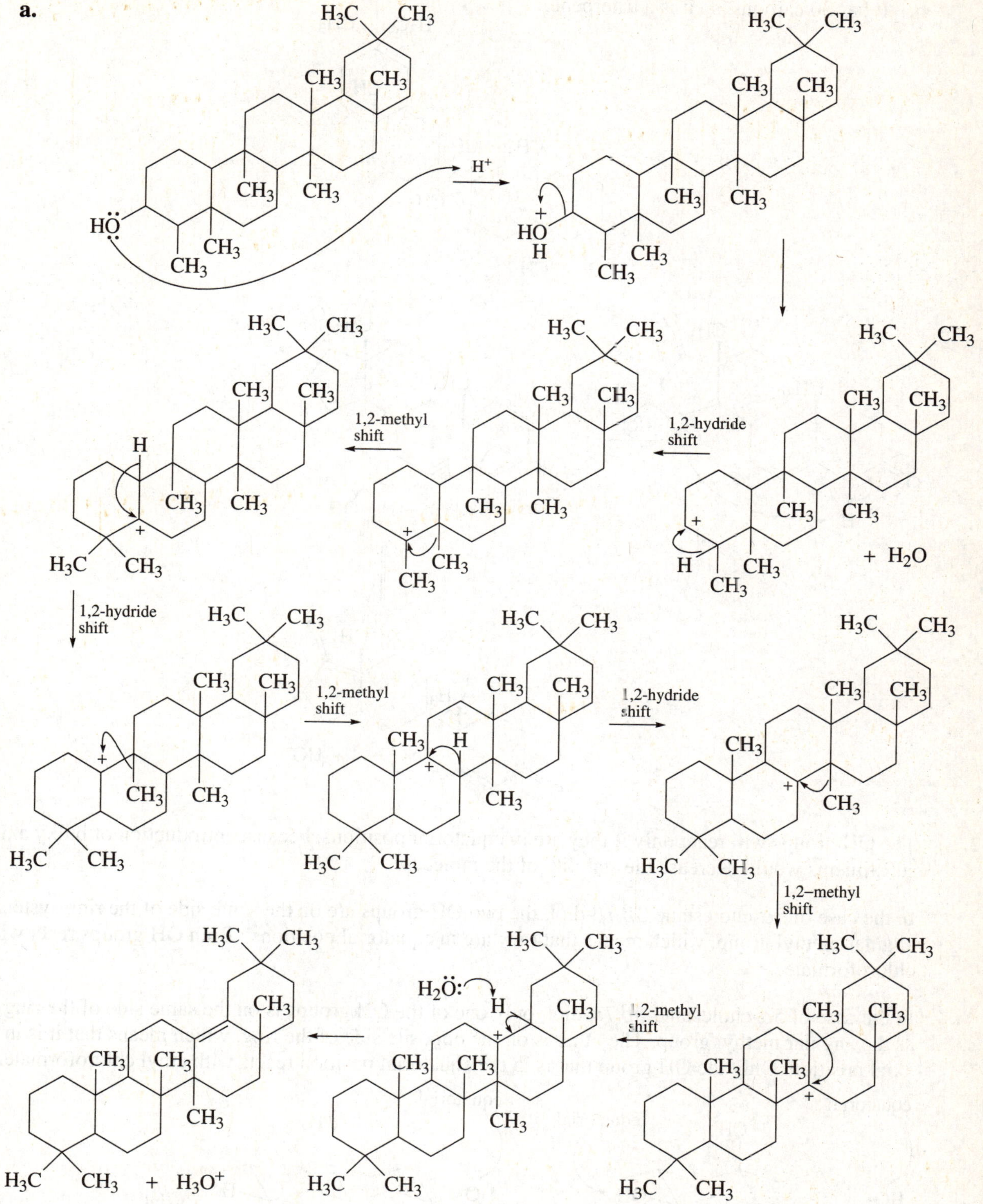
H⁺
1,2-hydride shift
1,2-methyl shift
+ H₂O
1,2-hydride shift
1,2-methyl shift
1,2-hydride shift
1,2-methyl shift
1,2-methyl shift
H₂Ö:
+ H₃O⁺

b. It has 30 carbons, so it is a triterpene.

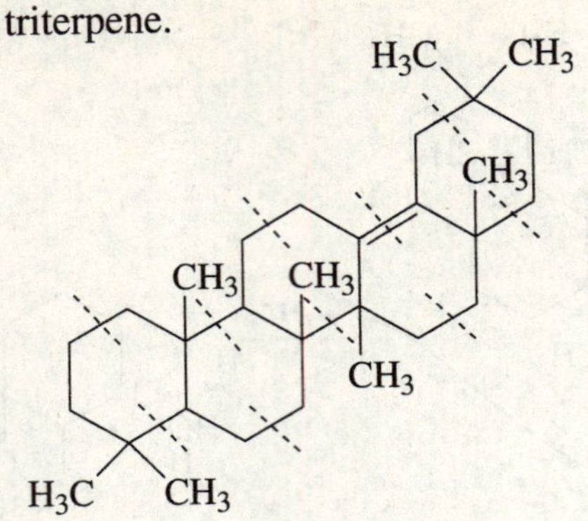

41.

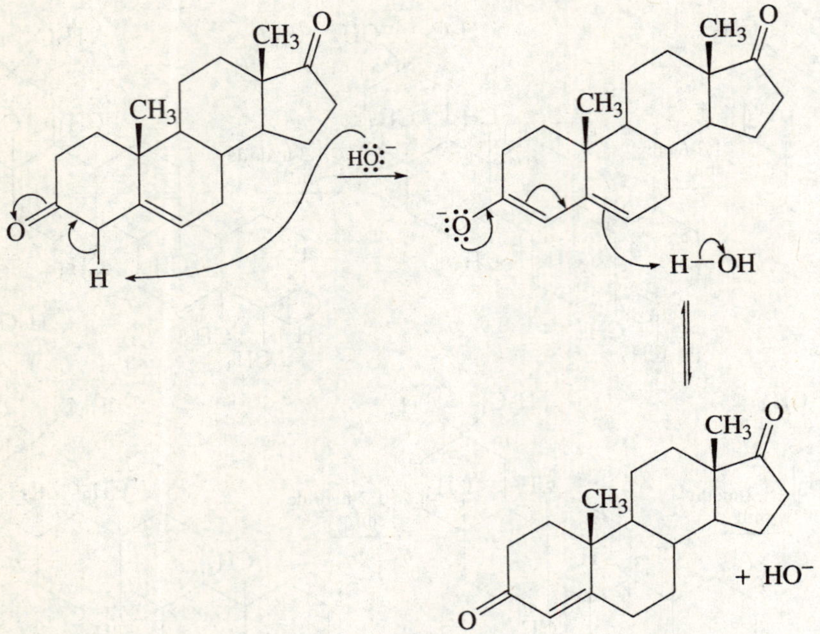

42. The OH groups will react only if they are in equatorial positions, because introduction of bulky axial substituents would decrease the stability of the molecule.

In the case of 5α-cholestane-3β,7β-diol, the two OH groups are on the same side of the ring system as the angular methyl group, which means that they are in equatorial positions. Both OH groups react with ethyl chloroformate.

In the case of 5α-cholestane-3β,7α-diol, only one of the OH groups is on the same side of the ring system as the angular methyl group. The other is on the opposite side of the ring, which means that it is in an axial position. Only the OH group that is in the equatorial position reacts with ethyl chloroformate.

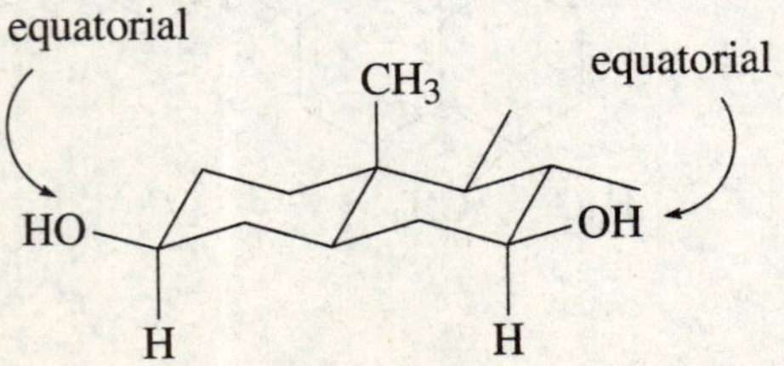

5α-cholestane-3β,7β-diol

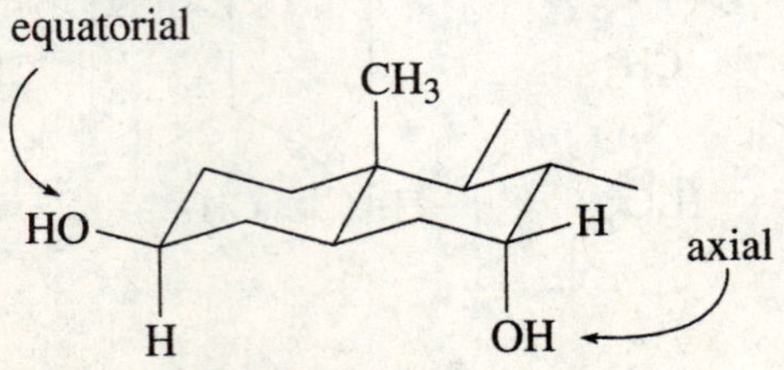

5α-cholestane-3β,7α-diol

43.

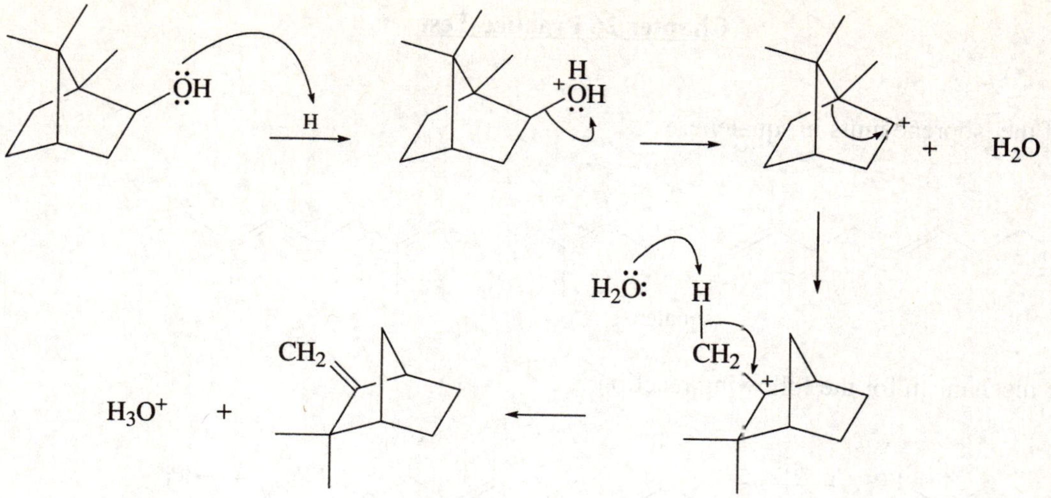

44.

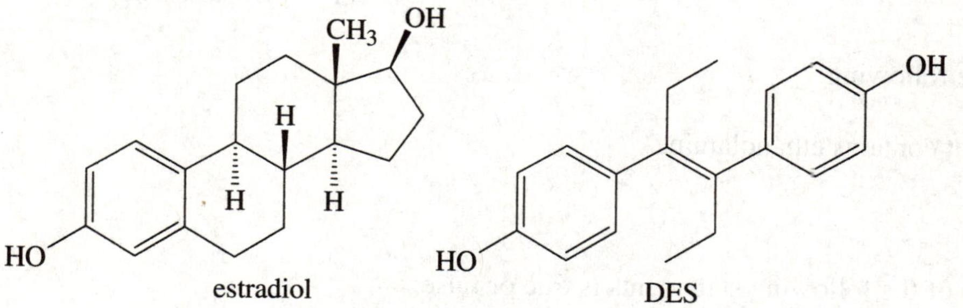

45. **a.** They are both sesquiterpenes.

b.

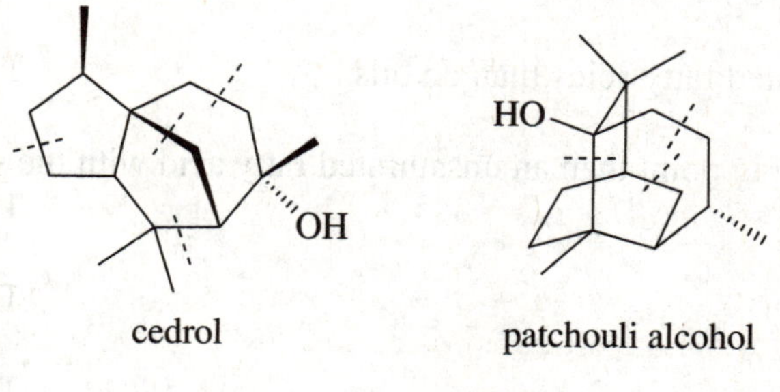

Chapter 26 Practice Test

1. Mark off the isoprene units in squalene.

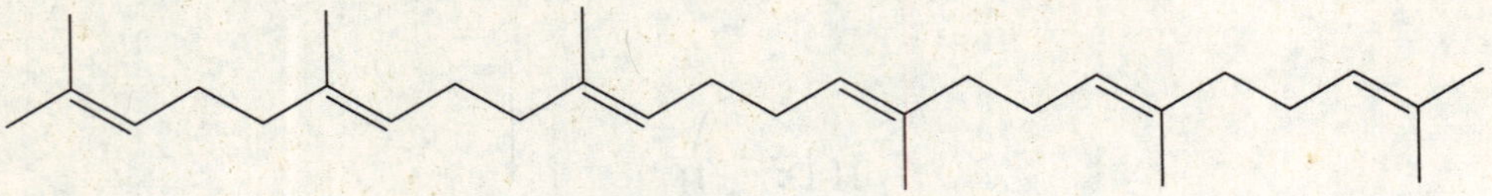

squalene

2. Give the mechanism for the following reaction:

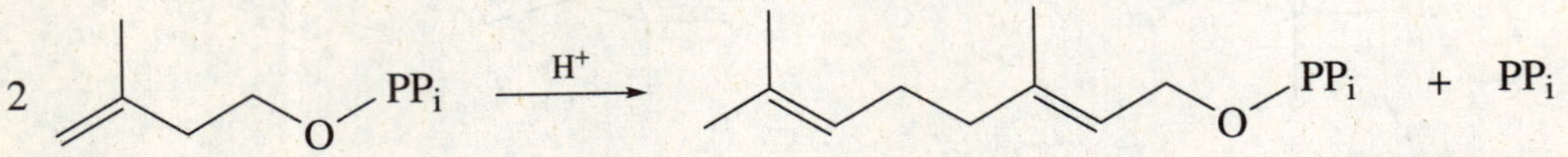

(PP$_i$ = pyrophosphate)

3. Give structures for the following:

a. a phospholipid that contains ethanolamine

b. a wax

4. Indicate whether each of the following statements is true or false:

a. Cholesterol is the precursor of all other steroids. T F

b. Saturated fatty acids have higher melting points than unsaturated fatty acids. T F

c. Fats have a higher percentage of saturated fatty acids than do oils. T F

d. A saturated fatty acid has a lower melting point than an unsaturated fatty acid with the same number of carbons. T F

e. Lipids are insoluble in water. T F

f. A diterpene contains 20 carbons. T F

g. Vitamin A is not a coenzyme. T F

5. What is the first step in the reaction that causes fatty acids to become rancid?

6. How many carbon atoms are there in a triterpene?

7. What size is the ring in a prostaglandin?

8. What is the starting material for the synthesis of prostaglandins?

CHAPTER 27
Nucleosides, Nucleotides, and Nucleic Acids

Important Terms

anticodon	the three bases at the bottom of the middle loop in a tRNA.
antisense strand (template strand)	the strand in DNA that is read during transcription.
autoradiograph (autorad)	the exposed photographic plate obtained in autoradiography.
autoradiography	a technique used to determine the base sequence of DNA.
base	a heterocyclic compound (a purine or a pyrimidine) in DNA and RNA.
codon	a sequence of three bases of mRNA that specifies the amino acid to be incorporated into a protein.
deamination	a hydrolysis reaction that results in removal of ammonia.
deoxyribonucleic acid (DNA)	a polymer of deoxyribonucleotides.
deoxyribonucleotide	a nucleotide where the sugar component is D-2-deoxyribose.
dideoxy method	a method used to determine the sequence of bases in DNA.
dinucleotide	two nucleotides linked by phosphodiester bonds.
double helix	the term used to describe the secondary structure of DNA.
exon	a stretch of bases in DNA that are a portion of a gene.
gene	a segment of DNA.
gene therapy	a technique that inserts a synthetic gene into the DNA of an organism defective in that gene.
genetic code	the amino acid specified by each three-base sequence of mRNA.
genetic engineering	the starting material for the biosynthesis of terpenes.
human genome	the total DNA of a human cell.
informational strand (sense strand)	the strand in DNA that is not read during transcription; it has the same sequence of bases as the synthesized mRNA strand (with Us in place of Ts).
intron	a stretch of bases in DNA that contain no genetic information.

major groove	the wider and deeper of the two alternating grooves in DNA.
minor groove	the narrower and more shallow of the two alternating grooves in DNA.
nucleic acid	the two kinds of nucleic acids are DNA and RNA.
nucleoside	a heterocyclic base (purine or pyrimidine) bonded to the anomeric carbon of a sugar (D-ribose or D-2-deoxyribose).
nucleotide	a nucleoside with one of its OH groups bonded to phosphoric acid in an ester linkage.
oligonucleotide	three to ten nucleotides linked by phosphodiester bonds.
polymerase chain reaction (PCR)	a method that amplifies segments of DNA.
polynucleotide	many nucleotides linked by phosphodiester bonds.
primary structure	the sequence of bases in the nucleic acid.
promoter site	a short sequence of bases at the beginning of a gene.
recombinant DNA	DNA that has been incorporated into a host cell.
replication	the synthesis of identical copies of DNA.
replication fork	the position on DNA where replication begins.
restriction endonuclease	an enzyme that cleaves DNA at a specific base sequence.
restriction fragment	a fragment that is formed when DNA is cleaved by a restriction endonuclease.
ribonucleic acid (RNA)	a polymer of ribonucleotides.
ribonucleotide	a nucleotide where the sugar component is D-ribose.
ribozyme	an RNA molecule that acts as a catalyst.
RNA splicing	the step in RNA processing that cuts out nonsense bases and splices informational pieces together.
semiconservative replication	the mode of replication that results in a daughter molecule of DNA having one of the original DNA strands plus a newly synthesized strand.
sense strand (informational strand)	the strand in DNA that is not read during transcription; it has the same sequence of bases as the synthesized mRNA strand (with Us in place of Ts).
site-specific recognition	recognition by a molecule of a specific site on another molecule.

stacking interactions	van der Waals interactions between the mutually induced dipoles of adjacent pairs of bases in DNA.
stop codon	a codon that says "stop protein synthesis here."
template strand	the strand in DNA that is read during transcription.
transcription	the synthesis of mRNA from a DNA blueprint.
transfer RNA (tRNA)	a single stranded RNA molecule that carries an amino acid to be incorporated into a protein.
translation	the synthesis of a protein from a mRNA blueprint.

Solutions to Problems

1. The ring is protonated at its most basic position. In the case of a purine, this is the 7-position. In the next step, the bond between the heterocyclic base and the sugar breaks, with the anomeric carbocation being stabilized by the ring oxygen's nonbonding electrons.

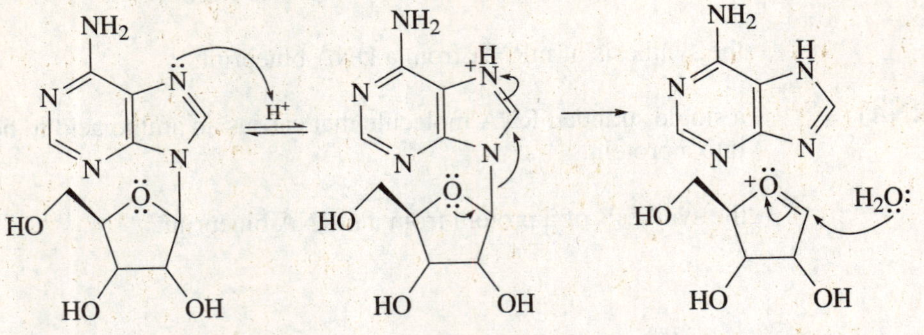

The mechanism is exactly the same for pyrimidines except that the initial protonation takes place at the 1-position.

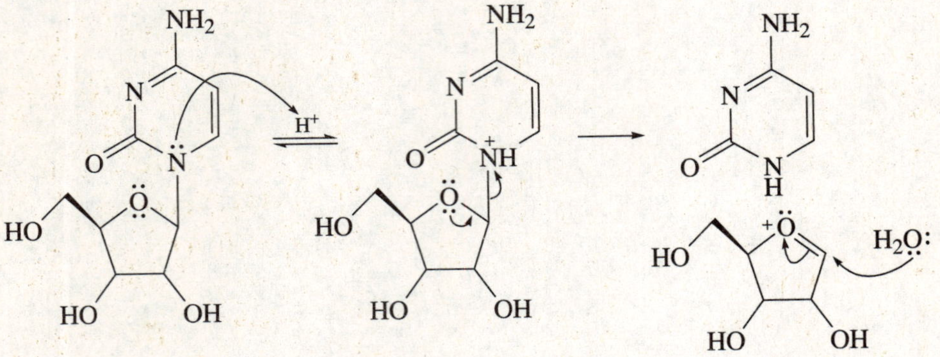

2. **a.**

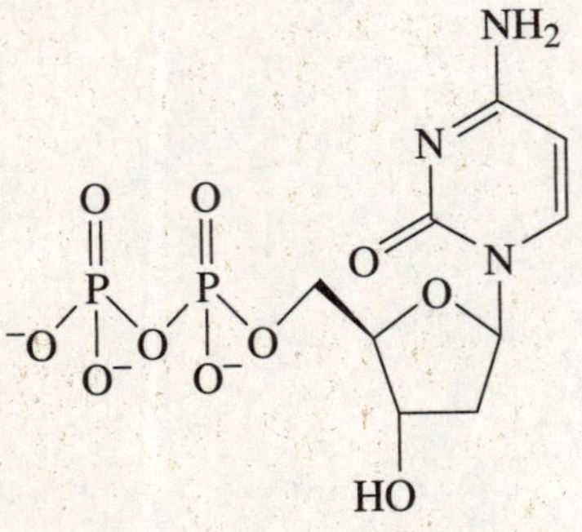

dCDP

b.

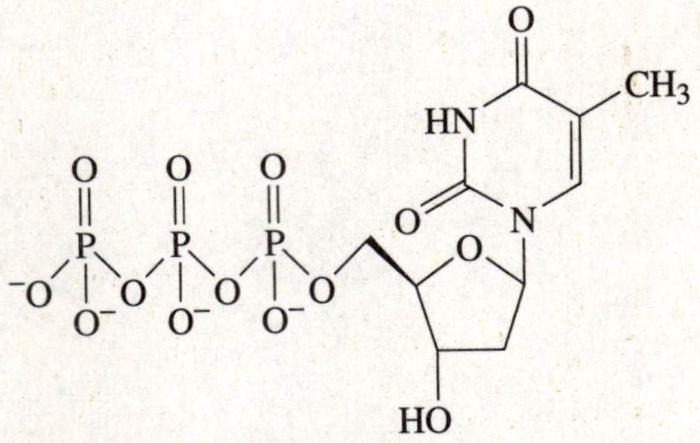

dTTP

c.

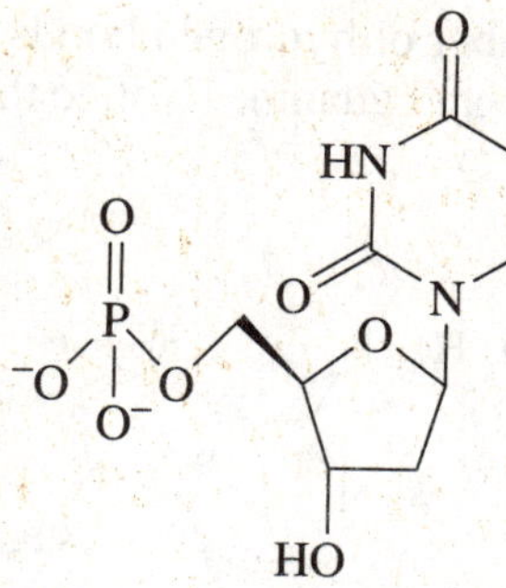

dUMP

e.

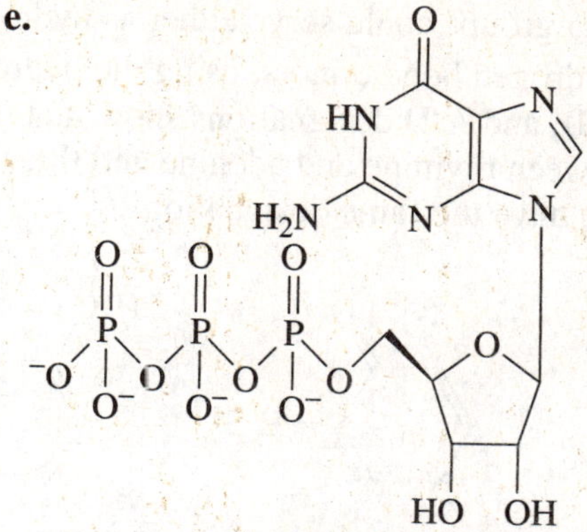

guanosine 5'-triphosphate
GTP

d.

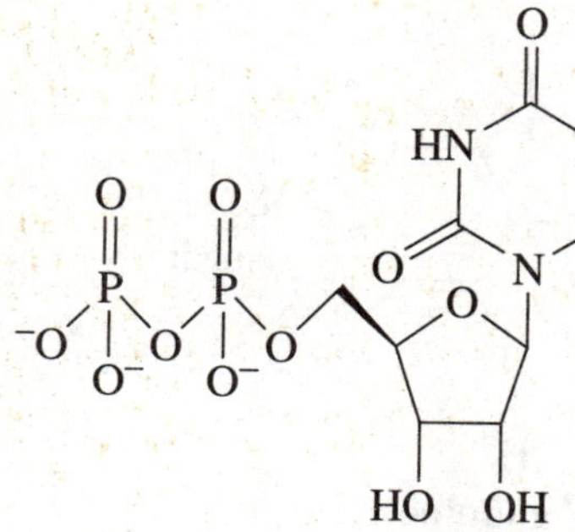

UDP

f.

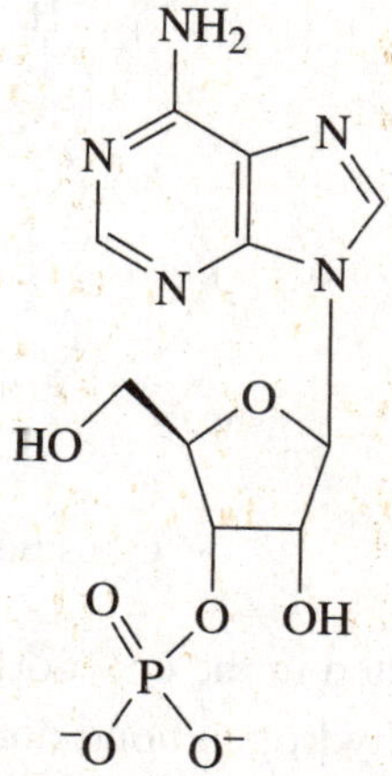

adenosine 3'-monophosphate

3. The hydrolysis of cyclic AMP forms adenosine 5'-phosphate and adenosine 3'-phosphate.

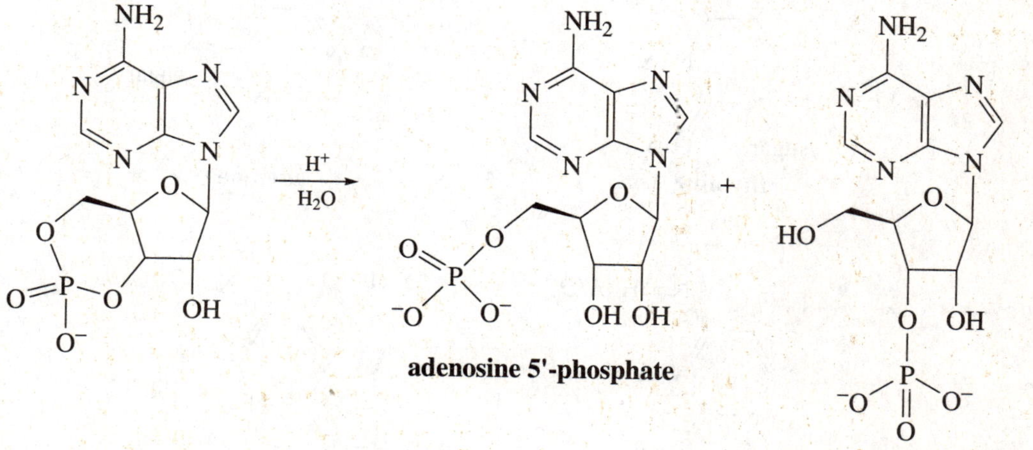

adenosine 5'-phosphate

adenosine 3'-phosphate

4. The NH_2 groups could serve either as hydrogen bond acceptors, using the lone pair electrons on nitrogen, or as hydrogen bond donors, using the hydrogen bonded to the nitrogen.
The **A**, **D**, and **A/D** designations show that the maximum number of hydrogen bonds that can form are two between thymine and adenine and three between cytosine and guanine. Notice that uracil and thymine have the same designations.

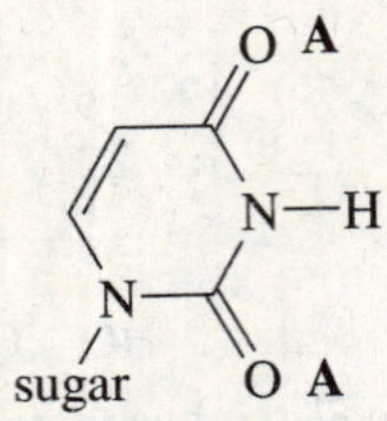

uracil

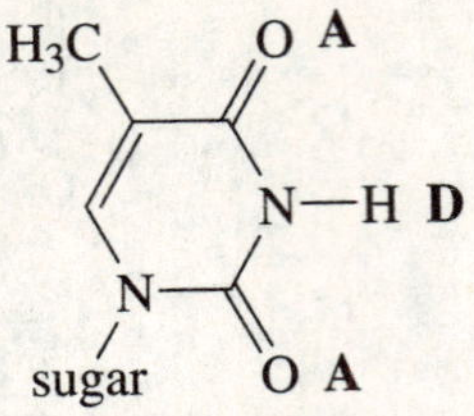

thymine

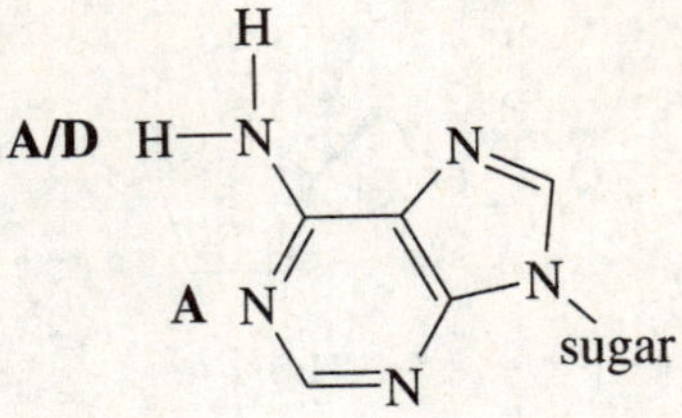

adenine

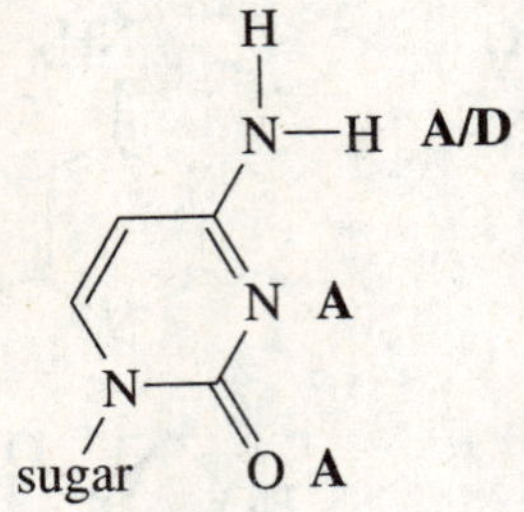

cytosine

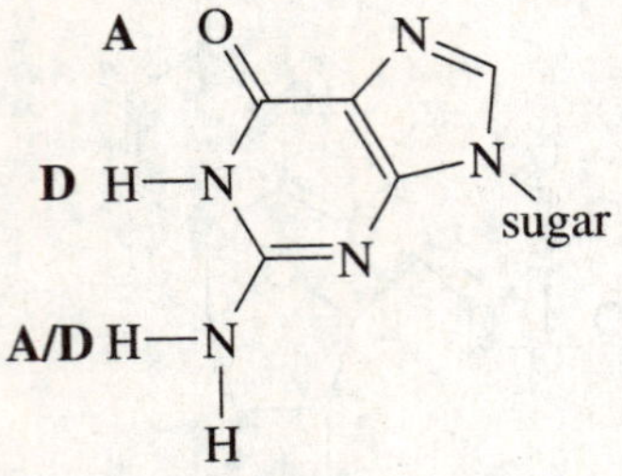

guanine

5. If the bases existed in the enol form, the OH groups and NH_2 groups could act either as hydrogen bond acceptors or as hydrogen bond donors. The maximum number of hydrogen bonds that could form is one between thymine and adenine and two between cytosine and guanine.

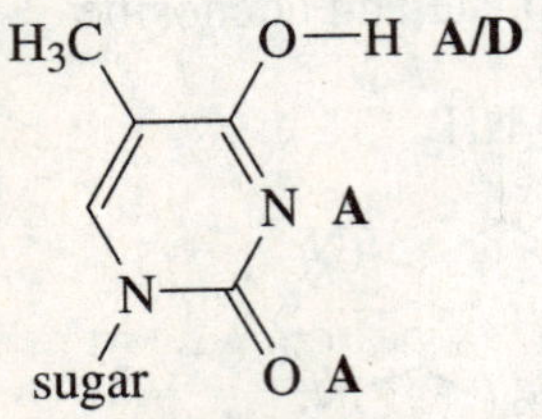

thymine

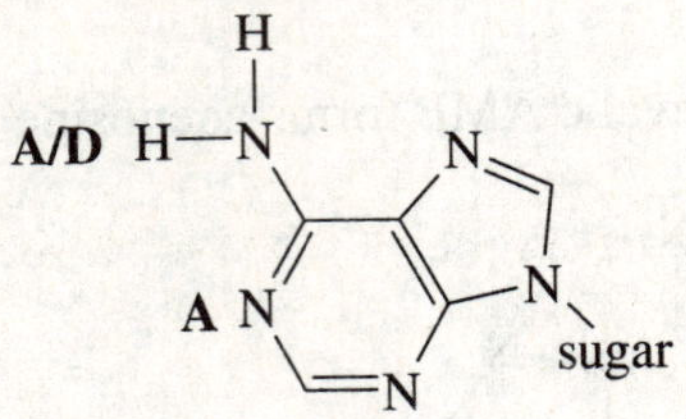

adenine

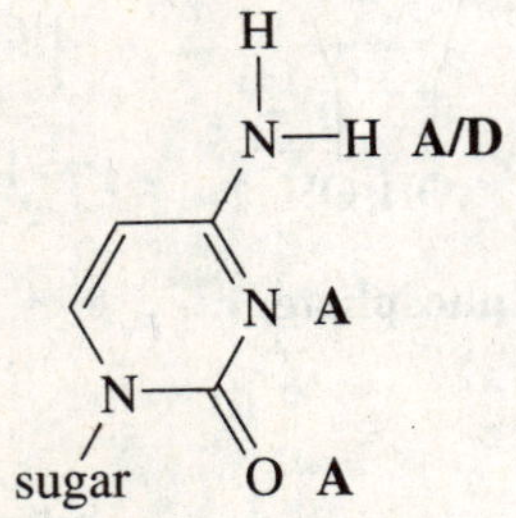

cytosine

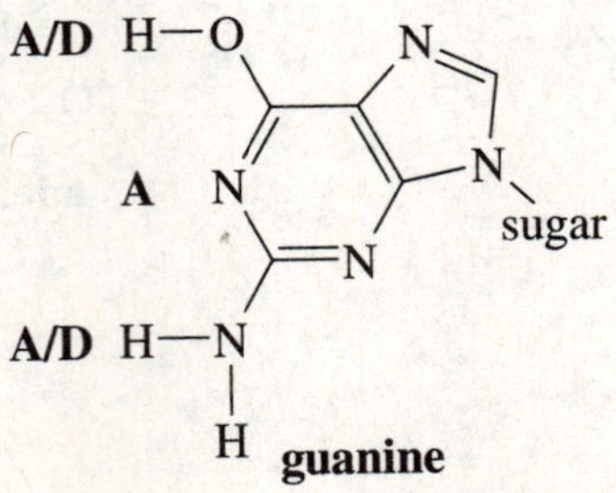

guanine

6. Notice that in the first step of the mechanism shown on page 1207, when the nucleophile attacks the phosphorus of the diester, the π bond breaks. However, when a nucleophile attacks the phosphorus of the anhydride intermediate, an anhydride bond breaks in preference to the π bond.

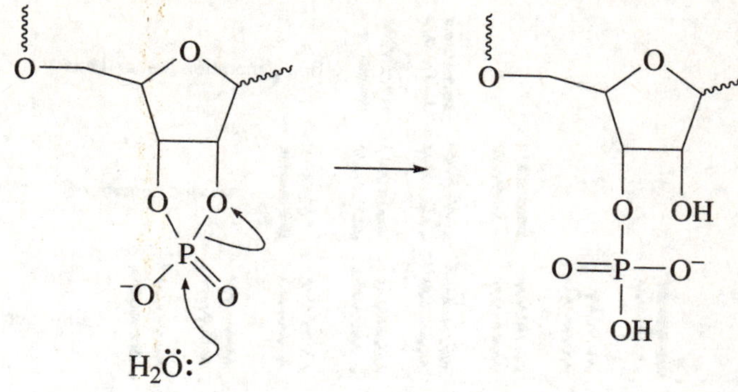

or

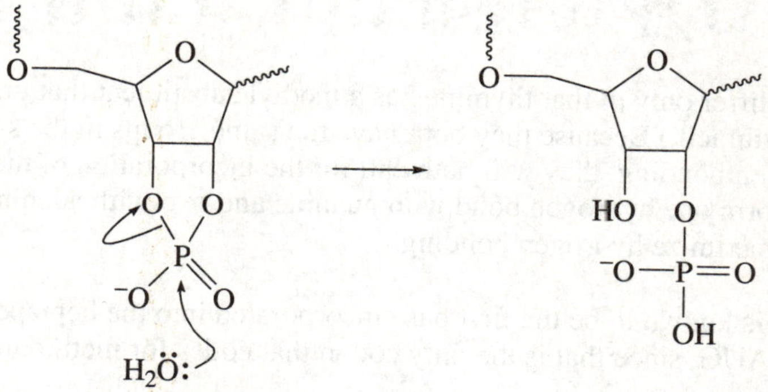

7. **a.** 3'—C—C—T—G—T—T—A—G—A—C—G—5'

b. guanine

8. 5-Bromouracil is incorporated into DNA in place of thymine because of their similar size. Thymine pairs with adenine via two hydrogen bonds. 5-Bromouracil exists primarily in the enol form. The enol can form only one hydrogen bond with adenine, but it can form three hydrogen bonds with guanine. Therefore, 5-bromouracil pairs with guanine. Because 5-bromouracil causes guanine to be incorporated instead of adenine into newly synthesized DNA strands, it causes mutations.

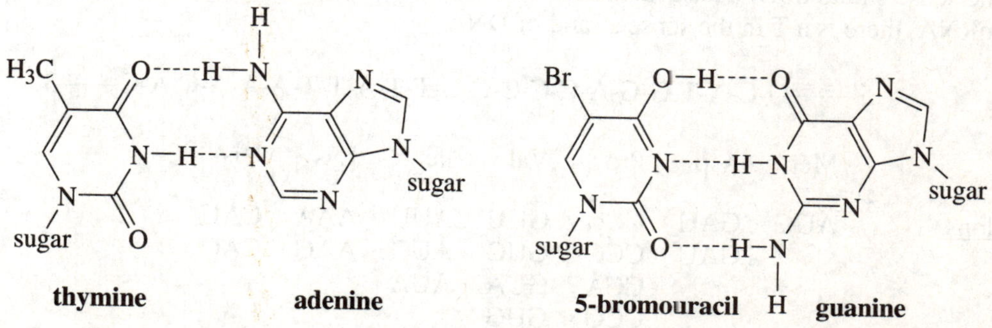

9.

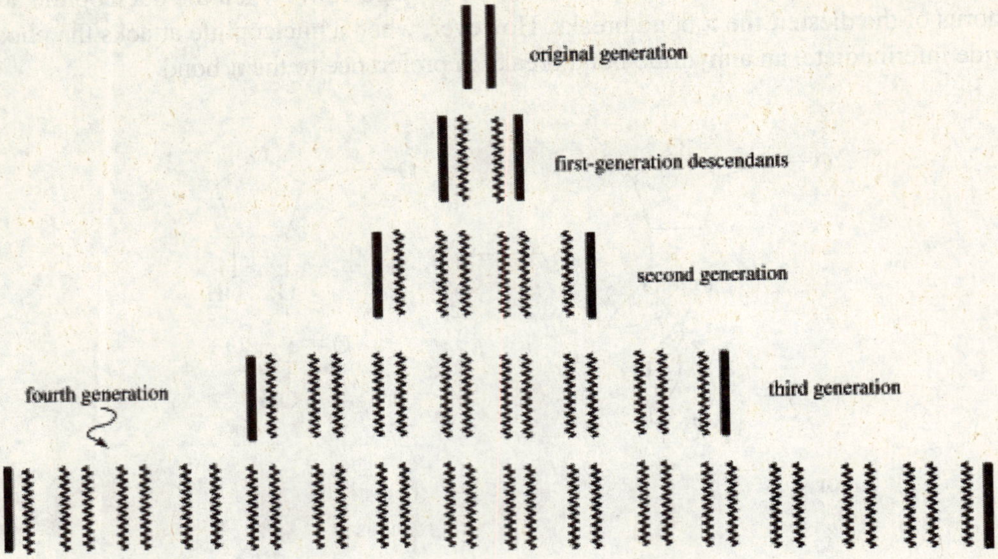

10. Thymine and uracil differ only in that thymine has a methyl substituent that uracil does not have. (Thymine is 5-methyluracil.) Because they both have the same groups in the same positions that can participate in hydrogen bonding, they will both call for the incorporation of the same purine. Because thymine and uracil form one hydrogen bond with guanine and two with adenine, they will incorporate adenine in order to maximize hydrogen bonding.

11. Because methionine is known to be the first base incorporated into the heptapeptide, the mRNA sequence is read beginning at AUG, since that is the only codon that codes for methionine.

Met-Asp-Pro-Val-Ile-Lys-His

12. Met-Asp-Pro-Leu-Leu-Asn

13. It does not cause protein synthesis to stop, because the sequence UAA does not occur within a triplet. The reading frame causes the triplets to be AUU and AAA.

14. The sequence of bases in the template strand of DNA specify the sequence of bases in mRNA, so the bases in the template strand and the bases in mRNA are complementary. Therefore, the sequence of bases in the sense strand of DNA are identical to the sequence of bases in mRNA, except wherever there is a U in mRNA, there is a T in the sense strand of DNA.

5'——G-C-A-T-G-G-A-C-C-C-C-G-T-T-A-T-T-A-A-A-C-A-C——3'

15.

	Met	Asp	Pro	Val	Ile	Lys	His
<u>codons</u>	AUG	GAU	CCU	GUU	AUU	AAA	CAU
		GAU	CCC	GUC	AUC	AAG	CAC
			CCA	GUA	AUA		
			CCG	GUG			

anticodons

Note that the anticodons are stated in the 5' → 3' direction. For example,the anticodon of AUG is stated as CAU (not UAC).

codon 5′ A U G 3′

anticodon 3′ U A C 5′

CAU	AUC	AGG	AAC	AAU	UUU	AUG
	GUC	GGG	GUC	GAU	CUU	GUG
		UGG	UAC	UAU		
		CGG	CAC			

16.

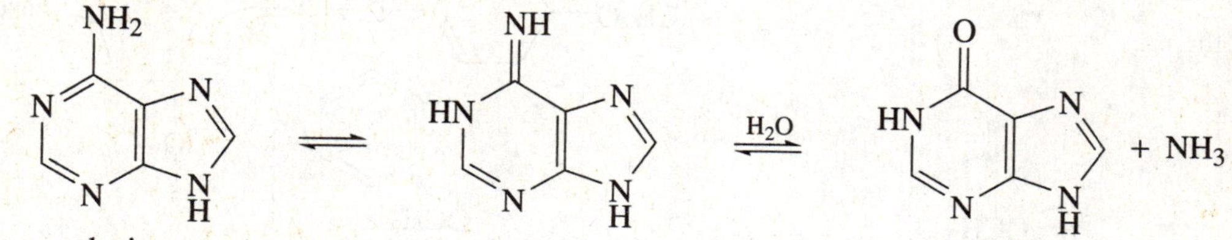

adenine hypoxanthine

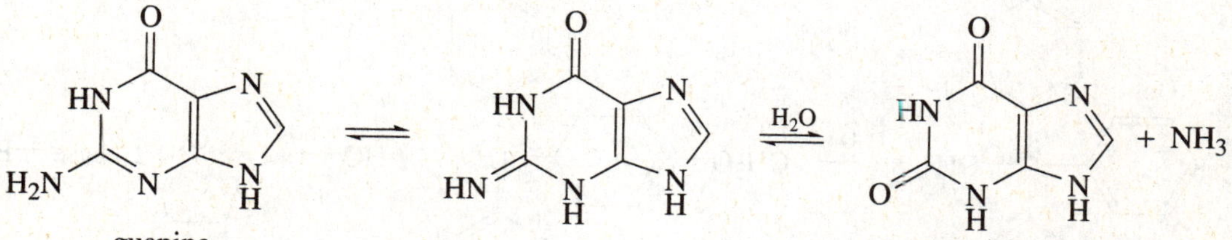

guanine xanthine

17. Thymine does not have an amino substituent on the ring, which means that it cannot form an imine. Deamination involves hydrolyzing an imine linkage to a carbonyl group and ammonia.

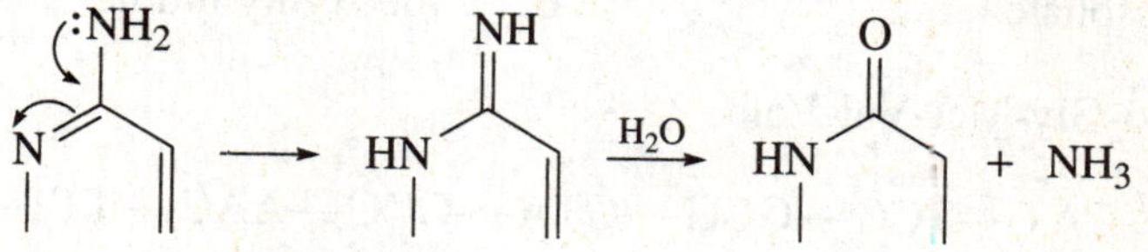

18. **a** is the only sequence that has a chance of being recognized by a restriction endonuclease because it is the only one that has the same sequence of bases in the 5' → 3' direction that the complementary strand has in the 5' → 3' direction.

ACGCGT

19. All the fragments will end in "G".

^{32}P—TCCGAGGTCACTAGG

^{32}P—TCCGAGGTCACTAG

^{32}P—TCCGAGG

^{32}P—TCCGAG

^{32}P—TCCG

20.

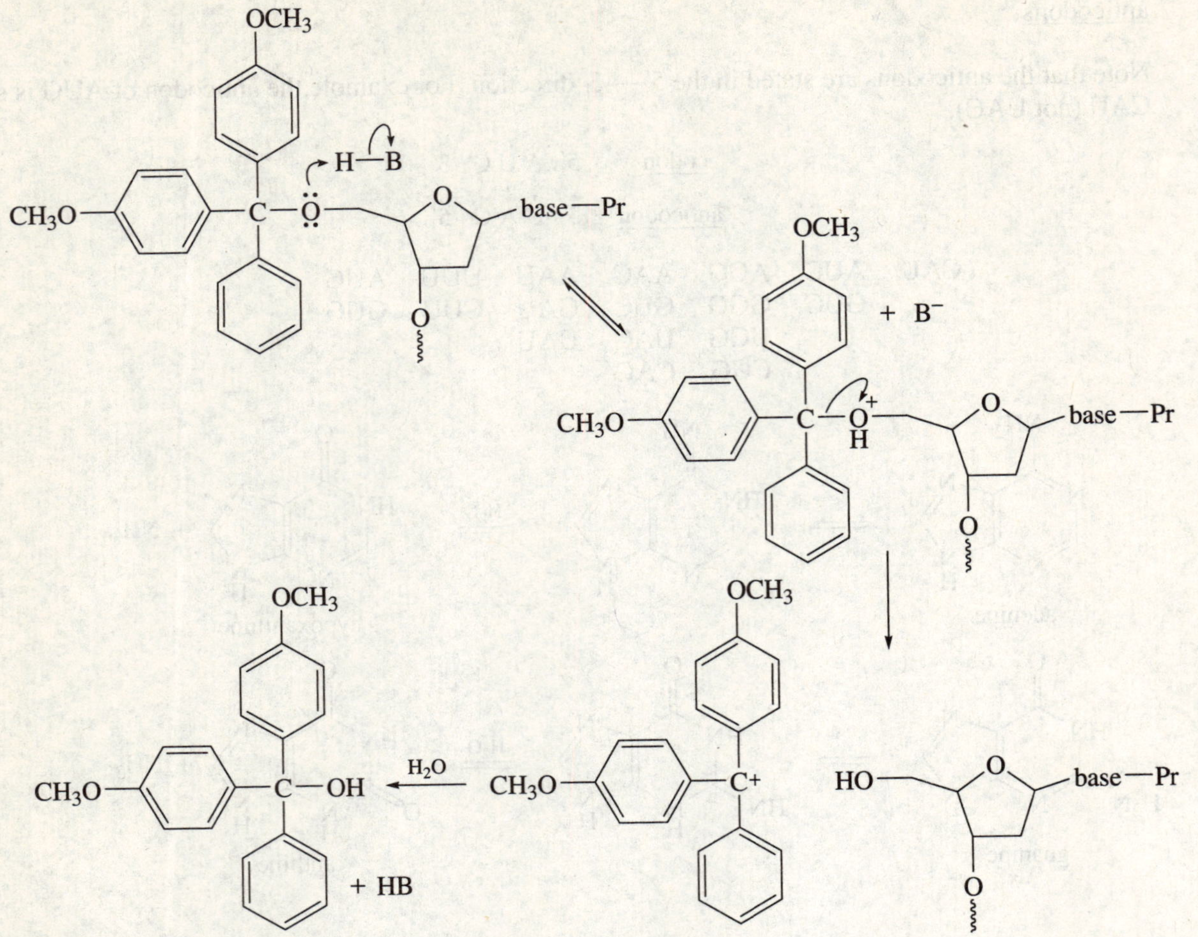

21. **a.** guanosine 3'-monophosphate **c.** 2'-deoxyadenosine 5'-monophosphate
b. cytidine 5'-diphosphate **d.** 2'-deoxythymidine

22. Lys-Val-Gly-Tyr-Pro-Gly-Met-Val-Val

23. 5'—GAC—CAC—CAT—TCC—GGG—GTA—GCC—AAC—TTT—3'

24. 5'—AAA—GTT—GGC—TAC—CCC—GGA—ATG—GTG—GTC—3'

25. **a.** Ile **b.** Asp **c.** Val **d.** Val

26. The third base in each codon has some variability.

mRNA 5'-GG(UCA or G)UC(UCA or G)CG(UCA or G)GU(UCA or G)CA(U or C)GA(A or G)-3'
or AG(U or C) AG(A or G)

DNA
template 3'-CC(AGT or C)AG(AGT or C)GC(AGT or C)CA(AGT or C)GT(A or G)CT(T or C)-5'
or TC(A or G) TC(T or C)
sense 5'-GG(TCA or G)TC(TCA or G)CG(TCA or G)GT(TCA or G)CA(T or C)GA(A or G)-3'
or AG(T or C) AG(A or G)

Notice that Ser and Arg are two of three amino acids that can be specified by six different codons.

27.

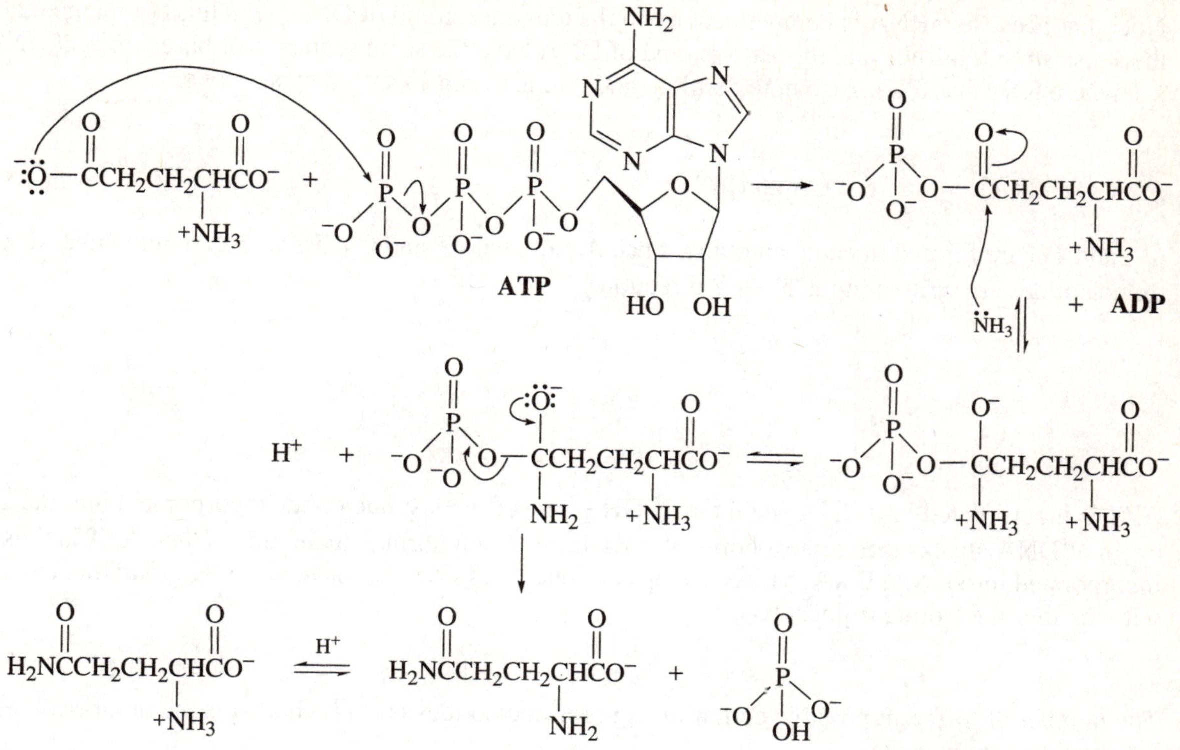

28.

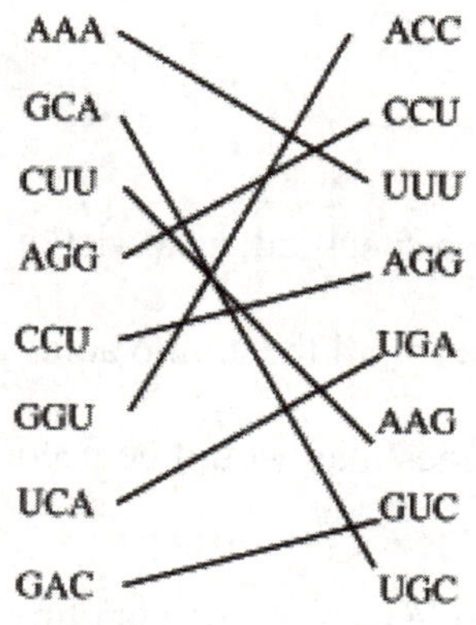

29.

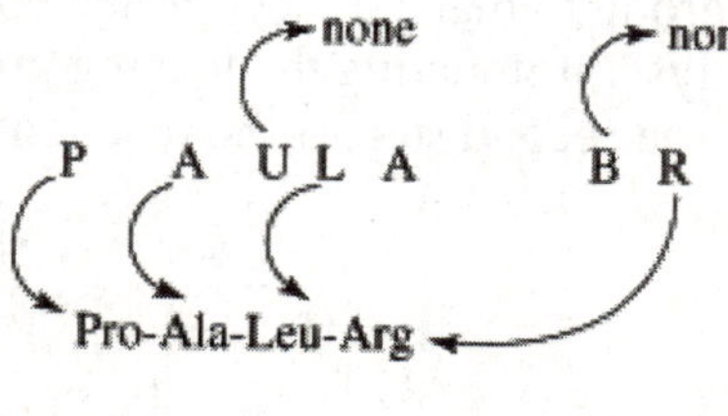

mRNA CC(UCA or G)GC(UCA or G)CU(UCA or G)CG(UCA or G)CG(UCA or G)
UU(A or G) AG(A or G)

DNA (sense strand) CC(TCA or G)GC(TCA or G)CT(TCA or G)CG(TCA or G)CG(TCA or G)
TT(A or G) AG(A or G)

Note that because mRNA is complementary to the template strand of DNA, which is complementary to the sense strand, mRNA and the sense strand of DNA have the same sequence of bases (expect DNA has a T where RNA has a U). Also note that Leu and Arg are each specified by six codons.

30. **a.** CC and GG **c.** CA and TG

CA and TG are formed in equal amounts, since A pairs with T and C pairs with G. (Remember that the dinucleotides are written in the 5' → 3' direction.)

5' C A 3'
3' G T 5'

31. AZT is incorporated into DNA when the 3´-OH group of the last nucleotide incorporated into the growing chain of DNA attacks the α-phosphorus of AZT instead of a normal nucleotide. When AZT is thus incorporated into DNA, DNA synthesis stops because AZT does not have a 3´-OH group that can react with another nucleotide triphosphate.

32. The number of different possible codons using four nucleotides is $(4)^n$ where *n* is the number of letters (nucleotides) in the code.

for a two-letter code: $(4)^2 = 16$
for a three-letter code: $(4)^3 = 64$
for a four-letter code: $(4)^4 = 256$

Since there are 20 amino acids that must be specified, a two-letter code would not provide an adequate number of codons.
A three-letter code provides enough codes for all the amino acids and also provides the necessary stop codons.
A four-letter code provides many more codes than would be needed.

33. In the first step of the reaction, the imidazole ring of one histidine acts as a general-base catalyst, removing a proton from the 2-OH group to make it a better nucleophile. The imidazole ring of the other histidine acts as a general-acid catalyst, protonating the leaving group to make it a weaker base and therefore a better leaving group. In the second step of the reaction the roles of the two imidazole rings are reversed.

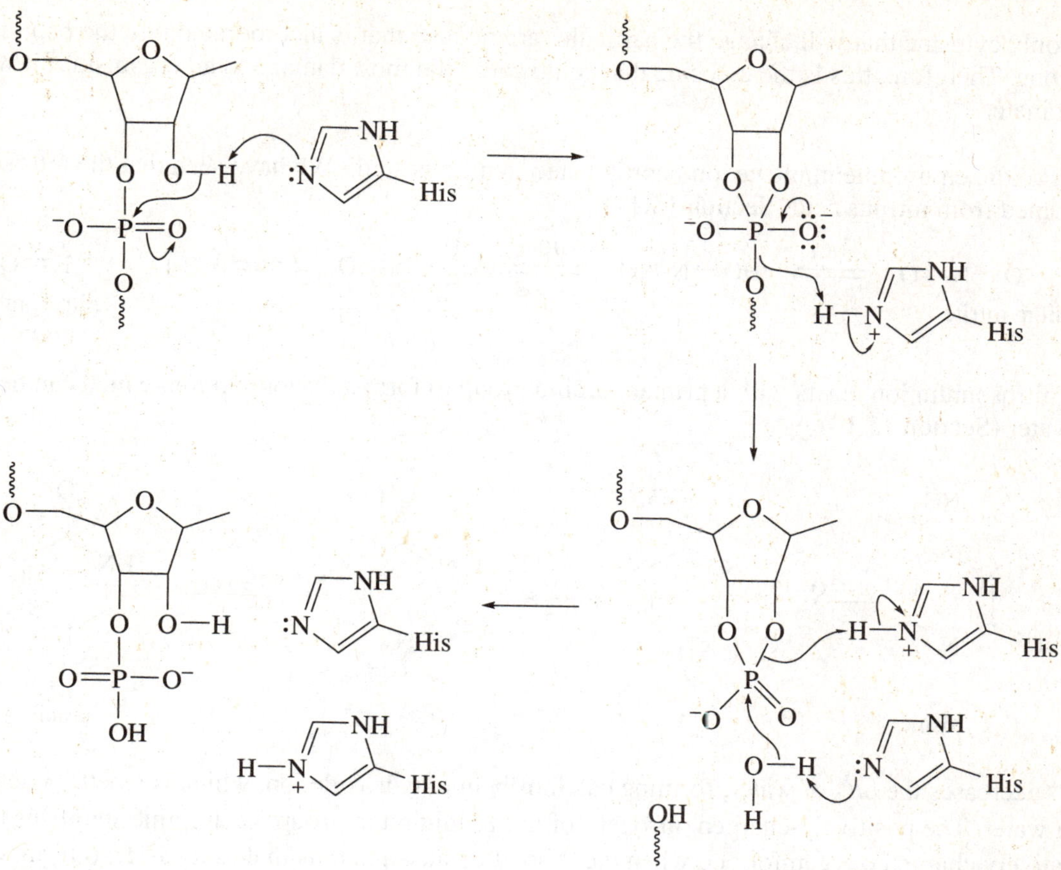

34. The normal and mutant peptides would have the following base sequence in their mRNA.

normal: CA(AG) UA(UC) GG(UCAG) AC(UCAG) CG(UCAG) UA(UC) GU(UCAG)

mutant: CA(AG) UC(UCAG) GA(AG) CC(UCGA) GG(UCGA) AC(UCAG)

a. The middle nucleotide (A) in the second triplet was deleted. This means that an A was deleted in the sense strand of DNA or a T was deleted in the template strand of DNA.

b. The mRNA for the mutant peptide has an unused 3'-terminal two-letter code, U(UCGA). The last amino acid in the octapeptide of the normal fragment is leucine, so its last triplet is UU(AG) or CU(UCAG).

This means that the triplet for the last amino acid in the mutant is U(UCGA)(UC) and that the last amino acid in the mutant is one of the following: Phe, Ser, Tyr, or Cys.

35. If deamination does not occur, the mRNA sequence will be:

AUG-UCG-CUA-AUC which will code for the following tetrapeptide
Met - Ser - Leu – Ile

Deamination of a cytosine results in a uracil.
If the cytosines are deaminated, the mRNA sequence will be:

AUG-UUG-UUA-AUU which will code for the following tetrapeptide
Met - Leu - Leu - Ile

The only cytosine that will change the particular amino acid that is incorporated into the peptide is the first one. Therefore, this is the cytosine that could cause the most damage to an organism if it were deaminated.

36. In an acidic environment, nitrite ion is protonated to nitrous acid. We have seen that the nitrosonium ion is formed from nitrous acid (Section 15.11).

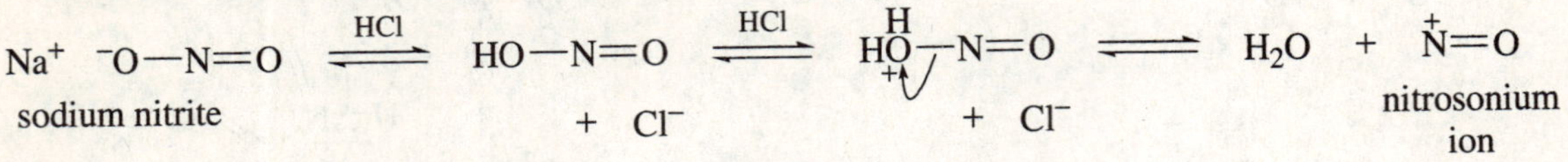

The nitrosonium ion reacts with a primary amino group to form a diazonium ion, which can be displaced by water (Section 15.10).

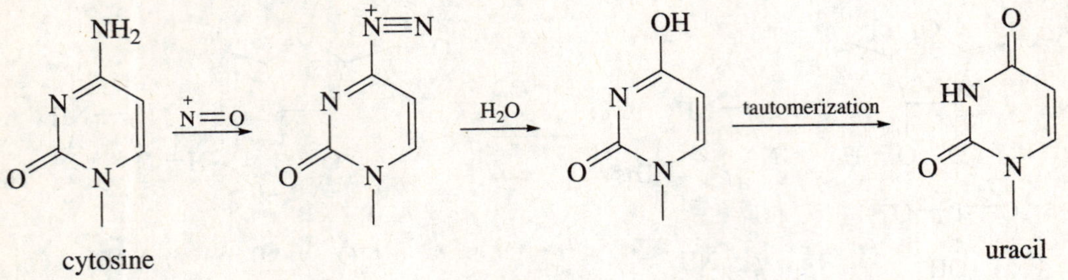

37. Ca^{2+} decreases the pK_a of water, forming calcium-bound hydroxide ion, which is a better nucleophile than water. The positively-charged nitrogen (of the guanidinium group) of arginine stabilizes the negatively charged oxygen formed when the π bond breaks, and thus makes it easier to form. Glutamic acid protonates the OR oxygen, thereby making it a weaker base and therefore a better leaving group when the π bond reforms.

38. The ribosome has a binding site for the growing peptide chain and a binding site for the next amino acid to be incorporated into the chain.

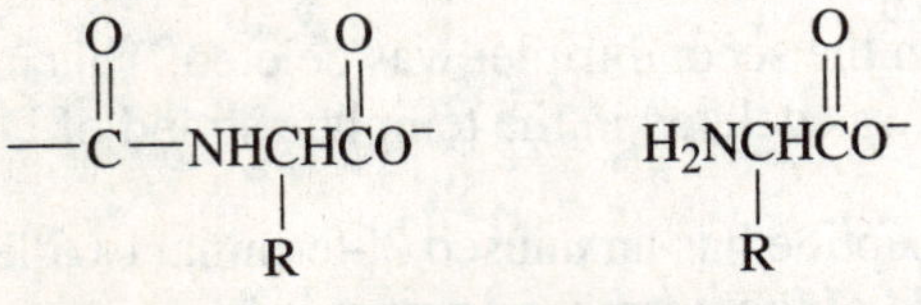

peptide binding site amino acid binding site

In protein synthesis, all peptide bonds are formed by the reaction of an amino acid with a peptide, except formation of the first peptide bond, which has to be formed by the reaction of two amino acids. Therefore, in the synthesis of the first peptide bond, the first (N-terminal) amino acid has to have a peptide bond that will fit into the peptide binding site.

The formyl group of *N*-formylmethionine will provide the peptide group that will be recognized by the peptide binding site, and the second amino acid will be bound in the amino acid binding site.

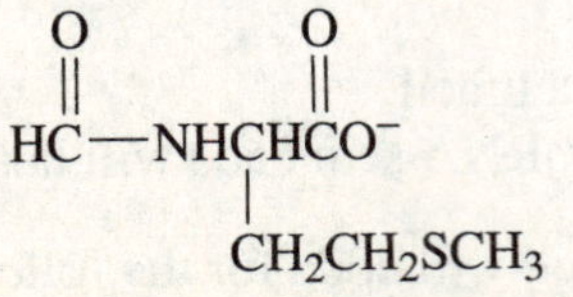

N-formylmethionine

39. It requires energy to break the hydrogen bonds that hold the two chains together, so an enormous amount of energy would be required to unravel the chain completely. However, as the new nucleotides that are incorporated into the growing chain form hydrogen bonds with the parent chain, energy is released, and this energy can be used to unwind the next part of the double helix.

Chapter 27 Practice Test

1. Is the following compound dTMP, UMP, dUMP, or dUTP?

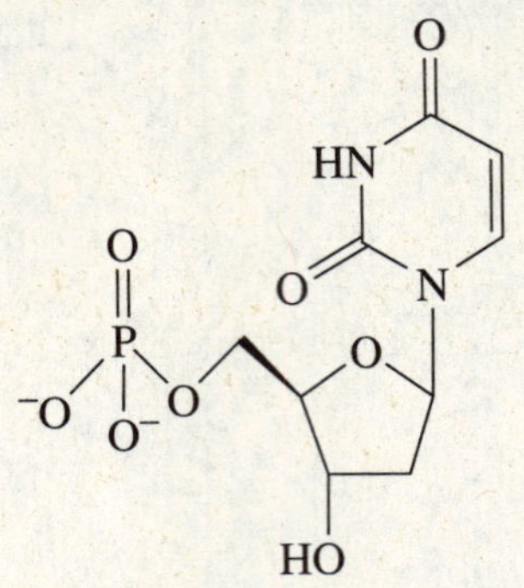

2. If one of the strands of DNA has the following sequence of bases running in the 5' → 3' direction, what is the sequence of bases in the complementary strand?

5'—A—C—T—T—G—C—A—T—3'

3. What base is closest to the 5′-end in the complementary strand?

4. Indicate whether each of the following statements is true or false:

a. Guanine and cytosine are purines. T F

b. The 3′-OH group allows RNA to be easily cleaved. T F

c. The number of A's in DNA is equal to the number of T's. T F

d. rRNA carries the amino acid that will be incorporated into a protein. T F

e. The template strand of DNA is the one transcribed to form RNA. T F

f. The 5′-end of DNA has a free OH group. T F

g. The synthesis of proteins from an RNA blueprint is called transcription. T F

h. A nucleotide consists of a base and a sugar. T F

i. RNA contains T's and DNA contains U's. T F

5. Which of the following base sequences would most likely be recognized by a restriction endonuclease?

1. ACGCGT 3. ACGGCA 5. ACATCGT

2. ACGGGT 4. ACACGT 6. CCAACC

6. What would be the sequence of bases in the m-RNA obtained from the following segment of DNA?

sense strand

5'—G-C-A-T-G-G-A-C-C-C-C-G-T—3'
3'—C-G-T-A-C-C-T-G-G-G-G-C-A—5'

template strand

7. Give the labeled fragments that would be obtained if the following strand of DNA were added to a mixture of the labeled primer, DNA polymerase, dATP, dGTP, dCTP, dTTP and 2′,3′-dideoxycytidine.

3'—CTTGAAGTATGAACTAC

^{32}P—GA

8. Why is NADPH needed in order to convert uridines into thymidines?

9. Which of the following pairs of dinucleotides will occur in equal amounts in DNA? (Remember nucleotides are always written in the 5' → 3' direction.)

CA and GT	CG and AT
CG and GG	CA and TG

CHAPTER 28
Synthetic Polymers

Important Terms

addition polymer (chain-growth polymer) made by adding monomers to the growing end of a chain.

alpha olefin a monosubstituted ethylene.

alternating copolymer a copolymer in which two monomers alternate.

anionic polymerization chain-growth polymerization where the initiator is a nucleophile; the propagation site, therefore, is an anion.

aramide an aromatic polyamide.

atactic polymer a polymer in which the substituents are randomly oriented on the extended carbon chain.

biodegradable polymer a polymer that can be broken into small segments by an enzyme-catalyzed reaction.

biopolymer a polymer that is synthesized in nature.

block copolymer a copolymer in which there are blocks of each kind of monomer.

cationic polymerization chain-growth polymerization where the initiator is an electrophile; the propagation site, therefore, is a cation.

chain-growth polymer (addition polymer) made by adding monomers to the growing end of a chain.

chain transfer a growing polymer chain reacts with a molecule XY in a manner that allows X· to terminate the chain, leaving behind Y· to initiate a new chain.

condensation polymer (step-growth polymer) made by combining two molecules while removing a small molecule (usually water or an alcohol).

conducting polymer a polymer that can conduct electricity down its backbone.

copolymer a polymer formed using two or more different monomers.

cross-linking connecting polymer chains by bond formation.

crystallites regions of a polymer in which the chains are highly ordered.

elastomer a polymer that can stretch and then revert back to its original shape.

epoxy resin	formed by mixing a low molecular weight prepolymer with a compound that forms a cross-linked polymer.
graft copolymer	a copolymer that contains branches of a polymer of one monomer grafted onto the backbone of a polymer made from another monomer.
head-to-tail addition	the head of one molecule is added to the tail of another molecule.
homopolymer	a polymer that contains only one kind of monomer.
isotactic polymer	a polymer in which all the substituents are on the same side of the fully extended carbon chain.
living polymer	a nonterminated chain-growth polymer that remains active. Therefore, the polymerization reaction can continue upon addition of more monomer.
materials science	the science of creating new materials to take the place of known materials such as metal, glass, wood, cardboard, and paper.
monomer	a repeating unit in a polymer.
oriented polymer	a polymer obtained by stretching out polymer chains and putting them back together in a parallel fashion.
plasticizer	an organic molecule that dissolves in a polymer and allows the polymer chains to slide by each other.
polyamide	a polymer with many amide groups.
polycarbonate	a step-growth polymer in which the dicarboxylic acid is carbonic acid.
polyester	a polymer with many ester groups.
polymer	a large molecule made by linking monomers together.
polymer chemistry	the field of chemistry that deals with synthetic polymers; part of the larger discipline known as materials science.
polymerization	the process of linking up monomers to form a polymer.
polyurethane	a polymer with many urethane groups.
propagating site	the reactive end of a chain-growth polymer.
radical polymerization	chain-growth polymerization where the initiator is a radical; the propagation site, therefore, is a radical.
random copolymer	a copolymer with a random distribution of monomers.

ring-opening polymerization	a chain-growth polymerization that involves opening the ring of the monomer.
step-growth polymer (condensation polymer)	made by combining two molecules while removing a small molecule (usually water or an alcohol).
syndiotactic polymer	a polymer in which the substituents regularly alternate on both sides of the fully extended carbon chain.
synthetic polymer	a polymer that is not synthesized in nature.
thermoplastic polymer	a polymer that has both ordered crystalline regions and amorphous non-crystalline regions.
thermosetting polymer	cross-linked polymers that, after they are hardened, cannot be re-melted by heating.
urethane	a compound with a carbonyl group that is both an amide and an ester.
vinyl polymer	a polymer in which the monomer is ethylene or a substituted ethylene.
vulcanization	increasing the flexibility of rubber by heating it with sulfur.
Ziegler-Natta catalyst	an aluminum-titanium initiator that controls the stereochemistry of a polymer.

Solutions to Problems

1. **a.** $CH_2{=}CHCl$ **b.** 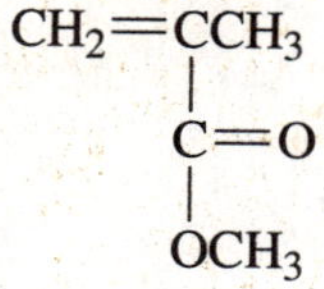 **c.** $CF_2{=}CF_2$

2. Poly(vinyl chloride) would be more apt to contain head-to-head linkages because a chloro substituent is less able (compared with a phenyl substituent) to stabilize the growing end of the polymer chain by electron delocalization.

3. 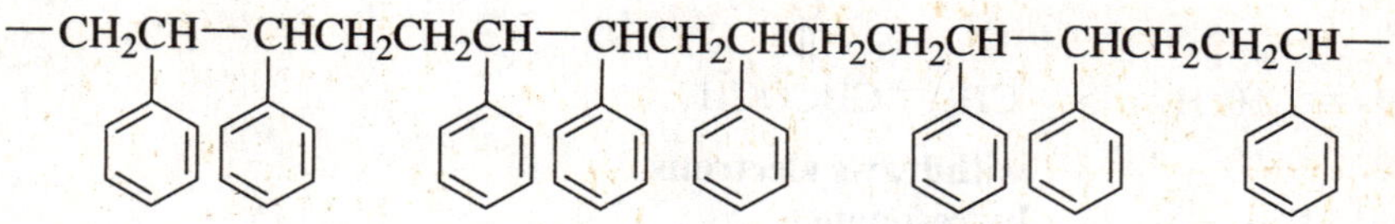

4. 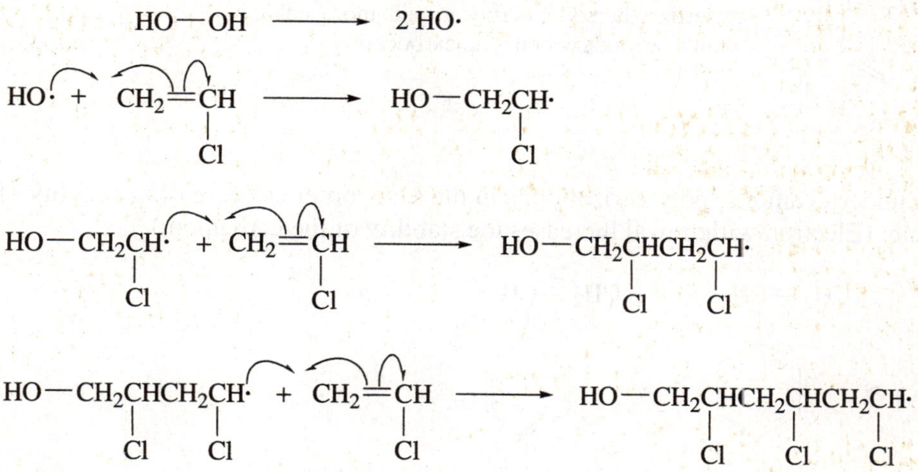

5. Since branching increases the flexibility of the polymer, beach balls are made from more highly branched polyethylene.

6. 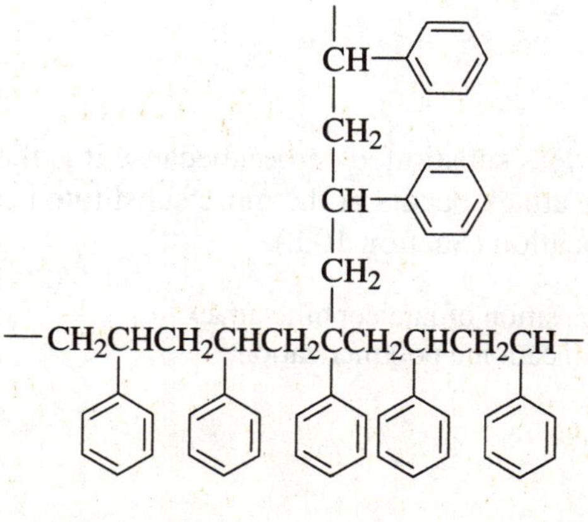

7. Decreasing ability to undergo cationic polymerization is in the same order as decreasing stability of the carbocation intermediate. (Electron donation increases the stability of the carbocation.)

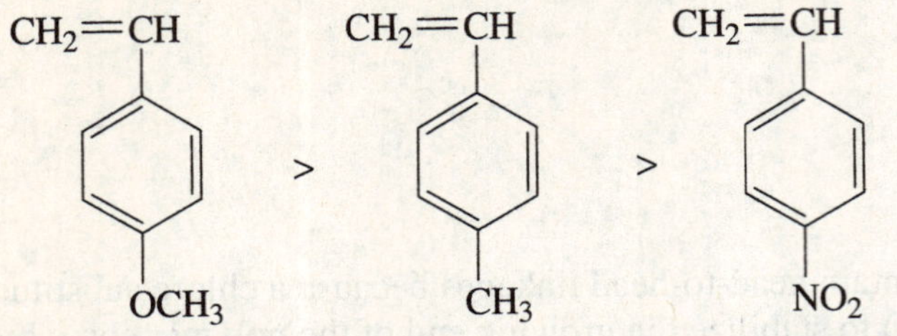

donates electrons by resonance **withdraws electrons by resonance**

$CH_2{=}CHOCCH_3$ (C=O) > $CH_2{=}CHCH_3$ > $CH_2{=}CHCOCH_3$ (C=O)

donates electrons by resonance **withdraws electrons by resonance**

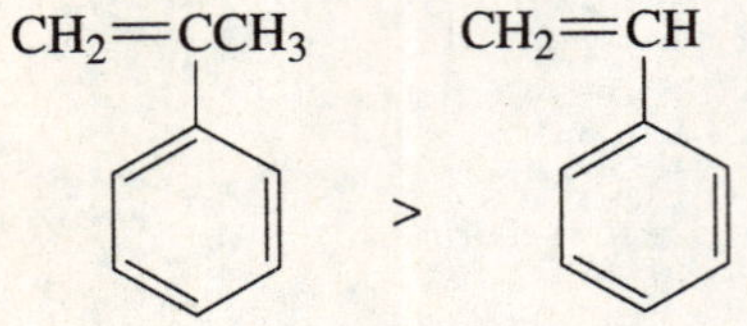

A tertiary benzylic carbocation is more stable than a secondary benzylic carbocation.

8. Decreasing ability to undergo anionic polymerization is in the same order as decreasing stability of the carbanion intermediate. (Electron withdrawal increases the stability of the carbanion.)

a.

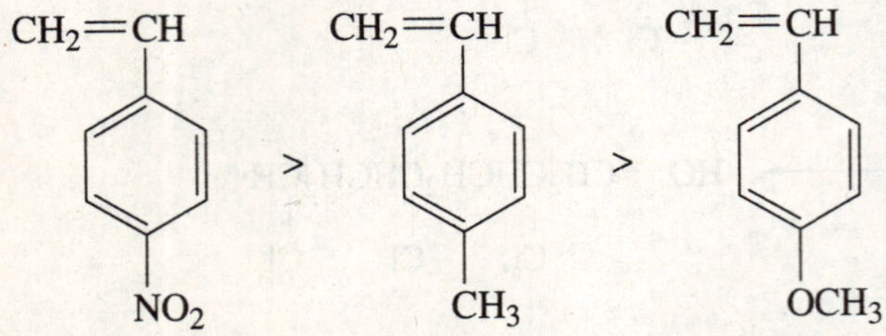

withdraws electrons by resonance **donates electrons by resonance**

b. $CH_2{=}CHC{\equiv}N$ > $CH_2{=}CHCl$ > $CH_2{=}CHCH_3$

withdraws electrons by resonance

9. In anionic polymerization, nucleophilic attack occurs at the less substituted carbon because it is the less sterically hindered; in cationic polymerization, nucleophilic attack occurs at the more substituted carbon, because the ring opens to give the more stable partial carbocation (Section 10.8).

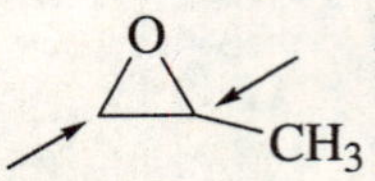

position of nucleophilic attack in cationic polymerization

position of nucleophilic attack in anionic polymerization

10. **a.**

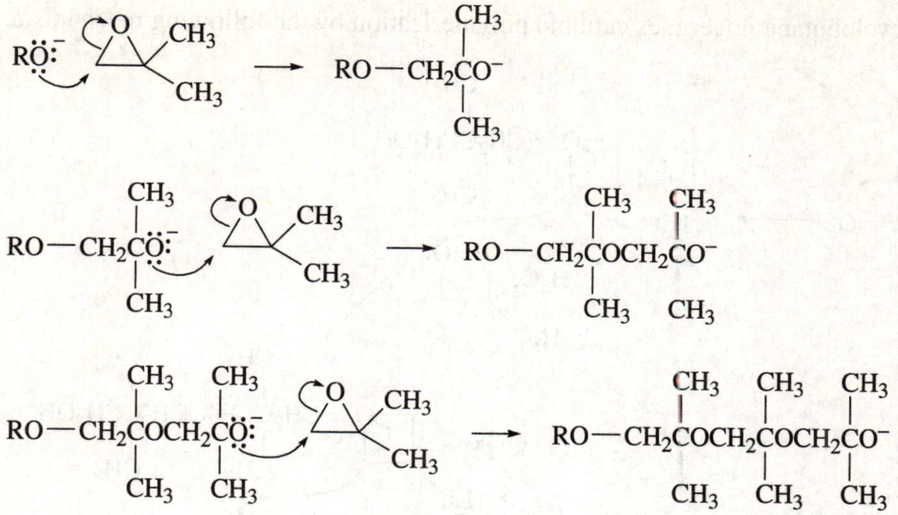

b.

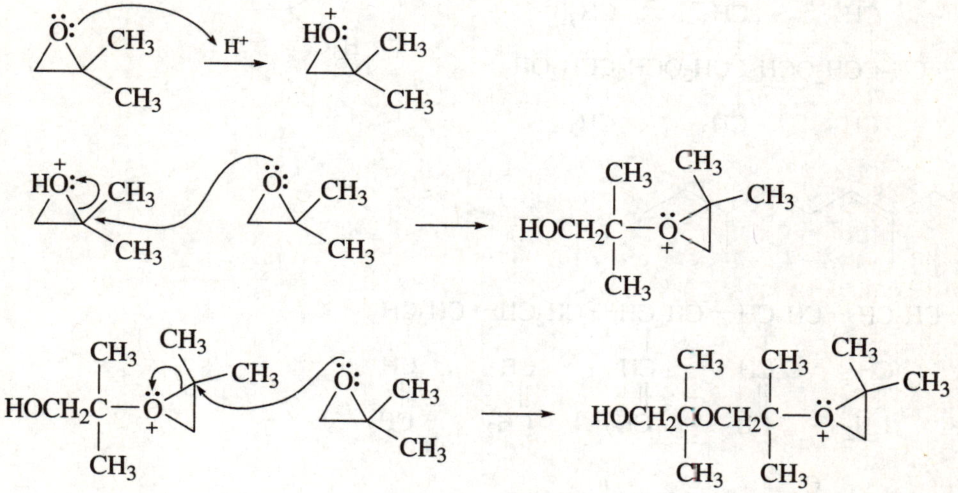

11. **a.**

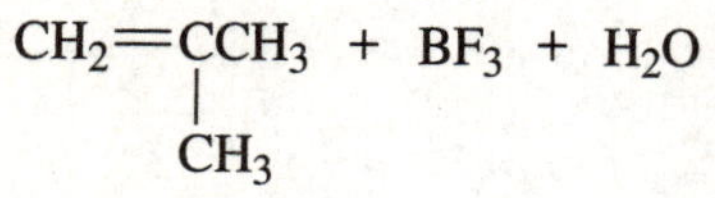

b.

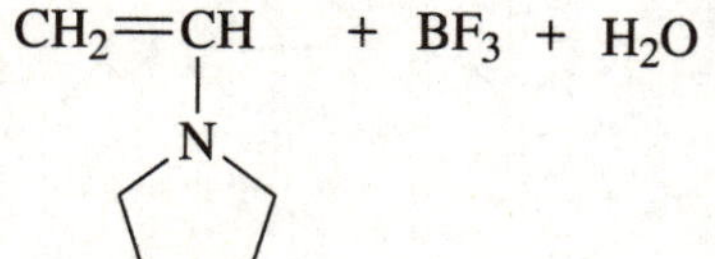

c.

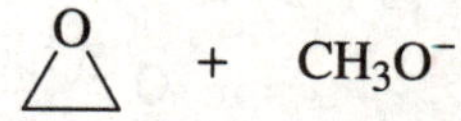

d.

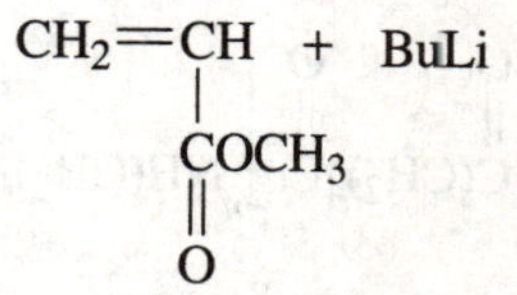

12. 3,3-Dimethyloxacyclobutane undergoes cationic polymerization by the following mechanism.

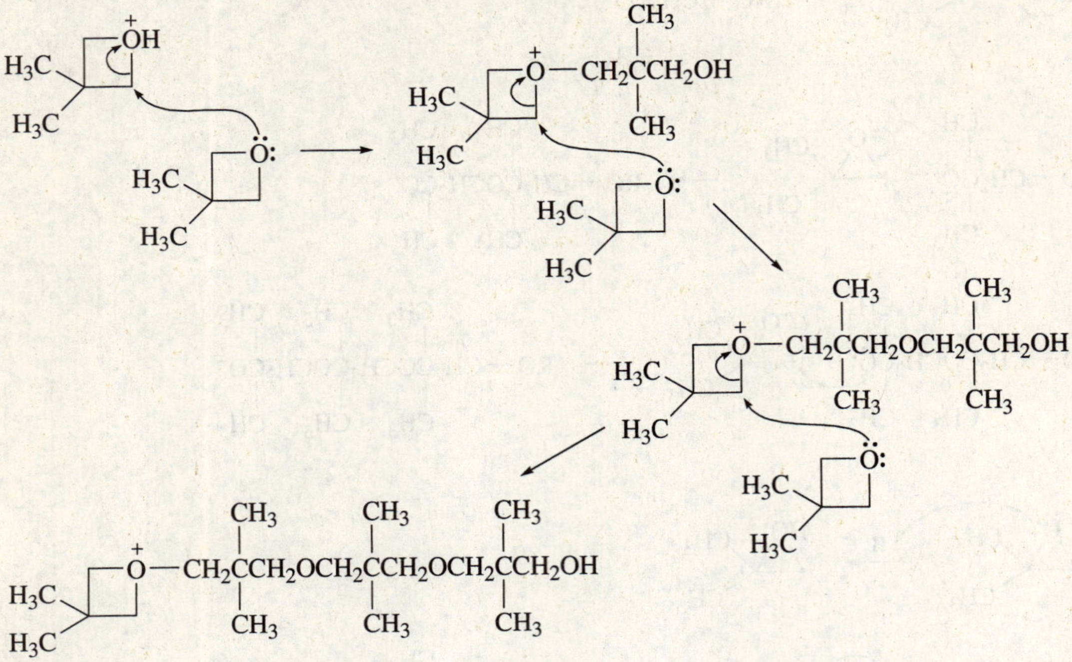

13.

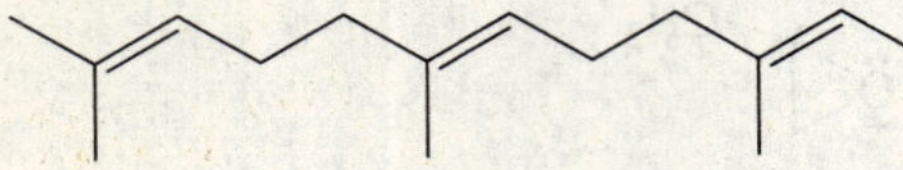

14.

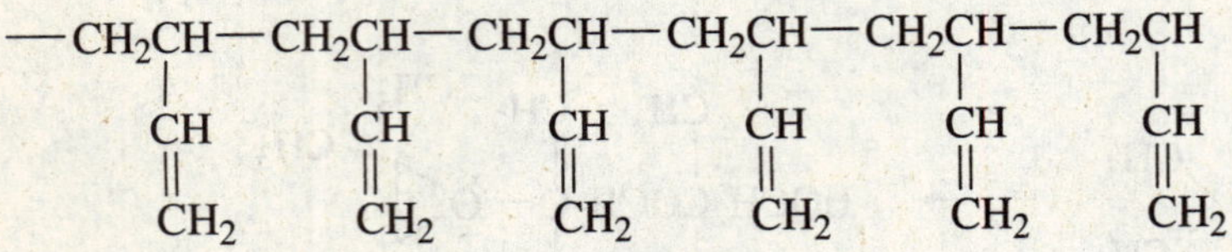

15. **a.**

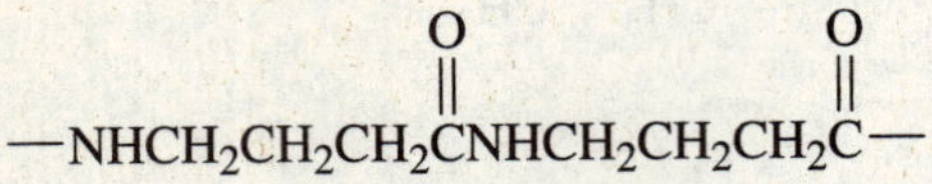

b.

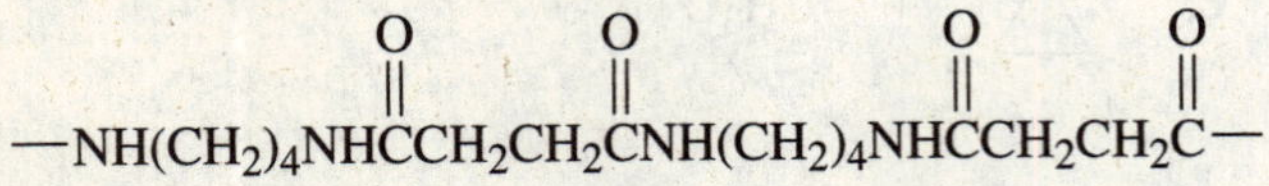

16.

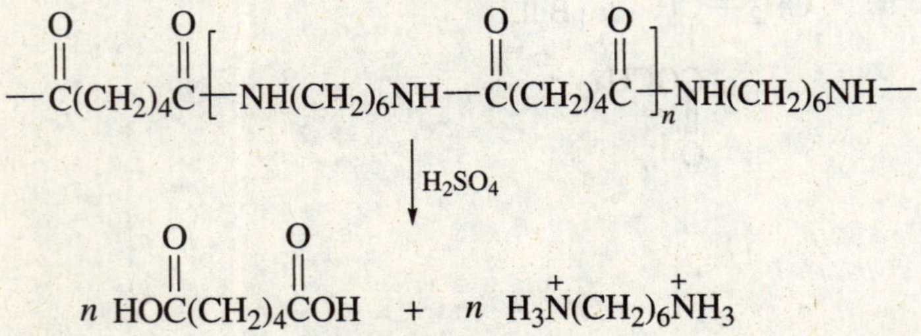

17.

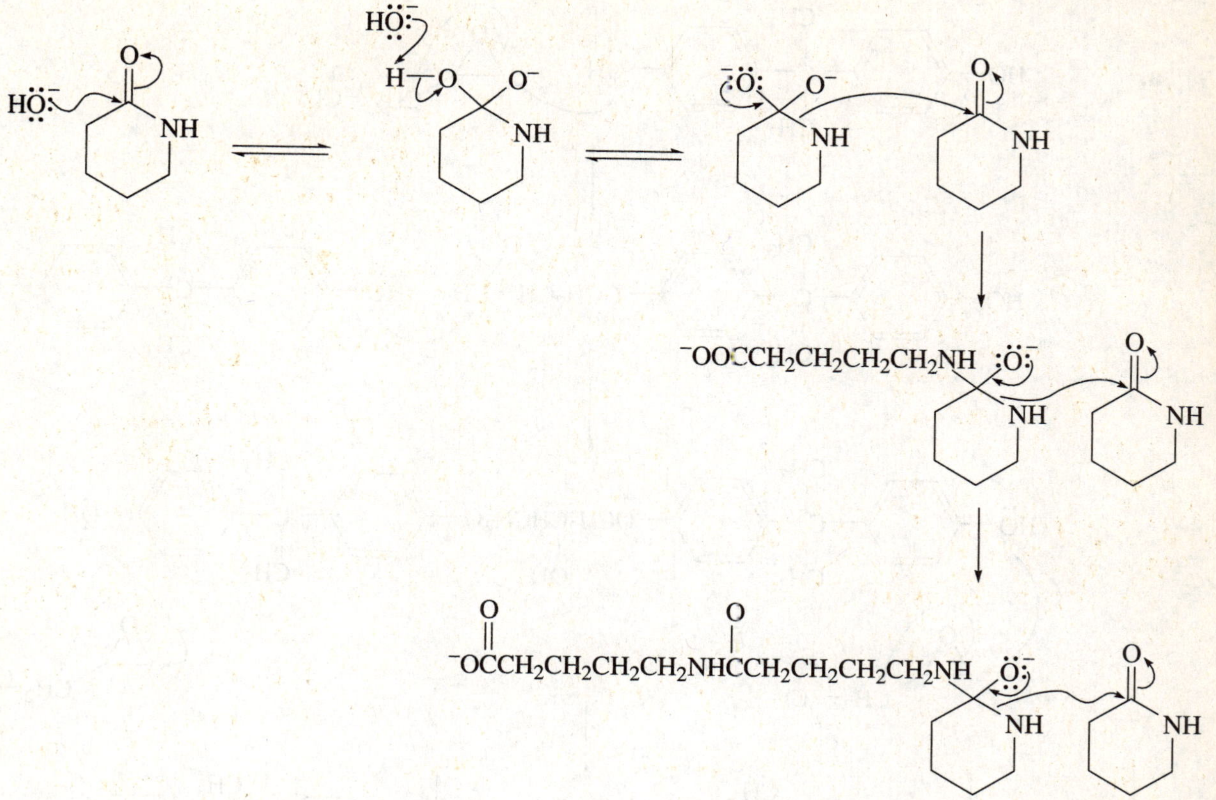

18. They hydrolyze to give salts of dicarboxylic acids and diols.

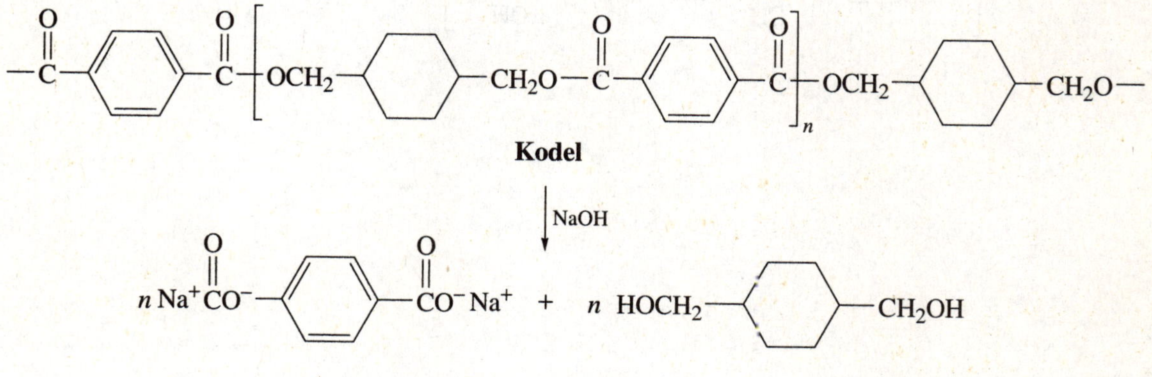

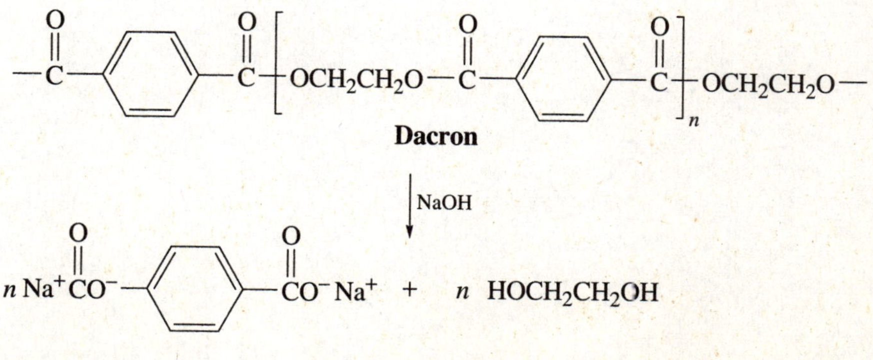

19. **a.**

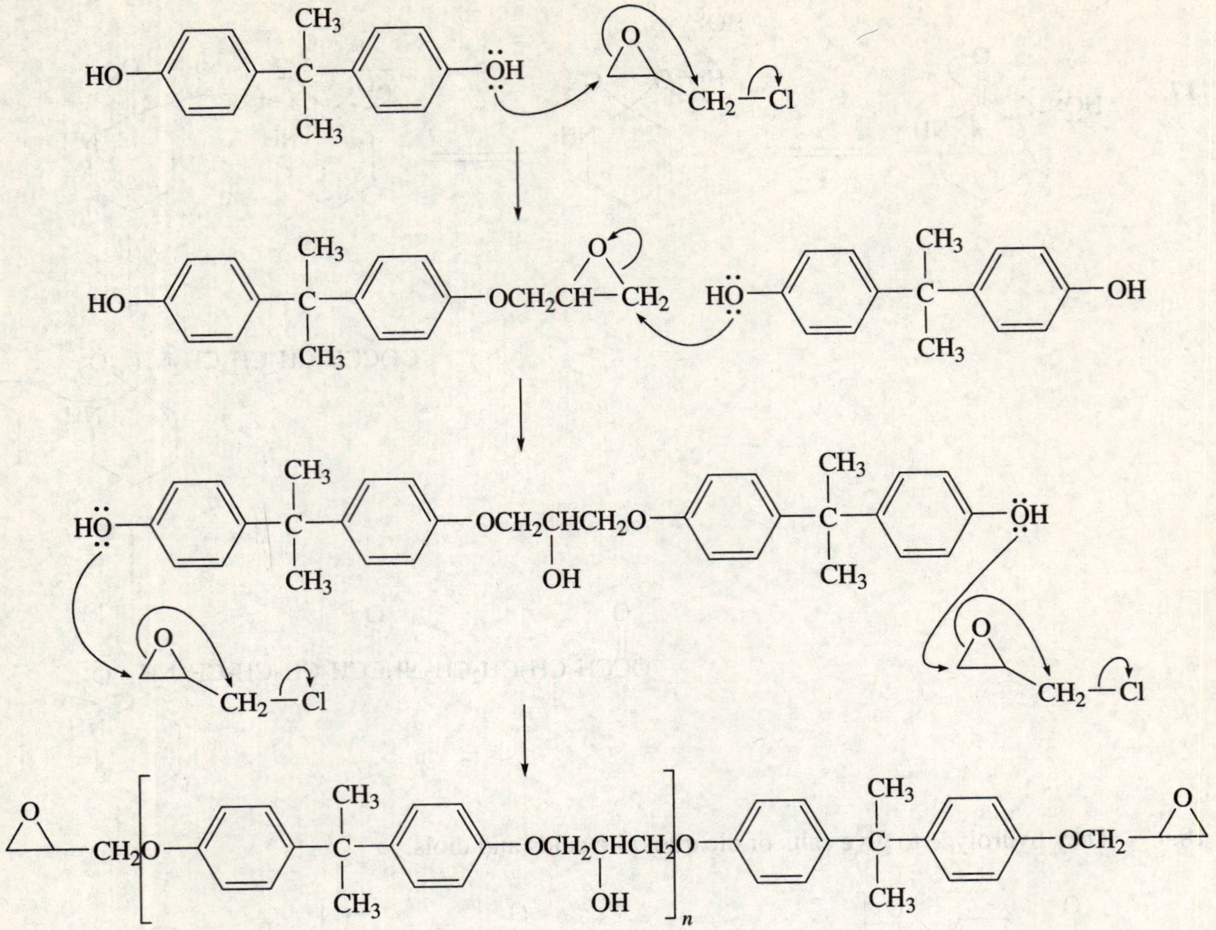

b.

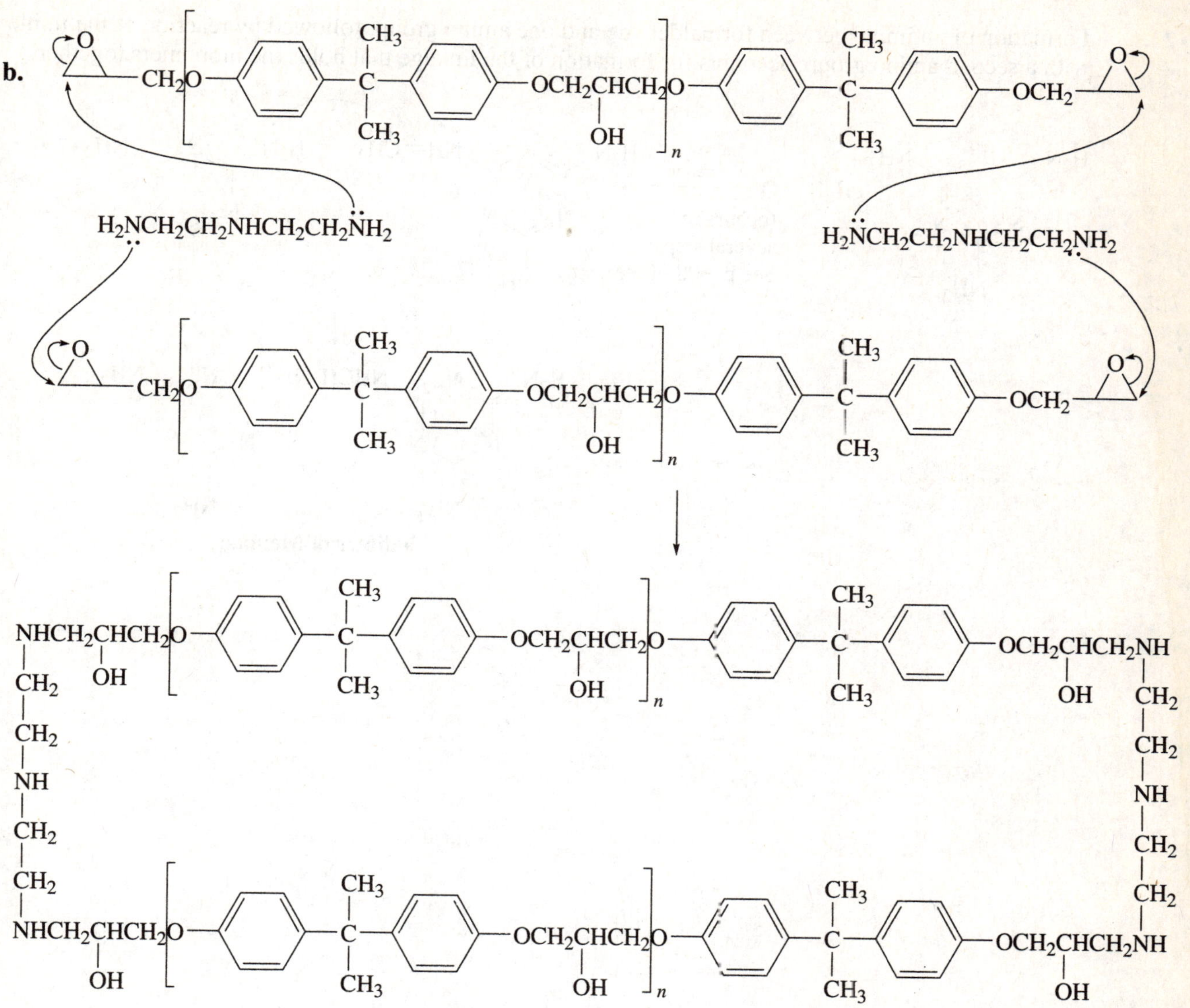

20. The polymer is stiffer because glycerol cross-links the polymer chains.

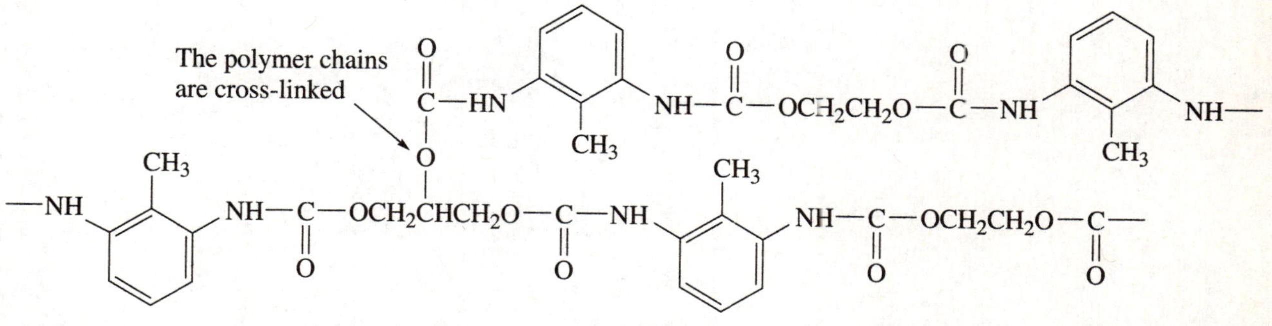

21. Formation of an imine between formaldehyde and one amino group, followed by reaction of the imine with a second amino group, accounts for formation of the linkage that holds the monomers together.

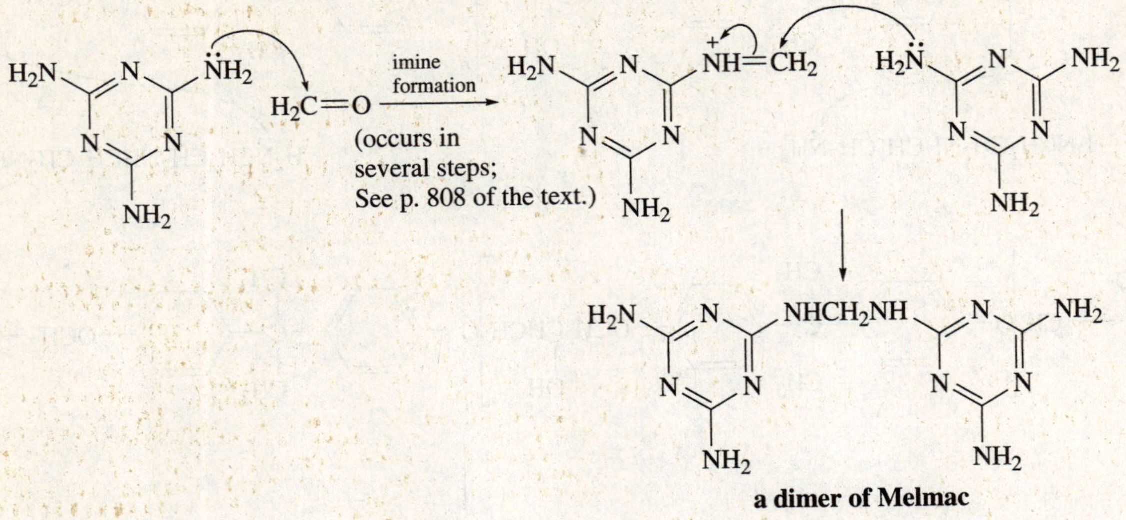

22.

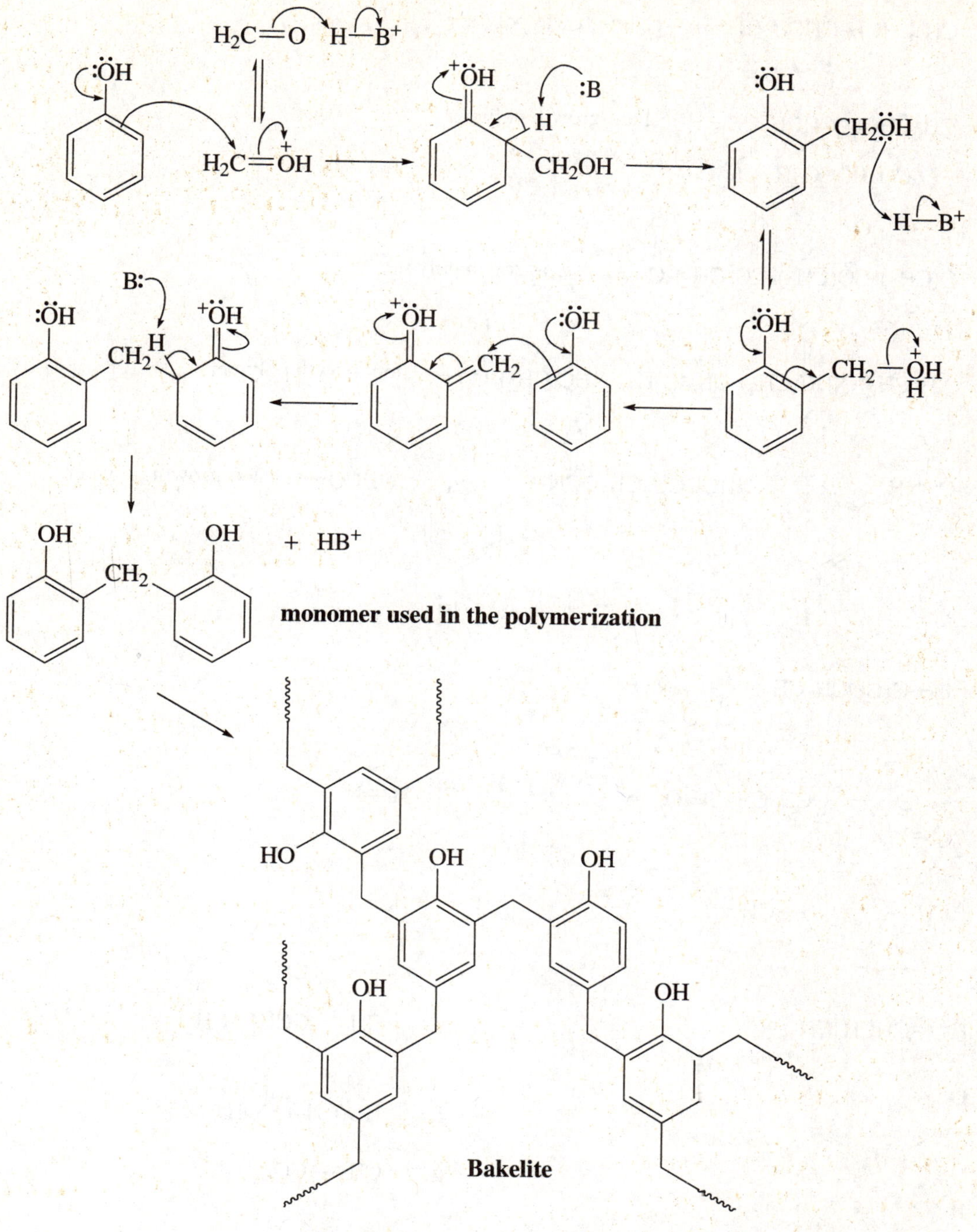

23. **a.** —$CH_2CH(F)CH_2CH(F)CH_2CH(F)$— chain-growth polymer

b. —$CH_2CH(CO_2H)CH_2CH(CO_2H)CH_2CH(CO_2H)$— chain-growth polymer

c. —$O(CH_2)_5C(=O)O(CH_2)_5C(=O)O(CH_2)_5C(=O)$— step-growth polymer

d. —$NH(CH_2)_5NHC(=O)(CH_2)_5C(=O)NH(CH_2)_5NHC(=O)(CH_2)_5C(=O)$— step-growth polymer

e. 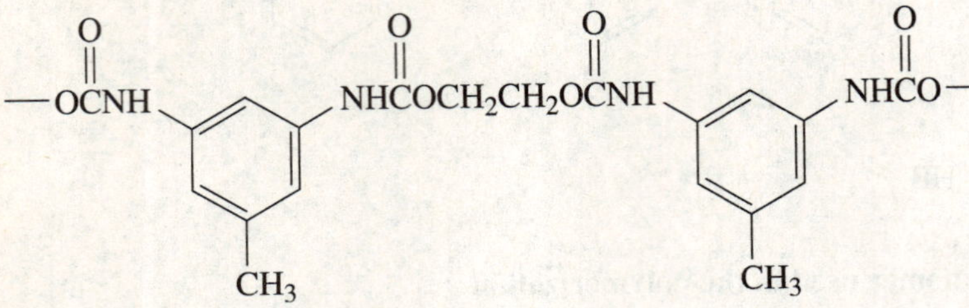 step-growth polymer

24. **a.** 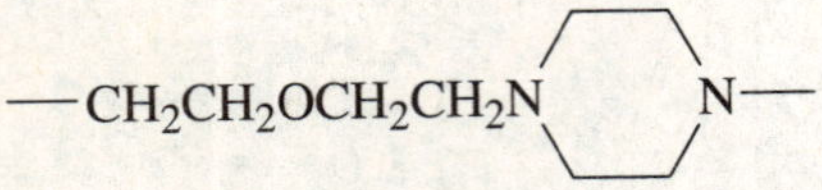

b. 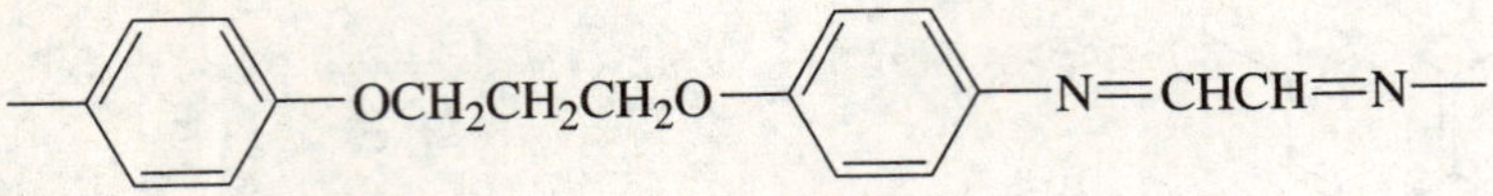

c. 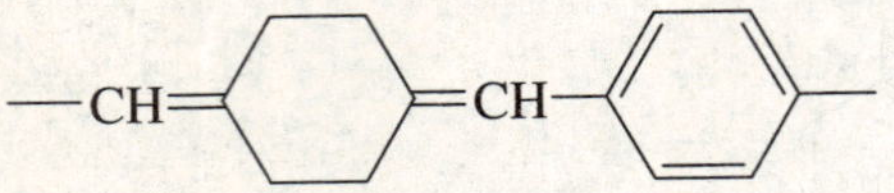

25. **a.** $CH_2{=}CHCH_2CH_3$

b. 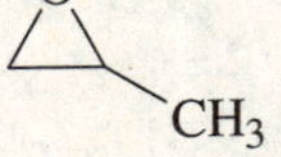

c. 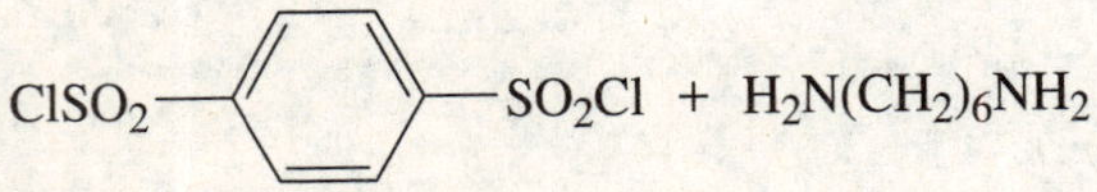

d. 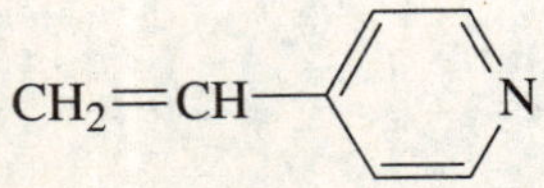

e. $CH_2{=}C(CH_3)CH{=}CH_2$

f. $HO(CH_2)_4C(=O)OH$

g. 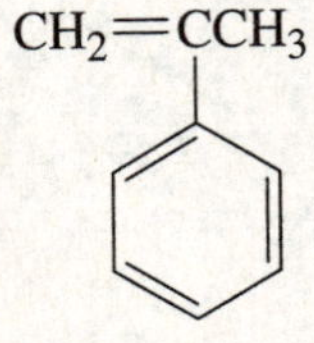

h. 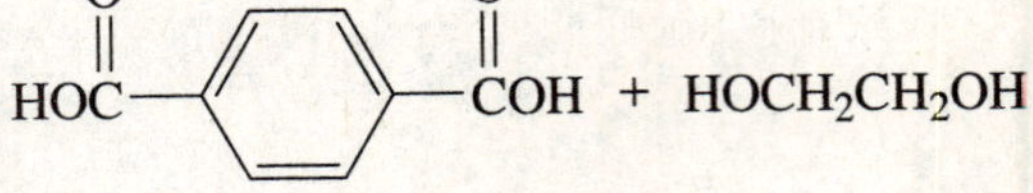

a, **b**, **d**, **e**, and **g** are chain-growth polymers.

c, **f**, and **h** are step-growth polymers.

26. Whether a polymer is isotactic, syndiotactic, or atactic depends on whether the substituents are all on one side of the carbon chain, alternate on both sides of the chain, or are random with respect to the chain. Because a polymer of isobutylene has two identical substituents at each carbon in the chain, different configurations are not possible.

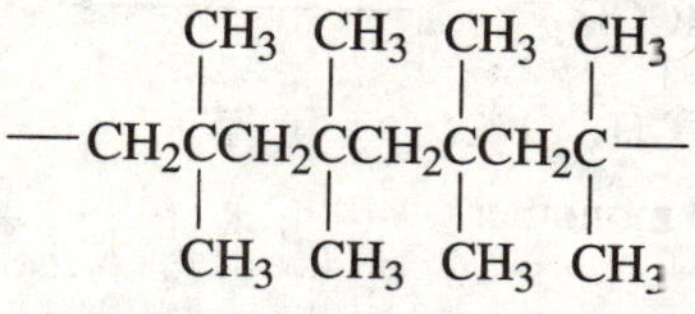

polymer of isobutylene

27. **a.**

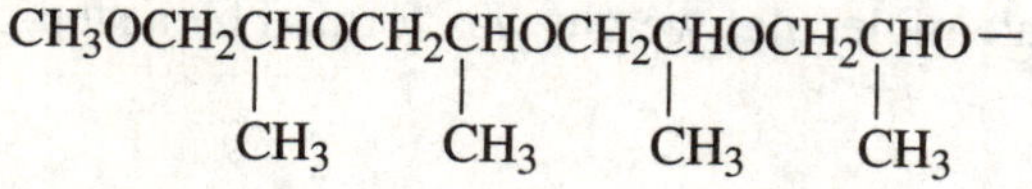

b.

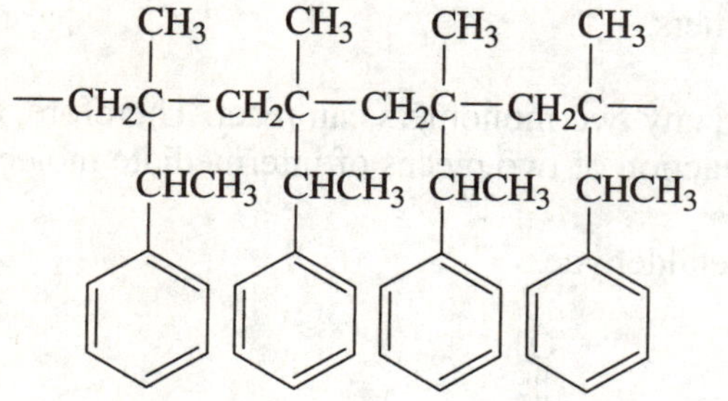

c.

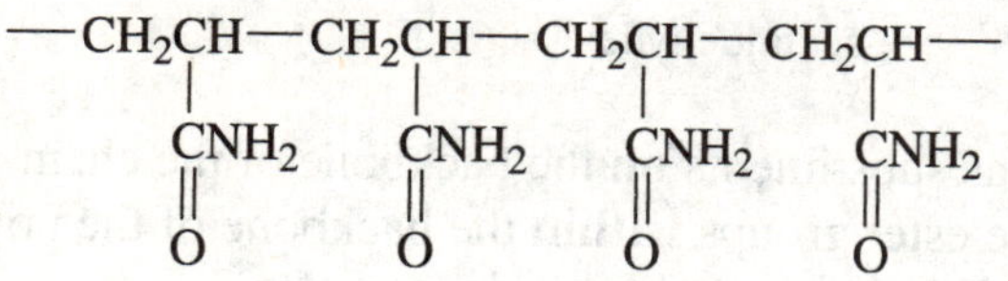

d. Depending on the particular Ziegler-Natta catalyst used, the double bond can be either cis or trans

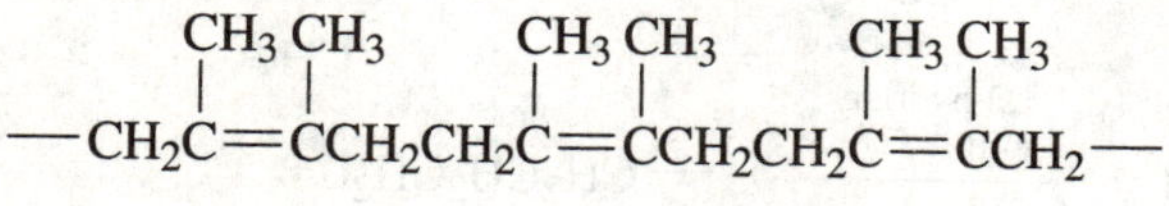

or

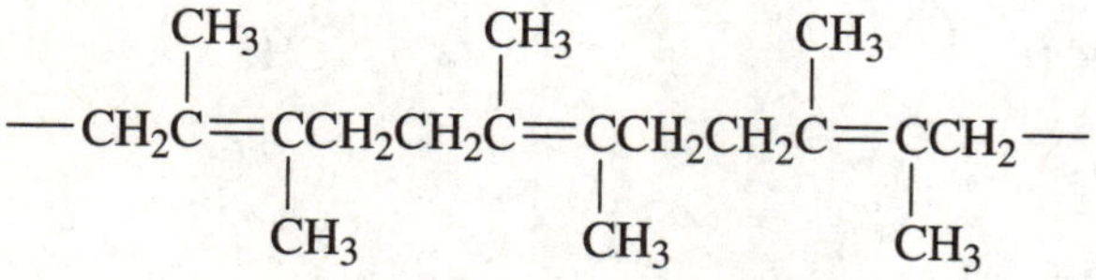

e.

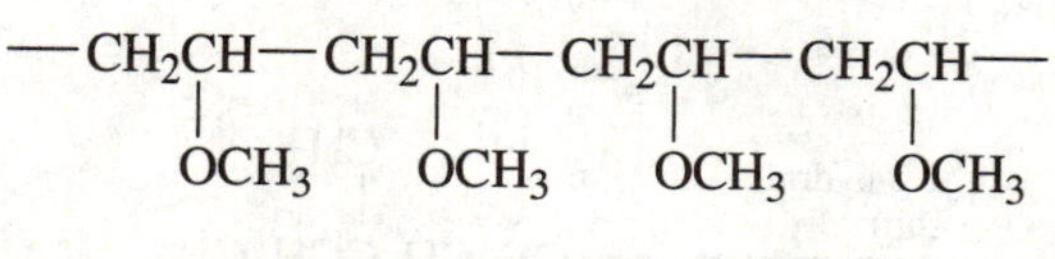

28. **a.** Because it is a polyamide (and not a polyester), it is a nylon.

b.

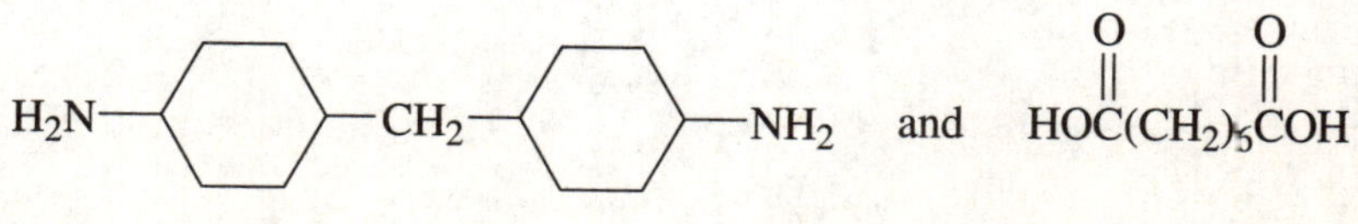

29. A copolymer is a polymer composed of more than one kind of monomer. Because the initially formed carbocation can rearrange, two different monomers are involved in formation of the polymer.

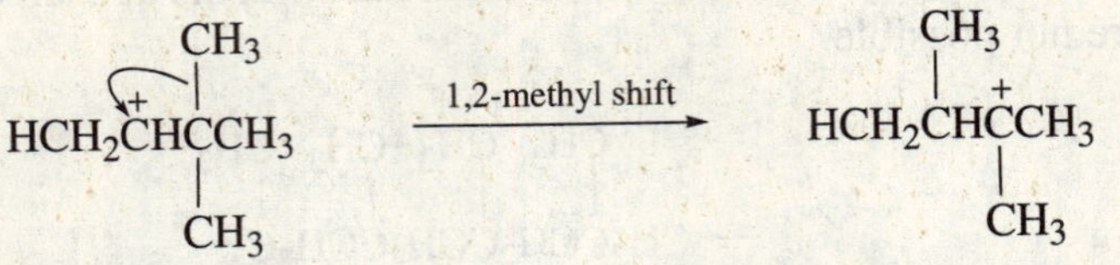

unrearranged monomer — rearranged monomer

30. The polymer in the flask that contained a high molecular weight polymer and little material of intermediate molecular weight was formed by a chain-growth mechanism, whereas the polymer in the flask that contained mainly material of intermediate molecular weight was formed by a step-growth mechanism.

In a chain-growth mechanism, monomers are added to the growing end of a chain. This means that at any one time there will be polymeric chains and monomers.

Step-growth polymerization is not a chain reaction; any two monomers can react. Therefore, high molecular weight material will be formed by the reaction of two pieces of intermediate molecular weight.

31. **a.** Vinyl alcohol is unstable, it tautomerizes to acetaldehyde.

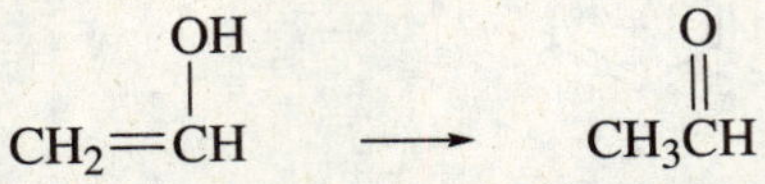

vinyl alchol — acetaldehyde

b. It is not a true polyester. It has ester groups as substituents **on** the backbone of the chain so it does have "polyester" groups, but it does not have ester groups **within** the backbone of the polymer chain. A true polyester has ester groups in the backbone of the polymer chain.

32. Each of the five carbocations shown below can add the growing end of the polymer chain.

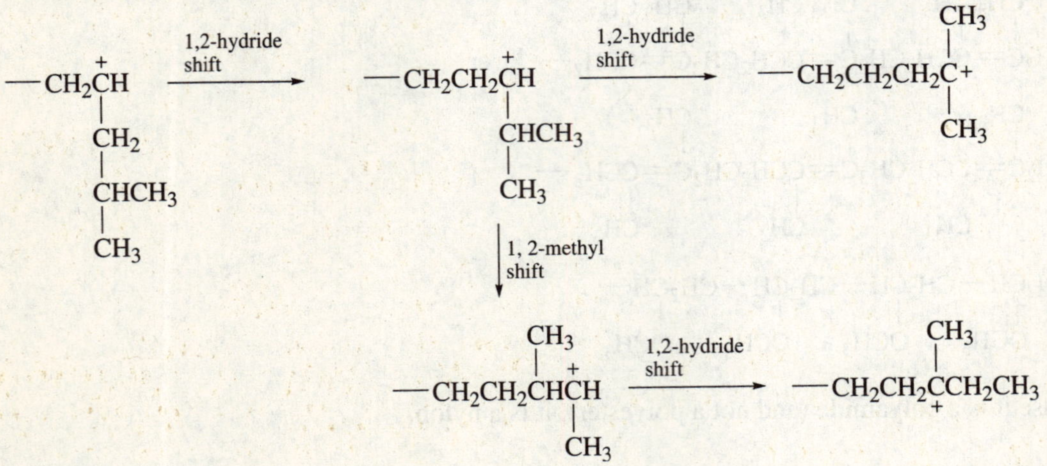

33. Because 1,4-divinylbenzene has substituents on both ends of the benzene ring that can engage in polymerization, the polymer chains can become cross-linked, which increases the rigidity of the polymer.

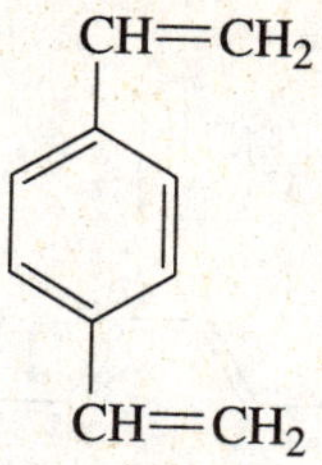

1,4-divinylbenzene

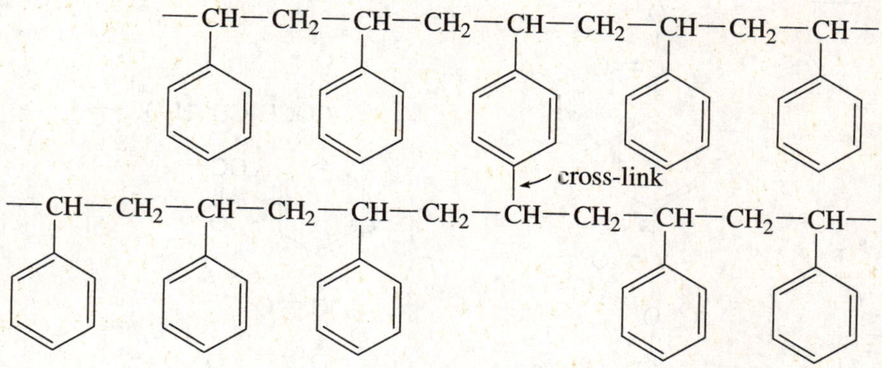

34. Glyptal gets its strength from cross-linking.

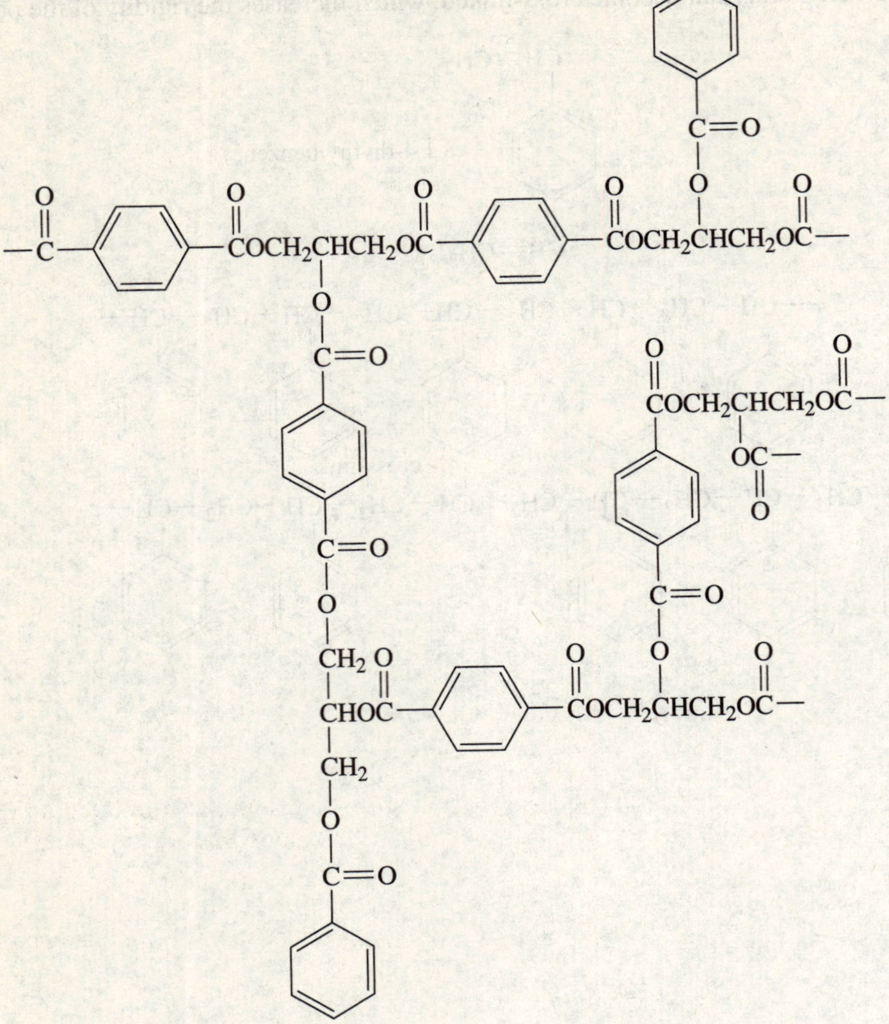

35.

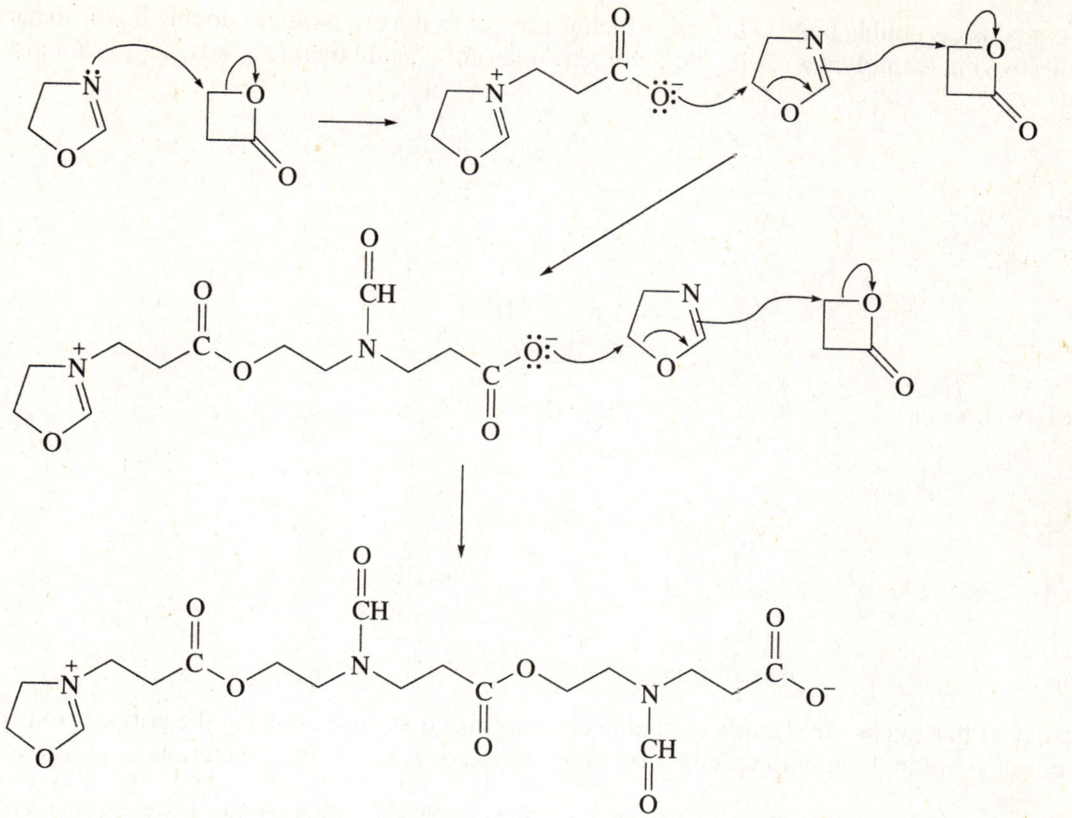

36. Both compounds can form esters via intramolecular or intermolecular reactions. The product of the intramolecular reaction is a lactone; the intermolecular reaction leads to a polymer.

5-Hydroxypentanoic acid reacts intramolecularly to form a six-membered-ring lactone, whereas 6-hydroxyhexanoic acid reacts intramolecularly to form a seven-membered-ring lactone.

$HOCH_2CH_2CH_2CH_2\overset{\overset{O}{\|}}{C}OH$ ⟶ 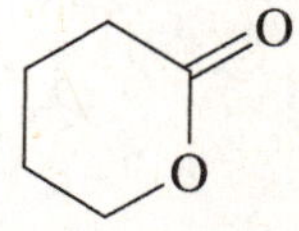

5-hydroxypentanoic acid

$HOCH_2CH_2CH_2CH_2CH_2\overset{\overset{O}{\|}}{C}OH$ ⟶ 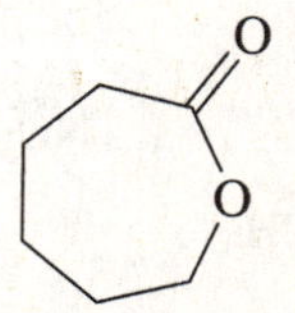

6-hydroxypentanoic acid

The compound that forms the most polymer will be the one that forms the least lactone, because the two reactions compete with one another.

The six-membered-ring lactone is more stable and so is the transition state for its formation, compared to a seven-membered ring lactone. Since it is easier for 5-hydroxypentanoic acid to form the six-membered ring lactone, than for 6-hydroxyhexanoic acid to form the seven-membered ring lactone, **6-hydroxypentanoic acid** will form more polymer.

37. Rubber contains cis double bonds. Ozone, which is present in the air, oxidizes double bonds to carbonyl groups, destroying the polymer chain. Polyethylene does not contain double bonds, so it is not air-oxidized.

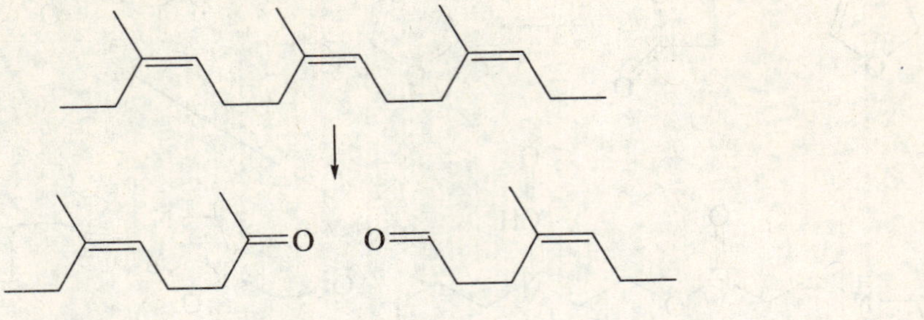

38.

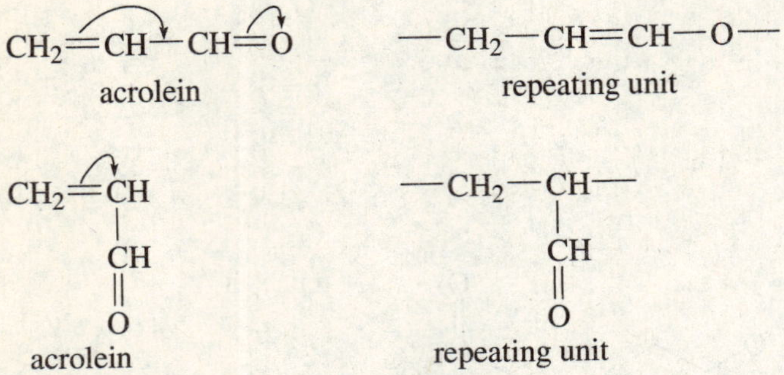

39. The plasticizer that keeps vinyl soft and pliable can vaporize over time, causing the polymer to become brittle. For this reason, high boiling materials are preferred over low boiling materials as plasticizers.

40. **a.** Because the negative charge on the propagation site can be delocalized onto the carbonyl oxygen, the polymer is best prepared by anionic polymerization.

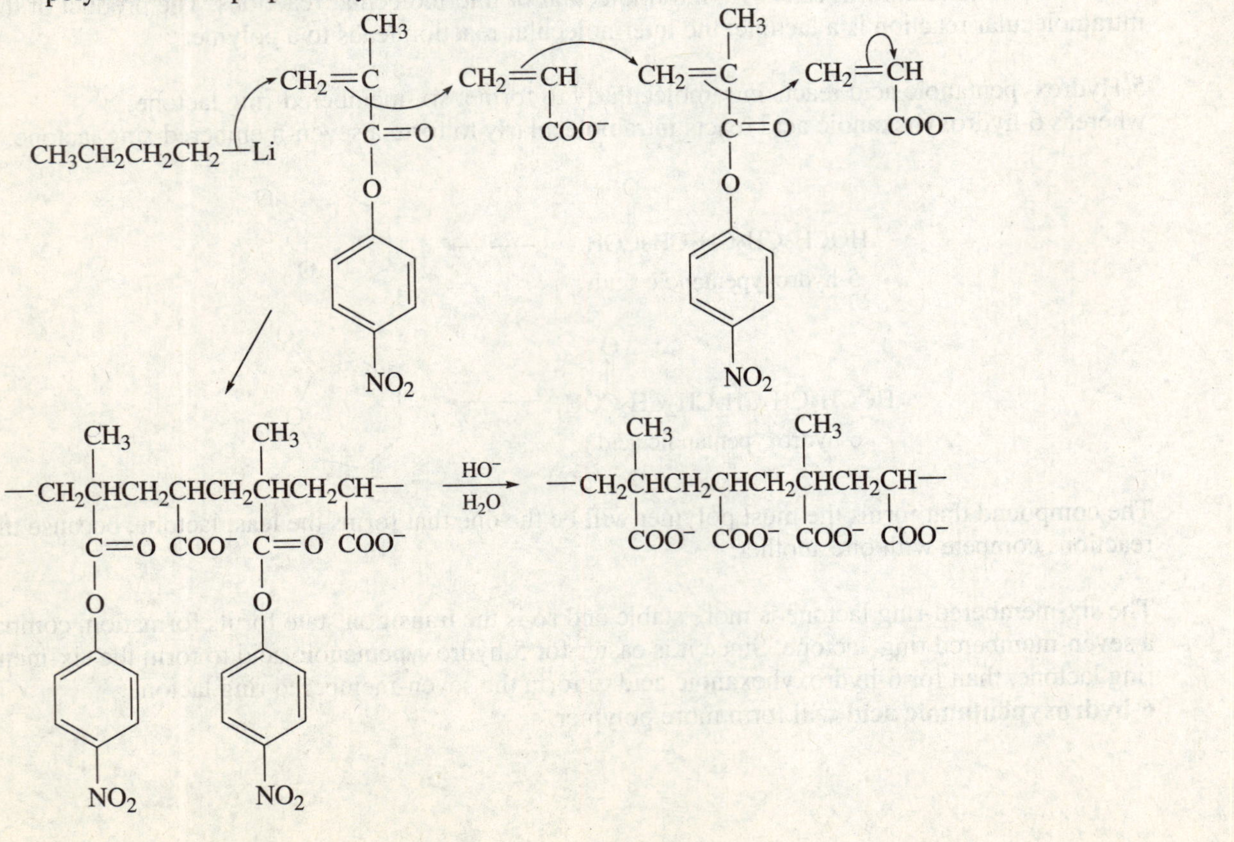

b. The carboxyl substituent is in position to remove a proton from water, making water a better nucleophile.

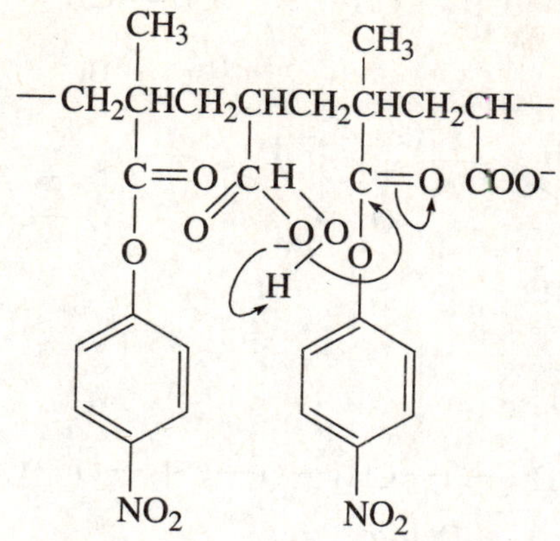

41. Hydrolysis converts the ester substituents into alcohol substituents, which can react with ethylene oxide to graft a polymer of ethylene oxide onto the backbone of the random polymer of styrene and vinyl acetate.

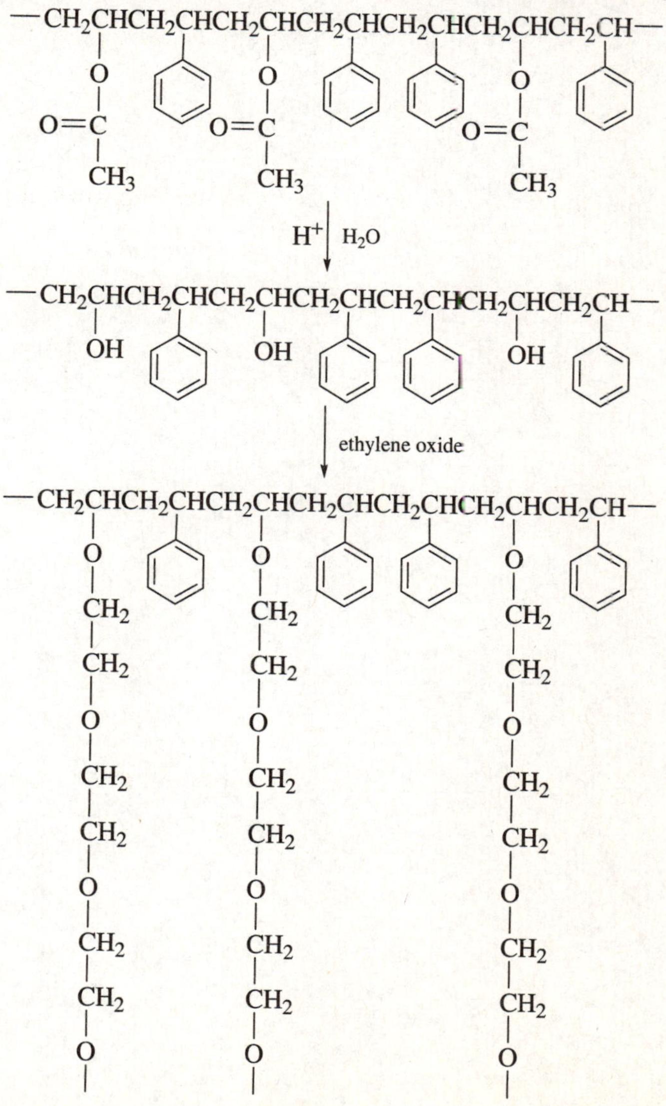

42. The desired polymer is an alternating copolymer of ethylene and 1,2-dibromoethylene.

43. **a.**

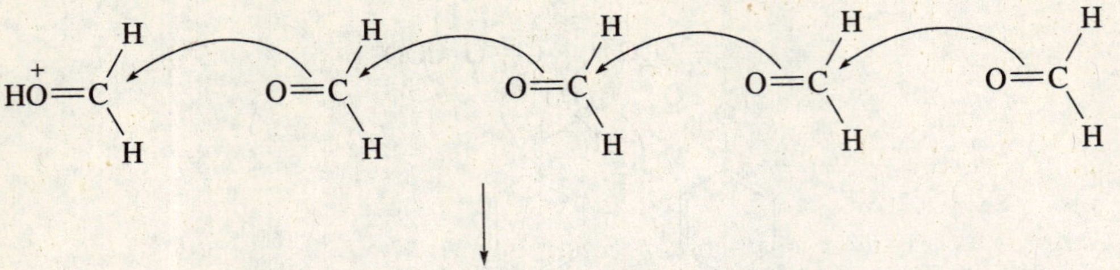

$$HO{-}CH_2{-}O{-}CH_2{-}O{-}CH_2{-}O{-}CH_2{-}O{-}CH_2{-}$$

Delrin

b. Delrin is a chain-growth polymer.

CHAPTER 29
Pericyclic Reactions

Important Terms

antarafacial bond formation formation of two σ bonds from opposite sides of the π system.

antarafacial rearrangement a rearrangement where the migrating group moves to the opposite face of the π system.

antibonding π molecular orbital a molecular orbital that results when two parallel atomic orbitals with opposite phases interact. Electrons in an antibonding orbital decrease bond strength.

antisymmetric molecular orbital a molecular orbital in which the left half is not a mirror image of the right half but would be if one half of the MO were turned upside down.

bonding π molecular orbital a molecular orbital that results when two parallel atomic orbitals with the same phase interact. Electrons in a bonding orbital increase bond strength.

Claisen rearrangement a [3,3] sigmatropic rearrangement of an allyl vinyl ether.

conrotatory ring closure achieves head-to-head overlap of π orbitals by rotating the orbitals in the same direction.

conservation of orbital symmetry theory a theory that explains the relationship between the structure and stereochemistry of the reactant, the conditions under which a pericyclic reaction takes place, and the stereochemistry of the product.

Cope rearrangement a [3,3] sigmatropic rearrangement of a 1,5-diene.

cycloaddition reaction a reaction in which two π-bond-containing molecules react to form a cyclic compound.

disrotatory ring closure achieves head-to-head overlap of π orbitals by rotating the orbitals in opposite directions.

electrocyclic reaction a reaction in which a π bond in the reactant is lost so that a cyclic compound with a new σ bond can be formed.

excited state a description of which orbitals the electrons of an atom or molecule occupy when an electron in the ground state has been moved to a higher energy orbital.

frontier orbital analysis determining the outcome of a pericyclic reaction using frontier orbitals.

frontier orbitals the HOMO and the LUMO.

frontier orbital theory a theory that, like the conservation of orbital symmetry, explains the relationship between reactant, product, and reaction conditions in a pericyclic reaction.

ground state	a description of which orbitals the electrons of an atom or molecule occupy when they are all in their lowest energy orbitals.
highest occupied molecular orbital (HOMO)	the molecular orbital of highest energy that contains an electron.
linear combination of atomic orbitals (LCAO)	the combination of atomic orbitals to produce a molecular orbital.
lowest unoccupied molecular orbital (LUMO)	the molecular orbital of lowest energy that does not contain an electron.
molecular orbital (MO) theory	describes a model in which the electrons occupy orbitals as they do in atoms but with the orbitals extending over the entire molecule.
pericyclic reaction	a reaction that occurs as a result of a cyclic reorganization of electrons.
photochemical reaction	a reaction that takes place when a reactant absorbs light.
polar reaction	the reaction between a nucleophile and an electrophile.
radical reaction	a reaction in which a new bond is formed using one electron from one reagent and one electron from another reagent.
selection rules	the rules that determine the outcome of a pericyclic reaction.
sigmatropic rearrangement	a reaction in which a σ bond is broken in the reactant, a new σ bond is formed in the product, and the π bonds rearrange.
suprafacial bond formation	formation of two σ bonds from the same side of the π system.
suprafacial rearrangement	a rearrangement where the migrating group remains on the same face of the π system.
symmetric molecular orbital	a molecular orbital in which the left half is a mirror image of the right half.
symmetry-allowed pathway	a pathway that leads to overlap of in-phase orbitals.
symmetry-forbidden pathway	a pathway that leads to overlap of out-of-phase orbitals.
thermal reaction	a reaction that takes place without the reactant having to absorb light.
Woodward-Hoffmann rules	a series of selection rules for pericyclic reactions.

Solutions to Problems

1. a. electrocyclic reaction
   b. sigmatropic rearrangement
   c. cycloaddition reaction
   d. cycloaddition reaction

2. a. bonding orbitals = ψ_1, ψ_2, ψ_3; antibonding orbitals = ψ_4, ψ_5, ψ_6
   b. ground-state HOMO = ψ_3; ground-state LUMO = ψ_4
   c. excited-state HOMO = ψ_4; excited-state LUMO = ψ_5
   d. symmetric orbitals = ψ_1, ψ_3, ψ_5; antisymmetric orbitals = ψ_2, ψ_4, ψ_6
   e. The HOMO and LUMO have opposite symmetries.

3. a. 8 π molecular orbitals
   b. ψ_4
   c. 7 nodes (there is also a node that passes through the nuclei, that is, through the centers of the *p* orbitals).

4. a. 1,3-Pentadiene has two conjugated π bonds, so it has the same π molecular orbital description as 1,3-butadiene, a compound that also has two conjugated π bonds. (See Figure 29.2 on page 1267 of the text.)
   b. The π bonds in 1,4-pentadiene are isolated, so its π molecular orbital description is the same as ethene, a compound with an isolated π bond. (See Figure 29.1 on page 1266 of the text.)
   c. 1,3,5-Heptatriene has three conjugated π bonds, so it has the same π molecular orbital description as 1,3,5-hexatriene, a compound that also has three conjugated π bonds. See Figure 29.3 on page 1268 of the text.)
   d. 1,3,5,8-Nonatetraene has three conjugated π bonds and an isolated π bond. The three conjugated π bonds are described by Figure 29.3 and the isolated π bond by Figure 29.1.

5. a. 2, 4, or 6 conjugated double bonds

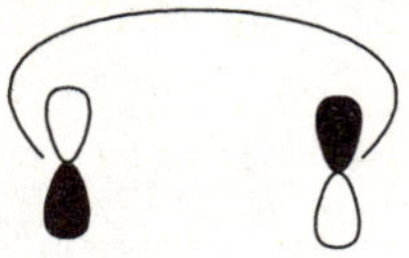

3, 5, or 7 conjugated double bonds

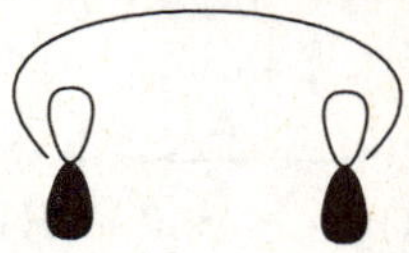

6. a. (2*E*,4*Z*,6*Z*,8*E*)-Decatetraene has an even number of conjugated π bonds (4). Therefore, under thermal conditions, ring closure will be conrotatory.
   b. The substituents point in opposite directions, and conrotatory ring closure of such substituents will cause them to be trans in the ring-closed product.
   c. Under photochemical conditions, a compound with four conjugated π bonds will undergo disrotatory ring closure.
   d. Because ring closure is disrotatory and the substituents point in opposite directions, the product will have the cis configuration.

7. a. correct b. correct c. correct

8. 1. a. conrotatory
      b. trans
   2. a. disrotatory
      b. cis

9. The reaction of maleic anhydride with 1,3-butadiene involves three π bonds in the reacting system, and such a reaction under thermal conditions involves suprafacial ring closure.

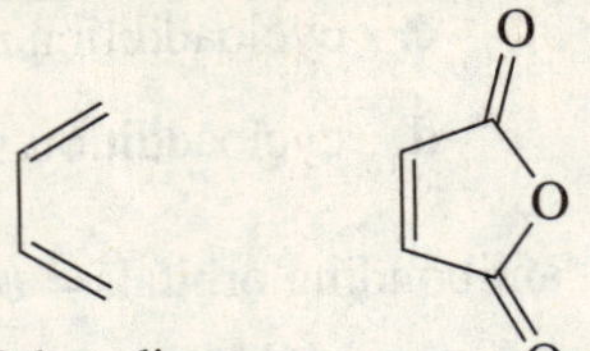

1,3-butadiene

The reaction of maleic anhydride with ethylene involves two π bonds in the reacting system, and such a reaction under thermal conditions involves antarafacial ring closure, which cannot occur with a four-membered ring.

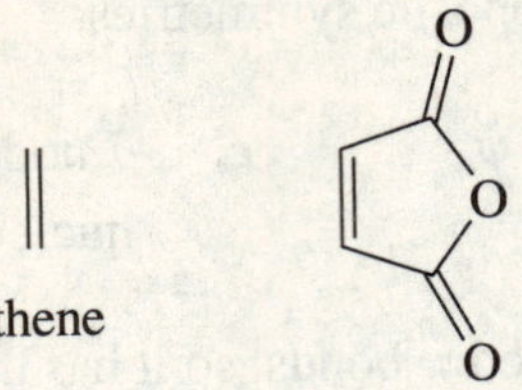

ethene

10. Solved in the text.

11. Because we are dealing with formation of a small ring, ring closure will have to be suprafacial. Under photochemical conditions, suprafacial ring closure requires an even number of π bonds in the reacting system. Thus, a concerted reaction will occur but it will use only one of the π bonds of 1,3-butadiene.

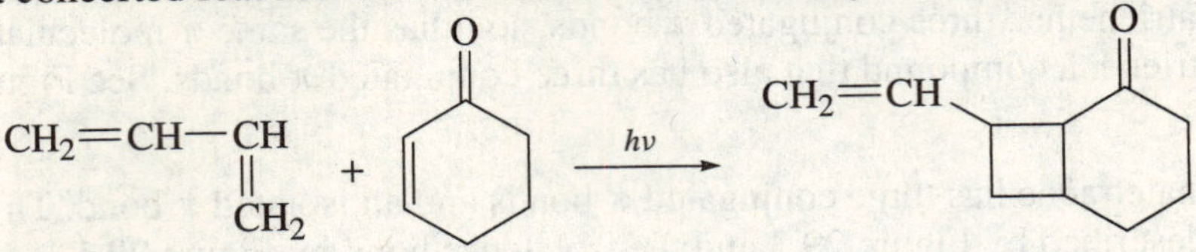

12. **a.** **1.** [1,7] sigmatropic rearrangement
2. [1,5] sigmatropic rearrangement
3. [5,5] sigmatropic rearrangement
4. [3,3] sigmatropic rearrangement

b. **1.**

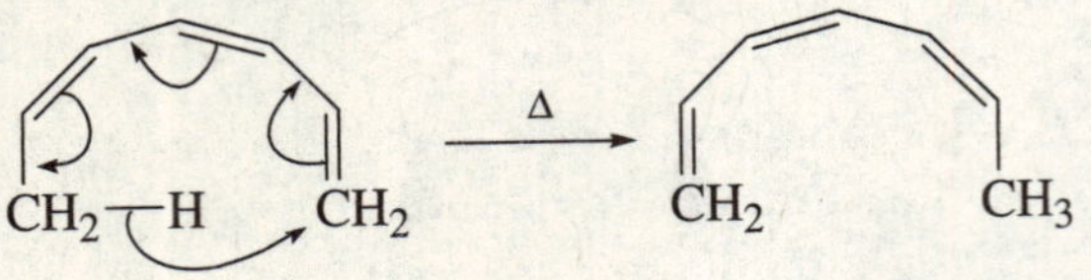

2.

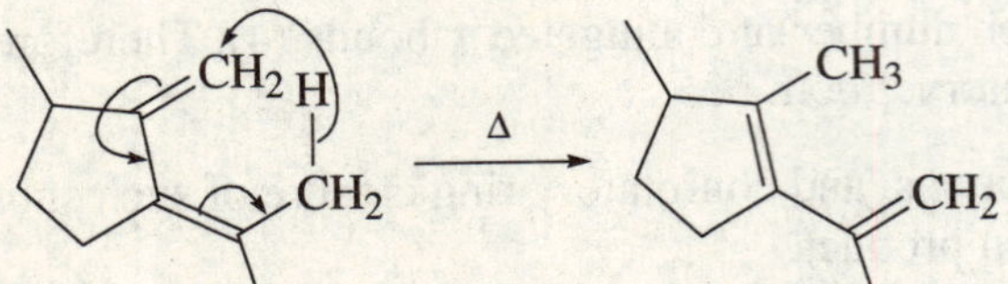

3.

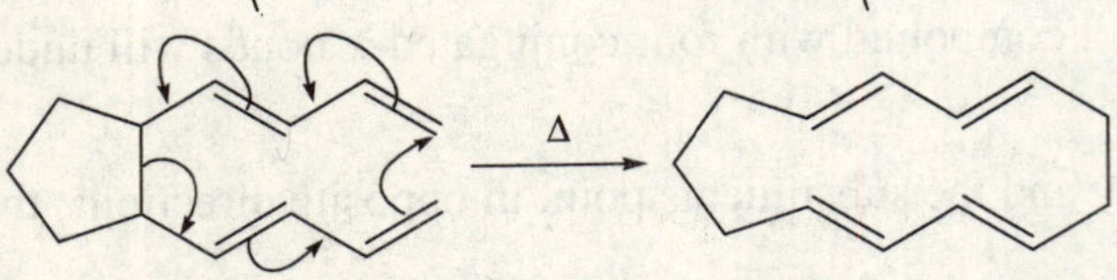

4.

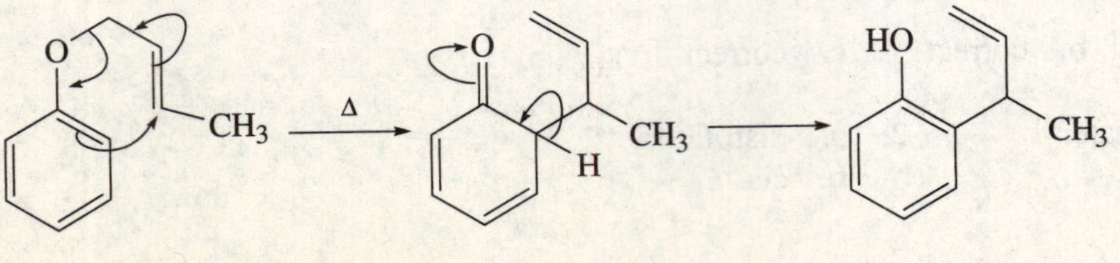

13. **a.** and **b.**

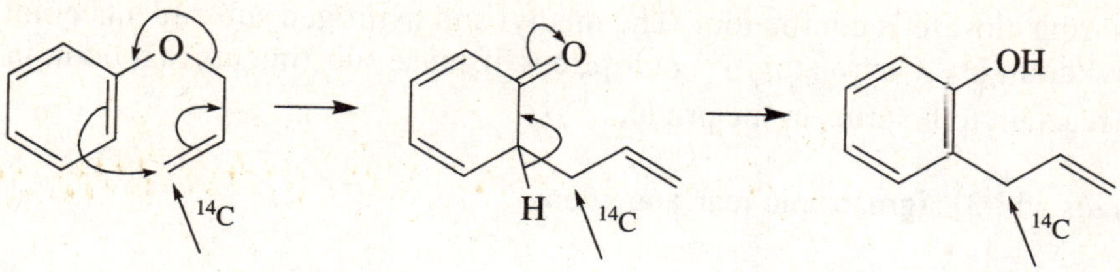

14. If a nondeuterated reactant had been used, the product would have been identical to the reactant. Therefore, the rearrangement would not have been detectable.

15. A suprafacial rearrangement can take place under photochemical conditions if there are an even number of electrons in the reacting system. Therefore, a 1,3-hydrogen shift occurs.

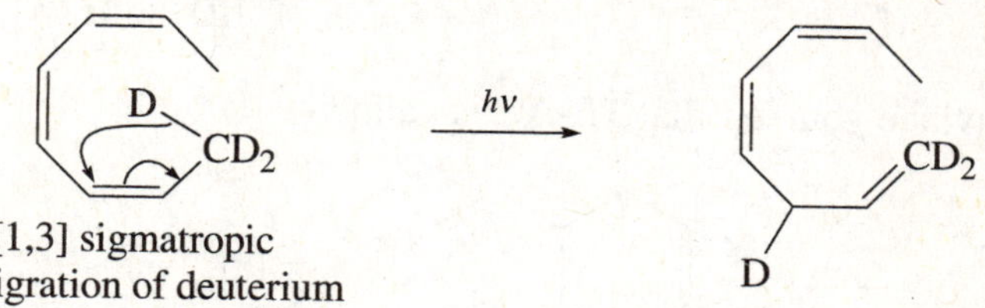

a [1,3] sigmatropic
migration of deuterium

A suprafacial rearrangement can take place under thermal conditions if there are an odd number of electrons in the reacting system. Therefore, a 1,5-hydrogen shift occurs.

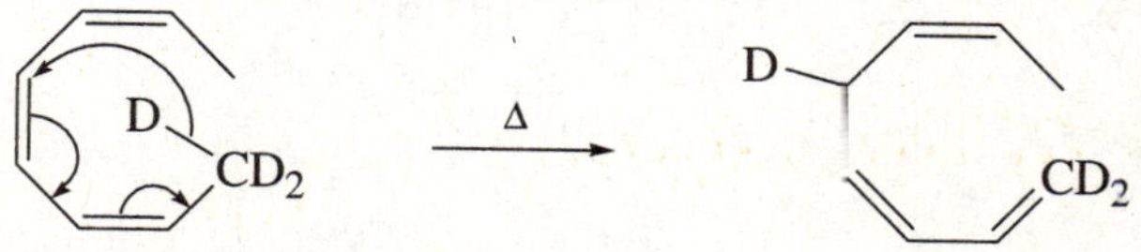

a [1,5] sigmatropic
migration of deuterium

16. Solved in the text.

17. [1,3] Sigmatropic migrations of hydrogen cannot occur under thermal conditions, because the four-membered transition state does not allow the required antarafacial rearrangement.

[1,3] Sigmatropic migrations of carbon can occur under thermal conditions because carbon can achieve the required antarafacial rearrangement by using both of its lobes when it migrates.

18. **a.** Because 1,3-migration of carbon requires carbon to migrate using both of its lobes (it involves an even number of pairs of electrons, so it takes place by an antarafacial pathway), migration will occur with inversion of configuration.

b. Because 1,5-migration of carbon requires carbon to migrate using only one of its lobes (it involves an odd number of pairs of electrons, so it takes place by a suprafacial pathway), migration will occur with retention of configuration.

19. Because the [1,7] sigmatropic rearrangement takes place under thermal conditions and involves an even number (4) of pairs of electrons, migration of hydrogen involves antarafacial rearrangement.

20. Because the reactant (provitamin D3) has an odd number (3) of conjugated π bonds and reacts under photochemical conditions, ring closure is conrotatory. The methyl and hydrogen substituents point in opposite directions in provitamin D3. Conrotatory ring closure will cause substituents that point in opposite directions in the reactant to be trans in the product.

21. Chorismate mutase catalyzes a [3,3] sigmatropic rearrangement.

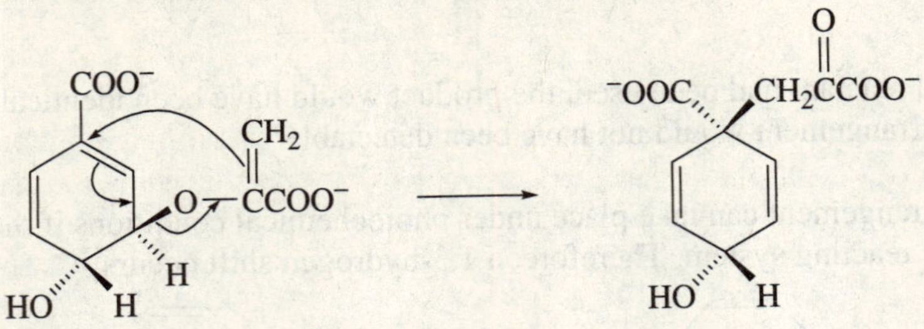

22. Were you able to convince yourself that TE-AC is valid?

23. **a.**

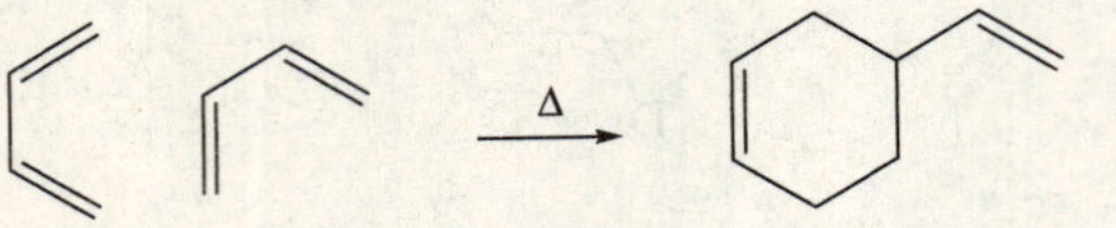

b.

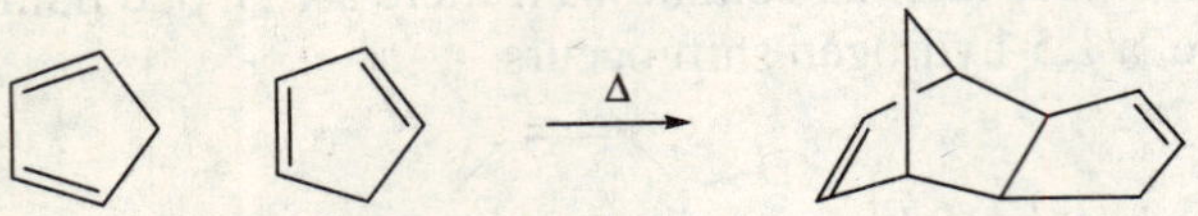

c.

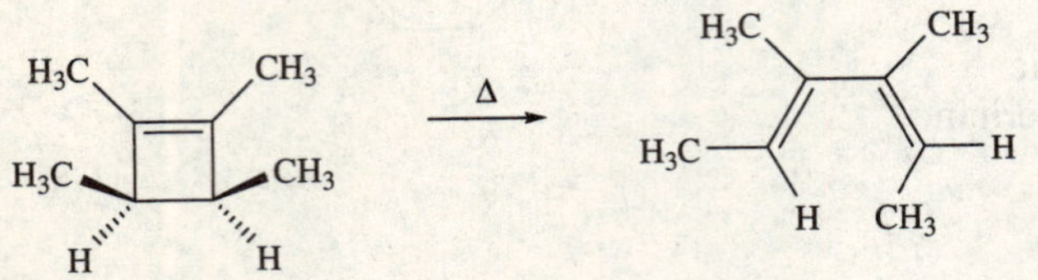

d.

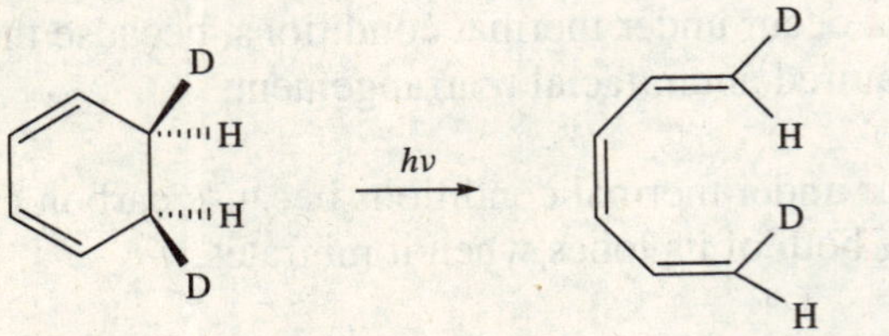

e.

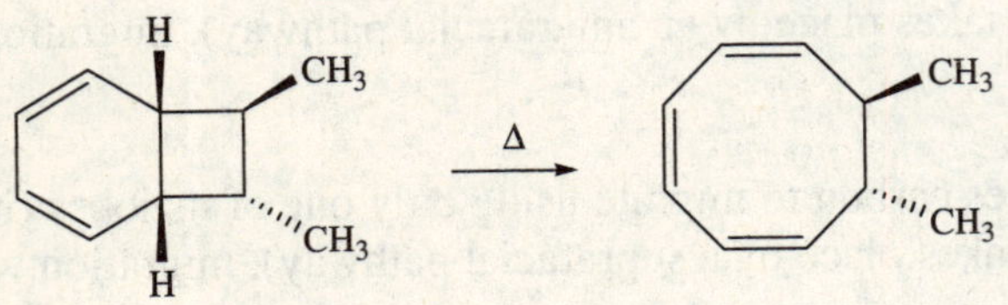

f.

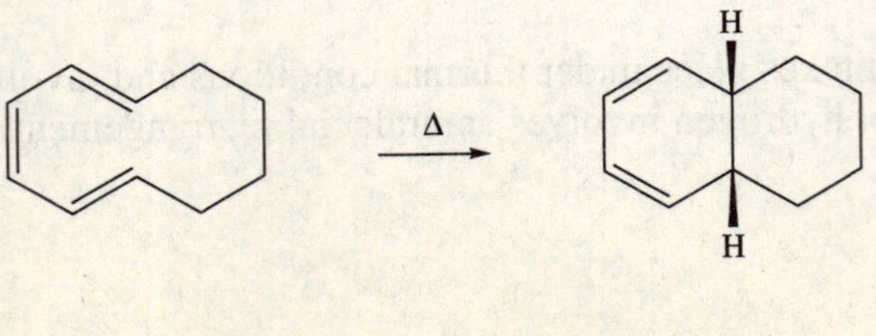

g. 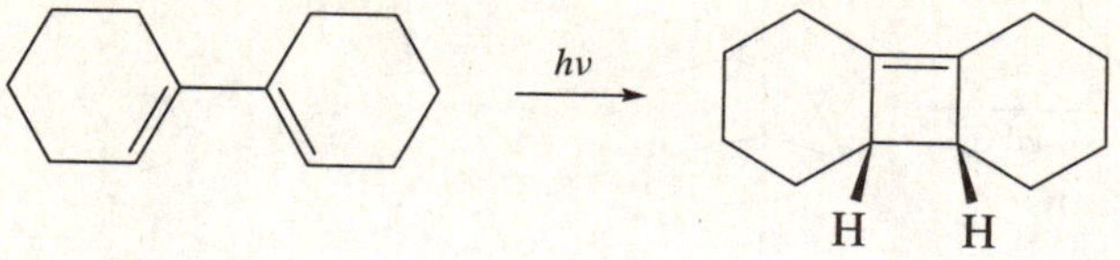

h. 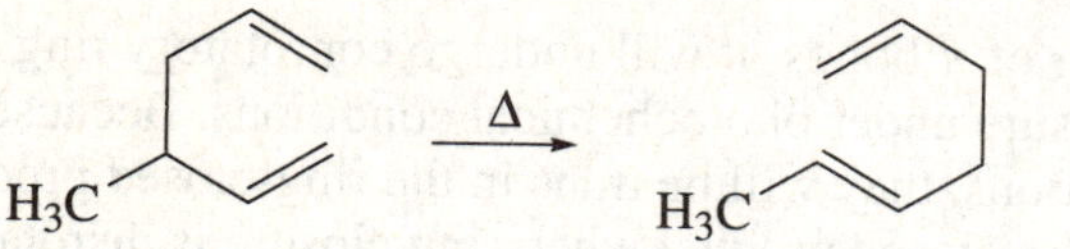

24. Because the compound has an odd number of π bonds, it will undergo disrotatory ring closure under thermal conditions and conrotatory ring closure under photochemical conditions.
In the compounds in which the two methyl substituents point in opposite directions, they will be cis in the ring-closed product when ring closure is disrotatory and trans in the ring-closed product when ring closure is conrotatory.
In the compounds in which the two methyl substituents point in the same direction, they will be trans in the ring-closed product when ring closure is disrotatory and cis in the ring-closed product when ring closure is conrotatory.

a. 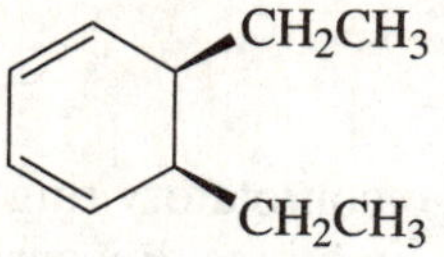

c. 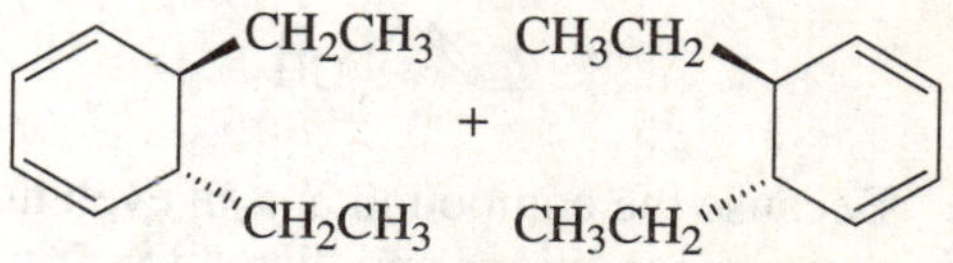

b. 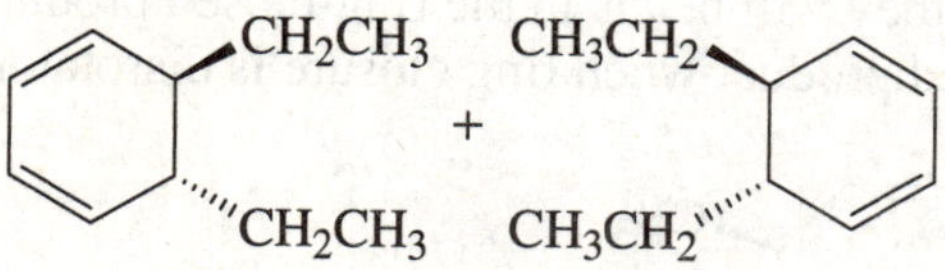

d. 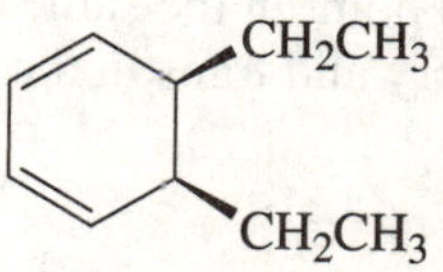

25. The hydrogens that end up at the ring juncture in the first reaction point in opposite directions in the reactant. Because ring closure is disrotatory (odd number of π bonds, thermal conditions), the hydrogens in the ring-closed product are cis. (See Table 29.2 on page 1275 of the text.)

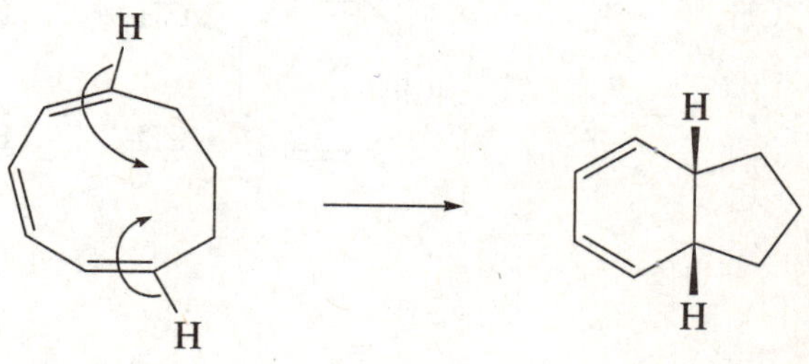

The hydrogens that end up at the ring juncture point in the same directions in the reactant. Ring closure is still disrotatory, so the hydrogens in the ring-closed product are trans.

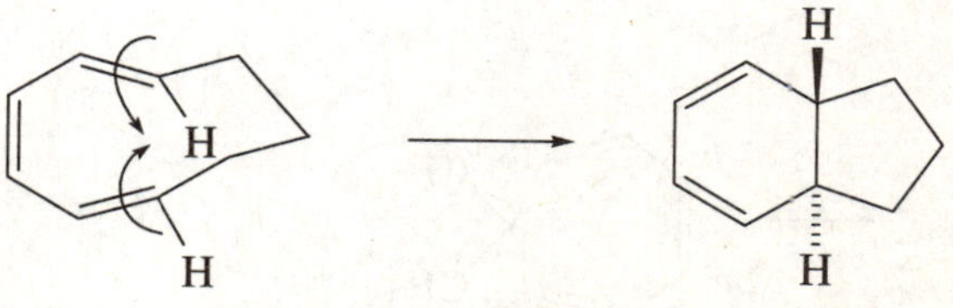

26.

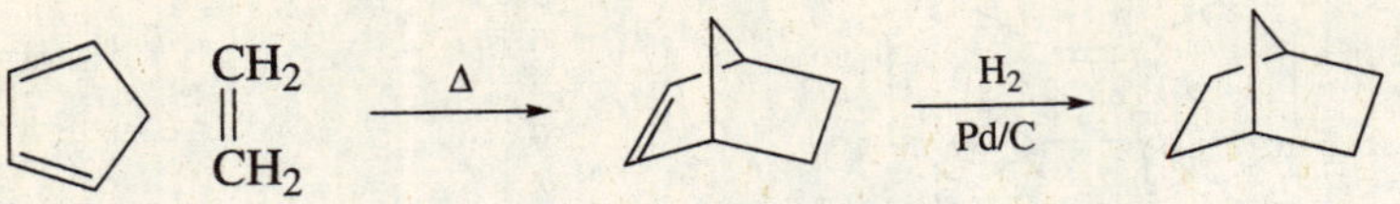

27. **1.** Because the compound has an even number of π bonds, it will undergo conrotatory ring closure under thermal conditions and disrotatory ring closure under photochemical conditions. Because the two methyl substituents point in opposite directions, they will be trans in the ring-closed product when ring closure is conrotatory and cis in the ring-closed product when ring closure is disrotatory.

a.

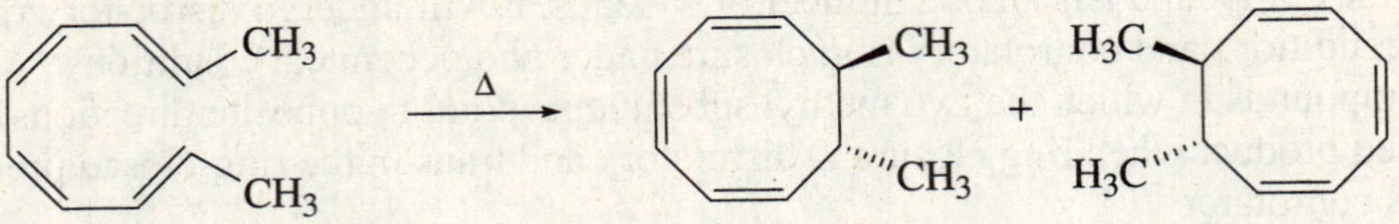

b.

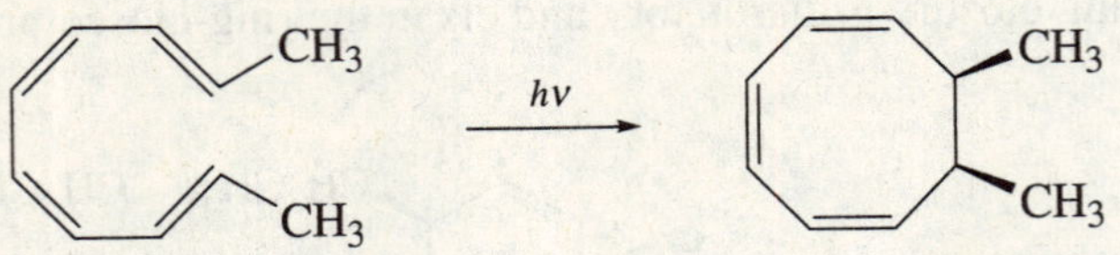

2. Because the compound has an even number of π bonds, it will undergo conrotatory ring closure under thermal conditions and disrotatory ring closure under photochemical conditions. Because the two methyl substituents point in the same direction, they will be cis in the ring-closed product when ring closure is conrotatory and trans in the ring-closed product when ring closure is disrotatory.

a.

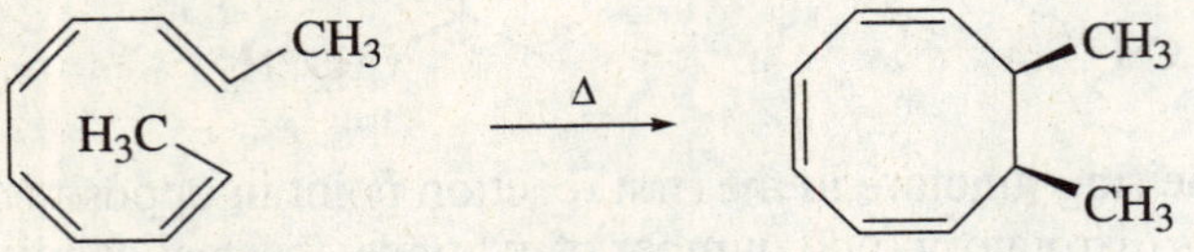

b.

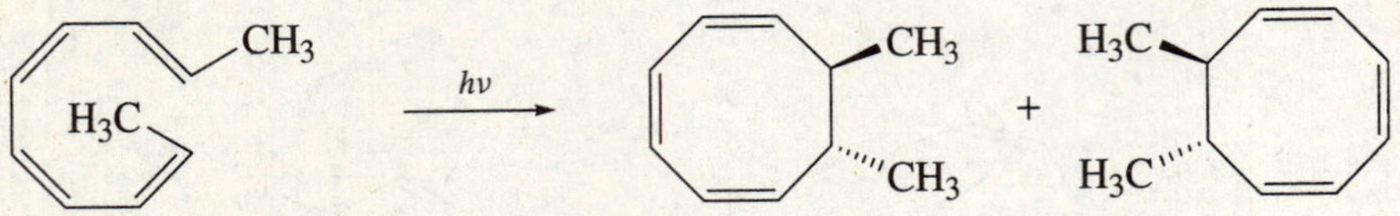

28. **a.**

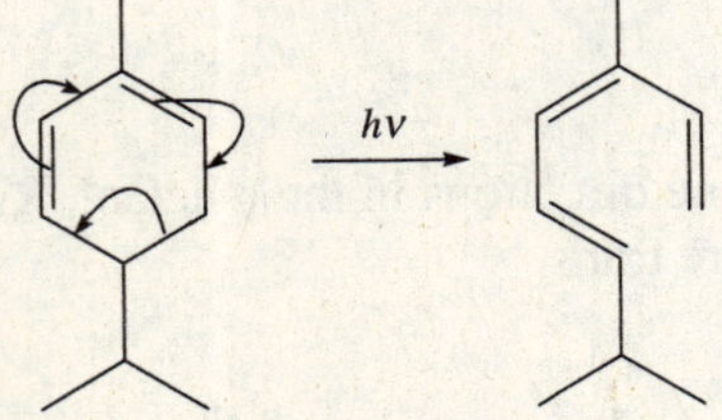

b.

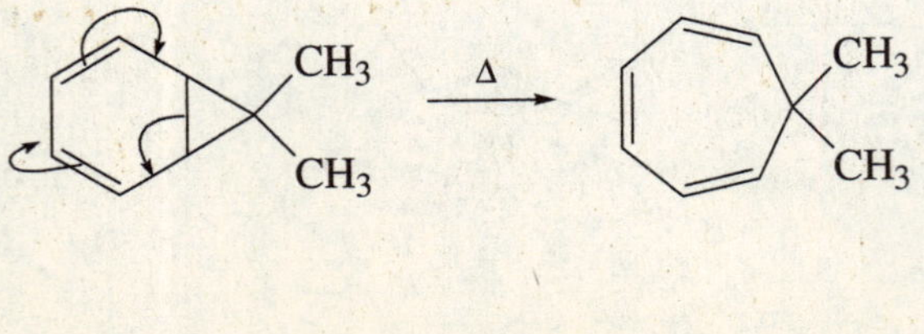

c.

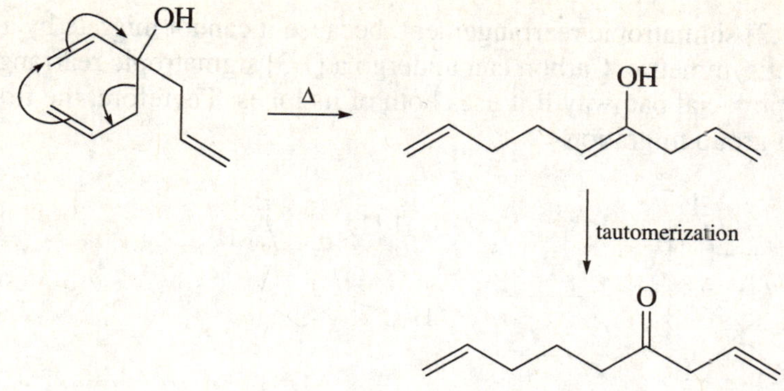

d.

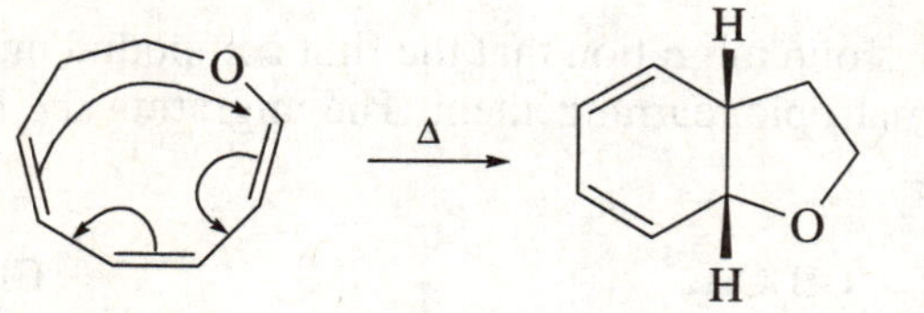

e.

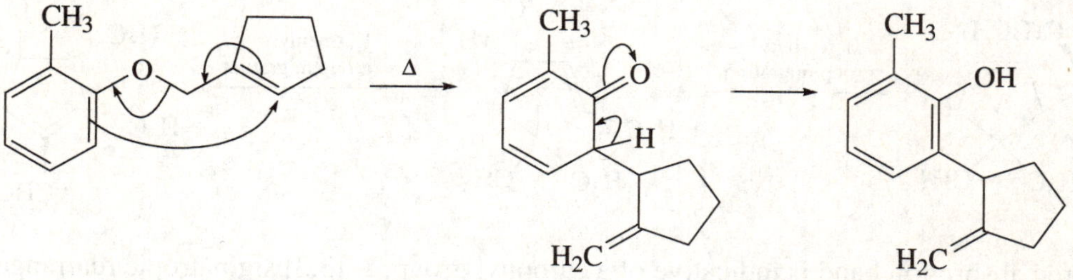

29. **B is the product. Because the reaction is a [1,3] sigmatropic rearrangement, antarafacial ring closure is required. Carbon, therefore, must migrate using both of its lobes. This means that the configuration of the migrating carbon will undergo inversion. The configuration of the migrating carbon has been inverted in B (the H's are cis to one another but were trans to one another in the reactant) and retained in A.**

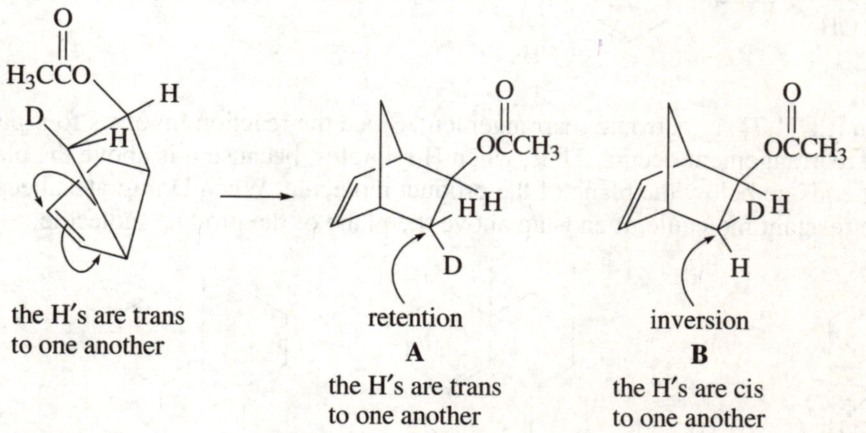

30. **At first glance it is surprising that the isomerization of Dewar benzene (a highly strained and unstable molecule) to benzene (a stable aromatic compound) is so slow. However, the isomerization requires conrotatory ring-opening, which is symmetry-forbidden under thermal conditions. The reaction, therefore, cannot take place by a concerted pathway and has to take place by a much slower stepwise process.**

31. Hydrogen cannot undergo a [1,3] sigmatropic rearrangement, because it cannot migrate by a suprafacial pathway since its HOMO is antisymmetric. Carbon can undergo a [1,3] sigmatropic rearrangement because it can migrate by a suprafacial pathway if it uses both of its lobes. Therefore, the first compound can undergo only a 1,3-methyl group migration.

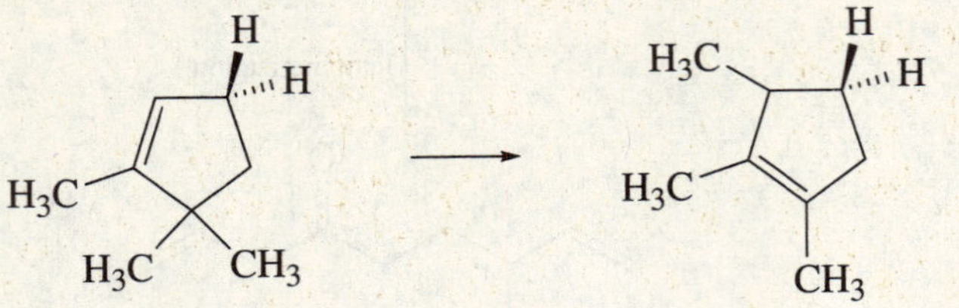

The second compound can under the 1,3-methyl group migration that the first compound undergoes, and the *sec*-butyl group can also undergo a [1,3] sigmatropic rearrangement. The migrating *sec*-butyl group will have its configuration inverted.

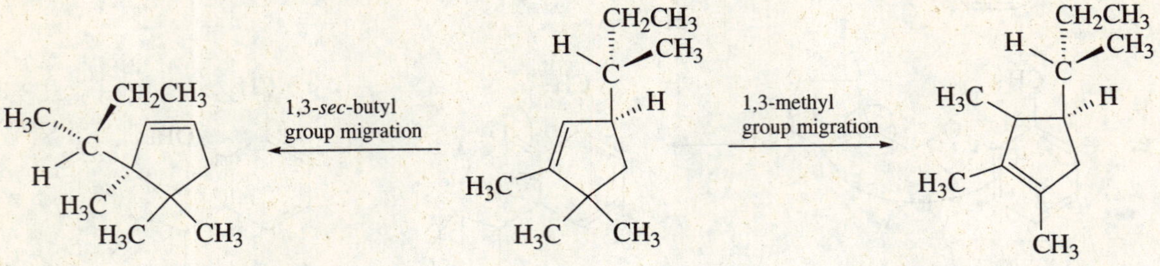

32. An infrared absorption band is indicative of a carbonyl group. A [3,3] sigmatropic rearrangement of the reactant leads to a compound with two enolic groups. Tautomerization of the enols results in keto groups. The ketone carbonyl groups give the absorbance at 1715 cm^{-1}.

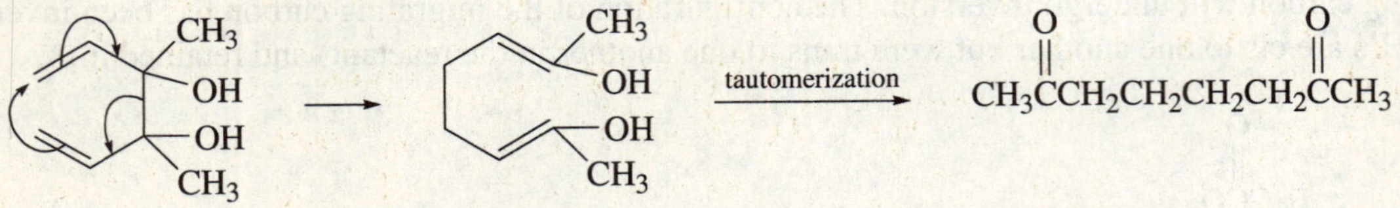

33. The reaction is a [1,7] sigmatropic rearrangement. Since the reaction involves four pairs of electrons, antarafacial rearrangement occurs. Thus, when H migrates, because it is above the plane of the reactant molecule, it ends up below the plane of the product molecule. When D migrates, because it is below the plane of the reactant molecule, it ends up above the plane of the product molecule.

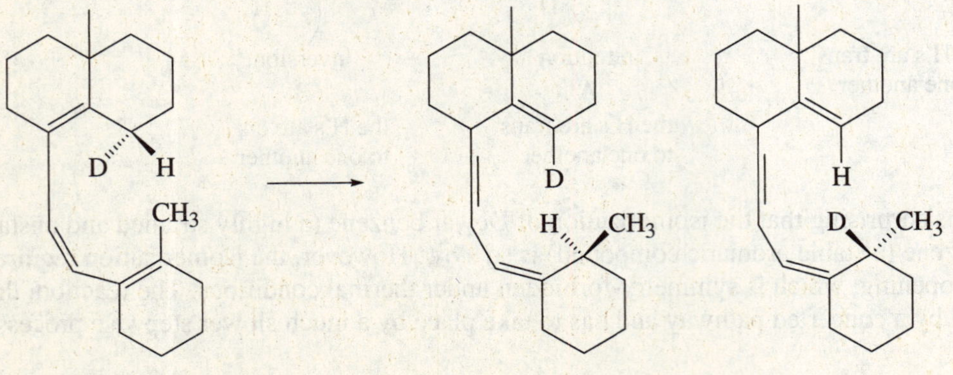

34. **a.**

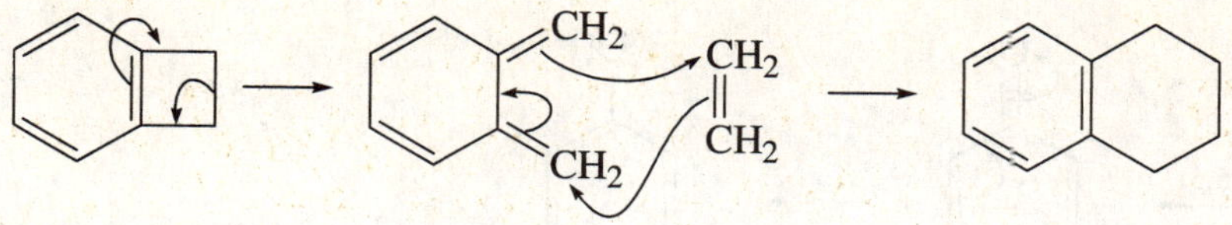

b.

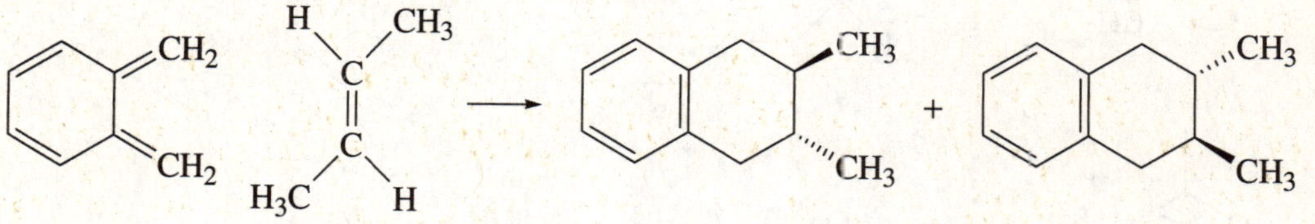

35. Disrotatory ring closure of (2*E*,4*Z*,6*Z*)-octatriene leads to the trans isomer, which can exist as a pair of enantiomers. One enantiomer is formed if the "top lobes" of the *p* orbitals rotate toward one other, and the other enantiomer is formed if the "bottom lobes" of the *p* orbitals rotate toward one other.

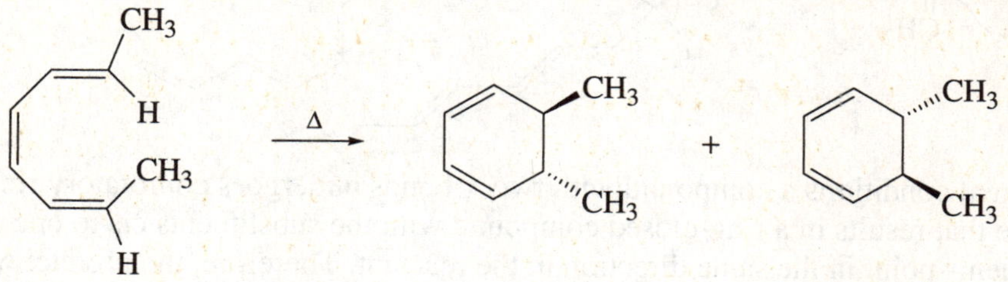

(2*E*,4*Z*,6*Z*)-octatriene

In contrast, disrotatory ring closure of (2*E*,4*Z*,6*E*)-octatriene leads to the cis isomer, which is a meso compound and, consequently, does not have a nonsuperimposable mirror image.Therefore, the same compound is formed from the "top lobes" of the *p* orbitals rotating toward one other and from the "bottom lobes" of the *p* orbitals rotating toward one other.

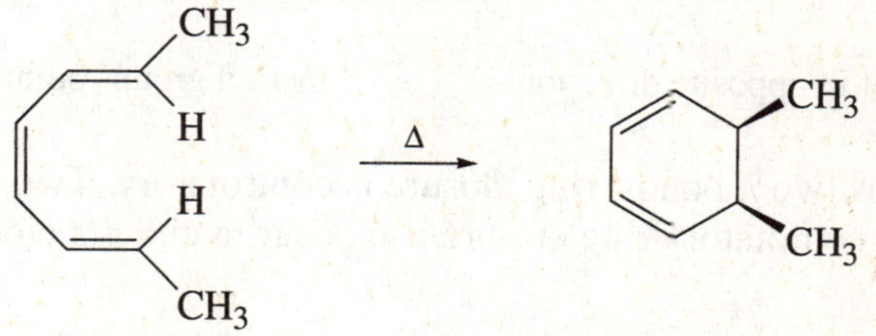

(2*E*,4*Z*,6*E*)-octatriene

36. **a.**

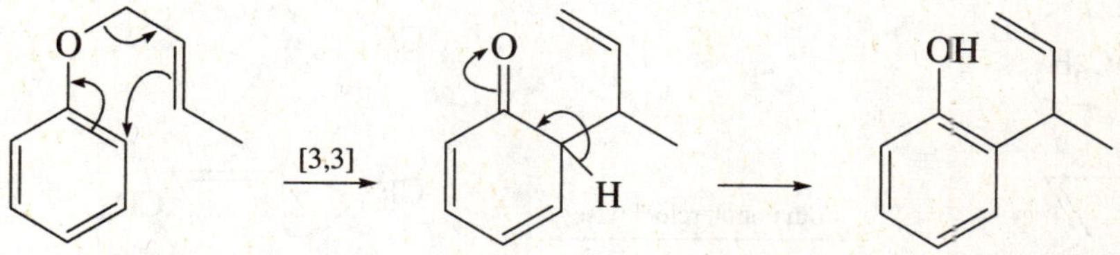

b.

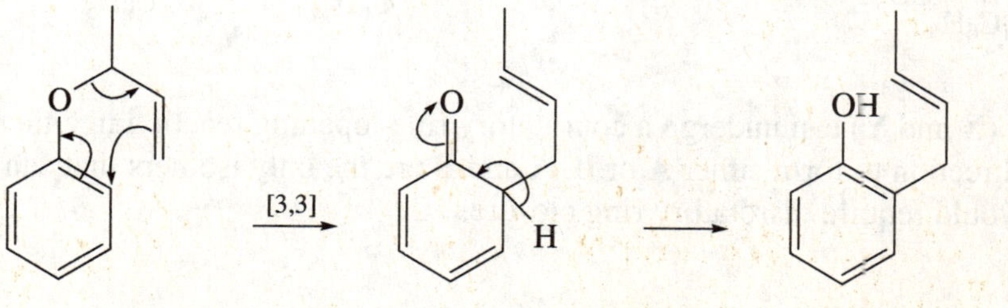

c.

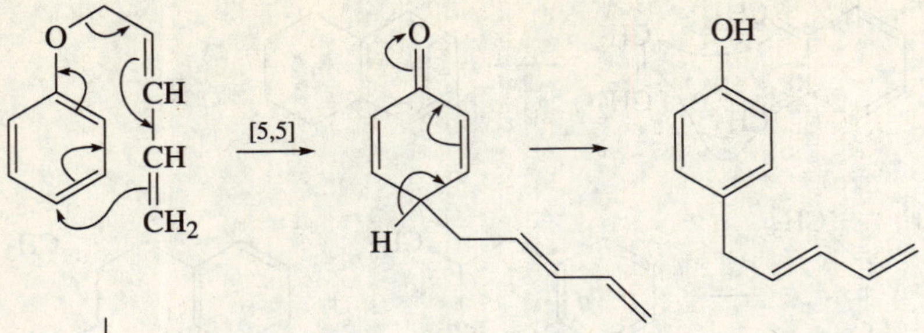

d.

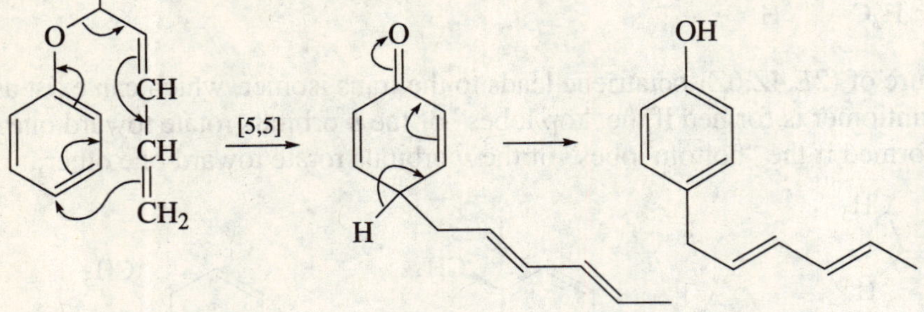

37. Under thermal conditions a compound with two π bonds undergoes conrotatory ring closure. Conrotatory ring closure that results in a ring-closed compound with the substituents cis to one another requires that the substituents point in the same direction in the reactant. Therefore, the product with the methyl substituents pointing in the same direction is obtained in 99% yield.

1% **99%**

methyl groups point in opposite directions methyl groups point in the same directions

38. Because the reactant has two π bonds, ring closure is controtatory. Two different compounds, **X** and **Y**, can be formed because conrotatory ring closure can occur in either a clockwise or a counterclockwise direction.

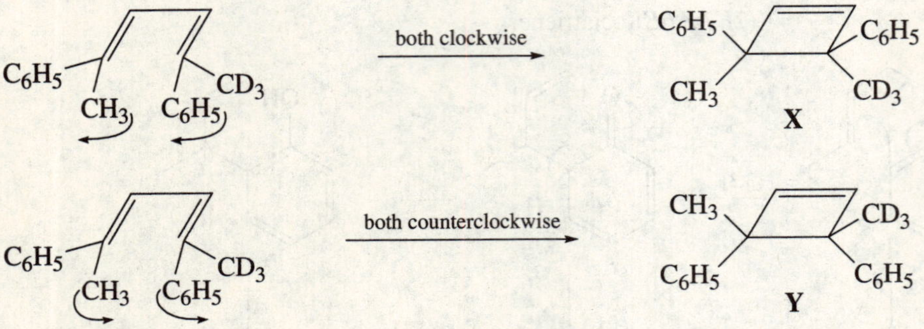

Each of the compounds (**X** and **Y**) can undergo a conrotatory ring–opening reaction in either a clockwise or a counterclockwise direction to form either **A** or **B**. **A** and **B** are the only isomers that can be formed; formation of **C** and **D** would require disrotatory ring closure.

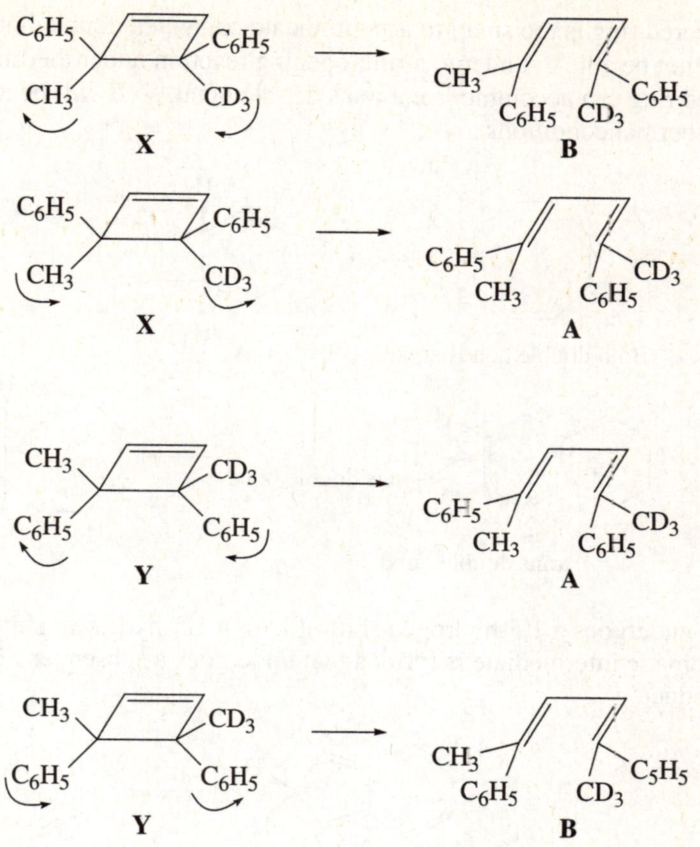

39.

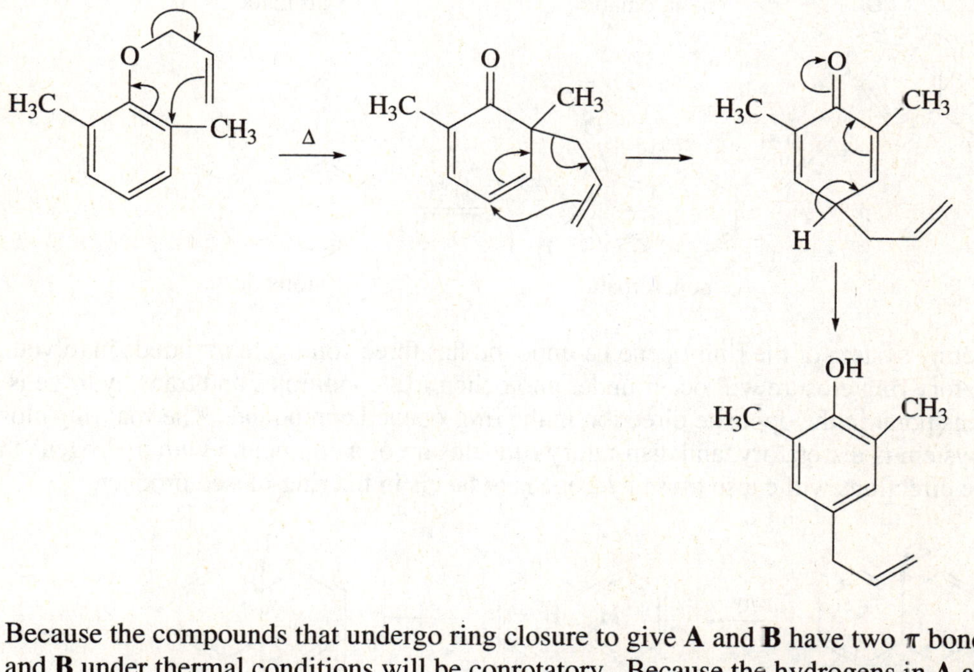

40. Because the compounds that undergo ring closure to give **A** and **B** have two π bonds, ring opening of **A** and **B** under thermal conditions will be conrotatory. Because the hydrogens in **A** and **B** are cis, they must point in the same direction in the ring-opened product. To have the two hydrogens pointing in the same direction, one of the double bonds in the ring-opened compound must be cis and the other must be trans.

An eight-membered ring is too small to accommodate cojugated double bonds with one cis and the other trans, so **A** will not be able to undergo a ring-opening reaction under thermal conditions.
A ten-membered ring can accommodate a trans double bond, so **B** will be able to undergo a ring-opening reaction under thermal conditions.

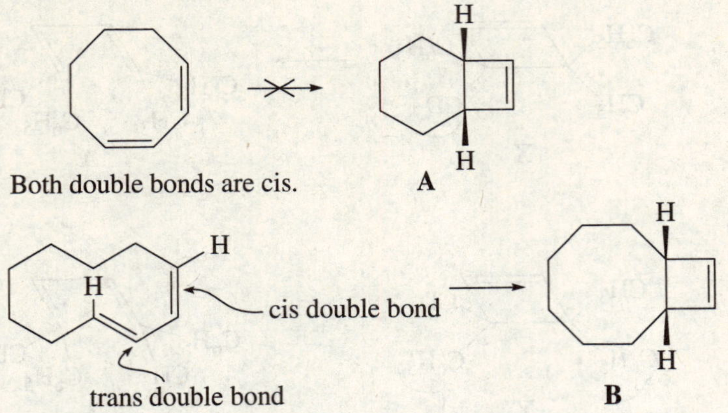

41. The compound undergoes a 1,5-hydrogen shift of D or a 1,5-hydrogen shift of H. In each case, an unstable nonaromatic intermediate is formed that undergoes a subsequent 1,5-hydrogen shift to form an aromatic product.

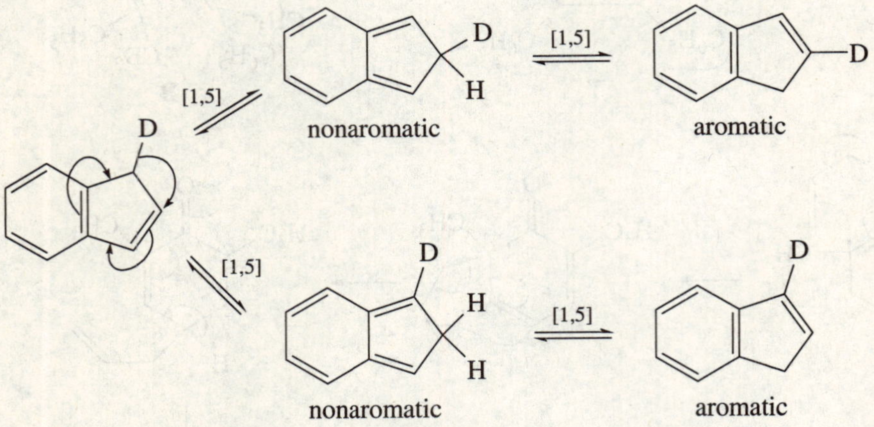

42. Because the reacting system of the ring-opened compound has three conjugated π bonds involved in the reaction, conrotatory ring closure will occur under photochemical conditions, and trans hydrogens require that the hydrogens point in the opposite direction in the ring-opened compound. Thermal ring closure of a three π bond system is disrotatory, and disrotatory ring closure of a compound with hydrogens that point in opposite directions will cause those hydrogens to be cis in the ring-closed product.

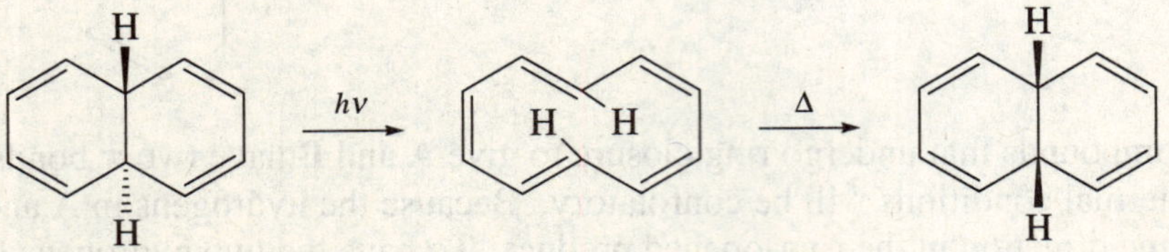

43. A Diels-Alder reaction is followed by an extrusion reaction that eliminates CO_2. An extrusion reaction is a reaction in which a neutral molecule is eliminated from a molecule.

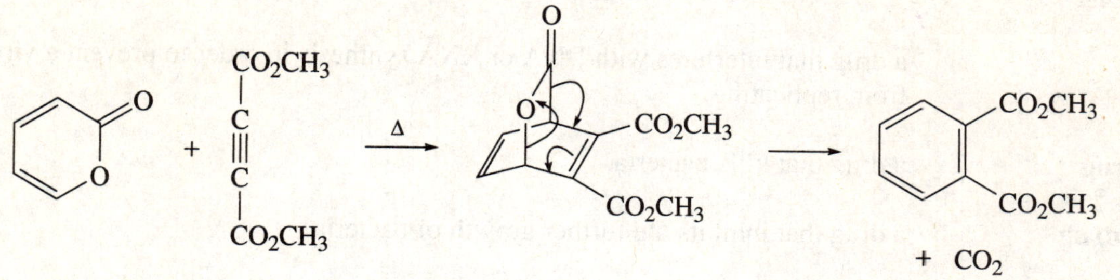

44.

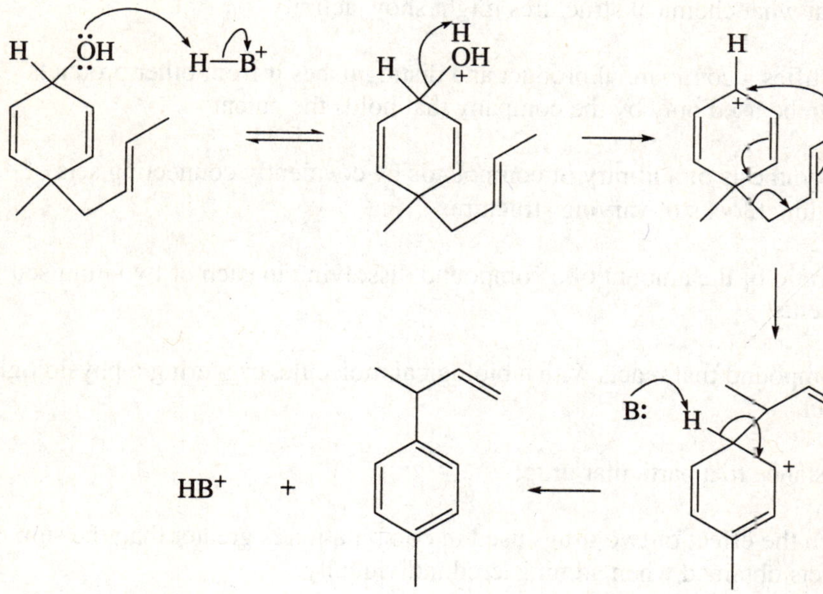

CHAPTER 30
The Organic Chemistry of Drugs: Discovery and Design

Important Terms

antiviral drug	a drug that interferes with DNA or RNA synthesis in order to prevent a virus from replicating.
bactericidal drug	a drug that kills bacteria.
bacteriostatic drug	a drug that inhibits the further growth of bacteria.
blind screen (random screen)	the search for a pharmacologically active compound without any information about what chemical structures might show activity.
brand name	identifies a commercial product and distinguishes it from other products. It can be used only by the company that holds the patent.
combinatorial organic synthesis	the synthesis of a library of compounds by covalently connecting sets of building blocks of varying structure.
distribution coefficient	the ratio of the amount of a compound dissolving in each of two immiscible solvents.
drug	a compound that reacts with a biological molecule, triggering a physiological effect.
drug resistance	resistance to a particular drug.
drug synergism	when the effect of two drugs used in combination is greater than the sum of the effects obtained when administered individually.
generic name	a name for a drug that anyone can manufacture.
lead compound	the prototype in a search for other biologically active compounds.
molecular modeling	computer-assisted design of a compound.
molecular modification	changing the structure of a lead compound.
orphan drug	a drug for a disease or condition that affects fewer than 200,000 people.
prodrug	a compound that does not become an effective drug until it undergoes a reaction in the body.
quantitative structure-reactivity relationship (QSAR)	the relation between a particular property of a series of compounds and their biological activity.

random screen (blind screen) — the search for a pharmacologically active compound without any information about what chemical structures might show activity.

rational drug design — designing drugs with a particular structure to achieve a specific purpose.

receptor — the site where a drug binds in order to exert its physiological effect.

suicide inhibitor — a compound that inactivates an enzyme by undergoing part of the normal catalytic mechanism.

therapeutic index — the ratio of the lethal dose of a drug to the therapeutic dose.

Solutions to Problems

1. **a.** ethyl *para*-aminobenzoate **b.** 2-(*N,N*-diethyl)ethyl *para*-aminobenzoate

 c. 2'-deoxy-5-iodouridine

2. There are many possibilities. The following are just a few of them.

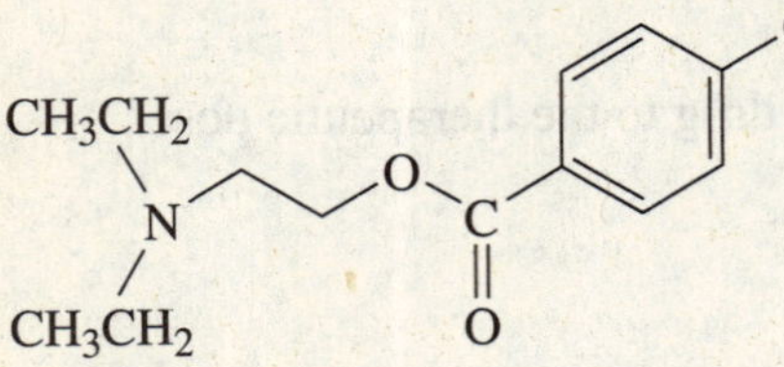

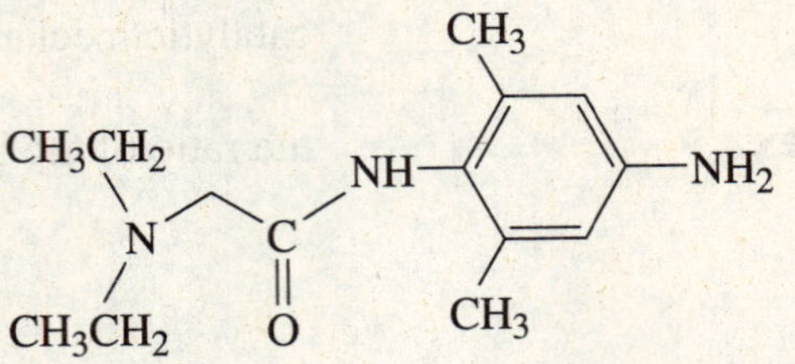

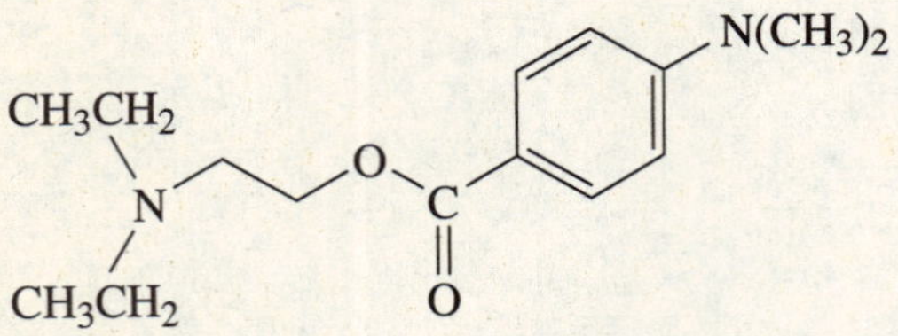

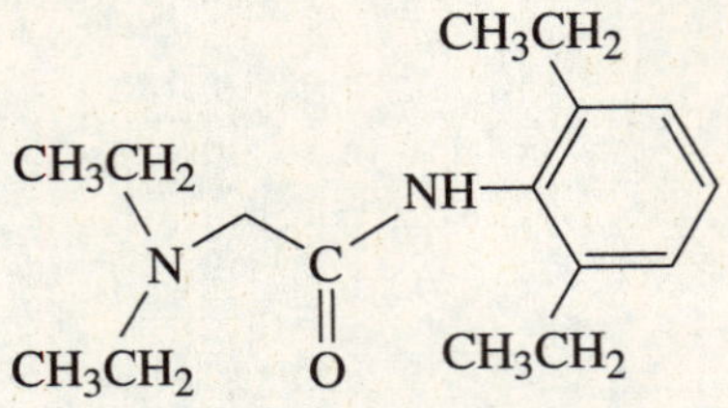

3. The compound on the left has an electron-donating substituent, whereas the compound on the right has an electron-withdrawing substituent in the same position. Because the compounds known to be effective tranquilizers have an electron-withdrawing substituent in that position (Cl or NO_2), the compound on the right is more likely to show activity as a tranquilizer.

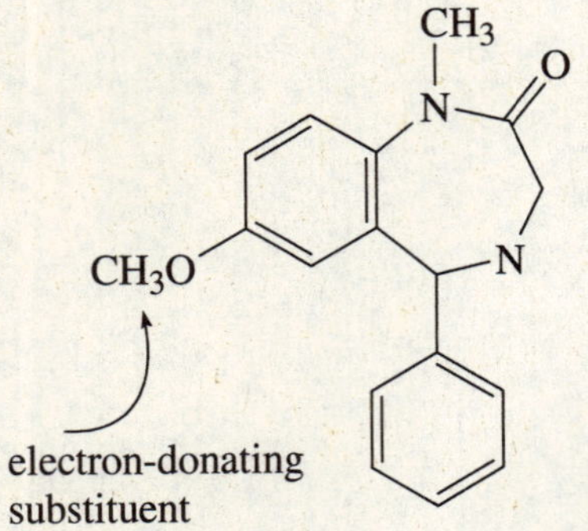

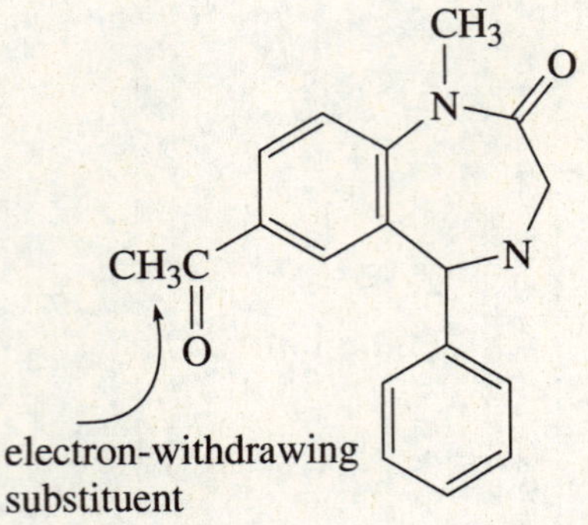

4. Anesthetics have been found to have similar distribution coefficients, which means that they have similar polarities. Since diethyl ether is a known anesthetic, ethyl methyl ether with a polarity similar to that of diethyl ether is more apt to be a general anesthetic than propanol, which is considerably more polar.

5. The electron-withdrawing SO_2 group is weakly basic and, therefore, an excellent leaving group. The weak carbon-sulfur bond aids imine formation.

6. **a.** The chloro-substituted compound would be expected to be a more potent inhibitor because potency increases with the ability of the substituents to donate electrons into the π system.

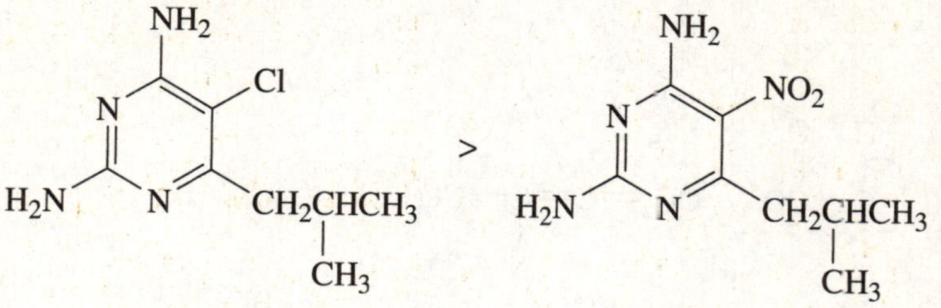

b. The compound with the pentyl substituent would be expected to be a more potent inhibitor because potency increases with increasing hydrophobicity of one of the substituents.

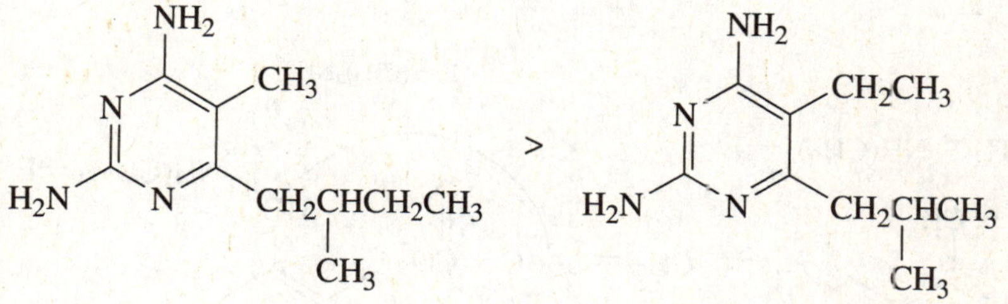

7. Tetrahydrocannabinol is a safer drug because it has a higher therapeutic index.

$$\text{therapeutic index} = \frac{\text{lethal dose}}{\text{therapeutic dose}}$$

$$\text{tetrahydrocannabinol} = \frac{2.0\ \text{g/kg}}{20\ \text{mg/kg}} = \frac{2000\ \text{mg/kg}}{20\ \text{mg/kg}} = 100$$

$$\text{sodium pentothal} = \frac{100\ \text{mg/kg}}{30\ \text{mg/kg}} = 3.3$$

8. Hydrolysis frees the coenzyme and forms a reactive α,β-unsaturated amino acid that can react with the coenzyme bound to its enzyme. This reaction causes the imine linkage to be converted into an amine linkage that cannot be hydrolyzed to release the enzyme. In this way the enzyme is deactivated.

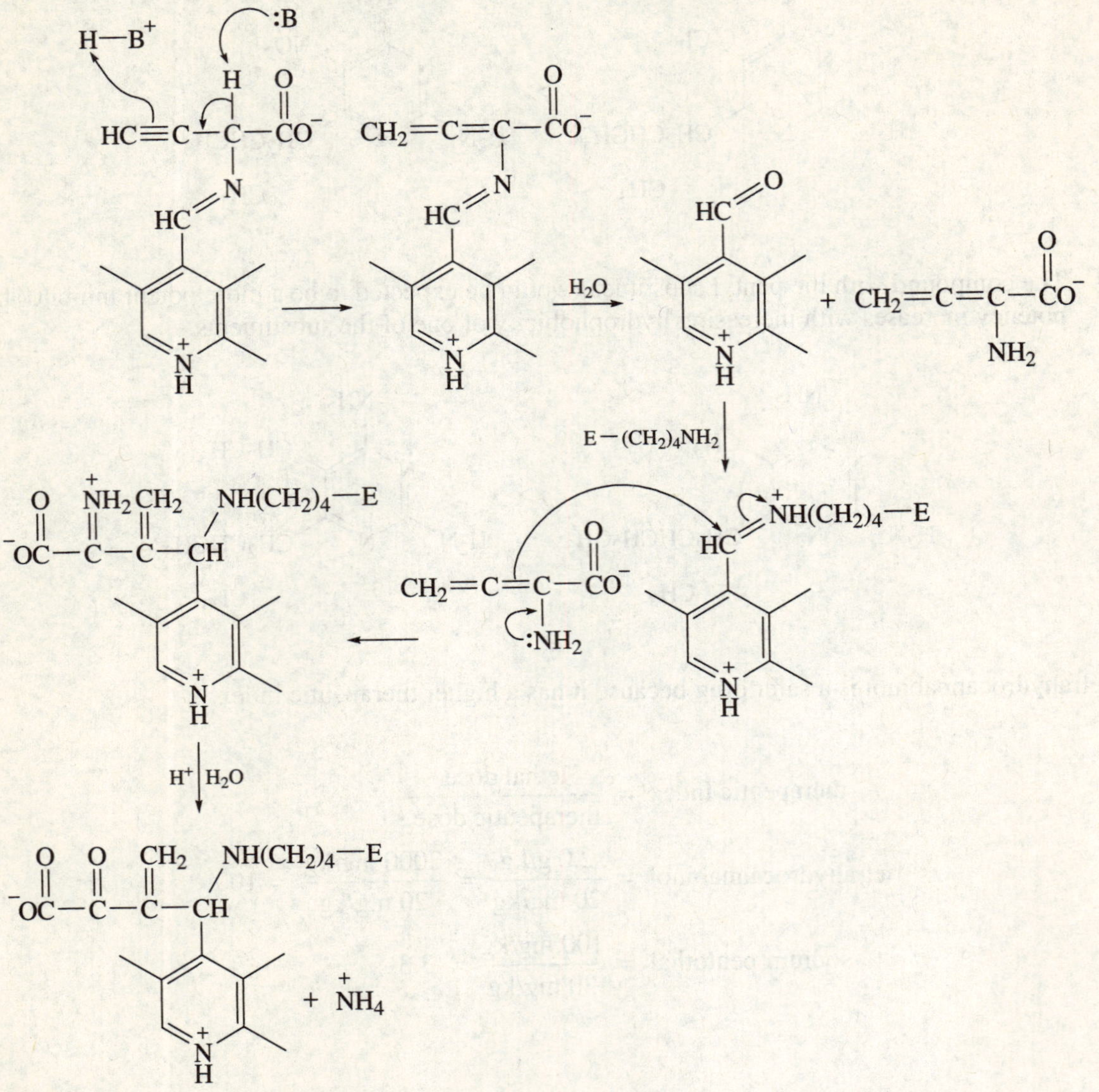

9. Aclovir differs from guanosine in that it has a hydroxy-ether substituent rather than a ribose.

Cytosar differs from cytidine in that it has the 2'-OH group in the β-position rather than in the α-position.

Viramid has an unusual heterocyclic base attached to ribose.

Herplex differs from 2'-deoxyuridine in that it has an iodo substituent in the 5-position.

10.

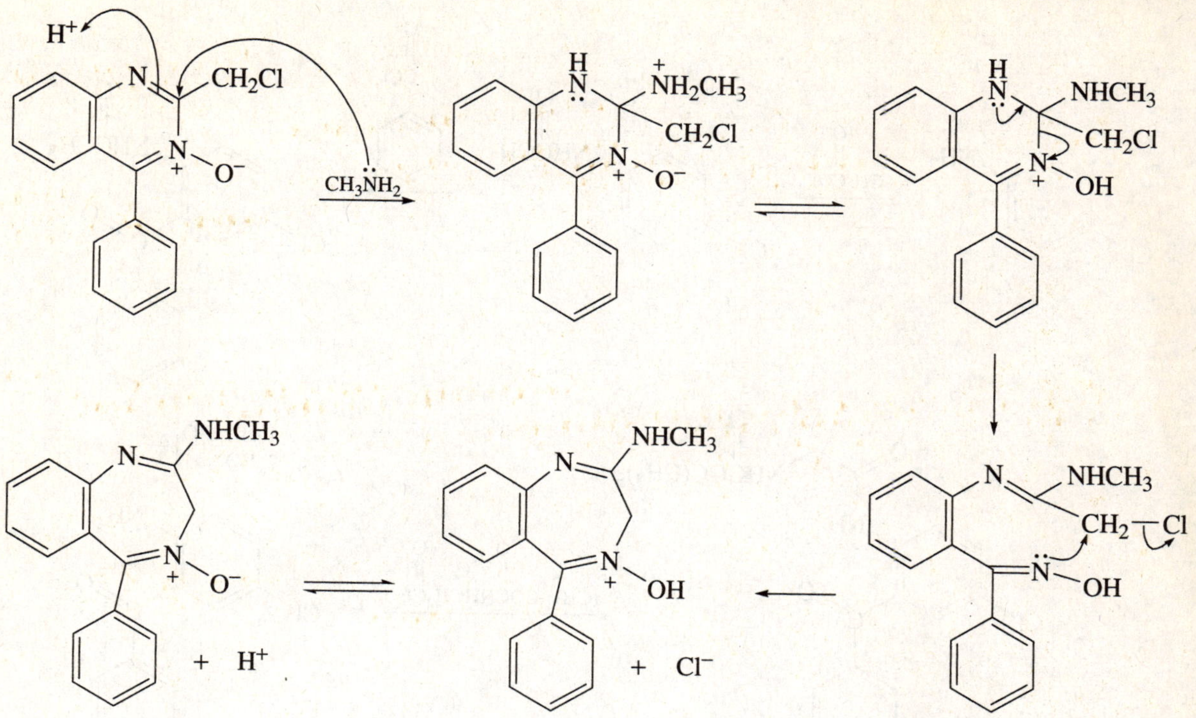

11.

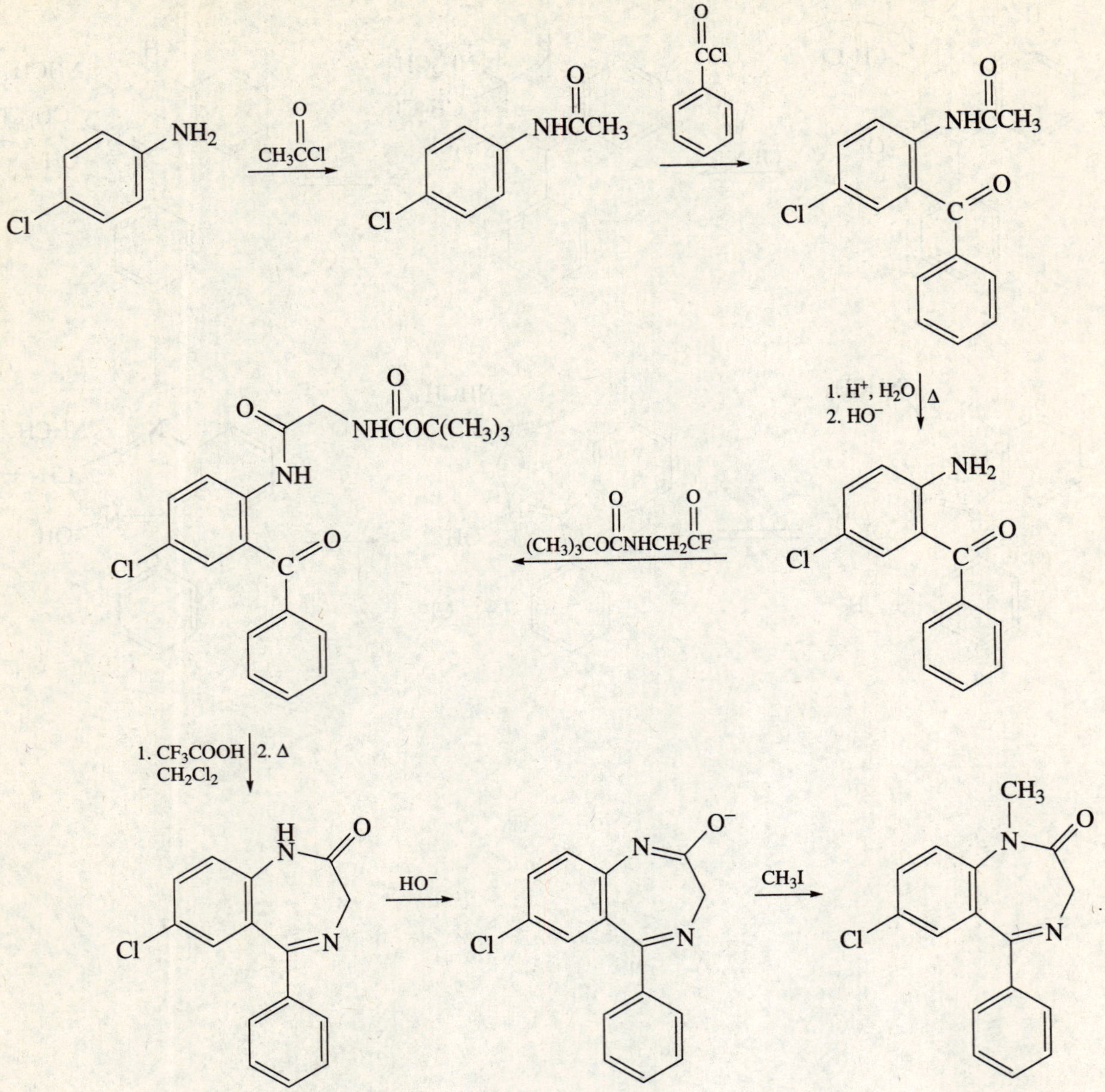
NH2
Cl
CH3CCl
NHCCH3
CCl
NH2
1. H+, H2O
2. HO−
Δ
(CH3)3COCNHCH2CF
NHCOC(CH3)3
NH
1. CF3COOH
CH2Cl2
2. Δ
HO−
CH3I
CH3
N
O−

12.

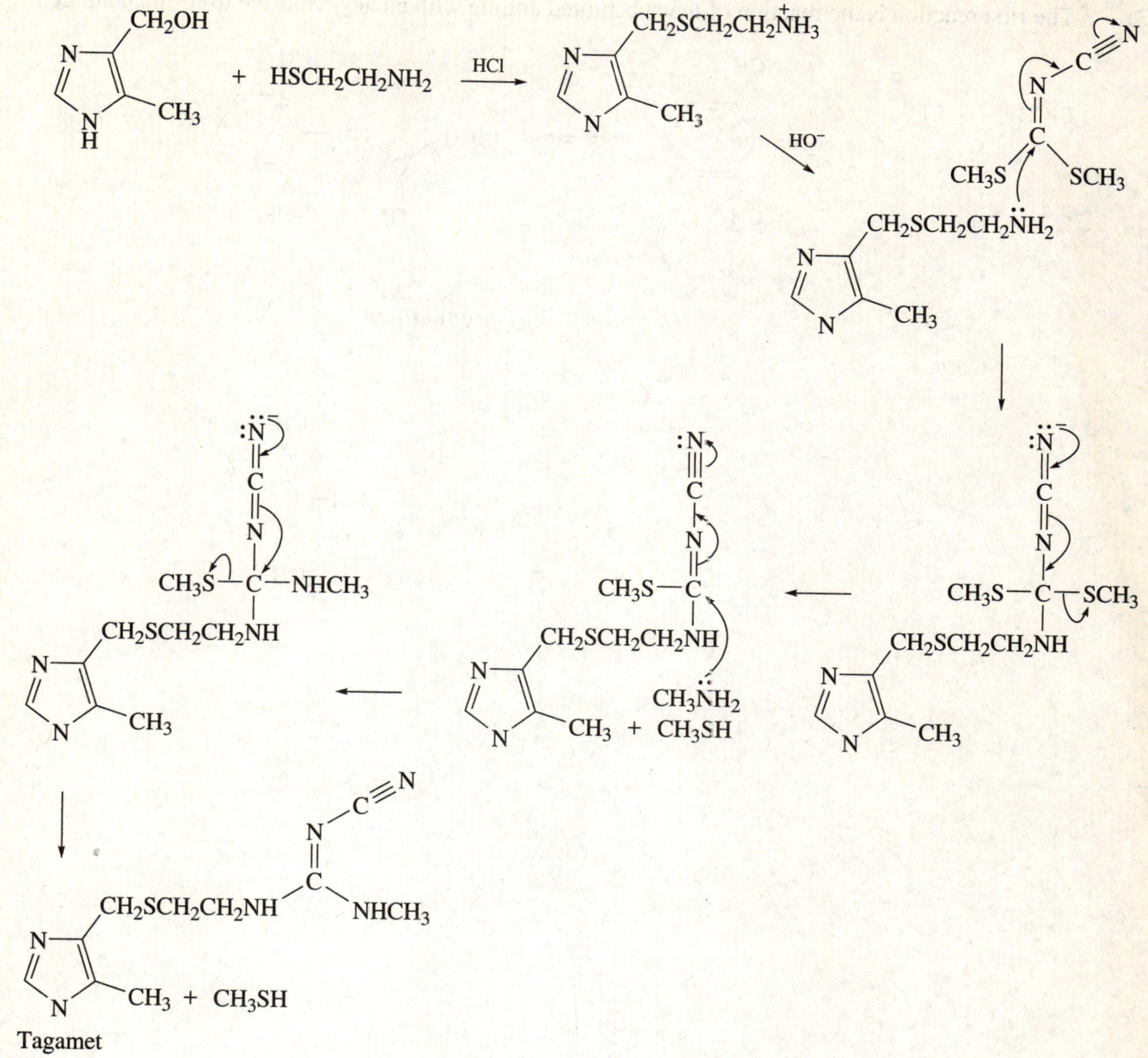
CH2OH
N
N
H
CH3
+ HSCH2CH2NH2
HCl
CH2SCH2CH2NH3
HO−
CH3S
SCH3
CH2SCH2CH2NH2
CH3S—C—SCH3
CH2SCH2CH2NH
CH3S—C
CH3NH2
CH3 + CH3SH
CH3S—C—NHCH3
NHCH3
CH3 + CH3SH
Tagamet

13. The first reaction is the reaction of the substituted aniline with an acyl chloride to form an amide.

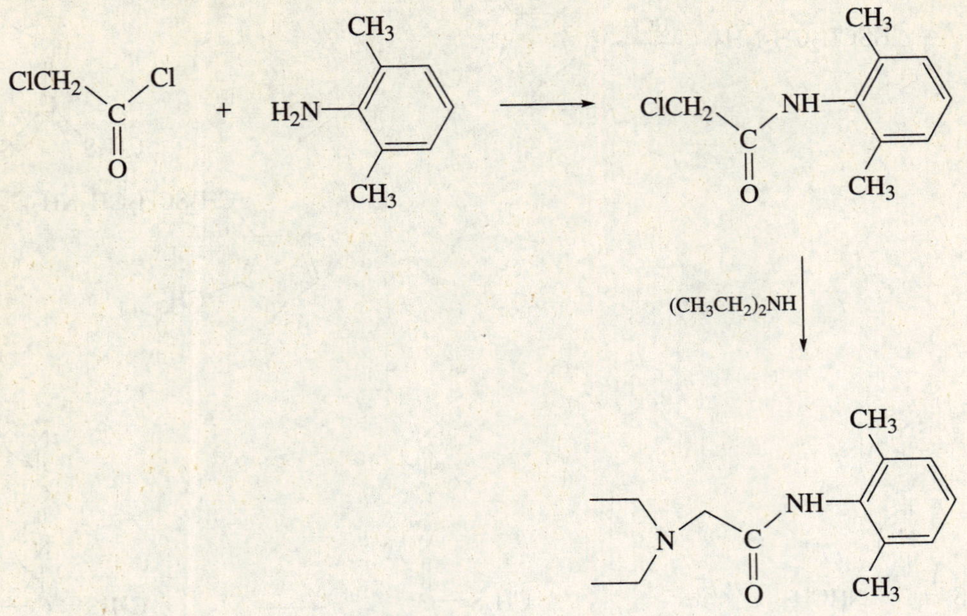

Answers to Chapter 1 Practice Test

1. **a.** HBr **b.** NH_3 **c.** a carbon-fluorine bond **d.** CCl_4

2. $^{+}CH_3$ $^{-}CH_3$ $^{\bullet}CH_3$

sp^2 sp^3 sp^2

3. $H:\ddot{O}:\overset{:O:}{C}:\ddot{O}:^{-}$

4. CH_3COO^- CH_3CH_2OH CH_3OH $CH_3CH_2\overset{+}{N}H_3$

5. $^{+}NH_4$ **6.** $^{-}NH_2$

7. $CH_3CH_2CH_2CH{=}CH_2$ **or** $CH_3CH_2CH{=}CHCH_3$ **or** $CH_3CH(CH_3)CH{=}CH_2$

8. **a.** $CH_3OH + \overset{+}{N}H_4 \rightleftharpoons CH_3\overset{+}{O}H_2 + NH_3$ **b.** reactants

9. **a.** $1s^2 2s^2 2p_x 2p_y$ **b.** $1s^2 2s\, 2p_x 2p_y 2p_z$

10. $CH_3CH_2CH(Cl)COOH$

11. O=C=O (sp) HC(=O)OH (sp^2) HC≡N (sp) CH_3OCH_3 (sp^3) $CH_3CH{=}CH_2$ (sp^3)

12.
- **a.** A pi bond is stronger than a sigma bond. F
- **b.** A triple bond is shorter than a double bond. T
- **c.** The oxygen-hydrogen bonds in water are formed by the overlap of an sp^2 orbital of oxygen with an *s* orbital of hydrogen. F
- **d.** HO^- is a stronger base than $^{-}NH_2$. F
- **e.** A double bond is stronger than a single bond. T
- **f.** A tetrahedral carbon has bond angles of 107.5°. F
- **g.** A Lewis acid is a compound that accepts a share in a pair of electrons. T
- **h.** CH_3CH_3 is more acidic than $H_2C{=}CH_2$. F

Answers to Chapter 2 Practice Test

1. 3-methyloctane

2. **a.** 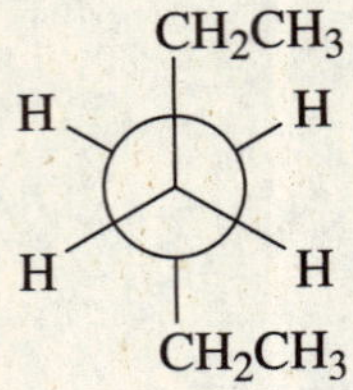

b. 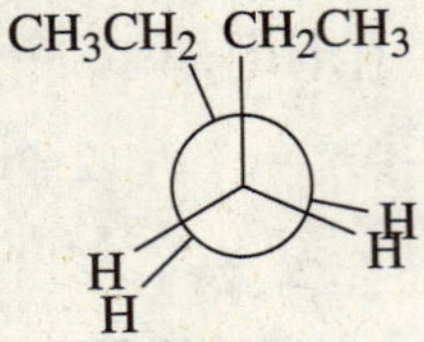

c. 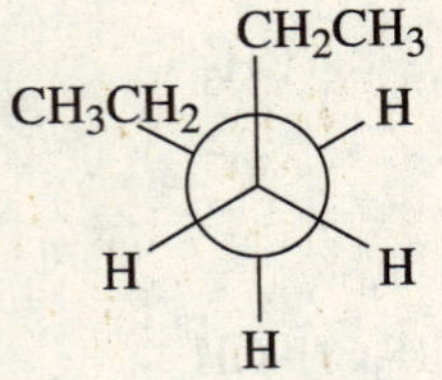

3. **a.** *sec*-butyl chloride, 2-chlorobutane
b. isohexyl alcohol, 4-methyl-1-pentanol
c. cyclopentyl bromide, bromocyclopentane

4. **a.** $CH_3CH_2CH_2CH_2CH_2Br$ **1** $CH_3CH_2CH_2Br$ **3** $CH_3CH_2CH_2CH_2Br$ **2**

b. $CH_3CH_2CH_2CH_2CH_3$ **3** $CH_3CH_2CH_2CH_2OH$ **1** $CH_3CH_2CH_2CH_2Cl$ **2**

c. 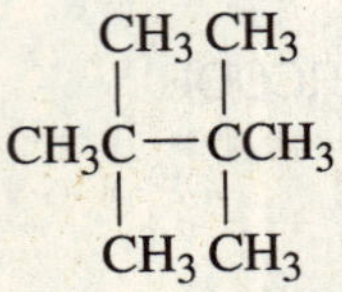 **3** $CH_3CH_2CH_2CH_2CH_2CH_2CH_2CH_3$ **1** 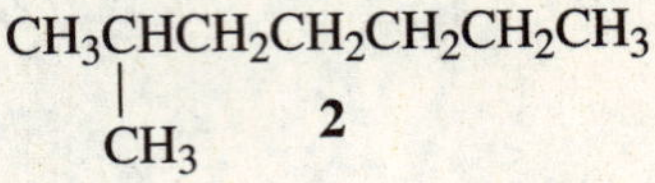 **2**

5. **a.** 6-methyl-3-heptanol
b. 3-ethoxyheptane
c. 1-bromo-3-methylcyclopentane
d. 1,4-dichloro-5-methylheptane

6. 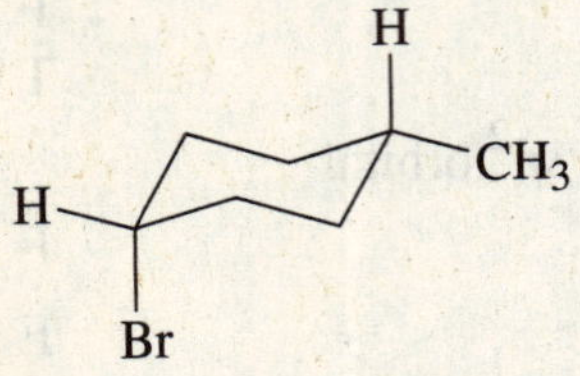

7. 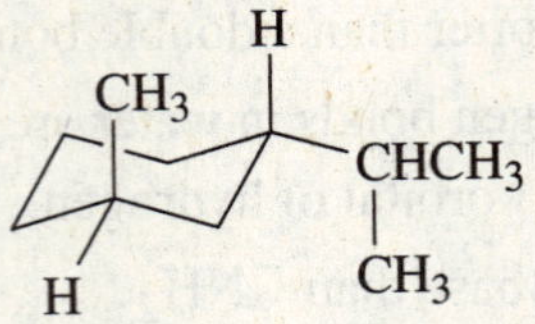

8. **a.** butyl alcohol
b. 1-butanol
c. hexane
d. pentylamine
e. ethyl alcohol

9. **a.** 1-bromo-3-methylbutane
isopentyl bromide
b. 3-methyl-1-butanol
isopentyl alcohol
c. 3-methyl-1-butanamine
isopentylamine

10. **a.** a staggered conformer
b. the chair conformer of methylcyclohexane with the methyl group in the equatorial position
c. cyclohexane

11. a. 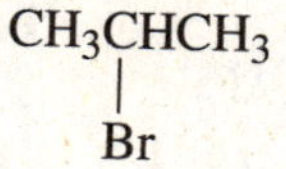

d. 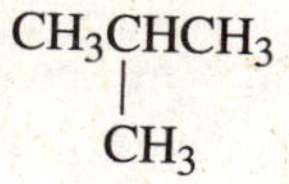

e. $CH_3CH_2CH_2OH$ 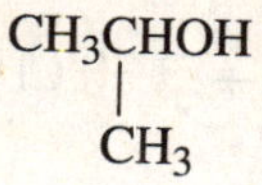 $CH_3CH_2OCH_3$

b. $CH_3CH_2NHCH_3$

c. 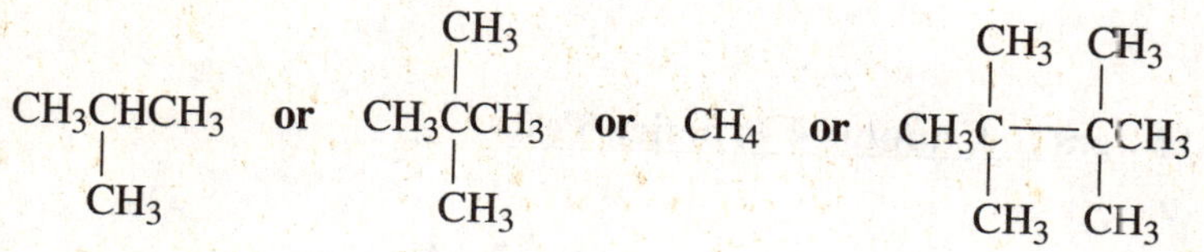

Answers to Chapter 3 Practice Test

1. a. 4-methyl-1-hexene
   b. 4-bromocyclopentene
   c. 7-methyl-3-nonene
   d. 4-chloro-3-methylcyclohexene

2. 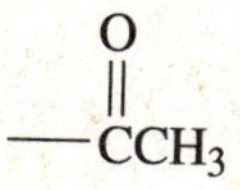 2 —$CH{=}CH_2$ 4 —Cl 1 —$C{\equiv}N$ 3

3. a. 2-pentene
   b. 3-methyl-1-hexene
   c. 3-methyl-2-pentene
   d. 1-methylcyclohexene

4. a. Increasing the energy of activation, increases the rate of the reaction. F
   b. Decreasing the entropy of the products compared to the entropy of the reactants makes the equilibrium constant more favorable. F
   c. An exergonic reaction is one with a $-\Delta G°$. T
   d. An alkene is an electrophile. F
   e. The higher the energy of activation, the more slowly the reaction will take place. T
   f. Another name for *trans*-2-butene is *Z*-2-butene. F
   g. *trans*-2-Butene has a dipole moment of zero. T
   h. A reaction with a negative $\Delta G°$ has an equilibrium constant greater than one. T
   i. Increasing the energy of the reactants, increases the rate of the reaction. T
   j Increasing the energy of the products, increases the rate of the reaction. F

5. a. *Z* b. *Z*

6. a. $CH_2{=}CHCH_2OH$
   c. 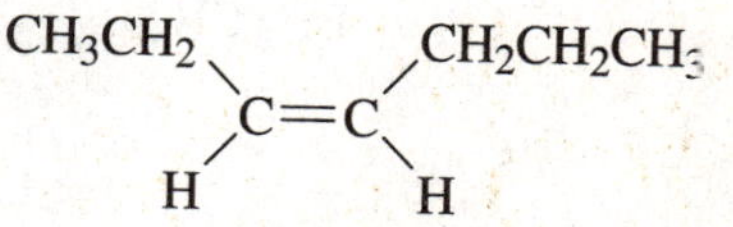
   b. 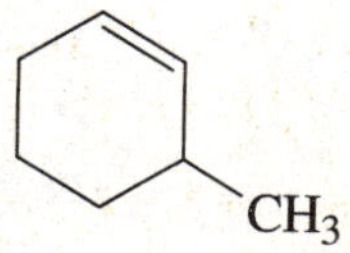
   d. $CH_2{=}CHBr$

7. 5

8. 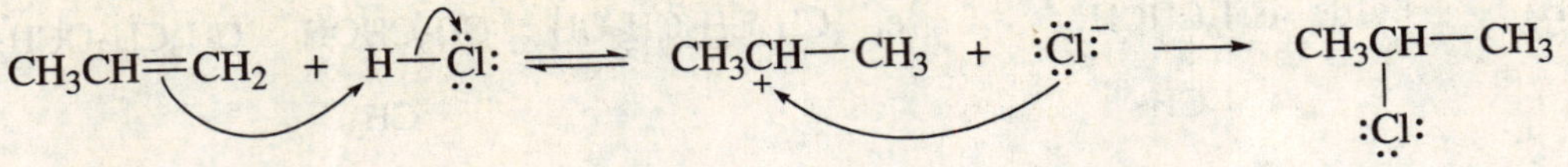

9. b and c **10. a.** 4 kcal/mol **b.** 35 °C **c.** one reactant forms two products

Answers to Chapter 4 Practice Test

1. **a.** 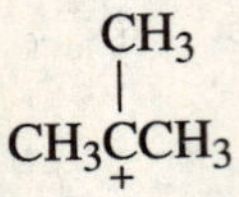

b. $CH_3CH_2\overset{+}{C}HCH_3$

c. $CH_3CH_2\overset{+}{C}H_2$

d. 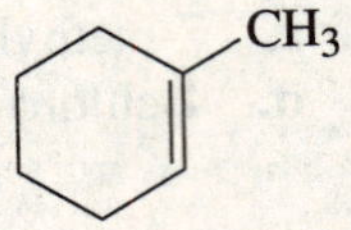

e. 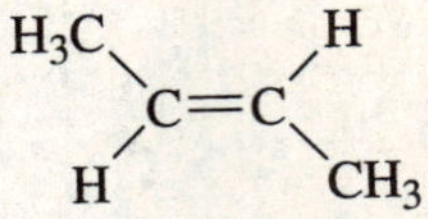

2. $CH_3CH_2CH_2CH{=}CH_2$

3. **B**

4. **a.** 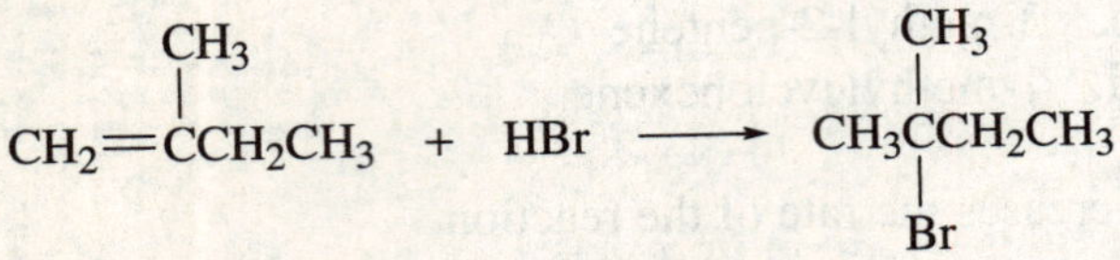

b. 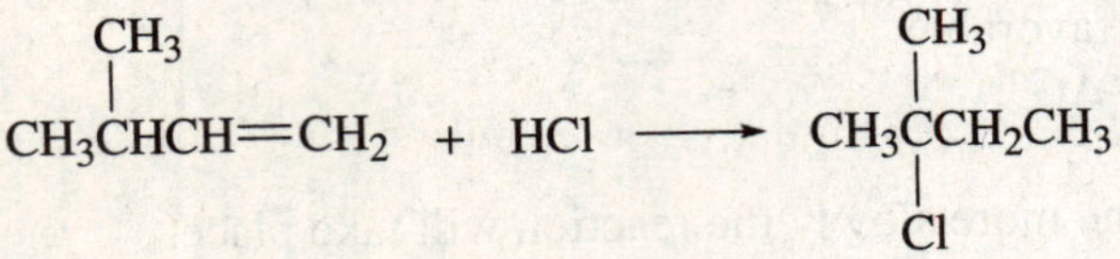

c. 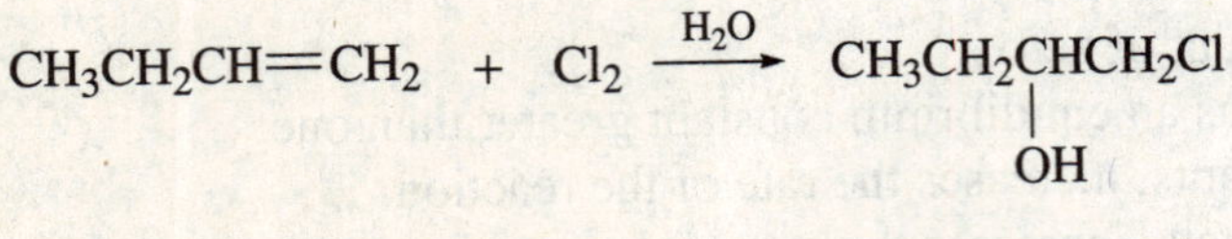

d. 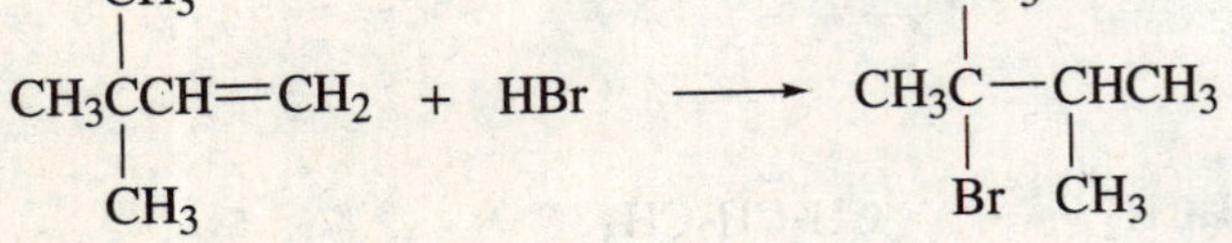

5. **a.** 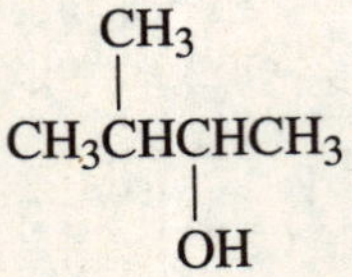

b. 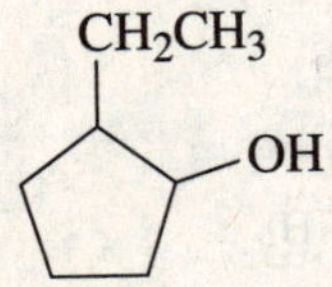

6. a. 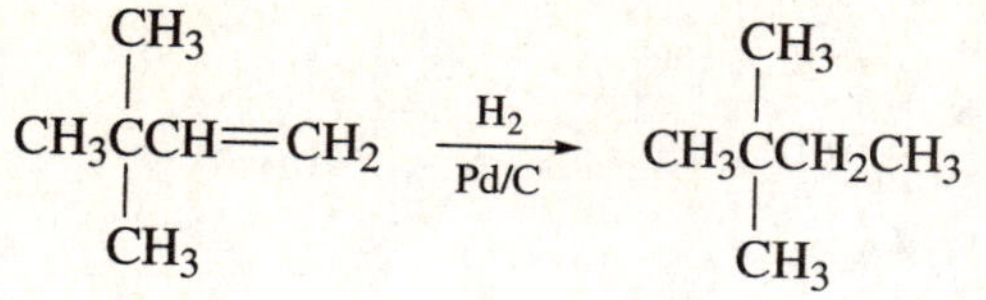

b. 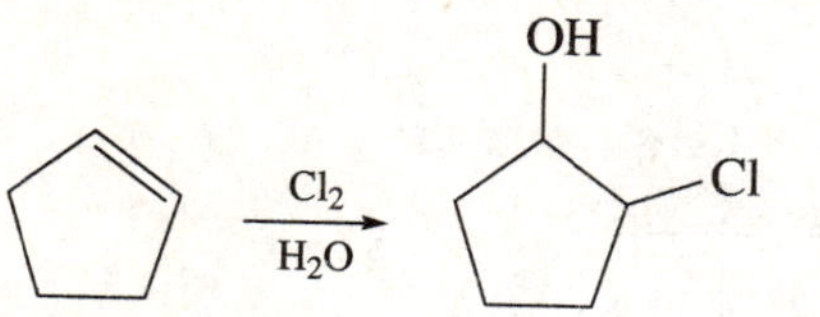

c. 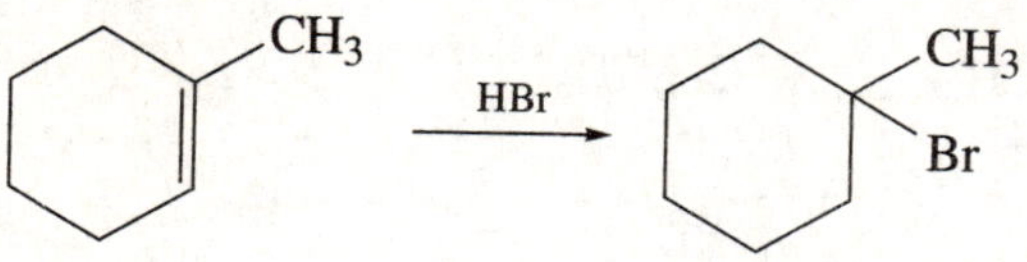

d. 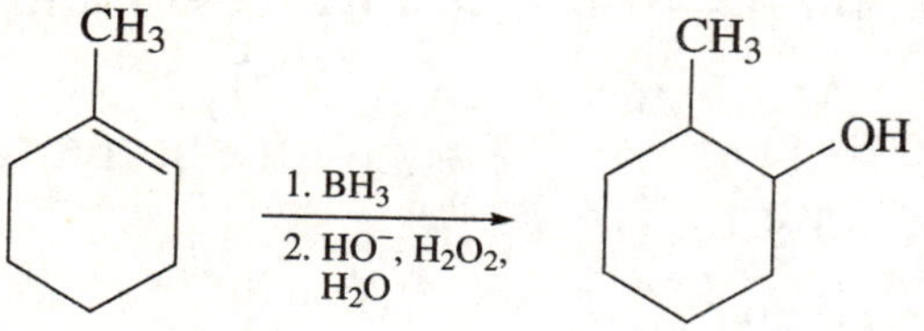

7. 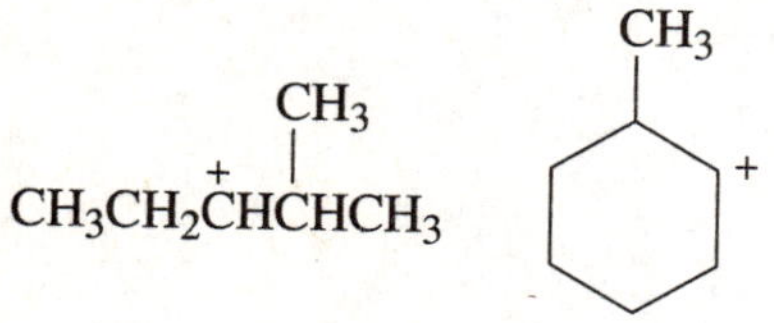

8.
a. The addition of Br_2 to 1-butene to form 1,2-dibromobutane is a concerted reaction. F
b. The reaction of 1-butene with HCl will form 1-chlorobutane as the major product. F
c. 2,3-Dimethyl-2-pentene is more stable than 3,4-dimethyl-2-pentene. T
d. The reaction of HBr with 3-methylcyclohexene is more highly regioselective than the reaction of HBr with 1-methylcyclohexene. F
e. The reaction of an alkene with a carboxylic acid forms an epoxide. F
f. A catalyst increases the equilibrium constant of a reaction. F

Answers to Chapter 5 Practice Test

1. a pair of enantiomers

2. -3.0

3. 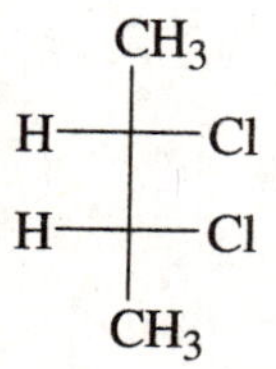

4. 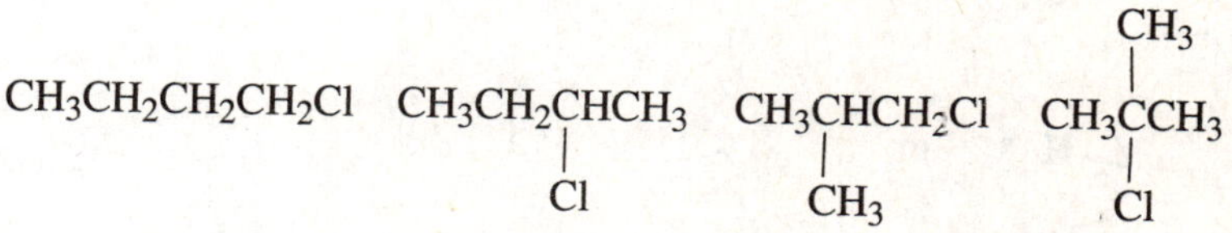

5. a. 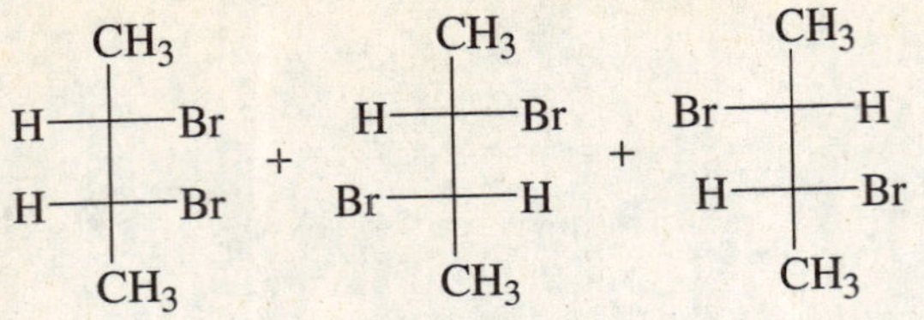

b. 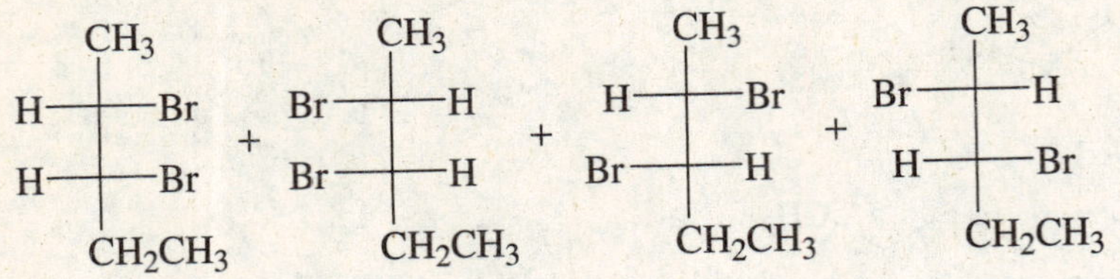

6. a. 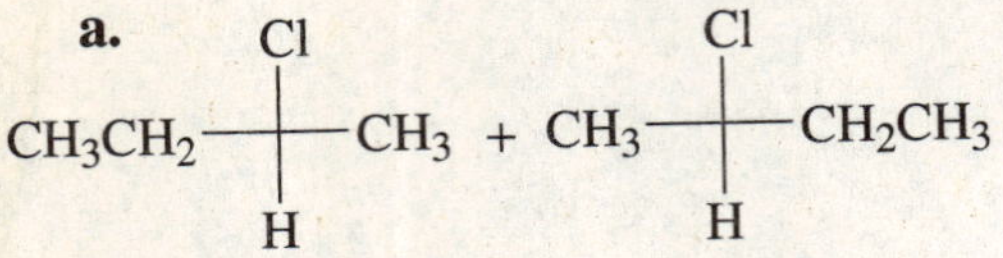

b. 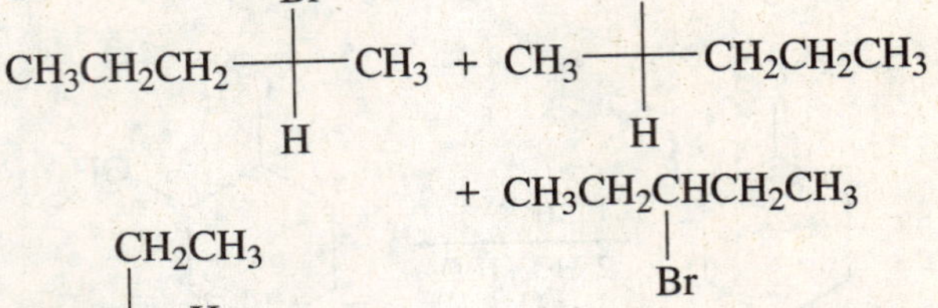

c. 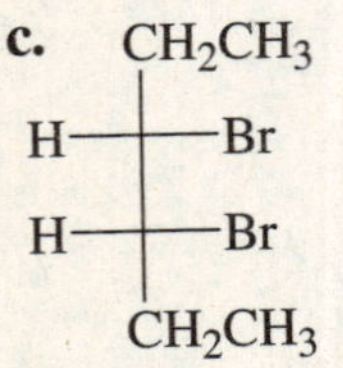

d. 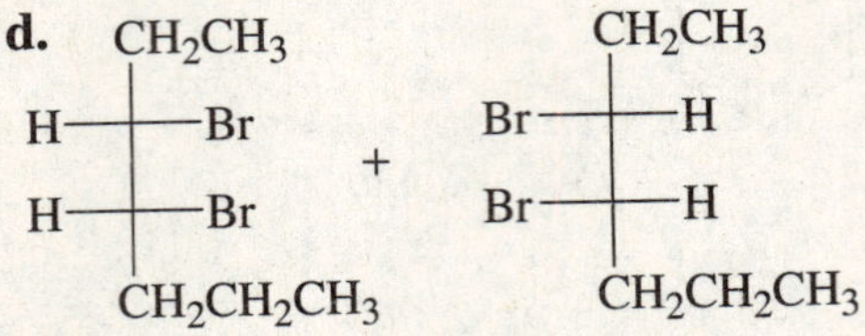

7. (–)-2-Methylbutanoic acid has the *S* configuration.

8. a. 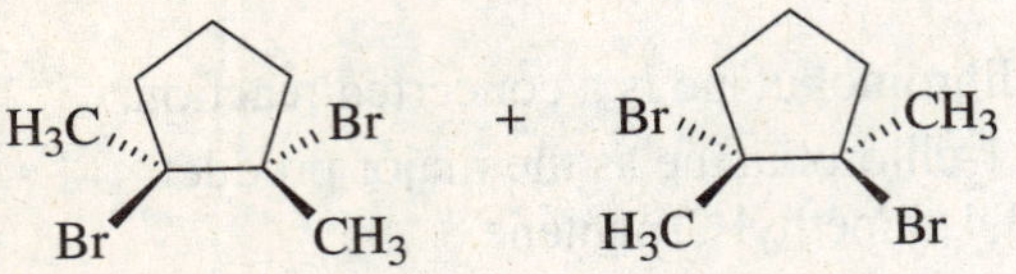

c. 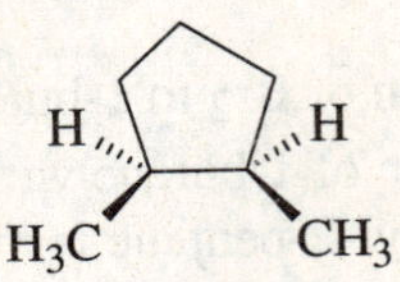

b. 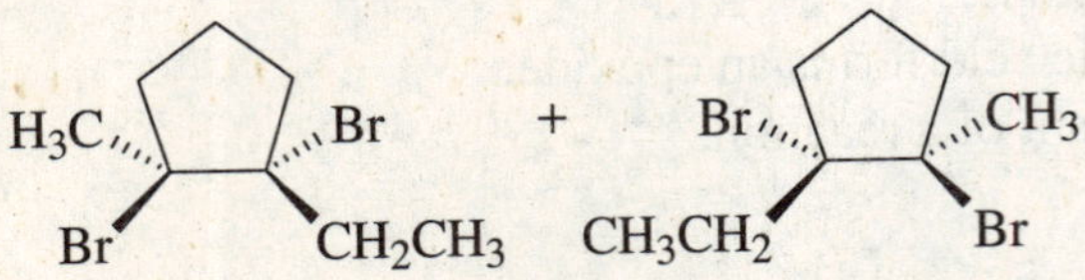

d. 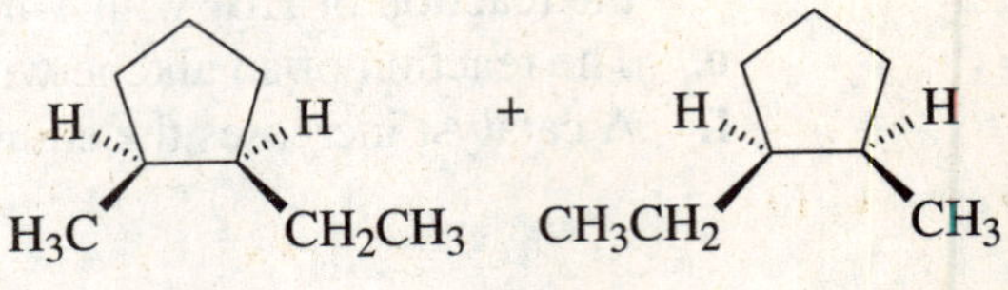

9. 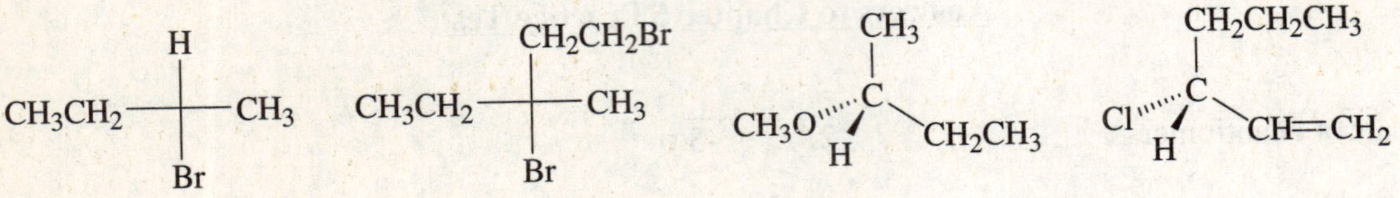

10. 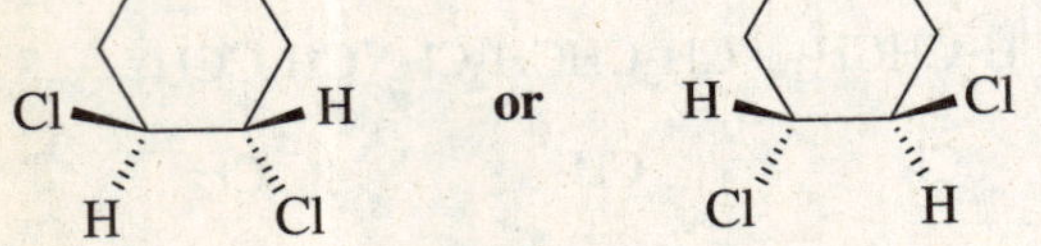

11. a. Diastereomers have the same melting points. F
b. The addition of HBr to 3-methyl-2-pentene is stereospecific. F
c. The addition of HBr to to 3-methyl-2-pentene is stereoselective. F

d. The addition of HBr to to 3-methyl-2-pentene is regioselective. T
e. Meso compounds do not rotate polarized light. T
f. 2,3-Dichloropentane has a stereoisomer that is a meso compound. F
g. All compounds with the *R* configuration are dextrorotatory. F
h. A compound with three asymmetric centers can have a maximum of nine stereoisomers. F

Answers to Chapter 6 Practice Test

1. **a.** 1. disiamylborane **2.** HO^-, H_2O_2, H_2O

b. H_2/Lindlar catalyst

2. 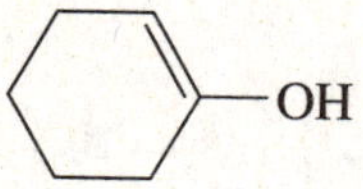

3. **a.** $CH_3CH_2CHC{\equiv}CCH_2CHCH_3$ (with CH_3 on C-3 and on C-7)

b. 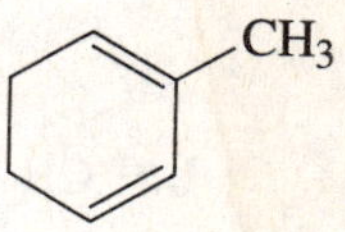

4. **a.** A terminal alkyne is more stable than an internal alkyne. F
b. Propyne is more reactive than propene toward reaction with HBr. F
c. 1-Butyne is more acidic than 1-butene. T
d. An sp^2 hybridized carbon is more electronegative than an sp^3 hybridized carbon. T

5. **a.** 1-bromo-5-methyl-3-hexyne **b.** 5-methyl-3-hexyn-1-ol

6. $CH_3CH_2CH_2C{\equiv}CH$

7.

NH_3	$CH_3C{\equiv}CH$	CH_3CH_3	H_2O	$CH_3CH{=}CH_2$
3	2	5	1	4

8. $CH_3CH_2CH_2\overset{O}{\overset{\|}{C}}CH_2CH_3$ 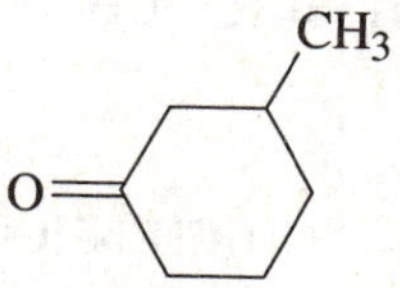

9. **a.** $CH_3CH_2C{\equiv}CH$ $\xrightarrow{^-NH_2}$ $CH_3CH_2C{\equiv}C^-$ $\xrightarrow{CH_3CH_2Br}$ $CH_3CH_2C{\equiv}CCH_2CH_3$ $\xrightarrow{H_2/Pt}$ $CH_3CH_2CH_2CH_2CH_2CH_3$

b. $CH_3CH_2C{\equiv}CH$ $\xrightarrow{1.\ ^-NH_2\quad 2.\ CH_3CH_2Br}$ $CH_3CH_2C{\equiv}CCH_2CH_3$ $\xrightarrow[\text{Lindlar catalyst}]{H_2}$ $CH_3CH_2CH{=}CHCH_2CH_3$ $\xrightarrow{HBr}$ $CH_3CH_2CH_2\underset{Br}{\underset{|}{C}}HCH_2CH_3$

c. $CH_3CH_2C{\equiv}CH \xrightarrow{^-NH_2} CH_3CH_2C{\equiv}C^- \xrightarrow{CH_3CH_2Br} CH_3CH_2C{\equiv}CCH_2CH_3 \xrightarrow{H_2O,\ H_2SO_4} CH_3CH_2\overset{O}{\overset{\|}{C}}CH_2CH_2CH_3$

Answers to Chapter 7 Practice Test

1. a. 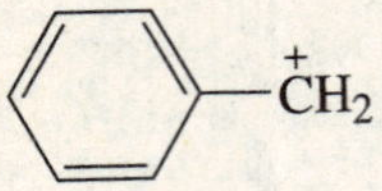

c. $CH_3\bar{C}H\overset{O}{\overset{\|}{C}}CH_3$

e. 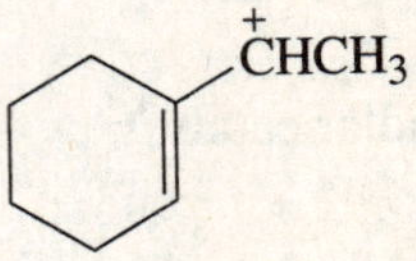

b. $CH_3\bar{C}HC{\equiv}CH$

d. $CH_2{=}CH\overset{+}{C}H_2$

2. a. $CH_3CH{=}CH{-}\ddot{O}CH_3 \longleftrightarrow CH_3\bar{C}H{-}CH{=}\overset{+}{O}CH_3$

b. $CH_3CH{=}CH{-}CH{=}CH{-}\overset{+}{C}H_2 \longleftrightarrow CH_3CH{=}CH{-}\overset{+}{C}H{-}CH{=}CH_2 \longleftrightarrow CH_3\overset{+}{C}H{-}CH{=}CH{-}CH{=}CH_2$

c. $^-CH_2{-}CH{=}CH{-}\overset{O}{\overset{\|}{C}H} \longleftrightarrow CH_2{=}CH{-}\bar{C}H{-}\overset{O}{\overset{\|}{C}H} \longleftrightarrow CH_2{=}CH{-}CH{=}\overset{O^-}{\overset{|}{C}H}$

3. $CH_3\overset{CH_3}{\overset{|}{\underset{+}{C}}}CH_2CH{=}CH_2$ $\quad CH_2{=}CHCH_2CH{=}CH_2$ $\quad CH_3CH_2NHCH_2CH{=}CHCH_3$ 

4. a. 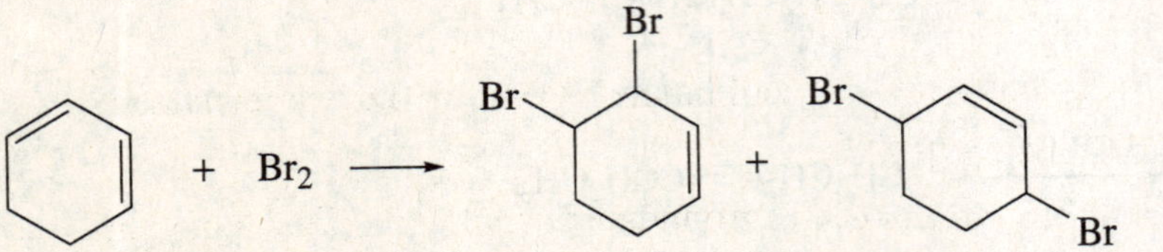

b. 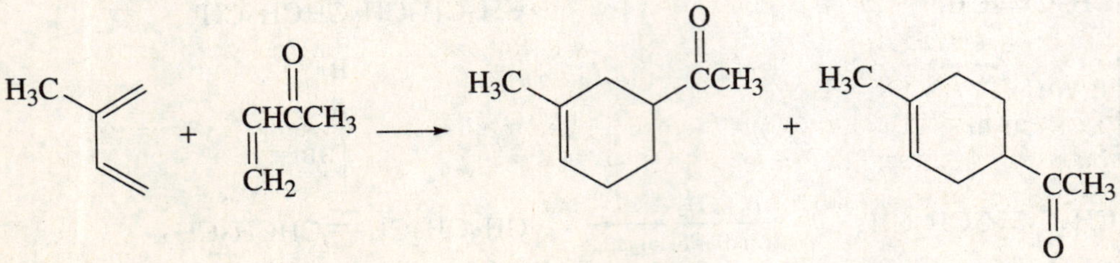

5. 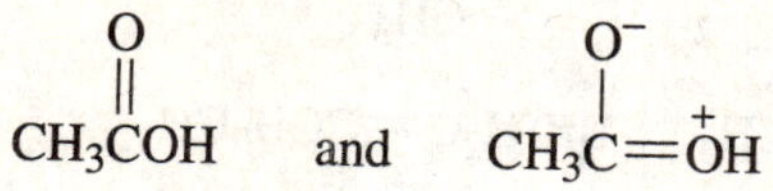

6. 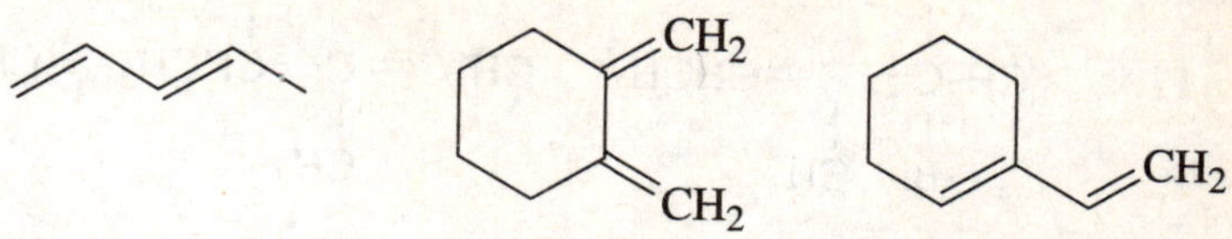

7. 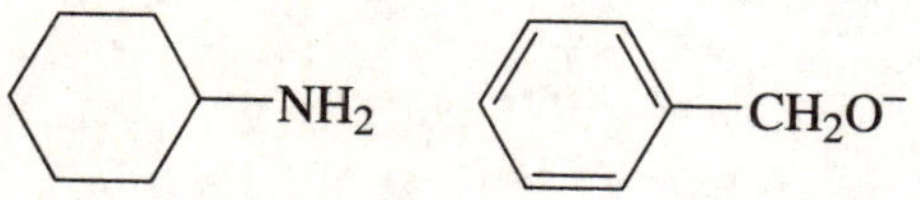

8. **a.** 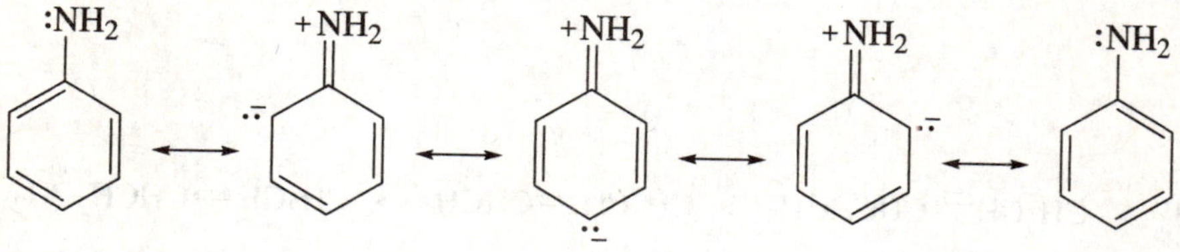

b. 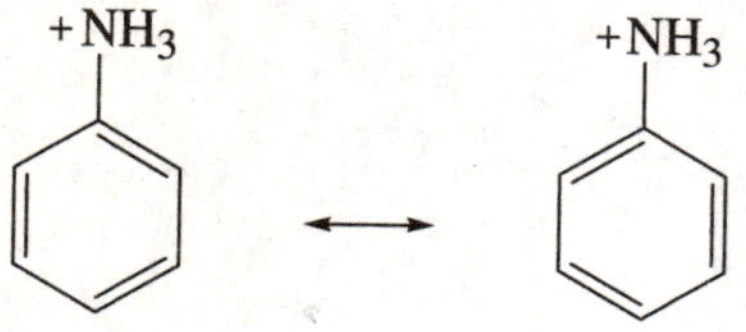

c. 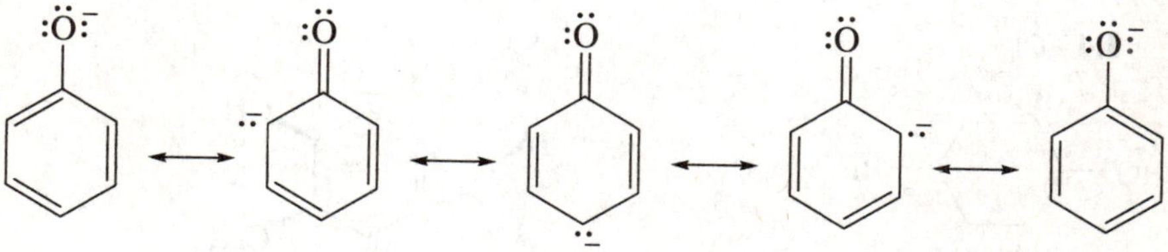

9. **a.** 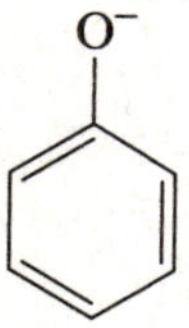 **b.** 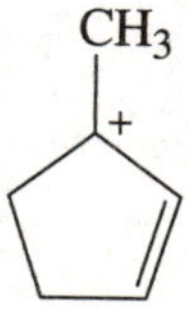

10. Indicate whether the following statements is true or false:

- **a.** A compound with four conjugated double bonds will have four molecular orbitals. F
- **b.** ψ_1 and ψ_2 are symmetric molecular orbitals. F
- **c.** If ψ_3 is the HOMO in the ground state, ψ_4 will be the HOMO in the excited state. T
- **d.** If ψ_3 is the LUMO, ψ_4 will be the HOMO. F
- **e.** If the ground-state HOMO is symmetric, the ground-state LUMO will be antisymmetric. T
- **f.** A conjugated diene is more stable than an isomeric isolated diene. T
- **g.** A single bond formed by an sp^2—sp^2 overlap is longer than a single bond formed by an sp^2—sp^3 overlap. F
- **h.** The thermodynamically controlled product is the major product obtained when the reaction is carried out under mild conditions. F
- **i.** 1,3-Hexadiene is more stable than 1,4-hexadiene. T

11. 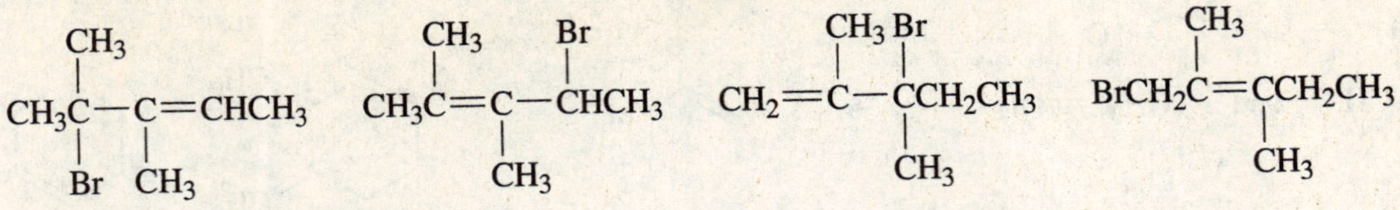

12. 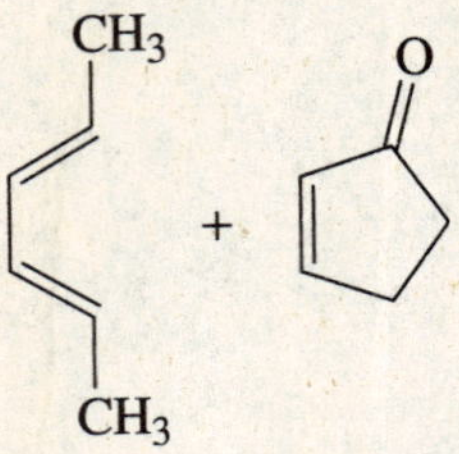

13. 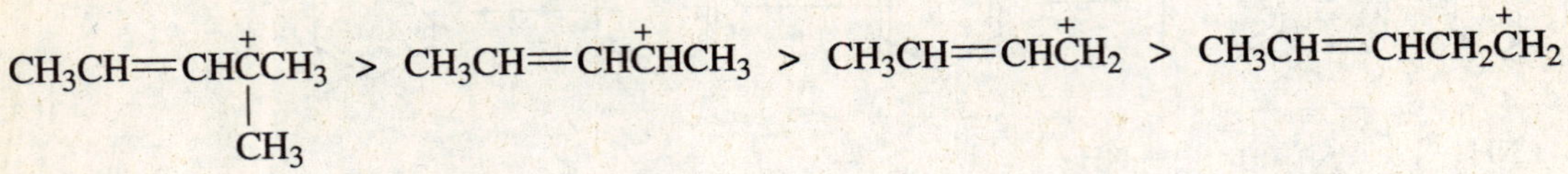

14. **a.** 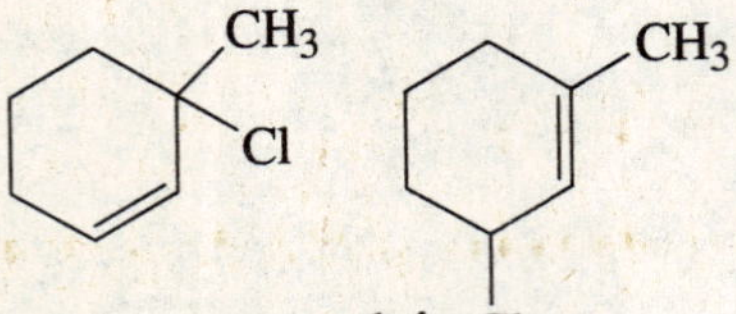

1,2 1,4

b. 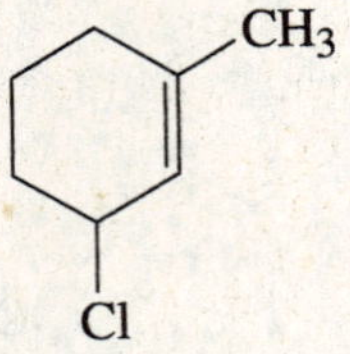

15. 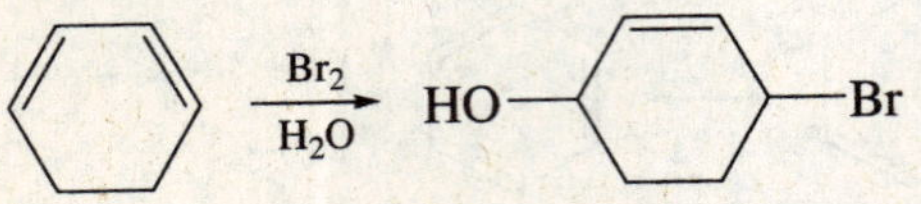

16. 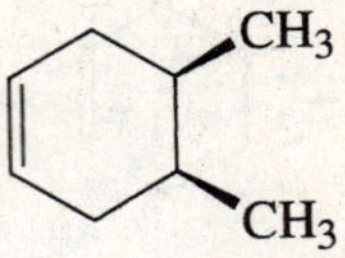

17. **a.** 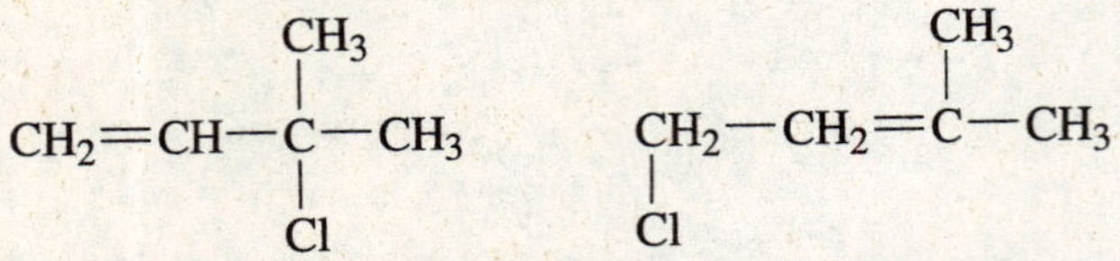

product of kinetic control product of thermodynamic control

b. 

product of kinetic control product of thermodynamic control

Answers to Chapter 8 Practice Test

1. a. 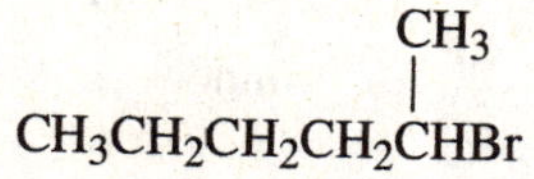

   b. 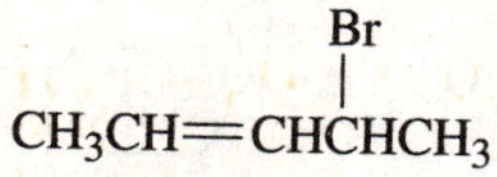

2. a. 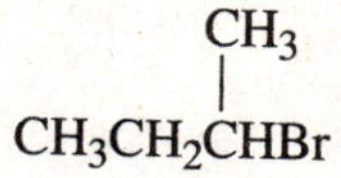

   b. 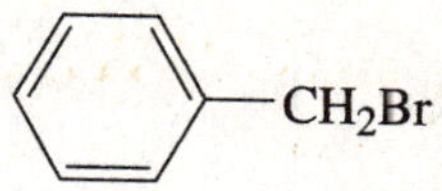

3. 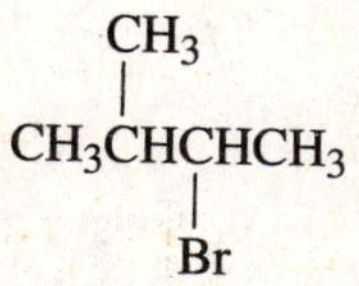

   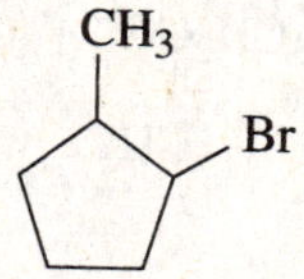

4. a. Increasing the concentration of the nucleophile favors an S_N1 reaction over an S_N2 reaction. F
   b. Ethyl iodide is more reactive than ethyl chloride in an S_N2 reaction. T
   c. In an S_N1 reaction, the product with the retained configuration is obtained in greater yield. F
   b. The rate of a substitution reaction in which none of the reactants is charged will increase if the polarity of the solvent is increased. T
   e. An S_N2 reaction is a two-step reaction. F
   f. The pK_a of a carboxylic acid is greater in water than it is in a mixture of dioxane and water. F
   g. 4-Bromo-1-butanol will form a cyclic ether faster than will 3-bromo-1-propanol. T

5. a. CH_3O^- b. CH_3S^-

6. a. $CH_3CH_2CH_2Cl + HO^-$
   b. $CH_3CH_2CH_2I + HO^-$
   c. $CH_3CH_2CH_2Br + HO^-$
   d. 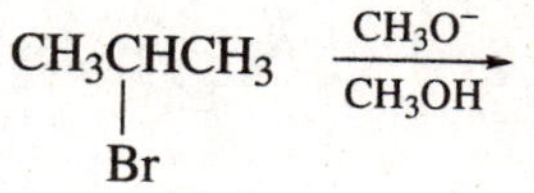
   e. $BrCH_2CH_2CH_2CH_2NHCH_3$

7. All are aprotic solvents except ethanol.

8. a. The rate of the reaction would increase.
   b. The rate of the reaction would increase.
   c. The rate of the reaction would decrease.
   d. The pK_a would decrease.
   e. The pK_a would decrease.

Answers to Chapter 9 Practice Test

1. a. 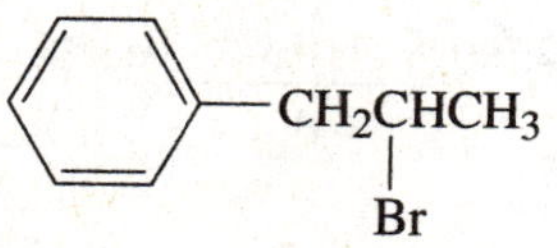

   b. 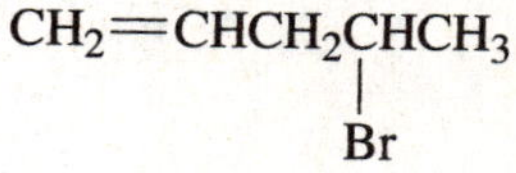

2. 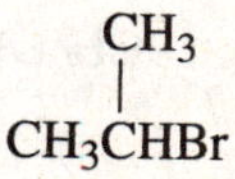

3. 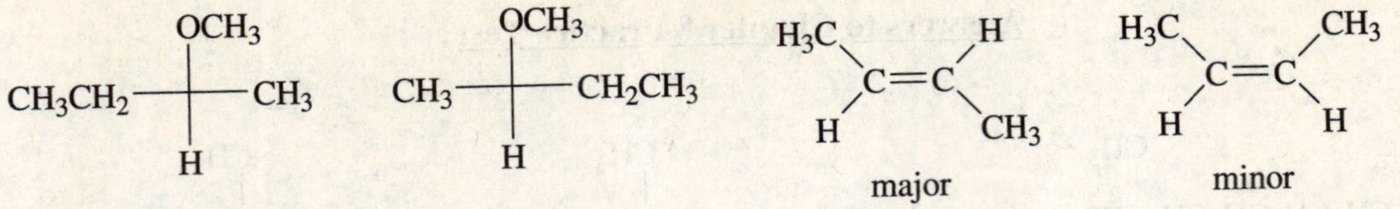

4. a. 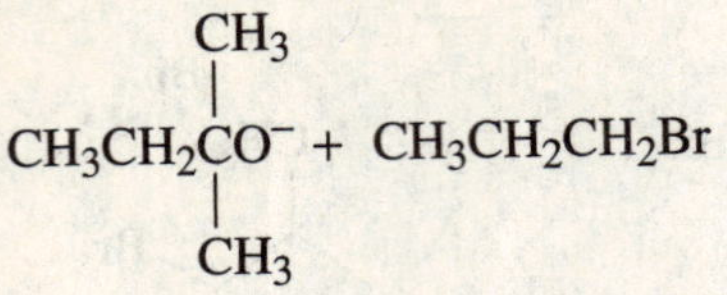 b. 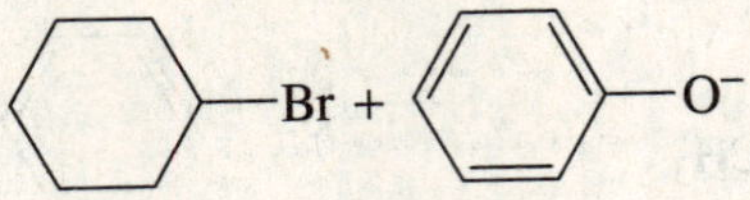 c. 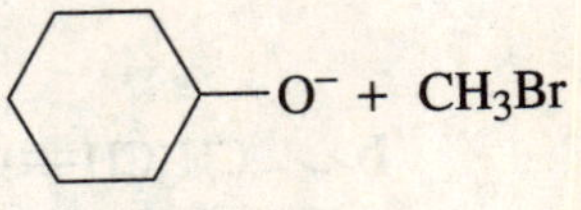

5. a. 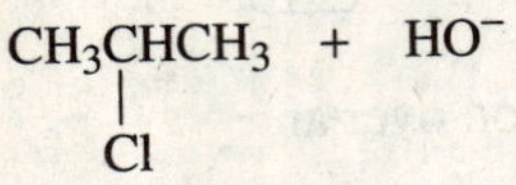 d. 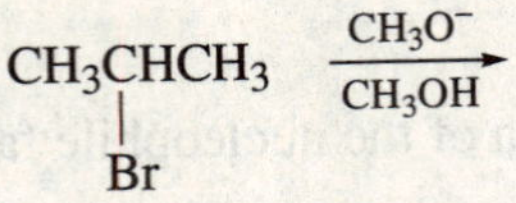

b. $CH_3CH_2CH_2I + HO^-$ e. 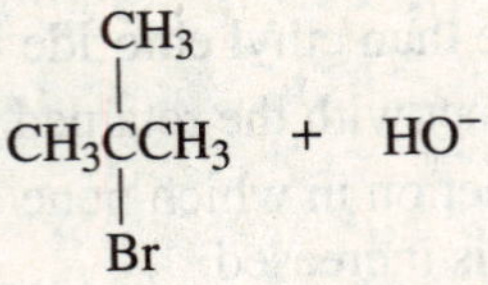

c. $CH_3CH_2CH_2Br + HO^-$

6. a. 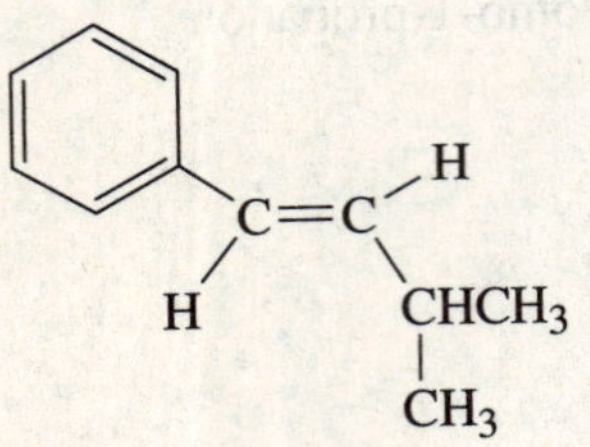 b. 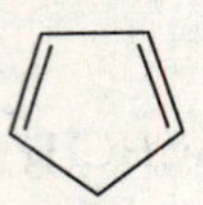 c. 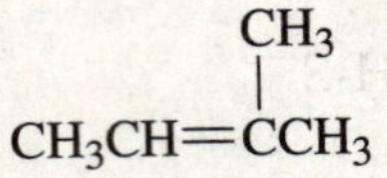 d. 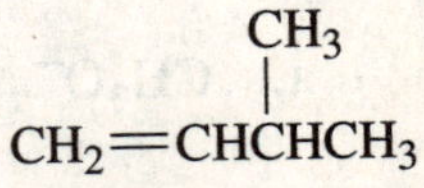

7. a. *cis*-1-bromo-2-methylcyclohexane
b. The cis and trans isomers react at about the same rate.

Answers to Chapter 10 Practice Test

1. HBr 2. $SOCl_2$

3. a. 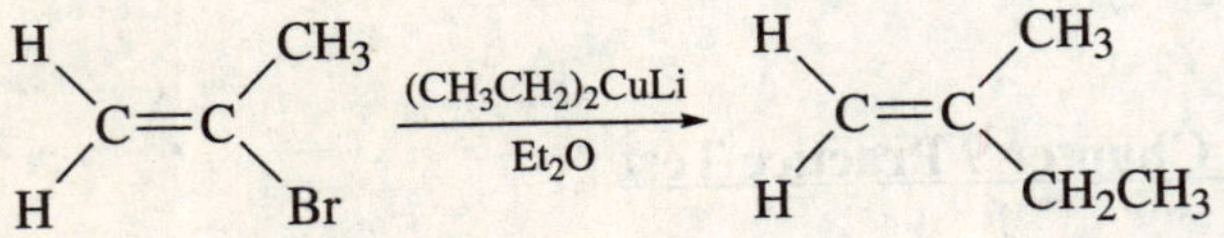

b. 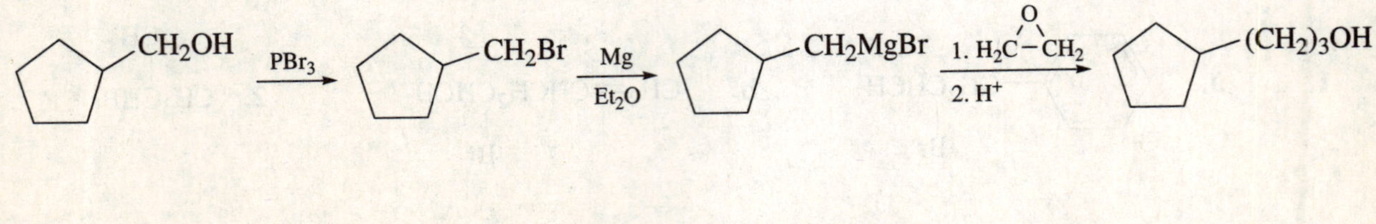

4. a. 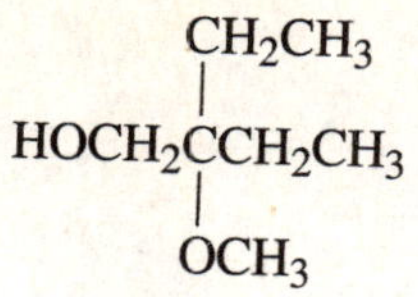 b. 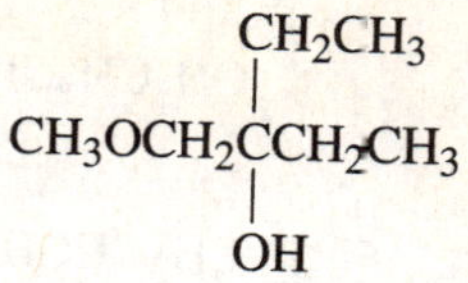

5. a. 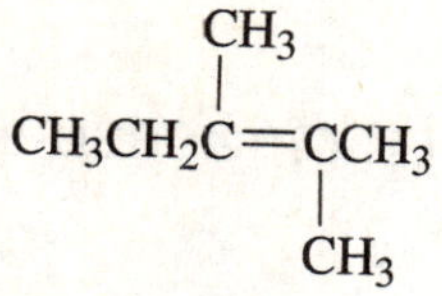 b. 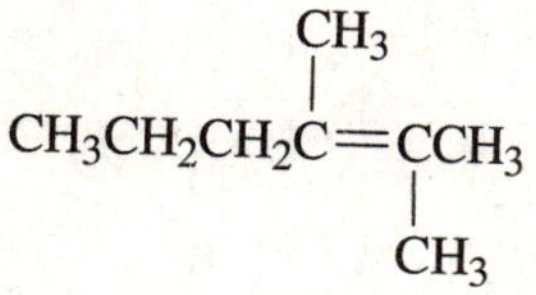 c. $CH_3CH_2CH{=}CHCH_3$ d. 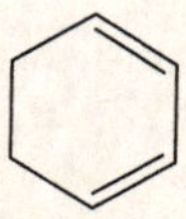

6. a. Tertiary alcohols are easier to dehydrate than secondary alcohols. T
   b. Alcohols are more acidic than thiols. F
   c. Alcohols have higher boiling points than thiols. T

7. a. 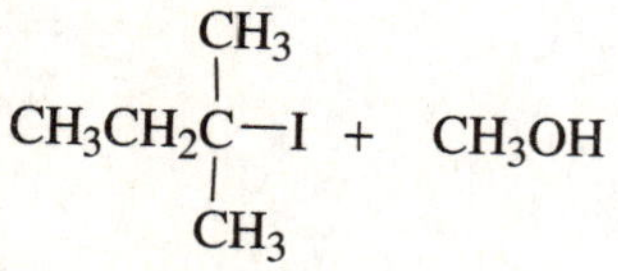 b. 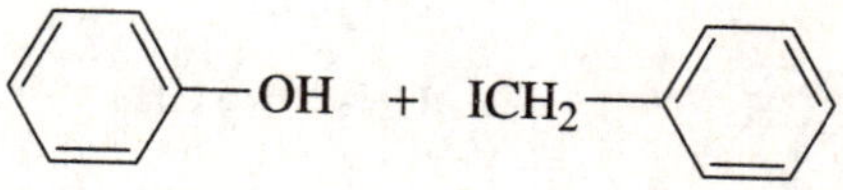

8. a. $CH_3CH_2CH_2CH_2CH_2OH$ b. 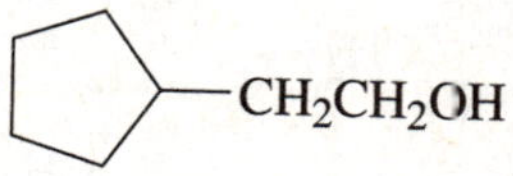

9. a. 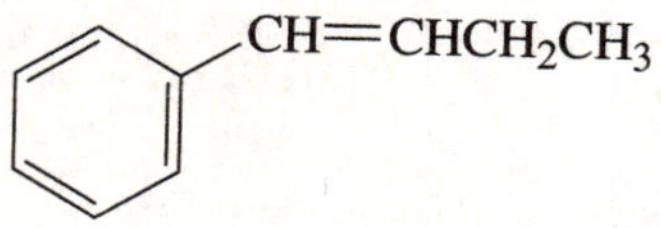 c. 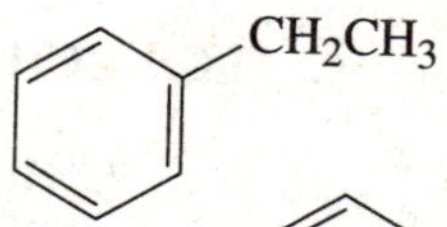

   b. 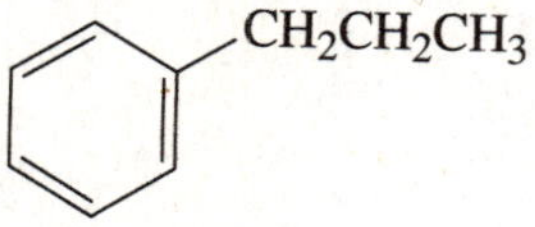 d. 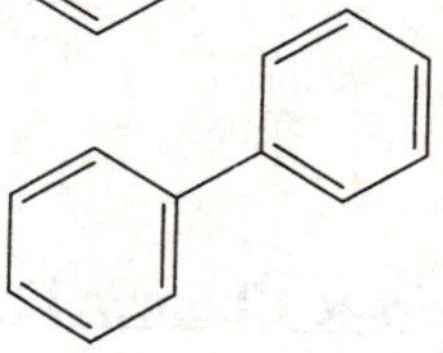

Answers to Chapter 11 Practice Test

1. 3

2. $CH_3CH_3 + \cdot Cl \longrightarrow CH_3\dot{C}H_2 + HCl$

3. 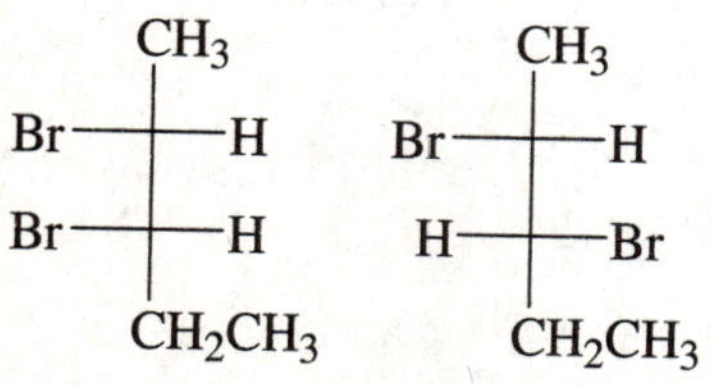

4. $CH_3CH_3 + Cl\cdot \longrightarrow CH_3\dot{C}H_2 + HCl \quad \Delta H° = 101 - 103 = -2$

$CH_3\dot{C}H_2 + Cl_2 \longrightarrow CH_3CH_2Cl + Cl\cdot \quad \Delta H° = 58 - 85 = -27$

5. $\dot{C}H_2CH_2CH_2CH{=}CH$ **3** $CH_3CH_2\dot{C}HCH{=}CH_2$ **1**

$CH_2CH_2CH_2CH{=}\dot{C}H$ **4** $CH_3\dot{C}HCH_2CH{=}CH_2$ **2**

6. **a.** 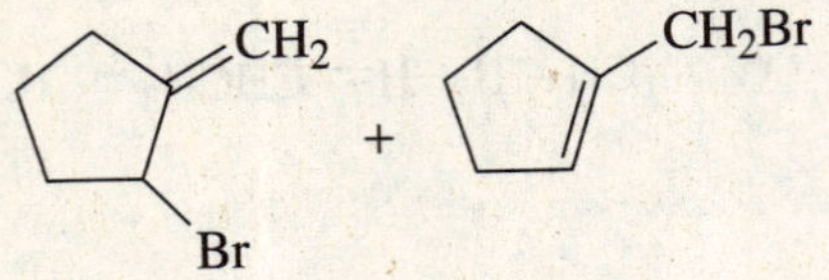

b. 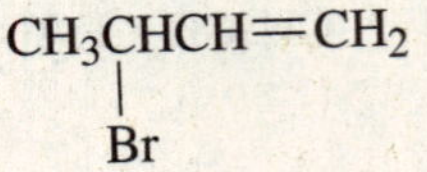 + $CH_3CH{=}CHCH_2Br$

7. **a.** 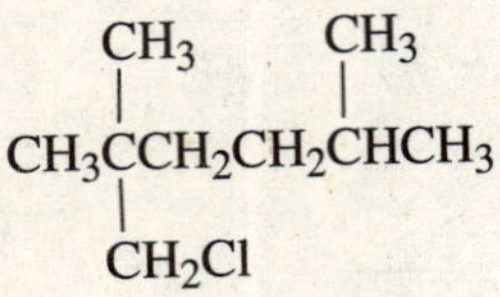 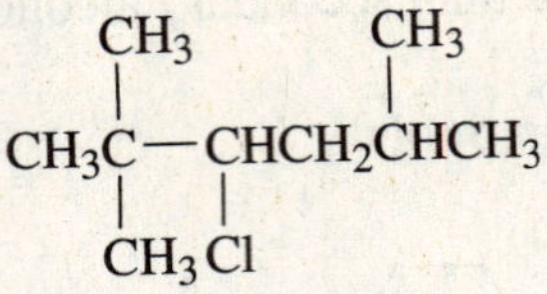 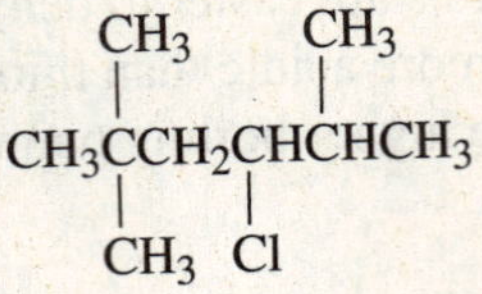

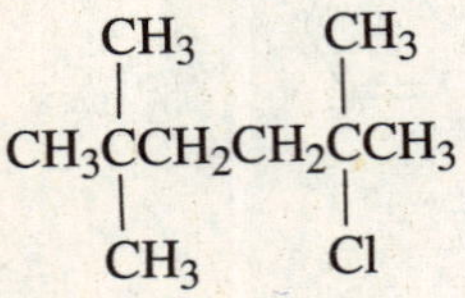 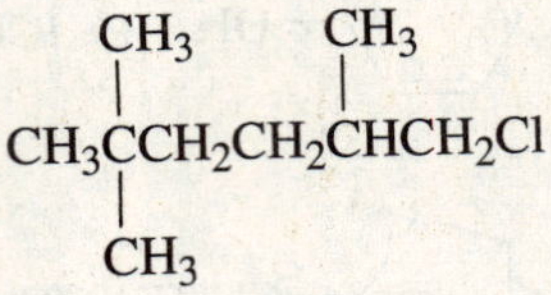

b. 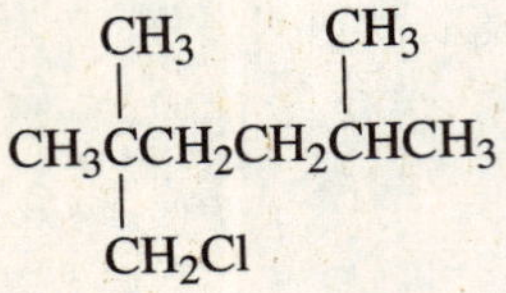

c. 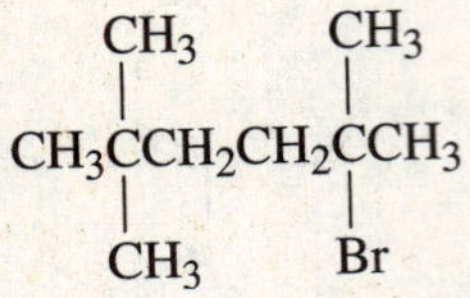

8. 23%

Answers to Chapter 12 Practice Test

1. **a.** 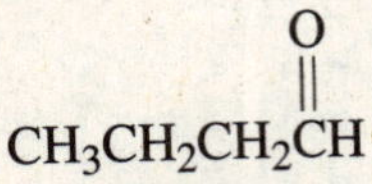
~2700 cm^{-1}

b. 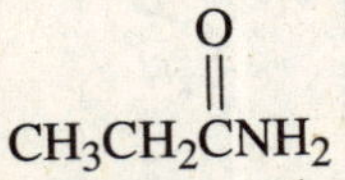
~3300 cm^{-1}

c. $CH_3CH_2CH_2CH_2OH$
~3600-3200 cm^{-1}

d. 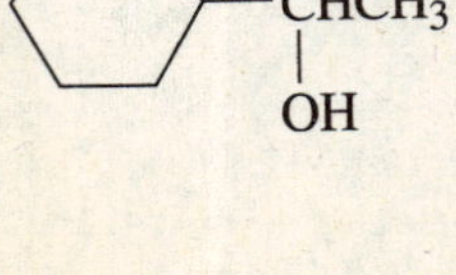 ~1380 cm^{-1}

e. 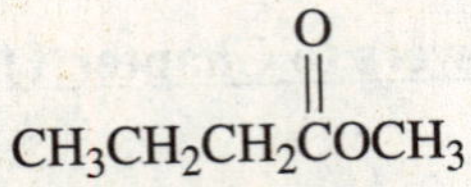
~1050 or ~1250 cm^{-1}

f. $CH_3CH_2CH{=}CHCH_3$
~ 1600 cm^{-1}
~ 3100 cm^{-1}

$CH_3CH_2C{\equiv}CCH_3$
~ 2100 cm^{-1}

g. $CH_3CH_2C{\equiv}CH$
~ 3300 cm^{-1}

2.

a. The O-H stretch of a concentrated solution of an alcohol occurs at a higher frequency than the O-H stretch of a dilute solution. F

b. Light of 2 μm is of higher energy than light of 3 μm. T

c. It takes more energy for a bending vibration than for a stretching vibration. F

d. Propyne will not have an absorption band at 3100 cm^{-1} because there is no change in the dipole moment. F

e. Light of 8 μm has the same energy as light of 1250 cm^{-1}. T

f. The M + 2 peak of an alkyl chloride is half the height of the M peak. F

g. A chromophore that exhibits both $n \rightarrow \pi^*$ and $\pi \rightarrow \pi^*$ transitions will have the $n \rightarrow \pi^*$ transition at a longer wavelength. T

3.

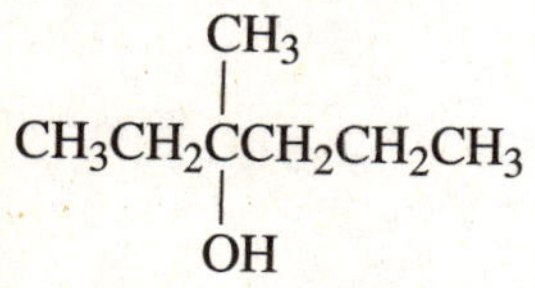

4.

Absorbance = molar absorbtivity × concentration × length of light path (in cm)

0.72 = molar absorbtivity $\times 3.8 \times 10^{-4} \times 1$

molar absorbtivity = 1,900 M^{-1} cm^{-1}

5.

a.

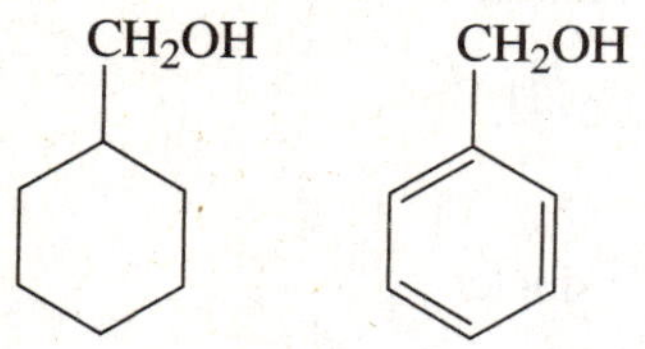

~3100 cm^{-1}
~1500 cm^{-1}

c.

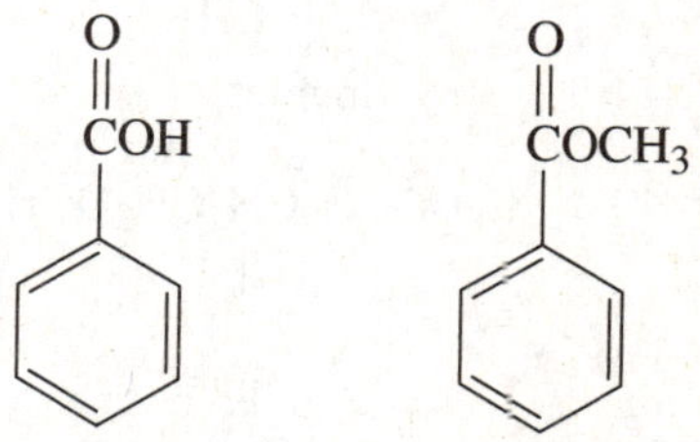

~3300-2500 cm^{-1}

~2900 cm^{-1}
~1380 cm^{-1}
~1050 cm^{-1}

d.

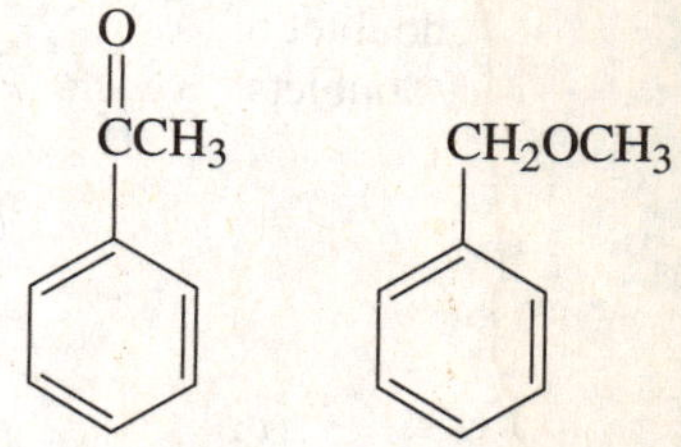

~1700 cm^{-1}

~1050 cm^{-1}

b.

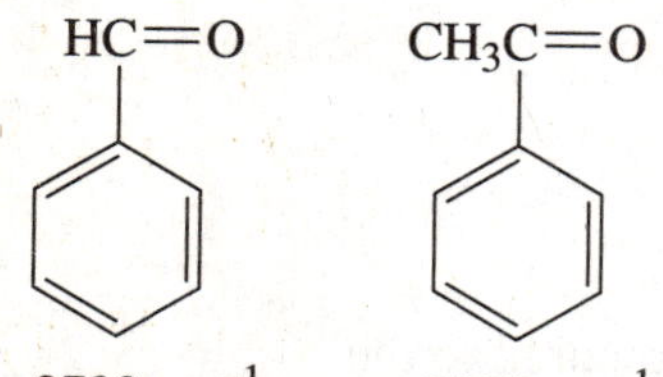

~2700 cm^{-1}

~2900 cm^{-1}
~1380 cm^{-1}

e.

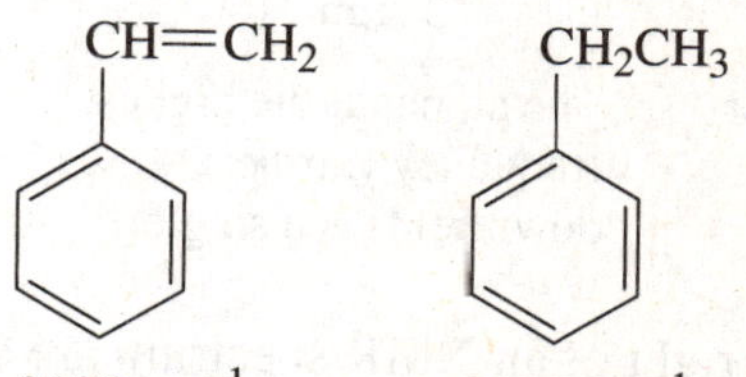

~3100 cm^{-1}

~2900 cm^{-1}
~1380 cm^{-1}

6.

a.

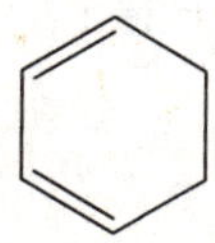

b.

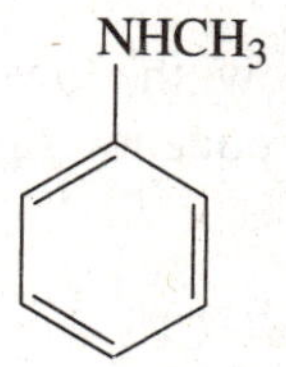

c.

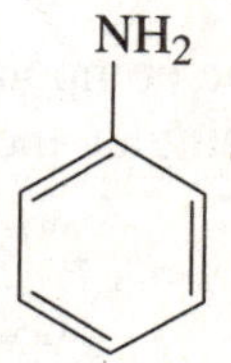

7.

Absorbance = molar absorbtivity × concentration × length of light path (in cm)

$0.76 = 1200 \times$ concentration

concentration = 6.3×10^{-4} M

Answers to Chapter 13 Practice Test

1. $CH_3CH_2CH_2\overset{\overset{O}{\|}}{C}CH_3$ **4** $CH_3CH_2\underset{\underset{Cl}{|}}{C}HCH_2CH_3$ **3** $CH_2{=}CH\overset{\overset{O}{\|}}{C}H$ **4**

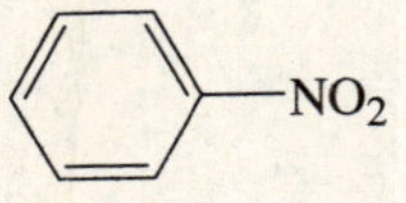

3

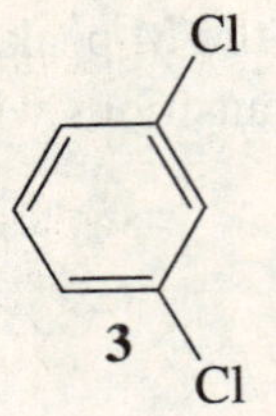

3

$CH_3\overset{\overset{CH_3}{|}}{C}HCH_2\overset{\overset{CH_3}{|}}{C}HCH_3$

3

2. $CH_3CH_2\overset{\overset{O}{\|}}{C}CH_3$ ← quartet

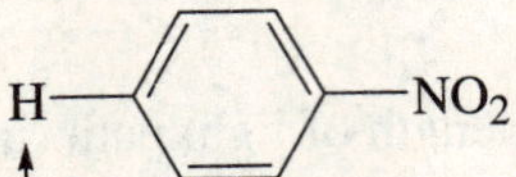

← triplet

$CH_3CH_2\overset{\overset{O}{\|}}{C}OCH_2CH_3$ ← triplet

$CH_3\underset{\underset{CH_3}{|}}{C}HCH_2Cl$ ← doublet

$ClCH_2CH_2CH_2OCH_3$ ← multiplet

doublet of doublets

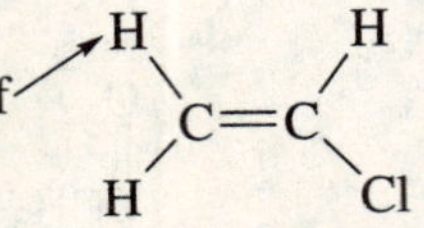

$CH_3OCH_2CH_2CH_2OCH_3$ ← quintet

$BrCH_2CH_2Br$ ← singlet

3. $CH_3\overset{\overset{O}{\|}}{C}OCH_2CH_3$ 3 signals

The signal at the highest frequency (farthest downfield) is a quartet.

$CH_3CH_2\overset{\overset{O}{\|}}{C}OCH_3$ 3 signals

The signal at the highest frequency (farthest downfield) is a singlet.

$H\overset{\overset{O}{\|}}{C}OCH_2CH_2CH_3$ 4 signals

4.
- **a.** The peaks on the right of an NMR spectrum are deshielded compared to the peaks on the left. F
- **b.** Dimethyl ketone has the same number of signals in its ^1H NMR spectrum as in its ^{13}C NMR spectrum. F
- **c.** In the ^1H NMR spectrum of the compound shown below, the lowest frequency signal (the one farthest upfield) is a singlet and the highest frequency signal (the one farthest downfield) is a doublet. T

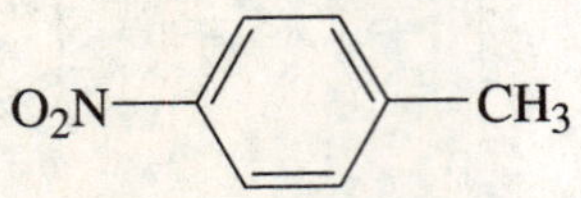

- **d.** The greater the frequency of the signal, the greater its chemical shift in ppm. T

5. a. $CH_3CH_2CH_2Cl$ triplet, 3 signals

b. $CH_3CH_2\overset{O}{\overset{\|}{C}}OCH_3$ singlet, 3 signals

c. $CH_3\underset{Br}{\underset{|}{C}}HCH_3$ septet, 2 signals

6 a. $CH_3CH_2CH_2Cl$ 3 signals, triplet

b. $CH_3CH_2\overset{O}{\overset{\|}{C}}OCH_3$ 4 signals, singlet

c. $CH_3\underset{Br}{\underset{|}{C}}HCH_3$ 2 signals, doublet

Answers to Chapter 14 Practice Test

1. 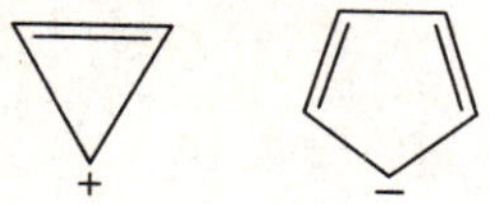

2. 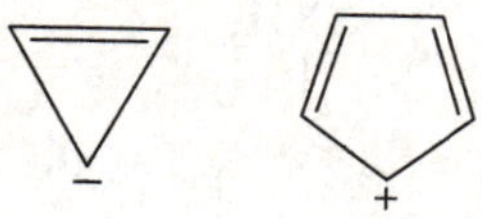

3. 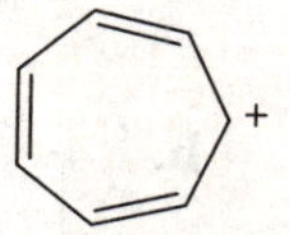

4. 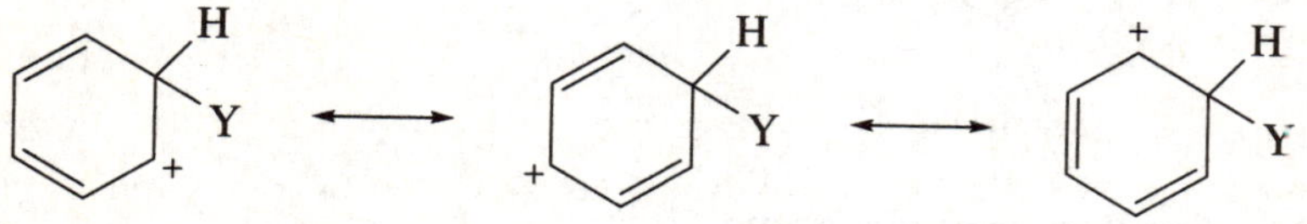

5. 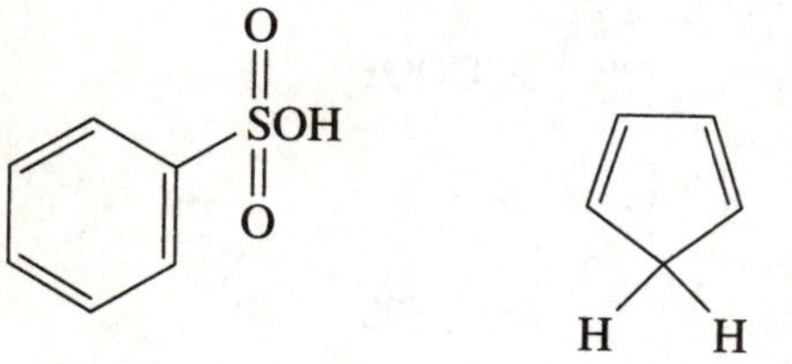

6. $CH_3CH_2CH_2CH_2Cl$ $CH_3CH{=}CHCH_3$ $CH_3CH_2CH_2CH_2OH$

$CH_3CH_2\underset{Cl}{\underset{|}{C}}HCH_3$ $CH_2{=}CHCH_2CH_3$ $CH_3CH_2\underset{OH}{\underset{|}{C}}HCH_3$

7. $CH_3CH_2\overset{O}{\overset{\|}{C}}O\overset{O}{\overset{\|}{C}}CH_2CH_3$

8. a. 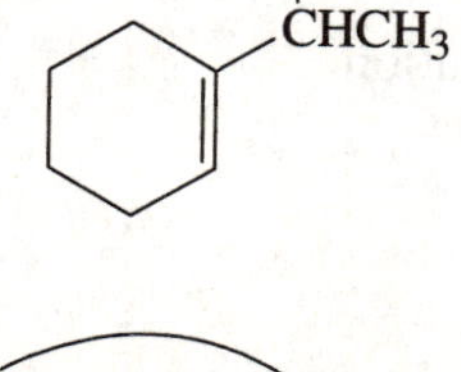

b. $CH_3\overset{+}{C}HCH{=}CH_2$

c. $CH_3\bar{C}H\overset{O}{\overset{\|}{C}}CH_3$

9. $H\ddot{O}{-}NO_2 + H{-}OSO_3H \rightleftharpoons H\overset{H}{\overset{+|}{\ddot{O}}}{-}NO_2 \rightleftharpoons H_2O + {}^{+}NO_2$

Answers to Chapter 15 Practice Test

1. a. *meta*-nitrotoluene
3-nitrotoluene
b. 1,2,4-tribromobenzene
c. *ortho*-ethylbenzoic acid
2- ethylbenzoic acid
d. *para*-chlorophenol
4-chlorophenol

2.

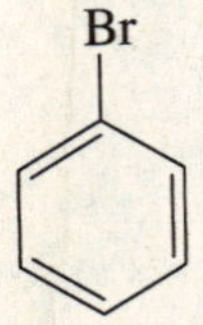

4

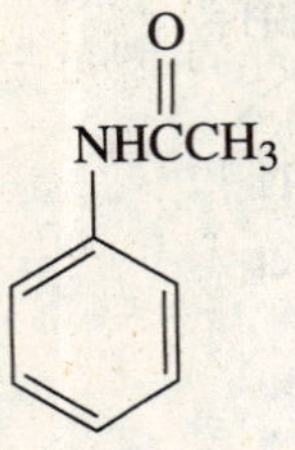

1

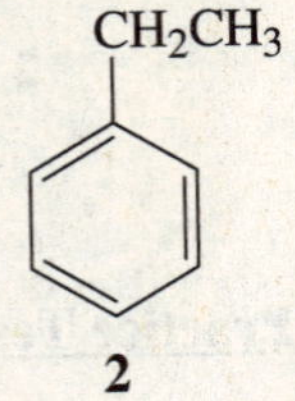

2

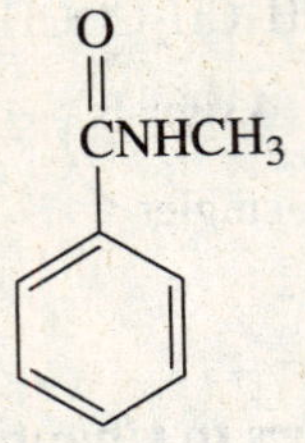

5

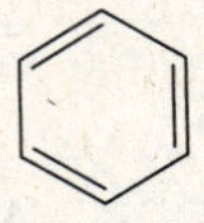
3

3. a.
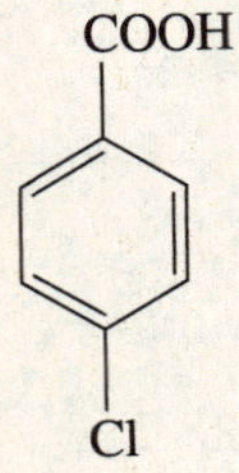

b.
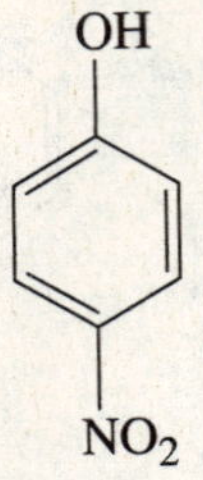

c.
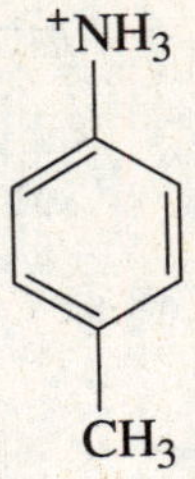

d.
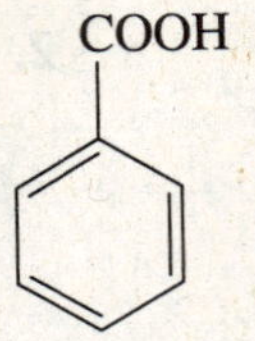

4. a. *para*-bromonitrobenzene b. *para*-bromoethylbenzene

5. a.
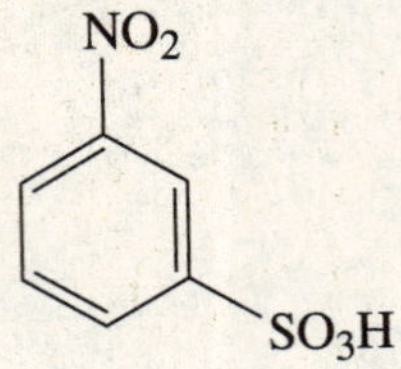

b.
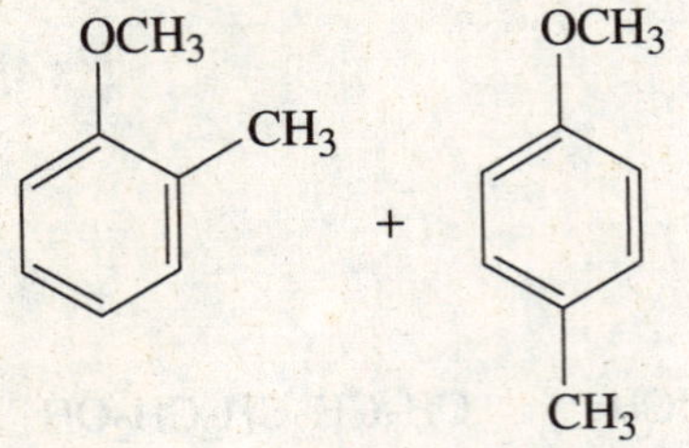

c.
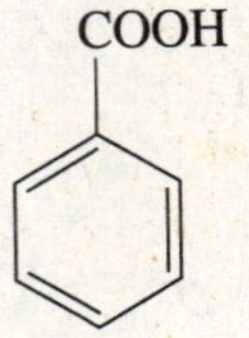

d.
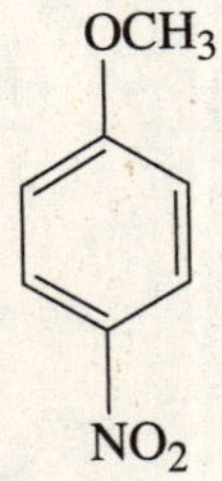

e.
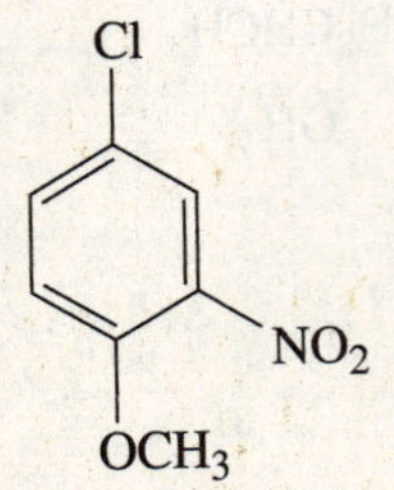

f.
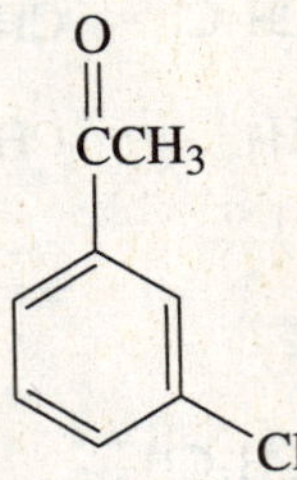

6. a. Benzoic acid is more reactive than benzene towards electrophilic substitution. F
b. *para*-Chlorobenzoic acid is more acidic than *para*-methoxybenzoic acid. T
c. A $CH=CH_2$ group is a meta director. F
d. *para*-Nitroaniline is more basic than *para*-chloroaniline. F

Answers to Chapter 16 Practice Test

1. a. $CH_3\overset{O}{\overset{\|}{C}}OCH_3$ b. 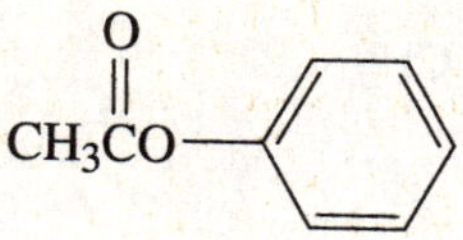 c. 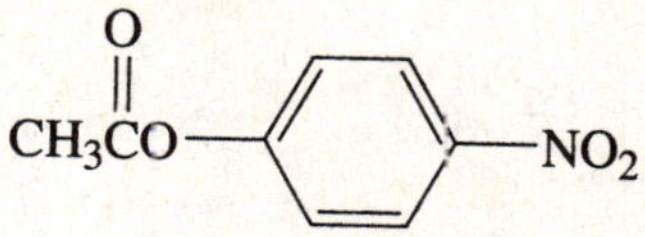 d. $CH_3\overset{O}{\overset{\|}{C}}Cl$

2. a. *N*-ethylpentanamide
   b. 3-methylpentanoic acid
   c. methyl 4-phenylbutanoate
   d. ethanoic propanoic anhydride

3. a. $CH_3CH_2\overset{O}{\overset{\|}{C}}O\overset{O}{\overset{\|}{C}}CH_2CH_3$

   b. $CH_3\overset{O}{\overset{\|}{C}}Cl + H_2O \longrightarrow CH_3\overset{O}{\overset{\|}{C}}OH + HCl$

   Any reaction in which a reactant is cleaved as a result of reaction with water.

   c. $CH_3\overset{O}{\overset{\|}{C}}OCH_3 + CH_3CH_2OH \xrightleftharpoons{H^+} CH_3\overset{O}{\overset{\|}{C}}OCH_2CH_3 + CH_3OH$

4. a. $CH_3{-}\overset{O}{\overset{\|}{C}}{-}OH$ b. $CH_3{-}\overset{O}{\overset{\|}{C}}{-}OCH_3$ c. $CH_3{-}\overset{O}{\overset{\|}{C}}{-}NH_2$ d. $CH_3{-}\overset{O}{\overset{\|}{C}}{-}OH$

5. a. $CH_3CH_2\overset{O}{\overset{\|}{C}}OH + \overset{+}{N}H_4$

   b. $CH_3CH_2\overset{O}{\overset{\|}{C}}O^- + CH_3CH_2CH_2OH$

   c. $CH_3CH_2CH_2\overset{O}{\overset{\|}{C}}OH + CH_3CH_2\overset{O}{\overset{\|}{C}}OH$

   d. $CH_3CH_2\overset{O}{\overset{\|}{C}}NHCH_2CH_3 + Cl^-$

   e. 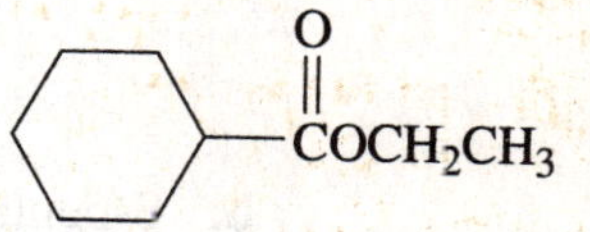 $+ CH_3OH$

   f. $CH_3CH_2CH_2\overset{O}{\overset{\|}{C}}OCH_2CH_2CH_3$

   g. $CH_3CH_2\overset{O}{\overset{\|}{C}}OH + CH_3\overset{+}{N}H_3$

   h. 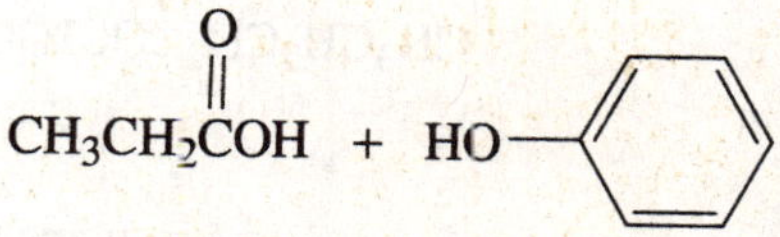

Answers to Chapter 17 Practice Test

1. a. 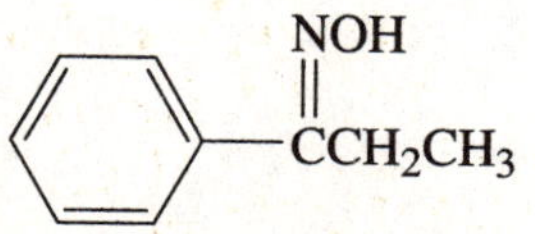 $+ H_2O$ d. 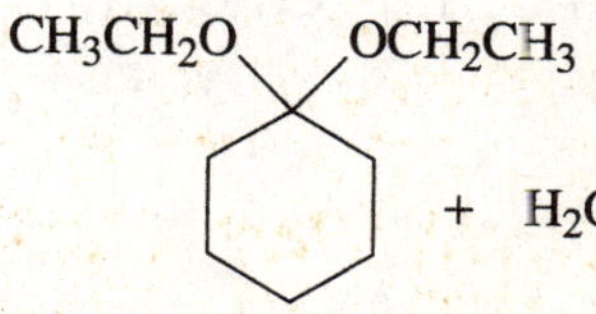 $+ H_2O$ g. $CH_3CH_2\underset{C\equiv N}{\overset{OH}{C}}CH_2CH_3$

b. 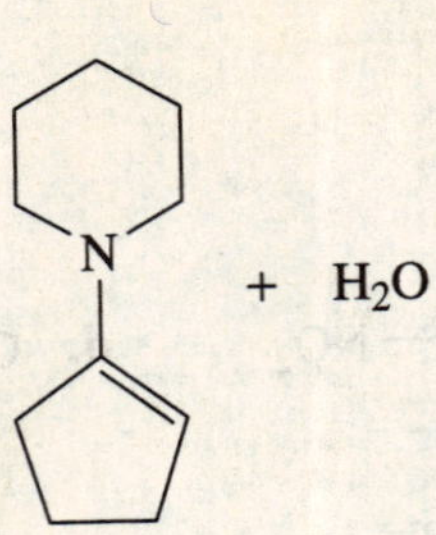

e. 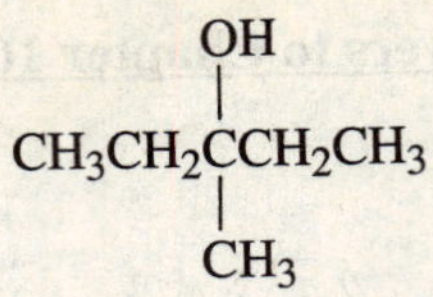

h. 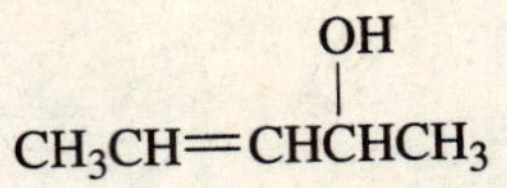

c. 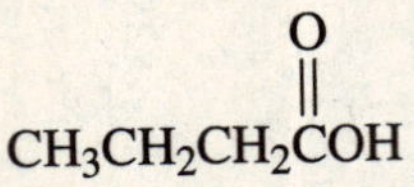

f. 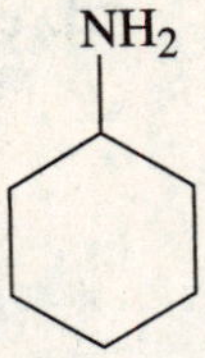

i. 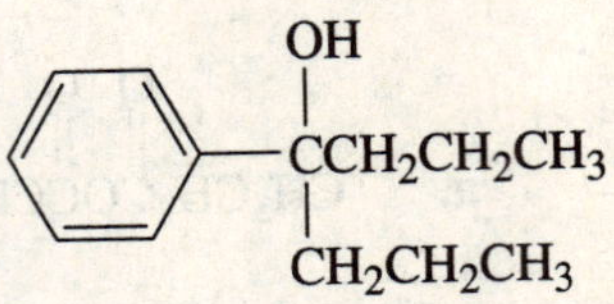

2. 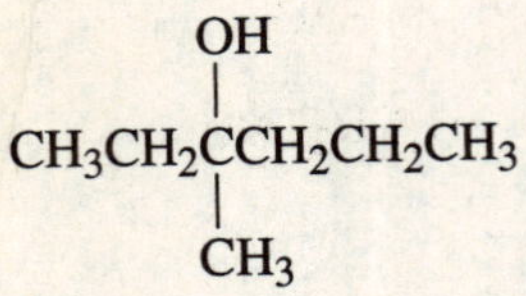

3. 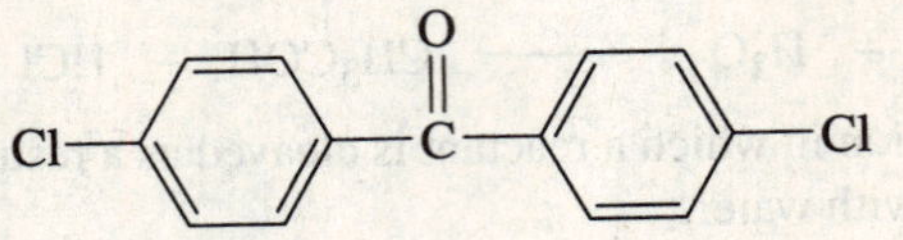

4. a. 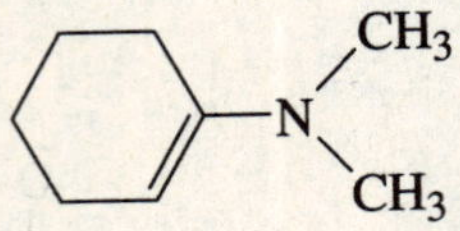

b. 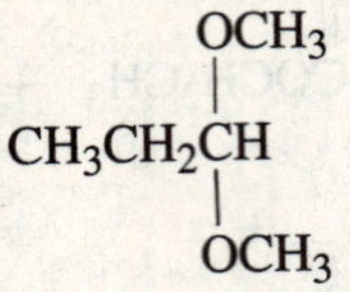

c. 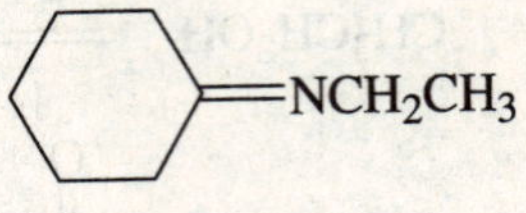

d. 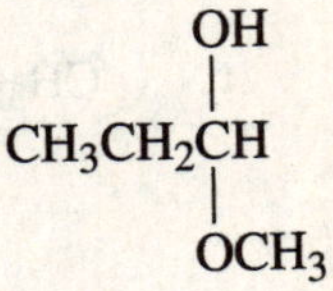

e. 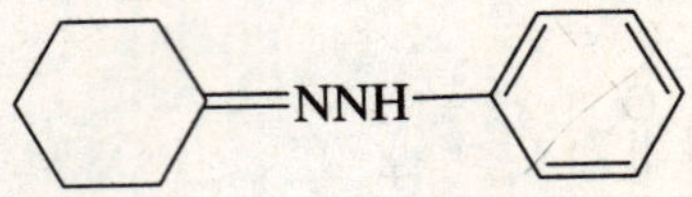

5. a. butanal b. 3-pentanone

6. 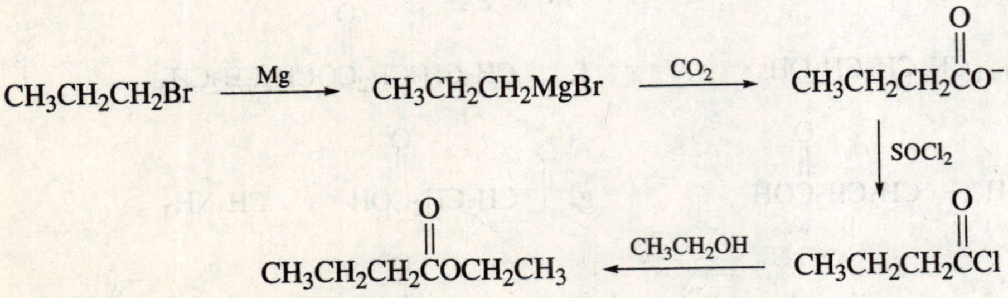

Answers to Chapter 18 Practice Test

1. 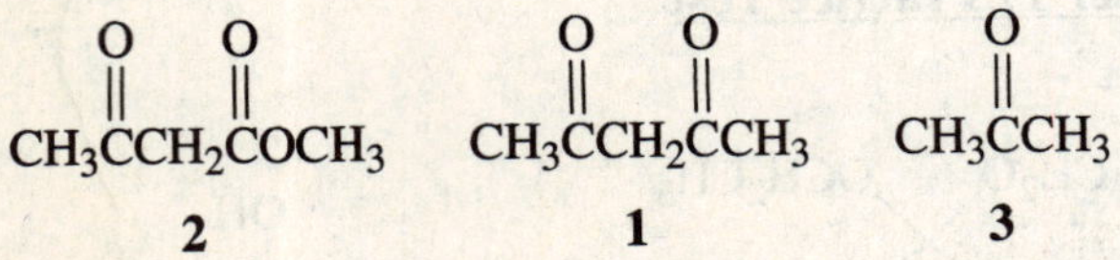

2. **a.** $CH_3C(OH){=}CHC(=O)CH_3$ **b.** $CH_3C(=O)CH_2C(=O)OCH_3$

3. **a.** $CH_3CH_2C(=O)CH_3 + CO_2$

b. $CH_3CH_2C(Br)_2{-}C(=O)H + 2\,Br^-$

c. $CH_3CH_2CH_2C(=O)CH(CH_2CH_3)C(=O)OCH_3$

d. $CH_3C(=O)CH(CH_2CH_2C(=O)CH_3)C(=O)CH_3$

e. 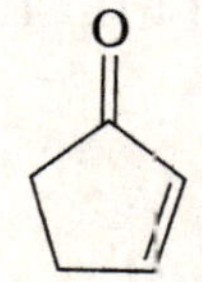

f. 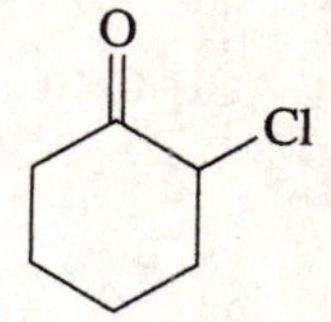

g. $CH_3CH_2CH_2CH_2C(=O)OH$

4. **a.** $2\,CH_3CH_2C(=O)H \xrightleftharpoons{HO^-} CH_3CH_2CH(OH)CH(CH_3)C(=O)H$

b. $2\,CH_3CH_2C(=O)H \xrightleftharpoons{HO^-} CH_3CH_2CH(OH)CH(CH_3)C(=O)H \xrightarrow[\Delta]{H_2SO_4} CH_3CH_2CH{=}C(CH_3)C(=O)H$

c. $2\,CH_3CH_2C(=O)OCH_3 \xrightarrow[2.\ H_3O^+]{1.\ CH_3O^-} CH_3CH_2C(=O)CH(CH_3)C(=O)OCH_3 + CH_3OH$

d. $CH_3OC(=O)CH_2CH_2CH_2CH_2CH_2C(=O)OCH_3 \xrightarrow[2.\ H_3O^+]{1.\ CH_3O^-}$ 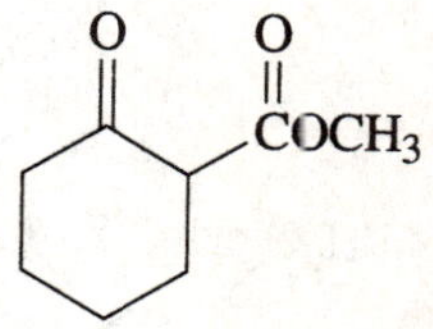

e. $CH_3CH_2OC(=O)CH_2C(=O)OCH_2CH_3 \xrightarrow[2.\ CH_3CH_2Br;\ 3.\ H^+, H_2O, \Delta]{1.\ CH_3CH_2O^-} CH_3CH_2CH_2C(=O)OH$

f. $CH_3C(=O)CH_2C(=O)OCH_2CH_3 \xrightarrow[2.\ CH_3CH_2Br;\ 3.\ H^+, H_2O, \Delta]{1.\ CH_3CH_2O^-} CH_3CH_2CH_2C(=O)CH_3$

5. 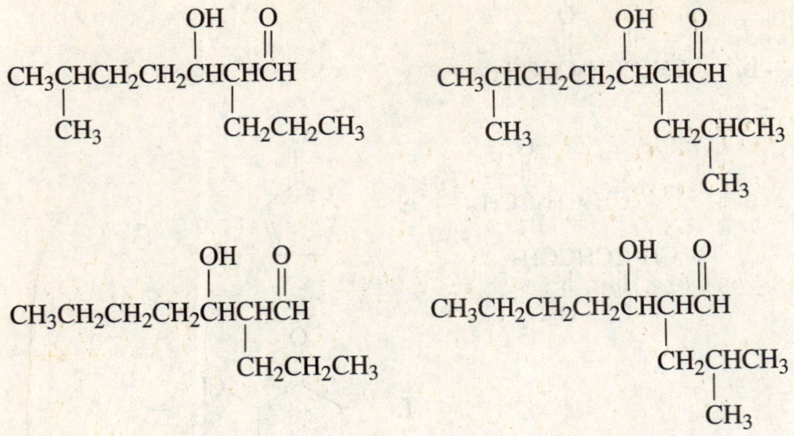

6. a. 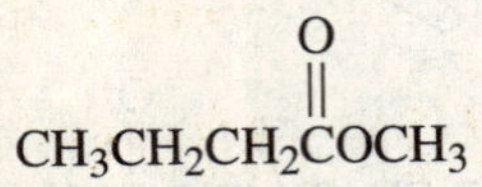 b. 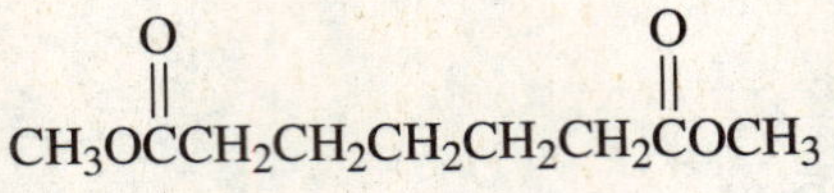

Answers to Chapter 19 Practice Test

1. a. reduction b. oxidation

2. a. 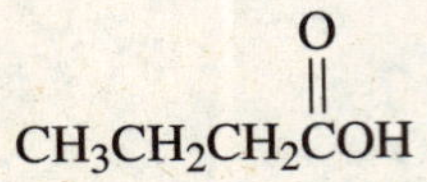

b. no reaction

c. $CH_3CH_2CH_2NHCH_2CH_3$

d. 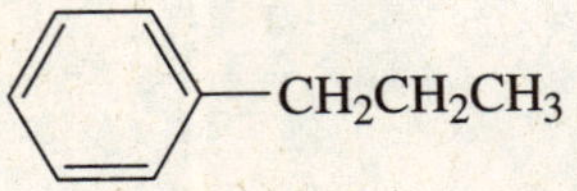

e. 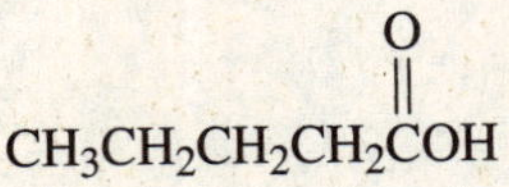

f. 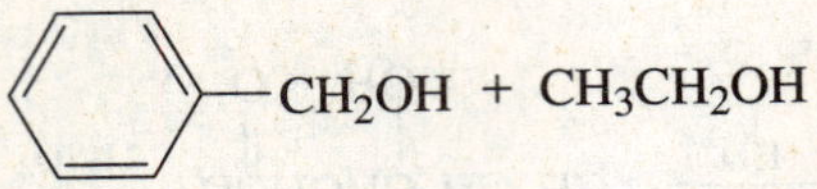

g. 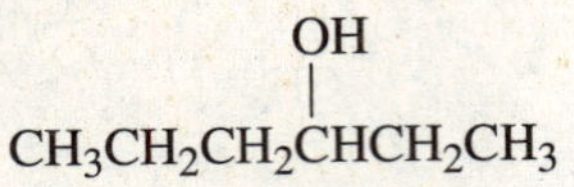

h. 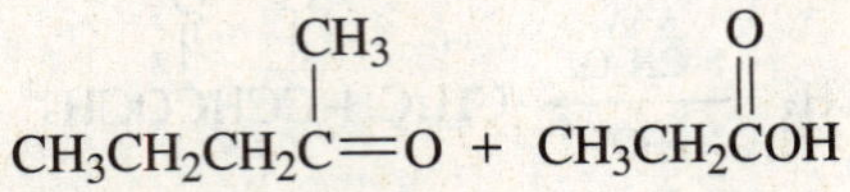

3. 1. Ag_2O, NH_3
2. H_3O^+

4. 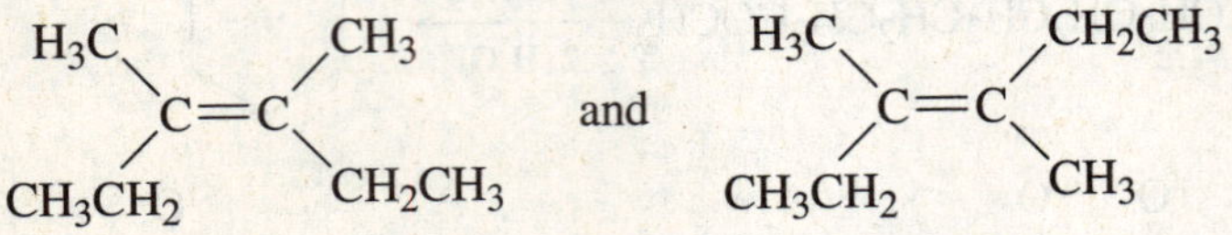

5.
a. NaBH4 is a weaker reducing agent than LiAlH4. T
b. Esters are easier to reduce than ketones. F
c. In an oxidation-reduction reaction the oxidizing agent is oxidized. F
d. Ketones are reduced to primary alcohols. F
e. Aldehydes are oxidized to carboxylic acids. T
f. Acyl halides are oxidized to aldehydes. F
g. Alkenes cannot be reduced with NaBH4. T

6. $CH_3CH_2CH_2CH_3 \xrightarrow[h\nu]{Br_2} CH_3CH_2CH(Br)CH_3 \xrightarrow{HO^-} CH_3CH_2CH(OH)CH_3 \xrightarrow{H_2CrO_4} CH_3CH_2C(=O)CH_3$

7. c. 1. O_3 2. Zn, H_2O or 1. O_3 2. $(CH_3)_2S$

Answers to Chapter 20 Practice Test

1. a. 2,4-dimethyl pyrrolidine
   2,4-dimethylazacyclopentane
   b. *N*-methylpiperidine
   N-azacyclohexane
   c. 3-ethyltetrahydrofuran
   3-ethyloxacyclopentane
   d. 3-bromopiperidine
   3-bromoazacyclohexane

2. a. $CH_2{=}CHCH_2CH(CH_3)CH_2N(CH_3)CH_3$

   b. $CH_3C(=O)NHCH_2CH_3$ + $CH_3CH_2\overset{+}{N}H_3$ Cl^-

   c. 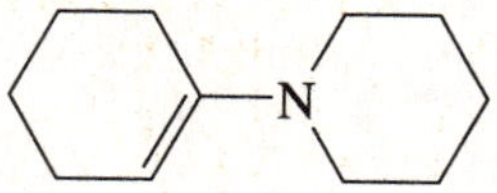 + H_2O

   d. $CH_3\overset{+}{N}(CH_3)(O^-)CH_3$

   e. 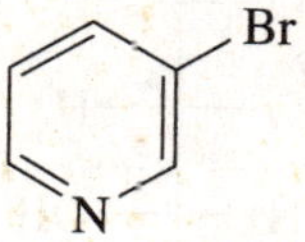

   f. 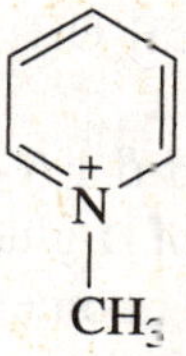

   g. 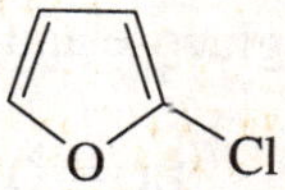

   h. $CH_2{=}CH_2$ + $N(CH_3)(OH)CH_2CH_3CH_3$

3. a. 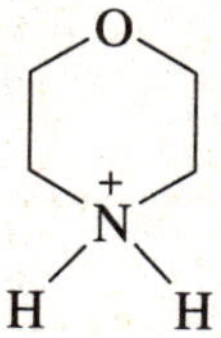
   b. 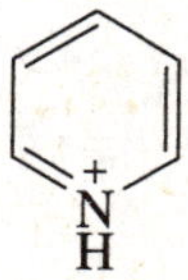
   c. 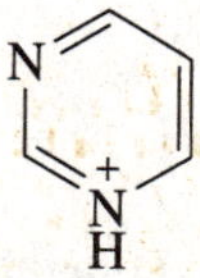
   d. 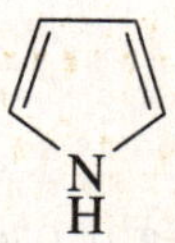

4. a. Pyridine is more reactive toward nucleophilic aromatic substitution than is pyrrole. T
   b. Pyrrole is more reactive toward electrophilic aromatic substitution than is furan. T
   c. Pyrrole is more reactive toward electrophilic aromatic substitution than is benzene. T
   d. Pyridine is more reactive toward electrophilic aromatic substitution than is benzene. F

Answers to Chapter 21 Practice Test

1. a. 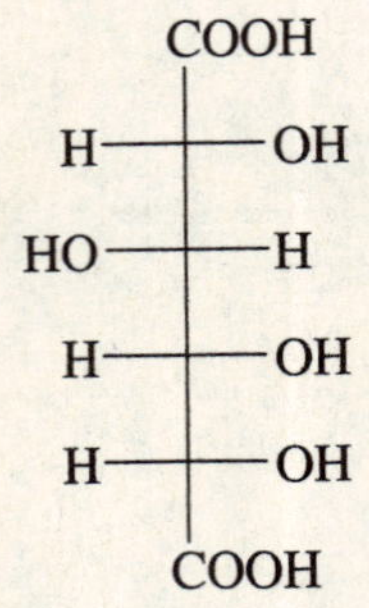

b. 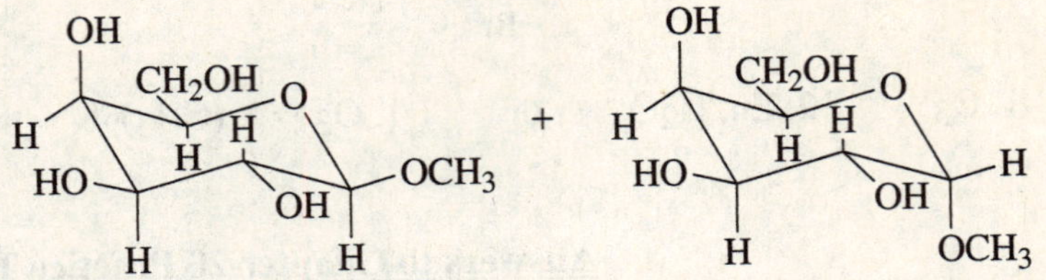

c. 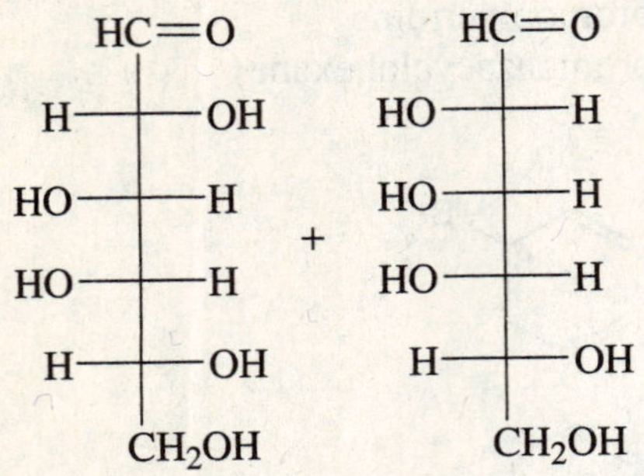

d. 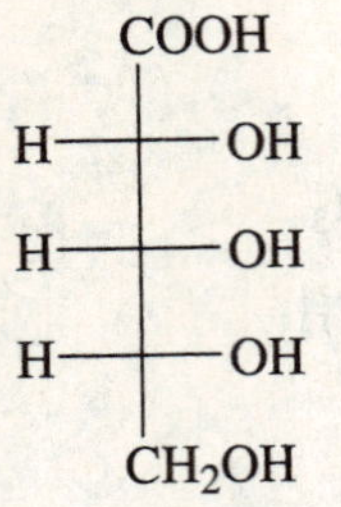

2. a. Glycogen contains α-1,4' and β-1,6'-glycosidic linkages. F
   b. D-Mannose is a C-1 epimer of D-glucose. F
   c. D-Glucose and L-glucose are anomers. F
   d. D-Erythrose and D-threose are diastereomers. T
   e. Wohl degradations of D-glucose and D-gulose form the same aldotetrose. F

3. 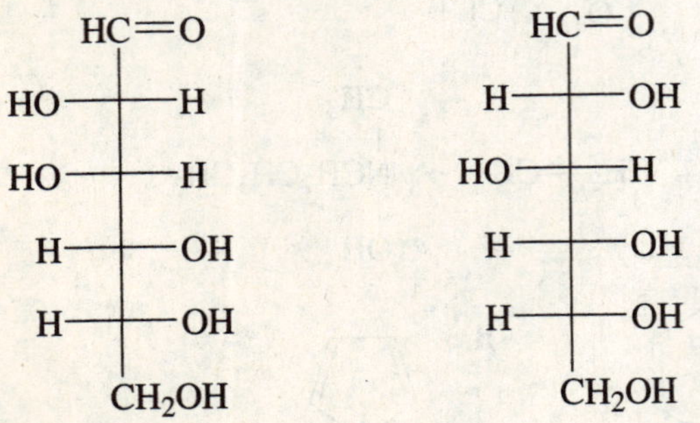

4. D-mannose and D-glucose
5. D-tagatose
6. D-altrose

7. Amylose has α-1,4'-glycosidic linkages, whereas cellulose has β-1,4'-glycosidic linkages.

8. D-gulose and D-idose
9. D-allose

10.

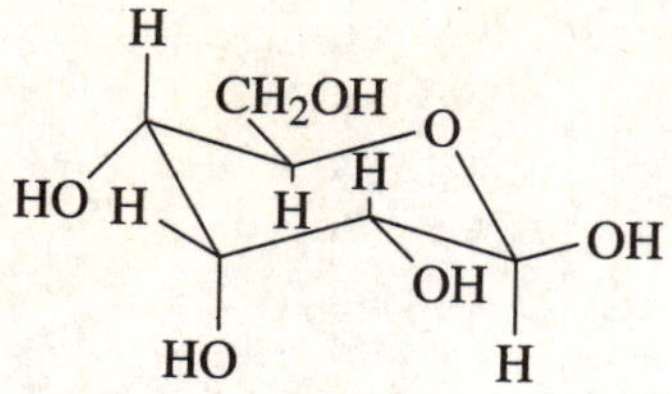

Answers to Chapter 22 Practice Test

1.

a.

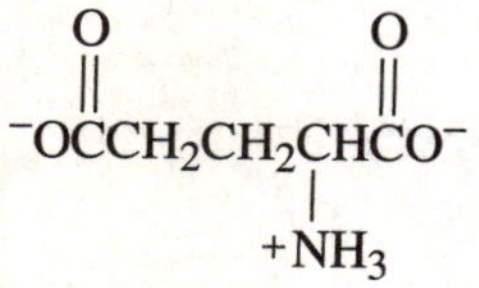

c.

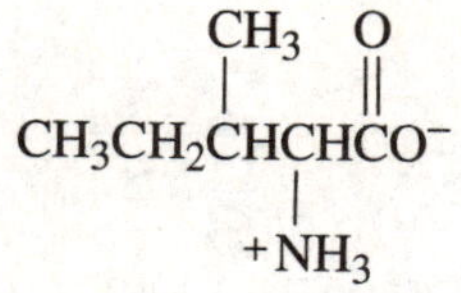

d.

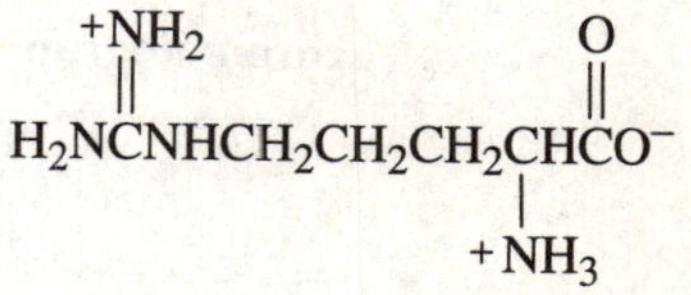

b.

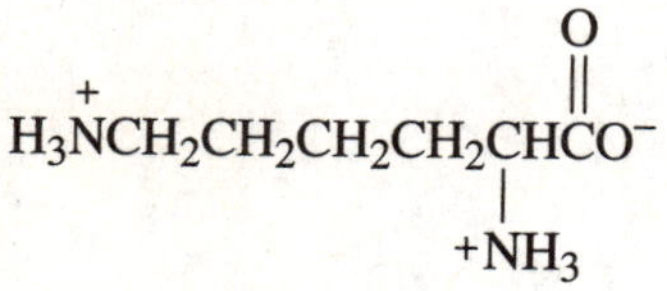

e.

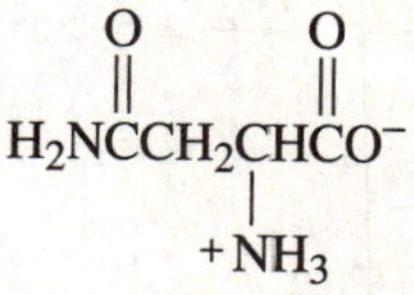

2.

a.

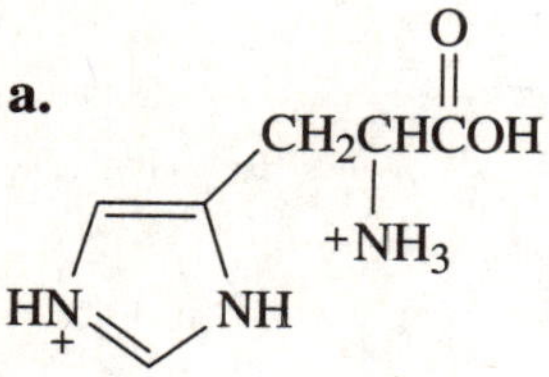

b.

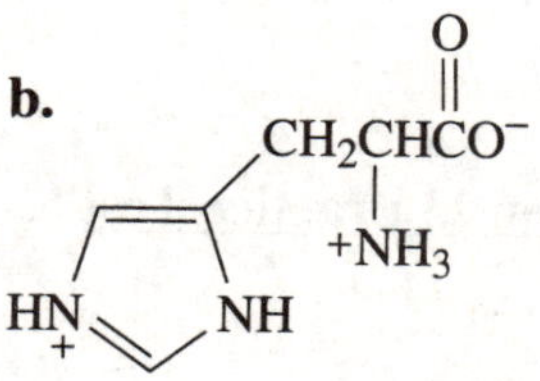

c.

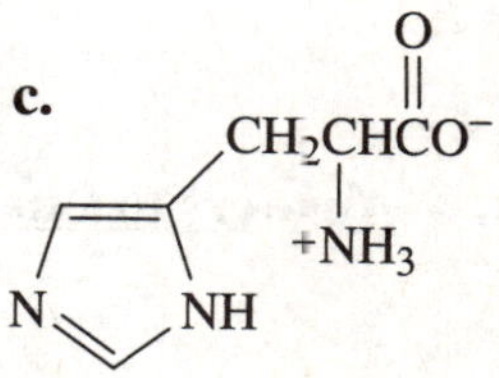

d.

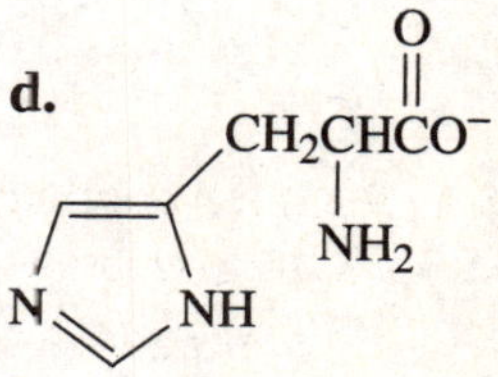

3.

a. Alanine, because it is farther away from its pI.
b. glycine
c. leucine and isoleucine
d. aspartic acid

4. The electron-withdrawing protonated amino group causes the carboxyl group of alanine to have a lower pK_a.

5.

a. A cigar shaped protein has a greater percentage of polar residues than a spherical protein. T
b. Naturally occurring amino acids have the L-configuration. T
c. There is free rotation about a peptide bond. F

6.

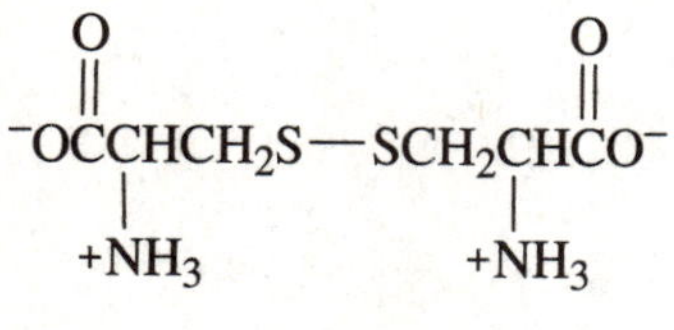

7.

a. The sequence of the amino acids in the protein chain.
b. The three-dimensional arrangement of all the atoms in the protein.
c. A description of the way the subunits of an oligomer are arranged in space.

8.

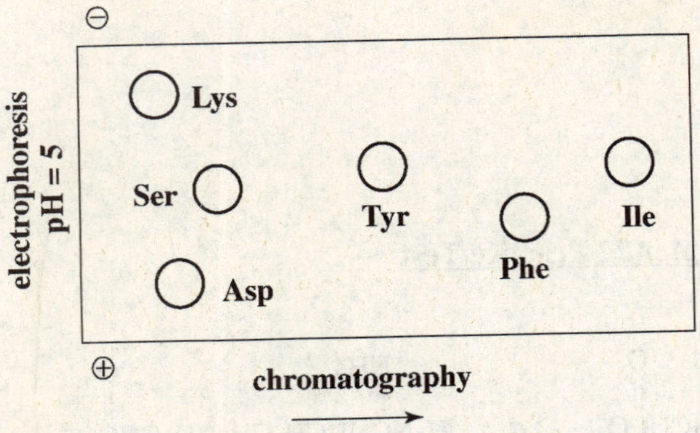

9. **a.** $\frac{2.16 + 9.18}{2} = \frac{11.34}{2} = 5.67$ **b.** $\frac{9.04 + 12.48}{2} = \frac{21.52}{2} = 10.76$

10. Ala Ser Arg Gly Arg Met His Phe Lys Ile

Answers to Chapter 23 Practice Test

1. **a.** A catalyst increases the equilibrium constant of a reaction. F
b. An acid catalyst donates a proton to the substrate and a base catalyst removes a proton from the substrate. T
c. The reactant of an enzyme-catalyzed reaction is called a substrate. T
d. Complexing with a metal ion increases the pK_a of water. F

2. It protonates the leaving group to make it a better leaving group.

3. **a.**

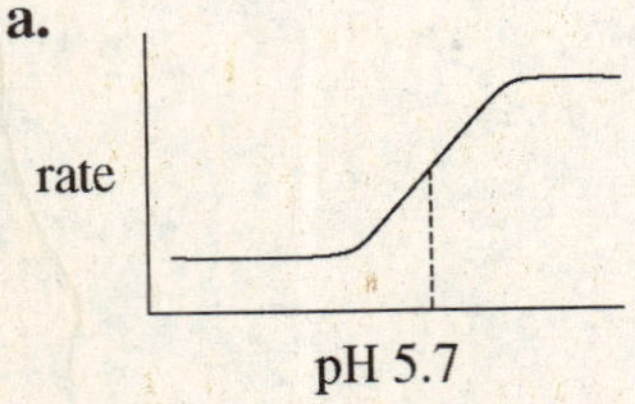

b.

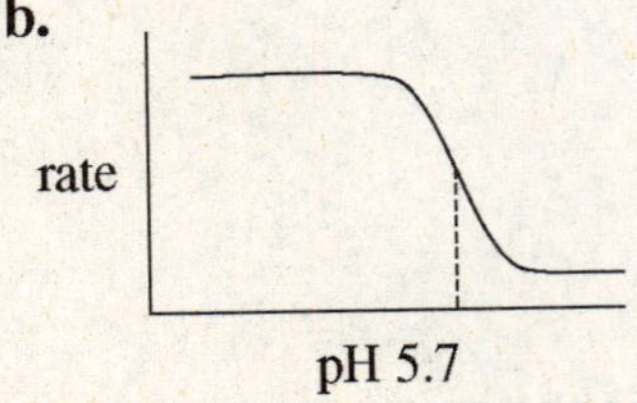

4.

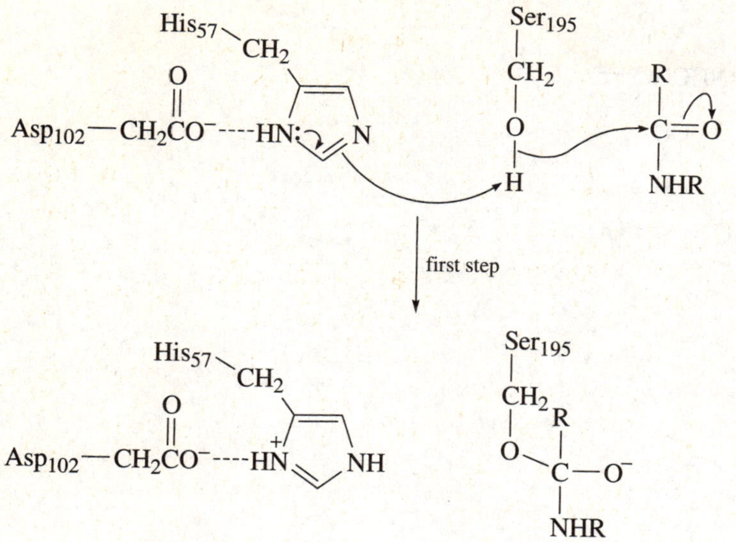

a. base catalyst
b. nucleophilic catalyst
c. it stabilizes the positive charge on histidine

Answers to Chapter 24 Practice Test

1. biotin and vitamin KH_2

2. methyl (CH_3), methylene (CH_2), formyl (HC=O)

3.

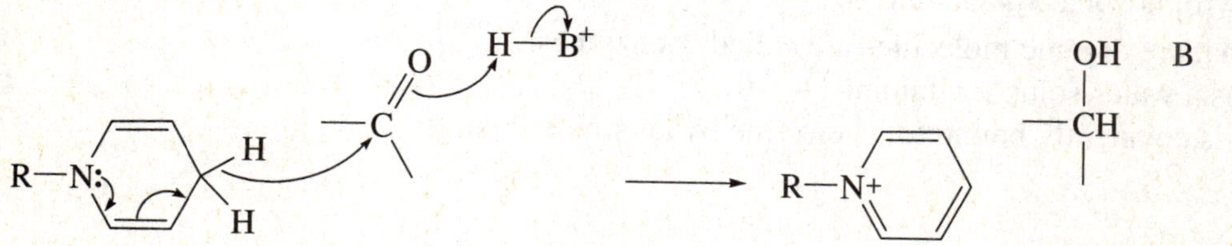

4. The first step is removing the hydrogen atom that is going to change places with a group on an adjacent carbon.

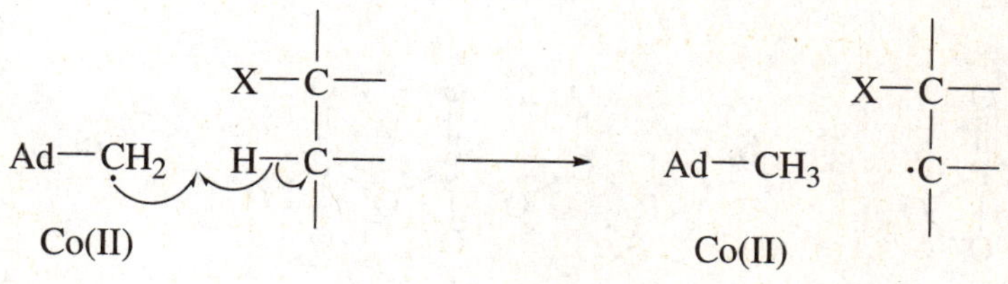

5. first coenzyme = biotin
second coenzyme = coenzyme B_{12}

6. 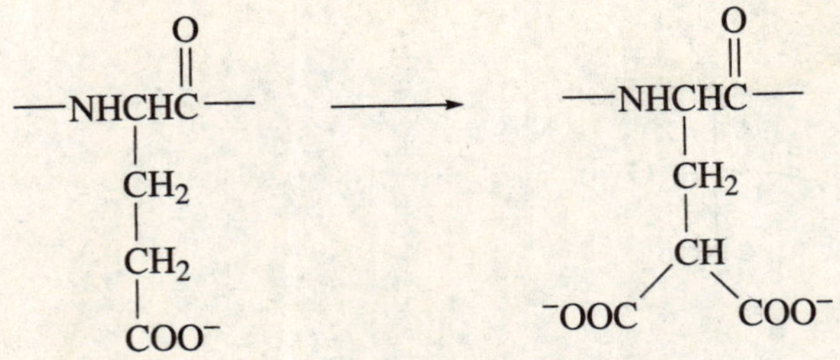

7. 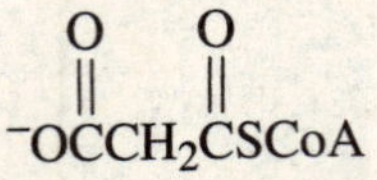

8. ATP, Mg^{2+}, HCO_3^-

9. FAD oxidizes dihydrolipoate to lipoate.

10. 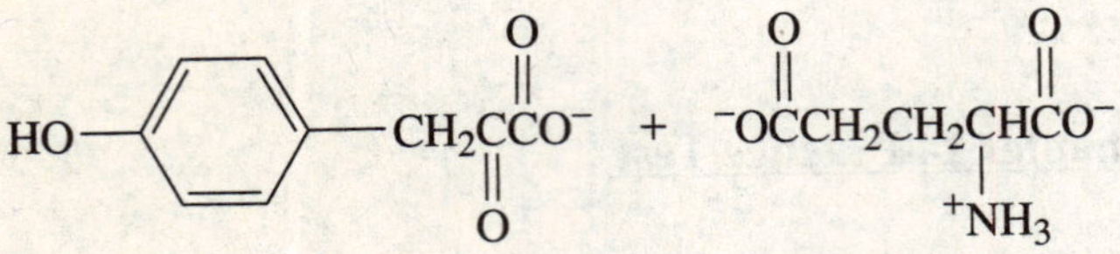

11.
a. Vitamin B_1 is the only water-soluble vitamin that has a coenzyme function. F
b. $FADH_2$ is a reducing agent. T
c. Thiamine pyrophosphate is vitamin B_6. F
d. Cofactors that are organic molecules are called coenzymes. T
e. Vitamin K is a water-soluble vitamin. F
f. Lipoic acid is covalently bound to its enzyme by an amide linkage. T

Answers to Chapter 25 Practice Test

1. a. 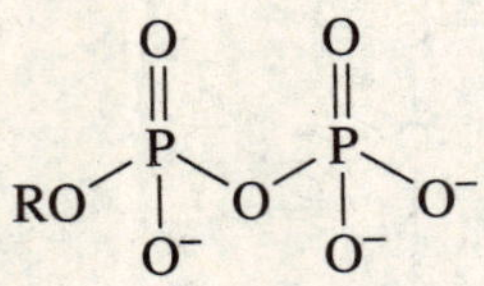

b. 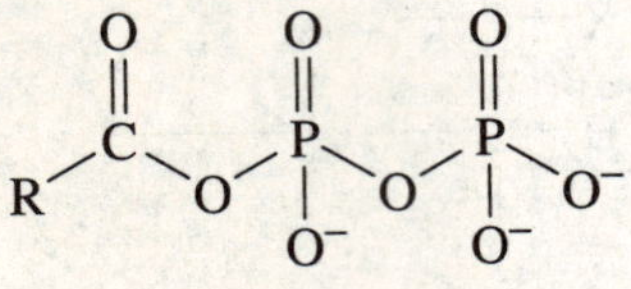

c. 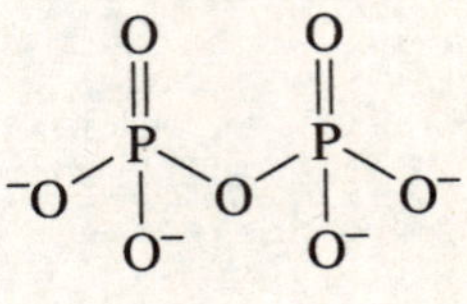

2.

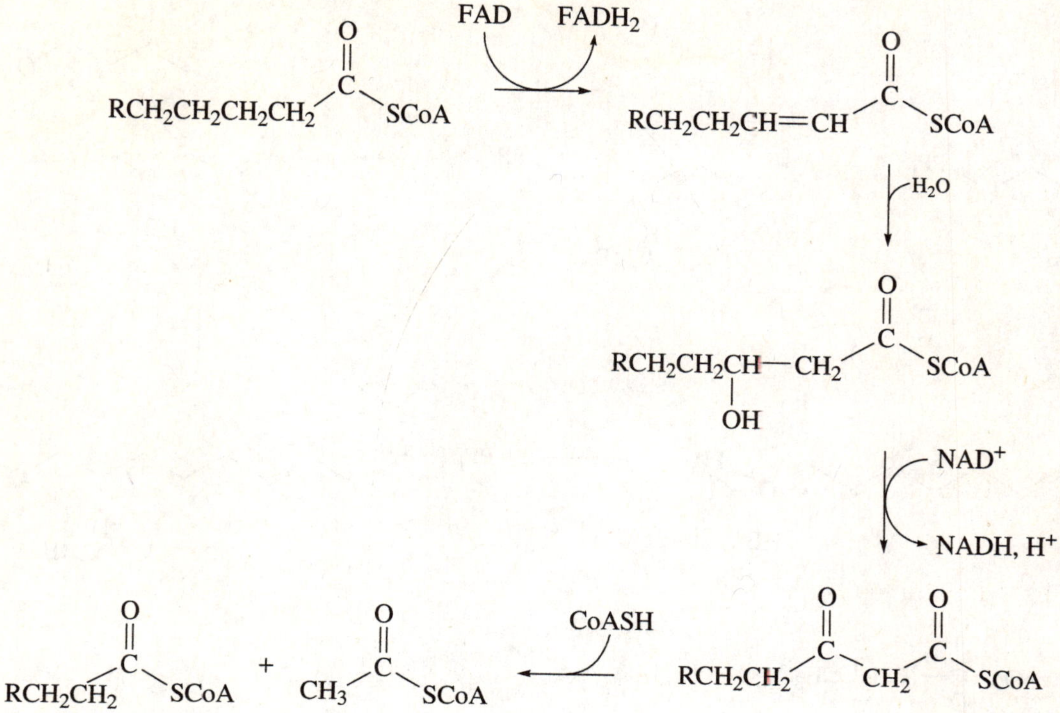

3. acetate

4. catabolic reactions

5. amino acids

6. acetyl-CoA

7.
- **a.** Each molecule of $FADH_2$ forms 3 molecules of ATP in the fourth stage of catabolism. F
- **b.** $FADH_2$ is oxidized to FAD. T
- **c.** NAD^+ is oxidized to NADH. F
- **d.** Acetyl-CoA is a citric acid cycle intermediate. F

Answers to Chapter 26 Practice Test

1.

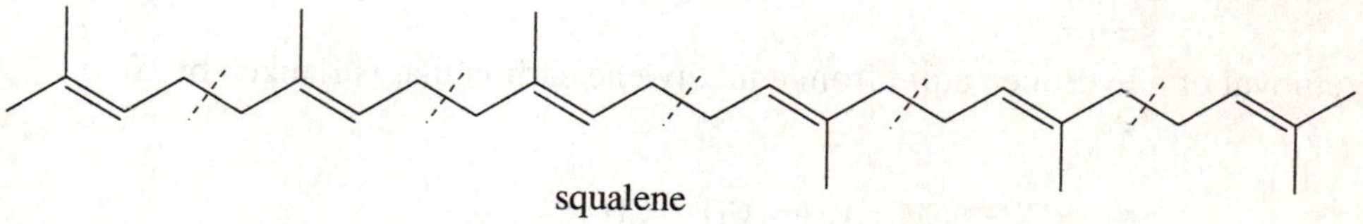

squalene

2.

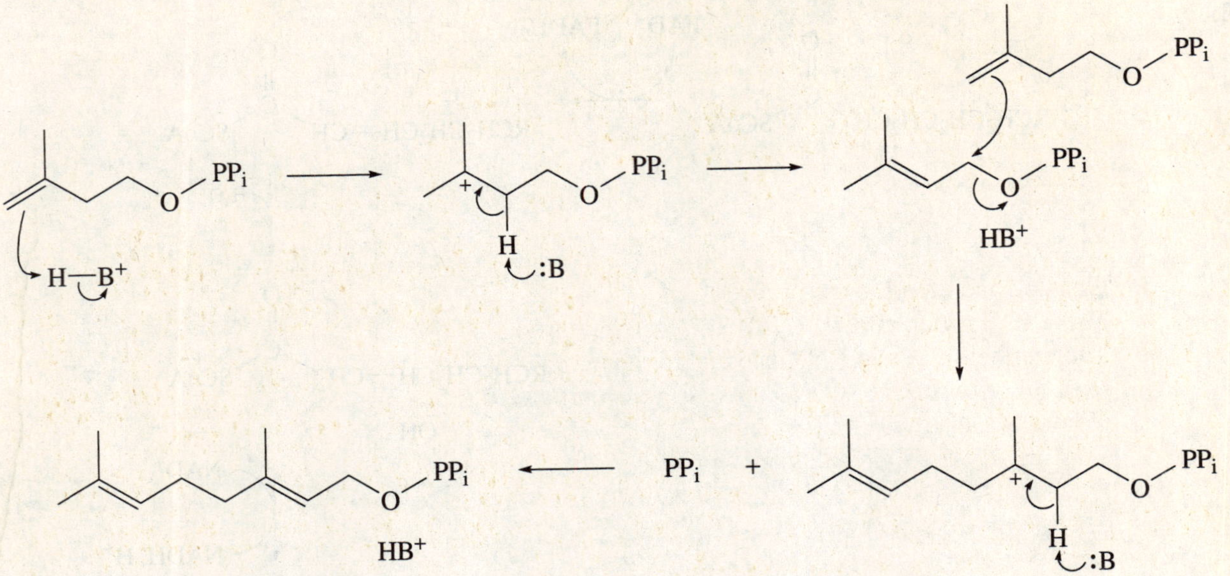

3. a.

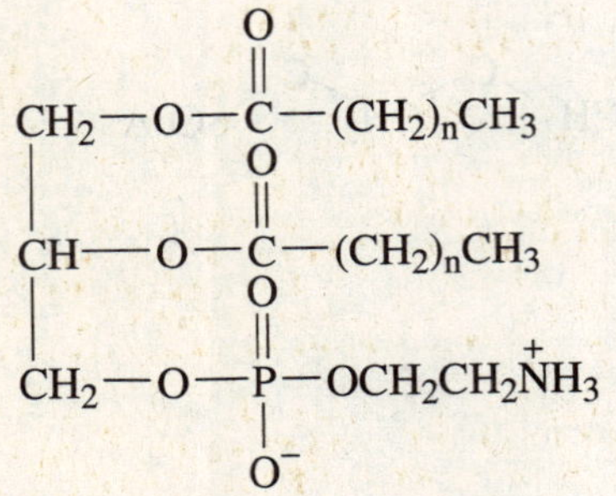

b. $CH_3(CH_2)_xCO(CH_2)_yCH_3$ (with C=O on the carbonyl carbon)

x = an even number equal to or greater than 14
y = an odd number equal to or greater than 15

4.
- **a.** Cholesterol is the precursor of all other steroids. T
- **b.** Saturated fatty acids have higher melting points than unsaturated fatty acids. T
- **c.** Fats have a higher percentage of saturated fatty acids than do oils. T
- **d.** A saturated fatty acid has a lower melting point than an unsaturated fatty acid with the same number of carbons. F
- **e.** Lipids are insoluble in water. T
- **f.** A diterpene contains 20 carbons. T
- **g.** Vitamin A is not a coenzyme. T

5. The first step is removal of a hydrogen atom from a methylene carbon that is flanked by two *sp*[2] carbons.

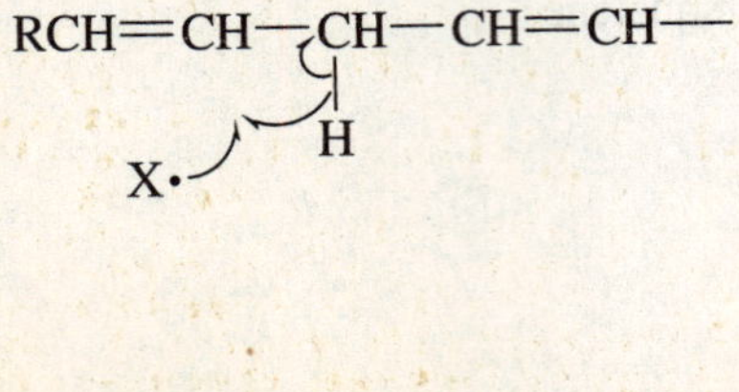

6. 6

Answers to Chapter 27 Practice Test

1. dUMP

2. 5'—A—T—G—C—A—A—G—T—3'

3. A

4.
   - **a.** Guanine and cytosine are purines. F
   - **b.** The 3'-OH group allows RNA to be easily cleaved. F
   - **c.** The number of A's in DNA is equal to the number of T's. T
   - **d.** rRNA carries the amino acid that will be incorporated into a protein. F
   - **e.** The template strand of DNA is the one transcribed to form RNA. T
   - **f.** The 5'-end of DNA has a free OH group. F
   - **g.** The synthesis of proteins from an RNA blueprint is called transcription. F
   - **h.** A nucleotide consists of a base and a sugar. F
   - **i.** RNA contains T's and DNA contains U's. F

5. only #1

6. 5'—G-C-A-U-G-G-A-C-C-C-C-G-U—3'

7. ^{32}P—GAAC

   ^{32}P—GAACTTC

   ^{32}P—GAACTTCATAG

8. In the process of converting uridines into thymidines, tetrahydrofolate is oxidized to dihydrofolate. NADPH is used to reduce dihydrofolate back to tetrahydrofolate.

9. **a.** CC and GG **c.** CA and TG